ICE Manual of Geotechnical Engineering

emerald PUBLISHING
ice Publishing

ICE Manual of Geotechnical Engineering

Second edition

Volume II
Geotechnical design, construction and verification

Edited by

Michael Brown
John Burland
Tim Chapman
Kelvin Higgins
Hilary Skinner
David Toll

ice | **manuals**

Published by Emerald Publishing Limited, Floor 5, Northspring, 21-23 Wellington Street, Leeds LS1 4DL.

ICE Publishing is an imprint of Emerald Publishing Limited

Other ICE Publishing titles:

UK Specification for Ground Investigation
The Association of Geotechnical and Geoenvironmental Specialists. ISBN 978-0-7277-6523-9

Environmental Geotechnics in Practice
Robert W Sarsby. ISBN 978-0-7277-6363-1

Disturbed Soil Properties and Geotechnical Design, Second edition
Andrew Schofield and Stuart Haigh. ISBN 978-0-7277-6155-2

A catalogue record for this book is available from the British Library

ISBN 978-0-7277-6681-6 (volume I)
ISBN 978-0-7277-6683-0 (volume II)
ISBN 978-0-7277-6685-4 (two-volume set)

Cover photo: Shutterstock/FWStudio

Commissioning Editor: Viktoria Hartl-Vida
Assistant Editor: Cathy Sellars
Production Editor: Sirli Manitski

Typeset by Nova Techset
Index created by Thomas Michael Allerton
Printed and bound in Great Britain by Henry Ling, Dorset

Contents

Foreword and endorsement

Geotechnical engineering quite literally underpins the construction industry.

Using a combination of geophysics, structural engineering and materials science, geotechnical engineers explore the interactions between soil, rock, water and structures, and develop solutions that enable buildings, tunnels, bridges, roads and other forms of infrastructure to carry the loads they are designed for, under all the conditions they may be expected to experience, over their planned lifetime and beyond.

Throughout my career, I have worked on numerous projects which have relied heavily on the skills of geotechnical engineering, so I am personally aware of the need to ensure the next generation of engineers continues to develop the knowledge and skills necessary to ensure our infrastructure continues to be built on safe, sustainable and long-lasting foundations.

I therefore commend this impressive and comprehensive work to all those who wish to develop and improve their knowledge of geotechnical engineering.

Keith Howells BSc, MBA, CEng, FICE, FCIWEM, FREng
ICE President, 2022–2023

The *ICE Manual of Geotechnical Engineering* is an essential reference document for experienced geo-professionals and non-specialists seeking to improve their knowledge of the subject. It covers historical aspects of how the subject evolved, the principles of soil mechanics, their application to designing and building in and on the ground, and how to manage and report geotechnical aspects of projects. The British Geotechnical Association is proud to remain affiliated with this updated edition of the document.

Dr Andrew Ridley
Chair of the British Geotechnical Association

Preface

We began to formulate the initial ideas for the first edition of this Manual as early as 2006. It had become apparent to us that civil and structural engineers not specialising in geotechnics face a daunting knowledge gap when they come up against a geotechnical problem. Most civil engineers leave university with very little grounding in geotechnical engineering. They will have a fair grasp of applied mechanics (mainly aimed at structural engineering). They will have had a basic introduction to geology and they will have studied the elements of soil mechanics and rock mechanics. But a recent graduate usually lacks a coherent understanding of the approach to, and methods of, geotechnical engineering and how these differ from other more widely practised branches of engineering. A survey carried out by ICE Publishing showed that information tends to be obtained from a wide range of sources through word of mouth, the internet and various publications. For the young practitioner this leads to a fragmented approach. Much of the geotechnical material is written by specialists for specialists, and its ad hoc application by a general practitioner is often inappropriate and can be extremely dangerous. We felt that it would be of great benefit to our profession to provide a single first-port-of-call authoritative reference source aimed at informing the less experienced engineer. To our delight this concept was endorsed by the ICE Best Practice Panel and the British Geotechnical Association and has offered a unique opportunity to provide authoritative guidance within a coherent framework of good geotechnical engineering.

It is most gratifying to note that the first edition of this Manual proved to be the most downloaded e-book of ICE Publishing's portfolio in 2022. Most of the chapters in this second edition have been extensively revised and updated to include the latest guidance documents and references. A major advance is the inclusion of carbon as a key metric in the choice of foundations and basement systems. In this regard, great emphasis is placed on avoiding unnecessary carbon emissions by avoiding over-conservatism in new designs and the application of codes sensitively in the remediation of old structures. Another important emerging challenge is the impact of climate change which will affect earthworks and slope engineering in particular.

As with the first edition, this second edition of the *ICE Manual of Geotechnical Engineering* has been a labour of love! The contribution of 99 contributors and ten section editors has made it possible to distil a great deal of experience from the profession into the chapters you see here. Don't imagine this will cover everything that a geotechnical engineer will face in their career – but it provides a 'starting point' from which to build experience while remaining grounded in robust fundamentals.

As mentioned previously, the Manual is aimed at people in the early stage of their careers who need a readily accessible source of information when working in new aspects of geotechnical engineering. However, it is expected that it should also prove valuable to all geotechnical engineering professionals. The aim has been to produce a manual that addresses the practice of geotechnical engineering in the twenty-first century including contemporary procurement, process and design standards and procedures. The grouping of chapters has been carefully chosen to facilitate a multidisciplinary and holistic approach to the solution of construction challenges. A key message is the importance of drawing on 'well-winnowed experience' for the smooth and reliable execution of projects. Such experience is best gained by working closely with a suitably experienced design or construction team.

It is hoped that this Manual will help in the training and development of the next generation of geotechnical engineers and will act as a useful source of reference to those with more experience.

The editors are grateful to all those contributors and section editors who have generously given so much of their time and knowledge in producing such a comprehensive book.

Michael Brown, John Burland, Tim Chapman, Kelvin Higgins, Hilary Skinner and David Toll

List of contributors

GENERAL EDITORS:
M Brown, University of Dundee, UK
JB Burland, Imperial College London, London, UK
T Chapman, Arup, London, UK
K Higgins, Geotechnical Consulting Group LLP, London, UK
HD Skinner, COWI UK, London, UK
DG Toll, Durham University, UK

SECTION EDITORS:
M Brown, University of Dundee, UK
JB Burland, Imperial College London, London, UK
JA Davis, Geotechnical Consulting Group, London, UK
M Devriendt, Arup, London, UK
S Hardy, Laing O'Rourke, Dartford, UK
I Jefferson, University of Birmingham, UK
T O'Brien, Mott MacDonald, UK
W Powrie, University of Southampton, UK

CONTRIBUTORS:
P Bailie, Arup, London, UK
J Ball, Roger Bullivant Piling, Swadlincote, UK
D Baxter, ArcelorMittal, UK
A Bell, Cementation Skanska Ltd, Doncaster, UK
A Berry, Arup, London, UK
D Bhuva, GSS Piling Ltd, Welwyn Garden City, UK
J Binns, Byland Engineering, York, UK
C Brewis, Arup, London, UK
M Brown, University of Dundee, UK
S Bullivant, Taranto Ltd, Craigavon, UK
JB Burland, Imperial College London, London, UK
F Chamley, Arup, London, UK
T Chapman, Arup, London, UK
J Chick, Expanded Ltd (Geotechnical), Dartford, UK
B Clarke, University of Leeds, UK
P Coney, Jacobs, UK
J Cowburn, Roger Bullivant Piling, Swadlincote, UK
JC Cripps, University of Sheffield, UK
MA Czerewko, AECOM Ltd, Chesterfield, UK
JA Davis, Geotechnical Consulting Group, London, UK
AM Daykin, Atkins, Leeds, UK
MH de Freitas, Imperial College London and First Steps Ltd, London, UK
M Devriendt, Arup, London, UK
C Doulala-Rigby, Tensar International Ltd, Blackburn, UK
A Duley, Jacobs, UK
PG Dumelow, Balfour Beatty plc, Thame, UK
C Edmonds, Stantec, Reading, UK
RD Essler, RD Geotech, Skipton, UK
I Farooq, Mott MacDonald, UK
ER Farrell, AGL Consulting, Dublin, Republic of Ireland
M Free, Arup, London, UK
S French, BES Group Ltd, Warrington, UK
R Garland, BAM Nuttall Ltd, UK
B Ghosh, Mott MacDonald, UK
P Gilbert, Jacobs, UK
S Glover, Arup, London, UK
S Hardy, Laing O'Rourke, Dartford, UK
A Harwood, Arup, London, UK
T Hayward, Laing O'Rourke, Dartford, UK
A Heathcote, Keller, Coventry, UK
G Herrmann, Keller, Coventry, UK
G Horgan, Huesker, Warrington, UK
C Hughes, Arup, London, UK
P Ingram, Arup, London, UK
I Jefferson, University of Birmingham, UK
LD Jones, British Geological Survey, Nottingham, UK
M Kemp, Atkins, Epsom, UK
V Kumar, Mott MacDonald, UK
M Large, Keller, Coventry, UK
R Lindsay, Atkins, Epsom, UK
S Markvukaj, Mott MacDonald, UK
WA Marr, Geocomp Corporation, Massachusetts, USA
P Marsden, Keller, Coventry, UK
BT McGinnity, retired formerly London Underground, London, UK
D Milo, Balfour Beatty, UK
P Morrison, Arup, London, UK
S Mortimore, Atkins, London, UK
D Mothersille, Geoserve Global Limited, London, UK
J Murray Reid, retired formerly TRL Limited, Wokingham, UK
R Nicholson, AECOM Ltd, Chesterfield, UK
PA Nowak, Atkins Ltd, Epsom, UK
T O'Brien, Mott MacDonald, UK
T Orr, Trinity College, Dublin, Republic of Ireland
H Pantelidou, Arup, London, UK
D Patel, Arup, London, UK
S Pennington, Arup, London, UK
G Pittaro, Mott MacDonald, Singapore
D Potts, Imperial College London, London, UK
JJM Powell, Geolabs Ltd, Watford, UK

W Powrie, University of Southampton, UK
D Preece, Laing O'Rourke, Manchester, UK
M Preene, Richter Associates, Huddersfield, UK
JA Priest, University of Calgary, Canada
D Puller, Geotechnical Experts, Sevenoaks, UK
N Pye, Mott MacDonald, UK
JM Reynolds, Reynolds Geo-Solutions Ltd, Mold, UK
C Robinson, Roger Bullivant Limited, Swadlincote, UK
CDF Rogers, University of Birmingham, UK
A Royal, University of Birmingham, UK
CS Russell, Geotechnical Consulting Group, London, UK
N Saffari, London Underground, London, UK
H Shields, Arup, London, UK
HD Skinner, COWI UK, London, UK
JA Skipper, Geotechnical Consulting Group, London, UK
P Sloan, COWI, London, UK
P Smith, Geotechnical Consulting Group, London, UK
L Sodano, Mott MacDonald, UK
J Standing, Imperial College London, London, UK
M Stanley, Bachy Soletanche Ltd, Camberley, UK
J Strange, Card Geotechnics Ltd, Godalming, UK
J Taggart, Arup, London, UK
R Talby, Mott MacDonald, UK
B Thompson, Keller, Coventry, UK
DG Toll, Durham University, UK
M Walsh, Arup, London, UK
MJ Whitbread, Wales, UK
A Willoner, Mott MacDonald, UK
D Woods, Huesker, Warrington, UK
L Zdravkovic, Imperial College London, London, UK

About the editors

Hilary Skinner (Editor)

Hilary Skinner graduated in Engineering from the University of Cambridge in 1990. After a period of research at BRE, she joined Whitby Bird (later Ramboll) as Associate Director in Geotechnics. Now a Director, she has been at COWI (previously Donaldson Associates) since 2010. She has authored or co-authored a number of papers and books for which she has been awarded the BGA Research Medal and the Halcrow Prize from the ICE. All have involved collaboration with UK and European consultants, contractors and researchers. She continues to collaborate on research into working platforms, piling and tunnelling with universities and industry bodies. Hilary is a Chartered Engineer, a Fellow of the ICE and a past Chair of the British Geotechnical Association (BGA). The importance of measurement and case history data has been critical to the development of our science, which marries theory and judgement, and this has been central to a lifelong pursuit of technical excellence in her work and our industry.

David Toll (Editor)

David Toll is Chair of the British Geotechnical Association (BGA) and Professor of Engineering and Co-Director of the Institute of Hazard, Risk and Resilience at Durham University. He is a member of ISSMGE's Technical Committee TC 107 on Tropical Soils, has been Chair of the ISSMGE's Technical Committee TC 106 on Unsaturated Soil and was the inaugural Chair of the Joint Technical Committee JTC2 of FedIGS on Geo-Engineering Data. David is a Fellow of the ICE and has been Chair of both ICE North East and the Northern Geotechnical Group. He is a Fellow of the Learned Society of Wales. He has been carrying out research into climate resilience of infrastructure and information technology applied to civil engineering for over 30 years and has published over 200 papers and 12 books. He has held Visiting Professor and Research Fellow posts at Tongji University, China, National University of Singapore, Nanyang Technological University, University of Sydney, University of Western Australia and University of Newcastle, Australia. He was the founding editor of the journal *Geotechnical and Geological Engineering* and has been a member of the editorial boards for *Géotechnique*, *Transportation Geotechnics* and the *Quarterly Journal of Engineering Geology and Hydrogeology*.

Kelvin Higgins BSc, MSc, DIC, CEng, FREng, FICE, FCIHT (Editor)

Kelvin Higgins is a Senior Partner at Geotechnical Consulting Group LLP (GCG) and a Visiting Professor at Imperial College London. In 2019 Kelvin became a Fellow of the Royal Academy of Engineering and since 2012 he has been a Visiting Professor at Imperial College London. Until recently he was Chair of the British Geotechnical Association and Chair of CIRIA's Geotechnical Advisory Panel. Since joining GCG in 1986 Kelvin has worked on the application of numerical methods to engineering problems, including the design of tunnels, retaining structures, highways, dams, embankments, cuttings and foundations (offshore and onshore). During his career he has gained broad practical and design experience of foundations, highway construction earthworks (cuttings, embankments and drainage schemes), deep basements and the interaction between structures. He has acted as an expert witness in legal disputes. In addition, he has directed and facilitated research in different aspects of geotechnical engineering. Kelvin has extensive experience of the assessment of the effects of underground construction, (excavations, shafts and tunnels) on adjacent structures. This includes developing alternative designs, construction techniques and sequencing of works to mitigate the effects of construction on adjacent structures and services by predicting the effects of construction on adjacent structures, services, operational tunnels and installations within these structures (e.g. escalators, permanent way, etc.). In many cases he has been able to demonstrate economic benefits to alternative sequences and designs. Kelvin has investigated the causes of the failures of foundations and slopes, designed remedial schemes and monitored construction. His research and publications cover earth-retaining structures, slopes, sheet piling, new monitoring technologies, soil behaviour, the performance and analysis of foundations, tunnels, embankment dams, slope stability, thermal piles, monitoring and the use of advanced numerical methods. Prior to joining GCG, he worked for Sir Alexander Gibb & Partners in the UK and internationally, where he gained experience in foundation design and the design of embankment dams. He oversaw the construction of highway schemes and other facilities in remote and challenging environments, having to adapt designs to accommodate local conditions.

Michael Brown BEng (Hons), PhD, GMICE (Editor and Section Editor)

Michael Brown is Professor of Geotechnical Engineering in the School of Science and Engineering at the University of Dundee, UK. Mike is a researcher, former practitioner and an educator at both undergraduate and post-graduate level. As Professor of Geotechnical Engineering his focus is on translating the complexities and specialist testing approaches of academia into practical outputs that can be easily deployed by practising engineers. Currently his research interests lie in offshore renewable energy, with the development of novel deep foundation systems and anchoring solutions for future floating wind. Onshore he is looking at efficiencies in deep foundation excavation and support and the development of efficient ground energy systems for deployment in diaphragm walls for heating and cooling in the UK and South America. Mike undertook his PhD at the University of Sheffield, UK, looking at the rapid load testing of piles in clay and soil rate effects in general. Based upon this work, the rapid load testing technique was made more accessible in the UK with inclusion in the *ICE Specification for Piling and Embedded Retaining Walls* and Federation of Piling Specialist (FPS) load testing guidance. Internationally he has contributed to national guidance on rapid load testing in the Netherlands and chaired the CEN working group that was responsible for the creation of ISO/CEN standards on rapid load pile testing and dynamic piles testing. These were subsequently adopted as British Standards. Based upon his expertise in deep foundation testing and developing novel foundation deployment, he is currently a UK representative on the ISSMGE Technical Committee 212 Deep Foundations, as well as a member of the International Press-in Association (IPA): IPA-TC4 concerned with press-in and rotary pile installation. Since joining the University of Dundee he has undertaken a significant amount of contract testing and research direct for industry based around specialist physical modelling and centrifuge testing. This type of work keeps Mike industrially relevant and allows him to appreciate the needs of both industry and academia. These specialist interests and skills have recently been recognised, with him being appointed as secretary of ISSMGE TC 104: Physical Modelling in Geotechnics.

Professor John Burland (Editor and Section Editor)

Professor John Burland is Emeritus Professor of Soil Mechanics and Senior Research Investigator at Imperial College London. Born in the UK, Professor Burland was educated in South Africa and studied Civil Engineering at the University of the Witwatersrand. He returned to England in 1961 and worked with Ove Arup and Partners for a few years. After studying for his PhD at Cambridge University, John Burland joined the UK Building Research Station in 1966, became Head of the Geotechnics Division in 1972 and Assistant Director in 1979. In 1980 he was appointed to the Chair of Soil Mechanics at Imperial College London, before becoming Emeritus Professor and Senior Research Investigator there. In addition to being very active in teaching and research, John Burland has been responsible for advising on the design of many large ground engineering projects worldwide, including the underground car park at the Palace of Westminster and the foundations of the Queen Elizabeth II Conference Centre in London. He specialises in problems relating to the interaction between the ground and masonry buildings. He was London Underground's expert witness for the Parliamentary Select Committees on the Jubilee Line Extension underground railway and has advised on many geotechnical aspects of that project, including ensuring the stability of the Big Ben Clock Tower. He was a member of the international board of consultants advising on the stabilisation of the Metropolitan Cathedral of Mexico City and was a member of the Italian Prime Minister's Commission for stabilising the Leaning Tower of Pisa. He has received many awards and medals, including the Gold Medal for Engineering Excellence of the World Federation of Engineering Organisations and the Gold Medals of the UK Institution of Structural Engineers and of the UK Institution of Civil Engineers. In 1994 he was awarded the Kevin Nash Gold Medal of the International Society of Soil Mechanics and Geotechnical Engineering 'In recognition of outstanding services to ISSMGE, to

International Goodwill and to International Geotechnical Practice and Education'. In 1996 he was awarded the Harry Seed Memorial Medal of the American Society of Civil Engineers 'for distinguished contributions as an engineer, scientist and teacher in soil mechanics'. He is a Fellow of both the UK Royal Academy of Engineering and of the Royal Society of London, a Member of the US National Academy of Engineering and was appointed Commander of the Most Excellent Order of the British Empire in 2005.

Tim Chapman BE, MSc, DIC, CEng, FICE, FIEI, FREng (Editor)

Tim Chapman is a Director at Arup, London, with deep expertise in all aspects of geotechnical engineering – his particular areas of expertise relate to ground risk minimisation for large projects and for the design of deep underground structures with and around tunnels. He has had extensive experience in the design of all types of infrastructure and devising how future infrastructure can best serve and shape future society. A current goal is helping clients to choose the right projects to meet their desired outcomes and setting up those projects to succeed far more fundamentally from the outset. He advocates that our infrastructure systems should increase national happiness by promoting prosperity and well-being as well as being fit for the future in terms of decarbonisation, social value, biodiversity and circular economy. Among other Arup roles, he led its London-based Infrastructure Design group for over a decade until April 2020 and was Office Leader for its London office, Arup's largest globally, through the COVID-19 pandemic and was also client for the creation of its new flagship head office, the smartest office space in the world and lowest carbon office building in London, designed for collaboration and inclusion, to attract staff back into a conducive space – delivered on time and under budget. Tim is a Fellow of the ICE and of the Royal Academy of Engineering. In 2011 he was awarded the ICE President's Medal for producing the institution's first significant thought piece on how to decarbonise the infrastructure sector. He is currently Co-Chair of the ICE Community Advisory Board for Fundamentals and Behaviours and Chair of the Association of Consulting and Engineering's (ACE) Risk Task Force, charged with reducing project risks for the benefit of all. For the RAEng he chairs the Sainsbury Management Fellowship helping to equip engineers with more business-facing skills and represents the RAEng on EuroCASE, the federation of European engineering academies, which seeks to influence the European Commission on key topics, one of which is 'sustainable materials for sustainability'.

Stuart Hardy MEng, ACGI, PhD, DIC, CEng, FICE (Section Editor)

Stuart Hardy joined Laing O'Rourke's central technical group as their Geotechnical Lead in 2021, with an overview of all geotechnical works for Laing O'Rourke, from tender through to completion and beyond. After completing an undergraduate degree in Civil Engineering and a PhD in geotechnical engineering at Imperial College London, Stuart joined Mott MacDonald in Croydon and worked on the design of the new Wembley Stadium and Heathrow T5. After leaving Mott MacDonald, Stuart joined the geotechnics department of Arup's office in London and stayed for nearly 16 years, working on a number of challenging basement and foundation projects located mainly in the Middle East and London. These included the Post Building, the Exhibition Road Quarter for the Victoria and Albert Museum, Elizabeth House, the ArcelorMittal Orbit sculpture in the Olympic Park and the redevelopment of the Shell Centre building in Waterloo. For the last five years at Arup Stuart led the geotechnical design of the S1 and S2 contracts on HS2, working with the Main Civils Contractor SCS JV. Stuart is one of the UK's representatives to CEN on the redrafting of the Eurocodes and chairs the parallel BSI committee. Stuart was one of the leading authors on the CIRIA C760 publication on the design of embedded retaining walls and is a RoGEP registered Advisor and a Fellow of the ICE.

Tony O'Brien FREng (Section Editor)

Tony O'Brien is a Fellow of the Royal Academy of Engineering and a Fellow of the ICE. He is the Global Practice Leader for Geotechnics at Mott MacDonald and a Visiting Professor at the University of Southampton. He has provided technical leadership on several major projects across Europe, North America and the Asia-Pacific region and worked as an expert during litigation/dispute resolution. Tony has contributed to best practice guidance, including CIRIA Guide C791 on advanced numerical modelling, and recently published a book, *The Observational Method in Civil Engineering*, and contributed to several international ISSMGE committees.

Mike Devriendt MSc, DIC, CEng, MICE (Section Editor)

Mike Devriendt is Associate Director at Arup, London, and is a geotechnical and tunnelling engineer with over 25 years of experience. He is responsible for leading large multidisciplinary design teams on urban infrastructure projects such as Crossrail 2, HS2, Tideway and Crossrail, and a range of commercial and residential building projects. Mike specialises in working collaboratively with clients and contracting organisations to develop cost-effective and buildable designs. Mike is author of over 25 journal or conference papers. As a recognised technical expert, Mike regularly leads and contributes to national and international steering committees. He has been appointed to a number of industry-wide ICE and CIRIA roles and has been an expert panel member or author of geotechnical and tunnelling standards for BSI, CIRIA and HS2 Ltd. Mike has particular research interests in the practical application of new technologies in the field of monitoring and asset management. He has led a number of consulting commission and collaborative research projects with industrial partners such as UK Power Networks, HS1, National Highways and with academic partners such as the University of Cambridge, University of Oxford and University College London. As part of these projects Arup have provided consultancy advice and developed new techniques for application of photogrammetry, machine learning, laser and thermal scanning to achieve health and safety benefits and operational cost and programme savings associated with infrastructure monitoring and maintenance.

William Powrie (Section Editor)

William Powrie is Professor of Geotechnical Engineering at the University of Southampton, where he served as Dean of the Faculty of Engineering and the Environment 2010–2018. His main technical expertise is in transportation geotechnics and waste/resource management, for which he was elected Fellow of the Royal Academy of Engineering in 2009. William's research on geotechnical transport infrastructure encompasses groundwater control, in-ground construction, understanding and mitigating vegetation and climate change effects, and fundamental soil behaviour. Major projects on which he has worked include the A55 Conwy Crossing, the Jubilee Line extension and HS1. He chairs HS2's Independent Geotechnical Expert Panel. He is Convenor of the 14-university UK Collaboratorium for Research on Infrastructure and Cities (UKCRIC), and leads the UK Rail Research and Innovation Network (UKRRIN) Centre of Excellence in Infrastructure and the Infrastructure for Port and Coastal Cities and Towns Network. William's research in waste/resource management focuses on landfill engineering and science-based policy and practice. He worked on the design and engineering risk assessment of low-level radioactive waste repositories at Drigg, and chaired the Technologies Advisory Committee for Defra's £30M research and demonstrator programme for new technologies for treating biodegradable waste. He is the author of the widely respected textbook *Soil Mechanics: Concepts and Applications*.

John A Davis BSc, MSc, DIC, CGeol, EurGeol (Section Editor)

John A Davis is a Senior Partner at Geotechnical Consulting Group LLP. John has worked for contractor, consultant, client and government organisations. He has had two extended periods of secondment, one into London Underground to work on the potential effects of rising groundwater on below-ground assets, and another to the Crossrail client organisation where he was responsible for geotechnical matters in the eastern half of the project. He has interests in commercial management of ground risk and quaternary engineering geology. John is currently Chair of the Engineering Group of the Geological Society.

Ian Jefferson BEng, DIS, PhD, FGS (Section Editor)

Ian Jefferson is Deputy Head at the College of Engineering and Physical Sciences, University of Birmingham, UK and Professor of Geotechnical Engineering. He has research interests that span geotechnical engineering and engineering geology, addressing issues associated with geohazards and problematic soils, from assessments through modelling to mitigation, using both traditional and novel areolation techniques. Past Vice President (Northern Europe) of the International Association of Engineering Geology (2011–2015), he has some 30 years' experience working on several problematic soils, most notably collapsible and expansive soils. More recently he has focused on sustainable geotechnics and urban near-subsurface, serving on the Technical Committee TC 307 on Sustainable Geotechnics of the International Society of Soil Mechanics and Geotechnical Engineering. Recently he has led European-funded work, ARLI (Alternative Resource of Low Impact) supporting small and medium enterprises to develop new materials from waste streams, supporting the net-zero agenda. Since 2017 he has been Co-Director of the National Buried Infrastructure Facility, where he leads the work on geostructures and works with several industrial partners both in the UK and internationally. He has served on regional, national and international technical committees across both geotechnical engineering and engineering geology, including as a member of the ASCE Geo-institute's Engineering Geology and Site Characterization committee.

Section 5: Design of foundations

Section editor: **Tony O'Brien**

Chapter 51
Introduction to Section 5

doi: 10.1680/icemge.66830.0759

Tony O'Brien Mott MacDonald, UK

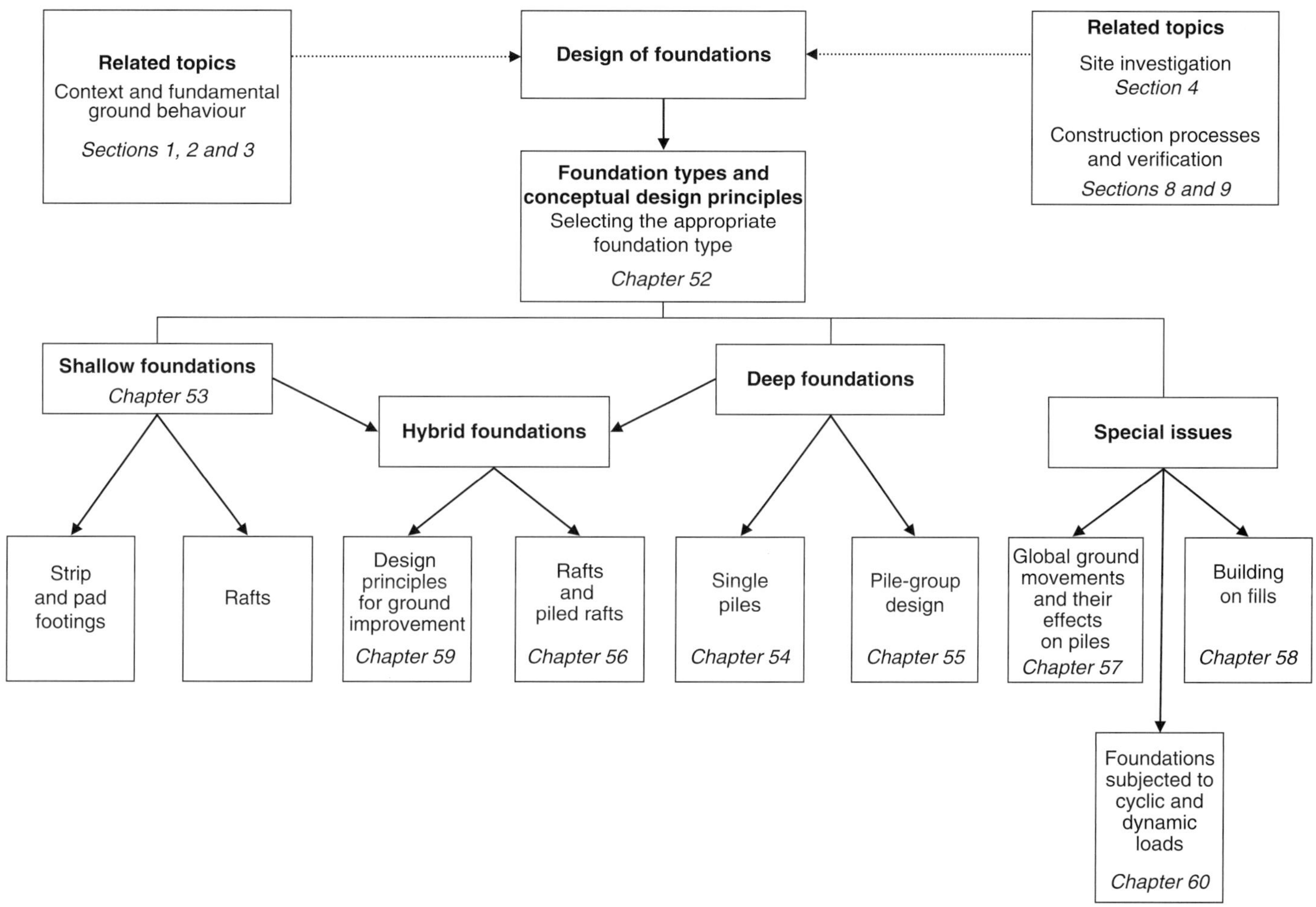

Figure 51.1 Layout of chapters in Section 5

Figure 51.1 outlines the layout and contents of Section 5 *Design of foundations.*

Foundation design is usually divided into two broad categories – shallow and deep – and these categories are also used here. In modern foundation engineering it is helpful to consider a third category: hybrid. Hybrid foundations have their own unique features, although their design usually requires an assessment of both shallow and deep foundation behaviour. Examples of hybrid foundations include deep ground improvement and piled rafts.

The final set of topics, special issues, include building on fills, the influence of global ground movements on deep foundations, and foundations subject to cyclic and dynamic loading, including earthquakes.

In common with other sections of this manual, this section is intended to provide guidance to practising engineers. Given the enormous breadth of the subject and the vast range of ground conditions and structures which a foundation designer may encounter, this section cannot, within the available space, be completely comprehensive. The intent is to outline

- the fundamental principles of good design practice
- the key mechanisms of ground and ground–structure interaction behaviour which need to be considered

- commonly used design methods
- the need to integrate a range of specialist disciplines at different stages of the design process;

and to provide

- references for detailed study
- some brief case histories which highlight key issues.

A quote from Terzaghi (1939) is particularly pertinent (and is still relevant!):

> Whoever expects from soil mechanics a set of simple, hard and fast rules for settlement computations will be deeply disappointed. … The nature of the problem strictly precludes such rules.

Hence, it is essential for the foundation designer to think carefully about the likely deformation and failure mechanisms which may occur both during and after construction, then compile a checklist of questions which need to be considered and answered.

There have been enormous developments in foundation engineering during the last few decades. Despite these developments, failures still regularly occur, and there are also numerous examples of grossly over-conservative design and poor construction practice. There are several commercial and technical factors that can cause these problems. Technically, the following are important.

- Recognise that foundation engineering is a 'process', and that success depends on a series of interlinked activities during *both* design and construction.
- Have a coherent approach to ground risk management; the 'geotechnical triangle' is a valuable framework in this context. An understanding of the site's history and its groundwater regime are critically important.
- Have good communication across different design/construction teams.
- Develop a good understanding of ground–structure interaction. It is vitally important for geotechnical and structural engineers to have early two-way discussions, so that the overall behaviour of the proposed works are understood. In particular, the best opportunity for economic foundation design is to set realistic (rather than arbitrary and usually over-conservative) limits on foundation movement.
- Have a good awareness of relevant case histories of past foundation performance, and to carefully assess the relevant parameters for assessing stability – and in particular, foundation deformation. Sophisticated analysis is not a substitute for a proper selection of design parameters, based on well-designed and supervised ground investigations.

Chapter 9 *Foundation design decisions* provides an introduction to these themes, and these are developed in more detail throughout Section 5.

Although it is a simplification, it is fair to state that routine ground investigation and analysis methods can lead to

- over-conservative design of foundations in heavily overconsolidated soils, such as stiff clays
- unsafe foundation design in normally and lightly overconsolidated soils, such as soft clays.

Modern developments in ground investigation and geotechnical analysis can avoid these problems, although these modern techniques are still under-used across the civil engineering industry. The chapters in Section 5 highlight some of the developments that are mature enough to be used more routinely. It is hoped that Section 5 will stimulate an improvement in foundation design practice.

The second edition

The topics covered and overall layout remain as for the first edition; however, all the content has been reviewed and updated where necessary.

The issues discussed above for the first edition remain as important as they were in 2012 and are covered throughout Section 5. Some topics and issues which were introduced in the first edition are now given greater prominence, including

- sustainability and carbon reduction
- the need to avoid over-conservative design, which will enable carbon reduction.

An unfortunate trend is that limit state design codes, such as Eurocode 7, have not in general stimulated greater investment in ground investigations. For example, it was hoped that there would be greater emphasis on obtaining higher-quality data on ground stiffness, especially small strain stiffness data.

The greater focus on serviceability (movement) checks and widespread use (more specifically misuse) of advanced finite element methods, combined with a lack of high-quality ground investigation data, can often result in over-conservative foundation design. This will not help the industry achieve its targets on carbon reduction.

It is hoped that this unfortunate trend will be reversed. To support this, Chapter 52 *Foundation types and conceptual design principles* has a revised section on parameter selection, including small strain stiffness. Chapter 53 *Shallow foundations* includes a new section on bearing capacity, since VHM methods should now be considered mature enough for general use; these methods can support the design of shallow foundations under complex load combinations. It is hoped that more practitioners will consider the use of piled rafts, addressed in Chapter 56 *Rafts and piled rafts*, which has been revamped to highlight simplified design approaches for different piled raft types.

There have been important developments in deep ground improvement. Chapter 59 *Design principles for ground improvement* describes recent advances, in particular for deep soil mixing and the characterisation of soilcrete strength and stiffness. Finally, Chapter 60 *Foundations subjected to cyclic and dynamic loads* has been rewritten to reflect lessons learned from the many developments in earthquake engineering, foundations for renewables (including offshore wind) and a new section on geotechnics associated with high-speed rail.

Reference

Terzaghi K (1939) Soil mechanics: a new chapter in engineering science. *Proceedings of the Institution of Civil Engineers* **7**: 106–141.

doi: 10.1680/icemge.66830.0763

CONTENTS

Chapter 52
Foundation types and conceptual design principles

Tony O'Brien Mott MacDonald, UK
Rob Talby Mott MacDonald, UK
Alan Willoner Mott MacDonald, UK
Stephanie Markvukaj Mott MacDonald, UK

The main types of foundation include: shallow foundations (pad, strip, raft); deep foundations (piles, caissons, barrettes) and hybrid foundations (deep ground improvement, piled rafts). Foundation engineering requires a broad range of skills, from an appreciation of geology and hydrogeology to structural engineering. The performance of foundations is dependent not only on how they are designed, but also on how they are constructed. The overall design process needs to be well managed, to ensure that there are good communications between different design teams and between design and construction. To provide a framework for selecting the most appropriate type of foundation, a useful mnemonic is the five Ss:

Soil, Structure, Site, Safety, Sustainability

The magnitude of allowable foundation movement is a key factor in determining the type, size and cost of foundations, and this chapter provides guidance on routine limits. Parameter selection is a common pitfall, and the critical information requirements are described.

52.1. Introduction

This chapter outlines the types of foundation that are commonly used, together with the key factors which need to be considered before selecting a particular foundation type.

Foundation engineering requires a working knowledge of

- geotechnics (including both engineering geology and geotechnical engineering)
- construction methods
- ground–structure interactions.

Professor Peck (1962) outlined the three areas of knowledge that are needed in geotechnics as

(*a*) a knowledge of precedents (i.e. a knowledge of case histories)

(*b*) a working knowledge of geology

(*c*) familiarity with soil mechanics.

In the vast majority of cases, a foundation design is developed on the basis of an appreciation of the site, any particular site constraints, the site geology and the nature of the proposed structure. The role of analysis is then to check that a proposed foundation design will be acceptable. Although analysis is important, it is just one part of the overall design process.

Many young civil engineers will have some knowledge of (*c*) above, but it is important that they endeavour to develop their knowledge of (*a*) and (*b*). A knowledge of geology (and hydrogeology) is critically important, since it enables many of the risks associated with foundation engineering at a particular site to be assessed. Simplifying assumptions have to be made before any analysis can be carried out (this includes sophisticated computer modelling). The site geology and hydrogeology have to be considered and understood as early as possible in the project. Discussions should be held with an experienced geologist. Relevant technical literature and, particularly, local case histories should be reviewed. Once this has been done then the appropriateness, or not, of certain assumptions inherent in a particular method of calculation can be judged, and the likely errors associated with the predictions from calculations can be assessed. A knowledge of case histories is probably the most important of the three attributes stated by Peck, since this enables the engineer to have an understanding of

- what has worked, or not, in the past
- the consequences of particular construction activities for subsequent performance
- past performance, against which the reliability of different methods of analysis can be compared
- when a proposed activity is going beyond what has been attempted before (and going beyond the empirical database) – this will then require a special effort and expert advice in order to develop a safe, buildable and economical design.

A foundation is a structural member and supports a superstructure. Therefore, an awareness of the sources and nature of structural loads and the structure's tolerance of foundation movements and an understanding of ground–structure interactions is also needed. Finally, the foundations must be built economically and safely. Hence, a designer needs to have an appreciation of construction methods and equipment in order to develop a design that is practical and safe to build.

This list of attributes, which a foundation designer needs to possess, is rather intimidating. The author has not yet come across the god-like creature who possesses a perfect knowledge of all these topics. Therefore, first and foremost, the foundation designer must be prepared to ask questions and discuss these with other design and construction specialists (e.g. geologists, structural and material engineers, specialist subcontractors). Good foundation design is usually the result of a multidisciplinary team effort, where two-way communications across the team occurs regularly and frequently.

52.2. Foundation types

There are two generic types of foundation: shallow and deep. Foundations developed on improved ground can be considered to be a hybrid of shallow and deep foundations, although ground improvement design requires additional considerations. The advantages and disadvantages of different foundation types are outlined in **Table 52.1**.

Foundation type	Advantages	Disadvantages
Pad/strip footing	(1) Simple to construct (2) Cheap and quick	(1) Competent soils,[a] need to be reasonably close to ground surface (typically less than 3–4 m) (2) If the groundwater table is above the base of the footing and there are permeable soils, then barriers or dewatering are needed during construction (3) Limited application if applied loads include large horizontal forces or overturning moments (4) Vulnerable to large differential settlements if applied loads vary significantly across structure or if ground conditions are heterogeneous (5) Interaction between adjacent footings, if closely spaced?
Raft foundations	(1) Simple to construct (2) Specialist equipment and subcontractors not required (3) If sufficient rigidity, then potential for differential settlement is significantly reduced (4) Can 'bridge' across local soft pockets of soil	(1) As noted above (2) As noted above (3) Large area of excavation required, hence large amounts of spoil generated (particular problem if excavated soils are contaminated) (4) Depth of influence of significant increases in stress will be considerable. Will it affect underground infrastructure? Are weak layers present at depth? (5) Raft deformation needs to be carefully checked. Is movement acceptably small?
Shallow foundations generally	(1) Simplicity (2) Cheap if appropriate conditions	(1) Inappropriate if site affected by 'global' ground movements, that may result in unacceptable differential movements[b] (2) Excessive depth may be needed to locate foundation below seasonal changes in moisture content or frost penetration[c] or, if close to rivers, potential scour depth (3) Inappropriate if desired footing depth would lead to undermining of existing foundations, utilities, roads, canals, railways, etc.
Piled foundations	(1) Large variety of pile types, diameters, depths are possible, so very large applied loads can be resisted (2) If near-surface soils are weak or of variable thickness then pile lengths can be readily modified to suit (3) If near-surface soils are contaminated then exposure can be minimised by appropriate choice of pile type (4) If site boundaries limit width of shallow foundations then piled foundations may be more appropriate (5) Horizontal and moment loads can be resisted (although a pile group may be more appropriate, rather than a single pile)	(1) Can be relatively expensive (2) Construction will require specialist contractor input, hence construction sequence may become complex (3) Piling construction can be adversely affected by local variations in ground and groundwater conditions; technical and commercial implications? (4) Pile capacity can be significantly affected by construction method and equipment and timing of key activities (5) Costs and time to carry out pile testing (6) Pile construction may adversely impact adjacent structures, utilities, etc.
Caissons/ shafts/ barrettes	(1) Can provide extremely high-capacity foundations for 'special' structures (2) Can provide relatively high stiffness/capacity to resist large horizontal loads or overturning moments (3) Shafts, in appropriate circumstances, can provide high-capacity foundations in sites with limited plan area or headroom	(1) May be complex or expensive to construct (2) Design and construction require expert input; these skills are usually in short supply (3) Verification of design assumptions may be challenging (4) Deformation behaviour requires careful assessment, may not be amenable to routine analysis methods

Table 52.1 Advantages and disadvantages of different foundation types (continued on next page)

Foundation type	Advantages	Disadvantages
Deep ground improvement	(1) Can create cost-effective foundations in poor ground conditions (2) Wide variety of techniques are available, which can allow a broad range of ground conditions to be improved (3) Techniques can optimise the behaviour of the in situ ground, hence is a 'sustainable' solution, with a relatively low carbon footprint	(1) The wide range of techniques can be difficult for non-specialists to understand, hence selection of appropriate method can be challenging (2) Effectiveness of ground improvement can be very sensitive to local changes in ground conditions and to changes in 'workmanship', installation method and equipment (3) A significant effort is needed to verify the effectiveness of ground improvement; this can often be underestimated (in terms of cost/time and complexity of interpretation)
Piled rafts	(1) More economic than conventionally designed pile groups (2) Since fewer piles to install, quicker to construct (3) Appropriately located piles, below large column loads, can significantly reduce structural forces in raft, hence raft can be thinner, with less steel reinforcement	(1) Complex ground–structure interaction analysis may be required; expert input will be required (2) Due to (1), more sophisticated ground investigation and interpretation of ground conditions may be required (3) Raft behaviour is important. Hence, as for conventional rafts, competent soils needed at reasonable depth below ground surface (4) Due to (3), may not be appropriate if adjacent construction activities lead to 'undermining' of raft

[a] The definition of 'competent' soils will depend upon the foundation loads and allowable movements. Typically, competent soils include firm to stiff clays, becoming very stiff at depth, or medium dense to very dense sands.
[b] Global ground movements may develop due to a wide range of causes which are independent of the proposed foundation loads; refer to Chapter 53 *Shallow foundations*. For example, the presence of old mine workings, solution features, adjacent slope movements, etc. may lend to 'global' ground movements.
[c] Minimum depths for frost protection and seasonal moisture content (pore water pressure) change are given in local building regulations and codes; also refer to Chapter 53 *Shallow foundations*. Seasonal moisture content changes are mainly affected by the type of vegetation present on site; if there are high-water-demand trees on high-plasticity clays then the effects can be significant and extensive (refer to NHBC Standards, 2023, and Chapter 53 *Shallow foundations*).

Table 52.1 Continued

(*a*) *Shallow foundations*: These typically comprise pad, strip or raft foundations; refer to **Figure 52.1**. Pad foundations (**Figure 52.1(a)**) may be square, circular or rectangular. A pad foundation usually only supports one or two columns. Strip foundations (**Figure 52.1(b)**) are commonly used to provide support to a load-bearing wall or for several closely spaced columns. The length of a strip foundation is much greater than its width. A raft foundation (**Figure 52.1(c)**) could support the entire structure or a substantial part of it. A 'compensated' raft

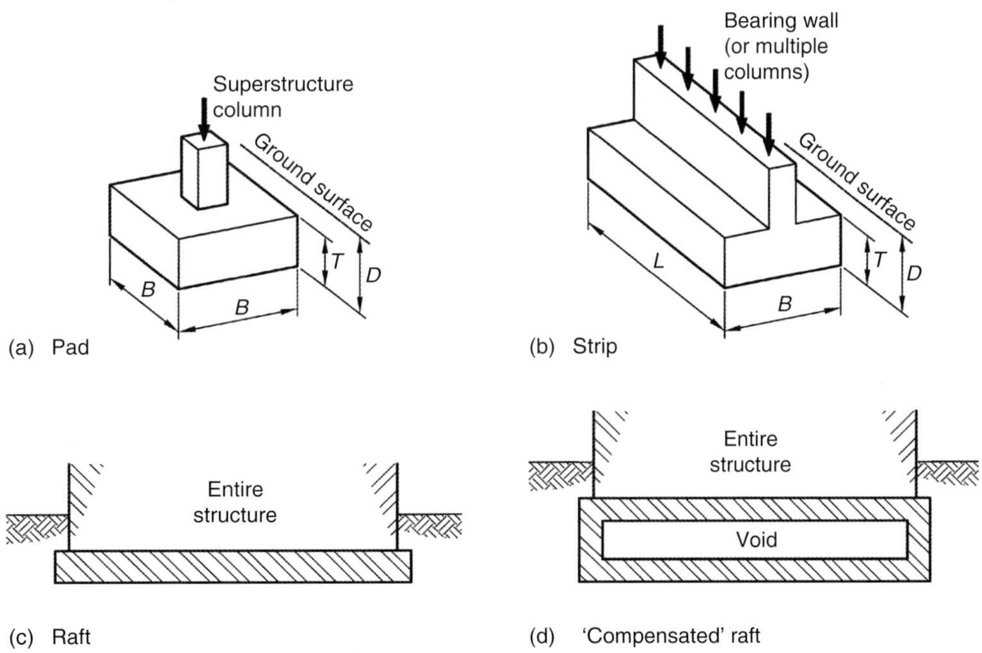

Figure 52.1 Types of shallow foundations

(also known as a 'buoyant' raft) is a special type of raft. It includes a void, thereby reducing the 'net' increase in bearing pressure (Section 52.5) on the ground below the foundation. This will increase the factor of safety against bearing capacity failure and reduce foundation settlement, compared with a conventional raft. The structural form of shallow foundations can vary, but pad and strip foundations could comprise either mass concrete or reinforced concrete, whereas raft foundations usually comprise reinforced concrete. The soil resistance to the loads applied by the structure is predominantly developed by the near-surface soil below the base of the foundation. Shallow foundations are usually constructed using simple excavation plant and general labour. Tomlinson (1987) gives useful advice on routine foundation design for low-rise buildings.

(*b*) *Deep foundations*: For the majority of routine structures, deep foundations comprise piles; refer to **Figure 52.2**. There are numerous different types of pile, discussed in Chapter 54 *Single piles*, although piles are usually classified as either driven or bored. Pile construction is a specialist activity and should only be carried out by contractors who have the appropriate equipment and experience.

The layout of pile foundations can vary substantially from a single pile below a load-bearing column, to a large number of piles in a group supporting the entire structure, through a pile cap (which transfers load from the structure directly into the piles). For a conventionally designed pile group, it is assumed that the piles carry the entire load.

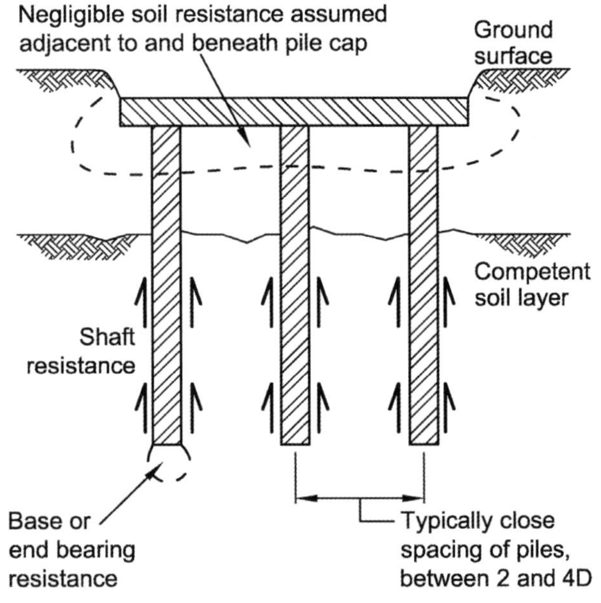

Figure 52.2 Deep foundation, conventional pile group

Deep foundations also include barrettes, shafts or caissons, which tend to be bespoke foundations for special circumstances. Caissons, barrettes and shafts typically have a large diameter or cross-sectional area relative to their depth in comparison with conventional piles. Barrettes are rectangular in plan, and are installed using diaphragm-wall construction techniques. Both shafts and caissons are circular in plan; however, they are usually constructed by different methods. Shafts could be hand-dug or excavated using back-actors, with excavation support installed top-down, using precast concrete rings or cast in situ concrete lining. Caissons are usually jacked or sunk into place using combinations of kentledge, excavation and perimeter lubrication (using bentonite or polymers). For both piled foundations and deep shafts the load transfer to the ground is through shear along the vertical sides of the shaft or pile and by end-bearing resistance.

Piled rafts are a hybrid type of foundation, with the structural load being transferred to the soil through the raft (as for a shallow foundation) and the piles (as for a deep foundation). The principal difference between a piled raft and a conventional pile group is that both the piles and raft provide load-bearing resistance. Chapter 56 *Rafts and piled rafts* provides clear definitions of the different types of piled raft and associated design concepts, which can facilitate practical design. Compared with a conventional pile group, a piled raft has a relatively small number of more widely spaced piles (**Figure 52.3**). Piled-raft design requires specialist input.

(*c*) *Foundations on improved ground*: In general, the purpose of ground improvement is to increase the strength and stiffness of the ground below a proposed structure,

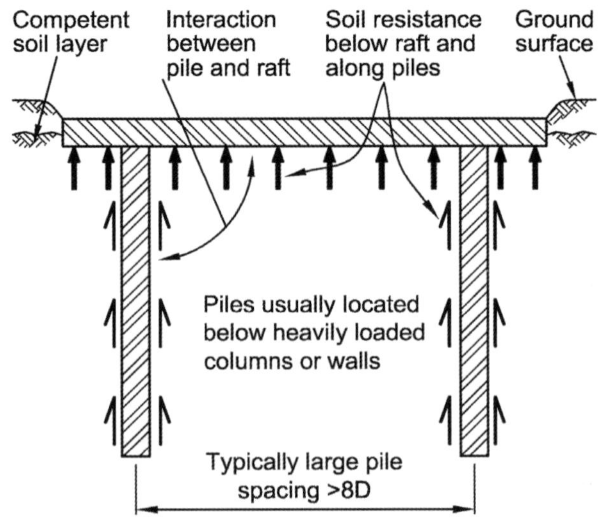

Figure 52.3 Hybrid foundation, piled raft

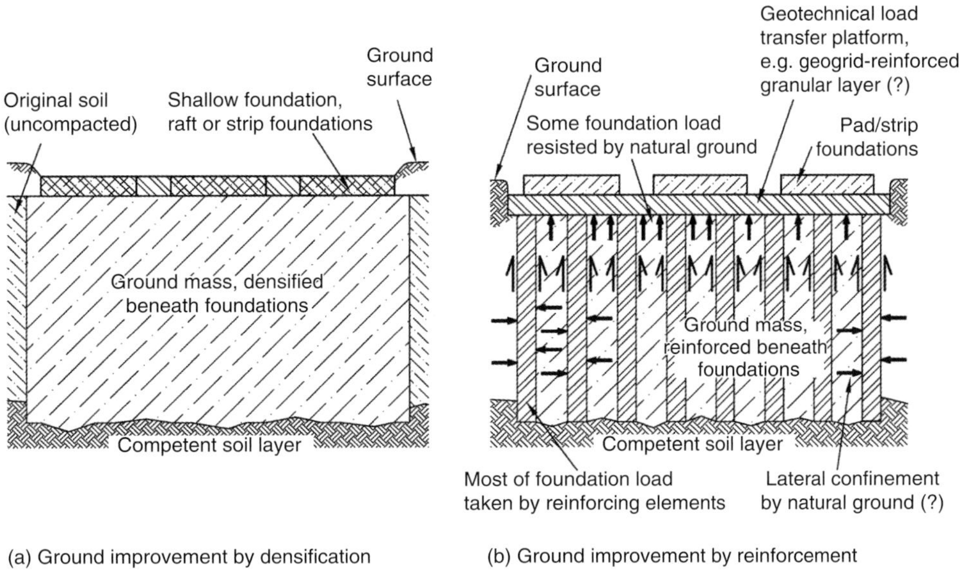

Figure 52.4 Hybrid foundation, deep ground improvement

and then shallow structural foundations are constructed on the surface of the improved ground (**Figure 52.4**). Below the structural foundations, a granular layer (sometimes reinforced with geogrids) is often used to transfer loads into the reinforcing elements (**Figure 52.4**). The mechanism of ground improvement is usually either

- reinforcement – usually for clays, silts or made ground; a large number of stiff elements are introduced into the compressible layer
- densification – usually for sands or gravels; a compaction process forces a rearrangement of the soil particles into a denser state.

Reinforcement: Techniques such as vibro stone columns or deep soil mixing facilitate the replacement of soft compressible materials by stronger materials introduced into the ground using specialist equipment. These stronger elements carry most of the foundation load. However, the remaining soft compressible materials will also carry some of the foundation loads, hence the term 'reinforcement'. It is important to note that the reinforcing elements usually have negligible tensile and bending resistance. Sometimes reinforcing elements such as stone columns are called 'stone piles'; however, this is quite incorrect and misleading. Foundations built upon improved ground will behave in a quite different manner to piled foundations (**Figure 52.5**).

Densification: Soils such as loose sand with a low silt or clay content can be compacted into a denser state. Once densified the sand will be a more competent stratum and be able to carry the foundation loads directly. Hence, conventional shallow foundations can be placed on the densified layer. Densification may be carried out using a wide range of specialist equipment. The key issue for densification projects is usually the relative

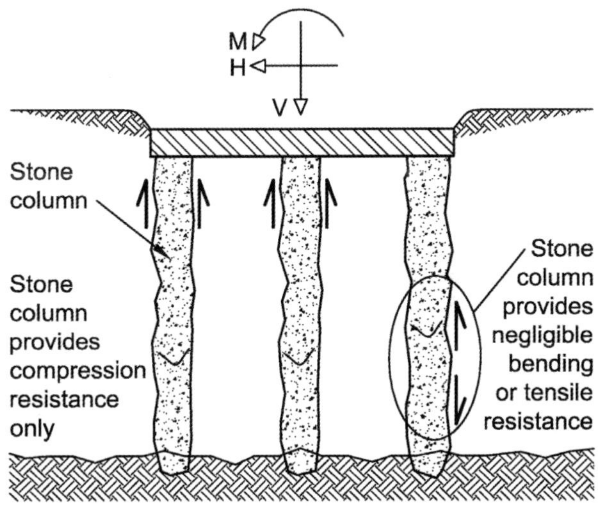

Figure 52.5 Ground improvement as a reinforcement method has limited bending and tensile resistance

density that can be achieved compared with the value which is needed to ensure that foundation settlement is acceptable.

The success of ground improvement (both by reinforcement or densification) is sensitive to the nature of the soil profile and the soil macrofabric. For example, densification by vibrocompaction may be straightforward if the sand is homogeneous and has a low silt or clay content. Compaction will be much more difficult and often less effective if there are numerous bands of silt or clay in the sand. These soil features can be easily missed by conventional boreholes and sampling at discrete depth intervals. Continuous sampling or profiling techniques (such as static cone penetration testing, or CPT) are usually very beneficial when designing ground improvement. Similar considerations can apply to reinforcement techniques. Quality control and field testing to verify the effectiveness of ground improvement are a significant component of any ground improvement project.

52.3. Foundation selection: conceptual design principles

52.3.1 Information requirements and the foundation design process

Figure 52.6 outlines the typical information requirements, the design deliverables (reports, etc.) and the project phases that are followed to design and construct a foundation. During the early stages the key questions are as follows.

- *Site geology, hydrogeology and history*: What are the geological age and the depositional environment of the main deposits under the site? What subsequent changes may have occurred (e.g. landslides, man's influence)?
- *What will be built?* Form of structure, likely construction methods, main site constraints, applied loads, principal constraints (e.g. allowable structural movements)?
- *Engineering knowledge*: What existing knowledge do we have? For example, relevant technical literature, good case histories, local customs and experience or expertise.

This collation of available data into a desk study report is a crucially important and cost-effective means of managing ground risks on a project. Sole reliance on boreholes in the absence of a proper desk study is potentially very dangerous. The main ground (and groundwater/gas) hazards should be identified from the desk study, and the key features of ground behaviour should then be assessed. This will inform the design of intrusive ground investigation (refer to Chapters 40 *The ground as a hazard*, 41 *Man-made hazards and obstructions*, 43 *Preliminary studies* and 44 *Planning, procurement and management*) and subsequent design decisions. It is often the case that designers 'inherit' ground investigation reports from previous or adjacent projects and studies. It is very important that the scope, adequacy and reliability of this information are assessed in the context of the current project requirements. If the scheme or client requirements change – for example, structure has been relocated, or more onerous loading or allowable movement criteria are required – or there are doubts about ground investigation data reliability, then an additional ground investigation may be needed (perhaps using more sophisticated investigation techniques). There can be confusion about geotechnical reporting requirements, and these may change depending on particular code or client requirements. For both the desk study and reporting of intrusive ground investigations it is important to differentiate between *factual* data and *interpreted* data. Interpretation of ground properties and of the groundwater conditions involves several stages, as noted in **Figure 52.6**. Initially a summary of reliable and representative data is required (this requires interpretation), and then depending on the specific design situation (and analysis methods) different parameters may be appropriate for use in design. These should be selected by the designer and reported in a design report.

Foundation design is an iterative process. Once relevant information has been obtained for the site, the key data should be summarised on a plan and vertical cross-sections drawn through *and* beyond the site boundaries. Drawn to scale, with levels referenced to an appropriate survey datum, as a minimum, the following features should be shown

- the previous and proposed ground surfaces
- any previous and neighbouring structures and underground services
- ground profile, strata boundaries (including selected ground properties from testing), the borehole log(s) and offset(s) from the section
- groundwater levels
- outline of the proposed structure and proposed locations for foundations
- the key site constraints and hazards (list as bullet points).

Based on a knowledge of the ground profile, the type of structure and key site constraints, an experienced engineer should be able to identify appropriate conceptual designs for the foundations. Analyses would then be carried out to check that this concept had an acceptable level of stability and that foundation movements were acceptable. If these are not acceptable, then the concept would be modified (say, pile foundations would be deepened or shallow foundations would be widened) and rechecked. It is common to consider a number of different options. Once technically acceptable foundation options are identified, then the relative safety, performance, cost, buildability, risk and sustainability of each option can be assessed. Throughout this conceptual design process, the way in which uncertainties can be best managed in order to minimise risks needs to be considered, together with the buildability of the concept (in essence, 'buildability' is a consideration of how simple, quick and safe the proposed design will be to construct). Once an acceptable foundation design has been identified, more detailed analysis is carried out to finalise the design and prepare drawings and specifications for construction. It is important to note that it is much more cost-effective to change or adapt a foundation option at the concept design stage, than at the detailed design stage or during construction. Usually, a change in foundation design has important effects on the superstructure design and construction planning (foundations are often on a project's critical path). The nature of the ground risks will usually inform the development of site-specific specification requirements, which may involve observations by experienced engineers or geologists at critical

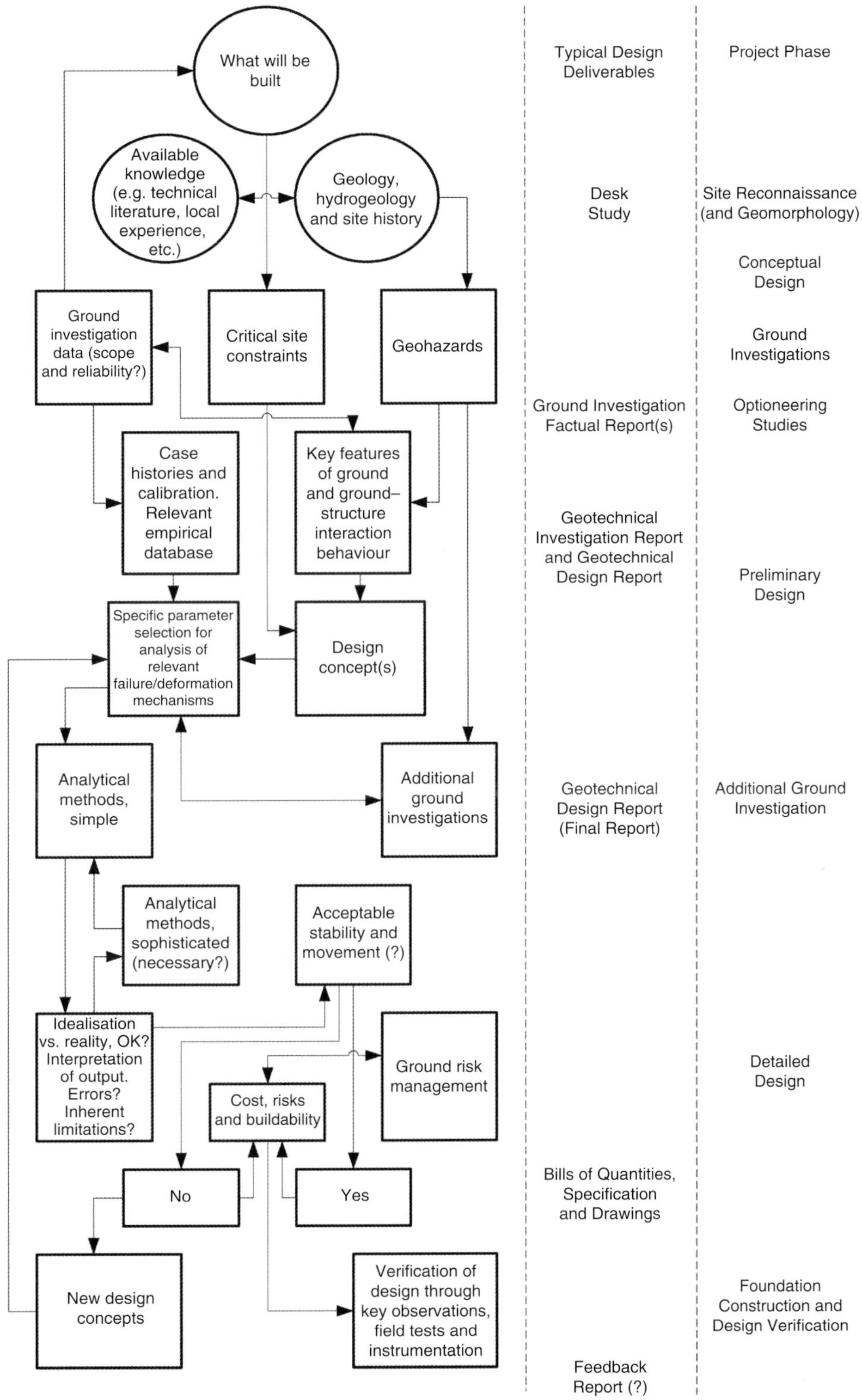

Note: These require the ground properties to be interpreted, which involves several stages: collation of factual data; checking correctness and identifying representative data (i.e. removing anomalously high/low values); using empirical correlations to cross-check relationships between parameters; finally selecting parameters for use in analysis – this final stage has to relate to specific design situations and analysis methods and therefore should be given in the design report.

Figure 52.6 Information requirements and the foundation design process construction stages and require specific field tests to confirm design assumptions.

The foundation design process does not end when design drawings are issued. Given the intrinsic uncertainties associated with variations in ground conditions, and the way in which different construction processes may affect subsequent ground behaviour, it is always important that the designer has the opportunity to verify critical design assumptions. However, with some forms of modern contracts and procurement practices (e.g. design and build), it can be challenging for a designer to get adequate site supervision opportunities. Nevertheless, it is important that this happens.

52.3.2 General considerations

Any foundation design has to include consideration of

- acceptable stability and deformation
- risk management and buildability
- sustainability, costs and programme for construction.

52.3.2.1 Acceptable stability and deformation

The foundations need to carry the foundation loads safely; hence, adequate factors of safety against collapse are required and excessive foundation movement must be prevented. There can be a wide range of different collapse or deformation mechanisms that may need to be considered (**Figure 52.7**). One pitfall can be to focus only on considering collapse or deformation mechanisms for the foundation operating in isolation, solely due to the structural loads imposed by the superstructure. Additional collapse and deformation modes may exist due to: the nature of the soil and groundwater regime; the site's location (e.g. if the site is adjacent to an unstable slope); the site's history (e.g. if it has been affected by mining or quarrying); or due to the nature of the regional hazards (e.g. earthquakes, flooding). The increasingly severe effects of climate change need to be considered – for example, the potential vulnerability to intense rainfall, site flooding and the effects of scour, especially for soils vulnerable to erosion and loss of fines (e.g. silts, fine sands, made ground containing ash deposits). In many cases detailed analysis is inappropriate, and eliminating the hazard (e.g. relocating foundations, stabilising the ground mass, preventing erosion) is the best approach. 'Global' ground movements across a site need to be considered. **Figure 52.7(g)** illustrates a common situation where a sloping site requires cut and fill earthworks to create a level platform in order to build the new structure. Strip foundations form the foundations for the new structure. In this situation, even if the earthworks are well engineered and properly constructed, significant global ground movements may subsequently develop (possibly resulting in unacceptable differential movement), because of the following.

(*a*) Beneath the zone of fill there will be deep-seated increases in the vertical stress, due to the weight of the fill, in the underlying natural ground; this will cause settlement (in addition to any settlement of the fill itself and settlement due to the bearing pressure on the strip foundation).

(*b*) Similar to (*a*), beneath the zone of excavation there will be deep-seated decreases in the vertical stress in the underlying natural ground, which will cause heave and swelling. This may be larger than any settlement induced by the foundation bearing pressure below the strip foundations, and may lead to a *net upward* movement of the foundations.

Any calculations of foundation movement would need to consider the stress changes and associated ground movements induced by the cut and fill activities, as well as those induced immediately below the strip footings. Potentially the foundations underlain by the area of fill could suffer substantial settlement, and the foundations in the area of excavation could suffer substantial upward movement. Consequently significant differential movement could develop across the interface between the cut and fill areas. The magnitude of the movements depends upon: the compressibility and stiffness of the natural ground; the magnitude and areal extent of the cut and fill required; the plan dimensions of the shallow foundations; and the magnitude of the foundation bearing pressure. If differential movements were excessive, then consideration could be given to: movement joints in the structure; relocation of the structure so it was either solely in an area of fill or solely in a cut area; use of a raft instead of isolated pad or strip footings; or piling. The most sustainable and economic solution would often be to undertake the earthworks for the cut and fill in a preliminary contract, allowing time for some of the global ground movements to occur. If piles are chosen the pile design needs to consider the negative skin friction below the fill area and the heave-induced tension below the excavation areas. Potential causes of movement, independent of foundation loads, associated with a range of different soils, rocks and fills are also discussed in Section 3 *Problematic soils and their issues* and Chapters 40 *The ground as a hazard* and 41 *Man-made hazards and obstructions*.

Throughout the design life of the structure, foundation deformations need to be acceptably small, and allowable deformation will usually be the governing design criterion. Historically, there has been little confidence in the reliability of deformation calculations. **Figure 52.8** provides a comparison between computed and measured settlements for several different foundations in a range of ground conditions. As shown, the actual settlement can be quite different to the predicted value. Hence, codes of practice used to recommend large factors of safety in order to limit foundation deformation. More recently, limit state codes of practice have been published (e.g. Eurocode, EC7), and typically their partial factors of safety against various collapse modes (e.g. ultimate limit state) are

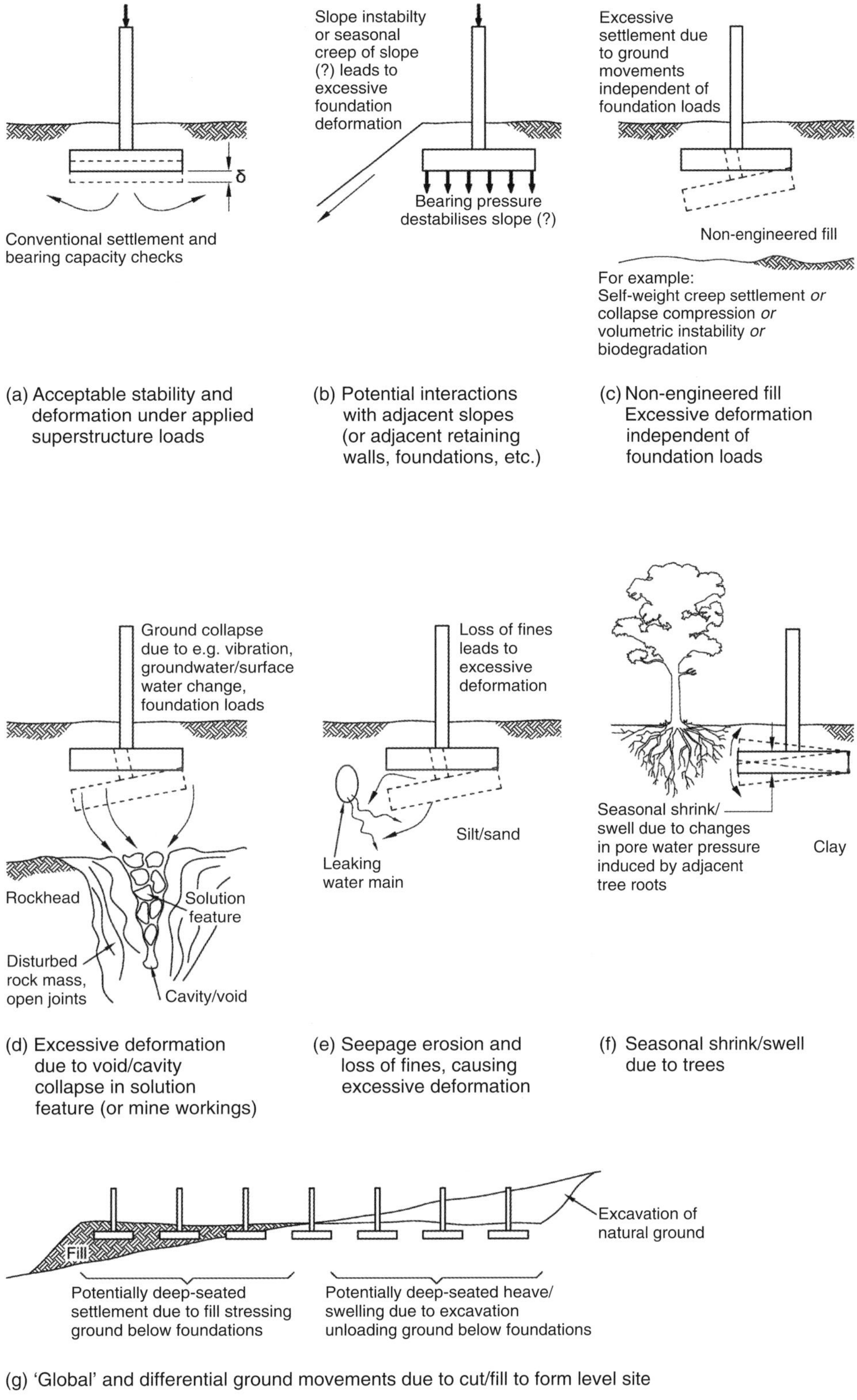

Note:
This is not intended to be a comprehensive summary of all potential deformation and collapse modes. Important to recognise that excessive deformation/collapse may occur due to a variety of causes, some of which are independent of the applied foundation loading.

Figure 52.7 Examples of various deformation and collapse modes

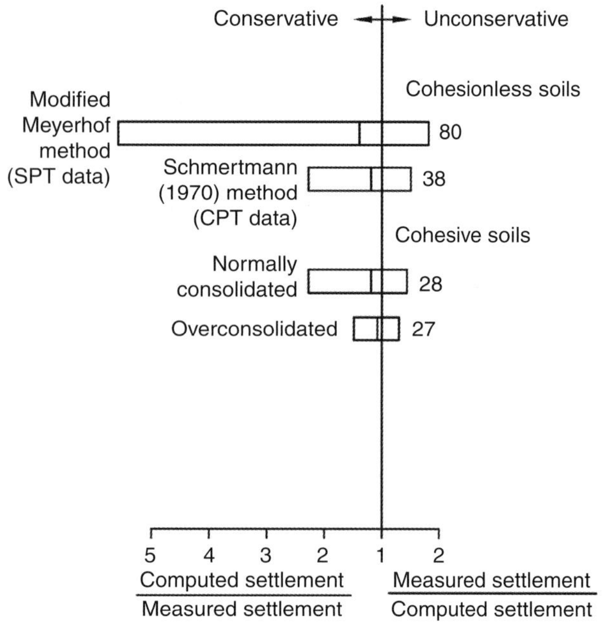

Figure 52.8 Reliability of deformation calculations. The figure gives a comparison between computed and measured settlements of spread footings. Each bar represents the 90% confidence interval (i.e. 90% of the settlement predictions will be within this range). The line in the middle of each bar represents the average prediction and the number to the right indicates the number of data points used to evaluate each method (Based on data from Burland and Burbidge, 1984; Butler, 1975; Schmertmann *et al.*, 1978; Wahls, 1981)

lower than those previously recommended – see **Table 52.2**. **Table 52.3** summarises the site-specific issues that should be considered before selecting a factor of safety. Factors of safety given in codes are minimum values, and there may be a need for higher factors of safety if there are significant uncertainties.

Failure mode	Foundation type	Historic practice: typical overall factor of safety	Limit state codes, e.g. EC7: partial factor on soil strength		
Base sliding	Shallow	1.5–2.0	tan ϕ	c'	S_u
Bearing capacity	Shallow	2.0–3.0	1.25	1.25	1.4
			1.25	1.25	1.4
Overall stability	Shallow	1.3–1.5	1.25	1.25	1.4
Shaft resistance	Deep	1.5–2.5	Pile resistance		
			Shaft		Base
			Varies depending on pile type and test regime		
Base resistance	Deep	2.0–3.5			

Table 52.2 Typical factors of safety for foundation design

Site-specific factors	Typical issues
Intrinsic complexity of geology and hydrogeology	Soil profile homogeneous? Artesian or sub-artesian groundwater pressures? Challenging soil behaviour (e.g. sensitive 'quick' clays)?
Onerous design criteria	Allowable settlement very low? Is horizontal displacement or rotation critical?
Consequences of failure	Potential loss of life? Major economic loss?
Nature of applied loads	Large and frequent cyclic loads? Occasional wind load? Or accidental impact load?
Availability of local experience and case history data	What has been published? What is known by local engineers from past experience?
Scope and reliability of ground investigation	Appropriate tests and locations investigated? Well supervised? Factual data checked?
Level of geotechnical risk	Ductile or brittle failure mechanism? Complexity of soil–structure interaction?
Construction uncertainties	Can design assumptions be verified? Level and quality of site supervision? Foundation behaviour intrinsically sensitive to workmanship?

Table 52.3 Some considerations for selecting a factor of safety

In favourable circumstances limit state codes can facilitate more economic designs, particularly if the ground investigations include high-quality measurements of ground stiffness at small strain (refer to Section 52.6). When using limit state codes and checking serviceability limit states, there is more onus on a designer to exercise judgement, both concerning the reliability of deformation calculations and the appropriate magnitude of 'allowable' deformation (refer to Section 52.4). For example, since the 1980s there has been a considerable improvement in understanding of the stiffness characteristics of overconsolidated clays subject to monotonic increases in stress. In contrast, there remains significant uncertainty about the deformation behaviour of soils subject to cyclically varying loads.

The overriding requirement for assessing deformation behaviour is an understanding of past observations of foundation performance. To be of practical benefit, this experience needs to be based on a careful analysis of foundation performance of similar foundation types under comparable loading, within similar geology and hydrogeology. Case history data can then be used to calibrate the proposed calculation model, and for practical applications the model needs to be as simple as possible while still capturing the key features of the soil–foundation interaction. Great care is required if extrapolation is necessary beyond the existing database of knowledge (for example, due to the magnitude of the loading, scale or complexity of the foundations). In these circumstances, expert advice, special investigations, field trials and so on are likely to be required. When utilising a database of past performance and an associated calculation model, it is vitally important to verify that the geology is relevant to the site under consideration. For example, Burland and Burbidge (1984) developed an empirical method for calculating the settlement of silica sands. It would be quite dangerous to use this method to calculate the settlement of sands with a different mineralogy, such as calcareous or micaceous sands.

52.3.2.2 Risk management

Risks, during and after foundation construction, need to be assessed and managed. There can be a wide range of different types of risk

- health and safety
- technical issues (e.g. ground, groundwater and foundation behaviour)
- performance of foundation contractor
- environmental
- commercial (costs and programme)
- design interfaces (communication).

Because of the broad range of issues that may need to be assessed, it can be beneficial for the project risks to be assessed formally in a risk workshop, attended by key members of the project team including the geotechnical and structural engineers. In addition, at an early stage of the project, it is helpful to carry out 'Peer Assist' (Powderham, 2010; SCOSS, 2009). Peer Assist involves one or two senior engineers carrying out a proactive independent review of the available project data (including desk study and ground investigation data), the project brief and constraints, and assessing the foundation engineering options identified by the project designers. A review of the risk register and challenging the assumptions and design methods used by the designers are important aspects of Peer Assist. Peer Assists are best carried out in a workshop environment with the project design team. The duration of a Peer Assist can vary from half an hour for simple projects to several days for complex projects.

The geotechnical triangle is a valuable framework for ground risk management. The geotechnical triangle is introduced in Chapter 4 *The geotechnical triangle* and practical applications are discussed in Chapter 9 *Foundation design decisions*. All aspects of the triangle must be considered. If one or more elements of the triangle are forgotten or poorly implemented, then foundation failures are much more likely. The identification of ground risks requires a deep understanding of ground behaviour (in the context of the site setting, history, geology and hydrogeology) and of the envisaged construction process.

Uncertainty is an intrinsic part of foundation engineering. Hence, there is a need to be aware of what could potentially go wrong and to assess the consequences. The most critical aspect of risk management is to identify risk mitigation measures that are practical to implement and verify on site. For the design to be robust the foundation's behaviour should be relatively insensitive to small changes in ground or groundwater conditions. Construction, particularly below ground, is seldom perfect. Therefore, the design should be simple to build and allow for normal construction tolerances and workmanship. The foundation designer will usually need to make several assumptions, and during foundation construction the designer will need to ensure that these key assumptions can be readily verified. This verification process may involve tests, monitoring and observations. The verification process could vary from a simple inspection and observation of the exposed ground at foundation level and verifying that the foundation will be placed on ground that is as strong and stiff as that assumed during design, through to complex instrumented load tests. These observations and tests are an important part of the foundation design process. If a design assumption cannot be readily verified and there are serious consequences if the assumption is not valid, then consideration should be given to changing the design to one which can be more easily verified. Foundation construction below the water table or in a maritime environment can be particularly challenging in this respect. For example, if soft soils are supposed to be dredged below water prior to foundation construction, how can the designer check that all the soft soils have been removed? If soft soils are left behind, instability may result.

52.3.2.3 Sustainability, costs and programme

The preferred foundation solution will usually be the most economical to build, provided that the long-term performance and associated risks are deemed to be acceptable. Sustainability is increasingly important, and as discussed in Chapter 9 *Foundation design decisions*, the embodied carbon for different foundation options should be assessed as part of the foundation selection process. Calculating the foundation costs may be based on assessing the quantities (e.g. volume of concrete for strip footing, or number, depth, etc. of piles) for the permanent works, and using unit rates for each of the main foundation elements (e.g. cost per m^3 for forming strip footing). A major problem with this approach is that it may not allow for

- cost of temporary works (e.g. lowering the groundwater if excavations are required below the water table)
- cost of supporting excavations if movement-sensitive structures or utilities are nearby
- cost of removing obstructions if present (e.g. large boulders in glacial till or old foundations from previous buildings on the site)
- costs of treating or removing contaminated soils, if present
- cost of verification tests (e.g. for ground improvement or piling this can be significant).

Hence, the normal methods used for assessing the costs of civil engineering construction may be quite unreliable when used for underground construction. In addition to assessing cost, the embodied carbon should also be assessed. Transport of materials and equipment to site is often a significant proportion of a foundation's carbon footprint. Reuse of materials (including existing foundations) and waste minimisation should also be considered and this will often benefit from a holistic appraisal of both the temporary and permanent works for foundation construction. To estimate foundation costs reliably it is important to take account of any necessary temporary works and to understand how the site's history and geology may impact upon foundation construction costs. Foundation construction is often on the critical path for a project. Hence, the preferred foundation solution will often be the one which is quickest to construct and has the least effect on other site activities, since overall project costs are often dominated by the overall duration for construction. 'Unforeseen' ground conditions are the most common cause of project delays and cost increases, with 40–50% of all delays due to 'unforseen' ground conditions (TRL, 1994; Chapter 7 *Geotechnical risks and their context for the whole project*). In particular, problems caused by groundwater are often the root cause of geotechnical problems on site. Therefore, reliable sustainability, cost and programme estimates for foundation construction will be heavily dependent on carrying out good project risk management and upon the advice of the geotechnical engineer.

52.3.3 Foundation selection: the five Ss

Table 52.4 provides a series of questions that a foundation engineer should ask themselves, before developing a conceptual design. They can be remembered as the five Ss: soil, structure, site, safety and sustainability. Each will be briefly discussed in turn.

52.3.3.1 Soil (and groundwater)

Once the soil profile and groundwater regime are known, and the local geology and hydrogeology are understood, then the decision between shallow or deep foundations or ground improvement will usually be straightforward. Construction planning will also be heavily dependent on an understanding of the soil profile and groundwater regime. On the basis of geological age, it is often helpful to consider the following four soil types.

(*a*) *Geologically old, pre-glacial deposits:* These are usually heavily overconsolidated 'competent' foundation materials. When they are encountered close to the ground

Soil
- What is soil profile?
- Depth to 'competent' soil
- What is depth to water table? (plus seasonal fluctuations)
- Verification of design assumptions?
- Effect of groundwater regime on foundation construction?
- Local experience of foundation construction and long-term performance?
- Likelihood of 'obstructions'? (natural or manmade)
- Do soils exhibit unusual behaviour, e.g. volumetric instability; highly sensitive?
- Are near-surface soils able to be treated/compacted to improve their engineering behaviour?

Structure
- Nature of structure (structural form, materials)?
- Magnitude of applied loads?
- Will loads vary with time (cyclic loads, large live loads, impact or dynamic loads)?
- Acceptable total and particularly differential settlement?
- Predominantly vertical load?
- Large horizontal, moment or torsion loads?
- Does structure have any special features or brittle finishes?

Site
- Space available for construction?
- Available headroom?
- Evidence of unstable ground in vicinity of site?
- Access for plant?
- Have historical mining/quarrying activities taken place, potential for voids/unstable ground at depth?
- Neighbouring structures and utilities, are they movement/vibration sensitive

Safety
- Foundation stable, short term and long term?
- Does foundation construction cause adverse effects on adjacent area?
- Acceptable risks during foundation construction?
- Is site contaminated due to past/current activities?
- Are near-surface soils sufficiently stable for plant access?
- Will construction involve significant temporary works? Stability issues?
- Will construction involve large fill embankments or large excavations? Stability or ground movement risks? Influence of ground/surface water?

Sustainability
- Reuse on-site materials?
- Can existing foundations (if present) be reused?
- Minimisation of waste due to construction; options?
- Specify low-carbon footprint materials?
- Use foundations as geothermal elements?

Table 52.4 Foundation selection, 'the five Ss'

surface (say within 3–4 m) then shallow foundations can often be utilised.

(*b*) *Glacial deposits:* These soils are often extremely heterogeneous – refer to Chapter 31 *Glacial soils*. Some glacial deposits, such as lodgement tills, can be stiff or very stiff clays or dense or very dense sands and be of very low compressibility. As for (*a*), if they are close to the ground surface then simple shallow foundations will often be appropriate. However, some glacial deposits (such as outwash or melt-out deposits) can be very soft and compressible. Changes can occur rapidly from one soil type to another (within lateral distances of 10–20 m or less). The interpretation of ground investigation data can be challenging, and input from experienced geologists will be important. Large, strong boulders can often be present, which can make piling problematic.

(*c*) *Post-glacial deposits:* This class of material is seldom suitable as a founding material. These are recent (less than 10 000 years old) deposits and often comprise soft compressible clays or peats or very loose sands or silts. Geologically they are usually classified as normally consolidated. However, many of these deposits can be lightly overconsolidated in a soil mechanics sense (i.e. the pseudo pre-consolidation pressure or yield stress is higher than the current vertical effective stress, but foundation bearing pressures may exceed the pre-consolidation pressure). This may mean that shallow foundations could be feasible for lightly loaded buildings, especially if 'compensated' foundations are used (**Figure 52.1**). However, expert advice and special ground investigations (to verify the magnitude of the pre-consolidation pressure profile) are required. More commonly, for lightly loaded structures, ground improvement is utilised. For heavily loaded structures piles are used to take the foundation loads to more competent strata at a depth below the post-glacial deposits.

(*d*) *Made ground and non-engineered fills:* Foundation construction is often carried out in areas that have been affected by previous human activities. These may have involved excavating the shallow natural soils and rocks and backfilling with any readily available materials, including construction arisings, domestic refuse and industrial waste materials. Unless the fill was properly engineered, using acceptable inert materials and well compacted, then it is unlikely to be a suitable founding stratum. Sometimes, lightly loaded foundations can be built on non-engineered fills, although special measures may be needed; foundation engineering in fills is discussed in more detail in Chapter 58 *Building on fills*.

For any of the above scenarios, and particularly for (*a*) and (*b*) when shallow foundations may be used, it will be important to check if subsequent events (either due to human activities or geological processes) have significantly affected the site, such as mining, quarrying or landslips.

Groundwater regime: It is always important to consider the groundwater regime. Its importance cannot be overstated, and is best summed up by the memorable statement from Terzaghi (1939):

> in engineering practice difficulties with soils are almost exclusively due, not to the soils themselves, but to the water contained in their voids. On a planet without any water there would be no need for soil mechanics.

Despite its importance, many ground investigations do not provide reliable data about a site's groundwater regime. No reliance should normally be placed on groundwater strike information during borehole drilling to assess if a particular soil layer is 'dry' or not. Reliable groundwater information requires careful installation of piezometers, and then monitoring the piezometers over a prolonged time period, preferably covering a period over the winter and spring, when groundwater pressures may be at a maximum. In geotechnical practice, the time period for piezometer monitoring is often inadequate. There is a high risk to any foundation engineering activity if there is an inadequate understanding of a site's groundwater regime. In addition to piezometric data, local experience of groundwater problems should be considered and discussions held with an experienced hydrogeologist (to assess if, for example, artesian groundwater conditions are present). Inputs by a hydrogeologist are also important when considering potential changes in the groundwater regime during the design life of a proposed scheme, including the potential effects of climate change.

As noted by Day (2000), 'probably more failures in geotechnical and foundation engineering are either directly or indirectly related to groundwater and moisture migration than any other factor'. Some simple examples in foundation engineering are given below.

- *Shallow foundations:* Forming a shallow excavation above the water table will be relatively quick and cheap, using conventional plant and general site labour. However, in permeable soils, if the water table is above the base of the foundation level then the works will be more complex, expensive and time consuming. The groundwater will usually need to be excluded, by, for example, forming an impermeable cut-off wall, or by lowering the groundwater table by pumping from wells. Both of these activities will often require specialist subcontractors. Groundwater lowering may lead to ground settlement concerns or concerns with disposal of the pumped water (especially if it is contaminated or if there is limited local discharge capacity). Installation of a cut-off wall by driven sheet piling may lead to concerns about noise and vibration. If the groundwater is not properly controlled, then the sides or base of the excavation may become unstable. The bearing capacity and long-term settlement performance would then be adversely affected.
- *Deep foundations:* Bored piles are commonly used in the UK. However, groundwater inflows and unbalanced groundwater pressures can seriously reduce the shaft friction and base resistance of a bored pile. If bored piles are to be drilled through

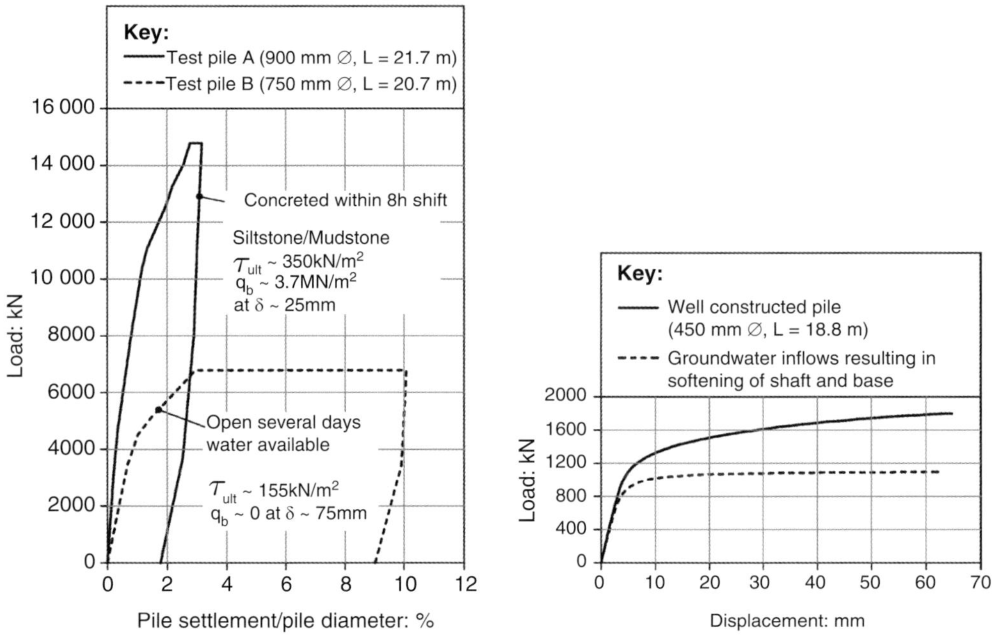

(a) Weak rock

(b) Interbedded stiff clays and dense sands

Figure 52.9 Influence of groundwater on pile capacity for weak rock, stiff clay and dense sand (Lambeth Group geology, south-east England and weak rock, Middle East)

cohesionless soils, weak fractured rocks, low-plasticity silts or clays below the water table, then some combination of temporary casing, water balance or support fluid (such as polymer or bentonite) will usually be required. Pile test data are shown in **Figure 52.9**, which illustrates the dramatic reduction in pile capacity due to unbalanced groundwater pressures in weak rock and in an interbedded sequence of overconsolidated clays and sands. A common alternative to conventional bored piles is continuous flight auger (CFA) piles. The decision to use either conventional bored or CFA piles is usually dependent on the risks associated with groundwater.

52.3.3.2 Structure

The size and height of the structure, its structural form and the layout of structural supports (spacing of columns) and the type, magnitude and direction of the applied loads will have a major impact on the choice of foundations. For example, the modern trend for multistorey building design is to create large internal spaces, which leads to relatively large column loads, and the consequential requirement for piled foundations. In addition to the magnitude and distribution of loads, the way in which the load varies over time is also important. The bulk of foundation design experience has been gained with structures that have live loads of less than 15–30% of the dead load and change infrequently. Foundation problems are often associated with tanks, silos and industrial units, where the ratio of live to dead load is high, and the live loads change frequently. The reliability of deformation predictions when cyclic loading is significant remains poor. These potential risks are exacerbated if large cyclic horizontal or moment loads are also applied.

One of the most important, but frequently overlooked, decisions is agreeing a sensible allowable settlement limit for the structure (with the structural engineer), based on a realistic appraisal of its purpose and vulnerability to damage. All too often over-conservative and arbitrary limits are placed on total settlement, which then leads to unnecessary increases in foundation costs. The important topic of allowable foundation deformation is discussed in more detail in Section 52.4. The main opportunity for a geotechnical engineer to design economic foundations is to agree realistic limits for foundation deformation.

52.3.3.3 Site

In most industrialised countries, such as the UK, foundation construction (for either buildings or rail and road links) often takes place adjacent to existing structures. Old foundations may lie buried across the site. In addition, structures may be located on the site and have to be demolished, prior to the start of new foundation construction. Therefore, space for the construction of new foundations is often severely limited, particularly for sites with previous generations of pile construction, which are often difficult to remove. The congestion of sites by existing piles increases our need to collect data on installed piles to increase the potential for reuse in the future. Space constraints might also mean there is inadequate space to create a large pad or strip footing, or there is limited headroom, which means certain piling or ground improvement equipment cannot be used.

A consequence of development in an urban environment is that there are often neighbouring structures or utilities, which may be sensitive to movement or vibration. If this is a concern it would preclude certain foundation types and methods of construction.

The site's history may mean that it is underlain by unstable ground due to past mining or quarrying activity; a thorough desk study (refer to Chapter 43 *Preliminary studies*) is essential. Because of the above considerations, it is usually important to visit the site and to understand the constraints that may affect foundation selection. When visiting the site it is also important to scrutinise neighbouring areas, since this may provide evidence of past problems due to unstable ground or long-term movements. Following completion of a desk study, the main environmental constraints (such as ecologically sensitive areas or areas with special environmental classifications) should be known. How these may affect foundation construction should also be considered.

52.3.3.4 Safety

Under CDM regulations a designer has a duty to formally assess the risks that may arise during construction (HSE, 2015). Given the uncertainties and the site-specific nature of ground hazards, the foundation designer needs to carefully review all the available information from

- desk studies
- ground investigations
- site visits
- local experience.

This must be considered in the context of

- planned construction activities at the site
- the condition of neighbouring land and structures and so on.

A particular concern when developing many sites across the UK (and other industrialised countries) is that past industrial activity has left a legacy of chemical or biological contamination. This contamination may have impacted the soil, groundwater and ground gas. In this case, foundation engineering will need to be coordinated and integrated into the strategy for minimising risks due to contaminated land. Often this will affect the choice of foundation type, depending on whether there will be large-scale removal and treatment of contaminated near-surface soils or if the site is to be capped off and the contaminated soils left in situ. For example, the use of a displacement piling system may be preferred to bored piles, so contaminated soils are not brought to the surface. Similarly, if ground improvement was being considered then a dry bottom-feed vibro system may be preferred to alternative wet vibro systems, which require subsequent treatment of the wet contaminated slurry. The chosen foundation or ground improvement solution should not create a pathway for the migration of contaminated soils, groundwater or ground gas that could increase the risks associated with contaminated land.

52.3.3.5 Sustainability

Sustainability comprises three main elements

- environmental
- social
- economic.

Environmental impacts can be wide-ranging, including the challenges and constraints associated with working adjacent to an environmentally sensitive area; the potential impact of contaminated land or groundwater; minimising the use of materials with a high embodied energy or carbon footprint.

Social impacts may include, for example, the effects of noise and vibration induced by pile driving in an urban area, or lorry movements associated with removing excavation arisings or spoil from bored piling, or spoil from a large excavation for a raft.

In highly sensitive locations, environmental and social issues may dictate the chosen foundation solution. In general, the objective is to minimise the environmental and social impacts of a project, while implementing it at a reasonable cost. An overview of sustainability in foundations is given in DFI (2015).

Current themes in foundation engineering are

- foundation reuse
- the use of foundations as geothermal elements (refer to Adam and Markiewicz, 2009)
- optimising geometry of foundation elements to minimise material quantities
- nonlinear ground stiffness to avoid over-conservatism when serviceability controls foundation design (see Section 52.6)
- consideration of sustainable materials, such as timber piles (Reynolds and Bates, 2009)
- advances in construction techniques, such as vacuum zero trim piling and HIPER Pile (Menteth, 2022).

Foundation reuse: It may be possible to reuse existing foundations of, say, an old bridge or building, for the new superstructure. For many practical applications the old foundations are either supplemented by additional new foundations or are strengthened. Foundation reuse can apply to both shallow and deep foundations. Foundation reuse is reasonably common for transportation projects, such as railway bridge refurbishment, and is increasingly being considered for many building projects. CIRIA Report C653 (Chapman *et al.*, 2007) provides an introduction to the topic, and Butcher *et al.* (2006) provide more detailed background information. The current technical literature mainly focuses on the issues associated with reusing piles. Similar issues also apply to reusing shallow foundations, although physical investigations and inspections of shallow footings are often easier than for piled foundations. The potential to reuse existing foundations is often limited by the lack of available reliable and comprehensive data on their construction, necessitating costly and time-consuming desk studies, investigations and technical evaluations as outlined below. The creation of digital models and records describing the design and construction of modern structures provides an opportunity to address this issue in the future. More comprehensive design and as-built data for an asset will help form the basis for better decision-making for the management and modification of new assets, increasing the potential for foundation reuse in the future. The key technical issues concerning foundation reuse are as follows.

- *Are the old foundations in the correct location?* What is the plan location of old foundations relative to the new superstructure loads – for example, a new building column grid may be offset from the old piles. Can the new substructures (e.g. new beams or rafts) be built economically to transfer loads from the superstructure into the old foundations?
- *Load-carrying capacity and settlement characteristics:* The ultimate geotechnical and structural capacity of the old foundations, and their likely movement under the new superstructure loads, needs to be assessed. Since the ground has consolidated (and strengthened) under the old foundation loads and will be subject to an unload–reload cycle (due to demolition of the old superstructure and reloading by the new superstructure), the settlement of old foundations would usually be relatively small until the old foundation load is exceeded. Once this load is exceeded much larger movements could develop. If, as is often the case, there is a mix of old and new foundations for the new superstructure then the different settlement characteristics of the different foundation types need to be carefully considered. Normally, the aim is to keep the net bearing pressures under the new superstructure to a smaller value than those under the old structure. Hence, an assessment of the loads that were applied by the old structure is fundamentally important.
- *Durability:* Over time the old foundations may have degraded – for example, steel may have corroded or concrete may have been subject to sulfate attack. Assessing the residual design life of the old foundations can be the most complex part of foundation reuse assessment. Geotechnical engineers will usually need to get additional advice from materials specialists.

52.4. Allowable foundation movement

The magnitude of the allowable settlement is critical in determining the type and size of foundations that will be required, and, therefore, the foundation cost (see Chapter 19 *Settlement and stress distributions* for an introduction to methods of settlement prediction). It is important that the foundation engineer is involved in discussions with the structural engineer before the allowable settlement is decided, and challenges values that are unreasonably small. Reference to Chapter 5 *Structural and geotechnical modelling* will be helpful in understanding the differences in approaches between structural and geotechnical engineers. As noted by Burland *et al.* (1977):

> Compared with the literature on the prediction of foundation movements, the influence of such movements on the function and serviceability of structures and buildings has received little attention. Yet major and costly decisions are frequently taken on the design of foundations purely on the basis of rather arbitrary limits on total and differential settlements.

Ideally, the foundation engineer would be able to predict the amount of differential settlement a structure can tolerate, and then predict the differential settlement that will actually occur due to the imposed structural loads and the response of the ground below the foundations. For most practical situations it is impossible to make this prediction. As noted by Fang (2002):

> It is an extremely complex indeterminate analytical problem.

He warns that:

> If an attempt is made to model analytically the structure and calculate the effect of differential settlements, one obtains ridiculously low allowable differential settlements because of the unrealistically large bending moments that will be calculated.

The factors that will affect real soil–foundation–superstructure interaction behaviour (Boone, 1996) include

- variations in soil properties
- variations in the construction sequence of the foundations and superstructure
- uncertainties regarding the rigidity of connections across the foundation and within the superstructure
- flexural and shear stiffness of the superstructure
- changes in stiffness of the superstructure during construction and how load redistribution to the foundations will be affected
- degree of movement (or 'slip') between the foundation and the ground
- overall shape of the ground-movement profile and the location of the structure within it
- building shape (both in plan and height)
- uncertainties in the way that the loads will redistribute in the long term as the structure settles differentially
- variable influence of time, both in the rate of settlement of the ground, and how creep and yield within the structure will develop.

Because of the above complexities there is a need for simple guidelines based on previous experience. The assessment of allowable settlements and the development of criteria for routine limits on allowable settlements have been established empirically on the basis of observations of settlement and damage in actual structures. The paragraphs below summarise these observations and provide simple guidelines. It should be emphasised that these are 'routine limits' for conventional structures and are *not* intended to be rigid rules. They will not apply to all structures; Burland *et al.* (1977) and Powderham and O'Brien (2020) describe case histories where the simple guidelines were *not* appropriate. This is an area of engineering where geotechnical and structural engineers must work closely together.

First, it is important to distinguish between

- total settlement – this may cause damage to services connecting into a structure or, occasionally, adjacent structures, but will not lead to damage to the structure itself
- differential settlement – due to rigid body rotation or tilt, which may be noticeable in high buildings and affect lifts, escalators and so on
- differential settlement due to relative displacements within the structure – this is this movement that can lead to structural damage (e.g. there may be differential settlement between adjacent building columns or bridge piers).

There is a wide range of terminology in the literature; the most commonly used terms are described in Chapter 26 *Building response to ground movements and tunnelling.*

There can be confusion about the criteria that need to be satisfied when assessing appropriate settlement or movement limits (see Chapter 26 *Building response to ground movements and tunnelling* for a full discussion of building responses to ground movements). In general, three basic criteria need to be considered.

(*a*) Visual appearance – this includes deviation of walls, columns or floors from the vertical or horizontal, respectively, and cracking of walls or cladding. These matters are highly subjective and vary between different people and across different regions of the world. Burland *et al.* (1977) have given a five-point classification system for visible damage, which is widely used. It should be noted that the system is not intended to be a direct measure of the degree of damage and relates specifically to the ease of repair of the brickwork, masonry and plaster. The two lowest categories of degree of damage are very slight (crack widths less than 1 mm) and slight (crack widths less than 5 mm). At these damage categories the building would be remote from structural instability. However, crack widths of 3–5 mm would be unsightly. Walls, columns and floors which deviate from the vertical or horizontal by more than 1 in 250 would usually be noticeable, and judged to be unacceptable. Whether these movements affect the functionality or serviceability of the structure depends on the nature and configuration of the structure.

(*b*) Serviceability limit state (SLS) – a loss of serviceability typically means that there is

- loss of weathertightness
- loss of fire resistance
- loss of thermal or sound resistance
- damage to connections to external pipework (water, gas, etc.)
- damage such that lifts, cranes or internal machinery cannot operate properly
- doors or windows sticking.

(*c*) Ultimate limit state (ULS) – this means that individual structural members may fail, or in an extreme case the whole structure may collapse. Structural distress is usually associated with excessive bending or shear stress in superstructure members. These limiting stresses are defined in the relevant structural design codes.

For the majority of buildings the allowable settlement will be governed by (*a*) or (*b*) above – that is, visual appearance or serviceability limits rather than by structural damage. For residential buildings (*a*) is likely to be most important, whereas for industrial buildings criterion (*b*) will be more relevant. For bridges, SLS for foundations are usually governed by avoiding SLS or ULS limits in the bridge deck and movement joints at piers and abutments.

52.4.1 Routine guides on limiting settlement

Published guidance on routine limits for allowable settlement consider sands separately from clay. The reason for this is that structures are more likely to suffer damage if the settlement occurs rapidly than if it develops slowly over many years. Hence, the settlement limits for sands are lower than for clays. **Figure 52.10(a)** plots maximum differential settlements against maximum settlement for buildings on raft foundations, and **Figure 52.10(b)** provides a similar plot for frame buildings on isolated foundations (e.g. pads, strips, footings).

Most of the data have been taken from studies by Skempton and MacDonald (1956) and Grant *et al.* (1974). The data for buildings founded directly on clayey soils are shown separately from those founded on a stiff layer (such as dense sand or gravel) overlying the clay. In a few extreme cases differential settlements (for both rafts and pads or strips) are nearly as large as the maximum settlements, and lie above the Bjerrum (1963) guidelines for the relationship between differential and maximum settlements. However, for the vast majority of situations the Bjerrum guidelines seem to be conservative. For rafts founded on overconsolidated London and Lambeth Group clays, Morton and Au (1974) report differential settlements equal to about 25% of the maximum settlement. Isolated foundations, such as pad or strip footings, would normally be considered to be 'flexible' when considering the overall interaction of the foundation and superstructure. Although, as individual elements they may be assumed to be rigid when calculating settlement by routine methods (refer to Chapter 53 *Shallow foundations*).

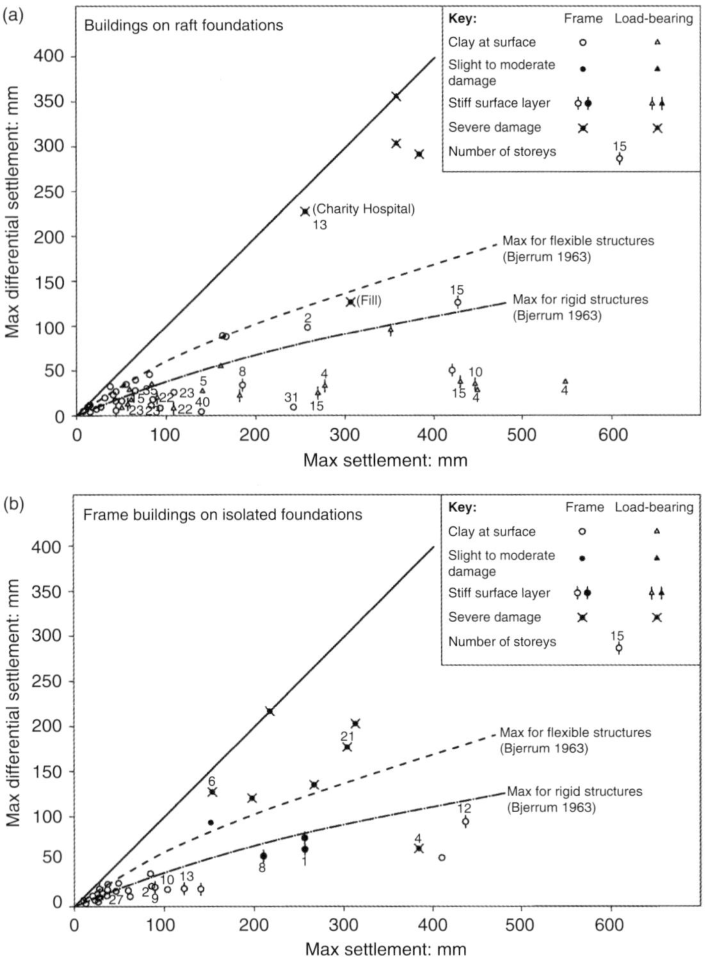

Figure 52.10 Observations of damage. Settlement and differential settlement. (a) Raft foundations; (b) isolated foundations, pad and strip footings (Data taken from Burland *et al.*, 1977)

The routine limits for the maximum total settlement for buildings are summarised in **Table 52.5**.

Foundations on sands seldom exhibit large settlements due to monotonic increases in stress; typically settlements will be less, often far less, than 75 mm (Bjerrum, 1963; Terzaghi, 1956). Most problems with foundations on sand are due to vibration or other forms of cyclic loading, or due to groundwater seepage and erosion. In areas of the world that are prone to seismic activity, earthquake shaking can cause large-scale instability and highly damaging movements of foundations on sands (refer to Chapter 60 *Foundations subjected to cyclic and dynamic loads*).

According to the data shown for frame buildings on isolated foundations, in **Figure 52.10(b)** damage usually occurs when differential settlements exceed about 50 mm, or maximum settlements exceed about 150 mm. Based on an analysis of 69 case studies, Ricceri and Soranzo (1985) suggested a settlement limit of 80 mm to avoid structural damage. For buildings on rafts (**Figure 52.10(a)**) significantly larger settlements seem

Foundation type	Soil type	Routine limits: maximum total settlement (mm) (S_{max})[f]	Typical differential settlement
'Isolated', pad, strip[a]	Clay	65	$< 0.66\ S_{max}$, if $S_{max} \leq 50$ mm $< 0.5\ S_{max}$, if $S_{max} > 200$ mm[b,c]
Raft	Clay	100	$< 0.33\ S_{max}$[d]
'Isolated', pad, strip[a]	Sand	40	$< 0.75\ S_{max}$[e]
Raft	Sand	65	

(a) Pad and strip footings are normally considered to be 'flexible' when assessing the overall interaction of the foundations and superstructure. As individual elements they may be rigid, depending on their thicknesses compared with their lengths or widths (see Chapter 53 *Shallow foundations*).
(b) These limits apply when the clay is immediately below the foundation. When a stiff layer is between the foundation and the clay, then the differential settlements can be substantially reduced and typically the differential settlement is $< 0.5\ S_{max}$. This stiff layer could be natural (say, due to a dense sand or gravel stratum) or be constructed as part of the foundation design.
(c) Assume linear interpolation to assess differential settlement when maximum settlements are between 50 and 200 mm.
(d) This is usually conservative; as the relative raft bending stiffness increases (refer to Chapter 53 *Shallow foundations*) the differential settlements can be substantially reduced to values of less than 0.15 S_{max}.
(e) Differential settlements on sands can exceed 0.75 S_{max} (Bjerrum, 1963).
(f) These routine limits are only intended for low-risk situations and simple structures; also refer to **Table 52.7**.

Table 52.5 Routine limits for the maximum total settlement of buildings

to be tolerable, with damage only being recorded for maximum settlements in excess of 250 mm. It is notable that some buildings founded on relatively rigid rafts have tolerated maximum settlements in excess of 400 mm without suffering significant damage. For these structures, the raft rigidity has limited the differential settlement to about 10% of the maximum settlement. More recently Zhang and Ng (2005) have collated a large database of building damage. This database comprises about 300 buildings, of which about 100 were in Hong Kong, and includes a variety of different building types (office blocks, warehouses, factories, hotels, hospitals, etc.), mainly on isolated foundations. Structural types include steel and reinforced concrete frame and load-bearing wall structures. Underlying ground conditions include sand, clay and alluvium. **Figure 52.11(a)** shows histograms of the settlement for 95 buildings, for both shallow and deep foundations. They classified damage as either intolerable (where repair work is needed to the structure) or tolerable, and these categories are, therefore, different to the damage categories shown in **Figure 52.10**. Most buildings with settlements of less than 100 mm are undamaged, and there were no reports of damage when the settlement is less than 50 mm. However, when the settlements exceed 200 mm intolerable damage becomes common; see **Figure 52.11(a)**. For angular distortion, Zhang and Ng report that values which are more severe than 1 in 400 are usually intolerable. Therefore, based on the case history data shown in **Figures 52.10** and

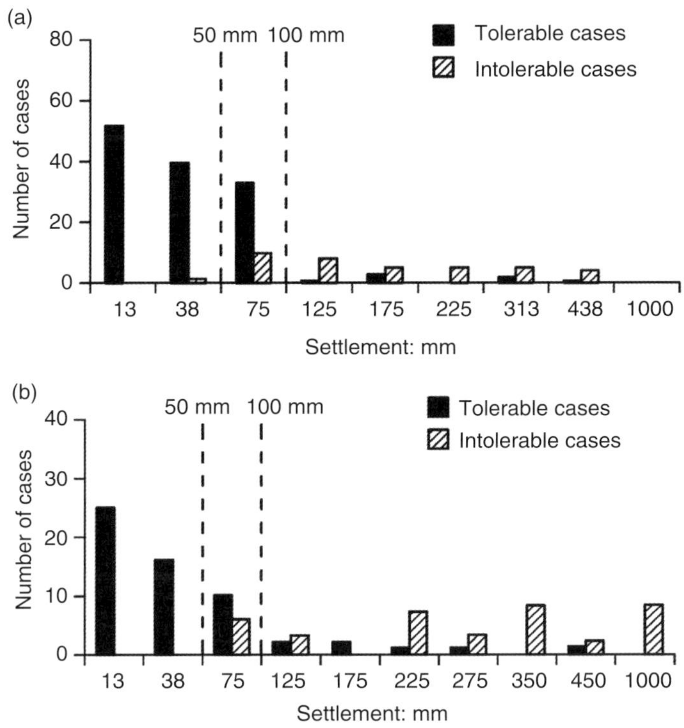

Figure 52.11 Observation of (a) building damage and settlement; (b) bridge damage and settlement (Reproduced from Zhang and Ng, 2005)

52.11(a), the routine limits for the maximum settlement given in **Table 52.5** will usually be conservative. This is particularly the case when raft foundations are used.

Moulton (1985) carried out a comprehensive study of tolerable movements of highway bridges in the USA and Canada. The bridge types included steel, concrete and composite, and were continuous spans or simply supported. They were small to medium span (less than 50 m span) bridges and of simple structural form (e.g. cable stay and suspension bridges were not considered). The bridges were built on a range of soils (clays, sands) and rocks. Of the 580 bridges which were studied, 439 experienced some form of significant movement. The primary causes were excessive movements of approach embankments, inadequate lateral resistance and large foundation displacements. **Figure 52.11(b)** provides a histogram of settlement for 171 bridges. From these data it can be seen that intolerable damage is usually observed if settlements exceed about 75–100 mm. Zhang and Ng (2005) also report that if angular distortion exceeds 1 in 250 then intolerable damage is usually caused. Movements are usually tolerable if settlements are less than 50 mm or if angular distortions are less severe than 1 in 500. Barker *et al.* (1991) note that settlements are more damaging when accompanied by horizontal movements (and if the structure has inadequate horizontal strength).

Table 52.6 summarises routine limits for allowable movements (generally as a limiting angular distortion) for a range of structures and associated infrastructure such as drains, lifts and so on.

52.4.2 Site-specific assessments

The routine limits given in **Table 52.5** should only be used for low-risk site conditions and for simple structures. For more complex situations (some examples are outlined in **Table 52.7**) a more detailed consideration of differential ground movements and allowable limits for foundation design may be necessary. As noted earlier, realistic analysis of stresses within the structure due to differential ground movements are seldom practical or necessary. If a more detailed assessment is necessary, then consideration should be given first to the following.

(*a*) Identify the component(s) of the structure that will be most sensitive to foundation movement, and when they will be built (or a likely construction sequence), and the relative weights of the main components of the structure.

(*b*) The sequence of construction and likely immediate settlements and differential settlements that may arise due to the weight of each of the principal components of the structure, assuming the structure is flexible.

(*c*) Each component of the structure will only be vulnerable to differential settlement and any resulting damage, which occurs *after* the component is built into the rest of the structure.

(*d*) Following (*b*) and (*c*), assess the relative rigidity of the foundation or substructure and apply corrections in order to derive the appropriate differential settlements (refer to Chapter 53 *Shallow foundations*).

(*e*) Compare the resulting differential settlements and angular distortion against the routine limits (given above) for the appropriate component of the structure.

Simple examples are given below for a frame building on a raft and a single-span bridge. The relevant settlements that need to be considered for a particular stage are those that are time dependent due to previous construction stages, together with those that are short term plus time dependent due to later construction stages. **Figure 52.12** shows a plot of settlement against time for different parts of a frame building. During excavation some heave occurs. The raft is constructed first and it is affected by subsequent differential settlements. As additional elements of the structure are built, short-term settlement occurs due to the incremental increases in weight of the structure. The parts of the structure that are built and tied together will experience differential settlement (and potentially be vulnerable to damage) due only to the settlements that occur after the element under consideration has been built. During construction the cladding will be added and internal components such as lifts and escalators will be built and connections to external utilities will be made. Finally the live load will be applied. Hence, the maximum settlements and differential settlements experienced by different parts of the building may be very different. The differential settlement of the

Type of structure	Type of damage	Criterion	Routine limits	Comments
Framed buildings and reinforced load-bearing walls	ULS structural damage	Angular distortion	1 in 150 to 1 in 250	ULS concerns at these limits
Framed buildings and reinforced load-bearing walls	SLS cracking of walls, cladding, partitions	Angular distortion	1 in 300 to 1 in 500	Typically SLS concerns at these limits
Unreinforced masonry walls	Visual onset of cracking	Deflection ratio	Sagging 1 in 2500 ($L/H = 1$) 1 in 1250 ($L/H = 5$) Hogging 1 in 5000 ($L/H = 1$) 1 in 2500 ($L/H = 5$)	At these limits there is only the onset of cracking; the damage is very slight. Tolerable movements are several times larger L = length of structure, H = height of structure
Steel, fluid storage tanks	SLS leakage	Angular distortion	1 in 300 to 1 in 500	
Utility connections	SLS	Maximum settlement	150 mm	Less for sensitive utilities, such as gas mains
Crane rails	SLS crane operation	Angular distortion	1 in 300	Depends on specific crane configuration
Floors, slabs	SLS drainage	Angular distortion	1 in 50 to 1 in 100	Depends on specific falls, alignment
Stacking of goods	ULS, collapse	Tilt	1 in 100	
Machinery	SLS, efficient operation	Angular distortion	1 in 300 to 1 in 5000	Depends on machine type and sensitivity. Bjerrum (1963) suggests 1 in 750 as 'typical'
Towers, tall buildings	Visual	Tilt	1 in 250	Tilts in excess of this will be noticeable and concerning, although possibly remote from collapse, depending on structure configuration. For the Leaning Tower of Pisa the tilt is 1 in 10
Towers, tall buildings	Lift and escalator operation	Tilt, after installation	1 in 1200 to 1 in 2000	Sequence of construction and timing of lift and escalator installation are important
Bridge	SLS	Angular distortion	1 in 250 to 1 in 500	Depends upon bridge-deck characteristics and articulation arrangements
Bridge	SLS	Maximum settlement	60 mm	Typical values
Bridge	SLS, bearing	Horizontal displacement	40 mm	Typical values

Table 52.6 Routine limits for allowable movements (Based on Barker *et al.*, 1991; Boone, 1996; Burland and Wroth, 1974; Day, 2000)

raft will be larger than for internal structural members, and much larger than for elements such as the cladding or lifts, which are usually the most sensitive to differential settlement. The risk of structural damage reduces as the ratio of live load to dead reduces, and the risk of visual or serviceability damage reduces if the construction of cladding and other sensitive components is delayed as much as possible.

Figure 52.13 shows a plot of settlement and horizontal displacement against time for a single-span bridge. Most of the weight of the bridge and resulting settlement will be due to the abutments and their backfill. The settlement, and particularly the differential settlement experienced by the bridge deck, will usually be the main concern, and the residual differential settlement experienced by the bridge deck may be a small fraction of the maximum settlement of the bridge abutments. It is also apparent that the bridge-deck settlement and the horizontal displacement of the bridge-deck bearings will be affected by the timing of backfilling to the abutments, relative to the bridge-deck construction. If the abutments are backfilled first, then the subsequent movements, which may affect the deck and bridge bearings, will be much smaller than if the deck is constructed before the abutment backfill.

Factor	Comments
Heterogeneous ground conditions	Significant variations in soil type, layer thickness or competence of the soil layers below different parts of the foundation
Variations in loading	Different loads applied across the foundation
Variations in construction duration	May induce increases in differential settlement, if the degree of consolidation of clay layers varies significantly
Variations in site history	Part of the site has been preloaded by old foundations, or part of the site has been excavated and the underlying clay has swelled and softened
Variations in site topography	If the foundation is on sidelong ground, then different amounts of excavation may lead to substantial changes in net foundation bearing pressure
Changes in superstructure stiffness across the foundation	Changes in configuration of the structure, connection joint stiffnesses or cladding details (e.g. height, thickness or spacing of columns, walls, etc.) could lead to local stress concentrations and increased sensitivity to differential ground movements
Form of structure, potential for 'brittle' behaviour	Certain forms of structure, for example short or thick columns and flexible floor slabs, are vulnerable to relatively small differential ground settlement (see Burland *et al.*, 1977)

Table 52.7 Factors that could increase the risk of damage due to differential ground movements

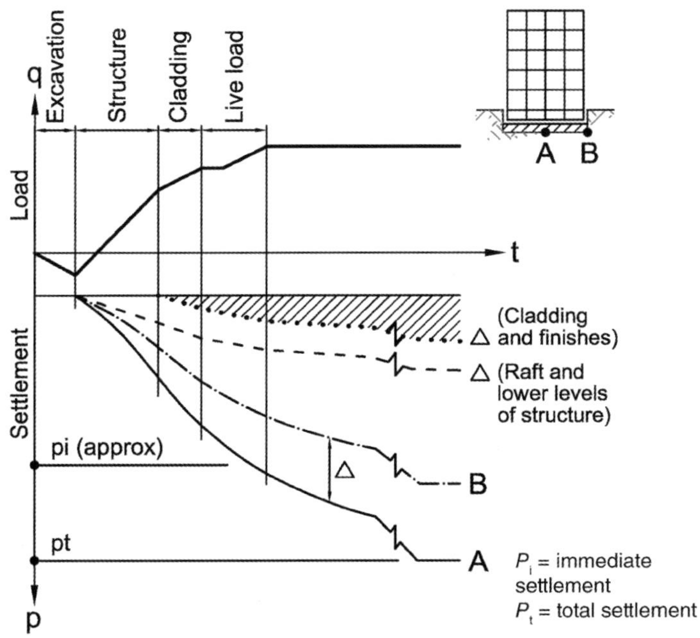

Figure 52.12 Frame building settlement against time

52.5. Design bearing pressures

The superstructure loads calculated by a structural engineer will enable the *gross* foundation bearing pressures to be determined (see Chapter 21 *Bearing capacity theory* for an introduction to bearing capacity theory). However, it is the *net* foundation bearing pressure that should be used for both bearing capacity and settlement calculations (**Figure 52.14**), *not* the gross bearing pressure. The *net* pressure is that part of the gross applied pressure that requires the shear strength of the ground to support it. Hence, a factor of safety should always be considered in the context of *net* pressures only. There can be confusion about the terminology that is used for foundation bearing pressures; the following definitions are important.

- q, *gross foundation pressure* – total applied load at the founding level (includes the weights of the superstructure, substructure and foundation) divided by plan area of foundations.
- q_n, *net foundation pressure* – gross pressure minus the overburden pressure at the founding level.
- q', *effective foundation pressure* – gross pressure minus the groundwater pressure.
- q_f or q_u, *ultimate bearing pressure* – value of bearing pressure at which the ground fails in shear.
- q_s, *maximum safe bearing pressure* – value of bearing pressure at which the risk of shear failure is acceptably low.
- q_a, *allowable bearing pressure* – value of bearing pressure at which the risk of excessive deformation and shear failure is acceptably low.
- q_w, *working pressure* – value of bearing pressure applied to foundations.

For limit state code calculations, q_s refers to the ULS and q_a to the SLS.

The application of transient loads should consider both the type of ground at founding level and whether ULS or SLS is being considered. The total settlement of clays would usually consider the dead load plus a fraction of the live load (typically about 25%), since clay consolidation will not be significantly affected by short-term transient loads (such as wind loads). In contrast, the stability of foundations on clays should consider both the dead and the full-live loads, and since the maximum bearing pressure will usually include short-term transient loads, the strength of clay should be based on undrained rather than drained (effective stress) parameters. For clays and other fine-grained, low-permeability soils, undrained bearing capacity is usually more critical than drained. If in doubt (especially for inclined loading conditions) both undrained and drained conditions should be checked. For coarse-grained, high-permeability soils, it is the drained bearing capacity which is relevant.

It is important to clearly define the factor of safety that is being used. Different codes use different definitions and they are not necessarily equivalent. For example, for coarse-grained soils (sands, gravels, etc.) two common definitions are as follows.

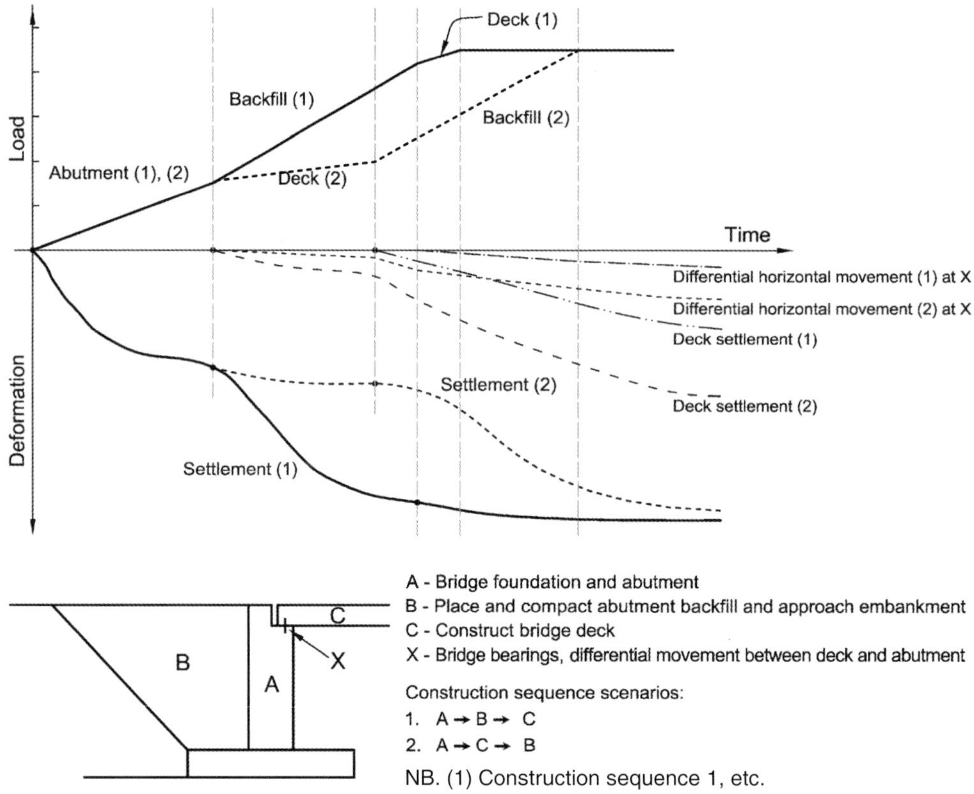

Figure 52.13 Bridge abutment movement against time

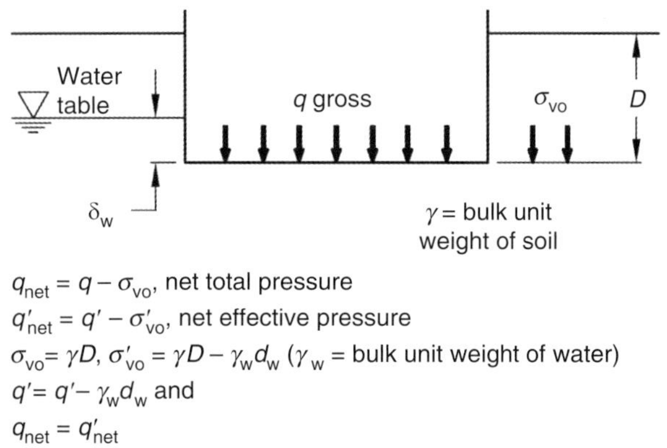

Figure 52.14 Definitions of net and gross bearing pressure

Factor of safety for strength (angle of friction, φ'),

$$F_{sd} = \frac{\tan \varphi'}{\tan \varphi'_{mob}} \tag{52.1}$$

Factor of safety for drained bearing capacity

$$F_{bd} = \frac{q'_f(\text{net})}{q'_w(\text{net})} \tag{52.2}$$

From **Figure 52.15(a)** it is clear that F_{bd} does *not* equal F_{sd}, and that F_{bd} is much greater than F_{sd}. In contrast, for the undrained bearing capacity, $F_{bu} = F_{su}$ (refer to **Figure 52.15(b)**), where

$$F_{bu} = \frac{q_f(\text{net})}{q_w(\text{net})} \text{ and } F_{su} = \frac{S_u}{S_{umob}} \tag{52.3}$$

and S_{umob} is the mobilised undrained shear strength under q_w (net).

52.6. Parameter selection

52.6.1 Introduction

Poor parameter selection is a common pitfall in foundation engineering, especially of deformation characteristics for assessing foundation movements. Parameter selection can appear to be a black art, both for young geotechnical engineers and to general civil and structural engineers. Parameters such as soil strength and compressibility are significantly affected by a wide range of factors, including the testing methods, rate and direction of loading and the volume of ground tested. Sampling disturbance affects the reliability of many laboratory tests and in situ tests can be difficult to interpret. Hence, parameters cannot be easily and accurately measured or used indiscriminately in a wide range of

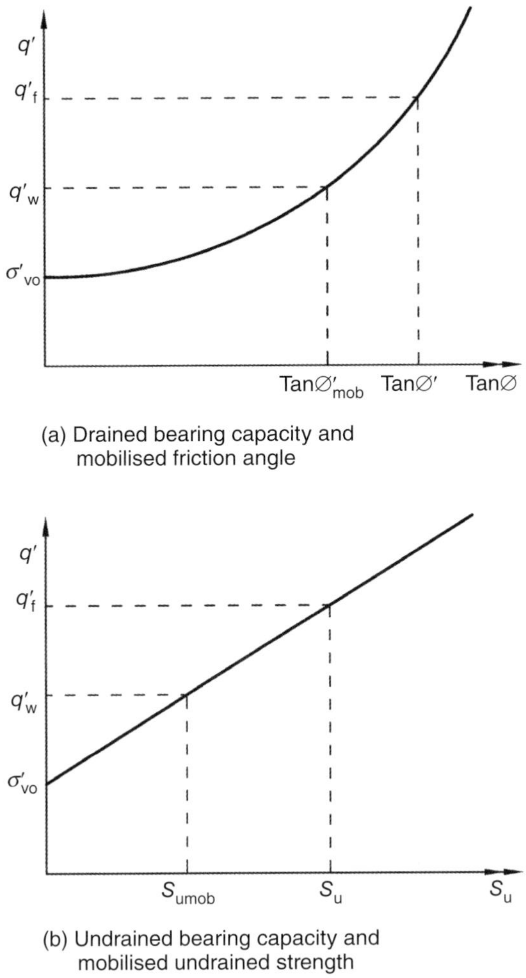

Figure 52.15 Bearing capacity and mobilised strength

different calculations. The ground investigation needs to, as a minimum, identify

- key soil characteristics, including strength, compressibility, the presence of water-bearing layers (sand or silt) and the presence of weak layers
- likely variability of properties: variation in soil strata thicknesses, relative density, strength and so on
- construction effects: influence of construction on ground properties (e.g. unloading and groundwater seepage causing strength/stiffness to be reduced), remoulding of soil at construction boundaries (e.g. pile shaft or base of footing).

Ground investigation needs to be designed based upon desk study, project requirements and the published literature for the site geology. For routine ground investigations strength can usually be more reliably established than compressibility or stiffness. Routine sampling often induces significant disturbance and, hence, variability in laboratory measurements of strength and stiffness. Therefore, variability in properties is usually best assessed using in situ test methods, such as CPT profiling or using continuous sampling and split/describe/photograph and high-frequency classification tests (e.g. moisture contents, Atterberg limits, particle grading). Understanding the nature of the soil fabric is important in terms of assessing overall behaviour – for example, in terms of rates of drainage or softening (Rowe, 1972).

In addition to ground investigation methods, the following factors will also need to be considered when selecting design parameters

- the soil type and its geological origins and subsequent history
- the nature of the chosen method of analysis
- the magnitude, direction and type of loading
- the scale and type of failure mechanism(s) – several may need to be considered
- construction influences
- code requirements and factor of safety selection.

Once all the factual data from a ground investigation have been assembled, data interpretation follows six main steps.

- Derive strength and compressibility from the factual information, usually directly from laboratory data, or indirectly using in situ tests (from theoretical analyses or empirical correlations).
- Identify representative values (and ranges of values) for each stratum, which usually involves deleting excessively high or low values (due to sampling or testing errors or local 'hard' layers).
- Cross-correlate different types of data to check that consistent and logical changes in strength and compressibility have been derived.
- Check – against desk study data, conceptual geological model and so on and published data – for strata under consideration.
- Once robust reliable datasets have been collated, trend lines can be determined (e.g. worst credible, mean or median, upper bound); this may involve some statistical analysis (especially of index tests or penetration tests), but is largely dependent on experienced judgement.
- Design parameter selection – **Table 52.8** gives comments on some of the most common issues that affect soil strength and compressibility parameter selection. These basic parameters always need to be assessed. Some common foundation design scenarios that lead to different strength or stiffness parameters being assumed for design are illustrated in **Figure 52.16.** Hence, in general strength and stiffness parameters for any soil layer should never be considered to be unique – that is, there is no such thing as *the* strength or *the* stiffness. A particular strength or stiffness should only be assumed to be relevant for a specific design situation and applied load.

Ground profile and site history: As discussed in Section 53.7, shallow foundations can be vulnerable to any misinterpretation of a site's history or geology, because this may mean that soft compressible soils are found at foundation level rather than competent low-compressibility soils. Deep foundations

Soil type, geological origins and subsequent history	Soil strength and stiffness are usually anisotropic due to bedding and stress history. If the yield stress is within the foundation bearing pressure, then substantial changes in compressibility will occur following yield (**Figure 52.17(b)**). The ground shear (or Young's) modulus can degrade by an order of magnitude with strain amplitude (**Figure 52.17(c)**). Thin low-strength layers may substantially affect the safe bearing pressure or overall stability.
Method of analysis	Ground investigation data typically provide simplified parameters: laboratory data are affected to some extent by disturbance, and in situ tests are difficult to correlate to fundamental behaviour. It is essential to check the relevance and reliability of the analytical method for the soil type considered and the design situation. Parameter selection for all methods of analysis (even Cat 3 methods – see **Table 52.9**) depends to an extent on careful calibration of selected output against good case history data.
Magnitude, direction and type of loading	Deformation estimates are critically dependent on selection of stiffness and compressibility parameters (e.g. Young's modulus). If these are linear elastic, then the modulus must be appropriate to the stress–strain range and loading direction. The mobilised strength and deformation modulus under different loading regimes, say sustained cyclic loads, impact loads or long-term monotonic 'static' loads, is quite different.
Scale and type of failure mechanism, and construction influences	A cautious approach is needed if failure could occur within a small zone of the soil, such as beneath a narrow pad, strip or small-diameter end-bearing pile. A less conservative approach is appropriate if failure will inevitably involve shearing across a large mass of ground, such as where there is large raft or shaft resistance along a friction pile. Softening, loosening and remoulding of soil may occur during construction, especially if exposed to surface water or groundwater.
Code requirements, magnitude of factor of safety and definition, ULS and SLS criteria and assessment methods	Many codes used to adopt large factors of safety (2–3 or more for bearing capacity) in order to limit deformation and to allow for crude ground investigation methods. Modern limit state codes typically place greater emphasis on SLS conditions (and therefore ground/foundation movements). This can lead to over-conservative design unless appropriate deformation moduli are selected, which may justify more sophisticated intrusive investigation and testing (e.g. sampling, field and lab testing) and often requiring nonlinear moduli to be determined.

Table 52.8 Factors which may influence the selection of soil strength and compressibility parameters for subsequent analysis

can also be vulnerable if a site's history and/or the engineering implications of its history are poorly understood. Examples of problems that can occur include: piles installed below or adjacent to old basements, with piles designed on the basis of total stress methods without allowance for softening of stiff clays in the vicinity of basements; end-bearing piles founded in an area affected by collapsed mine workings. Desk studies and ground investigations must be implemented with great care and thoroughly checked. Even if old investigation data are available for a site, it is prudent to carry out some additional investigations to check data reliability. It is important that investigations are fully supervised by experienced staff.

Groundwater conditions: As noted in Section 52.3.3, the most common cause of unforeseen difficulties is the influence of groundwater and its effects during and after construction. A desk study should include a consideration of groundwater conditions. This should start with an appraisal of the regional hydrogeology (e.g. presence of aquifers, likely groundwater flow directions, surface water bodies), and development of a conceptual groundwater model for the site. From this a series of questions should be asked about how groundwater may affect, for example, ground behaviour, construction means and methods, foundation performance. In addition, questions need to be asked about how the foundations may influence the groundwater regime. These issues can be particularly sensitive if a site has been affected by ground/groundwater contamination, or if foundations may penetrate into or close to aquifers used for potable water supply (this may, depending on the nature of foundation construction, create a migration pathway, albeit practical mitigation measures can usually be identified). The intrusive investigations should be designed to address these questions. A common pitfall with many ground investigations is inadequacies in terms of: number and location of piezometers; poor installation (leading to unreliable groundwater data); inadequate monitoring periods (groundwater levels tend to vary seasonally and with rainfall); and poor distribution across different strata with differing hydrogeological characteristics. Many ground investigations would benefit through a greater investment in understanding a site's groundwater regime. This investment is also likely to be repaid through a significant reduction in the risk of project delays and disputes through 'unforeseen' conditions claims.

Methods of analysis: The different categories of analysis are outlined in **Table 52.9**, and it is worth remembering that even the more theoretically sound analytical methods (Categories 2B or 3B) will depend on some empiricism (e.g. back-analysis and calibration against case histories). When apparently sophisticated finite element methods are used, if a simple linear elastic Mohr–Coulomb-type stress–strain model is used, then a high level of empiricism is still required to select the appropriate Young's modulus profile, and this stiffness profile will not necessarily be the same as that derived from simpler analytical methods. Methods of analysis (and their associated input parameters) will commonly only be appropriate for a

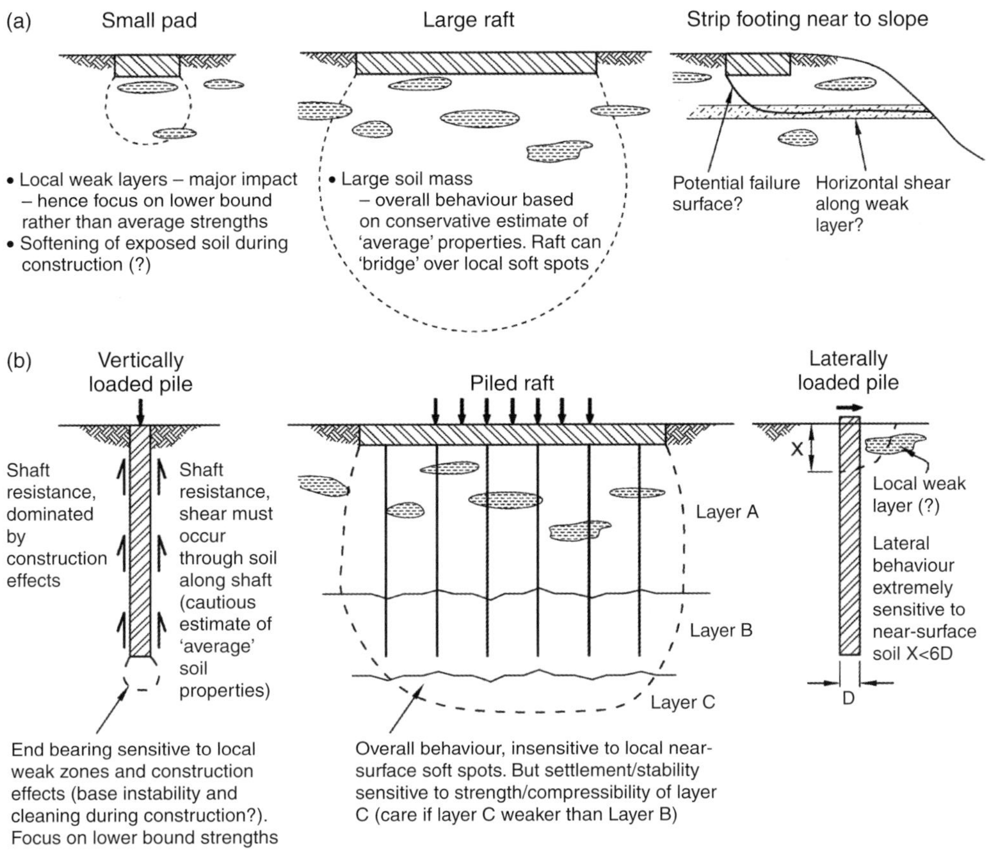

Figure 52.16 Parameter selection, general issues: (a) shallow foundations, (b) deep foundations

limited set of design situations, and it is vital that the geologist or engineer understands when a particular analysis method (and associated parameters) is relevant to the problem under consideration and how best to obtain the necessary input parameters for a particular analytical method.

Mobilised ground stiffness, introductory comments: Most foundation analyses involve Category 1 or 2A procedures. For simple deformation analyses, which usually assume linear elastic ground behaviour, the assessment of an elastic modulus, such as Young's modulus, is often required. Refer to Chapter 17 *Strength and deformation behaviour of soils* for a discussion of the properties of linear elastic porous materials. Research during the last few decades has clarified many of the factors that affect the mobilised stiffness and compressibility of soils (Atkinson, 2000). Competent heavily overconsolidated soils, commonly utilised as founding strata, can often be characterised as nonlinear elastic. It is important to differentiate between different elastic moduli (shear, Young's, bulk) and whether secant or tangent moduli are being considered (**Figure 52.17(a)**). Key features of nonlinear elastic behaviour are

- the stiffness at very small strains, less than 0.001% strain (characterised by small strain shear modulus, G_0) which is practically the same for static and dynamic loading
- G_0 applies for both drained and undrained conditions
- for overconsolidated soils G_0 will often be higher for horizontal loading than vertical loading.

G_0 provides an extremely important reference value (or 'anchor point') for assessing ground stiffness, and mobilised stiffness for many practical foundation design applications is closer to G_0 than is commonly assumed (as discussed by Burland, 1989; Clayton, 2011). Appropriate test methods are described in Chapter 47 *Field geotechnical testing.*

Prior to the 1980s there was considerable confusion concerning mobilised ground stiffness. It was only after the importance of strain amplitude on mobilised elastic moduli was appreciated (Jardine *et al.*, 1984) that a rational basis for selecting elastic moduli could be developed. The assumption that the ground behaves as a nonlinear elastic material is reasonable provided that, following application of the foundation bearing pressure, the effective stresses in the ground are remote from the soil's pre-consolidation pressure (or yield stress) (refer to **Figure 52.17(b)**). The 'elastic' modulus of the ground can reduce by more than an order of magnitude from G_0 (or equivalent E_0 values) with increasing strain amplitude to 'large strain' (**Figure 52.17(c)**). 'Large strain' moduli are typically measured through conventional tests such as

Category	Subdivision	Characteristics	Typical method of parameter selection	Level of empiricism
1	A	Empirical (direct)	Simple in situ, e.g. penetration tests (SPT 'N' or CPT 'q_c'). Direct application of in situ test data	3
	B	Empirical (indirect), plus some basic theory	In situ (as 1A) plus correlation to soil parameters (such as Young's modulus)	3
2	A	Simplified theory, uses soil mechanics principles, e.g. bearing capacity (plasticity) or settlement (elasticity)	Routine in situ and laboratory tests, together with empirical correlations	3
	B	As 2A, but theory allows for nonlinearity in a simplified manner	May use correlations with routine testing, more sophisticated testing or published data and back-analysis	2
3	A	'Basic' numerical modelling. Theory is isotropic, linear elastic–perfectly plastic. Allow for some interactions or topography variations	Rely on empirical correlations for soil stiffness and compressibility. Back-analysis of case histories. Reliability of foundation movement predictions mainly dependent on selection of 'elastic' (?) moduli. Ground movements remote from foundation likely to be unreliable, potential for differential movements to be underpredicted	3
	B	Nonlinearity or anisotropy allowed for in a simplified manner. More realistic assessment of interactions possible. Monotonic loading usually	Usually needs sophisticated test data (both in situ and laboratory), together with back-analysis of case histories	2
	C	Nonlinearity or anisotropy allowed for by way of proper constitutive models. Rarely used in current practice	Needs specialist tests (using research laboratories) and advice for selecting input. The interpretation of the output is likely to be difficult	1

Advanced methods of ground investigation and analysis tend to be more appropriate if there is: limited local experience or the case history data is poor; there is a high level of geotechnical risk; the design objectives are complex; or there is a high potential for cost savings. Level of required empiricism: 3 = high; 2 = intermediate; 1 = low.

Table 52.9 Categories of analysis

oedometer tests. Many empirical correlations are only relevant for a particular design situation or a particular geology (so the empirical basis for any correlation needs to be understood). Hence, the judgement required in selecting an appropriate modulus for linear elastic analyses is largely associated with understanding the likely strain amplitude induced by the foundation loads together with an appreciation of relevant case histories.

The assumption of linear elasticity can be acceptable when foundation behaviour can be considered in 'isolation' – for example, when calculating raft bending moments or the overall settlement of a strip footing on stiff clay. Linear elasticity is usually inappropriate if interactions between, say, a new foundation and adjacent infrastructure, or when ground movements outside and beneath a loaded area are important, or when the foundation behaviour is intrinsically sensitive to nonlinearity (e.g. interactions between piles within a pile group or the interaction between foundations and adjacent retaining walls or earthworks – say, new embankments or cuttings). For these situations nonlinear elasticity (or more complex stress–strain models) are likely to be required, and it will also be important to have early expert advice and guidance. If foundation design is being controlled by serviceability (i.e. movement) considerations, then often more economical designs can be achieved through a more sophisticated assessment of appropriate mobilised Young's (or shear) moduli.

Empirical correlations for G_o *in sands and clays:* Vardanega and Bolton (2013) collated G_o data for a wide range of overconsolidated clays and proposed

$$\frac{G_o}{p'_r} = \frac{B}{(1+e)^{2.4}}\left(\frac{p'}{p'_r}\right)^{0.5} \tag{52.4}$$

where e is the soil void ratio; p'_r is the reference pressure, taken as 1 kPa; p' is the mean effective stress; and B is the material coefficient, typically between 15 000 and 25 000. B typically increases with decreasing plasticity and decreases as degree of weathering increases. For clays which are lightly cemented, typically unweathered very stiff/hard clays (transitioning towards mudstones) at depth, B increases significantly towards values of 50 000.

Examples of G_o and shear wave velocity, V_s, profiles for a stiff high-plasticity fissured clay and for stiff low-plasticity lodgement tills are shown in **Figure 52.18**.

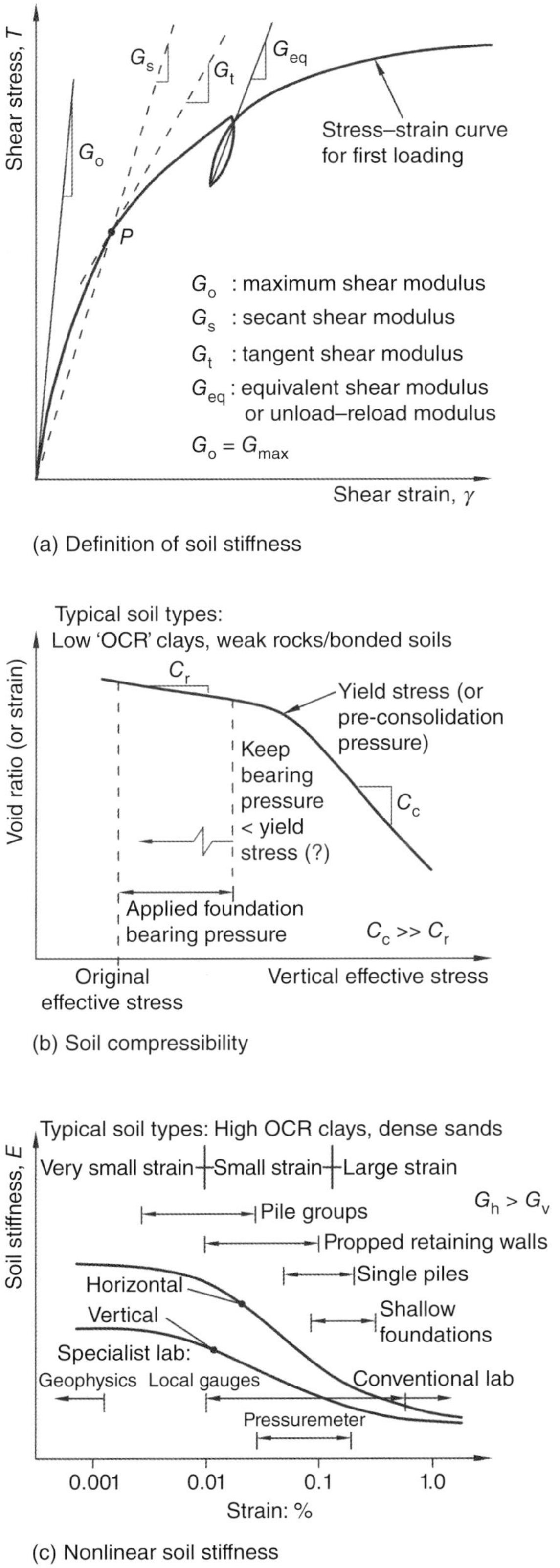

Figure 52.17 Ground stiffness and compressibility: fundamental considerations

For sands and silts (and sandy clays), Clayton (2011) proposed:

$$G_0 = C_p(1+e)^{-3}\left(\frac{p'}{p_{\text{atm}}}\right)^{0.5} \tag{52.5}$$

where e is the soil void ratio; p' is the mean effective stress; p_{atm} is atmospheric pressure, typically 101 kPa; and C_p is the material constant in MPa, typically 300–600 MPa, with higher values for clean sands and lower values for silts and clayey sands. Even modest amounts of interparticle bonding can significantly increase C_p. Clayton (2011) reports a C_p value of 1200 for Folkestone Beds, a Cretaceous-age dense sand. Jamiolkowski and Lo Presti (2003), also note the effects of interparticle bonding and ageing effects; for the Messina sands and gravels the deeper Pleistocene deposits had a C_p value several times larger than shallow Holocene-age deposits.

There are many published correlations relating penetration test data (both CPT and SPT) to G_o; one of the more reliable correlations is that by Robertson (2009) for CPTs. Correlations between SPT 'N' values and G_o should generally be treated with caution.

The above empirical correlations for G_o are very helpful, although as noted by Oztoprak and Bolton (2013), 'engineers should be encouraged to measure G_o by seismic methods in the field whenever possible'.

52.6.2 Strength and compressibility of clays

Classification data: A clay's strength and compressibility characteristics are fundamentally linked to its

- plasticity index
- overall macrofabric (presence of silt, sand layers; presence, spacing and orientation of fissures)
- stress history, in particular its overconsolidation ratio.

The liquidity index can be approximately correlated with a clay's overconsolidation ratio – for example, near-surface clays (within, say, 10–15 m depth) that have a liquidity index close to zero, or negative values, are likely to be heavily overconsolidated, whereas clays that have large positive values (say, in excess of + 0.3) are likely to be lightly overconsolidated or normally consolidated. Burland (1990) introduced a similar (but more robust) parameter termed the void index. A plot of the void index against the vertical effective stress (which can be readily derived from basic index tests) can provide a good indication of the clay's compressibility behaviour (**Figure 52.19**). Clays that plot well below the intrinsic compression line (ICL) are likely to be heavily overconsolidated and form competent founding strata. Clays that plot close to the ICL are likely to be lightly overconsolidated or normally consolidated and of moderate to

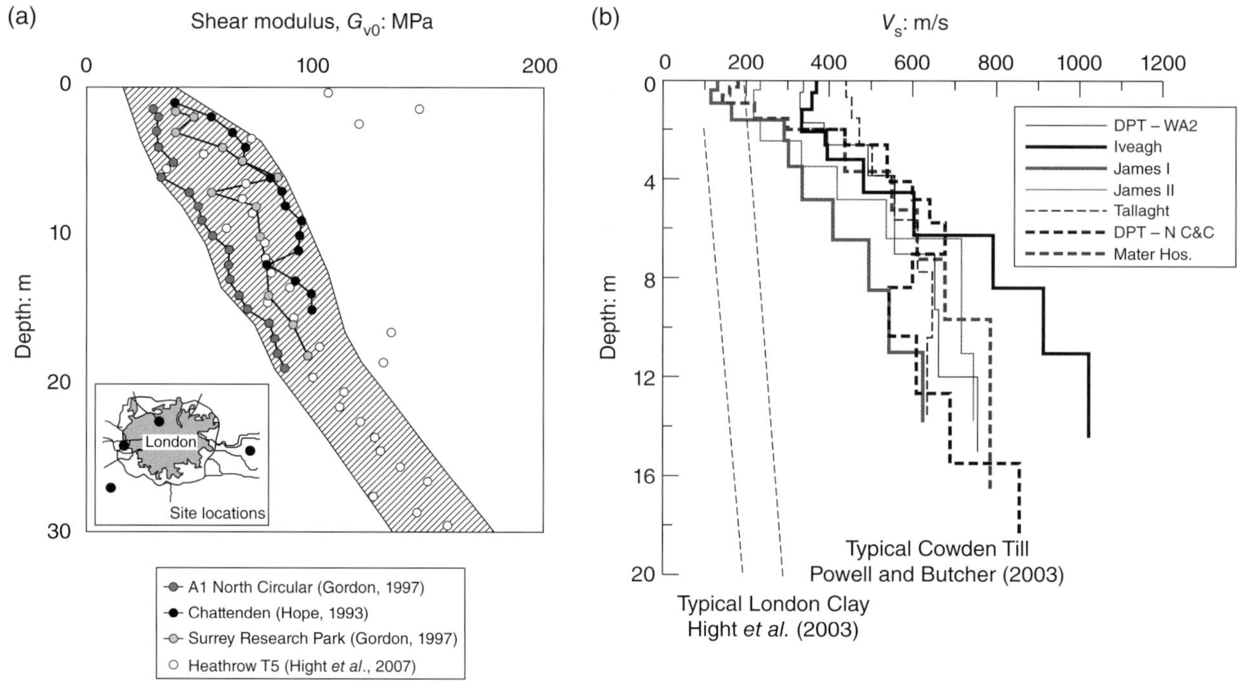

Figure 52.18 Shear modulus and shear wave velocity profiles for some stiff clays: (a) London Clay (After Clayton, 2011); (b) Dublin Boulder Clay (After Long and Menkiti, 2007)

low sensitivity (sensitivity is the ratio of peak to remoulded undrained shear strength). Clays which plot close to or above the sedimentation compression line (SCL) are also likely to be lightly overconsolidated or normally consolidated but will be of high sensitivity.

Any sites underlain by clays that plot above the SCL are likely to be extremely challenging; and these clays have been associated with significant geotechnical problems. Conventional empirical methods are likely to be unreliable, and specialist advice and more sophisticated investigation and analysis are likely to be necessary. The selection of parameters, such as permeability, coefficient of consolidation, coefficient of earth pressure and so on tends to be challenging. If required, then usually more sophisticated ground investigation methods and specialist engineering expertise are likely to be needed to measure and select appropriate values. Some common foundation design scenarios that lead to different strength or stiffness parameters being assumed for design are illustrated in **Figure 52.16**.

Undrained shear strength of overconsolidated clays: When selecting undrained strengths, the following should be noted (**Figure 52.20**).

- Low-plasticity clays – undrained strengths tend to be sensitive to small changes in natural moisture content, hence any poor handling of samples, leading to drying or wetting, will lead to large increases/decreases in undrained strength (relative to actual in situ strengths). Alluvial, estuarine and glacial clays often have a pronounced macrofabric (i.e. silt/sand seams within the clay) which often leads to unconsolidated undrained triaxial tests recording much lower strengths relative to actual in situ strengths – due to loss of moisture from silt/sand seams and then softening of adjacent clay.
- High-plasticity clays are often fissured when heavily overconsolidated – undrained strengths are sensitive to fissure spacing and orientation (are laboratory specimens large enough to be representative?) and mean effective stress. Hence, unconsolidated undrained triaxial tests often exhibit large scatter due to sampling disturbance leading to changes in mean effective stress within the sample, and due to variable effects of fissures within tested specimens.
- Field penetration testing – often give more reliable indications of actual in situ variability than routine laboratory tests. However, care is needed in correctly selecting the appropriate correlation factors (e.g. correct for specific geology, design situation, correlated against appropriate fundamental measurement of strength?). Clayton (1995) discusses the range of factors which can influence SPT results. For example, White *et al.* (2019) report that, because of changes in SPT equipment and methods, the correlation coefficient between SPT 'N' and UU triaxial undrained strengths (on driven-tube samples) should be increased by 25–35% from those proposed by Stroud (1990) based on practice during the 1980s. Hence, for high-plasticity fissured overconsolidated clay it is suggested that the correlation coefficient should be increased from an average of 4.4 to 5.7 (and similar increases should be expected for low-plasticity clays). Nevertheless, the limited accuracy of

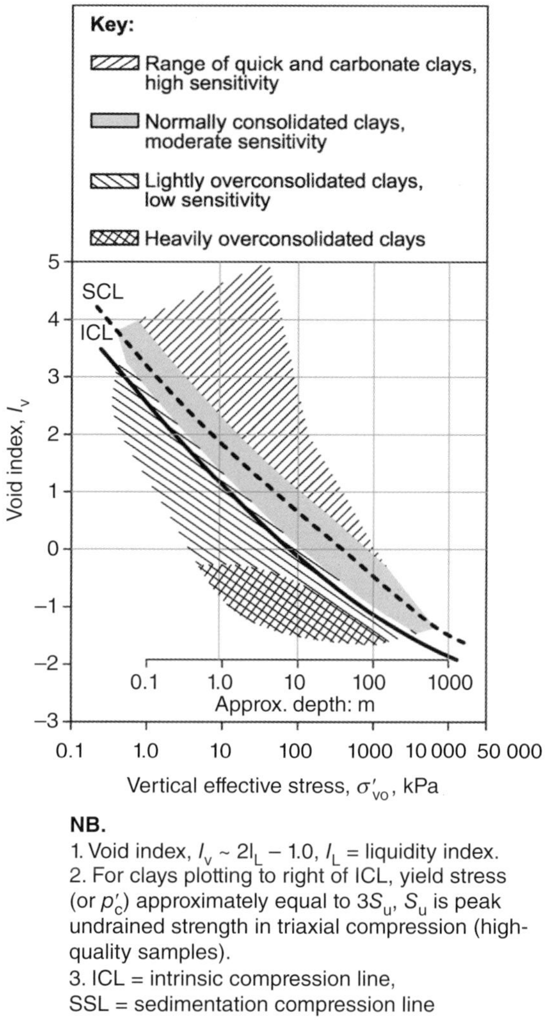

Figure 52.19 In situ stress states in terms of void index for different clay types (Data taken from Chandler *et al.*, 2004)

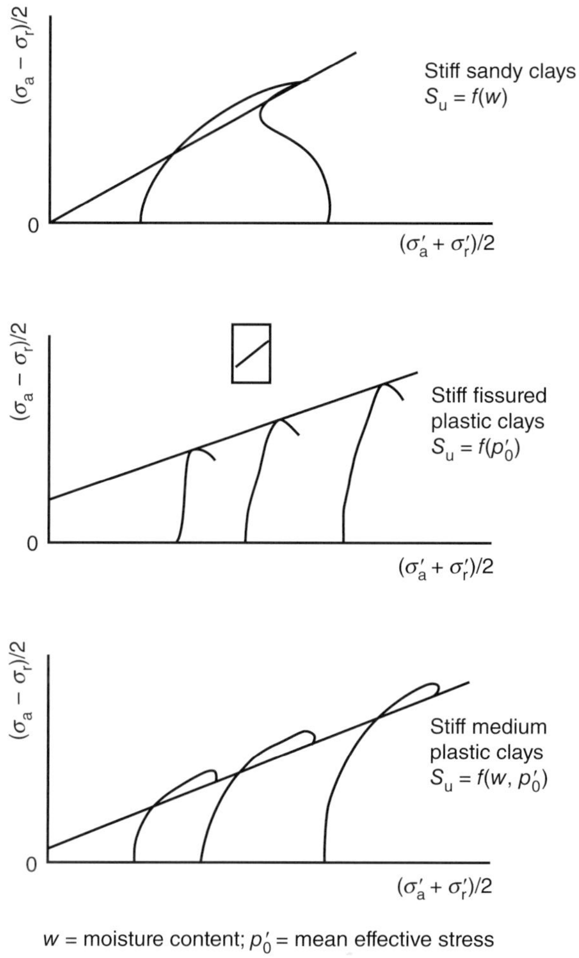

Figure 52.20 Undrained shear strength characteristics of overconsolidated clays (After Hight and Leroueil, 2003)

these sorts of empirical correlations (between two crude test methods – SPTs and UU triaxials) should be noted. Correlation factors varied between 4.5 and 7 across nine sites (for the same geology), and at one site the correlation factor increased from 5.1 to 6.6 with increasing depth. Correlations with CPT data (if installation is practical) tend to be more reliable since the CPT is carried out on a more consistent and repeatable basis than the SPT.

Stiffness and compressibility of overconsolidated clays: Poulos *et al.* (2001) provide some approximate relationships between normalised shear modulus and factor of safety; an example is shown in **Figure 52.21.** Based on this, the following can be concluded.

- The secant shear modulus decreases with a decreasing factor of safety.
- For a given factor of safety, the secant shear modulus for the shallow footing is lower than for a laterally loaded pile, which is lower than for a vertically loaded pile. This difference reflects differences in the induced strains around a particular type of foundation.

The relationships shown in **Figure 52.21** are intended mainly for preliminary assessment purposes, since many simplifying approximations have been made to develop the plotted curves. However, these relationships emphasise the point that different elastic moduli should be anticipated for different foundation applications and the mobilised secant modulus will drop rapidly as the factor of safety reduces. O'Brien (2015) discusses a practical framework for assessing the nonlinear stiffness characteristics for a wide range of soils and rocks.

One of the practical implications of nonlinear elastic behaviour is that for most foundation engineering applications the mobilised 'elastic' modulus will exhibit an increase with depth, even when the soil strength is judged to be reasonably constant (Jardine *et al.*, 1986). For many practical applications, the soil strength will also increase with depth. Hence, if using linear elastic analyses, then the appropriate

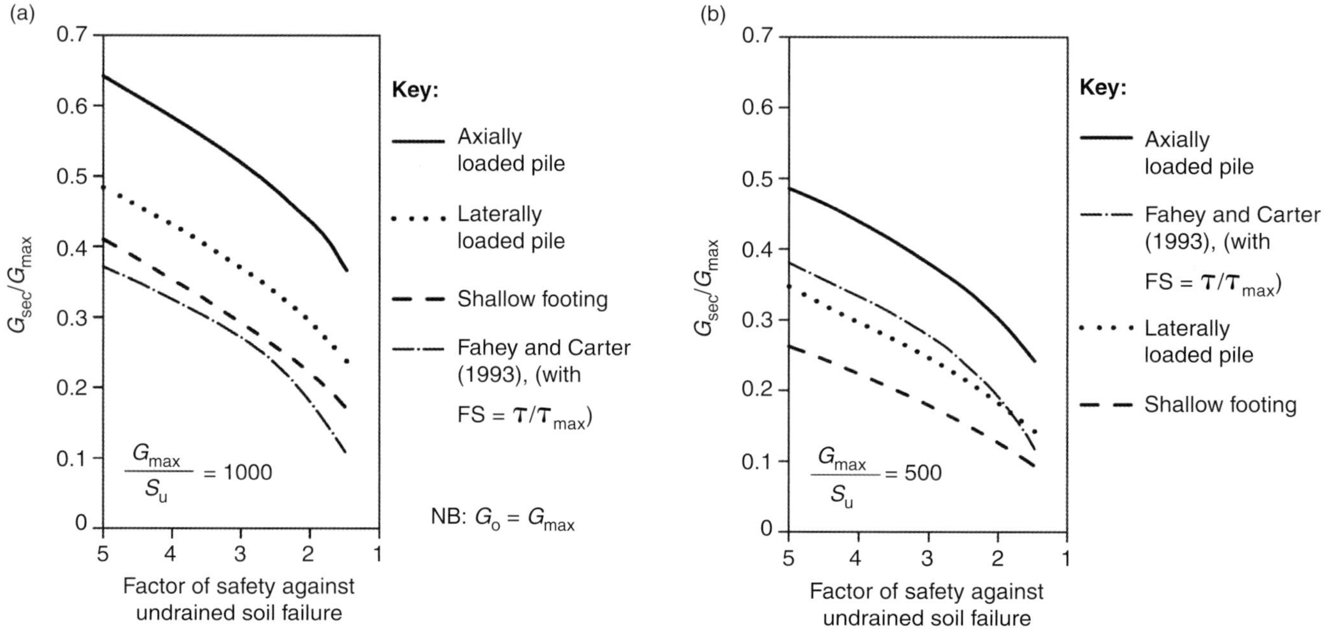

Figure 52.21 Mobilised secant shear modulus, various foundations on clay (Reproduced from Poulos *et al.*, 2001; all rights reserved)

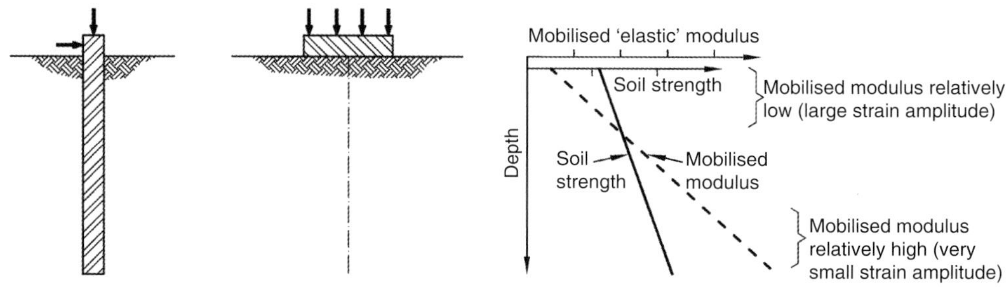

Figure 52.22 Mobilised 'elastic' modulus against depth, typical scenarios

profile of the elastic modulus would usually increase rapidly with depth, as in **Figure 52.22**. For example, at the QE2 Conference Centre in London (Burland and Kalra, 1986) the selected profile for the elastic moduli for the London Clay (based on a back-analysis of several case histories across Central London), for a raft of approximate plan dimensions 65 × 75 m, exhibited a variation in the drained Young's modulus for vertical loading, varying with depth from a value of about 100 times the clay's undrained shear strength, S_u, at the clay surface, to about 450 S_u at a depth of 20 m below the clay surface. These drained stiffnesses were, on average, several times higher than those derived from routine oedometer tests.

Vardanega and Bolton (2011) present a chart (**Figure 52.23**) relating normalised shear modulus against shear strain as a function of liquid limit for fine-grained soils (i.e. clays, silts), derived from an international laboratory testing database. Provided good estimates of G_o are available (e.g. derived from

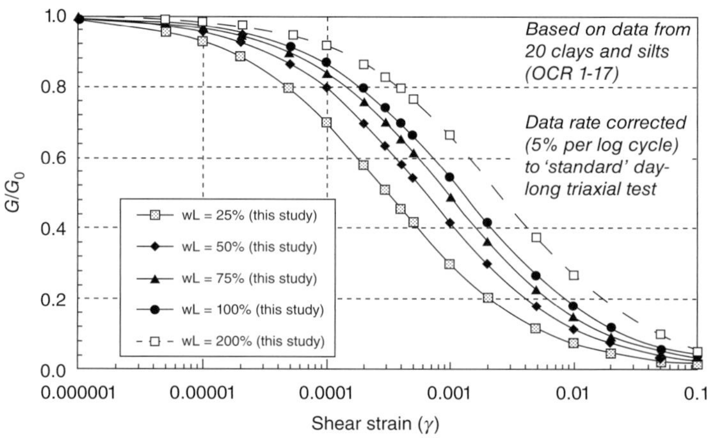

Note: Empirical curves may not be reliable for soils not represented in empirical database, such as lodgement tills

Figure 52.23 Normalised shear modulus against shear strain (After Vardanega and Bolton, 2011)

field measurements of shear wave velocity, V_s), then secant shear moduli can be estimated from small to large strain. Inevitably any empirical approach has limitations, associated with the scope of the test database. For example, stiff lodgement tills are not represented, and they tend to be more non-linear than indicated by their liquid limit. Nevertheless, this chart is simple and practical to use and should be useful for many clays and design situations.

52.6.3 Shear strength and shear modulus of sands

Shear strength: A simple and coherent approach to assessing the shear strength of sands is described by Bolton (1986)

$$\text{Peak angle of friction, } \Phi'_p = \Phi'_{cv} + \psi_d \quad (52.6)$$

$$\text{For triaxial conditions, } \psi_d = \Phi'_p - \Phi'_{cv} = 3I_R \quad (52.7)$$

$$\text{and } I_R = I_D(Q - \ln p') - 1 \quad (52.8)$$

where Φ'_{cv} is the critical state angle of friction, which is mainly dependent on particle shape, grading and mineralogy (refer to **Table 52.10**) and ψ is the dilatancy of the sand, which is mainly dependent on the relative density and the stress level at failure. I_R is a relative density index, with values between 0 and 4. I_D is the relative density, p' is the mean effective stress at failure (for $p' \geq 150$ kN/m^2) and Q depends on the strength or 'crushability' of the grains (refer to **Table 52.11**).

Dilatancy as a function of relative density and mean effective stress is given in **Figure 52.24** for quartz and feldspar sand. For micaceous and calcareous sands dilatancy will be much reduced, especially at high stress, since Q is lower (i.e. the particles are more easily crushed).

In general, making direct estimates of peak angles of friction from penetration tests is **not** recommended. An assessment of the critical state angle of friction should be made based on the sand mineralogy (e.g. from geological memoirs), visual description of particle angularity and particle grading and by reference to **Table 52.10**. The dilatancy component of the strength can be assessed from Equations 52.7 and 52.8 or **Figure 52.24**. An appropriate use of penetration test data is to assess the dilatancy component of the strength, from an assessment of relative density (i.e. I_D in Equation 52.8 and **Figure 52.24**). If necessary, appropriate laboratory tests can be carried out (Bolton, 1986). A cautious estimate of dilatancy is required. However, for dense sands it would be over-conservative to ignore dilatancy completely.

Sand type	Particle size[a]	Mineralogy[b]	Particle shape[c]	D50: mm	D10: mm	Unif coeff.	e_{max}	e_{min}	Φ'_{cv}
Ottawa	c	q	well rnd	0.75	0.65	1.2	0.8	0.49	29.5
Ottawa	m	q	rnd	0.53	0.35	1.7	0.79	0.49	30.0
Chattahoochee	m	q	s ang	0.37	0.17	2.5	1.10	0.61	32.5
Mol	f–m	q	s rnd	0.19	0.14	1.5	0.89	0.56	32.5
Ticino	c	q	s rnd	0.53	0.36	1.6	0.89	0.6	31.0
Sacramento	f–m	q+f	s ang/s rnd	0.22	0.15	1.5	1.03	0.61	33.3
Reid Bedford	f–m	q+sf	s ang	0.24	0.16	1.6	0.87	0.55	32.0
Hokksund	c	q+f	s ang	0.39	0.21	2.0	0.91	0.55	32.0
Toyoura	f	q	s ang	0.16	0.11	1.5	0.98	0.61	32.0
Mersey	f–m	q	s ang/s rnd		0.1	2.0	0.82	0.49	32.0
Milton Mines	f–m	q+f	ang	0.2	0.11	2.0	1.05	0.62	35.0
Southport	f–m		ang	0.2	0.12	1.8	0.88	0.53	35.0
Crushed quartz	f	q	v ang	0.12	0.07	2.0	1.15	0.55	36.4
Crushed feldspar	f	f	v ang	0.12	0.07	2.0	1.21	0.49	38.7
River sand and gravel	37 mm to f sand	f+q	s rnd/s ang	4.8	0.6	8			35.0
Glacial outwash sand	f–c		s ang	0.75	0.15	6	0.84	0.41	37.0
San Francisco	50 mm to fines	basalt	ang						38.0
Furnas Dam	10 mm to fines	quartzite							39.0
Granite gneiss	37 to 4 mm		ang						40.8

(a) Particle size; f = fine, m = medium, c = coarse.
(b) Mineralogy: f = feldspar, q = quartz, s = some.
(c) Particle shape: rnd = rounded, s ang = subangular, s rnd = subrounded, v ang = very angular.

Table 52.10 Critical state angle of friction, Φ'_{cv}, for some sands and gravels (Based on Bolton, 1986; Stroud, 1990)

Sand mineralogy	Q
Quartz and feldspar	10
Limestone	8
Quartz and glauconite	7.8
Calcareous	7.5
Chalk	5.5

From Bolton (1986) and Randolph *et al.* (2004); *Q* can be assumed equal to the natural logarithm of the particle-crushing strength in kN/m².

Table 52.11 Crushability *Q* for some sands

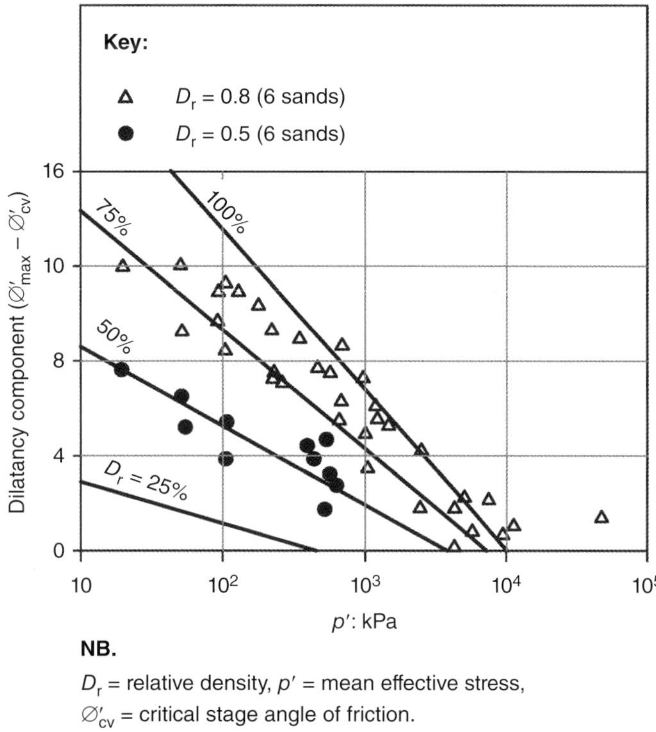

Figure 52.24 Dilation plotted against relative density and mean effective stress, coarse-grained soils (After Bolton, 1986)

Shear modulus: For sands many of the correlations between elastic moduli and penetration resistance are of extremely doubtful reliability, as discussed in Section 53.7. The ratio of Young's moduli to SPT 'N' can vary by more than an order of magnitude depending on the sand's relative density and the appropriate strain amplitude for the situation under consideration. Oztoprak and Bolton (2013) summarise a large international laboratory testing database for normalised shear modulus against shear strain (**Figure 52.25**). The spread of data is surprisingly small, given the wide range of sands and gravels in the database. Therefore, the main focus for site characterisation needs to be making reliable field measurements of shear modulus at small strain, G_o. If good G_o data are

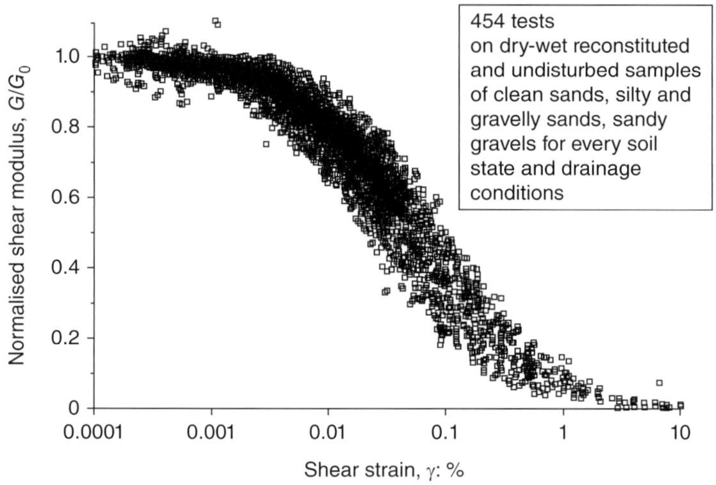

Figure 52.25 Normalised shear modulus plotted against shear strain, coarse-grained soils (sands and gravels)

available, then secant shear moduli can be readily estimated for larger strain amplitudes.

52.6.4 Strength and stiffness of weak rock

General considerations: Assessing the strength and stiffness of rocks is predominantly concerned with determining the discontinuity characteristics and degree of weathering of the rock. If the rockhead is close to the ground surface, then trial pits or trenches will be valuable for assessing the rock mass characteristics; for deeper deposits, large-diameter (P or S size core, i.e. 90–100 mm dia.) triple-tube coring is recommended (ideally combined with down-hole acoustic televiewer), together with logging by an experienced geologist. A common error is to carry out rotary coring with too small a diameter, which can be highly misleading due to excessive drilling-induced disturbance. The initial steps in assessing the rock mass properties typically comprise the following.

- Classify the rock mass with respect to its geological origin (i.e. igneous, metamorphic or sedimentary).
- Assess the rock mass quality by assessing the nature, orientation and spacing of all discontinuities and joints, including if the joints are open or closed, the strength and compressibility of discontinuities and any infill (which may include soft clay) in the joints. Practical applications normally utilise a rock mass classification system (such as the Geological Strength Index (GSI)); Hoek (1999).
- Assess the weathering grade of the rock; a number of weathering schemes have been developed following Chandler (1969), although no weathering scheme can be universally applied. In complex interlayered sequences (say of interbedded mudstone and sandstone) more weathered material may underlie less weathered material.

A general problem is that poorer zones of rock often cannot be practically sampled. Hence, reliance on laboratory tests can lead to an unsafe bias in the data on strength and compressibility, because only the higher-quality rock can be sampled and tested in the laboratory. Hence, in situ tests, such as pressuremeter and geophysics, can be invaluable.

Strength: The discontinuity characteristics, orientation and spacing need to be understood. Investigation techniques are discussed by Wyllie (1991). The Hoek–Brown failure criterion (Hoek and Brown, 1997) provides a coherent and practical approach to assessing appropriate strength parameters. However, it is only appropriate if the rock mass discontinuities are randomly orientated and spacing is small relative to the failure mechanism under consideration, otherwise the shear strength mobilised along specific discontinuities needs to be included in any analysis. CIRIA Reports C574 (Lord *et al.*, 2002) and C570 (Chandler and Forster, 2001) provide specific guidance on Chalk and Mercia Mudstone, respectively.

Rock mass stiffness: The stiffness of a rock mass is usually a fraction of the stiffness of intact rock core, depending on rock mass quality. Field geophysics, as discussed by Clayton (2011), is often the most practical means of making good estimates of mass stiffness (provided some allowance is made for changes of mobilised stiffness with strain amplitude). The ratio of field to laboratory geophysics wave velocities can also provide a good indication of rock mass quality.

52.7. Foundation selection: a brief case history

A single-span bridge was required to carry heavy rail freight traffic across an existing road. The new bridge was to be built adjacent to an old rail bridge, which was in poor condition. The original foundation design envisaged a conventional pile group (**Figure 52.26**) comprising 24 × 600 mm dia. bored piles (in a 6 × 4 group) about 6 m deep, socketed about 1 m deep into weak Coal Measures rocks. Because of the existing approach embankment a temporary excavation was required to construct the pile cap. The excavation was to be supported by a sheet-piled retaining wall. When work started on site it was discovered that this solution could not be built due to a 132 kV electricity cable passing beneath the proposed pile cap. Numerous other services were also identified, and the owners of these services refused to allow sheet piles to be driven. The original foundation design had to be abandoned. A new foundation designer was appointed to develop a revised foundation solution. The revised foundation design was carried out under a sub-consultancy agreement to the original designers, who specified the foundation loads and allowable movements. A review by the new designers indicated that the available geotechnical data were inadequate: a report indicated that coal mining was 'suspected' in the area; however, a comprehensive desk study had not been carried out; four boreholes had been drilled but these were too shallow and only penetrated a couple of metres into the rockhead; data on rock strength, groundwater and ground contamination were lacking, and the condition of the coal seams was unknown.

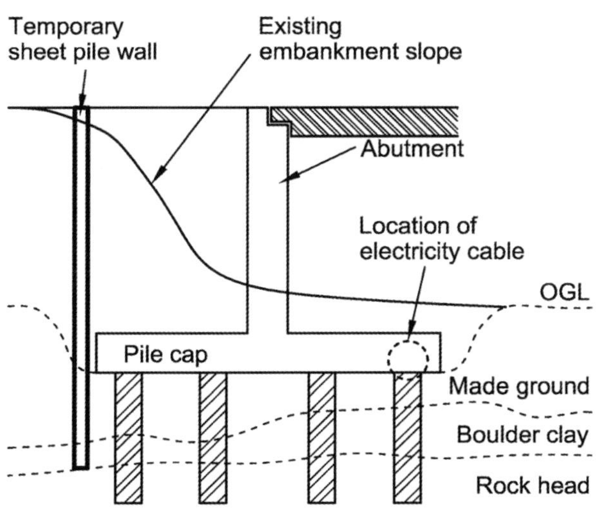

Figure 52.26 Foundation selection, case history: original unbuildable foundation design

52.7.1 Ground conditions and site history

A desk study was carried out, which indicated that coal mining had been carried out at several levels below the site and that a chemical manufacturing works had been active adjacent to the site in the late nineteenth and early twentieth centuries. Within the upper 35 m three coal seams were identified and it was expected that some or all had been extensively worked, and two shafts were identified adjacent to the site boundary. The subsequent investigation comprised four deep 100 mm dia. triple-tube rotary cored boreholes (two 25 m deep, two 40 m deep) and down-hole geophysical logging was used to supplement physical logging of the condition of the rock mass by a highly experienced geologist. The subsequent ground investigations confirmed the presence of two worked coal seams in the upper 20 m (numerous voids and soft mining backfill materials were recorded), and unstable abandoned shafts. The third coal seam, at 35 m depth, appeared to have been unworked. The Coal Measures rocks were predominantly mudstone, with variable and relatively low strengths (fluctuating between a very stiff clay and a very weak rock) and were heavily fractured (with low rock quality designation (RQD) values) in the upper zones overlying the upper two coal seams. In one area there was evidence that the coal workings had collapsed, causing significant disruption to the overlying rock mass. At greater depths, below the second coal seam, the mudstone strength increased markedly, becoming moderately weak to moderately strong, together with higher RQD values. A variable thickness of loose made ground (up to 3.5 m thick) and a discontinuous layer of firm to stiff glacial till up to 4 m thick were identified above the rockhead level. The made ground comprised industrial wastes and end-tipped clays and sands (probably from the old shaft excavations) and was chemically contaminated.

52.7.2 Site preparation

The site was in an unsafe and unstable condition. Therefore, before foundation construction could begin, it was essential to carry out preliminary works to infill and stabilise the old mine workings and remove heavily contaminated materials, which could harm the environment. Because of the scale of the works necessary (and the uncertain time period required to complete these works), this was let as a separate construction contract prior to the main foundation contract. The stabilisation works comprised mainly

- infilling the worked seams and overlying disturbed rock mass with grout of varying viscosity
- probing and uncovering the shafts, infilling with mass concrete and constructing new reinforced concrete caps.

The deep grouting involved both vertical and inclined rotary open-hole drilling (inclined holes were needed due to limited access in order to stabilise the area below the foundation and approach embankment). Low-pressure grouting was then carried out. Grout take volumes varied from about 0.2 tonne per hole to a maximum of 29 tonne per hole (introduced in three separate grout phases). These variable grout takes reflected the disturbance of the rock mass arising from local collapses of the worked coal seams.

Quality control of the grouting, including permeability testing of the grouted areas and high-quality rotary coring to check the post-grouted condition of the coal seams, was essential to verify the effectiveness of the ground stabilisation.

52.7.3 Selected foundation type, design verification and construction control

The foundation type was largely dictated by

- the limited available space along the site boundary (and existing utilities, which could not be diverted)
- an expectation that large obstructions were likely to be encountered in the made ground (demolition rubble, old concrete foundations, etc.), and the need to drill through the moderately strong mudstone at depth
- the need to minimise temporary works, in particular the avoidance of sheet piles.

The initial foundation solution was to construct at each abutment four 1200 mm dia. bored piles 26 m deep (with the pile toe below the second worked seam, but well above the third unworked coal seam), immediately below the plan location of the new bridge-deck bearings. These piles were formed as a load-bearing contiguous bored-pile retaining wall (**Figure 52.27(a)**). Unfortunately, the original designer insisted that changes to the bridge bearings would not be allowed and that these could not tolerate more than 6 mm of horizontal displacement. Despite several discussions and alternative bridge bearings (that were more tolerant of foundation movement) being proposed, this constraint was not removed. This allowable movement constraint had a significant impact on foundation costs, with the final design solution comprising 12 1200 mm dia. piles. The horizontal stiffness of the foundation was maximised by forming piled wing walls, U-shaped in plan, together with a 1.5 m deep pile cap, to ensure pile head

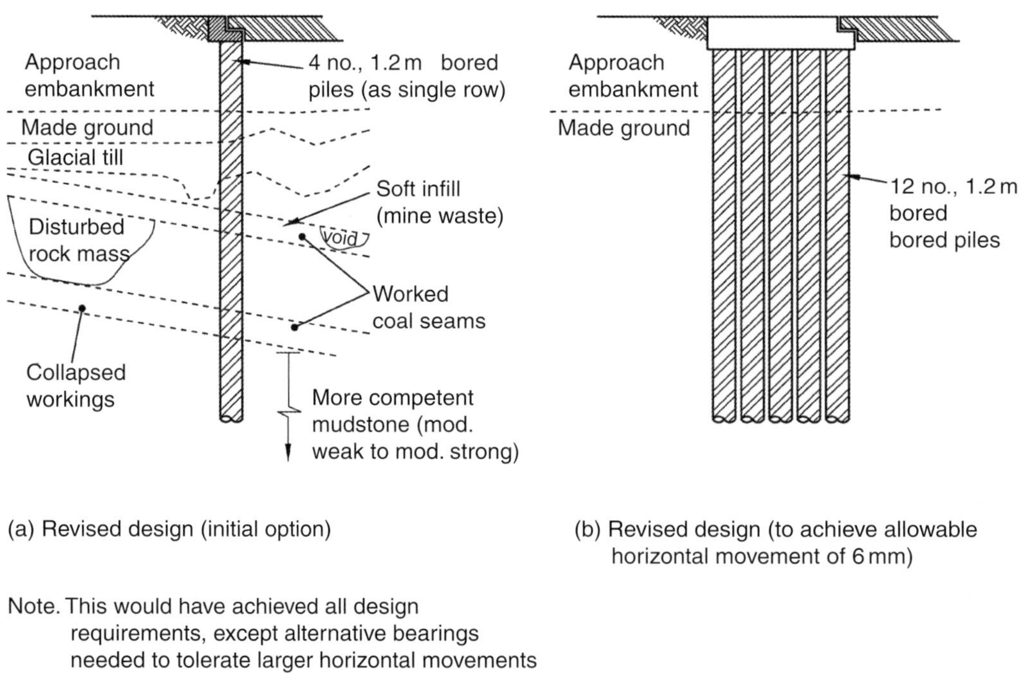

Figure 52.27 Foundation selection, case history: revised foundation design options

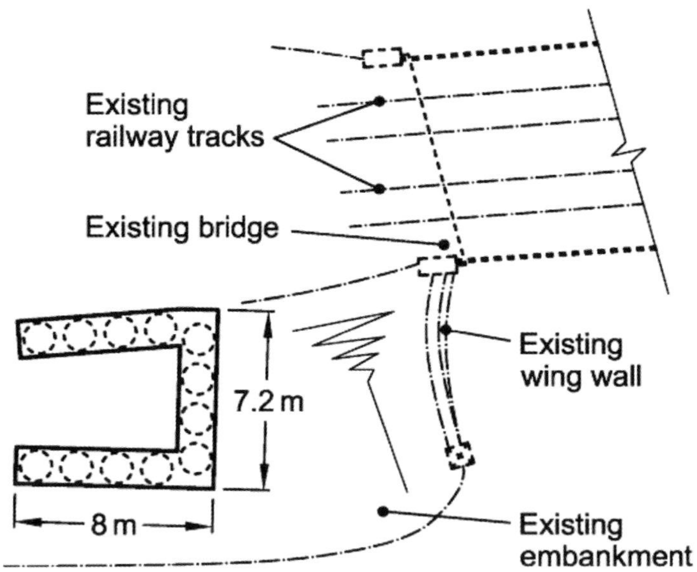

Figure 52.28 Foundation selection, case history: revised foundation design, plan

fixity (**Figures 52.27(b)** and **52.28**). The piling rig had to be sufficiently powerful to penetrate through obstructions. The contiguous bored-pile wall also acted as a retaining wall for the approach embankments.

The analyses used for design checks were relatively simple.

(*a*) Vertical capacity: the shaft capacity in the disturbed ground above the first coal seam was ignored. Below the first seam, the shaft capacity was limited to avoid overstressing the deeper worked coal seam, see (*e*) below. A maximum ultimate shaft resistance of 240 kN/m^2 was assumed in the competent mudstone below the second coal seam, and an ultimate shaft resistance of 70 kN/m^2 was assumed in the weathered mudstones and disturbed ground between the first and second coal seams. The allowable end-bearing pressure was limited to 2000 kN/m^2 in order to minimise pile settlement. The overall factor of safety was 3.0 for the shaft and the base resistance (to allow for the complex geology, site history and residual uncertainties). The group capacity, based on the plan area of the 'U-shaped' block, was more critical than the sum of the individual pile capacities.

(*b*) Vertical deformation: since limited end-bearing pressures were assumed, and a high factor of safety was being used, the total settlement was expected to be less than 10 mm overall (allowing for group effects) and less than 5 mm after deck construction.

(*c*) Horizontal capacity: this was based on the method of Broms (refer to Fleming *et al.*, 1992).

(*d*) Horizontal deformation: this was based on linear elastic pile-group analyses (refer to Fleming *et al.*, 1992).

(*e*) A maximum allowable bearing stress on the coal seams of less than 150 kN/m^2 was assumed. This was assessed by assuming an equivalent raft at two-thirds of the working shaft depth (below the first seam) and a 1 in 4 (horizontal to vertical) load spread above the 'raft' and 1 in 2 (horizontal to vertical) below.

A relatively small number of piles formed the bridge abutments; hence, it was not cost-effective to carry out preliminary or working pile tests. It was more economical to use a large factor of safety rather than carry out relatively expensive load tests on large piles (if the piling contract comprised a much larger number of piles, then the cost–benefit balance would move towards carrying out preliminary tests and reducing the factor of safety).

The main construction controls were therefore

- to log the pile bore arisings and verify that the depth was sufficient so as to be below the deepest worked coal seam, and that the piles were founded in competent mudstone at the specified depth
- to carry out cross-hole sonic logging to check the pile concrete integrity was adequate.

52.7.4 Concluding remarks

This case history highlights the importance of the following.

- Careful collation of all relevant information for a site into a desk study report, including geology, historical activities (coal mining in this case) and existing site constraints (such as existing services). The original foundation design was unbuildable due to the presence of a 132 kV electricity cable. The presence of unstable mine workings beneath the proposed pile group had not been identified by the original designers.
- High-quality ground investigation data. The initial investigation had been poorly designed and supervised, the boreholes were too short and important tests to identify the mechanical and chemical characteristics of the soils and rocks had not been scheduled. During the second ground investigation, high-quality logging of the rock mass characteristics by an experienced geologist was profoundly important in deciding upon the appropriate toe level for the piles.
- Risk management. The site was in an unsafe condition and the mine workings had to be stabilised by grouting prior to bringing large (100 tonne) piling rigs onto the site. There were several intrinsic uncertainties associated with the grouting works (timescale, grout volumes, etc.); hence, from a commercial perspective, it was best to implement these works as a preliminary contract, prior to the main bridge-construction contract.
- Allowable movement criterion. The onerous limit on the allowable horizontal movement meant that the final design had three times the number of piles than an alternative option (which would have been technically viable if the movement limit had been 15 mm rather than 6 mm).

52.8. Overall conclusions

- Foundation engineering requires a broad range of knowledge and skills in construction methods; ground–structure interaction; geology; soil and rock mechanics; and understanding relevant case histories. The foundation designer must be prepared to ask questions and discuss these with other specialists. Good foundation design is usually the result of a multidisciplinary team effort.

- The main types of foundations are shallow (pad, strip, raft) and deep foundations (piles, barrettes, caissons). Hybrid foundations include aspects of both shallow and deep foundation designs; examples include deep ground improvement and piled rafts.
- Design requires a consideration of: foundation stability and deformation; buildability and risk management; sustainability, costs and programme for construction. The 'geotechnical triangle' is a valuable framework for ground risk management, and all aspects of the triangle must be considered. Peer Assist carried out at an early stage can provide significant benefits.
- Foundation selection requires a consideration of the five Ss: **s**oil or rock types (and groundwater); **s**tructure type, including loads, allowable movements; **s**ite, including available space, headroom, neighbouring structures and sensitive environmental areas; **s**afety, in particular the site history may have left the site in an unstable or chemically contaminated state; **s**ustainability, as there are often major opportunities when designing foundations, including the reuse of excavated materials, foundation reuse and the use of the ground as an energy source.
- Major and costly decisions are often taken on foundation design, on the basis of arbitrary and over-conservative limits on total and differential settlements. Routine limits for total settlement vary between 40 and 100 mm, depending upon the soil and foundation type. The actual risk of damage is dependent on differential movements and the construction sequence. The timing of the installation of sensitive components, such as architectural cladding, internal machinery such as lifts and escalators and so on is particularly important.
- Selection of parameters, such as strength and, especially, deformation modulus, is challenging, because of the wide range of factors that may influence the value which is mobilised in a particular design situation, and the inherent limitations of analytical methods to model the complexity of real-ground and ground–structure interaction behaviour. Routine investigation methods induce significant disturbance to the ground, which exacerbates the difficulties associated with interpretation of factual data. Ground investigations should include both in situ and laboratory testing. Continuous profiling of soil properties, either by methods such as CPTs or continuous sampling and detailed logging, is vitally important to understand the soil fabric.
- In the last few decades there have been major improvements in understanding the deformation characteristics of soils. Stress–strain behaviour is highly nonlinear, and the stiffness at very small strain (often known as G_o or G_{max}) is an important reference or 'anchor' point. Whenever possible, field geophysics should be used to measure G_o values in situ. Stiffness degrades rapidly with increasing strain amplitude, so that at large strains it is often more than an order of magnitude smaller. Empirical correlations are available for the variation of normalised shear modulus with shear strain. These show consistent trends for many soils. Conventional tests, such as oedometer tests, measure stiffness (or compressibility) at large strains. Different foundation types (or different applied loads) induce different strain amplitudes in the ground, hence the *mobilised* deformation modulus can vary significantly (even for the same soil type). Commonly, the *mobilised* deformation modulus will increase rapidly with depth below the ground surface. Because foundations design is often controlled by SLS requirements, significant economies and sustainability benefits are often possible if nonlinear stiffness behaviour is taken into account.

References

Adam M and Markiewicz A (2009) Energy from earth-coupled structures, foundations, tunnels, and sewers. *Géotechnique* **59(3)**: 229–236.

Atkinson JH (2000) Non-linear soil stiffness in routine design. *Géotechnique* **50(5)**: 487–508.

Barker RM, Duncan JM, Rojiani O, Ooi PSK and Tan CK (1991) *Manuals for the Design of Bridge Foundations*. NCHRP Report 343. Transport Research Board, Washington, DC.

Bjerrum L (1963) Discussion. *Proceedings of the European Conference SMFE, Wiesbaden, Germany*, vol. 2, p. 135.

Bolton MD (1986) The strength and dilatancy of sands. *Géotechnique* **36**: 147–167.

Boone ST (1996) Ground movement related building damage. *Geotechnical and Geoenvironmental Engineering ASCE* **122(11)**: 886–896.

Burland JB (1989) Ninth Laurits Bjerrum Memorial Lecture: 'Small is beautiful' – the stiffness of soils at small strains. *Canadian Geotechnical Journal* **26(4)**: 499–516.

Burland JB (1990) On the compressibility and shear strength of natural clays. *Géotechnique* **40**: 329–378.

Burland JB and Burbidge MC (1984) Settlement of foundations on sand and gravel. *Proceedings of the Institution of Civil Engineers, Part 1* **78**: 1325–1381.

Burland JB and Kalra JC (1986) Queen Elizabeth II conference centre. *Proceedings of the Institution of Civil Engineers, Part 1* **80**: 1479–1503.

Burland JB and Wroth CP (1974) *Settlement of Buildings and Associated Damage: Settlement of Structures*. Pentech Press, Cambridge, UK.

Burland JB, Broms BB and De Mello VFB (1977) Behaviour of foundations and structures. *Proceedings of the 9th International Conference on Soil Mechanics and Foundation Engineering (ICSMFE), Tokyo, Japan*, vol. 1, pp. 495–546.

Butcher PS, Powell JJM and Skinner HD (2006) *Reuse of Foundations for Urban Sites: A Best Practice Handbook*. BRE Press, Watford, UK.

Butler FG (1975) Heavily overconsolidated materials: a state of the art review. *Proceedings of the Conference on Settlement of Structures, Cambridge, UK*. Pentech Press, Cambridge, UK, pp. 531–578.

Chapman T, Anderson S and Windle J (2007) *Reuse of Foundations*. CIRIA Report C653. CIRIA, London, UK.

Chandler RJ (1969) The effect of weathering on the shear strength properties of Keuper Marl. *Géotechnique* **19**: 321–334.

Chandler RJ and Forster A (2001) *Engineering in Mercia Mudstone*. CIRIA Report C570. CIRIA, London, UK.

Chandler RJ, de Freitas MH and Marinos D (2004) Geotechnical characterisation of soils and rocks: a geological perspective. *Proceedings of the Skempton Conference, London, UK*, vol. 1, pp. 67–102.

Clayton CRI (1995) *The Standard Penetration Test (SPT): Methods and Use*. CIRIA Report 143. CIRIA, London, UK.

Clayton CRI (2011) Stiffness at small strain: research and practice. *Géotechnique* **61(1)**: 5–37.

Day RW (2000) *Geotechnical Engineer's Portable Handbook*. McGraw-Hill, New York, NY.

DFI (2015) What is sustainability? *Deep Foundation Magazine*, Nov./Dec.

Fang HY (2002) *Foundations Engineering Handbook*, 2nd edn. Kluwer Academic, Dordrecht, the Netherlands.

Fleming GK, Weltman AJ and Randolph MF (1992) *Piling Engineering*, 2nd edn. Halsted Press, Ultimo, Australia.

Grant R, Christian JTC and Vanmarke EH (1974) Differential settlement of buildings. *Geotechnical and Geoenvironmental Engineering ASCE* **100(GT9)**: 973–991.

Hight DW, Leroueil S (2003) Characterisation of soils for engineering purposes. *Characterisation and Engineering Properties of Natural Soils* **1**: 255–362.

Hoek E (1999) Putting numbers into geology: an engineer's view point. Second Glossop lecture. *Quartely Journal of Engineering Geology* **32**: 1–19.

Hoek E and Brown ET (1997) Practical estimates of rock mass strength. *International Journal of Rock Mechanics, Mining Sciences and Geomechanics Abstracts* **34**: 1165–1186.

HSE (2015) *Managing Health and Safety in Construction. Construction (Design and Management) Regulations*. HSE, London, UK.

Jamiolkowski M and Lo Presti DCF (2003) Geotechnical characterisation of Holocene and Pleistocene Messina sand and gravel deposits. *Characterisation and Engineering Properties of Natural Soils* **2**: 1087–1120.

Jardine RJ, Symes MJ and Burland JB (1984) The measurement of soil stiffness in the triaxial apparatus. *Géotechnique* **34(3)**: 323–340.

Jardine RJ, Potts DM, Fourie AB and Burland JB (1986) Studies of the influence of nonlinear stress–strain characteristics in soil–structure interaction. *Géotechnique* **36(3)**: 377–396.

Long M and Menkiti CO (2007) Characterisation and engineering properties of Dublin Boulder clay. *Proceedings of the Second International Workshop of Characterisation and Engineering Properties of Natural Soils, Singapore*, pp. 2003–2045.

Lord A, Clayton C and Mortimore R (2002) *Engineering in Chalk*. CIRIA Report C574. CIRIA, London, UK.

Menteth T (2022) HS2: Keltbray trials 'revolutionary' low carbon pile at Euston. *Ground Engineering*, Feb.

Morton K and Au E (1974) Settlement observations on eight structures in London. *Proceedings of Conference Settlement and Structures, Cambridge, UK*. Pentech Press, Cambridge, UK, pp. 183–203.

Moulton LK (1985) *Tolerable Movement Criteria for Highway Bridges*. Report No. FHWA/RD-85/107. Federal Highway Administration, Washington, DC.

NHBC (2023) *Standards 4.2: Building Near Trees*. NHBC, Milton Keynes, UK.

O'Brien AS (2015) Geotechnical characterisation for underground structures: recent developments and applications. *Geotechnical Engineering for Infrastructure and Development: XVI European Conference on Soil Mechanics and Geotechnical Engineering, London, UK*, vol. 7, pp. 3729–3734.

Oztoprak S and Bolton MD (2013) Stiffness of sands through a laboratory test database. *Geotechnique* **63(1)**: 54–70.

Peck RB (1962) Art and science in subsurface engineering. *Géotechnique* **12**: 60–68.

Poulos HG, Carter JP and Small JC (2001) Foundations and retaining structures: research and practice. In *Proceedings of the 15th ISCMGE, Istanbul, Turkey*, vol. 4, pp. 2527–2606.

Powderham AJ (2010) Managing risk through safety driven innovation. *Proceedings of the Deep Foundations Institute, London, UK*.

Powderham AJ and O'Brien AS (2020) *The Observational Method in Civil Engineering, Minimising Risk, Maximising Economy*. CRC Press, London, UK.

Randolph MF, Jamiolkowski MB and Zdravkovic L (2004) Load carrying capacity of foundations. *Advances in Geotechnical Engineering, the Skempton Conference*. Thomas Telford, London, UK, vol. 1, pp. 207–240.

Reynolds T and Bates P (2009) The potential for timber piling in the UK. *Ground Engineering*, Jan.

Ricceri G and Soranzo M (1985) An analysis of allowable settlements of structures. *Rivista Italiana di Geotecnica* **19(4)**: 177–188.

Robertson PK (2009) Interpretation of cone penetration tests: a unified approach. *Canadian Geotechnical Journal* **46**: 1337–1355.

Rowe PW (1972) The relevance of soil fabric to site investigation practice. 12th Rankine Lecture. *Géotechnique* **22(2)**: 195–300.

Schmertmann JH, Hartman JP, Brown PR (1978) Improved strain influence factor diagrams. *Journal of Geotechnical Engineering, ASCE* **104(GT8)**: 1131–1135.

SCOSS (2009). Guidance note: independent review through Peer Assist. www.scoss.org.uk (accessed 19/05/2023).

Skempton AW and MacDonald DH (1956) The allowable settlements of buildings. *Proceedings of the Institution of Civil Engineers, Part 3* **5**: 727–768.

Stroud MJ (1990) The standard penetration test: its application and interpretation. *Proceedings of Penetration Testing in the UK, London, UK*. Thomas Telford, London, UK, pp. 29–49.

Terzaghi K (1939) Soil mechanics: a new chapter in engineering science. *Proceedings of the Institution of Civil Engineers* **7**: 106–141.

Terzaghi K (1956) Discussion. *Proceedings of the Institution of Civil Engineers, Part 3* **5**: 775.

Tomlinson DB (1987) *Foundation for Low Rise Buildings*. BRE, Watford, UK.

TRL (1994) *Study of the Efficiency of Site Investigation Practices*. TRL Project Report 60. TRL, Crowthorne, UK.

Vardanega PJ and Bolton MD (2013) Stiffness of clays and silts: normalizing shear modulus and shear strain. *Journal of Geotechnical and Geoenvironmental Engineering, ASCE* **139**: 1575–1589.

Vardanega PJ and Bolton MD (2011) Practical methods to estimate the non-linear stiffness of fine-grained soils. *Proceedings of the International Symposium on Deformation Characteristics of Geomaterials, Seoul, South Korea*, pp. 372–379.

Wahls HE (1981) Tolerable settlement of buildings. *Journal of Geotechnical Engineering, ASCE* **107(GT11)**: 1489–1504.

White F, Ingram P, Nicholson D, Stroud M and Betru M (2019) An update of the SPT-C_u relationship proposed by M. Stroud in 1974. *Proceedings of the XVII ECSMGE-2019, Reykjavik, Iceland*, pp. 1–8.

Wyllie DC (1991) *Foundations on Rock*. Spon Press, London, UK.

Zhang LM and Ng AMY (2005) Probabilistic limiting tolerable displacements for serviceability limit state design of foundations. *Géotechnique* **55(2)**: 151–161.

It is recommended this chapter is read in conjunction with

- Chapter 9 *Foundation design decisions*
- Chapter 19 *Settlement and stress distributions*

All chapters in this book rely on the guidance in Sections 1 *Context* and 2 *Fundamental principles.* A sound knowledge of ground investigation is required for all geotechnical works, as set out in Section 4 *Site investigation.*

Chapter 53
Shallow foundations

Tony O'Brien Mott MacDonald, UK
Luca Sodano Mott MacDonald, UK
Imran Farooq Mott MacDonald, UK

The design of shallow foundations is exceptionally wide ranging, and can vary from simple strip footings for a lightly loaded domestic property to the design of a large raft for a nuclear power station. The likelihood of geohazards affecting a site must always be assessed, irrespective of the scale of the project. This is best assessed by a comprehensive desk study and a site walkover by an appropriately experienced specialist. The prime requirement for successful design and construction is an understanding of the site history and the ground and groundwater conditions across the site. Sophisticated analysis will not compensate for a lack of knowledge of these issues. Prior to detailed analysis it is worthwhile considering the likely behaviour of the whole structure (which may comprise multiple closely spaced strip or pad footings) in the context of the overall geology, adjacent structures and the site topography. The potential risk of adverse interactions within, beneath and around the proposed structure should be considered. Foundation settlement will often be a more critical consideration than bearing capacity. Simple methods of analysis are usually adequate; the biggest challenge is to make reliable estimates of the compressibility of the ground. The bearing capacity and settlement of rocks will be sensitive to the orientation, spacing and characteristics of discontinuities in the rock mass, the degree of weathering and voids (either natural or artificial).

doi: 10.1680/icemge.66830.0801

CONTENTS

53.1. Introduction

The geotechnical design of shallow foundations can vary from very simple footings for small, lightly loaded domestic buildings to the complex requirements of a raft for a nuclear power station. **Table 53.1** poses some questions that should help the foundation designer identify the most appropriate level of sophistication for the ground investigation(s), analyses and subsequent supervision and monitoring. The table is based upon the geotechnical categories proposed in Eurocode 7. For Category 1 structures, site-specific calculations are often unnecessary, and design can be based on 'presumed bearing pressure values' from local codes or past experience. The geotechnical category should be assessed at the start of the project, and may need to be revised as more information becomes available. This process indicates the level of effort required during design and construction, together with the level of geotechnics expertise required for the project. It is particularly important to note that the geotechnical category is not solely related to the size and complexity of the superstructure. For example, a simple structure may be Category 3 if the site geology is particularly complex. For shallow foundation design, many textbooks focus on the analysis of bearing capacity and settlement; however, prior to these design checks, the overall site geology, site history and associated hazards must be carefully considered. As discussed in Chapter 52 *Foundation types and conceptual design principles*, the **five Ss** (**S**oil, **S**tructure, **S**ite, **S**afety, **S**ustainability) should be assessed for the site before deciding that shallow foundations are the most appropriate foundation type.

In order to facilitate these decisions, a comprehensive desk study should be produced for the site under consideration. This will then provide a coherent framework for planning ground investigations, and in some cases (if the geology is well known) allow a preliminary design to be carried out.

This chapter: considers causes of foundation movement (which can be independent of applied loads from the superstructure); introduces some of the construction processes that need to be considered; and gives an overview of foundation layout issues and potential interaction effects. Bearing capacity and settlement analysis methods are summarised in Sections 53.5 and 53.6. The key topic of parameter selection is discussed in Section 53.7.

Because of space constraints, the behaviour of soft clays (and associated analytical methods and parameters) is not discussed in detail. Normally, they would not be considered to be appropriate founding strata. However, if soft clays are present within the foundation's zone of influence, then expert advice should be sought.

53.2. Causes of foundation movements

It is important to recognise that foundations may settle excessively due to a variety of causes that are independent of the applied bearing pressures from the superstructure. **Table 53.2** provides a summary in the context of shallow foundations. Reference should also be made to Chapter 52 *Foundation types and conceptual design principles* (Section 52.2, **Figure 52.7**) and Chapter 9 *Foundation design decisions* (**Table 9.1**). The global movements that may be caused can be extremely

Questions	Category 1, simple	Category 2, intermediate	Category 3, complex
1	Is the structure small and simple?	If NO to Category 1 and 3 questions (refer to Notes below)	Is structure very large or unusual?
2	Is site geology well known and understood?		Is site geology complex or poorly understood?
3	Are ground and groundwater conditions known from comparable experience to be sufficiently straightforward that routine methods may be used for investigations, design and construction?		Does it involve unusual or exceptionally difficult ground or groundwater conditions?
4	If excavation below the water table is required, does comparable experience indicate that it will be straightforward?		Does it involve abnormal risks?
5	Is the site free from abnormal risks, such as onerous differential settlement criteria, unusual or significant cyclic loading or 'global' hazards such as earthquakes?		Can significant economies or reductions in construction programme be achieved by utilising Category 3 methods?
6			Can improvements in economy, safety or sustainability be achieved by utilising Category 3 methods?
7			Are the differential settlement criteria onerous (refer to Chapter 52 *Foundation types and conceptual design principles*)?
8			Are the loading conditions unusual or exceptional (e.g. high ratio of live to dead load, frequent large cyclic loads, large lateral loads)?
9			Is the site subject to significant global hazards, such as earthquakes?
Likely investigation requirements	Desk study, plus a few trial pits, penetration tests and index tests to confirm site geology and ground conditions at founding levels	Desk study plus full range of conventional investigation methods. Sufficient testing to characterise ground properties below foundations (with particular attention to the zone within two-thirds of foundation width), and assess construction methods and project risks	Desk study, plus comprehensive suite of ground investigations, including: highest-quality sampling techniques; advanced in situ and laboratory testing; may be a need for large-scale instrumented load tests
Likely analysis methods	If deemed necessary: clays – simple bearing capacity checks plus high FoS (3.0) to ensure low settlement (refer to Section 53.5); sands+rocks – simple empirical checks to verify low settlement (for sands, refer to Section 53.6)	Analytical methods, outlined in Sections 53.5 and 53.6	Nonlinear numerical modelling or more advanced analytical methods for bearing capacity and settlement assessment
Likely construction supervision requirements	Visual inspection of ground at the foundation level	Specialist supervision during key phases, local probing and testing may be needed to verify design assumptions; monitoring of, say, groundwater may be needed to control key project risks	Full-time specialist supervision plus subsurface instrumentation to monitor ground response

Category 1 vs 2: if the answer to one of the Category 1 questions is 'No', then it should be categorised as Category 2 unless previous experience clearly indicates that Category 1 is a suitable category. Category 3 vs 2: Category 3 is usually reserved for exceptionally demanding sites and projects, hence usually for Category 3 to apply the answer to more than one question would be 'Yes'. The selection of an appropriate category requires experienced judgement. The category may change if project constraints are changed (e.g. allowable movements for foundations, or state/sensitivity of adjacent infrastructure, environmental constraints, etc.).

Table 53.1 Shallow foundations: likely design issues and requirements related to geotechnical categories

Cause of movement	Comments	Typical mitigation measures
Non-engineered fills	Large movements may occur due to physical, chemical or biological degradation mechanisms. Refer to Chapters 29 *Arid soils* and 58 *Building on fills*. Examples include collapse settlement on wetting; expansion of iron and steel slag; decomposition of organic material; and gas or leachate generation	Ground improvement can be effective in loose fills (e.g. methods such as dynamic or vibrocompaction, preload/surcharge) provided organic materials are largely absent. If deformation mechanisms are due to chemical or biological factors, conventional compaction methods may be ineffective
Interaction effects	The construction of new structures, services, earthworks and so on may induce global ground movements, which can cause additional foundation movements	Design and construction of any new infrastructure need to assess likely interaction effects and mitigate if damaging movements are anticipated. Appropriate phasing of construction sequences may reduce adverse effects
Vibration	Loose or very loose coarse-grained soils (silts, sands and gravels) are vulnerable to large differential settlement when subjected to vibration from an external source (e.g. road traffic, machinery)	Ground improvement to densify the coarse-grained soils into a dense or very dense state (e.g. dynamic or vibrocompaction); refer to Chapter 59 *Design principles for ground improvement*
Erosion	Uniform fine sand or silt is particularly vulnerable to erosion from surface or groundwater flows	Variety of erosion-protection measures, from natural, well-graded, granular filter layers, through to propriety geotextile products
Vegetation and shrink–swell	Magnitude of movement will increase as plasticity index of clay increases, and 'water demand' of trees increases. Not relevant for non-plastic soils; refer to NHBC (2023)	For ground affected, or likely to be affected, by trees, then special foundation design measures will be required; refer to (NHBC 2023). This may include piled foundations (able to resist heave and swelling pressures) and suspended floors on void formers to avoid swelling forces and upward movement
Adjacent slopes	Sloping ground near foundations can increase the risk of bearing capacity failure. The slope may (without foundation loading) be close to failure, and near-surface soils on the slope may be vulnerable to seasonal movement and creep	Detailed inspection and assessment of slopes. Take into consideration for bearing capacity checks, and assess 'global' stability. If factors of safety are inadequate, then slope stabilisation will be necessary
Bulk earthworks, cut and fill	Large-scale filling or excavation will change the stresses in the underlying natural ground; hence, settlement or swelling (respectively) will result. The fill itself may also settle or swell if it is fine-grained	If possible, allow sufficient time for settlement or swelling to occur prior to building new foundations. If not feasible, then foundation deformation calculations will need to allow for additional ground movements due to cut and fill
Metastable or voided ground	Ground previously mined is likely to be unstable. Natural deposits, such as limestone and chalk, can contain voids and caves	Investigations need to be carefully designed and targeted. If voids are suspected, then large-scale grouting is normally required; refer to Chapter 59 *Design principles for ground improvement*
Frost	Silts, chalk and fine sands can be vulnerable to frost heave	Locate foundations below depth affected by frost. In the UK this is typically about 0.5 m

Groundwater changes are also a common cause of global ground movements; refer to Table 9.1 in Chapter 9 *Foundation design decisions*

Table 53.2 Potential causes of foundation movement independent of applied bearing pressures

large (some can be in excess of 1 m, and they are often in excess of 100 mm) and can lead to ultimate or serviceability limit state (ULS, SLS) failures. Identification of these hazards is critically important and is usually best achieved by a combination of a desk study (to understand the site's geology and history) and a site walkover by suitably experienced specialists. If the hazard is present it is usually best to eliminate it directly, or to relocate the foundations if feasible.

The most common cause of excessive foundation movement for shallow foundations in the UK is probably seasonal shrink–swell movement induced by trees. These differential movements are particularly significant in high-plasticity clay where trees are growing, or where previous tree growth has desiccated the clay (movements of circa 100–150 mm) and have caused structural damage. **Figure 53.1** gives an example of seasonal moisture-content change for a high-plasticity clay affected by high-water-demand trees. **Figure 53.2** shows a comparison between moisture content and soil suction measured at a high-plasticity clay site for a location remote from trees and in an area affected by trees, and illustrates the large stress (i.e. suction) changes which may be induced by trees. NHBC (2023) provides comprehensive guidance and simple rules for appropriate foundation depths, which will reduce the risk of structural damage due to the effects of trees. They define 'shrinkable' soils as soils that contain more than 35% fines (particle size <60 μm) and with a *modified* plasticity index (MPI) of more than 10% (see **Table 53.3**). The risk of excessive foundation movement will also depend on the type of trees growing close to the foundations, and NHBC provides water-demand categories

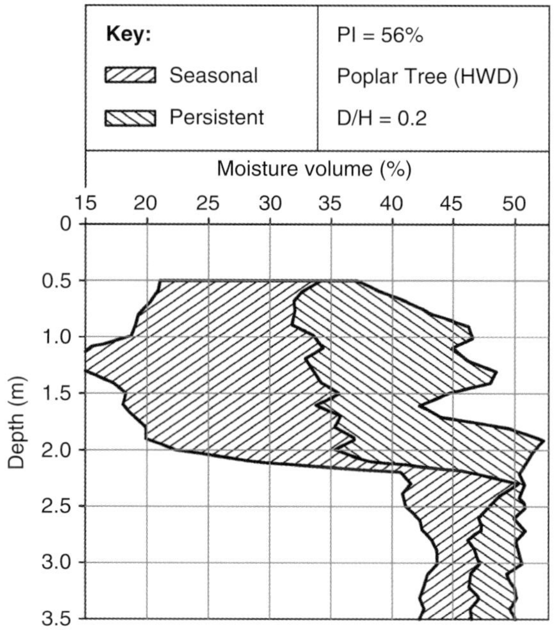

NB.
D = distance from tree
H = height of tree
PI = plasticity index of clay
HWD = high water demand
'Persistent' = reduction in moisture content due to tree, will only increase back to natural once tree dies or is cut down
'Seasonal' = change in moisture content between summer and winter

Figure 53.1 Example of changes in moisture content due to trees, Gault Clay

Modified plasticity index	Volume-change potential
≥40%	High
20–40%	Medium
<20%	Low

[a] MPI, Modified plasticity index = [(Plasticity index) (particles < 425 μm, %)]/100; that is, if PI = 30% and only 50% of particles pass through a 425 μm sieve then MPI = 15%.
[b] Shrinkable soils are vulnerable to volume change as their moisture content changes.

Table 53.3 Volume-change potential (Data taken from NHBC, 2023)

and mature tree heights for common species. The appropriate foundation depth is dependent on the MPI of the soil, the water demand and the mature tree height of existing trees or new trees that may be planted; **Figure 53.3** provides an example of one of the NHBC charts. Remote from significant vegetation, a foundation depth of 0.9 m is usually adequate for foundations on clay.

53.3. Construction processes and design considerations

53.3.1 Foundation excavations

Foundation excavations will usually be on the critical path for a project's construction programme, since many activities will not be able to start until they are completed. Hence, careful planning based on appropriate desk study and ground investigations is needed. **Figure 53.4** provides a flowchart that outlines the main decisions that need to be made. Although the temporary works will usually be a contractor's responsibility, the foundation designer needs to consider these matters to check that the proposed works are buildable, safe and do not endanger adjacent infrastructure, buildings and so on. Sometimes it is quicker and cheaper to excavate an area in one large operation than in several small independent excavations. In many soils and rocks, excavation faces can stand at steep angles (in excess of 45°) for a short period following excavation. With increasing time (perhaps varying from minutes to months) the steep temporary slopes will fail and move towards slope angles similar to those of 'permanent' slopes (refer to Chapter 72 *Slope stabilisation methods* and Chandler, 1984). The period of time before a temporary slope fails is not predictable solely by theoretical calculations. The use of local experience in selecting appropriate temporary slope angles is extremely important. **Figure 53.5** highlights some typical issues. Health and safety regulations require that all excavations deeper than 1.2 m should be designed, and inspected regularly, by a competent person. Contingency measures should be planned and implemented if adverse slope movements are detected.

53.3.2 Groundwater control

Groundwater control is an important consideration for any site that has a water table above the foundation excavation level. In most parts of the UK the groundwater table will be within a few metres of the ground surface. Poor groundwater control is probably the most common cause of ground-related problems during construction. The consequences of poor groundwater control can vary from project delays to catastrophic collapse of excavations. The adverse effects can be classified as follows.

- *Seepage effects.* If seepage is concentrated across a relatively small area, then steep hydraulic gradients may arise. Steep hydraulic gradients can lead to: excavation slope instability; erosion of silts and sands, leading to piping, which can cause excavation basal instability or loss of lateral support to retaining walls; rapid weakening of otherwise strong competent soils (such as stiff clays, weak rocks) – layered 'laminated' soils with alternating low- and high-permeability layers (such as stiff clay with thin silt or sand layers) are particularly vulnerable.
- *Hydraulic uplift.* In alternating layers of clays and sands instability will arise if the groundwater pressure in a sand layer (below the excavation base) exceeds the overburden stress due to the overlying clay layers between the excavation level and the sand layer.

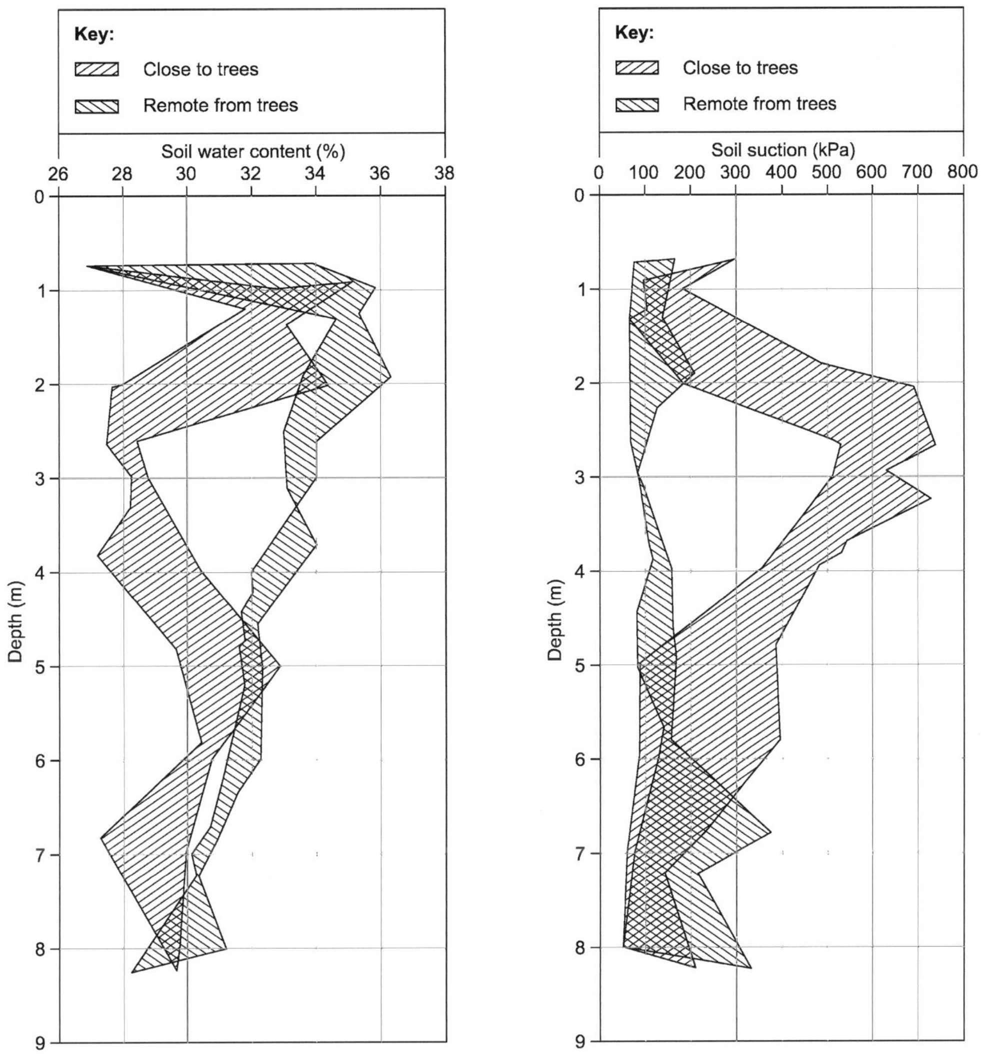

Figure 53.2 Water content and soil suction profiles remote from and close to trees (Modified from Crilly and Driscoll, 2000)

- *Excessive foundation settlement.* Instability may not develop. However, if competent foundation soils, such as dense sands, stiff clays or weak rocks are affected by groundwater seepage, then this can cause reductions in the relative density of the sands or gravels, softening of stiff clays or joint opening in fractured rocks. Hence, excessive settlement may then occur once the foundation loads are applied. The original ground stiffness parameters of the soils or rocks in their undisturbed state will no longer be relevant. An example of the foundation problems caused by groundwater is given in Chapter 9 *Foundation design decisions*.

Figure 53.6 provides a sketch of the range of problems and some possible solutions to strong groundwater seepage and hydraulic uplift. The two main methods of avoiding serious groundwater problems are

- lowering the groundwater pressure by installing pumped wells or well points around the perimeter of the excavation
- increasing the length of the groundwater flow path, either by installing deeper cut-off walls, providing more extensive grouted zones or loading the excavation base with appropriately graded filter material.

Passive relief, or bleed, wells (drilled holes, backfilled with permeable filter material) can reduce the risk of hydraulic uplift, since the groundwater pressure gradient will be reduced from sub-artesian to hydrostatic. In highly permeable soils passive relief wells may be 'swamped' and become ineffective.

For dewatering and depressurisation schemes, three main factors need to be considered.

- How much groundwater is present and how quickly it will be replenished – that is, what groundwater flow rates need to be managed. In turn, this depends on ground permeability, continuity of permeable layers, presence of free water sources (rivers and lakes) and their connectivity to permeable layers.

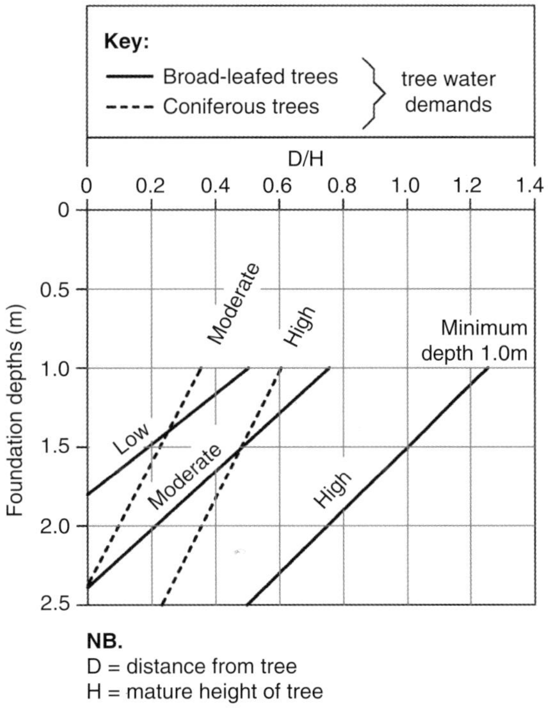

Figure 53.3 Example of NHBC foundation depth charts (soil with high volume-change potential, MPI ≥40%). Shrinkable soils are defined as soils with more than 35% fines (finer than 60 μm) with an MPI of 10% or greater (Reproduced from NHBC, 2023)

- How the properties of the soils within and around the boundaries of the excavation will be affected. For example, the risk of loss of fines or risk of settlement once pore water pressures are reduced.
- How much water will need to be disposed of and the ease of disposal. Groundwater disposal requires approval of the regulator. The state of the groundwater, particularly the degree to which it may be contaminated, will often be a critical consideration.

If the above factors are favourable, then dewatering can be relatively cheap, when pre-planned, compared with using cut-off walls or grouting. For complex situations, a combination of cut-off walls, grouting and depressurisation may be necessary. Chapter 80 *Groundwater control* provides more detailed guidance on groundwater control. A wide range of different dewatering systems are available, from discrete pumped wells to well points; the pros and cons are discussed in Chapter 80 *Groundwater control*. As a general rule, it is always preferable to locate the groundwater control systems so that groundwater is extracted *outside* the footprint of the excavation, since this positively reduces hydraulic gradients below the base of an excavation. Sump pumping from shallow holes at the base of an excavation is often used since it is relatively cheap and simple. However, it is not recommended except for in low-risk situations, such as the occasional removal of small volumes of seepage water from perched water contained in limited zones of sand or gravel. Continuous pumping from sumps located inside the excavation is high risk since hydraulic gradients are not reduced and there is a high probability of loss of fines and piping instability.

53.3.3 Buoyancy and flotation

For sites with a high groundwater table or those sites that are vulnerable to flooding, it is important to check that the structure has an acceptable factor of safety against buoyancy failure, both during construction and in its permanent condition. Basement-type structures and cellular rafts (compensated foundations) are particularly vulnerable. Groundwater exerts uplift pressure on any overlying non-vertical surface (**Figure 53.7**). There are four main methods of reducing the risk of buoyancy failure.

- Ensure that the substructure, once sealed against water inflow, always exerts a larger downward force than the uplift force from the water. The substructure must be structurally strong enough to withstand the water pressure acting on all external faces.
- Allow the substructure to flood once the water level goes beyond the critical level for buoyancy. This can be acceptable during construction, but it is not appropriate for permanent structures! A simple means of achieving this is to install 'bleed' or pressure relief wells through the base of the substructure (**Figure 53.7**) or install holes through the sides of the substructure. The 'spill-over' level for the relief wells would be set at the level which maintains the target factor of safety against buoyancy.
- Have dewatering or depressurisation systems available to keep the water level below the critical level. However, the reliability and maintainability of these systems needs to be carefully considered. Dewatering can be appropriate to deal with groundwater, but will not be effective against surface-water flooding.
- Ground anchors can be installed to anchor the structure to competent ground (**Figure 53.7**).

During construction, when the structure is incomplete, an additional stabilising load (e.g. granular fill, concrete blocks) can be used. The kentledge can be removed once the permanent structure has sufficient weight. To prevent overturning, the centre of the uplift force should be close to the centre of gravity of the structure.

53.3.4 Preparation of foundation formation

The settlement of shallow foundations can be sensitive to how carefully the excavated surface is prepared prior to pouring foundation concrete. In wet climates, such as the UK, in addition to properly dealing with groundwater, care is required when excavating in inclement weather to minimise softening of exposed soils to rainfall. A practical means of minimising softening is to carry out bulk excavation to within, say, 1 to 0.5 m of the foundation level. The final excavation stage is subsequently carried out as a final task, in short bay widths, and the exposed soil or rock is immediately (or within the same work shift) protected by structural-quality blinding concrete. When dewatering is carried out the groundwater surface level should be kept a minimum of 0.5 m, and preferably more than 1 m, below the foundation level.

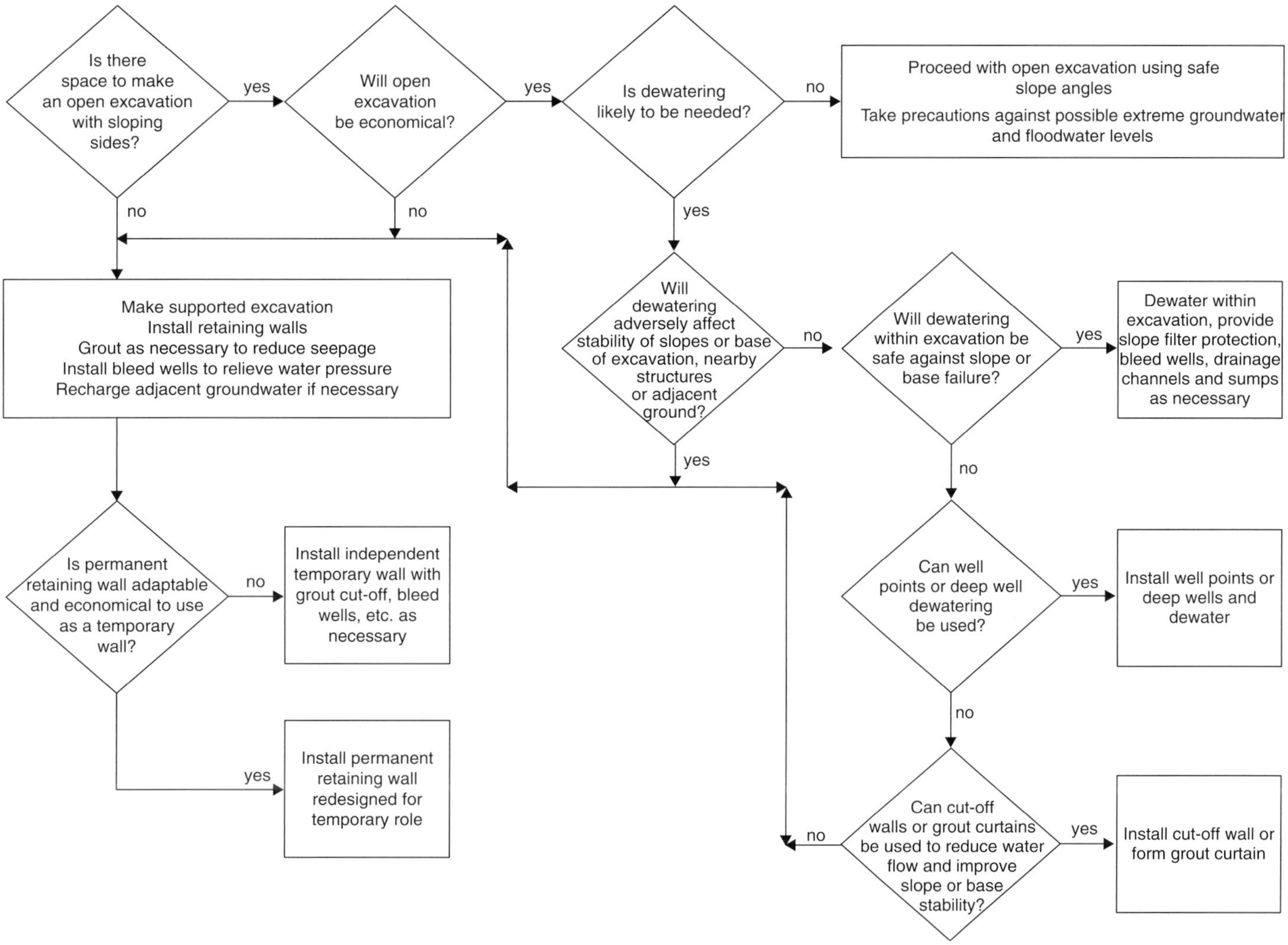

Figure 53.4 Excavation method decision diagram (Reproduced from Cole, 1988)

Foundation works for many sites may require the removal of old foundations, large boulders and 'soft spots'. This may result in an irregular profile, partly below the foundation level. A final trim should be made to remove disturbed material. Then voids below foundation level will need to be infilled with good-quality coarse-grained fill or mass concrete. Coarse-grained fill is acceptable only if it can be properly compacted. The possible presence of soft spots can be assessed by probing (Chapter 47 *Field geotechnical testing*) or by observing if ruts develop when large plant traverse the formation. At this stage, if it is apparent that the ground conditions are different to those assumed by the designer, then it is vital that the designer is informed and has the opportunity to modify the foundation design.

53.4. Applied bearing pressures, foundation layout and interaction effects

53.4.1 Applied bearing pressures

Gross and net bearing pressures are defined in Chapter 52 *Foundation types and conceptual design principles* (Section 52.5). The net bearing pressure is used to check the factor of safety against bearing capacity failure and to assess the total settlement of foundations. This is the principle that is used to design 'floating' or 'compensated' foundations (Chapter 52 *Foundation types and conceptual design principles*, **Figure 52.1**), where the weight of the structure is close to the weight of soil removed at the base of the foundations, so that the net bearing pressure on the soil is close to zero. D'Appolonia and Lambe (1971) summarise observed movements of compensated foundations on very soft clays, all of which exhibited very small settlements, which terminated shortly after construction.

The applied loads on a foundation will usually comprise a variety of different types of load, including dead loads (weight of substructure, superstructure) and transient (or live) loads, including wind or impact (or accidental) loads. It is important to consider the different types of loading in the context of the ground behaviour adjacent to the foundation. For example, it would be inappropriate to carry out a drained bearing

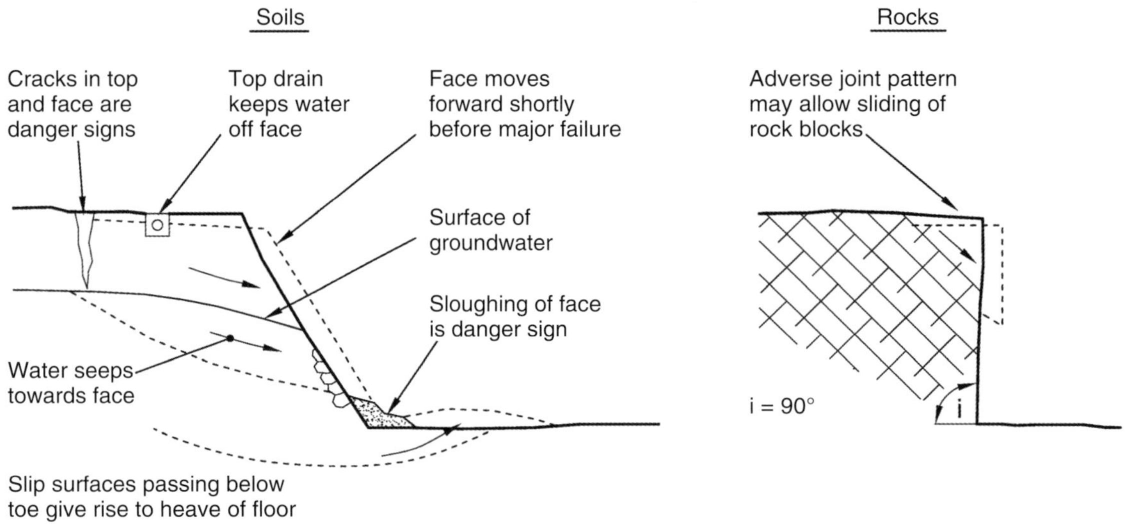

Typical short-term slope angles: stiff clays 35 to 60°; sandy clays 35 to 50°; these angles assume dry conditions. These slopes will degrade and become unstable over time, which may vary from minutes to months depending on mass permeability of ground. Laminated/varved clays are particularly vulnerable to rapid degradation. Good drainage, groundwater control and protection of slope from weather are critically important.

Beware adverse joint pattern, may allow sliding of blocks. Some rocks, notably mudstones/shales deteriorate rapidly once exposed to air/water (wetting/drying).

Figure 53.5 Excavation side-slope stability (Reproduced from Cole, 1988)

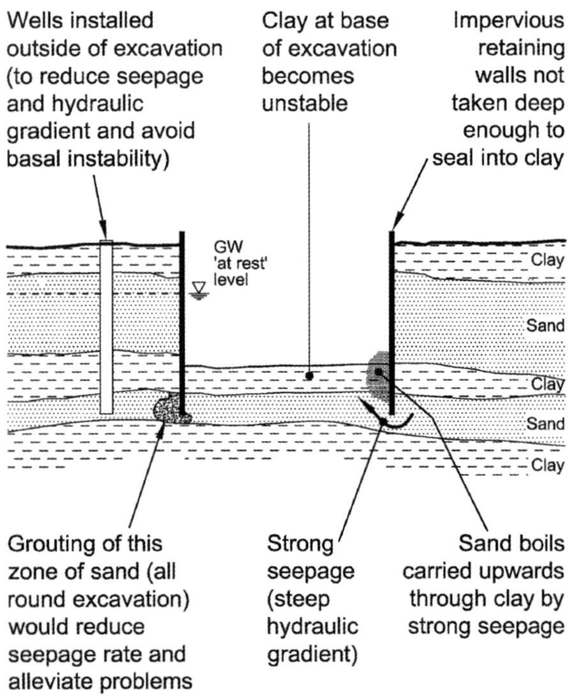

Figure 53.6 Excavation instability due to groundwater and common mitigation measures (Modified from Cole, 1988)

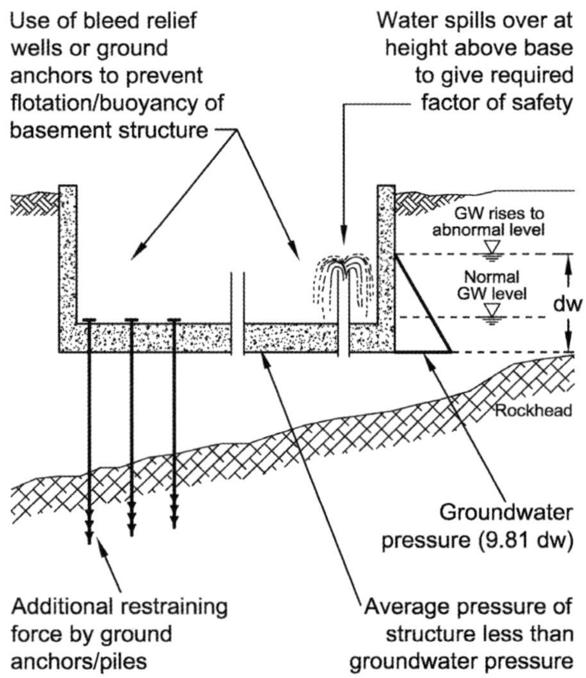

Figure 53.7 Buoyancy and flotation failure of substructure and common mitigation measures (Modified from Cole, 1988)

capacity analysis for clays for a load combination that included an impact load that lasts a fraction of a second. In contrast, foundations on sands can be vulnerable to significant incremental increases in settlement if subjected to regular transient loads, due to a ratcheting-type deformation mechanism (e.g. tall towers or chimneys subjected to high wind loads). When calculating the long-term settlement of clays, occasional transient loads can be ignored and usually only a fraction of the live load would be considered (say, 25–30%).

53.4.2 Foundation layout and interaction effects

When carrying out bearing capacity and settlement checks it is important to consider the behaviour of the whole structure, in addition to the individual foundation elements, within the context of the site and adjacent structures and topographical features. **Figures 53.8(a)** and **53.8(b)** highlight an example of shallow foundations for a building where the strip footings are located relatively close to one another. For this situation it would be wise to carry out two separate checks.

- Check the bearing capacity and settlement of an isolated footing.
- Check the bearing capacity and settlement of a hypothetical raft beneath the entire building, subjected to an equivalent net bearing pressure (based on the overall building weight and the width and length of the structure).

Particular care is required if a foundation is subjected to a lateral load, since this increases the risk of adverse interaction effects (**Figures 53.8(c)** and **53.8(d)**). **Figure 53.8(d)** illustrates a small single-span bridge founded on shallow pad footings. Shear keys can be designed to resist the applied horizontal loads. However, if the 'passive wedges' in front of the shear keys for each bridge abutment interact, then the available passive resistance will be much less than that for a single isolated foundation. Care is also required if forming shallow foundations that are adjacent to existing shallow foundations, adjacent to slopes or if drainage/utility trenches are constructed close to shallow foundations. For these situations the bearing capacity and settlement checks must take account of the potentially adverse interaction effects.

53.5. Bearing capacity

53.5.1 General comments

The bearing capacity can be defined as the maximum load per unit area that can be imposed on a soil at a given depth before failure occurs. It is a function of many parameters, including physical properties of the soil, foundation geometry, foundation depth, soil failure mode and load inclination or eccentricity. Some design scenarios of practical significance which can result in either unsafe or over-conservative estimates of bearing capacity will be highlighted. In addition, this section will

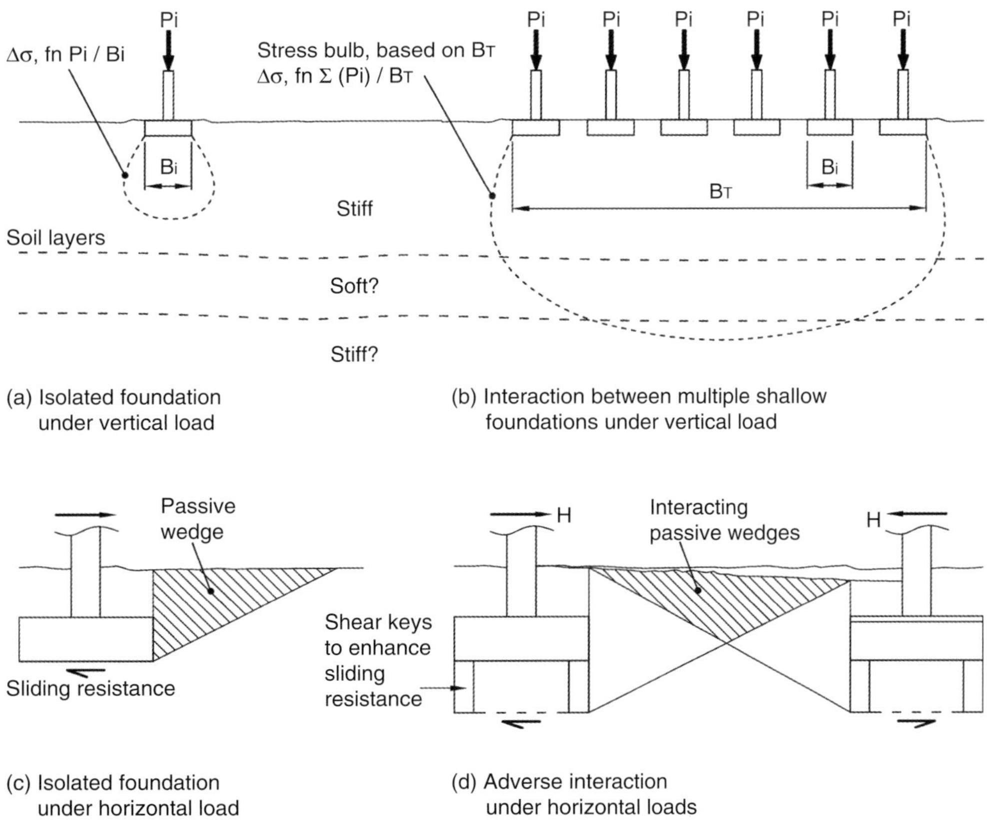

Figure 53.8 Multiple shallow foundations; examples of interactions under vertical and horizontal loads

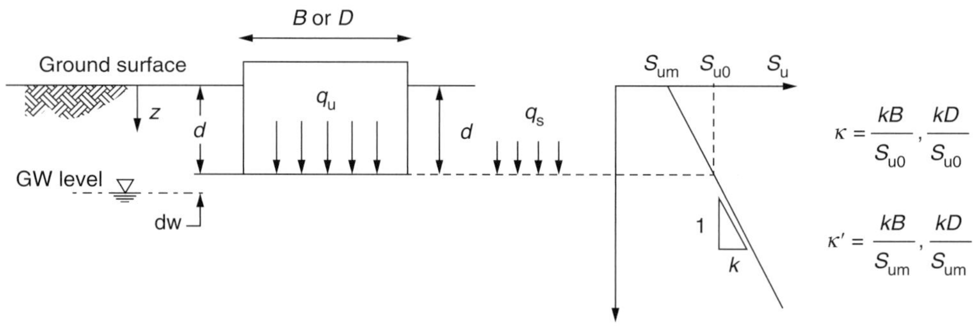

Figure 53.9 Geometry and parameters for ultimate bearing capacity; definition of degree of soil strength heterogeneity for embedded foundations (Modified from Randolph and Gourvenec, 2011)

briefly outline modern bearing capacity theory for undrained fine-grained soils, which is recommended instead of classical methods when

- undrained strength increases with depth (which commonly occurs)
- foundations are subject to combined horizontal, moment and vertical loads.

53.5.2 General bearing capacity formulae

Chapter 21 *Bearing capacity theory* should be studied for an introduction to the topic; the basic formulae for undrained saturated fine-grained soils (i.e. clays and silts) and drained soils (typically for coarse-grained soils) are given. For many practical situations the simple formulae given in Chapter 21 *Bearing capacity theory* have to be modified, and these are highlighted below and discussed in subsequent sections.

- *Undrained bearing capacity.* Increasing strength with depth (**Figure 53.9**). The classic bearing capacity formula assumes undrained strength is constant with depth, hence there are situations when modern theory should be used to check that the classic formula is not seriously in error (e.g. for wide raft foundations and/or when undrained strength increases rapidly with depth and the simple formula can then be unsafe). **Table 53.4** gives bearing capacity factors (derived by Houlsby and Martin, 2003) for surface and embedded foundations (both for uniform strength and strength increasing with depth). As the foundation shape changes and as embedment depth increases the failure mechanism is modified, and this is also influenced by the rate of strength increase with depth. As noted by Houlsby and Martin (2003), when strength increases with depth, the shape and depth correction factors can be quite different to those assumed in classical theory. The bearing capacity factors given in **Table 53.4** allow for depth effects. For a surface footing, the shape factor reduces below unity when strength increases rapidly with depth (Salencon and Matar, 1982).
- *Drained bearing capacity.* **Bearing capacity factor:** for coarse-grained soils, bearing capacity seldom controls design; more commonly foundation deformation is the critical consideration. Nevertheless, it should be noted that a bewildering range of values for the effective stress bearing capacity factor, N_γ, have been published. The formulae by Terzaghi and Vesic are now considered to be grossly unconservative (Poulos *et al.*, 2001). The more conservative equations by Brinch-Hansen (1970) or Davis and Booker (1971) are recommended. The Brinch-Hansen equation is given in Chapter 21 *Bearing capacity theory*. It should be noted that as the angle of friction increases beyond 30°, the bearing capacity factors increase rapidly, and even small increases in the angle of friction lead to dramatic increases in bearing capacity. **Soil compressibility**: important for relatively loose sands and sands with non-silica mineralogy, or sands with intermittent organic pockets and/or clay layers (refer to Section 53.5.4). This correction will lead to a reduction in bearing capacity. **Stress level**: as mean effective stress at failure increases, the ability of sands to dilate is reduced. Hence, as foundation depth or foundation width increases, the mobilised angle of friction will tend to reduce below the peak friction angle (Randolph *et al.*, 2004). **Groundwater table and groundwater pressures**: crucially important, because this can lead to a significant reduction in bearing capacity, especially if sub-artesian or artesian groundwater conditions are present (refer to Section 53.5.5).
- *Soil layering* (refer to Section 53.5.6). This can be a common scenario. Particular care is required if a more competent layer overlies a weaker layer (e.g. a sand over clay, or a stiff clay over softer clay). A punch-through failure can occur at a smaller load than indicated by the bearing capacity of the more competent upper layer, depending upon the foundation dimensions and thickness of the more competent layer.
- *Combined vertical, moment, horizontal loads.* Conventional theory should only be considered adequate if the resultant load inclination is small (i.e. within the middle third of the footing width) or for a load inclination of less than about 18° (Poulos *et al.*, 2001). For more severe load inclinations the simplifying assumptions and multiple correction factors become increasingly less reliable, and for clays can become grossly over-conservative. The ULS checks for shallow foundations often include complex load combinations due to the effects of various extreme events (such as storm wind loads, large vehicle impacts, etc.). A radically different approach to dealing with complex load combinations (through the use of VHM failure envelopes – see Randolph *et al.*, 2004) has been developed over the last 20 years or so (mainly stimulated by offshore engineering). These methods are briefly outlined in Chapter 21 *Bearing capacity*

k' (kD/S_{um})	d/D	N_{cv*} (based on roughness factor α)	
		$\alpha = 0.4$	$\alpha = 1.0$
0.0	0.0	5.97	6.05
	0.1	6.24	6.30
	0.25	6.57	6.61
	0.5	7.02	7.05
	1.0	7.70	7.71
	2.5	8.99	8.99
1.0	0.0	6.65	6.95
	0.1	6.87	7.14
	0.25	7.11	7.35
	0.5	7.39	7.60
	1.0	7.79	7.94
	2.5	8.53	8.62
2.0	0.0	7.20	7.63
	0.1	7.30	7.68
	0.25	7.39	7.73
	0.5	7.53	7.80
	1.0	7.78	7.98
	2.5	8.43	8.53

The vertical uniaxial foundation capacity (V_{ult}) for a circular foundation can be derived as

$V_{ult} = N_{cv*} \times A \times S_{u0}$

where N_{cv*} is the bearing capacity factor for surface and embedded circular foundations (both for uniform strength and strength increasing with depth) provided in the table; S_{u0} is the undrained strength of the soil at the underside of the foundation (see **Figure 53.9**).

For rectangular foundations the vertical capacity can be obtained by using a shape factor (s_c) as

$V_{ult,Rectangle} = V_{ult,circle}/s_c$

where s_c is the shape factor for rectangular foundations and can be derived from (Randolph *et al.*, 2004, based on Skempton, 1951):

$S_c = S_{circle} - (S_{circle} - 1)B/L$

where $B \leq L$ and S_{circle} is the ratio between the bearing capacity factor of circular and strip foundations and is a function of the non-dimensional factor k as summarised below (Martin, 2001).

$k = KB/S_{u0}$ or KD/S_{u0}	0	1	2
S_{circle}	1.18	1.05	1.00

Table 53.4 Bearing capacity factors $N_{cv*} = q_{ult}/S_{u0}$ for embedded circular smooth-sided foundations (derived by Houlsby and Martin, 2003) (Reproduced from Randolph and Gourvenec, 2011)

theory and discussed here in Section 53.5.3. Load inclination, especially for sands, can lead to a dramatic reduction in bearing capacity. A useful overview of recent developments in bearing capacity theory is given by Poulos *et al.* (2001), Randolph *et al.* (2004) and Randolph and Gourvenec (2011). Vulpe *et al.* (2014) provide a useful summary of solutions for a range of scenarios and associated references.

53.5.3 Combined vertical, moment and horizontal loads

53.5.3.1 'Effective width' method for a combined vertical and moment load

Meyerhof (1953) proposed a simple 'effective width' method, originally developed for strip foundations, to assess the impact of an overturning moment due to an eccentric, or offset, vertical load. For this situation (classified as 'one-way eccentricity'), the plan dimensions of the foundations should be reduced to

$$B' = B - 2e_x \tag{53.1}$$

and the equivalent gross bearing pressure will be

$$q_e = V/(B' \times L) \tag{53.2}$$

where B' is the corrected width, e_x is the eccentricity of the load in the width direction, V is the foundation load, B is the actual width and L is the length.

Based on q_e, the uniform net bearing pressure over the effective width for the eccentrically loaded foundation can be calculated, q_{ne}; q_{ne} should then be used for both bearing capacity and settlement checks. The alternative approach of assuming a linear variation of bearing pressure (of trapezoidal or triangular shape) is *not* recommended. If eccentricity is caused, or partly caused, by a sustained load, then the foundation can be designed more efficiently by placing the superstructure column off-centre, so that the resultant of the axial and lateral loads passes through (or closer to) the centroid of the foundation.

53.5.3.2 Undrained bearing capacity under combined vertical, horizontal and moment (VHM) loads

The original effective width method has been applied as an 'effective area' method for a range of foundation geometries and embedment ratios, and has been the most common means of assessing bearing capacity under complex loading. Rigorous analytical solutions (e.g. Randolph *et al.*, 2004) have shown that this approximate approach (together with its multiplicative correction factors) is inaccurate. **Figure 53.10** shows a rectangular foundation subjected to an eccentric inclined load (which is decomposed into a vertical and horizontal load intersecting the centre of an equivalent fictitious rectangle of smaller dimensions B′ and L′, for which the bearing capacity is calculated). The load inclination is included by further reducing the bearing capacity of the rectangle. Then the foundation capacity is defined as the vertical capacity of the equivalent rectangle. Feng *et al.* (2018) have investigated the reliability of the effective area method and concluded that

- under 'one-way eccentric loading' due to a vertical and moment load, the simplified 'effective width' method compares well with more rigorous solutions

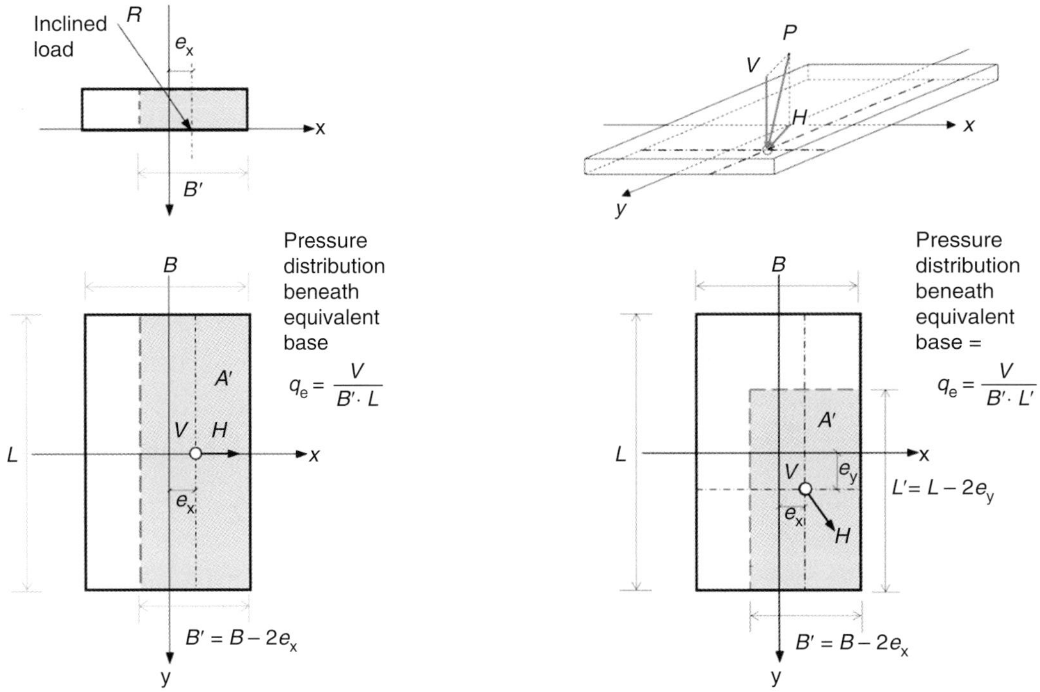

Figure 53.10 A shallow foundation subject to eccentric load and equivalent area carrying a uniform vertical pressure: (a) equivalent width under 'one-way eccentric loading'; (b) equivalent area under 'two-way eccentric loading' (Modified from Feng *et al.*, 2018)

- under 'two-way eccentric loading' (i.e. the resultant load is offset in both the x and y directions from the centroid of the foundation due to applied moments in x and y directions – **Figure 53.10**) the simplified method considerably underestimates the available bearing capacity; this is mainly because the effective contact area (between footing and soil) tends to be significantly underestimated by the simplified 'effective area' method
- if a horizontal load is applied to the foundation in addition to a moment, the available bearing capacity is again significantly underestimated.

Hence, it is recommended that the new bearing capacity methods are used to facilitate more realistic assessments of bearing capacity for shallow foundations under combined VHM loading. The procedure for developing a failure envelope for foundations of a specific shape, embedment and undrained strength increase with depth can be summarised as follows (following Randolph and Gourvenec, 2011).

- Calculate uniaxial bearing capacities for each individual load component (i.e. vertical, V_{ult}, horizontal, H_{ult} and moment M_{ult}, when $H = M = 0$, and $V = M = 0$ and $V = H = 0$, respectively) to define the apex points of the failure envelope.
- Define the shape of the failure envelope through a closed-form expression as a function of normalised loads (V/V_{ult}, H/H_{ult}, M/M_{ult}); see Vulpe *et al.* (2014) for a summary.
- Identify if the design load condition falls within the failure envelope; if so, the design is 'safe'. If not, the foundation capacity must be increased (i.e. area or embedment ratio increased) or the design loads reduced.

Table 53.5 provides a summary of relevant formulae to assess the uniaxial foundation capacities (V_{ult}, M_{ult} and H_{ult}), assuming that the undrained shear strength is constant with depth and zero tension is mobilised at the underside of the foundation. **Figure 53.11** provides a comparison between ULS under general VHM loading from classical theory and recently developed VHM failure envelopes. It is clear from **Figure 53.11** that even a modest embedment creates significant additional resistance to VHM loads which is ignored by simple theory (mainly because the interaction between horizontal and moment resistance is not represented in classical theory). Hence, there is an opportunity to avoid unnecessary use of piled foundations and use shallow pad foundations more often. Pad foundations are generally easier, cheaper and quicker to construct than piled foundations and avoid the mobilisation of a specialist contractor. The equations given in **Table 53.5** should be conservative for most applications since the critical ULS load event is often associated with a rapidly applied load (e.g. due to wind loads, vehicle impact or other extreme event). As discussed by Lehane *et al.* (2008), for these design scenarios allowing for rate effects (and associated increased undrained strength), and when appropriate tension effects on the underside of a foundation, can avoid over-conservatism in shallow foundation design; however, these aspects require project-specific specialist advice. A particular advantage of the modern failure envelope approach is that it is easy to assess the consequences of changes in individual load components on the real factor of safety against collapse.

Uniaxial foundation capacity	Bearing capacity factor	Shape factor	Depth factor
$V_{ult} = 0.85\ N_{cv^*} \times A \times S_{u0} \times s_{cv}$	N_{cv^*} from **Table 53.4** ($k' = 0$)	$s_{cv} = 1 + 0.214\frac{B}{L} - 0.067\left(\frac{B}{L}\right)^2$	Included in N_{cv^*}
$H_{ult} = N_{ch} \times A \times S_{u0} \times d_{ch}$	$N_{ch} = 1$ (fully rough interface)	n/a	$d_{ch} = 1 + 3.01\frac{d}{B} - 1.12\left(\frac{d}{B}\right)^2$
$M_{ult} = N_{cm} \times A \times B \times S_{u0} \times s_{cm} \times d_{cm}$	$N_{cm} = 0.64$	$s_{cm} = 1 + 0.075\frac{B}{L} - 0.005\left(\frac{B}{L}\right)^2$	$d_{cm} = 1 + 1.27\frac{d}{B} + 1.27\left(\frac{d}{B}\right)^2$

The depth factors are proposed by Gourvenec (2008) for a rough embedded strip foundation assuming uniform undrained shear strength with depth; S_{u0} is undrained strength between the underside of the foundation and a depth equal to two-thirds of the foundation width below the footing; d is the foundation embedment (see **Figure 53.9**); B is the foundation width; the shape factor, S_{cm}, for a rectangular foundation is proposed by Gourvenec (2007). Ultimate moment resistance is mobilised at a vertical load of 0.5 V_{ult}.

The bearing capacity assessment can be undertaken using the three-dimensional failure envelope proposed by Taiebat and Carter (2002) for a surface foundation with a zero-tension interface on a soil with uniform undrained shear strength with depth, expressed by the equation below (Gourvenec, 2007) and also shown in **Figure 53.11**

$$\left(\frac{h}{h^*}\right)^2 + \left(\frac{m}{m^*}\right)^2 = 1$$

where $h^* = 4(v - v^2)$ for $0.5 < v < 1$; $h^* = 1$ for $0 < v < 0.5$ and $m^* = 4(v - v^2)$, and v,h,m are the normalised loads (load/ultimate load):

$$v = \frac{V}{V_{ult}}, h = \frac{H}{H_{ult}} \text{ and } m = \frac{M}{M_{ult}}$$

Table 53.5 Summary of the relevant formulae to assess the uniaxial capacities and the VHM failure envelope (Randolph and Gourvenec, 2011) for rectangular footings (assuming uniform undrained shear strength with depth; zero mobilised tension at the underside of the foundation)

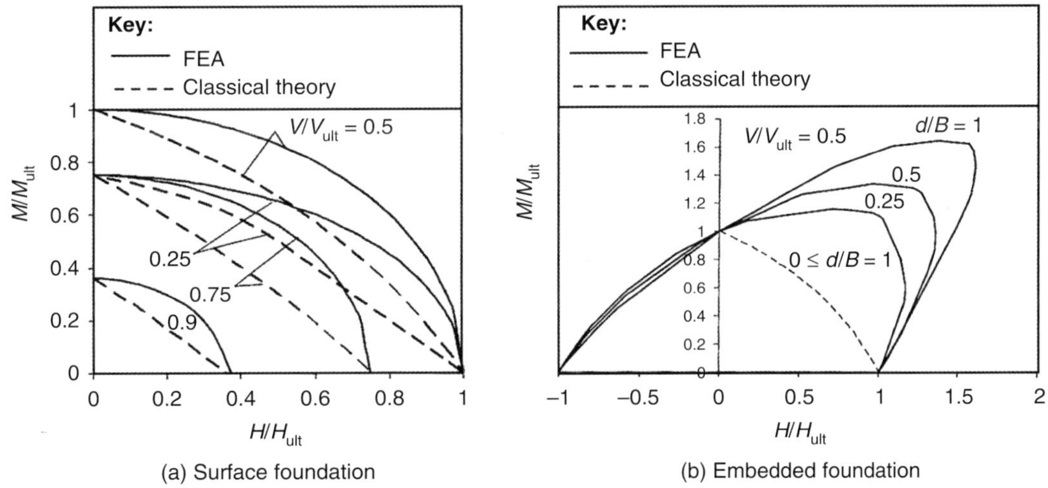

Figure 53.11 Comparison of ULS under general VHM loading from classical bearing capacity theory and finite element analysis (Reproduced from Randolph and Gourvenec, 2011)

53.5.4 Allowing for soil compressibility

It is often assumed that bearing capacity is dependent purely on the ground strength; in practice it is also a function of its compressibility. The compressibility of sand is a function of its relative density, its mineralogy and, if present, local lenses of clays/silts/organics and so on. The basis of conventional bearing capacity theory is 'general shear failure', which is inappropriate for the drained failure of medium-dense to very loose silica sands and any sand which is intrinsically compressible (such as calcareous or micaceous sands), or the drained failure of clays, and requires the bearing capacity factors for general shear failure to be corrected. A simple means of allowing for soil compressibility is to reduce the soil's friction angle in the formula for drained bearing capacity. For loose and very loose sands, the mobilised angle of friction is typically assumed to be two-thirds of the characteristic or moderately conservative value (interpreted from ground investigation data). More rigorous analytical methods are discussed by Randolph *et al.* (2004). For clays and silts, in the case of eccentric self-weight loading (such as for towers), low mobilised stiffness can lead to rotational buckling instability at significantly lower loads than indicated by classical theory (Potts and Burland, 2000). Lancellotta (1993) outlines a simplified equation for leaning instability.

53.5.5 Drained bearing capacity, influence of groundwater table level and groundwater pressure

The general bearing capacity equation is based on the assumption that the water table is located well below the foundation ($dw \geq B$). However, if – as is often the case – the water table is close to the foundation, then some modifications to the unit weight and surcharge terms (γ and q, respectively) are essential depending on the location of the water table.

- **Case 1:** If the water table is located above the bottom of the footing, then the effective surcharge is $q = q'$ (the vertical effective stress at the foundation level is $\gamma D - u$, where u is the groundwater pressure at the foundation level). The unit weight of the soil needs to be adjusted as

$$\gamma' = \gamma_{sat} - \gamma_w \tag{53.3}$$

where γ_{sat} is the saturated unit weight, γ_w is the unit weight of water and γ' is the buoyant unit weight.

- **Case 2:** If the water table is located at the foundation level (i.e. $dw = 0$), then q is the total vertical stress at the foundation level (since the groundwater pressure is zero).

$$\gamma' = \text{buoyant unit weight (as Case 1)} \tag{53.4}$$

- **Case 3:** If the depth to the water table $dw \geq B$, then q is the total vertical stress at the foundation level and

$$\gamma' = \text{saturated unit weight} \tag{53.5}$$

- **Case 4:** If, assuming no seepage forces, the water table is located at a depth $dw < B$ (i.e. the depth to water is within a depth less than the footing width) then interpolate between Cases 2 and 3:

$$q = \gamma D \text{ (as Case3)} \tag{53.6}$$

The unit weight in the first term of the bearing capacity equation must be replaced by the factor

$$\gamma_{mod} = \gamma' + d/B(\gamma - \gamma') \tag{53.7}$$

- **Case 5:** If the water table is at the foundation level, $dw = 0$, and there is uniform upward seepage (of gradient $du/dz > \gamma_w$), then $\gamma' = \gamma - du/dw$ and

$$\gamma' = \gamma - \gamma_w(1 + i) \tag{53.8}$$

where i is the hydraulic gradient (hence, a positive gradient can substantially reduce the bearing capacity).

53.5.6 Sand overlying clay

Various studies have been conducted into the ultimate bearing capacity of a footing on a layer of sand overlying clay (e.g. Hanna and Meyerhof, 1980; Okamura *et al.*, 1998). This is a common scenario, and 'punch-through' failure may be a genuine concern, particularly for relatively thin sand layers overlying clays.

Conventional analyses of the bearing capacity of sand over clay use limit equilibrium techniques and generally look at two proposed failure mechanisms, as shown in **Figures 53.12(a)** and **53.12(b)**. In each case the strength of the sand is analysed in terms of effective stress, using the effective unit weight (γ) and the friction angle (ϕ), while the analysis of the clay is in terms of total stress, characterised by its undrained shear strength (S_u). In the case of the 'projected area' method an additional assumption about the angle α (**Figure 53.12(a)**) is required, although a value of 30° is often assumed.

Okamura *et al.* (1998) assessed the validity of these two approaches and proposed an alternative failure mechanism; see **Figure 53.12(c)**.

53.5.7 Foundations on rock

The vast majority of rocks are not intact and have discontinuities, which, depending on their orientation and spacing, and the properties of the discontinuity infill material (if present), will have a significant impact on the bearing capacity of the rock. Franklin and Dusseault (1989) describe several different failure mechanisms.

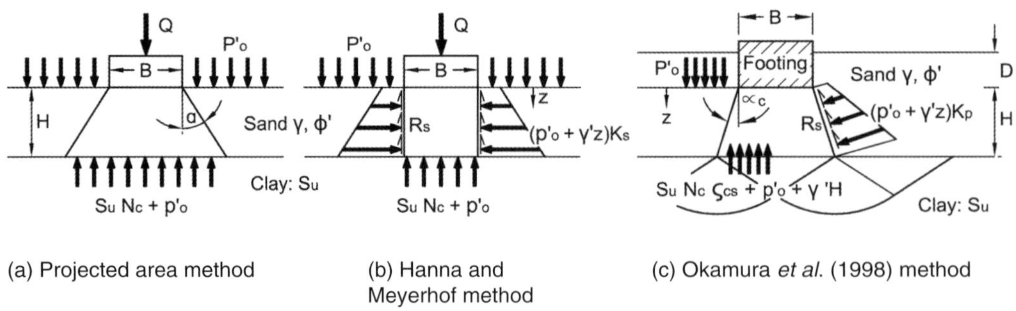

Figure 53.12 Alternative failure mechanisms, sand over clay

- General bearing capacity failures, similar to those which occur in soil, are uncommon in rock. These failures may occur beneath heavily loaded foundations on weathered, weak, highly fractured mudstone.
- 'Consolidation' failures in weathered, fractured rocks, where unweathered rock corestones fail due to a combination of low shear strength along clay-coated vertical joints and voids or compressible fillings along horizontal joints.
- 'Punching' failure in porous rocks, such as volcanic tuffs, limestone and chalk, due to local crushing of the rock adjacent to voids where the rock is heavily stressed.
- Wedge or sliding failure along unfavourably inclined joints (**Figure 53.13**). For this situation, the overall equilibrium of the foundation due to potential sliding along the discontinuity is checked by resolution of forces, and consideration of the specific shear strength characteristics of the critical discontinuities. If these are infilled with clay, fault gouge or are slick-sided, the shear resistance may be very low.
- Subsidence due to collapse of strata, undercut by subsurface voids, where the voids are natural or mining-induced. Natural voids can be formed by solution weathering of soluble rocks, such as gypsum, rock salt, chalk and limestone. These can form deep pipe- or cave-like features covered by metastable arches of soil. These may subsequently collapse due to changes in the stress or groundwater regimes.

The bearing capacity of a jointed rock mass is discussed by Wyllie (1991), who also discusses the stability assessment of foundations on rock slopes and on steeply bedded formations. If the rock mass contains numerous open joints, then the ultimate bearing capacity can be assumed to be equivalent to the unconfined compressive strength of the rock (**Figure 53.13(c)**).

For relatively weak and fractured rocks, Annex G of Eurocode 7 provides a simple set of charts with presumed design unit bearing resistance values of square pad foundations on rock (for settlements not exceeding 0.5% of the foundation width). These curves are only intended to be used for the design of simple, small structures. The charts assume joints are closed, and if open or infilled joints are suspected, then the allowable bearing pressures will need to be reduced, perhaps halved. For other structures, more detailed analysis is required (refer to Wyllie, 1991).

When the rock material is very weak, has very closely spaced discontinuities or is very heavily weathered or fragmented, the rock can be treated as a soil mass and designed on the basis of the conventional bearing capacity equations. For situations where there is a mix of strong and weak materials with complex discontinuity patterns, then the Hoek–Brown strength criterion (Hoek and Brown, 1997) should be used to evaluate the bearing capacity of the rock mass (Hoek, 1999; Hoek *et al*., 2002), **provided** the nature of the discontinuities is consistent with the Hoek–Brown criterion (e.g. spacing is small compared with foundation width, discontinuities do not have a dominant orientation, etc.). Specialist advice should be sought where inclined or eccentric loads are applied to footings on rock, and when voids are suspected in the rock (rocks such as limestones, chalk and carbonate-based rocks are susceptible to voiding or karstification). Argillaceous rocks (mudstone, shale and siltstones) are apt to swell once unloaded during foundation excavation, and exposure to water or cycles of wetting and drying can lead to considerable softening and may lead to degradation to a soil-like material (depending on their mineralogy, e.g. smectite-rich mudstone can be particularly susceptible).

53.6. Settlement

53.6.1 Introduction

The settlement of a foundation must be within tolerable or acceptable limits, as discussed in Chapter 52 *Foundation types and conceptual design principles*. For most shallow foundations the estimated settlement will be a more critical design consideration than bearing capacity. There are a wide range of methods available to estimate settlement; some of the commonly used methods will be discussed in this section. *However, for all methods the most challenging aspect is establishing the soil or rock properties and the associated deformation parameters. The use of more sophisticated analytical methods will not compensate for poor ground investigation*

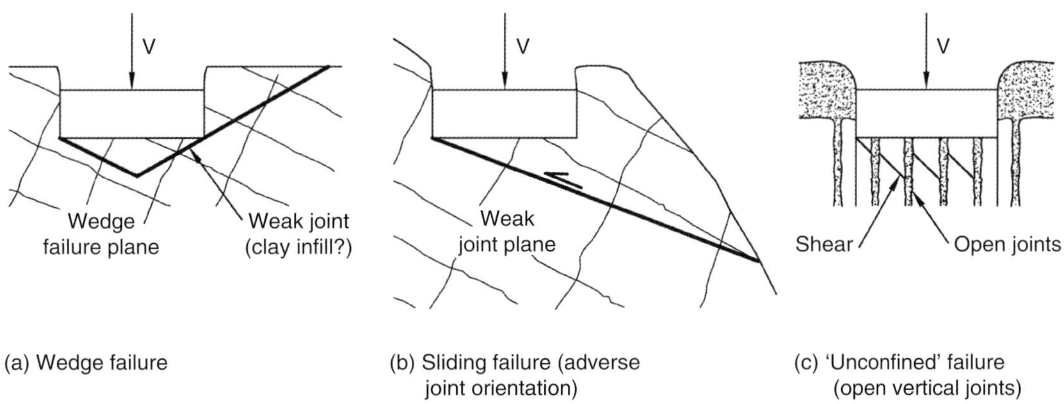

Figure 53.13 Typical failure mechanisms for jointed rock masses

data or an inappropriate interpretation of ground properties. If ground movements do not occur due to external sources (refer to Section 53.2), then the main components of settlement due to foundation loading are summarised in **Table 53.6**. The basic methods for calculating settlement in sands and clays are described in Chapter 19 *Settlement and stress distributions*.

Clays and silts experience both undrained shear and primary consolidation settlement. Creep settlement is usually assumed to be negligible for heavily overconsolidated clays and silts (which would usually be the founding materials), although can be significant for normally or lightly overconsolidated clays and silts (especially for highly organic materials, such as peat).

The commonly used methods of settlement analysis implicitly assume linear elastic ground behaviour, or simple nonlinear behaviour in the form of either a nonlinear elastic response or by distinguishing between the changes in compressibility, which occur between overconsolidated and normally consolidated states (or pre-yield and post-yield states for bonded soils and some weak rocks). As outlined by Burland *et al.* (1977), for the case of foundations that are predominantly vertically loaded, with a high factor of safety (typically more than three), the major principal stress increments remain practically vertical beneath the foundation (and stress and strain increment axes are mainly coincident). Hence, for this situation relatively simple stress–strain relationships are adequate. It should always be remembered that the main body of experience in shallow foundation behaviour is related to foundations subjected to predominantly vertical static loads. For other situations design will need to be directed by appropriately experienced geotechnics professionals and more detailed investigations and sophisticated analyses may be necessary.

If a foundation is subjected to an offset or eccentric vertical load, the Meyerhof method described in Section 53.5 (refer to **Figure 53.10**) can be used to calculate the equivalent uniform bearing pressure. Conventional methods can then be used to assess the overall foundation settlement.

53.6.2 Stress changes due to applied bearing pressures

Theoretical considerations are discussed in Chapter 19 *Settlement and stress distributions*. For practical applications, the changes in stress with depth are mainly based upon homogeneous, isotropic, linear elastic theory (e.g. Boussinesq) or similar. Burland *et al.* (1977) have discussed in detail the influence of a wide range of factors on the accuracy of predicted stress changes with depth. For many ground conditions the equations based upon linear elastic theory give a reasonably accurate estimate of *vertical* stress changes with depth. Poulos and Davis (1970) provide a comprehensive summary of elastic stress changes below shallow foundations for a wide range of situations.

A useful design chart for estimating vertical stress changes with depth has been produced by Janbu *et al.* (1956) and is shown in **Figure 53.14**. This chart gives the increase in vertical stress beneath the centre of a uniformly loaded flexible area of strip, rectangular or circular shape.

53.6.3 Elastic methods

Elastic methods should not be used when the final stress state in a layer exceeds the pre-consolidation or yield stress of the ground (i.e. $\sigma'_{voi} + \Delta\sigma_{vi}' > p'_{ci}$), where σ'_{voi} is the initial vertical effective stress and $\Delta\sigma_{vi}'$ is the vertical effective stress increment for the layer under consideration. p'_{ci} is the pre-consolidation pressure or yield stress.

Settlement component	Relevant ground conditions	Comments
1. Undrained shear distortion (no change in volume/void ratio)[a]	Clays, clayey silts, very weak and weak argillaceous rocks	Becomes more significant as the factor of safety reduces. Typically, EOC settlement is 10–20% of total settlement for soft clays and about 50–75% of total settlement for stiff clays[b]
2. Primary consolidation settlement due to a reduction in volume/void ratio, due to an increase in effective stress	All soils and rocks	Main component of settlement for most conditions. Case histories show significant primary consolidation settlement occurs during construction, for overconsolidated clays, sands and rocks
3. Creep, or secondary consolidation, settlement, due to a reduction in volume/void ratio, at constant effective stress	Soft lightly overconsolidated and normally consolidated clays and peats	Becomes significant once pre-consolidation pressure is exceeded. Creep increases as organic content increases, and as clay sensitivity increases
4. Deformation along or across joints, fissures or discontinuities	Rocks	At high stresses, creep can be significant – refer to text. Chalk can be vulnerable to creep once yield stress is exceeded. Manifested as a time-dependent 'creep' settlement. Swelling during excavation, or poorly controlled groundwater conditions, can exacerbate 'creep' during subsequent loading
5. Change in volume/void ratio due to cyclic loading or vibration	Sands, gravels and sandy silts	Observations (e.g. Burland and Burbidge, 1984) indicate that conventional foundations exhibit time-dependent settlement, even under 'static' load (implying 'creep' deformation, see point 3 above). But significant increases in time-dependent settlement occur if 'fluctuating' loads are applied

(a) For saturated soils.
(b) End of construction settlement (EOC) includes both undrained and a proportion of the primary consolidation settlement.

Table 53.6 Components of foundation settlement due to applied bearing pressure

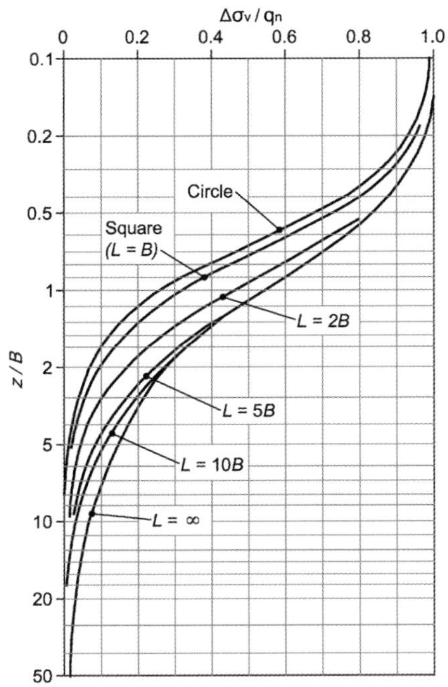

Figure 53.14 Vertical stress increase beneath the centre of flexible footings of varying geometry (Reproduced from Janbu *et al.*, 1956)

The undrained settlement of clays is often *mislabelled* as 'elastic settlement' because elastic theory has been used for calculation purposes; this erroneous terminology should not be used. It is now widely accepted that a wide range of overconsolidated soils and rocks can be treated as 'elastic' for predicting the *total* settlement (where total settlement is undrained *plus* time-dependent settlement) – see Burland *et al.* (1977) and Mayne and Poulos (1999).

Shallow foundation settlement can be assessed very quickly by using displacement influence factors derived from elastic theory. Unfortunately, many published solutions cannot be recommended because they assume the elastic material has a constant value of Young's modulus with depth and the elastic material is of infinite thickness, which is highly unrealistic for most practical situations. A recommended method is given by Mayne and Poulos (1999) for the settlement of a shallow foundation resting on an isotropic elastic material of finite thickness, whose Young's modulus increases linearly with depth (i.e. $E = E_0 + kZ$, where E_0 is Young's modulus directly beneath the foundation base at $Z = 0$ and kZ is the rate of increase of the modulus with depth); as noted in Chapter 52 *Foundation types and conceptual design principles*, this type of stiffness variation is relevant for many design situations

$$S = qDI_G(1 - \nu^2)f_R I_E / E_0 \qquad (53.9)$$

where S is the foundation settlement, q is the net bearing pressure, D is the diameter of the circular foundation, I_G is the influence factor for non-uniform ground stiffness, given in **Figure 53.15(a)**, ν is Poisson's ratio, E_0 is Young's modulus directly beneath the foundation, f_R is the correction factor for the foundation stiffness (refer to Section 53.6.7) and I_E is the correction factor for the foundation depth (**Figure 53.15(b)**).

The influence factor for non-uniform ground stiffness is plotted against the dimensionless term, $\beta = E_0/kD$, where k is the increase in Young's modulus with depth. Although developed for circular foundations, this method can be used for square and rectangular foundations (for $L \leq 3B$, where L is the foundation length and B is the breadth) and has been shown to compare well against more rigorous solutions. For

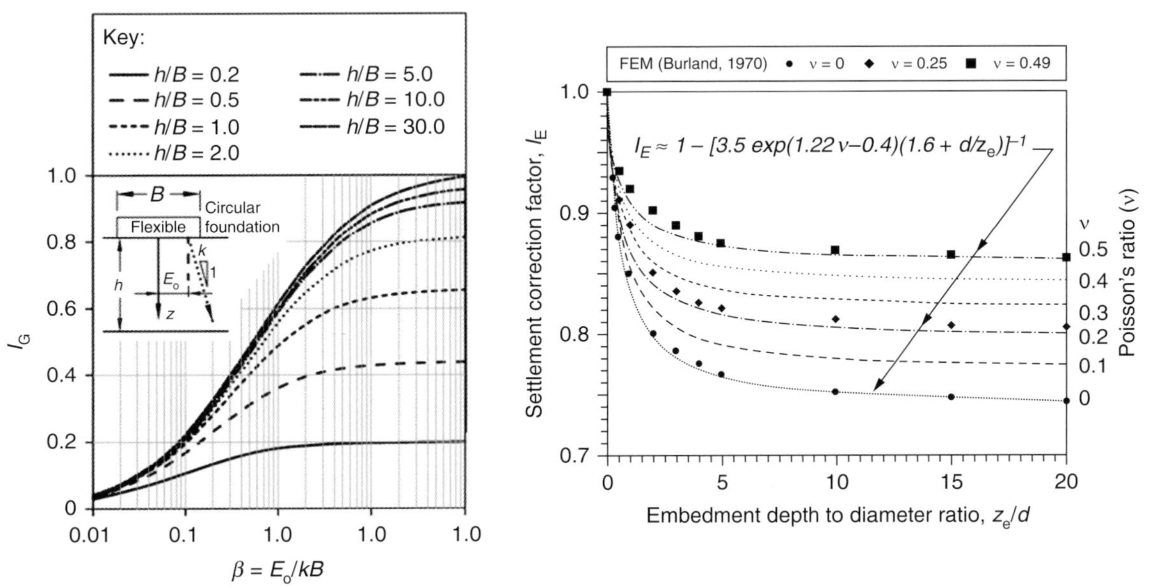

Figure 53.15 Displacement influence factor (I_G) and depth correction factor (I_E) for the settlement of a flexible circular foundation

rectangular foundations where $L > 3B$, Meigh (1976) has produced appropriate elastic solutions.

53.6.4 Settlement of clay

This section considers heavily overconsolidated clays only. The commonly used methods for estimating the settlement of clays are summarised in **Table 53.7**. The use of elastic methods has been outlined above. For the classical one-dimensional method the foundation settlement is calculated from

$$S_{\text{oed}} = \sum_{i=0}^{i=n} \left(m_{\text{v}i} \Delta\sigma'_{\text{v}i} H_i \right) \tag{53.10}$$

where m_{vi} is the coefficient of volume compressibility derived from oedometer testing. The one-dimensional method has been criticised as being too crude, and other methods, offering an apparently more sophisticated approach, have often been preferred. The accuracy of the various methods is discussed in detail in Chapter 19 *Settlement and stress distribution*, and by Burland *et al.* (1977). The main conclusion is that the one-dimensional method, for overconsolidated clays, provides results that are often more accurate than results from apparently more sophisticated methods *if the soil compressibility is appropriate for the stress change and strain amplitude imposed by the foundation.* Hence, for overconsolidated clays the one-dimensional method is sufficient for most practical purposes. The principal problem is that it has relied upon oedometer test data, which often gives excessively high m_{v} values (and hence gross overestimates of the total settlement) due to errors introduced during sampling and testing, and inherent shortcomings in the test method (the test only gives values of 'large strain' compressibility – refer to Section 52.6). These problems can be overcome by using a modified one-dimensional method, which enables the mobilised compressibility to be determined from a range of other tests (and specifically for the appropriate stress and strain amplitude for each particular layer within the foundation's influence zone)

$$S_{\text{od}} = \left[\sum_{i-0}^{i=n} \left(\Delta\sigma'_{\text{v}i} H_i \,/\, D'_i \right) \right] f_{\text{r}} f_{\text{d}} \tag{53.11}$$

where D'_{i} is the constrained modulus for layer i (for the relevant change in stress and mobilised strain amplitude), $\Delta\sigma'_{\text{v}i}$ is the change in the vertical effective stress for layer i and H_i is the thickness of layer i; f_{R} and f_{d} are correction factors for the foundation depth and rigidity (Section 53.6.7), respectively, and S_{od} is the foundation settlement.

The constrained modulus, D′, can be related to other isotropic elastic parameters

$$D' = 3K'(1 - v') \,/\, (1 + v') \tag{53.12}$$

$$D' = E'(1 - v') \,/\, (1 + v')(1 - 2v') \tag{53.13}$$

where K' is the bulk modulus, E' is the drained Young's modulus and ν' is the drained Poisson's ratio. For overconsolidated clays, ν' is close to 0.1 and, therefore, D' is approximately equal to $2.5K'$ or about equal to E'. The drained Poisson's ratio quoted in many text books is too high and for the strain amplitudes relevant for most design situations it can be assumed to be less than 0.2 (refer to Mayne and Poulos, 1999). For the recompression portion of the vertical strain against log σ'_{v} oedometer curve (i.e. where the clay is overconsolidated), it can be shown that:

$$D' = \frac{1}{m_v} = (1 + e_0)\sigma'_{\text{v}0}\ln(10)/C_{\text{r}} \tag{53.14}$$

where e_0 is the initial void ratio, $\sigma'_{\text{v}0}$ is the average of the initial and final vertical effective stress and C_{r} is the recompression index; as noted earlier, oedometer tests can only provide a 'large strain' compressibility. For anisotropic soils

$$D' = E'_{\text{v}} / \left[1 - 2v'^{2}_{\text{VH}} \left(n'/1 - v'_{\text{HH}} \right) \right] \tag{53.15}$$

where E'_{v} is Young's modulus in the vertical direction, v'_{VH} is Poisson's ratio for the effect of vertical strain on the horizontal strain, v'_{HH} is Poisson's ratio for the effect of horizontal strain on the complementary horizontal strain and $n' = E'_H/E'_V$, where E'_H is Young's modulus in the horizontal direction. The above relationships allow the one-dimensional approach to utilise data from a wide range of different test types in order to assess the ground's compressibility characteristics across a wide range of changes in stress and strain amplitude, from very small strain (e.g. from geophysics tests) through to large strain (from oedometer tests). For overconsolidated clays

$$S_{\text{OD}} = S_{\text{T}} \tag{53.16}$$

where S_{T} is the total settlement, and

$$S_{\text{PC}} = S_{\text{T}} - S_{\text{U}} \tag{53.17}$$

where S_{PC} is the primary consolidation settlement and S_{U} is the undrained settlement.

The ratio of the undrained to the total settlement has been studied for a variety of situations (e.g. Burland and Wroth, 1974). As the rate of increase of stiffness with depth increases, the ratio of the undrained to the total settlement decreases. At a typical stiffness increase for London Clay, the ratio of the undrained to the total settlement is about one-third. Observations of settlement for numerous structures on overconsolidated clays (London Clay, Lambeth Group and Gault Clay) indicate a typical ratio for the end of construction to the total settlement of about 0.6. Hence, a proportion of the primary consolidation settlement is likely to have taken place rapidly, and undoubtedly the end of construction settlement includes some consolidation settlement, as well as the undrained settlement. Independent analysis (described by Atkinson, 2000) also suggests the end of construction settlement includes some consolidation settlement. Hence, the long-term settlement, which occurs after construction is completed, will often be much lower than indicated by theoretical calculations that assume the clay is undrained during construction. **Table 53.8** summarises several case histories of shallow foundation settlement, and these data are plotted on **Figure 53.16** for both high- and low-plasticity overconsolidated clays. The influence of the plasticity index is clear, with high-plasticity clays exhibiting larger settlements than the low-plasticity Lodgement Tills. The influence of overconsolidation ratio is also significant. The New York Varved Clays are low-plasticity clays and silts, which are only moderately overconsolidated, with overconsolidation ratio (OCR) between 3 and 6, and they exhibit similar settlement to the more heavily overconsolidated clays (typical OCR > 10), which have a high-plasticity index (such as London, Frankfurt and Gault Clays). **Table 53.9** provides a summary of typical index properties for the clays included in **Table 53.8** and **Figure 53.16**.

53.6.5 Settlement of sand

Excessive settlement of sand is seldom a practical problem; most observations of foundation settlement on sands indicate that settlement is less than 40 mm. Large settlements are usually only associated with loose or very loose sands, or sands containing a high organic content or clay layers. In addition, the bulk of the settlement takes place during construction, so the post-construction settlement, which may lead to structural damage, is relatively small. Cases of structural damage for foundations on sand are extremely rare. Therefore, the first step for the practising engineer is to qualitatively check if there is likely to be any real settlement problem. **Table 53.10**, based on a study by Clayton *et al.* (1988), summarises those situations that may lead to significant settlement. For more common situations a simple approach to estimating settlement is often sufficient. **Figure 53.17** is based upon a compilation of field measurements of foundation settlement on silica sand, and three broad categories are given based on a visual description or the average of the standard penetration test (SPT) 'N' values: 'loose' (SPT 'N' < 10); 'medium-dense' (SPT 'N' = 10–30); and 'dense' (SPT 'N' > 30). Fairly well-defined upper limits are given for medium-dense and dense silica sand. There is much greater scatter for loose sands, and these soils would usually be considered as unsuitable as founding materials (without prior ground improvement to densify or stiffen the sand). Hence, for many

Method	Compressibility parameters	Advantages/disadvantages	Comments
(1) One-dimensional method, general comment	See below	Simple. Soil layering readily taken into account. Vertical stress changes reliable for many practical situations	Burland *et al.* (1977) showed that the one-dimensional method is more reliable than more complex methods *if* soil compressibility is appropriate for stress change and strain amplitude
(a) Oedometer	Coefficient of volume compressibility, m_v	Principal problem is reliance on m_v from oedometer tests, which are usually unreliable due to sample disturbance	For overconsolidated clays, settlement is grossly overpredicted due to deficiencies in conventional oedometer test data
(b) Modified	Constrained modulus, D'	Constrained modulus can be related to moduli obtained from a wide variety of tests, such as oedometer, triaxial and bender element (see Section 53.6.4)	**Most appropriate method for stiff overconsolidated clays**; modulus for each layer specific to stress level and strain amplitude, see Sections 53.7 and 53.8
(c) Compression indices	Compression indices, pre and post pre-consolidation pressure (C_r and C_c)	Compression indices can be correlated to carefully conducted oedometer tests, or, more commonly, empirical correlations	Most appropriate method for soft clays; pre-consolidation pressure is the critical parameter
(d) Skempton and Bjerrum	m_v and 'correction factor' U_g	Historically justified to 'correct' the settlement derived from routine oedometer tests. Calculates 'consolidation' settlement only; extra calculation needed for undrained settlement	**As noted by Burland *et al.* (1977), this method is obsolete** and does not improve reliability of predictions
(2) Linear elasticity (a) Classical equation	Young's modulus and Poisson's ratio	**Principal flaw is reliance on horizontal stress change,** which for many practical situations (involving anisotropy, increasing stiffness with depth and nonlinearity) is unreliable. Experience indicates mobilised Young's moduli will increase rapidly with increasing depth below foundation	**Inappropriate for soft clays**. Commonly used method and utilised in commercially available software and simple models in finite element codes. Can be acceptable when reliable moduli are available from relevant case history data (similar foundation size and stress changes)
(b) Displacement influence factors	Young's modulus and Poisson's ratio	Very simple and rapid method. Published charts often assume uniform modulus and infinite depth; these are unreliable for most practical situations	Methods that allow for increasing stiffness with depth are useful for preliminary design checks. Inappropriate if significant layering below the foundation

Table 53.7 Common methods for calculating the settlement of clays

Site[e]	Foundation dimensions: m				Soil profile		Net foundation pressure: kN/m²	Observed settlement: mm	
	D	B	L	Type	Type[a]	Thickness: m		EOC[b]	Final
Clapham Rd	1.5	17	28	Raft	SG	7	190	58	91
					LC	25			
					LG	15			
Hurley Rd	1.5	17	26	Raft	SG	5	190	68	109
					LC	24			
					LG	14			
King Edward	1.7	20	27	Raft	LC	4	210	32	52
					LG	15			
Addiscombe Rd	11	14	14	Raft	LG	5	226	48	59
Waterloo Bridge	7	8.3	36	Bridge piers	LC	25+	290	–	90–130[c]
Elstree	2.1	1.5	3.0	Pad footings	LC	15	85	–	14
Bracknell	1.0	3	3	Pad footing	LC	10+	110	–	17
Britannic House	18	27	70	Raft	LC	13	100	15–20	30–45
					LG	10			
Didcot	6–7.5	9	9	Pad footings	GC	30+	180 per footing	20–25	30–40[d]
New Wembley	1.5	50	50	Embankment	LC	–	170	–	110–120
					LG				
CN Tower	7.3	32	47	Raft	Till	19	370	20	30
					SG	9			
Toronto Hospital	5	18	70	Raft	Till	12	85	–	15
				Interglacial	Till	30+			
Gorlev Silos	4.0	22	22	Raft	Till	–	250	–	18–24
Scotia Centre	7.3	31	32	Raft	Till	12	225	32	49
					Shale	20+			
Dublin	<0.5	1.5	1.5	Pad footing	Till	>20	1010	–	16
AfE, Frankfurt	13	43	43	Raft	FC	100	350	150	230
Library, Frankfurt	8	32	52	Raft	FC	100	45	12	20
University, Frankfurt	7	14	96	Raft	FC	100	200	42	67
Bureau, Frankfurt	7	22	22	Raft	FC	100	250	65	100
Grant, New York	–	12.2	70.2	Raft	MS	9.2	140	43	51
					VSC	27.5			
Madison, New York	–	12.2	55	Raft	MS	4.6	140	71	97
					VSC	19			
Roosevelt, New York	–	16	48	Raft	MS	6.1	180	63	81
					VSC	15			

Note:
[a] SG is sand/gravel; LC is London Clay; LG is Lambeth Group; GC is Gault Clay; Till is Lodgement Till; FC is Frankfurt Clay; MS is medium sand; VSC is New York Varved Silt/Clay (Glacial Lake Deposits).
[b] End of construction (EOC) settlement as a percentage of total settlement varies between about 50 and 75%; typically it is about 60%. For EOC settlement, construction periods typically last 6 months to 2 years.
[c] Construction problems caused an increase in settlement at some piers.
[d] Multiple pad footings at close spacing.
[e] References: COSOS (1974), Clark (1997), De Jong and Harris (1971) and Parsons (1976).

Table 53.8 Stiff overconsolidated clays, observed shallow foundation settlement for various sites

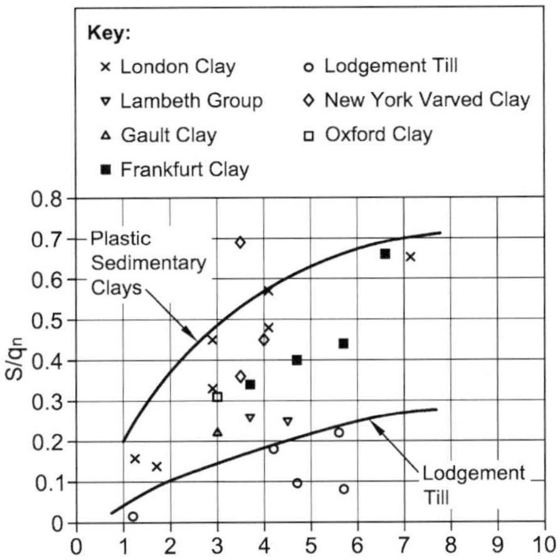

NB.
All clays are heavily overconsolidated, except New York varved clay with OCR ~ 3 to 6.
S = foundation settlement, q_n = net bearing pressure, B = foundation width.

Figure 53.16 Stiff clays, observed settlement ratios against the square root of foundation width

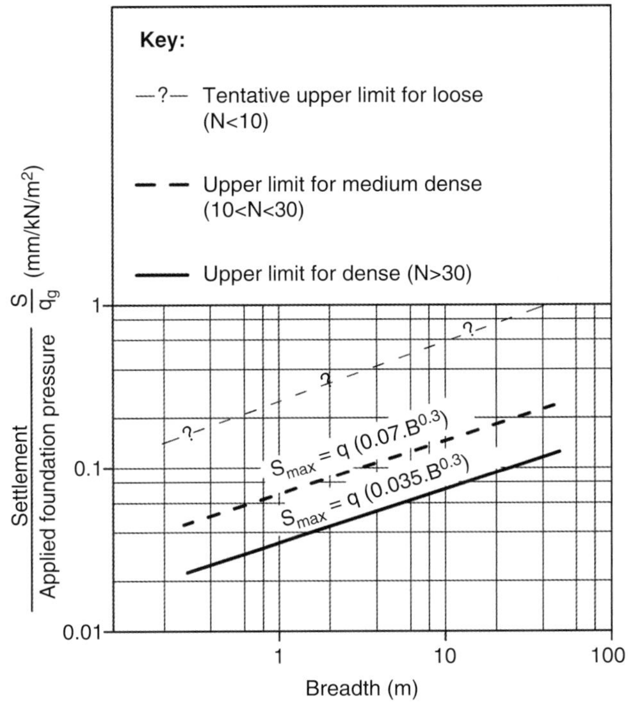

NB.
q_g = gross foundation pressure, S = settlement, N is average SPT 'N' over depth = 1.5B below foundation. Probable settlement is about one-half of upper limit. Settlement is 'short-term' settlement, refer to Burland *et al.* (1984) for assessment of 'creep'.

Figure 53.17 Sands, upper-limit settlement ratios (Reproduced from Burland *et al.*, 1977)

routine structures it is sufficient to use **Table 53.10** in conjunction with **Figure 53.17**, first to assess if any problems are likely, and if not use **Figure 53.17** to assess the settlement of medium-dense and dense silica sands. The 'probable' settlement is likely to be about half the upper-limit settlement. Based upon a review of settlement observations, Burland *et al.* (1977) stated that small plate tests (less than 1 m wide) are unlikely to be a reliable basis for assessing the settlement of much wider foundations. This should be borne in mind when assessing the type of testing needed to check the effectiveness of ground improvement methods. Small plate tests are still commonly used as field tests and, although useful as a quality control test, they will not be sufficient to assess full-scale foundation settlement.

When more detailed quantitative methods are needed to calculate the settlement of sands, then the following two methods are recommended

- Burland and Burbidge (1984)
- Schmertmann *et al.* (1978).

Burland and Burbidge (1984) is mainly based on empirical correlations with SPT 'N' values (although a simple correlation is provided between SPT 'N' and CPT q_c values). In

Soil type	Moisture content: %	Plasticity index: %	Liquidity index	Undrained strength: kN/m²	Description
Weathered London Clay	30–35	50	0 to 0.1	45–65	Firm to stiff
Unweathered London Clay	25–30	40–50	0 to −0.1	70–250	Stiff to very stiff
Gault Clay	20–30	50	0 to −0.2	100–300	Stiff to very stiff becoming hard
Lambeth Group	20–30	20–50	0 to −0.1	150–400	Very stiff to hard
Frankfurt Clay	30–35	35–55	+0.1 to −0.1	80–200	Stiff to very stiff
Lodgement Tills	10–15	10–20	0 to −0.5	150–500	Very stiff to hard
New York Varved Clays	30–45	5–40	0.1 to 0.3	95–140	Firm to stiff varved clay

For New York Varved Clays, the OCR is between 3 and 6; for other clays the typical OCR > 10 at shallow depth.

Table 53.9 Stiff clay case histories: typical index properties

Factor	Comment
1. Narrow (<1.5 m), heavily stressed foundations	Carefully check bearing capacity. High risk of failure or low factor of safety, especially if disturbance occurs to sands (reduction in density), high water table (refer to Section 53.5.5) or inclined loading. Good construction control needed
2. Very wide (>10 m) foundations	Careful assessment of variation of compressibility with depth (e.g. weak layers?)
3. Metastable ('collapsing' sands), e.g. dune sands and lightly cemented sands (e.g. sabkha)	Site geology needs to be well understood. If lightly cemented sands, penetration tests will be misleading. Large plate tests may be needed to assess 'yield' stress of bonded or cemented sand
4. Sands containing layers of organic or clayey material	Static cone penetration tests could identify layers. Sampling and laboratory testing of discrete layers required. SPTs and spaced sampling may miss critical layers
5. Non-silica sands (mineralogy)	Empirical penetration test-based settlement methods are only based on performance of silica sands. Micaceous or calcareous sands, for example, may exhibit far larger settlement. If in doubt, the sand mineralogy should be checked. Specialist technical literature and testing and analytical methods will need to be considered

Table 53.10 High-risk situations for excessive settlement of sands (Reproduced from Clayton, 1988)

contrast, Schmertmann *et al.* (1978) make direct use of CPT q_c data. Both of these methods are well described in several references (e.g. Clayton, 1995; Lunne *et al.*, 1997) and will not be repeated in detail here.

Comparative studies indicate that both methods have a similar level of reliability and accuracy. Many other methods of predicting sand settlement have been published. Tan and Duncan (1991) carried out an assessment of the reliability of 12 different methods and showed that some methods can exhibit a systematic bias in settlement predictions. More sophisticated analytical methods (including finite element analyses) do not necessarily improve the reliability of predictions. A prediction symposium (Briaud and Gibbons, 1994) involved making Class A predictions of a footing (3 × 3 m) loaded to 4000 kN (factor of safety of 2.5) for a sand layer with an average SPT 'N' value of 20. A settlement of 14 mm was measured, whereas a sophisticated finite element analysis gave a value of 75 mm.

53.6.6 Settlement of rock

Assessments of foundation settlement on rock routinely utilise linear elastic methods (refer to Section 53.6.3); however, the processes and concepts used to assess the deformation characteristics of the rock mass are fundamentally different to those used for soil. A soil mass is usually assumed to be a continuum, whereas for a rock mass the assessment of the nature, orientation and frequency of joints and other discontinuities is absolutely critical to assess the rock mass deformation properties. The orientation of the dominant discontinuity sets in relation to the loading direction can have a significant influence on the deformation behaviour of the rock mass. Based on in situ load tests, Barton (1986) defined three characteristic modes of deformation behaviour. Joint surfaces are usually irregular; hence, forces across joints will be transmitted by reduced contact areas. As foundation bearing pressures increase, the contact stresses across the joints may become extremely high, which may lead to local yield and time-dependent creep.

Engineers often erroneously assume that 'rock' will provide a very strong competent foundation; however, this is not always the case. When considering a rock, engineers often visualise strong igneous rocks, such as massive unweathered granite (with widely spaced joints). However, many rocks, especially of sedimentary origin, can be heterogeneous and of very poor quality. Weak rocks may need to be investigated and assessed far more carefully than soil deposits before an appropriate foundation design can be prepared. A commonly used index to assess rock mass quality is the RQD value, and Hobbs (1974) used a 'mass factor' *j* to relate the modulus for a rock mass, *Em*, to the intact properties of the rock (refer to Section 53.7.4) by way of RQD values. The modulus for a rock mass is then used in either the classic linear elastic (refer to Section 53.6.3) or the modified one-dimensional method (refer to Section 53.6.4) to assess settlement. For the vast majority of weak rocks, Young's modulus for a rock mass will increase rapidly with depth, especially beneath weathered material. Meigh (1976) used a linear elastic approach to back-analyse the observed settlement of foundations in weak rock.

Three weak rock types commonly encountered across the UK are:

- Triassic Series (Mercia Mudstone, previously known as 'Keuper Marl', and Sherwood Sandstone), which outcrops across much of the Midlands, north-west England and parts of south-west England and southern Wales
- Chalk, which outcrops across much of southern and eastern England and parts of the Yorkshire and Lincolnshire Wolds
- Carboniferous Coal Measures rocks, which are encountered across many parts of northern England.

For foundation design in these rock types, one of the main challenges is their variability in strength and compressibility, which can change from heavily weathered 'soil' to relatively strong unweathered 'rock'.

A summary of observed settlements for shallow foundations on a number of UK rocks is given in **Table 53.11**. **Figure 53.18** gives a plot of observed settlement ratio against foundation width for structures founded on Mercia Mudstone.

Site	Foundations B: m	L: m	Type	Ground profile	Foundation pressure: kN/m²	Observed settlement: mm
M1 bridges	6.1	11.6	Pad	Grade 1, Mercia Mudstone	240 (net)	14–15
	2.7	41	Strip	Grade 3, Mercia Mudstone	160 (net)	10
34-storey AGT Tower, Edmonton	18	19	Raft	Interbedded bentonitic mudstone (shale) and sandstone	160 (net) 430 (gross)	48[c]
20-storey building (single-level basement)	17	35	Closely spaced pads	Grade 3, Mercia Mudstone (Grade 2 to 1 at depth)	210 (net)	55
Second Severn Crossing Bridge	6	27	Caissons	Grade 2, locally 4, Mercia Mudstone	600 (gross)	<25
Oldbury Nuclear Power Station	26 dia.		Circular raft	Grade 1, Mercia Mudstone	1100 (net)	105–125[a]
4 silos, Bury St. Edmonds	22.9 dia.		4 circular rafts	Low-density chalk	350	45–75
South Abutment Thames Barrier	14.7	31.5	Pad	High-density chalk	480	20
4-storey building, Basingstoke	3.3	3.3	Pad	Low-density chalk	365	7
Bridge, Luton	8.5	20	Pad	Low-density chalk	300	9
Hinckley Point 'B' Nuclear Power Station	19 dia.		Circular raft	Interbedded mudstone and limestone, Lower Lias	1200 (net)	Reactor 3, 14–16[b] Reactor 4, 30–35[b]

(a) End of construction settlement (EOC), approximately 80–95 mm, subsequent time-dependent settlement (9-year period) described as creep.
(b) EOC settlement, approximately 7–8 mm (Reactor 3) and approximately 16–18 mm (Reactor 4). Time-dependent settlement took about 10 years to complete.
(c) The 13.7 m deep basement excavation caused 55–60 mm heave and swelling of mudstone. EOC settlement about 35 mm.

Table 53.11 Weak rocks, observed shallow foundation settlement for various sites

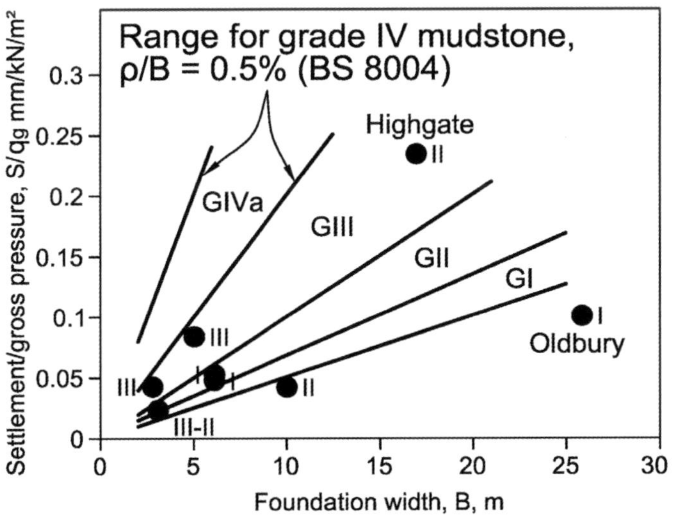

Figure 53.18 Mercia Mudstone, observed settlement ratios against foundation width (Modified from CIRIA C570, Chandler and Forster, 2001; www.ciria.org)

For Mercia Mudstone when bearing pressures and foundation widths are modest (pressures less than about 300 kN/m², widths less than about 10 m) the observed settlement of an isolated foundation is usually less than about 25 mm, provided weathering grade IV mudstone is largely absent. Settlement occurs mainly during construction, so differential settlement can be largely built out.

For chalk, foundation settlement increases rapidly once bearing stresses exceed the yield stress, q_y – refer to Section 53.7.4. Hence, it is prudent in order to limit settlement to keep the foundation bearing pressure below q_y. At stresses below q_y an isolated 5 m wide foundation would be expected to settle less than 10 mm (CIRIA C574, Lord *et al.*, 2002), provided weathering grade Dm chalk is absent. For larger foundations or higher foundation bearing pressure, CIRIA C574 provides more detailed guidance on appropriate methods of settlement assessment.

An interesting feature of the case history data is that significant time-dependent settlement can develop once foundation bearing pressures become relatively high. At Oldbury, on unweathered Keuper Sandstone and Siltstone, under a net bearing pressure of 1100 kN/m² (compared with UCS values mainly in excess of 10 MN/m²), post-construction settlement, after a period of about 10 years, had increased by about 50–60% compared with the settlement that occurred during construction. The time-dependent settlement was ascribed to creep in a 13 m thick 'leached' zone (where gypsum nodules had gone into solution and left a porous sandstone with numerous small cavities) (Meigh, 1976). At Hinkley Point, two large foundations were built on interbedded unweathered mudstone and limestone of the Lower Lias of Jurassic age.

The net bearing pressure was about 1200 kN/m^2 for both foundations (less than 10% of the UCS for the moderately strong rock). As noted in **Table 53.11**, the settlement for one of the reactor foundations was about double that experienced at the other reactor foundation, despite similar ground conditions. The much greater settlement was ascribed to ineffective dewatering of the high groundwater pressures, which caused joint opening in the fractured limestone and softening of the unloaded mudstones (COSOS, 1974). The settlement of Reactor 4 was about three times larger than that originally predicted.

53.6.7 Corrections for foundation rigidity and foundation depth

The classic one-dimensional and elasticity equations are based on a flexible foundation resting on the surface of a compressible medium. Hence, corrections are necessary for

- the actual rigidity of the foundation
- the actual depth of the foundation below the surface of competent ground.

The settlement of a rigid footing is usually assumed to be 0.8 times that of a flexible footing. Horikoshi and Randolph (1997) provide a rigidity correction factor that can be applied across varying foundation rigidity (from perfectly rigid to flexible) and geometry (circular, rectangular, etc.).

Figure 53.19 gives a plot of the normalised differential settlement against the relative raft–soil stiffness ratio K_{rs}

$$K_{rs} = 5.57(E_f/E_{Sav})[(1-v_s^2)/(1-v_f^2)](B/L)^{0.5}(t_f/L)^3 \quad (53.18)$$

$$\Delta S^* = (S_A - S_B)/S_{av} \quad (53.19)$$

where S_A is the settlement at the centre, S_B is the settlement at the middle of the short side of a footing or the settlement of a corner (**Figure 53.19**), S_{av} is the average settlement of a footing, E_f is Young's modulus for the footings or raft material (e.g. reinforced concrete), E_{Sav} is the representative Young's modulus for the ground beneath the foundation (i.e. the value of E_s at a depth equal to the radius of a circular footing or half the foundation width), t_f is the footing thickness, B is the footing width, L is the footing length, v_s is Poisson's ratio for the soil and v_f is Poisson's ratio for the footing material.

To use Equations 53.18 and 53.19 it is most convenient to assume that the foundation is initially rigid (i.e. apply a correction factor of 0.8) and then modify the results in accordance with **Figure 53.19** and Equations 53.18 and 53.19. It is apparent from the figure that foundation rigidity changes rapidly with changes in the ratio t_f/L. Hence, many foundations will either be practically flexible ($K_{rs} < 0.1$) or practically rigid ($K_{rs} > 5$).

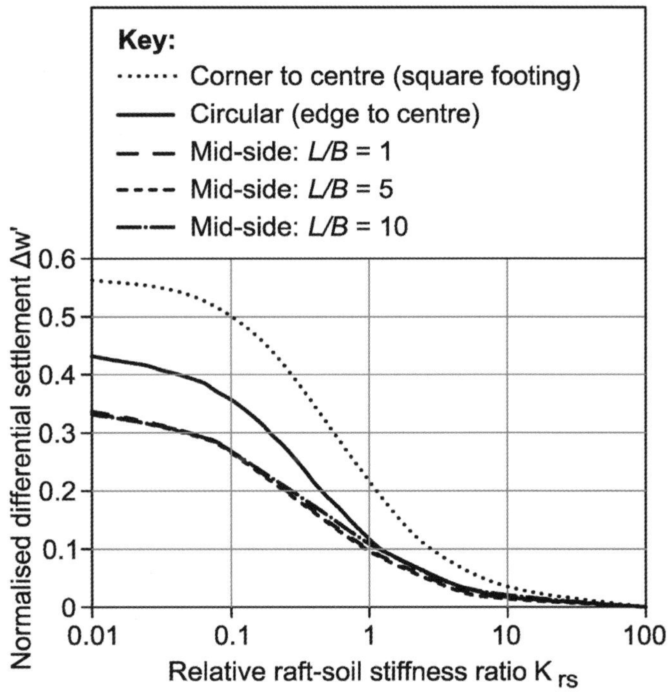

Figure 53.19 Normalised differential settlement as a function of raft stiffness ratio

The effect of foundation embedment on reducing settlement is often overestimated, in particular Fox's depth correction factor (given in many textbooks) significantly overpredicts the influence of embedment depth for shallow foundations. The depth correction factor given by Mayne and Poulos (1999) is more realistic (see **Figure 53.15(b)**).

53.6.8 Heave and swelling

Swelling has led to damaging movements of shallow foundations. Long-term swelling may develop in very stiff clays, mudstones/siltstones and shales due to one of the following mechanisms.

(*a*) *'Conventional', due to reduction of effective stress:* This is the 'textbook' process of undrained heave (due to reduction of total stress, for example, during excavations), followed by long-term swelling if the net change in effective stress is negative (i.e. a reduction) relative to the original effective stress state (therefore may arise from a combination of change in total stress and change in long-term groundwater pressure).

(*b*) *Oxidation of pyrite:* A chemical reaction triggered, for example, by disturbance and exposure to oxygen, producing sulfates (such as gypsum), leading to large volume expansion (~200%).

(*c*) *Hydration of anhydrite:* A change in groundwater regime may trigger a chemical reaction, leading to production of gypsum (with large volume expansion).

(*d*) *Change in groundwater chemistry (osmotic diffusion):* A change in the ambient groundwater regime may lead to an ion concentration gradient between pore water and 'new' groundwater, leading to volumetric expansion.

The most severe engineering problems tend to arise when chemical changes (mechanisms (*b*), (*c*) and (*d*)) cause large (and often unexpected) ground displacements. As outlined above, these chemical changes can arise when a chemically unstable mineral present in the ground is exposed to a new environment (typically as it is exposed during excavation works and different waters, levels of oxygen, etc.). The evaluation of this type of hazard requires specialist geological studies, and evaluation of mineralogy, ground and groundwater geochemistry; Terzaghi *et al.* (1996), Hawkins and Pinches (1987) and Czerewko *et al.* (2011) give some practical examples and further discussion.

Conventional swelling (mechanism (*a*)) tends to be less problematic since this mechanism is more easily anticipated. When the ground is unloaded beneath an excavation, undrained heave will develop, and additional upward movement can also develop due to time-dependent swelling. For sands and gravels, heave and swelling are practically negligible; however, for medium- to high-plasticity overconsolidated clays and argillaceous rocks (e.g. mudstone, shale) heave and swelling can be significant (even without chemically induced volumetric expansion). For example, vertical deformations due to heave and swelling of the order of 60 mm and 100 mm have been measured beneath 14 m and 20 m deep excavations, respectively, in bentonitic shale (e.g. Chan and Morgenstern, 1987; De Jong and Morgenstern, 1973). Interbedded layers of water-bearing limestone facilitated rapid swelling of Frankfurt Clay and movements of up to 140 mm were reported for 25 m deep excavations by Breth and Amman (1974).

The mechanisms associated with heave and swelling of overconsolidated fissured clays and mudstone or shales can be more complex than those associated with settlement. Some shales and mudstones that contain active clay minerals (e.g. montmorillonite or smectite) can swell considerably following stress release due to water absorption by the clay minerals. Swelling then can lead to progressive destructuration (as bond/cementing are progressively broken down during swelling). Cycles of wetting and drying can exacerbate this degradation process, which in some cases may lead to a weak rock deteriorating to a clay over a relatively short time period. This can mean that if the ground mass at the base of an excavation is exposed for prolonged periods of time, then it is likely to deteriorate. The subsequent settlement of shallow foundations will then be much larger than would be expected, based on the 'undisturbed' ground properties. Hence, it is extremely important that the ground surface at foundation level is protected as much as practical and as quickly as possible after initial exposure (refer to Section 53.3.4).

53.6.9 Simplified nonlinear methods and numerical modelling

For foundation design, sophisticated modelling is usually not required; however, there are situations when engineers may need to consider the influence of nonlinear ground stiffness behaviour in more detail, as discussed below. For foundations on stiff clays/weak rocks, a consideration of the nonlinear small strain stiffness characteristics of the ground can result in significant economies, especially (as is often the case) when the design is controlled by SLS (i.e. allowable movement). Simplified nonlinear methods are available, summarised in **Table 53.12**, which can provide valuable insights into the effects of stiffness nonlinearity. When the design situations are appropriate, they enable the effects of stiffness nonlinearity to be rapidly assessed without recourse to full nonlinear finite element analysis. Finite element methods (FEM) are now commonly available, although they can be easily misused. As stated in CIRIA C791 (O'Brien and Higgins 2020), FEM 'is not straightforward and if not used properly can be dangerous'. In particular, C791 states 'however simple the engineering challenge might be, numerical modelling is an intrinsically complex process'. There are many potential pitfalls, and it is recommended that the guidance given in C791 is followed if numerical modelling is used. **Table 53.13** provides a summary of the key steps involved in numerical modelling; checking by experienced specialists, model calibration and careful interpretation of outputs are always important. Hence, the methods outlined in **Table 53.12** provide an important alterative when foundation deformation requires more detailed assessment.

The modified one-dimensional method is beneficial for assessing total settlement and long-term swelling, whereas the MSD and Atkinson's methods are useful for assessing deformation under short-term loading, as described by Osman *et al.* (2007). However, these techniques need to be used under the guidance of a suitably experienced specialist. Nonlinear methods may be useful when the following are being assessed

- variation of settlement with changes in net bearing pressure
- potential interaction between adjacent structures
- differential settlement adjacent to, or beneath, a shallow foundation
- the potential for a more economic foundation design
- the potential to save time (since foundation construction is often on a project's critical path)
- the potential to reduce carbon and create more sustainable foundation solutions
- settlement and differential settlement of sensitive structures (say, a limiting settlement of less than 25 mm)

Method	Type of analysis	Nonlinear parameters	Foundation geometry	Soil layering	Comments
Osman and Bolton (2005)	Plasticity theory	Mobilised undrained strength against shear strain (linked to applied load and foundation settlement by way of scaling factor)	Isolated circular footing on clay	No. A single 'representative' stress–strain curve is used	Footing load–settlement behaviour is linked to triaxial stress–strain curve by way of bearing capacity theory and scaling factors to link triaxial shear strain to footing settlement. Can also be used to assess horizontal and rotational deformation of foundation under horizontal and moment loads. Undrained deformation of clays only
Atkinson (2000)	Elasticity theory	Normalised Young's modulus (E/E_0) against axial strain (linked to applied load and foundation settlement by way of scaling factor)	Isolated circular, strip or pad footings	No. A single 'representative' stress–strain curve is used	Footing load–settlement behaviour is linked to triaxial stress–strain curve by way of elasticity theory and scaling factors to link triaxial strain to footing settlement. Can also be used to assess horizontal and rotational deformation under horizontal and moment loads. Experience to date – undrained deformation of clays
O'Brien and Sharp (2001)	Modified one-dimensional method	Constrained modulus against axial (vertical) strain	Any geometry, both isolated and multiple closely spaced footings	Yes. Multiple layering and variations of stiffness with depth	Elastic theory used to calculate vertical stress changes. Constrained moduli at various stress and strain amplitudes assessed from oedometer, triaxial, geophysics data or empirically. Ideally field geophysics available to define 'anchor point' and stiffness variation with depth; more routine data can be used with empirical correlations. Total settlement and swelling of clays and silts

Table 53.12 Simplified methods, which incorporate nonlinear ground stiffness, for foundation settlement

Step	Description	Comment
1	Objectives	Specify why modelling is needed and what outputs are critical for design. Discuss with experienced modeller. Assess options and limitations.
2	Application category	May vary from simple modelling through to complex modelling which is safety critical. This will affect the necessary level of checking and seniority of staff involved. Past experience in type of modelling and specific application being considered is an important factor.
3	Design team organisation	Roles and responsibilities. Is the modeller also the designer? Information flow, etc. Who checks/approves? The time and budget available are also key factors.
4	Competencies and training	Necessary skills/experience? Future training needs?
5	Information for model	Surveys, ground investigations, design/site information, construction scenarios, etc. Adequate for scope/complexity of modelling?
6	Simplifications and assumptions	Some idealisations are always necessary, and need to be stated clearly. *All* assumptions and data sources must be described, communicated/agreed across project team and customer.
7	Calibration/testing	What types of calibration/testing will be made (e.g. element, component or case history checks)? Important to assess model output reliability.
8	Robustness	Some sensitivity studies are necessary to check for 'cliff-edge' effects.
9	Checking	Both inputs and outputs, independent of originator.
10	Reporting	Broken down into three parts: assumptions; factual reporting of the analysis; and finally interpretation of output (often this will be the most complex task!).

Table 53.13 Numerical modelling: the 'ten steps' for better numerical modelling (After CIRIA C791)

- negative skin friction and displacement of 'floating' piles in stiff clays (refer to Chapter 57 *Global ground movements and their effects on piles*)
- heave-induced tension, and displacement, of piles beneath excavations in stiff clay (refer to Chapter 57 *Global ground movements and their effects on piles*).

One of the important practical consequences of ground stiffness nonlinearity is that subsurface settlements and settlements adjacent to a foundation become concentrated much closer to the foundation than indicated by linear elastic calculations (Jardine *et al.*, 1986). The vast majority of foundation settlement is caused by subsurface settlement which occurs within $0.5B$ to $1B$ below a foundation (where B is the foundation width). This has important practical implications, both for planning ground investigations and control of works on site.

The assessment of shallow foundation SLS displacements can be carried out based on the methods outlined in Osman *et al.* (2007). Undrained deformations (horizontal

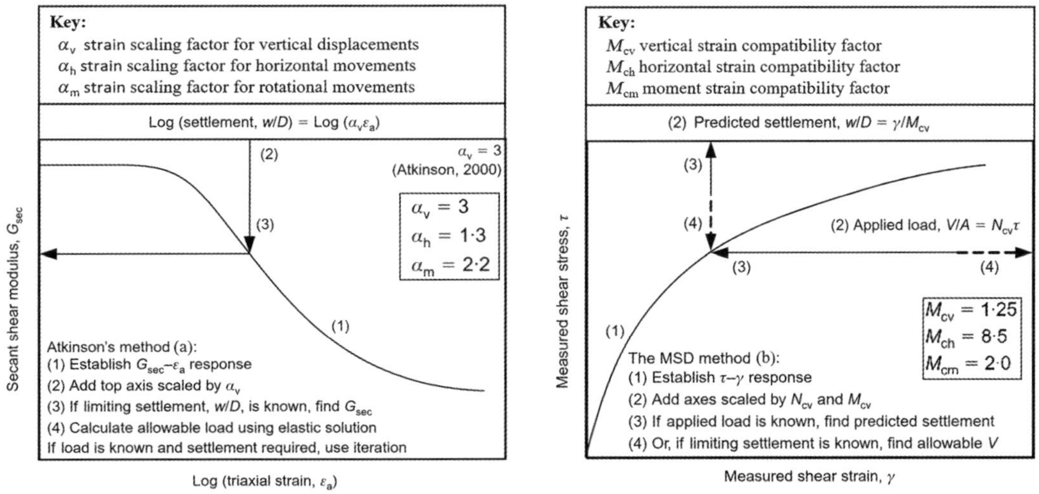

Figure 53.20 Illustration of Atkinson's method (a) and the mobilised strength design (MSD) method (b) (Reproduced from Osman *et al.*, 2007)

Box 53.1 Assessment of shallow foundation displacements: Atkinson's method (Atkinson, 2000)

Atkinson's method is a deformation calculation method based on elasticity that allows soil stiffness nonlinearity to be considered. The calculation method assumes that the decay in overall soil stiffness (beneath a shallow footing) with normalised footing settlement, w/D, takes the same form as the decay in selected soil stiffness with strain amplitude at a specific point below the foundation. The methodology, which links mobilised shear modulus to settlement, is briefly outlined in **Figure 53.20(a)**. Atkinson's original method specifically considers a circular surface foundation under pure vertical loading, with a proposed vertical scaling factor, α_v, equal to 3. Osman *et al.* (2007) extended the methodology to include proposed scaling factors for horizontal and moment loading, respectively $\alpha_h = 1.3$ and $\alpha_m = 2.2$, for rough circular shallow foundations on undrained clay. The effect of foundation embedment is accounted for using an elastic depth factor proposed by Mayne and Poulos (1999) – see **Figure 53.15(b)**. The shape of the foundation can be approximated by equating its base area to the area of an equivalent circle (acceptable if $L < 3B$). For vertical and moment resistance the secant shear modulus, G_{sec} against strain amplitude curve needs to be representative for the ground at a depth of 0.3B (where B is foundation depth), whereas for horizontal resistance it should be representative of the underside of the foundation. The curve(s) can be derived from: published literature; site-specific advanced testing; or Atkinson's (2000) nonlinear model (e.g. refer to Eadington and O'Brien, 2011). The G_{sec} values can then be linked to the applied load using elastic solutions proposed by Osman *et al.* (2007). The applied vertical (V), horizontal (H) and moment (M) loads are calculated as follows based on the dimensionless elastic coefficients, which are dependent on the Poisson's ratio, v.

Direction	Loads	Dimensionless elastic coefficients (Osman *et al.*, 2007)
Vertical	$V = K_v \times wGD$	$K_v = \dfrac{2\ln(3-4\nu)}{1-2\nu} = 4$ (undrained)
Horizontal	$H = K_h \times uG \times D$	$K_h = \dfrac{16(1-\nu)}{7-8\nu} = 2.67$ (undrained)
Moment	$M = K_m \times \theta GD^2$	$K_m = \dfrac{1}{3(1-\nu)} = 0.67$ (undrained)

The horizontal and moment load formulae above are based on the assumption that there is no cross coupling between horizontal and moment stiffness ($K_{mh} = 0$), which is conservative; w is the vertical displacement, u is the horizontal displacement and θ is the rotation; for undrained conditions ($v = 0.5$) the dimensionless elastic coefficients are $K_v = 4$, $K_m = 0.67$, and $K_h = 2.67$.

Atkinson's method is implemented in the following manner.

(a) Derive deformation: secant shear modulus behaviour for vertical, horizontal and moment loading using Atkinson's method over a defined movement range.

(b) Use elasticity constants to obtain vertical, horizontal and moment loading associated with relevant mobilised G_{sec} and deformation.

(c) Using (*a*) and (*b*), plot load–displacement behaviour of foundations and obtain deformations under relevant SLS foundation loading.

(d) Apply elastic depth correction factor to the calculated displacements to account for foundation embedment.

Note: calculating the load associated with an assumed deformation simplifies the calculations (by avoiding the need for iteration); if an appropriate range of assumed deformations are selected then a load–deformation curve can be calculated for each relevant load component.

displacement, rotation, settlement) can be calculated if combined VHM loads are applied. Hence, these methods complement the VHM method for ULS bearing capacity assessment (Section 53.5.3 and **Table 53.5**). Atkinson's (2000) method is based on elasticity, whereas Bolton's (1993) method is a plasticity-based mobilised strength design (MSD) approach. These methods are briefly illustrated in **Figure 53.20** and **Boxes 53.1** and **53.2**. Both approaches explicitly consider the nonlinear reduction of soil stiffness with strain and have been shown to provide good agreement with results from nonlinear finite element analysis (using advanced constitutive soil models) across a practical working load range (equivalent to factors of safety between 1.5 and 5.0) – refer to Osman and Bolton (2005). The methods outlined in Osman *et al.* (2007) specifically consider surface foundations on undrained materials; however, the methods can be adapted to account for the influence of foundation embedment using depth factors relevant to elasticity and plasticity methods for Atkinson's method and the MSD approach respectively. The Atkinson and Bolton methods can work well if ground conditions are relatively uniform for a depth of $\sim 1B$ (where B is foundation width) below a footing. However, they would be inappropriate if the ground comprises significantly layered strata of differing characteristics within this principal zone of influence.

The modified one-dimensional method enables total settlement under vertical loading to be calculated. The modified method is the same as the conventional one-dimensional method, the only modification is that the ground compressibility is adjusted to ensure the mobilised ground compressibility (for each sub-layer) is appropriate for the stress change and mobilised strain amplitude in each sub-layer. This can be readily implemented through a simple iterative routine. The nonlinear ground stiffness (based on constrained modulus against axial strain) curves can be derived from published literature, site-specific advanced testing, Atkinson's nonlinear ground stiffness model (e.g. refer to Eadington and O'Brien, 2011; O'Brien *et al.*, 2023) and Equations 53.12–53.15.

Box 53.2 Assessment of shallow foundation displacements: the mobilised strength design (MSD) approach (Bolton, 1993)

The MSD approach is a design approach proposed by Bolton (1993) which allows nonlinearity to be considered in a routine calculation based on plasticity. The approach scales the stress–strain response to a load–displacement response using scaling factors on both axes.

Similar to Atkinson (2000), Osman and Bolton (2005) originally demonstrated the use of MSD for a smooth circular foundation under pure vertical loading, assuming a vertical strain compatibility factor of $M_{cv} = 1.35$. Osman *et al.* (2007) extended the MSD approach for rough circular shallow foundations on undrained clay under combined loading by proposing $M_{cv} = 1.25$ and additional horizontal and moment compatibility factors, $M_{ch} = 8.5$ and $M_{cm} = 2.0$. The methodology is briefly outlined in **Figure 53.20(b)**. The effect of foundation embedment is accounted for by using a plasticity depth factor for vertical resistance (Skempton, 1951); $d_c = 1 + 0.2d/B \leq 1.5$) can be used, or factors indicated in **Table 53.4** (which include depth effects). For horizontal and moment resistance, depth factors are given in **Table 53.5**. The nonlinear shear stress–shear strain relationship can be obtained using: published data; site-specific advanced testing; power law relationships using Vardanega and Bolton (2011) or Atkinson's nonlinear model (e.g. Eadington and O'Brien, 2011). The nonlinear shear stress against shear strain relationship needs to be representative for the ground at a depth of about 0.3B below the footing (where B is the foundation width). Engineering shear strain γ_s is related to axial strain ε_a using $\gamma_s = 1.5\varepsilon_a$ for undrained loading.

The bearing capacity factors appropriate to vertical, horizontal and moment loading, N_{cv}, N_{ch} and N_{cm}, are used to scale the measured shear stress axis to obtain a relationship with the applied load. These factors can be assumed to be the following.

Parameter	Formulae	Notes
Bearing capacity factor for vertical loading	$N_{cv*} = \dfrac{V}{As_u}$	N_{cv*} is from **Table 53.4** (depth factor is included in N_{cv*})
Bearing capacity factor for horizontal loading	$N_{ch} = \dfrac{H}{As_u} = 1\,d_{ch}$	$N_{ch} = 1$ (for uniaxial loading assuming fully rough interface); d_{ch} from **Table 53.5**
Bearing capacity factor for moment loading	$N_{cm} = \dfrac{V}{ADs_u} = 0.61\,d_{cm}$	d_{cm} is from **Table 53.5**

The MSD method is implemented in the following manner.

(a) Derive the shear stress–shear strain relationship for vertical, horizontal and moment loading using the MSD method to obtain displacements from applied loads. The mobilised shear stress, (shear strength), τ is derived from the applied bearing pressure by way of the relevant bearing capacity factors.

(b) Calculate displacements with/without plasticity depth factor through changes in ultimate bearing capacity to account for changes in mobilised strength (and hence mobilised shear strain and displacement) to assess effects of foundation embedment.

The modified one-dimensional method can easily take account of layered strata and associated changes in stiffness nonlinearity.

A practical application of the modified one-dimensional method is outlined in a short case history in Section 53.8, which facilitated foundation reuse for a new settlement-sensitive structure (whereas conventional 'linear elastic' calculations indicated piles would be required).

53.6.10 Unsaturated soils and non-engineered fills

Deep deposits of unsaturated natural soils are commonly encountered across many parts of the world (e.g. Central Asia, South Africa, the Americas), although, in the context of foundation design, they are rarely a serious concern in temperate climate zones such as the UK. For shallow foundations, the principal concern is the metastable volume-change potential of unsaturated soils; see **Figure 53.21**. Below the threshold stress, if moisture content (or pore water pressure) increases then heave can develop, whereas above the threshold stress large collapse-type settlement will occur if moisture content (or pore water pressure) increases. These movements can develop quite rapidly as the moisture content increases, and have caused substantial structural damage. The assessment of unsaturated soil behaviour is a specialist task. Fredlund and Rahardjo (1993) provide detailed guidance on the theoretical and practical issues.

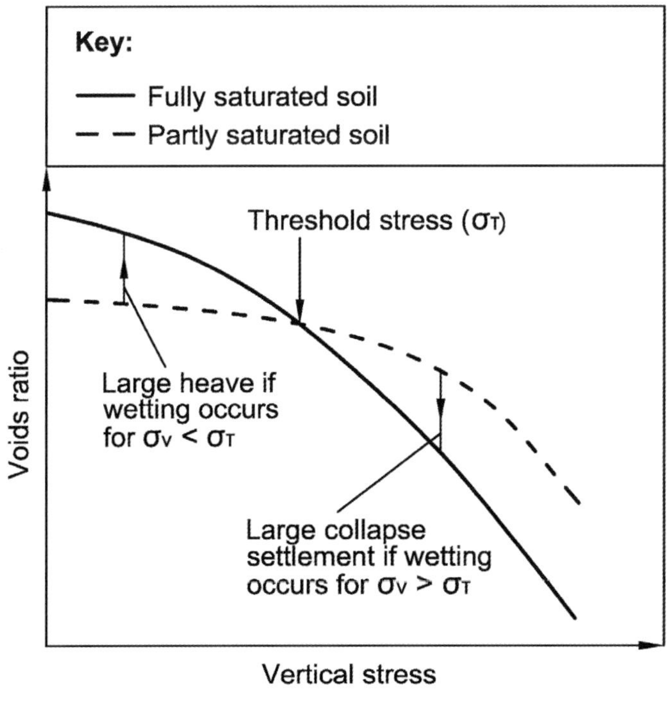

Figure 53.21 Metastable settlement and heave behaviour for partly saturated soils

In the UK, the practical foundation problems associated with unsaturated ground are usually associated with non-engineered fills (discussed in detail in Chapters 34 *Non-engineered fills* and 58 *Building on fills*) or with clays that have been desiccated by vegetation (discussed in Section 53.2).

53.7. Information requirements and parameter selection

53.7.1 Ground profile and site history

Many of the serious problems that occur with shallow foundations are due to a misinterpretation of the site's history or geology, which means that soft compressible soils are found at foundation level rather than competent low-compressibility soils. Two simple examples are given below.

- A low-rise building was to be founded on shallow strip footings. The local geology was Bagshot Beds (interbedded stiff clay and dense sands) over stiff London Clay. Boreholes had been sunk in the vicinity of the building. During initial excavations for the strip footings very soft to soft black sandy clay was encountered. Excavations down to 5 m depth did not find any improvement in strength. The original shallow foundation design had to be abandoned and replaced by piled foundations, with substantial cost increases and delays to the project. Subsequent investigations revealed that the area had been used as a brick pit; the natural competent soils had been excavated and after brick-making had finished the natural soils were end-tipped loosely into the pit. The borehole logs had proven to be completely unreliable.
- Several low-rise buildings were successfully designed and constructed, with shallow strip footings, and an additional building was then commissioned close to the site perimeter. The location for the proposed building was being used for agricultural purposes, and intrusive ground investigations could not be carried out due to access and land ownership problems. No problems were anticipated due to the positive experience from the previous works. Once access into the area was achieved, construction began straight away. Unfortunately, excavations for the strip footings only found very soft clays. Piled foundations were eventually used instead of the original strip footings. Aerial photographs of the area indicated the existence of an old pond, which had subsequently been infilled.

With the benefit of hindsight it may seem surprising that these problems occurred, especially since experienced designers had been involved in both sites. Desk studies and ground investigations must be implemented with great care and thoroughly checked. Even if old investigation data are available, it is prudent to carry out some additional investigations to check their reliability. Ground investigations need to be fully supervised by experienced staff and 'fresh' soil samples inspected and logged as soon as they are sampled. The scope and scale of ground investigations will vary depending upon the size and complexity of the proposed structures, local experience and the nature of the geomorphology, geology and hydrogeology in the vicinity of the site. A comprehensive desk study is always required. The objectives of any subsequent investigations will include

- confirmation of the conceptual ground and groundwater models derived from the desk study
- assessment of the local variability of the ground properties
- derivation of simple strength and compressibility parameters (often from empirical correlations with penetration tests or basic laboratory testing)
- assessment of likely construction problems (especially if groundwater is likely to be encountered).

For simple (Category 1) projects (with shallow pad or strip footings) in an area with well-known geology it may be sufficient to carry out trial pits, dynamic probing or static cone penetration tests. For Category 2 and 3 projects a broader range of more sophisticated in situ and laboratory tests will be necessary. In general, for sites underlain by clays, the emphasis for carrying out laboratory testing of strength and compressibility should be a small number of tests on high-quality samples (push-in thin-wall tubes, triple-tube rotary coring with foam flush for stiff clays or push-in piston tubes for soft clays), rather than carrying out a mass of routine triaxial and oedometer tests on relatively poor-quality driven tubes. The case history in Section 53.8 highlights the advantages of this approach.

For sands, if the foundation requirements are particularly onerous, or the sands have unusual mineralogy (micaceous or calcareous) or contain significant clay or organic content, then laboratory testing of reconstituted samples for strength and compressibility should be considered. As noted in Section 53.7.3, empirical correlations between penetration resistance and sand compressibility are of limited reliability. More sophisticated testing, including plate tests or pressuremeter tests, may need to be considered for complex situations.

The assessment of groundwater conditions will usually require piezometers to be installed in boreholes. The groundwater levels should be monitored over several months (preferably including a winter and spring period). In situ permeability tests should also be carried out in more permeable layers that may be the source of groundwater problems during construction.

For all projects, irrespective of their scale and complexity, the ground investigation should include in situ testing and sampling and laboratory testing. These are complementary techniques, with their own particular advantages. In situ testing is intrinsically better suited to assessing ground variability, whereas laboratory testing is better suited to assessing fundamental strength and compressibility characteristics provided the sample quality is sufficiently high. It is always important to sample and inspect the soil visually and describe it in a comprehensive and consistent manner using engineering descriptions consistent with appropriate codes (Rowe, 1972).

A major deficiency with conventional practice is that sampling is usually carried out at discrete depth intervals that are too large – typically 1.5 m depth intervals. Most foundation settlement will be due to settlement occurring within a depth of about two-thirds of the foundation width. Hence, *continuous* sampling of boreholes is recommended, down to a depth of about 1.0B, where B is the proposed foundation width. Laboratory testing and penetration tests should be particularly intensive within a depth of about one-half to two-thirds of foundation width (for small pad footings this may mean intensive testing of the upper 2 m below footing depth).

Much of the UK is underlain by glacial soils; these soils are often competent and suitable for shallow foundations – indeed, piling can often be far more problematic; refer to CIRIA C504 (Trenter, 1999); however, investigations can be challenging due to the potential presence of

- significant spatial variability in soil type (exhibited by large changes in strength, OCR, compressibility and consolidation properties)
- large variations in depth to rockhead and associated variations in the thickness of overlying compressible soils
- water-bearing layers of silt, sand, gravel of variable thickness and continuity
- artesian groundwater pressures
- cobbles and large boulders.

Figure 53.22 shows a profile through glacial deposits in north-east England, where a weaker, lightly overconsolidated, glaciolacustrine layer underlies a very stiff heavily

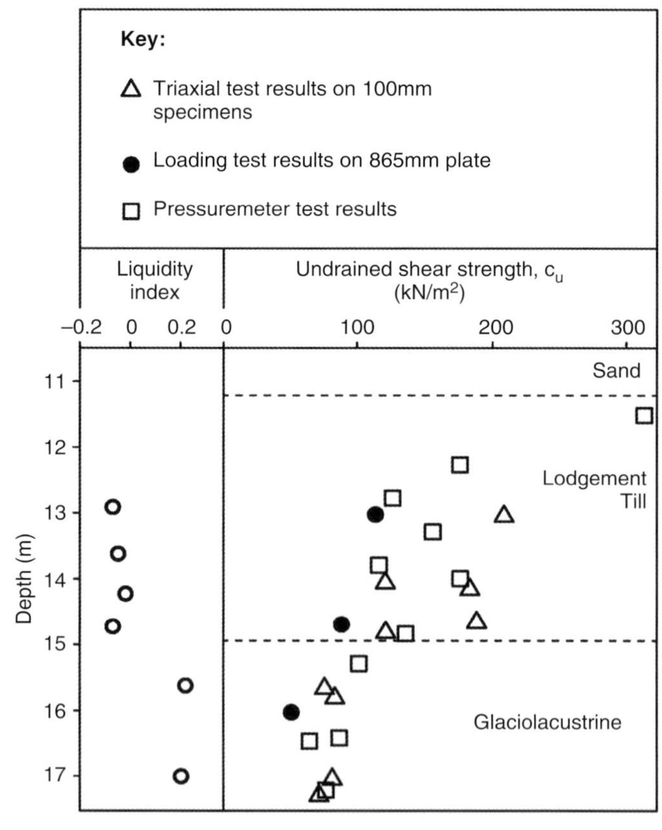

Figure 53.22 Glacial deposits, weaker layer beneath stiff clay (Reproduced from CIRIA C504, Trenter, 1999; www.ciria.org)

overconsolidated Lodgement Till. In these ground conditions the full range of ground investigation methods may be necessary in order to adequately characterise the ground conditions.

53.7.2 Strength and compressibility of clays

Classification: As noted in Section 53.6.4 (**Tables 53.6** and **53.7** and **Figure 53.17**), the observed settlement of shallow foundations is closely linked to the plasticity index (PI) and degree of overconsolidation (OCR) of the underlying clays. The shear strengths of clays are similarly dependent on PI and OCR, and are also significantly affected by their macrofabric and structure (i.e. the presence of silt and sand lenses, fissures and bonding). Classification test data can be used to derive void indices (Burland, 1990) and compare these against intrinsic and sedimentation compression lines (ICL and SCL, respectively); refer to Chapter 52 *Foundation types and conceptual design principles* and **Figure 52.21**. Void index plots can provide a robust framework for assessing settlement behaviour, especially for soft clays.

Reference should be made to Chapter 17 *Strength and deformation behaviour of soils* for an introduction to the topic and Chapter 52 *Foundation types and conceptual design principles* for foundation engineering-related considerations.

Heavily overconsolidated clays – undrained strength (S_u): The common method in the UK for measuring strength is to sample with driven thick-walled tubes and carry out quick undrained triaxial (QUT) tests. Data derived in this way usually exhibit significant scatter, much of which is induced because of the limitations associated with these basic sampling and testing procedures and does not necessarily reflect field behaviour. Variations in ground strength (either spatial variations across a site or depth-dependent changes) are usually best assessed through in situ testing. **Figure 53.23** gives examples of the differences between different test data. For Lodgement Till, the plate tests indicate an increase of strength with depth, whereas the U100 QUT data indicate a reduction of strength with depth. The misidentification of soft layers can be a common problem, especially in clays with silt or sand lenses. This is due to the clay softening after sampling as water migrates from the silt and sand to the adjacent clay. In situ penetration tests (such as CPT or SPT) can be a cost-effective means of identifying more realistic strength profiles. The SPT is a simple and robust method; its main value is that it can be used in a wide range of conditions. Approximate estimates of undrained strength can be derived through correlation factors; however, precision should not be expected (there is a wide scatter across any published correlation factors, in part due to the variations in the details and condition of the basic equipment). For example, Stroud (1990) is now considered obsolete due to changes in the SPT equipment.

Whenever practical, the CPT should be preferred to the SPT; the CPT provides detailed data on soil fabric and variations in soil type, and is far more repeatable and less prone to operator error and equipment-induced variability. Lunne *et al.*

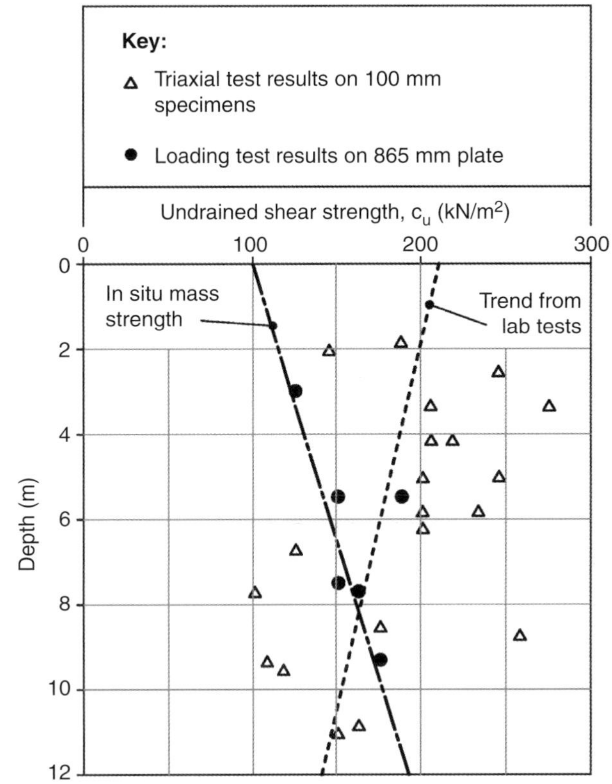

Figure 53.23 Undrained shear strength data for a stiff clay

(1997) give guidance on deriving the undrained strength from CPTs in stiff clays, where the correlation is dependent mainly on fissure spacing.

If the project is relatively simple and a high factor of safety (say, 3.0 or more) against bearing capacity failure is used, then the potential errors introduced by routine sampling and testing may be of little practical consequence. However, for some projects (e.g. settlement-sensitive structures, complex ground or groundwater conditions, onerous site-specific constraints, challenging construction timescales, opportunities to save time/costs) the potential errors associated with routine methods can become critical and then more sophisticated ground investigation and analysis should be seriously considered. The case history at the end of this chapter gives an example, for a fast-track design/build settlement-sensitive building project, of how more sophisticated sampling and testing enabled shallow foundations to be used, whereas a routine approach indicated piled foundations were required. This case history also demonstrates the value of CPTs, even for very stiff Lodgement Till. Although CPT penetration was limited, it was sufficient (given the limited depth of influence of the pad footings) to demonstrate the range of strength to be expected within the upper few metres of Till below the proposed footings.

Heavily overconsolidated clays – compressibility: Estimates of ground movements and foundation settlement are commonly based upon empirical correlations between simple parameters (such as quick undrained strength from UU tests and or SPT 'N' values). For example, Stroud (1990) suggested that vertical drained Young's modulus, E'_v varies from 200 S_{um} to 250 S_{um} for high- and low-plasticity clays respectively, where S_{um} is a cautious estimate of the **mean** undrained strength (i.e. not a characteristic strength). Empirical correlations, such as Stroud's, are valuable and helpful for practitioners; unfortunately, they are commonly misused – either by applying the correlations beyond the limits of the original database or being indiscriminately reused in different analytical methods. For example, if an empirical correlation is based on the use of simple elastic equations (with their implicit simplifying assumptions), the same correlations should not be reused in a finite element analysis (the resulting movement predictions are likely to be erroneous).

It is now well known that the deformation modulus (or compressibility) of soils is highly nonlinear, and as discussed in Chapter 52 *Foundation types and conceptual design principles*, it may change by more than an order of magnitude between very small strain and large strain amplitudes. Hence, the confusing range of soil stiffness correlations quoted in various technical publications. Research has shown that the degradation of deformation modulus, with increases in strain amplitude or shear stress, forms a consistent pattern – refer to Section 52.6 for further guidance. Atkinson (2000), Vardanega and Bolton (2013), Bolton and Oztoprak (2013) and Clayton (2011) give additional information on this topic. **Figure 53.24** plots drained Young's moduli derived from oedometer and geophysical tests (i.e. at large and very small strains, respectively) for a site in central London (QE2 Conference Centre; Burland and Kalra, 1986). Also shown are mobilised Young's moduli back-analysed from observations of foundation settlement. It can be seen that at shallow depths the mobilised stiffness tends towards large strain values (from oedometer tests), but rapidly increases with depth towards very small strain values (from geophysical tests). Also shown is the Young's modulus profile based on Stroud's correlation, which significantly underestimates the rate of increase of mobilised stiffness with depth.

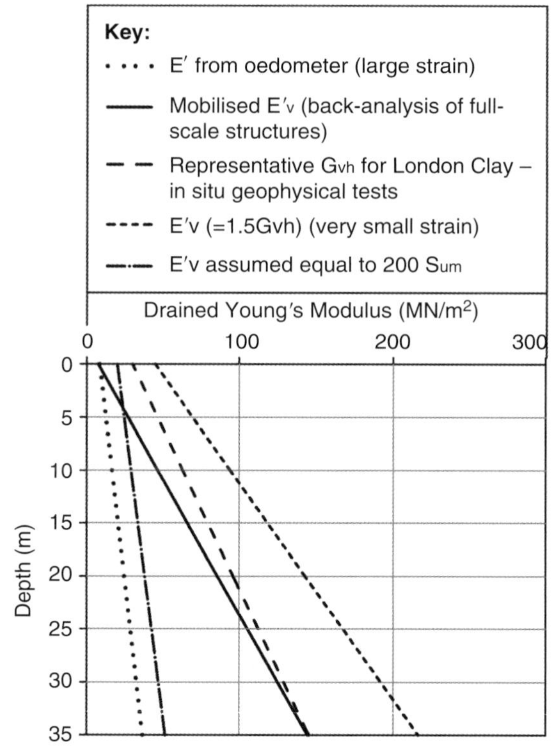

NB.
Mobilised E′v and oedometer E′v (after Burland and Karla, 1986).
Representative Gvh for London Clay (derived from Hight *et al.* 2003).
Gvh = vertical shear modulus;
Sum = mean undrained strength.

Figure 53.24 Drained Young's moduli against depth for a stiff plastic clay

53.7.3 Strength and compressibility of sand

Classification: To assess the strength and compressibility of sands the following basic information will be required

- geological age (e.g. post-glacial, glacial or pre-glacial) and likely OCR (overconsolidated or normally consolidated)
- particle grading and particle angularity
- mineralogy (silica or non-silica)
- relative density.

Clayton (1995) and Lunne *et al.* (1997) provide guidance on the derivation of relative density from SPT 'N' and CPT q_c values, respectively.

Chapter 17 *Strength and deformation behaviour of soils* gives an introduction and Chapter 52 *Foundation types and conceptual design principles* discusses foundation engineering-related considerations.

Strength: A simple and coherent approach to assessing the strength of sands has been described by Bolton (1986) and is summarised in Chapter 52 *Foundation types and conceptual design principles.*

Compressibility: It is recommended that the original paper by Burland and Burbidge (1984) is reviewed together with the subsequent published discussion (*Proceedings of the Institution of Civil Engineers*, 1986), which provide insights into the various issues that can affect the settlement of sand. The main practical conclusion is that there are seldom significant settlements associated with shallow foundations on sands which have a relative density of medium-dense or more. It is important for the engineer to appreciate the limitations of any method of settlement prediction that is based on the results of penetration tests; for example, a sand's stress history has a significant influence on its compressibility but has a negligible

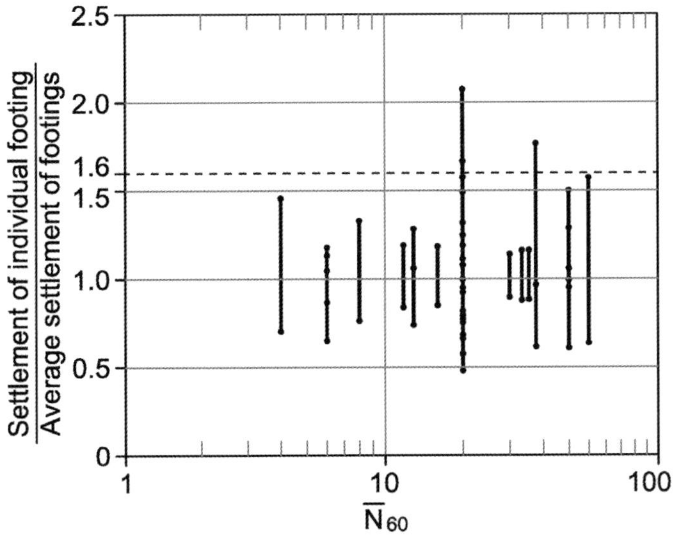

Figure 53.25 Variability of sand: observed variations of footing settlement at a site (Reproduced from Burland and Burbidge, 1984)

influence on penetration resistance. It is also important to note that the natural variability of sands tends to be high. **Figure 53.25** gives the results of a statistical study by Burland and Burbidge (1984) for 13 sites, which shows that the ratio of the maximum to the minimum settlement of practically identical footings (subject to the same loads) is large, and a ratio of maximum to average settlement of 1.6 is a reasonable bound to the data spread.

If it is necessary to derive Young's moduli from SPT 'N' values, then **Table 53.14** provides some guidance, and Lunne *et al.* (1997) also provide guidance for interpretation of CPT data (for E' mobilised at strain amplitudes of about 0.1%, and

Standard penetration value (blows/300 mm), N	E'/N (MN/m^2) at ...[a,c,d]		
	Lower limit (large strain)	Mean (small strain)	Upper limit[b] (very small strain)
<4	0.4–0.6	1.6–2.4	3.5–5.3
10	0.7–1.1	2.2–3.4	4.6–7.0
30	1.5–2.2	3.7–5.6	6.6–10.0
>60	2.3–3.5	4.6–7.0	8.9–13.5

(a) Based on back-analysis of Burland and Burbidge (1984) database and for initial guidance only.
(b) The upper-limit values should not be used in routine practice; site-specific verification is required by way of geophysical or other independent tests or correlations.
(c) E' = 'N' and E' = 2'N' are often used in UK practice for normally and overconsolidated sands respectively.
(d) For important projects, site-specific testing may be required.

Table 53.14 Correlation between SPT 'N' and Young's modulus (Data taken from Clayton, 1995)

the variation of Young's moduli with loading intensity). Many of the published correlations for relating penetration test values to Young's moduli are unreliable.

53.7.4 Strength and compressibility of rocks

Classification and strength: reference should be made to Chapter 52 *Foundation types and conceptual design principles* for discussion on these topics.

Compressibility: As indicated by **Figure 53.26**, the compressibility of a rock mass (due to the influence of discontinuities) can be far higher than the compressibility of the intact material. This figure compares the intact stiffness of chalk in the laboratory with the mass stiffness derived from surface-wave geophysics and large (1.8 m dia.) plate tests. At comparable strain amplitudes the intact lab stiffness is several times higher than the mass stiffness given by the field tests. Hobbs (1974) has related the rock mass quality by way of RQD to a rock mass factor, j, which reduces the intact Young's modulus (typically measured in the laboratory) to a 'mass' modulus, from

$$E_m = j(E_i) \qquad (53.20)$$

where E_m is the mass modulus, j is the mass factor and E_i is the intact modulus. The relationship between rock quality and j is given in **Table 53.15**. Some care is needed when using RQD since it takes no account of joint opening or infill materials

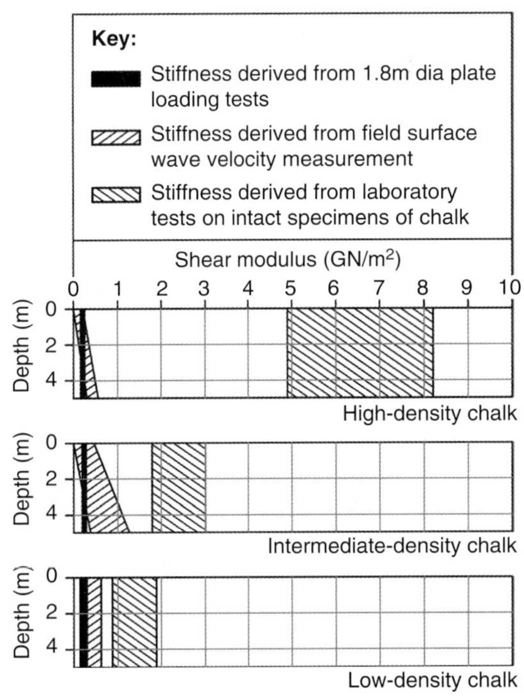

Figure 53.26 Comparison of initial modulus from 1.8 m dia. plate tests with stiffness measured on intact chalk, and derived from surface-wave geophysics (Reproduced from Clayton *et al.*, 1994)

Quality classification[a]	RQD: %	Fracture frequency per metre	Velocity index[b] $(V_f/V_l)^2$	Mass factor (j)
Very poor	0–25	15	0–0.2	0.2
Poor	25–50	15–8	0.2–0.4	0.2
Fair	50–75	8–5	0.4–0.6	0.2–0.5
Good	75–90	5–1	0.6–0.8	0.5–0.8
Excellent	90–100	1	0.8–1.0	0.8–1.0
	Rock type		**Modulus ratio: M_r**[c]	
Group 3	Very marly limestones, poorly cemented sandstones, cemented mudstones and shales, slates and schists (steep cleavage and foliation)		150	
Group 4	Uncemented mudstones and shales		75	

Notes:
(a) As BS 5930.
(b) V_f is the wave velocity in the field and V_l is the wave velocity in the laboratory.
(c) M_r is the modulus ratio, where $E_m = jM_r q_{uc}$; E_m is the modulus for a rock mass, j is the mass factor and q_{uc} is the unconfined compressive strength.
These approximate correlations are for preliminary settlement estimates only. Site-specific tests may be necessary depending on project requirements.

Table 53.15 Approximate correlations for rock mass stiffness

(which are assessed in more comprehensive rock mass classification systems such as the GSI or RMR systems). An open joint infilled with clay would lead to a much lower mass modulus than one with closed joints. Hoek and Diederichs (2006) outline a more sophisticated empirical approach based on GSI. For preliminary settlement assessments and routine design of simple structures it can be assumed that $E_i = Mq_{uc}$, where M is the ratio between the intact modulus E_i and the unconfined compressive strength q_{uc}. Values of M for different rock types are given in **Table 53.15**.

Perhaps the biggest challenge is that many weak rocks can rapidly transition between stiff clays/dense sands and weak rocks as the degree of weathering varies. From both theoretical and practical perspectives these 'intermediate geomaterials' can be problematic since both soil and rock mechanics skills may be required, and the investigation, interpretation and analytical tools required may need to be similarly varied. For weak rocks such as Mercia Mudstone it may not be practical to obtain samples of sufficient quality for reliable laboratory testing of strength, stiffness, compressibility and so on. Then more reliance will need to be placed on field tests, such as the pressuremeter and geophysics, together with high-quality logging by experienced geologists (describing the degree of weathering, nature of discontinuities, etc.). Penetration tests are not very reliable in weak rocks; for example, the influence of discontinuities and lithorelicts on the mass stiffness of a partially weathered rock mass will not necessarily be reflected by the recorded 'N' values. Mass compressibility

Density	Grade	Es[a]: MPa	q_y: kPa	E_y[b]: MPa	q_u[c]: MPa
Medium/high	A	1500–3000	>1600–2400	?[d]	>16
Medium/high	B	1500–2000	300–500	35–80	4.0–7.7
Medium/high	C	300–1500	300–500	35–80	4.0–7.7
Low	B and C	200–700	250–500	15–35	1.5–2.0
	Dc	200[e]	225–500	20–30	–
	Dm	6	?	–	–

(a) E_s is the initial modulus, but standardised at a net applied stress of 200 kPa; q_y is the yield stress.
(b) E_y is the post-yield modulus, assumed to be fully developed once ρ/D exceeds about 0.4%.
(c) q_u is the ultimate stress, measured at $\rho/D = 10–15\%$ of the plate diameter.
(d) Yield has not been observed for this grade of chalk.
(e) Values given in this table are short term, based upon plate loading tests. In the long term this figure reduces to 75 MPa.
The above values are for initial guidance only, site-specific tests may be necessary.

Table 53.16 Empirical mass compressibility parameters for the design of shallow foundations in chalk (Data taken from CIRIA C574, Lord *et al.*, 2002; www.ciria.org)

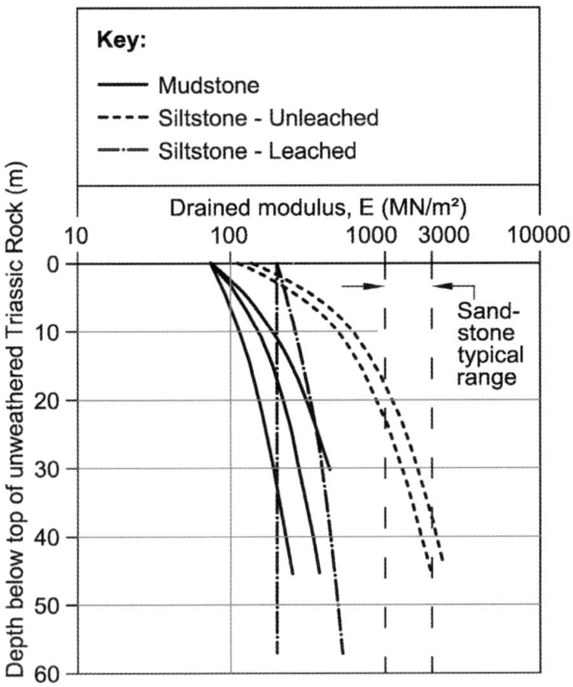

Figure 53.27 Measured secant drained Young's moduli in unweathered Triassic rocks (Data taken from Meigh, 1976)

parameters for the design of shallow foundations in chalk are given in **Table 53.16**. For Mercia Mudstone, profiles of rock mass Young's modulus against depth, based on the back-analysis of foundation settlement in unweathered mudstone, siltstone (typically Grades 1 and II) and sandstone, are given in **Figure 53.27**. Meigh (1976) suggested that weathering could change the mobilised Young's modulus by more than an order of magnitude in weak rocks, highlighting the importance of high-quality engineering geologist inputs.

53.8. Case history for a prestigious building on glacial tills

53.8.1 Introduction

A large public building was to be constructed in an urban area. An important architectural feature was a special glass frontage across one corner of the building. This meant that an onerous differential settlement criterion was imposed. The maximum net foundation bearing pressure was 250 kN/m^2.

A desk study and preliminary ground investigations indicated a surface layer of loose and variable made ground, about 2–3 m thick. Beneath this was a stiff to very stiff layer of glacial till. An existing old building constrained the site redevelopment, both for carrying out ground investigations and for construction of the new building (since demolition was on the overall project critical path).

The building was to be constructed under a 'design and build' contract, and there were very severe penalties under the contract if the building was completed late. The consultant responsible for the building design (including the foundations) was commissioned by the building contractor

53.8.2 Ground conditions and site history

Desk studies had shown that neither mining nor quarrying had affected the site. A key observation from the ground investigation was the presence of numerous boulders within the glacial till. The description of the glacial till indicated it was consistently stiff or very stiff and its liquidity index was about –0.2 (±0.1) in the uppermost 5 m of the till. These were indicators of the heavily overconsolidated nature of the till and, together with the PI, which was typically about 15%, indicated that compressibility of the till would be very low. The above observations were consistent with the relevant technical literature and local experience, which was collated during the desk study.

53.8.3 Foundation options and risks

The initial design criterion was for foundation settlement to be less than 5 mm. As a consequence, bored piles were initially selected for the new building foundations. There were serious risks with this foundation type, because the numerous boulders would form significant obstructions to pile boring. Hence, there was a major risk of large delays to the overall construction programme. Past experience had shown that shallow foundations on stiff glacial tills usually exhibit small settlement under the envisaged foundation bearing pressures. The old building foundations were pad foundations, and the decision was taken to supplement the existing shallow foundations, so mass concrete was used to broaden and lengthen the existing pads, which were then overlain by reinforced concrete pads. In addition, negotiations with the architectural finishes supplier enabled a less onerous differential settlement criterion of 10 mm to be agreed.

53.8.4 Additional ground investigations and geotechnical analysis

An additional ground investigation was carried out. This comprised three 4 m deep trial pits to expose the glacial till and enable high-quality block samples to be taken. Advanced laboratory tests were then carried out to obtain realistic drained stiffness parameters for the till. Glacial till is often heterogeneous and, because of access constraints (due to the existing building), there were several areas across the site where there were no geotechnical data. To mitigate this risk, static CPTs were carried out as soon as the existing building was demolished and before foundation construction for the new building began. The CPTs could be carried out and interpreted quickly. Cone penetration was limited, typically to about 7–8 m depth, due to boulders, but this was adequate given the limited depth of influence of the narrow, ~2 m wide, pad footings. These tests confirmed the till strength was adequate across the whole site for the proposed design.

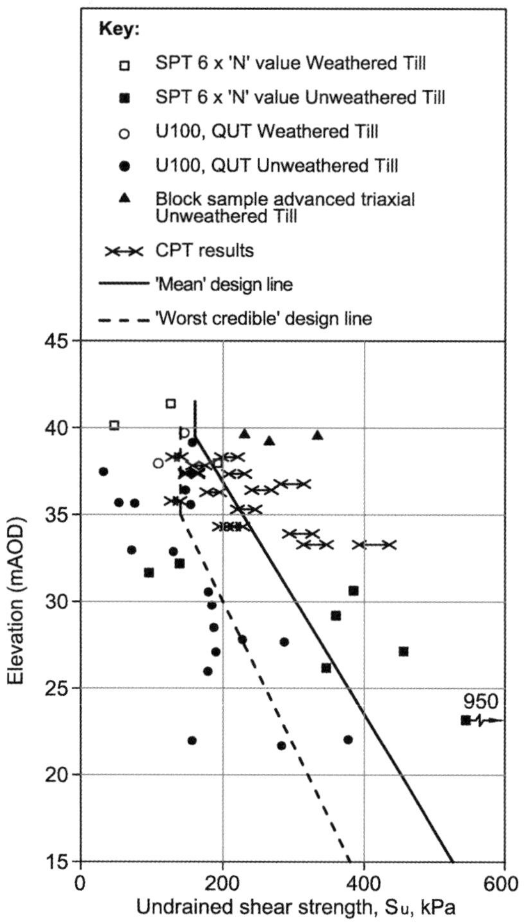

Figure 53.28 Case history, glacial till undrained shear strength data

The undrained strength data are plotted in **Figure 53.28**. The routine triaxial data (on U100 samples) obtained from the first investigation indicated relatively low undrained strengths compared with the block samples taken from the trial pits. The subsequent CPT profiles indicated that the till was reasonably uniform and that strength increased with depth. The design profiles were based on the block samples and the CPTs.

To assess the till's compressibility across a wide range of strain amplitudes, several different test techniques were specified: geophysical (bender element) tests for compressibility at very small strain; oedometer tests for large strain compressibility; and triaxial tests with local instrumentation to measure compressibility at intermediate strain amplitudes. The data derived from some of these tests are shown in **Figure 53.29**. It is worth noting that the oedometer tests (using relatively large load increments/decrements of 100 kN/m^2 plus) on U100 samples (taken during the first ground investigation) gave an average coefficient of volume compressibility m_v of 0.1 m^2/MN, whereas oedometer tests on the block samples (with load–unload–reload loops and small load increments on reloading, typically 20–30 kN/m^2) gave m_v values that were 2–4 times smaller.

Bearing capacity checks showed that the factor of safety was greater than three. In view of the onerous different settlement criterion, a careful analysis of the foundation settlement was required. The main method of settlement analysis was a modified one-dimensional method, which allowed for the variation of deformation modulus with both strain amplitude and mean effective stress due to foundation loading (O'Brien and Sharp, 2001). The configuration of the pad and strip footings beneath the building footprint was quite complex; however, this was easily taken into account by allowing for superposition of vertical stress increments (calculated from Boussinesq theory).

53.8.5 Design verification and construction control

A critical risk was the potential for the till to soften rapidly if it was exposed to water (either groundwater or rainfall). During trial pitting, no problems were encountered with water seepage. Nevertheless, the construction specification required the till at foundation level to be carefully protected. In particular, a protective layer of till 0.6 m thick was to be left above the foundation level, and once final excavation was carried out the till was to be covered with a structural-quality layer of blinding concrete 100 mm thick within the same shift.

The shallow foundations were constructed rapidly and without difficulties, and the project was completed ahead of schedule. Subsequent monitoring indicated foundation settlement of less than 10 mm, which was consistent with the predictions derived from the additional ground investigation data.

53.8.6 Conclusions

The building featured a special glass frontage with an onerous settlement criterion. It was built under a fast-track design and build form of contract. If piling had been attempted it is likely that severe delays would have adversely affected the project, due to large boulders within the glacial till. Local experience generally indicated that shallow foundations exhibited only small settlement when founded on the very stiff heavily overconsolidated till. The first ground investigation, using routine sampling and testing methods, indicated that shallow foundation settlement would be excessive. Hence, additional investigations were carried out using modern testing methods. These high-quality data were used together with a modified one-dimensional settlement analysis (Section 53.6.4), which took account of nonlinear elastic behaviour, to demonstrate that shallow foundations would perform satisfactorily. The old building foundations were reused and supplemented by new shallow foundations, and the new building was successfully completed.

53.9. Overall conclusions

- There can be several causes of shallow foundation movements, which are independent of the applied bearing pressures imposed by a superstructure (**Table 53.2**). The global ground movements caused by these factors can lead to SLS or ULS failure of structures. The seasonal shrink–swell movement of clay soils, induced

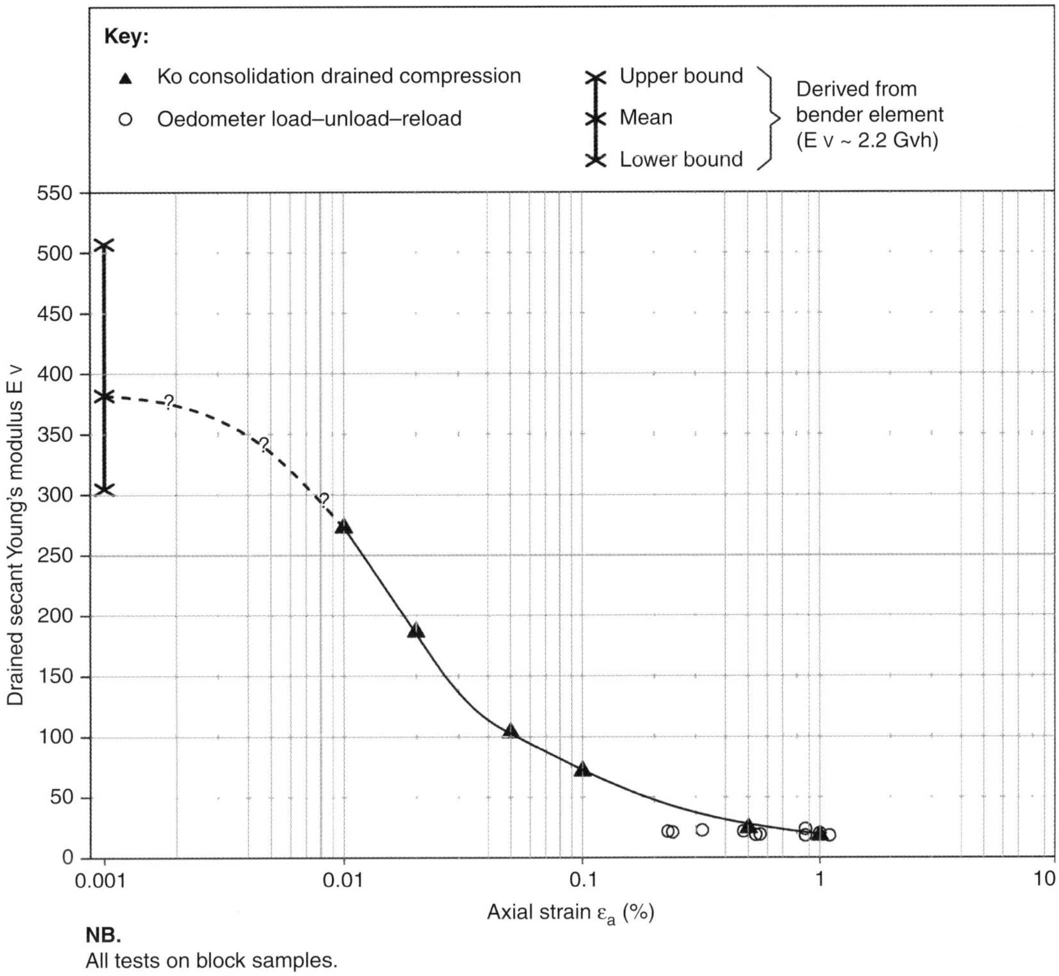

Figure 53.29 Case history, glacial till drained secant Young's modulus variation with strain amplitude

by trees, is a common cause of excessive shallow foundation movement.

- The temporary stability of excavations and the potential for groundwater inflow require careful consideration because of the safety and cost implications.
- The applied loads on a foundation usually comprise a variety of different types of load, including dead loads and transient loads (including wind, impact and accidental loads). It is important to understand the nature of the applied loads before carrying out bearing capacity and settlement calculations. For example, when calculating the long-term settlement of clays, occasional transient loads would usually be ignored.
- When carrying out bearing capacity and settlement checks, it is important to consider the behaviour of the whole structure, in addition to the individual foundation elements, within the context of the site, adjacent structures and topographical features. The risk of adverse interaction between adjacent structures tends to increase if foundations are subject to lateral loads.
- There have been significant developments in bearing capacity theory beyond the original equations proposed by Terzaghi and Skempton. A wide range of situations, including increasing strength with depth, anisotropy, soil layering and allowances for soil compressibility and stress level can be analysed (refer to Section 53.5.2). Foundations subjected to combined vertical, horizontal and moment loads have conventionally been assessed by applying inclination factors to the bearing capacity factors for vertical loading. However, the reliability of these correction factors reduces as the load inclination (*H*/*V* ratio) increases and the nature of the applied loads becomes more complex. Recently developed bearing capacity theories based on a *VHM* failure surface (Chapter 21 *Bearing capacity theory*) and discussed in Section 53.5.3 are more appropriate. A simple conclusion is that the 'equivalent width' method (**Figure 53.10(a)**) is reasonably accurate; however, the 'equivalent area' method (**Figure 53.10(b)**) is inaccurate because the actual soil/foundation contact area is poorly represented by the simplified method.
- The bearing capacity and settlement of rock will be sensitive to the orientation, spacing and characteristics of discontinuities within the rock mass (**Figure 53.13**), the degree of weathering and the presence of voids.
- Vertical stress changes beneath a foundation predicted by homogeneous, isotropic, linear elastic theory are reasonably accurate for many practical situations, which may involve anisotropy,

nonlinearity and increasing stiffness with depth. In contrast, horizontal stress changes predicted by elastic theory are often inaccurate for these situations.

- Foundation settlement can be assessed quickly by using displacement influence factors derived from elastic theory. Published solutions that assume the elastic material has constant Young's modulus with depth or is of infinite thickness are not recommended. Solutions that allow for finite thickness and with Young's modulus increasing with depth will generally provide more realistic predictions.
- For heavily overconsolidated clays, the simple one-dimensional method can be used to calculate the *total* settlement, and this method is as accurate as (and often more accurate than) many more sophisticated methods. The major difficulty is reliable measurements of compressibility, which are appropriate for the relevant stress changes and strain amplitudes. The modified one-dimensional method, which uses the constrained modulus D, is helpful in this respect since D can be related to a variety of measurable parameters that are relevant for a wide range of strain amplitudes, varying from very small strain geophysics to large strain oedometer data.
- The settlement of medium-dense and dense silica sands is usually small (some exceptions are summarised in **Table 53.8**), and for simple structures the use of the empirical correlations in **Figure 53.17** will often be sufficient to estimate the likely bounds of settlement. A variety of more complex methods are available; however, these usually depend on relating penetration resistance (from SPTs or CPTs) to sand compressibility and therefore are unlikely to improve the accuracy of predictions.
- Many of the serious problems that occur with shallow foundations are due to a lack of understanding of the site's history, geology or hydrogeology. Hence, comprehensive desk studies are always important, and ground investigations must be implemented with great care and thoroughly checked.
- For shallow foundations on heavily overconsolidated clays, settlement is concentrated immediately beneath the foundation and more than half of the settlement will generally occur over a depth of about half the foundation width. Hence, much more emphasis needs to be placed on measuring soil properties in this area. The *mobilised* soil stiffness generally increases rapidly with depth (typically from 'large strain' values immediately beneath the foundations to 'very small strain' values remote from the foundations).
- For soft clays and bonded materials (such as chalk) perhaps the most critical parameter for foundation design is the assessment of the yield stress (commonly known as the pre-consolidation pressure for soft clays). The yield stress for chalk is reasonably well established (**Table 53.16**) and at net bearing pressures below the yield stress settlement will be small. The compressibility of a rock mass will generally be far higher than the intact material because of the influence of discontinuities.

References

Atkinson JH (2000) Non-linear soil stiffness in routine design. *Géotechnique* **50(5)**: 487–508.

Barton N (1986) Deformation phenomena in jointed rock. *Géotechnique* **36(2)**: 147–167.

Bolton MD (1986) The strength and dilatancy of sands. *Géotechnique* **36(1)**: 65–78.

Bolton MD (1993) Design methods. In *Predictive Soil Mechanics: Proceedings of Wroth Memorial Symposium, Oxford, UK.* Thomas Telford, London, UK, pp. 50–72

Bolton M and Oztoprak S (2013) Stiffness of sands through a laboratory test database. *Géotechnique* **63**: 54–70, 10.1680/geot.10.P.078.

Breth H and Amman P (1974) Time settlement and settlement distribution with depth in Frankfurt Clay. *Proceedings of the Conference on Settlement of Structures, Cambridge, UK.* Pentech Press, London, UK, pp. 141–154.

Briaud JL and Gibbons RM (1994) Test and prediction results for five large spread footings on sand. *Geology Speciality Publication ASCE* **41**: 92–128.

Burland JB (1990) On the compressibility and shear strength of natural clays. *Géotechnique* **40**: 329–378.

Burland JB and Burbidge MC (1984) Settlement of foundations on sand and gravel. *Proceedings ICE* **78**: 1325–1381.

Burland JB and Kalra JC (1986) Queen Elizabeth II Conference Centre. *Proceedings ICE* **80**: 1479–1503.

Burland JB and Wroth CP (1974) *Settlement of Buildings and Associated Damage: Settlement of Structures.* Pentech Press, Cambridge, UK, pp. 611–654.

Burland JB, Broms BB and De Mello FB (1977). Behaviour of foundations and structures. *Proceedings of the 9th ICSMFE, Tokyo, Japan*, vol. 1, pp. 495–546.

Chan DH and Morgenstern NR (1987) Analysis of progressive deformation of the Edmonton Convention Centre excavation. *Canadian Geotechnical Journal* **24**: 430–440.

Chandler RJ (1984) Recent European experience of landslides in overconsolidated clays and soft rocks. *Proceedings of the 4th International Conference on Landslides, Toronto, ON*, vol. 1, pp. 61–81.

Chandler RJ and Forster A (2001) *Engineering in Mercia Mudstone.* Report C570. CIRIA, London, UK.

Clark JI (1997) The settlement and bearing capacity of very large foundations on strong soils. *Canadian Geotechnical Journal* **35**: 131–145.

Clayton CRI (1995) *The Standard Penetration Test (SPT): Methods and Use.* Report C143. CIRIA, London, UK.

Clayton CRI (2011) Stiffness at small strain: research and practice. *Géotechnique* **61(1)**: 5–37, 10.1680/geot.2011.61.1.5.

Clayton CRI, Simons NE and Instone SJ (1988) Research on dynamic penetration testing of sands. *Proceedings of the 1st International Symposium on Penetration Testing, Orlando, FL*, vol. 1, pp. 415–422.

Clayton CRI, Gordon MA and Matthews MC (1994) Measurement of stiffness of soils and weak rocks using small strain laboratory testing and field geophysics. *Proceedings of the 1st International Conference on Pre-failure Deformation Characteristics of Geomaterials, Sapporo, Japan*, vol. 1. Balkema, Rotterdam, the Netherlands, pp. 229–234.

Cole KW (1988) *Foundations.* Thomas Telford, London, UK.

COSOS (1974) *Proceedings of the Conference on Settlement of Structures*, Cambridge, UK.

Crilly MS and Driscoll RMC (2000) The behaviour of lightly loaded piles in swelling ground and implications for their design. *Proceedings of the Institution of Civil Engineers, Geotechnical Engineering* **143**: 3–16.

Czerewko MA, Cross SA, Dumelow PG and Saadvandi A (2011) Assessment of pyritic Lower Lias mudrocks for earthworks. *Geotechnical Engineering* **164(GE2)**: 59–77.

D'Appolonia DJ and Lambe TW (1971) Floating foundations for control of settlement. *Journal of the Soil Mechanics and Foundations Division, ASCE* **97(SM 6)**: 899–915.

Davis EH and Booker JR (1971) The bearing capacity of strip footings from the standpoint of plasticity theory. *Proceedings of the 1st Australia–New Zealand Conference on Geomechanics, Melbourne, Australia*, vol. 1, pp. 276–282.

De Jong J and Harris MC (1971) Settlement of two multi-storey buildings in Edmonton. *Canadian Geotechnical Journal* **8**: 217–235.

De Jong J and Morgenstern NR (1973) Heave and settlement of two tall building foundations in Edmonton, Alberta. *Canadian Geotechnical Journal* **10**: 261–281.

Eadington J and O'Brien AS (2011) Stiffness parameters for a deep tunnel: developing a robust parameter selection framework. *Proceedings of the 15th European Conference on Soil Mechanics and Geotechnical Engineering, Athens, Greece*. Millpress, Rotterdam, the Netherlands.

Feng X, Gourvenec S, Randolph MF and Teng R (2018) Investigation of effectiveness of effective area method for assessing undrained capacity of shallow rectangular foundations. *Journal of Geotechnical and Geoenvironmental Engineering*, **145(2)**. 10.1061/(ASCE)GT.1943-5606.0002009.

Franklin JA and Dusseault M (1989) *Rock Engineering*. McGraw-Hill, New York, NY.

Fredlund DG and Rahardjo H (1993) *Soil Mechanics for Unsaturated Soils*. Wiley, New York, NY.

Gourvenec S (2007) Shape effects on the capacity of rectangular footings under general loading. *Géotechnique* **57(8)**: 637–646.

Gourvenec S (2008) Undrained bearing capacity of embedded footings under general loading. *Géotechnique* **58(3)**: 177–185.

Hanna A and Meyerhof GG (1980) Design charts for ultimate bearing capacity of sand overlying soft clay. *Canadian Geotechnical Journal* **17**: 300–303.

Hawkins AB and Pinches GM (1987) Cause of significant heave at Llandough Hospital, Cardiff: a case study of ground floor heave due to gypsum growth. *Quarterly Journal of Engineering Geology and Hydrogeology* **20**: 41–57.

Hobbs NB (1974) The factors affecting the prediction of the settlement of structures on rock, State of the Art review. *Proceedings of the Conference on Settlement of Structures, Cambridge, UK*. Pentech Press, London, UK, pp. 579–610.

Hoek E (1999) Putting numbers into geology: an engineer's view point (Second Glossop Lecture). *Quarterly Journal of Engineering Geology* **32**: 1–19.

Hoek E and Brown ET (1997) Practical estimates of rock mass strength. *International Journal of Rock Mechanics, Mining Sciences and Geomechanics Abstracts* **34**: 1165–1186.

Hoek E and Diederichs MS (2006) Empirical estimation of rock mass modulus. *International Journal of Rock Mechanics, Mining Science* **43(2)**: 203–215.

Hoek E, Caranza-Torres CT and Corkum B (2002) Hoek-Brown failure criterion: 2002 edition. *Proceedings of the North American Rock Mechanics Society (NARMS-TAC 2002)* (Bawden HRW, Curran J and Telsenicki M (eds)). Mining Innovation and Technology, Toronto, ON, pp. 267–273.

Horikoshi K and Randolph MF (1997) On the definition of raft–soil stiffness ratio for rectangular rafts. *Géotechnique* **47(5)**: 1055–1061.

Houlsby G and Martin C (2003) Undrained bearing capacity factors for conical footings on clay. *Géotechnique* **53**: 513–520, 10.1680/geot.53.5.513.37507.

Janbu N, Bjerrum L and Kjaaernsli B (1956) *Soil Mechanics Applied to Some Engineering Problems*. Norwegian Geotechnical Institute, Oslo, Norway.

Jardine RJ, Potts DM, Fourie AB and Burland JB (1986) Studies of the influence of non-linear stress–strain characteristics in soil–structure interaction. *Géotechnique* **36(3)**: 377–396.

Lancellotta R (1993) Stability of a rigid column with non-linear restraint. *Géotechnique* **43(2)**: 331–332.

Lehane BM, Gaudin C, Richards DJ and Rattley MJ (2008) Rate effects on the vertical uplift capacity of footings founded in clay. *Géotechnique* **58(1)**: 13–21, 10.1680/geot.2008.58.1.13.

Lord JA, Clayton CRI and Mortimore RN (2002) *Engineering in Chalk*. Report C574. CIRIA, London, UK.

Lunne T, Robertson PK and Powell JJM (1997) *Cone Penetration Testing in Geotechnical Practice*. Spon Press, London, UK (reprinted in 2004).

Martin CM (2001) Vertical bearing capacity of skirted circular foundations on Tresca soil. *Proceedings of the 15th International Conference on Soil Mechanics and Geotechnical Engineering, Istanbul, Turkey*, vol. 1, pp. 743–746.

Mayne PW and Poulos HG (1999) Approximate displacement influence factors for elastic shallow foundations. *Journal of Geology and Geoenvironmental. Engineering ASCE* **125(6)**: 453–460.

Meigh AC (1976) The Triassic rocks, with particular reference to predicted and observed performance of some major foundations. *Géotechnique* **26(3)**: 391–452.

Meyerhof G (1953) The bearing capacity of foundations under eccentric and inclined loads. *Proceedings the 3rd International Conference on Soil Mechanics and Foundations Engineering, Zurich, Switzerland*, pp. 440–445.

NHBC (2023) *Building Near Trees*. NHBC, London, UK, Chapter 4.2.

O'Brien A and Higgins K (2020) *Management of Advanced Numerical Modelling in Geotechnical Engineering: Good Practice*. Report C791. CIRIA, London, UK.

O'Brien AS and Sharp P (2001) Settlement and heave of over-consolidated clays: a simplified non-linear method of calculation. *Ground Engineering*, Oct.: pp. 21–28; Nov.: pp. 48–53.

O'Brien T, Ho X and Tan R (2023) A practical framework for characterising the non-linear stiffness of overconsolidated soils and weak rock. *Proceedings of the Institution of Civil Engineers – Geotechnical Engineering*, 10.1680/jgeen.22.00210.

Okamura M, Takemura J and Kimura T (1998) Bearing capacity predictions of sand overlying clay based on limit equilibrium methods. *Soils and Foundations* **38**: 181–194.

Osman ASEK and Bolton MD (2005) Plasticity based method for predicting undrained settlement of shallow foundations on clay. *Géotechnique* **55(G)**: 435–447.

Osman AS, White DJ, Britto AM and Bolton MD (2007) Simple prediction of the undrained displacement of a circular surface foundation on non-linear soil. *Géotechnique* **57(9)**: 729–737.

Parsons JD (1976) New York's glacial lake formation of varved silt and clay. *ASCE Journal of Geotechnical Engineering* **102(GT6)**: 605–638.

Potts DM and Burland JB (2000) Development and application of a numerical model for simulating the stabilisation of the Leaning Tower of Pisa. In *Developments in Theoretical Geomechanics* (Smith DW and Carter JP (eds)). Balkema, Rotterdam, the Netherlands, pp. 737–758.

Poulos HG and Davis EH (1970) *Elastic Solutions for Soil and Rock Mechanics.* Wiley, New York, NY.

Poulos HG, Carter JP and Small JC (2001) Foundations and retaining structures: research and practice. *Proceedings of the 15th International Conference on Soil Mechanics and Geotechnical Engineering, Istanbul, Turkey.*

Proceedings of the Institution of Civil Engineers (1986) **80**.

Randolph MF and Gourvenec S (2011) *Offshore Geotechnical Engineering.* CRC Press, London, UK, pp. 272–292.

Randolph MF, Jamiolkowski MB and Zdravkovic L (2004) Load carrying capacity of foundations. *Advances in Geotechnical Engineering: The Skempton Conference.* Thomas Telford, London, UK, pp. 207–240.

Rowe PW (1972) The relevance of soil fabric to site investigation practice: 12th Rankine Lecture. *Géotechnique* **22(2)**: 195–300.

Salencon J and Matar M (1982) Bearing capacity of circular shallow foundations. *Journal of Mechanical Theory and Application* **1(1)**: 237–267.

Schmertmann JH, Hartmann R and Brown T (1978) Improved strain influence factor diagrams. *ASCE Journal, GE* **104(8)**: 1131–1135.

Skempton AW (1951) The bearing capacity of clays. *Building Research Congress, London, UK*, vol. 1, pp. 180–189.

Stroud MJ (1990) The Standard Penetration Test: its application and interpretation. *Proceedings of Penetration Testing in the UK.* Thomas Telford, London, UK, pp. 29–49.

Taiebat HA and Carter JP (2002) A failure surface for the bearing capacity of circular footings on saturated clays. *Proceedings of the 8th International Symposium on Numerical Models in Geomechanics, Rome, Italy*, pp. 457–462.

Tan CK and Duncan JM (1991) Settlement of footings on sands: accuracy and reliability. *Proceedings of the Geotechnical Engineering Congress, ASCE, Geotechnical Speciality Publications* **27(2)**: 446–455.

Terzaghi K, Peck RB and Mesri G (1996) *Soil Mechanics in Engineering Practice*, 3rd edn. Wiley, New York, NY.

Trenter NA (1999) *Engineering in Glacial Tills.* Report C504. CIRIA, London, UK.

Vardanega PJ and Bolton MD (2011) Practical methods to estimate the non-linear shear stiffness of clays and silts. *Proceedings of the 5th International Conference on the Deformation Characteristics of Geo-materials, Seoul, South Korea.* IOS Press, Amsterdam, the Netherlands, pp. 372–379.

Vardanega PJ and Bolton MD (2013) Stiffness of clays and silts: normalizing shear modulus and shear strain. *Journal of Geotechnical & Geoenvironmental Engineering* **139(9)**: 1575–1589.

Vulpe C, Gourvenec S, and Power M (2014) A generalised failure envelope for undrained capacity of circular shallow foundations under general loading. *Géotechnique Letters* **4(3)**: 187–196.

Wyllie DC (1991) *Foundations on Rock.* Spon Press, London, UK.

It is recommended this chapter is read in conjunction with

- Chapter 19 *Settlement and stress distributions*
- Chapter 21 *Bearing capacity theory*

All chapters in this book rely on the guidance in Sections 1 *Context* and 2 *Fundamental principles.* A sound knowledge of ground investigation is required for all geotechnical works, as set out in Section 4 *Site investigation.*

doi: 10.1680/icemge.66830.0841

Chapter 54
Single piles

Andrew Bell Cementation Skanska Ltd, Doncaster, UK
Christopher Robinson Roger Bullivant Limited, Swadlincote, UK
Alan Willoner Mott MacDonald, UK

This chapter considers the design of single piles, from selection of appropriate pile type(s) through to pile load testing strategies. Methods which are commonly used for estimating axial pile capacities in three principal ground types, as well as layered soils, are discussed, covering both empirical and theoretical methods. Other aspects of single-pile design covered by this chapter relate to factors of safety, pile settlements and pile response to lateral loading.

CONTENTS

54.1. Introduction

Piles are generally required to transfer load from a superstructure through weak or compressible strata, or through water, to stronger and less compressible soils or rock. Piles can also be required to reduce both overall and differential settlements of the supported structure and may be required to facilitate specific construction processes, such as top-down construction. They may be required to resist uplift loads as well as compressive loads when used, for example, to support tall structures which are subject to overturning forces caused by wind. Where trees are to be removed or basements are to be constructed, piles may also be subject to heave-induced tension forces. A more recent additional use of piles is within ground source heating and cooling systems as part of a development's carbon reduction strategy.

Piles used for marine structures are often subject to lateral loads from ship impact and wave forces. Many land-based piles are also subject to lateral forces; for example, in bridge works there are earth pressure forces, expansion and contraction forces and braking and traction forces. Combinations of vertical forces, horizontal forces and bending moments are common, and pile sections have then to be designed according to the relevant structural codes in addition to geotechnical considerations. Foundations beneath machinery may also be subject to cyclic and vibration forces.

Before embarking on the design of a piled foundation, it is essential to have a thorough understanding of all the relevant factors that might affect the performance of the piled foundation being considered.

The design of piled foundations has traditionally been based on a combination of simple empirical methods and local experience, but over the past few decades there has been a gradual change towards more soundly based theoretical methods. Poulos (1989) provided a summary of the range of analysis and design methods available for piled foundations, ranging from simple empirical methods to more sophisticated nonlinear numerical analysis methods. When selecting the appropriate design approach, the following factors should be considered

- the quality of the site investigation and extent of geotechnical data available
- the design stage (e.g. conceptual design or design for manufacturing and construction)
- the budget and timescale for foundation design
- the scale and sensitivity of the proposed structure, especially allowable total and differential settlements
- the complexity of the ground conditions and loading regime
- the potential impact of the piled foundations on adjacent structures, services or other assets.

54.2. Selection of pile type

Foundation selection is discussed in Chapters 9 *Foundation design decisions* and 52 *Foundation types and conceptual design principles.*

The key factors which often affect the choice of pile type and ultimately the design of the pile itself are

- ground conditions
  - strength or weakness of the overlying and founding strata
  - ground variability
  - potential for downdrag, heave loading or lateral loading
  - presence of obstructions
  - aggressive ground conditions

- construction constraints
  - location
  - access
  - working area (e.g. plan constraints or height restrictions)
  - sensitivity of adjacent assets
  - underground utilities and structures
- safety and environmental constraints
  - risks to operatives and the public
  - risks to adjacent structures and underground utilities/assets
  - risk of contamination to surrounding environment
  - effects of noise and/or vibration
- nature and magnitude of loads
  - direction of foundation loading (e.g. lateral, tensile, compression)
  - source and characteristics of loadings (e.g. heave, downdrag, cyclic, static)
- economic considerations
  - what it will cost and whether there is a more economic solution
- Local contractor experience
  - what solutions have worked successfully previously or, conversely, have encountered challenges.

Prior to embarking on the design of a pile, it is essential that the designer first assesses the type of piles and construction methods that are most appropriate for the particular site under consideration. There is clearly no sense in designing a driven pile in an area where excessive noise or vibration may present an insurmountable issue. Designing a continuous flight auger (CFA) pile 40 m deep when available plant will only reach 34 m is equally futile. Under current health and safety regulations (HSE, 2015) it is also essential for the designer to take into account the risks associated with the installation of the proposed pile solution during the design and specification process; to complete this satisfactorily the designer will need at least a general understanding of the various pile types available and the advantages, disadvantages and key safety considerations of each.

Before any design work is carried out it is essential that an assessment of pile buildability is carried out. Reference should be made to Chapters 52 *Foundation types and conceptual design principles*, 79 *Sequencing of geotechnical works* and 82 *Piling problems* in this respect.

It is important to recognise that each pile type and installation method disturbs the ground in different ways, and as such the design method and considerations may vary between the different systems. This disturbance can improve or reduce both skin friction and end-bearing, and as such will affect the bearing capacity of the complete pile. In addition to these factors, workmanship relating to each and every pile may influence pile performance under loading.

Piles of varying type and length may be used on a site provided a suitable assessment of likely differential settlement is undertaken. This should include

- pile size (diameter and length) – that is, geotechnical axial load capacity in compression and tension of a single pile
- displacement at working load of a single pile
- differential settlement between adjacent foundations
- influence of lateral loading, assessment of geotechnical and structural capacity and lateral displacement
- assessment of pile group behaviour under vertical and lateral loading (see Chapter 55 *Pile-group design*)
- purpose of the piles – if it is to reduce settlement/bending moments in a raft then consider a piled raft (see Chapter 56 *Rafts and piled rafts*)
- influence of any 'global' ground movements both during construction and in the long term (see Chapter 57 *Global ground movements and their effects on piles*)
- testing strategy for validating the integrity of piles
- testing strategy for verification of vertical and/or lateral load–settlement behaviour
- potential for foundation reuse.

54.3. Axial load capacity (ultimate limit state)

The approach described in this chapter to assess the resistance of piles to compressive loads is the static 'soil mechanics' method for piles loaded axially. Dynamic methods of design and analysis are available – a brief summary relating to these is included in Section 54.9. In general this section is focused on problems involving static or quasi-static axial loading of single piles only.

The general simplified formula for the design of single bearing piles typically takes the form

$$Q_{\text{ult}} = Q_{\text{b}} + Q_{\text{s}} - W_{\text{p}} \qquad (54.1)$$

where Q_{ult} is ultimate pile capacity, Q_{b} is base capacity, Q_{s} is shaft capacity and W_{p} is pile self-weight.

Furthermore, $Q_{\text{b}} = q_{\text{b}} A_{\text{b}}$ and $Q_{\text{s}} = q_{\text{s}} A_{\text{s}}$, where q_{s} is unit shaft resistance, q_{b} is unit base resistance, A_{b} is area of pile base and A_{s} is area of pile shaft. The diagram in **Figure 54.1** shows this relationship for circular piles.

The above approach, separating the evaluation of shaft and base capacity, forms the basis of all static pile capacity calculations. These components can be broken down further to resistance per unit shaft or base area, and the relative magnitude is dependent on the combination of pile geometry, pile installation method and ground conditions.

Typically, the shaft capacity of a pile is mobilised at much smaller displacements than the base capacity. Shaft capacity can be fully mobilised at movements of only 0.5–2% of

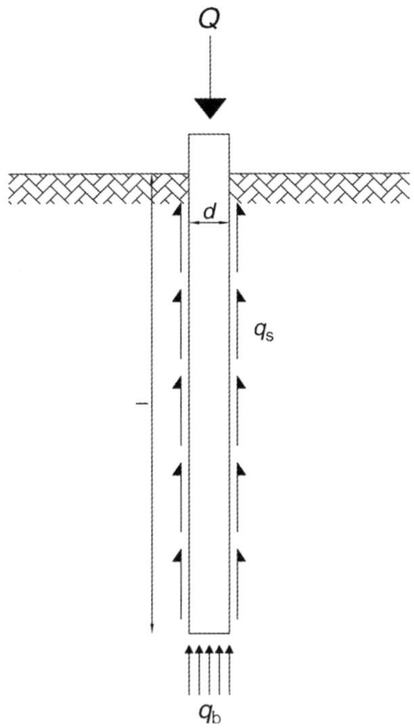

Figure 54.1 Pile under vertical load (Courtesy of Cementation Skanska)

pile diameter, whereas displacements of around 10–20% of pile diameter may be required to mobilise the base resistance, even more if the soil at the base of the pile is disturbed as part of the installation process. It is essential that these differences between the load–deformation characteristics of the pile shaft and base are considered early in the pile design process; **Figure 54.2** shows this as an idealised load–displacement response curve. Generally, long and slender piles are more efficient than short and wide piles with regard to settlement control, but limitations on construction methods may often dictate the pile geometry.

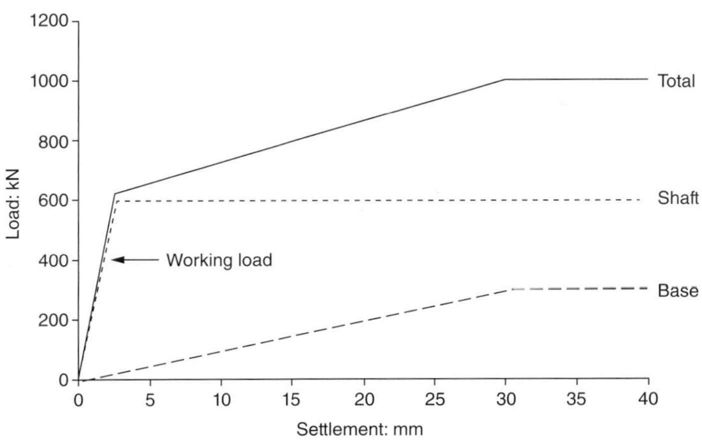

Figure 54.2 Idealised load–displacement response (Courtesy of Cementation Skanska)

There can be concerns about the axial pile capacity under tension loads. De Nicola and Randolph (1993) showed that shaft capacity under tension load is between about 70% and 90% of the shaft capacity under compression loads (depending on the relative soil/pile compressibility). However, most codes of practice require higher factors of safety for piles subject to tension loads than those for piles subject to compression. If these higher factors of safety are used, then there is no need to reduce the shaft capacity calculated by routine methods, since the potential adverse effects are taken into account by the use of a larger factor of safety. For piles subject to tension, the base resistance is ignored.

54.3.1 Fine-grained soils

54.3.1.1 Total stress end-bearing

The end-bearing of piles in fine-grained soils is generally considered using total stress methods, in terms of the undrained shear strength of the clay, S_u, and a bearing capacity factor, N_c, where

$$q_b = N_c S_u \tag{54.2}$$

The bearing capacity factor relevant to the depth of most piles is usually taken as 9, after Skempton (1951). Allowance should be made for the potential of a softened or disturbed base in some circumstances, particularly when considering under-reamed piles with enlarged bases. In such cases a reduction in S_u should be applied. A reduced value of N_c should also be considered when the pile tip only just penetrates the founding fine-grained strata. A value of $N_c = 6$ is recommended for founding in the top of such founding strata, increasing linearly to 9 after three pile diameters' penetration into the stratum.

Shaft friction

Piles in fine-grained soils generally derive most of their capacity in shaft friction, with the base resistance often not mobilised under working loads, particularly in the short term.

Shaft friction, in terms of total stress, is generally calculated using an empirical factor α, where

$$q_s = \alpha S_u \tag{54.3}$$

The value of α adopted for bored piles typically ranges from 0.3 to 0.6 for straight-shafted piles; higher values have been proven for ribbed or irregular pile shafts. Values of 0.45–0.6 are generally adopted for bored piles in stiff overconsolidated clays, but can be as high as 1.0 for large displacement piles. The value of α depends on a combination of the pile type and installation method and the nature of the fine-grained soil and also how undrained strength is determined. In the UK, the LDSA's (London District Surveyors Association) Guidance Note 1 (2017) gives clear recommendations on appropriate values for α for bored and CFA piles founded in London

Clay. It is also important for more complex and variable fine-grained soils, such as glacial till, that reference is made to specific guidance such as Weltman and Healy (1978) and Trenter (1999). **Figure 54.3** details the variation of α with undrained shear strength in glacial till, after Weltman and Healy (1978).

Poulos (1989) summarised some of the most commonly used methods in calculating α for driven piles, an extract of which is included in **Table 54.1**.

It is generally considered good practice to limit the *average* skin friction along the entire shaft in fine-grained soils to 100 kN/m^2 unless proven higher by pile load testing or by suitable experience of pile performance in similar ground condition.

The S_u value in Equation 54.3 is really an empirical index value, closely linked to historical sampling/lab test methods together with back-analysis of pile test data, rather than a 'fundamental' soil parameter. In the UK the method is mainly based on experience in London Clay, with the results of quick undrained triaxial (QUT) tests on U100 samples being compared with the results of maintained load tests (refer to Patel, 1992 for further background). The S_u value selected for design is required to be a 'characteristic' value (that is, a 'cautious estimate of the value affecting the occurrence of the limit state'). It is important to note that the empirical α factors were originally derived from comparing a 'mean' profile through measured S_u values with the results of pile load tests. Originally these were based on small-diameter (38 mm) triaxial tests and pile load tests which applied a higher rate of loading than is commonly used in current practice. The empirical factors were then adjusted for larger-diameter (100 mm) triaxial tests and modern (and slower) pile load testing rates. Hence, the 'characteristic' S_u should be close to a cautious estimate of the mean, rather than a lower-bound or moderately conservative (i.e. well below the mean) value.

54.3.1.2 Effective stress (fine-grained soils)

Historically, the use of effective stress calculations for the design of piles in fine-grained soils has been limited, but research and project experience over the last couple of decades have provided more accurate methods of using this approach. The Imperial College design methods (Jardine *et al.*, 2005) provide recommendations on appropriate calculation of both end-bearing and shaft resistance for driven steel piles in terms of effective stress. Chapter 22 *Behaviour of single piles under vertical loads* provides guidance on effective stress design to assess the capacity of single piles, both bored and driven, subjected to vertical loading. An important application of effective stress methods is for piles below or adjacent to deep excavations (say, in excess of 4 m deep), when total stress methods are likely to be *unsafe*.

Alpha factors	Reference
$\alpha = 1.0$ ($S_u < 25$ kN/m^2) $\alpha = 0.5$ ($S_u > 70$ kN/m^2) Linear variation in between	API (2008)
$\alpha = 1.0$ ($S_u < 35$ kN/m^2) $\alpha = 0.5$ ($S_u > 80$ kN/m^2) Linear variation in between Length factor applies for $L/d > 5$	Semple and Rigden (1984)
$\alpha = \left(\frac{c_u}{\sigma'_v}\right)_{nc}^{0.5}\left(\frac{c_u}{\sigma'_v}\right)^{-0.5}$ for $\left(\frac{c_u}{\sigma'_v} \le 1\right)$ $\alpha = \left(\frac{c_u}{\sigma'_v}\right)_{nc}^{0.5}\left(\frac{c_u}{\sigma'_v}\right)^{-0.25}$ for $\left(\frac{c_u}{\sigma'_v} \ge 1\right)$	Fleming *et al.* (2009)

Table 54.1 Summary of commonly used α factors for driven piles in fine-grained soils

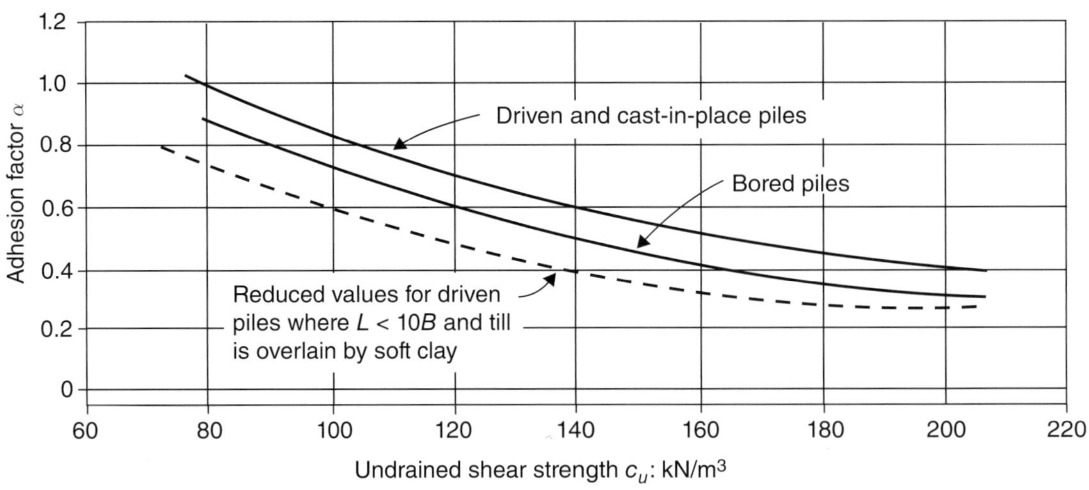

Figure 54.3 Adhesion factors, α, for piles in glacial till (Reproduced from CIRIA PG5, Weltman and Healy (1978); www.ciria.org)

Care is needed with such methods, as for total stress methods, to limit the calculated values of end-bearing and shaft friction to those achievable in the field for the pile type and pile construction methods being employed.

End-bearing

The use of effective stress methods to calculate the end-bearing of piles in clay soils is, in general, not recommended. It would only be appropriate to adopt such an approach based on the results of suitable pile load testing or appropriate experience in similar ground conditions.

Shaft friction

Assuming loading takes place under fully drained conditions, where remoulding and drainage at the pile–soil interface destroys all cohesion so that $c' = 0$, then the unit shaft friction can be calculated using

$$q_s = \sigma'_h \tan\delta \qquad (54.4)$$

where σ'_h is the horizontal effective stress following pile installation, taken as an average over the section of pile shaft being considered, and δ is the effective interface friction angle.

For bored piles δ would usually be assumed to equal the critical state angle of friction, φ'_{cv}, whereas for driven piles it would be assumed to equal the residual angle of friction, φ'_{res}. A simplification can be made if σ'_h is assumed equal to $\sigma'_v\ K$, in which case $q_s = \sigma'_v\ K \tan\delta$, further simplified, where $K \tan \delta = \beta$, to $q_s = \beta\sigma'_v$.

Values of β can vary significantly depending on consolidation history (i.e. normally consolidated or overconsolidated) and method of pile installation (driven or bored), as discussed in Chapter 22 *Behaviour of single piles under vertical loads*. Examples of typical values of K relative to K_0 for bored piles in overconsolidated clays are given in **Table 54.2**. Appropriate values for β are discussed in Chapter 22.

The main challenge in applying Equation 54.4 is to assess the appropriate values of σ'_h or K. As shown in **Figure 54.4**,

K value	Reference
K is the smaller of K_0 and $0.5(1 + K_0)$	Fleming *et al.* (2009)
$K/K_0 = 2/3$ to 1; K_0 is a function of OCR	Stas and Kulhawy (1984)

Table 54.2 Typical K values for bored piles in fine-grained soils

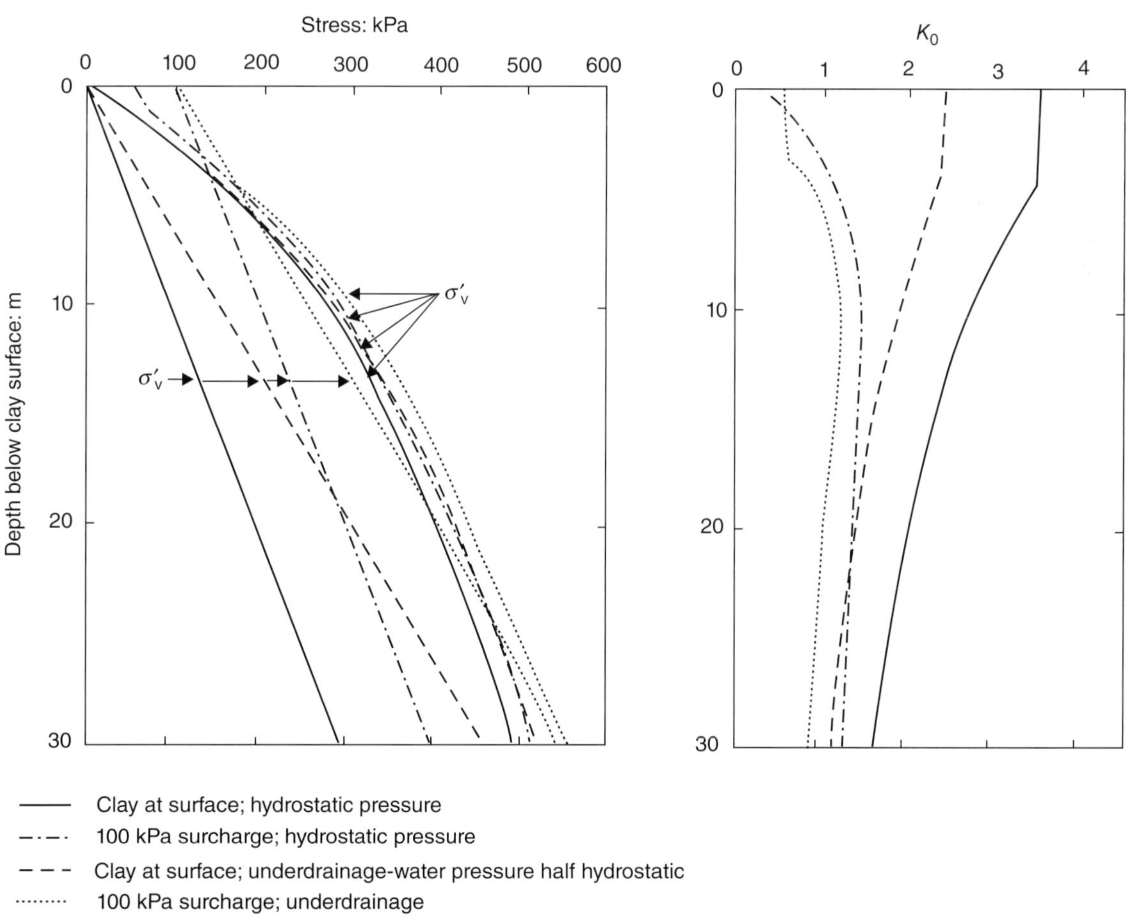

Figure 54.4 Influence of stress history on K_0 and σ'_h in overconsolidated clay (Reproduced from Burland *et al.*, 1979)

the value of K is sensitive to the site's stress history and also depends on stress changes during pile installation. However, the value of σ'_h is relatively insensitive to stress history, as also shown in **Figure 54.4**, and reductions in horizontal stress for well-constructed bored piles are relatively modest, as discussed by Bown and O'Brien (2008).

The β approach is particularly useful for normally consolidated or lightly overconsolidated soils, where K_0 can be assumed to be close to $1 - \sin\varphi'$. Then, for example, in the case of a driven pile, if the angle of effective friction is taken as, say, 21°, the ultimate friction becomes $0.25\sigma'_v$. For a very wide range of normally consolidated soils the value of β is likely to be between 0.2 and 0.3. This value is often used as the basis for calculation of downdrag or 'negative skin friction'; refer to Chapter 57 *Global ground movements and their effects on piles* for further discussion on this topic.

For high OCR clays, such as London Clay, β values can be far higher; refer to Chapter 22 *Behaviour of single piles under vertical loads.*

54.3.2 Coarse-grained soils

Approximate design methods for piles in coarse-grained soils are well documented, and reference should be made to Fleming *et al.* (2009) and Tomlinson (2014) for a wider understanding of the range of design methods available. Recent research has focused on steel-driven piles, aimed predominantly at the offshore industry, and reference should be made to the work of Randolph and White (2008) and Jardine *et al.* (2005).

54.3.2.1 End-bearing

The end-bearing capacity of piles in coarse-grained soils can be calculated by

$$q_b = \sigma'_{vB} N_q \tag{54.5}$$

where σ'_{vB} is effective vertical stress at the pile toe and N_q is bearing capacity factor.

N_q is often derived based on a correlation with φ'. A commonly used method is that by Berezantzev *et al.* (1961) (**Figure 54.5**). The main problem with this approach is that it is unduly sensitive to small changes in φ', and does not allow for the reduction in mobilised φ' as effective stress levels increase. It also implies that end-bearing pressure will continue to increase proportionally with depth. An alternative approach is that given in Fleming *et al.* (2009), which relates N_q to relative density, effective stress and the critical state angle of friction φ'_{cv}, shown graphically in **Figure 54.6** for $\varphi'_{cv} = 30°$; φ'_{cv} values for a range of sands are summarised in Chapter 52 *Foundation types and conceptual design principles* where I_D is the uncorrected relative density of the soil. Relative density can be assessed from standard penetration test (SPT) 'N' or cone penetration test (CPT) 'q_c' values (Section 54.3.4). Routinely used methods that can be used to calculate end-bearing are summarised in **Table 54.3**.

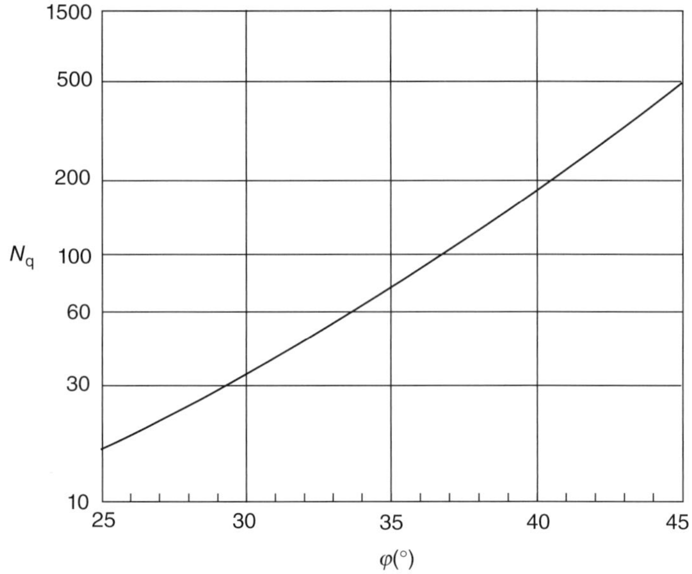

Figure 54.5 Variation of N_q with φ (Reproduced from Berezantzev *et al.*, 1961)

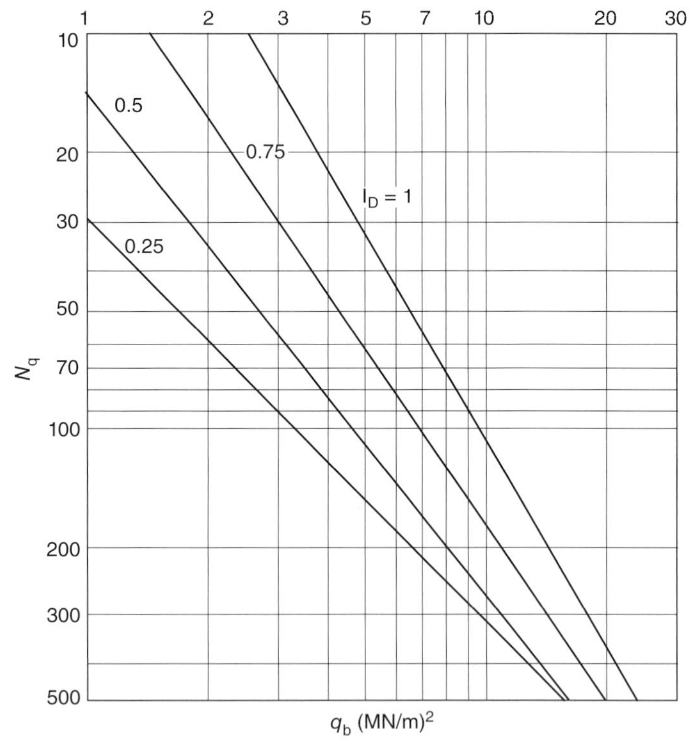

Figure 54.6 End-bearing capacity, coarse-grained soils, with $\varphi'_{cv} = 30°$ (Reproduced from Fleming *et al.*, 2009; Taylor & Francis Group)

With any design approach, it is very important to limit the calculated end-bearing to achievable values to ensure serviceability limits are not exceeded; this is discussed further in Sections 54.3.5 and 54.8. Code-based definitions of ultimate resistance usually adopt a limiting settlement criterion,

Material	N_q value	Reference
Silica sand	$N_q = 40$	API (2008)
	N_q plotted against φ'	Berezantzev *et al.* (1961)
	N_q related to φ', relative density and mean effective stress	Fleming *et al.* (2009)
	N_q from cavity expansion theory, as a function of φ' and volume compressibility	Vesic (1972)
Uncemented calcareous sand	$N_q = 20$	Datta *et al.* (1980)
	Typical range of N_q = 8–20	Poulos (1989)
	N_q determined for reduced value of φ'	Dutt and Ingram (1984)

Table 54.3 Methods of calculating end-bearing

typically 10% of pile diameter, and it is important to recognise that piles in coarse-grained soils will only mobilise a fraction of their theoretical end-bearing resistance at a settlement of 10% of the pile diameter. When adopting theoretical bearing capacity based methods it is possible to significantly overestimate the code-defined end-bearing resistance, particularly for pile construction methods that may disturb the soil at the pile toe, or for short piles that are highly dependent on end-bearing.

The end-bearing capacity of a pile in a uniform layer of coarse-grained soils will not continue to increase proportionately with depth – the Fleming *et al.* (2009) method replicates this nonlinear variation of q_b with depth, and is recommended for routine use. Simplified design methods (e.g. API, 2000) typically assume end-bearing pressure reaches a limiting value (based on Vesic, 1977), typically for pile lengths in excess of about 20*D* (where *D* is pile diameter). For this reason, end-bearing is often limited conservatively to 10–15 MN/m^2 for dense sands or gravels, depending on the pile installation method, and to about half these values for medium-dense sands or gravels. Higher values can be adopted based on the results of suitable pile load testing or appropriate experience in similar ground conditions. In general, for end-bearing piles, and especially those which need to mobilise end-bearing resistance at working loads, it is recommended that the Fleming (1992) method is used to assess the relationship between end-bearing resistance and settlement at both serviceability limit states (SLS) and ultimate limit states (ULS). The pile base stiffness strongly influences the mobilised end-bearing at both SLS and ULS – in the absence of site-specific tests, Fleming (1992) and England (1999) give practical guidance on base stiffness values for a range of soil types. For loose coarse-grained soils, very little end-bearing resistance may be mobilised at displacements which are small enough for normal foundation requirements.

54.3.2.2 Shaft friction

The ultimate shaft friction of piles founding in sands or gravels is generally calculated using

$$q_s = K_s \sigma'_v \tan\delta \tag{54.6}$$

where K_s is earth pressure coefficient after pile installation, σ'_v is effective vertical overburden pressure and δ is the angle effective interface friction between the pile and the ground.

δ is generally derived from φ'_{cv}, with the correlation dependent on the roughness of the pile–soil interface, resulting from the combination of pile installation method and soil characteristics, generally ranging from 0.75 to 1.0. For most bored piles the angle of skin friction may be taken as approximately equal to the critical state angle of friction of the soil. φ'_{cv} values are summarised in Chapter 52 *Foundation types and conceptual design principles* – also refer to Fleming *et al.* (2009) for further details. If support fluids are to be utilised during bored pile construction, it is important that any potential effects on pile capacity are considered; EFFC/DFI (2019) provides further guidance on this.

For full displacement driven piles, and when using simplified design methods, it is usually assumed that δ equals $0.75\varphi'_{cv}$. As discussed by Randolph and Gourvenec (2011), analysis of modern driven pile test databases reveals that the interface friction angle for coarse-grained soils mainly depends on the roughness of the interface *relative* to the size of the coarse-grained particles, and also by their mineralogy and shape (hence the interface friction tends to reduce as median particle size increases). Commonly used empirical design methods for driven piles in coarse-grained soils assume silica mineralogy; for calcareous sands much lower shaft friction has been observed.

For most bored piles, $K_s = 0.7$ is appropriate, but for CFA piles the value of K_s is more sensitive to the soil characteristics and appears to be dependent on technique. The following values are recommended for the design of CFA piles:

- clean medium/coarse sand: $K_s = 0.9$
- fine sand: $K_s = 0.7$–0.8
- silty sand: $K_s = 0.6$–0.7
- interlayered silt and sand: $K_s = 0.5$–0.6

For full displacement driven piles K_s can be estimated as $N_q/50$; this relationship gives a typical value of $K_s = 1.2$ for such piles, and limiting values between 0.7 and 2.0 are suggested.

For partial displacement piles K_s values will be lower than for full displacement, and it is usually reduced by 20%. For driven cast in situ piles some reduction in horizontal stress will occur when the casing is extracted, and normally a K_s value of 1.0 is assumed (if wet concrete is placed).

For auger displacement, rotary displacement and helical displacement piles, there are examples of K_s values derived from pile load testing well in excess of 1.2 (Bell, 2010). Such installation methods can produce significant improvements in shaft friction over conventional bored or driven piles, but such high values must be used with caution and in conjunction with a suitable pile load testing regime. The shaft friction of such piles is also highly dependent on the shape of the boring tool, of which there are numerous variants in the marketplace.

The basic formula for calculating q_s can be further simplified to $q_s = \beta\sigma'_v$. Guidance on β values for design are included in FHWA (2010), based on an extensive bored pile test database for a wide range of ground conditions. Mascarucci *et al.* (2016) have outlined the range of factors that can influence the mobilised shaft friction, including scale and depth effects (with unit shaft friction being higher for small-diameter piles compared with large-diameter piles, due to dilatancy effects at the pile–soil interface).

Mean unit shaft friction does not increase in proportion to shaft length. When using simplified design methods, such as API (2008), limits on shaft friction are assumed (with the limit value dependent on relative density and grading). Limiting values well in excess of the API limits can be achieved for some installation techniques, but should ideally be proven by suitable pile load testing or relevant experience in similar ground conditions. The ultimate shaft friction of driven piles in coarse-grained soils will approach a limiting value for pile lengths in excess of 20 pile diameters, a similar phenomenon to that for end-bearing. The limiting value is dependent on several factors (including installation methods, due to friction fatigue effects). White and Lehane (2004) provide a useful reference for this in addition to the work carried out by Jardine *et al.* (2005).

It is also important, particularly for bored and CFA piles in sands, especially silty sands, not to place too much reliance on the base capacity of the pile at working loads. It is common practice to design such piles with a factor of safety greater than 1.0 on the shaft capacity alone to limit settlements at SLS loads to acceptable levels as determined through specified serviceability limits

If base resistance needs to be mobilised at working loads (even for driven piles), then a separate serviceability check should be undertaken to assess if the working load can be mobilised at a tolerable settlement magnitude, as noted above.

54.3.3 Layered soils

From Sections 54.3.1 and 54.3.2 it should be clear that piles in coarse-grained soils may mobilise relatively high end-bearing capacity compared with fine-grained soils. Hence, when designing piles in interbedded sand and clays the location of the pile toe is of paramount importance. A common practical situation when designing driven piles would be to attempt to take advantage of the potentially high end-bearing resistance of coarse-grained layers. However, if the pile toe is too close to the interface with adjacent fine-grained layers then the pile will *not* develop its full end-bearing capacity (i.e. the end-bearing resistance which would be calculated by the methods outlined in Section 54.3.2).

Meyerhof (1976) first considered this important issue, and suggested that end-bearing capacity would be reduced if the pile toe was within 10*B* (where *B* is pile diameter) of the adjacent weaker layer (**Figure 54.7**). Subsequent work by Meyerhof and Sastry (1978) and Matsui (1993) indicates that Meyerhof's original guidance is, in general, over-conservative; nevertheless it remains a useful guide for preliminary design.

When the pile toe is within the zone of influence of the adjacent weaker layer, the end-bearing capacity can be assessed from **Figure 54.7** by interpolating between the capacity for the coarse-grained founding layer and the adjacent weaker layer.

A key practical issue for pile design in layered deposits is the reliable identification of the level, thickness and continuity of the proposed coarse-grained founding layer. Hence, a large number of boreholes/CPTs may be required to verify that the founding layer does not thin out and is not in the form of isolated lenses.

If there is significant uncertainty, which can be a particular issue in glacial deposits, then it may be appropriate to undertake pile design on the worst credible stratification unless proven otherwise by full-scale testing. A common approach is

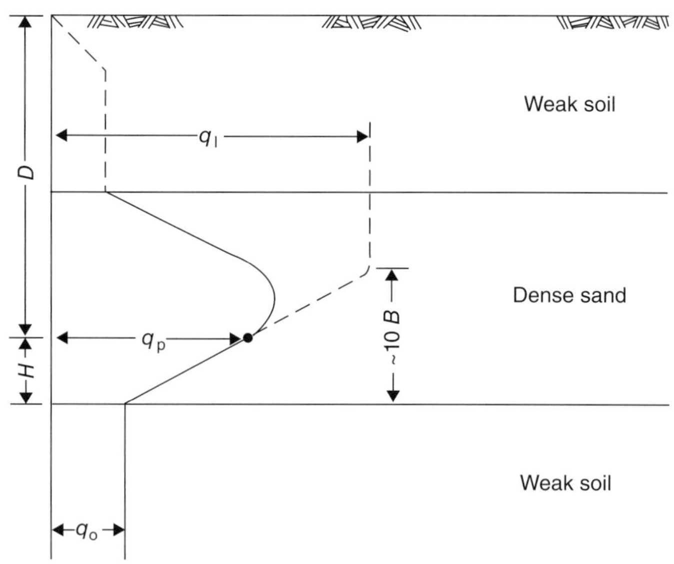

Figure 54.7 Relation between ultimate point resistance of pile depth in thin sand overlying weak soil (Reproduced from Meyerhof, 1976, with kind permission of ASCE)

to seek a factor of safety on shaft capacity alone in excess of 1.0, based upon the worst-case stratigraphy in terms of shaft capacity, and to place a limit on end-bearing (e.g. assuming a clay soil is present at the pile toe).

54.3.4 Correlations with SPT/CPT

In general, it is possible to use either 'indirect' or 'direct' correlations. In the 'indirect' approach the SPT or CPT is used to correlate the measured in situ test value with laboratory test parameters or other indices of soil behaviour (such as relative density for sands/gravels). The methods outlined in Sections 54.3.1 and 54.3.2 are then used. In the 'direct' approach, the measured in situ test value is directly correlated with shaft friction or end-bearing capacity. Both approaches can be of value, especially when there is a high level of uncertainty associated with predicting the ultimate pile capacity (e.g. driven piles in coarse-grained soils).

Many of the more recent driven pile design methods directly use CPT data to derive pile capacity, and the available evidence indicates that these methods are more reliable than the older empirical methods. Where driven piles are proposed, it is strongly recommended that CPTs are used to the maximum practical extent.

54.3.5 SPT

54.3.5.1 Fine-grained soils

A number of correlations between SPT results ('N' value) and in situ undrained strength, S_u, of silts and clays have been proposed. The most commonly adopted correlation within the UK for heavily overconsolidated clays is that proposed by Stroud and Butler (1975): $S_u = 4$ to $6 \times$ 'N' kN/m^2; recently this empirical correlation has been updated (reflecting changes to the SPT equipment), with the correlation coefficient increased by about 30% – refer to Schneider *et al.* (2019).

Other published correlations are available, but these can be unreliable. Some of the differences in correlations may result from variations in the methodology adopted in the field while undertaking SPTs or the types of interpretive methods used for pile tests rather than any significant differences in the soils being tested. It is to be noted that any correlations should be compared to all other available data (e.g. results from triaxial tests, field shear vane test results, etc.).

There are direct methods of calculating single-pile capacities from the results of SPTs undertaken in fine-grained soils, although there is limited experience with their use for UK conditions, and they should therefore be used cautiously.

Poulos (1989) has proposed the relationship

$$q_s = \alpha + \beta N \tag{54.7}$$

where q_s is the unit ultimate shaft friction, N is the SPT penetration resistance and α and β are constants depending upon soil and pile type.

The relationship between SPT N value and unit ultimate end-bearing proposed by Poulos is

$$q_b = KN \tag{54.8}$$

where K is a constant depending upon soil and pile type and N is the SPT penetration resistance.

The values of α, β and K are highly dependent on the type of pile and method of pile construction. CIRIA Report 143 (CIRIA, 1995) and Poulos (1989) give suggested values of α, β and K.

54.3.5.2 Coarse-grained soils

Correlations between SPT 'N' and peak friction angle are available, and the correlation by Peck *et al.* (1967) is often used and is usually conservative. This correlation is given in **Table 54.4**, and although fairly crude (it ignores many fundamental factors which govern mobilised friction in coarse-grained soils), it is useful for the preliminary assessment of the end-bearing resistance of small-diameter (say, less than 0.5 m) driven piles of moderate length (say, 10–15 m). The SPT 'N' values are uncorrected.

It is usually appropriate to also use SPT 'N' values to assess relative density, with

- overburden correction, $N_1 = C_N N$
- overburden correction factor, $C_N = 2/[1 + (\sigma'_v/100)]$
- relative density, $D_r = [(N_1)_{60}/60]^{0.5}$.

Here, N is measured as SPT 'N', σ'_v is vertical effective stress and $(N_1)_{60}$ is the SPT blow count normalised to a vertical effective stress of 100 kN/m^2 and corrected to 60% of free-fall energy. The above relationships are for uncemented, normally consolidated sands (with a median particle size of less than

SPT 'N'	φ' (degrees)
<5	28
5	29
10	30
15	31.5
20	33
25	34.5
30	36
35	37
40	38
>40	38
N.B. use $\varphi' > 38°$ with caution	

Table 54.4 SPT 'N' correlation to φ

0.5 mm). For soils of large particle size (i.e. gravels/cobbles) the above correlations are non-conservative. CIRIA Report 143 (CIRIA, 1995) provides further guidance.

Meyerhof (1976) proposed a set of direct correlations between SPT 'N' values and end-bearing resistance/shaft friction as follows

$$q_b = 0.4\ \text{'N'}(\text{MN/m}^2)\ \text{for sands, when } P > 6D \qquad (54.9)$$

$$q_b = 0.3\ \text{'N'}(\text{MN/m}^2)\ \text{for silts, when } P > 6D. \qquad (54.10)$$

Note: P is penetration of the pile toe beneath the surface of coarse-grained soil. The 'N' value is averaged over about three pile diameters above/below the pile toe. For shaft friction, Meyerhof recommends

$$\begin{aligned} q_s &= 2\ \text{'N'}(\text{MN/m}^2)\ \text{for driven piles} \\ q_s &= 1\ \text{'N'}(\text{MN/m}^2)\ \text{for bored piles.} \end{aligned} \qquad (54.11)$$

54.3.6 CPT

54.3.6.1 Indirect methods: fine-grained soils

As with the SPT, a number of correlations between in situ undrained shear strength, c_u, and CPT cone resistance, q_c, have been proposed. The general relationship is given by

$$c_u = q_c / N_k + p_0 \qquad (54.12)$$

where N_k is a cone factor, q_c is the unit cone end resistance (kN/m^2) and p_0 is the total overburden pressure (kN/m^2).

It is important to note that some correlations include the effect of overburden pressure, p_0, within the cone factor, N_k. Some correlations are specifically for use with the piezocone.

The cone factor, N_k, is a variable which is affected by many factors, including the rate of penetration, strength anisotropy, soil fabric (e.g. fissuring, laminations), the degree of overconsolidation of the clay deposit and the reference tests which the CPT is compared against. Overconsolidated clays generally tend to exhibit significantly higher values of N_k (especially if fissured) than normally consolidated clays.

Comprehensive guidance for the selection of appropriate values of N_k can be found in Lunne *et al.* (1997).

54.3.6.2 Direct methods: fine-grained soils

Fully mobilised unit end-bearing, q_{bu}, of a pile can be directly correlated to the unit cone resistance, q_c, determined by CPT testing. Effects of scale and rate of testing mean that the ratio q_{bu}/q_c is usually less than 1 (Poulos *et al.*, 2000).

The Imperial College Pile (ICP) design method (Jardine *et al.*, 2005) provides methods for calculating base capacity for both closed- and open-ended driven piles directly from CPT tests. The ICP design method does not, however, provide a direct correlation to pile shaft friction.

Estimates of pile shaft friction for driven piles can be made from CPT sleeve friction (e.g. $q_s = f_c$), but a high degree of caution is required since there may be large differences in the degree of consolidation of the soil around the pile when compared to the CPT test. Results of pile tests have indicated that long-term shaft friction values can be as much as 500% greater than those measured during driven pile installation (Fleming *et al.*, 2009).

Correlations between unit cone resistance, q_c, and unit shaft friction, q_s, generally yield more reliable results than those derived from CPT sleeve friction. These correlations themselves have a reasonably wide degree of variation – for example, $q_s = q_c/10$ (Fleming and Thorburn, 1983) and $q_s = q_c/40$ (Thorburn and McVicar, 1971).

A more reliable method for assessing pile shaft friction is usually attained from the approaches outlined in Section 54.3.1.

54.3.6.3 Coarse-grained soils

As noted in Chapter 22 *Behaviour of single piles under vertical loads*, where a pile toe is located within a sand layer, the CPT cone tip can be regarded as a model of the pile toe, and so the base capacity, q_b, will correlate with the cone resistance, q_c. Care needs to be taken to suitably derive an appropriate value of q_c given the typical variability of the magnitude of q_c. White and Bolton (2005) have carried out a careful comparison of ultimate base capacities measured in pile tests against measured values of CPT q_c. They found that

$$q_{bu} = 0.9\ q_{cm} \qquad (54.13)$$

where q_{bu} is ultimate, or 'plunging failure', capacity (refer to Section 54.8) and q_{cm} is the weighted average of q_c values within the zone of influence of the pile toe (up to 8D above and 4D below the toe, where D is pile diameter), based upon the 'Dutch' method.

Lehane *et al.* (2005) carried out further analysis on a relatively large pile test database and found that

$$q_{b0.1} = 0.6\ q_{cm} \qquad (54.14)$$

where $q_{b0.1}$ is bearing capacity mobilised at a pile settlement equal to 10% of the pile diameter and q_{cm} is as given in Equation 54.13.

In both of the above studies the important influence of adjacent weaker layers was emphasised, and this is readily taken into account by the selection of q_{cm} based upon the Dutch method. Shaft resistance is correlated to q_c in both the ICP design and UWA design methods (Lehane *et al.*, 2005).

54.3.7 Rock

The consideration of pile design in weak rocks differs widely for replacement piles (bored and CFA) and displacement piles (driven or bored). This is predominantly because it is possible to install significant rock sockets with replacement techniques, particularly when using modern high-torque piling rigs, whereas displacement systems typically meet refusal at the top of the weak rock strata or at a very short penetration into the rock. Generally strong rocks underlying a development site would not require bearing piles to penetrate them by any significant distance. There are of course occasional exceptions to this general rule – for example, where tension loads are high or in locations where adjacent tunnels are required to be isolated from bearing pile loads with slip liners.

54.3.7.1 Shaft capacity: rock sockets (generally applicable to replacement piles only)

Transfer of pile loads by shear transfer along the pile shaft–rock interface is relatively complex and depends on a number of factors, including

- rock strength
- rock socket roughness
- rock–concrete bond
- degree of 'contamination' of the rock socket with overlying superficial deposits.

Driven piles will generally not penetrate significant distances into rock of moderate strength. For driven piles in weak to medium strong rock, skin friction can be calculated as

$$q_s = 0.5 K_s \sigma'_{vo} \tan\delta \tag{54.15}$$

where K_s is an earth pressure coefficient, σ'_{vo} is the effective overburden pressure and δ is the angle of friction between the rock and pile shaft.

In general, pile driving into mudstone will create a 'skin' of remoulded material around the shaft, and mobilised shaft friction will only be a small fraction of that mobilised around bored (replacement) piles. Skin friction in weathered argillaceous rocks, particularly weathered mudstones/siltstones, can be calculated using methods described in Section 54.3.1, although care is needed to suitably account for remoulding of material at the pile–ground interface.

For bored cast in situ piles (replacement piles) a number of correlations between the unconfined compressive strength of the rock and rock socket skin friction have been proposed. The most common correlations are those proposed by Horvath (1978), Rosenberg and Journeaux (1976), Williams and Pells (1981) and Rowe and Armitage (1987a).

Whitworth and Turner (1989) summarised the various proposed formulae for calculating ultimate unit shaft friction (**Table 54.5**).

In **Table 54.5**, q_{uc} is unconfined compression strength, α is a reduction (adhesion) factor relating to unconfined compression strength and β is a correction (rock socket side friction) factor related to discontinuity spacing within the rock mass. Most of these formulae can be simplified into the following form where ψ is a dimensionless coefficient, ranging between 0.9 and 2 (Equation 54.16)

Pile design method	Ultimate unit shaft friction, q_s (kN/m²)
Horvath (1978)	$0.33(q_{uc})^{0.5}$
Horvath and Kenney (1979)	$0.20–0.25(q_{uc})^{0.5}$
Meigh and Wolski (1979)	$0.22(q_{uc})^{0.6}$
Rowe and Armitage (1987)	$0.45(q_{uc})^{0.5}$
Rosenberg and Journeaux (1976)	$0.375(q_{uc})^{0.515}$
Williams and Pells (1981)	$\alpha\beta(q_{uc})$

Table 54.5 Typical correlations between q_{wc} and ultimate shaft friction where UCS is in MPa

$$q_s = 220\psi\sqrt{q_c} \tag{54.16}$$

Clearly, the above correlations can yield a wide range of values for ultimate shaft friction. Some of these differences are due to the construction methods adopted for test piles (and the 'roughness' of the rock socket), different back-analysis methods to derive the correlations and different rock materials and/or rock masses in which back-analysed test piles were constructed. However, difficulty in collecting suitable test samples to determine q_c in extremely weak to weak rocks is often of concern, with drilling disturbance, poor recovery, short fracture spacing and the variability of the material leading to inherent sampling bias.

Kulhawy and Phoon (1993) have proposed values of ψ back-analysed from pile tests in clay and rock based on a relationship with UCS. This database has been added to by Markvukaj *et al.* (2022), who have developed a relationship between geological strength index (GSI) and ψ for extremely weak to very weak Mercia Mudstone to reduce reliance on empirical correlations with q_c.

Alternatively, use of the effective stress approach for Mercia Mudstone (described in Sections 22.4 and 54.3.2.2) has been advocated by Chandler and Davis (1973) with back-analysed values of q_c published in CIRIA 570 (Chandler and Forster, 2001) and Markvukaj *et al.* (2022), removing reliance on the determination of q_c completely.

It must be noted that particular care needs to be taken when calculating design values and supervising the construction of rock sockets in mudstones. Experience has shown that the calculated values for unit shaft friction based on mudstone rock strengths are often not achieved in practice, largely due to softening/polishing effects during rock socket construction (**Figure 54.8**). These adverse effects can be exacerbated if groundwater seepages into the socket are not properly

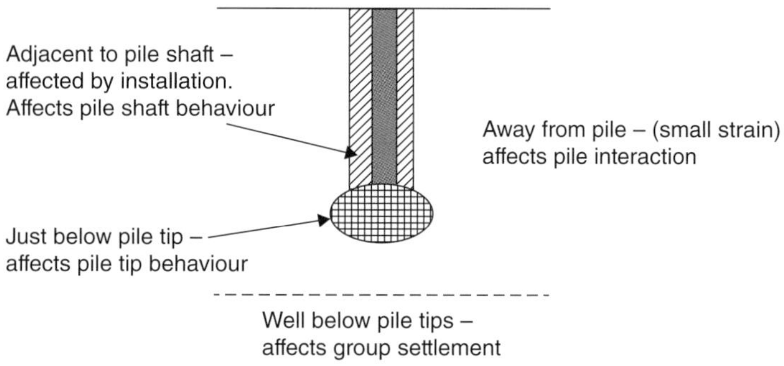

Figure 54.8 Zones around a pile and construction influences (After Poulos 2017)

controlled, and/or if an underpowered rig has the auger rotating excessively rather than penetrating into the rock.

The most recent published guidance for pile shaft friction in chalk is based on an effective stress approach. Reference should be made to CIRIA Report PR86 (Lord *et al.*, 2003) for specific guidance.

54.3.7.2 Base capacity (applicable to both displacement and replacement piling techniques)

Where piles are driven to practical refusal on strong rock, the limiting factor on pile capacity is usually the maximum allowable direct stress on the pile cross-section. In the UK this is typically taken as $0.25 \times f_{cu}$. Care should be taken when driving precast concrete piles to ensure that the piles are not overdriven, as this could result in damage to the pile section. When driving steel piles, care should be taken not to overdrive the pile, as this could cause shattering of the bearing stratum, resulting in a reduced pile capacity. Care should also be taken to ensure that the design assumptions relating to rock material (e.g. strength) are compatible with the rock mass (e.g. joint infilling, joint orientation, joint aperture, solution features, weaker underlying strata). Bored piles can also be constructed to practical refusal on strong rock. The definition of practical refusal should be agreed in advance with the engineer where it is anticipated this will occur. In certain circumstances the definition of practical refusal may be qualified subject to undertaking satisfactory maintained load tests.

For both driven and bored piles constructed within rock, the ultimate unit end-bearing capacity is often taken as 2.5–5.0 times the characteristic UCS value for the rock stratum for very weak rock. The ratio of characteristic UCS to q_b *reduces* as rock strength *increases* for a consistent level of jointing. Strong rocks often have their geotechnical behaviour largely controlled by the frequency and orientation of joints with a lower-bound q_b based on expected rock mass performance. Where rock is massive or compact and strong, the structural strength of the pile is likely to govern the ultimate design resistance of the pile. The achievable quality of the pile base to rock interface must also be considered. It is important that rock strength descriptions are understood in the context of the specific standard they refer to. In this context, it is important to recognise that engineering descriptions of rock strength have changed over time, meaning that the date of rock descriptions should be cross-referenced to the correct version of the relevant (site investigation) code of practice.

Where piles derive a significant proportion of their capacity from shaft friction, there may be less concern over consideration of the end-bearing component of capacity in terms of the pile load–settlement behaviour, particularly if the pile has a factor of safety on shaft capacity alone of 1.5 or greater.

Where a significant proportion of the pile capacity is being derived from the end-bearing component, care must be taken to ensure that the pile toe remains in intimate contact with the rock. Factors such as soil heave (from overburden removal or driving further displacement piles) and disturbance due to breaking down the piles can lead to lifting of the pile toe, which will result in greater settlements upon loading than would otherwise have been experienced. For driven piles it is often considered to be good practice to re-drive piles (subject to programme and access constraints) to ensure the toe of the pile is in good contact with the bearing stratum.

Sloping bedrock can pose particular risks to pile construction. Driven and bored piles may be deflected by steeply bedded rock surfaces. Driven piles may be installed to a refusal criterion yet upon loading cause instability of the inclined rock mass due to downward sliding of blocks of rock. Infilled quarries can pose a particular risk to piling: the position of the highwall is critical in assessing the required pile lengths and suitability of particular construction techniques. It may not be possible or practicable to construct piles in the zone where highwalls are located.

When constructing bored piles within rock, care should be taken to ensure that the base of the pile bore is thoroughly cleaned, so that the design end-bearing capacity can be achieved without adversely affecting the piles' load–settlement behaviour. This can only be realistically achieved when large-diameter bored piling techniques are adopted, facilitating the use of cleaning buckets. Typically, large-diameter bored piling ranges from 600 mm to 3 m dia.

Reference should be made to CIRIA Report PR86 (Lord *et al.*, 2003) for specific guidance on constructing piles within chalk.

54.4. Factors of safety

54.4.1 Global safety factors

Historically, the design of piles in the UK has adopted global factors of safety where the safe working load (SWL) is given as

$$\text{SWL} = Q_b/F_b + Q_s/F_s \qquad (54.17)$$

Although overall factors of safety are not necessarily the best way to govern pile performance, the numbers traditionally adopted are

- 2.0 overall minimum – usually with load test verification on expendable piles
- 2.5 overall or 1.5 (on shaft) with 3.0 (on base) – usually with working pile load tests
- 3.0 overall – may not require any load testing.

These factors of safety should only be used as a guide, and different values and combinations can be used in conjunction with suitable experience.

There are a number of ground conditions where different factors are used, placing more reliance on either end-bearing or shaft friction. Examples include piles in chalk, where large factors on end-bearing are adopted, as high as 10, to allow for the potential for soft spots and dissolution features beyond the pile toe. Similarly, the design of bored or CFA piles in weak rocks may often be dictated by ensuring a factor of safety as high as 1.5 on the shaft capacity alone to allow for the potential for machining/polishing or softening/remoulding of the pile shaft, or disturbance of the pile base, and to limit settlements at working load.

If a pile relies predominantly on end-bearing, then great care will be required both for design and construction, and more specialist techniques and expertise will normally be required. Site-specific preliminary pile tests may be needed unless there is significant previous experience in similar ground conditions and using similar piling techniques.

54.4.2 Partial safety factors

Partial factors are commonly used in design and these are applied to the particular parameters and loading conditions to reflect more accurately where the uncertainties lie. Most modern design codes, in particular the Eurocodes such as EC7 (BS EN 1997-1), recommend pile design based on the use of partial safety factors. This code does allow reduced factors of safety where there are several pile tests on a site. Such an approach encourages the use of preliminary pile testing.

The factor of safety, whether partial or a lump factor, is a factor on strength. However, as reported by Atkinson (2000), the key parameter that governs pile behaviour is stiffness of both the ground and the pile. The factor of safety is only of relevance in determining the likelihood of satisfactory performance beyond working load. For example, an end-bearing pile can fail to meet the required performance at working load because of low stiffness and still have the prescribed factor of safety. Conversely it can perform at working load and still not meet the required factor of safety. In either case it is necessary to determine whether the behaviour is of sufficient concern to be detrimental to the structure.

The main difficulty is in selecting criteria for settlement and factors of safety that are compatible and allow for the likely variability of the ground. This can only be done by making a prediction of pile behaviour based on the best estimate of the ground parameters and allowing a reasonable margin in the permitted test criteria to allow for natural variability in the ground.

54.5. Pile settlement

Structures do not generally suffer due to total settlement as such. It is differential settlement of, and the vulnerability of services connected to, structures that give rise to most concern. Refer to Chapter 52 *Foundation types and conceptual design principles*.

One of the reasons for introducing a global, or lumped, factor of safety when considering allowable pile capacity (such lumped safety factors applied to calculated ultimate pile capacities are most often in the range 2.0–3.0) is to ensure that settlement of single piles under working load (i.e. at SLS) is within tolerable limits for most structures.

A specified factor of safety does not guarantee a specific settlement and it is important to review the risk of excessive settlement as part of the pile design process. For most piling techniques in a range of ground conditions, except those that rely predominantly on end-bearing, it is good practice to ensure a suitable factor on the shaft capacity alone to limit settlements under working loads (i.e. to ensure suitable serviceability performance). UK practice commonly adopts a minimum factor of safety on ultimate shaft resistance > 1.0 to control pile serviceability performance and the LDSA *Guidance Notes for the Design of Straight Shafted Bored Piles in London Clay* requires a factor of safety > 1.2 on shaft alone. It is recommended that the settlement of end-bearing piles is always assessed. Group settlements will need to be subject to separate consideration to ensure that pile group settlement is tolerable for the structure (refer to Chapter 55 *Pile-group design*).

It is widely acknowledged that more settlement is required to mobilise end-bearing than shaft friction. It is also generally expected that piles which rely significantly on end-bearing (carrying a low proportion of the total applied load in shaft friction) will settle more under load than piles which are predominantly frictional piles (i.e. piles which carry the majority of applied load by shaft friction). Patel (1992) has summarised pile test data for friction piles in London Clay (**Figure 54.9**) which indicates that at typical working loads (factor of safety of 2 or more) the single-pile settlement would be expected to be less than about 1.0% of the pile diameter. Similar behaviour would be expected for comparable stiff overconsolidated fissured clays (Gault Clay, Weald Clay, etc.).

An additional factor which must be considered in any assessment of pile settlement is the magnitude of pile head displacement caused by elastic shortening of the pile under load. This can be a significant proportion of the total pile head settlement under load, but becomes increasingly significant when considering long and slender piles constructed through

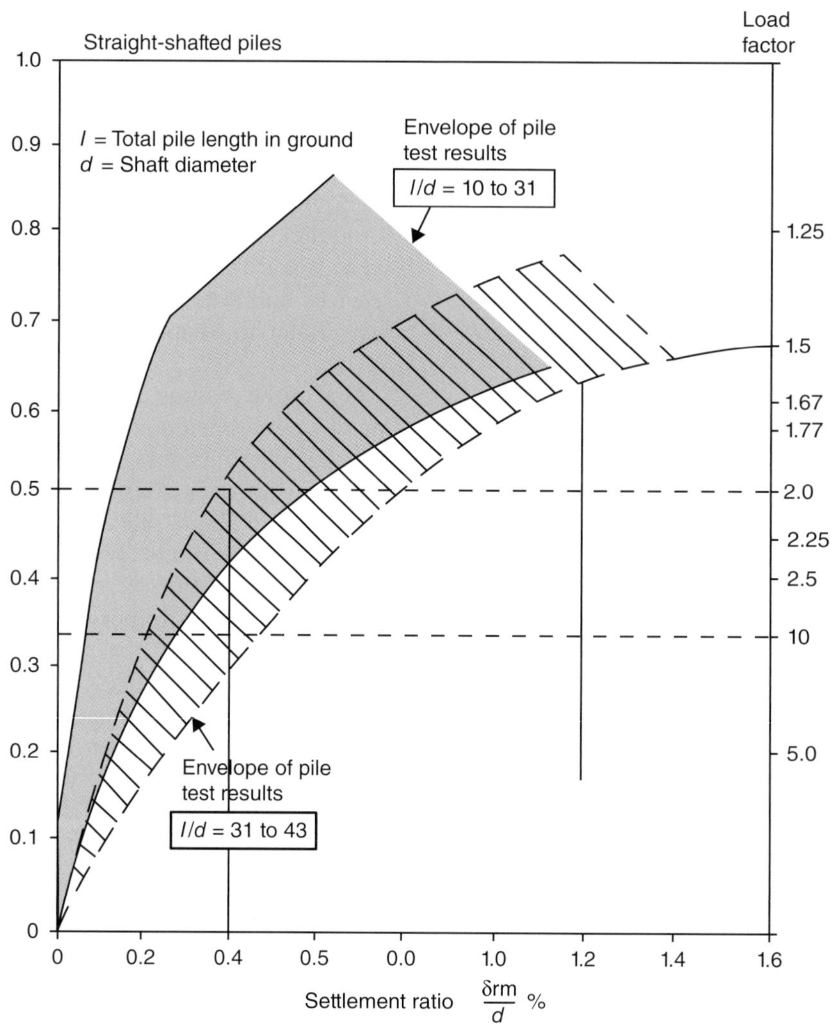

Figure 54.9 Results of loading tests on straight-shafted bored piles in London Clay – normalised load–settlement chart (Reproduced from Patel, 1992)

deep and weak deposits (i.e. transmitting a significant proportion of applied load to competent strata).

A number of different methods have been proposed over recent years: elastic approaches (e.g. Poulos and Davis, 1968; Tomlinson, 2014), simplified empirical methods (e.g. Burland *et al.*, 1966), finite element/boundary element analyses and hyperbolic analytic methods (e.g. Fleming, 1992).

Burland's proposed simplified empirical approach is specifically for the assessment of settlement of end-bearing under-reamed piles in London Clay (Burland *et al.*, 1966). Burland showed that where plate load tests are taken to failure, the normalised load–settlement curve can be plotted, and provided the base factor of safety exceeds 3, the settlement of the pile base can be obtained for any diameter of pile base. Alternatively, reference may be made to Tomlinson (2014).

In his paper, Fleming (1992) noted that sophisticated analytical techniques require sophisticated input data (e.g. certain geotechnical parameters) which are not usually available from routine site investigations. The method described by Fleming combines hyperbolic functions describing both the shaft and base components of pile capacity, together with the component of pile settlement derived from elastic shortening of the pile. The method takes advantage of the fact that the hyperbolic function adopted requires definition only of its origin and either its initial slope or a single point. The shaft flexibility parameter, M_s, does not vary significantly, with M_s values typically between 0.0010 and 0.0025 for many pile types and ground conditions. However, as expected, the base stiffness parameter, E_b, varies across a wide range for different pile and soil/rock types, and is particularly sensitive to pile construction methods and workmanship. E_b should not be confused with other 'elastic' soil parameters which are quoted in the literature: it is essentially an empirical parameter derived from back-analysis of pile tests and is as sensitive to pile construction method as to the fundamental ground properties. Guidance on appropriate parameters is given in Fleming (1992). The definition of axial capacity adopted within Fleming's method is the plunging failure (or asymptotic) definition (i.e. that load at which the soil resistance is fully mobilised and where settlement continues indefinitely for the load magnitude reached). Corke *et al.* (2001) provide interesting practical examples of Fleming's method in the context of developing pile testing performance criteria. Fleming's method is a powerful and simple method for single-pile settlement analysis; it is especially useful for large-diameter piles (say, in excess of 1.0 m dia.) and for end-bearing piles.

Fleming's method has a significant additional benefit, in that it can be used in reverse to back-analyse the results of maintained load tests to determine shaft friction, end-bearing and elastic shortening of the pile, subject to the pile test having moved the pile sufficiently to mobilise end-bearing sufficiently, typically more than about 5% of the pile diameter.

54.6. Pile behaviour under lateral load

54.6.1 Introduction

Piles can be subjected to lateral loads arising from different sources. These sources typically fall into two main categories: *active* loading (where external forces are applied to the pile) and *passive* loading (where movement of the soil immediately adjacent to the pile imposes bending stresses on the pile). A more refined classification of the sources of lateral load may involve (after Elson, 1984)

- static (e.g. structural reactions, lateral earth pressure)
- dynamic – cyclic (e.g. from operating reciprocating equipment, earthquake, cyclic wind loading (wind turbines), wave action)
- dynamic – transient (e.g. general wind loading, berthing, vehicular braking (impact) loads)
- other (undrained soil movements, soil consolidation adjacent to piles, thermal movements, soil creep, soil shrinkage/expansion – soil moisture changes).

Geotechnical capacity is rarely an issue even in weak/soft soils. The issues are usually either ones of serviceability (i.e. pile deflection/deformation) or structural capacity of the pile. Passive loading is a particularly high risk in soft soils, where large global ground movements may readily develop (refer to Chapter 57 *Global ground movements and their effects on piles*).

Piles subjected to cyclic loading require particularly close attention since cyclic loading will tend to modify soil behaviour and can lead to progressive failure. Consideration of dynamic cyclic and dynamic transient loads falls outside the scope of this chapter (refer to Chapter 60 *Foundations subjected to cyclic and dynamic loads*).

54.6.2 Deformation/failure modes due to 'active' loading

Piles subjected to lateral loads will experience an increase in normal stress in front of the pile (in the direction of the lateral load), and a decrease behind the pile. Lateral pile deformation is usually limited to a length equivalent to approximately 10 pile diameters below ground level, or less; the concept of a critical pile length (which is dependent upon the ratio of soil/pile stiffness) is particularly useful in this context (reference should be made to Fleming *et al.*, 2009).

There are two principal failure modes for unrestrained (or free-headed) single piles subjected to 'active' lateral loading (**Figure 54.10**) Unrestrained short rigid piles will fail by rotation when the passive resistance of the soil (in front of the pile above the point of rotation, and behind the pile below the point of rotation) are exceeded. Long piles will tend to rotate (and ultimately yield) at some depth down the pile shaft. For the purposes of analysis, a plastic hinge is assumed to develop which is capable of transmitting shear. The upper portion of the pile shaft above the plastic hinge will rotate while the lower pile shaft, below the hinge, will not move significantly.

For restrained (or fixed-headed) piles (**Figure 54.11**) there are three possible failure modes

- translation
- one-hinge failure (at underside of pile cap/restraint)
- two-hinge failure (at underside of pile cap/restraint and at some depth).

54.6.3 Fixed- and free-headed piles

Fixed-headed piles are restrained at head level, typically by a pile cap (pile groups of three or more) or ground beams (pairs of piles or single piles) or potentially by a structural slab (if so designed). Piles need to be restrained in two orthogonal directions to be considered truly fixed-headed. Free-headed piles are free to rotate and are not subject to orthogonal restraint.

In free-headed piles the bending moment in the pile at ground level is positive and acts in the same direction as the

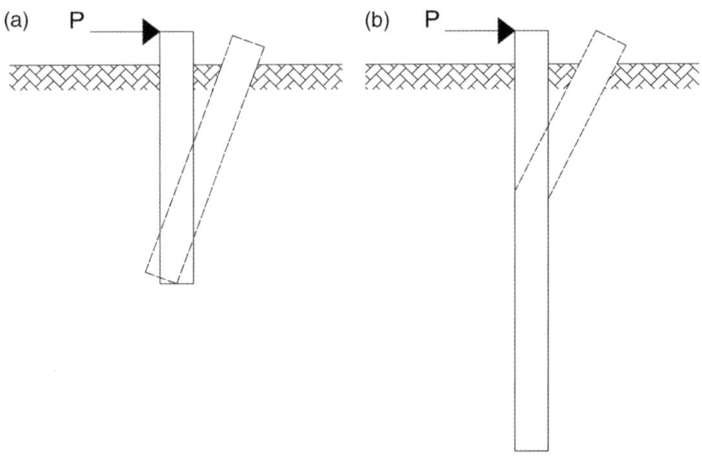

Figure 54.10 Possible failure modes for free-headed piles: (a) rotation; (b) hinge failure (Broms, 1964a, 1964b; Courtesy of Cementation Skanska)

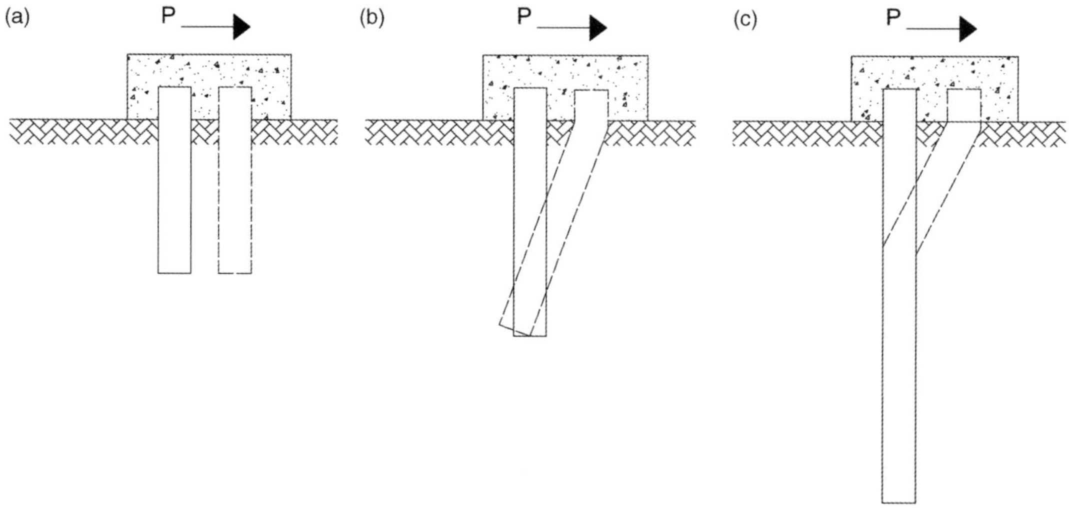

Figure 54.11 Possible failure modes for fixed-headed piles: (a) translation; (b) one-hinge failure; (c) two-hinge failure (Broms, 1964a, 1964b; Courtesy of Cementation Skanska)

applied lateral load. In fixed-headed piles the bending moment in the pile at ground surface is negative and acts in the opposite direction to the applied lateral load.

54.6.4 Lateral resistance of single piles

The ultimate lateral resistance of single piles can readily be estimated from the solution proposed by Broms (1964a, 1964b). This limit equilibrium method has the benefit of being straightforward, and only requires soil strength parameters.

However, the limitation in using simple limit equilibrium models is that real piles have flexural stiffness (EI) and do not act as a rigid body, therefore soil structure interaction analysis is recommended for the detailed design of laterally loaded piles.

Once the bending moment and shear force envelopes have been derived adopting appropriate analysis methods, the pile can be designed in accordance with relevant structural design codes of practice (e.g. BS EN 1992-1-1).

54.6.5 Lateral deformation

When considering the lateral deflection of laterally loaded piles, the near-surface soil parameters merit particular attention. This can often cause some problems since the very near-surface deposits (generally the upper 2–3 m) are often those about which very limited information is obtained during a ground investigation. This near-surface zone can also be affected by seasonal groundwater pressure changes and subsequent disturbance (e.g. installation of utilities).

Lateral pile loading tests are occasionally specified to be undertaken on piles. It is the authors' experience that these are often specified without full cognisance being taken of the head fixity of the piles in the final construction. The lateral load–deformation behaviour of a fixed-head pile is very different from that of a free-headed pile, and this must be taken into account when considering how the results of a lateral pile test (usually free-headed) will be analysed and used for the design of permanent piles (usually fixed-headed). The analysis method used for lateral pile test interpretation and permanent works design must be sophisticated enough to allow for

- potential changes in pile head fixity
- potential differences in load eccentricity between pile test and permanent works
- potential changes in near-surface ground conditions between pile test and permanent works.

If there is a coherent testing and analysis strategy, then lateral load testing of piles, particularly those whose load deflection behaviour is deemed to be of significant importance, or where particularly high lateral loads will be applied to the pile (relative to the ground conditions and the pile geometry), can provide a very useful check on performance of piles and the analysis techniques adopted for pile design.

The two analysis approaches commonly adopted for calculating the lateral displacement of single piles subjected to lateral loads are the subgrade reaction approach and the elastic continuum approach. Lateral load–deformation behaviour is strongly nonlinear even under moderate lateral loads, and linear elastic methods are usually inappropriate for piles subject to significant lateral loads. Ignoring nonlinear behaviour can lead to grossly inaccurate and unconservative predictions of lateral deformation.

The subgrade reaction approach models the ground as a series of discrete springs along the length of the pile shaft (i.e. as a Winkler idealisation of the soil mass). A refinement of the early models allows the spring stiffness to vary along the length of the pile (Reese and Matlock, 1956) and a further refinement is the *p–y* form of the Winkler soil model (Matlock, 1970; Reese *et al.*, 1974, 1975). The disadvantage of the *p–y* form of the model is that selection of reliable *p–y*

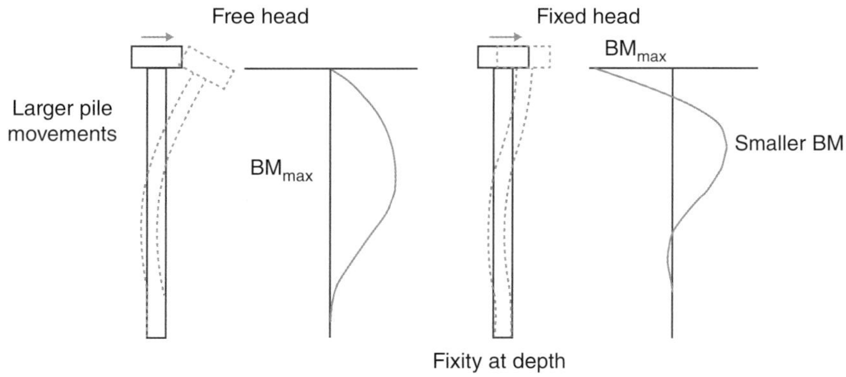

Figure 54.12 Fixity of single pile (After Raison and Egan 2016)

curves requires careful judgement, although typical forms of the curves have been presented by Reese *et al.* (1974) in sand and by Matlock (1970) in clays, and are summarised in the API code (American Petroleum Institute, 2000). In North America there is considerable experience in the successful practical application of p–y analysis for design. However, it needs to be remembered that the empirical database is predominantly for smaller-diameter piles (typically <750 mm dia.) and for large L/D ratios (>15).

Both Poulos (1971) and Randolph (1981) have published solutions relating to the behaviour of piles under lateral loads based on elastic continuum models. One principal advantage of the elastic continuum models over subgrade reaction models is that the method can be extended to pile groups, since interaction between adjacent piles can be modelled. Where large lateral loads are required to be resisted by piled foundations, for example marine/offshore structures or complex/unusual superstructures, it will be more appropriate to assess lateral displacements by adopting nonlinear elastic plastic methods. A useful practical procedure for estimating lateral load–deformation behaviour, proposed by Brettmann and Duncan (1996), is the 'characteristic load' method. Parameters are provided for sands and clays, and for free- and fixed-headed piles. The influence of pile head fixity on pile flexure and induced bending moment under lateral loading is illustrated in **Figure 54.12**.

Some geotechnical software assumes a fully fixed-head condition, which may be relevant for larger pile groups, but this can generate an excessively large negative bending moment at the top of the pile which does not occur in partially fixed small pile groups. Soil–structure interaction software (e.g. Oasys ALP, Rocscience RSPile or Geosolve Wallap) provides functionality to apply a moment restraint at the head of the pile for a calculated level of fixity. A method to calculate the rotational stiffness is illustrated in **Figure 55.35** and relies on a prediction of pile settlement or heave which could be made adopting the method described by Fleming or from measured settlements determined by maintained pile load tests.

54.7. Pile load testing strategy

Strategies for pile load testing often lack clear objectives. The requirements for testing are frequently driven by a desire to comply with regulations and to follow common practice. Testing is rarely seen as a part of a value engineering process, and therefore opportunities to optimise the foundation solution are often missed. Pile load testing, if suitably detailed and planned, can play an important part in producing the most sustainable deep foundation solution.

It is important that load testing of piles is considered as early as possible and included in the initial cost plan and programme stage. The programme must allow sufficient time for suitable tests to be carried out and for an objective appraisal of the test results and subsequent design revisions or value engineering to be carried out.

Without such an early review to set clear load testing objectives, combined with poorly specified requirements, avoidable problems will arise. Common problems include

- inappropriate test method specified and lack of flexibility in the testing regime
- final working conditions not modelled by the load tests
- piles not tested to failure or sufficient load to allow suitable evaluation
- insufficient time to carry out tests and to evaluate the results
- unrealistic/unachievable performance criteria specified
- no provision for value engineering
- unrealistic expectations of the possible benefits of load testing.

The use of suitably instrumented pile load tests, loaded well beyond the SLS, will provide credible data that will enable the state of the art with regard to pile design to progress and more innovative and low-carbon solutions to be developed. Such an approach will enable a performance-optimised design approach.

In recent years the use of fibre-optic instrumentation within test piles has proven to provide excellent data to help advance the understanding of pile performance and challenge current design approaches and parameters. Such instrumentation must be considered holistically as part of the overall load testing strategy.

54.7.1 Objectives

The strategy for pile testing needs to be established at the time the piles are designed. For most projects the main purpose of pile testing is to validate the design before construction and to check compliance with the specification during construction. However, in many cases there are considerable benefits in using testing for design development to identify the optimum solution. Testing strategies can therefore be divided into four main categories

- design development
- design validation
- quality control
- research.

The scope of testing will depend on the complexity of the foundation solution, the nature of the site and the risk of piles not meeting the specified requirements. The designer therefore needs to assess the risks and develop the testing regime accordingly. So what are the main considerations when assessing the risks?

- Site investigation data – are they sufficient?
- Previous knowledge – are there similar piles that have been tested on adjacent sites?
- Time – is there enough time to verify the design? If not, a conservative approach is needed.
- Cost – can the testing provide cost savings? What are the cost and/or construction programme implications of tests failing?

For simple structures on a site where the ground conditions are well understood and there are test data from adjacent sites which have used similar piling solutions, then the risks are low and the testing can usually be restricted to routine checks for compliance with the specification. For situations where the ground conditions or structural requirements are complex and there is little experience of similar piling work, then careful evaluation of the piling proposals is essential prior to embarking on the main piling works. The testing regime therefore needs to be considered in two phases, comprising preliminary pile testing before the main piling works and testing of working piles.

54.7.2 Preliminary pile tests

Preliminary testing is carried out in advance of the main piling works to verify that the design assumptions and construction method will achieve the required performance. The testing strategy for preliminary pile testing should address a specific set of stated objectives, which should include

- minimise risk – uncertainties about ground conditions, contractor's experience, novel techniques
- optimise the design – pile diameters, lengths and factors of safety
- confirm pile installation criteria – founding strata identification, pile set, refusal criteria
- assess buildability – site variability, pile integrity, pile uplift and displacement, soil remoulding
- check the pile performance meets the structural requirements – load/settlement behaviour, compression, tension and lateral loading
- confirm safety procedures – piling platform, working method, inspection, testing methods
- assess environmental impact – noise, vibration, pollution.

Where piles are required to carry very heavy loads, it may be impractical to carry out full-scale load tests. In such circumstances consideration should be given to carrying out tests on smaller-diameter piles using the same method of construction, provided the results of the tests can be extrapolated with some degree of confidence to predict the load–settlement characteristics of the larger piles. The test piles should be founded at the same level and in the same soil as the works piles.

For preliminary pile tests it is essential to have a borehole and/or CPT close to the test pile so that the test results can be reliably evaluated.

54.7.3 Working pile tests

Working pile tests are carried out during the main piling works to verify that the workmanship and materials meet the specified requirements. Testing on the working piles comprises a range of quality control checks. These checks will normally include load tests, integrity tests and materials testing.

The strategy for testing working piles should address

- load–settlement behaviour – if settlement criteria are specified then load tests should be carried out
- risk of integrity problems – the number of tests should reflect that certain ground conditions and pile systems may have increased risk
- installation criteria – use pile records, sets and dynamic testing where appropriate to confirm the installation criteria are being met.

The testing strategy for working piles should relate to the level of risk and the characteristics of the piling project. Test results and pile records should be continually reviewed during the works to reassess the risk level.

54.7.4 Number of tests

The need for preliminary testing in advance of the working piles, or in advance of new areas of piling on large projects, should be weighed against their cost and the time taken for installation, testing and analysis of results. It may be more economical and more logical to adopt a more conservative approach to design, for example by using larger or longer piles or reduced pile loads. Also, it may be more beneficial to undertake further site investigation to obtain information which is key to the success of the installation and hence achieve the necessary confidence in the piling. The value of preliminary testing undertaken to prove the design approach or pile performance is wasted unless the ground conditions are sufficiently well defined to allow the design to be applied to piles elsewhere on site.

The level of testing of working piles should consider the variability of the ground conditions from the point of view of the demands these place on the control of pile construction, area by area. Ensuring experienced supervision and monitoring, plus good pile records, can be preferable to increasing the level of control testing above the typical 1% of all piles installed.

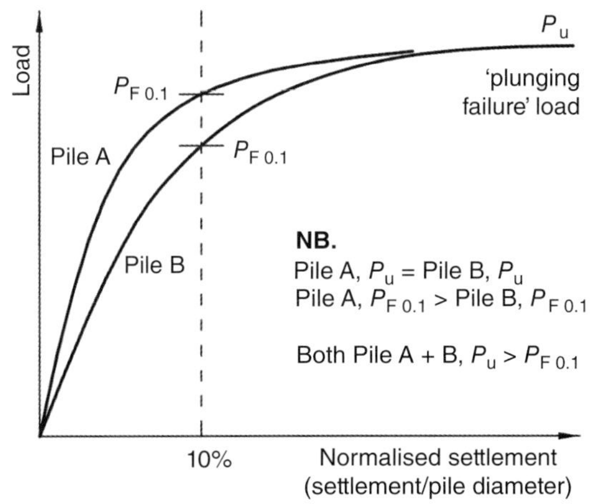

Figure 54.13 Definitions of pile failure

54.7.5 Methods of load tests

The various available methods of testing piles are best characterised by the duration that the force is applied to the pile and the strain induced in the pile. Tests involving large forces applied for several minutes or hours, such as static load tests (also known as maintained load tests), are used to assess pile load capacity, and small-energy low-strain tests are used to assess pile integrity. In dynamic and kinetic tests, although the force is comparable in magnitude to a static test, it is applied over a much shorter period than in a static load test. Careful consideration is therefore needed in the interpretation of the dynamic effects in order to derive static load capacities. The various test methods and their applications are detailed in Chapter 98 *Pile capacity testing.* There are a number of published documents dedicated solely to the subject of pile testing, such as Weltman (1980) and Handley *et al.* (2006), and these should be referred to for more detailed information on the subject of pile load testing. Chapter 98 *Pile capacity testing* focuses specifically on pile load testing.

54.8. Definition of pile failure

The definition of axial load capacity (or 'pile failure') needs to be clearly understood, since this can be the cause of confusion. Unambiguous definition becomes very important for piles which rely mainly on end-bearing resistance. Two definitions are commonly used; these are very different and must not be confused.

(*a*) Ultimate capacity, 'plunging failure' – failure is defined as the load at which settlement continues indefinitely, shown as P_u in **Figure 54.13**. This is often a theoretical concept rather than a practical reality as most pile tests are not taken to sufficient load to reach this value. Calculation methods based on bearing capacity theory will usually derive a 'plunging failure' load.

(*b*) Deformation controlled capacity, '10% failure load' – because the 'plunging failure' load cannot often be practically reached and measured during a pile test. Most codes/standards define failure as the load which is measured at a settlement equal to 10% of the pile base diameter, shown as $P_{F0.1}$ in **Figure 54.13.**

For friction piles there may be little practical difference between the above definitions. However, for end-bearing piles there can be a considerable difference between the capacities derived by (*a*) and (*b*). Referring to **Figure 54.13**, Pile A and B both have the same 'ultimate' capacity, but in terms of a deformation controlled capacity (consistent with most code definitions of failure) Pile A has a higher capacity than Pile B. For end-bearing piles, the deformation controlled capacity will be a function of the mobilised deformation modulus of the ground immediately beneath the pile toe.

54.9. Driven piles

Some additional commentary relating to dynamic design methods for driven piles is given here.

After the usual static geotechnical design stages, the design process for driven piles may differ from other pile types through adoption of wave equation analysis, with the following additional design steps checks being undertaken.

- Drivability checks by wave equation analysis are used to avoid excessive stresses or blow counts.
- Initial test programme – this should consider both maintain load static testing and dynamic load testing, to facilitate correlation between the two test methods (this is likely to be particularly important where ongoing reliance on dynamic testing is made in preference to further maintained static testing) and verifying any significant set-up or set-down.
- Establishing an installation criteria 'set' – the minimum penetration per blow for a number of blows and trend of set tightening – and permitted deviance from pile lengths derived from static formulae. This must take into account permissible driving stresses to avoid 'overdriving' of piles, which may lead to structural damage of the pile section.

The final length of working piles is generally determined at the time of pile installation. If 'set-down' (false sets) are identified by load testing at the end of the piling works, significant delays to the project may occur. Therefore, it may be beneficial to carry out a rigorous program of early pile testing (both maintained static load testing and dynamic testing) to determine appropriate acceptance criteria before working piles are installed. To mitigate against this potential, restrikes on working piles are usually undertaken at least 24 hours after initial pile installation.

Further details on the methods relating to wave equation analysis design for driven piles can be found in relevant software user manuals and publications – for example, FHWA (2013), Biddle (2005), BSI (2007, 2015).

54.10. New developments in piled foundations

New developments in terms of pile type and design approach will be required as innovative ideas are trialled or certain pile types gain rapid growth in use. This will include new methods and new materials. Examples of new pile types include helical screw piles, hollow piles and ribbed piles, all of which will have their own specific design approaches that allow determination of base and shaft capacity. Some of these will be mature enough to have published guidance – one such system is steel helical piles.

Helical piles have become more widely used in recent years. Useful references for this technology include IPENZ (2015), Perko (2009) and Clayton (2005).

Some of these new systems will be covered within current standards, with others needing to be considered within the envelope of existing standards even when not specifically referenced. It is important that standards do not constrain the development of new, innovative and low-carbon piling systems or design approaches. The pile designer must be encouraged to develop the optimum solution for a particular project, which may include new and innovative systems and materials with limited previous full-scale use.

With all of these new systems, construction-related limitations must be well understood prior to finalising any design to ensure it is practical to install. Chapter 81 *Types of bearing piles* provides guidance on a wide range of pile types and the selection process.

The development and adoption of new and innovative materials is an important consideration for the pile design process. New low-carbon concrete mixes and steel reinforcement alternatives such as basalt or glass fibre derivatives offer the potential to significantly reduce the embodied carbon within a piled foundation and can also produce safety and commercial benefits. The design process should consider the latest materials available in order to produce the optimum solution. Adoption of very new materials may require preliminary trials in order to assess their impact on the designed works, and this needs to be factored into the design process and construction programme.

It is also important to consider new materials that are not part of the permanent works, including support fluids used during the construction of bored piles. EFFC/DFI (2019) provides useful guidance on this subject.

References

API (American Petroleum Institute) (2000) *Recommended Practice for Planning, Designing and Constructing Fixed Offshore Platforms: Working Stress Design*. American Petroleum Institute, Washington, DC, API RP 2A-WSD.

API (2008). *Recommended Practice for Planning, Designing and Constructing Fixed Offshore Platforms: Working Stress Design.* API RP 2A-WSD. 21st edition, errata and supplement 3. API: Washington, DC.

Atkinson JH (2000) Non-linear soil stiffness in routine design. *Géotechnique* **50(5)**: 487–508.

Bell AG (2010) Foundation solutions for the urban regeneration of Glasgow city centre. *Proceedings of the DFI/EFFC 11th International Conference on Urban Regeneration, London, UK.*

Berezantzev VG, Khristoforov VS and Golubkov VN (1961) Load bearing capacity and deformation of piled foundations. *Proceedings of the 5th International Conference of the International Society for Soil Mechanics and Foundation Engineering, Paris, France*, vol. 2, pp. 11–15.

Biddle AR (2005) *SCI P335 H-Pile Design Guide.* SCI, London, UK.

Bown AS and O'Brien AS (2008) Shaft friction in London Clay: modified effective stress approach. *Foundations: Proceedings of the Second British Geotechnical Association International Conference on Foundations, Dundee, UK* (Brown MJ, Bransby MF, Brennan AJ and Knappett JA (eds)). IHS BRE Press, Watford, UK, vol. 1, pp. 91–100.

Brettmann T and Duncan JM (1996) Computer application of CLM lateral analysis to piles and drilled shafts. *ASCE, Journal of Geotechnical Engineering* **122(6)**: 496–498.

Broms B (1964a) Lateral resistance of piles in cohesive soils. *Journal of the Soil Mechanics and Foundations Division; Proceedings of the American Society of Civil Engineers* **90(SM2)**.

Broms B (1964b) Lateral resistance of piles in cohesionless soils. *Journal of the Soil Mechanics and Foundations Division; Proceedings of the American Society of Civil Engineers* **90(SM3)**.

BSI (British Standards Institution) (2007) BS EN 1993-5:2007. Eurocode 3: Design of steel structures – Part 5. BSI, London, UK.

BSI (2015) BS EN 12699:2015: Execution of special geotechnical work – Displacement piles. BSI, London, UK.

Burland JB, Butler FG and Dunican P (1966) The behaviour and design of large diameter bored piles in stiff clay. *Proceedings of the Symposium on Large Bored Piles, London, UK.* ICE, London, UK, pp. 51–71.

Burland JB, Simpson B and St John HD (1979) Movements around excavations in London Clay. *Proceedings of the 7th European Conference on Soil Mechanics and Foundation Engineering, Brighton, UK*, vol. 1, pp. 13–29.

Chandler RJ and Davis AG (1973) *Further Work on the Engineering Properties of Keuper Marl.* CIRIA Report 47. CIRIA, London, UK.

Chandler RJ and Forster A (2001) *Engineering in Mercia Mudstone.* CIRIA Report 570. CIRIA, London, UK.

CIRIA (1995) *The Standard Penetration Test (SPT): Methods and Use*. CIRIA Report 143. CIRIA, London, UK.

Clayton DJ (2005) *Basic Helical Screw Pile Design*. Earth Contract Products, Olathe, KS.

Corke DJ, Fleming WK and Troughton VM (2001) A new approach to specifying performance criteria for pile load tests. *Proceedings of the International Symposium Underground Construction, London, UK*, pp. 401–410.

Datta M, Gulhati SK and Rao GV (1980) Crushing of calcareous sands during drained shear. *SPE Journal* **20(2)**: 77–85.

Dutt RN and Ingram WB (1984) Jackup rig siting in calcareous soils. *Proceedings of Offshore Technology Conference, Houston, TX*, pp. 541–548.

EFFC/DFI (2019) *Guide to Support Fluids for Deep Foundations*. EFFC/DFI, Faversham, UK.

Elson WK (1984) *Design of Laterally Loaded Piles*. CIRIA Report 103. CIRIA, London, UK.

England MG (1999) *A Pile Behaviour Model*. PhD thesis, Imperial College London, London, UK.

FHWA (Federal Highway Administration) (2010) *Drilled Shafts, Construction Procedures and LRFD Design Methods*. FHWA, Washington, DC.

FHWA (2013) *Implementation of AASHTO LRFD Design Specifications for Driven Piles*. FHWA, Washington, DC,

Fleming WGK (1992) A new method for single pile settlement prediction and analysis. *Géotechnique* **42**: 411–425.

Fleming WGK and Thorburn S (1983) Recent piling advances: state of the art report. *Proceedings of the Conference on Advances in Piling and Ground Treatment for Foundations, London, UK*. ICE, London, UK.

Fleming K, Welman W, Randolph M and Elson K (2009) *Piling Engineering*, 3rd edn. Taylor & Francis, London, UK.

Handley B, Ball J, Bell A and Suckling T (2006) *Handbook on Pile Load Testing*. Federation of Piling Specialists, Beckenham, UK.

Horvarth RG (1978) *Field Load Test Data on Concrete to Rock Bond Strength for Drilled Pier Foundations*. University of Toronto, Toronto, OR.

Horvarth RG and Kenney TC (1979) Shaft resistance of rock socketed drilled piers. *Proceedings of the ASCE Annual Convention, Atlanta, GA*.

HSE (Health and Safety Executive) (2015) *The Construction (Design and Management) Regulations 2015 (CDM 2015)*. HSE, London, UK.

IPENZ (2005) *Practice Note 28: Screw Piles – Guidelines for Design Construction & Installation*. IPENZ, Wellington, New Zealand.

Jardine R, Chow F, Overy R and Standing J (2005) *ICP Design Methods for Driven Piles in Sand and Clays*. Thomas Telford, London, UK.

Kulhawy FH and Phoon K (1993) Drilled shaft side resistance in clay soil to rock. *Proceedings of the Design and Performance of Deep Foundations: Piles and Piers in Soil and Soft Rock, Dallas, TX*. ASCE, New York, NY, pp. 172–183

LDSA (London District Surveyors Association) (2017) *Foundations: Guidance Notes for the Design of Straight Shafted Bored Piles in London Clay*. LDSA, London, UK.

Lehane BM, Schneider JA and Xu X (2005) The UWA-05 method for prediction of axial capacity of driven piles in sand. *Proceedings of the International Symposium on Frontiers Offshore Geotechnics, Perth, Australia*, pp. 683–689.

Lord JA, Hayward T and Clayton CRI (2003) *Shaft Friction of CFA Piles in Chalk*. CIRIA Report PR86. CIRIA, London, UK.

Lunne T, Robertson PK and Powell JJM (1997) *Cone Penetration Testing in Geotechnical Practice*. Blackie, London, UK.

Markvukaj SA, O'Brien AS, Farooq I, Pye S and Vance M (2022) Shaft friction design for piles in extremely weak to weak Mercia Mudstone. https://learninglegacy.hs2.org.uk/document/shaft-friction-design-for-piles-in-extremely-weak-to-weak-mercia-mudstone (accessed 29/05/2023).

Mascarucci Y, Miliziano S and Mandolini A (2016) 3M analytical method: evaluation of shaft friction of bored piles in sand. *Journal ASCE* **143(3)**, 10.1061/(ASCE)GT.1943-5606.00013.

Matlock H (1970) Correlations for design of laterally loaded piles in soft clay. *Proceedings of the 2nd Offshore Technical Conference, Houston, TX*, vol. 1, pp. 577–594.

Matsui T (1993) Case studies on cast-in-place bored piles and some considerations for design. *Proceedings of BAPII, Ghent, Belgium*. Balkema, Rotterdam, the Netherlands, pp. 77–102.

Meigh AC and Wolski W (1979) Design parameters for weak rocks: design parameters in geotechnical engineering. *Proceedings of the 7th European Conference on Soil Mechanics and Foundation Engineering, Brighton, UK*. Thomas Telford, London, UK, vol. 5, pp. 59–79.

Meyerhof GG (1976) Bearing capacity and settlement of pile foundations. *Journal of the Geotechnical Engineering Division, ASCE*, **102(GT3)**: 197–228.

Meyerhof GG and Sastry VVRN (1978) Bearing capacity of piles in layered soils. Part 2. Sand overlying clay. *Canadian Geotechnical Journal* **15(2)**: 183–189, 10.1139/t78-018.

Nicola A and Randolph M (1993) Tensile and compressive shaft capacity of piles in sand. *Journal of Geotechnical Engineering* **119(2)**, 10.1061/(ASCE)0733-9410(1993)119:12(1952).

Patel DC (1992) Interpretation of results of pile tests in London Clay. In *Piling: European Practice and Worldwide Trends* (Sands MJ (ed)). Thomas Telford, London, pp. 100–110.

Peck RB, Hanson WE and Thornburn TH (1967) *Foundation Engineering*. Wiley, New York, NY.

Perko HA (2009) *Helical Piles: A Practical Guide to Design and Installation*. Wiley, New York, NY.

Poulos HG (1971) Behaviour of laterally loaded piles – I: single piles. *Journal of the Soil Mechanics and Foundations Division; Proceedings of the American Society of Civil Engineers* **97(SM5)**.

Poulos HG (1989) Pile behaviour – theory and application. 29th Rankine Lecture, Imperial College London. *Géotechnique* **39**: 363–416.

Poulos HG (2000) Pile testing: from the designer's viewpoint. In *STATNAMIC Loading Test '98, Tokyo, Japan* (Kusakabe O, Kuwabara F and Matsumoto T (eds)). Balkema, Rotterdam, the Netherlands, pp. 3–21.

Poulos HG (2017) *Tall Building Foundations*. CRC Press, London, UK.

Poulos HG and Davis EH (1968) The settlement behaviour of single axially loaded incompressible piles and piers. *Géotechnique* **18**: 351–371.

Raison C and Egan D (2016) Design of piles – UK practice. *Proceedings of the International Symposium on Design of Piles in Europe, Leuven, Belgium*.

Randolph MF (1981) Response of flexible piles to lateral loading. *Géotechnique* **31(2)** : 247–259.

Randolph MF and Gourvenec S (2011) *Offshore Geotechnical Engineering*. Spon Press, London, UK.
Randolph MF and White DJ (2008) Offshore foundation design: a moving target. *Foundations: Proceedings of the Second British Geotechnical Association International Conference on Foundations* (Brown MJ, Bransby MF, Brennan AJ and Knappett JA (eds)). IHS BRE Press, Watford, UK.
Reese LC and Matlock H (1956) Non-dimensional solutions for laterally loaded piles with soil modulus proportional to depth. *Proceedings of the 8th Texas Conference on Soil Mechanics and Foundation Engineering, Austin, TX*, pp. 1–41.
Reese LC, Cox WR and Koop FD (1974) Analysis of laterally loaded piles in sand. *Proceedings of the 6th Offshore Technical Conference, Houston, TX*, pp. 473–483.
Reese LC, Cox WR and Koop FD (1975) Field testing and analysis of laterally loaded piles in stiff clay. *Proceedings of the 7th Offshore Technical Conference, Houston, TX*, vol. 12, pp. 671–690.
Rosenberg P and Journeaux NL (1976) Friction and end bearing tests on bedrock for high capacity socket design. *Canadian Geotechnical Journal* **13**: 324–333.
Rowe RK and Armitage HH (1987a) A design method for drilled piers in soft rock. *Canadian Geotechnical Journal* **24**: 126–142.
Rowe RK and Armitage HH (1987b) Theoretical solutions for axial deformation of drilled shafts in rock. *Canadian Geotechnical Journal* **24**: 114–125.
Schneider AM, Stanier SA, White DJ, and Randolph MF (2019) Shallow penetrometer tests: theoretical and experimental modelling of penetration and dissipation stages. *Canadian Geotechnical Journal*, 10.1139/cgj-2018-0656.
Semple RM and Rigden WJ (1984) Shaft capacity of driven pipe piles in clay. *Proceedings of the Symposium on Analysis and Design of Deep Foundations, San Francisco, CA*, pp. 59–79.
Skempton AW (1951) The bearing capacity of clays. *Proceedings of the Building Research Congress, London, UK*. ICE, London, UK, vol. 1, pp. 180–189.
Stas CV and Kulhawy FH (1984) Critical evaluation of foundation design methods under axial uplift and compression loading. Electric Power Research Institute, Palo Alto, CA, report EL-3771.
Stroud MA and Butler FG (1975) The standard penetration test and the engineering properties of glacial materials. *Proceedings of the Symposium on the Engineering Behaviour of Glacial Materials, Birmingham, UK*, pp. 117–128.
Thorburn S and McVicar RSL (1971) Pile load tests to failure in the Clyde alluvium. *Proceedings of the Conference on Behaviour of Piles, London, UK*. ICE, London, UK, pp. 1–7, 53–54.
Tomlinson MJ (2014) *Pile Design and Construction Practice*, 6th edn. Spon Press, London, UK.
Trenter NA (1999) *Engineering in Glacial Tills*. CIRIA Report C504. CIRIA, London, UK.
Vesic AS (1972) Expansion of cavities in infinite soil mass. *Journal of the Soil Mechanics and Foundations Division, ASCE* **98**: 265–290.
Vesic AS (1977) *Design of Piled Foundations*. NCHRP Synthesis 43. Transport Research Board, Washington, DC.
Weltman AJ (1980) *Pile Load Testing Procedure*. CIRIA Report PG7. CIRIA, London, UK.
Weltman AJ and Healy PR (1978) *Piling in 'Boulder Clay' and Other Glacial Tills*. CIRIA Report PG5. CIRIA, London, UK.
White DJ and Bolton MD (2005) Comparing CPT and pile base resistance in sand. *Proceedings of the Institution of Civil Engineers* **158(GE1)**: 3–14.
White DJ and Lehane BM (2004) Friction fatigue on displacement piles in sand. *Géotechnique* **54(10)**: 645–658.
Whitworth LJ and Turner AJ (1989) Rock socketed piles in the Sherwood Sandstone of central Birmingham. *Proceedings of the International Conference on Piling and Deep Foundations, London, UK* (Burland JB and Mitchell JM (eds)). Balkema, Rotterdam, the Netherlands, pp. 327–334.
Williams AF and Pells PJN (1981) Side resistance rock sockets in sandstone, mudstone and shale. *Canadian Geotechnical Journal* **18**: 502–513.

It is recommended this chapter is read in conjunction with

- Chapter 19 *Settlement and stress distributions*
- Chapter 22 *Behaviour of single piles under vertical loads*
- Chapter 55 *Pile-group design*
- Chapter 82 *Piling problems*

All chapters in this book rely on the guidance in Sections 1 *Context* and 2 *Fundamental principles*. A sound knowledge of ground investigation is required for all geotechnical works, as set out in Section 4 *Site investigation*.

doi: 10.1680/icemge.66830.0863

Chapter 55
Pile-group design

Tony O'Brien Mott MacDonald, UK
Stephanie Markvukaj Mott MacDonald, UK
Alan Willoner Mott MacDonald, UK

Pile design often focuses on ultimate geotechnical capacity. When piles are located within a pile group, a wide range of additional factors need to be considered, including the deformation of the pile group and the structural forces induced in the piles and pile cap. Interaction between adjacent piles depends on several factors, including the direction of applied loads. When large horizontal loads are applied a complex range of interaction effects may need to be considered. Computer-based analytical methods are commonly used; nevertheless, it is important to use simplified methods of analysis both for preliminary design and to judge the reliability of output from computer-based methods. Pile-group deformation is usually more critical than group capacity, and, hence, the selection of appropriate deformation moduli is particularly important. Case history data and predictions from nonlinear modelling are compared with linear elastic methods. The delegation of design responsibility for pile-group design requires careful judgement and is also discussed.

CONTENTS

55.1. Introduction

The design of single piles is discussed in Chapter 54 *Single piles*, where the main focus is the assessment and verification of the ultimate geotechnical capacity of the pile. For pile groups, a wider range of factors will need to be considered. The design must consider the overall performance of the whole foundation system (**Figure 55.1**). It is often the case, especially for larger pile groups, that pile-group deformation and the structural forces (bending moments and shear forces) induced in the piles are the main design issues, rather than ultimate geotechnical capacity. A pile group will stress a much larger zone of soil than a single pile (**Figure 55.2**), and this will be extremely important when considering how to develop an appropriate analytical model and input parameters, and assessing potential failure and deformation mechanisms. For example, if there is a soft compressible layer at depth below the pile toe, it may have little effect on single-pile behaviour, but could cause excessive pile-group deformation or even collapse.

The overall behaviour of pile groups is affected by their geometry (length, width, etc.) and number of piles within a group, n. Pile groups have been classified by Basile (2020) by the number of piles in a group: 'small' ($n < 5$ piles), 'medium' (5–25 piles) and 'large' ($n > 25$ piles). This classification is particularly useful when considering pile-group capacity (i.e. ultimate limit state (ULS) design) – refer to

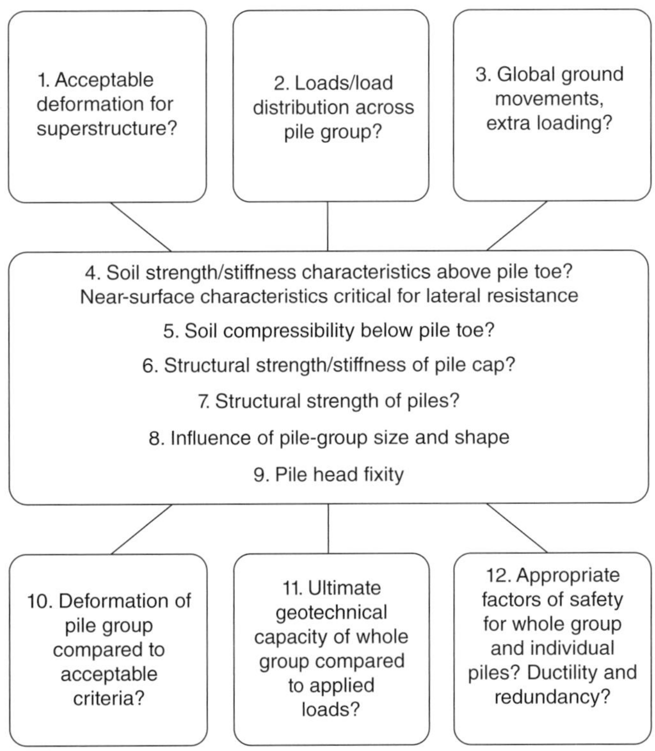

Figure 55.1 Design considerations for pile groups (Modified from O'Brien and Bown, 2010)

Section 55.2. The design of 'medium' and 'large' pile groups is typically controlled by deformation (i.e. serviceability limit state (SLS)) and structural design considerations – refer to Sections 55.3–55.6. A geometrical parameter which gives more general and powerful insights into group behaviour (both ULS and SLS) is the pile-group aspect ratio R (Randolph and Clancy, 1993), defined as $R = (ns/L)^{0.5}$, where n is the number of piles in the group and s and L are the pile spacing and length, respectively (both in metres). When the pile-group aspect ratio is large (**Figure 55.3(a)**), then most of the applied load is likely to be shed into the ground through compression of the deposits below the pile-toe level. In contrast, if R is small (**Figure 55.3(b)**), then most of the applied load is likely to be shed into the ground through shear around the pile-group perimeter. Randolph (1994) suggests that a pile group is 'large' if $R > 4$ and 'small' if $R < 2$ (therefore 'medium' applies if R is 2–4). Viggiani *et al.* (2012) define 'large' pile groups as those with a width, B, greater than pile length, L, whereas 'small' pile groups are those where $B < L$. Viggiani *et al.* provide several practical examples of how a simple appreciation of pile-group geometry (using the above classifications) can facilitate an appropriate choice of analytical methods and improved pile-group design.

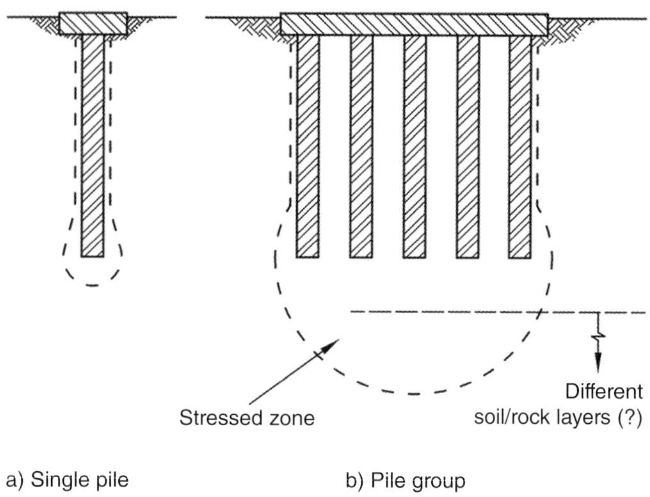

Figure 55.2 Comparison of influence zones beneath single piles and pile groups

In modern engineering practice, pile-group analysis usually relies upon commercially available software. However, most software has a number of limitations, namely

- soil behaviour is often assumed to be linear elastic
- the influence of soil layering below the pile-toe level may be poorly simulated.

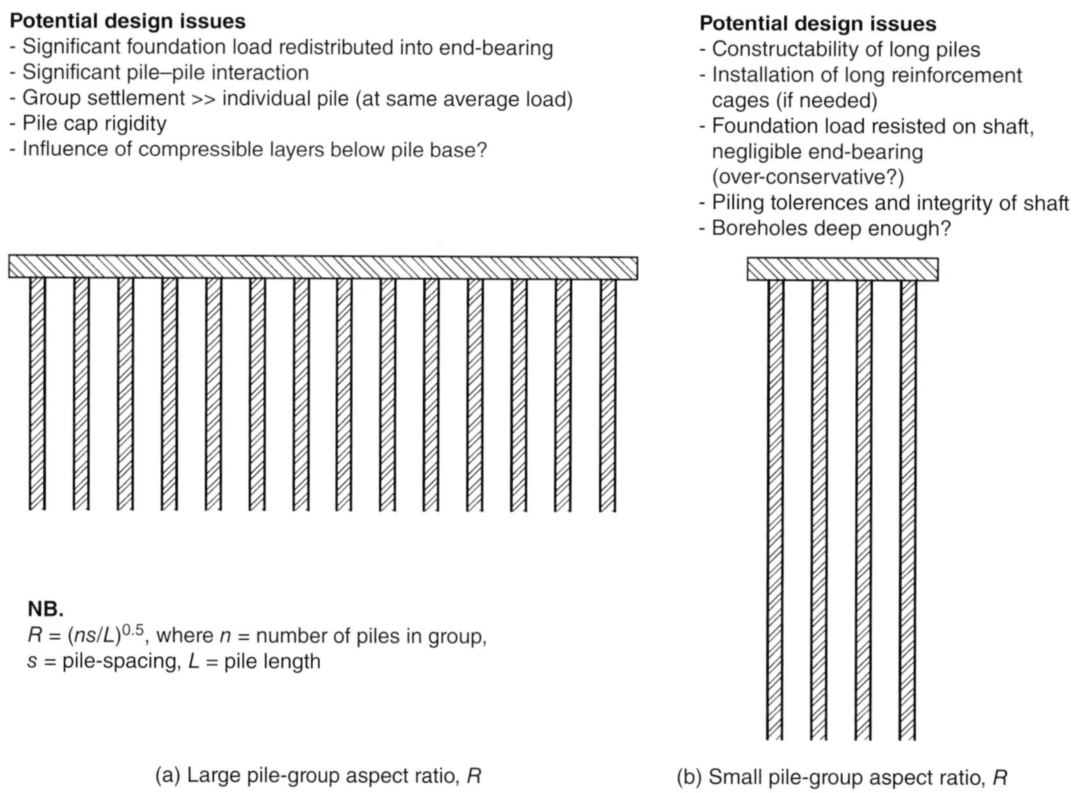

Figure 55.3 Pile-group aspect ratio, *R*, and potential design issues

Analytical methods for pile groups are often complex; therefore, the main intent of this chapter is to highlight

- the key factors that will affect pile-group behaviour
- several simplified analytical methods that can be useful either for preliminary design or to check computer output
- when linear elastic methods may give output that is appropriate for practical design, and when more complex methods are necessary
- how selected outputs, in particular axial load distributions, can be interpreted to facilitate more economic design.

55.2. Pile-group capacity

Independent calculations should be made of ultimate capacity based upon

(*a*) the sum of the individual pile capacities (based on capacity of isolated piles, see Chapter 54 *Single piles*) (**Figure 55.4(a)**)

(*b*) failure of the 'block' that encloses the perimeter of the pile group (**Figure 55.4(b)**)

(*c*) failure of rows of piles (**Figure 55.4(c)**).

For vertically loaded pile groups, usually only (*a*) and (*b*) will need to be checked; failure mode (*c*) becomes more important if the spacing between pile rows is variable or if the horizontal or moment loading is relatively large.

Most design codes offer little guidance on pile-group design, which often leads to over-conservative design. It is not necessary to provide the same geotechnical compressive ground resistance to peak axial load ratio for, say, a corner pile of a large pile group as is required for a single pile or small pile group (defined here as $n < 5$ piles). If the pile cap has adequate structural strength and stiffness, and the pile group has five or more piles (and, therefore, can be considered fully redundant (NCHRP, 2004)), then the designer should not be concerned about **local** geotechnical 'factors of safety' for individual piles within a group. As a highly loaded pile in a medium or large group attempts to settle more than adjacent piles, load redistribution occurs through the pile cap, which then reduces local peak loads and leads to a more uniform load distribution across a group. The important limit states to check are then

- the structural strength and stiffness of the pile cap and structural capacity of the single piles
- that the SLS pile-group deformation is acceptably small for the superstructure
- the durability of the foundations for the intended design life of the structure
- potential effects of global ground movements on the foundations (refer to Chapter 57 *Global ground movements and their effects on piles*)
- if relevant, earthquake effects such as liquefaction potential (refer to Chapter 60 *Foundations subjected to cyclic and dynamic loads*)
- if relevant, dynamic or cyclic response to wind or wave actions (refer to Chapter 60 *Foundations subjected to cyclic and dynamic loads*).

The failure of pile groups is rare; however, the following situations require particular care.

(*a*) When large global horizontal ground deformations develop around a pile group, usually due to out-of-balance loading on a soft clay or peat layer (**Figure 55.5(a)**), either due to placing fill or excavating adjacent to the pile group.

(*b*) When a layer of soft clay underlies a thin competent layer, and end-bearing piles are toed into the competent layer (**Figure 55.5(b)**).

(*c*) When end-bearing piles are toed in rock, which contains either dipping clay-filled discontinuities or voids (**Figure 55.5(c)**). Voids may arise in a rock mass either due to natural solution features or due to historic mining activity.

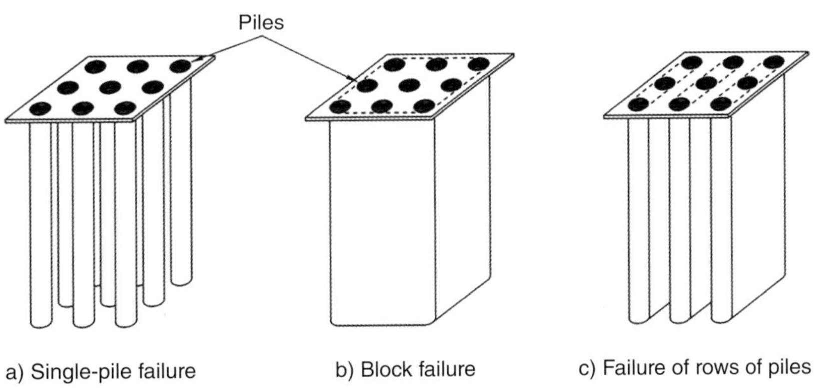

Figure 55.4 Failure modes for pile groups (Reproduced from Fleming *et al.*, 2009)

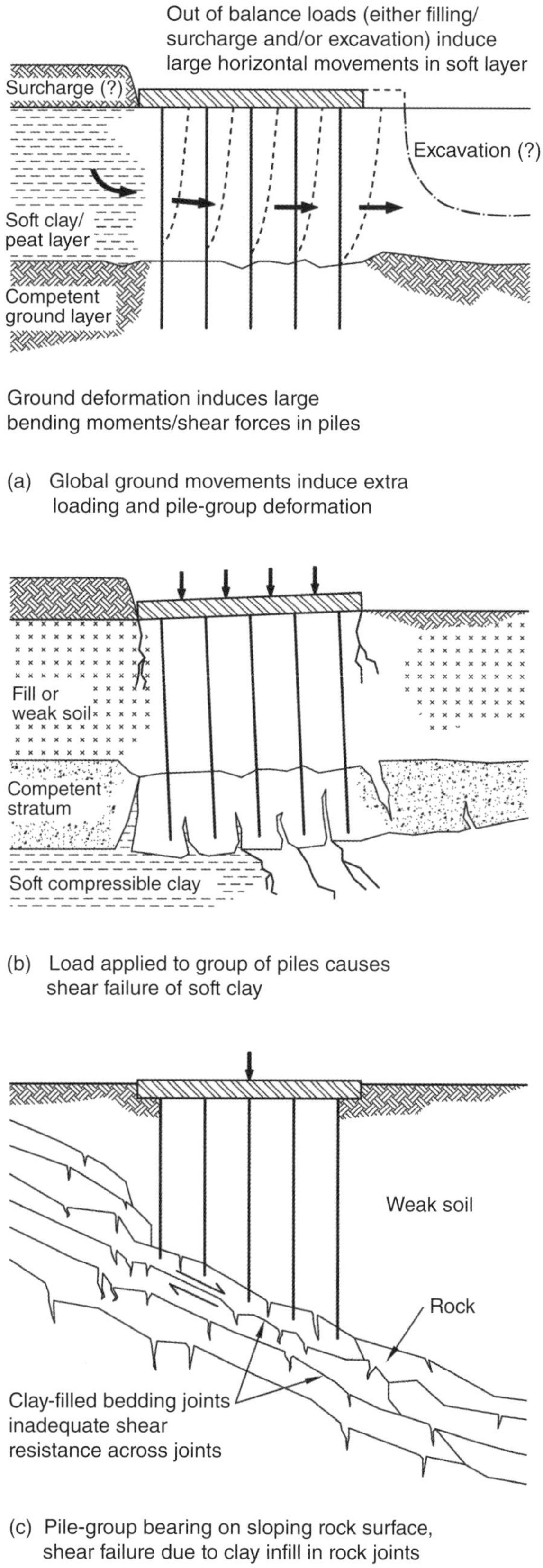

Figure 55.5 Pile-group failure, some high-risk situations

In the authors' experience, (*a*) is the most common cause of pile-group failure (usually due to inadequate structural pile strength). This has occurred when the pile designers or constructors did not appreciate the potential impact of global ground movements and this failure mode was completely overlooked. Chapter 57 *Global ground movements and their effects on piles* discusses the influence of global ground movements (both vertical and horizontal) on pile-group behaviour. This failure mode can often be avoided either by appropriate ground improvement (Chapter 59 *Design principles for ground improvement*) or by appropriate construction planning and sequencing. For example, installing the piles *after* the fill or excavation, adjacent to the pile group, has been completed. The analytical methods used to assess ultimate pile-group capacity are based on conventional bearing capacity theory (refer to Chapters 21 *Bearing capacity theory* and 53 *Shallow foundations*). Hence, for *vertical loading of a pile group in clay* the ultimate capacity of the block of soil encompassed by the group is

$$Q = 2L(B_1 + B_2)s_{u,av} + s_{u,b}N_c B_1 B_2 \tag{55.1}$$

where L is the pile length or embedded depth in competent founding strata, B_1 is the pile-group breadth, B_2 is the pile-group width (i.e. B_2 is the larger of the plan dimensions for a rectangular pile group), $s_{u,av}$ is the average undrained shear strength around the group perimeter, $s_{u,b}$ is the undrained shear strength beneath the group (the average between the pile-toe level and $0.5B_1$ below the toe level for homogeneous conditions), and N_c is the bearing capacity factor allowing for appropriate depth, width and shape (refer to **Table 55.1**, which summarises the appropriate bearing capacity and shape factors). Total stress methods are commonly used for clays. However, effective stress methods can be utilised (refer to Chapter 54 *Single piles*) and are recommended if deep

	N_c	
L/B_1	$B_2/B_1 = 1$	$B_2/B_1 \geq 10$
0.25	6.7	5.6
0.50	7.1	5.9
0.75	7.4	6.2
1.00	7.7	6.4
1.50	8.1	6.8
2.00	8.4	7.0
2.50	8.6	7.2
3.00	8.8	7.4
≥ 4	9.0	7.5

For intermediate B_2/B_1 ratios, use linear interpolation.

Table 55.1 Bearing capacity factors for pile groups (After Viggiani *et al.*, 2012)

excavations (say in excess of 4 m) are constructed above or adjacent to the pile group (since the excavation would lead to a reduction in the clay's mobilised strength in the long term).

The influence of an underlying compressible layer on the ultimate capacity of the group (acting as a block or equivalent pier, refer to Section 55.5.4) can be assessed (based on studies by Matsui, 1993) as

$$Q_b = Q_{u2} \quad \text{for} \frac{h_c}{d_{eq}} \leq 0.5 \tag{55.2}$$

$$Q_b = Q_{u1} \quad \text{for} \frac{h_c}{d_{eq}} \geq 3 \tag{55.3}$$

and for z_c/d_{eq} between 0.5 and 3

$$Q_b = Q_{u2} + \left(\frac{0.4h_c}{d_{eq}} - 0.2\right)(Q_{u1} - Q_{u2}) \tag{55.4}$$

where Q_b is the ultimate bearing capacity of the base, Q_{u1} is the ultimate bearing capacity of the upper layer, Q_{u2} is the ultimate bearing capacity of the underlying weak compressible layer, h_c is the depth of the underlying layer below the pile-group toe level and d_{eq} is the equivalent pier diameter (**Figure 55.6**).

If the upper layer is sand or gravel then Q_{u1} can be assessed by using conventional bearing capacity theory based on effective stress parameters (refer to Chapters 21 *Bearing capacity theory* and 53 *Shallow foundations*) together with appropriate shape and depth correction factors. For weak rocks, the nature and spacing of discontinuities in the rock mass need to be considered, as outlined in Chapter 53 *Shallow foundations*. For clays, conventional bearing capacity theory can be used, together with appropriate shape and depth correction factors (Equation 55.5)

$$\text{Ultimate bearing pressure } q_b = N_c s_{u,b} \tag{55.5}$$

where N_c is the appropriate bearing capacity factor, allowing for depth and shape correction factors (**Table 55.1**) and $s_{u,b}$ are defined above.

For pile groups subject to horizontal or moment loading and underlain by relatively weak layers, the potential for edge failure should be checked by assessing the bearing resistance beneath the outer (more heavily loaded) pile row. It is usually sufficient to calculate an average bearing pressure at the toe level beneath the row (based on the mobilised end-bearing pressure). The factor of safety against local failure can then be checked against Equations 55.2–55.5.

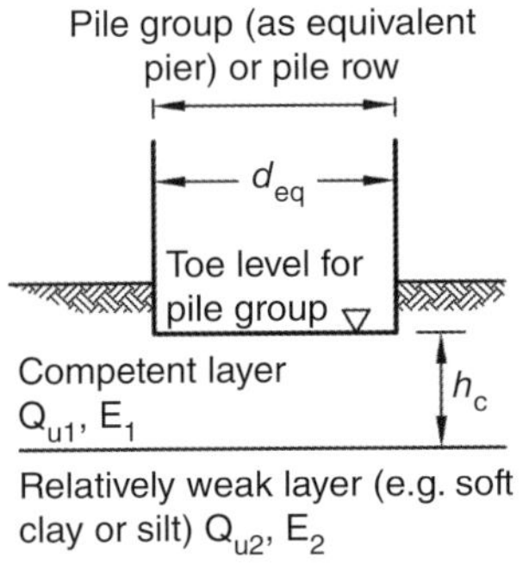

Figure 55.6 Pile group (or row) founded on a competent layer over a weak layer (After Matsui, 1993)

Historically, there was some concern about the capacity of a group of piles relative to the sum of individual pile capacities, and the term group 'efficiency' has been used in some textbooks. This concept has led to confusion. It should be noted that the efficiency factors quoted are based on a limited set of small-scale model tests, and general experience (together with full-scale tests, e.g. O'Neill *et al.*, 1982) indicates that there is little merit in using efficiency factors. For piles driven into sand, the group capacity is likely to be much higher than the sum of individual capacities. However, settlement considerations will usually be critical, and the actual ultimate capacity does not generally govern design. For bored piles (in sands or clays) it could be argued (with some justification) that there would be an overall relaxation of horizontal ground stress around the pile group (thus reducing the pile-group capacity). However, this will be counterbalanced by the fact that shear failure would occur mainly through an undisturbed soil mass (rather than through the highly disturbed or remoulded ground adjacent to an individual pile shaft). In summary, therefore, the use of group efficiency factors is *not* recommended for calculations of ultimate group capacity.

For pile groups subject to horizontal loading, the most likely geotechnical failure mode is shown in **Figure 55.4(c)**; commonly this will be more critical than the sum of individual pile capacities due to the 'shadowing' effect of adjacent piles in a row (**Figure 55.7**). The failure of pile rows under horizontal loading is discussed by Fleming *et al.* (2009).

For practical purposes the ultimate lateral capacity of a pile group can be estimated as the lesser of

- (*a*) the sum of the ultimate capacities of individual piles in the group (Chapter 54 *Single piles*)
- (*b*) the ultimate capacity of an equivalent block containing the piles and soil
- (*c*) the sum of the ultimate capacities of the pile rows.

For (*b*), this failure mode checks if the shear resistance of the soil between the piles is less than the limiting resistance of an individual pile (which then controls the limiting resistance provided by the soil in the wake of other piles in a group); only the 'short-pile' case is relevant (Fleming *et al.* 2009), and failure mode (*b*) is only likely to be critical when there are a large number of closely spaced and relatively short piles. For this scenario, the soil shear strength would only be ignored across a depth equal to 1.5 times the pile diameter. Pile-group failure under a lateral load will usually involve rotation as well as horizontal translation of the group. The ultimate geotechnical capacity will then be a function of both the axial and lateral

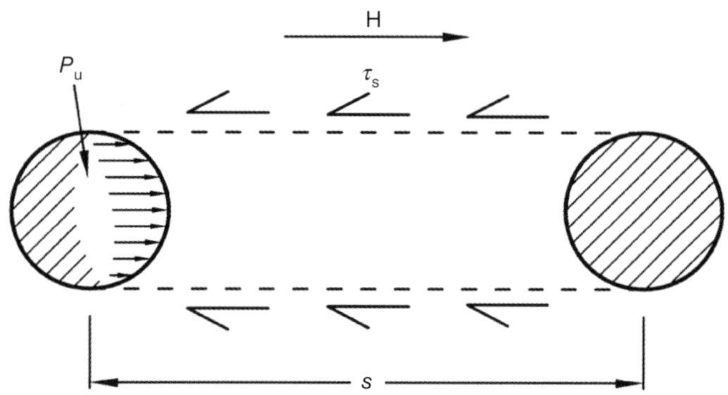

NB: The limiting force per unit length for piles in the wake of another pile is:

$$P_u < 2s\tau_s$$

where P_u is the limiting lateral resistance provided by the soil for an individual pile, τ_s is the ultimate shear resistance provided by the soil between piles in a group and s is the pile spacing.
P_u and τ_s are evaluated as appropriate for fine- or coarse-grained soils (see Chapter 54 *Single piles*).

Figure 55.7 Failure and deformation mechanism ('shadowing') for a pile row under a horizontal load

capacities, with piles behind the axis of rotation failing in tension and piles in front of the axis of rotation failing in compression. In situations involving large horizontal or moment loading, the induced structural forces (bending moment, shear force, axial forces, tension or compression) and horizontal or rotational movements will usually be more critical than the ultimate geotechnical capacity. Pile-group deformation and induced structural forces will be discussed later in this chapter.

55.3. Pile-to-pile interaction: vertical loading

When piles are constructed close to one another, their subsequent load–deformation behaviour is affected by the presence of the neighbouring piles. **Figure 55.8(a)** provides a simple example of these 'pile-to-pile' interaction effects; it shows four piles under equal vertical loading (by way of a flexible pile cap). Each pile settles under the load and a zone of soil settlement develops around each pile. These soil settlement zones overlap and interact; hence, the middle piles settle more than the outer piles (and all the 'group' piles settle more than an individual isolated pile). If the pile group has a rigid pile cap (**Figure 55.8(b)**) each pile has to settle by an equal amount (because the pile cap is rigid) but the axial loads in the piles will now vary across the group, with the outer piles developing higher axial loads than the middle piles (because the outer piles are stiffer than the inner piles and, hence, attract more load). The pile-to-pile interaction and how it affects both pile-group deformation and the redistribution of forces across piles (and into the pile cap) within a group is of fundamental importance. In practice, the pile-to-pile interaction depends on a large number of factors (**Table 55.2**). For vertical loading, many of the factors given in **Table 55.2** lead to a reduction in group interaction effects; this contrasts with horizontal loading, where group interaction effects may increase.

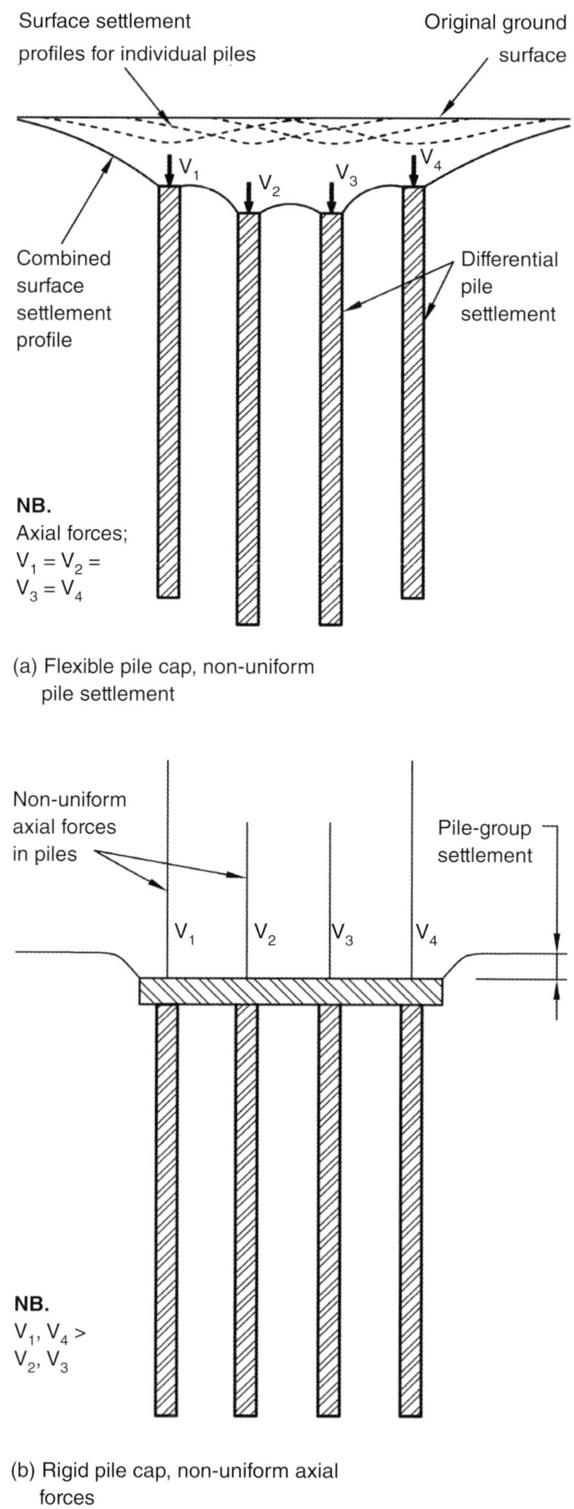

Figure 55.8 Effects of pile-to-pile interaction, vertical loading

Factor	Vertical loading		Horizontal loading	
	Group settlement ratio	Non-uniformity of axial loads	Group deformation	Stresses in piles
Increase of ground stiffness with depth	–	–	– (if $< L_c$)	– (if $< L_c$)
Ground layering, relatively stiff layer at depth below pile toe	–	–	NE	NE
End-bearing pile (compared with friction pile)	–	–	NE	NE
Ground layering, relatively compressible layer at depth below pile toe	+	+	NE	NE
Ground, nonlinear stress–strain behaviour	–[a]	–[a]	– or + [b]	– or + [b]
Pile installation[c]	–	–	NE	NE
Pile-group layout	NE	NE	+	+
Pile spacing ($s > 3d$)[d] ($s < 3d$)	– +	– +	– +	– +
Increasing displacement	–	–	+	+
Near-surface soils ($< 6d$) (if relatively weak layer)	NE	NE	+	+
Pile cap stiffness reduces	NE	–	NA	NA
Pile head fixity, fixed to free head condition	NA	NA	+	Depends on pile location in group

Key
– Leads to a reduction in interaction effects (e.g. smaller group settlement, more uniform axial loads); + leads to an increase in interaction effects (e.g. larger group deformation, more non-uniform axial loads and stresses in piles); L_c is the critical pile length for horizontal loading; NE, negligible effect; NA, not applicable.
[a] Particularly important for large pile groups;
[b] depends on selected 'elastic modulus', typical outcome. If the selected 'elastic' stiffness is an intermediate or large strain value, then the calculated bending moments and pile-group displacements are likely to be overestimated (refer to **Figure 55.29**);
[c] the locally disturbed zone around a pile reduces interaction effects;
[d] interaction effects reduce significantly as spacing increases, typically negligible at spacings $> 6d$.

Table 55.2 Pile interaction, influential factors

In the context of assessing pile-group settlement, two parameters are useful

$$\text{Settlement ratio, } R_s = \frac{\text{Average group settlement}}{\text{Settlement of single pile at same average load as a pile in the group}} \quad (55.6)$$

$$\text{Group reduction factor, } R_g = \frac{\text{Average group settlement}}{\text{Settlement of single pile at same total load as the group}} \quad (55.7)$$

Group reduction factor, R_g, is a conceptually difficult parameter; nevertheless, it is useful when comparing the performance of different pile groups. R_g is simply related to R_s by $R_s = nR_g$, where n is the number of piles in the group. The value of R_g varies between $1/n$ and 1.0. An important practical conclusion from various studies (e.g. Butterfield and Douglas, 1981) is that the precise layout of the piles has a negligible influence on the group settlement ratio R_s; for example, a square pile group has the same value of R_s as a rectangular or circular group, at the same average pile spacing.

Two of the most important factors which influence R_s are

- the variation of ground stiffness with depth (which may be expressed in terms of the shear modulus G or Young's modulus E)
- the influence of a rigid (or a relatively much stiffer or stronger) layer at some depth below the toe level of the pile group.

The influence of these factors is illustrated in **Figures 55.9** and **55.10**. **Figure 55.9** compares the values of R_s for a pile group in a soil layer that has constant stiffness with depth ($\rho = 1$) against a group in a soil layer that has stiffness increasing linearly with depth ($\rho = 0.5$). In both cases the soil layer is infinitely deep. It can be seen that the degree of interaction (and the value of R_s) is much less when the stiffness increases linearly with depth (which occurs commonly in practice, in many geologies, due to the combined effects of increasing effective stress and reductions in weathering with depth). **Figure 55.10(a)** plots a correction factor, F_h, against the ratio

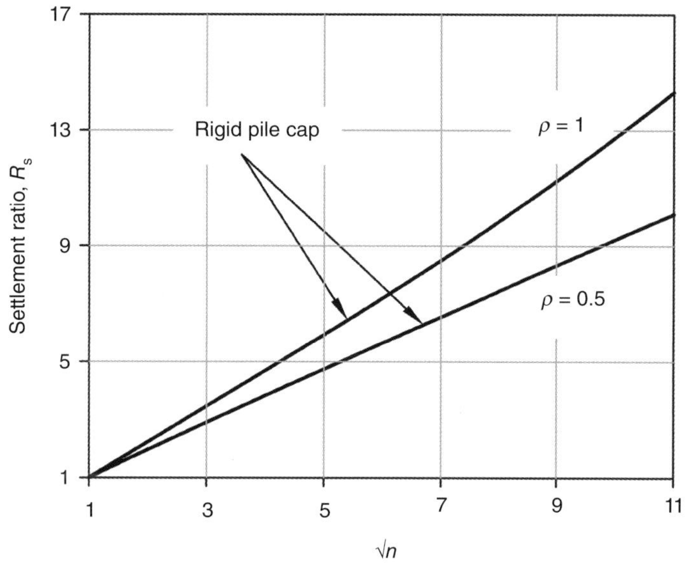

NB.
ρ = 1, Young's (or shear) modulus is constant with depth; ρ = 0.5, Young's (or shear) modulus increases linearly with depth; n = number of piles in group;
$R_s = W/W_s$, W = pile-group settlement, W_s = single-pile settlement (under same average load)

Figure 55.9 Influence of soil stiffness profile on pile-group settlement (Reproduced from Fleming *et al.*, 2009)

of compressible layer thickness, h, to the pile length, L. F_h is defined as

$$F_h = \frac{R_s \text{ for finite layer of depth } h}{R_s \text{ for infinitely deep layer}} \tag{55.8}$$

This effect becomes more significant for larger pile groups, with F_h values of 0.6–0.8 for a wide range of scenarios, reducing to 0.4 for some larger pile groups.

Figure 55.10(b) shows the potential difference in R_s value (indicated by the correction factor, F_b) for friction piles 'floating' in an infinitely deep homogeneous layer compared with end-bearing piles toed into a layer of varying stiffness, relative to the overlying strata (E_b/E_s). From **Figure 55.10(b)** it can be seen that R_s decreases (i.e. the pile-to-pile interaction decreases, and, similarly, the axial load distributions become more uniform) as the relative stiffness of the bearing layer increases. The effect is most significant for short stiff piles.

In practice, piles are often installed in geological sequences that exhibit marked variations in stiffness with depth, or 'layering'. Although this is a common situation, computer software is often not able to accurately model the influence of these layers on pile-group behaviour. **Figure 55.11** illustrates two different situations. First, **Figure 55.11(a)** shows the increase or decrease in pile-group settlement (for pile groups increasing in size up to 121 piles, pile spacing of 3d and pile length of 25 m) due to underlying layers of lower or greater stiffness, respectively. **Figure 55.11(b)** shows the increase in settlement due to an underlying layer of lower stiffness (for pile groups with up to 64 piles, pile spacing of 4d and pile length of 15 m). **Figure 55.11** shows that

- the effect of the underlying layer is more significant for larger pile groups
- for large pile groups, settlement may be increased by a factor of three or more when underlain by a more compressible layer
- if underlain by a relatively stiff layer, pile-group settlement may be more than halved.

Therefore, it is essential that ground investigations are deep enough to adequately characterise the ground beneath pile groups, and proper consideration is given to the effects of deeper layers on pile-group performance (noting that commercial software may not facilitate this, and some of the methods outlined in this chapter may need to be implemented). Axial load distributions across a pile group will be similarly affected: more compressible underlying layers will lead to more non-uniform axial loads across the group, whereas stiffer underlying layers will lead to more uniform axial loads across the group.

Pile-group analyses often assume that the soil behaviour is linear elastic and that pile installation has no effect on soil properties. Both of these assumptions, as noted in **Table 55.2**, lead to an overestimation of pile-to-pile interaction effects for vertical loading. **Figure 55.12** illustrates the influence of nonlinear elastic against linear elastic assumptions for soil behaviour; **Figure 55.12(a)** indicates that modelling the soil behaviour as linear elastic will tend to overpredict the soil settlement adjacent to a pile. **Figure 55.12(b)** plots contours of soil strain around a pile (derived from a sophisticated nonlinear soil model), which indicates that most of the soil adjacent to the pile experiences strains of less than 0.01%, whereas close to the pile there are small zones of soil experiencing larger strains. Hence, the mobilised soil stiffness close to the pile will be considerably lower than that mobilised further away from the pile, at any given depth below the ground surface. This effect will be exacerbated by the influence of pile installation. For either bored piles or driven piles there will be a zone of soil immediately adjacent to the pile that will be intensely disturbed and remoulded. **Figure 55.13** compares predictions from a range of different software with measured pile-group settlements (Poulos, 1989). **Figure 55.13(a)** indicates that although some software could accurately predict single-pile settlement, the group settlement was overpredicted – that is, interaction effects, or the equivalent R_s value, were overpredicted. **Figure 55.13(b)** plots R_s against the number of piles in the group. This indicates that conventional linear elastic analysis overpredicts R_s and this overprediction becomes worse as the pile-group size increases. Relatively good predictions are

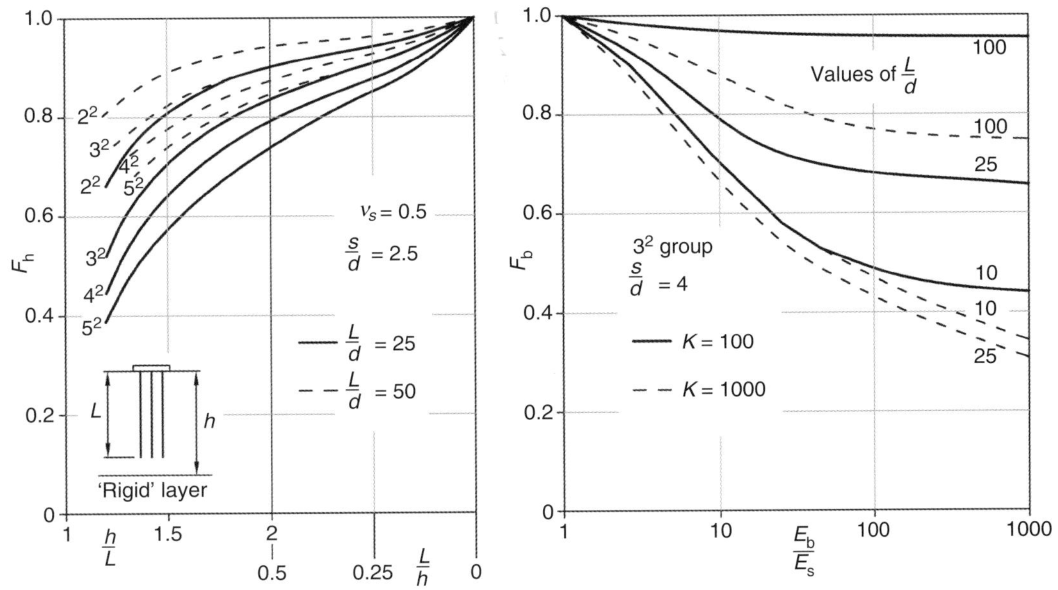

NB.
h = thickness of soil layer, L = pile length, d = pile diameter, s = pile spacing, 3^2 = 3x3 pile group

(a) Reduction coefficient F_h for effect of finite layer thickness, compared with infinite layer

NB.
K = relative pile/soil stiffness, $K = (E_p RA)/E_s$, E_p = pile Young's modulus, RA = area ration (= 1.0 for solid pile) = $A_p/(0.25\pi d^2)$; A_p = area of pile wall thickness

(b) Reduction coefficient F_b for effect of bearing stratum Young's modulus, E_b, to overlying soil modulus, E_s

Figure 55.10 Influence of finite layer thickness and bearing stratum stiffness on pile-group settlement (Reproduced from Poulos and Davis, 1980)

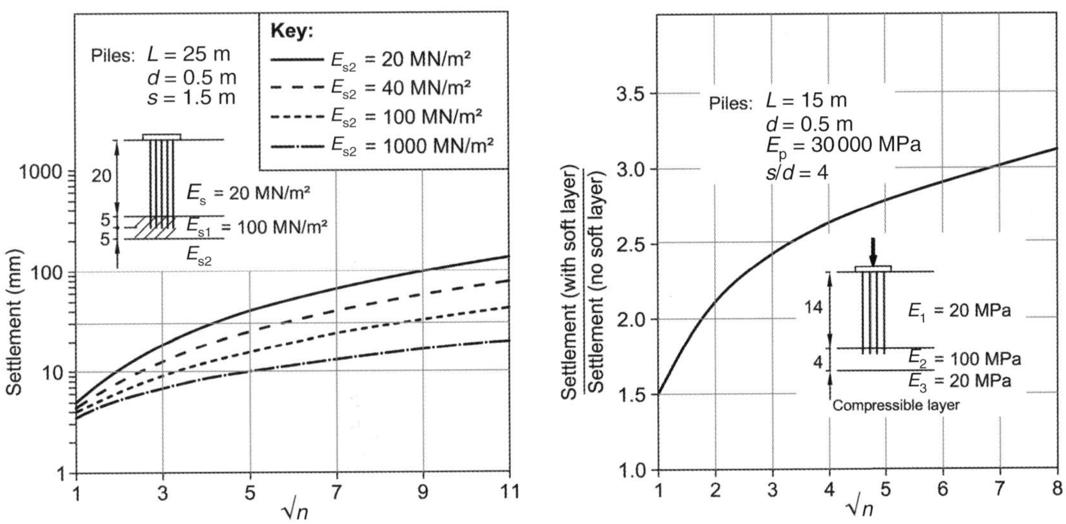

NB.
n = number of piles in group

(a) Effect of underlying compressible layer Young's Modulus, E_{s2}, on group settlement

(b) Effect of underlying compressible layer on group settlement

Figure 55.11 Influence of soil layering on pile-group settlement (Reproduced from (a) Poulos, 2005, with permission of ASCE; (b) Poulos *et al.*, 2001)

made for a modified analysis in which the soil between the piles was assumed to be stiffer than that immediately adjacent to the piles (**Figure 55.13(b)**) (O'Neill *et al.*, 1982); a Young's modulus of $750s_u$ was assumed for the thin zone of remoulded soil adjacent to the piles, and a Young's modulus of about $2500s_u$ for the soil beyond the remoulded zone (equivalent to the modulus at very small strain derived from cross-hole geophysics data).

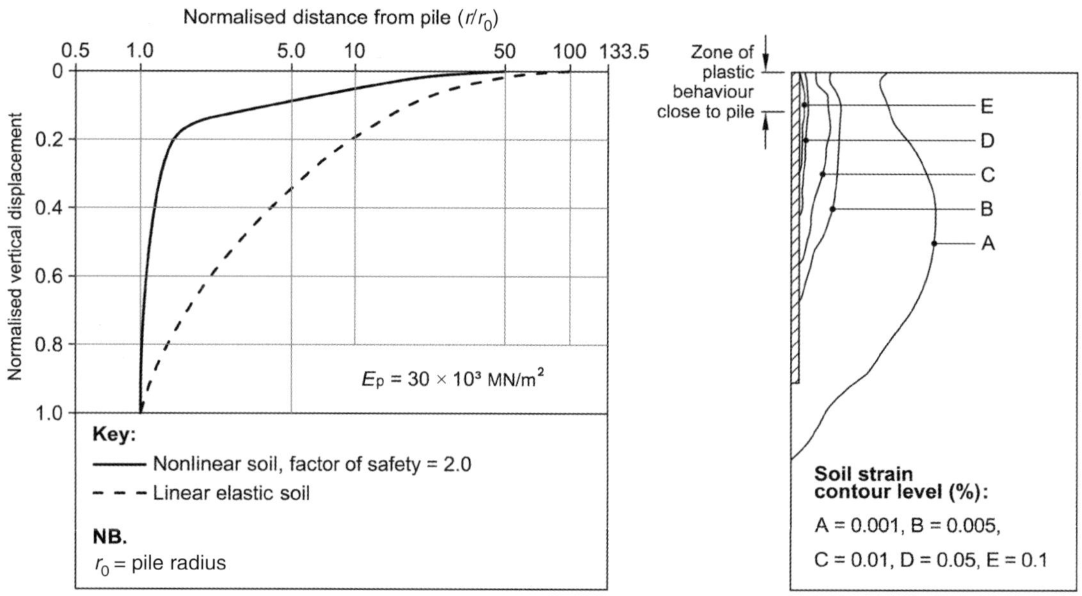

(a) Profiles of surface settlement adjacent to a pile 30 m long

(b) Contours of deviatoric strain around a pile 30 m long when factor of safety = 2.0

Figure 55.12 Influence of soil nonlinearity on pile-to-pile interaction (Reproduced from Jardine *et al.*, 1986)

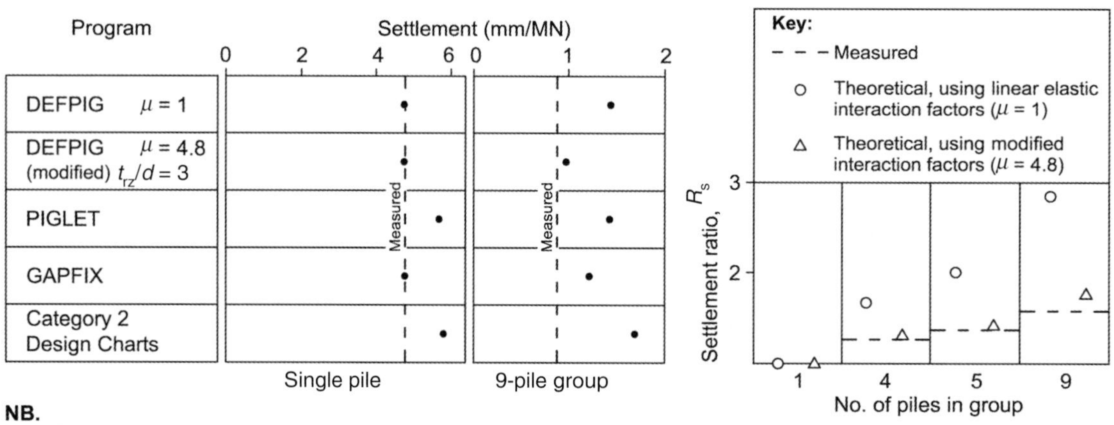

NB.

$\mu = \frac{E_{sm}}{E_{sr}}$; E_{sm} = Young's modulus of soil beyond t_{rz}; E_{sr} = Young's modulus of remoulded soil adjacent to pile; t_{rz} = thickness of remoulded zone; E_s varies linearly across t_{rz} between E_{sr} and E_{sm}; d = pile diameter.

(a) Effect of analysis method on single pile and group settlements

(b) Theoretical and measured group settlement behaviour (tests of O'Neill *et al.*, 1982)

Figure 55.13 Influence of analysis method on single-pile and pile-group settlement (Reproduced from Poulos, 1989)

For end-bearing piles toed into a relatively stiff layer, **Figure 55.14** shows that pile axial loads rapidly tend towards a more uniform distribution as the stiffness of the bearing stratum increases; with V/V_{av} less than 1.2 for E_b/E_s values in excess of 10, where V is the axial load on an individual pile and V_{av} is the total vertical load applied to the group divided by the number of piles. Measurements of axial loads in pile groups are rare; however, Mandolini *et al.* (2005) have collated the available data, shown in **Figure 55.15**. Typically, the ratio of corner to centre pile axial loads is between about 2.0 and 2.5 for common pile spacings of about 2.5–4 times the pile diameter. It is apparent that pile axial loads become more uniform with increasing pile spacing, and interaction effects become negligible for pile spacings of 6–8 times the pile diameter. **Table 55.3** gives measured axial load ratios (V_{max}/V_{min}) for a few specific case histories. At Stonebridge Park, given the large pile-group aspect ratio, a much larger value of V_{max}/V_{min} would have been expected. In this case, the pile cap was relatively flexible, which led to more uniform pile axial loads. For the Cargliano Bridge, the end-bearing piles led to a relatively low value of V_{max}/V_{min}. The pile-group test reported by Koizumi and Ito, with friction piles in a deep compressible layer and a rigid pile cap, led to relatively high values of V_{max}/V_{min}.

Figure 55.16(a) summarises the results of a study by O'Brien and Bown (2010), which plots the ratio of corner to centre pile axial loads against pile-group size. V_{max} is the maximum predicted axial load, which for square pile groups

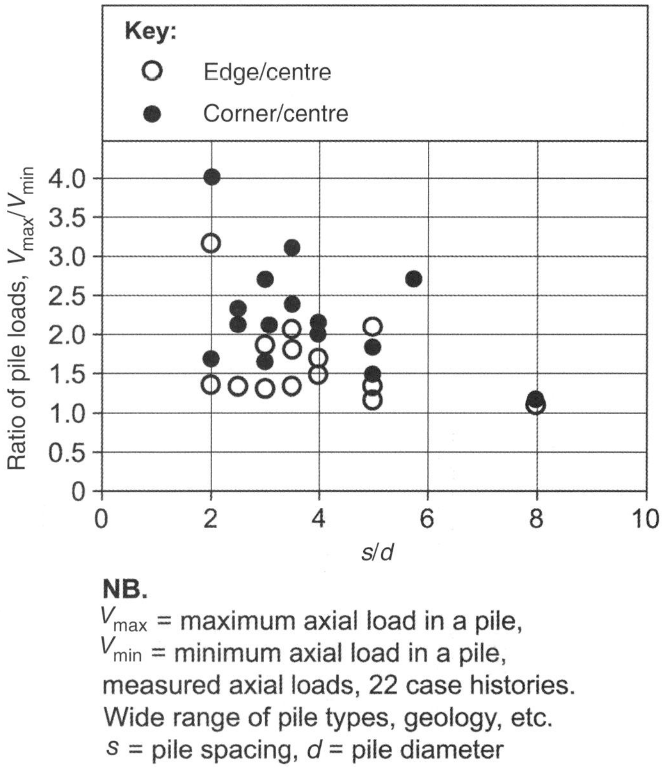

Figure 55.15 Measured axial load distributions, pile groups (Reproduced from Mandolini *et al.*, 2005)

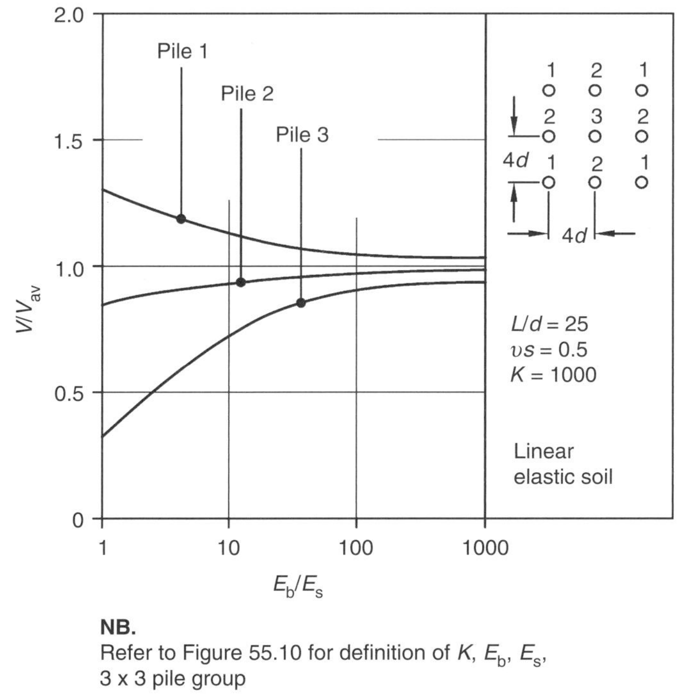

Figure 55.14 Influence of bearing stratum stiffness on axial load distribution (Reproduced from Poulos and Davis, 1980)

occurs in the corner piles. V_{min} is the minimum predicted axial load, which occurs in the piles close to the centre of the pile group. This shows that linear elastic analyses overestimate the load redistribution across the group when compared with nonlinear analyses and case history data, and that this discrepancy increases with increasing pile-group size.

For large groups, if pile axial loads are calculated by linear elastic methods then it is inevitable that corner piles will have an apparently low geotechnical 'factor of safety' even when the overall group factor of safety is two or more (**Figure 55.16(b)**). This can be a problematic issue when considering how to apply code requirements. This is discussed in Section 55.10.

55.4. Pile-to-pile interaction: horizontal loading

For pile groups subject to horizontal load the main design issue is usually the structural demands (bending moment, shear force, etc.) on the piles relative to their structural strength. Mandolini *et al.* (2005) summarise several studies on pile-group behaviour when horizontal loads are applied. Pile-to-pile interaction under a vertical load is fully developed at relatively small displacements, and group interaction effects tend to reduce at larger displacements (ground nonlinearity, including local yield, becomes more focused in a thin zone around the pile–soil interface). In contrast, under horizontal loading, group interaction effects amplify as the pile-group

Pile group	Ground profile	Pile type	Pile length: m	Pile diameter: m	Number of piles	Spacing: m	Pile-group aspect ratio	Axial load ratios		Pile cap stiffness
								C/I	E/I	
Stonebridge Park (Cooke *et al.*, 1981)	Deep layer of overconsolidated clay (London Clay)	Bored (friction)	13	0.45	351	1.61	6.6	2.2	1.7	Flexible
Cargliano Bridge (Mandolini *et al.* 2005)	Interbedded soft clays and silts over dense sand and gravel	Driven tubular (end-bearing)	48	0.38	144	1.14	1.1	1.6	1.3	Intermediate
Pile-group test (Koizumi and Ito, 1967)	Deep layer of overconsolidated clay	Driven closed-end (friction)	5.5	0.3	9	0.9	1.2	2.6	1.9	Rigid

C, corner pile axial load; *I*, internal pile axial load; *E*, edge pile axial load.

Table 55.3 Examples of measured axial load distributions

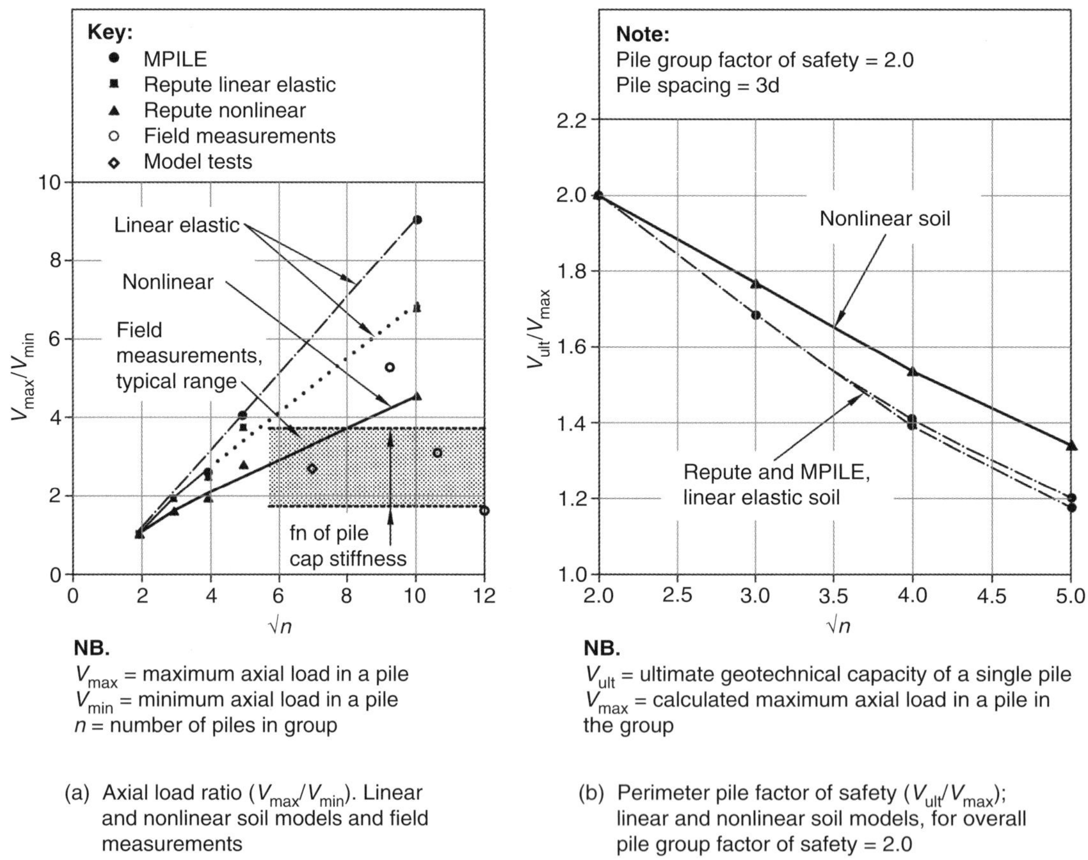

Figure 55.16 Axial load distribution for linear and nonlinear soil models against pile-group size (Reproduced from O'Brien and Bown, 2010)

displacement increases (and ground nonlinearity – e.g. zone of passive yield in front of a pile – increases as horizontal displacement increases). Hence, structural demands across a pile group tend to become more non-uniform. **Figure 55.17** provides some measured data from instrumented pile groups and centrifuge tests (Mandolini and Viggiani, 2005) for free head conditions. The ratio between the maximum bending moment in different piles in a group and in a single isolated pile is plotted against the pile-group displacement (the ratio is derived for the same average horizontal load per pile). The moments in the group piles are generally larger than the single pile, and the ratio increases with increasing displacement. Larger bending moments develop in the leading row (LR) piles compared to either the middle (MR) or trailing row (TR) piles (**Figure 55.17**); this phenomenon has been attributed to 'shielding' or 'shadowing' effects between adjacent piles in the same row (refer to **Figure 55.7**). This means that for horizontal loading, different group configurations (e.g. square, rectangular, circular) will exhibit different group interaction effects even if the number of piles and average spacing are the same (this contrasts with group interaction effects under a vertical load, which are practically unaffected by the group configuration for a given number of piles and similar average spacing). Hence, for piles under horizontal loading it is potentially unsafe to assess the bending moments in piles within a group from an assessment of single-pile behaviour; in addition the assumption of linear elastic behaviour is generally inappropriate (refer to **Figure 55.31**). The use of *p–y* curves is well established for predicting the behaviour of single piles under horizontal loads. However, the use of *p–y* multipliers to account for group interaction effects is of more doubtful validity because of the large number of complicating factors that can affect real pile-group behaviour under horizontal loading.

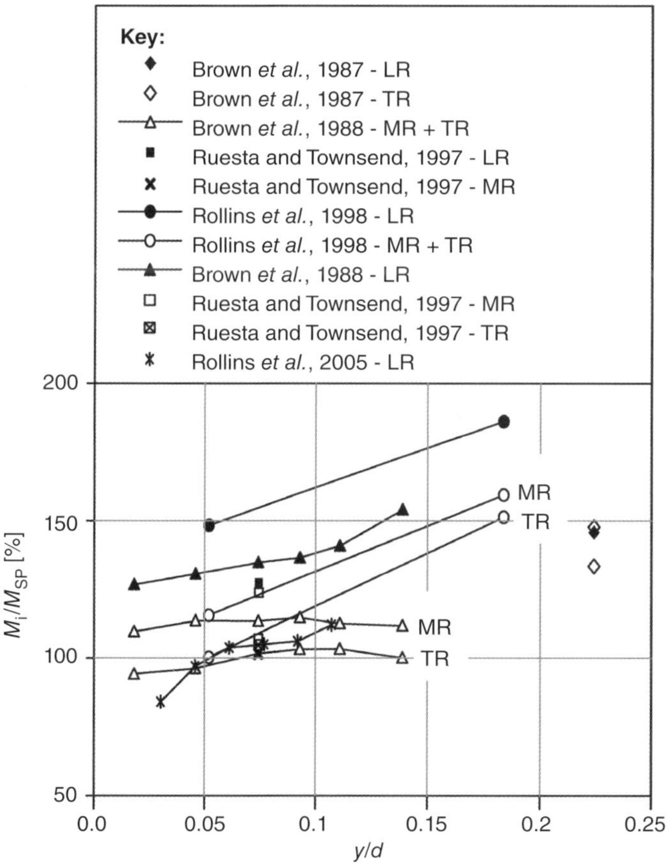

NB.
M_i = maximum bending moment in pile in a group; M_{sp} = maximum bending moment in a single isolated pile; LR = piles in a leading row MR = piles in middle of a row; TR = trailing piles

Figure 55.17 Ratio of bending moments for piles in a group and single isolated pile against displacement (Reproduced from Mandolini *et al.*, 2005)

55.5. Simplified methods of analysis

55.5.1 General comments

There are four simplified methods that may be used to assess pile-group settlement

- pile-group settlement ratio
- elastic interaction factors
- equivalent pier
- equivalent raft.

Each method has distinctive advantages and disadvantages (summarised in **Table 55.4**). Simplified analysis methods should always be used to check computer-based methods, which may have serious limitations in certain situations.

There are no simplified methods for checking axial load distributions across a group. However, **Figures 55.15** and **55.16** indicate the ranges of V_{max}/V_{min} that should be reasonably anticipated for friction piles (noting that linear elastic methods can lead to gross overestimates of axial loads in perimeter/corner piles; **Figure 55.16**). **Figure 55.14**, for piles toed into stiffer strata, indicates that V_{max}/V_{min} is close to 1. **Tables 55.2** and **55.3** indicate the factors that may influence axial load distribution.

It should be noted that the use of structural analysis software (which typically simulate pile behaviour as independent elastic springs) is not recommended, since pile–pile interaction is not modelled. Poulos and Davis (1980) give examples of some of the errors that can be produced.

55.5.2 Pile-group settlement ratio, R_{Se} (empirical correlations)

Mandolini *et al.* (2005) proposed an empirical correlation between the pile-group settlement ratio R_{Se} and the pile-group aspect ratio, based on an analysis of 63 case histories (for pile groups in varying geologies, and with varying pile length, type, diameter, etc.), which is shown in **Figure 55.18**. Equations have been developed for the upper-bound, best-estimate and lower-bound settlements.

$$\text{Best estimate } R_{Se} = \frac{a}{(R)^b} = \frac{0.29}{(R)^{1.35}}(n) \qquad (55.9)$$

where n is the number of piles in the group, the pile-group aspect ratio $R = (ns/L)^{0.5}$ and R_{Se} is the empirical pile-group settlement ratio.

$$\text{Pile-group settlement, } W = R_{Se} W_s \qquad (55.10)$$

Simplified method	Advantages	Disadvantages
Empirical settlement ratio	Very quick and simple. Best suited for friction piles, in deposits with increasing strength and stiffness with depth Useful as a 'sense' check of more complex methods	The influence of specific geological features, pile or ground properties cannot be assessed Potentially unsafe if weak or compressible strata underlie the bearing stratum. Potentially over-conservative if a relatively stiff layer underlies the bearing stratum
Elastic interaction factors	Influence of varying pile length, diameter and spacing can be quickly checked Best suited for friction piles in deposits with increasing strength and stiffness with depth	Cannot directly check influence of underlying strong or weak layers (refer to **Figures 55.10** and **55.11**) Care needed in amplifying single-pile settlement for large groups, **use initial tangent pile stiffness**, rather than secant; otherwise likely to be over-conservative
Equivalent pier	Well suited for pile groups with relatively small aspect ratio, $R < 3.0$ If using elastic solutions for single pile, then is quick to use Flexible, method can also be used within sophisticated numerical models, in axisymmetric mode	Not appropriate for large pile-group aspect ratios, $R > 3.0$ Inappropriate if pile lengths in group vary significantly
Equivalent raft	Most appropriate simplified method for checking influence of strong or weak layers at depth	Over-conservative for pile groups with a small aspect ratio, $R < 3.0$ Significant judgement needed to assess appropriate equivalent raft level and dimensions

Table 55.4 Pile-group settlement: advantages and disadvantages of simplified methods

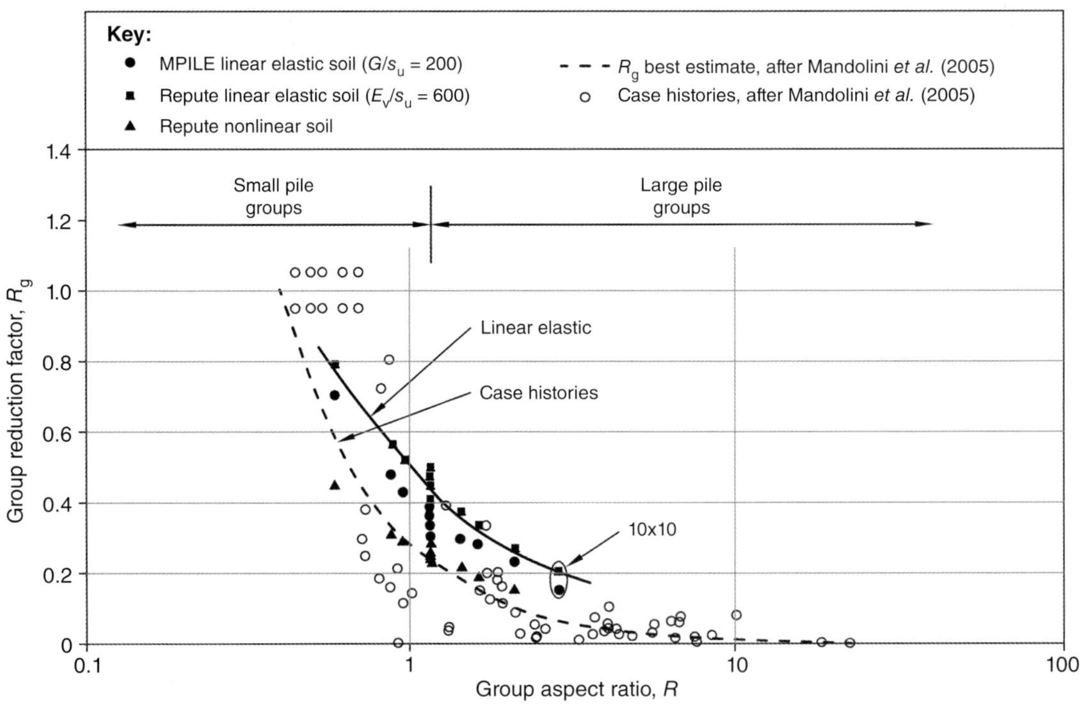

NB.
Settlement ratio, $R_s = nR_g$, n = number of piles in group, R_g = group reduction factor. Group aspect ratio, $R = (ns/L)^{0.5}$, s = pile spacing, L = pile length. G = soil shear modulus, E_v = Vertical Young's modulus of soil, s_u = undrained shear strength

Figure 55.18 Group reduction factor R_g against group aspect ratio R; field measurements and parametric study (Data taken from Mandolini *et al.*, 2005, and O'Brien and Bown, 2010)

where W_s is the single-pile settlement under the average working load of piles within the group (Q/n) and Q is the total vertical load applied to the pile group.

The single-pile settlement W_s can be calculated by using either relevant pile test data or Fleming's method for single-pile settlement (Fleming, 1992). The latter has been found to be reliable for a wide range of pile types and ground conditions (refer to Chapter 54 *Single piles*). The upper- and lower-bound values of R_{Se} are

$$\text{Upper-bound}\, R_{Se} = \frac{0.5}{R}\left[1 + \frac{1}{3R}\right]n \qquad (55.11)$$

$$\text{Lower-bound}\, R_{Se} = \frac{0.17}{(R)^{1.35}}(n) \qquad (55.12)$$

Hence, application of the above equations enables a rapid estimate to be made of the likely range of pile-group settlement.

Experience indicates that Equation 55.9 is generally appropriate for friction piles with soils with strength and stiffness increasing gradually with depth; Equation 55.11 applies if relatively deep and uniform soils underlie the pile group; and Equation 55.12 applies if the soil or rock beneath the group exhibits a rapid increase in strength and stiffness close to the toe level of the pile group.

55.5.3 Elastic interaction factors (design charts)

Elastic interaction factors have been widely published (for example Poulos and Davis, 1980). However, many of the early solutions relied on unrealistic assumptions (e.g. an infinite elastic medium with constant Young's modulus, rigid pile, etc.) and overestimated group interaction effects. Randolph (1994) has outlined a simple and practical approach to improve the reliability of elastic interaction factors, which is shown in **Figure 55.19**. The initial stiffness derived from the single-pile load–settlement response is given by the tangent line OA (and will be a function of the small strain stiffness of the soil). Plastic strain, which will develop at higher load levels, leads to the offset AB. This plastic strain will develop locally adjacent to the pile shaft. Group interaction effects under vertical load are predominantly due to elastic, rather than plastic, strains. Therefore, the pile-group settlement ratio (derived from linear elastic theory) R_{St} is only applied to the elastic component of the single-pile settlement. Hence, the elastic pile-group settlement is shown as OC in **Figure 55.19** where OC = R_{St}OA and the total pile-group settlement OD = OC + CD (with CD = AB). For most conventionally designed pile groups, which will have overall factors of safety well in excess of 2.0, the plastic component of settlement will be relatively minor. Back-analysis of a wide range of case histories by Mandolini and Viggiani (1997) confirms that Randolph's approach is usually far more realistic than applying the group settlement ratio R_{St} to the single-pile secant stiffness at the average working load of the piles within the group.

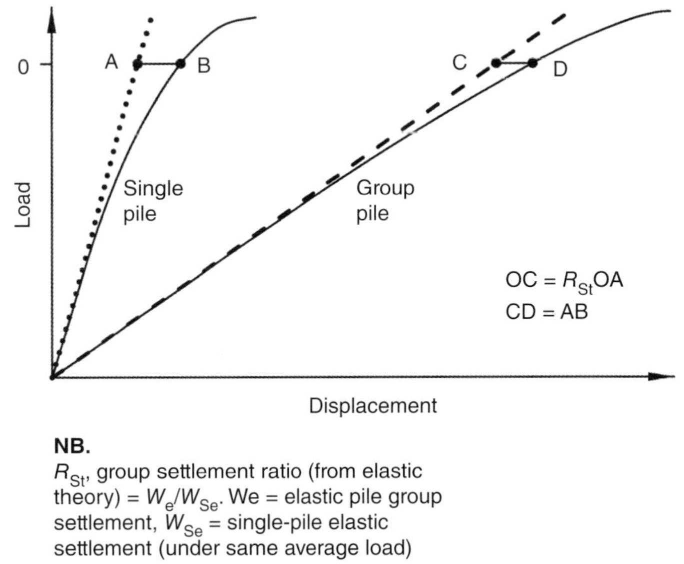

Figure 55.19 Applying elastic theory to pile interaction and nonlinear load–settlement behaviour for vertical loading (Reproduced from Randolph, 1994)

To a reasonable approximation, the group settlement ratio is (Fleming *et al.*, 2009)

$$R_{St} = n^e \qquad (55.13)$$

and

$$W = R_{St}W_{Se} + W_{Sp} \qquad (55.14)$$

where W is the pile-group settlement, R_{St} is the group settlement ratio derived from linear elastic theory, W_{Se} is the elastic settlement under the average working load of piles within the group (Q/n), n is the number of piles in the group, Q is the total vertical load applied to the group and W_{Sp} is the plastic settlement of a pile at Q/n (estimated from a pile test or from a Fleming (1992) analysis for single-pile settlement at the average load for a pile in the group).

For piles working mainly in shaft friction, e is typically between 0.3 and 0.5. Based on Randolph (1994), **Figure 55.20** gives a set of design charts; the exponent e is given by a base value, e_1 (depending on the pile slenderness ratio L/d) and four correction factors, c_1–c_4

$$e = e_1c_1c_2c_3c_4 \qquad (55.15)$$

where c_1 is a function of the pile stiffness to the ground stiffness, E_p/G_b, and E_p is Young's modulus for the pile. G_b is the ground shear modulus at the pile-toe level, c_2 is a function of the pile spacing, s, to the pile diameter d; c_3 is a function of the rate of increase of the ground shear modulus with depth,

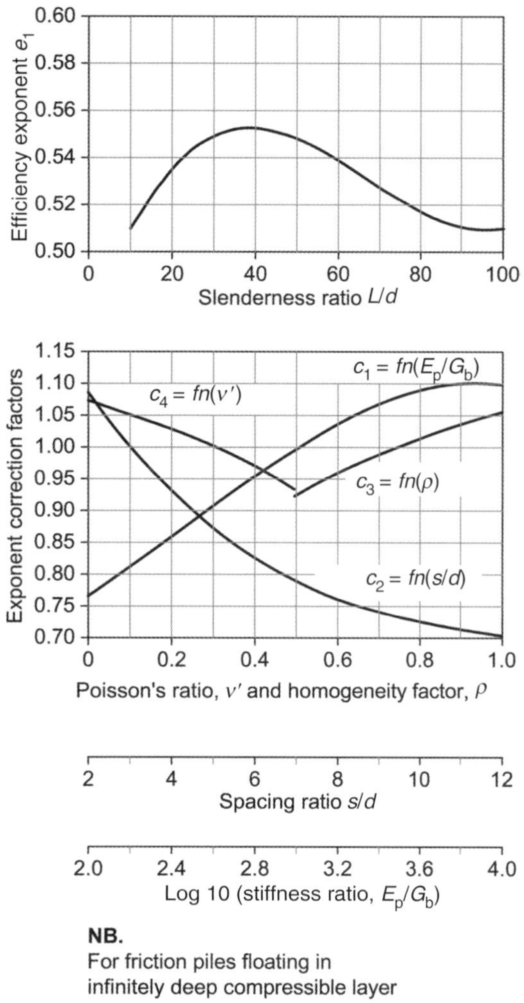

Figure 55.20 Design charts (linear elastic soil) for pile-group settlement ratio, R_{St} (Reproduced from Fleming *et al.*, 2009)

$\rho = G_{av}/G_b$; G_{av} is the ground shear modulus halfway down the pile; and c_4 is a function of Poisson's ratio, ν'. The value of e_1 is based on $E_p/G_b = 1000$, $s/d = 3$, $\rho = 0.75$ and $\nu' = 0.3$. These charts assume an infinitely deep elastic layer. Reference to **Figures 55.10** and **55.11** allows judgements to be made about the influence of a finite layer thickness and stiffer or weaker layers below the pile toe. Poulos (1989) indicates that for friction piles at typical working loads (factor of safety in excess of 2.0), plastic deformation would be expected to only increase single-pile settlement by 10–25% compared with the elastic response derived from the initial pile stiffness.

55.5.4 Equivalent pier and equivalent raft

Equivalent pier: the equivalent pier approach replaces the pile group by a buried 'cylinder' of appropriate dimensions and stiffness (**Figure 55.21**). Solutions for calculating the single-pile settlement can then be used to calculate the settlement of the equivalent pier (provided due account is taken of the compressibility of the cylinder, the stiffness of the soils below the base and the shear stiffness of the interface around the sides of the cylinder). Alternatively, the settlement of the equivalent pier can be simulated within an axisymmetric numerical model.

For a pile-group of area A_g, the diameter of the equivalent pier d_{eq} is

$$d_{eq} = \left(\frac{4}{\pi} A_g\right)^{0.5} = 1.13(A_g)^{0.5} \quad (55.16)$$

The equivalent Young's modulus E_{eq} of the pier is

$$E_{eq} = E_{s,av} + (E_p - E_{s,av})(A_p/A_g) \quad (55.17)$$

where E_p is Young's modulus of the pile, $E_{s,av}$ is the average Young's modulus of the soil penetrated by the piles, A_p is the total cross-section area of the piles in the group and A_g is the total plan area encompassed by the pile group.

The elastic solutions for settlement of a single pile presented by Randolph (1994), which allow for the relatively low aspect ratio of equivalent piers (for typical pile-group geometries) can be conveniently programmed in modern spreadsheets to facilitate rapid analyses of pile-group settlement. Alternatively, the nonlinear approach proposed by Fleming (1992) can be used (refer to Chapter 54 *Single piles*), although it should be noted that the strength and stiffness properties for the base and shaft resistance around the equivalent pier should mainly be representative of the undisturbed soil mass rather than the disturbed remoulded soil immediately adjacent to an individual pile.

The influence of an underlying compressible layer on the load–settlement behaviour of the equivalent pier can be assessed (based on studies by Poulos, 2005) from **Figure 55.22**. **Figure 55.22** provides a plot of the base stiffness reduction factor as a function of h_c/d_{eq} and E_2 and E_1 (where E_2 and E_1 are the Young's moduli of the upper and lower layers, respectively) – refer to **Figure 55.6**. The base stiffness reduction factor is defined as the ratio of the base stiffness with the underlying layer E_2 to the base stiffness of E_1 only (i.e. with no underlying layer present).

Equivalent raft: the equivalent raft method has been used extensively for estimating pile-group settlement and is described in many textbooks. The pile group is replaced by a raft foundation acting at a representative depth below the surface with some equivalent dimensions. Many variants are available; however, the approach by Tomlinson (1987) is relatively simple and widely used and is shown in **Figure 55.23**. As noted by Poulos *et al.* (2001), this method relies on considerable engineering judgement to assess the representative depth and appropriate raft dimensions, which depend on an understanding of the relevant load transfer mechanism for the pile group under consideration. Once the equivalent raft is established, then the settlement can be computed using the analysis methods used for shallow foundations.

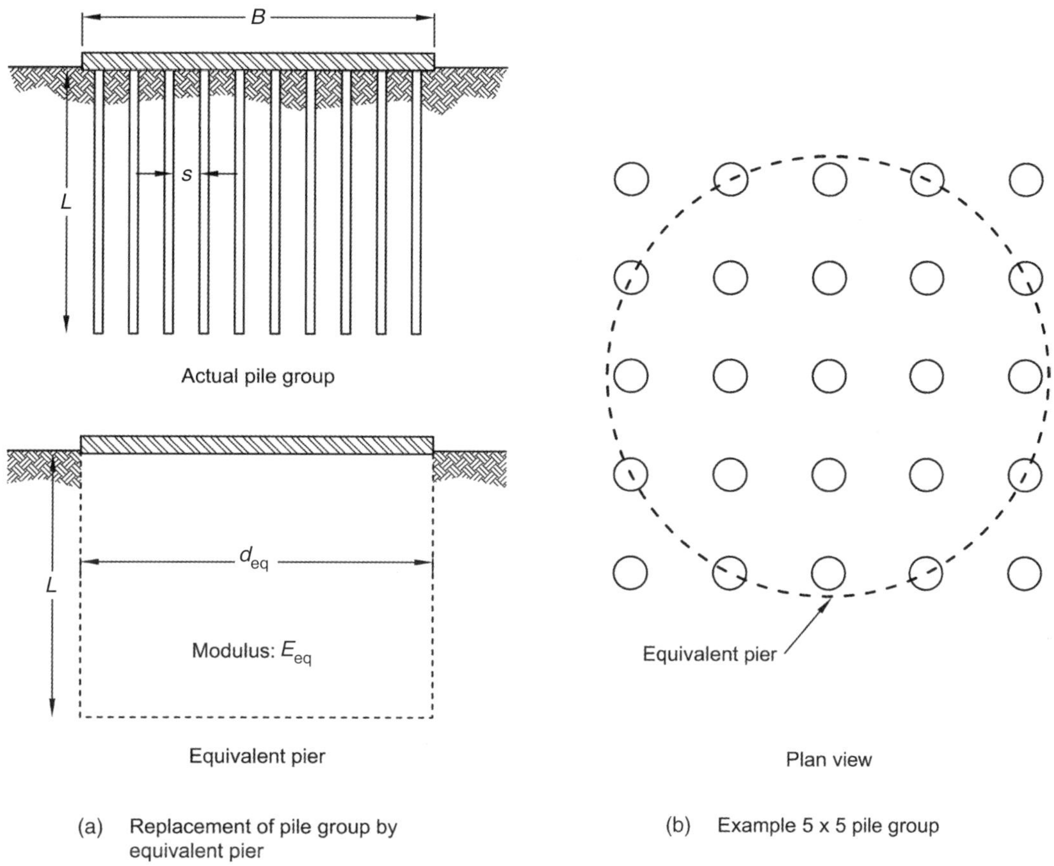

Figure 55.21 Replacement of pile group by an equivalent pier

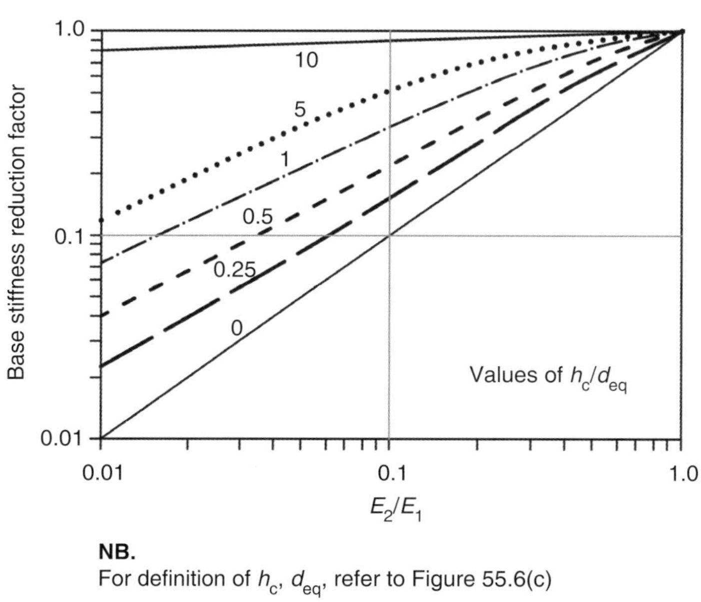

Figure 55.22 Base stiffness reduction factor for pile group (equivalent pier) underlain by compressible layers (Reproduced from Poulos, 2005, with permission of ASCE)

The average pile-group settlement W_{av} is then

$$W_{av} = W_{raft} + W_{ec} \tag{55.18}$$

where W_{ec} is the elastic compression of the piles above the equivalent raft level (i.e. acting as free-standing columns). Assuming elastic soil behaviour

$$W_{raft} = F_{av} F_D q \sum_{i=1}^{n} \left(\frac{I_E}{E_{sec}} \right) h_i \tag{55.19}$$

where W_{raft} is the equivalent raft settlement, q is the average pressure applied to the raft, I_E is an influence factor (refer to **Figure 55.24**) for the ith layer, h_i and E_{sec} are the layer thickness and secant Young's modulus, respectively, for the ith layer, F_D is a depth correction factor (as derived by Fox, refer to **Figure 55.25**) and F_{av} is equal to 0.8 (to correct from centre line settlement to average raft settlement). The embedment depth of the raft should be assumed to be the depth below the competent bearing stratum, rather than the ground surface (**Figure 55.23**).

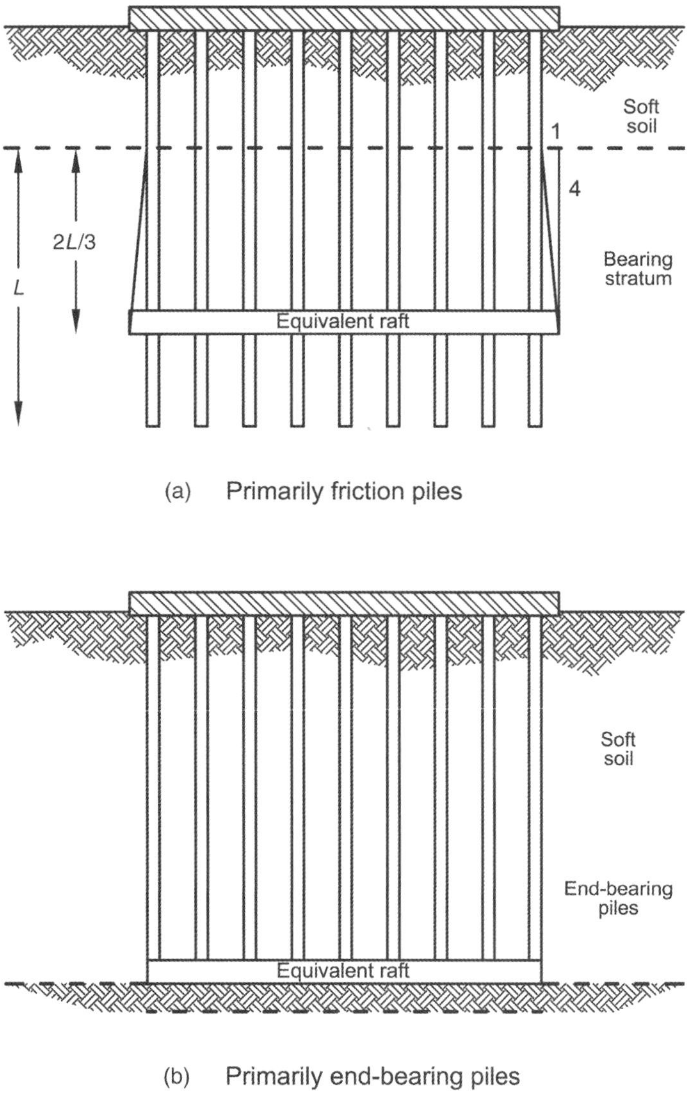

Figure 55.23 Replacement of pile group by an equivalent raft

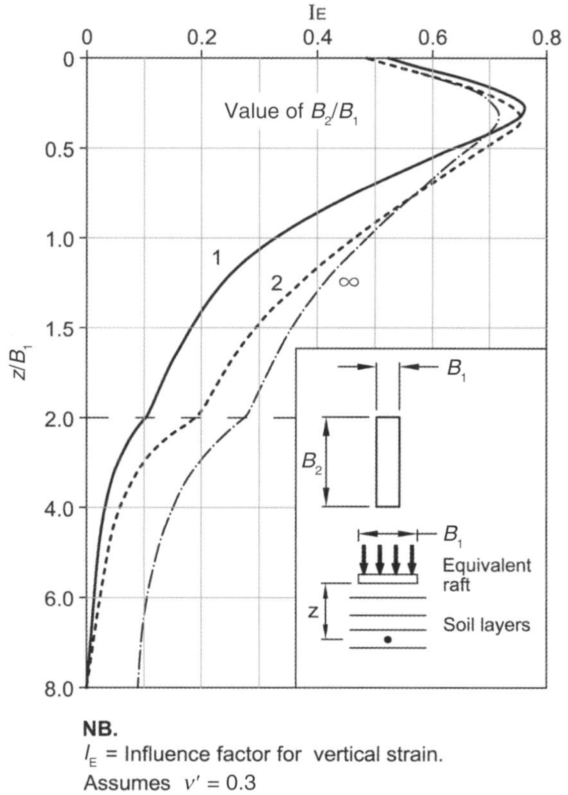

Figure 55.24 Equivalent raft settlement using elastic theory; strain influence factor against depth (Data taken from Randolph, 1994)

When using the equivalent raft analogy for calculating pile-group settlement, care is required to ensure input parameters and the calculation method are relevant; for example, using parameters derived from observations of shallow foundation settlement is likely to be inappropriate. In general, the use of elastic methods, such as Equation 55.19 (e.g. Poulos, 1993), is likely to be suitable provided the strata at, and below, the pile-toe level are competent (i.e. stiff heavily overconsolidated clays, dense sands, etc.). Elastic analyses will *not* be appropriate if a layer of normally or lightly overconsolidated clay is present at depth below the pile-toe level, *and* the pile-group bearing pressures below pile-toe level exceed the clay's pre-consolidation pressure.

The main advantage of the equivalent raft method is that it enables the influence of softer or stiffer soil layers below the toe level of a pile group to be assessed in a simple and transparent way. This is particularly important in situations where softer soil layers may exist at some level below the pile group; as noted above, some commercially available software cannot analyse the situation reliably and **a separate check is essential**.

55.5.5 Which is appropriate, equivalent raft or equivalent pier?

If the piles are mainly dependent on shaft friction and the pile group is relatively deep and narrow, the resistance mobilised around the perimeter of the equivalent pier is likely to substantially exceed the applied working load (i.e. SLS load). Hence, the load reaching the base of the pile group is likely to be a small proportion of the applied load, which will be mainly dissipated in shear around the pile-group perimeter. For this situation, an equivalent pier will be appropriate. In contrast, if the piles are mainly end-bearing or the pile group is relatively wide, then a substantial proportion of the applied working load will need to be resisted at the base of the pile group. In this case the equivalent raft will be an appropriate analogue for analysis purposes.

The choice between using either an equivalent raft or equivalent pier is mainly dependent on the overall pile-group aspect ratio or normalised width: **Figure 55.26** illustrates this

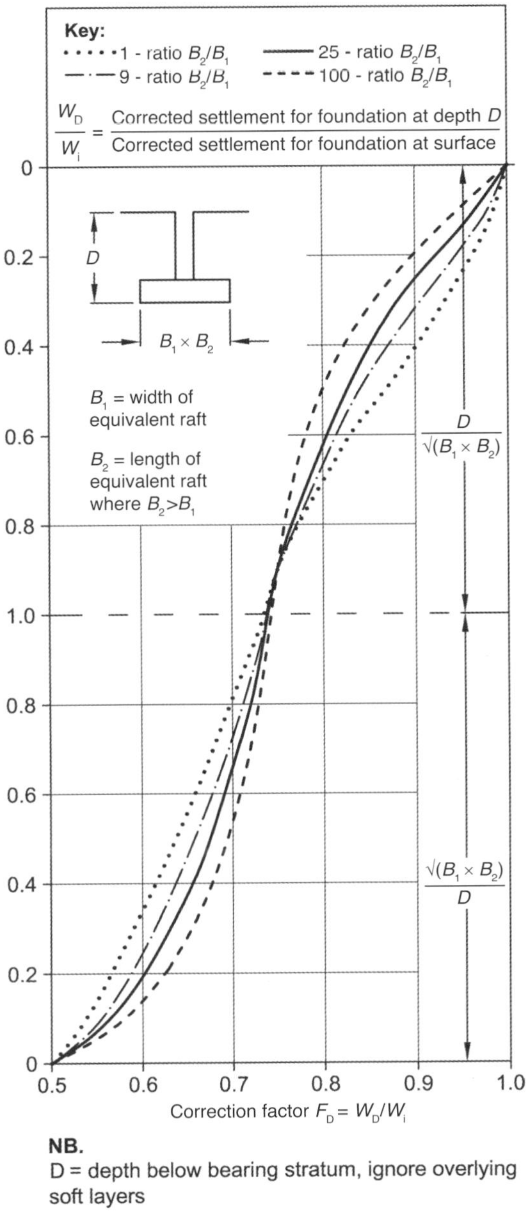

Figure 55.25 Equivalent raft settlement using elastic theory; Fox's depth correction factor (Data taken from Randolph, 1994)

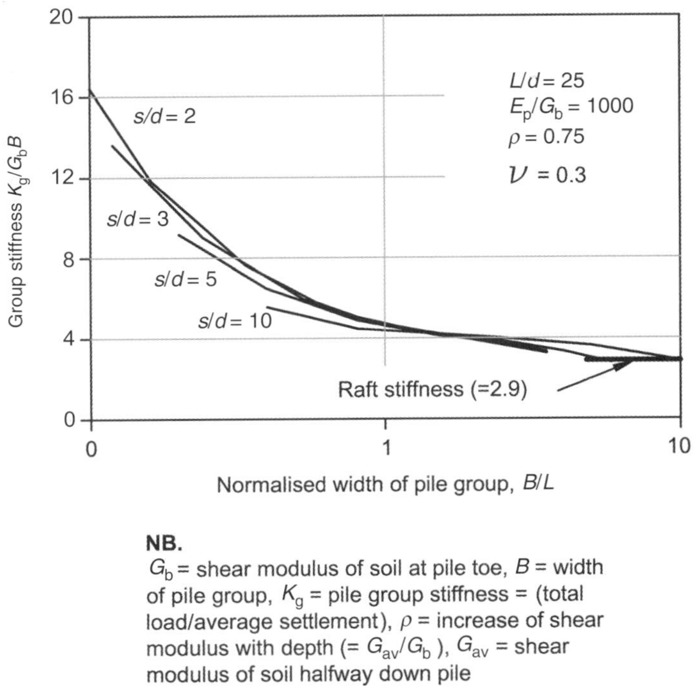

Figure 55.26 Normalised stiffness of pile groups and equivalent raft against normalised width (Reproduced from Randolph, 2003)

quite clearly. When the normalised pile-group width (B/L; B is the group width and L is the pile length) is less than 1.0, the normalised pile-group stiffness is much higher than that for the equivalent raft. Hence, the equivalent raft approach would tend to be over-conservative and the equivalent pier should be used. Once the normalised width exceeds 1.0, then the pile-group stiffness tends towards that of an equivalent raft. Randolph (1994) has suggested that, for friction piles, if the pile-group aspect ratio exceeds about 3 or 4, then an equivalent raft is appropriate, whereas for lower values of R an equivalent pier is more appropriate (and then the accuracy of the equivalent pier is likely to be within 20% of more rigorous solutions (Randolph, 2003), which is perfectly satisfactory for practical design purposes).

55.6. Pile cap analysis and differential settlement

Pile-group analysis commonly assumes that the pile cap is perfectly rigid or, occasionally, that it is perfectly flexible. Most pile-group analysis software cannot carry out analyses with a pile cap of intermediate stiffness. While this may be a reasonable assumption for small groups of piles, it becomes increasingly inaccurate as the size of the group increases. The flexibility of the pile cap has a significant influence on the computed load distribution to individual piles.

Basile (2020) states that for small to medium-sized groups (up to 25 piles), the assumption of a rigid pile cap is generally reasonable for practical purposes. However, this assumption becomes increasingly inaccurate for larger pile groups. Pile cap stiffness can be assessed by calculating the raft–soil stiffness ratio (K_{rs} – see Section 53.6.7 and **Figure 55.27**). Basile (2020) provides further guidance on selecting an appropriate value for ground stiffness and dealing with layered soils. Limits for K_{rs} equal to 0.1 and 1.5 are suggested for design purposes for which pile caps may be considered as 'practically flexible' and 'practically rigid', respectively (**Figure 55.27**). Axial loads for pile groups with a 'practically flexible' pile cap can be assumed to be uniform (assuming a uniform applied load).

The decision to use pile groups is often driven by the wish to reduce settlement, in particular to reduce differential settlement, in order to reduce the risk of superstructure damage

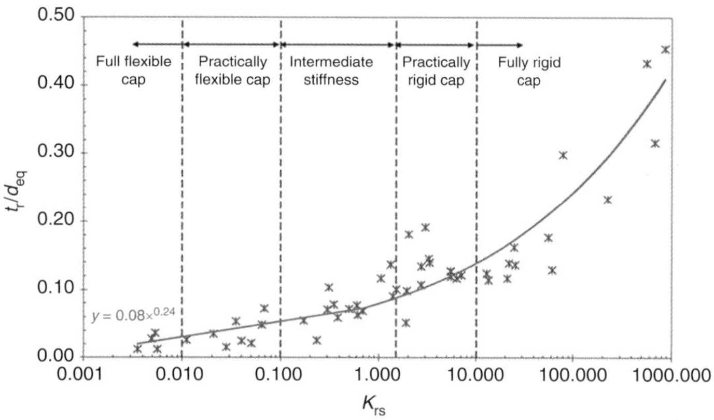

Figure 55.27 Pile cap–soil stiffness ratio K_{rs} and relationship with pile cap thickness t_r/equivalent diameter d_{eq} (After Basile 2020)

(refer to Chapter 52 *Foundation types and conceptual design principles*). Hence, a check on likely differential settlement will be important, especially for large pile groups or sensitive structures.

Pile-group differential settlement depends on the pile-group aspect ratio and on the pile cap stiffness. Randolph and Clancy (1993) have shown that the normalised differential settlement of pile groups, with a fully flexible pile cap, is mainly a function of the pile-group aspect ratio R. Normalised differential settlement is reasonably independent of the precise number of piles, spacing and so on. Hence, for a flexible pile cap

$$\Delta W_{flex} = f(R/4)W_{av} \quad (\text{for } R \leq 4) \tag{55.20}$$

$$\Delta W_{flex} = fW_{av} \quad (\text{for } R > 4) \tag{55.21}$$

with $f = 0.3$ for differential settlement between the centre and mid-side, $f = 0.5$ for centre to corner; ΔW_{flex} is the differential settlement across a flexible pile cap, R is the pile-group aspect ratio and W_{av} is the average pile-group settlement. The pile-group differential settlement ΔW is

$$\Delta W = F_R \Delta W_{flex} \tag{55.22}$$

where ΔW_{flex} is derived from Equations 55.20 or 55.21, and F_R is a correction factor dependent on the pile cap rigidity (varying from 1.0 for fully flexible to 0 for fully rigid), which can be assessed from solutions given in Chapter 53 *Shallow foundations* (Section 53.6.7).

The pile cap rigidity also significantly influences the bending moments induced in the pile cap, especially for medium and large pile groups, as indicated in **Figure 55.28** (for a 5 × 5 pile group). The type and location of applied loads onto the pile cap also affect the induced bending moment (comparing **Figures 55.28(a)** for a uniformly distributed load with **55.28(b)** for a load applied at the centre of the pile cap). Viggiani *et al.* (2012) have shown that simple analytical methods and the 'strut and tie' (STM) analogy are adequate for structural design of pile caps for 'small' pile groups ($n < 5$ piles); however, these methods will usually be inadequate for 'large' pile groups. For large pile groups, more sophisticated analysis (e.g. a 'shell and beam model') of interactions between the pile cap, piles and the ground may facilitate more economic design. Significant economies may be possible if large pile groups are designed as piled rafts (refer to Chapter 56 *Rafts and piled rafts*), with the ground resistance below the pile cap (or raft) taken into account.

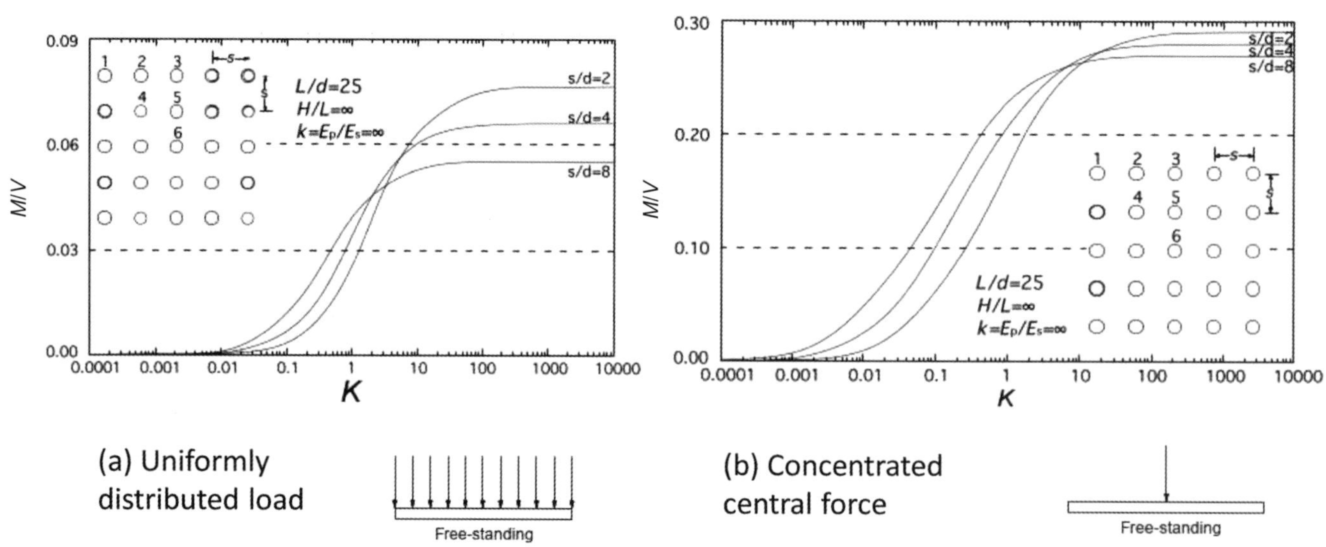

NB: where M is the maximum bending moments at the centre of the raft (bending moments are positive when the lower edge of the raft is subject to tension), V is the total vertical load applied, K is relative raft–soil stiffness, E_p is the Young's modulus of the pile, E_s is the Young's modulus of the soil, H is the depth to rigid layer, L is the pile length, d is the pile diameter and s is the pile spacing.

Figure 55.28 Pile cap bending moments as a function of pile cap rigidity (Reproduced from Viggiani *et al.*, 2012)

55.7. Time-dependent settlement

Time-dependent settlement of pile groups under foundation loading can occur due to

- primary consolidation, due to dissipation of excess pore pressure, in clays and silts (Hooper, 1979; Katzenbach *et al.*, 2000)
- settlement due to creep, in sands, gravels and weak rock (Mandolini and Viggiani, 1997).

The pile-group aspect ratio and load transfer mechanism are important factors in controlling the amount of time-dependent settlement. For small aspect ratios when load transfer is primarily due to shear, the time-dependent settlement at working load would be expected to be small. Studies by Poulos (1993) indicate that consolidation settlement would be less than about 15% of total settlement. In contrast, compression of the ground below the pile-group base may become increasingly important when the group aspect ratio increases to values in excess of 3 (or the B/L value exceeds 1). Mandolini *et al.* (2005) compared observations of 'end of construction' settlement with observations of time-dependent movements for rafts and piled rafts in Frankfurt and London Clay (both stiff overconsolidated plastic clays). End of construction settlement (which probably includes some primary consolidation settlement) varied between 50% and 75% of the total, with the ratio increasing as B/L reduced (from 2.5 to 1.0). Poulos *et al.* (2001) suggest that creep settlement is unlikely to be significant at working loads (when the group factor of safety exceeds 1.4).

A separate issue that can cause relatively large time-dependent settlement is when large groups of full displacement (closed-end) piles are driven at close spacings into soft clays, silts or peats. Significant heave and lateral displacement can occur during pile driving. The shearing and remoulding of soft soils can generate substantial excess pore pressures, which will lead to large settlement of the heaved clay as the pore pressures dissipate. If the pile group is end-bearing, then negative skin friction forces will result; if it is acting in friction, then large pile-group settlements can develop (Adams and Hanna, 1970; Bjerrum, 1967; Brzezinski *et al.*, 1973). This will be independent of, and additional to, any settlement due to foundation or structure loading.

55.8. Optimising the pile-group configuration

Once the decision has been made to use pile groups, some thought will be needed on how to optimise the layout of the pile group in terms of the length, diameter and spacing of the piles. This will be dependent on a number of factors.

- *A shallow foundation would have adequate bearing capacity, but is it likely to suffer excessive settlement?* If a shallow foundation such as a raft has adequate bearing capacity and the requirement is to control settlement, differential settlement or raft bending moment, then a piled raft is likely to be most cost-effective; refer to Chapter 56 *Rafts and piled rafts*. A piled raft would usually comprise a small number of widely spaced piles, normally located beneath the superstructure columns.
- *A shallow foundation would have inadequate bearing capacity?* A conventional pile design approach should be used. A single pile may be adequate (Chapter 54 *Single piles*), although for heavily loaded foundations a deep large-diameter pile may be needed. This may occasionally be advantageous – for example, landing an underbridge to a railway during a short blockage, for designing out pile caps in a floodplain or for some offshore applications. An alternative may be to form a small pile group by installing several smaller piles and a pile cap. This is usually advantageous since the mobilisation costs for a smaller piling rig would be lower. The overall cost–benefit of different piling options will be dependent on several site-specific issues (refer to Chapter 52 *Foundation types and conceptual design principles*), in particular site access and headroom. A holistic approach to site development is needed in terms of the requirements for all foundations across the site and matters such as overall construction sequence and programme, and the available space for foundation construction plant and for building pile caps.
- *Site geology*: if there is a distinctive competent bearing stratum at depth and the piles act predominantly as end-bearing piles, then the appropriate pile length will be obvious, and the main design variables will be diameter and spacing. Alternatively, if the bearing stratum exhibits a gradual increase of strength or stiffness then the benefit of increasing the pile length, rather than the number or diameter of piles, may not be immediately obvious; this is discussed in more detail below.
- *Applied loads*: if high horizontal loads will be applied, then increasing the pile diameter will generally be more effective than increasing the pile length. A critical pile length L_C can be defined (Fleming *et al.*, 2009) for lateral loading. L_C/d depends on the relative ground to pile stiffness, but is typically between $6d$ and $10d$ (where d is the pile diameter). Piles longer than L_C will not reduce lateral pile deflection. Structural strength will often be the critical issue for laterally loaded pile groups, and structural strength rapidly increases with increasing pile diameter. For friction piles, if vertical loading and associated pile-group settlement is the key consideration, then increasing the pile length will usually be more cost-effective than increasing the pile diameter. For friction piles there is a critical pile length L_t beyond which further increases in length will not reduce settlement (Fleming *et al.*, 2009). L_t is approximately $1.5(E_p/G_b)^{0.5}d$, where E_p is Young's modulus for the pile concrete, G_b is the shear modulus at the pile toe and d is the pile diameter. For friction piles, an equally important consideration is the pile spacing. Provided ultimate group capacity is adequate, a relatively small number of widely spaced piles can provide similar pile-group stiffness to a larger number of more closely spaced piles. The design charts, shown in **Figure 55.20**, can facilitate a rapid comparison between the effectiveness of varying any of pile-group length, diameter and spacing on pile-group settlement performance. Pile spacing is often chosen to be $3d$ for conventionally designed pile groups; increasing the pile spacing to $4d$ or $5d$ may be worthy of consideration since pile interaction effects will be substantially reduced. For friction piles, spacings below $3d$ are usually inappropriate because pile interaction effects become far more significant and the potential benefits (in terms of bearing capacity and settlement) are likely to be negligible relative to the increased costs. For end-bearing piles minimum pile spacing is less of a concern for loss of efficiency.

In general, on the basis of observed performance (e.g. Mandolini *et al.*, 2005) it is reasonable to assume that pile-to-pile interaction is negligible for practical purposes at pile spacings of between 6*d* and 8*d* for both vertical and horizontal loading. At these spacings, the pile design can be based on single-pile behaviour; refer to Chapter 54 *Single piles*.

Raking piles: raking piles are sometimes used to support foundations subject to horizontal loads, and they can provide a very stiff support system. Although horizontal deformation will be relatively small compared with vertical piles of the same diameter (under an equivalent horizontal load), the use of vertical piles is usually favoured for several practical and technical reasons.

- Construction tolerances tend to be far poorer for raking piles than for vertical piles. The relevant tolerances are the position and alignment tolerances – that is, comparing the intended toe position with its actual position.
- Construction problems, such as installation damage for driven piles or potential integrity problems for bored cast in situ piles. Any potential problems associated with pile installation, such as groundwater inflows, pile bore instability, penetration into or through hard strata or obstructions, will all be far more challenging for raking than vertical piles. Because of this, raking piles will often be more expensive and slower to install than vertical piles.
- Raking piles should *not* be used in situations where there are likely to be large global ground movements, such as for bridge foundations on soft soils when adjacent approach embankments are likely to induce large vertical and horizontal ground movements (or, for similar reasons, adjacent to deep excavations); nor for heavily loaded foundations that are likely to undergo large settlement. In both situations the relative soil–pile movements are likely to induce bending stresses in the raking piles. In these circumstances, raking piles are vulnerable to being overstressed.

Raking piles were often used historically because of the lack of adequate analytical tools to assess the lateral load–deformation behaviour of vertical piles. Nowadays, there is a wide range of pile-group analysis software available to practitioners, which allows lateral load behaviour to be readily analysed. Therefore, for most circumstances, groups of vertical piles should be adequate. Nevertheless, there are some situations when raking piles can be cost-effective; one example is maritime applications when vertical loads may be negligible but there are substantial horizontal loads, such as for ship berthing facilities. Marine applications also tend to favour the use of driven piles, which are generally better suited to being installed at a rake than bored piles.

55.9. Information requirements for design and parameter selection

55.9.1 Ground profile

Golder and Osler (1968) described the performance of five pile groups (each with 32 piles, 20 m long), which were constructed to support an industrial facility. The piles were founded at a high level in a layer of dense sand. A pile test indicated a settlement of about 1 mm at the working load and pile-group settlement was expected to be less than 10 mm. Fifteen years after construction, the observed settlements were more than 70 mm and continuing to increase. The settlement of the pile groups at each end of the row of five groups showed significant tilt due to interaction with the adjacent pile groups. Subsequent analysis indicated that layers of compressible clay (about 3 m below the pile-toe level) were mainly responsible for the excessive pile-group settlements, which were predicted to reach 90 mm. About 80 mm, out of the total of 90 mm, was due to settlement of the underlying clay layers. This (and other) experience indicates that supporting a structure on a large pile group does not necessarily eradicate all risk of excessive foundation movement. It also highlights the importance of a proper understanding of the overall ground conditions and ground–structure interaction. Clearly, if considered in isolation, the results of a pile load test can be highly misleading.

The requirements for a proper desk study and comprehensive ground investigations are just as important for piled foundations as they are for other aspects of geotechnical engineering. For pile groups, an appreciation of the relative strength and compressibility of strata below the pile group is particularly important. A minimum of the pile-group width and, depending on the overall geology, up to 2–3 times the width should be investigated below the pile-toe level. Hence, deep boreholes may be necessary, together with careful logging and index testing at close centres, to check for weak layers within and below the anticipated bearing stratum. If lateral loading is significant, then the strength and compressibility of near-surface soils (typically the uppermost 6*d*, where *d* is the pile diameter) will be particularly important. This is challenging since near-surface soils can be very variable. Cost-effective investigations of near-surface soils will probably include some form of penetration or probing tests (refer to Chapter 47 *Field geotechnical testing*) to obtain rapid and cheap coverage across the site, together with several trial pits to facilitate interpretation.

55.9.2 Applied loading

The critical load case for pile-group design (especially for bridges and maritime structures) will often include a significant transient live load (due to, for example, vehicle impact or wind loading). For this type of short-term loading appropriate parameters (e.g. undrained strength and stiffness parameters for clays) should be used for pile-group analysis.

55.9.3 Mobilised ground stiffness, vertical loading

Randolph (1994) discusses the implications of nonlinear ground stiffness behaviour on both single-pile and pile-group settlement. **Figure 55.29** summarises the output from these analyses, plotted as normalised secant ground stiffness against factor of safety for a large pile group. The pile-group settlement remains relatively linear until the overall group factor of safety reduces below about 1.6. At a factor of safety of 2.0, the secant stiffness for the soil within the pile group is more than 80% of the initial tangent stiffness (which is close to the modulus mobilised at a strain amplitude of about 0.01% for many soil types). The relationship between choice of elastic modulus and pile-group size is shown schematically in **Figure 55.30**. **Table 55.5** provides guidance on the selection of the appropriate linear elastic moduli for single piles and pile groups of varying size at different factors of safety. Mandolini *et al.* (2005) report back-analyses of 48 pile groups across a range of geologies. Linear elastic analyses assuming very small strain stiffness (initial tangent stiffness from a pile test or derived from field geophysics G_{max} or G_0) generally gave estimated pile-group settlements within 20% of observed settlements. In contrast, a 'secant' stiffness approach (consistent with conventional practice, assuming a stiffness from a pile test at the working load, or using conventional empirical correlations for large strain stiffness) generally gave gross overestimates of pile-group settlement. Chapter 52 *Foundation types and conceptual design principles* gives some guidance on selection of G_{max}.

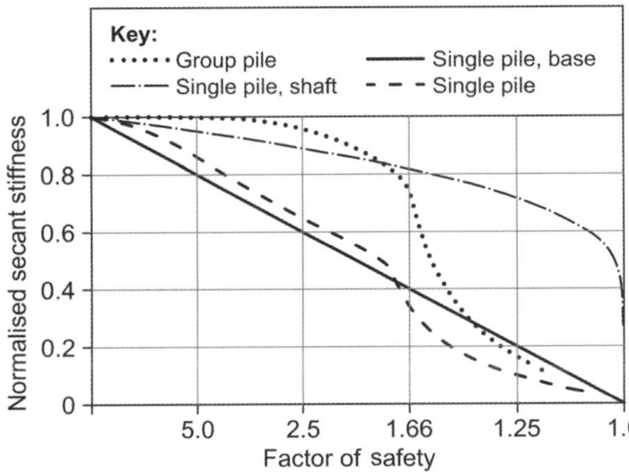

NB.
Normalised secant stiffness = 1.0, when G_{mob} (E_{mob}) = G_o (E_o); G_{mob} = mobilsed shear modulus, G_o = shear modulus at very small strain. E = Young's modulus

Figure 55.29 Normalised secant stiffness against factor of safety. Nonlinear settlement for a single pile and a pile group (Modified from Randolph, 1994)

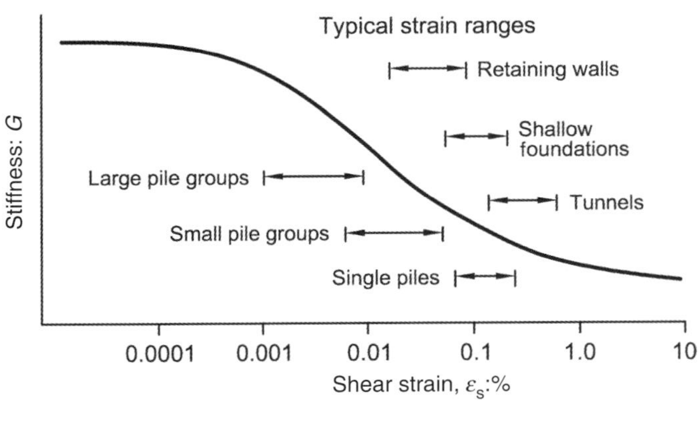

Figure 55.30 Influence of pile-group size on the selection of the appropriate 'elastic' modulus

Factor of safety	G/G_{max} Single pile	Medium pile group	Large pile group
>5.0	>0.6	>0.8	>0.9
3.0	0.45–0.6	0.7–0.9	>0.8
2.0	0.3–0.45	0.6–0.8	>0.8

Above values are for vertical loading and intended as preliminary guidance. A medium pile group has 5–25 piles; large pile group >25 piles. G_{max} is the shear modulus at very small strain for vertical shear. The values are relevant for linear elastic analysis and they assume the pile spacing is $3d$; as spacing increases the guide values will be increasingly conservative.

Table 55.5 Tentative values of shear modulus ratio (G/G_{max}), linear elastic analyses

55.9.4 Mobilised ground stiffness, lateral loading

For lateral loading, pile-group performance (in terms of deformation and structural demands) can be far more strongly influenced by nonlinear material behaviours than that for vertical loading. For example, both nonlinear ground stiffness reduction with strain amplitude and reduction of structural bending stiffness (for reinforced concrete piles) can become significant as pile-group deformation increases. The potential importance of both of these nonlinear behaviours will depend on the pile head fixity and the overall shape and size of the pile group (**Figure 55.31**). For example, at one extreme if pile head fixity is close to a 'pinned' condition, or for small pile groups (or a single row of piles), then performance is likely to be strongly influenced by nonlinear material behaviours. Modern software (e.g. LPILE or RSPile) can readily analyse the effects of nonlinear ground behaviour through the use of p–y curves,

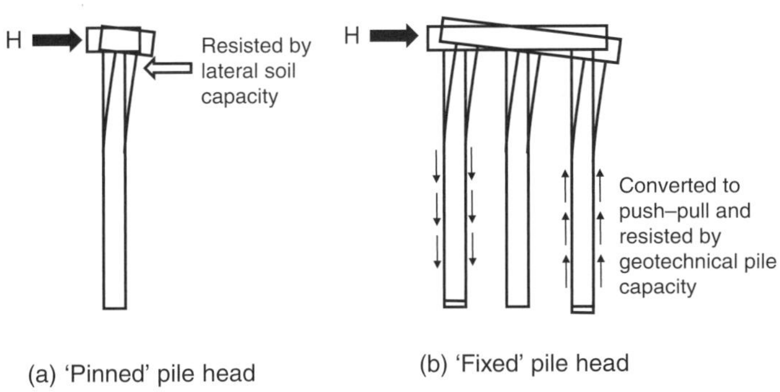

Figure 55.31 Comparison of pile head fixity on group behaviour

for which there is a significant empirical database (e.g. refer to Parkes *et al.*, 2018)).

At the other extreme, if the pile head is approaching a 'fixed' condition, then most of the applied lateral loading tends to be converted into a 'push–pull' action (that is, the applied lateral load is resisted by vertical pile resistance (compression and tension) across the group), which then reduces the practical effects of nonlinearity. If linear elastic analyses are being used, then it is recommended that a simple set of checks are implemented: first, the sensitivity of pile bending moments and group deformation to variations in shallow ground stiffness is checked (i.e. within a depth equal to the 'critical pile length' for lateral loading, L_c, typically the upper $6d$ to $10d$, where d is the pile diameter), perhaps by doubling and halving ground stiffness compared with the design assumption; second, the structural demands are considered in the context of local yield of the pile material (e.g. concrete 'cracking moment').

For critical situations, specialist advice should be sought and more sophisticated nonlinear analyses may be required. Hardy and O'Brien (2006) discussed the influence of ground stiffness nonlinearity on the predictions of bending moment in pile groups subjected to horizontal loads. For single piles, back-analysis of vertical and lateral load tests highlighted the significant influence of nonlinearity on lateral pile performance (across the working load range, 'elastic' secant stiffness reduced to about half its initial value for vertical loading, whereas for horizontal loading it reduced to less than 15% of its initial value). **Figure 55.32** shows that for a medium-size pile group, the pile bending moments can be sensitive to the soil constitutive model (hyperbolic or linear elastic) and, for the linear elastic model, the magnitude of the selected moduli. If the design is substantially influenced by lateral load effects, then nonlinear analyses should be implemented (Mandolini *et al.*, 2005; O'Brien, 2007). These analyses may require specialist input and guidance, although software such as REPUTE includes a hyperbolic model which is quick to use for suitably experienced practitioners.

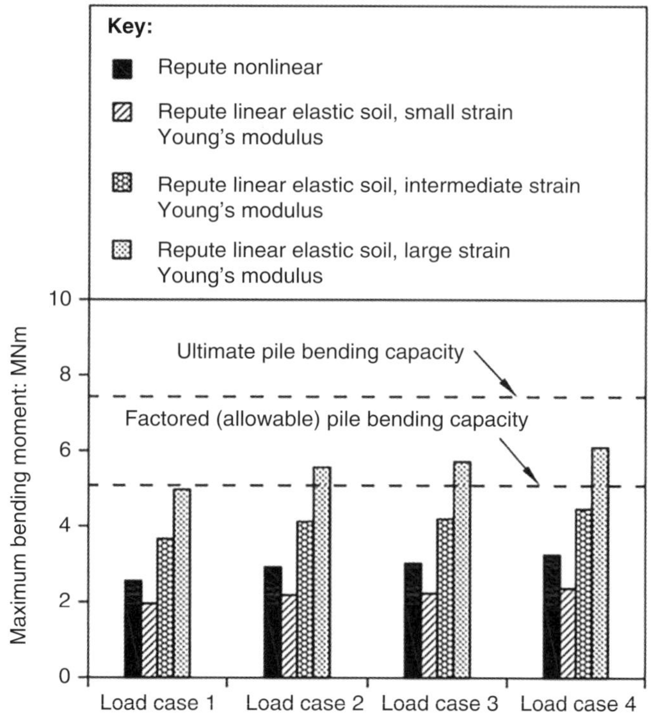

Figure 55.32 Influence of soil model and selected Young's modulus on the peak bending moment under horizontal loading (Reproduced from Hardy and O'Brien, 2006)

55.9.5 Ground stiffness characterisation for pile groups

When considering the selection of input parameters for pile-group analysis, and, in particular, comparing single-pile with group performance, it is helpful to consider the different zones of ground around a pile or pile group that will influence behaviour (**Figures 55.33** and **55.34**), and the relevant information that will be needed for making decisions. Five separate zones, and associated ground stiffness moduli, can affect behaviour.

- The stiffness of the soil or rock close to the pile shaft: this zone will strongly influence the settlement of a single pile at typical working loads. It will be strongly affected by the construction process (remoulding and horizontal stress changes), either due to pile boring or pile driving.
- The stiffness of the soil or rock at the pile toe: this zone is dominated by the construction process and is generally far more sensitive to the construction process than shaft stiffness. Base stiffness will be crucially important for end-bearing piles; however, for many piles (which often gain significant shaft resistance) the base stiffness only becomes influential at low factors of safety. For driven piles the base stiffness tends to be relatively high compared with the original ground stiffness. For bored piles, the base stiffness will tend to be lower, sometimes much lower, than the original ground stiffness and will be highly dependent on factors such as groundwater inflows, debris at the pile toe and the effectiveness of any base-cleaning activities (Fleming, 1992).
- The stiffness of the ground close to the ground surface: this can strongly affect pile behaviour under horizontal loading, especially for single piles and small pile groups. Ground investigations frequently overlook the characterisation of this zone, and it can be extremely variable. Also, site redevelopment (excavation and filling) will often change the nature of this zone. Pile concrete stiffness under lateral loads will tend to be lower (due to yield, creep and cracking) than that for compression.
- Ground stiffness between the piles: this zone has a dominant influence on pile-group settlement. This zone will not be significantly affected by construction activities. As noted earlier, the relevant modulus, for linear elastic analysis, will be close to the very small strain stiffness of the soil or rock, especially for medium and large pile groups.

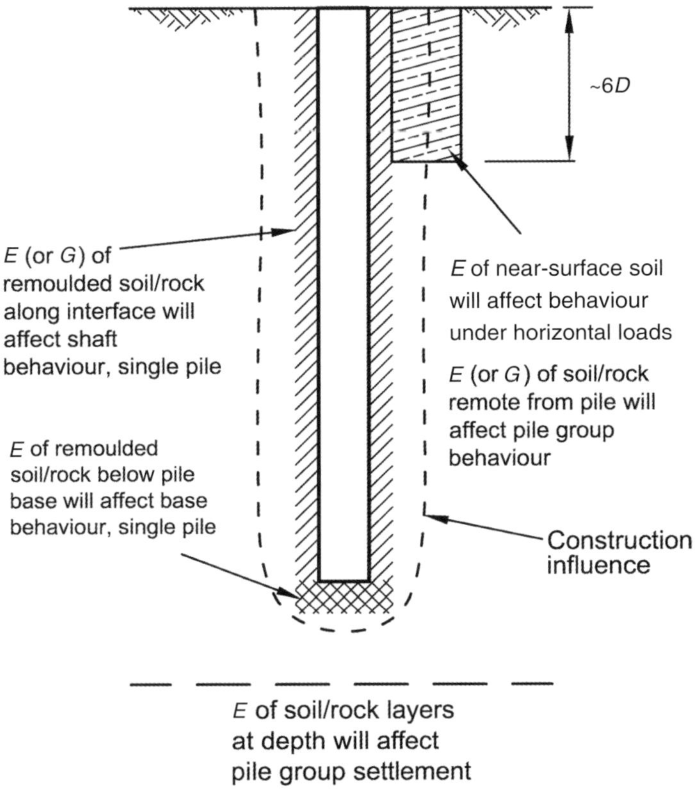

Figure 55.33 Different ground moduli that may influence single-pile and pile-group behaviour

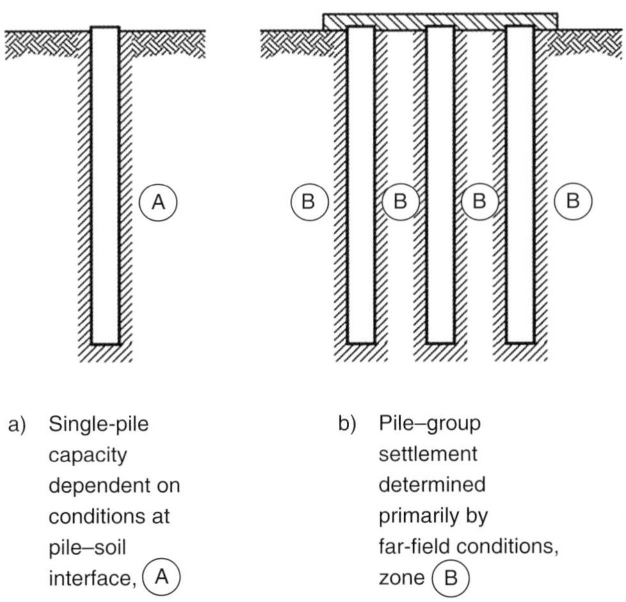

Figure 55.34 Influence of (a) the pile–soil interface on single-pile capacity and (b) far-field conditions on pile-group settlement (Reproduced from Randolph, 2003)

- Ground stiffness well below the pile-toe level: this zone will be unaffected by construction activities. It will not affect the behaviour of single piles, but will influence pile-group behaviour, especially for pile groups with a large group aspect ratio.

Hence, the information required will involve a consideration of the following.

- Construction effects and the back-analysis of single-pile behaviour (e.g. England, 1999; Fleming, 1992). For laterally loaded piles, tests on single piles will usually be free-headed, whereas piles in a group will generally be fixed-headed. This different head fixity must be allowed for when comparing isolated piles and pile-group behaviour.
- Data from relevant case histories (e.g. Mandolini *et al.*, 2005). Case histories are particularly useful for calibration of software. Software calibration is particularly important for large pile groups or more sophisticated nonlinear analytical methods.
- Site-specific ground investigation data. Useful additional data (although seldom provided by routine investigations) would be measurements of G_0 or G_{max}. Modern empirical correlations (e.g. Atkinson, 2000) are helpful for the selection of the appropriate moduli. Chapter 47 *Field geotechnical testing* provides details of field methods for direct measurement of soil stiffness. Modern empirical correlations are helpful for the selection of the appropriate moduli and further guidance is given in Chapter 52 *Foundation types and conceptual design principles*.

55.9.6 Selecting nonlinear ground stiffness parameters for pile-group analysis

Some commercially available pile-group analysis software can simulate nonlinear ground stiffness. The mathematical formulations may vary (although simple power law or hyperbolic models are commonly used) and the inputs cannot necessarily be linked to fundamental or commonly measured ground properties. The ideal means of calibrating these software models is to curve-fit the software load–displacement curve against full-scale pile test data (either site-specific or published data for the relevant geology) for vertical or lateral pile tests as appropriate. In the authors' experience, in the absence of relevant pile test data, the following approaches can also be successful, if implemented with care.

- For vertical loading: use the Fleming (1992) method (for which a good empirical database is available) to generate an appropriate nonlinear load–settlement curve for a single pile, and then calibrate the software's nonlinear model against this. The calibrated model inputs can then be used for pile-group analysis.
- For horizontal loading: use relevant empirical *p–y* curves and calibrate against these for a single pile, prior to use for a pile-group analysis.

When developing the pile-group model, the issues highlighted in **Figures 55.33** and **55.34** also need careful consideration, especially for end-bearing piles, so realistic group behaviour is simulated (e.g. appropriate layering below the toe level of the pile group).

55.9.7 Single-pile capacity compared to pile-group deformation

Ultimate geotechnical capacity of a single pile is mainly dependent on local conditions at the pile–soil interface, which are strongly affected by construction methods and can exhibit significant variability (e.g. NCHRP, 2004). In contrast, pile-group deformation is determined primarily by far-field conditions (i.e. ground stiffness between and remote from the piles), and is relatively unaffected by construction processes (**Figure 55.34**). Hence, pile-group deformation can often be more reliably predicted than the ultimate capacity of a single pile (Randolph, 2003). This is in stark contrast to many other aspects of geotechnical engineering (such as predicting movements around retaining walls or shallow foundation movements), where predicting movement is perceived as more difficult than predicting ultimate capacity.

55.10. Some practical design considerations

55.10.1 Factors of safety

Burland (2006) highlighted the profound importance of understanding the mode of foundation failure mechanisms. Routine geotechnical and structural engineering practice is based on the assumption of ductile behaviour. The load–settlement behaviour of pile foundations is usually ductile. However, this is not always the case; for example, experience indicates that some rock socket piles can exhibit brittle behaviour, such as piles socketed into some rocks (e.g. limestone and sandstone) or end-bearing piles toed into thin competent strata and underlain by soft clays. If brittle behaviour is anticipated, then a cautious approach is essential. Rather than summing shaft friction and end-bearing capacity, it would be more appropriate to consider a deformation-based capacity (i.e. discounting some fraction of either shaft or end-bearing resistance). Preliminary pile tests to failure would clearly be helpful. Pile-group design would need to limit pile-group deformation to ensure brittle failure was not induced.

For more normal situations, when pile load–settlement behaviour is ductile, the selected factors of safety will be influenced by

- pile cap and substructure stiffness
- number of piles in the group (i.e. level of redundancy)
- code requirements
- nature and direction of applied loading (e.g. cyclic or monotonic loads, vertical or horizontal loads)
- analysis methods
- reliability and scope of ground investigations.

Table 55.6 gives a commentary on these issues.

Many codes and standards do not make specific statements about factors of safety for pile-group design. However, EC7 provides the following, in Clause 7.6.2.1 (the code uses the term 'stiffness' in the context of a structure's ability to redistribute loads, whereas the term 'rigidity' is generally used in this chapter).

- The stiffness and strength of the structure connecting the piles in the group **shall** be considered when deriving the design resistance of the foundation.
- If the piles support a stiff structure, advantage may be taken of the ability of the structure to redistribute load between the piles. A limit state will occur only if a significant number of piles fail

Factor	Comment
Pile cap and substructure stiffness	A stiff pile cap or substructure can redistribute axial loads; hence, the individual pile factor of safety is not significant. If the pile cap is flexible then differential settlement needs to be considered. If the pile group is small, individual pile factors of safety need to be considered.
Number of piles in the group	If more than five piles, then there is redundancy, and 'failure' of a pile within the group does not imply failure of the group (NCHRP, 2004). For large pile groups there is considerable redundancy.
Code requirements	Many codes do not discuss pile groups in detail and mainly focus on single piles. EC7 provides some guidance, see text. AASHTO (and NCHRP, 2004) gives guidance on reduced risk of failure associated with varying levels of redundancy.
Direction of loading	For horizontal and moment loading, carefully check structural strength of piles, pile cap and pile–pile cap connections. For large long-term moment loading, nonlinear analysis should be considered and risk of excessive group rotation in the long term (creep) assessed, based on holistic assessment of overall group behaviour.
Analysis method	For nonlinear methods reliable calibration of the model is important. Use simple methods to check factor of safety against failure. Computer-based methods are more appropriate for assessing deformation and structural demands on piles.
Reliability and scope of ground investigations	The most important factor to consider. It is especially important to verify the strength and stiffness of layers below the pile group. Near-surface materials are important for laterally loaded pile groups. The greatest uncertainty lies with establishing the geological model, the idealisation of the ground profile for analysis and the selection of appropriate geotechnical parameters.
Nature of loading	Guidance in this chapter is solely for pile groups with predominantly static, monotonic loading. Under prolonged cyclic loading, significant degradation of shaft resistance can occur, with associated substantial increases in deformation and reductions in ultimate capacity (Jardine, 1991).

Table 55.6 Factors that may influence the factor of safety for a pile group

together; therefore a failure mode involving only one pile need not be considered.

- Special attention should be given to possible failure of edge piles caused by inclined or eccentric loads from the supported structure.

Hence, EC7 allows piles within a pile group to have ratios of ultimate geotechnical capacity to axial load to be below the partial factors required for a single isolated pile (for GEO ULS) provided appropriate checks are made to verify that the overall structural strength/stiffness of the pile cap, pile–pile cap connection and piles are satisfactory (to ensure load redistribution can occur). Hence, it is necessary to check that the pile cap can be considered to be stiff (or 'rigid') rather than flexible. Referring to **Figure 55.27**, the pile cap may be considered 'stiff' if $K_{rs} > 1.5$, and then if the structural strength of the pile cap, pile–pile cap connections and the piles are sufficient, geotechnical failure of the pile group can only develop if either the 'block' resistance is inadequate or the sum of individual pile capacities is inadequate (refer to **Figures 55.4(a)** and **55.4(b)**). In general, there are three options for verifying GEO ULS capacity.

(*a*) Sum the design resistance of all the piles, or the equivalent block, and check it exceeds the applied design loads. This is relatively simple if the pile group is subject to vertical loads, but more complex for combined loading.

(*b*) For pile groups subject to combined vertical, moment and lateral loads, input the **design resistance** of the piles and loads (i.e. relevant resistance and load factors applied) into a **nonlinear** pile-group analysis (i.e. using a hyperbolic model or similar) and check that there are piles with available 'spare' capacity to avoid an overall failure mechanism.

(*c*) For pile groups subject to combined vertical, moment and lateral loads, check that the design resistance of the leading row exceeds the sum of the mobilised axial loads.

Option (*c*) is quite conservative, since the 'spare' capacity available in the other rows in a group is ignored and the ability of the stiff pile cap to redistribute loads is also ignored (refer to **Figure 55.4(c)**). It will be particularly conservative if based on a linear elastic analysis, since the axial loads in the perimeter piles will be overpredicted. Nevertheless, the check is quick and simple to undertake. Option (*b*) can enable more economic pile groups to be designed. As noted in the case history example (and by O'Brien and Bown, 2010), there are major structures where some of the piles within the group have factors of safety of less than 1.0 (based on linear elastic analyses). However, when the pile cap is stiff and the pile response is ductile, then the overall pile-group stability is entirely satisfactory and pile-group deformation is relatively small. In practice, group GEO ULS capacity is seldom critical, and SLS and structural strength are usually the key criteria for pile groups, especially for medium to large pile groups. Therefore, it is important to consider the structural demands (e.g. axial loads, bending moments) across a pile group, together with pile group deformation.

55.10.2 Structural checks and pile head fixity

The structural strength of piles, pile cap and pile–pile cap connections should always be checked and be sufficient to resist the induced axial forces, bending moments and shear forces. This is particularly important for pile groups subject to large moment or horizontal forces, since inadequate structural strength in any part of the overall foundation system could lead to a brittle and progressive collapse of the structure.

The concept of free and fixed head piles in Section 54.6.5 is explained further by considering partial fixity provided by the pile cap in small pile groups. Small pile groups (< 5 piles) are very common for supports to single columns in buildings and partial fixity or moment restraint at the pile head can be considered to improve the economy of the structural design of the piles. When designing for large accidental impact actions the need to consider partial fixity to achieve a practical solution is particularly important.

Some geotechnical software assumes a fully fixed head condition which may be relevant for larger pile groups, but this can generate an excessively large negative bending moment at the top of the pile which does not occur in partially fixed small pile groups. Soil–structure interaction software including, but not limited to, LPILE and RSPile provide functionality to apply a moment restraint at the head of the pile for the predicted level of fixity. The method to calculate the rotational stiffness is illustrated in **Figure 55.35** and relies on a prediction of pile settlement or heave which could be made by use of Fleming's method for single-pile settlement (Fleming, 1992) or from pile load testing. A case history to illustrate this method is included in Section 55.12.

55.11. Pile-group design responsibility

Two options are normally considered: contractor's design or engineer's design.

(*a*) *Contractor's design:* this is commonly used, especially for small building projects. The engineer employed by the client as the overall project designer should provide the piling tenderers with relevant information, including site history, ground conditions, applied structural loads and acceptance criteria for pile loading tests. The contractor is responsible for detailing the appropriate pile length, diameter and reinforcement to resist the specified loads. The pile cap is usually designed by the overall project designer.

(*b*) *Engineer's design:* under this arrangement design responsibility lies with the overall project designer, who

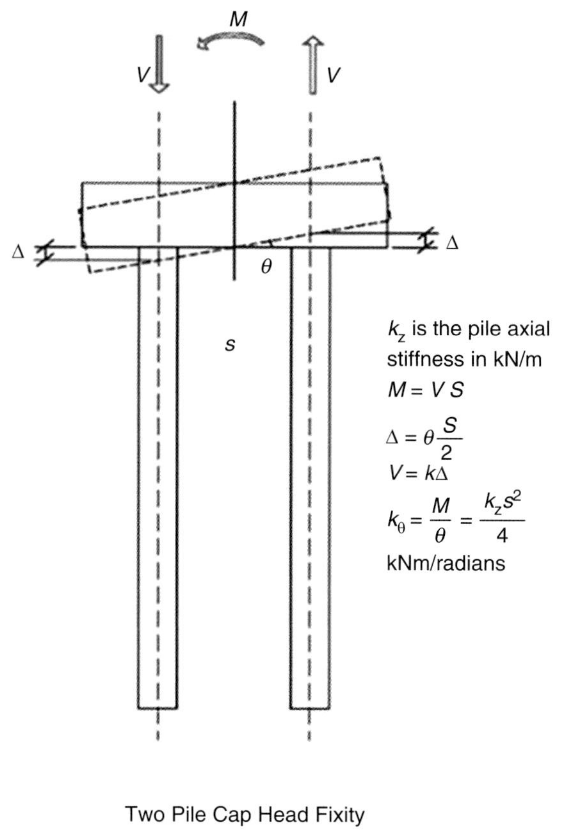

Figure 55.35 Rotational fixity of twin piles by pile cap (Reproduced from Raison and Egan 2016)

would design and detail the pile lengths, reinforcement and so on. This arrangement is often followed for major civil engineering projects, such as bridges or large building projects. The contractor is responsible for workmanship and method of construction.

There are advantages and disadvantages with both options. The specialist piling contractor will have a lot of expertise on how to best utilise his materials, plant and labour. He may also have local experience on the likely pile capacities that can be achieved in particular geological sequences. Therefore option (*a*) will favour those situations where the geotechnical capacity of a single pile is the dominant consideration, where buildability is very challenging (due to poor access, etc.) or if particular proprietary piling systems are well suited to local conditions. Option (*a*) is less appropriate if pile-group deformation is critical or if there are broader ground–structure interaction problems (due to complex loading or if global ground movements are a concern). A particular problem with option (*a*) for pile-group design is the appropriate specification of pile loads and pile test acceptance criteria. For example, axial loads will not be uniform across the group, and the stresses induced in the piles and pile cap will be a function of the pile length, diameter and so on and the pile cap bending stiffness. The deformation of the pile group is dependent on several factors and not just the settlement characteristics of an isolated pile. Hence, the overall designer needs to have a good understanding of these ground–structure interaction issues before realistic and sensible specification requirements for pile capacity and settlement (for an isolated pile) can be developed. In many cases, if design responsibility is to be split then it is more appropriate to split responsibility at the top of the pile cap level rather than at the top of the pile level. The overall designer then only needs to specify the appropriate loads at the pile cap level, and the acceptable limits for overall pile-group deformation and differential settlement across the pile cap. Option (*b*) will also be more appropriate for large pile groups or for pile groups underlain by compressible layers (where single-pile performance cannot be easily correlated with pile-group performance). Option (*b*) would also facilitate more economic solutions, such as piled rafts (refer to Chapter 56 *Rafts and piled rafts*).

55.12. Case histories

Two case histories are discussed: first, pile-group analyses and observed performance are discussed for a large pile group; second, analysis for a small pile group is outlined which facilitated more economic design.

55.12.1 Large pile group

This case history is for the Emirates Twin Towers in Dubai (which has been described in detail by Poulos and Davids, 2005). The ground conditions mainly comprise interbedded calcareous sandstone and siltstone. The measured unconfined compressive strength is variable and typically between about 0.5 MN/m^2 and 1.5 MN/m^2. The stiffness at very small strain was measured by cross-hole geophysical tests, and shear moduli varied between 2 and 3.5 GN/m^2. A summary of the ground profile is given in **Table 55.7**. The twin towers were founded on two triangular-shaped pile groups comprising 92 and 102 1.2 m dia. piles, 40 m long. The estimated ultimate geotechnical capacity of a single pile was about 42 MN.

Based on the published data, two different analyses have been carried out, the first using a computer-based method and the other using a simple empirical method. The computer-based analysis, using the software REPUTE, was used with two different models (linear elastic and nonlinear hyperbolic) to simulate the load–displacement characteristics of the piles. The input parameters are summarised in **Table 55.8**. The load–settlement behaviour of a single pile was modelled initially and calibrated against preliminary pile test data for a 0.9 m dia., 40 m long pile (**Figure 55.36**) prior to modelling the entire pile group. For the single-pile calibration, Fleming's method (Fleming, 1992) was used to derive reasonable nonlinear parameters for use in the hyperbolic model.

The calculated pile-group axial load distributions and settlement given by the two different analytical models are

Strata number	Strata designation	Typical depth of strata base: mbgl	Description
1	Silty sand	5	Uncemented calcareous silty sand, loose to moderately dense
2	Sand	10	Variably and weakly cemented calcareous silty sand
3	Calcareous sandstone	29	Calcareous sandstone, slightly to highly weathered, well cemented
4	Cemented sand (?)	35	Calcareous silty sand, variably cemented, with localised well-cemented bands
5	Calcareous siltstone and conglomerate	56	Variably weathered, very weakly to moderately well cemented
6	As Unit 5	70	As Unit 5

mbgl, metres below ground

Table 55.7 Case history: ground profile

	Linear elastic[a]	Nonlinear[b]	
Strata	Secant Young's modulus: MN/m^2	Tangent Young's modulus (small strain): MN/m^2	Interface strength τ: kN/m^2
Silty sand	15	$20 + 8Z$	$20 + 5.5Z$
Sand	50		
Calcareous sandstone	250	$1500 + 40Z$	$150 + 12Z$
Cemented sand (?)	50		
Calcareous siltstone and conglomerate	200	$3000 + 50Z$	500

[a] Linear elastic model, pile base stiffness, $E' = 40$ MN/m^2.
[b] Nonlinear model, hyperbolic constants, $R_{shaft} = 0.65$, $R_{base} = 0.99$, maximum shaft friction = 500 kN/m^2, maximum base pressure = 2700 kN/m^2 and pile base modulus $E_0' = 750$ MN/m^2 (E_0' = small strain tangent stiffness).
Z is the depth below the ground surface.

Table 55.8 Linear elastic and nonlinear input parameters used in the REPUTE analyses of the case history

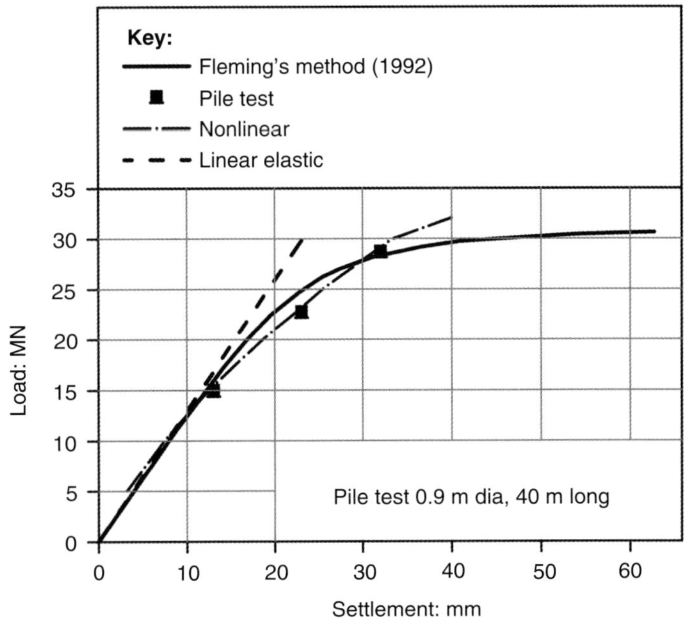

At a load of 15 MN, Geo factor of safety ~ 2.0	
Analysis/data	Settlement: mm
Linear elastic	11.6
Nonlinear	12.3
Pile test	12.9

Figure 55.36 Case history: pile test in weak rock

summarised in **Figure 55.37**. The linear elastic analysis predicts a highly non-uniform distribution of axial load; in particular, axial loads in the corners are especially large. The linear elastic analysis predicts maximum axial loads of 43 MN compared with a maximum of 31 MN from the nonlinear analysis. The ratio of corner to central pile axial load from the linear elastic analysis is about 5, and for the nonlinear analysis it is about 2. Hence, the local factor of safety for a pile in the group (based on linear elastic analysis) is less than 1.0, although the overall factor of safety for the pile group is about 2.0. This is similar to the results from boundary element analysis reported by Poulos and Davids (2005), where they state that 'a number of the piles reached their full geotechnical design resistance (i.e. the factor of safety for some piles was about 1.0), but the foundation as a whole could still support the ultimate design loads'. The pile-group settlement was observed to be about 8–10 mm after 70% of the load had been applied, and the estimated final settlement is 20–40 mm. The nonlinear analysis indicates a settlement of 25 mm, compared with 55 mm from the linear elastic analysis (**Figure 55.37**).

The pile-group aspect ratio is about 2.9 and Mandolini's empirical correlations for pile-group settlement (refer to Section 55.5.2) gives settlement ratios R_{Se} between 4 and 7. Based on the site geology (weak rock with stiffness increasing with depth) the best-estimate and lower-bound pile-group settlement ratios are relevant (Equations 55.9 and 55.12). The single-pile settlement at the average working load of 21 MN is about 10–11 mm. Hence, based on Mandolini's empirical method, the pile-group settlement would be expected to be about 40–75 mm. Although conservative relative to

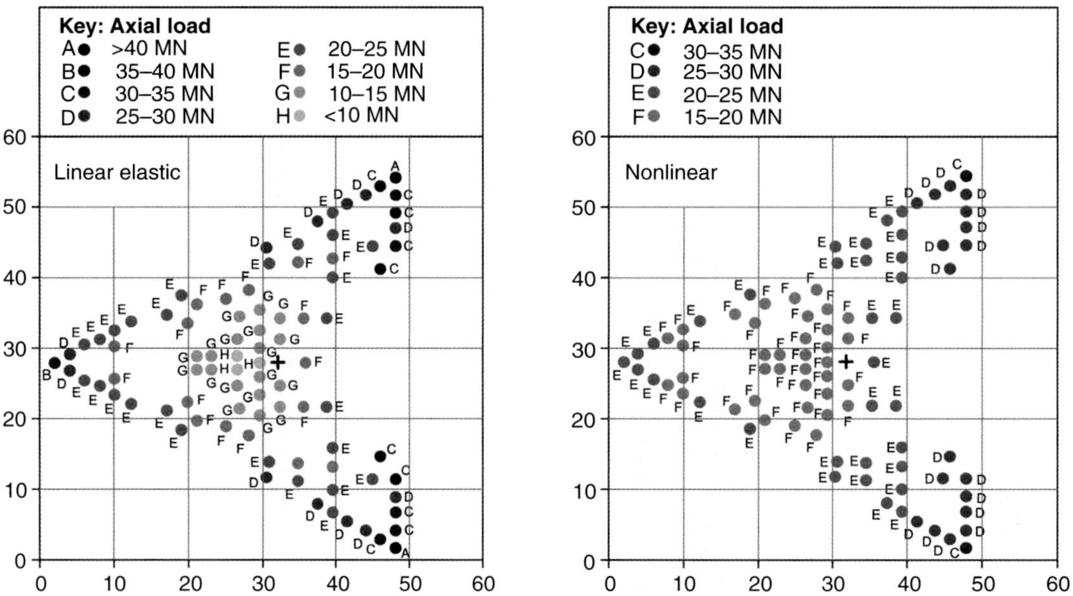

1.2m dia. piles, ultimate capacity ~42 MN, Geo facor of safety for group ~2.0		
	Peak axial load	Group settlement: mm
Linear elastic	43 MN	55 mm
Nonlinear elastic	31 MN	25 mm
Estimated settlement (in situ measurements) ~20 to 40 mm		

Figure 55.37 Case history: pile group in weak rock; axial load distributions and settlement for the linear elastic and nonlinear models

observations and the estimated final value, it is more accurate than the designer's original estimate of about 90–140 mm (based on a linear elastic boundary element analysis). Poulos and Davids describe the two main reasons for the initial overprediction using boundary element analysis

- overestimation of pile–pile interaction
- underestimation of the ground stiffness for Unit 5 (**Table 55.7**).

Poulos and Davids produced revised analyses where the stiffness varied radially from each pile, so that the ground stiffness between the piles was five times stiffer than that adjacent to the piles (to values of about 2–2.5 GN/m^2 – that is, close to the geophysics, G_0, values), and the ground stiffness for Unit 5 was increased from 80 to 600 MN/m^2. This revised analysis indicated a settlement of 23–40 mm, (allowing for differential settlement across the pile cap). With the benefit of hindsight, it seems likely that the laboratory-measured stiffness for Unit 5 was far too low because of sample disturbance.

This case history emphasises the challenges in deriving realistic ground stiffness values, especially in weak rocks, and the potential errors even when sophisticated analyses are utilised. It also highlights the value of simple empirical methods to check computer output. It is likely that the axial loads around the perimeter of the pile group were high and the factor of safety against geotechnical failure for the perimeter piles was much lower than code values (nonlinear analyses predict 'local' factors of safety of about 1.3–1.4 for the perimeter piles). Nevertheless, the overall group factor of safety is code compliant (equal to about 2.0), the piles and pile cap have adequate strength and stiffness to redistribute forces across the group and therefore the overall pile-group performance is satisfactory.

55.12.2 Small pile group

Pile-group analyses are briefly described for a small pile group on the Werrington Grade Separation scheme, near Peterborough, UK. A new footbridge was constructed with one abutment very close to the live track and the risk assessment resulted in a need to design the foundations for an accidental horizontal impact force of 2000 kN acting at the top of the pile cap. Due to restricted space in the area of the foundation, a four-pile group was chosen with 900 mm dia. bored piles at $2D$ spacing, 11 m in length penetrating made ground over Oxford Clay, where strength improved with depth from a firm to a very stiff clay or extremely weak mudstone (**Figure 55.38**).

The worst-case orientation of impact for the pile design was 45° to the group, where the combined compressive and horizontal actions resulted in 2009 kN compression on the pile furthest from impact, a tension of −1134 kN on the nearest pile and +438 kN compression on the two middle piles. The vertical pile stiffness was based on a load of 2000 kN divided by an estimated pile settlement of 0.01 m, giving 'best estimates' for a vertical pile spring of k_z of 200 000 kN/m and a rotational spring k_θ of 162 000 kNm/radian, derived using the formula in **Figure 55.35**. **Figure 55.39** illustrates the differences in pile bending moment and deflection for the full range of pile head fixity.

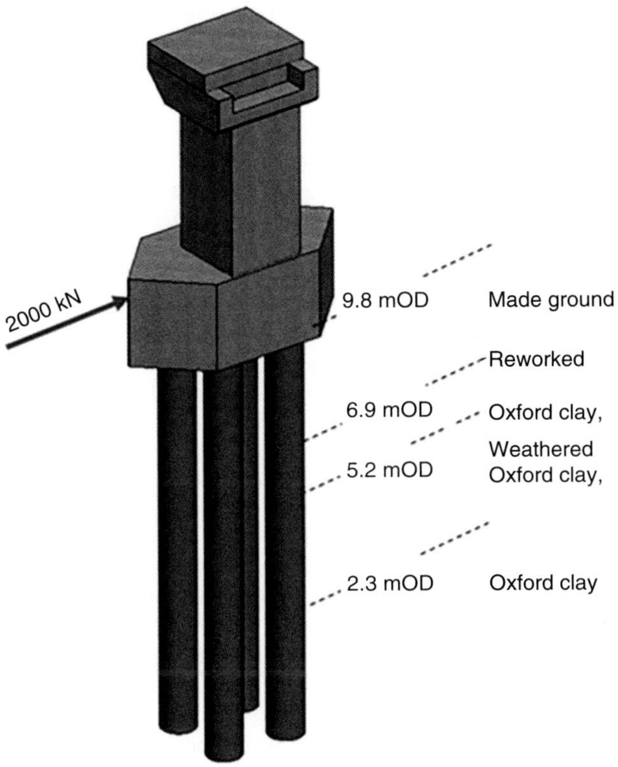

Figure 55.38 3D view of the pile group to the west abutment of Cock Lane Footbridge, subject to accidental derailment action

Due to the lack of pile load test data to verify vertical stiffness, a sensitivity analysis was carried out assuming half and double the best estimate of applied moment restraint at the pile head in order to calculate the required reinforcement. The partially fixed laterally loaded pile analysis resulted in a 25% reduction in pile reinforcement and avoided steel congestion in the pile cap (all reinforcement had a clear spacing greater than 100 mm). Hence, buildability was improved and the design was safer to build.

55.13. Overall conclusions

- The overall behaviour of pile groups is affected by their geometry. An important parameter is the pile-group aspect ratio (**Figure 55.3**). Where R is large, the group settlement will be many times larger than the settlement of an isolated pile (at the same average load). For pile groups with a large aspect ratio, $R > 3$, an 'equivalent raft' analogy is appropriate; where $R < 3$ an 'equivalent pier' analogy is more appropriate.
- The failure of pile groups is rare; the risk of failure increases when global movements develop around the group (typically due to out-of-balance loads on soft clays and peats); end-bearing piles toed into a thin competent layer are underlain by a weaker underlying strata; or piles are toed into steeply dipping fractured rock or rock containing voids.
- Pile–pile interaction, under vertical or horizontal loads, depends on many factors (**Table 55.2**), including pile spacing and the

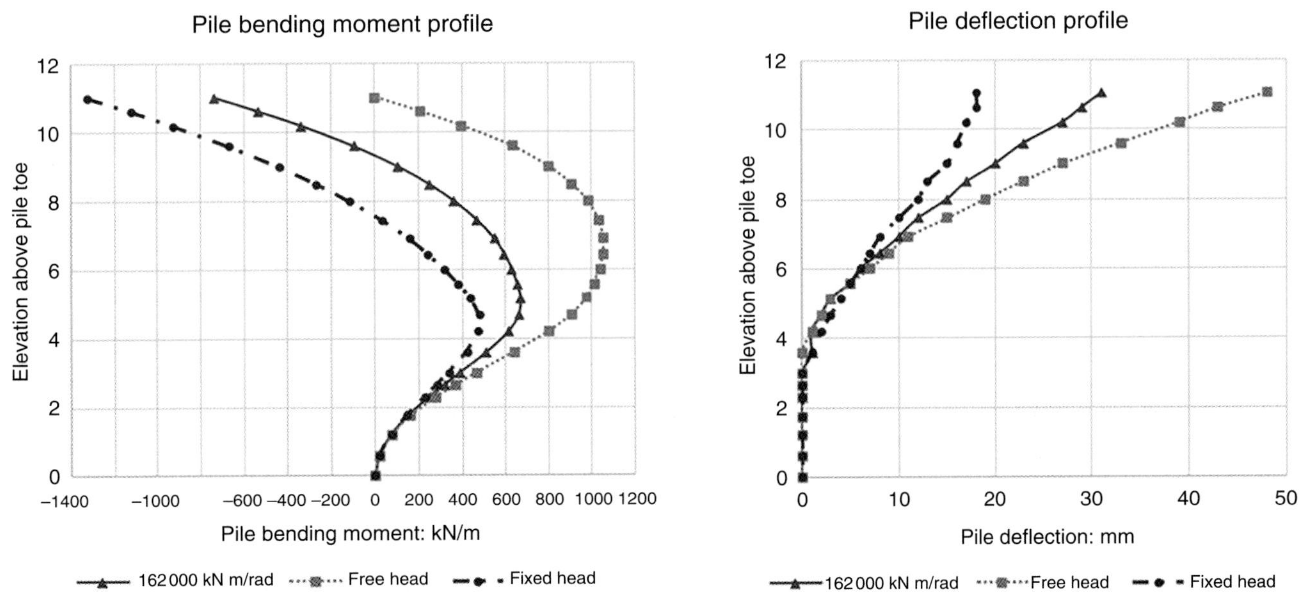

Figure 55.39 WALLAP single-pile analysis results for bending moment and deflection of a pile under different head fixity conditions

variation of ground stiffness with depth, especially 'hard' or 'soft' layers beneath the pile toe (for vertical loads) or adjacent to the pile head (horizontal loads). For horizontal loads, the pile head fixity is important, whereas for large pile groups under vertical loads the pile cap stiffness is important.

- For small- to medium-sized vertically loaded pile groups linear elastic analysis may be adequate for a practical design, although care is required to avoid over-conservatism. However, for larger groups or pile groups subject to large horizontal loads, linear elastic analyses are likely to be inappropriate. Simplified methods of analysis (based on empirical settlement ratios, elastic interaction factors or an equivalent pier) will often be sufficient to calculate the group settlement and should always be carried out as a check on more sophisticated methods.
- When considering the selection of input parameters for pile-group analysis, it is helpful to consider the five separate zones (and associated deformation moduli) that can affect behaviour (**Figures 55.33** and **55.34**): (*a*) the stiffness of the ground adjacent to the pile shaft; (*b*) the zone immediately below the pile toe; (*c*) the stiffness of the ground close to the pile head; (*d*) the ground stiffness between the piles; and (*e*) the ground stiffness at depth below the pile-toe level.
- The appropriate values for factors of safety selected for pile-group design depend on several factors – refer to **Table 55.6**. These include pile cap and substructure rigidity, the number of piles in a group (redundancy), code requirements, the nature and direction of loading, analysis methods and the reliability and scope of ground investigations. In addition, the ductility or brittleness of pile load–settlement behaviour is an important consideration.
- If design responsibility is split for pile-group design, then careful thought is required regarding the most appropriate point to divide responsibility (pile head or pile cap level). When pile-group deformation is critical, or if global ground movements or complex loads are applied, then pile and pile cap design should be considered together so that the soil–structure interaction is properly assessed.

References

Adams JI and Hanna TH (1970) Ground movements due to pile driving. *Proceedings of the Conference on the Behaviour of Piles, London, UK*. ICE Publishing, London, UK, pp. 127–133.

Atkinson JH (2000) Non-linear soil stiffness in routine design. *Géotechnique* **50(5)**: 487–508.

Basile F (2020) Influence of cap rigidity on pile group design. *Proceedings of the Piling 2020 Conference*. ICE Publishing, London, UK.

Bjerrum L (1967) Engineering geology of normally-consolidated marine clays as related to the settlement of buildings. *Géotechnique* **17(2)**: 83–117.

Brzezinski LS, Shector L, MacPhie HL and Vander Noot HJ (1973) An experience with heave of cast *in situ* expanded base piles. *Canadian Geotechnical Journal* **10(2)**: 246–260.

Burland JB (2006) Interaction between structural and geotechnical engineers. *The Structural Engineer*, 18 Apr.: pp. 29–37.

Butterfield R and Douglas RA (1981) *Flexibility Coefficients for the Design of Piles and Pile Groups. CIRIA Technical Note 108.* CIRIA, London, UK.

Cooke RW, Bryden Smith DW, Gooch MN and Sillet DF (1981) Some observations of the foundation loading and settlement of a multi-storey building on a pile raft foundation in London Clay. *Proceedings of the ICE, Part 1* **70**: 433–460.

England M (1999) *A Pile Behaviour Model*. PhD thesis, Imperial College, University of London, London, UK.

Fleming WGK (1992) A new method for single pile settlement prediction and analysis. *Géotechnique* **42(3)**: 411–425.

Fleming WGK, Weltman AJ, Randolph MF and Elson WK (2009) *Piling Engineering*, 3rd edn. Taylor & Francis, London, UK.

Golder HQ and Osler JC (1968) Settlement of a furnace foundation. *Canadian Geotechnical Journal* **5(1)**: 46–56.

Hardy S and O'Brien AS (2006) Nonlinear analysis of large pile groups for the New Wembley Stadium. *Proceedings of the 10th International Conference on Piling and Deep Foundations, Amsterdam, the Netherlands* (Bottiau LM and VanTol AF (eds)). Deep Foundations Institute, Hawthorne, NJ, pp. 303–311.

Hooper JA (1979) *Review of Behaviour of Piled Raft Foundations*. CIRIA Report 83. CIRIA, London, UK.

Jardine RJ (1991) The cyclic behaviour of large piles with special reference to offshore structures. In *Cyclic Loading of Soils* (O'Reilly MP and Brown SF (eds)). Blackie, Glasgow, pp. 174–248.

Jardine RJ, Potts DM, Fourie AB and Burland JB (1986) Studies of the influence of non-linear stress–strain characteristics in soil structure interaction. *Géotechnique* **36(3)**: 377–396.

Katzenbach R, Arslan U and Moormann C (2000) Piled raft foundation projects in Germany. In *Design Applications of Raft Foundations* (Hemsley JA (ed)). Thomas Telford, London, UK, pp. 323–391.

Koizumi Y and Ito K (1967) Field tests with regard to pile driving and bearing capacity of piled foundation. *Soil and Foundations* **3**: 30–53.

Mandolini A and Viggiani C (1997) Settlement of piled foundations. *Géotechnique* **47(3)**: 791–816.

Mandolini A, Russo G and Viggiani C (2005) Pile foundations: experimental investigations, analysis and design. *Proceedings of the 16th ICSMGE, Osaka, Japan*, vol. 1, pp. 177–213.

Matsui T (1993) Case studies on cast-in-place bored piles and some considerations for design. *Proceedings of BAP II, Ghent, Belgium*. Balkema, Rotterdam, the Netherlands, pp. 77–102.

NCHRP (National Cooperative Highway Research Program) (2004) *Load and Resistance Factor Design (LRFD) for Deep Foundations. Report 507.* Transportation Research Board, Washington, DC.

O'Brien AS (2007) Raising the 133 m high triumphal arch at the New Wembley Stadium, risk management via the observational method. *Proceedings of the 14th European Conference on Soil Mechanics and Geotechnical Engineering, Madrid, Spain*, vol. 2, pp. 365–370.

O'Brien AS and Bown A (2010) Pile group design for major structures. *Proceedings of the 11th International Conference on Piling and Deep Foundations, London, UK*. DFI, London, UK.

O'Neill MW, Hawkins RA and Mahar LJ (1982) Load transfer mechanisms in piles and pile groups. *Journal of the Geotechnical Engineering Division* **108(GT12)**: 1605–1623.

Parkes J, Castelli R, Zelenko B, *et al.* (2018) *Design, Analysis, and Testing of Laterally Loaded Deep Foundations that Support Transport Facilities. Geotechnical Engineering Circular FHWA-HIF-18-031.* Department of Transport, Washington, DC.

Poulos HG (1989) Pile behaviour: theory and application. *Géotechnique* **39(3)**: 365–415.
Poulos HG (1993) Settlement prediction for bored pile groups. In *Deep Foundations on Bored and Auger Piles* (van Impe WF (ed)). Balkema, Rotterdam, the Netherlands, pp. 103–189.
Poulos HG (2005) Pile behaviour: consequences of geological and construction imperfections. *Journal of Geotechnical and Geoenvironmental Engineering* **131(5)**: 538–563.
Poulos HG and Davids AJ (2005) Foundation design for the Emirates Twin Towers, Dubai. *Canadian Geotechnical Journal* **42**: 716–730.
Poulos HG and Davis EH (1980) *Pile Foundation Analysis and Design*. Wiley, New York, NY.
Poulos HG, Carter JP and Small JC (2001) Foundations and retaining structures: research and practice. *Proceedings of the XV International Conference on Soil Mechanics and Foundation Engineering, Istanbul, Turkey*, vol. 4, pp. 2527–2606.
Raison C and Egan D (2016) Design of piles: United Kingdom practice. *International Symposium on Design of Piles in Europe, Leuven, Belgium*, pp. 1–24.
Randolph MF (1994) Design methods for pile groups and piled rafts. *Proceedings of the 13th ICSMFE, New Delhi, India*, vol. 5, pp. 61–82.
Randolph MF (2003) Science and empiricism in pile foundations design. The 43rd Rankine Lecture. *Géotechnique* **53(10)**: 847–875.
Randolph MF and Clancy P (1993) Efficient design of piled rafts. *Proceedings of the 2nd International Geotechnical Seminar on Deep Foundations on Bored and Auger Piles, Ghent, Belgium*, pp. 119–130.
Tomlinson MJ (1987) *Pile Design and Construction Practice*, 3rd edn. Palladian, Farnham, UK.
Viggiani C, Mandolini A and Russo G (2012) *Piles and Pile Foundations*. Spon Press, London, UK.

It is recommended this chapter is read in conjunction with

- Chapter 19 *Settlement and stress distributions*
- Chapter 22 *Behaviour of single piles under vertical loads*
- Chapter 54 *Single piles*

All chapters in this book rely on the guidance in Sections 1 *Context* and 2 *Fundamental principles.* A sound knowledge of ground investigation is required for all geotechnical works, as set out in Section 4 *Site investigation.*

Chapter 56
Rafts and piled rafts

Tony O'Brien Mott MacDonald, UK
John B Burland Imperial College London, London, UK
Tim Chapman Arup, London, UK

Rafts can provide a very cost-effective foundation solution. Their design requires a careful assessment of the soil–structure interaction. Many factors, including soil and raft stiffness, superstructure and substructure stiffness, local yield and soil nonlinearity and time-dependent effects, can influence differential settlement and structural forces in the raft. Errors associated with assuming that a raft sits on a bed of uniform soil 'springs' are discussed, and a practical alternative is outlined. Piled rafts, which are a hybrid foundation comprising a raft and piles, can offer many benefits. Currently there is little guidance in the codes and there is confusion about appropriate design methods. Piled-raft behaviour is potentially very complex. Significant simplifications are possible, which facilitate practical design and greater confidence in analysis output. A key aspect is a clear definition of two different types of piled raft: the raft-enhanced pile group and the pile-enhanced raft (previously known as a raft with settlement-reducing piles). The design concepts for each type of piled raft are very different. The intermediate zone between these two types should be avoided.

doi: 10.1680/icemge.66830.0897

CONTENTS

56.1. Introduction

This chapter provides guidance on the design of rafts and piled rafts.

Chapter 52 *Foundation types and conceptual design principles* describes the main foundation types, which include shallow foundations such as pads or strip footings and deep foundations such as piles. If individual pad or strip footings are relatively large or closely spaced it may be more practical and cost-effective to create a single reinforced-concrete raft beneath the structure. The rule of thumb that is commonly used is that if the pad or strip footings occupy more than about half of the superstructure footprint, then it is usually more appropriate to use a raft. However, site-specific issues may change this simple criterion. There are four main reasons to use raft foundations.

- The individual excavations for numerous pad or strip footings are relatively close to one another, and for construction purposes it is easier to create a single large excavation.
- The differential settlement between individual footings is excessive, and the use of a relatively rigid raft is needed to minimise differential settlement to a tolerable value. As noted in Chapter 53 *Shallow foundations* (**Figure 53.8**), it is necessary to check the influences of the interaction between pads and strips when assessing the overall settlement and differential settlement of a structure. Buildings on raft foundations have withstood large total settlements (in excess of 100 mm) with negligible damage, because the bending stiffness of the raft has been sufficient to reduce differential settlement to tolerable levels – refer to Chapter 52 *Foundation types and conceptual design principles* and **Figure 52.10**.
- There is a reduced risk of damaging differential settlements in heterogeneous ground conditions: in some situations it may be difficult to confidently assess the likely total and differential settlement, and past experience has shown that rafts can provide effective and economic foundations that mitigate this risk. An example is the use of rafts for house foundations on deep layers of non-engineered fill (refer to Chapter 58 *Building on fills*), sometimes with the building weight balanced by partial excavation to reduce the net imposed load. There may be other situations where it is deemed appropriate to use rafts in order to minimise the risk of damage due to site-specific geohazards (refer to Chapter 52 *Foundation types and conceptual design principles*).
- There is a reduced risk of bearing capacity failure and excessive settlement or differential settlement through using cellular rafts in soft clays: the raft is placed at the base of an excavation and the building weight is compensated for by the weight of soil that is excavated.

Piled foundations are often used because there is a concern about excessive settlement of shallow foundations, even though a raft would have an acceptable factor of safety (FoS) against bearing capacity failure. Conventional design of the pile group would then ignore the resistance provided by the pile cap or raft, and solely consider the resistance provided by the piles (refer to Chapters 54 *Single piles* and 55 *Pile-group design*); that is, all the design load is transferred through the piles. In the past three decades there has been increasing recognition that this conventional assumption is overly conservative when shallow foundations can be founded on competent soils (such as stiff clays). It is now known that a hybrid foundation can be designed, which makes use of both a raft and piles, and allows for load-sharing between the raft and piles. This foundation type is usually known as a piled raft. Unfortunately, current codes and standards do not provide specific guidance on piled-raft design, and there are still some misunderstandings about the design of piled rafts. The main intent of this chapter is to provide clear and coherent guidance on the key concepts that should be used for piled-raft design.

There are several reasons why a piled raft may be used

(*a*) to minimise total settlement, and, in particular, differential settlement

(*b*) to reduce bending moments and shear forces in the raft (especially effective where there are a small number of heavily loaded columns)

(*c*) a further practical advantage from (*b*) is that the raft thickness can be minimised, overcongested reinforcement can be reduced and, perhaps, excavation below the water table avoided.

This chapter builds upon many of the design issues described in Chapters 53 *Shallow foundations*, 54 *Single piles* and 55 *Pile-group design*, and it is assumed that these will be referred to by the reader. This chapter is split into the following sections:

(56.2) *Analysis of raft behaviour:* Describes the main soil–structure interaction issues that affect raft design and particularly the pitfalls associated with a 'raft on springs' design approach.
(56.3) *Structural design of rafts:* Introduces the key practical design and detailing issues for raft design.
(56.4) *Design of a real raft:* Reviews the main steps in the design process and buildability considerations, and includes a mini case history example.
(56.5) *Piled rafts, conceptual design principles:* Outlines fundamental differences in behaviour between different types of piled raft, provides a new terminology for piled rafts, gives guidance on appropriate ground conditions and outlines the main steps in design.
(56.6) *Piled-raft ground–structure interaction, general considerations*: Provides an introduction to the complex interactions which can develop in a piled raft.
(56.7) *Raft-enhanced pile groups:* Summarises the main design issues and outlines appropriate design approaches and how undue complexity can be avoided.
(56.8) *Compensated piled rafts:* Piled rafts located at the base of deep excavations need to take account of additional design and construction issues, and this section explains the main issues and gives some practical examples.
(56.9) *Pile-enhanced rafts:* Introduces a simple design approach and explains how a highly efficient piled raft can be designed.
(56.10) *Lateral loads on piled rafts:* A piled raft can be far more effective at resisting lateral loads than a conventionally designed pile group; the main differences in lateral load resistance between different types of piled raft are outlined.
(56.11) *A case history of a pile-enhanced raft: the Queen Elizabeth II Conference Centre.*
(56.12) *Key points:* A summary of the main design issues.

56.2. Analysis of raft behaviour

56.2.1 Design requirements

Raft design requires an assessment of

(*a*) total settlement and, particularly, differential settlement

(*b*) bending moment and shear forces across the raft, either due to uniformly distributed loads or concentrated loads from superstructure columns

(*c*) practical issues related to the thickness of the raft, the density of reinforcement, location of the groundwater table, size of excavation and so on.

Total settlement can be assessed by the methods of analysis outlined in Chapter 53 *Shallow foundations*; (*c*) will be discussed in more detail in Section 56.4. The next section focuses on the assessment of differential settlement, bending moment and shear force. Initially some of the key soil–structure interaction issues will be discussed together with the pitfalls associated with commonly used 'raft on spring' methods. Finally, a practical approach to raft–soil interaction analysis will be described.

56.2.2 Raft–soil interaction

Several different factors can influence raft differential settlement, bending moments and shear forces; these include

- the relative soil–raft stiffness (**Figure 56.1** and **56.2**)
- the influence of the superstructure or substructure stiffness on the 'effective' raft stiffness (**Figure 56.3**)
- the influence of variations in the soil stiffness and the thickness of the compressible layer
- the influence of soil nonlinearity and local yield (**Figure 56.4**).

Figure 56.1 illustrates the fundamental influence of the raft stiffness on differential settlement and bending moment. For a raft that is practically rigid, **Figure 56.1(a)**, the differential settlement is negligible, but the maximum bending moment is very high. In contrast, for a raft that is practically flexible, **Figure 56.1(b)**, the differential settlement is large and the

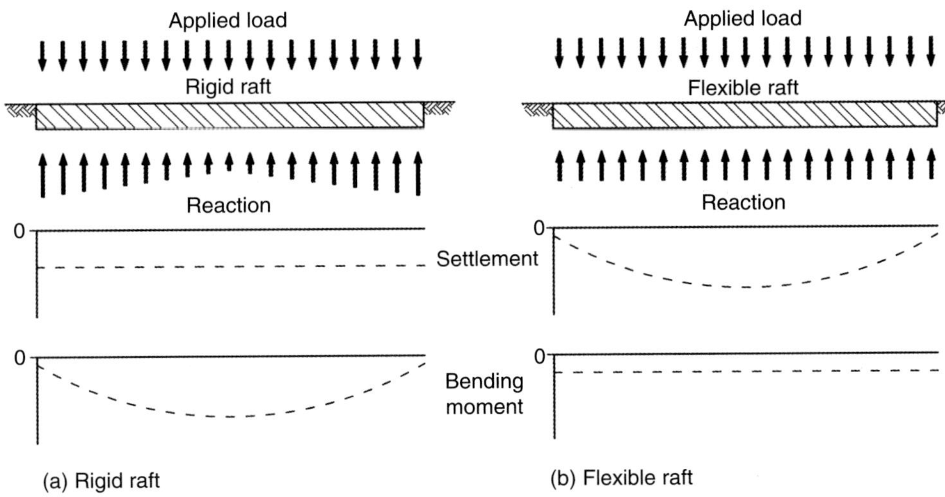

Figure 56.1 Influence of raft stiffness on behaviour

bending moment is zero. The stiffness of many real rafts will be between these two extremes, and then the relative soil–raft stiffness and soil–structure interaction need to be considered.

Figure 56.2 shows the variation of differential settlement and maximum raft bending moment for a uniformly loaded circular raft on a deep elastic soil layer, based on studies by Brown (1969) and Hooper (1974). This indicates that for the relative soil–raft stiffness $K < 0.1$ the raft is effectively 'flexible' (and therefore ineffective in reducing differential settlement) and for $K > 5$ the raft is practically rigid (and therefore differential settlement is negligible). For the zone $0.1 < K < 5$ the relationship between the raft bending moment and differential settlement changes rapidly. The practical value of these sorts of charts is that

- a simple calculation of K can indicate if the raft is sufficiently stiff to limit differential settlement to acceptable values
- for rafts at either extreme, very simple models can be used to assess both structural forces and foundation settlement, since the soil–structure interaction has a limited effect
- for rafts of intermediate stiffness ($0.1 < K < 5$) a more sophisticated method of analysis will be necessary, since the soil–structure interaction will have a dominant effect on raft behaviour.

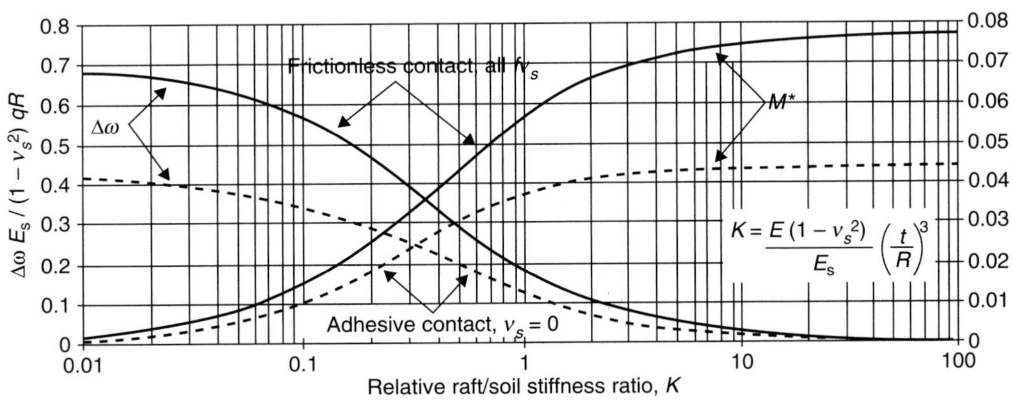

NB.
'Adhesive' contact means full soil strength mobilised at raft–soil interface, this is not an appropriate assumption for design purposes.
M^* = maximum raft bending moment, $\Delta\omega$ = differential settlement
E, v = Young's modulus and Poisson ratio respectively, for raft
E_s, v_s = Young's modulus and Poisson ratio respectively, for soil
t = raft thickness, R = raft radius

Figure 56.2 Maximum differential settlement ($\Delta\omega$) and bending moment (M^*) for a uniformly loaded circular raft on a deep elastic layer (Data taken from Brown, 1969, and Hooper, 1974)

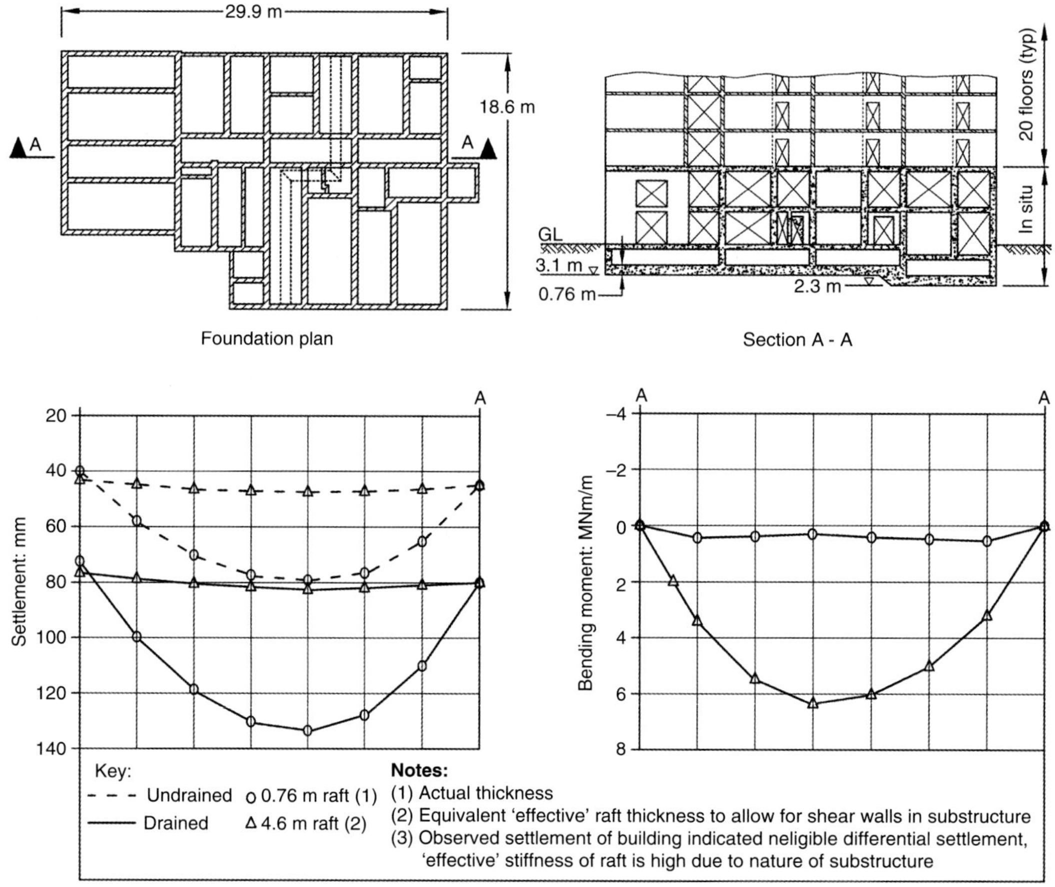

Figure 56.3 Influence of substructure stiffness on 'effective' raft stiffness (Modified from Hooper and Wood, 1977; all rights reserved)

A general approach to calculating K for rectangular rafts of varying aspect ratios is given in Chapter 53 *Shallow foundations.*

Figure 56.3 illustrates the importance of properly assessing how the substructure or superstructure may contribute to the 'effective' stiffness of the raft. The raft is only 0.76 m thick, and would be expected to be flexible given its small t/B ratio (t is raft thickness and B is its width or length). However, agreement between measured and computed differential settlement was only obtained by allowing for the high bending stiffness of the lowermost two storeys of the structure, which comprised stiff continuous shear walls. The actual differential settlement was negligible, but only at the expense of relatively high bending moments, which would be underpredicted if the substructure stiffness had been ignored. The influence of the superstructure or substructure on raft behaviour will depend on their relative stiffnesses, as discussed by Brown and Yu (1986), Lee and Brown (1972) and Fraser and Wardle (1975), among others. Hence, as for the case above, the stiffening effect of rigid reinforced-concrete shear walls on the raft is significant, whereas for a flexible light steel-framed structure the effect on raft behaviour will be small. This example illustrates the advantages that shear walls can have in reducing differential settlements of a raft.

In general, as the soil stiffness increases, the raft settlement and differential settlement will tend to reduce, and, as a consequence, the raft bending moments will also reduce. Hence, the compressibility of the soil, especially close to the raft base (typically within half the raft width) must be carefully assessed. For wide rafts, the influence of relatively stiff soils (say, dense sands or weak rock) at depth can be important (especially if within a depth of one times the raft width). These stiff layers will reduce raft settlement, differential settlement and bending moment. **Figure 56.4** illustrates the important influence of local yield on the distribution of contact stress along the underside of a raft. For an idealised linear elastic soil, the contact stresses are relatively low across the centre of the raft, and then increase towards the edge of the raft to extremely high values of p/q (where p is the local contact stress and q is the vertical pressure applied to the raft). For a real soil (which will fail *locally* once the contact stress fully mobilises the soil strength) the distribution of contact stress differs from the idealised elastic profile. As the applied vertical pressure increases (varying from a pressure equivalent to an undrained FoS against bearing capacity failure of about 10 to a value of 1.1) the contact stress becomes more uniform, particularly for drained conditions. The rule of thumb

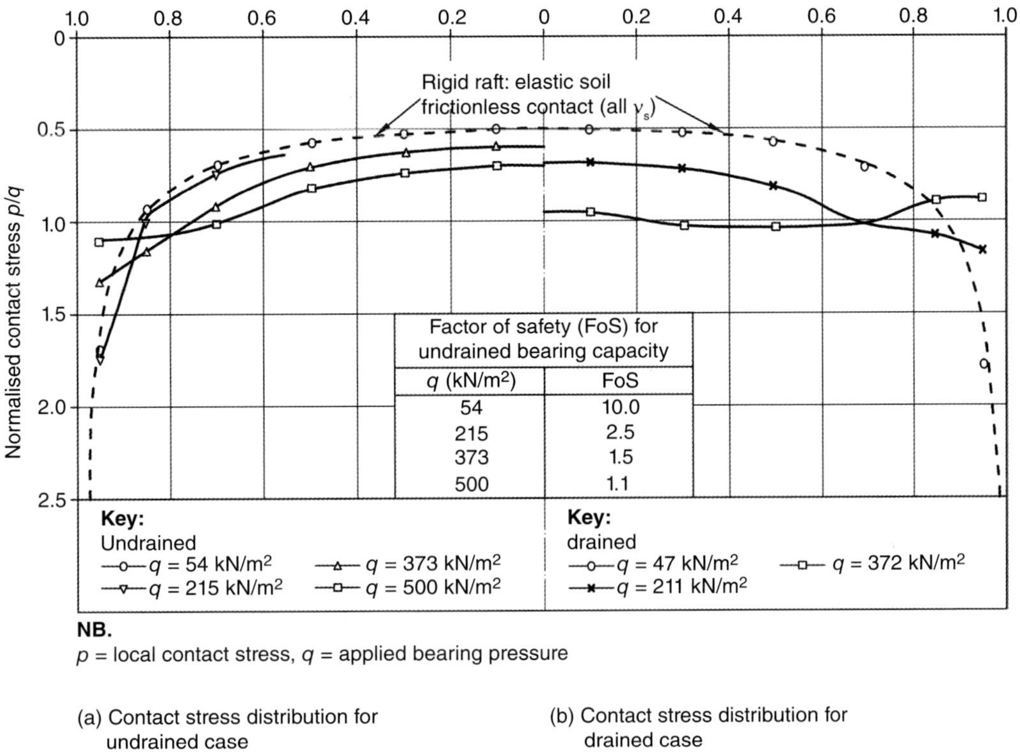

Figure 56.4 Influence of local yield on raft contact stress distribution (Modified from Hooper, 1983)

commonly used by designers, limiting the contact stress to 50–100% above the average applied pressure from the structure, appears reasonable.

Referring back to **Figure 56.2**, it can be seen that if it is assumed that the full soil shear strength can be mobilised at the soil–raft interface, then the predicted maximum bending moment is substantially reduced. However, as noted by Hooper (1983) it is usually very unwise to rely on significant raft–soil interface strength for the purposes of structural design. Usually, a low interface strength should be assumed to avoid unsafe provision of raft reinforcement. This should be remembered if numerical models (finite elements, etc.) are used for raft design, and it will be essential to include special interface elements along the underside of the raft model, with appropriately low values of strength specified.

56.2.3 Soil models for analysis: springs vs continuum

Structural engineers usually wish to model the soil as a series of springs, since this allows conventional structural analysis software to be used. Traditionally the soil springs have a uniform stiffness equivalent to the soil's modulus of subgrade reaction. There are problems with this approach, which are well known in the geotechnical engineering community; for example, the spring stiffness cannot be directly related to measurable soil parameters, since parameters such as the subgrade reaction moduli are also affected by the foundation width. Also the spring stiffness cannot directly allow for the effects of soil layering. However, the more fundamental problems with 'spring-type' models are often less understood. **Figure 56.5** illustrates a fundamental flaw in spring-type models: if a raft of uniform stiffness is subject to uniform loading and is analysed using soil 'springs', then it will settle a uniform amount (**Figure 56.5(a)**). This is inconsistent with real raft behaviour: a raft of finite stiffness will settle more in the centre than at the edges (**Figure 56.5(b)**). For a spring model to generate this curvature, the soil springs at the centre must be less stiff than those at the edges. Hence, appropriate soil spring stiffnesses cannot be uniform and are dependent on a range of factors, which will be dependent on the raft and the structure, the soil and the distribution of applied loads. The serious errors that can result from a non-interactive approach are illustrated in **Figures 56.6** and **56.7**. For a raft subject to edge loads (**Figure 56.6**) the error associated with a spring model is dependent on the relative soil–raft stiffness and on the rotational stiffness of the edge of the raft (e.g. it may be 'clamped' by a stiff perimeter wall). In this case the spring model is conservative. However, it is not possible to provide any generalised set of 'correction' factors, since spring models can lead to significant overestimates or underestimates of bending moment depending on the magnitude and distribution of the applied load (Hemsley, 1987). This is illustrated in **Figure 56.7**, where the raft differential settlement and bending moments predicted by the spring model are wildly different (and of opposite sign) to the more realistic continuum model.

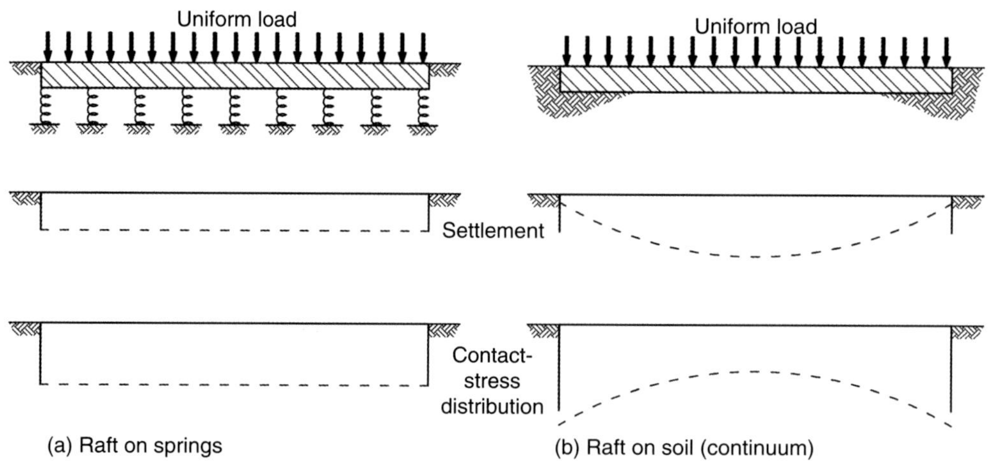

Figure 56.5 Raft on springs compared with raft on soil

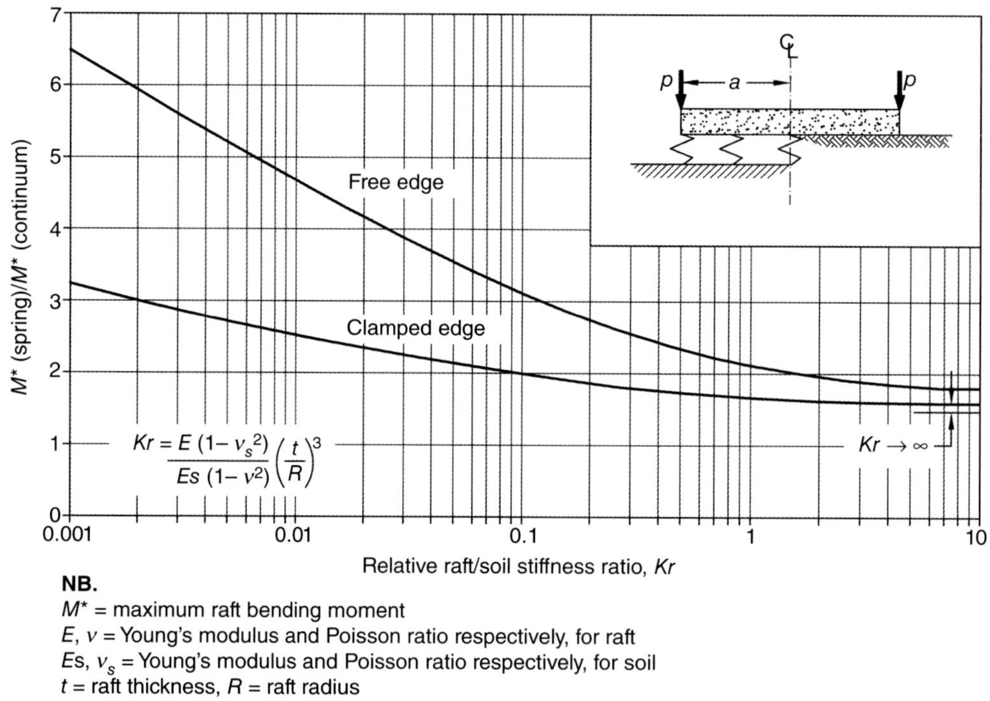

Figure 56.6 Ratio of maximum bending moments for an edge-loaded circular raft of varying stiffness (*Kr*), spring vs continuum model (Modified from Hemsley, 1987)

The key practical conclusion is that the traditional approach of providing subgrade reaction moduli for 'uniform' soil springs and then running an independent structural analysis can lead to serious errors and should *not* be used for raft design.

56.2.4 Simplified methods for raft–soil interaction analysis

For the simple situation of a rectangular raft that is uniformly loaded, the maximum bending moment and differential settlement can be assessed from charts developed by Horikoshi and Randolph (1997). The estimation of differential settlement is discussed in Chapter 53 *Shallow foundations.* Hemsley (1998) also provides charts that are useful for preliminary assessment purposes. Unfortunately, the various published charts are usually too idealised to be sufficient for detailed structural design. Usually the structure-specific loading distribution, spatial arrangement of walls, columns and so on differs substantially from those assumed in published solutions. Nowadays, it is possible to carry out 3D numerical modelling of soil–raft systems, although it is usually inappropriate (due to complexity, cost, time and lack of

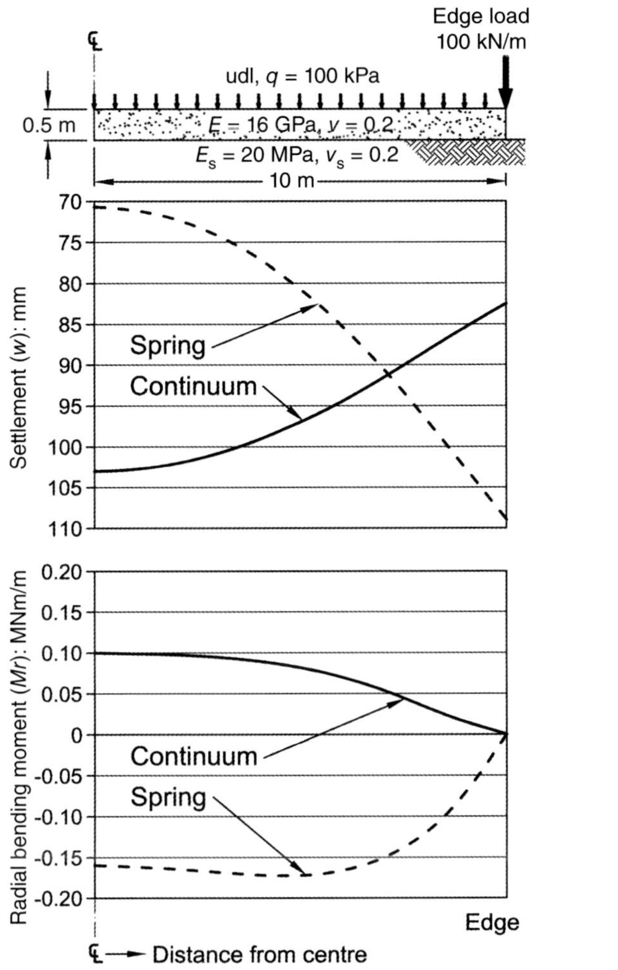

Figure 56.7 Settlement and bending moment for a circular raft subject to uniform pressure and edge load, continuum model compared to spring model (Data taken from Hemsley, 1987)

necessary specialist input) to implement a highly complex analysis for the majority of structures. A practical approach, which avoids the pitfalls associated with a non-interactive spring-type model (discussed above), is summarised in the flowchart in **Figure 56.8**. This relies upon a structural spring model and a soil continuum model being run in tandem, and both are iterated several times until there is sufficient convergence between the spring stiffness values (say, 95% convergence). Depending upon the design phase (preliminary or detailed design), the geotechnical category for the site, the structure and so on, the structural and soil models can be of varying levels of sophistication. The 'interactive spring stiffness', S_i, is defined as

$$S_i = \frac{\text{Structural spring reaction force}}{\text{Local soil continuum displacement at spring location}} \tag{56.1}$$

The soil continuum model assumes the raft is flexible.

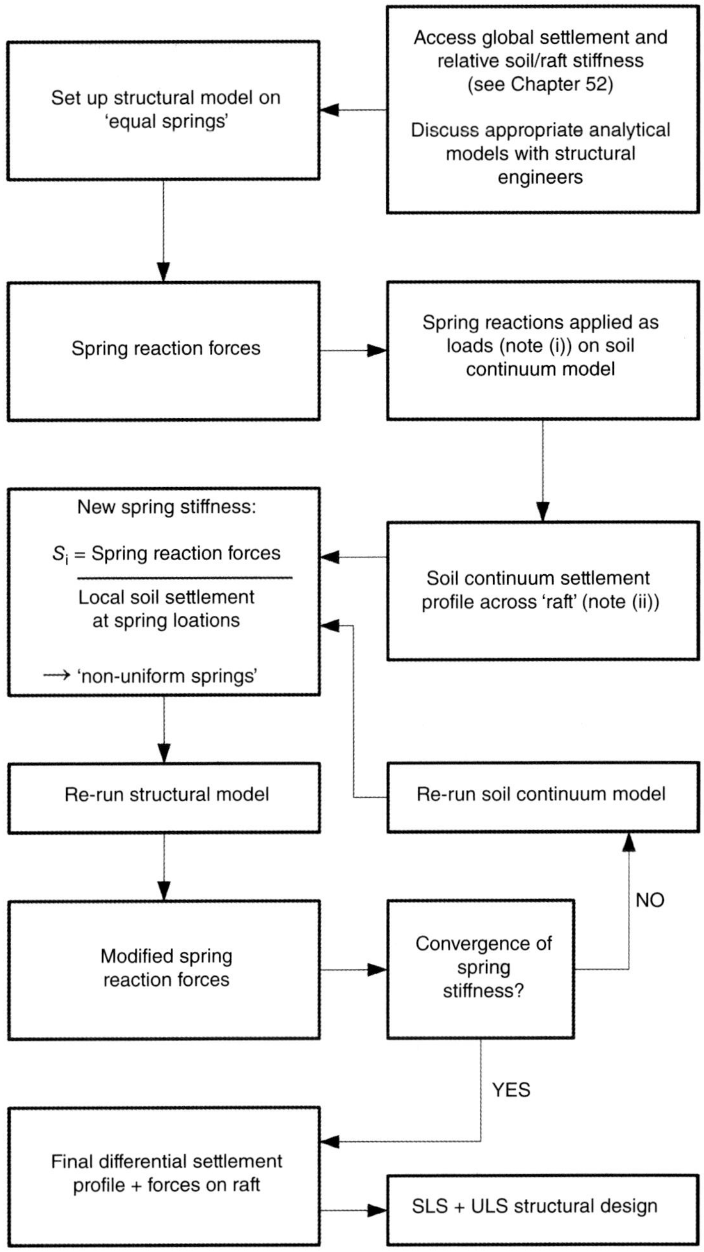

Notes
(i) Raft assumed 'flexible' in soil continuum.
(ii) Consider gross and 'net' forces for short-term and long-term settlement respectively; refer to text.

Figure 56.8 Flowchart for raft–soil–structure interaction analyses

The iterative analysis is usually started by estimating the spring stiffness from the overall raft settlement (calculated by conventional soil mechanics methods; Chapter 53 *Shallow foundations*). These spring stiffnesses are then used in a structural model (with appropriate raft and substructure stiffnesses, applied loading for columns, walls, etc.). The spring reaction forces are then applied as loads in the soil continuum model, with appropriate characterisation of the soil behaviour (stiffness, layering, etc.). This will provide a non-uniform settlement profile along the 'raft' base, which is then used to recalculate the new spring stiffnesses of Equation 56.1. These values are re-input into the structural model and re-run. This process is repeated until the deflections from the soil continuum and structural models are similar at all spring locations. The stiffness of the soil springs will usually vary significantly beneath the raft, typically with softer springs in the centre and stiffer springs around the edge of the raft.

Some care is required when the raft is located at the base of an excavation, and there is a significant difference between the gross and net bearing pressures (refer to Chapter 52 *Foundation types and conceptual design principles* for definitions). The structural analysis will require gross pressures; however, the long-term soil settlement will be dependent on net bearing pressures (based on the changes in effective stress). For rafts on clay, gross bearing pressures would be used for undrained settlement and net bearing pressures for drained (total) settlement (or swelling, if the net effective bearing pressure is negative, allowing for overburden removal and any changes in the water table level).

56.3. Structural design of rafts

As for all foundations, rafts must remain structurally integral and not deflect excessively and cause unacceptable angular distortions in the structure above. There are two main modes of potential structural failure

- bending
- shear.

These are illustrated in **Figure 56.9**.

While unreinforced concrete has some capacity to resist both of these sets of forces, generally steel reinforcement is used to augment capacity and provide adequate resistance. In both cases, structural codes (such as BS 8110 or Eurocode 2) provide detailed guidance on how to design reinforced concrete. The purpose of this part of the chapter is to outline the basic principles.

Simple beam theory states

$$\frac{f}{y} = \frac{E}{R} = \frac{M}{I} \tag{56.2}$$

where f is the peak stress in the extreme fibre of the section at distance y from the neutral axis; E is Young's modulus for the concrete and R is the induced radius of curvature (note that for small curvatures the radius of curvature is high); M is the corresponding applied moment in the concrete section and I is the moment of inertia for the section, which is (1/12) bd^3 for a rectangular section, where d is the depth and b is the breadth, equal to 1 m per metre run.

The equation is often simplified to

$$M = EI/R \tag{56.3}$$

so that the product of the moment and the radius of the induced curvature is equal to the section's flexural rigidity EI, and from this it can be deduced for any section that the bending moment and radius of curvature are inversely related. Hence, adding more steel, so that the section can resist greater moments, will reduce the radius of curvature – that is, it will increase the actual curvature and hence differential settlements. However, a stiff section may attract load, so the benefit of continually adding more steel may reduce.

Excessive bending in a structural element also causes cracks in the extreme tension fibres of the concrete, which can reduce durability. The significance of cracks in structural elements that are exposed to saturated soil with no free oxygen is a topic of debate, but structural codes provide firm guidance.

Bending resistance is provided by steel on the tension face, and sometimes also by steel on the compression face. Thus a raft often includes two distinct layers of steel, in two perpendicular directions. This is shown in **Figure 56.10**. For lightly reinforced rafts, mesh reinforcement is sometimes used.

A critical consequence of bending is the induced curvatures, which lead to differential settlements between different parts of the raft. A key issue in raft design is to control these differential settlements so that the supported structure above remains fully functional – that is, angular distortions in the structure are acceptable, finishes are undamaged and the building function is not impaired by excessive tilt.

Shear reinforcement in a concrete raft is needed when either a column or a pile may 'punch' through the element, as shown in **Figure 56.11(a)**. If a base plate is added beneath the column, the plane of potential rupture is forced further out and a greater perimeter of concrete is mobilised to resist rupture

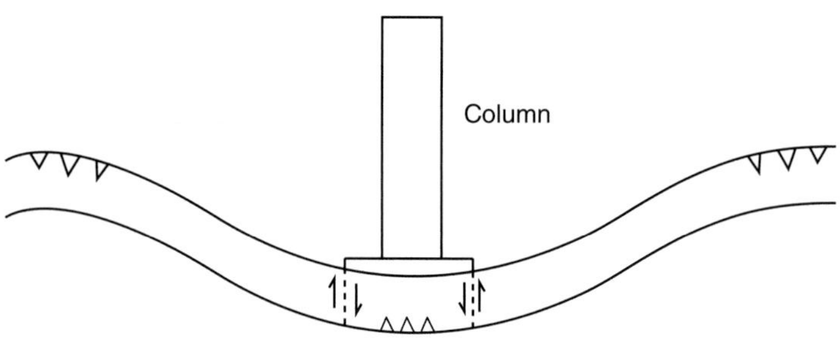

Figure 56.9 Modes of failure of rafts

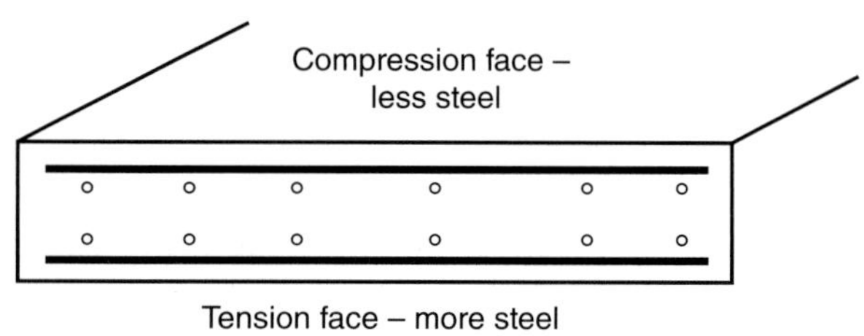

Figure 56.10 Typical bending steel in a raft designed for sagging

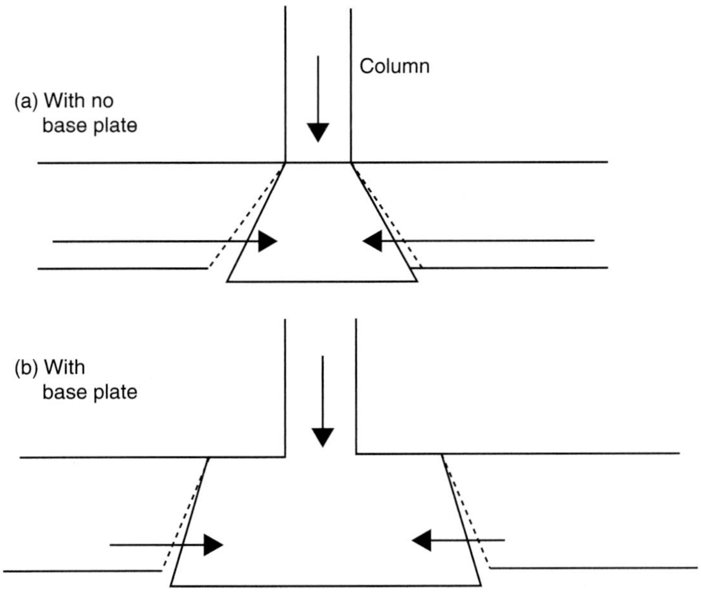

Figure 56.11 Punching of a column through a raft slab

(see **Figure 56.11(b)**). **Figure 56.12** shows a simplified shear stress example. The shear force down is 5000 kN – 200 kN/m^2 × 1 m^2 = 4800 kN, and the area of concrete resistance is 1.2 × 1 × 4 = 4.8 m^2. Therefore, the shear stress is 4800 kN/4.8 m^2 = 1000 kN/m^2 = 1 N/mm^2. Since unreinforced concrete can resist a shear stress of up to 1 N/mm^2, normally no special shear reinforcement is needed. For higher induced shear stresses, extra reinforcement may be needed to resist these shear stresses.

In the case shown in **Figure 56.12**, the induced shear stress in the concrete v_c is 1 N/mm^2, which is at the limit where shear reinforcement might usually be needed. As shear

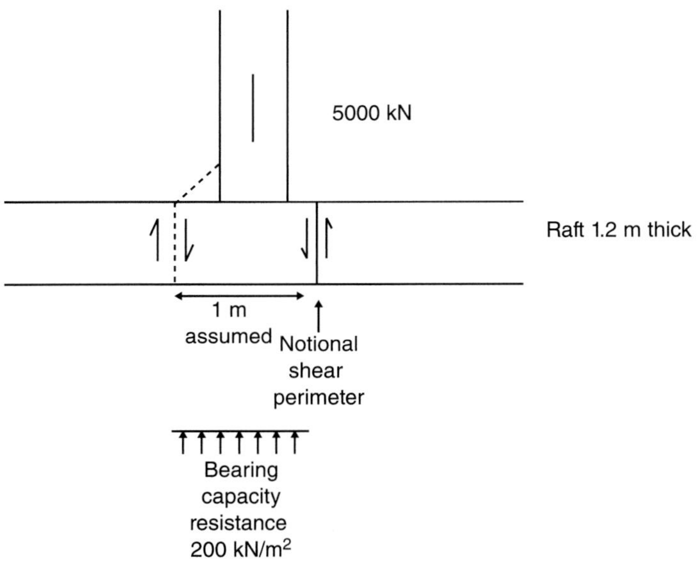

Figure 56.12 Simplified shear stress example

reinforcement is difficult to fabricate, other options to avoid or limit its use might include

- widening the column base to extend the shear perimeter
- thickening the raft, although the extra excavation and concrete will be expensive, and in a basement may necessitate also a deeper retaining wall
- adding a 'shear-reducing' pile beneath the column location to increase the soil resistance locally and hence reduce the net load that needs to be shed into the rest of the raft.

56.4. Design of a real raft

56.4.1 Bracketing by load cases

Raft design needs to accommodate a wide range of design cases and the induced forces and deflections are very sensitive to those load cases. The design cases include loads from the structure and loads and reactions from the ground. Loads from the structure range from a minimum of the dead load only to a maximum of the full dead and live loads. The ground loads include water pressures and heave pressures (which can be maxima or minima) from under a basement and a range of ground reactions based on upper- and lower-bound ground stiffnesses. These various combinations give a wide range of curvatures, often with sagging or hogging at the same location for different cases, and hence the steel needed in upper and lower reinforcement layers can originate from apparently odd load combinations. An example of a set of load combinations for a raft design is given in **Table 56.1**.

56.4.2 Design steps

Raft design, almost more than the design of any other geotechnical element, depends on mutual comprehension between geotechnical and structural engineers. As the raft design process is more complex, often piles are chosen ahead of rafts, even though the raft design may be more efficient in terms of cost, ease of construction, programme, risk reduction and even carbon saved.

The design complexity depends on

- the magnitude of loads, connected to the building height and column grid layout
- whether there is a basement, and if there will be water or heave pressures
- the number of load cases that should be considered
- the magnitude of variation in loads, particularly between adjacent columns
- whether the ground is relatively soft
- the drive for extreme economy in the raft.

The complexity of analysis needed crudely depends on the compatibility between the structural and the geotechnical analyses. A typical process is shown in **Figure 56.13**. Where

Basic load cases[a]	Additional load cases[d]
ULS	Wind and SLS
1.4A1 + 1.4A2 + 1.6A3 + 1.4A4 + 1.4A5 + 1.6A6	DL + SDL + IL + Water tank + Earth + Surcharge
	DL + SDL + IL + Water tank + Earth + Surcharge + Wind (X)
	DL + SDL + IL + Water tank + Earth + Surcharge + Wind (–X)
	DL + SDL + IL + Water tank + Earth + Surcharge + Wind (Y)
	DL + SDL + IL + Water tank + Earth + Surcharge + Wind (–Y) found to be most critical
SLS	SLS – DL[b]
A1 + A2 + A3 + A4 + A5 + A6	SDL + IL + Water tank + Earth + Surcharge
	SDL + IL + Water tank + Earth + Surcharge + Wind (X)
	SDL + IL + Water tank + Earth + Surcharge + Wind (–X)
	SDL + IL + Water tank + Earth + Surcharge + Wind (Y)
	SDL + IL + Water tank + Earth + Surcharge + Wind (–Y) found to be most critical
	Critical wind[c]
	SDL + IL + Water tank (1,2) + Earth + Surcharge + Wind (–Y)
	SDL + IL + Water tank (1,3) + Earth + Surcharge + Wind (–Y)
	SDL + IL + Water tank (2,3) + Earth + Surcharge + Wind (–Y)
	SDL + IL + Water tank (3,4) + Earth + Surcharge + Wind (–Y)

[a] A1: Dead load (DL); A2: superimposed dead load (SDL); A3: imposed load (IL); A4: water tank (e.g. the primary treatment works building has a few tanks and different combinations of fully loaded tanks were considered in the differential settlement check); A5: earth pressure; A6: surcharge (from soil above the roof); A7: wind X; A8: wind –X; A9: wind Y; A10: wind –Y.
[b] DL is omitted for settlement and differential settlement checks on the assumption that the proposed utilities and pipeline connections will be installed after the DL has taken effect.
[c] Based on the –Y wind direction being the most critical for the different tank load combination effects on the utilities and pipeline connections.
[d] The load cases were for a sewage treatment facility.

Table 56.1 Example of a set of load cases for a raft design

the two sets of analyses yield relatively close agreement at first iteration, then a less rigorous design process can be followed.

As stated earlier, the structural analyses usually assume the ground reaction can be modelled by simple springs while the geotechnical analyses model the ground response following Boussinesq or similar principles that more accurately model settlement interactions between adjacent loaded areas, such as the OASYS PDISP analysis program. The errors associated with spring models are described in Section 56.2.3. Geotechnical analyses are usually seriously limited in their ability to model structural behaviour.

Sometimes a sufficiently accurate simplification can be achieved by performing a second structural iteration, refining the springs to better approximate the likely soil displacements. Otherwise a longer series of iterations between structural and geotechnical analyses will yield a better approximation, as discussed in Section 56.2.4 and shown in **Figure 56.8**. The structural analysis is initially run to give a load disposition over a number of loaded areas; those loads are then fed into a geotechnical analysis which gives a new disposition of settlements, taking into account adjacent effects; those settlements are used to refine the springs assumed in the structural analysis and it is again re-run. The process is continued until a reasonable state of convergence is achieved between the structural and geotechnical analyses.

Finite element (FE) techniques can be used to give even more precise results. An FE approach combines the structural and geotechnical modelling into a single analysis. But even 2D FE modelling can require too much simplification as both building layouts and settlements are very 3D by their nature, so often a 3D analysis is required. And once 3D analyses are applied to a wide range of load combinations, the whole design process can become very time-consuming and expensive.

In addition to these interactions crude checks are needed at an early stage for global and local bearing capacity failure. It is rare to find a raft that could exhibit a kinematically valid bearing capacity failure mode, as the more likely modes are excessive displacement or a structural failure, but this relatively rapid check remains important as an indicator of potential design problems.

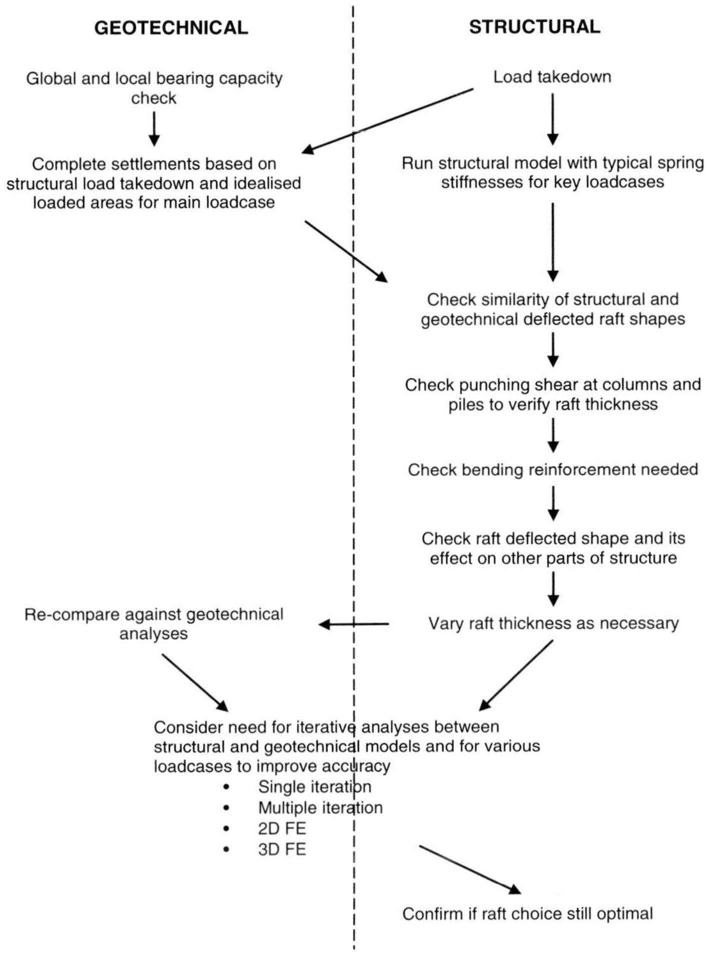

Figure 56.13 Simplified raft design process

Whenever there are significant water loads, a check for net buoyancy should also be made for the critical load stages, which could include a future partial demolition of the building above.

56.4.3 Factor of safety considerations

Usually an FoS calculation is less critical in raft design than in other types of geotechnical design, as usually movements (serviceability) are more critical than ultimate failure considerations. As for other aspects of design, it is easy to muddle geotechnical and structural safety factors and design codes. Both British Standards and Eurocodes offer little guidance. Hence, careful thought is needed to identify and resolve these potential inconsistencies. Generally, as the magnitude of the resisting forces is very large, a lumped FoS of 2 is assumed to be sufficient margin against bearing capacity failure.

Simplistically for a pure raft, the structural analysis considers the deflected shape and external forces acting on the element while the geotechnical analysis considers movement and the global bearing capacity failure of the whole raft foundation. Local bearing capacity calculations can sometimes be relevant, but are usually obviated as the structural check against failure prevents a local failure of a discrete part of the raft from being kinematically admissible.

56.4.4 Buildability considerations for raft construction

The first set of buildability choices relate to the raft's formation level, which corresponds to the chosen raft thickness. The formation layer should ideally be dry, well drained and above the water table. Often it will be necessary to thin a raft to bring its formation level above a likely water table level. From a structural perspective, it is expedient to thicken a raft beneath columns to resist the increased shear forces, but such stepped formations are difficult and potentially less safe to build.

Steel is relatively expensive and slow to place in cages, especially significant quantities of shear steel. There is always a trade-off between a thicker raft with less steel and a thinner raft with more steel. If a raft has too much steel it can be difficult to compact the concrete and so the raft may be more prone to defects, long-term durability and water-penetration issues.

If the ground contains some contamination or there is a risk that it may, a thinner raft needing less excavation may be advantageous. Ground gas penetration can be an issue, especially at joints – for instance, with a retaining wall or at any penetrations through the raft.

If the design calls for an under-slab drainage layer to reduce the design water pressures, the layer needs to be considered carefully. Consideration of price alone suggests gravel-filled grip drains, while broader buildability and construction-safety considerations will probably result in a no-fines concrete drainage layer that can also act as a blinding layer. Care must be taken to ensure that the no-fines concrete has high permeability, rather than just porosity, and will not be prone to clogging. Often a roddable pipe network is included as a precaution, connected to manhole covers that can lift if water pressures exceed the design levels.

56.5. Piled rafts, conceptual design principles

56.5.1 Advantages

Compared with conventionally designed pile groups and rafts, piled-raft design offers significant advantages when used in suitable ground conditions

(*a*) minimise total settlement

(*b*) minimise differential settlement

(*c*) reduce bending moment and shear force in the raft

(*d*) a further practical advantage from (*c*) is that the raft thickness can be minimised, overcongested reinforcement can be reduced and, perhaps, excavation below the water table avoided (*e*) costs can be significantly reduced (e.g. number of piles can be significantly reduced compared with conventionally designed pile groups – for many piled rafts the number of piles can be less than half of a conventionally designed pile group)

(*f*) construction can be simplified and time periods for foundation construction can be reduced.

There are often safety and sustainability benefits associated with (*d*), (*e*) and (*f*).

Viggiani (1998) back-analysed the conventionally designed pile group for the Stonebridge Park building in London (comprising 350 13 m piles of 450 mm dia.). The use of a piled raft would have enabled a reduction in the number of piles to one-third of the original number (i.e. to 117 piles), and this would have only caused an increase in settlement of about 5–10 mm, with the raft carrying 20–25% of the load in the long term. Hence, the use of piled rafts can create considerable cost and carbon savings. **Table 56.2** provides a selection of piled raft case histories.

56.5.2 Conceptual design scenarios

Usually in foundation design the first option is to consider a shallow foundation to support the superstructure and, if not adequate, a conventional piled foundation is then adopted. It follows that, from a conceptual standpoint, piles are utilised for helping the raft in satisfying ultimate limit state (ULS; i.e. adequate safety against a bearing capacity failure) and/or serviceability limit state (SLS; i.e. acceptable settlements under working loads) requirements. Consider the simple case of a piled-raft foundation loaded by only a vertical load V_{PR}. The load will be shared between piles (V_P) and raft (V_R). Consider the load-sharing ratio α_{pr} (**Figure 56.14(a)**), defined as

$$\alpha_{pr} = \frac{V_P}{V_{PR}} = 1 - \frac{V_R}{V_{PR}} \tag{56.4}$$

A load-sharing ratio $\alpha_{pr} = 0$ represents a shallow foundation with no piles, while $\alpha_{pr} = 1$ represents a conventionally designed pile group (with the pile cap, or raft, assumed to be isolated from the ground). Piled-raft foundations cover the range $0 < \alpha_{pr} < 1$.

Figure 56.14(b) shows a schematic chart for selecting the appropriate design approach (Viggiani *et al.*, 2011). Point A represents an optimum condition for an unpiled raft which has a global FoS *FS* equal to some minimum acceptable value and also an overall 'average' settlement w_{UR} equal to some allowable value ($w_{UR}/w_{all} = 1$). Conceptually, three design situations may occur.

Reference/structure	Ground conditions	Pile spacing: s/d	Pile length: m	Raft width: m	Pile-group area raft area	Settlement: mm	Raft load: %
Urawa Building (Yamashita *et al.*, 1994)	Interbedded sands and clays	8	15.8	23	0.9	15	51
Stonebridge Park Building (Cooke *et al.*, 1981)	London Clay	3.6	13	21	0.95	12	23
Carigliano Bridge (Russo and Viggiani, 1995)	Interbedded clays or sands and gravels	3.0	50	10.6	0.88	52	20
MesseTurm (Sommer *et al.*, 1991)	Frankfurt Clay	3.5–6.0	27–35	59	0.85	140	45
Japan Centre (Katzenbach *et al.*, 2000)	Frankfurt Clay	5.5	22	43	0.45	60	60
DG Bank (Katzenbach *et al.*, 2000)	Frankfurt Clay	4–6	30	54	0.52	110	50
Hyde Park Barracks (Hooper, 1979)	London Clay	4.3	25		0.72	20–25	40
QE2 Conference Centre[a] (Burland and Kalra, 1986)	London Clay	N/A	16	33.5	N/A	16–20	>90
Napoli Tanks (Mandolini *et al.*, 2005)	Medium-dense sand	5 and 5.8	13.7 and 9.7	12.5 and 10.5	0.82	10–15 and 20–25	46 and 50
Canary Wharf South (Nicholson *et al.*, 2002)	Stiff clay (raft level), very dense sand (pile toe level)	4	25.5	25	0.9	45	16

[a] The sole purpose of the piles was to reduce the shear stresses and bending moments in the raft, beneath a few heavily loaded columns.

Table 56.2 A selection of piled-raft case histories

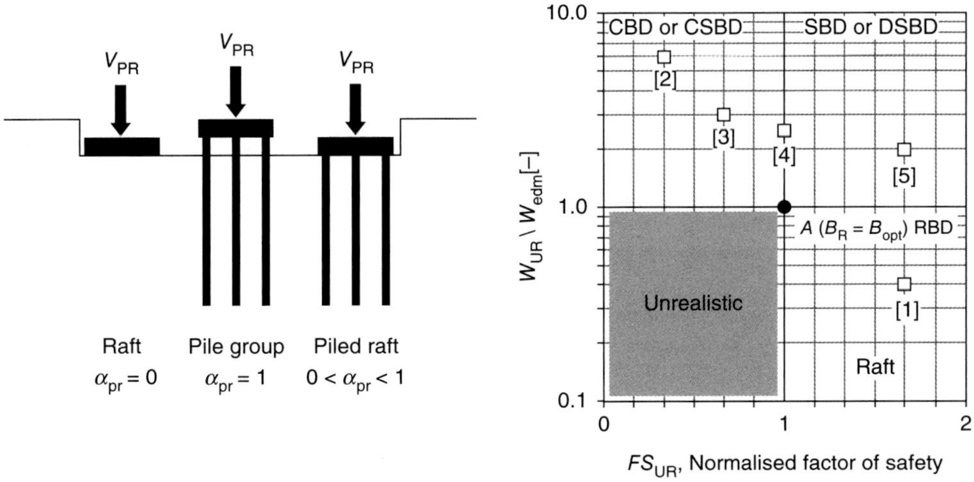

(a) Foundation systems

(b) Chart for selection of the design approach

Figure 56.14 Foundation systems and conceptual design principles

- Both the estimated values of FS_{UR} and w_{UR} are acceptable (point [1] in **Figure 56.14(b)**): the adoption of an unpiled raft is possible (i.e. raft-based design), 'RBD zone'; piles may be required to reduce the differential settlement and associated structural stresses in the raft – see 'pile-enhanced raft' below.
- Both the estimated values of FS_{UR} and w_{UR} are not acceptable (points [2] and [3]); piles have to be added to increase the value of FS_{UR} and to reduce the settlement w_{UR} (termed 'capacity and serviceability-based design' (i.e. CSBD zone)) to satisfy both ULS and SLS requirements. In general, this will require a 'raft-enhanced pile group'.
- Although the FoS of the unpiled raft is equal (point [4]) or larger (point [5]) than the minimum admissible value, the predicted displacement is larger than w_{all}. Piles have thus to be added with the aim of reducing total (and often also differential) settlement (termed 'serviceability-based design'; SBD zone) to satisfy only SLS requirements. For this situation, either a 'pile-enhanced raft' or a 'raft-enhanced pile group' may be acceptable.

In general, piled-raft behaviour can be very complex; see Section 56.6. Significant simplifications are possible which will greatly facilitate practical design and create greater confidence in analysis output. To achieve these simplifications, it is essential to have clear definitions (see Section 56.5.3) of two different types of piled raft: a 'raft-enhanced pile group' and 'a pile-enhanced raft' (previously known as a raft with 'settlement-reducing piles'). The design concepts for each type of piled raft are very different.

56.5.3 Different types of piled raft, key definitions and design principles

The following definitions will be used.

- *Raft-enhanced pile group:* Both the piles and raft are working within a pseudo-elastic range of behaviour. The ultimate pile-group capacity is *not* fully mobilised at the anticipated working load (and based on the expected raft–pile load-share). Referring to **Figure 56.14(b)**, raft-enhanced pile groups can be used for both CSBD and SBD zones (and can also be used in the RBD zone to minimise raft differential settlement).
- *Pile-enhanced raft:* The piles must be designed to mobilise all their ultimate geotechnical capacity and must exhibit ductile load–settlement characteristics. The piles will usually be located beneath heavily loaded superstructure columns. Referring to **Figure 56.14(b)**, pile-enhanced rafts should only be used within the SBD and RBD zones (i.e. to minimise raft differential settlement).

The different modes of behaviour and design principles are also illustrated in **Figure 56.15**, which shows a load–settlement curve for

- a raft (curve 1)
- a conventionally designed pile group (curve 2)
- a raft-enhanced pile group (curve 3)
- a pile-enhanced raft (curve 4).

In **Figure 56.15** the raft has a substantial ultimate bearing capacity, but the settlement at working load is deemed to be excessive. The conventionally designed pile group is very stiff and its settlement at working load is far smaller than the 'allowable' settlement. The raft-enhanced pile group (curve 3) is also relatively stiff, and this is achieved with a smaller number of piles than the conventionally designed pile group because the piles are operating more efficiently. The pile-enhanced raft (curve 4) exhibits more settlement than the conventional pile group and raft-enhanced pile group, but is slightly stiffer than the raft and can achieve the allowable settlement for this structure. The load–settlement curve for the pile-enhanced raft is within the nonlinear range. It is apparent that if the allowable settlement is set at too low a value, then the opportunity to

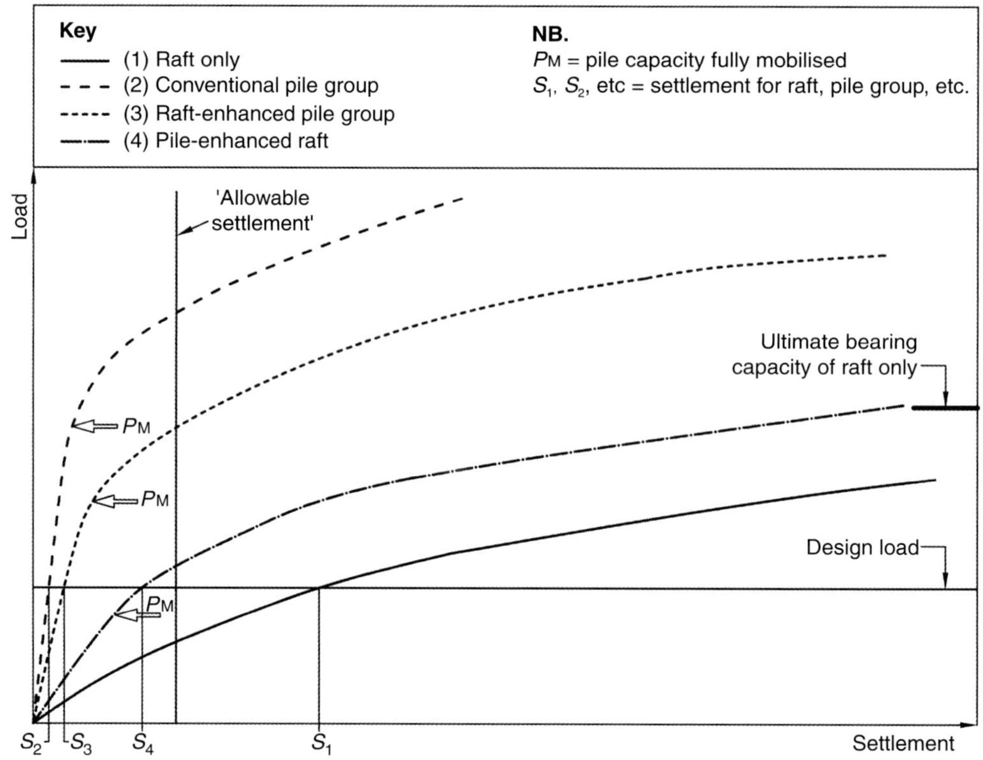

Figure 56.15 Load–settlement behaviour of rafts, conventional pile groups and different types of piled raft

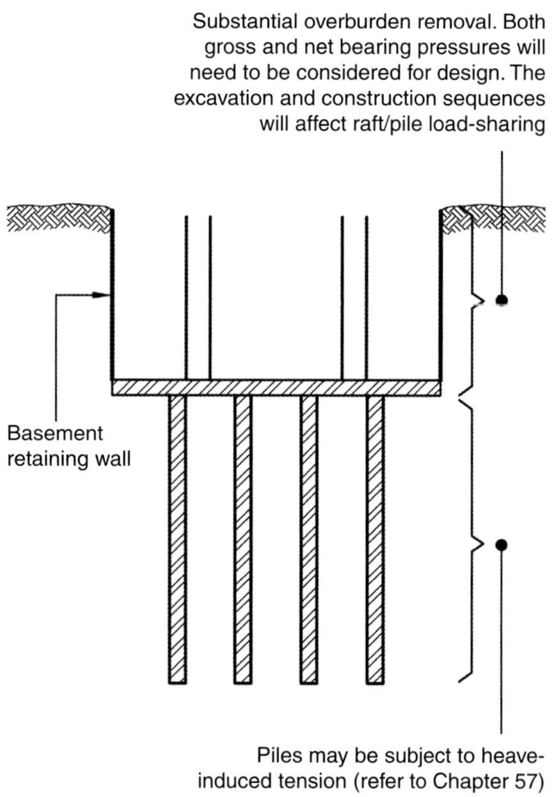

Figure 56.16 Compensated piled rafts

use piled rafts, especially 'pile-enhanced rafts', will usually be lost. Hence, defining a realistic value of allowable settlement is important; refer to Chapter 52 *Foundation types and conceptual design principles.*

An additional type of piled raft is the compensated piled raft (**Figure 56.16**).

- *Compensated piled raft*: This foundation is situated within a deep excavation, and because of the large reduction in overburden stress and, usually, a large groundwater pressure acting on the underside of the raft, this type of piled raft has additional design considerations (discussed in Section 56.8) beyond those of a piled raft close to the ground surface.

For the purposes of practical design, it is also useful to subdivide piled raft (in particular raft-enhanced pile groups) foundations into two broad categories: 'small' and 'large' piled rafts (**Figure 56.17**), defined as follows.

- *Small piled rafts:* Both piles and raft are required to meet ULS requirements (i.e. to achieve the required overall FoS). The foundation width, B, is relatively small compared with the pile length, L, with $B/L < 1$. In practical terms this will usually mean that B is about 15 m or less. Referring to **Figure 56.14(b)**, small piled rafts will usually be located in the CSBD zone. Small piled rafts should be designed as raft-enhanced pile groups; the primary benefit of the raft is to enhance the overall ultimate capacity of the foundation system (compared with a conventional pile group). The pile group can be designed more efficiently than a conventionally designed pile group.

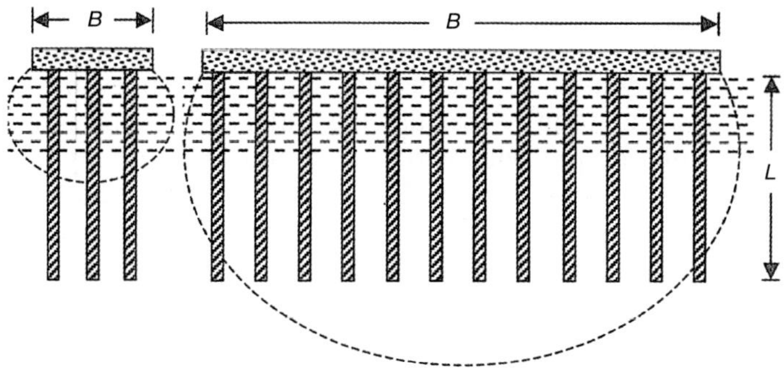

Figure 56.17 Small and large piled rafts

- *Large piled rafts:* The bearing capacity of the raft (on its own) is usually adequate. Hence the primary benefit of the piles is to reduce total and differential settlement and structural forces in the raft (bending moment, shear force, etc.). The width of the raft is large compared with pile length, with $B/L > 1$. Referring to **Figure 56.14(b)**, large piled rafts will usually be located in the SBD or RBD zones. Large piled rafts could be designed as either pile-enhanced rafts or raft-enhanced pile groups (selection is discussed below).

Pile-group stiffness (the ratio of total applied load to average foundation settlement) has been compared with the stiffness of a surface raft in a normalised plot of stiffness against pile-group width by Randolph (2003) (**Figure 56.18**). The stiffness envelope of pile groups and the surface raft can be matched by using an equivalent pier approximation (Randolph, 1994). This plot is useful for conceptual design since it illustrates the key influence of the ratio of pile-group width to pile length (B/L) on the reduction in settlement that may be achieved relative to a surface raft. Small piled rafts ($B/L < 1$) are usually far stiffer than a surface raft, whereas for a large piled raft ($B/L > 1$) the settlement reduction is more modest and often the focus is on reducing differential settlement rather than total settlement. Pile-group settlement can be calculated using approximate analytical models

- an 'equivalent pier' is most suitable when the pile-group aspect ratio $R < 2$
- an 'equivalent raft' is most suitable when the pile-group aspect ratio $R > 4$

where $R = (ns/L)^{0.5}$.

These approximate analytical models have been shown to be robust and reliable for a wide range of situations; refer to Chapter 55, *Pile-group design*.

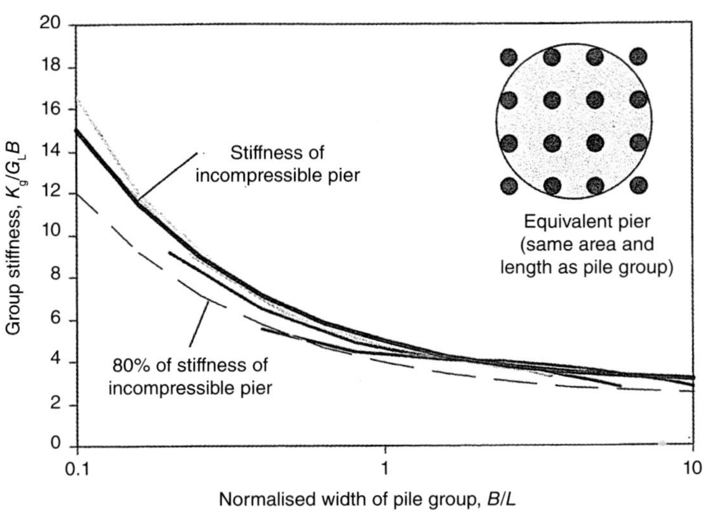

Figure 56.18 Normalised stiffness of pile groups, equivalent pier and surface raft (After Randolph, 2003)

56.5.4 Selection of an appropriate type of piled raft

Selection of a piled raft rather than a conventionally designed pile group generally depends upon appropriate ground conditions being present at the underside of a raft. The key questions that need to be considered by designers are summarised in **Table 56.3**.

Ground conditions	Are the overall geology and hydrogeology well known and understood? Is there competent ground (stiff clay or medium-dense or dense sand) at raft level?
Global ground movement	Are significant ground movements likely to develop adjacent to the raft in the future? Could these lead to a loss of support to the raft?
Raft bearing capacity and settlement	Would the raft have an adequate bearing capacity (say FoS in excess of 2) under anticipated loads? What is the likely upper- and lower-bound settlement?
Allowable settlement and differential settlement	Is the raft differential settlement acceptable? Can the allowable settlement be increased?
Pile type	Given the nature of the site, what types of piles are appropriate: friction or end-bearing? Is there potential for significant variability in pile stiffness and capacity? Is the behaviour ductile?
Piled raft	If the ground is competent at raft level, could a piled raft be used if the raft settlement is excessive, or if the bending moment or shear force in the raft is excessive? What type of piled raft is appropriate: a raft-enhanced pile group or a pile-enhanced raft?
Analysis requirements and design competency	Are appropriate analytical tools available? Are specialist staff available to deliver design?
Additional field tests	Are additional ground investigations required (e.g. to verify ground conditions at the raft or ground stiffness characteristics)? What additional pile tests are required?

Table 56.3 Piled rafts: key questions to consider

There are some ground conditions which are unsuitable for the use of piled rafts.

- Soil profiles containing heterogeneous non-engineered fill, soft clays or loose sands close to the underside of the raft, and which are underlain by more competent soils at depth in which the piles will be located. In this case the raft will not provide significant stiffness or bearing capacity compared with the piles. Even if some load-sharing is developed in the short term, long-term load-sharing between the raft and the piles cannot be relied upon. The piles should then be designed conventionally.
- Site locations and soil profiles which may be vulnerable to significant settlement due to external causes, (i.e. those situations where significant ground movements may develop independently of structural loads, such as due to water table fluctuations or soil creep or earthquake shaking). These must be carefully assessed. Piled rafts may be possible, but expert guidance will be necessary.

Selection of a pile-enhanced raft rather than a raft-enhanced pile group typically depends upon five factors: appropriate geology; foundation geometry; pile load–settlement characteristics; magnitude of allowable settlement; preliminary pile load tests.

A raft-enhanced pile group can be used for a broader range of design situations and is more tolerant of variations in ground conditions than a pile-enhanced raft. However, if feasible to use, a pile-enhanced raft will be far more economic, easier to design and build and will have less embodied carbon.

Potential variability in the ultimate geotechnical capacity of a pile is a key consideration for pile-enhanced rafts. Variations in ultimate pile capacity and pile stiffness are sometimes observed at an individual site. These are usually due to variations in pile construction methods (or quality problems) and/or the nature of the site geology and associated variations in ground properties. Some examples are shown in **Figure 56.19**. **Figure 56.19(a)** illustrates the behaviour of CFA piles in alluvial soils, and it can be seen that the initial stiffness is reasonably similar despite wide variations in ultimate capacity (due to changes in operation of the CFA piling rig). **Figure 56.19(b)** illustrates the wide variation in pile stiffness and capacity of driven piles in a soil, which varies between sandy clay and clayey sand. The variation is mainly due to the large changes in end-bearing resistance when the piles are toed into 'sand' rather than 'clay'. These ground conditions and associated variability in pile ultimate capacity would usually be unsuitable for pile-enhanced rafts. Deep deposits of homogeneous stiff heavily overconsolidated clay provide far more suitable conditions for pile-enhanced rafts. In this type of geology, pile load–settlement characteristics and ultimate geotechnical capacity will tend to be more consistent, provided pile installation is carefully controlled.

Many rafts are designed for a wide range of load cases and so there can be a risk of excessive differential settlement between adjacent columns. **Figure 56.20** indicates that where the *shaft* FoS is more than about 1.2–1.3, the pile settlement becomes much smaller and more consistent. Hence, it is not the overall FoS (i.e. shaft plus base), but the *shaft* FoS which mainly controls its stiffness. For end-bearing piles, both their stiffness and ultimate capacity tend to be more variable than for friction piles, because end-bearing behaviour is more sensitive to local variations in ground conditions and construction method than shaft behaviour. Therefore, before deciding upon the use of a raft-enhanced pile group or a pile-enhanced raft, it is important to consider the type of piles that can be used, the nature of the ground conditions beneath the site and

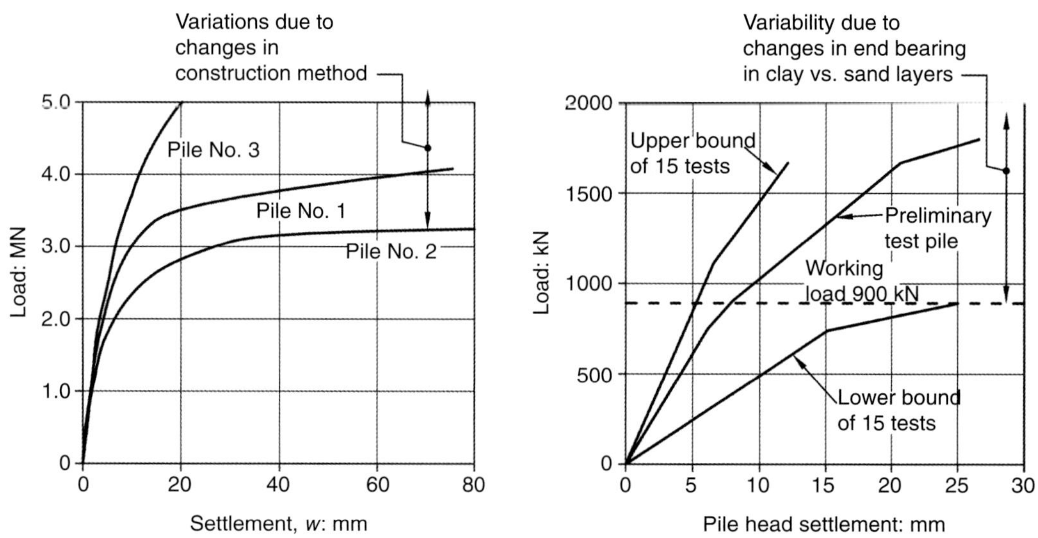

(a) CFA piles in alluvial soils (after Mandolini *et al.*, 2005)

(b) Driven piles in glacial till (after Corke *et al.*, 2001)

Figure 56.19 Examples of load–settlement curves for CFA and driven piles

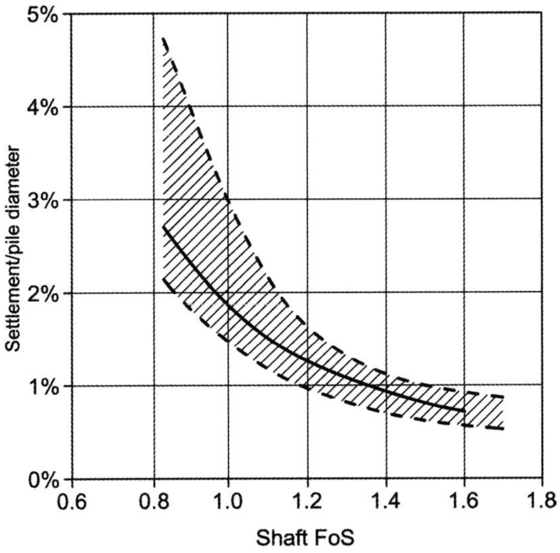

Figure 56.20 Pile settlement plotted against FoS on shaft capacity (Modified from Corke *et al.*, 2001; all rights reserved)

the potential variations in pile capacity (especially for pile-enhanced rafts) and stiffness together with the methods that can be used to minimise these variations.

Figure 56.21 illustrates a raft-enhanced pile group and a pile-enhanced raft, together with guidance on the ground profiles which would be most appropriate for their application. For a pile-enhanced raft, friction piles (i.e. piles which mobilise their resistance mainly in shaft friction) are suitable pile types, and generally bored piles have been used. For raft-enhanced pile groups, a wide range of pile types have been used, acting both as friction and end-bearing piles. The starting point for raft-enhanced pile groups and pile-enhanced rafts will often be very different (**Figure 56.21**). The raft-enhanced pile group is usually a pile group which is value-engineered to make use of the pile cap for resisting the applied loads and reducing the required number of piles. In contrast, a pile-enhanced raft is usually a raft which does not meet all the serviceability criteria, where a few piles are used to resolve (typically local) non-compliances.

56.5.5 Piled rafts: main steps in the design process

Poulos (2001) has outlined the main steps in piled-raft design.

(*a*) *'Conceptual' design:* Assess feasibility of using a piled raft and the required number of piles to broadly satisfy design requirements (settlement, bearing capacity).

(*b*) *'Preliminary' design:* Assess where piles are required and pile characteristics (length, diameter, etc.), differential settlement of raft.

(*c*) *'Detailed' design:* Finalise optimum number, location and configuration of the piles, and calculate detailed distribution of settlement, bending moment, shear force in raft and the pile loads, moments and so on.

Stages (*a*) and (*b*) should be based on an appraisal of the ground conditions, the type of pile which will be suitable (based on the site conditions and locally available piling technology) and on relatively simple calculations (outlined below). The detailed design stage will require appropriate analytical software, which accounts for interaction between piles, raft, soil and superstructure. This can vary from a modified version of the iterative and interactive spring approach commonly used for assessing raft behaviour (refer to Section 56.2.4 and **Figure 56.8**), through to 3D nonlinear numerical modelling. Even when sophisticated numerical modelling is used, simplified methods should always be used to 'sense check' sophisticated analysis outputs. In addition, there would normally be two separate analytical models, one focusing mainly on structural behaviour and the other focusing mainly on soil behaviour.

For conceptual design it is essential to consider

- raft behaviour without the piles (bearing capacity, settlement, differential settlement)
- pile-group behaviour without the raft (ultimate capacity, settlement, differential settlement).

For both cases relatively simple methods (commonly used for assessing the bearing capacity and settlement of shallow and deep foundations) are appropriate (refer to Chapters 53 *Shallow foundations* and 55 *Pile-group design*). If the ultimate capacity of the raft on its own is a small fraction of the required ultimate capacity or would be adversely affected by global ground movements which are independent of the applied structural loads, then a conventional pile-group design would usually be appropriate. However, if the raft can be founded on reasonably competent ground which can provide reliable long-term resistance, then the raft can be used to significantly reduce the piling requirements necessary to achieve the design criteria (ultimate capacity, settlement, etc.).

The next steps will then be to assess the ultimate capacity and settlement of the combined piled-raft system. Poulos (2001) suggested a simplified trilinear load–settlement curve for preliminary analysis, termed here the 'PDR method' (**Figure 56.22**). The ultimate capacity of all the piles below the raft (based on the sum of all the individual pile capacities or the overall group capacity) is mobilised at a total load of P_1 (**Figure 56.22**). As noted in the introduction, considerable simplifications are possible in analysis and design of piled rafts if the definitions recommended in Section 56.5.3 are adhered to. Raft-enhanced pile groups operate in the initial stiff zone prior to P_1 being mobilised. In contrast, pile-enhanced rafts operate in the 'softer' second zone after P_1. *It is very important (in order to avoid complex analysis of load-sharing between pile and raft) to ensure that for* ***all*** *load cases the piled raft does* ***not*** *traverse P_1 and enter the intermediate zone between raft-enhanced pile groups and pile-enhanced rafts.*

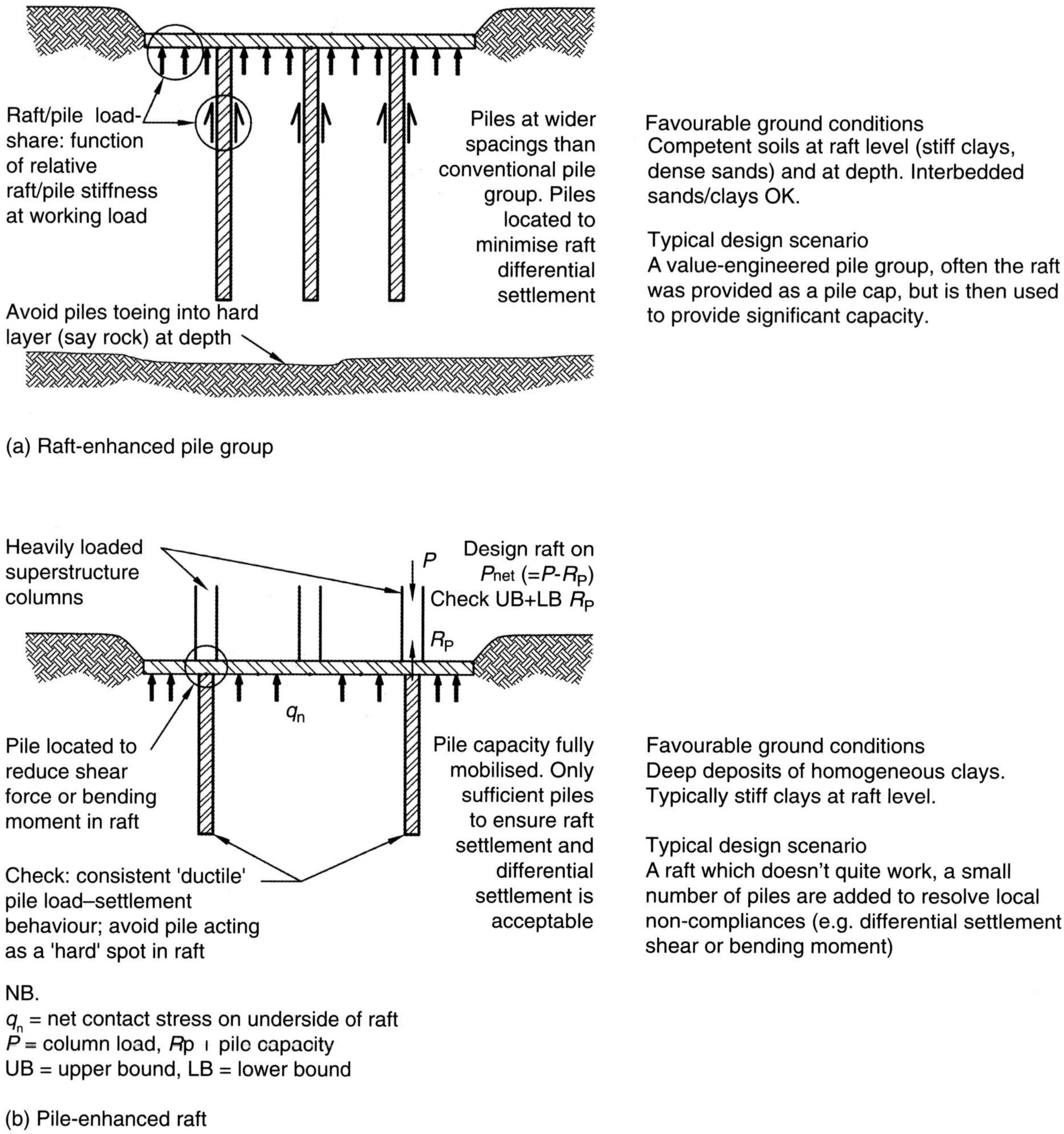

Figure 56.21 A raft-enhanced pile group compared to a pile-enhanced raft

Katzenbach *et al.* (2000) report nonlinear numerical modelling studies for piled rafts (**Figure 56.23**). Load-sharing between piles and raft is shown in **Figure 56.23(c)**: for model M1 the pile-group capacity exceeds the applied load, and then the load-share remains practically constant across a wide range of applied loads; for model M2 the pile group ultimate capacity is less than the applied load and the load-share between piles and raft becomes strongly nonlinear once the pile-group capacity is exceeded. *It is recommended that the situation illustrated by model M2 should be avoided for practical design purposes, in order to avoid excessive complexity.*

It is strongly recommended that the simplified overall load–settlement response is calculated for any piled raft. At later design stages the component behaviour of the raft and piles can be computed by assuming nonlinear elastic behaviour. A raft-enhanced pile group should be operating within the initial, relatively stiff, part of the trilinear load–settlement curve (**Figure 56.22**), with the design load $\leq P_2$, whereas a pile-enhanced raft will be operating in the 'softer' second part of the load–settlement curve with the design load $> P_3$. For the plausible range of applied loads, it is important that the raft-enhanced pile group does not traverse P_1 in **Figure 56.22**.

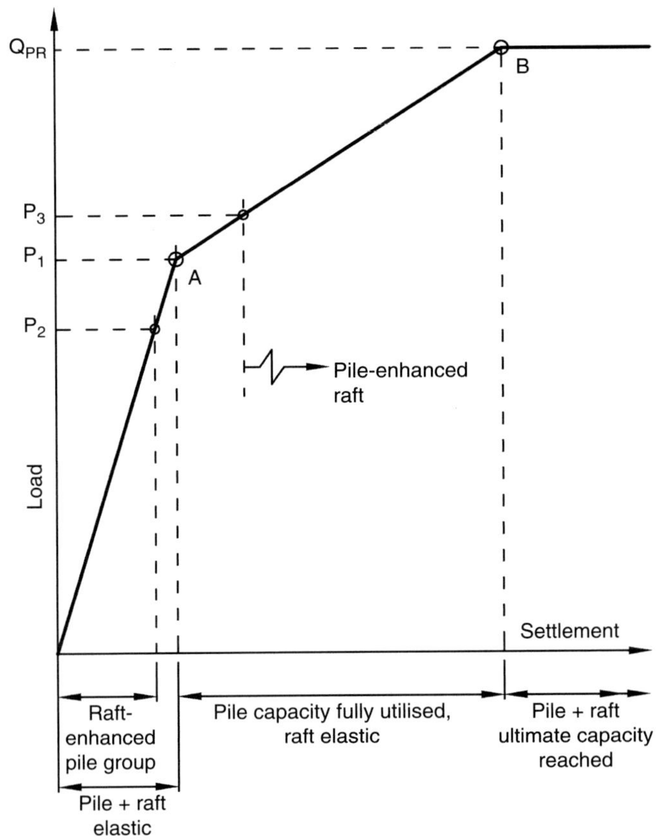

Figure 56.22 Simplified load–settlement curve for a piled raft

To reduce the risk of traversing P_1, a margin should be applied to the ultimate pile-group capacity, Q_P. The margin to be allowed is a matter of judgement depending on the confidence associated with estimating the pile-group capacity; in turn this will depend on the level of understanding of the geology, ground properties, type of piling and precedent experience and so on. The margin should not be confused with an FoS: it is simply a pragmatic means of ensuring that the raft or pile load-sharing (and associated structural forces) is constrained within reasonably predictable bounds.

56.6. Piled-raft ground–structure interaction, general considerations

Parametric studies by Randolph (1994), Poulos (2001) and Katzenbach *et al.* (1998) have provided useful insights into piled-raft behaviour, and some key aspects of behaviour are summarised in **Table 56.4**. In the general case of a piled raft, very complex interactions take place between the various elements. These are summarised in **Figure 56.24**, which shows a piled raft with non-uniform applied loading, in contact with the underlying soil. The following interactions can take place.

(*a*) *Raft–soil interaction:* The contact stresses between the raft and soil are transmitted into the soil and settlement of the raft takes place.

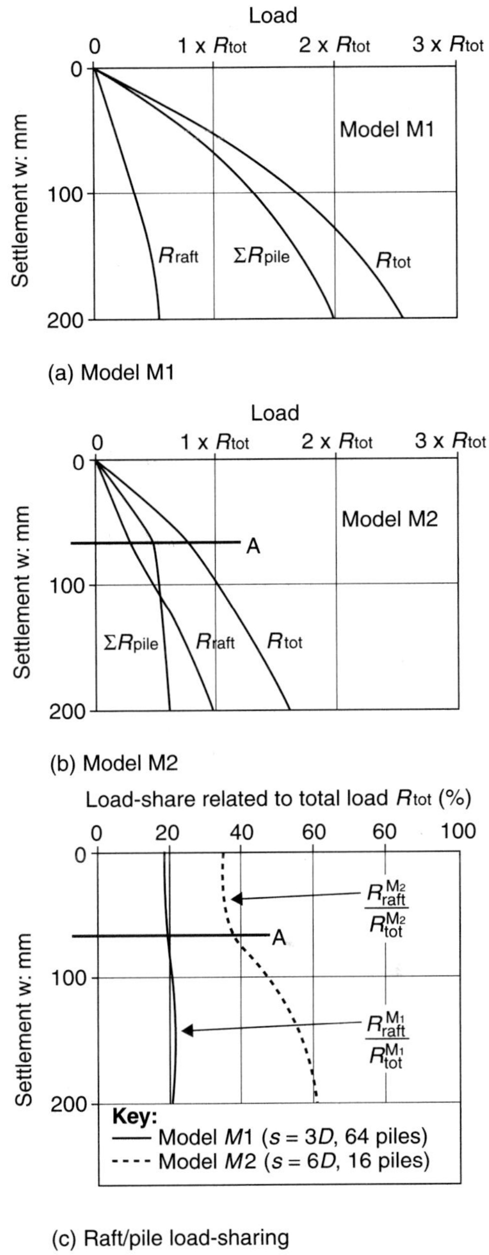

Figure 56.23 The influence of pile-group capacity on load-sharing between the raft and pile group (Modified from Katzenbach *et al.*, 2000)

(*b*) *Raft–soil–raft interaction:* Interaction takes place through the soil with other parts of the raft.

(*c*) *Raft–soil–pile interaction:* The raft contact stresses are also transmitted through the soil and interact with the piles.

(*d*) *Raft–pile interaction:* Loads are transmitted into the piles directly by the raft.

(*e*) *Pile–soil interaction:* The pile loads disperse into the ground surrounding the piles.

Design issues	Comment
Number of piles	Increasing the number of piles will be of limited benefit once a threshold number is exceeded
Pile location	Typically, for optimum effectiveness focus on the area beneath the most heavily loaded columns and across the central half of the raft
Pile length and spacing	A smaller number of longer piles will be more effective than a large number of short piles. Keep the pile spacing as wide as possible, typically in excess of 4*D*
Raft thickness	A thicker raft will reduce differential settlement at the expense of higher bending moments. Raft thickness has a negligible effect on load-sharing or maximum settlement
Distribution of applied loads	Important for assessment of differential settlement and raft bending moment, negligible effect on maximum settlement or raft/pile load-sharing
Soil–structure interaction analysis	Most methods tend to overpredict the proportion of load carried by the piles, and therefore underpredict the load carried by the raft. Hence, care is needed in the structural design of a raft. Sophisticated 2D analyses are likely to be less accurate than approximate methods that allow for the 3D nature of the problem

Table 56.4 Piled rafts: some factors influencing behaviour

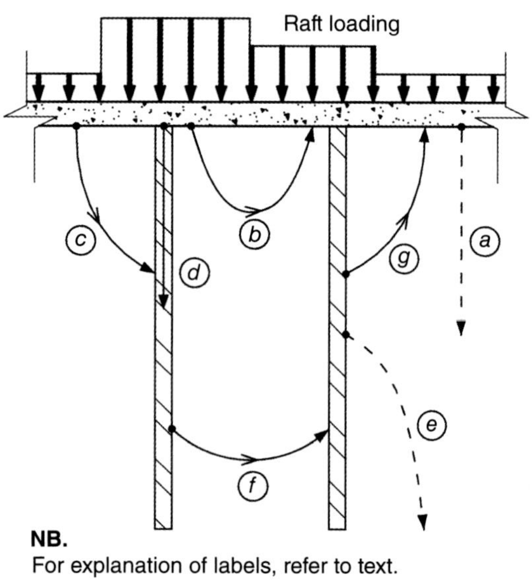

Figure 56.24 Interaction between different components of a piled raft

(*f*) *Pile–soil–pile interaction:* Interaction takes place between each pile through the soil to other piles.

(*g*) *Pile–soil–raft interaction:* Interaction also takes place between each pile through the soil to the underside of the raft.

A comprehensive analysis of the behaviour of a raft-enhanced pile group would require sophisticated 3D numerical modelling, which would need to include the following.

- A nonlinear stress–strain model for the soil; as noted in Chapter 55 *Pile-group design*, modelling of small-strain stiffness nonlinearity can lead to more economic pile-group design because interaction effects between piles are more realistically simulated (and will often be far lower than predicted by a linear elastic–plastic soil model). Similarly, interaction effects between piles and raft will also be more realistically predicted.
- Appropriate modelling of substructure stiffness. The 'effective' stiffness of a raft is often relatively high due to the influence of the substructure and/or superstructure (refer to **Figure 56.3**).
- Interface elements between soil–raft and soil–pile elements. Appropriate strength and stiffness characteristics need to be selected otherwise erroneous and potentially unsafe predictions of structural forces can be predicted.

The above requirements are challenging and require significant input from appropriately qualified specialists. Careful calibration of these complex models is extremely important and should include:

- checking that the load–settlement characteristics of raft only and single piles are realistically simulated
- checking that the load–settlement characteristics of pile group only (without a contribution from the raft) is realistic
- back-analysis of a relevant case history for a similar raft-enhanced pile group.

Management and calibration of advanced numerical modelling is discussed by O'Brien and Higgins (2020).

Studies by Burland (1995) and Katzenbach *et al.* (2000) indicate how raft–pile interaction modifies the behaviour of piles below a raft, compared with 'conventional' piles (i.e. without a raft in contact with competent soil) and how the contact stress beneath a raft is modified by the piles. Burland (1995) describes a study of the behaviour of a single free-standing pile, 1 m dia. and 20 m long, compared with the behaviour of an identical pile beneath a very wide rigid raft foundation which is loaded such that the shaft resistance of the pile is fully mobilised (**Figure 56.25(a)**). Three cases were studied: (*a*) homogeneous, linear elastic–perfectly plastic; (*b*) increasing stiffness with depth, linear elastic–perfectly plastic; (*c*) nonlinear elastic–perfectly plastic. Cases (*b*) and (*c*) are representative of most realistic situations. Fully drained conditions were assumed and the maximum shaft friction was specified as 50 kPa, the equivalent of half the undrained strength ($\alpha = 0.5$). **Figure 56.25(b)** shows the load–settlement behaviour of the free-standing single pile compared with that of the pile beneath the raft for each of the above cases. It can be seen that for cases (*b*) and (*c*) the load carried by the pile mobilises a little less rapidly than for the equivalent free-standing pile. **Figure 56.25(c)** shows a graph of normalised settlement against depth beneath

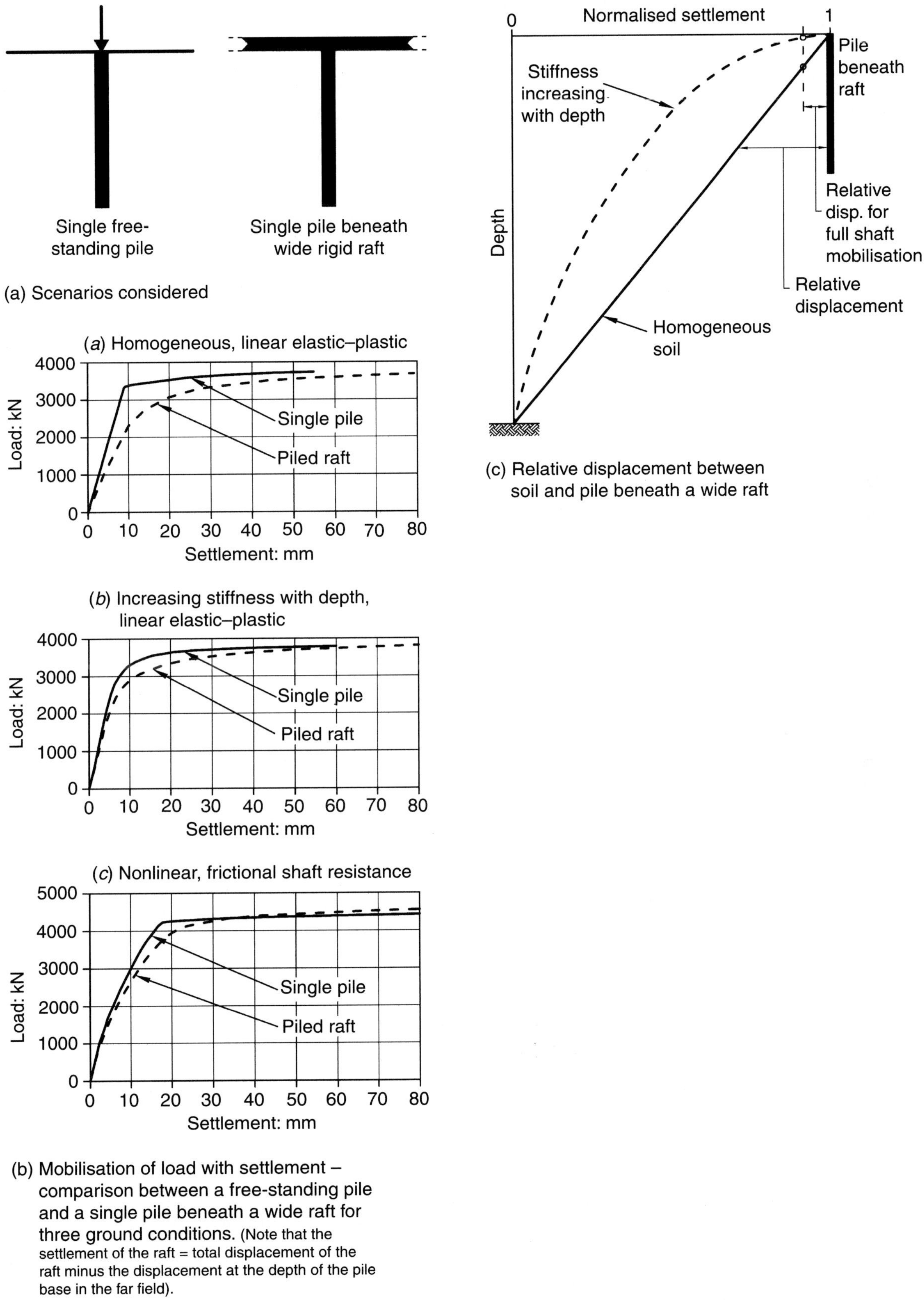

Figure 56.25 Influence of raft–soil–pile interaction on pile settlement characteristics (Modified from Burland, 1995)

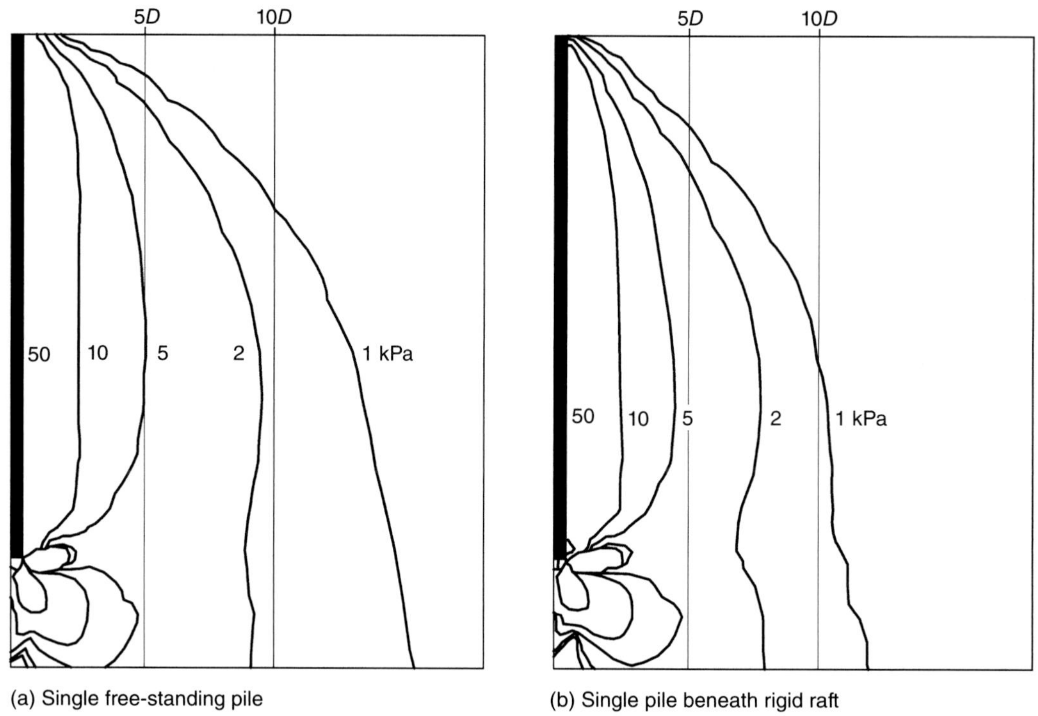

Figure 56.26 Contours of vertical shear stress corresponding to 50 mm settlement for a linear elastic–perfectly plastic soil with stiffness increasing with depth (Modified from Burland, 1995)

the raft and the relative displacement (between pile and soil) of a rigid pile beneath the raft. Full mobilisation of shaft friction requires a finite relative displacement between the soil and pile (typically about 5–10 mm). For a pile beneath a raft the relative displacement varies from zero at the top of the pile to a maximum at the pile toe. As raft settlement increases, shaft friction is mobilised at the pile toe and progressively moves upwards with increasing raft settlement. The rate at which shaft friction is mobilised will depend significantly on the variation of soil stiffness with depth. When soil stiffness increases rapidly with depth (a common occurrence in practice) the shaft friction will then be mobilised more rapidly than for the case when stiffness is constant with depth. However, there will always be a small region close to the pile-head, where the shaft friction is not fully mobilised.

Figure 56.26 (for the case (*b*) conditions shown in **Figure 56.25(b)**) shows the distribution of vertical shear stress around a single free-standing pile and the corresponding pile beneath a raft at a settlement of 50 mm. It is evident that the differences between the two piles are small but the stresses are slightly more concentrated around the pile beneath the raft. At a radius of about six times the pile diameter the vertical shear stresses have reduced to less than about 5% of the shaft friction. It is worth noting that near the top of the pile beneath the raft, the shaft resistance is not fully mobilised due to raft–soil–pile interaction.

The pile beneath a raft also influences the raft (Katzenbach *et al.*, 2000). The contact stress on the underside of a raft is locally reduced around the pile shaft (within about 2*D* to 2.5*D* of the pile centre-line, where *D* is the pile diameter).

56.7. Raft-enhanced pile groups

56.7.1 Bearing capacity

The first step is the assessment of the ultimate capacity of the piled raft. The checks outlined below can be used for both 'small' and 'large' piled rafts. However, the bearing capacity check of the combined system will usually be more important for 'small' piled rafts than 'large' piled rafts because large piled rafts will often have adequate bearing capacity based on the raft alone. The vertical bearing capacity can be taken as the lesser of

- the capacity of the block containing piles and raft, plus the portion of the raft beyond the pile-group perimeter
- the factored ultimate capacity of the raft plus the piles, as indicated in Equations 56.5 and 56.6.

Mandolini *et al.* (2005) indicate that pile-group block failure is likely at small values of pile spacing s, below s_{crit}, with s_{crit}/d varying between 2.5 (for 3 × 3 pile groups) and 3.5 (for 9 × 9 pile groups) and d equal to pile diameter. For pile spacings larger than s_{crit} the piled raft ultimate geotechnical capacity, Q_{PR}, can be taken as the lesser of

$$Q_{PR} = \alpha_R Q_R + \alpha_P Q_P \tag{56.5}$$

where Q_R and Q_P are the ultimate capacity of raft and pile group alone; α_R varies as shown in **Figure 56.27** and $\alpha_P = 1.0$.

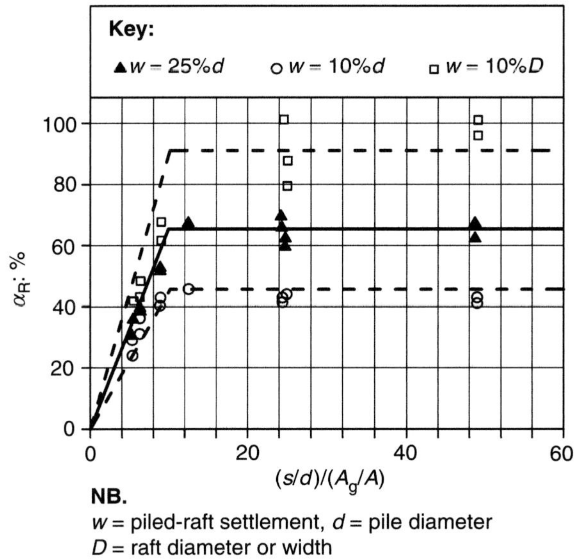

Figure 56.27 Piled-raft ultimate capacity factor, α_R, plotted against pile spacing and area (Modified from Mandolini *et al.*, 2005; all rights reserved)

In **Figure 56.27** A_g is area of the pile group (plan area of the pile-group 'block', i.e. not the sum of individual pile cross-sections) and A is the area of the raft. The value of α_R is dependent on the definition of failure (i.e. the magnitude of piled-raft settlement at 'failure'). A limiting α_R of 0.65 for a piled-raft settlement equal to 25% of d is generally appropriate for this bearing capacity check, although at typical pile spacings and group geometries α_R varies between 0.3 and 0.5, or (Mandolini *et al.*, 2005)

$$Q_{PR} = 0.8(Q_R + Q_P) \tag{56.6}$$

The lesser of Equations 56.5 or 56.6 should be taken as the ultimate geotechnical capacity of the foundation and used for ULS checks. The ultimate geotechnical capacity of individual components of the foundation (i.e. raft, pile group or individual piles) are not relevant for the ULS check of a piled raft.

56.7.2 Small raft-enhanced pile groups: settlement and differential settlement

Viggiani *et al.* (2012) summarise analytical studies of 'small' piled rafts for a range of pile diameters, spacings and lengths. The only parameters which significantly influenced the average piled-raft settlement were the *B*/*L* and A_g/A ratios, which is consistent with **Figure 56.18**. Factors such as pile diameter, pile spacing and number of piles had a negligible influence. The study also indicated that at working loads, for A_g/A ratios of practical interest (typically >0.8 for small piled rafts), the piles attract the vast majority of the applied load. The raft may attract a greater load-share (e.g. refer to the Napoli tanks case history; Mandolini *et al.*, 2005) in certain situations; however, for preliminary design of small piled rafts it will often be adequate to assume that

- piled-raft settlement is only controlled by the pile-group stiffness (i.e. the piled-raft settlement is based upon an equivalent pier method; refer to Chapter 55 *Pile-group design* and Randolph, 2003)
- all the working load is carried by the pile group.

For most small piled rafts the differential settlement will be practically insignificant, since the total settlement of the piled raft is likely to be small (given a large *L*/*B* ratio) and the raft itself is likely to be relatively stiff (given a practical range of dimensions, raft thickness and width).

56.7.3 Large raft-enhanced pile groups: load-sharing, settlement and differential settlement

56.7.3.1 Load-sharing and average settlement

At working loads the interaction between raft and pile will be of practical significance, and for preliminary design the PDR method can be used to assess the average settlement of the piled raft and load-sharing between raft and pile (**Figure 56.22**). The PDR method assumes: piles and raft are linear elastic; the raft is rigid; and a vertical load is applied at the raft centroid. Using the stiffness and ultimate capacity of pile group and raft considered as individual elements, the PDR method combines them to predict the trilinear load–settlement behaviour of the piled-raft system, as shown in **Figure 56.22**. The expression for the piled-raft stiffness and load-sharing are

$$K_{pr} = \frac{1 - \dfrac{K_r}{K_p}(2\alpha_{rp} - 1)}{1 - \dfrac{K_r}{K_p} \times \alpha_{rp}^2} K_p \tag{56.7}$$

where K_r is stiffness of the raft; K_p is stiffness of the pile group; K_{pr} is stiffness of the piled raft (= total load divided by average settlement); and α_{rp} is the interaction factor between raft and piles; and

$$\frac{P_r}{P_t} = \frac{P_r}{P_r + P_p} = \frac{(1 - \alpha_{rp})K_r}{K_p + (1 - 2\alpha_{rp})K_r} \tag{56.8}$$

where P_r is load carried by the raft; P_p is load carried by the piles; and P_t is total load.

The stiffness of the raft and the pile group can be assessed by conventional methods typically used for shallow and deep foundation design, although care is required in assessing appropriate ground stiffness values at the relevant mobilised stress/strain levels. **Figure 56.28** (based on elastic theory) summarises the change of α_{rp} for large variations in pile spacing, pile-group size and soil–pile stiffnesses; it should be noted that α_{rp} only varies across a small range of values of 0.65–0.8 (Randolph, 1994).

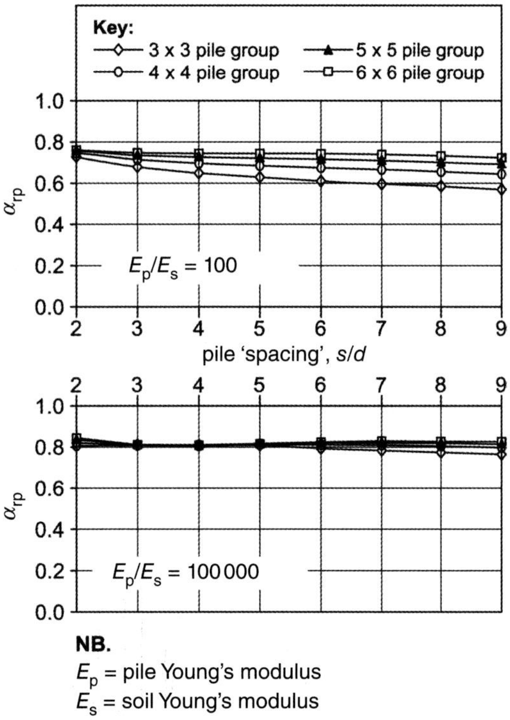

Figure 56.28 Pile–raft interaction factor (α_{rp}) plotted against pile spacing and pile-group size (Data taken from Clancy and Randolph, 1993)

A key consideration is the relative stiffness of the raft to the pile group. It will be important to assess upper- and lower-bound stiffnesses for the raft and pile group (i.e. the ratio of K_r to K_p) (**Figure 56.29**). The relative raft/pile stiffnesses will be sensitive to variations in ground stiffness with depth. Poulos (2001) observed that the load carried by the raft is often underestimated. This is probably due to the influence of small-strain stiffness nonlinearity (e.g. Jardine *et al.*, 1986), which will lead to a relatively high initial raft stiffness. Commonly used ground stiffness correlations used for shallow foundation design are relevant for large strain amplitudes; more

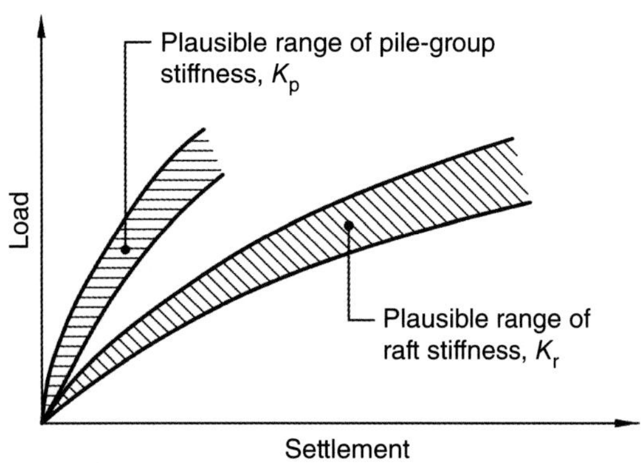

Figure 56.29 Raft-enhanced pile groups: key design check against relative raft and pile stiffnesses

relevant ground stiffnesses at small to intermediate strain amplitudes can be more than double these large strain values (**Figure 56.30**).

It should also be noted that the load-sharing between raft and piles will change over time due to the influence of

- changes in ground stiffness due to transition from undrained to drained conditions for clay soils, and influence of creep in sandy soils – typically this will lead to the load carried by the raft reducing with time
- influence of changes in groundwater pressure over time.

Hence, analyses of pile/raft load-sharing will need to take account of the changes in ground stiffness between short-term and long-term conditions. For example, at a site in London Clay (Cooke *et al.*, 1981) the raft initially carried 40% of the total load during construction, and this reduced to less than 25% in the long term as the clay consolidated. For situations where the proportions of load carried by raft and pile components varies over time, each must be designed for their greater percentage of load.

The pile-group capacity is fully mobilised at a total applied load of P_1 (**Figure 56.22**), given by

$$P_1 = (Q_P)/(1 - P_r/P_t) \tag{56.9}$$

where P_r/P_t is calculated from Equation 56.8 and Q_P is the ultimate capacity of the pile group.

For loads beyond P_1 the stiffness of the piled raft is simply equal to the raft alone (K_r) until the ultimate capacity of

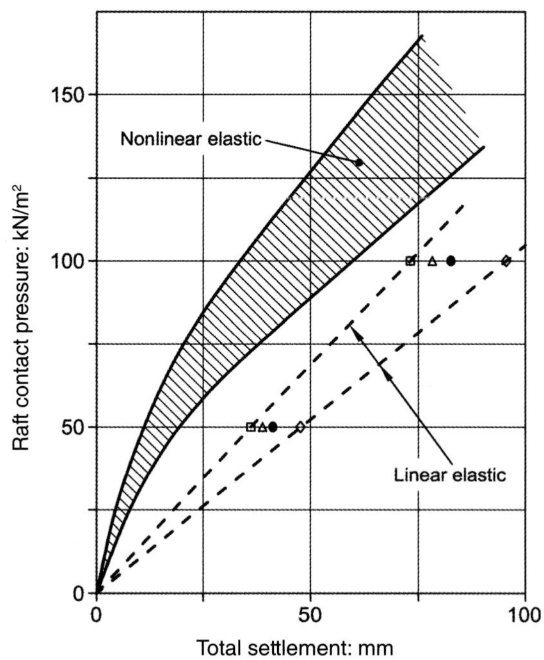

Figure 56.30 Influence of small-strain nonlinearity on raft settlement (Modified from O'Brien and Sharp, 2001; Emap)

the piled raft (from Equations 56.5 and 56.6) is reached. The relative stiffness between raft and pile group controls load-sharing, the overall stiffness of the raft-enhanced pile group and the magnitude of P_1 on **Figure 56.22.**

Based on observations of load-sharing for 22 large piled rafts (acting as raft-enhanced pile groups), Mandolini *et al.* (2005) provide a useful empirical plot of raft load (%) compared to the dimensionless parameter $(s/d)/(A_g/A_r)$ (**Figure 56.31**). Typically for the rafts within the database (mainly in stiff London and Frankfurt Clays) the raft takes 20–60% of the total load.

Figure 56.32 shows the results for average piled-raft settlement from a parametric study by Viggiani *et al.* (2012) for a 45 m wide piled raft, founded in a deep deposit of stiff clay, for pile groups of varying: size (9–225 piles), spacings (3*d*–15*d*) and pile length. The piled-raft settlement is only affected (to a moderate extent) by increasing pile length (i.e. larger *L*/*B* ratio); this result is consistent with the trends illustrated in **Figure 56.18** (when $B/L > 1$). Increasing the number of piles from 49 to 225 has a negligible effect, whereas an increase in *L*/*B* ratio from 0.4 to 1.0 has a more significant effect. It should also be noted that increasing the raft stiffness from $K_{rs} = 0.01$ to 0.1 only provides a minor reduction in settlement relative to a considerable increase in concrete volume. **Figure 56.33** (based on a separate study by Katzenbach *et al.*, 1998) shows that the addition of a small number of relatively long piles is far more effective at reducing settlement than the installation of a large number of short piles. For example, 10 piles with $L/d = 30$ gives a settlement of about half that for 30 piles with $L/d = 10$ (and about 25% of an unpiled raft). Also, adding more piles becomes progressively less efficient (in this example, more than 20 piles provides little extra benefit).

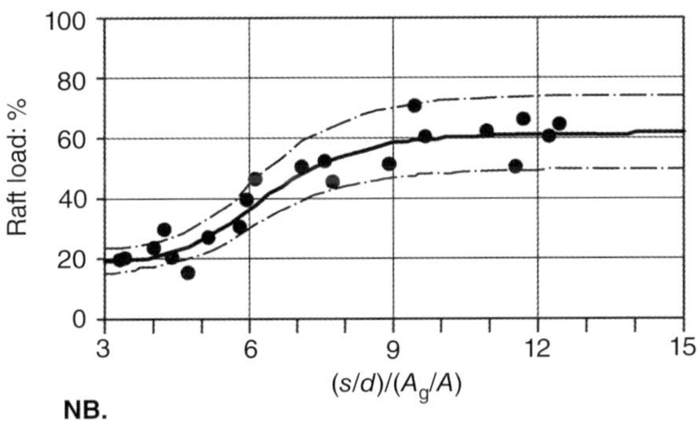

Figure 56.31 Observed load-sharing between raft and piles, 22 case histories (Modified from Mandolini *et al.*, 2005; all rights reserved)

56.7.3.2 Differential settlement

Addition of a small number of piles can provide a considerable reduction in differential settlement. **Figure 56.34** plots differential settlement against A_g/A_r (from the parametric study by Viggiani *et al.*, 2012, outlined above). Also summarised is the effect of changing raft stiffness, K_{rs}, from 0.01 to 0.1 (and an associated increase in raft thickness from 1.1 to 2.3 m), which is a relatively inefficient means of reducing differential settlement. A small number of relatively long piles located within the central one-third of the raft (A_g/A_r about 0.35 and $L/B = 1$) is far more effective at minimising differential settlement than a large number of short piles distributed across the whole raft ($A_g/A_r > 0.8$ and $L/B = 0.4$). The optimum location of piles in a piled raft will depend on the nature of the applied loading (uniformly distributed or a series of concentrated loads beneath heavily loaded columns or shear cores). For uniformly distributed loads the principles for differential settlement control are outlined in **Figure 56.35**. Buildings often apply large loads around the outer edge of a raft; Reul and Randolph (2004) discuss the results of a study where 50% of the total load was applied to the edge of the raft, with the remaining 50% distributed uniformly across the central 25% of the raft. In this case central pile support reduces differential settlement by a factor of 3.5 compared with the unpiled raft (differential settlement of the

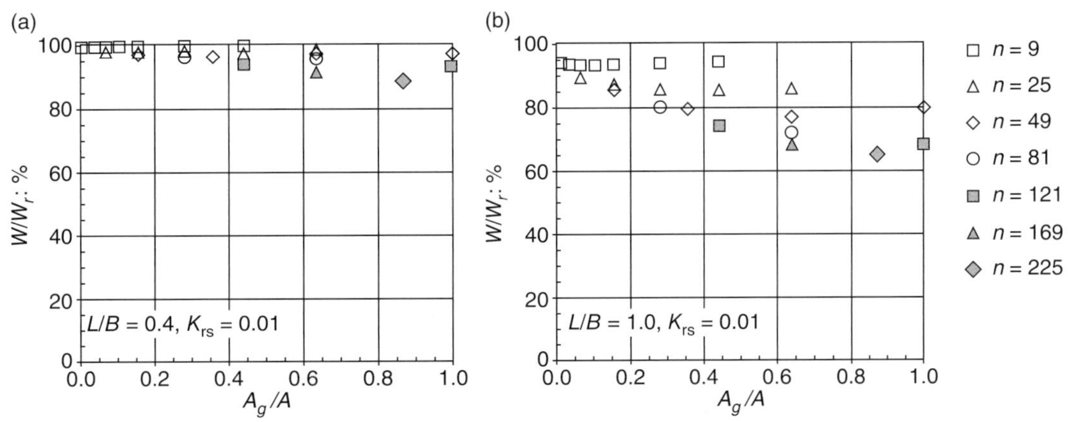

Figure 56.32 Settlement of a large raft-enhanced pile group compared to pile filling ratio, A_g/A_r

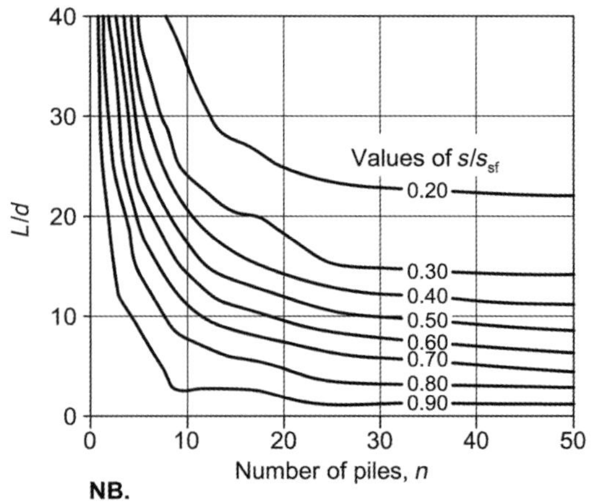

NB.
s = settlement of piled raft, s_{sf} = settlement of raft alone, L = pile length, d = pile diameter

Figure 56.33 Piled-raft interaction diagram showing settlement reduction compared to pile length and number of piles

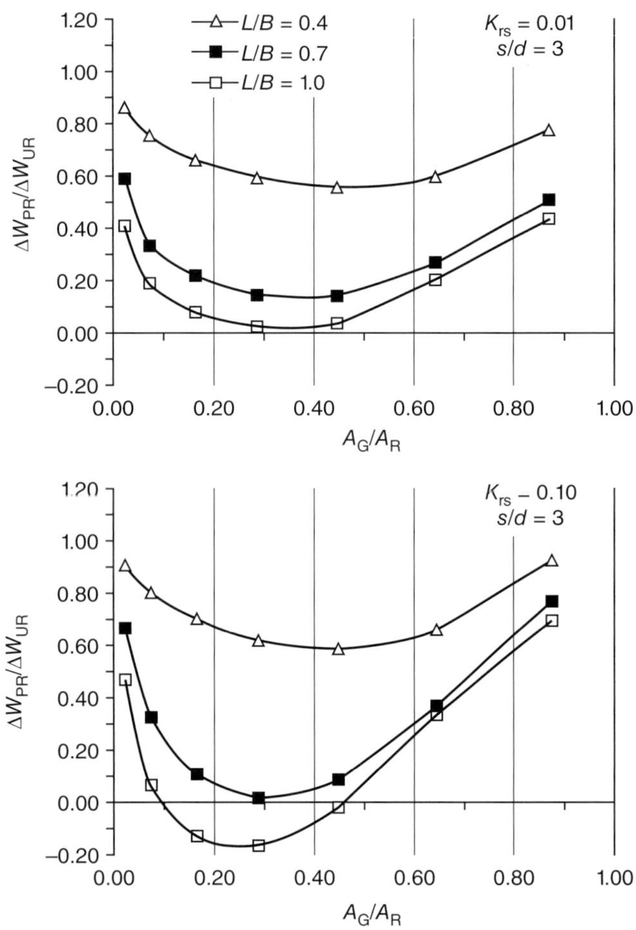

Figure 56.34 Differential settlement of a raft-enhanced pile group: results of numerical analysis for piles at $s/d = 3$ (After Viggiani *et al.*, 2012)

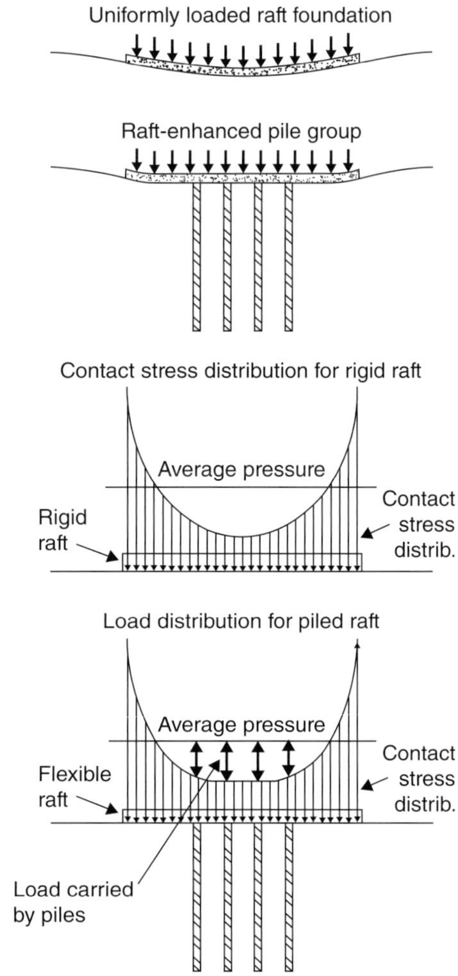

Figure 56.35 Optimising the pile locations for a uniformly loaded piled raft (Modified from Randolph, 1994; all rights reserved)

piled raft is only 24% of the average settlement). In general, the principle is to locate pile support beneath a raft in such a way as to minimise raft differential settlement and bending moment. This will minimise foundation costs without compromising performance.

56.8. Compensated piled rafts

56.8.1 Introduction

When piled rafts are located at the base of a deep excavation (known as 'compensated piled rafts') their behaviour is different to piled rafts located close to the ground surface. Because of the large reduction in overburden stress and the buoyancy force applied by water pressure, there are several additional factors which need to be carefully considered during design (e.g. O'Brien, 2019; Sales *et al.*, 2010). Compensated piled rafts are increasingly important for many urban metro projects because the deep underground station boxes are often also used to support 'oversite developments' (OSDs). The OSDs are typically large tower superstructures, with the underground structural box (and its retaining walls) acting as a stiff substructure.

- Pile behaviour and load-sharing between the raft and piles is affected by the excavation and pile installation sequence.
- The excavation is usually supported by deep embedded retaining walls, which will also pick up loads (hence, loads will be shared between three different elements: piles, retaining walls and raft).
- The presence and magnitude of the buoyant force (i.e. groundwater pressure) will have a significant effect on overall behaviour and load-sharing. The mass permeability of the ground will control the timing of the application of the full buoyant force (i.e. 'long-term' groundwater pressure), hence at intermediate time periods load-sharing may be different to that in the long term. Any future changes in the groundwater regime will also influence overall behaviour and load-sharing.
- Piles are usually subject to uplift forces and heave-induced tension at some point during and after construction, depending on the sequence of construction and the timing of excavation, pile installation and application of OSD loads (above the 'box' excavation).
- OSD construction is often part of a separate development (the metro owner will install the box and foundations and the OSD owner will build the OSD afterwards) and therefore the timing of OSD construction can be uncertain, which generates several additional load case scenarios which must be considered in the piled-raft design.
- Large changes in effective stress (both negative and positive) may occur at different stages during the design life of the piled raft. These will control the capacity and stiffness of the different components of the piled raft, in particular any piles or walls acting as vertical load-bearing elements. Hence, empirically based formula (in particular, total stress methods for clays) are often inappropriate for compensated piled-raft design. More fundamental methods based on effective stress (e.g. Bown and O'Brien, 2008; Burland and Twine 1988) need to be used.
- Upward movement (i.e. 'heave') may be as much of an issue as settlement, and differential movements will often be the main concern, especially if the underground box is an asymmetrical shape or the OSD applies an eccentric load to the piled raft.
- The piles and retaining walls may often take a smaller proportion of the load and the raft a larger proportion than for a conventional (near-surface) piled raft.
- The embedded retaining walls potentially provide a high stiffness/capacity component of the overall foundation system. The capacity/stiffness of the walls will be influenced by the means/methods for installation and subsequent excavation support (e.g. bracing or ground anchors, top-down or bottom-up construction sequences, etc.) since this will influence the horizontal effective stress and interface friction mobilised along the retaining walls. Oversimplified assumptions can lead to uneconomic and difficult-to-build designs. For example, interface friction (between wall and retained soil) is likely to undergo several reversals during and after construction, and interface friction will often be mobilised at small displacements (which should be considered when checking geotechnical ULS load cases).

56.8.2 Net change in effective stress, buoyancy force and construction stages

Table 56.5 shows an example of the building loads compared with the weight of excavated soil and the buoyant force from groundwater pressure for the Hyde Park Barracks in London. This case history was described by Hooper (1973) and back-analysed by Sales *et al.* (2010). The basement is nearly 9 m deep, and the groundwater table is about 4 m below ground level. For this project, the weight of excavated soil is about 32% of the total load from the structure, and the weight of excavated soil plus the buoyancy force is about 44% of the total load from the structure. Clearly, for a compensated piled raft, any analysis ignoring the influence of the weight of excavated soil and the buoyancy force will give erroneous predictions

Construction stage (refer to text)	Time: months	Soil condition	Load: MN				
			Structure			Soil and groundwater	
			Dead load		Live load	Excavated soil	Buoyant force
			Raft	Structure			
1A	2.5	Undrained	22	–	–	−77.3	–
1B	4.4	Undrained	22	37.7	–	−77.3	−0.99
1B	8.5	Undrained	22	47.3	–	−77.3	−11.14
1B	10.25	Undrained	22	70.7	–	−77.3	−15.42
2	12.7	Undrained	22	103.5	–	−77.3	−21.5
2	15.7	Undrained	22	159.3	–	−77.3	−28.96
2	18.8	Undrained	22	168.8	–	−77.3	−29.7
2	24.8	Undrained	22	205	–	−77.3	−29.7
3	>24.8	Drained	22	206	17	−77.3	−29.7

Table 56.5 Compensated piled raft – example for Hyde Park Barracks (Data taken from Sales *et al.* 2010)

of the piled-raft behaviour. The modelling of the construction process should be split into several key stages.

- *Stage 0:* Excavate to raft formation level.
- *Stage 1A:* Raft concreting: the wet concrete will apply a load to the base of the excavation. At this stage the raft is flexible and the 'piled raft' is ineffective. The settlement of sandy soils may be largely completed before the concrete has hardened, whereas for clayey soils some undrained settlement will occur and time-dependent consolidation settlement will occur after raft hardening (and hence lead to some raft–pile load-sharing).
- *Stage 1B:* Subsequent construction of the structure will take place with an effective piled raft after the raft has hardened, and there will be load-sharing between the raft and piles. At this stage the weight of the structure is less than the weight of the excavated soil and the buoyancy force; only towards the end of 1B will the net effective weight be positive.
- *Stage 2:* Throughout this stage the structural weight exceeds the weight of excavated soil plus buoyancy force. The increment of soil settlement that occurs should be based on the net loads (i.e. the weight of structure minus (the weight of the excavated soil plus buoyancy force)). In contrast, the loads that are transferred through the raft and piles should be based on the weight of the structure only.
- *Stage 3:* This is the long-term condition, after consolidation of the soil.

The axial loads in the piles will be directly affected by the construction sequence. If the piles are installed before the bulk excavations for the basement, then heave-induced tensile forces will be induced in the piles due to short-term heave of the piles (this is discussed in Chapter 57 *Global ground movements and their effects on piles*). If the piles are constructed after basement excavation, then short-term heave-induced tensile forces will not be generated in the piles. During Stage 1, when the structure weight is less than the weight of the excavated soils and the buoyancy force, there will be two components of vertical soil deformation: undrained settlement due to increments of load from the construction of the structure, and time-dependent swelling (i.e. upward movement) due to the net reductions in vertical effective stress at the base of the excavation. The overall 'net' effect (i.e. the overall downward or upward movement) will be dependent on the soil permeability compared with the speed of construction. If the soil is relatively permeable there will be a net upward movement, whereas for relatively low-permeability soils there will be a net downward movement. For the back-analysis of Hyde Park Barracks, Sales *et al.* (2010) used a continuum model (finite element for the raft and boundary element for soil and piles). The soil below the raft is assumed to behave as a linear elastic–perfectly plastic material. In order to match the observed behaviour, the following were important.

- An increase in the 'elastic' modulus for the soil immediately below the raft (from a value of about 57 MN/m^2 to 284 MN/m^2 – that is, five times higher and close to the very small strain modulus, E_o or E_{max} (refer to Chapter 52 *Foundation types and conceptual design principles*)).
- An increase in the flexural stiffness of the raft to replicate the stiffening effect of the substructure shear walls; the actual raft thickness was 1.5 m, but locally in the model this was increased to 10 m to simulate the much higher effective flexural stiffness of the basement structure.

The change in groundwater pressure over time was replicated by assuming a linear increase in the buoyancy force with time from month 4 (after excavation) to the full buoyancy force after month 16. This is a simplified way of allowing for the dissipation of excess pore water pressures in the stiff clay with time. For other (more permeable) soils this could occur more rapidly.

56.8.3 Changes in load-sharing, settlement and differential settlement

Figure 56.36 gives the cross-section through a metro station box constructed in a very stiff clay, underlain by interbedded dense sands and very stiff clays. The compensated piled raft is located at a depth of 22 m below ground level, the retaining walls for the metro station box are formed by 1.2 m thick diaphragm walls (D-walls) and the piles were 1.8 m and 2.1 m dia. bored piles. The D-wall is about 30 m deep, but alternate D-wall panels are extended as deep barrettes to provide OSD load-carrying capacity.

Preliminary design used simplified methods to assess settlement/heave and load-sharing, as outlined in Section 56.7.3 for large raft-enhanced pile groups. Generally, the nonlinear settlement/swelling behaviour of the raft needs to be taken into account (**Figure 56.30**). For detailed design a 3D coupled effective stress FE analysis (calibrated against relevant case history data) was used to simulate several different construction scenarios. Because of uncertainty about the timing of OSD construction, the barrettes and piles also needed to act as tension elements to resist buoyancy forces (assuming OSD construction was delayed for several years). **Figure 56.37** illustrates the predicted load-share between the raft and the deep foundation elements (piles and barrettes) if the OSD construction followed immediately after completion of the underground station (with groundwater pressures equilibrating to long-term values after OSD construction). It can be seen that load-sharing changes markedly once the buoyancy force develops (note the 'remaining' load-sharing elements are the retaining walls for the metro box). An important benefit of the numerical modelling is to assess several 'what-if' scenarios and the impact of various design changes and different project development scenarios.

A key design question was the differential settlement across the metro box and the connecting metro tunnels. **Figure 56.38** shows the impact of shortening the deep foundation elements from 62 m to 41 m, which led to an increase in differential settlement across the box from 1 in 1700 to

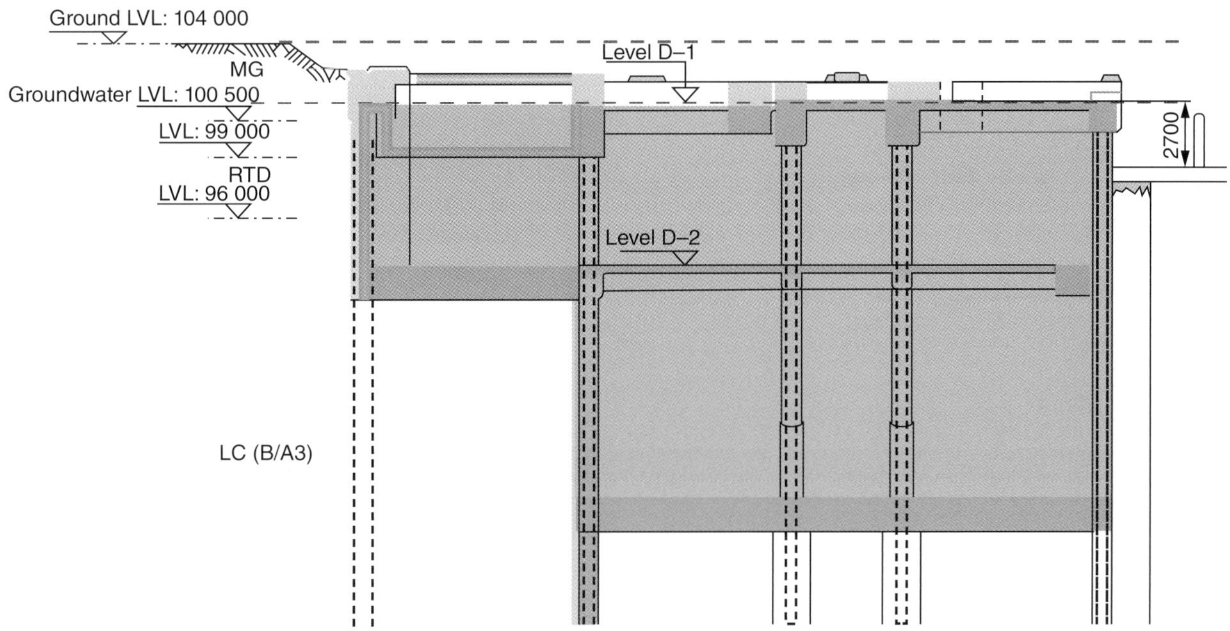

Figure 56.36 A compensated piled raft for a metro station and subsequent OSD development

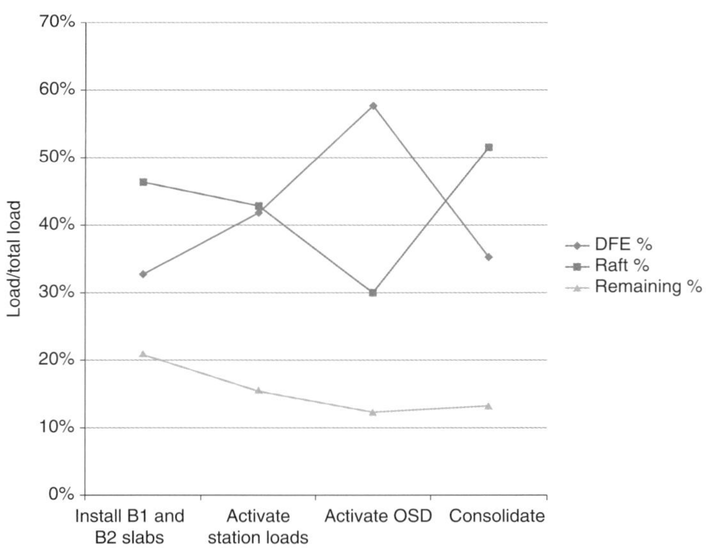

Figure 56.37 Compensated piled raft with OSD, example of changes in load-sharing for a design scenario

1 in 900. The shorter foundation elements had significant benefits, including: avoiding penetration of sand layers at depth (and associated use of bentonite slurry support), health/safety and buildability challenges (mainly related to handling large and heavy reinforcement cages within a constrained site), together with cost and carbon savings. These benefits need to be balanced against the increase in differential settlement, hence consideration of what is deemed to be an allowable maximum differential settlement is often a critical issue. This needs effective communication between structural and geotechnical engineers and realistic limits set for allowable differential settlement (these issues are also discussed by Burland, 2006).

56.9. Pile-enhanced rafts

56.9.1 Introduction

Sections 56.2 and 56.3 describe approaches for the analysis of settlement and the structural performance of raft foundations, respectively. Usually, for a raft foundation the bearing capacity is not a limiting factor. When analysis indicates that the settlement (total or differential) is excessive or the structural resistance of the raft is exceeded locally, consideration has to be given to how to improve the design. The structural performance of a raft can be modified by increasing its thickness. However, this can prove costly. Moreover, local thickening is often undesirable as changes in the excavation level are best avoided and the associated reinforcement details can be difficult and expensive to implement.

Reducing settlement will usually require piles. The conventional methods of pile design are essentially based on providing an adequate FoS against bearing capacity failure. As pointed out in Section 56.1 this leads to the question ‘How many piles are required to carry the weight of the building?’ Thus, what is essentially a limiting settlement problem is converted into a bearing capacity problem, which can result in a very expensive solution. When settlement is the controlling factor, it is more appropriate to pose the question ‘How many piles are required to reduce the settlement to an acceptable amount?’ The approach of providing a sufficient number of piles to limit settlement can also be used to control differential

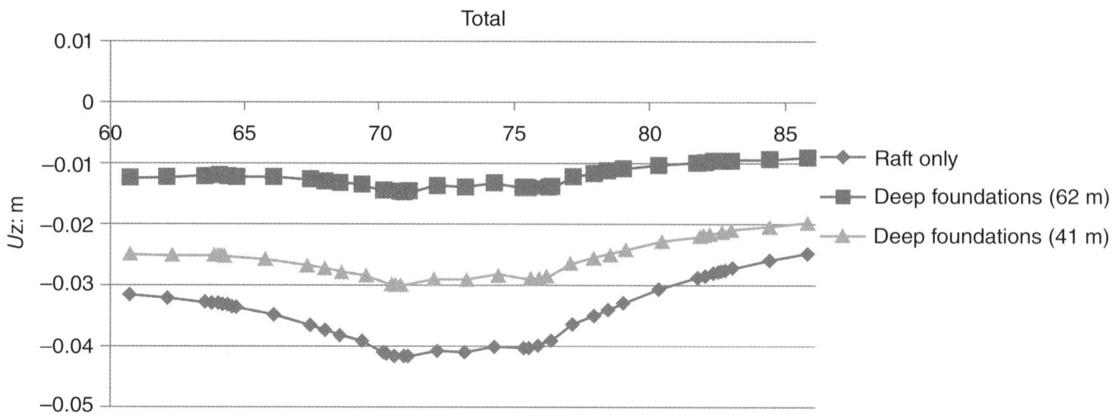

Figure 56.38 Compensated piled raft with OSD, example of changes in settlement with varying length of deep foundations for a design scenario

settlement, bending moments and shear stresses in the raft. Rafts that are designed with these objectives are called *pile-enhanced rafts*. Note that this term is more appropriate than *settlement-reducing piles*, which was originally proposed by Burland *et al.* (1977) and Burland (1995).

56.9.2 Ductile load–settlement behaviour of piles

Straight-shafted piles are used for pile-enhanced rafts, and an essential requirement is that their load–settlement behaviour should be ductile. This is usually the case for straight-shafted piles in clayey soils. As pointed out in Chapter 22 *Behaviour of single piles under vertical loads*, shaft resistance is essentially a frictional phenomenon and once it is fully mobilised it usually does not change significantly with further displacement. Some examples of pile tests on a wide variety of clay soils are given in the following.

Reese and O'Neill (1988) present numerous load–settlement results for straight-shafted bored piles in clay soils. These show that the carrying capacity is fully mobilised at settlements of about 10 mm and thereafter remains approximately constant for large settlements. The results of the classic research by Whitaker and Cooke (1966) on bored piles in London Clay are of particular relevance for pile-enhanced rafts. **Figure 56.39** shows the results of a test on an under-reamed bored pile in London Clay in which a load cell was used to separate the base resistance from the shaft resistance. For the first part of the test the loads on the pile were applied in increments, with the load at each stage being maintained constant until settlement had practically ceased. In some cases the load was held constant for 15 or more days. The second part of the test consisted of carrying out a constant rate of penetration test on the pile so as to determine its ultimate carrying capacity. The full line in **Figure 56.39(a)** shows the load–settlement behaviour of the pile as a whole.

The broken line labelled 'shaft' in **Figures 56.39(a)** and **56.39(b)** shows the measured load–settlement behaviour of the shaft. It is clear that the carrying capacity of the shaft is fully mobilised at very small settlements – less than 10 mm.

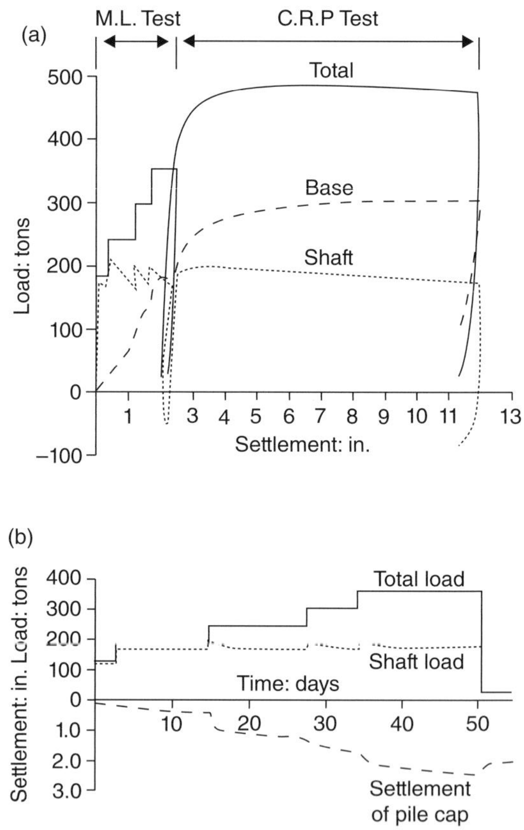

Figure 56.39 Load test on instrumented under-reamed bored pile in London Clay: length is 12.2 m, shaft diameter is 0.79 m and base diameter is 1.68 m (Whitaker and Cooke, 1966)

The jagged 'saw tooth' shape of the load–settlement curve is of interest. When an additional increment of load is applied to the pile, the load carried by the shaft initially increases and then reduces as settlement takes place, as can be seen from **Figure 56.39(b)**. The temporary increase of load carried by the shaft is a 'rate effect' due to the initial high rate of settlement with time (Burland and Twine, 1988). As the rate of settlement steadily reduces, the load carried by the shaft reduces back to its previous value, with the reduction in shaft load being shed to the base. It can be seen that the equilibrium fully mobilised carrying capacity of the shaft remains remarkably constant with increasing settlement, confirming the ductile nature of the shaft behaviour. All of the pile test results reported by Whitaker and Cooke show similar behaviour. As pointed out in Chapter 22 *Behaviour of single piles under vertical loads*, it is important to note in **Figure 56.39(a)** that, for working loads in excess of about 170 tons, the pile is operating with full mobilisation of the shaft carrying capacity. There must be tens of thousands of large-diameter bored piles operating satisfactorily in this way worldwide.

Figure 56.40(a) shows the load–settlement behaviour of a 762 mm dia. 55 m open-ended tubular pile driven into a soft low-plasticity clay at Pentre, England (Cox *et al.*, 1992). The dotted line shows the measured elastic shortening of the pile. It can be seen that the peak resistance develops at a relative settlement of about 10 mm (total settlement minus elastic shortening). Thereafter there is a small reduction in capacity for an additional settlement of about 60 mm. The reduction in capacity is less than 10%. **Figure 56.40(b)** shows the load–settlement behaviour of a 762 mm dia. 33.5 m open-ended tubular pile driven into hard till overlying Oxford Clay at Tilbrook Grange, England (Cox *et al.*, 1992). The behaviour is remarkably similar to the pile in the soft Pentre clay. Peak resistance develops at a relative settlement of about 10 mm and thereafter there is a small reduction of carrying capacity with increasing settlement up to more than 70 mm.

In conclusion, shaft resistance is essentially a frictional phenomenon and as such it is permanent and reliable provided the effective stresses acting normal to the pile shaft do not change significantly. Numerous test results on straight-shafted driven and bored piles in clayey soils show that full mobilisation takes place at settlements of about 5–10 mm and that at larger settlements little or no reduction or increase in shaft resistance takes place. If a pile-enhanced raft is proposed it is prudent to carry out loading tests to large settlements on the proposed type of pile to confirm that the load–settlement behaviour is sufficiently ductile.

56.9.3 Basic mechanisms of behaviour of pile-enhanced raft foundations

We return to the question posed previously but expand it a little.

> *How many piles are required, together with their arrangement, to reduce the settlement (both total and differential) or the stresses in the raft to an acceptable level?*

This question lies at the heart of designing a pile-enhanced raft.

We deal first with the question of the number of piles required to limit the settlement to an acceptable amount. **Figures 56.41(a)** and **(b)** show, respectively, a raft foundation on a deep layer of clay and a graph of settlement plotted against the number of piles. With no piles the calculated maximum final long-term settlement is given by point A in **Figure 56.41(b)**. After a consideration of the structural and

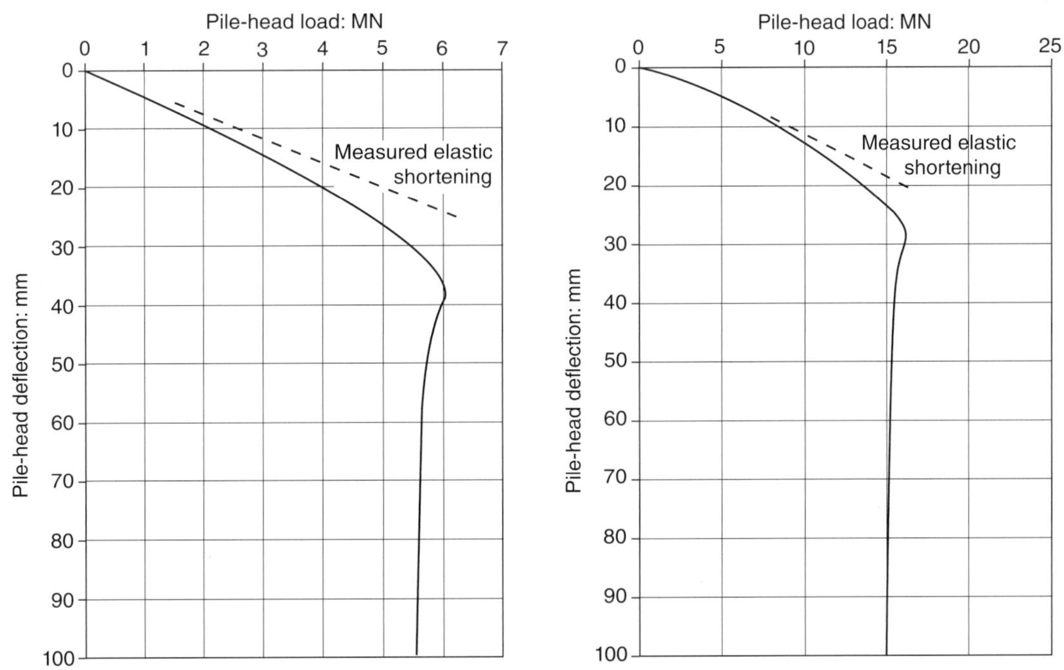

Figure 56.40 (a) Load test on 762 mm dia. 55 m open-ended pipe pile driven into normally consolidated low PI clay at Pentre; (b) load test on 762 mm dia. 33.5 m open-ended pipe pile driven into hard overconsolidated Oxford Clay at Tilbrook Grange

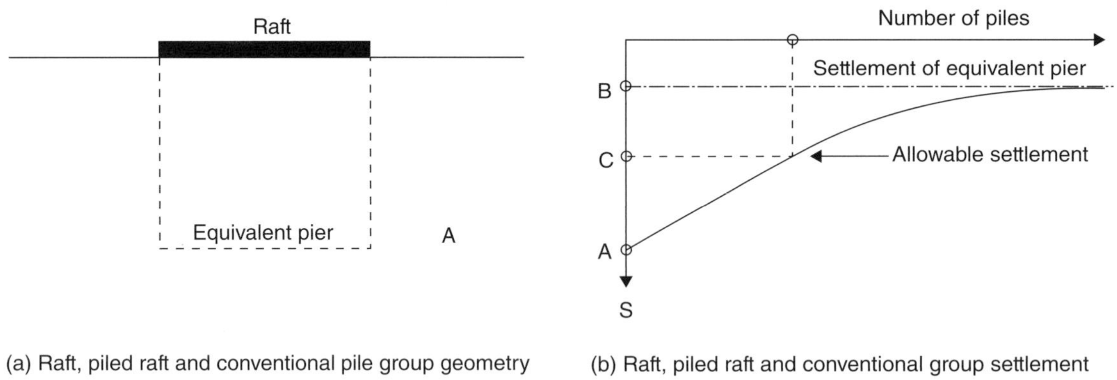

Figure 56.41 Relationship between number of piles and settlement of a piled raft: (a) raft, piled raft and conventional pile group geometry; (b) raft, piled raft and conventional group settlement

architectural requirements of the building to be supported by the raft, it is concluded that the settlement (or, usually, the differential settlement) is unacceptably large so that piles are required to reduce the settlement. In this situation a full pile group designed by conventional methods can be represented by an equivalent pier as shown in **Figure 56.41(a)** – see Chapter 55 *Pile-group design* (Section 55.4.5.5). A settlement calculation of the equivalent pier will give a much-reduced settlement represented by point B in **Figure 56.41(b)**, but at considerable cost.

If, instead of the conventional approach, the decision is taken to adopt a pile-enhanced raft solution, it will first be necessary to decide on an acceptable maximum settlement – see Chapter 52 *Foundation types and conceptual design principles* (refer to Section 52.4). Point C in **Figure 56.41(b)** represents the chosen maximum allowable settlement for this case. Straight-shafted (ductile) piles are used and the load required to fully mobilise the shaft resistance of each is calculated. If a small number of piles are used, the settlement of the raft will be sufficient to fully mobilise their shaft resistances. The upward resistance of each pile acting on the raft foundation will reduce the load transmitted to the ground through the raft by the mobilised shaft resistance. If we proceed to add one pile at a time it will be found that the settlement of the raft will initially reduce in inverse proportion to the number of piles, as shown in **Figure 56.41(b)**. This is because the spacing between the piles is large and there is little or no interaction between them (typically the case for a pile-enhanced raft). As the number of piles is increased the spacing between them reduces until there comes a point where the interaction between the piles and the raft foundation begins to become significant (typically the case for a raft-enhanced pile group). The relationship between the settlement and the number of piles then becomes nonlinear and approaches the settlement of the equivalent pier foundation asymptotically. Provided the chosen allowable total settlement is large enough to be consistent with the guidance summarised in Chapter 52 *Foundation types and conceptual design principles*, then the number of piles is likely to be small enough (and the spacing between them large enough) for there to be negligible interaction between them.

Burland (1995) drew the following conclusions from a study of pile–raft interaction (described in Section 56.6).

- The raft–soil–pile interaction results in the pile beneath the raft mobilising its shaft resistance less rapidly with settlement than a free-standing pile. However, for raft settlements greater than 20 mm, more than 90% of the total shaft resistance of the equivalent free-standing pile is mobilised.
- Computed contours of vertical shear stress around the pile beneath the raft show that the pile–soil–pile interaction can be ignored for spacing factors greater than about 6, where the soil stiffness increases rapidly with depth (this often applies).
- The pile–soil–raft interaction and the raft–soil–pile interaction can be simply accounted for in design by reducing the fully mobilised capacity of the piles by about 10%.

For the critical spacing factor for the pile–soil–pile interaction to be practically negligible depends on the variation of ground stiffness with depth, varying between about 6*d* and 8*d* (for stiffness increasing with depth and constant stiffness with depth, respectively) (Mandolini *et al.*, 2005). If piles are spaced more widely than this critical value across the raft, then the piles can be assumed to act as single piles in terms of their load–settlement characteristics.

56.9.4 A simple design approach

The practical importance of these conclusions is that the analysis of a pile-enhanced raft can be carried out by a standard raft computer program (see Section 56.2) in which the raft is modelled as plate elements and the deflections of the ground are derived from elastic half-space theory or the 1D method (see Chapter 53 *Shallow foundations*). The actions of the fully mobilised piles are simply modelled as appropriate upward forces applied at the chosen pile locations. For a given raft geometry and loading distribution, the locations of the piles can be successively adjusted to minimise the differential settlements, bending moments and shear stresses in the raft.

The above approach is illustrated in **Figure 56.42(a)**, which shows a raft foundation on a relatively uniform deep layer of clay and subjected to local high column loads. As shown in **Figure 56.42(b)**, the raft foundation will undergo significant settlement in a sagging mode and there will be high local bending moments beneath the columns, as shown in **Figure 56.42(c)**. By placing a suitably designed ductile pile under each column it is possible not only to reduce the bending moments in the raft as shown in **Figure 56.42(c)**, but also the total and, in particular, the differential settlement as shown in **Figure 56.42(b)**.

Figures 56.34 and **56.35** illustrate how pile locations can be optimised to minimise differential settlements for raft-enhanced pile groups. Similar approaches can be adopted for pile-enhanced rafts, but the analysis will be much simpler since pile–raft and pile–pile interactions can be ignored. If the pile spacing ratios turn out to be less than about 6*d* to 8*d* (depending on the variation of soil stiffness with depth), then a pile-enhanced raft will not be appropriate. However, it is worth revisiting the reasons for the choice of the maximum allowable settlement. Increasing the allowable total settlement will reduce the number of piles, but this does not necessarily lead to increases in differential settlement or raft stresses, provided the piles are located appropriately.

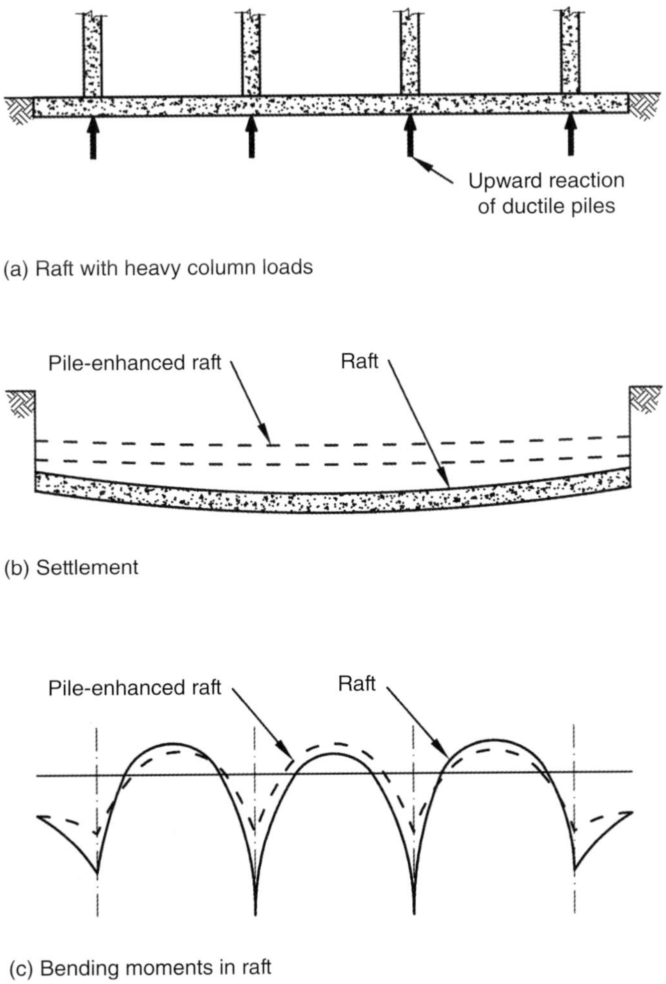

Figure 56.42 Illustrative example of a pile-enhanced raft foundation showing how the use of ductile piles beneath heavily loaded columns can reduce the total and differential settlement and the bending moments

Love (2003) describes a case history of the design of a pile-enhanced raft foundation on London Clay for which there was a considerable saving both in the number of piles and in the raft thickness when compared with a conventional design of a fully piled rigid raft. The building is a reinforced-concrete framed structure, seven storeys high, with a basement. Love stresses the following design requirements.

- The load–displacement behaviour after full mobilisation should be as near as possible constant.
- In practice the actual load deflection relationship will not be perfectly constant – see, for example, **Figure 56.40**. Unlike conventional pile designs, overcapacity can be as much of a problem as undercapacity (**Figure 56.43**).
- Raft settlement should always be greater than the amount required to mobilise full shaft friction, generally more than 20 mm.
- Reasonable upper and lower bounds should be set for the fully mobilised pile loads (**Figure 56.44**), and the raft design checked for both extremes. **Figure 56.44** is taken from Love's paper and illustrates the designed upper- and lower-bound settlement profiles across the raft at various sections. It should be noted that a pile was not needed beneath every column.
- In order to reduce the potential variation between the lower- and upper-bound pile capacities, the end-bearing resistance was eliminated. A variety of different methods can be used to eliminate or reduce end-bearing if thought necessary.
- Three pile tests were carried out and the tests were required to pass the design upper- and lower-bound load–settlement criteria

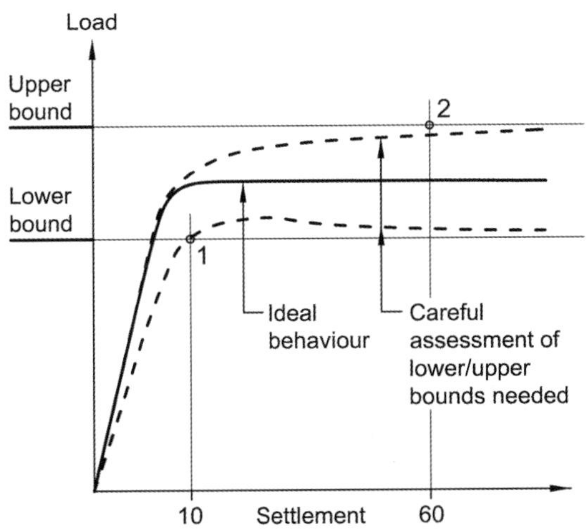

Figure 56.43 Pile-enhanced rafts: required pile load–settlement response

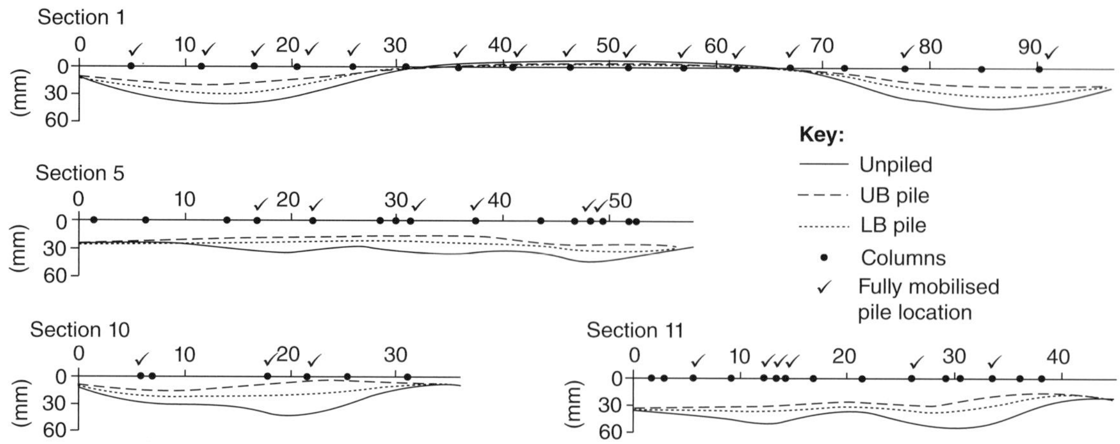

Figure 56.44 Upper- and lower-bound settlement profiles for a pile-enhanced raft compared to an unpiled raft (Love, 2003)

set for the design (points 1 and 2 in **Figure 56.43**). In order to capture the full load–settlement behaviour of the pile, the load tests were run as load controlled up to yield and as displacement controlled thereafter. As mentioned in Section 56.9.2, it should be noted that the shaft resistance is rate dependent and a constant rate of displacement of 2 mm/min is likely to give shaft resistances about 15% higher than the static resistance (Burland and Twine, 1988).

56.10. Lateral loads on piled rafts

56.10.1 Introductory comments

Horizontal load on a piled raft is potentially resisted by four components: piles; passive resistance in front of the raft; frictional resistance along the sides of the raft; and frictional resistance along the underside of the raft. This is in a similar way to pile shaft resistance (primarily a frictional phenomenon) being mobilised at relatively small displacement and end-bearing resistance being mobilised at larger displacement. The frictional resistance along the sides and along the underside of a raft are mobilised at relatively small displacement, whereas the passive resistance in front of the raft and the lateral resistance of the piles typically requires larger displacements to be fully mobilised. The potential for friction to be mobilised along the sides and along the underside of a raft means that lateral loads can be shared between the raft and piles in a similar manner to vertical loads. When assessing the potential for lateral load-sharing, a careful assessment of the contact stress on the raft is very important. This requires an assessment of plausible upper- and lower-bound vertical load-sharing between raft and piles, which in turn depends on the **relative stiffness** of the raft and piles. Once the potential variations in vertical load-share are calculated, then the lower-bound contact stress can be derived and then the lateral frictional resistance provided by the raft can be easily calculated. This frictional resistance is mobilised at relatively small displacements and well before the passive resistance in front of the raft and lateral resistance of vertical piles is mobilised. Therefore, **Figure 56.22** together with **Figures 56.28–56.30** and implementation of the associated simple analytical methods are important for conceptual and preliminary design. It will also be important to avoid the piles being overly stiff for horizontal loading (since the piles will then attract most of the applied load). Hence, the use of relatively flexible piles would be appropriate for this application – for example, steel piles or smaller-diameter reinforced-concrete piles (with reinforcement design focused on ensuring ductility).

56.10.2 Lateral loads on raft-enhanced pile groups

As stated above, the potential for friction to be mobilised on the underside of a raft means that lateral loads can be shared between the raft and piles in a similar manner to vertical loads. However, in general for raft-enhanced pile groups a cautious approach is necessary to ensure that the piles are structurally strong enough to resist most of the applied horizontal load. The critical scenario for lateral loading is a careful assessment of the lower-bound contact stress on the raft. This is likely to occur for the lower-bound soil stiffness close to the underside of the raft (hence a relatively 'soft' raft stiffness), and an upper-bound soil stiffness at depth (and hence a relatively stiff pile-group stiffness). The behaviour of pile groups under a horizontal load is introduced in Chapter 55 *Pile-group design*.

Mokwa (1999) carried out horizontal loading experiments on a small square piled raft (using steel H-piles). These experiments confirm that the stiffness and capacity of a piled raft under horizontal load is much greater than a free-standing pile group. Centrifuge tests by Katzenbach and Turek (2005) indicated that at small horizontal displacement most of the horizontal load was resisted by friction along the underside of the raft (this raft load-share reduced as horizontal displacements increased – that is, as the piles began to pick up the horizontal loads).

For small piled rafts, refer to Section 56.5.3 and **Figure 56.17**, it is recommended that frictional resistance on the underside of the raft is ignored (at least for the purposes of conceptual and preliminary design). In general, only large piled rafts (**Figure 56.17**) would reliably develop sufficient contact stress for this component of resistance to be useful for the purposes of practical design.

56.10.3 Lateral loads on pile-enhanced rafts

The use of pile enhancement for rafts carrying lateral loads has the potential to achieve very significant economies. In the traditional approach to pile-group design most of the vertical load is assumed to be carried by the piles. This leads to difficulties in carrying lateral loads since no reliance can be placed on frictional resistance between the base of the raft and the underlying ground, as would be the case for an unpiled foundation. Therefore, either the pile group has to be designed to carry the horizontal loads or raking piles must be introduced. Both of these solutions are expensive and unsatisfactory for a variety of reasons. It is evident that a 'vicious circle' has developed in which an excessive number of vertical piles inhibits the ability of the foundation to efficiently transfer horizontal loads into the underlying ground.

However, if a pile-enhanced raft design is adopted, the number of piles can be substantially reduced and a significant contact stress between the raft and the ground is retained. For a pile-enhanced raft, the raft will usually carry more than half the vertical load (often substantially more than half). Hence the frictional resistance between the underside of the foundation and the ground can be confidently utilised to transmit the horizontal shear stresses into the ground and the need for additional piles, either vertical or raking, is eliminated. This approach requires a careful assessment of the likely horizontal displacements of the raft together with the associated induced bending moments at the top of the piles. As noted above for raft-enhanced pile groups it is important that the relative stiffness of the piles and raft under lateral loads needs to be considered (and *not* just their ultimate capacity), in order to assess the load-share between the raft and piles. The reader is referred to Chapter 54 *Single piles* for an introduction to the design of single piles subject to lateral loads.

56.10.4 Raft foundations supported by pile-reinforced ground

An alternative design concept for piled rafts, when the primary goal is settlement and differential settlement reduction (i.e. the SBD and RBD zones in **Figure 56.14(b)**), is to use piles solely as ground reinforcement. This design concept (outlined by Wong *et al.*, 2000) uses piles which are *not* structurally connected to the raft and a gap is left between the underside of the raft (supported by a granular mattress) and the top of the piles. The mechanism of vertical load transfer is then similar to using ground improvement (such as deep soil mix columns acting as 'stiff inclusions' in the ground below a foundation) where load is transferred into the pile reinforcing elements by negative skin friction acting through the upper part of the pile shaft. This design concept has particular merit when a foundation is subject to rare but very high horizontal loads (either due to earthquake or ship impact loads), since the shear force and bending moment transferred into the piles is substantially reduced. Conceptually this has benefit because the load path is more reliable (lateral resistance is provided through friction around and beneath the raft, and from passive resistance in front of the raft). For horizontal design loads which are infrequent short-duration extreme events (such as earthquake or impact loads) this design concept provides the potential for a more flexible system which can absorb and dissipate the energy and avoids the risk of brittle or progressive failure. Some major structures, such as the Venice tidal barriers (Jamiolkowski *et al.*, 2009) and the Rio–Antirrio Bridge (Teyssandier *et al.*, 2003), have used this design concept.

56.11. A case history of a pile-enhanced raft: the Queen Elizabeth II Conference Centre

An early application of pile-enhanced rafts was the design of the foundations of the Queen Elizabeth II Conference Centre in Westminster, London (Burland and Kalra, 1986; **Figure 56.45**). The conference centre is founded on a 2 m thick raft in a relatively shallow excavation. The calculated maximum total settlement was in excess of about 20 mm. Over half of the total weight of the superstructure is transmitted to the raft by means of a few columns, which are located at the edges of the raft and carry loads of up to 26 MN. The

Figure 56.45 Queen Elizabeth II Conference Centre

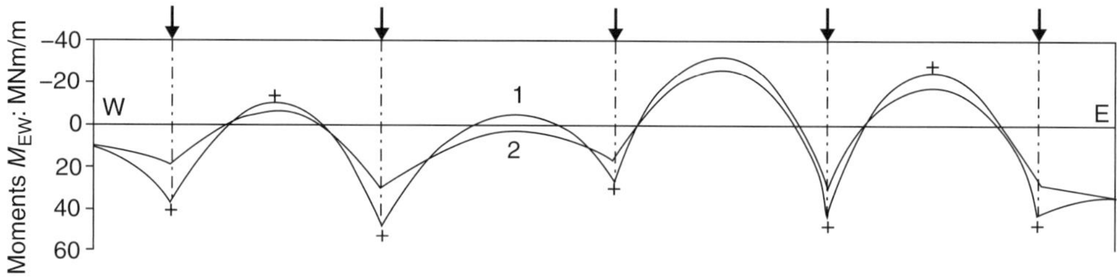

Figure 56.46 Queen Elizabeth II Conference Centre: calculated bending moments on a 3 m wide strip at the edge of the raft: (1) is for the raft alone and (2) shows the effects of the ductile piles (Modified from Burland and Kalra, 1986)

preliminary analysis of the raft identified local areas of high bending moments and shear forces, and concern was expressed that the bearing pressures under some of these columns might induce local yielding and excessive settlement in the underlying London Clay.

One method of dealing with the problem would have been to thicken the raft locally. This would have been a costly and time-consuming approach with the potential for causing delays and difficulties in the construction of the major raft foundation. The local thickenings would have been at the edges of the raft and there were significant 'knock-on' costs in the design of the retaining walls.

A pile-enhanced raft appeared to be a suitable solution. Single straight-shafted piles were placed directly beneath the most heavily loaded columns in the knowledge that the settlement of the raft would be sufficient to fully mobilise the shaft resistance of the piles. The effect was to apply a constant upward force beneath each column. By this means the resultant loads transferred from the columns into the raft were significantly reduced. The piles are each 16 m long and 1.8 m dia. The calculated ultimate shaft resistance was 6.4 MN, giving effective reductions in the resultant column loads of 25–46%.

Figure 56.46 shows the calculated bending moments along a 3 m wide edge strip, with the locations of the columns indicated by arrows. The bending moment labelled 1 relates to the undrained case without piles. The crosses refer to the worst values for some other cases considered, including the fully drained long-term case and partially drained cases involving some heave. The bending moment labelled 2 is the undrained case with stress-reducing ductile piles in place beneath the columns. This proved to be the worst case considered using such piles. It can be seen that the presence of the ductile stress-reducing piles beneath each column results in a considerable reduction in the bending moments, particularly beneath the columns.

In view of the novelty of this approach the Building Research Establishment was invited to instrument one of the piles to measure the magnitude of the load transferred to it through the raft foundation. A full description of the investigation is given by Price and Wardle (1986). **Figure 56.47** shows the results of the measurement. The full line and the broken line represent the measured pile-head loads and the calculated column loads, respectively, noting that they are plotted to different scales. Early on in construction the pile-head loads showed some fluctuations due to intermittent load applications, with a tendency for the underlying clay to swell during intervening periods. The main construction began in the middle of 1982, as can be seen by the rapid increase in the column loads. The measured pile-head loads reflect the loads on the columns. Construction was completed in 1984. It can be seen that by 1986 the measured pile-head load was in good agreement with the calculated value of 6.4 MN for the fully mobilised shaft resistance. Measurements continued through to 1996 and it is clear that no significant change in the load carried by the pile has taken place over a period of about 10 years.

56.12. Key points

- *Rafts:* The differential settlement and bending moment induced in a raft is fundamentally affected by its stiffness. Raft stiffness is a function of soil and concrete stiffness, and, particularly, the ratio of raft thickness to raft width (or diameter).
- *Raft on springs vs soil continuum:* A structural design often assumes the soil beneath a raft acts as a bed of uniform springs. There are many pitfalls and associated errors with this approach, and it is *not* recommended. A practical and more realistic approach is to have 'interactive' spring stiffnesses, with a structural spring model run in tandem with a soil continuum model. This will result in non-uniform springs, typically stiffer towards the raft edge and softer at the centre.
- *Piled rafts:* Compared with conventionally designed rafts and pile groups, a piled raft can provide many benefits when used in suitable ground conditions, such as reduced costs, lower carbon and easier, quicker and safer construction. There are two main types of piled raft: the *raft-enhanced pile group* and the *pile-enhanced raft*. The design concepts are very different and must *not* be mixed. The application of these concepts enables considerable simplifications for practical design.
- *Raft-enhanced pile group:* Both the raft and pile group behave in a pseudo-elastic manner. A key issue is to assess the load-sharing between the pile group and raft, which is dependent on the relative stiffness of the pile group and raft. The upper- and lower-bound stiffnesses of pile group and raft need to be assessed. This type of piled raft can be thought of as a 'value-engineered pile group', with piles omitted to the extent that raft resistance can be

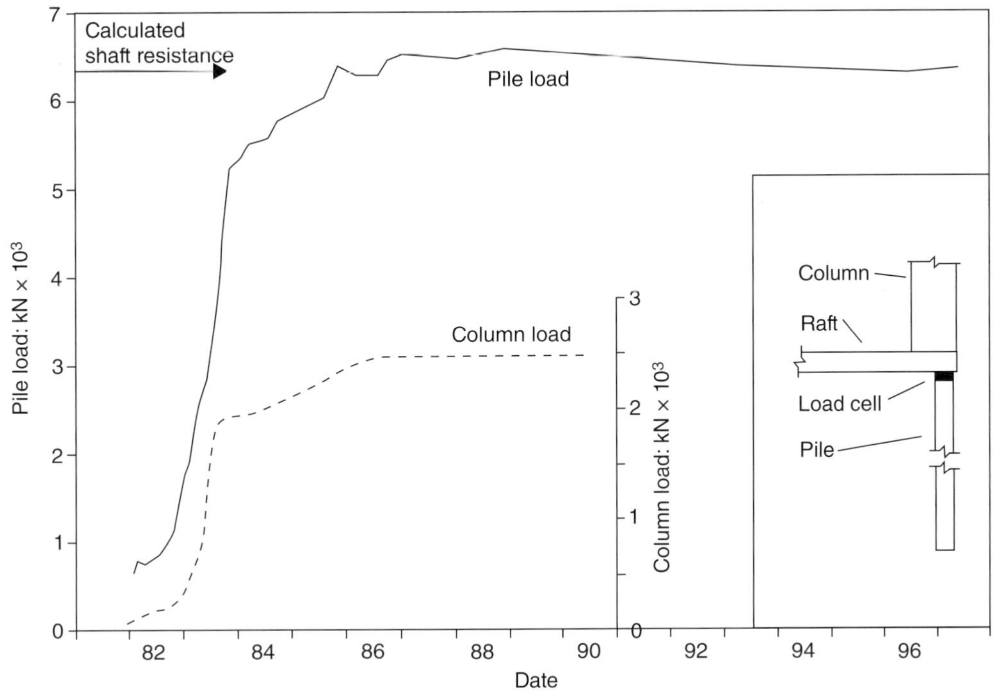

Figure 56.47 Queen Elizabeth II Conference Centre: load against time response of instrumented pile beneath the edge of the raft foundation carrying a heavily loaded column (Data taken from Burland and Kalra, 1986)

relied upon. There are two subtypes – 'small' and 'large' – and their behavioural trends are different. For the former assessment of enhanced ultimate capacity is important, while for the latter it is load-sharing, settlement and differential settlement.

- *Compensated piled raft:* A piled raft located at the base of a deep excavation is known as a *compensated piled raft*. Behaviour tends to be close to that of raft-enhanced pile groups; however, several additional issues become important, such as: large changes in effective stress; typically large buoyancy force; increased importance of construction sequence and timing on movements and load-sharing and so on. Because the *net change in effective stress* at the base of the excavation may be negative (as a minimum for some interim construction stages), then heave and associated effects become important (including changes in pile capacity with time). An increasingly important application is the use of compensated piled rafts for large underground metro station boxes and their use for OSD foundation support.
- *Pile-enhanced raft:* The piles will fully mobilise their ultimate geotechnical capacity. It is essential that the pile load–settlement behaviour is ductile. A key issue is to assess the upper- and lower-bound pile ultimate geotechnical capacities. This type of piled raft acts predominantly as a raft, with additional local support by piles beneath heavily loaded columns. The term 'pile-enhanced raft' is considered more appropriate than the term 'settlement-reducing piles', because the main benefits may be broader than settlement reduction alone. A pile-enhanced raft is generally chosen when a raft by itself just fails to work.
- *Lateral load-sharing:* Lateral loads can be shared between piles and raft; this is most relevant for either large raft-enhanced pile groups or pile-enhanced rafts. Pile-enhanced rafts are particularly efficient means of resisting lateral loads. A consideration of the relative stiffness of piles and raft is important, together with ensuring the piles can cope with imposed lateral movements, and pile reinforcement is detailed to ensure a ductile response.

References

Bown, A and O'Brien, AS (2008) Shaft friction in London clay: modified effective stress approach. *Proceedings of 2nd BGA International Conference on Foundations, Dundee, UK*. HIS BRE Press, Watford, UK, pp. 91–100.

Brown, PT (1969) Numerical analysis of uniformly loaded circular rafts on elastic layers of finite depth. *Géotechnique* **19(2)**: 301.

Brown, PT and Yu, SKR (1986) Load sequence and structure foundation interaction. *Journal of Structural Engineering, ASCE* **112(1)**: 481–488.

Burland, JB (1995) Invited special lecture: Piles as settlement reducers. *Proceedings of the 19th National Conference on Geotechnics, Associazione Geotecnica Italiana, Pavia, Italy*, pp. 21–34.

Burland, JB (2006) Interaction between structural and geotechnical engineers. *The Structural Engineer* **84(8)**: 29–37.

Burland, JB and Kalra, JC (1986) Queen Elizabeth II Conference Centre: geotechnical aspects. *Proceedings of the Institution of Civil Engineers* **1(80)**: 1479–1503.

Burland, JB and Twine, D (1988) The shaft friction of bored piles in terms of effective stress. *Conference on Deep Foundations on Bored and Augered Piles, Ghent, Belgium*. Balkema, Rotterdam, the Netherlands, pp. 411–420.

Burland, JB, Broms, B and de Mello, VFB (1977) Behaviour of foundations and structures. *Proceedings of the 7th International Conference on Soil Mechanics and Foundation Engineering, Tokyo, Japan*, vol. 1, pp. 495–548.

Clancy, P and Randolph, MF (1993) An approximate analysis procedure for piled raft foundations. *International Journal for Numerical and Analytical Methods in Geomechanics* **17(12)**: 849–869.

Cooke, RW, Bryden Smith, DW, Gooch, MN and Sillet, DF (1981) Some observations on the foundation loading and settlement of a multi-storey building on a piled raft foundation in London. *Proceedings of the Institution of Civil Engineers* **107(1)**: 433–460.

Corke, DJ, Fleming, WK and Troughton, VM (2001) A new approach to specifying performance criteria for pile load tests. *Symposium Proceedings of Underground Construction, London, UK*, pp. 401–410.

Cox, WR, Cameron, K and Clarke, J (1992) Static and cyclic axial load tests on two 762 mm diameter pipe piles in clay. *Proceedings of the Conference on Large-Scale Pile Tests in Clay, London, UK*. Thomas Telford, London, UK, pp. 268–284.

Fraser, RA and Wardle, LJ (1975) A rational analysis of shallow footings considering soil–structure interaction. *Australian Geomechanics Journal* **1**: 20–25.

Hemsley, JA (1987) Elastic solutions for axisymmetrically loaded circular raft with free or clamped edges founded on Winkler springs or a half space. *Proceedings of the Institution of Civil Engineers* **2(83)**: 61–90.

Hemsley, JA (1998) *Elastic Analysis of Raft Foundations*. Thomas Telford, London, UK.

Hooper, JA (1973) Observations on the behaviour of a piled-raft foundation on London Clay. *Proceedings of the Institution of Civil Engineers* **55(2)**: 855–877.

Hooper, JA (1974) Analysis of a circular raft in adhesive contact with a thick elastic layer. *Géotechnique* **24(4)**: 561.

Hooper, JA (1979) *Review of Behaviour of Pile Raft Foundations*. CIRIA Report 83. CIRIA, London, UK.

Hooper, JA (1983) Nonlinear analysis of a circular raft on clay. *Géotechnique* **33(1)**: 1–20.

Hooper, JA and Wood, LA (1977) Comparative behaviour of raft and piled foundations. *Proceedings of the 9th International Conference on Soil Mechanics, Tokyo, Japan*, vol. 1, pp. 545–550.

Horikoshi, K and Randolph, MF (1997) On the definition of raft–soil stiffness ratio for rectangular rafts. *Géotechnique* **47(5)**: 1055–1061.

Jardine, RJ, Potts, DM, Fourie, AB and Burland, JB (1986) Studies of the influence of nonlinear stress–strain characteristics in soil–structure interaction. *Géotechnique* **36(3)**: 377–396.

Katzenbach, R and Turek, J (2005) Combined pile–raft foundation subjected to lateral load. *Proceedings of the 16th International Conference on Soil Mechanics and Geotechnical Engineering, Osaka, Japan*. Millpress, Rotterdam, the Netherlands, vol. 4, pp. 2001–2004.

Katzenbach, R, Arslan, U, Moorman, C and Reul, O (1998) Piled raft foundation: interaction between piles and raft. *Darmstadt Geo-technics* **4**: 279–296.

Katzenbach, R, Arslan, U and Moorman, C (2000) Piled raft foundation projects in Germany. In *Design Applications of Raft Foundations* (Hemsley, JA (ed)). Thomas Telford, London, pp. 323–391.

Lee, IK and Brown, PT (1972) Structure–foundation interaction analysis. *Journal of the Structural Division, ASCE* **98(ST11)**: 2413–2430.

Love, JP (2003) Use of settlement reducing piles to support a raft structure. *Proceedings of ICE, Geotechnical Engineering* **156(GE4)**: 177–181.

Mandolini, A, Russo, G and Viggiani, C (2005) Pile foundations: experimental investigations, analysis and design. *Proceedings of the 16th International Conference on Soil Mechanics and Geotechnical Engineering, Osaka, Japan*. Millpress, Rotterdam, the Netherlands, vol. 1, pp. 177–213.

Mokwa, RL (1999) An investigation of the resistance of pile caps to lateral loading. PhD thesis, Virginia University, Charlottesville, VA.

Nicholson, DP, Morrison, PRJ and Pillai, AK (2002) Piled raft design for high rise buildings in East London, UK. *Proceedings of the 9th International Conference on Piling and Deep Foundations, Nice, France*.

O'Brien, AS (2019) The evolution of deep foundations for transportation projects: from minor to major. Invited Speaker, TC212 International Conference, Santa Cruz, CA.

O'Brien, AS and Higgins, KG (2020) *The Management of Advanced Numerical Modelling in Geotechnical Engineering: Good Practice*. CIRIA Report C791. CIRIA, London, UK.

O'Brien, AS and Sharp, P (2001) Settlement and heave of overconsolidated clays: a simplified nonlinear method of calculation. *Ground Engineering*, Oct.–Nov.: pp. 21–28, 48–53.

Poulos, HG (2001) Piled-raft foundation: design and applications. *Géotechnique* **51(2)**: 95–113.

Price, G and Wardle, IF (1986) Queen Elizabeth II Conference Centre: monitoring of load sharing between piles and raft. *Proceedings of the Institution of Civil Engineers* **1(80)**: 1505–1518.

Randolph, MF (1994) Design methods for pile groups and piled rafts. *Proceedings of the 13th International Conference on Soil Mechanics and Geotechnical Engineering, New Delhi, India*. Balkema, Rotterdam, the Netherlands, vol. 5, pp. 61–82

Randolph, MF (2003) Science and empiricism in pile foundations design. The 43rd Rankine Lecture. *Géotechnique* **53(10)**: 847–875.

Reese, LC and O'Neill, MW (1988) Field load tests on drilled shafts. *Conference on Deep Foundations on Bored and Augered Piles, Ghent, Belgium*. Balkema, Rotterdam, the Netherlands, pp. 145–191.

Reul, O and Randolph, MF (2004) Design strategies for piled rafts subjected to nonuniform vertical loading. *Journal of Geotechnical and Geoenvironmental Engineering* **130(1)**: 1–13.

Russo, G and Viggiani, C (1995) Long term monitoring of a piled foundation. *4th International Symposium on Field Measurements in Geomechanics, Bergamo, Italy*, pp. 283–290.

Sales, MM, Small, JC and Poulos, HG (2010) Compensated piled rafts in clayey soils: behaviour, measurements, and predictions. *Canadian Geotechnical Journal* **47**: 327–345.

Sommer, H, Tamaro, G and DeBenedittis, C (1991) Messe Turm, foundations for the tallest building in Europe. *Proceedings of the 4th DFI Conference on Piling and Deep Foundations, Stresa, Italy*, pp. 139–145.

Teyssandier, JP, Combault, J, Morand, P (2003) The Rion–Antirion Bridge, design and construction. *Proceedings of the 12th World Conference on Earthquake Engineering, Auckland, New Zealand*, paper 1115.

Viggiani, C (1998) Pile groups and piled rafts behaviour. In *Deep Foundations on Bored and Auger Piles* (van Impe, WF and Haegman, W (eds)). Balkema, Rotterdam, the Netherlands, pp. 77–90.

Viggiani, C, Mandolini, A and Russo, G (2012) *Piles and Pile Foundations*. CRC Press, Boca Raton, FL.

Whitaker, T and Cooke, RW (1966) An investigation of the shaft and base resistance of large bored piles in London Clay. *Proceedings of the Conference on Large Bored Piles, London, UK.* ICE, London, UK, pp. 7–49.

Wong, IH, Chang, MF and Cao, XD (2000) Raft foundations with disconnected settlement reducing piles. In *Design Applications of Raft Foundations* (Hemsley, JA (ed)). Thomas Telford, London, UK.

Yamashita, K, Kakurai, M and Yamada, T (1994) Investigation of a piled raft foundation on stiff clay. *Proceedings of the International Conference on Soil Mechanics of Foundation Engineering, New Delhi, India*, vol. 2, pp. 543–546.

It is recommended this chapter is read in conjunction with

- Chapter 19 *Settlement and stress distributions*
- Chapter 21 *Bearing capacity theory*
- Chapter 22 *Behaviour of single piles under vertical loads*
- Chapter 53 *Shallow foundations*

All chapters in this book rely on the guidance in Sections 1 *Context* and 2 *Fundamental principles.* A sound knowledge of ground investigation is required for all geotechnical works, as set out in Section 4 *Site investigation.*

doi: 10.1680/icemge.66830.0937

Chapter 57
Global ground movements and their effects on piles

Tony O'Brien Mott MacDonald, UK
Alan Willoner Mott MacDonald, UK

CONTENTS

Global ground movements may occur for a variety of reasons, but are most commonly associated with cut-and-fill operations, particularly in clay soils. The impact on structural elements such as piles depends strongly on construction sequence, and normally the presence of soft clay or peat is most problematic, since large movements may occur over several months or years. The chapter considers negative skin friction, heave-induced tension and lateral interaction. Lateral ground movements are of particular concern since they have led to the structural failure of piles and pile groups on many occasions.

57.1. Introduction

Global ground movements may occur for a variety of reasons, but are most commonly associated with cut-and-fill operations, particularly in clay soils – for example, adjacent to embankments or excavations. The impact on structural elements such as piles depends strongly on construction sequence, and normally the presence of soft clay or peat is most problematic since large movements may occur over several months or years. The chapter considers

- negative skin friction (in soil which is settling, e.g. beneath an embankment) – this increases the axial load on piles, and will tend to increase pile settlement significantly
- heave-induced tension (in soil which is heaving upwards, e.g. beneath an excavation) – this may cause tensile failure of the piles or pile cap (especially at the pile–pile cap connection)
- lateral interaction (in soil which is moving horizontally, e.g. adjacent to an embankment or excavation, or in an unstable slope) – this can be particularly damaging since the piles may fail in bending or shear. This is a common cause of pile or pile-group failure if this mode of ground movement is overlooked. A specific application is when the piles are installed as a remedial measure to stabilise a slope.

Poulos (2017) illustrates a wide range of circumstances where piled foundations may be subject to loading from soil movement (**Figure 57.1**). An important cause of global ground movements is earthquake-induced ground movements in seismically active areas of the world (e.g. Ashford *et al.*, 2011); this is discussed further in Chapter 60 *Foundations subjected to cyclic and dynamic loads*. Note that driven piling-induced heave can be a problem when working close to sensitive receptors – for example, sheet piles being installed near a rail track at night to avoid the risk of pile-induced heave derailing a train.

In each section the response of a single (individual) pile is initially considered, before proceeding to additional considerations for the group response. In general, piles at the centre of a group tend to be 'sheltered' or 'shielded' from interaction effects, and therefore the assessment of a single-pile response tends to be conservative for pile groups. When lateral interaction develops, progressive structural failure may be an issue (because the structural failure mode may be brittle, rather than ductile) and then a cautious approach is necessary.

Owing to the importance of construction sequence, close communication and coordination are likely to be required between the contractor or project manager and the designer. Close control of activities on site, particularly where subcontractors may not be aware of or prefer to ignore the implications of their actions, is also important.

Owing to the complexity of many of the problems and their relation to subtle aspects of ground behaviour, input from a geotechnical specialist will be essential. The text below highlights the main points and indicates key references for more detailed information. However, most of the references are published in specialist geotechnical journals, which will assume prior knowledge of geotechnics.

Often the most pragmatic approach may be a practical 'engineering solution' which substantially 'removes' the problem. For instance

(*a*) inherently reduce soil movement, such as by carrying out ground improvement

(*b*) wait for significant soil movement to be substantially complete before constructing the piles and pile caps

(*c*) isolate the piles from the soil movement.

Approach (*a*) normally requires that the cause of movement is largely removed, for instance by constructing an embankment from lightweight fill or by carrying out deep ground improvement. Approach (*b*) highlights the difficulty

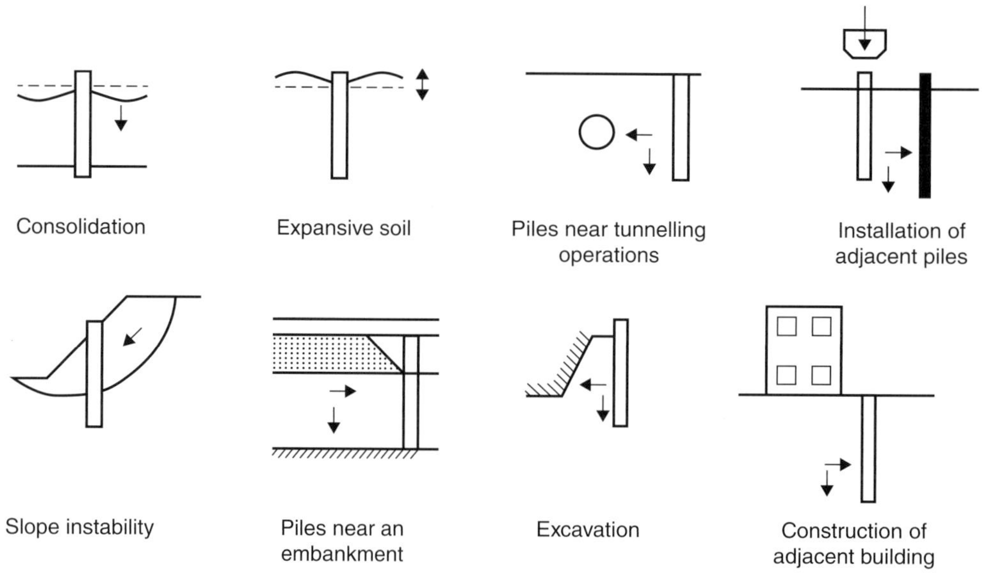

Figure 57.1 Some sources of global ground movements which may affect piled foundations (After Poulos, 2017)

of low-permeability clay soils, which are likely to require months or years to fully consolidate. It may be undesirable to wait this long. Time-dependent ground movements tend to be predominantly vertical, whereas significant lateral ground movements mainly occur during construction (unless there is ongoing instability). Hence, when lateral ground movements are the main concern point (*b*) can be effective. An example of (*c*) is the use of double casings for bored cast in situ piles. The annulus between the two casings can isolate the pile from the outer casing, which moves with the soil. Point (*c*) is generally appropriate if settlement and negative skin friction is a concern; it is unlikely to be appropriate if lateral ground movements are an issue.

The alternative to 'removing' the cause of the problem is to select a pile type and/or design the piled foundations in such a way as to be able to withstand the applied loads from the global ground movements. For example, use of driven steel H piles where the section size is generally controlled by temporary driving stresses and there is often redundancy in flexural capacity that can accommodate additional stress from lateral soil movement. Nevertheless, this latter approach will require an estimate to be made of the additional loads due to global ground movement, which will require an understanding of the ground movements and the ground–structure interaction between piles and the ground.

Figure 57.2 shows an example of the extreme consequences of global ground movements on foundations. In this situation it appears that lateral interaction (due to excavation on one side and stockpiling on the other) was a significant contributory factor, which led to a catastrophic collapse (*China Daily*, 2009). This failure illustrates a few key principles.

Figure 57.2 Building collapse due to failure of the piles (Shanghai, 2009) (*China Daily*; www.chinadaily.com.cn)

- It is vital to have an awareness of all the construction processes and their relationships with site topography, existing structures and infrastructure, both across a site and beyond the site boundaries.
- The sequence and timing of activities can be important.
- Piles which are designed solely for vertical loads (common for buildings) can be extremely vulnerable to lateral ground movements.

This means that temporary works can have a significant impact on the permanent works. Temporary and permanent works are often designed by separate organisations, and the commercial and procurement environment will also be a factor, as well as technical issues. The avoidance of these major hazards, although often obvious with the benefit of hindsight, can require a major coordination effort both during design and construction.

57.2. Negative skin friction

57.2.1 Introduction

As shown in **Figure 57.3(a)**, ordinarily for piles the vertical load applied at the head (Q_h) is resisted by a combination of the resistance on the shaft (Q_s) and the base (Q_b); see Chapters 10 *Codes and standards and their relevance* and 54 *Single piles* for further details. Hence

$$Q_h = Q_s + Q_b \tag{57.1}$$

The maximum axial load in the pile is Q_h.

It is assumed that the pile displaces downwards, while the surrounding soil is static except near the pile, where it will tend to move with it. However, as shown in **Figure 57.3(b)**, if the soil inherently tends to settle more than the pile over at least part of the length, then there will be both 'positive' and 'negative' skin (or shaft) friction (Q_{s+} and Q_{s-}, respectively). The transition from positive to negative skin friction is the point where settlement of the pile and soil are equal, and will be referred to as the neutral point. When there is negative skin friction (NSF) the maximum axial load in the pile (Q_{max}) occurs at the neutral point and exceeds Q_h

$$Q_{max} = Q_h + Q_{s-} = Q_{s+} + Q_b \tag{57.2}$$

Thus, any NSF on the pile increases the maximum axial load, which may have implications for the structural design of the pile, by exceeding the permitted compressive stress in the concrete or by inducing bending stresses in raking piles. However, settlement of the pile will also be increased, and this may be the most significant practical issue that may govern design (Poulos, 2008). In general, NSF does not cause geotechnical ultimate limit state (ULS) failure, hence the designer's focus should generally be on the potential impact of NSF on pile settlement (serviceability limit state (SLS)). For friction piles, note that shaft friction (positive or negative) is mobilised at small relative displacement (between soil and pile), typically less than 10 mm. For end-bearing piles (and noting that ULS may be defined as a settlement equal to 10% of pile diameter), for small-diameter driven piles NSF may affect ULS, depending on the magnitude of NSF relative to the stiffness and ultimate resistance of the founding stratum for end-bearing.

Settlement of the soil layer may often be due to imposition of a surcharge (e.g. fill material) at the surface, although lowering of the water table or dissipation of excess pore water pressure after pile driving may also cause significant settlement without any change in load at the surface (especially for soft clays or peat). Furthermore, vegetation-induced changes in effective stress may cause both settlement and heave (depending on seasonal climate effects); refer to Chapter 53 *Shallow foundations*. The settling material may include fill material, which has been recently placed and is settling with the underlying material that it has loaded, as well as potentially settling within its own thickness. Where the soil is clay it may take significant time for consolidation settlement to occur. If the thickness of compressible soil is variable, say soft clay on a steeply dipping rockhead, then lateral movement may also develop in addition to settlement. This would then require a consideration of lateral interaction effects; refer to Section 57.4.

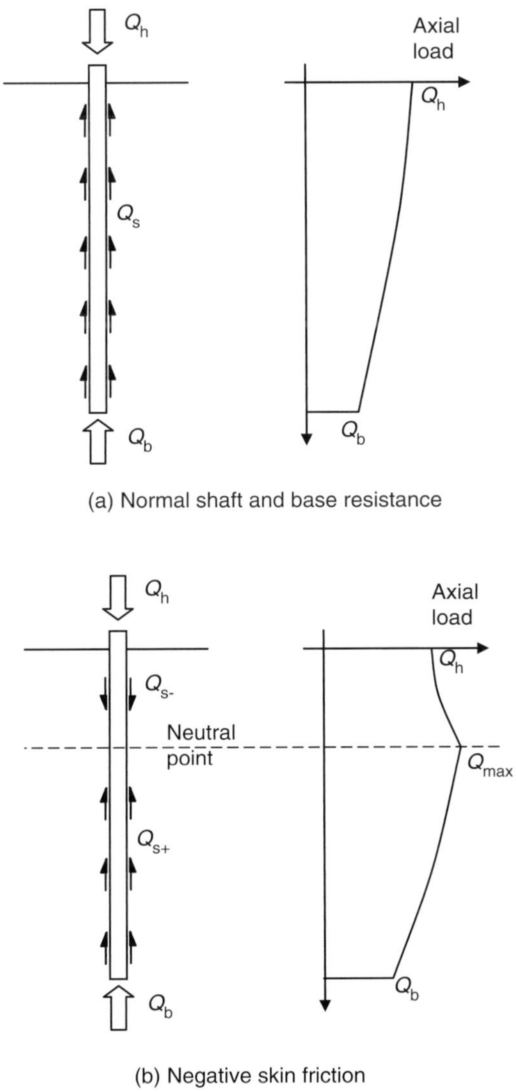

Figure 57.3 Effect of negative skin friction compared to normal pile action

57.2.2 Single piles: end-bearing and 'floating'

57.2.2.1 End-bearing piles

As shown in **Figure 57.4(a)**, the simplest form of NSF loading is for a pile that derives nearly all of its capacity from end-bearing in an underlying stratum, which is significantly stiffer or stronger than the overlying soil (e.g. rock). The neutral point can then be assumed to be close to the bottom of the settling soil layer, and it will be slightly conservative to assume

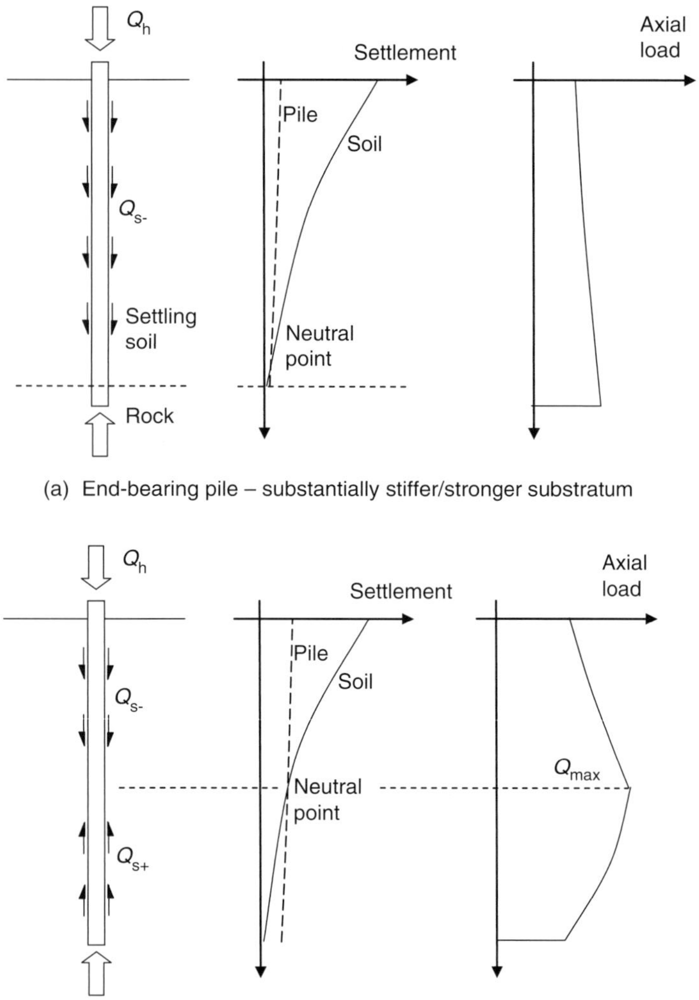

(a) End-bearing pile – substantially stiffer/stronger substratum

(b) Floating pile – gradual increase in soil strength/stiffness with depth

Figure 57.4 End-bearing and floating piles

that NSF exists over the entire length of the shaft ($Q_{s-} = Q_s$, $Q_{s+} = 0$). Equation 57.2 then becomes

$$Q_{max} = Q_h + Q_{s-} = Q_b \tag{57.3}$$

If the pile also derives a significant shaft capacity in the underlying stratum (i.e. $Q_{s+} > 0$), then the original version of Equation 57.2 may be appropriate. However, the neutral point is still assumed to occur at the boundary between the two strata and, hence, determination of Q_{s-} and Q_{s+} is relatively straightforward.

When NSF is induced by soft clays or peats it is common to assess this by using total-stress methods. However, there are a number of problems: first, the empirical alpha values quoted in the literature are not relevant for NSF (since they have only been derived from conventional load tests to measure positive shaft friction); second, there is usually a wide spread of alpha and c_u values, so selection of appropriate design values is problematic; and finally, depending upon the magnitude of the surcharge, the total-stress method may be either unsafe or over-conservative.

For soft clays or peats it is strongly recommended that NSF is calculated by using an effective-stress (β) approach (refer to Chapter 22 *Behaviour of single piles under vertical loads* and Chapter 54 *Single piles*).

57.2.2.2 'Floating' piles

As shown in **Figure 57.4(b)**, where there is a more gradual increase of soil strength or stiffness with depth the situation becomes more complicated, since the depth of the neutral point will not be immediately apparent. Rather, it depends upon a comparison of estimates of soil and pile settlement with depth.

57.2.2.3 End-bearing piles and brittle response

The design of end-bearing piles is generally more straightforward, since the location of the neutral point can be readily assumed. However, despite the complication of establishing the depth of the neutral point for floating piles, the response will be relatively ductile (or 'self-correcting'), since if Q_{s-} is larger than anticipated, the pile settlement will be increased, but this will in turn move the neutral point upward, tending to reduce Q_{s-}. End-bearing piles are not 'self-correcting' in the way that floating piles are, although they normally have ductile load–settlement characteristics. Nevertheless, it will be important to check that the base resistance has sufficient reserve capacity to cope with uncertainties in the assessment of NSF. For piles which obtain most of their resistance from shaft friction in rock sockets, and also have limited (or uncertain) end-bearing (either due to base cleaning not being practical, or due to potential voids in the rock mass), then particular care may be required if they are subjected to NSF. For example, the shaft friction that develops in some rocks (sandstones, limestones) can exhibit brittle behaviour (i.e. the shaft friction reduces after a limiting displacement has developed). The designer may consider that some 'margin' needs to be applied to estimates of NSF; this requires engineering judgement and depends upon a consideration of several factors, including

- specific application – for example, is the pile used as a single pile or within a pile group (with adequate redundancy and a stiff pile cap to facilitate load redistribution – see Chapter 55 *Pile-group design*)
- precedent experience in specific geology under consideration, and uncertainties associated with estimating shaft and end-bearing resistance
- consequences associated with exceeding an SLS (e.g. is the supported structure relatively tolerant of movement or not? See Chapter 52, *Foundation types and conceptual design principles*)

- whether a larger NSF would lead to a structural or geotechnical limit state being exceeded
- whether the geotechnical resistance characteristics are potentially 'brittle' or not, as noted above.

The 'margin' discussed above is any allowance which a designer may consider necessary to allow for uncertainties, if Equations 57.2 or 57.3 are used, in excess of a code factor of safety (or partial factor). In general, if ground–structure interaction is being carried out (with a consideration of relative stiffness of the system components), then the analysis should be unfactored, and any factors only applied to analysis outputs.

57.2.3 Design considerations for single piles

Even for a large pile group, a consideration of the behaviour of a single pile is likely to be a logical starting point for design.

57.2.3.1 Determination of skin friction load

Determination of Q_{s+} and Q_{s-} (and Q_b) is likely to be based on routine methods, as outlined in Chapter 54 *Single piles.* Shaft capacity is normally assessed using effective-stress methods in free-draining soils or a total-stress approach in clay. However, in general, an effective-stress approach is more logical and far more appropriate when assessing NSF, for both soft and stiff clays. NSF increases gradually as the clay consolidates and vertical and horizontal effective stresses increase with time (as excess pore water pressures dissipate). Hence, the original measured undrained strength is no longer relevant once the clay starts to consolidate. Chapter 22 *Behaviour of single piles under vertical loads* discusses the shortcomings of total-stress methods; for both soft and stiff clays a simple design approach using the β method is discussed. For soft clays, Burland and Starke (1994) analysed case records in which NSF has been measured (for monitoring periods of up to 17 years). The data covered a wide range of conditions; nevertheless the β values fell within narrow limits of 0.15 and 0.35 (for low-plasticity marine clays β was between 0.15 and 0.25, whereas for high-plasticity clays and silty clays it was between 0.2 and 0.35). For stiff clays a β method is also generally more reliable than total-stress methods; the β value will vary depending upon the nature of the overconsolidated clay, as discussed in Chapter 22 *Behaviour of single piles under vertical loads.* Bown and O'Brien (2008) reviewed case records for the shaft friction of bored piles in a stiff fissured clay and related the shaft friction to in situ horizontal effective stress. This method was used for piles at the new Wembley Stadium, including those for the foundation of the Wembley Arch, which were subject to NSF (Cheong, 2007).

Since skin friction is mobilised at relatively small pile–soil displacement it may normally, and conservatively, be assumed that the 'switch' from Q_{s+} to Q_{s-} at the neutral point occurs instantaneously. The estimation of base capacity as settlement of the pile increases is likely to be more important, as discussed below.

57.2.3.2 Determination of the soil settlement profile with depth

The profile of soil settlement is an important element of any NSF design. It will require knowledge of the increment in vertical effective stress and the one-dimensional stiffness with depth (see Chapters 19 *Settlement and stress distributions* and 53 *Shallow foundations*). The distribution of strain and, hence, settlement with depth can then be derived. The resulting profile is likely to have a curvature as shown in **Figure 57.4(b)**, representing reducing strain with depth. The increment of vertical stress will reduce with depth to some extent, particularly as the depth becomes large compared to the plan area of the load. Furthermore, the stiffness of the soil is likely to inherently increase with depth. This effect is compounded since ground stiffness also increases significantly as strain reduces (see Chapter 53 *Shallow foundations*), so that the relevant 'operational' stiffness may increase very significantly as strains reduce with depth. Ignoring the curvature of the settlement profile (**Figure 57.4(b)**) can lead to significant overestimation of the thickness and, hence, magnitude of NSF loading (Cheong, 2007).

The interdependence of strain and stiffness can be accounted for in a finite element analysis with a suitably sophisticated soil constitutive model, but care will also be required that the nonlinearity of the response is correctly modelled by conducting the analysis in sufficiently small increments (Potts, 2003). An iterative analytical method, which can be undertaken using a spreadsheet, is described by O'Brien and Sharp (2001a, 2001b), and this simpler approach will typically be more appropriate for most practical design applications.

Once the 'free-field' soil settlement profile (i.e. not specifically accounting for the pile) has been determined, it becomes an input into the subsequent analysis, and can be assumed to remain constant. This will be conservative, particularly for large pile groups (see below), since the presence of the piles will tend to locally reduce settlement of the soil. A finite element analysis including the piles can potentially be used to overcome this assumption, inherently incorporating the soil–pile interaction. However, implementation of the analysis will require an experienced user, and careful checking (see O'Brien and Higgins 2020). Furthermore, an appropriate soil constitutive model will be required, and modelling of the pile–soil interface properties will require careful consideration.

57.2.3.3 Determination of the pile settlement profile with depth

In general, the vertical strain of concrete in the pile will be one or two orders of magnitude lower than the strain in the surrounding soil if the soil is settling significantly. Hence, an assumption that the pile is rigid will often be appropriate, leading to a profile of settlement with depth, which is a vertical line on a plot such as **Figure 57.4**.

Settlement at the toe of the pile may be significant unless there is a very stiff bearing layer. Fleming (1992) presents a relatively simple method based on hyperbolic functions, which can be used for estimation of the base response. Alternatively, a suitable factor on the ultimate bearing capacity can be used to limit settlement at the toe of the pile (see Chapter 54 *Single piles*).

57.2.3.4 Determination of pile length

Routine design of a pile (**Figure 57.4(a)**) is likely to require some iteration in the sense that even for a given pile diameter the required pile length will initially be unknown. It will be dictated either by the factor of safety against ULS failure or by the allowable settlement at the head. Normally the pile will be made longer until these criteria are achieved.

As described above, for an end-bearing pile (**Figure 57.3(a)**) the design is likely to follow this approach. The total load ($Q_h + Q_{s-}$) must be resisted in the stronger substratum. Note that the pile will need to extend a few diameters into the stronger substratum to develop the full bearing capacity of this layer in any case (see Chapters 54 *Single piles* and 55 *Pile-group design*). Any weathering of this material near the top of the layer will also be an issue. It may be decided to extend the pile a significant distance into the stronger layer so that a significant Q_{s+} can also be exploited. However, the ability of piling plant to extend (drive or bore) the pile into this layer may be a constraint, particularly as diameter increases.

For a floating pile the design will be governed by serviceability. A practical starting point for design is to assume that the pile is rigid and settles uniformly by an amount corresponding to the maximum allowable settlement at the pile head. Using the soil settlement profile this then allows determination of Q_{s-}, and hence ($Q_{s+} + Q_b$) required for equilibrium. Since the calculation is a serviceability calculation, a conventional ULS partial factor (or factor of safety) is *not* needed on Q_{s+}; however, as mentioned above, Q_b will require more careful consideration or a larger factor to account for potentially significant settlement at the toe. For piles in stiff clay, Q_b is typically quite small compared with Q_{s+}; hence, it could be conservatively ignored for a simple assessment. The importance of an accurate estimation of the distribution of settlement with depth is stressed, since this can have a far more significant impact on Q_{s-} for a given settlement. *It should be emphasised that conventional linear elastic calculations of the variation of settlement with depth will usually lead to grossly conservative assessments of NSF in stiff clays.*

Poulos (2008) gives guidance on the relevant margin (or factor of safety) required in the 'stable' (non-settling zone), which should limit settlements to acceptable values. In the discussion of that paper Bourne-Webb (2009) highlights the potential importance of settlement at the toe, arguing that this approach should not rely heavily on Q_b unless this can be justified. The value of Q_b will be very important if the pile is toed into cohesionless soil or rock (when Q_b will be relatively large compared with Q_{s+}). For these situations the Fleming (1992) method is recommended, so that the variation of Q_b as a function of settlement can be more carefully assessed.

57.2.4 Pile groups

Lee *et al.* (2002) report finite element modelling of pile-group effects on NSF loading. As might be expected, they found that piles near the centre of a large group are 'shielded' from the effect of NSF. Thus, the estimation of the effect on a single pile as above is conservative (and for large pile groups likely to be highly conservative).

Lee *et al.* (2006) furthermore consider the effect of a pile cap on group behaviour. Again, piles at the edge of the group are most prone to NSF, and tend to cause settlement of the group, which is resisted by piles in the centre. In the absence of a significant external vertical load on the pile group this can lead to tension at the head of the outer piles as they pull the cap down. This would need to be carefully considered in the structural design of the connection of the pile head to the pile cap.

Global ground movements will affect the distribution of axial forces in a pile group. For example, observations of axial forces in a piled bridge abutment at Newhaven (Reddaway and Elson, 1982) indicated a relatively uniform load distribution across the group, whereas a conventional linear elastic analysis (which ignored the influence of adjacent embankment settlement) gave axial forces in the front row of piles that were nearly three times larger than the rear row.

57.2.5 Methods to reduce NSF

For precast driven piles the pile may be coated with a low-friction material (e.g. bitumen) over the depth where NSF is anticipated, thus reducing Q_{s-}. However, there is a risk that the coating will be removed as the pile is driven. Even if the coating is undamaged, the NSF over this length will not be zero (e.g. residual friction angles for bentonitic materials may be of the order of 10°).

For bored piles a double casing can be used over the upper portion of the pile. The outer casing is directly in contact with the settling soil, but not the inner casing or pile. The inner casing allows the pile to be cast in situ through the full depth. However, the piles are then laterally unsupported over this length, which may have implications for buckling under an axial load or there may be a significant loss of capacity to carry a lateral load. A 'double casing' approach is also an expensive and time-consuming solution.

57.3. Heave-induced tension

57.3.1 Single piles

Heave-induced tension (HIT) occurs when piles are constructed in clay soil beneath a deep excavation, where upward movement is ongoing, or in an expansive swelling soil (e.g. due to changes in the water table or saturation) – see Section 57.3.3.

Here, the main issue is that if there is insufficient load at the head of the pile, tension will occur, with consequent cracking of the concrete if there is insufficient tensile reinforcement. **Figure 57.5(a)** shows the situation with no load at the pile head. The top portion of the pile is 'pulled' upwards by what would conventionally be 'positive' skin friction, and vertical equilibrium is preserved by an equal amount of 'negative' skin friction anchoring the pile into the soil at greater depth, which is heaving less. There is no load at the base of the pile which is being 'lifted'. Since soil strength tends to increase with depth, the neutral point is likely to be approximately at a depth of between half and two-thirds of the pile length. For the situation shown in **Figure 57.5(a)**

$$Q_{s+} = Q_{s-} = \frac{Q_s}{2} \tag{57.4}$$

Therefore, a pragmatic safeguard against tensile failure of piles due to HIT is to provide sufficient tensile reinforcement to carry a load equal to half the shaft capacity through the full length.

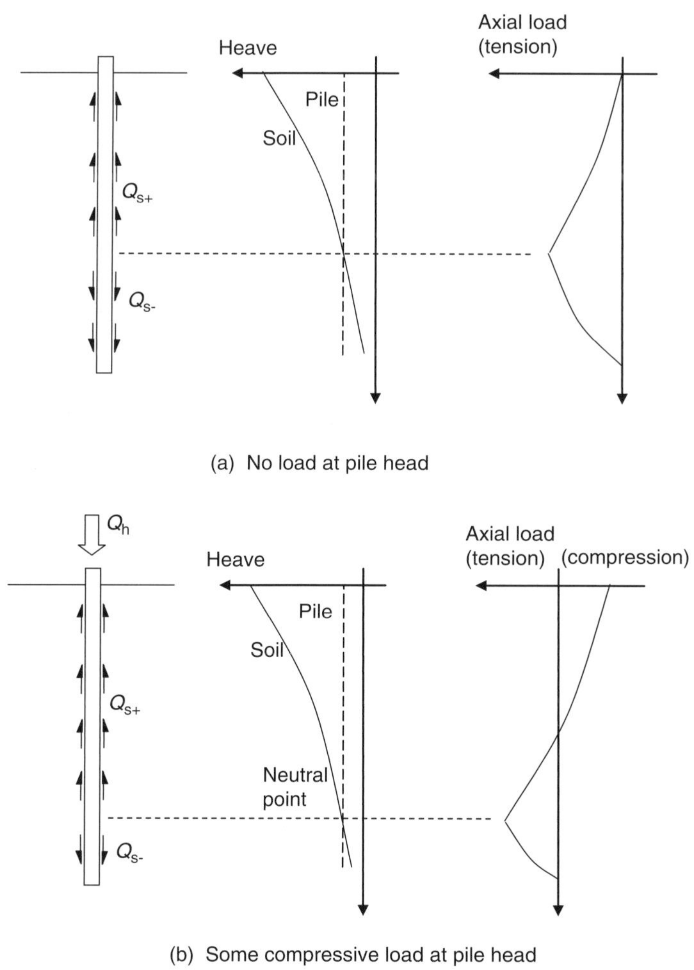

Figure 57.5 Heave-induced tension

As shown in **Figure 57.5(b)**, if there is some compressive load Q_h at the head of the pile, tension will still occur near the bottom of the pile unless $Q_h > Q_s$. However, the position of the neutral point is lower and the heave of the pile is reduced.

For clay soils beneath an excavation the strength and effective stresses will reduce as swelling occurs. Hence, if the pile is installed before excavation then there should be a check on the HIT for both short- and long-term conditions to ensure the critical condition is identified. Typically, short-term conditions will be critical for structural design, and either short- or long-term conditions may be critical for serviceability (i.e. upward movement) of the pile, depending upon the magnitude of the compressive load at the pile head. If the pile is installed after excavation the structural design checks would be similar to those above; however, upward movement would solely be due to time-dependent movement (the total minus undrained movement).

A free-field profile of soil heave against depth can be calculated, as in Section 57.2, based on a reduction in the effective stress. *The soil swelling curve will tend to be strongly nonlinear, and this is an important consideration for making reasonable estimates of the soil–structure interaction.*

57.3.2 Pile groups and base slabs

Where a base slab (or basement or floor slab) is cast near ground level, it may be designed to be suspended from the pile caps, and cast onto a void former, which collapses at a small load on the underside of the floor slab as the soil heaves beneath. Where the slab is not isolated from movement and uplift pressure of the soil beneath, the connection to the piles would be required to carry tension, and the slab would need to be designed for large hogging bending moments induced by the soil uplift pressure. Where a pile cap is used for a pile group, the underside of the pile cap will also be subject to uplift pressures unless it is isolated from the underlying soil using a void former. Piles near the centre of the group will probably be 'sheltered' from the effect of soil heave to some extent, whereas HIT on the outer piles will tend to cause uplift of the pile cap. If there is an insufficient external load on the pile cap (including its weight), there may be significant tension acting at the heads of the central piles as the cap is lifted by HIT on the outer piles. This is essentially a reversal of the situation described above for pile groups subject to NSF, where the outer piles are potentially subject to tension. An important difference between piles subject to unrestrained swelling (i.e. a suspended base slab) and restrained swelling (a base slab or pile cap in contact with the ground) is the development of shaft friction with depth. For unrestrained swelling, the relative pile–soil displacement (and shaft resistance) can develop near the pile head and move progressively downwards. In contrast, for restrained swelling, shaft resistance will develop at depth and move upwards, since the base slab or pile cap will inhibit the relative soil–pile displacement near the pile head (the same principle, working in reverse, holds for piles beneath piled rafts – refer to Chapter 56 *Rafts and piled rafts* (Section 56.6)).

Soil–structure interaction will develop between the slab, piles and swelling soil in contact with the floor or base slab. The swelling pressure will have two components: the soil 'effective heave (or swelling)' pressure plus the groundwater pressure. The magnitude of the soil 'effective' swelling pressure, and the tensile forces and displacement of the piles, will depend on a wide range of factors, including: nonlinear swelling characteristics, permeability and strength of the ground, the thickness of swelling ground, the stiffness of the structure (both the bending stiffness of the base slab and stiffness of any restraints due to piles or embedded retaining walls), excavation geometry, the pile length and the timing of excavation and casting of the base slab (especially once the clay is exposed at final excavation level) and any sources of water (including those within silt/sand lenses within the clay). Probably the most critical and most uncertain parameter is the ability of the clay to retain the suction which will develop below the base of the excavation due to the excavation process. Some indication of the ability of a clay to retain suction can be assessed from specialist suction measurements carried out on high-quality samples (Burland and Maswoswe, 1982). As far as the authors are aware, currently there are no reliable field measurements of suction below the base of deep excavations that have been reported in the technical literature. Hence, it is necessary to adopt a cautious approach to estimating the effective heave pressure on the underside of base slabs.

Figure 57.6 illustrates the interrelationship between some of these factors. The swelling pressure will lead to an increase in the tensile force in the pile (**Figure 57.6(a)**), the neutral point will move upwards and the upward displacement of the pile will increase. As the upward displacement of the structure increases, the swelling pressure on the slab will reduce (**Figure 57.6(b)**). Therefore, a relatively stiff structure (B) will move less, S_B, but larger swelling pressures P_B, will be 'locked in', compared with a more flexible structure (A) (**Figure 57.6(b)**). Hence, an interactive analysis is needed to assess the equilibrium values of the swelling pressure, the ground and structure displacements and the forces induced in the piles and structure.

Irrespective of the 'internal' structural stiffness of the structure, some global upward displacement, S_g, will develop as friction develops on the basement walls and piles. Typically, this situation will occur for deep basement structures and the 'global' stiffness and displacement of the whole structure needs to be considered, as well as the local interactions between the pile, slab and soil (**Figure 57.6(c)**). It is particularly important to check that the structural strength of connections between the pile, slab and basement walls is adequate. The groundwater pressure is unaffected by the soil–structure interaction and needs to be added to the effective swelling pressure from the soil. Simpson (2018) provides a theoretical analysis of the effective heave pressure which may develop below the base of a perfectly rigid and rigidly restrained base

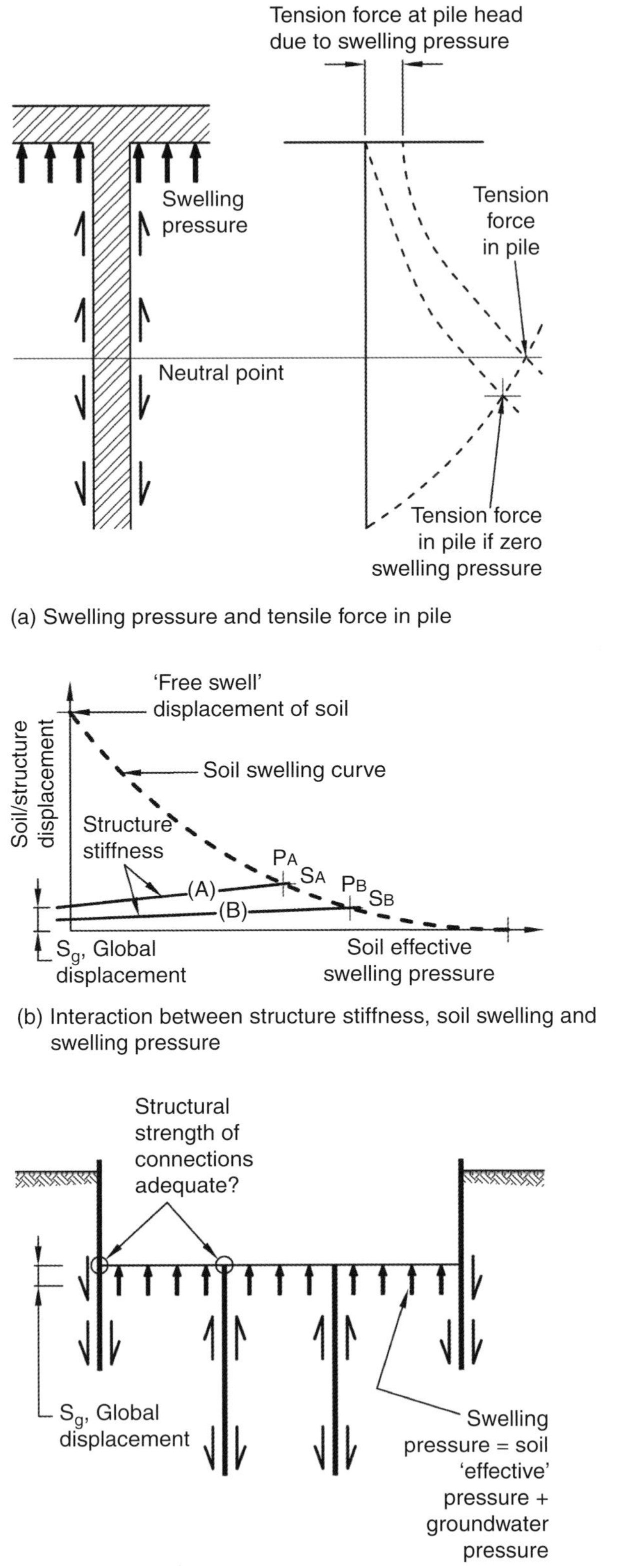

Figure 57.6 Heave-induced tension, floor/base slab in contact with swelling soil

slab due to swelling of an elastic material. It is shown that a wide range of effective heave pressures may develop depending on the assumptions made. It is concluded that the process outlined in **Figure 57.6** should generally give safe predictions of effective heave pressures. This concurs with the author's experience; the simplified process outlined in **Figure 57.6** usually provides reasonable estimates of effective heave pressure for most practical design scenarios when compared with nonlinear finite element analysis, provided the flexibility of the base slab and its restraints is taken into account and the nonlinear swelling characteristics of the ground are properly characterised (assuming linear elastic behaviour is inappropriate). Centrifuge modelling and nonlinear finite element analyses reported by Chan and Madabhushi (2021) and Chan *et al.* (2022) provide some important insights into effective heave pressures and the influence of base slab flexibility. A 'flexible' and 'stiff' base slab were modelled in the centrifuge and simulated numerically for a 15 m deep excavation (plan area 30 × 15 m) in heavily overconsolidated clay; the effective heave pressure below the centre of a 'flexible' base slab was less than 30% of the value for the 'stiff' base slab. For the 'stiff' base slab the effective heave pressure at the centre of the slab was measured at about 150 kPa during the centrifuge test compared with the original vertical effective stress of about 240 kPa (i.e. about 60% of initial vertical effective stress). A refinement of the process outlined in **Figure 57.6** is to assume a parabolic contact stress at the underside of the base slab, which improved the match with centrifuge and nonlinear finite element modelling derived effective heave pressure and base slab displacement, while still providing a suitably cautious estimate for design.

The incorporation of drainage beneath the slab can potentially eliminate (or substantially reduce) the groundwater pressure acting directly on the underside of the slab and reduce the soil swelling displacement, swelling pressures and forces induced in the structure. However, the drainage must be easy to maintain in order to remain effective throughout the structure's design life (often asset owners are uncomfortable with the long-term maintenance liability associated with base slab drainage of underground structures, unless there is a positive business case and a simple robust means of managing maintenance and longer-term risks).

57.3.3 Expansive and collapsible soils

Collapsible and expansive soils are considered in Chapters 32 *Collapsible soils* and 33 *Expansive soils*, respectively. The behaviour of such soils is very complex, and in certain scenarios it may not be clear which situation will arise (i.e. NSF or HIT) – see Chapter 53 *Shallow foundations*. It may then be necessary to adopt a robust structural design for the piles, which will cope with both eventualities. Furthermore, the piles will be required to penetrate to sufficient depth where the soil is known to be unaffected by tendencies for expansion or collapse.

57.4. Piles subject to lateral ground movements

57.4.1 Single piles

This form of interaction is most often considered using a horizontal plane normal to the axis of the pile (which is normally vertical), as shown in **Figure 57.7(a)**. The simplest situation is for an 'isolated' single pile where the boundaries are at a large distance compared with the diameter of the pile. The relative pile–soil displacement in the horizontal plane will be referred to as δ. Relative movement most commonly occurs when the pile carries a horizontal load applied at the head and it moves relative to the stationary soil, which resists this movement – see Chapter 54 *Single piles*. However, this chapter considers the reverse situation – where the soil moves and loads the pile. Examples of this type of interaction are considered in Section 57.4.2.

In the most fundamental consideration of this interaction, it is immaterial how Δ originates, and a horizontal load acting on the pile (often expressed as an equivalent pressure), p, will result as shown in **Figure 57.7(b)**. As is often the case, the interaction may initially be idealised as elastic, becoming plastic as the strength of the ground is fully mobilised. However, the actual response shows a more gradual transition between these bounds of behaviour.

Such loading can be particularly damaging if the designer is unaware of the risk of lateral ground movements loading the pile. If the pile is only structurally designed to accommodate loading applied at the pile head, then there is a high

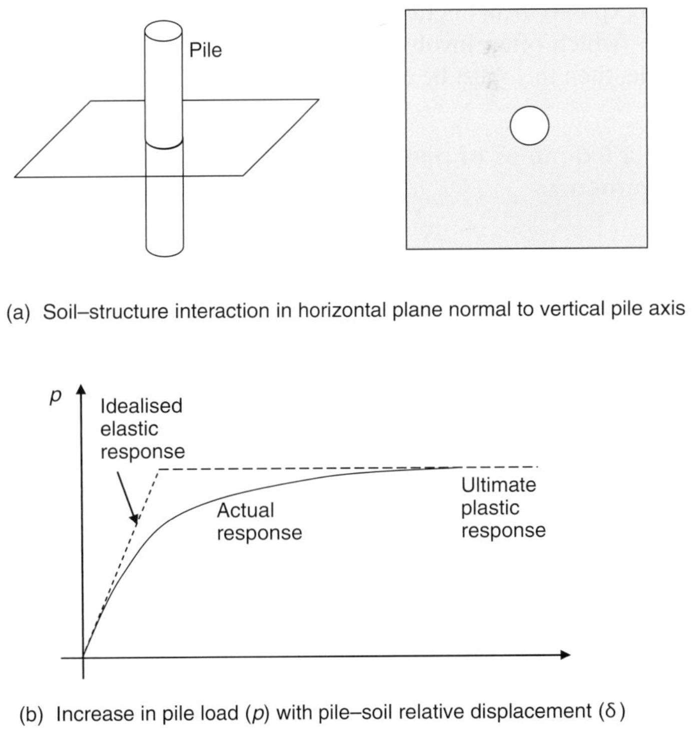

Figure 57.7 Fundamental concepts of lateral pile–soil interaction

probability that the bending and shear capacity of the pile will be inadequate (the reinforcement of vertically loaded piles is often relatively short, whereas piles exposed to global lateral ground movements will often require full-length reinforcement). Historically semi-empirical methods have been used to allow estimation of the soil loading or maximum moment directly. However, such methods tend to rely on a very general consideration of behaviour and are inherently limited in their ability to accurately represent specific circumstances. Finite element analysis is now likely to be considered appropriate where a complex and potentially damaging soil–structure interaction problem is identified. However, the complexities of such an approach should not be underestimated (see Section 57.4.3). A preliminary set of analyses using one, or preferably two, of the available simplified methods should always be carried out first before undertaking complex finite element analyses.

The text below will first consider situations where this type of interaction may occur. As described in Section 57.1, the construction sequence and so on may have a significant impact and 'engineering solutions' may exist to avoid significant lateral interaction effects. This is generally the best course of action, rather than attempting complex analysis. Key references are listed, which may be of use in making simplified preliminary calculations. The incorporation of this type of interaction in a finite element analysis is then briefly considered.

The piles will often be in rows or groups. Some of the references below specifically consider group interaction effects. Progressive structural failure may be an issue for multiple pile rows exposed to an interaction; if the structural failure mechanism (which often involves combined bending and shear) is brittle, then this must be carefully considered and avoided.

57.4.2 Examples of piles subjected to lateral ground movements

57.4.2.1 Piles adjacent to embankments on soft clay

When an embankment is constructed on soft clay or peat the soil beneath tends to move horizontally outwards, rather like toothpaste being squeezed from a tube. As shown in **Figure 57.8**, where piles are situated at the edge of an embankment they will be horizontally loaded by the soft clay, in addition to any structural loading at their head. The effect of a pile-head restraint should be considered – where the pile head is restrained against rotation this reduces the displacement of the pile but increases the interaction loading from the soft clay, leading to larger structural demands (bending moment and shear force) on the pile. Where an abutment structure is supported by the piles there may be additional interaction issues associated with the embankment fill 'arching' onto the abutment due to loss of support from the soft clay (Ellis and Springman, 2001).

In this situation there will be an immediate undrained movement, which is specifically associated with construction

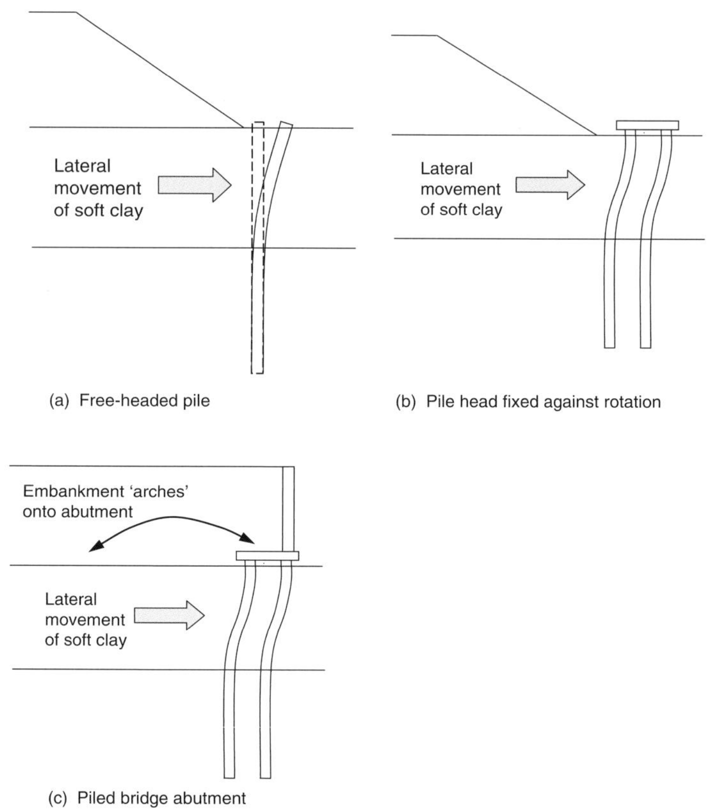

Figure 57.8 Piles adjacent to an embankment on soft clay

of the embankment and, thus, the construction sequence is of significant importance. Using a lightweight fill (e.g. expanded polystyrene blocks) or supporting the embankment itself on piles or improved ground (refer to Chapter 59 *Design principles for ground improvement*) may be considered to reduce soil movement. It may be possible to isolate the piles from the moving soil, such as by using an annulus of soft bentonite clay, or inside an oversized liner. However, the structural design of the pile must then consider the pile to be laterally unsupported by the soil over this height. In general it is preferable to reduce the problem at source – that is, to construct the piles after the embankment has been constructed and allowed to settle (usually the most economic option) or use lightweight fill or deep ground improvement of the underlying soil.

Key references

DeBeer and Wallays (1972) (also summarised in Fleming *et al.*, 2008) give an example of a 'pressure-based method' for the direct estimation of the lateral pressure on piles near an embankment as a fraction of the nominal increase in vertical stress from embankment loading (q). The ratio of q to the undrained shear strength of the soft soil (c_u) is also considered, and many other later methods (see below) also use this ratio. Ultimate bearing failure occurs at a value of $(q/c_u) = 5$, but the risk of damage to the piles is generally considered to become significant when a value of about 3 is reached.

Springman and Bolton (1990) proposed a method for the simple calculation of soil deformation and, hence, pressure loading due to adjacent loading on a soft clay layer. The method also considers the resulting bending moment distribution in the piles.

Stewart *et al.* (1994) give a useful summary of existing analytical methods at that date and is strongly recommended for preliminary design. A method for direct but approximate estimation of the maximum pile moment loading is proposed, and the method proposed by Springman and Bolton (1990) is reviewed and modifications are proposed.

Goh *et al.* (1997) give a numerical procedure for the determination of pile pressure and moment loading resulting from known 'free-field' soil movement (which would occur in the absence of the piles). They compare the results with some case studies, and propose an equation for the simple direct estimation of the maximum moment in the piles.

Chen and Poulos (1997) give a boundary element procedure for consideration of pile loading, again from known 'free-field' soil movement. They use this approach to produce design charts, which potentially can be used for the 'preliminary design' of piles near embankments or piles used to stabilise slopes (see below). Leroueil *et al.* (1985) present case study information regarding the lateral displacement associated with the construction of an embankment on soft clay.

Ellis and Springman (2001) describe physical model testing and plane strain finite element analyses of full-height piled bridge abutments on soft clay. A mechanism of arching onto the pile cap (**Figure 57.8(c)**) was highlighted for this type of structure. The authors conclude that empirical methods which do not account for specific boundary conditions of the piles are likely to be inadequate for a direct estimate of the pile moment loading in complex situations. However, a simple structural analysis of the pile under the actual boundary conditions and pile pressure load obtained through one (or more) of the methods above may be adequate for preliminary design. For detailed design the paper demonstrates that a plane strain finite element analysis can be used provided there is some facility for incorporating lateral interaction between the soil and pile row (see Section 57.4.3).

57.4.2.2 Piles near excavations

It is helpful to consider two different situations

(*a*) relatively long piles near a shallow open excavation or cantilever retaining wall

(*b*) relatively short piles near a deep multi-propped retaining wall.

For (*a*) the piles are likely to experience some horizontal movement, whereas for (*b*) the piles are likely to experience settlement.

Where an excavation (e.g. an open excavation or a cantilevered retaining wall) is constructed near a pile, the ground will tend to move towards the excavation, causing interaction loads on the pile and a corresponding horizontal deformation (**Figure 57.9**). The deformation and structural demands on the pile will be affected if the pile is restrained against rotation or movement at the head. For a simple cantilevered wall the zone affected by ground deformation may extend to at least two times the retained height behind the retaining wall (refer to Long, 2001), and within the 'active wedge' (**Figure 57.9**) relatively large movements are likely. As discussed by Long (2001), construction means and methods, as well as the ground and groundwater conditions, will influence the magnitude of ground movements which may develop behind a retaining wall (and therefore the associated interaction effects on piles located behind a wall).

The most effective means of reducing the retaining wall movement, ground movements behind the wall and adverse interaction effects on any piles behind a wall will be to install a prop at the top of a retaining wall as soon as feasible during the excavation. It is recommended that initial estimates of 'free-field' soil deformations are made based on relevant case history data, such as that collated by Long (2001) and Clough and O'Rourke (1990). The structural demands on the pile can then be assessed by imposing the 'free-field movements' on the pile by use of '*p*–*y*' springs (e.g. software such as LPILE or RSPile). For excavations in soft clays the magnitude of the ground movement is very sensitive to the factor of safety against basal instability. Typically this is best minimised through deep ground improvement to ensure the base is stable. Movement of the pile tends to reduce the relative pile–soil displacement and hence the interaction pressure; thus, it

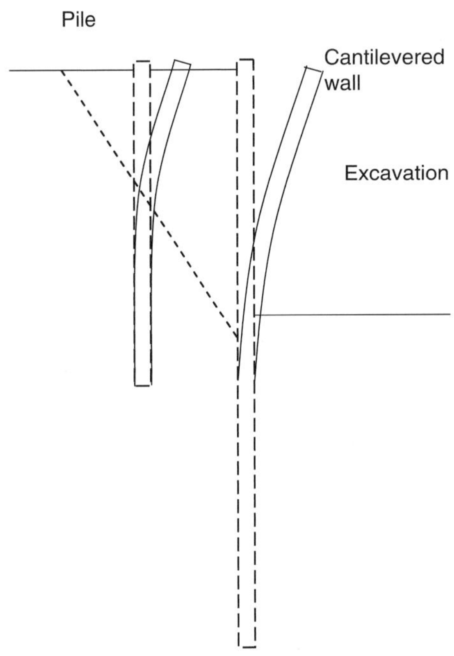

Figure 57.9 Piles near an excavation

is conservative if it is assumed that the pile does not move. It may be possible to isolate the piles from the moving soil, such as inside an oversized liner. However, the structural design of the pile must then consider it to be laterally unsupported by the soil over this height.

For (*b*) above – short piles adjacent to a deep multi-propped excavation – experience in Amsterdam has shown that piles are more likely to settle rather than experience significant bending moments. **Figure 57.10** shows a cross-section through one of the deep excavations for the Amsterdam Metro (Korff and Mair, 2013) and the adjacent piled building. The building foundations comprise timber piles driven through 12 m of soft clay and into the first sand layer. **Figure 57.11** shows the building displacement ('LevellingS') compared with the ground surface ('GroundSurface') settlement and the ground settlement at the pile-toe level ('ExtensoNAP-12m'). The building settlement is equal to the ground settlement at a depth below ground surface of about $0.3L_p$ to $0.5L_p$ (where L_p is pile length). For a building founded on modern steel driven piles, the building settlement was equal to the ground settlement at a depth of between about $0.8L_p$ and $1.0L_p$ (Korff and Mair, 2013); most importantly the differential settlement of the piled building was significantly less than that indicated by the 'free-field' settlement profile for the deep excavation.

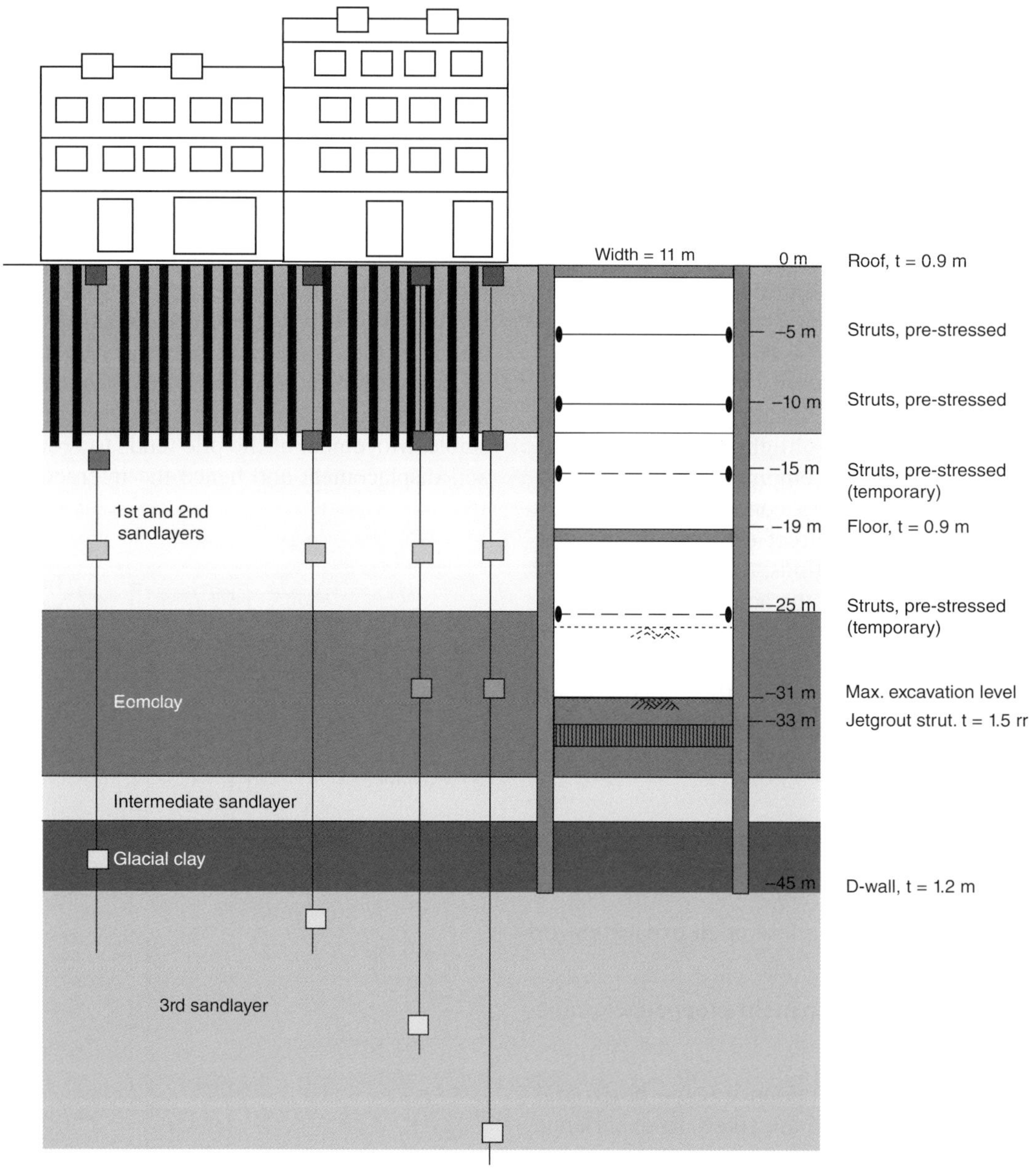

Figure 57.10 Amsterdam Metro, cross-section through Ceintuurbaan station (After Korff and Mair, 2013)

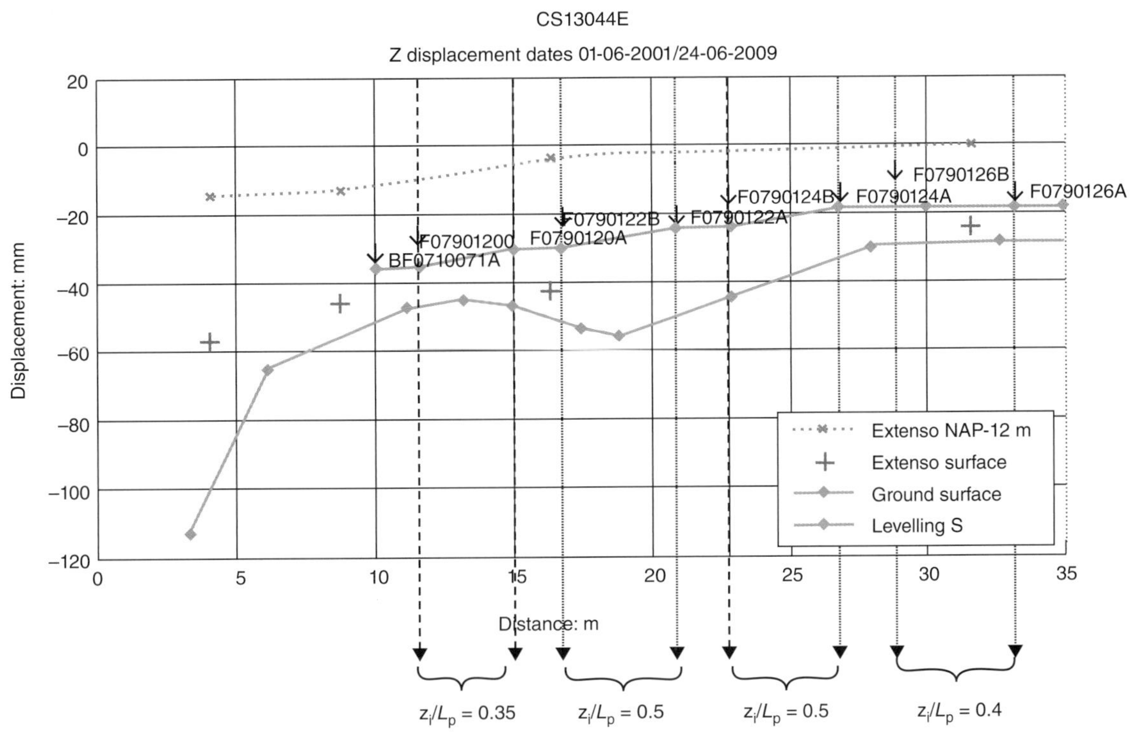

Figure 57.11 Amsterdam Metro, observed settlement of piled building (driven timber piles), soil settlement profiles also shown for ground surface and at the pile-toe level (After Korff and Mair, 2013)

Key references

Poulos and Chen (1997) present a method for analysing the response of piles to a nearby excavation, focusing on braced excavations in clay. Design charts for maximum bending moment and deflection of the piles are presented. Korff and Mair (2013) present a summary of case history data and guidance on practical design issues.

57.4.2.3 Piles in unstable slopes

Piles are often used to stabilise unstable or potentially unstable slopes. The piles are designed to act as 'dowels' across potential slip surfaces (**Figure 57.12(a)**). For reasons of economy and sustainability 'discrete' piles are typically spaced at a distance s along the row, of 3–6 times the pile diameter (d). However, it is important to check that the unstable soil loading the piles 'arches' across the gap between adjacent piles and does not 'flow' through it (**Figure 57.12(b)**). Piles constructed in an unstable slope for any other reason will attract similar loading. For piles installed in firm to stiff clays the above range of pile spacings have been found to be effective (with no evidence of long-term soil 'flow' between the piles), although the geometry of the 'arch' which develops behind the row of piles needs to be considered relative to any movement-sensitive infrastructure (i.e. the pile row needs to be located more than half the pile spacing beyond movement-sensitive infrastructure).

There are a number of aspects of any design.

- Establishing the critical slip (where there is not a well-defined existing slip) and required stabilising force, F_x, to give the desired factor of safety for the slope. The nature and scale of the deformation mechanism needs to be carefully considered, including how the mobilised field strength may vary with depth to avoid excessively deep failure mechanisms being produced by circular searches using slope stability software. Non-circular failure mechanisms also need to be considered. The F_x provides the disturbing force, which needs to be provided by the pile 'back into the slope' in order to stabilise the slope, and geotechnical resistance, supporting the pile, is provided by the net available passive resistance acting below the critical slip surface (taking account of the effective slope at this level when deriving available passive resistance). These disturbing and resisting forces can then be used to produce the structural design of the pile.
- The location of the piles in the slope, which is likely to be affected by the practicality of construction and the desire to protect infrastructure either at the crest or toe of the slope. TRL Reports 466 (Carder and Temporal, 2000) and 493 (Carder and Easton 2001) recommend that the position of a piled wall to stabilise a slope should be approximately one-third to one-half of the distance up from the toe, to avoid failures above or below the row of piles. However, when stiff support is required at the crest, slope-stabilising piles close to the crest together with earthworks berms have been used to provide cost-effective solutions.
- Choice of spacing ratio (s/d).

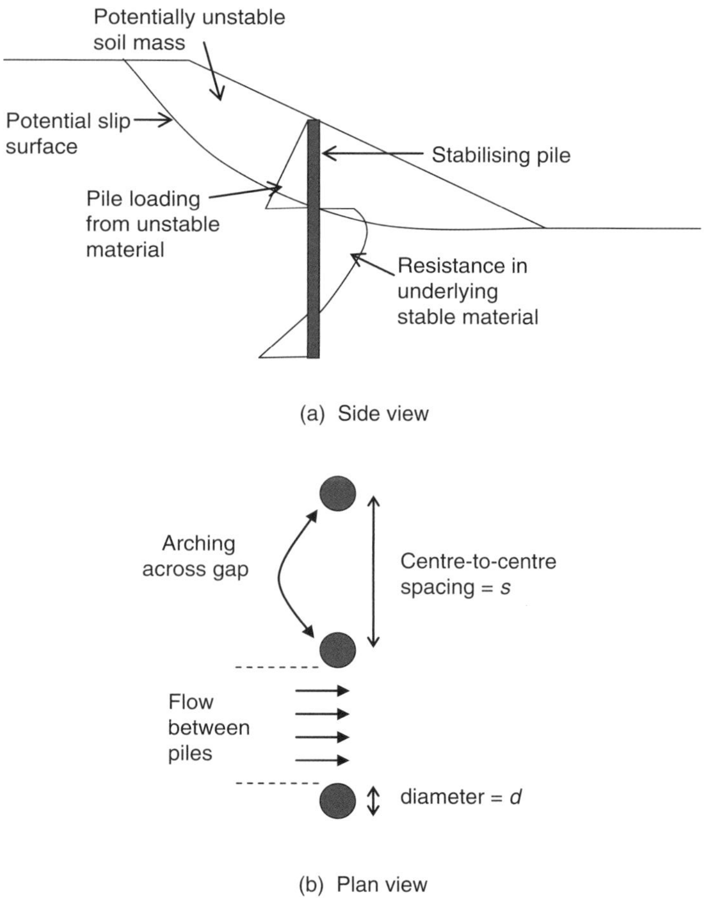

Figure 57.12 Discrete pile row stabilising slope

Factor	Comment
Operational railway or highway	Access may be restricted to slope and toe
	Construction plant height may be restricted particularly if there are overhead electric cables
	High levels of site management required
Environmental constraints	Minimise noise and vibration
	Ecology may impose significant constraints for investigation, design and construction
Site access	Often limited, hence small specialist plant and sequencing of operations required
Temporary works	Usually significant and often the principal cost item
Existing structures (bridges/retaining walls/culverts)	The structures may be in poor condition and hence need to minimise stress increase close to structures
Historic stabilisation or widening	Grouting, counterfort drains, etc., may provide significant obstructions
	Local zones or more extensive areas of demolition rubble, etc. may be encountered
	'Reservoirs' of trapped water may be present
Space along slope toe	Key for stabilisation method selection
Permanent works, material costs	Usually a minor consideration, costs often dictated by programme, access/temporary works
Lineside cables (signals/power) and/or existing utilities	May require special temporary works or impose constraints on stabilisation measures
	Cable diversions are excessively expensive
	If in poor condition, then potentially high risk of disturbance

Table 57.1 Use of piles for slope stabilisation: factors which may influence design and construction (Modified after O'Brien, 2007)

- Provision of adequate moment capacity in the piles (noting that the load per pile increases in proportion with *s*).
- Sufficient embedment of the piles in the underlying stable material to give the required passive resistance, and to ensure that a deep failure cannot pass beneath the piles.
- Horizontal restraint may be provided at the pile head using a capping beam, which itself may be restrained by ground anchors or raking piles. This reduces displacement of the pile head and the maximum moment in the pile.
- The minimisation of temporary works and integration of temporary works into the final permanent works is often key, since most of the time and cost may be associated with temporary works.
- The temporary stability of the piling rigs working on an unstable slope. This may include using temporary toe berms and keeping the rig and hence pile diameter to a small size and/or starting work from a stable location and using the recently constructed piles to contribute to the stability of the working platform as piling progresses into the slip zone.

Table 57.1 provides a summary of the factors which normally need to be considered when stabilising unstable slopes (O'Brien, 2007). The stabilisation of ageing infrastructure slopes (rail and highway) is a common activity, and for slope heights of 4–9 m the use of mini-piling has often been a cost-effective solution (with pile diameters of 300–450 mm). Typically the piling is combined with earthworks berms (in the temporary stage these are used as access ramps for piling, then regraded for the permanent design). Field monitoring carried out by Smethurst and Powrie (2007) over a period of nearly 4 years has provided some insights into the long-term soil–structure interaction behaviour of mid-slope discrete piles (600 mm dia.), and **Figures 57.13** and **57.14** summarise the key results.

As noted above, there can be concerns that gaps between discrete piles lead to long-term slope deformation, or 'squeezing' or 'flow' of the soil through the gaps between adjacent piles (i.e. if a stable 'arch' does not develop between the spaced piles). The piles are spaced at 4*D*, where *D* is the pile diameter, with a clear gap of 1.8 m between the piles. Following installation of the piles, further lateral soil movements are relatively small and there is no evidence of unstable squeezing or soil flow between the piles. The bending moment profile developed along the pile is significantly affected, and reduced,

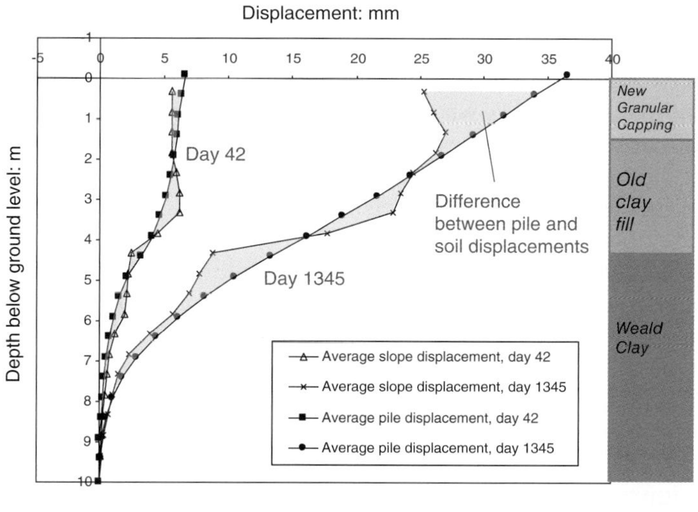

Figure 57.13 Embankment stabilisation: mid-slope discrete bored piles, pile and slope displacement (After Smethurst and Powrie, 2007)

by the influence of the granular regrade/toe berm, which acts as a stiff restraint at the pile head (**Figure 57.14**), although this would not normally be relied upon for permanent works design. The granular layer and toe berm, by acting as a capillary break, minimises seasonal deformation and surface creep. If measures are not taken to control seasonal deformation and surface creep, then long-term ground movements may continue even after the slope is 'stabilised'. High-plasticity clay (especially dumped clay fills found in old railway embankments) often experiences large vegetation-induced seasonal deformation, especially if high-water-demand trees are present on a slope. O'Brien (2012) summarises field measurements of vegetation-induced seasonal deformations; in general, if ecological considerations allow, it is best to minimise these effects by replacing high-water-demand trees with lower-water-demand vegetation and/or regrading with surface cover of coarse-grained fills to form a capillary break layer. Smethurst *et al.* (2020) summarise long-term monitoring of bending moments developed in slope-stabilising piles.

Note that the use of a 'free-field' soil displacement to determine the loading on the piles (e.g. Chen and Poulos, 1997) is not that common in design and not recommended for general use due to the difficulties in reliably estimating long-term deformations in unstable, or potentially unstable, slopes. Instead, the required stabilising force, derived from limit equilibrium analysis, is normally used to determine the disturbing force, as outlined above, on the piles above the slip.

Key references

Viggiani (1981) gives a simple method for the derivation of the moment in piles by assuming that the loading from unstable and underlying stable material is uniform with depth.

Chen and Poulos (1997) give a boundary element procedure for consideration of pile loading from known 'free-field' soil movement. They use this approach to produce design charts, which potentially can be used for the 'preliminary design' of piles in unstable slopes.

Hayward *et al.* (2000) examine the effect of pile spacing for the stabilisation of a clay cutting using the results of geotechnical centrifuge tests.

Carder and Temporal (2000) give a good overview of the topic and associated references at the time of publication.

Smethurst and Powrie (2007) and Smethurst *et al.* (2020) give recent case studies with monitoring data.

O'Brien (2007) provides a summary of the practical design issues which need to be considered when selecting piles for slope stabilisation.

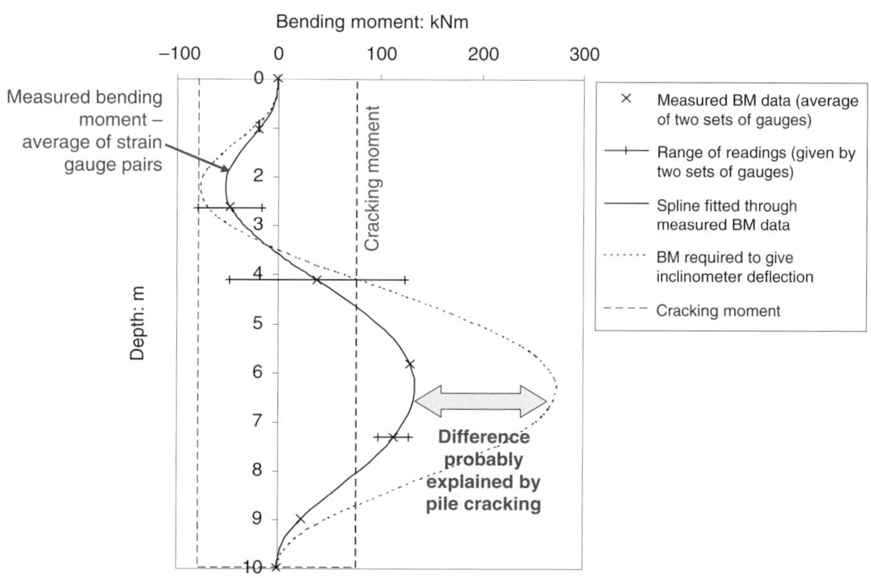

Figure 57.14 Embankment stabilisation: mid-slope discrete bored piles, bending moment profile (After Smethurst and Powrie, 2007)

Ellis *et al.* (2010) refer to many recent numerical studies, and give guidance on various generic aspects of design, including critical pile spacing.

57.4.2.4 Piles near tunnels

Tunnelling can have a significant effect on the surrounding ground, depending on the tunnelling method, and the effects will increase as the tunnelling volume loss increases. Where existing piled foundations are in the vicinity of a tunnel that is to be constructed there should be an assessment of the effect on the piles (**Figure 57.15**). In general, construction of a tunnel will cause movement of the soil towards the tunnel, with a consequent relaxation and redistribution of both vertical and horizontal stresses in the soil in the vicinity of a tunnel (**Figure 57.15**). It is also important to remember that ground settlement will *increase* with increasing depth below the ground surface towards the tunnel. There used to be concerns that tunnelling would lead to a significant loss of pile capacity. Recent research and field monitoring have provided evidence that this risk is far lower than previously thought (Mair and Williamson, 2014).

For a pile located adjacent to a tunnel, lateral movement of the soil towards the tunnel may cause bending and shear forces to be induced in the pile. Tunnelling adjacent to piles is discussed by Pang *et al.* (2005). In general the experience is that the effects on piles are relatively small, with some increase in bending moment and settlement. Analytical assessments may overpredict the effects, unless care is taken to account for 3D effects and nonlinear behaviour at the pile–soil interface.

There is greater concern about tunnelling effects on piles when tunnelling directly beneath piles. It is important to consider piles working primarily as end-bearing piles (i.e. end-bearing is mobilised at working loads) separately from piles working primarily as friction piles.

Research and field monitoring is summarised by Mair and Williamson (2014), who conclude that: 'there is little or no loss of capacity of the piles. Hence, the concept of "loss of capacity" is not generally applicable. In cases where the end-bearing component is a significant element of the working load of the pile there may be some reduction of this end-bearing component associated with the tunnel volume loss; however, the consequence is likely to be settlement of the pile in order to re-mobilise this end-bearing component.' Full-scale field trials of tunnelling effects on driven end-bearing piles, summarised by Kaalberg *et al.* (2005), indicate that significant pile settlements due to tunnelling stress relief only occur at distances of less than $0.25D_t$ (between piles and tunnel, where D_t is tunnel diameter). **Figure 57.16** shows that, for driven end-bearing piles (with a factor of safety of 1.5–2.0), the influence of pile-toe location on the induced pile settlement can be summarised as follows.

- Zone A, the induced pile settlement can be assumed to be slightly larger than the ground surface settlement (see below).
- Zone B, the induced pile settlement can be assumed to be equal to the ground surface settlement.
- Zone C, the induced pile settlement can be assumed to be less than the ground surface settlement.

For driven end-bearing piles located directly above a tunnel – that is, within Zone A of **Figure 57.16** – it is generally assumed that the induced pile settlement is similar to the 'free-field' (or 'greenfield') settlement of the ground at the level of the pile toe.

Guidance for the effects of tunnelling on bored friction piles in stiff clays is summarised in **Figure 57.17**, which shows (for bored friction piles, with a factor of safety of 2.0) the influence of pile location on the induced pile settlement. Mair and Williamson (2014) note that: 'pile failure does not occur, even at high tunnel volume loss. The pile simply experiences increasing settlement so that sufficient positive friction is generated to maintain equilibrium.' Because shaft friction is mobilised after small pile settlements (typically 5–10 mm or less) the induced pile settlements, for friction piles, will often be very small, as indicated by case history data. Hence, the

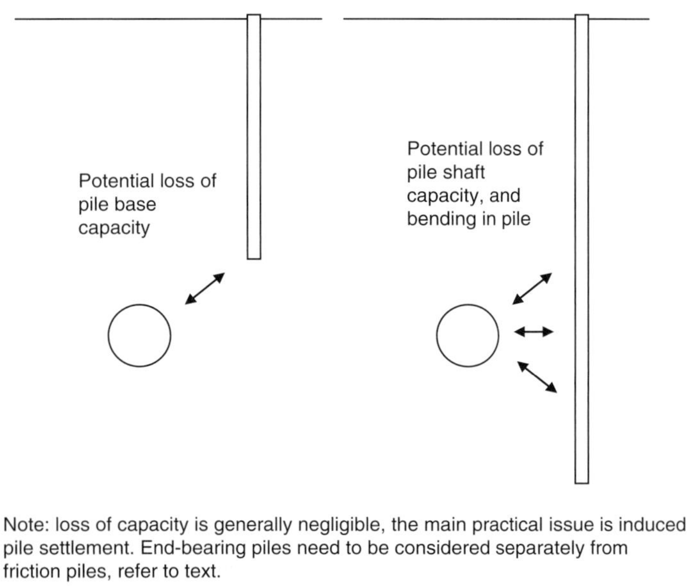

Note: loss of capacity is generally negligible, the main practical issue is induced pile settlement. End-bearing piles need to be considered separately from friction piles, refer to text.

Figure 57.15 Tunnelling near piles

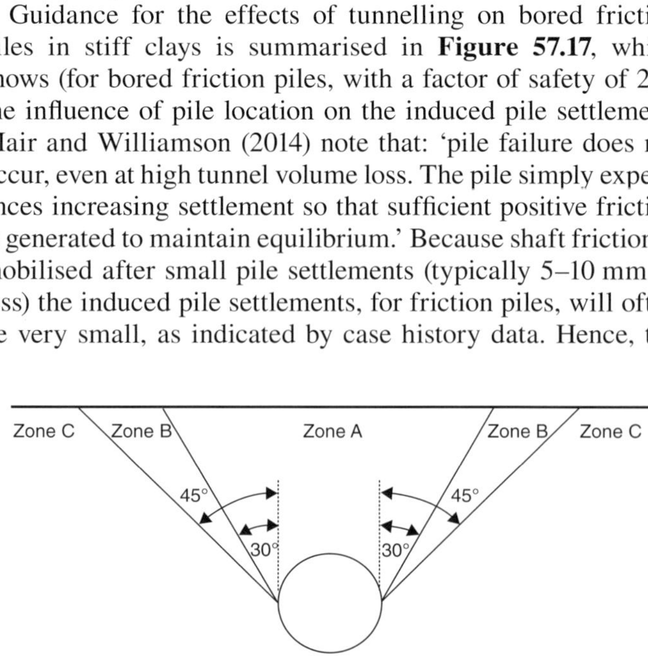

Figure 57.16 End-bearing piles: influence of pile-toe locations on induced pile settlement due to tunnelling (After Kaalberg *et al.*, 2005)

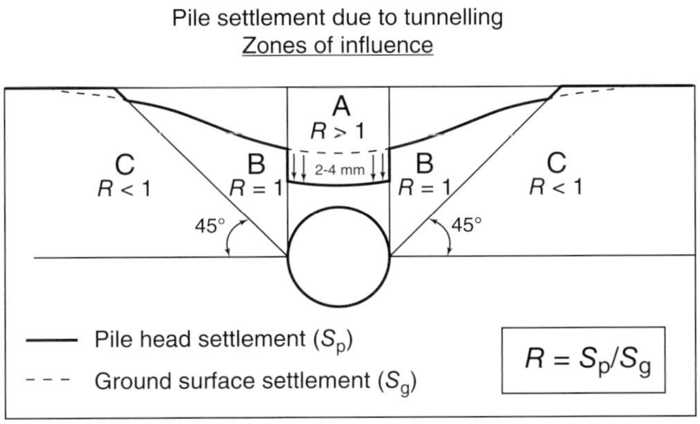

Figure 57.17 Friction piles: influence of pile-toe locations on induced pile settlement due to tunnelling (After Selemetas, 2005)

effects of tunnelling on friction piles in stiff clay tend to be far smaller than those for end-bearing piles, with the induced pile settlements (for friction piles directly above a tunnel) being similar to the greenfield settlement at the *ground surface*.

Key references

As noted above, Mair and Williamson (2014) provide a summary of relevant research and practical guidance, together with a comprehensive reference list. Other useful references include: Chen *et al.* (1999), Chung *et al.* (2006), Jacobsz *et al.* (2003), Korff *et al.* (2016) and Selematis and Standing (2017).

57.4.3 Numerical analysis

Early computational methods used bespoke software based on specific analytical approaches. Such methods often rely upon an estimate of the 'free-field' ground movement (in the absence of piles) to derive the interaction pressure on the pile, or estimate this directly. Once this has been done it is still necessary to combine this loading with other conditions of restraint on the pile (both structural and from soil that is not moving) to derive a bending moment distribution. Consequent deformation of the pile tends to reduce the interaction, and thus iteration may be required, or it is conservative to assume that the pile does not move.

More recently finite element analysis is likely to be considered appropriate, and can inherently consider the simultaneous compatibility of pile–soil relative displacement and restraint and the structural loading of the pile. Plane strain modelling will require the use of 'link' elements to model the interaction, and there are other considerations for 'scaling' the three-dimensional pile row to the plane strain equivalent (e.g. Ellis and Springman, 2001). Three-dimensional analysis offers the benefit of removing these complications. However, other significant complications are introduced beyond the increased computation effort required to run the analysis (e.g. Ellis *et al.*, 2010). O'Brien and Higgins (2020) provide further guidance on the management of advanced numerical modelling.

It should be emphasised that this type of soil–structure interaction analysis requires extremely complex numerical modelling, and it should not be attempted in routine geotechnical practice, nor for preliminary design assessments (even when appropriate at a detailed design stage).

57.5. Conclusions

Pile-supported structures are often considered to be 'safe', and therefore may not receive the attention they deserve. When piles fail, it is often because they were loaded in unexpected ways. The impact of global ground movements can be severe, and numerous failures have occurred, including the catastrophic collapse of buildings, as illustrated in **Figure 57.2**. Usually these failures are due to a lack of awareness on the part of designers and constructors that global ground movements could occur. Fragmentation across different design and construction teams can exacerbate the technical challenges, and therefore the procurement process, managed by a client, needs to facilitate proper coordination across any design or construction interfaces and the definition of clear roles and responsibilities. It is strongly recommended that a single competent senior geotechnical practitioner oversees the ground–structure interaction issues, and temporary and permanent works design interfaces (which is not common practice, despite the obvious technical and risk-reduction benefits). Therefore, the following are particularly important.

- Awareness of the situations when significant global ground movements may develop, either during construction or in the long term. This will usually require input by a senior geotechnical engineering specialist at an early stage of the project, and this oversight by a 'controlling mind' should be maintained throughout the project design and construction.
- An appreciation of the interaction that may develop between temporary and permanent works. This may need the temporary works to be carefully controlled or the permanent works design to be modified. Hence, the loading scenarios for design or specifications may need to be tailored to suit specific circumstances.
- An appreciation of the interactions that may develop between existing and new structures or which may develop due to intrinsic site hazards (e.g. non-engineered fills (see Chapter 58 *Building on fills*) or due to the site's history or geology).
- As stated earlier, it is usually wise to avoid adverse interactions developing due to global ground movements (e.g. by selecting an appropriate construction sequence (which would then need specifying on drawings, CDM registers, etc.)) rather than attempting a complex analysis.

References

Ashford S, Boulanger RW and Brandenberg SJ (2011) *Recommended Design Practice For Pile Foundations in Laterally Spreading Ground*. PEER report 2011/04, University of California, Berkeley, CA.

Bourne-Webb P (2009) Discussion: a practical design approach for piles with negative skin friction. *Geotechnical Engineering* **162(GE3)**: 187–188.

Bown A and O'Brien AS (2008) Shaft friction in London clay: modified effective stress approach. *Proceedings of the Second BGA International Conference on Foundations*, Dundee, UK, pp. 91–100.

Burland JB and Maswoswe J (1982) In situ measurement of horizontal stress in overconsolidated clay using push-in spade-shaped pressure cells: discussion. *Geotechnique* **32**: 285–286.

Burland JB and Starke W (1994) Review of negative pile friction in terms of effective stress. *Proceedings of the 13th International Conference on SMFE*, New Delhi, India, pp. 493–496

Carder DR and Easton MR (2001) *Analysis of Performance of Spaced Piles to Stabilize Embankment and Cutting Slopes*. TRL Report 493. TRL, Wokingham, UK.

Carder DR and Temporal J (2000) *A Review of the Use of Spaced Piles to Stabilise Embankment and Cutting Slopes*. TRL Report 466. TRL, Wokingham, UK.

Chan D and Madabhushi M (2021) Centrifuge and numerical modelling of the influence of structural stiffness on basement heave in overconsolidated clay. *Proceedings of the 10th International Symposium on Geotechnical Aspects of Underground Construction in Soft Ground, Cambridge, UK*. CRC Press, Abingdon, UK, pp. 259–265 .

Chan D, Madabhushi M, Hsu Y and Tan R (2022) Improving semi-analytical predictions of long-term basement heave in over-consolidated clay. *Journal of Geotechnical and Geoenvironmental Engineering, ASCE* **148(2)**: 04021179.

Chen LT and Poulos HG (1997) Piles subjected to lateral soil movements. *Journal of Geotechnical and Geoenvironmental Engineering, ASCE* **123(9)**: 802–811.

Chen LT, Poulos HG and Loganathan N (1999) Pile responses caused by tunnelling. *Journal of Geotechnical and Geoenvironmental Engineering, ASCE* **125(3)**: 207–215.

Cheong MT (2007) Case study: negative skin friction development on large pile groups for the New Wembley Stadium. *Ground Engineering*, Nov.: pp. 33–37.

China Daily (2009). Pressure difference cause of Shanghai building collapse. www.chinadaily.com.cn/china/2009-07/03/content_8376126.htm (accessed 31/05/2023).

Chung KH, Mair RJ and Choy CK (2006) Centrifuge modelling of pile–tunnel interaction. *Physical Modelling in Geotechnics: 6th ICPMG, Hong Kong* (Ng CWW, Wang YH and Zhang LM (eds)). CRC Press, London, UK, pp. 1151–1156.

Clough GW and O'Rourke TD (1990) Construction induced movements of *in situ* walls. In *Proceedings of Design and Performance of Earth Retaining Structures, New York, NY* (Lambe P and Hansen L (eds)). ASCE, New York, NY, pp. 439–470.

DeBeer EE and Wallays M (1972) Forces induced in piles by unsymmetrical surcharges on the soil around piles. *Proceedings of 5th European Conference on Soil Mechanics and Foundation Engineering*, Madrid, Spain, vol. 1, pp. 325–332.

Ellis EA and Springman SM (2001) Full-height piled bridge abutments constructed on soft clay. *Géotechnique* **51(1)**: 3–14.

Ellis EA, Durrani IKD and Reddish DJ (2010) Numerical modelling of discrete pile rows for slope stability and generic guidance for design. *Géotechnique* **60(3)**: 185–195.

Fleming WGK (1992) A new method for single pile settlement prediction and analysis. *Géotechnique* **42(3)**: 411–425.

Fleming WGK, Weltman AJ, Randolph MF and Elsong WK (2008) *Piling Engineering*, 3rd edn. Blackie, London, UK.

Goh ATC, The CI and Wong KS (1997) Analysis of piles subjected to embankment induced lateral soil movements. *Journal of Geotechnical and Geoenvironmental Engineering, ASCE* **123(9)**: 792–801.

Hayward T, Lees AS, Powrie W, Richards DJ and Smethurst J (2000) *Centrifuge Modelling of a Cutting Slope Stabilised by Discrete Piles*. TRL Report 471. TRL, Wokingham, UK.

Jacobsz SW, Standing JR, Mair RJ, *et al.* (2003) Tunnelling effects on driven piles. *Proceedings of the International Conference on Response of Buildings to Excavation-Induced Ground Movements, London, UK* (Jardine FM (ed)). CIRIA, London, UK, pp. 337–348.

Kaalberg FJ, Teunissen EA, van Toi AF and Bosch JW (2005) Dutch research on the impact of shield tunnelling on pile foundation. *Proceedings of the 5th International Symposium on Geotechnical Aspects of Underground Construction in Soft Ground, Amsterdam, the Netherlands* (Bakker KJ, Bezuijen A, Broere W and Kwast EA (eds)). Balkema, Rotterdam, the Netherlands, pp. 123–131.

Korff M and Mair RJ (2013) Response of piled buildings to deep excavations in soft soils. *Proceedings of the 18th International Conference on Soil Mechanics and Geotechnical Engineering*, Paris, France, pp. 2035–2038.

Korff M, Mair RJ and Tol F (2016) Pile–soil interaction and settlement effects induced by deep excavations. *Journal of Geotechnical and Geoenvironmental Engineering, ASCE* **142(8)**: 04016034.

Lee CJ, Bolton MD and Al-Tabbaa A (2002) Numerical modelling of group effects on the distribution of dragloads in pile foundations. *Géotechnique* **52(5)**: 325–335.

Lee CJ, Lee JH and Jeong S (2006) The influence of soil slip on negative skin friction in pile groups connected to a cap. *Géotechnique* **56(1)**: 53–56.

Leroueil S, Magnan J and Tavenas F (1985) *Remblais sur argiles molles* [Embankments on Soft Clays]. Ellis Horwood, Chichester.

Long M (2001) Database for retaining wall and ground movements due to deep excavations. *Journal of Geotechnical and Geoenvironmental Engineering, ASCE* **127(3)**: 203–224.

Mair RJ and Williamson MG (2014) The influence of tunnelling and deep excavation on piled foundations, Keynote Lecture. *Proceedings of the 8th International Conference on Geotechnical Aspects of Underground Construction in Soft Ground, Cambridge, UK*, pp. 21–30.

O'Brien AS (2007) Rehabilitation of urban railway embankments: research, analysis and stabilisation. *Proceedings of the 14th European Conference on Soil Mechanics and Engineering, Madrid, Spain*, vol. 1, pp. 125–143.

O'Brien AS (2012) The assessment of old railway embankments: time for a change? Partial saturation in compacted soils. *Géotechnique Symposium in Print*. ICE, London, UK, pp. 19–34.

O'Brien AS and Higgins K (2020) *The Management of Advanced Numerical Modelling in Geotechnical Engineering: Good Practice*. CIRIA Report C791. CIRIA, London, UK.

O'Brien AS and Sharp P (2001a) Settlement and heave of over-consolidated clays: a simplified non-linear method of calculation – part 1 of 2. *Ground Engineering*, Oct.: pp. 28–21.

O'Brien AS and Sharp P (2001b) Settlement and heave of over-consolidated clays: a simplified non-linear method of calculation – part 2 of 2. *Ground Engineering*, Nov.: pp. 48–53.

Pang CH, Yong KY, Chow YK and Wang J (2005) The response of pile foundations subjected to shield tunnelling. *Proceedings of the 5th International Symposium on Geotechnical Aspects of Underground Construction in Soft Ground, Amsterdam, the Netherlands* (Bakker KJ, Bezuijen A, Broere W and Kwast Ea (eds)). Balkema, Rotterdam, the Netherlands, pp. 737–744.

Potts DM (2003) Numerical analysis: a virtual dream or practical reality? *Géotechnique* **53(6)**: 535–573.

Poulos HG (2008) A practical design approach for piles with negative skin friction. *Geotechnical Engineering* **161(GE1)**: 19–27.

Poulos HG (2017) Design for ground movements. In *Tall Building Foundation Design*. CRC Press, London, UK, pp. 213–256.

Poulos HG and Chen LT (1997) Pile response due to excavation-induced lateral soil movement. *Journal of Geotechnical and Geoenvironmental Engineering, ASCE* **123(2)**: 94–99.

Reddaway AL and Elson WK (1982) *The Performance of a Piled Bridge Abutment at Newhaven*. CIRIA Technical Note 109. CIRIA, London, UK.

Selemetas D (2005) *The Response of Full-Scale Piles and Piled Structures to Tunnelling*. PhD thesis, University of Cambridge, UK.

Selemetas D and Standing JR (2017) The response of full-scale piles to EPBM tunnelling in London Clay. *Géotechnique* **67(9)**: 823–836.

Simpson B (2018) Effective heave pressures beneath restrained basement slabs. *Proceedings of the Institution of Civil Engineers* **171(1)**: 28–36.

Smethurst J and Powrie W (2007) Monitoring and analysis of the bending behaviour of discrete piles used to stabilise a railway embankment. *Géotechnique* **57(8)**: 663–677.

Smethurst J, Bicocchi N, Powrie W and O'Brien AS (2020) Behaviour of discrete piles used to stabilize a tree covered railway embankment. *Géotechnique* **70(9)**: 774–790.

Springman SM and Bolton MD (1990) *The Effect of Surcharge Loading Adjacent to Piled Foundations*. UK Transport and Road Research Laboratory Contractor Report 196. TRL, Wokingham, UK.

Stewart DP, Jewell RJ and Randolph MF (1994) Design of piled bridge abutments on soft clay for loading from lateral soil movements. *Géotechnique* **44(2)**: 277–296.

Viggiani C (1981) Ultimate lateral load on piles used to stabilise landslides. *Proceedings of the 10th International Conference on Soil Mechanics and Foundation Engineering*. Stockholm, Sweden, vol. 3, pp. 555–560.

It is recommended this chapter is read in conjunction with

- Chapter 53 *Shallow foundations*
- Chapter 55 *Pile-group design*
- Chapter 94 *Principles of geotechnical monitoring*

All chapters in this book rely on the guidance in Sections 1 *Context* and 2 *Fundamental principles*. A sound knowledge of ground investigation is required for all geotechnical works, as set out in Section 4 *Site investigation*.

doi: 10.1680/icemge.66830.0957

Chapter 58
Building on fills

Hilary D Skinner COWI UK, London, UK

This chapter outlines the main issues to consider when building on filled ground, and the references include further reading relevant to each topic. Successful construction depends not only on the load-carrying characteristics of the fill, but also on the sensitivity of the structure to ground movement. Due to the variability inherent in filled ground, the application of the geotechnical triangle is particularly important.

CONTENTS

58.1. Introduction

It is often necessary to construct buildings on areas covered by fill, for which the engineering properties are almost always poorer than the parent material. While this chapter will consider the issues related to construction on fills, other issues considered in the design, such as sloping ground, should not be neglected (see Chapter 9 *Foundation design decisions*).

Generally, little or no control was exercised over the disposal of non-engineered fills and so they may be highly variable in composition, placement and geometry. All of these provide potential risks to development. This means that the engineering behaviour of non-engineered fills is difficult to predict and that they are likely to afford very variable support to any building or structure constructed upon them.

When construction is carried out on an engineered fill placed for a different development, the investigation must identify whether the fill properties and engineering behaviour are suitable for the development being undertaken. If an engineered fill is to be placed for the development, the choice of materials and placement methodology should take into consideration the requirements of the structure to be supported. The behaviour of the materials below the fill should also be considered.

Because the settlement distribution over the loaded area of a fill, even under uniform loading conditions, may be very irregular, it is often necessary to design so that the total, as well as the differential, movements are restricted or can be withstood without damage. Hence, the structure should either be sufficiently rigid to redistribute the loads and thereby reduce relative settlement, or should be relatively flexible to accommodate them without cracks appearing in the structure. Foundation design, therefore, has to consider the predicted fill–structure interaction. Long, continuous structures should be avoided by dividing them into sections with flexible joints passing completely through the structure and foundation. Ordinary pad and strip footings are rarely adequate. Wide reinforced strip footings may be adequate, otherwise a reinforced raft is necessary.

If a fill is relatively thin and it can be penetrated and overlies a suitable stratum, then a beam and pier or beam and pile foundation may prove an alternative.

This chapter will outline the main issues to consider when building on filled ground, from investigation, characterisation of fill properties, causes of volume change and foundation design.

All construction requires appropriate investigation allied to careful design. In the subsequent sections only those issues which are different, or require a different approach or interpretation from the 'normal' design process, are highlighted. When building housing, additional requirements may be placed on the design and construction.

Section 3 *Problematic soils and their issues* discusses the behaviour and characterisation of fills in depth. A summary is given here.

58.2. Engineering characteristics of fill deposits

Some of the characteristics of a fill relevant to the design of foundations are related to the nature and extent of the deposit, and not just to the material of which it is composed. These should be identified as far as possible by desk studies and confirmed by investigation (see Section 4 *Site investigation* for more on this).

58.2.1 Surface extent and depth

The boundaries of the filled area need to be established reliably. With well-defined infilled excavations, such as old docks, this may be simple, but with other infilled excavations it may be more difficult. The depth of the fill is important in assessing its behaviour, and in evaluating the applicability of various ground treatment techniques and foundation designs.

Abrupt changes in depth may lead to differential settlement. Some old excavations, such as docks, have vertical sides, and it may be necessary to excavate the sides to a flatter slope to reduce subsequent differential settlement.

58.2.2 Age

The age of a fill deposit can have several different effects on its engineering behaviour.

- *Environment.* An old fill may have already been subjected to many changes in moisture content, water level and loading or chemical conditions, which may make it less vulnerable to large movements under future changes.
- *Fill material.* The length of time that has elapsed since fill placement is of major importance for fills containing a significant proportion of biodegradable materials, where settlement is largely related to decomposition.
- *Deposition.* The rate of long-term settlement under self-weight is related to the time that has elapsed since fill placement. Other time-related phenomena affecting a fill deposit include desiccation and weathering leading to the formation of a surface crust.

58.2.3 Method of placement

The mode of formation will have a major influence on subsequent behaviour. Fills may have been placed in thin layers and heavily compacted or they may have been end-tipped in high lifts in dry conditions or into standing water. The method of placement affects the density of the fill and the homogeneity of the deposit. Many of the problems with non-engineered fills are related to their heterogeneity. Depending on the method of placement and the degree of control exercised during placement, there may be variability in materials, density and age. Much of the usefulness of ground treatment (see Chapter 59 *Design principles for ground improvement*) techniques is related to reducing this variability. **Figure 58.1** shows an example of some control exercised in fill placement.

Hydraulic placement of materials generally produces fills with a high moisture content in a relatively loose or soft condition. Segregation of different particle sizes is likely.

Figure 58.1 Placement of fill in deep lifts

58.2.4 Groundwater level

The groundwater level within a fill is controlled by a variety of factors, such as the surrounding hydrogeology, whether there is an impermeable crust on the surface of the fill, the permeability of the fill or whether there is fill drainage.

The current level of the water table, the amount by which it has fluctuated in the past and the possibility of future fluctuations may have important consequences. A rise in the water level that inundates a fill for the first time may cause collapse compression in the fill. Lowering the water table can cause settlement due to the increase in effective stress within the fill.

58.3. Investigation of fills

Non-engineered fills are difficult to investigate – their variability is a key characteristic that can be difficult to quantify. Engineered fills should have lesser variability, but this should be confirmed by records and intrusive investigation. In addition, even when fills can be penetrated, fill properties are often altered dramatically by the investigation process, and standard correlations based on natural soils are rarely appropriate. Given the challenges in identification and characterisation, it is vital that a good desk study identifies what might be present and that the engineer sees the resulting material, as descriptions may not convey the characteristic of the material. As investigation necessarily only tests properties in the short term, it does not give a great deal of indication of how properties or settlements may change over time (see also Section 4 *Site investigation*).

58.3.1 Desk studies and walkovers

A key component in the investigation of fills is the desk study. Possible areas of filled ground can be identified by previous site or local land use and from British Geological Survey mapping (see Chapter 32 *Collapsible soils*). As it is unlikely that fill has been transported a significant distance, local arisings can be identified. Types of fill generated by different processes are discussed in Chapter 34 *Non-engineered fills.* Possible contamination arising from processes should also be identified; a good source is the Department of the Environment industry profiles, available online at the CL:AIRE website. The depth of infilled ground may be discerned from the evolution of the site, and also from knowledge of the processes that may result in a large depth of non-engineered ground (or from deep holes!) – that is, opencast mining, deep mining, quarrying or landfilling. Processes such as landscaping usually result in shallower depths of fill.

A walkover may identify regular patterns of hummocks or hills indicating a man-made origin, variations in vegetation or significant areas of 'empty space' near built-up areas. A combination of desk study and walkover may also help identify areas of infilled basements.

The investigation of the site and local area is a key component in identifying the likely presence, type and history of the fill, as well as other site issues requiring investigation and design. This allows key risks to the development to be identified early and the scope of intrusive investigation to be defined.

58.3.2 Non-intrusive investigation

Some geophysical techniques may aid the identification of the depth of fill where significant differences in stiffness, density or water content of the filled and natural ground exist (see Chapter 45 *Geophysical exploration and remote sensing*). Before specifying the technique, it is worth noting the change in properties that is detectable by each – a change in stiffness, for example – and finding out whether there may be a sufficient change in the property to enable the technique to locate the features required. In each case some form of intrusive confirmation is recommended.

58.3.3 Intrusive investigation

The techniques available to the engineer are likely to be the following.

58.3.3.1 Trial pits

For shallow fills a key element is simply digging trial pits. This will enable a good visual description (the number of trial pits required depends on how variable the fill may be and the type of construction required). Bulk disturbed samples will enable basic characterisation of the components by description, particle size, moisture content and Atterberg limits as appropriate. The visual characteristics of deposition may also be visible, such as *layered* or *very variable*. Some chemical testing may be required.

Depending on construction requirements, shallow in situ tests such as plate or skip tests may be a useful way of characterising the bearing capacity and immediate settlement. Care must be exercised that a sufficient volume of fill will be tested, and a number of tests may be necessary to assess variability. In situ CBR (California bearing ratio) tests must be analysed with particular care, as the size of test is small. If the fill is amenable to treatment, further testing on samples is helpful; if mixing is a possible treatment then consider compaction testing and chemical testing.

58.3.3.2 Window samples (WS) and dynamic probing (DP)

Dynamically driven window samplers may penetrate fills to a few metres, although they will refuse on large or flat hard obstructions. The samples obtained will be small; hence, they may not be representative of larger particles, and will be disturbed by the driving process. It is likely that a higher density will be retrieved and some particle breakdown will be caused.

Information from such samples, whether descriptive or a classification, should be interpreted with care.

Dynamic probing may be one of the only tests available. A dynamically driven cone is used to penetrate the fill and the number of blows per 100 mm step is recorded. Some indication of density, with a similar caveat on obstructions, can be obtained. A number of test profiles can give an idea of the variability and help refine additional test locations.

58.3.3.3 Deep in situ tests

Deep in situ tests can be divided into those involving pushing (CPT, CPTU), drilling (self-boring pressuremeter) or carried out from a borehole (DMT, PMT, SPT). For more information, see Chapter 48 *Geo-environmental testing*.

Few CPT correlations are relevant for fills, so this form of profiling and testing should be used with care. CPTs are also liable to refuse on or be damaged by obstructions and so are rarely used. Where the fill is similar to a natural soil (particularly some granular fills) it is, however, a useful tool and can readily identify both variability and improvement due to treatments such as dynamic compaction.

Pressuremeter and dilatometer testing is also rarely carried out, primarily because the sample size tested is not representative of a highly variable deposit. The cost of a large number of tests required allied to the potential for remoulding caused by the drilling process means they are often inappropriate.

SPT tests are often carried out where boreholes can be drilled, and give some indication of the strength of the fill as altered by the drilling and testing process. Due to the small sample size and lack of appropriate correlations these tests should be analysed with care. However, SPT testing may be the only data that can be readily obtained.

58.3.3.4 Drilling and sampling

Cable percussion (CP) drilling is likely to break down and densify fills; where water is added during the drilling process this will often aid this breakdown. Samples retrieved from CP boreholes should be analysed and described with care and are likely to be disturbed.

Rotary drilling may retrieve little material; where it does these will have been altered by the drilling and the drilling fluid.

Samples are very unlikely to be suitable for testing for strength or stiffness.

Sonic drilling may assist penetration and sample recovery in granular or mixed fills, but samples are disturbed.

58.3.3.5 Monitoring

Monitoring of water levels, gas and also long-term settlement are important tools for understanding the current state and future behaviour of a fill (refer to Section 58.6.4).

58.3.3.6 Trials

Where treatment is to be carried out that addresses deep-fill behaviour an initial trial area in which a greater density of monitoring is carried out is worthwhile. This may well lead to optimisation in the treatment method or increased confidence in its efficacy.

Property	Disturbed samples				In situ testing		Shallow load testing
	Trial pits	WS	CP	Rotary/sonic	DP	SPT	Plate
Density	x	x	x	x		Some indication	Inferred
Compactive state	Visual indication	x	x	x	x	x	Inferred
Water content	y	y	May be altered by drilling		x	x	x
Particle size	y	y			x	x	x
Plasticity	y		y – fines content may be altered by drilling		x	x	x
Strength	x	x	x	x		Some indication	y depending on load
Stiffness	x	x	x	x	x	x	y
Overconsolidation	x	x	x	x	x	x	x
Variability	Depends on number of locations tested						

y, technique likely to give some indication of the property; x, property unlikely to be determined by the technique

Table 58.1 Properties likely to be identified by intrusive investigation

In summary, when investigating a fill, routine methods for drilling and testing are often not useful in determining the behaviour or design parameters. A trial pit to determine the nature of the material (and care is needed to ensure that the material sampled and viewed is representative), backed up by a good desk study and monitoring information, is often the most useful source of information. Some information on boundaries can usually be supplemented by direct investigation. The properties that can be identified by intrusive investigation are listed in **Table 58.1**.

Information from case histories on similar fills or large-scale tests may be very useful. Again, care should be taken that the test is representative of the scale and type of behaviour anticipated. A small short-term load test on a fill where self-weight settlement or creep is still occurring is not likely to reveal the performance that a large structure built on the fill may experience over a number of years!

58.4. Fill properties

As fills may be difficult to investigate, information from tests on similar materials is vital for the designer.

58.4.1 Index and classification properties

Simple index and classification of properties enables fills to be classified in ways that are relevant to their engineering characteristics. BS 1377-2:1990 (BSI, 1990) describes classification tests for soils.

58.4.1.1 Particle-size distribution

Particle size has a strong effect on soil behaviour and provides a useful method of classification. Sieving and sedimentation can be used to determine the distribution by particle size. There are basic differences in behaviour between coarse 'granular' soils and fine 'cohesive' soils. Coarse soils tend to have high shear strength and permeability, whereas fine soils generally have lower strength and permeability. The percentage of silt- and clay-sized particles (i.e. finer than 0.06 mm) is important; when this percentage is high (usually taken as $Fc > 35\%$), the soil will cease to behave as a coarse soil. In reality, the distinction between 'granular' and 'cohesive' behaviour is a function of other properties in addition to particle size. With a well-graded soil it may not be immediately obvious whether the behaviour will be that of a coarse or a fine soil. 'BS 5930:1999: Code of practice for site investigations' gives helpful advice on this transition (BSI, 1999).

An uncompacted clay fill composed of lumps of stiff clay may behave more like a granular fill than a cohesive fill when loaded in the absence of water, but if inundated it may undergo collapse compression and subsequently behave like soft, saturated clay fill.

58.4.1.2 Moisture content and degree of saturation

Water is invariably present in fill materials. The moisture content for saturated fills is typically 20–80%. The placement moisture content affects the properties of fills, and the behaviour of clayey fills is very moisture-dependent. Moisture content changes can cause volume changes in fills.

58.4.1.3 Density

The density or 'compactness' of a fill has a major influence on its behaviour. This is a function of the method of placement of the fill and its subsequent stress history. The term 'compaction' is used to describe processes in which equipment is used to compress the fill into a smaller volume, increasing its density. Usually, engineering properties are improved with increased density. As solid soil particles and water are virtually incompressible, compaction reduces the percentage of air voids.

58.4.1.4 Coarse fills

For a fairly clean coarse hydraulic fill, the dry density is typically 1.6 Mg/m^3. A well-compacted rockfill might be used in an embankment dam and its porosity will typically be in the range 20–25%. In contrast, an uncompacted opencast backfill of similar material has a much higher porosity of about 40%.

58.4.1.5 Fine fills

The amount of densification that can be achieved for a given compactive effort on a clayey soil is a function of moisture content. Laboratory compaction test results are usually plotted as the variation of dry density with moisture content. Such plots are a useful way of representing the condition of a partially saturated fill. The density that can be achieved by a particular compactive effort on a clay fill depends on the moisture content of the fill. For a given compactive effort there is a maximum dry density, which is achieved at the optimum moisture content. The expressions 'maximum dry density' and 'optimum moisture content' refer to a specified compaction procedure and can be misleading if taken out of the context of that procedure. With a moisture content that is less dry than the optimum, the specified compaction procedure will result in a fill with larger air voids. For a moisture content that is significantly wetter than the optimum, the specified compaction procedure will produce a fill with a minimum of air voids, typically between 2% and 4% (**Figure 58.2**).

The fundamental factors listed above demonstrate that the moisture content during placement is key to the density – and hence engineering performance – to be expected.

58.4.1.6 Liquidity and plasticity indices

In this context the term plasticity describes the response of fine fills (or the finer fractions) to changes in moisture content and is the same property as for natural soils.

Care should be taken when assessing index tests related to the matrix, as the impact of this will be different for fills compared with natural materials.

58.4.1.7 Compressibility and stiffness

A wide range of compressibility is possible. Different relationships can be used below and above the optimum moisture content, and the relationships will also depend on the nature of the fill material and whether the grading envelope means that the voids are 'filled' or 'unfilled'. When fills are not close to failure it is often convenient to describe their behaviour using elastic parameters even though fills generally exhibit nonlinear non-recoverable stress-dependent behaviour (see Charles and Skinner, 2001b).

Conventionally Young's modulus, E, and Poisson's ratio, ν, are used to describe linear elastic behaviour. However, these parameters have little direct relevance to the stress conditions that generally pertain to fills. In practice, the settlement for many fills is closely related to one-dimensional compression, and the constrained modulus, D, is of more direct application. Typically for non-engineered fills it may be assumed that $D = 1.3E$ to $1.8E$.

Moisture content and dry density relationship test

In accordance with BS 1377-4:1990

Hole no/sample ref: R2 — Test procedure: 2.5 kg hand hammer
Depth (m): 1.00 — Mould type: 1 litre
Soil visual description: Firm to stiff (medium to high strength) orange patched grey slightly sandy CLAY with frequent flint gravel (8.3% retained on 20 mm sieve)

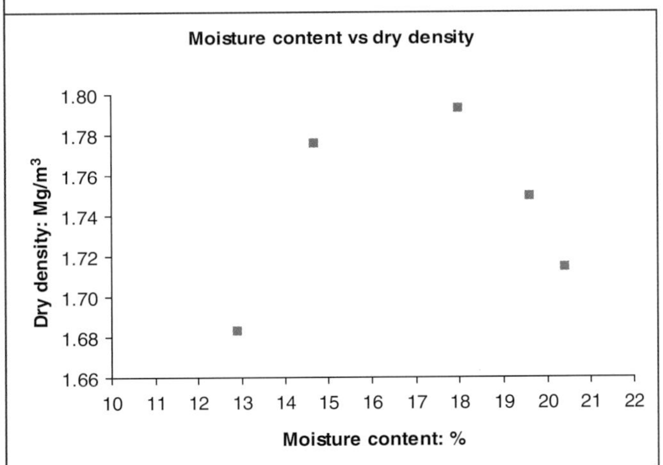

Natural moisture content (%)	18.50

Moisture content (%)	Dry density (Mg/m^3)
12.90	1.683
14.70	1.776
18.00	1.793
19.60	1.749
20.40	1.714

Optimum moisture content (%)	Maximum dry density (Mg/m^3)
16.41	1.794

Note:

Figure 58.2 Proctor test results for a fine-grained fill

When a uniform load is applied over a wide area of the surface of a fill, the resulting compression is largely one-dimensional. An application of load will produce an immediate compression in a partially saturated fill. Effective stress is not a simple concept for partially saturated soil and, for most practical purposes, compressibility can best be described by a constrained modulus expressed in terms of total stress. The constrained modulus is the ratio of the increment of applied total vertical stress, $\Delta\varepsilon_v$, to the increment of vertical strain, $\Delta\sigma_v$, which it produces under drained conditions

$$D = \Delta\sigma_v / \Delta\varepsilon_v \tag{58.1}$$

It should be noted that $D = 1/m_v$.

The constrained modulus can be measured in oedometer tests, but is usually not representative of in situ behaviour because of the small test size (**Table 58.2**).

58.4.2 Time dependency for fine fills

A fundamental difference between the deformation behaviour of fine fills, such as a clay fill, and coarse fills, such as a sand fill or a rockfill, is due to permeability.

Time-dependent movement depends on the nature and thickness of the fill, the method of placing and the nature of the underlying ground, especially the groundwater conditions. The best materials in this respect are obviously well-graded, hard and granular. Fills containing a large proportion of fine material, by contrast, may take a long time to settle. Similarly, old fills and those placed over low-lying areas of compressible or weak strata should be considered unsuitable unless tests demonstrate otherwise or the structure can be designed for low bearing capacity and irregular settlement. Frequently, poorly compacted old fills continue to settle for years due to secondary consolidation. Mixed fills that contain materials liable to decay, which may leave voids or involve a risk of spontaneous combustion, afford very variable support, and such sites should generally be avoided. Some materials, such as ashes and industrial wastes, may contain sulfate and other products that are potentially aggressive to concrete.

Generally, for uncompacted materials, rockfills will settle about 2.5% of their thickness, sandy fills about 5% and cohesive materials around 10%. The rate of settlement decreases with time, but in some cases it may take 10–20 years before movements are reduced to within tolerable limits for building foundations. In coarse-grained fills most of the movement generally occurs in the early years after the completion of filling, and after five years settlements are often small.

Fill	D_{sec}: MPa	ε_v: %
Sandstone rockfill ($I_D = 0.8$)	12	0.83
Sandstone rockfill ($I_D = 0.5$)	6	1.67
Sandstone rockfill (preloaded 0.5)	40	0.25
Colliery spoil (compacted)	6	1.67
Colliery spoil (uncompacted)	3	3.3
Clay fill ($I_p = 15\%$, $I_L = 0.1$)	5	20
Sandy gravel fill ($I_D = 0.8$)	50	0.2
Sandy gravel fill ($I_D = 0.5$)	25	0.4
Sandy gravel fill (preloaded 0.5)	200	0.05

Table 58.2 Typical values of the constrained modulus (D_{sec}) and strain for vertical stress increments of 100 kPa (from 30 kPa) (Data reproduced from Watts and Charles, 2015)

58.4.3 Strength

58.4.3.1 Undrained shear strength

The undrained behaviour is of practical importance for saturated clay fills when applied loads change faster than the induced pore pressure can dissipate. The undrained shear strength (c_u) is a function of moisture content, fabric and effective stresses. Plastic silt and clay hydraulic fills are cohesive and can be characterised by a profile of undrained shear strength against depth. With a high water table, effective stresses in a hydraulic fill will be low. Some desiccation is likely to occur at the surface of the fill, forming a stiff crust.

58.4.3.2 Drained shear strength

Non-engineered fills are likely to be in a loose condition. Consequently, their strength may be close to the constant volume angle of shearing resistance, φ'_{cv}. For a granular fill this angle of shearing resistance is usually similar to the angle of repose. If instability should occur, loose fill materials may be vulnerable to liquefaction and flow slides.

Box 58.1 Example: colliery spoil

The shear strength parameters of a coarse discard do not exhibit any systematic variation with depth in a spoil heap, and so are not related to age. This suggests that a coarse discard is not seriously affected by weathering. As far as the effective shear strength of a coarse discard is concerned, the angle of shearing resistance usually varies from 25° to 45°. The angle of shearing resistance and, therefore, the strength, increases in spoil that has been burned. With an increasing content of fine coal, the angle of shearing resistance is reduced. Also, as the clay mineral content in the spoil increases, so its shear strength decreases.

The shear strength of a discard within a spoil heap and, therefore, its stability, is dependent upon the pore water pressure developed within it. Pore water pressures in spoil heaps may develop as a result of the increasing weight of material added during construction or by seepage through the heap by natural drainage. A high pore water pressure is usually associated with fine-grained materials that have a low permeability and high moisture content. Thus, the relationship between permeability and the build-up of pore water pressure is crucial. In fact, in soils with a coefficient of permeability of less than 5×10^{-9} m/s there is little significant dissipation of the pore water pressure, while above 5×10^{-7} m/s the pore water pressure is generally completely dissipated. The permeability of a colliery discard depends primarily upon its grading and its degree of compaction. It tends to vary between 1×10^{-4} and 5×10^{-8} m/s, depending upon the amount of degradation in size that has occurred.

58.4.4 Permeability

A number of important effects and processes are influenced by water movement and permeability, including

- the rate of dissipation of excess pore pressure and the associated primary consolidation of saturated clay fills
- the influence of undrained or drained behaviour
- collapse compression of loose unsaturated fills
- the liquefaction of saturated fills
- the loss of ground due to erosion of fills.

Uncontrolled seepage can cause the migration of fine particles out of the fill or into the pores of a coarser soil. Where water flows out of a fill local instability may occur at the exit point. If water is flowing through a fill which is susceptible to erosion, then erosion can be prevented or controlled by protection of the fill with adequate filters, which will halt the loss of material.

58.5. Volume changes in fills

One of the most important assessments of the suitability of any fill for building purposes is of the likely volume changes within the fill, which could be observed as settlement, heave or lateral movements. Changes can occur as a result of the passage of time or in response to changes in stress or environmental conditions.

58.5.1 Self-weight

Stress changes caused by placement of fill over a wide area differ from the stress changes due to building foundations, which usually apply loads to relatively small areas. The wide extent of fill placement has the following consequences.

- Loading and compression behaviour of the fill is largely one-dimensional.
- The full depth of the existing fill may be affected by the loading caused by the extra fill.
- Underlying natural strata may be significantly loaded by the fill.

Similar conclusions hold for removal of layers of fill, although the stiffness properties may be very different on unloading, and stiffness is often significantly greater.

Movement caused by settlement or heave of the fill occurs in two stages

- 'immediate' movement as the applied stress increases or decreases
- time-dependent movement after the change in applied stress.

58.5.1.1 Movement during and shortly after fill placement

Any immediate compression occurring during an earthmoving operation before construction on the fill has no practical effect on the structures.

As extra fill is placed, most of the resulting compression of a coarse fill occurs immediately. The one-dimensional compression behaviour of coarse granular fills can be described in terms of the constrained modulus, *D*, which depends on the fill type, placement conditions including initial density and moisture content, the stress level and stress history. In contrast, the behaviour of a saturated or almost saturated clay fill, when extra fill is placed, differs considerably. The load will initially induce an excess pore pressure in the clay fill, and there is little immediate settlement. There may be some immediate settlement resulting from compression of any remaining air voids. This is followed by primary consolidation as the pore pressure dissipates. Any investigation should identify whether primary consolidation is still occurring. Often this is identified by changes in pore pressure.

58.5.1.2 Long-term movement

When movements due to the self-weight of the fill occur subsequent to building on the fill, damage to structures may result. The magnitude and rate of movement subsequent to fill placement are therefore of major interest.

For most fills there is an approximately linear relationship between creep compression and the logarithm of time. This is often expressed as

$$\alpha = \Delta s/[H \log(t_2/t_1)] \tag{58.2}$$

where compression $\Delta s/H$ of a fill of height H occurs between times t_2 and t_1.

Some of the results reported in Watts and Charles (2015) are given in **Table 58.3**.

58.5.2 Change in applied stress due to weight of buildings

Whereas fill placement is usually over a wide area, the weight of buildings will generally be concentrated onto relatively

Fill type	Compaction	α: %[a]	Case history[b]
Sandy gravel fill	Heavy vibrating roller	$0.04\sigma'_v$	Megget
Mudstone fill	Heavy vibrating roller	$0.12\sigma'_v$	Brianne
Sandstone or mudstone rockfill	Heavy vibrating roller	$0.13\sigma'_v$	Scammonden
Sandstone or mudstone rockfill	No systematic compaction	0.9–1.5	8
Stiff clay fill	Heavy systematic compaction	0.5	2

[a] σ'_{cv} in MPa. [b] The first three examples refer to case histories of embankment dam construction: Megget (Penman and Charles, 1985), Brianne (Penman and Charles, 1973) and Scammonden (Penman *et al.*, 1971).

Table 58.3 Creep compression rate parameter α for engineered and non-engineered fills (Data reproduced from Watts and Charles, 2015) small foundation areas. Where the foundation size is small, stresses will be significantly increased over only a relatively shallow depth immediately beneath the footing. In any evaluation of ground deformations caused by the weight of a building, it is important to have some knowledge of the stress distribution below the foundation, because in predicting settlement it is necessary to know to what depth the foundation will effectively stress the fill. The stress distribution below foundations is discussed in Chapter 53 *Shallow foundations*. Furthermore, for a foundation load of 50 kN/m, the ground is significantly stressed only to a depth of 2–3 m.

Apart from the immediate settlement of clay fills under undrained conditions, most movement due to the weight of structures is attributable to a volume reduction resulting from an increase in effective stress. Movement occurs both during and after construction. If movements are very large, structural damage can result from movements during construction, although these can often be built out and may cause little damage. However, significant settlement occurring after construction can seriously damage a structure.

In addition, some time-dependent strains will occur under a constant applied foundation load.

- For coarse fills, settlement will be caused by creep movements occurring at constant effective stress.
- For fine fills, settlement will be caused by primary consolidation as the pore pressure induced by construction loading slowly dissipates, followed by secondary compression at constant effective stress.

Long-term settlement due to constant imposed foundation loads often shows a linear relationship with the logarithm of time since construction, analogous to creep settlement arising from self-weight of the fill.

58.5.3 Change in water content

58.5.3.1 Change in groundwater level

The effective stress within a fill deposit depends strongly on the groundwater level. A rise in the groundwater level reduces effective stress, and a fall in the groundwater level increases effective stress. The smaller the particle density of the fill, the greater the percentage change in effective stress caused by a given change in the groundwater level. The magnitude of the movements caused by these changes in effective stress will depend on the stiffness of the fill. The changes will occur over large areas and so deformations will be largely one-dimensional. The movements can be described by a constrained modulus appropriate to the particular stress change. As the change in water level will usually result in unloading or reloading of the fill, the stiffness may be much larger than that operative during the first loading of the fill. The rate at which the groundwater level within a fill adjusts to any changes in external constraints is a function of the fill permeability. When an unsaturated clay fill is subjected to a rise in groundwater, the change may be rapid at first. Following softening of the clay lumps, often accompanied by large settlement, further changes are slowed by the significantly lowered permeability.

58.5.3.2 Swelling and shrinkage

The volume change of saturated clay fill, particularly high-plasticity clay, is moisture-content dependent. Drying may cause considerable shrinkage. As the fill dries and shrinks, the soil moisture exerts more suction. If water comes into contact with the fill, it is absorbed and the clay swells. These types of movement have caused damage to houses with shallow foundations on natural clay soils, and clay fills are vulnerable to the same processes. Shrinkage can occur through

- movement of moisture near the ground surface, due to evaporation and recharge
- the action of the roots of vegetation: the demand for moisture in dry weather increases with the amount of vegetation
- heating of the ground.

58.5.3.3 Collapse compression

Partially saturated fills may undergo a reduction in volume, known as collapse compression, when inundated. Inundation may occur owing to a rise in groundwater level or by downward migration of water. Most types of partially saturated fill are susceptible to collapse compression under a wide range of applied stress when inundated or wetted for the first time, if they have been placed in a sufficiently loose or dry condition. This phenomenon can occur without any change in applied total stress. When construction is about to take place on such a fill, susceptibility to collapse compression may be the most significant hazard because, where collapse compression occurs after construction has taken place, buildings may be seriously damaged (**Figure 58.3**). This most important facet of fill behaviour is examined in some detail in Charles and Watts (2001).

58.5.4 Biodegradation

The age of refuse landfill is important in evaluating the potential for further settlement. Not only has older refuse landfill had more time for organic matter to decompose, it is often an inherently better material with a much higher ash content than more recent refuse. Settlement behaviour is also affected by a number of other factors

- composition of the waste
- initial bulk density
- initial moisture content
- level of leachate within the fill
- the time over which the fill was placed.

Long-term movement can occur as biodegradation causes a reduction in volume, which may be in addition to movement

Figure 58.3 Damage to a building caused by collapse compression

caused by physical creep compression (both can be described by logarithmic compression rate parameters – see Watts and Charles, 1999).

58.5.5 Chemical action

Volume changes, usually expansion, may occur due to chemical reactions within the fill. Loss of material due to dissolution of carbonates could occur in some fills. In addition to a volume change of the waste fill, there may be an interaction between the construction materials and chemically aggressive ground.

Chemical reactions may have been occurring slowly since the waste fill was deposited. However, building development on the site could lead to an acceleration of these reactions as a result of effects connected with construction, such as the following.

- Earthmoving may lead to mixing of wastes, which brings different substances into contact and may introduce oxygen into the waste fill.
- Leaking drains may cause water to infiltrate the waste fill.

Fills that are particularly vulnerable include

- sulfate-bearing waste material
- iron and steel slag
- pyritic shale
- alkali waste.

Expert advice will be required for these materials.

58.6. Design issues

In foundation design, it is important to distinguish between the settlement attributable to the weight of the building and settlement attributable to other causes such as self-weight of the fill or collapse compression. The concept of bearing capacity is based on the premise that the settlement resulting from the weight of the building is the critical factor. With small structures on deep fills, almost invariably settlement attributable to causes such as self-weight of the fill or collapse compression will dominate and, consequently, the bearing capacity can be a misleading concept. Foundation design should be based on an assessment of the magnitude of differential movements of the fill subsequent to construction. It is necessary to identify first the cause of settlement. The small scale of many developments makes it difficult to ensure that there is an adequate investigation of the fill.

58.6.1 Bearing capacity

While the performance of building foundations on fills is rarely determined by bearing capacity, it is a design issue. For less settlement-sensitive structures (embankments or gravity retaining walls, for example) the bearing capacity is of key importance. The maximum stresses that can be applied to and by fills is also of relevance when considering the pressures on walls (compaction pressure, heave), and also in slope stability calculations.

Standard bearing capacity calculations can be used, but care should be taken with the fill properties used and a range of probable values must be considered (refer to Chapter 53 *Shallow foundations*).

58.6.2 Settlement

Settlement under applied loads, or due to fill settlement caused by other factors, is covered earlier in this chapter. The design must assess which type of settlement effects will apply in the situation under consideration. Following this, the anticipated amount and likely variability of settlement and its distribution (or heave) must be calculated.

Calculation approaches are often driven by the quantity and quality of data on fill properties on the site, and often may take the form of simple estimates. These may be based on elastic analogies (Boussinesq-type stress distributions), on semi-empirical approaches (such as Charles, 1996) or be fully empirical (Burland and Burbidge, 1985).

58.6.3 Geometry

Distortion and damage of buildings are related to the magnitude of the differential settlement across the building rather than the total settlement. Variation in the depth of fill under a building can be a cause of differential settlement and may also result in damaging horizontal movements. There are particularly acute problems associated with differential settlement at the edges of a filled area.

In a backfilled excavation, the severity of the differential settlement will depend on the steepness of the face of the excavation, and many different situations are commonly encountered

- an old dock with vertical sides
- a quarry in hard rock with a near-vertical face
- gravel and clay pits with slopes that are not so steep
- buried changes of the gradient of a slope
- the sloping base of an excavation.

There is a need to define the areas from which buildings should be excluded because the rate of change in fill thickness gives the potential for unacceptable long-term differential settlement. If unsafe decisions are made and exclusion zones are too small, there may be damage to the building and consequently legal action. On the other hand, an over-conservative approach will needlessly sterilise large areas of land and could make many sites uneconomic to develop. In determining the area of land where ground surface movements may be unacceptably large owing to variations in the depth of the fill, consideration needs to be given to both the properties of the fill and the geometry of the fill. Further consideration and prediction of the settlement pattern due to changes in fill depth can be found in Charles and Watts (2001) and Charles and Skinner (2001a).

58.6.4 Monitoring

In assessing the condition of a fill, the performance of a treatment or the condition of a structure it is often helpful to carry out monitoring, which may include

- surface levelling stations to measure the settlement of the fill surface
- extensometers (magnet or rod) to measure the settlement of incremental depths of fill
- standpipe piezometers to measure groundwater levels within the fill (local weather observations are also useful)
- load tests for direct estimation of the settlement produced by a load (**Figure 58.4**).

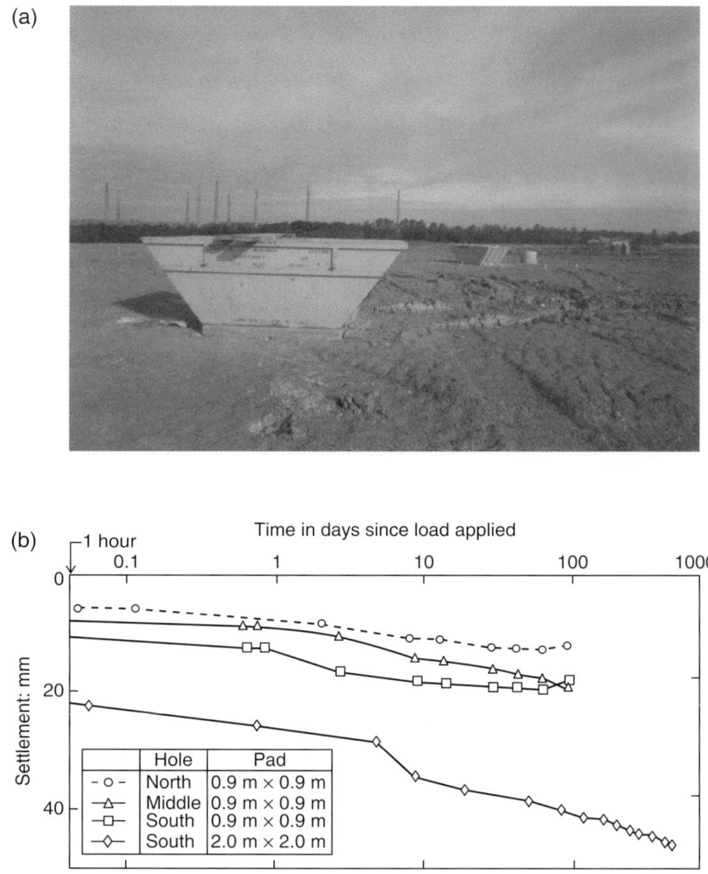

Figure 58.4 (a) Load testing using a skip; (b) load settlement results (Reproduced from Watts and Charles, 2015)

Such measurements can form one of the most important parts of an investigation of a filled site.

Surface settlement can be monitored by precise optical levelling. The settlement of the levelling stations is due to the vertical compression of the fill in which the station is installed, together with any movement of the underlying natural ground. Surface levelling stations may be very simple, and are easy to install. The levelling stations should be sufficiently robust to resist damage due to construction traffic, and where the stations protrude above the ground surface it is usually necessary to provide protection such as concrete manhole rings.

Although filled sites can be subject to large settlements, monitoring often takes place over a short period or may be intermittent. Accurate measurements are required to establish the rate of settlement reliably over a relatively short period. Measurements should be made to an accuracy of at least 1 mm. It can be helpful to locate levelling stations in a number of traverses with the stations at quite close intervals. In this way an indication of the likely differential settlement over the area of a building can be obtained.

Settlement at different depths within a deep fill can be measured by installing a magnet extensometer, whereas rod extensometers may be appropriate for shallow fills.

It is also important to measure water levels within a fill. Simple standpipe piezometers can be sealed into boreholes. In low-permeability fills the response time of a standpipe piezometer to a change in piezometric pressure may be too long. Pneumatic or vibrating wire piezometers can be installed in boreholes and will give a much more rapid response.

The period for monitoring depends on the situation. For example, to establish trends in settlement and groundwater levels reliably on a site a minimum period of at least a year will be required, and several years would be preferable. Where a site is being investigated for building development it will rarely be possible to monitor over such long periods. However, even where the period is quite short, it may be very advantageous to perform some monitoring. Monitoring may be used to identify when treatment, or the rate of settlement, has reached some design level. In each case the interval, period and type of monitoring should be linked to the parameter being measured. Further guidance on monitoring is given in Chapters 94 *Principles of geotechnical monitoring* and 95 *Types of geotechnical instrumentation and their usage*.

Post-construction settlement of buildings on fill should be monitored where possible, and BRE Digests 343 and 344 (BRE, 1989; BRE and Longworth 1989) provide guidance on measuring the movements of low-rise buildings.

58.6.5 Shallow foundations

Shallow foundations (see Chapter 53 *Shallow foundations*) are generally less than 3 m deep, so in a shallow fill (<3 m deep) a shallow foundation is all that is required to found the structure on the underlying natural stratum. With medium-depth (3–10 m) and deep (>10 m) fills, either deep foundations into the underlying natural stratum or a shallow foundation is adopted and the structure is founded on the fill. Where small structures are to be built on medium-depth or deep fills, the only economic solution may involve founding the structure on the fill. It should be ensured that

- the bearing capacity of the fill is adequate
- settlement under the working load will not damage the structure
- settlement due to causes other than building loads will not damage the structure.

Where fill is vulnerable to significant movement, ground improvement prior to construction on the fill should be considered. Ground treatment methods, when appropriately used, should limit and control settlement; they will not normally eliminate settlement (refer to Chapter 59 *Design principles for ground improvement*). Many types of ground treatment will not improve the full depth of a deep (>10 m) fill, but they may stiffen the upper part of the fill sufficiently to prevent excessive differential settlement.

The following basic approaches to shallow foundation design may be adopted.

- Construct shallow foundations on fill that has been sufficiently improved by ground treatment so that deformations are acceptably small.
- Design shallow foundations that are sufficiently stiff to withstand the anticipated deformations in the fill; this may not be difficult with small buildings, but tilt may still be a problem.
- Design shallow foundations and structures that are sufficiently flexible to tolerate the anticipated deformations in the fill.

In practice, solutions may comprise both ground treatment and special foundation design.

Fill that is susceptible to collapse compression on wetting is of particular concern. Several ways of dealing with the problem are listed below and a combination of these approaches may often be the most suitable.

- Eliminate susceptibility to collapse by pre-inundation (not generally feasible).
- Eliminate or greatly reduce susceptibility to collapse by some other form of ground treatment (depth of effectiveness may be limited).
- Prevent inundation of the fill during the life of structure (this often cannot be guaranteed).
- Design the foundations and structure to survive collapse movements (difficult if potential movements are large).

Reinforced concrete rafts with edge beams, sometimes termed semi-rafts or stiffened edge rafts, have commonly been used for housing on non-engineered fills. They have usually been designed so that some of the length will act as a cantilever over a potential void. However, where large differential movement may occur, very substantial foundations may be required and tilt may still be unacceptably large. Atkinson (1993) has described typical designs.

The problems associated with shallow foundations for low-rise buildings on filled ground can be minimised by

- avoiding building across the edges of filled areas where the structure would be partly founded on fill and partly on undisturbed natural ground
- restricting construction to small units and not building long terraces of houses.

Relative movement between the building and any services entering it, and between various sections of these services, merits careful consideration. If the movements are not likely to be large, the use of short lengths of pipes with flexible connections may be sufficient. In more severe cases it may be necessary to use flexible pipes or to carry services on piles bearing on a firm stratum beneath the fill. Falls given to drains should be adequate to reduce the risk that settlement will cause backfalls.

58.6.6 Deep foundations

Where a large structure is to be built on shallow or medium-depth fill, the poor load-carrying characteristics of the fill can be circumvented by using piled foundations and a suspended floor. The piles will be designed to transmit the loads applied by the weight of the building down through the fill to a competent stratum underlying the fill (refer to Chapters 54 *Single piles* and 55 *Pile-group design*).

There may be problems where piles hit obstructions. If the depth of fill is not accurately known at a pile location, it may be incorrectly assumed that the pile has reached natural ground.

The piles should be designed for downdrag (negative skin friction) caused by settlement of the fill. The downdrag can be calculated in terms of the effective stress (Burland, 1973; see Chapter 55 *Pile-group design*). In some situations it may be considered that the downdrag will cause too large a settlement of the pile. The use of a slip coating or a permanent sleeve can reduce its effect, but careful detailing (to ensure effectiveness) is required and complicates pile construction. Often it will be more cost-effective to increase the pile length.

Special attention is needed in the design of services that span from the filled ground into buildings founded on piles.

In a fill in which methane gas or other dangerous gases are generated by the decay and decomposition of organic matter, piles could form paths for the escape of the gas.

58.7. Construction on engineered fills

In some cases a fill is placed to support a building or other structure. The characteristics of the site should be considered with regard to the impact of the fill and other applied loads. Ideally the fill should be placed such that simple foundations can be utilised. Sometimes this is not the case, and in these circumstances the nature of the fill should be designed so it has the required properties (e.g. it can be penetrated by piles but will support the ground slab, or it can be compacted after placement).

Where suitable fill material is placed to an appropriate specification under controlled conditions, most of the problems associated with construction on fills should be greatly reduced if not eliminated. The two major considerations are, first, the preparation of an appropriate specification, and, second, the provision of adequate site supervision to ensure compliance with it.

The required behaviour of the fill should be well defined and it should be possible to make reliable estimates of settlement under working loads, including long-term settlement and bearing capacity. However, sometimes problems have been experienced because inappropriate specification or inadequate control of the placement of engineered fills has led to excessive settlement or heave in service.

While the specified fill may be adequate to support the building, other causes of settlement or ground failure must be considered

- the stability of the natural ground
- settlement of the underlying natural ground due to the weight of a fill
- settlement of an engineered fill due to self-weight
- movement in an engineered fill due to changes in groundwater level, pore pressure or seasonal changes in moisture content.

The specification of the engineered fill should define acceptable materials for the fill and the method of placement and compaction. Specifications for highway construction have often been used for the specification of fills to support buildings or other structures, but may not be adequate. Trenter and Charles (1998) have produced a model specification for fills for building purposes.

58.8. Summary

Construction on fills is often more challenging than for similar structures on natural soils. However, additional risks can be minimised provided that

- there is an understanding of the problems associated with the various types of fill and fill deposits
- the construction is appropriate
- thorough investigations are carried out
- the design recognises that movement may not just be related to construction
- adequate specifications for construction and monitoring are produced
- construction is carried out to the specifications, with appropriate quality control and feedback.

References

Atkinson, MF (1993) *Structural Foundations Manual for Low-Rise Buildings*. Spon, London, UK.

BRE (1989) *Simple Measuring and Monitoring of Movement in Low-Rise Buildings. Part 1: Cracks*. BRE Digest 343. BRE Press, Watford, UK.

BRE and Longworth, I (1989) *Simple Measuring and Monitoring of Movement in Low-Rise Buildings. Part 2: Settlement, Heave and Out of Plumb*. BRE Digest 344. BRE Press, Watford, UK.

BSI (British Standards Institution) (1990) BS 1377-2:1990: Methods of test for soils for civil engineering purposes – Classification tests. BSI, London, UK.

BSI (1999) BS 5930:1999: Code of practice for site investigations. BSI, London, UK.

Burland, JB (1973) Shaft friction of piles in clay: a simple fundamental approach. *Ground Engineering* **6(3)**: 30, 32, 37, 38, 41, 42.

Burland, JB and Burbidge, MC (1985) Settlement of foundations on sand and gravel. *Institution of Civil Engineers – Proceedings Part 1: Design and Construction* **78**: S1325–S1381.

Charles, JA (1996) Depth of influence of loaded areas. *Géotechnique* **46(1)**: S51–S61.

Charles, JA and Skinner, HD (2001a) The delineation of building exclusion zones over high walls. *Ground Engineering* **34(2)**: 28–33.

Charles, JA and Skinner, HD (2001b) Compressibility of foundation fills. *Proceedings of the Institution of Civil Engineers, Geotechnical Engineering* **149(3)**: 145–157.

Charles, JA and Watts, KS (2001) *Building on Fill: Geotechnical Aspects*, 2nd edn. BRE Press, Watford, UK.

Penman, ADM and Charles, JA (1973) Constructional deformations in rockfill dam. *ASCE Journal of Soil Mechanics and Foundations Division* **99(SM2)**: 139–163.

Penman, ADM and Charles, JA (1985) A comparison between observed and predicted deformations of an embankment dam with a central asphaltic core. *Transactions of 15th International Congress on Large Dams, Lausanne, Switzerland*, vol. 1, pp 1373–1389.

Penman, ADM, Burland, JB and Charles, JA (1971) Observed and predicted deformations in a large embankment dam during construction. *Proceedings of Institution of Civil Engineers* **49**: 1–21.

Trenter, NA and Charles, JA (1998) A model specification for engineered fills for building purposes. *Proceedings of the Institution of Civil Engineers, Geotechnical Engineering* **119**: 219–230.

Watts, KS and Charles, JA (1999) Settlement characteristics of landfill wastes. *Proceedings of the Institution of Civil Engineers, Geotechnical Engineering* **137**: 225–233.

Watts, KS and Charles, JA (2015) *Building on Fill: Geotechnical Aspects*, 3rd edn. IHS BRE Press, London, UK.

Useful websites

British Geological Survey: http://www.bgs.ac.uk

Department of the Environment Industry Profiles: http://www.claire.co.uk/useful-government-legislation-and-guidance-by-country/198-doe-industry-profiles

It is recommended this chapter is read in conjunction with

- Chapter 26 *Building response to ground movements and tunnelling*
- Chapter 25 *The role of ground improvement*
- Chapter 34 *Non-engineered fills*

All chapters in this book rely on the guidance in Sections 1 *Context* and 2 *Fundamental principles*. A sound knowledge of ground investigation is required for all geotechnical works, as set out in Section 4 *Site investigation*.

Chapter 59
Design principles for ground improvement

Robert D Essler RD Geotech, Skipton, UK
Nigel Pye Mott MacDonald, UK
Gerardo Pittaro Mott MacDonald, Singapore

The use of ground improvement on projects has increased worldwide as its design and application become more common. Many major projects now routinely adopt ground improvement, and this experience is then disseminated through technical publications, which in turn increases knowledge. Hence, there is a better understanding of how and when such techniques can be incorporated into projects and what associated risks may be attached. In the UK there remains a perception by some that ground improvement techniques are difficult to design securely, that the design knowledge is restricted to a small number of engineers or that the techniques are only to be used when there is an unexpected problem on site. It is, therefore, important that young practising engineers have a basic grounding in the design of ground improvement techniques so that they can recognise at an early stage when such processes could be used to advantage on their projects and have the confidence to propose such schemes. It is equally important that when such a situation arises the engineer is able to progress the design knowledgeably, whether by discussion with contracting specialists or by referral to experienced engineers within their organisations or other companies.

doi: 10.1680/icemge.66830.0971

CONTENTS

59.1. Introduction

There are many ground improvement processes with different methodologies and applications. Techniques can be categorised into either ‘creating inclusions’ (i.e. introducing new elements into the ground) or ‘not creating inclusions’ (whereby ground improvement is undertaken principally by densification or grouting). It is a difficult task to summarise the design principles of so many varied processes with different methodologies and applications; it is therefore hoped that the techniques set out below and described in outline will allow the engineer to understand which processes might be suitable and where to look for further insights to better inform the decision-making process. **Tables 59.1** and **59.2** summarise the principal ground improvement techniques with insight into their suitability in terms of applicability and performance for ‘creating inclusions’ and ‘not creating inclusions’. In some instances (e.g. concrete columns and soil replacement with geosynthetics) discussion on their design is covered elsewhere in the manual and reference is given to the relevant chapter in the tables.

59.1.1 Background

The improvement of ground by injection of grout dates back to at least 1802 (Littlejohn, 2003). This would suggest, then, that in the twenty-first century it should be a well-practised, well-understood technique. It is, therefore, surprising that ground improvement, particularly by grouting, is still perceived as a process that is not very well understood and can be difficult to incorporate within a project to provide definable performance. There is no doubt that until the 1970s or 1980s there was very little technical knowledge in the public domain, but from this time onwards the technical literature has increased substantially. In the USA the first specialist grouting conference was in New Orleans in 1982 and attracted a total of 64 papers. By the third conference (New Orleans, 2003) this number had risen to 127 from 21 countries. Because the UK construction market is significantly smaller than that in Europe or the USA, the development of these more experience-driven techniques is not as advanced and is limited to specialists who, generally, belong to larger groups working worldwide. In 2008 the probable size of the European ground treatment (grouting) market was in excess of £500 million annually. This should be compared to the UK where the market is probably on average £10–20 million annually and is distorted when large projects stretch the available resources, such as the jet grouting works undertaken for the Victoria Station upgrade (Rutty *et al.*, 2021). Therefore, it is critical to the development of these techniques that they become better understood and more available as tools for the engineer to employ in design.

59.1.2 Definition of ground improvement

Ground improvement can be defined as the introduction of materials or energy to soils to affect a change in performance of the ground such that it performs more reliably and can be incorporated within the design process.

Method	Technique	Depth capability: m					Suitable ground					Cement-based grouts			Chemical grouts		Chapter
		0–2	2–5	5–10	10–20	20+	Clay	Silt	Sand	Gravel	Rock	Cement PFA	Cement bentonite	Other cement grouts	Silicate	Resin others	
Steel/ wood columns	Vibrated	✓	✓	✓	✓			✓	✓	✓							54–56
	Driven	✓	✓	✓	✓	✓		✓	✓	✓							54–56
	Bored	✓	✓	✓	✓	✓		✓	✓	✓							54–56
Concrete columns	Vibrated	✓	✓	✓				✓	✓	✓							54–56
	Driven	✓	✓	✓	✓			✓	✓	✓							54–56
	Bored	✓	✓	✓	✓	✓		✓	✓	✓							54–56
Granular columns	Stone columns/ vibrocompaction	✓	✓	✓	✓	✓			✓								59
	Sand columns / sand compaction piles	✓	✓	✓	✓	✓	✓ S_u >15 kPa	✓	✓								–
	Stone columns/ vibroreplacement	✓	✓	✓	✓		✓ S_u >15 kPa	✓	✓								59
	Geosynthetic encased columns	✓	✓	✓	✓		✓	✓	✓								–
Deep mixing	Dry methods		✓	✓	✓		✓ S_u <50 kPa	✓	✓ (loose)								59
	Wet methods		✓	✓	✓	✓	✓	✓	✓	✓							
Jet grouting	Single system	✓	✓	✓	✓	✓	✓	✓	✓	✓	✓	✓	✓	Cement/ slag	Not usual	Not usual	59
	Double system	✓	✓	✓	✓	✓	✓	✓	✓	✓	✓	✓	✓	Cement/ slag	Not usual	Not usual	59
	Triple system	✓	✓	✓	✓	✓	✓	✓	✓	✓	✓	✓	✓	Cement/ slag	Not usual	Not usual	59

Table 59.1 Selection of ground improvement techniques by 'creating inclusions'

Method	Technique	Depth capability: m					Suitable ground					Cement-based grouts			Chemical grouts		Chapter
		0–2	2–5	5–10	10–20	20 +	Clay	Silt	Sand	Gravel	Rock	Cement PFA	Cement bentonite	Other cement grouts	Silicate	Resin others	
Compaction methods	Impact roller compaction	✓	✓						✓	✓							75
	Rapid impact compaction	✓	✓	✓					✓	✓							75
	Dynamic compaction	✓	✓						✓	✓							59
	Deep vibration								✓	✓							59
	Micro-blasting								✓	✓							–
Consolidation methods	Surcharge	✓	✓	✓	✓		✓	✓									–
	Vertical drains + surcharge	✓	✓	✓	✓		✓	✓									–
	Vacuum consolidation	✓	✓	✓	✓		✓	✓	✓								–
	Dewatering	✓	✓	✓	✓			✓	✓	✓							–
Thermal methods	Ground freezing	✓	✓	✓	✓	✓	✓	✓	✓	✓							–
	Ground heating	✓	✓	✓	✓	✓	✓	✓	✓	✓							–
Grouting methods	Permeation grouting	✓	✓	✓	✓	✓	Nil	Nil	✓	✓	✓	✓	✓		✓	✓	59
	Compaction grouting						✓	✓	✓	✓		✓					59
Soil replacement	Dig-out and replace	✓					✓	✓									70
	With geosynthetic/ metallic reinforcement	✓					✓	✓									73

Table 59.2 Selection of ground improvement techniques by 'not creating inclusions'

Soil type	Issues	Investigation techniques	Ground improvement techniques
Sands and gravels	Fines content and collapse potential	Obtain representative samples for laboratory testing, carry out cone penetration testing (CPT)	All forms of ground improvement
Soft clays	Long-term settlement	Obtain high-quality samples, carry out CPT testing, carefully supervised field vane tests	Compaction grouting. Soil mixing and vibroreplacement (insensitive clays, S_u > 15 kPa)
Peat and organic clays	Acidic ground, secondary consolidation	Chemical testing and if possible large-diameter, high-quality sampling	Jet grouting and soil mixing
Karstic rock	Presence of solution features	Geophysics and intrusive surveys	Void filling and compaction grouting
Fills	Presence of organic materials, obstructions and chemicals	Chemical testing, geophysics and intrusive surveys, CPT testing	Vibrocompaction or dynamic compaction, compaction grouting

Table 59.3 Ground-related issues for ground improvement

59.1.3. Types of ground improvement

The forms of ground improvement discussed within this chapter are

- void filling
- grouting (permeation, compaction, compensation, rock and jet grouting)
- improvement by densification or replacement (vibrocompaction, dynamic compaction and stone columns)
- soil mixing.

There are other forms of ground improvement, but the processes set out above are the major techniques used for a very high proportion of the ground improvement carried out. Of all these techniques, jet grouting, soil mixing, vibrocompaction and vibroreplacement probably represent most of the ground improvement carried out and should be considered carefully. Other forms of ground improvement are detailed elsewhere in the manual (refer to **Tables 59.1** and **59.2**).

59.2. General design principles for ground improvement

The following general principles apply to ground improvement and need to be followed to ensure that the design is appropriate and the application successful.

59.2.1 Ground conditions

Because most ground improvement techniques require some interaction with the ground, the correct identification of the soils present is of vital importance. It is therefore important that adequate ground investigation is carried out prior to any ground improvement to ensure that the correct decisions are taken as to which technique is most suitable.

Problem soils to consider carefully would be very soft organic clays, peats and uniformly graded sands and silts.

For the various deep mixing and grouting methods, as well as carrying out normal laboratory testing, chemical testing is also important as some chemicals can affect the strength development of some binders.

Table 59.3 sets out some soils that need careful investigation to ensure that appropriate ground improvement can be designed. It should be stressed that adequate, carefully designed and well supervised thorough ground investigation is always a prerequisite for successful ground improvement.

59.2.2 Laboratory and site trials

It is inevitable that for some applications there will be some doubt as to what degree of improvement might be obtained. Laboratory and site trials are invaluable for informing the design on a project-specific basis, and it is always recommended that some form of trial, whether an initial period of more intensive monitoring, validation at the start of the works or a completely separate advance trial, is carried out. The exact form of a trial will depend on both familiarity with the process within the soils under consideration, and equally the criticality of the degree of improvement to the design. The works programme should be developed assuming some period of initial trialling so that due allowance can be made for any resultant changes. Where the ground improvement performance is critical then advance trials are preferred with sufficient programme time not only for the actual site works but also for the on-site and off-site testing and evaluation.

A good example of this was the jet grouting carried out for the construction of the Victoria Station upgrade works, London (Rutty *et al.*, 2021), where the proposed jet grouting was critical to the stabilisation of water-bearing River Terrace Deposits through which sprayed concrete-lined tunnels were excavated under shallow overburden. Another example is the construction of a sandwich wall in Amsterdam, where jet grouting was critical to the support of a historic building, and a series of trials were carried out well in advance of the actual works (De Wit *et al.*, 2007).

For deep soil mixing (dry or wet methods), laboratory trials are recommended in advance of site trials.

59.2.3 Performance

It is always important when designing ground improvement works that the performance required is documented to such a degree that it can be validated on site. It is imperative that whatever design parameters are required from the ground improvement scheme, that they are correctly translated into field performance parameters that can be measured and validated so that the design can be assured. For example, a designer must decide whether the design parameters need to be the average of field results, the minimum, the maximum or even to have some other relationship. Evaluating the performance of ground improvement based upon a statistical assessment of quality assurance/quality control data (e.g. the target needs to be greater than a specified percentile of measurements) generally provides a more pragmatic rationale for assessing the quality of improvement rather than reliance on attainment of either minimum or maximum criteria (which may not be practically achievable). It is also important that the designer is involved in the type and location of testing to be undertaken to validate the ground improvement works and confirm design assumptions.

59.2.4 Environmental issues

Environmental issues resulting from ground improvement can either be due to polluting the ground with the cements or chemicals used, or equally as a result of changes to the local groundwater hydrogeology. It is important to review these potential effects when considering ground improvement design.

59.2.5 Existing infrastructure

The presence of existing infrastructure can have a significant bearing on the choice of ground improvement technique, its practical implementation and the measures required to avoid and mitigate impacts on third parties. Consideration could include vibrations from dynamic compaction on adjacent buildings, the presence of utilities obstructing the boring of holes for jet grouting and compensation grouting next to basements or tunnels. Designers specifying ground improvement should ensure that third-party interfaces are adequately considered and that risks to third parties' infrastructure can be managed.

59.3. Design principles for void filling

59.3.1 Methods and key issues

Void filling is most commonly employed to deal with abandoned mines or redundant underground structures and involves the drilling of holes on a regular grid and the injection of either a fluid grout or paste.

Generally, the design is to first create a perimeter barrier around the area to be treated to economise the operation. This is most often formed by drilling larger-diameter holes (100–250 mm dia.) and introducing gravel-sized material washed in with a grout to bind the gravel together.

Figure 59.1 shows how perimeter holes are executed using a mix of gravel and grout, which will not travel and, thus, when set forms a barrier to the more fluid infill grout. Typical perimeter holes would be installed on a 2–3 m grid.

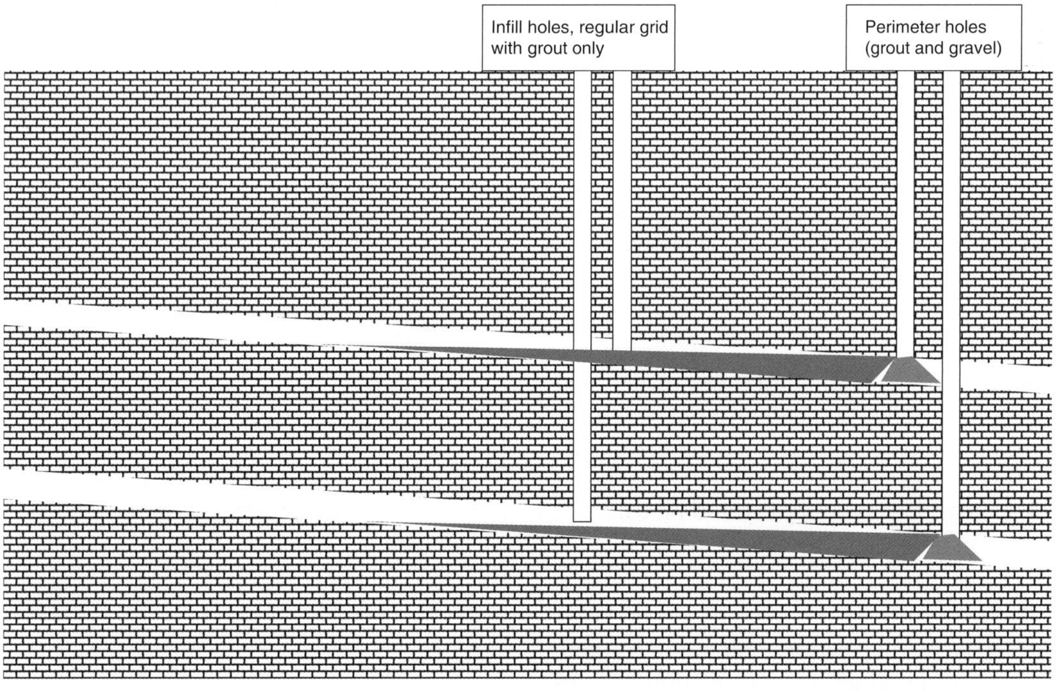

Figure 59.1 Schematic of perimeter and infill grouting for void filling

The design of the main works is then executed by drilling holes (50–75 mm dia.) on a regular grid, typically ranging from 3 to 6 m, and injecting a cement-based grout or paste. The design of the grid to adopt is based on how open the void is and, thus, how easy it will be for the grout to travel. Usually, the works are carried out following the recommendations of CIRIA Special Publication 32 *Construction over Abandoned Mineworkings* (Healy and Head, 1984), which sets out the extent of treatment beyond the structures of interest.

Key design issues relate to the control of any groundwater trapped within the voids, which may need to be pumped out as the void is filled, ensuring that access shafts are located and treated as these can be a source of future settlement. Groundwater pumped from these voids could be contaminated and should be disposed of appropriately.

Quite often there are underground records of mines, and these can be used to design the works and assess the amount of infilling required, as well as expected drilling depths. Historically a cement and pulverised fuel ash (PFA) mix was used in the proportion of 1 part cement to 10–20 parts PFA. The Environment Agency is now reluctant to sanction the use of PFA as it contains heavy metals and, thus, can introduce contaminants into the ground. Replacement grouts can be foam concrete or simple cement and sand mixes.

59.3.2 Design principles

Designing for void filling is essentially observational, with probe holes being drilled on a regular grid to locate any voids and subsequently grouting them. The design process should ensure that the depth of drilling is sufficient to investigate all known voids and the plan area treated should include the potential collapse zone of voids at the boundary of the site. Adequate site investigation is thus very important. As the resultant void is infilled, the grout is contained; a low strength of less than 5 MPa is usually sufficient.

It is essential good practice when developing a new site to check available coal mining records. These are available from the Mining Records Office in Mansfield.

59.3.3 Execution controls

Execution is controlled by documenting all holes drilled and using this information to develop the underground void profile. The grout or paste injected is controlled by testing its properties against a testing and inspection plan, which sets out limits on properties such as density, viscosity and bleed.

59.3.4 Validation

Validation is usually by the use of the ongoing probe holes within secondary or tertiary grid positions and proof grouted. Where complete filling is important, then validation by secondary grouting can be carried out using a more fluid grout and higher pressures.

59.3.5 Typical performance obtained

Generally, about 95% of the void can be expected to be filled, as some in situ bleed can take place. Filler strengths have a wide range depending on individual project design. Historically, void filling of mine workings is regarded as a low-tech solution and there are very few instances of problems arising from this type of work. Care should be taken, however, when the void being infilled is of a different type such as an abandoned tunnel within an urban environment, as performance here can be more critical and post-pressure grouting may be required. Where the infilled void is likely to be loaded by subsequent construction, the grout properties need to be considered in more detail as higher strengths may be required.

59.4. Design principles for compaction grouting

59.4.1 Methods and key issues

Compaction grouting, probably originating in the USA, is most commonly employed to increase the relative density of granular materials to improve stiffness or reduce future settlement potential. Compaction grouting is essentially the injection of a low-slump (typically 25–100 mm) grout such that an expanding frictional bulb forms. **Figure 59.2** shows the schematic effect of the technique. This expansion causes deformation and densification around it and ultimately improves the ground. This expansion will continue until the ground has densified sufficiently to resist the expanding pressure. This is observed during the process of grouting through the increasing injection pressure and also by ground heave. If the slump is relatively high, then a hybrid form of grouting is used where both expansion and hydrofracture occur.

The design is thus based on expanding cavities within the ground until a maximum relative density is achieved. The design process generally considers the improvement midway between injection points as this will be the least improved. In

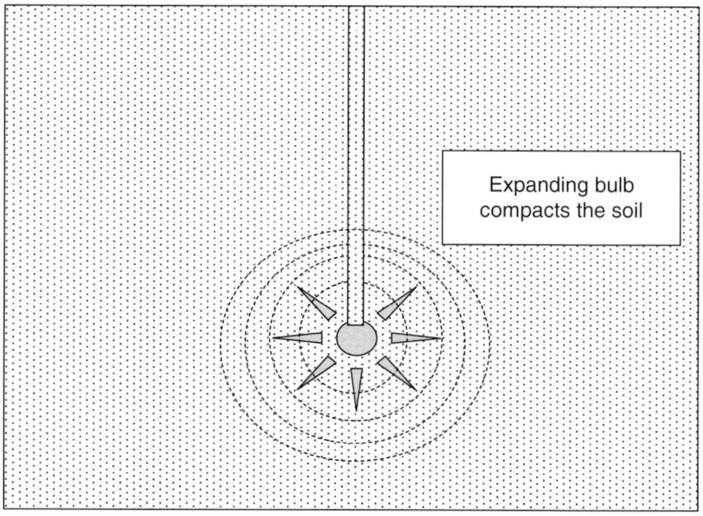

Figure 59.2 Compaction grouting

the UK, compaction grouting is mainly restricted to ground improvement where neither waterproofing or underpinning is required. It is sometimes used for settlement control, but as control is more difficult than the use of fracture grouting it is not favoured. Compaction grouting, on the other hand, is commonly used in the USA for settlement correction, ground improvement and seismic risk mitigation, where it has a long history of success. Typical uses in the UK are restricted to grouting to improve made ground or solution features formed within limestone or chalk. Compaction grouting has the advantage that the compactive effort coupled with the associated high pressures mean that a wider hole spacing can be used than for permeation grouting and, thus, it can be more economic in improving granular soils.

A secondary benefit when using compaction grouting to stabilise solution features is that the improvement can be achieved without the introduction of significant volumes of fluid grout or water, which could further destabilise the feature. For historic or domestic properties there is also an advantage that the original foundation support can be left in place without significant rebuilding as the soil itself is improved and no new load-carrying capacity is required. Therefore, this type of ground improvement is preferred when improving specific locations in proximity to an existing structure.

The technique is most appropriate for coarse-grained soils where the compactive effort is greatest. In silts the technique appears to work well and assists in forcing water from the silt and increasing strength. In clays the process is likely to cause significant pore water pressure response, which although causing some localised improvement to the clay in the long term can also result in triggering time-dependent settlement. It is unusual to apply compaction grouting in clays unless they exist as thin layers or bands within the main body of coarse-grained soils.

Historically, compaction grout is a mix of cement, bentonite and sand with an approximate water–solids ratio of 0.1 to 0.2 by weight, designed to be pumpable but having a low slump. Where available, PFA can be added to increase pumpability, but environmental issues may preclude its use in the future. There are now proprietary mixes available in Europe, such as Blitzdammer, manufactured by Heidelberg Cement, which are supplied ready mixed in dry form and which allow a much smaller site set-up.

59.4.2 Design principles

Generally, compaction grouting is designed on the basis of an expanding bulb compacting the ground around it. The degree of improvement depends on the soil type and existing overburden, as compactive effort is lost when the ground starts to heave. When designing compaction grouting, the hole spacing, which usually ranges from 1.5 to 3.5 m, is related to the depth of the compaction grouting and the zone of improvement required. A closer spacing is used where confinement is least and also where the treatment is at shallow depths. Treatment up to ground level is not possible as the compactive effort is minimal, and in this case the shallow zone can be excavated and replaced in layers while applying compactive effort or treatment by another form of ground improvement.

Soil type	Depth of treatment: m	Hole spacing: m	Improvement range: %[a]
Loose sands and sandy gravels	<3	1	150–200
	5–10	1–2	150–300
	>10	2–3	<300
Medium-dense sands	<3	1	125–150
	5–10	1–2	150–175
	>10	2–3	<200
Silts	<3	1	<150
	5–10	1–2	<200
	>10	2–3	<200

(a) Degree of improvement highly sensitive to silt/clay content; improvement reduces as silt/clay content increases.

Table 59.4 Hole spacing for compaction grouting

Table 59.4 gives an approximate indication of hole spacing and the likely degree of improvement. Care needs to be taken when dealing with fills as the variable composition can lead to unpredictable results.

Generally, the density of the ground can be improved such that standard penetration test (SPT) 'N' or CPT q_c values are increased by a factor of up to three, although the degree of improvement is highly sensitive to the fines content. The closer the spacing, the higher the degree of improvement, but overburden also plays an important part in providing the necessary confinement for the expanding bulb.

Design grouting pressures range from 40 bar at depths greater than 15 m to no more than 10 bar at depths less than 5 m. Holes are drilled to depth and the casing withdrawn in 0.5 to 1 m stages while injecting grout. Typically stage volumes are controlled by pressure or resultant movement: the grout is injected until either the target pressure is reached or surface movement is triggered.

Grout strength is not usually an issue as the cement in the mix is sufficient to produce cube strengths in excess of 10 MPa. Low-strength compaction grout can be produced by substitution of the cement with non-cementitious material.

A good example of the use of compaction grouting was the Castle Mall shopping centre at Norwich. The site was underlain by chalk at shallow depth, containing numerous solution features of variable size and depth. The infill material to these features was also variable but generally very weak. The development consisted of an 18 m deep excavation within a perimeter piled retaining wall, and the presence of solution features within the site which complicated the foundation design. Because the site investigation had located some features (8 out of 68 boreholes) but other features were likely to be present, the foundation design had to take account of the random location and nature of the features. The foundation design considered three solutions

- piled foundations, permanently cased through the features
- raft, spanning over any feature
- pad foundations combined with compaction grouting to improve the infill material.

The chosen foundation solution was pad foundations with ground improvement, as this allowed the chalk bedrock to be utilised for the majority of the foundations and compaction grouting would then allow the remaining foundations to be safely supported.

The engineer developed a strategy for dealing with the presence of features according to location and size, which is shown in **Figure 59.3**. This strategy was followed for the site works and was either a simple probe and bulk fill with low-pressure grouting or high-pressure compaction grouting.

The performance of the compaction grouting was based on post-testing using the SPT. This was set at an average of 15 and a minimum of 10. About 7% of the site area was discovered to contain features; there were 60 in number with a maximum size of 12 m dia. The largest feature had a plan area of 200 m^2. Grout volumes per metre of hole ranged from 0.26 to 1.0 m^3.

59.4.3 Execution controls

As with most ground improvement, execution controls are twofold: first, quality control of the materials employed; and second, the resultant effect on the ground. Compaction grouting is controlled by ensuring a consistent low slump for the grout while monitoring for excessive line pressure or resultant surface movement.

59.4.4 Validation

The most common validation for compaction grouting is to compare in situ testing before and after treatment, usually using a CPT or SPT. The designer normally sets a post-treatment minimum value to be obtained based on the required degree of improvement. On some projects, where justified, zone testing can be carried out, although the additional information gained may not justify the significant cost of the testing. **Figures 59.4** and **59.5** show typical results for CPT before and after treatment.

59.5. Design principles for permeation grouting

59.5.1 Methods and key issues

Permeation grouting by definition requires that grout is injected into the interstitial pores of the ground to replace air or water and, hence, bind the individual particles together.

A number of techniques are available, which are described in Section 59.5.2.

59.5.1.1 Grout selection

With permeation grouting the design of the grout is important, as it must penetrate the interstitial pores with minimum disturbance. If the grout viscosity is too high or the grout particulate size is too large then complete uniform permeation will not be possible and this will cause problems with both treatment quality and resultant movement.

The design of the grout to be used depends both on the ground conditions and the technique being used. Ultimately, when using permeation grouting only limited types of grout can be used for lancing and end-of-casing grouting due to the relatively crude nature of the delivery system, as discussed below. Tube-à-manchette is a more sophisticated delivery system and can use a wide variety of grouts and the selection of grout for this system can greatly influence the quality of the end product.

Figure 59.6 shows the range of penetrability for the various grouts; the most commonly used grouts are either cement bentonite or silicate-based grouts. As an example, Thames gravels are commonly treated with either cement bentonite or silicate grouts or a combination of the two.

59.5.1.2 Cement bentonite

When used as an installation and injection grout, a typical mix would have a water/cement ratio of 2 with 10% bentonite to cement. The unconfined compressive strength is in the range 2–4 N/mm^2 and the bleed <5%.

Sometimes the cement is replaced by a cement/PFA preblend where lower strength is required. An example might be where piling occurs after the grouting and there are concerns about penetrating grouted ground.

59.5.1.3 Silicate

In the UK, for grouting the silicate is usually a proprietary brand, and the brands have different specific gravities. The silicate solution is mixed with a hardener based on organic esters, which react and cause the mix to gel. The gelling time depends on the type (specific gravity) of the silicate being used and the percentage of hardener added. The setting time is around 20–45 minutes and depends, broadly, on the hardener percentage, but is also temperature-sensitive.

It is important to know the specific gravity when detailing the mix as differing dilutions will require differing amounts of hardener. A typical mix will contain 35–45% silicate solution, 5–7% hardener and 50–60% water.

59.5.1.4 Microfine cement

Microfine cement is available in the UK in a number of grades. The fineness of cement is given by the Blaine value, which is the surface area of particles per unit weight, usually expressed

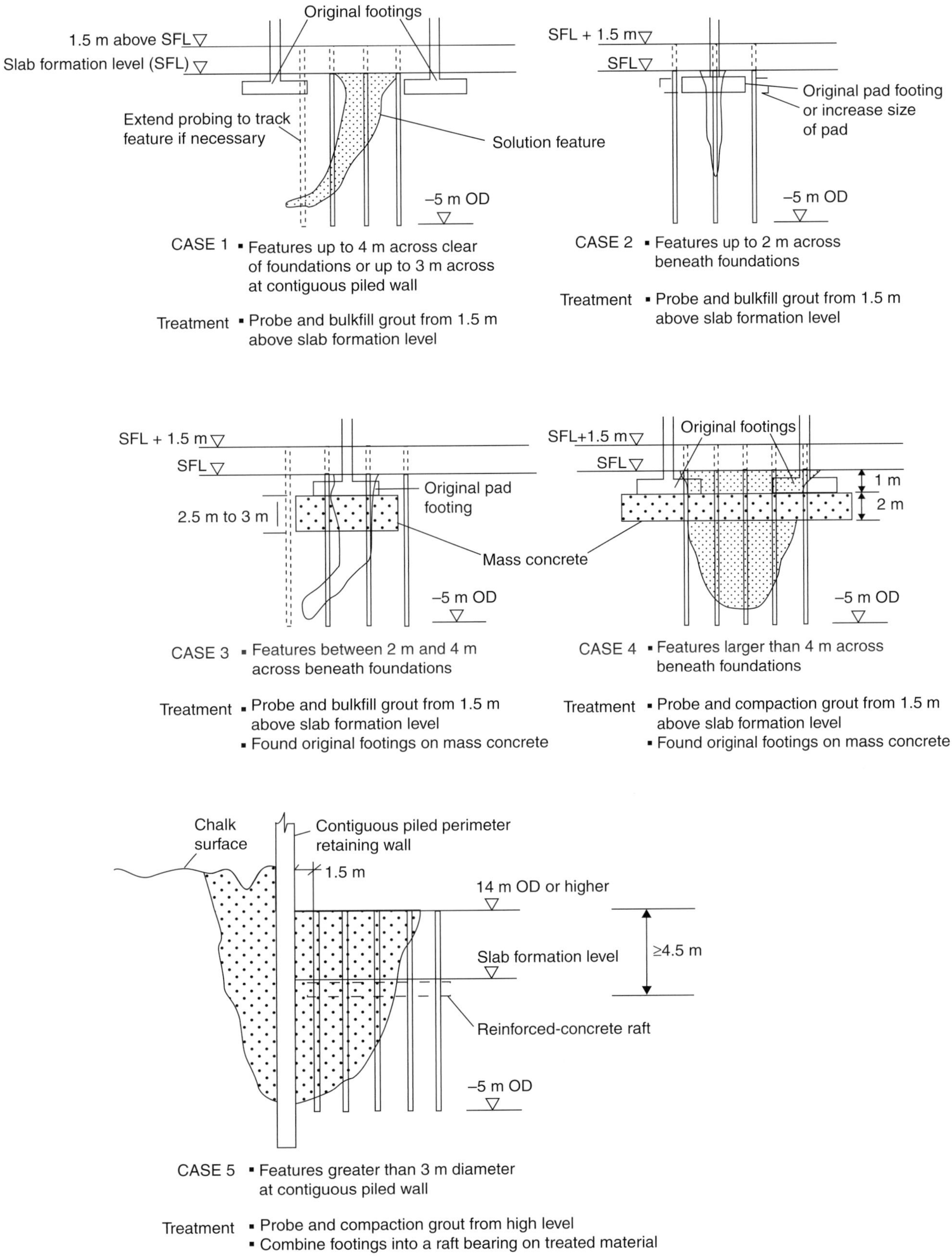

Figure 59.3 Cross-sections of solution features (Reproduced from Francescon and Twine, 1992)

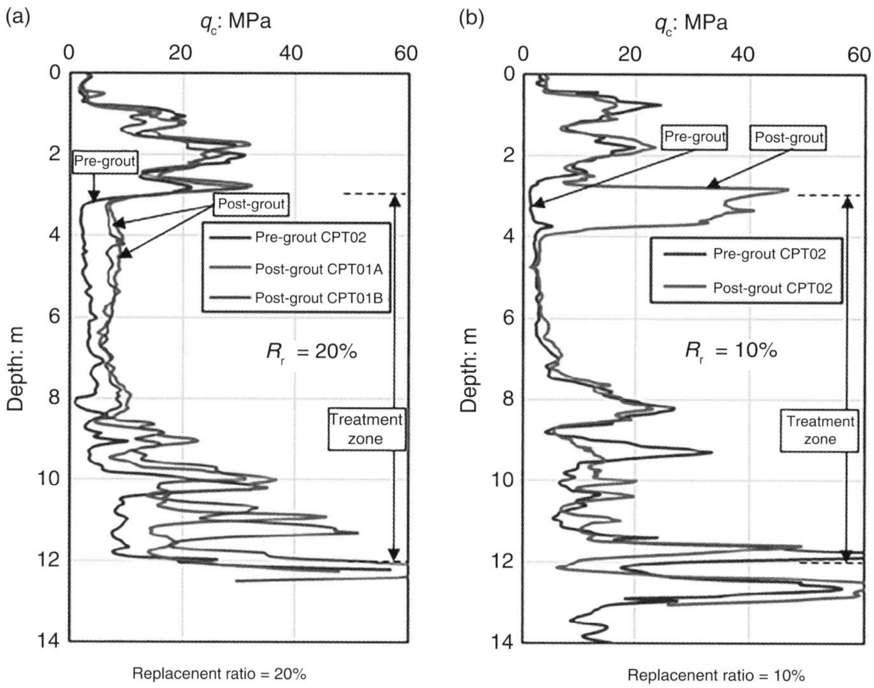

Figure 59.4 Example of pre- and post-testing for compaction grouting using CPT from two replacement ratios (i.e. 10% and 20%) (Reproduced from Yi and Johnson, 2021)

as m²/kg. The larger the value the finer the particles. As a rough guide the Blaine values for some cement-based materials are

- Ordinary Portland cement: 290–390 m^2/kg
- Rheocem 650: 625 m^2/kg
- Rheocem 900: 900 m^2/kg.

Microfine cement is normally utilised at relatively high water/cement ratios (1–3) and must be used with a plasticiser and bleed inhibitor to reduce the risk of blockage and to reduce bleed. The finer the material the faster the setting time; with Rheocem 900, for example, penetration will be impaired after 1 hour from mixing.

Microfine cements are somewhat peculiar in that cubes taken for strength testing will bleed severely and may not set properly, but if mixed with sand will set quickly and produce a high strength. This can sometimes lead to problems on site if the engineer is not familiar with this behaviour.

59.5.1.5 Colloidal silica

This is a relatively recent grout and is a 'two-component' injection system, based on a nanometric colloidal silica suspension. The working time can be adjusted with an accelerator. The product may be used between + 5°C and + 40°C and contains neither solvents nor toxic components.

The grout will penetrate quite fine materials (i.e. silty sands) such as the Thanet, Harwich or Upnor Formation and has been used for the Channel Tunnel Rail Link for this purpose.

59.5.2 Design principles

59.5.2.1 Lancing

Lancing is one of the simplest techniques but can be effective if carried out in the right circumstances. It is commonly used for temporary groundwater control for excavations below the water table but with very little head. **Figure 59.7** shows how an angled hole can be used to insert grout beneath a footing and improve bearing capacity. Lanced underpins are possible but risky as quality can be variable and, hence, can lead to settlement or even collapse. The most common use is to proof grout beneath slabs when settlement has taken place, loadings are increased or voids are present under slabs. Lancing is the process of driving a hollow tube or lance, at close spacings (usually less than 1 m), and injecting grout through the end as the lance is withdrawn. The lances are typically 25–38 mm dia. The tubes have a rivet blocking the drive end. A lance is driven in using a hand-held pneumatic drill, and on reaching the desired depth it is withdrawn slightly so that the rivet can be knocked out by inserting a strand into the lance. Normally the depth is limited to 1–2 m, but on occasions this has been increased to 3 or even 4 m in the right ground conditions. The grout injected is either a cement bentonite or silicate grout.

Lancing is, therefore, restricted to the most open types of ground such as clean gravels and some forms of made ground. It can also be used to prove ground for voids, where it is very cost-effective.

Currently hand lancing is not favoured and should not be considered at the design stage, as prolonged use of the

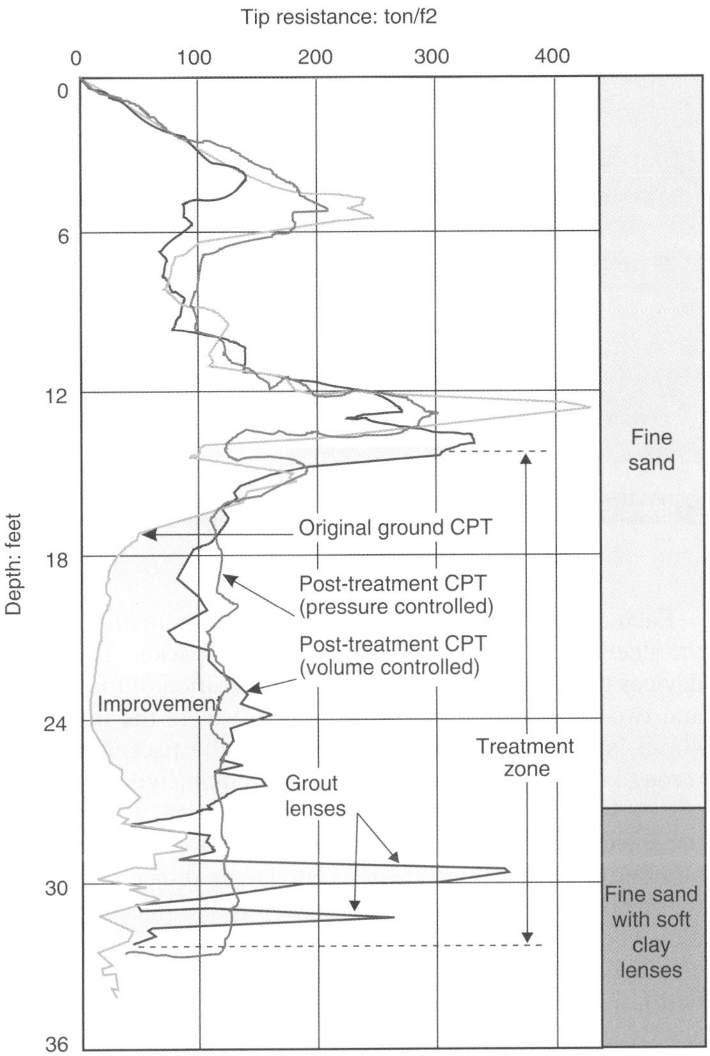

Figure 59.5 Example of pre- and post-testing for compaction grouting using CPT (Reproduced courtesy of Hayward Baker Inc © 2004)

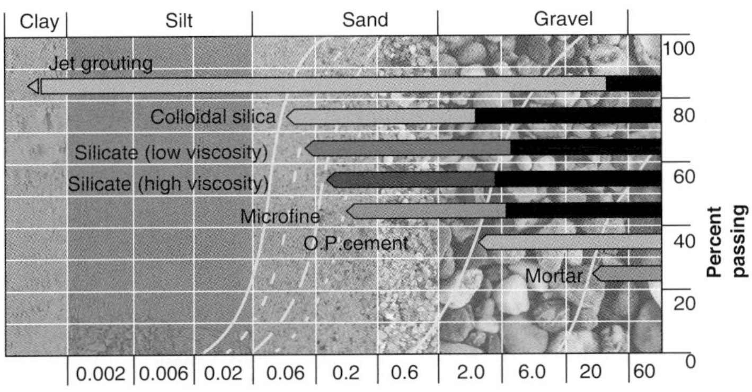

Figure 59.6 Range of penetrability for various grouts (Based on a figure from Keller brochure 67-03 E: Soilcrete® (jet grouting process); Keller Group plc)

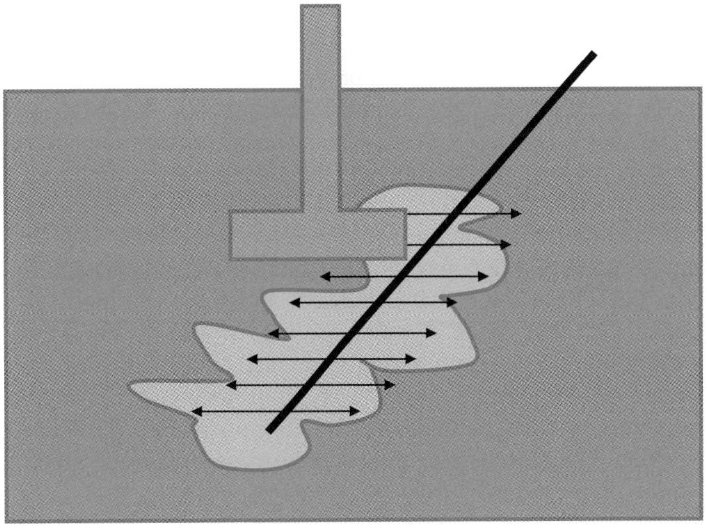

Figure 59.7 Example of hand lancing to improve the bearing capacity of a soil

pneumatic tool leads to a condition called 'white finger' where feeling is lost in the fingers, which then leads to compensation claims. Health and safety issues mean that this work is very limited and the wearing of anti-vibration gloves is recommended. This makes the work more costly and its limited penetrability means that it is used less and less.

Because of the limited penetration, the hole spacing should be designed accordingly. Typically, the hole centres are 0.5–0.75 m apart, except when carrying out under-slab proof grouting, when the spacing can be increased to 1 or 1.2 m. As a guide, expect to inject only 10–15% of the ground by volume unless the ground is very open or voided, when injected volumes could rise to 25%.

59.5.2.2 End-of-casing grouting

Typically, end-of-casing (EOC) grouting is used as a first-stage grouting in open or voided ground in advance of a second-stage tube-à-manchette (TAM) or compaction grouting scheme where use of the TAM or compaction grouting to fill voids and open ground would be slower and more costly. This type of grouting is used for probing and filling voids or open ground at depth. For example, it can be used for grouting around sheet piles when they have been jetted in and there is concern about toe restraint, canal lock and gate structures that are subjected to voided ground, and as a precursor to more sophisticated grouting techniques.

EOC grouting is a technique quite commonly used in open ground. As the name implies, the system comprises the drilling in or driving of a 75–114 mm casing to depth, the withdrawal of the drill rods or removal of the drive point and the injection of grout. It can be used to quite considerable depth but is usually limited to about 10 m. It is limited by breakout of grout around the casing. It is now commonly used at shallow depths due to the health and safety issues of hand lancing, as described above.

It is limited to relatively open ground as the injection pressures are relatively low. Cement bentonite or silicate-based grouts are commonly used. It has the advantage over hand lancing in that the casing can be advanced through most ground conditions. All rig sizes can be used as the casing can be handled by the smallest of rigs. Generally, EOC grouting can be used at wider spacings than hand lancing as the confining pressure is somewhat increased, and some sealing of the casing can take place. Typical spacings would be 1.0–1.5 m, depending on soil type. Grout types are either cement-based (common) or silicate (rare). The grout mix can vary from cement/bentonite in open ground to cement/PFA in voided ground. The selection of mix is determined from the size of void and, therefore, the ease of injection.

59.5.2.3 Tube-à-manchette grouting

TAM grouting, or sleeve port pipe grouting as it is known in the USA, is one of the more sophisticated permeation grouting techniques. It is probably the most commonly used technique, along with jet grouting, for most ground treatment applications. In its conventional form it is used for permeation grouting, but can also be used for compensation grouting where it is used effectively as a displacement grouting technique. Because the pipes are sealed into the ground, the problems associated with EOC grouting are overcome and both the quality and range of application are increased. It can be used for groundwater control, underpinning, excavation retention, slope stability and bearing capacity enhancement.

It is widely used throughout the world and is a technique common to most countries and, hence, can be expected to be available from local specialists. The technique is sensitive to operator control and quality can be variable, so supervision and checking of implementation is very important. Independent checks and audits are recommended. It is a grouting technique often adopted where there is a requirement to undertake repeated phases of grouting from the same bore and where ground conditions dictate that a stable bore cannot be maintained (i.e. when EOC grouting is ineffective). Common applications that utilise TAM grouting include the construction of grout curtains beneath dam foundations (CIRIA C774, Hughes *et al.*, 2018) and compensation grouting in response to tunnelling where repeated phases of grouting are required.

Figure 59.8 shows the principle of operation of a TAM. The system comprises a 40–60 mm dia. steel or plastic pipe with openings at predetermined spacings, usually two or three per metre length, through which grout can be injected. The openings are covered with flexible sleeves which open when pressurised by the grout and seal again when the pressure is removed. The injection or TAM pipe is installed in 90–150 mm dia. boreholes and sealed in with a cement/bentonite grout with typical unconfined compressive strength (UCS) in the range 3–5 N/mm^2.

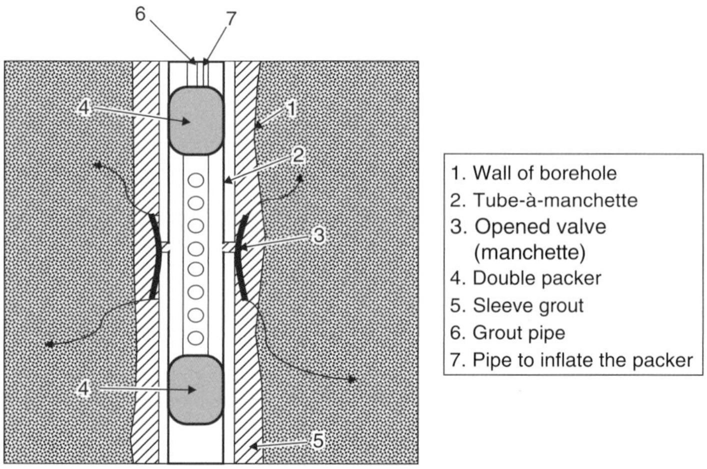

Figure 59.8 Operation of tube-à-manchette

Each injection position can be isolated from the rest of the sleeves in a pipe by the use of a double packer. These are devices that are lowered to straddle the position of the sleeve, and two flexible sleeves are inflated to isolate the injection. Grout is introduced down the centre of the packer and out through the sleeve. Sometimes, for small diameters, mechanical packers with leather washers can be used. In this way the injection volume, pressure and location of injection can be controlled to a high degree. For larger projects some of the major specialists use computer-controlled injection modules where a high degree of control and verification can be obtained. The experience of the site supervisor and engineer will have a large influence on the quality of work carried out and the reliance on such modules should be limited to an ability to translate the designer's scheme into practice.

TAM grouting has been used for many decades. **Figure 59.9** shows the layout of TAMs from a project in the 1960s. **Figures 59.10** and **59.11** show typical layouts for tunnelling and dam construction.

With the sealing in of the injection pipes there is effectively no path for the grout except into the ground. In certain cases this will inevitably lead to either injection of grout into ground that it cannot permeate or reinjection into ground where there is no available pore space. Either of these situations will eventually lead to ground or structural heave or large pressures being imposed on adjacent infrastructure (if ground movements are restrained). These potential risks should be taken into account in both design and specification. An assessment of services or structures at risk should be carried out and an appropriate monitoring regime implemented. Heave of 10–20 mm can occur when injecting silicate grouts into gravels without careful control and can be higher with cement-based grouts; similarly, pressures in excess of 100 kPa can develop local to grouting if it is poorly controlled and ground movements are restrained (e.g. by retaining walls or similar).

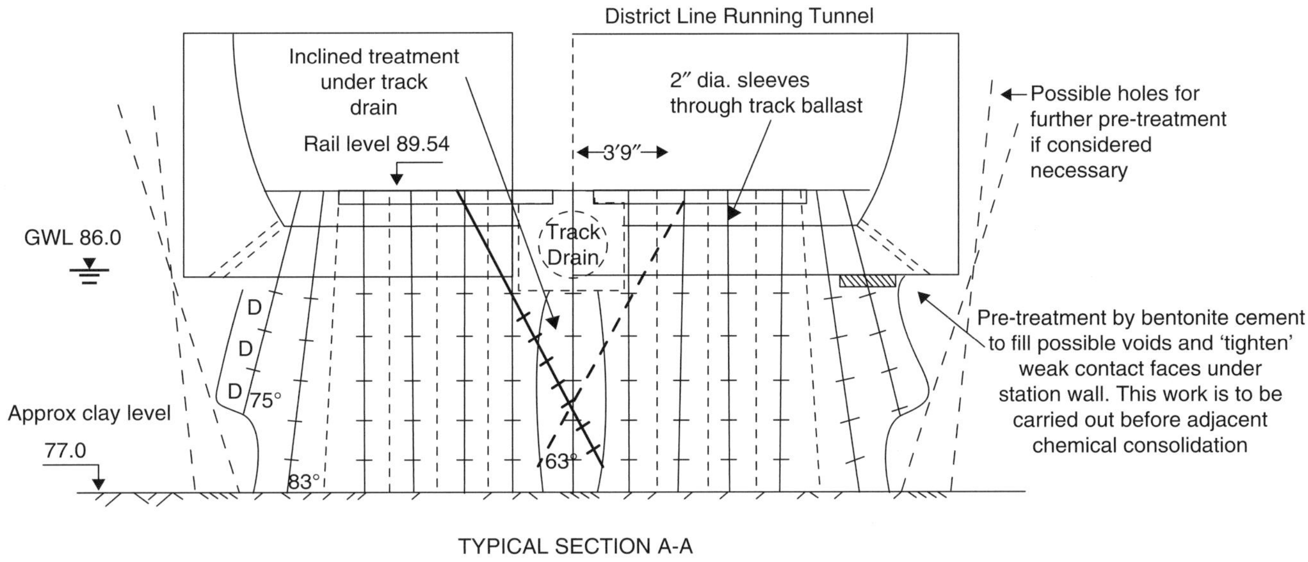

Figure 59.9 Example of historical layout of injection pipes (Courtesy of London Underground Ltd)

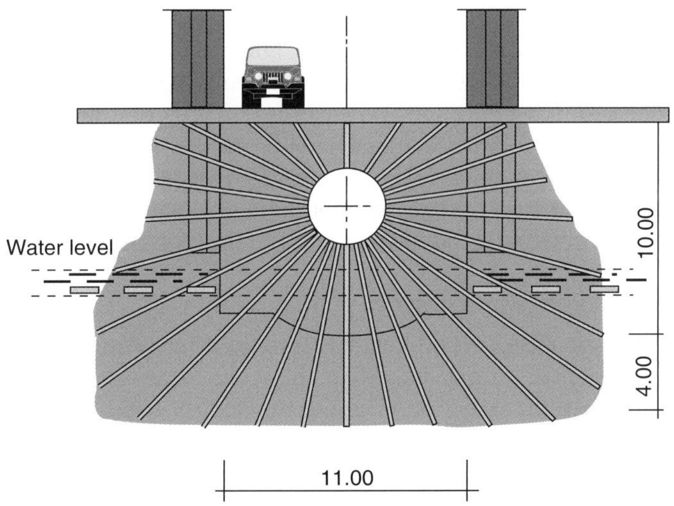

Figure 59.10 Example of injection layout for construction of tunnel using pilot tunnel

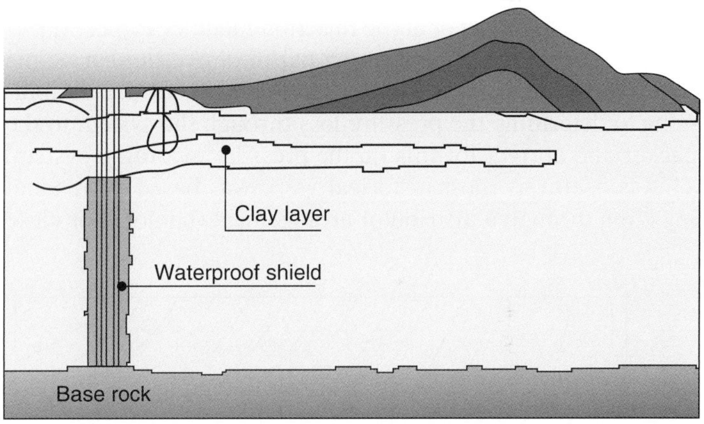

Figure 59.11 Example of use of permeation grouting to provide cut-off for a dam

It is possible to eliminate most heave and excessive pressure by careful sequencing and volume control. This heave can be of use in compensation grouting, which is discussed below.

During design, careful attention should be paid to the location of any service or underground structure, as the grout can end up in services, especially old sewers where the lining is not competent. It is essential to identify such services prior to grouting and consider the effects of the grouting on the service. In particular, for sensitive services (i.e. gas or chemical pipelines) additional monitoring and control will be required. For grouting adjacent to tunnels, due consideration to the effects of the injection pressure need to be considered especially for segmental construction.

The sequence of injection needs to be carefully designed to avoid the risk of trapped groundwater. If the sequence is such that as the grouting programme proceeds there is no flow path for groundwater (being displaced by the grout), then it can emerge into structures or above ground, or if unable to be displaced it can prevent ground treatment taking place as the grout cannot penetrate the ground. An escape path (such as pressure relief standpipes) needs to be allowed for when considering the injection sequence and the scheme monitored for excess groundwater pressure.

The hole spacing is generally in the range 1.2–1.7 m, usually with a triangular grid. Wider spacing is rarely used as the ability to permeate such distances becomes more problematic. The effect of hole deviation is discussed below for jet grouting, where it is considered in detail, but is also applicable to permeation grouting.

The design must consider hole spacings, grout type and target grout injection volumes.

For most applications the holes are installed on a triangular grid and the target injection is derived from

$$Q = \frac{D^2 \sin 60^\circ n}{N} 1000 \qquad (59.1)$$

where Q is the target injection volume per sleeve (litres), D is the hole grid spacing (metres), n is the ground porosity (%) and N is the number of sleeves per metre.

Typically, for a 1.5 m grid in ground similar to Thames gravel, the target injection volume will be about 150 litres. **Figure 59.12** shows the required injection volumes per metre for varying ground porosity and hole spacing.

The rate of injection is also important; injection rates of 8–12 l/min are typically used, leading to injection times of 10–20 minutes for each sleeve. Silicate grout should be injected at the lower end of these rates as penetration can be more difficult in finer ground.

Injection pressures are usually based on pressures measured at the top of the holes or at the injection module. Typical injection pressures are 0.25 bar to 0.5 bar/m of overburden. Some of the more sophisticated modules can carry out line calibrations to determine the pressure loss through the system to the packer and correct for this on the pressure recording system. This is useful as most historical work was based on manual injection through a distributor arrangement (banjo) at or close to the borehole mouth. Hence, the line loss between the pump and borehole was effectively eliminated and the 0.25–0.5 bar injection pressures work on this basis. Overcorrecting could lead to excessive pressures in the ground, which could encourage hydrofracture rather than permeation.

The use of silicate grouts will lead to lower ground strengths and lower deformation moduli than when cement-based grouts are utilised. Typically, the strength of treated ground will have a UCS of around 1–3 MPa.

An example of the use of permeation grouting was on the Channel Tunnel Rail Link (CTRL), where the new railway lines passed under the existing London–Southend railway lines. **Figure 59.13** shows the section at the crossing. The resulting cofferdam was predicted to cause too much movement to the adjacent viaduct, and it was decided to install a grouted plug to reduce movements. Because of the presence of the alluvium, the plug did not have to provide waterproofing and, hence, it was decided that permeation grouting would be acceptable. **Figure 59.14** shows the TAM layout and also the distribution of grout injected. Both silicate- and cement-based grout were utilised and **Figure 59.15** shows the results from an in situ permeability test obtained after TAM grouting was applied in sandy soil below a tunnel invert in Australia (Campbell *et al.*, 2023).

TAM grouting can also be undertaken in combination with other groundwater control measures, such as dewatering (to reduce water volumes to manageable volumes) or cut-off walls. It has also been undertaken as a remedial measure, such as to seal joints between adjacent diaphragm wall panels to mitigate against excessive groundwater leakage through the joints.

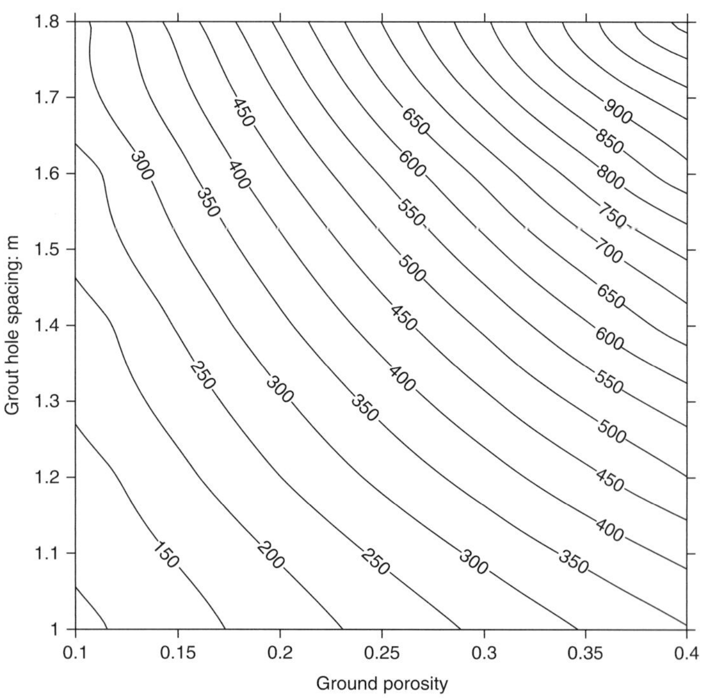

Figure 59.12 Required grout volumes per metre for differing ground porosities and hole spacing

59.5.3 Execution controls

Most execution control relates to initially ensuring that the injection holes are installed in the right locations and then monitoring the injection of grout. Secondary controls can be evaluating movements caused by the grouting.

For the simpler systems such as lancing or EOC, execution control is limited and generally will be paper-based simple records of hole numbers and volumes of grout injected.

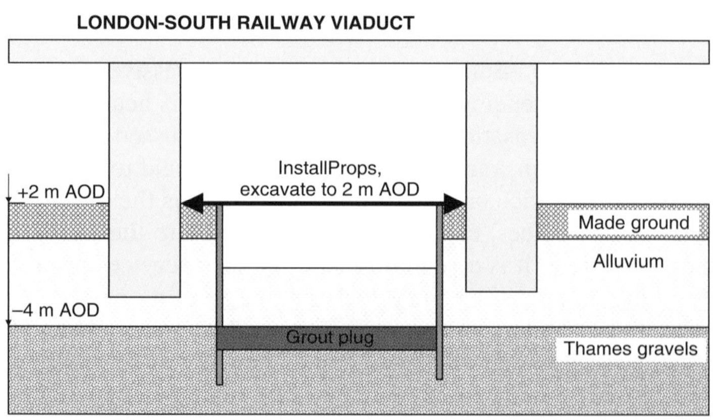

Figure 59.13 Section at CTRL railway crossing

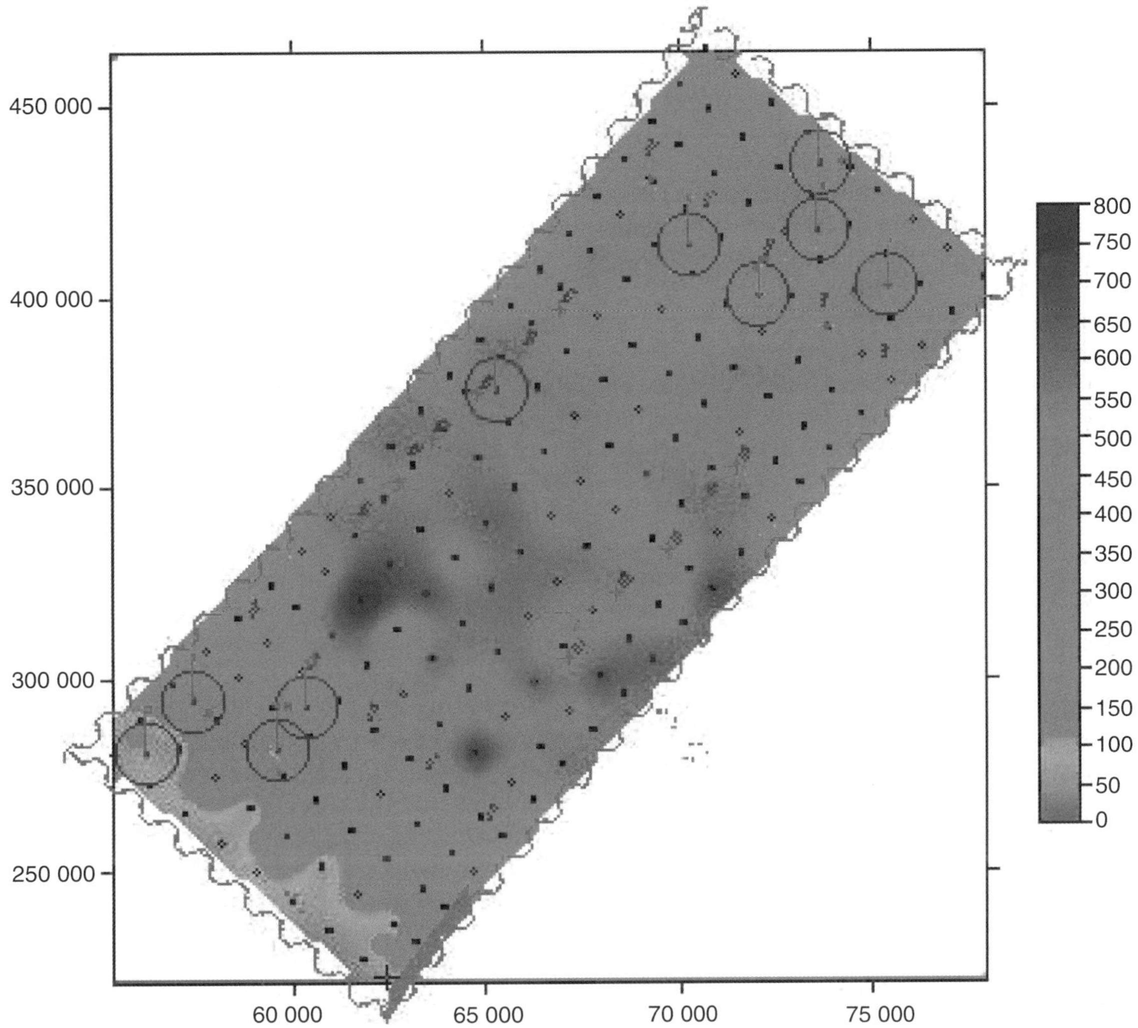

Figure 59.14 Example of validation of permeation grouting by reviewing injection data

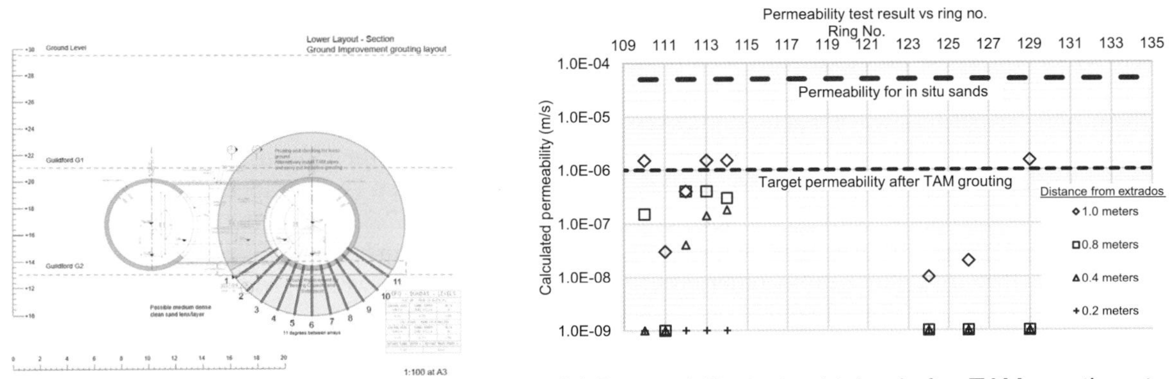

(a) Typical tunnel invert TAM installation pattern.

(b) Permeability tests obtained after TAM grouting at several longitudinal locations (ring numbers) and distance to the tunnel extrados (see legend)

Figure 59.15 In situ permeability tests obtained after TAM grouting was applied in sandy soil below a tunnel invert (Reproduced from Campbell *et al.*, 2023)

Additional refinements can be to record volumes injected for each stage of injection.

TAM grouting tends to push the limits of capability of ground treatment and, hence, needs careful control to avoid mistakes on site. In particular, the large number of injections requires a high degree of site control to ensure each sleeve is injected correctly. With the advent of site computers this is now more easily and efficiently accomplished, and their use should be regarded as a specified requirement for most projects.

59.5.4 Validation

Validation for hand lancing is not generally carried out except by the drilling and grouting of holes at intermediate positions. Where lancing is used for proof grouting under a slab or void filling, then additional intermediate holes are useful to assess the degree of coverage.

Validation for EOC grouting is generally restricted to either falling or rising head tests within the casing installed for testing purposes, or through piezometers installed following ground treatment. A further test could be the drilling and grouting of intermediate holes, the results being judged on the grout take. If strength or deformation are an issue, then either coring or pressuremeter testing could be carried out, but test cell formation can be a problem and core recovery poor due to the fact that the material consists of relatively strong aggregate grouted with a relatively low-strength grout.

Validation for TAM grouting is essentially as described for EOC grouting above. Additionally, it is possible to re-inject through sleeves to assess grout take and refusal pressure. When assessing the success of a TAM grouting project it is common practice to review the primary and secondary grout injection volumes. Normally one would expect to see a reduction in grout take from primary to secondary grouting coupled with some increase in refusal pressures. **Figure 59.14** shows such a validation following completion of grouting in terms of grout volumes; **Figure 59.15** shows a validation in terms of in situ testing. In situ permeability tests can be used to measure the hydraulic conductivity of the TAM grouting. This testing is critically important when TAM grouting is used to reduce the permeability of the in situ soil, such as when it is used to cover gaps in ERSS walls (such as utility gaps) or when it is used to reduce groundwater ingress into deep excavations. Permeability testing is not straightforward and requires careful consideration by appropriately experienced specialists; in general before-and-after testing is required and the target permeability (after grouting) needs to be judged to be practically achievable (e.g. it is unrealistic to assume the grouted zone is 'impermeable'!).

59.6. Design principles for rock grouting

59.6.1 Methods and key issues

Rock grouting is commonly employed to reduce the mass permeability of the ground where the control of groundwater or the contact/consolidation of a jointed rock mass is required. Rock grouting is typically associated with foundation treatment beneath dams, cut-off walls to dam foundations, shafts or in unlined adits or tunnels in rock where groundwater control is required. Rock grouting may also be implemented as a remedial measure during the operational life of an asset where groundwater penetration through tunnel linings or leakage through a dam foundation cut-off wall requires repair.

Best cut-off practice currently is to install two rows of opposing inclined drillholes at angles of around 15° to the vertical. This provides more opportunity for the grout holes to intersect vertically orientated discontinuities. This is standard practice for installing grout curtains around shafts.

Rock grouting is essentially the injection of grout to penetrate and infill open joints and discontinuities within a rock mass, displacing air and water. The injection of the grout results in a reduction in the mass permeability of the rock. Typically, grouting will continue until a fixed volume of grout has been injected, or the grout injection pressures reach a prescribed value above which hydrofracturing of the ground or uplift could occur or capped relative to the overburden pressure.

Typically, for rock grouting a neat water–cement grout is used, with a water–cement ratio by weight varying in the order of <0.67 through to 2. Higher water–cement ratios are adopted for microfine and ultrafine cement. The grout is designed to be pumpable and also have a high viscosity to allow for penetration into the jointed rock mass. A cement grout can penetrate discontinuities greater than 0.5 mm wide (Houlsby, 1990). In particular circumstances, additives can be introduced into the grout mix, including sand where open joints are to be sealed, or microfine and ultrafine cements where penetration through finer discontinuities is required. Additives may also be added to delay setting time, improve grout penetration and, in the case of bentonite, reduce bleed of cement-based grouts.

Rock grouting is usually implemented by adopting one of three approaches: *descending stage grouting*, advancing a bore to a defined length and grouting in stages as the bore is advanced with a packer installed at the top of the stage. After each stage the bore is redrilled and advanced to the base of the next stage. *Ascending stage grouting*, drilling the grout bore to the full depth and then installing a single packer and grouting beneath the packer. Using a specially adapted TAM as detailed in Section 59.5.2.3, where both ascending and descending stage grouting can be undertaken using a double packer to isolate the zone into which grout is to be injected. If required, multi-stage grouting can also be undertaken with TAMs.

A decision to undertake ascending or descending stage grouting should be based upon an understanding of the rock mass and orientation of discontinuities. Descending stage grouting is suited to rock with significant discontinuities, in situations where inclined bores are required and where there is the potential for bore instability. In ascending stage grouting, packers can be vulnerable to being grouted into the ground where an extensive network of discontinuities exist and high grout pressures are used. Grout can flow into the ground and can then enter the bore above the packer, leading to the packer being grouted into the hole. Ascending stage grouting is not suitable for ground where a stable bore cannot be formed.

In advance of undertaking grouting, a grout trial should be undertaken. The trial could encompass two elements: laboratory grout mix design trials to confirm the initial mix design for grouting and subsequent grout mixtures; and a field trial to confirm the requirement for grout hole spacing, grouting stage lengths and the suitability of grout mixes for actual ground conditions. The laboratory trials should be used to establish the index and rheological properties (e.g. density, viscosity, yield point, cohesion, bleed filtration and set time) of the grout mix for use to establish quality control criteria during the production works. Field trials should encompass the drilling and injection of a number of grout positions to confirm the suitability of the grouting methodology (e.g. ascending or descending stage, grout section length), grout mix design and operation of the grouting plant and equipment. On completion of the grouting trial operations, permeability testing should be undertaken to demonstrate the suitability of the method being employed and to allow for optimisation for the permanent works construction.

59.6.2 Design principles

The design of a successful rock grouting scheme is reliant on an understanding of the geological (ground and groundwater) conditions present at the site. Site investigations should be undertaken to quantify the rock, its discontinuities and rock mass permeability. Particular focus of the ground investigations include: joint spacing, width, infill material and orientation along with a spatial understanding of ground variability in the grouting zone. In situ permeability testing should be specified to provide a reference against which the reduction in ground permeability after grouting can be compared. Inclined boreholes may be required to understand discontinuity orientations.

The design of a grouting scheme for a dam foundation cut-off wall requires consideration of the depth and length of ground treatment, the number of grout curtain rows, the hole spacing and the grouting sequencing. Development of the design typically requires a hydrogeological model and seepage analysis, as the cut-off wall requirements can be significantly influenced by the site-specific geological conditions. For initial assessments (only), the United States Bureau of Reclamation proposed the following empirical relationship for assessment of the depth of grout curtain for dam foundations

$$D = 1/3H + C \tag{59.2}$$

where D is the depth of the grout hole from the proximal end of the grout curtain, H is the height of the dam above the proximal end of the grout bore and C is a constant varying as a function of the ground conditions. Guidance on the value of C is given by Thomas (1976), Grundy (1955) and Houlsby (1982).

While the length of the grout curtain is determined by ground conditions, on the abutment side slopes it is often linked to the reservoir impounding level, with the grout curtain extended up the abutment slopes to avoid groundwater flow through the rock mass, bypassing the grout curtain and resulting in leakage, washout or uplift along the dam abutment side slopes.

Consideration should be given to the number of grout curtains, whether a single curtain or multiple curtain rows. With a single row an acceptable reduction in ground permeability might be achieved; however, two or more rows is usually more robust, more cost-effective and provides a more durable design against unknown anomalies in the grout curtain which may only become apparent during the operation of a dam when leakage is observed. A multiple-row curtain is likely to be more robust in intercepting flow paths through a rock mass, which are often of irregular geometry. A wider multiple-row grout curtain tends to be more cost-effective than a single row with a lower permeability requirement.

Houlsby (1982) proposed a separation distance between grout curtains of no less than approximately twice the anticipated grout penetration to allow for overlap of grout penetration between successive rows within the grout curtain.

Deere and Lombardi (1985) proposed that the radius of penetration of grout be estimated as

$$R_{max} = P_{max} \times (T/C) \tag{59.3}$$

where R_{max} is the maximum radius of treatment penetration (m), P_{max} is the maximum injection pressure (MPa), T is the crack thickness/2 (m) and C is the grout cohesion (MPa).

The arrangement of the grout holes within a grout curtain should be established as a function of the spacing of discontinuities within the rock mass, their orientation and their persistence. Grout curtains may be inclined both parallel and perpendicular to the axis of a dam in order to intersect the maximum number of discontinuities. Grouting would typically be undertaken in a defined sequence of primary holes, followed by secondary and tertiary holes. The spacing of primary holes would be in the order of 6–12 m, with the corresponding secondaries at 3 and 6 m and tertiaries at 1.5 and 3 m, respectively. The requirement and depth of secondary and tertiary bores is a function of the grout takes from the primary bores; secondary and tertiary bores may be shallower than the primary holes. Grout takes in secondary holes should be lower than those from the primary holes, with a reduction ratio in the order of 0.25–0.75 between successive phases of grouting.

59.6.3 Execution controls

Initial grout injection pressures in the order of 0.25 bar/m below ground should be considered for grouting, taking account of hole inclination and the condition of the rock. In poorer-quality rock lower grouting pressures are likely to be appropriate, whereas in blocky rock higher pressures may be necessary. Section lengths for grouting would typically be no greater than 5 m, reducing to shorter lengths in poorer-quality ground or where bore stability issues are encountered.

Grouting intensity number (GIN) grouting, developed by Lombardi and Deere in the 1970s and 1980s (Lombardi and Deere, 1993), is a particular execution approach applicable to the grouting of fissured rock masses with a reduced likelihood of hydrofracture and heave using a single, stable and low water–cement ratio grout. The method is based upon grouting to a 'GIN' number that relates the grout injection pressure to grout injection volume. A precis of the GIN grouting methodology is presented in CIRIA C774 (Hughes *et al.*, 2018). For all stages of grouting and for all boreholes within a given grouting stage, grouting to a constant GIN number is undertaken. Grout injection is undertaken to a maximum pressure, maximum volume and at a uniform injection rate until the point at which the pressure/volume GIN curve intersects the GIN envelope boundary curve. Injection of grout is largely automated by the use of a computerised control system with a single grout mix.

Irrespective of the execution method employed, the grouting operations should be connected to instrumentation that continuously records all aspects of the grouting operations (e.g. grout injection pressures and volumes) for each grouting stage. Likewise, records should be maintained of packer installation depths and inflation pressures. Routine quality assurance and control testing of the grout mix should be undertaken to ensure consistency of the grout mix being injected.

59.6.4 Validation

The validation of rock grouting should typically be undertaken in two stages: during the implementation of the grouting works and following completion of the grouting works. During implementation, records of the grouting works should be maintained (e.g. grout hole injection layout, the grout mix design and injection pressures and volumes) along with index testing on the grout mix and strength testing on grout cubes/cylinders. The construction records thereafter form part of a progressive validation that the works have been implemented in compliance with the execution specification.

On completion of the grouting works, further validation should be undertaken principally by measuring the permeability of the grouted rock mass and undertaking a comparison with pre-grouted values. Water testing may be undertaken using rising or falling head tests, although more commonly where long-term groundwater pressures are liable to increase (e.g. associated with a dam cut-off wall), packer testing with Lugeon permeability measurement should be undertaken. Lugeon testing comprises the injection of water into the ground at a constant pressure and measurement of the volume of water injected. In the case of dam foundation treatment works where upstream and downstream grout curtains are installed, testing would be undertaken between the two grout curtains. Evidence of grouting quality can also be obtained from coring through the grouted rock mass and using downhole optical or acoustic televiewers to record the in situ ground conditions.

59.7. Design principles for jet grouting

59.7.1 Methods and key issues

Jet grouting differs from permeation grouting in that the intent with the latter is not to disturb the fabric of the ground, whereas with the former this has to be the case for successful application. Because the product of jet grouting tends to be more structural in nature, it has a great range of applications. When installed beneath buildings it can extend foundations through poor ground or support them while excavations, often unsupported, are carried out next to them. It can provide support underground where openings are necessary, such as for tunnel boring machine (TBM) breakout or break-in, cross passages or equally to support tunnel faces. It can provide groundwater control at the base of excavations and can prop retaining walls at the same time. It is a multi-application tool for support and groundwater control, and simultaneously increases the efficiency of site usage. Nevertheless, as with other geotechnical works, it requires experienced, site-specific design and diligent fieldwork supervision/checking. The three basic systems are

- single system
- double system
- triple system.

These are illustrated in **Figure 59.16**.

The single system is the injection of only grout at high pressure. This was the first system to be used. The column diameter

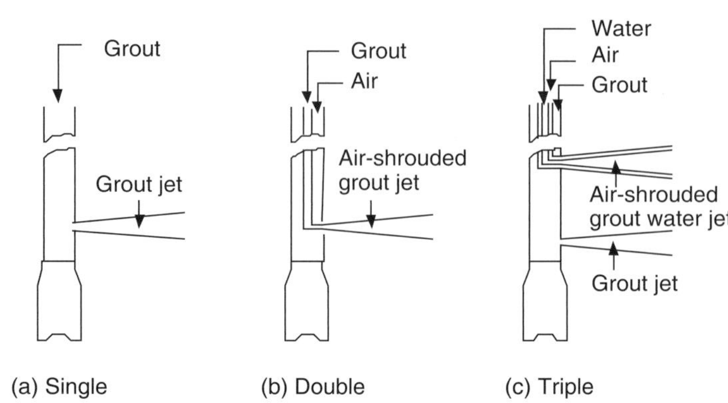

Figure 59.16 Jet grouting systems (Reproduced from Essler and Yoshida, 2005; Taylor & Francis Group)

is limited, and the borehole has a tendency to become blocked, resulting in ground heave. The design column diameters are small, usually up to 1.2 m in diameter. Some designers take the view that the single system provides the highest-strength columns and that air bubbles trapped in the columns from the other systems reduce strength.

The double system is effectively the single system with the addition of an air shroud to the nozzle. The addition of the air shroud increases jetting efficiency dramatically, typically giving a 30% increase in design diameter for the same jetting energy. These are very versatile and powerful systems with a design column diameter up to 4–5 m in most soils, although such large diameters are only practical on specific projects with no site constraints. Typical column diameters in normal use would range from 1 to 3 m.

The triple system differs from the single and double systems in that erosion of the ground is carried out by a high-pressure water jet shrouded with air with an additional low-pressure grout line. The design column diameters achieved with this system are greater than those of the single system, but as the energy is relatively low compared to the double systems, columns are usually only up to 1.7–2 m dia. (in exceptional ground). In the USA an advanced super triple system can deliver much higher energies and as a result design diameters of 3–4 m are possible.

The range of design applications for the systems are similar and selection is a site-specific requirement. The single system is most commonly used for horizontal applications, where air may cause problems. For most other applications the choice normally rests between the use of the double or triple systems. Generally, the triple system will produce more spoil but is usually less viscous and, hence, offers less risk for blockage and associated structural or ground movement. Moreover, when organic soils are encountered, the use of the triple system can be beneficial because the organic material can be flushed out during jetting with water, therefore improving the strength of the final admixture.

Jet grouting, historically, has been considered a risky ground treatment process due to the high pressures involved and the spoil generation. With good experience, site operatives, site control and design, jet grouting is no riskier than other processes. The main risks associated with the process are the use of high pressures and the potential generation of excessive movement. When jet grouting beneath buildings, care must be taken to ensure that heave of the walls or floors is maintained within agreed limits. It is absolutely critical that the construction details of the floors and walls are understood so that the effect of jet grouting can be considered. Where floor construction is poor and unsealed then it is advisable to maintain visual inspection and continual monitoring during jet grouting. Wall movement during jet grouting can be controlled in the long term by precise levelling and in the short term by the use of robotic survey systems and rotating laser targets fixed to selected building walls.

Design of the grouting sequence is extremely important for underpinning operations as there is the possibility of loss of support if too large an area is jetted. Where the structural foundation condition is poor (i.e. with poorly cemented masonry or brick) then a pre-grouting exercise using lancing or EOC can be considered with the intent to reinforce the available bonding and reduce the risk of partial wall collapse into a freshly completed column. For underpinning works, it is usual to restrict the jetting of adjacent columns to intervals of at least 24 or 48 hours. For isolated foundations, the jetting sequence must be designed to prevent loss of support; in extreme cases, temporary load transfer should be provided during the jet grouting operation. There is usually excellent bonding between the foundation and the jet grout columns as subsequent adjacent columns always treat this zone and ensure good structural contact. Grouting close to an underground structure needs careful consideration and any grouting carried out within 3–5 m of any underground (new or existing) structure will require a specific risk assessment. Pressure relief systems should generally be considered, as noted above for permeation grouting. When executing jet grouting in clays, the hole diameter should be large enough to allow the spoil to be ejected, typically a minimum 250–300 mm dia. hole is required, depending on rod size in use. Wherever possible, consultation with the original designer is advisable to agree the specifics of the ground treatment.

The most common application for jet grouting is either underpinning or to restrict movements associated with excavations. **Figure 59.17** shows the use of jet grouting to underpin the Reading Room of the British Museum (Scott and Essler, 2001, 2002). This is a Grade I listed building and the jet grouting provides permanent support. **Figure 59.18** shows the effect of the installation of a deep jet grout strut within an excavation to reduce retaining wall movements (Essler, personal communication from North–South Project Team, 2006). **Figure 59.19** presents an instrumented excavation where jet grouted struts below excavation level were installed in deep deposits of very soft marine clays in Singapore (Pittaro and Murphy, 2018). The figure shows that the 8 m thick jet grouted strut reduced wall movements to <0.2% of excavation depth. This type of application is commonly used when deep deposits of soft soils are encountered – it has had little use to date in the UK.

Jet grouting was successfully applied in temporary works, as exemplified with the ground improvement works for the Victoria Station upgrade (Rutty *et al.*, 2021). Jet grouting was used to stabilise water-bearing River Terrace Deposits, creating a stabilised zone of ground with a jet grout annulus some 2 m thick through which sprayed concrete-lined tunnels were subsequently excavated. In all, some 2500 jet grout columns were constructed using a double jet system forming columns through made ground, alluvium, River Terrace Deposits and London Clay, with ~1.6 m dia. columns and soilcrete strengths in the order of 4–6 MPa.

The strength of jet grout soilcretes depends on several factors, which can be classified into two categories: intrinsic soil

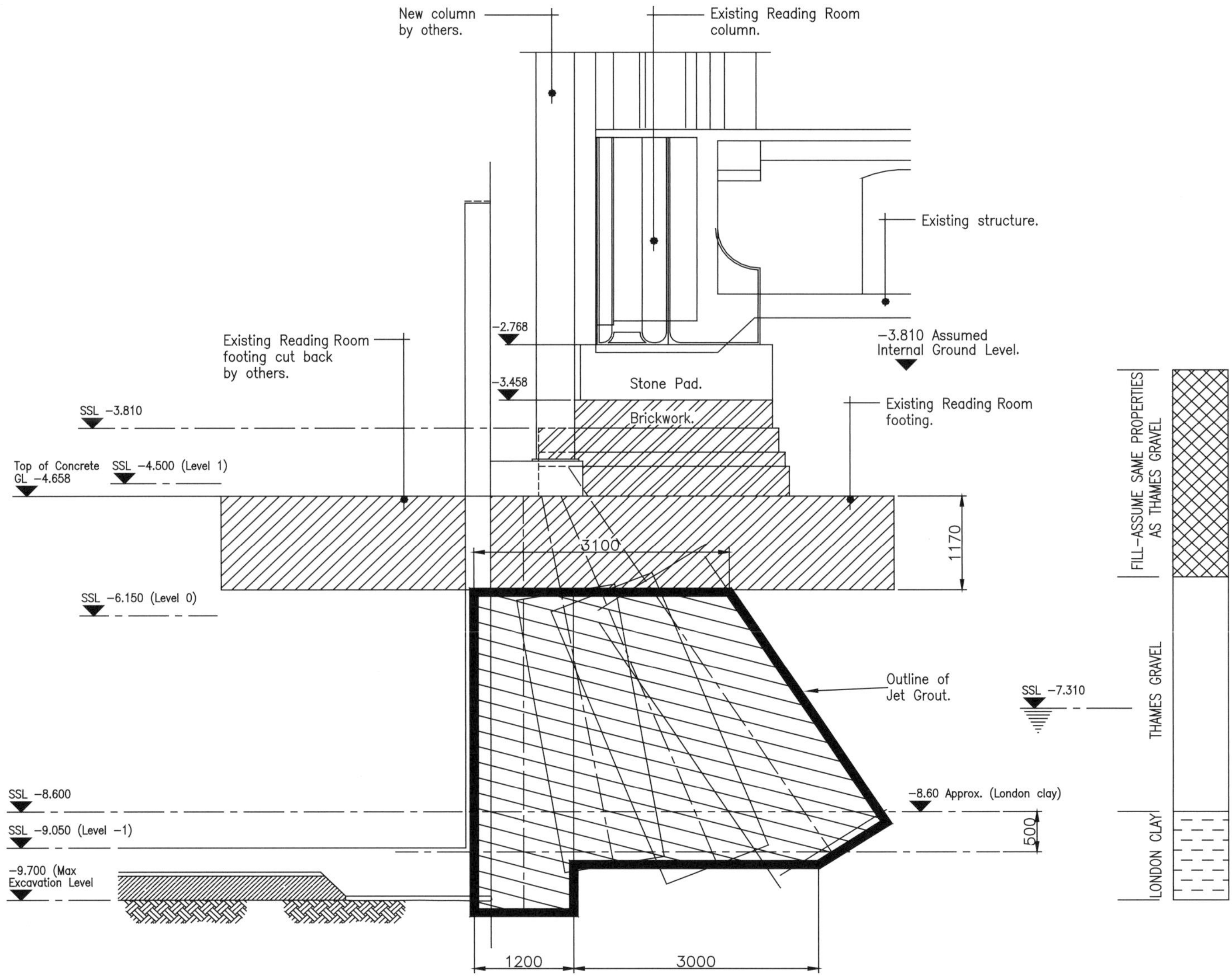

Figure 59.17 Cross-section of British Museum jet grouted underpin (Reproduced with permission from CIRIA SP199, Scott and Essler, 2003; www.ciria.org)

factors (i.e. soil properties) and external factors (binder, mixing, curing and loading conditions), summarised in **Table 59.5**. **Table 59.5** is actually prepared for deep soil mixing material; however, the factors affecting the strength for both admixtures are similar. Although the treated ground strength is expected to be much higher compared to that of the in situ soil, the strength variability is also expected to be considerably higher compared to that of the in situ soil. Generally, the scatter of values is measured through the coefficient of variability (COV), which is the ratio between the standard deviation and the mean of sample data.

The measured COV for jet grouting and deep soil mixing usually ranges from 0.20 to 0.99, with most of the past data falling between 0.30 and 0.60 (Pittaro, 2022). The higher the COV for jet grouting (and other types of deep cement mixing), the lower the values for the characteristic strength (to be used for design after appropriate partial factors are applied). UCS values are typically quoted for a 28-day strength; however, there are usually significant longer-term strength gains (e.g. Filz *et al.*, 2012). Both jet grout and deep soil-mix soilcretes share conceptually similar strength and stiffness characteristics, and Bruce *et al.*, (2013) and Kitazume and Terashi (2012) discuss soilcrete stress–strain behaviour in greater detail, including aspects such as tensile and residual strength (although these issues are not directly assessed for routine design purposes).

59.7.2 Design principles

Because of the ability of jet grouting to bond ground and create relatively strong bodies, known as ‘soilcrete’ (e.g. UCS range

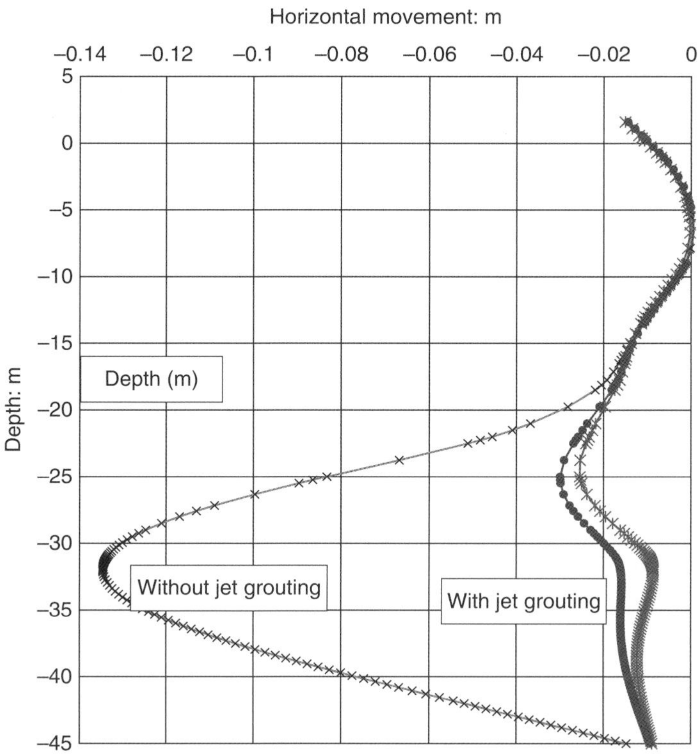

Figure 59.18 Effect of installation of jet grouted strut at 33 m

Factor type	Category	Factors
Intrinsic	Soil properties	• Physical, chemical and soil mineralogy • Organic content • pH of pore water • Water content
External	Characteristics of binder	• Type of binder(s) • Quality • Mixing water and additives
	Mixing conditions	• Amount of binder • Mixing efficiency • Timing of mixing/remixing
	Curing conditions	• Temperature • Curing time • Humidity • Wetting and drying, freezing and thawing, etc.
	Loading conditions	• Loading rate • Confining pressure • Stress path (e.g. compression, tension and simple shear)

Table 59.5 Factors affecting strength of treated ground (Reproduced from Bruce *et al*., 2013)

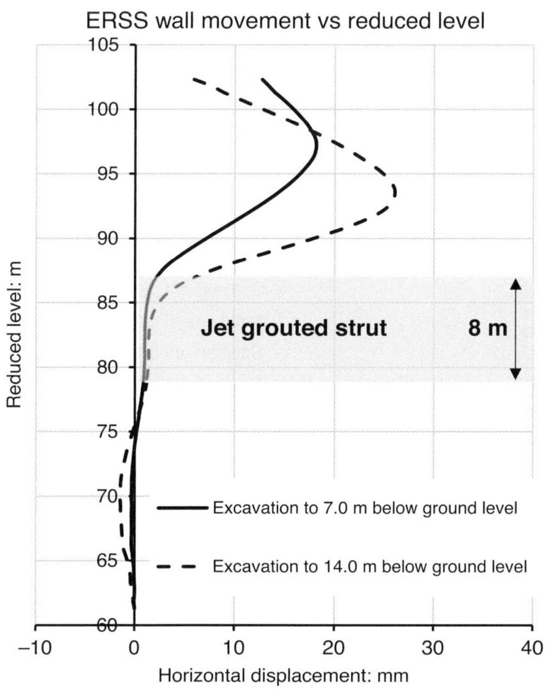

(a) Measured wall deflections for two excavation stages (see legend).

(b) Photo of the actual excavation after reaching 7 m below ground level.

Figure 59.19 Excavation in Singapore where jet grouted strut was effectively used below the final excavation level by Pittaro and Murphy (2018)

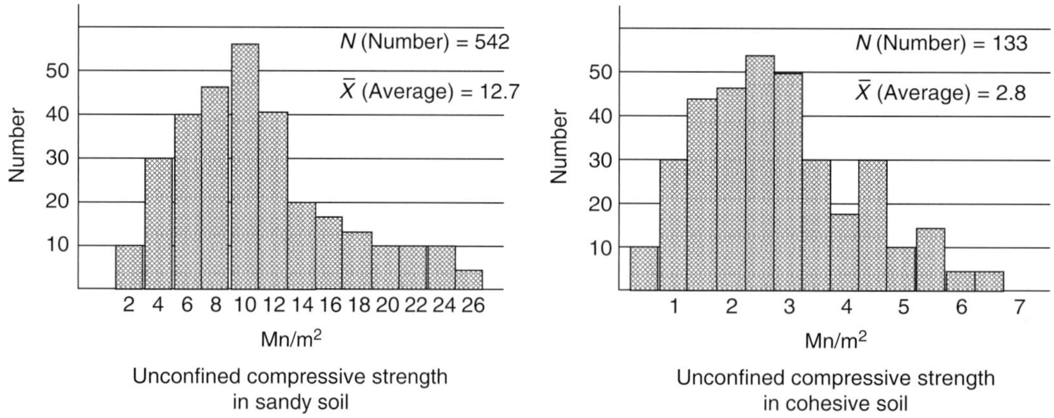

Figure 59.20 Typical strength distribution for jet grouted soils for sandy and cohesive soils (Reproduced from Shibazaki, 2003)

Soil type	Q_u: unconfined compressive strength: MN/m²	C: Cohesive strength: MN/m²	f: Bond strength: MN/m²	σ_t: Bending tensile strength: MN/m²
Fine-grained	1	0.3	0.1	0.2
Coarse-grained	3	0.5	0.17	0.33

Table 59.6 Typical design values for jet grouted materials

2–10 MPa can be achieved in many soil types), the design process is similar to the design of brick or masonry. The strength of jet grouted ground is usually assessed on the basis of UCS tests on samples obtained by wet grab sampling and/or coring. To define the appropriate strength for a specific project it is relevant to investigate past cases where jet grouting was installed in similar ground conditions and with similar installation parameters. For example, the histograms shown in **Figure 59.20** indicate the range of UCS measured in coarse- and fine-grained soils, and **Table 59.6** provides an indication of strength parameters which are useful for conceptual assessment purposes. The Japan Jet Grouting Association has adopted these simplistic distribution charts (i.e. **Figure 59.20**), defining the design UCS as a minimum safe value which should be adopted as within the lower 1–3% range of these charts. It is noted that **Table 59.6** presents values of tensile strength; however, to date this parameter is usually not relied upon in design – for more information about this property refer to Kitazume and Terashi (2012), EuroSoilStab (2002) and Pittaro (2016). **Table 59.5** outlines the influence of various factors on UCS and **Figure 59.21** indicates the trends to be expected between soil type, cement content, water–cement ratio and UCS (based on data from He, 2018; Lee *et al.*, 2005; Pittaro, 2022; Wanik *et al.*, 2017).

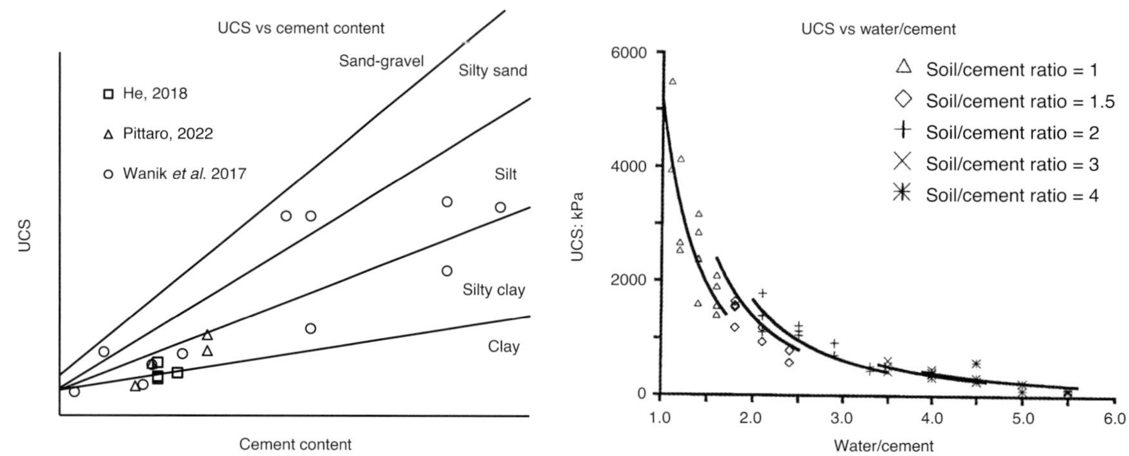

(a) UCS vs Cement content trend for different in situ soil conditions. Plot reproduced from Wanik et al. (2017) with additional information by He (2018)

(b) UCS vs Water/ Cement ratio obtained for laboratory prepared samples by Lee (2005).

Figure 59.21 UCS trend for cement content and water–cement ratio

Selecting a minimum safe value for design strength was standard practice historically; however, more rational, cost-effective and robust processes for selecting the design strength of jet grout zones are now available (note: deep soil-mix strength selection adopts similar procedures), which are consistent with the requirements of limit state codes. Typically these processes involve a consideration of the statistical variation in strength, then selecting a suitable fractile (or percentage value for which measured UCS is higher than the chosen fractile value) as a characteristic strength. For preliminary design, relevant published data should be used, and during design development and construction this strength data should be supplemented by site-specific data (through laboratory and/or field trials) and from quality control testing during full-scale implementation.

For ULS checks a characteristic UCS strength ranging from 10% to 20% fractile was proposed from experiences in Europe (Topolnicki, 2016). A 10% fractile of the UCS is recommended in Japan (Kitazume and Terashi, 2012), whereas in the USA a 20% fractile is often used (Filz *et al.*, 2012).

In the second generation of Eurocode (currently under preparation), Equation 59.4 is proposed to obtain the characteristic strength of the ground improvement

$$X_k = e^{Y_{Mean}(1 \pm k_n V_Y)} \tag{59.4}$$

where X_k is the characteristic value of the ground property X; Y_{Mean} is the mean from a number (n) of sample-derived log values; k_n is a coefficient that depends on the confidence level and the number of samples (n) used to calculate Y_{Mean}; V_Y is the coefficient of variation of the log values of the ground property X [$V_Y = s_Y/Y_{Mean}$], s_Y being the standard deviation of the sample-derived log values; $\pm$ denotes that $k_n V_Y$ should be subtracted when a lower value of X_k is critical and added when an upper value is critical. Design strength is then derived from this characteristic by applying a partial factor (for SLS checks the partial factor is unity).

Equation 59.4 can be used if a significant number of samples are tested (>10) and the recommended confidence level is 90% (10% fractile).

Typical values of UCS which have been adopted for design in the UK would be in the range 2–5 MN/m² for coarse-grained materials; far lower values would be appropriate for fine-grained materials as indicated in **Figure 59.20**. Strength is usually only an issue for retaining wall/shaft support struts, base slabs (for deep excavations or tunnelling projects) and shaft breakout or break-in. In these cases, strength is important for the design as the jet grouting is required to span across openings and resist significant earth and water pressures, and so has to be designed with adequate structural integrity. When the structural behaviour of the jet grout soilcrete is important, it is good practice to assess the mobilised strain in the jet grout slabs/blocks to ensure mobilised strain is small, relative to failure strain, and tensile strain is not significant.

Recent research has provided insights into the relationships between soilcrete volume, failure mode and risk of failure. This enables a coherent framework for parameter selection to be developed, taking account of the failure mode and the statistical and spatial variability of the ground improvement (Pittaro, 2022). The application of this framework results in higher characteristic strength when large volumes of cement-treated soil are used; in contrast, when a relatively small volume of ground improvement is used, more conservative design strength values should be selected. This approach highlights the greater robustness of large jet grouted zones and that far greater care is needed if, for example, reliance is to be placed on thin jet grouted struts. This proposed framework is described in detail by Pittaro (2022). **Figure 59.22,** based on research in Singapore Marine clay, shows the ratio between the characteristic to mean strength and the normalised volume of soilcrete (measured by the ratio of the thickness of the ground improvement (l) divided by its scale of fluctuation (δ_{xy})). The plot shows that as the volume of ground improvement increases, higher characteristic strength (and therefore fractile values) can be selected. This plot was proposed for ground improvement use for excavation support, but its principle is also relevant for other applications, although care is required until further experience is gained in this type of approach.

The soilcrete stiffness (usually defined as a secant Young's modulus) is another important design parameter, especially for ground–structure interaction analysis (to assess movements and the distribution of stresses through an underground structure),

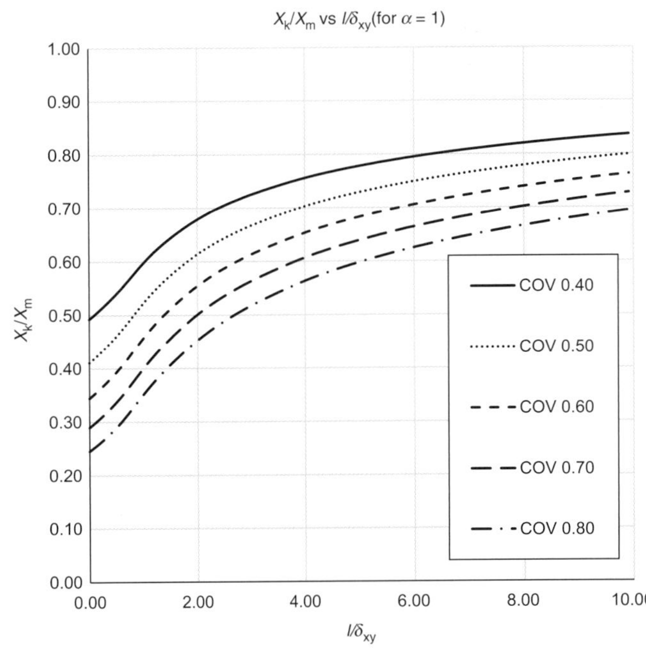

Figure 59.22 Plot to obtain the characteristic strength (X_k) from mean strength (X_m), thickness of the ground improvement (l), scale of fluctuation (δ_{xy}) and variability (COV) (Reproduced from Pittaro, 2022)

noting that it is not necessarily conservative to assume a lower-bound soilcrete stiffness (which can lead to underestimation of stress concentrations through a system). Usually, this parameter is obtained by multiplying the characteristic strength by a constant factor typically ranging between 150 and 300 (Bruce *et al.*, 2013; Pittaro 2022). Although such factors are commonly applied, other authors have proposed power law equations when plotting E_{50}/UCS as presented in **Figure 59.23(b)**. Van der Stoel (2001) suggested stiffness relationships as power law equations of UCS, which varied for fine- and coarse-grained soils. The data presented in **Figure 59.23** were obtained from jet grouting and deep soil mixing installed in soft clay, and highlight the large scatter in Young's moduli values, reflecting the limitations associated with routine sampling and lab testing of these materials. For important applications, field measurements of stiffness, using pressuremeter or geophysics, may need to be considered (e.g. Ng *et al.*, 2019). Back-analysis of full-scale retained excavations indicates that the selected design Young's modulus should be close to the mean or median of measured values, for SLS checks, rather than some lower fractile value (Pittaro and Murphy, 2018).

At detailed design the arrangement of the jet grout column layout should consider the interrelationships between setting out, the drilling verticality tolerances and likely as-built column diameters. Third-party constraints (e.g. the presence of utilities or buried infrastructure) that can impact on column layouts also need to be integrated into design and addressed in the column layout as far as reasonably practicable prior to starting construction.

As part of the detailed design a statistical assessment of jet grout column interlock should be undertaken, defining the locations and diameters of the grout columns and ensuring that there is continuity of contact (and overlap) between adjacent columns. Where the jet grouting works are intended to provide a cut-off to groundwater flow, gaps between adjacent columns or insufficient thickness of cut-off would adversely compromise the suitability of the jet grouting works, resulting in leakage, with **Figure 59.24** illustrating the flow paths between a single- and double-row jet grout cut-off. Likewise, where jet grouting is used to provide temporary propping in the base of a deep excavation, gaps in the continuity of the columns could compromise the continuity of strutting. Essler (2012, 2023) details the requirements for the design of jet grouting.

Spreadsheets can be used to evaluate the likelihood of an opening existing within a jet grouted block with consideration of the setting out tolerance (e.g. 75 mm); mast inclination tolerance (e.g. 2%); drilling deviation tolerance (e.g. 2%); and column diameter variability (e.g. 20%). Note, tolerances should be discussed and agreed with specialist jet grouting contractors to ensure that the design layout is aligned with what is practical for their construction. By randomly varying these parameters, a statistical analysis can be undertaken to confirm the absence of gaps in the ground improvement and the degree of overlap. If a minimum overlap is required, the spreadsheet could be written to verify

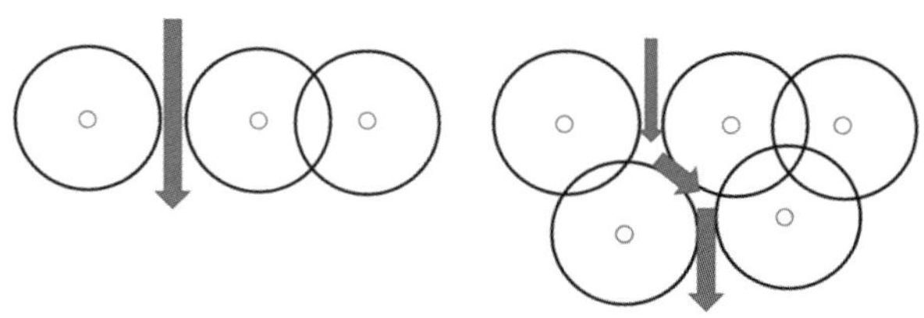

Figure 59.24 Visualisation of the flow paths for a single- and double-row jet grout cut-off block (Reproduced from Essler, 2023; Deep Foundations Institute)

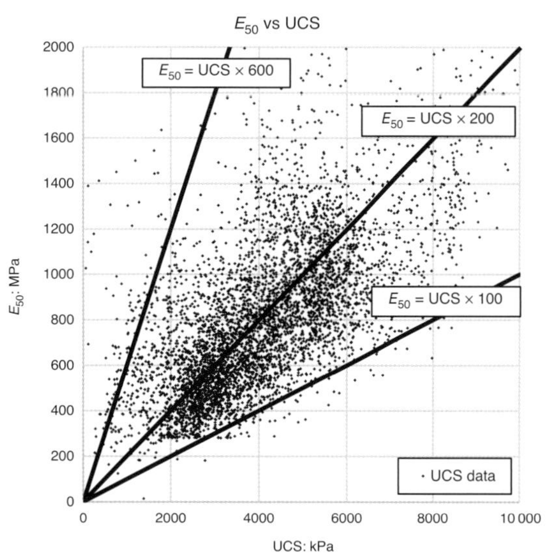

(a) Young's modulus vs UCS. The plot presents three linear relationships between E_{50} and UCS.

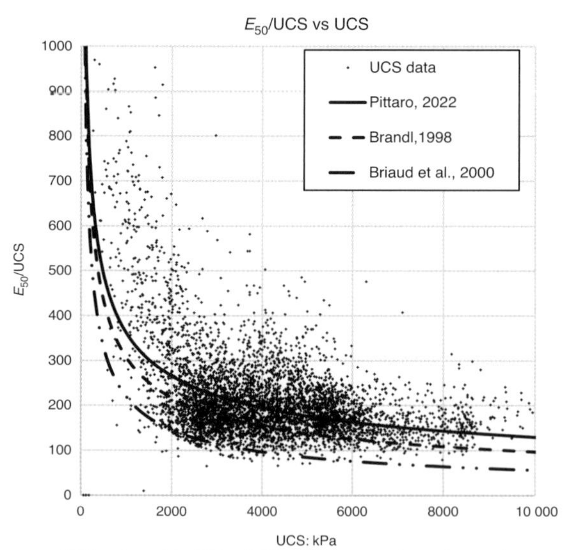

(b) Young's modulus/UCS vs UCS. The plot presents three references for power law equations between E_{50}/UCS and UCS.

Figure 59.23 Relationship between Young's modulus and UCS that the minimum overlap is achieved. An iterative process is then implemented to vary the jet grout column spacing and column diameter to achieve the performance requirements of the design.

For a two-row jet grout cut-off wall, groundwater flow through the grouted block would be possible where there are three or fewer interlocks between adjacent columns. This equates to a ~20% probability of open interlocks between columns. To mitigate the risk of an open interlock, the column spacing would need to be reduced, the column diameter increased or a combination of the two. In the case of a three-row cut-off, groundwater flow through the ground improvement would require five or more gaps to permit groundwater flow. Hence, with an increasing number of rows, the likelihood of leakage statistically decreases. Conversely, with a single row of columns there is significant likelihood of an open overlap existing within the cut-off wall.

A further consideration with respect to the design of jet grout columns is the depth of treatment and influence of construction tolerances. With increasing depth, hole deviation can become significant and avoiding open interlocks requires either larger-diameter columns or closer-spaced columns. This has also led to a drive to increase column diameter as this not only reduces cost but provides more security at depth as the column interlock is less dependent on drillhole deviation. Therefore, it is important to understand that depth plays a significant role in determining cost and potentially the layout of a jet grout system.

As an example, considering the design for a base slab 20 × 20 m at a depth of 10 m below ground level (Case A) and 20 m below ground level (Case B), the difference in number of columns for column diameters of 1.5 m and 3 m is illustrated in **Table 59.7**. Clearly there is a difference in that with 1.5 m dia. columns the number of columns increases by 132 from Case A to B, but for 3 m dia. columns this increase is only 18 columns.

The other factor that influences cost and design is the actual column volume cut compared to the volume required. For the example above, at a depth of 10 m below ground level (Case A) and 20 m below ground level (Case B) the volume of column cut per metre is respectively 1.77 m^3 and 7.07 m^3 for 1.5 m and 3 m dia. columns. **Table 59.8** illustrates the comparison.

Case	Depth: m	Column diameter: m	Nominal design spacing: m	Number of columns	Percentage increase with depth
A	10	1.5	1.2	418	
B	20	1.5	1.2	550	32
A	10	3.0	2.7	72	
B	20	3.0	2.7	90	25

Table 59.7 Effect of depth and diameter on column numbers required

The volume jetted will affect both cost and spoil production as they vary linearly in proportion to the meterage jetted.

Good examples of the correct design of jet grouting are the base struts incorporated into the design of the station boxes (Driesse *et al*., 2008a, 2008b) and the sandwich wall at Central Station (De Wit *et al*., 2006, 2007) on the North–South project in Amsterdam. **Figure 59.18** shows that without these struts the wall deflection would have been excessive, with resultant damage to the adjacent historical buildings. The initial design of the jet grouting is shown in **Figure 59.25** and

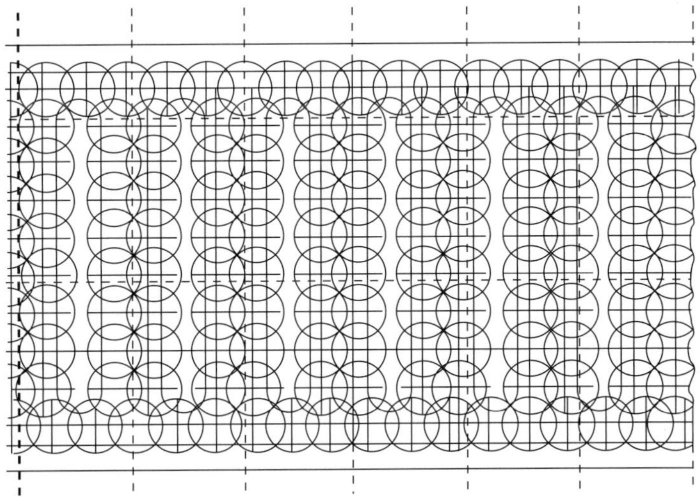

Figure 59.25 Initial column design layout for the North–South Line, Amsterdam

Case	Depth	Column diameter: m	Nominal spacing: m	Number of columns	Column volume jetted per metre: m^3/m	Column volume as % of slab volume: 400 m^3/m
A	10	1.5	1.2	418	739	185
B	20	1.5	1.2	550	972	243
A	10	3.0	2.7	72	509	127
B	20	3.0	2.7	90	636	159

Table 59.8 Effect of column diameter and depth on column volume required

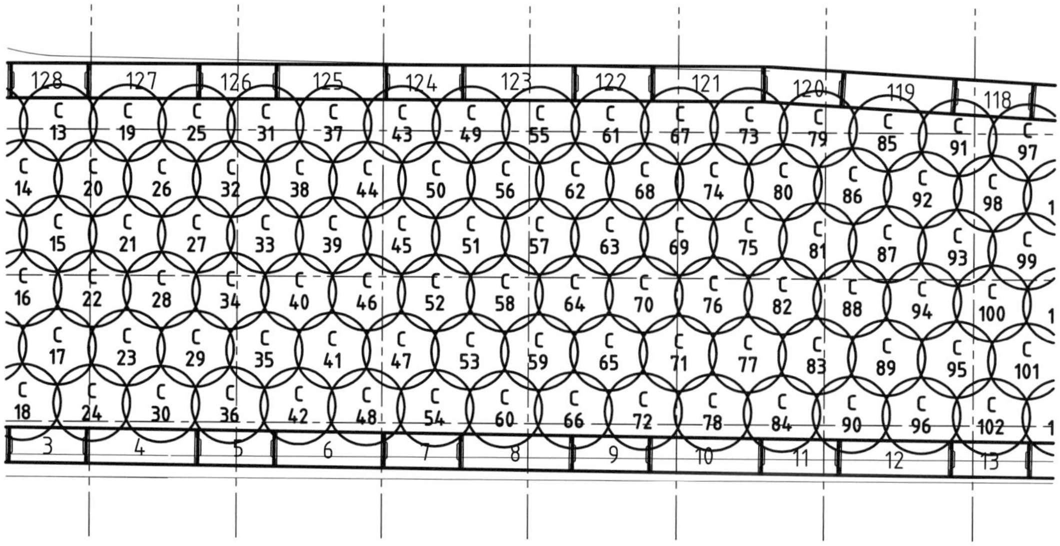

Figure 59.26 Revised column layout for the North–South Line, Amsterdam

essentially mimicked conventional strutting; however, as the strut was some 33 m below ground level, it ignored the difficulties of achieving positional accuracy without which the strut would not transmit the required loads. The resultant redesign assumed a uniform distribution of the columns within the station box with a nominal overlap. Extensive finite element analysis of over 1000 layouts of columns randomly positioned based on assumed diameter and positional accuracy demonstrated that the strut would behave as designed. **Figure 59.26** shows the revised layout.

Essler (2023) presents further statistical analysis of the percentage of area not grouted for a base-plugging jet grout scheme. Illustrated in **Figure 59.27** is the output from the analysis in terms of column diameter and spacing for a set of 1.5 m dia. columns on a triangular spacing of 1.05 m, corresponding to 70% of the column diameter. The figure shows the impact of jet grout column diameter variation of 0% and 20% (D) and column deviation of 0.5%, 1% and 2%, assuming no variation in column spacing. Essler (2023) illustrated that the theoretical maximum spacing assuming no variation in diameter or drilling deviation is 86.7%, and that a spacing equivalent to 80% of column diameter is close to the limit and results in significant generation of gap area.

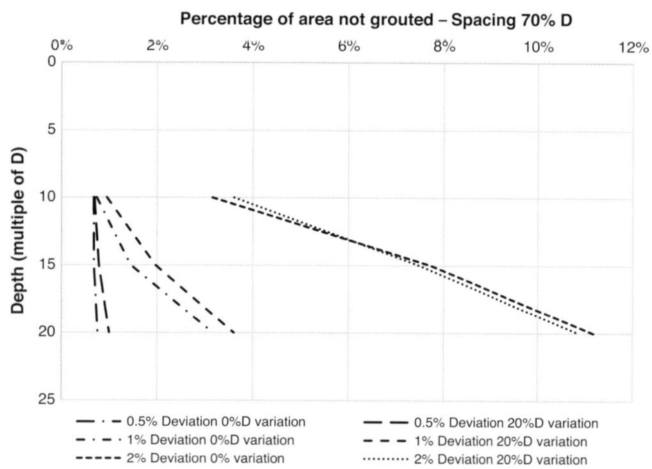

Figure 59.27 Analysis to determine the percentage area not grouted for a column spacing of 70% diameter (Reproduced from Essler, 2023; Deep Foundations Institute)

Designing jet grouting based on a statistical investigation of the column diameter and position is important and is recommended for any design where the depth and soil conditions could have significant effects. It is especially relevant for designs where the jet grouting is required for waterproofing, as column variations – and hence gaps between columns – actually control the water flow.

59.7.3 Execution controls

Execution controls comprise the quality assurance of the grout materials and the grouting process. Generally, most jet grouting rigs are fitted with sophisticated instrumentation that records all the required process parameters. This can give the designer confidence that the columns have been constructed reliably, provided the instrumentation is working correctly. There are some systems that include a built-in borehole deviation system that can be used to actually monitor the as-built locations of individual columns, or conversely the use of borehole deviation measurement systems such as the Shape Accel Array or DeviGyro systems through the drilling rods. As the rigs are instrumented, the drilling energy can be easily plotted against depth and this can give the designer valuable insights into ground variability and can drive process adjustment.

59.7.4 Validation

The validation process consists of quality control of the grout materials and the column jetting parameters followed by post-jet grouting testing. Because jet grouting generally is structurally sounder, coring can be more successful if care is taken, particularly when sonic-coring techniques are used that can be more effective than conventional rotary coring in coarse-grained soilcrete columns. When jet grouting in fine-grained soils, rotary coring can be more challenging due to the weaker soilcrete. Generally, high-quality rotary coring, typically used for sampling very weak rock, is required and site coring trials may be needed to optimise sampling. In addition, the sonic-coring process lends itself to trial exposure whereby a number of initial columns are jetted on site and then exposed after 24–48 hours to determine size and uniformity. With jet grouting, the column diameter tends to be the most important parameter for validation as this is fundamental to the construction of gravity or underpinning blocks. Column diameters can be checked using a number of proprietary methods such as Cyljet, hydrophones or heat of hydration. Strength is not usually an issue in coarse-grained ground.

Laboratory testing of cores can be carried out and the designer needs to be careful in relating potentially solid core results to the mass properties. **Figure 59.28** illustrates results of destructive drilling, which is relatively quick and economical. It is a powerful validation tool during design and can be calibrated against coring and subsequent laboratory testing. Destructive drilling must be instrumented and the drilling energy then plotted against depth. There is a correlation with the core strength, which can be used to identify zones of low strength. This methodology should be considered on all major projects where repetitive testing of large numbers of columns is required.

59.8. Design principles for vibrocompaction and vibroreplacement

59.8.1 Methods and key issues

Kirsch (1993) describes the techniques in detail, but typically vibrocompaction is carried out by inserting a vibrating closed-end steel tube of 300–500 mm dia., commonly known as a

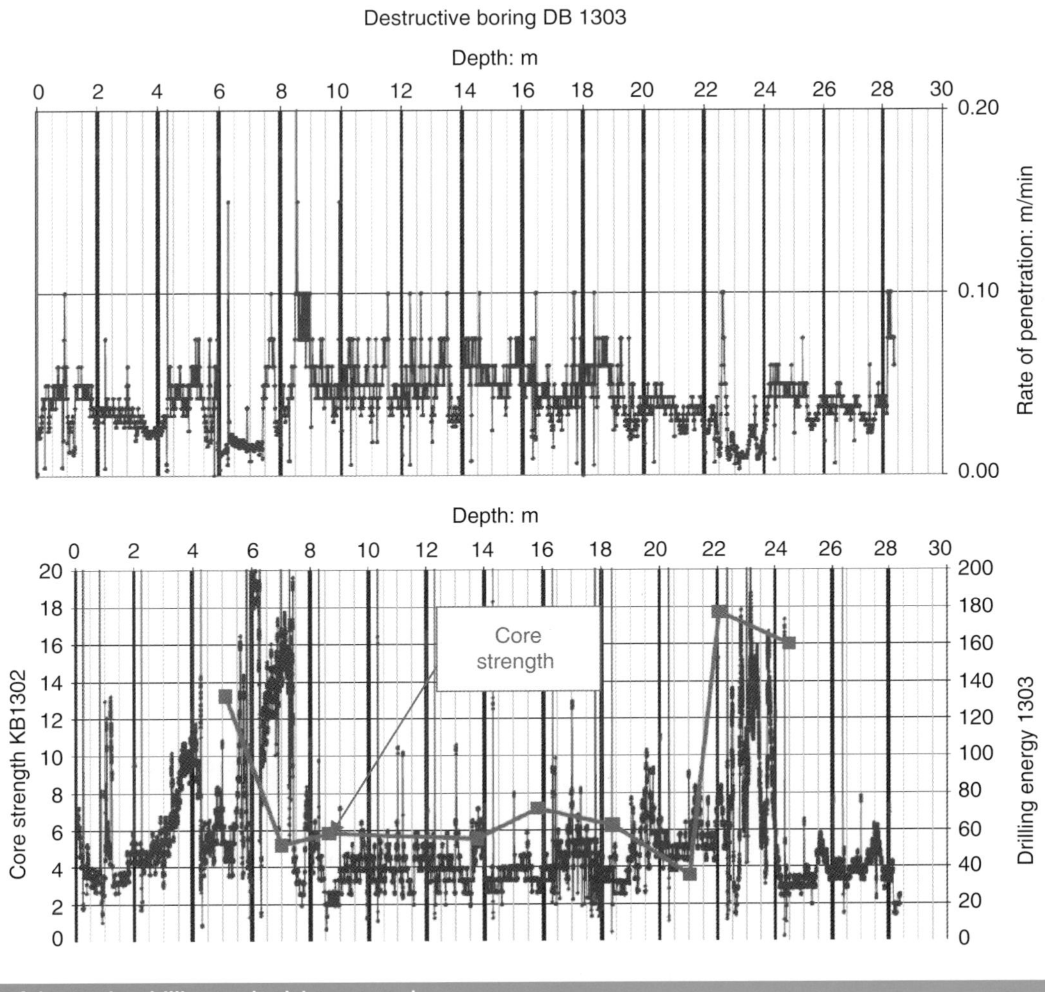

Figure 59.28 Use of destructive drilling to check jet grout columns

poker, into the ground and causing the surrounding ground to be densified by the ground vibrations. The design of individual pokers is commercially sensitive, but in principle they operate by hydraulically rotating an eccentric weight horizontally within the poker. In this way the vibrations have a significant horizontal amplitude and cause ground densification.

The vibrations cause the soil grains to be rearranged into a more dense packing and thus improve the soil density. Water jetting can be used in association with the poker to assist penetration of the poker and flush out soft finer material. It also has the effect of locally reducing the effective stresses and aiding ground restructuring and compaction.

Generally, the poker is lowered to depth and then raised in stages as the ground is compacted. The applied energy can be monitored during compaction as higher energy is needed when the surrounding ground becomes densified. The poker can be systematically raised and lowered in stages, with the lowering phase used to judge the compaction of the underlying ground.

Care needs to be taken with this process to ensure that the ground can be densified solely by compaction. Problems occur if the content of silt and clay is too high. If bands or layers of silt or clay are present, these are unlikely to be compacted and residual settlement problems may arise. CPT testing is an ideal means of assessing site suitability for vibrocompaction, Since problematic zones or layers can be readily identified. The layout of penetration points is usually triangular, with the spacing generally following from the soil type and available poker power. Generally, the more powerful the poker, the wider the spacing. Very powerful pokers have been developed over the years to deal with large areas of marine sands and reclaimed materials where wide spacings can be economically employed.

Vibroreplacement, or stone columns, used for fine-grained soils, is a similar process to vibrocompaction but it relies on creating columns of compacted stone (with adequate lateral confinement by the natural soil), which will transmit load or resist shear. On reaching depth, stone is introduced either at the top of the hole or at the bottom by way of a special tube built into the poker. These methods are known as top feed and bottom feed, respectively. The stone is compacted as the poker is withdrawn and thereby improves the ground. Bottom feed is used where there are weak soils with a high water table and thus the column may be unstable.

Great care is needed when assessing fine-grained soils, especially deep deposits of alluvial, estuarine or marine clays, below the water table. In general, stone (or sand) columns should only be considered suitable if the clays/silts are insensitive (i.e. the clay strength does not reduce significantly when remoulded) and remoulded undrained strength is in excess of 15 kPa (due to a lack of lateral stone column confinement). Similarly, stone columns are unsuitable if there are thick layers of organic materials or peat soils, generally less than about half the installed stone column diameter. For sensitive or weaker soils alternatives, such as geotextile-encased sand columns (Raithel *et al.*, 2005) or deep soil mixing, should be considered. Many UK stone column applications are for improving the properties of mixed layers of urban fill, for which there are many examples of cost-effective treatment (e.g. Watts and Charles, 2015).

59.8.2 Vibrocompaction and vibroreplacement design principles

Vibrocompaction is commonly utilised to provide reduced total or differential foundation settlements or to provide seismic improvement to coarse-grained soils that can be improved by compaction. **Figure 59.29** indicates the range of soils suitable for treatment. Design methods are generally based on the principle of achieving an overall increase in relative density. For fine-grained soils the design principle is very different, and relies upon a reduction of the original (unimproved) settlement. Typical improvement ratios for suitable fine-grained soils are between two and three (i.e. post-treatment settlement may be between one-half and one-third of the original value), depending on the stone column area ratio. Typical expected settlements are as set out in **Table 59.9**.

Priebe (1995) is commonly used for assessing the settlement reduction that may be achieved by vibroreplacement, indicated in **Figure 59.30.** The area ratio is defined as the area of a column (A_c) as a percentage of the area supported by the column (A). The improvement factor is effectively the settlement reduction compared with the original soil. Thus, if around 20–25% of the area were replaced with stone columns, this would provide around 200% improvement in settlement reduction. The use of area ratio gives a guide to the column spacing, assuming that column diameters range from 0.5 to 0.8 m. The degree of improvement is highly sensitive to the minimum column diameter (in practice the columns will not be of uniform diameter).

The applied load is then shared between a composite structure of stone column and soil on an areal basis, and hence performance relies more on the overall 'group' effect than individual columns. This can be expressed as

$$Aq = Q_c + (A - A_c)\sigma_s \tag{59.5}$$

where A is the area supported per stone column, q is the applied pressure, Q_c is the safe capacity of the stone column, A_c is the cross-sectional area of the stone column and σ_s is the safe capacity of the surrounding soil.

In this formula the safe capacity (assumed strength) of the soil is sometimes taken as zero for better post-treatment settlement performance.

There are three main design procedures used in the UK

- Hughes and Withers (1974)
- Baumann and Bauer (1974)
- Priebe (1995).

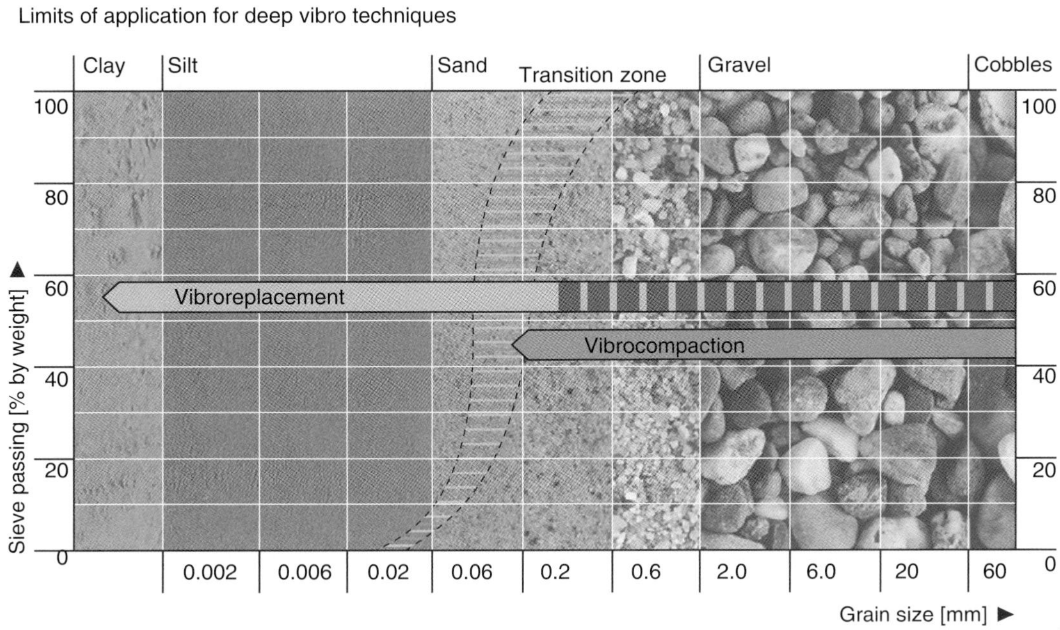

Figure 59.29 Range of application of vibroreplacement and vibrocompaction (Reproduced from Keller brochure 10-02 E: Deep Vibro Techniques; Keller Group plc)

Soil type	Bearing pressure: kPa	Settlement: mm
Made ground: mixed cohesive and granular	100–165	5–25
Made ground: granular fill, ash, brick rubble, etc.	100–215	5–20
Natural sands or sands and gravels	165–500	5–25
Soft to firm alluvial clays	50–100	a

(a) Post-treatment settlement depends on area ratio and is typically between one-third and one-half of unimproved values

Table 59.9 Bearing pressures and settlements for vibro-improved ground

The most commonly used method for an initial calculation of the safe capacity of a stone column is that of Hughes and Withers (1974)

$$\sigma'_v = (1 - \sin\phi'_c)(\gamma_b h_c + 4S_u + P) / (1 + \sin\phi'_c) \quad (59.6)$$

where σ'_v is the ultimate vertical effective stress in the soil (kN/m^2), ϕ'_c is the friction angle for the stone column material (typically 45°), γ_b is the bulk density of the soil (kN/m^3), h_c is the critical depth (m), S_u is the undrained shear strength of the soil (kN/m^2) and P is the surcharge (kN/m^2). The critical depth h_c is often taken as the depth from ground level to the

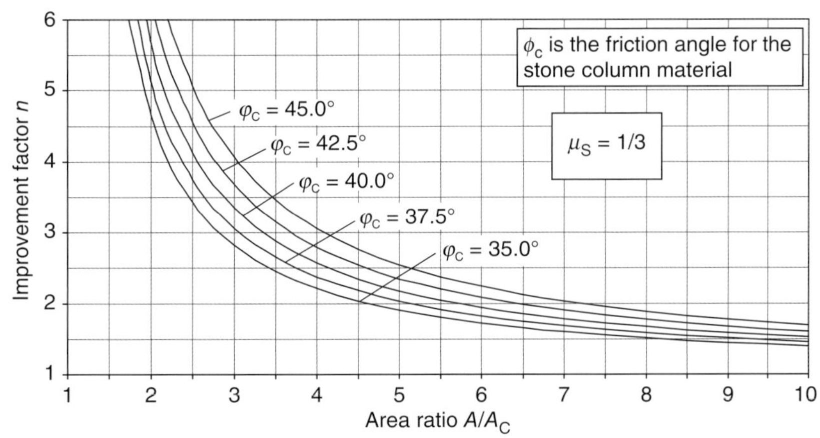

Figure 59.30 Design of stone columns (Reproduced from Priebe, 1995)

base of the foundation plus one stone column diameter. The safe capacity of the stone column is then given by

$$Q_c = \sigma_v' A_c F \tag{59.7}$$

where Q_c is the safe capacity of the stone column (kN), A_c is the area of the stone column (m^2) and F is the factor of safety.

Figure 59.31 shows how the capacity of a stone column varies with depth and assumed design undrained strength (note, the assumption of 'zero' soil strength is simply a means of developing a more cautious design; for stone columns to be effective a minimum remoulded undrained strength of 15 kPa is required, as noted above).

When loads are applied rapidly – such as when building road embankments on soft ground, silos and coal stockpiles – an allowance should be made for the development of excess pore water pressures in the soils surrounding the stone columns. This is often taken as $-u$ within the bracket of Equation 59.6 or as an r_u percentage. Thus, in Equation 59.6 P is replaced by $P - u$ or $\gamma_b h_c$ replaced by $r_u \gamma_b h_c$. Although stone columns may act as deep drains and facilitate more rapid consolidation of fine-grained soils, their drainage effectiveness can be significantly reduced by contamination of stone columns by fines and irregular column diameters.

The majority of stone column designs now use the Priebe (1995) method, where the area ratio of **Figure 59.30** can be used to assess the initial settlement reduction factor that is applied to the untreated settlement within the treatment depth. There will then be further potential settlements due to the imposed loads beneath the treatment depths, as well as ongoing self-weight settlements within fills and made ground, secondary consolidation and so on.

Stone columns have been used to enhance shear resistance (for slopes, gravity structures); however, routine methods (based on elasticity) can significantly overestimate the available shear resistance from stone columns. Failures have occurred due to inappropriate assumed column shear resistance under combined vertical and lateral loads. The stone column shear resistance may be close to fully mobilised under vertical loads, with little 'extra' resistance available to resist lateral or shear loads.

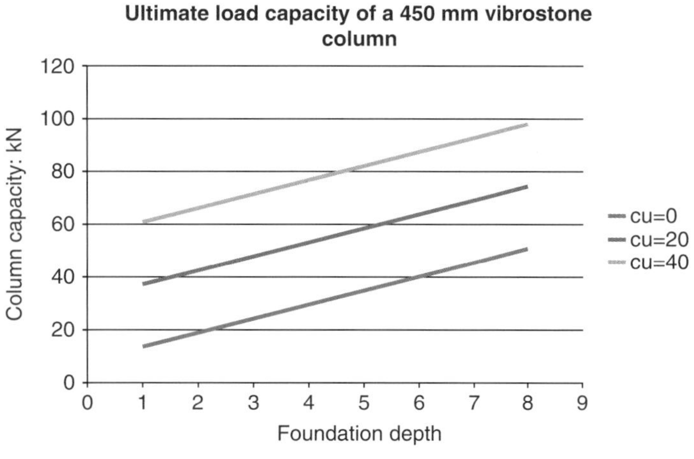

Figure 59.31 Stone column capacity for various depths and assumed clay strength

The treatment depth is a fundamental factor in the design, which will also dictate the vibro-equipment that is to be used to construct the stone columns (see Chapter 84 *Ground improvement*).

One of the largest vibroreplacement contracts was carried out in Barrow-in-Furness in 1991 for a proposed onshore gas terminal (Raison, 1999). Before development the majority of the site comprised three former settlement lagoons containing saturated PFA, a waste product from the adjacent coal-fired Roosecote power station. Site investigations revealed ground conditions beneath the PFA as comprising very loose silty gravelly alluvial sand over more dense glacial sands, with glacial till and sandstone at depth. Ground conditions were locally variable, and simplifications were necessary for design purposes. A typical geological section and variation of SPT with depth is shown in **Figure 59.32**. Particle size grading of the PFA indicated a sandy silt, which was often clayey. It was concluded during the project design stage that protection from seismic episodes – although small in the UK – was required, and a combination of driven piles and vibroreplacement was selected by the contractor. While the vibroreplacement was not deemed suitable for improvement of the very soft clayey PFA, which being a sandy silt was deemed too fine, it could improve the underlying soils and, thus, provide an acceptable foundation for the driven piles, which would be considerably shorter as a result. **Figure 59.33** shows the selected combination of driven piles and vibroreplacement columns. Driven piles were installed first, generally at 2 m centres, followed by the vibro stone columns, installed at centres ranging from 2.75 to 3 m.

59.8.3 Execution controls

Execution control is limited to setting out and depth, measurement of stone consumption or the compaction energy (usually measured as hydraulic or electric power consumed), water jetting (if used) and compaction lift increments. The above parameters can usually be measured by the rig instrumentation and provide an execution record for each column. The vibrocompaction case history in Chapter 9 *Foundation design decisions* highlights the considerable difference that a change in methodology (vibrolift increment, compaction duration per increment and use of water jets) made to post-compaction CPT results (50–100% increases).

59.8.4 Validation

Validation is typically in the form of SPT or CPT prior to and post-testing in association with zone testing on a group of columns (Greenwood, 1991). **Figure 59.34** shows the validation

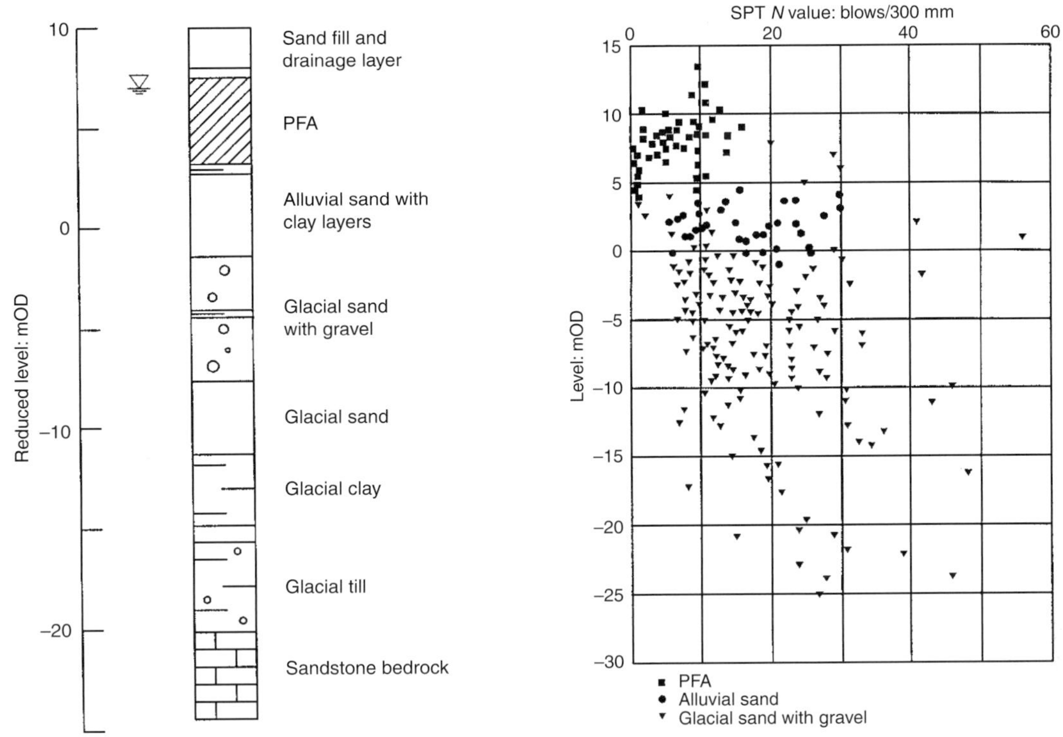

Figure 59.32 Ground conditions at Barrow (Reproduced from Raison, 1999)

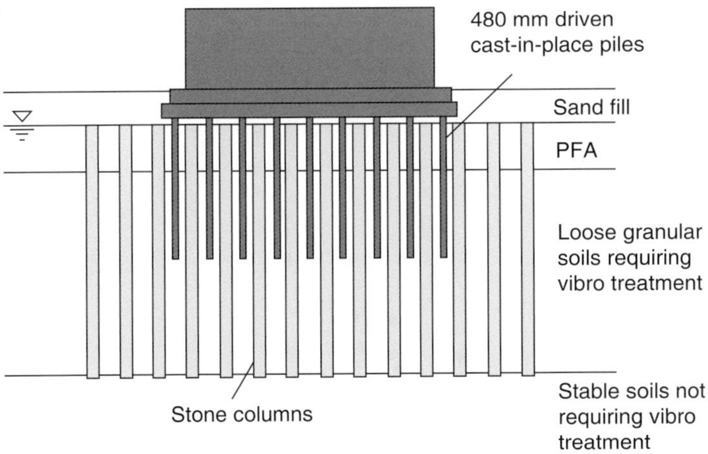

Figure 59.33 Proposed foundation improvement for Barrow (Reproduced from Raison *et al.*, 1995)

for the Barrow project in terms of CPT. Plate bearing on individual columns may be undertaken, noting they are unlikely to replicate the group performance of the ground improvement system.

59.9. Design principles for dynamic compaction

Dynamic compaction is similar to vibrocompaction in that the intent is to compact the ground and, thus, increase stiffness and create a more uniform competent material. It is more effective in coarse-grained materials although it has been used in finer materials. In the UK one of the main applications of dynamic compaction has been for the improvement of deep deposits of non-engineered fill, located above the water table (Watts and Charles, 2015). Dynamic compaction involves the repetitive dropping of heavy weights at fixed grid points to compress the ground. Historically, Menard owned a patent for this work and carried out pioneering works in Europe and the Far East. The weight of their dynamic tampers was as heavy as 40 tonnes and was supported on very high, specially constructed tripods.

The design of dynamic compaction is based on the input of a minimum dynamic energy per volume of ground and is carried out in a series of phases or passes. The maximum depth of treatment depends both on the weight of tamper and height of drop and is typically given by

$$\text{Depth of treatment} = 0.5 \times (\text{height of drop} \times \text{weight})0.5 \tag{59.8}$$

Thus, considering the use of a 20-tonne weight dropping 20 m, the depth of treatment would be limited to around 10 m. Fines content has a significant effect on compactive effort and care also needs to be taken if there are bands or layers of clay or silt, which will not be improved.

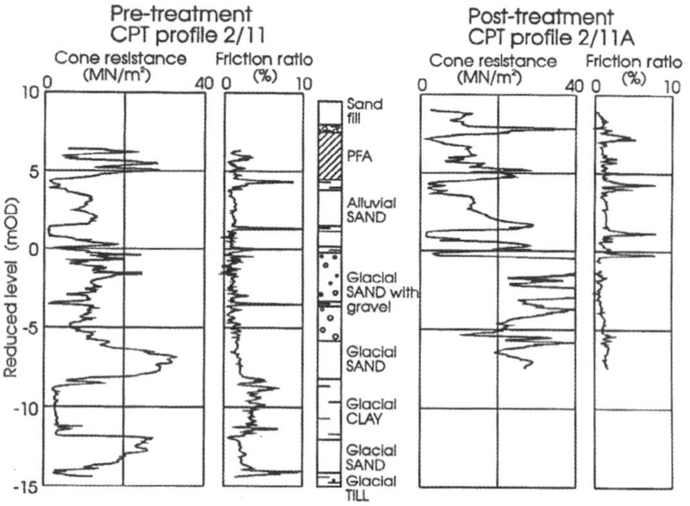

Figure 59.34 Results of pre- and post-CPT testing at Barrow (Reproduced from Raison *et al.*, 1995)

Typically, depending on the depth of treatment the compaction is carried out in three phases to ensure the full depth of the soil is treated.

- *Pass 1*: This will be based on widely spaced points and is designed to compact the deepest soils first. Spacing might be 7–10 m for a 10–12 m depth of treatment.
- *Pass 2*: This might be carried out on a 4–5 m spacing and is designed to compact the intermediate depths of ground below around 2 m.
- *Pass 3*: This is a final pass on a close grid, which is almost the plan area of the tamper, and is designed to compact 1–3 m depth. Sometimes the tamper configuration will be changed to provide a larger area.

Often a final surface compaction is necessary to complete the densification. As an approximate guide the application of around 25–30 tonne/m energy per cubic metre would provide a significant compaction of up to 60–70% relative density for 'clean' coarse-grained soils (say, less than 10% clay or silt content). **Table 59.10** is a guideline for the energies that might be required.

Type of deposit	D_{50}: mm	PI	Permeability range: m/s	Total energy: tonne/m/m³
Pervious coarse-grained, e.g. sand	>0.1	0	$>1 \times 10^{-4}$	<20
Semi-pervious, e.g. silt	0.01–0.1	<8	1×10^{-4}–10^{-8}	<30
Impervious above the water table, e.g. silt or clay	<0.01	>8	$<1 \times 10^{-8}$	<30
Landfill				<50

Table 59.10 Applied dynamic compaction total energy guidelines

59.9.1 Methods and key issues

Key issues associated with dynamic compaction are the high vibration levels associated with the process, which can be detrimental to nearby structures. This means that dynamic compaction must occur at least 20–30 m from buildings (or perhaps 50 m from relatively sensitive buildings), although this would need to be established on a project-specific basis. Results of ground-borne vibrations are shown in **Figure 59.35**.

59.9.2 Execution control

Execution control is limited to ensuring that the correct energy is applied. This requires control of the height and the number of drops for each grid point. The craters formed by the process are usually dozed level and a further guide to the improvement made can be seen by surveying the reduction in ground level.

59.9.3 Validation

Validation is similar to vibrocompaction and replacement and typically consists of CPT or other in situ testing. Zone testing

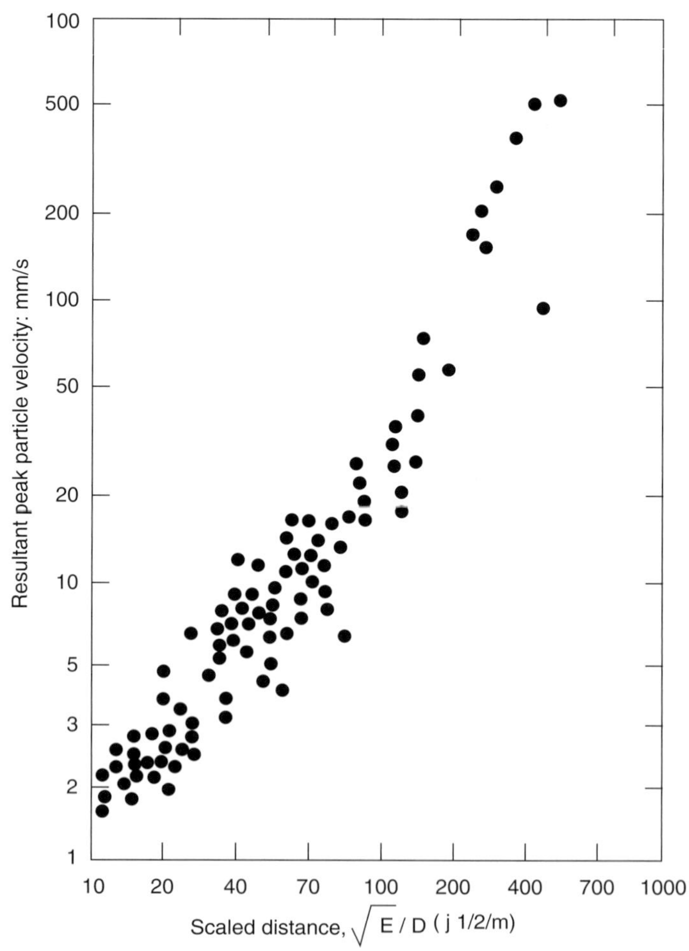

Figure 59.35 Ground-borne vibration levels caused by dynamic compaction (Reproduced from CIRIA C573, Mitchell and Jardine, 2002; www.ciria.org)

can also be carried out, although the limited depth of influence of the zone tests needs to be recognised compared with the proposed foundations.

59.10. Design principles for deep soil mixing

Deep soil mixing (DSM) was first established by Intrusion-Prepakt in the mid-1950s and then further developed in Japan in the 1960s. The level of research and development activity in Japan in relation to DSM remains the highest in the world, and annual DSM construction in Japan alone exceeds 1 million cubic metres of ground treated, with an accumulated total in excess of 25 million cubic metres.

The two generic forms of DSM are dry-soil mixing and wet-soil mixing. Dry-soil mixing consists of injecting a powder binder while mechanically disaggregating the soils. Wet-soil mixing consists of injecting a fluid binder while mechanically disaggregating the soils.

Dry-soil mixing is principally used in very soft to soft clays, silts and peats because the equipment was historically used for these soils, whereas wet-soil mixing can be used in more heterogeneous and stiffer soils.

Dry-soil mixing tends to remould the material while injecting the binder, while wet-soil mixing is designed more to mix in the binder to create a homogeneous material.

59.10.1 Laboratory and site trials

With the natural variability in the chemical and physical properties of the ground, the use of case study data to optimise binder dosage may not reliably predict the strength of the improved ground. A programme of laboratory testing is recommended to identify and optimise binder dosage for construction. The EuroSoilStab *Design Guide: Soft Soil Stabilisation* (2002) details a methodology for the preparation of soilcrete samples for laboratory strength and stiffness testing whereby samples with differing binder dosages are made and subsequently tested at varying ages after sample preparation (typically 7–90 days).

In advance, or at the latest at start of the ground improvement works, site trials should be undertaken. For treated organic clay and peat soils it can be expected that field soilcrete strengths are lower and sometimes significantly lower than laboratory strength. This is in response to factors including the differences in uniformity of material mixed between the laboratory and site conditions, sample curing conditions and variabilities on the materials between laboratory-tested samples and site conditions. Field strengths can be in the order of 20–50% of the strength of laboratory-treated samples (EuroSoilStab, 2002).

However, Pye *et al.* (2012) reported comparisons between laboratory and site trials where site strengths for both organic/peat soils and inorganic soils were both higher than the laboratory samples, in the order of 1.35 and 2.0, respectively. This illustrates the value and importance of undertaking both laboratory and site trials supporting design and to optimise binder contents for construction.

59.10.2 Design methods and key issues

The design of soil mixing is generally based on the required composite shear strength of the cement-stabilised soils (i.e. soilcrete) to achieve an appropriate factor of safety using limit equilibrium analysis. The Federal Highway Administration design manual *Deep Mixing for Embankment and Foundation Support* (FHWA, 2013) outlines a design methodology for routine applications, and the need to check a wide range of potential failure mechanisms.

For vertical support, columns can be utilised in interlocking patterns or as isolated columns. Where applied to resist shear, isolated columns are not recommended as their bending resistance is low and, thus, are typically utilised as panels of soil-mixed columns (typically 600 mm dia. at 500 mm centres perpendicular to the shear direction). Some configurations are shown in **Figure 59.36**.

The composite design shear strength is based on

$$S_u = a \times S_{\mathrm{u(column)}} + (1 - a)S_{\mathrm{u(soil)}} \qquad (59.9)$$

where a is the area replacement ratio of the treated ground. Hence, if the strength of the soil-mixed columns $S_{\mathrm{u(column)}} = 150$ kPa, the strength of the untreated material $S_{\mathrm{u(soil)}} = 20$ kPa and a composite shear strength of around 50–60 kPa was needed, then an area replacement ratio of about 25–30% would be required. This would correspond to panels with 600 mm dia. columns (at 500 mm centres) spaced about 1.6–2 m apart.

The detailed design should establish the area replacement ratio and soil-mixed column strengths required.

For drained soils analysis typical parameters are

$$c'_{\mathrm{(column)}} = bC_{\mathrm{u(column)}} \text{ and } \phi'_{\mathrm{(column)}} = 30° \qquad (59.10)$$

For initial design, b may be assumed to be between 0.1 and 0.3, but should be established for individual site conditions if critical for design.

As with all ground improvement techniques, both laboratory and field trials are extremely important to confirm the design requirements. Peats and organic clays are especially difficult to improve due to the high sulfate and acidic water content, which can depress cement hydration and lower the strengths of the mixed materials.

Most soil mixing is designed on the basis of consideration of the final in situ water–cement ratio as this ultimately affects the strength development as defined classically by Abrams in the 1920s. **Figure 59.37** shows data obtained by Swedish researchers for various soils and water–cement ratios. To estimate the required binder content, the initial

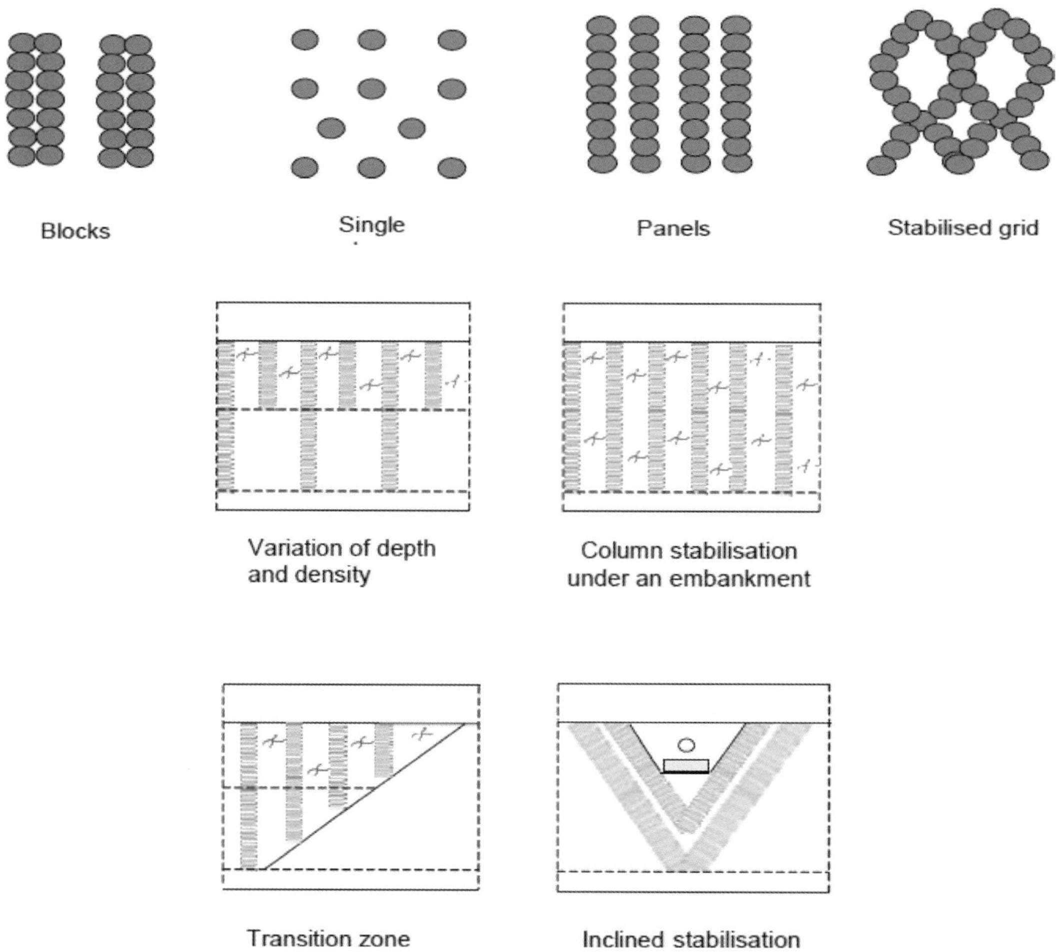

Figure 59.36 Examples of soil-mix column layout (Reproduced from EuroSoil Stab, 2002; IHSBRE Press)

moisture content of the soil is required together with any water added from the binder (if fluid) and the overall water–binder ratio is calculated based on a design addition of binder. For peats and organic clays, a binder containing ground granulated blast furnace slag (GGBS) is preferred as this binder tolerates sulfates and acids, and provides a higher strength. The factors (both 'intrinsic' and 'external', refer to **Table 59.5**) which affect the strength of DSM and jet grout soilcrete are conceptually similar in many respects (refer to Sections 59.7.1 and 59.7.2). The strength and stiffness of DSM soilcrete is discussed in further detail by Bruce *et al.* (2013) and Kitazume and Terashi (2012), including its tensile strength and residual strength. When the DSM soilcrete is performing an important structural support function (e.g. for deep excavations or sea walls on soft clays), then it is good practice to check the mobilised soilcrete strain (to check it is smaller than its failure strain) and that zones of tensile strain are insignificant.

One of the first successful large-scale uses of dry-soil mixing in the UK was carried out within the Port of Tilbury. Overloading of an aggregate stockpile caused a failure within the underlying soft clays and resultant failure of the quay wall. **Figure 59.38** shows the cross-section of the mode of failure. A number of solutions were considered based on piling, vibroreplacement and soil mixing, and it was decided that dry-soil mixing offered the most cost-effective and technically acceptable solution. Dry-soil mixing in panels was carried out to improve the soft clays and provide support to the aggregate stockpile. To create panels, 800 mm dia. columns were installed to the top of the underlying gravels with columns spaced at 700 mm. Individual panels were spaced at 2.3–2.8 m centres. The replacement volume for the soil-mixed columns varied from 25 to 33%, assuming a column shear strength of 150 kN/m^2. Over 3000 columns were installed, and the design was verified using an initial trial where columns were tested in situ and also exhumed. Extensive production testing was carried out using both CPT and pull-out testing and this is described by Lawson *et al.* (2005). **Figure 59.39** shows a comparison between CPT, pull-out testing and laboratory testing. In general, the field tests gave similar results, whereas the laboratory testing gave much lower results, which was attributed to sample disturbance.

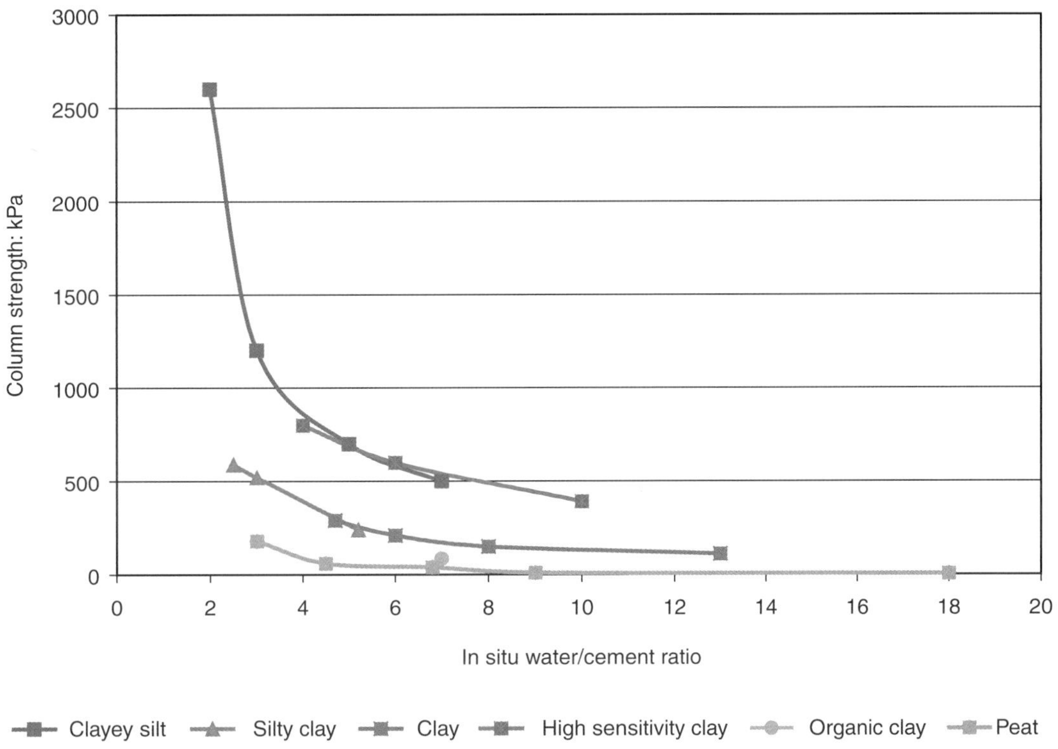

Figure 59.37 Expected soil-mix column strengths based on in situ water–cement ratio (Data taken from Åhnberg *et al.*, 1995)

59.10.3 Execution controls

Execution controls consist of controlling the binder content during mixing, and most specialists provide instrumentation on the rigs that record all important parameters. It is important for this instrumentation to be checked to ensure it is working correctly. For wet-soil mixing, additional controls are required for grout batching as for other forms of grout mixing.

59.10.4 Validation

Validation is generally by in situ testing. This can be CPT, plate testing or zone testing. Additionally, for dry-soil mixing a special vane penetration and pull-out test has been developed. The vane, which is around 0.45 m wide, is inserted below the mixing head, with a cable running up the kelly bar. The mixing head is inserted as usual and the column is mixed,

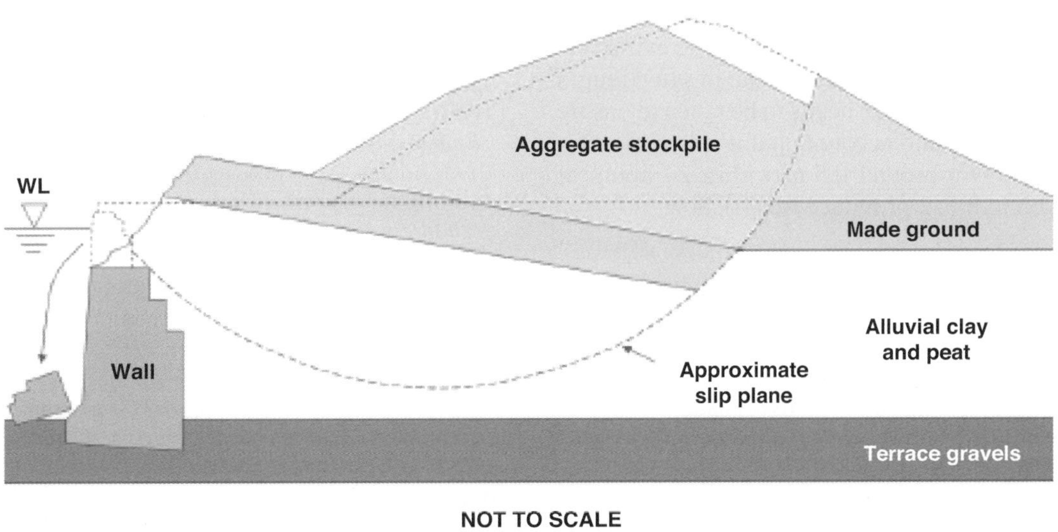

Figure 59.38 Failure of the quay wall at Tilbury (Reproduced from Lawson *et al.*, 2005)

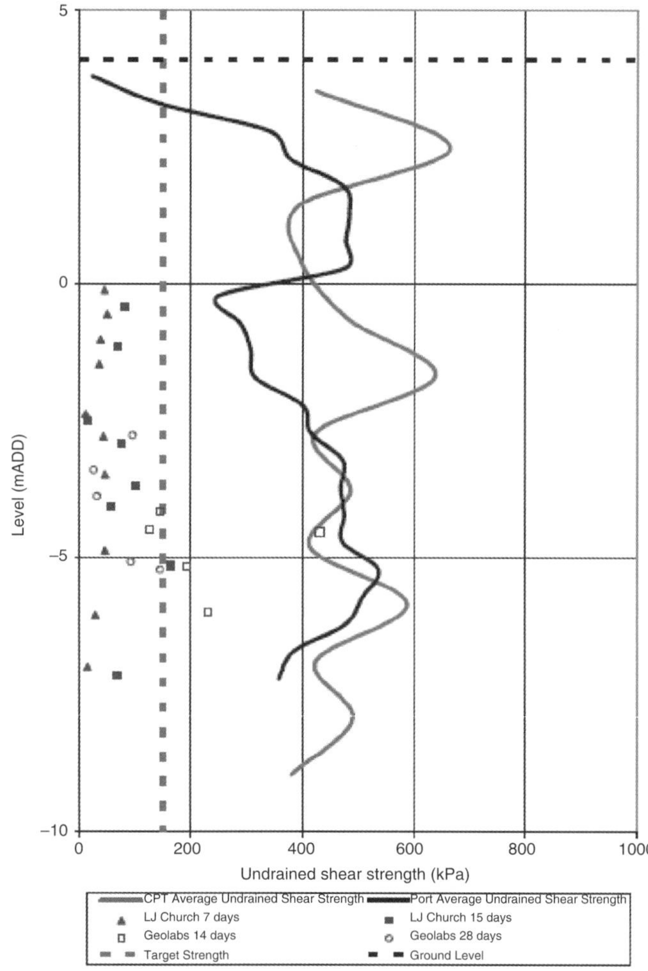

Figure 59.39 Results of testing for dry-soil mixing, Tilbury (Reproduced from Lawson *et al.*, 2005)

leaving the vane in place centrally and immediately below the column. After a set period of time, normally 3–7 days, the vane is withdrawn at a fixed rate of 2 cm/s and the extraction load monitored. This can then be related to the in situ strength using empirical relationships. Care needs to be taken to ensure that cable friction is taken into account, and it is common for the vane to be withdrawn around 0.5 m within 24 hours of column construction to break or reduce this friction.

References

Abrams DA (1920) *Design of concrete mixtures. Bulletin.* Structural Research Material Laboratory, Chicago, IL.

Åhnberg H, Johansson S-E, Retelius A, *et al.* (1995). *Cement och kalk för djupstabilisering av jord: En kemisk fysikalisk studie av stabiliseringseffekter. Report 48.* Statens Geotekniska Institut, Linköping, Sweden.

Baumann B and Bauer GEA (1974) The performance of foundations on various soils stabilised by the vibrocompaction method. *Canadian Geotechnical Journal* **11**: 509–530.

Bruce C, Collin J, Berg R, *et al.* (2013) *Federal Highway Administration Design Manual: Deep Mixing for Embankment and Foundation Support.* FHWA, Washington, DC.

Campbell A, Moradalizadeh M, Anders A and Napoli D (2023) Forrestfield Airport Link: Ground stabilization prior to reinstatement of damaged section of tunnel at Dundas Rd. *Proceedings of the 14th Australia and New Zealand Conference on Geomechanics, Cairns, Australia.*

Deere D and Lombardi G (1985) Grout slurries thick or thin? Issues in dam grouting. *Proceedings of the Sessions Sponsored by the Geotechnical Division of the American Society of Civil Engineers in Conjunction with the ASCE Conference, Reston, VA*, pp. 156–164.

De Wit J, Bogaards J, Essler RD, *et al.* (2006) Sandwich wall under Amsterdam Central Station, an innovative approach for jet grouting under difficult circumstances. *DFI Conference, Amsterdam, the Netherlands*, pp 1–13.

De Wit J, Bogaards J, Essler RD, *et al.* (2007) The design of the sandwich wall under Amsterdam Central Station: an innovative approach for jet grouting under difficult conditions. *Proceedings of the 14th European Conference, Madrid, Spain*, pp. 1–13.

Driesse A, Essler RD and Salet TAM (2008a) Grout struts for deep station boxes North–South Line Amsterdam, *Design. Proceedings of the 2nd BGA International Conference on Foundations, ICOF, Dundee, UK.*

Driesse A, Essler RD and Salet TAM (2008b) Grout struts for deep station boxes North–South Line Amsterdam, quality control and execution. *Proceedings of the 2nd BGA International Conference on Foundations, ICOF, Dundee, UK.*

Essler RD (2012) The design of jet grouting from concept to execution. *Proceedings of the Fourth International Conference on Grouting and Deep Mixing, New Orleans, LA*, pp. 2071–2081.

Essler RD (2023) Theoretical and practical aspects of jet grout cut-off and plug design. *Proceedings of the Sixth International Conference on Grouting and Deep Mixing, New Orleans, LA.*

Essler RD and Yoshida H (2005) Jet grouting. In *Ground Improvement*, 2nd edn (Moseley MP and Kirsch K (eds)). Spon Press, London, UK, Chapter 5.

EuroSoilStab (2002). *Design Guide: Soft Soil Stabilisation: Development of Design and Construction Methods to Stabilise Soft Organic Soils.* BRE, Watford, UK.

FHWA (Federal Highway Administration) (2013) *Deep Mixing for Embankment and Foundation Support.* FHWA, Washington, DC.

Filz G, Adams T, Navin M and Templeton AE (2012) Design of deep mixing for support of levees and floodwalls. *Proceedings of the 4th International Conference on Grouting and Deep Mixing, New Orleans, LA*, pp. 89–33.

Francescon M and Twine D (1992) Treatment of solution features in upper chalk by compaction grouting. *In Grouting in the Ground: Proceedings of the November 1992 Conference* (Bell AL (ed)). Thomas Telford, London, UK, pp. 327–348.

Greenwood DA (1991) Load tests on stone columns. In *Deep Foundation Improvements: Design, Construction, and Testing.* ASTM, West Conshohocken, PA.

Grundy CF (1955) The treatment by grouting of permeable foundations of dams. In *Proceedings of the Fifth ICOLD Congess, Paris, France.*

He L (2018) *Spatial Variability in Deep Soil Mixing – Centrifuge Model and Field Data Study.* PhD thesis, National University of Singapore, Singapore.

Healy PR and Head JM (1984) *Construction over Abandoned Mine Workings.* CIRIA Special Publication 32. CIRIA, London, UK.

Houlsby AC (1982) Cement grouting for dams. *Proceedings of ASCE Conference on Grouting in Geotechnical Engineering, New Orleans, LA.*

Houlsby AC (1990) *Construction and Design of Cement Grouting.* Wiley, London, UK.

Hughes JMO and Withers NJ (1974) Reinforcing soft cohesive soils using stone columns. *Ground Engineering* **7(3)**: 42–49.

Hughes AK, Bruggemann DA and Kettle C (2018) *Grouting for Reservoir Dams: A Guide to Good Practice. Report C774.* CIRIA, London, UK.

Kirsch K (1993) *Die Baugrundverbesserung mit Tiefenrüttlern, 40 Jahre Spezialtiefbau: 1953–1993. Festschrift*, Werner-Verlag, Düsseldorf, Germany.

Kitazume M and Terashi M (2012) *The Deep Mixing Method.* CRC Press, London, UK.

Lawson CH, Spink TW, Crawshaw JS and Essler RD (2005) Verification of soil mixing at Port of Tilbury, UK. *Proceedings of the International Conference on Deep Mixing Best Practice and Recent Advances, Stockholm, Sweden*, vol. 1, pp. 453–462.

Lee FH, Lee Y, Chew SH and Yong KY (2005) Strength and modulus of marine clay-cement mixes. *Journal of Geotechnical and Geoenvironmental Engineering* **131(2)**: 178–186.

Littlejohn SJ (2003) The development of practice in permeation and compensation grouting: a historical review (1802–2002). Part 1: permeation grouting. *Proceedings of the 3rd International Conference, New Orleans, LA.*

Lombardi G and Deere D (1993) Grouting design and control using GIN principle. *International Water Power and Dam Construction, Sidcup, UK*, pp. 15–22.

Mitchell J and Jardine F (2002) *A Guide to Ground Treatment.* CIRIA Report C573. CIRIA, London, UK.

Ng YCH, Danovan W and Ku T (2019) The potential of seismic cross-hole tomography for geotechnical site investigation. *E3S Web of Conferences* **92**: 1–5.

Pittaro GA (2016) Use of pressure relief wells to check ground improvement tensile strains in deep excavations. *19th Southeast Asian Geotechnical Conference & 2nd AGSSEA Conference (19SEAGC & 2AGSSEA), Kuala Lumpur, Malaysia.*

Pittaro GA (2022) *A Sustainable Design Framework Incorporating the Statistical and Spatial Variability of Cement-Treated Soils for Use in Excavation Support.* PhD thesis, National University of Singapore, Singapore.

Pittaro GA and Murphy J (2018) Deep excavations with ground improvement: a study using statistics and back analysis from a monitored excavation. *Underground Singapore Conference, Singapore.*

Priebe HJ (1995) The design of vibro-replacement. *Ground Engineering*, Dec.: pp. 31–37.

Pye N, O'Brien A, Essler R and Adams D (2012) Laboratory and field trials for deep dry soil mixing to stabilise a live railway embankment across Thrandeston Bog. *Proceedings of the Fourth International Conference on Grouting and Deep Mixing, New Orleans, LA*, pp. 554–564.

Raison CA (1999) North Morecambe Terminal, Barrow: pile design for seismic conditions. *Proceedings of the Institution of Civil Engineers and Geotechnical. Engineering* **137**: 149–163.

Raison CA, Slocombe BC, Bell AL and Baez JI (1995) North Morecambe Terminal, Barrow, ground stabilisation and pile foundations. *Proceedings of the 3rd International Conference on Recent Advances in Geotechnical Earthquake Engineering and Soil Dynamics, St Louis, MO*, pp. 187–192.

Raithel M, Kirchner A, Schade C and Leusink E (2005) Foundation of constructions on very soft soils and geotextile encased columns: state of the art. *Innovations in Grouting and Soil Improvement.* Geotechnical Special Publication 136. ASCE, Reston, VA, pp. 1–11.

Rutty P, Prangley C, Heath I and Mimnagh F (2021) Jet grouting for tunnelling at London Victoria station. In *Geotechnical Aspects of Underground Construction in Soft Ground: Proceedings of the Tenth International Symposium on Geotechnical Aspects of Underground Construction in Soft Ground, Cambridge, UK* (Elshafie M, Viggiani G and Mair R (eds)). Taylor & Francis, London, UK.

Scott P and Essler R (2001) Maintaining the integrity of the Reading Room during basement excavation at the British Museum. *Response of Buildings to Excavation-Induced Ground Movements: Proceedings of the International Conference, London, UK* (Jardine FM (ed)). CIRIA, London, UK, pp. 513–525.

Scott P and Essler RD (2002) Predicting and controlling movement of the Reading Room during basement construction at the British Museum, London. *Christian Veder Kolloquium, Institut für Bodenmechanik und Grundbau, TU Graz, Graz.*

Shibazaki M (2003) State of practice of jet grouting. *Grouting and Ground Treatment: Proceedings of 3rd International Specialty Conference on Grouting and Ground Treatment, New Orleans, LA*, pp. 198–217

Thomas HH (1976) *The Engineering of Large Dams.* Wiley, London, UK.

Topolnicki M (2016) General overview and advances in deep soil mixing. *XXIV Geotechnical Conference of Torino, Italy*, pp. 1–30.

Van der Stoel A (2001) *Grouting to Improve Pile Foundations.* Doctoral dissertation, Delft University of Technology, the Netherlands.

Wanik L, Mascolo MC, Bzówka J, Modoni G and Shen JSL (2017) Experimental evidence on the strength of soil treated with single and double fluid jet grouting. *Proceedings of the Grouting 2017 Conference: Grouting, Deep Mixing, and Diaphragm Walls, Honolulu, Hawaii.*

Watts K and Charles A (2015) *Building on Fill, Geotechnical Aspects*, 3rd edn. BRE, Watford, UK.

Yi F and Johnson RJ (2021) Case study of liquefaction mitigation by compaction grouting. *6th International Conference on Geotechnical and Geophysical Site Characterization, Budapest, Hungary*, pp. 1–9.

Further reading and useful websites

Xanthakos PP, Abramson LW and Bruce DA (1994) *Ground Control and Improvement.* Wiley, New York, NY. An excellent book which covers most forms of ground improvement in considerable detail; although it may not cover the most up-to-date innovations, it remains a source of significant knowledge.

Design principles for void filling

BRE (2006) *Stabilising Mine Workings with PFA Grouts: Environmental Code of Practice. Report 488*. BRE, Bracknell, UK.

Mining Records Office: http://www.coal.gov.uk/services/history/index.cfm

Design principles for compaction grouting

There are many specialist conference proceedings published in the USA, and these are the primary source of information. The American Society of Civil Engineers (ASCE) has published a number of special publications.

Johnsen LF, Bruce DA and Byle MJ (eds) (2003) *Grouting and Ground Treatment: Proceedings of the Third International Conference, New Orleans*, LA. ASCE, Reston, VA.

Krizek RJ and Sharp K (eds) (2000) *Advances in Grouting and Ground Modification (GeoDenver 2000)*. ASCE, Reston, VA.

Design principles for permeation grouting and jet grouting

BSI (2000) BS EN 12715:2000 Execution of geotechnical work – Grouting. BSI, London, UK.

Hughes AK, Bruggeman DA, Kettle C and Justino C (2018) *Grouting for Reservoir Dams: A Guide to Good Practice. Report C774*. CIRIA, London, UK.

Mitchell JM and Jardine FM (2002) *A Guide to Ground Treatment. Report C573*. CIRIA, London, UK.

Rawlings CG, Hellawell EE and Kilkenny WM (2000) *Grouting for Ground Engineering. Report C514*. CIRIA, London, UK.

Design principles for vibrocompaction and vibroreplacement and dynamic compaction

Mitchell JM and Jardine FM (2002) *A Guide to Ground Treatment. Report C573*. CIRIA, London, UK.

US Department of Transportation (2001) *Demonstration Project 116. Publication FHWA-SA-98-086R*. FHWA, Washington, DC.

Watts K and Charles A (2015) *Building on Fill, Geotechnical Aspects*, 3rd edn. BRE, Watford, UK.

Design principles for deep soil mixing

BSI (2005) BS EN 14679:2005 Execution of special geotechnical works – Deep soil mixing. BSI, London, UK.

Essler RD (2006) Personal communication from North–South Project team.

EuroSoilStab (2002) *Design Guide: Soft Soil Stabilisation*. BRE Press, Watford, UK.

Kitazume M and Terashi M (2012) *The Deep Mixing Method*. CRC Press, London, UK.

Swedish Geotechnical Institute (1995) *Design of Lime-Cement Columns. Report 4/95*. Swedish Geotechnical Institute, Stockholm, Sweden.

It is recommended this chapter is read in conjunction with

- Chapter 25 *The role of ground improvement*
- Chapter 84 *Ground improvement*
- Chapter 100 *Observational method*

All chapters in this book rely on the guidance in Sections 1 *Context* and 2 *Fundamental principles*. A sound knowledge of ground investigation is required for all geotechnical works, as set out in Section 4 *Site investigation*.

Chapter 60
Foundations subjected to cyclic and dynamic loads

Barnali Ghosh Mott MacDonald, UK
Vipul Kumar Mott MacDonald, UK

The basic concepts and references concerning shallow and deep foundation response to earthquake, wave and machinery effects are introduced. The information provided in this chapter will be useful to many readers because available recommendations in the codes and standards concerning the response of foundations subjected to cyclic and dynamic loads is limited with regard to the requirements encountered in practice. Soils tend to respond nonlinearly even at rather small strain. Nonlinear dynamic analysis and testing can produce chaotic results, and the best approach to a problem solution is to understand the fundamental issues associated with the problem, have an idea about the possible range of results and know the limitations of analysis methods used for its solution.

doi: 10.1680/icemge.66830.1009

CONTENTS

60.1. Introduction

Soil dynamics deals with the behaviour of soil and foundations under different dynamic loads. Operation of rotary machines or hammers, wave forces and earthquake ground motions constitute a class of dynamic loads that usually challenge engineers in their design of different kinds of foundations. The main goal of this chapter is to provide a better understanding to practising engineers by giving a brief introduction to the topic and providing a list of technical references that can be consulted for detailed information. The design of foundations which are subjected to cyclic/dynamic loads is a complex matter and should only be carried out by appropriately qualified specialists. The types of dynamic loading discussed in this chapter include

(*a*) seismic loads triggered by earthquakes
(*b*) wave forces imposed on offshore (marine) structures
(*c*) foundations which support vibrating machines.

Cyclic load effects may also need to be considered for tall slender structures subjected to wind loading, although the cyclic loads will tend to be relatively minor compared with the situations for (*a*) to (*c*) above. The adverse effects of cyclic/dynamic loading include

- degradation of ground strength, which can lead to failure at foundation loads below those expected based on 'static' strength
- degradation of ground stiffness and/or 'ratcheting'-type effects, leading to an accumulation of permanent foundation displacements
- dynamic loading can induce site-/structure-specific amplification of loads/movements due to resonance-type phenomena
- earthquake loads can induce large-scale ground instability which will impose large additional forces on foundations or cause catastrophic collapse.

It is noted in Chapters 52 *Foundation types and conceptual design principles* to 55 *Pile-group design*, inclusive, that the main body of experience is for foundations which are loaded vertically. A feature of cyclic/dynamic loading is that it commonly involves large horizontal loads, and therefore it is far more difficult to predict foundation behaviour. Hence, a more cautious approach to foundation design is required.

60.2. Cyclic loading

Vibration of foundations occurs when loads on them (from machinery) or their base displacements (from earthquakes) vary in time, or a structure is forced out of its equilibrium position (by blast) and left to oscillate freely. In the case of dynamic loads, the inertial forces are significant in comparison with the acting static loads. More details about dynamic and seismic loading of soil are given in Chapter 24 *Dynamic and seismic loading of soils.* The main parameters of cyclic behaviour are as follows.

- **Amplitude of vibration** indicates how much a cyclic state differs from the steady-state condition of a body (**Figure 60.1**). Acceleration amplitude is directly proportional to the inertia force acting on a body. Square of velocity amplitude is directly proportional to the kinetic energy of a vibrating body. Displacement amplitude of a vibration is proportional to the amount of deformation (strain) in cyclic condition (strain is the ratio between induced displacement and length over which such displacement has been achieved). The ratio between particle and wave propagation velocity equals the strain induced by unidirectional vibration within a body. Product of unit density, particle velocity and wave propagation velocity equals the stress acting within a body during its unidirectional vibration (e.g. Kramer, 1996; Timoshenko and Goodier, 1970) (stress is the ratio between applied force and the area over which it has been applied).
- **Vibration period** is the time difference between two equal states of body vibrations. The product of vibration period and wave

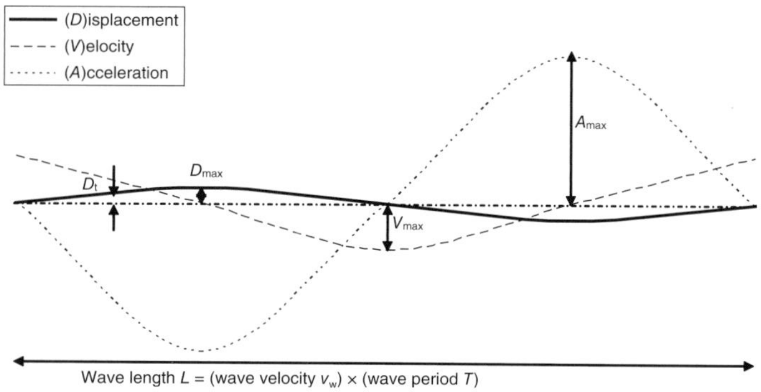

Shear strain γ_t at time t = (displacement D_t at time t)/($v_w \times t$) = (particle velocity v_{pt} at time t)/(wave velocity v_w)
Shear force T_t at time t = (mass m) × (acceleration A_t at time t)
Shear stress τ_t at time t = T_t /(cross sectional area) = (unit density ρ) × v_{pt} x v_w
Kinetic energy E_t at time t = (mass m) × (particle velocity V_{pt}^2 at time t)/2

Figure 60.1 A sketch of sinusoidal transversal wave with amplitude-related values

propagation velocity equals the wavelength. The reciprocal of vibration period is called vibration frequency. If the vibration period of free oscillation of a body corresponds to the vibration period of vibration excitation, then resonance could occur which can lead to the increase of the amplitudes of the source of vibration and damage or destruction of the oscillating body. In the case of harmonic (sinusoidal) vibration caused by machinery, the maximum amplitudes of displacement D, velocity V and acceleration A are related to each other as $V = \omega D$, $A = \omega V = \omega^2 D$, where ω is the so-called circular frequency $= 2\pi f$, where f is vibration frequency. Earthquakes are chaotic and the predominant periods (frequencies) of acceleration, velocity and displacement are different.

- **Number of cycles (i.e. duration of vibration)** is important because longer vibration duration can cause greater damage to materials (due to the so-called fatigue effect) than a few vibration cycles if the amplitudes are the same. Numerous definitions have been proposed for the measurement of both of these parameters (Bommer and Martinez-Pereira, 1999; Hancock and Bommer, 2004). The commonly used measure of duration is called the *significant duration* and it is based on the accumulation of Arias intensity, defined as the time elapsed between reaching 5% and 75% or 95% of the total energy. Seed *et al.* (1975) developed the concept of an equivalent number of significant stress cycles (caused by the horizontal ground acceleration) to represent an irregular time history of shear stresses by a uniform series of harmonic (sinusoidal) stress cycles. Seed *et al.*'s (1975) data can be approximated by a simple formula Neqv = 0.0008 M_L 4.88, where M_L is local (Richter) earthquake magnitude. Other researchers related Neqv to different magnitude types (e.g. Hancock and Bommer, 2004) and showed that Neqv depends also on the site-to-source distance and depth below ground level (e.g. Green and Terri, 2005) – that is, the soil type.

60.3. Earthquake and geoseismic hazard

Earthquakes are sudden ruptures of the Earth's crust caused by accumulating stresses (elastic strain energy) resulting from internal processes of the planet. Ruptures propagate over approximately planar surfaces called faults, releasing large amounts of strain energy. Energy radiates from the rupture as seismic waves. These waves are attenuated, refracted and reflected as they travel through the Earth, eventually reaching the surface where they cause ground shaking. Macroseismic intensity is a parameter that reflects the strength of the ground shaking at a given location, inferred from observations rather than instrumental measurements. There are several scales of intensity, the most widely used defining 12 degrees of intensity (Musson *et al.*, 2010), such as the European Macroseismic Scale, or EMS (Grünthal, 1998). A more objective measure of the size of an earthquake can be obtained by estimating the total amount of elastic strain energy released by the fault rupture. Different magnitude scales have been proposed depending on the quantities used for the determination of the energy released by an earthquake. In most historical earthquake databases, earthquake size is expressed in terms of the Richter local magnitude scale, denoted by M_L, or the surface wave magnitude, denoted by M_s. These are the most cited magnitudes for older earthquakes, but because of their deficiency to saturate at magnitudes higher than 6 they have been superseded by the moment magnitude scale, denoted by M_w, which does not suffer from saturation. It is worth noting that there are no significant differences between the different magnitudes for values ranging from 5.0 to 7.5 (Kanamori, 1977).

Ambraseys (1990) derived the following relationships between various common earthquake magnitude scales

$$\begin{aligned} 0.77 \times m_b - 0.64 \times M_L &= 0.73 \\ 0.86 \times m_b - 0.49 \times M_s &= 1.94 \\ 0.80 \times M_L - 0.60 \times M_S &= 1.04 \end{aligned} \qquad (60.1)$$

Figure 60.2 illustrates the different geotechnical hazards that can result from the occurrence of an earthquake, which the designer needs to incorporate in the design. There are two important hazards directly associated with the fault rupture: surface fault rupture and tsunami. For shallow earthquakes, the fault rupture may extend to the ground surface, generating scarps and lateral offsets of up to several metres. The extent of surface deformation is dependent on the type of fault and the depth and nature of surface soils. These deformations may be very damaging to buried services, roads, dams and railways. If a tectonic fault causes a rupture of ground surface, it results in destruction of structures crossing the fault and therefore placement of structures above active faults must be avoided wherever such faults exist.

Most earthquake destruction is caused by the strong shaking of the ground associated with the passage of seismic waves, and this shaking is also the trigger for the geotechnical hazards associated with earthquakes. The focus of most seismic hazard assessments is to quantify possible levels of ground shaking, which provides the basis for earthquake-resistant structural design. These characteristics of the ground motion at a given site are affected by numerous complex factors, such

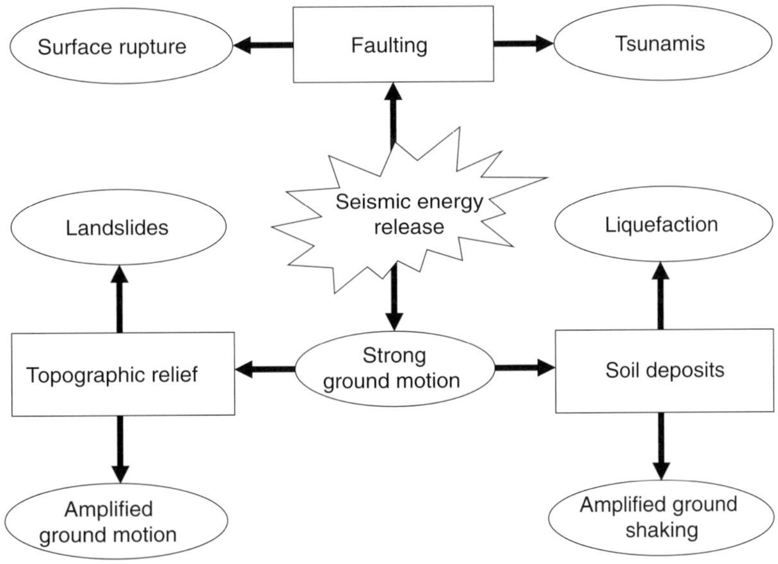

Figure 60.2 Earthquake processes and their interaction with the natural environment (ellipses) and the resulting seismic hazard (rectangles) (adapted from Bommer, 2022)

as the source-to-site distance, earthquake magnitude, effects of local soil and rock conditions, rupture directivity, topographic and basin effects, source mechanism and propagation path of seismic waves.

Liquefaction is associated with a significant loss of stiffness and strength in the liquefied soil due to development of large excess pore water pressures within the soil, leading to large ground deformations. Particularly damaging for engineering structures are cyclic ground movements during the period of shaking and excessive residual deformations such as settlements of the ground and lateral spreads. An example of the effect of soil liquefaction on loss of bearing capacity of shallow foundations and tilting of the whole buildings is shown in **Figure 24.12** of Chapter 24 *Dynamic and seismic loading of soils*. In other cases, buildings have sunk vertically or pipes and shafts have floated due to soil liquefaction. Piles can be damaged or destroyed in liquefied soil when a significant lateral movement of ground occurs due to flow-type failures. A devastating effect on the piled foundation caused not by soil liquefaction but by the 'double resonance' condition (amplification of bedrock motion by the soil deposit and amplification of the soil motion by the structure) is shown in **Figure 60.3**. The peak horizontal bedrock acceleration was only 0.03–0.04 g, but has been amplified several times (Kramer, 1996).

Earthquake-induced landslides usually affect large areas in the source zone, or even greater areas beyond the immediate source zone in the case of large-magnitude earthquakes. As demonstrated in several earthquakes, rockfalls, slope instabilities and associated hazards are very difficult to deal with, and are particularly challenging in an urban setting.

60.4. Global failure/deformation mechanisms

During earthquakes the propagation of seismic waves causes cracking and elevation differences even on level ground. However, the greatest ground deformations are caused by slope failures, rupture of ground surface by tectonic faults and soil liquefaction followed by flow-type failures.

Keefer (1984) studied the effects of 40 historical earthquakes on slope failures. The types of failures, their relative frequency and minimum triggering earthquake magnitudes are given in **Table 60.1**.

Keefer (1984) and Rodriguez *et al.* (1999) plotted the graphs of the maximum recorded distances of occurrence of landslides, depending on their type and earthquake magnitudes. Their graphs indicate that no slope failures are expected beyond 120 km epicentral distance, even from earthquakes with magnitudes up to 8, and that disrupted landslides are more prone to being triggered by earthquakes than are coherent landslides, spreads and flows. More information on slope stability issues is provided in Chapter 23 *Slope stability* and on slope stabilisation methods in Chapter 72 *Slope stabilisation methods*.

If liquefaction occurs, then the consequences can vary enormously from modest ground settlement to catastrophic large-scale ground instability. A range of factors will influence the nature of any consequences, including: the earthquake characteristics; site stratigraphy and topography; nature of structures (and their foundations) and earthworks on the site. Three important consequences are

- reduction of soil shear strength, leading to global instability of the ground (e.g. embankment collapse leading to failure of adjacent pile foundations) and increase in lateral pressure on retaining structures

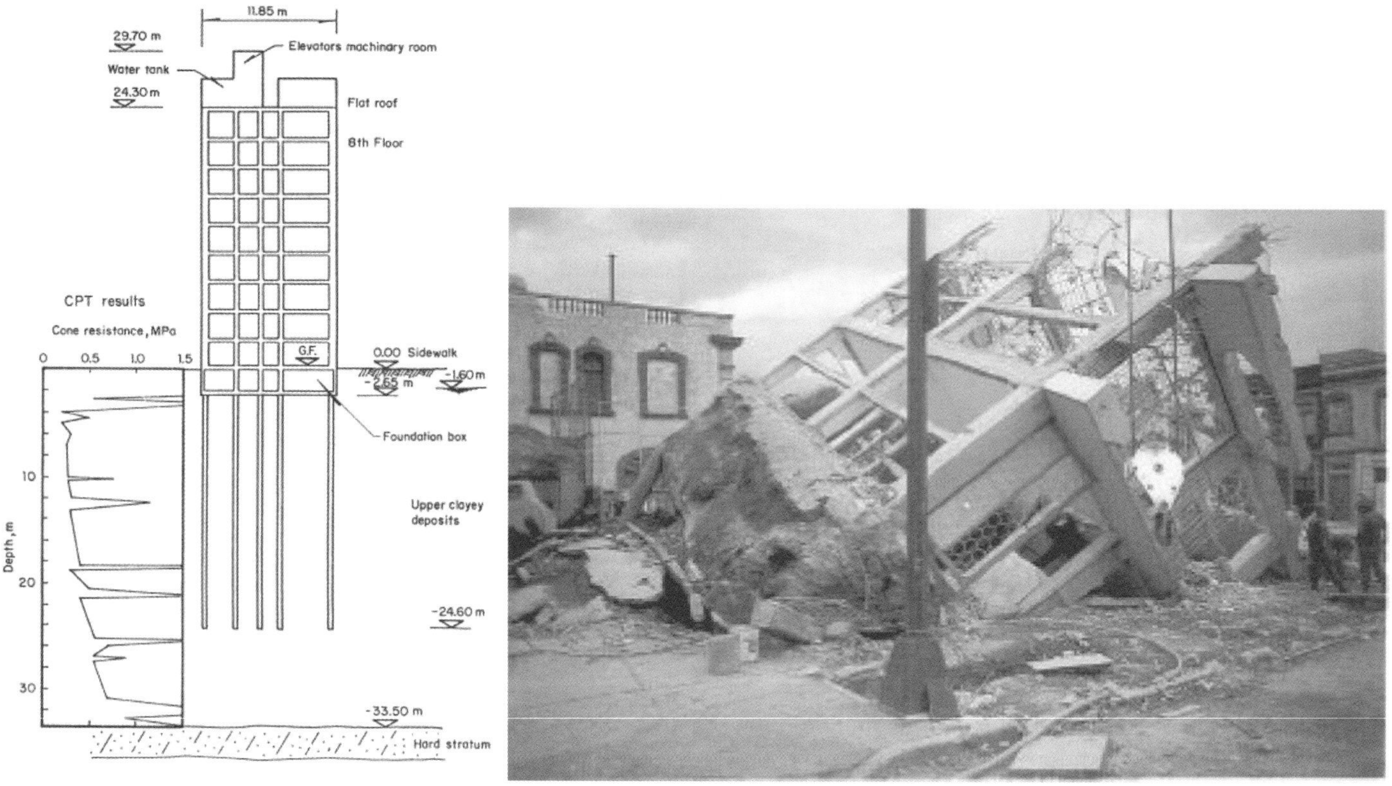

Figure 60.3 Elevation of a 10-storey building in Mexico City and after toppling (Reproduced from Meymand, 1998)

Type of slope failure	Frequency of occurrence during earthquakes	Minimum triggering earthquake magnitude M_L
Rock falls, disrupted soil slides	Very frequent	4.0
Rock slides		4.0
Soil slumps, soil block slides	Frequent	4.5
Soil lateral spreads		5.0
Soil avalanches		6.5
Soil falls	Moderately frequent	4.0
Rapid soil flows, rock slumps		5.0
Sub-aqueous landslides	Uncommon	5.0
Slow earth flows, rock block slides		5.0
Rock avalanches		6.0

Table 60.1 Types of slope failures, relative frequency and minimum triggering earthquake magnitudes

- settlement caused by reconsolidation of liquefied soil
- lateral deformation of gently sloping ground.

Figure 60.4 indicates how site-specific features, and the spatial distribution of liquefied zones, can affect the consequences of liquefaction.

- **Figure 60.4(a)** shows a continuous layer of liquefied soil which daylights in a riverbank, leading to post-earthquake instability and collapse of the building.
- **Figure 60.4(b)** shows a liquefied layer at depth, with no potential for lateral instability, which leads to post-earthquake settlement, which does not cause serious damage to the building but may cause damage to connections into the building (e.g. gas mains, water pipes) if they have insufficient flexibility. Ishihara and Yoshimine (1992) and Cetin *et al.* (2004) provide guidance on methods to assess reconsolidation settlement.
- **Figure 60.4(c)** shows shallow pockets of sand cause differential settlement and associated damage to the building (but not collapse).
- **Figure 60.4(d)** shows a zone of liquefied soils leads to post-earthquake instability of the embankment (termed 'flow' failure) because the overall soil shear resistance is less than the 'static' shear stress imposed by the embankment, as conventionally assessed by slope stability analyses. The key difference is that the 'residual' shear strength of the liquefied layer needs to be assessed. The lateral deformation of the ground will lead to additional forces on the piles, which could lead to structural failure of the piles if they are inadequately designed.

Soil liquefaction can cause not only flow-type failure of slopes but also loss of bearing capacity of foundations and their excessive settlement and horizontal movement. Shallow

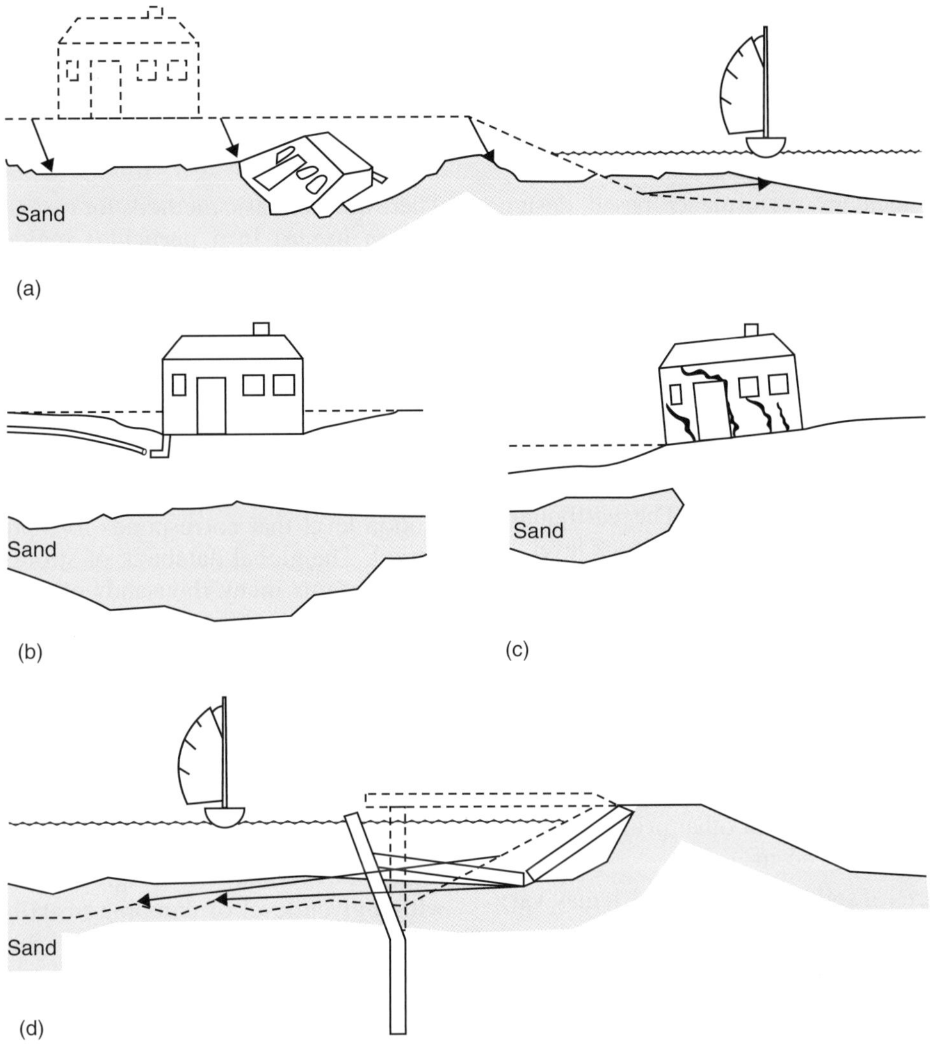

Figure 60.4 Effects of soil liquefaction: (a) flow-type slope failure leading to a building collapse; (b) deep soil layer and excessive building settlement with breakage of a pipe; (c) differential settlement and severe damage to a building; (d) flow-type slope failure of an embankment and damage to the deck and supporting piles

foundations can tilt or sink (e.g. Liu and Dobry, 1997). Piled foundations can be sheared in the case of flow-type slope failure (e.g. Hamada, 1992) because the top non-liquefied crust exerts significant lateral force, which equals the passive lateral resistance of the layer. Liquefaction in stratified soils needs careful consideration, and when the top non-liquefied soil layer is thick enough to prevent formation of sand volcanoes then sills (intrusions) are formed within non-liquefied layers. It also means that such layers are subjected to uplift pressure, which decreases their bearing capacity.

60.5. Hazard identification and mitigation

60.5.1 Overall design philosophy

The fundamental objective of earthquake engineering is to protect lives and livelihoods through the reduction of seismic risk using our knowledge of earthquakes, wave propagation and their effect on the built environment. Directly or indirectly, this generally requires quantification of the risk, for which quantification of the seismic hazard is required as a basic input. The fundamental assumption in seismic design is that the design codes provide a safe design which meets the limit states of collapse. The majority of seismic codes are written to prevent loss of life. This is achieved by designing a structure that is ductile, (i.e. one that absorbs energy through cracking and plastic hinging but without collapse). However, this means that significant damage can occur after a very large earthquake, which might make the structure unusable. If this is unacceptable to the client, a more stringent performance level can be achieved either by using a higher importance factor (I) than is required by the code, or by using a performance-based approach such as that given in FEMA368, which will be incorporated in the Eurocode updates. However, it can be argued that in certain types of structures one is not interested in collapse prevention for frequent earthquakes. This places emphasis on the idea that the fundamental cornerstone of geotechnical seismic design should be the management of performance of the structure to various hazard levels. The fundamental requirements shall refer to the state of damage in the structure, herein described through the *limit states*.

The design process includes performance-based design, which usually includes the following steps (**Figure 60.5**).

(*a*) Assess the earthquake magnitude and source location. Often several 'design' earthquake scenarios will be considered, varying from a relatively small event (for this the foundation would be expected to remain serviceable with minimal repair requirements), often called the OBE (operating basis earthquake), to a large event (for which there would be a 'no collapse' requirement), often called the SSE (safe shutdown earthquake). The earthquake properties will be defined for reference bedrock level.

(*b*) Assess the overall topography in the vicinity of the proposed structure, particularly the location of slopes, embankments and cuttings, ditches, riverbanks and so on to isolate any slope or basin effects.

(*c*) Assess local amplification or attenuation of bedrock earthquake motion through overlying soil deposits by including site response effects.

(*d*) Assess whether soil liquefaction or other primary or secondary seismic effects may occur.

(*e*) Assess consequences of soil liquefaction which may vary from local settlement to global failure.

(*f*) Assess dynamic soil–structure interaction effects.

(*g*) Depending upon (*e*) or (*f*), assess the type and extent of deep ground improvement to mitigate the potential adverse effects.

60.5.2 Code- and site-specific spectra

There are two basic methods for assessing the seismic ground-motion hazard in a particular region or at a specific site: deterministic methods and probabilistic methods. The commonly used approach for determining seismic design loads for engineering projects is probabilistic seismic hazard analysis (PSHA). The primary output from a PSHA is a hazard curve showing the variation of a selected ground-motion parameter, such as peak ground acceleration (PGA) or spectral acceleration (SA), against the annual frequency of exceedance (or its reciprocal, return period). The design value is the ground-motion level that corresponds to a preselected design return period. The global databank of strong motion accelerograms now contains many thousands of records, and these provide invaluable insight into the nature of earthquake ground shaking. More detailed information about deriving site-specific hazard curves can be seen in Bommer (2022). The methodology of probabilistic hazard assessment is very robust and adds value to the client by reducing the uncertainties associated with the definition of the hazard and can help in managing performance at different return periods.

The response spectrum is a representation of the maximum response experienced by single-degree-of-freedom oscillators with a given level of damping (usually assumed to be 5% of critical) to a specific earthquake motion. **Figure 60.6** compares the unified hazard response spectra (UHRS, bedrock)

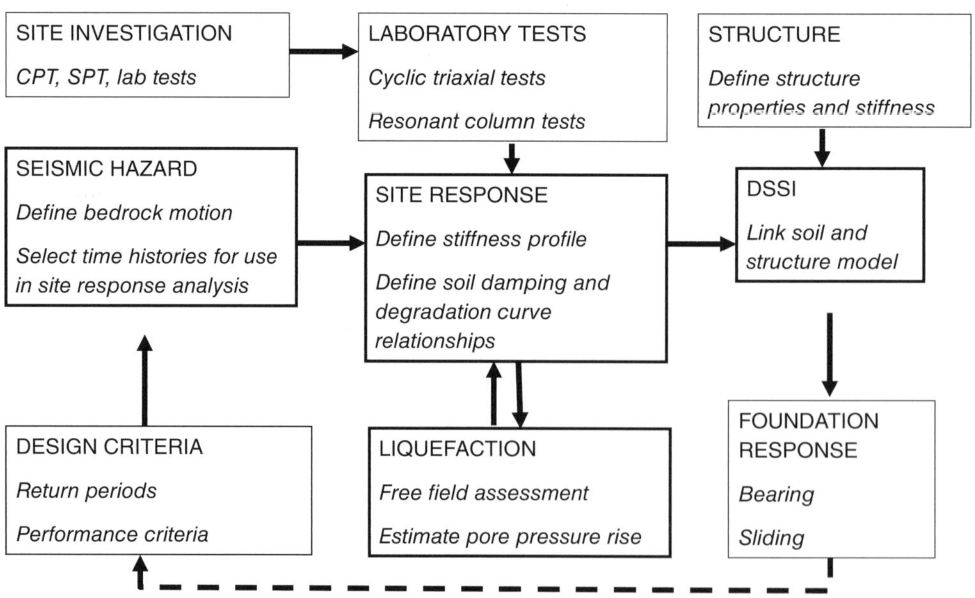

Figure 60.5 Typical design methodology followed for a project

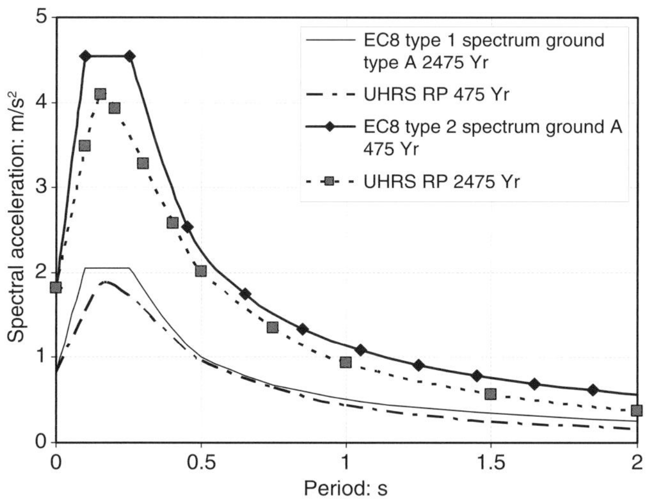

Figure 60.6 Comparison of bedrock spectra with PSHA analysis results

obtained for moderately seismic sites in North Africa by PSHA analysis. The UHRS spectra are compared with the Eurocode type 2 specified spectra for ground type A. This spectrum can be used if the earthquake that contributes to the seismic hazard defined for the site for the purpose of probabilistic hazard assessment has a surface wave magnitude (M_s) greater than 5.5. It can be seen that the code-specified spectra is often conservative, especially for long periods and more so for longer return periods. This issue becomes significant in highly seismic areas where the cost of effective seismic protection can influence the project design basis. Thus, it is important to offer a site-specific probabilistic seismic hazard assessment as a value-enhancing tool for the client for projects which are above a minimum threshold. The surface spectra can be obtained by following several routes, including by multiplying the bedrock spectra with the soil factors specified in the code. In Eurocode 8 the same S factor is used for all periods, whereas the International Building Code (IBC) 2021 specifies that different S factors can be used for long and short periods. In the second case, site-specific spectra can be determined by performing site response analysis (SRA).

60.5.3 Soil behaviour under cyclic load

It is found that for very small strain amplitudes the soil deforms in an elastic manner. That is, the strains are recoverable and the amount of damping is very small. These properties are usually measured with special testing devices, such as the resonant column or free torsional vibration. The shear modulus so determined is called G_o. It provides an upper limit on the stiffness of the soil and is a very important property when considering the earthquake response of a site.

A detailed description of soil behaviour in cyclic conditions is given in Chapter 24 *Dynamic and seismic loading of soils*. In summary, both loose to medium-dense sand and silt and soft to firm clay can exhibit a decrease of shear strength and stiffness with increase in the number of cycles and cyclic amplitudes. The stiffness of both coarse- and fine-grained soils is highly nonlinear, with stiffness reducing with increasing strain once a threshold strain has been exceeded. While the stiffness of soils reduces with increasing strain, the hysteretic or material damping increases. It should also be noted that the stiffness and the hysteretic damping of soils are generally not dependent on the frequency of the loading. In the seismic design situation, the soil should be considered to behave under undrained conditions, and the soil undrained behaviour may be studied in terms of total stresses, or in terms of effective stresses with due account of the excess pore water pressure generation.

60.5.4 Site response analysis

The presence of soil deposits above bedrock results in modification of the earthquake ground-motion amplitudes as well as frequency content and duration of the motion otherwise measured on a rock outcrop. This is known as the 'site response effect' and plays an important role in the seismic-resistance design. The key element of SRA is to capture the attenuation or amplification of vertically propagating shear waves as they pass through an idealised soil column. The elements needed for such an analysis include the dynamic soil properties, modulus degradation curves, soil density and earthquake time history that suit the tectonic regime of the site. Comprehensive technical advice is provided in several publications (e.g. Hashash *et al.*, 2020) which can be referred to for further information about this topic. The outputs from the site response analyses include the PGA at the surface, variation of shear stress/shear strain with depth, peak ground velocity (PGV), peak ground displacement (PGD) and surface response spectra.

Dynamic soil–structure interaction effects are dependent on the variation of soil shear modulus with depth and with strain amplitude. Empirical correlations are available for G_{max} (or G_o) (refer to Chapter 52 *Foundation types and conceptual design principles*); however, whenever feasible site-specific in situ measurements of G_{max} should be carried out (using, for example, seismic cone or down-hole geophysics tests). The variation of shear modulus with strain amplitude is usually based on cyclic triaxial tests (ASTM D3999-91) or resonant column tests (ASTM D4015-92). When the results of tests of soil stiffness are not available, BS EN 1998-5:2004 provides a table of average soil damping ratios and average reduction factors (plus one standard deviation) for shear wave velocity V_s and shear modulus G (for the upper 20 m depth range) as a function of peak horizontal ground acceleration.

In terms of total stresses, the value of the soil strength parameters applicable under static undrained conditions may be used. For fine-grained soils (such as silts and clays) the appropriate strength parameter should be the undrained shear strength S_u; S_u should consider cyclic degradation effects under long-duration earthquake actions. For cohesionless soil the appropriate strength parameter should be the cyclic undrained shear strength $\tau_{cy,u}$. When the results of tests of shear strength of sandy soil in the cyclic condition are not available, the frictional angles of sandy soil ϕ can be inferred from the critical cyclic stress ratios τ/σ' (=tan(ϕ)) which were initially proposed by Seed and Idriss (1971) as the boundaries between liquefied and non-liquefied sandy soil and later included in BS EN 1998-5:2004. For dense sand the friction angle in the cyclic condition is assumed to be equal to the angle in the static condition, which can be inferred from, for example, Peck *et al.* (1974).

Srbulov (2003) plotted the ratios between the horizontal peak accelerations at the surface and at depth (equivalent to 'bedrock'), shown in **Figure 60.7**, from 66 recorded case histories. This shows that peak ground surface acceleration is attenuated for peak horizontal accelerations at depth (i.e. bedrock) greater than 0.3–0.4 g. Attenuation of large bedrock accelerations is most likely because of soil yield at greater accelerations. Idriss (1990) also indicated attenuation of bedrock acceleration on soil sites when the acceleration on rock sites exceeded about 0.3–0.4 g (but Idriss' relationship is based on calculations rather than recorded data). However, special precautions should be taken when using this curve because it represents the ground response of several soft, sensitive and highly plastic clay deposits, which tend to have relatively lower damping values under seismic loading than those of low- to medium-plasticity clays.

60.5.5 Simplified liquefaction assessment

Seismic liquefaction refers to a sudden loss in shear stiffness and strength of soil due to cyclic loading effects of an earthquake. The post-liquefaction shear strength is commonly referred to as the residual shear strength and may be 1–10 times lower than the static shear strength. If the residual shear strength is sufficient, it will prevent a bearing capacity failure for level ground conditions, but large settlements may occur due to distortion during the earthquake and dissipation of excess pore water pressures after shaking. For sloping ground conditions, if the residual strength is sufficient, it will prevent a flow slide, but displacements commonly referred to as lateral spreading could occur along with settlements. These ground or foundation movements may be excessive relative to acceptable limits for the structure. Considerable engineering judgement is required to understand the implications and mechanism of liquefaction. Soils that should be evaluated for liquefaction susceptibility include sands, gravelly sands, silts, mine tailings and fine-grained soils with plasticity index not greater than 10. In soil layers with a clay fraction greater than 15% evaluation of liquefaction susceptibility may be omitted. The factors which influence the risk of liquefaction are reasonably well understood and include the following.

- *Soil type and groundwater table level:* Saturated coarse-grained soils (particularly fine/medium sands and coarse silts) in areas with a high water table level are at most risk. In general, the younger and looser the deposit the more vulnerable it will be to liquefaction. In general, deposits which are of Pleistocene age or older are usually considered to be at low risk. Dense coarse-grained soils, with a relative density greater than 80%, are not likely to liquefy. Post-glacial or Holocene soils of fluvial and alluvial depositional origin and loose man-made fill materials are highly vulnerable to soil liquefaction and form the class of soils for which the design guidelines are usually applicable.
- *Depth:* With increasing depth the confining pressure increases and liquefaction resistance increases. Commonly observed liquefaction depths are less than about 15 m. There are no recognised 'maximum' depths for liquefaction, although in general the risk of liquefaction below 20–25 m depth is believed to be low. There is some evidence that liquefaction resistance may improve with ageing and overconsolidation (preloading), but these effects

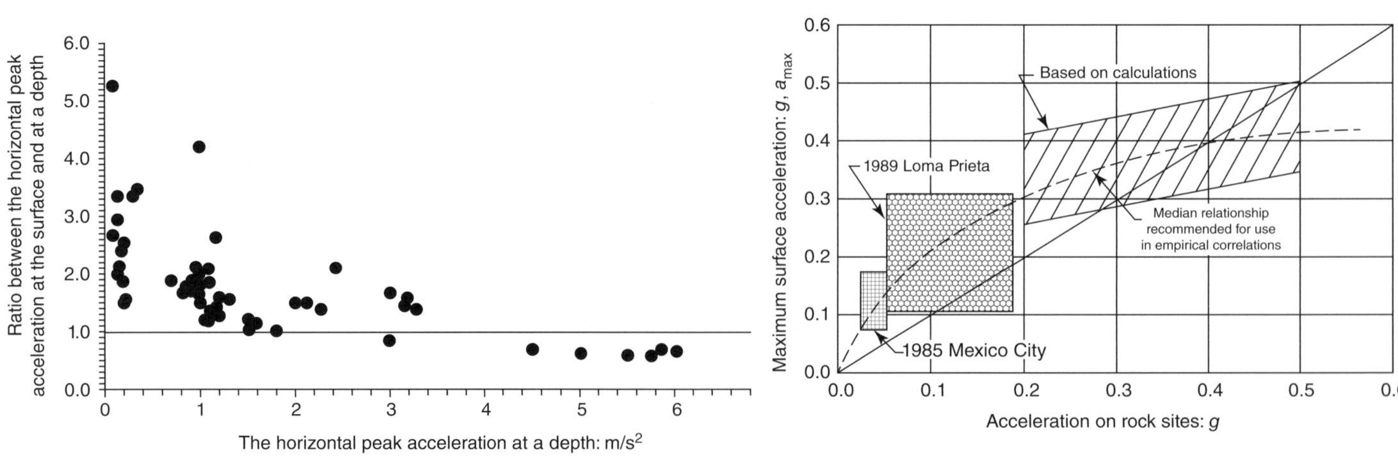

Figure 60.7 Ratios between the recorded horizontal peak accelerations at the surface and at depths (soil or rock) plotted against the horizontal peak accelerations at depths

are difficult to quantify and are usually not directly addressed in design procedures. However, such effects should be reflected in terms of increased penetration resistance and/or shear strength.

- *Drainage conditions and soil fabric:* Poor drainage allows excess pore pressure build-up and increases the risk of liquefaction; if loose sands are sandwiched between clay layers the loss of strength may be more serious due to upward seepage effects.
- *Earthquake-induced shear stress:* As the intensity of ground shaking increases the shear stress increases and liquefaction is more likely.
- *Earthquake duration:* As duration increases (generally a function of earthquake magnitude) the number of loading cycles increases, leading to an increase in excess pore water pressure generation.

Simplified empirical analysis procedures have been developed and codified, and the factor of safety against liquefaction is CRR/CSR, where CRR is the cyclic resistance ratio of the soil at a particular depth and CSR is the cyclic shear stress ratio imposed by the earthquake at a particular depth (e.g. Youd *et al.*, 2001; BS EN 1998-5:2004). The CSR is usually calculated by using Seed and Idriss' simplified equation. If the soil deposit is strongly inhomogeneous (or 'layered') with depth, the maximum shear stress should be calculated from an SRA, carried out by seismic experts. Seed and Idriss' simplified equation is

$$CSR = 0.65 \times a_{max} \times \frac{\sigma_v}{\sigma_v'} r_d \qquad (60.2)$$

where a_{max} is the peak ground surface acceleration for the design earthquake, σ_v and σ_v' are total and effective vertical stress, respectively, and r_d is a stress reduction factor which is depth-dependent and can be calculated using the equations provide by Idriss and Boulanger (2006). For simple projects the value of a_{max} at ground surface can be estimated from the local seismic codes, which provide the peak horizontal accelerations at bedrock but also amplification factors for different soil types. In order to use the simplified equation, reasonable values for a_{max} and r_d are required. These values vary significantly (Cetin *et al.*, 2004) and therefore it is recommended that a site-specific ground response analysis should be conducted for most projects.

The liquefaction resistance can be estimated empirically from in situ tests (commonly standard penetration tests (SPTs) and cone penetration tests (CPTs)) – see Figure B.1 of BS EN 1998-5:2004 – or from laboratory tests. The use of laboratory tests to assess CRR is challenging and requires very high-quality samples and specialist laboratory testing (e.g. cyclic simple shear tests). In situ methods are usually preferred over laboratory methods for evaluating liquefaction susceptibility because of the difficulty and high cost associated with securing high-quality undisturbed samples for laboratory testing which satisfactorily represent in situ conditions. For sandy soils with high fines content (e.g. sandy silts or very silty sands) or sands/silts which are suspected to be overconsolidated, and where empirical SPT-/CPT-based methods may be overly conservative, the use of specialist laboratory testing can be particularly valuable (and may avoid over-conservative liquefaction assessments and associated mitigation measures). Comprehensive guidance on assessing liquefaction risks is given in Youd *et al.* (2001) and Idriss and Boulanger (2008), which should be consulted for more specific information.

BS EN 1998-5:2004, Clause 4.1.4 provides guidance on assessing liquefaction potential. This states that liquefaction hazard may be neglected when the seismic action on the surface is less than 0.15 g (1.5 m/s^2) and either the soil is sufficiently dense (SPT $N_1(60) > 30$) or it has a clay content greater than 20% and plasticity index greater than 10%. Liquefaction of saturated loose coarse-grained soils is always accompanied by ground settlements as the excess pore pressures generated during the earthquake shaking dissipate. The settlement is typically completed within seconds to about 1 day after the shaking, depending on the characteristics of the soil and the length of the drainage path. Post-liquefaction settlements as high as 2–3% of the sand layer thickness for loose to medium sands and more than 5% for very loose sands have been reported.

Clays and clayey silts can suffer a loss of strength/stiffness during earthquake shaking, usually termed 'cyclic mobility'. The current understanding of the seismic behaviour of silts is limited due to lack of laboratory and field data on the cyclic performance of natural silts. Fundamentally the overconsolidation ratio and strength sensitivity (ratio of peak to remoulded strength) are important factors. Bray *et al.* (2004) have suggested the following classification based on moisture content (w), liquid limit (w_L) and plasticity index (I_p)

- $w/w_L > 0.85$ and $I_P < 12$: susceptible to liquefaction or cyclic mobility*
- $w/w_L > 0.8$ and $12 < I_P < 20$: moderately susceptible to liquefaction or cyclic mobility*
- $w/w_L < 0.8$ and $I_P > 20$: no liquefaction or cyclic mobility but may undergo significant deformation if cyclic shear stress exceeds static undrained shear strength (S_u).

* This classification may be revised on a site-specific basis using data from laboratory cyclic shear testing of good-quality field samples (e.g. samples obtained using thin-walled tube samples with sharpened (i.e. >5°) cutting edge and no inside clearance).

Boulanger and Idriss (2007) have recommended that fine-grained soils be classified as 'sand-like' (susceptible to liquefaction) if PI (plasticity index) <7, and 'clay-like' if PI ≥ 7. Bray *et al.* (2004) suggests that the residual strength (S_r) for silt and clay zones be determined as

- $w/w_L > 0.85$ and $I_P < 12$: S_r = remoulded shear strength ($S_{remoulded}$) unless appropriate testing of undisturbed samples can show greater strength

- $w/w_L > 0.8$ and $12 < I_P < 20$: $S_r = 0.85\ S_u$ where S_u = static undrained shear strength
- $w/w_L < 0.8$ and $I_P > 20$: $S_r = S_u$.

Based on case studies, coarse-grained soils like gravel can also liquefy in certain circumstances (e.g. Kokusho *et al.*, 1995). Dense to very dense coarse granular soil and overconsolidated fine-grained soil tend to dilate initially during shearing (i.e. after initial cycles) and exhibit greater shear resistance and stiffness than normally consolidated soil.

60.6. Dynamic soil–structure interaction

The overall performance of a structure during seismic shaking is influenced by the interaction between three interlinked systems: the structure, the foundation and the complex geological media underlying and surrounding the foundation. SSI (soil–structure interaction) effects may have a significant influence on the seismic behaviour of a building and the way in which some mechanisms might develop in the structure. Accordingly, possible SSI effects should be considered as part of an assessment and a decision made on how detailed and complex the inclusion needs to be. A simplified SSI analysis can be undertaken with upper- and lower-bound geotechnical parameters to determine the most adverse consequences from the probable range of deformations resulting from ground behaviour (e.g. range of foundation flexibility due to pile tension uplift) and step change scenarios (e.g. differential settlements due to liquefaction occurring or not occurring).

A seismic SSI analysis evaluates the collective response of the three systems to the specified free-field ground motion. Interaction effects are classified as inertial and kinematic and are usually handled by either the impedance approach or the direct approach. In both approaches the mathematical complexity is enough to persuade most design engineers to ignore these effects. The motion recorded at the base of a structure or in the immediate vicinity is different from that which would have been recorded in the absence of the structure. However, very few instrumented cases studies exist which quantify this difference; numerical modifications are usually used to account for such effects. In most cases the free-field surface-level ground motion is selected as the control motion used at the foundation level, neglecting the kinematic interaction effects. This assumption is based on the perceived beneficial role of SSI in reducing the seismic response. But there are numerous documented case histories where ignoring SSI has led to oversimplification in the design, leading to an unsafe design in foundation and superstructure (Gazetas and Mylonakis, 1998). The impedance method is more popular; in this method the founding medium is represented by homogeneous, isotropic, linear elastic or visco-elastic half-space extending to an infinite depth. Most of the foundation impedances have been derived assuming homogeneous half-space conditions for the soil which overpredict the damping for structures on actual soil profiles.

Engineers should note that it is important to consider the potential for the soil to be stronger/stiffer or weaker/softer and for this variability to be non-uniform in distribution. Similarly, imposed displacements or loads may be uniform or differential. **Figure 60.8** illustrates a simple example of the range of structural responses as a consequence of the soil strength/stiffness adopted. Assuming unrealistically stiff soil/foundations (e.g. fixed base assumptions) could result in an unrealistically low natural period of shaking for the structure (unrealistically high seismic loads) or underestimation of structural deformations. The converse also applies. SSI effects are complex but can often be simplified for assessment, particularly initial screening to assess sensitivity of behaviour. For example, this could be as simple as recognising that the soil support for a footing may not be rigid and reflecting on what this means for the rigidity of a supported column and its ability to receive flexural resistance/restraint at the base. This may influence the possible actions in the column and mechanisms that are possible in the structure. For this example it may be appropriate to at least consider

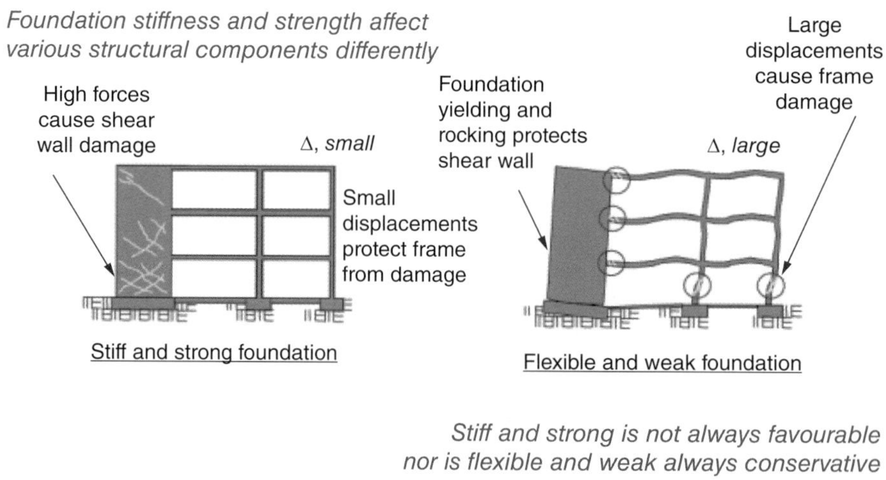

Figure 60.8 Influence of SSI on structural performance (Adapted from Mahoney, 2005 and www.buildings.nz)

the possibility of varying restraint, within appropriate bounds, when assessing the structure. Simple hand checks can be undertaken collaboratively with the structural engineer to assess if the building is likely to be sensitive to the deformation demands from foundation flexibility. The amount of acceptable deformation for foundations generally depends primarily on the effect of the ground-induced lateral deformation on the structure and ultimately on the life safety hazard that can develop.

SSI generally increases the period of vibration and damping ratio of structures (ATC, 1978; BSSC, 1995). The increase in the structural period may appear to be beneficial according to smoothed design spectra, whose amplitude decreases with the period increase after a threshold period, usually in the range of 0.5–1.0 s. However, site-specific acceleration spectra like the ones in Mexico City (Kramer, 1996) and in Kobe (Gazetas and Mylonakis, 1998) indicate that an increase in structural vibration period can be responsible for structural damage due to resonance effects between the vibration of a structure and soil layers when the period of vibration of the soil deposit is relatively large (e.g. 2 s in Mexico City and 1 s or more in Kobe). Stewart *et al.* (1999) evaluated SSI effects for 77 strong motion datasets at 57 building sites and found a pronounced influence of structure-to-soil stiffness ratio on inertial interaction, as well as secondary influences from structure aspect ratio and foundation embedment, shape and flexibility. They found that kinematic interaction effects on the 'input' motion at the bases of structures are relatively modest in many cases, whereas inertial interaction effects on the structural response to these motions can be significant. Wolf (1994) provides simplified expressions for the calculation of an equivalent period T_e of soil–foundation–structure systems. Some codes also provide simplified expressions. These can be useful to assess resonance-type effects.

The analysis of SSI effects should consider two effects.

- Inertial effects modify the dynamic response of the structure, caused by the change of the analysed system which affects its fundamental period and increases its damping, due to both radiation and material damping in the soil. A rule-of-thumb test to determine if inertial SSI effects are likely to be significant was developed in NIST (2012). This test combines three simple parameters that are relatively easy to obtain – building height, fundamental period of the structure and shear wave velocity – into a term called the structure-to-soil stiffness ratio, and correlates the ratio with the likelihood that SSI inertial effects will be significant.
- Kinematic effects that modify the seismic excitation at the base of the structure, and the loading of foundation elements with respect to the free-field. It should be recognised that kinematic effects remain controversial, and many designers would not consider them except for very large earthquakes. Modification of the foundation input motion should usually be taken into account for
  - deep foundations
  - foundations embedded at a depth of at least two basements, or a depth larger than $L/4$, where L is the largest dimension of the foundation in plan, with the foundation vertical surfaces in full contact with the surrounding soil
  - abutments of bridges with large embankments, or integral bridges
  - foundations with one or two large dimensions in plan ($L >$ 50 m) consisting of a slab, a single box foundation or footings interconnected with tie beams, where kinematic interaction should be considered.
  - Kinematic interaction may be neglected for the vertical component of the seismic action.

SSI may be neglected for an embedded foundation founded on rock with a shear wave velocity larger than 800 m/s and for low-seismicity regions. The effects of SSI should be considered if at least one of the following conditions applies as specified in several technical documents.

- Soil–structure systems in which an increase in the fundamental period due to SSI effects results in an increase of SAs.
- When the displacement of the structure controls the width of joints separating nearby buildings (existing or planned) or other performance criteria.
- Structures supported on soft soils in which the shear velocity compared to the averaged velocity over a depth equal to 3.0 times the maximum foundation width in the case of footings or to the maximum width in the case of a raft foundation is less than 250 m/s.
- Structures in which geometrical nonlinearity, known as the P–Δ effect, plays a significant role.

SSI effects can be assessed by a range of techniques with varying degrees of complexity (FEMA, 2020). These can be classified as

- simplified hand analysis to evaluate influence of ground
- simplified flexible base model using linear Winkler springs
- flexible base model using equivalent linear springs
- simplified flexible base model using compression-only or tension-only Winkler springs
- nonlinear pseudostatic analysis with explicit nonlinear soil springs
- direct finite element modelling.

For most assessments only the simplified techniques will be required. The more complex methods should only be used if

- a more simplified method has been applied first to develop an understanding of the likely SSI effects
- the assessment of the simplified analysis indicates that more complex analysis will be beneficial in better understanding the structure's behaviour and meeting the overall objectives of the project
- adequate investigation and assessment have been undertaken to define geotechnical and structural input parameters to a level of detail consistent with that of the analysis.

For the direct method usually state-of-the-art finite element packages like LS DYNA, FLAC, SWANDYNE and ANSYS are used. The analysis is often based upon section 5 of ASCE 4–16, together with supporting information from other sections of this standard (**Figure 60.9**). The direct method described in section 5 of ASCE 4-16 is used for analysing the combined soil–structure system in a single step without invoking superposition for the structure or the soil separately. The direct method will solve the SSI problem in the time domain. The direct method is implemented in a nonlinear time-history analysis. Essentially this analysis needs to capture nonlinear behaviour during larger earthquakes, which includes gapping and sliding and material nonlinearity. Dynamic time-history analysis of the whole structure–foundation–soil system may be used to capture both kinematic and inertial effects in a single step.

Complex analysis including direct nonlinear modelling of the soil and its interaction with the structure is possible and may be warranted in some situations. **Table 60.2** provides some further guidance on when to use the next level of sophistication of SSI modelling. Irrespective of the SSI modelling approach adopted, sanity checks of complex model situations by approximate calculation and a simplified ground model are essential. The variable nature of the soil and the way in which the building interacts with it means that analysis runs to investigate the sensitivity of the results to the modelling parameter will almost certainly be required. If SSI behaviour provides a beneficial influence on the structural performance (e.g. period elongation) the SSI analysis and geotechnical considerations should be cautiously appraised and also subjected to appropriate peer review.

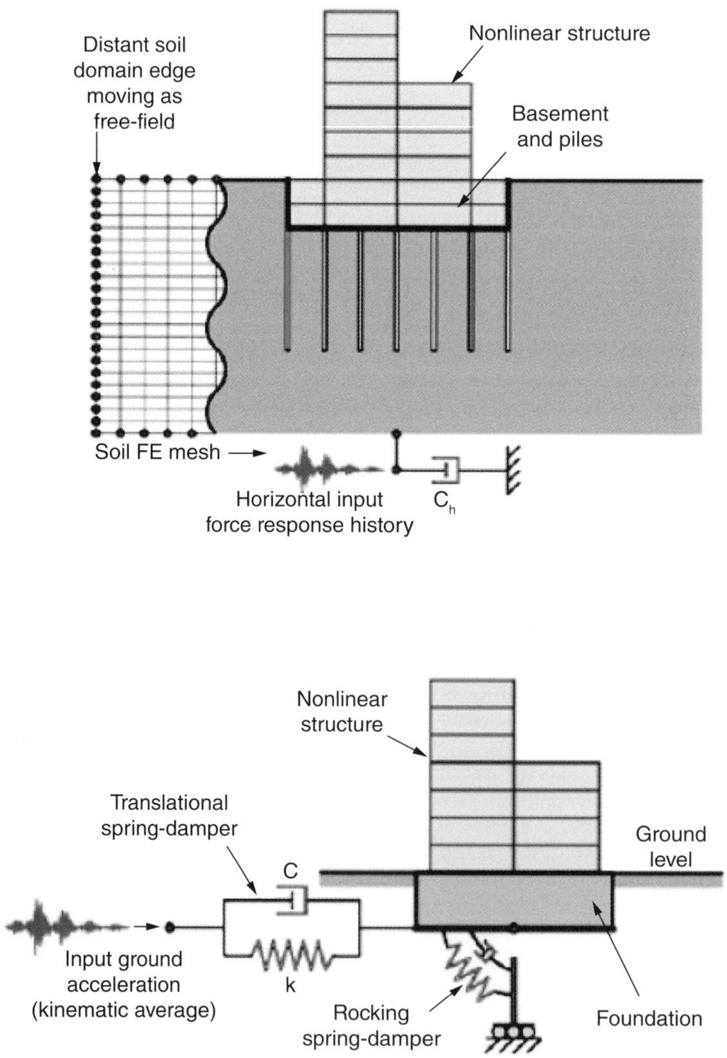

Figure 60.9 Direct and indirect SSI modelling (Deierlein *et al.*, 2010)

SSI analysis option	When to use/not to use	Comments
Fixed base model – no SSI consideration	This should not be used for high-rise buildings on piles or slender wall systems with shallow foundations	The foundation structure will still need to be assessed by hand • global overturning stability • yielding at the soil–foundation interface
Simplified flexible base model using linear Winkler springs	Shallow foundations Core walls Basement/part basements	The superstructure needs to be assessed for a fundamental period considering both fixed base and flexible base (i.e. building period) Shift due to foundation flexibility is to be considered
Flexible base model using nonlinear soil springs (either explicit nonlinear or equivalent linear springs) and site response analysis	Shallow foundations Core walls Basement/part basements	Equivalent linear springs need iteration between structural analysis and geotechnical *P–Y* curve analysis The use of rotational springs or multi-axial springs will need careful consideration of the assumed effective damping and equivalent linearisation of the nonlinear system
Flexible base – nonlinear dynamic history (e.g. nonlinear time-history analysis)	Irregular system on complex soils and foundations Soil and foundation could potentially result in catastrophic step change behaviour	The shape of the hysteresis curve should be realistic and reflective of the ground conditions No additional damping should be included for foundation radiation damping, etc. Horizontal springs can artificially damp out ground acceleration – these should be used with care
Advanced geotechnical SSI analyses (e.g. nonlinear finite element analyses)	Where ground deformations are potentially critical and significant (e.g. behaviour of high-rise buildings adjacent to a tunnel or steep slope)	There needs to be a robust process for interlinking the advanced/complex finite element ground model behaviour with the global structural models

Table 60.2 Different SSI analysis options

60.6.1 Shallow foundations

The failure mechanisms for shallow foundations in cyclic conditions are the same as those for static conditions, which are considered in Chapter 53 *Shallow foundations*. The engineer has a variety of approaches which may be applicable to the seismic design of shallow foundations. Indeed, it is quite likely that on any major project several of the different classes of analysis will be used to assess different aspects of the foundation behaviour as the project proceeds. An overview of the types of analysis which may typically be used is illustrated in **Figure 60.10** (May, 2015). The designer of shallow foundations subject to seismic loading ideally seeks solutions to the following questions.

- What transient and permanent foundation deflections will occur? Knowledge of the seismic bearing capacity will assist in answering this question, but will not provide the full solution.
- What transient accelerations, forces and/or displacements will be transmitted through the foundation and into the structure?
- Will the foundation continue to support the structure following the seismic event?
- How can economic foundation solutions be provided for new and existing structures?

The foundation of a structure in a seismic zone shall transfer the relevant forces from the superstructure to the ground without structurally unacceptable residual displacements. In a displacement-based approach, the calculated displacements due to sliding are acceptable to the superstructure and do not adversely affect the performance of any lifelines (e.g. water, gas, access or telecommunication lines) connected to the structure. Soil bearing capacity, foundation sliding and overturning resistances are usually checked using static methods with addition of an inertia force from the foundation and structural acceleration. Soil properties in cyclic conditions need to be used with the static methods. Also, the effect of inertial force on soil under the foundation needs to be considered, for example using an equivalent inclined foundation or inclined ground surface. The angle of inclination may correspond to the angle necessary to rotate the resultant of gravity and inertia forces acting on the soil under the foundation into the vertical. BS EN 1998-5:2004 (BSI, 2004b) provides an informative formula for assessment of seismic bearing capacity of shallow foundations which allows the load inclination and eccentricity arising from the inertia forces in the structure as well as the possible effects of the inertia forces in the supporting soil itself to be taken into account.

Bearing capacity calculations by means of conventional bearing capacity expressions have considerable limitations. Conventional bearing capacity expressions are inaccurate under two-way loading conditions (which occur under earthquake loading) and for embedded foundations, including foundations with skirts. Chapter 53 *Shallow foundations* provides an introduction to modern bearing capacity theory, which is generally more appropriate for shallow foundation design under earthquake (or other dynamic/cyclic) loading conditions. The advantage of using this type of VHM formulation is that the application of a load factor to the H and M terms provides an assessment of the factor of safety for a given value of V, whereas this is more difficult to determine from classical bearing capacity theory.

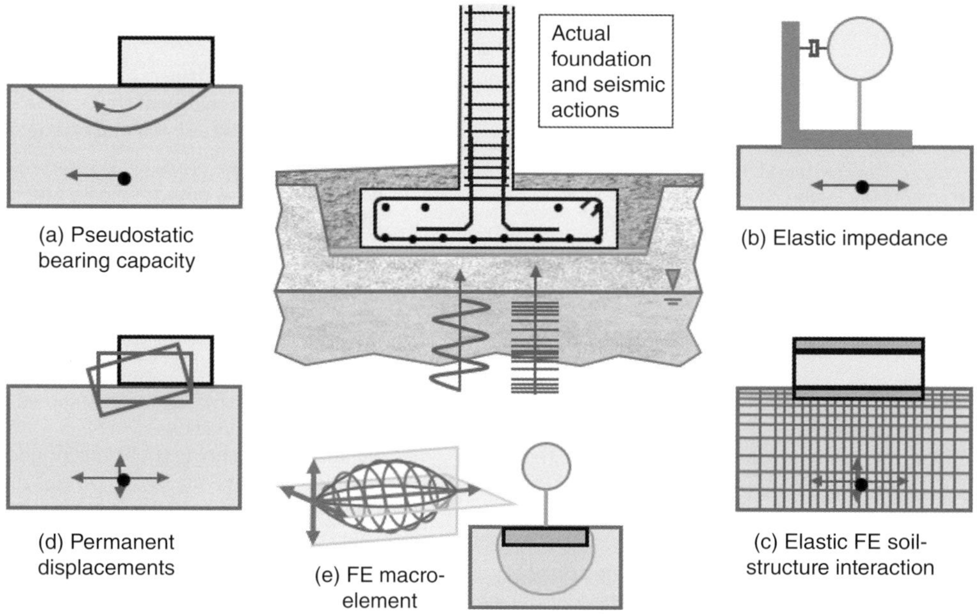

Figure 60.10 Analytical methods for seismic design of shallow foundations (May, 2015)

The formulae for static bearing capacity that use the coefficient $N\gamma = 2(N_q + 1)\tan(\varphi)$ are not considered safe to use. It should be noted that rocking and uplift of foundations can reduce the inertial forces entering the superstructure and therefore protect it with regard to these forces. Rocking can be allowed provided the induced residual rotation and settlement of the foundation are compatible with the performance criteria of the superstructure at the relevant limit state.

Settlement of shallow foundations during earthquakes is caused by soil densification – that is, decrease in soil porosity due to ground shaking (soil porosity is the ratio between the volume of voids and the total volume of soil). The foundation system should be verified against soil densification and settlements, calculated under free-field conditions. The Eurocode 8-5 formulae for seismic bearing capacity permit a limited amount of sliding displacement to be considered and implicitly allow for some permanent deformation in the partial model factors (γ_{Rd}) suggested for seismic bearing capacity analysis – see Pecker and Pender (2000). Methods for undertaking such displacement assessments are still under development. Richards *et al.* (1993) provided an approximate method of assessment based on a two-wedge idealisation of the bearing capacity failure surfaces leading to a seismic settlement calculated on the basis of a Newmark sliding block analogy similar to that used for assessing the displacements of gravity retaining walls.

In cases where the transient seismic loadings exceed the available foundation resistance, permanent displacements will occur. The accelerations at which this displacement begins are termed threshold accelerations. In many cases the peak earthquake accelerations can exceed the threshold values by a substantial margin with minimal foundation displacement occurring. Although EN 1998-5 generally requires that foundations remain elastic, for foundations above the water table, where the soil properties remain unaltered and the sliding will not affect the performance of any lifelines connected to the structure, a limited amount of sliding may be tolerated (section 5.4.1.1(7)). This requires careful consideration by a specialist.

Seismic codes contain limited recommendations for analyses of shallow foundation vibration. For simplified SSI analyses, shallow foundations are represented by equivalent vertical, horizontal and rotational springs (K) and dampers (C). Ground motion can be applied at the ends of the springs and dampers or on a structure. Vertical springs and dampers are needed even if only the horizontal ground-motion component is considered because of coupled horizontal and rocking modes of vibration, with the vertical component at the edge of the foundations. The formulae for calculation of values of the dynamic stiffness and damping coefficients are tabulated in the relevant textbooks (Gazetas, 1991; Wolf, 1994). If foundation vibration is considered excessive for the intended use of the structure, then foundation vibration isolation methods can be used.

60.6.2 Deep foundations

Piles are most frequently used for deep foundation types, although vertical shafts and caissons are used occasionally. It is not an overstatement to say that piles are used in most heavily loaded structures such as multistorey buildings, bridges, flyovers and oil platforms. Piles are also used in seismic-prone zones having loose to medium-dense sandy soil or soft clay. Deep foundations are subjected to kinematic and inertial interactions. Deep foundations can also be subjected to lateral loading from surrounding soils or kinematic interaction from induced ground deformations. Site response to earthquake shaking can cause amplification of the arriving shear waves and surface waves near the ground surface, resulting in shear deformations of the ground which lead to interactions between embedded piles and surrounding soil. Generally, such ground deformations are not damaging for well-designed piles except in the case of soft soils, liquefied soils, localised severe ground deformation and/or large relative displacements between soil and stiff piles caused by severe earthquake excitations. Significant research has been done in the area of soil–pile–superstructure interaction in liquefiable soil, and the reader is referred to specialist books on this subject (Bhattacharya and Madabhushi, 2008). Deep foundations extending through liquefiable soils will require the following special considerations.

- The ability of the foundation to resist lateral loads applied by the structure to the foundation may be reduced.
- The bearing (and uplift) capacity of the deep foundation may be reduced due to reduction in soil strength.
- Settlement of the liquefied soil may cause down-drag or negative friction forces on the deep foundation.
- Liquefaction-induced soil movements may cause lateral loading that will displace or fail the deep foundation.

The following approach can be adapted/modified from the generalised approach to gravity design of deep foundations as discussed in Chapters 53 *Shallow foundations* and 54 *Single piles*. Further information can be obtained from Building performance, Module 4 from the New Zealand Geotechnical Society's Earthquake Geotechnical Engineering Practice Guidelines.

- *Calculation of ULS axial capacity:* When evaluating the axial compression capacity of the pile using standard equations, values for unit side resistance and unit end bearing should be reduced for layers where pore water pressure generation or cyclic softening is expected. For cohesionless soils the effective stress and thus friction may be reduced by a factor of $1/(1 - R_u)$ (Boulanger and Brandenburg, 2004). R_u may be evaluated from **Figure 60.11** after determining the factor of safety against liquefaction. For clay soils, a reduction in undrained shear strength S_u may be estimated using the procedure of Boulanger and Idriss (2007).
- *SLS settlement check:* For cases where the SLS axial load is able to be resisted entirely by the mechanism of side resistance, settlements should remain small (i.e. less than 12 mm; Brown *et al.*, 2010). For cases where the mechanism of side resistance is likely to be mobilised, and where down-drag is possible, then some transfer of load to the pile base will occur, causing partial mobilisation of the end-bearing mechanism of the pile. Additional pile settlement is required for mobilisation and might be significant, especially for bored piles. A crude estimate of the additional settlement may be made by assuming that the calculated end-bearing capacity of the pile will be fully mobilised at settlements of 5–10% of pile diameter, depending on the assumptions of the calculation procedure used to determine capacity.
- *Pile buckling:* The strength and stiffness of liquefied layers may be reduced to very low values and the soil may provide inadequate stabilising support to slender piles through thick liquefied layers, resulting in pile buckling. Further information is given by Bhattacharya and Bolton (2004).

Three methods of analysis are usually suggested, depending on the level of complexity and importance of the case:

- *Method 1:* Limiting equilibrium calculation with factor of safety is appropriate for simple cases where site soils are expected to remain firm and where the site is stable. In this phase, the building is responding dynamically to the earthquake shaking and the resulting inertial loads (base shear) are applied to the foundations and there is no liquefaction. The available resistance to lateral load should be calculated including passive soil pressure acting against the embedded structural elements and the lateral load capacity of the piles. The ultimate lateral resistance of the piles below the pile head may be calculated using the conventional limiting equilibrium procedures.

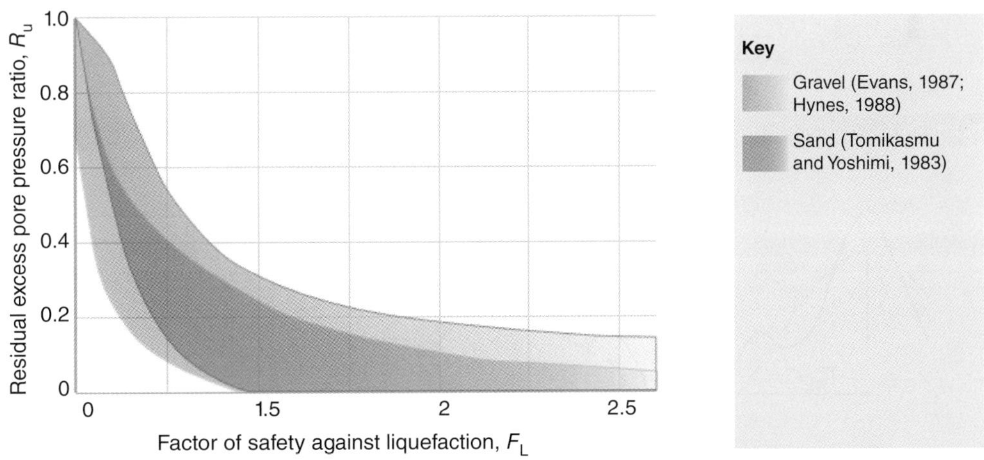

Figure 60.11 Excess pore water pressure ratio ($R_u = \Delta U / \sigma'_v$) plotted against factor of safety and liquefaction triggering under level ground conditions (Marcuson *et al.*, 1990)

- *Method 2:* Quasi-static beam-spring analysis. Method 2 is appropriate for most cases and can be adapted to analysing piles for most scenarios. In this phase, liquefaction has triggered in susceptible layers with resulting significant reduction in soil stiffness and strength. Other soil layers may also be softened by elevated pore water pressures or cyclic softening effects. Beam-spring analysis (e.g. '*P–Y*' analysis, (Reese *et al.*, 1974, and subsequent commercial software packages)) is used to calculate lateral displacements and pile bending strains under the ground deformations and building inertial loads (**Figure 60.12**).
- *Method 3:* Numerical time-history analysis. For more complex cases (e.g. piled waterfront structures) or buildings of high importance it may be more appropriate to carry out numerical time-history analyses. Such analyses should include the soils, foundations and building structure in a realistic way so that the complex interactions among them can be better understood.

Inertial interaction occurs because of structural mass, which together with structural stiffness causes a different vibration frequency (period) for a structure with respect to the adjacent ground. Different structural periods cause different periods of pile vibration, which tend to oscillate asynchronously with the ground. A superposition of incoming ground waves to a pile and outgoing ground waves from a pile vibration can cause superposition of wave amplitudes and new wave patterns as sketched in **Figure 60.13**. Guidance on combining inertial and free-field kinematic ground displacement loading for pseudo-static design procedures can be found in Idriss and Boulanger (2008).

Seismic codes contain limited recommendations for deep foundations during earthquakes. For simplified SSI analyses the formulae for calculation of values of the dynamic stiffness and damping coefficients of individual piles are tabulated in the relevant textbooks (Gazetas, 1991) for different variations of soil stiffness modulus with depth. If large lateral ground displacement occurs in the post-seismic stage, then the effect of lateral forces imposed on piles can be analysed using the methods described in Chapter 57 *Global ground movements and their effects on piles.*

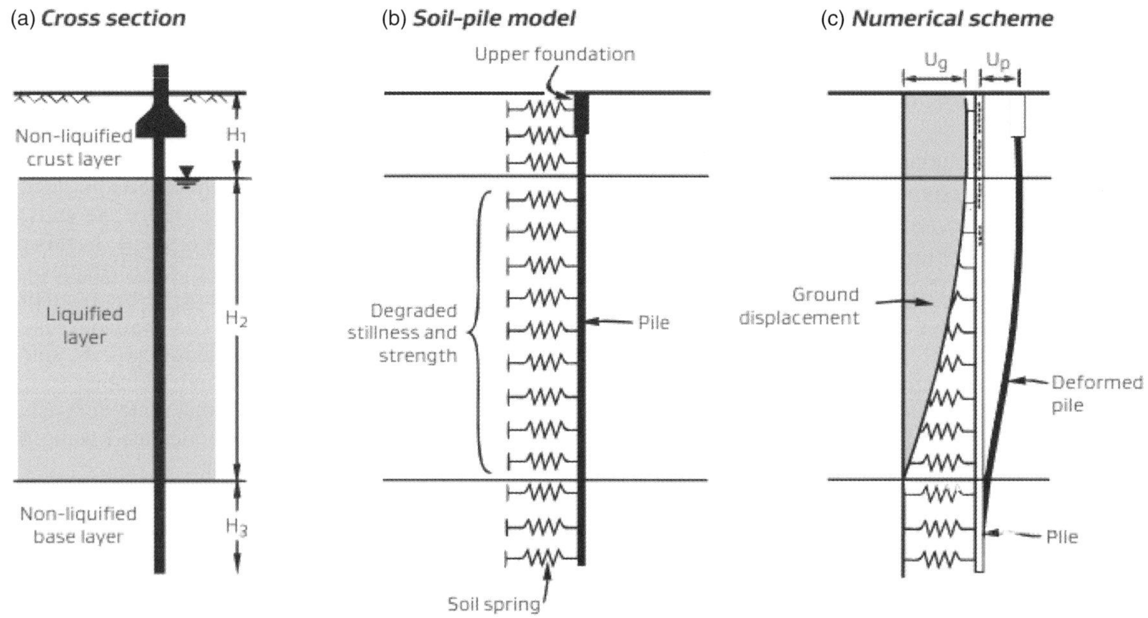

Figure 60.12 Typical beam-spring model for pseudo-static analysis of pile (Cubrinovski *et al.*, 2009)

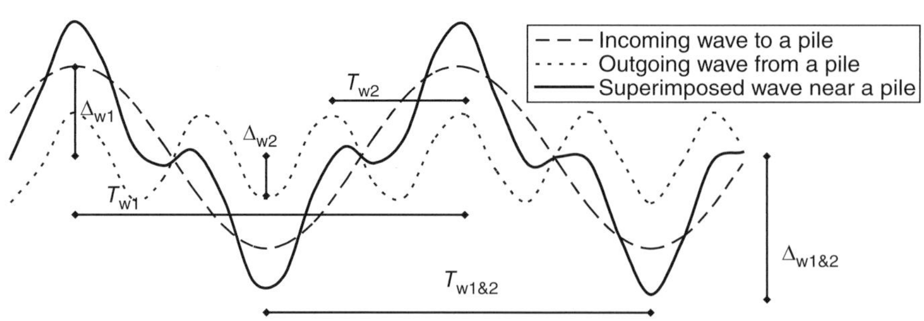

Figure 60.13 Sketch of the superposition of wave amplitudes near a pile due to different vibration properties of the adjacent ground and piles

Examination of dynamic behaviour of pile foundations in seismically liquefiable soils led to the following practical recommendations (Rostami *et al.*, 2020).

- In order to consider a possible boundary for safe design and avoid failure of the pile, it is suggested to consider the buckling mechanism together with the effect of lateral load. Pile foundations can collapse due to bending, buckling and combined action of bending of pile foundations in the seismic region.
- The review of results from various physical modelling techniques indicated that the dynamic behaviour of pile foundations depends not only on the site-specific characteristics (e.g. frequency of input motion, amplitude, etc.), but also on the dynamic (modal) characteristics of the system (fixity of the pile head, the pile tip, pile position within the group, thickness of liquefying soil layer, etc.), which considerably affect pile performance in liquefied grounds.
- The three methods outlined earlier can be used for design, depending on the complexity of the overall system.

60.6.3 Risk reduction measures

Risk reduction measures are dependent on the scale of the hazard and the consequence of failure. When foundation response is a source of hazard in competent ground but during strong earthquakes, then an increase in number, depth or diameter of piles can be helpful. A number of densification techniques (e.g. deep dynamic compaction, vibro-compaction, compaction piles) have been widely studied because these techniques are relatively simple and practical, and the resulting remediation success can be easily verified using in situ penetration techniques. For earthquake-prone areas the use of raking piles to support horizontal load requires careful consideration. Raking piles will provide a horizontally stiff foundation; however, they can be vulnerable to large adverse loads if significant global ground movements develop during or after an earthquake (e.g. if liquefaction occurs). The structural detailing of raking piles, and pile–pile cap connections, needs to ensure they behave in a ductile manner. If a pile or pile–pile cap connection fails in a brittle manner this could lead to progressive failure of the overall foundation. When behaviour of weak ground is a source of hazard, then the following measures can be considered.

- Effective ground improvement may include methods such as densification, grouting, gravel piers, chemical injection and vertical drains (refer to Chapter 59 *Design principles for ground improvement*).
- Solidification methods (e.g. deep soil mixing, jet grouting, walls, deep soil-mixed, sheet piles or lattice-shaped walls) are promising ground improvement methods which have proven to be effective in stabilising potentially liquefiable soil at several sites during earthquakes not only in controlling lateral spread but also in preventing liquefaction (refer to Chapter 59 *Design principles for ground improvement*).
- Slope stabilisation methods are described in Chapter 72 *Slope stabilisation methods*.
- When the fines content does not exceed about 15–20%, stone columns can be efficiently used to increase ground density, permeability and shear strength, which all decrease the potential for soil liquefaction.
- Ground drains and trenches can be used to speed up the dissipation of excess pore water pressure. Ground drains may become clogged in time by siltation, when groundwater level varies and by bacteria and algae growth.
- Soil mixing with binders can be used for the increase in stiffness of soil and decrease of amplitudes of its vibration when vibration and lateral deformations caused by stone column installation are not acceptable and when the spacing is limited, such as among existing piles and anchors (Kramer, 1996; Port and Harbour Research Institute, 1997). More details on ground improvement are provided in Chapters 59 *Design principles for ground improvement* and 84 *Ground improvement*. Mitchell and Wentz (1991) provides a useful summary of the effectiveness of ground improvement following the Loma Prieta earthquake in California, USA. Other guidance and recommendation can be seen in Madabhushi (2007).

60.7. Offshore foundation design

By 2030 it is expected that there will be 205 GW of power generated by offshore wind farms (OWFs) globally. For wind turbine foundation designers, some of the biggest challenges over the next decade will be the shift from shallower sites to deeper water, and a move into new markets where supply chains are not yet established. Of the 35 GW of offshore wind power installed around the world at the end of 2020, over two-thirds are in wind farms off the UK, Germany and China. The overwhelming majority of these wind farms use monopile foundations. More comprehensive guidance for the design of foundations for offshore structures can be seen in Bhattacharya (2019).

Typical hazards that may be present on a site for an offshore renewable project include, but are not limited to

- areas of soft soils (e.g. channel infill), the presence of which may affect foundation placement and installation depths and may also restrict the selection of installation vessels
- areas of mobile seabed, the presence of which may affect foundation behaviour, loads and installation depths and may also affect cable routing, installation and long-term burial/protection
- very hard soils or bedrock, the presence of which may affect foundation installation methods, installation depths as well as cable routing and burial/protection options and methods
- rapid change in foundation conditions that may determine the selection of more than one foundation type for a development area
- surface or buried obstructions, boulders, unexploded ordnance (UXO) and so on

- shallow gas, the presence of which may impact foundation stability and the safe drilling of geotechnical soil borings
- seismic risk and the potential for soil liquefaction.

60.7.1 Typical condition and wave loading

Wind and wave loading differ from an earthquake in the duration of an event. Whereas a large earthquake is over in a minute or so, a wind or wave storm may continue for hours and even, in a very extreme event, days. Wave heights are determined from meteorological observations at sites of interest. The periods of sea waves range from about 2 to 27 s (Poulos, 1988b) and therefore the inertia forces arising from wave load are not significant, but the effects of cyclic loads are significant and the equivalent static analyses, which are usually performed, need to consider soil properties in cyclic conditions. Detailed description of pile drivability analyses is provided by Dean (2009). The number of hammer blow counts from drivability analyses is used to calculate fatigue damage to pile steel and remaining useful lives in years based on so-called S–N curves (American Petroleum Institute, 2014; HSE, 2001). The effect of cyclic wave loading on fatigue damage to pile steel is also considered using S–N curves.

For most offshore foundations the allowable deformations are relatively large, and usually the critical design criterion is an acceptable factor of safety against collapse under a severe storm event. The geotechnical conditions at a wind farm site are critical to foundation design decisions. Weak soils can lead to excessive pile penetration and hard layers can limit pile drivability.

The fundamental mode of vibration of the structure highly depends on the soil stiffness. Unlike many offshore geotechnical problems, assessment of the soil stiffness is a high priority for offshore wind foundations. Also, underestimation of the soil stiffness involves stiffer and larger support structure design and is therefore one of the primary reasons for increased use of steel, meaning higher costs of the foundation and restriction of the use of current foundations to shallow water depths. The dynamic response, and in particular the amplitude of the structural vibrations, is then governed by the damping of the structure and foundation.

Finally, serviceability and fatigue design must also ensure limited displacement of the infrastructure over time. The maximum tilt rotation of the support structure of fixed foundations is generally specified by the turbine manufacturer to guarantee good operation of the turbine. Typically, the maximum tolerance for monopile foundation tilt over the wind turbine lifetime (including installation tolerance) is 0.5° (DNV, 2014), with a maximum requirement of 0.25° rotation induced by SLS and also a maximum of 0.25° rotation tolerated following installation.

In conclusion, for sites where solely wind and wave loadings are considered, the design of the offshore foundation needs to address

- accurate determination of foundation capacity during extreme events
- accurate estimation of the foundation stiffness during extreme events and operational conditions (and foundation damping)
- prediction of the accumulation of permanent deformations of the foundation.

60.7.2 Typical site investigation for offshore foundations

A reliable model of the seabed stratigraphy including quantification of engineering parameters for relevant layers is essential for engineering design of an offshore foundation. Often, preliminary design studies will be called for prior to any site investigation at the particular site, and estimation of the seabed characteristics must rely on regional knowledge. In any case, it is advisable to establish a site model that extends beyond the immediate location, as this allows for subsequent adjustments in the positions of facilities and indicates the spatial uniformity of seabed conditions across the region. Critical features such as faults, buried channels or other localised non-uniformities will need to be mapped.

Complete site characterisation involves three stages

- desk study, including preliminary metocean data
- geophysical investigation and
- geotechnical investigation.

Integration of these studies (**Figure 60.14**) will lead to the engineering properties needed for the foundation design of the planned structures. Note that the metocean data are not part of the geotechnical characterisation of seabed conditions but will be needed during preliminary foundation studies to determine loads. Knowledge of wave and current conditions will also help in assessing potential seabed features such as sand waves, storm-induced mud slides or ice gouges.

60.7.3 Soil behaviour

Soil behaviour in cyclic condition is described in Chapter 24 *Dynamic and seismic loading of soils.* Offshore marine clays can have very soft consistency or be underconsolidated and prone to cyclic degradation. Specific ground types encountered offshore are calcareous sand and corals. Calcareous sand has fragile grains that are crushed under impact load (Kolk, 2000) and can exhibit very small shear resistance, such as very soft clay. Coral structures may be collapsible because of the presence of cemented skeletons and numerous voids between the parts of skeletons (Touma and Sadiq ,1999). Global and local scour could reach significant depths offshore in loose top sand. Local scour causes gapping, which can significantly decrease the lateral pile resistance and affect vibration period (frequency). For this reason, scour protection is often used. Global scour decreases vertical effective stress around piles. Even submerged, soil offshore may be partially saturated due to the presence of natural gas.

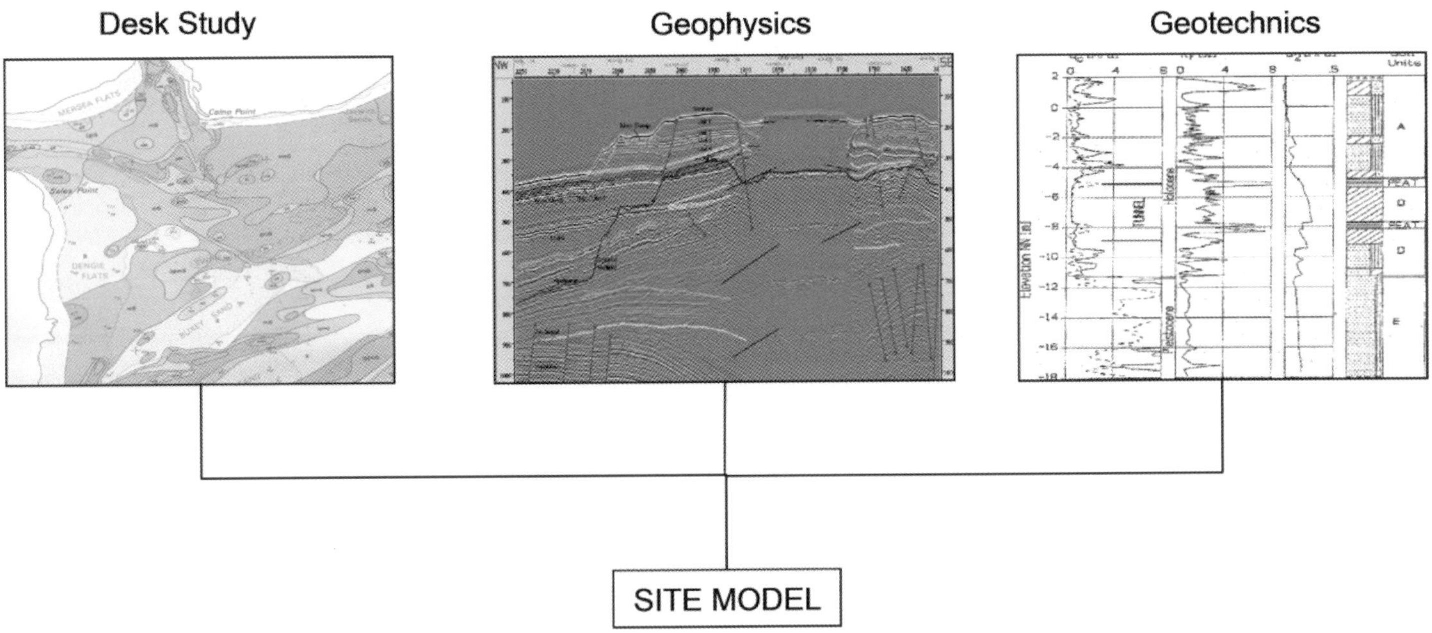

Figure 60.14 Integration of desk study, geophysics and geotechnics for developing an offshore ground model

60.7.4 Contour diagrams

The effect of the cyclic shear stress amplitude and the number of cycles can be expressed compactly by developing cyclic resistance curves, also known as contour diagrams. Cyclic resistance curves are used to define the shear stress τ_{cyc} to reach a given value of shear strain γ after a certain number of cycles N. Shear stress is typically normalised by the consolidation stress τ_{cyc}/σ_{vc} or by the monotonic shear strength (e.g. τ_{cyc}/s_u).

The results of cyclic loading tests (whether triaxial or simple shear tests) can be used to construct cyclic resistance curves of strain and pore pressure contours, which can then be used in design. For the strain contour diagram in **Figure 60.15** each dot represents the number of cycles resisted prior to the shear strain exceeding the magnitude, indicated by each contour line at each shear stress ratio tested. The single point at a shear stress ratio $\tau_{cyc}/\sigma_{vc} = 0.25$ is derived from a monotonic simple shear test. A pore pressure contour diagram looks very similar, except that the contours are for constant pore pressure ratio u_e/σ'_v, with zero indicating zero pore pressure and unity indicating liquefaction.

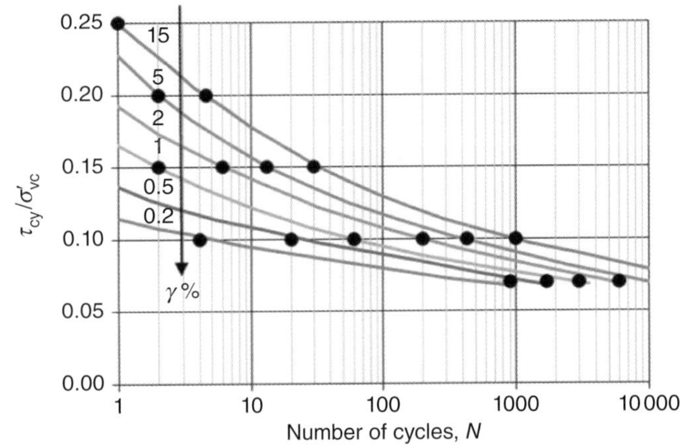

Figure 60.15 A typical contour diagram, constructed by plotting the number of cycles required to reach a particular shear strain magnitude, for shear stress ratios of 0.1, 0.15, 0.2 and 0.25

60.7.5 Shallow foundations

Typical foundation types are shown schematically in **Figure 60.16**.

A number of recommendations exist for design of offshore shallow foundations, such as API RP 2A-WSD (American Petroleum Institute, 2014), DNV-OS-J101 (DNV, 2014), DNV-RP-C212 (DNV, 2021), De Groot *et al.* (1996) and BS 6349-7:2010, among others. The issues concerning shallow foundations onshore, which are described in Chapter 53 *Shallow foundations*, are also relevant for offshore foundations.

As noted earlier, the soil shear strength and stiffness used for design must allow for degradation due to cyclic loading (**Figure 60.17**). Another common feature of offshore design is the high horizontal loads, which need to be resisted (due to large waves during storm events). Hence, the correction for inclined loads and load eccentricity become significant (refer to Chapters 21 *Bearing capacity theory* and 53 *Shallow foundations*). Modern developments in bearing capacity theory, using VHM failure envelopes, have been stimulated by the

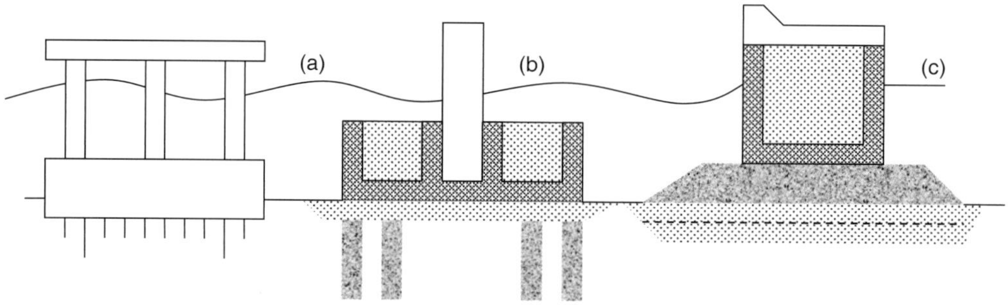

Figure 60.16 Sketches of (a) foundation with skirts and dowels under a gravity platform, (b) foundation of a pylon of a wind turbine with stone columns for ground improvement, (c) caisson breakwater on rubble mound over a reinforced sand layer

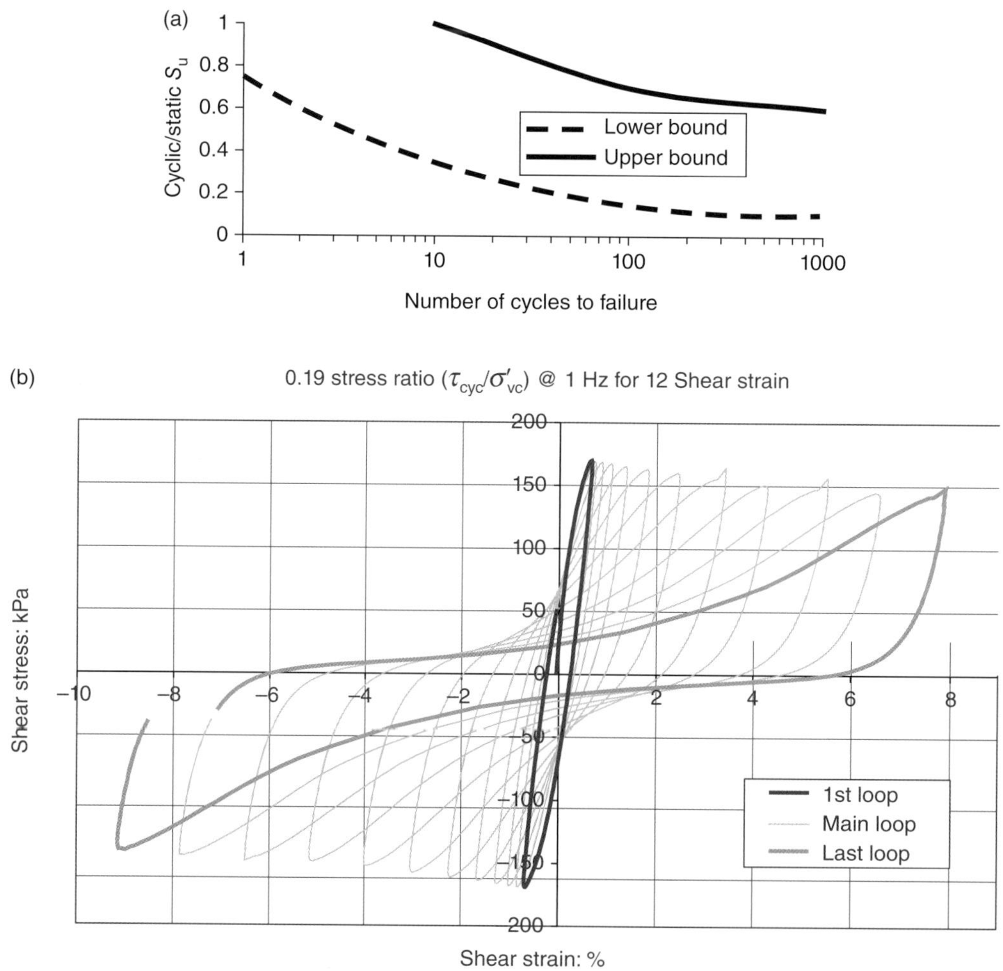

Figure 60.17 Effects of number of cycles on (a) the ratio of cyclic to static undrained shear strength S_u of clays (From data by Lee and Focht, 1976), (b) degradation of stiffness (the ratio between shear stress and strain) of a non-plastic silt specimen in a cyclic simple shear test

inadequacies of conventional theory (with its plethora of correction factors) for offshore design requirements. Offshore foundations are often very large and issues such as soil layering, stress level (for coarse-grained soil) and strength variations with depth (for fine-grained soil) can be important.

Besides ground bearing capacity at the ULS, foundation settlement in the SLS needs to be considered. A large number of cycles generated during storms induce excess pore water pressures in the ground under shallow foundations. Dissipation of such pressure causes additional settlement in loose to medium-dense sandy soils and soft clay/silts. The additional settlement may affect connecting tubes between an offshore structure and fluid-exporting pipelines if such settlement has not been considered in design. The design considerations frequently lead to recommendations for ground improvement, which are sketched in **Figure 60.16(b)** and **(c)**.

60.7.6 Deep foundations

Typical deep foundations offshore are shown in **Figure 60.18**. Offshore foundations typically need to resist larger loads (especially horizontal) and relatively larger cyclic loads than onshore foundations, and the methods of construction (e.g. installation of caissons using suction) will usually be quite different. However, in other respects the general design principles for offshore foundations are similar to deep foundations onshore. Modern monopiles that are used for OWFs are several metres in diameter and have a mass of several hundred tonnes. The dimensions between monopile foundations for offshore wind turbines and jacket piles are noticeably distinct. Offshore wind monopiles are substantially larger in diameter and stiffer than jacket piles, to resist the large overturning moment transmitted from the upper structure. Monopiles commonly have an outer diameter of 5.5–8 m and an embedded depth of 20–40 m. Their aspect ratio (also called embedment ratio, or length over diameter ratio *L*/*D*) is usually 4–8. In comparison, typical jacket piles used in oil and gas are 1.2–2.4 m dia. (1.8–2.4 m in the North Sea), of typical length 40–100 m and pile aspect ratio of 20–50. This means conventional empirical design methods do not work well for monopiles. Comprehensive guidance for the design of foundations for offshore structures can be seen in Bhattacharya (2019). As mentioned before for shallow foundations offshore, it is important to consider the effect of cyclic loads on soil properties but not inertia forces for the periods of vibrations of a few seconds. Dean (2009) describes in detail the foundations, methods for their design and installation, applicable codes and the available literature. A number of recommendations exist for design of offshore deep foundations, such as API RP 2A-WSD (American Petroleum Institute, 2014), DNV-OS-J101 (DNV, 2014) and SNAME TR-5A (2002), among others.

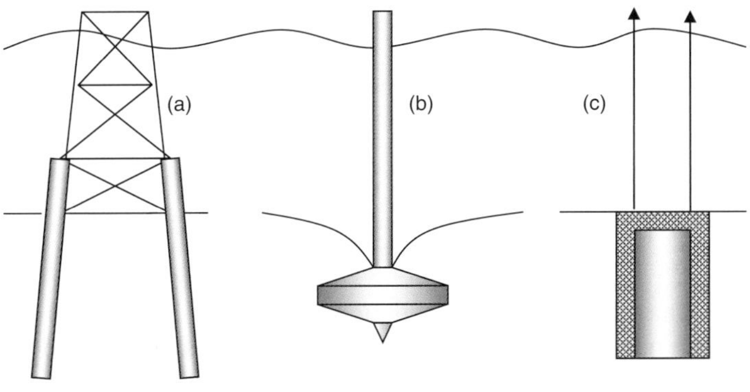

Figure 60.18 Sketches of (a) piled foundation of a jacket, (b) spud-can foundation of a jack-up (mobile platform), (c) caisson foundation of a tension leg or a deep-water anchor

Cyclic wave loading during storms causes degradation of ground shear strength and stiffness, which in turn causes decrease of pile shaft friction and increase of lateral deformation of piles. The degradation of shaft resistance under axial loading is a function of the cyclic displacement and the number of load cycles. A cyclic stability diagram is an important concept (Poulos, 1988a), and three main regions need to be identified: a cyclically stable region; a cyclically metastable region (in which cyclic axial loading causes a reduction of shaft capacity, but the pile does not fail within the specified number of cycles); and a cyclically unstable region, in which cyclic loading leads to failure within the specified number of axial load cycles. Randolph (1983) has provided guidance on the boundary between stable and metastable regions. Reduced soil shear strength and *P*–*Y* curves in cyclic conditions are often used to account for lateral load effects, although several important phenomena are still the subject of ongoing research. A summary of the PISA project and resulting method is provided by Byrne *et al.* (2015) and Taborda *et al.* (2020).

60.8. Machine foundations

Machine foundations are typically subjected to a large number of loading cycles with relatively small amplitudes. The operational conditions given below are an example of the requirements specified by the manufacturer of a particular machine.

- Large compressor (MAN) foundation velocity <2.8 mm/s in operational condition and <6 mm/s in accidental case between frequencies from 25 to 190 Hz.
- Gas turbine (EGT) foundation velocity <2 mm/s and where a peak-to-peak amplitude of any part of the foundation is less than 50 μm at the operating frequency of 250 Hz.

The specified frequency (i.e. its range) must be avoided in order to avoid resonance effects between the machinery and its foundation, which will damage the machine unless it has been switched off. The vibration resonance is indicated as the increase in vibration amplitude ratio at resonant frequency in **Figure 60.19** for harmonic vibration (Clough and Penzien, 1993), where $\beta_t = f_d/f_o$ is the tuning ratio, f_d is the frequency of machine vibration, f_o is the frequency of foundation free vibration, ξ is the damping ratio, a_o is the acceleration amplitude of foundation and a_i is the acceleration amplitude of machine vibration.

Foundations can be subjected to vibration with the origin located at a distance. For example, BS 5228-2:2009 (BSI, 2009) specifies the limits of peak particle velocities to which retaining walls are allowed to be subjected for minor or cosmetic damage (**Table 60.3**). The same limits are applicable to foundations.

Machine-induced vibrations usually only cause very small strains in the ground and therefore elastic soil properties are

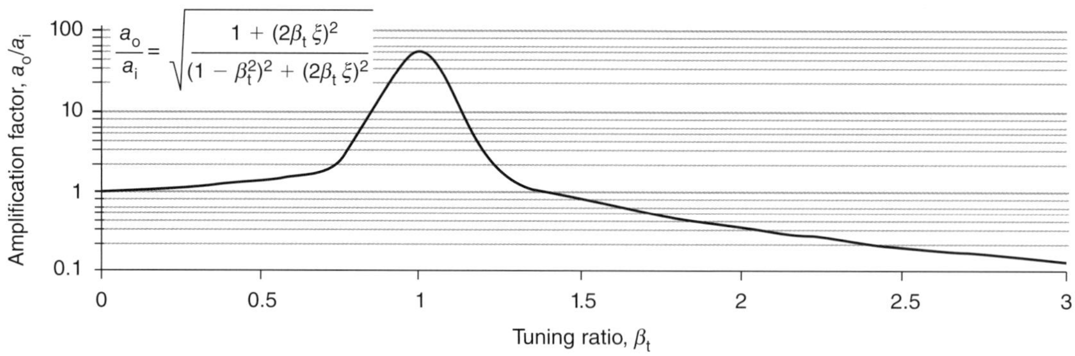

Figure 60.19 Amplification factor for a harmonic vibration with 1% damping ratio

Vibration type	Slender and potentially sensitive masonry walls	Propped or tied walls or mass gravity walls
Intermittent vibration	10 at the toe 40 at the crest	50–100% greater than for slender and potentially sensitive masonry walls
Continuous vibration	1.5–2.5 times smaller than the intermittent vibration limits	

Table 60.3 Threshold peak particle velocities in mm/s for minor or cosmetic damage according to BS 5228-2:2009 + A1:2014

used for dynamic analyses. Also, the soil damping ratio is small, in the order of 1%. The energy dissipation is mainly achieved by radiation damping.

Vertical vibration of machine foundations is usually considered separately from the horizontal and rocking vibrations, which are coupled. The basic equation of foundation motion arises from Newton's second law

$$m \times a + c \times v + k \times d = F \times \sin(2 \times \pi \times f \times t) \tag{60.3}$$

where m is foundation and machine mass, a is acceleration of foundation, c is damping coefficient (which is a frequency-dependent value), $c = \xi k/(\pi f)$, ξ is damping ratio (about 0.01), f is vibration frequency, v is velocity of foundation, k is stiffness coefficient, d is displacement of foundation, F is maximum amplitude of dynamic unbalanced force and t is time. The formulae for calculation of values of the dynamic stiffness k and damping coefficients c are tabulated in the relevant textbooks (Gazetas, 1991; Wolf, 1994). The formulae are provided for homogeneous soil and a soil layer over bedrock and should not be used for layered soil. If foundation vibration is considered excessive according to the requirements set by the manufacturer then foundation vibration isolation can be used.

The horizontal motion is coupled with the rotational motion, which is

$$I_a \times \theta_a + c_\theta \times \theta_v + k_\theta \times \theta - e \times (c \times v + k \times d) = M \times \sin(2 \times \pi \times f \times t) \tag{60.4}$$

where I is foundation and machine mass moment of inertia with respect to the centre of the masses, θ_a is rotational acceleration, $c\theta$ is rotation damping coefficient, θ_v is rotational velocity, $k\theta$is rotational stiffness coefficient, θ is rotation of foundation, e is the eccentricity of the base of foundation with respect to the centre of mass of foundation and machine and M is the maximum amplitude of rotational moment from unbalanced force with respect to the centre of mass of foundation and machine. The formulae for calculation of values of the rotational stiffness $k\theta$ and damping coefficients $c\theta$ are tabulated in the relevant textbooks (Gazetas, 1991; Wolf, 1994).

Simple guidance exists for determination of minimum foundation mass for vibration control (Anyaegbunam, 2011). The key performance criteria are as follows.

- The foundation should be safe from shear failure caused by superimposed loads and settlement should be within acceptable limits.
- Soil pressure should not exceed 80% of the allowable bearing pressure in static conditions.
- There should be no possibility of resonance. The natural frequency of the foundation shall be either greater or smaller than the operating frequency of the machine. Usually a value less than 0.5 times and more than 1.5 is accepted.
- The amplitude-under-service condition shall be within the permissible limit specified by the manufacturer.
- Machine foundations should be separated from adjacent building components by using expansion joints.
- The foundation should be so planned that subsequent alternation of the natural frequency can be managed by changing the base area or mass of the foundation.

Older publications contain examples of foundation vibration analyses by hand calculations (Irish and Walker, 1969), which may be used for rough checking of results of modern numerical analyses by computers. CP 2012-1 (1974) exists for foundations for reciprocating machinery that generates harmonic loads (e.g. steam engines, internal combustion engines, piston-type compressors and pumps), with normal rotational frequency range of 5–25 Hz, and is relevant for machines placed on a rigid block. DIN 4024-1 (1988) and DIN 4024-2 (1991) contain useful recommendations. A frequently cited book is Richart *et al.* (1970). Not only theoretical but also experimental methods have been used for analyses of foundation vibration.

60.9. Groundborne vibrations from high-speed trains

Groundborne vibrations from high-speed trains is a resonance-like condition which develops when a train reaches the critical speed of the track system (inclusive of geotechnical structures such as embankments). At resonance-like state, excessive dynamic amplification of vertical vibrations and track displacement can occur. These excessive vibrations could result in imposition of lower operating speeds, increased maintenance of tracks and catenary structures and, if there are sensitive buildings adjacent to the tracks, these vibrations can also transmit through the ground and cause excessive floor vibration. Such observations have been made in Sweden, the UK, France and Germany (Madshus and Kaynia, 2000; Woldringh and New 1999). **Figure 60.20** shows dynamic amplification in vertical deflections at train speeds approaching critical velocity in sites in the Netherlands, the UK and Sweden (Woldringh and New, 1999).

The critical speed of the track system is governed by the minimum phase velocity of the first Rayleigh mode of the soil and embankment profile (Madshus and Kaynia, 2000). With train speeds upwards of 150 km/h, either on purpose-built new high-speed train tracks or upgrading speeds on existing tracks, it is now possible for resonance-like conditions to occur if the route passes over regions of soft soils with low shear wave velocity (110–180 km/h) such as peats, organics and marine clays. For a homogeneous half-space the Rayleigh wave velocity can be derived from the shear wave velocity

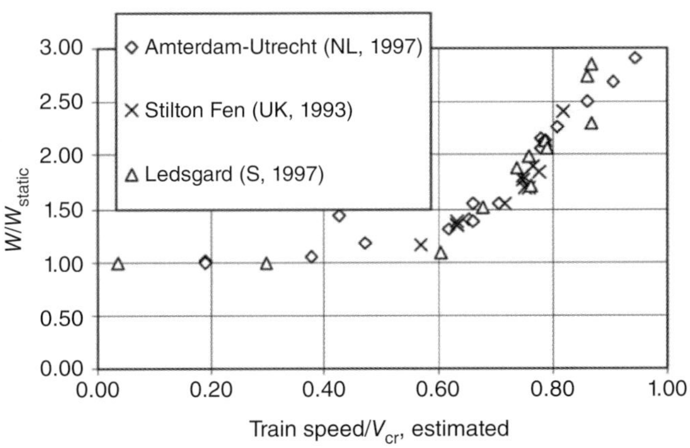

Figure 60.20 Field observations showing dynamic amplification in rail deflection as a function of normalised train speed (Woldringh and New, 1999)

and Poisson's ratio of half-space using the analytical equation proposed by Bergmann (1948). For more complex layered soil profiles, dispersion analysis based on surface geophysics is required to identify the first Rayleigh mode and minimum phase velocity in the free-field (i.e. in the absence of any track structures).

The stiffness contribution of track structures should be considered when assessing critical velocity. The Winkler analytical model (Kenny, 1954; Kneifati, 1985; Timoshenko, 1926) can be used to represent the track–subgrade system where an infinite beam is supported on discrete vertical springs without any shear coupling; superposition can be used to derive the behaviour of the beam for a multi-axle train. The critical speed of bending waves (V_{cr}) is expressed as

$$V_{cr} = \left(\frac{4kEI}{\mu^2}\right)^{\frac{1}{4}} = \left(\left(\frac{2.6EI}{\mu^2}\right)\left(\frac{E_s}{(1-\nu^2)}\right)\right)^{\frac{1}{4}} \sqrt[48]{\frac{(E_s B^4)}{EI}} \tag{60.5}$$

where k is the stiffness coefficient of foundation, EI is the flexural rigidity of the beam, μ is mass per unit length of the beam, E_s is the subgrade elastic modulus and ν is Poisson's ratio of the medium. Fryba (1972) provided closed-form solutions for the lightly damped case to determine vertical displacement of the beam. The presence of stiff embankments including ballast and sub-ballast can be considered in analytical solutions by idealising them as elastic beams on the visco-elastic subgrade. The stiffness coefficient k of the foundation can be derived from E_s of the visco-elastic foundation either using analytical solutions (Biot, 1937; Vesic, 1963) or with the help of finite element or finite difference numerical models.

More detailed and accurate predictions of critical velocity for site-specific geology and track properties can be obtained with numerical models. These numerical models are broadly classified in two categories: time domain solution (finite element and finite difference) and wave number domain (semi-analytical methods). Two-dimensional models in both categories offer limited benefit as they are unable to capture wave propagation along the length of track. Three-dimensional numerical models are computationally expensive due to requirements of high-fidelity mesh and small time steps for accurate simulation, but offer the advantage of considering complex soil profiles, geotechnical structures, adjacent structures and soil models (linear and nonlinear stiffness). Groundborne vibration problems have generally been modelled as an elastic medium as it is considered to be a small-strain problem, but at some soft soil locations (Madshus and Kaynia, 2000) strains in soft soil layers under train loads can be significant enough to induce nonlinearity. If nonlinearity is expected, then the soil model in analysis should be able to capture reduction in small-strain stiffness with increasing shear strains and generation of associated hysteretic damping. Boundaries in 3D numerical models

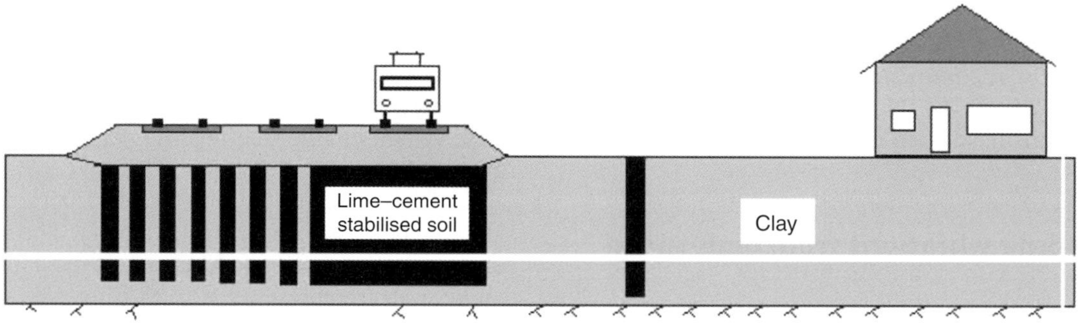

Figure 60.21 Schematic representation of dry deep mixing at Ledsgård for vibration mitigation (Holm *et al.*, 2002)

also need special consideration to minimise reflection of incident waves back into the main model domain. In the analyses of groundborne vibrations, absorbing boundary elements, infinite elements or special elements such as perfectly matched layer (PML) elements can be used.

Semi-analytical models based on Green's function offer speed compared to 3D numerical models but must assume that ground and track properties are invariant along the track. The model is solved in the wave number domain assuming a visco-elastic layered soil profile, and then Fourier transforms are used to extract output in the time domain. To overcome the limitations of semi-analytical models in modelling irregular cross-sections such as embankments, or to consider nonlinear stiffness of the soil (as an iterated equivalent linear stiffness), 2.5D coupled BEM–FEM models are also used in prediction of track response for train speeds approaching critical velocity. In these hybrid models the embankment–track system is modelled with finite elements, the layered soil profile is modelled with boundary elements and the coupling between the two systems is provided with finite elements.

Prediction, monitoring and remediation of groundborne vibrations from high-speed trains is an area of active research. Full-scale field tests, such as in Belgium (Kouroussis *et al.*, 2016), have been carried out in other European countries and China to gather field data for track performance, vibration generation and propagation for train speeds approaching expected critical velocity, but these studies have generally been limited to gathering data (such as peak particle velocity, PPV) to determine dynamic amplification in vibration of track structures instead of track displacements, which might be a more important parameter for geotechnical design. Vibration control measures have also been actively researched as it is not always economical to have profitable train operations when imposing a reduced train speed in areas where issues due to critical velocity are expected. Some of the common solutions include considering rigid inclusions within embankments, piled embankments which transfer dynamic loads to the stiff subgrade level, mixed-in-place walls and, where economics permit, concrete slab track. Vibration radiating away from tracks can also be attenuated with the help of active and passive wave barriers. Based on site-specific geology and groundborne vibration propagation in the free-field, dimensions (particularly depth) and material stiffness (open trench, EPS geofoam, bentonite, etc.) can be optimised to reduce vibration propagating from the tracks to adjacent structures. **Figure 60.21** shows a schematic representation of dry deep mixing undertaken at a Ledsgård site to reduce groundborne vibrations below embankments and as a vibration mitigation barrier.

References

Ambraseys NN (1990) Uniform magnitude re-evaluation of European earthquakes associated with strong motion records. *Earthquake Engineering and Structural Dynamics* **19**: 1–20.

American Petroleum Institute (2014) *API RP 2A-WSD: Recommended Practice for Planning, Designing and Constructing Fixed Offshore Platforms – Working Stress Design*. American Petroleum Institute, Washington, DC.

Anyaegbunam AJ (2011) Minimum foundation mass for vibration control. *Journal of Geotechnical and Geoenvironmental Engineering, ASCE* **137(2)**: 190–195.

ASTM (American Society for Testing Materials) (2000) D4015-92: Standard test methods for modulus and damping of soils by the resonant column method. ASTM, West Conshohocken, PA.

ASTM (2003) D3999-91: Standard test methods for the determination of the modulus and damping properties of soil using the cyclic triaxial apparatus. ASTM, West Conshohocken, PA.

ATC (1978) *Tentative Provisions for the Development of Seismic Regulations for Buildings*. Applied Technology Council, US Department of Commerce, Washington, DC, Report ATC 3-06.

Bergmann L (1948) *Ultrasonics and their Scientific and Technical Applications*. Wiley, New York, NY.

Bhattacharya S (2019) *Design of Foundations for Offshore Wind Turbines*. Wiley, New York, NY.

Bhattacharya S and Bolton M (2004) Buckling of piles during earthquake liquefaction. *Proceedings of the 13th World Conference on Earthquake Engineering, Vancouver, BC*.

Bhattacharya S and Madabhushi SPG (2008) A critical review of methods for pile design in seismically liquefiable soils. *Bulletin of Earthquake Engineering* **6**: 407–446.

Biot AM (1937) Bending of an infinite beam on an elastic foundation. *Journal of Applied Mechanics* **4**: A1–A7.

Bommer J (2022) Earthquake hazard and risk analysis for natural and induced seismicity: towards objective assessments in the face of uncertainty. *Bulletin of Earthquake Engineering* **20**: 2825–3069.

Bommer JJ and Martinez-Pereira A (1999) The effective duration of earthquake strong motion. *Journal of Earthquake Engineering* **3(2)**: 127–172.

Boulanger RW and Brandenberg SJ (2004) Neutral plane solution for liquefaction-induced down-drag on vertical piles. In *Geotechnical Engineering for Transportation Projects* (Yegian MK and Kavazanjian E (eds)). ASCE, Reston, VA, pp. 403–410.

Boulanger RW and Idriss IM (2007) Evaluation of cyclic softening in silts and clays. *Journal of Geotechnical and Geoenvironmental Engineering, ASCE* **133(6)**: 641–652.

Bray JD, Sancio RB, Durgunoglu T, *et al.* (2004) Subsurface characterisation at ground failure sites in Adapazari, Turkey. *Journal of Geotechnical and Geoenvironmental Engineering, ASCE* **130**: 673–685.

Brown DA, Turner JP and Castelli RJ (2010) *Drilled Shafts: Construction Procedures and LRFD Design Methods.* Federal Highway Administration, Washington, DC.

BSI (British Standards Institution) (1974) CP 2012-1:1974: Code of practice for foundations for machinery – Part 1: Foundations for reciprocating machines. BSI, London, UK.

BSI (2004a) BS EN 1998-1:2004: Eurocode 8: Design of structures for earthquake resistance – Part 1: General rules, seismic actions and rules for buildings. BSI, London, UK.

BSI (2004b) BS EN 1998-5:2004: Eurocode 8: Design of structures for earthquake resistance – Part 5: Foundations, retaining structures and geotechnical aspects. BSI, London, UK.

BSI (2009) BS 5228-2:2009: Code of practice for noise and vibration control on construction and open sites – Part 2: Vibration. BSI, London, UK.

BSI (2010) BS 6349-7:2010: Maritime structures – Part 7: Guide to the design and construction of breakwaters. BSI, London, UK.

BSSC (Building Seismic Safety Council) (1995) *NEHRP Recommended Provisions for Seismic Regulations for New Buildings, Part 1 – Provisions and Part 2 – Commentary.* BSSC, Washington, DC, Report No. FEMA 222A.

Byrne BW, McAdam RA, Burd HJ, *et al.* (2015) New design methods for large diameter piles under lateral loading for offshore wind applications. *Third International Symposium on Frontiers in Offshore Geotechnics, Oslo, Norway.*

Cetin KO, Seed RB, Kiureghian AD, *et al.* (2004) Standard penetration test-based probabilistic and deterministic assessment of seismic liquefaction potential. *Journal of Geotechnical and Geoenvironmental Engineering, ASCE* **130(12)**: 1314–1340.

Clough RW and Penzien J (1993) *Dynamics of Structures*, 2nd edn. McGraw-Hill, New York, NY.

Cubrinovski M, Ishihara K and Poulos H (2009) Pseudo-static analysis of piles subjected to lateral spreading. *Bulletin of the New Zealand Society for Earthquake Engineering* **42(1)**: 28–38.

De Groot MB, Andersen KH, Burcharth HF, *et al.* (1996) *Foundation Design of Caisson Breakwaters.* Norwegian Geotechnical Institute, Oslo, Norway.

Dean ETR (2009) *Offshore Geotechnical Engineering: Principles and Practice.* Thomas Telford, London, UK.

Deierlein GG, Reinhorn AM and Willford MR (2010) *Nonlinear Structural Analysis for Seismic Design, NEHRP Seismic Design Technical Brief No. 4.* NIST, Gaithersburg, MD.

DIN (Deutsches Institut für Normung) (1988) DIN 4024-1: Maschinenfundamente; Elastische Stützkonstruktionen für Maschinen mit rotierenden Massen. DIN, Berlin, Germany.

DIN (1991) DIN 4024-2: Maschinenfundamente; Steife (starre) Stützkonstruktionen für Maschinen mit periodischer Erregung. DIN, Berlin, Germany.

DNV (Det Norske Veritas) (2014) DNV-OS-J101: Design of offshore wind turbine structures. Det Norske Veritas, Hovik, Norway.

DNV (2021) DNV-RP-C212: Offshore soil mechanics and geotechnical engineering. Det Norske Veritas, Hovik, Norway

FEMA (Federal Emergency Management Agency) (2021) *P-2091: A Practical Guide to Soil–Structure Interaction.* FEMA, Washington, DC.

Fryba L (1972) Beams with various boundary conditions subjected to a moving load. In *Vibration of Solids and Structures under Moving Loads.* Springer, New York, NY, pp. 83–93.

Gazetas G (1991) Foundation vibration. In *Foundation Engineering Handbook* (Fang H-Y (ed)), 2nd edn. Chapman & Hall, New York, NY, pp. 553–593.

Gazetas G and Mylonakis G (1998) Seismic soil–structure interaction: new evidence and emerging issues. In *Geotechnical Earthquake Engineering and Soil Dynamics III* (Dakouls P, Yegian M and Holtz B (eds)). University of Washington Press, Seattle, WA, vol. 2, pp. 461–468.

Green RA and Terri GA (2005) Number of equivalent cycles concept for liquefaction evaluations – revisited. *Journal of Geotechnical and Geoenvironmental Engineering ASCE* **131**: 477–488.

Grünthal G (ed) (1998) *European Macroseismic Scale 1998.* Conseil de l'Europe, Luxembourg, Luxembourg.

Hamada M (1992) Large ground deformations and their effects on lifelines: 1964 Niigata earthquake. In *Case Studies of Liquefaction and Lifeline Performance during Past Earthquakes.* Vol. 1: *Japanese Case Studies* (Hamada M and O'Rourke T (eds)). National Centre for Earthquake Engineering Research, State University of New York at Buffalo, Buffalo, NY, pp. 1–123.

Hancock J and Bommer JJ (2004) Predicting the number of cycles of ground motion. *Proceedings of the 13th World Conference on Earthquake Engineering, Vancouver, BC,* paper 1989.

Hashash YMA, Musgrove MI, Harmon JA, *et al.* (2020) *DEEPSOIL 7, User Manual.* Board of Trustees of University of Illinois at Urbana-Champaign, Urbana, IL.

Holm H, Andréasson B, Bengtsson P, Bodare A and Eriksson H (2002) *Mitigation of Track and Ground Vibrations by High-Speed Trains at Ledsgard, Sweden.* Swedish Deep Stabilization Research Centre, Linköping, Sweden.

HSE (2001) *A Study of Pile Fatigue during Driving and In-Service and of Pile Tip Integrity.* HSE Books, Norwich, UK.

IBC (International Building Code) (2021) https://codes.iccsafe.org/content/IBC2021P2/chapter-16-structural-design (accessed 22/05/2023).

Idriss IM (1990) Response of soft soil sites during earthquakes. *H. Bolton Seed Memorial Symposium, Vancouver, BC* (Duncan JM (ed)). BiTech Publishers, Richmond, BC, vol. 2, pp. 273–289.

Idriss IM and Boulanger RW (2006) Semi-empirical procedures for evaluating liquefaction potential during earthquakes. *Journal of Soil Dynamics and Earthquake Engineering* **26**: 115–130.

Idriss IM and Boulanger RW (2008) *Soil Liquefaction during Earthquakes*. Earthquake Engineering Research Institute, Oakland, CA, EERI Publication, Monograph MNO-12.

Irish K and Walker WP (1969) *Foundations for Reciprocating Machines*. Concrete Publications, London, UK.

Ishihara K and Yoshimine M (1992) Evaluation of settlements in sand deposits following liquefaction during earthquakes. *Soils and Foundations* **32(1)**: 173–188.

Kanamori H (1977) The energy released in great earthquakes. *Journal of Geophysical Research* **82**: 2981–2987.

Keefer DK (1984) Landslides caused by earthquakes. *Bulletin of the Geological Society of America* **95**: 406–421.

Kenney JT (1954) Steady-state vibrations of beam on elastic foundation for moving load. *Journal of Applied Mechanics–Transactions of the ASME* **21(4)**: 359–364.

Kneifati MC (1985) Analysis of plates in a Kerr foundation. *Journal of Engineering Mechanics, ASCE* **11**: 1325–1342.

Kokusho T, Tanaka Y, Kawai T, *et al.* (1995) Case study of rock debris avalanche gravel liquefied during 1993 Hokkaido-Nansei-Oki earthquake. *Soils and Foundations* **35(3)**: 83–95.

Kolk HJ (2000) Deep foundations in calcareous sediments. In *Engineering for Calcareous Sediments* (Al-Shafei KA (ed)). Balkema, Rotterdam, the Netherlands, pp. 313–344.

Kouroussis G, Connolly DP, Olivier B, Laghrouche O and Costa PA (2016) Railway cuttings and embankments: experimental and numerical studies of ground vibration. *Science of the Total Environment* 557–558: 110–112

Kramer S (1996) *Geotechnical Earthquake Engineering*. Prentice Hall, Englewood Cliffs, NY.

Lee KL and Focht JA (1976) Strength of clay subjected to cyclic loading. *Marine Georesources and Geotechnology* **1(3)**: 165–168.

Liu L and Dobry R (1997) Seismic response of shallow foundations on liquefiable sand. *Journal of Geotechnical and Geoenvironmental Engineering, ASCE* **123(6)**: 557–567.

Madabhushi SPG (2007) Ground improvement methods for liquefaction remediation. *Proceedings of the Institution of Civil Engineers – Ground Improvement* **11(4)**: 195–206.

Madshus C and Kaynia AM (2000) High-speed railway lines on soft ground: dynamic behaviour at critical train speed. *Journal of Sound and Vibration* **231**: 689–701.

Mahoney M (2005) Performance-based earthquake engineering for geotechnical and structural engineers. *Proceedings of Geotechnical Earthquake Engineering and Soil Dynamics IV, Sacramento, CA*, pp. 1–32.

Marcuson WF, Hynes ME and Franklin AG (1990) Evaluation and use of residual strength in seismic safety analysis of embankments. *Earthquake Spectra* **6(3)**: 529–572.

May R (2015) The seismic design of shallow foundations: a state-of-the-art exploration. SECED Conference, Cambridge, UK.

Meymand P (1998) *Shaking Table Scale Model Tests of Nonlinear Soil–Pile–Superstructure Interaction in Soft Clay*. PhD dissertation, University of California, Berkeley, CA.

Mitchell JK and Wentz FL (1991) *Performance of Improved Ground during the Loma Prieta Earthquake*. EERC, University of California at Berkeley, Berkeley, CA, Report UCB/EERC-91/12.

Musson RM, Grünthal G and Stucchi M (2010) The comparison of macroseismic intensity scales. *Journal of Seismology* **14(2)**: 413–428.

NIST (National Institute of Standards and Technology) (2012) *Soil–Structure Interaction for Building Structures*. NIST, Gaithersburg, MD, Report NIST GCR 12-917-21.

Peck RB, Hanson WE and Thornburn TH (1974) *Foundation Engineering*, 2nd edn. Wiley, New York, NY.

Pecker A and Pender MJ (2000) Earthquake resistant design of foundations: new construction. *Proceedings of GeoEng 2000, Melbourne, Australia*, vol. 1.

Port and Harbour Research Institute (1997) *Handbook on Liquefaction Remediation of Reclaimed Land. Ministry of Transport, Japan. Translated by Waterways Experimental Station, USA*. Balkema, Rotterdam, the Netherlands.

Poulos HG (1988a) Cyclic stability diagram for axially loaded piles. *Journal Geotechnical Engineering, ASCE* **114(8)**: 877–895.

Poulos HG (1988b) *Marine Geotechnics*. Unwin Hyman, London, UK.

Randolph MF (1983) Design considerations for offshore piles. *Proceedings of the ASCE Speciality Conference of Geotechnical Practice in Offshore Engineering, Austin, TX* (Wright SG (ed)). ASCE, Baltimore, MD, pp. 422–439.

Reese LC, Cox WR and Coop FD (1974) Analysis of laterally loaded piles in sand. *Proceedings of the 6th Offshore Technology Conference, Houston, TX*, vol. 2, pp. 473–483.

Richards R, Elms DG and Budhu M (1993) Seismic bearing capacity and settlement of foundations, *ASCE Journal of Geotechnical Engineering* **119(4)**: 662–674.

Richart FE, Hall JR and Woods RD (1970) *Vibrations of Soils and Foundations*. Prentice-Hall, Englewood Cliffs, NY.

Rodriguez CE, Bommer JJ and Chandler RJ (1999) Earthquake-induced landslides: 1980–1997. *Soil Dynamics and Earthquake Engineering* **18**: 325–346.

Rostami R, Mickovski BS, Hytiris N and Bhattacharya S (2020) The dynamic behaviour of pile foundations in seismically liquefiable soils: failure mechanisms, analysis, re-qualification. In *Earthquakes* (Salazar W (ed)). InTech, London, UK.

Seed HB and Idriss IM (1971) Simplified procedure for evaluating liquefaction potential. *Journal of the Soil Mechanics and Foundations Division, ASCE* **107(SM9)**: 1249–1274.

Seed HB, Idriss IM, Makdisi F and Banerjee N (1975) *Representation of Irregular Stress Time Histories by Equivalent Uniform Stress Series in Liquefaction Analyses*. EERC, University of California at Berkeley, Berkeley, CA, Report EERC 75-29

SNAME (Society for Naval Architects and Marine Engineers) (2002) *SNAME TR-5A: Recommended practice for site specific assessment of mobile jackup units*. Society for Naval Architects and Marine Engineers, Washington DC.

Srbulov M (2003) An estimation of the ratio between the horizontal peak accelerations at the ground surface and at depth. *European Earthquake Engineering* **17(1)**: 59–67.

Stewart JP, Seed RB and Fenves GL (1999) Seismic soil–structure interaction in buildings. II: empirical findings. *Journal of Geotechnical and Geoenvironmental Engineering, ASCE* **125(1)**: 38–48.

Taborda D, Zdravković L, Potts F and Burd H (2020) Finite-element modelling of laterally loaded piles in a dense marine sand at Dunkirk. *Géotechnique* **70(11)**: 1014–1029.

Timoshenko S (1926) Method of analysis of statical and dynamical stresses in rail. *Proceedings of the International Congress for Applied Mechanics, Zurich, Switzerland.*

Timoshenko SP and Goodier JN (1970) *Theory of Elasticity*, 3rd edn. McGraw-Hill, New York, NY.

Touma FT and Sadiq MI (1999) Foundations on the Red Sea coastal coral in Jeddah area. In *Engineering for Calacareous Sediments* (Al-Shafei KA (ed)). Balkema, Rotterdam, the Netherlands, pp. 167–177.

Vesic AB (1963) Beams on elastic subgrade and Winkler hypothesis. *Proceedings of the 5th International Conference on Soil Mechanics Foundations and Engineering, Paris, France*, vol. 1, pp. 845–850.

Woldringh RF and New BM (1999) Embankment design for high-speed trains on soft soils: geotechnical engineering for transportation infrastructure. *Proceedings of the 12th European Conference on Soil Mechanics and Geotechnical Engineering, Amsterdam the Netherlands*, vol. 3, pp. 1703–1712.

Wolf JP (1994) *Foundation Vibration Analysis Using Simple Physical Models*. Prentice Hall, Upper Saddle River, NJ.

Youd TL, Idriss IM, Andrus RD, *et al.* (2001) Liquefaction resistance of soils: summary report from the 1996 NCEER and 1998 NCEER/NSF workshops on evaluation of liquefaction resistance of soils. *Journal of Geotechnical and Geoenvironmental Engineering, ASCE* **127(10)**: 817–833.

It is recommended this chapter is read in conjunction with

- Chapter 24 *Dynamic and seismic loading of soils*
- Chapter 45 *Geophysical exploration and remote sensing*

All chapters in this book rely on the guidance in Sections 1 *Context* and 2 *Fundamental principles.* A sound knowledge of ground investigation is required for all geotechnical works, as set out in Section 4 *Site investigation.*

Section 6: Design of retaining structures

Section editor: **Stuart Hardy**

doi: 10.1680/icemge.66830.1039

Chapter 61
Introduction to Section 6

Stuart Hardy Laing O'Rourke, Dartford, UK

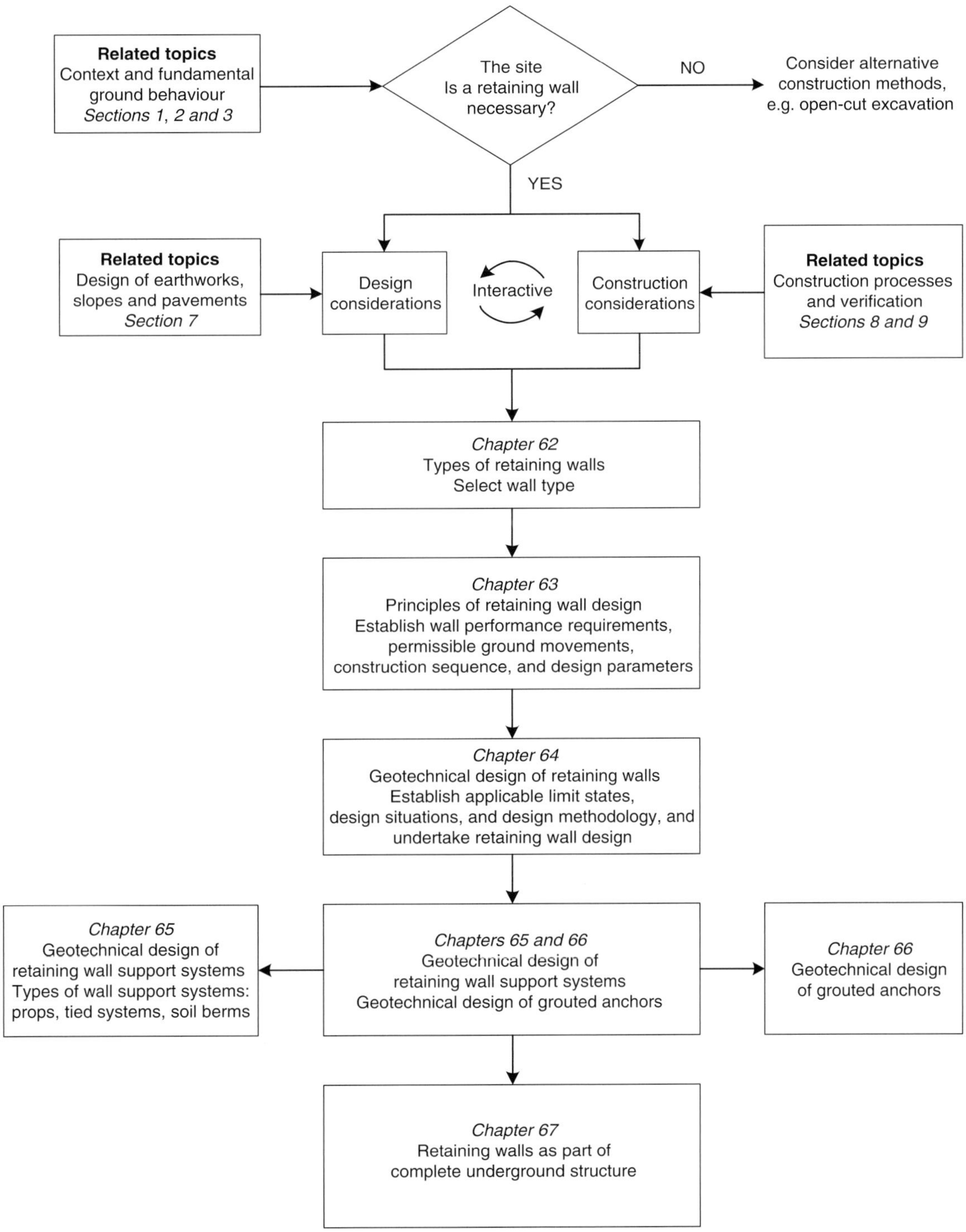

Figure 61.1 Relationships between the chapters in Section 6

Figure 61.1 shows the relationships between the chapters in Section 6.

Retaining walls are typically used to retain the ground at a steeper angle than that which can be sustained by the ground acting unsupported on its own. Common applications vary from modest terracing of sloping ground to perimeter walls for multistorey basements and underground structures for large infrastructure projects.

Retaining walls are usually divided into two broad categories: 'gravity' retaining walls and 'embedded' retaining walls, and this distinction is also made in this section. However, there is a third category – 'hybrid' walls – which have their own particular characteristics, and their design may demand an approach that requires consideration and assessment of both gravity and embedded wall behaviour. An example of a hybrid wall would be a gravity wall supported partially by piles embedded into the ground beneath the gravity structure.

In common with other sections of this manual, this section is intended to provide guidance to practising engineers on the design of retaining walls. Given the considerable breadth of the subject and the vast range of ground conditions which a retaining wall designer may encounter, this section cannot, within the available space, be completely comprehensive. The intent is therefore to provide outline guidance to designers on

- considerations of wall performance criteria, wall selection, construction methods and associated ground movements
- fundamental principles of good design practice in relation to the selection, design and construction of retaining walls and their support systems
- key mechanisms of ground and ground–structure interaction behaviour, which need to be considered
- commonly used design methods for gravity and embedded retaining walls and their support systems
- additional references for detailed study.

The chapters are arranged in a sequence to guide the reader in a systematic approach that considers the key issues that relate to the design of retaining walls while recognising that retaining walls may form part of an overall structure. It is important not to lose sight of the fact that the client's priority is the entire structure and not just the retaining walls.

Chapter 62 *Types of retaining walls* identifies different types of retaining wall along with the particular characteristics of each. The advantages and limitations of each wall type are compared and guidance is provided on the selection of the most appropriate wall type to satisfy the particular site and project requirements.

Chapter 63 *Principles of retaining wall design* provides guidance on the determination of key wall performance and design criteria, setting permissible limits for the associated ground movements and identifying the appropriate construction sequence for particular applications. The chapter also provides guidance on the assessment of drained and undrained ground behaviour and the selection of parameters appropriate for temporary and permanent works design.

Chapter 64 *Geotechnical design of retaining walls* offers good practice guidance on the geotechnical design of retaining walls. Using the approach outlined in this chapter, the reader will be able to size retaining walls to ensure adequate stability and calculate load effects to be used in the structural design of walls.

Chapters 65 *Geotechnical design of retaining wall support systems* and 66 *Geotechnical design of grouted anchors* give the reader an overview of a wide range of different methods of providing lateral support to both gravity and embedded retaining walls for temporary and permanent situations. Chapter 65 provides guidance on the appropriate selection and design of propping systems and berms for the lateral support of a wall. Chapter 66 is a complementary and comprehensive chapter that deals with the design of grouted anchors for the support of retaining walls and should be read in conjunction with Chapters 74 *Design of soil nails*, 88 *Soil nailing construction* and 89 *Grouted anchor construction* of this manual.

Chapter 67 *Retaining walls as part of complete underground structure* deals with the situation where the retaining walls are part of an overall structure. It provides holistic considerations of the ground–structure interaction, retaining wall watertightness requirements and the importance of communications between the retaining wall designer and other design and construction professionals.

Following this theme, it is important to recognise that a retaining wall designer cannot function in isolation from other members of the project team. To overcome the potential fragmentation of design and construction that can accompany the process of retaining wall design, it is important that the designer considers all parts of the construction sequence and, in doing so, communicates effectively with other complementary design and construction professionals. The client, designer and constructor and, where appropriate, the architect and the quantity surveyor, should be involved as early as possible to

- optimise the temporary and permanent use of a retaining structure (such as adopting one wall instead of two to serve both the temporary and permanent requirements), which is also compatible with long-term maintenance requirements
- establish appropriate design and performance criteria for the retaining structure, such as acceptable limits for wall deflection and associated ground movement
- consider an appropriate wall type
- consider an appropriate method and sequence of construction to ensure buildability.

As urban sites become more space-constrained the imperative to go deeper below ground to provide the required floor space becomes stronger and the use of engineered retaining walls becomes more commonplace, including for domestic buildings. As retaining walls have become more commonplace, their design, specification and construction have become more fragmented, with often multiple parties involved that change and evolve through the process. It is therefore imperative that a controlling mind is present to ensure the different systems work together and that there are no gaps in responsibility.

doi: 10.1680/icemge.66830.1041

Chapter 62
Types of retaining walls

Hilary Shields Arup, London, UK
Sara Anderson Arup, London, UK

CONTENTS

The term 'retaining wall' applies to structures that retain the ground at a steeper angle than the ground can sustain acting on its own. Common applications vary in scope, from modest terracing of sloping landscaping to perimeter walls for multistorey basements and underground spaces. Several different forms of retaining wall exist with different performance characteristics in terms of water retention, strength and stiffness, as well as constructability characteristics such as land-take and requirements for specialist plant and equipment. The different characteristics of retaining walls must be fully appreciated as part of the process of selecting a wall for a particular application.

This introductory chapter to Section 6 identifies the different forms of retaining wall along with their particular characteristics.

62.1. Introduction

Retaining walls exist in many shapes and forms but can be considered in terms of three different groups

- gravity walls
- embedded walls
- hybrid walls.

These are discussed in detail below.

62.2. Gravity walls

A gravity wall gains resistance to sliding and overturning through the weight of the wall and friction on its underside. Gravity walls are generally constructed on a flat surface prior to backfilling behind the wall to create the height difference that the wall retains. As such, they are commonly used where the ground profile is being raised locally, such as to retain a highway or railway embankment.

Gravity walls can also be used to support excavations below ground level. For these scenarios the excavation is typically an open-cut and backfill is provided behind the wall. This requires additional excavation and backfilling as well as temporary land-take during construction.

Although the distinction is not mentioned further here, BS 8002 (2015) subdivides gravity walls into gravity and semi-gravity retaining walls. Semi-gravity retaining walls support the retained ground by a combination of the wall's weight and that of the backfill on its base. The stem of a semi-gravity wall resists the horizontal thrust of the retained ground in bending.

Drainage provision in the form of interceptor drains behind the wall is a common feature of gravity walls to reduce the water-pressure load acting on the wall (e.g. figure 9 of BS 8002 (2015)).

Key references for gravity retaining walls include

- CIRIA C516 (Chapman *et al.*, 2000) – section 2 and section 4 mostly
- BS 8002 (BSI, 2015) – sections 5 and 6
- Eurocode EC7 (BS EN 1997-1:2004 + A1:2013, incorporating corrigendum February 2009 (BSI, 2004)).

62.2.1 Modular and non-modular walls

Modular walls may be constructed using a wide variety of techniques and materials. They share the common feature that their construction involves the erection of a number of mainly similar modular elements.

Examples of modular walls include

- precast reinforced-concrete stem walls
- preformed masonry walls
- dry-stack masonry walls (e.g. porcupine walls)
- crib walls
- gabion walls.

For modular walls, the modules are constructed off site and can be assembled on site with a minimum of specialist site operations.

In contrast, non-modular walls are predominantly constructed on site (e.g. in situ reinforced-concrete walls). As such, they require a greater level of site activity and more significant specialist activities such as shuttering, steel fixing and in situ concreting.

62.2.2 Reinforced-concrete stem walls

Reinforced-concrete stem walls are constructed from in situ or precast reinforced concrete. The walls are L-shaped or

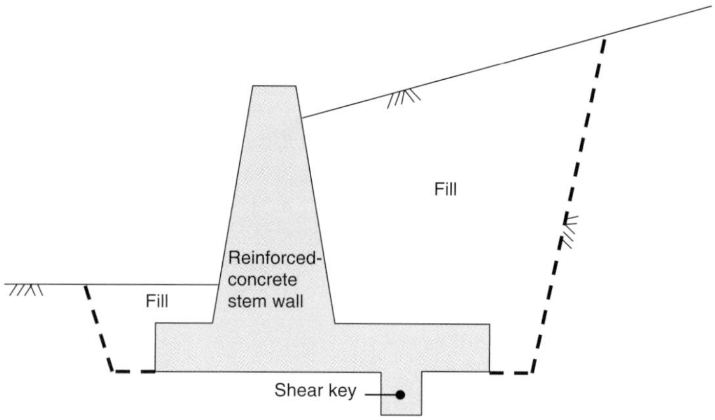

Figure 62.1 T-shaped reinforced-concrete stem wall with shear key

inverted T-shaped, where the concrete stem cantilevers from the wall base to support the retained material behind the wall. A shear key may be used to increase the shear resistance (**Figure 62.1**). The walls are not reliable as water-retaining structures, and are usually designed and constructed with suitable drainage behind the wall to limit the build-up of water at the back of the wall.

Precast walls are typically 1–3 m in height.

62.2.3 Masonry walls

Masonry walls are made with brick, block, natural or manufactured stone, conventionally bedded with mortar and founded on mass or reinforced-concrete foundations (**Figure 62.2**).

The walls should be provided with a coping layer to avoid water saturation and possible frost damage. Details for the use of copings, drainage, damp-proof courses and waterproofing to prevent saturation of the wall and improve durability are given in BS EN 1996-2 (BSI, 2006, incorporating corrigendum 2009). The walls should be constructed in panels, up to 10–15 m in length, with movement joints at the ends.

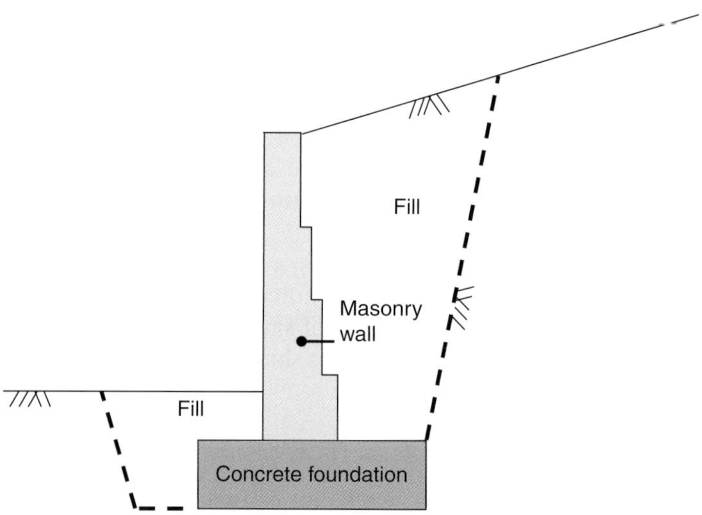

Figure 62.2 Typical masonry wall

62.2.4 Dry-stack masonry walls

Dry-stack masonry walls consist mainly of precast concrete special blocks and occasionally bricks that are designed to interlock with each other to produce a solid wall face (**Figure 62.3**). A dry-stack wall relies upon gravity to support the retained material and will have an interlock shear resistance between each layer of blocks. The presence of the interlock assists with the accurate placing of successive layers of blocks.

The wall foundations may be mass, reinforced or possibly precast concrete with interlocks compatible with the wall units.

Most of the walls are relatively free-draining, but some may require granular backfill material and a drain to avoid the build-up of water behind the wall.

The walls can be constructed vertically but are typically inclined or battered to provide better stability for greater heights. The coping blocks at the top of the wall, often attached to the rest of the wall with adhesive, are to prevent damage to the wall through vandalism.

62.2.5 Crib walls

Crib walls comprise a grillage of header and stretcher units placed on firm foundations, usually of mass or reinforced concrete. The header and stretcher elements may be made of

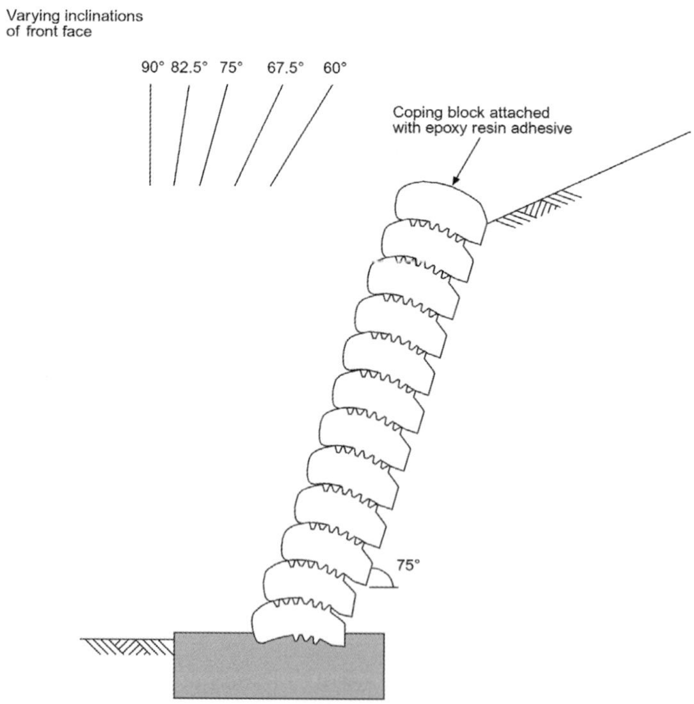

Figure 62.3 Porcupine wall: typical single-width dry-stack masonry wall (Reproduced with permission from CIRIA C516, Chapman *et al.*, 2000; www.ciria.org)

reinforced concrete or timber, and are designed to be interlocking to give the wall continuity. Walls can be made with single or multiple rows of cribs at the base, with single rows used for low walls and multiple rows for higher walls (**Figure 62.4**).

The spaces between the crib grillages are filled with a free-draining, non-aggressive coarse granular material. The spaces can also be filled with lean-mix concrete, making it more akin to a masonry wall.

Even without interceptor drains, the open structure of the granular fill can allow water to flow through the wall. Interceptor drains behind the wall will be required with a lean-mix concrete fill to prevent the build-up of water behind the wall.

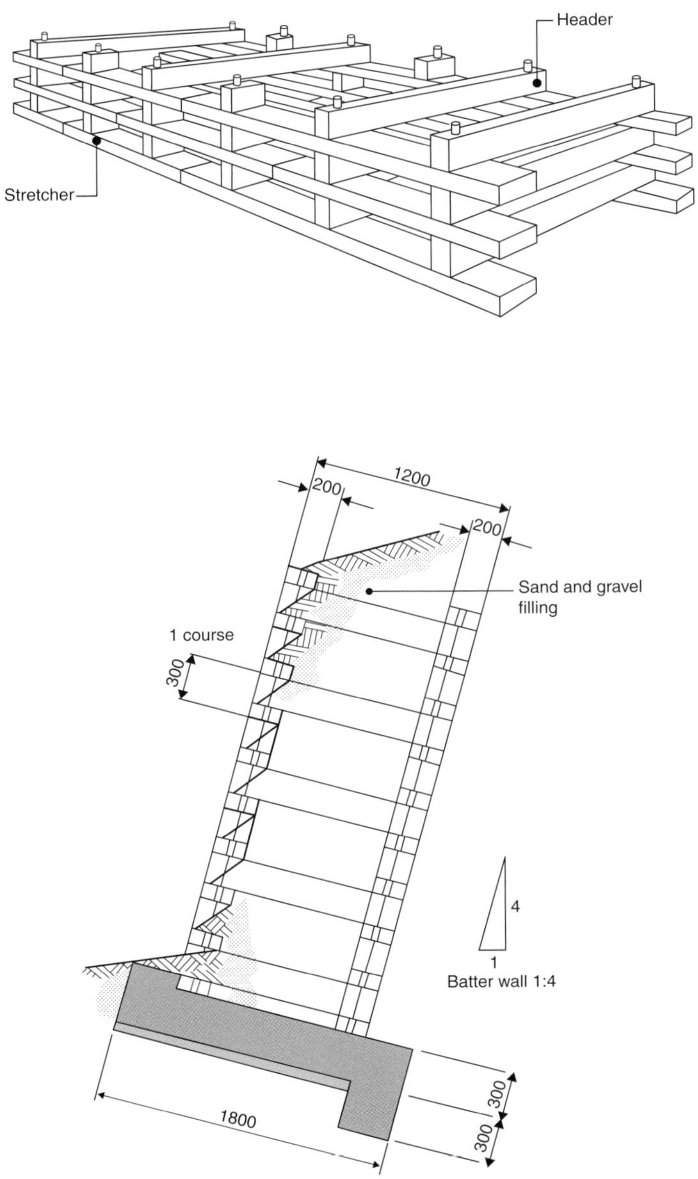

Figure 62.4 Crib wall (Reproduced with permission from CIRIA C516, Chapman *et al.*, 2000; www.ciria.org)

Crib walls are often laid to a batter; an angle corresponding to 1 horizontal to 12 vertical is typical; however, for shorter walls – generally those less than 2 m – the walls are sometimes built with a vertical face provided their width exceeds their height.

62.2.6 Gabion walls

Gabion walls are free-draining walls constructed by filling rectangular mesh cages with broken stone. The cages are made from sheet mesh strips or woven wires. Differing degrees of corrosion resistance of the steel mesh can be provided by galvanising (zinc or zinc–aluminium coating) or coating with a bonded plastic, thermoplastic or epoxy resin polymer coating. They are placed side by side and laced together in a stepped course to form a gravity structure (**Figure 62.5**). The walls are typically constructed with either a front stepped face or a back stepped face to incline the wall by at least 6–8° from the vertical towards the retained material.

The walls are typically constructed in 0.5–1.0 m heights, corresponding to the standard size of gabion-basket units. Two forms of cage are common: basket and mattress. Mattress cages tend to be long in relation to the height and are often used as a lower level to a gabion wall.

The flexibility of gabion walls makes them more suitable where the subgrade soil is poor, as the walls can accommodate larger total and differential settlements than other wall types.

The permeability of the walls also makes them well suited to saturated conditions, such as near to rivers. In such conditions, the grading of the fill behind the wall and in the wall needs to be carefully chosen so that the fines in the fill are not washed out by flowing water.

A variation of the gabion wall is a bastion. Bastion walls are collapsible wire or geotextile mesh multicellular structure systems which are designed for rapid erection from a flat-pack delivery. When assembled, the bastion forms a series of individual cells, which are lined with a non-woven geotextile. The cells are filled with granular material, such as sand, ballast, earth, stone or concrete. They are most often used for emergency works such as flood protection.

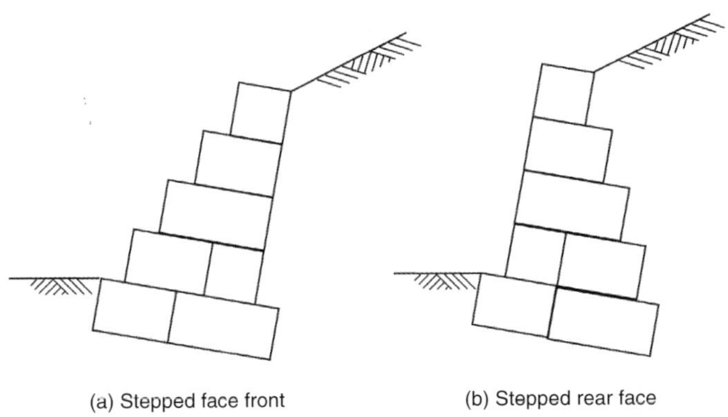

Figure 62.5 Gabion wall (Reproduced with permission from CIRIA C516, Chapman *et al.*, 2000; www.ciria.org)

To be used for permanent works, measures need to be incorporated to prevent the loss of fill material through the mesh, and the use of a facing to protect any geotextile mesh against ultraviolet degradation.

62.2.7 Reinforced soil

Reinforced soil can be considered as a gravity wall. In contrast to the gravity walls described above, the gravity structure is provided by reinforcing the ground itself rather than providing a separate structure (**Figure 62.6**).

Reinforced soil walls are described in Chapters 63 *Principles of retaining wall design* and 73 *Design of soil reinforced slopes and structures.*

62.3. Embedded walls

An embedded retaining wall penetrates the ground at its base to gain lateral support from the passive resistance of the ground in front of the lower part of the wall. This resistance may be used as a sole resistance to provide stability for the wall or may act in conjunction with anchorages, props or other sources of lateral support for stability (**Figure 62.7**).

Embedded retaining walls are often used in conjunction with lateral support. Methods of lateral support are described in Chapter 65 *Geotechnical design of retaining wall support systems.*

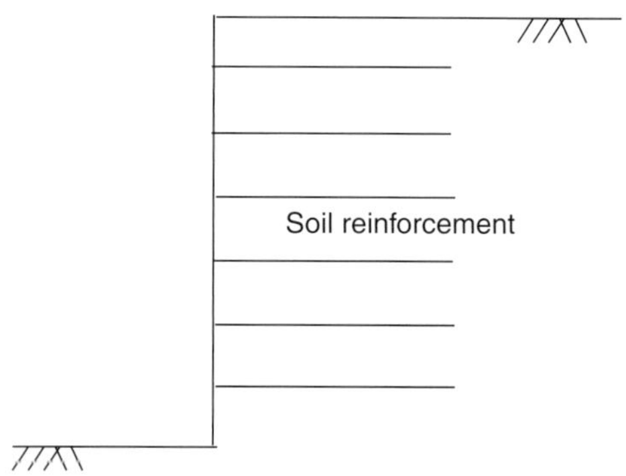

Figure 62.6 Reinforced soil wall

In contrast to gravity walls, embedded walls require specialist plant and operations on site. Embedded walls are often adopted in preference to gravity walls when

- deep excavations are required
- temporary land-take is not available for a temporary open-cut excavation
- buildings or other structures are in close proximity to the excavation and need to be supported or protected
- high groundwater conditions require excessive dewatering for a temporary open-cut excavation.

Key references for embedded retaining walls include

- CIRIA C760 (Gaba *et al.*, 2017)
- BS 8002 (BSI, 2015) – section 7
- *Piling Handbook*, 9th edition (ArcelorMittal, 2022)
- *Steel Intensive Basements* (Yandzio and Biddle, 2001).

62.3.1 Sheet pile walls

The key references for sheet pile walls are the *Piling Handbook* (ArcelorMittal, 2022) and *Steel Intensive Basements* (Yandzio and Biddle, 2001).

Steel sheet pile walls are constructed by driving, vibrating or pressing steel sheets into the ground without any material being removed. They are traditionally installed by means of driving methods using heavy hammers, which can result in significant noise and vibration that are often unacceptable in urban locations.

Sheet pile walls can be installed in the majority of soils. However, they can be unsuited to ground conditions with hard inclusions such as boulders or man-made obstructions, as these can lead to refusal of the pile driving or pushing. Preboring for the sheet piles or jetting along the toe of the sheet pile during installation may improve the penetration of sheet piles; however, these methods are often associated with additional settlement.

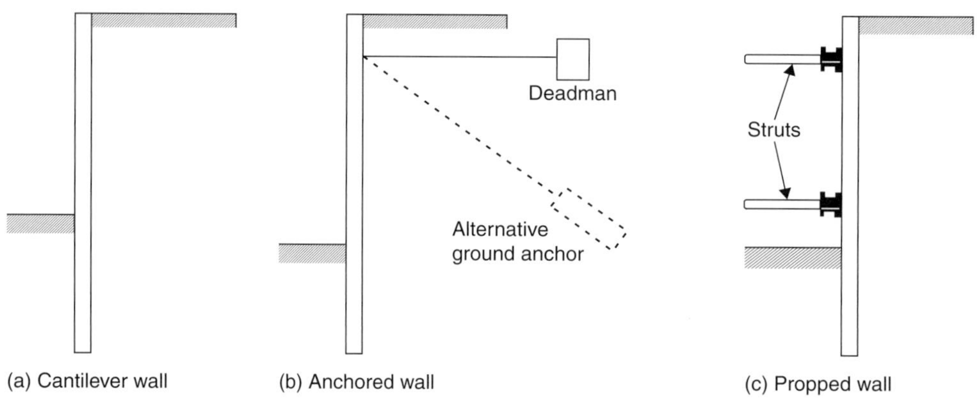

Figure 62.7 Embedded retaining wall (Reproduced with permission from BS 8002 © BSI (2015))

There are a number of standard shape profiles for sheet piles, as shown in **Figure 62.8**. These can be combined with circular or beam steel sections to create a number of sheet pile combinations with improved strength and stiffness performance. A selection of these combinations is shown in **Figure 62.9**.

If seepage through the sheet pile interlock is controlled, a sheet pile wall can be used for water-retaining applications. Sealing systems include non-swelling sealants, hydrophilic (water-swelling) sealants, combination systems and welding of the interlocks. Details of typical construction processes are provided in Section 8 *Construction processes*.

Advantages of sheet pile walls include

- provides an economic wall with a predictable surface finish
- no arisings
- suitable as water-retaining walls
- can be used as both a temporary and permanent wall.

Disadvantages of sheet pile walls include

- maximum pile length is approximately 30 m
- corrosion can lead to a reduced section size
- potential declutching in coarse-grained soils or in hard driving conditions.

62.3.2 King post walls

King post walls are also known as soldier piles with lagging or Berlin walls.

Isolated steel beams or posts are installed along the line of the wall, either driven into position or placed in bored

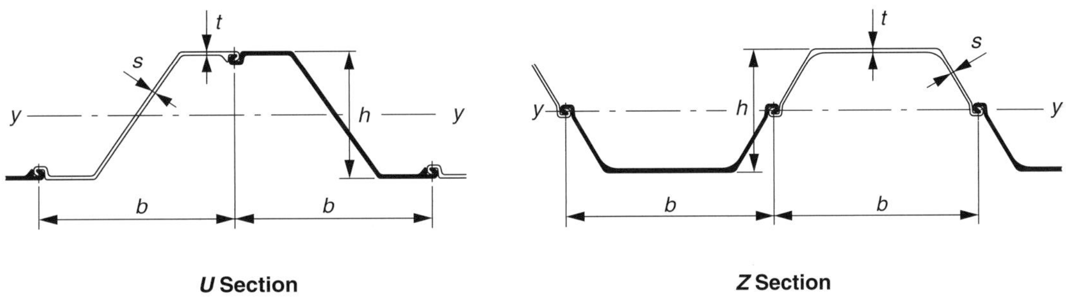

Figure 62.8 Standard sheet pile sections (Reproduced with permission from ArcelorMittal, 2022); all rights reserved)

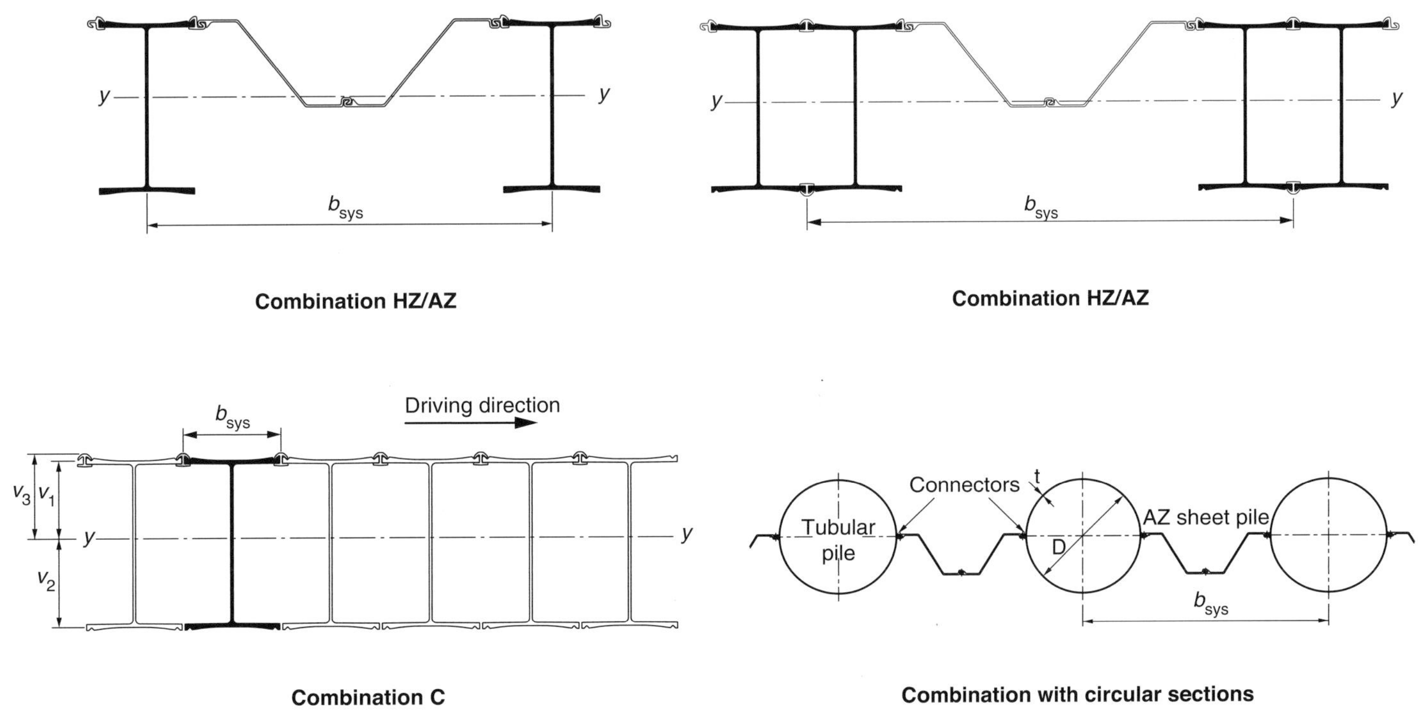

Figure 62.9 Sheet pile combinations (Reproduced with permission from ArcelorMittal, 2022; all rights reserved)

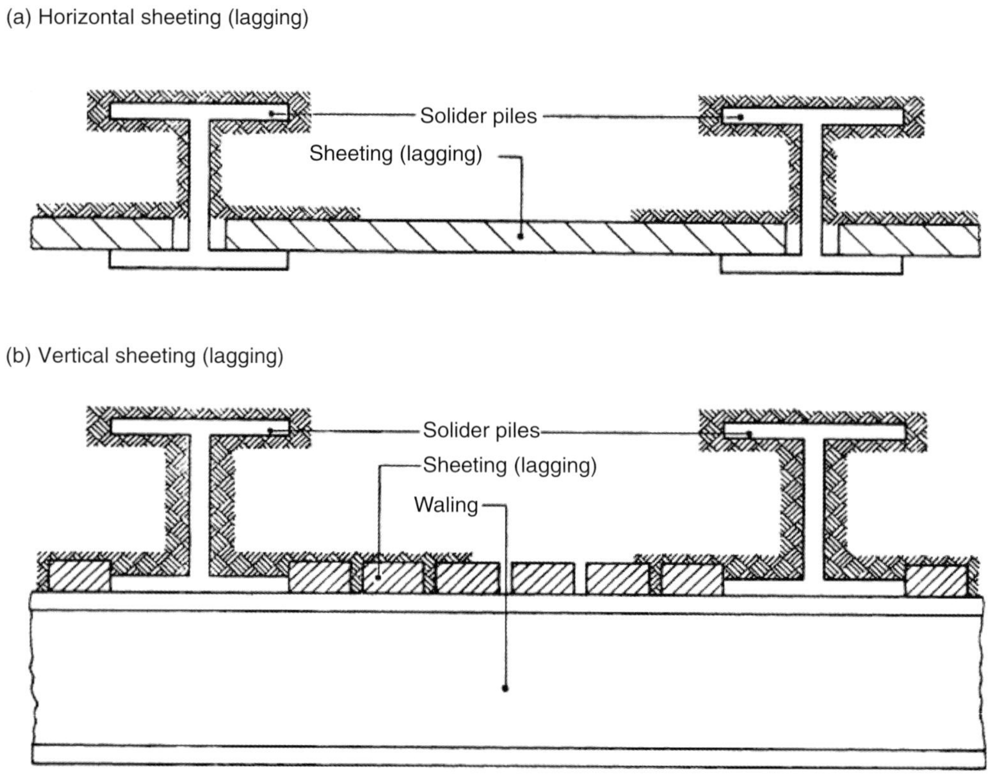

Figure 62.10 Soldier pile wall (Reproduced with permission from BS 8002.© BSI (2015))

cast-in-place piles. As the excavation proceeds the space between the posts is filled using either timber railway sleepers, precast concrete elements or in situ or sprayed concrete. As such, the wall is not suitable for excavation below the water table in coarse-grained soils without significant dewatering outside the excavation. The posts are typically spaced 1–3 m apart and can be varied, making it easier to lay out the wall to avoid localised obstructions (**Figure 62.10**).

Propped excavations up to 20 m can be achieved; however, shallower depths of excavation are common due to the requirements for groundwater control.

62.3.3 Contiguous bored pile walls

A contiguous pile wall consists of bored cast-in-place piles installed along the line of the wall. The piles are installed with a gap between individual piles. The size of the gap is chosen on the basis of the site dimensions and also the particular ground conditions, but is typically 50–150 mm (**Figure 62.11**).

The piles can be installed as either continuous flight auger (CFA) or traditional bored piles, using casings where required. The choice of pile type will depend on the ground conditions and depth of the wall. Details of the design and construction of piles are provided in Sections 5 *Design of foundations* and 8 *Construction processes*.

Due to the gap between individual piles, the wall is not water-retaining and is, therefore, not suited to construction below the water table in coarse-grained ground conditions. Traditional bored piles with thick-walled casings may be required to overcome harder strata and obstructions.

Contiguous piled walls cannot be used as a permanent wall due to the gaps between individual piles; however, they can be made into a permanent solution through the use of a structural facing wall.

For CFA piles, the depth of retaining wall will be limited by the maximum depth of the CFA piles (refer to Section 8 *Construction processes*). Propped excavations up to 20 m can be achieved using contiguous pile walls.

62.3.4 Secant bored pile walls

A secant pile wall consists of intersecting bored cast-in-place piles installed along the line of the wall (**Figure 62.12**). The pile construction may either use traditional piling rigs or CFA rigs. Refer to Chapters 79 *Sequencing of geotechnical works*, 81 *Types of bearing piles* and 82 *Piling problems* for more details on piling methods.

The primary (female) piles are installed first, followed by the secondary (male) piles, which cut through the primary piles during installation. This overlap or secanting of individual piles provides a barrier to groundwater flow across the wall. The secondary piles are constructed with full-strength reinforcement and provide the structural strength of the wall.

Three varieties of secant pile wall can be considered and are defined by the strength of the primary piles as follows

- hard/soft – a soft pile mix, typically cement and bentonite or bentonite and sand with a characteristic compressive strength of 1–3 N/mm^2
- hard/firm – a firm mix, with a characteristic compressive strength of 10–20 N/mm^2, and a retardant to reduce the strength of the piles during the drilling of the secondary piles
- hard/hard – a full-strength concrete, which can also be reinforced (**Figure 62.13**).

The size of the piles is based on standard pile diameters.

The size of the overlap and pile spacing is chosen on the basis of the site dimensions and consideration of the installation tolerance of the piles and the depth of overlap or secanting required.

The hard/soft wall is not a preferred solution for permanent water-retaining walls due to shrinkage and cracking of the soft pile mix. To create permanent water retention a structural internal lining wall can be cast against the face of the wall.

For hard/firm and hard/hard basement walls, the need for an additional facing wall, such as a drained cavity and blockwork wall, will depend on the basement grade that is specified (refer to BS 8102 (BSI, 2022)).

As a water-retaining wall, secant pile walls are suitable for permeable ground conditions where excavation is to extend below the water table. They are also suited to coarse-grained materials above the water table where the gaps in a contiguous piled wall could result in a loss of fines from the material retained behind the wall. Where permeable ground conditions are present above low-permeability strata (e.g. gravels above clay), the primary piles may be terminated with only limited penetration into the clays, with the secondary piles extending lower to form a contiguous wall within the clays.

Secant pile walls are suitable for most ground conditions.

For CFA piles, the depth of the retaining wall will be limited by the maximum depth of the CFA piles (refer to Section 8 *Construction processes*). The maximum depth of a secant wall may be dominated by the installation tolerance of the wall and, therefore, the depth at which effective overlap of the piles can be achieved to provide a barrier to groundwater flow. Propped excavations up to 20 m can be achieved or up to 25 m for cased hard/hard solutions.

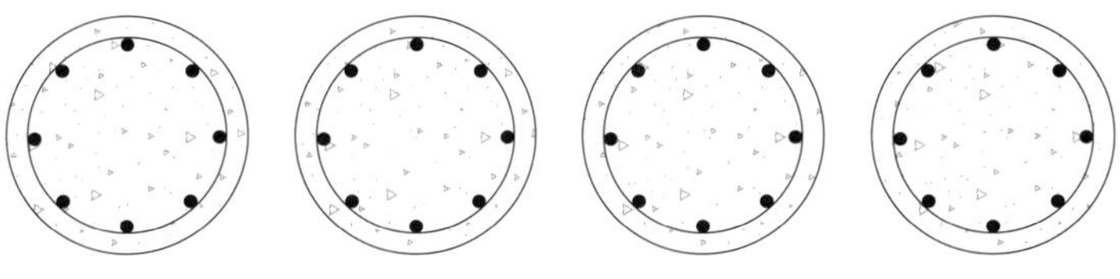

Figure 62.11 Contiguous pile wall (Reproduced with permission from CIRIA C760, Gaba *et al.*, 2017; www.ciria.org)

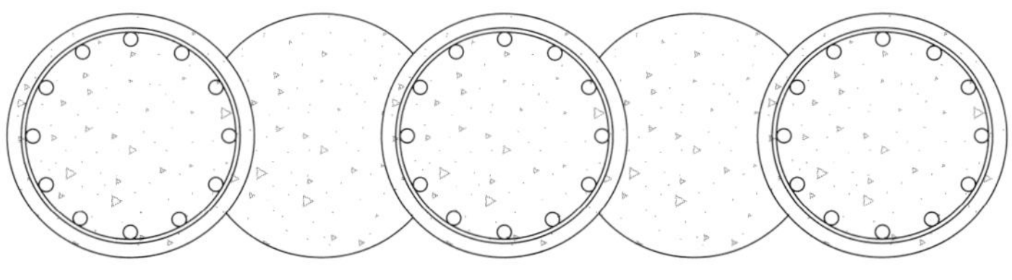

Figure 62.12 Hard/soft or firm secant pile wall (Reproduced with permission from CIRIA C760, Gaba *et al.*, 2017; www.ciria.org)

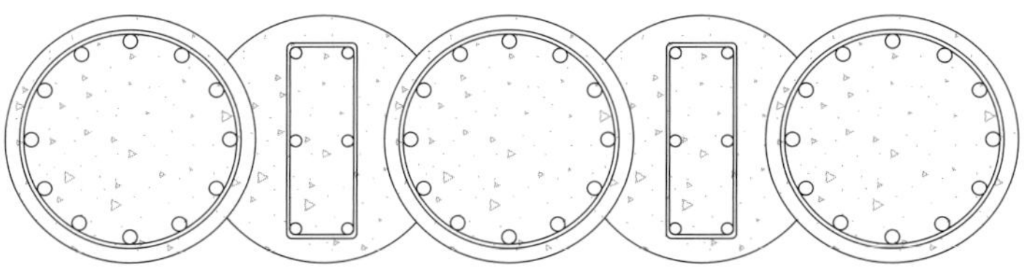

Figure 62.13 Hard/hard secant pile wall (Reproduced with permission from CIRIA C760, Gaba *et al.*, 2017; www.ciria.org)

62.3.5 Diaphragm walls

Diaphragm walls are interlocking reinforced-concrete panels (**Figure 62.14**).

The panels are formed by excavation under fluid to support the surrounding ground and lowering a prefabricated reinforcement cage into the trench excavation. Concreting is undertaken from the base of the pile using a tremie tube to displace the support fluid. Due to the requirements for a support fluid and the large volume of individual panels, diaphragm wall construction requires a larger laydown construction space than other forms of wall construction.

Panel widths and lengths are a function of the construction equipment available. Typical widths are 600, 800, 1000, 1200 or 1500 mm, which corresponds to the width of the grab or cutter used to form the hole. The typical grab length is 2.8 m, but may be reduced to 2.2 m. The lengths are a function of the number of bites used to form the panel (**Table 62.1**). For larger panel lengths, the ability of the support fluid to act in conjunction with the ground to arch around the large excavation to maintain a stable excavation should be considered.

Diaphragm walls are suitable for most ground conditions.

The depth of the wall is limited by the reach of the excavation machinery. Walls have been constructed to depths of up to 120 m; however, at this depth stringent installation tolerances are needed to provide an effective overlap of the panel width to the full depth of the walls. For such deep panels, detailed planning is required in relation to the practical difficulties of splicing cage sections together; the crane capacity to lift the cage; and the overall construction duration for the panels.

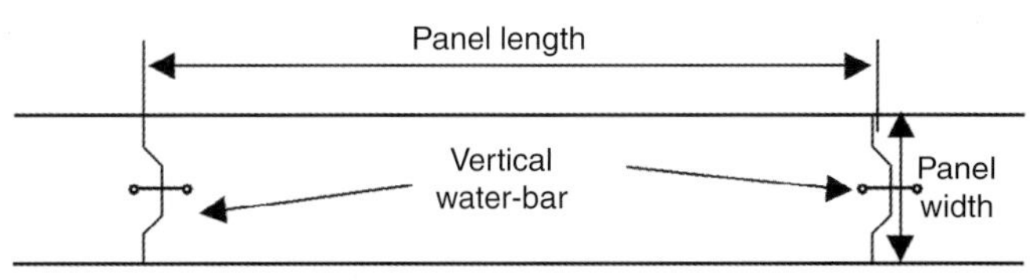

Figure 62.14 Diaphragm wall panel and joint (Reproduced with permission from CIRIA C580, Gaba *et al.*, 2003; www.ciria.org)

Number of bites	Panel length: m Minimum	Maximum	Excavation sequence
1	W	W	[1]
2	$W + T$	$2 \times W - T$	[1][ ][2]
3	$2 \times W + T$	$3 \times W - 2 \times T$	[1][ ][3][ ][2]

W is the grab length available (typically 2.8 m); T is the grab width: 600, 800, 1000, 1200 or 1500 mm.

Table 62.1 Diaphragm wall: typical panel widths (Data taken from CIRIA C580, Gaba et al., 2003; www.ciria.org)

Due to the size of the individual panels, diaphragm walls are less flexible in dealing with any changes in the alignment of the wall and are, therefore, more suited to long, straight walls as opposed to more complicated or changing geometries.

Water-bars constructed across the joints between individual panels make a diaphragm wall into a water-retaining wall.

Diaphragm walls are also known as slurry walls in some countries; however, care should be taken with this terminology as in the UK a slurry wall is an unreinforced wall used as a cut-off or containment wall rather than an embedded retaining wall.

Variations can include

- T-shaped panels for increased strength and stiffness
- post-tensioned panels
- precast panels, with or without pre-tensioning.

Precast panels are constructed by lowering the panel into the support fluid-filled trench and sealed into the ground with in situ concrete or grout placed by means of a tremie pipe. The size of the panels for precast walls is limited practically by the ability to lift the panels and place them into the trench.

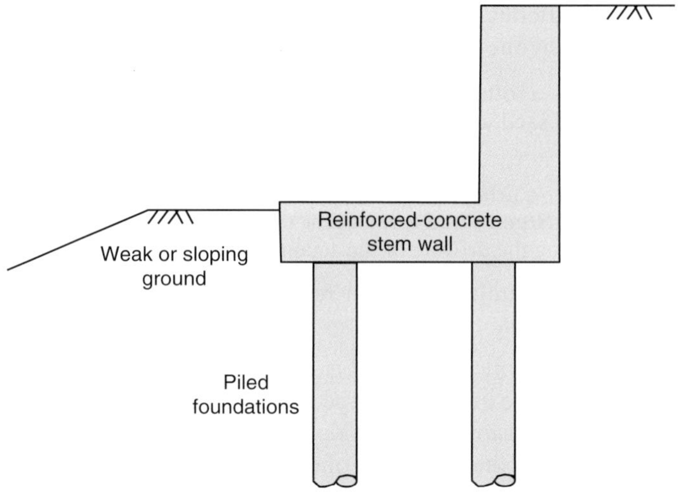

Figure 62.15 Example of a hybrid retaining wall

62.4. Hybrid walls

Hybrid walls combine aspects of gravity and embedded elements to retain the ground. An example of a hybrid gravity wall is a concrete stem wall, L-shaped or T-shaped, founded on individual piles (see **Figure 62.15**). Such a hybrid solution is suited to circumstances where there is insufficient sliding or bearing capacity for a gravity wall but sufficient capacity can be achieved using piled foundations for the wall.

62.5. Comparison of walls

A comparison of the advantages and disadvantages of the different retaining wall types is given in **Table 62.2**. Chapter 63 *Principles of retaining wall design* gives further information on factors affecting the choice of retaining wall types.

Wall type	Advantages	Disadvantages
Gravity walls – general	• Minimises construction activities on site by using precast or modular systems • Cost-effective retaining wall solution for supporting raised ground levels	• Requires site dewatering for construction below the water table • Requires temporary open-cut excavation for retaining wall construction below existing ground level
Precast reinforced-concrete stem walls	• Provides a wall with a predictable surface finish	• Height of wall may be limited practically by transporting and lifting requirements; 1–3 m typical retained height • Changes in retained height or plan alignment need to be planned in detail for prefabrication of units • Non-draining, appropriate drainage required
Masonry walls	• Can provide a wall with a predictable surface finish • Can be installed around obstructions at isolated points • Can be built to a batter	• Construction requires bricklaying skills • Foundation slab required • Non-draining, appropriate drainage required

Table 62.2 Comparison of retaining walls (Reproduced with permission from Gaba *et al.*, 2003, and Chapman *et al.*, 2000) (continued on next page)

Wall type	Advantages	Disadvantages
Dry-stack masonry walls	• Simple manual construction • Distinctive course effect • May be used with planting • Can produce curves in plan	• Foundation slab required • Non-draining, appropriate drainage required
Crib walls	• Simple manual construction • Distinctive 'honeycomb' appearance • May be used with planting • Can produce curves in plan	• Foundation slab required • Free-draining, although back drain may also be required • Backfill in compacted layers
Gabion walls	• Simple construction • Foundation slab may not be required • Fully draining, no back drain required unless they are filled with non-draining materials • Can be used with planting • Can produce curves in plan	• A flexible system which can result in an undulating wall profile
Embedded retaining walls – general	• Minimise volume of excavation • Enable deep excavation adjacent to existing structures and utilities • Enable deep excavation below the water table (water-retaining embedded walls only)	• Specialised construction plant and operation on site • Costly in comparison with open-cut construction methods
Sheet pile walls	• Provides an economic embedded wall with a predictable surface finish • No arisings to be removed • Suitable as a water-retaining wall • Can be used as both a temporary and a permanent wall	• Maximum pile length approximately 30 m • Potential declutching in coarse-grained soils
King post walls	• Can be installed around obstructions at isolated points	• Not suitable for retaining water in the long term • Cannot be used for excavation below the groundwater table in coarse-grained soils
Contiguous bored pile walls	• The cheapest form of concrete piled wall	• Not a water-retaining solution • Not a permanent solution in any soil due to the gaps between piles, unless a structural facing is applied
Hard/soft secant bored pile walls	• Acts as a water-retaining temporary wall • The use of soft piles enables hard piles to be formed using lower-torque rigs than for hard/hard secant piles	• Not usually a permanent solution for retaining water • The soft pile mix is not significantly cheaper than concrete. Local concrete plant is often unable to batch the soft material, so site batching is required • Depth is limited by the verticality tolerance, which may determine the depth of secanting
Hard/firm secant bored pile walls	• A permanent water-retaining wall • The firm material for the primary (female) piles is either a standard concrete mix, retarded to reduce its strength when the secondary (male) piles are constructed or a reduced-strength concrete mix	• Depth is limited by the verticality tolerance, which may determine the depth of secanting
Hard/hard secant bored pile walls	• A permanent water-retaining wall • Installed using standard piling plant with high-torque rigs	• The cutting of the hard primary (female) piles requires high-torque rigs or oscillators • Depth is limited by the verticality tolerance, which may determine the depth of secanting
Diaphragm walls	• A permanent water-retaining wall • Can be installed to great depths provided the verticality tolerances can be accepted • In some circumstances the face of the diaphragm wall can form the final finish subject to some surface cleaning and removal of protuberances • Fewer joints compared with piled walls	• Horizontal continuity is difficult to achieve between panels • Cannot follow intricate plan outlines • The installation equipment is extensive, requiring a large site area for accommodation of the support fluid plant, reinforcement cages and the excavation plant • Disposal of the support fluid is costly
Hybrid walls	• Provide a practical retaining wall solution where the site conditions will not allow a gravity wall but do not require an embedded retaining wall	• Combination of different wall and foundation elements results in site- and solution-specific wall type

Table 62.2 Continued

References

ArcelorMittal (2022) *Piling Handbook*, 9th edn. ArcelorMittal, London, UK. See http://www.arcelormittal.com/sheetpiling/page/index/name/arcelor-piling-handbook (accessed 27/03/2023).

BSI (British Standards Institution) (2004) BS EN 1997-1:2004 + A1:2013: Eurocode 7: Geotechnical design – Part 1: general Rules. BSI, London, UK.

BSI (2006) BS EN 1996-2:2006: Eurocode 6 – Design of masonry structures – Part 2: Design considerations, selection of materials and execution of masonry. BSI, London, UK.

BSI (2015) BS 8002:2015: Code of practice for earth retaining structures. BSI, London, UK.

BSI (2022) BS 8102:2022: Code of practice for protection of below ground structures against water from the ground. BSI, London, UK.

Chapman T, Taylor H and Nicholson D (2000) *Modular Gravity Retaining Walls: Design Guidance. Report C516.* CIRIA, London, UK.

Gaba A, Simpson B, Powrie W and Beadman DR (2003) *Embedded Retaining Walls: Guidance for Economic Design. Report C580.* CIRIA, London, UK.

Gaba A, Hardy S, Doughty L, Powrie W and Selemetas D (2017) *Guidance on Embedded Retaining Wall Design. Report C760.* CIRIA, London, UK.

Yandzio E and Biddle AR (2001) *Steel Intensive Basements. Report P275.* Steel Construction Institute, Ascot, UK.

Further reading

Geoguide 1 (2020) *Guide to Retaining Wall Design.* Hong Kong Government Geotechnical Control Office, Hong Kong.

NAVFAC (1986) *Foundations and Earth Structures. NAVFAC Design Manual 7.02. Naval Facilities Engineering Command (NAVAC)*, US Department of the Navy, Washington, DC.

Useful websites

CIRIA: http://www.ciria.org

Federation of Piling Specialists: http://www.fps.org.uk/index.php

Hong Kong Geo Publications: http://www.cedd.gov.hk/eng/publications/manuals/geo_publications.htm

Piling Handbook: http://www.arcelormittal.com/sheetpiling/page/index/name/arcelor-piling-handbook

It is recommended this chapter is read in conjunction with

- Chapter 85 *Embedded walls*
- Chapter 91 *Modular foundations and retaining walls*

All chapters in this book rely on the guidance in Sections 1 *Context* and 2 *Fundamental principles.* A sound knowledge of ground investigation is required for all geotechnical works, as set out in Section 4 *Site investigation.*

doi: 10.1680/icemge.66830.1051

Chapter 63
Principles of retaining wall design

Michael Devriendt Arup, London, UK

CONTENTS

Prior to carrying out the design of a retaining wall, basic concepts relating to the design should be understood by the designer. Considerations such as the characterisation of the wall, the construction sequence to form the retaining wall, performance criteria and the selection of an appropriate wall type should be understood. Following this appreciation of basic concepts, it is common for a designer to select design parameters for analysis in accordance with appropriate standards. The designer should also understand serviceability issues relating to the magnitude of ground movements and the potential for damage to nearby structures.

This chapter provides the reader with a basic understanding of these conceptual issues relating to the design of retaining walls and provides references to other material where more detailed descriptions may be found.

63.1. Introduction

This chapter provides a description of the basic concepts that govern retaining wall design. The concepts described should be appreciated by designers before carrying out the design of a wall as outlined in Chapters 64 *Geotechnical design of retaining walls* to 67 *Retaining walls as part of complete underground structure*. This chapter, therefore, serves as a basis for describing many of the conceptual issues that need to be addressed when planning the design of a retaining wall.

63.2. Design concepts

63.2.1 Geotechnical characterisation of retaining walls

When designing a retaining structure, it is important to understand how the retaining wall can be characterised. As described in Chapter 62 *Types of retaining walls*, retaining structures can broadly be considered to be characterised as gravity or embedded walls or a combination of both. Methods of analysis of the wall types are given in Chapters 64 *Geotechnical design of retaining walls* to 67 *Retaining walls as part of complete underground structure*.

For analysis or back-analysis of a retaining structure it is fundamental to obtain an understanding of ground and groundwater conditions. The use of soil parameters and adopting different groundwater conditions in design is discussed in greater detail later in this chapter and in Chapters 64 *Geotechnical design of retaining walls* to 67 *Retaining walls as part of complete underground structure*.

63.2.2 Construction sequence

63.2.2.1 General

The construction sequence that is followed to construct a retaining wall can have a significant effect on the design, cost and time spent constructing the wall. This section discusses the conceptual decisions that influence the selection of an appropriate construction sequence.

63.2.2.2 Design

Displacement of the retaining wall influences the in situ stress state in the ground around the retaining wall. Earth pressure theory was introduced in Chapter 20 *Earth pressure theory*. During installation of the wall and as displacement of the wall occurs, earth pressures will change from an in situ K_0 condition to active and passive states depending on whether relaxation or compression of the soil occurs behind or in front of the wall, respectively.

Figure 63.1 illustrates this concept of the amount of displacement (or strain) governing the stress state on the active and passive sides of the wall. An active state (K_a) is reached

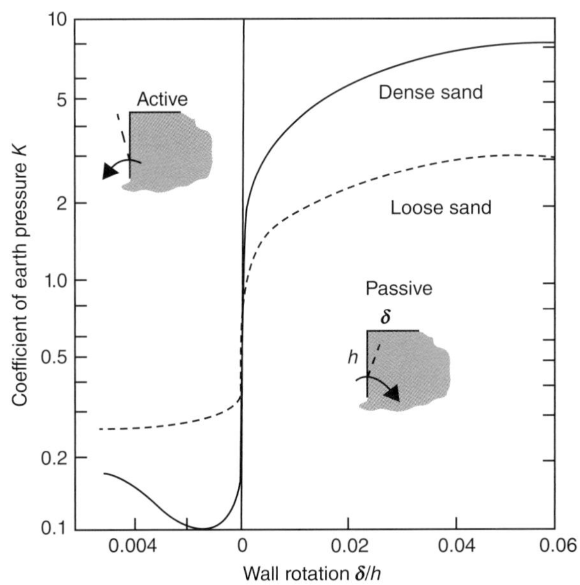

Figure 63.1 Relationship between earth pressure and horizontal strain by Terzaghi (1954) for normally consolidated sand (different scales for active and passive) (Reproduced from Simpson, 1992; all rights reserved)

at relatively small strain, while larger strains are required to reach a passive state (K_p). The earth pressures in front of or behind a retaining structure will influence the bending moment, shear forces and any prop forces supporting the wall. The construction sequence that is followed can also influence the displacement of a retaining structure and, therefore, affect the design of the retaining structure.

To appreciate the differences caused by construction sequence, an analysis that considers staged construction is necessary (i.e. the limit equilibrium method of analysis described in Chapter 64 *Geotechnical design of retaining walls* is not suitable). The use of more complex analysis methods such as pseudo- and full finite element programs is required to accommodate the effects of construction sequence. The use of such programs is discussed in Chapter 6 *Computer analysis principles in geotechnical engineering*, and their use for retaining wall design in particular is also covered in Chapters 64 *Geotechnical design of retaining walls* to 67 *Retaining walls as part of complete underground structure*.

Displacements of the wall and those caused from construction of the wall itself will influence ground movements around the retaining wall. This is discussed in further detail in Section 63.3.

63.2.2.3 Basement construction

The excavation of basements, particularly in confined urban sites, is a situation where the construction sequence has a significant impact on the successful outcome of the project. In particular, careful consideration of the construction sequence is required where there is an existing basement that occupies the site and the retention of existing basement retaining walls is necessary.

63.2.2.4 Top-down and bottom-up

Where retaining walls are used for excavation of basements and confined spaces such as station box structures, two generic forms of construction sequence are commonly used to form the basement space: 'top-down' and 'bottom-up'. Variants of these generic sequences are also commonly adopted to suit particular site constraints.

Top-down sequences involve constructing sequential permanent basement slabs to support the retaining walls as the excavation progresses. Bottom-up sequences use temporary propping to support the basement space during the excavation, then construct the permanent basement slabs after the excavation is complete.

The two generic sequences are compared in **Table 63.1** through reference to some common considerations. Photographs of construction carried out in top-down and bottom-up sequences are shown in **Figures 63.2** and **63.3**.

The construction sequence followed can have a significant effect on the programme and the cost of the proposed development. In particular, the programme for development in urban sites can have an effect on the overall successful completion of a project. The effect of delay on costs is discussed in greater detail in Chapter 7 *Geotechnical risks and their context for the whole project*.

63.2.3 Design requirements and performance criteria

63.2.3.1 General

Prior to the design of a retaining structure, the designer must first gather the generic performance criteria for the structure. The performance criteria will vary depending on the requirements of the structure during its design life (for instance, taking into account that vertical loading may be applied to the retaining structure at some time in the future).

63.2.3.2 Design life

The design life of a structure may influence several issues relating to the design of a retaining wall.

The forces exerted on a retaining wall may also be influenced by variable loads such as wave or seismic loading. The magnitude of these forces is commonly related to a return period and probabilistic approach of exceedance for a given structural design life; refer to Eurocode 8 Parts 1 (BSI, 2004b) and 5 (BSI, 2004c) and BS 6349-1-3-2012 (BSI, 2012) on how

Property	Top-down	Bottom-up
Construction of superstructure	Allows concurrent construction of basement and superstructure	Generally basement must be constructed prior to construction of superstructure
Bearing piles used to support superstructure within plan area of basement space	Piles installed with low cut-off and columns plunged into piles	Piles can be installed before or after basement excavation. If before excavation, the use of plunged tubes for reception of columns may be necessary
Spoil removal	Spoil needs to be removed from moling hole in slab	Allows 'clear space' or 'blue sky' excavation of spoil, generally allowing quicker excavation of spoil provided lorry movements at the surface are not restricted
Ground movements	Generally a greater stiffness basement space is created, therefore minimising ground movements	Temporary propping tends to be less stiff than permanent propping, resulting in greater movements; however, this can be mitigated by using additional temporary propping

Table 63.1 Top-down and bottom-up considerations

Figure 63.2 Example of top-down excavation (One New Change, London) (Photo courtesy of Arup)

Figure 63.3 Example of bottom-up excavation (Ropemaker, London) (Photo courtesy of Arup)

to select return periods. The proposed design life of a structure can influence the return period that is considered in the design and, therefore, the magnitude of loads. In particular, for the design of retaining walls in seismic regions, the size of the active and passive zones may be affected by seismic loading. Consequently, the positions of support features such as tie-backs should be reviewed with respect to the seismic design.

The integrity of the overall structural form of the retaining wall and structural support also needs to be considered. Where a structure is exposed to adverse conditions (for instance sheet piles exposed to corrosion in a saline environment) the reduction in thickness of section needs to be accounted for in the design – see the *Piling Handbook* (ArcelorMittal, 2022).

63.2.3.3 Temporary and permanent works

The responsibility for design of the permanent and temporary works is often split between companies or groups of designers. As identified in Section 63.2.2, the construction sequence that is followed can have a significant effect on the design of the retaining wall. It is, therefore, important that the temporary and permanent works designers share design input information and the construction sequence that each respective designer anticipates is the same. In particular, consideration to respective prop and retaining wall properties and wall stiffness parameters should be considered. An analysis should be carried out to check the entire sequence (both temporary and permanent works).

63.2.3.4 Drained or undrained

When determining whether drained or undrained soil conditions apply, account should be taken of the rapidity with which drained conditions are achieved. For fully saturated soil, undrained conditions are characterised by there being no changes in volume, although a change in shape can occur. It is possible for such conditions to prevail in the short term in low-permeability soils such as high-plasticity sedimentary clays, where excess water pressure dissipates slowly and suction is generated by loading and unloading, respectively. The detailed fabric of the soil dominates the magnitude of the water pressure at equilibrium. The importance of obtaining an accurate record and gaining a good understanding of the soil fabric cannot be overemphasised (refer to Chapters 13 *The ground profile and its genesis* to 15 *Groundwater profiles and effective stresses* and 46 *Ground exploration*).

Stiff overconsolidated clays are generally jointed, fissured and layered as a result of their depositional history. Although the permeability of the clay between the joints and discontinuities may be low, the presence of small sandy or silty partings can have a disproportionate effect on the magnitude of the coefficient of permeability of the soil. This influences the rate at which water may flow from one part of the soil mass to another and the rate at which water may be drawn into the soil. To assess whether undrained conditions are likely to prevail for a significant time, the following factors should be considered.

- Soil stratigraphy and fabric. There is usually no dilation of presheared surfaces, which means that shearing can occur at constant volume under drained residual conditions. This should not be confused with undrained behaviour.
- The mass in situ permeability of the soil (in the vertical and horizontal directions).
- The proximity and likelihood of available water.
- The soil stiffness, which affects the value of the coefficient of consolidation, c_v. This might be particularly high following a change in the direction of the stress path.
- Previous experience of construction in similar ground conditions.

The designer should evaluate drainage paths and assess the duration over which drained conditions may be restored.

As a general guide, undrained conditions may be assumed in the short term where the mass in situ permeability of the ground is low and the construction period is sufficiently short (i.e. a coefficient of permeability of the order of 10^{-8} m/s or less and a construction period of 6 months or less; see CIRIA C760 (Gaba *et al.*, 2017)). Where the mass permeability is not low, in coarse-grained soils or where extended construction periods are anticipated, drained conditions should be assumed in design.

63.2.3.5 Cycling of loads on integral bridges

For the design of bridges, it is common to have an earth-retaining structure forming part of the bridge abutment supporting either side of the bridge deck; see **Figure 63.4**. To avoid the use of bearings to support the deck structure, which can reduce long-term maintenance costs, it is common to make the abutment integral with the deck (an integral bridge).

For integral bridges, temperature variations will induce cyclic movement of the abutments and the soil retained behind them. The effect of these cyclic movements on the earth pressure influences the design of the retaining wall. In addition, in granular soils the cyclic behaviour may result in settlement occurring behind the retaining wall. This may have an effect on the design of the road or railway approach structure to the bridge.

Various design approaches have been proposed to account for the increased earth pressures due to cyclic loading; see, for instance, the UK's Highways Agency's 'Highway structures: Approval procedures and general design' (Highways Agency, 2003). An appraisal of design approaches is given by Lehane (2006) through reference to centrifuge laboratory model testing.

63.2.3.6 Performance criteria for horizontal and vertical support

Although the primary function of a retaining wall is, in many instances, to provide lateral support and retention of the ground and groundwater, some retaining walls, particularly

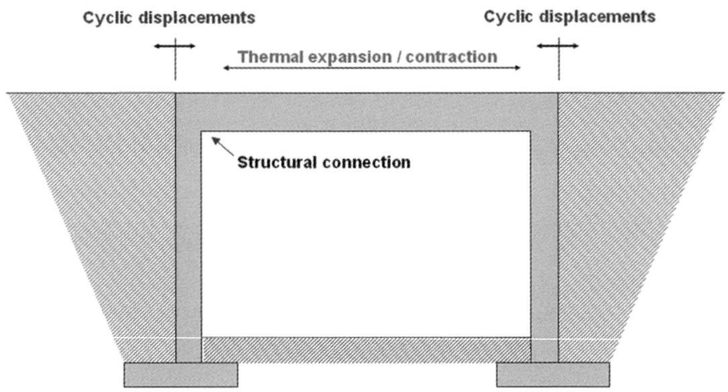

Figure 63.4 Schematic of an integral bridge (Image courtesy of Arup)

Factor	Description or comment
Availability of space around perimeter of site for retaining wall construction	Where space is constrained and the client wants to build up to the property boundary an embedded retaining wall is generally favourable over a gravity wall. The thickness of an embedded retaining wall section is also a consideration. Greater wall thickness generally increases the stiffness and reduces movement; however, these walls will take up more space.
Unsupported height of retaining wall	Where large unsupported heights of a retaining wall are desirable, generally a thicker retaining wall section will be required to have sufficient moment capacity and limit displacements of the wall to acceptable magnitudes.
Water retention	The permeability of the ground and requirements for the basement space influence the selection of wall. Generally, the fewer the connections in the wall, the lower the leakage rate and the greater the water retention. Improved long-term basement space may be obtained by using drained cavities or inner lining walls. Guidance on how to achieve watertightness requirements for bored piles, diaphragm walls and sheet pile walls is provided in CIRIA Report 139 (Johnson, 1995) and BS 8102 (BSI, 2022).
Stiffness	Higher system stiffness (i.e. the wall and its support system) leads to reduced wall and ground movement. This is achieved by increasing the thickness of the wall or increasing the number of props or ties supporting the wall.
Cost	Cost savings may be obtained where continuous support is not required for the purpose of water retention (for instance, selection of a contiguous bored pile wall rather than a secant wall). The relative costs of materials (for instance steel for sheet piles) will also fluctuate with time.
Programme	Similar to the cost section above, savings in the programme can be made if contiguous bored pile walls are feasible rather than a secant wall. Savings can also be made by reducing the overlap for secant walls or increasing the diameter of the piles (therefore reducing the number of piles to install).
Design life	The design life of a retaining wall is a primary consideration where the imposed forces on a retaining wall are governed by duration of exposure (for instance seismic or wave loading) or where the wall degrades or corrodes with time (for instance sheet piling in a maritime environment).
Embodied carbon	The total carbon used in the construction and operational life of the retaining wall. Further details of these considerations are given in Chapter 11 *Sustainable geotechnics*.

Table 63.2 Factors influencing the selection of retaining walls

those used for basement construction, may also be required to provide vertical support. For these scenarios both the allowable vertical and horizontal displacement should be considered. Further discussion regarding the design for horizontal loading is given in Chapter 64 *Geotechnical design of retaining walls* and for designing for vertical loading in Chapter 67 *Retaining walls as part of complete underground structure.*

63.2.3.7 Design of retaining structures in a marine environment

Further consideration, when designing structures in a marine environment, are the effects of tide levels, swell, storm surge, wave impact loads and sea-level rise over the design life of the retaining structure. By considering these effects, overtopping of the retaining wall can be minimised. Allowance should also be made for the water level difference between the active and passive sides of the retaining wall. BS 6349-1-3:2021 (BSI, 2012) should be referred to for further guidance on appropriate considerations for design in a marine environment.

63.2.4 Consideration for wall selection

Chapter 62 *Types of retaining walls* describes the main types of retaining wall that are commonly constructed to retain soil and groundwater. **Table 63.2** provides a summary of the factors that influence the selection of retaining walls.

63.3. Selection of design parameters

63.3.1 Design philosophy and approach (limit states, moderately conservative, most probable)

63.3.1.1 Limit states

For the design of retaining walls, appropriate soil and groundwater parameters need to be selected for the calculations. The objective of design calculations is to demonstrate that a limit state will not be exceeded. Various limit states are generally considered. Current guidance (BS EN 1997-1:2004: Eurocode 7; BSI, 2004a) identifies that a designer must check that ultimate, serviceability or accidental limit states are not exceeded.

A serviceability limit state (SLS) defines the limit of satisfactory functioning of the structure or structural members under normal use, the comfort of those affected by the design and the appearance of the construction works. Exceedance of an SLS will typically result in inconvenience, disappointment and less manageable costs.

An ultimate limit state (ULS) is breached where one or more of the following occurs

- loss of equilibrium of the structure or any part of it considered as a rigid body
- failure by excessive deformation or transformation
- deformation resulting in a mechanism or rupture
- loss of stability of the structure or any part of it, including supports and foundations
- failure caused by fatigue or other time-dependent effects.

Exceedance of a ULS will result in serious failure involving risk of injury or major cost. An appropriate design must render this very unlikely or make this an 'unrealistic' possibility.

63.3.1.2 Selection of characteristic soil parameters

The zone of ground governing the behaviour of a geotechnical structure at a limit state is much larger than a test sample or the zone of ground affected in an in situ test. Consequently the value of the governing parameter often requires careful consideration of a range of values covering a large surface or volume of the ground.

The characteristic value should be a cautious estimate. Therefore, the selection of the characteristic value should be more conservative than selecting a most probable value.

If statistical methods are employed in the selection of characteristic values for ground properties, the methods used should differentiate between local and regional sampling and should also consider any experience in using comparable ground properties. In accordance with Eurocode 7-1 Clause 2.4.5.2(11), the characteristic value should be calculated 'such that the calculated probability of a worse value governing the occurrence of the limit state under consideration is not greater than five percent'. Refer to CIRIA C760 (Gaba *et al.*, 2017) for more guidance.

Reference should be made to Chapters 17 *Strength and deformation behaviour of soils* and 27 *Geotechnical parameters and safety factors* for further discussion on the selection of design parameters.

Examples of the selection of characteristic soil parameters for design of retaining walls is given in Chapter A7 of CIRIA C760 (Gaba *et al.*, 2017).

The examples describe the process that engineers typically follow when trying to interpret ground investigation data.

63.3.1.3 Selection of characteristic groundwater parameters

A conservative estimate of the active and passive groundwater conditions should be taken when selecting the characteristic levels and groundwater pressure profiles on the active and passive sides of a retaining wall. The designer should check that the following have been considered

- the proximity of sources of free water and the likelihood of such sources becoming available over the design life of the wall
- the effects on the local hydrogeology of the site due to the construction of the wall – for example, the potential damming of natural groundwater flows or a long-term rise in aquifer groundwater levels
- the effects of wall toe levels not reaching the target design levels due, for example, to obstructions or hard driving resistance

- the effects of drainage or dewatering during construction and during the lifetime of the wall
- changes to water pressures due to the growth or removal of vegetation
- changes to water pressures due to long-term climatic variations.

Based on the above considerations, the designer should determine the following.

(*a*) The water pressure and seepage forces that represent the most unfavourable values that could occur in extreme or accidental circumstances at each stage of the wall's construction sequence and throughout its design life. An example of an extreme or accidental event may be a burst water main in close proximity to the wall.

(*b*) The water pressures and seepage forces that represent the most unfavourable values likely to occur in normal circumstances at each stage of the wall's construction sequence and throughout its design life. Extreme events such as a nearby burst water main may be excluded, unless the designer considers that such an event may reasonably occur in normal circumstances.

The groundwater pressure corresponding to (*a*) should be compared to the ULS, while (*b*) should be compared to the SLS.

Undrained conditions

Where undrained conditions are considered to prevail during the construction of the wall, total stress considerations will apply. In such circumstances, there may be a tendency for tension cracks to develop in the retained soil. These tension cracks can be considered to have the potential to flood, depending upon the availability of water. Tension cracks may be less likely behind a propped wall compared with a cantilever wall. The depths of tension cracks and the groundwater pressure to assume are provided in CIRIA C760 (Gaba *et al.*, 2017) and Chapter 64 *Geotechnical design of retaining walls.*

In the case of an embedded cantilever wall or where water is expected to be present – for example, from water-bearing deposits overlying stiff clay – hydrostatic pressure should be used on the retained side of the wall.

CIRIA C760 (Gaba *et al.*, 2017) proposes that where water is not expected, the total pressure acting on the retained side of the wall at any depth below the ground surface should be assumed to be not less than a minimum effective fluid pressure (MEFP) of $5z$ kN/m^2 (where z is the depth below the ground surface).

Drained conditions

Where the ground fabric, permeability and construction duration indicate that drained conditions should be designed for, effective stress considerations will apply. The designer should evaluate the water pressure over the whole of the wall, assuming steady-state conditions at relevant stages of construction and over the lifetime of the wall. Flow nets are commonly used to establish the groundwater pressure distribution around the retaining wall. It should be noted that, in reality, the ground is likely to be anisotropic with a non-uniform coefficient of permeability, which may greatly affect the pore water pressure distribution.

A commonly used simplified method of evaluating water pressure around an embedded retaining wall excavation in uniform isotropic conditions is shown in **Figure 63.5**.

This simplified method applies where the differential head of water across the wall dissipates uniformly along the length of the flow path adjacent to the wall, and is sometimes referred to as linear seepage. This assumption tends to underestimate water pressures beneath narrow excavations. In general, the expressions in **Figure 63.5** should not be applied to excavations of widths that are less than four times the differential water head across the wall (BSI, 2015). At greater excavation widths, the assumption of linear seepage provides a good approximation to the pore water pressure distribution around the wall.

In anisotropic ground conditions and in situations where the design of the wall is sensitive to small changes in earth and water pressures around the wall, water pressures should be evaluated by computer-based analytical methods or by flow nets.

Mixed undrained and drained conditions

It is possible for the retained side of the wall to be considered drained (e.g. due to unloading laterally and the likelihood of open fissures allowing water into the clay and the potential for consequential softening or due to the presence of water-bearing silt and sand partings in the clay) while the restraining soil is considered to be undrained (e.g. due to an impermeable wall extending into impermeable ground thereby providing a cut-off to the groundwater recharge beneath the excavation on the restraining side). There is less likelihood of fissures opening and allowing water into the restraining clay due to its lateral confinement. In such circumstances, effective stress conditions may apply on the retained (active) side. Total stress conditions may apply in the restraining soil on the passive side and, possibly also, on the retained (active) side below the depth of the tension cracks.

The above is a very specific example used to highlight a scenario where mixed undrained and drained conditions may apply. The designer should carefully consider the likely ground behaviour on each side of the wall before finalising the design assumptions.

63.3.1.4 Other considerations

The designer should include consideration of the following prior to carrying out the retaining wall design.

- Adequate consideration must be given to the dead and live loading that the retaining wall may be subjected to throughout the design life.

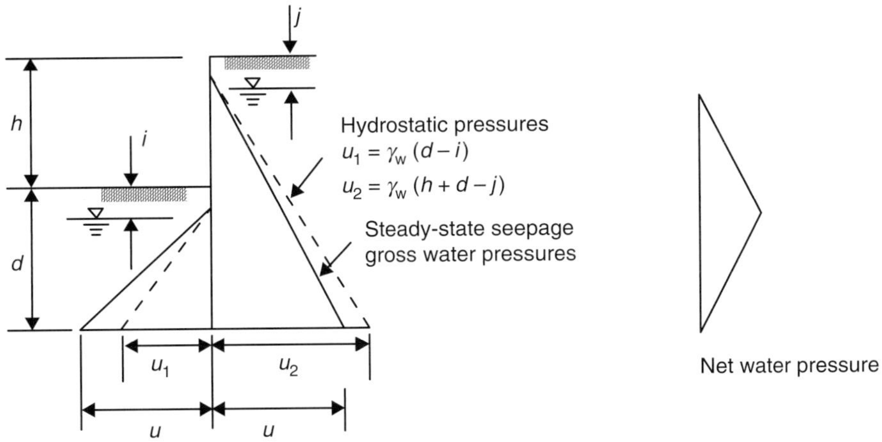

= Assumption (1)
Uniform dissipation of differential head along flow path adjacent to the wall

$$u = \frac{2(d + h - j)(d - i)}{2d + h - i - j} \gamma_w$$

= Assumption (2)
Average hydrostatic pressure at wall toe

$$u = \frac{u_1 + u_2}{2} = \gamma_w \left| d + \left(\frac{h - i - j}{2} \right) \right|$$

Figure 63.5 Uniform ground: simplifying assumptions for calculating water pressures for steady-state seepage conditions (Reproduced from Gaba *et al.*, 2017)

- Unplanned excavation on the passive side of the retaining wall is generally assumed to be the lesser of 0.5 m or 10% of the total height retained for cantilever walls, or the height retained below the lowest support level for propped or anchored walls.
- There is potential for softening of the soil on the passive side of the retaining wall.
- Vertical forces applied to the wall through wall friction will have an effect.

Further guidance on the selection of the above is given in CIRIA C760 (Gaba *et al.*, 2017) and Chapters 64 *Geotechnical design of retaining walls* to 67 *Retaining walls as part of complete underground structure*.

63.3.2 Temporary works design

It is very important that the roles and responsibilities of each of the parties (client, owner, designer, contractor and subcontractor) are clearly defined and fully understood by all parties in respect of the design and construction of the temporary and permanent works. Please refer to Chapters 42 *Roles and responsibilities*, 78 *Procurement and specification* and 96 *Technical supervision of site works*.

The design of temporary works should be demonstrably robust. This robustness should be provided by identifying the hazards and risks and ensuring that these are adequately addressed by the temporary works design. The principal risks lie in predicting ground behaviour and dealing with parameter uncertainty.

- Will the ground respond in a drained or undrained manner, or a combination of both during construction?
- If undrained, will tension cracks develop? Are these likely to be dry or full of water?
- What groundwater regime will apply during construction?
- What provision should be made to allow for the deterioration in soil strength due to the installation of the wall and its support system and the disturbance due to excavation and partial drainage during the period of construction?
- What measures or controls will be applied to confirm that
  - the loading conditions assumed in design will not be exceeded
  - unplanned excavation of the formation will not occur (see Section 63.3.1.4)
  - the actual support (props, anchors, berms, etc.) to the wall will be as assumed in design
  - the actual wall performance (deflection, watertightness, etc.) will remain within predefined limits?

The assumptions made in temporary works design in respect of the above depend upon how the associated risks are mitigated during construction. Risk mitigation is undertaken through close control of the process of design and the construction of the temporary works on site. This requires an integrated interactive approach between design and construction on site with the ability to adapt the design quickly to suit changes in construction methods and vice versa. Appropriate contractual arrangements should be in place to permit this.

Special consideration should be given where there is uncertainty over whether undrained or drained conditions may be assumed for the duration of the temporary works (see Section 63.3.1.3). In terms of risk, three possible sets of assumptions are possible for the design of temporary works in stiff clays

- undrained conditions on both sides of the wall
- mixed undrained and drained conditions
- drained conditions on both sides of the wall with a steady-state groundwater regime.

The assumption of undrained conditions on both sides of the wall may be associated with significant risk, while least risk is attached to the assumption of drained conditions. Further consideration of these assumptions is given in CIRIA C760 (Gaba *et al.*, 2017).

63.4. Ground movements and their prediction

63.4.1 General

For the design and construction of retaining walls, it is inevitable that some movement will occur both in front of and behind the retaining wall.

The amount of movement that is acceptable will depend upon the structures, utilities and infrastructure that are located within the vicinity of the new retaining structure and their tolerance of displacement.

Figure 63.6 shows the general type and form of movement that can be expected to occur.

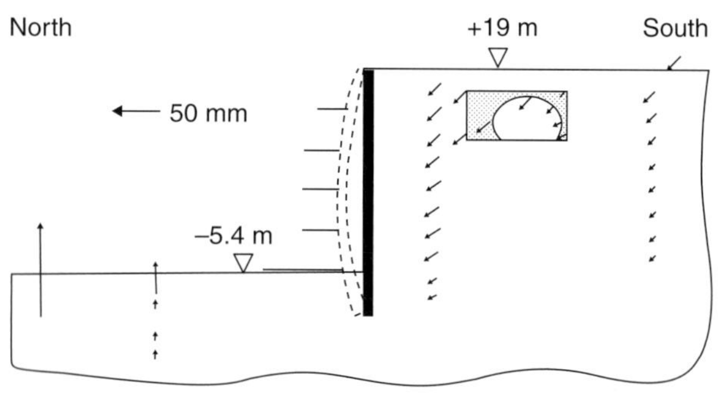

Figure 63.6 Typical ground movements around a deep excavation (Reproduced from Simpson, 1992; all rights reserved)

Movements can be generally split into the following types.

- *Installation of the wall.* These could be due to a relaxation of the ground during the formation of piles or panels of an embedded retaining wall. Compression of the ground may occur, resulting in heave where sheet piles are driven into the ground. For gravity walls, ground movements may occur due to the relaxation of the ground arising from forming the excavation where the gravity structure will be located or from settlements from loading the ground by the gravity structure.
- *Excavation related.* The excavation in front of a retaining wall will result in the retaining wall displacing laterally. This will lead to horizontal displacements and settlements behind the wall; see **Figures 63.7** and **63.8**.
- *Heave.* The removal of soil on the passive side of the wall will result in unloading of the ground. This will lead to heave of the ground both on the passive and active sides of the retaining wall; see **Figure 63.7**.
- *Long term.* In fine-grained soils, the effects described above will result in long-term displacements arising from equalisation of the pore water pressures in the ground; see **Figure 63.7**.

Ground movements tend to be localised and aggravated by construction problems, such as unacceptable vibration or

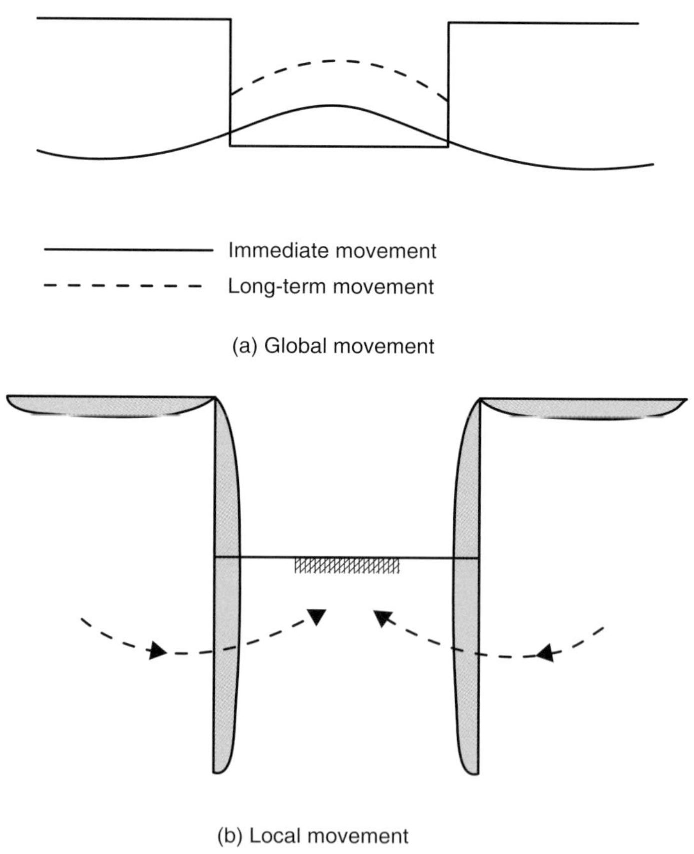

Figure 63.7 Typical ground movement pattern associated with excavation stress relief (Reproduced with permission from CIRIA C760 (Gaba *et al.*, 2017); www.ciria.org)

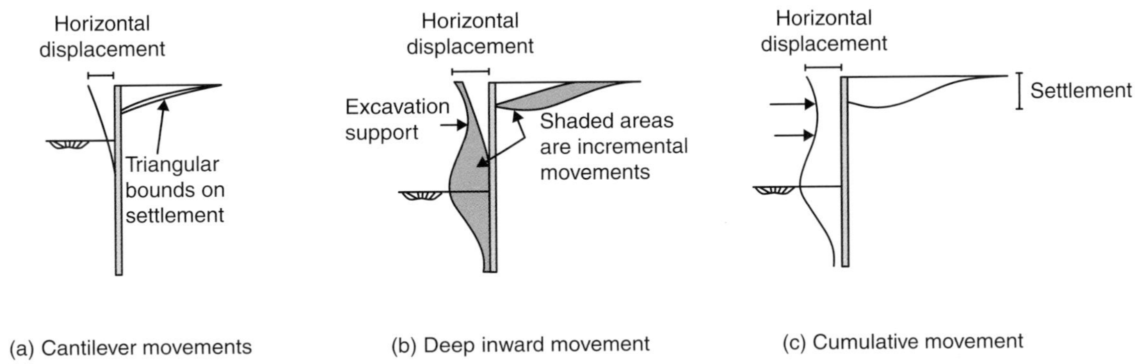

Figure 63.8 General pattern of wall movement and ground displacement (Reproduced with permission from CIRIA C760 (Gaba *et al.*, 2017); www.ciria.org)

over-break. The removal of obstructions and the excavation of guide trenches before wall construction may also cause as much or more movement as wall installation itself. Further details of issues relating to construction may be found in Chapter 85 *Embedded walls*.

There are various techniques that can be used to assess the magnitude of ground deformation. The techniques broadly split into the categories of empirical approaches, numerical methods and physical models. For conventional design, empirical approaches and numerical methods are generally used. Further details of these techniques are given in Section 63.4.2.

Many of the methods described in this section consider only plane strain displacements (i.e. a two-dimensional assumption). It should be realised that where retaining walls are used to form shafts or basements, the displacements will be significantly modified by the increased stiffness offered by the presence of corners around excavations. A basic consideration of how to account for these effects is given in CIRIA C760 (Gaba *et al.*, 2017).

63.4.2 Calculation-based methods

63.4.2.1 Introduction

Calculation-based methods can be broadly split into empirical and numerical methods. If designing in accordance with Eurocodes, calculations are carried out using characteristic soil parameters selected by the designer. As discussed in Section 63.3, any characteristic value should be a cautious estimate. As a consequence, calculation-based methods that involve the selection of characteristic soil parameters will also result in a conservative estimate of the magnitude of ground movements.

63.4.2.2 Empirical methods

Empirical methods rely on the collation of case history data to provide an estimation of ground movement. Numerous authors have provided empirical correlations for ground movements around retaining structures. In particular, the following references are often considered when calculating surface displacements: Clough and O'Rourke (1990), Gaba *et al.* (2017), Moorman (2004) and Long (2001). Displacements are commonly calculated as a function of the retaining wall excavation depth and are grouped with respect to the stiffness of the excavation and ground conditions where they are carried out; see **Figure 63.9**.

CIRIA C760 (Gaba *et al.*, 2017) also provides an empirical method for relating wall displacements from pseudo-finite element methods, such as those calculated by FREW or WALLAP, to the ground surface displacement behind the retaining wall.

For use of these methods the case study database should be considered and compared with the particular conditions being analysed. Further assumption or correlations with case study movements are required if estimates of subsurface ground displacements are to be made without resorting to more rigorous methods of analysis.

63.4.2.3 Numerical methods

Numerical methods such as finite element or finite difference programs are commonly used for the design of retaining walls. The analysis methodology generally involves discretising the retaining wall and surrounding ground into elements within a mesh and specifying boundary conditions distant from the retaining wall. With the exception of discrete element analysis methods, the ground is modelled as a continuum. Further details on the use of numerical techniques may be found in Chapter 6 *Computer analysis principles in geotechnical engineering*. The primary reasons for using numerical methods include the following.

- They overcome the limitations and assumptions made by more simplistic analyses (e.g. limit equilibrium or beam spring, pseudo-finite element programs).
- The use of numerical techniques can be readily extended to more complex problems (including berms).

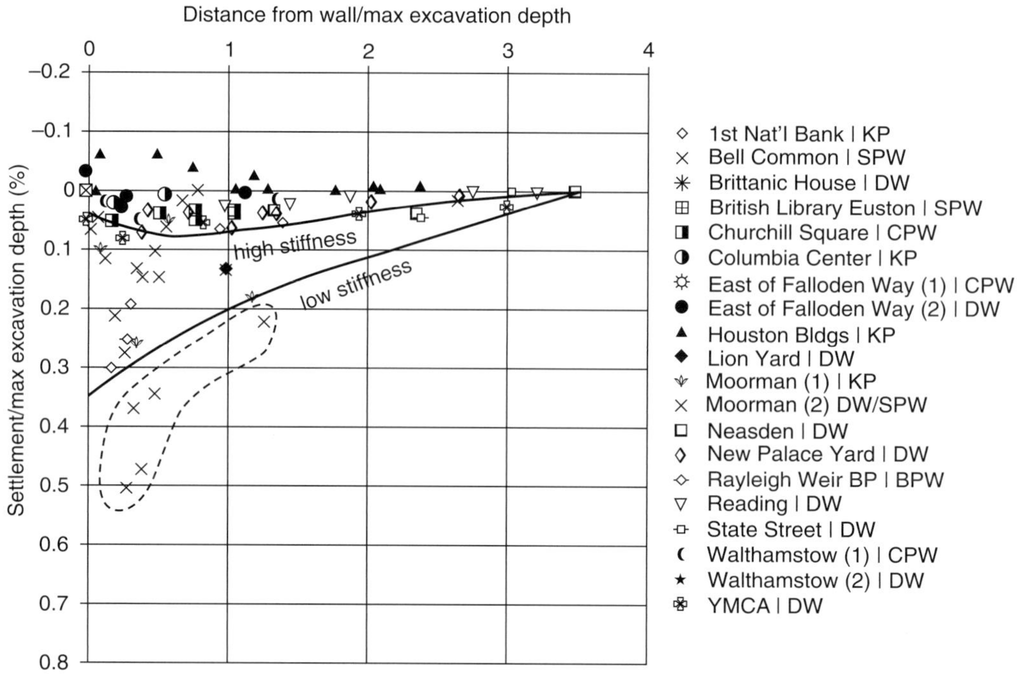

Figure 63.9 Ground surface movements due to excavation in front of a wall in stiff clay (Reproduced with permission from CIRIA C760 (Gaba *et al.*, 2017); www.ciria.org)

- Three-dimensional (3D) analysis can be carried out if the geometry of the retaining wall is not readily approximated to behave in a two-dimensional plane strain manner.
- The analysis can incorporate time-dependent consolidation.

Alongside this list of benefits, the following should also be borne in mind when considering numerical analysis.

- Interpretation of the results of numerical analysis requires more knowledge and appropriate practical experience by the analyst.
- The active and passive limits are not easily computed.
- Data entry is more complex than for other forms of computational analysis.
- Full numerical analyses carried out by a computer take longer than limit equilibrium or pseudo-finite element or finite difference analyses.
- Interpretation of results takes longer and is often not as clearly understood by non-specialists.

63.5. Principles of ground movement and impact assessment

Further to carrying out an assessment of the magnitude of ground movements arising from retaining wall excavations, it is important to assess whether the movements may have an impact upon adjacent structures and utilities. The principles of assessing the effect of ground movement on structures due to excavation are described in Chapter 26 *Building response to ground movements and tunnelling.*

It should be appreciated that lateral displacements and the resulting ground strains should also be considered as they may form a significant contribution to an affected structure's deformation. For many structures it should be appreciated that the interaction between the ground and structural deformation is complex. As a consequence, either conservative assumptions or advanced forms of analysis may be required to adequately model this interaction.

References

ArcelorMittal (2022). *Piling Handbook*, 9th edn. ArcelorMittal, London, UK. See https://sheetpiling.arcelormittal.com/news/piling-handbook-9th-edition-revision-2022 (accessed 10/03/2023).

BSI (British Standards Institution) (2004a) BS EN 1997-1:2004. Eurocode 7: Geotechnical design – Part 1: General rules. BSI, London, UK.

BSI (2004b) BS EN 1998-1:2004. Eurocode 8: Design of structures for earthquake resistance – Part 1: General rules, seismic actions and rules for buildings. BSI, London, UK.

BSI (2004c) BS EN 1998-5:2004. Eurocode 8: Design of structures for earthquake resistance – Part 5: Foundations, retaining structures and geotechnical aspects. BSI, London, UK.

BSI (2012) BS 6349-1-3:2012. Maritime works – Code of practice for geotechnical design. BSI, London, UK.

BSI (2015) BS 8002:2015. Code of practice for earth retaining structures. BSI, London, UK.

BSI (2022) BS 8102:2022. Protection of below ground structures against water ingress – Code of practice. BSI, London, UK.

Burland JB, Potts DM and Walsh NM (1981) The overall stability of free and propped embedded cantilever retaining walls. *Ground Engineering* **14(5)**: 28–38.

Clough GW and O'Rourke TD (1990) Construction induced movements of insitu walls. Geotechnical Special Publication, ASCE No. 25. ASCE, Reston, VA, pp. 439–470.

Gaba AR, Hardy S, Doughty L, Powrie W and Selemetas D (2017) *Guidance on Embedded Retaining Wall Design*. CIRIA C760. CIRIA, London, UK.

Highways Agency, UK (2003) DMRB Volume 1 Section 3 Part 12 (BA 42/96) Highway structures: Approval procedures and general design. General design. Design of integral bridges (including Amendment No. 1 dated May 2003). The Stationery Office, London, UK.

Johnson RA (1995) *Water Resisting Basements: A Guide. Safeguarding New and Existing Basements against Water and Dampness*. CIRIA 139. CIRIA, London, UK.

Lehane BM (2006) Geotechnical considerations for integral bridge abutments. *3rd National Symposium on Bridge and Infrastructure Research in Ireland*. Department of Civil, Structural & Environmental Engineering, Trinity College, Dublin, Ireland, vol. 1, pp. 385–392.

Long M (2001) Database for retaining wall and ground movements due to deep excavations. *Journal of Geotechnical and Geoenvironmental Engineering* **127(3)**: 203–224.

Moorman C (2004) Analysis of wall and ground movements due to deep excavations in soft soil based on a new worldwide database. *Soils and Foundations* **44(1)**: 87–98.

Simpson BS (1992) Retaining structures: displacement and design. 32nd Rankine lecture. *Géotechnique* **42(4)**: 541–576.

Terzaghi K (1954) Anchored bulkheads. *Transactions of the American Society of Civil Engineers* **119**: 1243–1281.

Further reading

Chapman T, Taylor H and Nicholson D (2000) *Modular Gravity Retaining Walls: Design Guidance*. CIRIA C516. CIRIA, London, UK.

Zdravkovic L, Potts DM and St John HD (2005) Modelling of a 3D excavation in finite element analysis. *Géotechnique* **55(7)**: 497–513.

It is recommended this chapter is read in conjunction with

- Chapter 20 *Earth pressure theory*
- Chapter 64 *Geotechnical design of retaining walls*
- Chapter 85 *Embedded walls*
- Chapter 100 *Observational method*

All chapters in this book rely on the guidance in Sections 1 *Context* and 2 *Fundamental principles*. A sound knowledge of ground investigation is required for all geotechnical works, as set out in Section 4 *Site investigation*.

doi: 10.1680/icemge.66830.1063

Chapter 64
Geotechnical design of retaining walls

Jack Taggart Arup, London, UK

CONTENTS

This chapter outlines the design approach for retaining walls, following from the principles of retaining wall design outlined in Chapter 63 *Principles of retaining wall design*. The mechanisms that typically require consideration in the design of gravity, reinforced soil and embedded retaining walls are identified and the design process for each type of wall is explained. The design basis is in accordance with BS EN 1997-1:2004 (Eurocode 7). Using the methods outlined in this chapter the reader will be able to size retaining walls to ensure adequate stability and calculate load effects for the structural design of the wall.

64.1. Introduction

This chapter outlines the design method for gravity retaining walls, reinforced soil walls and embedded retaining walls. Design concepts outlined in Chapter 63 *Principles of retaining wall design* are developed, and the limit states that should be typically considered when designing a retaining wall are identified. The chapter describes both traditional overall factors of safety and appropriate application of the partial factors incorporated within BS EN 1997-1:2004 (BSI, 2004). Limit-equilibrium analyses that can be calculated by hand are described, in addition to a design using more complex numerical methods to take account of soil–structure interaction effects.

64.2. Gravity walls

64.2.1 Introduction

Gravity walls derive the majority of their stability from their self-weight, or soil placed on the toe or heel of the wall. This differs from an embedded wall, which derives a significant proportion of its support from resistance of the ground in front of the wall for its stability. There are a variety of different gravity wall types, as described in Chapter 62 *Types of retaining walls*, and the selection of a particular wall type will depend upon a variety of factors, as discussed in that chapter. There are broadly two types of gravity wall, one in which internal stability is provided by structural resistance of the wall either through shear or bending (e.g. reinforced-concrete stems), and another where gravity action and the self-weight of the wall provide stability without mobilising tension in the structure (e.g. mass concrete or gabion walls). These are illustrated in **Figure 64.1**.

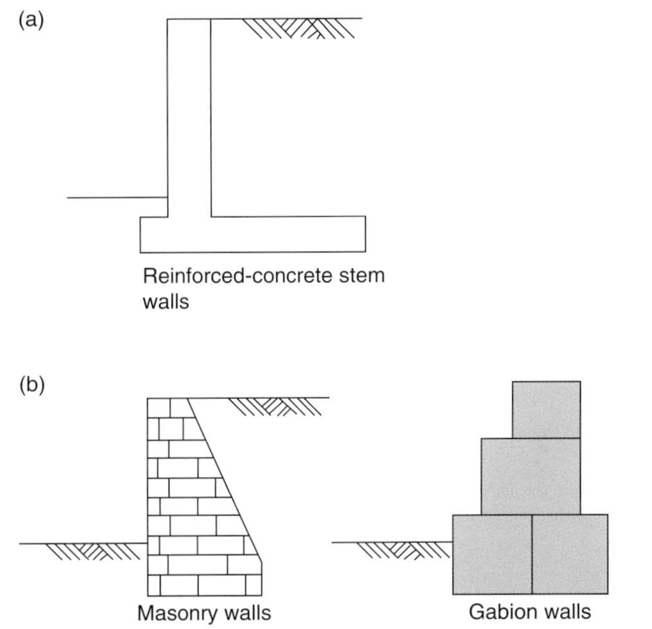

Figure 64.1 Types of gravity wall: (a) internal stability maintained by structural strength of reinforced-concrete stem wall; (b) internal stability maintained by gravity action – no (or minimal) development of tension

64.2.2 Limit states

BS EN 1997-1:2004 prescribes a limit state design approach in which the structure should be designed to sustain all loads liable to occur during its construction, perform adequately in normal use and have adequate durability. The limit states relevant to the geotechnical design of gravity walls are discussed below. These are listed in sections 9.2, 9.7 and 9.8 of BS EN 1997-1:2004.

64.2.2.1 Ultimate limit states

BS EN 1997-1:2004 requires the following ultimate limit states (ULS) to be considered for gravity retaining walls

- bearing resistance failure of the soil below the base
- failure by sliding at the base of the wall
- failure by toppling of the wall

- loss of overall stability
- failure of a structural element such as a wall, anchor, waler or strut, or failure of the connection between such elements
- combined failure of the ground and a structural element
- movement of the retaining structure that may cause collapse or affect the appearance or efficient use of the structure, nearby structures or services that rely on it
- unacceptable leakage through or beneath the wall
- unacceptable transport of soil grains through or beneath the wall
- unacceptable change to the flow of groundwater.

Some of these limit states are illustrated in **Figure 64.2**.

In addition to the limit states listed above, the designer should consider whether any additional limit states should be taken into account in respect of specific site conditions or other requirements not covered by this manual.

64.2.2.2 Serviceability limit states

The following serviceability limit states (SLS) should be considered

- movement that may cause damage to structures or services which rely on the support of the wall
- excessive tilting, deflection or settlement of the wall – the acceptable values for these deformations should be assessed on a project-specific basis.

64.2.3 Factors of safety

In traditional practice, gravity walls have been designed by applying lumped factors of safety to the mechanisms analysed. In simple terms, the traditional methods used unfactored characteristic soil parameters and analyses were undertaken to assess factors of safety against overall stability; overturning about the toe of the wall, sliding along the base of the wall and bearing capacity failure beneath the wall. Typical values of factors of safety adopted using such traditional design methods for designing permanent structures are given in **Table 64.1**.

Mechanism	Target factor of safety
Overall stability	1.3
Overturning	2.0
Sliding	1.5
Bearing capacity	3.0

Table 64.1 Traditional lumped factors of safety for gravity wall design

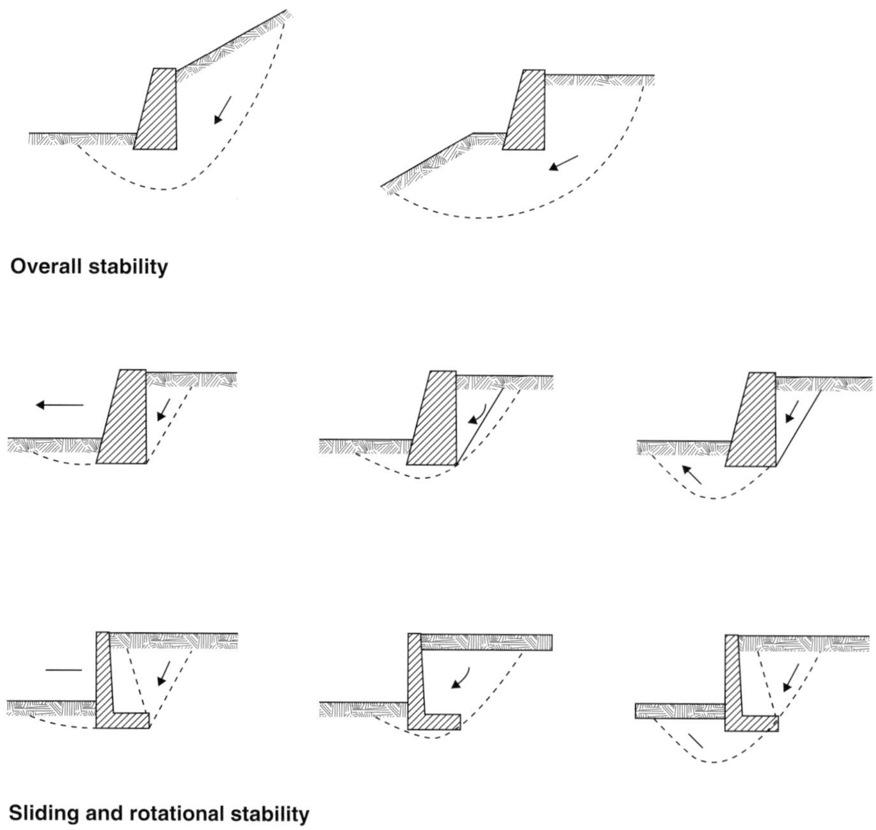

Figure 64.2 Gravity wall: ultimate limit states

In a change from traditional practice, the design approach prescribed in BS EN 1997-1:2004 and described in this manual requires application of partial factors to characteristic values of soil strength, geometry and actions to determine design values. Rather than seeking an overall lumped factor of safety, partial factors are applied to different inputs, such as material strength, actions, wall geometry and water pressure. The size of the wall is then selected to ensure that the design actions, or effects of design actions, are less than the design resistances for each of the limit states described above.

The design approach and partial factors to be used in design are stated in the relevant National Annex of BS EN 1997-1:2004. In the UK, Design Approach 1 is applicable for gravity walls; this approach comprises two combinations

- combination 1 (DA1-1), where the material partial factor is unity and actions (or action effects) are factored by a value greater than unity
- combination 2 (DA1-2), where partial factors are applied to material strengths and variable actions, but the factor on the permanent actions (or action effects) is unity.

The reader should refer to the appropriate National Annex for confirmation of these partial factor values.

It may be necessary to undertake an additional SLS analysis where all partial factors are unity. Traditionally, large lumped factors acted to reduce the proportion of strength mobilised, and hence control movement. The partial factors in BS EN 1997-1:2004 may lead to lower equivalent overall factors of safety and therefore there is potential for more movement. As a consequence, an explicit check is needed for the SLS case.

64.2.4 Design situations, earth pressure, design geometry, groundwater conditions

64.2.4.1 Introduction

An illustration of the typical loads acting on a gravity retaining wall are summarised in **Figure 64.3**.

From a stability perspective, some actions are unfavourable as they tend to reduce the stability of the structure, while others are favourable and tend to improve stability. Unfavourable actions are often referred to as disturbing actions and favourable actions are referred to as restoring actions. The objective is to ensure that the design resistance exceeds the design action effects – that is, the sum of the restoring actions is greater than the disturbing actions. The following considerations should be addressed.

- For *bearing capacity*: sum the vertical actions, horizontal actions and moments about the centre of the base, then undertake a check as a shallow footing – the design approach described in Chapter 63 *Principles of retaining wall design* should be followed.
- For *sliding*: calculate the net horizontal load and check that this is smaller than the sliding resistance. For effective stress-based calculations this is a function of the vertical load.

The traditional check on *overturning* is not generally necessary. If the wall was on the limit of toppling (or even past the limit of toppling), then the effective footing width from the bearing capacity calculation above would be very small or the line of action would fall outside the base of the wall. This would therefore be identified as a bearing capacity failure. The only case where this may not be correct is for walls bearing on strong rock. In such cases the bearing capacity may be

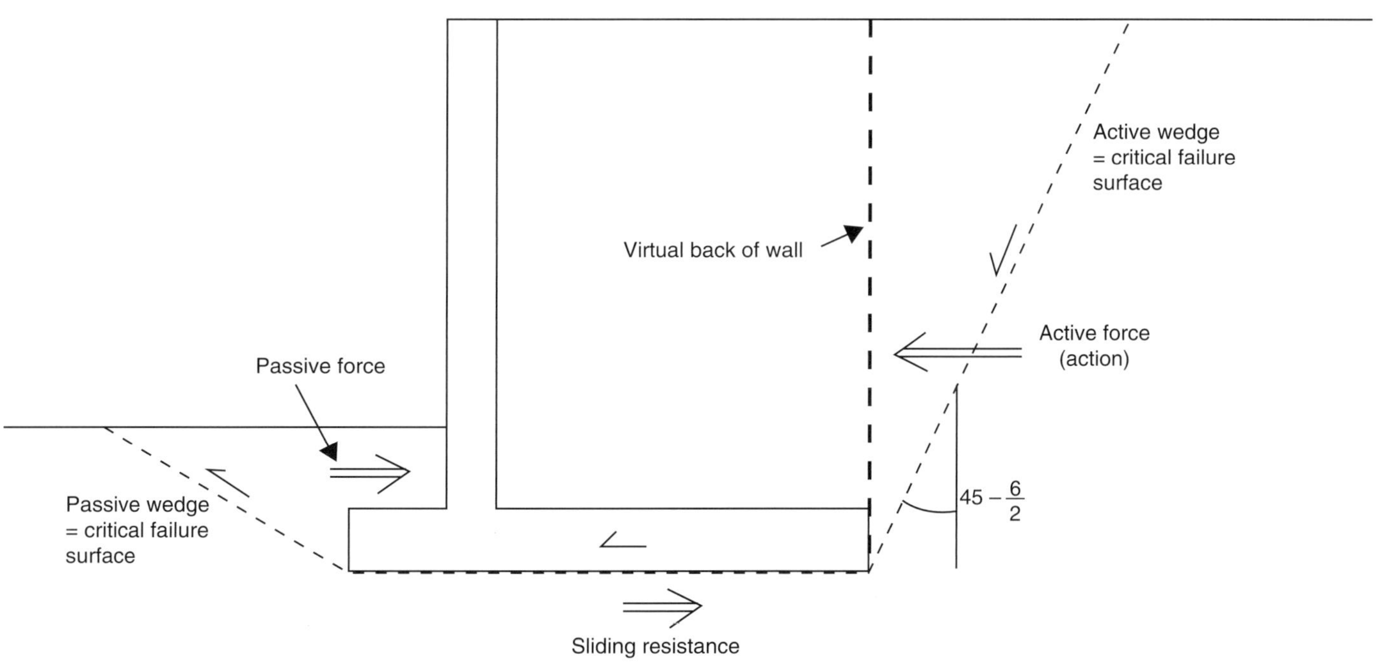

Figure 64.3 Idealised actions on a gravity wall

sufficient to resist very high load eccentricities and so overturning stability must be checked explicitly.

The sections below provide guidance on calculating the individual components of the load acting on a gravity wall.

64.2.4.2 Design situations

Before any analysis is undertaken, it is necessary to establish the appropriate design situations. Issues to consider include the following.

- *Design soil stratigraphy.* In variable ground conditions it may be necessary to consider more than one design soil stratigraphy. Each one would comprise a separate design situation and analysis case.
- *Soil parameters.* Characteristic values should be assessed based on the approach outlined in Chapter 27 *Geotechnical parameters and safety factors*.
- *Temporal effects.* For example, whether the analysis will be short term (i.e. during construction) or long term (i.e. permanent condition). If necessary, both conditions may need to be checked using the appropriate soil parameters as separate design situations.
- *Construction sequence.* It is necessary to ensure that all potentially critical stages are considered in the design.
- *Design geometry.* Where this varies for the structure – for example, because of changes in the overall retained height – consider how the wall will be divided up for design purposes and identify the critical section for each. Each geometry will then be considered as a separate design situation and analysis case.
- *Imposed loading.* Are there any externally applied loads, such as traffic surcharges, impact loads or loads from adjacent structures that need to be considered in the analysis?

While it may be necessary to calculate more than one design situation, by careful consideration of the relevant combinations that are likely to prove critical, it is possible that once a solution for the critical design situation has been determined, alternative situations may be confirmed as acceptable either by inspection, or through a reduced number of analyses and calculations.

64.2.4.3 Earth pressure

The principal actions on a gravity retaining wall are usually earth and water pressures. The earth pressure is derived from the self-weight of the soil and possibly from loads applied to the ground surface. Additionally, there may be some direct loading of the wall – for example, from structures bearing onto the wall or from impact loads applied to parapets connected to the wall.

Typically, a gravity wall analysis is carried out by assuming the limit-equilibrium earth pressure acts on the structure, and it is relatively unusual for a gravity wall to be designed using advanced numerical methods. Soil structure–interaction analysis may be appropriate where the behaviour of the gravity wall is integral to the overall performance of a more complex structural system or where it is important to understand the behaviour of the gravity structure in response to complex applied loading, such as due to seismic events. While the general design principles outlined in this chapter may be applied to a soil–structure interaction analysis of a gravity wall, the focus of this chapter will remain primarily on the limit-equilibrium approach.

Gravity walls rarely have a vertical back in contact with the retained ground, so the first step in identifying the earth pressure acting on a wall is to select a virtual back (see **Figure 64.3**). Soil within this line is considered to act with the structural components of the wall as a rigid body. Normal stresses and shear stresses must then be evaluated along the virtual back. For many situations the virtual back is taken as a vertical line upwards from the heel of the wall. In this case it is assumed that a Rankine active wedge forms both within the soil mass over the heel of the wall and in the retained ground behind the virtual back. This means the force acting normal to the virtual back can be calculated using Rankine earth-pressure theory. Because the soil within the active wedge over the heel of the wall and in the retained fill are both moving downwards together, zero shear stress is mobilised along the virtual back (see **Figure 64.4**).

An alternative approach is to assume the virtual back lies on a plane joining the heel of the wall to the crest of the wall (see **Figure 64.5**).

For either of these two cases the active forces acting on the virtual back should then be evaluated for the chosen geometry, using either tabulated values of earth-pressure coefficients (e.g. Kerisel and Absi, 1990), the charts in BS EN 1997-1:2004 or the analytical method of calculating earth-pressure coefficients given in BS EN 1997-1:2004. The *design* material strength should be used in this calculation.

It should be noted that since DA1-1 and DA1-2 in BS EN 1997-1:2004 give different values for the partial factor on

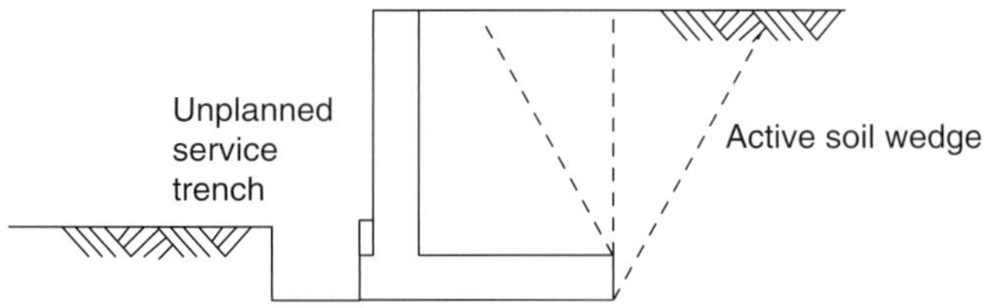

Figure 64.4 Rankine active wedge

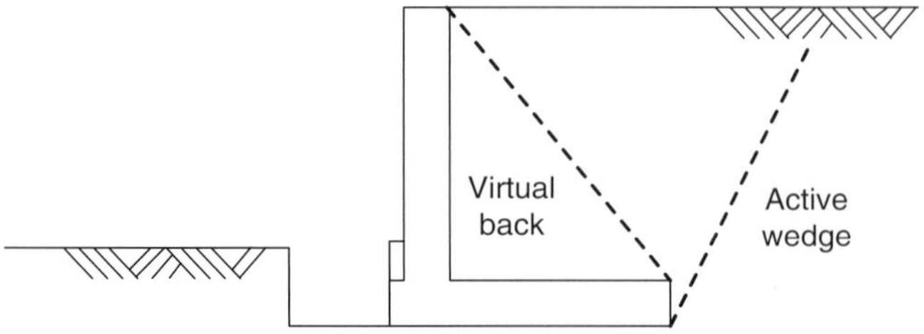

Figure 64.5 Inclined virtual back

the material strength, then the earth-pressure coefficients and hence the limiting earth pressures used for each design approach will also be different.

Where the base of the wall is founded below the ground level in front of the wall, passive earth pressures may be considered to act on the front of the wall. Careful consideration needs to be given as to the reliability of this pressure being mobilised. In order to form the wall base, the material in front of the wall is likely to have been excavated and then backfilled once the wall has been constructed. The type of this material, the method of placement and the level of compaction used to replace the material may therefore influence the magnitude of passive resistance that can be mobilised and relied upon. The passive resistance may also be affected by unplanned excavation (see Section 64.2.4.4). If there is any uncertainty regarding the reliability of the passive resistance, it is recommended that it be omitted from the analysis.

Where sliding is shown to be a critical factor in design, the stability of the wall may be improved by the construction of a shear key or the use of an inclined base. Further details regarding the inclusion of the effects of shear keys in a gravity wall analysis can be found in CIRIA C516 (Chapman *et al.*, 2000).

As required by BS EN 1997-1:2004, it is important to ensure that the assumed earth pressures for a given limit state are compatible with the anticipated movements at that limit state. For geotechnical stability (i.e. sliding, toppling, bearing) it is reasonable to assume sufficient movement will occur to mobilise active earth pressures. However, for the structural design of the wall it is possible that the wall deformation under in situ conditions may be insufficient to mobilise active pressures. For structural design it will therefore often be appropriate to use the 'at-rest' earth pressures. BS EN 1997-1:2004 includes information on the magnitudes of any movement that are required to mobilise active and passive earth pressures and these are also discussed in Chapter 63 *Principles of retaining wall design*. The enhanced in situ pressures may also be applicable to the SLS case.

Since gravity walls typically retain placed fill, the in situ pressures must take into account the construction processes that formed the backfill. Typically some form of compaction will be used behind the wall to achieve the required backfill strength and to limit the potential for future settlement of the backfill. Compaction tends to lock in stresses in the ground, which can exceed the values calculated based on a typical in situ earth-pressure coefficient, K_0. The horizontal pressure acting along the back of the wall due to compaction is given in **Figure 64.6**.

The values of h_c, σ'_{hrm} and z_{cr} can be determined as

$$h_c = \frac{1}{K_0}\sqrt{\frac{2P}{\pi\gamma}} \tag{64.1}$$

where P is the effective line load per metre of the roller and γ is the density of the material undergoing compaction.

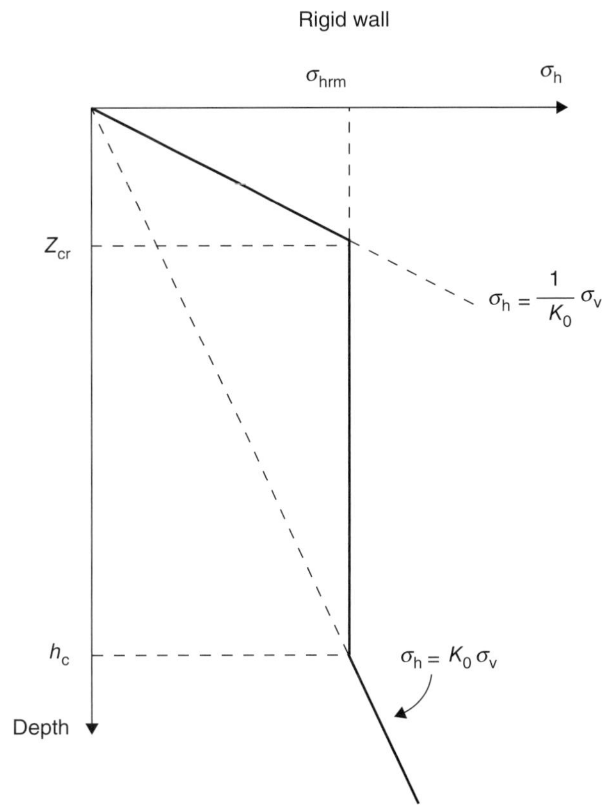

Figure 64.6 Calculation of compaction pressures

The maximum pressure σ'_{hrm} can be calculated from

$$\sigma'_{hrm} = \sqrt{\frac{2P\gamma}{\pi}} \tag{64.2}$$

and this occurs below a depth of z_{cr}, where

$$z_{cr} = K_0\sqrt{\frac{2P}{\pi\gamma}} \tag{64.3}$$

64.2.4.4 Unplanned excavation

When assessing the geometry for a given design situation for a wall where the passive resistance of the ground is relied upon for stability, it is important to take into account the potential for unplanned excavation in front of the wall. BS EN 1997-1:2004 requires that for normal site controls 10% of the retained height, up to a maximum of 0.5 m, should be assumed for an unplanned excavation.

64.2.4.5 Surcharges and direct loads

The design of a retaining wall must take account of any surcharge that may be applied to the ground retained by the wall. Such a surcharge may be derived, for example, from highway traffic in the permanent condition, plant loading during the construction phase or from the placement of stockpiled materials behind the wall. The effect of such surcharges must be incorporated into the assessment of the earth pressure acting on the wall.

A minimum surcharge of 10 kPa on the retained surface has traditionally been assumed in UK practice for the design of gravity retaining walls. This is intended to allow for any unplanned additional vehicle loading, stockpiled materials, goods or containers. There is, however, no explicit requirement in BS EN 1997-1:2004 for a minimum surcharge, but for most walls some allowance should be made and a 10 kPa uniform distributed load is considered to be a reasonable minimum. Larger surcharges should be assumed if appropriate, for example due to special vehicle traffic or rail traffic. Useful guidance in selecting appropriate uniformly distributed loads is given in section 4.5.1.2 of BS 8002 (BSI, 2015).

Typically, such surcharges would be considered as a variable action. In these cases consideration should be given to whether the effect is beneficial to the stability of the wall. In particular, if a live load surcharge acts on the soil directly over the heel of the wall, this would contribute additional load to resist overturning and sliding (for free-draining granular materials), which cannot be relied upon. However, for bearing capacity checks and the structural design of the wall, the additional vertical load from the surcharge over the heel may prove to be adverse. As a rule, live load surcharges are typically only applied to the ground outside the virtual back. In this way the destabilising effect of the load is considered, but no beneficial effects are relied upon (**Figure 64.7**).

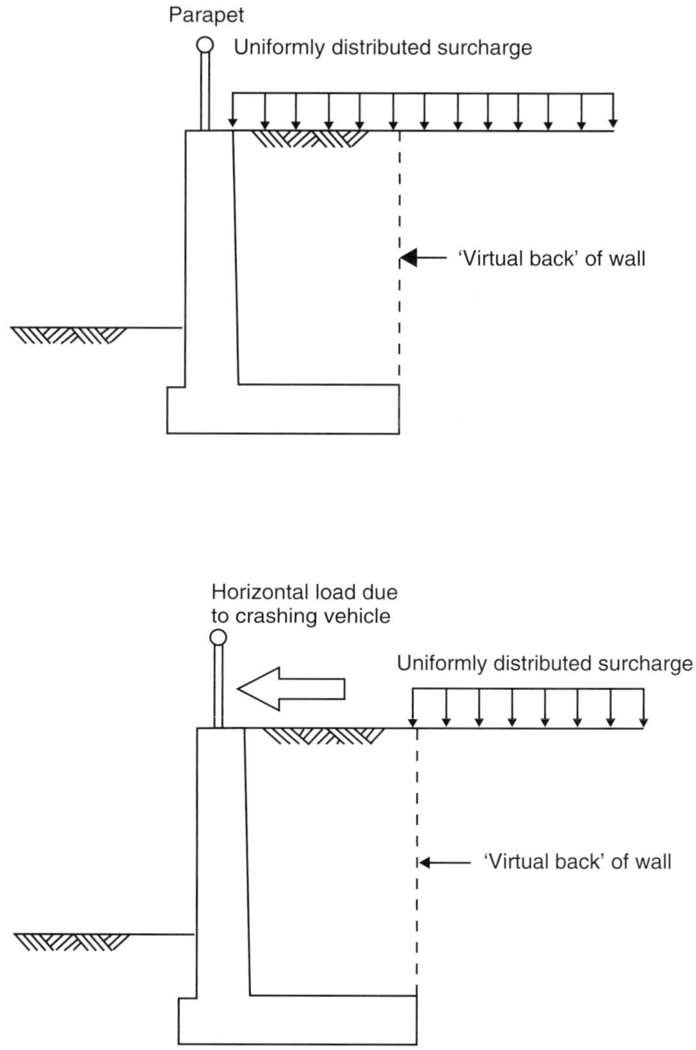

Figure 64.7 Surcharge assumptions for different design cases. (a) Load case 1: often critical for bearing pressure calculations and design for internal stability. (b) Load case 2: often critical for sliding stability and bearing pressure calculations

64.2.4.6 Water pressure

As is commonly the case for geotechnical design, water pressure can be a significant component of load on a gravity wall. Since gravity walls retain fill material, there is potential to install a drainage system behind the wall and hence control the water pressure. However, such a system would require ongoing maintenance to ensure that it continues to perform.

In the absence of a drainage system, BS EN 1997-1:2004 requires that the water level is assumed to be at the surface of the retained ground level for low-permeability backfills. Consideration also needs to be given to the occurrence of seepage around a wall. In the absence of seepage analysis it may be appropriate to assume a linear variation in pore pressure across the base, from the pressure behind the wall to the pressure in front of the wall.

For the design groundwater conditions, the following should be determined.

(*a*) Water pressure and seepage forces that represent the most unfavourable values that could occur in extreme or accidental circumstances at each stage of the wall's construction sequence and throughout its design life. An example of an extreme or accidental event may be a burst water main in close proximity to the wall.

(*b*) Water pressures and seepage forces that represent the most unfavourable values likely to occur in normal circumstances at each stage of the wall's construction sequence and throughout its design life. Extreme events such as a nearby burst water main may be excluded, unless the designer considers that such an event may reasonably occur in normal circumstances.

Different approaches to groundwater are required for DA1-1 and DA1-2. In DA1-1 the actions of the water are factored by a value greater than unity, therefore it is recommended that water condition (*b*) is assumed, as the factor on the action of the water will give a margin of safety. For DA1-2, the design actions are derived from design material parameters, with a factor on permanent actions of unity. It is therefore recommended that the design water pressure is derived from a directly assessed design groundwater level, which is defined as in Equation 64.1.

64.2.4.7 Drainage systems and fill materials

There is a need to have consistency between the design groundwater assumptions and the specification of backfill or drainage. It is becoming more common to use site-won material. This may have lower strength and lower permeability than traditional backfill material so the designer needs to know in advance what is available. Without drainage, an assumption of hydrostatic conditions from the ground level is required by BS EN 1997-1:2004 for medium- and low-permeability soils.

64.2.5 Design of structural elements

As noted above, it is necessary to ensure compatibility between the movement of the wall at a given limit state and the value of the earth pressure. This means that the magnitude of the pressure for structural strength design may be higher than for geotechnical stability. For reinforced-concrete walls, the load effects derived from the analysis are design effects to be used in the structural calculation.

For mass gravity-type walls (e.g. mass concrete or gabion walls) it will be necessary to undertake checks at different points down the height of the wall to ensure that tension is not generated where it cannot be resisted, and that the shear forces do not exceed the sliding resistance – for example, along the boundary between two levels of gabion baskets.

64.2.6 Design method

The preceding sections outline how the various loads acting on a wall can be assessed. Once these actions have been identified, it is necessary to analyse the wall to verify that the limit states identified in Section 64.2.2 are not exceeded. Typically this analysis is iterative, where an initial estimate of the wall geometry is made and stability against each failure mode is checked. The geometry of the wall may then be adjusted to optimise the solution while still ensuring stability. Such analysis and iteration may be calculated by hand, calculated using proprietary software or by using a fully checked spreadsheet.

The design stages for a gravity wall are shown in the flowchart in **Figure 64.8** and described below.

(*a*) Confirm the relevant limit states for the problem, as identified in Section 64.2.2. For SLS agree the movement limits or other serviceability criteria that are to be adopted.

(*b*) Identify the design situations to be considered in the analyses, considering the issues described in Section 64.4. Select an initial situation to analyse, taking account of any historic construction to ensure a correct initial stress state is considered.

(*c*) For the selected design situation, estimate the initial geometry of the wall.

(*d*) Verify the global stability of the structure, in accordance with Chapter 23 *Slope stability*.

(*e*) Calculate the individual components of the actions acting on the wall for DA1-1 and DA1-2, using the guidance of Section 64.4.

(*f*) For DA1-2, apply the relevant partial factors to the characteristic actions to determine the design actions. On the basis of these design actions, determine the design bearing resistance of the ground.

(*g*) Check whether the design actions exceed the design bearing resistance. If they do, return to step (*c*) and revise the wall geometry.

(*h*) Once sufficient bearing resistance has been demonstrated, verify that the sliding and overturning limit states are also acceptable. If not, return to step (*c*).

(*i*) Verify the geotechnical ULS for DA1-1. Return to step (*c*) if insufficient capacity is identified.

(*j*) Calculate the ULS in situ earth pressure acting on the wall, including the compaction pressure where appropriate, for DA1-1 and DA1-2.

(*k*) Determine the design actions on the wall for the structural design for DA1-1 and DA1-2. Verify the structural capacity of the wall is acceptable.

(*l*) Calculate the SLS in situ earth pressure acting on the wall, including the compaction pressure, where appropriate. Estimate the anticipated movement of the wall and check against the agreed SLS. If acceptable, the design situation has been verified as acceptable, otherwise return to step (*c*).

(*m*) Review other design situations: start at step (*b*) for each design situation.

(*n*) Prepare the geotechnical report, including details of maintenance requirements associated with any drainage systems included in the design.

When carrying out the different ULS checks, it is important to consider whether the actions are beneficial as this will influence the partial factor that is applied.

The procedure for verifying the bearing resistance should be undertaken in accordance with Chapter 53 *Shallow foundations*.

64.3. Reinforced soil walls

64.3.1 Introduction

Reinforced soil walls are essentially a specific type of gravity wall. Two different checks need to be made, one for external stability and the other for internal stability. The size of the block of reinforced soil is selected to provide overall stability, with the internal check being made to ensure that the reinforcement holds the block together. The external stability check should follow the procedure set out in Section 64.2.

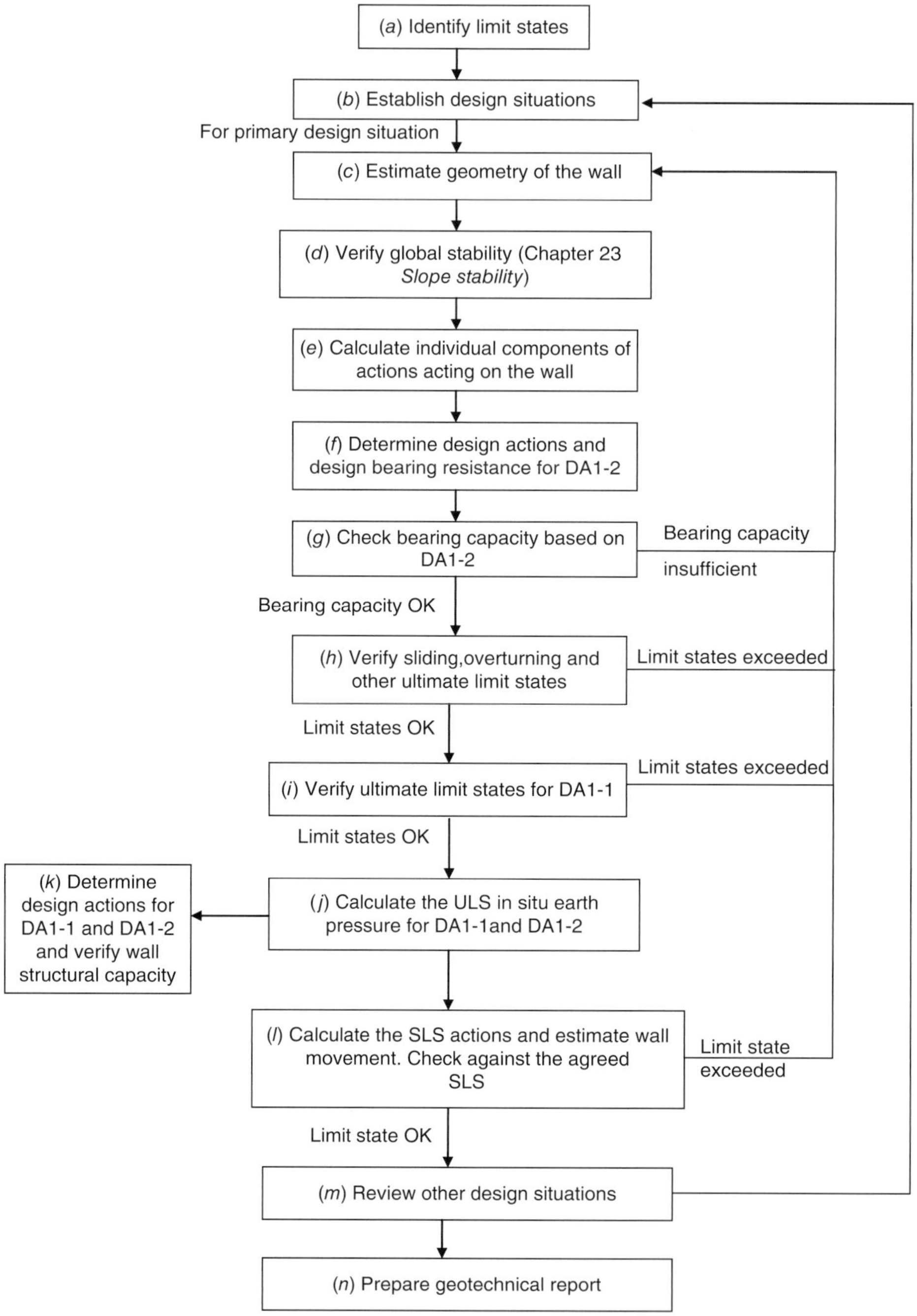

Figure 64.8 Gravity wall design process

A discussion of the available types of soil reinforcement and a detailed explanation of the design of gravity walls is given in Chapter 73 *Design of soil reinforced slopes and structures*, with a brief overview below.

It is noted that BS EN 1997-1:2004 is not applicable to reinforced soil walls and their design should be undertaken in accordance with the current British Standard BS 8006 (BSI, 2010) and all associated annexes.

64.3.2 Design for external stability

As with any gravity wall, the overall block of reinforced soil must be selected to ensure the geotechnical limit states are met.

Slope stability, bearing capacity, sliding and overturning limit states must be checked.

For many projects it is common to design the soil block for external stability, which is typically presented as a minimum length of reinforcement.

64.3.3 Design for internal stability

Once the overall size of the reinforced soil block has been selected, the reinforcement must be designed. This typically includes parameters such as the gauge of reinforcement and the spacing of layers. On occasion it may also be necessary for the straps to be longer than the minimum required for external stability to ensure that pull-out failure is avoided.

There are a number of different proprietary soil reinforcement systems available, and while they generally work on the same principles, the detail of the design will vary. In general, the design for internal stability must ensure that a soil block remains intact. Typically this requires checking the pull-out resistance and whether the reinforcement ruptures, which are described in Chapter 73 *Design of soil reinforced slopes and structures.*

Since the various systems differ, it is common for a specialist reinforced soil subcontractor to undertake the detailed design.

64.4. Embedded walls

64.4.1 Introduction

As in Chapter 62 *Types of retaining walls*, an embedded retaining wall penetrates the ground at its base to gain lateral support from the resistance of the ground in front of the lower part of the wall. This resistance may be used as the sole resistance to provide stability for the wall or it may act in conjunction with anchorages, props or other sources of lateral support for stability. The structural resistance of the wall in bending and shear is usually critical.

Embedded walls can be used to retain backfilled material, much like gravity walls, with typical applications including construction of new marine geo-structures or land reclamation. Unlike gravity walls, embedded walls can be installed into natural ground, which is then excavated in front of the wall. This means there is no control over the quality of the ground that the wall is supporting nor is there an opportunity to install drainage.

In general, embedded walls can be categorised as cantilever (fixed earth) or propped (free earth). While the overall principles for design are comparable between these two categories, there are differences in the specifics of the design approach, and these are identified in the following sections.

64.4.2 Limit states

As with gravity walls, BS EN 1997-1:2004 adopts the principle of a limit state design. The limit states relevant to embedded walls are discussed below. These are listed in sections 9.2, 9.7 and 9.8 of BS EN 1997-1:2004.

64.4.2.1 Ultimate limit states

BS EN 1997-1:2004 lists the following ULS as relevant to embedded walls

- loss of overall stability
- failure by rotation or translation of the wall or parts thereof
- failure of a structural element such as a wall, anchorage, wale or strut or failure of the connection between such elements
- failure by lack of vertical equilibrium
- combined failure in the ground and in the structural element
- failure by hydraulic heave and piping
- movement of the retaining structure, which may cause collapse or affect the appearance or efficient use of the structure or nearby structures or services which rely on it
- unacceptable leakage through or beneath the wall
- unacceptable transport of soil particles through or beneath the wall
- unacceptable change in the groundwater regime.

Some of these limit states are illustrated in **Figure 64.9**.

The design of an embedded wall must ensure that these limit states are avoided. In addition to the limit states listed above, the designer should consider whether any additional limit states should be taken into account in respect of specific site conditions or other requirements not covered by this manual.

64.4.2.2 Serviceability limit states

The primary SLS noted by BS EN 1997-1:2004 for embedded walls is movement of the retaining structure, which may cause collapse or affect the appearance or efficient use of the structure or nearby structures or services which rely on it. This may involve both lateral and vertical movement of either the wall or the soil around the wall.

64.4.3 Factors of safety: partial factors and lumped factors

As discussed in Section 64.2.3, BS EN 1997-1:2004 adopts a partial factor approach to design. For embedded walls this concept has been in common use for a number of years, and should be readily adopted in design. There are, however, a number of traditional methods of embedded wall design that use a lumped factor approach. Typically these either sought a sufficiently large ratio of the restoring moment to the overturning moment in a rotational stability check, or applied a single factor to a measure of the passive resistance of the soil.

The ULS identified above typically relate to either the BS EN 1997-1:2004 GEO limit state ('failure or excessive deformation of the ground, in which the strength of soil or rock is significant in providing resistance') or STR ('internal failure or excessive deformation of the structure or structural elements in which the strength of structural materials is significant in providing resistance'). Partial factors to be used in the design of embedded walls are provided in BS EN 1997-1:2004 for these limit states. DA1 has been adopted in the UK: walls must be checked against both DA1-1 and DA1-2.

In DA1-2 factors greater than unity are applied to the characteristic soil parameters to derive the design values to be used

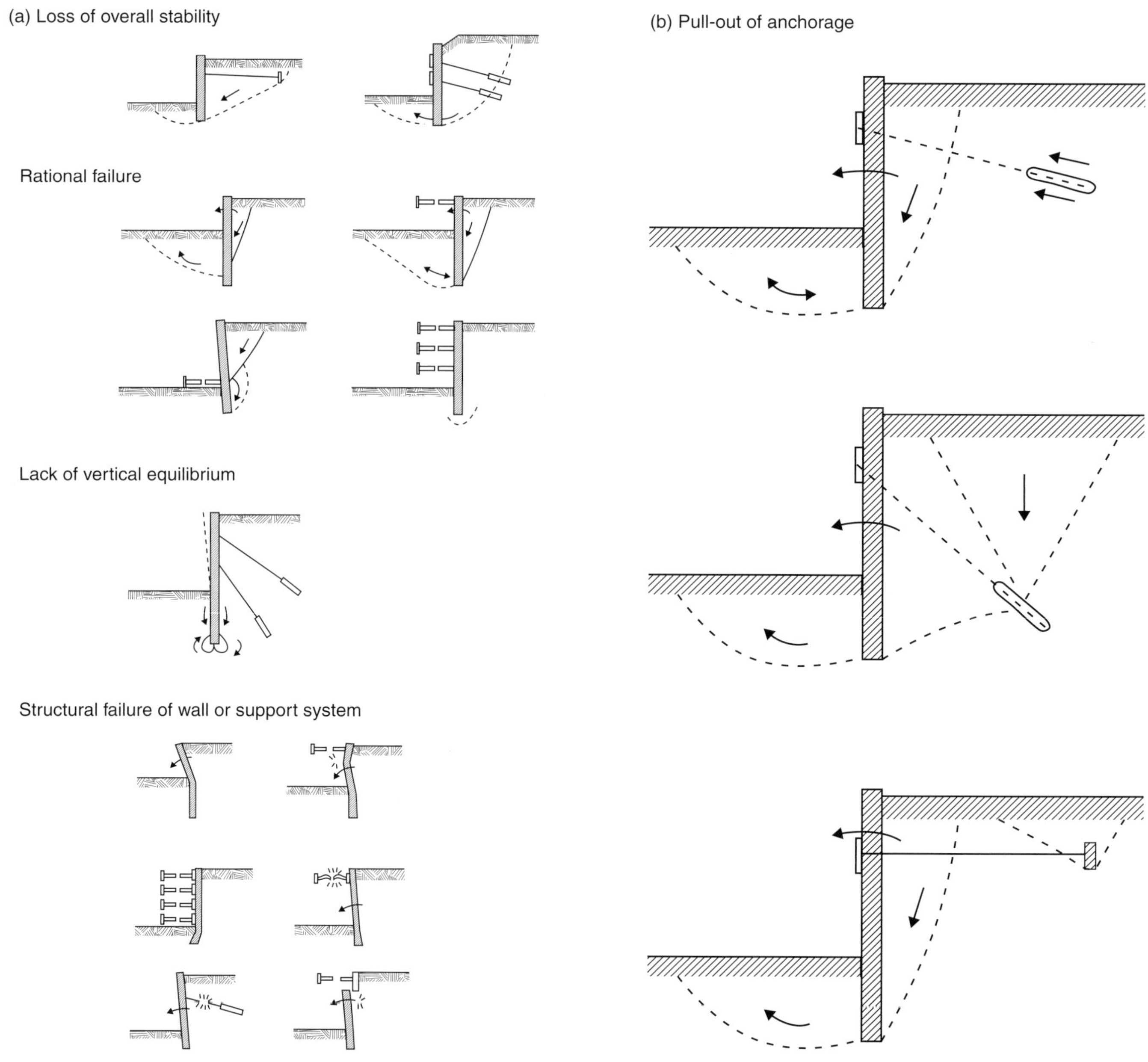

Figure 64.9 Ultimate limit state failure modes for embedded walls

in the analysis. This is usually the critical case for stability and will establish the required wall geometry. The load effects in the wall from this analysis are used to check the structural capacity of the system, along with the output from DA1-1.

For the SLS check, partial factors of unity are used.

64.4.4 Design situations, design geometry, groundwater conditions

64.4.4.1 Introduction

The general aim of an embedded wall design is to select a wall length that is sufficient to ensure stability and to derive load effects so that the adequate structural performance of the wall and any support system is ensured. In addition it is common that some form of movement analysis is undertaken to ensure that the structure meets the design performance requirements and to ensure nearby structures and utilities are not adversely affected.

The following sections outline the key considerations that the designer should address in designing an embedded retaining wall.

64.4.4.2 Design situations

Design situations are discussed in BS EN 1997-1:2004. When designing an embedded wall it is important to consider whether more than one design situation is applicable. For example, where construction is staged, such as for multiple-propped walls, it is necessary to consider each stage of the construction sequence individually to ensure that the relevant limit states are avoided.

For each design situation, the designer must consider the following.

- *Design soil stratigraphy.* In variable ground conditions it may be necessary to consider more than one design soil stratigraphy. Each would comprise a separate design situation.
- *Soil parameters and ground model.* Characteristic soil parameters should be assessed based on the approach outlined in Chapter 27 *Geotechnical parameters and safety factors.* Walls are linear structures and therefore have the potential to encounter variations in the design stratigraphy and soil parameters. The designer must consider this variability and if necessary analyse the different ground models.
- *Temporal effects.* For example, whether the analysis will be short term (i.e. construction stages) or long term (i.e. permanent condition). If necessary, both conditions may need to be checked as separate design situations in combination with the appropriate set of soil parameters.
- *Construction method.* The designer should consider the intended method of construction and make an allowance for any construction-related effects that may affect performance of the embedded retaining structure (e.g. pre-augering of driven sheet piles or construction duration for rotary bored cast-in-place piles).
- *Construction sequence.* It is necessary to ensure that all potentially critical stages are considered in the design.
- *Design geometry.* Where ground levels or retained heights vary along the length of the wall, it may be appropriate to divide the wall into sections and select a representative geometry for each section. Each geometry will then be considered as a separate design situation.
- *Groundwater and drainage.* This is a key consideration when developing the design situations for a retaining wall. Depending upon the soil type, the construction sequence and the rate of construction, it may be necessary to analyse situations which consider that the ground is undrained in some cases and drained in others. This is discussed in detail in Chapter 63 *Principles of retaining wall design.*
- *Imposed loading.* Consider surcharges due to building loads or traffic and their effect.

While it may be necessary to calculate more than one design situation, by careful consideration of the relevant combinations that are likely to prove critical, it is possible that once a solution for the critical design situation has been determined, alternative situations may be confirmed as acceptable either by inspection, or through a reduced number of analyses and calculations.

64.4.5 Earth pressure

In order to verify the stability and movement of a wall, the earth pressure acting upon it must be calculated. There are two general analysis methods that are typically used to evaluate this pressure

- limit equilibrium
- soil–structure interaction.

Limit-equilibrium methods of calculation are based on conditions at collapse, when the full strength of the soil is mobilised around the retaining wall based on assumed failure mechanisms. Calculations are typically based on simple assumed linear lateral stress distributions, which represent a particular collapse mechanism. Collapse mechanisms are well developed for cantilever and single-propped walls, but less so for other forms of embedded retaining walls, such as multi-propped walls and walls propped significantly below the top. Limit-equilibrium analyses may be calculated using computer software or by hand.

Soil–structure interaction analyses require the use of specialist software. In such analyses the initial stresses in the ground are established, and the sequence of construction is modelled. At each stage changes are made to the model to disrupt the equilibrium of the system – for example, soil is excavated or a support removed, and the calculation models the deformation of the system and the mobilisation of stresses in the ground and the structure as the analysis attempts to restore equilibrium. These analyses therefore directly calculate the earth pressure that is generated around the wall and the associated wall and soil movement.

A discussion of the relative merits of limit-equilibrium and soil–structure interaction analyses is given in CIRIA C760 (Gaba *et al.*, 2017).

64.4.5.1 Limit-equilibrium analysis

For a limit-equilibrium analysis, earth-pressure profiles are based on established mechanisms, and the selection of the appropriate earth-pressure distribution depends upon the design situation. The established mechanisms for two common situations – cantilever and propped walls – are illustrated in **Figure 64.10**.

For the fixed-earth mechanism the wall is assumed to rotate about some pivot point in the ground. Above this point, the earth pressure on the retained side is assumed to be at the active limit and the earth pressure on the excavated side is assumed to be at the passive limit. To ensure rotational and lateral stability can be achieved, below the pivot point this is reversed so that on the retained side the pressure is at the passive limit and the pressure on the excavated side is at the active limit. The stability of the wall is verified by considering the moment equilibrium about the assumed pivot point, and the equilibrium of the assumed horizontal forces. To achieve equilibrium an iterative calculation is required, where the toe and pivot points are varied until equilibrium is achieved. This process is best performed by specialist analysis software but can be undertaken by hand.

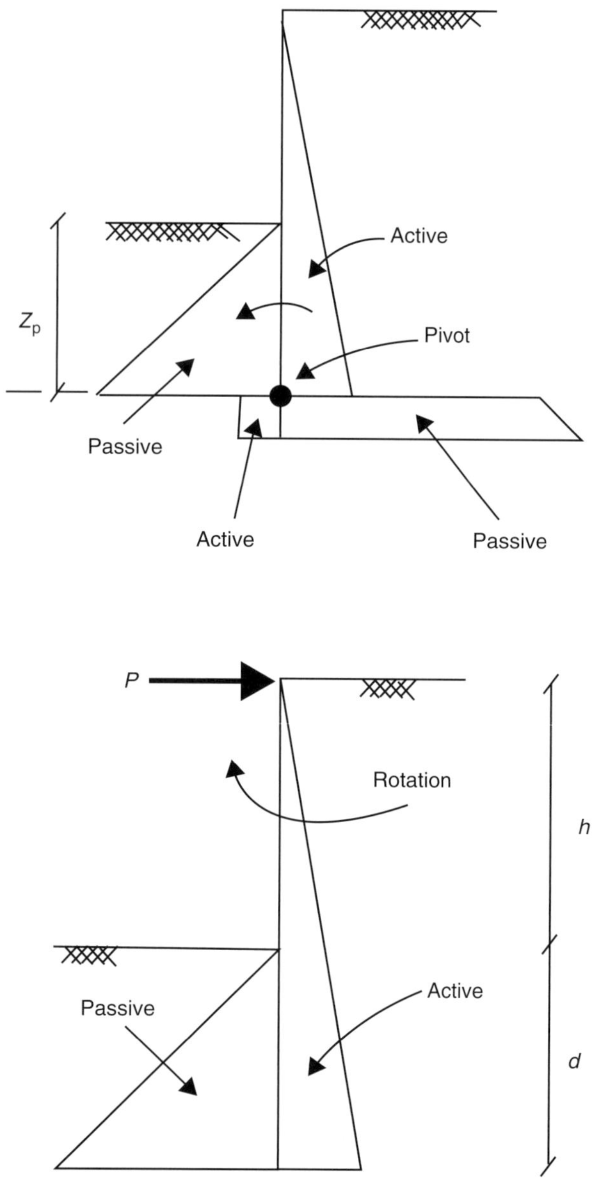

Figure 64.10 Limit-equilibrium earth-pressure distributions. (a) Fixed earth – cantilever. (b) Free earth – propped wall

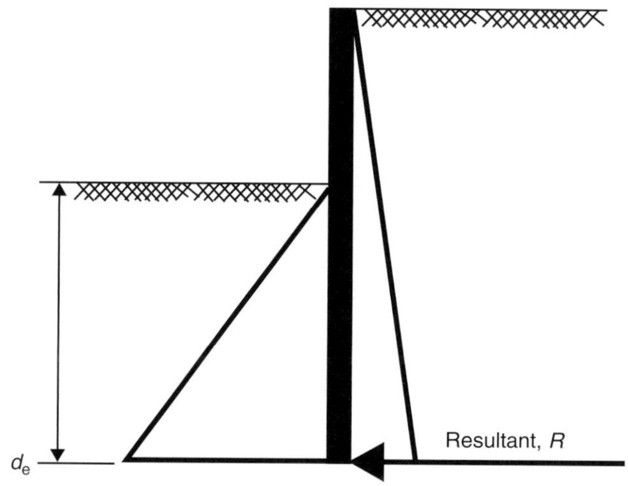

Figure 64.11 Simplification of the fixed-earth mechanism

A simplification of the calculation is often adopted for hand calculations. In this case the length of wall below the pivot point is replaced by an assumed resultant force, R, as illustrated in **Figure 64.11**.

The depth to the pivot point, d_0, is then varied until moment equilibrium is achieved and the magnitude of the resultant, R, that is required for horizontal stability can be calculated. To ensure the wall is long enough to ensure stability, the toe of the wall is generally assumed to be 20% deeper than the pivot point (i.e. $1.2d_0$). It is recommended that this length of wall below the pivot point is sufficient to mobilise a reaction that is greater than the assumed resultant force, R.

The free-earth condition is simpler to assess as it is assumed that the wall rotates about the level of the prop. The earth pressure on the retained side is therefore assumed to be at the active limit and the earth pressure on the excavated side is assumed to be at the passive limit. To verify stability, moments are taken about the assumed pivot point and the toe of the wall varied until equilibrium is achieved. Any imbalance in the lateral pressure acting on the wall is taken up as a reaction in the prop.

For either mechanism, once the stability of the system has been verified, the earth pressure arising from that analysis is used to derive the action effects (e.g. bending moments and shear forces) in the wall.

Design values of soil parameters should be used when determining the values of the limiting earth-pressure coefficients that are inputs into the calculation of DA1-1 and DA1-2 combinations; these will yield different results for the wall length, pivot points and prop forces. This is because the different partial factors on material properties and actions will lead to different limiting earth-pressure coefficients and imposed load effects. In terms of wall length, the longer wall from the two combinations should be adopted in the design.

The use of limit-equilibrium methods to derive design action effects where the length of the wall is fixed by criteria other than rotation or lateral stability, such as to achieve water cut-off or for vertical load capacity, requires further consideration. The toe level required for the limit-equilibrium mechanisms to achieve an equilibrium condition will be different to the wall toe level required to satisfy the other criteria. Load effects will therefore not be calculated for the full length of the wall. This may be overcome by deriving the equilibrium earth pressure and load effects in the wall based on the limit-equilibrium analysis, and then extrapolating the load effects down to the actual required toe level, as illustrated in **Figure 64.12**. This approach also applies to the SLS condition, where the toe

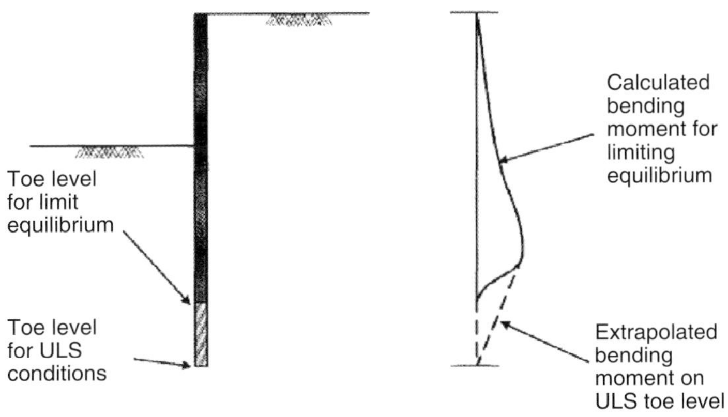

Figure 64.12 Extrapolation of load effects from limit-equilibrium analysis

level from the limit-equilibrium analysis will be shorter than from the ULS analysis, therefore the SLS load effects need extrapolating to the ULS toe level.

If this extrapolation of the load proves to be a critical consideration in the wall design, it may be appropriate to consider using soil–structure interaction methods as described below.

A further limitation of the limit-equilibrium method is the lack of arching or earth-pressure redistribution in the assumed earth-pressure profile. This is particularly relevant for a wall propped at or near the top, where there is a risk of underestimating the prop loads. This is discussed further in Section 64.4.7.

64.4.5.2 Soil–structure interaction analysis

As noted above, in a soil–structure interaction analysis the earth pressure acting on a wall is calculated using computer software. A number of different methods of soil–structure interaction analysis are available

- subgrade reaction (e.g. Geosolve WALLAP)
- pseudo-finite element (e.g. Oasys FREW, Geosolve WALLAP)
- finite element and finite difference (e.g. Plaxis, Oasys Gofer, FLAC).

This chapter does not go into the detailed theory behind the different methods and the method of software operation, as these are dependent upon the program used. However, the main considerations to be taken into account for a soil–structure interaction analysis are presented below.

Unlike a limit-equilibrium analysis, the in situ stresses are an important input for a soil–structure interaction analysis. Typically the software will require a value of the at-rest earth-pressure coefficient, K_0, for each soil layer.

One benefit of a soil–structure interaction analysis is the potential for modelling the sequence of construction. The sequence in which a wall is constructed will affect the earth pressure that is generated in the soil adjacent to the structure and therefore the action effects acting on the wall. The stages of the construction sequence that are to be modelled should therefore be established when the design situations are considered.

During analysis of the construction sequence, temporal effects can also be taken into account. This would include allowing fine-grained materials to behave in an undrained manner in the short term, but as drained materials in the longer term. Creep and corrosion of structural components may also be modelled, and it is important that such effects are represented in the software following the recommendations in the specific user manual.

It is important to read and understand the technical basis upon which any software carries out calculations to ensure that the correct assumptions and inputs are made for the analysis. It is also crucial that the output is checked carefully. With a complex analysis it can be difficult to understand its theoretical basis, and there is a danger that output is accepted without being critically reviewed. Before accepting the results of a soil–structure interaction analysis, the user should consider whether the results are as expected – does the earth pressure look reasonable? How much of the soil is at the active or passive limit? Does the mode of deformation look reasonable? It is useful to relate the results back to a limit-equilibrium analysis of the same problem, where possible.

One particular difficulty with a soil–structure interaction analysis is the determination of the limiting stability, such as would be required when shortening a wall to find the minimum length that is stable. Whether or not an analysis converges will depend upon the number of iterations specified and the magnitude of any convergence tolerances specified in a particular software program. Often an analysis may indicate that it has converged, which could lead the user into thinking that the wall is stable; however, if excessive deformation has occurred in order to reach convergence, it may be considered that the wall has actually failed, despite reaching convergence. Again this should be picked up when checking the detailed output of an analysis.

64.4.5.3 Other influences on earth pressure

The preceding sections outlined the basic failure mechanisms assumed in limit-equilibrium analysis, and the general approach to soil–structure interaction analysis necessary to calculate the earth pressure acting on a wall. However, there are a number of additional factors that will influence the basic approach that need to be taken into account when assessing the actions acting on an embedded wall. These are discussed below.

64.4.5.4 Wall–soil friction

The interface friction angle between the soil and the wall structure will affect the limiting earth pressure that is mobilised adjacent to a wall. The magnitude of the soil–wall interface friction angle will depend upon the ground conditions and the wall type. The direction of the wall–soil interface friction

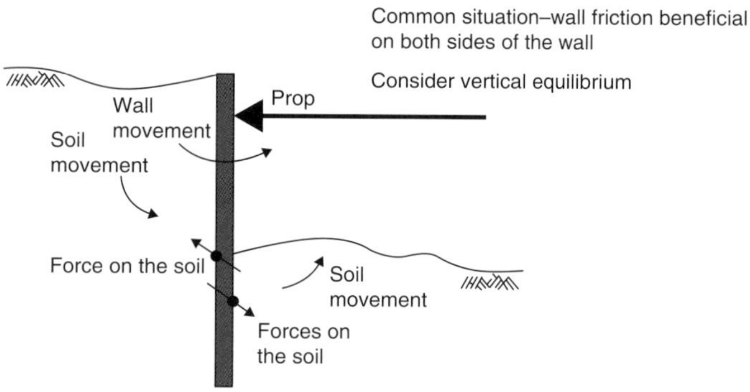

Figure 64.13 A common situation for wall friction

angle will depend upon the direction of relative movement between soil and wall, which can be affected if the wall is carrying vertical loads. For the conventional case, on the retained side the soil moves downward relative to the wall, resulting in an upward reaction on the wall–soil interface, whereas on the excavated side the soil heaves upward relative to the wall, giving a downward reaction (**Figure 64.13**). Helpful guidance on the use and interpretation of a soil–structure interaction analysis is given in Chapter 6 *Computer analysis principles in geotechnical engineering.*

If, however, the wall is carrying a vertical load, upward shear stresses need to be mobilised on the wall–soil interface in order to support the applied loads. Over the retained height of the wall it may be considered that the soil and wall move downwards together, generating no net friction at the interface. The resistance to the applied load therefore comes from the embedded portion. On the excavated side this corresponds to a reversal in the direction of the interface friction in relation to the conventional case. This is important as it can significantly reduce the limiting values of the passive resistance and may therefore increase the length of wall required for stability. Further details of walls carrying vertical loads are given in Chapter 62 *Types of retaining walls* as part of complete underground structures.

64.4.5.5 Passive softening

During excavation of the ground in front of a wall, the soil providing lateral stability for the wall is unloaded vertically. If this soil is a fine-grained material that is assumed to act in an undrained manner during construction there is a risk that the unloading will lead to a softening of the soil near the surface of the excavation as the excess pore pressure is dissipated. Any such softening will reduce the passive resistance available for wall stability and may therefore increase the required wall length. It is recommended that the value of the undrained shear strength at the surface of an excavation in fine-grained soils is taken as zero, rising to the design value of undrained shear strength at some depth, L, within the soil block, as illustrated in **Figure 64.14**. Where there is no potential for groundwater recharge this depth may be taken as 0.5 m. In other situations, where recharge may occur at excavation level but where there is no recharge within the soil, it may be taken as

$$L = \sqrt{(12c_v t)} \qquad (64.4)$$

where c_v is the coefficient of consolidation and t is the elapsed time.

64.4.5.6 Tension cracks

In fine-grained materials modelled with undrained shear strengths in the short term there is a theoretical potential that on the retained side the active limit will be negative. Since soil cannot sustain tensile stresses, there is the potential for a crack to open between the soil and the wall. In this zone the pressure upon the wall would be zero. Consideration must be given to such tension cracks becoming flooded with water – for example due to perched water on top of a clay layer. If a tension crack becomes flooded the pressure acting on the wall will be equal to the hydrostatic pressure of the water in the crack. Even if there is no obvious source of water, it may not be appropriate to rely on a tension crack exerting zero pressure

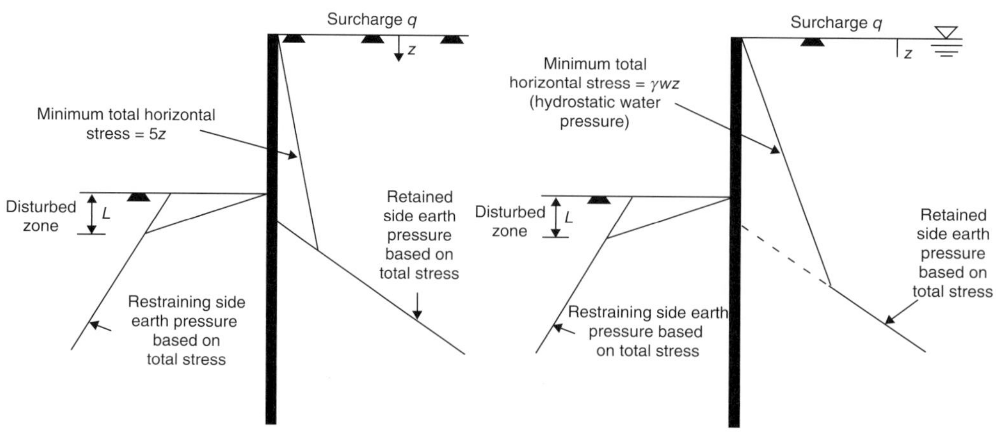

Figure 64.14 Passive softening and tension cracks. (a) Undrained conditions: dry tension cracks. (b) Undrained conditions: flooded tension cracks

on the wall. In such cases it is common to limit the minimum pressure that is assumed to act on a wall to a minimum equivalent fluid pressure (MEFP). In dry conditions the MEFP is typically taken as $5z$, where z is the depth below the retained ground surface. Water-filled tension cracks and the assumption of MEFP are illustrated along with passive softening of the ground in **Figure 64.14**.

64.4.5.7 Sloping ground

Sloping ground, either in front of or behind a wall, can significantly affect the pressure acting on the wall. BS EN 1997-1:2004 gives analytical and graphical methods for calculating the earth-pressure coefficients for sloping ground. A difficulty arises when the slope is uneven or of limited extent. In such situations, if approximating the profile to an infinite slope is unreasonably conservative, one of the following methods may be considered.

- Carry out a Coulomb wedge analysis to determine the active lateral thrust at different depths down the wall.
- Model the effect of the slope as a series of surcharges. It is noted that this method does not model the active thrust within the slope above the top of the wall. The active thrust should therefore be quantified separately and applied as an imposed force at the top of the wall.
- Undertake a finite element or finite difference analysis of the proposed slope geometry and model its effect explicitly.

64.4.5.8 Reverse passive

During excavation of the ground in front of a wall propped close to the top, the section of wall above the prop tends to rotate backwards into the retained soil. In a soil–structure interaction analysis this can mobilise an earth pressure up to the passive limit close to the top of the wall. In this zone, however, the soil is generally moving down relative to the wall (giving a negative value for the wall–soil interface friction, based on the standard sign convention for K_p); therefore it is important that the value of K_p applied over the soil zone near the top of the wall takes this into account. It is generally found that a value of $K_p = 1$ is appropriate.

A limit-equilibrium analysis will not incorporate the reverse passive pressure near the top of the wall as the assumed pressure profile does not include it. The absence of this increased pressure is the main reason that limit-equilibrium analyses underestimate prop forces in a wall support system. A model factor should be applied to the calculated limit-equilibrium prop force to ensure the design is robust. Further details are given in Section 64.4.7.

64.4.5.9 Multi-propped walls

Where walls are propped by more than one level of support, instability of the wall due to rotation or translation is not a concern, provided the structural design of the wall and the support system is adequate. However, to design the wall and the support system it is still necessary to calculate the earth pressure acting on the wall. Limit-equilibrium analysis is not particularly well suited to multi-propped situations, although some approaches are discussed in CIRIA C760 (Gaba *et al.*, 2017). In general, it is recommended that for multi-propped walls a soil–structure interaction analysis is undertaken, with the sequence of prop installation and excavation being modelled.

64.4.5.10 Wall and soil stiffness

While not required for a limit-equilibrium analysis, wall and soil stiffness are important inputs in a soil–structure interaction analysis. It is important that the stiffness adopted is appropriate for the level of strain anticipated in the soil. Generally for retaining wall design this strain will be lower than for foundation design; therefore higher values of stiffness may be justified. In all cases the resulting wall and soil movements arising from the analysis should be verified against comparable experience, such as the wall and soil movement case histories presented in CIRIA C760 (Gaba *et al.*, 2017) or other technical papers, to ensure that the assumptions are reasonable.

Modelling wall and prop stiffnesses should take into account any time-related effects. For example, the bending stiffness of a reinforced-concrete wall is given as $E_o I$, where E_o is the short-term Young's modulus of the reinforced concrete and I is the second moment of area. However, as soon as a wall is loaded by bending, cracking will occur on the side of the wall that is in tension, which acts to reduce the stiffness of the wall that is bending. In the absence of a detailed calculation it is recommended that a value of $0.7E_o I$ is taken as the initial stiffness of the wall. In the long term, reinforced-concrete walls have the potential to creep due to strain at constant stress. Typically the long-term bending stiffness is taken as $0.5E_o I$.

In steel retaining walls a reduction in stiffness may occur due to corrosion of the steel over the lifetime of the structure. The reader should refer to BS EN 1993-5:2004 and the appropriate National Annex for guidance on corrosion rates to be applied in design.

Modelling changes in wall stiffness between the short and long term needs to be appropriate to the software being used, and the user should consult the specific guidance for the software. In general it is not acceptable to simply change the wall stiffness at a particular stage of the analysis.

64.4.5.11 Thermal effects

Temperature variations can lead to expansion and contraction of support systems, which may induce additional loads on the structure. In general this is more likely to be significant where the support system is exposed to atmospheric conditions, such as during construction, especially where temporary steel props are used. A method for calculating thermal effects on props is given in Section 64.4.7.

Thermal effects may be significant for integral bridges (bridges without movement joints between the deck and the abutments). In these cases the abutment walls are subject to cyclic expansion and contraction that can lead to strain ratcheting – an effect that increases the earth pressure behind the wall. Specific guidance for the analysis and design of integral bridges is given in the *Design Manual for Roads and Bridges* (Highways Agency, 2000).

64.4.5.12 King post walls

King post walls are a specific type of embedded wall. It is generally assumed that retaining walls are a plane-strain problem and therefore the calculations are generally carried out on the basis of a 'unit length' of wall. King post walls, however, derive support from discrete elements embedded at a fixed spacing, *s*. Where *s* is less than three times the diameter of the support element it may be appropriate to assume plane-strain conditions; however, for wider-spaced supports a different approach is required, which is set out in CIRIA C760 (Gaba *et al.*, 2017) and summarised below.

The individual support elements should be designed individually to resist the calculated lateral loads. Full active conditions should be assumed behind the retaining wall. The passive resistance in front of the wall may be taken as follows (taken from Gaba *et al.*, 2017: Equations 4.16–4.20).

The king posts should be designed as piles in lateral loading, with an ultimate net effective resisting force P_a' per metre length of

$$P_a' = K_p \times b \times p'\,v/S \text{ at embedment depths } z \leq 1.5b \qquad (64.5)$$

and

$$P_a' = K_p^2 \times b \times p'\,v/S \text{ at embedment depths } z \leq 1.5b \qquad (64.6)$$

where b is the king post width, s is the spacing of the king posts ($s > 3b$), K_p is the passive earth-pressure coefficient ($K_p > 3$) defined as $(1 + \sin\varphi')/(1 - \sin\varphi')$ and p_v' is the effective overburden pressure at depth (Fleming *et al.*, 2008).

The above expressions for P_a' are based on the work of Barton (1982) as reported by Fleming *et al.* (2008), and are considered applicable for K_p values of 3.0–5.3 (i.e. $30° \leq \varphi' \leq 43°$).

In a total stress analysis, the ultimate net lateral resisting force per metre length P_u is given by Fleming *et al.* (2008) as

$$P_u = \left[2 + (7z/3b) \times s_u \times b/s\right] \text{ at embedment depths } z \leq 3b \qquad (64.7)$$

and

$$P_u = 9 \times s_u \times b/s \text{ at embedment depths } z \leq 3b \qquad (64.8)$$

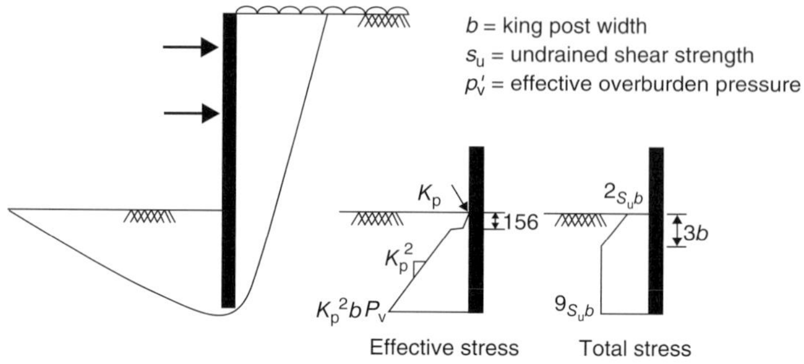

Figure 64.15 King post wall design. (a) Overall stability. (b) Lateral loading of individual king posts

where b is the king post width, s is the spacing of the king posts ($s > 3b$) and s_u is the undrained shear strength at embedment depth z.

This is illustrated in **Figure 64.15**.

64.4.5.13 Unplanned excavation

When assessing the geometry for a given design situation, it is important to take into account the possibility of an unplanned excavation in front of the wall. This is particularly critical for an embedded wall, where the passive resistance of the ground is necessary for stability. BS EN 1997-1:2004 requires that for normal site controls 10% of the retained height, up to a maximum of 0.5 m, should be assumed for unplanned excavations.

64.4.5.14 Surcharges and direct loads

The design of a retaining wall should consider the possibility of surcharges being applied to the ground retained by the wall. Such surcharges may arise, for example, from traffic in the permanent condition, plant loading during the construction phase or from the placement of stockpiled materials behind the wall. The effect of such surcharges must be incorporated into the assessment of the earth pressure acting on the wall.

The method of modelling a surcharge will depend on how the wall is analysed. Where specialist software is used, the user manual should be consulted to ensure the surcharges are properly modelled. Uniformly and non-uniformly distributed loading, such as loading over a limited area, line load and point loads, can be modelled directly in a limit-equilibrium analysis. CIRIA C760 (Gaba *et al*., 2017) gives different methods for determining the limiting lateral pressure due to non-uniformly distributed loads.

Traditionally in UK practice a minimum 10 kPa surcharge has been applied behind a wall. Although this is not an explicit requirement in BS EN 1997-1:2004, for most walls some allowance should be made and 10 kPa is considered a reasonable minimum value. Larger surcharges may be required for particularly heavy road traffic or for rail traffic. Guidance is provided in CIRIA C760 (Gaba *et al*., 2017).

64.4.5.15 Groundwater

Understanding the groundwater conditions is crucial to embedded retaining wall design. Since the water pressure directly influences the effective stresses, and hence strength of the ground, applying a partial factor on the water pressure may lead to unreasonable design situations. In order to ensure a margin of safety is achieved, it is recommended that a margin is applied to the groundwater levels upon which the pore pressures are based. The approach for selecting the groundwater level differs in DA1-1 and DA1-2: in DA1-1 the actions of water on the structure are factored by a value greater than unity, whereas in DA1-2 the factor on permanent actions is unity. This means that for DA1-2 no factors are applied to the effect of water, so it is recommended that a suitably cautious design water pressure is derived from a directly assessed design groundwater level, defined as

> (1) Water pressure and seepage forces that represent the most unfavourable values, which could occur in extreme or accidental circumstances at each stage of the wall's construction sequence and throughout its design life. An example of an extreme or accidental event may be a burst water main in close proximity to the wall.

For DA1-1 a factor greater than unity is applied to water effects, so it is considered that a less cautious assessment of the groundwater level may be used than for DA1-2. The recommended approach is defined as

> (2) Water pressures and seepage forces that represent the most unfavourable values likely to occur in normal circumstances at each stage of the wall's construction sequence and throughout its design life. Extreme events such as a nearby burst water main may be excluded, unless the designer considers that such an event may reasonably occur in normal circumstances.

When selecting the design groundwater levels, consideration must be given to any potential effect the construction of the embedded wall may have on the groundwater regime. For example, if the wall penetrates a layer of flowing groundwater, there may be a 'dam' effect that leads to a rise in groundwater levels adjacent to the wall. Conversely, the installation of drainage within a basement may lead to a lowering of the equilibrium groundwater level. Such effects need to be considered in the design of the wall, and possible effects on adjacent structures must also be considered.

Where differing groundwater levels are identified on each side of a retaining wall, the potential for seepage around the wall must be investigated, as this will have an effect on the pore pressures assumed to act along the length of the wall. Where an impermeable wall penetrates an impermeable stratum, the assumption of unbalanced pore pressures at the toe of the wall may be acceptable. However, in many situations, especially over the design lifetime of a structure, since the wall and the soil have a finite permeability there will be seepage around the wall and generation of non-hydrostatic pressure profiles.

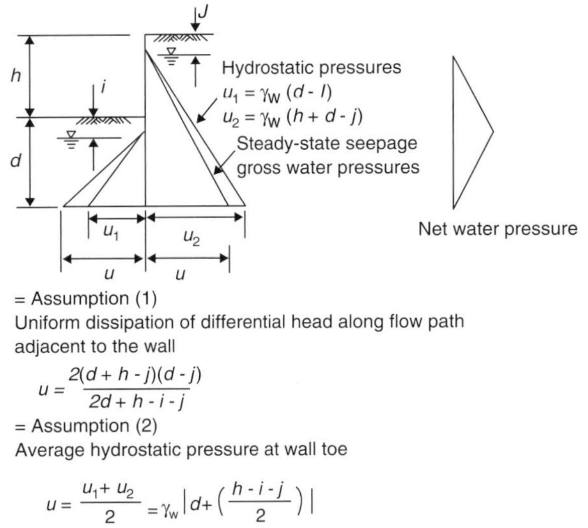

Figure 64.16 Simplified approach for calculating the linear steady-state seepage water pressure

In order to assess the pore pressures acting on the wall accurately, it may be necessary to undertake a seepage analysis. However, there is also a commonly adopted simplification to the seepage problem that has been developed on the basis of isotropic permeability and wide excavations. For this to be applied, the design must verify that these simplifying assumptions are reasonable. The method of calculating this simplified linear seepage profile is given in **Figure 64.16**. Further details of the effect of seepage on retaining wall design are discussed in CIRIA Special Publication 95 (Williams and Waite, 1993).

64.4.6 Design of structural elements

The load effects from either the limit-equilibrium or soil–structure interaction analyses should be used in the structural design of a wall.

64.4.7 Design of propping systems

The design of propping systems is covered in Chapters 65 *Geotechnical design of retaining wall support systems* and 66 *Geotechnical design of grouted anchors.* Careful consideration must be given to determining the design load effects that are to be used in the propping design. In general these will be derived from the reactions calculated in the retaining wall analysis. However, there may be additional factors that need to be taken into account to ensure a robust solution is achieved.

64.4.7.1 Limit-equilibrium prop loads

As noted in Section 64.4.5, the limit-equilibrium method for propped walls does not consider soil arching or redistribution of the earth pressure. This can lead to a significant underestimate of the prop loads that may be appropriate for use in design. While this is conservative for the wall design, it is not conservative for prop design. CIRIA C760 (Gaba *et al.*, 2017) recommends that a model factor of 1.15 be applied to the support reactions derived from a limit-equilibrium analysis to ensure a robust design is achieved. It is recommended that this model factor is increased to 1.3 when the wall is propped by a single prop.

64.4.7.2 Distributed prop load method

A method for deriving design actions for a temporary prop design is the distributed prop load (DPL) method. This method is outlined in CIRIA C517 (Twine and Roscoe, 1999) and is based on extensive field measurements of prop loads for flexible and stiff walls and for the range of ground conditions commonly encountered in the UK. Where the DPL method is adopted, the results should be compared with the findings of the wall analysis for the comparable temporary case. Where the prop loads differ between the two methods the designer should consider the reasons for the difference and adopt appropriate values in the design of the temporary props.

64.4.7.3 Thermal effects

As noted in Section 64.4.5, thermal effects can increase the design load effects for props and retaining walls. The effect on props due to thermal variations can be calculated as follows.

An increase or decrease in the temperature of a prop from its installation temperature will cause the prop to expand or contract according to the relationship

$$\Delta L = \alpha \times \Delta t \times L \qquad (64.9)$$

where ΔL is the change in prop length, α is the thermal coefficient of expansion for the prop material, Δt is the change in prop temperature from the installation temperature and L is the prop length.

If the prop is restricted or prevented from expanding freely, an additional load is generated in the prop. The magnitude of this addition load is

$$\Delta P_{temp} = \alpha \times \Delta t \times E \times A \times (\beta/100) \qquad (64.10)$$

where E is Young's modulus of the prop material, A is the cross-sectional area of the prop and β is the percentage *degree of restraint* of the prop.

CIRIA C760 (Gaba *et al.*, 2017) recommends values of 50% for stiff walls in stiff ground and 30% for flexible walls in stiff ground for β; however, the designer should select the appropriate value of β to suit particular project circumstances. Values other than those generally recommended above may be applied where the designer is confident that such values can be justified (e.g. on the basis of comparable experience).

64.4.8 Ground movement

Excavation in front of an embedded wall will lead to movement of the wall and the ground around the excavation. Where sensitive infrastructure or buildings are in close proximity to the wall, there is a risk of damage if movement is not limited. A check of the ground movement caused by the construction of an embedded wall is often required as part of the verification that SLS criteria have not been exceeded.

Methods for predicting ground movement are discussed in detail in Chapter 63 *Principles of retaining wall design*.

64.4.9 Design method

The preceding sections outlined how the various loads that may act on a wall can be assessed. Once these actions have been identified, it is necessary to analyse the wall to verify that the limit states identified in Section 64.4.2 will be avoided. For a soil–structure interaction analysis, an iterative approach is necessary, where an initial estimate of the wall geometry is made and stability against each failure mode is checked. The geometry of the wall may then be adjusted to optimise the solution while still ensuring stability. Such analysis and iteration is likely to require proprietary software.

This section outlines the stages in the design process for an embedded wall. These stages are illustrated by the flowchart in **Figure 64.17** and described below.

(*a*) Confirm the relevant limit states for the problem, as identified in Section 64.4.2. For SLS agree the deflection limits or other criteria that are to be adopted.

(*b*) Identify the design situations to be considered in the analysis, considering the issues described in Section 64.4.4, and make an assumption regarding the anticipated support system for the wall. Select an initial situation to analyse.

(*c*) Carry out a ULS analysis of the wall stability (DA1-2). For a soil–structure interaction analysis, an initial size for the wall must be assumed. In such cases it may often be appropriate to carry out a limit-equilibrium analysis to make an initial estimate of the required wall length.

(*d*) On the basis of (*c*), select the design length of the wall. For a soil–structure interaction analysis, step (*c*) may need to be iterated to optimise the solution. Consider whether other factors, such as global stability, vertical stability or water cut-off requirements, will be critical for the wall length.

(*e*) Undertake a further ULS analysis (DA1-1) based on the geometry from (*d*).

(*f*) Consider the load effects from (*c*) and (*e*) and determine the design load effects in the wall and the support system.

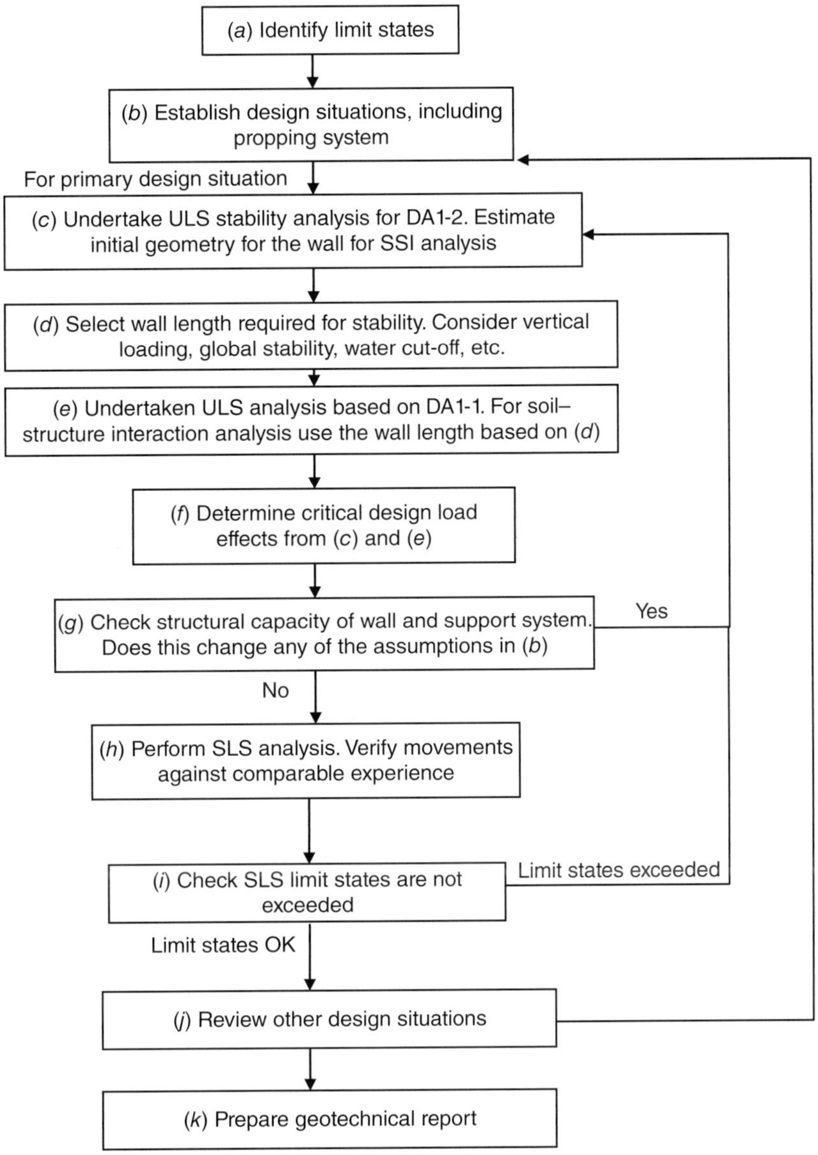

Figure 64.17 Embedded wall design process

(*g*) Undertake the structural design of the wall and support system. Consider whether this changes any of the initial assumptions (e.g. wall stiffness, support levels or support stiffness). If so, return to step (*c*) and re-analyse.

(*h*) Undertake SLS analysis to estimate wall movements. Validate against comparable experience or case history data, such as that presented in CIRIA 760 (Gaba *et al.*, 2017).

(*i*) Check SLS limits are not exceeded. If they are, return to step (*c*).

(*j*) Review other design situations.

(*k*) Prepare the geotechnical report.

References

Barton YO (1982) *Laterally Loaded Model Piles in Sand: Centrifuge Tests and Finite Element Analyses.* PhD thesis, University of Cambridge, Cambridge, UK.

BSI (British Standards Institution) (2004) BS EN 1997-1:2004: Eurocode 7: Geotechnical design – Part 1: General rules. BSI, London, UK.

BSI (2010). BS 8006:2010: Reinforced earth walls. BSI, London, UK.

BSI (2015). BS 8002:2015: Codes of practice for earth retaining structures. BSI, London, UK.

Chapman T, Taylor H and Nicholson D (2000) *Modular Gravity Retaining Walls: Design Guidance.* CIRIA Publication C516. CIRIA, London, UK.

Fleming K, Weltman A, Randolph M and Elson K (2008) *Piling Engineering*, 3rd edn. CRC Press, London, UK.

Gaba AR, Hardy S, Doughty L, Powrie W and Selemetas D (2017) *Guidance on Embedded Retaining Wall Design*. CIRIA Publication C760. CIRIA, London, UK.

Highways Agency (2000) *Design Manual for Roads and Bridges (DMRB) BD 42/00*. The Stationery Office, London, UK.

Kerisel J and Absi E (1990) *Active and Passive Earth Pressure Tables*. Balkema, Rotterdam, the Netherlands.

Twine D and Roscoe H (1999) *Temporary Propping of Deep Excavations: Guidance on Design*. CIRIA Publication C517. CIRIA, London, UK.

Williams BP and Waite D (1993) *The Design and Construction of Sheet-Piled Cofferdams*. CIRIA Special Publication 95. CIRIA in conjunction with Thomas Telford, London, UK.

Further reading

ArcelorMittal (2022) *Piling Handbook*, 9th edn. ArcelorMittal, London, UK. See http://www.arcelormittal.com/sheetpiling/page/index/name/arcelor-piling-handbook (accessed 27/03/2023).

Bond A and Harris A (2008) *Decoding Eurocode 7*. Taylor & Francis, London, UK.

Padfield CJ and Mair RJ (1991) *Design of Embedded Retaining Walls in Stiff Clay*. CIRIA Publication R104. CIRIA, London, UK.

Puller M (2003) *Deep Excavations: A Practical Manual*, 2nd edn. Thomas Telford, London, UK.

Simpson B and Driscoll R (1998) *Eurocode 7: A Commentary*. BRE, Garston, UK.

Useful websites

Highways Agency's *Design Manual for Roads and Bridges*: www.standardsforhighways.co.uk/dmrb/

It is recommended this chapter is read in conjunction with

- Chapter 20 *Earth pressure theory*
- Chapter 64 *Geotechnical design of retaining walls*
- Chapter 85 *Embedded walls*

All chapters in this book rely on the guidance in Sections 1 *Context* and 2 *Fundamental principles.* A sound knowledge of ground investigation is required for all geotechnical works, as set out in Section 4 *Site investigation.*

doi: 10.1680/icemge.66830.1083

CONTENTS

Chapter 65
Geotechnical design of retaining wall support systems

Hilary Shields Arup, London, UK
Sara Anderson Arup, London, UK

As identified in Chapter 62 *Types of retaining walls*, embedded retaining walls gain lateral support from passive resistance of the soil in front of the embedded section of the wall, while gravity walls gain lateral support from passive resistance and friction of the soil on the underside of the wall. These walls may also rely on anchorages, props or other sources of additional lateral support for stability.

Many forms of support systems are available with various structural properties and will invariably impose different constraints to working conditions within the excavation. An understanding of these different properties as well as the site-specific constraints and characteristics is required to determine the most suitable support system for a project.

This chapter provides the reader with an overview of different methods of providing lateral support to both gravity and embedded retaining walls.

65.1. Introduction

Embedded retaining walls gain lateral support from the passive resistance of the soil in front of the embedded section of the wall, while gravity walls gain lateral support from a combination of the passive resistance in front of the wall and the friction of the soil on the underside of the wall. Under many circumstances it is necessary to provide additional lateral support to the wall to restrict the lateral movement of the wall, or to provide a more economical retaining wall solution.

This chapter presents the design and performance criteria to be considered by the designer as well as the advantages and disadvantages of adopting different methods of lateral support.

Key references for retaining wall support systems include

- CIRIA C517 (Twine and Roscoe, 1999)
- CIRIA C760 (Gaba *et al.*, 2017)
- *Piling Handbook* (ArcelorMittal, 2022)
- BS 8002 (2015)
- Eurocode EC7, BS EN 1997-1:2004+A1:2013, incorporating corrigendum February 2009 (BSI, 2004).

65.2. Design requirements and performance criteria

Support to a retaining wall must be considered during all stages of the life of the retaining wall. In many cases, the temporary conditions during construction, which are dependent upon the sequence of construction adopted, can be more onerous than the permanent conditions, and therefore critical for the design of the wall.

65.2.1 Temporary situations

Temporary situations predominantly occur during construction, but can also relate to later activities such as major remedial or refurbishment works. It is common for the stability of a basement embedded retaining wall to be dominated by the temporary condition, as the excavation reaches its maximum depth – but before the lowest permanent basement slab is constructed. Similarly the construction or maintenance of drainage in front of gravity retaining walls can be a dominant condition for stability.

65.2.2 Permanent situations

In the permanent condition, retaining walls may be provided with permanent support systems – such as slabs or permanent anchors – which will need to be designed to resist the lateral loading appropriate to the full design life of the wall.

65.2.3 Bottom-up/top-down construction sequences

For basement excavations using embedded retaining walls, two different sequences of construction may be adopted, as identified in Chapter 63 *Principles of retaining wall design*.

The *bottom-up construction sequence*, in its simplest form, consists of a cantilever retaining wall supporting the ground surrounding the excavation, followed by construction of the permanent works within the cantilever excavation. However, there are many situations when additional temporary support, often in the form of temporary props or ground anchors, are utilised to supplement the earth pressures on the passive side of the wall during excavation. Following excavation, the permanent works are constructed within the excavation; the temporary support is removed once the permanent works are in place to provide sufficient lateral restraint.

Top-down construction is defined by the use of the permanent internal structure of a basement to prop the retaining wall during excavation without the reliance on temporary propping systems. To achieve this, the higher-level slabs are constructed before the excavation progresses below the slab level and before casting the lower-level slabs.

65.2.4 Control of ground movements

Ground movements result from both the installation of the retaining wall and due to excavation of the ground. Control of ground movements is important in urban settings where buildings and other third-party assets (e.g. utilities and transport infrastructure) are in close proximity to an excavation and will incur damage from ground movements. The magnitude of ground movements due to excavation in front of a retaining wall is dependent upon the stiffness of the wall and support systems. As such, the requirements for the control of ground movements may be critical in the selection and design of retaining wall support systems.

65.2.5 Accidental conditions

Accidental conditions – for example, the potential for the loss of a support element such as a prop or anchor – should be considered in the design. The loss of one support element will shed loading into adjacent support elements. If these are not designed to resist this additional loading, they may also fail, leading to progressive loss of support elements and therefore excessive deflection if not progressive collapse of the retaining wall.

65.2.6 Design responsibility

The responsibility for the design of both the retaining wall and any lateral support will not always lie with the same party, and therefore must be clearly defined in the contract documents and understood by all parties. Situations where there is likely to be more than one design party involved include the following.

- *Temporary retaining wall.* A temporary wall may be used to form an excavation, into which the permanent structure will be constructed. Performance criteria, such as the maximum wall deflection to ensure that the wall does not impinge on the permanent works, need to be clearly defined for the designer of the temporary wall. The temporary wall design will also need to consider the proposed construction sequencing to minimise constraints placed by the temporary works on the construction of the permanent works. The staging of the temporary works, for example the removal of temporary propping, can affect the permanent works. As such, the design of both the temporary and permanent works must fully account for any restrictions or reliance placed on one system by the other.
- *Permanent retaining wall with temporary support.* For basement excavations it is not uncommon to utilise the permanent embedded retaining wall with temporary propping or anchoring during excavation and construction of the basement. The design loading acting on the retaining wall during construction will be dependent on the strength, stiffness and sequence of installation and removal of the temporary support system. In this situation, the designer of the permanent wall will need to inform the designer of the temporary support system of the assumptions made regarding the strength, stiffness and staging of the temporary works in the design of the permanent works. If the temporary support designer is unable to satisfy these requirements, then the design of the permanent works, temporary works or construction staging will need to be modified to achieve a coordinated temporary and permanent works design solution.
- *Specialist lateral support.* The permanent works designer may not possess the relevant specialist design skills for some of the methods of lateral support. Alternatively, a specialist contractor may provide additional value to the project by undertaking the design of the permanent lateral support to a wall. For this situation, key design assumptions of strength, stiffness, maximum wall deflection, construction staging and design life must be clearly communicated by the permanent works designer to the specialist designer to achieve a coordinated design solution.

65.3. Types of wall support systems

Many different forms of wall support systems can be used for retaining walls. Some of the more common types of wall support systems are briefly described below.

- *Cantilever.* The retaining wall relies only on the passive ground resistance or sliding resistance of the wall and therefore has no additional wall support system.
- *Single support.* A single level of support is provided to the retaining wall. The introduction of additional support will reduce the bending moment in the wall and therefore reduce its size, compared to a cantilever wall. It will also increase the stiffness of the excavation support system and therefore reduce ground movements behind the wall. As such, support is typically adopted either where the excavation depth is too deep for an economic cantilever wall or where a cantilever wall would result in unacceptably large wall deflections and associated ground movements, or where the introduction of support results in an overall saving for the project.
- *Multiple support.* For deeper excavations, a single level of support may not be sufficient to provide the strength and stiffness required to retain the ground behind the wall with acceptable wall deflections and associated ground movements; therefore, several levels of support are required.
- *Circular construction.* Retaining walls, constructed in a circular layout, derive support from hoop action due to the shape of the wall layout and excavation. For walls which provide a solid cross-section – such as secant walls or diaphragm walls – support is likely to be provided by hoop action through the wall itself. In other situations, a continuous ring waling beam will be required to provide the hoop action.
- *Buttressed walls.* A buttress is a local thickening of a gravity retaining wall. It acts to increase the stiffness of the wall and also to improve the resistance of the wall, particularly to overturning. Buttresses are more commonly provided to masonry walls and form an architectural feature to many historical masonry buildings and walls.

- *Temporary props.* Props consist of steel sections, typically (but not always) orientated perpendicular to the line of a retaining wall. They provide lateral support by acting in compression, transmitting loading from the retaining wall on one side of the excavation onto a retaining wall on the opposite side, or onto another source of resistance.
- *Ties.* Support to a retaining wall can also be provided by tying it back to another structure using a steel rod. The ties can be connected back to an object such as a mass concrete block which relies on friction and passive resistance of the ground between the block and the wall to resist the tie force. They can also be connected back to embedded elements such as piles or other nearby embedded walls which utilise their stiffness and the passive resistance of the ground in front of the piles/embedded wall section to resist the tie loads.
- *Berms.* The use of an earth berm in front of, and immediately adjacent to, a retaining wall reduces the effective depth of excavation experienced by the retaining wall while allowing a deeper excavation within the centre of the site. Raking props from the top of the berm to a thrust block at the bottom of the excavation in the central area of the site can then be used to replace lateral support provided by the berm when excavation proceeds to full depth adjacent to the wall. Alternatively, removal of the berm down to the final excavation level and construction of the lowest slab level can be carried out in isolated sections or bays. Although there is a reduction in the lateral support provided to the retaining wall, arching in the soil and distribution of loading through the retaining wall capping beam reduces the loading and deflection that occurs on the short section or bay of wall.
- *Ground improvement.* Strengthening or stiffening of the ground, particularly within the passive zone below the excavation level, can be utilised to provide support. Ground improvement methods include the following.
  - *Soil mixing, jet grouting or other strengthening measures.* Such measures can be used to create a 'slab' of improved soil spanning between retaining walls (retaining an excavation) to provide enhanced passive restraint.
  - *Crosswalls.* These are constructed using unreinforced slurry or diaphragm wall trenches and can enhance the passive strength and stiffness of the main retaining wall. Crosswalls provide enhanced passive performance at their locations, but little improvement between crosswalls. As such, they are most effectively used with diaphragm walls where a single crosswall can be provided for each diaphragm wall panel.
- *Ground anchorages.* Anchors provide support to a retaining wall by transferring tension loading to a load-bearing stratum behind the wall through friction over a grouted length of the anchor. Ground anchors are discussed further in Chapter 66 *Geotechnical design of grouted anchors.*

65.4. Props

65.4.1 General

Temporary props are typically formed from steel sections but can also be formed from reinforced concrete. Ideally the props should be located at regular intervals along the length of the excavation using a waling beam to distribute the restraint between the individual props. Both the strength and stiffness of the propping system will be dependent on the properties of the waling beam, as well as the size and spacing of the props.

Temporary props will span across an excavation and provide a physical restriction on the working methods that can be adopted within the excavation. From a programme and ease of construction point of view, there is a desire to have propping at larger intervals to minimise the restrictions on working methods. This desire needs to be balanced not only with the strength and stiffness of the propping system, but also with the practical issues related to installing and supporting large prop and waling sections.

65.4.2 Propping layout

The layout of propping will depend on the shape of the excavation (**Figure 65.1**). For long, thin trough excavations, propping at regular intervals can be provided across the excavation. For rectangular basement excavation, propping will be required in two directions. This can be achieved using cross-bracing and diagonal propping in the corners.

Diagonal bracing can also be utilised to provide effective propping with an enlarged opening between props, which is often used as the location for mucking out of the excavation (**Figure 65.2**).

65.4.3 Inclined struts

Propping is typically installed horizontally. However, inclined struts can be used as raking props in conjunction with a berm solution, as described in Section 65.3 (**Figure 65.3**). They can also be used for unbalanced excavations where the excavation depth and loading are different on the opposite sides of an excavation.

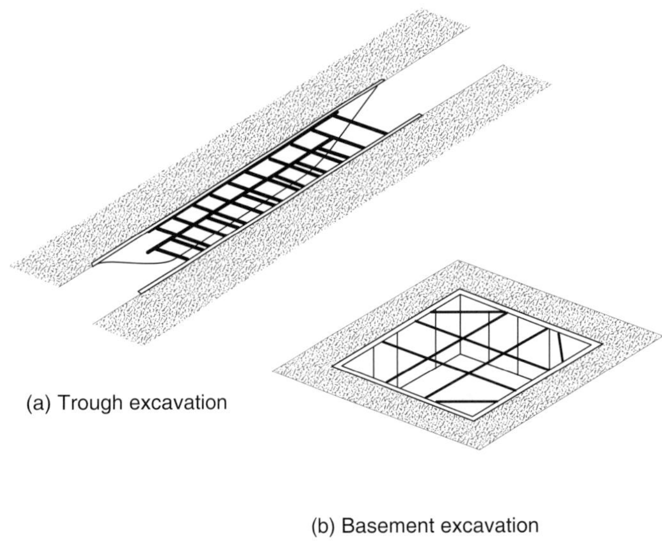

Figure 65.1 Typical excavations with temporary propping (Reproduced with permission from CIRIA C517, Twine and Roscoe, 1999; www.ciria.org)

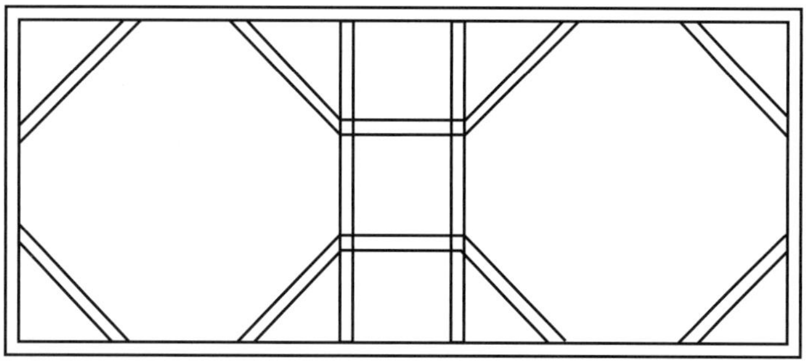

Figure 65.2 Use of diagonal bracing to provide enlarged opening for access (Reproduced with permission from CIRIA C517, Twine and Roscoe, 1999; www.ciria.org)

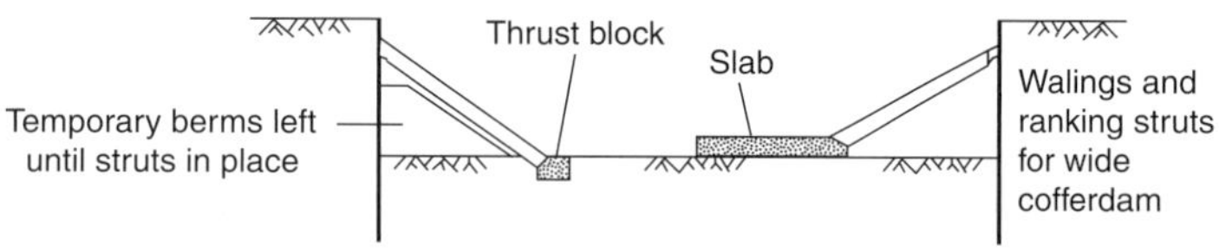

Figure 65.3 Use of inclined struts with temporary berm (Reproduced with permission from CIRIA C517, Twine and Roscoe, 1999; www.ciria.org)

65.4.4 Prop stiffness/preload/prestress

The stiffness of a propping system per metre length of wall (k) is dependent on the Young's modulus of the propping member (E), the cross-sectional area of the member (A), the effective length of the prop (L) and the spacing of the individual props (s), and is given by

$$k = A \times E/(s \times L) \tag{65.1}$$

Any inclination of the props relative to the face of the retaining wall should be resolved trigonometrically to give an equivalent stiffness perpendicular to the face of the wall.

For 'balanced' excavations, where excavation depth and loading on both ends of the prop are similar, L can be taken as half of the excavation width. For unbalanced or sway excavations, where the excavation depth and loading are much larger on one end of the prop, the length should be taken as the full excavation width.

Although this assessment of propping stiffness is commonly used, it is an approximation as it does not consider the influence of the waling beam deflection (δ) on the effective stiffness of the propping system (**Figure 65.4**). For further guidance, refer to chapter 8 of CIRIA C760 (Gaba *et al.*, 2017).

The application of preload or prestress into the propping system – typically around 10% of the design (working) load – can be used to take up the slack in the system. Higher preloading

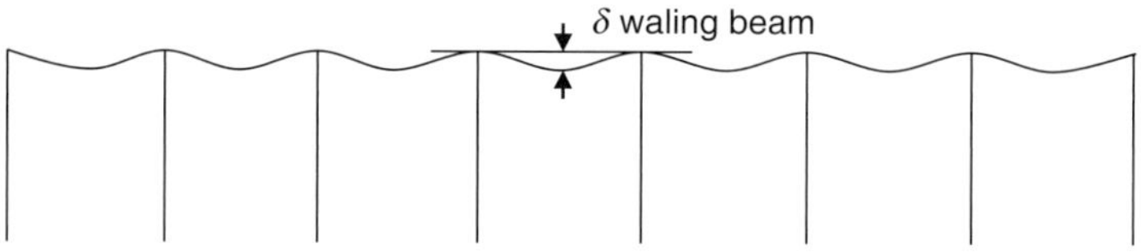

Figure 65.4 Waling beam deflection

can be used to stiffen the support system to reduce wall deflections and hence movements behind the wall.

65.4.5 System stiffness

Available case history data indicate that the magnitude of horizontal wall deflection of embedded retaining walls is dependent on the effectiveness of the support system.

Clough *et al.* (1989) define system stiffness as $EI/(\gamma_w h^4)$, where I is the second moment of area of the wall section, γ_w is the bulk unit weight of water and h is the average vertical prop spacing of a multi-propped support system. Case history data indicate that wall deflections and associated ground movements are relatively insensitive to variation in wall thickness and stiffness, provided the overall system stiffness is not significantly reduced. Because of this, CIRIA C760 (Gaba *et al.*, 2017) comments that economies in wall type and size can be achieved by adopting flexible walls (e.g. sheet pile walls) in stiff soils, without greatly increasing ground movements. Thus, in general, flexible walls with many props (smaller h) will give similar displacements to stiff walls with fewer props (larger h). However, it should be noted that the cost of additional propping should be offset against the benefit of using a more flexible wall. More extensive propping may also be counterproductive, as it may slow the rate of construction and thereby increase ground movements due to consolidation of the ground.

65.4.6 Temperature effects

The change in length of a prop due to temperature effects depends upon the thermal coefficient of expansion (α), increase or decrease in the temperature of a prop from its installation temperature (Δt) and prop length (L)

$$\Delta L = \alpha \Delta t L \tag{65.2}$$

In practice, the waling beam, wall and retained ground at the end of the props will restrict or prevent free expansion, leading to an increase in loading within the props. The effects of temperature will be more significant for steel props in comparison to reinforced concrete due to higher values of coefficient of expansion in steel. Refer to CIRIA C760 (Gaba *et al.*, 2017) for further guidance. Thermal effects will be more pronounced for props in direct sunlight and less pronounced for more shaded lower levels of propping.

65.4.7 Accidental loading

The design of individual props should be robust and include consideration of accidental loading and the possible loss of a prop. The design needs to

- incorporate the loss of a prop in the design of the wall and support systems or
- provide adequate mitigation of the risk of accidental damage or loss of a prop through design changes and robust construction management strategy.

65.4.8 Buckling

Props are slender, axially loaded elements. Therefore the design of props needs to consider buckling due to excessive or eccentric axial loading. For wider excavations or highly loaded props it may be necessary to provide mid-span restraint to the props. This can be provided using temporary king post piles within the centre of the excavation.

65.4.9 Determination of prop load for design

Refer to Chapter 64 *Geotechnical design of retaining walls* for the determination of serviceability limit state and ultimate limit state prop loads using limit-equilibrium and soil–structure interaction methods.

Designers of temporary props may also make use of the distributed prop load (DPL) method presented in CIRIA C517 (Twine and Roscoe, 1999). The DPL method for calculating prop loads is based on the back-analysis of field measurements of prop loads relating to 81 case histories, of which 60 are for flexible walls (steel sheet pile, king post walls) and 21 are for stiff walls (contiguous, secant, diaphragm walls). The case histories relate to excavations ranging in depth from 4 to 27 m in stiff to very stiff clays, and 10 to 20 m in coarse-grained soils.

Based on the case history data, a family of DPL diagrams have been determined and are presented in CIRIA C517 (Twine and Roscoe, 1999) and summarised in CIRIA C760 (Gaba *et al.*, 2017) (**Figure 65.5**). Further reference should be made to these publications for the details of the method as well as the considerations to be satisfied when using it.

65.4.10 Permanent structure propping

Permanent propping of embedded retaining walls is typically provided by permanent reinforced-concrete slabs. Although the Young's modulus of reinforced concrete is lower than that of steel, the cross-sectional area of a solid slab is significantly larger than for temporary steel propping. Therefore, a reinforced-concrete slab provides a stiffer support solution than temporary steel propping. This enhanced stiffness is one of the benefits of a top-down construction sequence to minimise wall deflections and hence ground movements.

The stiffness (k) of the slab to axial loading is dependent upon the Young's modulus of concrete (E), thickness of the slab (t) and effective length of the slab perpendicular to the wall (L)

$$k = Et/L \text{ per m width of slab} \qquad (65.3)$$

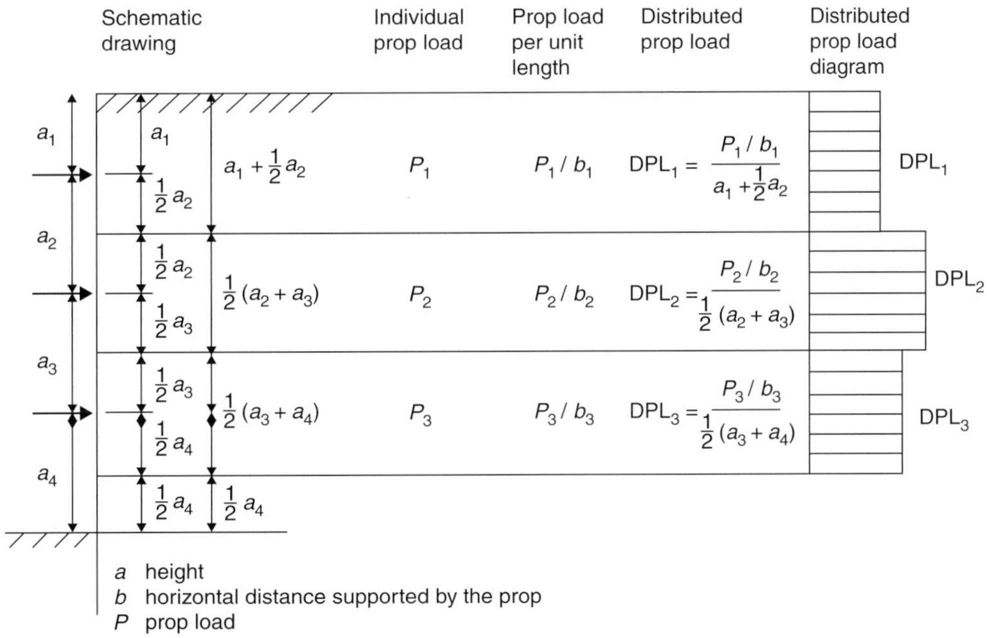

Figure 65.5 Method for calculating the DPL. In order to compute prop loads from a distributed load diagram, the procedure is reversed (Reproduced with permission from CIRIA C517, Twine and Roscoe, 1999; www.ciria.org)

As for the stiffness of temporary props, the effective length is dependent upon whether the excavation is balanced or subject to sway loading – see Section 65.4.4.

Large holes in the slab, such as for services, can locally influence the stiffness of the slab in supporting the retaining wall. The effect is most significant for holes adjacent to the retaining wall.

65.5. Tied systems

65.5.1 Ties and deadman anchors

Deadman anchors can be formed of either a plate section (e.g. sheet pile section) or a mass concrete block, and provide restraint to a wall by means of passive resistance in front of the plate section or mass concrete block (**Figure 65.6**).

Where the passive soil zone in front of the anchor does not interfere with the active soil zone behind the main wall, the design of the main wall and the anchor can be undertaken independently of each other (**Figure 65.7**).

Ties can also be used to connect or tie a wall back to embedded elements for support; for example, raking tension piles (**Figure 65.8**). Embedded elements are able to gain passive resistance over a deeper extent than the plates or mass concrete plates used in deadman anchors. For raking elements, the resistance is provided by tension within the piles.

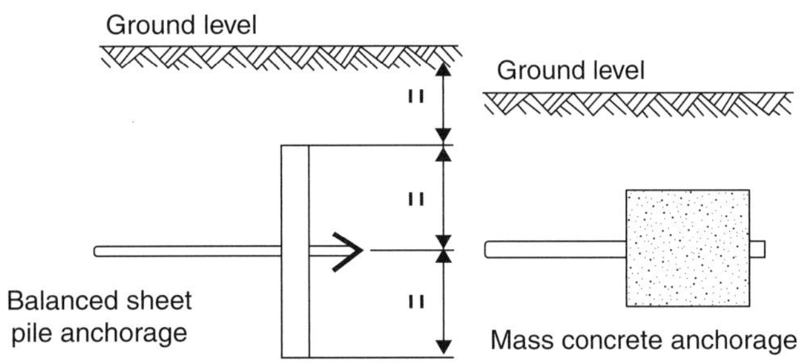

Figure 65.6 Deadman anchors (Reproduced with permission from BS 8002 © BSI (2015))

The effectiveness of a tied system will depend on the depth to which the ties and deadman anchors can be installed. This will be affected by practical considerations of excavation for the anchors. As such, tied systems are unlikely to be feasible where multiple levels of support are required; for instance, where the ground rises behind the retaining wall or where the groundwater level is high.

65.5.2 Earth-filled cofferdams

Earth-filled double-wall or cellular cofferdams are another form of tied structure. They are formed from two parallel lines of sheet pile walls connected to each other by means of waling beams and tie rods. The gap between the two lines of piling is backfilled with sand, gravel, broken rock or other coarse-grained material. Cellular cofferdams are formed by interlocking sheet pile sections to form cells which are then backfilled with coarse-grained material (**Figure 65.9**).

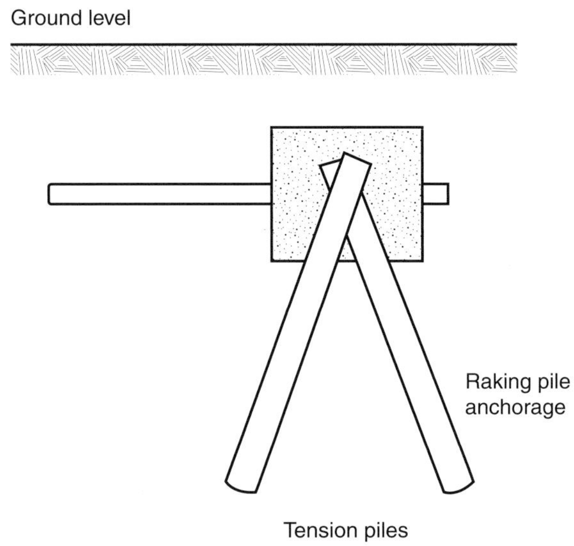

Figure 65.8 Ties and raking pile anchorage (Reproduced with permission from BS 8002 © BSI (2015))

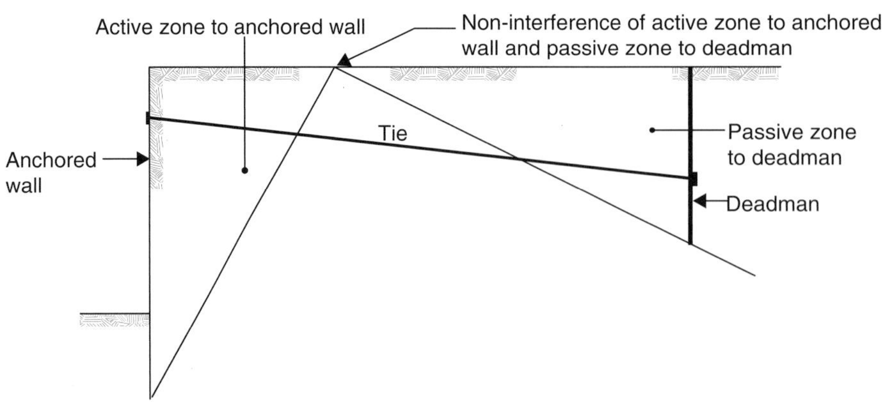

Figure 65.7 Non-interference zone for anchors (Reproduced with permission from BS 8002 © BSI (2015))

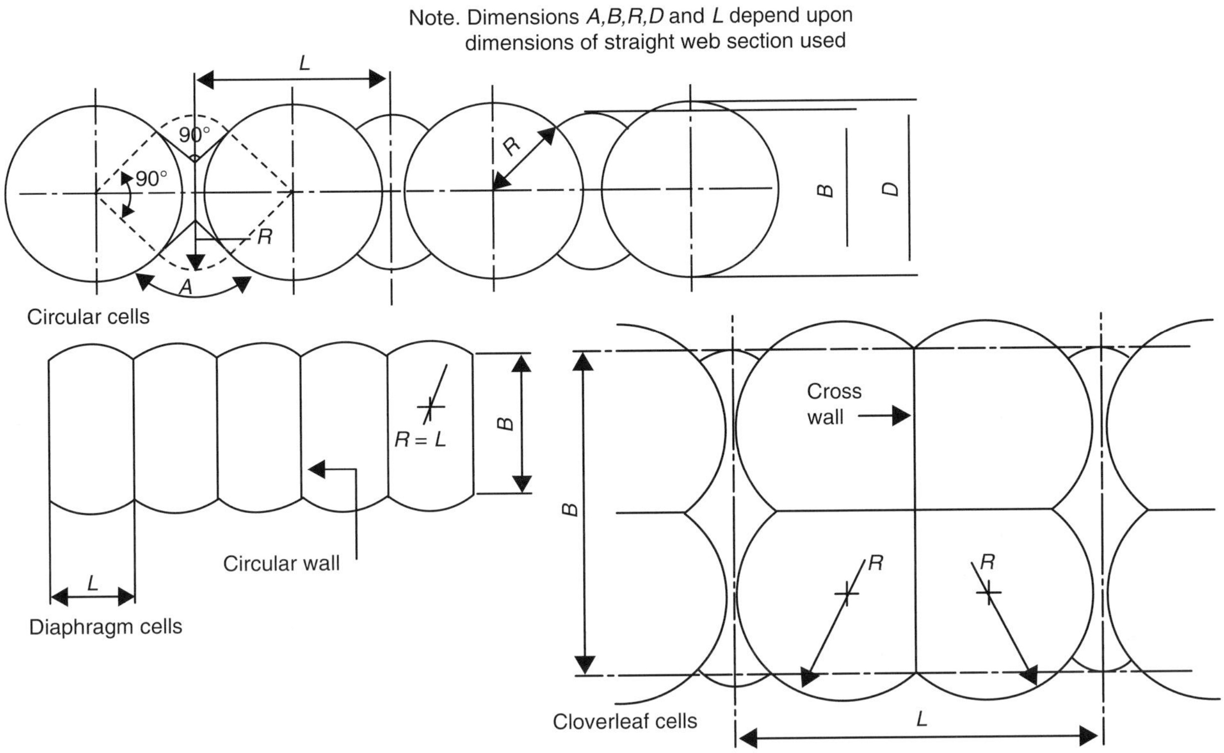

Figure 65.9 Types of cellular cofferdam (Reproduced with permission from BS 8002 © BSI (1994))

Earth-filled cofferdams are typically used in marine applications. The sheet piles are installed through the water and into the underlying ground before placing backfill between the rows or within the cells to above the surrounding water level. The cofferdams are designed to allow dewatering and excavation from within the area they enclose.

Further design consideration of tied systems including earth-filled cofferdams is given in BS 8002 (BSI, 2015), the *Piling Handbook* (ArcelorMittal, 2022) and Williams and Waite (1993); for ground anchors, guidance is provided in Chapter 66 *Geotechnical design of grouted anchors.*

65.6. Soil berms

Berms can be used to provide support to an embedded retaining wall by reducing the effective depth of the excavation.

The degree of support provided by a berm will depend on the ground conditions, the height of the berm (H), the width of the bench of the berm (B) and the angle of the berm slope (S) (**Figure 65.10**).

The slope of the berm will be governed by the soil and groundwater conditions, and for soils of low permeability will depend on the duration for which the berm is to be relied upon to support the wall. The height of the berm, bench width and also the slope of the berm will be influenced by the practical considerations of space and access. As such, the design of a berm will need to seek a balance between providing one of a suitably large size to provide meaningful support to the excavation, and yet be suitably small to minimise the restrictions on space and access within the excavation.

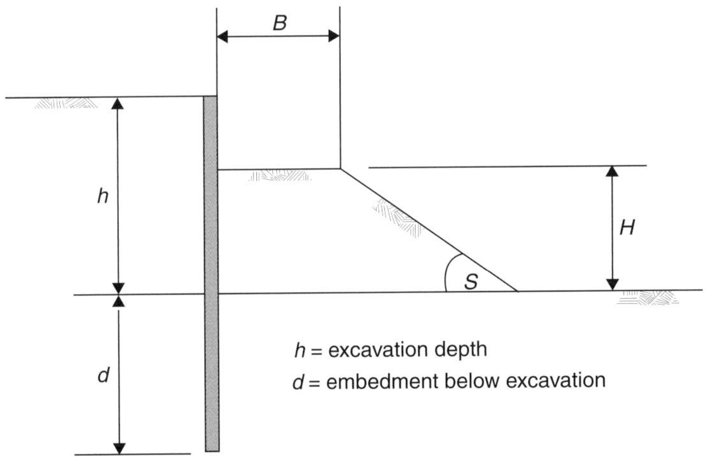

Figure 65.10 Definition of berm geometry (Reproduced with permission from CIRIA C760, Gaba *et al.*, 2017; www.ciria.org)

CIRIA C760 (Gaba *et al.*, 2017) comments that most methods of representing the effects of a berm in a limit-equilibrium or soil–structure analysis, using subgrade reaction and pseudo-finite element methods, are semi-empirical and that

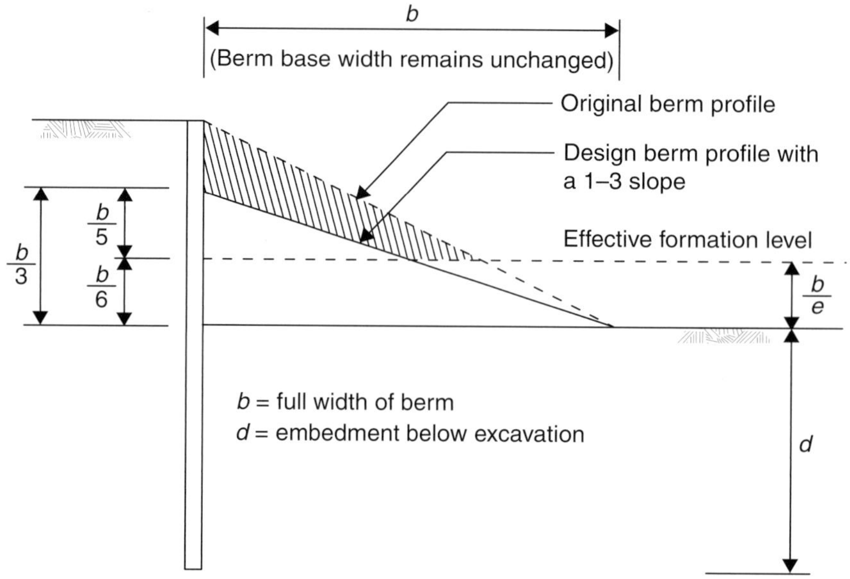

Figure 65.11 Representation of a berm by means of a raised effective formation level (After Fleming *et al.*, 2008; reproduced with permission from CIRIA C760, Gaba *et al.*, 2017; www.ciria.org)

this difficulty of analysis may explain why berms have often been used in conjunction with the observational method (Chapter 100 *Observational method*).

Several methods are commonly used to represent an earth berm in limit-equilibrium or soil–structure interaction analysis methods. CIRIA C760 (Gaba *et al.*, 2017) recommends the adoption of a raised effective formation level (after Fleming *et al.*, 2008) for routine analyses.

The raised effective formation level proposed by Fleming considers the berm as having a slope of 1:3 without any bench and with the full width of the berm. The effective formation level to be used in design analyses is determined as half of the reduced berm height, which is equivalent to one-sixth (*b*/6) of the full berm width. The benefits of berm structure beyond the assumed 1:3 berm profile and the effective formation level (shown as shaded in **Figure 65.11**) are incorporated into the analyses as a surcharge representing the weight of this additional berm structure applied at the effective formation level.

For information on other methods of modelling earth berms for consideration of wall stability or wall deflections, refer to CIRIA C760 (Gaba *et al.*, 2017: chapter 8).

With the increased use of finite element and finite difference programs in routine design, it is possible to explicitly model the geometry of a berm in the analysis and therefore not require any of the approximations outlined above that are necessary when using limit-equilibrium, subgrade reaction or pseudo-finite element methods. The general requirements of Chapter 6 *Computer analysis principles in geotechnical engineering* should be considered if this approach is adopted.

65.7. Other systems of wall support

65.7.1 Circular construction

For circular excavations where the wall is not suitable to provide sufficient support through hoop action, support is typically provided in the form of a continuous ring beam. Due to the shape, a reinforced-concrete ring beam is typically adopted.

For circular excavations, BS 8002 (BSI, 2015) identifies that earth pressures should be calculated as for straight-sided excavations; refer to Chapter 64 *Geotechnical design of retaining walls.*

In practice, the profile of the excavation and ring beam will deviate from a true circle. Williams and Waite in CIRIA Special Publication 95 (1993) explain a check for buckling in the ring beam with a radial waling load, W (kN/m), which is dependent upon the Young's modulus of the waling material, E (N/mm^2) and the moment of inertia about the x–x axis, I (cm^4), and the radius of the circular excavation, R (m), and is defined by

$$W = 1.5\frac{EI}{R^3 \times 10^5} \quad (65.4)$$

65.7.2 Buttressed walls

The design of masonry walls to resist horizontal loading is covered by BS 5628 Parts 1 and 2 (BSI, 2005a, 2005b). Buttresses act to improve the internal stability of the masonry wall structure as well as to increase the effective width of the wall footing (**Figure 65.12**). As such, they improve the

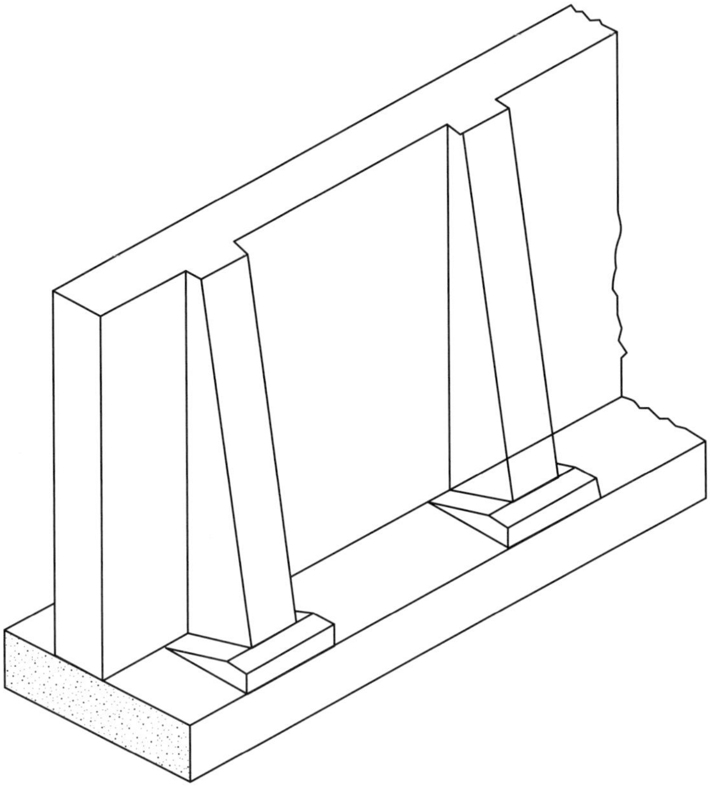

Figure 65.12 Buttressed masonry retaining wall on concrete footing (Reproduced with permission from BS 8002 © BSI (2015))

ability of the walls to transmit horizontal pressures acting on the back of the wall down to the foundation level, where sliding resistance is provided through friction on the underside of the wall, and overturning resistance is provided through eccentric bearing resistance on the underside of the wall footing.

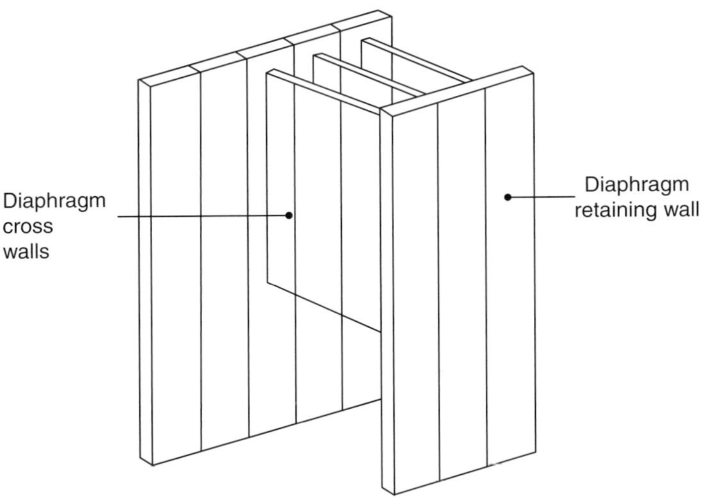

Figure 65.13 Use of crosswalls

65.7.3 Ground improvement

Ground improvement methods such as soil mixing, jet grouting and crosswalls act by increasing the lateral support to an embedded retaining wall; they are more specialised than the other techniques discussed in this chapter (**Figure 65.13**). Such methods are generally only considered for excavation within soft soils where the benefit of improving the ground before excavation can outweigh the costs and time required for such ground improvement.

Guidance on the design and implementation of deep soil mixing and jet grouting is included in

- CIRIA C573 (Mitchell and Jardine, 2002)
- the EuroSoilStab Design Guide (2002)
- BS EN 14679 (BSI, 2005a)
- BS EN 12716 (BSI, 2018).

References

ArcelorMittal (2022) *Piling Handbook*, 9th edn. ArcelorMittal, London, UK. See http://www.arcelormittal.com/sheetpiling/page/index/name/arcelor-piling-handbook (accessed 29/03/2023).

BSI (British Standards Institution) (1994) BS 8002:1994: Code of practice for earth retaining structures. BSI, London, UK.

BSI (2004, incorporating corrigendum February 2009) BS EN 1997-1:2004+A1:2013: Eurocode 7: Geotechnical design – Part 1: General rules. BSI, London, UK.

BSI (2005a) BS EN 14679:2005: Execution of special geotechnical works – Deep mixing. BSI, London, UK.

BSI (2005b) BS 5628-1:2005: Code of practice for the use of masonry: Structural use of unreinforced masonry. BSI, London, UK.

BSI (2005c) BS 5628-2:2005: Code of practice for the use of masonry: Structural use of reinforced and prestressed masonry. BSI, London, UK.

BSI (2015) BS 8002:2015: Code of practice for earth retaining structures. BSI, London, UK.

BSI (2018) BS EN 12716:2018: Execution of special geotechnical works – Jet grouting. BSI, London, UK.

Clough GW, Smith EM and Sweeney BP (1989) Movement control of excavation support systems by iterative design procedure. *ASCE Foundation Engineering: Current Principles and Practices*, vol. 1, pp. 869–884.

EuroSoilStab (2002) *Development of Design and Construction Methods to Stabilise Soft Organic Soils.* IHS BRE Press, Bracknell, UK.

Fleming WGK, Weltman AJ, Randolph MF and Elson WK (2008). *Piling Engineering*, 3rd edn. Blackie, Glasgow, UK.

Gaba A, Hardy S, Doughty L, Powrie W, and Selemetas D (2017). *Guidance on Embedded Retaining Wall Design.* CIRIA Publication C760. CIRIA, London, UK.

Mitchell JM and Jardine FM (2002). *A Guide to Ground Treatment.* CIRIA Publication C573. CIRIA, London, UK.

Twine D and Roscoe H (1999). *Temporary Propping of Deep Excavations: Guidance on Design.* CIRIA Publication C517. CIRIA, London, UK.

Williams BP and Waite D (1993). *The Design and Construction of Sheet Piled Cofferdams*. CIRIA Special Publication SP95. CIRIA, London, UK.

Further reading

Geoguide 1 (2020). *Guide to Retaining Wall Design*. Hong Kong: Hong Kong Government Geotechnical Control Office.

Useful websites

Construction Industry Research and Information Association (CIRIA): http://www.ciria.org

Hong Kong GEO Publications: http://www.cedd.gov.hk/eng/publications/manuals/geo_publications.htm

Piling Handbook: http://www.arcelormittal.com/sheetpiling/page/index/name/arcelor-piling-handbook

It is recommended this chapter is read in conjunction with

- Chapter 67 *Retaining walls as part of complete underground structure*
- Chapter 85 *Embedded walls*

All chapters in this book rely on the guidance in Sections 1 *Context* and 2 *Fundamental principles.* A sound knowledge of ground investigation is required for all geotechnical works, as set out in Section 4 *Site investigation.*

Chapter 66
Geotechnical design of grouted anchors

Devon Mothersille Geoserve Global Limited, London, UK

doi: 10.1680/icemge.66830.1093

CONTENTS

A grouted anchor is a form of ground anchor that relies on a grouted fixed length to generate resistance to applied prestressing forces. Grouted anchor technology can be employed to solve various problems in geotechnical and civil engineering. This includes anchors designed to resist sliding, overturning, dynamic loading, ground prestressing and direct tension.

Successful applications worldwide include structures subject to hydrostatic uplift such as tunnels, dock structures, stormwater tanks, tension loads imposed by suspension bridges, tension roof structures and cantilevered sports stadia. Grouted anchors can resist tension loads such as those exerted by transmission towers and wind farm turbines or destabilising forces such as sliding and overturning in concrete gravity dams. In addition, grouted anchors are commonly employed for the stabilisation of soil and rock slopes, landslides and cliffs. This chapter deals with the design of grouted anchors that provide restraint to many forms of retaining structures, including embedded retaining walls, gravity walls, deep excavation support incorporating diaphragm walls and even some forms of hybrid walls. It also covers general features concerning temporary and permanent anchors, load transfer mechanisms, tendon design and overall stability considerations.

66.1. Introduction

This chapter should be read in conjunction with Chapter 89 *Grouted anchor construction* because the design aspects of ground anchors are inextricably linked with their construction.

66.1.1 Definitions

66.1.1.1 BS EN 1997-1+A1:2013

The current geotechnical design standard for grouted anchors, BS EN 1997+A1:2013, defines an anchor as an ‘installation capable of transmitting an applied tensile load through a free length to a load-bearing stratum’. A grouted anchor is defined as an ‘anchor that uses a bonded length formed of cement grout, resin or similar material to transmit the tensile force to the ground’, where the term ‘ground’ includes both soil and rock. Grouted anchors are normally prestressed to a predetermined tensile load, which once locked-off, provides restraint to the structure. The loads are applied to the anchor by tensioning the anchor tendon at the anchor head, usually by means of a hydraulic jack. The applied load is transferred from the anchor head, along the tendon, through the debonded free tendon length and into the ground through the fixed anchor length.

The anchor head transfers the applied anchor force onto the structure, the fixed anchor length transfers the anchor force into the ground and the free anchor length spans the gap between the anchor head and the fixed anchor length.

For reference, these terms are illustrated in **Figure 66.1**.

66.1.1.2 Temporary anchors

The design standard defines a temporary anchor as one with a design life of 2 years or less and a permanent anchor as having a design life in excess of 2 years.

The current execution standard (BS EN 1537:2013) requires that for a temporary anchor the steel components of the anchor shall be provided with protection that will inhibit or prevent corrosion over a minimum design life of 2 years. The standard further requires that the exposed steel in the tendon bond length shall be provided with a minimum grout cover of 10 mm from the borehole wall, and in aggressive ground conditions it may be appropriate to enhance the protection by surrounding the tendon with a single corrugated-plastic duct.

‘BS 8081:2015+A2:2018: Code of practice for grouted anchors’ draws attention to the fact that for temporary anchors the anticipated service life of the anchor should include an allowance for any unforeseeable delay. If the service life exceeds two years, then the anchor should be classified as permanent and the appropriate protection provisions provided. Various degrees of protection may be applied, but unless there is a positive demonstration that anchors are not at risk within their service life, protection graded according to the severity of the environment should be provided. In general, for corrosion resistance, the anchor should be protected overall, as partial protection of the tendon might only induce more severe corrosion on the unprotected part. Thus, the least protected zone of a grouted anchor defines the class of protection provided.

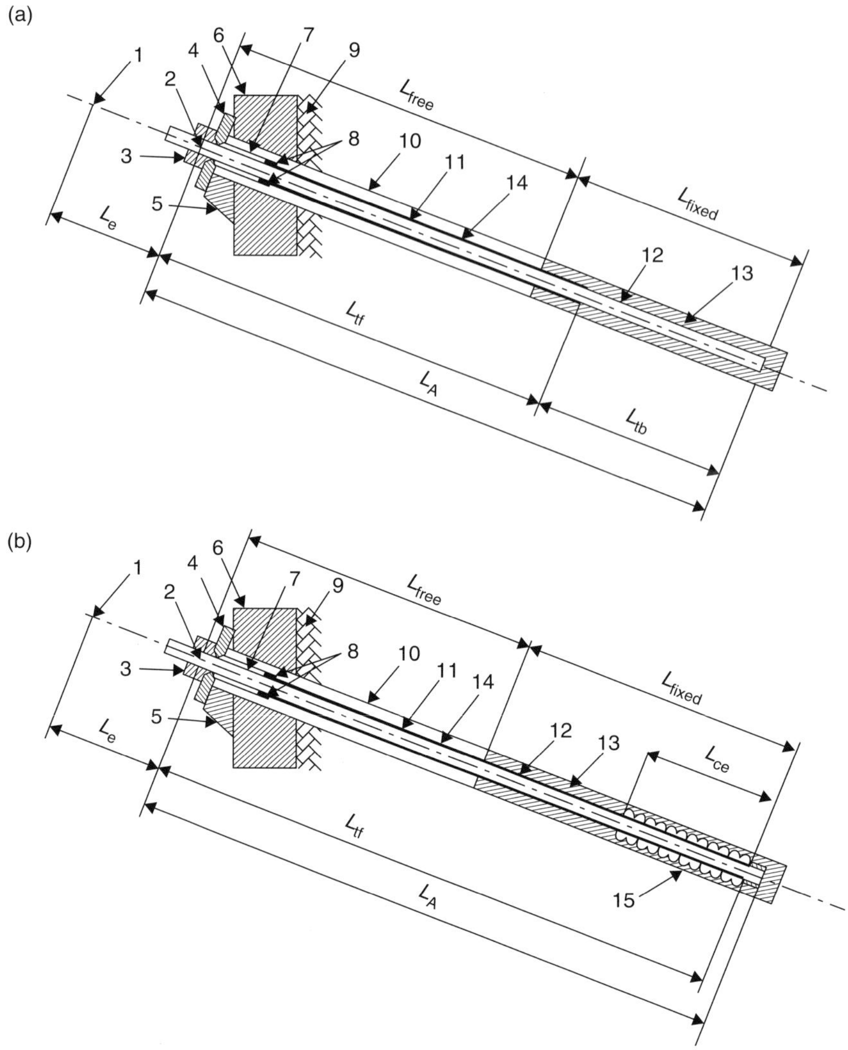

Key

1	Anchorage point at jack during stressing	9	Soil/rock
2	Anchorage point at anchor head in service	10	Borehole
3	Tensioning element at anchor head	11	Debonding sleeve
4	Bearing plate	12	Tendon
5	Load transfer block	13	Fixed-length grout body
6	Structural element	14	Free-length filling, where appropriate
7	Trumpet or anchor head tube	15	Compression element
8	O-ring		

L_{free} Free anchor length

L_{tb} Tendon bond length

L_{tf} Tendon free length

L_A Anchor length

L_{app} Apparent tendon free length

L_e External length of tendon measured from the tendon anchorage in the anchor head to the anchorage point in the stressing jack

L_{ce} Length of compression element

L_{fixed} Fixed anchor length

Figure 66.1 Typical grouted anchor nomenclature: (a) bond type; (b) compression type (Reproduced with permission from BS 8081; © BSI)

66.1.1.3 Permanent anchors

For a permanent anchor, BS EN 1537 requires that corrosion protection to the steel tendon should consist of at least a single continuous layer of corrosion protective material that will not degrade during the design life of the anchor. Acceptable systems stipulated within BS EN 1537 are

(*a*) a single protective barrier to corrosion, the integrity of which shall be proven by testing each anchor in situ (e.g. by falling head water test)

(*b*) two protective barriers to corrosion such that if one barrier is damaged during installation or loading, the second barrier remains intact.

Examples of corrosion protection that may be considered to satisfy the above principles of protection for permanent anchors are described in Annex C of BS EN 1537.

66.1.2 Nomenclature

The importance of the use of standardised nomenclature cannot be overemphasised since it avoids confusion and potentially costly misunderstandings among relevant parties. BS EN 1537 illustrates two types of grouted anchor: the bond-type anchor and the compression-type anchor, shown in **Figures 66.1(a)** and **66.1(b)**, respectively.

In UK practice, a permanent anchor will consist of an encapsulated tendon bond length (which transfers the prestress load into the fixed anchor length), a free tendon length and an anchor head. It should be noted that the fixed anchor length is not necessarily the same as either the tendon bond length or the encapsulation bond length. Correspondingly, the free anchor length is not necessarily the same as the free tendon length. This differentiation is intentional within BS EN 1537 and BS 8081, and allows flexibility of design by the designer.

In the case of a temporary anchor, the tendon bond length and the fixed anchor length are usually, but not always, the same.

66.1.3 Classification of grouted anchors

As detailed in Chapter 89 *Grouted anchor construction*, the sequence of constructing a grouted anchor is usually

- drill borehole
- install the tendon
- place grout (extracting temporary drill casing if used)
- apply prestress.

It is generally acknowledged that the performance of a grouted anchor, and particularly the magnitude of the bond mobilised at the ground–grout interface, is highly sensitive to the construction process. Most practitioners accept that drilling and, to an even greater extent, grouting techniques are the major factors in this regard. BS 8081:2015+A2:2018 proposed a general classification of cement-injected grouted anchors into four groups, based primarily upon the type and pressure of the grouting applied. BS 8081 termed these as anchor types A–D, described below and illustrated in **Figure 66.2**.

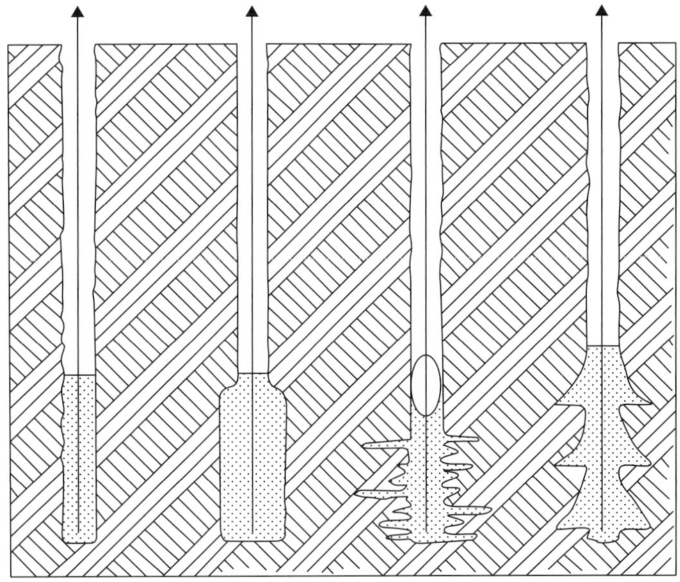

Figure 66.2 Main types of cement-injected grouted anchor

- Type A anchor: Grout is placed in the borehole under gravity (usually by tremie pipe) in straight-shafted boreholes which may be temporarily lined or unlined, depending on hole stability.
- Type B anchor: Grout is injected into the borehole under low pressure, typically less than 10 bar (1000 kN/m^2), using a temporary or permanent casing, or an in situ packer at the top of the fixed anchor.
- Type C anchor: Grout is injected into the borehole under high pressure, typically in excess of 20 bar (2000 kN/m^2) using a lining tube or in situ packer. Typically, the injection process uses a tube-à-manchette system, or similar, which allows multiple episodes of grout to be introduced if required.
- Type D anchor: Grout is placed under gravity, but the borehole is formed with a series of enlargements or under-reams.

66.2. Review of design responsibilities

BS 8081:2015+A2:2018 introduces the roles of the supported structure designer and that of the specialist anchor designer/contractor and emphasises that since grouted anchors form one part of an anchored structure the design responsibilities of the supported structure designer and specialist anchor designer/contractor should be defined. BS 8081 also includes a guide identifying an appropriate separation of design and construction duties, as shown in **Table 66.1**.

Supported structure designer[a]	Specialist anchor designer/contractor
1. Site investigation data for grouted anchors (borings at the location and depth of the fixed anchor locations which can be outside the site working area)	1. Anchor components and details
2. Decision to use grouted anchors, required trials and testing and provision of a specification	2. Determination of fixed anchor dimensions
3. Overall design of the anchored structure, and calculations of anchor force required. Definition of partial factors to be employed	3. Anchor spacing, anchor orientation and individual anchor loads if not specified by the supported structure designer (i.e. if only horizontal capping or waling beam load per metre is specified)
4. Definition of anchor life (permanent/temporary) and requirement for corrosion protection	4. Detailing the corrosion protection system for the grouted anchor
5. Anchor spacing and orientation. Waling beam capacity	5. Supply and installation of the grouted anchor system
6. Free anchor length and anchor characteristic serviceability limit state load, $F_{Serv;k}$ and design ultimate limit state load, $F_{ULS;d}$	6. Where relevant, supply and installation of the anchor load monitoring system
7. Anchor behaviour monitoring system (structure/anchors) and interpretation of results	7. Quality control of the anchor works
8. Supervision of the works (inspection by testing and sampling) structure	8. Execution and assessment of anchor tests
9. Maintenance specification for anchors	9. Evaluation of on-site anchor tests
10. Instruction of all contracting parties regarding key items within design philosophy to which special attention should be directed	10. Anchor maintenance as directed by the specialist anchor designer/contractor

[a]The supported structure designer may be employed by the client, the main contractor, a specialist subcontractor or a consultant.

Table 66.1 Recommended design and construction duties (based on BS 8081:2015+A2:2018)

From **Table 66.1**, the detailed 'design' of a grouted anchor falls within the ambit of activities 1–4 of the specialist anchor designer/contractor duties. In the context of anchors supporting a retaining wall, two important points emerge from the identified roles.

(*a*) The wall designer, who is responsible for the design of the retaining wall itself, would be expected to identify the requirement for anchors to support the structure, but also the details of individual anchor loads, their spacing and orientation, and their minimum length to ensure the overall stability of the structure (although, in practice, these design decisions may be undertaken in conjunction with the specialist anchor designer/contractor).

(*b*) The specialist anchor designer/contractor is responsible for the selection and design of the anchor type and its component parts, and is also responsible for the overall design of the grouted anchor system, taking full account of the ground conditions anticipated at the site.

It should be borne in mind that it is not unusual for the wall designer to identify the magnitude of the horizontal restraint required to be provided to the wall, and its level or levels of action, leaving the specialist anchor designer/contractor to determine the detail of individual anchor loads and their spacing and orientation. It should, however, remain the wall designer's responsibility to identify a minimum free anchor length or the extent of the zone beyond which the designed fixed anchor length must be placed to ensure the overall stability of the structure. This is on the basis that only the wall designer should be able to identify the full range, combination of forces and effects affecting the design, construction and performance of the retaining structure.

BS 8081 advocates that the overall supported structure designer activities 1–10 in **Table 66.1** are undertaken by the wall designer, albeit in possible conjunction with the specialist anchor designer/contractor in an advisory or support capacity. The specialist anchor designer/contractor in turn is responsible for specialist execution activities 1–4, where they affect 'design' matters. The assessment of site investigation data, for instance, is not limited to a 'design' issue but can also be part of the execution phase in determining the means to be adopted to install an anchor.

As examples of the above division of the duties undertaken by the various parties responsible for the design, it is not uncommon in a project containing an anchored retaining wall that the design responsibilities are divided in one of the following ways.

(*a*) A project designer employed by either the client or the principal or main contractor will identify the need for such a retaining structure within the overall project scheme. A specialist foundation engineering contractor may be employed by the principal contractor, or directly by the client, to be responsible for the design and installation of the retaining wall, which would typically be an embedded retaining wall. The foundation engineering contractor may subcontract the design and execution of the anchors to a specialist geotechnical contractor, who may also design the anchors, or employ a specialist consultant to design them on their behalf.

(*b*) Alternatively, the supported structure designer may undertake the design and specification of the retaining wall and identify the need for grouted anchors as part of the applied restraint to the wall at levels and spacings identified in the design. The construction of the retaining wall might be undertaken by a foundation engineering contractor to a specification prepared by the supported structure designer, with the anchors being designed and installed by the foundation engineering contractor, or by an anchoring specialist. The anchors must provide the loadings specified by the project designer.

It will be appreciated that there are many such possible variations of the division of responsibility for the design and execution of an anchored structure, which underscores the necessity of ensuring that those responsibilities are clearly understood and accepted by all parties.

66.3. The design of grouted anchors for the support of retaining walls

66.3.1 Introduction

As an essential precursor to design, a substantial retaining wall that requires the use of grouted anchors for support should be accompanied by an adequate ground investigation. It is often forgotten, or not recognised, that such grouted anchors may extend beyond the confines or ownership of the site, and that in such circumstances any ground investigation should be expected to extend beyond the site boundaries. In addition, questions of wayleaves or permissions to install the anchors beyond the site boundary must be addressed.

As a principle, any site investigation should ideally extend beyond and below the anticipated limits of the anchors. Where this is not possible or practicable the design involves additional risk by extrapolating geotechnical and geological data beyond known boundaries to formulate a predictive ground model.

While the requirement for adequate ground information is similar in essence to the requirements that would be expected for the ground investigation for the design of the wall itself, it is not unusual in the design process that the need for a retaining structure is identified at some time before the requirement for anchors emerges. It might be only at this later stage that the need for ground information at some distance behind the wall becomes clear, and this might have to be obtained as a second investigation phase.

The ground information should be sufficient to allow the specialist anchor designer/contractor to understand the range, sequence and properties of the strata behind the retaining wall into which, or through which, the anchors will be installed. It should allow the specialist anchor designer/contractor to illustrate this understanding with adequate and sufficient cross-sections through the wall and its attendant anchors. This appreciation of the three-dimensional geological picture, including groundwater information and the potential variation across the site, is vital to the successful design and execution of the anchors – as it is for any other geotechnical design.

It should be appreciated, however, that the scope of the ground investigation may depend upon the size and type of project, and the practicalities of gaining access for an adequate investigation. It may also depend upon a degree of local knowledge and experience of ground conditions. It is possible, and permissible, to take an informed view of the degree of risk and uncertainty implied by the relative lack of ground information when undertaking design and construction. At the least, however, this must imply that the robustness of the ground model developed by the specialist anchor designer/contractor must be adequate and flexible for the task in hand, and any potential shortcomings are sufficiently understood and capable of being addressed within the design and execution of the anchors. This recommendation is underpinned by the recognised fact that inadequate ground investigation is a hazard.

66.3.2 Matters that should be addressed at the design stage of the retaining wall

In the context of the above discussion, it is useful to review how these matters equate within the framework of the activities outlined in **Table 66.1**. With reference to the supported structure designer activities, activities 1–4 encompass both the site investigation phase and the design of the retaining wall, which will have led to the decision to use grouted anchors, including decisions on their design life.

The next phase of the design, activities 5 and 6, are to review the anticipated layout of the anchor system in relation to the structural design of the wall and the ability of the ground within any potential fixed anchor length zone to transfer the projected design loads, and the ability of the wall to accept them. An example would be the consideration of whether it was preferable to use a large number of lower-capacity anchors, or to adopt a smaller number of anchors with higher individual loads.

As a guide, for an 'average' embedded retaining wall, grouted anchor design working loads would typically be 200–600 kN, with a mean typically in the range 300–400 kN, and with horizontal and vertical anchor spacings of 2–4 m. Designs requiring anchors outside this range would often be concerned with larger, deeper retaining walls and more onerous conditions.

At this point, it is important to point out that although still commonly used, the term 'anchor working load' is not recognised in Eurocode 7. Instead, designers are required to derive the characteristic value of the maximum anchor force, $F_{Serv;k}$. BS 8081:2015+A2:2018 provides a useful explanation of the relationship between working load and $F_{Serv;k}$.

> Regarding the relationship between the 'working load of anchorage' T_w (BS 8081:1989, Clause 3) and the characteristic value of the maximum anchor force, $F_{Serv;k}$, the design and

testing of grouted anchors in the UK was traditionally based on a specified 'working load' termed T_w. BS 8081:1989, 2.21.7 defines the working load as 'the safe load (T_w) of the anchor', the magnitude of which is routinely derived from limit equilibrium stability analysis of the anchored structure.

In order to allow for load losses (a large component of which was historically due to relaxation of prestressing steel), it has been routine practice that anchors have been locked-off at 110% T_w. This took account of the fact that relaxation in prestressing steels could account for up to approximately 9% prestress loss. In situations where the lock-off load is set at 110% T_w, and no greater load is reasonably expected during the service life of the anchor, then T_w can be taken as equivalent to $F_{Serv;k}/1.1$, since, by definition, $F_{Serv;k}$ includes the effect of the lock-off load.

However, as it is currently acknowledged that modern prestressing steels account for prestress losses of less than 1.5%, a lock-off load of about 102% T_w might be more appropriate. In the context of this code of practice, it is therefore adopted that T_w approximates to the maximum characteristic anchor force $F_{Serv;k}$.

Anchor inclinations typically vary between 20 and 45° below the horizontal, although inclinations outside these limits are not uncommon. Both the wall designer and the specialist anchor designer/contractor need to bear in mind that the more steeply an anchor is inclined, the greater the vertical component of the anchor force exerted on the wall. This can be particularly relevant for steel sheet piles, with their relatively low cross-sectional area.

It is not unusual for an anchored retaining wall to have corners or bends, to suit site constraints, ownership boundaries or the like. When corners become re-entrant (i.e. at an angle of greater than 180° relative to the site), there is an increasing tendency for anchors to clash with one another. A typical situation is illustrated in **Figure 66.3**. Measures should be identified during design to ensure that in such circumstances anchors are positively designed to avoid intersecting one another, and that sufficient clearance is left between passing anchors. This is usually done by varying both the inclination and skew (horizontal angle to the wall) for some or all the affected anchors. Even relatively simple intersection problems often become sufficiently complex that they are best solved by simple modelling of the layout, either by computer or by constructing a physical model.

66.3.3 Notes on the use of BS 8081

BS 8081:1989 was, and still is, regarded by many practitioners as the most comprehensive guide to anchor design, construction and testing. In fact, even though the document has been updated, the original version is still specified in many parts of the world (e.g. South Africa and parts of the Middle East). In the UK, the adoption of Eurocodes necessitated that the original document be revised so that it is Eurocode-compatible. The revised version was published in 2015 and was subsequently subjected to two corrigenda published in 2018, leaving the current document entitled 'BS 8081:2015+A2:2018: Code of practice for grouted anchors'. The relationship between this document and the suite of related documents in Eurocode 7 is important, and its publication was based on the condition that it provides non-contradictory, complementary information in relation to the following Eurocode 7 documents

- BS EN 1997-1:2004+A1:2013: Geotechnical design – General rules (and its UK National Annexes)
- BS EN 1537:2013: Execution of special geotechnical works – Ground anchors
- BS EN 22477-5:2018: Geotechnical investigation and testing – Testing of geotechnical structures – Part 5: Testing of grouted anchors.

The retention of a revised form of BS 8081, despite the existence of the Eurocodes, was driven by the fact that it contains recommendations and guidance on best practices and the development and use of design parameters, particularly in terms of bond values for the various interfaces involved in

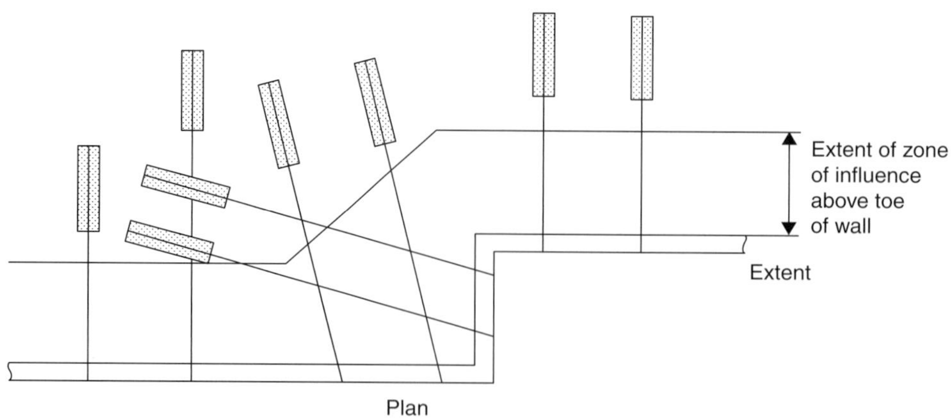

Figure 66.3 Typical intersection of anchors at re-entrant corners (Reproduced with permission from BS 8081; © BSI 1989)

grouted anchor design. It should also be understood, however, that where there is any potential conflict between any recommendations within BS 8081 and Eurocode 7, the latter would normally take precedence. It is further emphasised that the principal feature of Eurocode 7 is that its requirements are *mandatory* principles, which must be followed.

66.4. Detailed design of grouted anchors

66.4.1 Introduction

As previously explained, BS 8081:2015+A2:2018 replaced BS 8081:1989 in the UK, although the latter is still implemented in parts of the world that do not use Eurocodes and is regarded by many practitioners as the most systematic and comprehensive coverage of grouted anchors. However, the design philosophy associated with grouted anchors as individual structural elements is comprehensively covered in the 2018 revision and, importantly, has been tailored for Eurocode compatibility. Based on this, the 2018 revision will be used as the basis for the following guide on the detailed design of grouted anchors.

Four main design items should be considered by the specialist anchor designer/contractor

(*a*) the type and strength of the tendon

(*b*) load transfer into the ground

(*c*) the location of the fixed anchor zone, together with uplift or pull-out capacity

(*d*) load transfer into the structure.

Items (*b*) and (*c*) depend upon knowledge of the characteristics of the ground, as revealed by the ground investigation.

66.4.2 Anchor tendon design

Until recently, grouted anchor design in the UK was based on a global factor of safety or permissible stress approach. However, as explained in Section 66.3.3, the introduction of Eurocodes has required that engineers alter their approach to accommodate a partial factor limit state design basis. This document focuses on the design or 'sizing' of a grouted anchor as an individual structural element and does not address the derivation of forces used to support the structure.

The following values, all derived from serviceability and ultimate limit state analyses in accordance with BS EN 1997-1:2004+A1:2013, should be provided by the supported structure designer and used for the dimensioning of the fixed anchor length and sizing of the anchor tendon.

- $F_{\text{Serv;k}}$: the characteristic value of the maximum anchor force, including the effect of the lock-off load, and sufficient to prevent a serviceability limit state in the supported structure.
- $F_{\text{ULS;d}}$: the design value of the force required to prevent any ultimate limit state in the supported structure.

The ultimate limit state design resistance of a prestressing steel tendon $R_{\text{T;d}}$ should be determined in accordance with BS EN 1992-1-1 (Clause 3.3) where prestressing steel is used, and BS EN 1992-1-1 (Clause 3.2) where reinforcing steel is used (material with a characteristic yield strength typically in the range 400–600 MPa).

The following inequalities should be satisfied by calculation when establishing the capacity of a tendon

$$R_{\text{T;d}} \geq \max(F_{\text{ULS;d}}; P_{\text{p}}; \gamma_{\text{T}} F_{\text{Serv;k}}) \qquad (66.1)$$

where P_{p} is the proof load and γ_{T} is a resistance factor (see **Table 66.2**).

In the case where prestressing steel is used

$$R_{\text{T;d}} = f_{\text{pd}} \times A_{\text{T}} \qquad (66.2)$$

where f_{pd} is the design strength of prestressing steel and A_{T} is the cross-sectional area of the tendon.

In the case where reinforcing steel is used

$$R_{\text{T;d}} = f_{\text{yd}} \times A_{\text{T}} \qquad (66.3)$$

	Minimum resistance factor			
Anchorage category	Ground–grout interface,[c,d] γ_{GG}	Ground–encapsulation interface,[d] γ_{GE}	Grout–tendon interface,[d] γ_{GT}	Tendon, γ_{T}
Permanent anchors and temporary anchors	3.0[b]	3.0[a]	3.0	1.5

[a] Minimum value of 2.0 may be used if investigation tests are undertaken but not less than $F_{\text{ULS;d}}/F_{\text{Serv;k}}$. This value should be increased in cases where $F_{\text{ULS;d}}/F_{\text{Serv;k}} > 2$.
[b] May need to be raised to 4.0 to limit ground creep.
[c] Minimum resistance factors for the ground–grout interface generally lie between 2.5 and 4. However, these may vary, where investigation tests provide sufficient additional information to permit a reduction.
[d] The resistance factors applied to the ground–grout interface will invariably be higher compared with the grout/tendon or grout encapsulation interface values, the additional magnitude representing a margin of uncertainty in the magnitude of the ultimate resistance of this interface. The magnitudes of the minimum resistance factors are recommended where previous experience provides insufficient confidence that the anchor will ultimately satisfy the test criteria in BS EN 1997-1:2004+A1:2013.

Table 66.2 Minimum resistance factors recommended for calculation of the size of individual fixed anchor lengths (Reproduced from BS 8081; © BSI 2018)

Where prestressing steel is used

$$f_{pd} = f_{p0.1k}/\gamma_T \tag{66.4}$$

and where reinforcing steel is used

$$f_{yd} = f_{yk}/\gamma_s \tag{66.5}$$

'BS EN 1992-1-1:2004: Eurocode 2: Design of concrete structures – Part 1' defines the following

- $f_{p0.1k}$: characteristic 0.1% proof-stress of stressing steel
- f_{yk}: characteristic yield strength of reinforcement
- f_{yd}: design yield strength of reinforcement
- γ_s: partial factor for reinforcing or prestressing steel.

66.4.3 Load transfer to the ground

The load transfer between the grouted anchor tendon and the soil or rock within which it is embedded depends upon the design of three boundaries

- anchor grout to ground (fixed anchor length)
- encapsulation to anchor grout, if appropriate (encapsulation length)
- tendon to anchor grout or encapsulation grout (tendon bond length).

The fixed anchor length should be designed to provide resistance at load levels $F_{Serv;k}$ and $F_{ULS;d}$ in accordance with the acceptance criteria set out in BS EN 1997-1:2004+A1:2013 and NA+A1:2014 to BS EN 1997-1:2004+A1:2013.

66.4.3.1 Guidance on fixed anchor length design

Fixed anchor length design generally focuses on three convenient 'groups' of anchoring material: coarse soils, fine soils and rocks. It should be realised that there can be some blurring of the boundaries between these idealised groups, as intact rocks weather towards 'soils', for instance.

As a rule (e.g. BS 8081) the fixed anchor length should:

(*a*) Not exceed 10 m long (because of a trend of decreasing capacity per metre run, as discerned from back-analyses of load testing). However, multi-stage anchors are able to employ longer load transfer lengths by means of 'stacking' fixed anchor lengths (see Section 66.4.3.3).

(*b*) Not less than 3.0 m; although for design characteristic serviceability limit state loads below 200 kN, a minimum fixed anchor length of 2 m may be acceptable in suitable circumstances.

66.4.3.2 Design of fixed anchor length by consideration of ground–grout interface resistance

The magnitude of the ultimate ground–grout interface resistance in a grouted anchor is highly sensitive to the construction process. This limits the ability of the specialist anchor designer/contractor to make accurate predictions solely based on empirical rules. Consequently, the calculated values are not used exclusively in every case since they often merely provide the experienced designer with a means to compare values. BS 8081 points out that in anchor technology, practical knowledge is just as essential to a good design as the ability to make calculations. Based on the work of Littlejohn and Bruce (1977) and others, BS 8081 presents rock–grout bond values that have been employed in practice, those that have been recommended for design and those derived from anchor tests in rock. It is important to emphasise, however, that the values presented are indicative and represent values derived from specific site and construction conditions. The specialist anchor designer/contractor is advised, therefore, to consider the need to supplement or validate the values by site-specific investigation tests.

To establish the dimensions of the fixed anchor diameter and length, D and L_{fixed}, the following procedure should be adopted. The characteristic ultimate ground–grout interface resistance, $R_{GG;k}$, should be determined by

(*a*) empirical or analytical calculation methods, the validity of which have been demonstrated by load tests in comparable situations, or by previous experience, and/or

(*b*) the results of investigation tests, which have been demonstrated, by means of calculations or otherwise, to be consistent with other relevant experience.

It is normal practice to designate the diameter prior to the derivation of the fixed anchor length based on experience, the feasibility of construction in the site conditions and the required number of strands/encapsulations.

Where the size of the fixed anchor length is determined by calculation, the calculated ultimate ground–grout interface resistance, $R_{GG;calc}$, should be found using characteristic values of geotechnical parameters in accordance with BS EN 1997-1:2004+A1:2013.

Where the size of the fixed anchor length is determined by investigation tests, the measured ultimate ground–grout interface resistance, $R_{GG;m}$, should be derived from the investigation tests.

The characteristic ultimate ground–grout interface resistance, $R_{GG;k}$, should be derived from

$$R_{GG;k} = \min(R_{GG;calc}; R_{GG;m}) \tag{66.6}$$

The following inequalities should be satisfied when establishing the fixed anchor length by calculation or by results of investigation tests

$$F_{Serv;k} \leq R_{GG;k}/\gamma_{GG} \tag{66.7}$$

and

$$F_{ULS;d} \leq R_{GG;k} \tag{66.8}$$

66.4.3.3 Incorporating an efficiency factor when determining $R_{GG;calc}$

It has been acknowledged by numerous researchers for over 70 years that when a tensile load is applied to a steel tendon in the grout, whether founded in rock or soil, the load distribution within the fixed anchor length is non-uniform due to the action of progressive debonding. Progressive debonding can occur at either or both grout–tendon and grout–ground interfaces because of the different elastic properties of the tendon, the grout and the surrounding ground.

This mechanism, which influences the load transfer, inhibits the development of uniform stress distribution on these interfaces. Field tests have shown that the magnitude of the mobilised and measured ultimate ground–grout interface resistance ($R_{GG;m}$) does not increase linearly with an increase in fixed anchor length and is less than that assuming a uniform stress distribution along the fixed anchor length (Barley, 1997).

In order to accommodate the effect of progressive debonding, a factor f_{eff} is applied in the evaluation of the calculated ultimate ground–grout interface resistance ($R_{GG;calc}$), where it is assumed that the stress distribution is uniform.

In clays (silty clays, sandy clays, boulder clay and glacial till)

$$f_{eff} = 1.6L_{fixed}^{-0.57} \quad (66.9)$$

and in fine sands

$$f_{eff} = 1.91L_{fixed}^{\tan\varphi} \quad (66.10)$$

where L_{fixed} is the fixed anchor length in metres.

66.4.3.4 Design of anchors in coarse soil

Type A Anchors

Type A anchors are not used generally in coarse soils, except where they are cemented. Where the coarse soil is cemented, the soil is treated as a weak rock, and the ground–grout interface design is undertaken accordingly.

Type B Anchors

For low-pressure grouted anchors of type B, design equations for the estimation of the calculated ultimate ground–grout interface resistance of the anchor, $R_{GG;calc}$, are based primarily on piling design technology (Bassett, 1970; Littlejohn, 1970; Lundahl and Adding, 1966; Oosterbaan and Gifford, 1972; Robinson, 1969).

Equation (66.11) is more generally used, where the calculated ultimate ground–grout interface resistance of the anchor is related to anchor dimensions, soil properties and local soil overburden stresses (Littlejohn, 1970, 1979)

$$R_{GG;calc} = f_{eff}A\sigma'_v\pi DL_{fixed}\varphi' + B\gamma h\frac{\pi}{4}(D^2 - d^2)$$
$$= (\text{side shear}) + (\text{end-bearing}) \quad (66.11)$$

where

- A is the ratio of contact pressure at the fixed anchor–soil interface to the average effective overburden pressure
- σ'_v is the average effective overburden pressure adjacent to the fixed anchor equivalent to $\gamma[h + (L_{fixed}/2)]$ for a vertical anchor (in kPa) (Littlejohn, 1970)
- B is the bearing capacity factor equivalent to $N_q/1.4$
- N_q is the bearing capacity factor (Berezantzev *et al.*, 1961)
- f_{eff} is the resistance efficiency factor
- D is the diameter of the fixed anchor (in metres)
- d is the diameter of the grout shaft above the fixed anchor (in metres).

In very dense, cemented, fine sands, the behaviour of the soil might be like a weak rock, and it might be appropriate to establish a ground–grout interface resistance value for use in Equation 66.22.

In Equation 66.11, the value of B depends on the angle of shearing resistance of the soil adjacent to the top of the fixed anchor and the slenderness ratio (h/D).

Table 66.3 shows the variation in N_q with the variation in both slenderness ratio, h/D, and ϕ'. A complementary study (Trofimenkov and Mariupolskii, 1965) has also indicated that N_q/B equals 1.4, and this combined information is used to estimate B.

For compact sandy gravel ($\phi' = 40°$) at Vauxhall Bridge, London, and compact dune sand ($\phi' = 35°$) at Ardeer, Scotland, values of B equal to 101 and 31, respectively, have been measured in the field (Littlejohn, 1970) and are in good agreement with the respective values of 99 and 35 estimated from Berezantzev *et al.* (1961) (**Figure 66.4**).

The value of A is influenced by the construction technique for the type B anchor described by Equation 66.11. Values of 1.7 and 1.4 have been recorded in compact sandy gravel ($\phi' = 40°$) and compact dune sand ($\phi' = 35°$), respectively (Littlejohn, 1970).

Littlejohn (1970) estimated that the values of A, applied in Equation 66.11, lay in the range 1–2, but that, if the soil was not compacted or displaced during the casing installation and no residual grout pressure was left at the fixed anchor–soil–grout interface on completion of the injection stage, the value

Slenderness ratio	Bearing capacity Value of ϕ'	Factor N_q			
h/D	26°	30°	34°	37°	40°
15	11	20	43	75	143
20	9	19	41	74	140
25	8	18	40	73	139

Table 66.3 Approximate relationship between bearing capacity factor N_q and slenderness ratio

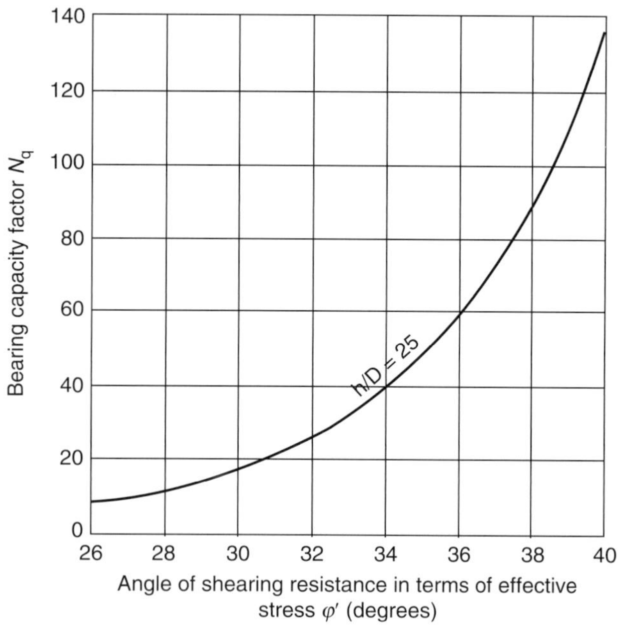

Figure 66.4 Relationship between bearing capacity factor N_q and angle of shearing resistance in terms of effective shear stress (After Berezantzev *et al.*, 1961)

of *A* might reduce to a value approximating to K_0. In the light of experience, this reduction is now considered unduly pessimistic since, even with tremie grouting, the full hydrostatic head of the grout is applied at the fixed anchor–grout interface, which creates a contact pressure greater than $K_0\sigma_v'$ in normally consolidated ground. Therefore, even for the tremie grouting method, it is difficult to envisage a value of *A* less than unity for design purposes.

An accurate assessment of the fixed anchor diameter (*D*) in Equation 66.11 is not generally possible, but estimates can be made from grout taken in conjunction with ground porosity. For boreholes of 0.1–0.15 m, *D* values of 0.4–0.5 m can be attained in coarse sands and gravels, say 3–4 *d*. Where grout permeation is not possible and only local compaction is achieved, for the above borehole diameters and an applied pressure of up to 1000 kN/m^2, effective values of *D* might range from 0.2 to 0.25 m for medium dense sand (1.5–2 *d*). For very dense sand (1.2–1.5 *d*), *D* values of 0.18–0.2 m have been attained (Littlejohn, 1979).

An alternative approach to Equation 66.11 is to neglect the end-bearing component and rely on side shear (Robinson, 1969) to produce Equation 66.12

$$R_{\text{GG;calc}} = f_{\text{eff}} K \pi D L_{\text{fixed}} \sigma_v' \tan \phi' \qquad (66.12)$$

where: *K* is a coefficient of earth pressure.

For medium to dense sandy gravel, values of *K* ranging from 1.4 to 2.3 have been adopted where no grout injection pressure is used. For fine sands and silts, values of *K* of 1.0 and 0.5 have been adopted for high and low relative densities, respectively, while for dense sands a value of 1.4 has been obtained.

Equation 66.12 can be modified to take account of the grout injection pressure, which can lock in a residual pressure in the ground, to give Equation 66.13

$$R_{\text{GG;calc}} = f_{\text{eff}} p_i \pi D L_{\text{fixed}} \tan \phi' \qquad (66.13)$$

where p_i is the grout injection pressure (but see also Barley and Windsor, 2000); the residual pressure might range from $p_i/3$ to $2p_i/3$.

$R_{\text{GG;calc}}$, in kN, may be estimated from Equation 66.14

$$R_{\text{GG;calc}} = L_{\text{fix}} n \tan \phi' \qquad (66.14)$$

It should be noted that *n* is valid in the grouting pressure range of 30–1000 kPa and f_{eff} is not applied to this equation as the evaluation of $R_{\text{GG;calc}}$ is based on field data.

Field experience (Littlejohn, 1970) indicates that for coarse sands and gravels (permeability of the ground $k_w > 10^{-4}$ m/s), *n* ranges from 400 to 600 kN/m, while in fine to medium sands ($k_w = 10^{-4}$ to 10^{-6} m/s) *n* reduces to 130 to 165 kN/m. These figures were initially measured in normally consolidated materials for borehole anchor diameters of approximately 0.1 m. Where the enlargement diameter *D* varies significantly, *n* is modified in the same proportion.

Equation (66.14) is simple but crude and is used mainly by specialist contractors familiar with their anchor system. The rule tends to be conservative in view of the limited use of information concerning fixed anchor dimensions and ground parameters, and an underestimate can be significant if the rule is applied to overconsolidated alluvium where the *n* values (see Equation 66.14) were initially established in normally consolidated materials at a density lower than the critical value for dilation.

Type C Anchors

It is a feature of type C anchors that calculations are based on design curves created from field experience in a range of soils, rather than relying on a theoretical or empirical equation using the mechanical properties of a particular soil. In alluvium, for example, test results (Jorge, 1969) have indicated that for 0.1–0.15 m dia. boreholes, calculated ground–grout interface resistances ($R_{\text{GG;calc}}$) of 90–130 kN/m of fixed anchor length at a grouting pressure p_i of 1000 kN/m^2, and 190–240 kN/m at a p_i of 2500 kN/m^2, can be expected, where p_i is the grout injection pressure.

For sandy gravels and gravelly sands, **Figure 66.5** shows that the ultimate load-holding capacity increases with density and the uniformity coefficient.

The effect of high-pressure grouting in medium loose and dense sands has been demonstrated by Egan and Merrifield (1997) by centrifuge modelling. In medium loose sands the measured ultimate ground–grout interface resistance was enhanced by 180%, while in medium dense sands an increase

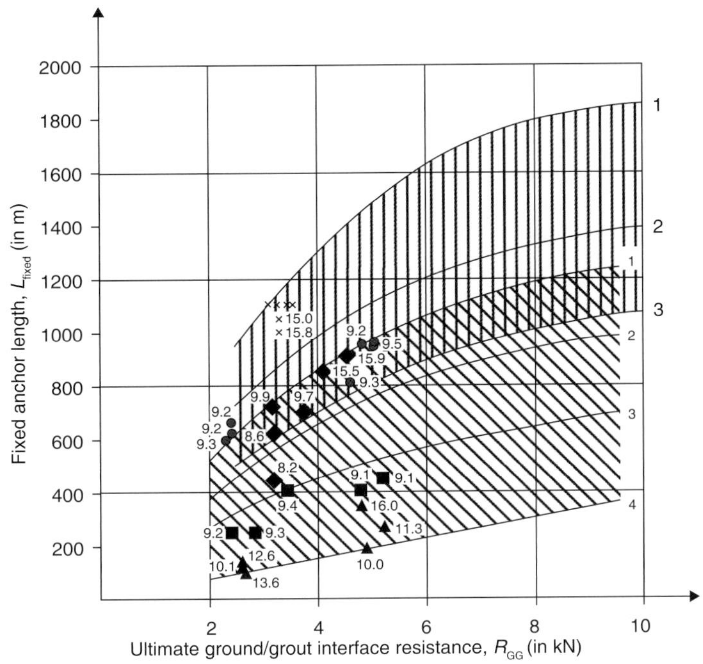

	Type of soil	Density	SPT (*N* = Number of blows per 0.3 m)
●	Gravelly sand $U = 10$ ($D_{60} = 1.6$; $D_{10} = 0.16$)	Very dense	$N = 120$
◆	Gravelly sand $U = 10$ ($D_{60} = 1.6$; $D_{10} = 0.16$)	Dense	$N = 60$
■	Gravelly sand $U = 10$ ($D_{60} = 1.6$; $D_{10} = 0.16$)	Medium dense	$N = 43$
▲	Gravelly sand $U = 10$ ($D_{60} = 1.6$; $D_{10} = 0.16$)	Loose	$N = 11$
×	Sandy gravel $U = 5$ ($D_{60} = 1.5$; $D_{10} = 0.3$)	Very dense	$N > 130$

where:

U is the uniformity coefficient (D_{60}/D_{10})

D_{60} is the maximum particle size of the smallest 60%

D_{10} is the maximum particle size of the smallest 10%

Sandy gravel
$U = 5$ to 10

Gravelly sand
$U = 8$ to 10
and
Medium to coarse sand (with gravel)
$U = 3.5$ to 4.5

NOTE 1 Field evidence is limited to a fixed anchor range of 2 m.

NOTE 2 Diameter of grouted bodies in the range $D = 0.1$ m to 0.15 m

NOTE 3 The relationships between soil density and SPT values are not in accordance with BS 5930

Figure 66.5 Ultimate load-holding capacity of anchors in sandy gravels and gravelly sands, showing influence of soil type, density and fixed anchor length for type C anchors (After Ostermayer and Scheele, 1978)

of 200% was observed. The failure mechanism associated with high-pressure grouting was found to be brittle, resulting in a decrease in anchor pull-out capacity with further increment of displacement. An increase in radial or normal stress at the ground–grout interface to values of 2–10 times the effective overburden pressure has also been noted (Wernick, 1978).

In practice, density is commonly measured indirectly by in situ penetrometer tests. **Figure 66.6** illustrates how penetration resistance can be used to provide a rough estimation of ultimate load-holding capacity for 3, 6 and 9 m fixed anchor lengths. It is emphasised, however, that certain fluctuations in test results are possible due to soil inhomogeneity, even when anchors have been properly installed.

Bustamante and Doix (1985) published empirical correlations of ultimate skin friction in coarse and fine soils as a function of both standard penetration testing (SPT) and pressuremeter test values. These relate the ultimate skin friction, τ_{ult}, to the pressure limit p_L, obtained from the pressuremeter test, and the SPT N-value.

Type D Anchors

If the enlargements are required to take all the load in the fixed anchor length, then shearing across the interface of the nominal borehole diameter and the enlargement is an important consideration. This is in respect of the number of enlargements that are required and the subsequent effect this has on the transfer of the load into the fixed anchor length (Barley, 1978).

Distribution of Skin Friction in Coarse Soils

Designs are normally based on the assumption of an equivalent, uniform skin friction. Field values of the distribution of skin friction (Oosterbaan and Gifford, 1972; Ostermayer and Scheele, 1978) are rare and have been derived from the measured distribution of load in the tendon and estimation of the geometry of the anchor fixed length enlargements.

The strong influence of relative density, particle size distribution and grouting pressure on the shear resistance along the fixed anchor has also been highlighted by Petrasovits (1981).

Limit of Fixed Anchor Length in Coarse Soils

Taken to the point of failure, there exists a limit to the effective fixed anchor length beyond which there is no evident increase in capacity. For dense coarse soil (SPT N = 50), Fujita *et al.* (1978) indicate very small capacity increases for *L* greater than 6.7 m, which supports Ostermayer (1974), who concluded that 6–7 m was optimal from an economic point of view. It is noteworthy that the fixed anchor length is not normally less than 2 m.

66.4.3.5 Anchors in fine soil

General

The application of low grouting pressure and the penetration of drill casing into the soil may be beneficial to the capacity of the anchor when compared to the case when no grout pressure is applied. This is particularly true in fine soil strata interbedded with weak mudstones or siltstones. Even without the presence of hydrofracture, the fissure penetration of the grout can enhance the skin friction values. Adhesion factors in excess of 0.45 can be mobilised and can be verified by investigation tests.

Type A Anchors

For tremie- (or gravity displacement-) grouted, straight shaft anchors of type A, the design rules are similar to

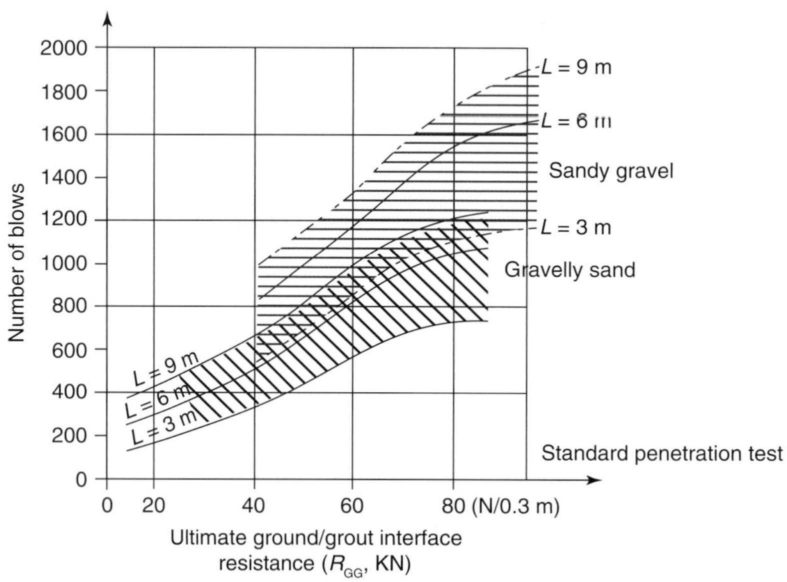

Figure 66.6 Relationship between ultimate load-holding capacity, fixed anchor length and dynamic penetration for two types of coarse soil (After Ostermayer and Scheele, 1978)

those developed for bored piles (Littlejohn, 1970; Neely and Montague-Jones, 1974; Sapio, 1975), and are based on the use of the undrained shear strength of the soil.

For guidance, the calculated geotechnical ultimate ground–grout interface resistance of the anchor, $R_{GG;calc}$ in kN, may be estimated from Equation 66.15

$$R_{GG;calc} = f_{eff}\pi DL_{fixed}c_u \tag{66.15}$$

where

- c_u is the average undrained shear strength of the ground over the fixed anchor length (in kPa)
- D is the diameter of the fixed anchor length (borehole diameter; in metres)
- L_{fixed} is the fixed anchor length (in metres).

The actions of drilling and grouting cause stress changes within the ground that cannot be accurately modelled by either an effective stress analysis or a total stress analysis. It has been shown that an effective stress analysis as in Equation 66.11 results in a calculated ultimate ground–grout interface resistance, $R_{GG;calc}$, higher than that derived from a total stress analysis. Furthermore, a total stress analysis, considering the efficiency factor, has been shown to yield results more closely resembling ultimate ground–grout interface resistances measured from tests. Therefore, total stress analysis is considered more appropriate to derive the ultimate ground–grout interface resistance.

Type C Anchors

The load-carrying capacity of anchors in suitable soils or weak rocks can also be enhanced by under-reaming or by post-grouting techniques. In the latter technique, additional grout is injected into the fixed anchor length at high pressure after the initial anchor grout has set.

In addition, higher anchor loads can often be developed using 'multi-stage' anchor tendons (also known by the proprietary name of Single Bore Multiple Anchors), a technique whereby a number of unit anchors are placed within the same borehole. Each unit anchor is designed to transfer load over a separate section of the anchor hole. Hence, with such anchors, the free tendon length of each strand is different, as illustrated in **Figure 66.7**. Mothersille *et al.* (2014) describe the application of the technology when used with a diaphragm wall to support a 25 m deep excavation in alluvial deposits.

For further details on such techniques and systems, see Chapter 89 *Grouted anchor construction*.

Where high grout pressures can be safely permitted, type C anchors, with or without post-grouting, are often used.

The results of many fundamental tests are shown in **Figure 66.8**, which can be used as a design guide for borehole diameters of 0.08–0.16 m. Skin friction (τ_M) increases with an increasing consistency index and decreasing plasticity.

The consistency index I_c (shown in **Figure 66.8**) is given by $I_c = (w_L - w)/(w_L - w_p)$. In stiff clays (i.e. a consistency index, $I_c = 0.8$–1.0) with medium to high plasticity, skin frictions of 30–80 kPa are the lowest recorded, while the highest values ($\tau_M > 400$ kPa) are obtained in sandy silts of medium plasticity and very stiff to hard consistency ($I_c = 1.25$). The technique of post-grouting is also shown to generally increase the skin friction of very stiff clays by some 25–50%, although greater improvements (from 120 kPa up to about 300 kPa) are claimed for stiff clay of medium to high plasticity. In **Figure 66.9**, the influence of post-grouting pressure on skin friction is quantified, showing a linear increase in τ_M with an increase in post-grouting pressure, to a maximum post-grouting pressure of approximately 3000 kPa.

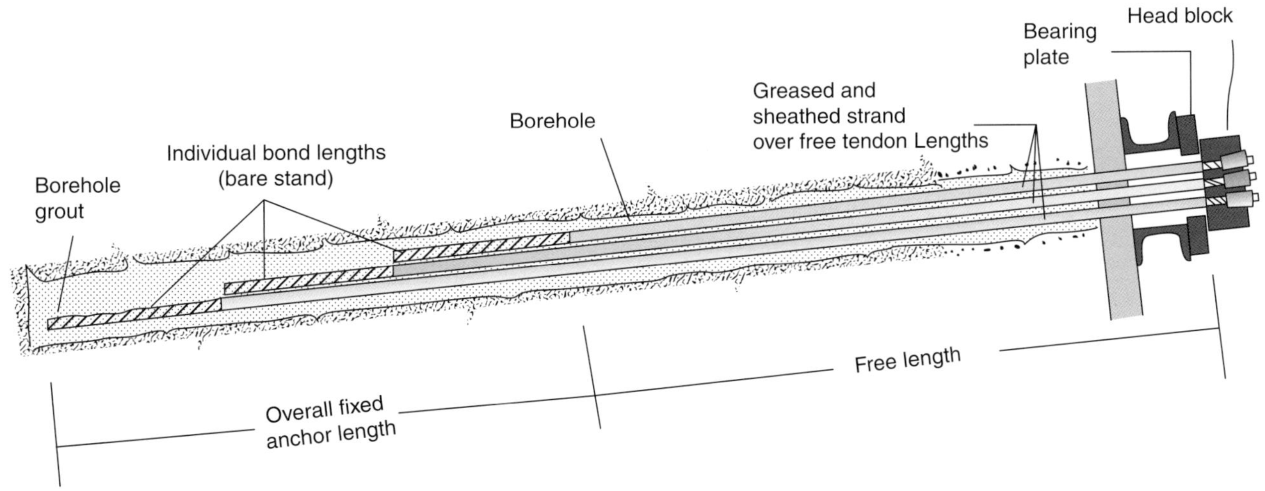

Figure 66.7 Characteristic features of a multi-stage ground anchor (temporary anchor shown)

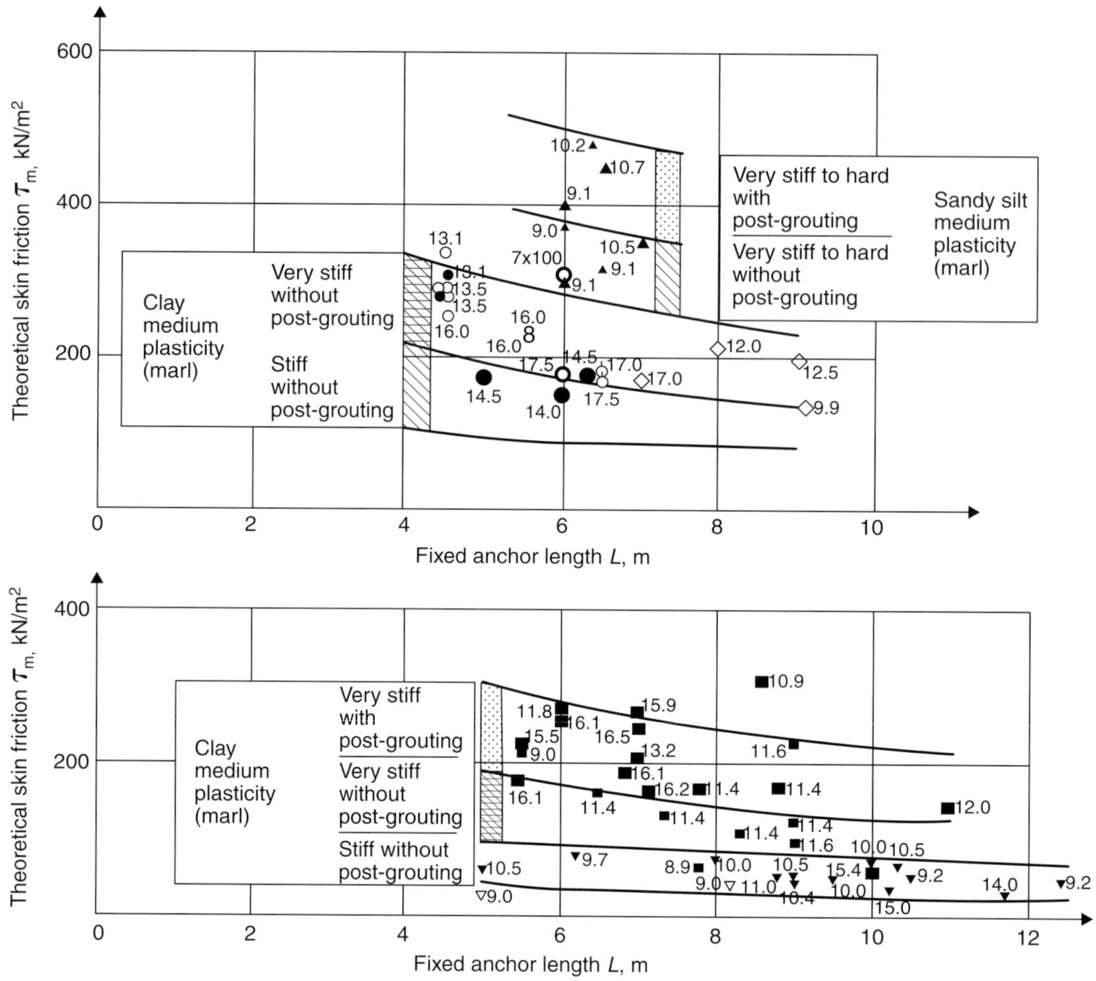

Failure load was reached	Failure load was not reached	Post-grouting	Type of soil	L_L %	P_L %	I_c %
▴	▵	Without	Silt, very sandy (marl) medium plasticity	~ 45	~ 22	~ 1.25
▲	△	With				
•	○	Without	Clay (marl) medium plasticity	32 to 45	14 to 25	1.03 to 1.14
○	◯	With				
● (with tick)	○ (with tick)	Without		36 to 45	14 to 17	1.3 to 1.5
● (with tick)	**○** (with tick)	With				
	◇	Without	Silt medium plasticity	23 to 28	5 to 11	0.7 to 0.85
▪		Without	Clay medium to high plasticity	48 to 58	23 to 35	1.1 to 1.2
■		With				
▼		Without		45 to 59	16 to 32	0.8 to 1.0

Figure 66.8 Skin friction in fine soils for various fixed anchor lengths, with and without post-grouting (After Ostermayer, 1974)

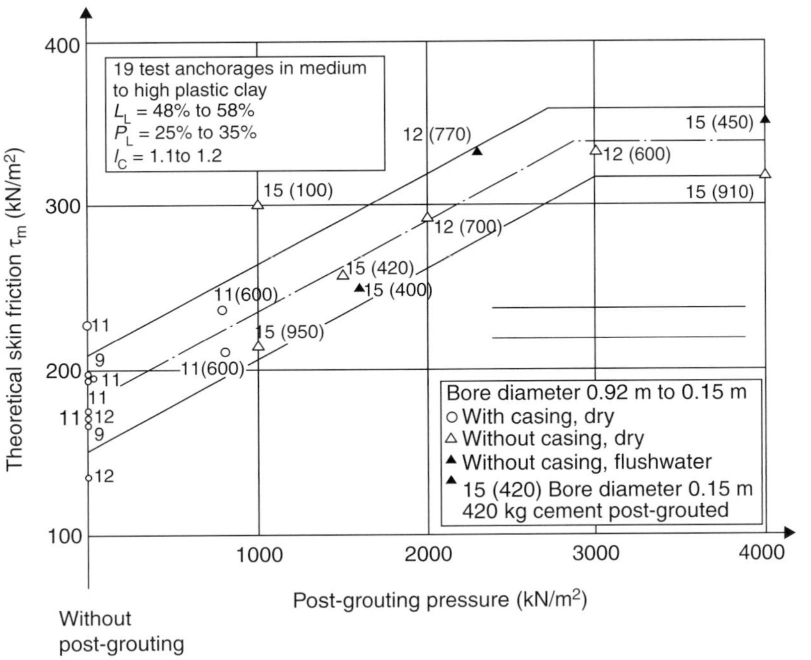

NOTE. The theoretical skin friction is calculated from the borehole diameter and designed fixed anchor length

Figure 66.9 Influence of post-grouting pressure on skin friction in fine soils (After Ostermayer, 1974)

Type D Anchors

Design rules are available for multi-under-reamed anchors of type D (Basset, 1970; Bastable, 1974; Littlejohn, 1970; Neely and Montague-Johns, 1974). For guidance, the calculated geotechnical ultimate ground–grout interface resistance of the anchor, $R_{GG;calc}$ (in kN), may be estimated from Equation 66.16:

$$\begin{aligned} R_{GG;calc} &= \pi D L_{fixed(U)} C_s + \pi/4(D^2 + d^2) N_c C_b \\ &\quad + \pi d L_{fixed(NU)} C_a \\ &= [\text{side shear} + \text{end-bearing} \\ &\quad + \text{side shear (no under-ream)}] \qquad (66.16) \end{aligned}$$

where

- $L_{fixed(U)}$ is that part of the fixed anchor length which has under-reams
- $L_{fixed(NU)}$ is that part of the fixed anchor length which has no under-reams.

A value for C_a of $0.3c_u$ to $0.35c_u$ is commonly adopted.

In the absence of results from investigation tests on trial anchors in the field, empirical multiplier reduction coefficients ranging from 0.75 to 0.95 are sometimes applied to the side shear and end-bearing components of Equation 66.16, to allow for factors such as construction techniques and under-ream geometry (Bassett, 1970; Buttling, 1977).

In the particular case where the clay adjacent to the fixed anchor contains open or sand-filled fissures, a reduction coefficient of 0.5 is generally applied for the side shear and end-bearing components (Littlejohn, 1970).

Of vital importance in fine soil deposits is the time during which drilling, under-reaming and grouting take place. These processes are kept to a minimum in view of the softening effect of water on the clay. The consequences of delays of only a few hours include reduced load-holding capacity and significant short-term prestress losses. In the case of sand-filled fissures, for example, where water flushing is employed, a period of only 3–4 hours might be sufficient to reduce the average undrained shear strength of the ground over the fixed anchor length, c_u, to near the fully softened value.

Under-reaming is ideally suited to clays of c_u greater than 90 kPa. Some difficulties, for example local collapse or breakdown of the neck portion between the under-reams, can occur where c_u values of 60–70 kPa are recorded. Under-reaming is impracticable where the clay has a c_u of 50 kN/m² or less. Under-reaming is also difficult in soils of low plasticity, such as those with a plasticity index <20.

For type D anchors, Equations 66.17 and 66.18 can be used to estimate the maximum allowable spacing of under-reams, δ_u, to give failure along a cylindrical surface, which helps to minimise fixed anchor displacement to develop load. Where fixed anchor displacement is unimportant, more widely spaced under-reams may be employed, when they will act independently.

$$\pi D\delta_u c_u < \pi/4(D^2 + d^2)N_c c_u \quad (66.17)$$

Where

$$\delta_u < [(D^2 + d^2)4D]N_c \quad (66.18)$$

In London Clay at Orford Ness ($c_u = 54$–72 kPa), where type D anchors were constructed using a mechanically expanded double flight under-reamer, Bassett (1970) allowed for the contribution to resistance from the free anchor length in back-analysis using Equation 66.19

$$\text{Calculated pull-out resistance} = \pi DL_{\text{fixed(U)}}f_u c_u + \pi/4(D^2 + d^2)(N_c c_b + \sigma'_E) + \pi dlf_s c_u \quad (66.19)$$

where

- σ'_E is the effective stress normal to the proximal end (in kPa)
- c_b is the undrained shear strength at the proximal end of the fixed anchor length (in kPa)
- c_u is the average undrained shear strength over the anchor shaft length (in kPa).

The following design values were used

- $f_u = 0.75$–0.95
- $f_s = 0.3$–0.6
- $N_c = 6.5$ (range 6–13).

The on-site anchor dimensions were

- $D = 0.46$–0.53 m
- $d = 0.14$ m
- $L_{\text{fixed}} = 3$ m
- $l = 7.6$ m.

With regard to the maximum allowable spacing of under-reams, it is stipulated that $\delta_u < 1.5$–$2D$ and $d < 0.6$–$0.7D$, in order to ensure cylindrical shear failure. For stiff to very stiff, fissured, silty clay ($c_u = 130$–290 kPa) at Neasden Underpass, London, with a mean value of 175 kPa assumed for design, test results (Bastable, 1974) for a multi-flight, mechanical under-reamer ($D = 0.54$ m and $d = 0.175$ m) have indicated an efficiency factor for undrained shear strength along the under-reamed fixed anchor of $f_u = 0.75$.

The success of multi-under-reamed anchors over straight shafts is illustrated by reference to Ostermayer and Barley (2003).

Range for Fixed Anchor Length in Fine Soil

The practical range for fixed anchor lengths in fine soil lies between 2 and 10 m.

66.4.3.6 Anchors in rock

General

While anchor types A–D can all be applied in rock, dependent on strength, the tremie- or packer-grouted, straight shaft borehole type A anchor is the most popular in current practice, based on cost and simplicity of construction.

In type B anchors, low-pressure grouting is used to enhance rock–grout bonding and to increase the rock–grout interface area in weak or fissured rock. When enhanced bonds are employed, investigation and suitability tests are necessary to verify the design assumptions.

For type C anchors, investigation and suitability tests are used to verify the calculated ultimate ground–grout interface resistance of the anchor.

In type D anchors, under-reaming is used to increase the diameter of the fixed anchor in weak rock. Investigation and suitability tests are used to verify the design method.

Distribution of Skin Friction in Rock

Based on the work of Berardi (1967) and Coates and Yu (1970), it can be concluded that the distribution of the bond mobilised at the rock–grout interface is unlikely to be uniform unless the rock is weak. It appears that non-uniformity applies to most rocks where E_C/E_R is less than 10, where E_C is the Young's modulus of concrete or grout and E_R is the Young's modulus of rock.

Although it would appear from the available evidence that the assumptions made in relation to uniform bond distribution are not strictly accurate, it is noted that few failures are encountered at the rock–grout interface and new designs are often based on the successful comparable experience.

Rock/Grout Shear Resistance Values

For strong rocks where there is an absence of shear strength data or results from investigation tests to failure, the ultimate skin friction τ_{ult} is taken as 10% of the unconfined compressive strength of the rock, up to a maximum value of 4.0 MPa. This applies to any rock assuming that the design unconfined compressive strength of the grout is equal to or greater than 40 MPa. Increasing grout strength beyond 40 MPa does not lead to a significant increase in rock–grout bond. Having said this, it is common that minimum grout strengths of 60 MPa are specified for ultra-high-capacity anchors with characteristic serviceability limit state loads exceeding 8.5 MN.

In some rocks, particularly granular, weathered varieties with a relatively low tan ϕ' value, the assumption that τ_{ult} equals 10% unconfined compressive strength of the rock can lead to an artificially low estimate of shear strength. In such cases, the assumption that τ_{ult} equals 20–35% unconfined compressive strength of the rock may be justified.

In weak rocks where the unconfined compressive strength is less than 7 MPa, shear tests on representative samples are normally carried out to measure the maximum shear resistance to be used in the design.

In chalk and clay-infilled fissured rock, the value of skin friction might be affected by the presence of borehole smear.

The degree of weathering of the rock is a major factor that affects not only the ultimate bond but also the load-deflection characteristics. The degree of weathering is seldom quantified, but for designing in soft or weathered rocks the SPT N-value can be used to derive the ultimate bond or side shear resistance and the rock–grout interface resistance. In highly weathered or completely decomposed granite the magnitude of the ultimate bond or skin friction at the rock–grout interface, τ_{ult} (in kPa), has been determined from Equation 66.20

$$\tau_{ult} = 7N + 120 \tag{66.20}$$

Similarly, Equation 66.21 has been established for stiff/hard chalk (weathering grades A to C)

$$\tau_{ult} = 10N \text{ kPa} \tag{66.21}$$

where N is the number of blows per 0.3 m (SPT). Note that SPT results are not generally used for chalk classification.

Barley (1988), however, provides field case histories in chalk (grades I to V), which indicate that pressure grouting can create high skin friction (e.g. τ_{ult}=20–30N kPa). However, it is to be noted that N values in chalk are subject to considerable scatter (Dennehy, 1975), and where this is found, investigation tests are undertaken to verify design assumptions.

Fixed Anchor Lengths in Rock

The fixed anchor length is not normally less than 3 m (2 m if the characteristic serviceability limit state load <200 kN).

It is recognised that fixed anchor lengths less than 3 m can suffice, even after the application of an adequate reduction factor on the calculated ultimate ground–grout interface resistance. However, for a very short fixed anchor length, any sudden reduction in rock quality along the fixed anchor length can induce a large decrease in the ultimate ground–grout interface resistance of the anchor. In general, the fixed anchor length can be derived from Equation 66.22.

The possibility of rock blasting in the vicinity of anchors does not appear to require any increase in fixed anchor length (Mothersille and Weerasinghe, 2000). However, where limited information is available, investigation tests are undertaken, especially in weak, deformable rocks.

The practical limit of fixed anchor lengths in rock does not normally exceed 10 m.

Type A Anchors

For type A anchors, design assumes that the mobilised bond stress is uniformly distributed along the fixed anchor length (Coates, 1970; Fargeot, 1972; Mascardi, 1972; White, 1973). Thus, the calculated ultimate ground–grout interface resistance, $R_{GG;calc}$, of the fixed anchor (in kN) is estimated from Equation 66.22

$$R_{GG;calc} = \pi D L_{fixed} \tau_{ult} \tag{66.22}$$

where

- τ_{ult} is the ultimate bond or skin friction at the rock–grout interface (in kPa)
- D is the diameter of the fixed anchor (in metres)
- L_{fixed} is the fixed anchor length (in metres).

BS 8081 provides indicative values of τ_{ult}.

Equation (66.22) is based on the following assumptions.

(*a*) The transfer of the load from the fixed anchor to the rock is by an average uniformly distributed bond stress, τ_{ult}, acting over the whole of the perimeter of the fixed anchor.

(*b*) The diameters of the borehole and the fixed anchor length are identical and there is no reduction in borehole diameter in the anchor free length, creating no end-bearing effect at the proximal end of the fixed anchor length.

(*c*) Failure takes place by shearing at the rock–grout interface (smooth borehole) or by shearing adjacent to the rock–grout interface, in the weaker medium (rough borehole).

(*d*) There are no discontinuities or inherent weakness planes along which failure can be induced.

The assumption of uniformly distributed stress along the fixed anchor length might require careful consideration in terms of the likely stress concentrations at the proximal end of the fixed anchor in weak, deformable rock. As a result of stress concentrations at the grout–tendon interface, failure in the rock–grout interface could be initiated, even though average stress–strength conditions have been satisfied. Under such conditions, it might be necessary to base the design directly on investigation test results.

66.4.3.7 Encapsulation to anchor grout bond

The design of the corrugated-plastic sheath forming this interface is typically required to be such that a mechanical interlock will be formed between the external and internal grout on either side of the boundary. BS 8081 recommends a maximum ultimate bond stress of 3.0 N/mm^2 at this interface where cement grouts are used, assuming that the bond is uniform over the surface. Higher values can be adopted if adequately proven. Bond tests have confirmed this is a reasonable but conservative low bound, and in most cases this is not a critical interface in designs.

It should be remembered that, where two concentric sheaths are used, the inner sheath will be the critical interface.

66.4.3.8 Tendon to encapsulation grout or anchor grout bond

For temporary anchors, where the steel tendon is embedded directly in the anchor grout, the tendon bond length is usually the same as the fixed anchor length. The length necessary for the grout–ground bond is usually more than adequate to ensure against failure of the tendon–ground bond. In the case of permanent anchors using strands, however, the encapsulation length is often shorter than the fixed anchor length (because a shorter encapsulation length is more convenient for handling the tendon at installation). In such a case the tendon–grout interface can become critical.

BS 8081 recommends that the value for the ultimate bond stress at this interface when using cement grout should be limited to 2.0 N/mm^2 for a clean prestressing strand or deformed bar, rising to 3.0 N/mm^2 for strands that have been locally noded (where the outer wires of each strand are unravelled and then rewound around small collars placed onto the king-wire at intervals to form small nodes). Tests described by Turner (1980) suggested that a value of 2.0 N/mm^2 for a plain strand was reasonable, but that deformed bars could develop in excess of the 5.0 N/mm^2 ultimate bond stress.

66.4.4 The location of the fixed anchor zone (overall stability)

Table 66.1 indicates that the responsibility for the specification of a minimum distance from the structure to the fixed length, to ensure the overall stability of the structure, lies with the supported structure designer. However, the specialist anchor designer/contractor should ensure that the anchors are deeply enough embedded beyond the zone identified by the wall designer to avoid an overall pull-out failure of the body of soil or rock surrounding the anchors.

The methodology generally adopted in design is to determine the location of a notional failure surface behind the wall that has acceptable stability against general failure. The fixed anchor lengths of the ground anchors should then be located beyond this failure surface. It is not uncommon that the midpoint of the fixed anchor length is located on this surface, although in UK practice it is more usual to locate the whole of the fixed anchor beyond this surface (e.g. Littlejohn *et al.*, 1971).

In contemporary design practice, the use of finite element analyses is common in refining the location and extent of predicted failure zones. The advantage of this approach is that geotechnical parameters can be varied to test their sensitivity, and the strength of materials can be factored to facilitate both serviceability and ultimate limit state analyses. However, regardless of the approach adopted in analyses, the fundamentals associated with the estimation of the depth of failure zones behind walls should be appreciated.

66.4.4.1 For cohesionless coarse soils

The design methods used to determine the location of an acceptably stable surface within coarse soils vary from 'planar' wedge or 'sliding block' analysis techniques to conventional circular or non-circular stability analyses

The planar wedge method is illustrated in **Figure 66.10**. It has been used for design in both coarse and fine soils, and for multiple rows of anchors. Values for the parameters β and x in this figure, which have been used in practice, vary between 27° and 45° to the vertical for β, with x varying between 0 and 6.0 m. However, it is a commonly held view (e.g. Hanna, 1982) that justification of the method is difficult because the mode of failure is unlikely to be planar.

A typical 'sliding block' analysis is illustrated in **Figure 66.11**. It is suitable for analysing a single row of anchors. The method illustrated is based upon that developed by Kranz (1953) and modified by Locher (1969) and Littlejohn (1970, 1977). Other methods are also described in BS 8081.

As described by Littlejohn *et al.* (1971), the lateral earth pressure, P_n, on the vertical surface b–c drawn through the proximal end of the fixed anchor length is calculated using a 'nominal' friction angle ϕ'_n, which is less than the effective angle of internal friction of the soil such that

$$\tan \phi'_n = (\tan \phi'_d)/F \qquad (66.23)$$

where F is a safety factor, and ϕ'_d is the design value of the angle of internal friction of the soil. The resultant R_n on the inclined sliding plane c–d will also be inclined at ϕ'_n to the normal to c–d.

By iteration, when the disturbing and restoring forces are in equilibrium, by revising the geometry of the soil block a–b–c–d, then the desired safety factor has been achieved. The additional stabilising force due to the passive resistance at the toe of the wall is ignored in the calculation, which is a conservative assumption.

The dashed lines in **Figure 66.11** represent the typical shape of the failure curve, derived from experimental tests on plate or 'deadman' anchors. The polygon a–b–c–d is a simplification of this shape for calculation purposes.

It should be noted that the examples in **Figures 66.10** and **66.11** both show the minimum distance to the proximal end of the fixed anchor length from the face of the wall as being at the boundary of the calculated 'failure' surface.

It should also be noted that the anchors must have a sufficient depth of vertical cover above the fixed anchor length to avoid localised passive failure of the soil, such as the failure condition associated with shallow 'deadman' anchors. A typical rule of thumb cited by Littlejohn (1972) and based upon early experience of ground anchors in UK alluvial sands and gravels suggests that a minimum depth of cover of 5–6 m is normally considered sufficient to guarantee a deep-seated failure condition at pull-out. It can also be seen that Equation 66.12 for the calculation of the load capacity of an anchor is sensitive to the overburden pressure around the fixed anchor length. Hence, by implication, it would be expected that the

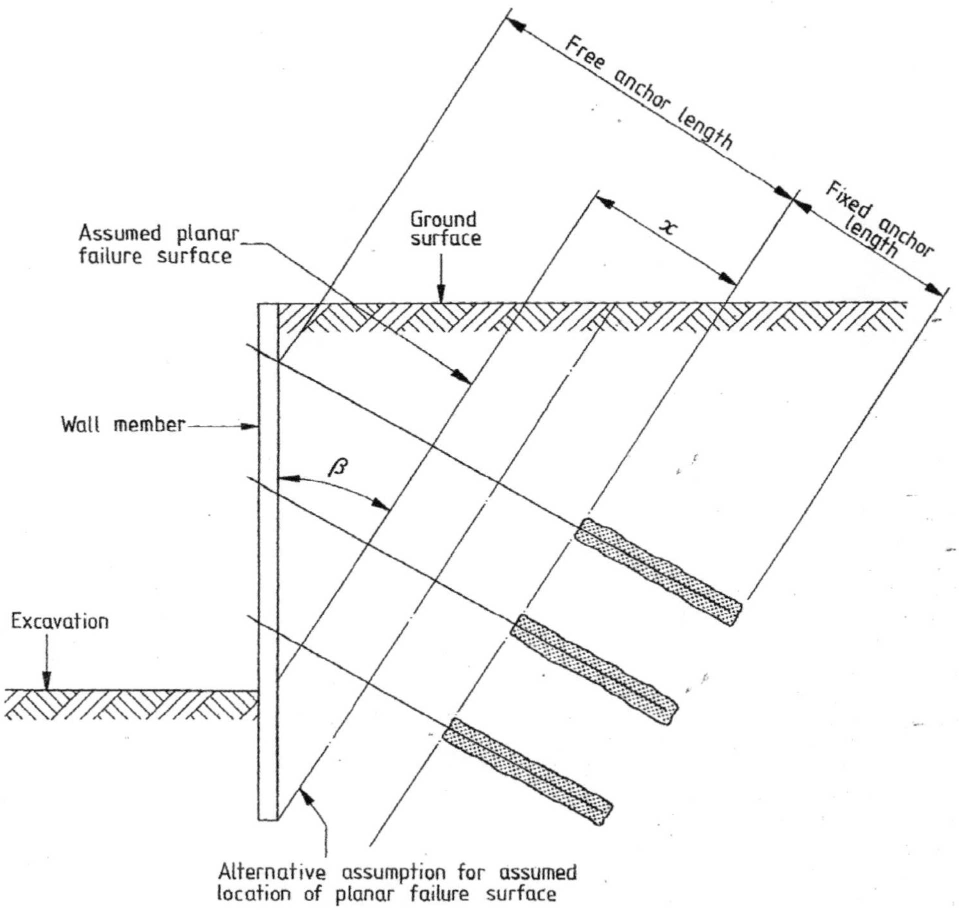

Figure 66.10 Typical planar wedge method of analysis (Reproduced with permission from BS 8081; © BSI 1989)

smaller the overburden pressure above a given fixed anchor length, the lower the capacity of the ground anchor. If there is any doubt about the performance of a near-surface anchor, its performance should be confirmed either by investigation or suitability testing to BS EN ISO 22477-5:2018.

For multi-tied walls, a circular type of stability analysis is usually recommended to determine the location of the acceptably stable failure surface. This is because the shape of the sliding surface is not clear experimentally with multiple rows of anchors. Littlejohn *et al.* (1971) recommended a logarithmic spiral method for defining this surface, as illustrated in **Figure 66.12**. The reason for this is that a log spiral has the property that the radius from the spiral centre to any point on the curve forms a constant angle ϕ_s with the line drawn normal to the curve. Hence, if a nominal friction angle, ϕ'_n, for the soil is selected for ϕ_s, then the line of action of the resulting forces on each part of the sliding surface will pass through the centre point of the spiral. Hence, none of the forces along the failure surface will create a moment around the centre point and they can be neglected when equating moments about this point.

In a similar manner to the sliding block method, the value for F is satisfied when the value for ϕ'_n is such that the moments of the remaining forces sum to zero, as shown in **Figure 66.12**.

It is again usual to ignore the stabilising effect of the passive resistance of the soil beyond the toe, so that when the moments W_s and W_o are equal, a conservative value is obtained of $F = \tan \phi'_d / \tan \phi'_n$.

The definition $F = \tan\phi'_d / \tan\phi'_n$ can be considered as a partial factor of safety against failure of the soil body. The chosen value of F used in practice can vary between 1.2 and 1.5 for a sliding block design, and $\geq$1 for a spiral-shaped sliding surface design. The higher value would be selected for more critical applications, or where conditions are less well defined.

All the analysis techniques illustrated in **Figures 66.11–66.13** make an underlying assumption that the prestress exerted by the post-tensioned anchors increases the shear strength of the soil sufficiently to ensure the potential failure plane is beyond the proximal end or the midpoint of the fixed anchor length, depending on the design philosophy noted above.

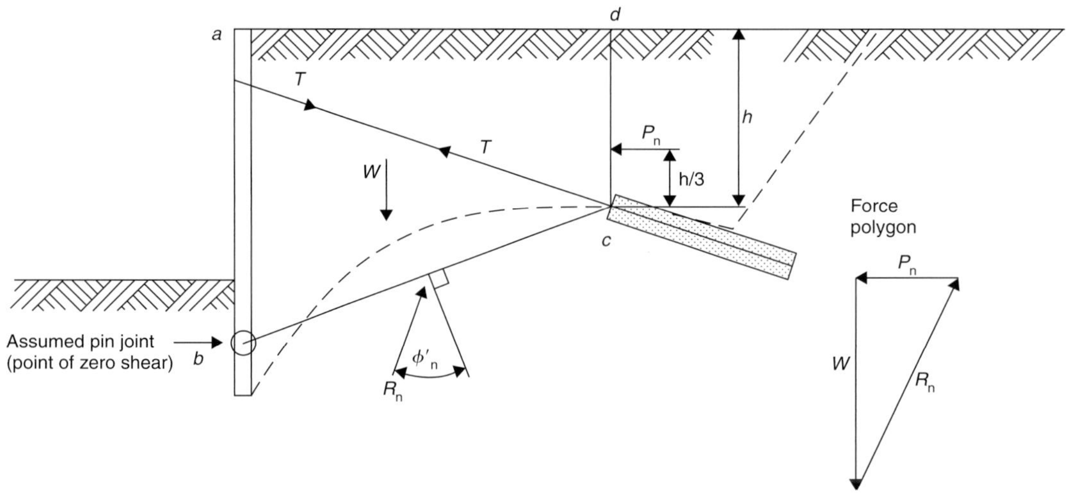

Factor of safety F is given by:

$$F = \frac{\tan \phi'_d}{\tan \phi'_n} \geq 1.5$$

where

ϕ'_n is nominal angle of shearing resistance in terms of effective stress (in degrees).

NOTE: If ϕ'_n has been correctly assumed, the weight W and the forces R_n, and P_n are in equilibrium. If this is not the case ϕ'_n has to be altered.

Figure 66.11 Typical sliding block method of analysis (Reproduced from Littlejohn, 1972)

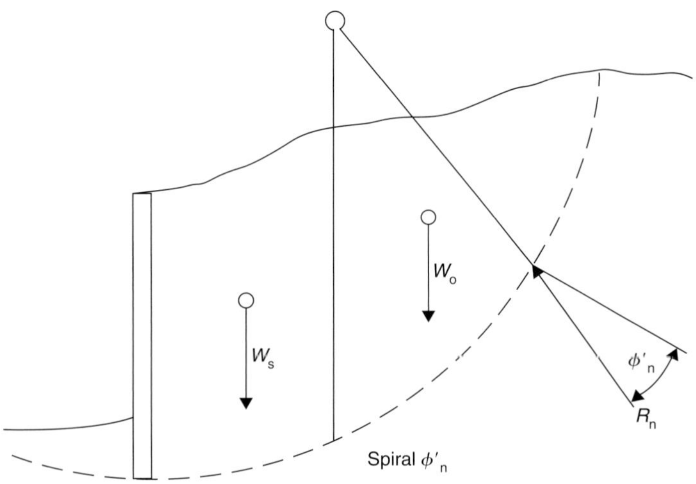

$$F = \frac{\tan \phi'_d}{\tan \phi'_n} \quad \text{For equilibrium} \quad \frac{\text{Moment due to } W_s}{\text{Moment due to } W_o} \geq 1$$

Figure 66.12 Typical stability analysis using a log spiral-shaped sliding surface (Reproduced from Littlejohn, 1972)

66.4.4.2 Fine soils

For fine soils a circular form of failure surface would normally be preferentially analysed, and, again, as a general recommendation, the whole of the fixed anchor length should be located beyond the failure surface with the required value for the factor of safety, unless the designers agree otherwise.

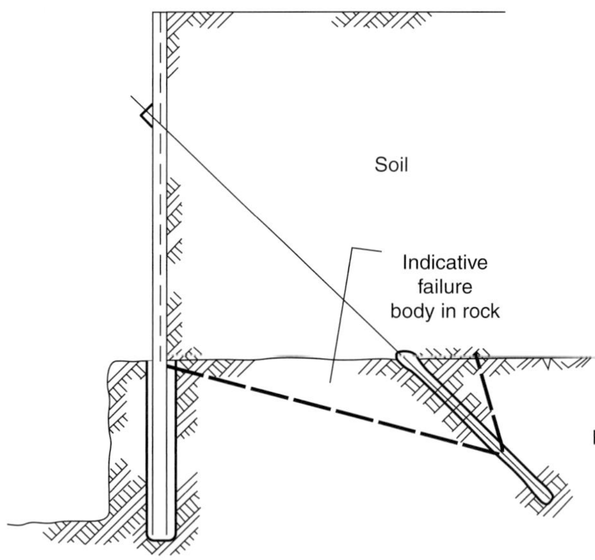

Figure 66.13 Indicative pull-out failure body within rock stratum

Reference has been made to the use of a planar wedge design method described above for coarse soils, being used also for fine soils, but this method has to be used with caution, and also perhaps previous experience. It has been used successfully in the past on such projects as the Neasden Underpass in London (Sills *et al.*, 1977).

Compared with the design assumptions for coarse soils, it can be appreciated that anchor prestress will only slowly increase the shear stress of fine soil over time. Therefore, it is usual in such soils that a conventional analysis of overall stability of the wall be undertaken, ignoring the effect of the proposed ground anchors, and identifying the location of the failure surface that has an acceptable factor of safety, in a similar manner to that described above for granular soils. The fixed anchor lengths of the anchors should then be located beyond this calculated surface.

For mixed soil and rock conditions, where, typically, anchors might be in a rock stratum beneath the overlying soils, as illustrated in **Figure 66.13**, it should be appreciated that rock anchors generally develop higher unit bond strengths than soil anchors. In some circumstances, there could, therefore, be a risk of pull-out failure of a conical or wedge-shaped mass of rock by the tensile force imposed by the wall if the anchor is not embedded sufficiently deeply into the rock stratum. This possibility needs to be checked by the anchor designer.

66.4.5 Load transfer into the structure

The head assembly for a ground anchor typically consists of a stressing head or anchor block onto (or into) which the tendon is anchored, and a bearing plate, which transfers the anchor force onto the structure. In the case of an embedded retaining wall, the bearing plate can often bear onto steel or concrete waling beams to distribute the imposed anchor forces more evenly into the wall.

An important factor in such load transfer and distribution systems is that the vertical component of the anchor force should be accommodated within the design.

In steel sheet piles and king post walls, with steel waling beams, some form of angled bracket assembly must be provided. Typical solutions to this problem are illustrated in **Figure 66.14**.

In contiguous or secant piled walls, very often reinforced concrete walings or load transfer assemblies would typically be employed, as in **Figure 66.15**.

For diaphragm walls, a suitable angled pocket is usually cast into the wall panel, together with a guide tube through the wall section to locate the drill string clear of the reinforcement.

With reference to secant and diaphragm walls, it may be necessary to make provision for the anchor head system to withstand hydrostatic pressures. This is often more difficult to achieve than might be expected, and needs close attention to design, installation and construction procedures.

BS 8081:2015+A2:2018 outlines the design and tolerance requirements of the anchor head. One requirement is that the anchor head should be designed to tolerate an angular deviation of up to ±5° from the axial direction of the tendon. While this is a normal requirement for post-tensioning systems, it can cause difficulties with the protective systems for permanent ground anchors if attention is not paid to this detail. Specific requirements should be highlighted, discussed and clarified with the suppliers of such systems.

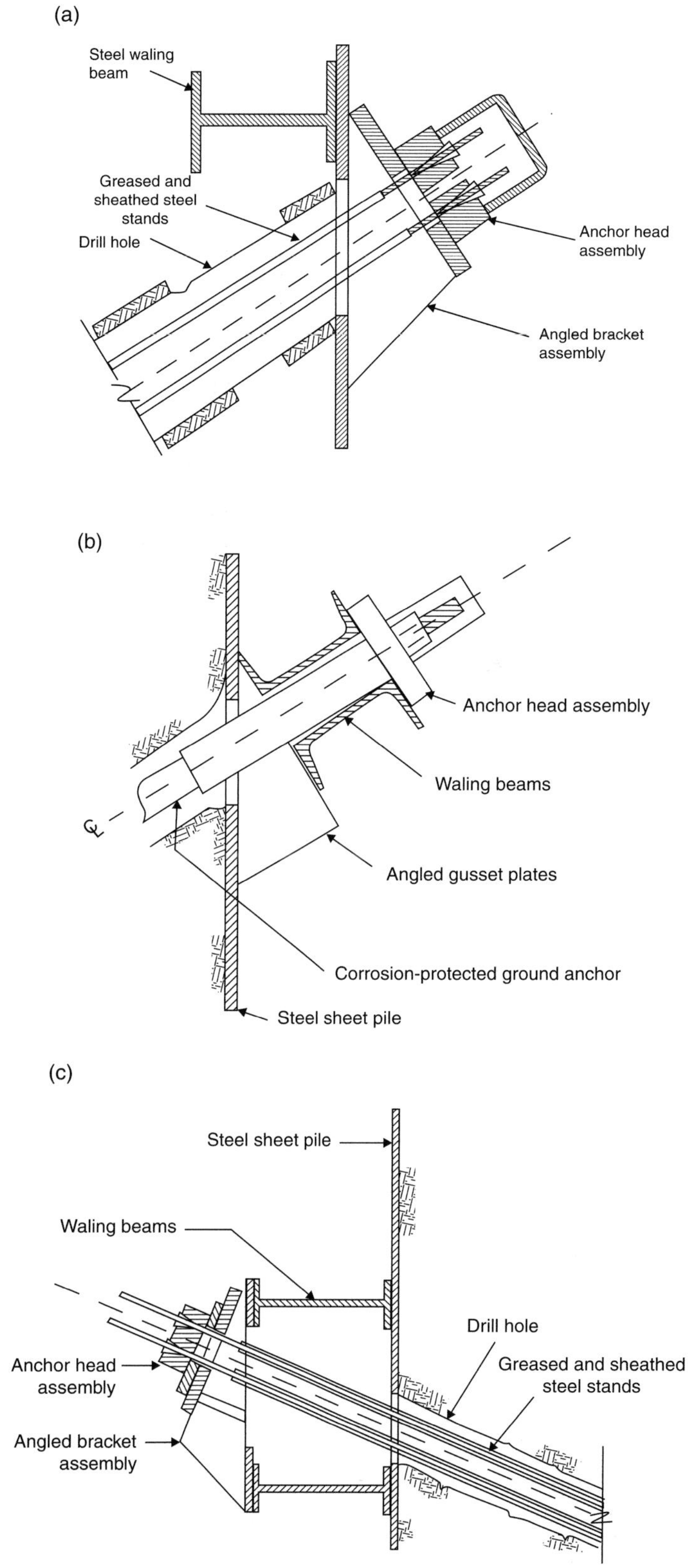

Figure 66.14 Typical use of ground anchors with steel walings: (a) internal waling, with external angled bracket assembly; (b) external angled waling; (c) external waling with angle bracket assembly

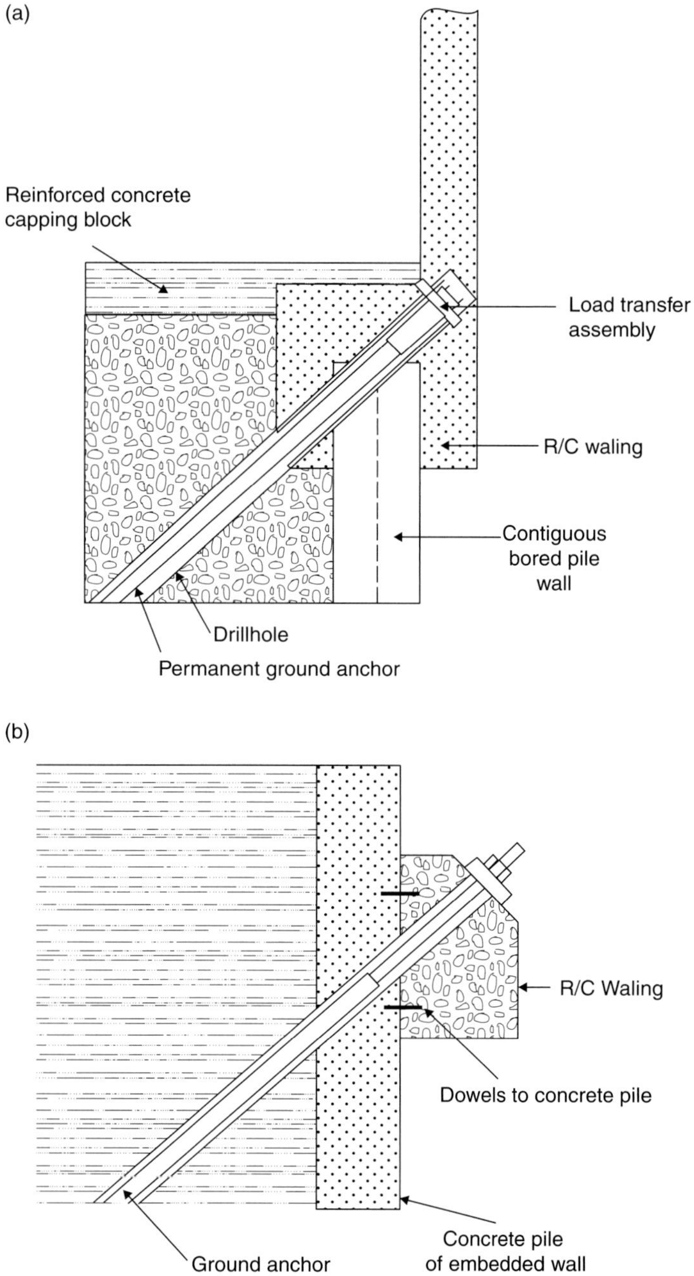

Figure 66.15 Typical use of ground anchors with reinforced concrete walings or load transfer assemblies

BS 8081 also makes specific reference to the design requirements of anchors in terms of the facility for the anchor to be restressed or even detensioned during its service life. It defines the 'restressable' and 'detensionable' anchor heads as follows.

- A restressable anchor head is an anchor head that permits the tendon load, throughout the service life of the anchor, to be measured by check-lifting, and enables small losses of up to 10% of the service state load to be recovered by shimming or thread-turning
- A detensionable anchor head is an anchor head that has all the properties of the restressable anchor head and, in addition, permits the tendon to be detensioned in a controlled way at any time during the service life of the anchor

It is important that the wall designer, as well as the anchor designer, is aware of these different facilities and is clear what level of 'adjustability' is required for their case.

BS 8081 also recommends that a degree of redundancy should be allowed within the anchored system to a retaining wall. A structure should be designed so that, in the limit, any one anchor can fail without the load on the other anchors exceeding the proof load to which they were initially tested.

66.4.6 Testing and load monitoring

The grouted anchors supporting retaining walls should be subjected to the regime of investigation, suitability and acceptance testing recommended by BS EN ISO 22477-5:2018.

As part of the routine monitoring of the supported excavation, it is prudent to consider monitoring loads in selected anchors, especially for deep, multi-tied excavations. It is important to confirm that the performance of a wall is maintained within its design envelope, and monitoring anchor loads is one part of this regime. Such load-checking can be undertaken using load cells or by routine check-lifting operations (Littlejohn and Mothersille, 2008a, 2008b).

It is also not unknown for anchor loads to decrease or fluctuate during the life of the structure, especially for sheet-piled walls where new fill has been placed behind the wall, in the construction of new river walls, for instance. Turner and Richards (2007) discuss load changes and lateral movements recorded in an anchored sheet pile wall constructed along the River Thames in 1979, where the site was redeveloped some 20 years later. Horizontal movements of the piled wall, supported by a single row of inclined ground anchors, were found to vary in a roughly cyclic manner by 25–30 mm over a year.

References

Barley AD (1978) *A Study and Investigation of Under Reamed Anchors, and Associated Load Transfer Mechanisms*. MSc thesis, University of Aberdeen, UK.

Barley AD (1988) Ten thousand anchorages in rock. *Ground Engineering*, Sept.–Nov.

Barley AD (1997) The single bore multiple anchor system. *International Conference on Ground Anchorages and Anchored Structures, London, UK*. Thomas Telford, London, UK, pp. 65–75.

Barley AD and Windsor CR (2000) Recent advances in ground anchor and ground reinforcement technology with reference to the development of the art. *Proceedings of the International Conference on Geotechnical and Geological Engineering, Melbourne, Australia*.

Bassett RH (1970) Discussion to paper on soil anchors. *ICE Conference on Ground Engineering, London, UK*, pp. 89–94.

Bastable AD (1974) Multibell ground anchors in London Clay. *7th FIP Congress, Session on Prestressed Concrete Foundations and Ground Anchors, New York, NY*, pp. 33–37.

Beradi G (1967) *Sul Comportamento degli Ancoraggi lmmersi in Terreni Diversi.* Publication series III, no. 60, Institute of Construction Science, University of Genoa, Genoa, Italy.

Berezantzev VG, Khristoforov VS and Golubkov VN (1961) Load bearing capacity and deformation of piled foundations. *Proceedings of the 5th International Conference on Soil Mechanics and Foundation Engineering, Paris, France*, vol. 2, pp. 11–15.

BSI (British Standards Institution) (1989) BS 8081:1989: Code of practice for ground anchorages. BSI, London, UK.

BSI (2000) BS EN 1537:2000: Execution of special geotechnical work – Ground anchors. BSI, London, UK.

BSI (2004a) BS EN 1997-1:2004: Eurocode 7 – Geotechnical design – Part 1 – General rules. BSI, London, UK.

BSI (2004b) BS EN 1992-1-1:2004: Design of concrete structures – General – Common rules for building and civil engineering structures. BSI, London, UK.

BSI (2013) BS EN 1537:2013: Execution of special geotechnical works – Ground anchors. BSI, London, UK.

BSI (2014) NA+A1:2014 – National Annex to BS EN 1997-1:2004: Eurocode 7 – Geotechnical design – Part 1 – General rules. BSI, London, UK.

BSI (2018) BS EN ISO 22477-5:2018: Geotechnical investigation and testing – Testing of geotechnical structures – Part 5: Testing of grouted anchors. BSI, London, UK.

Bustamante M and Doix B (1985) Une méthode pour le calcul des tirants et des micropieux injectés. *Bulletin de Liaison des Laboratoires de Ponts et Chaussées, Paris, France*, pp. 75–92.

Buttling S (1977) Report on discussion to session IV by C Truman-Davies. In *A Review of Diaphragm Walls.* ICE, London, UK, p. 76.

Coates DF (1970) *Rock Mechanics Principles.* Department of Energy, Mines and Resources Mines, Ottawa, ON.

Coates DF and Yu YS (1970) Three-dimensional stress distributions around a cylindrical hole and anchor. *Proceedings of the 2nd International Conference on Rock Mechanics, Belgrade, Serbia*, vol 2, pp. 176–182.

Dennehy JP (1975) Correlating the SPT N value with chalk grade for some zones of the Upper Chalk. *Géotechnique* **25(3)**: 610–614.

Egan D and Merrifield C (1997) Modelling of high-pressure grouted compaction anchors. *Proceedings of International Conference on Ground Anchorages and Anchored Structures.* ICE, London, UK, pp. 460–469.

Fargeot M (1972) Reply to FIP questionnaire. In: Draft of the recommendations and replies to FIP questionnaire, FIP sub-committee on prestressed ground anchors (mimeo).

Fujita K, Ueda K and Kusabuka MA (1978). Method to predict the load displacement relationship of ground anchors. *Revue Française de Géotechnique*, **3**: 58–62.

Hanna TH (1982) *Foundations in Tension: Ground Anchors.* Trans Tech Publications, Clausthal-Zellerfeld, Germany.

Jorge GR (1969) The regroutable IRP anchorage for soft soils, low capacity or karstic rocks. *Proceedings of the 7th International Conference on Soil Mechanics and Foundation Engineering, Mexico City, Mexico*, pp. 159–163.

Kranz F (1953) *Über die Verankerung von Spundwänden.* Wilhelm Ernst & Sohn, Berlin, Germany.

Littlejohn GS (1970) Soil anchors. *ICE Conference on Ground Engineering, London, UK*, pp. 33–44.

Littlejohn GS (1972) Anchored diaphragm walls in sand. *Ground Engineering*, **5(1)**: 12–17.

Littlejohn GS (1977) Ground anchors: installation techniques and testing procedures. In *Review of Diaphragm Walls.* ICE Publishing, London, UK, pp. 93–97; discussion pp. 98–116.

Littlejohn GS (1979) Ground anchors: state-of-the-art. *Symposium on Prestressed Ground Anchors.* The Concrete Society of South Africa, Johannesburg, South Africa.

Littlejohn GS and Bruce DA (1977) *Rock Anchors: State of the Art.* Foundation, Brentwood, UK.

Littlejohn GS and Mothersille DKV (2008a) Maintenance and monitoring of anchorages: guidelines. *Geotechnical Engineering* **161**: 93–106.

Littlejohn GS and Mothersille DKV (2008b) Maintenance and monitoring of anchorages: case histories. *Geotechnical Engineering* **161**: 107–114.

Littlejohn GS, Jack BJ and Sliwinski ZJ (1971) Anchored diaphragm walls in sand: some design and construction considerations. *Journal of the Institute of Highway Engineers* **18(4)**: 15–29.

Locher HG (1969) Anchored retaining walls and cut-off walls (unpublished).

Mascardi C (1972) Reply to questionnaire on rock anchor design. Geotechnics Research Group, Department of Engineering, University of Aberdeen (unpublished).

Mothersille DKV and Weerasinghe RB (2000) The response of low to medium capacity rock anchors to close proximity blasting. *Proceedings of the International Conference on Geotechnical and Geological Engineering, Melbourne, Australia*, vol. 2.

Mothersille DKV, Duzceer R, Gokalp A and Okumusoglu B (2014) Support of 25m deep excavation using ground anchors in Russia. *Proceedings of the Institution of Civil Engineers, Geotechnical Engineering* **168(4)**: 281–295.

Neely WJ and Montague-Jones M (1974) Pull-out capacity of straight-shafted and under-reamed ground anchors. *Die Siviele lngenieur in Suid-Africa* **16(4)**: 131–134.

Oosterbaan MD and Gifford DG (1972) A case study of the Bauer earth anchor. *Proceedings of the Speciality Conference on Performance of Earth and Earth Supported Structures – Part 2, West Lafayette, IN*, pp. 1391–1401.

Ostermayer H (1974) Construction carrying behaviour and creep characteristics of ground anchors. *Conference on Diaphragm Walls and Anchorages.* ICE, London, UK, pp. 141–151.

Ostermayer H and Barley AD (2003) Anchor design guidelines. In *Geotechnical Engineering Handbook*, vol. 2 (Smoltczyk U (ed)). Ernst & Sohn, Berlin, Germany, pp. 189–205.

Ostermayer H and Scheele F (1978) Research and ground anchors in non-cohesive soils. *Revue Française de Géotechnique* **3**: 92–97.

Petrasovits G (1981) Interaction between soil and grouted anchor. *Proceedings of the 10th International Conference on Soil Mechanics and Foundation Engineering, Stockholm, Sweden*, vol. 2, pp. 213–217.

Robinson KE (1969) Grouted rod and multi-helix anchors. *Proceedings of the 7th Conference on Soil Mechanics and Foundation Engineering, Mexico City, Mexico*, Speciality Session 15, pp. 126–130.

Sapio G (1975) Comportamento di tiranti de ancoraggio in formazioni de argille preconsolidate. *Atti XII Convegno Nazionale de Geotecnica, Cosenze, Italy*, pp. 39–47.

Sills GC, Burland JB and Czechowski, MK (1977) Behaviour of an anchored diaphragm wall in stiff clay. *Proceedings of the 9th International Conference of Soil Mechanics and Foundation Engineering, Tokyo, Japan*, vol. 2, pp. 147–154.

Trofimenkov JG and Mariupolskii (1965) LG screw piles used for mast and tower foundations. *Proceedings of the 6th International Conference on Soil Mechanics and Foundation Engineering, Montreal, QC*, vol. 2, pp. 328–332.

Turner MJ (1980) Rock anchors: an outline of some current design, construction and testing practices in the United Kingdom. *International Conference on Structural Foundations in Rock, Sydney, Australia*. Balkema, Leiden, the Netherlands, pp. 87–103.

Turner MJ and Richards D (2007) The long-term performance of permanent ground anchors forming part of the Thames Barrier Project. *International Conference on Ground Anchorages and Anchored Structures in Service, London, UK*. ICE, London, pp. 72–81.

Wernick E (1978) Stress and strains on the surface of anchors. *Revue Française de Géotechnique* **3**: 113–119.

White RE (1973) Reply to questionnaire on rock anchor design. Geotechnics Research Group, Department of Engineering, University of Aberdeen (unpublished).

It is recommended this chapter is read in conjunction with

- Chapter 89 *Grouted anchor construction*
- Chapter 94 *Principles of geotechnical monitoring*

All chapters in this book rely on the guidance in Sections 1 *Context* and 2 *Fundamental principles*. A sound knowledge of ground investigation is required for all geotechnical works, as set out in Section 4 *Site investigation*.

Chapter 67
Retaining walls as part of complete underground structure

Peter Ingram Arup, London, UK

Retaining walls usually form part of an overall structure. They are often used both to create basement space and to act as part of the overall foundation system. As a geotechnical designer it is important not to lose sight of the fact that the function of the entire structure is the priority for the client, rather than the function of the walls and foundations alone.

This chapter is intended to highlight how, as part of designing an overall underground structure, it is necessary for the retaining wall designer to consider not only how the various elements forming the underground structure will interact and react as a whole structure, but also how the geotechnical design must be consistent with the work of other disciplines.

The following discussions provide the reader with a basic understanding of the issues relating to the holistic design of the whole basement/underground structure and interaction with other disciplines.

doi: 10.1680/icemge.66830.1117

CONTENTS

67.1. Introduction

This chapter provides a description of some of the wider issues which need consideration for holistic basement design in addition to basic retaining wall design calculations. The whole structure should be considered by designers before and during the design of the wall, as described here and in Chapters 64 *Geotechnical design of retaining walls* to 66 *Geotechnical design of grouted anchors*. Successful application of the concepts contained in this chapter can require a high degree of interaction between the various engineering disciplines involved in basement design.

67.2. Interfaces with structural design and other disciplines

Underground structures require input from a variety of disciplines, often with conflicting requirements. In order to safely design for the whole structure, it is important that the various engineering disciplines and the architect understand clearly the pinch points in each other's designs. For example, floor slab (e.g. propping) levels may be constrained by stair heights, and may therefore not be at the optimum level to suit the geotechnical design. For small projects it is relatively easy to get a consensus as, typically, a small number of individuals are involved. For larger projects, this is most likely to be successfully achieved by

- having the team in one location
- regular multidisciplinary design reviews
- production of formal design guidance, setting out clearly how the different disciplines will design their respective elements of the structure, and how their designs will interact with others.

The following sections deal principally with common design interfaces with the structural engineer.

67.2.1 Design life and durability

BS EN 1990:2002 requires the designer to specify the design working life for the structure, and outlines indicative time periods for typical structures (BSI, 2002). Building structures and other common structures are to be designed for a suggested working life of 50 years, with monumental building structures, bridges and other civil engineering structures having a suggested design working life of 100 years.

Durability in concrete structures is typically provided by three means

(*a*) having an appropriate concrete mix to take account of the method of placement, and exposure to aggressive ground conditions – determined with reference to BRE (2005)

(*b*) appropriate reinforcement detailing to limit cracking, usually incorporating a continuous layer of reinforcement in each face, for walls and slabs in contact with the ground

(*c*) appropriate concrete cover to reinforcement in accordance with the requirements of BS EN 1992-1-1 (BSI, 2004a).

67.2.2 Design philosophy for buried structures (crack width control, construction joints, watertightness, etc.)

Understanding the client's requirements for end use, and hence watertightness, is key to determining appropriate wall details.

BS 8102 (BSI, 2022) defines grades of performance for below-ground spaces, and methods of achieving them. **Table 67.1** is reproduced from the standard.

The 1990 edition of BS 8102 (BSI, 1990) referred to grade 4 (or special) environments. The 2009 and 2022 editions have not retained this grade, stating that the only difference between grades 3 and 4 is the performance level related to ventilation, dehumidification or air conditioning, and suggests referring to

Grade	Performance definition
1a	Seepage and damp areas from internal and external sources are tolerable where this does not impact on the proposed use of below-ground structure. Internal drainage might be necessary to deal with seepage.
1b	No seepage. Damp areas from internal and external sources are tolerable.
2	No seepage is acceptable. Damp areas as a result of internal air moisture/condensation are tolerable; measures might be required to manage water vapour/condensation.
3	No water ingress or damp areas are acceptable. Ventilation, dehumidification or air conditioning are necessary, appropriate to the intended use.

Table 67.1 Grades of waterproofing protection (Data taken from BS 8102 (BSI, 2022) – refer to the standard for definitions)

BS 5454 (BSI, 2000) for recommendations for the storage and exhibition of archival documents.

The *Specification for Piling and Embedded Retaining Walls* (ICE, 2017) provides guidance on levels of acceptable wall seepage for the grades defined in BS 8102.

- *For grade 1 structures*: beading and limited damp patches are tolerable, but no weeping.
- *For grade 2 structures*: beading and limited damp patches are tolerable.
- *For grade 3 structures*: no weeping. Other components will be needed in addition to the retaining wall component to achieve the required watertightness of the whole system.

BS 8102 states that resistance to water ingress is usually provided by one or a combination of the following methods.

- *Type A (barrier) protection*: protection against water ingress which is dependent on a separate barrier system applied to the structure.
- *Type B (structurally integral) protection*: protection against water ingress, provided by the structure.
- *Type C (drained) protection*: protection against water ingress into usable spaces, provided by the incorporation of an appropriate internal water management system.

Watertightness provided by the structure (types A and B) is affected by cracking, construction joints, differential movements, openings for services and so on, so design for an appropriate watertightness requires an understanding of how to control water ingress at these points.

BS EN 1992-1-1 (BSI, 2004a) provides requirements for crack control in reinforced concrete structures. Gaba *et al.* (2017) state that 'Cost savings are possible if a pragmatic approach is taken to crack width control'. Consideration of appropriate limits of cracking for buried retaining wall structures can lead to a significant reduction in the reinforcement required to control cracks. Useful guidance on this is given in IStructE (2004) and Gaba *et al.* (2017).

In addition to consideration of appropriate limits on flexural cracking in the retaining wall section, and the cracking of wall structural elements, the following are areas where particular care is required.

- Interfaces between structural wall elements and capping beams or slabs. These require careful detailing if water flow paths are to be restricted.
- Points of gross change in construction (e.g. where new and existing constructions join) result in higher risk of entry for water.
- Discrete movement joints should be avoided where possible in the outer envelope of basements as they are a frequent point of weakness and leakage. Where such movement joints cannot be avoided they should be protected against water ingress.
- Construction joint spacing in water-resistant concrete should be determined in accordance with requirements to limit shrinkage and early age cracking.
- Applied barrier system waterproofing must be able to accommodate the predicted lateral wall movements following installation of the system.

Further discussion on watertightness of basement structures, and guidance for clients, can be found in ICE (2009).

67.2.3 Design codes and standards

Chapter 10 *Codes and standards and their relevance* provides a comprehensive list of the main codes and standards to be used in the UK for design. The chapter draws attention to the first point-of-call design guidance for underground structures in the UK, although these are likely to be broadly applicable to other geographic locations. The primary sources of geotechnical and structural design guidance for underground structures in the UK are shown in **Figure 67.1**.

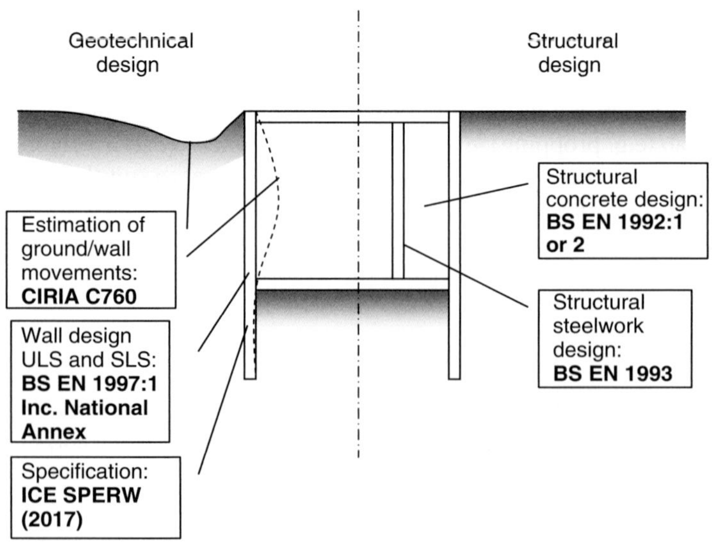

Figure 67.1 UK design guidance

In addition to the guidance highlighted in **Figure 67.1**, a series of execution standards are available to complement the Eurocodes. Most relevant to geotechnical construction of basements are:

- BS EN 1536:2010 Execution of special geotechnical works – Bored piles (BSI, 2010)
- BS EN 1537:2013 Execution of special geotechnical works – Ground anchors (BSI, 2013)
- BS EN 12063:1999 Execution of special geotechnical works – Sheet pile walls (BSI, 1999).

67.2.4 Load combinations (lateral and vertical)

BS EN 1997-1 Eurocode 7 (BSI, 2004b) defines the following load factors for use in retaining wall design. Design is to be carried out using Design Approach 1 (DA1) in the UK, using both Combination 1 and Combination 2 (DA1-1 and DA1-2, respectively) factors (**Figure 67.2**). Both of these load combinations are ultimate limit state (ULS) conditions. Serviceability must be considered separately.

67.3. Resistance to lateral actions

As part of the whole underground structure, the retaining wall needs to fulfil a variety of functions, and must

- be able to resist lateral forces in bending or in overall horizontal stability, to provide an adequate factor of safety
- be able to support the applied vertical loading at all stages of construction, and in the permanent condition, including the need to maintain a satisfactory factor of safety against flotation and soil heave
- provide hydraulic cut-off and/or maintain satisfactory basal stability
- be of sufficient stiffness to restrain ground movements outside the structure to within tolerable limits.

67.3.1 Determination of retaining wall toe level for lateral stability and associated load effects

Chapter 64 *Geotechnical design of retaining walls* provides design guidance for considering retaining wall lateral stability and loading, highlighting the need for compatibility with the requirement for the wall to support vertical loading, and the importance of designing for both.

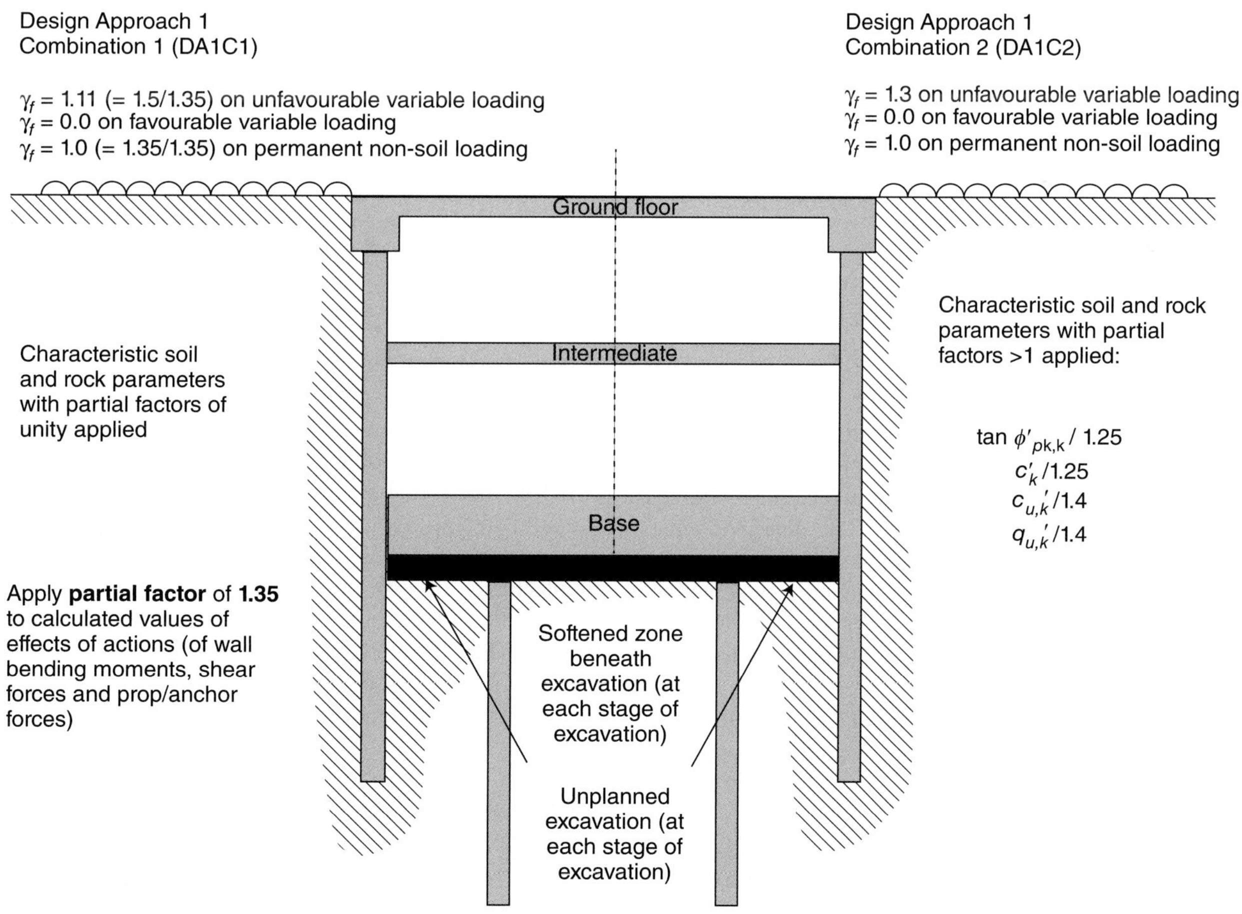

Figure 67.2 Design factors to BS EN 1997-1:2004, Eurocode (Reproduced with permission from CIRIA C760, Gaba *et al.*, 2017; www.ciria.org)

The design of underground structures requires a holistic and consistent approach to the design of the retaining walls, internal piles and base slabs. The following is a step-by-step approach to considering the interaction between the designs of the various elements.

(*a*) Determine the magnitude of wall frictions for vertical/horizontal compatibility of wall movements.

(*b*) Analyse the walls to determine the minimum toe depth for lateral stability; see Chapter 64 *Geotechnical design of retaining walls*.

(*c*) Consider the ability of the walls to carry the vertical loads during excavation, and increase the toe depth if necessary; see Chapter 64 *Geotechnical design of retaining walls*.

(*d*) Determine the additional foundation dimensions to carry vertical loading (e.g. pile foundations).

(*e*) Adopt a design wall toe level and review the direction and magnitude of the assumed wall friction and adhesion at the soil–wall interface. Redo lateral analyses if appropriate to confirm wall design toe level.

(*f*) Consider vertical uplift forces (these can be reduced where deflection of the structure is acceptable).

(*g*) Consider any heave void or under-slab drainage.

(*h*) Increase the foundation element/wall depths to ensure a sufficient factor of safety against uplift, where necessary.

(*i*) Consider the design of the base slab to accommodate any net uplift forces.

(*j*) Consider the base slab/piles/walls acting as an overall foundation system to resist vertical loads.

67.3.2 Accidental loading

Accidental loss of a temporary prop. The designer should allow for the accidental loss of a temporary prop during construction, which could happen for a variety of reasons; for instance, if something were dropped from a crane. The designer must ensure an adequate load factor, and deformations must remain within repairable limits. If the design is particularly sensitive to the loss of a temporary prop, the designer may opt to specify redundancy within the propping system to account for such an eventuality.

Loss of a permanent prop. Most basement walls are unlikely to be able to survive this scenario without serious damage, and would most likely totally collapse. If this is recognised by the designer as a viable load case (for instance, in the design of metros which may be vulnerable to terrorist activity), then the designer needs to ensure that the slab/prop is structurally able to tolerate the foreseeable forces to which it may be subjected in such a scenario.

Flooding. The designer should consider whether flooding is likely, to what extent and for what duration. Flooding can increase loads on the structure in two main ways: (*a*) it may increase lateral water pressures behind the walls; and (*b*) it may apply an additional surcharge on the slabs and the soil retained by the walls.

67.3.3 Out-of-balance forces

Underground structures are often subjected to out-of-balance lateral forces for a variety of reasons. It is important to consider the interaction of out-of-balance forces in design, rather than just designing the wall with the higher lateral forces and not considering how this will affect (or be affected by) the opposing wall.

Typical causes of permanent asymmetric loads are

- ground or groundwater conditions (e.g. a sloping site, heavily variable stratigraphy or a basement adjacent to water on one side)
- surcharges adjacent to the basement (e.g. from nearby structures, roads or railways)
- superstructure of varying building heights over the basement causing an imbalance in vertical loading of walls
- substructure stiffness (e.g. different response of opposing walls formed in different manners)
- adjacent tunnels or other underground infrastructure
- pressures from adjacent ground treatment works.

In addition, it is important that the designer considers temporary imbalances such as

- loading from adjacent construction plant
- construction storage or stockpiles
- asymmetrical construction sequences.

When the design is being carried out using finite element modelling (FEM), catering for these imbalances once they have been established is relatively straightforward. For underground structures subject to significant out-of-balance forces, FEM is likely to be the most expedient way of understanding and designing for these forces. It is common to design retaining walls using limit-equilibrium or pseudo-finite element (PFE) analyses which consider a single wall only. For simpler force imbalances it is possible to iterate between two separate PFE models, each representing opposite sides of the excavation, with the prop prestress forces adjusted at each iteration until deflection and prop force are balanced across the excavation.

67.4. Resistance to vertical actions

For a basement structure it is likely (and usually economical) that the retaining wall will be used to resist vertical loading. Vertical loading on the wall is likely to be different in the temporary and permanent conditions. A typical basement structure carried out using a bottom-up methodology may be subject to the following variations in loading over its construction and use (**Figure 67.3**).

(*a*) Excavation of the basement using temporary props – walls lightly loaded.

(*b*) Completion of substructure. Vertical downwards loads – increased from (*a*) due to the slabs – are now supported by a combination of walls and base slab bearing pressure. Uplift or heave forces on the structure are to be resisted by a combination of structure dead weight and wall frictional resistance (or controlled by a heave void/drainage layer).

(*c*) Completion of the superstructure. Increased resistance to uplift/heave forces are provided by the superstructure weight. This also increases vertical downwards forces into walls/base slab.

(*d*) Redevelopment of the superstructure while reusing the basement structure – this is similar to condition (*b*).

The magnitude and direction of the vertical loading affect the direction of the friction forces acting on the wall, and hence the lateral pressure applied to the wall by the surrounding ground. The design of embedded retaining walls in the UK is often carried out using limit-equilibrium or PFE computer programs such as FREW or WALLAP, where the designer must select the magnitude and direction of wall friction when determining active and passive earth pressure coefficients. It is therefore important to consider the variation of vertical loading in the temporary and permanent cases when determining appropriate wall frictions for use in lateral design.

67.4.1 Design of retaining walls to support/resist vertical loading

The design of embedded retaining walls in the UK is described in the previous section. Typical directions of wall friction adopted for basement wall design are shown in **Figure 67.4**, which has been reproduced from CIRIA C760 (Gaba *et al.*, 2017). These assumptions of wall friction are appropriate for excavations where vertical loading of the retaining wall is minor, such as in **Figure 67.5(a)**.

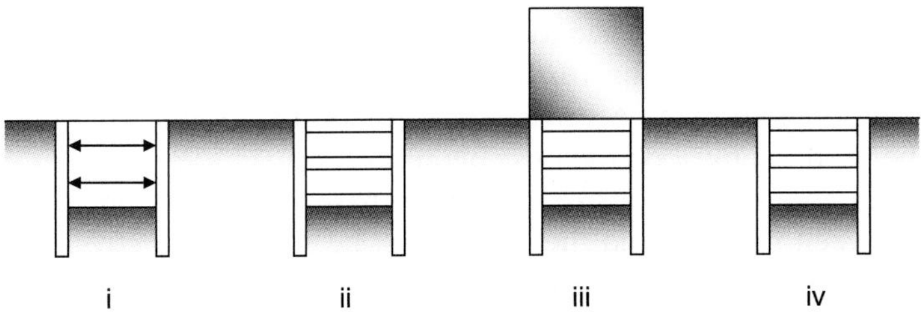

Figure 67.3 Typical construction conditions for basement wall loading

Diaphragm or piled retaining walls for deep excavations, particularly in the top-down method of construction or anchored walls, may be subject to significant vertical compressive loading during construction, in particular just prior to the casting of the base slab (**Figure 67.5(b)**). The designer must consider whether the magnitude of applied vertical loading is such that the assumed applied sense of the wall frictions reverses, as this will influence the applied horizontal pressures (see, for example, Ingram *et al.*, 2009).

If the design is carried out using appropriate finite element analysis, then compatibility of friction assumptions can be modelled explicitly.

For excavations below the water table or in heaving soils, the retaining wall may become a tension element resisting the upward pressures acting on the underside of the base slab. Under these conditions, the friction directions shown in **Figure 67.6** may be appropriate.

The change in wall frictions caused by uplift conditions will reduce the passive resistance of the ground, increasing the horizontal load in the slabs or props and modifying the bending moment and shear forces in the retaining wall. This should be considered as part of the wall design, but is often overlooked.

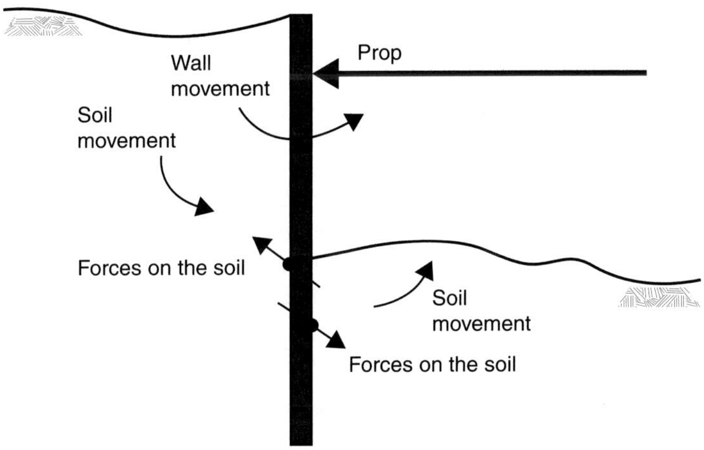

Figure 67.4 Effect of wall friction (Reproduced with permission from CIRIA C760, Gaba *et al.*, 2017; www.ciria.org)

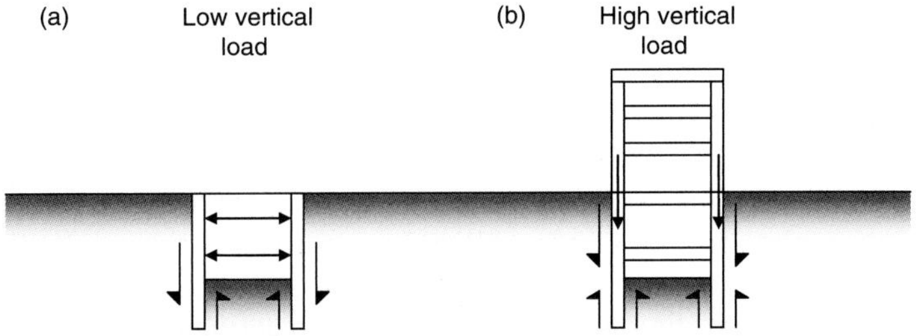

Figure 67.5 Possible wall frictions around excavations with low and high vertical loads

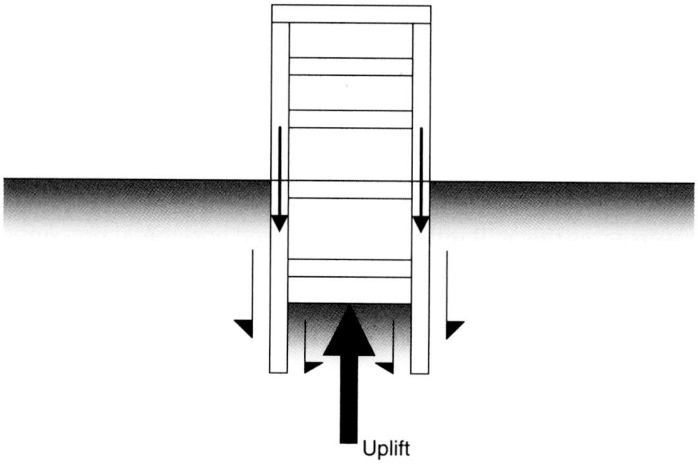

Figure 67.6 Possible wall frictions around excavations subject to long-term uplift

67.4.2 Flotation and heave

Uplift actions on underground structures can be caused by pore water pressure acting under the base slab and, in over-consolidated fine-grained soils, by soil heave pressure. The approach to dealing with the uplift pressures associated with these two phenomena should be considered early in the design process. For both processes, it is possible to either resist or dissipate the forces on the substructure.

Flotation can be accommodated in design by either dissipating under-slab pore water pressures by under-slab drainage, or by resisting the imposed forces by a combination of the building weight and soil friction acting on the walls and any tension piles. Under-slab drainage can introduce other factors such as increased maintenance and operation costs and may result in widespread settlement as a result of consolidation and/or loss of fines which could affect both the basement structure and adjacent third parties.

Possible methods of providing under-slab drainage are

- a no-fines concrete drainage blanket beneath the base slab blinding
- a series of shallow French drains beneath the base slab blinding
- a number of passive wells, discharged by artesian pressure into the base slab drainage system
- a number of active wells, discharged by pumping.

Soil heave pressure will dissipate if upward movement of the ground is allowed (see **Figure 67.7**). Vertical movement of the ground can be achieved either by specifying a heave void or by allowing the structure to rise upward (which may not be appropriate from a serviceability perspective). If required, all or some of the heave pressure can be resisted using a combination of the building weight and soil friction acting on the walls and any tension piles.

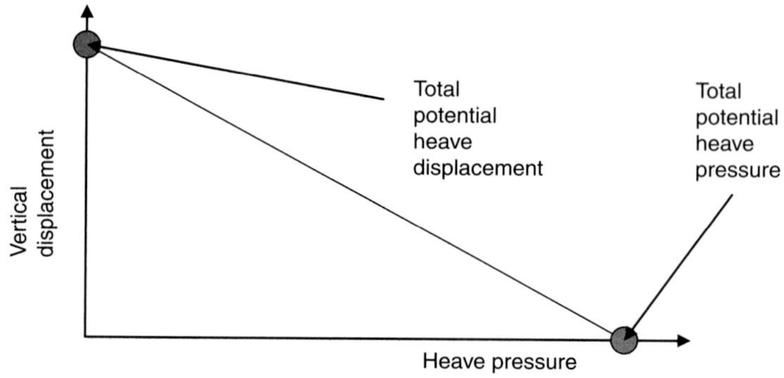

Figure 67.7 Dissipation of soil heave pressure with vertical displacement. Note: assumes linear elastic behaviour for simplicity

It should be noted that where the designer opts to dissipate the uplift/heave forces using a drainage blanket or heave void, the effective stress under the base slab can be reduced, resulting in reduced vertical load capacity of the walls and piles.

67.4.3 Design of ground-bearing base slab (for vertical downward loading and to resist heave and water pressures)

Aside from the geotechnical design of the base slab as a foundation, it is important to consider how it functions as part of the overall structure. The designer should consider the robustness of the whole structure, including connections. The base slab connects the walls to the ground below, and also to any additional piled foundations. Interaction of these various elements must be considered in design.

The base slab must be strong enough to resist the upwards flotation and heave forces, restrained by the self-weight of the box structure together with any tension forces mobilised through shaft friction in the piles and walls. It must also be able to distribute any concentrated applied loads to the supporting ground.

Provided that the appropriate overall ULS and serviceability limit state considerations are satisfied, an economic foundation which just satisfies global resistance to uplift (or downwards) loading may well have individual pile or wall elements with a lower factor of safety.

The base slab must be strong enough to safely redistribute the loading in the scenario where the individual pile or wall elements only just achieve (or less obviously, achieve more than) their theoretical working load resistance. For example, piles that perform significantly in excess of predictions may settle less under applied loads, and therefore act as stiff points, potentially attracting further load. A robust consideration of the potential relative stiffnesses of the foundation system can be used to envelope the possible responses.

In addition to the strength of the base slab, the connection between the slab and the walls must be sufficiently strong to transfer all of the possible loads between the walls and the foundation structure.

67.5. Design of bored piles and barrettes to support/resist vertical loading beneath the base slab

Where piles or barrettes are used as a foundation solution within a basement, their design has additional considerations compared to the design of piles with a cut-off level at or close to the ground surface. Excavation of the basement causes two phenomena which must be allowed for in pile design: (*a*) a reduction in overburden stress and hence ground strength; and (*b*) heave-induced ground movements.

The change in the horizontal stress state ($\Delta\sigma_h'$) in the ground following a reduction in overburden stress can be calculated by considering an equation of the type

$$\Delta\sigma_h' = \left(\frac{v}{1-v}\right)\Delta\sigma_v' \tag{67.1}$$

where v is Poisson's ratio and $\Delta\sigma_v'$ is the change in vertical effective stress due to excavation.

Piles beneath deep basements should be designed using effective stress parameters, as total stress design does not account for strength changes (unless, for example, c_u was measured after basement excavation). A detailed discussion of the effective stress approach to pile behaviour and design is given in Chapter 22 *Behaviour of single piles under vertical loads.*

Where piles are installed within a deep basement, pile reinforcement typically should extend full depth to resist heave forces. Consideration of the relative displacement and capacity of the pile shaft can be used to reduce the length of cages where the piles are not required to provide additional tensile capacity.

References

BRE (Building Research Establishment) (2005) *Concrete in the Ground.* BRE, Watford, UK.

BSI (British Standards Institution) (1990) BS 8102:1990: Code of practice for protection of structures against water from the ground. BSI, London, UK.

BSI (1999) BS EN 12063:1999: Execution of special geotechnical works – Sheet pile walls. BSI, London, UK.

BSI (2000) BS 5454:2000: Recommendations for the storage and exhibition of archival documents. BSI, London, UK.

BSI (2002) BS EN 1990:2002 + A1:2005. Eurocode: Basis of structural design (incorporating corrigendum December 2008 and April 2010). BSI, London, UK.

BSI (2004a) BS EN 1992-1-1:2004 + A1:2014: Eurocode 2: Design of concrete structures – Part 1: General rules and rules for buildings (incorporating corrigenda January 2008, November 2010 and January 2014). BSI, London, UK.

BSI (2004b) BS EN 1997-1 + A1:2013: Eurocode 7: Geotechnical design – Part 1: General rules (incorporating corrigendum February 2009). BSI, London, UK.

BSI (2009) BS 8102:2009: Code of practice for protection of below ground structures against water from the ground. BSI, London, UK.

BSI (2010) BS EN 1536:2010: Execution of special geotechnical works – Bored piles. BSI, London, UK.

BSI (2013) BS EN 1537:2013: Execution of special geotechnical works – Ground anchors. BSI, London, UK.

BSI (2022) BS 8102:2022: Protection of below ground structures against water ingress – Code of practice. BSI, London, UK.

Gaba AR, Hardy S, Doughty L, Powrie W and Selemetas D (2017) *Guidance on Embedded Retaining Wall Design.* CIRIA Publication C760. CIRIA, London, UK.

ICE (Institution of Civil Engineers) (2009) *Reducing the Risk of Leaking Substructure: A Clients' Guide.* Thomas Telford, London, UK.

ICE (2017) *Specification for Piling and Embedded Retaining Walls,* 3rd edn. Thomas Telford, London, UK.

Ingram PJ, Chodorowski AR, Anderson SE and Gaba AR (2009) Design methodology for retaining walls for deep excavations in London using pseudo finite element methods. *Proceedings of the 17th International Conference on Soil Mechanics and Geotechnical Engineering, Alexandria, Egypt,* vol. 2, pp. 1437–1440.

IStructE (Institution of Structural Engineers) (2004) *Design and Construction of Deep Basements Including Cut and Cover Structures.* IStructE, London, UK.

Further reading

Johnson RA (1995) *Water Resisting Basements: A Guide – Safeguarding New and Existing Basements against Water and Dampness. CIRIA Report R139.* CIRIA, London, UK.

Useful websites

CIRIA: http://www.ciria.org

Eurocodes: http://www.eurocodes.co.uk and http://www.bsigroup.com/templates/FourTabContent.aspx?id=147778&epslanguage=EN

ICE Specification for Piling and Embedded Retaining Walls: http://www.thomastelford.com/books/bookshop_main.asp?ISBN=9780727733580

It is recommended this chapter is read in conjunction with

- Chapter 2 *Foundations and other geotechnical elements in context – their role*
- Chapter 26 *Building response to ground movements*

All chapters in this book rely on the guidance in Sections 1 *Context* and 2 *Fundamental principles.* A sound knowledge of ground investigation is required for all geotechnical works, as set out in Section 4 *Site investigation.*

Section 7: Design of earthworks, slopes and pavements

Chapter 68
Introduction to Section 7

doi: 10.1680/icemge.66830.1127

Paul A Nowak Atkins Ltd, Epsom, UK

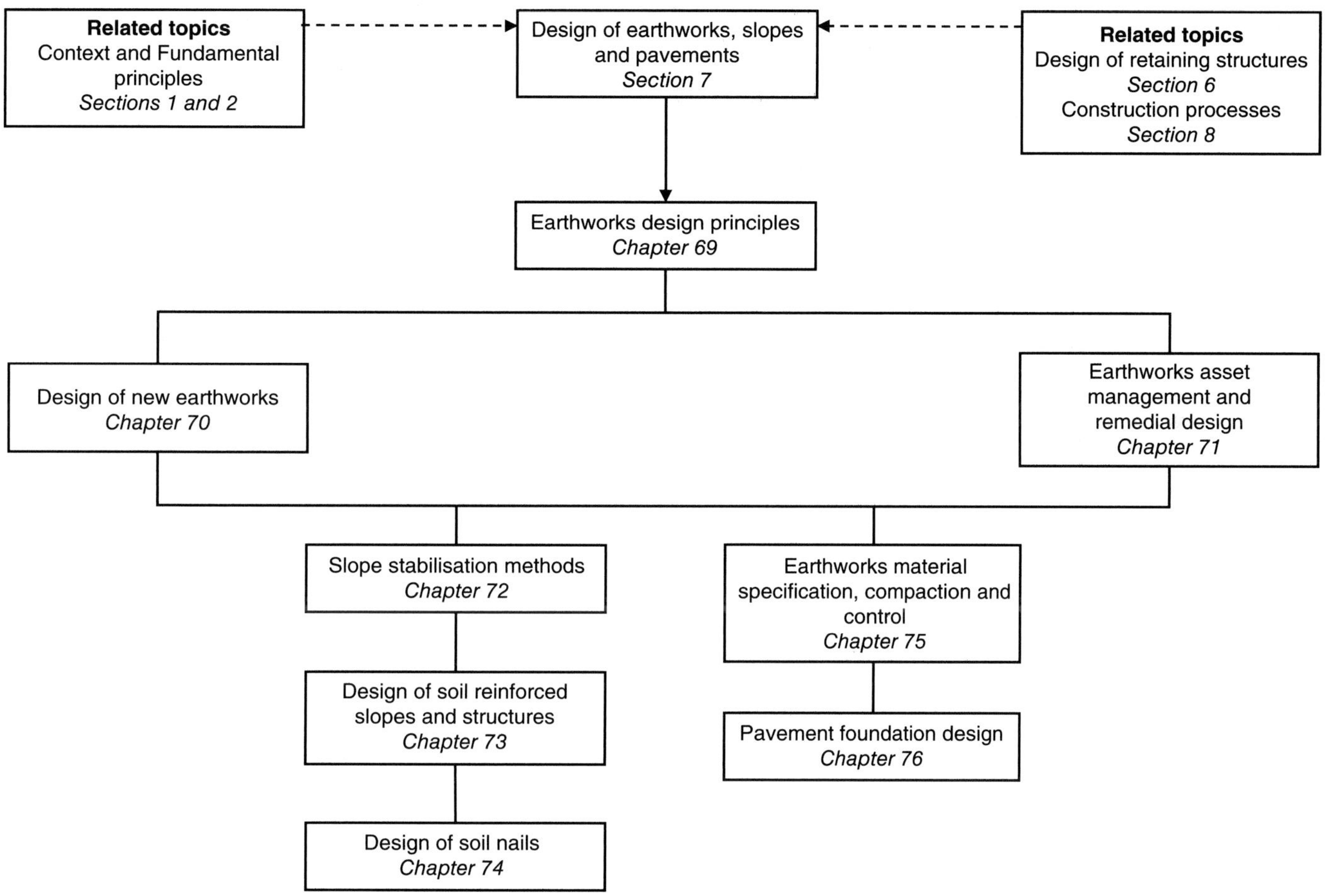

Figure 68.1 Relationships between the chapters in Section 7

Mankind has constructed earthworks in Europe since the Stone Age, but a major expansion of activity occurred in the nineteenth century with the development of the railways, in order to meet operational maximum gradient requirements.

The development of major road networks in the twentieth century continued this expansion and heralded the extensive use of earthworks plant and formal methods of compaction.

In both cases the earthworks were constructed to provide a balance between excavated and deposited material and to provide a stable platform on which the infrastructure can operate.

Current earthworks engineering can be considered to have three major streams

- design and construction of new-build routes
- alteration or widening of existing earthworks
- assessment and maintenance of existing assets.

The design of earthworks, although predominantly engineering, has always demanded a feel for material that is essentially heterogeneous and of varying properties, and the balance has changed over time. Much of earthworks construction occurred before the formal advent of soil mechanics theory in relation to earthworks engineering. Its development in the UK through the 1950s to the 1970s forms the basis of the concepts used in modern-day design. It grew in great part due to failures of constructed slopes on the railway network and slopes under construction on the trunk road network, and the need to understand failure mechanisms to prevent further earthworks failures. This development was in parallel with that in the USA, which was driven mainly by failures of embankment dams such as Teton Dam.

Although the failure mechanisms were generally different in the UK (the presence of relict slip features in new-build failures and pore water pressures in existing earthworks slopes) compared with the USA (internal erosion of embankment dam shoulders and core), the same theories were established which have allowed the development of more rigorous analysis as soil mechanics has evolved. It is interesting to note that, in both cases, the presence of water generally had a great influence on the instability.

Traditional earthworks design, where input assumptions have been tested against a varying final factor of safety against failure, has largely been replaced in the UK by the design approach of Eurocode 7, the use of which was written into EU law in April 2010. This adopts the concept of partial factors on input parameters from structural engineering and represents a change in thinking, where the choice of input parameters demands more focus, although it does not detract from previous practice. Its introduction has provoked extensive use of this new approach but has not fully replaced the use of the final factor of safety approach.

Overall earthworks design usually considers ultimate limit state failure but, increasingly, in the assessment of existing assets, serviceability limit state behaviour is an important consideration, as 'failure' can be defined as loss of service through unacceptable movement of the earthworks formation. This is particularly true in relation to secondary consolidation of normally consolidated cohesive material below embankment construction where, with concession periods of up to 30 years, the concession company rather than the overall asset owner now has to allow for additional maintenance.

Chapter 23 *Slope stability* provides an introduction to slope stability design. **Figure 68.1** outlines the layout and contents of Section 7 *Design of earthworks, slopes and pavements*. This section covers the design of both unreinforced and reinforced earthworks, the assessment of existing assets, the specification of earthworks materials and consideration of pavement design.

The future holds an interesting mix of potential earthworks problems. New-build schemes will demand a potentially differing approach from that of the maintenance and operation of existing, ageing assets and their potential replacement. The effects of climate change are likely to produce changes in rainfall patterns and a rise in sea levels. This will demand greater resilience of both existing and new-build earthworks and earth structures, and is likely to generate the need for earthworks engineering in the provision of flood defence and sea erosion protection measures.

doi: 10.1680/icemge.66830.1129

Chapter 69
Earthworks design principles

Paul A Nowak Atkins Ltd, Epsom, UK

Earthworks, in the form of cuttings and embankments, have been constructed for some 4000 years. This chapter presents a brief history of development of analysis methods and the development of factors of safety to the current use of Eurocode 7.

CONTENTS

69.1. Historical perspective

The construction of earthworks in the UK can be traced back some 4000 years to when earthworks took the form of burial mounds and mounded features such as Silbury Hill in Wiltshire (Charles, 2008).

The tradition was carried on by the Romans, whose roads tended to follow natural contours but contributed with earthworks in the form of barriers like Hadrian's Wall to keep invading tribes out of the Empire.

The bulk movement and placing of earthworks coincided with the Industrial Revolution in the mid-eighteenth century, initially in the construction of the canal network. This comprised the excavation of channels with gradients being overcome by the construction of locks – the most spectacular of which is probably the system on the Leeds–Liverpool Canal to take it over the Pennines.

The advent of railways from the 1830s required the construction of cuttings and embankments, as locomotives could not negotiate gradients of greater than 1 in 50. Initially, all bulk earthmoving was carried out using horse-drawn wagons, with the materials being end-tipped into embankments, then cut and placed at their angle of natural repose. This form of technology did not detract from large volumes of construction, and Skempton (1995) reports the movement of two to three million cubic yards of material per year over the period 1834–1841. In the late nineteenth century steam-driven equipment was introduced but the method of placing of the fill was largely the same.

Embankment settlements could be large both during and subsequent to construction and were controlled by re-ballasting of the track and the imposition of speed restrictions. McGinnity and Russell (1995) describe London Clay embankment construction on the London Underground in the late nineteenth and early twentieth centuries, where settlement during construction was topped up with London Clay and steam locomotive ash.

The development of the roads network in the UK from the 1920s required earthworks gradients of less than 1 in 10 and coincided with large-scale mechanisation of earthmoving plant in the USA. The development of mechanised plant allowed the control of fill quantity and the compaction of soil in layers for embankment construction.

69.2. Fundamental requirements of earthworks

The fundamental requirement of an earthwork is to form a stable foundation for the proposed end use over its design life. This usually represents minimal settlement for a road pavement, railway track bed or platform for industrial development.

With respect to cuttings, the fundamental requirement is the stability of the cutting side slopes which, if they fail within the design life of the earthwork, may cause the deposition of debris at the toe of the cutting or failure of material at its base, causing damage to the road surface or railway track or damage to structures at the base of a cutting.

With respect to embankments, failure of side slopes is similarly important as loss of slope could result in undercutting of the asset that the embankment supports. Additionally, the settlement of the embankment over its design life, while not necessarily contributing to catastrophic failure, may result in the loss of service of the asset due to unacceptable differential movement. This is particularly the case with respect to railways, where differential settlement between the running rails is usually set at less than 5 mm, or buildings with settlement-sensitive plant. Acceptable differential settlement may also vary along the earthwork. The most critical area is usually where an earthwork adjoins a structure. The structure abutment is designed to minimise settlement of the overall structure, which will be less than the likely settlement of the embankment adjoining the abutment. This can lead to eventual cracking of the road pavement or loss of horizontal rail alignment. In the short term it can lead to a reduction in ride quality.

In planning earthworks for a linear transportation scheme, it is usual, where practicable, to endeavour to set the vertical alignment such that the scheme is 'in balance'. This means that the volume of material excavated from cuttings is nearly the same as that required to form embankments. The calculation of overall volume will take account of the amount of cut material that is unsuitable for use as embankment fill historically. This material was likely to have been removed from the site. Common practice is now, however, to incorporate unacceptable material into landscape features, or to render it acceptable for use in the works by conditioning. This is dealt with in more detail in Chapter 75 *Earthworks material specification, compaction and control*. In calculating the overall volume of earthworks, bulking of the material from cutting to placing into embankment is sometimes considered. For most earthworks materials the volume placed in embankments is up to 5% less than the volume won from cuttings. This is due to the specification of formal compaction and the fact that mechanical placement and compaction in an embankment do not fully replicate the geological process under which the cut material was first deposited. The notable exception to this rule is unweathered chalk, which shows up to 5% shrinkage in volume from cut to fill. This is due to the fact that the material occurs naturally with a high voids ratio and open jointing – which are destroyed by the compaction process.

It is common practice to ignore the volume increase due to bulking from cut to fill when calculating earthworks quantities, or to include it but balance it with the volume of earthwork embankment lost due to settlement of the foundation material. Overall guidance to the excavation of materials and their compaction into earthworks embankments for highways schemes in the UK was given in document HA44/91 (1995) produced by the Highways Agency. More up-to-date guidance is provided by the set of European Earthworks Standards BS EN 16907-1 to 7 (BSI, 2018a–f, 2021)

69.3. Development of analysis methods

It should be noted that most of the earthworks built in the nineteenth and early twentieth centuries were constructed when soil mechanics was in its infancy. Collin (1846) produced a slip circle stability analysis, but it was not until the work of Terzaghi (1925) on the influence of pore water pressure on the deformation and strength of soils that stability and prediction of failure started to be understood.

In the 1930s the principle of effective stress for earthworks materials began to be understood and a move from empirical to numerate determination of slope stability began to be developed. Over time a number of methods have been developed to ascertain the factor of safety of new and existing earthworks slopes. The rigour of the analysis was, however, tied to the development of analysis technology, and hence in the 1950s and 1960s (before the development of computer analysis programs) more traditional methods of mathematical computation were used. The main methods of analysis can be summarised as

- infinite slope method
- stability charts
- limit equilibrium
- numerical modelling
- observational method.

The infinite slope method is a quick computational method developed for failures that occur at shallow depth in cuttings and embankments where the failure surface is assumed to be a plane parallel to the slope surface. The method is described in Skempton and DeLory (1952) and Trenter (2001).

Stability charts, like those by Taylor (1948) and Bishop and Morgenstern (1960), were developed from more rigorous analysis to produce a series of charts and tables where knowledge of shear strength, angle of friction and water conditions allowed a factor of safety to be computed for varying slope angles without further rigorous computation. These charts have been updated by Bond and Harris (2008) for use with Eurocode 7.

The limit equilibrium method was developed as the more rigorous analysis of a series of slices through the proposed slope, where interactions between the slices were computed. Bishop (1955) developed the method for analysing circular potential failure surfaces. The method for analysing non-circular modes of failure was developed by Morgenstern and Price (1965) and Janbu (1972).

Computational limitations in the 1950s and 1960s restricted the number of potential failure surfaces that could be practically analysed, and work carried out by Skempton (1964) and Chandler (1972), among others, tended to concentrate on back-analysis of slope failures, where the form of failure surface was already known. The advent of computer analysis in the 1980s has allowed these analysis methods to be applied to everyday earthworks analysis, with many failure surfaces analysed in the same computational run.

The development of finite element and finite difference analysis techniques in the 1990s has allowed more rigorous analysis of earthworks slopes to be computed from the limit equilibrium methods described above. These techniques model elements bounded by nodes and allow both ultimate failure conditions and movement of slopes, during their design life, to be determined.

The observational method of earthworks slope analysis, summarised by Nicholson *et al.* (1999), represents an approach in which design assumptions are reviewed during the construction and design life of an earthwork by use of instrumentation. The principles of the observational methods can be summarised as

- establishment of acceptable limits of behaviour
- assessment of the range of possible behaviour showing that there is an acceptable probability that the actual behaviour will be within acceptable limits – a plan of monitoring is devised which will determine whether actual behaviour lies within acceptable limits
- the response time of the instrumentation and proposals for analysing the results shall be sufficiently rapid in relation to the possible evolution of a failure mechanism
- adoption of a plan of contingency actions which may be enacted if the monitoring reveals behaviour outside acceptable limits.

More information on the analysis and stability of natural and artificial earthworks can be found in Chapter 23 *Slope stability*.

69.4. Factors of safety and limit states

69.4.1 Factors of safety

Traditionally the factor of safety against failure of an earthworks slope has been considered as a target 'lump' factor which should represent the minimum value that an analysis of failure can achieve. Trenter (2001) commented that the chosen minimum factor of safety will depend on two sets of circumstances

- technical assessment of the geotechnical data collected for a potential slip
- judgement on the safety, environmental and economic costs of any failure.

BS 6031 (BSI, 1981) gives minimum factors of safety of 1.3–1.4 for first-time failures, and 1.2 for failures involving ancient shear surfaces. BS 6031 (BSI, 2009) covers design to BS EN 1997-1.

Trenter, considering environmental, economic and safety risks, suggests the minimum factors of safety for first-time and reactivation failures of cuttings and embankments, which are summarised in **Table 69.1**.

The factors of safety for new dam construction are shown in **Table 69.2** (Johnston *et al.*, 1999).

Other studies have considered a variable factor of safety for earthworks stability considering design soil parameters and groundwater/loading conditions. Egan (2005) considered the acceptable factors of safety for Network Rail earthworks, shown in **Table 69.3**.

Perry *et al.* (2003) considered the variable minimum factors of safety for infrastructure embankments (**Table 69.4**).

	Factor of safety (first-time failure)	Factor of safety (reactivation failure)
Cuttings		
Permanent	1.30–1.50	1.10–1.30
Temporary	1.10–1.30	>1.00–1.20
Embankments		
Permanent	1.40–1.60	1.30–1.50
Temporary	1.20–1.40	1.10–1.30

Table 69.1 Typical factors of safety for cuttings and embankments (Data taken from Trenter, 2001)

Loading condition	Typical minimum acceptable factor of safety
End of construction	1.3–1.5
Steady seepage with reservoir full	1.5
Rapid drawdown	1.2

Table 69.2 Typical factors of safety for new dam construction (Data taken from Johnston *et al.*, 1999)

As can be seen from **Table 69.4**, the choice of minimum factor of safety lies with the designer and generally reflects the choice of design soil parameters, the application of loading and groundwater conditions and the influence of failure on the infrastructure asset. These aspects are discussed in greater detail in Chapter 70 *Design of new earthworks*.

The implementation of BS EN 1997-1, Eurocode 7 (BSI, 2004), introduced the concept of partial safety factors to input parameters, rather than the use of a final target 'lump' factor for the design of earthworks slopes.

The required condition to be met by the analysis is

$$E_d \leq R_d \tag{69.1}$$

where E_d is the design value of the effect of actions, and R_d is the design value of the resistance to an action.

The National Annex to BS EN 1997-1 for the UK (BSI, 2007) adopts Design Approach 1, which requires the computation of two load combinations

- Combination 1 = $A1 + M1 + R1$
- Combination 2 = $A2 + M2 + R1$

where A = actions, M = materials, and R = resistance. The partial safety factors to be applied are given in **Table 69.5**.

Analysis to Eurocode 7 is usually reported as satisfying the condition above – that is, $E_d \leq R_d$.

Bond and Harris (2008) introduce the concept of a degree of utilisation, where the analysis result is reported as a percentage of the design value of the resistance to an action, rather than a factor of safety.

Failure class	Factor of safety: Moderately conservative peak parameters	Factor of safety: Moderately conservative residual soil parameters
1. Affecting track and lineside services	1.3	1.1
2. Affecting earthwork	1.2	1.1
3. Deeper failures	Not less than preworks condition	1.1

Table 69.3 Typical factors of safety for Network Rail embankments (Data taken from Egan, 2005)

	Worst credible	Moderately conservative
Shallow failure	1.05	1.15
Deep failure	1.10	1.30

Table 69.4 Typical factors of safety for infrastructure embankments (Data taken from Perry *et al.*, 2003)

Action	Symbol	Set A1	Set A2
Permanent – unfavourable	γ_G	1.35	1.0
Variable – unfavourable	γ_Q	1.5	1.3
Variable – favourable	$\gamma_{Q,fav}$	0	0
Soil parameter		**M1**	**M2**
Angle of shearing resistance (tan ϕ)	γ_ϕ	1.0	1.25
Effective cohesion (c')	$\gamma_{c'}$	1.0	1.25
Undrained shear strength (c_u)	γ_{qu}	1.0	1.4
Unconfined strength (q)	γ	1.0	1.4
Weight density (γ)	γ_γ	1.0	1.0
Resistance		**R1**	**R2**
Earth resistance	γ_{Re}	1.0	1.0

Table 69.5 Partial safety factors for analysis to Eurocode 7

Nowak and Gilbert (2015) consider that in many cases it is advisable to design slopes using a traditional global factor of safety as this gives more flexibility in the selection and modification of input parameters to an analysis method more suited to geotechnical engineering. A check on the final proposed slope can then be made for Eurocode 7 compliance.

Additionally, it is common practice to consider that failure will occur if the analysis degree of utilisation is >100%, as it is incorrectly associated with a traditional factor of safety of 1.0. This is not the case, as a partial safety factor has been applied in the design input and a degree of utilisation of 100% would likely result in an equivalent factor of safety in the range 1.25–1.3 if characteristic design values mirror best-estimate design values from the dataset.

69.4.2 Limit states

The calculation of a satisfactory factor of safety against failure, or satisfactory achievement of design value of resistance to Eurocode 7, should result in a stability state where overall failure of the slope is unlikely to occur. This can be considered to be the ultimate limit state failure of the slope.

Conditions can, however, occur where the earthwork does not reach its ultimate failure state – movement occurs which affects its performance. This can be considered to be a serviceability limit state condition and is controlled by the performance of the infrastructure that the earthwork supports. Common examples of this condition are

- creep of cutting slopes, resulting in unacceptable movement of services (e.g. drainage, electricity or communications cables)
- settlement of embankment foundations or creep of embankment slopes, resulting in unacceptable deflections of railway tracks or highway surfacing/services.

BS 6031 (BSI, 2009) recommends that when considering deformation of earthworks, it is appreciated that they can undergo large deformations without detriment to their own serviceability, although the effect of deformations on shear strength may be sufficient to cause failures at ultimate limit state as shear surfaces develop. It is important to consider the effect of deformation of structures adjacent to or supported by earthworks, as these may control the overall earthworks design.

Deformation of earthworks over their design life can be determined by numerical methods of finite element or finite difference modelling, outlined in Section 69.2.

References

Bishop AW (1955) The use of the slip circle in the stability of slopes. *Géotechnique* **5**: 7–17.

Bishop AW and Morgenstern NR (1960) Stability coefficients for earth slopes. *Géotechnique* **10**: 129–150.

Bond A and Harris A (2008) *Decoding Eurocode 7.* Taylor & Francis, Abingdon, UK.

BSI (British Standards Institution) (1981) BS 6031: Code of practice for earthworks. BSI, London, UK.

BSI (2004) BS EN 1997-1: Eurocode 7: Geotechnical design – General rules. BSI, London, UK.

BSI (2007) UK NA to BS EN 1997-1:2004: UK National Annex to Eurocode 7: Geotechnical design. BSI, London, UK.

BSI (2009) BS 6031: Code of practice for earthworks. BSI, London, UK.

BSI (2018a) BS EN 16907-1: Earthworks Part 1: Principles and general rules. BSI, London, UK.

BSI (2018b) BS EN 16907-2: Earthworks Part 2: Classification of earthworks. BSI, London, UK.

BSI (2018c) BS EN 16907-3: Earthworks Part 3: Construction procedures. BSI, London, UK.

BSI (2018d) BS EN 16907-4: Earthworks Part 4: Soil treatment with lime and/or hydraulic binders. BSI, London, UK.

BSI (2018e) BS EN 16907-5: Earthworks Part 5: Quality control. BSI, London, UK.

BSI (2018f) BS EN 16907-6: Earthworks Part 6: Land reclamation earthworks using dredged hydraulic fills. BSI, London, UK.

BSI (2021) BS EN 16907-7: Earthworks Part 7: Hydraulic placement of extractive waste. BSI, London, UK.

Chandler RJ (1972) Lias Clay: weathering processes and their effect on shear strength. *Géotechnique* **22**: 403–431.

Charles JA (2008) The engineering behaviour of fill materials: the use, misuse and disuse of case histories. *Géotechnique* **58**: 541–570.

Collin A (1846) *Recherches Expérimentales sur les Glissements Spontanés des Terrains Argileux.* Caralian-Goeury et Dalmont, Paris, France.

Egan D (2005) Earthworks management: have we got our designs right? *Proceedings of the Conference on Earthworks Stabilisation Techniques and Innovations, Birmingham, UK*, pp. 1–13.

Highways Agency (1995) *Design Manual for Roads and Bridges: Volume 4, Section 1, HA44, Earthworks – Design and Preparation of Contract Documents.* The Stationery Office, London, UK.

Janbu N (1972) Slope stability computations. In *Embankment Dam Engineering* (Hirschfield RC and Poulos SJ (eds)). Wiley, New York, NY, pp. 47–86.

Johnston TA, Millmore JP, Charles JA and Tedd P (1999) *An Engineering Guide to the Safety of Embankment Dams in the United Kingdom*, 2nd edn. BRE, Watford, UK.
McGinnity BT and Russell D (1995) Investigation of London Underground earth structures. *Proceedings of the International Conference on Advances in Site Investigation Practice, London, UK*. Thomas Telford, London, UK, pp. 230–242.
Morgenstern NR and Price VE (1965) The analysis of the stability of general slip surfaces. *Géotechnique* **15(1)**: 79–93.
Nicholson D, Tse CM and Penny C (1999) *The Observational Method in Ground Engineering: Principles and Applications.* CIRIA Report 185. CIRIA, London, UK.
Nowak PA and Gilbert PJ (2015) *Earthworks: A Guide*, 2nd edn. ICE Publishing, London, UK
Perry J, Pedley M and Reid M (2003) *Infrastructure Embankments: Condition, Appraisal and Remedial Treatment.* CIRIA Report C592. CIRIA, London, UK.
Skempton AW (1964) Long term stability of clay slopes. *Géotechnique* **14**: 77–101.
Skempton AW (1995) Embankments and cuttings on the early railways. *Construction History* **11**: 33–49.
Skempton AW and DeLory FA (1952) Stability of natural slopes in London Clay. *Proceedings of the 4th International Conference on Soil Mechanics and Foundation Engineering, London, UK*, vol. 2, pp. 378–381.
Taylor DW (1948) *Fundamentals of Soil Mechanics.* Wiley, New York, NY.
Terzaghi K (1925) *Erdbaumechanik*. Franz Deuticke, Vienna, Austria.
Trenter NA (2001) *Earthworks: A Guide*. Thomas Telford, London, UK.

It is recommended this chapter is read in conjunction with

- Chapter 70 *Design of new earthworks*
- Chapter 71 *Earthworks asset management and remedial design*
- Chapter 94 *Principles of geotechnical monitoring*

All chapters in this book rely on the guidance in Sections 1 *Context* and 2 *Fundamental principles.* A sound knowledge of ground investigation is required for all geotechnical works, as set out in Section 4 *Site investigation*.

doi: 10.1680/icemge.66830.1135

Chapter 70
Design of new earthworks

Paul A Nowak Atkins Ltd, Epsom, UK

The design of earthworks needs to consider the interaction of a number of factors to understand their stability and performance over time. These include soil design parameters, groundwater conditions, external loading and the effects of vegetation. Additionally, for embankment design the nature of fill material and the competence of the bearing stratum need to be considered. This chapter introduces these concepts in the framework of design of cutting slopes and embankments.

An introduction to the mechanics of natural and earthworks slopes is presented in Chapter 23 *Slope stability* and an introduction to earthworks slopes can be found in Chapter 69 *Earthworks design principles*.

CONTENTS

70.1. Failure modes

70.1.1 Introduction

The design of earthworks is predominantly driven by the possibility of failure of the completed earthworks slope. The failure of earthworks can generally be considered to fall into two categories

- failure during or soon after construction
- failure during operation through the design life of the earthwork.

Although a new earthwork would be analysed in the same way to take account of both types of failure, the mechanisms of failure are fundamentally different as they are the result of a different set of circumstances.

Most modern earthworks are constructed to fulfil a 60-year design life (design life = period without significant maintenance rather than failure after the period of the design life) but throughout the UK infrastructure many earthworks have been in place for more than 150 years with no significant signs of distress, particularly on the railway network.

70.1.2 Types of slope failure

The main types of potential slope failure are summarised in **Table 70.1**. The failure mechanisms are shown in **Figure 70.1**.

The design against failure of a new earthworks slope is generally by limit equilibrium methods, described in Chapter 69 *Earthworks design principles*, and tends to concentrate on rotational circular failure, after Bishop, or rotational non-circular/translational failure, after Morgenstern and Price, as these are the analysis methods commonly available in computer analysis programs.

Compound failure, flow slides and slab slides are more difficult to predict as a failure mechanism for new earthworks. The analysis of compound failures tends to be the subject of back-analysis of failed or partially failed slopes. The factor of safety against failure by flow slides and slab slides can, however, be evaluated by the infinite slope method or by limit equilibrium methods where shallow circular failures close to the slope surface are similar in shape to shallow flow slides.

Progressive failure is discussed further in Section 70.1.4. It is difficult to analyse using limit equilibrium as the mechanism is strain-dependent and empirical adjustments are required to the measured or estimated material strength. A limit equilibrium idealisation of progressive collapse for analysis is included in the TfL *Guide for Slope Stability Analysis* (2014) and is presented as **Figure 71.2** in Chapter 71. The analysis is more suited, however, to finite element analysis methods which can be used to predict the behaviour of strain softening soils. It should be noted that calculation of progressive collapse has not predominantly been developed as a predictive failure tool but as a method of explaining the failure mechanism of failed slopes.

70.1.3 Failure during construction

The failure of earthworks slopes during and within a short period after construction is rare in modern earthworks, but when they occur they can be attributed to a number of factors. The failure of cutting slopes can usually be attributed to heterogeneity of the material in which the cutting has been formed, such as

- the presence of geological structures, bedding/laminations/fissures
- the presence of zones of contrasting permeability (e.g. interbedded sands and clays)
- the presence of historical slip surfaces.

The presence of geological structures is more common in rock slopes, but where they occur in soil slopes in the form of bedding or laminations these can form planes of preferential sliding if these surfaces are inclined detrimental to the inclination of the cut face.

Failure mode	Definition	Comments
Rotational slide	Rotation of mass of soil along curved surface	Not usually differentiated between mechanisms
Circular slide	Rotational slide on a slip surface that is approximately circular	
Rotational non-circular	Rotational slide on a slip surface that is not wholly circular	
Translational	Movement of a shallow mass of soil in a plane roughly parallel to the slope due to a weakness on a plane	
Compound	Movement of soil mass that combines the characteristics of a rotational and translational slide	
Mud flow	Translational slide in saturated soil, caused by a sudden increase in pore pressure, in which the soil flows as a viscous liquid	
Debris slide	Translational slide of debris, forming a mantle on a slope or the disturbed material at the toe of a rotational slide when rainfall or surface water causes downward movement of debris	
Slab slide	Translational slide in which the sliding mass remains more or less intact	Usually occurs in weathered surface of a slope
Progressive failure	Brittle failure of a soil mass by development of a rupture surface which migrates, causing failure before the surface fully develops	Not specifically covered by BS EN 1997-1 (2004)

Table 70.1 Main types of failure of earthworks slopes

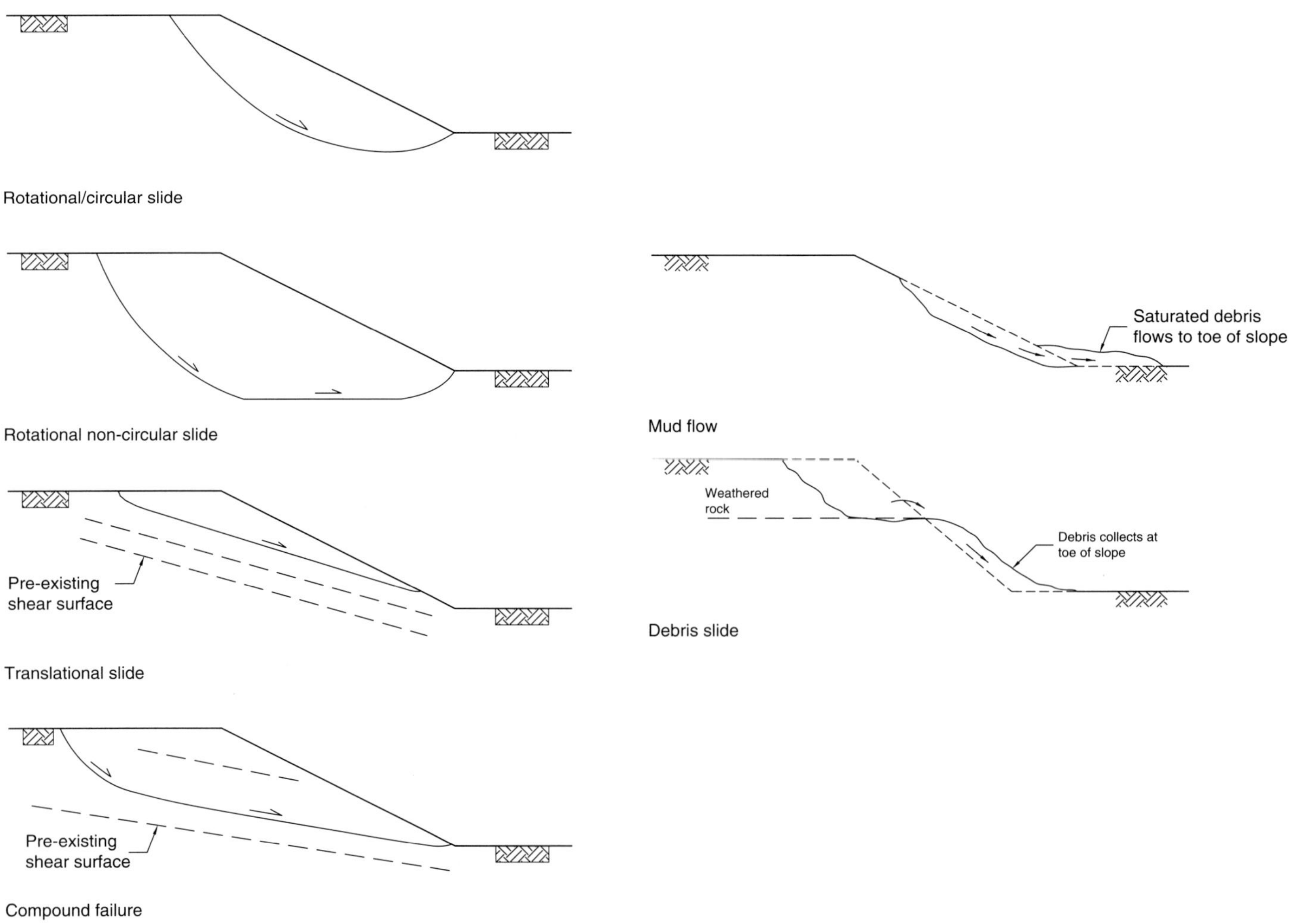

Figure 70.1 Types of failure of earthwork slopes

The presence of fissures in cohesive materials can reduce material strength over time, particularly in overconsolidated clays when pore pressures within a saturated soil mass readjust to new stress conditions.

The presence of zones of contrasting permeability can lead to large movements and slope failures on construction where the groundwater table is high in relation to depth of cutting. Where granular materials are interbedded with cohesive layers, drainage is impaired and flow of groundwater is possible only through the daylighting of the granular layers. The presence of groundwater in these layers considerably reduces their strength and flow can occur, undermining the surrounding cohesive soils and causing instability of the slope.

The M3 motorway at junction 3 is underlain by Eocene Bracklesham Beds material comprising interbedded silty fine sands and clays. The final construction earthworks slopes were designed as 1V:3H. During construction in the 1970s failure of the slope occurred due to flow of groundwater and instability of the sand layers. Stable slopes were eventually formed at 1V:10H following the employment of extensive drainage measures.

The presence of historical slip surfaces is common in areas where cohesive strata are present which have been subject to peri-glacial activity and freeze–thaw action has caused the historical movement of material on shallow slope angles. The presence of the slip surfaces considerably reduces the material strength, as described in Section 70.2.2.

Failure of cutting slopes during construction occurred in Weald Clay on the A21 Sevenoaks bypass in the 1960s, reported by Symons and Booth (1971), and in Gault Clay on the M25 motorway at Godstone in the 1970s. In both cases original cutting slopes were designed without taking reduced strength of relict slip surfaces into account. On the M25 motorway west of London shear surfaces were recognised in London Clay and Reading Beds material (Spink, 1991), and cutting slopes of 1V:6H were designed to prevent failure.

Hughes in Vasilikos (2009) reports a failure of a slope at Dromore on the M1 Belfast–Dublin motorway where a pre-existing shear zone in the Glacial Till caused failure of a 1V:2H cutting slope during a period of heavy rainfall.

The failure of embankments during construction can usually be attributed to

- slope height, geometry and angle
- foundation inadequacy
- the presence of pre-existing shear surfaces in the embankment foundation material
- change in the nature of embankment construction materials.

Failure due to slope geometry and angle are rare in current construction as adequate tools are available for the design of safe slope angles.

Similarly the adequacy of foundations can be controlled in order to prevent failure of the overall embankment slope, and these are discussed further in Section 70.2.6.

Historical failures of this type can be identified by lower slopes of low gradient, representing failures during construction. The 12-metre-high Roding Valley embankment on the London Underground Central Line was built in 1903 and comprises 4 m of London Clay at the base at side slopes of 1V:6H overlain by 8 m of ash. Contemporary accounts of its construction describe failure of the lower levels of embankment construction, sometimes overnight, due to failure of the underlying alluvial clay foundation.

Change in material type for embankment construction can present similar problems to those encountered in cuttings formed in interbedded materials described above, but usually manifest themselves as shallow translational slides close to the surface of the embankment.

Construction of the M25 motorway between Egham and the M3 in Surrey in the 1970s comprised embankments up to 10 m high built of Eocene Bagshot Beds silty fine sand. Some layers of the embankment material were laid at high moisture content which on further embankment loading caused build-up of pore pressure and subsequent loss of strength in these layers, leading to sloughing of the embankment slopes, with regrading required. Similar failures were experienced during construction of the M65 motorway in Lancashire in the 1990s, where an embankment 19 m high was constructed of layered Glacial Till and fine-grained Glacial Sand.

70.1.4 Failure during operation

The failure of an earthworks slope during operation can be generally subdivided into two modes

- shallow failure by translation or shallow circular failure
- deep failure, usually circular in nature by progressive collapse.

Shallow failures, particularly with respect to the UK highways network, have been studied over the last 30 years in an attempt to understand the mechanisms causing failure in order to minimise maintenance requirements over the design life of the earthwork.

Perry (1989) surveyed 570 km of the then constructed 2700 km of UK motorways. His study relates the angle of constructed slope for cuttings and embankments with the incidence of failure in particular geological strata. It presents a probabilistic and risk-management approach in which varying slope angles in the same material were examined for percentage of failure over the total lengths of earthwork. Stable slope angles were proposed for the various materials in cutting and embankment where incidence of failure was recorded as less than 5% of total earthwork length. Perry concluded that the failure surface rarely exceeded a depth of 1.5 m, with a minimum depth of 0.2 m and a maximum depth of 2.5 m.

Crabb and Atkinson (1991) studied failures in road embankments and cuttings in the south of England and recorded failures to 1–2 m below the slope surface. Reid and Clark (2000) observed similar depths of failure on slopes generally greater than 4.0 m in height. This concurs with the work of Perry, where shallow failures were more critical in slopes greater than 5.0 m in height.

Investigation of these failures indicates that the most likely mechanism for failure is one of moisture content or pore pressure increase in the near-surface of the embankment slope together with a potential reduction in strength properties of the near-surface material. This increase is usually due to infiltration by surface water during and after precipitation and could generally be attributed to the lack or non-performance of slope drainage.

Regular inspection of earthworks of the UK major highway network over the last 25 years has revealed not only shallow failures as described by Perry, but also signs of creep and tension cracking on existing slopes. While these mechanisms may not pose an immediate danger to infrastructure users, any lack of maintenance or remediation may cause prolongation of the potential failures in surfaces, with more significant volumes of failed material in the long term.

Perry *et al.* (2003b) record that operational serviceability failure of railway embankments is generally slow and insidious, associated with excessive movement rather than overall instability failure. Ultimate limit failure is less frequent but develops if serviceability failure is not addressed. Coppin and Richards (2007) report shallow embankment failures generally do not exceed 2.0 m in depth and are commonly translational, although some shallow circular slips can occur.

Operational failure of embankment slopes can usually be attributed to

- shrink–swell induced due to seasonal moisture changes and vegetation
- deterioration of drainage
- presence of water in the embankment over time (discussed further in Section 70.2.3).

Work by Skempton (1964) after the Second World War on failures to London Underground cuttings, predominantly in London Clay, considered failure some time after construction. The failure mechanism and type were different to that described above, involving deep-seated circular failures which caused a significant hazard to the operation of the railway. Work by Potts (1997, 2000) has postulated a theory of 'progressive collapse' for these types of failure where the fissured, overconsolidated cohesive soil acts in a brittle failure mode with an increase in pore pressure along a failure surface propagating from the toe of the slope. Failure occurs when the strength of the soil on the remainder of the surface above which the increased pore pressure is acting is less than the overall weight of material, causing it to fail as shown in **Figure 70.2**.

Ellis and O'Brien (2007) have attempted to extend this work as a predictive tool for safe slope angle and slope height where progressive collapse will not occur. Their findings show that progressive failure is unlikely for cutting slopes in London Clay shallower than 1V:3H and less than 8 m in height. For steeper slopes or those greater than 8 m in height progressive collapse may occur up to 125 years after construction.

70.2. Typical design parameters

70.2.1 Derivation of design parameters

Traditionally, for slope stability analysis to a target factor of safety 'best estimate' or 'worst credible' parameters are usually chosen from a dataset of laboratory or in situ test results or from back-analysis of failed slopes. These generally equate to characteristic and worst credible values in Eurocode 7.

Selection may be based on statistical analysis. Statistical analysis is applied to man-made construction materials, steel and concrete, which are the products of a 'controlled' industrial process and elastic perfectly plastic in behaviour. The application of statistical analysis to soils and rocks that have formed by natural processes and are heterogeneous and elasto-plastic in behaviour is questionable. Application of engineering judgement usually yields parameters of sufficient accuracy for design. A typical example is shown in **Figure 70.3**.

The selection of best estimate or worst credible parameters will lead to a different target factor of safety being chosen as an acceptable minimum output from the analysis. From the discussion in Chapter 69 *Earthworks design principles* a value in the range 1.3–1.4 is usually considered acceptable where best estimate parameters are used. Where worst credible parameters are used a minimum value in the range 1.05–1.15 would be acceptable. Worst credible parameters can also be derived from the back-analysis of failed slopes as these parameters equate to a factor of safety of just less than 1.0 immediately prior to the failure of the slope.

In deriving design values for slope analysis, the variation in parameters in relation to displacement should be considered as shown in **Figure 70.4**.

Values of c'_{pk}, φ'_{pk} reduce with greater displacement to c'_{cv}, φ'_{cv} and, in the case of cohesive soil where the plasticity index (PI) is greater than 25%, to c'_{r}, φ'_{r} when movement is sufficient to develop shear surfaces.

Best estimate parameters tend to equate to values between c'_{ck}, φ'_{pk} and c'_{cv}, φ'_{cv}, whereas worst credible parameters tend to equate to c'_{cv}, φ'_{cv} or, in the case of cohesive soils with existing shear surfaces, c'_{r}, φ'_{r}.

The adoption of Eurocode 7 (BSI, 2004) introduces the concept of a set of characteristic values for analysis which are then factored prior to analysis, as described in Chapter 69 *Earthworks design principles*.

Clause 2.4.5.2 (P) of BS EN 1997-1 defines the characteristic value as 'a cautious estimate of the value affecting the occurrence of the limit state'. BS 6031 (2009) suggests the factors affecting the choice of a characteristic design value are

- geological, historical and other background data
- the amount of measured data relating to the parameters under consideration

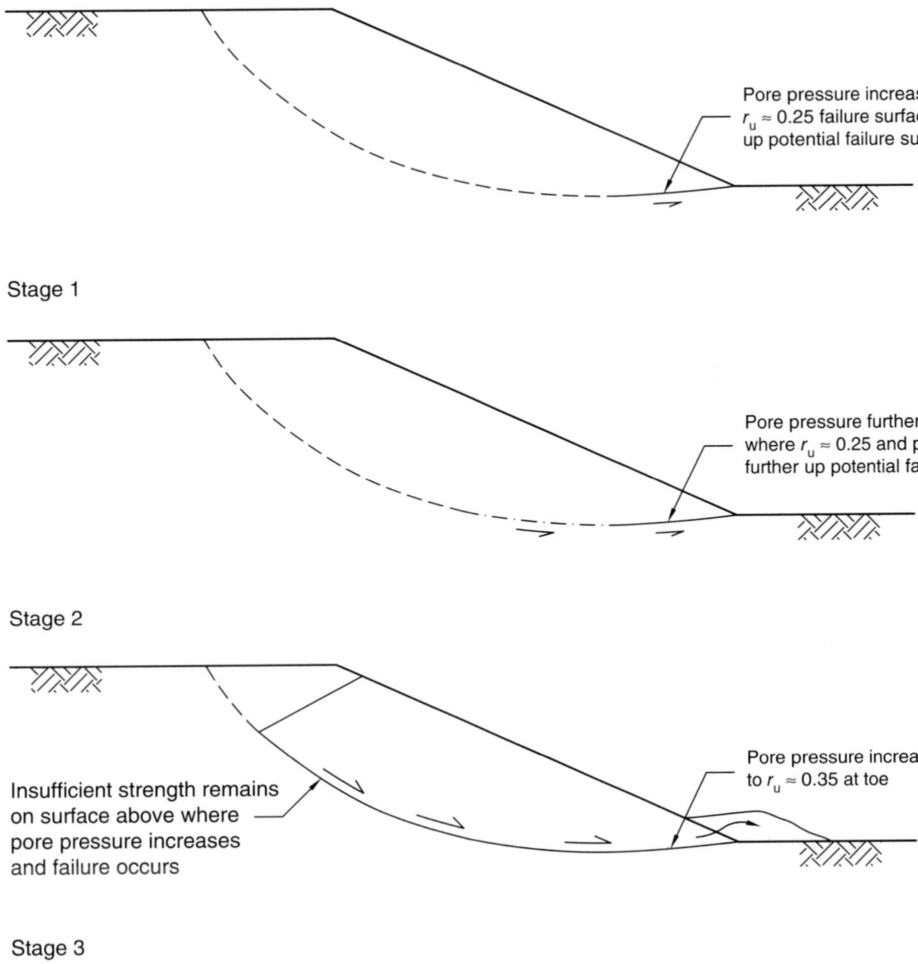

Figure 70.2 Simplified mechanism of progressive collapse

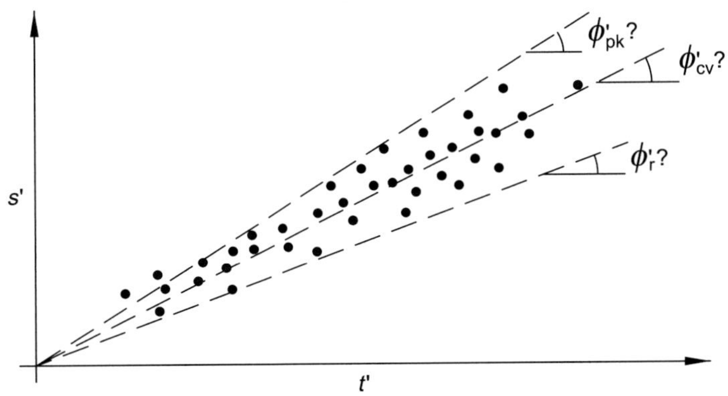

Figure 70.3 Possible selection of design parameters

- variability and degree of confidence in the measured data
- extent of the zone of ground governing the limit state under consideration
- ability of the ground to transfer load from weak to strong zones
- the consequences of failure at the limit state.

The characteristic value is, therefore, a choice through engineering judgement and although this is likely to equate to a value close to the best estimate value, it may tend to worst credible value if the consequences of failure are deemed to be exceptionally serious.

The choice of a value tending to worst credible value must be viewed in the light of the likely failure mechanism and the fact that design values will have been assigned a subconscious partial safety factor prior to those applied through Eurocode 7. Additionally, if worst case parameters are used it should be recognised that Eurocode 7 partial factors are not applied.

It may be viewed as conservative to generally use lower design parameters for long-term stability of earthworks slopes when, as described in Section 70.2.1, likely failure surfaces are shallow with no immediate effect on the overall failure of the earthwork immediately subsequent to failure.

Bond and Harris (2008) equate characteristic design values to the 'moderately conservative' value used for design of retaining walls (Gaba *et al.* 2017). This equates to a low cautious average of the design dataset. They state that 'The difference between CIRIA 104's moderately conservative value and

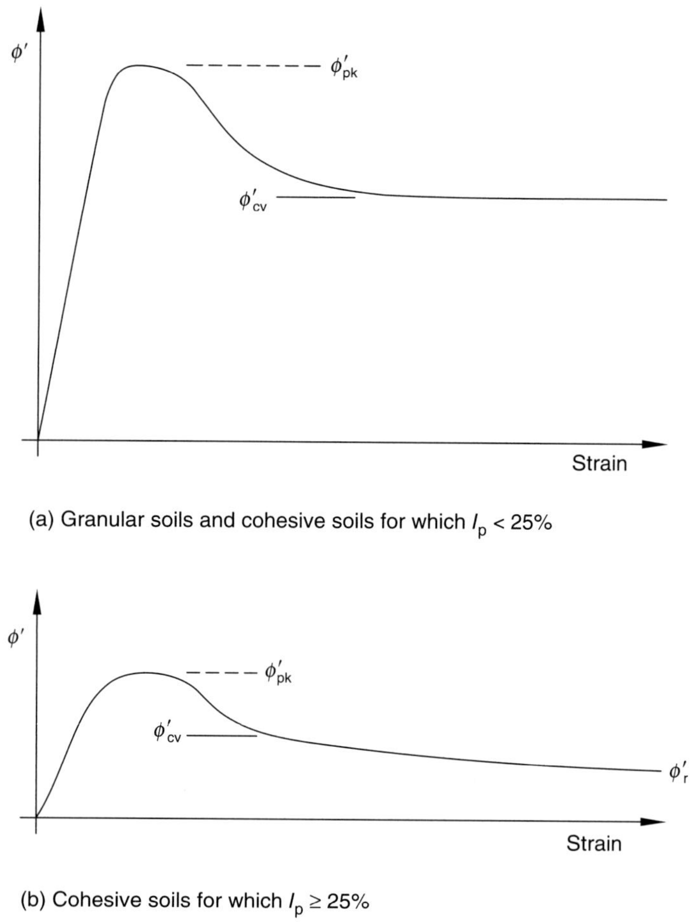

Figure 70.4 Variations of φ' with displacement (Modified with permission from BS 6031 © BSI 2009)

Eurocode 7's cautious estimate is … merely one of semantics' (Bond and Harris, 2008: p. 139). This should be treated with caution in light of the discussion above.

Bond and Harris also develop the theory of an upper and lower characteristic value from a dataset from Eurocode 7. It is considered that the derivation of a characteristic value is unique to a set of ground conditions and the impact of failure of an earthwork as defined in BS 6031, and that a single set of values only should be derived.

70.2.2 Typical design parameters

Guidance on design values where no existing data are available, or where comparison of a dataset with existing data and relationships is desirable, is available from a large number of sources, some of which are described below.

For cohesive soils, BS 8002 (2015) presents a relationship between φ'_{cv} and plasticity index (I_p). Terzaghi *et al.* (1990) present a similar relationship between φ'_{pk} and I_p. These are presented in **Table 70.2**.

Skempton (1977) reports the following design values from testing and analysis of cutting slopes in London Clay.

I_p (%)	15	30	50	80
φ'_{cv} (°)	30	25	20	15
φ'_{pk} (°)	34	28	25	22

Table 70.2 Comparison of the relationship between I_p and $\varphi'_{cv}/\varphi'_{pk}$ for cohesive soils

$\varphi'_{pk} = 20°$, $c'_{pk} = 14$ kN/m², 38 mm samples
$\varphi'_{pk} = 20°$, $c'_{pk} = 7$ kN/m², 250 mm samples
$\varphi'_{cv} = 20°$, $c'_{cv} = 1$ kN/m², back-analysis critical state
$\varphi'_{r} = 13°$, $c'_{r} = 1$ kN/m².

It can be seen that φ'_{pk} and φ'_{cv} remain the same and that the difference in design value results from a reduction in cohesion (c'). Nowak and Gilbert (2015) recommend a similar variation in the c' relationship with maximum c' of 2 kN/m² for fissured and structured overconsolidated clays and maximum c' of 10 kN/m² for non-structured clay for first time slides.

Crabb and Atkinson (1991) report φ'_{cv} values from a study of shallow failures of embankments and cuttings in south-east England, where failure was recorded at a depth of 1–2 m below the slope surface. These are summarised in **Table 70.3**.

It should be noted that the results reported by Crabb and Atkinson do not entirely correlate with the results reported by Skempton or the I_p to φ'_{cv} relationship in BS 8002. This indicates a scatter of results which is reported by Terzaghi *et al.* (1990). By way of illustration, **Table 70.4** presents a range of values reported for London Clay.

Soil design parameters, derived from previous analysis and testing, are available for selected strata in the literature, listed in **Table 70.5**.

Undisturbed samples of granular soils are more difficult to obtain from ground investigation, and design parameters tend to be taken from in situ testing, predominantly the standard penetration test (SPT) or cone penetration testing (CPT). Guidance on derivation of design parameters from in situ test results is provided in the literature from sources such as Clayton (1995) and Lunne *et al.* (2002). Design values can be derived from shear box testing (see **Figure 70.6**), but only if the in situ density of the material is replicated by the test.

Stratum	I_p (%)	φ'_{cv} (°)
Gault Clay	21–22	23
Oxford Clay	32	25
London Clay	26	25
Reading Beds	34	19
Kimmeridge Clay	27	22
Weald Clay	26	24

Table 70.3 Comparison of I_p and φ'_{cv} (Data taken from Crabb and Atkinson, 1991)

Source	c': kN/m²	φ': °
Cripps and Taylor (1981) (weathered/unweathered)	12–18/31–252	17–23/20–29
Potts *et al.* (1977)	0–20	20
LUL Standard 1-054	2	21
BAA Terminal 5, Heathrow	0–8	23
BAA M11/A120 Junction	0–5	25

Table 70.4 Comparison of c'/φ' for London Clay

Reference	Stratum
Cripps and Taylor (1981)	Mudrocks
Davis and Chandler (1973)	Mercia Mudstone
Chandler and Forster (2001)	Mercia Mudstone
Chandler (1972)	Lias Clay
Hight *et al.* (2004)	Lambeth Group
Lord *et al.* (2002)	Chalk

Table 70.5 Useful literature sources for design values of cohesive strata

On the A483 Newtown Bypass project, measured shear box values of φ' varied by up to 7° on the same material due to differences in test density and different manufacturers of test equipment.

The relationship between φ' and SPT from commonly used relationships like Peck *et al.* (1974) will generally give φ' values in the range 28–43°. These are usually acceptable for derivation of design values for cutting stability.

It should be noted, however, as described in Section 70.1.3, that the density of granular materials will be influenced by the presence of groundwater. This is particularly the case for silty fine-grained granular soils where removal of overburden below the groundwater table will effect considerable loosening of the soil structure. During the construction of the M25 motorway, Bagshot Beds strata, comprising silty fine sand with an in situ SPT N-value of greater than 50, were seen to 'flow' when excavated below the water table. This is a particular concern for the short-term stability of cutting faces in fine-grained granular soils before the groundwater table has reached its new equilibrium level subsequent to the formation of the cutting face and may require dewatering measures to be installed during construction to prevent short-term failure.

Where embankment construction comprises predominantly granular materials it is not normally common practice to test the density of the compacted material in situ and normally a conservative estimate of the φ' design value is taken. The London Underground Standard 1-054 gives a design value φ' of 35° for granular embankments composed of well-graded sandy Alluvial Terrace gravel. The implementation by the Highways Agency in the UK of BA 42/96 (Highways Agency, 1996) for the design of backfill behind integral bridge abutments has focused the minds of materials engineers to provide a low φ' material. Recent experience of shear box testing on well-graded granular materials for this purpose indicates that, when compacted to the required density, design φ'_{peak} is in the range 40–45°. The approach of BA 42/96 has now been largely superseded by British Standard Document PD6694-1:2011 (BSI, 2020), which advocates the use of φ'_{max}, after Bolton (1986).

70.3. Pore pressures and groundwater

Pore pressures within an earthwork will vary with time and can influence the factor of safety against failure. This is shown in **Figure 70.5**. At a point A beneath an embankment foundation pore pressure increases initially on loading and then reduces as dissipation occurs over time, increasing the factor of safety against failure over the design life of the earthwork. This is particularly marked in embankment foundations comprising cohesive materials. At points B and C in the slope of an earthwork, pore pressure suctions act initially and dissipate with time, resulting in a decrease in the factor of safety.

The consideration of groundwater in slope design is usually modelled as

- adoption of a pore pressure ratio, r_u
- modelling of a defined groundwater table.

The concept of r_u was developed in conjunction with the investigation of failures of existing earthworks slopes in order to simulate pore pressure conditions on a known failure surface by back-analysis. Early soil mechanics, prior to the development of the computer, investigated conditions on a single or a small number of potential failure surfaces generally by hand calculation.

It must be borne in mind that the adoption of an r_u value is unique to a given failure surface and its use in computer analysis programs will generate a variable groundwater table with each potential failure surface generated. This can lead to conservative modelling of groundwater, particularly when a free-draining material overlies a material of low permeability.

Exploratory hole data will usually allow the modelling of a defined groundwater table. The recorded groundwater level will, however, be influenced by the permeability of the soil. Whereas higher-permeability sands and gravels are likely to indicate equilibrium groundwater level as flow into an exploratory hole during its construction, the natural groundwater level in cohesive soils may not equilibrate for some time and accurate prediction will only be achieved from readings taken from piezometer installations.

In the modelling of the groundwater profile for long-term stability of earthworks slopes, the effect of the earthworks construction should be considered with respect to long-term groundwater conditions.

Where cuttings are formed in materials where the original groundwater profile is above the final finished road level this

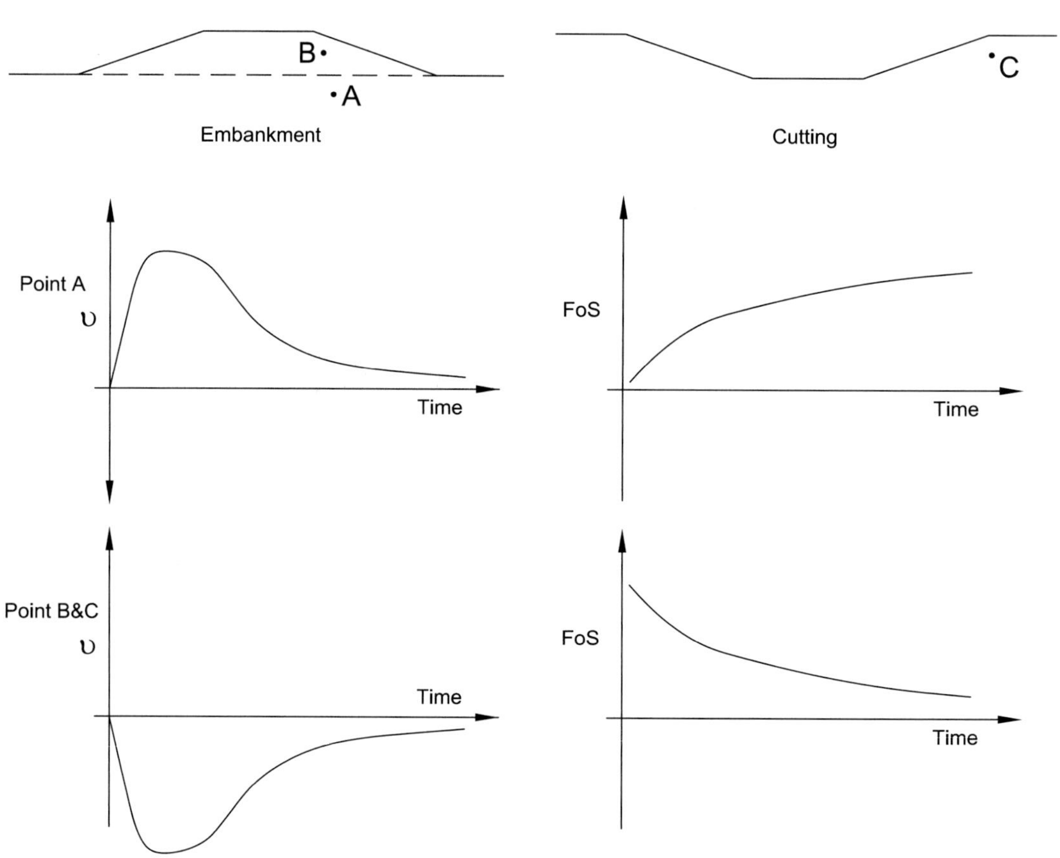

Figure 70.5 Change in pore pressure with time

will be influenced by drainage installed at the base of the cutting to develop a revised profile over time, depending on material permeability. This is illustrated in **Figure 70.6**.

Where the groundwater level is below the base of the cutting it is common to allow for a presence of groundwater in the cutting slope above the groundwater table. Farrar (1978) suggests a value of r_u in the range 0.1–0.3 should be used for slopes in cohesive materials. London Underground (2000) also suggest a similar approach for London Clay slopes with a base value of 0.25 within a range 0.15–0.35. The lower value is employed where slopes are underdrained and the higher value where no slope vegetation is present.

Groundwater conditions for embankment design differ from those in cuttings as the groundwater table will generally occur in natural ground below the embankment construction. It is usually accepted that some percolation of precipitation onto the embankment will occur over time and a nominal r_u value in the range 0.05–0.1 is adopted throughout the embankment material.

In adopting any groundwater conditions within the embankment the nature of the constructed embankment material and the likely infiltration of precipitation should be considered.

Where the embankment construction comprises granular material the permeability of the material is high and any percolating water will drain relatively easily. Where the embankment construction comprises cohesive material percolation will be slower, but once saturated the embankment material will take much longer to drain.

When considering infiltration into an embankment its use should be taken into account. A highway embankment will be capped by an impermeable surfacing and possibly verge drainage which will preclude the infiltration of precipitation, apart from on the shoulders of the embankment. A railway embankment is normally capped with permeable ballast to support the tracks and, hence, infiltration can occur across the whole area of the embankment. This is shown in **Figure 70.7**.

It is common to simulate the increase of moisture content close to the slope surface, described in Section 70.1.4, by an increase in r_u value in the upper 1.5–2.0 m of the slope surface. A typical value of 0.2 is suggested by Jewell (1996), which is similar to the range 0.1–0.3 suggested by Farrar (1978). On recent infrastructure projects designers have applied $c' = 0$ and $r_u = 0.2$ to a 'vadose' zone up to 2.0 m into the face of an earthworks slope to represent worst case pore water conditions over the design life where the surface material progressively softens. Eurocode 7 partial factors have then been applied, modelling a condition that cannot credibly exist.

Design of earthworks should consider the provision of drainage both to prevent the flow of surface water onto or below the newly formed earthwork and also to control the flow of groundwater.

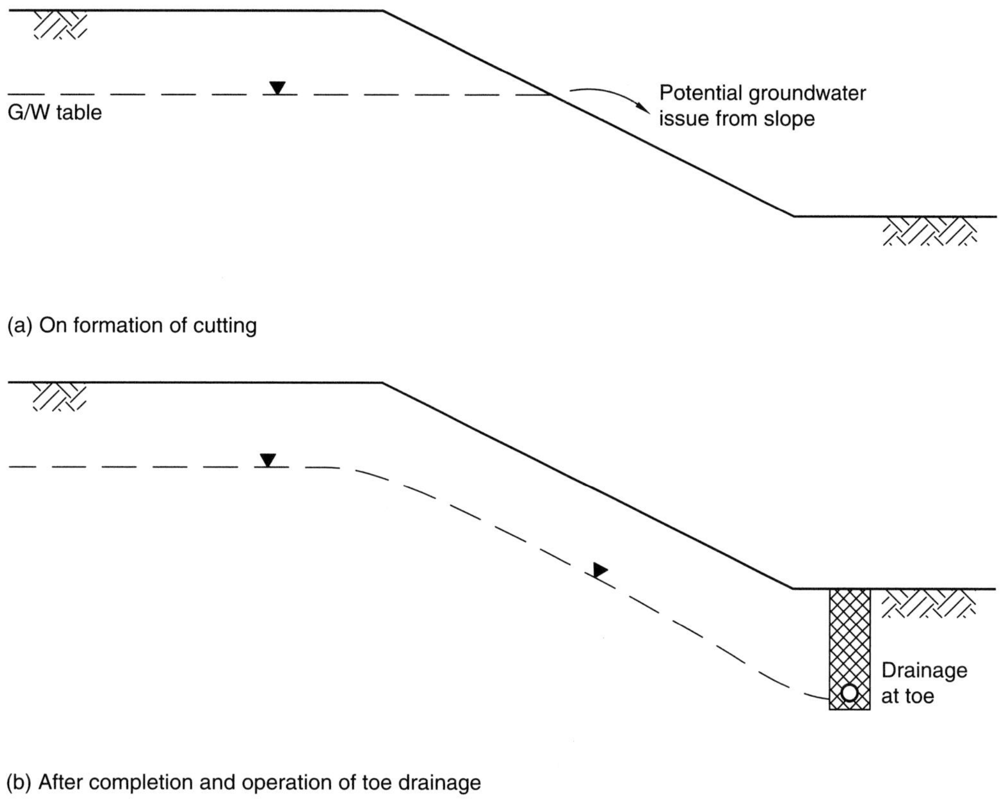

Figure 70.6 Influence of drainage of long-term groundwater level in cutting slope

The prevention of surface water flow is usually critical to cutting slopes. It takes the form of pre-earthworks drainage as it is usually constructed prior to the main earthwork slope and normally comprises the construction of a cut-off ditch at the crest of the cutting. This both collects run-off water from adjacent areas and can also collect any field drainage that may be intercepted as a result of the new cutting.

Pre-earthworks drainage in areas of embankment construction usually comprises measures to deal with existing watercourses and ditches in order that the earthwork can be constructed without detriment to the existing groundwater regime. This normally takes the form of new collector drains or culvert construction.

Interception of groundwater in a cutting face and its subsequent lowering may be critical to the short-term (construction) life of the earthwork or its long-term stability. Construction of permanent dewatering usually takes the form of gravel-filled trench drains in parallel or herring-bone pattern. Positive pumping from well points or deep wells is commonly employed for dewatering of temporary earthworks slopes and excavations. Long-term adoption of these measures is, however, normally impractical as a permanent measure due to the cost of operation and maintenance.

More details on the aspects of drainage and dewatering design are given in BS 6031 (2009), Hutchinson (1977), Preene *et al.* (2016) and Sommerville (1986).

Any drainage design should consider the discharge of collected water. This can be into existing watercourses or pipework, possibly through ponds or attenuation facilities. Alternatively, a sustainable drainage system (SUDS) can be designed. Details of SUDS systems can be found in CIRIA C697 (Woods and Kellagher, 2007). In all cases, discharge should be agreed with the relevant authorities (e.g. the Environment Agency and water companies).

70.4. Loadings

Two types of loading should be considered in the design of earthworks slopes, namely

- permanent loading
- transient loading.

Permanent loading will be applied from buildings adjacent to cutting slopes and abutments of railway and highway structures that are constructed mid-height on the cutting slope. Transient loads can be applied to earthworks in the form of

- construction loads
- stored materials/maintenance loads
- road/rail loading on embankments.

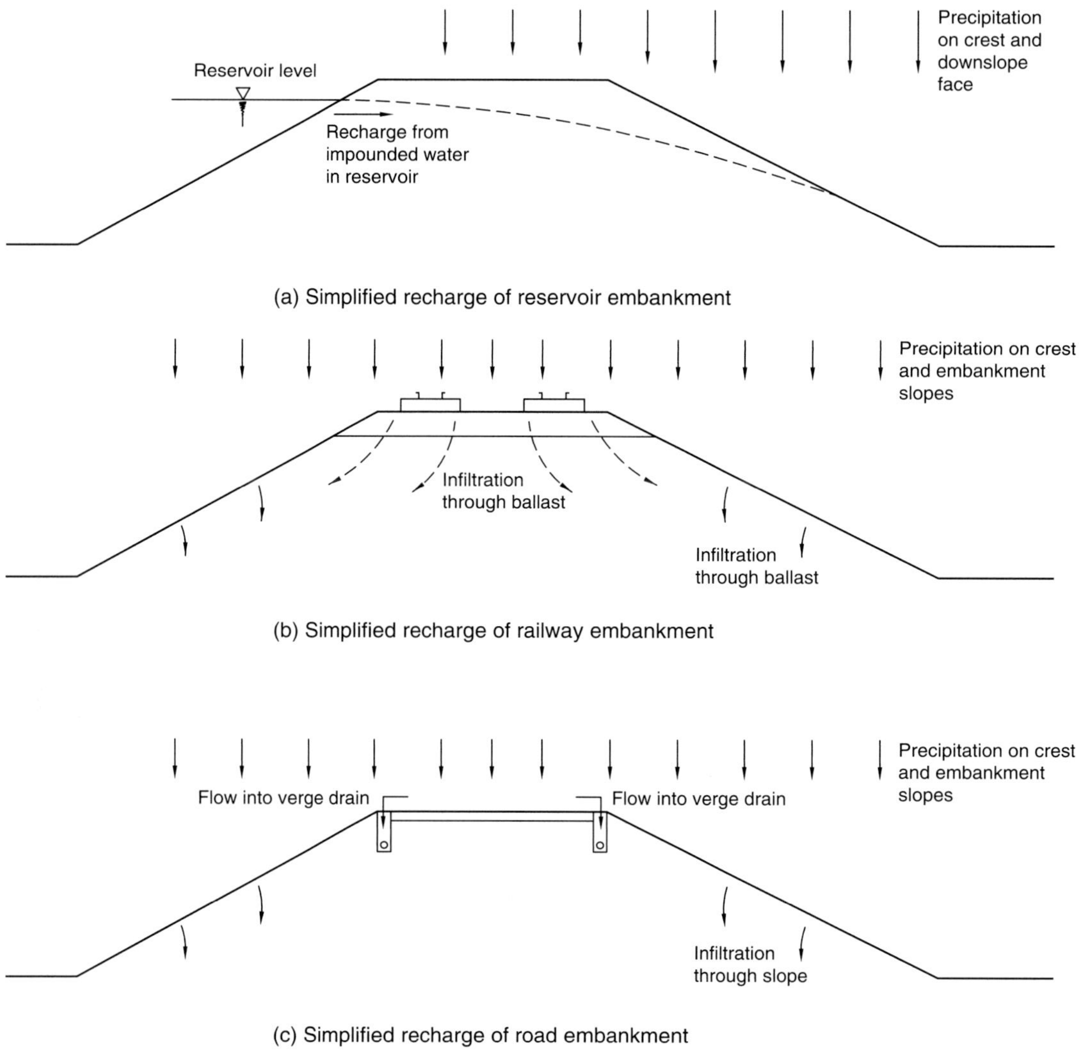

Figure 70.7 Recharge into different types of embankment

It is usual to apply transient loads only where they contribute to the potential failure of an earthworks slope. Where transient loads may be of benefit to the stability of the slope they should be ignored.

Construction loads will influence the stability of a slope in the short term and are generally applied as a uniformly distributed load (UDL) at the top of cuttings and embankments. It is common practice to set loading from construction plant at 20 kN/m^2 for normal loading conditions. Where large items of construction plant are working at the top of an earthwork, individual calculation of the applied load should be based on the weight of the plant from manufacturers' data. It should be borne in mind that large items of plant will generally operate on a construction mat comprised of granular material which will distribute the applied plant load as a UDL at the base of the mat but will also impart load in the short term.

Where materials may be stored or maintenance carried out to the earthwork over time it is usual to apply a 10 kN/m^2 UDL to take account of this eventuality. As stated above, this load would only be applied in areas where it may contribute to the failure of the slope (e.g. at the top of a cutting slope). The application of this transient load is not normally applied to cutting or embankment slope faces as significant maintenance activities or material storage is unlikely to occur at these locations.

Design of infrastructure earthworks require the application of traffic or rail loading to the top of embankment construction, as detailed in **Table 70.6**. Information on road and rail loading is also contained in Eurocode 1 (BSI, 2008, 2010).

In applying these loads for the calculation of stability of new earthworks, the following should be noted.

- The loads are derived from Eurocode 1 and the UK National Annex. In the application of the loads to structures they are in direct contact with the structure and are applied to ensure that there is sufficient redundancy in the structure design.

Standard load	UDL: kN/m²	Application
Highways Agency HA	10	Normal roads
Highways Agency HB	20	Trunk roads and motorways
Highways Agency Exceptional	37.5	Routes with exceptional traffic
Rail RL	30	London Underground and light rail
Rail RU	50	UK railway standard

Table 70.6 Typical design loads applied to structures

- CIRIA C592 (Perry *et al.*, 2003b) comments that loading should be applied only to the worst credible analysis case and not to an analysis using moderately conservative values. The loadings above are transient, and a cohesive soil is acting in its undrained condition during loading, so applying load with drained strength parameters is only suitable for the worst credible case.
- BS 6031 (2009) comments that traffic loads are normally applied to the surface of an earthwork and usually contribute only a small proportion of the total earthwork load, and thus may generally be ignored in settlement assessments unless there is a particular reason to take them into account.

On the A14 Cambridge to Huntingdon Improvement project an approach to foundation loading considering the Frequent approach in Eurocode 0 (BSI, 2002) was accepted by Highways England, which reduced the applied design loads. It was also applied to the design traffic loading on earthworks. This approach is described by Nowak *et al.* (2021).

Additionally, measurement of settlement and pore pressure increase during traffic loading was measured in a dynamic trial on a 45 km haul road embankment constructed for Tenghiz Oilfield Growth Project in Kazakhstan. The results are discussed in papers by Nowak and Barr (2022a, 2022b). Pore pressures less than 10% of the applied load were recorded in the normally consolidated clay underlying the 1.5 m high embankment construction and settlements under traffic loading were some 15% of those calculated using finite element modelling.

70.5. Vegetation

The effect of vegetation on long-term stability is not extensively considered in the design of new earthworks slopes. Perry *et al.* (2003a) detailed the main benefits of vegetated slopes as

- control of erosion by shielding against the impact of rain, wind and water flow
- enhancing shear strength through reinforcement by plant roots
- removal of water by transpiration.

These are indicated in **Figure 70.8**.

Perry *et al.* indicate that most root reinforcement extends generally to 50 mm depth below the slope surface, with tree roots penetrating up to 3.0 m depth.

Norris *et al.* (2008) build on the work of Greenwood (2004) and Norris and Greenwood (2006) in developing the concept

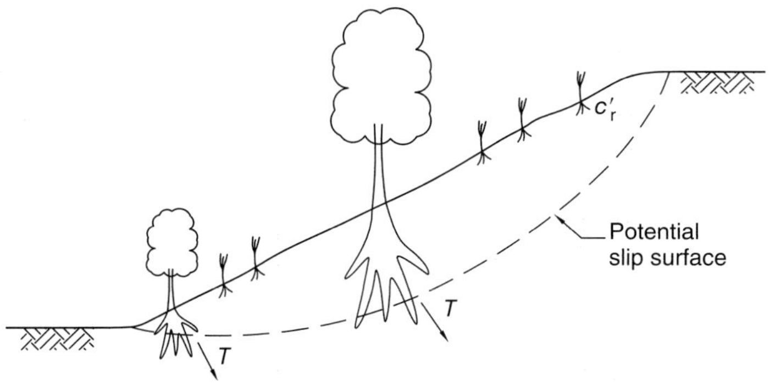

Figure 70.8 Forces exerted on a slope by vegetation (Data taken from Norris *et al.*, 2008)

of increase in soil cohesion due to root growth (c'_r). Typical values of c'_r from historical data after Norris *et al.* are summarised in **Table 70.7**.

A similar approach is developed by Coppin and Richards (2007), who indicate that an enhanced c' value of up to 20 kN/m² can be taken in the upper surface of a slope where root structures penetrate.

A practical observation of enhanced c' due to plant growth is provided by the Longham Wood Cutting Trial in the Gault Clay on the A20, reported by Greenwood *et al.* (2001).

Norris *et al.* (2008) do not attribute any specific effect of vegetation on the groundwater table, but note that pore water suctions can develop close to the surface of vegetated slopes. These can be evaluated by piezometers such as those developed by Ridley *et al.* (2002).

The benefits of this suction are difficult to quantify on a definitive basis as they will vary with vegetation pattern and geology. They are likely to be beneficial to the stability of cutting slopes, and the London Underground Standard for Earth Structures Assessment (2000) allows a reduction in design r_u of 0.05 where cutting slopes are vegetated. However, Vaughan *et al.* (2004), from work on London Underground embankment slopes, report that pore water suctions generate a seasonal pore water gradient through the embankment construction which can lead to seasonal shrink–swell movement in the form of

	c'_r: kN/m²	
Vegetation type	Maximum	General range
Grasses and shrubs	60.0	2.0–6.0
Deciduous trees	63.0	3.0–10.0
Coniferous trees	94.3	3.0–6.0

Table 70.7 Shear strength contribution from vegetation (Data taken from Norris *et al.*, 2008)

creep. This movement can eventually lead to unacceptable movement of line-side services, and London Underground have instituted a managed vegetation policy over the last few years to handle this problem. The difference between rail and highway embankments in the regime of potential recharge, described in Section 70.3, should be noted, this problem being more likely on rail embankments.

70.6. Embankment construction

Embankment construction to the early twentieth century was carried out with no formal compaction, with slopes generally formed at the angle of repose of the deposited material.

The mechanisation of earthmoving equipment, particularly in the USA in the 1920s and 1930s, created the need for control of earthworks compaction by a rapid means of determination. The work of Proctor (1933) developed the concept of 'optimum moisture content' and 'maximum dry density' for compacted material, with laboratory testing carried out prior to placement being controlled on site by the use of moisture content determination. The use of optimum moisture content for control of compacted embankment fill is described more fully in Chapter 75 *Earthworks material specification, compaction and control.*

The Casagrande system of soil classification, published in 1942, drew on the concept of the plasticity index for cohesive soil. These classifications were developed in the UK for control of compacted earthworks material by the Road Research Laboratory, leading to the development of the Department of Transport's *Specification for Highway Works* (Highways Agency, 2009). As the classifications were predominantly developed for quality control they do not translate directly into design values for earthworks materials, but design values can be derived when the ranges of acceptability set are considered.

The control of moisture content for granular materials based on optimum moisture content or cohesive materials based on plastic limit achieves a compacted density of greater than 90% of optimum and a measured density similar to that of the naturally occurring material from which they were derived. Hence, design values derived material from samples of natural material (as described in Section 70.2) will be similar when compacted into earthworks embankments.

For most granular materials an acceptable moisture content of –3% to +2% of optimum achieves a compacted density of greater than 90% of optimum, but this should be tested for all proposed materials. Care must be taken when specifying the upper limit for fine-grained granular materials, particularly those with a high silt content. These materials develop high pore pressures on compaction at moisture contents above optimum, with a resulting loss of strength, and limiting acceptable moisture content to optimum is sometimes necessary.

For cohesive materials the limiting of moisture content to around the plastic limit was not only to produce an acceptable compacted density but also to control the range of undrained shear strength of the placed material. Whyte (1982) showed that the shear strength of remoulded compacted clay is 1.5 kN/m^2 at its liquid limit and 100–130 kN/m^2 at its plastic limit (PL). The lower bound of moisture content for acceptability was set at PL –4% by the UK *Specification for Highway Works*, and this reflects a maximum undrained shear strength of approximately 200 kN/m^2, at which normal compaction plant could achieve greater than 90% of maximum dry density.

Clayton (1979) notes that the purpose of the upper limit of moisture content was set to ensure embankment stability in the short term, keep self-settlement within acceptable limits and avoid problems due to loss of trafficability of earthmoving plant on the freshly placed fill. The latter is particularly important to the economics of earthmoving, and Arrowsmith (1978) quoted minimum undrained shear strength of 35 kN/m^2 for caterpillar tractors and scrapers and 50 kN/m^2 for large rubber-tyred scrapers.

Dennehy (1978) relates remoulded undrained shear strength to tyre pressures of earthworks plant, where he indicates a limiting rut depth of 275 mm for efficient plant working. This equates to a minimum shear strength in the range 40–60 kN/m^2 for low-tyre-pressure plant and 60–80 kN/m^2 for high-tyre-pressure plant.

The Transport and Road Research Laboratory (1975) record that for scrapers with a 16–18 m^3 struck capacity, rut depth increased from 30 mm to 110 mm from a moisture content at PL to one at 1.1 × PL. This reduced average speed from 15 km/h to 5 km/h and resulted in a 75% increase in earthworks costs. The depth of rutting of recently compacted material by plant tipping the layers above increases with moisture content, and at undrained shear strength of less than 50 kN/m^2 it is likely that rutting could penetrate the previously compacted layer, where compacted layer thickness is normally of the order of 250–300 mm. Further information is also provided by Farrar and Darley (1975).

The range of multipliers on PL for upper-bound acceptable moisture content varied with material, and usually PI, but accepted limits were in the range PL × 1.1 and PL × 1.3. Arrowsmith (1978) presents a range between PL × 1.1 and PL × 1.2 for glacial clays in the north-west of England, from which a value of PL × 1.2 was chosen to produce a minimum compacted remoulded shear strength of 70 kN/m^2.

The use of moisture content, with optimum moisture content or PL, as a compliance tool does present a problem with respect to the reporting of results, particularly for cohesive materials. To comply with BS 1377, samples have to be dried for 24 hours before a moisture content can be determined. In the case of cohesive samples the dried sample is used to determine PL and a further 24 hours is required to determine the corresponding moisture content at PL. This means that in excess of 48 hours could pass between sampling and determination of compliance or non-compliance, over which time further earthworks layers will have been placed and remedial measures are difficult to effect if non-compliance is determined.

In order to overcome this time lag between sampling and confirmation of compliance, Parsons (1976; Parsons and Boden, 1979) developed the moisture condition test. This uses site-based laboratory compaction equipment to determine an immediate value which can be compared with a predetermined acceptability range from pre-earthworks testing. The pre-earthworks testing is accompanied by the determination of undrained shear strength; a correlation curve of the moisture condition value (MCV) against undrained shear strength can be plotted and MCV limits set based on the minimum and maximum compacted shear strength measured. This method was developed further by Parsons and Darley (1982) in relation to the operation of earthmoving plant.

It is usual to set the minimum MCV value at 8, which generally corresponds to a minimum undrained shear strength of 50 kN/m^2 for most cohesive materials. Correlation testing should, however, be undertaken on all materials prior to incorporation in embankment earthworks as the relationship between MCV and remoulded undrained shear strength is unique. On the M65 Blackburn bypass Glacial Till material achieved a minimum 50 kN/m^2 at MCV as low as 6.5.

The upper MCV limit is usually set at 16, but for most materials the equivalent remoulded shear strength is in excess of 300 kN/m^2 at this value. As stated above, minimum moisture content in terms of PL was set to achieve a maximum remoulded shear strength of some 200 kN/m^2. Earthmoving plant has developed since this criterion was set and a remoulded shear strength of 250 kN/m^2 will still allow greater than 90% of maximum dry density to be achieved on compaction. This is likely to equate to an MCV value in the range 13–14 for most cohesive earthworks materials.

70.7. Embankment settlement and foundation treatment

70.7.1 Failure of embankment foundation

In considering the performance of new-build embankments, other factors apart from the stability of the embankment slopes should also be taken into account. These are

- failure of the embankment foundation
- settlement of the foundation material
- self-settlement of the embankment fill.

The potential failure of embankment foundation by failure surfaces passing below the level of the embankment fill should be determined as part of the overall assessment of the stability of the embankment slopes. This failure mechanism is unlikely to occur where the embankment is underlain by granular material or overconsolidated clay. It is, however, a potential failure mechanism where the embankment is underlain by lightly overconsolidated and normally consolidated silts and clays or organic rich materials such as peat. These materials have lower strength properties in short-term undrained conditions rather than long-term drained conditions, and failures have, historically, occurred during construction, as mentioned in Chapter 69 *Earthworks design principles*. Analysis of the critical failure case should, therefore, consider undrained strength of these materials in slope stability analysis.

It is common where these materials are shallow, up to 2 m thickness, that they are removed and replaced with acceptable embankment fill, usually granular material, prior to the main embankment construction. Excavation to a greater depth can be considered if economically viable and there are no short-term stability problems with the excavated slopes. This was the case on the N8 Fermoy bypass in Ireland, where 6 m of soft material was removed prior to the construction of a 9 m high embankment and replaced with Old Red Sandstone rockfill.

Where soft material cannot be wholly removed below embankment construction and the stability of the overall embankment slope is an issue, a number of options are available to improve stability. The main options are

- geogrids or geocells at the base of embankment construction
- vibro stone columns or mix-in-place columns below embankment construction
- vibro concrete columns, rigid inclusions or driven precast piles.

The use of geogrids or geocells provides additional tensile strength along a potential failure plane in a manner similar to their use in steepened soil slopes. Design of this type of basal reinforcement is discussed in Chapter 73 *Design of soil reinforced slopes and structures.*

The use of vibro stone columns or soil mix columns produces a block of soil of greater strength than the existing ground and they are usually installed on a grid pattern over the footprint of the embankment. They perform, however, in different ways. Vibro stone columns introduce gravel size material as a column, usually up to 600 mm dia., and displace the surrounding soft soil, effecting a degree of compaction. The process produces a composite soil of bearing capacity up to 150 kN/m^2 but is generally limited to a depth of 10 m. Vibro stone columns are not very efficient in cohesive soil of undrained shear strength of less than 25 kN/m^2 as the untreated soil provides insufficient lateral restraint.

Soil mix columns introduce lime and cement into weak soils by initially mixing the column of soft soil then mixing in lime and/or cement by a dry mix process to produce a column of greater strength. The soil mix columns, normally up to 600 mm dia., are usually installed as a series of rows. The use of soil mix columns is not as cost effective in soils of high organic content as the required cement content increases significantly.

Outside the UK, jet grouting is more popular for treating weak soils. Columns of diameter up to 1.5 m can be formed to produce an overlapping grid of stiffened columns by a wet mix process.

A different mechanism of load transfer is employed if vibro concrete columns or driven precast piles are used. These methods transfer the embankment load onto a stiffer stratum through end-bearing on the column or pile.

Vibro concrete columns are installed by the same process as that to install vibro stone columns, with the same depth restrictions, but the vibrated material is a wet concrete mix. Most of the vibro concrete columns installed in the UK have been constructed to transfer load to either granular strata, weak rock (e.g. Chalk or Mercia Mudstone) or more competent bedrock.

Rigid inclusions are installed using a continuous flight auger (CFA) or a displacement auger piling rig with the bore filled with concrete and nominal reinforcement installed. Applied load from the overlying earthworks is transferred by skin friction and/or end-bearing in the underlying competent ground.

Driven precast piling is used particularly in treating ground on the approach to structures, the process for which is described in Section 70.7.2. It was extensively used as a technique on the Channel Tunnel Rail Link in areas of soft sediments in the Thames Estuary. The same contract also experimented with screw piles, a product used extensively in Belgium (Lovelace, 2002) for the same purpose. On the M74 completion project in Glasgow, driven precast piles up to 35 m long were used for all approaches to bridge abutments to transfer the embankment loading through the lightly overconsolidated Clyde Alluvium onto the Carboniferous Coal Measures bedrock below.

Typical load transfer mechanisms are shown in **Figure 70.9**.

70.7.2. Embankment foundation settlement

In addition to overall embankment failure, settlement of the founding material should also be considered as part of new embankment design.

Settlement below embankment construction, like that below structural ground-bearing foundations, can be considered to comprise three phases

- immediate elastic settlement on loading
- primary consolidation settlement
- secondary consolidation settlement.

Settlement in granular materials below embankment construction with a permeability of greater than 1×10^{-5} m/s can be considered and calculated using elastic theory such as

$$\Delta h = \Delta p \times I \times \frac{H}{E'} \tag{70.1}$$

where Δp = applied pressure (kN/m^2), I = influence factor taking account of depth below embankment foundation, H = thickness of layer in metres and E' = long-term Young's modulus of soil (kN/m^2).

For cohesive soils, settlement will occur on applied load as partly immediate elastic and partly primary consolidation. For overconsolidated soils this is usually calculated using

$$\Delta h = \delta p \times h \times m_v \tag{70.2}$$

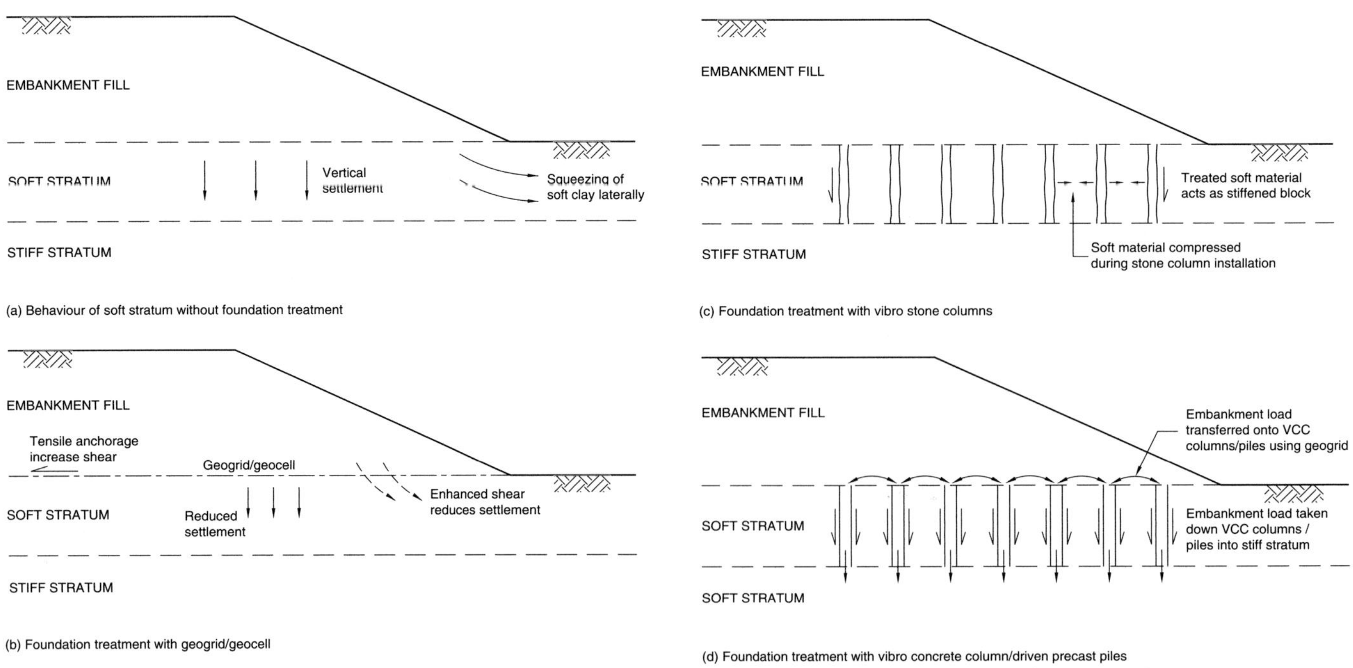

Figure 70.9 Embankment foundation treatment

where δp = applied pressure (kN/m^2), h = thickness of layer in metres and m_v = coefficient of compressibility (m^2/MN). For normally consolidated soils it is usually calculated using

$$\Delta h = \frac{C_c}{1 + e_o} \times H \times \log \frac{p_o + \Delta p}{p_c} \tag{70.3}$$

where C_c = compression index, e_o = initial void ratio, H = thickness of layer in metres, p_o = initial overburden pressure at centre of layer (kN/m^2), p_c = pre-consolidation pressure (kN/m^2) and Δp = applied pressure (kN/m^2). The parameters m_v, C_c, e_o and p_c can usually be obtained from the results of laboratory oedometer testing. m_v should be taken in the range $p_o + \Delta p$ to obtain a realistic value with regard to embankment height. C_c is taken as the gradient of the straight portion of the consolidation curve and p_c is taken from construction from the change of gradient of the consolidation curve.

The reduction in applied pressure in the soil beneath the embankment construction can be determined using elastic theory such as Boussinesq or Newmark (1942). The most appropriate is possibly that developed by Oesterberg and presented by Leroueil (1990).

It should be noted that the equations above calculate both immediate elastic and primary consolidation settlement. The proportion of each component will depend on the stress history of the cohesive soil. For an overconsolidated clay the two components are usually equal, but for normally consolidated materials the proportion of primary consolidation is greater. Further guidance is provided by Padfield and Sharrock (1983).

The rate of primary consolidation settlement is governed by the coefficient of consolidation, C_v, which can be obtained from oedometer testing and is expressed as

$$C_v = \frac{K}{m_v} \times \gamma_w \tag{70.4}$$

where k = coefficient of permeability (m/s) and γ_w = unit weight of water. C_v values from oedometer testing should be taken over a representative pressure range, as for m_v.

The time over which primary consolidation occurs subsequent to embankment construction is expressed as

$$t = T_v \times \frac{d^2}{C_v} \tag{70.5}$$

where T_v = time factor and d = length of drainage path (m).

The length of the drainage path is critical to the time required for primary consolidation to occur. If permeable materials such as granular strata both overlie and underlie the cohesive layer, two-way vertical drainage can be considered and the length of the drainage path is halved with a subsequent reduction in the time over which primary consolidation is achieved.

In layered soils horizontal drainage should also be considered, as although the drainage path to the sides of the embankment may be longer, C_h may be considerably greater than C_v and the resulting combination with a longer horizontal drainage path may result in a shorter consolidation period than the shorter vertical drainage path combined with a lower C_v value.

In some cases large consolidation settlements may be theoretically calculated, but the length of the drainage path is such that these will be realised over a long period of time. This was the case for a 12 m high embankment construction for Heathrow Airport Terminal 5 spur which was underlain by up to 30 m of London Clay. Although 400 mm of total settlement was calculated, the time to realise primary consolidation was over 200 years.

In soft compressible soils, where the pre-consolidation pressure is usually exceeded by the applied pressure from embankment loading, settlements will be large and normally unacceptable if the time for primary consolidation exceeds the construction period. A number of solutions are available to mitigate this situation, namely

- use of lightweight fill
- preloading or surcharging
- drainage
- structural elements.

The use of lightweight fill such as polystyrene, pulverised fuel ash (PFA) or lightweight aggregate will reduce the initial applied load from embankment construction and, hence, the amount of settlement, but will not influence the rate of consolidation.

Preloading, with embankment construction at the start of and prior to main construction works, will lengthen the period over which primary consolidation can occur. As the rate of consolidation settlement with time is hyperbolic, this additional time may allow sufficient primary consolidation to occur such that any remaining consolidation subsequent to completion of construction will have minimal impact on the finished works.

The use of surcharge, where greater embankment height and load are applied, results in primary consolidation occurring in response to the greater applied load, and a smaller percentage of total consolidation settlement is required over time to achieve an acceptable percentage under the original embankment loading.

Drainage usually comprises the construction of vertical drains either as vertical plastic strips or vertical sand drains. These are installed at close centres (1.0–2.5 m) in order to significantly reduce the drainage path and produce two-way drainage. A similar effect is produced by the use of vibro stone columns described in Section 70.7.1.

Structural elements generally comprise driven precast piles or vibro concrete columns and act in the same way as that described in Section 70.7.1 by carrying the applied embankment load through the soft compressible stratum to a competent bearing stratum below.

The design of embankments on soft compressible soil next to structures should also take account of the influence of embankment loading on the structure abutments, which are likely to be on piled foundations. Leroueil (1990) reports for an applied load greater than pre-consolidation pressure that horizontal movement is 20% of vertical deflection under load. This lateral squeezing of the soil will generate additional shear and bending moments which will have to be taken into account in the pile reinforcement, in addition to the requirement from structural loading. An approach to calculate additional load on structural abutment foundations is presented by Springman and Bolton (1990).

One approach to reduce any additional lateral load is by reducing applied vertical load through the use of lightweight aggregate. Another alternative is to support the earthworks approach to the abutment on a structural platform using piles, vibro stone columns, vibro concrete columns or soil mix columns. This has the advantage of not only limiting settlement, but also reducing lateral stress transferred from the soil below embankment construction onto abutment foundations. Driven precast piles have been used extensively in this application, and guidance on their use is given by Reid and Buchanan (1984). Structural support to the embankment construction behind a structure abutment also minimises the differential settlement between the earthwork and the structure, which without treatment creates a hard spot requiring ongoing maintenance.

Embankment construction on soft compressible soils will also result in an increase in pore pressure if the drainage path is not sufficiently short that it can dissipate in a reasonable period subsequent to application of load.

In these conditions, where the drainage path is not reduced by the installation of vertical drainage described above, the increase in pore pressure can be such that failure of the soft material below the embankment can occur as the pore pressure is greater than the strength of the material.

In order to prevent failure it is usual to limit the construction rate so pore pressure can dissipate during embankment lifts. The build-up of pore pressure and its dissipation would usually be observed with a piezometer installed in the soft material below the embankment construction. Additionally, staged construction may be applied where hold periods are introduced subsequent to lifts of embankment construction. This allows pore pressure to dissipate prior to construction of the next lift. Examples of staged embankment construction are presented by Leroueil (1990).

The advent of PPP (public–private partnership) contracts with longer maintenance periods has demanded that the designer consider more closely the effects of secondary consolidation with time subsequent to embankment construction. Historically, secondary consolidation has been considered mainly in peat and organic soils, where further breakdown and deterioration of the organic structure occurs subsequent to realisation of primary consolidation.

Secondary consolidation, which is load-independent, can also occur in cohesive materials, particularly normally and lightly overconsolidated clays where applied load is greater than the pre-consolidation pressure in the form of creep. For overconsolidated clays the pre-consolidation pressure is not usually exceeded and loading is applied on the shallow gradient of the consolidation curve, with the result of very small settlements.

Secondary consolidation can be determined as

$$p_c' = C_\alpha \times H \times \log\left(\frac{t_1}{t_0}\right) \tag{70.6}$$

where C_α = coefficient of secondary consolidation, H = thickness of layer (m), t_0 = time for 95% primary consolidation and t_1 = time after start of embankment construction. C_α can be derived from natural moisture content after the approach of Mesri (1973). It can also be derived as a ratio with C_c, the compression index. Mesri and Feng (1991) and Mesri and Godlewski (1977) report C_c/C_α relationships in the range 0.024–0.055 and 0.025–0.085, respectively, the higher values being reported in peats and organic soils. Andersen *et al.* (2004) report C_α values in the range 0.019–0.026 for alluvial silt and 0.055–0.108 for alluvial peat from back-analysis.

The time to reach primary consolidation is important in the calculation of secondary settlement, which may increase in the medium term if measures are taken during construction to accelerate primary consolidation. The use of vertical drainage will reduce the time to realisation of 95% primary consolidation settlement to, say, 6 months rather than 5 years without drainage. Although solving construction problems, the onset of secondary consolidation occurs at an earlier stage in the maintenance period after construction rather than at a later stage, and is of greater magnitude.

A method for reducing the amount, and onset, of secondary consolidation is the use of surcharge kept in place subsequent to achievement of the required amount of primary consolidation settlement within the construction period. This method is described by Mesri and Feng (1991) and Lambrechts *et al.* (2004).

70.7.3 Embankment self-settlement

The self-settlement of embankment fill, like that of the founding material, is dependent on fill type. Trenter (2001) reports the total self-weight settlement of fill can be calculated from

$$p_i = 0.5 \times \frac{\gamma H}{D*} \tag{70.7}$$

where $\gamma =$ bulk density of fill material (kN/m^3), $H =$ total height of embankment fill (m) and $D^* =$ equivalent constrained modulus. Watts and Charles (2015) give values of D^* for a 10 m high embankment of 50 for well-compacted sandy gravel and 6 for low-plasticity clay.

Charles (2008) reports settlements of less than 0.5% embankment height for well-graded granular embankments and less than 1% for clay embankments.

70.8. Instrumentation

Instrumentation is commonly used to confirm earthworks design assumptions and to monitor the performance of earthworks during construction. **Table 70.8** shows commonly used types of instrumentation. Instrumentation can be installed during a ground investigation, but its location should be considered as it may be destroyed by earthmoving plant during construction.

The installation of embankment monitoring equipment is usually carried out as a pre-construction activity and, in the case of piezometers and rod and plate settlement gauges the reading tubes are extended as embankment construction proceeds. Survey monitoring has the disadvantage that survey stations cannot usually be installed until earthworks are complete, and movement during construction may have been missed.

The use of piezometers should be cognisant of the time over which readings are required and the permeability of the stratum into which they are installed. Whereas a rapid response can be expected in granular materials, the low permeability of cohesive materials is such that a period of time is required for the groundwater level in normal standpipe piezometers to stabilise. If readings are required rapidly after installation, pneumatic or vibrating wire piezometers are preferable.

The observational method (Nicholson *et al.*, 1999) is now commonly used to monitor existing assets, particularly in the railway environment, where instrumentation monitors the medium- and long-term performance of an earthwork, rather than employing immediate remediation, where initial analysis of stability indicates that the current factor of safety against failure is lower than would be normally acceptable.

The observational method is also used in earthworks construction. In this case a suitable contingency measure needs to be determined prior to construction to consider the situation where the installed instrumentation is showing the onset of potential failure. The contingency measure needs to be commensurate with the nature of the earthwork and the time for its instigation. An example is the temporary stockpiling of soil to act as a rapid toe-weight in the case of earthworks cutting failure.

Instrumentation	Earthwork	Purpose
Piezometer	Cutting	Confirmation of in situ pore water pressure
	Embankment	Increase in pore pressure of founding material during construction
Inclinometer	Cutting	Movement of cutting slope
	Embankment	Movement of founding material beyond embankment toe
Settlement loop	Embankment	Settlement of founding material
Rod and plate	Embankment	Settlement of founding material
Survey monitoring	Embankment and cutting	Movement of earthwork during and after construction

Table 70.8 Commonly used instrumentation

References

Andersen EO, Balanko LA, Lem JM and Davis DH (2004) Field monitoring of the compressibility of municipal solid waste and soft alluvium. *Proceedings of the Fifth International Conference on Case Histories in Geotechnical Engineering, New York, NY*, paper 8.08.

Arrowsmith EJ (1978) Roadworks fills: a materials engineer's viewpoint. *Proceedings of Clay Fills Conference, London, UK.* ICE, London, UK, pp. 25–36.

Bolton MD (1986) The strength and dilatency of sands. *Géotechnique* **36**: 65–78.

Bond A and Harris A (2008) *Decoding Eurocode 7.* Taylor & Francis, Abingdon, UK.

BSI (British Standards Institution) (2002) BS EN 1990: Eurocode: Basis of structural design. BSI, London, UK.

BSI (2004) BS EN 1997-1: Eurocode 7: Geotechnical design. BSI, London, UK.

BSI (2008) BS EN 1991:2003: UK National Annex to Eurocode 1: Actions on structures – Part 2: Traffic loads on bridges. BSI, London, UK.

BSI (2009) BS 6031: Code of practice for earthworks. BSI, London, UK.

BSI (2010) BS EN 1991-2:2003: Eurocode 1: Action on structures – Part 2: Traffic loads on bridges. BSI, London, UK.

BSI (2015) BS 8002: Code of practice for earth retaining structures. BSI, London, UK.

BSI (2020) PD6694-1:2011+A1:2020: Recommendations for the design of structures subject to traffic loading to BS EN 1997-1:2004+A1:2013. BSI, London, UK.

Chandler RJ (1972) Lias Clay: weathering processes and their effect on shear strength. *Géotechnique* **22(3)**: 403–431.

Chandler RJ and Forster A (2001) *Engineering in Mercia Mudstone.* CIRIA Report C570. CIRIA, London, UK.

Charles JA (2008) The engineering behaviour of fill materials: the use, misuse and disuse of case histories. *Géotechnique* **58**: 541–570.

Clayton CRI (1979) Two aspects of the use of the moisture condition apparatus. *Ground Engineering* **12(2)**: 44–48.

Clayton CRI (1995) *The Standard Penetration Test (SPT): Methods and Use.* CIRIA Report R143. CIRIA, London, UK.

Coppin NJ and Richards IG (2007) *Use of Vegetation in Civil Engineering.* CIRIA Report C708. CIRIA, London, UK.

Crabb GI and Atkinson JH (1991) Determination of soil strength parameters for the analysis of highway slope failures. *Proceedings of the International Conference on Slope Stability, London, UK*. Thomas Telford, London, UK, pp. 13–18.

Cripps JC and Taylor RK (1981) The engineering properties of mudrocks. *Quarterly Journal of Engineering Geology* **14**: 325–346.

Davis AG and Chandler RJ (1973) *Further Work on the Engineering Properties of Keuper Marl*. CIRIA Report 47. CIRIA, London, UK.

Dennehy JP (1978) The remoulded shear strength of cohesive soils and its influence on the suitability of embankment fill. *Proceedings of Clay Fills Conference, London, UK*. ICE, London, UK, pp. 87–94.

Ellis EA and O'Brien AS (2007) Effect of height on delayed collapse of cuttings in stiff clay. *Geotechnical Engineering* **160**: 73–84.

Farrar DM (1978) Settlement and pore-water pressure dissipation within an embankment built of London Clay. *Proceedings of Clay Fills Conference, London, UK*. ICE, London, UK, pp. 101–106.

Farrar DM and Darley P (1975) *The Operation of Earthmoving Plant on Wet Fill*. Special Report SR351. Transport and Road Research Laboratory, Crowthorne, UK.

Gaba A, Hardy S, Doughty L, Powrie W and Selemetas D (2017) *Guidance on Embedded Retaining Wall Design*. CIRIA Report C760. CIRIA, London, UK.

Greenwood JR, Vickers AW, Morgan RPC, Coppin NJ and Norris JE (2001) *Bioengineering, The Longham Wood Cutting Field Trial*. CIRIA Project Report 81. CIRIA, London, UK.

Greenwood JR, Norris JE and Wint J (2004) Assessing the contribution of vegetation to slope stability. *Geotechnical Engineering* **157**: 199–208.

Hight DW, Ellison RA and Page DP (2004) *Engineering in the Lambeth Group*. CIRIA Report C583. CIRIA, London, UK.

Highways Agency (1996) Section 3, BA 42/96: The design of integral bridges. In *Design Manual for Roads and Bridges*, vol. 1: *Highway Structures: Approval Procedures and General Design*. The Stationery Office, London, UK.

Highways Agency (2009) *Specification for Highway Works*. The Stationery Office, London, UK.

Hutchinson JN (1977) Assessment of the effectiveness of corrective measures in relation to geological conditions and types of slope movement. *Bulletin of the International Association of Engineering Geology* **16**: 133–155.

Jewell RA (1996) *Soil Reinforcement with Geotextiles*. Special Publication 123. CIRIA, London, UK.

Lambrechts JR, Layhee CA and Straub NA (2004) Analyzing surcharge needs to reduce secondary compression at embankment interfaces. *Proceedings of the GeoTrans Conference, Los Angeles, CA*, pp. 2048–2057.

Leroueil S (1990) *Embankments on Soft Clays*. Ellis Horwood, Chichester, UK.

London Underground (2010) *Civil Engineering: Earth Structures*. Engineering Standard E1-054, A3. London Underground, London, UK.

Lord JA, Clayton CRI and Mortimore RN (2002) *Engineering in Chalk*. CIRIA Report C574. CIRIA, London, UK.

Lovelace N (2002) Screw piles to be used on CTRL. *New Civil Engineer*, 7 March: p. 7.

Lunne T, Robertson PK and Powell JJM (2002) *Cone Penetration Testing in Geotechnical Practice*. Spon Press, London, UK.

Mesri G (1973) Coefficient of secondary consolidation. *Journal of the Soil Mechanics and Foundations Division* **99(1)**: 123–137.

Mesri G and Feng TW (1991) Surcharging to reduce secondary settlements. *Proceedings of the International Conference on Geotechnical Engineering for Coastal Development, Yokohama, Japan*, vol. 1, pp. 359–364.

Mesri G and Godlewski PM (1977) Time and stress compressibility inter-relationships. *Journal of the Geotechnical Engineering Division ASCE* **103(GT5)**: 417–430.

Newmark NM (1942) Influence charts for computation of stress in elastic foundations. University of Illinois Bulletin 338. University of Illinois, Urbana, IL.

Nicholson D, Tse C-M and Perry C (1999) *The Observational Method in Ground Engineering: Principles and Applications*. CIRIA Report R185. CIRIA, London, UK.

Norris JE and Greenwood JR (2006) Assessing the role of vegetation on soil slopes in urban areas. Geological Society of London Paper 744. Geological Society of London, London, UK.

Norris JE, Stokes A, Mickovski SB, *et al.* (eds) (2008) *Slope Stability and Erosion Control: Ecotechnological Solutions*. Springer, Dordrecht, the Netherlands.

Nowak P and Barr J (2022a) TengizChevroil future growth project: haul road design, construction and operation, *Proceedings of 5th International Seminar on Earthworks in Europe, Prague, Czech Republic*, pp. 281–290.

Nowak P and Barr J (2022b) TengizChevroil future growth project: the design of temporary haul roads. *Proceedings of the Institution of Civil Engineers: Geotechnical Engineering*, 10.1680/jgeen.22.00090.

Nowak PA and Gilbert PJ (2015) *Earthworks: A Guide*, 2nd edn. ICE Publishing, London, UK.

Nowak PA, Patel B and Sandberg J (2021) Variable loads for foundation design: experience from the A14 project. *Proceedings of the Piling 2020 Conference* (Higgins K, Ainsworth Y, Toll PG and Osman A (eds)). ICE Publishing, London, UK, pp. 79–84.

Padfield CJ and Sharrock MJ (1983) *Settlement of Structures on Clay Soils*. CIRIA Special Publication 27. CIRIA, London, UK.

Parsons AW (1976) *The Rapid Measurement of the Moisture Condition of Earthworks Materials*. Report LR750. Transport and Road Research Laboratory, Crowthorne, UK.

Parsons AW and Boden JB (1979) *The Moisture Condition Test and Its Potential Applications in Earthworks*. Report SR522. Transport and Road Research Laboratory, Crowthorne, UK.

Parsons AW and Darley P (1982) *The Effect of Soil Conditions on the Operation of Earthmoving Plant*. Laboratory Report LR1034. Transport and Road Research Laboratory, Crowthorne, UK.

Peck RB, Hanson WE and Thorburn TH (1974) *Foundation Engineering*. Wiley, New York, NY.

Perry J (1989) *A Survey of Slope Condition on Motorway Earthworks in England and Wales*. Report RR199. Transport and Road Research Laboratory, Crowthorne, UK.

Perry J, Pedley M and Brady K (2003a) *Infrastructure Cuttings: Condition Appraisal and Remedial Treatment*. CIRIA Report C591. CIRIA, London, UK.

Perry J, Pedley M and Reid M (2003b) *Infrastructure Embankments: Condition Appraisal and Remedial Treatment*. CIRIA Report C592. CIRIA, London, UK.

Potts DM, Kovacevic N and Vaughan PR (1997) Delayed collapse of cut slopes in stiff clay. *Géotechnique* **47(5)**: 953–982.

Potts DM, Kovacevic N and Vaughan PR (2000) Delayed collapse of cut slopes in stiff clay. Discussion by Bromhead and Dixon and authors' reply. *Géotechnique* **50(2)**: 203–205.

Preene M, Roberts TOL and Powrie W (2016) *Groundwater Control: Design and Practice*, 2nd edn. CIRIA Report C750. CIRIA, London, UK.

Proctor RR (1933) The design and construction of rolled earth dams. *Engineering News Record* **111(9)**: 245–248; **111(10)**: 216–219; **111(12)**: 348–351; **111(13)**: 372–376.

Reid JM and Clark GT (2000) *A Whole Life Cost Model for Earthworks Slopes*. Report 430. Transport Research Laboratory, Crowthorne, UK.

Reid WM and Buchanan NW (1984) Bridge support piling. *Piling and Ground Treatment: Proceedings of the International Conference on Advances in Piling and Ground Treatment for Foundations, London, UK*. Thomas Telford, London, UK, pp. 267–274.

Ridley A, Brady KC and Vaughan P (2002) *Field Measurement of Pore Water Pressures*. Report 555. Transport Research Laboratory, Crowthorne, UK.

Skempton AW (1964) Long term stability of clay slopes. *Géotechnique* **14**: 77–101.

Skempton AW (1977) Slope stability of cuttings in brown London Clay. *Proceedings of the 7th International Conference on Soil Mechanics and Foundation Engineering, Tokyo, Japan*, vol. 3, pp. 261–270.

Sommerville SH (1986) *Control of Groundwater for Temporary Works*. CIRIA Report R113. CIRIA, London, UK.

Spink TW (1991) Periglacial discontinuities in Eocene clays near Denham, Buckinghamshire. In *Engineering Geology Special Publication 7*. Geological Society, London, UK, pp. 389–396.

Springman SM and Bolton MD (1990) *The Effect of Surcharge Loading Adjacent to Piles*. Contractor Report 196. Transport and Road Research Laboratory, Crowthorne, UK.

Symons IF and Booth AI (1971) *Investigation of the Stability of Earthwork Construction on the Original Line of the Sevenoaks Bypass, Kent*. Report LR393. Transport and Road Research Laboratory, Crowthorne, UK.

Terzaghi K, Peck RB and Mesri G (1990) *Soil Mechanics in Engineering Practice*, 3rd edn. Wiley, New York, NY.

Transport and Road Research Laboratory (1975) The effect of soil conditions on the productivity of earthmoving plant. Leaflet LF510. Transport and Road Research Laboratory, Crowthorne, UK.

Transport for London (2014) *Earth Structures: Guide for Slope Stability Analysis*. Guidance Document G0054B. TfL, London, UK.

Trenter NA (2001) *Earthworks: A Guide*. Thomas Telford, London, UK.

Vasilikos P (2009) Report on BGA meeting on Irish Glacial Till. *Ground Engineering* **42(9)**: 11–12.

Vaughan PR, Kovacevic N and Potts DM (2004) Then and now: some comments on the design and analysis of slopes and embankments. *Advances in Geotechnical Engineering: The Skempton Conference, London, UK*. Thomas Telford, London, UK, pp. 241–290.

Watts K and Charles A (2015) *Building on Fill: Geotechnical Aspects*, 3rd edn. BRE, Watford, UK.

Whyte IL (1982) Soil plasticity and strength: a new approach using extrusion. *Ground Engineering* **15(1)**: 16–24.

Woods B and Kellagher R (2007) *The SUDS Manual*. CIRIA Report C697. CIRIA, London, UK.

It is recommended this chapter is read in conjunction with

- Chapter 69 *Earthworks design principles*
- Chapter 94 *Principles of geotechnical monitoring*

All chapters in this book rely on the guidance in Sections 1 *Context* and 2 *Fundamental principles*. A sound knowledge of ground investigation is required for all geotechnical works, as set out in Section 4 *Site investigation*.

doi: 10.1680/icemge.66830.1155

Chapter 71
Earthworks asset management and remedial design

Brian T McGinnity retired, London Underground, London, UK
Nader Saffari London Underground, London, UK

CONTENTS

Earthworks are the most common product of civil engineering activities and perform a crucial function in the efficient operation of civil infrastructure.

This chapter provides guidance on the asset management, stability, condition appraisal and stabilisation of existing earthworks.

71.1. Introduction

Earthworks or earth structures are embankments and cuttings that allow transport infrastructure, whether it be a railway, highway or waterway, to maintain a certain line, level and grade through the landscape. Embankments are made from materials placed on natural ground and are commonly composed of soil or rock excavated from elsewhere. A cutting is constructed through high ground to maintain vertical alignment. The material excavated from a cutting is often used to build adjacent sections of embankment.

Asset management is a specialist field in its own right, and this chapter does not attempt to cover it in detail other than to comment on some of the principal considerations of asset management where they relate to earthworks and to guide the reader to other references where a wider or more detailed appreciation of the subject is required (Hooper *et al.*, 2009).

The key texts on earthworks asset management to which reference should be made are Perry *et al.* (2003a) and Perry *et al.* (2003b). The recommendations of these CIRIA publications are relevant to all earthworks asset managers in the UK.

71.1.1 History

Embankments and cuttings form civil engineering structures known as earthworks or earth structures. They are the most common product of civil engineering activities. Little can be constructed without some excavation and transfer of soil or rock. The total length of embankments in the UK is considerably longer than that of bridges.

It was not until relatively recently that earthworks were designed. Earthworks for roads, canals and railways in the eighteenth and nineteenth centuries were built at slope angles that were based on experience, and essentially by trial and error. If failure occurred, which it often did, the slope was slackened or other remedial works implemented (Skempton, 1996). These earthworks are generally operating at lower safety factors and steeper slopes than would be considered prudent today.

The quality of construction of these older earthworks was often poor in comparison to modern engineering practice. Before the 1930s nearly all embankments were constructed of relatively uncompacted material as the process of compaction was poorly understood, and due to the lack of modern construction plant. Large settlement commonly occurred soon after construction. The legacy of these construction methods is reflected in the performance of these historic embankments and, hence, in the degree of current maintenance.

Since the 1930s the development of construction plant and a greater understanding of the discipline of geotechnical engineering have resulted in modern earthworks being better compacted and built at generally less steep slopes than historic earthworks. As a result, modern earthworks in the UK suffer less from settlement due to their better compaction. However, slope failures still occur (Perry, 1989).

71.1.2 Asset management systems

All earthworks should be managed by an appropriate asset management system to ensure that acceptable performance is achieved and that the earthworks do not present a risk to users. CIRIA reports C591 and C592 (Perry *et al.*, 2003a, 2003b) outline the requirements for cost-effective earthworks asset management, which are very different to conventional engineering design, as the process is mainly one of comparing relative risks across the network – for example, does earthwork X need more urgent attention than earthwork Y?

Cost-effective asset management allocates sufficient maintenance resources, within budgetary constraints, for efficient performance. Until fairly recently in the UK a reactive approach to earthwork management has frequently prevailed, but this is disruptive, inefficient and uneconomic, and inconsistent with long-term asset management objectives. Engineering performance has consequently suffered as a result. More recently there has been an increasing trend to adopt a more proactive asset management approach. This requires the implementation of a reliable system of condition appraisal, maintenance and repair or renewal so that existing earthworks can be kept in good condition, avoiding service loss and minimising expensive unplanned works.

In the UK, the formal documentation of good asset management practices was aided by the publication of the Publicly Available Specification document PAS 55 in 2004, which was substantially revised in 2008 (BSI, 2008). This was followed by the International Standard ISO 55000, developed under the leadership of the Institute of Asset Management and first published in 2014 (BSI, 2014).

The nature of the asset management system will vary to reflect the use of the earthworks and the risks they pose. An asset management system should as a minimum include

- an asset catalogue – an inventory and summary of condition
- performance requirements, service level and required duty
- an asset strategy and plan, including consideration of whole-life costs and deterioration models
- a risk register.

Typical performance requirements for earthworks include

- safety and reliability
- satisfying level of service and cost of service requirements
- operational efficiency
- satisfying statutory and regulatory obligations
- value for money and business improvement
- minimising environmental impact
- maximising environmental value.

Figure 71.1 illustrates a simplified earthworks asset management cycle, which follows a continuous process of condition appraisal, maintenance and improvement, repair or renewal, leading to a constant awareness of and improvement in asset

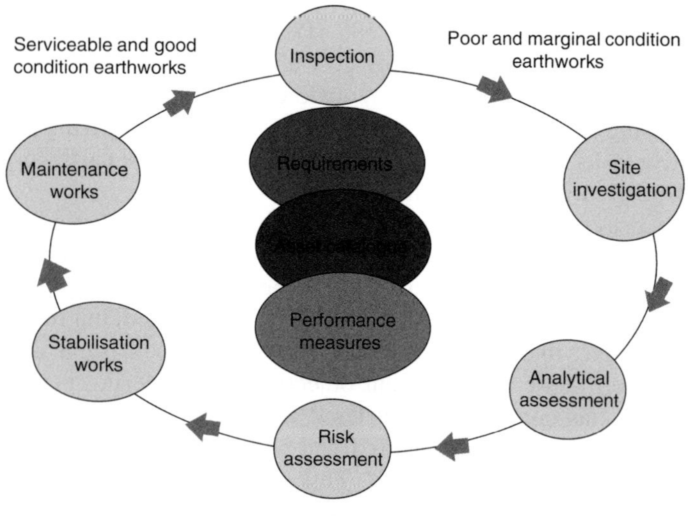

Figure 71.1 Simplified earthwork asset management cycle

condition and performance. An integral part of this process is the management of asset data and the provision of information links between condition appraisal and intervention planning.

An earthworks asset management system (EAMS) comprises a framework that provides asset life cycle management from conception and design, construction, the work management required during the asset's operation and the planning for renewal, repair and replacement. An EAMS can support all condition appraisal and maintenance work management processes from the creation of work orders, through issuing of work, to the recording of work done. It can be a powerful support tool assisting asset managers to ensure

- an earthwork's condition is improved to reduce the risk to safety and service loss
- an earthwork achieves a life expectancy that meets the performance requirements
- asset management decisions are made on an economic whole-life basis
- the knowledge and understanding of an earthwork is constantly improved.

The level of complexity and sophistication required of such a system will depend on the size and character of the earthwork asset base. Historically, earthworks management has relied upon written records, including card-indexing systems and paper-based files. Although such systems may still be adequate for the management of very small numbers of assets, the demands of managing large numbers of earthworks and dealing with an ever-increasing quantity of information mean that computer-based database systems, relying on digital data storage and retrieval, are now the norm.

Like any other structure, earthworks will have a finite life before significant renewal or replacement works are required. This lifespan can be extended by systematic, continued maintenance and repair.

It is very common for earthwork assets to be in the operate-and-maintain phase of their life cycle, with no likelihood of closure or decommissioning. In the UK, much of the earthwork asset base has already been in service for longer than the design life of equivalent modern assets.

71.1.3 Whole-life asset management

Whole-life asset management balances maintenance, repair, refurbishment, renewal, replacement and upgrade activities to optimise the long-term value of an asset. The concept of whole-life costing is described by Perry *et al.* in CIRIA Report C592 (2003a). It can be a useful tool when comparing the capital cost of renewal schemes against longer-term maintenance costs. Although whole-life asset management is a potentially useful tool, it does have its limitations, particularly for existing earthworks that are expected to have very long service lives. These limitations should be understood and the process used with care to ensure sensible results. In practice it is difficult to set up a reliable model for the management of existing earthworks because the performance requirements, the availability of expenditure and an appropriate discount rate over the likely service life are very difficult to estimate. There is a risk that whole-life asset management models can become immensely complicated; however, if they are too simplistic this defeats the whole object of the exercise and their results may be misleading.

The characteristics of transport infrastructure earthwork assets bring their own particular challenges with respect to whole-life asset management, including

- managing a portfolio of largely existing assets that cannot economically be replaced, but must be maintained and in many cases see their economic lives extended
- the implications of practical asset life extension from current engineering design codes and dealing with uncertainties associated with inherently variable earthworks.

71.1.4 Risk assessment

A unique aspect of earthwork assets, as opposed to highways, tracks or bridges, for example, is their inherent variability. Even if our information and knowledge of future performance were perfect (which it never is), there would still be a variability associated with our future predictions, related to the uncertain behaviour of geological materials in earthworks. The behaviour of earthwork assets is also significantly affected by environmental conditions, notably surface water and groundwater (including rainfall events), which are also uncertain, particularly as it is likely the UK will experience more extreme weather conditions with ongoing climate change. A risk-based approach is essential to understand this variability, and to assign probabilities to future behaviour rather than presenting future behaviour as if it were certain. The primary output of a risk-based approach will be the ability to make informed budget allocation decisions with a proper understanding of the associated risk. A full approach will also include appreciation of future changes in environment and asset duty. For the earthwork asset this would include further uncertainty, such as the impact of climate change.

The need for earthwork risk assessment also arises in the UK to satisfy statutory safety obligations. The Management of Health and Safety at Work Regulations (HMG, 1999; HSE, 1999) require employers to put in place arrangements to control health and safety risks, including a statutory duty to conduct regular risk assessments.

Under UK safety law, the Health and Safety at Work etc Act (1974), asset management of existing earthworks has to follow the approach of reducing risk levels to ‘as low as reasonably practicable’ (ALARP), giving consideration to cost–benefit analysis of whether spending on interventions will result in a cost benefit over a realistic service life.

The ALARP principle allows that safety improvements should not be pursued at any cost, but only if the cost of averting a risk is not grossly disproportionate to the risk averted. The risk assessment process, therefore, permits a logical analysis of the safety risks; how these can be mitigated or controlled; and the associated funding needs and financial forecasts. Minimal extension of the scope of the risk analysis can have secondary benefits, such as identifying the exposure to non-safety-related business risks and helping to formulate business cases and overall spending forecasts and plans. As described by ISO 55000 for asset management (BSI, 2014), decision-making requires understanding and resolution of cost, risk and performance aspects. Risk assessment is, therefore, a tool that earthworks asset managers apply to ensure that safety objectives are met within a business framework, and that funds are justified and allocated in response to safety and business needs.

The management of risk should also be a key aspect of earthwork repair or renewal projects, as in all construction activity (Godfrey, 1996), and should include

- programme, quality and financial risks to ensure the successful delivery of the project
- health and safety and environmental risks to satisfy statutory requirements.

All UK earthwork repair or renewal projects will come under the requirements of the Construction (Design and Management) Regulations (2015), known as CDM. All those involved in design and construction activities associated with earthwork repair and renewal should consider the requirements of the project and seek to follow the ‘spirit of CDM’ to ensure the approach taken is appropriate for the project.

A project risk register should be established and include all risks identified by those involved with the design and construction of earthworks, and those involved in repair and renewal.

71.1.5 Sustainability

Geotechnical engineers and earthworks asset managers need to consider not just the technical aspects of the work, but how it can be undertaken in a way that will enhance the environment and maximise sustainability – for example, by maximising the use of existing fill materials in earthwork remedial works’ design and construction, and with a sensitivity to the surrounding environment. The project should aim to minimise its carbon footprint by minimising the use of concrete and steel and opt for softer solutions such as regrading, soil nailing and similar solutions. This can also be achieved through innovation in material used and efficiency in design methodology and the design parameters in order to reduce the scope of structural inclusions. The fact that many of the organisations responsible for earthworks have environmental policies and have set specific targets relating to environmental issues is testimony to their importance.

However, it should be appreciated that this is a rapidly developing field and further changes are likely. All those involved in earthworks asset management will need to be well informed and vigilant in understanding the consequences of changing legislation on environmental issues (also see Chapter 11 *Sustainable geotechnics*).

71.1.6 Earthwork asset interfaces

Earthworks can carry a number of different infrastructure elements, such as structures, services, drainage or signage. It is important to coordinate all maintenance activities so that best use is made of time and that the works to all elements are integrated to prevent damage by piecemeal working. Asset managers should be kept informed of and approve all activities to be undertaken involving earthworks, as it is possible for overall performance to be adversely affected by the activities of others (e.g. an excavation for a service at the toe of the slope could trigger slope movement).

Conversely, the designer should assess the potential impacts of earthwork repair or renewal on existing adjacent infrastructure. In all these cases earthwork activities can change the loading on the existing infrastructure or modify the surface water and groundwater flow paths, both of which could be detrimental to the stability of the adjacent assets and, therefore, require consideration within the design. Where the existing slope is identified as being of poor stability, then the design of the temporary works will require special consideration to ensure that stability is maintained throughout construction.

Typical activities that could interface with an earthwork are

- vegetation management
- track maintenance and renewal work
- drainage maintenance and renewal work
- cable route management system
- new construction within, above or below the earthworks.

A management system should be set up such that any activities that might impact the future access, performance and safety of the earthworks is coordinated and reviewed in order to ensure the impact is safely and appropriately managed.

71.2. Earthwork slope stability and performance

Earthwork slope stability depends on the capability of the soil or rock mass to withstand its gravitational forces, the additional loads acting on the slope and potential dynamic loads (such as that of an earthquake). Methods of analysis for soil slope stability are well established (Powrie, 2018) and after determining soil properties and groundwater conditions an appropriate slope angle for long-term stability can be calculated. Slope movement under the influence of gravity can be subdivided into three major classes: slides, flows and falls (Bromhead, 1992). The major differences between these three are in the way in which movement takes place. In a slide, the moving material remains largely in contact with the parent or underlying rocks during the movement, which takes place along a discrete boundary shear surface. The term flow is used when the material becomes disaggregated and can move without the concentration of displacement at the boundary shear. Although a flow can remain in contact with the surface of the ground it travels over, this is by no means always the case. Lastly, falls normally take place from steep faces in soil or rock, and involve immediate separation of the falling material from the parent rock or soil mass, with movement involving only infrequent or intermittent contact thereafter, until the debris comes finally to rest.

71.2.1 Mechanisms

71.2.1.1 Shallow instability

Shallow instability is usually confined to the sloping part of the cutting or embankment and the failure surface does not pass through the transport corridor (railway or highway) or the associated services usually located at the toe of the cuttings or on the crest of embankments. The failure surface is generally translational (parallel to the slope surface) and has typically a depth of 1–1.5 m.

In granular materials this type of failure manifests itself as ravelling of material down the slope and collecting at the toe. In cuttings this will result in debris in the margins of the transport corridor at the toe and will require to be cleared through routine maintenance. On embankment slopes, ravelling often undermines support to routeway equipment such as overhead masts, gantries, signals and cable routes.

In cohesive soils this mechanism is generally controlled by the surface saturation within the zone of seasonal variation caused by near-surface rainfall infiltration, in particular following a prolonged dry season.

Vegetation on the slope surface generally has a positive effect on this type of instability through the reinforcement provided by the roots (Coppin and Richards, 2007) and the control of the pore pressures within the slope surface. In cohesive soils seasonal shrink and swell will cause surface cracking of the slope during the dry period, leading to increased permeability.

71.2.1.2 Deep-seated instability

For deep-seated instability the failure surface is more extensive, covering the whole length of the slope, and may pass through the soil layer beneath the embankment or cutting. The slip surface is generally deeper than 1–2 m and affects the transport corridor or the associated services located on the crest of embankments or the toe of cuttings. This type of failure, when it happens, is more disruptive and can have serious consequences, such as the derailment of trains.

For free-draining granular soils deep-seated instability is not generally a major concern and is generally easier to predict. The instability in this type of material mostly manifests itself as surface ravelling of materials downslope, leading to maintenance problems.

Deep-seated instability in cohesive soils can be categorised as first-time slides and slides along old failure surfaces.

First-time slides

First-time slides happen to slopes that have not experienced previous failures and, hence, do not involve pre-existing shear surfaces. First-time slides have been studied extensively, in particular in relation to cuttings in stiff clays (Chandler, 1984; Chandler and Skempton, 1974; DeLory, 1957; Skempton, 1948, 1964, 1970, 1977).

The results of these studies indicated that failure generally occurs many years or decades after construction due to a very slow rate of pore pressure equilibration (delayed failure). The results also indicated that the average strength at failure was considerably smaller than the peak soil strength measured in the laboratory, but greater than the residual strength. The results of back-analysis of long-term failures in five cuttings in brown weathered London Clay by Skempton (1977) indicated the following operating effective strength parameters at failure

$$\phi' = 20°, \qquad c' = 1\ \text{kN/m}^2 \tag{71.1}$$

with the average pore pressure ratio r_u typically ranging between 0.25 and 0.35.

The peak and residual strengths measured in the laboratory were:

$$\phi'_p = 20°, c'_p = 7\ \text{kN/m}^2\ \text{(Sandroni, 1977)} \tag{71.2}$$

$$\phi'_r = 13°, c'_r = 1\ \text{kN/m}^2\ \text{(Skempton and Petley, 1967)} \tag{71.3}$$

It is now well established that the most likely reason for the difference in the peak strength and the operating strength at failure is due to the mechanism of progressive failure (Ellis and O'Brien, 2007; Potts *et al.*, 1997).

The mechanisms of delayed failure and progressive failure referred to above are defined as follows.

Delayed failure

The deep-seated delayed failure of UK cutting slopes in stiff clays has been the subject of extensive study, with a significant body of published research (e.g. Potts *et al.*, 1997; Skempton, 1964). When a cutting is excavated in stiff overconsolidated clays of high plasticity due to unloading effects, the pore pressures in the soil are reduced sometimes even to negative values. With time these pore pressures will start to increase towards the long-term equilibrium values and the soil will start to swell. This process will lead to a reduction in the mean effective stress and the onset of delayed failure. The length of time to this failure will depend on the permeability of the soil and it may take several decades (Potts *et al.*, 1997; Vaughan and Walbancke, 1973).

Progressive failure

Progressive failure occurs as a result of the non-uniform development of strains along a failure surface, leading to non-uniform mobilisation of shear strength. This phenomenon is specific to clays that exhibit highly brittle behaviour, defined as the magnitude of the drop from peak to residual strength (Ellis and O'Brien, 2007; Potts *et al.*, 1997; Vaughan, 1994).

In a cutting or embankment formed in such a clay material, a non-uniform loading due to swelling of the soil near the toe will result in the mobilisation of the peak strength in an element of the soil while the other parts are still in the pre-peak state of stress. This is accompanied by the development of a slip surface at this location. With further increases in pore pressure the strength at this location is reduced from the peak towards the residual strength. Further increases in pore pressures will lead to the development of the rupture surface further into the slope. This will in turn lead to larger displacements and the final failure of the slope. At this time the strength along the rupture surface will be close to residual at some points, post-peak at some other points and close to peak along the rest. Hence, the average strength of the soil along the failure surface will be somewhere between the peak and residual strength of the soil.

Based on case studies and applied research (Chandler, 1984; Cooper *et al.*, 1998; Take and Bolton, 2004), conceptual models for progressive failure have been developed in both embankments and cuttings. **Figure 71.2** shows a conceptual model for a typical embankment on the London Underground network. In this model the rupture surface developed at the toe is shown as a weakened basal layer with shear strength at or close to the residual strength. These models can be set up in a slope stability limit equilibrium analysis and using appropriate soil parameters and pore pressures the safety factor for this failure mechanism can be determined. Further details on the progressive failure and the use of a conceptual model in limit equilibrium slope stability analysis can be found in London Underground Guidance Document G0054B (2019).

Slides along previous failure surfaces

This type of failure is generally caused by failure along pre-existing shear surfaces or partially pre-sheared surfaces.

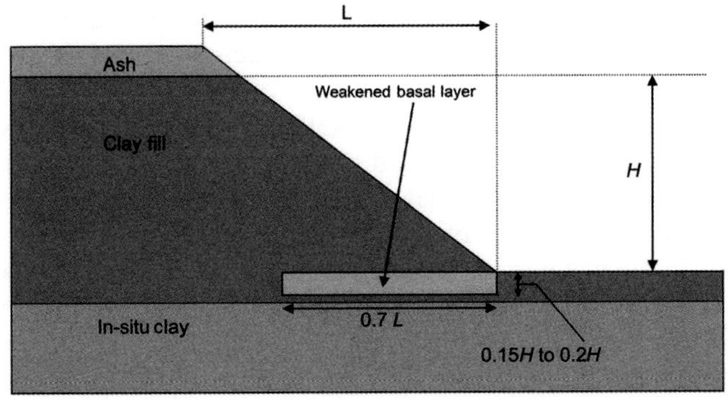

Figure 71.2 Conceptual embankment model for progressive failure

These surfaces may have developed as the result of past instability, commonly during construction or by landslides or other geological processes. Repair work does not always remove the shear surfaces fully. Shear surfaces are sometimes present within the softer alluvial deposits underlying embankments and, hence, may have been left in place during earthworks construction.

The stability analysis for this type of failure is carried out using residual effective strength parameters or parameters close to residual strength where partially pre-sheared surfaces are suspected.

71.2.1.3 Seasonal shrink and swell

Seasonal shrink–swell movements occur in embankments and cuttings containing intermediate and high-plasticity clay soils due to seasonal changes in the soil moisture content. Field observations and numerical modelling have shown that the deformation and delayed failure of old rail embankments, composed of dumped clay fills, is significantly affected by seasonal climate variations and by different types of vegetation. These embankments are particularly vulnerable to seasonal shrink–swell deformation, which ultimately can lead to progressive failure (O'Brien, 2013).

The seasonal movement is strongly influenced by the presence of high-water-demand mature trees, which leads to increased changes in the soil moisture content in the proximity of the trees. In the dry summer months, due to lack of rainfall and the presence of mature vegetation, moisture is removed from the soil, resulting in shrinkage. This causes shrinkage cracking. In the wet autumn and winter season the soil is recharged with moisture. This wetting also combines with less demand for moisture from the trees, leading to significant increases in the soil moisture content, which in turn causes swelling to occur. The resulting movements will influence the performance of the transport corridor located on the crest of an embankment and possibly at the toe of a cutting (e.g. excessive track deformation; **Figure 71.3**).

Shrinkage cracking of the soil in the dry periods leads to the formation of water pathways and increased permeability of the soil. This in turn leads to a more rapid inflow of water into the soil and a deeper penetration depth. This allows the rainwater to reach significant depth in the following wet season, thus, increasing the zone of influence of shrink–swell effects.

In intermediate and high-plasticity clay fills where increases in strain beyond the peak will result in a post-peak reduction in the shear strength, cycles of shrink–swell can lead to a loss of strength and onset of progressive failure. The swelling of soils in the wet period is associated with upward and outward movements, whereas the dry season is associated with mainly downward movement of the soil. This leads to a phenomenon called 'ratcheting', where the horizontal swelling that occurs in the wet periods does not fully recover in the dry summer months. These effective stress changes cause steady increases in shear strain (Kovacevic *et al.*, 2001; O'Brien, 2007), eventually leading to progressive slope failure. The risk of failure will be mainly dependent on the maximum pore water pressure that may develop, which is predominantly controlled by the permeability variations in the embankment fill and foundation. The above factors that control behaviour, together with the unsaturated nature of railway embankments, needs to be considered by practising engineers. The variation of pore water pressure with depth will seldom be hydrostatic, except perhaps within a near-surface layer, typically affected by desiccation cracking (O'Brien, 2013).

Figure 71.3 Effect of trees on track deformation

Further information on seasonal shrink–swell movements and the effects of vegetation that may lead to progressive failure and how the risk of failure may increase in the future can be obtained from McGinnity *et al.* (1998), Kovacevic *et al.* (2001), Loveridge and Anderson (2007), Butcher and Pearson (2007), Take and Bolton (2004), Hudacsek and Bransby (2008), Nyambayo *et al.* (2004), O'Brien (2013), Briggs *et al.* (2017) and Stirling *et al.* (2021).

71.2.1.4 Embankment shoulder instability

Shoulder instability is a shallow type of instability that is more specific to embankments where, due to various factors described below, the shoulder of an embankment cannot provide the required support to the transport corridor on the crest (see **Figure 71.12(a)** and **71.12(b)**). The main factors responsible for this type of instability are

- steep upper-slope angle, which can cause gradual ravelling of material down the slope, leading to narrowing of the shoulder
- dynamic loading of trains or road vehicles – this loading can travel laterally and mobilise the soil particles, and with time these soil particles will move down the slope under gravity
- embankment material
- seasonal effects including changes in embankment fill moisture content.

Fine granular materials such as ash, which is commonly used in old railway embankments, are particularly prone to this type of instability. During drier periods of the year, the ash becomes mobile particularly under dynamic loading from track vehicles, and the greatest magnitude of movement occurs at this time. As the track and ballast lie directly on the ash, this leads to track level deterioration (Standing *et al.*, 2021).

71.2.1.5 Flooding and scour

Flooding and scour at the toe of embankments can occur as the result of rivers or streams flowing parallel or through an embankment.

Increased water levels during wet seasons can cause flooding, resulting in increased water pressure at the toe of the embankment. This can result in reduced safety factors against deep-seated failure, leading potentially to the onset of instability, depending on the duration of flood.

Figure 71.4 Scour erosion at the toe of a railway embankment

Figure 71.5 Concrete channels provide a more durable solution

Flooding can also cause scour erosion of the toe of the embankment, which again results in reduced safety factors against deep-seated failure due to the reduced weight at the toe of the embankment (**Figures 71.4** and **71.5**).

The important factors to consider for the safe management of the slope are

- direction of flow of water in the river
- extent of the flooding regulatory authority (e.g. Environment Agency) flood zone and capacity
- topography and land use adjacent to the toe
- presence, condition and capacity of drainage at the toe.

71.2.1.6 Flow failure

There are a number of categories of flow failure in the classification of landslides, developed by Varnes (1978) and updated by Hungr *et al.* (2014). Those relevant to UK earthworks include the following, among others.

- *Debris flow:* Rapid soil mass movements that typically require relatively steep slopes and high topographic variance. Field observations of Scottish debris flows over a 20-year period suggest that the particle sizes of materials forming debris flows may range from coarse granular (including boulders) to fine cohesive, with most sizes between being potentially represented (Winter *et al.*, 2010). They often occur during intense rainfall when the soils at the surface of the ground become saturated by infiltration and/or by water seeping upward from more pervious soil or rock beneath. The saturated soil has little strength, and flows like a wet paste. Debris-flow events can occur in open, steep-sided hillsides, or be focused along stream channels and previous gulley tracks. Debris flows often originate on a hillside above transport infrastructure – for example, the passenger train derailment on the Fort William to Mallaig line at Loch Eilt in January 2018 following the collision with material from a debris flow (RAIB, 2018). Originating from the adjacent mountainside the debris flow mobilised rapidly and was encountered by a passenger train, with the driver unable to stop in advance of the failed material. Long-term antecedent rainfall followed by short intense storm rainfall is believed to characterise many debris-flow events in Scotland. Analysis of a series of debris flows sampled from six study sites across upland Scotland provides insights into key topographic and material controls on debris-flow activity, with implications for effective management of the associated geohazard both locally and in similar areas of high relief (Milne *et al.*, 2015). It seems most likely that the nature of debris-flow events in Scotland may alter with climate change. Whereas at present such events seem to be mainly driven by storms affecting ground that has already been well saturated over a long period of time, it seems possible that in time they will become increasingly driven by the very rapid ingress of water to drier materials, particularly during the summer months, when many of the source materials may be in a dry state and vulnerable to rapid inundation and erosion (Winter *et al.*, 2010).

- *Earthflow:* Flow in fine-grained soils or clayey rocks under saturated conditions. Hungr *et al.* (2014) define an earth flow as a 'Rapid or slower, intermittent flow-like movement of plastic clayey soil, facilitated by a combination of sliding along multiple discrete shear surfaces and internal shear strains'. Liquefaction of the failed material often leads to the formation of a bowl at the head of the slope and creates a unique 'hourglass' effect.
- *Mudflow:* A rapid flow consisting of at least 50% clay, silt and sand particles. Mudflows are characterised by a high water content and are sometimes referred to as mudslides.
- *Creep:* A subtle and steady type of downward flow caused by shear deformation. Shear stresses are high enough to cause displacements but not adequate to result in shear failure. A common feature that reveals creeping flow is the presence of tilted tree trunks.
- *Washouts:* Slope failures as a result of uncontrolled concentrated surface water flow eroding and scouring the ground over which it flows are often termed washouts in the UK. The term 'washout' overlaps a number of categories of flow failure defined by Varnes (1978), such as debris flow, earthflow and others. For example, in June 2019 a passenger train travelling at 40 mph (64 km/h) collided with debris around 1 mile (1.6 km) north of Corby station. The debris comprised sandy gravel and soil which had been washed onto the track from the adjacent cutting slope by flood water (RAIB, 2020). Washout failure is most likely to occur in high-permeability granular materials as a result of rapid saturation caused by short bursts of high-intensity rainfall or a storm event. The saturation is caused by infiltration of rain water into the slope surface but also, more significantly, from concentrated surface run-off in areas of poor drainage. Hence, this type of failure is much more likely to occur in cuttings where the topography, land use and poor drainage condition at the crest of the cutting allow a large volume of water to be directed towards the slope in a very short period of time. The high permeability of the soil within the cutting allows this water to be absorbed within the slope rapidly, leading to a mass flow of material.

From the above it can be seen that the key parameters for flow failure to develop are

- short-duration, high-intensity rainfall events
- high-permeability soil

Figure 71.6 A83 Argyll and Bute, Scotland, Rest and Be Thankful debris flow, August 2012 (Courtesy of the British Geological Survey)

- slope geometry
- topography
- land use
- vegetation
- drainage.

Low-intensity, long-duration rainfall events are unlikely to lead to a flow failure as, in a granular material, there will be adequate time for the excess water to drain. Flow failures are unlikely to develop in low-permeability soils, where rapid saturation cannot take place. Finally, the right geometric and topographic conditions must exist to generate a concentrated flow of water. Hence, embankments are less likely to be at risk from this type of failure.

The consequences of this type of failure can be very significant. In a cutting a large volume of soil can flow onto the transport corridor at the toe of the cutting, causing derailment of trains or accidents on the roads (**Figure 71.6**).

71.2.1.7 Rock cuttings

Generally, rock cutting slopes are steeper than soil cutting slopes. Rock detaching from cutting faces is a common phenomenon ranging from small-scale spalling and ravelling to major rockfalls that can result in large volumes of material falling onto the infrastructure below (Network Rail, 2021a). Of critical importance to the understanding of failures in rock cuttings is the presence of discontinuities within the rock mass. In general, rock materials will rarely fail as a material – almost always the failure will be of the rock as a mass. In understanding and analysing such failures, critical importance needs to be placed on the discontinuities within the rock mass, their orientation, persistence and strength parameters. There are several modes of failure that can affect rock cutting slopes

- rockfalls including toppling failure
- planar failure on a dominant discontinuity set

- wedge failure on a combination of discontinuities
- mass failure of slopes containing closely spaced discontinuities
- failure of mixed soil and rock slopes.

Rock cutting failures can occur with little evidence of the early development of failure. Whereas soil slopes may show slope bulging, tension cracks and disturbed vegetation, such precursors are less common in rock slopes. Frost action can lead to gradual degradation of some rock cutting slopes, depending on the permeability, porosity and the joint system within the rock. Ice crystals that form within the pores and between the joints in a slope face induce strains, which result in splitting and shattering of the slope surface.

The volume change caused by the growth of ice crystals can result in an increase in the joint width and depth. Hence, repeated freeze–thaw actions can lead to an increase in the permeability of the rock mass, which in turn increases the depth of influence of the frost action.

One of the rock types particularly susceptible to frost action is chalk. Shattering or loosening of the surface of cutting slopes in chalk results in the generation of debris at the toe of the cutting, which affects the serviceability of the transport corridor located at the toe (Lord *et al.*, 2002) (**Figure 71.7**).

Frost shattering can have more significant consequences in the transport corridor at the toe when loosening of the surface leads to destabilising of the vegetation and tree roots on the surface of the slope, which can fall onto the rail or road at the toe.

71.2.2 Limit states

71.2.2.1 Ultimate limit state

The ultimate limit state (ULS) is reached when the disturbing forces acting on a slope exceed the resisting forces. Under this condition collapse, slope instability or other forms of structural failure will occur which could significantly affect the safety of the services or the structures associated with the earthworks (Perry *et al.*, 2003a, 2003b).

For an embankment a ULS failure will lead to significant disruption of the safe operation of the railway or highway transport running on the crest of the embankment (**Figure 71.8**).

In cuttings, the slope failure, depending on the size and extent of the rupture surface, may affect the operation and the safety of the transport infrastructure and the associated services running at the toe of the cutting. The failure could also affect the structures on the crest of the cutting, such as buildings, roads, car parks and their associated services such as drains, cables and so on.

In order to prevent such a mode of failure, a margin of safety against the disturbing forces has to be introduced into the design and maintained during the life of the structure. For existing slopes this can be achieved by performing slope stability analyses. Where the stability of slopes is found to be marginal, remedial measures should be implemented to increase the margin against instability.

71.2.2.2 Serviceability limit state

A serviceability limit state (SLS) is reached when deformation within an earthwork exceeds the desired or prescribed limits such that the structure can no longer perform its required functions satisfactorily.

Deformations within an embankment can affect the performance of the transport infrastructures running along the top, leading to excessive settlement of roads or railway tracks or the associated services such as cable runs and signals.

In cuttings, deformations within the slope can affect the buildings, roads, car parks, drains and other structures supported on the crest.

It is not common practice to carry out analysis of a slope to predict deformation. It is generally assumed that adequate safety factors against deep-seated slope instability would ensure the satisfactory performance of the slope during its design life.

However, where required, numerical analysis can be carried out to estimate slope deformation. For existing slopes,

Figure 71.7 Frost shattering on a chalk railway cutting

Figure 71.8 Embankment failure as a result of increased pore water pressures after a prolonged rainfall event (1994)

however, any such analysis would need to model the loading history experienced by the slope since construction. The analysis would also require appropriate parameters such as strength, stiffness, permeability of the soil and the pore pressure regime within the slope in order to simulate the field conditions leading to realistic predictions.

The serviceability of slopes is affected by the following factors, some of which were discussed in the preceding sections

- seasonal shrink–swell
- vegetation
- dynamic loading
- animal burrowing
- heavy rainfall
- flooding
- frost.

71.3. Weather and climate change

It has long been understood that performance of earthworks is influenced by weather and climate, with particularly wet periods resulting in greater numbers of earthworks failures. (Department of Transport, 2014). Slope failures generally occur after prolonged periods of rainfall in winter and after intense rainfall in summer. In August 2004, long-lasting and intense rainfall led to a large number of landslides in the form of debris flows, which affected important parts of the Scottish major road network (Winter *et al.*, 2010).

The seasonal nature of evapotranspiration and rainfall in the UK, with much higher evapotranspiration and generally lower rainfall during summer than winter, causes annual cycles of soil drying and wetting, which can lead to earthwork serviceability problems, most critically track settlement for railway infrastructure (Scott *et al.*, 2007). The effective stress changes caused by the seasonal changes in pore water pressures can cause steady increases in shear strain in earthwork clay slopes (Kovacevic *et al.*, 2001; O'Brien, 2007), eventually leading to progressive failure.

Powrie and Smethurst (2019) describe both observations and numerical modelling carried out over a number of years to understand changes in water content, pore water pressure and ground movements in earthworks caused by climate and vegetation to start to develop a quantitative understanding of the effects on earthworks, and their implications for the management of earthworks in the future.

As well as dealing with the highly variable UK weather and climate, infrastructure owners now have to factor in the influence of anthropogenic climate change. It is postulated that as the climate changes, the behaviour of the UK's earthworks will require modified asset management approaches to be implemented by infrastructure owners (Loveridge *et al.*, 2010). These issues have been further brought to national attention by the fatal railway derailment at Carmont, Aberdeenshire, in August 2020 (RAIB, 2022), where a passenger train collided with debris washed from a cutting slope drain onto the track following very heavy rainfall. This tragedy prompted the UK Government Secretary of State for Transport to request a wide-ranging review of the impact of climate change on the resilience and safe performance of the railway network, and in particular the management of railway earthworks, drainage and weather to minimise the risk of such an event happening again. In response, Network Rail decided to commission two taskforces to advise on the ways it could improve its understanding of earthworks management and potential improvements to its mitigation measures. Lord Robert Mair CBE, FREng, FRS, a geotechnical expert, led an earthworks management taskforce to advise Network Rail on how it could improve the management of its earthwork portfolio. Dame Julia Slingo FRS, former chief scientist at the Meteorological Office, led a weather action taskforce with the objective of better equipping Network Rail to understand the risk of rainfall to its infrastructure. The taskforces' report findings and recommendations (Network Rail, 2021a, 2021b) are guiding Network Rail's ongoing earthworks asset management and operational mitigation strategies.

There is no doubt that climate change is affecting the UK and shifting the odds of certain high-impact weather events. It is likely the UK will experience more extreme weather conditions with reference to past climatic trends due to climate change (Lowe *et al.*, 2019). This means that infrastructure that was previously resilient to the UK's past weather and climate may no longer be so. The threats to earthwork stability posed by climate change may be summarised (Network Rail, 2021a) as

- longer periods of prolonged rainfall in winter months leading to rising groundwater levels and higher pore water pressures in earthwork slopes, which could trigger first-time failures and reactivation of previous failures in late winter
- more frequent periods of more intense rainfall which will increase surface run-off which could trigger washouts and debris flows
- hotter and drier summers that will increase serviceability problems on clay embankments and will increase the amplitude of cyclic pore pressure changes experienced on the slopes, promoting additional progressive failures
- intense rainfall falling on a desiccated cracked earthworks slope will result in a rapid increase in pore water pressure and the risk of shallow slope failures
- intense rainfall could surcharge the local drainage network and direct flow towards the adjacent earthworks leading to flooding and surface erosion.

71.4. Earthwork condition appraisal, risk mitigation and control

71.4.1 Inspection

For an EAMS to be successfully delivered it is necessary to develop an appropriate earthwork inspection system. Its main purpose is to

- appraise structural safety
- detect incipient defects at an early stage and monitor the development of defects to determine the urgency for, and the nature of, corrective action
- classify the condition of the earthworks according to a defined system
- identify the need for preventive action.

The inspection approach varies between UK infrastructure asset owners: in all cases it involves an observation and recording of the earthwork condition at the time of the inspection. Some UK asset owners (e.g. Network Rail) use the term 'examination' in lieu of 'inspection'.

The following factors should be considered in developing an inspection regime

- defect definition, classification and condition rating system
- inspection frequency
- inspection types (cyclical or ad hoc)
- data-collection techniques
- risk.

For high and steep slopes where access for inspection is considered unsafe, close inspection and examination of the slope face may be carried out by abseiling technique by trained and qualified professionals.

The frequency of cyclical inspections should be set considering the history and condition of the earthwork together with the likelihood and consequences of failure.

Wherever possible a record should be made of the drainage condition and water management associated with earthwork slopes as an integral part of the inspection process. The inspection should also establish the presence and overall condition of any additional stabilising structures (retaining walls, soil nails, etc.) within the earthwork. Some of the stabilising structures might require inspection by other disciplines and so harmonised inspection programmes should be considered.

The inspection should not only focus on the specific asset being inspected but should be watchful of nearby activities, including those outside the transport property boundary, that may adversely affect the earthwork. It is acknowledged that collection of information about slope instability features outside the transport boundary may be restricted. In particular, the identification of deficient or inadequate water management from catchments beyond the boundary fence is likely to be limited, especially if the inspections are not undertaken during or following heavy rainfall. The information collected from inspections should be used to identify defects and to monitor the asset condition and degradation rate.

Remote inspection from airborne platforms is becoming an increasingly powerful tool to support traditional site inspection and to aid strategic planning. It can offer a safer and more cost-effective method for identifying problem areas than visits conducted on foot. Duffell *et al.* (2005) described the use of airborne remote-sensing techniques to assist in prioritising detailed ground-based inspections of earthworks. Helicopter flights are of considerable value, especially in hilly or mountainous terrain, for inspection and provision of visual evidence, particularly immediately after an extreme weather event that may have resulted in a slope failure or instability. Unmanned aerial vehicles (UAVs or drones) have reduced in price and developed rapidly from a technological perspective, and offer considerable potential to significantly enhance the earthwork inspection process. A review of the use of UAVs for the assessment of hazards is provided by Antoine *et al.* (2020), while a more general review of the use of UAVs in engineering geology is provided by Giordan *et al.* (2020). While these reviews are more general, they have close applicability to earthworks. UAVs are primarily a mechanism for the deployment of other technologies, such as photogrammetry, multispectral data capture, thermal imagery, LiDAR (light detection and ranging) and potentially SAR (synthetic aperture radar), providing a flexible platform. The scale of data collection can range from the local, in which the operator collects data with a small UAV flown within line of sight, to the regional, with the UAV operating autonomously (noting that there are substantial UK civil aviation regulatory hurdles in this respect at present). Large areas can be surveyed quickly and effectively, and response times can be short.

However, for slopes which are densely vegetated remote inspection from airborne platforms may not prove to be effective in identifying the critical slope instability features.

71.4.2 Analytical assessment

The objectives of an earthwork analytical assessment are more rigorous and detailed than an inspection and are to

- provide data from a detailed desk study, ground investigation and topographical survey to define the earthwork history, slope geometry and the ground model for slope stability analysis
- perform slope stability analysis to accurately determine the operating safety factor for the earthwork
- carry out a detailed assessment of the condition and stability of specific earthworks based on the calculated safety factor as well as the physical condition of the slope
- classify the condition of the earthwork in accordance with a classification system defined in the asset management system (the condition classification of an earthwork typically spans between very good to extremely poor or life expired)
- provide data to carry out a risk assessment to calculate the safety and performance risks to the operation of the transport system
- identify those earthworks that need priority attention when formulating asset management plans
- plan renewal and maintenance budgets and programmes, based on risk prioritisation
- anticipate future problems and identify a rate of deterioration
- provide information for the design of any required stabilisation.

An earthwork assessment typically consists of the stages outlined below.

71.4.2.1 Desk study and walkover survey

A desk study typically includes a review of all relevant data obtained from inspection and maintenance records, and a detailed site inspection (walkover survey) to identify visually any areas of distress. In addition to geological and geotechnical data the desk study would also typically consider drainage, environmental, ecological, services and access information.

71.4.2.2 Ground investigation

A ground investigation includes exploratory holes, sampling, in situ and laboratory testing in order to define the ground and groundwater condition at a specific earthwork location. The ground investigation may include instrumentation and monitoring to obtain information on pore pressure regime as well as any slope deformation.

71.4.2.3 Interpretation and stability analysis

Information obtained from the ground investigation and desk study, including any monitoring data, must be interpreted to establish the ground model and geotechnical parameters. Likely failure and deformation mechanisms must be considered and slope stability analysis performed to determine the operating safety factors against deep-seated and shallow instability.

71.4.2.4 Reporting and prioritisation

Results from the desk study, ground investigations, their interpretation and the stability analysis are presented, with an explanation of the condition and criticality of the earthworks, leading to a condition classification. Recommendations for mitigation and intervention works should be provided with the associated cost estimates.

71.4.3 Earthwork risk mitigation and control

The principal responsibility of the asset manager is to ensure that earthworks are maintained in a condition such that the safety of users and the public is not compromised. Factors such as age, increased traffic loading, inadequate or poor maintenance and deferred repairs reduce the performance of earthworks and may ultimately compromise operational safety. Specific earthwork performance requirements are set by infrastructure owners (e.g. railway embankment deformation limits relate to maintaining track quality for a specified line speed).

When an earthwork experiences stability problems it is often necessary to impose temporary restrictions on speed or loading, or close traffic lanes to maintain operational safety. When its structural capacity is inadequate and there is a risk of collapse, public safety is jeopardised and complete withdrawal of service may be necessary.

If instability or other loss of functionality is predicted, the following mitigation options may be adopted individually or in combination in increasing order of impact

- increased inspection frequency
- monitoring
- routine maintenance (e.g. unblocking drains)
- temporary support measures
- service restriction, either while mitigation is put in place or stabilisation is undertaken
- renewal works
- withdrawal from service.

71.4.4 Monitoring existing earthworks

The purpose of monitoring earthworks includes

- mitigation of risk until repairs are implemented
- determining the rate of movement and establishing if it is constant, accelerating or decelerating
- establish in situ pore water pressures and any variation over time
- where movements have occurred, gathering information on the depth, area and lateral extent
- to establish the effects of construction adjacent to earthworks, both during construction and in the long term
- to obtain quantitative performance data
- to validate the output from analytical assessment, particularly where it is considered to have produced conservative results, and to confirm design assumptions.

It is essential to recognise that monitoring on its own cannot and will not predict the timing of failure in a slope. However, neither will analysis (either simple or advanced) with a degree of accuracy that could be used in a practical sense to avoid the problem. The key to the effective assessment of existing earthwork slopes is the combination of analysis and monitoring (i.e. the observational approach) to determine if the earthwork slope has the conditions required for a future failure to occur.

It is important to distinguish between two principal objectives of monitoring in the context of existing earthwork asset management (Network Rail 2021a).

- *Monitoring objective A:* Slope movement detection (i.e. detection of rapid loss of functionality that may have a direct consequence on the safety of the transport infrastructure) and reaction (e.g. stop or slow trains using alert alarm systems).
- *Monitoring objective B:* Provide data on the performance and condition of a slope or embankment, and possible precursors to failure.

The designer of the monitoring system should define the following aspects in advance of implementation of the scheme

- objectives
- techniques
- required accuracy
- frequency
- trigger levels
- data collection and reporting
- action plans.

There is a vast range of instrumentation available and further guidance can be obtained from Dunnicliff (1998). Smethurst *et al.* (2017) provide a comprehensive review of current and future technologies for monitoring the performance of transport infrastructure slopes. Recent developments in this field include web-based, cable-free or fibre-optic systems, which offer remote, real-time monitoring and have the potential to be integrated to automate alert alarms. Owing to the rate of technological development in this field, discussions should always be held with specialist instrument manufacturers and installers.

Instrumentation for an earthwork monitoring system typically includes

- inclinometers to measure displacement within slopes or buried structures such as piles or retaining walls
- shape arrays, also known as shape accel arrays, which are a type of in situ inclinometer to measure subsurface displacement within ground and structures
- standpipe or vibrating wire piezometers to measure pore pressures
- hydraulic (flushable) piezometers to measure negative pore pressures
- strain gauges to measure strains within structural members
- tiltmeters to measure near-ground surface displacement and/or tilt in structural members
- extensometers to measure vertical displacement in the ground
- fibre-optic systems to measure strain within the ground and structures
- rain gauges to measure rainfall intensity and precipitation.

It is generally beneficial to review and assess the monitoring data in combination with the data on rainfall and in situ state of soil moisture in order to establish if there is any correlation with changes in weather condition.

In establishing monitoring trigger levels, the likely magnitude of normal baseline values should be established. For example, seasonal vertical movements of up to 50 mm have been recorded on some London Underground embankments due to the effects of vegetation on high-plasticity clay. Therefore, the monitoring system should be put in place sufficiently in advance of control periods so that the range and pattern of background movements can be established.

A review of the use of remote sensing for the monitoring and management of railway infrastructure is provided by Donzelli (2019) in work that has been undertaken in the EU Horizon 2020 Multi-scale Observation and Monitoring of Railway Infrastructure Threats (MOMIT) project.

For new or modified earthworks, monitoring can aid in finalising the design either through the approach recommended in BS EN 1997-1:2004 (BSI, 2004) or through observation itself (Nicholson *et al.*, 1999).

71.5. Maintenance and remedial works

Many older UK transport infrastructure earthworks have now exceeded the typical engineering design life used in modern design and present a range of ULS and SLS defects. However, the majority of these older transport infrastructure earthworks are still used intensively, and it is both impractical and too expensive to try to replace them with new assets. Therefore, interventions are required once cyclical inspection is no longer considered to be acceptable for controlling earthwork asset risk at an ALARP level and to maintain the earthwork in a serviceable condition. Interventions include

- maintenance (preventive and corrective)
- refurbishment (repair)
- renewal (strengthening).

71.5.1 Preventive and corrective maintenance regimes

As discussed in Section 71.1.3, maintenance forms a part of the asset management regime and is undertaken in order to maintain the assets in their current condition and to minimise asset deterioration and degradation. Preventive maintenance is carried out on a routine basis, such as through vegetation management, removal of toe debris and cleaning drainage systems. The frequency of preventive maintenance is governed by the severity of any potential problems and their consequences. Corrective maintenance is undertaken where defects are observed during inspection or site walkovers. Prioritisation based on risk is used to determine the maintenance programme. Maintenance for earthworks could include

- vegetation management
- debris clearance
- drainage cleaning
- infilling and ongoing management of animal burrows
- erosion control measures
- embankment shoulder support system
- construction of a toe-retaining structure to capture debris
- construction of a mesh and anchor system to contain and capture rockfalls.

71.5.2 Refurbishment (repair) and renewal (strengthening) works

Refurbishment (repair) or renewal (strengthening) interventions are required

- where the earthworks can no longer provide the required function against the performance criteria
- where the asset does not meet the minimum requirements of the standards set by the regulatory organisations
- where a full or partial slope failure has occurred or is considered to be imminent.

The aim of these interventions is to improve the stability of the earthworks against the ULS and SLS. However, during the design, other safety and environmental features are considered and included where appropriate in the works in order to upgrade the asset to modern engineering standards.

The performance criteria for earthwork repair and strengthening design generally consist of

- improving the stability of the earthworks in order to meet the requirements for the minimum safety factor against deep-seated instability
- improving the stability of the earthworks in order to meet the requirements for the minimum safety factor against shallow instability
- limiting the subsequent deformation such that the relevant maintenance target levels as defined in relevant infrastructure owner standards are not exceeded – for example, London Underground Standard S1159 (2015) requires the differential settlement of the outer rail to be limited to 1:500 along any 10 m length of track with a limit of 1:300 across the rails
- designing remedial work such that it will have no detrimental effect on the stability of the existing structures within or adjacent to the earthworks
- ensuring permanent earthworks should achieve a specified design life that varies between 60 and 120 years, depending on which design standards are being followed.

The stability of slopes can be improved by increasing restoring forces or reducing disturbing forces. The following options can be considered for the design of earthwork repair and strengthening works.

71.5.2.1 Soil cuttings

Regrading

Cutting slopes can be stabilised by regrading to a shallower angle, provided sufficient room exists at the crest of the cutting. Regrading can also be combined with excavation of the cutting face and replacement with granular material.

In general, this option requires a significant amount of excavation and off-site disposal of large volumes of material. It also requires the complete removal of all vegetation and replacement after completion of construction. Plastic clay earthworks can also present difficult working conditions during wet periods.

Spaced bored piles

This solution consists of installing bored piles around the mid-slope to pin potentially unstable or slipping material into underlying stable ground. This method can also be used in the stabilisation of embankments. The piles are installed discretely a short distance apart (along the line of the piles) for economy of construction. The system relies on the soil arching between the piles to resist the active pressure, hence the gap between piles cannot be too large, otherwise the soil will be able to flow through it. It has been found that piles work satisfactorily at spacing up to four pile diameters centre to centre for embankments and six pile diameters centre to centre for cuttings (Carder and Barker, 2005; Davies *et al.*, 2003; Powrie and Smethurst 2015; Smethurst, 2003; Smethurst and Powrie, 2007; Smethurst *et al.*, 2020). However, the spacing between the piles is a critical design parameter and must be considered carefully for each specific site condition for the piles to be effective in stabilising the slope.

The spacing of the bored piles in the longitudinal direction is dependent on the ground conditions and individual pile capacity, and is selected so as to provide the required safety factor against deep-seated instability. The selection of the pile diameter is affected by the sizes of the available piling rigs and the associated temporary works required. Safety issues associated with the construction of piles above an operational highway or railway are important when selecting the appropriate construction plant.

The location, pile diameter and spacing will need to be optimised to provide the most efficient system, as well as allowing for buildability issues such as the size of the piling rig, temporary works and the clearance from existing services and structures.

Construction of bored piles can be carried out using a mini-piling rig from a temporary bench excavated into the slope of the cutting or using a scaffolding platform. Small piling rigs, such as the TD610 for pile diameters up to 450 mm, can be used. Where the bending moment and shear forces require larger-diameter piles, piling rigs such as the Klemm KR 709 or similar can be used. This type of rig has a larger size and weight and a higher mast when fully extended. Hence, the requirements for a suitable piling platform and the associated temporary works will need to be considered as well as the safety concerns associated with tall plant working adjacent to a live transport corridor (**Figure 71.9**).

The advantages of this system are

- relatively low cost
- much of the vegetation on the upper slope can remain
- minimised earthworks
- access for plant and equipment

Figure 71.9 Construction of spaced bored piles on a London Clay cutting: (a) vegetation removal and site preparation; (b) construction of the mid-slope bench for piling; (c) different size piling rigs (TD610 and Klemm 709R) to suit the pile diameters; (d) lowering of the reinforcement cage; (e) slope after completion of construction (continued on next page)

Figure 71.9 Continued

- low environmental impact
- ease of construction
- design life.

Soil nailing

Soil nailing comprises the installation of nails into the body of the slope at shallow angles of 10° to 30° below the horizontal, penetrating the soil mass beyond the critical slip circle in the passive zone. Various types of soil nails are available and the choice is dependent on the particular application and the availability of space and access for installation plant, and the required design life and cost (Phear *et al.*, 2005). Soil nails have been used to stabilise embankments and cuttings (Payne *et al.*, 2020).

Hollow self-drilling soil nails are quick and easy to install in some granular soils. These are installed by simultaneous drilling and grouting, using a sacrificial bit attached to the advancing end of the hollow soil nail. Soil nailing a slope will require the complete removal of vegetation from the slope where nails are to be installed.

For sites where access is difficult, an abseiling technique can be adopted for installation of the nails. However, only a limited length of nail and depth of penetration can be achieved using this system. A soil-nailing scheme constructed under difficult access conditions using an abseiling rig is shown in **Figure 71.10**.

Design life and corrosion protection will significantly affect the financial viability of the scheme. A risk-based approach should be adopted where corrosion protection is related to the possibility of failure (Phear *et al.*, 2005). When designing the soil nails, the length and the angle of the soil nails should be optimised to ensure that the nails will remain within the boundary of the land ownership, otherwise, an easement for the extension into third-party land will be required.

Toe-retaining walls

Where space is available, the construction of a wall such as a mass-concrete gravity wall or a gabion wall at the toe of the cutting combined with backfilling with granular fill will

Figure 71.10 Soil nailing at West Kensington cutting, before, during and after construction: (a) slope before remediation; (b) slope during construction – due to space constraints only hand excavation was possible and a conveyor system was used to transport material away from the slope; (c) slope after completion of soil nailing; (d) final slope profile after vegetation growth

provide additional weight at the toe to increase the safety factor against deep-seated instability. The depth to the base of the wall should be designed such that any failure surface passing below the wall has the required safety factor. This may require deep excavations at the toe, which may not be practical given the space constraints often existing at the toe. Excavations at the toe could lead to slope instability and need careful considerations as part of the temporary works design. Where excavation is impractical at the cutting toe, a piled wall with a capping beam may provide a better solution.

Drainage

Where it has been established that the instability of a slope is due to excessive pore pressures within the slope, the installation of a drainage system to reduce the pore pressures may be a viable solution.

The following drainage systems may be considered

- cut-off drains at the crest in order to reduce the flow of water into the slope
- counterfort drains in order to drain the body of the slope
- toe drains in order to keep the water level down at the toe
- a combination of the above.

However, when designing a drainage system the following must be considered

- need for reliable pore water pressure data for design
- requirement for regular long-term pore water pressure monitoring and maintenance of any slope drainage system in order to ensure continued operation during the intended design life
- disposal of drainage discharge, which can require connection works to the existing track drainage
- this solution may require to be combined with other remedial options to meet the safety factor requirement.

Relying solely on a drainage solution for slope stability over a long service life (e.g. 120 years is required by some UK infrastructure owners) may not be practical as this would impose a heavy reliance on the inspection and maintenance regime to ensure the drainage remains functional. Also, for such a long service life the longevity of the drainage type and the materials used would need to be carefully considered.

71.5.2.2 Rock cuttings

Methods of rock cutting slope stabilisation fall into three categories (Wyllie, 2017)

- scaling and rock removal
- reinforcement
- containment and protection methods.

It is important that the appropriate stabilisation method be used for the particular conditions at each site. For example, where the cutting slope is steep and the toe is close to the transport infrastructure, there will be no space to excavate a catch ditch or construct a fence. Alternative stabilisation measures may be to remove loose rock, secure it in place with bolts, or cover the slope with mesh. It is generally preferable to remove loose rock and eliminate the hazard, but only if this will form a stable face and not undermine other potentially loose rock. In general, scaling and rock removal is a preferred method of stabilisation because the work eliminates the hazard. However, removal should only be used where it is certain that the new face will be stable and where construction on live transport routes can be carried out efficiently and safely.

Scaling and removal of loose blocks

Scaling describes the removal of loose rock, soil and vegetation on the face of a slope, generally using hand tools. On steep slopes workers are usually supported by ropes anchored at the crest of the slope and tied to belts and carabiners around their waists. It is often employed prior to or in combination with other containment and protection methods. When carrying out this method, it is essential that the works are supervised by an experienced person as excessive or incorrect scaling and removal could be detrimental to overall slope stability.

Reinforcement

There are a number of reinforcement techniques that may be implemented to secure potentially loose rock on the face of a rock cutting. Tensioned rock bolts can be installed across potential failure surfaces and anchored in sound rock beyond the surface. The application of a tensile force in the bolt, which is transmitted into the rock by a reaction plate at the rock surface, produces compression in the rock mass and modifies the normal and shear stresses across the failure surface. Loosening and failure of small blocks of rock on the cutting slope face can be prevented by the installation of passive dowels at the toe of the block. Dowels are composed of lengths of reinforcing steel grouted into holes drilled in the underlying, stable rock, with a cap of reinforced concrete encasing the exposed steel. Zones or beds of closely fractured or degradable rock can be protected by applying a layer of shotcrete to the rock face; the shotcrete controls both the fall of small blocks of rock and progressive ravelling that will produce large, unstable overhangs on the face. However, shotcrete provides little support against sliding of the overall slope; its primary function is surface protection. When a rockfall has occurred that forms a cavity in the slope face, it may be necessary to construct a concrete buttress in the cavity to prevent further falls. The buttress serves two functions: to retain and protect areas of weak rock and to support the overhang.

Containment and protection methods

An effective method of minimising the hazard of rockfalls is to let the falls occur and to control their distance and direction of travel. Methods of rockfall control and protection of transport infrastructure include netting and meshing, catch fences, ditches and barriers. A common feature of all these protection structures is their energy-absorbing characteristics, which either stop the rockfall over some distance or deflect it away from the infrastructure that is being protected.

Netting and meshing

The design, installation and whole-life management of rock netting is described in CIRIA C775 (Koe *et al.*, 2018). Wire netting or mesh hung on the face of a rock slope (draped netting) can be an effective method of containing rockfalls close to the face and preventing them from falling onto the transport infrastructure, as shown in **Figure 71.11**. When used in combination with bolting it can also be effective in improving rock mass stability (stabilisation netting). The process involves netting the unstable cutting face with either double twist steel mesh or polymer geogrid. The mesh has to be securely fastened at the top of the cutting. For steep slopes the mesh is allowed to hang under its own weight to just above the base of the cutting, or on higher cuttings to a bench, where the rockfall debris can accumulate. It is important to consider how debris accumulating can be safely removed during maintenance. On shallower and irregular slopes the mesh may be fixed tightly against the cutting face using dowels or face cables, reducing both visual impact and the ability for rockfall to occur.

When deciding on the type of rock netting, careful consideration should be given to access for construction as well as future inspection and maintenance of the netting system, including removing the accumulated debris and rock blocks.

Figure 71.11 Netting used to stabilise a rock cutting at Lambhill on the Glasgow-Anniesland, by way of Maryhill, railway line following a failure in 2017

Catch fences

Where it is not practical to provide rock netting, but there is a risk of boulders or loose blocks of rock falling, catch fences can be considered to absorb the impact before damage is caused to the infrastructure. Alarm systems, such as a mechanical rockfall trip wire that is linked to railway signalling or highway communication systems, provide some reduction in the consequence by slowing or stopping the traffic in the event of a failure. Catch fences are typically 2–3 m high and should be located an adequate distance from the toe of the cutting, or area of instability on a slope, to intersect bouncing or rolling rockfall of diameter up to 1 m. They are designed as flexible structures to help absorb impact, often comprising vertical posts securely bolted to the rock surface, with horizontal cables and netting spanning between then. Where lateral impact forces on posts are high, the posts may need to be restrained laterally by ties. The catch fence is normally dimensioned to ensure only minimal damage arises due to impact from the anticipated size of rockfall. However, it may not be possible to prevent all damage, and therefore regular inspection and repair are needed to ensure the catch fence offers the necessary degree of protection.

71.5.2.3 Soil embankments

Regrading

Embankment slopes can be stabilised by placing a berm against the slope and regrading to a shallower angle, provided sufficient room exists at the toe.

The berm solution can be used to widen the crest and provide more support to the shoulder of the embankment. The regraded embankment will typically have a uniform gradient. For embankment widening and regrading, a Class 1A granular material is generally specified. Recycled material such as crushed concrete can be utilised for this purpose, provided that chemical composition of the fill is found to be acceptable for the intended purpose.

This material will be benched into the existing slope, removing any existing local over-steepening or slope bulges. The fill will be placed and compacted in accordance with the contract specification. A layer of seeded topsoil is usually placed on top of the new fill to provide vegetation (**Figure 71.12**). A steel mesh can be incorporated at the surface to provide erosion protection for the steeper slopes or to provide protection against animal burrowing.

Ruglei verge protection system

In areas where the slopes of embankments have a satisfactory safety factor against deep-seated and shallow slips but widening of the cess is required to provide a stable shoulder, the Ruglei verge protection system can be considered.

This system consists of an angular galvanised-steel mesh and a geogrid inset to contain and support the fill (coarse granular fill or ballast) that is required for the widening of the cess. Vertical and raked piles are used as necessary to stabilise the system. Driven, old rail sections or steel H-piles can be used as vertical piles. The type, spacing and length of the vertical and raked piles depend on the slope geometry and ground conditions. The coarse granular fill or ballast provides a free-draining medium, which would also act as a natural barrier against vegetation. Hence, the lateral drainage of the track is improved and the application of weedkillers can largely be eliminated. A rapid rate of construction per shift can be achieved, in particular if rail sections are used as vertical dowels. This system is economical and is considered suitable for areas where the only requirement is to widen the cess and to provide shoulder stability (Saffari and Smith, 2005) (**Figure 71.13**).

Figure 71.12 Regrading of an embankment to increase stability, incorporating a walkway at the toe and crest

Anchored bored-pile wall

For this option a system of embedded vertical small-diameter (of the order of 300 mm) bored piles with raking bored piles and a capping beam can be used to improve the stability of slopes and/or provide support for the widening of the cess at the crest. The capping beam acts as a small retaining structure to support the widened cess. The vertical piles act as a foundation for the capping beam and provide overturning stability, and the raking piles control lateral deformation.

The design and construction of this retaining wall system should ensure that the slope in front of the wall will remain stable and not subject to future erosion or movement. Any such movements in the future will lead to exposing the bored piles and increasing the design retained height of the wall, leading to potential pile damage and instability. The space between the exposed piles will potentially be used by burrowing animals, leading to further instability.

Toe wall

Where space at the toe of an embankment is too limited to regrade the slope to stable angles, a combination of a wall

Figure 71.13 Shoulder instability on a railway embankment and construction of a Ruglei shoulder stabilisation system: (a) and (b) poor embankment shoulder before remediation; (c) driving of old rail section for Ruglei support; (d) backfilling of Ruglei basket with granular fill; (e) completed Ruglei shoulder stabilisation system

at the toe of slope and regrading by placing additional fill between the top of the toe wall and the edge of the widened crest can be used. Issues regarding the design and construction of this type of wall are generally similar to those for cuttings. Space at the toe of an embankment may be more readily available to accommodate a retaining wall compared to cuttings.

King post wall

Where the safety factor against deep-seated instability is satisfactory and the main problem is to provide a solution to mitigate shoulder instability and widen the embankment shoulder, king post walls can be used.

In principle a king post retaining wall consists of embedded vertical structural members acting as king posts, with horizontal structural members spanning between them to form a retaining function. Backfilling with granular fill can be carried out behind the wall to provide the widened shoulder.

The vertical members can be bored or driven concrete or steel piles. Along the shoulder of an embankment, space and access for construction are very limited and, hence, driven steel piles such as universal bearing piles using a track-mounted rig would provide a more practical solution. This arrangement eliminates the need for a piling platform and the associated temporary works. The horizontal members can be precast concrete panels fitted between the flanges of the steel pile or against the web. An example of this type of wall is shown in **Figure 71.14**.

Grundomat pile wall

A Grundomat pile wall is a lighter form of a king post wall comprising a 3 mm thick, 100–150 mm outer diameter

Figure 71.14 An example of a king post wall installed along the shoulder of a railway embankment. The wall comprises steel universal bearing piles and precast concrete panels. A handrail is fixed to the steel piles after installation of the wall and backfilling was completed: (a) installation of the piles using a track-mounted 13 t excavator over the cable run; (b) the guide rails for pile alignment; (c) installed piles and the temporary platform alongside the embankment for operative access; (d) and (e) completed wall

steel casing which is bottom-driven to the required depth using compressed air. Consecutive sections of tubes can be welded together and driven in until the design pile depth is achieved. The pile reinforcement typically comprises a minimum 70 × 70 × 8 mm square hollow section (SHS) which is placed 0.5 m above the base of the casing which forms the driving head for the casing. The SHS will protrude from the Grundomat casing to the required height above ground level to support the horizontal concrete planks which provide the retaining function. The SHS post will then be grouted inside the casing with concrete or grout. Backfilling with granular fill can be carried out behind the concrete planks to provide the widened shoulder. The Grundomat piles can be driven into the ground using very small plant, which makes this system more suited for restricted-access sites where larger plant cannot be utilised.

Sheet piling

Sheet piles can be installed at the toe or crest of an embankment slope to form a retaining wall (Baker, 2015; Payne *et al.*, 2020). Backfilling behind the sheet piles will then be carried out to widen the shoulder and also increase the toe weight, which contributes to improving stability. The depth of the sheet piles will need to be designed to resist lateral loading as well as the rotational slope instability. A capping beam, steel or concrete, can be constructed to connect the sheet piles together. The sheet piles will need to be designed against corrosion to provide the required design life. The design should consider and ensure that the sheet pile wall does not adversely affect the drainage of the embankment by damming the water behind. This can be achieved through introduction of weep holes and/or intermittent shorter sheet piles to allow drainage through or beneath the wall.

Installation of the sheet piles using a standard crane-mounted vibrating pile-driving method can be unsuitable at some sites due to restrictions on the use of crane-mounted rigs and on lifting long sections of sheet piles adjacent to railways.

An alternative installation method using the proprietary Giken GRB non-staging system can be considered under these circumstances. This is a silent press method using installed piles to provide reaction for a hydraulic pressing system. The piling press, power pack, pitching crane and pile transport equipment are all mounted on and travel along the installed piles; hence, no temporary access or staging is required. As the pile lifting equipment is clamped to the installed piles, there is a reduced risk to the railway from toppling plant. An example of this type of sheet piling is shown in **Figures 71.15** and **71.16**.

(a)
(b)

Figure 71.15 Construction of Giken sheet piling at the toe of a railway embankment within a tight space close to third-party buildings: (a) components of the Giken sheet piling rig; (b) piling rig during operation resting on adjacent piles for support

71.5.3 Remedial and strengthening work design

For existing slopes, remedial and strengthening work often needs to be carried out while the transport system running above or below the earthworks remains operational. Hence, the main challenges for existing slopes can be

- very tight and limited space
- limited access
- safe operation of the railway, highway and associated services
- environmental constraints.

Hence, innovative techniques are required in order to arrive at cost-effective, efficient and safe solutions for remedial and strengthening work. Many innovative solutions such as the Ruglei shoulder protection system, Giken sheet piling, soil

Figure 71.16 Construction of Giken sheet piling at the toe of a railway embankment within a tight space close to third-party buildings: (a) installed piles close to third-party boundary; (b) view along finished wall with capping beam; (c) view of finished embankment from above, next to the railway

nailing and various types of piling have successfully been implemented. **Figure 71.17** shows the replacement of an old and worn-out king post wall in a difficult access area adjacent to a congested network of cables.

In some cases, due to site and access constraints, a solution can only be achieved by providing protection rather than remediation. **Figure 71.18** shows the protection provided for the steep toe of a cutting, which could not otherwise be regraded due to site constraints, by the placing of Armortec revetment blocks.

71.5.4 Environmental considerations

Vegetation on an earthworks slope is generally the most aesthetically pleasing form of surface protection and is fundamental to enhancing the environment. The need to maintain and enhance environmental value has become a major factor in overall earthworks asset management.

One of the key findings of the independent review to examine Network Rail's approach to lineside vegetation management (Varley, 2018) was that the approach of the organisation was 'reactive and inconsistent', with little thought given to the environmental impact. In response to Varley, the UK Department of Transport has set out a clear strategy for how it expects Network Rail to protect and enhance the UK's lineside environment, while ensuring the safety of passengers and services. *Enhancing Biodiversity and Wildlife on the Lineside* was published (Department of Transport, 2019) with the headline challenge for Network Rail to achieve no net loss in biodiversity by 2024 and biodiversity net gain on each route by 2040.

London Underground's activities are fully aligned with the Mayor of London Transport Strategy (2018) that aims for London to be a National Park City where more than half of its area is green, where the natural environment is protected and the network of green infrastructure is managed to benefit all Londoners. Within this context, protecting and enhancing green infrastructure on transport land will help to improve the natural environment and contribute to London's overall resilience to climate change. Accordingly, London Underground operates a proactive vegetation management system and sets out its requirements for landscaping and vegetation management in the off-track areas in the London Underground Standard S1165 (2017).

Environmental factors could have an impact on the performance of an existing earthwork or on the maintenance or construction of the remedial works. Environmental factors could also affect the type of design solution adopted for the remedial works in order to comply with regulations.

These aspects, when identified, have the potential to significantly delay a project until approval from the relevant authorities is obtained. Hence, a consideration of environmental aspects should be made from the start of any project in order to ensure that all risks are taken into account. An environmental risk assessment can be carried out at the start of a project and reviewed throughout the development of the project. A list of environmental factors that could have an effect on an earthwork is discussed below.

Figure 71.17 Replacement of an old king post wall adjacent to a congested network of cables: (a) the old king post wall next to the cables; (b) temporary soil-nailed wall with timber facing; (c) constructed mass-concrete wall with cable ducts and drainage pipes

71.5.4.1 Vegetation

Performance

The effect of vegetation on the performance of earthworks has been discussed in the preceding sections. While the presence of mature trees reduces the pore pressures within a slope, leading to increased stability, on clay slopes it can cause seasonal swelling and shrinkage, leading to serviceability problems. 'The influence of vegetation on the railway and earthwork slopes in particular is rarely, if ever, wholly beneficial or wholly detrimental but almost always manifests as some combination of the two extremes' (Gellatley *et al.*, 1995).

The presence of vegetation may serve to reinforce and strengthen a slope through the root system. However, depending on the type of vegetation and the slope geometry, the root system may be quite shallow and ineffective and become unstable during storm events and fall onto the transport corridor below. Many aspects of the long-term behaviour of existing earthwork slopes are affected by vegetation (Glendinning *et al.*, 2009). It is clear that the role of vegetation and its part in the effective management of earthwork slopes is a complex issue, with many conflicting effects. The latest research (Smethurst *et al.*, 2015) indicates that on the whole the presence of mature vegetation on slopes is generally beneficial and the vegetation management strategy should strike a balance between the positive and negative impacts.

Vegetation management

Earthworks vegetation management needs to balance the negative impacts of vegetation (blocked ditches and pipes, leaf fall, tree fall, ground desiccation and shrinkage) against its positive impacts (reducing surface erosion, providing root reinforcement, avoiding channelling of flows, maintaining surface pore

(a)

(b)

Figure 71.18 Steep toe-erosion protection where access to regrade is not available: (a) installation of the Armortec revetment blocks; (b) a steep toe after installation of the Armortec system

water suction). The potential negative impacts of devegetation should be carefully considered. Vegetation should be treated as an asset and the transport earthworks regarded as a wildlife corridor. Management of vegetation on earthwork slopes should maximise the beneficial actions while minimising the detrimental ones. The necessity for vegetation clearance, bearing in mind its cost and potential negative impact on slope stability, needs careful consideration on an asset-specific basis. However, the management of existing vegetation and replacement of vegetation with new species can result in improvements of slope stability and reduced erosion. It can also result in managing soil moisture fluctuations in the near-surface zone to reduce the potential for desiccation cracks to form and to prevent significant shrinkage during droughts. The use of managed vegetation and bioengineering to stabilise earthwork slopes is a cost-effective technique with the potential to be used more extensively on earthwork slopes as a preventive and remedial measure. However, further work is required to quantify the effectiveness of managed vegetation and bioengineering for earthworks slopes and to adapt the approach for routine practical applications (Network Rail, 2021a).

Tree preservation order (TPO)

Trees which carry a TPO cannot be removed from a construction site. There is usually an exclusion zone associated with these trees, which could be quite significant depending on the type and size of the trees and the root system. This can significantly affect the design of the remedial works in the vicinity of the tree. A preservation order on hedgerows on the crest of a cutting can significantly affect access to the cutting during construction. Hence, it is important that any such orders are investigated at the start of a project in order to avoid any design changes and delays to the project.

Third party

Vegetation and trees that may have a significant effect on the stability of a cutting or embankment could be located on third-party land. In this case the vegetation may be outside the control of the owner of the infrastructure. Hence, caution must be applied during the design of the works regarding the long-term pore pressure regime within the structure.

71.5.4.2 Environment Agency (EA)

The EA is a non-departmental public body, established in 1995 with responsibilities relating to the protection and enhancement of the environment in England (and until 2013 also Wales). In 2013 that part of the EA covering Wales was merged into Natural Resources Wales, a separate body managing the Welsh environment. The Scottish Environment Protection Agency (SEPA) is Scotland's environmental regulator and national flood forecasting, flood warning and strategic flood risk management authority. Streams and rivers located adjacent to an embankment or passing beneath it may be affected by construction activities. Drilling of exploratory holes for the design or drilling of piles may cause pollution in the river. Increasing an embankment's volume as the result of the construction of a berm may affect the EA flood capacity in the vicinity of a river. In this case, consultation with the EA during the design phase would be required. The design of a drainage scheme at the toe of an embankment may include a discharge of the collected water into the river. Prior approval from the EA would be required, including an outfall design that complies with EA requirements.

71.5.4.3 Ecological

For any investigation or remedial work, the following ecological aspects should be considered

- nesting birds
- badger setts
- great crested newts
- slow-worms.

An ecological survey would be required well in advance of the work so that appropriate mitigation measures are undertaken in order to comply with the relevant standards.

References

Antoine R, López T, Tanguy M, *et al.* (2020) Geoscientists in the sky: un-manned aerial vehicles responding to geohazards. *Surveys in Geophysics* **41(6)**: 1285–1321.

Baker DA (2015) King sheet piling (KSP): a major advance in sheet pile retaining wall design and installation. *Proceedings of the XVI ECSMGE Geotechnical Engineering for Infrastructure and Development, Edinburgh, UK*, pp. 1035–1040.

Briggs KM, Loveridge FA and Glendinning S (2017) Failures in transport infrastructure embankments. *Engineering Geology* **219**: 107–117.

Bromhead EN (1992) *The Stability of Slopes.* Taylor & Francis, London, UK.

BSI (British Standards Institution) (2004) BS EN 1997-1:2004: Eurocode 7: Geotechnical design – Part 1: General rules. BSI, London, UK.

BSI (2008) PAS55:2008-1:2008: Specification for the optimized management of physical assets. BSI, London, UK.

BSI (2014) BS ISO 55000:2014: Asset management – Overview, principles and terminology. BSI, London, UK.

Butcher D and Pearson A (2007) Serviceability of London Clay embankments. *Engineering Geology of London Clay, Joint Bicentennial Conference on the Thames Valley Regional Group and the Engineering Geology of the Geological Society, London, UK.*

Carder DR and Barker KJ (2005) *The Performance of a Single Row of Spaced Bored Piles to Stabilise a Gault Clay Slope on the M25.* TRL Report 627. TRL, Crowthorne, UK.

Chandler RJ (1984) Delayed failure and observed strengths of first-time slides in stiff clays: a review. *Proceedings of the 4th International Conference Landslides, Toronto, ON*, vol. 2, pp. 19–25.

Chandler RJ and Skempton AW (1974) The design of permanent cutting slopes in stiff fissured clays. *Géotechnique* **24(4)**: 457–464.

Cooper MR, Bromhead EN, Petley DJ and Grant DI (1998) The Selborne cutting stability experiment. *Géotechnique* **48(1)**: 83–101.

Coppin NJ and Richards IG (2007) *Use of Vegetation in Civil Engineering.* CIRIA Report C708. CIRIA, London, UK.

Davies JP, Loveridge FA, Perry J, Patterson D and Carder D (2003) Stabilization of a landslide on the M25 freeway London's main artery. *12th Pan-American Conference on Soil Mechanics and Geotechnical Engineering, Boston, MA.*

DeLory FA (1957) *Long-Term Stability of Slopes in Overconsolidated Clays.* PhD thesis, University of London, London, UK.

Department for Transport (2014) *Transport Resilience Review: A Review of the Resilience of the Transport Network to Extreme Weather Events.* Department for Transport, London, UK.

Department for Transport (2019) *Enhancing Biodiversity and Wildlife on the Lineside.* Department for Transport, London, UK.

Donzelli V (2019) Satellites and drones to support railway infrastructure maintenance. *Global Railway Review.* https://www.globalrailwayreview.com/article/82771/momit-drones-and-satellites-railway (accessed 25/11/2022).

Duffell CG, Rudrum DM and Willis MR (2005) Remote sensing techniques for highway earthworks assessment. *Proceedings of ASCE Geo-Frontiers, Austin, TX*, 10.1061/40785(164)3.

Dunnicliff J (1988) *Geotechnical Instrumentation for Monitoring Field Performance.* Wiley, New York, NY.

Ellis EA and O'Brien AS (2007) Effect of height on delayed collapse of cuttings in stiff clay. *Proceedings of the Institution of Civil Engineers, Geotechnical Engineering* **160(GE2)**: 73–84.

Gellatley M, McGinnity BT, Barker DH and Rankin WJ (1995) Interaction of vegetation with the LUL surface railway system. *Vegetation and Slopes: Stabilisation, Protection and Ecology: Proceedings of the International Conference Oxford, UK*, 10.1680/vasspae.20313.0007.

Giordan D, Adams MS, Aicardi I, *et al.* (2020) The use of unmanned aerial vehicles (UAVs) for engineering geology applications. *Bulletin of Engineering Geology and the Environment* **79(7)**: 3437–3481.

Glendinning S, Loveridge FA, Starr-Keddle RE, Bransby MF and Hughes PN (2009) Role of vegetation in sustainability of infrastructure slopes. *Proceedings of the Institution of Civil Engineers, Engineering Sustainability* **162(ES2)**: 101–110.

Godfrey PS (1996) *Control of Risks: A Guide to the Systematic Management of Risk from Construction.* CIRIA Special Publication 125. CIRIA, London, UK.

HMG (Her Majesty's Government) (1974) Health and Safety at Work Act, etc. *1974.* HMSO, London, UK.

HMG (1999) Management of Health and Safety at Work Regulations 1999. HMSO, London, UK. Statutory Instrument 1999 No. 3242.

HMG (2015) Construction (Design and Management) Regulations (CDM) 2015. HMSO, London, UK. Statutory Instrument 2015 No. 51.

Hooper R, Armitage R, Gallagher A and Osorio T (2009) *Whole-Life Infrastructure Asset Management: Good Practice Guide for Civil Infrastructure.* CIRIA Report C677. CIRIA, London, UK.

HSE (1999) *Management of Health and Safety at Work Regulations 1999: Approved Code of Practice and Guidance L21*, 2nd edn. HMSO, London, UK.

Hudacsek P and Bransby M (2008) *Centrifuge Modelling of Embankments Subject to Seasonal Moisture Changes.* BIONICS research project, University of Dundee, Dundee, UK.

Hungr O, Leroueil S and Picarelli L (2014) The Varnes classification of landslide types, an update. *Landslides* **11(2)**: 167–194.

Koe A, Murphy W and Nicholson R (2018) *Rock Netting Systems: Design, Installation and Whole-Life Management.* CIRIA Report C775. CIRIA, London, UK.

Kovacevic N, Potts DM and Vaughan PR (2001) Progressive failure in clay embankments due to seasonal climate changes. *Proceedings of the 15th International Conference in Soil Mechanics and Geotechnical Engineering, Istanbul, Turkey*, pp. 2127–2130.

London Underground (2015) Standard S1159: Track-dimensions and tolerances, A4. London Underground, London, UK.

London Underground (2017) Standard S1165: Landscaping and vegetation, A4. London Underground, London, UK.

London Underground (2019) Guidance G0054B: Earth structures: guide for slope stability analysis. London Underground, London, UK.

Lord JA, Clayton CRI and Mortimore RN (2002) *Engineering in Chalk.* CIRIA Report C574. CIRIA, London, UK.

Loveridge F and Anderson D (2007) What to do with a vegetated clay embankment. *Slope Engineering Conference, London, UK.* Thomas Telford, London, UK.

Loveridge FA, Spink TW and O'Brien AS (2010) The impact of climate and climate change on infrastructure slopes, with particular reference to southern England. *Quarterly Journal of Engineering Geology and Hydrogeology* **43**: 461–472.

Lowe J, Bernie D, Bett P, *et al.* (2019) *UKCP 18 Science Overview Report November 2018* (Updated March 2019). https://www.metoffice.gov.uk/pub/data/weather/uk/ukcp18/science-reports/UKCP18-Overview-report.pdf (accessed 06/05/2023).

Mayor of London (2018) Mayor's Transport Strategy. Mayor of London, London, UK.

McGinnity BT, Fitch T and Rankin WJ (1998) A systemic and cost-effective approach to inspecting, prioritising and upgrading London Underground's earth structures. *ICE Proceedings of the Seminar Value of Geotechnics in Construction, London, UK.* ICE, London, UK, pp. 309–322.

Milne FD, Brown MJ, Davies MCR and Cameron G (2015) Some key topographic and material controls on debris flows in Scotland. *Quarterly Journal of Engineering Geology and Hydrogeology* **48(3–4)**: 212–223.

Network Rail (2021a) *A Review of Earthworks Management. Prepared by a Task Force Led by Professor Lord Robert Mair.* Network Rail, London, UK.

Network Rail (2021b) *Weather Advisory Task Force Final Report February 2021. Prepared by Prof. Dame Julia Slingo, Paul Davies and Prof. Hayley Fowler.* Network Rail, London, UK.

Nicholson D, Tse C and Penny C (1999) *The Observational Method in Ground Engineering: Principles and Applications.* CIRIA Report C185. CIRIA, London, UK.

Nyambayo VP, Potts DM and Addenbrooke TI (2004) The influence of permeability on the stability of embankments experiencing seasonal cyclic pore water pressure changes. *Proceedings of Advances in Geotechnical Engineering: The Skempton Conference, London, UK*, vol. 2, pp. 993–1004.

O'Brien A (2007) Rehabilitation of urban railway embankments: investigation, analysis and stabilisation. In *Proceedings of the 14th European Conference on Soil Mechanics and Geotechnical Engineering, Madrid* (Cuellar V, Dapena E, Alonso E, *et al.* (eds)). Millpress, Amsterdam, the Netherlands, pp. 125–143.

O'Brien AS (2013) The assessment of old railway embankments: time for a change? *Partial Saturation in Compacted Soils: Géotechnique Symposium in Print, London, UK*, pp. 19–32

Payne I, Clifton L, Holt S, Griffiths I and Wadesmith D (2020) Ground risk and rail asset management in East Anglia. *Proceedings of the Institution of Civil Engineers, Transport* **173(4)**: 232–244.

Perry J (1989) *A Survey of Slope Condition on Motorway Embankments in England and Wales.* TRL Research Report RR199. TRL, Crowthorne, UK.

Perry J, Pedley M and Brady K (2003a) *Infrastructure Cuttings: Condition Appraisal and Remedial Treatment.* CIRIA Report C591. CIRIA, London, UK.

Perry J, Pedley M and Reid M (2003b) *Infrastructure Embankments: Condition Appraisal and Remedial Treatment*, 2nd edn. CIRIA Report C592. CIRIA, London, UK.

Phear A, Dew C, Ozsoy B, *et al.* (2005) *Soil Nailing: Best Practice Guidance.* CIRIA Report C637. CIRIA, London, UK.

Potts DM, Kovacevic N and Vaughan PR (1997) Delayed collapse of cut slopes in stiff clay. *Géotechnique* **47(5)**: 953–982.

Powrie W (2018) *Soil Mechanics: Concepts and Applications*, 3rd edn. CRC Press, Boca Raton, FL.

Powrie W and Smethurst JA (2015) Discrete passive piles for infrastructure slope stabilization. *International Conference on Soft Ground Engineering, Singapore, Singapore*, pp. 1–16.

Powrie W and Smethurst J (2019) Climate and vegetation impacts on infrastructure cuttings and embankments. *Proceedings of the 8th International Congress on Environmental Geotechnics: Towards a Sustainable Geoenvironment* (Zhan L, Chen Y and Bouazza A (eds)). Springer, New York, NY, vol. 1, pp. 128–144.

RAIB (Rail Accident Investigation Branch) (2018) Landslip and derailment at Loch Eilt, north-west Scotland. 22 January 2018. Report 10/2018. RAIB, Derby, UK.

RAIB (2020) Train collision with material washed out from a cutting slope at Corby, Northamptonshire. 13 June 2019. Report 04/2020. RAIB, Derby, UK.

RAIB (2022) Derailment of a passenger train at Carmont, Aberdeenshire. 12 August 2020. Report 02/2022. RAIB, Derby, UK.

Saffari N and Smith R (2005) Stabilisation of embankment shoulders on the London Underground. *Conference Proceedings of Railway Engineering 2005, Earthworks Stabilisation,* London, UK.

Sandroni S (1977) *The Strength of London Clay in Total Effective Stress Terms.* PhD thesis, University of London, London, UK.

Scott JM, Loveridge F and O'Brien AS (2007) Influence of climate and vegetation on railway embankments. *Proceedings of the 14th European Conference on Soil Mechanics and Geotechnical Engineering, Madrid, Spain* (Cuellar V, Dapena E, Alonso E, *et al.* (eds)). Millpress, Amsterdam, the Netherlands, pp. 659–664.

Skempton AW (1948) The rate of softening in stiff fissured clays, with special reference to London Clay. *Proceedings of the 2nd International Conference on Soil Mechanics and Foundation Engineering, Rotterdam, the Netherlands*, vol. 2, pp. 50–53.

Skempton AW (1964) Long-term stability of clay slopes. *Géotechnique* **14(2)**: 77–101.

Skempton AW (1970) First-time slides in overconsolidated clays. *Géotechnique* **20(3)**: 320–324.

Skempton AW (1977) Slope stability of cuttings in brown London Clay. *Proceedings of the 9th International Conference on Soil Mechanics and Foundation Engineering, Tokyo Japan*, vol. 3, pp. 261–271.

Skempton AW (1996) Embankments and cuttings on the early railways. *Construction History* **11**: 33–39.

Skempton AW and Petley DJ (1967) The shear strength along structural discontinuities in stiff clays. *Proceedings of the Geotechnical Conference, Oslo, Norway*, vol. 2, pp. 29–46.

Smethurst JA (2003) *The Use of Discrete Piles for Infrastructure Slope Stabilisation.* PhD thesis, University of Southampton, Southampton, UK.

Smethurst JA and Powrie W (2007) Monitoring and analysis of the bending behaviour of discrete piles used to stabilise a railway embankment. *Géotechnique* **57(8)**: 663–667.

Smethurst JA, Briggs KM, Powrie W, Ridley A and Butcher DJE (2015) Mechanical and hydrological impacts of tree removal on a clay fill railway embankment. *Géotechnique* **65(11)**: 869–882.

Smethurst JA, Smith A, Uhlemann S, *et al.* (2017) Current and future role of instrumentation and monitoring in the performance of transport infrastructure slopes. *Quarterly Journal of Engineering Geology and Hydrogeology* **50**: 271–286.

Smethurst JA, Bicocchi N, Powrie W and O'Brien AS (2020) Behaviour of discrete piles used to stabilise a tree-covered railway embankment. *Géotechnique* **70(9)**: 774–790.

Standing JR, Vaughan PR, Charles-Jones S and McGinnity BT (2021) Observed behaviour of old railway embankments formed of ash and dumped clay fill. *Géotechnique* **71(11)**: 938–956.

Stirling RA, Toll DG, Glendinning S, *et al.* (2021) Weather-driven deterioration processes affecting the performance of embankment slopes. *Géotechnique* **71(11)**: 957–969.

Take WA and Bolton MD (2004) Identification of seasonal slope behaviour mechanism from centrifuge case studies. *Proceedings of Advances in Geotechnical Engineering: The Skempton Conference, London, UK*, vol. 2, pp. 993–1004.

Varley J (2018) *Network Rail Vegetation Management Review: Valuing Nature, a Railway for People and Wildlife. The Network Rail Vegetation Management Review.* Department of Transport, London, UK.

Varnes DJ (1978) Slope movement types and processes. In *Landslides, Analysis and Control* (Schuster RL and Krizek RJ (eds)). Special Report 176. Transportation Research Board, National Academy of Sciences, Washington, DC, pp. 11–33.

Vaughan PR (1994) Assumptions, predictions and reality in geotechnical engineering. *Géotechnique* **44(4)**: 573–603.

Vaughan PR and Walbancke HJ (1973) Pore pressure changes and delayed failure of cutting slopes in overconsolidated clay. *Géotechnique* **23(4)**: 531–539.

Winter MG, Dent J, Macgregor F, *et al.* (2010) Debris flow, rainfall and climate change in Scotland. *Quarterly Journal of Engineering Geology and Hydrogeology* **43(4)**: 429–446.

Wyllie DC (2017) *Rock Slope Engineering: Civil Applications*, 5th edn. CRC Press, London, UK.

It is recommended this chapter is read in conjunction with

- Chapter 94 *Principles of geotechnical monitoring*
- Chapter 100 *Observational method*
- Chapter 101 *Close-out reports*

All chapters in this book rely on the guidance in Sections 1 *Context* and 2 *Fundamental principles.* A sound knowledge of ground investigation is required for all geotechnical works, as set out in Section 4 *Site investigation.*

doi: 10.1680/icemge.66830.1183

Chapter 72
Slope stabilisation methods

Paul A Nowak Atkins Ltd, Epsom, UK

Earthworks slopes may require the use of structural solutions where geometric constraints or the proximity of settlement-sensitive structures prevent the design of a full-height earthworks solution. Structural solutions may also be used for road widening where further land purchase is not possible, and for the stabilisation of existing earthworks slopes. This chapter introduces the major forms of structural slope solutions.

CONTENTS

72.1. Introduction

Stabilised slopes incorporating structural components, nailing or reinforcing elements in new earthworks are usually employed either

- where long-term performance of a conventional slope may impact adjacent structures (e.g. adjacent buildings) or
- where geometric constraints do not allow the construction of a conventional slope that is stable over its design life.

Stabilised slopes are, however, common in stabilising existing assets and the widening of existing motorways where measures cannot extend outside the existing asset boundary.

The selection of a stabilisation method can depend on the balance of the capital expenditure to construct and maintain compared to the adoption of a less expensive option that may require additional measures to be applied at a later date. This is usually the decision of the asset owner and can depend on the consequences of failure. A useful approach outlining responsibilities is given in Charles and Watts (2002).

This chapter describes briefly methods to stabilise slopes, which can be divided into

- embedded solutions
- gravity solutions
- reinforced/nailed solutions
- slope drainage.

In some cases, solutions comprise more than one of these categories.

72.2. Embedded solutions

Embedded solutions generally comprise

- concrete pile walls
- sheet pile walls.

Concrete pile walls usually comprise bored cast in situ or continuous flight auger piles. Pile diameter and spacing will depend on the size of plant that requires access to the location and the requirements of the design with respect to bending moment and shear in the pile shaft.

Piled walls are usually designed in order that potential failure surfaces passing below the toe of the wall achieve a satisfactory factor of safety/degree of utilisation as described in Chapter 69 *Earthworks design principles*. Wall design should also consider the action of the piles to act as shear keys on potential failure surfaces passing through the wall and should be designed as a conventional embedded retaining wall after Gaba *et al.* (2017).

Where the design of the wall cannot satisfy bending moment and shear requirements of the pile shaft, or deflection causing unsatisfactory settlement of the ground behind the wall, additional support can be considered. This would usually comprise (*a*) cantilever walls; (*b*) ground anchors; (*c*) raking piles; and (*d*) tie bars (see **Figure 72.1**).

Ground anchors, installed at the head of the pile, apply a restraining force in an active manner directly after construction; the active load in the anchor is applied immediately after installation. The design of the anchor should be after BS 8081 (BSI, 2015a).

Raking piles, installed in a similar manner to anchors, apply no initial restraining force and act passively, whereby forward movement of the wall generates friction on the raking pile shaft. This initiates an anchoring action at the vertical pile head. The raking piles should be designed as conventional piles in tension.

Tie bars act in the same way as raking piles, but can be post-tensioned to act similarly to ground anchors. They require a restraint mechanism and are most practicably used in the stabilisation of existing earthworks embankments where there is a piled wall on the opposite side of the embankment. This solution has been employed in the stabilisation of London Underground embankments.

Sheet pile walls are designed in the same manner as bored pile walls, but may be precluded where vibration could affect adjacent structures if a driven method of installation is employed. This can be overcome by the use of a hydraulic jacking installation method. The use of a sheet pile solution for the remediation of a failed slope may be attractive as it can be implemented rapidly after the failure occurs.

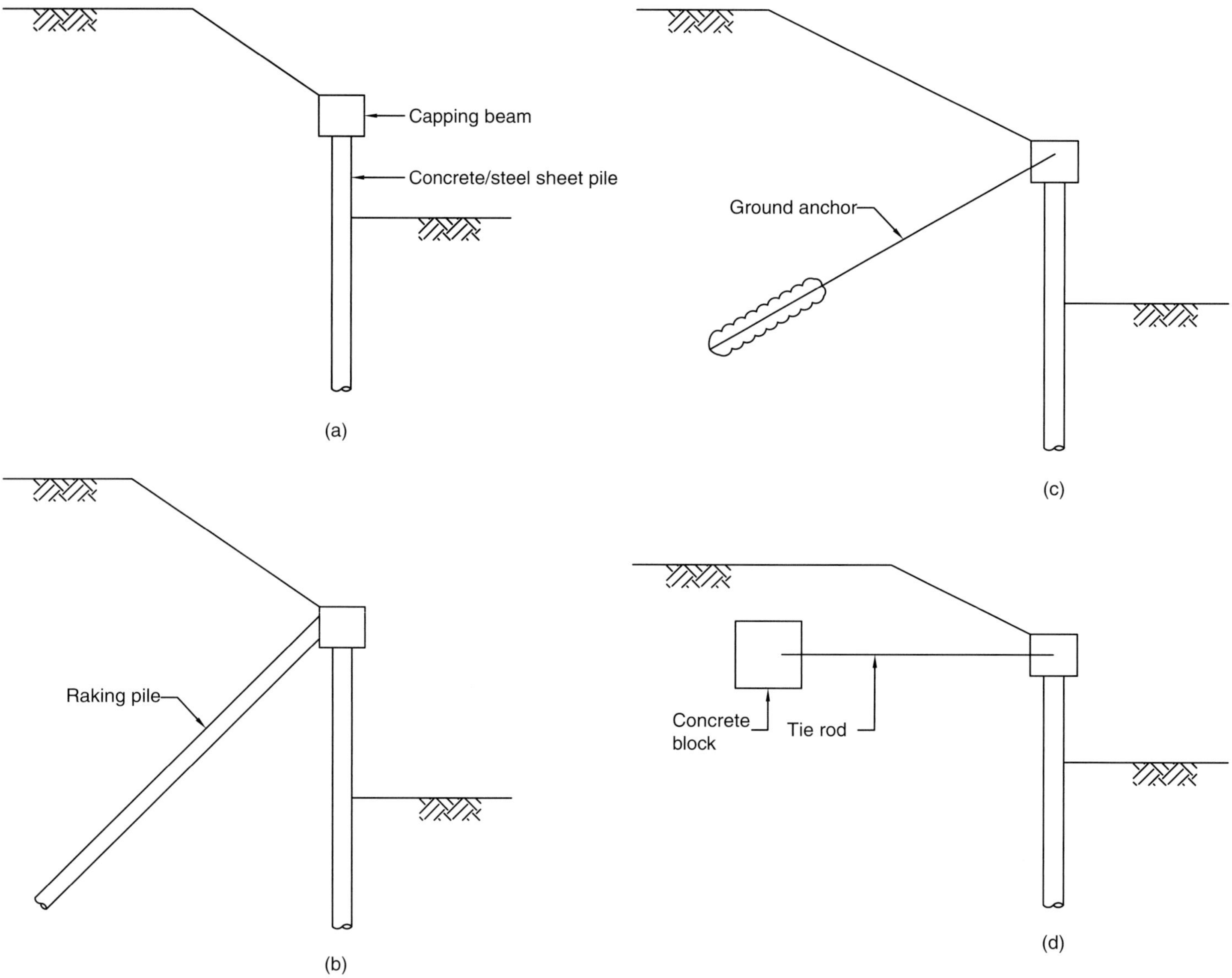

Figure 72.1 Typical embedded solutions: (a) cantilever wall; (b) wall with ground anchor; (c) wall with raking pile; (d) wall with tie and deadman block

Care should be taken in the design of sheet pile walls in that the design section chosen should be capable of being driven in the prevailing ground conditions. Guidance is given in Williams and Waite (1993).

Structural design of bored pile walls should be undertaken to BS EN 1992-1 (BSI, 2004a), BS8110-1 (BSI, 1997) and BS 8500 (BSI, 2015c). Structural design of steel sheet piles should be undertaken to BS EN 1993-5 (BSI, 2007). More advice on the design of embedded walls is given in Section 6 *Design of retaining structures.*

72.3. Gravity solutions

Gravity structures would usually comprise the following solutions: (*a*) reinforced concrete wall; (*b*) gabion wall; (*c*) dry block wall; (*d*) crib wall; (*e*) reinforced soil wall; and (*f*) reinforced earth wall. Typical solutions, without soil reinforcement, are illustrated in **Figure 72.2**.

The gravity structure performs under its own stability in terms of its resistance to sliding, overturning and bearing capacity failure. They can be designed in accordance with Chapman (2000). Reinforced earth and reinforced soil walls should be designed in accordance with BS 8006-1 (BSI, 2010) as described in Chapter 73 *Design of soil reinforced slopes and structures.* More advice on the design of reinforced gravity walls is given in Section 6 *Design of retaining structures.*

In addition to the stability of the gravity structure, the designer should also consider the stability of the slope that it retains. It is common for walls greater than 3 m which retain

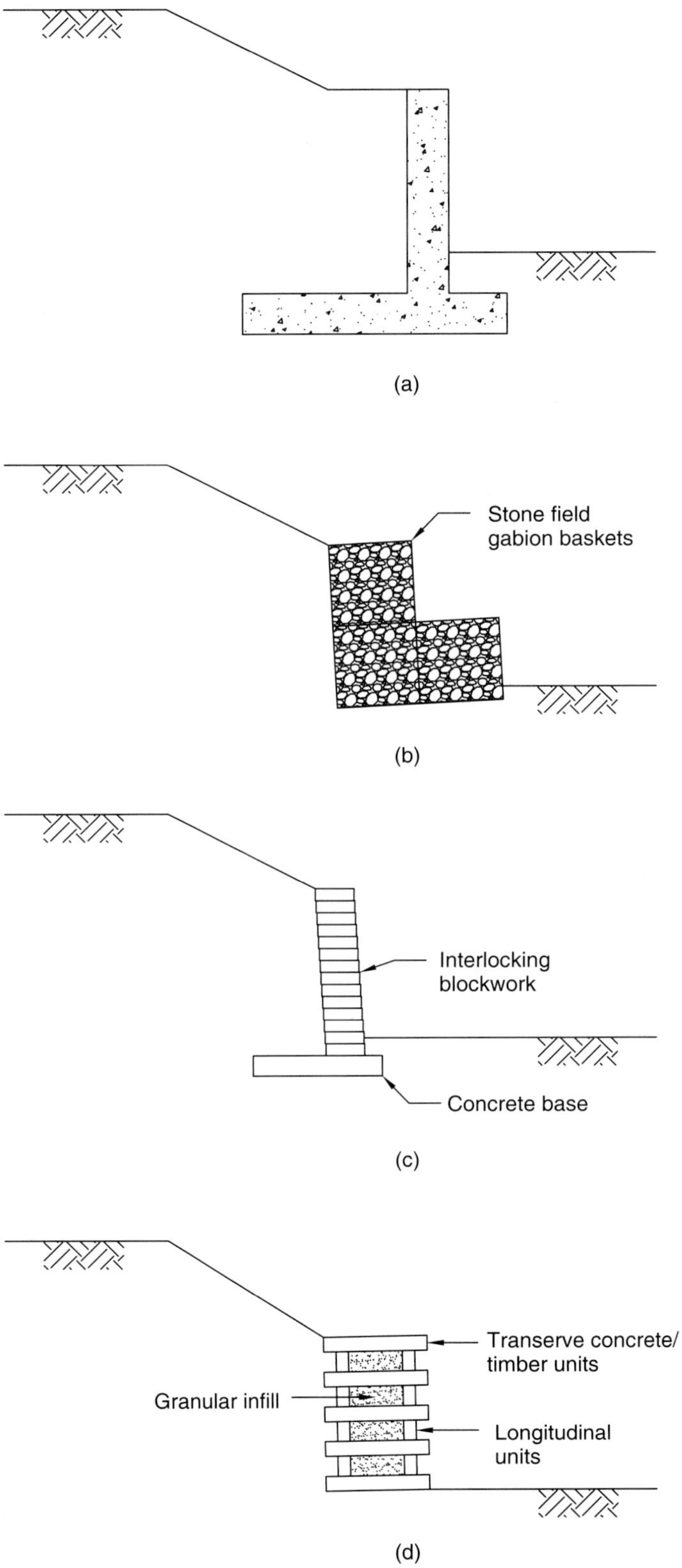

Figure 72.2 Typical gravity solutions: (a) reinforced concrete wall; (b) gabion wall; (c) dry block wall; (d) crib wall

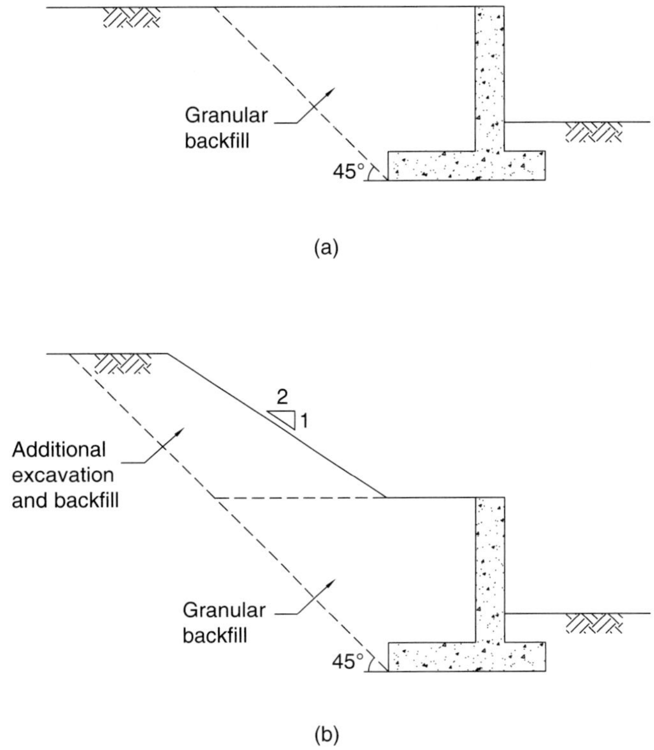

Figure 72.3 Effect of sloping ground on excavation and backfill volume: (a) gravity wall with horizontal ground behind; (b) gravity wall with sloping ground behind

a slope behind them to require embedment of the wall below ground surface on the passive side in order to prevent overall failure of the composite slope and wall.

It is also usual that gravity structures are designed with granular backfill in the area of the active wedge behind the wall. Care should be taken where a gravity structure is specified on existing or widened infrastructure, as it may result in significant excavation of the slope behind the wall, as shown in **Figure 72.3**.

It is common that gravity structures, particularly masonry and brick walls, form part of an existing infrastructure asset. Where an assessment is required as part of an asset condition survey this will be carried out to current best practice – that is, BS EN 1997 (BSI, 2004b) and/or BS 8002 (BSI, 2015b). It should be borne in mind, particularly with respect to railway infrastructure, that the structure may be up to 100 years old and would not have been constructed to current best practice. Assessment of the structure to BS EN 1997 or BS 8002 is likely to indicate failure of the structure or earthwork which shows no visible structural distress. In this case, the asset owner should be consulted and a sensible approach to the management of the asset agreed.

Additional commentary on the practicalities of gravity block walls and gabion walls is provided by Winter *et al.* (2022a) and Duffy-Turner *et al.* (2022b), respectively.

72.4. Reinforced/nailed solutions

Reinforced or nailed slopes will generally comprise the use of geogrid reinforcement or a grid of soil nails to provide additional tensile strength along likely failure surfaces. Typical examples are shown in **Figure 72.4**.

Design should be carried out in accordance with BS 8006-2 (BSI, 2011); further guidance is provided in Chapters 73 *Design of soil reinforced slopes and structures* and 74 *Design of soil nails*. The use of soil reinforcement and nailing is not confined to steep-faced (approximately 60°) slopes; it can be applied to slopes of any angle where additional tensile strength of the reinforcement or nailing allows the adoption of a stable slope at a steeper angle than if no reinforcement were employed. On the Terminal 5 spur road project at Heathrow Airport, geogrid reinforcement was employed to allow the use of 1V:2H slopes in London Clay embankments where geometric constraints prevented the adoption of 1V:4H slopes used for the majority of the embankment construction.

Geogrids have also been used to repair slope failures, particularly cutting slopes. They were extensively used on the M4 motorway in Berkshire in the 1970s and were also used in a London Clay cutting on London Underground at Hendon. The slipped material is excavated back beyond the slip surface, stockpiled and recompacted with layers of geogrid to provide additional tensile strength to the original failed material, as shown in **Figure 72.5**.

This solution may not always be practicable if there is insufficient facility to temporarily store the slipped material locally. If storage of excavated material is not available local to the failed area it is common practice for the slip repair to be effected using imported granular material. Additional commentary on the practicalities of reinforced soil walls and soil nailed slopes is provided by Winter *et al.* (2022b) and Duffy-Turner *et al.* (2022a), respectively.

72.5. Slope drainage

Stability of constructed slopes can be influenced by water either as groundwater (see Chapter 70 *Design of new earthworks*) or as precipitation.

The former may result in deep-seated slope failure if uncontrolled and not in line with design assumptions, particularly in interbedded cohesive and non-cohesive strata; groundwater flow in the non-cohesive strata causes softening of the surrounding cohesive material.

The ingress of precipitation can cause softening of the surface layers of cohesive earthworks, which leads to reduction in any effective cohesion in the slope material and the development of shallower failure surfaces. These are commonly seen on major road networks and require remediation.

It is common practice to counteract these problems with slope drainage, either as part of the original slope construction or as a remedial works measure.

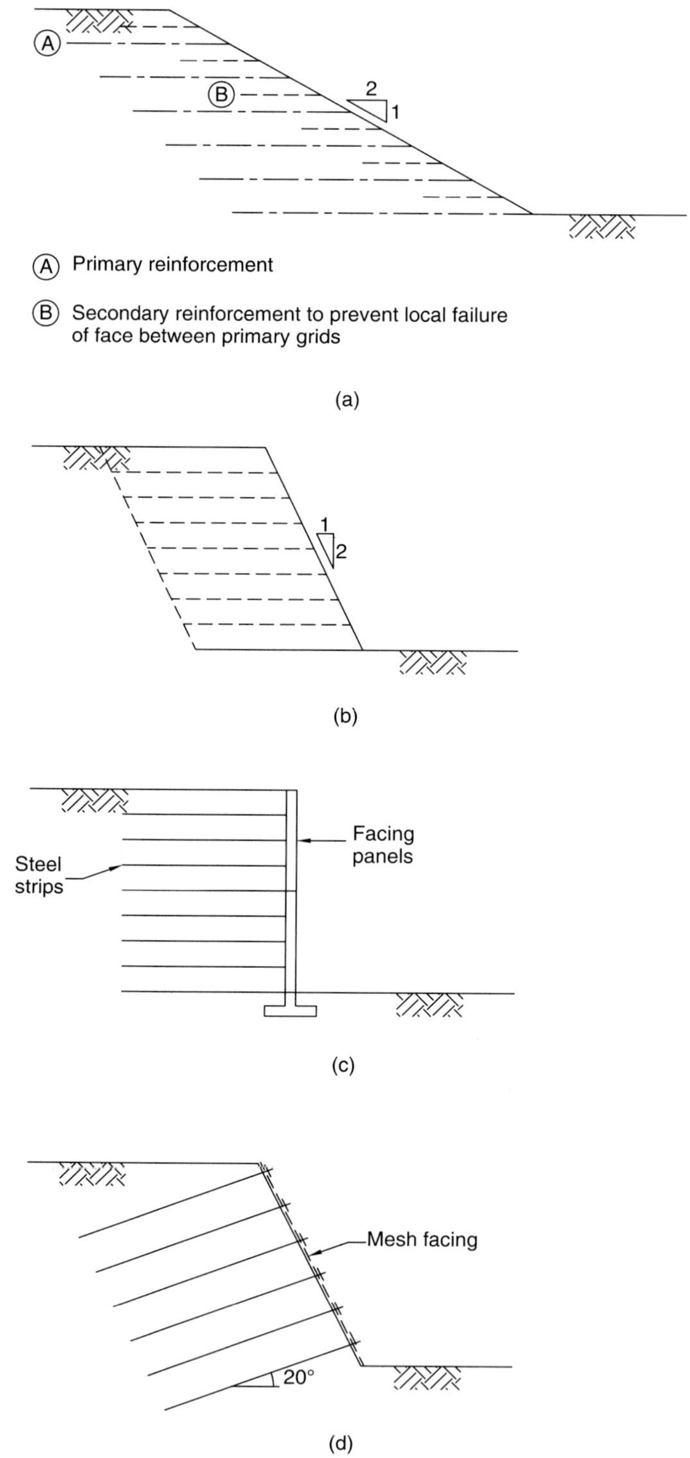

Figure 72.4 Typical reinforced/nailed solutions: (a) reinforced soil slope; (b) reinforced soil block; (c) reinforced earth slope; (d) soil nail slope

Slope drainage usually takes the form of gravel-filled drains orthogonal to the slope (counterforts) or as a 'herring bone' pattern. The drains are commonly constructed with a carrier pipe and geotextile wrap to prevent the ingress of fine-grained material which would impact on their long-term performance. Drains are usually spaced at intervals of 5–10 m along the

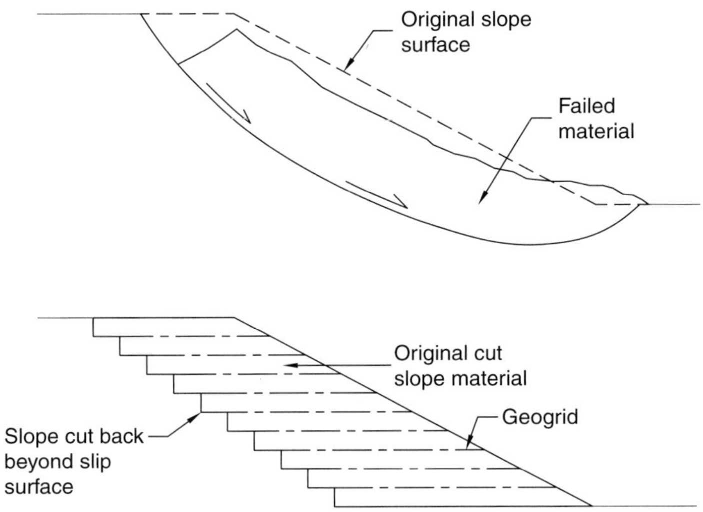

Figure 72.5 Repair of slope failure using geogrids

slope, depending on the composition of the slope material. Useful guidance on their design and the philosophy behind slope drainage is provided by Hutchinson (1977).

In designing slope drainage, a suitable outfall has to be considered to prevent ponding of drainage water that will occur at the base of the earthwork, potentially resulting in the softening of the toe and increasing the potential for failure in this area.

Slope drains can be connected to a drain constructed parallel to the toe of an embankment slope, or to verge drainage in a cutting. Outfall of the system can be by means of positive flow, such as an existing river course or existing public sewers, or to a sustainable drainage system (Woods and Kellagher, 2007).

Additional commentary on the practicalities of slope drainage is provided by Nettleton *et al.* (2022).

References

BSI (British Standards Institution) (1997) BS 8110-1: Structural use of concrete: Code of practice for design and construction. BSI, London, UK.

BSI (2004a) BS EN 1992-1: Eurocode 2: Design of concrete structures – General rules and rules for buildings (including National Annex). BSI, London, UK.

BSI (2004b) BS EN 1997-1: Eurocode 7: Geotechnical design (including National Annex). BSI, London, UK.

BSI (2007) BS EN 1993-5: Eurocode 3: Design of steel structures – Piling (including National Annex). BSI, London, UK.

BSI (2010) BS 8006-1: Code of Practice for strengthened/ reinforced soils and other fills. BSI, London, UK.

BSI (2011) BS 8006-2: Code of practice for strengthened/ reinforced soils – Part 2: Soil nail design. BSI, London, UK.

BSI (2015a) BS 8081: Code of practice for grouted anchors. BSI, London, UK.

BSI (2015b) BS 8002: Code of practice for earth retaining structures. BSI, London, UK.

BSI (2015c) BS 8500: Concrete: Complementary standard to BS EN 206-1. BSI, London, UK.

Chapman T (2000) *Modular Gravity Walls – Design Guidance.* CIRIA Report C516. CIRIA, London, UK.

Charles JA and Watts KS (2002) *Treated Ground, Engineering Properties and Performance.* CIRIA Report C572. CIRIA, London, UK.

Duffy-Turner M, Nettleton IM, Winter MG and Webber I (2022a) *Forensic Examination of Critical Special Geotechnical Measures: Soil Nails.* Information Note PPR1033. TRL, Crowthorne, UK.

Duffy-Turner M, Nettleton IM and Winter MG (2022b) *Forensic Examination of Critical Special Geotechnical Measures: Gabion Walls.* Information Note PPR1036. TRL, Crowthorne, UK.

Gaba A, Hardy S, Doughty L, Powrie W and Selemetas D (2017) *Guidance on Embedded Retaining Wall Design.* CIRIA Report C760. CIRIA, London, UK.

Hutchinson JN (1977) Assessment of the effectiveness of corrective measures in relation to geological conditions and types of slope movement. *Bulletin of the International Association of Engineering Geology* **16**: 133–155.

Nettleton IM, Winter MG and Duffy-Turner M (2022) *Forensic Examination of Critical Special Geotechnical Measures: Counterfort Drains.* Information Note PPR1034. TRL, Crowthorne, UK.

Williams BP and Waite D (1993) *The Design and Construction of Sheet Piled Cofferdams.* CIRIA Special Publication SP95. CIRIA, London, UK.

Winter MG, Duffy-Turner M and Nettleton IM (2022a) *Forensic Examination of Critical Special Geotechnical Measures: Gravity Block Walls.* Information Note PPR1035. TRL, Crowthorne, UK.

Winter MG, Duffy-Turner M and Nettleton IM (2022b) *Forensic Examination of Critical Special Geotechnical Measures: Reinforced Soil.* Information Note PPR1037. TRL, Crowthorne, UK.

Woods B and Kellagher R (2007) *The SUDS Manual.* CIRIA Report C697. CIRIA, London, UK.

Further reading

BSI (2009) BS 6031: Code of practice for earthworks. BSI, London, UK.

Highways Agency (2009) *Specification for Highway Works.* The Stationery Office, London, UK.

It is recommended this chapter is read in conjunction with

- Chapter 23 *Slope stability*
- Chapter 69 *Earthworks design principles*
- Section 6 *Design of retaining structures*

All chapters in this book rely on the guidance in Sections 1 *Context* and 2 *Fundamental principles.* A sound knowledge of ground investigation is required for all geotechnical works, as set out in Section 4 *Site investigation.*

doi: 10.1680/icemge.66830.1189

Chapter 73
Design of soil reinforced slopes and structures

Stuart Mortimore Atkins, London, UK
Graham Horgan Huesker, Warrington, UK
Dave Woods Huesker, Warrington, UK
Sebastien Manceau Kent, Glasgow, UK
Colin Macdiarmid GeoSolve Ltd, Otago, New Zealand

Reinforced soil structures are composite constructions involving some form of reinforcement (usually geosynthetic or metallic), generally installed in horizontal layers within a soil mass. The reinforcement layers extend beyond the potential failure surface, absorb the tensile strains that would otherwise cause failure of the non-reinforced soil and redistribute them in the soil beyond the failure plane. This chapter aims to introduce the general concepts of soil reinforcement, the various materials and the general principles involved in the design of reinforced soil structures. Reinforced soil walls and abutments, reinforced soil slopes and basal reinforcement and reinforced working platforms are discussed.

CONTENTS

73.1. Introduction and scope

The concept of reinforced soil structures consists of placing (typically horizontal) layered reinforcing elements with sufficient axial tensile stiffness within a soil mass to improve its tensile and shear capacities. This allows the construction of reinforced soil slopes and reinforced soil walls at significantly steeper angles – which would be unstable without the reinforcement. When used as basal reinforcement this also allows the construction of embankments over poor ground or areas prone to subsidence that would otherwise fail without the reinforcement in their foundations.

The idea of introducing reinforcing elements to improve the strength of fill can be traced back to the earliest human times, when primitive people used sticks and branches for the reinforcement of mud dwellings. The earliest remaining example of reinforced fill is the Aqar Quf ziggurat built in modern-day Iraq by the Babylonians around 3000 BC, using clay bricks reinforced with woven mats of reed laid horizontally on layers of sand and gravel, with plaited ropes of reed passing through the structure. Parts of the Great Wall of China built more than 2000 years ago also adopted a form of reinforced fill where tamarisk branches were used to reinforce a mixture of clay and gravel.

In the early 1960s Henri Vidal introduced and developed the modern form of reinforced fill using flat metallic reinforced strips laid horizontally on a frictional fill. In the late 1960s and 1970s extensive studies of reinforced soil structures sponsored by national bodies, notably the Laboratoire central des ponts et chaussées (LCPC) in France, led to a fuller understanding of the concepts involved and a better acceptance of the method. More or less simultaneously, advances in synthetic fabrics led to the development of geotextiles and geogrids. These new materials were soon put to good use in the construction of geosynthetic-reinforced soil structures and as basal reinforcements. It should be noted that reinforced soil is the general term for reinforcing elements placed within a soil mass and covers both metallic and polymeric reinforcement. The term 'reinforced earth' is the trademark for the Reinforced Earth Company – using its founder's (Henri Vidal) 'Terre Armée' concept.

This chapter introduces the various types of reinforcements and their properties, discusses the general principles of the reinforcement action and provides guidance for the design of reinforced soil structures. Reinforced soil walls and abutments, reinforced soil slopes, hybrid reinforced soil combined with soil nailing applications and basal reinforcement, including load transfer platforms and temporary works geosynthetic reinforced platforms are then covered in more detail. Particular forms of reinforcement such as anchored earth (Chapter 89 *Grouted anchor construction*) and soil nails (Chapter 74 *Design of soil nails*), and particular applications of support to load transfer platforms (Chapter 70 *Design of new earthworks*), are discussed elsewhere in this manual.

73.2. Reinforcement types and properties

73.2.1 Types of reinforcement

There are essentially two main types of reinforcement, defined by their extensibility.

- *Extensible reinforcement:* Defined in BS 8006-1+A1:2016 (BSI, 2010) as reinforcement that sustains the design loads at strains greater than 1% and which are usually polymeric.
- *Inextensible reinforcement:* Defined in BS 8006 as reinforcement that sustains the design loads at strains less than or equal to 1% and which are usually metallic.

The design life of the reinforcement can vary from a few months (e.g. temporary working basal reinforcement for an embankment on soft ground) to up to 120 years (e.g. walls, abutments and slopes). The principal function of the reinforcement is to withstand tensile loads; the ability of the reinforcement to perform its primary function, both initially and throughout its design life, will largely depend on the material from which it is produced and also its strain compatibility.

73.2.2 Properties of polymeric reinforcement

Polymeric reinforcement can take a variety of forms, such as grids, meshes, strips (commonly used to reinforce slopes and walls) and geotextile sheets (commonly used for basal reinforcement).

All polymers are essentially nonlinear viscoelastic materials, and as such are load rate dependent. In addition, polymers are subject to creep and their behaviour is therefore time dependent.

When subjected to a constant load all materials will increase in strain over time. This phenomenon is known as creep and is significant in polymers at ambient temperatures. The tensile rupture strength of polymeric reinforcement will reduce over time, predominantly due to creep from an initial short-term ultimate tensile rupture strength (UTS) to a tensile creep rupture strength at the end of the selected design life. The propensity of polymeric reinforcement to creep will be primarily dependent on the particular polymer used to produce the geosynthetic reinforcement (e.g. polypropylenes tend to have a greater creep propensity than polyesters).

Isochronous curves plot the tensile load against the strain for a given time (during which the reinforcement has been subjected to the tensile force). The curves are derived from extensive testing and interpolations following guidelines in PD ISO/TR 20432 (BSI, 2007) and are unique to each proprietary geosynthetic reinforcement. The stress axis is generally normalised by expressing the tensile load as a percentage of the initial short-term UTS of the reinforcement.

These curves enable the designer to establish both the initial strains and post-construction strains for polymeric reinforcement at a given load/stress level; strains must remain below prescribed design limits. BS 8006 imposes varying serviceability limits on post-construction strain depending on the type of structure considered. Post-construction strains are limited to 0.5% for bridge abutments and retaining walls with permanent structural loads, and to 1.0% for retaining walls with no applied structural loading. For slopes where deformations are not critical, post-construction strains in the order of 5% may be acceptable. There usually is no limit on post-construction strain for basal reinforcement (used for embankments on soft ground), but there are limits usually imposed when the basal reinforcement is used over areas prone to subsidence and for load transfer platforms.

Additionally, the curves allow the designer to determine the variations of the reinforcement stiffness with time. A typical set of isochronous curves is shown in **Figure 73.1**.

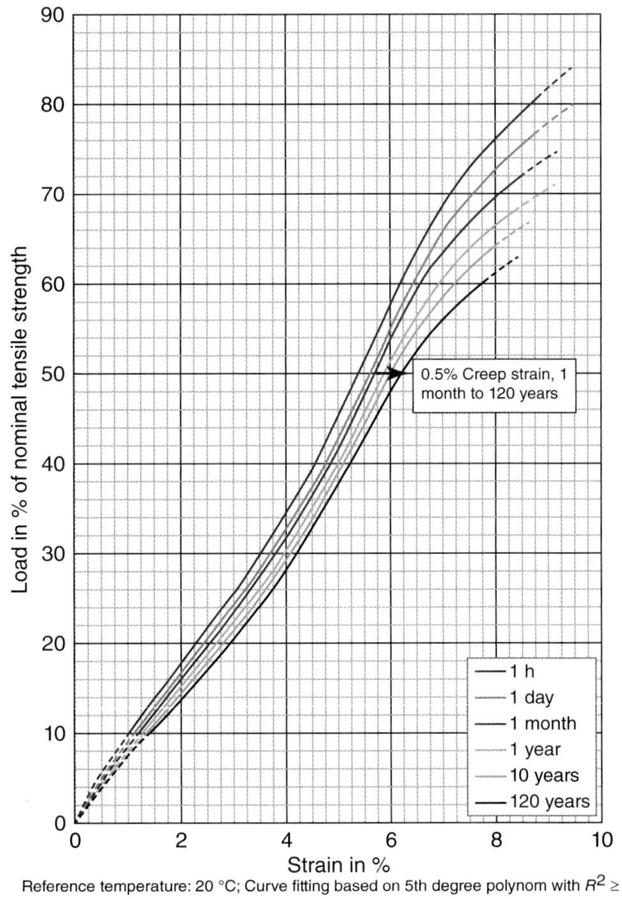

Figure 73.1 Typical isochronous curves for polyester-based reinforcement (Courtesy of Huesker (Isochrones for Stabilenka product))

73.2.3 Properties of metallic reinforcement

Metallic reinforcement can take a variety of forms, such as grids, meshes, strips, bars or rods (commonly used to reinforce walls and abutments).

Creep in metallic reinforcement is generally negligible in ambient temperatures. Metallic reinforcement can therefore be assumed to exhibit linear elastic behaviour to yield, independent of time. Hence, the tensile yield strength of metallic reinforcement will reduce over time, almost uniquely due to corrosion.

Corrosion in metallic reinforcement is essentially a process of oxidation; this forms a protective oxide layer which tends to retard further corrosion. The effects of corrosion are considered in design by allowing for a loss of the cross-sectional area of the reinforcement. BS 8006 provides guidance on the prescribed sacrificial thickness to allow for at the design stage, which varies depending on

- particular metal or alloy and corrosion protection used
- design service life of the reinforcement

- corrosivity of the fill, soil or environment in which the reinforcement is placed (e.g. high-acidity soil, which would be highly corrosive).

73.3. General principles of reinforcement action

73.3.1 Effects of introducing reinforcements in soil

When an inclined or vertical load (due to soil self-weight or surcharge) is applied to a soil, this will generate axial compressive strain in the soil and a corresponding lateral tensile strain. The maximum load that can be applied to a soil is limited by its internal shear strength. The introduction of reinforcement into the soil (of adequate stiffness) improves the shearing resistance of the soil (shear strength of reinforced soil = mobilised shearing in the soil + mobilised tensile force in the reinforcement) and has the effect of reducing both the axial compression and lateral deformations.

Deformation of the soil along a potential failure plane causes shear forces to develop in the soil and tensile forces to develop in the reinforcement intercepting the failure plane. The reinforcement is most effective when placed in the direction in which the tensile/lateral strains develop. However, in most practical reinforced soil applications the reinforcement is installed horizontally (with the exception of soil nails, which are commonly inclined at an angle of 10–20° from horizontal).

73.3.2 Strain compatibility

The relative magnitude of the shear force mobilised in the soil and the tensile force mobilised in the reinforcement will be a function of deformations and the relative stiffness properties of both soil and reinforcement (stiffer reinforcement will carry a higher percentage of the mobilised force). The magnitude of mobilised soil shearing resistance and the mobilised reinforcement strain need to be considered to provide equilibrium.

Inextensible (metallic) reinforcement will absorb, through frictional contact, the disturbing forces in the active zone at low soil deformations and transfer these forces beyond the failure plane. For more extensible polymeric reinforcements the issue of strain compatibility becomes increasingly important – to ensure design reinforcement strain limits remain compatible with mobilised soil strains.

Figure 73.2 shows the mobilised shear resistance of a typical granular soil and the typical mobilised reinforcement force (time constant). From the graph in **Figure 73.2(a)** it can be seen that as strains develop, the mobilised shear resistance in the soil increases (and the required reinforcement force to satisfy equilibrium decreases) up to a point where the mobilised shear strength reaches a maximum or peak strength φ'_p. Beyond this point (φ_m), increasing deformations result in reducing shear strength towards a minimum value φ'_{cv} (shear strength at constant volume) independent of strain, with a corresponding increase in the force required to satisfy equilibrium. At the same time, as strains develop, the force mobilised in the reinforcement corresponding to a specific loading period (t_d) and temperature (T_d) increases (**Figure 73.2(b)**).

Strains develop until a strain level is reached (in both the soil and the reinforcement) whereby the force mobilised in the reinforcement is that required to satisfy equilibrium (**Figure 73.3**). At this point, further deformations will only occur as a result of additional loading, or through stress relaxation, or creep, of the reinforcement over time.

The stress–strain response of soil will vary depending on a number of factors, including mineralogy and stress history; similarly, the stress–strain response of reinforcement will vary depending on a variety of factors, including its form, manufacturing process and raw material. It is generally not practical to develop a set of compatibility curves for every soil type and proprietary reinforcement product available. BS 8006 provides practical recommendations for addressing strain compatibility by adopting certain values depending on the type of reinforced soil application. For example, φ'_p would be adopted for walls and steep slopes where mobilised strains and deformations are considered more critical to the overall

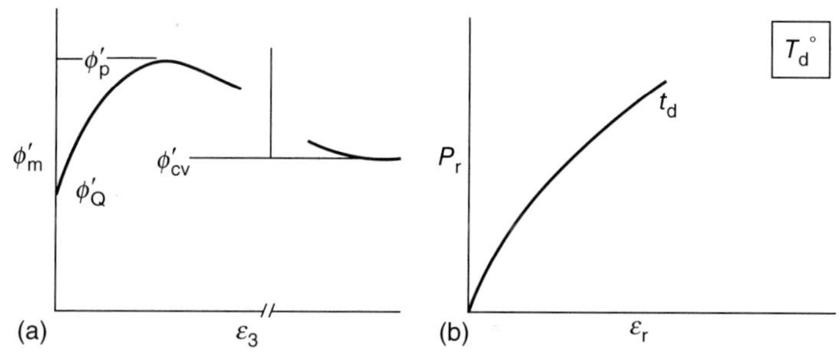

Figure 73.2 Relations for strain compatibility: (a) mobilised soil shearing resistance, (b) mobilised reinforcement force. P_r = mobilised reinforcement force; t_d = loading period; T_d = temperature; ε_3 = maximum tensile strain (Reproduced, with permission, from CIRIA SP123, Jewell, 1996; www.ciria.org)

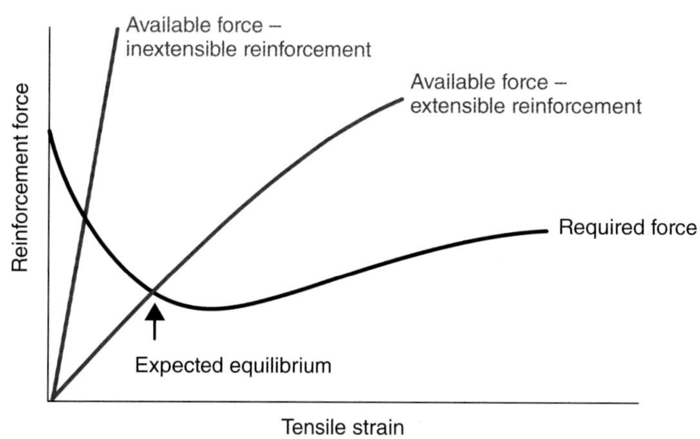

Figure 73.3 Typical compatibility curve for determining equilibrium in reinforced soil (Adapted, with permission, from CIRIA SP123, Jewell, 1996; www.ciria.org)

performance, and φ'_{cv} for shallow slopes and basal reinforcement over soft soils where deformations are considered less critical.

73.3.3 Interaction between soil and reinforcement

Reinforcement develops a bond with the soil through either friction (for granular soils) or adhesion (for cohesive soils). There are two interaction modes

- direct sliding, in which a block of soil slides over a layer of reinforcement
- pull-out (bond), in which a layer of reinforcement pulls out of the soil after mobilising the maximum available bond stress.

73.3.3.1 Direct sliding

The sliding resistance of geotextile reinforcement is generated over the full plan area of contact. The sliding resistance of geogrids and strips is generated from both direct contact between the reinforcement and the soil, and from soil-to-soil contact through the apertures or adjacent to the reinforcement.

Modified direct shear tests are suitable for measuring the coefficient of direct sliding between soil and any type of reinforcement.

73.3.3.2 Pull-out

The pull-out resistance of geotextile reinforcement is generated from surface shear over the full plan area. Geogrids develop pull-out resistance partly from surface shear and partly through bearing resistance from the transverse members of the geogrids. It is usually sufficient for the design to calculate the bond coefficient from theoretically derived values for these two components.

For narrow, rough strip reinforcement embedded in dense, cohesionless fill, the shear stresses developed during pull-out lead to dilatancy in the fill, which causes the vertical effective stress to locally rise above the overburden pressure. This gives rise to an enhanced pull-out resistance. Although it is generally recommended that field pull-out tests be carried out to verify the bond strength of these types of reinforcement (BS 8006), pull-out tests are relatively difficult to perform and can be significantly influenced by the boundary conditions of the test. They can, therefore, be unreliable.

73.4. General principles of design

The design of reinforced soil structures and slopes is not currently covered by the Eurocodes. The Eurocodes are, however, in the process of being fully revised, and it is anticipated that reinforced soil will be covered within the revised version of BS EN 1997:2004+A1:2013. The current draft for public comment has 'Reinforced Fill Structures' as section 9 of prEN 1997-3.

In the UK, design of reinforced soil structures follows 'BS 8006: Code of practice for strengthened/reinforced soils and other fills'. (Earlier recommendations in 'HA68/94: Design methods for the reinforcement of highway slopes by reinforced soil and soil nailing techniques' (National Highways, 1994) have been incorporated into the later versions of BS 8006.) BS 8006 follows a limit state approach and provides partial factors of safety calibrated for reinforced soil structure design through the assessment of historical data. The standard was first published in 1995, and its current version was published in 2010, with amendments made in 2016.

73.4.1 Limit state approach

Reinforced soil structures are designed against the occurrence of ultimate limit states (ULS) and serviceability limit states (SLS). ULS are associated with collapse, rupture or major damage, and SLS with excessive settlement or deformations.

73.4.2 Partial factors of safety approach

BS 8006 provides various partial factors of safety which are used to generate the overall safety of reinforced soil structures. Although BS 8006 does not specifically group them in this fashion, it is convenient to think of the partial factors of safety as being introduced to reduce material properties and resistances, and to increase loads (a similar approach to that used in the Eurocodes).

A partial factor for ramification of internal failure (f_n) is also introduced to account for the ramifications (consequences) of internal failure. The value of f_n depends on the class of risk for the structure and introduces additional safety for structures where the consequences of failure are most onerous. This factor is applied both as an additional material factor and as an additional resistance factor on the reinforcement strength and fill reinforcement interactions, respectively.

73.4.2.1 Partial material factors

A partial material factor of safety on reinforcement (f_m) is applied on the reinforcement capacity to account for uncertainty on the reinforcement properties (manufacture and extrapolation of test data), susceptibility to damage during construction (e.g. due to shape and size of fill used) and environmental effects (e.g. temperature, exposure to ultraviolet radiation).

The partial factor for ramification of internal failure (f_n), as discussed above, is also applied to further reduce the reinforcement design strength and therefore acts as an additional material factor.

A partial material factor of safety is applied on soil strength parameters (f_{ms}) to account for uncertainty on the characteristic value of the soil parameters considered.

73.4.2.2 Partial load factors

Partial load factors are applied to increase the soil self-weight (f_{fs}), the external dead loads (f_f) and the external live loads (f_q). For walls and abutments the worst load combination for the various design criteria must be identified and considered in the design.

BS 8006 uses high values for partial load factors on fill or soil self-weight, which is one of the parameters that is likely to be well known and not subject to large variations. However, it should be borne in mind that the values of partial factors of safety provided in BS 8006 have been specifically calibrated for reinforced soil structures.

73.4.2.3 Partial resistance factors

For the internal failure mechanisms (i.e. within the soil/reinforcement block) involving fill/reinforcement interaction, a partial factor of safety on pull-out resistance (f_p) is applied to the resistance generated by fill/reinforcement interaction behind the potential failure surface when this intersects a layer of reinforcement. A partial factor of safety on sliding resistance along reinforcement (f_s) is also applied to the resistance generated by fill/reinforcement interaction when the base of a potential failure surface coincides with a layer of reinforcement.

The partial factor for ramification of internal failure (f_n) is also applied to further reduce the fill/reinforcement resistances (pull-out and direct sliding) and therefore acts as an additional resistance factor.

For the external mechanisms (e.g. sliding or bearing capacity) a partial factor on sliding resistance along soil (f_s) is applied to the resistance generated by soil–soil interaction on any horizontal plan, either within or at the base of the reinforced soil structure where there is soil–soil contact. A partial factor on bearing capacity resistance (f_{ms}) is also applied to the resistance generated by the ultimate bearing capacity. Please note that BS 8006 considers this partial factor of safety as a material factor (hence the notation f_{ms}); however, bearing capacity is not an intrinsic soil property and, therefore, in the following the partial factor on bearing capacity is considered as a resistance factor.

73.5. Reinforced soil walls and abutments

73.5.1 General

Reinforced soil walls and abutments are composite structures consisting of fill, reinforcement and a facing, as illustrated in **Figure 73.4**. Provided the reinforcement layout and properties are adequate to prevent an internal failure of the reinforced soil block, the block of reinforced soil retains the soil at its back as a mass gravity structure, and transmits and spreads the effects of surcharges and earth pressures over its wide base. Reinforced soil walls are cost-effective soil-retaining structures that can tolerate large settlements. Reinforced soil structures with a face within 20° from the vertical can be analysed as reinforced soil walls.

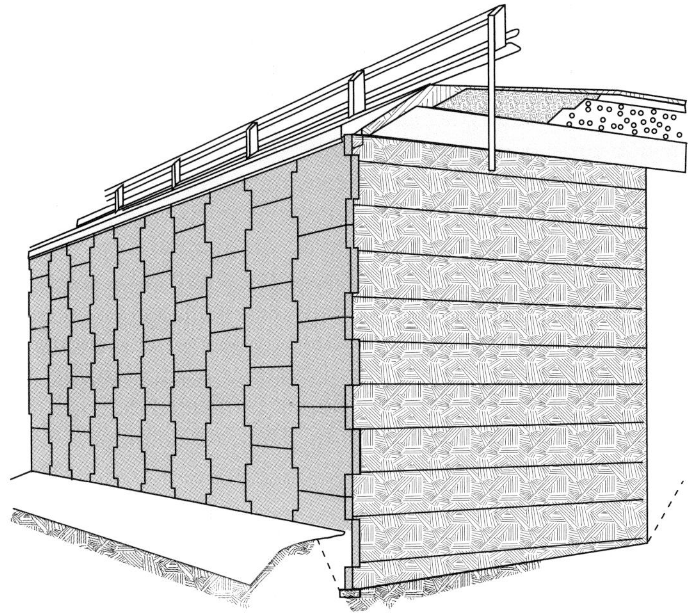

Figure 73.4 3D section through reinforced soil structure (Courtesy of Terre Armée Internationale)

73.5.2 Materials

73.5.2.1 Fill

Both the mechanical and chemical properties of the fill need to be assessed. Generally, frictional class 6I/6J fill of the *Specification for Highways Works* (National Highways, 2016) is used in the construction of reinforced soil walls and abutments. When steel metallic reinforcements are used the electrochemical properties of the fill also need to be considered.

Polymeric reinforcements are not influenced by electrochemical properties, but the influence of chemicals on the proprietary polymeric reinforcements needs to be assessed.

In certain circumstances, provided durability and compatibility with the reinforcement have been considered, recycled aggregate, cohesive fills, pulverised fuel ash, colliery spoil, and chalk can be used (for details, refer to BS 8006, section 3). The selection and specification of fill for reinforced soil structure should be considered as early as possible in the project, and aspects such as sustainability, availability, impacts on design and costs should be borne in mind. The use of material other than class 6I/6J fill (provided durability and compatibility with the reinforcement have been considered) may lead to a design requiring longer, more closely spaced and stronger reinforcements, and to increased earthworks fill testing requirements during construction. However, the use of alternative material could also lead to a significant reduction in the quantities of selected fill to be imported if site-won material can be reused (thus reducing the quantity of material to be disposed of off-site). It could also reduce haulage costs if sources of acceptable fills are identified in the vicinity.

73.5.2.2 Reinforcement

The reinforcement in reinforced soil walls and abutments typically consists of metallic strips, polymeric geogrids or geocomposites. The reinforcement properties and the principles of their interaction with the fill are discussed in Sections 73.2 and 73.3.

73.5.2.3 Facing

The facing provides local support to the fill between reinforcement layers, prevention against weathering of the exposed soil and an aesthetically acceptable finish. Common facing types include full-height panels, discrete panels, segmental blocks and wrap-arounds. They are constructed of concrete (reinforced or mass), steel, wood or polymers. Discrete panels (in hexagonal, square or cross patterns) are the most commonly used with metallic reinforcement; segmental blocks and rigid gabion mesh facings are commonly used with polymeric reinforcements for permanent structures where a hard facing is acceptable. Wrap-arounds for polymeric reinforcement can be utilised but are more susceptible to ultraviolet degradation, accidental damage and vandalism. They are commonly used in temporary structures.

73.5.3 Common geometries and typical dimensions

Figure 73.5, in which L is the reinforcement length, illustrates common reinforced soil-retaining wall and abutment geometries that can be used for assessing options at conceptual design stage and for preliminary sizing. These dimensions are given relative to the mechanical height H. For a wall the mechanical height is derived as the vertical distance between the base of the reinforced soil block and the intersection of the finished profile and a line at $1V{:}0.3H$ from the front bottom corner of the reinforced soil block. For an abutment, H is the vertical height between the base of the reinforced soil block and the finished profile at the back of the abutment. Dimensions chosen for preliminary sizing may need amending following detailed design. Other geometries including trapezoidal walls can be designed to suit particular site requirements.

73.5.4 Elements of design

The design of reinforced soil walls considers two distinct aspects.

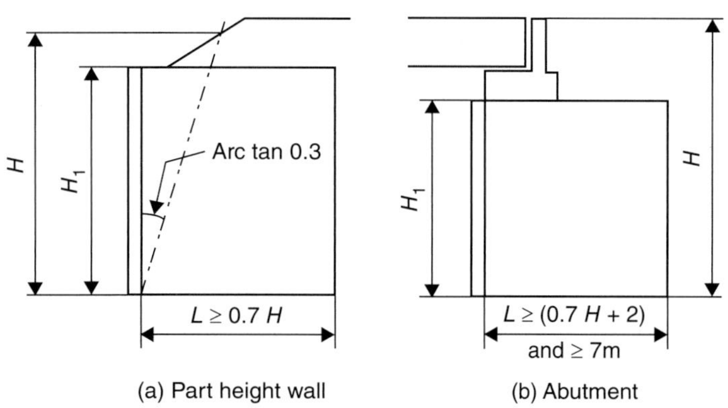

Figure 73.5 Common geometries and typical dimensions for (a) a reinforced soil wall; and (b) an abutment (Reproduced with permission from BS 8006-1 © BSI 2010+A1:2016)

- External stability covers the design of the reinforced soil structure as a unit and follows the principles used for a general mass gravity retaining structure.
- Internal stability covers the internal mechanisms within the reinforced soil block and deals with considerations of stress distribution and reinforcement layout.

73.5.4.1 External stability

For external stability the following ULS and SLS should be considered

- bearing and tilt failure – ULS
- sliding – ULS
- slip failure – ULS
- settlement – SLS.

The factored applied bearing pressure at the base of the reinforced soil block (q_r) is derived using a Meyerhof distribution (equivalent rectangular distribution on a reduced area) as

$$q_r = \frac{R_v}{L - 2e} \qquad (73.1)$$

where R_v is the resultant of all factored vertical load components; L is the reinforcement length at the base of the wall; and e is the eccentricity of resultant load R_v about the centre-line of the base of width L.

The bearing capacity of the foundation soil derived using traditional bearing capacity equations must be greater than the applied bearing pressure (refer to Chapter 53 *Shallow foundations* for further details on bearing capacity).

At the base of the reinforced soil block, sliding should be checked either on a soil–soil interface or, if a layer of reinforcement is provided at the base of the reinforced soil structure, on a soil–reinforcement interface. Any potential slip failure passing at the back of the reinforced soil structure should also be checked using traditional slope stability analysis software.

For the settlement check, both the settlement of the foundation soil and the internal settlement of the reinforced fill should be considered (refer to Chapter 70 *Design of new earthworks* for further details on fill self-settlement). However, when properly compacted the selected fill used in reinforced soil walls and abutments will produce small internal settlements, and this component is often neglected.

In current UK practice the internal stability of the reinforced structure is often designed by suppliers who do not generally assess the external stability. Instead, they provide target values for allowable bearing capacity of the foundation soil to prevent bearing capacity failure; target values for the shear strength of the foundation soil to prevent sliding; and SLS applied bearing pressures at the base of the reinforced soil block to allow an assessment of the settlements to be undertaken. It is critical to understand the limitations of such suppliers' designs and to ensure the following.

- The external stability is adequately assessed, based on the actual ground investigation information and that, where required, ground improvement is applied to the foundation soil (or the geometry of the reinforced soil block is amended, or a combination of both) to ensure external stability is satisfied.
- The responsibilities of the various parties and the objectives/specification of the design are clearly defined (for most common applications BS 8006 will provide a safe basis for design). Furthermore, notwithstanding the expertise of the suppliers, suppliers' designs should not be taken at face value.
- All other aspects such as drainage, services, fencing and aesthetics that could have an impact on the design are considered and communicated between the parties involved, to ensure an integrated overall design is achieved.

The court case that followed subsidence on a load transfer platform in Enniskillen, Northern Ireland (High Court of Justice in Northern Ireland, Queen's Bench Division, 2005) illustrates what can go wrong when the principles above are not followed.

73.5.4.2 Internal stability

Within the reinforced mass there are two zones delimited by a potential failure surface. In the active zone (between the face of the wall and a potential failure surface) the fill sheds shear stress into the reinforcement and induces a tensile force in it. The reinforcement transfers this tensile force in the resistant zone (between the potential failure surface and the back of the reinforced soil zone) where the reinforcement sheds shear stress into the fill (**Figure 73.6**).

For the internal stability, the following ULS and SLS are considered

- local stability of a layer of reinforcement – ULS
  - rupture of the reinforcement (or/and connection)
  - breakdown of the interaction between the fill and the reinforcement (pull-out or/and direct sliding)
- instability of a wedge of any size or shape assumed to behave as a rigid body – ULS
- deformations – SLS.

Extensible reinforcements in which the design load is sustained at an axial tensile strain greater than 1% (polymeric reinforcement) are designed using the tie-back wedge method, which is based on a theoretical approach.

Inextensible reinforcements in which the design load is sustained at an axial tensile strain less than 1% (metallic reinforcement) are designed using the coherent gravity method. This method is mainly empirical and is based on observations on instrumented reinforced soil structures.

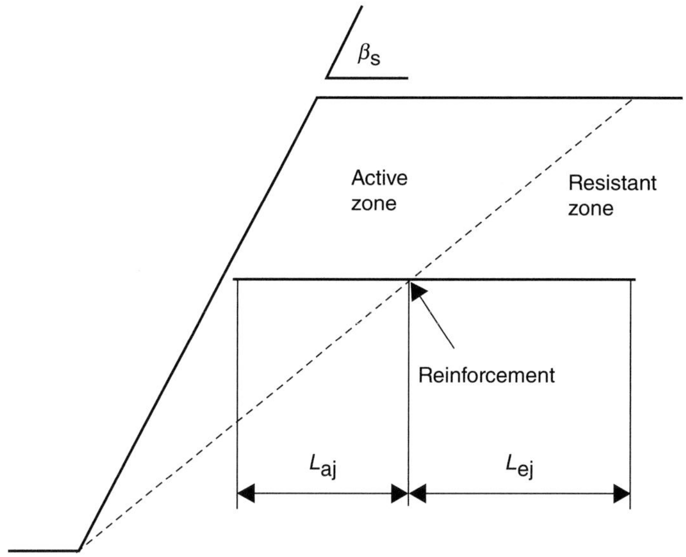

Figure 73.6 Internal stability of a reinforced soil wall. L_{aj} = length of reinforcement within the active zone, L_{ej} = length of reinforcement within the resistant zone, β_s = slope angle (Reproduced with permission from BS 8006-1 © BSI 2010+A1:2016)

Both methods follow broadly similar steps which are summarised below.

Tensile force in reinforcement

Both the tie-back wedge and the coherent gravity methods calculate the maximum factored tensile force in the *j*th layer of reinforcement T_j, which is given as

$$T_j = Tp_j + Ts_j + Tf_j \tag{73.2}$$

where

- Tp_j is the tensile force at the *j*th layer of reinforcement generated by self-weight of fill, surcharge and retained backfill
- Ts_j is the tensile force at the *j*th layer of reinforcement generated by vertical load applied to a strip
- Tf_j is the tensile force at the *j*th layer of reinforcement generated by horizontal load applied to a strip.

The component due to the fill self-weight, surcharge and retained backfill is given as

$$Tp_j = K\,\sigma v_j\, sv_j \tag{73.3}$$

where

- K is the earth pressure coefficient. For extensible reinforcement (tie-back wedge method) active earth pressure coefficient K_a shall be used throughout the structure. For inextensible reinforcement (coherent gravity method) values varying between at rest earth K_0 at the top of the structure and active K_a lower down the structure shall be used

- σv_j is the vertical stress acting on the jth layer of reinforcement assuming the Meyerhof distribution. The derivation of σv_j varies slightly between the tie-back wedge and the coherent gravity method (refer to BS 8006 for details)
- Sv_j is the vertical spacing of the reinforcement at the jth level in the wall.

The component due to any vertical load applied to a strip is given as

$$Ts_j = K\,\Delta\sigma v_j Sv_j \tag{73.4}$$

where $\Delta\sigma v_j$ is the increase in vertical stress generated by strip vertical loading at the level of the jth layer of reinforcement. The tie-back wedge method assumes a simple $2V{:}1H$ dispersion of the strip load through the reinforced fill. The coherent gravity considers the dispersal based on the Boussinesq equation for strips with an edge on the front face of the wall, and applies the principle of superposition to represent the strip at a distance from the front face of the wall. The mathematics involved in the original coherent gravity method can be cumbersome and the 2010 revision of BS 8006 allows the use of the simple $2V{:}1H$ load dispersal assumption in conjunction with the coherent gravity method. The component due to any horizontal load (F_L) applied to a strip is given as

$$Tf_j = a_j\,F_L\,sv_j \tag{73.5}$$

where $\alpha_j\,F_L$ is the horizontal stress generated by the strip horizontal loading at the level of the jth layer of reinforcement. The tie-back wedge method assumes a dispersion of the strip load at $45° + \varphi'/2$ from the back of the strip through the reinforced fill. The coherent gravity assumes a dispersion of the strip load at 45° from the back of the strip through the reinforced fill.

The various components of the tensile force in the jth layer of reinforcement (for the tie-back wedge method) are illustrated in **Figure 73.7**.

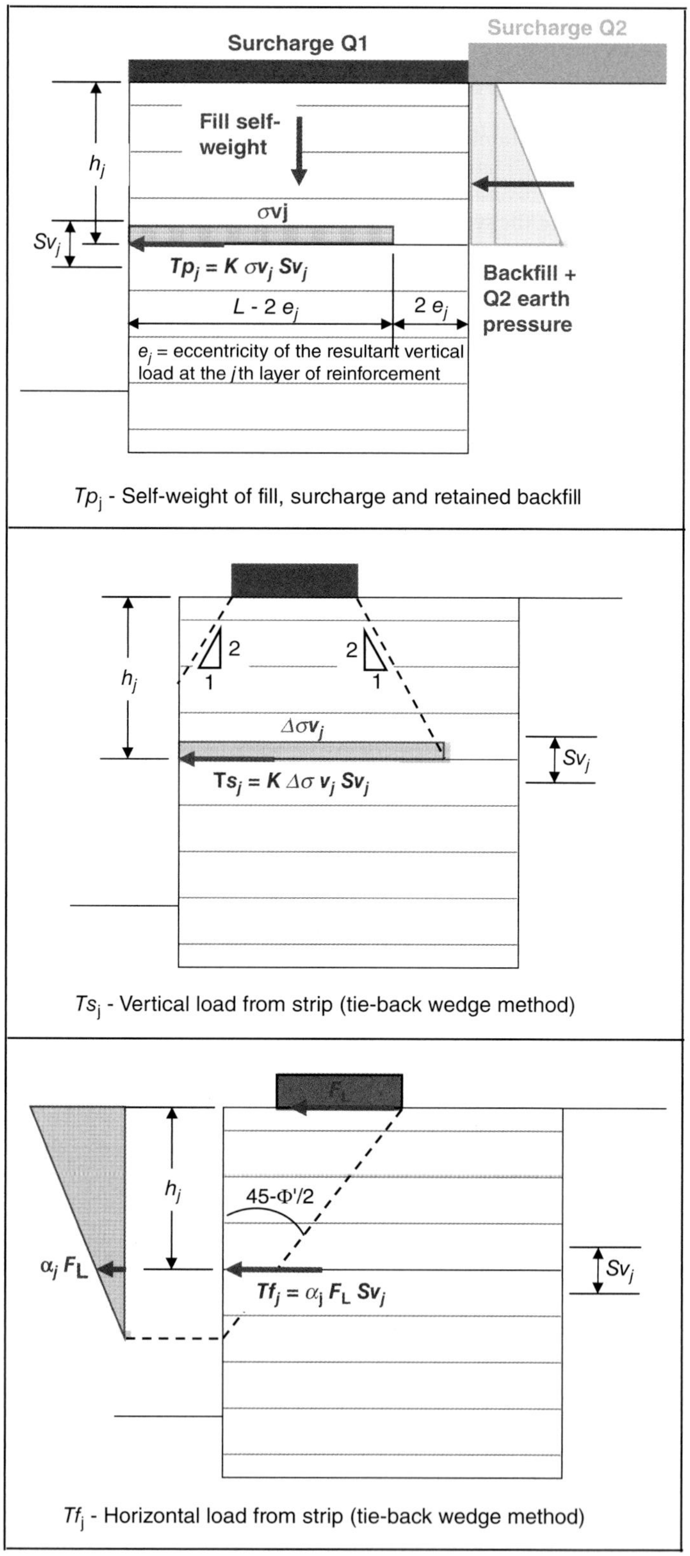

Figure 73.7 Tensile force in the jth layer of reinforcement

Local stability checks

For each layer of reinforcement, check that the design tensile strength of the reinforcement is greater than the factored tensile force to prevent rupture of the reinforcement. Also, for each layer of reinforcement, check that the design adherence capacity of the reinforcement behind the failure wedge being considered is greater than the factored tensile force to prevent pull-out of the reinforcement.

For extensible polymeric reinforcement subject to creep, the serviceability criterion is generally more critical than the ULS local stability check (see ‘Serviceability’ on p. 1197).

Surface failure/lines of maximum tension

In the tie-back wedge method the failure plane with the most tensile force needs to be determined by considering the stability of wedges, of any size or shape, assumed to behave as rigid bodies. The friction along the potential failure plane and the tensile resistance of reinforcement beyond the failure plane (rupture capacity or pull-out capacity, whichever is the smallest) must be greater than the disturbing applied loads.

In the coherent gravity method the results of monitoring installed along reinforcements have shown that the tensile force varies along the reinforcement. Lines of maximum tension are defined (the shape of the line considered depends on whether strip loadings are applied or not) and the tensile force in the reinforcement is calculated on these defined failure surfaces. Provided the local stability checks (rupture and pull-out) are satisfied on these defined failure surfaces, there is no requirement to carry any further wedge stability analysis (unless the geometry or loadings are unusual).

The failure plane in the tie-back wedge method and the lines of maximum tension in the coherent gravity method are illustrated in **Figure 73.8**.

Vehicle restraint barriers on reinforced soil walls

Examples of strip loads that require local wedge stability checks to be undertaken in the design of the reinforced soil supporting layers may include a vehicle restraint barrier. The supporting reinforced concrete base slab should be designed to resist all the loads which the parapet is capable of transmitting, up to and including failure in any mode that might be induced by a vehicle collision, without damage to the base slab or the structure upon which it is founded. The layout and strength of the reinforcing layers below the base slab may need alteration to provide the necessary local wedge stability support. The design requirements are covered in Appendix E of BS 8006 and a typical example of a precast coping unit with stitch beam detail to be cast into the supporting cast in situ base slab to connect the two units together, is shown in **Figures 73.9** and **73.10**.

Connections

The tensile force in the connection between the facing and the reinforcement depends on their properties. Unless the facing is stiff (e.g. full-height panels with no movement capacity at connections), the tensile force in the connection will be less than the maximum tensile force calculated. However, the tensile force at connection is significant and the connection has to be designed to prevent rupture under this load throughout the design life. When metallic reinforcements are used, the connection generally involves a bolting arrangement that can lead to the tensile capacity at the connection being less than that of the full reinforcement strip and for some details this can be particularly vulnerable to corrosion.

Facings

The facings are designed structurally to resist the connection tensile loads, any externally applied loads (e.g. impact loadings from traffic) and the effect of any construction tolerance. In the case of segmental block facings, sliding and movements between blocks and overtopping of the blocks above the uppermost layer of reinforcement must be checked.

Serviceability

Deformations post-construction can happen as the result of creep of polymeric reinforcement, where the tensile stiffness decreases with time from an initial short-term value. In order to limit movements due to creep, the capacity of the reinforcement should be considered in conjunction with a limiting post-construction value of strain (typically 0.5% for abutments and 1% for walls) at the end of the design life, using isochronous load–strain curves. For metallic reinforcement, creep is negligible.

73.5.5 Construction

Factors affecting the construction of reinforced soil walls and abutments include the

- nature and properties of the soil at the base of the reinforced soil block
- site controls and placement and compaction requirements for the specified fill
- safe storage and handling of the reinforcement
- facings
- simplicity and robustness of the connection detail
- requirements for drainage
- site constraints
- end use
- erection rate.

Buildability is discussed in more detail in BS 8006 and in Chapter 86 *Soil reinforcement construction*. Recommended construction tolerances are also presented in BS 8006.

73.6. Reinforced soil slopes

73.6.1 General

Reinforced soil slopes are composite structures consisting of fill, reinforcement and facing. The front face is usually grassed to provide a green slope finish. A distinction is made between 'shallow slopes', defined as no steeper than 45° from the horizontal, and 'steep slopes', defined as being between 45° and 70° from the horizontal.

73.6.2 Materials

73.6.2.1 Fill

Generally, frictional fill will be used as fill for reinforced soil slopes, but the range of fill that can be used is wider than for reinforced soil walls and abutments. As discussed in Section 73.5.2.1 for reinforced soil walls and abutments, the selection and specification of fill for reinforced soil structures should be considered as early as possible in the project. Aspects such

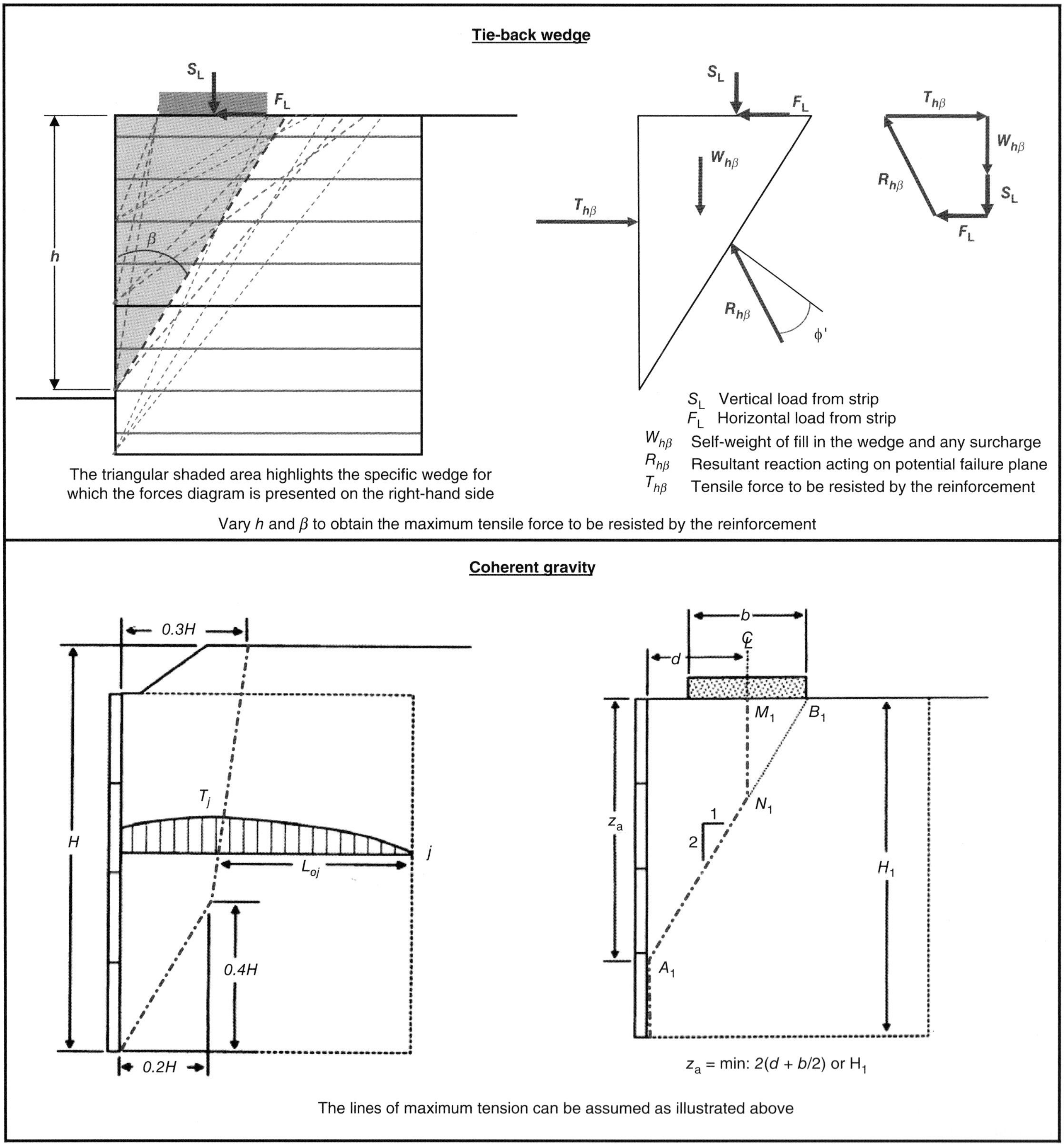

Figure 73.8 Surface failure/lines of maximum tension: coherent gravity (Reproduced with permission from BS 8006-1 © BSI 2010+A1:2016)

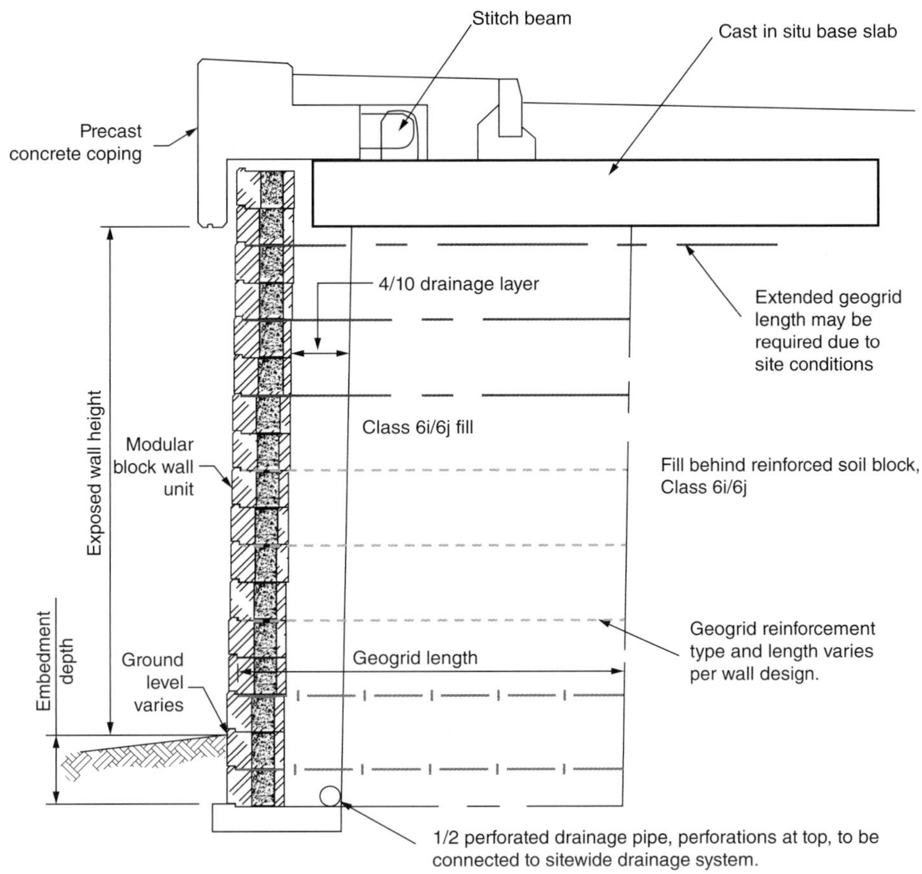

Figure 73.9 Typical detail of precast coping on modular block reinforced soil wall

Figure 73.10 Photograph of the Allan Wall system with precast coping unit awaiting connection to cast in situ support slab at the top of a reinforced soil wall

as sustainability, availability, impacts on design and costs should be borne in mind. Site-won material is often used as an economic and sustainable fill for reinforced soil slopes. Properties of particular significance are the drained shear strength parameters (φ' and occasionally c', although effective cohesion should be used with great care) which will govern the soil reinforcement pull-out capacity, and the maximum particle size which will govern the risk of mechanical damage during construction – and hence the geosynthetic reinforcement rupture capacity and durability.

73.6.2.2 Reinforcement

For reinforced soil slopes, geogrids are predominantly used. The reinforcement properties and the principles of their interaction with the fill are discussed in Sections 73.2 and 73.3.

73.6.2.3 Facing

For shallow slopes (no steeper than 45° from the horizontal) it is generally possible to place the fill without permanent or temporary supports. Intermediate short secondary geogrids are often placed between layers of main reinforcements to prevent small shallow-seated failures in the front face. Some form of erosion mat is often provided to ensure that the topsoil placed does not erode in the short term, and to facilitate the establishment of vegetation. Health and safety issues associated with cutting grass on relatively steep slopes should be considered.

For steep slopes (steeper than 45° from the horizontal) the facing is generally built using either wrap-around or sacrificial or permanent steel mesh. In a wrap-around detail the reinforcing geogrid or geotextile is folded through 180° to form the face and is anchored back into the fill or connected to the next layer of reinforcement above. Temporary formwork is often required to allow placement and compaction of the fill. In a steel mesh facing arrangement the reinforcing geogrid or geotextile can be connected directly to the steel mesh, or continued behind the face of the steel mesh and returned into the fill to allow for the future deterioration of the sacrificial steel facing. In both cases topsoil is placed just behind the facing to facilitate the establishment of vegetation. In addition to health and safety issues associated with cutting grass on such steep slopes, the difficulties of establishing, irrigating and maintaining vegetation on them requires careful consideration.

Typical reinforced soil slope configurations are presented in **Figure 73.11**.

73.6.3 Elements of design

Similar to the design of reinforced soil walls and abutments, the design of reinforced soil slopes considers both external and internal stability. In current UK practice, the stability of reinforced soil slopes is often designed by suppliers who, unless they are provided with adequate information, tend to concentrate on internal stability, and the principles addressed in Section 73.5.4.1 apply. In addition, compound stability, where the potential failure surface is partly within the reinforced soil zone and partly behind it, should also be considered.

73.6.3.1 External stability

For external stability the following ULSs and SLSs are considered

- bearing and tilt failure – ULS
- sliding – ULS
- slip failure – ULS
- settlement – SLS.

73.6.3.2 Internal stability

Similar to reinforced soil walls and abutments, a potential failure surface delimits an active zone and a restraint zone. For the internal stability, the following ULSs and SLSs are considered

- local stability of a layer of reinforcement – ULS
  - rupture of the reinforcement (or/and connection)
  - insufficient interaction between the fill and the reinforcement (pull-out or/and direct sliding)
- deformations – SLS.

Internal stability assessment is generally undertaken using one of many limit equilibrium methods available. The most commonly used are the two-part wedge mechanism and the method of slices, which are discussed below. Other methods, such as a non-circular analysis, a log spiral or a coherent gravity method, could be used.

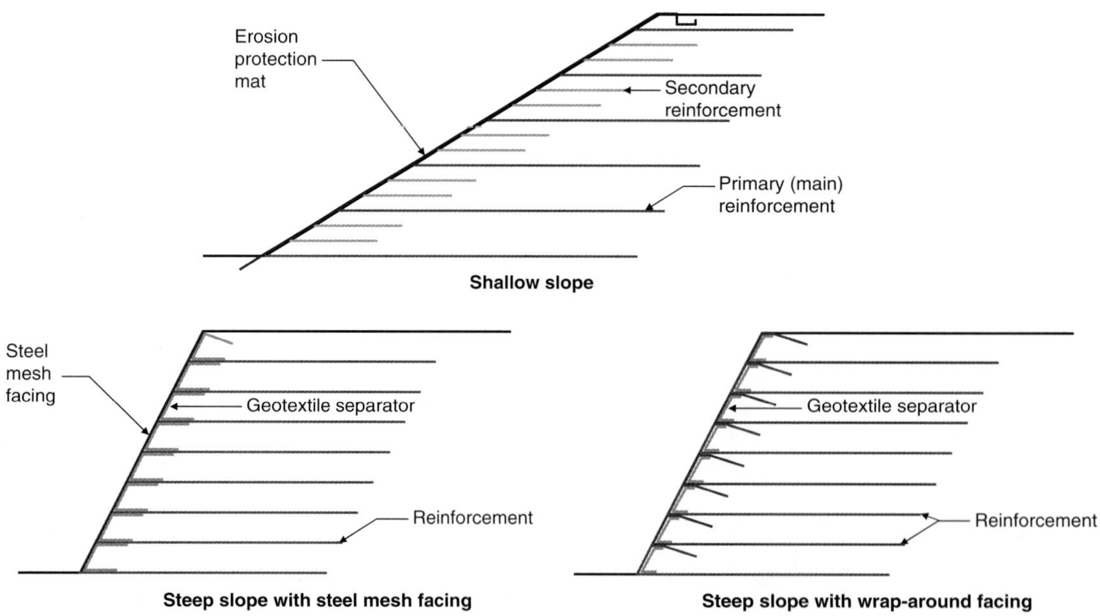

Figure 73.11 Typical reinforced soil slope configurations

Two-part wedge mechanism

The two-part wedge mechanism is an extension of the Coulomb wedge approach as illustrated in **Figure 73.12**. The mechanism may take any form (i.e. θ_1, θ_2 and X can vary) provided the mechanism outcrops at the toe of the slope and the inter-wedge boundary is vertical. By resolving the forces acting on the wedges (which requires assumptions on the inter-wedge boundary conditions), the out-of-balance force that needs to be taken by the reinforcement can be assessed for each mechanism. For each slope there is a unique critical two-part wedge mechanism for which the out-of-balance force that needs to be taken by the reinforcement is maximum. For more details on the two-part wedge mechanism, refer to BS 8006.

Considering the simple case of a slope with flat ground at the top and no surcharge, the maximum factored total force to be taken by the reinforcement T_{max} can be expressed as

$$T_{max} = 0.5\,K\,\gamma\,H^2 \quad (73.6)$$

where K is a non-dimensional parameter equivalent (but not equal) to an earth pressure coefficient; γ is the fill unit weight (kN/m^3); and H is the fill height (m).

The factored tensile force in the jth layer of reinforcement T_j is given as

$$T_j = K\,\gamma\,z_j\,Sv_j \quad (73.7)$$

where z_j is the depth below crest level to the jth layer of reinforcement; and Sv_j is the vertical spacing at the jth layer of reinforcement.

For each layer of reinforcement a local stability check should be undertaken. The design tensile strength of the reinforcement should be greater than the calculated factored tensile force to prevent rupture of the reinforcement. The design adherence capacity of the length of reinforcement (L_{ej}) behind the wedge considered should be greater than the factored tensile force to prevent pull-out of the reinforcement.

For each layer of reinforcement the smallest of the design tensile strength and the design adherence capacity can be referred to as the design capacity of the jth layer of reinforcement: $Tdes_j$. For the local stability check, it should be confirmed that for each layer of reinforcement

$$Tdes_j \geq T_j \quad (73.8)$$

The overall reinforcement provided should be sufficient to prevent rupture of the two-part wedge mechanism, and it should be checked that

$$\sum_{j=1}^{n} Tdes_j \geq T_{max} \quad (73.9)$$

Slice method for circular slip mechanism

The method of slices is traditionally used for unreinforced soil slopes and is adapted to take into account the effects of introducing the reinforcement layers. The additional restoring forces and moments provided by the reinforcing elements beyond the slip failure are considered in the calculations of the slope stability factor of safety. BS 8006 only considers moment equilibrium and suggests that inter-slice forces are ignored in the calculations. However, most current slope stability software allows for analyses using both force and moment equilibrium, and will consider some form of inter-slice interaction.

For each layer of reinforcement a local stability check (**Figure 73.13**) should be undertaken, and it should be verified that the design capacity of the reinforcing layer (either the pull-out resistance or the rupture strength of the reinforcement) is greater than the tensile load generated in the layer.

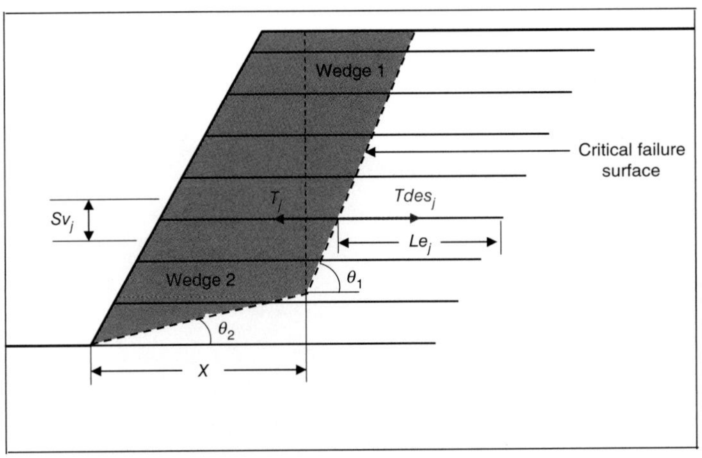

Figure 73.12 Two-part wedge mechanism

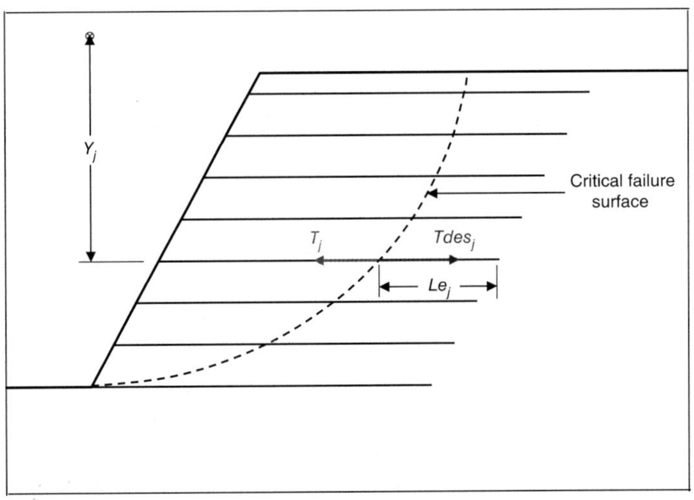

Figure 73.13 Circular slip failure mechanism

The overall reinforcement provided shall be sufficient to prevent slope failure and it shall be verified that

$$F_{RS} + \sum_{j=1}^{n} Tdes_j \geq F_D \qquad \text{(force equilibrium)} \qquad (73.10)$$

$$M_{RS} + \sum_{j=1}^{n} (Tdes_j \times Y_j) \geq M_D \qquad \text{(moment equilibrium)} \qquad (73.11)$$

where Y_j is the vertical distance from the centre of rotation of slip to the reinforcement layer under consideration; F_{RS} (M_{RS}) is the restoring force (moment) due to shear strength of soil; F_D (M_D) is the disturbing force (moment) due to the weight of soil and surcharge.

73.6.3.3 Compound stability

Compound stability considers failure mechanisms where the potential failure surface is partly within the reinforced soil zone and partly behind it. The principles for analysing compound stability are similar to those for analysing internal stability.

73.7. Hybrid reinforced soil and soil nail slopes

Reinforced soil (this chapter) and soil nailing (see Chapter 74 *Design of soil nails*) are both well understood as standalone systems of improving soil stability through the inclusion of tensile reinforcing elements. Reinforced soil is an active reinforcement system where the tensile strength in the reinforcing elements is mobilised during the placement and compaction of the fill material. Since the reinforcement lengths can be significant, reinforced soil is better suited to building new embankment structures.

Where an existing soil mass needs stabilising and its removal and replacement are impractical, soil nail reinforcement (typically steel bars, but can also be carbon or glass fibre rods) can be introduced into the soil, either through drilling and grouting or through dynamic insertion. In principle, soil nails differ little from reinforced soil although they are generally inclined downwards from the horizontal. Soil nails are a passive reinforcement system where the tensile strength in the reinforcement elements is generated through skin friction along the nail/soil or grout/soil interface as the soil begins to move.

The main constraint on the feasibility of reinforced soil solutions in cuttings is insufficient space to excavate and place the reinforced fill mass. This is particularly evident when founding reinforced soil slopes or walls on existing shallow embankment slopes to facilitate highway widenings where it may be preferable to soil nail the existing slope rather than excavate significant volumes of fill to place the reinforcement. In such situations it is possible to combine a permanent passive means of stabilising the cut face, such as soil nailing or rock bolting, with a narrow reinforced soil facing to provide the additional width required as a hybrid solution. Similarly, a hybrid solution may be preferable for existing walls and slopes requiring stability enhancement that are supporting services or access routes, in close proximity to site boundaries (where easement agreements are in place), or on sites seeking an aesthetic facing (**Figure 73.14**).

Design of both the active and passive reinforcement is typically undertaken by considering the reinforcement effect beyond a series of postulated failure surfaces defined by single or multi-part wedges, slip circles or log spirals with partial factors applied to soil and reinforcement properties, interface friction coefficients and applied loads. In the UK this design is generally undertaken in accordance with 'BS 8006-1:2010. Code of practice for strengthened/reinforced soils and other fills' and 'BS 8006-2:2011: Code of practice for strengthened/reinforced soils – Part 2: Soil nail design'.

(a) (b)

Figure 73.14 Failed wall repair with soil-nailed hybrid reinforced soil wall under construction and finished wall with original masonry reattached (Huesker/Dywidag)

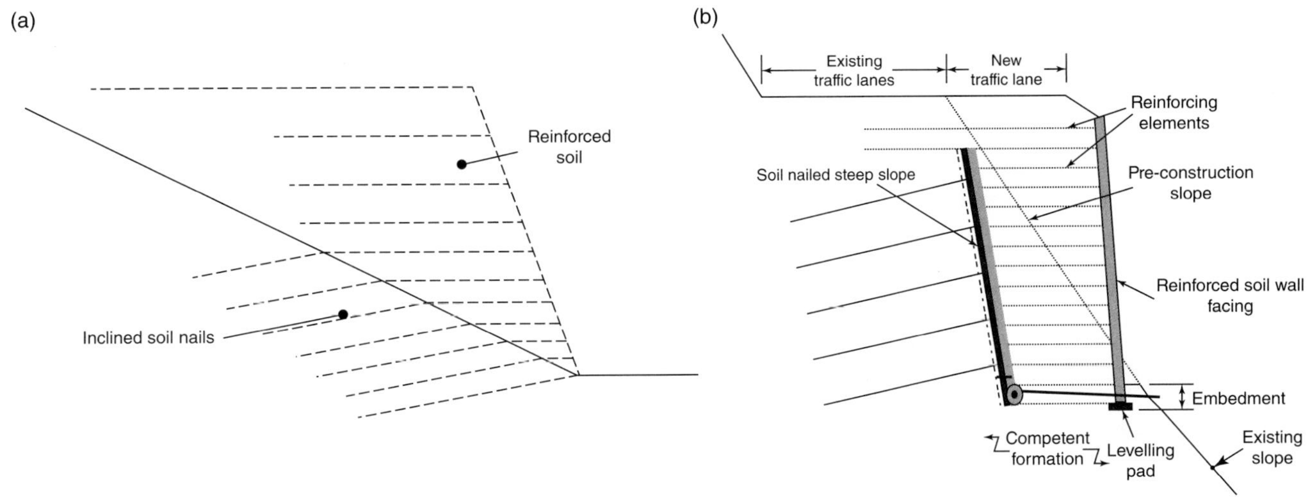

Figure 73.15 National Highways HA68/94 hybrid (a) and FHW modular shored walls (b)

While the two design documents explain the design methods for their respective types of reinforcement, there is limited guidance available for when the two approaches are combined to form a hybrid solution. The National Highways' 'HA68/94: Design methods for the reinforcement of highway slopes by reinforced soil and soil nailing techniques', published in 1994, does provide the user with an approach for hybrid slopes with a finished face of less than 70° degrees. The US Department of Transportation Federal Highway Administration published a design guide for 'Shored mechanically stabilised earth wall systems' in 2006; this provides the user with guidance on when reinforced soil walls are to be constructed to form new or widened roads in mountainous terrain, taking the benefit of the soil nails used to stabilise the back face of the platform cut to form the wall (**Figure 73.15**).

The approach suggested in HA68/94 consisted of mechanically joining or overlapping the soil nails and geosynthetic reinforcement at the active/passive interface and matching the vertical spacing of the two systems (**Figure 73.16**) for simplification of the two-part wedge method analysis, which only applies to reinforced soil faces shallower than 70°.

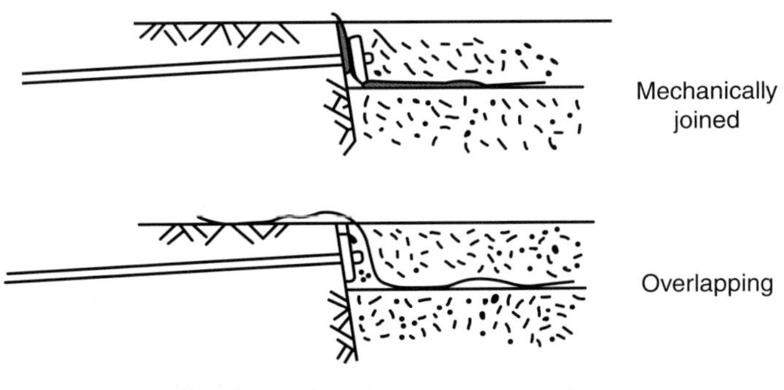

Figure 73.16 National Highways HA68/94 showing connection methods

The matching of soil nail and reinforcement layers is generally impractical due to the significant differences in strength between the soil nail steel bars and the geosynthetic reinforcement used for the reinforced soil facing. The spacings for soil nails are typically in excess of 1.0 m, while the vertical spacings of geosynthetic reinforcement are typically 0.4–0.7 m.

The typical application area for hybrid solutions is usually against steep top-down nailed or rock bolt faces. This steeper face lends itself well to a vertical or near-vertical bar system fixed to the soil nail heads, which can then be connected to the reinforcement of the reinforced soil slope facing (see **Figure 73.17**).

Since the active zone of the reinforced soil block will undergo some vertical compression during compaction relative to the passive soil-nailed face, the physical connection should be designed to permit sufficient vertical movement to minimise moments being induced at the soil nail head interface.

Designs are typically undertaken using standard single or multi-wedge analysis or slip circle analysis for the reinforced soil facing. Similarly, the soil nail design should be undertaken independently to stabilise the required cut face. A supplementary check should then be undertaken to ensure the total design load from the reinforced soil facing element does not exceed the total design capacity for the soil nails. A more detailed global stability analysis should then be carried out using staged construction for both the top-down soil nailing process and for the hybrid soil nails and reinforced soil facing elements together, to ensure that the overall solution meets the design requirements.

In creating a hybrid facing, the narrow reinforced soil zone is typically at least 0.3 × height or a minimum 1.5 m width to facilitate compaction of the fill and a safe working zone.

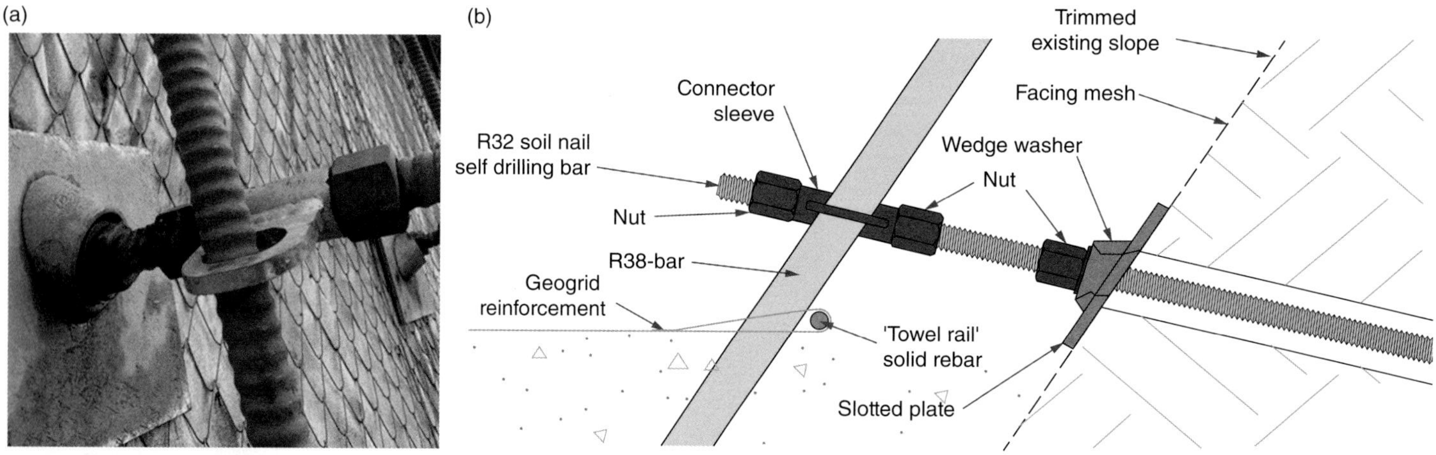

Figure 73.17 Typical detail of soil nail head connection to 'towel rail' geosynthetic reinforcement

73.8. Basal reinforcement

73.8.1 General

Reinforcement can be introduced at the base of embankments over poor ground to avoid ULS failure through shear in the foundation of the embankment and/or SLS failure through excessive deformations. It should be noted that BS 8006 limits the applications of basal reinforcement to foundations for earthworks. The common applications are

- basal reinforcement over soft to very soft foundation soil to control initial ULS stability of the embankment without controlling the settlement
- basal reinforcement combined with another form of ground improvement technique over soft to very soft foundation soils (typically piled embankment with basal reinforcement acting as a load transfer platform) to control both the initial ULS stability and long-term SLS settlements of the embankment
- basal reinforcement over foundation soils prone to subsidence (e.g. mining) where the reinforcement is designed to span the void (ULS stability) and limit the deformations at the top of the embankment.

Basal or confining geosynthetic reinforcement can also be introduced within a layer of granular fill to improve the stability of haul roads or working platforms to enable safe access for construction plant.

The products used in basal reinforcement applications generally consist of geotextiles or geogrids, but products such as steel meshes are sometimes considered.

73.8.2 Basal reinforcement for an embankment on soft to very soft ground

When constructing an embankment on soft soil, shear failures are likely to develop through the foundation soil. The problem is always more acute in the short term (during construction) since soft soils will consolidate and strengthen over time due to the dissipation of the excess pore water pressures generated in the soil by the construction of the embankment. Once consolidation is complete, the basal reinforcement is often redundant. Basal reinforcement is often used in conjunction with other construction techniques (e.g. staged construction discussed in Chapter 70 *Design of new earthworks*) to limit the capacity required from the reinforcement. Strain compatibility between the reinforcement and the soft foundation soil should be satisfied in order to achieve a maximum bond coefficient. This is particularly the case when dealing with sensitive foundation soils that exhibit rapid strength loss post-peak, hence the strain in the reinforcement (SLS case) should not exceed the strain at which the peak shear strength is mobilised in the foundation soil (see Section 73.3.2).

The ULS limit equilibrium mechanisms to consider are

- deep-seated failure
- lateral sliding
- extrusion.

73.8.2.1 Deep-seated failure

The analysis of deep-seated failure is generally performed using the method of slices for a circular failure mechanism, described in Section 73.6.3. The reinforcement provided should have a sufficient design capacity *Tdes* (against rupture and pull-out beyond the failure mechanism) to provide moments and forces equilibrium. In the case of basal reinforcement, it is also necessary to ensure that the length of reinforcement within the failure surface mechanism is sufficient to prevent pull-out (**Figure 73.18**).

73.8.2.2 Lateral sliding

The reinforcement provided shall have a sufficient design capacity *Tdes* (against rupture and pull-out beyond the failure mechanism) to resist the outward thrust (active earth pressures) from the embankment fill and any surcharge which is at

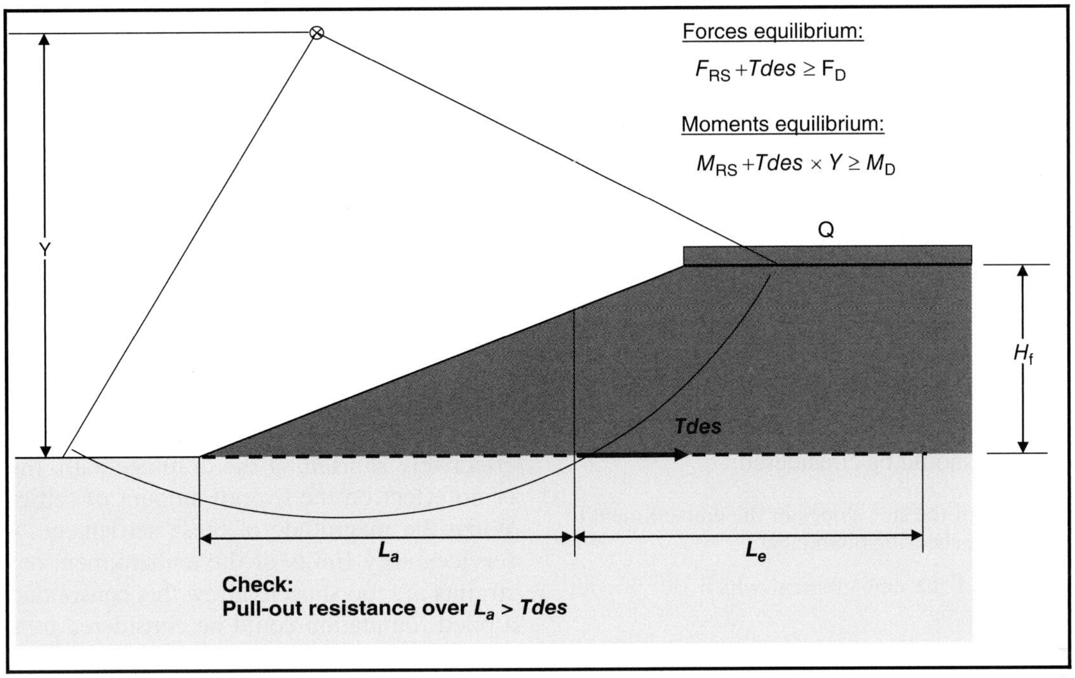

Figure 73.18 Basal reinforcement – deep-seated failure. Q = surcharge loading, H_f = height of embankment, L_a = length of reinforcement within active zone, L_e = length of reinforcement within resistant zone

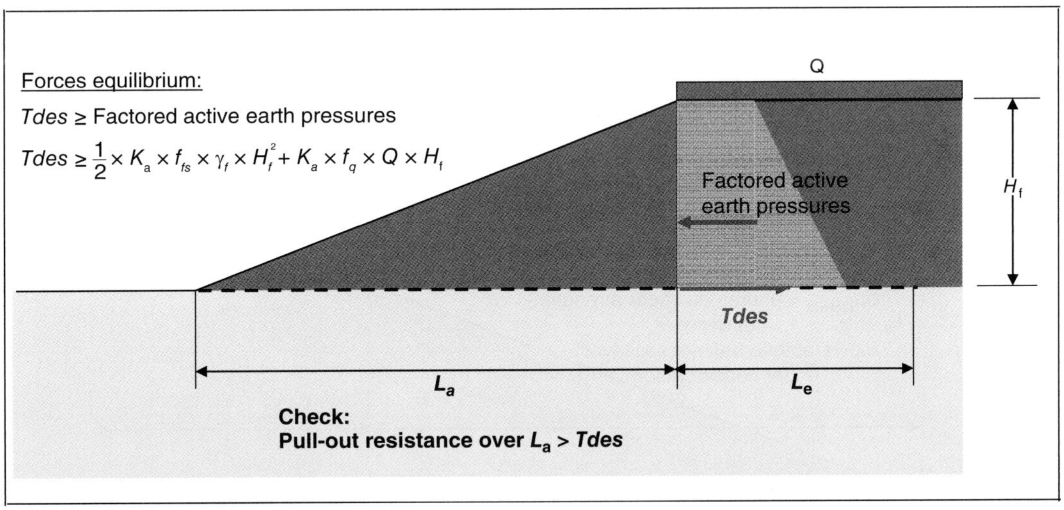

Figure 73.19 Basal reinforcement – lateral sliding. K_a = active earth pressure coefficient, γ_f = fill unit weight

a maximum at the top edge of the embankment slope. Again, it is also necessary to ensure that the length of reinforcement within the failure surface mechanism is sufficient to prevent pull-out (**Figure 73.19**).

For cohesive foundation soils, subject to undrained loading, the bond stress developed between the soil and the reinforcement is directly related to the undrained shear strength of the soil by an adhesion factor, which will normally have a value less than unity when considering the direct sliding mechanism.

73.8.2.3 Extrusion

Extrusion is particularly problematic for heavily loaded embankments over weak soil layers, where the imposed loading from the embankment can cause extrusion of the weak underlying soils, which have insufficient strength to resist the out-of-balance active earth pressures. This squeezing failure is comparable to a toothpaste tube under pressure – the soft soil undergoes plastic deformation out from underneath the embankment, causing the embankment to settle by displacement.

Extrusion assumes the outward lateral displacement of a rectangular block of soft soil under the embankment fill. The reinforcement provided shall have a sufficient design capacity *Tdes* (against rupture and pull-out beyond the failure mechanism) to resist the tensile load generated by the outward foundation shear. In addition, it is also necessary to ensure that at any depth H_s within the reinforced soil the rectangular block of soil is in equilibrium. The restoring forces are the adhesion along the base of the soil block, the adhesion between the reinforcement and the soil at the top of the block and the passive earth pressures on the outward side of the block (**Figure 73.20**).

If the factor of safety against extrusion failure cannot be satisfied by the stabilising forces indicated above, then additional stabilising options should be considered

- Reduce the batter angle of the side slopes of the embankment to increase the length of the shearing boundary.
- Use a berm at the toe of the embankment which can provide counter-pressure to extrusion.
- Remove the soft soil and replace it with component engineered fill.
- Incorporate a pile cut off wall and retain the soft soils.
- Use a shear key of granular fill placed under the shoulders of the embankment.

A shear key/trench detail (see **Figure 73.21**) should be checked for shear planes through it, where the active earth pressure is resisted by the shear key, taking into consideration the frictional shear strength (ϕ') of the trench fill (Scotland *et al.*, 2019). The creation of a shear key is a well-established earthworks technique used to disrupt potential weak slip-planes in the foundation soil (Giffen 2015).

73.8.3 Basal reinforced platforms over pile foundations

Chapter 70 *Design of new earthworks* (Section 70.7) discusses the issue of settlement and failure of new embankment foundations built over soft soils. It highlights foundation treatment options and the use of geogrids to control embankment instability for both short- and long-term conditions. While additional measures such as vertical drains can be installed in the foundation soil to speed up the consolidation process by effectively shortening the drainage path, they will have little or no effect on the overall amount of settlement that occurs. When the magnitude of these settlements would exceed the serviceability limits of the embankment or programme constraints are too short to allow this consolidation to occur, then a piled foundation could be considered to essentially bypass the soft stratum and transfer the embankment load to a firm stratum below, in combination with a geosynthetic reinforcement formed over the piles to transfer the embankment load onto the pile caps.

The design of the geosynthetic reinforcement incorporated to aid distribution of the embankment load onto the piles is covered in BS 8006 (2016), Clause 8.3.3.2. One of the principal assumptions underpinning the design is that all the embankment loading is ultimately carried by the piles, either directly by arching of the soil concentrating vertical stresses onto the

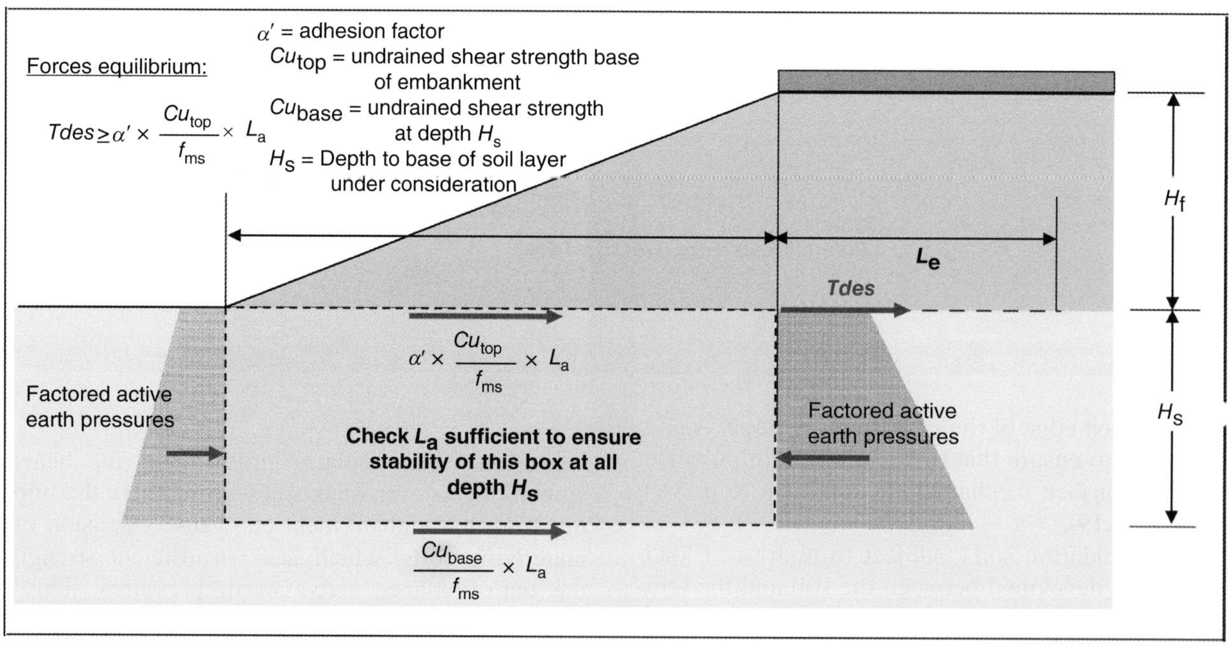

Figure 73.20 Basal reinforcement – extrusion. α' = adhesion factor, Cu_{top} = undrained shear strength base of embankment, Cu_{base} = undrained shear strength at depth H_s, H_s = depth to base of soil layer under consideration

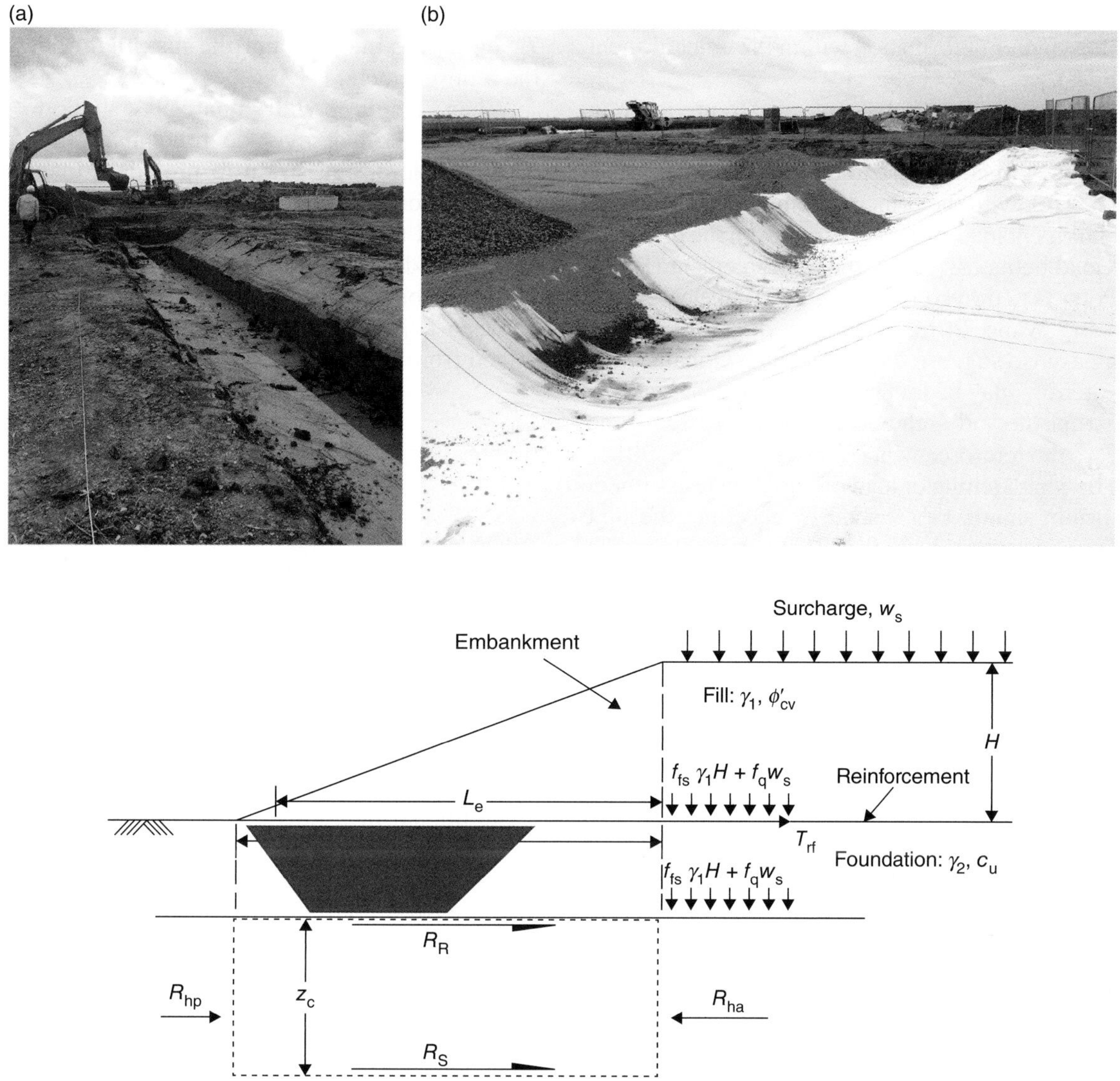

Figure 73.21 Shear key/trench detail (Courtesy of Huesker)

pile caps, or indirectly by the geosynthetic reinforcement acting in catenary between the piles distributing the weight of the embankment soil (below the arch) onto the pile caps (see **Figure 73.22**).

This principal assumption, that the entire load from the overlying embankment is taken by the piles, may seem unduly onerous, and indeed other ground improvement options exist that introduce stiffer vertical elements into the foundation soil but utilise some of the adjacent soil to support part of the embankment loading (i.e. rigid inclusions).

However, the consideration of long-term partial support offered by the ground between piles is not recommended for routine design unless it can be demonstrated that this partial contribution will be provided for the entire service life of the embankment. The effective stress conditions within the existing ground can alter irrespective of the embankment loading (e.g. seasonal fluctuations in the groundwater level which can increase the effective stress of the in situ soil, resulting in further consolidation and thereby further embankment stress

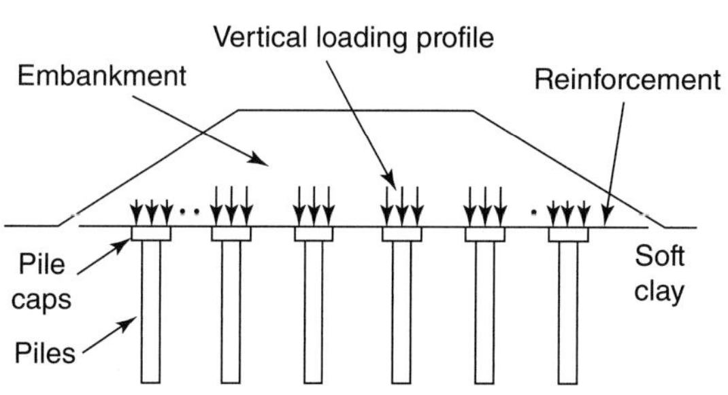

Figure 73.22 Vertical load shedding (Reproduced with permission from BS 8006, figure 76c)

redistribution or arching changes, increasing the direct loads acting on the piles and/or deformation within the embankment fill above the piles).

BS 8006 considers a number of different arching approaches to determine the stress acting across the reinforcement. There has been ongoing research since the publication of BS 8006-1 in 2010, with the principal aim to determine the amount of embankment loading acting directly on the piles and conversely the load being carried by the reinforcement (Blanc *et al.*, 2013; Potts & Zdravkovic, 2010; van Eekelen *et al.*, 2013; Zhuang and Wang, 2018; Zhuang *et al.*, 2010, 2012). Some research has also suggested that when the embankment is shallow, the arching may be influenced by transient loading, and as such disrupt the soil arch and effectively increase the load acting across the reinforcement (van Eekelen *et al.*, 2012).

Irrespective of what arching or load distribution is assumed, vertical equilibrium should be satisfied by checking the unit area of embankment carried by the single pile. Once the amount of load to be carried by the reinforcement (W_T) is quantified then the tensile load in the reinforcement can be determined from

$$T_{rp} = \frac{W_T(s-a)}{2a}\sqrt{1+\frac{1}{6\varepsilon}} \qquad (73.12)$$

where T_{rp} is the tensile load in the reinforcement (**Figure 73.23**); ε is the strain in the reinforcement; a is the size (or equivalent size) of the pile caps; and s is the pile spacing between adjacent piles.

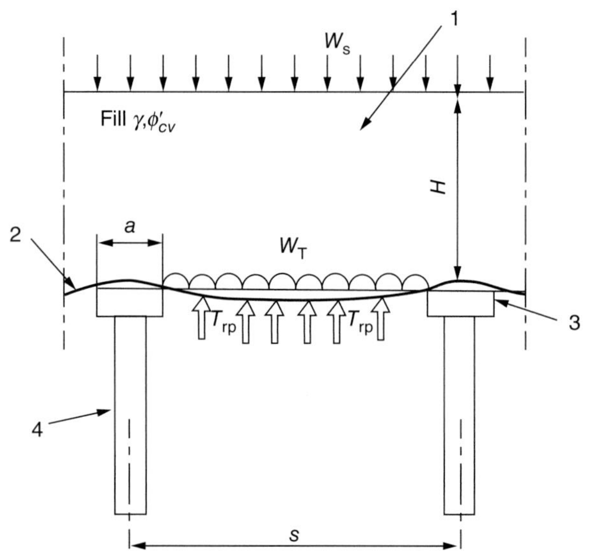

Key
1 Embankment
2 Reinforcement
3 Pile cap
4 Pile

Figure 73.23 Variables used in determination of T_{rp} (Reproduced with permission from BS8006 figure 79)

Equation 73.12 has two unknowns, T_{rp} and ε. It is solved for T_{rp} by taking into account the maximum allowable strain in the reinforcement.

When dealing with relatively shallow embankments [$h < 1.4(s - a)$], it may be prudent to limit the allowable geosynthetic design strain to provide a maximum mid-span deflection of the supporting reinforcement. BS 8006 recommends that the strain should be limited to 3% or a practical maximum deflection of 300 mm.

The amount of deflection of the reinforcement spanning between adjacent piles for an assumed strain can be determined from

$$y = (s-a)\sqrt{\frac{3\varepsilon}{8}} \qquad (73.13)$$

where a is the size of the pile caps (assuming full support can be generated at the edges of the caps); s is the spacing between adjacent piles; and ε is the strain in the reinforcement.

It should also be ensured that most of the design strain is mobilised during the construction period and that any post-construction reinforcement strain and associated deflections at the surface of the embankment are minimised. The risk of associated deflections at the surface of the embankment is increased with shallow embankments.

Additional measures that can be considered to encourage the deflection of the supporting reinforcement during construction are discussed further in Chapter 86 *Soil reinforcement construction*.

73.8.4 Basal reinforcement over foundation soils prone to subsidence

Basal reinforcement can also be used to support embankments to be formed over a void developing/existing in the formation below. Man-made voids are discussed in more detail in Chapter 41 *Man-made hazards and obstructions*. Allowable serviceability limits are prescribed for the surface of the embankment and hence the mobilised strain in the supporting reinforcement is limited to ensure the allowable surface deflections are not exceeded.

This design approach is often dominated by the serviceability design controlled by the allowable surface deflections (some infrastructure, such as railways, has tight allowable cross-rail deflection to ensure a twist fault does not develop). This can result in very high tensile stiffness reinforcements being considered to quickly mobilise the required tensile capacity at low deformations as a void develops.

The method described in BS 8006 (2016) assumes a constant volume of soil within the zone of depression of the soil influenced by the underlying void, assumed to be an inverted cone/wedge of soil extending up from the edges of the void (**Figure 73.24**). It does not consider any bulking of the fill in the zone of depression. Further conservatism is provided in the

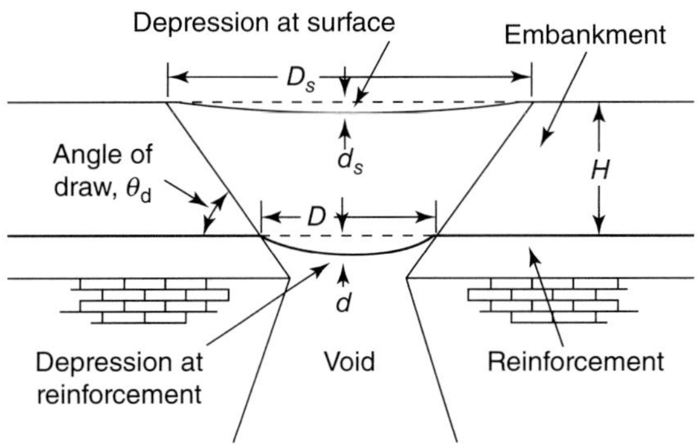

Figure 73.24 Conceptual role of basal reinforcement in limiting surface deformations due to subsidence (Reproduced with permission from BS 8006:1995)

design assumption in that no arching occurs in the overlying fill. However, a modified methodology for predicting surface settlements based on an increase in the volume of soil in the 'zone of depression' due to dilatancy or bulking has been presented (Bruhier and Sobolewski, 2000). The methodology is very similar to the British Standard method described earlier. However, it allows for a volume increase in the zone of depression such that the volume of fill occupying the shape of the deflected membrane is larger than the volume of the surface depression due to the dilatancy or bulking of the soil in the 'zone of depression'.

By considering bulking of the fill in the collapse zone one can effectively design a larger allowable reinforcement deflection or strain for a given or prescribed allowable surface deflection.

Consideration of arching of an embankment will effectively reduce the zone of soil to be supported by the geosynthetic and may result in minimal deflections at the surface. There are a number of arching theories; the most common theory associated with void spanning was described by Terzaghi (1943). Terzaghi assumes a vertical shear boundary extending up from the edge of the void. Shear along this boundary helps to prevent the downward movement of the soil overlying the void. This effectively reduces the vertical stress previously acting across the area of the void and increases the stress acting on the side of the void. This arching theory is also attributed to the crown hole migration where stable soil/rock at the apex of the arch is disturbed and collapses and migrates towards the ground surface.

Increasingly, basal reinforcement is being incorporated for a risk-based approach to void mitigation measures. This can be coupled with monitoring layers incorporated below the supporting reinforcement layer that provide an earlier warning system of ground movements and migrating voids. Recent research and site installations have focused on incorporating optical fibres into monitoring layers, enabling continuous measurement along those fibres. This can provide detailed data on ground movement and void development in real time (Xu *et al.*, 2022).

73.8.5 Reinforced haul roads and working platforms

In the temporary condition, sites over poor soil will require adequate haul roads and working platforms to facilitate site access and the safe movement and operation of site machinery including heavy plant such as piling rigs and cranes. There are a number of empirical and analytical approaches to the design of granular haul roads and working platforms, as either unreinforced or reinforced with geosynthetic elements typically in the form of geogrids or geotextiles, incorporated at, or close to, formation level.

Where geosynthetics are laid beneath or within a granular platform, they will act to improve the structure of the platform and minimise deformations during trafficking and loading through either, or a combination, of

- tension membrane effect
- lateral restraint by way of friction and or interlock/fill confinement.

Geotextiles are primarily used as a separation layer to reduce contamination of the platform fill material from the underlying ground, but will also provide a degree of stabilisation. Several geosynthetic manufacturers have developed design charts, nomographs and software which determines the thickness of the haul road or working platform and type of geosynthetic to be incorporated. The principal benefit of incorporating a geosynthetic element is to reduce the thickness of granular fill required to provide an equivalent or even enhanced level of trafficability/operation. This reduction in required fill depth (**Figure 73.25**) often provides both commercial and sustainability benefits, particularly on sites where the granular fill needs to be imported.

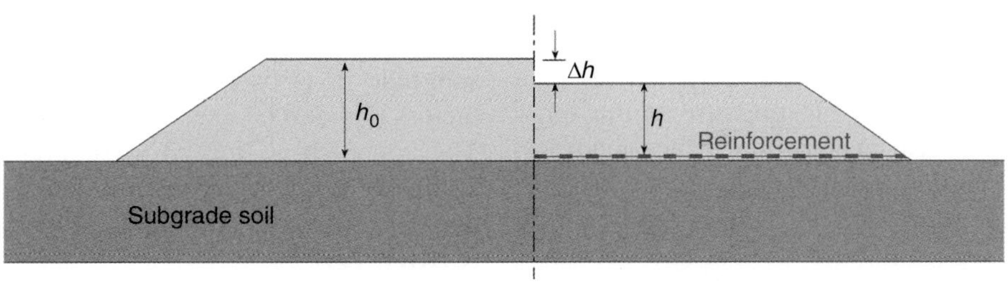

Figure 73.25 Unreinforced and geosynthetic reinforced haul road (Courtesy of Huesker)

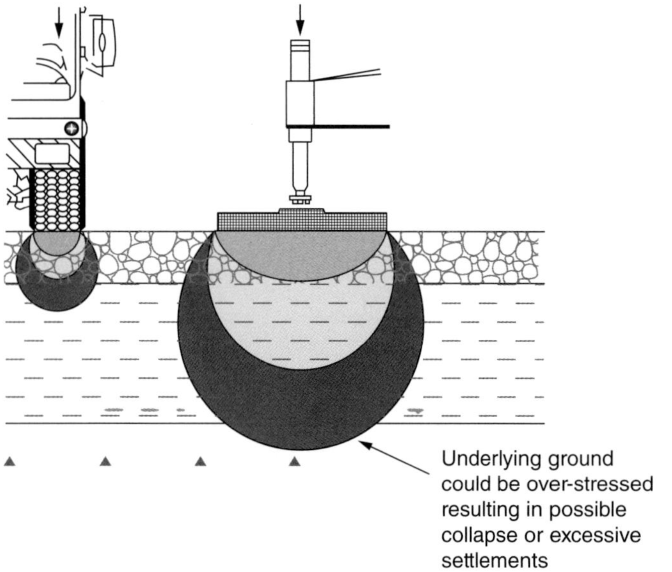

Figure 73.26 Depth of influence for bearing defined by pressure bulbs of plant loading (Courtesy of Huesker)

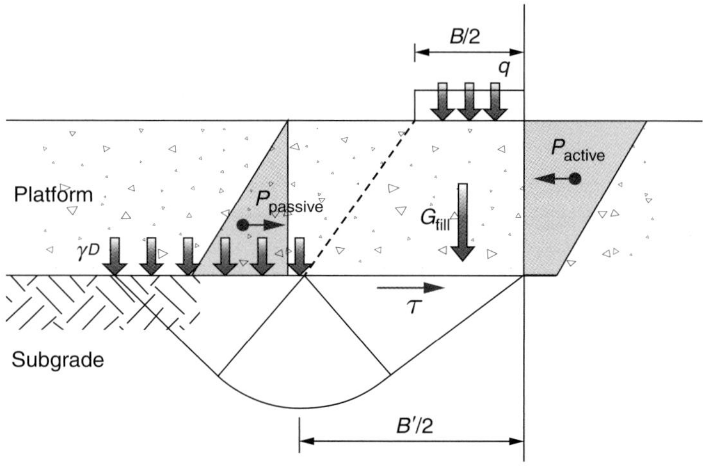

Figure 73.27 CIRIA SP123 lateral shear model (Reproduced with permission from TWf, 2019)

There is, however, a key distinction that must be drawn between the design of haul roads, which are typically relatively lightly loaded, and the design for working platforms, where the loads generated from the operational construction plant can be significant, with a larger depth of influence below the working platform (**Figure 73.26**). For this reason the reliance on geosynthetic manufacturers' developed design charts, nomographs and software should be treated with caution, particularly for the design of working platforms, and more rigorous approaches should be adopted such as those provided by BRE BR470 (BRE, 2004), CIRIA SP123 (Jewell, 1996) and TRRL LR1132 (Powell *et al.*, 1984) publications.

The 2019 publication by the Temporary Works forum 'Working platforms: design of granular working platforms for construction plant – a guide to good practice' provides a good collation of the available UK guidance, extracts of which are summarised in the following sections.

73.8.5.1 CIRIA SP123: Soil reinforcement with geotextiles

The special report published by CIRIA in 1996 features a predominantly analytical approach considering the bearing capacity of a subgrade limited by outward shear stress. The complex equations used require iteration to find the resulting platform thickness, hence calculation spreadsheets are required.

The method determines the bearing pressure at the formation of the working platform, where the surface loading (tracked plant or wheel) is dispersed through the fill to the base of the platform (**Figure 73.27**). A load distribution angle (β) is assumed to determine the vertical and outward stress applied. It is typical in geotechnical engineering to use a load spreading angle of 26.6° (2:1, vertical to horizontal). However, many studies show load dispersal in reinforced platforms can be significantly higher. For geotextiles this has been observed at around 40°, while for geogrid reinforcement this has been recorded to be greater than 45° (Palmeira and Antunes, 2010).

The underlying subgrade's bearing capacity factor N_c varies nonlinearly with mobilised shear stress τ up to a maximum of $\pi + 2$, where no outward support is required. The average horizontal earth pressure resulting from this distribution is used to determine the horizontal force to be resisted by the geosynthetic.

For reinforced platforms the document suggests the full bearing capacity of the soil can be mobilised when the reinforcement can resist all the lateral shear stresses. This can require high strength reinforcements, unless the platform thickness is increased, to reduce the net outward stress.

73.8.5.2 BRE BR470: Working platforms for tracked plant

Published in 2004, this guidance document is probably still the most widely used for unreinforced working platform design in the UK and is increasingly used internationally. This document was developed in conjunction with the UK Federation of Piling Specialists (FPS) to improve safety for tracked plant on construction sites, following a number of high-profile failures.

The document introduced a straightforward semi-empirical design method. Its methodology is based on a punching shear failure of a loaded track through a platform onto the subgrade. It is based largely on empirical bearing capacity factors that can be used with cohesive or granular subgrades. The approach is limited to subgrades of strengths between $20\ kN/m^2 < Cu < 80\ kN/m^2$. Outside this range designers are advised to seek alternative guidance, such as SP123 (CIRIA, 1996).

The BRE method is extremely sensitive to the shear strength of the platform fill. Corke and Gannon (2010) showed only a small 12% increase in platform material angle of friction from 40° to 45° led to a 27% reduction in platform thickness. While the benefits of adopting a high design friction angle in an attempt to minimise the platform thickness could be adopted, the designer should also consider the ability of actually achieving this in practice. Influences that could reduce the design platform fill friction angle may be from: not achieving the required compacted density of the platform fill; potential contamination from fines and water forming a slurry within the voids of the granular fill during the site operations; and deterioration of the fill itself, particularly recycled fills.

Jefferis (2019) urges caution in using recycled crushed materials: 'testing recycled materials for environmental acceptability alone is not sufficient. Recycled materials must be tested for in-service behaviour, in in-service environments.'

Within the BR470 approach, geosynthetic reinforcements can be included to reduce the thickness of the platform by providing additional punching resistance across the vertical 'punching' boundary (**Figure 73.28**).

73.8.5.3 Geosynthetic tensile strength–strain relationship within the platform

As the geosynthetic acts in tension under the load, it restrains the outward movement of the fill material (from under the load) and thus reduces or eliminates horizontal shear on the formation, thereby improving the bearing capacity. In addition, by improving interlock/confinement, the geosynthetic acts to increase the load spread angle, thereby increasing the effective bearing area and the overall bearing resistance.

In order to mobilise the design geosynthetic tensile strength for the tensioned membrane effect, the reinforcement needs to strain and the deformation needed to mobilise this mechanism could exceed the serviceability requirements of the working platform. Thus, when reinforcing geosynthetics are selected for use, the tensile strength used should be for small strains, typically in the range of 2–5%, in order to keep the mobilising deformations within acceptable limits.

For stabilisation geogrids the mechanism is lateral restraint and interlock/confinement. The radial stiffness of the geogrid limits the lateral strains, typically in the order of 0.5%, thereby restricting lateral aggregate particle movement, increasing the load spread, and consequently improving the bearing resistance.

All geosynthetics used for working platforms should be configured to provide an equal degree of restraint longitudinally and transversely, or radially to any load. This may be achieved through the use of multi-axial, biaxial or multiple cross-laid uniaxial geosynthetics.

It is important to note that sufficient deadweight is required over a geosynthetic to prevent it from pulling laterally through the ground and to allow it to strain adequately to work when loaded. A minimum thickness of fill (typically 300 mm or half the plant track width) and an anchorage length should be specified.

73.8.5.4 Alternative platform design methods

Since the publication of BR470 in 2004 and a subsequent publication in 2011 by BRE, 'Use of "structural geosynthetic reinforcement": a BRE review seven years on', geosynthetic manufacturers have undertaken further product development which has resulted in the development of various design methodologies that are bespoke to their products. Such alternative methods may be deemed acceptable for the design of temporary working platforms, provided that the manufacturers can demonstrate that their proprietary methods have been validated by appropriate theoretical design and experimental testing of an appropriate scale, with sufficient evidence of past experience for its intended application. It should also be recognised

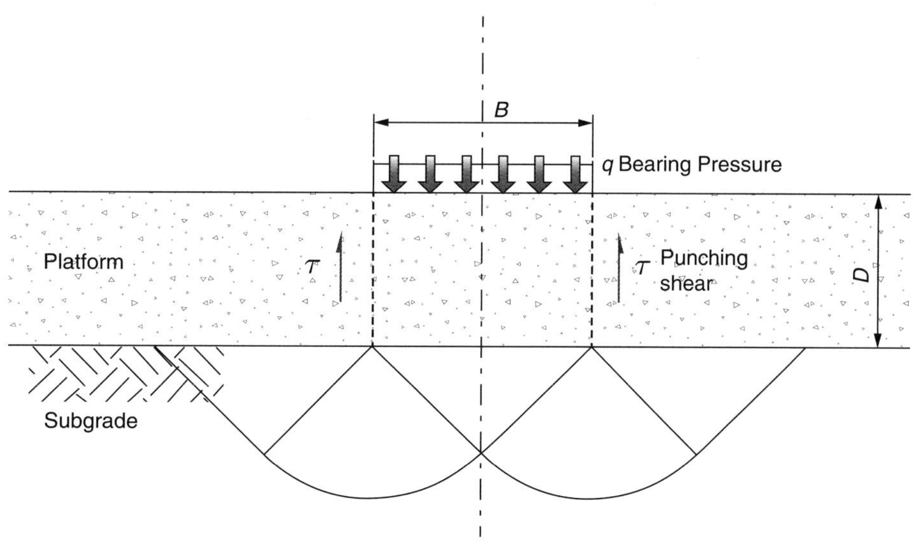

Figure 73.28 BR470 punching shear mechanism (Reproduced with permission from TWf, 2019)

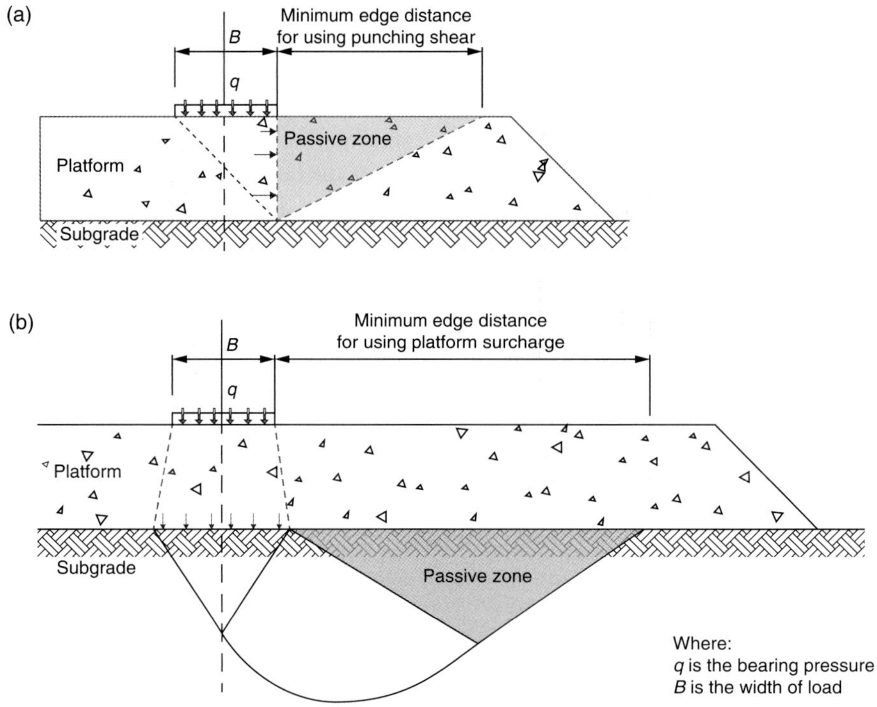

Figure 73.29 Minimum distance of effective working area from platform edge (Reproduced with permission from TWf, 2019)

that the substitution of another manufacturers' geosynthetic product into a proprietary working platform design must be avoided due to the specific differences in product development and potential impact to the performance of the granular working platform.

73.8.5.5 Edge restraint and effective working platform area

Irrespective of what design approach is adopted for the load distribution within the working platform, failures are often initiated close to the edge of the working platform; hence, this should be checked as a shallow embankment for both rotational (**Figure 73.18**) and extrusion failure (**Figure 73.20**). Edge detailing for all working platforms must be carefully considered, and the design must cater for each site's individual ground conditions, anticipated loading conditions and the selected geosynthetic and platform fill materials. The edge of the operational working area must also be clearly marked to define an exclusion zone between the effective working area and the edge of the platform (**Figure 73.29**).

The minimum edge distance is needed to ensure

- mobilisation of punching shear resistance within the platform
- mobilisation of ground bearing support due to platform surcharge (if used)
- pull-out resistance of geosynthetic reinforcement (for the BR470 method)
- acceptable surcharge condition near embankments or retaining walls.

Additional advice relating to the specification and construction of working platforms is provided in Chapter 83 *Underpinning*.

References

Blanc M, Rault G, Thorel L and Almeida M (2013) Centrifuge investigation of load transfer mechanisms in a granular mattress above a rigid inclusions network. *Geotextiles and Geomembranes* **36(1)**: 92–105.

BRE (2004) *BR470 Working Platforms for Tracked Plant: Good Practice Guide to the Design, Installation, Maintenance and Repair of Ground Supported Working Platforms.* BRE, Watford, UK.

BRE (2011) *BR470 Working Platforms for Tracked Plant: Use of 'Structural Geosynthetic Reinforcement' – a BRE Review Seven Years On.* BRE, Watford, UK.

Bruhier J and Sobolewski J (2000) Protection of road and railways embankments against collapse involved by sinkholes. *Proceedings of the Second European Geosynthetics Conference, Bologna, Italy*, vol. 1, pp. 299–304.

BSI (British Standards Institution) (1995) BS 8006: Code of practice for strengthened/reinforced soils and other fills. BSI, London, UK.

BSI (2004) BS EN 1997-1: Eurocode 7 – Geotechnical design – Part 1: General rules: BSI, London, UK.

BSI (2007) PD ISO/TR 20432: Guidelines for the determination of the long-term strength of geosynthetics for soil reinforcement. BSI, London, UK.

BSI (2010) BS 8006-1+A1:2016: Code of practice for strengthened/reinforced soils and other fills. BSI, London, UK.

Corke D and Gannon J (2010) Economic design of working platforms for tracked plant. *Ground Engineering*, Feb.: pp. 29–31.

Giffen AD (2015) Design and construction of a strengthened embankment for an intermodal transfer facility. *Proceedings of the XVI ECSMGE, Edinburgh, UK*, pp. 583–588.

High Court of Justice in Northern Ireland, Queen's Bench Division (Commercial List) (2005) Enniskillen Case on Load Transfer Platform as a Good Warning Story – Neutral Citation No [2005]. Belfast: NIQB 68, delivered 24 October.

Jefferis SA (2019) Recycled crushed concrete: engineering and environmental issues. *Proceedings of the XVII European Conference on Soil Mechanics and Geotechnical Engineering, Reykjavik, Iceland*, pp. 1–9.

Jewell RA (1996) *Soil Reinforcement with Geotextiles*. CIRIA Special Publication 123. CIRIA, London, UK.

National Highways (1994) *Design Methods for the Reinforcement of Highway Slopes by Reinforced Soil and Soil Nailing Techniques. Design Manual for Roads and Bridges. HA 68/94*. London: National Highways.

National Highways (2016) *Specification for Highway Works: Manual of Contract Documents for Highway Works*. National Highways, London, UK.

Palmeira EM and Antunes LGS (2010) Large scale tests on geosynthetic reinforced unpaved roads subjected to surface maintenance. *Geotextiles and Geomembranes* **28**: 547–558.

Potts VJ and Zdravkovic L (2010) Finite-element study of arching behaviour in reinforced fills. *Proceedings of the Institution of Civil Engineers – Ground Improvement*, **163**: 217–229.

Powell WD, Potter JF, Mayhew HC and Nunn ME (1984) *LR1132: The Structural Design of Bituminous Roads*. TRRL, Crowthorne, UK.

Scotland I, Poberezhniy V and Tatari A (2019) Reinforced soil shear key to mitigate extrusion failure in soft soils under working platforms. *Proceedings of the XVII European Conference on Soil Mechanics and Geotechnical Engineering, Reykjavik, Iceland*, pp. 1–8.

Terzaghi K (1943) *Theoretical Soil Mechanics*, Wiley, New York, NY.

TWf (Temporary Works forum) (2019) *TWf2019:02: Working Platforms: Design of Granular Working Platforms for Construction Plant – a Guide to Good Practice*. TWf, London, UK.

van Eekelen SJM, Bezuijen A and van Duijnen PG (2012) Does a piled embankment 'feel' the passage of a heavy truck? Field measurements. *Proceedings of 5th European Geosynthetics Congress, Valencia, Spain*, pp. 162–166.

van Eekelen SJM, Bezuijen A and van Tol AF (2013) An analytical model for arching in piled embankments. *Geotextiles and Geomembranes* **39(4)**: 78–102.

Xu X, Kechavarzi C, Wright D, *et al.* (2022) Fibre optic instrumented geogrid for ground movement detection. *Proceedings of 11th International Symposium on Field Monitoring in Geomechanics, London, UK.*

Zhuang Y and Wang KY (2018) Finite-element analysis of arching in highway piled embankments subjected to moving vehicle loads. *Géotechnique* **68(10)**: 857–868.

Zhuang Y, Ellis EA and Yu HS (2010) Plane strain FE analysis of arching in a piled embankment. *Proceedings of the Institution of Civil Engineers – Ground Improvement* **163(4)**: 207–215.

Zhuang Y, Ellis EA and Yu HS (2012) Three-dimensional finite-element analysis of arching in a piled embankment. *Géotechnique* **62(12)**: 1127–1131.

Further reading

Association Française de Normalisation (AFNOR) (2009) Calcul Géotechnique, Ouvrages de Soutènement, Remblais Renforcés et Massifs en Sol Cloué. Norme Française NF P 94-270.

BSI (2006) BS EN 14475: Execution of special geotechnical works – Reinforced fill. BSI, London, UK.

Geotechnical Engineering Office, Civil Engineering Department, Government of the Hong Kong Special Administrative Region (2002) *Guide to Reinforced Fill Structure and Slope Design: Geoguide 6*. HKSAR, Hong Kong.

Smith CC and Tatari A (2016) Limit analysis of reinforced embankments on soft soil. *Geotextiles and Geomembranes* **44**: 504–514.

It is recommended this chapter is read in conjunction with

- Chapter 23 *Slope stability*
- Chapter 69 *Earthworks design principles*
- Chapter 72 *Slope stabilisation methods*
- Chapter 86 *Soil reinforcement construction*

All chapters in this book rely on the guidance in Sections 1 *Context* and 2 *Fundamental principles*. A sound knowledge of ground investigation is required for all geotechnical works, as set out in Section 4 *Site investigation*.

doi: 10.1680/icemge.66830.1215

CONTENTS

Chapter 74
Design of soil nails

Adam M Daykin Atkins, Leeds, UK
Martin J Whitbread Haverfordwest, UK

Soil nails are a system for reinforcing in situ soil. They are used in the construction of steep slopes, to strengthen failing slopes and to tie back existing retaining walls. Soil nails are installed in groups and form an integrated system in which the forces in the soil nails, load plates, facings and surrounding soil all interact. Soil nails need to be designed with careful consideration of the ground and groundwater conditions, the surrounding environment and the construction process. The pull-out resistance of individual nails, the stability of the soil nail group and the global stability of the slope are considered by the designer. Various slope facings may be adopted for soil-nailed slopes, ranging from a hard finish such as shotcrete to a green finish with live vegetation.

74.1. Introduction

Soil nailing is a method of reinforcing the ground, comprising the installation of slender tension-carrying elements known as soil nails into the in situ ground. They are used to strengthen soil slopes usually in the following circumstances.

- To improve the stability of existing, potentially unstable slopes.
- To reinforce soil such that new slopes can be formed at steeper slopes than the soil properties would otherwise allow.
- To improve the stability of existing retaining structures.

The composition of a typical soil nail is illustrated in **Figure 74.1**.

74.2. History of soil nail design

Soil nails were first used in the early 1970s, notably in France and Brazil. The Clouterre Recommendations (Clouterre, 1991) facilitated a significant step in the development of soil nail design. These were based on a French research project in which soil nails were used to tie back a vertical wall. In the UK during the early 1990s there were numerous research projects at universities and the Transport Research Laboratory which generated debate about and improved understanding of the behaviour of soil nails. This led to greater consensus about appropriate methods of design and construction. Since the mid-1990s design has become increasingly standardised and soil nails are now in common use on major government infrastructure projects.

'BS 8006-2: Soil nailing' was first issued in 1995. This was followed by further technical guidance documents. In 2005 CIRIA C637 *Soil Nailing, Best Practice Guide* (Phear *et al.*, 2005) was issued. This was a comprehensive and detailed guide which remains extremely useful today. The current British Standard, 'BS 8006-2:2011+A1:2017: Code of practice for strengthened/reinforced soils: Part 2 – Soil nail design' was issued in 2017. The second generation of Eurocodes are going to be issued soon. This second generation will introduce a new document to complement 'Part 1: General rules' and 'Part 2: Ground investigation'. It will be called 'Part 3: Geotechnical structures', and its annexes will provide specific rules to be applied for design and verification of certain types of geotechnical structure. This will include soil nails. Further information will be provided in 'Annex F: Reinforced fill, soil nailed, and rock bolted structures'.

74.3. Behaviour of soil nails

Soil nails are constructed in groups that act together as a system. A schematic section through a soil nail system is illustrated in **Figure 74.2**. Soil nails are designed on the basis that steep slopes comprise an active outer zone which is prone to sliding and a more stable inner zone known as the resistant zone. The extent of the active zone is defined by a critical slip plane. The critical slip plane is usually not an active slip; it is generally a theoretical potential slip which the soil nails are designed to stabilise.

Soil nails are typically spaced around 1.5–2.0 m apart. The part of each nail that lies in the resistant zone is known as the distal end of the nail. If the active zone starts to move, the soil nails are pulled into tension. The tension in the distal ends of the nails provide a restoring force which balances the destabilising forces acting on the critical slip. This restoring

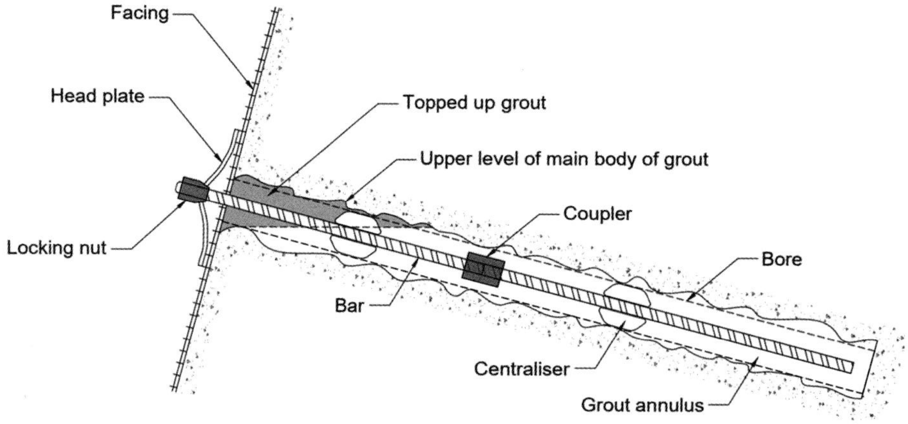

Figure 74.1 General composition of a soil nail

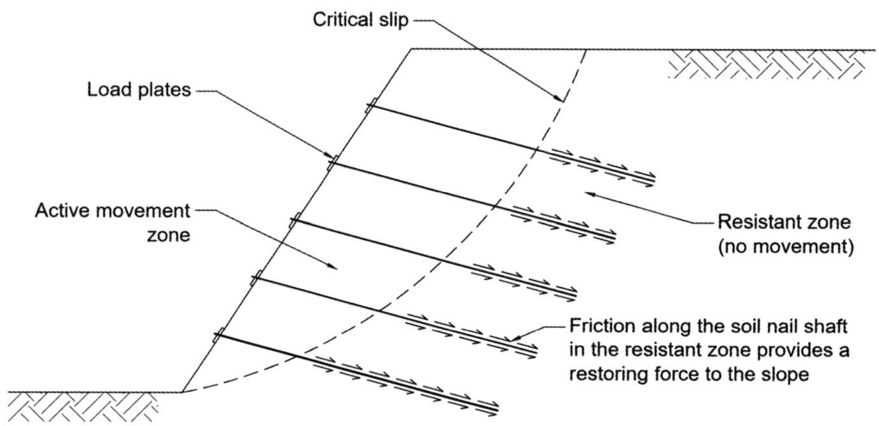

Figure 74.2 Schematic section through a soil nail system

force is generated by the bond strength between the outside of the grouted annulus around the nail and the surrounding soil. Soil nails are a passive system; movement needs to occur in the active zone for the tensile forces to be generated. This is fundamentally different to ground anchors, which are tensioned during construction and provide an active rather than a passive system. The tensile forces are generated by relatively small movements. During the experimental period of the early 1990s consideration was given to using soil nails as dowels to provide shear strength across potential movement planes. However, it was found that in practice excessive deformation is required before a significant proportion of the potentially available shear strength can be mobilised to support the soil (Jewell and Pedley 1992). Therefore, only the tensile forces are used in the design of soil nails.

The ground in the active zone is prevented from slipping off the soil nails by a combination of the bond strength between the soil nail and the surrounding ground in the active zone and the action of the load plate, which provides a bearing force on the ground around the nail. This bearing force may be enhanced by the action of a facing.

74.4. Suitability of ground conditions

A wide range of soil types are suitable for soil nailing. Considerations of different soil types are provided in BS 8006. Some general points for guidance are provided below.

Soil nailing is generally suitable for most cohesive soils with an undrained shear strength of 50 kN/m^2 or greater. If boulders are present in glacial till these will present an obstruction to drilling. This can usually be overcome by relocating soil nails if problems are encountered. Clays with an undrained shear strength of less than 50 kN/m^2 may not be sufficiently self-supporting to remain stable during construction. High-plasticity clays tend to be prone to shrinking and swelling and may not be suitable. Granular soils provide good long-term strength but there may be difficulties with temporary stability. Self-drilling nails are likely to be required unless bores are cased. Also, subject to the amount of fine material present in the soil it may be difficult to construct stable temporary slopes. Weak rocks are generally suitable, although careful consideration of bedding planes will be required. Drilling through intermittent hard and soft layers can cause difficulties.

Groundwater is a very important consideration. Problems can arise where there are elevated groundwater levels, including perched water tables and spring lines. In these situations, the designer should pay close attention to the incorporation of temporary and permanent drainage, including slope drainage, toe drains and crest drains.

74.5. Types of soil nail

Soil nails are installed using one of the following methods.

- Drilling of a sub-horizontal hole followed by the insertion of a bar and grouting of the hole using a tremie tube, starting from the base of the hole and working outwards. Spacers are placed around the bars to ensure that the bar is placed centrally within the bore. This method, which is generally known as 'bored and grouted', is the most economic where the ground conditions are suitable – for example, within firm to stiff cohesive soils in which collapse of the hole is generally unlikely. However, care must be taken with this method when there is a delay between the drilling and grouting of a hole resulting in the undetected collapse and blockage of the hole, leading to only partial grouting of the hole.
- Simultaneous drilling and grouting whereby a hollow high-yield steel bar is used as the drill rod with a sacrificial drill bit to form the hole by percussive drilling. During the drilling process the grout is pumped through the hollow centre of the bar and the hole drilled to the required depth. Grouting continues until clean grout emerges at the surface. These nails are generally referred to as 'self-drilling'. This method is very rapid and can be used in most types of soil. The grout supports the bore at all times, preventing collapse of the hole.
- Driving a soil nail into a slope using ballistic methods. This method has become very rare and is not discussed further.

74.6. The design of soil nail systems

Soil nails are usually designed following the limit state principles of BS EN 1997-1:2004. Further recommendations and guidance are provided in BS 8006:2011+A1:2017. This guidance will be supplemented in the near future by BS EN 1997-3.

The modes of failure and serviceability limit states for stabilised slopes normally considered are described and illustrated in BS 8006. They are

- rotational failure through the stabilised soil block
- deep-seated rotational failure exploiting the soil behind the soil nails
- sliding of the reinforced soil block
- failure at the interface between the grouted annulus and the surrounding ground, resulting in pull-out of soil nails from the resistant zone
- failure at the interface between the bar and the surrounding grout
- failure of the facing
- failure of the soil bar.

Analysis is usually undertaken using 2D, called limit state analysis. Numerical analysis is sometimes used to model complex situations or where it is necessary to utilise a more rigorous approach to estimating deformations. Generally, most of the analysis is undertaken using proprietary slope stability software. The designer will usually compare different combinations of soil nail spacing and length before the most efficient solution is found. Figure 13 in BS 8006 provides some indicative soil nail spacings and lengths, together with suggested types of facing for slopes with a range of steepnesses. This guidance can provide a suitable starting point.

The design approach for improving the stability of soil-nailed walls is similar to that for soil-nailed slopes. However, careful consideration is required with regard to the connection and load transfer between the wall and the soil nails. Also, the fact that movement is required before resistant tensile forces are generated in the nails could be a more significant factor for designs relating to walls than it is for slope design. Therefore, stabilisation of walls is best suited to walls where there is a pragmatic acceptance that a degree of movement will occur. For example, this approach was used successfully on the stone wall supporting the N71 'Ring of Kerry' National Road, close to Moll's Gap in Ireland, as can be seen in **Figures 74.3** and **74.4**.

74.6.1 Selection of soil strength parameters

When selecting soil strength and groundwater parameters, the designer must consider the likely long-term future ground conditions. Selection of parameters for the overall stability of the slope should be considered in much the same way as a typical earthworks slope. Effective stress parameters are usually

Figure 74.3 Soil nailing at Moll's Gap, Co. Kerry (With kind permission of Atkins and Transport Infrastructure Ireland)

Figure 74.4 Soil-nailed retaining wall at Moll's Gap, Co. Kerry (With kind permission of Atkins and Transport Infrastructure Ireland)

used to model the long-term stability of the slope. Undrained parameters are generally only appropriate for temporary works where the duration is short enough for the soil suction which gives rise to the selected undrained strength to be relied upon.

BS 8006 gives the designer the freedom to choose drained or undrained parameters for determining the ultimate stress in the bond between the grouted nail and the surrounding soil. Empirical methods and test data are also allowed. In contrast to long-term soil strength associated with slope stability at shallow depths, the choice of undrained parameters may be a reasonable proposition when evaluating bond strength. This is because the distal part of the nail will be deep in the ground, where factors such as stress relief and exposure – which lead to reduction of strength – are likely to be less prevalent than in the shallow ground near the surface of the slope. Similarly, where effective stress parameters are used there may be a stronger argument for using effective cohesion as a component of bond strength than there would be for shallow soil strength. The use of effective cohesion in earthworks design is discussed in the *Earthworks Guide* (Nowak and Gilbert, 2015). Whichever parameters are used, the designer must ensure they are reliable long-term values.

74.6.2 Inclination of soil nails

There are a number of factors that influence the optimum soil nail angle. A detailed discussion is provided in TRL 537 (Johnson *et al.*, 2002), and the conclusion is that for most slopes the optimum inclination is generally around 10–15° to the horizontal. As noted in TRL 537 and in BS 8006, the resistance forces of the soil nail diminish significantly as the downward angle increases beyond 15°.

74.6.3 Practical considerations and buildability

Careful consideration of the site constraints and how the soil nails will be constructed should be an integral part of the design process. This will facilitate safe and cost-effective construction. In some situations, it may be advantageous to obtain advice from a soil nailing contractor during the design process. Chapter 88 *Soil nailing construction* describes how soil nails are constructed. Some construction-related issues which are frequently important factors for the designer to consider are discussed herein.

Ease of access may influence the construction methods. In some circumstances this could influence the length and diameter of bore that can be achieved. For example, if access constraints and slope geometry are such that rope access and suspended rigs are required, then the length and diameter is likely to be less than what could be achieved with standard drilling rigs or using a drilling mast mounted on an excavator. The cost of construction is also likely to be affected. Arrangements to facilitate access, such as purchasing land, negotiating wayleaves or clearing vegetation are best considered early. If the designer is involved in the early stages of the project, they may be able to influence appropriate early action by the client to secure good access and achieve better value for money.

Generally, newly cut soil-nailed slopes are constructed in rows from the top down. The first step is excavation of the top section of slope, typically to a depth of 1–1.5 m to form a platform from which the nails can be installed. Alternatively, the nails can be installed using long-reach plant from either the crest or toe of the slope. Once the first row of nails has been installed and the grout has reached its required strength, excavation of the next layer of the slope is carried out and the second row of nails is installed. The process is then repeated until the whole excavation is complete. The required facing and securing plates may be placed over the full height, or may be placed in sections from the top down such that the period during which the ground between the nails is unsupported is minimised. When this top-down approach is used, the soil nails in the upper part of the slope are brought into tension as the lower sections are excavated. Thus, on completion of the works the nails are likely to be in tension. This will reduce the potential for deformation of the slope face after construction.

Sometimes, as an alternative to top-down construction, the slope is trimmed to its finished profile over its full height. This approach generally requires the soil to be self-supporting over a greater height. It is only appropriate where the soil is suitably stiff and the period for which the ground is unsupported is short. Where this approach is used, the benefit of putting the nails into tension during construction will not be gained and so there is a greater propensity for long-term displacement of the slope face than where top-down construction is used. Furthermore, the risk of slope movement during construction is higher than for the top-down method. Excessive movement will have an adverse effect on the quality of the finished works. Examples of problems with temporary slope instability are provided in TRL PPR1033 (Duffy-Turner *et al.*, 2022). This approach is generally only suited to low-height slopes, and the work is usually carried out in short bays to control the risk of deformation. The designer should opt for top-down construction unless there is a good reason not to.

The soil must be sufficiently self-supporting that it can stand up unsupported until it is nailed and the facing is placed. The designer should satisfy themselves that this short-term stability requirement is a reasonable expectation before choosing a soil nail-based solution. In addition to concerns about safety and buildability the designer should consider whether the potential for movement of the face during construction might have an adverse effect on the long-term quality of the work. The designer may wish to specify a maximum bench height. Practical factors may influence the maximum bench height. For example, where sheets of steel mesh are used as a permanent facing, it would facilitate efficient construction if the temporarily unsupported face height were a function of the smallest dimension of the mesh sheets.

Construction plant will require a safe working platform. The allocation of responsibility for designing these platforms may vary between projects and the detailed design of the platforms may only be possible when the contractor has decided upon the plant to be used. However, in general the designer should at least satisfy themselves that plant will be able to work safely.

The height of the lowest row of soil nails should be considered in relation to drilling equipment. Tracked drilling rigs generally require around 0.5 m clearance from nail entry point to the top of the construction platform, although this may be less for some rigs. Further clearance from the base of the slope may also be required to allow room for the construction platform unless this is embedded into the adjacent ground.

The designer must allow for construction tolerances. This is generally done by considering the most adverse values allowed within the construction tolerances in the design calculations. Recommendations on positional tolerance and nail inclination are provided in BS 8006. Further guidance, based on CIRIA C637 (Phear *et al.*, 2005), is provided in Chapter 88 *Soil nailing construction*. As noted in Chapter 88, these values are intended for guidance, and the designer should consider the way in which tolerances might influence the costs of construction and whether or not there is a justifiable benefit to the client. Obstructions can result in a requirement for changing nail positions or adding extra nails such that the design spacing is not exceeded. The designer may have a preference as to how this is dealt with. This can be communicated through the construction documents.

The designer should consider the potential influence of the construction process on the position of the bar within the bore, the size and shape of the grouted annulus and the condition of any protective coatings. When soil nails are installed using bored and grouted techniques, the process of ensuring a central position of the bar within the grouted bore is relatively straightforward to control and to verify. When soil nails are installed using self-drilling techniques the process is more difficult to control and to verify. Part of the process of constructing self-drilling nails is to surge the hollow bar backwards and forwards while injecting grout through the hollow bar. The aim of this action is to flush out and replace the grout, which has served its temporary purpose of supporting the bore but has become mixed with soil during drilling and is not suitable for the permanent works. The surging process can cause attrition damage to any protective coating on the nails, particularly if the centralisers rub up and down the bar. Also, surging can result in loss of centrality of the bar, particularly if the centralisers dig into the sides of the bore. Noting that self-drilling nails tend to be used in ground which is not self-supporting, there may be uncertainty about the shape of the bore and the position of the bar within it at the end of this process. There is no practical way of checking the finished construction other than testing the completed nails.

While centralisers have obvious benefits, they can impede the flow of the grout. When designing the nails and specifying centralisers the designer will need to weigh the importance of centralising the bar against potentially increasing the difficulty flushing the bore. TRL PPR1033 (Duffy-Turner *et al.*, 2022) provides examples of severe problems where spacers have not been used.

There are actions the designer can take to mitigate the design risks associated with these practical issues.

- Avoid relying on protective coatings if it is likely that they will be damaged during construction. Alternative measures such as use of a larger-diameter bar may be more practical than protective coatings.
- Increasing the diameter of the bar will make it less prone to sagging between spacers, and so a thicker bar may be advantageous where there are fewer centralisers. However, sagging is not the only factor that affects centralisation. Use the specification to ensure that spacers are properly secured.
- Consider limiting coatings to the proximate end of the nail. It is common for higher-quality protection to be required at the proximate end of the nail, which is more exposed to the atmosphere. The centraliser at the proximate end is usually added after insertion of the nail and thus does not impede grout. The coating on this important section will generally be less prone to damage.
- Limit the degree of reliance on the grout for corrosion protection if there are doubts about the quality of encapsulation.
- If there are doubts about the competency of the grouted annulus or the encapsulation of the bar within it, consider reducing the expected size of the grouted annulus when calculating the pull-out resistance of the soil nail.
- Contractor engagement may help to ensure that the specified requirements are not excessively difficult to achieve.
- Provide a suitably rigorous testing regime.
- Try to ensure provision of good auditing of construction.

74.6.4 Load plates

Load plates have three purposes: they secure the facing; they provide a bearing force which, together with friction along the part of the nail in the active zone, prevents the soil from falling away from the soil nails; and they encourage soil arching across the spaces between soil nails, which reduces surface instability.

BS 8006 provides a method for calculating the required plate size. This is based on the method contained in Appendix E of HA 68/94 (Highways Agency, 1994). This method is likely to be conservative for a few reasons. The section of the nail away from the slope surface tends to experience the highest load and the nail head itself may not experience the full load except in the unlikely event of approaching slope failure. The load transferred to the nail head generally increases as the slope angle increases, and the use of larger plates for shallower slopes, as in the Highways Agency guide, is likely to be conservative.

Load plates may be mounted on the surface of the slope or cut into the slope. In any case, it is important that they are in good contact with the face of the slope such that they can provide a bearing force. To obtain good contact it is good practice to fill up the void that invariably occurs at the outside of the bore where the level grout meets the slope, as illustrated in **Figure 74.1**.

Generally, plates of dimensions up to 300 × 300 mm can be easily handled. The use of larger plates poses problems for installation due to the weight of the plates, such as manual-handling issues. Designing to allow for contribution from the facing as described below can help to limit the required plate size.

74.6.5 Slope facing

Facing supports or protects the soil between the soil nails. BS 8006 categorises facings into three types; soft, hard and flexible. Guidance on the design of facings is provided in BS 8006 and CIRIA C637 (Phear *et al.*, 2005). Figure 13 in BS 8006 provides an indication of the range of slope angles over which these facing types are typically used. The worked examples in the appendices to CIRIA C637 include facing design.

Soft facings may be used for various reasons, such as erosion protection or promoting the growth of vegetation. They are not considered to provide structural support and should only be used on slopes which can be relied upon to be self-supporting. BS 8006 recommends that the upper end of the range of slope inclination for soft facings is 45°. Care should be taken towards the upper end of this range because in the long term cohesive soils are not likely to be self-supporting at such steep gradients. Enlarged plates might help or it may be appropriate to consider a flexible facing. The slope may be covered in topsoil for gradients up to around 30° or hydroseeded for steeper slopes. However, where the slope is too steep and topsoil cannot be applied directly to the slope, a cellular topsoil retention system can be included in the design. This consists of a cellular geogrid, which is pinned to the slope. The cells are filled with seeded topsoil of sufficient depth to bury the nail heads. This approach will prevent the potential hazard of the nail heads protruding from the slope.

Hard facings provide a rigid structural restraint between the soil nails. They are usually used on slopes steeper than 70° and may be used on slopes up to 90°. They may comprise reinforced concrete, cribwork or gabions. Sprayed concrete (shotcrete) has been the most common hard facing material in the past. However, the appearance may be a disadvantage. It usually incorporates one or two layers of mesh, with the lowest layer of mesh secured by the load plates. While load plates may provide a practical way of securing the mesh, they are not necessarily needed for the permanent works. Provision of weep holes and back-of-wall drainage should be considered.

Flexible facings provide a structural restraint between the soil nails and allow a degree of movement. They are typically used on slopes with angles of up to 70°. Flexible facings can be designed in two ways. They can be designed as 'simple flexible facings' in which the facing is designed to support the soil between the soil nails, but the load plate is designed with no contribution from the facing. Alternatively, they can be designed as 'complex flexible facings' in which the load plate is designed with allowance for some contribution towards its capacity from the action of the facing. BS 8006 recommends that this approach is limited to slopes with a maximum gradient of 60°.

Flexible facings are usually constructed from woven or welded meshes. Welded meshes are generally much stiffer than woven meshes and deflect less. Caution is required when designing facings using woven meshes or any facing that is very flexible. Invariably slopes requiring soil nailing will be at gradients for which the soil requires support. The soil is likely to deform until it pushes the facing into tension. This could result in the front face deforming and taking on a 'studded settee' appearance, as can be seen in **Figure 74.5**. The amount of deflection will be a function of the flexibility of the mesh and how securely it is fixed against the face of the slope. Excessive deflection could result in softening of the soil at the slope face or could cause high stresses in the wire around the edge of the load plate. BS 8006 recommends that horizontal deflection remains within serviceability limits and does not exceed 2% of the diagonal distance between soil nails – that is, around 40 mm for a soil nail on a 1.5 m square grid. In practice this value may be hard to achieve with a woven wire mesh. To avoid ambiguity regarding what deflection is allowable it would be good practice for the maximum deflection to be stated in the works information. The designer should also check that the proposed use of the product is within the manufacturer's expected parameters, which are usually provided on the manufacturer's certification. This could affect the applicability of manufacturer certifications or warranties.

Welded mesh facings typically comprise one or two layers of square aperture mesh. This is the type of mesh that is typically used to reinforce concrete slabs. The mesh can be up to around 12 mm dia. and thus offers a high level of rigidity, and can also be designed with an allowance for loss of section due to corrosion. A backing geotextile is sometimes

Figure 74.5 Slope face deformation

used to prevent loss of soil through the apertures. It can be faced with systems which improve the appearance and cover the load plates and nail heads. This can comprise gabion-type baskets filled with coarse granular material such as can be seen in **Figure 74.6**. Another popular option is to use a green facing comprising durable bags infilled with soil. The bags can be impregnated with grass seed prior to placing or can

Figure 74.6 Steel mesh basket facing under construction

Figure 74.7 Bagwork facing on River Leven (With kind permission of Atkins and Stockton-on-Tees Borough Council)

be hydroseeded retrospectively. Specialist advice should be sought when considering green solutions with due consideration of environmental factors such as exposure to sunlight, moisture and slope gradient. **Figure 74.7** shows this type of facing.

74.7. Drainage

If not properly understood and controlled, groundwater has the potential to adversely affect the construction or long-term performance of the soil nail system. A good understanding of groundwater conditions and an appropriate drainage system is a critical part of the design of any earthworks or earth-retaining system. It is particularly important for soil-nailed slopes because the soil nail system comprises several elements which are sensitive to water ingress. For example, the soil nail bores can provide a conduit for water. This could cause instability of unsupported bores or washout of grout during construction, or it could adversely affect the soil nail–soil bond in the long term. Water can cause deformation of flexible facings due to softening or washout of soil. Water trapped behind concrete facings can cause a build-up of pressure, leading to cracking of the facing; it can also cause damp patches.

As with the design of any slope, consideration should be given to the provision of surface water interceptors if there is potential for flow onto the site from the adjacent ground. Toe drains may be useful for drawing down the groundwater level and/or providing a means to take away water collected from other parts of the drainage system. Various proprietary systems are available for providing drainage behind hard facings. These generally comprise a void former encapsulated in a geotextile and are usually connected to the toe drain. These can be seen, rolled up, during construction in **Figure 74.3**.

Raking drains are very well suited to soil nail construction. These typically comprise a slotted pipe installed within a bored hole sloping at 5–10° down from the horizontal and connected to a weep hole in the facing. They are usually constructed by the same drilling plant that is used to bore the soil nails. Therefore, the activity may be easily arranged at relatively low cost. The drains can be drilled to a similar length to the soil nails and thus improve the whole of the soil-nailed block of ground. They can be evenly spaced or can be targeted at suspected wet areas. Some indicative details of drainage systems are provided in CIRIA 637 (Phear *et al.*, 2005) and BS 8006.

Consideration should be given to the longevity of the drainage system and any requirements for long-term maintenance. Any drainage system concealed behind a hard facing should include a means of maintenance (e.g. access points where the inclined drains could be rodded). Such requirements should be incorporated in the safety file and/or operating manual. If there is no way of inspecting or maintaining any elements of the drainage there is a risk that they could fail to function properly in the future. In such circumstances it would be prudent not to rely on these elements for long-term stability assessments, although they could potentially be incorporated to provide a general benefit. The risk of drains deforming or becoming blocked and having an adverse effect on stability in the future should be accounted for. This risk is particularly important on sites where ground movement has been a historical problem.

74.8. Corrosion of soil nails

The design of corrosion protection for soil nails depends on several factors, including the specified design life, the aggressivity of the ground and the soil nail installation method. The typical design life for most soil-nailed slopes is 60 years; however, where soil nails are incorporated into other structures, such as retaining walls, a design life of 120 years is sometimes specified. Guidance on the aggressivity of the ground is given in TRL Report 380 (Murray, 1993), which also recommends thicknesses for sacrificial steel that may be incorporated as a method of corrosion protection. This allows the diameter of the steel to be designed so as to give sufficient corrosion protection over the design life and also to provide sufficient structural strength. However, BS 8006 recommends that for aggressive ground conditions or where corrosion is considered to be a significant risk, loss of section should not be the only means of addressing corrosion. For aggressive ground conditions a further layer of protection may be provided, such as galvanisation or epoxy coatings. Alternatively, a plastic duct may be placed within the upper part of the grout column around the steel bar, as the corrosion potential is likely to be greatest in the outer part of the soil nail – that is, within the zone that is most exposed to the environment and likely to be subjected to seasonal variations in the groundwater level.

74.9. Construction specifications

Clear and detailed construction drawings and a construction specification are crucial to ensuring that the design is constructed according to the designer's intentions. There is no standard published 'off the shelf' specification for the construction of soil nails. The designer is therefore required to produce their own bespoke specification. For projects which are to be constructed in accordance with the *Specification for Highway Works* this specification would sit within Appendix 6.10. BS EN 14490:2010 is a useful guide to designers producing soil nailing specifications. It provides useful guidance on the information needed to construct the works and it specifies various requirements relating to construction, materials and testing. The designer's specification would normally cross-reference this document and add further project-specific requirements.

74.10. Auditing of construction

Soil-nailed slopes are relatively complex systems comprising several integrated parts. Technical auditing of the construction for compliance with the design and production of construction records – including compliance checks, testing and records relating to any departures from the design – are important. It is essential that this role is allocated to a suitable party in advance of construction. A representative of the designing organisation is usually appropriate. It is a good idea to produce a check-sheet which will facilitate systematic checks during construction. BS EN 14490:2010 provides a helpful list of suggested quality control checks.

74.11. Testing soil nails

Three principal types of soil nail tests are described in BS EN 14490:2010. They are categorised according to their purpose.

- *Design investigation tests* are an option for obtaining information ahead of design. They are carried out by testing sacrificial nails to failure. They are used to investigate the ultimate bond strength between the soil nail and the surrounding ground. They are not a mandatory requirement for any structure.
- *Suitability tests* are undertaken to verify the bond strength used in the design. They are carried out by testing sacrificial nails to failure. They are mandatory for category 2 structures, except where directly applicable experience exists, and they are mandatory for category 3 structures in all circumstances. They may be undertaken ahead of or during the construction works.
- *Acceptance tests* are undertaken to demonstrate satisfactory performance at the acceptance load. They may be carried out on specially installed sacrificial nails or on working nails. They are mandatory for category 2 and 3 structures.

BS EN 14490:2010 indicates the minimum numbers of tests for various situations. Soil nail tests do not provide an exact representation of the failure of a soil-nailed block of soil, and careful interpretation is required. One particularly significant factor to consider when evaluating test results is that the bond strength may be designed using effective stress parameters to model likely future conditions. However, the pull-out test will generate an undrained response which will give a different, usually higher, result in clay soils. Commentary on the interpretation of soil nail testing is provided in TRL Report 537 (Johnson *et al.*, 2002) and CIRA Report C637 (Phear *et al.*, 2005). CIRIA C637 recommends undertaking separate tests to measure the bond strength in the active zone and the resistant zone. This can be achieved by testing short nails to represent the bond strength in the active zone and full-length nails, debonded through the active zone, to represent the bond strength in the resistant zone.

74.12. Inspection and maintenance

Soil-nailed structures generally require little maintenance. The nails are passive and are not reliant on maintaining a specific amount of tension. Therefore, routine load testing after construction is not usually required.

Sometimes, for complex structures, structures which are at high risk or if long-term checks on expected performance are required, a system of testing and/or inspection could be put in place. The requirements would depend on the particular risks and concerns but could include inspection/checks of the slope facing to check for any signs of bulging, drainage, abnormal vegetation growth; and checks for loose components such as load plates. Where nail heads are visible, monitoring by precise surveying may be carried out. Guidance is provided in CIRIA Report C794 (O'Donovan *et al.*, 2020).

If load tests are required to check long-term bond strength, sacrificial nails could be installed adjacent to the structure, and these can be tested at intervals throughout the required design life.

References

BSI (British Standards Institution) (2004) BS EN 1997-1:2004: Eurocode 7: Geotechnical design. London: BSI, London, UK.

BSI (2010) BS EN 14490:2010: Execution of special geotechnical works – Soil nailing. BSI, London, UK.

BSI (2017) BS 8006-2:2011 + A1:2017: Code of practice for strengthened/reinforced soils: Part 2 – Soil nail design. BSI, London, UK.

Clouterre (1991) *French National Research Project Clouterre – Recommendations.* English translation, FHWA-SA-93-026. Federal Highway Administration, Washington, DC.

Duffy-Turner M, Nettleton IM, Winter MG and Webber I (2022) *Forensic Examination of Critical Special Geotechnical Measures: Soil Nails Information Note.* Report PPR1033. TRL, Wokingham, UK.

Highways Agency (1994) *HA 68/94 Design Methods for the Reinforcement of Highway Slopes by Reinforced Soil and Soil Nailing Techniques.* HMSO, London, UK.

Jewell RA and Pedley MJ (1992) Analysis for soil reinforcement with bending stiffness. *ASCE Journal of Geotechnical Engineering* **118(10)**: 1505–1528.

Johnson PE, Card GB, Darley P (2002) *Soil Nailing for Slopes.* Report 537. TRL, Wokingham, UK.

Murray RT (1993) *The Development of Specifications for Soil Nailing.* Research Report 380. TRL, Wokingham, UK.

Nowak P and Gilbert P (2015) *Earthworks: A Guide.* ICE Publishing, London, UK.

O'Donovan J, Feudale FE, Turner M, Irvin C and Mothersille D (2020) *Grouted Anchors and Soil Nails: Inspection, Condition Assessment and Remediation.* CIRIA Report C794. CIRIA, London, UK.

Phear A, Dew C, Ozsoy B, *et al.* (2005) *Soil Nailing Best Practice Guide.* CIRIA Report C637. CIRIA, London, UK.

It is recommended this chapter is read in conjunction with

- Chapter 69 *Earthworks design principles*
- Chapter 72 *Slope stabilisation methods*
- Chapter 88 *Soil nailing construction*

All chapters in this book rely on the guidance in Sections 1 *Context* and 2 *Fundamental principles.* A sound knowledge of ground investigation is required for all geotechnical works, as set out in Section 4 *Site investigation.*

Chapter 75
Earthworks material specification, compaction and control

Philip G Dumelow Balfour Beatty, London, UK

This chapter describes how an effective earthworks specification can be produced for a project. It will define permitted material types, placement methods and appropriate means of control. The objective is earthworks which are safe during construction and use, durable and economically viable. Sustainable construction is a significant consideration.

doi: 10.1680/icemge.66830.1225

CONTENTS

75.1. The earthworks specification

75.1.1 Necessity and purpose

It is essential that a materials specification is produced for any earthworks contract, the core requirements being defined by the client. Whenever possible, all parties in the contract should be involved in preparing and developing the specification at the earliest stage. This will include the contractor and the designer, whether appointed by the client or the contractor. The nature of certain contracts encourages such a collaborative approach

- design and build (D&B)
- design, build, fund and operate (DBFO)
- early contractor involvement (ECI).

Client, designer and contractor collaboration utilises the broadest possible experience and knowledge of materials, geotechnics and construction methods to guide how the specification is written. The purpose is to deliver a project built to best practice, economically and safely.

The completed specification will define

- the client's core requirements for the works (e.g. levels, tolerances)
- methodologies for carrying out the works (e.g. excavating cuts, filling embankments and preparing formation)
- materials suitable for constructing the works (e.g. classes of material for specific purposes)
- material factors for physical, mechanical and chemical suitability (e.g. material parameters and limits for the required performance).

75.1.2 Current UK specifications

The earthworks specifications in widest use are produced by the Highways Agency (HA), and are discussed in the following.

1. **Manual of Contract Documents for Highway Works (MCHW)**

- Volume 1: *Specification for Highway Works* (SHW)
- Series 600: Earthworks (**Figure 75.1**)

Although this has been produced by the HA for their network of trunk roads and motorways, it is used by other clients. These include those in other infrastructures and disciplines. Use of the SHW is either in full or as the basis from which a specification is developed. Throughout this chapter, SHW and related HA publications will be used for reference; engineers working in other infrastructure disciplines must ascertain what specification is applicable to them.

The SHW has been developed from the Ministry of Transport's *Specification for Road and Bridge Works*, in use from the 1960s to 1983, when it was replaced by the SHW. It is freely available, in all its parts, Series 000 to 3000 (generally incrementing by 100 for each series) plus appendices, as a download from the HA's website.

The SHW refers to British Standards (BS) and the European Standards (BS EN) which replace them. These are not numbered the same as the superseded BS, and the requirements are often very different. Prior to beginning any earthworks project and planning any of the works, an engineer must know which specifications and standards apply to the contract. These must be acquired and the requirements understood. Not to do so is inviting failure.

In addition, SHW references publications other than BS and BS EN; these are listed in Appendix F of SHW, 'Publications Referred to in the Specification'. It is the responsibility of the engineer to be in possession of the relevant publications.

Contract-specific requirements are defined in the numbered appendices prepared by the specification's compilers.

Accompanying and supporting the SHW are the following HA publications.

2. **Manual of Contract Documents for Highway Works (MCHW)**

- Volume 2: *Notes for Guidance on the Specification for Highway Works* (NG)
- Series 600: Earthworks (**Figure 75.2**)

The clauses mirror those in SHW. The purpose of the NG is to assist in preparation of the contract; this includes commenting on

MANUAL OF CONTRACT DOCUMENTS FOR HIGHWAY WORKS
VOLUME 1 SPECIFICATION FOR HIGHWAY WORKS

SERIES 600
EARTHWORKS

Contents

NATIONAL ALTERATIONS OF THE OVERSEEING ORGANISATIONS OF SCOTLAND, WALES AND NORTHERN IRELAND

Scotland

Northern Ireland

\# denotes a Clause or Sample Appendix which has a substitute National Clause or Sample Appendix for one or more of the Overseeing Organisations of Scotland, Wales or Northern Ireland.

Amendment - November 2006 1

Figure 75.1 Highways Agency SHW 600 Series (Reproduced from Highways Agency, *Manual of Contract Documents for Highway Works*, Volume 1: *Specification for Highway Works*, Series 600. © Crown Copyright 2017)

MANUAL OF CONTRACT DOCUMENTS FOR HIGHWAY WORKS
VOLUME 2 NOTES FOR GUIDANCE ON THE SPECIFICATION FOR HIGHWAY WORKS

SERIES NG 600
EARTHWORKS

Contents

NATIONAL ALTERATIONS OF THE OVERSEEING ORGANISATIONS OF SCOTLAND, WALES AND NORTHERN IRELAND

\# denotes a Clause or Sample Appendix which has a substitute National Clause or Sample Appendix for one or more of the Overseeing Organisations of Scotland, Wales or Northern Ireland.

Amendment - November 2005 1

Figure 75.2 Highways Agency NG 600 Series (Reproduced from Highways Agency, *Manual of Contract Documents for Highway Works*, Volume 2: *Notes for Guidance on the Specification for Highway Works*, Series NG 600. © Crown Copyright 2017) the technical background underlying the SHW requirements and listing the appendices which must be produced for the contract (compare with 'Scope of the specification' below). Engineers beginning an earthworks contract must read the NG in conjunction with SHW for a full understanding of the specification.

3. **Manual of Contract Documents for Highway Works (MCHW)**

- Volume 1: *Specification for Highway Works* (SHW)
- Series 100: Preliminaries
- Series 500: Drainage and service ducts

The engineer must be aware that other series within the SHW are related to earthworks activities (e.g. Series 500 specifies the requirements for drainage material).

The most significant of these is Series 100, clause 105. This requires the production of Appendix 1/5, which includes Table 1/5, defining the frequency at which all classes and types of material used in the works are to be sampled and tested. This includes all classes of earthworks materials.

75.1.3 Scope of the specification

The specification compilers will prepare the numbered appendices defining earthworks requirements for the contract.

App No.	Appendix title
6/1	Requirements for acceptability and testing, etc. of earthworks materials
6/2	Requirements for dealing with Class U1B and Class U2 unacceptable materials
6/3	Requirements for excavation, deposition, compaction (other than dynamic compaction)
6/4	Requirements for Class 3 material
6/5	Geotextiles used to separate earthworks materials
6/6	Fill to structures and fill above structural foundations
6/7	Sub-formation and capping and preparation and surface treatment of formation
6/8	Topsoiling
6/9	Earthwork environmental bunds, landscape areas, strengthened embankments
6/10	Ground anchorages, crib walling and gabions
6/11	Swallow holes and other naturally occurring cavities and disused mine workings
6/12	Instrumentation and monitoring
6/13	Ground improvement
6/14	Limiting values for pollution of controlled waters
6/15	Limiting values for harm to human health and the environment
1/5	Testing to be carried out by the contractor

Table 75.1 Numbered appendices required for an earthworks specification

Sample appendices for an earthworks specification are listed in NG 600 Series (see **Table 75.1**). These inform the compiler what should be included in each appendix. The specification compiler must also prepare an Appendix 1/5 and Table 1/5. Advice on test frequency is provided by Table 1/1 in SHW NG. Similarly to the overall specification, this compilation team should include client, designer and contractor to include the broadest knowledge and experience.

Some of the appendices are associated with tables with the same numbering. For site control of earthworks, the most important of these are

- Table 6/1: Materials permitted in the works, acceptability limits, compaction method
- Table 6/2: Grading limits for each material class
- Table 6/4: Method compaction for earthworks materials: plant and methods
- Table 1/5: Testing to be carried out by the contractor.

Extracts from a completed Table 6/1 are shown in **Figures 75.3(a)** and **75.3(b)**, and from Table 1/5 in **Figure 75.3(c)**.

The main material classes in Table 6/1 are shown in **Table 75.2**. Each class may be subdivided by a lettered suffix to describe differing material types (wet, dry, coarse, fine) or end use, especially for Class 6 materials. Class U1 material may be reverted to an acceptable class by treatment to modify its physical characteristics. Class U2 is unacceptable due to chemical contamination. This class may also be subdivided into material which can be treated to be acceptable and material which, either by the nature or concentration of contamination, must be removed from site to an appropriately licensed facility.

75.1.4 Key requirements for an effective specification

To produce the specification, the compilers must refer to and draw data and information from different sources. These will include the following.

- The client's core requirements: for an SHW type, this is SHW & NG 600 Series.
- Ground investigations (GI) carried out for the project: The engineer must be aware that other series within the SHW are related to earthworks activities (e.g. Series 500 specifies the requirements for drainage material)
- Reports by other bodies: generally, these are by the Transport Research Laboratory (TRL) or the Building Research Establishment (BRE) for specific material types and conditions.
- Personal knowledge and experience: these are especially relevant if local materials merit specific consideration, including local knowledge of recycling and secondary materials sources.

The complete earthworks specification, whether using the format of SHW or any other, must deliver

- clear, unambiguous statements on how the earthworks will be carried out

APPENDIX 6/1 REQUIREMENTS FOR ACCEPTABILTY AND TESTING ETC. OF EARTHWORKS MATERIALS

(a)

CLASS	GENERAL MATERIAL DESCRIPTION	TYPICAL USE	PERMITTED CONSTITUENTS (All subject to requirements of Clause 601 (SHW) and Appendix 6/1)	MATERIAL PROPERTIES REQUIRED FOR ACCEPTABILITY (in addition to requirements on use of fill materials in Clause 601 (SHW) and testing in Clause 631 (SHW))				COMPACTION REQUIREMENTS IN CLAUSE 612 (SHW)
				PROPERTY (See exception in previous column)	DEFINED AND TESTED IN ACCORDANCE WITH:-	ACCEPTABLE LIMITS WITHIN:		
						Lower	Upper	
2A	Wet Cohesive material	General fill	Any material, or combination of materials, other than chalk For details of properties (iv) and (v) in the next column, refer to Notes 12 and 13.	(i) grading	BS 1377 Part 2	Tab 6/2	Tab 6/2	Table 6/4 Method 1 except for materials with liquid limit greater than 50, determined by BS1377: Part 2, only deadweight tamping or vibratory tamping rollers or grid rollers shall be used.
				(ii) plastic limit (PL)	BS 1377 Part 2	-	-	
				(iii) mc	BS 1377 Part 2	-	-	
				(iv) MCV	Clause 632	8	12	
				(v) Undrained shear strength	Clause 633	50 kN/m^2	100 kN/m^2	
2B	Dry Cohesive material	General Fill	Any material, or combination of materials, other than chalk. For details of properties (iv) and (v) in the next column, refer to Notes 12 and 13. For the upper limit of property (iv) refer to note 14.	(i) grading	BS 1377 Part 2	Tab 6/2	Tab 6/2	Table 6/4 Method 2
				(ii) plastic limit (PL)	BS 1377 Part 2	-	-	
				(iii) mc	BS 1377 Part 2	-	-	
				(iv) MCV	Clause 632	12	18	
				(v) Undrained shear strength	Clause 633	100 kN/m^2	-	

Version 4 Status: Construction

Page 3 of 20 Issue Date: July 2009

Figure 75.3 Extracts from (a) a typical Table 6/1 for cohesive material compiled for a specific project; (b) a typical Table 6/1 for granular material compiled for a specific project; (c) a table in a typical Appendix 1/5 compiled for a specific project (Extracts from Highways Agency, *Manual of Contract Documents for Highway Works*, Volume 1: *Specification for Highway Works*, Series 600 © Crown Copyright 2017) (continued on next page)

APPENDIX 6/1 REQUIREMENTS FOR ACCEPTABILTY AND TESTING ETC. OF EARTHWORKS MATERIALS

(b)

CLASS	GENERAL MATERIAL DESCRIPTION	TYPICAL USE	PERMITTED CONSTITUENTS (All subject to requirements of Clause 601 (SHW) and Appendix 6/1)	MATERIAL PROPERTIES REQUIRED FOR ACCEPTABILITY (in addition to requirements on use of fill materials in Clause 601 (SHW) and testing in Clause 631 (SHW))				COMPACTION REQUIREMENTS IN CLAUSE 612 (SHW)
				PROPERTY (See exception in previous column)	DEFINED AND TESTED IN ACCORDANCE WITH:-	ACCEPTABLE LIMITS WITHIN:		
						Lower	Upper	
6N	Selected well graded granular material	Fill to structures	Natural gravel, natural sand, crushed gravel, crushed rock, crushed concrete, slag, well-burnt colliery spoil or any combination thereof. None of these constituents shall include any argillaceous rock. Recycled aggregate except recycled asphalt	(i) grading	BS 1377 Part 2 (On site)	Tab 6/2	Tab 6/2	End product 95% of maximum dry density of BS 1377: Part 4 (vibrating hammer method)
					BS EN 933-2 (Off site)	Tab 6/5	Tab 6/5	
				(ii) uniformity coefficient	See Note 5	10	-	
				(iii) Los Angeles coefficient	Clause 635	-	40	
				(iv) Undrained shear parameters (c and φ)	Clause 633	-	-	
				(v) effective angle of internal friction (φ') and effective cohesion (c')	Clause 636	35	45	
				(vi) permeability	Clause 640	$5x10^{-5}$ m/s	-	
				(vii) mc	BS 1377: Part 2	OMC - 2.5%	OMC +1%	
				(viii) MCV	Clause 632	-	-	

Version 4

Page 12 of 20

Status: Construction

Issue Date: July 2009

Figure 75.3 Continued

APPENDIX 1/5 TESTING TO BE CARRIED OUT BY THE CONTRACTOR

(c)

Clause	Work, Goods or Material		Test	Frequency of Testing	Test Certificate	Comments
Series 600						
601, 631 to 637 & 640	Acceptable material		For recycled aggregate, see sub-clauses 601.12 and 601.18. Cross-reference should be made to any requirements in Appendix 6/1			
	Class	General Description				
	1	General granular fill	Grading / uniformity coefficient	Twice a week	Required	
			MC (N)	2 per 1000m^3 up to max of 5 per day per source		
			OMC (N)	Once per week per source		
		1C only	Los Angeles coefficient (N)	Weekly		
	2	General cohesive fill	Grading	Twice a week	Required	
			MC (N)	2 per 1000m^3 up to max of 5 per day per source		
			PL/LL (N)	Weekly		
			MCV (N)	2 per 1000m^3 up to max of 5 per day per source		
			Undrained shear strength – Hand Shear Vane	2 per 1000m^3 up to max of of 5 per day per source		See Note 13 to Table 6/1
			Undrained shear strength of remoulded material (N)	2 per 5000m^3 up to max of 2 per week per source		See Note 13 to Table 6/1
	2E	Reclaimed pulverised fuel ash cohesive material	MC (N)		Required	
			Bulk Density (N)	-		
	4	Fill to Landscape Areas	Grading	Daily	Required	
			MCV (N)	Daily		
	5	Topsoil	Grading	Daily	Required	

Version 5 Status: C – Status

Page 9 of 46 Issue Date: September 2010

Figure 75.3 Continued

Class	General material description	Typical use
1	Granular material	General fill
2	Cohesive material	General fill
3	Chalk	General fill
4	Various materials	Landscape fill
5	Topsoil	Topsoiling
6	Selected granular material	Specific uses (e.g. starter layer, backfill, capping)
7	Selected cohesive materials	Specific uses (e.g. backfill, material for stabilisation)
8	Class 1, 2 or 3 material	Lower trench fill
9	Lime or cement stabilised granular or cohesive material	Capping (synonymous with 'foundation' in pavement design)
U1	Any type of material	Materials physically unacceptable for the works
U2	Any type of material	Materials chemically unacceptable for the works

Table 75.2 The main material classes in Table 6/1 (materials permitted in the works, acceptability limits, compaction method) of SHW

- the means by which the works can be carried out safely
- a durable end product to the client's satisfaction
- characteristics and limits for the acceptability of materials, without overdesigning
- methodologies for dealing effectively with specific materials and conditions
- the most sustainable earthworks solution.

From these derive the limits applicable to the test used for quality control of the project. Sustainable earthworks must be a key consideration. This can be addressed by the specification compilers aiming to

- minimise waste by balancing the cut with the embankment/landscape fill volumes
- maximise the reuse of the excavated materials to minimise import
- create opportunity for the optimum use of site-won material
- utilise locally available recycled and secondary materials
- improve poor-quality materials by treatment (e.g. modification or geosynthetics)
- involve the contractor in additional GI and materials trials.

Additionally, specification compliers must ensure, when producing Table 6/1 and 1/5, that artificial constraints are not imposed. This can occur when

- the contractor is not involved in specification compilation
- the contractor is required to design part of the permanent works.

The specification must not unnecessarily restrict the materials allowed. The contractor may have experience of the successful use of alternatives which deliver the same performance. These may be more sustainable or economic or the methodology may have a programme benefit, yet their use is not permitted.

Amending the specification to permit materials not included in the specification takes time, resources and cost. Furthermore, there may be contractual implications. A 'Departure from Standards' may need to be formally requested through the design protocol. Engineers must be aware that adoption of a Departure may affect the ownership of risk associated with the material or method concerned.

75.1.5 Specifying test frequency

As referred to by item (*h*) in Section 75.4, SHW NG advises on frequency of testing. For many materials – including most earthworks materials – the test frequency in NG Table 1/1 is only advisory. The reason for this is defined by the notes in the key at the foot of the table:

> Where materials are known to be marginal or if initial test results show them to be such, the frequency of testing should be increased. Conversely where material properties are consistently in excess of specified minimum requirements or well below specified maximum limits, then the frequency of testing should be reduced.

As discussed in Section 75.5, it is essential that the specification compilers understand the

- likelihood of the occurrence of non-conformance with specification
- implications of non-conformance on the structural integrity of the works.

Two examples:

- Primary aggregate Type I subbase is supplied from a major quarry, subjected to a mature quality assurance system and CE marked.
- 6N/6P backfill to a structure under a motorway is being supplied from multiple sources, processing to produce recycled aggregates.

In the first example the likelihood of occasional non-conformance is low. In this situation the deleterious effect on the works of, say, the product's grading analysis being 2–3% non-compliant on one of the coarser sieves is insignificant.

In the second example it is not the author's intent to imply that the use of recycled aggregate is less suitable – quite the contrary, recycled and secondary aggregates should be used wherever possible for sustainable construction – but a specification compiler potentially in this situation must consider:

- multiple sources are being used
- recycled material is produced from a variable feedstock

- possible structural failure from frequent non-conformance
- the nature and criticality of failure (e.g. slope instability, settlement).

Both compliant materials may be completely acceptable in performance terms and suitable for use at the locations, but to manage risk effectively it would be reasonable to expect material being used in the second example to be sampled and tested more frequently than that in the first.

75.2. Compaction

Compaction is the final action (other than trimming of the uppermost layer) in the sequence of earthworks events which began with excavation or import, followed by haulage to the point of deposition, tipping/discharge and spreading. Throughout these processes the material has been disturbed and loosened, has high air voids and is mechanically unstable.

Compaction is achieved by the application of a force to the surface of the spread material. On major earthworks projects this is usually in the form of a heavy roller, either towed or self-propelled and deadweight or vibratory. Where space is very restricted, or on small projects, point load plant may be used instead.

Compaction does not necessarily produce a 'dense' material; the term is subjective. The purpose of compaction is to produce a material which has, by packing the soil particles more closely together

- air voids lower than a required maximum

and/or

- dry density higher than a required minimum
- predictable settlement and consolidation characteristics within prescribed limits
- increased mechanical stability under load.

Further consequences may be, depending on the material type

- increased impermeability
- improved resistance to erosion
- greater strength or stiffness.

75.2.1 The mechanics of compaction

75.2.1.1 Terminology

- Moisture content: the mass of water in a soil as a percentage of the mass of dry soil
- bulk density: the mass per unit volume of soil including water
- dry density: the mass per unit volume of the soil excluding water, expressed by

$$\text{Dry density (Mg/m}^3\text{)} = \frac{[\text{Bulk density (Mg/m}^3\text{)} \times 100]}{[\text{Moisture content (\%)} + 100]} \quad (75.1)$$

Any earthworks material comprises solid particles, water and air voids. Compaction increases the packing of the solid particles and reduces air voids: water lubricates the soil particles and displaces air. From a dry condition, increasing water (or moisture) content increases the particle packing – measured as the dry density – and reduces air voids when a soil is compacted.

For any soil and applied compactive effort there is a condition when the water in the soil is just sufficient for the particle packing to be maximised and air voids minimised. This condition is the maximum dry density (MDD). The water content at which MDD is achieved is optimum moisture content (OMC).

At moisture contents below OMC there is neither sufficient water to lubricate the particles for maximum density nor to fill all the voids between the particles. Above OMC water displaces the solid particles. As a result, dry density decreases and, with increasing moisture content, the material becomes increasingly unstable (e.g. deforming more under load).

The relationship between moisture content, dry density and air voids is demonstrated by drawing a compaction curve. This is from a series of tests measuring the dry density, at varying moisture contents, achieved by compacting each specimen in an identical manner, to the method defined in BS 1377 (**Figure 75.4**).

MDD and OMC are a function of the compactive effort applied to any soil. Increasing the compactive effort raises MDD and lowers OMC. The compaction curve, while maintaining approximately the same shape, moves up and left. It does not approach the 0% air voids line more closely. A state in which air voids equal zero is not physically attainable with practical earthworks compaction methodologies.

MDD, OMC and the shape of the curve are influenced by physical properties of the soil:

- particle shape
- particle size
- range of particle sizes.

Generally, angular or fine soils have a higher water demand than rounded or coarse soils. Well-graded materials, where particle size distribution is over a wide range of sizes, tend to have steeper curves than uniform materials, with particle sizes over a small range. It would appear from this that the permutations for OMC/MDD would be infinite; however, general ranges can be outlined for different material types.

Material	OMC range (%)
Clay	25–35
Silty, sandy clay	20–30
Pulverised fuel ash	15–25
Fine sand	10–20
Coarse sand	5–15
Sand and gravel	5–10
Crushed rock	5–10

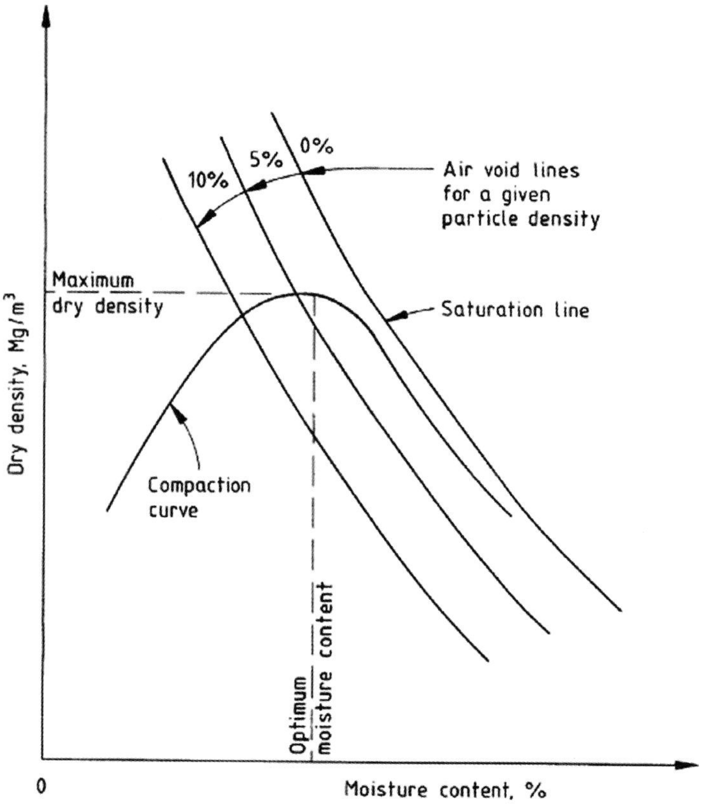

Figure 75.4 Dry density, moisture content and air voids relationship (Reproduced with permission from BS 1377-2. © BSI, 2022)

The compaction curve is used by the specification compiler to set the moisture content acceptability limits for a soil type or class. The lower limit should ensure that air voids are less than the maximum acceptable for the design. The upper limit should be before increasing moisture causes any significant instability.

75.2.2 The compaction specification

The compaction specification must define the requirements for each class of material allowed in the project, as defined in Table 6/1.

75.2.2.1 Method compaction

For each type of compaction plant and the amount of compactive effort imparted (e.g. mass per metre width of roll), the requirements are defined for

- the maximum thickness of the layer being compacted, when compaction is complete
- the number of passes (roller) or applications (point load plant).

Table 6/4 lists the permitted permutations for each material class and method number in Table 6/1 – see **Figure 75.5(a)** and **75.5(b)** for methods 1–6 (Method 7 is omitted). SHW permits plant and methods other than those in this table if the effectiveness of the methodology is proven by site trials.

Method compaction is generally used for bulk earthworks operations. The methods specified in Table 6/4 assume a conservative approach and have been determined to deliver maximum air voids of 5–10%, depending on the type of material and material class.

75.2.2.2 End product compaction

The required level of compaction is specified. In earthworks this is usually a minimum in situ density, expressed as percentage of the MDD of that material. This percentage is 90–100%, depending on the material type and class. The MDD is obtained from the laboratory test defined in Table 6/1; either

- BS 1377-2 (2.5 kg rammer method) – usually used for cohesive materials, or
- BS 1377-2 (vibrating hammer method) – usually used for granular materials; also imparts a greater compactive effort than the 2.5 kg rammer test.

The BS 1377-2 (4.5 kg rammer method) test is not used by SHW Table 6/1.

The type of plant used, the thickness of the final compacted layer and the number of passes are not specified. Selection of these is the responsibility of the contractor carrying out the compaction operation, guided by the compaction plant manufacturer's recommendations.

Table 6/1 also defines the maximum and minimum moisture contents at which a material is acceptable to be placed in the works and compacted.

Recent changes to SHW have incorporated BS EN standards and greater emphasis on suppliers to provide conformity compliance control. Performance-based specifications allow a wider range of materials (e.g. recycled and secondary aggregates) to be more widely used.

75.3. Compaction plant

Many types and sizes of compaction plant are available. The method compaction options in Table 6/4 permit many of these for each material class, although experience has proved some to be more effective than others. For end product compaction not all of these are suitable for all materials, locations and conditions. Selection of plant for end product compaction should be made with careful reference to the plant manufacturer's recommendations and, more importantly, an experienced earthworks manager.

The drum roller depicted in **Figure 75.6** is towed behind a tractor dozer, often the same plant used for spreading the tipped material. This plant can be obtained either as a deadweight roller or with a vibratory mechanism fitted.

This plant is normally deployed as the compaction plant where large volumes of Class 2 material (cohesive general fill) are to be placed and compacted to a method compaction specification.

Table 6/4: method compaction for earthworks materials: plant and methods (method 1 to method 6) (this table is to be read in conjunction with sub-clause 612.10)

(a)

Type of compaction plant	Ref no.	Category	Method 1		Method 2		Method 3		Method 4		Method 5		Method 6		
			d	n#	d	n#	d	n#	d	n	d	n	n for d = 110 mm	n for d = 150 mm	n for d = 250 mm
Smoothed wheeled roller (or vibratory roller operating without vibration)		mass per metre width of roll:													
	1	over 2100 kg up to 2700 kg	125	8	125	10	125	10*	175	4	unsuitable		unsuitable	unsuitable	unsuitable
	2	over 2700 kg up to 5400 kg	125	6	125	8	125	8*	200	4	unsuitable		16	unsuitable	unsuitable
	3	over 5400 kg	150	4	150	8	unsuitable		300	4	unsuitable		8	16	unsuitable
Grid roller		mass per metre width of roll:													
	1	over 2700 kg up to 5400 kg	150	10	unsuitable		150	10	250	4	unsuitable		unsuitable	unsuitable	unsuitable
	2	over 5400 kg up to 8000 kg	150	8	125	12	unsuitable		325	4	unsuitable		20	unsuitable	unsuitable
	3	over 8000 kg	150	4	150	12	unsuitable		400	4	unsuitable		12	20	unsuitable
Deadweight tamping roller		mass per metre width of roll:													
	1	over 4000 kg up to 6000 kg	225	4	150	12	250	4	350	4	unsuitable		12	20	unsuitable
	2	over 6000 kg	300	5	200	12	300	3	400	4	unsuitable		8	12	20
Pneumatic-tyred roller		mass per wheel:													
	1	over 1000 kg up to 1500 kg	125	6	unsuitable		150	10*	240	4	unsuitable		unsuitable	unsuitable	unsuitable
	2	over 1500 kg up to 2000 kg	150	5	unsuitable		unsuitable		300	4	unsuitable		unsuitable	unsuitable	unsuitable
	3	over 2000 kg up to 2500 kg	175	4	125	12	unsuitable		350	4	unsuitable		unsuitable	unsuitable	unsuitable
	4	over 2500 kg up to 4000 kg	225	4	125	10	unsuitable		400	4	unsuitable		unsuitable	unsuitable	unsuitable
	5	over 4000 kg up to 6000 kg	300	4	125	10	unsuitable		unsuitable		unsuitable		12	unsuitable	unsuitable
	6	over 6000 kg up to 8000 kg	350	4	150	8	unsuitable		unsuitable		unsuitable		12	unsuitable	unsuitable
	7	over 8000 kg up to 12000 kg	400	4	150	8	unsuitable		unsuitable		unsuitable		10	16	unsuitable
	8	over 12000 kg	450	4	175	6	unsuitable		unsuitable		unsuitable		8	12	unsuitable
Vibratory tamping roller		mass per metre width of a vibrating roll:													
	1	over 700 kg up to 1300 kg	100	12	100	12	150	12	100	10	unsuitable		unsuitable	unsuitable	unsuitable
	2	over 1300 kg up to 1800 kg	125	12	125	12	175	12*	175	8	unsuitable		12	unsuitable	unsuitable
	3	over 1800 kg up to 2300 kg	150	12	150	12	200	12*	unsuitable		unsuitable		8	12	unsuitable
	4	over 2300 kg up to 2900 kg	150	9	150	9	250	12*	unsuitable		400	5	6	10	unsuitable
	5	over 2900 kg up to 3600 kg	200	9	200	9	275	12*	unsuitable		500	6	6	10	unsuitable
	6	over 3600 kg up to 4300 kg	225	9	225	9	300	12*	unsuitable		600	6	4	8	unsuitable
	7	over 4300 kg up to 5000 kg	250	9	250	9	300	9*	unsuitable		700	6	3	7	12
	8	over 5000 kg	275	9	275	9	300	7*	unsuitable		800	6	3	6	10

Figure 75.5 Extracts from Highways Agency SHW 600 Series, Table 6/4 (Reproduced from Highways Agency, *Manual of Contract Documents for Highway Works*, Volume 1: *Specification for Highway Works*, Series 600. © Crown Copyright 2017) (continued on next page)

Table 6/4: method compaction for earthworks materials: plant and methods (method 1 to method 6) (this table is to be read in conjunction with sub-clause 612.10)
(b)

Type of Compaction Plant	Ref No.	Category	Method 1		Method 2		Method 3		Method 4		Method 5		Method 6		
			D	N#	D	N#	D	N#	D	N	D	N	N for D = 110 mm	N for D = 150 mm	N for D = 250 mm
Vibratory roller		Mass per metre width of a vibratory roll:													
	1	over 270 kg up to 450 kg	unsuitable		75	16	150	16	unsuitable		unsuitable		unsuitable	unsuitable	unsuitable
	2	over 450 kg up to 700 kg	unsuitable		75	12	150	12	unsuitable		unsuitable		unsuitable	unsuitable	unsuitable
	3	over 700 kg up to 1300 kg	100	12	125	10	150	6	125	10	unsuitable		16	unsuitable	unsuitable
	4	over 1300 kg up to 1800 kg	125	8	150	8	200	10*	175	4	unsuitable		6	16	unsuitable
	5	over 1800 kg up to 2300 kg	150	4	150	4	225	12*	unsuitable		unsuitable		4	6	12
	6	over 2300 kg up to 2900 kg	175	4	175	4	250	10*	unsuitable		400	5	3	5	11
	7	over 2900 kg up to 3600 kg	200	4	200	4	275	8*	unsuitable		500	5	3	5	10
	8	over 3600 kg up to 4300 kg	225	4	225	4	300	8*	unsuitable		600	5	2	4	8
	9	over 4300 kg up to 5000 kg	250	4	250	4	300	6*	unsuitable		700	5	2	4	7
	10	over 5000 kg	275	4	275	4	300	4*	unsuitable		800	5	2	3	6
Vibrating plate compactor		Mass per m^2 of base plate													
	1	over 880 kg up to 1100 kg	unsuitable		unsuitable		75	6	unsuitable		unsuitable		unsuitable	unsuitable	unsuitable
	2	over 1100 kg up to 1200 kg	unsuitable		75	10	100	6	75	10	unsuitable		unsuitable	unsuitable	unsuitable
	3	over 1200 kg up to 1400 kg	unsuitable		75	6	150	6	150	8	unsuitable		unsuitable	unsuitable	unsuitable
	4	over 1400 kg up to 1800 kg	100	6	125	6	150	4	unsuitable		unsuitable		8	unsuitable	unsuitable
	5	over 1800 kg up to 2100 kg	150	6	150	5	200	4	unsuitable		unsuitable		5	8	unsuitable
	6	over 2100 kg	200	6	200	5	250	4	unsuitable		unsuitable		3	6	12
Vibro-tamper		Mass:													
	1	over 50 kg up to 65 kg	100	3	100	3	150	3	125	3	unsuitable		4	8	unsuitable
	2	over 65 kg up to 75 kg	125	3	125	3	200	3	150	3	unsuitable		3	6	12
	3	over 75 kg up to 100 kg	150	3	150	3	225	3	175	3	unsuitable		2	4	10
	4	over 100 kg	225	3	200	3	225	3	250	3	unsuitable		2	4	10
Power rammer		Mass:													
	1	100 kg up to 500 kg	150	4	150	6	unsuitable		200	4	unsuitable		5	8	unsuitable
	2	over 500 kg	275	8	275	12	unsuitable		400	4	unsuitable		5	8	14
Dropping-weight compactor		Mass of rammer over 500 kg weight drop:													
	1	over 1 m up to 2 m	600	4	600	8	450	8	unsuitable		unsuitable		unsuitable	unsuitable	unsuitable
	2	over 2 m	600	2	600	8	unsuitable		unsuitable		unsuitable		unsuitable	unsuitable	unsuitable

Figure 75.5 Continued

Figure 75.6 Smooth single-drum roller – towed (Courtesy of Balfour Beatty)

As it is towed by a tracked machine, this plant can be used on a wide range of soft materials where a self-propelled roller might become bogged or cause rutting. A single-drum roller, towed by a tractor, may have greater productivity than a self-propelled roller (see **Figure 75.7**), but is less manoeuvrable.

The self-propelled roller is effective in compacting most material types in many construction operations, including earthworks. The roller is normally fitted with a vibratory mechanism, which can be selected to work at various amplitudes and frequencies to suit the material. The manufacturer will provide data on the correct selection of these, and an experienced earthworks manager should also be consulted.

Table 6/1 precludes the use of vibratory rollers on cohesive materials above a certain liquid limit. This plant may be operated as a deadweight roller to satisfy this requirement. The driving wheels of this plant are pneumatic tyres with deep treads. These provide reasonable traction on soft cohesive materials. The self-propelled smooth drum roller is also available as a twin-drum roller.

Figure 75.7 Smooth single-drum roller – self-propelled (Courtesy of Balfour Beatty)

Available in deadweight or vibratory form, the tamping roller, commonly referred to as a sheepsfoot roller (**Figure 75.8**), is very effective at compacting high plasticity index clays. It can also be effective on sands and gravels, although usually those with a high silt and clay content. It is rarely used on granular fills, where a smooth drum vibratory roller is more effective.

A drawback of tamping rollers is the depressions left in the surface, which can trap water, leading to a softening of the material. If rain is expected, it is good practice to compact the surface with a smooth drum roller or, if this does not remove the depressions, blade off the surface and recompact with a smooth drum.

As with the drum roller, Table 6/1 precludes using this plant in vibratory mode on clays above a specific liquid limit to avoid

(a) (b)

Figure 75.8 Tamping roller (sheepsfoot): (a) towed and (b) self-propelled (Courtesy of Balfour Beatty)

Figure 75.9 Grid roller – towed (Courtesy of Broons)

Figure 75.11 Single- (a) and twin- (b) smooth drum – self-propelled, pedestrian controlled ((a) Courtesy of Terex; (b) Courtesy of Balfour Beatty)

damage to the structure of the clay. The grid roller (**Figure 75.9**) is normally a towed item of plant. It can be obtained in a range of roll widths and masses. It is not usually available to operate as a vibratory action. Grid rollers are particularly effective on coarse rockfill (e.g. starter layers to SHW Class 6B and 6C). They can be effective on sands and gravels, although other types of plant are more often used on this material.

The roller depicted in **Figure 75.10** is also available as a heavier item of plant with wider drums. This is rarely used in major earthworks/bulk fill operations due to the lack of traction from two smooth drums. The smaller unit pictured here is used where the area to be compacted precludes the use of larger plant. This will typically be used in small general fill activities, trench backfill and backfill to structures. The size of this plant restricts the permitted layer thickness, constraining productivity where large areas are to be compacted.

The self-propelled, pedestrian-controlled single- and twin-drum rollers depicted in **Figure 75.11** often have lifting hitches (visible in both **Figures 75.10** and **75.11**) for plant to be craned into otherwise inaccessible locations (e.g. for compacting backfill to structures, where they are widely used). However, the compactive effort of these items of plant is limited: care has to be taken to ensure that the limit for layer thickness is not exceeded. This plant is available as remote controlled, for compacting material in a trench without operator access.

Figure 75.10 Smooth twin-drum roller – self-propelled (Courtesy of Balfour Beatty)

Along with the vibro-tamper (not pictured), the vibrating plate and power rammer depicted in **Figure 75.12(a)** and **75.12(b)** are the lightest items of compaction plant in use in UK construction. Use of these is generally confined to narrow, shallow, trench backfill. Neither is recommended for compacting material to an end product compaction specification.

Although a permitted option in Table 6/4, the pneumatic tyred roller (PTR) (**Figure 75.13**) is rarely encountered in UK earthworks. It is frequently used to compact cement or bituminous bound pavement layers, where the kneading action of the tyres is effective in compacting stiff materials.

(a)

(b)

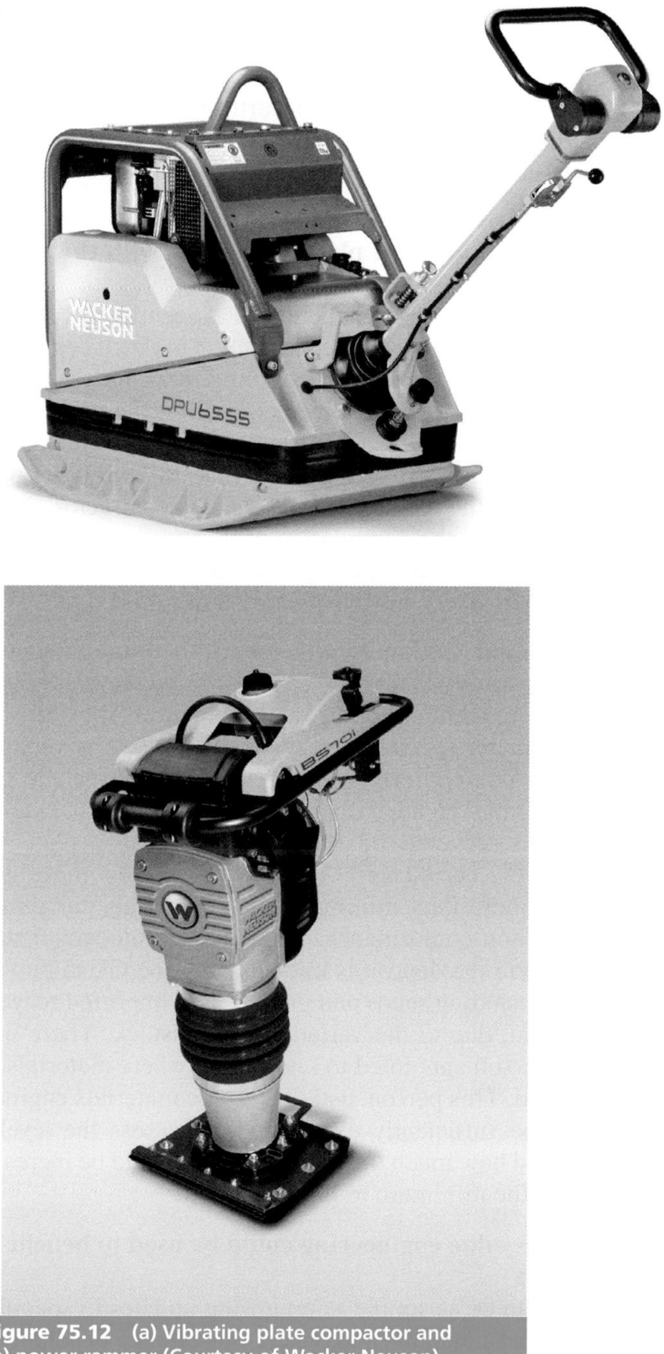

Figure 75.12 (a) Vibrating plate compactor and (b) power rammer (Courtesy of Wacker Neuson)

Figure 75.13 Pneumatic tyred roller (PTR) (Courtesy of Balfour Beatty)

75.4. Control of earthworks

Without a strategy to control an earthworks operation in place prior to the works beginning, non-conformances with the specification are inevitable. These may take the form of

- unacceptable material placed in the works
- the wrong class of material placed in the works
- inadequate compaction.

Any of these occurring can result in some form of physical failure in the works. This might result in

- physical injury to workforce or the public
- prosecution
- poor perception by the public of the works
- deterioration in the client–contractor relationship
- damage to both client and contractor reputations
- increased cost and delays.

Protocols and processes must be established. For the quality control of the earthworks these must identify

(*a*) the client's specific/core requirements

(*b*) which specifications, and which revision of these, are relevant to the contract

(*c*) whether the drawings have the right information, and there is no ambiguity

(*d*) who contractually is responsible for risk

(*e*) management and supervision

(*f*) a clearly defined chain of command from site operatives to the earthworks manager

(*g*) the need for a UKAS-accredited site laboratory

(*h*) inspection and test plans to comply with the client's requirements

(*i*) additional testing requirements to manage specific risks

(*j*) where value engineering could be used to benefit the project (*k*) a system for rapidly reviewing test results to monitor for trends

(*l*) transparency of test results between contractor, designer and client

(*m*) a system for reporting, addressing and closing non-conformances

(*n*) a no-blame culture, especially to allow early warnings to be acted upon.

Here the above are considered in detail.

(*a*) **The client's specific/core requirements**
A proposed method or material, an alternative to criteria here, may require a formal departure from standard to be approved by the client. All parties must be fully cognisant of how adopting such a departure alters the balance of risk ownership (see (*d*)).

(*b*) **Which specifications, and which revision of these, are relevant to the contract**
Every engineer's office on a construction project should have a hard copy of the client's overall specification and the contract-specific documents. These should be issued by one person on the contract – a document controller – responsible for ensuring all documents are kept up to date. This also applies to external specifications.

(*c*) **Whether the drawings have the right information, and there is no ambiguity**
Not simply a task delegated to the document controller, there should be a central register of dated current drawing revisions to which any engineer can refer to confirm the right one is being used.

(*d*) **Who contractually is responsible for risk**
A risk register compiled before the works begin must identify this issue. It must not be seen as a buck-passing exercise, but so that the parties responsible can be prepared and plan the mitigation.

(*e*) **Management and supervision**
A senior manager must be appointed to oversee the planning, management or coordination of the earthworks operations. On most projects, a works manager for earthworks will be appointed to supervise site operations.

(*f*) **A clearly defined chain of command from site operatives to the earthworks manager**
If an earthworks subcontractor is used, this must include how its managers, staff and workforce collaborate with the main contractor. An organogram is essential. It should name staff and provide job titles for all parties, client, designer, contractor and subcontractor. The inclusion of mobile phone numbers is useful, as are photos of staff at the start of a contract. The organogram should be displayed in a part of the office accessed by all personnel.

(*g*) **The need for a UKAS-accredited site laboratory**
The specification will state whether or not this is a core requirement of the client. Most contracts require the contractor to carry out all earthworks testing; on any medium to large project adequate control is not feasible without a laboratory presence on site, if purely for reaction time to any situation.

(*h*) **Inspection and test plans to comply with the client's requirement**
SHW requires the production of Appendix 1/5: Testing to be carried out by the contractor. This includes tests to be carried out, at defined sampling rates, for the materials, including bulk earthworks, used in the project. The compilers take advice on test rates from SHW NG and HA44/91. It is strongly advised that, as with other parts of the earthworks specification, the contractor should contribute to the compilation of this appendix. This will draw on the broadest experience for the highest-risk materials to be controlled and risk minimised.

(*i*) **Additional testing requirements to manage specific risks**
This is in addition to test rates defined in Appendix 1/5. Some materials can be realistically viewed as very low risk of non-conformance. These typically include processed and blended material imported from external quarries, covered by the supplier's quality plan and, often, EC certificated. The test regime for these need be no more than minimal. Other materials carry more risk of non-conformance. These include site-won materials where the deposit is known, from the GI, to be variable (including sands and gravels) and imported recycled material, due to the variation of feedstock. There must be a person appointed to know from where materials are sourced. This person, usually the site materials engineer, must be sufficiently experienced to assess the level of risk and how much the sampling rate should be increased above the minimum to mitigate this.

(*j*) **Where value engineering could be used to benefit the project**
This can be as simple as reviewing updates in specifications and standards and, where this offers a benefit to the project, for client, designer and contractor collaborating for their implementation. It can also be as complex as a major redesign of an element in the works. In any case, it is incumbent on all parties to encourage creative thought among staff.

(*k*) **A system for rapidly reviewing test results to monitor for trends**
A large earthworks project will generate a lot of test data, whether from a site or external laboratory. This is useless unless subjected to frequent, regular review. A UKAS-accredited laboratory will, for each test carried

out, produce a hard copy test report. This is a requirement of accreditation, but does not make review of multiple results realistic. A simple spreadsheet on which earthworks testing – in situ and in the laboratory – can be entered allows rapid review of all testing, for whichever material or property is filtered. Non-conformances can be quickly identified and, more importantly, trends identified, so early action can be taken to avoid non-conformance.

(*l*) **Transparency of test results between contractor, designer and client**
Having created a results database, it should be reviewed by all parties with a stakeholding in the success of the project. If a project has a local electronic network it should be stored so that the client, client's agent, designer and contractor have open access to the data. Security measures must prevent alteration by those other than appointed persons.

(*m*) **A system for reporting, addressing and closing non-conformances**
A system should exist for openly addressing non-conformances to allow closure to the satisfaction of the client and designer. A non-conformance must be seen as an opportunity to improve, not as a means to apportion blame for a failure. This can only work effectively in…

(*n*) **A no-blame culture, especially to allow early warnings to be acted upon.**

75.5. Compliance testing of earthworks

All testing, whether for compliance of the material or to measure the level of compaction, must have a minimum frequency specified as stated above. The frequency of testing must take into account all risks associated with the material.

- Likelihood of structural failure through non-conformance: some elements are critical (e.g. backfill to a structure under a motorway pavement).
- Potential variability of material composition: recycled aggregate will vary more than a quarry product from primary aggregate.
- Potential variability of material properties (e.g. moisture content): recycled aggregate is likely to be more variable than primary aggregate.

75.5.1 Material compliance

All earthworks materials, whether reused, site-won or imported, must be used within the limits defined for various parameters by the client's specification and the designer's geotechnical design report. For projects using the SHW these are specified in the 600 Series and associated appendices and tables.

There is a wide range of physical, mechanical and chemical soil characteristics for which tests are carried out to ascertain compliance with specified suitability limits. It is outside the scope of this chapter to list all these, but **Table 75.3** gives tests which may be used to control the physical suitability of cohesive material for earthworks general fill (i.e. SHW Class 2).

75.5.1.1 Compaction compliance: method compaction

Although the tests quoted in **Table 75.4** may be carried out on method compaction earthworks, this is usually to confirm the conclusions of a geotechnical design, rather than as a control. The key control of method compaction is to ensure that the methodology is correctly carried out. This entails checking

- that the method is suitable for the material class (SHW Table 6/4)
- that the type of compaction plant is suitable
- the compactive effort (mass/m roll width, etc.) of the plant.

These can be obtained by reference to the contract specification and to the plant manufacturer's performance data sheets.

In SHW, for each method and category of compactive effort, there are limits for the maximum thickness of compacted layer and minimum number of passes. These should be ascertained by direct measurement. Daily records must be kept of all observations and measurements.

75.5.1.2 Compaction compliance: end product compaction

Where a minimum level of compaction is specified for a material the compacted material must be tested in situ to ascertain compliance with the specification. It is strongly recommended that material is not overlaid with the next layer until it has been confirmed that the layer under test complies with the specification.

The test methods in regular use in the UK are listed in **Table 75.4**. For each test method the moisture content of the soil must be determined in laboratory conditions to calculate the dry density. This is compared to the MDD obtained for that material (by the method specified) to be expressed as % of MDD.

Testing with a nuclear density meter (NDM) has been used in the UK since at least the early 1980s and is a methodology accepted by specifiers, designers and contractors. The technology is proven for bulk density measurement but not moisture measurement; dry density must be calculated from the moisture content, tested in the laboratory, of a soil sample from the NDM test location.

The legislation regarding the ownership and storage of NDM is extensive and it is not within the scope of this chapter to discuss this in detail. Anyone considering the acquisition of NDM is strongly recommended to seek the advice of a radiation protection adviser in the first instance, as soon as possible. Registration with the Environment Agency (EA) may take 4 months. NDM site testing may be carried out by a hired third-party operator registered with the EA; this is less onerous.

Test	Advantages of test	Constraints of test	Typical acceptable limits
Moisture condition value (MCV)	Rapid – test takes about 15 min. Can be carried out on site.	Must be correlated against fundamental property tests to define limits. Not especially effective for granular soils, or very silty or sandy clays.	8–16
Hand shear vane (HSV)	Rapid – test takes 1 min. Can be carried out on site. Direct measure of cohesion/undrained shear strength C_u.	Not suitable for clay containing much sand or gravel.	>70 kPa
Plastic limit (PL)	Used to define material properties at design stage – not often used for earthworks control.	Must be carried out in laboratory. Test can take 24 hrs from sampling to result.	See MC
Optimum moisture content (OMC)	Effective for granular soils, or very granular clays (e.g. glacial till) where MCV or HSV may not be suitable.	Must be carried out in laboratory. Test can take >48 hrs from sampling to result.	See MC
Quick undrained triaxial	Used to define material properties at design stage – rarely used for earthworks control. Direct measure of cohesion/undrained shear strength C_u.	Must be carried out in laboratory. Test can take >48 hrs from sampling to result.	>70 kPa
Moisture content (MC)	Fundamental soil property which should be measured for every soil sample taken. Can be rapid if microwave method used (correlated against oven dried).	Requires another test, e.g. OMC or PL, to provide control reference.	<PL×1.2 >OMC-2 <OMC+3
Particle size distribution (PSD) – also known as grading	Not frequently carried out on general fills (SHW Class 1 or 2) as other parameters of these materials affect suitability more. More often carried out on selected fills (SHW Class 6), where PSD has a greater effect on performance.	Generally, 48 hrs from sampling to result as two drying cycles are required for materials which have to be washed of clay and silt. A sample size of at least 80 kg is needed for material of 90 mm nominal size.	See SHW Table 6/2
California bearing ratio (CBR)	CBR is not a routine acceptability test, but engineers must be aware of the test. CBR is an empirical measure of material shear strength. It is used to design the thickness of the foundation layer of a pavement. It also identifies soft areas which require treatment (replacement or improvement) prior to placing the foundation. A rapid assessment of CBR can be made using a hand-held MEXE probe.		>2.5%
Falling-/lightweight deflectometer (FWD/LWD)	Like CBR, this is not a routine acceptability test but is used to measure the in situ stiffness of soils. This provides data for pavement foundation design and design verification testing.		–

Table 75.3 Tests which may be used to control the physical suitability of cohesive material for earthworks general fill (SHW Class 2)

Test	Advantages of test	Constraints of test
Nuclear density meter (NDM) (**Figure 75.14**)	Rapid – a result from a single test point takes less than a minute. Speed of test enables many test points to be taken. Suitable for most soil types, fine and coarse grained; unbound and hydraulically bound.	Significant administrative input required to manage issues of: health and safety licensing security. NDM apparatus requires regular correlation for each material against actual density. Moisture measurement by NDM is not reliable
Core cutter (**Figure 75.15**)	Quick and simple to carry out on site. Can be rapid if microwave method used for MC (correlated against oven dried).	Not suitable for other than fine cohesive soils.
Sand replacement density	Suitable for coarse-grained soils.	Slow and labour-intensive.

Table 75.4 Compaction test methods in regular use in the UK

Figure 75.14 In situ NDM testing (Courtesy of Balfour Beatty)

Figure 75.15 In situ core cutter testing (Courtesy of Balfour Beatty)

75.5.2 Selection of compliance parameters and limits

The specification compiler must select the test methods to be used on the project to control the earthworks. The compiler must take into account

- suitability of the method for the materials to be encountered
- urgency of the result to determine compliance with specification
- avoiding conflicting parameters/limits.

Tables 75.3 and **75.4** provide guidance on the first two of these points. It is worth stressing the importance of selecting the most rapid method appropriate to the nature of the contract and earthworks: the triaxial test will certainly define whether a clay is a suitable fill or not; specifying it as the means to control the earthworks on a project where 1 000 000 m^3 of clay has to be moved in a season is not practical.

The MCV test was developed by TRL (then TRRL) in the 1970s as a rapid method of controlling earthworks for material acceptability. The method is ideal for medium- to high-plasticity 'heavy' clays (e.g. Lias, London, Gault), but can be effective on very silty or sandy clays if correlation testing against fundamental properties (e.g. shear strength) is carried out prior to the works beginning.

The hand shear vane, when calibrated, is an effective method for cohesive materials of varying plasticity. It is less suitable if the clay is expected to include fractions larger than medium sand: coarser particles engaging with the vanes can produce erroneous results.

The third point needs careful consideration. In allowing more than one test method as a means of control, the specification compiler can introduce a paradox where a material tested to one method may comply but, tested to the other, may not. This may be a result of limit selection being imperfect, variation in the material from that used in correlation testing or just that natural material does not necessarily conform precisely to the models made for it.

In the example specification provided (**Figure 75.3(a)**), this possibility was addressed by a footnote to the table. This allowed earthworks control by hand shear vane *or* MCV, so long as the Appendix 1/5 requirement for the minimum number of MCV was satisfied.

Selection of limits for a material must take account of

- the material properties
- results of any correlation testing
- understanding how material will perform when subjected to the expected load
- other physical conditions that may affect this performance
- the acceptable level of risk/factors of safety.

Determining these limits is the role of geotechnical specialists of the designer and the client and, where possible, the contractor.

75.5.3 Dealing with non-conformance

This chapter has used 'non-conformance' and 'failure' as distinct and non-interchangeable terms. Non-conformance, where one or more of the tested properties of a material lies outside the specified limits, does not necessarily imply that the latter, structural failure, is inevitable or even possible.

While all non-conformances must be investigated, it must be noted that, in statistical analysis of a table of test results – the normal distribution curve (bell graph) – a small number of 'defectives' are expected. About 4% of results of any test might be expected to be marginal non-conformances. It is not within the scope of this chapter to discuss how far outlying the specified limits constitutes a marginal non-conformance, or the statistical analysis of data.

When a test result lies outside the acceptable limits for that parameter, most quality control systems require the production of a non-conformance report (NCR). This will describe the

- material type and class (if appropriate)
- location used
- specified limits
- test result obtained.

This will allow the team assessing the NCR to make a judgement on the action required. This may vary and include

- the criticality of the non-compliance, by severity and location
- consideration that the factor of safety has not been affected and no action is necessary
- resampling and retesting – the non-conformance may result from sampling bias
- treatment to modify the characteristics of the material to comply with the specification
- removal and replacement with compliant material.

As with all matters involving earthworks and the specification, the review of the NCR and the decision-making process should include client, designer and contractor. A solution that is without negative effect on the safety of the project, but is practical and commensurate with the situation, is to be sought.

75.6. Managing and controlling specific materials

If not correctly managed, all materials may create problems for the earthworks operations. Some, in certain conditions, may be considered by the engineer as 'difficult' (i.e. more prone to failure) even though there may have been no non-conformance with the specification.

75.6.1 Clay

Although not necessarily a difficult material to manage, working with clay merits consideration and planning that will benefit an engineer controlling earthworks.

Clay is moisture-sensitive, more so with increasing silt content. While light rainfall might not necessitate a temporary halt in earthworks, an experienced earthworks manager must be consulted on when work should cease. Continuing to traffic clay with earthworks plant, especially on narrow, channelled haul roads, during rainfall has several negative effects.

- Safety: on clay, particularly firm–stiff high-PI clays, light rain will reduce traction/grip for wheeled construction plant.
- Softening of acceptable material, so that it becomes unacceptable. This is increased by rutting.
- Rutting. Once started this can lead to a rapid deterioration of the material, as water is trapped in ruts, softening the clay and leading to deeper ruts (**Figure 75.16**).

(a)
(b)

Figure 75.16 Earthworks after heavy rain: (a) not well drained; (b) rolled beforehand and well drained (Courtesy of Balfour Beatty)

Rutting is a serious problem as trapped water will have to be drained away. On an embankment this may be done relatively easily, but in a cutting this may need pumping and long pipe runs. This has a negative effect on the programme and entails additional cost.

Planning, especially if rain is forecast, can mitigate or eliminate the effects of rain, mainly by managing the earthworks operations to create drainage paths.

- *Crossfall:* Manage deposition and compaction so that there is always a fall to the outer edge of the works.
- *Longitudinal fall:* In a fill, crossfall is easily created and is effective. In cuts there is no over-the-edge drainage. In this case, while a crossfall will take the water initially away from the works and haul road, drainage requires a longitudinal fall to the open end of the cutting. Care must be taken to ensure there is no accidental damming.

- *Grips:* Surface drainage channels to carry water away from low spots that will inevitably occur no matter how well the works are planned.
- *Proof rolling:* Although rolling is concurrent with deposition, with rainfall imminent additional attention should be taken to close the surface of the material, to prevent ponding. A smooth surface will also shed water along the crossfall faster.

Any action to drain rainwater must take into account that a discharge offsite may require licensing by the EA. A limit on suspended solids in any such discharge may be imposed.

Haul roads on clay must be maintained to avoid rutting. If ruts are allowed to develop, haulage times are increased, more fuel is burned by the plant and, in the worst cases, plant may become stuck and need pulling or pushing out. All these reduce the efficiency of the operation, increasing cost and adding delays. Rainwater ponding in ruts accelerates any deterioration.

Allowance must be made for a grader to trim the haul road so rutting does not develop. A roller may also be required.

Wet or soft unacceptable clay, either naturally occurring or through deterioration, can be reverted, but this needs planning and has programme and cost implications.

- *Natural drying:* Wet clay is excavated and loosely spread in windrows where sun and wind can effect natural drying. This process requires confidence for a prolonged spell of fair weather; it is not an option for winter. The process can take from a few days to a week or more, depending on the condition of the clay and the weather.
- *Lime modification:* Quicklime reduces moisture content by chemical combination and by the exothermic reaction evaporating some of the water. Quicklime is spread and rotavated into the clay. It should not be compacted immediately; the steam generated must be allowed channels to escape (**Figures 75.17** and **75.18**).

Figure 75.17 Lime being added to saturated earthworks – the cloud is steam from the exothermic reaction between the quicklime and water, not lime dust (Courtesy of Balfour Beatty)

Figure 75.18 Improvement with lime addition – dissipating steam (Courtesy of Balfour Beatty)

Lime modification must not be confused with stabilisation. Modification does not change the material from Class 7 to Class 9 and does not require the testing for those classes of material. However, the testing required for Class 2 must be carried out after treatment to confirm the material complies with the acceptability limits; further treatment is necessary if not.

Care should be taken if the clay is suspected to contain sulfate-bearing minerals such as pyrite or gypsum. Although modification uses a lower addition of quicklime (typically 1–2%) than stabilisation, there is still the potential for long-term deleterious expansion from reaction with the sulfates.

The GI should be thoroughly assessed for

- unacceptable clay, for which the earthworks strategy may need to be modified
- gypsum or pyrites described in the BH or TP logs
- elevated sulfates
- any testing carried out to TRL 447 (Reid *et al.*, 2001).

If there is any doubt, samples should be obtained from trial pits and tests 1, 2 and 4 in TRL Report 447 must be carried out. Advice should be sought from a geotechnical specialist to interpret the results.

The earthworks programme must be reviewed. Is there a programme-critical operation from late autumn to early spring when wetter weather is more likely and natural drying not feasible? Assessment and testing are required as above.

These processes take time and must be planned and undertaken well in advance of earthworks in that area of the project.

75.6.2 Silt

All the caveats for clay apply to silt, but more so. Silt is highly sensitive to variations in moisture content. As silt is more permeable than clay, water can penetrate the material more readily.

If, during earthworks involving silt – or a very silty clay – it starts to rain, it is prudent to cease operations sooner than one would on clay. A small change in moisture content can render an acceptable fill material into slurry. Trafficking silt during rain will rapidly deteriorate the material to significant depths.

To revert unacceptable silt, lime modification can be used, but it is slightly less effective than with clay. As with clay, quicklime will reduce moisture by exothermic reaction with the free water. However, silt differs chemically from clay, and further reaction between the clay and the lime (as occurs in stabilisation) does not occur.

Natural reversion, by excavating and windrowing to dry, is just as effective for silt as it is for clay.

Draining the earthworks is even more important for silt than clay. It must be considered that silt erodes much more easily than clay and care must be taken to limit the fall of any grip to minimise this. Likewise, over-the-edge drainage of embankments must be carefully thought out to prevent the formation of deep washout rills in the batterslope and top of batter. Repairing these can be more problematic, and cost more, than the damage the drainage was intended to avoid.

Maintenance of haul roads over silt is critical to avoid rutting developing. Due to the sensitivity of the material, consideration must be made for providing a granular haul road to permit haulage in all weathers.

Haul roads can be formed from locally available material. Crushed rock, or crushed concrete/masonry is better than sand and gravel. This is due to the greater mechanical interlock of the particles providing more stability – less tendency to deformation – under the load of construction plant.

75.6.3 Fine, uniform sand

SHW Table 6/1 identifies this material as Class 1B. Two physical properties of this material characterise it.

- *Fineness:* Particle sizes are, typically, 100% <0.6 mm, 50% <0.3 mm, 10% <0.06 mm.
- *Uniformity:* The distribution of particles is over a small size range.

As a result of the effects of these characteristics, this material has

- restricted acceptable moisture range
- high mobility – readily moved by wind or water
- increased risk of washout in heavy rain
- poor trafficability by wheeled construction plant.

Techniques for managing this material generally require the close monitoring and control of moisture content.

- *Drier than OMC:* Water should be added during deposition. Water bowsers should be employed for this, and used during dry weather to prevent desiccation of the surface.
- *Wetter than OMC:* This material drains reasonably quickly. Only when excavation is below the water table is encountering this likely. If excavated and windrowed this material will drain and dry; it is important to monitor this to avoid overdrying.
- *Drainage runs:* Run-offs must be carefully planned to avoid significant falls and the risk of erosion.
- *Over-the edge drainage:* This is to be avoided, as the risk of washout is too great. To avoid the worst effects of run-off erosion it may be better to allow rainwater to pond on the earthworks surface.

75.6.4 Chalk

Even the strongest, blockiest chalk can be reduced to a structureless mass by incorrect handling. Taking the right actions, most grades of chalk can be used in construction without increased risk:

- *Season:* Determined by the specification compilers in Appendix 6/4. This is decided by reference to rainfall and temperature records for the area and knowledge of the chalk. The usual period is March to November. However, if early spring is much wetter or colder than usual the start should be delayed.
- *Control:* While intact dry density and chalk crushing value are used, the basic test for controlling chalk earthworks is the moisture content test. Microwave drying should be correlated to BS oven drying methods so rapid assessment can be made.
- *Excavation:* this – in fact, every part of the chalk operation – must be managed to retain as much of the chalk structure as possible. Chalk should only be excavated as a face. This should be as deep as possible; 5 m is the usual minimum, although site conditions and depth of cut may constrain this (**Figure 75.19**).
- *Haulage:* Chalk can lose about 2% moisture in haulage, in fair weather conditions and on an average haul distance. Chalk that is slightly wetter than the maximum limit may revert in haulage.

Figure 75.19 Chalk cutting excavated as a single deep face (Courtesy of Balfour Beatty)

Figure 75.20 Chalk compaction with a deadweight roller (Courtesy of Balfour Beatty)

This should be taken into account when the sampling and test plan is written.

- *Deposition:* Chalk should never be rehandled. It should be discharged from the dump truck at the point of final deposition. The lightest plant commensurate with efficiency should be used in spreading. Chalk should be spread in the thickest layer possible to maintain the structure; 0.5 m is not unreasonable.
- *Compaction:* SHW Table 6/4 permits light vibratory rollers. Unless the chalk is very strong it is normal practice to use deadweight rollers, again to minimise damage to the material. A rule-of-thumb guide is that, when the chalk starts to adhere to the drum, rolling should cease (**Figure 75.20**).

Any chalk fill must undergo a settlement period before any pavement layers are added. This period is defined by the designer. It may be shortened by surcharging, which is removed prior to the addition of further layers of the permanent works.

75.6.5 Recycled and secondary materials

Engineers are likely to encounter recycled and secondary materials as fills and aggregates; there is a mature UK marketplace to supply these materials. Recycled and secondary aggregates are not interchangeable terms. Neither are sourced from primary aggregate – that is, naturally occurring sand and gravel (land or marine sourced) or crushed rock. Generally accepted definitions of each are

- secondary aggregate: produced as by-products of other processes
- recycled aggregate: produced from the processing of inorganic material used in a construction process.

Use of recycled and secondary aggregates is encouraged by government initiatives for sustainable construction. However, legislation and regulation require certain procedures to be carried out: there is a duty of care to ensure that contaminating substances, at concentrations liable to cause harm to human health or the environment, are not spread about.

The regulations covering the use of non-primary aggregates are still evolving. It would be unwise to provide definitive instructions in this chapter as they may easily be superseded in the future. Recommendation can be given on sources of reference, the most significant of these being the Waste and Resources Action Programme (WRAP) which, in conjunction with the EA, has produced 'The Quality Protocol for the Production of Aggregates from Inert Waste'. Aggregates produced in accordance with this protocol are considered to be fully recovered and thus are not waste.

When recycled aggregates are not produced to the WRAP protocol they are probably still considered to be waste by the regulators. Producers must then comply with the Environmental Permitting Regulations: a permit must be acquired from the EA, a process that may take many weeks to complete.

Potential users of recycled aggregate are strongly advised that, prior to importing any material to their works, they should ascertain that the supplier has produced the material in accordance with the WRAP protocol and can prove this and be subjected to audit by the purchaser. The supplier must provide test results which confirm the material complies with the protocol. Contaminated Land: Applications in Real Environments (CL:AIRE) has, in conjunction with the EA, published 'The Definition of Waste: Development Industry Code of Practice' Version 2 in March 2011. Anyone intending to reuse or recycle material is strongly recommended to acquire this freely downloadable document.

Specification compilers and engineers are recommended to consult

- their in-house environmental manager
- external waste/environmental advisers
- WRAP
- CL:AIRE
- the EA.

In addition to compliance with waste regulations, the use of non-primary materials may be restricted by the client's core requirements. The HA define their requirements for recycled aggregate in SHW Clause 601.12, and Table 6/1 includes specific classes of material where recycled material is permitted.

In addition, HA DMRB Vol. 1, Sec. 1, Part 2: HD 35/04 'Conservation and the Use of Secondary and Recycled Materials' (now withdrawn) provides detailed advice on using recycled materials to specifiers. Table 2.1 from this document is in part reproduced in **Table 75.5**.

It must be noted that clients other than HA may have significantly different requirements for using recycled aggregates.

Material	Application and series						
	Pipe bedding	Embankment and fill	Capping	Unbound mixtures for subbase	Hydraulically bound mixtures for subbase and base	Bitumen bound layers	PQ Concrete
	500	600	600	800	800	900	1000
Blast furnace slag	✓	✓	✓	✓	✓	✓	✓
Burnt colliery spoil	×	✓	✓	✓	✓	×	×
China clay sand/stent	✓	✓	✓	✓	✓	✓	✓
Coal fly ash/pulverised fuel ash (CFA/PFA)	✓	✓	✓	×	✓	✓	✓
Foundry sand	✓	✓	✓	✓	✓	✓	✓
Furnace bottom ash (FBA)	✓	✓	✓	×	✓	×	×
Incinerator bottom ash aggregate (IBAA)	✓	✓	✓	✓	✓	✓	✓
Phosphoric slag	✓	✓	✓	✓	✓	✓	✓
Recycled aggregate	✓	✓	✓	✓	✓	✓	✓
Recycled asphalt	✓	✓	✓	✓	✓	✓	✓
Recycled concrete	✓	✓	✓	✓	✓	✓	✓
Recycled glass	✓	✓	✓	✓	✓	✓	×
Slate aggregate	✓	✓	✓	✓	✓	✓	✓
Spent oil shale/blaise	×	✓	✓	✓	✓	×	×
Steel slag	✓	✓	✓	✓	✓	✓	×
Unburnt colliery spoil	×	✓	×	×	✓	×	×

KEY:
✓ Specific (permitted as a constituent if the material complies with the Specification (MCHW 1)) or
General Provision (permitted as a constituent if the material complies with the Specification (MCHW 1) requirements but not named within the Specification (MCHW 1)).
× Not permitted.

IMPORTANT NOTES:

1. Table 2.1 is for guidance only and reference must be made to the accompanying text and the Specification (MCHW 1). Materials indicated as complying with the Specification (MCHW 1) for a particular application may not necessarily comply with all the requirements of the series listed, only particular clauses. For example in the 600 Series, Unburnt Colliery Spoil can satisfy the Specification (MCHW 1) as a general fill, but is excluded as a selected fill; and in Series 1000 recycled or secondary materials are not permitted within the running surface of PQ concrete. Reference should also be made to the Specification (MCHW 1) for any maximum constituent percentages of specific recycled or secondary aggregates. For example in the 1000 Series, the maximum by mass constituent of Recycled Asphalt is given under the limits for 'other material' (Table 10/2) within the Specification (MCHW 1).
2. There is no Specific or General Provision for the use of recycled glass as an aggregate in PQ concrete or Hydraulically Bound Mixtures due to the potential for deleterious alkali-silica reaction (ASR). However, its use may be permitted by the Overseeing Organisation if sufficient provisions to minimise the risk of deleterious ASR are included in the mixture design.
3. There is no Specific or General Provision for the use of steel slag as an aggregate in PQ concrete or Hydraulically Bound Mixtures due to the potential for volume instability. However, its use may be permitted by the Overseeing Organisation if sufficient assurance of volume stability is provided.

Table 75.5 ***Specification for Highway Works*** **(MCHW 1): application of secondary and recycled aggregates (Reproduced from Highways Agency,** ***Design Manual for Roads and Bridges*****, HD 35/04. © Crown Copyright 2004)**

Before proceeding, specifiers or users must ensure they understand what their client permits to be used and what may require a departure from standards, if this is possible.

Specification compilers and users should be aware that a client's core requirements may not include current industry best practice. For example, tyre bales are being increasingly used as lightweight embankment fill, but are not identified in **Table 75.5**.

Discussion between client, designer/specifier and contractor is, as before in this chapter, strongly recommended to maximise the use of recycled and secondary material for sustainable construction.

75.6.6 Manufactured aggregate

The most likely material of the kind that will be encountered is lightweight aggregate. This is manufactured from sintered PFA or expanded clay and marketed under certain brand names. The materials are not included in the SHW. They are only likely to be used to resolve specific geotechnical problems, and designers should seek advice from the manufacturers on the material properties and performance.

75.6.7 Peat

Although significant volumes of peat are uncommon in England and Wales, engineers expecting to work in Scotland

and Ireland should be prepared to encounter this material (**Figure 75.21**). Peat characteristics are

- low cohesion/shear strength – values of C_u are unlikely to exceed 10 kPa
- high moisture content – values of >300% are common
- high compressibility – when loaded, water is driven from the structure leading to a large volume change
- unpredictable settlement – because peat is organic and not homogeneous, settlement is uneven over any unit of area. Actual settlement is very difficult to predict from laboratory testing.

Peat is not acceptable, in any condition, as general fill for an earthworks project. Classification of this material, according to SHW terminology, will always be Class U1. Neither is it suitable as a load-bearing material for an embankment. It must be dealt with or removed.

A strategy for dealing with peat first must consider how thick the layer is and, consequently, the volume to be removed. As it is not suitable even for landscaping, all the material may have to be removed from site. This may involve significant cost.

- There may not be a suitably licensed landfill locally.
- Any landfill may have limits on the annual permitted intake of this material.
- It may attract the higher level of landfill tax, if this applies.

If peat is to be removed, it must be done with consideration for the stability of the remaining mass to each side of the excavation.

It may be preferable to leave the material in place. In this case the load of the overlying structure must be transferred to the underlying strata, usually by a piled raft. Any project encountering peat needs to be carefully planned in advance of the works.

Figure 75.21 Typical low-lying peat landscape (Courtesy of Balfour Beatty)

75.6.8 Topsoil

Topsoil is not used as a structural engineering soil. This is due to the organic content of the soil, which can create stability problems and, if used close to the surface, regrowth. However, topsoil is part of the earthworks and, as it is generally encountered as the first and final operations, must be considered and included in the programme.

75.6.8.1 Topsoil removal (topsoil strip)

- In cut or at grade areas: topsoil is removed before other earthworks can begin.
- In fill areas: topsoil may be removed or, when an embankment is above a minimum height, left in place. The advantages of this are
  - reduced disposal if there is a surplus of topsoil
  - reduced imported fill if there is a deficit of suitable material in the works
  - preservation of archaeology
  - programme and cost benefits.

  The minimum height for this must be agreed between designer and contractor. Below this, topsoil must be removed.

Topsoil which will be reused in the works must be stored in topsoil bunds. To avoid damage to the soil there is a height restriction to these, which engineers must confirm with the designer before proceeding.

Topsoil strip is usually carried out by a scraper box towed behind a track-type tractor. This is effective if the topsoil bund is close to the area being stripped. However, if these are further away stripping with a 360° excavator and haulage by articulated dump truck (ADT) are more efficient.

75.6.8.2 Topsoil replacement (resoiling)

After completion of structural earthworks topsoil is replaced where the contract requires. Areas to be resoiled will include verges, batterslopes, landscaping and areas not otherwise overlaid by, for example, bound/block pavement, bound or unbound decorative finishes, surface treatments or over filter drains.

Topsoil is spread by a wheeled or tracked 360° excavator or a light track-type tractor. Topsoil is not compacted.

Engineers should note that SHW has different material classes for topsoil existing on site (5A) and imported topsoil (5B). Different parameters and limits may apply to each. Each of these classes may be subdivided so the composition may varied to suit the ecology of a specific area.

References

BSI (British Standards Institution) (2022) BS 1377-2: Methods of test for soils for civil engineering purposes: classification tests and determination of geotechnical parameters. BSI, London, UK.

CL:AIRE (2011) *The Definition of Waste: Development Industry Code of Practice Version 2*. http://www.claire.co.uk (accessed 12/06/2023).

Highways Agency (2004) *Design Manual for Roads and Bridges, Volume 7, Section 1, Part 2, HD 35, Conservation and the Use of Secondary and Recycled Materials*. The Stationery Office, London, UK (withdrawn 2018).

Highways Agency (2017) *Specification for Highway Works*. The Stationery Office, London, UK.

Reid JM, Czerewko MA and Cripps JC (2001) *Sulfate Specification for Structural Backfills*. TRL Report 447. Transport Research Laboratory, Crowthorne, UK.

Useful websites

British Standards Institution (BSI): http://www.bsigroup.com
Building Research Establishment (BRE): http://www.bre.co.uk
CL:AIRE Code of Practice: http://www.claire.co.uk/index.php?option=com_content&view=article&id=210&Itemid=82
Environment Agency: http://www.environment-agency.gov.uk
Highways Agency standards: http://www.dft.gov.uk/ha/standards
Transport Research Laboratory (TRL): http://www.trl.co.uk
WRAP: http://www.wrap.org.uk/index.html

It is recommended this chapter is read in conjunction with

- Chapter 37 *Sulfates and sulfides in soils and rocks*
- Chapter 69 *Earthworks design principles*
- Chapter 78 *Procurement and specification*

All chapters in this book rely on the guidance in Sections 1 *Context* and 2 *Fundamental principles*. A sound knowledge of ground investigation is required for all geotechnical works, as set out in Section 4 *Site investigation*.

doi: 10.1680/icemge.66830.1251

Chapter 76
Pavement foundation design

Paul Coney Jacobs, UK
Peter Gilbert Jacobs, UK
Alice Duley Jacobs, UK
Reviewed by **Professor Paul Fleming**, Loughborough University, UK

CONTENTS

There have been significant developments in the design of pavements and trackbed for roads and railways in recent years. The latest design specifications take a more rational approach than the traditional recipe method based on experience and offer potential rewards including the wider use of more sustainable materials and reduction in construction thicknesses. While the aims of these developments are valid, it is considered an increasing focus on analysis has also coincided with a reduction in emphasis on the local geology, groundwater conditions and using appropriate construction materials for the prevailing conditions. This is considered a significant failing, and the authors have experienced many examples of design and construction errors resulting from poor understanding of the ground conditions, with some resulting in multi-million-pound legal disputes. This chapter attempts to address this trend and emphasises the importance of properly assessing the subgrade and developing a design that suits the ground conditions present. It gives a brief outline of good current design practice for UK roads and railways, and references are given to some of the key literature available for further reading. The focus is on the design methods developed for motorways and trunk roads (produced by National Highways) and railways (Network Rail) as the majority of research carried out in the UK has been related to these. A brief description of the important considerations for high-speed rail pavement foundations is provided as well as reference materials for pavement applications not specifically covered in this chapter, such as ports, airfields and unsurfaced roads.

76.1. Introduction

Transportation is a fundamental requirement of our lives, and in the UK one is never far from a road or railway. Considering this, it may seem surprising that pavement engineering, and pavement foundation design in particular, was relatively slow to develop as a technical aspect within civil engineering. Traditionally, pavements were built based on experience or precedent, and therefore construction has tended to be conservative or in some cases inadequate. Recent years have seen significant developments of a more theoretical framework for design. This approach has been driven largely by the change in the procurement method of many civil engineering projects towards public–private partnership, design and build, and design, finance and operate type contracts. These contracts, where the construction risk is transferred from the public to the private sector, have generated a greater desire for innovation to provide cost and environmental benefits by the use of alternative cheaper construction materials, and to minimise construction thicknesses while considering whole-life costing.

Pavement design practice involves several civil engineering disciplines, including materials and structural engineering, hydraulics and geotechnics. The geotechnical engineer is required to assess the ground on which pavement will be constructed (subgrade). The pavement engineer (for highways) or permanent-way engineer (for railways) will be required to determine requirements for the materials directly subject to traffic loading. The boundary between the responsibilities of the two disciplines varies in practice. Like building structures, the founding soil and the pavement foundation should be considered in unison (see Chapter 3 *A brief history of the development of geotechnical engineering*). Traditionally, for roads design, the boundary has been taken as the top of the capping; the geotechnical engineer is responsible for a suitable platform for the construction of the upper layers. The latest developments towards performance-based design have potentially changed the boundaries as the analytical design (which may be carried out by the pavement engineer) will require knowledge of the properties of each pavement layer. It is clear, however, that the geotechnical engineer's involvement should not be limited to the assessment of the subgrade conditions. In particular, a geotechnical engineer is often best placed to judge the practicality of different foundation solutions – which is vital for successful construction. Design works best when the pavement engineer, geotechnical engineer, alignment engineer and drainage engineer work together to produce a suitable design solution.

Unfortunately, poor practice is not uncommon, and examples include a failure to understand and address the ground and groundwater conditions, designs that do not work for the short (during construction) and long term and the use of inappropriate materials for the site conditions present. These failings have resulted in delays, re-work and multi-million pound construction claims. To reap the benefits from a more analytical approach the industry must not forget to address the various factors required to reach a successful design and construction outcome.

This chapter concerns the design of pavement foundations. For the purpose of this chapter the term 'pavement foundation' is used to refer to the materials below the base course for roads (i.e. usually subbase, capping and subgrade) and materials below the sleepers for railways (i.e. usually the ballast, subgrade and any intermediate layers). A variety of terminology is used to refer to the construction layers, with different terms being used for roads and railways. An attempt has been made to use consistent terminology: as examples, 'pavement foundation' is also used to mean 'trackbed', and 'trafficking' is used to refer to road or rail loading, as appropriate.

The focus is on the design methods developed for motorways and trunk roads (produced by National Highways) and railways (Network Rail), as the majority of research carried out in the UK has been related to these. Pavement construction for other purposes (e.g. ports and airports) is not specifically discussed; however, many of the principles apply to any pavement foundations, and some key reference materials are provided in Section 76.5.3. Section 76.9 focuses on the particular considerations when designing pavement foundations for high-speed rail, including the differences from the design of more traditional railways on which the majority of this chapter focuses. Thus, reference to 'railways' in the majority of this chapter refers to more traditional non-high-speed lines.

The aim here is to provide an outline of good design practice and give references to some key texts for further information. It is not possible to provide a detailed description of pavement engineering theory – references are provided to some key literature. The recent work on developing design approaches for high-speed rail in the UK has provided a good opportunity to demonstrate how pavement engineering should be done correctly, starting with a thorough identification of the ground conditions and working up to provide a suitable platform for the overlying construction.

76.2. Purpose of pavement foundation

The basic terminology used for the various construction layers used in road and rail pavements is given in **Figures 76.1** and **76.2.**

Each pavement layer has different functions which are described in various sources (for roads, see Thom, 2008; for railways, see Network Rail, 2020; Selig and Waters, 1994). Trafficking applies a dynamic load to the pavement. This load will be distributed through the (upper) pavement to the pavement foundation. The fundamental roles of the pavement foundation are as follows.

- Provide a stable platform during construction to enable placement and compaction of the overlying pavement layers.
- Reduce the stress from traffic loading to an acceptable level for the subgrade and prevent excessive deformation (in the short and long term).

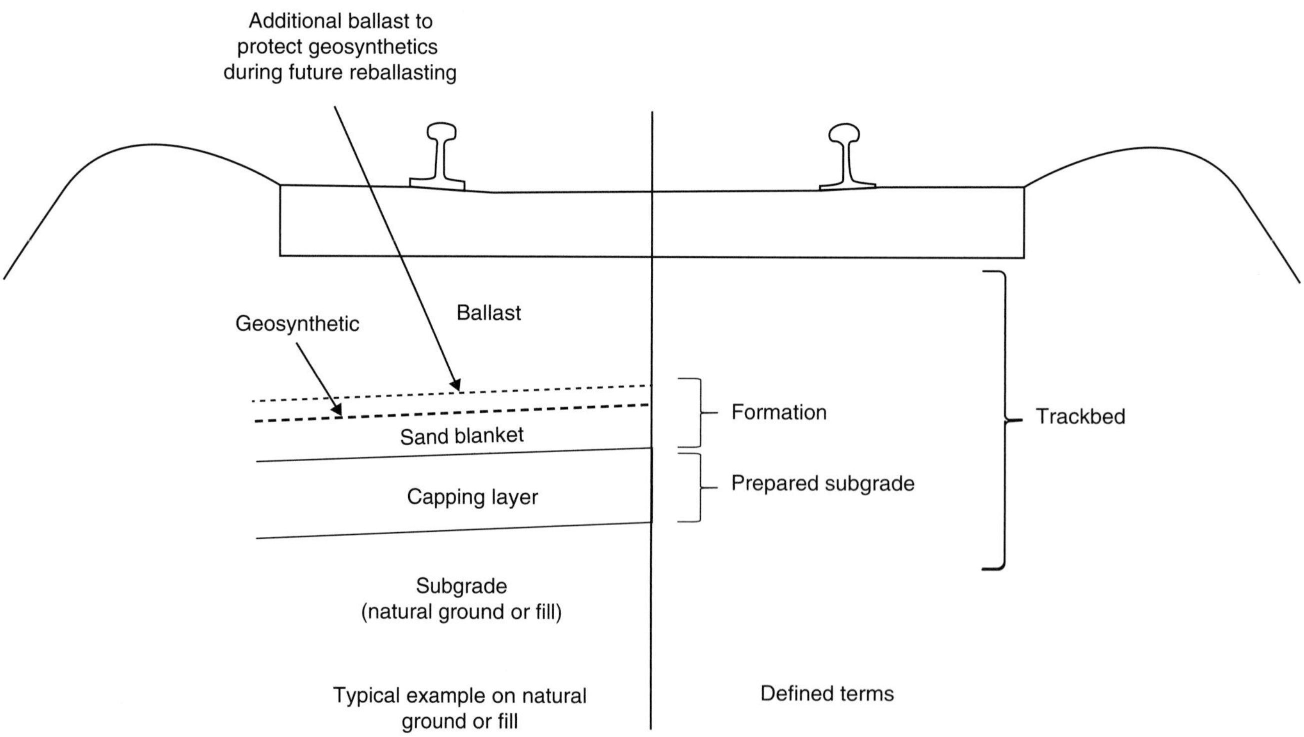

Figure 76.1 Railway trackbed terminology (Reproduced from NR/L2/TRK/4239 Copyright holder © Network Rail 2020)

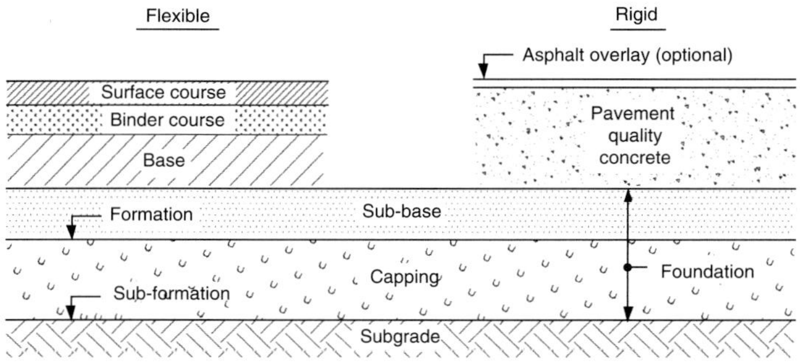

Figure 76.2 Road pavement layer terminology

- Maintain material properties under traffic loading. Worsening of material properties can occur either by attrition of the material particles or by migration of fines from adjacent layers.
- Resist lateral and longitudinal forces from traffic loading. This is most important for ballast, which is required to maintain support to the sleepers.
- Allow drainage of water from the pavement layers to the drainage system. A separate but related topic is the drainage of the subgrade which, in cuttings, is delivered by filter drains to avoid groundwater entering the pavement foundation.

These functions are required both during construction and in service. For a pavement during construction, stresses in the foundation due to construction traffic are relatively high, although the number of stress repetitions is relatively low compared to in-service loading conditions. The extreme consequence of inadequate pavement foundations is rutting of the pavement and the underlying subgrade. For the pavement foundation, rutting is more likely during construction and should be remedied appropriately (usually by excavation and replacement of the failed material), otherwise inadequate pavement performance is likely in the long term. Poor pavement foundation performance commonly results in significantly higher long-term maintenance costs.

76.3. Pavement foundation theory

There is insufficient opportunity to explain the theory of pavement design in this chapter, and sources such as Thom (2008) for roads and Selig and Waters (1994) for railways should be referred to for further detail. For road design, the Transport Research Laboratory (TRL) research of the 1960s to 1980s still provides much of the underlying logic that is followed today; therefore it is also advisable to read TRL LR1132 (Powell *et al.*, 1984) and refer to TRL LR889 (Black and Lister, 1979) for further details.

The upper-bound layers (e.g. asphalt or concrete surfacing for a road, or the rails and sleepers of a railway) are designed to spread the traffic load, reducing it to a level that is tolerable for the pavement foundation. In turn, the foundation materials must be selected to perform under the design loading and reduce the stress passed to the subgrade.

As described by Brown (1996), when a dynamic load is applied to an unbound material (soil), it will deform initially elastically (resilient deformation) and then plastically (permanent deformation), dependent on the stress path. The unloading action allows for only the elastic deformation to be recovered. The load cycles therefore cause an accumulation of permanent deformation, the magnitude of which depends on the size of the stresses applied, the number of load cycles and the stiffness and strength of the material. A typical shear stress–shear strain plot for an unbound material under dynamic loading is shown in **Figure 76.3**.

As shown in **Figure 76.4**, with increasing load applications at low levels of applied shear stress the tendency is for plastic deformation to occur at an ever-decreasing rate or actually cease altogether. At higher levels of shear stress, plastic deformation will increase at an increasing rate, leading to large deformation or failure (Thom, 2008). The shear stress

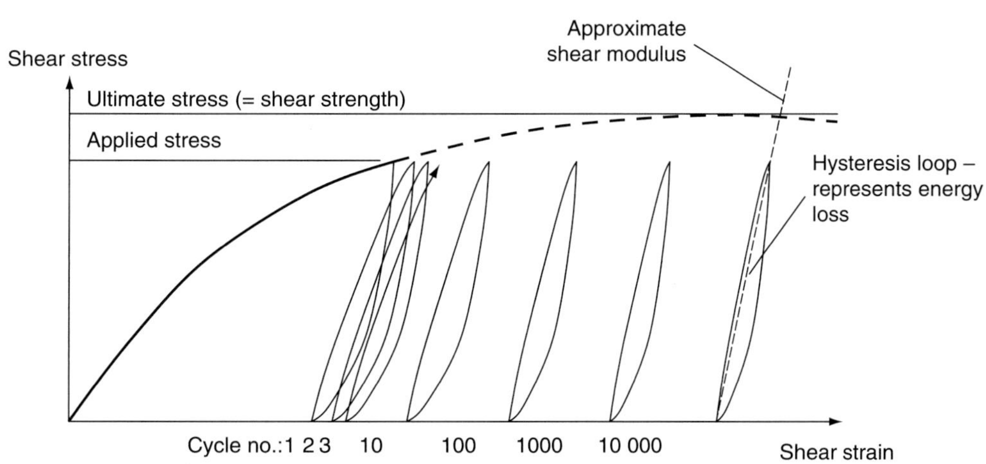

Figure 76.3 Idealised stress–strain behaviour of an unbound material under dynamic loading (Reproduced from Thom, 2008)

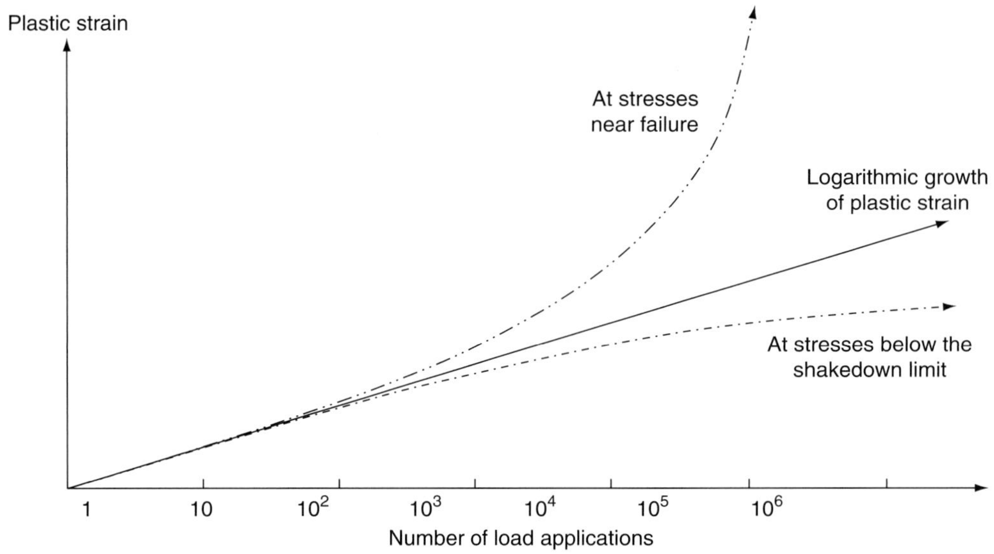

Figure 76.4 Plastic strain accumulation (Reproduced from Thom, 2008)

dividing these two conditions is called the threshold stress, which as a rough guide has been suggested as around half the undrained shear strength for a cohesive soil (Brown, 1996). The term 'shakedown limit' is also sometimes used in the available literature; it has a similar meaning to threshold stress but is arguably more complex.

The key point for the practising engineer is: if the stress applied remains below the threshold limit, then the effect of the applied stress will be to compact the soil rather than to soften it. Perhaps this logic is seen in nature by the evolution of chickens – whose feet are large enough to impart low stresses to the soil. Consequently, chickens stay clean even during rainy periods, and if you dig the ground below a chicken coop you will find the upper layer to be extremely compact.

The important parameters for pavement foundation design are therefore the elastic stiffness, some measure of plastic deformation and shear strength under repeated load. These parameters are difficult to determine in practice (and the ground will vary along the route, so many tests are required) which has led to significant simplifications in both the testing undertaken and the design.

76.4. Brief recent history of pavement foundation design

In the 1930s the California State Highways Department developed the California bearing ratio (CBR) as an index test for soils, specifically for the purpose of pavement design. The CBR test involves pushing a small plunger into soil at a constant rate and measuring the displacement. The CBR values obtained are a ratio determined by comparing the results to a 'standard' soil. The plunger penetration is due partly to elastic, plastic and shear deformation, and the CBR value obtained is an empirical value with no fundamental meaning. Despite these limitations, it has the advantages of using low-technology equipment, can (unlike more advanced tests) be performed in the laboratory and in situ, is quick to carry out and has become widely used. Various empirical design approaches have been developed around the world to allow for local conditions using CBR values as a measure of stiffness and strength of a material.

The last few decades have brought important developments in the understanding of soil mechanics, testing procedures, materials research and design tools. There are now many possible test techniques which can help determine soil parameters, and design can be undertaken analytically using computer models. However, the difficulty of determining the required soil parameters remains a problem; a practical testing technique has yet to be developed (see Section 76.6 and Fleming *et al.*, 2000). Design standards remain partially empirical and CBR testing and reference to CBR values as an indirect measure of the stiffness of the subgrade are commonplace due to their ease of use.

An underlying problem for pavement foundation design is that the stiffness of the soil prior to construction (i.e. during design) will in many soils be adversely affected by the construction process. The final stiffness of the soil will depend on many factors that the designer has only limited control over; these relate to the construction method employed, and how variations in ground and groundwater conditions are managed. Consequently, soil mechanics theory is only part of what goes into a pavement foundation design; potentially more important is engineering experience and for this reason the empirical CBR approach continues to have a place in practice.

76.5. Current design standards

Sections 76.2 and 76.3 describe how the pavement foundation must reduce the stress from traffic loading to an acceptable level (i.e. below the threshold stress) for the subgrade. The design of the pavement foundation to perform this task is achieved by following industry standards and specifications, and more recently analytical design methods can also be used.

For UK roads and railways, several of the authorities responsible for the construction and maintenance of pavements have their own design standards. The two main standards are those produced by National Highways and Network Rail, both of which are described in this section. Several local councils also have their own design standards for roads, although commonly they are based on a version of the National Highways standards such as the now superseded National Highways HD 25/94 standard (National Highways, 2007) (or even its predecessor, Road Note 29), which remain very informative documents.

76.5.1 National Highways standards

National Highways, responsible for motorways and trunk roads in England, present their design standards in their *Design Manual for Roads and Bridges*. This design standard is also used by the roads authorities in Northern Ireland, Scotland and Wales.

Until recently the National Highways design standard for pavement foundations was HD 25/94, which followed a method or recipe approach to ensure adequate performance. The standard was based on large-scale trials and long-term experience, which is described in LR1132 (Powell *et al.*, 1984). HD 25/94 included a table which relates the subgrade design CBR value to the required thickness of subbase and capping (see **Figure 76.5**). Options are given for a combined subbase and capping layer, or a subbase-only construction – that is, a thinner but higher-stiffness foundation layer. The standard also provides good practical advice, such as restricting changes in foundation design along the route to prevent inconsistent ride quality. The standard was successfully used on many schemes; its main disadvantage is that no methodology was provided to allow for the different properties of foundation materials or traffic loading. An analytical approach which would take account of the engineering properties of the proposed foundation materials would potentially allow a wider range of materials to be used – and foundation thicknesses to be reduced.

In response to this need, in the early 2000s National Highways introduced a new standard superseding HD 25/94: Interim Advice Note (IAN) 73/06 (National Highways, 2006), which has now become CD 225: *Design for New Pavement Foundations* (National Highways, 2020). As well as allowing analytical design for the pavement foundations, CD 225 also links to the latest guidance from National Highways on the design of the upper road layers in CD 226 (National Highways, 2021a), which permits a thinning of the upper layers depending on the stiffness of the foundation (for Classes 3 and 4). However, the old HD 25/94 and previous references remain of value for the design theory and for good practical advice.

CD 225 specifies four permissible foundation classes, defined by their long-term minimum stiffness at the top of the foundation level (termed foundation surface modulus)

- Class 1: 50 MPa
- Class 2: 100 MPa (see example below)
- Class 3: 200 MPa
- Class 4: 400 MPa.

Class 1 is acceptable only for minor roads, Class 2 usually represents a granular pavement foundation, while Classes 3 and 4 represent lime- or cement-stabilised (stiffer) foundations.

The construction (short-term) target values are different to the long-term values, and allow for subsequent confinement for granular materials, or curing rate and cracking in the case of bound materials.

The foundation surface modulus is not to be confused with the stiffness of an individual layer (termed layer modulus). The foundation surface modulus is a 'composite value representing all the foundation layers under the completed pavement' and is affected by the stiffness and thickness of the foundation layers and the underlying subgrade stiffness. Hence, an unbound subbase foundation of sufficient thickness and with a layer modulus of 150 MPa overlying a subgrade with a layer modulus of 50 MPa (approximate CBR of 5%) would provide a Class 2 foundation with a foundation surface modulus of 100 MPa.

Traditionally, before the introduction of CD 225 and IAN 73/06, for an unbound pavement foundation the CBR at the top of the capping was assumed to be in excess of 15% and the CBR at the top of the subbase was assumed to be in excess of 30%. These values corresponded to the CBR of these individual layers and should not be compared to the foundation surface modulus, which, as stated above, is a composite behaviour of the various layers under a 300 mm dia. loaded plate. If the individual foundation layers were sufficiently thick (much thicker than in normal practice) then the loading from the plate would be distributed entirely within the individual foundation layer. In this case, the foundation surface modulus measured by the plate would be the same as the layer modulus of that layer measured using the same test method.

It is important to clarify that capping cannot easily be tested for CBR due to the coarse grading of most capping materials. Static plate load testing can be used to determine an equivalent CBR, but is rarely carried out.

As a rough guide, a CBR of 15% for capping may correspond to a layer modulus of around 100 MPa, and a CBR of 30% may correspond to a layer modulus of around 150 MPa. However, a note of caution is required as correlation of CBR to stiffness is not well defined for higher-stiffness granular or stabilised materials. It is also of note that CD 225 seems to suggest that capping may achieve a stiffness of only 75 MPa, which may correspond to a CBR of less than 15% (depending on the correlation used). This does not correlate well with

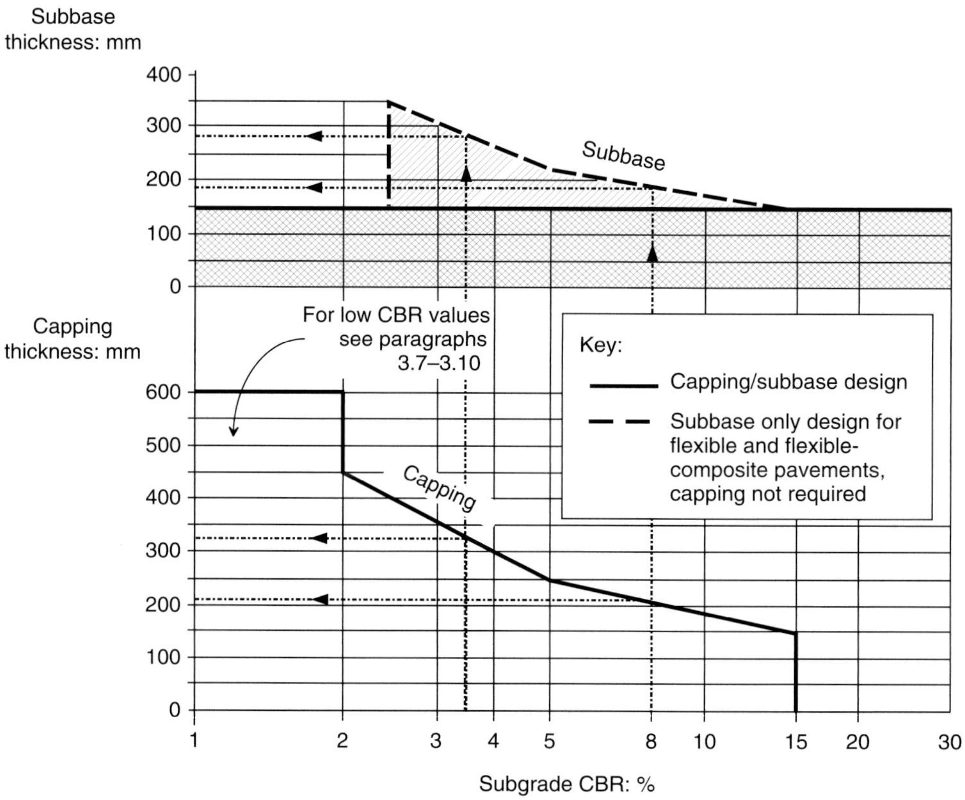

Figure 76.5 Capping and subbase thickness design (Reproduced from National Highways, *Design Manual for Roads and Bridges, HD 25/94.* © Crown Copyright 1994)

traditionally used values; however, it may reflect that capping can range in quality and consistency and a design layer stiffness of 75 MPa is appropriate to represent the wide range of potential sources, including lower-cost locally available and recycled materials, together with higher-quality quarried products that were traditionally used. CD 225 does give the option of carrying out a performance design using a higher stiffness. Experience shows that it is important that the material selected to use as a capping has a low fines content and is from a source that can deliver a layer of reasonably consistent nature. If this is achieved, then the result should be a layer of stiffness suitable for a capping layer.

Two design processes are encompassed in CD 225: a 'performance' approach and a more traditional 'restricted' approach. The restricted design involves determination of the pavement foundation thicknesses based on the subgrade stiffness using a set of charts. The 'performance approach' design methodology is presented in the flow chart shown in **Figure 76.6**. This allows many more theoretical permutations and combinations of materials and design thickness than using the restricted method or the previous HD 25/94 standard. Note that **Figure 76.6** is reproduced from the now superseded IAN 73/06 and is not contained in CD 225, but is retained here as it provides a useful summary of the intended process.

For the performance design, the designer initially chooses appropriate layer stiffness values for the subgrade and foundation layers, and determines the design of the layer thicknesses based either on charts provided for some limited design permutations or, more usually, from the equations provided with the aid of a computer model. The analytical model is based on multilayer linear elastic analysis, which involves a number of simplifications and assumptions from the actual conditions which are described in Thom (2008). Testing and assessment of both the subgrade and foundations materials are required to determine appropriate design stiffness values.

After the design has been completed, a series of demonstration trials are required. The philosophy is that the demonstration trials are situated on various subgrade conditions, which reflect the range of site conditions that will be encountered. The trials are to be constructed using the methods and materials to be used in the main works. Testing and monitoring of the trials are carried out to validate the design, primarily relating to the density and stiffness of the placed layers, and satisfactory resistance to rutting under repeated traffic loading. A key development in the performance design approach is inclusion within the trials and main works of the use of in situ dynamic plate tests (lightweight deflectometer and falling-weight deflectometer, described in Section 76.6) to evaluate the 'as built' stiffness to prove compliance with the required

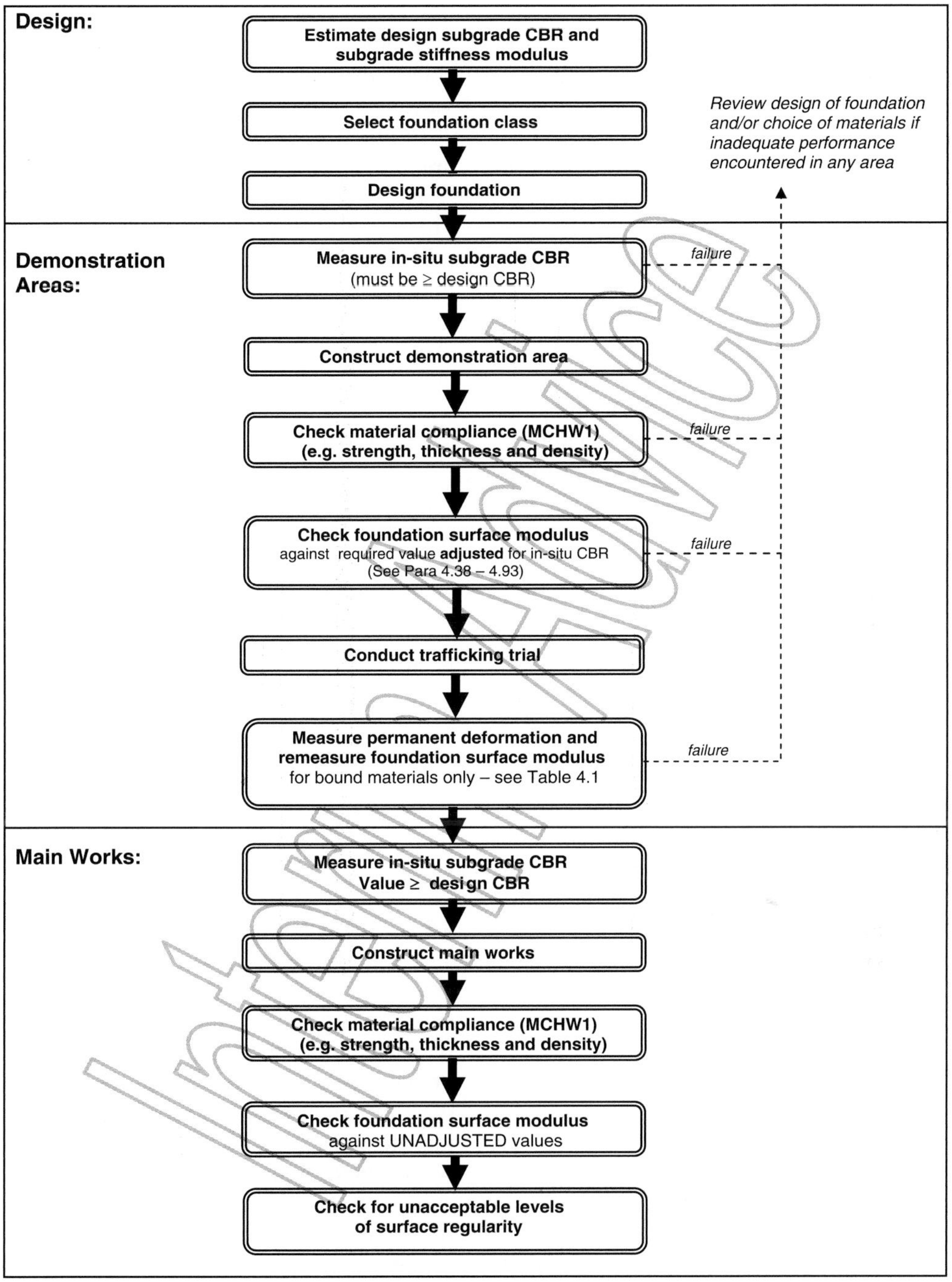

Figure 76.6 Summary flowchart of IAN 73/06 performance design (Reproduced from The National Highways, IAN 73/06 Rev.1 © Crown Copyright 2009)

foundation surface modulus for the intended foundation class. The foundation surface modulus values to be achieved during the trials and works are those expected in the short term and are not a direct method for determining the stiffness in the long term once the foundation is covered by the overlying pavement layers (Edwards and Fleming, 2009).

There has been a fair amount of debate on the benefits of the design approach described in CD 225, and some merits and limitations are provided in Fleming *et al.* (2008). As stated above, the merits are the potential for the use of a wider range of materials – including the reuse of materials giving environmental and cost benefits. The demonstration trials and testing during the works also offer better assurance of performance. However, a number of limitations have been suggested, including very conservative thickness requirements for the 'restricted design' and extensive costly testing for the performance design approach, which for some designs would appear inappropriate. A fundamental assumption remains that short-term behaviour will represent long-term performance. Carrying out trials which represent all the appropriate subgrade materials in all conditions which will influence performance (such as weather wetting/drying effects) in advance of the works will often not be realistic and a degree of judgement will be required. Encountering alternative conditions on site may require additional trials, which are likely to have a knock-on effect to the programme and to the cost.

While the move to a more satisfactory analytical design-based method is welcome, significant development is required to achieve a practical design approach. It should also not be forgotten that the most important factor in pavement foundation design (whichever approach is taken) is to undertake an appropriate assessment of the ground and groundwater conditions, and to develop a design which works for these conditions and the likely method of construction (see Section 76.6). In this regard it is useful to refer to HA 44/91 *Earthworks – Design and Preparation of Contract Documents* (National Highways, 1991), where chapter 10 provides some useful background on the design and specification of pavement foundations. HA 44/91 was produced prior to HD 25/94 and therefore some of the information has now been superseded. However, several of the key principles of pavement foundation design are explained briefly (with cross-reference to LR1132) (Powell *et al.*, 1984) and remain relevant – for example, the purpose of foundation layers (10.1 and 10.2) and discussion on subgrade CBR values (10.9–10.23). Note that HA 44/91 no longer forms part of the *Design Manual for Roads and Bridges*, but remains a useful reference.

76.5.2 Network Rail standards

The most relevant Network Rail standards concerning trackbed foundation design are

- NR/L2/TRK/2102 (Network Rail, 2021) Design and construction of track
- NR/L2/TRK/4239 (Network Rail, 2020) Track bed investigation, design and installation.

NR/L2/TRK/2102 gives minimum ballast depths depending on the category of the track (based on train speed and track usage), rail and sleeper types, and range between 150 and 300 mm. This is based on experience and on minimum depths to allow maintenance, including tamping. For the design of the subgrade and formation, the specification refers to NR/L2/TRK/4239.

NR/L2/TRK/4239 gives recommendations for the design of trackbed treatments. It describes the various potential trackbed problems and suggests possible investigation techniques. Advice is provided on trackbed thickness, and a chart illustrates the required thickness of trackbed layers, which varies depending on the subgrade stiffness and railway type. This is based on a combination of empirical data and multilayer elastic theory, and is in general agreement with German railway standards. Sketches of standard solutions are also provided for various criteria.

In practice, the Network Rail standards for trackbed design are likely to continue to develop and change with further research to encompass a broader range of design cases and ground conditions. At present, the designer needs to have a good understanding of the underlying theory and the research undertaken, particularly by British Rail in the 1950s to 1980s (as presented by Selig and Waters, 1994) in order to determine an overall design in conjunction with the advice given in the current standards.

76.5.3 Other applications

Pavement construction for other purposes (e.g. ports and airports) is not covered in this chapter; however, many of the principles apply to any pavement foundations. Some key reference materials are

- ports: Uniclass L534 (Knapton, 2007)
- airfield pavements: DMG27 (Ministry of Defence, 2011)*
- unsurfaced tracks (TWf, 2019)

* Note: DMG27 was produced by the Ministry of Defence for government airfields but has also been used for commercial airport pavements. Several commercial airports also have their own bespoke standards and design guidance.

76.6. Subgrade assessment

The most important factor in pavement and track foundation design is to undertake an appropriate assessment of the subgrade and groundwater conditions, and to develop a design which is appropriate for these conditions. For the purposes of the assessment, the subgrade should be considered to be the layer or layers of ground below the foundation and not just the sub-formation surface on which the foundation will be founded. This is because the material properties of the subgrade may change with depth and along the route, and in particular the stiffness may reduce. The depth of subgrade which needs to be considered depends on the depth to which the majority of the stress from traffic will be dissipated during construction and in service. The influence of traffic-induced stress can extend to as much as 5 m below the bottom of the sleepers for railways (Brown, 1996). However, in the majority of circumstances stress is usually considered to be distributed to around 1.5 m below the base of the foundation. This is supported by CD 225, which suggests a rigid layer is modelled for analytical design 1.5 m below the foundation.

Unlike the other foundation layers, the subgrade is variable and its performance is influenced by many factors. The factors that apply are site-specific, but are likely to include some of the following, and ground investigations should be designed to address these issues

- soil type, grading (especially for borderline cohesive/granular soils)
- variability within the soil type
- permeability and stiffness
- horizontal and vertical geological variability
- soil fabric (e.g. lamination)
- likely presence of hard/soft spots
- groundwater conditions and drainage (prior to excavation)
- conditions that encourage elevation of pore water pressures during construction (e.g. silts/fine sand or a thin clay layer over sand)
- topography (e.g. transition zones from cut to fill)
- construction procedures adopted and skill of the site foreman to implement these
- construction season, timing of drainage installation, exposure time of subgrade
- quality control systems.

It is a common misconception that the subgrade stiffness is the only factor which needs to be considered; it is just one item on a long list. To illustrate this issue, consider projects where the subgrade consists of glacial till which typically comprises a sandy gravelly clay/silt with cobbles and boulders, and has a relatively low fines content. The subgrade usually has a relatively high stiffness with design CBR values typically varying between 3% and 6%. However, due to the grading of the material, which includes a low but dominant, fines fraction, the engineering properties of the material can be very sensitive to changes in moisture content. The permeability of the material can also be reasonably high. If groundwater is high (such as in cuttings) and is not adequately controlled, the subgrade stiffness can deteriorate rapidly on excavation.

In determining the design subgrade stiffness, consideration should be given to both the short-term (during construction) and long-term (in service) conditions. The stiffness of the subgrade is dependent on the moisture content, particularly in fine-grained soils. If an excavation for a road pavement is undertaken during the summer in dry conditions, the stiffness recorded in situ may be high, but following construction of the road in service, the moisture content is likely to increase to the equilibrium moisture content – resulting in a reduction in stiffness. Conversely, construction in wet conditions in winter is likely to find lower initial stiffness subgrade conditions during construction and these may then deteriorate to below equilibrium moisture content stiffness. In this case the material will never recover to a significantly higher stiffness in service.

There are many possible exploratory and testing methods to investigate the subgrade, many of which are discussed in Thom (2008) and Network Rail (2020) and are not repeated here. As with all ground investigations, the methods should be designed specifically for the ground conditions present. Practical experience shows that it is better for the ground investigation to obtain a large number of relatively simple tests along the route (such as index properties, grading and a simple measure of shear strength) than simply to focus on a small number of advanced laboratory tests to determine stiffness.

Despite significant work carried out in recent years, there is no one simple method to determine subgrade stiffness, either in situ or in the laboratory. CD 225 specifies use of the falling-weight deflectometer (FWD) and/or the lightweight deflectometer (LWD) during construction for performance-based design. However, a review of a wide body of data from various LWD development trials on capping materials has shown that the stiffness values determined can vary significantly. This is complicated further by the large number of factors from the material properties and material states that can influence this variability (Lambert, 2007). Similar variability is typical for other in situ tests and laboratory testing. It is recommended that a range of testing is carried out to allow correlation between the various methods and a degree of engineering judgement, based on experience, is usually required to determine design stiffnesses.

The requirement to determine the long-term stiffness is also problematic. The long-term stiffness should be based on the equilibrium moisture content – which is the maximum moisture content that is likely to be present during the operation of the pavement. It is possible to undertake testing in the laboratory over a range of moisture contents to reflect the conditions which are likely to be experienced during construction and operation of the pavement. The equilibrium moisture content can also be determined from measurements on a suction plate (Black and Lister, 1979). However, this is not widely undertaken in practice due to uncertainty in the equilibrium moisture content and difficulties of the testing. As a worst case, testing can be carried out on saturated samples, with a common (but unsatisfactory) example being soaked CBR testing. However, this may be too severe and can result in very low results which may be unrealistic. In response to these difficulties, National Highways produced a table that provides an estimation of the subgrade CBR value at equilibrium moisture content for various soils (see **Table 76.1**). These estimates are mainly used for clay soils for which the estimated CBR depends on the plasticity index, the water table, the construction conditions and the thickness of the pavement.

Table 76.1 was originally published in LR1132 (Powell *et al.*, 1984) and has been subsequently included in HD 25/94 and IAN 73/06 revision 0. CD 225, which supersedes IAN 73/06, does not contain the table, but reference is given to LR1132, which remains an important document for designers to understand the underlying logic. The table is widely used in practice to estimate the long-term CBR values for clay subgrades and has been found to give reasonable results. It needs to be used with caution as there are a number of complicating factors.

Soil	Plasticity index	High water table[a] construction conditions						Low water table[b] construction conditions					
		Poor		Average		Good		Poor		Average		Good	
	%	Thin[d]	Thick[c]	Thin	Thick	Thin	Thick	Thin	Thick	Thin	Thick	Thin	Thick
Heavy clay	70	1.5	2	2	2	2	2	1.5	2	2	2	2	2.5
	60	1.5	2	2	2	2	2.5	1.5	2	2	2	2	2.5
	50	1.5	2	2	2.5	2	2.5	2	2	2	2.5	2	2.5
	40	2	2.5	2.5	3	2.5	3	2.5	2.5	3	3	3	3.5
Silty clay	30	2.5	3.5	3	4	3.5	5	3	3.5	4	4	4	6
Sandy clay	20	2.5	4	4	5	4.5	7	3	4	5	6	6	8
	10	1.5	3.5	3	6	3.5	7	2.5	4	4.5	7	3	> 8
Silt*	–	1	1	1	1	2	2	1	1	2	2	2	2
Sand (poorly graded)	–	20											
Sand (well graded)	–	40											
Sandy gravel (well graded)	–	60											

*Estimated, assuming some probability of material saturating.
[a] A high water table is 300 mm below formation or sub-formation.
[b] A low water table is 1000 mm below formation or sub-formation.
[c] A thick layered construction is a depth to subgrade of 1200 mm.
[d] A thin layered construction is a depth to subgrade of 300 mm.

Table 76.1 Equilibrium subgrade CBR estimation (Reproduced from National Highways, Interim Advice Note IAN 73/06. © Crown Copyright 2006)

- Plasticity index is determined using samples of material with a grading of less than 425 μm. For well-graded soils that contain both a granular and cohesive fraction such as glacial till, the stiffness is affected by both fractions and the CBR values determined from **Table 76.1** may not be representative for material with a low fines content.
- Plasticity index results are often variable, and the designer must choose a representative value or range of values.
- For low-plasticity soils with a plasticity index of 5–10%, the designer must judge if the subgrade will act as a silt (CBR of 1–2%) or a clay (CBR of 2.5–8%).
- It is also important to realise that if construction conditions are poor and groundwater is not controlled, the stiffness of fine-grained soil can easily fall below the values in **Table 76.1** and will not fully recover.
- The table also gives CBR values for granular materials which are relatively high and the actual stiffness will depend on the in situ density and pore water pressure. However, in practice, the requirement is simply to check if the subgrade remains above a CBR of 15%.

To obtain confirmation of the proposed design, a CBR investigation of an existing pavement of similar construction could be carried out. This approach is not often followed, and the few results obtained tend to show that the standard values are reasonable.

It is important for the designer to check that the original shear strength of the soil is sufficient to deliver the expected equilibrium moisture content CBR value, as construction will usually lead only to a softening below the undisturbed condition. In this regard, a commonly used approach is to apply an approximation for clay soils of CBR = $C_u/23$ (Black and Lister, 1979), where C_u is the undrained shear strength.

Recognising the importance of equilibrium moisture content and determining the design based on an assessment of long-term and short-term conditions cannot be underestimated. There have been several construction disputes where a contractor and designer have naively assumed they can simply carry out some in situ testing at the time of construction and adjust the pavement thickness accordingly, with no consideration of the long-term subgrade stiffness.

76.7. Other design issues

Once the ground conditions are confirmed, the majority of the design effort is often focused on providing a sufficient construction thickness to prevent (*a*) subgrade strength failure; and (*b*) excessive subgrade deformation. While these two failure mechanisms are obviously important, the design should also consider a range of issues as follows.

76.7.1 Materials

The design should carefully consider and specify the materials to be used in the pavement foundation. In particular, the materials specified should suit the ground and groundwater conditions present at the site. This is usually achieved by reference to material types given in standard specifications, such as the *Manual of Contract Documents for Highway Works* (MCHW) (National Highways, 2009), with further advice given in 'BS 6031: Code of practice for earthworks' (BSI,

2009). The following examples illustrate a few of the most common issues to be considered.

CD 225 offers a reduction in foundation thickness, should a single subbase layer be used rather than a two-layer subbase on a capping due to the higher stiffness of the higher-quality subbase layer. Contractors will generally prefer a subbase-only solution based on an economic assessment of material costs. In practice, however, for many sites the use of capping is advisable to manage the subgrade conditions, as it is coarser and quicker to construct. This is particularly true where subgrade conditions are susceptible to changes in moisture content and the capping provides some protection and a platform for the construction of the subbase layer (which is finer and requires more precise construction). CD 225 also offers benefits from providing stiffer lime- or cement-stabilised foundations. In practice, it is important to consider whether the ground conditions are well suited to the use of thin subbase only or stabilised foundation solutions and whether any additional measures (e.g. temporary drainage or modification of the stabilisation mix) may be necessary to avoid problems with moisture-susceptible subgrades.

CD 225 performance-based design, together with recent revisions of the MCHW, has opened up the possibility of a wide range of materials to be used in the foundation. This allows a greater potential for the reuse of locally available materials. The capping layer, in particular, is intended to be a locally available material, and the new specification provides greater potential for this to be realised.

The materials need to be selected to ensure that they do not deteriorate under trafficking. This is a common problem for railways, as the ballast is highly loaded due to its position directly under the sleeper, the heavy loading from trains and mechanical maintenance. The ballast will break down with time and eventually become clogged with fines. Once pore water pressures cannot dissipate under loading, track geometry will be affected and the occurrence of wet spots will become a common feature. This is described in NR/L2/TRK/4239 (National Rail, 2020).

Geotextiles are often incorporated in pavement foundations. The Network Rail standards (see NR/L2/TRK/4239) specify the use of a separating geotextile or geocomposite where there are subgrade erosion (pumping) issues, which may be used in conjunction with blanketing sand. The use of separating geotextiles is relatively uncommon for road pavements, as subgrade pumping is usually less of an issue (due to the lower traffic loading and usually better drainage), although it may be considered for particular conditions. Geogrids may be incorporated in pavement foundations, particularly where the subgrade has a very low stiffness.

Near-surface materials should not be susceptible to frost, and guidance on frost protection is provided in Series 800 of the MCHW (National Highways, 2021b). For routine cases, all material within 450 mm of the road surface should be non-frost-susceptible, although this may be reduced if the frost index is low. The frost index is a measure of the severity of a period of cold weather and provides a means of assessing likely penetration of frost into a road.

Environmental issues should be carefully considered when designing pavement foundations. Considering that some 25–30% of all materials extracted from the ground finish up in highways pavements (Thom, 2008), the environmental impact of pavements is clearly significant. Innovation to limit the environmental impact and reduce cost is a key aspect of design. The design life is an important consideration for the upper-bound layers, but is not an easy consideration for pavement foundations. For these, it is usually desirable to use locally available materials to reduce transport costs during construction. Recycled materials are also often used, particularly for capping layers in road pavements, and National Highways has made significant steps in recent years to encourage recycling (e.g. the addition of a wider range of recycled aggregates to table 6/1 of the MCHW, and CD 225, which permits a wide range of construction materials).

76.7.2 Drainage

Adequate drainage is of vital importance to the performance of the pavement and should be provided during both the construction and operation of the pavement.

During construction, the subgrade should be protected to avoid deterioration due to surface water and groundwater. Failure to adopt suitable measures can result in the requirement for extensive additional excavation and replacement below sub-formation, which can significantly add to the construction costs. For moisture-susceptible subgrades and cutting areas where groundwater is high, deep drainage is often required before the verge to draw down groundwater prior to excavation to sub-formation level. Additional drainage in the centre of the pavement or cross-drains may also be required in some cases. Open excavations may need to be limited, particularly in moisture-susceptible soils. If a drainage layer is present at the base of the pavement foundation, the sub-formation should be sloped so that water can be shed to the sides – to prevent ponding during construction, or entrapment of water in the completed pavement. If the upper pavement foundation layers are also intended to act as drainage layers (e.g. ballast or type 1 subbase), their base may also need to be sloped.

Drainage during the operation of the pavement should be designed to maintain groundwater level below sub-formation level (earthworks drainage) and to drain water entering the pavement foundation by seepage through the upper pavement layers (pavement drainage). The pavement foundation layers should be constructed with a fall towards the drainage. This is a particularly important consideration for widening projects and a fall should be maintained through the existing and proposed pavement foundation layers to the drains. For these schemes, this can be the governing factor that determines the thickness of the pavement foundation layers.

76.8. Construction specification

The design needs to be adequately presented using drawings and/or a specification to allow construction. Unless the scheme is small and the ground conditions are very well understood, it is likely that the conditions in some areas of the site will be different to that indicated by the ground investigation information. This presents a risk of delays on site while the design is amended, or the risk that the construction will be inadequate for the ground conditions present. This risk can be addressed by adopting a conservative design which will be adequate for most conditions likely to be present, preferably by providing flexibility in the design. The flexibility, where variable site conditions are encountered during construction (or limited pre-construction site data are available), is embodied in the 'observational approach' (Nicholson *et al.*, 1999). A similar methodology has been developed for a number of design and build schemes (Gilbert, 2004); it has proved appropriate for large highway foundation areas and allows swift construction. This was achieved by including full details of the design assumptions on the drawings and specification, including

- pavement foundation layer specifications and thicknesses
- design subgrade CBR value or stiffness
- design subgrade ground conditions (e.g. clay, sand)
- design groundwater level, if relevant.

A defined regime was also specified, comprising site inspection coupled with rapid in situ testing methods to confirm the design assumptions and identify locations where the conditions differ from those expected. Clear procedures were also developed to enable changes to be made on site, based on the actual conditions encountered. This allows the designer to make realistic predictions of a range of pavement foundation requirements. However, for this approach to work, trust and cooperation are required between all parties (client, contractor and designer) to allow the optimum foundation for the scheme to be built

The approach given in CD 225 for performance-based design is similar to Gilbert's methodology in that the design assumptions must be clearly stated and rapid in situ testing methods are used during construction to confirm the design. The potential drawback is that there is no simple, quick procedure to enable changes to be made on site based on the actual conditions encountered, and as stated in Section 76.5.1, additional trials could be required which are likely to have a knock-on effect to programme and cost.

In addition to the issues mentioned above, the drawings and specification should address a number of other issues for the construction, including

- permanent and temporary drainage
- measures to protect the subgrade during construction

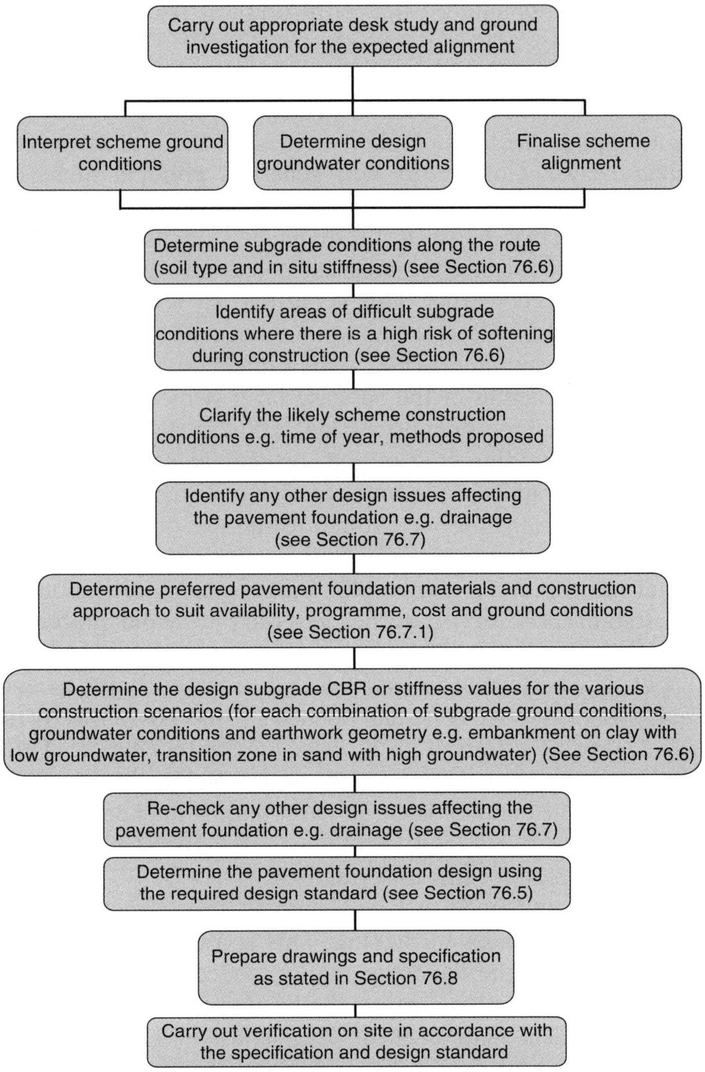

Figure 76.7 Pavement foundation design summary flow chart

- transition zones, particularly cut/fill intersections, should be given special attention to avoid significant variations in pavement stiffness
- consideration of consistent ride quality (i.e. minimise changes in foundation along the route).

A flow chart summarising the steps required in the design of pavement foundations is given in **Figure 76.7**.

76.9. Considerations for high-speed rail pavement foundations

High-speed rail is playing a growing role in improving the capacity, availability and carbon cost of national infrastructure, through both the construction of new routes and the upgrading of existing routes. High-speed rail line speeds are typically 250–400 km/h ('conventional rail' line speeds are up to 200 km/h), requiring stricter tolerances on ground engineering performance than any other form of transport infrastructure. Currently in the UK design speeds for high-speed trains are 300 km/h on High Speed 1 (2007) and 360 km/h on High Speed 2 (expected 2029).

76.9.1 Design approach for high-speed rail

The design approach for a project should be driven by the client's vision for how the entire track and earthworks system is to be managed through design, construction and railway operation. The geotechnical engineer is typically responsible up to the formation level (the top of the protection layer, which caps the embankment fill), above which sits the trackbed/track system (sub-ballast, slab or ballast, sleepers and rails) (**Figure 76.8**). The earthworks design team (geotechnical, drainage and structural engineers) must work together to deliver a stable earthworks platform of consistent stiffness upon which the track system can be constructed.

The track form for speeds of up to 320 km/h is commonly ballasted track, which has the benefit of being compatible with conventional track maintenance methods to maintain alignment. For faster line speeds slab track becomes the only practical option. Ballast track is cheaper to construct, but the additional construction cost of slab (ballastless) track is often considered to be offset by lower maintenance costs within the first 10–20 years of operation. However, rigid track slab systems can only accommodate very limited deflections in the underlying ground, which necessitates an even greater focus on the earthworks design and construction. Changes between the two track forms should be kept to a minimum as these transitions add design complexity.

For both track forms high train speeds demand tight controls on track deflection; to achieve this the earthworks design should satisfy the following requirements.

- Settlement affecting the track is minimised.
- Heave in deep cuttings within overconsolidated clay soils is minimised.
- Fill materials are selected and compacted to achieve low air voids and high stiffness.

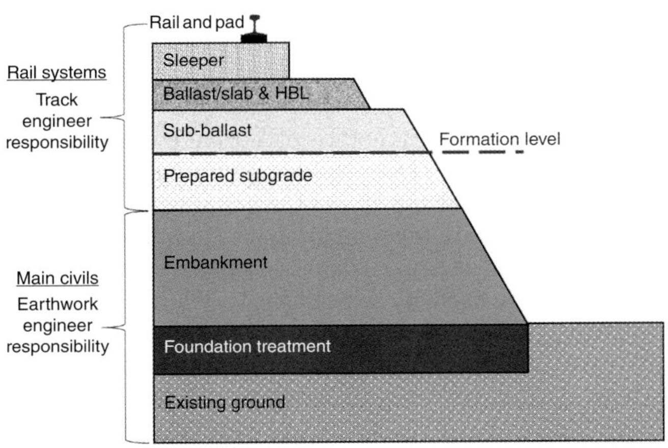

Figure 76.8 Typical contract responsibilities

- Variation in stiffness along the trackbed is minimised.
- The earthwork and subgrade will not be at risk of dynamic response (critical velocity effects) under high-speed train loading.

To achieve these requirements the design of appropriate 'foundation treatment' (see **Figure 76.8**) becomes an essential part of the overall earthworks and trackbed design. The design must resolve problems in the natural ground in order to deliver acceptable performance at the top of the earthwork. This requires robust ground risk management through the design and construction phases, including: a phased ground investigation, detailed engineering geology ground modelling and checks during construction to address any unexpected and/or unacceptable ground conditions. The design should determine foundation treatment requirements under embankments and through earthworks sections at-grade and in cuttings, as shown in **Figure 76.9**. This approach should be considered for all infrastructure, but it becomes central to high-speed rail earthworks design due to the extremely tight deflection criteria for settlement at the top of the earthwork when in operation.

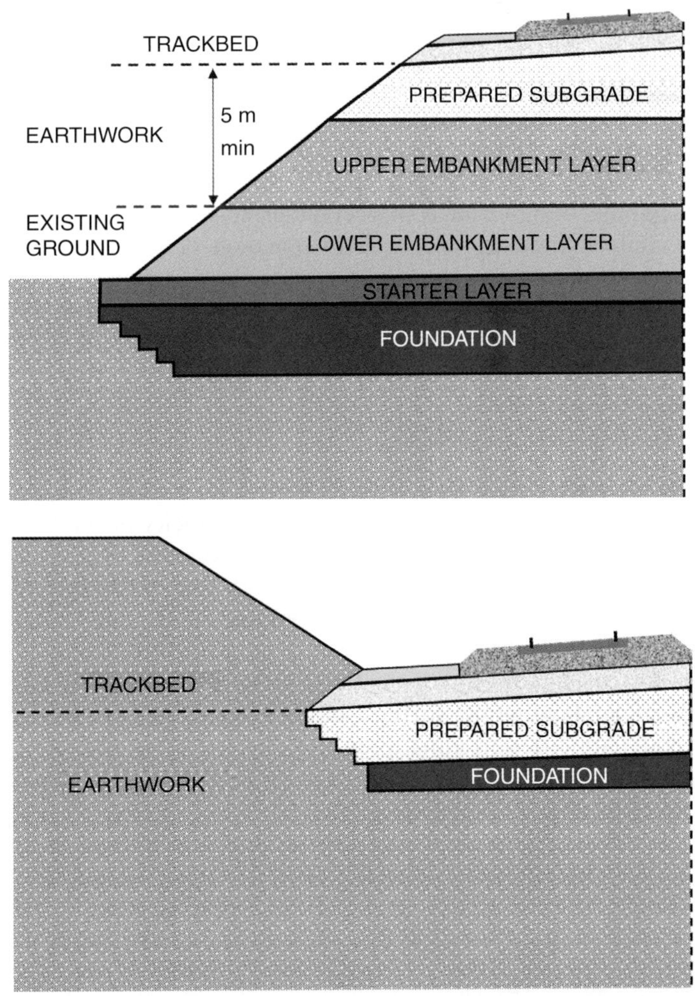

Figure 76.9 Typical earthworks cross-sections for embankments and cuttings, to meet HS2 Technical Standard (2019)

Settlement of the natural ground beneath the railway earthworks tends to be the biggest single driver that determines the foundation treatment requirements, particularly in the case of embankments. The management of settlement requires the greatest attention at transitions, such as earthwork to structures, cutting to embankment, major changes in geology or changes in track form.

The one design criterion unique to high-speed rail is the requirement to address dynamic response issues. The geotechnical design should include a geodynamic assessment to ensure that the combination of earthwork, foundation treatment and underlying natural ground will be of sufficient overall stiffness to avoid critical velocity effects (a phenomenon seldom considered in conventional rail design). This topic is discussed further in Section 76.9.3.

76.9.2 High-speed rail design standards

The design of high-speed rail earthworks and trackbed is covered by HS2's Earthworks Technical Standard (HS2, 2019), which was developed considering experience from other countries. The HS2 project required a significant increase in line speed for UK rail, which necessitated an approach to address dynamic response issues within the earthworks design.

There are no Eurocodes specifically addressing high-speed rail design, although the 2020 UIC document *IRS70719-1: Earthworks and Trackbed Layers for Railway Lines* covers some design and construction principles, and the HS2 Technical Standard approach to selection of prepared subgrade thickness uses the UIC subgrade classification system.

Section 76.9 reflects the overall requirements of the HS2 Technical Standard, and summarises international project experience of how to resolve high-speed rail earthworks design requirements.

76.9.3 Critical velocity phenomena

Critical velocity effects occur where the velocity of the train approaches or exceeds the Rayleigh wave velocity (V_R) of the combined underlying ground, earthwork and track system. The resulting displacements and vibration can be much larger than usual, resulting in significantly increased track maintenance costs, as well as potentially reducing running speeds. By understanding the stiffness of the ground and earthworks, geodynamic design aims to ensure that the V_R of the site is greater than the train speed so as to prevent the occurrence of any detrimental critical velocity effects.

While problems are more likely for high-speed trains, critical velocity effects have been recorded around the world for a range of train speeds – as the phenomenon is caused by the ratio of the train speed to the site, critical velocity problems can be experienced on very soft ground at relatively low train speeds. **Figure 76.10** is a dynamic amplification curve, based upon data from real critical velocity sites. Track displacement, normalised by static load values, is plotted against train speed, normalised by the site critical velocity. A strong trend can be seen, with the curve divided into three indicative zones.

If the train speed is approaching or equal to the site critical velocity then significant dynamic amplification occurs and displacements are generally unacceptably high, sitting within the red zone. Conventional railways did not specifically consider critical velocity in their design, but most operate in the green zone, with displacements during train passage similar to those under static loading.

It is not practical to restrict high-speed rail to the green zone – the amber zone sets an upper limit of acceptable performance, which is normally taken as a ratio of train speed to V_{crit} of the railway system (equivalent to V_R at the track level). For contractual certainty this is simplified to the ratio of train speed to V_R at the top of the earthwork (the HS2 Technical Standard sets this ratio at 0.625).

It is also possible to define deflection or acceleration limits for high-speed rail; however, the use of such criteria in a large-scale design process is not practicable. As the acceptable limits would be extremely small, the accuracy of any prediction model would need to be extremely high, requiring significant amounts of accurate input data and processing power. Such models may have a role at discrete locations, but for the bulk design of linear infrastructure a critical velocity criterion remains the most functional option.

76.9.4 Geodynamic design

Geodynamic design is the approach used to avoid problems associated with critical velocity. While critical velocity effects are generally measured at the track level (as in **Figure 76.10**), contractual practicalities mean earthworks are typically designed to a target V_R at the formation level (**Figure 76.8**). This can only be measured using geophysics, once earthworks construction is complete. During design V_R must be predicted based upon knowledge of the natural ground, foundation materials and earthworks. Rayleigh waves are a ground-borne surface wave, which are strongly related to the shear wave velocity (V_S) of the underlying soil layers. This enables the V_S of individual components of the ground and the earthworks to be used as a proxy for the overall V_R. Richart (1962) provides more detail on quantifying the relationship, but for a typical soil V_S is approximately 90–95% of V_R.

The closer a soil layer is to the surface, the greater the impact it has on the overall V_R. Design is commonly based on effective characterisation of the top 15 m of material below the foundation level. Historically, problem locations include softer or more organic soils, such as alluvial deposits, marine clays, peaty soils and other surface deposits, which have very low V_S due to their lower shear modulus. At high train speeds more 'normal' soil types which may be acceptable from other engineering perspectives can also be problematic. Strong weathering or low levels of overburden can also result in a soil with a comparatively low V_S – for example, the V_S of a weathered mudstone may be significantly lower than that of competent mudstone.

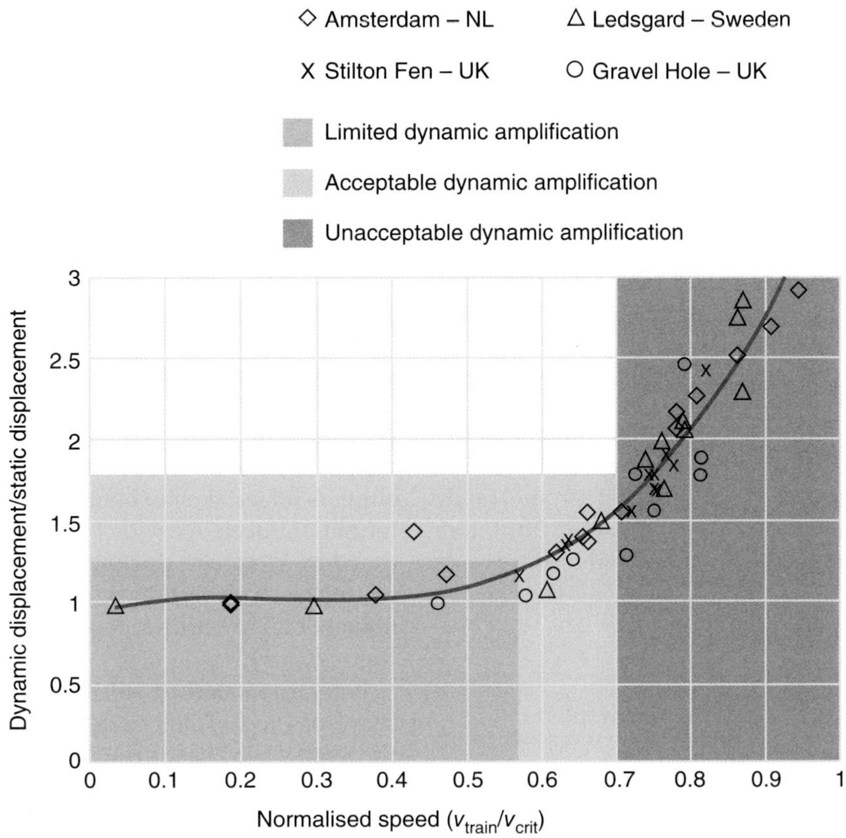

Data from Woldringh and New (1999) and Duley (2018)

Figure 76.10 Dynamic amplification curve from real critical velocity sites (Reproduced from Duley, 2021, with permission from the copyright holder © Ground Engineering 2021)

Ground tends to fall into three categories for geodynamic assessment: very poor ground, where treatment for settlement or support stiffness normally resolves the geodynamics; very stiff ground, where it is clear no geodynamic treatment is required; and the 'middle' ground, where foundation treatment may need to be increased to address geodynamic issues.

The geodynamic design requires specific soil properties obtained by special ground investigation methods for both the natural ground and the proposed fill materials. In general terms, the design uses numerical or analytical models to predict the V_R at finished earthwork formation level. Ideally the design predictions should be confirmed and further developed through full-scale earthworks site trials, which clarify the geodynamic properties of the treated ground and fill materials. The designer needs to judge which locations should be tested to enable an overall understanding of the site conditions. This is complicated by the reality that geodynamic earthworks design is a developing art.

76.9.5 Ground investigation for geodynamics

Conventional ground investigation methods are used to develop the overall ground model for the site and to understand the range of soil properties in terms of density, strength and stiffness. Specific geophysical testing is required to measure the geodynamic properties of the various soil types at sufficient locations to give an overall understanding of the soil profiles for geodynamic design. Comparison of the geophysics and the conventional ground investigation is important.

Indirect estimates of V_S using the standard penetration test (SPT) correlation (see e.g. Wair *et al.*, 2012) are not reliable, but can be useful for initial estimation purposes.

A range of geophysical techniques can be used to measure or estimate the V_S profile of the existing ground. Surface survey techniques include multi-channel analysis of surface waves (MASW), continuous surface wave (CSW) and refraction microtremor (ReMi), which typically utilise a linear array of geophones on the ground surface and an impulse source. Depending on setup and ground conditions, these techniques can estimate the V_S profile in sufficient detail down to approximately 30 m below existing ground level. For higher granularity of results, measurements covering the top 15 m are most common. In locations of future cuttings, it is important to ensure the survey design will span below the level of the future formation.

Invasive geophysical techniques include seismic cone penetration tests (sCPT) and down-hole (DH) and crosshole

(XH) surveys. Generally, invasive techniques offer the most accurate measurements of V_S. However, surface surveys such as MASW and CSW are the most practical methods for managing earthworks projects because they allow significant linear lengths of route to be tested for design purposes, and the same test method can be used for post-construction verification.

The designer should use all available information to make their best prediction of typical V_S profiles for each material and location. Combining geophysical measurements with ground models allows calibration of the soil type to V_S, enabling any data gaps in the geophysics to be covered and an understanding of the V_S profile along the entire route to be estimated. It is important to recognise the limitations of the test methods available, and the inherent variabilities in geophysical measurements due to equipment, ground and human data processing elements (see e.g. Foti *et al.*, 2018, for magnitudes). Such uncertainties are commonly addressed through the application of a partial factor during the design process.

76.9.6 Fill properties and construction control for high-speed rail

While general considerations for earthworks and embankment fill are similar to those in conventional earthworks, a higher magnitude and more consistent earthworks stiffness is required in high-speed rail. Furthermore, compacted fill material must remain stable under cyclic loading over decades. This dictates that the fill materials selected be compacted to reduce air voids, and many general fill materials will require stabilisation.

Quality assurance (QA) testing includes plate load testing to measure the deformation modulus on the second load cycle (E_{V2}). On HS2, for slab track with a design speed of 360 km/h, the E_{V2} requirements (based on a 600 mm dia. plate) are 60 MPa in the upper embankment fill and foundation treatment (see **Figure 76.9**). Lower stiffness is permissible in the lower embankment fill, but the project must be satisfied that the completed earthwork will deliver the required V_R.

The fill design process follows normal good practice of laboratory relationship testing under high compactive effort to identify the acceptable range of fill material moisture content that will achieve low air voids and adequate stiffness (CBR testing of the compacted samples is used as an indicator of stiffness).

Trial embankments are then required to check whether the fill materials compacted under site conditions behave as predicted. Geophysics is undertaken on the compacted fill to identify V_S for use in design. E_{V2} testing is also undertaken on the compacted fills to check that the QA stiffness criteria can be satisfied (E_{V2} testing will be undertaken throughout construction), and if achieved should deliver the required V_S (which will only be tested on the completed earthwork). The combined information from the laboratory relationship testing and the trial embankments is used to develop the earthwork fill specification.

76.9.7 Practical geodynamic assessment method

A practical approach to geodynamic design can be to determine the ground model at a location and assign a V_S to each layer (measured for natural ground and predicted for fill materials), and then judge acceptability based on V_S relative to a target value (e.g. $V_{S\,\text{target}} = 1.05 \times V_R$). At locations where some of the ground does not meet the target value, then forward modelling is undertaken to obtain a prediction of V_R at earthworks formation level. As with V_R, the forward modelling is influenced by V_S values, thickness of the layer and depth below the top of the earthwork (near-surface layers have greater overall influence). The earthworks design then progresses as shown in **Figure 76.11**.

Forward modelling software predicts the overall dispersion curve of the finished design. A dispersion curve is a plot of Rayleigh wave phase velocity against either wavelength or frequency – this is what surface wave geophysics directly measures. The lowest phase velocity within a certain frequency range (relating to the train loading) is taken as the predicted V_R at formation level. The forward modelling method is well established for most ground conditions; research continues to explore the accuracy of these predictions for stiff embankments over softer ground.

More complex design solutions such as 4D modelling (time-domain 3D numerical modelling) are generally only used for research purposes because these models are time-consuming, and typically require advanced laboratory testing to source strain-dependent parameters. The research continues to inform practical design for projects in this emerging field.

The most common form of foundation treatment is to excavate and replace in the zone illustrated in **Figure 76.9**, and this can be modelled by forward modelling. Across very soft ground more expensive treatments such as piled embankments are sometimes required, but it remains unclear how effective geophysics or forward modelling will be for such earthworks.

76.10. Conclusion

There have been significant recent developments in pavement foundation design, particularly for roads and high-speed rail. Whatever the application or method used for design, it should be remembered that the most important factor is to undertake an appropriate assessment of the subgrade and groundwater conditions and to develop a design which works for these conditions.

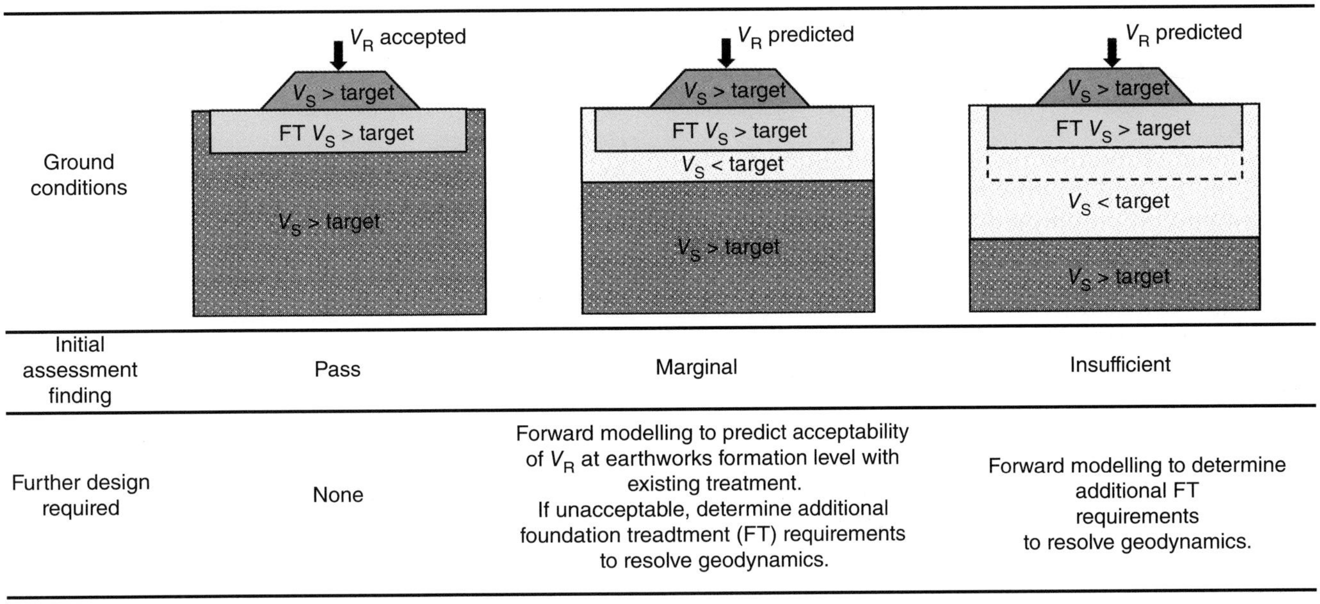

Figure 76.11 Overview of a practical geodynamic assessment method

References

Black WPM and Lister NW (1979) *The Strength of Clay Subgrades: Its Measurement by a Penetrometer (LR889).* Transport Research Laboratory, Crowthorne, UK.

Brown SF (1996) Soil mechanics in pavement engineering. 36th Rankine Lecture of the British Geotechnical Society. *Géotechnique* **46(3)**: 383–426.

BSI (British Standards Institution) (2009) 'BS 6031: Code of practice for earthworks'. BSI, London, UK.

Duley A (2021) Practical geo-dynamic assessment of high speed rail earthworks on the Align contract (High Speed 2). *Ground Engineering*, Nov.

Edwards JP and Fleming PR (2009) *LWD Good Practice Guide Version 9.* Working Draft. Department of Civil and Building Engineering, Loughborough University, Loughborough, UK.

Fleming PR, Frost MW and Rogers CD (2000) A comparison of devices for measuring stiffness in situ. *Proceedings of the 5th International Conference on Unbound Aggregates in Roads, Nottingham, UK*, pp. 193–200.

Fleming PR, Frost MW, Gilbert PJ and Coney P (2008) Performance related design and construction of road foundations: review of the recent changes to UK practice. *Advances in Transportation Geotechnics, Proceedings of the International Conference, Nottingham, UK.* Taylor & Francis, Abingdon, UK, pp. 135–142.

Foti S, Hollender F, Garofalo F, *et al.*. (2018) Guidelines for the good practice of surface wave analysis: a product of the InterPACIFIC project. *Bulletin of Earthquake Engineering* **16**: 2367–2420.

Gilbert PJ (2004) Practical developments for pavement foundation specification. *Proceedings of the International Seminar on Geotechnics in Pavement and Railway Design and Construction, Athens, Greece* (Gomes C and Loizos A (eds)). Millpress, Rotterdam, the Netherlands.

HS2 (2019) HS2-HS2-GT-STD-000-000001 Technical standard: Earthworks, Rev P08. HS2, London, UK.

Knapton J (2007) *Uniclass L534: Heavy Duty Pavements*, 4th edn. Interpave, Leicester, UK.

Lambert JP (2007) *Novel Assessment Test for Granular Road Foundation Materials.* Engineering doctorate thesis, CICE Loughborough University, Loughborough, UK.

Ministry of Defence (2011) *DMG 27: Design & Maintenance Guide 27: A Guide to Airfield Pavement Design and Evaluation.* Ministry of Defence, London, UK.

National Highways (1991) *Design Manual for Roads and Bridges. Vol. 4 Geotechnics and Drainage: Section 1 Earthworks, Part 1 HA44/91.* The Stationery Office, London, UK.

National Highways (2006) *Design Guidance for Road Pavement Foundations (Draft HD 25). Interim Advice Note IAN 73/06.* The Stationery Office, London, UK.

National Highways (2007) *Design Manual for Roads and Bridges. Vol. 7 Pavement Design and Maintenance, Section 2 Pavement Design and Construction, Part 2 HD 25/94.* The Stationery Office, London, UK.

National Highways (2009) *Design Guidance for Road Pavement Foundations (Draft HD 25). Interim Advice Note IAN 73/06 Revision 1.* The Stationery Office, London, UK.

National Highways (2020) *CD 225 Design for New Pavement Foundations.* The Stationery Office, London, UK.

National Highways (2021a) *CD 226 Design for New Pavement Construction.* The Stationery Office, London, UK.

National Highways (2021b) *Manual of Contract Documents for Highway Works (MCHW).* Series 800. The Stationery Office, London, UK.

Network Rail (2020) NR/L2/TRK/4239: Track bed investigation, design and installation. Network Rail, London, UK.

Network Rail (2021) NR/L2/TRK/2102: Design and construction of track. Network Rail, London, UK.

Nicholson D, Tse CM and Penny C (1999) *The Observational Method in Ground Engineering.* CIRIA Report R185. CIRIA, London, UK.

Powell WD, Potter JF, Mayhew HC and Nunn ME (1984) *The Structural Design of Bituminous Roads (LR1132).* Transport Research Laboratory, Crowthorne, UK.

Richart FE (1962) Foundation vibrations. *Transactions of the ASCE* **127(1)**: 863–898.

Selig ET and Waters JM (1994) *Track Geotechnology and Substructure Management.* Thomas Telford, London, UK.

Thom NH (2008) *Principles of Pavement Engineering.* Thomas Telford, London, UK.

TWf (Temporary Works Forum) (2019) *Working Platforms: Design of Granular Working Platforms for Construction Plant. A Guide to Good Practice.* Temporary Works Forum, Institution of Civil Engineers, London, UK

UIC (2020) *IRS70719-1: Earthworks and Trackbed Layers for Railway Lines.* UIC, Paris, France.

Wair BR, DeJong JT and Shantz T (2012) *Guidelines for Estimation of VS Profiles: PEER Report 2012/08. Pacific Earthquake Engineering Research Center*, University of California, Berkeley, CA.

Useful websites

National Highways Standards: https://standardsforhighways.co.uk/ha/standards/mchw/vol1/index.htm

It is recommended this chapter is read in conjunction with

- Chapter 75 *Earthworks material specification, compaction and control*
- Chapter 99 *Materials and material testing for foundations*

All chapters in this book rely on the guidance in Sections 1 *Context* and 2 *Fundamental principles.* A sound knowledge of ground investigation is required for all geotechnical works, as set out in Section 4 *Site investigation.*

Section 8: Construction processes

Section editor: **Stuart Hardy**

Chapter 77
Introduction to Section 8

doi: 10.1680/icemge.66830.1271

Stuart Hardy Laing O'Rourke, Dartford, UK

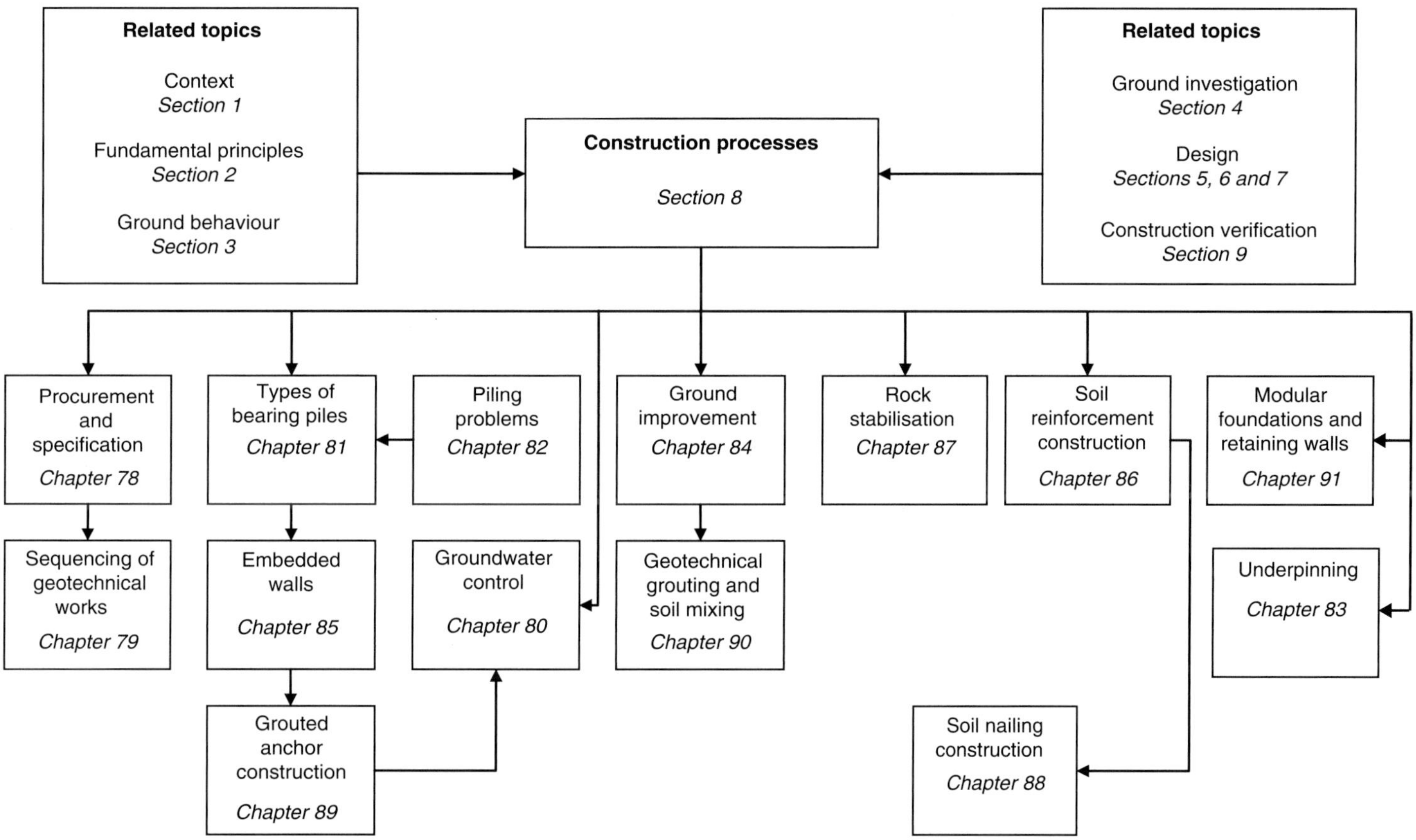

Figure 77.1 Layout of chapters in Section 8

Figure 77.1 outlines the layout and contents of Section 8 *Construction processes.*

Geotechnical engineering differs from most other forms of civil engineering in that it does not deal with man-made materials whose properties and behaviour can be reliably dictated and predicted. Man-made materials such as concrete and steel have been rigorously tested in the past, and their engineering properties perfected away from the site environment. A concrete or steel member will perform the same way on different construction projects, provided the same materials are used and the same level of construction workmanship applied.

The ground is not normally man-made, and so every construction project will have a unique set of ground conditions, including the groundwater regime. It is this unique combination of ground conditions and the applied load from buildings and structures that makes geotechnical engineering so fascinating to those who work in the industry, and so frustrating to those who do not. It is no surprise that the biggest risks for new developments are often associated with the ground.

The physical characteristics of the ground need to be investigated for every new development. Geotechnical engineers use a combination of experience, empiricism and idealised engineering models to try to best predict the ground behaviour and its unique interaction with a new building or structure. This is as important for the practical aspects associated with construction as it is for design.

Ground behaviour is implicitly linked to the geotechnical construction process – for example, a bored pile of a certain length will have a different load capacity to a driven pile of the same length. Consequently, in geotechnical engineering the design cannot and must not be divorced from the construction process. Because of the intrinsic link between design and construction it is common in geotechnical engineering for specialist subcontractors to offer both as a single package. If the design is carried out by a party different from that undertaking the construction, then the geotechnical designer must communicate the needs of the design to the contractor, and the contractor must communicate the construction process to the designer, so that both parties can ensure that design and construction are compatible with each other. This lack of compatibility is frequently the cause of disputes.

There are many different geotechnical construction processes, and even the same process can be undertaken in many different ways. A process may be considered safe on one site but have a high level of risk on another. A process may also not be technically or practically possible in some ground conditions or site environments. It is recommended that during design development and the tender process for any new development, experienced geotechnical designers and experienced geotechnical contractors are brought together to ensure that the proposals are both viable and technically sound. It is essential that the final process is monitored during construction so that the impact on the design of any unforeseen changes in ground conditions can be appropriately managed.

For those who are not experienced in geotechnical engineering the best advice is to get specialist advice. There are many geotechnical consultancies, specialist contractors and trade organisations available to provide expert advice and services. A failure to take specialist advice is fraught with danger due to the reasons described above.

Section 8 *Construction processes* describes the most common geotechnical processes in use today. Note that there are some specialist processes that are not covered here, such as ground freezing. The processes that are covered are introductions for the reader, but they are absolutely not a replacement for specialist advice. This section does not consider design, which is covered in other sections. The exception is Chapter 80 *Groundwater control*, as the design of such systems is not described elsewhere in this manual.

doi: 10.1680/icemge.66830.1273

Chapter 78
Procurement and specification

John Chick Expanded Ltd (Geotechnical), Dartford, UK
Tony P Suckling A-squared Studio Engineers Ltd, London, UK

CONTENTS

Each construction project will have a unique set of ground conditions to be considered, and it is this unique combination of ground conditions and future applied loadings which presents risk for new developments. Consequently the procurement process which defines the scope of work, allocates responsibilities and risk and chooses the contractor to do the work is critically important. The specification that defines the minimum standard of the work is a key tool in ensuring a successful outcome. Tender documents and submissions must be clear and non-contradictory, and the tender negotiation process must be managed to ensure that the chosen contractor has the necessary skills and tools to undertake the work in the prevailing ground conditions without unduly affecting the design. There are many forms of contract and subcontract used in geotechnical construction, but none has been designed specifically for such specialist works. A common problem with geotechnical construction is identifying who is responsible at interfaces between different elements of work. Works at such interfaces must be clearly allocated in the contract documents, especially when one party will be relying upon another to supply a service. In this way, enabling works can have a significant impact on geotechnical works.

78.1. Introduction

Each construction project will have a unique set of ground conditions to be considered, including the groundwater regime. It is this unique combination of ground conditions and future applied loadings (from the building or structure) which present the most risk for new developments. Consequently, the procurement process which defines the scope of work, allocates responsibilities and risk and chooses the contractor to do the work is critically important. The specification which defines the minimum requirements for the work is a key tool in ensuring a successful outcome.

78.2. Procurement

Ground behaviour is implicitly linked to the geotechnical construction process being used. Consequently, in geotechnical engineering, design cannot (and must not) be divorced from the construction process being used. Because of this, it is common in geotechnical engineering for specialist subcontractors to offer both a design and construction service. If the design and the construction are being carried out by different parties, the geotechnical designer must communicate the needs of the design to the contractor, and the contractor must communicate the construction process being used to the designer. This is to ensure design and construction are mutually compatible. Both geotechnical designer and geotechnical contractor must be clear what their technical and pricing assumptions are based upon in terms of the ground investigation information they are using, and what their design and planned construction processes rely upon in terms of construction sequence. The assumed temporary works support to surrounding ground and structures at all stages is of particular importance and must be clearly communicated to and understood between affected parties. During procurement the role of the design coordinator is of great importance.

There are various forms of procurement, and employers have their own preferences. In general, the forms of procurement are

- competitive tender – cheapest price
- competitive tender – best value
- preferred supplier
- partnering arrangement.

After a competitive tender process, ‘cheapest price wins’ is the most common method of procurement in construction. Most experienced geotechnical engineers would not advocate this procurement method; they can foresee that ground-related problems will necessarily increase their costs, and so are unlikely to offer the cheapest price and be successful.

Additionally, seeking the cheapest price alone can unintentionally invite the submission of the cheapest ‘defensible’ price – that is, a heavily qualified price based purely on the information included in the enquiry package and failing to allow for previous geotechnical experience of the area or possible risks and cost influences. Such a bid will bear little resemblance to the outturn cost. ‘Cheapest price wins’ is a valid method of procurement for only the simplest of geotechnical projects, such as a repeat phase of simply loaded piles on a well-understood site.

The disadvantages of cheapest-price competitive tendering for geotechnical works can be partially overcome if the tenders are not assessed on cost alone. Other considerations may include construction experience in the appropriate ground conditions, design experience in the appropriate ground conditions, safety records, sustainability performance and capability, quality systems and the ability to manage risk and change. This is known as 'best value'.

Some employers have their own preferred supply chain, comprising organisations which meet specific important criteria. These criteria vary and may include safety standards, quality assurance (see Chapter 93 *Quality assurance*), relevant experience, design capability and sustainability performance. The scope of the works is identified and the preferred supplier provides the works. The costs for doing the work may be agreed by a competitive tendering process, direct negotiation or reference to a pre-negotiated schedule of rates.

Some employers may enter into a partnering arrangement with a preferred supplier. Again, the scope of the works is identified and the preferred supplier provides the works. The costs for doing the work are likely to have been agreed by negotiation, and if the scope of the work changes the cost implications and associated responsibilities will already have been identified in the arrangement. The advantage of this form of procurement is that the risks are normally identified and allocated early in the process.

Chapman (2008) states that geotechnical mitigation is better than the management of risks by contingency, and hence argues strongly against awarding a tender on cost rather than on the quality of the design and construction.

A modern approach for large projects has been to assess tenders on a declared split between cost and quality. This may be, say, 80% on cost and 20% on quality, but for complex projects it has been known to be 20% on cost and 80% on quality. This assessment method is seldom used to make the final award and is commonly seen as a useful way of reducing tendering contractors to the best-scoring two or three. The final award will then be based on cost, programme and approach to risk.

Some degree of unconscious bias in scoring is almost impossible to avoid. Previous trading history and contractor relationships influence the initial scoring. For large, complex or publicly funded schemes, reducing the list to a valid set of competitors is cost-effective. This scoring method focuses tender resources and business costs for all parties in the most productive place. It should be understood by all, however, that due to unconscious bias, an assessment process with a cost quotient greater than 25–30% means cost will be the most important driver from the outset.

78.2.1 Tender process

The employer (i.e. the body funding the work) normally will appoint a professional team comprising someone to manage the technical aspects of the work (normally the engineer) and someone to manage the commercial aspects of the work (normally the quantity surveyor), as a minimum. A document describing the employer's requirements is produced and normally the engineer then designs the building or structure to meet these requirements. It is common and advisable that the engineer seeks expert advice regarding geotechnical solutions at this early stage from geotechnical specialist contractors. A range of advice from different contractors should be sought rather than approaching a single source, and this will inform a more efficient and competitive design solution. It should be highlighted when specific advice relates to rare or even unique equipment or techniques.

Bills of quantities are then prepared, normally in conformity with a standard method of measurement, approved for use with the relevant form of contract. However, some of the provisions of these standard methods are not well suited to geotechnical processes and need to be amended. The type and number of amendments vary according to the specific requirements of the geotechnical work, the ground conditions and for each type of construction process. Quantities of geotechnical work should generally be billed as 'provisional' and measured and valued as 'executed', since it is unlikely the ground will be homogeneous enough that design quantities will be precise. There will normally be variations or additions to the design quantities, which should be anticipated and allowed for.

The employer's requirements are provided to prospective contractors with an invitation to tender, and typically comprise a bill of quantities, a description of the works (normally as drawings), a specification describing the minimum standard of work that will be acceptable, contract conditions and information on the expected ground conditions. The tender invitation may include a request that the tenderer provide a 'lump sum' for the works, including taking all of the risk on quantities – for example, dealing with voids in the ground and obstructions. This request will seldom result in any compliant bids as the risk profile of such an undertaking without very specific and detailed background investigation is impossible to assess.

The contractor will often submit a tender which is not fully in accordance with the original documents. Technically this counter-offer may be for a different design or a different construction process, or to suit a specific type of equipment. Commercially, this counter-offer may be for a different level of risk or responsibility. Whatever the basis of the counter-offer, it is key that the professional team understands what is different, the reasons for the differences and the impact these may have. Post-tender discussions and negotiation are essential for geotechnical works as every project will be dealing with a unique set of ground conditions. If the professional team does not include a geotechnical adviser, as defined by the ICE site investigation steering group, then one should be considered. The consultant must have a very clear brief, as procurement is confused by consultants who appear to be issuing instructions or modifying specifications with no construction responsibility.

It is important that the professional team takes full account of risk items excluded, and opportunities highlighted, by tenderers. Risk items are commonly excluded from tenders where the nature of the risk does not rest with the party best equipped to deal with it. The professional team should understand where a simplistic 'risk roll down' approach is potentially failing to provide a 'best for project' outcome. Tenderers may highlight alternatives and opportunities which may give them a cost or programme advantage, but which are also better for the project. Provided they *are* actually better for the whole project, in the round, such proposals can make a significant difference to the assessment outcome.

78.2.2 Tender documents and submissions

The list of documents which would normally form part of a contract is given below, to give the reader an indication of what is likely to be needed for complex geotechnical works where specialist advice must be taken.

- The details of the client, contractor, designer, principal contractor, principal designer and design coordinator.
- The form of tender (a formal bid for the work) including the period for which the tender is valid, who is engaged to supervise the works and the period required for any relevant on-site testing.
- General and project-specific specifications, including either a tender schedule or the performance criteria the design needs to achieve. If the contractor is required to design the works, this should be expressly stated.
- Drawings should include details of underground services (and planned diversions), existing structures and other known obstructions and infrastructure. Site-specific information and ground investigation data should include existing ground levels, ground levels at borehole and other test positions, and a drawing showing the positions of such boreholes relative to the outline location of the works. Caution should be applied where the professional team are relying on 'inherited' ground investigation information and data commissioned for a superseded scheme. Such inherited information can be inadequate in detail, and in the wrong locations or of the wrong scope for the new scheme. Advice from a geotechnical specialist should be sought regarding the benefits of procuring additional, project-specific ground investigations.
- The conditions of contract should include the location of the site and access to it, any limitation on working hours, any special conditions limiting noise and vibration or other environmental constraints, any phasing of the works and the available working and storage areas.
- In the case of subcontract documents, any known special conditions or restrictions should be described, and information on such matters as insurances, retention, liquidated damages, programme requirements and bonds should be provided. The bill of quantities and the form of acceptance are also required.
- The tender offer should include a method statement giving a description of the proposed type of equipment to be used for the execution of the works, the method of construction and any assumptions made, including attendance from others and ideally the assumed programme and construction sequence. Also needed are an estimate of the time required for completion of the design, the period of notice required before commencement and the contract period for working.

Part A of the Institution of Civil Engineers' *Specification for Piling and Embedded Retaining Walls* (ICE, 2017) gives details which are relevant to deep foundation works.

78.2.3 Common forms of contract

Contractors may be invited to tender to an employer using one of the established forms of contract, or a contract document drawn up especially for management-type contracts. There are four basic types of contractual arrangement under which construction work may be undertaken.

(*a*) Civil engineering works with an engineer responsible to an employer for design and supervision.

(*b*) Building works with an architect responsible to an employer for design and supervision, advised by an engineer who may also be responsible to the employer for structural engineering elements of the works but who has no formal status under the conditions of the contract. Under such a contract, it is desirable that the architect authorises an experienced geotechnical engineer to act as his representative in connection with the geotechnical works.

(*c*) Building or civil engineering works with a contractor responsible to an employer for design and construction. The contractor may appoint an engineer to undertake the engineering duties relating to the geotechnical works, or may entrust those duties to a suitably qualified engineer of their own.

(*d*) Building works with an architect responsible to an employer for design and supervision, but having no engineering adviser. The architect should consider the responsibilities that devolve upon them in these circumstances and should recognise that such an arrangement may not be in the best interests of either the employer or themselves. This is not recommended for complex geotechnical works.

The cost consultant or quantity surveyor can advise the most effective form of contract under which to procure the geotechnical work. Consideration should include the time required to negotiate contract terms with a preferred bidder, where a simple form of contract has been extensively amended in favour of the employer. Resource requirements are an important consideration in the case of some forms of target cost contract which are programme-driven and practically require all parties to assign programmes to the project.

Part A of the *Specification for Piling and Embedded Retaining Walls* (ICE, 2017) gives further details relevant to deep foundation works.

78.2.4 Common forms of subcontract

The use of specialist subcontractors is very common for geotechnical works, and such organisations often offer both a design and construction service. Subcontract tenders for geotechnical works may be invited in the following ways.

(*a*) Tenders may be invited from specialist contractors for nomination as a subcontractor; in such cases it is normal for a prime cost item to be included in the main contract bills of quantities. It is necessary to ensure that the specification and other relevant items from the subcontract documents are included in the main contract tender documents – so that tenderers for the main contract are aware of, and can price for, their responsibilities to and their attendance on the subcontractor. Equally, the subcontract tender documents should include information regarding the form of the main contract, including the appendix to the contract, and sufficient detail to ensure that the subcontractor is aware of their responsibilities and liabilities at the time of preparing the tender.

(*b*) The geotechnical works may be measured and included in the bills of quantities for the main contract, and the document may stipulate that these works shall be executed by any one organisation from an approved list of contractors. The contractor may have the option of proposing further companies for approval.

(*c*) The geotechnical works may be measured and included in the bills of quantities for the main contract, but without any list of approved subcontractors. In such cases, the contractor will seek prices from their own selection of subcontractors, for whom it is normal practice to seek approval before they are employed.

(*d*) Following appointment, the contractor may be instructed to carry out additional works not included in the contract. Should this additional work include geotechnical work, tenders for such works will normally be invited from a list of companies agreed by the contractor and the engineer. It is most desirable that the tender document should be approved by the engineer before enquiries are issued.

With procedures (*b*), (*c*) and (*d*), the selected subcontractor is normally appointed as a domestic (direct) subcontractor of the contractor.

Part A of the *Specification for Piling and Embedded Retaining Walls* (ICE, 2017) gives further details relevant to deep foundation works.

78.3. Specifications

78.3.1 Model specifications

There are many specifications published for use with specialist geotechnical works, and most are updated on a regular basis. Some examples follow.

78.3.1.1 Execution codes

- BS EN 1536+A1:2015 Execution of special geotechnical work – Bored piles (BSI, 2010b)
- BS EN 1537 Execution of special geotechnical work – Ground anchors (BSI, 2013)
- BS EN 1538+A1:2015 Execution of special geotechnical work – Diaphragm walls (BSI, 2010a)
- BS EN 12063 Execution of special geotechnical work – Sheet pile walls (BSI, 1999)
- BS EN 12699 Execution of special geotechnical work – Displacement piles (BSI, 2015a)
- BS EN 12715 Execution of special geotechnical work – Grouting (BSI, 2020)
- BS EN 14199 Execution of special geotechnical work – Micropiles (BSI, 2015b)
- BS EN 14731 Execution of special geotechnical work – Ground treatment by deep vibration (BSI, 2005b)

78.3.1.2 Specifications/codes of practice

- American Petroleum Institute *Specification for Line Pipes 5L* (American Petroleum Institute, 2020)
- BS EN 12794 Precast concrete products, foundation piles (BSI, 2005a)
- BR 391 *Specifying Vibro-Stone Columns* (BRE, 2000)
- BR 458 *Specifying Dynamic Compaction* (BRE, 2003a)
- BR 470 *Working Platforms for Tracked Plant* (BRE, 2004)
- BRE 479 *Timber Pile and Foundations* (BRE, 2003b)
- ICE *Specification for Piling and Embedded Retaining Walls* (ICE, 2017)
- ICE *Specification for Ground Treatment* (ICE, 1987)

While model specifications provide a useful starting point and guide, most geotechnical projects will incorporate project-specific aspects which are either not covered or are unintentionally excluded by use of such a specification.

Sustainability, technological and testing advances tend to drive retrospective specification changes. Specifications and codes tend to lag development. For sustainability benefits to be incorporated into designs and tenders, specifications should leave room for alternative sustainable materials, techniques and technologies to be offered. Such alternatives will generally have been proven and be reasonable rather than wishful.

78.3.2 Design responsibility

The model specifications described above are written to be used in conjunction with all forms of contract. Hence they do not always state who should have design responsibility and therefore if they do not, this must be clearly stated in the contract documents.

In most cases, the engineer will design the superstructure and the part of the substructure which is not embedded in the ground – which may include spread footings, pile caps and ground beams. A specialist subcontractor or trade contractor will normally be asked to design the specialist geotechnical element of the works, such as the piling, ground treatment, grouting or embedded retaining wall. The engineer will nevertheless be expected to take overall responsibility for the substructure design and, in some cases, may be required to adopt any contractor design elements. Where the design roles are thus split, clear communication of all the design requirements is essential. For example, the client's structural engineer may design the pile caps and groundbeams with piles of certain diameters shown in the pile caps. The specialist could be free to redesign numbers of piles under the caps and the piles per cap of different diameters and type. The specialist will often design the pile length and diameter. Opportunities to change pile layout should be enabled early in the procurement process or the specialist may propose good ideas for which there is no time or designer's budget to exploit.

In certain circumstances the engineer may consider that it is preferable to design the complete foundation. This is usually the case in more complex circumstances, such as slope stabilisation, or where the geotechnical works will interact with other structures such as existing tunnels or where it may be difficult to define the constraints adequately.

A good example is *Specification for Piling and Embedded Retaining Walls* (ICE, 2017), which requires an explicit award of responsibility for the design of the piles and foundation, and for the construction method.

78.4. Technical issues

78.4.1 Common engineer–main contractor issues

Below are some examples of the issues that regularly cause dispute between the engineer and the main contractor. These are best prevented by early appointment of the contractor so they can work with the engineer and influence the design during its development. In the absence of construction advice during the design process, the engineer may find that

- their design and/or specification is not practically buildable
- their design is not safe to build
- their design does not consider the latest technologies
- instrumentation is not installed correctly or at the right time, or is not adequately protected
- monitoring of instrumentation is not carried out
- access requirements are not understood or accounted for.

78.4.2 Common main contractor–subcontractor issues

Below are some examples of the issues that regularly cause disputes between the main contractor and the subcontractor. These are best prevented by early appointment of the subcontractor and clear discussions regarding responsibilities and reliance

- site not ready for subcontractor upon arrival
- working platform not adequate for the specialist plant (refer to Section 78.4.4)
- working platform not adequately maintained during the geotechnical works
- below-ground obstructions either not removed or removed causing disturbance to the ground or adjacent structures
- temporary works such as props or shores designed without reference to the geotechnical contractor's working space requirements
- construction sequence changed, requiring different performance from the geotechnical works
- main or principal contractor fails to plan, manage and control other subcontractors, comprising the subcontractors design
- agreed attendances not provided.

78.4.3 Common subcontractor–subcontractor issues

Below are some examples that regularly cause dispute between the subcontractors or trade contractors on a site. These are best prevented by getting all interfacing subcontractors together soon after their appointments to discuss and agree interfaces and responsibilities. The contractor needs to manage these subcontractors as they are likely to be relying on the services of others. Leaving subcontractors working alone on a site is not acceptable, and disputes often arise when this happens. The list of issues is

- one subcontractor is late, thereby delaying another subcontractor
- several subcontractors need to use the same part of the site at the same time (including deliveries, use of tower cranes)
- assumed sharing of site facilities, access or plant. These must be defined before award
- one subcontractor's design is not compatible with another's design
- one subcontractor's product is not compatible with another's product
- when something goes wrong, all the subcontractors blame each other and none can be proven to be responsible
- follow-on subcontractor fails to protect geotechnical works or instrumentation from damage
- follow-on subcontractor damages geotechnical works through use of inappropriate techniques
- follow-on subcontractor responsible for attendance to the geotechnical contractor monopolises attendance resources for their own activities
- scope of obstruction removal. This activity is often split by removal to a certain depth by one party and the risk then passes to another party. This is not generally best for a project.

78.4.4 Preparation for construction

Specialist geotechnical works often require the use of specialist plant, and the loadings these may apply to the ground may be considerable. The working platform is a key item of temporary works, and responsibility for its design, specification, construction, maintenance and repair must be clearly allocated. There have been fatal consequences from an inadequate working platform in the past. For piling platforms, the Federation of Piling Specialists members have developed, maintained and updated a piling platform design program. Specific piling rig loads and geometry can be used in the program to derive the actual design requirements of the working platform. This design can then be constructed by an appropriate contracted party.

Responsibility for identifying and removing obstructions in the ground needs careful consideration. Normally, shallow obstructions can be safely removed unless there are sensitive structures nearby. It may not be possible to remove deeper obstructions practically or safely, and hence these will have an impact on the design and construction of the geotechnical works. It may also not be possible to identify all obstructions in the ground, and the risks associated with these should be considered carefully.

Where obstructions have been removed from the ground, the geotechnical designer must take account of the methods used and any impact on their design or on adjacent structures both above and below ground. Backfilling of removed obstructions needs to be a controlled procedure using materials and methods that do not unduly affect other works.

Unexploded ordnance in the ground is often a real hazard and should be investigated on most sites. Expert advice is provided in CIRIA report C785 (CIRIA, 2019).

References

American Petroleum Institute (2020) *Specification for Line Pipes 5L*, 46th edn. API, Washington, DC.

BRE (Building Research Establishment) (2000) *Specifying Vibro-Stone Columns*. BRE, Watford, UK.

BRE (2003a) *Specifying Dynamic Compaction*. BRE, Watford, UK.

BRE (2003b) *Timber Pile and Foundations*. BRE, Watford, UK.

BRE (2004) *Working Platforms for Tracked Plant*. BRE, Watford, UK.

BSI (British Standards Institution) (1999) BS EN 12063:1999. Execution of specialist geotechnical work – Sheet pile walls. BSI, London, UK.

BSI (2005a) BS EN 12794:2000. Precast concrete products, foundation piles. BSI, London, UK.

BSI (2005b) BS EN 14731:2005. Execution of specialist geotechnical work – Ground treatment by deep vibration. BSI, London, UK.

BSI (2010a) BS EN 1538:2010+A1:2015. Execution of specialist geotechnical work – Diaphragm walls. BSI, London, UK.

BSI (2010b) BS EN 1536:2010+A1:2015. Execution of specialist geotechnical work – Bored piles. BSI, London, UK.

BSI (2013) BS EN 1537:2013. Execution of specialist geotechnical work – Ground anchors. BSI, London, UK.

BSI (2015a) BS EN 12699:2015. Execution of specialist geotechnical work – Displacement piles. BSI, London, UK.

BSI (2015b) BS EN 14199:2015. Execution of specialist geotechnical work – Micropiles. BSI, London, UK.

BSI (2020) BS EN 12715:2020. Execution of specialist geotechnical work – Grouting. BSI, London, UK.

Chapman TJP (2008) The relevance of developer costs in geotechnical risk management. *Proceedings of the 2nd BGA International Conference on Foundations, ICOF 2008* (Brown MJ, Bransby MF, Brennan AJ and Knappett JA (eds)). IHS BRE Press, Watford, UK.

CIRIA (2019) *Unexploded Ordnance (UXO) Risk Management Guide for Land-Based Projects*. CIRIA, London, UK.

ICE (Institution of Civil Engineers) (1987) *ICE Specification for Ground Treatment*. Thomas Telford, London, UK.

ICE (2017) *ICE Specification for Piling and Embedded Retaining Walls*, 3rd edn. Thomas Telford, London, UK.

It is recommended this chapter is read in conjunction with

- Chapter 93 *Quality assurance*
- Chapter 96 *Technical supervision of site works*

All chapters in this book rely on the guidance in Sections 1 *Context* and 2 *Fundamental principles*. A sound knowledge of ground investigation is required for all geotechnical works, as set out in Section 4 *Site investigation*.

doi: 10.1680/icemge.66830.1279

Chapter 79
Sequencing of geotechnical works

Toby Hayward Laing O'Rourke, Dartford, UK
Mark Pennington Balfour Beatty Ground Engineering, Basingstoke, UK
Tony P Suckling A-squared Studio Engineers Ltd, London, UK

CONTENTS

To achieve the safety and technical requirements of geotechnical works, it is imperative that a sequence of working is agreed before beginning, and is subsequently adhered to by all the organisations working on a project at the same time. The roles for planning, managing and coordinating the construction works are set out under the Construction Design and Management Regulations 2015 (CDM 2015 – see Chapter 8 *Health and safety in ground engineering*). When the construction works involve more than one contractor, a CDM client must appoint an organisation (or individual) to plan, manage, monitor and coordinate each of the phases of the work. Under the CDM Regulations these roles are referred to as the principal designer and the principal contractor. There is a key link between the principal designer, who plans, manages, monitors and coordinates health and safety in the pre-construction phase, and the principal contractor, who executes these items in the construction phase. When agreeing the sequencing of the geotechnical works, the role of the principal contractor is crucial - they are likely to be the only party with an understanding of the overall construction sequence and the roles and responsibilities of each organisation on site. The principal contractor needs to understand each organisation's ideal method of working to achieve their own particular safety and technical requirements. If, as is common, many organisations on a site interact with each other or are reliant on one another, the principal contractor must help these organisations to modify their ideal method of working without compromising safety or technical requirements. Geotechnical design often requires a particular construction sequence to be followed or else the design is invalidated and the geotechnical works may not progress to plan; this would introduce another complication which would need to be managed by the principal contractor.

79.1. Introduction

Every construction project will have a unique combination of particular contractors, subcontractors, processes, products and materials on site. The roles and responsibilities for planning, managing, and coordinating the construction works are set out under the Construction Design and Management regulations 2015 (CDM, 2015). The client has overall responsibility for the CDM coordination on a project. When the construction works involve more than one contractor, a CDM client must appoint an organisation (or individual) to plan, manage, monitor and coordinate each of the phases of the work. These roles are the CDM principal designer and principal contractor. If the client does not make these appointments, they must fulfil the duties of these roles themselves. There is a key link between the principal designer, who plans, manages, monitors and coordinates health and safety in the pre-construction phase, and the principal contractor, who implements these items in the construction phase. The principal designer will advise the client in bringing together pre-construction information and provide the information designers and contractors need to carry out their duties. The principal designer should keep the principal contractor informed of any risks that need to be controlled during the construction phase, including any sequencing assumptions used in the design. The principal contractor should liaise with the client and principal designer for the duration of the project to ensure that all risks are effectively managed during the delivery of the project. The principal contractor should control the construction sequencing of projects that involve more than one contractor. This includes a written construction phase plan prepared before the construction phase begins, which is implemented and then regularly reviewed to make sure it remains fit for purpose.

Geotechnical engineering is a specialist aspect of civil engineering and so is normally undertaken by specialist organisations. The geotechnical works need to be undertaken safely and achieve all the technical requirements. The safety of on-site personnel and the public is paramount, and the integrity of works by others (existing, new or proposed) must not be compromised.

It is common for the geotechnical works to be undertaken by one or more specialist organisations under the control of a principal contractor. This leads to two separate construction sequencing issues: the overall project construction sequence and the specific task construction sequence. To achieve the safety and technical requirements it is imperative that a sequence of working is agreed prior to commencement and then adhered to by all the organisations working on a project at the same time. When agreeing the sequencing, the role of the principal contractor is crucial – they will have responsibility for the overall project construction sequence. They are likely to be the only party on site with an understanding of the overall construction sequence and the roles and responsibilities of each organisation on the site. The principal contractor needs to understand the ideal method of working for each of these organisations to achieve their own particular safety and technical requirements, and to investigate if any of the organisation's methods of working will impact on one another. If, as is common, many organisations on a site interact with each other or are reliant on one another, the principal contractor must help these organisations to modify their ideal method of working without compromising safety or technical requirements. As the principal contractor has an important role in managing health and safety risks during the construction phase, they must have the skills, knowledge, experience and, where relevant, organisational capability to carry out this work.

Specific task construction sequences are generally better defined and controlled. In geotechnical works, each specific task sequence is generally the responsibility of one contractor who is experienced in the activity. Detailed method statements and risk assessments are carried out. It is hard to control the interaction and impact of specific task construction sequences on each other. A relatively minor change in the sequence of one task may have a significant impact on the overall process.

Geotechnical design often requires a particular construction sequence to be followed or else the design is invalidated and the geotechnical works may not progress to plan; this would introduce another complication which would need to be managed by the principal contractor. The geotechnical designer needs to appreciate the constructability constraints in the design. If this is not the case and the contractor finds that the particular construction sequence cannot be followed due to practical or technical reasons, or due to safety concerns, then the work must cease, and the sequence and design must be suitably modified before the works recommence.

79.2. Design construction sequence

Geotechnical design is dependent on a number of different factors; many of these are not exact and are based on the best available knowledge. Examples include assumptions about the ground conditions, the previous usage of the site, the groundwater regime and the type of construction methodology. It is important that all concerned understand any assumptions made in the design construction sequence so that they are in a position to assess the impact of any changes during construction.

In order to accommodate unknowns or variables (due to the very nature of working with the ground), the geotechnical design would normally require one or more of the following to be carried out on site: observations, monitoring, investigation, testing and reporting back to the designer to check the validity of the design. If the contractor chooses to deviate from the construction sequence used in the design, either by choice or due to a lack of site control, it is likely that the design will be invalidated and the works will not progress to plan.

The geotechnical designer should have sufficient appreciation of the capabilities and limitations of the construction process, products and materials so that it is not unreasonable to expect the particular construction sequence assumed in the geotechnical design to be closely followed by the contractor. In reality, this can actually be difficult to achieve as many specialist geotechnical contractors use proprietary systems, and if the procurement process and contractual arrangements do not appoint the contractor before the design is finalised, then the design requirements may not be compatible with the contractor's capabilities or requirements. If the contractor finds that the particular construction sequence used in the design cannot be followed due to practical or technical reasons, or due to safety concerns, then the sequence and the design must be suitably modified before the works begin.

To best communicate the construction sequence between designer and contractor, a detailed drawing is recommended (see the example in **Figure 79.1**), which can then be incorporated into the construction phase plan. Geotechnical design is generally concerned with excavation into the ground or building on top of the ground. To this end, design construction sequences are generally two-dimensional drawings showing the maximum excavation or construction level for each phase. It is important to not overlook the plan constraints of the site works.

79.3. Site logistics

Site logistics are crucial in determining a successful construction project. There would be constraints on a very small site where plant movement is very limited, and on a large site where plant movement is a significant element in the programming of the project. The type of specialist plant available and how the plant can be moved may be an integral part of the overall solution. Contractors may organise work around available plant, which is difficult to plan during the design phase of contracts. However, it is important that allowance is made during design to cater for the most likely option. On particularly complex projects, advice should be taken from various specialist contractors to establish likely construction methodology.

Site logistics need to be planned to follow the specific design requirements of each element. For example, it is not appropriate to design a complicated conveyor system to move spoil from a tunnel excavation if this does not fit around the propping system in the excavation at the tunnel portal.

Specific task sequences have to fit within the overall project sequence. The site logistics for specific activities are easier to plan than those for the overall project. Specific activities generally lie with one contractor, who is fully aware of their

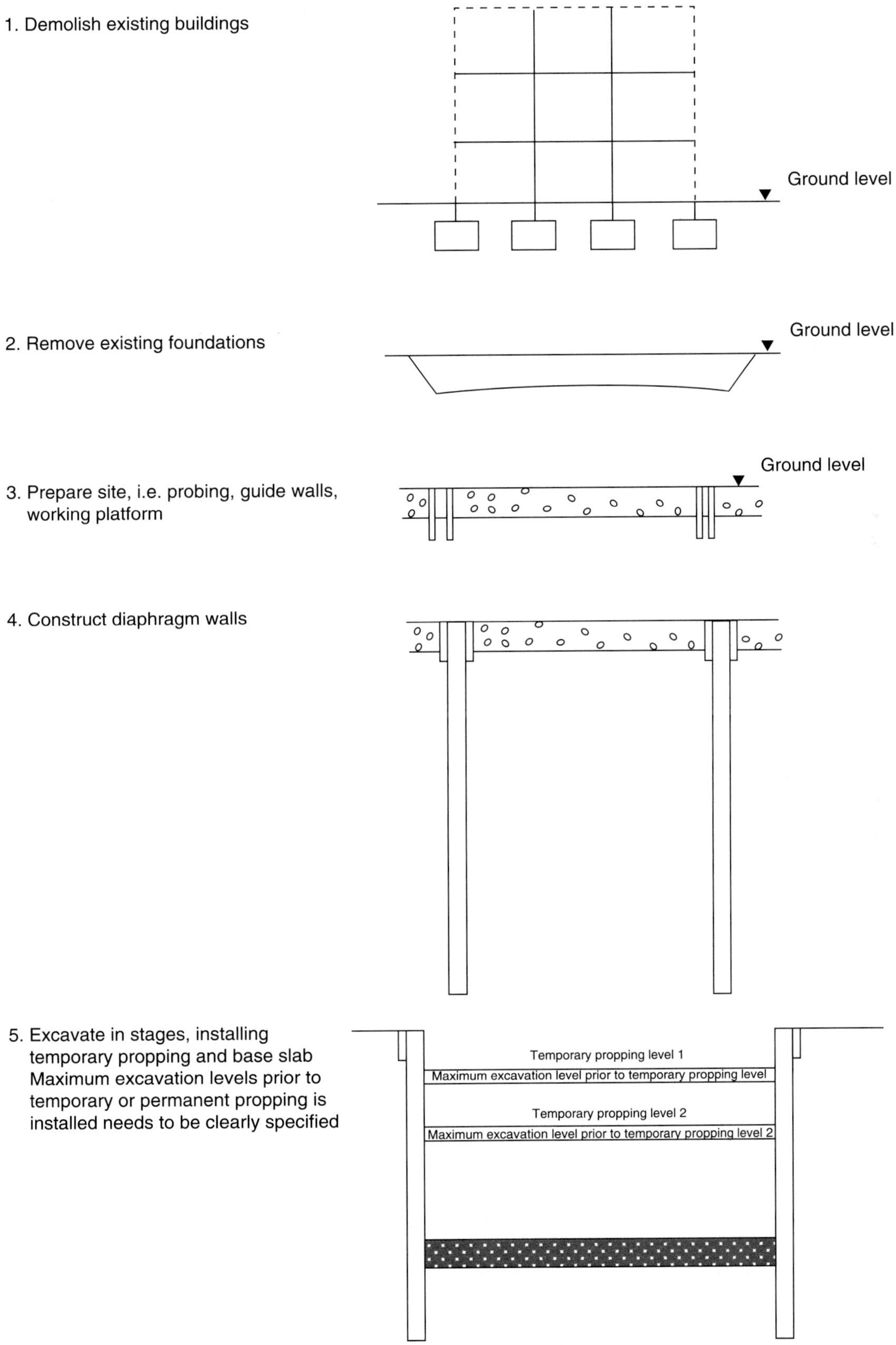

Figure 79.1 Example of a construction sequence for a tunnel boring machine launch chamber (continued on next page)

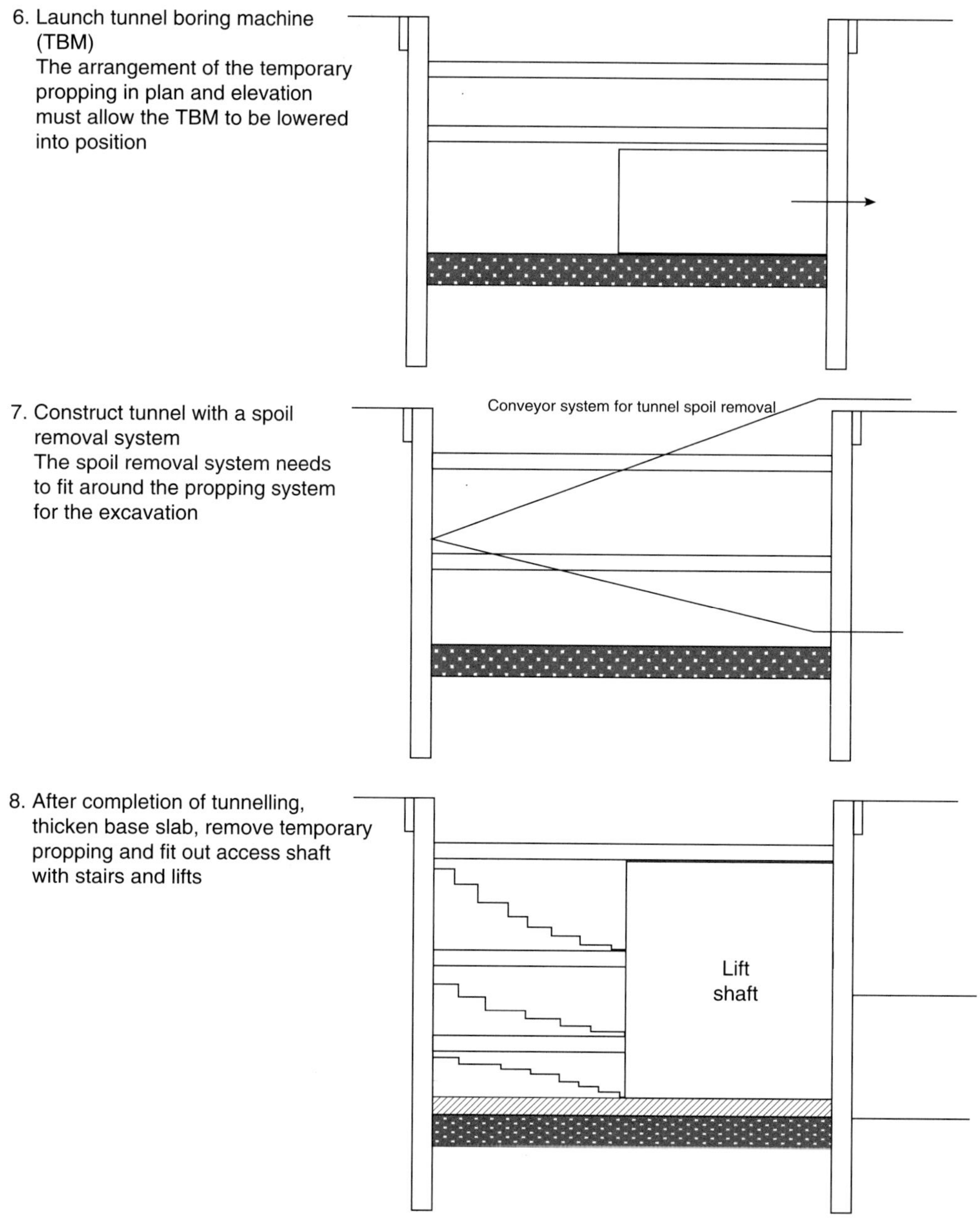

Figure 79.1 Continued

own plant requirements and limitations. For example, following the construction of a section of retaining wall, excavation may be required to proceed as quickly as possible in front of the wall. One aspect to consider in how quickly this can happen is the movement of plant around the site. One contractor may be responsible for the wall construction and another for the excavation. The full sequence of working can only be developed when the plant and site logistics of each element are understood. If these logistics require a change to the designed sequence, it would necessitate a redesign and may impact many activities.

Minimum working areas must be defined for each contractor working on the site. In addition, access to each work site area has to be clearly agreed between all parties. Construction sites are, by definition, constantly changing and developing. It is important that drawings showing the working space requirements for each contractor are prepared and agreed by *all* parties involved. These will have to be prepared on a regular basis to reflect the changing nature of the site (see **Figure 79.2**). It is common for site logistics to be drawn in plan, but the vertical excavation or construction should not be ignored in these drawings.

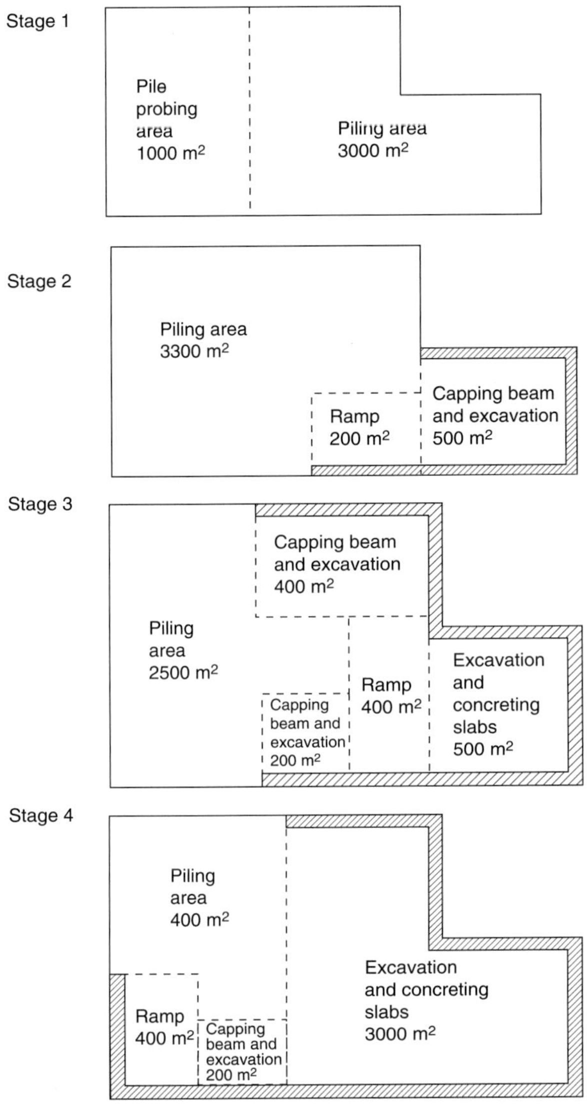

Figure 79.2 Example of typical working areas for a basement construction

79.4. Safe construction

Health and safety in geotechnical engineering is described in Chapter 8 *Health and safety in geotechnical engineering.*

Generic and specific risks and hazards need to be identified for both the specific and project construction sequences. Each contractor is responsible for establishing the hazards and risks for the activities they are undertaking, and also for the residual risks they find on arrival, and those they hand on to any follow-on contractor.

Geotechnical works are increasingly being carried out on brownfield sites which have had one or more previous uses. Any construction sequence undertaken needs to consider this previous usage.

Every contractor needs to assess the generic hazards and their control or mitigation measures for the specific characteristics of the site, and also to identify site-specific hazards and carry out a risk assessment for these (see **Table 79.1**).

Part of the site-specific risk assessment process is to identify if the proposed construction works or their sequence cause a hazard to others. It is common for the geotechnical works to be undertaken by one or more specialist organisations under the control of a principal contractor. To achieve the safety requirements, it is imperative that a sequence of working is agreed to by all prior to commencement on site, and then adhered to by everyone working on a project at any one time. When agreeing this sequencing, the role of the principal contractor is crucial; they are likely to be the only party on site with an understanding of the overall construction sequence and the roles and responsibilities of each organisation on the site. The principal contractor needs to understand the ideal method of working for each of these organisations and then to investigate if any of the organisation's methods of working will impact on one another. If, as is common, many organisations on a site interact with each other or are reliant on one another, the principal contractor must help these organisations to modify their ideal method of working to achieve a maximum level of safety.

Hazard	Likely consequences if hazard arises	Impact rating	Likelihood	Severity rating before mitigation (impact× likelihood)	Mitigation measures	Impact rating	Likelihood	Severity rating after mitigation (impact× likelihood)
	Risk of injury to either contractor's personnel	5	3	15		5	1	5
One contractor working in close proximity to another contractor	Risk of damage to plant	3	3	9	Provide barriers between work areas, exclusion zones within work areas and pedestrian access around the work areas	3	1	3
	Risk of damage to work done	3	3	9		3	1	3

Table 79.1 Example of how a generic risk assessment has been modified to take account of others on site concurrently. Ratings and likelihoods vary from 1 (low) to 5 (high)

79.5. Achieving the technical requirements

Geotechnical engineering is complex. The following technical requirements are often key components of the design and construction process.

- The installation method must not cause significant ground disturbance or movement and so invalidate the design (see **Figure 79.3**).
- The construction method must not be a nuisance (e.g. too noisy).
- The construction method must not cause damage to existing buildings or structures, either above or below ground (e.g. by causing too much vibration).
- Plant must not be too heavy or work in a way that could cause damage to underground structures or services, or to newly constructed works.
- Sequencing of plant within temporary works needs to be carefully planned so that plant has sufficient and safe working areas, and temporary works continuously fulfil their function.
- Stockpiling of materials is done in a way that ground movements resulting from the applied surcharge to the ground do not damage existing buildings/structures or newly constructed works.
- Construction method may cause softening or instability of the ground if work progresses too slowly.
- Construction method may cause premature failure of the ground if work progresses too fast.
- Follow-on works are planned in a way not to cause damage to the geotechnical works.

The relevance of each of these depends on the actual ground conditions, the geotechnical solution being adopted, the construction method being used and the design philosophy. Hence, geotechnical specialists are needed to assess each of these and communicate their findings to the professional team as an integral part of the design and construction process. The impact of one element of works on another, either generally or at interfaces, needs to be carefully considered.

Figure 79.3 Piles which have been affected by moving ground

79.5.1 Piling works

The following list is not exhaustive and there are many more examples of achieving the technical requirements for piles. Similar lists could be produced for all geotechnical processes.

- In ground conditions containing hard layers, the rotation of continuous flight augers (CFAs) can cause significant smearing of the bore, or induce ground movements invalidating the pile design.
- The driving of piles may be too noisy at the site boundary.
- Piles may be too close together: during construction of a new pile, a recently constructed one is damaged.
- Bearing pressures from piling plant on a working platform overlying a soft subgrade may damage recently concreted piles.
- Piling plant working within temporary propping to basement walls has to be carefully designed and planned, with the props being removed and replaced in a controlled sequence.
- Stockpiling of fill material to form a piled embankment over soft clay is done in a limited way, so that ground movements resulting from the applied surcharge to the ground do not damage the piles.
- The time between boring and concreting a pile is to be minimised.
- Loading of a test pile in chalk should be undertaken in small increments.
- Trimming piles is to be carried out in a way so as not to cause damage to the piles.

79.6. Monitoring

It is often necessary for a geotechnical solution to be monitored during construction and in some cases the long term. The monitoring can be to confirm design assumptions or can be an integral part of the design – for example, when filling for an embankment or when the observational method (see Chapter 100 *Observational method*) is being used.

The instrumentation should be installed prior to construction so that baseline readings can be established. The instruments and their cabling must be protected during construction, and the construction methods and sequencing should take this into consideration (unfortunately this is often not the case). The monitoring data should be augmented with details of the construction sequence followed (e.g. site photos, levels and dates of excavation, dates of prop installation/removal), so this information can be used to give context when reviewing the site monitoring date.

79.7. Managing changes

Ground conditions are rarely homogeneous and are seldom investigated in enough detail prior to construction. Consequently it is common for the ground conditions to be slightly (or sometimes significantly) different to that expected. When this occurs, everyone involved comes under commercial pressure to provide an appropriate solution. In these circumstances, it is important to ensure that the changed sequencing of the geotechnical works still achieves the aforementioned safety and technical requirements.

It is common for an accident or technical problem to be related back to a change in circumstances on the site, which may be a change in ground conditions or in something else. When on site, one should be aware that not all parties may follow the previously agreed method of working, either for good reason or for no reason at all.

There are always changes on site and it is difficult to ensure that all those involved have reassessed their works and their impact on others. However, it is imperative to do so. A common problem occurs when a change is made to a specific task sequence, making it more efficient and cost-effective. Checks may be made to show that the change in sequence has no impact on the design of that specific task. However, if this necessitates significant changes to the follow-on construction sequence (to ensure the design is valid), then the sequence should not be changed without the agreement of all the parties involved.

79.8. Common problems

Below are some specific problems that commonly occur on complex urban geotechnical projects.

(*a*) Poor-quality material is used to backfill old foundations that have been removed, leading to excessive concrete over-break on a new embedded wall installed along the line of the old foundations, which then has to be laboriously chiselled back to meet the requirements of the specification.

(*b*) A ramp into an existing basement or excavation to allow a piling rig to access and egress to undertake piling from the base of the excavation normally needs to be at a slope of less than 1 in 10. Hence such a ramp needs to be over 100 m long for a 10 m deep excavation. It is often more practical to temporarily backfill an existing basement to allow new piling, but this may then require coring through any basement slab, or the breaking out of holes in the slab prior to backfilling.

(*c*) Top-down basement construction can often be beneficial on congested sites, as one can then win back new laydown areas for materials and space for plant. For tall structures, top-down construction provides a shorter programme; for smaller structures it may provide the only practical and safe method of construction.

It is recommended this chapter is read in conjunction with

- Chapter 8 *Health and safety in geotechnical engineering*
- Chapter 67 *Retaining walls as part of complete underground structure*
- Chapter 94 *Principles of geotechnical monitoring*
- Chapter 96 *Technical supervision of site works*

All chapters in this book rely on the guidance in Sections 1 *Context* and 2 *Fundamental principles.* A sound knowledge of ground investigation is required for all geotechnical works, as set out in Section 4 *Site investigation.*

doi: 10.1680/icemge.66830.1287

Chapter 80
Groundwater control

Martin Preene Richter Associates, Huddersfield, UK

Groundwater control encompasses the range of temporary works techniques used to allow below-ground construction projects to be carried out in dry and stable conditions. Two principal approaches can be used: groundwater control by pumping (also known as construction dewatering), which involves pumping from an array of wells or sumps to lower groundwater levels in the vicinity of an excavation; or groundwater control by exclusion, which relies on low-permeability cut-off walls around the excavation to prevent or reduce groundwater inflows.

CONTENTS

80.1. Introduction

Groundwater control encompasses the range of temporary works techniques used to allow below-ground construction projects to be carried out in dry and stable conditions. Two principal approaches can be used: groundwater control by pumping (also known as construction dewatering), which involves pumping from an array of wells or sumps to lower groundwater levels in the vicinity of an excavation; or groundwater control by exclusion, which relies on low-permeability cut-off walls around the excavation to prevent or reduce groundwater inflows.

80.2. Objectives of groundwater control

The occurrence and control of groundwater are fundamental to geotechnical engineering (see Chapter 16 *Groundwater flow*). Where groundwater is encountered during excavation, problems can occur either in terms of flooding of the excavation or in the form of instability induced by the presence of groundwater. Groundwater control can be defined as the range of activities used to allow construction work to proceed at depths below groundwater level.

The most obvious (but not necessarily the most important) objective of groundwater control is to prevent an excavation below the natural groundwater level from flooding.

However, especially in fine-grained soils of low to moderate permeability (such as silts or silty sands), another objective is to avoid groundwater-induced instability of the excavation by controlling pore water pressures and hence effective stresses around the excavation. In these soils the pore water pressures associated with even very small seepages into the excavation can cause serious instability. It is not always understood that in fine soils the objective of groundwater control is to control pore water pressures to ensure stability, and not to literally 'dewater' the soil.

Basic soil mechanics theory (see Chapter 15 *Groundwater profiles and effective stresses*) states that soil behaviour is controlled by the effective stress σ' which is related to total stress σ (due to external loads) and the pore water pressure u by Terzaghi's equation

$$\sigma' = \sigma - u \tag{80.1}$$

The shear strength τ_f of a soil depends on the normal effective stress, according to the Mohr–Coulomb failure criterion

$$\tau_f = \sigma' \tan \phi' \tag{80.2}$$

where ϕ' is the effective angle of soil friction. Equations 80.1 and 80.2 illustrate that a reduction in pore water pressure at constant effective stress (for example, as may result from a drawdown in groundwater level) increases the normal effective stress and enhances the ability of the soil to resist shear, thus improving stability. Conversely, the positive pore water pressures associated with seepage into the excavation have a destabilising effect and result in slumping of side slopes and softening of the base, as shown in **Figure 80.1(a)**. Base heave in an excavation can also occur due to unrelieved pore water pressures beneath very low-permeability layers below excavation formation level (**Figure 80.1(b)**). Both forms of instability can be avoided by the use of a suitable groundwater control system.

Groundwater control alone should not be expected to achieve a totally dry excavation. Groundwater control can only give a totally dry excavation in favourable ground conditions when used with an effective surface water control system. In other circumstances groundwater control can give a stable and workably dry excavation, but some residual seepages may remain.

80.2.1 Permanent groundwater control

Although the predominant application of pumped groundwater control systems is as part of temporary works during construction, occasionally such systems are used on a 'permanent' basis for the entire design life of a structure. The most common application of permanent groundwater control is to reduce piezometric pressures below large underground structures. Reducing piezometric pressures will reduce the uplift buoyancy forces on the structure and may allow reduction in

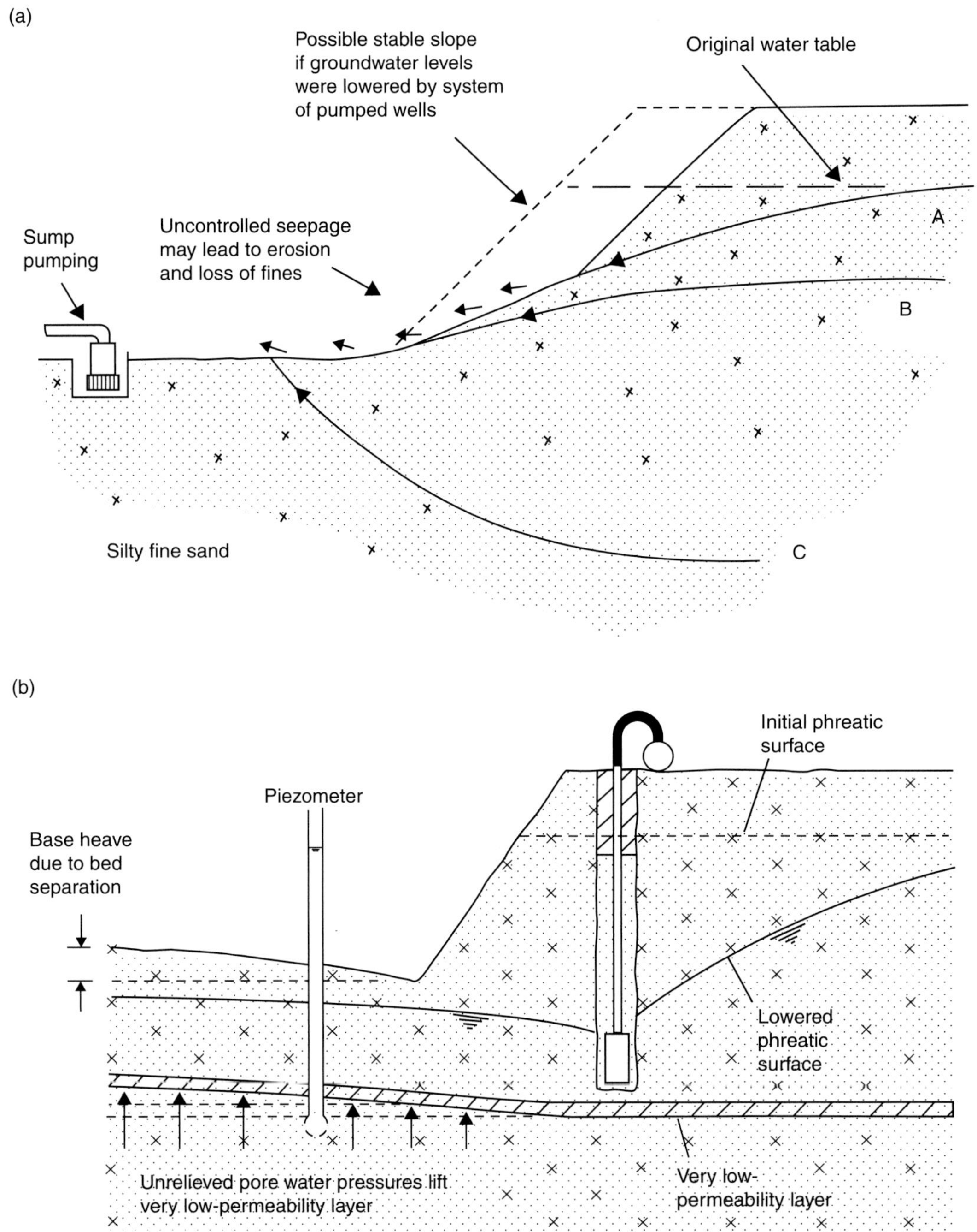

Figure 80.1 (a) Groundwater-induced instability of excavations; (b) instability of the base due to unrelieved pore water pressure ((a) Reproduced with permission from Cashman and Preene, 2021, © Taylor & Francis Group; (b) reproduced with permission from CIRIA C750, Preene *et al.*, 2016; www.ciria.org)

deadweight of the permanent structure. In such cases the savings in capital cost of the structure may outweigh the costs of long-term operation of the pumping system (Whitaker, 2004).

80.2.2 Control of surface water

Groundwater control is separate from the control of surface water such as rainfall run-off. Many contractual disputes have arisen when a groundwater control system has been operating but puddles of water remain in the excavation. If layers of clay exist in the soil, then these puddles may simply be ponded rainwater. Any excavation, including those above the water table, should have a system for surface water control, typically consisting of sump pumps and French drains. Surface water can come from a variety of sources, including rainfall, direct seepage from nearby rivers or lakes, leaking sewers and water mains or the construction operations themselves. Whatever its source, surface water will need to be controlled to allow efficient construction operations. Guidance on surface water control can be found in Cashman and Preene (2021) and Powers *et al.* (2007).

80.3. Methods of groundwater control

The requirement to pump groundwater can be reduced or even avoided by installing a physical cut-off wall or cofferdam (to form a very low-permeability barrier) around the excavation to exclude groundwater.

If a very low-permeability barrier exists at shallow depth beneath the excavation then the wall may be able to penetrate down into that stratum to create a complete cut-off (**Figure 80.2(a)**). The only pumping requirement will be to drain the water trapped within the cut-off and to deal with leakages through the wall and through the very low-permeability stratum.

On the other hand, if a very low-permeability stratum does not exist at a convenient depth, only a partial cut-off can be formed (**Figure 80.2(b)**); groundwater will still enter the excavation, but the cut-off increases the seepage path length and reduces the flow rate compared to the case when there is no cut-off at all. The cut-off should be designed to be of adequate penetration to prevent piping failure of granular soils (see Chapter 16 *Groundwater flow*).

Alternatively, it may be possible to form a horizontal barrier or 'floor' to the cut-off structure to prevent vertical seepage (**Figure 80.2(c)**). The construction of horizontal barriers is relatively rare, but has been carried out using methods such as jet grouting (see Chapter 90 *Geotechnical grouting and soil mixing*) (Newman *et al.*, 1994) and artificial ground-freezing techniques (Harris, 1995).

80.4. Groundwater control by exclusion

A wide range of methods can be used to exclude groundwater from excavations (Bell and Mitchell, 1986). Key attributes of more commonly used cut-off methods are described in **Table 80.1**. Some methods used to form groundwater cut-off barriers are described elsewhere in this book, including diaphragm walls, secant pile walls, sheet pile walls, permeation grouting, jet grouting and mix-in-place walls (Chapter 85 *Embedded walls* and Chapter 90 *Geotechnical grouting and soil mixing*).

Some of the methods are temporary; for example, the groundwater will thaw out when artificial ground freezing is discontinued, or steel sheet piles can be extracted at the end of the job. These temporary methods should not have a significant effect on groundwater conditions at the site once the project is completed. However, methods that permanently affect soil permeability (e.g. grouting) can permanently alter groundwater flow regimes at the site – it is essential that the potential impact of this is assessed at design stage (see Section 80.6.5).

Even when a cut-off is used, some pumping will be required to cope with

- groundwater trapped within the cut-off area
- rainfall and precipitation
- seepage through the wall and through the ground.

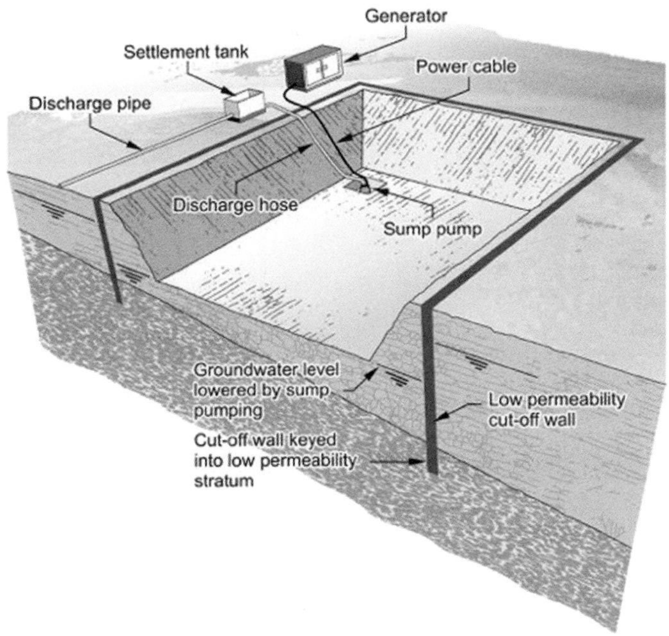

(a) penetrating into very low-permeability stratum

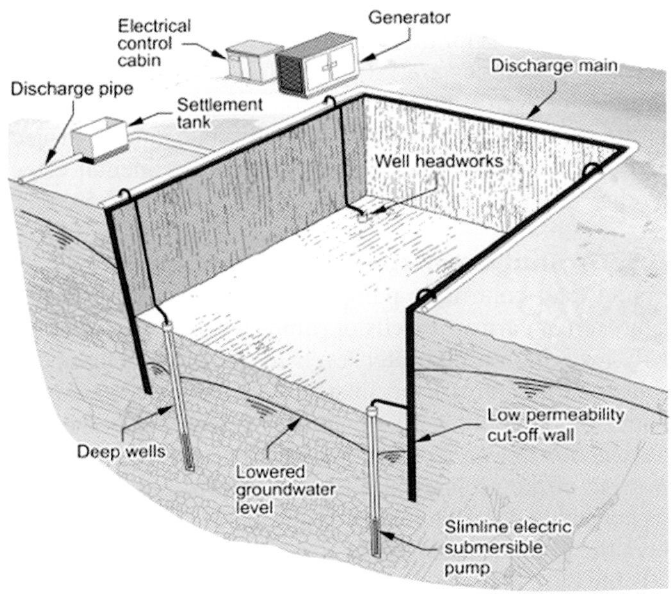

(b) used in combination with dewatering methods

Figure 80.2 Physical cut-off walls (Reproduced with permission from Cashman and Preene, 2021, © Taylor & Francis Group) (continued on next page)

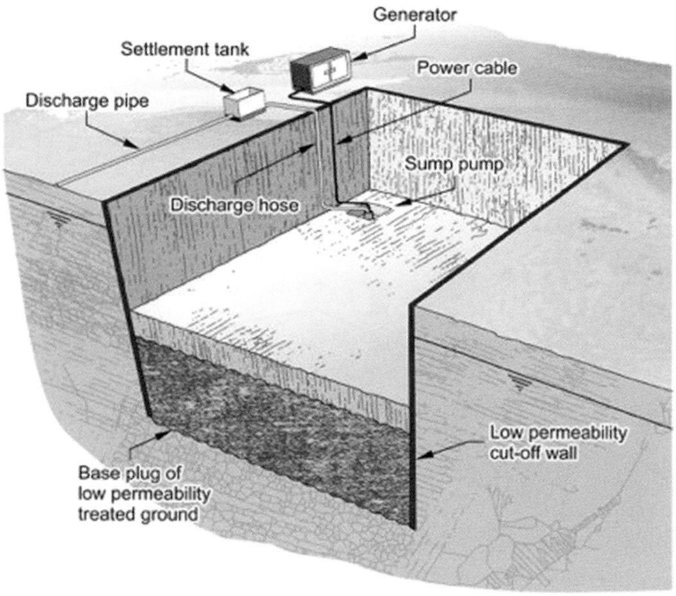

(c) used with horizontal barrier to seal base

Figure 80.2 Continued

It is important to recognise that most cut-off walls will leak. For the common example of steel sheet piles, clutches may leak and declutching of piles may occur in difficult driving conditions. The presence of cobbles and boulders will have a detrimental effect on the integrity of many cut-off methods.

Despite the potential problems, groundwater control by exclusion is widely used in the UK and internationally. Perhaps one of the reasons for this is that, if carried out effectively and if ground conditions are favourable, an exclusion system can minimise or even eliminate any groundwater lowering outside the site area. This can be a very attractive outcome on projects where settlement of nearby structures or detrimental effects on other water users are a concern (see Section 80.6.5).

80.5. Groundwater control by pumping

Groundwater control by pumping involves pumping groundwater from an array of wells or sumps with the aim of temporarily lowering groundwater levels to allow civil engineering projects to be constructed in stable conditions. Groundwater control by pumping is also known as *groundwater lowering*, *construction dewatering* or simply *dewatering*. The amount of lowering of the groundwater level is known as drawdown, so it is normal to describe the way a dewatering system will draw down the groundwater levels.

Table 80.2 lists the various techniques available. However, in the UK the vast majority of projects are carried out using just four main conventional dewatering techniques

- sump pumping
- wellpoints
- deepwells
- ejector wells.

Each of the main pumping techniques will be described briefly in the following sections. **Figure 80.3** shows the range of application of each method relative to the two key parameters: drawdown required and soil permeability (the shaded areas show zones where more than one technique may be suitable).

A variation on groundwater control by pumping is 'passive drainage', where groundwater flow from an excavation or slope is facilitated by gravity flow – that is, without the requirement for direct pumping. Indirect pumping of the water, after it has emerged from the ground, may be necessary to prevent the water causing problems elsewhere on site. Relief wells (see Section 80.5.5.1) are one of the commonly used methods of passive drainage.

An important distinction within the pumping methods is between open pumping and pre-drainage methods. Open pumping, most commonly carried out by sump pumping, involves pumping from within the excavation to remove water that has already entered the excavation. While simple in practice, open pumping has the disadvantage that groundwater levels cannot be lowered in advance of excavation. Open pumping typically requires water to enter the excavation before it can be pumped away, and localised instability of the excavation may result.

In contrast, pre-drainage methods (which include wellpoints, deepwells and ejector wells) work on the principle of lowering groundwater levels in advance of excavation works. This group of methods has the advantage that groundwater can be managed so that water does not enter the excavation, reducing the risk of groundwater-induced instability in the excavation.

80.5.1 Sump pumping

When sump pumping is implemented, water is allowed to enter the excavation, where it is collected in drains and fed to special pits called sumps, from where it is pumped away (**Figure 80.4**). The major limitations of sump pumping are that the sumps and collector drains take up space within the excavation and that it is not possible to lower groundwater levels much below formation level; the sump must be continually deepened as excavation proceeds.

Under favourable conditions sump pumping systems can be a simple and cost-effective means of controlling groundwater inflows to an excavation. Under unfavourable conditions a sump pumping approach can result in delays, cost overruns and, occasionally, catastrophic failure. The primary limitation on sump pumping is the instability of the soil under the action of the seepage forces generated by the groundwater entering the excavation (**Figure 80.1**). This is commonly referred to as 'running sand conditions' or 'boiling' and can cause rapid loss of base and side slope stability, leading to a risk of undermining and settlement to adjacent structures.

Method	Typical applications	Comments
Displacement barriers		
1. Steel sheet piling	Open excavations in most soils, but obstructions such as boulders may impede installation	Temporary or long-term. Rapid installation. Can support the sides of the excavation with suitable propping. Vibration and noise of driving may be unacceptable on some sites, but 'silent' methods are available
2. Vibrated beam wall	Open excavations in silts and sands. Will not support the soil	A vibrating H-pile is driven into the ground and then removed. As it is removed, grout is injected through nozzles at the toe of the pile to form a thin, low-permeability membrane. Relatively cheap
Excavated barriers		
3. Slurry trench cut-off wall using bentonite or native clay	Open excavations in silts, sands and gravels up to a permeability of about 5×10^{-3} m/s	The slurry trench forms a low-permeability curtain wall around the excavation. Quickly installed and relatively cheap, but cost increases rapidly with depth
4. Structural concrete diaphragm walls	Side walls of excavations and shafts in most soils and weak rocks, but presence of boulders may cause problems	A trench is constructed under bentonite slurry as a series of panels by grab or hydromill cutter. A reinforcing cage is placed in the trench and the bentonite slurry displaced by concrete. A continuous reinforced concrete wall is formed by the interconnecting panels, providing a permanent barrier. Often used to form the below-ground structural retaining walls of the permanent structure
5. Concrete bored pile walls (secant and contiguous bored piles)	Side walls of excavations and shafts in most soils and weak rocks, but presence of boulders may cause problems	Secant pile walls have similar attributes to concrete diaphragm walls, but are more likely to be economic for temporary works use Contiguous pile walls do not form an effective groundwater cut-off. Sealing between contiguous piles can be difficult, and additional grouting or sealing of gaps between piles may be necessary
Injection barriers		
6. Permeation grouting using cementitious grouts	Tunnels, shafts and relatively small excavations in gravels and coarse sands, and fractured rocks	The grout comprises a suspension of cement particles (and other admixtures) in water. The cement can be ordinary Portland cement or microfine or ultrafine types. The grout fills the pore spaces in soil and fractures in rock, preventing the flow of water through the soil. Equipment is simple and can be used in confined spaces
7. Permeation grouting using chemical and solution grouts	Tunnels, shafts and relatively small excavations in medium sands (chemical grouts), fine sands and silts (resin grouts) and fractured rocks	The grouts comprise either solutions or colloidal suspensions of very fine particles; this allows the grout to penetrate smaller pores or fractures. Materials (chemicals and resin) can be expensive. Silty soils are difficult and treatment may be incomplete, particularly if more permeable laminations or lenses are present
8. Jet grouting	Tunnels, shafts and relatively small excavations in most soils and very weak rocks	Typically forms a series of overlapping columns of soil/grout mixture. Can be deployed in inclined or horizontal applications. Can be messy and create large volumes of waste slurry. Risk of ground heave if not carried out with care
9. Mix-in-place methods	Side walls of excavations and shafts in most soils and very weak rocks	In situ mixing of soil and injected grout is used to form overlapping columns or panels. Columns formed using auger-based equipment, panels formed using cutter soil mixing equipment. Produces little spoil. Less flexible than jet grouting
Other types		
10. Artificial ground freezing using brine or liquid nitrogen	Tunnels, shafts and relatively small excavations. Will not work if groundwater flow velocities are excessive (>2 m/day for brine or >20 m/day for liquid nitrogen)	Temporary. A 'wall' of frozen ground (a freezewall) is formed, which can support the side of the excavation as well as excluding groundwater. Can be deployed in inclined or horizontal applications. Plant costs are relatively high. Ground freezing using liquid nitrogen is expensive but quick; brine freezing is cheaper but slower
11. Compressed air	Confined chambers such as tunnels, sealed shafts and caissons in most soils and weak rocks	Temporary. Compressed air (up to 3.5 bar) applies a counter-pressure to the exposed ground and temporarily achieves an approximate balance between the tunnel pressure and pore water pressure in the ground, reducing the hydraulic gradient and limiting groundwater inflow. Potential health hazards to workers in compressed air, and specialist medical supervision is required. Air losses may be significant in strata of high permeability. High set-up and running costs
12. Closed-face tunnel boring machines (TBM) using earth pressure balance (EPB) or slurry methods	Tunnels in most soils and weak rocks	A fluid within the head of the TBM applies a counter-pressure to the face, stabilising the soil/rock and limiting groundwater inflow. For EPB TBMs the fluid is a paste formed of a mixture of soil cuttings, groundwater and conditioning agents (such as polymer or bentonite muds). For slurry TBMs the fluid is a bentonite or polymer-based slurry circulated in a closed circuit and used to carry cuttings from the face. TBMs need to be carefully selected to deal with prevailing ground conditions; set-up and running costs may be high
13. Low-permeability surface barriers	Open excavations and shafts in most soils and rocks	Low-permeability materials (including geosynthetic membranes, blinding concrete, sprayed concrete) are used to temporarily seal exposed soil/rock surfaces or faces. The objective is to reduce infiltration of surface water and/or softening and degradation of exposed material

Table 80.1 Physical cut-off techniques for exclusion of groundwater (Data taken from Preene *et al.*, 2016, and Cashman and Preene, 2021)

Method	Typical applications	Comments
Open pumping methods		
Drainage pipes or ditches (e.g. French drains)	Control of surface water and shallow groundwater (including overbleed and perched water) in rock and coarse-grained soils	May obstruct construction traffic, and will not control groundwater at depth. Unlikely to be effective in reducing pore water pressures in fine-grained soils. May generate silt- or sediment-laden discharge water, causing environmental problems if pumped water is not adequately treated prior to discharge
Sump pumping	Shallow excavations in clean coarse-grained soils or stable fractured rock for control of groundwater and surface water	Cheap and simple. May not give sufficient drawdown to prevent seepage from emerging on the cut face of a slope, possibly leading to instability. May generate silt- or sediment-laden discharge water, causing environmental problems if pumped water is not adequately treated prior to discharge
Pre-drainage methods		
Wellpoints	Generally shallow, open excavations in sandy gravels down to fine sands and possibly silty sands. Deeper excavations (requiring >5–6 m drawdown) will require multiple stages of wellpoints to be installed	Relatively cheap and flexible. Quick and easy to install in sands. Difficult to install in ground containing cobbles or boulders. Maximum drawdown is $\sim$6 m for a single stage in sandy gravels and fine sands, but may only be $\sim$4 m in silty sands
Horizontal wellpoints laid by trenching machine	Generally shallow trench or pipeline excavations or large open excavations in sands and possibly silty sands	Horizontal wellpointing is suitable for trench excavations outside urban areas, where very rapid installation is possible
Deepwells with electric submersible pumps	Deep excavations in sandy gravels to fine sands and water-bearing fractured rocks	No limit on drawdown. Expensive to install, but fewer wells may be required compared with most other methods. Close control can be exercised over well screen and filter
Deepwells with electric submersible pumps and vacuum	Deep excavations in silty fine sands or rock of low permeability, where drainage from the ground into the well may be slow	No limit on drawdown. More expensive than ordinary deepwells because of the separate vacuum system. Number of wells may be dictated by the requirement to achieve an adequate drawdown between wells, rather than the flow rate, and an ejector system may be more economical
Shallow bored wells with suction pumps	Shallow excavations in sandy gravels to silty fine sands and water-bearing fractured rocks	Particularly suitable for coarse, high-permeability materials where flow rates are likely to be high. Closer control can be exercised over the well filter than with wellpoints. Drawdowns limited to $\sim$4–8 m depending on soil conditions
Ejector system	Excavations in silty fine sands, silts or laminated or fissured clays, or rock of low permeability in which pore water pressure control is required	Low-capacity wells are pumped by a nozzle and venturi system. In practice drawdowns generally limited to 30–50 m. Low energy efficiency, but this is not a problem if flow rates are low. In sealed wells a vacuum is applied to the soil, promoting drainage
Passive relief wells	Relief of pore water pressure in confined aquifers (including fractured rocks) or sand lenses below the base of the excavation to ensure basal stability	Cheap and simple. Relief wells create a vertical flowpath for water into the excavation. Water overflowing from relief wells must then be directed to sumps and pumped away. Used for both temporary and permanent applications
Vertical drains	Drainage of perched or shallow groundwater downward into strata with lower groundwater levels, to reduce seepages into excavation slopes	Cheap and simple. May be applied in combination with active drainage methods used to lower groundwater levels in deeper strata
Collector wells	Tunnels or deep excavations in relatively permeable soils (such as sands and gravels and sands) and rock, where surface access does not allow the installation of large numbers of wells	Sub-horizontal wells (laterals) are drilled radially outwards from the central pumping shaft. Each collector well is expensive to install, but relatively few wells may produce large flow rates and be able to dewater large areas
Horizontal directionally drilled (HDD) wells	Groundwater control for linear infrastructure projects or for lowering groundwater levels in areas where surface access for vertical wells is restricted	Specialist HDD rigs are used to drill steered sub-horizontal wells, within which screens and filters are installed. Wells can be pumped or be allowed to flow passively if topography allows
Artificial recharge	Soils of moderate to high permeability and fractured rocks, where lowering of groundwater is to be controlled so that environmental impacts can be mitigated. Can also be used as a means to dispose of dewatering water, where other discharge routes are not available	Pumped groundwater is reinjected into the ground, under careful control, by way of trenches or wells. Typically complex to operate and maintain. Recharge wells and trenches often suffer from clogging due to water chemistry and bacterial growth effects and may require periodic backflushing and cleaning
Electro-osmosis	Soils of very low permeability (e.g. clays, silts and some peats)	Very specialised technique using a system of anodes and cathodes to apply electrical potential gradients to promote groundwater flow in materials of very low permeability. Only generally used for pore water pressure control or ground improvement when considered as an alternative to artificial ground freezing. Installation and running costs are comparatively high

Table 80.2 Summary of principal pumped well groundwater control methods (Data taken from Preene *et al.*, 2016, and Cashman and Preene, 2021)

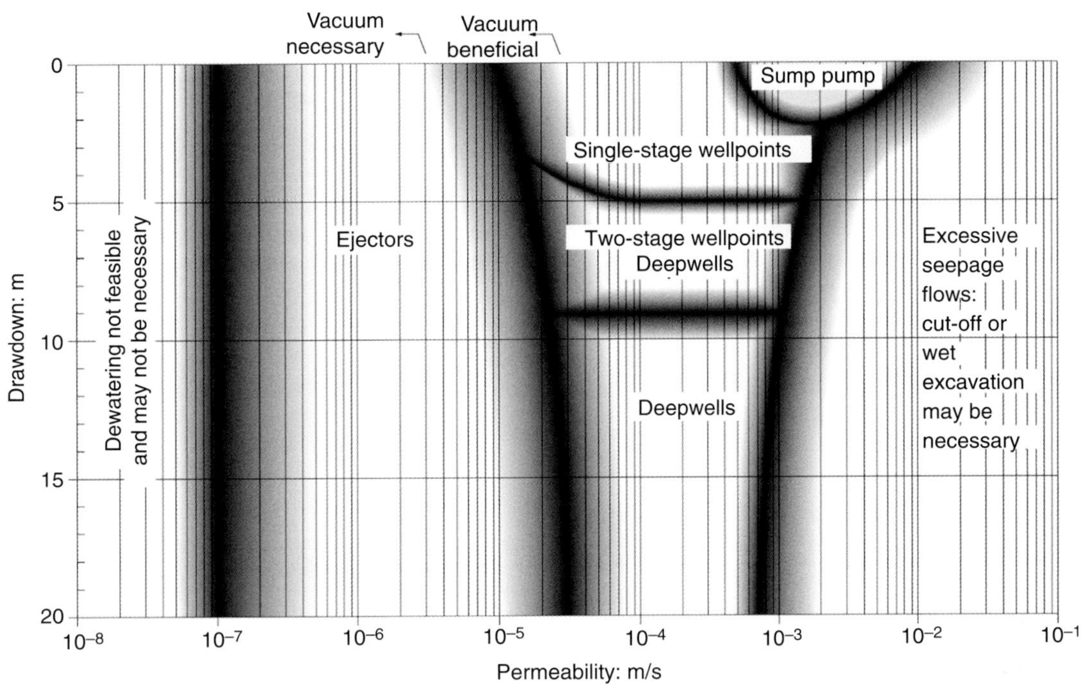

Figure 80.3 Range of applications of pumped well groundwater control techniques (Reproduced with permission from CIRIA C750, Preene *et al.*, 2016; www.ciria.org)

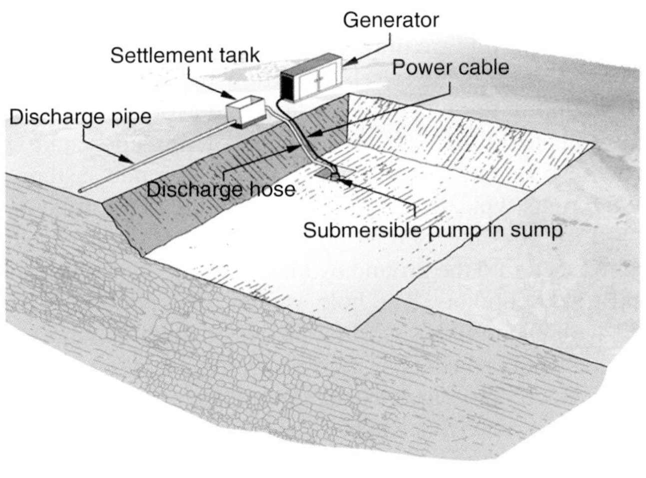

Figure 80.4 Sump pumping system (Reproduced with permission from Cashman and Preene, 2021, © Taylor & Francis Group)

The following are requirements for a sump.

- Depth: the sump should be deep enough to drain the excavation and drainage network, allowing for the pump intake level and some accumulation of sediment.
- Size: the sump should be substantially larger than the size of the pump to allow space for sediment and cleaning.
- Filter: the sump lining should be perforated or slotted, typically with a hole size or slot width of 10–15 mm. The sump should be surrounded with coarse gravel (20–40 mm).
- Access: good access is required to allow removal of the pumps for maintenance and cleaning of the sumps to remove any accumulation of sediment.

When carrying out sump pumping operations, some of the sand and fines fraction in the soil will initially be removed in the immediate vicinity of the sump and drainage network. It is good practice to pass the discharge water through a settlement tank to allow the situation to be monitored and to remove the solids that settle readily prior to discharge. Settlement ponds or lagoons may be needed to remove any silt or clay fraction present to meet discharge consent requirements. If persistent movement of fines occurs, leading to ground loss and settlement, or if an excavation shows signs of instability, sump pumping should be stopped and supplementary or other methods adopted. If the ground loss or instability is serious, it may be necessary to flood the excavation to maintain stability while the situation is reassessed.

80.5.2 Wellpoints

Wellpointing involves closely spaced small-diameter shallow wells (known as wellpoints) in a ring around the excavation (**Figure 80.5**), or in lines alongside long trench excavations (**Figure 80.6**). The intention is to form a curtain of wellpoints around the excavation in order to intercept the groundwater. If a trench excavation is narrow and shallow, wellpoints may be required on one side of the trench only (single-sided system), but for wider or deeper trenches wellpoints will be required on both sides (double-sided system).

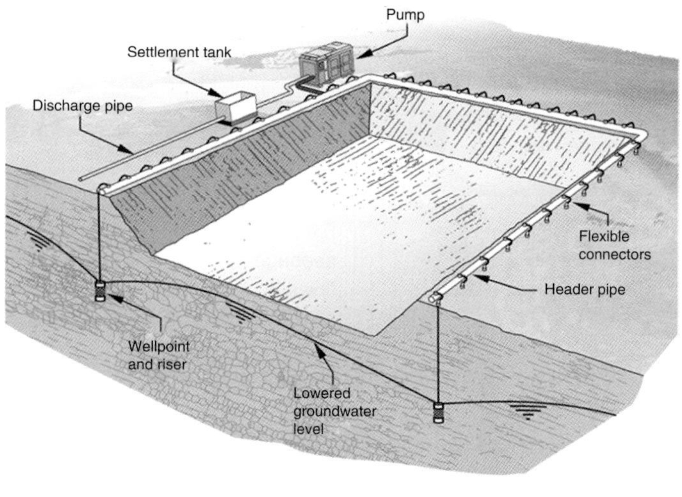

Figure 80.5 Wellpoint system (Reproduced with permission from Cashman and Preene, 2021, © Taylor & Francis Group)

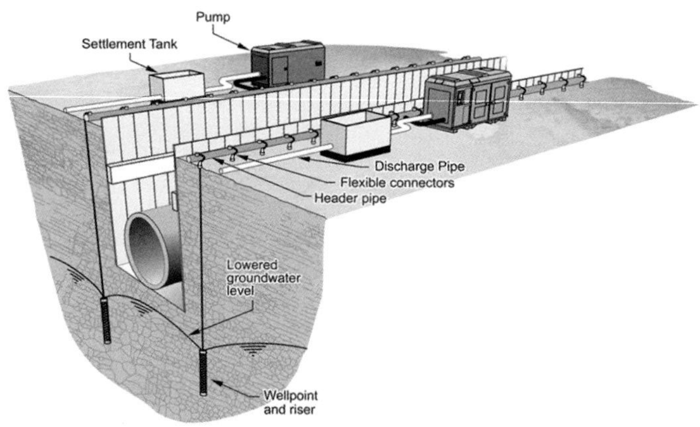

Figure 80.6 Double-sided wellpoint system for pipeline construction (Reproduced with permission from Cashman and Preene, 2021, © Taylor & Francis Group)

Wellpoints are typically installed at spacings of 1.5–3 m between wellpoints and linked using flexible 'swing' connectors to a 150 mm dia. header pipe laid alongside the excavation. The header acts as a manifold allowing one pump to act on many wellpoints (typically one 150 mm wellpoint pump can handle 60–80 wellpoints). The wellpoint pump is located at ground level and applies a vacuum to the header pipe to draw groundwater up the wellpoints, along the header pipe to the pump, from where the water is discharged. Pumps may be of a piston variety or may be a centrifugal pump with an ancillary vacuum pump.

Because the pump draws water up out of the wellpoints by suction there is a practical limit to the drawdown that can be achieved using the wellpoint technique. The suction limit of the pump means that drawdowns of more than 5–6 m below the level of the pump and header pipe are rarely feasible. If greater drawdowns are required a second stage of wellpoints can be installed at a lower level after the first stage has achieved its drawdown (**Figure 80.7**). Alternatively, the first stage wellpoints and pump could be installed at a lower level after a site strip has been carried out.

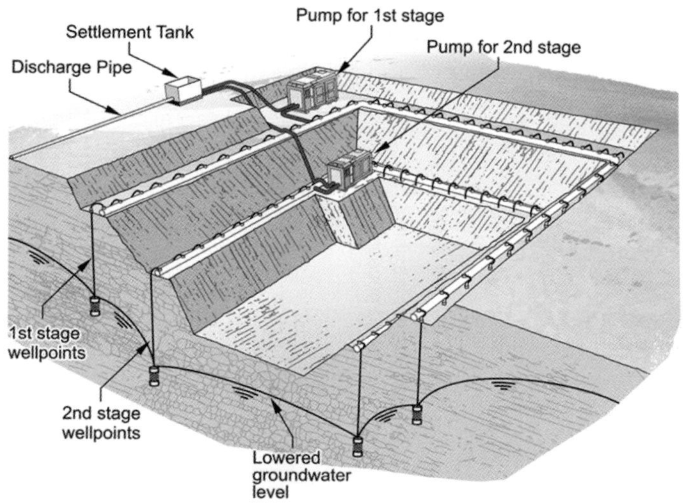

Figure 80.7 Multistage wellpoint system (Reproduced with permission from Cashman and Preene, 2021, © Taylor & Francis Group)

During operation, care must also be taken that the wellpoints are carefully adjusted or trimmed. Trimming is needed when wellpoints begin to pump excessive quantities of air when groundwater levels are drawn down to the top of the wellpoint screens. Trimming involves throttling the control valve on each wellpoint to reduce the air flow and avoid exceeding the pump's air-handling capacity. A poorly trimmed system will not achieve as much drawdown as a well-trimmed system.

Wellpoints typically consist of 38 mm dia. uPVC tubes with a 0.5–1 m long filter screen on the lower end. A steel jetting tube, washed into the ground by high-pressure water, is used to install the wellpoints. The hole around the wellpoint is backfilled with clean filter sand in a process known as sanding-in. The filter sand prevents fine particles being drawn into the wellpoints, which could lead to clogging of the screens. Jetting is very quick in sands, but can be slow in gravels or where cobbles are present. In difficult soils a 'holepuncher' with a hammer action can be used together with the addition of compressed air to the high-pressure water supply. Hydraulic augers are also sometimes used to penetrate strata of stiff clay. Wellpoints are typically installed to 6–7 m below groundwater level (mbgl) or occasionally deeper for pressure-relief purposes.

For long, shallow trench excavations horizontal perforated tubes are sometimes installed by a special trenching machine in a technique known as horizontal wellpoint dewatering (**Figure 80.8**). One end of the pipe is pumped on by a wellpoint pump every 100 m or so.

80.5.3 Deepwells with submersible pumps

A deepwell system involves widely spaced deepwells around an excavation, each pumped by an electric submersible pump

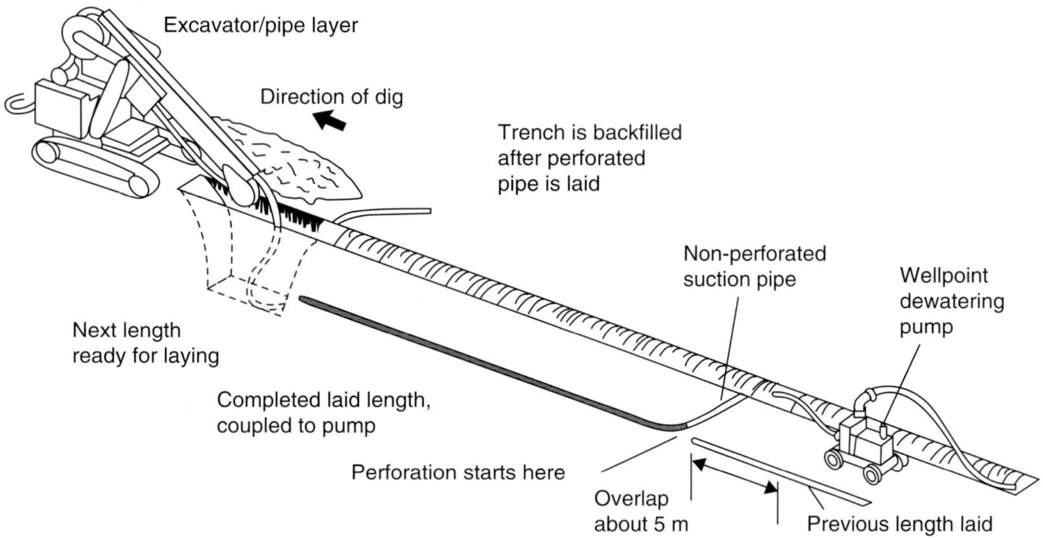

Figure 80.8 Horizontal wellpoint installation (Reproduced with permission from CIRIA C750, Preene *et al.*, 2016; www.ciria.org)

at depth. Because the submersible pump lifts the water from the base of the well there is no suction limit to the drawdown that can be achieved; theoretically, drawdown is only limited by the depth of the wells or by soil stratification. Unlike wellpointing, where the intention is to create a 'curtain' around the excavation, widely spaced deepwells rely on action at a distance. All the wells act together to cumulatively lower groundwater levels over a wide area.

A typical deepwell (**Figure 80.9**) involves drilling a borehole (typically 250–350 mm dia.) to 1.5–2 times the depth of the proposed excavation. A well liner (uPVC or HDPE tube) is then installed in the borehole. The lower section of the liner will be slotted or perforated to allow water to enter; this section is known as the well screen. The annulus between the borehole walls and the well screen is backfilled with filter gravel (known as the gravel pack). The annulus above the gravel pack may be filled with bentonite or grout to prevent surface water percolating down into the groundwater. The well itself may be drilled by a cable percussion rig or by a rotary drilling rig. Typically wells are installed at spacings of 15–60 m between wells and are linked at ground level by electrical supply cables and water collection pipes.

The most common layout for a deepwell system is to space the wells evenly around the perimeter of the area where control of groundwater is required (**Figure 80.10**). The number of wells required for a scheme may be flexible. A few high-capacity wells or a greater number of smaller wells may give a similar extraction flow and drawdown. A few high-capacity

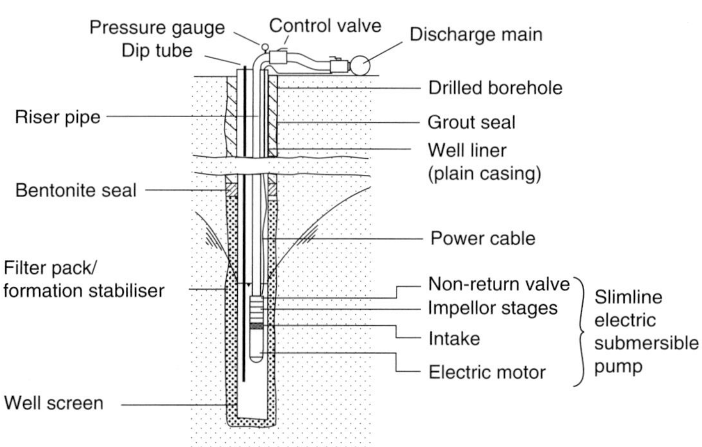

Figure 80.9 Typical deepwell (Reproduced with permission from CIRIA C750, Preene *et al.*, 2016; www.ciria.org)

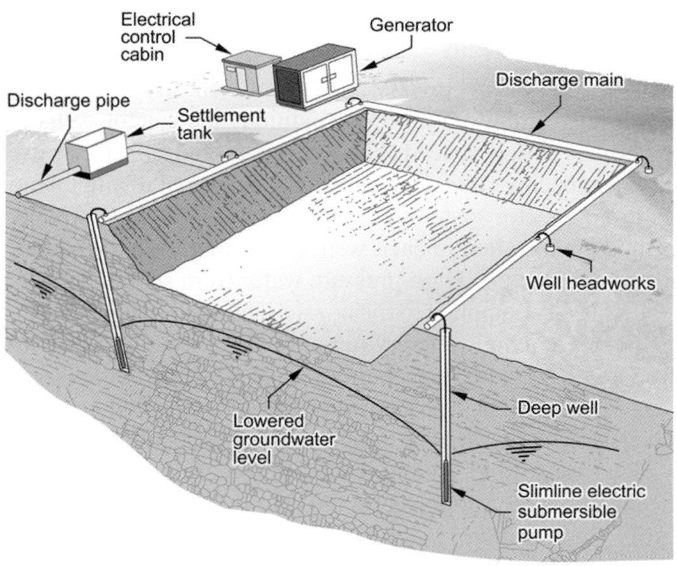

Figure 80.10 Deepwell system (Reproduced with permission from Cashman and Preene, 2021, © Taylor & Francis Group)

wells may seem more cost-effective, but if there are uncertainties in the ground investigation information or a possibility of perched water, a larger number of smaller wells may give better control of the groundwater. Also, a scheme with too few wells may be unacceptable if the stoppage of a single pump could cause flooding or even catastrophic failure. Standby electric power supply facilities can be readily provided, but standby pumping plant ready for immediate start up is rarely provided for economic reasons. Typically, the solution is to make sure that there is sufficient redundancy in the pumping capacity and that the system is not highly dependent on any one well. This can be a problem for schemes which comprise fewer than three or four wells. There have to be sufficient wells to draw the water table down. Keeping the groundwater level down may require a reduced flow rate compared with the initial period of pumping.

One potential problem that can occur when deepwells are run for long periods is the build-up of a sticky red-brown gelatinous slime in the well and pump. This process is called biofouling and the slime is biomass produced by the iron-reducing aerobic bacteria *Gallionella*, which may occur naturally in groundwater. The lowering of the water level in the well provides the bacteria with an abundant source of oxygen. The *Gallionella* thrive in such conditions and adsorb soluble iron (Fe) present in the groundwater and excrete the resulting insoluble iron oxide and oxyhydroxide compounds. If not removed, the biomass can clog well screens and increase wear on pumps dramatically. Cleaning requires the pumps to be removed and washed down; the well can be cleaned by surging with compressed air. Other chemical and bacterial fouling problems can occur: calcium carbonate can leave a thick white paste in the wells; anaerobic sulfate-reducing bacteria create a black slime at the bottom of pumps leading to acidic conditions which may eat into metal pump casings.

80.5.4 Ejector wells

Ejector well systems involve an array of wells around the excavation. Depending on ground conditions the ejector wells are sometimes closely spaced (like wellpoints) and sometimes used at wider spacing (like deepwells). The principal distinguishing feature of ejector systems is the method of pumping. An ejector body is installed in each well, containing a small-diameter nozzle and venturi. High-pressure water from supply pumps at ground level is supplied to the nozzle. The water passes through the nozzle at a high velocity (up to 30 m/s), creating a pressure drop and generating a vacuum of up to 0.95 bar at the ejector. This vacuum draws groundwater through the well screen to the ejector body, where it joins the supply stream of water and is piped back up to ground level. The water that comes out of the ejector is the supply water plus the groundwater; in simple terms you get more water out of an ejector than you put in. The water from the ejectors goes to a tank feeding the supply pumps and is recirculated back to the ejectors (**Figure 80.11**). The excess water abstracted from the ground will start to fill up

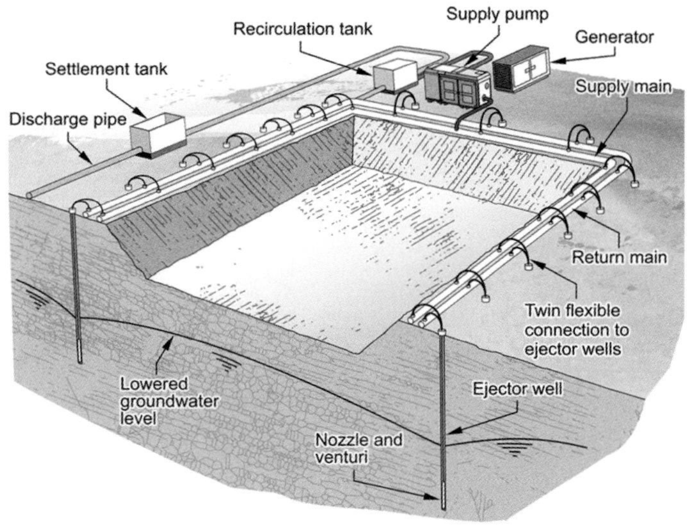

Figure 80.11 Ejector well system (Reproduced with permission from Cashman and Preene, 2021, © Taylor & Francis Group)

the tank and a V-notch weir is normally provided to allow the excess to overflow to a suitable disposal point. **Figure 80.11** shows that ejector systems have two header pipes: a supply header containing the high-pressure feed to each well and a low-pressure return header to carry the recirculated water back to the supply pumps. The ejector body is fitted with a non-return valve which is intended to prevent high-pressure water being inadvertently injected into the ground.

Ejector wells are generally only used in lower-permeability strata such as silty sand or silts where groundwater flow rates are likely to be low. This is because commercially available ejector bodies have a relatively small pumping capacity. A typical ejector might only be capable of pumping 0.3–0.8 l/s of groundwater (the supply flow is in addition to this). Higher-capacity ejector bodies are theoretically possible but are not available because of the low mechanical efficiency of ejectors. For every kilowatt of energy input into an ejector system, less than 25% is used to pump groundwater (the rest is used to circulate supply water and to overcome friction losses). This is a much lower efficiency than either deepwells or wellpoints, so for high groundwater flow rates, instead of designing larger ejectors, it is usually more economic to use either deepwells or wellpoints instead.

A typical ejector well is similar to a deepwell but, depending on the type of ejector used, can often be of smaller diameter than deepwells. Single-pipe ejectors (where the supply and return flows are carried in concentric pipes) can fit down a 50 mm dia. well liner (**Figure 80.12(a)**). Twin-pipe ejectors (where the supply and return flows are carried in separate pipes) can generally fit down a 105 mm dia. liner (**Figure 80.12(b)**). In practice, for the supply pressures normally used of 100–160 psi (690–1100 kPa), the drawdown achievable is limited to about 35 m below the level of the supply pump.

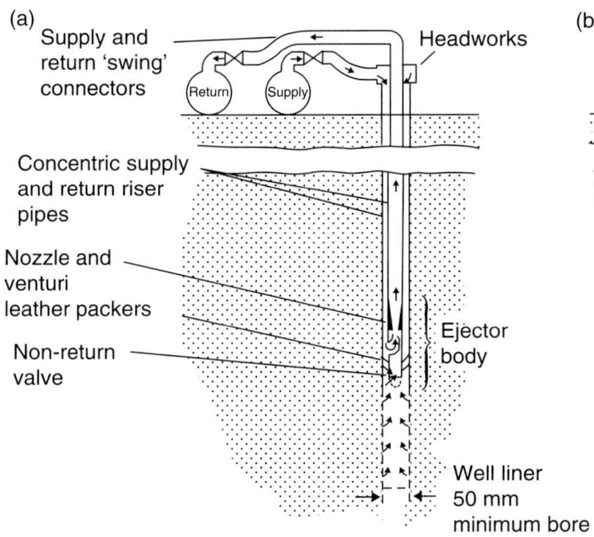

Figure 80.12 Single-pipe and twin-pipe ejector bodies (Reproduced with permission from CIRIA C750, Preene *et al.*, 2016; www.ciria.org)

Like wellpoint and deepwell systems, ejector wells are generally laid out in a ring configuration around the area to be dewatered. Spacing of ejector wells will be controlled by the flow rate and the capacity of the ejectors used. If the soil stratification indicates the possibility of perched water or overbleed seepage, the well spacing may have to be reduced. In practice, ejector well spacings generally fall between those used for wellpoint systems (i.e. 1.5–3 m) and those used for deepwells (i.e. 10 m or more).

In a similar way to deepwells, ejector wells are prone to biofouling by bacterial action. Due to the smaller size of the flow paths inside the ejector, where the biomass can build up, it is possible for an ejector body to become totally clogged and cease functioning altogether. Regular monitoring of supply and return flow rates and drawdown levels should allow the problem to be identified at an early stage. The ejectors can be cleaned by compressed air surging in a similar way to deepwells.

80.5.5 Other groundwater pumping techniques

80.5.5.1 Relief wells

Relief wells (also known as pressure-relief wells) can be used to prevent base heave of excavations (see Chapter 16: *Groundwater flow*) by allowing groundwater to flow upward from pressurised strata beneath an excavation (**Figure 80.13**). Relief wells are an alternative approach to pumped wells (typically located outside the excavation) and comprise a gravel-filled borehole (with or without a perforated liner) within the excavation. The relief well forms a vertical high-permeability pathway. As the excavation progresses deeper, groundwater will flow up the relief well into the excavation, thereby reducing pore water pressures below the base of the excavation and improving base stability. The water entering the excavation is typically disposed of by sump pumping. A simple example of the application of relief wells is given in Ward (1957).

80.5.5.2 Collector wells

A collector well (sometimes known as a Ranney well after a proprietary system) typically comprises a vertical shaft (approximately 5 m dia.), sunk as a caisson, from which

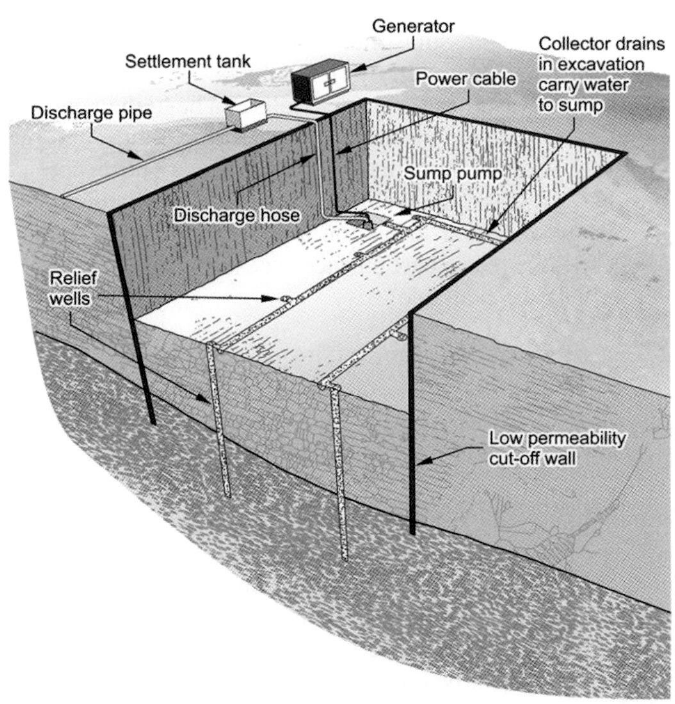

Figure 80.13 Relief wells linked to a sump pumping system (Reproduced with permission from Cashman and Preene, 2021, © Taylor & Francis Group)

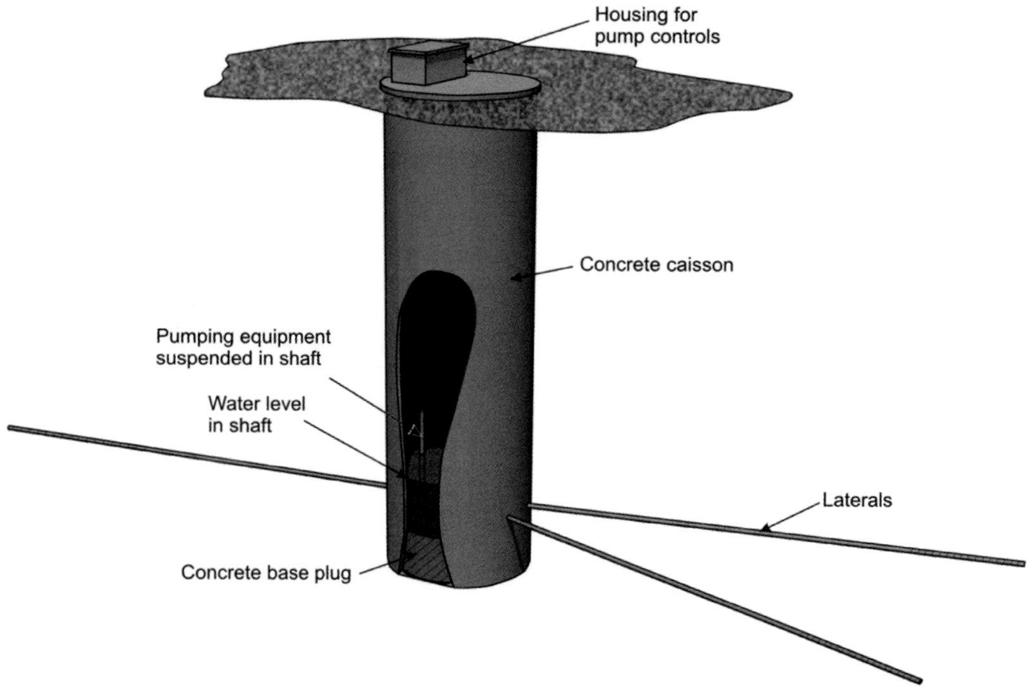

Figure 80.14 Collector wells

small-diameter horizontal or sub-horizontal screened wells (termed laterals) are drilled or jacked radially outward (**Figure 80.14**). Typically a single duty pump is used to pump from the shaft, causing a pressure gradient along the laterals, causing them to discharge into the shaft.

The cost of installing collector wells can be high in comparison to other pumped groundwater control methods, and they are rarely used in temporary groundwater control applications. Nevertheless, the technique has occasionally been used to lower groundwater levels (on a temporary and permanent basis) beneath areas where surface access to drill conventional deepwells is not available.

80.5.5.3 Artificial recharge systems

Artificial recharge involves injecting (or recharging) water into the ground in a controlled manner by way of recharge wells or recharge trenches. The intention is to mitigate the potential environmental impacts (including ground settlement) caused by pumping. Such systems need to be designed and operated with care as they must achieve their objective of mitigating the impacts of pumping, without preventing groundwater control being effective in the excavation itself.

Water may be recharged by way of shallow recharge trenches or specially constructed recharge wells. Recharge trenches can be quick and cheap to install, but, because of their shallow depth, are only suitable to recharge into unconfined aquifers with water tables near ground level. In contrast, recharge wells can be designed to inject water into specific aquifers, including (if appropriate) confined aquifers at depth beneath a site. A recharge well is essentially similar to a pumping well, but with the following features.

- The annulus around the well casing is typically sealed at the surface with a grout or concrete seal. This is to prevent injected water short-circuiting up to ground level by way of the filter pack.
- A downspout is used to prevent the recharge water from cascading into the well and becoming aerated. Aeration of the water may promote clogging processes in the well.
- Air vents are used to allow air to be purged from the system.
- A control valve and flowmeter are used to regulate flow.

A fundamental issue with the operation of artificial recharge systems (wells and trenches) is that they are very vulnerable to loss of performance due to clogging. Clogging may be caused by low levels of suspended solids (sand, silt or clay particles) in the recharge water, or by chemical or biological processes (such as the build-up of iron deposits or bacteriological slimes). It is normal to allow for regular cleaning of recharge wells by backflushing, and to allow for additional recharge capacity to account for long-term deterioration in performance of the wells. Further information on artificial recharge to mitigate the impacts of pumping can be found in Powers (1985). An example of the application of a recharge system is given in Powrie and Roberts (1995). Examples of the use of recharge trenches are given in Cliff and Smart (1998).

80.5.5.4 Electro-osmosis

Pumped groundwater control systems are often ineffective in very low-permeability soils such as silts and clays. This is because groundwater movement under the influence of hydraulic gradients caused by pumping will be excessively slow. In such circumstances, groundwater flow may be more rapid under the influence of electrical potential gradients created by the technique of electro-osmosis.

Electro-osmosis is a very specialised technique, whereby a direct current is passed through the soil between an array of anodes and cathodes (**Figure 80.15**). The electrical potential gradient causes positively charged ions and water molecules around the soil particles to migrate from the anode to the cathode. The small volumes of water generated can be pumped away from the cathode by wellpoints or ejectors.

The main application of electro-osmosis is for the stabilisation of excavations in very soft silts or clays. Further information can be found in Casagrande (1952) and Casagrande *et al.* (1981).

80.5.5.5 Pumping and exclusion methods used in combination

Pumped groundwater control methods are sometimes used in combination. Common examples include the following.

- *Pumping to dewater cofferdams.* Even where the soil stratification allows a cut-off wall to form a complete cofferdam (**Figure 80.2(a)**) to exclude groundwater, some residual groundwater will remain trapped in the excavation. Groundwater pumping from within the excavation will be necessary to manage this water, to prevent it from interfering with construction operations.
- *Cut-off walls used to reduce pumped flow rates.* In very high-permeability soils pumped flow rates from excavations will be very large. It may be appropriate to use cut-off walls penetrating below excavation level (**Figure 80.2(b)**). If the abstraction points (wells or sumps) are located within the cut-off walls, the pumped flow rate will be reduced to some degree, potentially reducing pumping costs and making the pumped discharge easier to dispose of.
- *Groundwater lowering to reduce loading on cut-off structures.* Where the sides of an excavation are supported by cut-off walls such as sheet piles or diaphragm walls, a significant proportion of the lateral loading on the walls may be due to external groundwater pressures (see Chapter 63 *Principles of retaining wall design*). In some circumstances pumped dewatering wells are located outside the excavation to lower groundwater levels and reduce loading on the retaining structure. This can allow overall cost savings to be made by reducing the necessary wall stiffness and propping requirements (Roscoe and Twine, 2001).

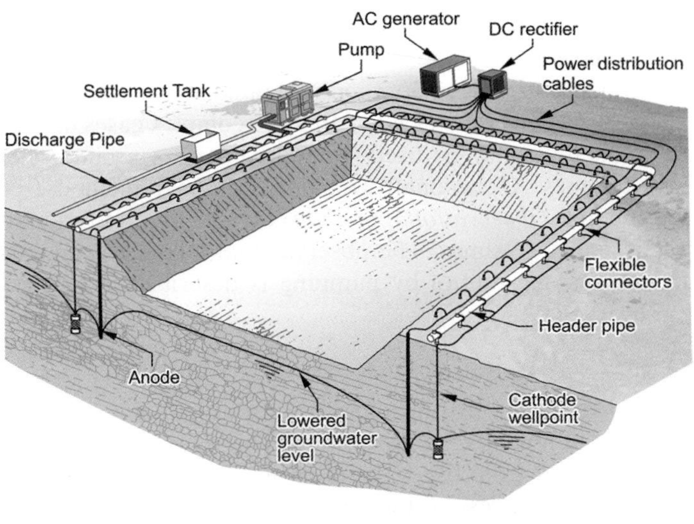

Figure 80.15 Electro-osmosis system for stabilisation of an excavation (Reproduced with permission from Cashman and Preene, 2021, © Taylor & Francis Group)

80.6. Design issues

The design process needs to address hydrogeological factors in relation to the calculation of pumped flow rates, environmental impacts and so on, but must also address the performance and selection of suitable technologies. In all but the simplest of groundwater control problems, the following steps should be followed as part of the design process

- definition of problems and constraints
- development of the hydrogeological conceptual model
- selection of the method of groundwater control
- design calculations
- assessment of the environmental impacts
- review of the design.

Each of these stages is briefly discussed in the following sections. On large and complex projects, the observational method is sometimes used (see Chapter 100 *Observational method*). Further details of design methods are given in Preene *et al.* (2016) and Cashman and Preene (2021).

80.6.1 Definition of problems and constraints

To allow a groundwater control system to be designed, three aspects must be addressed.

- *Define the objectives of groundwater control.* The objectives must be established at the start. Is there a need to prevent inundation of excavations in permeable soils, or is pore water pressure control needed to ensure slope stability in fine-grained soils, or is there a need to reduce groundwater uplift pressures to prevent base heave? Such differing objectives may drive the design down different avenues.
- *Identify key constraints.* Groundwater control is not carried out in isolation; it must integrate with the rest of the construction project. Are there space and time constraints for the site works? Is the site in a sensitive environmental setting, where prolonged pumping may cause problems (see Section 80.6.5)?

- *Gather and review available information.* It is important to collate information at an early stage to understand what is known about the site, and hence what is not known or is uncertain. Often, even if recent site investigations have been carried out there may be uncertainty over the permeability of soil and rock at the site.

80.6.2 Hydrogeological conceptual model

A key stage in the design process is the development of a hydrogeological conceptual model relevant to the groundwater control problem, to provide the information necessary for design. If the conceptual model is a poor match for actual conditions, any subsequent design work is likely to be of little value. If there are insufficient reliable data to formulate a convincing model, this could be a sign that further site investigation is needed. Some key factors in the conceptual model are outlined below.

80.6.2.1 Aquifer types and properties

The aquifer(s) (water-bearing strata in which groundwater levels are to be lowered) must be defined and their nature identified. On major projects there may be a requirement to involve expert hydrogeologists to characterise groundwater conditions. Further details on the characteristics of aquifers can be found in hydrogeological texts such as Younger (2007).

80.6.2.2 System geometry

The depth and size of the excavation will have a key influence on the design of a groundwater control scheme. In particular the depth of the deepest part of the excavation below groundwater level is a key parameter, which must be determined.

80.6.2.3 Soil permeability

The permeability of the aquifer is a critical parameter in the design of groundwater control systems, and must be assessed as part of the design process. Methods of estimation of permeability are described in Chapter 47 *Field geotechnical testing*. Permeability can be very difficult to determine accurately, either from laboratory tests or in situ tests, and uncertainties in permeability can lead to major variations in calculated flow rates. It is often necessary to carry out sensitivity analyses to assess the impact of different permeability values on design.

80.6.2.4 Aquifer boundary conditions

Aquifers are rarely uniform and will typically have physical or hydraulic boundaries that influence their behaviours. Any significant boundary conditions need to be identified and accounted for in the design. Examples of boundary conditions include low-permeability clay layers that may impede vertical water flow within sandy aquifers, or bodies of surface water (such as rivers, lakes or the sea) which may act as large local sources of water for an aquifer.

80.6.3 Selection of method of groundwater control

Correct selection of the method of groundwater control is essential for successful design. It is likely to be very expensive and disruptive to change methods and technologies at any stage after the initial design.

A key step is to identify whether the most appropriate approach is groundwater control by pumping or groundwater control by exclusion, or a combination of the two. This decision is likely to be influenced by a range of factors, including

- technical issues such as order of magnitude of pumped flow rates, likely depth of cut-off to low-permeability layers (some scoping calculations may be needed at this stage)
- working space or programme issues for the main construction works
- requirement to minimise external groundwater impacts (which would tend to favour exclusion options)
- availability of water-disposal routes of adequate capacity (if no disposal routes are available, pumping options are unlikely to be feasible)
- cost and availability of experienced contractors for the various methods.

Once the method (exclusion or pumping) has been selected, the technology must be initially selected. The choice of technology will affect the design process, and so must be selected early, and then reviewed periodically during the design process to confirm it is appropriate. For groundwater control by pumping, initial technology selection can be made from **Figure 80.3**, which relates the range of applications of pumped dewatering methods to the required drawdown and estimated soil or rock permeability. For groundwater control by exclusion, initial technology selection can be made from **Figure 80.16**, which relates cut-off methods to estimated soil or rock permeability.

80.6.4 Design calculations

Design calculations for pumped groundwater control systems are described in Preene *et al.* (2016) and Cashman and Preene (2021). In simple cases the design will be carried out using seepage calculations derived from classic Darcy's Law flow models applied to appropriate boundary conditions (see Chapter 16 *Groundwater flow*). In more complex cases numerical modelling may be used to model complex groundwater flow regimes. Occasionally the observational method (see Chapter 100 *Observational method*) may be used for groundwater control schemes. The observational method as applied to groundwater control by pumping is described in Roberts and Preene (1994).

80.6.5 Assessment of environmental impacts

Dependent on the hydrogeological setting, groundwater control has the potential to have measurable effects at considerable distances from the excavation itself. These effects need to be assessed at design stage so that any necessary mitigation measures can be identified.

Pumping of groundwater can lower groundwater levels around a site. Occasionally groundwater levels may be

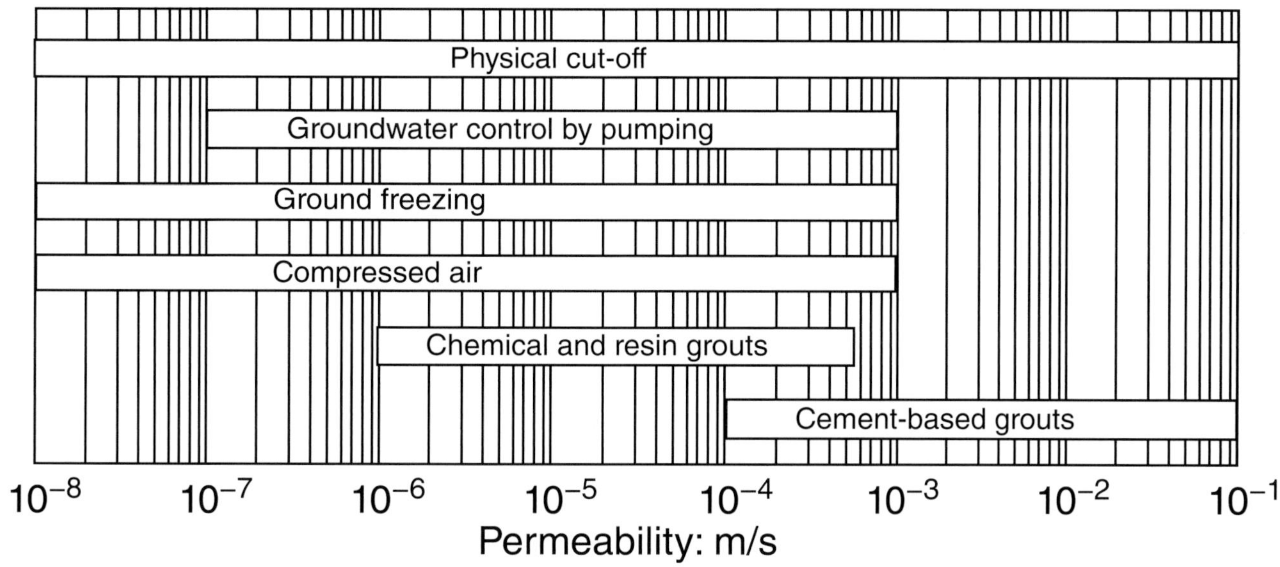

Figure 80.16 Approximate range of application of groundwater exclusion techniques in soils (Reproduced with permission from CIRIA C750, Preene *et al.*, 2016; www.ciria.org)

lowered by small amounts up to several hundred metres from the site. Cut-off walls installed as part of groundwater exclusion schemes may act as underground dams and may cause local changes in groundwater levels.

Table 80.3 summarises the range of potential impacts from groundwater control works. The more significant impacts include

- reduction of yield from nearby water supply wells as a result of dewatering pumping
- settlement of nearby structures as a result of effective stress increases caused by groundwater lowering (Preene, 2000)
- increased movement of contaminated or saline groundwater as a result of dewatering pumping
- increases in groundwater level as a result of the installation of cut-off walls.

Guidance prepared by UK regulators is available on methodologies for hydrogeological impact appraisals (HIA) to be applied to dewatering abstractions (Boak *et al.*, 2007). The HIA approach is often 'tiered', with different levels of complexity of assessment, dependent on the nature of the project. Simple HIAs may be little more than a desk study to confirm there are no sensitive receptors nearby. However, a small number of large and complex projects may require extensive numerical modelling and field monitoring.

Further details of potential environmental impacts and associated mitigation measures related to groundwater control activities are given in Preene and Brassington (2003).

80.6.6 Review of design

It should be recognised that even when a thorough site investigation has been carried out, and when design has been carried out in a rational manner, uncertainty may remain, perhaps in relation to values of soil or rock permeability or the magnitude of potential environmental impacts. It is important that a brief review is carried out at the end of the design process to consider if the remaining uncertainty is acceptable in the context of the proposed project. If necessary, the design may need to be revised or additional ground investigation information gathered.

80.7. Regulatory issues

80.7.1 Abstraction licensing and discharge consents

In the UK, groundwater control by pumping is subject to specific government legislation and is overseen by regulatory authorities. These authorities are

- the Environment Agency (England)
- Natural Resources Wales (Wales)
- the Scottish Environmental Protection Agency (Scotland)
- the Environment Agency Northern Ireland (Northern Ireland).

Publications are available which outline the regulatory approach (Environment Agency, 2018).

There are two facets to the legal requirements. The first deals with pumping of groundwater (termed abstraction) and is intended to make sure the regulators can control groundwater use so that groundwater control systems do not cause nearby groundwater users to lose their supplies. The second deals with disposal of groundwater (termed discharge) and is intended to ensure that the pumped water does not itself cause pollution. Up-to-date guidance on legal requirements can be found on the regulators' websites.

Category		Potential impacts	Duration	Relevant construction activities
1	Abstraction	Ground settlement Derogation of individual sources Effect on aquifer – groundwater levels Effect on aquifer – groundwater quality Depletion of groundwater-dependent features	Temporary	Dewatering of excavations and tunnels using wells, wellpoints and sumps Drainage of shallow excavations or waterlogged land by gravity flow
			Permanent	Permanent drainage of basements, tunnels, road and rail cuttings, both from pumping and from gravity flow
2	Pathways for groundwater flow	Risk of pollution from near-surface activities Change in groundwater levels and quality	Temporary	Vertical pathways created by site investigation and dewatering boreholes, open excavations, trench drains, etc. Horizontal pathways created by trenches, tunnels and excavations
			Permanent	Vertical pathways created by inadequate backfilling and sealing of site investigation and dewatering boreholes and excavations and by permanent foundations, piles and ground improvement processes Horizontal pathways created by trenches, tunnels and excavations
3	Barriers to groundwater flow	Change in groundwater levels and quality	Temporary	Barriers created by temporary or removable physical cut-off walls such as sheet piles or artificial ground freezing
			Permanent	Barriers created by permanent physical cut-off walls or groups of piles forming part of the foundation or structure or by linear constructions such as tunnels and pipelines Barriers created by reduction in aquifer hydraulic conductivity (e.g. by grouting or compaction)
4	Discharge to groundwaters	Discharge of polluting substances from construction activities	Temporary	Leakage and run-off from construction activities (e.g. fuelling of plant) Artificial recharge (if used as part of the dewatering works)
			Permanent	Leakage and run-off from permanent structures Discharge by way of drainage soakaways
5	Discharge to surface waters	Effect on surface waters due to discharge water chemistry, temperature or sediment load	Temporary	Discharge from dewatering systems
			Permanent	Discharge from permanent drainage systems

Table 80.3 Impacts on groundwater conditions from civil engineering works (Reproduced from Preene and Brassington, 2003, with permission of John Wiley & Sons Ltd)

80.7.2 Management of discharge water quality

Adequate management of the discharged water is essential for any groundwater control scheme. Discharge consents or permissions are required for all groundwater discharges (see Section 80.7.1).

Disposal options for discharge water include

- to surface waters (e.g. river, watercourse, lake, sea) – consent is required from the regulators depending on the site location
- to groundwater (e.g. soakaways, recharge wells, recharge trenches) – consent is required from the regulators depending on the site location
- to an existing sewerage network – permission is required from the sewer authority (e.g. water utilities or their agents).

Silt pollution caused by poorly managed pumped discharges from groundwater control operations is a common problem, typically resulting from poorly planned and executed sump pumping operations. Silt pollution can harm the aquatic environment in several ways, including

- injuring fish by its abrasive action
- clogging the gills of fish, causing them to die by suffocation
- destroying spawning sites and insect habitats on the riverbed, removing the source of food for fish
- coating the leaves of aquatic plants, limiting their growth.

The ideal way to manage suspended solids in discharges is to tackle the cause of the problem and design and specify the groundwater control system with adequate filters to minimise sediment in the discharge water. Provided that suitable filters are installed, wellpoint, deepwell and ejector well systems do not normally produce discharges with high sediment contents, except during the initial periods of pumping and development, when dirty water may be produced for short periods.

The method which most commonly produces sediment-laden water is sump pumping (Section 80.5.1). Installation of adequate filters around sumps can be difficult and, as a result, clay, silt and sand-size particles can be drawn to the pump and entrained in the discharge water. Whenever sump pumping is carried out, arrangements should be made, before final discharge, to remove any suspended solids (for example by using appropriately designed and operated settlement tanks or ponds) to below the maximum levels set in the discharge consent. If this is not possible, it may be necessary to change to another groundwater control method, such as wellpoints, with adequate filters.

References

Bell FG and Mitchell JK (1986) Control of groundwater by exclusion. *Groundwater in Engineering Geology* (Cripps JC, Bell FG and Culshaw MG (eds)), Geological Society, London, UK, pp. 429–443.

Boak R, Bellis L, Low R, *et al.* (2007) *Hydrogeological Impact Appraisal for Dewatering Abstractions.* Science Report Sc040020/SR. Environment Agency Science, Bristol, UK.

Casagrande L (1952) Electro-osmotic stabilisation of soils. *Journal of the Boston Society of Civil Engineers* **39**: 51–83.

Casagrande L, Wade N, Wakely M and Loughney R (1981) Electro-osmosis projects, British Columbia, Canada. *Proceedings of the 10th International Conference on Soil Mechanics and Foundation Engineering, Stockholm, Sweden*, pp. 607–610.

Cashman PM and Preene M (2021) *Groundwater Lowering in Construction: A Practical Guide to Dewatering*, 3rd edn. CRC Press, Boca Raton, FL.

Cliff ML and Smart PC (1998) The use of recharge trenches to maintain groundwater levels. *Quarterly Journal of Engineering Geology* **31**: 137–145.

Environment Agency (2018) *The Environment Agency's Approach to Groundwater Protection*, Version 1.2. Environment Agency, Bristol, UK.

Harris JS (1995) *Ground Freezing in Practice.* Thomas Telford, London, UK.

Newman RL, Essler RD and Covil CS (1994) Jet grouting to enable basement construction in difficult ground conditions. *Grouting in the Ground* (Bell AL (ed)). Thomas Telford, London, UK, pp. 385–402.

Powers JP (1985) *Dewatering: Avoiding Its Unwanted Side Effects.* ASCE, New York, NY.

Powers JP, Corwin AB, Schmall PC and Kaeck WE (2007) *Construction Dewatering and Groundwater Control: New Methods and Applications*, 3rd edn. Wiley, New York, NY.

Powrie W and Roberts TOL (1995) Case history of a dewatering and recharge system in chalk. *Géotechnique* **45(3)**: 599–609.

Preene M (2000) Assessment of settlements caused by groundwater control. *Proceedings of the Institution of Civil Engineers, Geotechnical Engineering* **143:** 177–190.

Preene M and Brassington FC (2003) Potential groundwater impacts from civil engineering works. *Water and Environmental Management Journal* **17(1)**: 59–64.

Preene M, Roberts TOL and Powrie W (2016) *Groundwater Control: Design and Practice.* CIRIA Report C750. CIRIA, London, UK.

Roberts TOL and Preene M (1994) The design of groundwater control systems using the observational method. *Géotechnique* **44(4)**: 727–734.

Roscoe H and Twine D (2001) Design collaboration speeds Ashford tunnels. *World Tunnelling* **14(5)**: 237–242.

Ward WH (1957) The use of simple relief walls in reducing water pressure beneath a trench excavation. *Géotechnique* **7(3)**: 134–139.

Whitaker D (2004) Groundwater control for the Stratford CTRL station box. *Proceedings of the Institution of Civil Engineers, Geotechnical Engineering* **157**: 183–191.

Younger PL (2007) *Groundwater in the Environment: An Introduction.* Blackwell, Oxford, UK.

Useful websites

Environment Agency (England): www.environment-agency.gov.uk

Environment Agency Northern Ireland (Northern Ireland): https://www.daera-ni.gov.uk/northern-ireland-environment-agency

Natural Resources Wales (Wales): https://naturalresources.wales/?lang=en

Scottish Environmental Protection Agency (Scotland): www.sepa.org.uk

It is recommended this chapter is read in conjunction with

- Chapter 15 *Groundwater profi les and effective stresses*
- Chapter 16 *Groundwater flow*
- Chapter 46 *Ground exploration*
- Chapter 100 *Observational method*

All chapters in this book rely on the guidance in Sections 1 *Context* and 2 *Fundamental principles.* A sound knowledge of ground investigation is required for all geotechnical works, as set out in Section 4 *Site investigation.*

doi: 10.1680/icemge.66830.1305

CONTENTS

Chapter 81
Types of bearing piles

David Preece Laing O'Rourke, Sale, Manchester, UK
Andrew Heathcote Keller Limited, Ryton-on-Dunsmore, UK
James Binns Byland Engineering, York, UK

Piles are slender, columnar foundation elements. They are most commonly designed as discrete bearing piles to transfer part or all of the vertical loading imposed by a structure through weak and compressible soils to competent and stable soils, or rock. This chapter describes the most commonly available types of bearing pile and discusses the methods of construction and the advantages and constraints of individual techniques. Other applications of piling are discussed in Chapters 83 *Underpinning* and 85 *Embedded walls*. Section 81.2 describes the plant and a range of methods used in the construction of bored piles, including rotary cast-in-place, continuous flight auger and displacement auger techniques. Brief discussion of some specialised techniques, including under-reaming, base grouting, embedded columns and geothermal piles, completes this section. Driven piling techniques including timber, segmental precast, prestressed precast and driven cast in situ concrete are described in Section 81.3, along with steel H and box section piles and steel tube piles. A discussion of modern driven piling plant is set in a historical context. The impact of ground conditions, environmental aspects and particular construction issues such as drivability, installation control and the effects of pile driving on the ground are also discussed. Finally, Section 81.4 discusses the wide-ranging application of micro-piling techniques. Practical aspects of various construction methods are discussed, covering rotary case and auger, rotary duplex, rotary down-the-hole-hammer, rotary percussive, hollow-segmental-auger and driven micro-piling. The section is completed with discussions of the grouts used in micro-piling and the particular environmental, quality assurance and health and safety issues that apply.

81.1. Introduction

The use of piles as a foundation solution is in itself a key decision in the design process (see Section 5 *Design of foundations*). Once the use of piles has been decided, it is important to consider which type of pile and method of construction is most appropriate for the project. The selection of appropriate pile type should take place early in the foundation design if time is not to be lost pursuing uneconomical or impracticable solutions. The constraints of a particular project may dictate that only a narrow range of pile types is applicable, or may leave room for a wider choice of solution, which can then be based on more environmentally sustainable and economic grounds. The factors affecting the selection of pile type are many and varied, and are often interrelated. They include those set out in **Table 81.1**.

These factors are commonly in conflict, and a degree of engineering judgement is required to identify the most appropriate solution. A commercial and residential tower development, for instance, may call for relatively few piles with correspondingly high axial and lateral loading. On a site with reasonably open access, large-diameter bored piles (see Section 81.2.1) would be suitable. On the other hand, if the site is subject to severe access constraints and headroom limits, the installation of large-diameter piles may be uneconomical or even impracticable. This may therefore dictate that a micro-piled solution (see Section 81.4) is adopted. Such conflicts can have profound implications for the design of the substructure, and in some cases the overall feasibility of the development.

Performance	Environmental	Site constraints
• Bearing capacity • Uplift capacity • Lateral load capacity • Durability • Settlement	• Noise • Vibration • Spoil disposal • Contamination • Carbon efficiency	• Restricted access • Restricted headroom • Restricted working hours • Existing assets and structures

Safety	Geotechnical
• Work next to operational railways • Airside work at airports • Work adjacent to sensitive assets • Work on sloping sites	• Very weak or compressible strata (e.g. peat) • Deep unstable strata • Water-bearing strata • Obstructions • Rock

Table 81.1 Factors affecting selection of pile type

In another case, a relatively open site may be underlain by a significant volume of contaminated soil. Here the use of displacement piles (which avoid the need for off-site disposal of spoil) would present itself as an attractive solution. On a site where vibration is not an issue this might be most

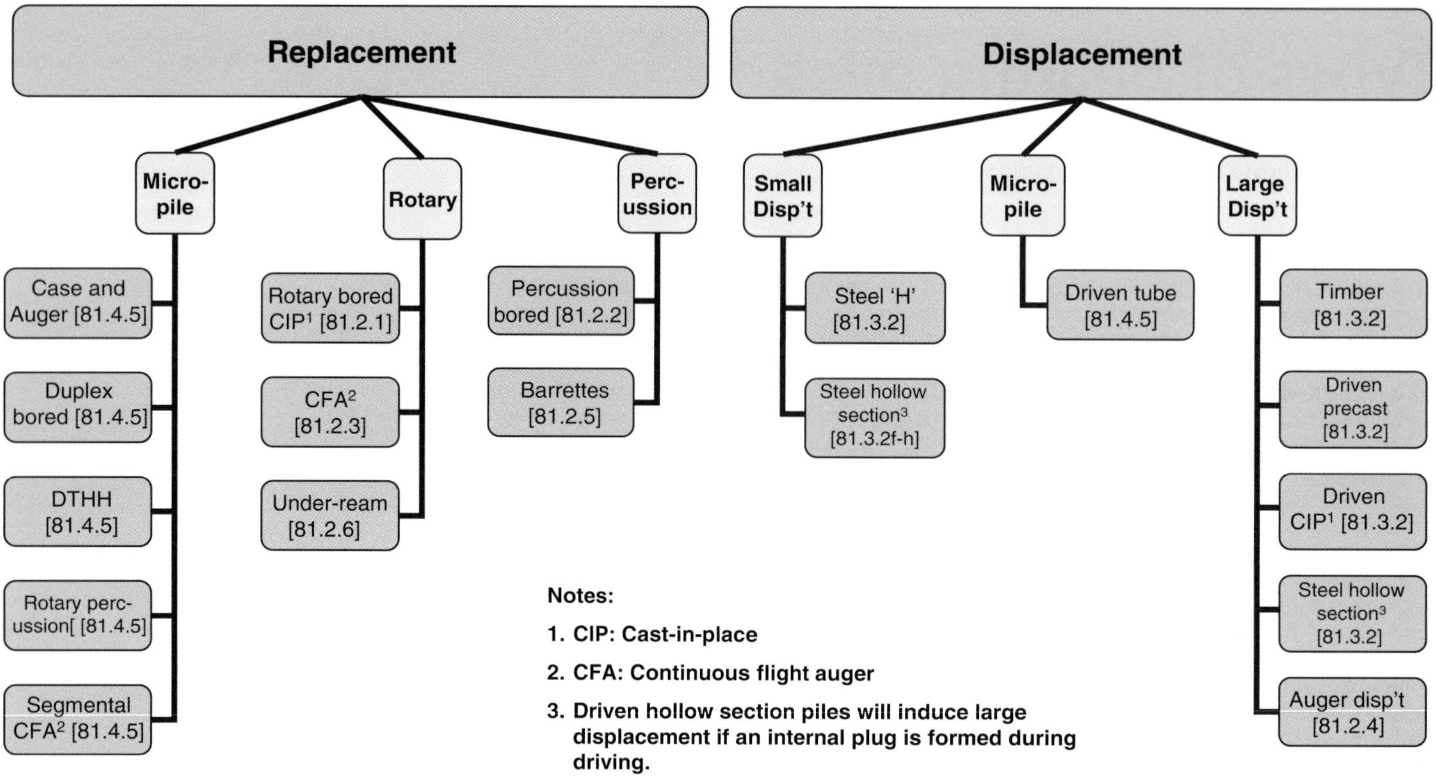

Figure 81.1 Broad classification of common bearing pile types (with reference to the relevant sections of this chapter)

economically achieved with a form of driven pile (see Section 81.3). However, adjacent to a vibration-sensitive installation, or in a residential location, a vibration-free displacement auger pile solution (see Section 81.2.4) may be more appropriate.

In practice it is found that, however similar, no two projects are quite the same in regard to the founding solution. An extensive residential development, for example, may deploy a driven pile solution in the first instance. As the development grows and some sectors are populated, even though the performance requirements and ground conditions remain the same, later adjacent plots may require an alternative solution such as continuous flight auger (CFA) piling (see Section 81.2.3) to keep vibration or noise to acceptable levels.

Figure 81.1 presents a simple classification of pile types, categorising each according to whether installation involves replacement or displacement of the soil (as originally proposed in BS 8004:1986 – now superseded). The figure includes references to the relevant sections in this chapter, where each pile type is briefly discussed.

81.2. Bored piles

Bored pile construction in the UK is generally carried out in accordance with the *Specification for Piling and Embedded Retaining Walls* (SPERW) (ICE, 2017) and is subject to the requirements of the European execution code, BS EN 1536 (BSI, 2010). A bored pile is defined in BS EN 1536 as 'a pile formed, with or without casing, by excavating or boring a hole in the ground and filling with plain or reinforced concrete'. This section describes the processes involved in constructing bored piles and gives illustrations of the plant most commonly used in the UK for this purpose. Some more specialised methods of construction are also discussed. Summaries of typical construction tolerances and the most commonly available pile sizes and depth ranges are given. The particular case of bored piles installed with micro-piling equipment is described in Section 81.4.

Bearing piles are most commonly used to transfer structural loads through weaker near-surface soils, to more competent strata at depth. Unsupported bores and excavations through weak and unstable strata, such as very soft clays, highly organic soils, highly fractured rock or non-cohesive soils (silts, sands, gravels, especially if they are water-bearing) can result in uncontrolled inflows of soil and water. This may create unstable cavities outside the pile shaft, undermine adjacent footings and pose a risk to the ultimate integrity of the concreted shaft (see Chapter 82 *Piling problems*). In unstable soils and beneath the water table it is therefore necessary to provide a means of temporary or permanent support. The means of support may take the form of

- temporary or permanent casings
- drilling/stabilising fluid
- soil-filled auger flights.

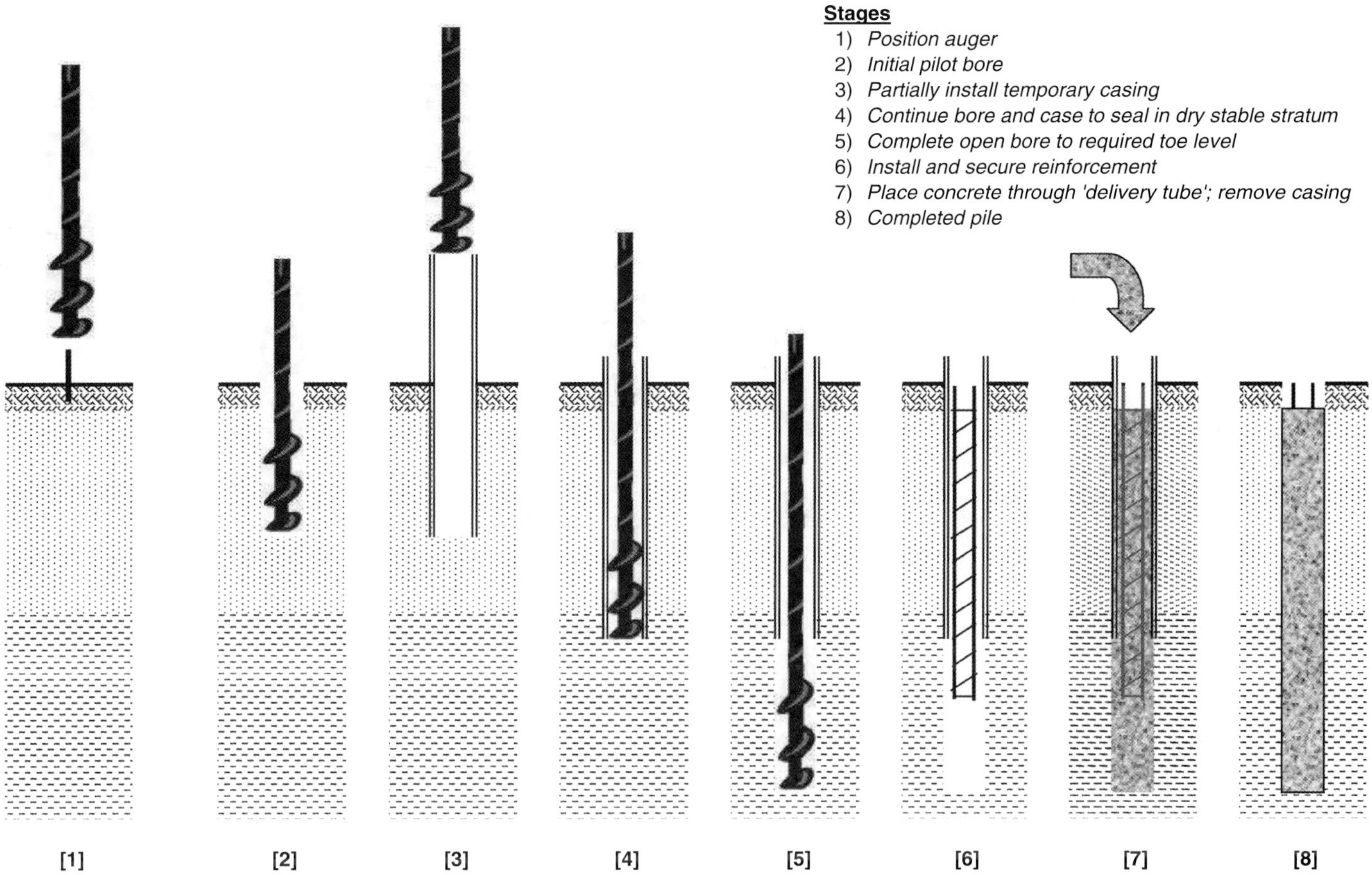

Notes: i Temporary casing may be installed in a single length (as shown) or segmentally as the bore proceeds.
ii Reinforcement generally extends beyond casing to prevent cage uplift during casing extraction.
iii Concrete 'delivery tube' should be long enough to direct concrete and avoid segregation by impact on steel (ICE SPERW B3.5.2.3).
iv Concrete should be cast to a level that ensures that a sound concrete connection can be made after trimming to the specified level (ICE SPERW table B1.5).

Figure 81.2 Basic construction sequence for a rotary bored cast-in-place pile

These stabilisation methods, coupled with the types of tools and processes used to excavate the piles, provide the key distinctions between the different forms of bored piles that are described below.

The majority of bored piles installed in the UK are constructed using either rotary bored and cased methods or CFA techniques. Variations of these methods, such as tripod bored piles, circular clamshell grabs and down-the-hole hammer (DTHH) techniques, are sometimes used to suit particular subsoil and environmental conditions. Increasing use is also being made of cast-in-place displacement piles, particularly to support relatively light structural loads. These distinct bored piling methods are discussed below. High vertical loads, and particularly high lateral loads, may justify the use of barrettes, constructed using rectangular clamshell grabs and diaphragm walling techniques (see Chapter 85 *Embedded walls*). Chapters 79 *Sequencing of geotechnical works* and 82 *Piling problems* include discussion of the need for stable working platforms, the sequencing of pile installation and the problems that can arise during and following pile construction.

81.2.1 Rotary bored cast-in-place piles

The basic process of constructing a rotary bored cast-in-place pile is shown in **Figure 81.2**.

After setting the rig up over the pile location [1], an initial pilot hole is bored [2] using a short, flighted auger, mounted on a kelly bar. A temporary casing is then inserted [3] and the bore and casing are progressed together until a seal into a stable, dry stratum is achieved [4]. The bore is then completed to the required depth [5], the auger is withdrawn and the reinforcement cage is lowered into position and secured [6]. Concrete is placed in the bore to the required level [7] and the temporary casing is withdrawn. Other types of bored cast-in-place pile generally follow this basic process, but with

(a) (b) (c)

Figure 81.3 Rotary bored cast-in-place piling rigs: (a) crane-mounted rotary piling rig; (b) light-/medium-duty hydraulic rig; (c) heavy-duty hydraulic rig (Photos courtesy of Cementation Skanska Ltd)

Pile type	Rig characteristic	Range of diameters: mm[a,b]	Maximum depth: m[a,b]	Typical axial load range[c]
Rotary bored	Light rotary (<35 t)	300–1200	18–35	<5 MN
	Medium rotary (40–80 t)	<1800	40–60	<20 MN
	Heavy rotary (>100 t)	>2400	<90	>30 MN
Percussion bored	Tripod	300–600	<30	<600 kN
CFA	Light CFA (<35 t)	300–750	18–24	<3 MN
	Medium CFA (40–80 t)	450–1200	24–28	<7.5 MN
	Heavy CFA (>100 t)	450–1200	<32	<10 MN
Cased CFA	Twin-drive (>100 t)	600–1200	Casing: 18 m; bore: 25 m	<10 MN
Auger displacement	Rotary (>60 t)	400–600	25	<2 MN

The above figures are given as a preliminary guide only. The advice of specialists should be sought before selecting the particular pile type, plant type and method of construction.
(a) Maximum achievable pile diameters and depths vary according to the plant manufacturer. Figures shown are for guidance only.
(b) In addition to the rig size and power, the available pile diameters and depths may also be limited by the ground conditions.
(c) Irrespective of pile diameter and depth, the achievable axial loads are dependent on ground conditions.

Table 81.2 Typical bored pile characteristics

variations to one or more aspects of construction to meet particular project-specific constraints.

The particular case of CFA piling is discussed in Section 81.2.3.

81.2.1.1 Bored piling rigs

Bored piling rigs have developed considerably over the past 30 years in response to the changing and increasing demands made by the construction industry. The crane-mounted rigs that dominated rotary piling until the 1980s (**Figure 81.3(a)**) have been steadily replaced by purpose-built, self-erecting, vertically masted hydraulic units (**Figure 81.3(b)**). Simultaneously, as the demands of the commercial and infrastructure sectors have risen, increasingly powerful rigs have been developed, capable of installing large-diameter piles to depths in excess of 60 m (**Figure 81.3(c)**). Tomlinson and Woodward (2014) provide a useful summary of the characteristics of self-erecting hydraulic rigs from various suppliers. Commonly available pile diameters and depths are listed in **Table 81.2**.

81.2.1.2 Excavation tools

A wide range of tools are available for use with rotary piling rigs to excavate pile shafts through the variable ground conditions found in the UK (**Figure 81.4**). One or more of these tools may be used during the construction of an individual pile.

Augers are generally used to excavate soft ground (sands, clays, gravels, etc.) and some weak rocks (including chalk and marl). In hard ground a variety of chisels may also be deployed to break up the material at the base of the bore before removal by auger. Specially designed toothed augers can also be used to excavate stronger rocks, although production rates fall significantly at strengths above 10–15 MN/m^2, depending on the type of rig and tool design being used.

Drilling buckets are designed to recover materials that are not easily retained on an auger during retrieval from the bore. This can arise in water-bearing soils and fractured rocks. Specially designed buckets may be used when drilling under

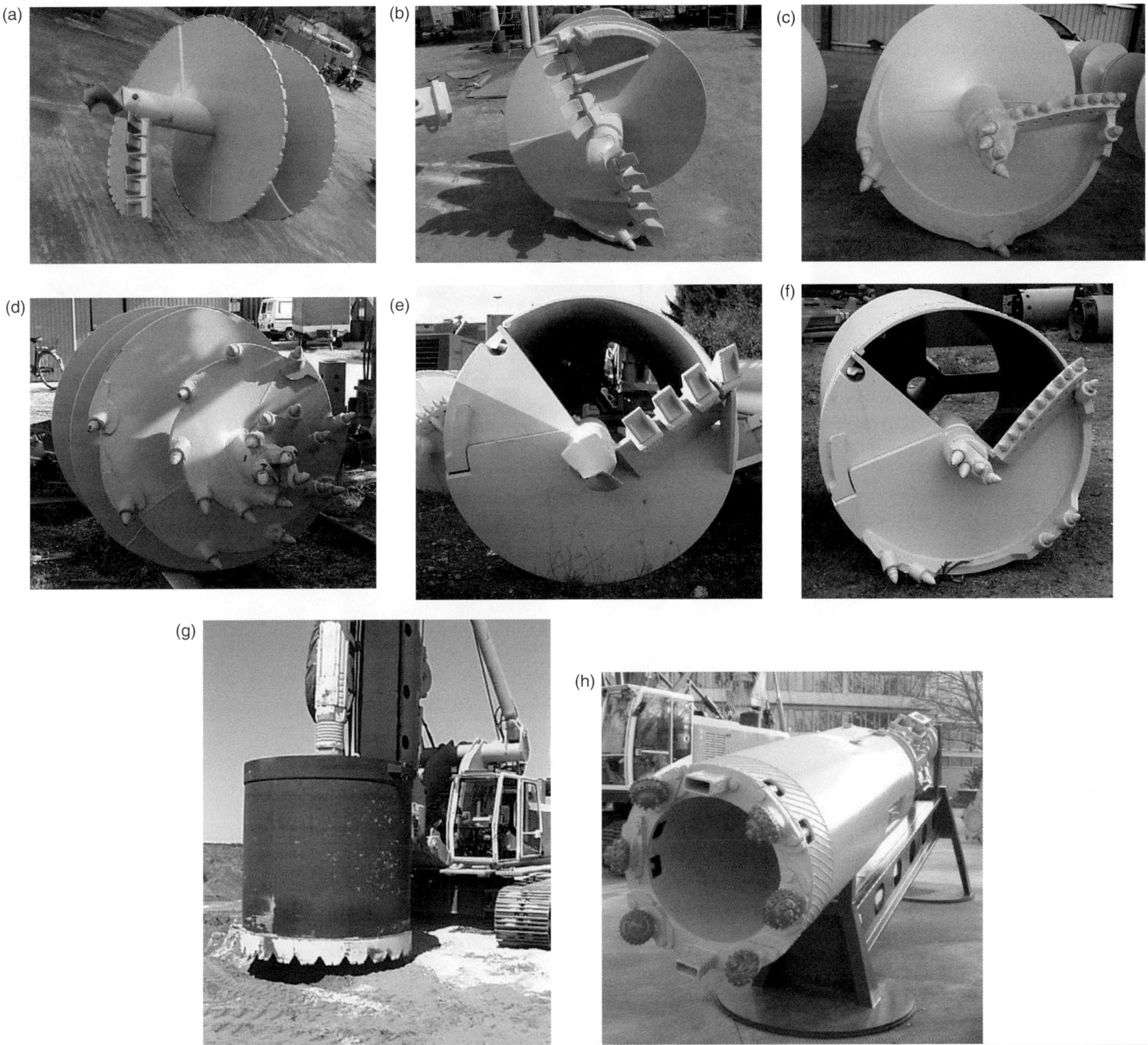

Figure 81.4 Rotary bored piling – excavation tools: (a) single start soil auger; (b) double start soil auger; (c) single start rock auger; (d) single start progressive rock auger; (e) single cut digging bucket; (f) single start rock digging bucket; (g) large-diameter coring barrel; and (h) heavy-duty coring barrel with air flush (Photos courtesy of Bauer Equipment UK Ltd)

stabilising fluid, to minimise turbulence during insertion and retrieval of the tool, and avoid the resulting shaft and base instability. Where appropriate, a drilling bucket may also be used to clean the base of the pile before concreting. Some caution is required in cohesive soils, however, since this can create a softened layer at the pile toe, thereby diminishing the end-bearing response of the pile.

An alternative means of excavating pile shafts, not commonly used in the UK, is to employ a circular-section, clamshell grab or hammer grab. These tools are rope-suspended from a tracked crane and can be combined with segmental casing techniques (see below) to excavate through most soils, and form sockets in weak rocks (using chisel attachments).

Harder and more massive rocks call for alternative methods

- coring barrel
- rock roller assemblies coupled with air flush or reverse-circulation systems to remove the cuttings (described in greater detail by Tomlinson and Woodward, 2014).

81.2.1.3 Casings

Temporary casings most commonly take the form of single-wall steel tubes, installed in one piece as illustrated in **Figure 81.5(a)**. These may be installed and removed from the bore using the piling rig or an attendant crane, with or without the assistance of a casing vibrator.

Figure 81.5 Rotary pile casings: (a) single-wall temporary casing; (b) slip-coated permanent lining; (c) segmental casing; (d) cutting head; (e) casing oscillator (Photos (a) to (d) courtesy of Cementation Skanska Ltd; (e) courtesy of Bauer Equipment UK Ltd)

Where closer control of the boring process is required, segmental casings may be used. These are stiff double-walled casings, typically 40–60 mm thick, mechanically jointed in 1–6 m lengths, with a cutting head at the leading edge (**Figures 81.5(b)–(c)**). Segmental casings can be installed to significant depths using the rotary drive of medium- and heavy-duty hydraulic rigs. Where greater installation/extraction capability is required, a casing oscillator (**Figure 81.5(e)**) or specially designed casing extractor may be used.

Segmental casings, installed as the bore proceeds, offer greater control over pile verticality and enable virtually continuous support of the bore through unstable deposits. With the right combination of power and cutting head they can also be used in conjunction with heavy-duty coring tools to overcome substantial obstructions (**Figure 81.6**), including, in some cases, old piles.

Drill casings can be left in place as permanent linings. These linings may be used to prevent washout of concrete fines by flowing groundwater or to support the concreted shaft in very soft soils. They may also be slip-coated to reduce negative skin friction or to reduce the load shed by the pile in the vicinity of sensitive structures such as tunnels and basement walls (**Figure 81.5(b)**). Lightweight permanent linings may alternatively be installed and cast within the pile bore, allowing recovery of the more expensive drill casings for reuse.

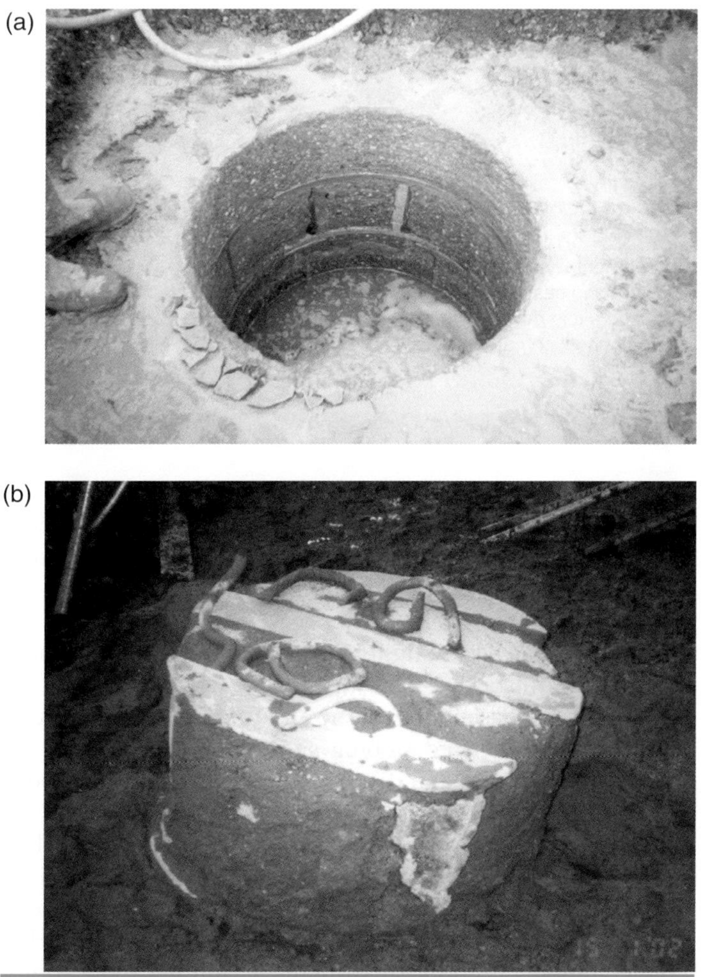

Figure 81.6 Rotary piling – obstruction removal: (a) core hole through concrete; (b) removal of reinforced concrete obstruction (Photos courtesy of Cementation Skanska Ltd)

81.2.1.4 Stabilising fluids

Stabilising fluids are used to support pile bores where they pass through unstable soils at depths beyond the economical use of temporary casing. A typical pile construction sequence using stabilising fluid is shown in **Figure 81.7**. Stabilising fluids take the form of a bentonite clay suspension or a polymer fluid. The composition, properties and operation of bentonite and the various polymer fluids are distinct from one another and call for understanding and experience on the part of the practitioner. A few practical observations are made here.

- Drilling with fluid support is only feasible from a working level at least 1.5–2.0 m above the effective water table since support of the bore wall is provided by maintaining a positive pressure differential between the fluid and the surrounding groundwater.
- Polymer fluids generally hold less detritus in suspension, for less time, than bentonite, making base cleaning more critical with polymer fluids.
- Polymer fluids suit smaller sites since they require less space for the mixing and treatment plant.
- Waste bentonite must be disposed of by tanker to licensed tips. Polymer fluids may be broken down to water and polymer, reducing the difficulty and cost of disposal.

More detailed accounts of the use of bentonite and polymer stabilising fluids may be found in Fleming *et al.* (2009).

81.2.1.5 Placing concrete

In a dry bore, SPERW (ICE, 2017) requires that concrete is placed in a single continuous operation by way of a hopper and sufficient rigid delivery tube to avoid a fall of more than 10 m within a reinforcement cage. This limits segregation of the mix caused by impact on the reinforcement during the fall. In a wet bore (filled with water or stabilising fluid) concrete must be placed with a full-length watertight tremie tube (**Figure 81.7**). The initial charge of concrete is separated from the fluid by a disposable plug to avoid segregation or contamination of the mix, and delivered to the pile toe through the tremie tip. This charge displaces the fluid upward, covering the pile base and engulfing the tip of the tremie. Further concrete is delivered through the tremie, which is progressively shortened as the concrete level rises, until the required concrete level is reached. The displaced fluid is pumped away at the pile head. Throughout this operation the fluid and concrete levels must be maintained to counteract the effects of any groundwater pressures along the pile shaft. The tremie tip must remain embedded between 3 m and 6 m into the rising concrete at all times.

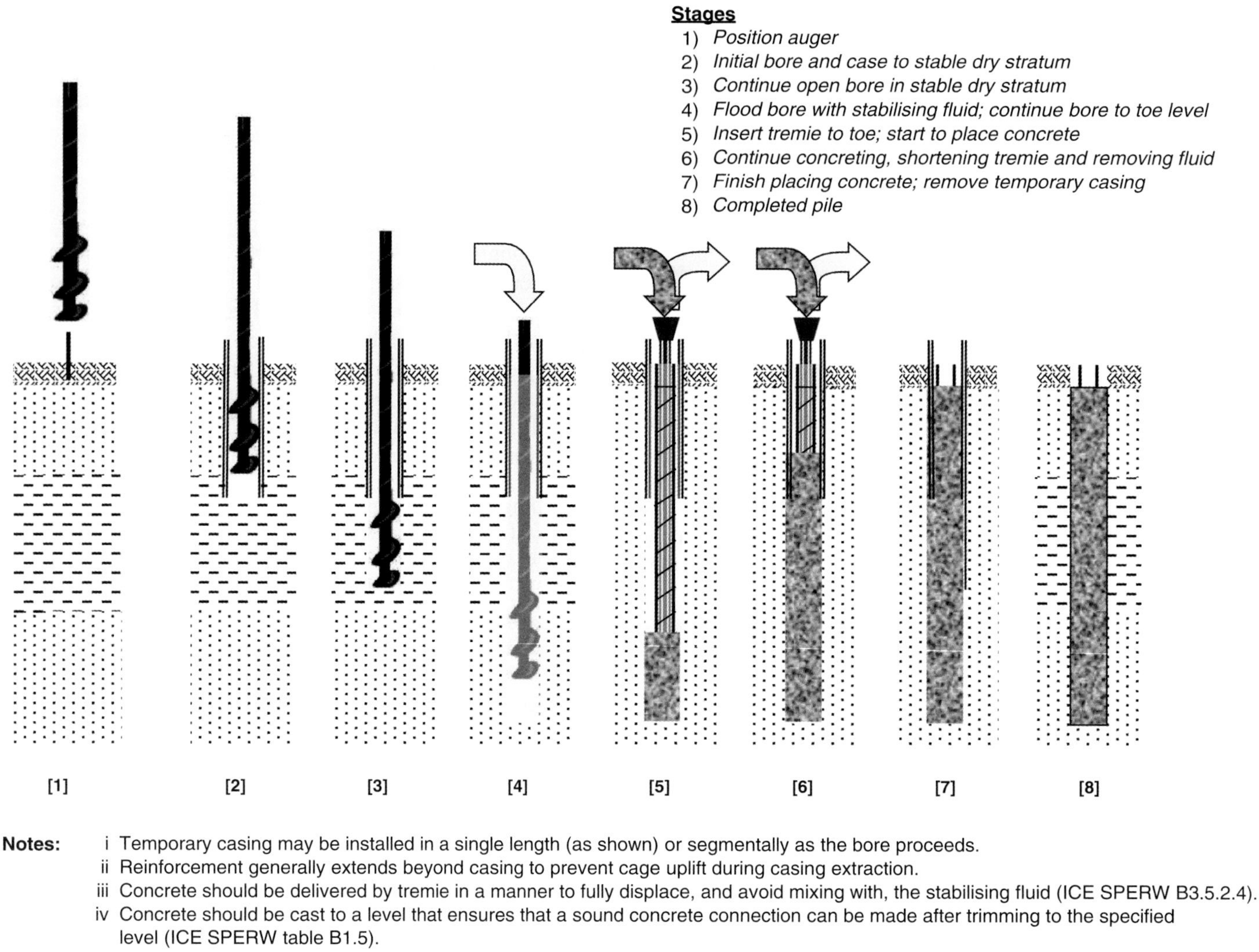

Figure 81.7 Typical construction sequence for bored cast-in-place pile under stabilising fluid

81.2.2 Percussion bored cast-in-place piling

Tripod-mounted cable percussive boring equipment continues to be available for piling in the UK, although it is increasingly replaced by the growing capabilities of hydraulic micro-piling rigs. The method involves much manual handling of heavy tools and casings, and achieves relatively slow production. This makes it less attractive than other methods on both health and safety and commercial grounds. The boring tools, including clay cutters, 'shells' for sand and gravel and chisels, are suspended by rope from the rig and sequentially lifted and dropped in the bore to cut or pump soils up from the base of the bore (**Figure 81.8**). The tool is raised and spoil is cut away or emptied out on the working platform for disposal. Considerable experience is needed to ensure that this process does not significantly soften or loosen the surrounding soils. Temporary casings, which are hammered or surged into place, are handled in short threaded sections, screwed together and installed as the bore proceeds. The reinforcement and concrete processes are similar to rotary piling except that concrete is commonly delivered by dumper, through a chute at the pile head, or by pump.

Tripod piling rigs are light and easily transported, and have been used for installing piles where access and headroom are limited (within basements and in the vaults beneath existing buildings, benched into sloping sites, etc.). Diameters up to 600 mm are possible and depths in excess of 30 m have been successfully achieved.

Earlier percussive processes, such as the Benoto method (Puller, 2003), have been used to install larger-diameter piles. These methods have died out in the UK, in the face of more economical techniques. In recent years, however, DTHH, more commonly used for small-diameter mining applications, has become available in sizes suited to bored piling. Single-bit hammers up to 750 mm dia. are available. Larger diameters

(a)
(b)

Figure 81.8 Percussion bored piles – tripod piling using: (a) 'shell' for boring in granular soils; (b) 'clay cutter' for boring in cohesive soils (Photos courtesy of Cementation Skanska Ltd)

can be drilled using cluster drills (multiple hammer units). These units can be used in combination with cased rotary piling to form deep rock sockets (Fleming *et al.*, 2009). The technique requires large quantities of compressed air and casing for borehole stability. Hydraulically driven hammers are also becoming available. These can reduce the need for temporary casing.

81.2.3 Continuous flight auger

The CFA process (**Figure 81.9**) allows economical and rapid pile construction in water-bearing strata without the need for casing or stabilising fluid. It is suitable for constructing piles through sands, gravels, silts and clays, and for limited penetration of weak rocks. In comparison to rotary piling, greater power is required to drill in and extract the auger in a single pass. It is particularly important that the CFA piling equipment is sufficiently powerful to achieve the required pile depth without significant flighting of material from the bore (note: 'flighting' is the upward displacement of spoil along the auger flights to ground surface). The available depth and diameter range, and the ability to penetrate hard strata with CFA piling, are therefore more limited than with conventional rotary piling (**Table 81.2**). CFA piling is not appropriate in ground containing frequent and substantial obstructions, including old foundations and boulders, without due consideration in the foundation design.

After setting the rig up over the pile location [1] the hollow-stemmed, continuously flighted auger is drilled in a single stroke to the full pile depth [2]. The disturbed spoil remains on the flights at this stage [3]. Concrete, delivered by pump, is injected through the hollow stem of the auger to the base of the pile [4]. The auger, which is still rotating slowly, is lifted a short distance (<100 mm) to allow entry of the concrete into the bore so that the initial charge engulfs the tip, and is then returned to the full bore depth. The spoil-laden auger, still rotating slowly, is then gradually retracted as further concrete is injected to form a continuous shaft [5]. Once the shaft is fully concreted to ground level the auger is withdrawn, the pile head is scraped clean and the reinforcement cage is inserted into the concrete [6] leaving the top of the steel at ground level [7].

81.2.3.1 CFA piling rigs

CFA piling became established in the UK piling market during the 1980s, coinciding with the development of purpose-built hydraulic piling rigs (**Figure 81.10**). Many hydraulic rigs are capable of modification to work in either rotary or CFA mode. The distinguishing features of CFA rigs are the single 'string' of the continuously flighted auger, and the location of the drive table at the top of the auger.

Many CFA piling rigs are fitted with a winch-driven or hydraulic 'crowd' facility designed to increase the downward thrust of the auger while drilling through harder strata. Extraction capability is enhanced by providing a 'foot' at the base of the mast which bears on the working platform to stabilise the rig during withdrawal of the auger. Mechanical auger cleaners are fitted to CFA rigs to prevent spoil from rising with the auger string during extraction. The rising spoil would otherwise present a hazard to operatives working below, and potentially destabilise the rig.

81.2.3.2 Excavation tools

As the name implies, the excavation tools are restricted to augers. In common with rotary piling, the lead augers are mounted with a wide range of cutting tools designed and arranged to suit the expected ground conditions (**Figure 81.11**).

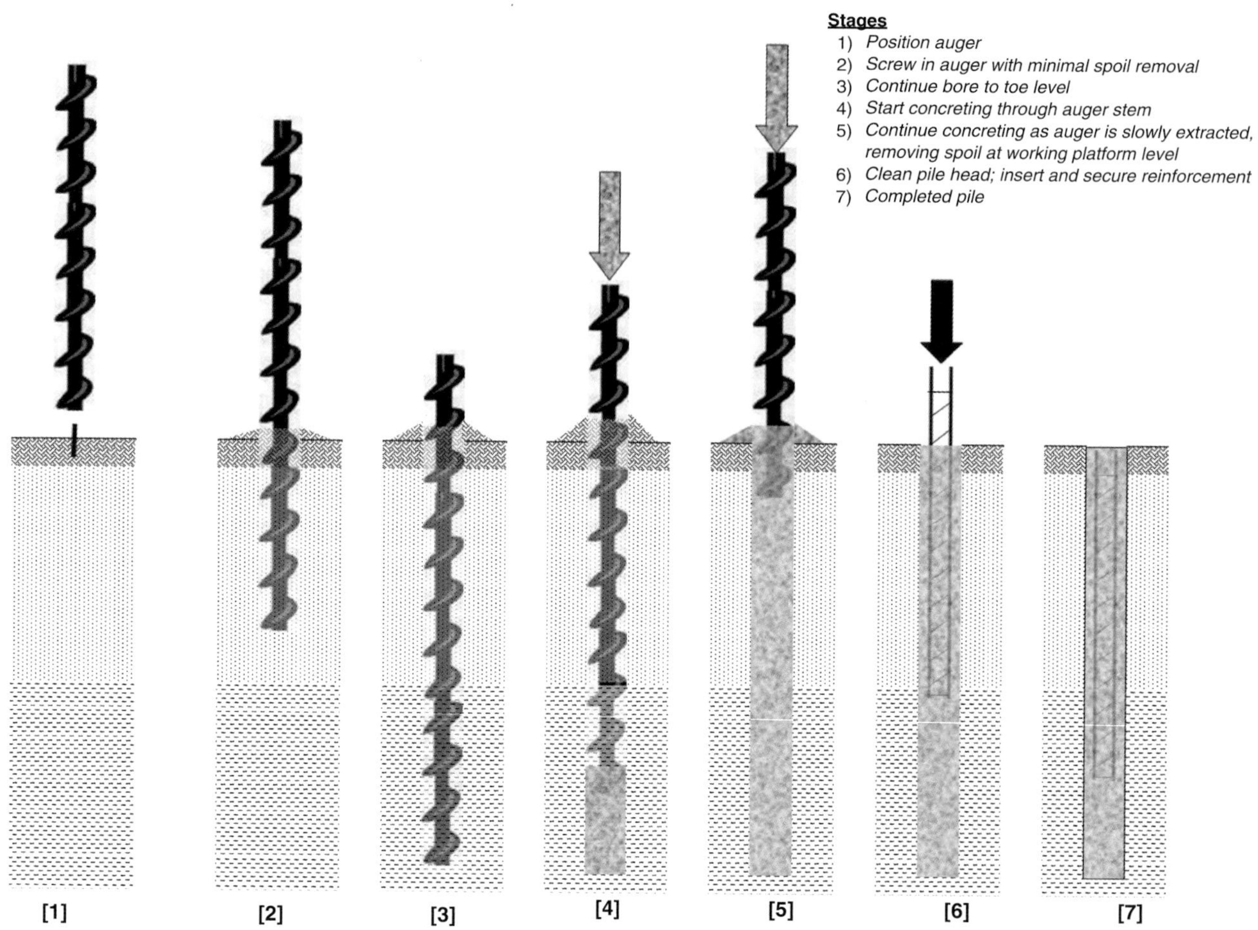

Notes:
i Pile boring and concreting stages should generally be subject to automatic real-time monitoring (ICE SPERW B4.4.8).
ii Concrete should be cast to the commencing surface level to allow removal of contamination before installing reinforcement.
iii Reinforcement cage must be robust enough to resist the stresses imposed by insertion through fresh concrete (ICE SPERW B4.4.7).

Figure 81.9 Basic construction sequence for a CFA pile

Figure 81.10 CFA piling rig (Photo courtesy of Expanded Piling)

The various configurations of flat teeth, picks, bullet teeth and so on are suitable for boring through most soils and into weak rocks. CFA piling is not generally capable of penetrating stronger rocks or hard obstructions.

81.2.3.3 Placing concrete

CFA piling relies on a steady delivery of concrete to the pile. Concrete is therefore often delivered to site by ready mix wagon and held in an agitator ready for delivery by pump to the piling rig. Once the auger has been drilled to depth the pump lines and hollow stem are pre-charged with concrete. The auger, still rotating slowly, is lifted a short distance (<100 mm) to allow entry of the concrete into the bore (through the hollow auger stem and an end or side entry valve at the auger tip). The auger is returned to the full bore depth so that the initial charge engulfs the tip, and then steadily extracted with the spoil as more concrete is delivered to fill the resulting pile shaft. It is essential that a positive pressure of concrete is

Figure 81.11 CFA piling: typical lead auger and cleaner configuration (Photo courtesy of Cementation Skanska Ltd)

maintained at the auger tip throughout the concreting process and that the auger tip remains embedded within the rising column of concrete. Failure to control and coordinate the delivery of concrete and extraction of the auger can lead to pile integrity problems (see Chapter 82 *Piling problems*).

81.2.3.4 Rig instrumentation

During the early development of the CFA technique in the UK, a number of projects encountered serious pile integrity problems. Others suffered ground loss and an unexpected variability in load-bearing performance. These problems arose in part because the CFA process is largely hidden from the eye and cannot be directly verified. Fleming (1995) discussed these issues and highlighted the need for close control of the boring and concreting phases of pile construction. It is now a standard requirement (ICE, 2017) that automated instrumentation is provided to monitor critical parameters throughout pile construction.

Data from built-in instrumentation is displayed in real time on a computer mounted in the operator's cab. The data include critical boring and concreting parameters such as auger depth and rates of rotation and penetration, applied torque, concrete pressure and rate of delivery (**Figure 81.12**). This enables the operator to control flighting of material and over-rotation during boring, and to prevent over- or undersupply of concrete during auger extraction. The data also provide a permanent construction record for each pile. These detailed records, coupled with integrity test results (Chapter 97 *Pile integrity testing*), provide essential tools with which to assure the quality of the CFA operation.

81.2.3.5 Placing reinforcement

It is necessary to cast the pile to the starting level to allow removal of contaminated concrete before insertion of the reinforcement cage. Cages must be robust enough to tolerate insertion into wet concrete to the required depth without distortion. Insertion depths up to 12 m are common. Greater depths are feasible (depths over 20 m have been achieved) but commonly require additional measures which can include careful cage detailing, special concrete mix designs and the use of a cage vibrator. Particular caution is required when piling in dry silts and sands. These soils can draw moisture from the concrete, causing an early set, and thus prevent full insertion of the cage.

81.2.3.6 Cased CFA piling

Cased CFA piling combines the advantages of CFA piling with the additional capabilities of a cased system. The process is the same as the CFA method except that a single length of thick-wall temporary casing is inserted and removed simultaneously with the auger. The casing has a cutting head and is counter-rotated against the auger, creating a higher torque to enable penetration into harder strata. The stiff temporary casing helps to prevent flighting of superficial deposits when drilling in hard strata, and provides greater verticality control. The system requires a very high-powered rig with either a single drive head coupled with a torque-multiplier or two independent rotary drive heads (**Figure 81.13**). The former system has a typical depth limit of 16–18 m, while with the latter cased depths of 21 m and bored depths of 27 m are claimed. When using the twin-drive system it may be feasible to cast the piles with a cut-off level below the starting level, in the same way as cased rotary piles.

The simultaneous extraction of auger and casing results in spoil being discharged from a point high above the working platform. The additional weight of spoil and equipment at height can have a destabilising effect on the rig. Particular attention is therefore required in the design of the working platform and in assessing the rig stability during operation. The safe discharge of the spoil to platform level must be controlled by purpose-built chutes (**Figure 81.13**).

81.2.4 Cast-in-place displacement auger piling

Cast-in-place displacement auger piles are installed using high-torque rotary equipment with drilling tools designed to form the bore by displacement rather than removal of the soil.

The system is relatively quiet and vibration-free, minimises the volume of spoil handling and concrete supply and can improve the compaction of granular deposits. In the simplest form (e.g. the 'Omega' pile, Fundex system) a tapered or 'bullet-headed' auger is screwed into the ground, displacing the soil, and extracted as concrete is placed through the hollow stem, to leave a straight-shafted pile. A typical construction sequence is illustrated in **Figure 81.14**.

(a)

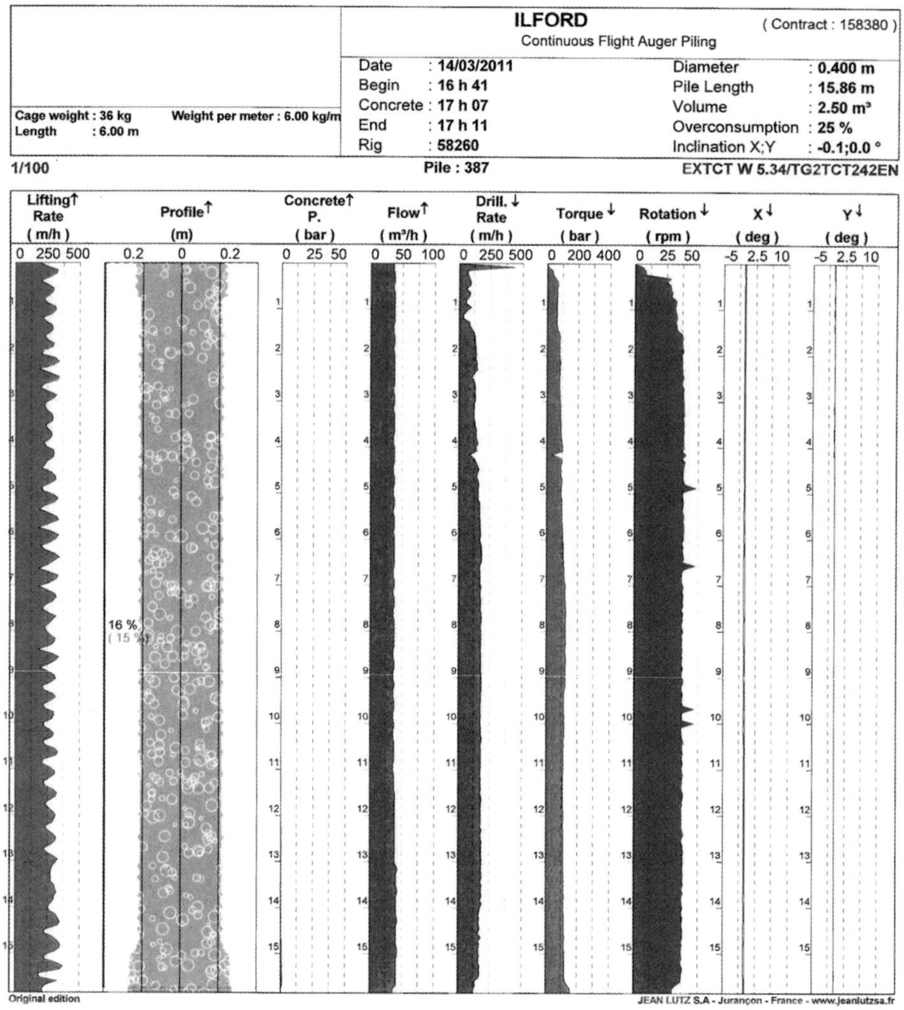

(b)

Figure 81.12 CFA piling rig instrumentation (Photo courtesy of Cementation Skanska Ltd)

Alternative methods (e.g. Atlas pile, continuous helical displacement (CHD) pile and the 'Screwsol' system) employ a short, flighted, bullet-headed auger, screwed into the ground in one pass and carefully back-screwed along the same trajectory during concreting, to leave a screw-form pile (**Figure 81.15**). The construction of screw piles is subject to the requirements of the European execution code, BS EN 12699:2015 (BSI, 2015a).

All of these methods require comprehensive instrumentation to confirm successful construction of the pile at the required diameter. It is vital to confirm that each pile has reached the design bearing stratum and, in the case of screw-form piles, that the flights are robust enough to prevent premature failure. As with CFA piling, the reinforcement is installed after completion of the concrete column to ground level.

When considering the use of this technique, it should be borne in mind that in common with driven displacement piling, cast-in-place displacement piling is likely to cause heave of the surrounding ground.

81.2.5 Barrettes

Barrettes may be used to carry exceptionally large vertical loads and high lateral loads. They are constructed using diaphragm walling methods and equipment (Chapter 85 *Embedded walls*). They are most commonly rectangular but may take a variety of forms (**Figure 81.16**).

In soft ground, and some weak rocks, barrettes may be excavated using rope-suspended grabs. In stronger rocks, rock cutters or milling machines may be required, although deployment of such equipment is only occasionally economical compared to rotary piling.

81.2.6 Specialised techniques

81.2.6.1 Under-reaming

Under-reaming is an extension of the rotary piling technique which can significantly enhance the capacity of relatively slender and short piles. It is only practicable in stable

Figure 81.13 Cased-CFA piling rig (Photo courtesy of Cementation Skanska Ltd)

strata, typically stiff clays. After forming a straight-shafted bore to the required level, a specialised under-reaming tool (**Figure 81.17**) is lowered in the closed position to the toe of the bore. As the tool is rotated the reaming arms are slowly opened to form the under-ream. Cuttings are removed by a collecting bucket at the bottom of the tool.

Under-reaming tools are designed to form a maximum under-ream diameter of approximately three times the pile shaft diameter, with a roof angle commonly limited to about 35° to vertical (ICE, 2017). Given the dependence of such piles on end-bearing it is essential that the drilling process is precisely controlled to ensure that no remoulded material or crumbs of soil remain on the base of the ream. This is usually proven by the use of CCTV, and testing or sampling apparatus mounted on the tool (for safety reasons physical descent and inspection is no longer an accepted process).

81.2.6.2 Base grouting

Base grouting is a secondary process applied to rotary piles. Its purpose is to mobilise very high base capacities in dense sands without excessive pile settlement. The effect is twofold

- consolidation of sand loosened by the pile bore
- pre-compression of the shaft, reversing the direction of applied shaft friction.

This gives the pile a very stiff response under load, a necessary requirement for particularly tall structures, and can result in economies in the pile design (see Chapter 54 *Single piles*). Base grouting is generally only economical in large-diameter piles and requires the pile to have sufficient shaft capacity to resist the uplift caused by grouting.

It is achieved by casting grout circuits into the pile. Each circuit comprises injection and exit tubes linked by a 'tube-à-manchette' (Chapter 86 *Soil reinforcement construction*) across the pile toe. The number of circuits depends on the pile diameter (**Figure 81.18**). After casting, and before the pile concrete gains significant strength, each circuit is pressurised with water to 'crack' the grout valves and the base concrete, in preparation for grouting. At a later stage, cementitious grout is injected under pressure at the pile base in a controlled process that upwardly compresses the pile shaft. The effect of the grouting is confirmed by monitoring the pile head for uplift (a minimum of 1–2 mm is typically specified) or by instrumentation installed close to the toe of deeper piles, where the pile head may not respond.

81.2.6.3 Embedded columns

Embedded (plunge) column technology has been developed to enable accurate installation of permanent columns within basements constructed using top-down techniques. The columns are embedded in large-diameter rotary cast-in-place piles using specially designed placing equipment. These may be simple mechanically adjustable guides, or more sophisticated hydraulically operated frames that can install columns to tolerances consistent with BS EN 1993 (BSI, 2005a) in dry or wet bores. **Figure 81.19** illustrates the typical embedded column installation sequence. The columns are generally rolled steel column sections, although other forms such as precast concrete may be accommodated. After installation and removal of temporary casings the open bore around the column must be temporarily backfilled to stabilise the column and the surrounding ground and to prevent accidents.

81.2.6.4 Geothermal piles

Piled foundations can be used to exploit geothermal energy exchange between the supported structure and the ground. Coupled with heat exchanger technology, geothermal piles can be used to heat and cool buildings in an environmentally friendly manner. Flexible plastic pipework is incorporated in the reinforcement cages (**Figure 81.20**) of rotary cast-in-place piles or plunged into CFA piles. These pipes are connected by way of a manifold in the substructure to a heat pump, which allows a reversible exchange of energy between the structure and the ground (Bourne Webb *et al.*, 2009).

81.3. Driven piles

81.3.1 Introduction

Driven piles have historically been described by the means of installation or construction, driven into the ground using

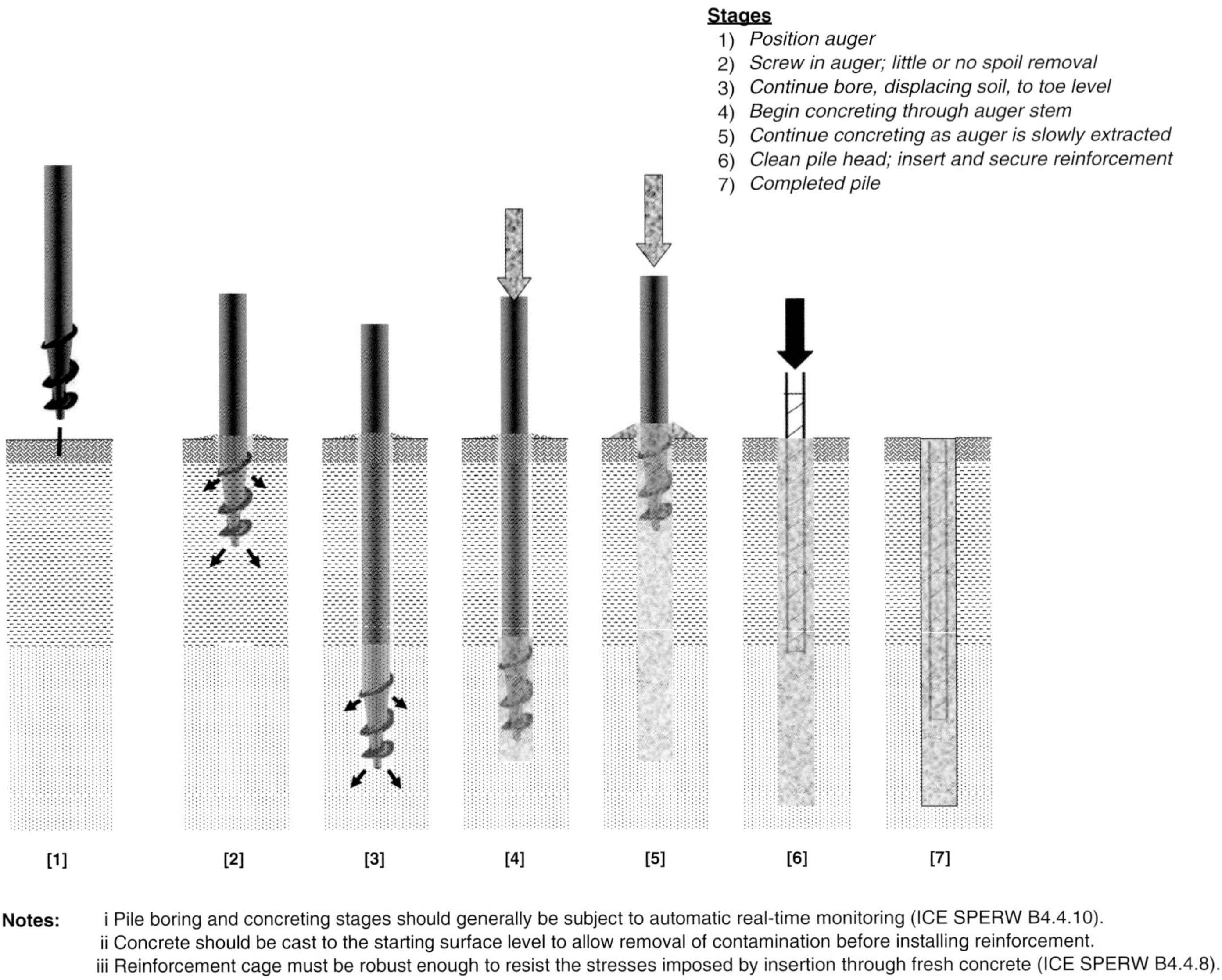

Figure 81.14 Displacement auger piling: typical construction sequence

a piling hammer. A better term would be displacement piles, describing the manner in which the pile interacts with the soil during driving. The soil is displaced, predominantly in a lateral direction, rather than replaced as with a bored pile. A driven pile may be preformed or cast in place, and may be a large or small displacement type, as explained below.

In the UK, driven piles are generally installed or constructed in accordance with SPERW (ICE, 2017) and the process is subject to the requirements of the execution code for displacement piles, BS EN 12699:2015 (BSI, 2015a).

81.3.2 Types of displacement pile used in the UK

(*a*) The concept of driving a pile for construction purposes is credited to a Neolithic tribe called the 'Swiss Lake Dwellers' who lived in what is now Switzerland about 6000 years ago. They used timber piles, not for supporting heavy loads, but for elevation to protect against attack from wild animals. The Romans built the first bridge across London's River Thames in the year AD 60 using timber piles. The modern age of wood-preserving began in England in 1832 with the concept of pressure-injecting chemicals into wood.

In more recent times timber piles have been used almost exclusively in sea defence works (groynes), installed as the vertical posts supporting timber planks which act together to reduce longshore drift on beaches. Considerations of durability and drivability require that they are invariably produced from hardwood species, with section sizes of 225–350 mm^2.

It is necessary to protect the lower end of the timber pile with a pointed cast iron shoe, as shown in **Figure 81.21**,

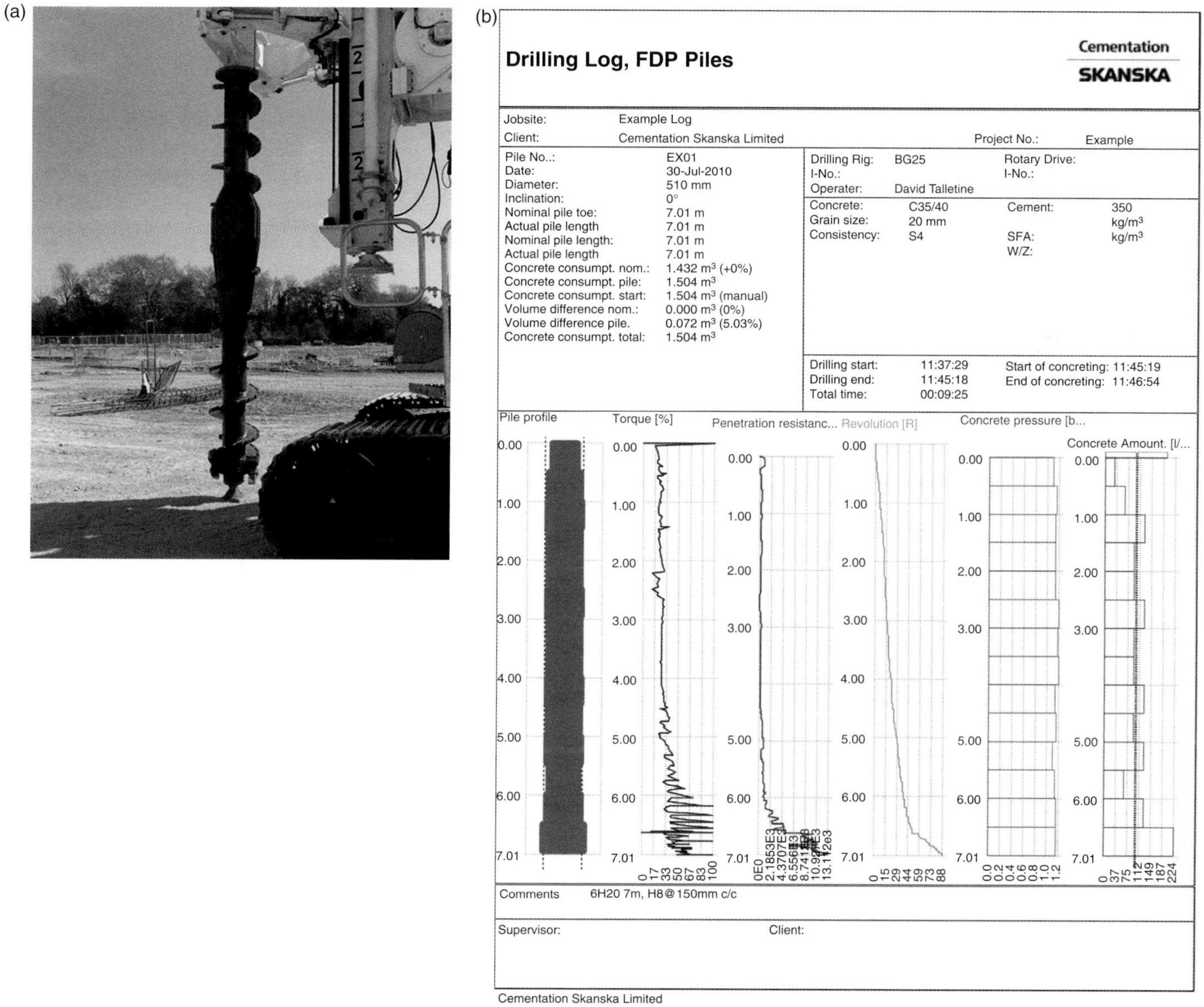

Figure 81.15 Displacement auger piling: (a) typical displacement auger; (b) typical displacement auger pile log (Photos courtesy of Cementation Skanska Ltd)

to aid penetration through the granular beach deposits without undue damage or deviation from plan position or vertical alignment. To prevent 'brooming' or splitting of the pile head a steel band is fitted (see **Figure 81.22**).

(*b*) Segmental precast concrete piles were introduced in the 1970s into the UK from Scandinavia. These large displacement piles are now universally square in section and proprietary section sizes range from 150 to 400 mm^2, with working loads usually in the range 200–1400 kN, although higher loads are achievable in favourable soil conditions.
Individual segment lengths normally range from 3 to 15 m, although some manufacturers are able to supply up to 17 m in a single length. Where soil conditions require, the pile can be extended almost indefinitely (the longest to date in the UK is in excess of 70 m) using pile joints. BS EN 12794 (BSI, 2005b) is the regulatory standard for both pile segments and pile joints, which are classified according to structural capability and robustness.
Piles are cast in moulds and removed to a storage area for curing the day after casting (**Figure 81.23**). Once the required characteristic concrete strength has been reached, usually 8–10 days after casting, the piles can be delivered to site for installation. All pile segments should be labelled to provide traceability (**Figure 81.24**).

(*c*) While segmental precast concrete piles are generally reinforced with straight high-tensile main bars and a square

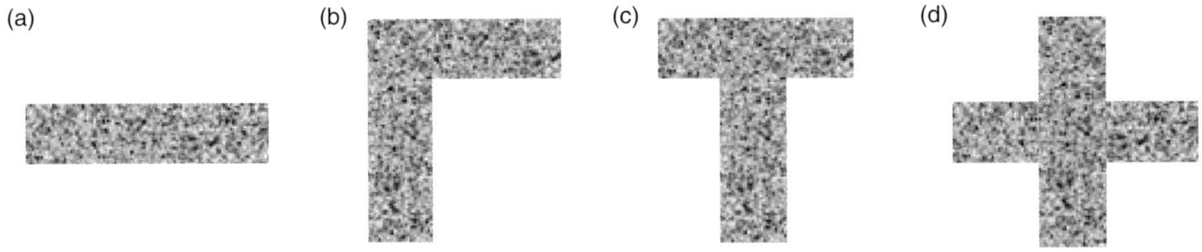

Figure 81.16 Typical forms of barrette

Figure 81.17 Under-reaming tool (Photo courtesy of Cementation Skanska Ltd)

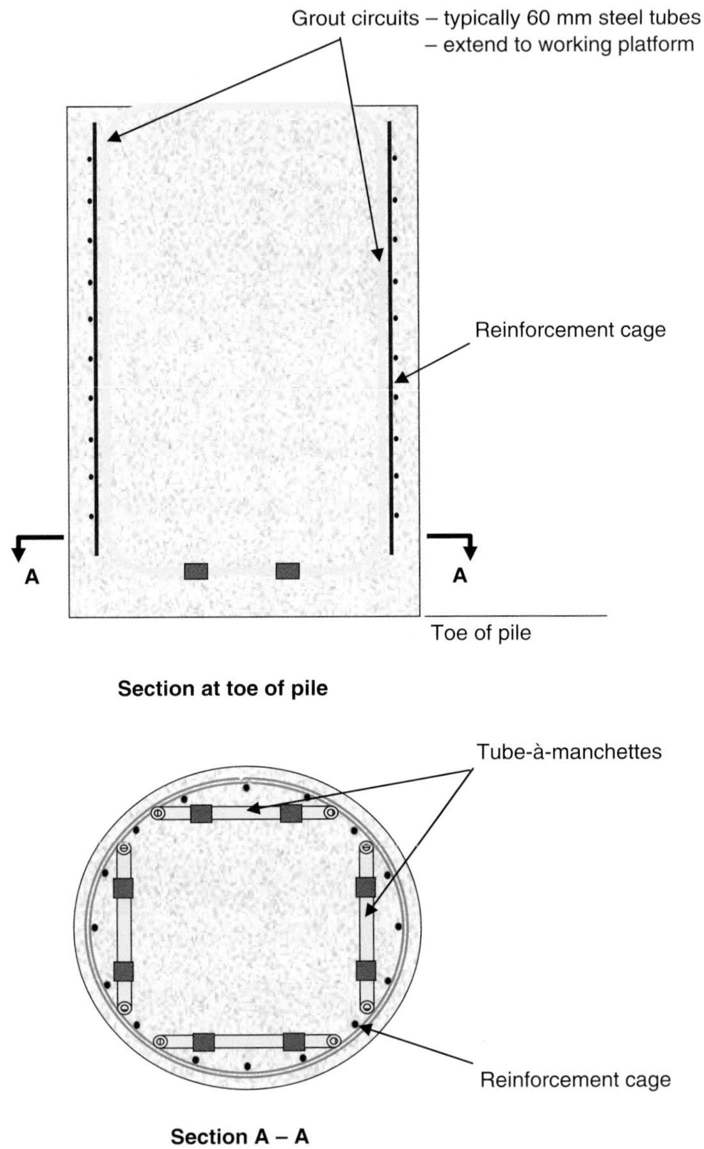

Figure 81.18 Typical base grouting circuit configuration

helical binding, prestressed concrete piles manufactured using strand reinforcement are becoming more common within the UK market. This type of pile permits longer segment lengths and can be more economic in terms of manufacturing and materials costs in some instances.

Where the piles are required to resist larger lateral forces or moments, reinforcement can be incorporated into the top section of the pile to provide extra structural capacity and bond steel into the pile cap (**Figure 81.25**).

(*d*) Driven cast-in-place (DCIP) concrete piles are, as the term suggests, a large displacement pile formed in the

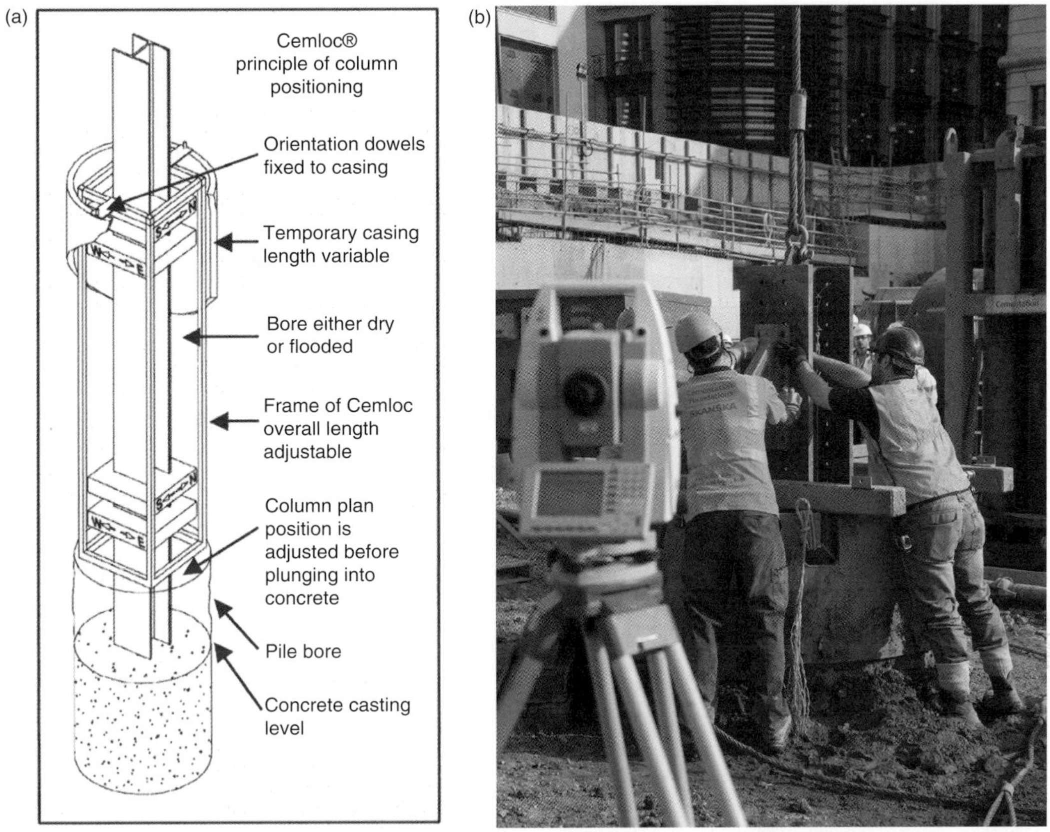

Figure 81.19 Plunge column piles: (a) plunge column installation; (b) precise installation of plunge column (Photos courtesy of Cementation Skanska Ltd)

Figure 81.20 Geothermal piles: (a) typical pipe installation; (b) completed geothermal pile with plunge column (Photo courtesy of Cementation Skanska Ltd)

Figure 81.21 Timber piles with cast iron shoes (Photo courtesy of Aarsleff Piling)

Figure 81.22 Timber pile head protection (Photo courtesy of Aarsleff Piling)

ground using a temporary steel casing. The lower end of the casing is closed off during driving with an expendable steel shoe, and the casing is then driven to the design toe level, embedment or required resistance, generally with a hydraulic drop hammer striking the head of the casing.

Traditionally, prior to placing high-slump concrete, a steel reinforcing cage bound with circular helical wire is introduced into the casing. The casing is then withdrawn using either a vibrator or rapid blows of the piling hammer to assist with extraction and concrete compaction. This sequence of construction is illustrated in **Figure 81.26**. More recent developments include pumping the concrete into the tube during driving and plunging

Figure 81.23 Precast concrete piles being removed from casting moulds (Photo courtesy of Aarsleff Piling)

the reinforcement into the wet concrete after withdrawal of the tube. This vastly reduces the cycle time per pile.

At all times it is necessary to maintain sufficient head of concrete in the casing to minimise the risk of 'necking' of the shaft, and piles must be cast above the water table and cut-off level, generally to piling platform level. These precautions are designed to prevent loss of pile section and/or washing out of fines from the concrete.

DCIP piles are readily adaptable to changes in founding level and can incorporate flared heads as an extension of the shaft when used under concrete floor slabs or embankments.

Proprietary sizes range from 340 to 600 mm dia., with safe working loads between 600 and 1500 kN, depending on the ground conditions.

In recent years the use of DCIP piling in the UK has declined in the face of fierce cost competition from cheap mass-produced precast piles.

(*e*) Steel H piles take their name from the cross-section of the profile used. These may be universal column sections or universal bearing piles where the web and flanges are of the same dimensions and thickness. H piles are small displacement piles and are well suited to hard driving

Figure 81.24 'CE' marked precast concrete piles being installed on a power station project in Kent (Photo courtesy of Aarsleff Piling)

Figure 81.25 Prestressed concrete piles with starter bars cast into the pile head

through ground where less robust piles might be damaged by artificial or natural obstructions.

They are able to resist higher levels of combined axial and lateral loads than the equivalent sizes of concrete pile and can be installed at a rake to enhance this capability further. This high combined load-bearing capacity, together with their low soil displacement characteristics, mean that they have been used frequently on highways projects (**Figure 81.27**).

H piles can be delivered to site in single lengths of up to 30 m, but this would require the use of very large rigs. Alternatively piles can be extended in situ by butt-welding, although this is time-consuming and expensive.

Section sizes vary from 203 × 203 mm × 45 kg/m to 356 × 368 mm × 174 kg/m, capable of supporting maximum axial loads ranging from 600 to 2300 kN in compression when manufactured from high yield strength steel.

(*f*) Steel box piles are formed by welding two U-section sheet piles or two pairs of Z-section sheet piles back to back, forming a symmetrical box profile (**Figure 81.28**). The clutches of the piles can be used to introduce the box piles into a line of sheet piling where high vertical and horizontal loads can be resisted without adversely affecting the appearance of the wall.

They can also be used as individual bearing piles in open jetty or marine dolphin construction as they have geometrical properties that permit long lengths with little or no lateral support.

They are generally driven open-ended and are, therefore, small displacement piles unless they plug with soil during driving (see the following section on large-diameter steel tube piles for an explanation of plugging). They can be supplied in similar lengths to H piles and extended in the same way.

(*g*) Large-diameter steel tube piles are commonly used in marine piling such as for jetties, piers and mooring dolphins, and utilise their high buckling strength as columns and energy-absorbing capacity in bending to resist loads where the piles project some distance above the seabed (**Figure 81.29**). Tubes with wall thicknesses greater than 10 mm are suitable for driving open-ended and so can be considered small displacement piles unless they plug with soil during driving.

Plugging is likely to occur once the cumulative shaft friction from the trapped soil on the inside of the tube wall exceeds the gross end-bearing resistance of the tube when considered as closed.

Large-diameter steel tubes can also be driven with closed ends. These piles range in size from 273 to 2020 mm dia.

(*h*) Small-diameter steel casings are an alternative to precast concrete piles in situations where artificial or natural obstructions could deflect or damage the concrete pile. Commonly known as Aberdeen casings, they are a recycled product from the offshore oil industry, originally manufactured for use as drill casing.

Available as non-prime or second-hand material, they are manufactured to an American Petroleum Institute (API) specification, with steel having yield strength of the order 550 to 600 N/mm^2. However, being uncertified

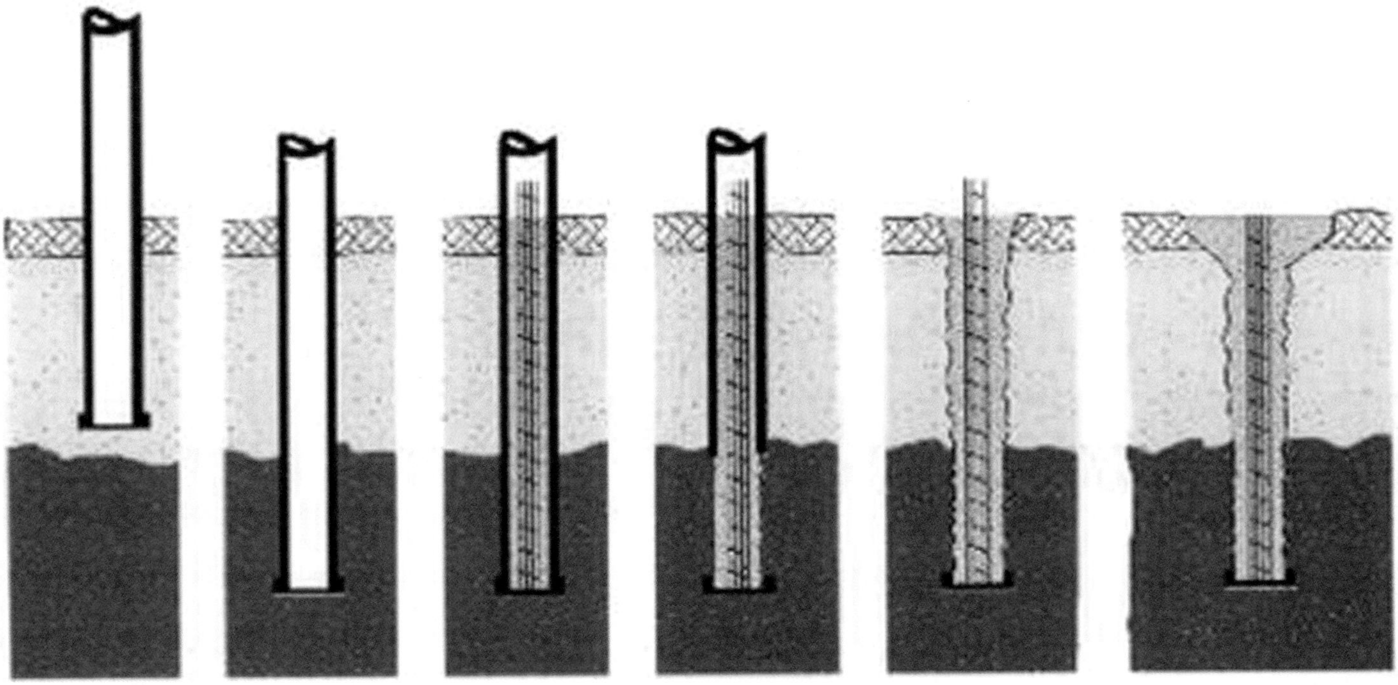

Figure 81.26 DCIP construction sequence (flared head shown at the right side of the schematic) (Courtesy of Cementation Skanska Ltd)

Figure 81.27 Steel H piles being installed to support a major road crossing a railway line (Photo courtesy of Aarsleff Piling)

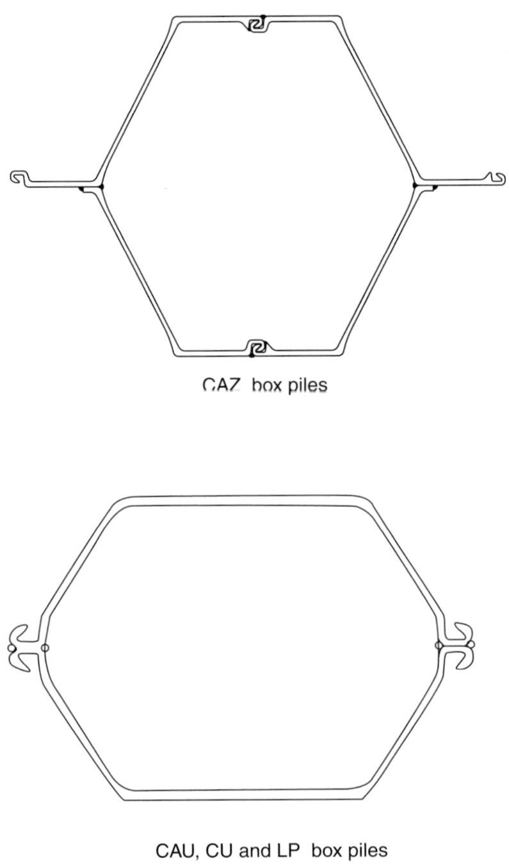

Figure 81.28 CAZ and CU steel box piles (Reproduced with permission from ArcelorMittal; all rights reserved)

Figure 81.29 Pairs of 23 m long, 610 mm dia. raking steel tube piles installed for a quay wall in Glasgow (Photo courtesy of Aarsleff Piling)

Figure 81.31 244 mm dia. Aberdeen casing collar (Photo courtesy of Aarsleff Piling)

Figure 81.30 339 mm dia. Aberdeen casings to support a wind turbine foundation near Inverness (Photo courtesy of Aarsleff Piling)

material they are usually treated as being at least equivalent to grade S355JH. Independent batch testing of larger quantities can be arranged if deemed appropriate for particular project requirements.

Driven open-ended (and so qualifying as small displacement piles) their robust nature and drivability means that safe working loads of 400–1200 kN are achievable using the same installation equipment as is required for precast concrete piling (**Figure 81.30**).

Diameters range from 178 to 339 mm, and stock lengths are in the range 11–14 m. The piles can be extended by driving the plain end of a second tube into the collar at the head of the first length driven (see **Figure 81.31** for detail of the collar).

81.3.3 Types of displacement pile used outside the UK

Within mainland Europe, driven piling has generally been restricted to Scandinavia, the Netherlands, Belgium, Germany and Spain.

As stated in Section 81.3.2(b), segmental precast concrete piles originated in Scandinavia and are still widely used in those countries. They are often used as the upper part of a composite pile, with a timber shaft below the water table where the risk of timber decay is minimal. Segmental precast concrete piles are also common in northern Germany and are now being regularly used in Poland and some former Eastern bloc countries.

In addition to installing (generally) prestressed concrete piles, contractors in the Netherlands and Belgium employ many variations on the DCIP piling technique, known locally as vibro-piling. In this context a combi-pile, as the name suggests, involves the introduction of a preformed element as the shaft of the vibro-pile. In areas of deep compressible soil the annulus can be filled with bentonite to reduce downdrag, or otherwise filled with grout to improve positive shaft friction.

The term 'combi-pile' is also used to describe tubular steel piles, usually driven open-ended to form the primary elements of an embedded steel pile wall. Clutches welded on either side of the tubular piles allow one or more sheet pile sections to form the secondary elements between the tubular piles.

In Belgium and Germany the Frankipile is still used regularly, having become practically obsolete in the UK. DCIP piles, including the Frankipile, are occasionally used in France and Italy but to a far lesser extent than bored or CFA piles.

In Spain, driven precast concrete piles, both segmental and prestressed, will be encountered on projects in the southern regions of the country.

In Austria, Germany and Switzerland, small-diameter ductile cast iron driven piles, installed with high-frequency hydraulic hammers, have established themselves as a low-vibration displacement piling system.

In a wider geographical context, small-diameter steel 'pipe piles' are common in North America, along with proprietary systems such as the 'Tapertube' pile. Precast concrete piles are also used, but section sizes are larger than those generally encountered in the UK. In many areas of the Far East, where deposits of soft compressible materials are encountered, pre-stressed spun concrete 'pipe piles' are commonplace.

81.3.4 Practically extinct types of driven pile (UK only)

Towards the end of the twentieth century the UK piling industry witnessed the decline and/or demise of two versatile driven pile techniques, which lost their significant market share to faster and/or less labour-intensive methods. Improvements in integrity testing of piles and rig instrumentation, which boosted confidence in DCIP and CFA piles, also contributed to the demise of the first of these to be discussed here.

(*a*) Developed and patented by Russian marine engineer Alexander Rottinoff, the West's shell pile was constructed using precast, fibre-reinforced concrete tubes, 900 mm long, threaded onto a steel mandrel and driven into the ground with a solid concrete shoe placed at the bottom of the mandrel. The shells were held together with circular steel bands to prevent ingress of groundwater or silt and driven into the ground using a rope-operated drop hammer rig, as shown in **Figure 81.32**. Once the specified depth or resistance had been achieved, the mandrel was withdrawn and concrete placed in the core, with a reinforcing cage to suit the diameter and loading conditions. Diameters ranged from 380 to 600 mm. The same company also produced the West's segmental pile, an early form of solid precast concrete pile 280 mm dia., with 900 mm long segments being joined together during driving with a simple spigot and socket joint. On completion of driving, a full-length single reinforcing bar was grouted into a central longitudinal hole. This system was only suitable for light loads in soft soil conditions and became obsolete on the introduction of West's Hardrive precast concrete pile.

Figure 81.32 Westpile 30RB based shell piling rig

(*b*) The Frankipile, which is still used worldwide, was first developed by the Belgian engineer Edgar Frankignoul, who registered an international patent for the system in 1902. It is a cast-in-place concrete pile with an enlarged base and a cylindrical shaft which can penetrate stiff soils and readily reach depths of 20 m.

The pile is constructed using a temporary steel casing and an internal drop hammer, the sequence being as illustrated in **Figure 81.33**. A plug of gravel is placed in the bottom of this tube and compacted by the hammer, which then enables the casing to penetrate the ground by virtue of the friction between the inner wall and the gravel plug.

On reaching the design depth, the casing is lifted slightly and the gravel plug broken out. Dry concrete is then introduced and expelled by the hammer to form an expanded base, thereby significantly improving the original bearing capacity of the founding stratum.

After inserting a reinforcing cage, the pile shaft was traditionally formed in the same way using a semi-dry concrete, thus producing a pile with excellent end-bearing resistance and improved shaft resistance due to the ribbed profile caused by successive lifting of the tube and hammering out of the concrete.

Such a construction process is slow and labour-intensive, with production rates being only 20–25% of alternative DCIP or driven precast concrete piles. The technique was modified in the mid-1970s by using high-slump wet concrete to form the pile shaft, and where particular circumstances such as aggressive ground conditions require a protective coating, a precast concrete pile can be used to form the shaft.

The Frankipile system is now obsolete in the UK.

81.3.5 Driven piling plant

Driven piling frames provide a guide for the alignment of the pile and hammer during driving, and usually have at least two winches. Early piling frames comprised open platforms on which the winches were mounted (often steam-powered with a large boiler) and a vertical lattice or tubular mast that supported the piling leader used to guide the pile and hammer, as shown in **Figure 81.34**.

On land these frames usually moved on railway lines or by using a system of rollers; some Frankipile machines were able to 'walk' using an ingenious system of hydraulic jacks. These large

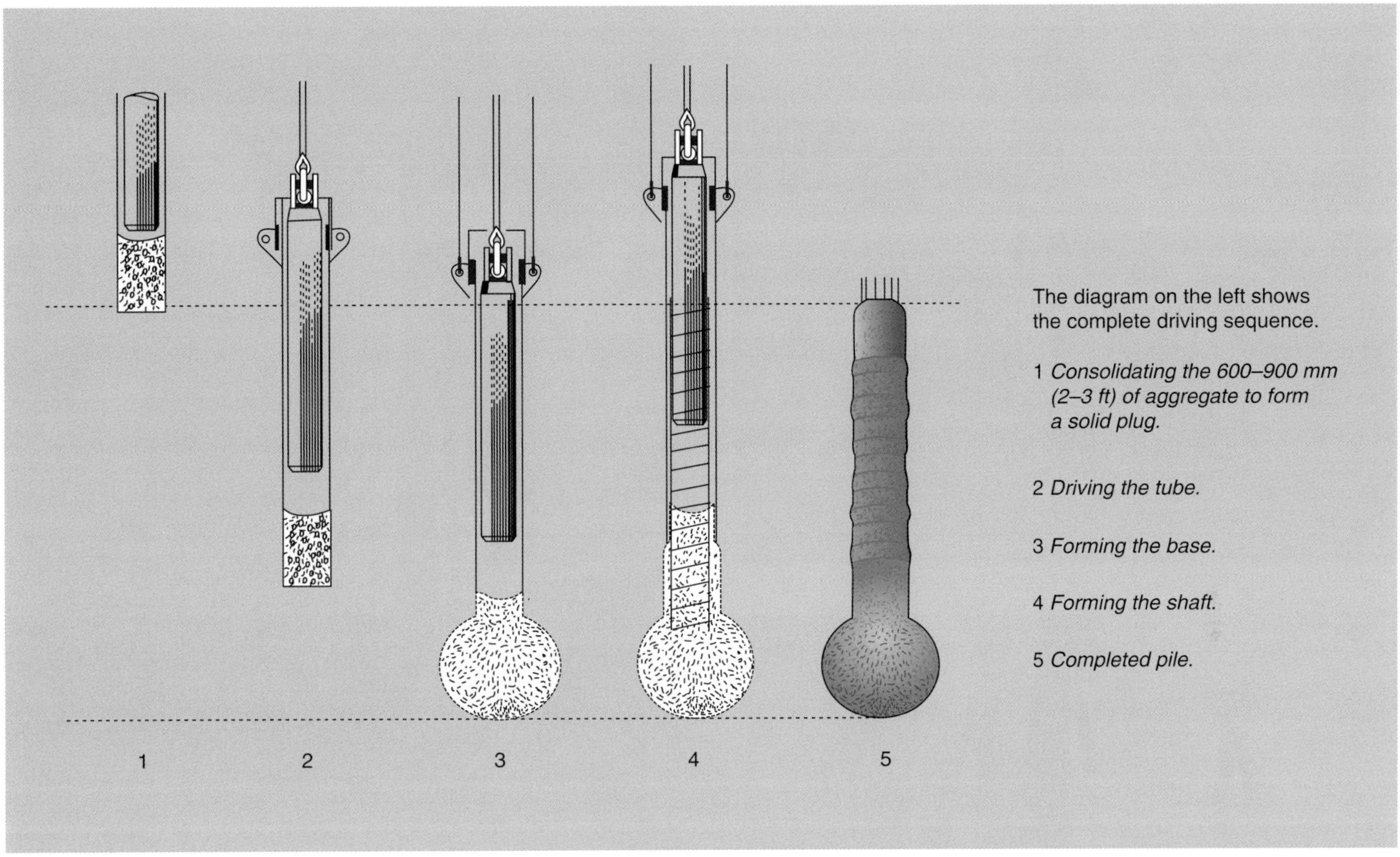

Figure 81.33 Frankipile construction sequence (Reproduced from FRANKI® pile)

frames provide the rigidity and stability needed to control the alignment of very large piles, and more recent versions can still be found working in marine applications. However, conditions on most onshore projects favour the use of smaller equipment.

Initially the move away from large fixed piling frames involved the use of hanging leaders suspended from the jib of a crawler crane. This type of rig was successfully used by West's Piling and Construction (later Westpile) for many years to install shell piles using a rope-operated drop hammer (see **Figure 81.32**), and the same type of equipment, modified by the addition of an adjustable foot boom and foot pad, was marketed by BSP International Foundations for the installation of steel piles using their own diesel hammers. At about the same time, piling contractors such as Expanded Piling, Keller (then GKN) and Cementation adapted the BSP style of equipment to suit the rigours of DCIP piling, predominantly by strengthening the leaders. Typical examples of these largely obsolete piling rigs can be seen in **Figure 81.35**.

The arrival of precast concrete piles in the 1970s from Scandinavia brought with it the compact fixed leader rig mounted on an Akermann hydraulic excavator, manufactured by Banut in Sweden. Although hanging leader rigs were used for some considerable time to install precast concrete piles in the UK, the Banut rig allowed the development of a hydraulic piling hammer where the drop weight is lifted using a hydraulic ram instead of a rope and winch, the result of which is a better-controlled driving operation. Banut and Junttan (from Finland) have developed the fixed leader rigs to a level where they are suitable for all forms of driven piling. Junttan have further developed a hydraulic hammer with an accelerated fall which increases the efficiency rating of the hammer blow by up to 35%. A modern Junttan PM20 rig is illustrated in **Figure 81.36**. A Banut 700 rig is featured in **Figure 81.30**, while a smaller Banut 500 rig can be seen in **Figure 81.37**.

Box piles and steel tube piles are generally installed using a combination of hydraulic/electric vibrator and/or hydraulic drop hammer suspended from a crawler crane, although fixed leader rigs can be used given satisfactory access and working space. Thin-walled steel tubes can be bottom-driven using an internal hammer falling on a plug of semi-dry concrete, placed in the pinched end of the tube, in the same manner as a Frankipile.

81.3.6 Suitable and unsuitable ground conditions

Driven concrete piles can be used in a wide range of ground conditions. Often considered to be only suited to end-bearing conditions in granular soils and weak rock, they can be used in

Figure 81.34 Rail-mounted Frankipile rig c.1930

Figure 81.35 Historical photograph of crane-based DCIP drop hammer rigs

cohesive soils provided that the limitations of driving stresses (for precast concrete piles) and casing extraction forces (for DCIP piles) are considered. DCIP piles are not normally suitable for pile depths much in excess of 20 m because of the effort required to extract the casing. Boulders and massive obstructions will preclude the use of driven piles unless they can be economically removed prior to piling. Steel tube and H piles are particularly suited to hard driving situations or where small obstructions are anticipated. In the latter case it would be prudent to use Aberdeen casings as these are more readily available than ex-mill materials if additional piles are required. Durability in aggressive ground conditions will influence the choice of pile type and material. Steel tube piles can incorporate a sacrificial element of wall thickness, and both precast concrete and steel piles can be given a protective coating to improve resistance to corrosion.

Figure 81.36 Junttan PM20 piling rig with 5 tonne accelerated hydraulic hammer (Photo courtesy of Keller Ltd)

Figure 81.37 Banut 500 rig with shrouded 3 tonne hydraulic hammer unloading piles on a housing site (Photo courtesy of Aarsleff Piling)

81.3.7 Environmental considerations

A common reaction to a proposal to use driven piling is that it will create undesirable levels of noise and vibration.

Figure 81.38 Junttan PM20 rig fitted with CFA auger string and rotary head for pre-boring (Photo courtesy of Aarsleff Piling)

However, most construction processes create detectable levels of vibration and noise, and tolerance is greatly increased when the origin, likely levels and duration of noise or vibration are made known in advance and those affected are reassured that consideration has been given to these factors.

It may not be possible, for technical reasons, to replace driven piling by a 'quieter' piling technique. Even if it is feasible, the operation of the alternative method may prolong the piling duration so that the overall disturbance to the community may not actually be reduced. Piling hammers are available that are completely shrouded (**Figure 81.37**), and most hydraulic driven piling rigs can be adapted to pre-bore pile positions prior to driving to reduce the level of vibration (**Figure 81.38**).

Factors affecting the acceptability of site noise include site location and layout, pre-existing ambient noise levels and the daily hours of work. In the case of both noise and vibration, measured levels will reduce with increasing distance from the source. Annex F of BS 5228-1 + A1 (BSI, 2009a) gives guidance on predicting the expected levels of noise from construction sites, and Annex E of BS 5228-2 + A1 (BSI, 2009b) contains empirical expressions derived by Hiller and Crabb (2000) for the prediction of vibration levels which take into account driving energy, soil conditions and distance between the source and the point of measurement.

Further considerations which play a part in the overall environmental impact of a piling scheme are traffic movements in and out of the site, the efficient use of natural materials, and the embodied energy/CO_2 for the complete process.

A positive environmental consideration is that driven piling produces no spoil, making it particularly suitable for sites with contaminated made ground that would otherwise have to be disposed of safely and at great cost.

Extensive research at the University of Sheffield, the results of which have been endorsed by the Environment Agency, has concluded that solid square or circular piles pose no additional threat to an underlying aquifer when they are driven through contaminated ground that is separated from the aquifer by a clay layer of sufficient thickness.

81.3.8 Drivability

For driven piles, the length required to satisfy the design must be achievable using available equipment and without damaging the piles or driving equipment in the process. This drivability assessment can be made based on experience, and a responsible piling specialist will always be prepared to comment on the viability of a scheme at an early stage of the design.

In addition to experience, computer software is available that can model the soil, pile and driving system to predict resistance to driving and stresses in the pile during installation.

81.3.9 Installation control

A driven pile provides valuable feedback on the strength of the ground through which it is being installed. This resistance to penetration will usually be recorded on the first piles driven (and thereafter at regular intervals) as the number of blows per unit of penetration (usually 250 mm in the UK) for a given hammer weight and drop.

'Sets' will be recorded at varying intervals and at the end of driving three consecutive sets will usually be recorded. The precise definition of a set is the permanent downward displacement of the pile through the soil per hammer blow; however, measurements of set are generally recorded as the total movement for ten successive blows. Set calculations provide a useful guide to the resistance that should be encountered for a given load-bearing capacity, but should not be used in isolation. The relevant ground investigation information must also be reviewed to verify that the level at which the set is being achieved is reasonable.

Additional measurements of temporary compression (sometimes referred to as 'quake') of the pile and soil under individual hammer blows will also provide an indication of whether the pile toe is in competent material or is, for instance, sitting on a thin, hard band underlain by weaker soil. Driven piles can be designed and installed to an embedded length in cohesive soils, in which case the set achieved is not necessarily indicative of bearing capacity, although comparing penetration records will highlight any potential variation in soil strengths and pile performance.

In weak rocks, the depth of embedment will be as relevant as the set achieved, and in those materials prone to relaxation (such as Coal Measure mudstones) it will generally be necessary to overdrive the pile to achieve a satisfactory outcome. Driven preformed piles have the advantage of permitting restrike sets to be measured at some time after installation to determine whether the ground is relaxing or conversely setting up and improving. These restrike measurements can be combined with dynamic load testing where appropriate.

81.3.10 Ground-related aspects of pile installation

As stated in Section 81.3.1, driven piles displace soil as they are installed. Although the displacement is predominantly in a lateral direction, this can result in ground heaving upwards in closely spaced groups of piles. In the case of short piles end-bearing on rock, this can be enough to unseat the piles, but should be detectable by routine restrikes on preformed piles, and any risen piles can then be reseated by redriving. Ground heave can also affect the accuracy of setting out in large pile groups by moving marker pins. In the case of longer DCIP piles, heave forces can compromise the integrity of freshly placed concrete if piles are installed consecutively too close together. In confined working areas such as inside cofferdams or basements, ground displacement can have adverse effects on adjacent temporary or permanent structures.

Closely spaced DCIP piles should not be installed consecutively even where ground heave is not an issue. The driving of the casing can jeopardise the integrity of an adjacent pile shaft in which the concrete is still workable or not set. The minimum permissible spacing between adjacent piles formed in the same day will be a function of casing diameter and soil strength. Guidance is given in BS EN 12699:2015 (BSI, 2015a). The sequencing of operations should, therefore, take account of this risk, and plant movements should also be planned to avoid tracking over constructed piles.

Silty soils that are prone to dilation can give rise to 'false sets' where apparently high dynamic resistance does not result in the anticipated static resistance under a load test.

Driven piles installed through soft clays and silts can generate excess pore water pressures during driving, which will dissipate with time. As a result the ground will eventually reconsolidate (the same can happen if ground heave occurs) and induce an additional load on the shaft of the pile which should be accounted for in the pile design.

When precast concrete piles are installed first through soft or loose materials and then require prolonged driving through more competent soil, the pile will experience tensile forces caused by the stress wave travelling from the head of the pile to the toe and back to the head. In extreme cases these forces can be large enough to cause yielding of the main reinforcement in the pile. The computer software referred to in Section 81.3.8 can be used to predict the tensile forces in the pile, and the hammer weight, fall height and main reinforcement in the pile can be adjusted to suit the anticipated conditions. Dynamic test measurements taken during installation can subsequently be used to verify the actual driving stresses in the pile.

Driven piles installed in cohesive soils will usually exhibit a degree of 'set-up' after a period of time has elapsed after installation, this effect being caused by the clay remoulding around the shaft as the excess pore water pressures created in the soil during driving dissipate. The same effect will also be found to occur in some grades of chalk. With preformed piles this effect can be demonstrated by the tightening of the set at restrike compared to that measured at the end of drive. In the case of DCIP piles restrike sets are not possible, and in both cases load testing the pile too soon after installation will result in the pile underperforming compared to its long-term capability. However, as the ground may take several weeks to recover to a state where the pile performance can be optimised, considerations of programme rarely allow for this benefit to be maximised.

81.4. Micro-piles

81.4.1 Definition

Bored micro-piles are defined as bored piles with a shaft diameter no greater than 300 mm, and driven micro-piles are defined as driven thin-walled steel piles with a shaft diameter no greater than 150 mm.

81.4.2 History and introduction

Bored micro-piles were originally developed by Fernando Lizzi and Fondedile during the 1950s in Italy. Lizzi employed the Pali Radice ('root piles') micro-pile system for strengthening existing foundations which were settling or required upgrading (**Figure 81.39**) (Lizzi, 1982). Bored micro-piles have since become increasingly popular throughout Europe, North America and the rest of the world, and are used in many different applications. Driven micro-piles were developed during the 1980s, and they are primarily used to provide a low-cost foundation solution for the domestic building market. There are other types of micro-pile, such as jacked, jet grouted and post-grouted micro-piles; however, they are less common and as such they are not covered here. The installation of micro-piles is typically labour-intensive, expensive and time-consuming, and they should only be used when the more conventional piling techniques have been discounted. However, when used correctly and in the right circumstances they can bring real value to a project. There are a multitude of micro-pile types, and this chapter can only provide a brief introduction to them. Specialist micro-pile advice should be sought when the preliminary micro-pile scheme is being developed.

81.4.3 Where to use micro-piles

Micro-piles are extremely versatile and they can be used in a wide variety of compression and tension loading applications.

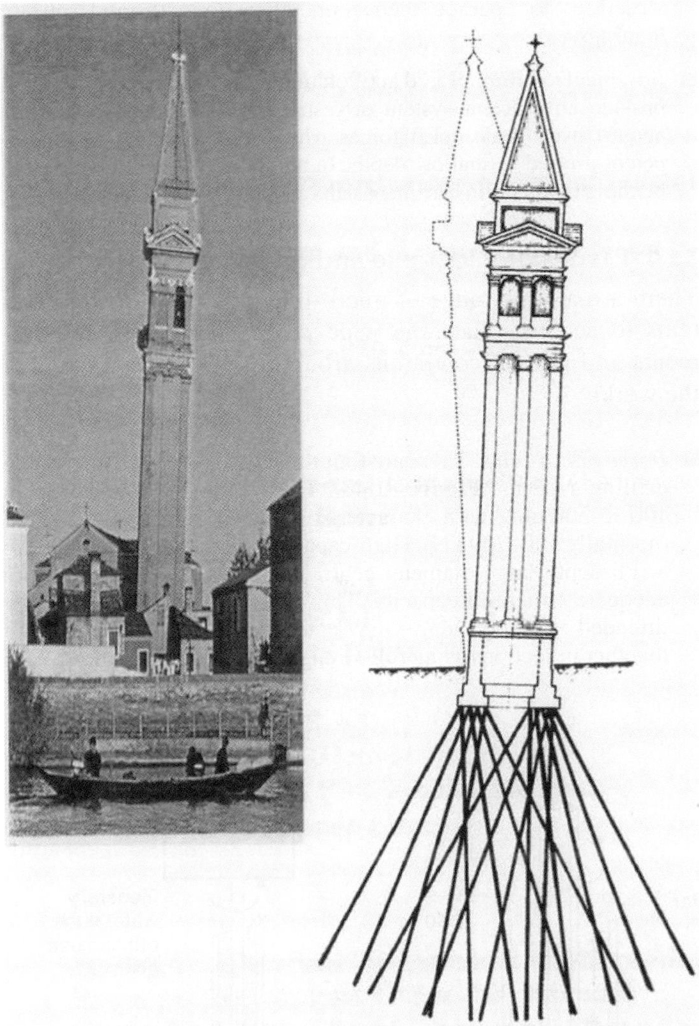

Figure 81.39 Burano Tower, Venice (Reproduced from Lizzi (1982); photo courtesy of Aarsleff Piling)

Figure 81.40 2 tonne drill rig

Figure 81.41 15 tonne drill rig (DTHH)

Figure 81.42 Grundomat driven micro-pile

- *Foundation upgrading and strengthening.* Pali Radice micro-piles are formed through existing foundations and develop an intimate bond with the structure; they are founded in competent ground, thus providing the foundation with 'roots' – hence their name.

- *Restricted access and working room.* There are a wide range of micro-pile drill rigs, from small 1 tonne rigs that can drive through doorways and work within 2 m of headroom (**Figure 81.40**), through to larger 15 tonne rigs with 9 + m high masts (**Figure 81.41**). The capabilities, production rates and costs of these micro-pile rigs vary considerably.
- *Lightly loaded structures.* Driven micro-piles typically used on new and existing (underpinning) housing projects where there is poor ground over firmer strata, with pile characteristic loadings in the range 100–300 kN. Driven micro-piles can be installed in restricted access and low headroom working locations using small drill rigs or even a hand-held grundomat hammer (**Figure 81.42**).
- *Difficult ground and hard rock.* Bored micro-piles can be installed using an extensive range of drilling techniques to penetrate almost any ground conditions and/or obstructions, except possibly steel and strong timber, where high-speed coring techniques may need to be employed.
- *Permanent casings.* Bored micro-piles can install permanent steel casings through very soft ground or voids (e.g. caves or solution features) as part of the construction process.
- *Seismic loadings.* An increasingly important international use of bored micro-piles is their ability to be retrofitted to existing structures to upgrade their foundations to withstand seismic loadings.
- *Basement restraint.* Bored antiflotation tension micro-piles (ATMs) provide an efficient system of restraining large deep basements against hydrostatic uplift forces where there is reasonably competent ground at shallow depth. In addition, they can help reduce bending moments in basement slabs and thus reduce their thickness.

81.4.4 Types of micro-pile and drilling systems

There are a large range of micro-pile types and drilling systems to accommodate the wide range of working environments and ground conditions around the UK and the rest of the world.

- *Bored micro-piles.* There are four main types of bored micro-pile drilling systems (**Figure 81.43**). These range in diameter from 100 to 300 mm, with characteristic pile loads of 100–1500 kN (normally 300–700 kN). High-capacity micro-piles often require a full-depth large-diameter central reinforcement bar to ensure adequate structural capacity. This central rebar is usually fully threaded (e.g. Dywidag GEWI) to enable it to be connected together in sections using full-strength couplers for low headroom

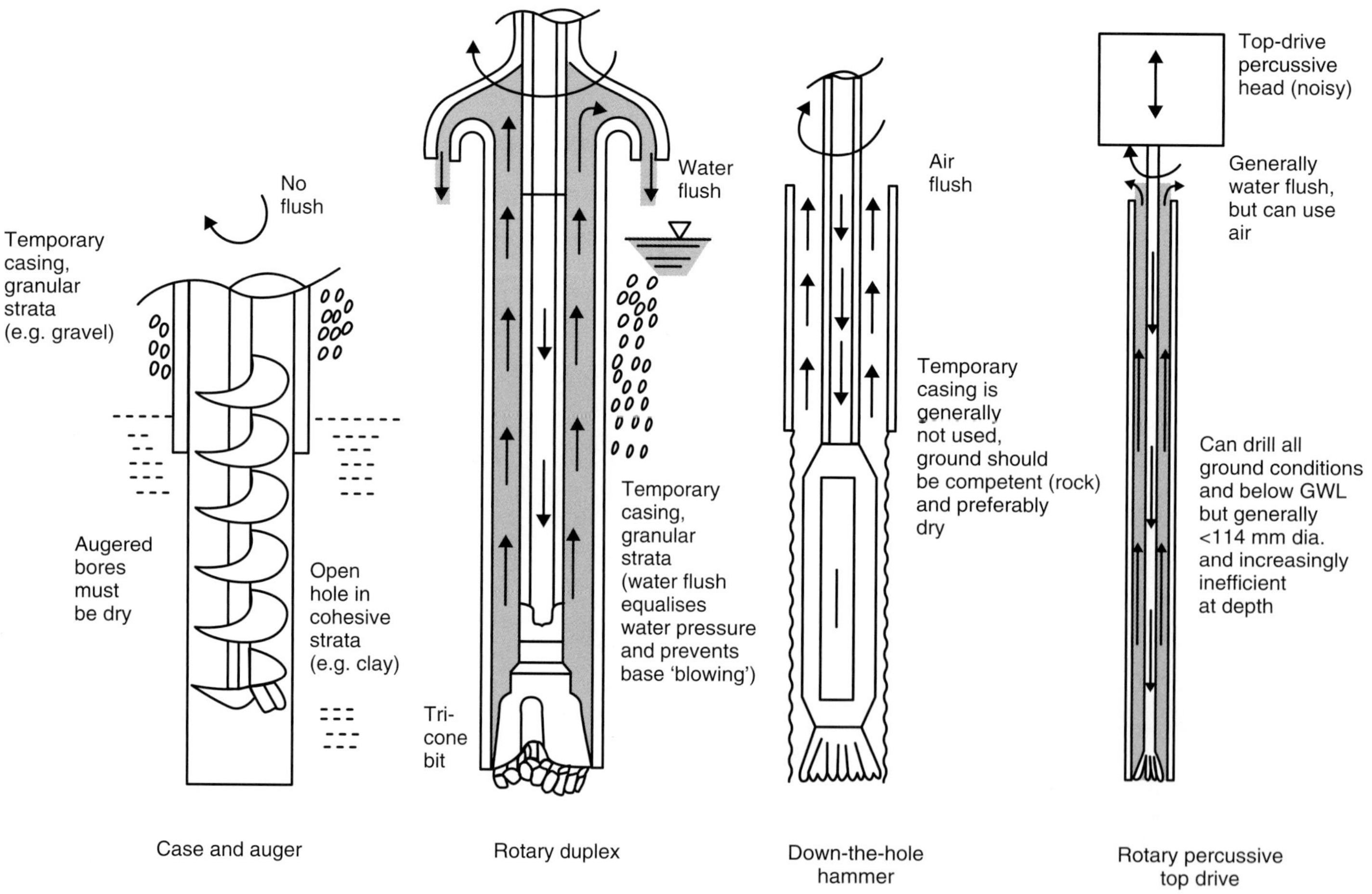

Figure 81.43 Bored micro-pile drilling systems

or deep pile applications. In addition, a steel head plate may also be required to facilitate an efficient force transference mechanism into the pile cap. If a steel head plate is used then this should be located in the middle of the pile cap and secured above and below with full-strength nuts. Bored micro-piles generate their geotechnical restraint primarily from shaft friction because the shaft area is considerably larger than the end-bearing area. They can provide high tension capacities as well as high compression capacities when they are founded deep into strong bedrock and incorporate a full-depth large-diameter central rebar (e.g. 50–75 mm dia.). Micro-piles are bored using 0.5–1.0 m lengths of drill rods/augers, depending upon the pile diameter and the headroom available, and can extend up to 50–70 m deep. It is unusual for micro-piles to be deeper than 30 m and particular allowance should be made in the calculations for the elastic compression behaviour of the pile shaft if they are. Bored micro-piles can be installed vertically or at any rake up to 45°.

- *Rotary case and auger bored micro-piles***.** These are the simplest and most popular form of bored micro-piles, and are constructed by sealing temporary steel casing into competent strata and then dry open-hole auger drilling thereafter. The depth of the temporary casing, if required, varies according to the ground conditions. These micro-piles are popular in London, where open-hole auger drilling in dry homogeneous London Clay is possible. If groundwater and/or obstructions are encountered then case and auger drilling may have to be stopped. The most popular temporary casing sizes range from 140 to 343 mm, with respective open-hole auger sizes from 102 to 300 mm. Temporary casing depths should be limited to ~15 m wherever possible; however, this depth is dictated by the type and power of the drill rig and the ground conditions encountered.

- *Rotary duplex bored micro-piles.* The rotary duplex drilling technique is useful in poor and/or granular strata with a high groundwater level, where there is a risk of borehole or basal instability. The temporary casing and drill rod with tri-cone drill bit ('rock roller') are simultaneously rotary drilled into the ground using water flush. The drill water passes down the centre of the drill rods and up the inside of the temporary casing, before exiting through the duplex head. The hydrostatic forces therefore remain positive on the inside of the temporary steel casing, which will prevent 'blowing' conditions in the base. The most popular size is 220 mm dia. temporary casing with a 187 mm dia. tri-cone bit, although a variety of sizes down to 140 mm are available. A wide range of ground conditions can be drilled with this system, including weak rocks, which can be open-hole drilled with a tri-cone bit below the temporary casing. However, penetration rates will become progressively slower in stronger rocks, leading to an increased risk of a 'smooth' bore, which can lead to lower grout-to-ground bond values. Large volumes of drilling water are required (~3 litres per second for a 220 mm dia. micro-pile) to ensure there is sufficient uphole velocity to transport the drill cuttings out of the bore. An efficient drill water recirculation system is necessary, which may comprise sandbag bunds and recirculation/settling tanks. This type of micro-pile is becoming less common due to the environmental implications of using and disposing of large volumes of drilling water. It is also possible to use air flush; however, this is not commonly undertaken because of dust, lack of lubrication, bit wear and overheating problems.

- *Rotary DTHH bored micro-piles.* This type of percussively drilled micro-pile has become increasingly popular in recent years, primarily due to the technological advancements in drilling equipment. Air-flushed DTHHs drill quickly through masonry, concrete and strong bedrock, and typical sizes range from 100 to 300 mm dia. There is also a water-flushed DTHH; however, this is expensive and not commonly used. Conventional DTHH micro-piles require temporary steel casing rotating into the ground until sealed into bedrock. The DTHH can then open-hole drill an efficient and 'rough' bore into the bedrock which can generate high bond values. DTHHs can drill strong bedrock quickly because the percussive hammer remains immediately behind the drill bit and it does not lose any energy as the bore gets deeper. However, there are risks with this type of micro-pile, primarily the failure to obtain an adequate casing seal into bedrock, which can lead to large volumes of air being pumped into the ground and causing problems, and borehole collapses behind the DTHH, which can lead to the loss of an expensive drilling system. DTHHs perform best in competent rocks (> 10 MPa) due to the percussive drilling action and the bore remaining stable and uniform; production rates are directly proportional to the pressure and volume of air provided. A 200 mm dia. DTHH requires a large (750 cfm/170 psi), expensive-to-operate compressor; however, the equivalent micro-pile cost per metre

Figure 81.44 Cased DTHH micro-piles

will be considerably less than rotary drilling in rock. The air-flushed DTHHs require substantially dry bores; otherwise, large volumes of water will be blown into the air. In recent years DTHH overburden drilling systems (essentially a 'cased DTHH') have been developed, such as Odex, Symmetrix, Maxbit and Super Jaws. These systems allow full-face DTHH percussive drilling with simultaneous advancement of the temporary steel casing to keep the bore open. Therefore, difficult granular strata with large boulders can be drilled efficiently using cased DTHH systems (**Figure 81.44**).

- *Rotary percussive micro-piles (including self-drilled hollow bar).* These micro-piles are installed using top-drive rotary percussive techniques and can achieve fast production rates. The percussive drill heads ('drifter') can be mounted on a variety of micro-pile rigs or powerful excavator base units. The top-drive percussive heads are noisy, and the drill string loses energy into the ground as the bore extends deeper. There are two main types of rotary percussive micro-piles. The first is rotary percussive duplex drilled. This is similar to normal rotary duplex drilling, except that there is additional percussion which speeds up drilling through difficult ground containing cobbles and boulders. Unfortunately, large percussive drill heads are required to drive the heavy percussive drill casing/bit and therefore standard micro-pile sizes are 76 mm and 114 mm dia. There is a 178 mm dia. system available but this requires an even larger, heavier and more powerful drifter, which requires a minimum 25 tonne base machine. The second system is the self-drilled hollow bar micro-pile, which is now very popular because of its rapid installation speed and the resultant low cost per metre. These micro-piles comprise a hollow reinforcement threadbar with an expendable drill bit which is percussively drilled into the ground using grout or water flush. It is preferable to use grout flush in difficult ground; however, if water flush is used during drilling to reduce grout waste, then it must be changed to grout flush immediately upon reaching the hole base. Grouting must be continued until there is a full column of clean grout around the hollow bar. Production rates of up to 400 m per shift can be achieved with this system. The drill flush keeps the bore open during installation, and there is no need for any temporary drill casing. This type of micro-pile can be installed into very difficult ground conditions, and typical sizes range from 75 to 150 mm, with larger bits in the range 200–300 mm also now available. The main risks with this type of micro-pile are the hollow bar deflecting off natural or man-made obstructions and/or loss of flush. This system is also commonly used for soil nails and passive tension micro-pile applications.
- *Hollow segmental auger (HSA) micro-piles (segmental CFA).* This type of micro-pile is similar to a CFA pile, except that the auger is segmental. Typical diameters are 300 mm and above, although the larger diameters are not strictly micro-piles but are commonly installed using micro-pile drill rigs. The hollow stem augers are rotated into the ground in nominal 1.0 m sections to the required depth and then a sand–cement grout or ready mix concrete is pumped down the central hollow stem while the auger is rotated slowly. This is continued until grout can be seen exiting around the augers at the surface. The augers are then carefully extracted while continuing to pump grout and maintaining a positive grout pressure. The grout level in the central stem should be visible at the surface as each section of auger is removed, and if not then it should be topped up as necessary. It is not possible to have a continuous monitoring record of the micro-pile construction as with CFA piles; however, maintaining a positive head of grout at all times will ensure good pile integrity. HSA micro-piles are popular in mixed or granular strata with high groundwater levels because the augers provide temporary support to overcome borehole instability problems.
- *Driven micro-piles.* These are generally bottom-driven vertical 150–200 mm diameter thin-walled 'closed-end' steel tubes with characteristic pile loads ranging from 100 to 300 kN. They are installed quickly and efficiently using a small drill rig with internal drop hammer onto a dry concrete plug with no spoil produced. Alternatively, smaller tubes can be driven into the ground using an internal air-powered grundomat hammer, without any drill rig (see **Figure 81.42**). These micro-piles may have typical characteristic pile loads of 100–200 kN. The tubes can be driven in various lengths and welded together as the installation proceeds. The depth of the micro-piles is typically dictated by the greater of a static calculation and a dynamic 'set' calculation; alternatively testing may be carried out to justify the design capacity. Once the design depth is reached, the steel tube is filled with high-slump concrete or grout and reinforcement as required for the applied loads or connection to the substructure.

81.4.5 Micro-pile drill rigs and coring

There is a wide range of micro-pile drill rigs and capabilities, from small ~1 tonne rigs with 2 m masts which can track through house doorways, through to 15–20 tonne rigs with 9 + m masts. The smaller rigs have a more limited range of drilling techniques and are generally slower (and therefore they install more expensive micro-piles), while the larger rigs can use any drilling system and are faster. Careful thought should always be given to choosing the appropriate micro-pile rig and drilling system, and the advice of specialist micro-pile contractors should be sought at an early stage. Micro-pile drill rigs have low rotation speeds and high torque, and can rotary core drill through concrete, masonry and other obstructions with standard tungsten-tipped steel casing; however, coring rates can be slow. DTHHs can drill quickly through unreinforced mass concrete. If there are significant amounts of coring and steel reinforcement then consideration should be given to using a specialist coring company (which uses high rotation speed and low torque diamond-tipped coring barrels).

81.4.6 Micro-pile construction techniques

One of the key issues with micro-piles is to choose the most appropriate drilling technique to ensure that the micro-pile can be bored safely and efficiently to the required depth. Guidance is given in BS EN 14199:2015 (BSI, 2015b). Detailed consideration of the borehole logs is required, taking careful note of the stratum type, the groundwater regime, obstruction drilling/chiselling and whether there are any cobbles or boulders. Micro-piles are normally formed with cementitious grout; however, ready mixed concrete can be used with larger 300 mm dia. micro-piles in HSA bores or dry open

Figure 81.45 Colloidal high-shear grout mixer

bores (checked by shining a light down). If an open bore is 'wet' then a measured small-diameter (25 or 38 mm dia.) plastic tremie pipe must be inserted completely to the base. The grout mix is generally a neat cement mix or a 1:1 sand:cement mix, with the neat mix used for the smaller diameters and the sanded mix more popular for the larger diameters. The sand should be a well-graded washed sharp concreting sand or similar (i.e. not a 'soft' sand) and it should preferably be used in 25 kg bags. Bulk sand can be used, but it must be measured using calibrated containers on site (e.g. 6 level buckets = 100 kg). The grout is mixed in either a paddle mixer or a colloidal (high-shear) mixer, which is more efficient and 'wets' more of the cement particles, thus leading to a superior low-bleed–high-strength mix (**Figure 81.45**). The correct amount of clean, potable water should always be added first, followed by the cement and finally the sand (if required). The standard cementitious grout mixes are

- neat cement grout (this 0.40 water:cement ratio mix will produce 72 litres of grout)
  - 40 litres of clean potable water
  - 100 kg of cement
- 1:1 sand:cement grout (this 0.45 water:cement ratio mix will produce 113 litres of grout)
  - 45 litres of clean potable water
  - 100 kg of cement
  - 100 kg of sharp sand

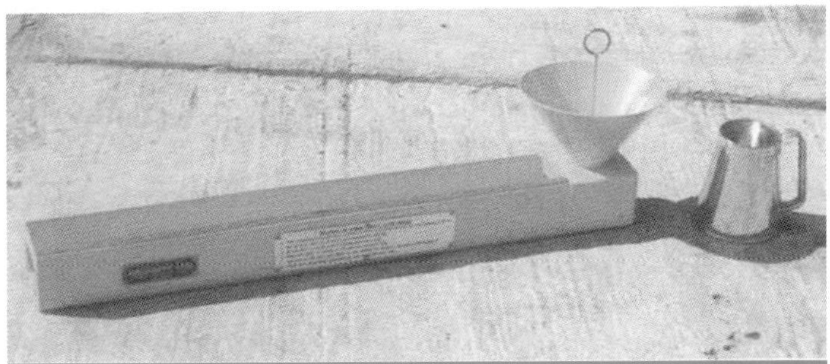

Figure 81.46 Colcrete flowmeter

Both of these mixes will produce high-quality cementitious grouts with unconfined compressive strengths of >40 MPa at 28 days. The cement should preferably be grade 42N or 52N. There are a wide range of sands available and difficulties may be experienced adding the full 100 kg of some 'softer' sands. If this occurs, *do not add more water*, which would only weaken the final grout strength; it is better not to add all of the sand (but record the amount added). If higher grout strengths are necessary then plasticiser can be added and the amount of water reduced accordingly. Grout fluidity can be checked using the Colcrete flowmeter test if required (**Figure 81.46**). This involves pouring 1 litre of grout into the cone and then releasing it and allowing it to flow along the level 'wetted' channel. A flowmeter reading of 375 mm (±50 mm) is normal for a 1:1 sand:cement grout and 450 mm (±50 mm) for a neat cement grout.

Micro-pile grouting should be carried out in a continuous operation by pumping clean grout into the base of the micro-pile until all the debris has been flushed out of the bore and a column of clean grout remains. Where temporary casing is used it can then be removed in sections, ensuring that the internal grout level is fully topped up to ground level each time. Finally, the clean reinforcement cage or bar should be carefully plunged into the grouted pile with appropriate centralisers (~3 m centres). If the micro-pile is very deep (or in dry sandstone, which may cause a flash set) and there is concern over inserting the reinforcement after grouting, then it can be installed first; however, this does increase the risk of smearing the rebar on the side of the bore. This is the case for self-drill micro-piles where the structural reinforcement is also the drill string, and careful consideration is often required as to the clearance around the couplers joining sections of hollow bar together to ensure that the bore can be flushed adequately. Upon completion of grouting, the grout level should be carefully monitored for a minimum of 1 hour and topped up as necessary until it has completely stabilised. In granular strata it is common for the grout level to continue to drop as the pressure head forces the grout to permeate into the surrounding ground. Additional pressure of 3–4 bar (43–58 psi) in such strata can be applied during casing extraction using a pressure head coupling to further enhance grout take, effective pile diameter and micro-pile capacity.

81.4.7 Micro-pile environmental, quality assurance and health and safety issues

Micro-pile construction has several environmental issues which need to be carefully managed

- disposal of water flush, which may have high silt contents and be harmful to fish
- disposal of air flush, which will contain fine dust particles
- noise suppression, especially from top-drive rotary percussive drilling systems
- disposal of grout overflush and washout.

Micro-pile quality assurance is similar to other piling techniques and requires the checking of the plan position, verticality, borehole depth, diameter and reinforcement. In addition, special care needs to be taken with site-mixed grout with regard to quantities of water, cement and sand. Grout cube sizes should be 100 × 100 × 100 mm.

Micro-pile construction is labour-intensive, with several health and safety issues

- manual lifting of casings, drill rods and reinforcement
- slippage with water and grout flush
- working near rotating augers (auger guarding during boring is now a statutory requirement)
- grout mixing on site
- working in restricted and noisy locations.

References

Bourne Webb PJ, Amatya B, Soga K, *et al.* (2009) Energy pile tests at Lambeth College. *Géotechnique* **59(3)**: 277–298.

BSI (British Standards Institution) (2005a) BS EN 1993-1-1:2005 + A1:2014: Eurocode 3: Design of steel structures – General rules and rules for buildings. BSI, London, UK.

BSI (2005b) BS EN 12794:2005 + A1: Precast concrete products – Foundation piles. BSI, London, UK.

BSI (2009a) BS 5228-1:2009 + A1:2014: Code of practice for noise and vibration control on construction and open sites – Part 1: Noise. BSI, London, UK.

BSI (2009b) BS 5228-2:2009 + A1:2014: Code of practice for noise and vibration control on construction and open sites – Part 2: Vibration. BSI, London, UK.

BSI (2010) BS EN1536:2010 + A1:2015: Execution of special geotechnical work – Bored piles. BSI, London, UK.

BSI (2015a) BS EN 12699:2015: Execution of special geotechnical work – Displacement piles. BSI, London, UK.

BSI (2015b). BS EN 14199:2015: Execution of special geotechnical work – Micropiles. BSI, London, UK.

Fleming WGK (1995) The understanding of CFA piling, its monitoring and control. *Proceedings of the Institution of Civil Engineering* **113**: 157–165.

Fleming WGK, Weltman A, Randolph M and Elson K (2009) *Piling Engineering*, 3rd edn. Taylor & Francis, Abingdon, UK.

Hiller DM and Crabb GI (2000) *Groundborne Vibration Caused by Mechanised Construction Works.* Report TRL 429. Transport Research Laboratory, Bracknell, UK.

ICE (Institution of Civil Engineers) (2017) *ICE Specification for Piling and Embedded Retaining Walls*, 3rd edn. Thomas Telford, London, UK.

Lizzi F (1982) *The Static Restoration of Monuments: Basic Criteria – Case Histories, Strengthening of Buildings Damaged by Earthquakes.* SAGEP Publisher, Genoa, Italy.

Puller MP (2003) *Deep Excavations: A Practical Manual*, 2nd edn. Thomas Telford, London, UK.

Tomlinson M and Woodward J (2014) *Pile Design and Construction Practice*, 6th edn. Taylor & Francis, Abingdon, UK.

Further reading

ArcelorMittal (2022) *Piling Handbook*, 9th edn. ArcelorMittal, London, UK.

Australian Drilling Industry Training Committee Limited (1997) *Drilling: The Manual of Methods, Applications and Management.* Lewis Publishers, Chicago, IL.

Barley AD (1988) Ten thousand anchors in rock. *Ground Engineering* **21(6)**: 20–29; **21(7)**: 24–35; **21(8)**: 35–39.

Barley AD and Woodward MA (1992) High loading of long slender micropiles. *Proceedings of the ICE Conference on Piling European Practice and Worldwide Trends, London, UK.* Thomas Telford, London, UK, pp. 131–136.

Biddle AR (1997) *Steel Bearing Piles Guide.* Steel Construction Institute, Ascot, UK.

Broms B (1978) *Precast Piling Practice.* Royal Institute of Technology, Stockholm, Sweden.

Bruce DA, Ingle JD and Jones MR (1985) Recent examples of underpinning using micropiles. *Proceedings of the 2nd International Conference on Structural Faults and Repairs, London, UK*, pp. 13–28.

EFFC/DFI Concrete Task Group (2018) *Guide to Tremie Concrete for Deep Foundations*, 2nd edn. EFFC, Faversham, UK. See www.effc.org/content/uploads/2018/08/EFFC_DFI_Tremie_Concrete_Guide_2nd-Edition_2018_Final_rev2_28-08-18.pdf (accessed 30/03/2023).

EFFC/DFI Support Fluids Task Group (2019) *Guide to Support Fluids for Deep Foundations*, 1st edn. EFFC, Faversham, UK. See www.fps.org.uk/content/uploads/2019/04/EFFC_Support_Fluids_Guide_FINAL.pdf (accessed 30/03/2023).

FPS (Federation of Piling Specialists) (2006) *Bentonite Support Fluids in Civil Engineering*, 2nd edn. FPS, Faversham, UK.

FPS (2017) *Introduction to the ICE Specification for Piling and Embedded Retaining Walls 2017. Presentation.* See https://www.fps.org.uk/guidance/piling-specification/#accordion1.

Hird CC, Emmett KB and Davies G (2006) *Piling in Layered Ground: Risks to Groundwater and Archaeology. Environment Agency Science Report SC020074/SR.* Environment Agency, Bristol, UK.

Littlejohn GS (1982) Design of cement based grouts. *Proceedings of the Conference on Grouting in Geotechnical Engineering, New Orleans, LA.* ASCE, New York, pp. 35–48.

Martin J (1994) The design & installation of bridge strengthening schemes using micropiles. *Proceedings of Structural Faults and Repair Conference, London, UK*, pp. 1–7.

Martin J (1999) Horsfall Tunnel, Todmorden: the design, construction and performance of a temporary reticulated micropile retaining wall. *Proceedings of Piling and Tunnelling 99 Conference, London, UK*, pp. 1–10.
Martin J (2000) Stabilisation of an existing railway embankment using small diameter micropiles and large diameter pin piles at Kitson Wood, Todmorden. *Proceedings of Railway Engineering 2000 Conference, London, UK*, pp. 1–12.
Martin J (2007) The design, installation and monitoring of high capacity antiflotation micropiles to restrain deep basements in Dublin. *Proceedings of the International Conference on Ground Anchorages and Anchored Structures in Service 2007, London, UK*, pp. 1–12.
Martin J (2008) Retrofit micropile system to increase the capacity of existing foundations. *Proceedings of the 2nd BGA International Conference on Foundations, ICOF, Dundee, UK.*
Martin J (2009) High capacity micropile groups for the Cannon Place redevelopment in London. *Proceedings of the 7th International Conference on Micropiles, London, UK.*

Useful websites

American Petroleum Institute: http://www.api.org
Deep Foundations Institute: http://www.dfi.ork/publications
Federation of Piling Specialists: http://www.fps.org.uk
Geosystems digital library: http://www.geosystemsbruce.com/v20/html/per_DonaldBruce2.html
International Society of Micropiles: http://www.ismicropiles.org
Technical papers from Byland Engineering: http://www.bylandengineering.com
Wave equation analysis of pile driving: http://www.pile.com

It is recommended this chapter is read in conjunction with

- Chapter 22 *Behaviour of single piles under vertical loads*
- Chapter 54 *Single piles*
- Chapter 82 *Piling problems*

All chapters in this book rely on the guidance in Sections 1 *Context* and 2 *Fundamental principles.* A sound knowledge of ground investigation is required for all geotechnical works, as set out in Section 4 *Site investigation.*

Chapter 82
Piling problems

Jon Ball Roger Bullivant Piling, Swadlincote, UK
James Cowburn Roger Bullivant Piling, Swadlincote, UK
Viv Troughton formerly at Arup, London, UK

doi: 10.1680/icemge.66830.1339

CONTENTS

A successful project requires adequate management of the inherent risks that piling is engineered to diminish. If piled foundations are correctly installed and designed to adequately address the risks of the inherently unpredictable ground conditions, then the successful management of those risks will lead to a successful project.

Where incorrect choices are made further risks can be introduced, be it from the piling methods chosen, a poor geotechnical model of the ground conditions encountered, or poor execution of the technique employed.

Cast in situ piles, CFA bored and auger displacement for example, can significantly alter their performance dependent upon the execution of the installation technique and ground support. The structural integrity is highly dependent on the sequence of works, concrete placement and the detailing of the reinforcement and the execution of installation.

In the case of driven piles the technique chosen, material used, degree of displacement and speed of installation will influence geotechnical performance of the technique and reliability of the integrity of the piles.

Environmental considerations may also be influential, either through pathway protection and inhibition, noise and vibration considerations or installation aids such as pre-boring and jetting.

This chapter provides guidance on how problems are commonly identified, how they are assessed and the ways in which they can be resolved.

82.1. Introduction

There are several key references that provide a comprehensive background to the potential problems that can arise in piling. The more common problems for driven and bored piles are described by Fleming *et al.* (2009) in their handbook *Piling Engineering.* Specific sections are also included covering driven cast in situ piles, continuous flight auger (CFA) piles and cast in situ screw piles. Thorburn and Thorburn (1977; CIRIA Report PG2) provide a review of piling problems associated with the construction of cast in situ concrete piles. Healy and Weltman (1980; CIRIA Report PG8) provide a survey of the problems associated with the installation of displacement piles, and although some of the techniques have changed, much of the guidance is still valid. Tomlinson and Woodward (2015) describe the potential problems associated with piles in specific situations. These include piling for machinery foundations and underpinning, piling in mining subsidence areas, in frozen ground, for foundations for bridges on land and over water and energy piles.

Specifications and codes are also an important source of information, and many of the requirements contained in these documents have been incorporated as a result of problems that have been identified in piling projects. The ICE *Specification for Piling and Embedded Retaining Walls* (SPERW) (ICE, 2017) covers most of the common types of piling and includes guidance as well as specification clauses. The sections cover bearing piles and piles incorporated in retaining walls as contiguous piled walls, secant and sheet pile walls and king post walls. The publication includes three parts, with Part C providing guidance for the use of the specification and essential background information for specifiers and contractors alike.

Execution of bored piles is covered in BS EN 1536:2010 + A1:2015 (BSI, 2015a) and execution of displacement piles in BS EN 12699:2015 (BSI, 2015b). The different types of bored and driven piles and their methods of installation are also described in Chapter 81 *Types of bearing piles.*

Noise and vibration control in piling is comprehensively covered in BS 5228:2009 + A1:2014. Part 1 (BSI, 2014a) covers noise, including methods of noise control, guidance on methods for predicting and measuring noise and assessing its impact. Part 2 (BSI, 2014b) covers vibration and means of vibration control, as well as guidance on methods of measuring and assessing vibration. Measured levels of vibration for piling are included in an appendix.

The durability of steel piling is covered in BS EN 1995 Part 5 (BSI, 2007a) and in the *Piling Handbook* (ArcelorMittal, 2022), which also contains useful practical guidance on pile driving. It details corrosion performance of steel piles and describes protective measures that can be adopted to increase their effective life. The durability of bored piles is detailed in both BS EN 1992 Part 1 (BSI, 2015c) and BS EN 206 (BSI, 2016), which describe the exposure classes affecting durability and the requirements for concrete cover to reinforcement.

Many of the common problems in piling can be avoided by selecting the right piling system for the site conditions, and then ensuring that the installation is properly controlled and executed. A comprehensive site investigation is an essential component in selecting the appropriate method, and the Association of Geotechnical and Geoenvironmental Specialists (AGS, 2016) guidelines are a very useful starting point and aide-mémoire.

With a good understanding of the ground conditions and the risks that can arise with different pile systems, the designer can select an appropriate method and control measures that will ensure successful construction of the piles. This is all part of the pile design process, which is described in Chapter 54 *Single piles*. The following sections cover the common problems that can arise with the installation of bored and driven piles.

82.2. Bored piles

82.2.1 Ground conditions

It is important to ensure that the equipment to be used for boring is adequate for the ground conditions. This may seem obvious, but the greatest unknown is still the ground. There may be unproven obstructions, both natural and man-made, that could prevent a 'less heavy' item of plant from completing the boring. Delays may occur if break-out tools have to be mobilised to site. Sometimes, the specialist contractor may know if natural obstructions are likely in a given area. There is also the possibility that, due to some unforeseen ground condition, the piles may have to be redesigned with a different diameter. While this should not be anticipated by the specialist contractor, their plant should ideally be able to switch to the alternative geometry.

By far the most common problem is the poor understanding of the ground conditions, usually due to inadequate site investigation. When this occurs, the client usually ends up paying for the cost and time lost caused by the 'unknown' conditions. There is also the common problem of a varying depth of founding stratum. This is more acute for CFA piling as the boring is essentially blind. Probing using the piling auger is possible, but is only a crude identifier of such depths.

Boring tools should be checked to ensure that the design diameter is to be achieved. The SPERW (ICE, 2017) recommends a tolerance of up to 5%. Depth indicators and other instrumentation should be regularly calibrated, particularly if drilling tools or auger strings are changed. Other issues that can be of concern are smear in clays and weak rocks, bores left open too long before concreting and bases not adequately cleaned. This list is not exhaustive.

82.2.2 Boring

The correct use of temporary casings is essential for the success of good pile construction. Casings should always be driven or screwed in advance of drilling tools through unstable soils to provide a safe working environment and to avoid over-excavation. A basic problem is found when, due to variations in the depth of unstable soils, the casing lengths used are inadequate and attempts to push them further into the ground risk the safety of personnel working around the bore (a minimum of 1 m should be left above the starting level as a safety barrier). With the less frequent use of high-speed rotation, the practice of 'mudding-in' has become less common and the prevalent method is to screw/crowd (i.e. push down) the casing, sometimes as the auger is rotating within the casing. Mudding-in has the inherent problem of creating a weak zone around the casing which, if adequate concrete head were not utilised during subsequent casing extraction during concreting, could lead to a reduction in the concrete cross-section (known as necking) as the ground slumps into the bore.

This phenomenon of necking can also occur when withdrawing temporary casings if there is insufficient head of concrete to resist the external groundwater pressure. This is illustrated in **Figure 82.1**. To avoid this, the SPERW (ICE, 2017) has specified minimum casting levels for concrete, which are included in **Table 82.1**. This type of problem is

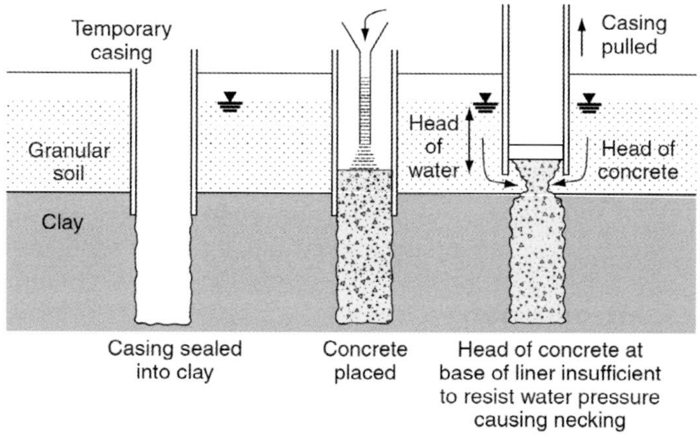

Figure 82.1 Necking in bored piles when the casing is pulled

Cut-off level below starting surface, H^a: m	Casting tolerance above cut-off level: m	Condition
0.15 to any depth	$0.3 + H/10$	Piles cast in dry bore within permanent casing or cut-off level in stable ground below base of casing
0.15–10.00	$0.3 + H/12 + C/8$	Piles cast in dry bore using temporary casing other than above
0.15–10.00	$1.0 + H/12 + C/8$ where C = length of temporary casing below the starting surface	Piles or walls cast under water or support fluid[b]

[a] Beyond $H = 10$ m, the casting tolerance applying to $H = 10$ m shall apply.
[b] In cases where a pile is cast so that the cut-off level is within a permanent lining tube, or for a wall, the appropriate tolerance is given by deleting the casing term $C/8$.

Table 82.1 Concrete casting level tolerance above cut-off levels for specified conditions (Data taken from ICE, 2017) most acute with long casings. The fall in concrete within the casing must be carefully observed as the casing is extracted in order to top up the bore to prevent such pressure differential affecting the concrete placement. Vibrating casings is an effective method for the larger-diameter bores, but care must be exercised as noted above and disturbance to previously concreted piles or other structures prevented.

The sequence of bored pile construction, whether for discrete bearing piles or wall piles, and whether by conventional rotary bored piling or CFA methods, is of importance to pile integrity. The problems usually occur in weak or soft soils where the shear strength of the soils is inadequate to resist the lateral pressures of the fluid concrete. If the concrete is poured to a level significantly above groundwater level (it is essential for all bored piles to be concreted to at least nominally 1 m above groundwater level), the excess hydrostatic pressure can be significantly greater than that in the ground. Casing may be required to ensure migration of fines from the concrete into any aquifer is minimised; special concrete mixes can also be formulated to address this concern.

During boring, the fluid concrete in an adjacent pile can be disturbed and flow laterally if the piles are close together and the ground between them insufficiently strong to resist the lateral pressures. During concreting, the pressurised column of fluid concrete could exceed the in situ pressures in a recently cast pile and cause ground movement towards the previously concreted pile, thus possibly causing a reduction in pile cross-sectional area. It is normal practice to leave spacing equivalent to at least three pile diameters between successive piles in any one working day. This spacing may have to be increased when using CFA methodology, which uses positive injection pressures to assist sound concrete placement.

When boring into aquifers under drilling fluid or water, it is important to raise and lower drilling tools slowly to prevent suction or pressurisation destabilising the bore, as shown in **Figure 82.2**. Drilling buckets should have a sufficient bypass for the fluid to pass through or around the bucket, such as the flat-sided bucket shown in **Figure 82.3**, in order to reduce the suction or pressurisation effects. Care must be taken to check the properties of the drilling fluid regularly. If the fluid density is too low this can lead to bore collapse; if the fluid is dense due to suspended solids it can lead to clean-up problems. Bores must be cleaned and the drilling fluid conditioned and recirculated after boring and cleaning to ensure that no 'sludge' remains at the base, which could lead to concreting problems and poor base loading of the pile. Sampling of the drilling fluid should be undertaken with special tools.

One of the most common problems encountered in CFA piling is the tendency to 'flight' material up the bore and cause over-excavation of the weaker soils, as illustrated in **Figure 82.4**. Guidance is given in SPERW (ICE, 2017), which states that care has to be taken particularly in silts and fine sands to ensure an adequate rate of penetration of the auger against the number of revolutions. Excessive rotation of the auger without penetration should not be permitted.

Auger displacement piles are a form of piling where the auger displaces the ground without removing spoil. There are

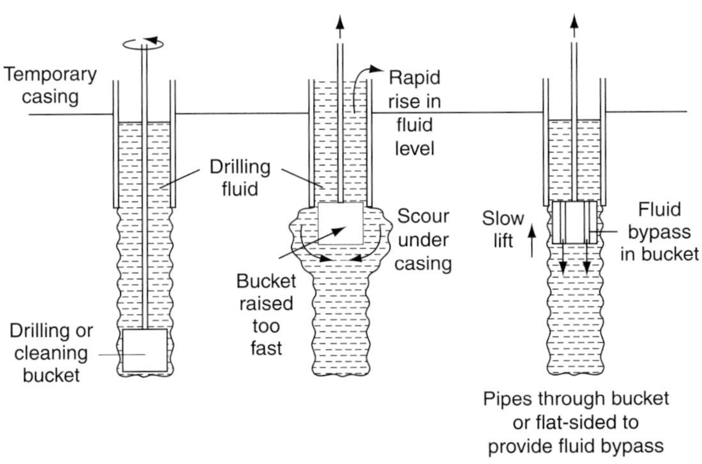

Figure 82.2 The effect of suction and scour on borehole stability when using drilling buckets

Figure 82.3 Flat-sided digging bucket to allow fluid to bypass (Courtesy of Bachy Soletanche)

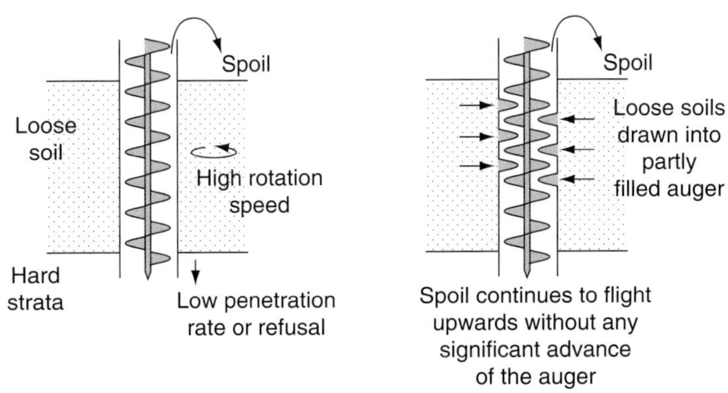

Figure 82.4 Flighting in CFA piles

two types of displacement piles: the helical (screw) and the soil displacement. For the screw displacement pile, the rate of penetration is controlled during insertion and removal of the auger to ensure that the auger follows the same path of flights – thus forming concrete ribs for the pile. Accurate monitoring and control of the rotation and penetration rate of the auger is essential to ensure integrity of the concrete ribs. For the soil displacement pile type, the soil is displaced without forming ribs. Substantial torque and crowd may be needed to ensure the auger penetrates the bearing stratum to a sufficient depth to achieve the required bearing capacity.

Other problems that may occur are discussed here. One problem is the meeting of natural or man-made obstructions. Even with the best desk study/collation of services information, the chances of hitting services are real. Smear to the bore walls, which can reduce the shaft friction, can occur in clays and weak mudstone due to over-rotation of the drilling tools. This is illustrated in **Figure 82.5** for a rotary bored pile shaft and in **Figure 82.6** for a CFA pile in mudstone, which was deliberately over-rotated. If bores are left open too long, clays

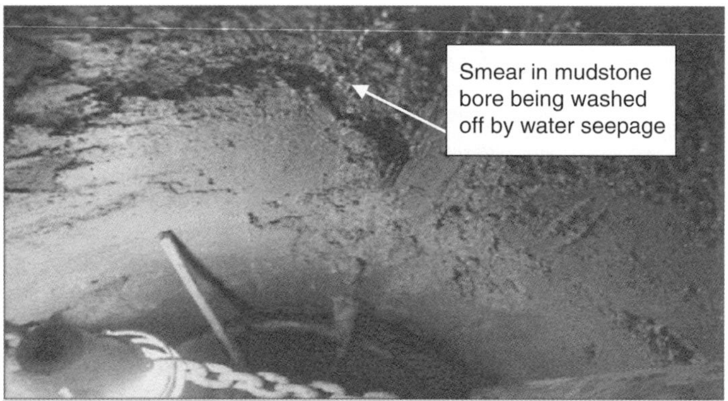

Figure 82.5 Smear in rotary bored shaft in weak mudstone

Figure 82.6 Smear on shaft of CFA bored pile in Mercia Mudstone (Courtesy of Balfour Beatty Ground Engineering)

can soften and again result in lower shaft friction being mobilised. Pile end bearing can also be affected by allowing loose material to collect on the base. In a dry bore, a bladed auger or cleaning bucket is used to remove the loose debris. In piles that are bored under water or drilling fluid, suspended soil particles can settle onto the pile base and a cleaning bucket, air lift or submersible pump should be used to clean the base.

82.2.3 Concreting

Some of the problems associated with concrete placement have been noted in Section 82.2.2, but this is by no means an exhaustive list. Since the most common method of concrete supply is ready-mix suppliers, basic checks of the concrete delivery ticket, however mundane a process, can save problems with incorrect batching and distribution. The next most important check is of slump; as with all types of bored piling, high-slump mixes are essential. Concrete that is too stiff is unsuitable for the self-compacting nature of the mixes employed and can cause blockages in tremie tubes and CFA delivery pipes. Placement pipes are desirable as they limit the segregation of concrete poured into bores, especially those that have pre-placed reinforcement cages. SPERW (ICE, 2017) recommends the use of a rigid delivery pipe placed centrally in the bore to prevent the concrete hitting the reinforcement and so avoid segregation.

When concrete is placed inside a temporary casing, sufficient concrete needs to be provided to fill the extra volume created by the removal of the temporary casing. It is essential, especially for long casings, to monitor the internal concrete level and allow for topping up as casings are extracted.

There is the thorny subject for conventional bored piling of 'how wet can the bore be' before normal concrete placement has to be replaced by tremie placement. This is usually a problem encountered in otherwise dry cohesive ground, as the water table level in granular soils will normally be known before construction and the appropriate concrete placement method selected. Clearly, water entering the bore, whether from some way down the bore walls or from the base, influences the decision. On occasions where the amount of water entry is minimal, it is impractical to use a tremie to place the concrete. In this situation it is quite adequate to either 'dry' the bore by the placement of dry cement or leave the final metres of boring until concrete is known to be ready on site for placement. If there is any doubt, the answer is to change concreting methodology and possibly pile design so as to avoid the problem.

While unlikely, the joints of tremie pipe assemblies can cause blockage if not adequately sealed due to the squeezing of water out of the fluid concrete by the excess pressure head in the pipe. Vigilance is important, particularly when removing and breaking such assemblies, and renewal of the rubber seal rings must be undertaken to prevent reoccurrence.

The removal of the tremie must be carefully controlled by logging the tremie depth as each volume of concrete is placed. Over-extraction of the tremie will cause a break in concrete quality in the bore, with a zone of trapped detritus. To avoid over-extraction, SPERW (ICE, 2017) requires a minimum 3 m embedment for the tremie pipe in the concrete. The EFFC/DFI *Best Practice Guide to Tremie Concrete for Deep Foundations* (EFFC/DFI, 2016) determines that many problems in bored piles were caused by (or in part due to) the use of inadequate concrete mixes with inadequate workability, or insufficient stability or robustness. It further identified the primary causes as inadequate concrete specifications and inadequate testing procedures. The publication provides guidance on the characteristic performance of fresh concrete and its placement using tremie methods in bored piles and diaphragm walls.

In CFA piling, a problem can occur either above or below the groundwater table due to the bleeding of water and possibly fines from the fluid concrete after placement. The problem is more prevalent above the groundwater table in dry granular soils and is caused by the squeezing of water/fines from the concrete due to the significantly higher pressure in the concrete than in the ground. Usually this problem can be foreseen and changes to concrete mix design (such as over-sanding and a higher dosage of plasticiser) are made. Problems also occur below the groundwater table for similar reasons but are not always evident before the start of work. Such problems are usually only detected when it becomes difficult or impossible to place reinforcement cages into the supposedly fluid concrete. The concrete in the affected depth zone will have lost its workability and present an obstruction to the smooth passage of the reinforcement cage. The remedy in these situations is to extract the reinforcement cage, re-bore and re-concrete the bore so as to have workable concrete throughout the pile depth.

Care needs to be taken with regard to the temperature of concrete during placement at air temperatures less than 5°C. SPERW (ICE, 2017) provides cold weather precautions. Concrete with cement replacement or chemical admixtures may have extended curing time at low temperatures.

Another 'golden rule' is: 'once the concrete is placed, leave it alone'. That is to say, there is risk of contamination of the concrete in the bore if overemphasis is placed on cleaning up the top of the pile. However, with care, clean up and reduction of concrete level is achievable, even to the extent of preparing/casting pile caps integral with the piles.

82.2.4 Reinforcement

Congestion of reinforcement in heavy cages can lead to poor concrete placement with the associated problems of design integrity of reinforcement bond. The problem is usually found when lapping multiple cages. BS EN 1536 (BSI, 2015a) requires a minimum of 100 mm between main bars or 80 mm when using aggregate of 20 mm or less. On occasion it may be necessary to limit aggregate size to conform to these requirements, or use double circles of main reinforcement or bent bars to crank radially in the pile.

Poor handling and coupling of cages can be avoided by cage welding, which is now becoming more prevalent, or by using stiffeners, particularly for larger-diameter cages. Pre-welded cages are usually more accurately produced than hand-tied cages and allow better matching of bars at laps. Correct lifting of cages should be observed to avoid collapse. Cages should be stored so as to remain essentially clean, and must not become covered with mud or other debris.

Spacers to ensure adequate cover for the reinforcement to the bore walls do not always get the attention they deserve. After the expensive mobilisation of specialist equipment and manpower and the procurement of high-quality materials, do not allow the final product to be ruined for the sake of inadequate spacers – either by quality or number. Often the inadequacy of spacers only comes to light when the hardened pile is excavated or cut down – to reveal either no spacers or bent or twisted ones that have allowed the cage to move eccentrically in the bore before the concrete has set and hence degraded the finished product.

Lifting rings should be designed and installed at appropriate points on the cage to allow ease of lifting and installation; again cage congestion should be taken into account to allow ease of concrete flow. The level at which reinforcement is placed should be done with care so as to avoid subsequent problems with inadequate projection when follow-on trades take over the adoption of the piles into the works.

82.2.5 Ancillary works

After the completion of any one pile, it is important to ensure it is protected until the concrete has attained strength. Plant must not be allowed to travel near such piles as the fluid or partially set concrete cannot offer adequate passive resistance to the lateral pressures so induced (**Figure 82.7**). This is often

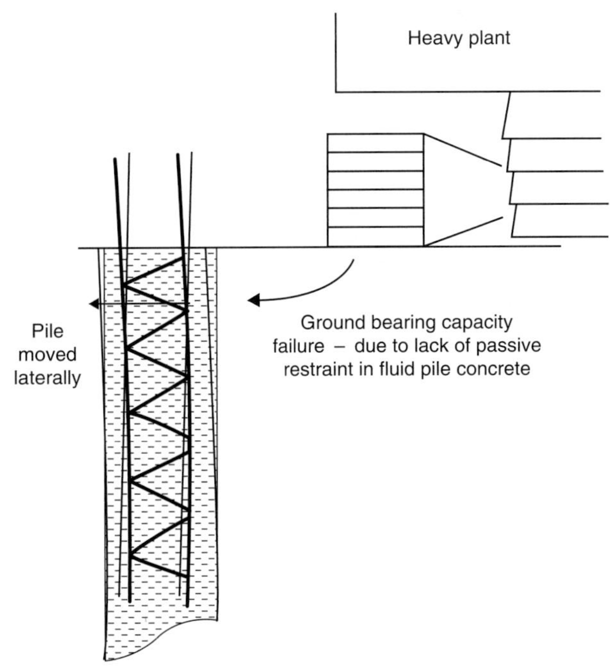

Figure 82.7 Damage to piles from nearby plant movement

the reason for inadequate concrete cover or misaligned reinforcement cages. The clear identification of piles should also be a basic item of housekeeping on any site.

Other good housekeeping should include the removal of spoil and general area cleaning to create a safe working environment and to prevent debris from entering the fluid concrete. This should be undertaken using appropriate reach plant to avoid surcharging the piles.

Cutting down piles must be done with care. The use of laterally applied breakers, unless of a multi-circular configuration, must not be permitted. Ideally, proprietary multi-teeth breakers should be utilised. The removal of the top concrete core is assisted by the use of foam sleeves affixed to the reinforcement above cut-off level. FPS (2008) have produced a guidance note covering this topic and there is also a CIRIA paper on the subject (Cox, 2009).

82.2.6 Structural problems

Structural problems arise when reinforcement is placed at the incorrect level, or the cage becomes displaced due to poor use of spacers or mechanical displacement. The proving of concrete strength can be fraught with practical difficulties such as poor or dirty cube moulds, inadequate labelling, poor storage, curing tanks not being temperature-controlled and poor off-site testing. Fortunately, the industry is moving to a self-certification system that is the province of the concrete supplier. The ready-mix industry has taken it upon itself to overcome the known shortcomings of this facet of construction and now collects samples from sites and produces records of the set concrete properties.

Pile eccentricity can cause problems, particularly if the designer has not taken account of the tolerance latitude that is permissible.

82.3. Driven piles

82.3.1 Installation methods and ground conditions

BS EN 12699:2015 (BSI, 2015b) covers the execution of concrete, steel, timber and cast iron displacement piles. Driven piles include low-displacement piles such as open-ended steel tubes, steel sheets and steel H-sections, and high-displacement piles such as pre-cast concrete sections, closed-ended steel tubes, timber piles and driven cast in situ concrete piles. Piles are installed by using drop hammers, vibratory techniques or by hydraulic press-in methods.

Hydraulically operated drop hammers are most commonly used to drive pre-cast concrete sections, and vibratory methods and drop hammers are both commonly used for steel piles. Hydraulic press-in methods are used to install steel sheets and some tube sections in areas where noise and vibration are to be avoided.

Adequate site investigation information is critical for making an assessment of driveability. The presence of obstructions and hard layers can affect driveability and additional measures such as jetting and pre-boring may be needed for the pile installation.

82.3.2 Pile manufacture

The manufacture of pre-cast concrete piles is covered by BS EN 12794:2007 (BSI, 2007b). The piles should be designed to withstand handling and driving stresses, which are often considerably greater than the stresses they will experience in service. Pre-cast concrete piles are cast in moulds; the concrete strength needs to be sufficient for their removal from the moulds, handling and transportation and then for final driving. The piles are cast in lengths that will subsequently be jointed on site. The ends of the sections therefore need to be cast square to the pile to ensure they fit tightly together and avoid eccentricity of loading (**Figure 82.8**).

Steel and timber pile manufacture is unlikely to present any particular issues. Straightness of the section is important to avoid induced bending during driving.

82.3.3 Pile installation

Piles installed using drop hammer or vibratory techniques are held in leaders or guide frames to guide the hammer or vibrator, pitch the pile and hold it during the driving process. For hydraulic press-in methods, the hydraulic jaws clamp onto the pile to guide and feed the sections into the ground.

Over-driving, described by Healy and Weltman (1980), can often occur when attempting to drive piles to an unreasonable

Figure 82.8 Ends of pre-cast piles, not cast square

design depth. This can result in permanent damage to the pile head which is visible, or to the pile toe where it is unseen. In the case of toe damage, a pile may appear to be driving normally but the set may not build up as expected due to crushing or distortion of the pile toe. Where it is important to reach a particular depth, jetting or pre-boring should be considered.

Jetting and pre-boring need particular care to ensure that they do not cause disturbance to the ground that would affect adjacent structures such as utilities and buildings, or affect the performance of the piles. For jetting, a high-pressure water jet is used to erode or loosen the soil at the pile tip. Water returns and pressures should be monitored to ensure the water is not finding an alternative flow path away from the pile tip. If the returns are blocked, pressure can rapidly build in the ground and cause heave. In pre-boring, an auger is used to loosen the soil prior to pile installation. Spoil removal should be limited to the minimum necessary to aid pile installation without affecting the geotechnical performance of the pile. Pile behaviour is best checked by dynamic or static test loading.

Damage during driving can occur, particularly at the pile head where the stresses are highest. For pre-cast concrete this will take the form of cracking and spalling of the concrete, as shown in **Figure 82.9**, and on steel sections the pile head may bend or buckle. In order to prevent spalling, pre-cast concrete pile suppliers normally provide a protective steel band at the pile head and keep the concrete cover to a minimum while still complying with durability requirements for protecting the reinforcement (Healy and Weltman, 1980, CIRIA Report PG8). Damage can also occur within the pile due to tensile stresses. This can occur anywhere in the pile shaft when piles are driven using excessive hammer energy in soft ground and the damage is not noticed until harder driving occurs later towards the end of the drive. The cracking induced earlier in the pile shaft then leads to spalling and compressive failure at depth.

Figure 82.9 Damaged pre-cast concrete pile head (Courtesy of Roger Bullivant Limited)

Pre-cast piles are usually reinforced to resist these driving stresses, and breakages are relatively uncommon. They tend to happen when obstructions are encountered. Although piles are relatively slender, buckling is unlikely because of the restraint provided by the ground. Even soft ground will usually provide adequate support to prevent buckling during driving.

Where steel pile heads are damaged during installation, they can be cut back and extension pieces welded on. Requirements for structural welds can be difficult to achieve under site conditions and usually require verification testing, which can delay the works. Procedures and testing requirements for both off-site and site welding are detailed in SPERW (ICE, 2017) and general guidance for arc welding is covered in BS EN 1011 Part 1 (BSI, 2009). Repairs to a concrete section can be made, but delays are likely to be considerable due to the time required for the new section to cure.

Particular care is needed where driven piles are to be founded in a dense layer of variable thickness that is underlain by weaker material. Piles may punch through the dense layer in some locations before reaching the required set. In these circumstances, detailed site investigation information regarding the variability of strength and thickness of the founding stratum is critical. This information may need to be backed up by driveability trials, selecting areas close to boreholes where the founding stratum thickness and strength are the most critical. Further caution is required in the driving process where a precast pile is driven through a dense layer into a weaker layer below, as this increases the risk of a phase change in the compressive driving stress wave as the resulting tensile wave may cause the pile to become damaged. This difficulty can be overcome through changes in the input energy of the hammer blows.

During driving, steel tube piles can plug. This is where the soil inside the tube moves down instead of up as the pile continues to be driven. The mechanics of whether plugging is likely to occur have been investigated by White *et al.* (2002), who showed how this affects the pile capacity and described measures for reducing the effects by fitting a pile shoe. Piles can also be made to deliberately plug by welding an internal plate to prevent upward travel of soil inside the pile. Plugging in H-pile sections is very rare according to Biddle (1997), and should not be assumed unless it is demonstrated during driving.

Particular care is needed with founding driven piles onto buried inclined rock surfaces and rock head which contains boulders. These conditions can potentially deflect the pile off-line, inducing bending and damage to the pile, and can lead to uncertainty in establishing a reliable set. Steep-sided rock faces, such as the edges of backfilled quarries, present a particular risk.

82.3.4 Final set

Driven piles are normally driven depending on the ground conditions, to a final set which is usually defined as the measured penetration for ten blows, a geotechnical design length or in some circumstances a combination of both. False sets can occur where there are obstructions or boulders in the ground, or where thin rock bands may overlie weaker layers and where the adverse effects of negative driving-induced pore water pressures are manifest. Trial drives at various locations around the site are a way of verifying consistent founding conditions and compatibility with the expected ground conditions and design. Pre-boring can be used where the pile driving is likely to be impeded by obstructions.

In very stiff clays and chalk the pile resistance can increase with time; this is known as set-up. A delay in the driving can then lead to extra energy being required to get the pile moving again. The increase in pore water pressure developed during driving reduces the ground resistance to driving. As this excess pressure dissipates, the resistance increases. This is beneficial to the long-term capacity, but it is important not to test the pile too soon after installation – otherwise the pile capacity will be underestimated. If the problem is assessed as potentially significant, piezometers can be used to monitor the dissipation of the excess pore water pressures.

In dense silts, some clays and some weathered rocks, false sets can occur as a result of negative pore water pressures being induced by the pile driving. The dissipation of the negative pore water pressure can lead to a relaxation of the pile set. It is therefore necessary to carry out re-drive checks in these circumstances to check for any relaxation or reduction in pile resistance. The problem of relaxation can also occur in some mudstones, where it has been suggested that the clay in the joints squeezes out or consolidates when load is applied after driving.

82.3.5 Downdrag

In soft clays and silts the driving of displacement piles can induce excess pore pressures, which will dissipate over time. As the excess water dissipates, the soil will consolidate, leading to an additional downdrag load developing in the pile. Pre-boring and sometimes slip coatings can be used to reduce these effects. Generally, slip coatings have not proved very effective and their integrity after driving is uncertain. The additional downdrag load will usually be assessed and included as an additional load in the pile design. Downdrag loads can also occur in loose granular soils where the driving causes compaction of the soils.

82.3.6 Ground displacement

The process of driving a pile into the ground causes displacement of the ground, which can lead to heave and lateral soil displacement in incompressible soils, or compaction and settlement in loose granular soils. The effects are greatest for high-displacement piles in incompressible ground. In a confined situation such as a cofferdam, the cumulative effect of driving several piles can lead to increased ground displacement and lateral pressures. These can affect the cofferdam structure or make further driving of piles increasingly difficult as the ground tightens up. Use of low-displacement piles, pre-boring and careful sequencing of the pile installation can all help to mitigate these effects. For large structures such as oil tanks it is normal practice to drive piles from the centre working outwards. When working near to retaining walls, piles should be installed by working progressively away from the wall.

Preformed piles, such as pre-cast concrete and steel sections, can all be re-driven if there is a suspicion that they have been uplifted by driving adjacent piles. This is not the case for cast in situ driven piles, where an assessment would need to be made of any potential effect on performance.

82.3.7 Pile trimming

Poorly controlled pile trimming is probably the most common cause of damage to pre-cast concrete piles. 'Breaking down of piles' (FPS, 2008) provides guidance on various methods, and states that heavy impact breakers should not be used on small-diameter and lightly reinforced piles, or piles in soft ground. Cox (2009) also describes the different methods of pile trimming and the failure of the pile reinforcement that can occur with some systems. In particular, Cox (2009) notes the specific benefit of using debonding foam to avoid pile damage.

Trimming of steel piles should be carried out using a disc cutter or steel saw, and not with burning equipment which will reduce the strength of the steel and may produce noxious fumes with certain types of steel.

82.3.8 Jointing piles

Pre-cast concrete piles are usually supplied in sections which are connected by joints. There are two types of joints in general use. The first is a full-strength mechanical joint which provides a compression, tension and moment connection. The second non-mechanical joint provides a compression connection only. Failure of joints is rare, but mechanical joints should be used where a lot of obstructions are known to be present. Guidance as to the classification of these joint types can be found in EN 12794:2007 (BSI, 2007b).

For steel-cased piles, the problems of jointing casing lengths are associated with the practical aspects of site welding, including preparation and inspection. Again, the recommendations provided in SPERW (ICE, 2017) are highly relevant. Occasionally, threaded joints from second-hand or non-prime drill casing are employed as a means of avoiding the delays associated with welding. The casing sections are joined by driving the end of the casing with the male thread into the female internally threaded collar. This provides an interference fit that is only suitable for compressive loads (not tensile loads). For bottom-driven thin-wall steel-cased piles, the casing sections are usually provided with simple interference fit spigot and socket type connections that can be joined by tack welding. Alternatively, riveted connections can be used as a temporary expedient for the casing where the connection is only required to resist tension forces during driving. Thin-walled steel casings are also employed for Odex drilled-in piles, and in these cases tack welding is also satisfactory for jointing as the casings, left in permanently, are not considered structural in the pile design process.

82.3.9 Noise and vibration

Noise is a potential problem for driven piling operations, particularly systems that are top-driven. Noise levels are generally less for bottom-driven methods, where the hammer is contained within a driving tube or where vibratory methods are used. Consideration needs to be given to the proximity to noise- and vibration-sensitive areas. Measures can be taken to reduce the noise using acoustic shrouds and by restricting working hours. Alternatively, hydraulic push-in systems can be considered which have very low noise levels. Field measurements using this system have been compared to the existing recommended limits by White *et al.* (2002).

BS 5228 (BSI, 2014a, 2014b) is the code of practice for noise and vibration control on construction sites; Part 1 (BSI, 2014a) covers noise. It provides guidance concerning methods of predicting and measuring noise and assessing the impact on those exposed to it. Noise control from piling sites is specifically addressed in section 8.5, and types of piling are discussed in Appendix H of the code. For top-driven systems, noise reduction can be achieved by introducing a non-metallic dolly (such as timber) between the hammer and the driving helmet. Acoustic shrouds have also been used to enclose the driving equipment. Data will normally be available from piling contractors for their particular system.

The driving of piles into the ground creates stress waves and vibrations that can affect the environment. The effects of vibration can be reduced by modifying the impact or frequency of the pile-driving equipment. Vibratory piling methods also induce vibration in the ground. On some occasions, isolation trenches or pre-boring can prove useful in protecting neighbouring structures as they act as an inherent barrier to such ground-borne vibrations. However, unpredictable passages of vibration, possibly through susceptible strata, can also occur. As an alternative to steel piling, hydraulic push-in piling methods can be adopted which result in very low levels of ground vibration. White *et al.* (2002) have provided prediction curves showing that this system has low vibration levels that enable it to be used much closer to residential buildings than can impact or vibratory methods.

The level of vibration depends on the source and distance. Even very low levels of vibration can be sensed by humans, but much higher levels are necessary to cause damage to structures. BS 5228-2 (BSI, 2014b) covers vibration. It provides guidance on methods of measuring vibration and assessing its impact on the environment. Annex C provides current measured vibration levels for piling and Annex D provides historic data.

The threshold of perception of vibration for humans occurs at very low levels of peak particle velocity (PPV) in the range 0.15–0.3 mm/s. The threshold criteria for damage to buildings occur at much higher levels and are presented in BS 7385-2 (BSI, 1993), and are summarised in Annex B of BS 5228-2 (BSI, 2014b). These depend on the type of structure, the PPV of the vibration and the frequency range. In general, for reinforced or framed structures and industrial and heavy commercial buildings, the threshold level for cosmetic damage from transient vibration is 50 mm/s. For unreinforced or light framed and for residential or light commercial buildings the threshold is 50 mm/s for frequencies above 40 Hz, but reduces to 20 mm/s at 15 Hz and down to 15 mm/s at 4 Hz. Minor damage can occur at twice these values and major damage can occur to building structures at four times these values.

82.3.10 Durability

For concrete piles the durability depends on the concrete quality and the concrete cover to the reinforcement. Healy and Weltman (1980) note that fine cracks of up to 0.2 mm in width may occur in high-quality piles, and these are not generally regarded as being of concern because they close up under the dead load from the structure. Larger cracks may occur as a result of improper handling, pitching or driving. The seriousness of excessive cracking depends on the aggressiveness of the ground into which the pile is being installed and the durability requirements. For concrete structures these are covered in BS EN 1992-1 (BSI, 2015c).

Corrosion of steel piles is most likely to occur either in disturbed ground or in a marine environment. Corrosion rates for steel piling in natural environments are given in BS EN 1993 Part 5 (BSI, 2007a), as shown in **Table 82.2**. Various

Water	Zone[a]	Required design working life (years)[b]				
		5	25	50	75	100
Common freshwater (river, ship, canal, etc.)	High attack (water line)	0.15	0.55	0.90	1.15	1.40
Very polluted freshwater (sewage, industrial effluent, etc.)	High attack (water line)	0.30	1.30	2.30	3.30	4.30
Sea water in temperate climate	High attack (low water and splash zones)	0.55	1.90	3.75	5.60	7.50
Sea water in temperate climate	Permanent immersion or in the intertidal zone	0.25	0.90	1.75	2.60	3.50

[a] The highest corrosion rate is usually found in the splash zone or at the low water level in tidal waters. However, in most cases the highest bending stresses occur in the permanent immersion zone.
[b] The values given for 5 and 25 years are based on measurement, the other values are extrapolated.

Table 82.2 Recommended values for the loss of thickness (mm) due to corrosion for piles and sheet piles in freshwater or in sea water (Data taken from BSI, 2007a)

methods are available to combat corrosion, and these are described by Fleming *et al.* (2009). Guidance is also provided by ArcelorMittal (2022) in their *Piling Handbook*.

82.4. Identifying and resolving problems

82.4.1 General guidance

The following sections describe the way in which piling problems are often identified, how they can be assessed and the ways in which they can be resolved. Quality management systems and the process of identifying piling non-conformance are described in Chapter 93 *Quality assurance*. Problems often arise from the results of pile tests. Useful guidance on pile testing and the interpretation and assessment of tests is provided in the *Handbook on Pile Testing* (FPS, 2006) and by Tomlinson and Woodward (2015). Poulos (2005) also provides a range of case studies of piling problems and comments on the particular issues involved in investigating and analysing them.

82.4.2 Identification

There are a range of tests and records that can be used to identify potential problems. These include

- piling records – inconsistencies in ground conditions, driving records, out-of-position piles, uplift and lateral displacement, anomalies in concreting or reinforcement records
- integrity tests – acoustic anomalies in integrity traces indicating the possibility of cracks, changes in section or inclusions in the pile; guidance on different types of integrity tests and their interpretation is provided in Chapter 97 *Pile integrity testing* and by Turner (1997) in CIRIA Report 144
- load tests – failure to meet the specified load/settlement behaviour, unexpected settlement behaviour indicating structural failure or low shaft resistance or a soft toe, sometimes identified by using back-analysis of load test results using methods such as those of Chin (1970) or Fleming (1992)
- materials tests – concrete cube tests not meeting the specified strength, failure of weld tests on steel connections
- inspection of trimmed piles – damaged pile head, weak concrete, buckled steel section
- observations from the resident engineer or inspection staff detecting non-conformances with the specification that lead to further investigations that highlight problems.

Piling problems do not always arise from non-conformance with the specification and may become apparent from observed inconsistencies. Sometimes the difficulty is to realise when there is a significant issue that needs further investigation. For example, a successful preliminary test on a bored pile in dry conditions in one part of a site may be found to be unrepresentative of wet conditions that become apparent in other areas. Another example is of driven piles which may all reach the required set but some are founded at a much higher level. Where this is unexpected and could affect the design, further investigation and design review or testing may be needed.

82.4.3 Diagnosis

Test results should be viewed critically to ensure they are accurate and reliable, and they should be considered along with all the other information from pile records and knowledge of the site conditions, as well as the anticipated geotechnical design. Further investigation may be necessary in the form of further testing or physical investigations before determining what remedial action, if any, is required. In some cases, the test itself may be limited in what it can reveal. For example, integrity testing in the form of sonic echo tests or cross-hole sonic testing relies on the interpretation of acoustic waves to identify potential anomalies in piles. While these tests can indicate acoustic inconsistencies within the pile and between piles, this should be verified by physical investigation or load testing and a review of the pile records before deciding whether the anomalies are structurally significant. Similarly, with load testing, unexpected pile load/settlement behaviour may be the result of variations in ground conditions, poor installation or an inadequately performed test. Further evaluation and testing may be required to confirm the cause of unexpected performance.

In some cases, the anomalous test results may be the result of follow-on works where piles have been damaged by trimming or ground movements due to other works. It is therefore important that the testing is timely and carried out during the piling works or immediately on completion rather than at a later stage – otherwise the cause of the problem may be difficult to attribute to a specific contractor. It is advisable that testing is carried out as an ongoing process during the piling works so that potential problems can be identified during the work and remedial action taken while the works are in progress.

Digital rig instrumentation can provide an additional source of piling records, including as shown in **Figure 82.10**:

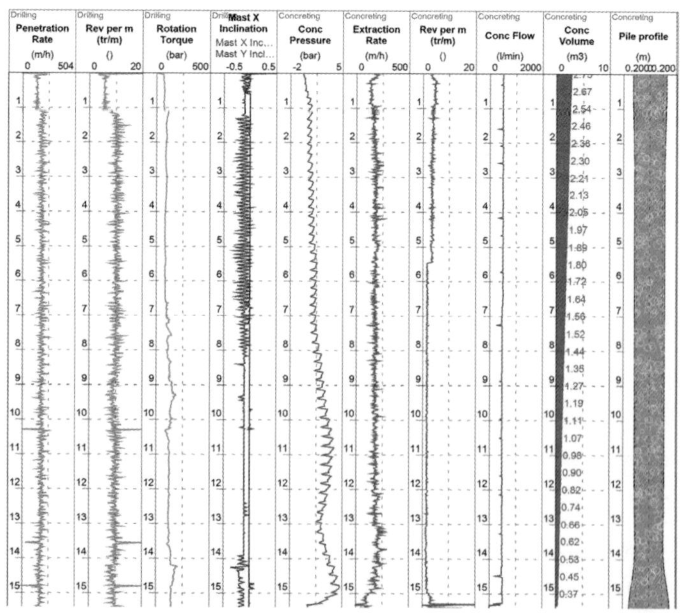

Figure 82.10 Typical rig instrumentation printout for a CFA pile

the inclination of the mast, drilling depth and speed, the torque and the speed of rotation of the auger, concrete flow, average consumption, the profile of the pile as formed, the pressure at the bottom of the auger and the deviation of the pile.

82.4.4 Resolving problems

Where tests or records indicate a potential problem with the piling or ground conditions, it is important to understand the impact of the problem for the structure. The issue may affect the durability or the structural performance.

In some cases, the problem may be easily accommodated by the structure without any remedial action, or with only minor remedial works. For example, with out-of-position piles, the pile reinforcement or the pile cap or ground beam arrangement may be sufficient to accommodate the eccentricity without any remedial action. Also, a pile load test that just exceeds the specified settlement limit for the serviceability state may not be structurally significant; the structure may be able to accommodate slightly larger settlements than were originally specified without detriment to the overall structural performance. In these cases, while there may be a non-compliance with the specification, it is important not to overreact as it may be easily resolved by further assessment and analysis. It is not in the client's best interests to conduct a major investigation into apparent anomalies which may have no real significance for the structure.

Some problems may indicate a more fundamental issue with the piling which may affect the performance, safety and long-term durability of the structure. This will require a clearly defined strategy of investigation and assessment. The process should determine

- the structural significance of the problem
- whether other piles may be affected
- any further testing that is considered necessary
- remedial works if required.

The procedures for dealing with non-conformance should be defined in the contractor's quality plan. The actual procedure will depend on the contractual arrangements and whether the contractor's workmanship will be verified by an external party or whether the contractor will self-certify their work. The different quality management systems are described in Chapter 93 *Quality assurance*.

A significant area of misunderstanding can arise from the process of exploring the anomaly. Tests to verify that a pile can be safely incorporated into the works are conventionally paid for by the client (as the pile is ultimately acceptable), while tests that reveal defects that need remedying may be paid for by the contractor if the process is at fault. Payment for consequential costs is often contentious, so it is in everyone's interest that the time for resolution is minimised.

The important underlying principle is that the lines of communication and project responsibilities should be clearly defined so that any non-conformance can be reported, and remedial action proposed, assessed and approved as quickly as possible. Where this is likely to impact on cost, programme or safety, the client needs to be kept informed of the solution.

Disclaimer

The example non-conformances included in this chapter are for illustrative purposes only and are not associated with or reflective of the authors or their employers.

References

AGS (2016) *Guidelines for Good Practice in Site Investigation*. Association of Geotechnical and Geoenvironmental Specialists, Beckenham, UK.

ArcelorMittal (2022) *Piling Handbook*, 9th edn. ArcelorMittal, London, UK.

Biddle AR (1997) *Steel Bearing Pile Guide*. Publication P156. The Steel Construction Institute, Ascot, UK.

BSI (British Standards Institution) (1993) BS 7385-2:1993: Evaluation and measurement for vibration in buildings – Part 2: Guide to damage levels from groundborne vibration. BSI, London, UK.

BSI (2007a) BS EN 1993-5:2007: Eurocode 3 Part 5: Design of steel structures – Piling. BSI, London, UK.

BSI (2007b) BS EN 12794:2007: Precast concrete products – Foundation piles. BSI, London, UK.

BSI (2009) BS EN 1011-1:2009: Welding – Recommendations for welding of metallic materials – Part 1: General guidance for arc welding. BSI, London, UK.

BSI (2014a) BS 5228-1:2009 + A1:2014: Code of practice for noise and vibration control on construction and open sites – Part 1: Noise. BSI, London, UK.

BSI (2014b) BS 5228-2:2009 + A1:2014: Code of practice for noise and vibration control on construction and open sites – Part 2: Vibration. BSI, London, UK.

BSI (2015a) BS EN 1536:2010 + A1:2015: Execution of special geotechnical work – Bored piling. BSI, London, UK.

BSI (2015b) BS EN 12699:2015: Execution of special geotechnical work – Displacement piles. BSI, London, UK.

BSI (2015c) BS EN 1992-1-1:2004 + A1:2014: Eurocode 2 Part 1-1: Design of concrete structures – General rules and rules for buildings. BSI, London, UK.

BSI (2016) BS EN 206:2013 + A1:2016: Specification, performance, production and conformity. BSI, London, UK.

Chin FK (1970) Estimation of the ultimate load of piles from tests not carried to failure. *Proceedings of the Second Southeast Asian Conference on Soil Engineering, Singapore*, pp. 83–91.

Cox D (2009). *Breaking Down Piles and the Significance of Debonding*. CIRIA, London, UK.

EFFC/DFI (2016) *EFFC/DFI Best Practice Guide to Tremie Concrete for Deep Foundations*, 1st edn. EFFC/DFI, Faversham, UK.

Fleming WGK (1992) A new method for single pile settlement prediction and analysis. *Géotechnique* **42(3)**: 411–425.

Fleming K, Weltman A, Randolph M and Elson K (2009) *Piling Engineering*, 3rd edn. Taylor & Francis: Abingdon, UK

FPS (Federation of Piling Specialists) (2006) *Handbook on Pile Testing*. Federation of Piling Specialists, Faversham, UK.

FPS (2008) *Breaking Down of Piles*. Federation of Piling Specialists, Faversham, UK.

Healy PR and Weltman AJ (1980) *Survey of Problems Associated with the Installation of Displacement Piles*. Report PG8 1980. CIRIA, London, UK.

ICE (Institution of Civil Engineers) (2017) *Specification for Piling and Embedded Retaining Walls*. Thomas Telford, London, UK.

Poulos HG (2005) Pile behavior: consequences of geological and construction imperfections. *Journal of Geotechnical and Environmental Engineering, ASCE* **131(5)**: 538–563.

Thorburn S and Thorburn JQ (1977) *Review of Piling Problems Associated with the Construction of Cast-in-Place Concrete Piles*. DOE and CIRIA Piling Development Group Report PG2. CIRIA, London, UK.

Tomlinson MJ and Woodward JC (2015) *Pile Design and Construction Practice*, 6th edn. E&FN Spon, London, UK.

Turner MJ (1997) *Integrity Testing in Piling Practice*. CIRIA Report 144. CIRIA, London, UK.

White D, Finlay T, Bolton M and Bearss G (2002) Press-in piling ground vibration and noise during pile installation. *Proceedings of International Deep Foundation Congress, Orlando, USA*, pp. 363–371.

Useful website

Association of Geotechnical and Geoenvironmental Specialists (AGS): www.ags.org.uk

It is recommended this chapter is read in conjunction with

- Chapter 96 *Technical supervision of site works*
- Chapter 97 *Pile integrity testing*
- Chapter 98 *Pile capacity testing*
- Chapter 101 *Close-out reports*
- Section 5 *Design of foundations*

All chapters in this book rely on the guidance in Sections 1 *Context* and 2 *Fundamental principles.* A sound knowledge of ground investigation is required for all geotechnical works, as set out in Section 4 *Site investigation.*

Chapter 83
Underpinning

Devji Bhuva GSS Piling Ltd, Welwyn Garden City, UK
Tim Jolley formerly at Geostructural Solutions Ltd, Welwyn Garden City, UK

Before a final decision is made on the choice of underpinning for a particular application, there are several aspects to be considered. The various types of underpinning available are discussed in order to provoke thought on which type may be worth further consideration for the project at hand. Some factors are given which may influence the choice of underpinning to enable an informed decision to be made. Finally some specific points are covered in relation to groundwater, subsidence settlement, safety and the financial aspects of underpinning.

doi: 10.1680/icemge.66830.1351

CONTENTS

83.1. Introduction

83.1.1 Where underpinning is carried out and why

Underpinning is carried out in three main areas

- rectification of distress
- extending foundations downwards to enable building and civil engineering works to be carried out
- additional loading on an existing building.

83.1.2 Objective of this chapter

Irrespective of the reason for which underpinning is required, there are some common fundamental principles worthy of consideration prior to design or installation of an underpinning scheme. Some points are highlighted that may need to be borne in mind prior to embarking on an underpinning scheme, whether it is at design stage or at construction stage. Information is therefore provided to provoke thought in order to facilitate correct decision-making at an early stage.

83.2. Types of underpinning

83.2.1 Mass concrete

Mass concrete underpinning is used as an extension of an existing masonry wall or footing. This may be necessary as a result of a requirement to lower the ground floor of a building where there is insufficient depth of existing foundation.

Another common use of mass concrete underpinning is to transfer the load of a building to stable ground, as in the case of a building having suffered the effects of subsidence-related settlement. Where this is the case, it is important to have sufficient information about the nature of the ground at an early stage to ascertain the likely depth at which stable ground will be found. This will enable a decision to be made on whether mass concrete underpinning is an appropriate solution in this case.

Where an existing building is being considered for the addition of further storeys, resulting in an increased loading on the foundations, mass concrete can be used to provide the required increase in bearing capacity of the soil upon which the building is founded. The proposed underpinning can be designed to increase the width of the existing footing or, alternatively, to deepen the existing footings in order to found onto soil having greater bearing capacity.

If a new basement construction is proposed below or adjacent to an existing building, then mass concrete may be used to create a new wall to retain the soil behind. Generally a reinforced concrete lining wall is used in conjunction with the underpinning to provide permanent support. To maximise the space available in the proposed new basement room, the underpinning profile is often maintained in line with that of the existing walls. The underpinning profile is then widened at a point below the level of the new floor to achieve the same width as that of the original footing. In order to determine the degree of propping required in the temporary case, the proposed underpinning should be designed as a mass concrete gravity retaining wall. The proposed reinforced concrete lining wall will be designed to provide permanent restraint.

83.2.2 Hit and miss underpinning

Where appropriate, for reasons of economy, the mass concrete underpinning bases can be spaced apart in a ‘hit and miss’ pattern. As the loading from the structure above will now be concentrated onto a smaller area it is particularly important to check that the soil upon which the underpinning is founded does not become overstressed.

It is also important to consider the integrity of the structure between the bases. In some cases it may be appropriate to introduce a lintel or series of lintels placed side by side to span between each adjacent base.

83.2.3 Reinforced concrete underpinning

Reinforced concrete underpinning bases are used most commonly when creating a new basement space below an existing building. They should be designed as a series of vertically spanning beams having a typical width of an individual underpinning base (say, 1.2 m). Continuity of reinforcement between the individual bases is therefore restricted to a series of dowel bars in order to ensure the new retaining wall acts as a homogeneous mass.

A reinforced wall is designed as the primary means of soil retention. In the permanent case, the wall will span as a beam between the various floor levels; in the temporary case, the underpinning may be required to resist the possibility of overturning and forward sliding. Depending on the circumstances surrounding each individual case, an 'L'-shaped underpinning profile may need to be adopted. The sectional dimensions of the horizontal leg will be determined from design analysis. Given the conditions in which underpinning bases are formed, it is advisable to provide a tolerance for construction of the base slab in relation to the horizontal leg. **Figure 83.1** shows a means of providing an economic design while allowing construction tolerance.

For reinforced concrete underpinning, the waterproofing is normally provided internally in the form of a cavity drain membrane or a flexible coating system. Waterproof concrete for construction of the reinforced underpinning bases can be considered, but depending on the circumstances for a particular project it is sometimes found to be both impractical and uneconomical.

83.2.4 Piled underpinning

If an entire property is to be underpinned for reasons of subsidence, then a piled raft scheme could be considered. This is especially so if the ground conditions dictate that the depth of mass concrete underpinning is likely to be in excess of, say, 2.5 m. This figure is based largely on economic and practical considerations, which are set out in Section 83.10.

A system commonly used to underpin an entire building for subsidence is formation of the raft by use of piles situated

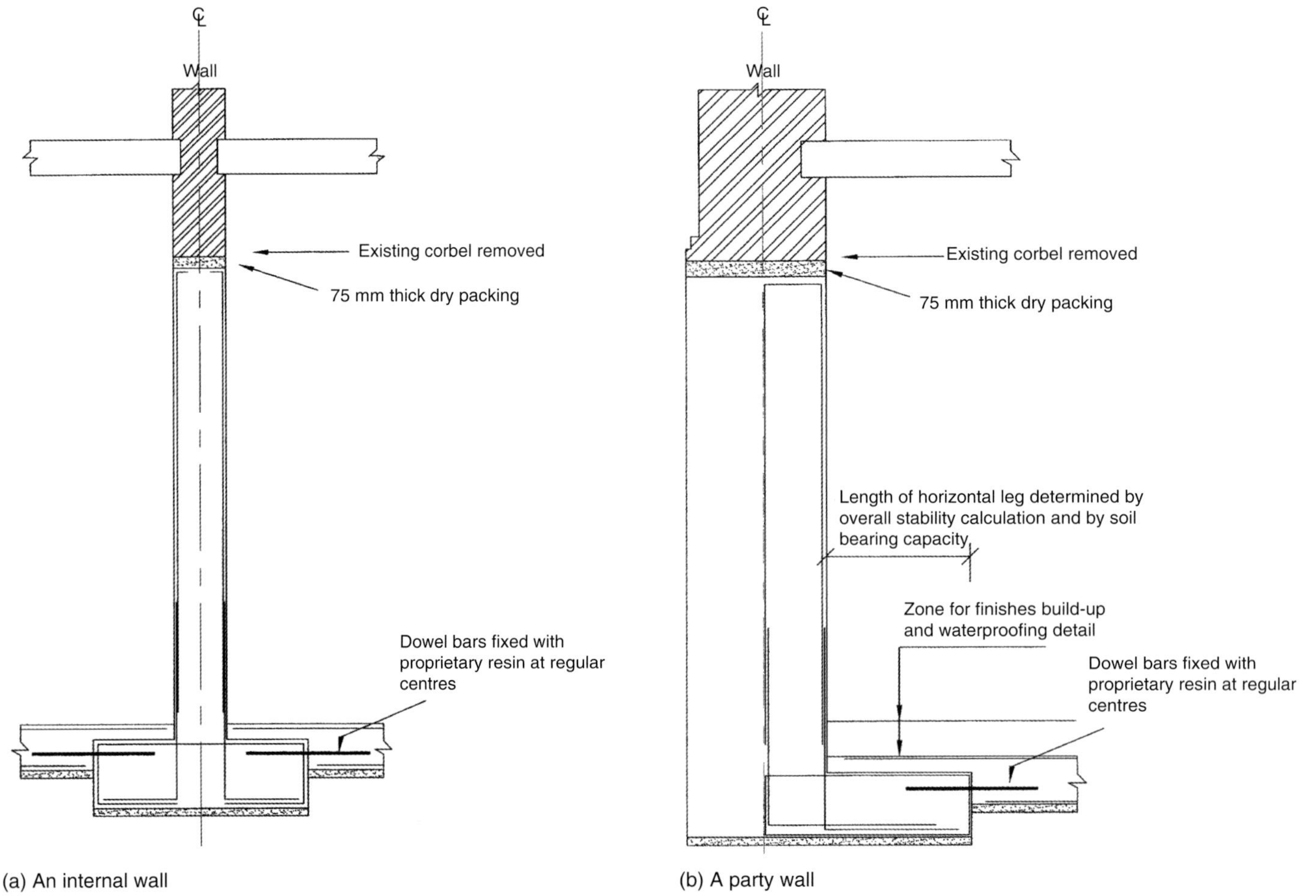

Figure 83.1 Underpinning detail for (a) an internal wall; (b) a party wall

entirely within the building. A raft slab is formed to support the external walls by cantilevering out from piles placed just inside the building. In this way, external drainage and other services around the building are not affected by the piles. Internal walls are supported by piles placed either side. Sacrificial stools are used to temporarily prop the walls during installation of the reinforced concrete raft. Typically, an entire building would be supported on a raft foundation formed from, say, three pours. In this way, stability to the building is ensured at all times.

For partial underpinning, a 'needle pile' system can be used. A reinforced concrete beam is formed under the wall or, preferably, within the depth of the existing foundation. If conditions permit, piles are placed each side of the wall at regular centres (say, 3 m), with reinforced concrete needle beams connecting them together. **Figure 83.2** illustrates this technique,

Proposed beam spanning between needle

Stools placed in pairs and jacked into position

300 mm dia. piles at needle locations

Completed beams and needles prior to dry-packing

Finished works ready for landscaping

Figure 83.2 Piled underpinning, St John's Church, Wembley used at St John's Church, Wembley. The brief was to underpin the external walls in order to overcome the effects of differential settlement caused by clay shrinkage within the prevailing London Clay soil. Sacrificial stools were used to form a reinforced concrete beam within the depth of the existing masonry structure, and needle beams connected through to bear on piles installed either side of the wall at regular centres.

When access to the interior of the building is not possible or is not preferred, then a variation of the needle system described above can be used. Pairs of piles are placed on the external side of the building to act as a tension-and-compression couple.

Alternatively, a single pile can be used on the external side of the building. In this case, the pile is designed to support the vertical load but is combined with a design bending moment resulting from the lever arm between the pile centre and that of the wall. A reinforced concrete beam can be placed under the wall and extended at regular centres to form a series of pile caps. Depending on the loading conditions, the centres of the piles can be closed up so that the individual pile caps penetrate into the wall to form a series of castellations on a hit and miss basis.

A system of underpinning which avoids the requirement for a reinforced concrete beam using sacrificial stools is called 'Pali Radice', meaning 'root piles'. Typically a vertical pile is formed externally. This pile is coupled with another pile drilled at an angle through the existing foundation of the building. The two piles are designed to act in unison to support the building.

Irrespective of the type of piling used, it is essential to ensure that there is sufficient bond between the old structure and the new pile to get the required load into the new pile. This is particularly important for the Pali Radice system as, unlike the other examples, there is no transition beam or slab to connect the existing structure to the piles. For the Pali Radice system each proposal would need to be assessed on its merits and a decision taken on whether sufficient bond can be achieved between the new pile and the old structure.

For the other cases, it is important to maintain adequate load transfer from the pile into the slab or beam system. The steel reinforcement protruding from the pile should achieve a full bond into the reinforced concrete support system and, where appropriate, a check should be made that the pile does not tend to punch through the slab through excessive load concentration.

83.2.5 Beam and pad underpinning

For subsidence-related underpinning the sacrificial stool system can be used to create a reinforced concrete beam below or within the depth of an existing foundation, as described above. Isolated mass concrete pads are then formed at regular centres (say, 3 m). This form of underpinning is often more economic to build than the equivalent mass concrete system. This is especially so if, due to the prevailing ground conditions, the bases are anticipated to be formed at considerable depth.

Reference should be made to the comments in Section 83.2.1 in relation to the bearing capacity of the soil below the isolated underpinning bases.

83.2.6 Jacked underpinning

Where a building has settled differentially it is often possible to jack the structure back to eradicate or at least reduce the difference in levels. These are usually independent solutions appropriate to each case, but as a general rule the principle involves forming a solid base from which to jack and to provide a 'cradle' under the building in order to contain it during the jacking operation. The cradle will then be used to support the building once the jacking is complete.

In the simple case where an extension has rotated downwards and away from the main building, it is important to provide a horizontal tie at foundation level. This is to allow the extension to rotate upwards during jacking but to prevent it from moving outwards at the base.

In all cases the jacking will only be effective if the existing cracks in the fabric of the structure are clear and free of previous repair or debris. If the building is allowed to 'pinch' on any material within the cracks, then it cannot rotate back to its original position.

83.3. Factors influencing the choice of underpinning type

83.3.1 Reason for carrying out the underpinning works

Where the underpinning is not scheduled to be used to retain soil – for example, to counter the effect of subsidence – then the choice of underpinning type can be determined by consideration of the most economical engineering solution given any access restrictions and taking into account the prevailing ground conditions. Where the underpinning is proposed to be used to retain soil, then mass concrete or reinforced concrete bases become applicable. The use of mass concrete or reinforced concrete will then depend on such aspects of a project as the requirements for space creation, and the design conditions relating to the ability for the wall to act as an independent gravity retaining wall without an undue amount of propping.

83.3.2 Study of the site investigation

Save for the most basic of requirements, an appropriate site investigation is essential to make an informed choice regarding the type and depth of underpinning. For example, clay soils may have suffered the effect of shrinkage through tree root activity. In certain circumstances, it is more economical to provide piled underpinning as an alternative to traditional underpinning. If the site investigation shows the soil to be desiccated to a depth in excess of 2.5–3 m, depending on the prevailing circumstances, it may well be more economical to consider a piled scheme.

If a clay soil is encountered, hand- or machine-excavated underpinning is relatively easily installed as the shoring can be safely placed at intervals, leaving sections of exposed ground between the props. On the other hand, if loose uniformly graded sand is encountered, then hand or machine excavation becomes much more difficult and therefore expensive as it is likely that a form of close boarded shoring will be required to make the excavation safe. For this form of shoring, horizontal and vertical trench sheeting is often used to fully retain the soil.

The site investigation data should be used as a means to establish the founding depth of the proposed underpinning. In certain cases – for example, dense gravel over soft clay – the ground upon which the existing foundations are formed can have a higher bearing capacity than that of the ground below. This will influence the depth and width of mass or reinforced underpinning.

For ground conditions such as peat or where there is no stable ground to be encountered at reliable depth, a piled solution should be considered. Where the water table is above the proposed founding level for the underpinning, a piled solution should be considered wherever possible.

When commissioning a site investigation, prior knowledge should be gained of the type of ground likely to be encountered. This will enable an initial assessment to be made of the type of underpinning likely to be required. The investigation can then be commissioned to provide adequate information to enable a detailed and informed design to be carried out. A 5 m deep borehole may be sufficient for excavated underpinning but will not be adequate with respect to a piled solution. Standard penetration testing (SPT) should be undertaken in granular soils, with U100 samples wherever possible in clay soils. The level and inflow rate of any water ingress should be noted and recorded. The final level of the water table should be noted and a piezometer should be installed in order to monitor any fluctuations in the level with time.

Soil suction tests should be undertaken when investigating the requirement for underpinning in the case of subsidence. Specialist advice should be sought from an appropriate ground investigation company to ascertain the type and degree of testing to be carried out.

83.3.3 Settlement potential

In general a piled underpinning solution will settle less than an excavated solution. If, however, due to the required function of the underpinning, an excavated technique must be used, then the likely settlement is an important factor for consideration.

It is possible to provide hydraulic jacking to alternate bases in order to pre-consolidate the soil as the bases are installed. In some cases, small-diameter steel-cased piles can be placed in the excavated base prior to pouring concrete. This will create a load-sharing mechanism to minimise the effect of settlement.

If, for example, a dense granular soil underlies soft clays or loose granular materials, it may be effective to deepen the underpinning over and above the required depth in order to found on the material having a lesser potential for settlement.

In order to avoid the effect of multi-stage settlement at each phase of underpinning, single-stage underpinning should be considered wherever possible.

83.3.4 Gravity retaining wall design

For basement design the underpinning should be designed as a gravity retaining wall in terms of overall performance. A second analysis should then be undertaken to check its performance as a masonry wall or alternatively as a reinforced wall in bending and shear.

83.3.5 Party wall matters

If possible, it is advisable to open discussion with party wall surveys and engineers at an early stage. This will minimise the risk of abortive design work and delay to the progress of the project as a whole. Reinforced concrete underpinning proposed along the party wall line may be considered as a 'special foundation' and can cause difficulty in achieving an award. The main principle, however, is not to encroach into the space below the plan area of the adjacent property and the choice of the underpinning system must reflect this requirement.

83.4. Bearing capacity of underpinning and adjacent footings

83.4.1 Bearing capacity theory

The shearing resistance of a foundation may be divided into three parts for ease of explanation (see **Figure 83.3**)

- frictional resistance resulting from the surcharge due to the depth of overburden
- frictional resistance resulting from the weight of soil within the rupture zone BCD
- cohesive resistance along the line ABC.

83.4.2 Excavation adjacent to an existing footing

Reference to **Figure 83.3** shows the effect of trench excavation to the level of an existing footing. As the depth of overburden has an influence on the bearing capacity of the soil below the existing foundation, its removal can only serve to reduce the

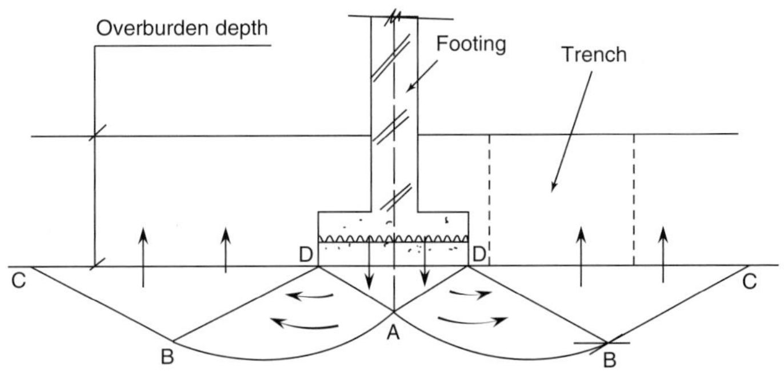

Figure 83.3 Failure of a foundation through inadequate shear resistance (Scott, 1980)

overall bearing capacity. This is especially so for granular soil as there is also no appreciable cohesive resistance along the rupture zone ABC shown in the diagram, and the ground is much more reliant on the presence of an overburden to achieve adequate bearing capacity.

83.4.3 Bearing capacity of the proposed underpinning

The mode of failure described above is a catastrophic collapse of the soil beneath the foundation through inadequate shear strength. Failures of this type are not very common, but when they do occur can lead to large distortions of the structure above. A more common mode of failure is through excessive settlement of the foundation.

For sands and gravels, the settlement takes place almost immediately and the cycle is usually complete by the end of the construction programme. This is because the settlement is caused through *compaction* of the soil through the increased load applied by the new foundation.

For clay soils, the settlement is caused through *consolidation* rather than compaction. In this case, settlement occurs in diminishing amounts with time and can continue for a significant part of the life of the structure. The settlement comes by consolidation of the particles in the tight soil matrix. This process is delayed, however, by gradual dissipation of the excess pore water pressure caused through loading the soil with the weight of the new foundation.

When underpinning in a clay soil it is therefore important to consider that the soil upon which the new foundation sits will settle with time by the process described above. It is advisable to leave a newly underpinned structure as long as possible to allow the greater part of the settlement process to take place before applying the final decorative finishes.

Any new foundation should have a calculable capacity and appropriate factor of safety. Wherever possible a site investigation should be carried out. Advice can be sought through the site investigation company on the most appropriate form of foundation to use, along with information on allowable bearing capacity. It is of course necessary to brief the site investigation company beforehand on the information required for a particular project so that they may carry out the appropriate testing with samples gained at the time of the investigation.

83.5. Shoring

83.5.1 Types of shoring

Timber, steel or precast concrete are often used. For a typical underpinning base the shoring supporting the face behind the wall will become sacrificial. For this reason, it is advisable to use steel or concrete to support this face, even if timber is used elsewhere.

83.5.2 Design of shoring

Safety for installation is a key factor in the design of shoring. While precast concrete planks or lintels are relatively low in cost and readily available, they are heavy to lift and are prone to progressive collapse when placed horizontally over the full height of an underpinning face. Steel trench sheets are equally robust but do not carry with them the same degree of safety concern.

For some soils, intermittent horizontal propping may be sufficient; for other soils – for example, loose sand – full sheeting may be required. The size (and type) of props will be designed in accordance with the calculated soil pressure due to the overburden and any surcharging that may need to be taken into account. **Figure 83.4** shows an example of underpinning carried out in loose sands and gravels.

(a) Excavation (b) Blinding (c) Steel fixing (d) Cast the base (e) First wall pour (f) Second wall pour

Figure 83.4 Illustration of reinforced concrete underpinning in sands and gravel

83.5.3 Minimising the risk of settlement

Any shoring technique will result in varying degrees of voiding behind the props due to the lack of fit and local decay of the soil face prior to concreting. Consideration should be given to dry-packing behind the face of the shoring. For basement construction, the shoring to each underpinning base can be left in place prior to excavation of the entire basement. Depending on the soil type, retaining the shoring can be more effective in reducing the risk of settlement than backfilling each base, as the excavated material cannot always be readily compacted.

83.6. Underpinning in sands and gravel

Figure 83.4 shows the construction techniques used to create reinforced underpinning bases in a single stage to a depth of 6 m below ground level.

83.7. Dealing with groundwater

83.7.1 Pre-grouting

Figure 83.5 shows a technique for consolidating a granular soil prior to underpinning below the water table. The method relies on the interstices between the soil particles being capable of being filled with an injected material. In many cases, the interstices are too fine to permit cementitious grout to be used and therefore a chemical is required. Often a gel is used which is biodegradable so that the ground below adjacent properties is not affected in the long term.

This form of prevention for water ingress during underpinning is by no means guaranteed. The majority of injected material is excavated when carrying out the underpinning and therefore the most densely compacted areas are removed. The diagrams showing the angle of drilling indicate that the effectiveness of injection decreases with the distance from the drill location. This method of water prevention has been shown to be effective in certain locations, but careful advice from specialist contractors should be sought if considering this method for a particular application.

83.7.2 Well pointing

The system typically comprises a series of perforated plastic tubes placed into boreholes and surrounded with granular

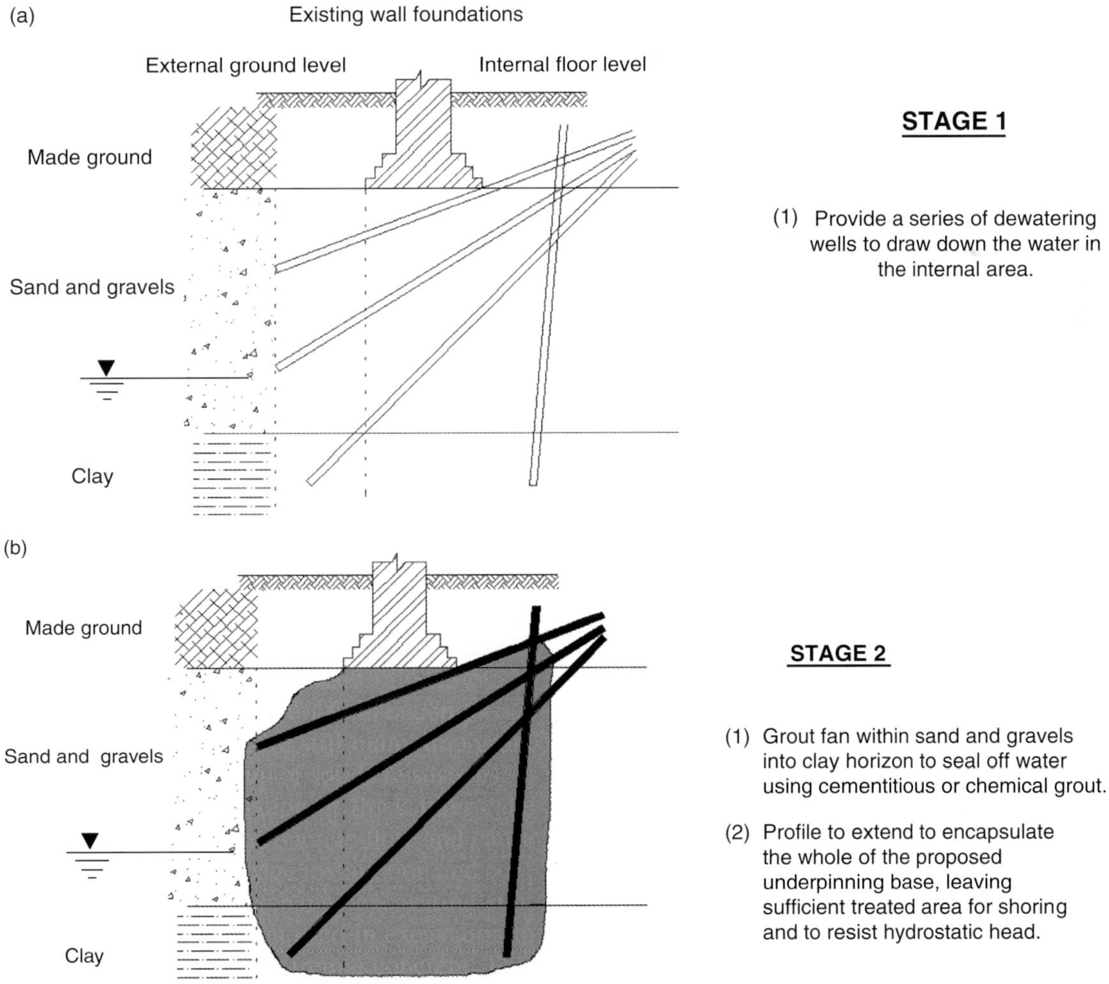

Figure 83.5 Illustration of the pre-grouting technique (continued on next page)

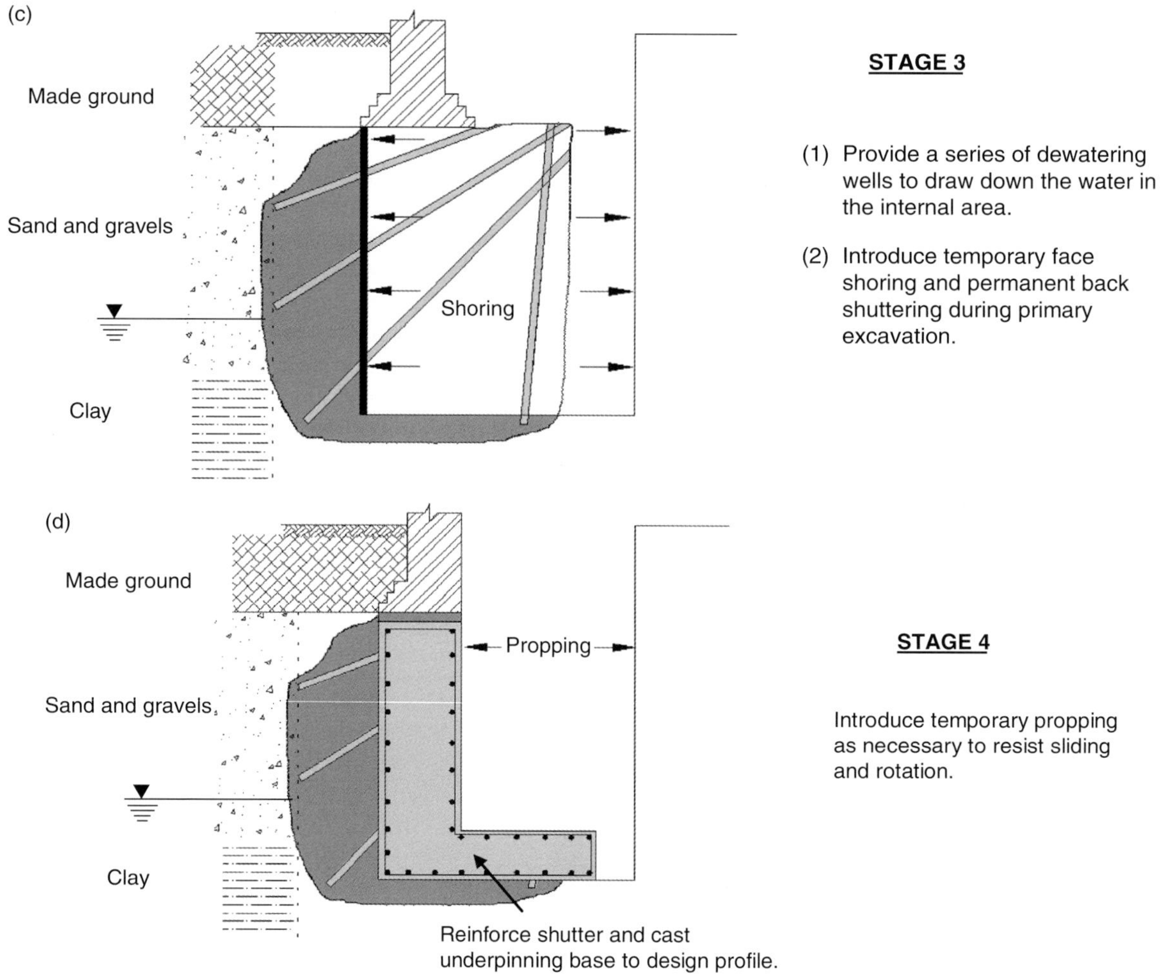

Figure 83.5 Continued

soil with filter membrane. As water drains into the tubing it is pumped out, thus lowering the water table locally. Depending on the soil type and the flow characteristics through it, the water table can be locally drawn down to permit the underpinning to be carried out in relatively dry conditions.

Lowering the water table can create an increased possibility for settlement of the adjacent foundations. The movement of water through the ground can also cause settlement through migration of fines. In the right circumstances, however, local dewatering through well pointing can be a very effective way of enabling underpinning works to be carried out below the water table.

Consideration should be given to where the water is to be discharged. If it is to be discharged into drains, then permission should be sought from the local authority. Water which is discharged locally and allowed to soak into the ground will eventually find its way back into the system.

83.7.3 Local pumping

Depending on the nature of water inflow, local pumping can be used to enable an underpinning base to be formed in wet conditions. The use of local pumping should, however, be restricted to conditions which are safe in terms of the personnel involved in carrying out the work and in terms of potential settlement to the surrounding structures. If seepage is allowed to continue unabated during the progress of the works to install the base then fines can be drawn into the excavation. This will compromise personnel safety and will increase the possibility of settlement occurring in the adjacent structures.

If the underpinning is required to penetrate the groundwater table, then depending on the depth of penetration and the prevailing ground conditions local pumping can be effectively used. The point discussed in Section 83.7.2 with regard to discharge of the pumped water is relevant here. For local pumping below the water table, it is usually difficult to control the amount of fines being pumped away along with the removal of water. The potential for settlement as a result of the pumping operation is correspondingly increased.

For both well pointing and local pumping the removal of fines within the soil will cause settlement to adjacent structures. The use of filters on the pumps is essential to minimise the possibility of this occurring.

83.8. Underpinning in relation to subsidence settlement

83.8.1 Choice of underpinning scheme

If the underpinning is not generally to be exposed or used as a soil-retaining structure, the choice of underpinning is governed by the requirement to satisfy the design brief while achieving the best economy of construction. While a performance specification or design brief can be developed by a consulting practice or surveyor, the actual choice of underpinning scheme may be left to the contractor's experience and specialist skills to conceive the best system in view of the prevailing ground conditions, access situation, safety aspects and overall economy.

83.8.2 Association of Specialist Underpinning Contractors

The Association of Specialist Underpinning Contractors (ASUC) is an organisation which has influence on activities related to subsidence settlement and other aspects of underpinning. Information concerning its initiatives, documentation and a list of approved contractors can be found on its website (www.asuc.org.uk).

83.8.3 Other subsidence-related stabilisation methods

In conjunction with underpinning works related to subsidence, a programme of crack repair and refurbishment is usually undertaken. Crack repair can take the form of resin injection, 'brick stitching' by cutting out and replacement of the cracked bricks or insertion of steel reinforcing bars into the horizontal joints.

When the cause of the subsidence settlement is clearly defined by the presence of a tree or trees then, depending on the prevailing circumstances, these can be removed or pollarded. This can have the result of halting or even reversing the effects of subsidence settlement. A programme for tree removal/pollarding with subsequent monitoring is often undertaken prior to making a decision to carry out underpinning works.

If tree root activity is the cause of subsidence, it is often important to implement and maintain a programme of continual tree maintenance to ensure that once the underpinning has been carried out further settlement does not take place. Continual tree maintenance is carried out with the objective of containing the level of root activity to its current state. In this way the shrinkable clay neither shrinks further nor recovers moisture, with resultant ground heave.

83.8.4 Extent of underpinning

For subsidence-related underpinning, the question of whether to underpin at all or how much of the structure to underpin is always a difficult one. In order to find an appropriate solution it will usually be necessary to monitor the structure to establish the pattern of movement or indeed to establish whether the movement is still active and ongoing.

Study of the cracking pattern alone can be deceptive as it only indicates the degree of differential settlement. The whole structure could be settling but at different rates. Partially underpinning the worst area will serve to 'lock' this part of the building but then cracking may continue as the rest of the building settles further.

If, for economic reasons or for any other reason, it is not possible to underpin the whole of the property, then appropriate measures should be taken to ensure that an adequate transition is made between the underpinned and non-underpinned structure. This will of course depend on the type of underpinning chosen for the main structure. Typically this would involve stepped underpinning, reducing the depth of the bases upwards towards the original foundation. To work effectively, however, each transition base should be separated from the adjacent one with polystyrene strips or similar to permit independent movement to take place. Otherwise the transition bases will tend to act as a corbel and therefore will not permit independent movement.

83.8.5 Underpinning of piers

Where existing piers or walls with concentrated loads are required to be underpinned, the additional load must be taken into consideration during the design and construction stage. The sequence and method of supporting the pier must be established prior to excavation.

83.8.6 Contractual matters in relation to subsidence settlement

A contract in relation to subsidence settlement will probably contain clauses concerning the following points.

(*a*) A statement of the depth of underpinning/piling allowed and the practical/financial arrangements should the depth vary upwards or downwards.
(*b*) For partial underpinning the extent of underpinning should be contained within the contract, a statement regarding any transition sections and arrangements concerning responsibility for the sections of the building not currently scheduled to be underpinned.
(*c*) The extent of the design brief. Is the specialist contractor expected to design a total solution regarding the type and extent of underpinning required, taking into account the damage that can be observed? Will the contractor take responsibility for variation of ground conditions? It will be necessary to accurately define the responsibility for both of these questions.
(*d*) An appropriate form of guarantee/warranty will usually be required. ASUC provides an insurance-backed guarantee, more details of which can be found on their website.

83.9. Safety aspects of underpinning

Underpinning works are considered by the Health and Safety Executive (HSE) as high risk, and there is a requirement to produce method statements and risk assessments. The historical rule that only excavations over 1.2 m deep require shoring has been changed to include all excavations, based on risk assessment.

HSE (2006) contains a chapter on groundworks. The publication warns of the danger of excavation in unsupported trenches and states the following:

> 341 Before digging any trenches, pits, tunnels, or other excavations, decide what temporary support will be required and plan the precautions that are going to be taken against:
>
> - collapse of sides;
> - people and vehicles falling into the excavations;
> - materials falling onto people working in the excavation;
> - undermining nearby structures;
> - underground and overhead services; and
> - the inflow of ground and serviced water.

Sections 338–381 provide information on safe practice with respect to the above subject headings.

Save for relatively shallow underpinning, the Confined Spaces Regulations 1997 will usually apply. As there is a risk of personnel falling into the excavations, the Work at Height Regulations 2005 apply, especially with regard to the provision of handrails around all excavations, irrespective of depth.

Under the current Construction (Design and Management) Regulations (CDM Regulations 2015), it is required that clients provide accurate information to enable the structural stability of the building and those adjacent to be assessed before work begins. A health and safety plan should be produced.

83.10. Financial aspects

As a general statement, deep hand-excavated underpinning is expensive; where appropriate, consideration should be given to a piled scheme. Piles can be considered as a total solution in the case of Pali Radice or more often in conjunction with a reinforced concrete beam/slab arrangement.

Reinforced underpinning is of course relatively more expensive to install than a mass concrete equivalent, but other considerations should be taken into account such as temporary works support and overall space creation. In general, reinforced underpinning will occupy less space than the equivalent mass concrete system with reinforced lining walls.

In consideration of space creation construction, tolerance should be allowed in the calculation. Due to the nature of construction, it is difficult to achieve tight construction tolerance both in terms of verticality and precise formation level.

83.11. Conclusion

The process for finding the most appropriate form of underpinning to suit a particular set of circumstances is involved and can become complex. Specialist advice should be sought at all stages of the process.

It is particularly important to produce a site investigation report that provides information in order to enable an informed and rational decision to be made on the correct type of underpinning to suit a particular set of circumstances. Advice from both specialist underpinning companies and site investigation companies should again be taken.

References

HSE (1997) The Confined Spaces Regulations 1997. The Stationery Office, London, UK.
HSE (2005) The Work at Height Regulations 2005. The Stationery Office, London, UK.
HSE (2006) *Health and Safety in Construction*, 3rd edn. HSE Books, London, UK.
HSE (2015) The Construction (Design and Management) Regulations 2015. The Stationery Office, London, UK.
Scott CR (1980) *An Introduction to Soil Mechanics and Foundations*. Spon, Abingdon, UK.

Useful websites

ASUC: www.asuc.org.uk
Geostructural Solutions: www.geostructuralsolutions.co.uk
GSS Piling Ltd: www.gsspiling.co.uk
HSE: www.hse.gov.uk
Oxford Hydrotechnics: www.h2ox.net

It is recommended this chapter is read in conjunction with

- Chapter 26 *Building response to ground movements and tunnelling*
- Section 5 Design of foundations

All chapters in this book rely on the guidance in Sections 1 Context and 2 Fundamental principles. A sound knowledge of ground investigation is required for all geotechnical works, as set out in Section 4 Site investigation.

doi: 10.1680/icemge.66830.1361

CONTENTS

Chapter 84
Ground improvement

Georg Herrmann Keller Limited, Ryton-on-Dunsmore, UK
Darren Milo Balfour Beatty Ground Engineering, Raynesway, UK
Colin Serridge Tritech Ground Engineering Ltd, UK and Edge Hill University, UK

Ground improvement techniques can often be used as an economical alternative to piled and deep foundation solutions for a wide range of made ground, fill materials and natural soils to support houses, offices, industrial units, tanks, road embankments and similar low-rise developments. Techniques used include in situ compaction of clean sands using depth vibrators, adding stone or concrete during compaction to form vibro stone or vibro concrete columns and dynamic compaction.

As the soil conditions have a large influence on the result, ground improvement techniques require an appropriate level of site and ground investigation to permit satisfactory geotechnical characterisation of the soil profile. Ground improvement also requires an appropriate level of understanding of where the differing techniques work and how to ensure correct and appropriate application. Quality control and monitoring procedures during execution of ground improvement techniques are essential to ensure successful implementation and performance.

Ground improvement techniques permit the adoption of relatively simple shallow foundations and groundbearing warehouse floor slabs. They can also provide significant sustainability advantages in comparison to more traditional deep foundation methods.

84.1. Introduction

Ground improvement involves the enhancement of ground properties, principally through the use of strengthening or stiffening processes and compaction or densification mechanisms, to achieve a specific geotechnical performance.

Of the various ground improvement methods available, the most commonly used are vibro techniques employing vibroflots (also referred to as vibrating pokers or depth vibrators) and dynamic methods incorporating the dropping of heavy steel tampers or weights on predetermined grid patterns. These form the main focus of the discussion in this chapter.

It is important that ground engineering practitioners involved with the design, procurement and supervision of ground improvement techniques have an appropriate level of understanding of the various treatment methods being considered, the likely benefits, including selection of the most appropriate technique, and any drawbacks or limitations of the technique selected for the prevailing combination of the ground conditions and applications being considered.

Choice of method will be influenced by the geotechnical requirements and engineering performance objectives of the ground improvement. Appropriate site supervision, quality control and monitoring procedures during execution of ground improvement techniques are essential to ensure successful implementation, as is an appropriate level of post-treatment testing to ensure satisfactory performance will be achieved.

Design principles for these techniques are covered elsewhere (Chapter 59 *Design principles for ground improvement*). Further guidance on these and a wider range of ground improvement techniques can be found in the CIRIA guides and Building Research Establishment (BRE) specifications referenced in this chapter. The ground improvement techniques of grouting and soil mixing are covered in Chapter 90 *Geotechnical grouting and soil mixing.*

Where any doubts or uncertainties exist, specialist ground improvement contractors should always be consulted. Such advice is normally free of charge, although advance trials may sometimes be recommended.

84.2. Vibro techniques (vibrocompaction and vibro stone columns)

84.2.1 Introduction and history

The modern origins of vibro techniques were conceived in Germany in the mid-1930s. Vibrocompaction was the first technique to be developed, with its first application being for the in situ densification of loose sands for a government building in Berlin in 1937. Further development of the technique proceeded in parallel in Germany and the USA in the 1940s. The development of the vibro stone column method in the 1950s permitted the application of vibro techniques to a much wider range of soil types, most notably finer-grained soils (cohesive and mixed cohesive and granular soils), thus increasing the range of soils which could be treated by vibro techniques. Vibro stone column techniques were introduced into Great Britain and France in the late 1950s and have been used extensively worldwide.

84.2.2 Essential features of the vibroflot/vibrator equipment

The principal piece of equipment used to carry out vibro ground improvement techniques (whether for in situ densification or construction of stone columns) is the vibroflot (also referred to as a vibrating poker or depth vibrator), which is either suspended from a crawler crane or mounted on a leader attached to a base machine, dependent upon the specific application. The essential features of the vibroflot are presented in **Figure 84.1**. The vibroflot equipment comprises three main components: the vibrator, the isolator and the extension tubes. The main component used to achieve compaction is the vibrator. Vibrations are produced in the lower part of the vibrator. These are induced by a series of rotating internal eccentric weights mounted on a shaft driven by a hydraulically or electrically powered motor located in the upper part of the vibrator casing. The rotating eccentric weights within the vibrator impart to the body of the vibrator a gyratory motion in a horizontal plane. Follower or extension tubes of similar or smaller diameter to the vibrator unit are attached to permit treatment of soils to varying depths. An elastic coupling isolates the vibration from the extension tubes and therefore prevents vibration travelling up the extension tubes to the supporting crane or base machines. Stabilising fins are often added to provide stability of the vibroflot in the bore to prevent rotation due to the application of torque when the vibroflot is working hard in the ground. Otherwise, such rotations could lead to twisting or snagging of any external hydraulic or electric cables.

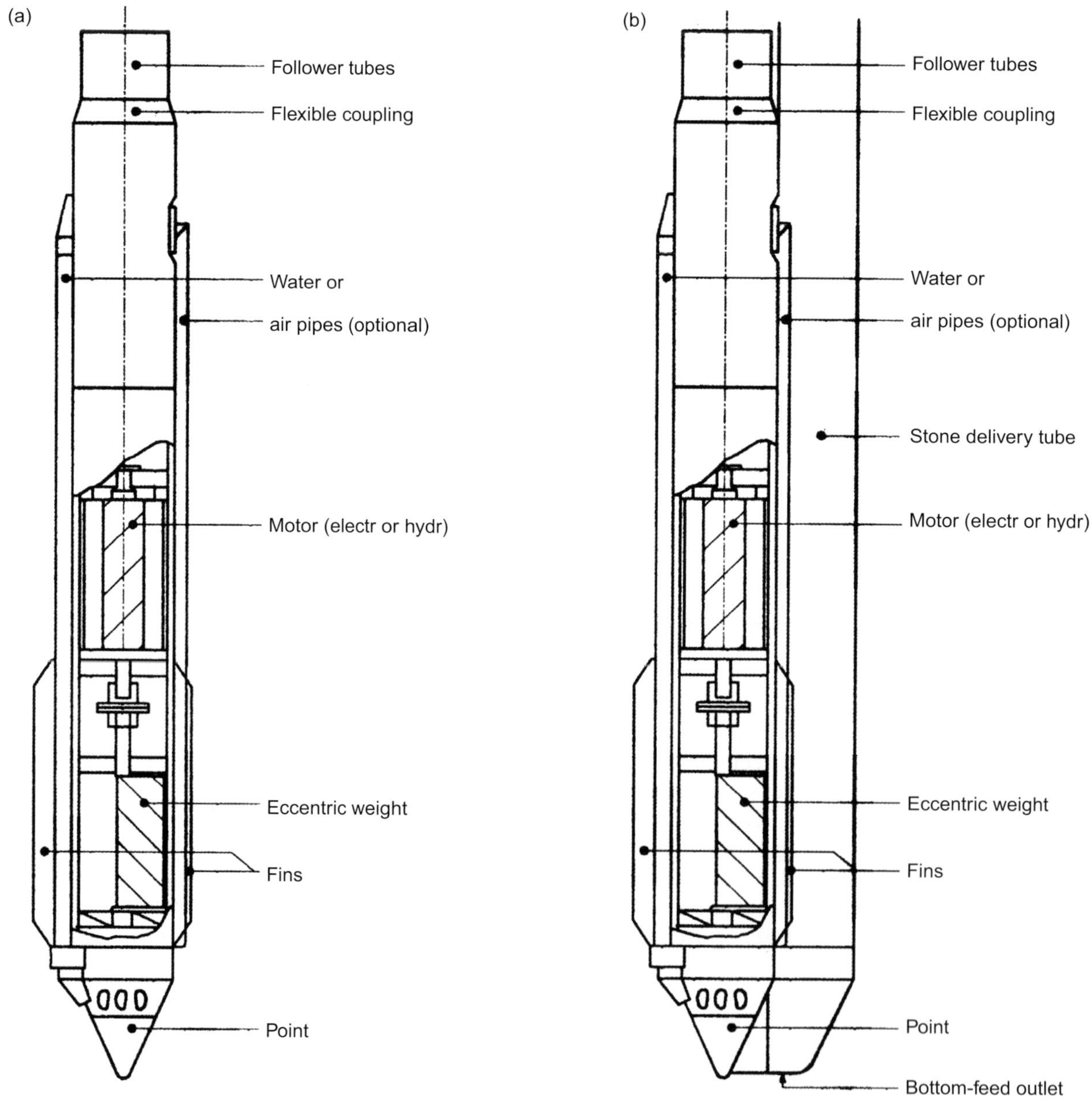

Figure 84.1 (a) In situ densification/top-feed and (b) bottom-feed vibroflot (vibrating poker) units (Reproduced from Slocombe *et al.*, 2000; all rights reserved)

While the basic components of the equipment have changed very little over the years, there have been significant developments in the reliability (with extended lifespans and reduced maintenance) and power ratings of the equipment, with the objective of achieving greater efficiency in densification and stone column production.

84.2.3 Applicability of vibro techniques

The vibroflot may be used for in situ compaction of cohesionless soils. Stone columns are constructed in both cohesionless and fine-grained soils. The vibro techniques are applicable to a range of soils, as shown in **Figure 84.2**. The actual mechanism of improvement is a function of whether the soils are essentially granular (coarse-grained) or cohesive (fine-grained).

It is generally recognised that in situ vibrocompaction is appropriate for granular soils with a total fines (particles finer than 0. 06 mm) content of up to 15%, of which the clay and fine silt content (particles smaller than 5 μm) should be less than 2% (Slocombe *et al.*, 2000).

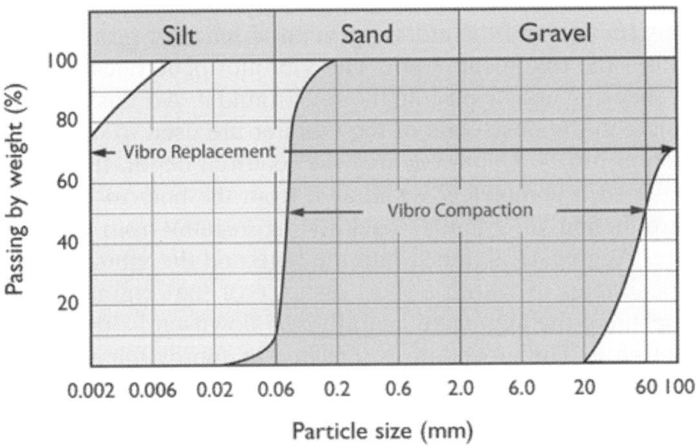

Figure 84.2 Range of soils treatable by vibrocompaction and vibro stone column techniques

84.2.3.1 Granular (coarse-grained) soils

The principle of vibrocompaction in granular soils is based upon particles of coarse-grained (non-cohesive) soil being rearranged into a denser state by means of dominantly horizontal vibration from the vibroflot. The resultant reduced void ratio and compressibility and corresponding increased angle of shearing resistance then permit the adoption of higher imposed design loadings at lesser settlement and increased seismic resistance.

84.2.3.2 Cohesive (fine-grained) soils

With fine-grained soils, the cohesion between the soil particles dampens the vibrations and prevents rearrangement and compaction occurring. Improvement is therefore achieved by 'reinforcing' the soil with relatively rigid stone columns. During the stone column construction, the coarse granular column material is pressed radially into the soil so that it is displaced beyond the diameter of the vibrator. The column of compacted dense granular material forms, together with the surrounding soil, a composite stone column – soil mass, with enhanced shear strength and bearing capacity, together with a corresponding reduction in settlements (attributed to the 'stiffening' effect of the stone columns). Stone columns in fine-grained soils also assist in the dissipation of excess pore water pressure under applied load or surcharge, which accelerates the consolidation process, together with providing enhanced slope stability parameters.

84.2.4 Vibrocompaction

For the vibrocompaction technique (**Figure 84.3**; see also **Figure 84.7(a)**), the vibroflot unit is positioned over the required compaction point, high-pressure water jets in the vibrator nose cone are activated and the vibroflot is slowly lowered into the ground assisted by a combination of the weight of the vibroflot and the jetting action of the water jets. In the immediate vicinity of the vibrator, the saturated soil

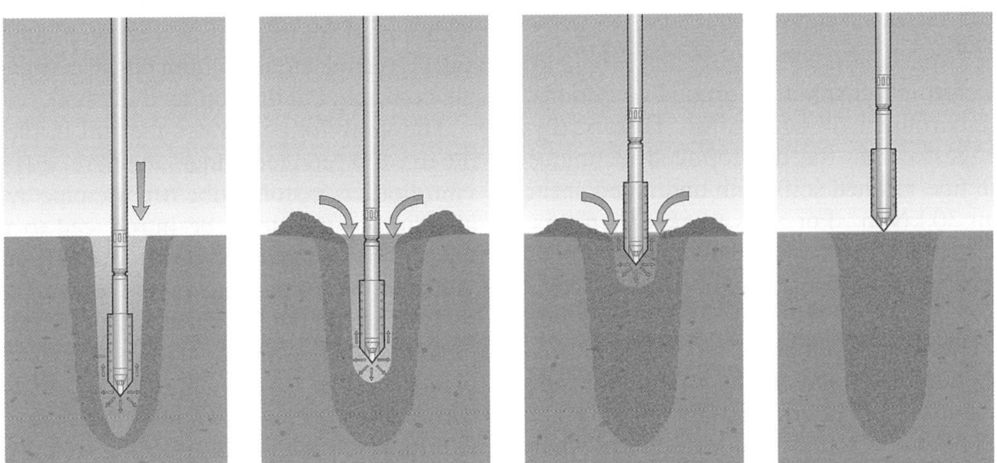

Figure 84.3 In situ vibrocompaction (sequence)

liquefies locally and temporarily under the influence of the vibrations. The vibroflot proceeds to sink rapidly into the soil under its own weight. When the required depth of penetration is reached, the vibrator is halted and the water supply reduced (or sometimes stopped). Under the action of the vibrations, the granular soil particles surrounding the vibrator are rearranged into a denser state. The effect of this compaction becomes visible at the ground surface in the form of a cone-shaped depression which is continually infilled with sand during the vibrocompaction process. When the required degree of compaction has been achieved (normally by monitoring the power consumption) at the depth of the vibrator, it is raised a short distance (typically 300–500 mm) from the base of the bore and the cycle repeated. Insertion of the vibroflot on a grid pattern will enable overlapping of the compacted cylinders and produce a homogeneous mass of compacted soil. Preliminary trials with pre- and post-treatment penetration testing may be employed to optimise the compaction spacings.

Sands of single-size grading or with a higher fines content are more difficult to compact than well-graded sands, as are sands that occur in relatively thin layers trapped between layers of clay and which require particular attention. Carbonate sands are susceptible to particle crushing and typically require closer grid spacings than silica sands.

84.2.5 Vibro stone column terminology and techniques

Terminology for vibro stone column techniques has not always been applied in a consistent manner, and some attempt at addressing this can be found in now-archived BRE Document BR 391, *Specifying Vibro Stone Columns* (BRE 2000). This specification employs terms that reflect the fundamental principles of top-feed or bottom-feed to describe the method of stone aggregate supply, and wet or dry to describe the 'jetting' medium. This has given rise to the following terminology

- dry top-feed technique
- dry bottom-feed technique
- wet top-feed technique.

The technique adopted is a function of ground conditions, bore stability and environmental constraints. Historically, the literature has suggested that the dry top-feed technique should not be used in fine-grained soils with undrained shear strengths of less than 30 kN/m^2. For soils weaker than this, the dry bottom-feed technique (or historically the wet top-feed technique) must be used. A lower-bound undrained shear strength of 15 kN/m^2 has typically been used for vibro stone column applications. However, developments in vibroflot technology (together with monitoring and quality control systems) have allowed weaker soils to be treated in certain applications. Consideration also needs to be given to soil sensitivity when selecting the appropriate technique, as saturated sensitive soils can undergo significant remoulding when exposed to vibrations. For soft soils, settlement tends to be more of an issue than bearing capacity.

All three vibro stone column techniques described above use similar types of vibroflots, normally hydraulically or electrically driven, the main difference being the dry bottom-feed equipment which has a stone (aggregate) feed tube attached to the vibroflot to permit the stone aggregate to be introduced to the tip of the vibroflot without it having to be removed from the bore, thus overcoming any instability issues in weak ground or treatment to beneath the water table. The stone aggregate is typically handled by front-end loaders with a side-tip facility.

All the vibro stone column techniques are aimed at constructing well-compacted granular columns, typically of crushed stone or gravel aggregate in the void formed in the ground by the vibrating poker, thus improving bearing capacity and drainage while reducing total and differential settlements.

84.2.5.1 Dry top-feed technique

The dry top-feed technique can be used in a range of soil types, but relies on the bore remaining stable during column construction and therefore a general absence of groundwater within the treatment depth. The vibroflot penetrates the soil by shearing and displacing the soil around it. Air flush by way of jets in the nose cone of the vibrator are used to overcome suction forces. Upon reaching the required depth, the vibroflot is then completely withdrawn from the bore to facilitate introduction of granular backfill (aggregate) from the surface. Aggregate is tipped into the bore and the vibroflot then repenetrates to within a short distance of the original depth, displacing the aggregate laterally and downwards, thus compacting it. The procedure is repeated in approximately 0.5 m lifts until the stone column construction is completed to the surface (**Figure 84.4**; see also **Figure 84.7(b)**). Stone column diameters of about 0.45–0.60 m are typically achieved, dependent upon soil properties.

84.2.5.2 Dry bottom-feed technique

The dry bottom-feed technique is used in weak soil profiles with a high water table, where bore stability cannot be guaranteed during stone column construction with the adoption of the conventional dry top-feed method.

The vibroflot penetrates the soil in the same manner as in the dry top-feed technique, the main difference being that the vibroflot has a stone tube (tremie pipe system) attached to it, allowing aggregate to be introduced to the tip of the vibroflot without it having to be removed from the bore. Hence the vibroflot, by remaining in the ground, supports the potentially unstable bore. An airlock in the stone tube allows compressed air to assist discharge of aggregate. The vibroflot unit is mounted on leaders attached to a dedicated base machine which provides pulldown resistance during penetration of the vibroflot into the ground. During initial penetration the stone tube is charged with stone aggregate, and when the required

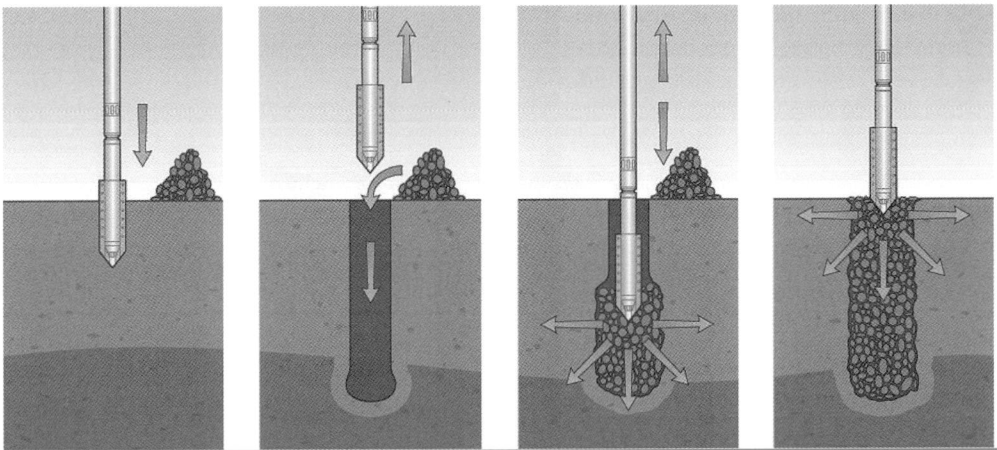

Figure 84.4 Installation sequence for dry top-feed vibro stone columns

depth is reached the vibroflot is withdrawn a short distance from the base of the bore and a charge of stone aggregate is discharged (with air pressure assistance) at the tip of the vibroflot, which is subsequently compacted on repenetration of the vibroflot. This cycle is repeated (in approximately 0.5 m lifts) until a compact stone column is constructed to the surface (**Figure 84.5**; see also **Figure 84.7(c)**).

84.2.5.3 Wet top-feed technique

The wet top-feed technique is employed in saturated fine-grained deposits and saturated granular soils deemed too silty for in situ vibrocompaction. The technique uses a similar technique to that used in vibrocompaction. When the required depth is reached, the vibroflot is sometimes surged up and down to flush out the bore. The vibroflot is then held a short distance off the bottom of the bore and the water pressure reduced sufficiently to allow a nominal outflow of water (and suspended fines) at the surface such that excess hydrostatic pressure and outward seepage forces support the uncased bore sufficiently long enough to form an annular space surrounding the vibroflot and permit construction of a stone column. A charge of aggregate is then introduced while the vibroflot is still in the bore, down the annulus against the continuing low pressure upflow of water. This is subsequently compacted by repenetration of the vibroflot into the aggregate. The vibroflot is then lifted sufficiently to allow introduction of further charges of aggregate, which are then compacted and the cycle repeated until a continuous column of compacted stone is formed to the surface (**Figure 84.6**; see also **Figure 84.7(d)**).

The wet top-feed technique requires consideration of provision of water supply, drainage ditches, settlement lagoons and contamination of the site surface with fines, as well as disposal of effluent.

While the dry bottom-feed technique has largely superseded the wet top-feed technique on environmental grounds (there is a high water demand and effluent is generated), there are still certain ground types (where contamination is not an issue) where the wet technique may be more suitable and often necessary to achieve satisfactory results. In coastal or estuarine environments where soft clays and alluvial soils persist, the wet technique may be preferred – for example, for bulk storage tanks or stockpiles of raw materials. The difference

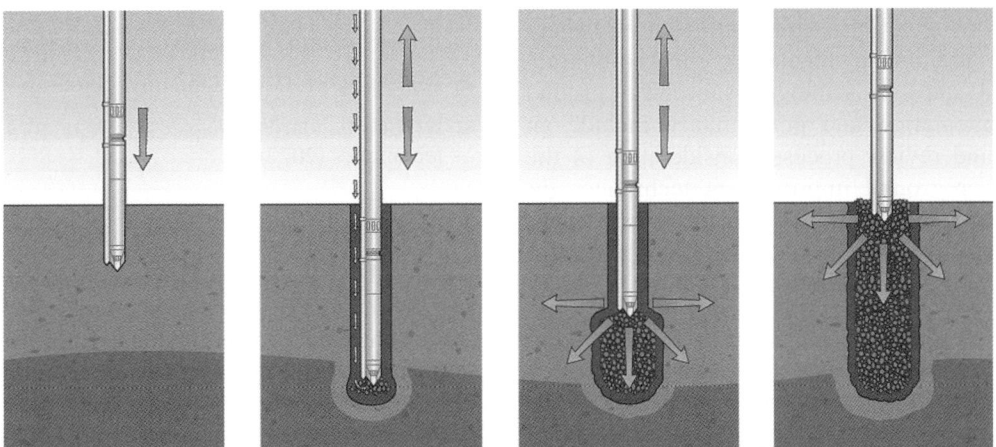

Figure 84.5 Installation sequence for dry bottom-feed vibro stone columns

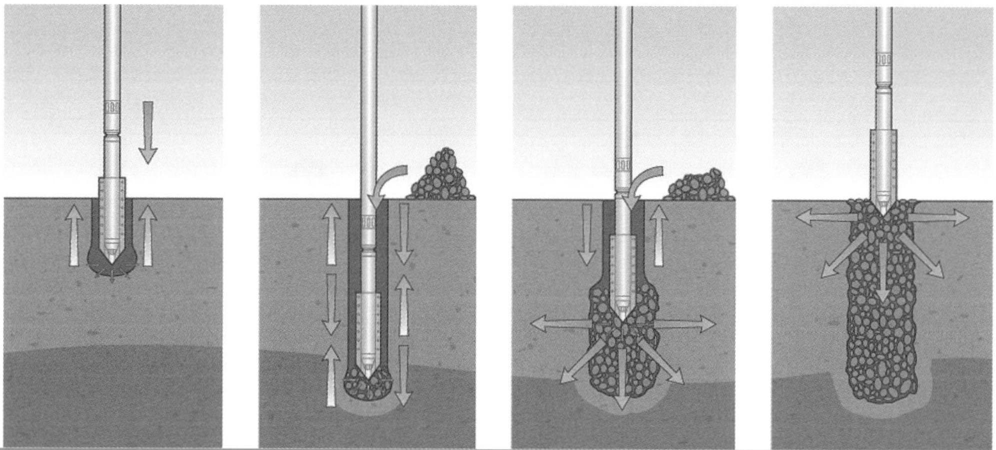

Figure 84.6 Installation sequence for wet top-feed vibro stone columns

in construction technique between the wet method and dry methods typically results in the diameter of the stone column formed by the dry method being smaller than that formed by the wet method.

Vibro techniques are increasingly being used in near-shore marine applications, working off barges or pontoons, sometimes using larger cranes reaching out from existing quays.

84.2.6 Soil geotechnical properties and characterisation

Because of the large influence of soil conditions on performance, ground improvement using vibro techniques requires a more extensive site and ground investigation programme compared with more conventional deep foundations solutions (Serridge, 2008). Site investigations are often designed to obtain soil parameters for deep foundation pile design. Ground improvement schemes often require greater detail and laboratory testing on the near-surface soils than is typically provided. The most common types of ground-related problems encountered concern soil strata boundaries (i.e. geometry not as anticipated) and the geotechnical properties and characterisation of the soil profile. Site and ground investigation requirements for vibro techniques are dealt with elsewhere (e.g. BRE 2000). In addition to satisfactory site geo-characterisation, important steps in achieving successful ground improvement implementation and performance include an appropriate design and review process, consideration of the interfacing with other ground improvement techniques (or deep foundation techniques), preliminary trials (where applicable), monitoring, quality control and testing. Management of geotechnical risk is discussed by Clayton (2001).

84.2.7 Vibro stone column design

Within the UK, design is typically based on either Priebe (1995) or Hughes and Withers (1974) for determination of stone column length. The guidance of Baumann and Bauer (1974) is frequently adopted for analysis of stress distribution between column and soil. Foundation settlement (without ground improvement) is first estimated using available geotechnical parameters and then applying appropriate settlement reduction factors (within the treated depth), according to Priebe (1995), to allow for the 'reinforcing' effect of the stone columns.

Although primarily applicable to embankment loadings over soft fine-grained/organic soils, it is important that secondary consolidation is considered, as in some instances it may contribute to a significant part of the total settlement.

Some vibro design principles are provided in Chapter 59 *Design principles for ground improvement*.

84.2.8 Specifications

Within the UK the main specification documents and guidance notes covering vibro stone column techniques include

- Institution of Civil Engineers (ICE), *Specification for Ground Treatment* (1987)
- BRE, *Specifying Vibro Stone Columns* (2000), BR 391 (archived)
- Charles and Watts (2002) *Treated Ground – Engineering Properties and Performance*, CIRIA Report C572
- BS EN 14731: Ground treatment by deep vibration (BSI, 2005)
- NHBC Standards, Chapter 4.5: Vibratory ground improvement techniques (2022).

84.2.9 Stone aggregate requirements

All the above-mentioned specifications convey a similar message with regard to stone aggregate requirements for vibro stone columns – that aggregate used in stone column construction should be sufficiently hard (i.e. must withstand the vibratory impact of the vibroflot); inert; free-draining (typically with less than 5% fines); and of appropriate grading, shape and angularity to form dense granular columns with high angles of

Figure 84.7 (a) Vibrocompaction technique ; (b) dry top-feed technique ; (c) dry bottom-feed technique; (d) wet top-feed technique (Courtesy of Keller (a, c) and Balfour Beatty Ground Engineering Ltd (b))

internal friction. Acceptable particle shapes for stone column material include rounded, angular and irregular. Unacceptable materials would be flaky and elongated.

Historically, a maximum aggregate crushing value (ACV) (BS 812 (BSI, 1990c)) of 30 or a minimum 10% fines value (TFV) (BS 812 (BSI, 1990c)) of 50–100 kN (saturated) has been deemed appropriate for vibro stone column aggregate. For the more powerful and bottom-feed vibroflot units, a minimum TFV of around 120 kN is typically required. However, with the advent of the new European Standards, the above test methods are no longer recognised and have been replaced by the Los Angeles abrasion value (LAA) (BS EN 1097-2 (BSI, 2020)). While no recognised correlation exists between ACV, TFV and LAA, an LAA value of 30–40 is recognised within

Installation technique	Typical aggregate grading (mm)
Dry top-feed process	40–75
Wet top-feed process	25–75
Dry bottom-feed process	8–50

Table 84.1 Stone column aggregate grading for different vibro techniques (Data taken from BS EN 14731:2005)

the industry as being acceptable for use in vibro stone columns, dependent upon specific applications. Typical grading requirements for stone column aggregate as outlined in 'BS EN 14731: Ground treatment by deep vibration' (BSI, 2005) are summarised in **Table 84.1**.

Smaller aggregate sizes are used in the bottom-feed delivery system to prevent blockages in the stone feed pipe. Rounded aggregates tend to flow better in the stone tube and it is imperative that the fines content is kept below 5%, also to prevent blockages.

Natural (primary) aggregate resources are not unlimited within the UK, and their extraction causes increasingly unacceptable geo-environmental impacts. As part of achieving environmentally sustainable development within the ground improvement sector, there is an increasing desire to utilise recycled (and secondary) aggregates in vibro stone columns. However, where such materials are considered for use, it is important that there are appropriate specifications and quality control/assurance procedures in place to ensure 'fitness-for-purpose'. Spent railway track ballast and crushed concrete currently have the greatest potential for this application in the UK, and other materials have been investigated. Further information on these aspects can be found in Slocombe (2003b) and Serridge (2005), among others.

84.2.10 Environmental considerations

84.2.10.1 Noise and vibration

While it is rare for noise issues to be a problem with vibro techniques, vibration levels need to be considered when working close to existing structures and services such that these are not adversely affected. The safe working distance will depend on a number of factors including type of ground, the vibroflot power rating, the nature and state of repair of the structure and also plant access, particularly if any basement structures are present. As a guide, minimum distances of 1–3 m are often appropriate. However, each situation must be assessed on its own merits and dilapidation or precondition surveys should be considered, in addition to appropriate risk assessments and vibration monitoring.

84.2.10.2 Contaminants

Appropriate risk assessment and precautions will be required on contaminated sites and to avoid exposure to the atmosphere of chemicals and materials such as asbestos.

84.2.10.3 Vibro concrete plugs

Penetrative ground improvement techniques such as vibro stone columns can generate potential pathways for contaminant migration on brownfield contaminated sites. This raises environmental concerns, particularly if sensitive groundwaters or underlying aquifers are present, due to potential pollutant linkages (i.e. source–pathway–receptor linkages). An innovative approach to address this issue has been the development of 'vibro concrete plug' technology, incorporating the introduction of lean mix concrete into the basal section (toe) of the stone column, thereby isolating any pathways for downward migration of contamination by way of the stone columns. The above technology has also been utilised to bridge thin peaty deposits within the treated depth.

84.2.11 Monitoring, quality control and testing

The ground improvement sector exhibits a high frequency of ingenious equipment and solutions employed in difficult ground conditions. Close monitoring of ground response, quality control and testing throughout the duration of the vibro ground improvement works is therefore essential.

Bell (2004) describes an example of a range of quality (control) issues possibly arising from insufficient attention to construction detail on what he considers should have been a routine vibro stone column ground improvement project in mixed (heterogeneous) soils in the UK (**Figure 84.8**). Poor (discontinuous) column construction was attributed to the lack of constructing the columns in discrete lifts, each of which is required to be compacted to a predetermined limit such as target hydraulic pressure or ammeter reading, dependent upon whether the vibroflot is hydraulically or electrically driven. It is also indicated that accurate records of the amount of stone aggregate consumed by each column would have revealed the deficiencies of construction during installation.

The majority of vibro stone column projects in the UK involve the support of structures. Therefore, in addition to the monitoring and recording of installation parameters, some form of validation testing using load testing and/or penetration testing, dependent upon soil type, is normally required. Load tests typically include

- short-term plate load tests
- zone load tests or skip tests
- embankment load tests (surcharge load tests).

84.2.11.1 Short-term plate load tests

Plate load tests are normally of less than two hours' duration with plate diameter comparable to the design stone column diameter, normally 450 or 600 mm dia. Loading is quickly applied to the test column to between 1.5 times and 3.0 times the design pressure over the plate area, using a crane or vibro rig as reaction. The tests are principally regarded as a quality

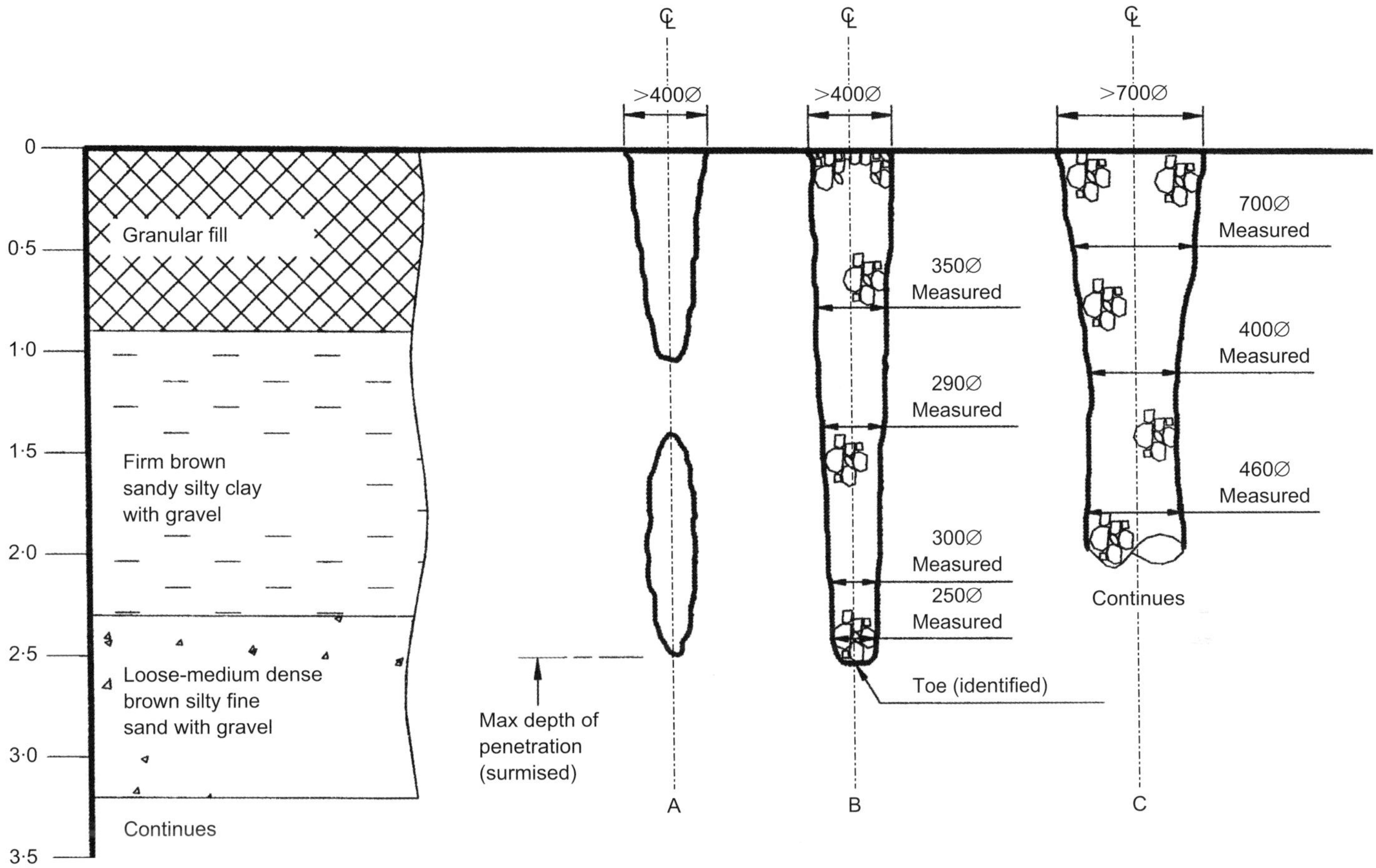

Figure 84.8 Example of poor quality control with vibro stone columns (Reproduced from Bell 2004; all rights reserved)

control test to assess the level of workmanship with regard to stone column installation.

84.2.11.2 Zone load tests or skip tests/dummy foundation tests

Zone load tests are normally of similar size to the project foundations and are performed over a group of columns – either by direct loading or against a system of grillage and kentledge for a period of between 24 hours and one week. These tests are more meaningful than the plate load tests since they apply load to the composite structure of both stone columns and soil, but are far more expensive because of the cost of importing kentledge to site.

Skip tests/dummy foundation tests provide an intermediate test between zone load tests and plate load tests, and may be more practical and economic on smaller projects (see BRE, 2000; NHBC Standards, 2022: chapter 4.5).

84.2.11.3 Embankment loading tests (surcharge load tests)

These tests can impose loads over a larger area for a longer period of time. They are particularly useful for road embankments on soft ground where time-dependent performance may be important (to ensure post-road construction settlements fall within acceptable limits). As with untreated soft soils, the rate of loading should be carefully controlled and monitored by way of appropriate field instrumentation. This facilitates better assessment of the performance of widespread loads. Where such tests are not possible, emphasis should be placed on monitoring and observing ground response to stone column installation and ensuring the design column diameters are achieved, supplemented by good quality control.

84.2.11.4 Penetration tests

In situ tests are typically employed to assess the technical success of in situ vibrocompaction, and performance is measured by the level of densification achieved against a specified target. The densification can be readily checked using standard penetration tests (SPTs) or cone penetration tests (CPTs). Comparisons can be made between pre- and post-treatment tests and care should be taken to ensure the same testing techniques are used in each situation. They are not normally employed on stone column projects, particularly where the soils contain a significant fines component.

With both load tests and penetration tests, it is important that there is an appropriate recovery period following completion of vibro ground improvement to allow a satisfactory level of pore water pressure dissipation to obtain meaningful results. A suitability rating for different testing methods for vibro techniques in presented in **Table 84.2**.

A modulus of subgrade reaction or California bearing ratio (CBR) (BS 1377-9 (BSI, 1990c)) value may sometimes be specified for a ground-bearing floor slab. It is important to recognise that while there may be some improvement in relative density between stone columns in granular soils, this will not be the case in cohesive soils (where the stone columns are acting as 'reinforcing elements'). In this instance the main objectives of stone columns are for control of total and differential settlements. Floor slab design would need to be based upon values obtained at subgrade between stone columns.

Test method	Granular	Cohesive	Comments
McIntosh Probe	*	*	Pre-/post-treatment essential. Can locate obstructions prior to treatment.
Dynamic cone	**	*	Too insensitive to reveal clay fraction. Can locate dense layers and buried features.
Mechanical cone	***	*	Rarely used in the UK.
Electric cone (CPT)	****	**	Particle size is important. Can be affected by lateral earth pressures generated by treatment. Most appropriate test for seismic liquefaction evaluation.
Boreholes and SPT	***	**	Efficiency and repeatability of test are important. Recovers samples (e.g. split spoon sampler).
Dilatometer	***	*	Rarely used.
Pressuremeter	***	*	Rarely used.
Small plate test	*	*	Does not adequately confine the stone column. Limited stress depth. Affected by pore water pressures.
Large plate test	**	**	Better confining action.
Skip test	**	**	Can maintain test for extended periods.
Zone loading test	****	****	Best test for realistic comparison with foundations.
Full-scale (e.g. surcharge load test)	*****	*****	Not commonly used. Tends to be confined to highway embankment projects.

* Least suitable; ***** Most suitable

Table 84.2 Suitability rating for test methods applied to vibro techniques (Data taken from Moseley and Priebe, 1993)

84.2.12 Practical issues

There are a number of practical issues to consider in the context of vibro stone column techniques. The more pertinent ones are discussed below.

84.2.12.1 Dealing with obstructions

Obstructions, dense granular made ground and remnant foundations can be an issue in terms of preventing vibroflot penetration and potential for forming hard spots. These should be suitably remediated (e.g. pre-loosening/excavation with removal and/or screening/crushing of oversize granular material before returning to excavation) and with provision of an appropriate cushion of granular material between the underside of the foundation or ground-bearing floor slab and the top of the buried remnant foundation.

84.2.12.2 Working platforms and stabilised platforms

Provision of a safe granular working platform is essential for safe access and execution of vibro ground improvement. Stabilised platforms (lime/cement) can be used, but due consideration needs to be given to implications of 'spragging' of rig tracks and ponding of water in inclement weather, which may necessitate placement of a surface granular layer.

84.2.12.3 Pre-boring

Many historic heterogeneous fills in the UK are quite competent, particularly in their uppermost layers. This, together with placement of any engineered/stabilised upfill in advance of vibro ground improvement, can impede vibroflot penetration, necessitating the use of pre-boring. Arisings will be generated which then need to be dealt with.

84.2.12.4 Slopes and interceptor drains

For construction at the base of a slope, it is important that appropriate drainage is provided at the toe to intercept water draining off the slope, to avoid any detrimental effects on the treated ground.

84.2.12.5 Lateral forces

Vibrocompaction and vibro stone column construction, as with driven piling, can generate lateral forces that can act on existing retaining structures, sheet piles and quay walls. Consideration of these effects requires specialist input.

84.2.12.6 Gas venting

On historic landfill sites, stone columns can actively vent any gases trapped in the fill. Therefore consideration of provision of gas venting measures may be required following appropriate risk assessment.

84.2.12.7 Installation tolerances

Accepted positional tolerances for vibro stone columns are more flexible than for piles and are typically quoted as 150 mm.

84.2.13 Conditions where caution is to be exercised for vibro stone columns

There are recognised ground conditions within the ground improvement industry which are either deemed not suitable for vibro stone column techniques, or where caution should be exercised. Some of the more significant ones are discussed briefly below.

(*a*) *Excessive degradable materials.* On sites where the made ground contains excessive quantities of degradable domestic refuse which would decay with time and cause loss of support to the stone columns and hence settlement. Generally, where the degradable refuse content is less than about 10% (by volume) and fairly evenly distributed, sites can be considered for treatment, but the design approach has to be conservative and whether or not vibro is adopted is dependent upon the building, its function and the client's (performance) requirements.

(*b*) *Non-engineered recent clay fills.* This is a situation where self-weight movement will not be complete. While introduction of stone columns may well improve the situation, it is very difficult to control self-weight movements. Typically, significant depths of clay fill (up to 5–8 m) should not be considered suitable for treatment if placed within the last 5–8 years. Each site must be judged on its own merits and with due consideration of settlement performance requirements.

(*c*) *Opencast mineworkings.* A variant on the backfilled clay pit is the backfilled opencast mineworking, where in addition to differential settlement implications for structures spanning the pit edge or buried high wall, there is the issue of restoration of the groundwater table (inundation), the timing of which is totally unpredictable. When the groundwater does re-establish itself, there is a breakdown of the clay fill and further settlement, which can be in excess of the self-weight movements. Vibro can be considered if careful attention is given to sealing the tops of the stone columns with clay immediately following their installation and during subsequent foundation excavation by concrete blinding, thus inhibiting surface water ingress due to, for example, heavy rain.

(*d*) *Thick peat deposits.* Thin peat layers can be accommodated (see BRE, 2000). Thickness, depth and lateral variations have to be considered very carefully in relation to the size of foundation and its loading. On some sites covered by up to 2 m of peat, the peat has been excavated along the lines of the foundation over a 2–3 m width and replaced with granular fill, vibratory stabilisation then being carried out in the normal way.

84.2.14 Design of foundations after vibro ground improvement

The bearing pressures achievable with vibro techniques are a function of the engineering properties of the soils prior to ground improvement. **Table 84.3** provides a guide as to what can be achieved in terms of load-bearing capacity and settlement performance for a range of soil types and applications after vibro ground improvement. Each site has to be assessed on its own merits.

Following adoption of vibro ground improvement techniques, conventional shallow pad/strip foundations and ground-bearing floor slabs can be adopted. Because the stone columns are formed by working against the overburden pressure, the stone aggregate at the top of the constructed columns is not as compact as at depth. For this reason, the main load-bearing foundations should be placed at a minimum depth of 600 mm below the level from which the ground improvement was carried out to fully realise the enhanced performance. In cohesive soils there should be appropriate allowance for minimum top and bottom mesh reinforcement. Levelling off and proof rolling of the subgrade following completion of vibro ground improvement is normally acceptable prior to construction of ground-bearing floor slabs.

84.2.15 Case histories

84.2.15.1 Introduction

Vibro techniques have been applied successfully on many projects and there are many well-documented case histories in the literature. Cases of unsatisfactory performance in vibro ground improvement inevitably occur but are not commonplace. They are usually not well publicised, principally

Soil type	Bearing pressure (kN/m^2)	Settlement range after treatment (mm)
Made ground: cohesive and mixed granular and cohesive	100–165	5–25
Made ground: granular	100–215	5–20
Natural sands or sands and gravels	165–500	5–25
Soft alluvial clays	50–100	15–75

Table 84.3 Typical improvements achievable in terms of load bearing and settlement after vibratory stabilisation (Data taken from Slocombe, 2001)

because of legal and contractual ramifications. Nevertheless, there is much to be learned from situations where problems have arisen. Examples of unsatisfactory performance include those relating to ground risk and insufficient site and ground investigation data; those relating to lack of understanding of vibro ground improvement techniques and their application; lack of appreciation of the implications of changes in site levels; issues such as inundation; those relating to quality control and workmanship issues (see Section 84.2.11); plus later drainage excavations undermining the constructed foundations on treated ground.

84.2.15.2 Successful case histories using vibro stone columns

Case study 1

For highway schemes, the approaches to bridges (or the transition between rigid bridge abutments and the consolidating soil behind the abutment) are a challenge for geotechnical engineers who are exploring cost-effective ground improvement solutions, to ensure a smooth settlement profile and vehicle ride quality and to reduce maintenance. Dependent upon the soil profile and geometry, this can be achieved in critical zones using a combination of ground improvement techniques such as vibro stone columns, vibro concrete columns and vertical band drains, and also by addressing issues such as lateral loads on abutment piles due to lateral 'squeezing' of the soil during soil consolidation beneath embankments. One application (see Serridge and Synac, 2007) is summarised in **Figure 84.9**, where vibro stone columns have been used in conjunction with other ground improvement techniques to support highway embankments over soft ground.

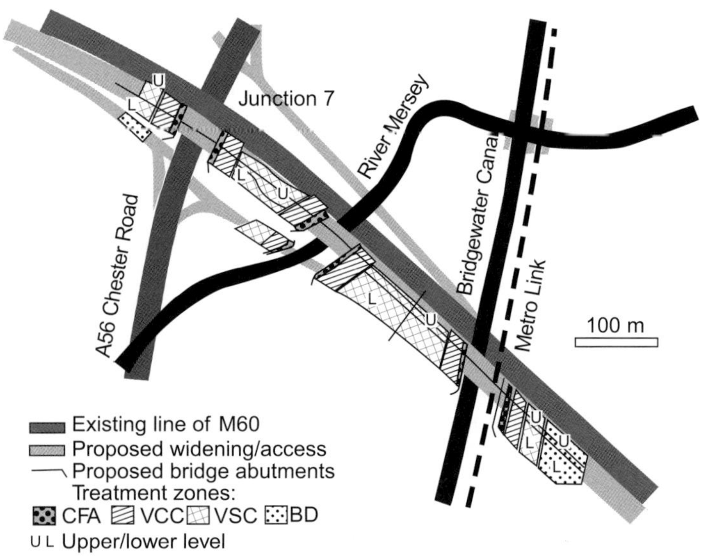

Figure 84.9 Ground improvement applications to support highway embankment construction over soft ground – M60 motorway widening, South Manchester (Reproduced from Serridge and Synac, 2007; all rights reserved)

A similar application using a combination of vibro stone columns and vibro concrete columns is described by Cooper and Rose (1999) for a project in Bristol.

Case study 2

Long-term renewable energy targets have led to the construction of increasing numbers of wind farms and biomass power plants. Vibro stone columns have been used to support their foundations, one example being for a 44 MW annual power production plant in Scotland with output sufficient to supply 70 000 homes.

The project involved a number of structures of various sizes and loadings, some also including heavy plant that was very sensitive to movement, particularly under dynamic loadings during operation. Settlements were required to be limited to no greater than 20 mm under loadings of up to 250 kN/m^2. The near-surface ground conditions were highly variable, comprising recent alluvial deposits that were present in some areas but absent in others, variable depth of water table and interbedded lenses and layers of saturated sands and silts, locally clays, all underlain by sandstone at depth. Some areas were considered suitable for dry top-feed vibro, some vibrocompaction without stone aggregate and others requiring bottom-feed treatment. The ground was so variable, however, that all treatment was performed using the dry bottom-feed technique to construct around 2000 stone columns to depths of up to 8 m. With such tightly specified post-treatment performance and variable soils additional design control checks were carried out using the monitoring systems on the bottom-feed installation rig to correlate the actual energy, depths of penetration and stone consumption as the work progressed. This effectively provided highly intensive investigation of the treatment areas, mapping areas to compare to soil properties reported in the site investigation.

One of the 'worst' areas was then subjected to a zone loading test on a 2.5 × 2.5 m reinforced-concrete base using a hydraulic jack reacting against kentledge blocks placed on grillage for maximum applied loading of 234 t or 1.5 times the design load. Settlements of 8 and 16 mm were respectively recorded at 1.0 and 1.5 times the working load and were well within the specified 20 mm limit. The results are presented in **Figure 84.10**.

84.2.15.3 Case history of poor settlement performance

Wilde and Crook (1992) describe the settlement of a steel portal frame factory unit within the River Mersey floodplain in Warrington, Cheshire. The site was shown to be underlain by soft alluvial soils varying in thickness from 5 to 10 m. Prior to implementation of vibro stone column ground improvement techniques, the site was brought up to the required plateau levels by addition of up to 1.5 m of fill coincident with the maximum 10 m thickness of soft alluvial deposits. The subsequently installed vibro stone columns did not fully penetrate the soft alluvial soils, and upon completion of ground

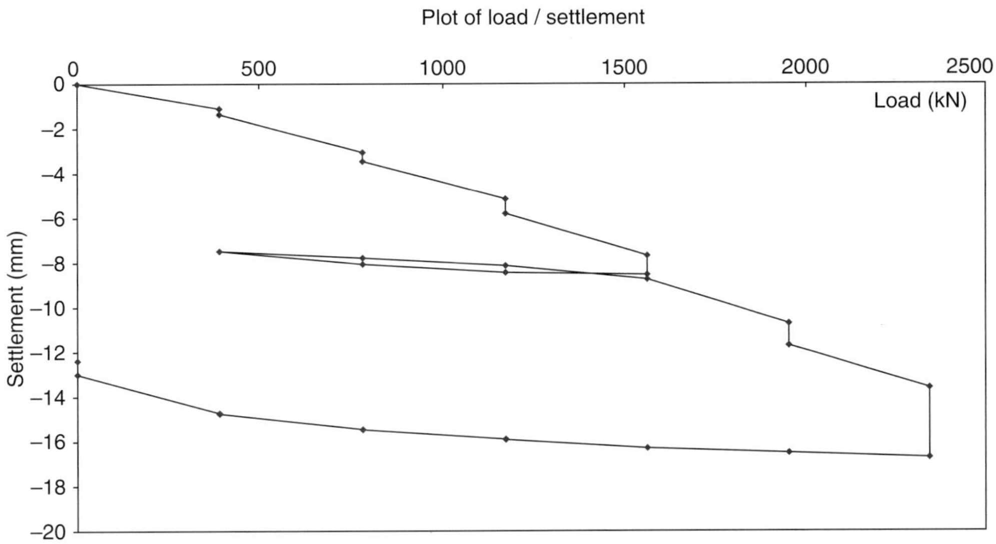

Figure 84.10 Zone load test on vibro stone columns for biomass plant (Data courtesy of Keller)

improvement the factory unit was constructed on simple pad and strip footings and with adoption of a ground-bearing floor slab. Subsequent monitoring of the portal frame structure showed that over a 6-year period 120 mm of total settlement occurred, with maximum differential settlement of 100 mm along the length of the structure. It was estimated that there was probably another 50 mm of settlement during the construction period, but which was not recorded. Levelling of the floor slab revealed similar movements to those experienced by the main foundations.

It is clear that the majority of recorded settlement is attributed to the surcharge effect of the fill associated with raising of site levels rather than the relatively small weight of the industrial unit. This brief example clearly demonstrates the need for careful consideration of changes in site levels in the context of vibro stone column techniques when considering treatment to thick layers of soft soils.

This example illustrates some of the difficulties of defining unsatisfactory performance. The building in this instance performed satisfactorily despite the high total settlements and without apparent structural distress or serviceability problems, and suggests that some settlement tolerances specified for low-rise lightweight buildings are unnecessarily stringent.

84.2.16 Conclusions

Vibro techniques are adaptable to a wide range of soil conditions and permit the adoption of simple foundation solutions. Vibrocompaction will increase the density of clean sands (and mitigate liquefaction potential) and vibro stone columns will reinforce fine-grained soils. Both techniques will improve the bearing capacity of the ground and result in reduced total and differential settlements.

Vibro design relies heavily on good-quality site and ground investigation data for accurate prediction of performance.

Good quality control procedures are necessary to ensure the successful implementation and performance of the vibro ground improvement techniques.

Where knowledge of a particular vibro technique is lacking, or there is uncertainty as to the applicability of vibro techniques to a specific set of site circumstances and performance requirements, then specialist advice should be sought to ensure appropriate vibro ground improvement selection and specification.

84.3. Vibro concrete columns

84.3.1 Introduction and history

Vibro concrete columns (VCCs) were first developed in Germany by injecting grout into already constructed stone columns. The aim was to overcome the lack of confining action to stone columns from very soft soils and peat. The grouting operation was then replaced by pumping suitable concrete directly down the bottom-feed delivery tubes, the first use being in the mid-1970s. VCCs were introduced into the UK in 1982.

84.3.2 VCC plant and equipment

VCCs are constructed using the specialist instrumented bottom-feed vibro stone column equipment and vibrators, but without the stone skip that normally runs up and down the leaders (see **Figure 84.11**). The concrete is then delivered to the top of the vibrator by way of an external pump.

84.3.3 VCC concrete, reinforcement and technique

In Germany, both high- and low-slump concretes with high cement contents are used. However, in the UK the technique generally uses low-slump concrete (typically less than around 60 mm) to be able to use vibration and ramming of

Figure 84.11 Vibro concrete column equipment (Courtesy of Keller)

their bases to densify basal granular soils to enhance end-bearing and generating enlarged 'end bulbs' for higher capacity at shorter length than for conventional pile design. Higher slumps have been used for some applications in the UK, depending upon site-specific circumstances. Once the end bulb is formed the vibrator is withdrawn at a controlled rate from the ground while maintaining positive pressure of concrete. The construction process also tends to be faster than normal piling when performed in suitable ground conditions.

VCCs are normally designed to support vertical axial loads. However, they are generally not constructed using reinforcement cages, sometimes simply with a central large section bar, sometimes with no reinforcement when beneath floor slab areas and road embankments, but normally with tie bars to tie into foundations. As a result individual VCCs have limited capacity to resist horizontal, shear and tension forces.

84.3.4 Applications and limitations

The VCC construction process allows a 'ground improvement' approach that is especially useful when considering the control of settlements of road embankments that are approaching piled bridges and in their superior settlement performance in comparison to vibro stone columns in very soft soils (see Cooper and Rose, 1999; Maddison *et al.*, 1996; Serridge, 2006). Enlarged heads are commonly constructed to assist in the transfer of embankment loadings by way of granular geogrid-reinforced load transfer platforms, as well as more efficient industrial floor slab design.

The construction of VCCs with end bulbs and tight interlock of concrete to soil can also prevent migration of contamination to underlying aquifers. Also, as the VCCs can develop safe loads of up to 600 kN with only short penetration of a sand/gravel stratum, they are useful where conventional piling would normally have to penetrate deeper. This shorter 'pile length' is particularly useful for the prevailing ground conditions to the east and west of London, where there is generally a suitable layer of sand and gravel underlain by either London Clay or Chalk aquifer.

The system is relatively quiet with low vibration in comparison to driven piles and does not require the removal of spoil that would be required for continuous flight auger (CFA) piling.

As with all displacement piles, care has to be exercised when installing VCCs at closer than about three times their diameter. It is also common practice to install single full-length central large section bars for spacings of up to about six times the VCC diameter (typically about 400–500 mm) to resist the effects of installing adjacent columns. Significantly larger diameters are generated in weak peat layers.

84.3.5 Site investigation (ground model)

The safe working capacities of VCCs are predominantly derived from end-bearing minus the assessed negative skin friction along the length of the pile shaft. As a result the depth to and penetration into suitable founding stratum has to be established over the extent of the proposed development. In addition, where VCCs are founded within granular soils that are underlain by cohesive strata, the capacity may have to be limited where the granular strata are relatively thin. Where there are thin clay layers within the granular founding strata the VCCs should be designed on the basis of a clay toe.

Such details are often best investigated by CPTs since these provide greater detail than SPTs in boreholes. It is relatively common for VCC quotations to include the performance of 1 or 2 days of CPT investigation to better refine the VCC design.

The soil chemistry must also be investigated to provide suitable buried concrete design, as per conventional piles.

84.3.6 Monitoring and testing

The specialist VCC rigs include computer-generated instrumentation, comparable to CFA piling rigs, to provide continuous records of the depths, concrete pressures, extraction rates and so on, of every column. Static and dynamic loading tests are often performed to confirm capacity. However, integrity tests are difficult to interpret due to the sometimes irregular shape of the pile shaft, particularly in the presence of thick peat layers.

84.3.7 Concluding remarks

VCCs are effectively an alternative type of pile that is constructed using vibro equipment, often in conjunction with stone columns for better settlement performance in softer and peaty soils. They are perhaps best considered as settlement-reducing piles.

84.4. Dynamic compaction

84.4.1 Introduction

Dynamic compaction (DC) improves weak soils by *controlled* high-energy tamping. The reaction of soils during DC treatment varies with soil type and energy input. A comprehensive understanding of soil behaviour, combined with experience of the technique, is vital to successful improvement of the ground. Dynamic compaction is capable of achieving significant improvement to substantial depth, often with considerable economy when compared to other geotechnical solutions.

The principle of dropping heavy weights on the ground surface to improve soils at depth has attracted many claims for its earliest use. However, the advent of large crawler cranes led to the current high-energy tamping levels first being performed on a regular basis in France in 1970 and subsequently in Britain in 1973 and in North America in 1975.

An extension of the concept of weights dropped onto the ground – rapid-impact compaction (RIC) – was developed in the late 1970s for the rapid repair of explosion damage to military airfield runways using modified (BSP) hydraulic piling hammers acting on a steel foot that remains in contact with the ground. This process is now used for civilian applications.

84.4.2 Processes

The original concept for DC was to collapse voids, particularly for the treatment of natural sands plus granular, mixed and cohesive made ground and fills. This was then extended to finer natural soils where the high impact energy effectively provided localised surcharge to squeeze water out of silts and clays, this being termed dynamic consolidation. Dynamic replacement was then developed to drive large-diameter columns of coarse imported materials through soft near-surface soils, particularly for peat and sabkha strata.

The worldwide use of DC has resulted in a large number of important terms, some of which can have different meanings in different countries or could be confused with other geotechnical descriptions. The following are the main terms that have been adopted in Britain (see Slocombe, 2003a, for a fuller listing).

- 'Effective depth of treatment' is the maximum depth at which significant improvement is measurable.
- 'Drop energy' is the energy per blow – that is, mass multiplied by drop height (tonne metres).
- 'Tamping pass' is the performance of each pattern of treatment over the whole treatment area.
- 'Total energy' is the summation of the energy of each tamping pass – that is, the number of drops multiplied by drop energy divided by respective grid areas (normally expressed in tonne-metre/m^2).
- 'Recovery' is the period of time allowed between tamping passes to permit the excess pore pressures to dissipate to a low enough level for the next pass.
- 'Induced settlement' is the average reduction in general site level as a result of the treatment.

84.4.3 Techniques

DC, whether applied to shallow or deep layers, improves the ground to the basal layers first and then progressively upwards to the shallower layers in a series of tamping passes (i.e. a bottom-up process) (see Section 84.4.6). In contrast, RIC improves the soils by first creating a 'plug' or layer of denser ground and then progressively driving this plug/layer to greater depth (i.e. a top-down process). The response of the ground to these two approaches is fundamentally different. Whereas the relatively lower number of high-energy impacts at wide grid centres of DC tends to initially bypass the upper layers and then by subsequent progressive treatment builds up the strength of the near-surface soils, the higher number of lower-energy impacts of RIC requires consideration of the possible generation of pore water pressures that could inhibit the required ground improvement in finer-grained soils.

84.4.4 Plant and equipment

At first sight, the physical performance of DC would appear to be simple: a crane of sufficient capacity to drop a suitable size of weight in virtual free fall from a certain height. Most contracts are performed with standard crawler cranes, albeit slightly modified for safety reasons and productivity, with a single lifting rope attached to the top of the weight (**Figure 84.12**). The operation must be performed safely; as a result, the Health and Safety Executive in Britain requires that a crane should operate at not more than 80% of its safe working load.

Recent crane developments allow automation of the whole work cycle. This is controlled by a data processing unit which plots for each compaction point its location, number, weight size, drop height, number of blows and measurement of imprint achieved. Some cranes include the ability for synchronous operation of two winches to lift larger weights than the conventional crane rating.

The majority of DC contracts in Britain have utilised weights within the range 6–20 tonnes dropped from heights of up to 20 m. However, the majority of UK work is now performed using 8 tonne weights dropped from heights of up to 12 m. Standard crawler cranes have also been used in America and Europe for weights of up to 33 tonnes and 30 m height. Specialist lifting frames with quick-release mechanisms have been utilised to drop weights of up to 170 tonnes in France.

Weights are typically constructed using toughened steel plate or box-steel and concrete. The effect of different sizes and shapes of the weight has also been extensively researched, with narrower weights being generally used for dynamic replacement columns. Treatment has also been performed below water using barge-mounted cranes and more streamlined weights with holes cut out to reduce water resistance and increase impact velocity on the seabed.

Within the UK, RIC typically employs a 7 tonne weight dropped repeatedly through a 1.2 m height onto a 1.5 m dia. steel articulated compaction foot which remains in contact with the ground (**Figure 84.13**). While the energy per blow is not large (typically 8.4 tonne-metre), the equipment permits a large number of impacts to be applied at a rate of about 40 blows per minute. Weights ranging from 5 to 12 tonnes are used worldwide.

84.4.5 Applications and limitations

DC can be applied to many of the filled sites that are considered to be suitable for treatment by vibro methods, except where weak natural clays and silts are the main problem layer (see Section 84.4.6.2). Many more UK sites are treated by vibro due to the noise and vibration generated by DC and the ability of the dry vibro methods to permit closer follow-on operations. DC can, however, provide superior performance in comparison to vibro for treatment to more recent deep made ground and in the presence of degradable constituents within landfills.

Figure 84.12 Dynamic compaction equipment (Courtesy of Keller)

Figure 84.13 Rapid-impact compaction equipment (Courtesy of Balfour Beatty Ground Engineering Ltd)

84.4.6 How dynamic compaction works

In contrast to having constructed a vibro stone or concrete column, the treatment at that location being then completed, DC is generally performed as a series of tamping passes over the treatment area with different combinations of energy levels designed to achieve improvement to specific layers within the depth requiring improvement. The most common approach is to consider the ground in three layers. The first tamping pass is aimed at treating the deepest layer by adopting a relatively wide grid pattern and a suitable number of drops from the full-height capability of the crane. The middle layer is then treated by an intermediate grid, often the midpoint of the first pass or half the initial grid, with fewer drops and reduced drop height. The surface layers then receive a continual (contiguous) tamp of a small number of drops from low height on a continuous pattern to compact the uppermost soil layers disturbed during the earlier higher-energy tamping passes.

It is sometimes feasible to combine, and sometimes necessary to subdivide, the three basic tamping passes for various reasons. The performance of increasing correctly controlled total energy input will normally lead to better engineering performance of the treated ground. This, however, is not a linear relationship, and the post-treatment parameters are heavily dependent upon the characteristics of the soil. As a general rule, similar total energies, whether per m^2 of area or m^3 of treatment depth, provide better performance to granular soils than mixed soils. Mixed soils are then better than cohesive, with refuse-contaminated soils generally offering the least performance.

Some DC design principles are provided in Chapter 59 *Design principles for ground improvement.*

For treatment using RIC, the operator monitors and can record the number of impacts, the total energy input applied, the foot penetration per blow and the cumulative penetration. When a specified parameter is reached – for example, foot penetration or set per blow – the RIC equipment is moved and positioned at the next treatment point. This primary treatment pass is normally performed on a closely spaced grid pattern, typically 1.5–2.5 m. Additional offset and/or lower-energy passes, or conventional proof roller compaction, are occasionally performed to achieve better coverage.

84.4.6.1 Granular soils

In dry granular materials, such as sand, gravel, ash, brick, rock slag and so on, it is easy to understand how heavy tamping improves engineering properties. Physical displacement of particles and, to a lesser extent, low-frequency excitation will reduce void ratio and increase relative density to provide improved load bearing and enhanced settlement characteristics.

When granular materials extend below the water table, a high proportion of the energy impulse is transferred to the pore water which, after a suitable number of surface impacts, eventually rises in pressure to sufficient level to induce liquefaction in a similar manner to vibrocompaction. Low-frequency vibrations caused by further stress impulses will then reorganise the particles into a denser state.

Dissipation of the pore water pressures, in conjunction with the effective surcharge of the liquefied layer by the soils above, results in further increase in relative density over a relatively short period of time. This can vary from 1 to 2 days for well-graded sand and gravel, to 1 to 2 weeks for sandy silts. The testing programme should therefore recognise the time-dependent response for soils that are normally considered to be free-draining.

Laboratory and in situ tests have consistently shown that to achieve maximum density, the lowest number of stress impulses to attain the required energy input will provide the optimum result. Saturated granular soils will normally require higher treatment energy overall, in a larger number of tamping passes, than if the soils were essentially dry.

When the individual sand particles are weak, such as the calcareous sands of the Middle East, 'sugar' sands of North America or the Thanet Sands of Britain, crushing tends to occur during the treatment. A similar response affects ash, clinker and weak aerated slags. When these soils are dry, the effect of such particle breakdown is not particularly significant. However, below the water table, the higher proportion of fines developing with increasing energy input results in a rapid change from a granular to a pseudo-cohesive soil response.

In summary, excellent engineering performance can easily be achieved in dry granular soils using both DC and RIC equipment. However, care must be exercised for the treatment of granular soils with significant silt content, particularly below the water table.

84.4.6.2 Cohesive soils

The response of clays is more complex than that of granular soils. There is again the distinction between above and below the water table.

With conventional consolidation theory, a static surcharge loading will collapse voids within clay fills and expel water to induce consolidation and increase strength. In contrast, DC applies a virtually instantaneous localised surcharge that collapses voids and transfers energy to the pore water. This creates zones of positive water pressure gradient which induce water to drain rapidly from the soils matrix. Consolidation can therefore occur more rapidly than would be the case with static loading. However, as with staged construction, the application of too high energy too soon can lead to problems.

Where the soils occur above the water table, the clays tend to be of relatively low moisture content, generally less than their plastic limit, where even a small reduction in water content can result in significant improvement in bearing capacity. Treatment is relatively straightforward and is mainly the collapse of voids to provide a more intact soil structure. However, care must be applied for the treatment to higher plasticity clays.

Where the clays occur below the water table, a much larger reduction in moisture content is generally required in the presence of a smaller available pore pressure gradient and a longer drainage path. These conditions can, if not properly controlled, result in localised failure of the clay matrix. Control is achieved by using greater numbers of tamping passes of lower energy input, requiring greatly extended contract periods in comparison to normal productivity.

To date, only nominal degrees of improvement have been achieved for thick layers of relatively weak saturated alluvial clays and silts, even with additional measures such as sand or wick drains. It is, however, more common in the UK to adopt vibro stone columns in such soils and then perform DC to preload the ground (see Slocombe, 1989), with the benefit of the stiffer columns also acting as drains to control excess pore water pressure.

It cannot be emphasised too strongly that the treatment of clayey made ground and fills together with natural clay soils requires experienced control on site. Particular care has to be exercised in the timing of successive tamping passes to permit adequate recovery of pore pressures to avoid excessive remoulding of the soils. Efficient treatment is achieved by attempting to provide as much improvement as quickly as possible while recognising that the response of the soils will dictate the speed of the treatment operations.

In summary, dry cohesive made ground and fills respond well to DC. Care must be exercised in the treatment of weak natural clayey soils or clayey made ground and fill below the water table (i.e. when saturated). The prior performance of vibro stone columns to both stiffen the ground and enhance drainage has been successfully combined with DC to clayey soils.

84.4.6.3 Landfills

The capability of DC to treat every square metre of road and parking areas is increasingly used in the development of former landfill sites where, depending on their age, the original degradable constituents have decayed to create extensive voids. It has also been performed to reduce ground levels to avoid costly removal to specialist tips to permit development at the desired site level.

As a general rule, the older the landfill the less the residual presence of matter susceptible to long-term decay; some older landfills, particularly those of high ash content, have been compacted to support structures that would normally be piled. However, more recent landfills generally contain significant proportions of organic matter, and structures would normally be piled.

There is as yet little documented proof that dynamically compacted landfill can affect the rate of decay of residual degradable constituents, although a paper by Sharma and Anirban (2007) clearly records far better post-treatment performance at a creep rate of 2% per log cycle than for static surcharge over a monitoring period of about 15 years. As there will be ongoing decay, when this technique is combined with piled structures increasing differential settlements will become apparent with time and a degree of maintenance may be required at some future time.

The principle of treatment of landfills is comparable to the treatment of mixed clayey made ground and fills but with generally higher energy input than for inert made ground and fills. This is to collapse near-surface voids and to 'overcompact' the remaining inert constituents. If a void then starts to develop due to localised long-term decay, then the inert materials will tend to ravel into the void, bulk up and spread the void effect rather than have a localised sharp deformity in the finished surface. Geogrids have also been used for a number of sites where the movements of heavy goods vehicles were critical to the development operations.

84.4.7 Site investigation

As with vibro designs, the extent of the site investigation should be appropriate to the type of development. For deep fill sites, desk studies to establish the locations of buried high walls plus the age and degree of control of placement are essential data. Water contents for comparison with liquid and plastic limits should be performed for clayey constituents. Densities as revealed by SPTs and CPTs plus the basal soils, whether clay, sand and gravel or rock are also required. The presence of any overhead wires, buried services or nearby structures should also be established.

84.4.8 Depth of treatment

Menard and Broise (1976) originally proposed that the effective depth of treatment was related to the metric energy input expression of $(WH)^{0.5}$ where W is the weight in tonnes and H the drop height in metres. This has been modified by a general factor of 0.5, reducing to 0.3 for heavier weights and greater drop heights. The range of treatment depths can also vary with initial strength, soil type and energy input, as illustrated in **Figure 84.14**, as well as the depth to the groundwater table.

High-speed photography has shown typical impact speeds of about 35 and 50 mph for 8 and 12 tonne equipment to achieve effective depths of treatment of about 5–6 m and 6–8 m, respectively. Within these depths, for similar volumetric energy input, the 12 tonne treatment will generally achieve better engineering performance than the 8 tonne weight, albeit at greater cost in view of the larger cranes required.

The shape of improvement in the ground tends to be similar to the Boussinesq distribution of stresses for a square foundation. Modification of energy levels for each tamping pass can be used to custom-design the treatment scheme to the specific soils profile and engineering requirements.

For RIC, BRE Report 458 (2003) records depths of treatment of about 2–4 m at total energy inputs ranging from 80 to 190 tonne-metre/m^2 for granular fills and silty sands. Greater treatment depths using higher energy inputs in favourable conditions, together with a range of applications, have been reported by Serridge and Synac (2006).

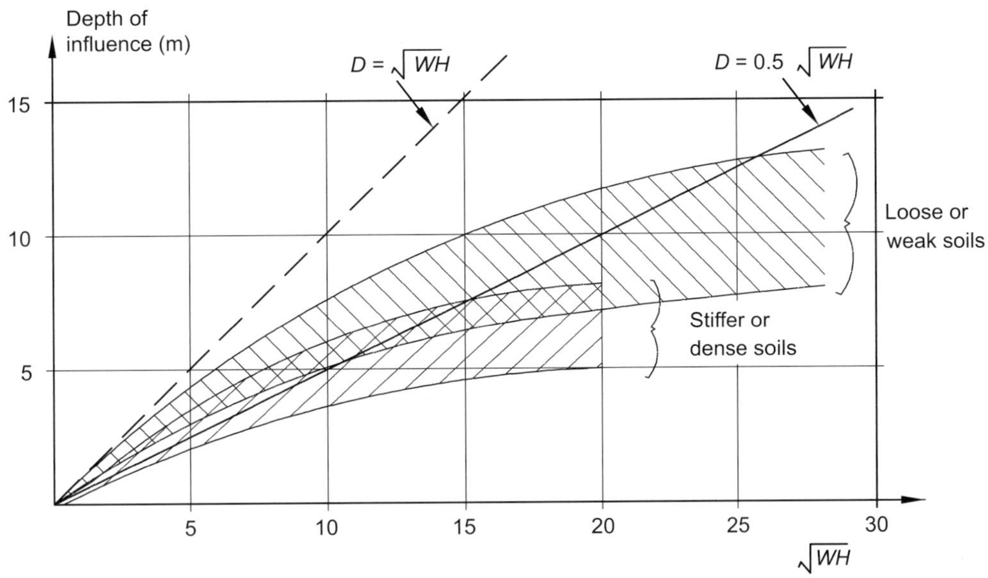

Figure 84.14 Depth of treatment

84.4.9 Practical issues

There are a number of practical factors that must be taken into account when performing DC contracts. The large crawler crane must be safely supported by a free-draining working surface, the thickness of which will depend upon the type of ground being treated. If the surface 1 m layer is granular, no imported working carpet is generally required. However, when working from a sandy surface, particularly during wet weather, fly-debris has been seen to be ejected through the air up to 60 m from the point of impact. If work is to be carried out near roads, railways or property, a movable screen is often used to intercept such fly-debris, albeit these affect the productivity. Alternatively, the programme should contain sufficient flexibility to permit treatment to be performed within, say, 50 m of such features only when the surface conditions permit its safe operation. As the RIC foot remains in contact with the ground there are no such issues with this equipment.

Where cohesive surface conditions exist, a free-draining granular working carpet is normally required. The thickness can be as little as 150 mm for light-energy treatment in reasonably competent soils to up to 1.5 m when treating heavily voided refuse fills. Winter working will place more onerous requirements on the adequacy of the working surface. The general rule is to increase the depth of the granular working carpet by 25% in comparison to summer thickness. When working in arid climates there is often no need for any working surface, even for clayey soils.

The free-draining granular surface materials are also used to infill the localised deep imprints generated by the high impact energy treatment operations by blading in with a large dozer. Alternatively, suitable on-site granular materials (e.g. crushed concrete) can be placed directly into the imprints. In such cases it is useful to compare the imported quantities to the assessed volume of the directly infilled imprints.

As the performance of DC tends to induce increases in water pressures, a pre-existing groundwater table within about 1.5–2 m of the working level can inhibit the productivity of the technique. In such cases bottom-feed vibro stone columns may be the preferred approach.

If more than one rig unit is to be used, they should be separated by at least 30 m. Similarly, subsequent operations by the main contractor may have to be delayed until the treatment operations are sufficiently remote. Furthermore, while DC can be performed over areas of vibro stone columns, it has to be performed before any adjacent piling operations to avoid possibly damaging the constructed piles.

84.4.10 Environmental considerations

DC utilises large, highly visible equipment. The process creates noise and vibration, both of which must be considered in the UK under the Control of Pollution Act 1974. The standards listed in the reference section provide further details (BS 5228-1 and 2: 2009; BS 7385 Part 1, 1990, and Part 2, 1993; BRE Digest 403, 1995).

Airborne noise levels are generated by a number of factors. Of these, the point of impact is by far the highest noise level, at typically 110–120 dB at source. However, its duration only occupies about 0.5% of the lifting cycle. The considerably lower noise values during lifting and idling when combined with the impact noise using the LAeq calculation method will normally meet most environmental limitations at distances of greater than 50 m from the treatment operations.

By far the most important consideration, however, is ground vibration. In addition to the magnitude of the vibration, the typical frequency of about 5–15 Hz is potentially damaging to structures and services, and particularly noticeable to human beings. It is suggested that there are three vibration levels that will influence the design of the treatment scheme. Guide values of resultant peak particle velocity at foundation level for buildings in good condition are

- structural damage: 40 mm/s
- minor architectural damage: 15 mm/s
- annoyance to occupants: 2.5 mm/s.

Lower values must be adopted for buildings in poor condition or environmentally sensitive situations such as schools, hospitals and data centres. Services and utilities must be considered on an individual basis depending upon their age, condition and importance, with values of 15–20 mm/s normally being considered acceptable, except for higher-pressure gas mains.

The level of vibration transmitted through the ground is an imprecise science because of the variable nature of the characteristics of soils. Field measurements of vibrations at ground level have revealed a number of trends which are illustrated in **Figure 84.15**. The upper DC limits tend to occur in the presence of granular or refuse-type soils, with the lower limits in cohesive strata.

Careful assessment is required where the soil being treated is directly underlain by relatively dense sand, gravel or rock, which will tend to transmit vibrations over larger than normal distances with comparatively little attenuation. Pre-existing dense surfaces or buried layers can have a similar effect of causing the transmission of higher than anticipated vibration levels.

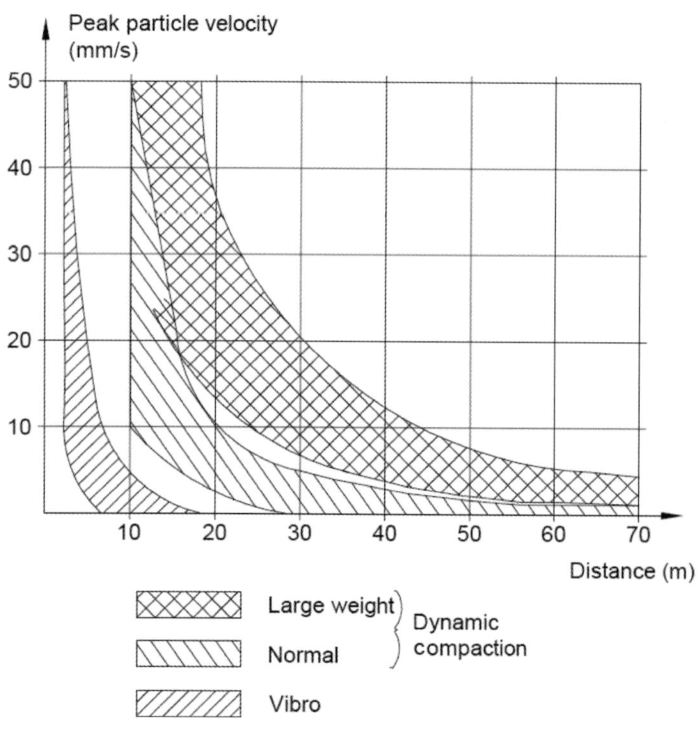

Figure 84.15 Vibrations

As the RIC equipment is based on lower impact energy but greater numbers of drops, this method has been employed as close as 10–15 m from an existing structure that was to be extended.

When vibrations become a problem, the main methods of reducing their effect are to simply reduce the drop height, to reduce both the impact energy and penetration of the stress impulse that may have attained an underlying dense stratum, and/or to excavate a cut-off trench to sufficient depth to intercept the surface wave.

People are particularly sensitive to vibrations and have a psychological reaction in believing that damage is caused even though the values are far below the well-established damage threshold levels. A public relations exercise can sometimes help to overcome concern among local residents. Building condition surveys prior to the commencement of treatment are often advisable.

DC is a highly sustainable technique since it does not use cement or quarried stone, normally only requiring suitable inert free-draining granular waste as a working platform and to infill localised deep imprints. It is also an area treatment technique that permits changes in foundation layouts and localised loadings – for example, mezzanine support foundations – anywhere within the treatment area. A number of DC contracts have permitted the rebuilding of developments where fire destroyed the original by simply performing loading tests upon the treated ground.

84.4.11 Monitoring and testing

Many contracts have simply involved the measurement of depth of first-pass imprints and monitoring of site levels. Post-treatment in situ (SPT and CPT) and loading tests are also often performed in a similar manner to vibro testing (see Section 84.2.11 and **Table 84.2**). In clayey soils, as with the performance of the treatment, it is essential that a sufficient recovery period be allowed to avoid ambiguous results.

It is common for DC to be performed for sites underlain by coarse made ground or including obstructions that would cause penetration problems for vibro and piling methods or in situ testing. Surface loading tests only are more normally performed in this situation.

Tests normally reveal superior performance being achieved when treating granular materials in comparison to clayey soils.

84.4.12 Additional comments

The general densification and collapse of voids will induce general reduction in site levels, the settlement being dependent on the total energy input and the manner in which it is applied. A convenient simple approach is to adopt approximate percentages (see **Table 84.4**) of the target treatment

Soil type	Percentage of treatment depth
Natural clays	1–3
Clay fills	3–5
Natural sands	3–10
Granular fills	5–15
Refuse and peat	7–20

Table 84.4 Guide to induced settlement

depth for 8 tonne equipment (50–100 tonne-metre/m^2) and 12 tonne (100–200 tonne-metre/m^2), the total energies with the 12 tonne energy applying to greater depth of treatment.

Higher percentages can be induced. However, the increase in energy input will not be a linear relationship with induced settlement, as during the treatment the material becomes progressively stronger and there is less and less potential void reduction available. Care has to be exercised to avoid over-treatment and possible loss in strength in these situations. Also, many sites, especially refuse tips, tend to be capped by clay soils.

Loose materials will obviously be induced to settle more than denser soils. Ash and certain types of slag also tend to break down during treatment to produce induced movements towards the higher value for granular fills given in **Table 84.4**.

Many sites of former heavy industry are now being reclaimed in the UK. These often present the designer with the problems of deep made ground, uncontrolled fills and massive obstructions from old basements and foundations. Any technique that has to make a hole in the ground will experience difficulty in gaining adequate penetration, and large excavations often have to be performed to remove the obstructions. However, with DC such features can be left in place provided they occur at sufficient depth to avoid excessive differential performance. The sequence of operation would be for advance earthworks to remove all known features down to a specific level, then perform treatment and normal construction operations.

Brownfield sites with minor contamination are well suited to treatment since the technique does not create a bore that could permit the migration of leachate. However, care should be exercised to ensure that the impact energies do not shear any underlying clay layer that may prevent the downward migration of contaminants into an aquifer. DC can also be controlled to avoid exposure of hazardous material to the atmosphere while still compacting soils at depth in the presence of, say, asbestos contamination.

Sites of former quarries are prime situations for treatment in view of the potential for piles to glance off the buried sub-vertical face between made ground and rock, and the possibility of constructing piles to inadequate depth as a result of false readings from boulders or inaccurate historical information on the depth of the quarry.

An area of increasing importance is the treatment of partially saturated made ground and natural soils that are susceptible to collapse compression due to inundation for the first time. DC reduces such potential by generally reducing voids, inhibiting surface water from penetrating the ground and by creating a ‘stiff crust’ over the effect of deeper movements, should they occur.

84.4.13 Case histories

84.4.13.1 Case study 1

A former colliery site was to be developed for several very large industrial and retail units, including three-storey offices. The existing fills and mounds of colliery spoil were first excavated and then recompacted in layers to minimum 95% maximum dry density. However, better settlement performance of maximum 35 mm total and 1 in 1000 differential over 25 years was required irrespective of the depth of fill.

On the first phase of 92 000 m^2 development, the depths of fill varied between about 5 and 8.5 m, with one area being up to 12.5 m depth beneath the proposed levels. The deeper zone was treated from the lower level during the filling operations and then from the final level using 12 tonne weights dropped from heights of up to 15 m.

The second phase of 45 000 m^2 was performed after the first phase had been constructed. This area was underlain by 1.5–5.5 m depth of fill and was also located near an electricity substation. This work was performed using 8 tonne weights and drop heights of up to 12 m.

The two phases of treatment were performed using similar volumetric energy input (tonne-metre/m^3) for the full depths of fill. They were then tested by zone loading tests on 2 × 2 m test bases to twice working load of 96 tonnes. In addition, a number of tests were performed on conventionally compacted fills (i.e. untreated ground). Comparison of their results revealed equivalent moduli as follows.

Untreated	E = 16–18 MPa
8 tonne DC	E = 30–50 MPa
12 tonne DC	E = 50–60 MPa

84.4.13.2 Case study 2

DC was performed in several contracts to a large area of landfill that was to be developed for industrial and office use. The first contract was to the deepest and newest landfill, parts of the site being up to 11.5 m deep and being filled with predominantly paper and cardboard waste only the year before treatment. Groundwater was typically at about 2 m depth.

All structures were supported on driven piles founded in the underlying competent clays, with DC being performed for the roads, parking and service areas. Treatment was performed using 12 tonne weights with cranes operating from nominal 1 m thickness of free-draining working platform.

The treatment area extended near to existing industrial units where the landfill was much shallower. This required a reduction in drop heights, while still capable of improving the full landfill depth within agreed vibration limits. The site was also near to a local railway line, where safety precautions were put in place.

Induced settlements of over 1 m were recorded within the deepest landfill area. Zone loading tests were performed to confirm that the required short-term performance had been achieved. The site was then monitored for six years with maximum settlement of 75 mm. This was assessed as being equivalent to rates of creep movements of about 1–1.5% per log cycle.

84.4.14 Concluding remarks

The DC method is a powerful tool when applied to suitable sites. A large database has been collated over the years to define its limitations and, more importantly, capabilities which can be used with confidence. As with every specialist technique, the designer and contractor performing this method of ground improvement must understand these capabilities and limitations. Such understanding can only be gained by experience.

References

Baumann V and Bauer GEA (1974) The performance of foundations on various soils stabilised by the vibro-compaction method. *Canadian Geotechnical Journal* **11**: 509–530.

Bell A (2004) The development and importance of construction technique in deep vibratory ground improvement. In *Ground and Soil Improvement* (Raison CA (ed)). Thomas Telford, London, UK, pp. 103–111.

BRE (Building Research Establishment) (1995) *Damage to Structures from Ground-Borne Vibration*. BRE Digest 403. BRE, Watford, UK.

BRE (2000) *Specifying Vibro Stone Columns*. BRE Report BR 391. CRC, Garston, UK (archived).

BRE (2003) *Specifying Dynamic Compaction*. BRE Report BR 458. CRC, Garston, UK (archived).

BSI (British Standards Institution) (1990a) BS 1377-9:1990, 4.3: Determination of the in situ California bearing ratio (CBR). BSI, London, UK.

BSI (1990b) BS 7385-1:1990 and BS 7385-2:1993: Evaluation and measurement for vibration in buildings. BSI, London, UK.

BSI (1990c) BS 812:1990: Methods for sampling and testing of mineral aggregates and fillers – Part 110: Method for determination of aggregate crushing value (ACV); Part 111: Method for determination of ten per cent fines value (TFV). BSI, London, UK.

BSI (2005) BS EN 14731:2005: Execution of special geotechnical works – Ground treatment by deep vibration. BSI, London, UK.

BSI (2009) BS 5228-1:2009 and BS 5228-2:2009: Code of practice for noise and vibration control on construction and open sites. BSI, London, UK.

BSI (2020) BS EN 1097-2:2020: Tests for mechanical and physical properties of aggregates – Part 2: Methods for the determination of resistance to fragmentation, Los Angeles abrasion (LAA) test. BSI, London, UK.

Charles JA and Watts KS (2002) *Treated Ground: Engineering Properties and Performance*. CIRIA Report C572. CIRIA: London, UK.

Clayton CRI (2001) *Managing Geotechnical Risk*. Thomas Telford, London, UK.

Cooper MR and Rose AN (1999) Stone column support for an embankment on deep alluvial deposits. *Proceedings of the Institution of Civil Engineers, Geotechnical Engineering* **137**: 15–25.

Hughes JM and Withers NJ (1974) Reinforcing of soft cohesive soils with stone columns. *Ground Engineering* **7(3)**: 42–49.

ICE (Institution of Civil Engineers) (1987) *Specification for Ground Treatment*. Thomas Telford, London, UK.

Maddison JD, Jones DB, Bell AL and Jenner CG (1996) Design and performance of an embankment supported using low strength geogrids and vibro concrete columns. *Geosynthetics: Applications, Design and Construction: Proceedings of the First European Geosynthetics Conference, Maastricht, Netherlands* (De Groot MB, Den Hoedt G and Termaat RJ (eds)). Balkema, Maastricht, the Netherlands, pp. 325–332.

Menard L and Broise Y (1976) Theoretical and practical aspects of dynamic consolidation. *Proceedings, Ground Treatment by Deep Compaction, London, UK*. ICE, London, UK, pp. 3–18.

Moseley MP and Priebe HJ (1993) Vibro techniques. In *Ground Improvement* (Moseley MP (ed)). Blackie, Glasgow, UK, pp. 1–19.

NHBC Standards (2022) Chapter 4.5: Vibratory ground improvement techniques. NHBC, Milton Keynes, UK.

Priebe HJ (1995) The design of vibro-replacement. *Ground Engineering* **28**: 31–37.

Serridge CJ (2005) Achieving sustainability in vibro stone column techniques. *Engineering Sustainability* **158(ES4)**: 211–222

Serridge CJ (2006) Some applications of ground improvement techniques in the urban environment. In *Engineering Geology for Tomorrow's Cities* (Culshaw MG, Reeves HJ, Jefferson I and Spink TW (eds)). Geological Society of London, London, UK.

Serridge CJ (2008) Site characterization and ground improvement applications for embankment construction over soft ground. *Proceedings of the BGA International Conference on Foundations, Dundee, Scotland*. IHS BRE Press, Watford, UK, pp. 1403–1414.

Serridge CJ and Synac O (2006) Application of the rapid impact compaction (RIC) technique for risk mitigation in problematic soils. In *Engineering Geology for Tomorrow's Cities* (Culshaw MG, Reeves HJ, Jefferson I and Spink TW (eds)). Geological Society of London, London, UK.

Serridge CJ and Synac O (2007) Ground improvement solutions for motorway widening schemes and new highway embankment construction over soft ground. *Ground Improvement* **11(4)**: 219–228.

Sharma HD and Anirban D. (2007) Municipal solid waste landfill settlement: postclosure perspectives. *Journal of Geotechnical and Geoenvironmental Engineering, American Society of Civil Engineers* **133(6)**: 619–629.

Slocombe BC (1989) Thornton Road, Listerhills, Bradford. *Proceedings, International Conference on Piling and Deep Foundations, London*, pp. 131–142.

Slocombe BC (2001) Deep compaction of problematic soils. In *Problematic Soils* (Jefferson I, Murray EJ, Faragher E and Fleming PR (eds)). Thomas Telford, London, UK, pp. 163–181.

Slocombe BC (2003a) Dynamic compaction. In *Ground Improvement*, 2nd edn (Moseley MP and Kirsch K (eds)). Spon Press, London, UK, pp. 93–118.

Slocombe BC (2003b) Ground improvement (nature versus nurture). *Ground Engineering* **36**(**5**): 20–23.

Slocombe BC, Bell AL and Baez JI (2000) The densification of granular soils using vibro methods. *Géotechnique* **50**(**6**): 715–725.

Wilde PM and Crook JM (1992) The monitoring of ground movements and their effects on surface structures: a series of case histories. In *Ground Movements and Structures* (Geddes JD (ed)). Pentech Press, London, pp. 182–189.

It is recommended this chapter is read in conjunction with

- Chapter 25 *The role of ground improvement*
- Chapter 59 *Design principles for ground improvement*
- Chapter 100 *Observational method*

All chapters in this book rely on the guidance in Sections 1 *Context* and 2 *Fundamental principles.* A sound knowledge of ground investigation is required for all geotechnical works, as set out in Section 4 *Site investigation.*

Chapter 85
Embedded walls

Andrew Bell Skanska UK Plc, Rickmansworth, UK
David Puller Geotechnical Experts, Sevenoaks, UK
David Baxter ArcelorMittal, Solihull, UK
Alec Courts Geotechnical Consultant, UK

This chapter considers the various forms of embedded wall and in particular focuses on the selection of – and construction issues associated with – the different types. It considers the economics of different wall types and gives guidance on issues such as adjacent structures, groundwater, temporary works, ground conditions and construction methods and materials. Systems covered include steel piled walls, concrete piled walls and diaphragm walling.

doi: 10.1680/icemge.66830.1385

CONTENTS

85.1. Introduction

The principal function of an embedded wall is to safely retain ground and groundwater to an agreed watertightness, although some embedded walls are used simply as groundwater or leachate barriers. The UK, with its range of geology (and its industrial heritage), employs and has contributed to the development of almost all embedded wall types.

The choice of embedded walling type is dependent upon a number of factors, including

- ground
- groundwater
- proposed depth
- site logistics
- noise and vibration restrictions
- economy (including programme economy)
- wall finish and watertightness
- performance criteria (e.g. allowable movement, installation tolerance)
- project size
- sustainability.

The 'generic' sketch (**Figure 85.1**) indicates the major issues that need to be considered when choosing a wall type. For further guidance, see Gaba *et al.* (2017) and ICE (2017) (CIRIA Report C760 and the *Specification for Piling and Embedded Retaining Walls*, or SPERW, respectively).

Walls in the UK seldom fail structurally – a great deal of discussion, debate and cost is spent on their aesthetics above dredge level, their contribution to local ground deformation, the effect of their placement and exposure on adjacent structures and particularly their 'watertightness'.

It is often the issue of watertightness that governs the ultimate judgement of the success of the wall, yet to date no agreed qualitative method of assessing this issue exists between users and providers. The publication *Reducing the Risk of Leaking Substructure: A Client's Guide* (Maloney *et al.*, 2009) and the further references within it are recommended reading for those involved in any aspect of embedded walling below groundwater.

If walls are judged post-construction on aesthetics and watertightness they appear to be procured almost purely on the basis of economics; **Figure 85.2** tries to indicate 'relative' costs for the various wall types and schematically illustrates where wall types are employed. The schematic indicates how difficult it is to generalise likely costs; at the polarised ends, small schemes for canals and temporary works for motorway cuttings are economically provided for by sheet pile walls, but deep shafts in difficult ground require diaphragm walls.

This chapter covers embedded walling by considering the following wall types

- diaphragm walls
- piled walls
- steel walls
- hybrid and other walls.

Each section addresses the generic issues raised in **Figure 85.1** and indicates how the walls are constructed, the range of their applications, their strengths and which particular interfaces need closest attention.

85.2. Diaphragm walls

Diaphragm walls are embedded concrete walls, usually reinforced, formed using the ground as a shutter. The process is illustrated in **Figure 85.3**. To ensure that the ground does not collapse, the panel is supported by a fluid suspension – normally a bentonite, polymer or hybrid suspension. Discrete 'panels' are excavated and supported by the fluid, and at their ends by 'stop-ends'. The bentonite support fluid suspends materials not removed by the excavation and cutting process

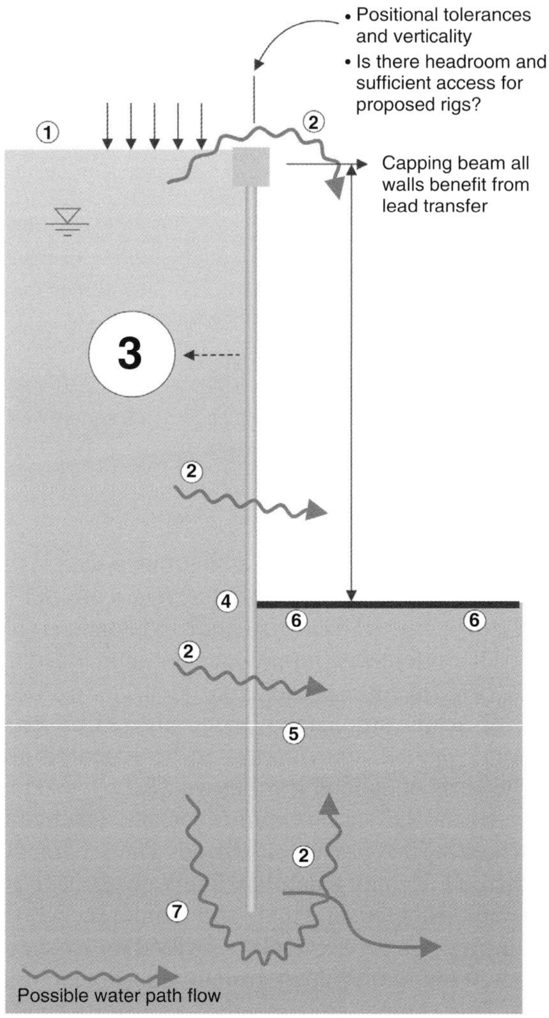

Some major issues

① Do adjacent services or structures inhibit access or apply constraints?

② Groundwater leakage possible by way of paths shown. The greater the number of interfaces/joints, the greater the problem. Walls are often ultimately judged on perceived watertightness.

③ The ground is the major influence on the choice of wall type. Can you drive, bore, drill or jack in wall elements? Is the ground stable? A soil or rock or a melange?

④ How easily are joints effected? How necessary are walings, props, anchors, plates, head details and any connections?

⑤ Depth to dredge dictates whether can be cantilevered or whether it needs props or berms. Depth below dredge can depend on cut-off, bearing and hold down requirements as well as stability.

⑥ Ground at dredge critical to stability, movements, water considerations and connections from wall to base slab.

⑦ Can wall depth be achieved by 'standard' construction methods? Need to be pinned/castellated/curtailed?

General

- How do temporary and permanent works interface and impact on the wall elements? Is this fully appreciated by all parties?
- Can predicted movement be accommodated or sensibly influenced?
- How important are finishes and aesthetics?

Figure 85.1 The major issues that need to be considered when choosing a wall type

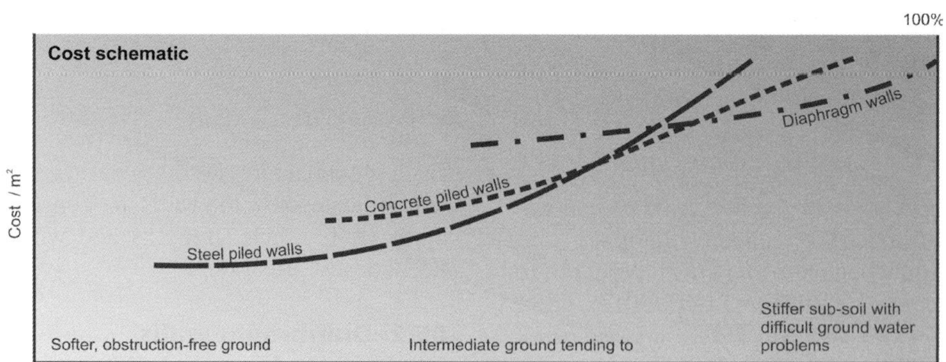

Total depth up to 20 m	Depth 20 to 40 m	Total depth > 40 m
Tendency for sheet pile walls	Choice dependent on local economics, contractual programme and construction requirements	Tendency for diaphragm walls
← Sheets	can be retrievable →	

Figure 85.2 'Relative' costs for the various wall types and where wall types are employed

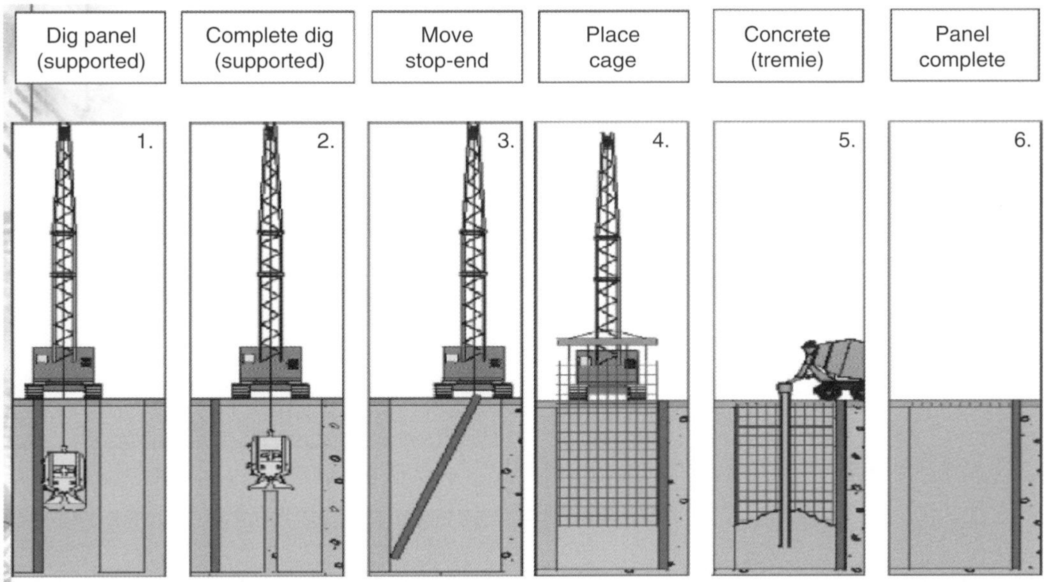

Figure 85.3 Diaphragm wall construction

and it requires cleaning by way of a replacement or recycling process. Reinforcement cages are then placed and the fluid suspension is displaced by specially designed concrete mixes with sufficient consistency to flow to the panel extrados and ‘scour off’ any bentonite from any element (stop-ends, reinforcement, box-outs, etc.) within the panel.

As illustrated in **Figure 85.2**, diaphragm walls tend to be used on larger, deeper walling contracts in more difficult ground – walls have been constructed to 150 m depth and up to 2.5 m thickness, although more commonly wall thicknesses of 800, 1000, 1200 and 1500 mm are used. They can be built in nearly all ground conditions, but special works may be necessary in very soft ground ($S_u < 20$ kPa) and hard ground (lightly fissured rock with an unconfined compressive strength greater than 20 MPa). Construction techniques can be readily configured to produce circular or near-circular shafts of diameter greater than 8 m, and panels in various shapes can efficiently carry large imposed loads. Introduced to the UK in the 1970s by ICOS, diaphragm walling has grown with the tendency for deeper basements and deeper shafts. Extended motorway and metro troughs of up to 1 km in length, shafts up to 120 m in depth and basements 30 m in depth have been constructed in the UK in recent years. Since diaphragm wall panels are typically 6.5 m in length (range 2.8–7.5 m), they have fewer joints (which can also be fitted with water bars) than other embedded walls and are favoured in ground where the control of groundwater, seepage and damp is necessary. It must be noted, however, that no better than Grade 1 watertightness in accordance with BS 8102 (see Chapter 64 *Geotechnical design of retaining walls*) should be assumed possible for any concrete embedded wall. In Western Europe and its influence zones, diaphragm walling seldom provides structural continuity across panel joints. Such continuity is more common in the Far East, at a considerable cost premium.

As with most embedded walling expertise, quality workmanship and close cooperation at the interfaces is necessary for successful construction. There is useful guidance in SPERW (ICE, 2017) and BS EN 1538 (BSI, 2015).

Excavation for diaphragm wall panels is usually carried out using grabs or hydromills (cutters or hydrofraise); see **Figure 85.4**. These are dedicated pieces of plant which experienced diaphragm wall specialists are reluctant to modify since any change affects the process. Designers for economy are encouraged to work with what is available rather than what is optimum. Grab and cutter widths are typically 2.8 m. Grabs (cable or hydraulic) were used in early developments and were used particularly for softer cohesive ground. A typical excavation rate for a grab is 100 m^2/shift. Modern grabs are fitted with instrumentation so that depth and deviations can be monitored in real time and recorded (particularly where rotating head kelly bars are fitted). Specification requirements of 1 in 100 verticality tolerance for cable grabs and 1 in 150 for hydraulic grabs can be bettered using such techniques. Grabs are usually employed to start excavation works when mills are to be used.

Mills (cutters or hydrofraises) are hydraulically powered drum wheel excavators which remove soil suspended in the slurry, which also acts as trench support. The process uses reverse circulation, with the spoil transport medium in a continuous circuit abstracted by a mud pump, fed by pipework to a separation plant (desanders, desilters) where the solid materials are removed, to be pumped back to the panel again (**Figure 85.5**). Developed initially for cohesionless soils and weak rocks, excavation rates were typically up to

Figure 85.4 (a) Hydraulic grab and (b) hydrofraise

Figure 85.5 (a) Desanding equipment and (b) cage lifting

five times that of a grab. However, at this rate of excavation in a soft cohesive soil, it is difficult to provide an efficient cleaning or replacement cycle to remove the finely graded soil material from the support fluid. There has been much use of mills in recent times in stiffer, carbonate and over-consolidated clays and in weak rocks. Verticality tolerances of better than 1 in 1000 have been achieved using mills, but contractors tend to limit their contracted tolerances to 1 in 300. In addition, contracts usually allow for protrusions of not greater than 100 mm from the 'true wall face'. This criterion is based on the experience of many contracts; however, it is generally controlled by the ground rather than workmanship. In loose cohesionless or soft cohesive soils, the over-break could be much greater, contrasting with very stiff clays where the finish can be smooth.

With the correct tooth configuration on the cutter wheels, mills can cut into the concrete of the adjacent primary panel, forming a 'continuous' concrete member (still without rebar continuity). This can be particularly useful for deep shafts (circular or ellipsoidal) where depth precludes the use of other methods of achieving good-quality joints between panels. However, this form of overcut joint has been found to have rather dubious watertightness and, hence, its use should be restricted to structures where panels tend to be in hoop compression or where watertightness is not considered to be critical. The choice between using a mill or a grab is usually a straightforward comparison between productivity and the difference in set-up costs (mills being more expensive), although site access, logistics, ground conditions and environmental considerations can be relevant.

Diaphragm wall ancillary equipment includes heavy chisels to aid excavation through rock and boulders, brushes to clean panel joints, service cranes for cage installation and concreting, and support fluid mixing and separation plants.

As noted, diaphragm walls are often employed where groundwater and soft soils are present and, hence, trench stability is critical. In these circumstances stability may be maintained by limiting panel length – thereby encouraging arching effects and by ensuring a positive head in the support fluid with respect to the groundwater level (usually not less than 1.5 m for bentonite and 2 m for polymer). This positive head has to be maintained throughout the process and is ensured by harnessing a reservoir or the containment effect of the guide walls. The panel size will also be important in terms of ground movement, with the excavation of larger panels tending to develop larger ground movements; however, they can be easier and more efficient to prop. The converse is true of smaller panels, which cause less ground movement but may require propping at closer spacings. Guide walls have several uses: they act as positional guides for the excavation tools, they indicate the positive fluid support level that must be maintained, they act as a reservoir for this level and they act as a support for the (often heavy) down-hole insertions such as stop-ends, tremie pipes and reinforcement cages. Guide walls should be designed for their purpose and constructed with care. They are seldom less than 1 m deep and 300 mm thick. Usually guide walls are constructed in sections which are structurally continuous from cast in situ reinforced concrete in advance of the diaphragm walling. Occasionally existing adjacent basement walls are employed as one side of a guide wall. Pre-fabricated guide walls have been used successfully.

The specific gravity of freshly mixed bentonite slurry is about 1.02–1.04, compared with water at 1.0; it is not surprising that water has occasionally been used as a support medium, but should only be considered as a water pressure stabilising fluid. Bentonite has the added property of thixotropy and can form a filter cake on the trench wall in many soils, thus enhancing panel stability by preventing side sloughing. Experience of using bentonite support fluid encompasses at least 50 years in the UK but there are still many important aspects to consider. Excessive build-up of solids in suspension during digging can lead to accumulation of thick filter cake in the cover zone, which can impede concrete flow and cause defects such as mattressing. Prior to cage installation and concreting it is essential to de-sand or exchange excavating support fluid with fluid which has not been used in excavation in order to promote good displacement by tremie concrete. This process of desanding (which should deploy a suction pump across the full base of the panel) assists in the removal of heavy mud and sediments from the base of the panel, which can otherwise be trapped at the base or elsewhere in cage box-outs or areas of congested reinforcement. Site tests on support fluid include density, viscosity, fluid loss, pH and sand content. ICE (2017: table B20.1) sets out which tests should be carried out at each stage of the construction process (freshly mixed, ready for reuse, during excavation and prior to concreting) and provides typical compliance values (ICE, 2017: table C20.1).

Polymers, where the long polymeric chains clog up the ground pores, are often the support fluid of choice for bored piling due to their easier disposal (in the public sewer) and the smaller plant set up required, which is extremely advantageous on congested inner-city sites. However, they have not been used widely in diaphragm walling in the UK, although they have been used successfully internationally. Increasingly, bentonite support fluids with polymer additives are being utilised to improve resilience to chemical attack (from cement and/or saline groundwater). The EFFC/DFI *Guide to Support Fluids for Deep Foundation* (EFFC/DFI, 2019) provides more detailed guidance.

Early diaphragm walling employed circular pipe stop-ends, which were extracted clear of the 'green' concrete as part of the panel formation process. Circular pipes gave way to 'organ pipe' stop-ends, which were still pulled several hours after the last concreting. Where panels were very deep or working hours restrictive, permanent steel (H-sections) or precast concrete stop-ends were employed.

Currently, most practitioners use 'peel off' stop-ends. These are prised off the end of the hardened concrete panel, usually by action of the grab, once it has been exposed during excavation of the adjacent panel. Stop-ends can be used to incorporate a water stop, which will extend into adjacent panels. This system has greatly facilitated diaphragm wall programming. However, experience of successfully placing and removing such stop-ends is limited to about 50 m depth. Accurate and vertical placing of stop-ends is a vital part of the process, and if high standards are not achieved then subsequent panel joints and exposed wall surfaces can be seriously compromised.

Diaphragm wall tool sizes are such that multi-returns and complex wall shapes are not encouraged. Panel dimensions are dependent on plant size, and there are stability considerations both for the panel and for adjacent structures. Logistics are also important since excavated panels soften; common practice is to concrete quickly, and handling much more than 200–300 m^3 of concrete within a normal working day is often difficult. However, panel volumes up to 1400 m^3 have been cast over an extended shift in the UK. Wherever possible, practitioners will promote simple rectangular panels, although corner panels must be accepted. T-, Z- and I-shaped panels can offer significant structural benefits in terms of stiffness and moment capacity, but it must be recognised that they are more difficult to construct. Here, as in the case of corner panels, continuity of reinforcement across changes in panel direction should be avoided if at all possible. However, in the case of re-entrant corners, continuous cages may be necessary for structural reasons.

Two examples of diaphragm walls are shown in **Figure 85.6**. The tendency for deeper walls, carrying bigger loads over larger spans, is leading to heavier densities of reinforcement in diaphragm wall panels. The need to incorporate connection box-outs and wall cut-outs can exacerbate problems. Panels have construction requirements (tremie space, clearance to stop-end and water bars), which constrain reinforcement cages. These tendencies act against the need for open, accessible panel faces to facilitate concrete movement, to ensure penetration, to promote scour of slurry or filter cake deposits and to ensure dense concrete cover. Although BS EN 1538 indicates a minimum clear vertical bar space of 80 mm may be allowed and 150 mm for horizontal reinforcement in special cases, the authors of this chapter believe that 100 mm and 150 mm should always be provided as a minimum to the vertical and horizontal reinforcement, respectively. ICE (2017: C10.4.4) provides more detailed recommendations and differentiates between vertical and horizontal reinforcement. It is extremely important that discussions take place between the wall constructor and designer well in advance of cage fabrication (ideally prior to the start of detailing) to ensure that buildability is not compromised.

(a)

(b)

Figure 85.6 (a) Deep shaft (France) and (b) 'peanut-shaped' (UK) diaphragm walls

The challenges faced by the constructor regarding the lifting and handling of reinforcement cages must not be overlooked. Issues such as transportation to site, safe lifting, on-site welding of reinforcement and cage splicing must all be discussed at an early stage and the appropriate risk assessments and temporary works design carried out. The Federation of Piling Specialists provides useful guidance regarding splicing and lifting of prefabricated reinforcement cages.

All embedded concrete walls require special concrete mixes with high workability, which can be placed reliably using tremie pipes and which self-compact without the need for vibrators or mechanical means. It should be noted that such concrete has evolved over decades to meet the specialist demands of deep-foundation contractors. This has resulted in more complex concrete mix designs which employ a number of admixtures to improve workability and useful life. As a result and following a number of problem projects, the testing of tremie concrete has moved beyond simply taking cubes and checking slump or flow. It is now industry practice to assess the visual stability index and to carry out other tests on fresh concrete, such as pressure bleed (Bauer filtration test) and/or ASTM gravity bleed testing to ensure that the mix will be stable and will not bleed excessively. Flow retention is also tested on samples of fresh concrete at time intervals after arrival on site in order to ensure that workability will be sufficiently maintained. Further guidance on concrete mix design, production and testing, as well as good tremie practice and reinforcement detailing, is provided in the EFFC/DFI *Guide to Tremie Concrete for Deep Foundations* (EFFC/DFI, 2016).

Diaphragm walls, with their flat surfaces, can easily be 'joined' or connected to adjacent structural elements (props, walers, beams, slabs). Box-outs are preferred in deep walling for connections and can be placed to within +150/–50 mm of the design elevation. The presence of a box-out can create significant resistance to concrete flow, especially when combined with a dense reinforcement cage. In recent years the advent of hydro-demolition and its use after the wall has been excavated can enable effective removal of box-outs and should be considered. Whereas simple pull-out bars will only provide very limited moment fixity and are used generally to aid transfer of shear, the use of couplers can provide a high degree of moment fixity. Alternatively, connection bars can be drilled and grouted into the wall face on exposure during bulk excavation. The joint between a diaphragm wall and the base slab (or roof slab in the case of cut-and-cover tunnels) is most likely to be subject to groundwater or water infiltration and needs particular and serious attention in its detailing. These are also likely to be regions of particularly high reinforcement congestion, and the 3D modelling of bars and inserts as solid objects is highly recommended to minimise the risk of construction problems later.

Diaphragm walls may be lengthened (deepened) where necessary to increase the flow path for groundwater beneath the toe of the wall or where they are employed as vertical load-carrying elements as well as acting as retaining walls. It is not commonly required to reinforce walls below the depth required for embedded wall stability in such cases. There are also circumstances where the ground is too hard and works cannot be economically progressed to the required 'design' depth. Such cases are often resolved by taking a series of pins through the panel into the hard ground below, ensuring lateral support at the toe. Provided there is adequate site investigation data, this is usually foreseeable at the design stage, and steel reservation tubes can be provided in the cage for this purpose as well as providing locations at which toe grouting can be carried out to significantly reduce rock permeability.

Diaphragm walling contracts need large, dedicated plant, which usually necessitates sole occupancy of the site. Bentonite plant (silos, mud pumps and cleaning or desanding equipment) typically requires 150 m^2. As noted above, it is often the logistics of the mud plant that controls overall planning of the diaphragm walling activity. Joints are inevitably 'leaky' and even with best practice can be areas where bentonite or ground inclusions are trapped. Contractors will make good areas with inclusions and will contract to exclude 'running leaks'. This particular issue needs upfront discussion with the client and project team and a clear understanding of what is likely to be achieved in a particular ground.

Diaphragm walling continues to be pushed to deeper and longer spans and to incorporate higher densities of reinforcement and ancillary inclusions (box-outs, soft eyes for tunnel connections, geothermal pipes, instrumentation tubes, sonic-coring tubing, etc.). It is a technique that is likely to grow in application.

Another example of the construction of diaphragm walls is shown in **Figure 85.7**.

85.3. Secant pile walls

85.3.1 Description and use

Secant pile walls are formed from a series of intersecting bored cast in situ concrete piles. In most cases alternate 'male' (secondary) piles are reinforced to provide the structural capacity of the wall, with the intervening 'female' (primary) piles acting as unreinforced barriers to groundwater flow (**Figure 85.8**). As well as providing retention of soil and groundwater pressures, secant walls can be used to carry superstructure loads. Considerable flexibility in plan shape is offered by secant pile walls (**Figure 85.9**), which leads to efficient use of the basement space. Hence, secant pile walls are a popular form of embedded retaining wall in building projects as well as civil engineering structures such as metro stations, underpasses, cut-and-cover tunnels and pumping stations. Circular and elliptical shafts (**Figure 85.10**) have also been constructed successfully using secant pile walls.

Secant piled walls constructed in the 1970s using mechanically configured rigs were relatively expensive and slow to install. However, the advent of powerful hydraulic piling rigs increased both the range of applications and the rates of production. By the mid-1980s, deep secant pile walls were being constructed as permanent walls. Later that decade, hard/hard secant pile walls were constructed at a number of sites in the UK using the continuous flight auger (CFA) method. The most recent developments have seen the use of cased CFAs, which are able to combine the benefits of high productivity through using a CFA with the tight construction tolerances obtained using temporary casing.

(a)

(b)

Figure 85.7 Station box for Channel Tunnel Rail Link (UK)

Some confusion exists in the terminology used to differentiate between the types of secant pile wall. The key constant in all walls is that the secondary piles are nearly always reinforced structural concrete, and the variable is usually the strength of the primary piles, which will typically vary as follows.

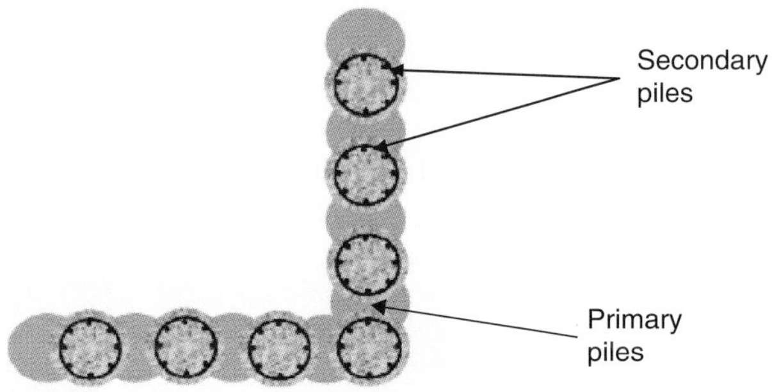

Figure 85.8 Primary–secondary pile sequence

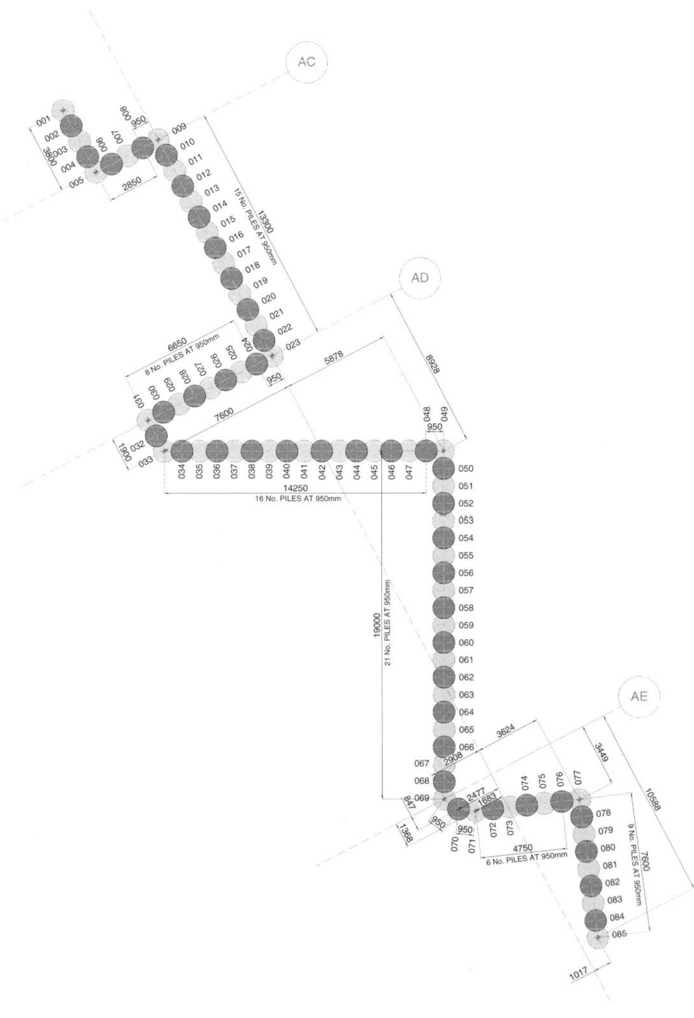

Figure 85.9 Flexibility in plan shape

- *Hard/soft walls.* This wall type is often used in temporary works or where the long-term durability of the primary pile can be neglected as its function is taken over by some other form of construction, such as an internal structural liner wall. Primary piles are designed simply as short-term water blockers using a cement/bentonite mix or similar (**Figure 85.11**).
- *Hard/firm walls.* This wall type is often used for basement walls for building projects and constructed usually using either rotary piling with segmental casings, CFA or cased CFA technique. Primary piles are designed to suit the design life of the structure but with detailed consideration of constructability. There is sometimes a perceived conflict here between the minimum cement requirements to ensure durability and the difficulty encountered when overcutting primary piles of excessive strength. Primary piles are often designed to achieve a strength of C8/10 at 56 days.
- *Hard/hard walls.* These walls are usually constructed using segmental casing and intended to act as permanent works with equal durability of both secondary and primary piles. Primary piles are usually unreinforced, but are constructed with a full structural concrete mix, typically either C28/35 or C32/40, with the mix designed to meet the required durability requirements.

Figure 85.10 Ellipsoidal shaft

Figure 85.11 Hard/soft secant pile wall

85.3.2 Range

The increased use of secant piles has been driven by the development of ever more powerful hydraulically powered rotary drilling rigs with torque in excess of 400 kN/m. However, the range of application varies significantly depending on the installation method.

It is difficult to provide hard definitive limits on the depth achievable with each technique since this is dependent on many factors, including rig power, pile diameter and ground conditions. However, in general, while it is possible to construct CFA piles to depths in excess of 30 m, cage depths are usually limited to about 20 m depending on ground conditions. Cased CFA pile depths are currently limited to about 25 m and rotary cased secant piles are rarely constructed beyond 30 m depth. It is important to remember that the depth over which the piles will actually be secanted will be determined by tolerances and not by the depth that the piling rig can drill (see **Table 85.1**).

85.3.3 Installation and materials

Piles are constructed at a predetermined spacing designed to ensure they will intersect over the required height (usually the depth required to achieve a barrier to groundwater flow), allowing for the combined positional and verticality tolerances of both piles. The provision of reinforced concrete guide walls (**Figure 85.12**) is essential to maintain good positional tolerances at the piling platform level. The scalloped profile of the guide walls is usually achieved using special formwork or pods, which are adjusted to achieve the desired pile layout.

Pile diameter: mm	Bore diameter below casing: mm	Minimum overcut: mm	Maximum overcut: mm
Secant pile walls constructed using the CFA method			
600	–	100[a]	250[a]
750	–	100[a]	250[a]
900	–	100[a]	250[a]
1050	–	100[a]	250[a]
Secant pile walls constructed using the cased CFA method			
750	–	100[a]	250[a]
880	–	100[a]	250[a]
Secant walls constructed using segmental casing			
620	520	100[a]	250[a]
750	670	100[a]	250[a]
880	760	100[a]	250[a]
1000	900	100[a]	250[a]
1180	1060	100[a]	250[a]

[a] Typical values that depend on the concrete mix, ground conditions and retained height.

Table 85.1 Range of pile diameters

Primary piles are constructed first at an initial spacing such that there is no risk of interference between recently constructed piles as a result of plant track or concreting pressures. For example, this may require a pile construction sequence of pile numbers 1, 5, 9, 13, 17, then 3, 7, 11, 15. For hard/hard and hard/firm walls, secondary piles are ideally constructed within 3–7 days of construction of the adjacent primary piles. This is particularly important for walls constructed using the CFA method, where the excessive concrete strength of the primary piles can cause severe verticality tolerance issues during secondary pile installation. Care must be taken not to open up too much of the wall by constructing an excessive number of primary piles without the intervening secondary piles. In addition, in the case of phased construction, it may prove necessary to install a gravel pile at the end of a construction phase by overcutting the last primary pile while it is still green and backfilling it with gravel; the overcut can be achieved several weeks later if necessary as the later phase of the works begins.

The productivity of CFA and cased CFA rigs can exceed ten piles per shift and, therefore, it is necessary for continuous sections of wall in excess of 50 m in length to be available for these rigs to work efficiently.

While excavation of CFA piles takes place under the support of the soil on the flights of the auger, cased CFA utilises a stiff temporary casing to support the ground and overcut adjacent primary piles. Air is often supplied down to the auger tip by a nozzle in the concreting pipe by way of the hollow auger stem to assist flighting of the drilling spoil up the auger. At the top, the spoil is directed into a series of telescopic muck bins, which transfer the drilling arisings safely to the ground (**Figure 85.13**). The casing and CFA are powered independently by separate rotary drives.

The guidelines for the reinforcement of diaphragm walls cited in Section 85.2 are in many areas equally applicable to secant pile walls. Particularly relevant is observing minimum

Figure 85.12 Guide walls

clear bar spacings of 100 mm for longitudinal and 150 mm for transverse reinforcement. Consideration to the pairing of bars should be given to ease congestion. Special attention should be paid to congestion at overlaps and the use of couplers may be required if there are two layers of reinforcement. Furthermore, the cranking of larger-diameter bars, such as 40 mm or 50 mm, may not be feasible.

Additional cover (100 mm rather than 75 mm) may be provided to facilitate the installation of long reinforcement cages in CFA and cased CFA secant piles. The slight resulting loss of structural capacity is normally recouped by the improvements in productivity (or lack of re-drilled piles due to stuck cages).

Concreting of CFA pile walls follows similar practices to those required for deep CFA bearing piles. Special care should be paid to concrete workability and the longevity of fresh mix properties for cased secant piles to avoid common faults such as cage displacement, stuck temporary casings and damage to green concrete during casing removal.

85.3.4 Tolerances

There are a number of essential ingredients to achieve good construction tolerances in secant pile walling

- design and construction of robust guide walls
- selection of appropriate plant and experienced personnel
- careful sequencing of the pile installation
- appropriate materials, especially primary pile concrete mix design
- attention to detail in designing the pile layout.

Wall installation tolerances are heavily dependent on the construction method. ICE (2017: table B1.4) gives some useful guidelines, and ICE (2017: section B11.4) gives further guidance on guide wall and reinforcement cage tolerances.

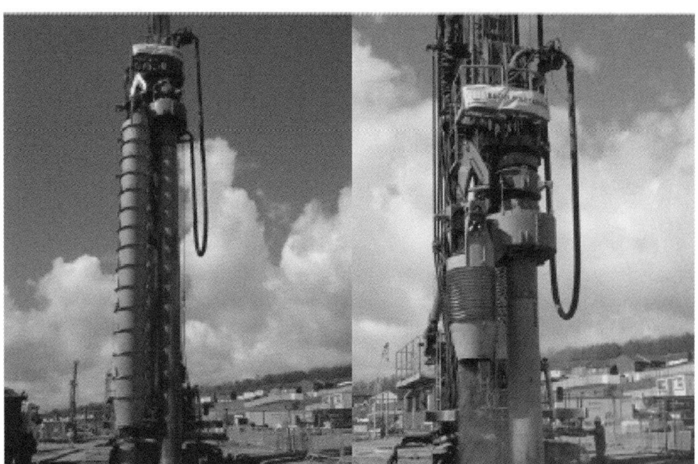

Figure 85.13 Cased CFA wall piling with spoil management system

85.3.5 Interfaces

Unlike diaphragm walls, it is difficult to install slab connection reinforcement or box-outs in secant pile walls without excessive risk that they will not be located within the required tolerances. Reinforcement cages installed either by plunging into fresh concrete or by installing prior to concreting within a temporary casing will usually become significantly twisted out of position, particularly at depth. As a result, it is usually preferable to achieve slab-to-pile connections by drilling and grouting in bars at the required level.

Secant pile walls that are supported by temporary props will require waling beams, which must be properly packed to achieve contact between the waler and every pile to ensure effectiveness. In many cases temporary props are arranged to support the wall by way of a top capping beam, which may be required anyway, hence, saving the need for a temporary waling beam. When temporary anchors are used, the provision of a waling beam can be avoided if every primary pile is restrained by an anchor using an anchor head block (**Figure 85.14**).

Waterproofing details for secant pile walls require particular attention given their geometrical shape at the slab-to-wall connections and also at the capping level. The provision of re-injectable trays for permanent structures can be beneficial, provided access is available in the long term.

85.4. Contiguous pile walls

Contiguous pile walls consist of a series of cast in situ bored piles, which do not intersect with each other. The gaps between the piles are usually sized so that, at a minimum, the piles do not clash at depth although they can be larger, dependent upon the ability of the ground to arch between the piles; typical gaps are around 100–200 mm, but can be bigger.

These walls are designed to resist earth pressure but are generally not designed to be groundwater retaining, which means some form of drainage detail is necessary at the formation level of the proposed structure. For aesthetics, the walls

Figure 85.14 Hard/hard cased secant pile wall, Royal Opera House, London

Figure 85.15 Contiguous bored piled wall

are often faced with a variety of finishes (e.g. panels, tiles, bricks or blocks). To protect both the drains and facing from ground sloughing in the gap, it is good practice to infill this area with a free-draining material, such as no-fines concrete or a propriety geofabric or plastic drain.

Contiguous bored piled walls are most often employed in low-permeability ground. Despite being less structurally efficient than diaphragm walls and suffering from the interface problems common to secant pile walling, they can offer a very economical solution, and walls have been constructed in a range of diameters, typically 600–2400 mm (**Figure 85.15**).

85.5. Sheet pile walls

85.5.1 Description and use

Sheet pile walls consist of a series of interlocked steel sections, which when installed are capable of withstanding horizontal and vertical loads and bending moments. Sheet pile walls can be used for both permanent and temporary works (for example, to an excavation or to exclude water). Common uses for sheet pile retaining walls include the following.

- *Quay walls.* Sheet pile walls are used in ports and harbours as well as rivers and inland waterways to provide berthing for vessels, embankments and river walls. To accommodate increasing vessel size and higher retained heights, it is possible to construct high-modulus walls, or 'combi-walls' where stiffer primary elements (H-sections, tubes or box piles) are placed with sheet pile infills between. This technique is covered in further detail in the next section.
- *Flood defences.* Sheet piles can be used in a variety of flood defence systems, providing an impermeable barrier above and below ground level. Seepage through the pile interlocks can be further reduced by highly effective sealing systems.

Figure 85.16 Basement construction utilising steel sheet piles and temporary propping

- *Retaining walls.* Sheet piles are commonly used along linear infrastructure (railways and highways) and in other situations where cuttings, embankments or slopes need to be supported. Piles may act in cantilever or can be anchored using tiebacks or anchors; alternatively, if space is available in front of the wall then waling beams and props can be installed.
- *Basements and underground construction.* Sheet piles are used as the permanent retaining structure and can also carry vertical perimeter structural loads. The exposed face of the wall is often left visible and in these cases the steel can be cleaned and painted to provide an aesthetically acceptable product (**Figure 85.16**). Further details on steel basements can be found in *Steel Intensive Basements* (Yanzio and Biddle, 2001).
- *Hydraulic and environmental barriers.* The impermeable barrier of a sheet pile wall can be used as an environmental barrier to prevent the flow of pollutants and contamination.
- *Cofferdams.* Temporary works can be constructed to retain soil and water, both on land or in water (e.g. for construction of shafts or bridge pier foundations).

85.5.2 Materials

Sheet piles can be manufactured as either hot-rolled or cold-rolled sections. For structural applications (especially permanent works) hot-rolled sheet piles are preferred as they have a defined closely fitting interlock which results in a continuous wall that acts as a single unit. The use of cold-rolled sections is generally limited to lightly loaded situations, as they are generally of a lighter section and do not interlock as tightly as hot-rolled sections.

The tolerances for the manufacture of cold-rolled sheet piles are covered in BS EN 10249 (BSI, 2023a). The inherent limitations of cold-rolled piles mean they are more susceptible to declutching if the pile is installed inaccurately or in denser, more difficult soils. The manufacture of hot-rolled sheet piles is covered in BS EN 10248 (BSI, 2023b). Hot-rolled sheet piling of high-grade, non-alloy steels and the manufacture of highly efficient sections to tight tolerances represent the state of the art in pile manufacture. The provision of a tight interlock allows for their use in more heavily loaded situations with more difficult ground conditions. There are a large range of sheet piles available on the market in both U- and Z-shaped profiles. Profile selection depends upon the design requirements. However, from an installation point of view it is important that the design takes account of the driving conditions. It may be that a design and selected pile section can be adequate for the permanent applied loads, but would fail during installation due to the temporary loads applied by driving. In these cases, a heavier pile section should be selected to accommodate the driving forces. The drivability of a sheet pile section is related to its cross-section, stiffness, length, steel grade, quality and the chosen method of installation. Further details on the factors affecting drivability can be found in the *Piling Handbook* (ArcelorMittal, 2022), along with detailed product information regarding their manufactured sections.

85.5.3 Installation

The method of installation will vary depending upon the type of wall, the selection of pile section and physical site characteristics. However, the method will generally fall into one of the following categories.

85.5.3.1 Pitch and drive

This methodology utilises specialist equipment to pick up and pitch and drive single piles to the required depth. Originally this technique was limited to situations where the soil was loose and the piles short. In recent years the equipment has become more powerful, allowing the technique to be used on more complex sites with more difficult soils. Care needs to be taken to ensure that the piles remain vertical during installation as the lack of any guidance frame can allow the piles to lean in the direction of driving, or rotate, particularly in stiffer soils or where driving longer piles.

85.5.3.2 Panel driving

With this methodology a number of piles are inserted into a preconstructed frame located along the required wall line.

85.5.4 Installation techniques

Three basic driving techniques are used.

85.5.4.1 Impact driving

The pile is repeatedly hit with a falling mass travelling over a controlled stroke length. The method can generate high levels of low-frequency vibrations and high noise levels. The advantage is that, given a variety of hammer weights and stroke lengths, high levels of energy can be put into the piles, overcoming most soil conditions.

85.5.4.2 Vibrodriving

Piles are installed using high-frequency vibration from sophisticated and powerful hammers, which can overcome most problematic soil types. However, the hammers are expensive to operate and repair compared with impact hammers.

85.5.4.3 Pressing

If pressing is used, installation into granular soils can be limited unless assisted by other methods such as water jetting or pre-boring. The major advantage is that the equipment generates low levels of noise and vibration and can be used near existing structures.

85.5.5 Plant

The plant used to install sheet piles will depend upon the method of installation. For impact driving a number of different types of hammer are available, employing various means of imparting a blow to the pile head. They may be hung from a crane or mounted on a hydraulic piling rig for additional manoeuvrability and control. Where freely suspended, additional stability and control of direction of force is improved by the presence of leg guides. The blow may be delivered freely under gravity or may be an accelerated fall of the ram (a double-acting hammer). The method of raising (and/or accelerating) may typically be by diesel hammer or hydraulics.

Vibratory hammers are available in a wide range of sizes and may operate in different frequency modes. Variable (resonance-free) high-frequency vibratory hammers allow the eccentricity to be adjusted at start up and shut down to eliminate unwanted vibrations. The rate of progress and the suitability of the method to changing ground conditions can be controlled through adjustments to the frequency and/or amplitude of the vibrations.

Press-in piling equipment relies on the principle of developing a reaction force to allow the pile to be pushed into the ground. This is primarily achieved by using neighbouring piles and by employing one or more hydraulic rams. Press-in piling equipment may be rig-mounted utilising multiple rams (**Figure 85.17**), or may be mounted on the already installed piles (**Figure 85.18**), from which they generate a reaction in a self-supporting manner (often referred to as the 'Japanese' or 'walking' press).

85.5.6 Tolerances

Installation tolerances have been established for many years and are summarised in a number of documents, including SPERW (ICE, 2017), the *Piling Handbook* (ArcelorMittal, 2022) and in BS EN 12063 (BSI, 2023c). Where there is a structural need for piles to be installed with increased accuracy, this should be possible by using the correct installation techniques for the ground conditions. A competent contractor should be able to achieve tolerances of 25 mm of plan position as opposed to the standard specification of within 75 mm. A similar improvement on verticality should also be feasible. The production tolerances of the sheet pile (**Figure 85.19**)

Figure 85.17 Sheet pile installation utilising leader rig with hydraulic pile press

Figure 85.18 Walking hydraulic pile press

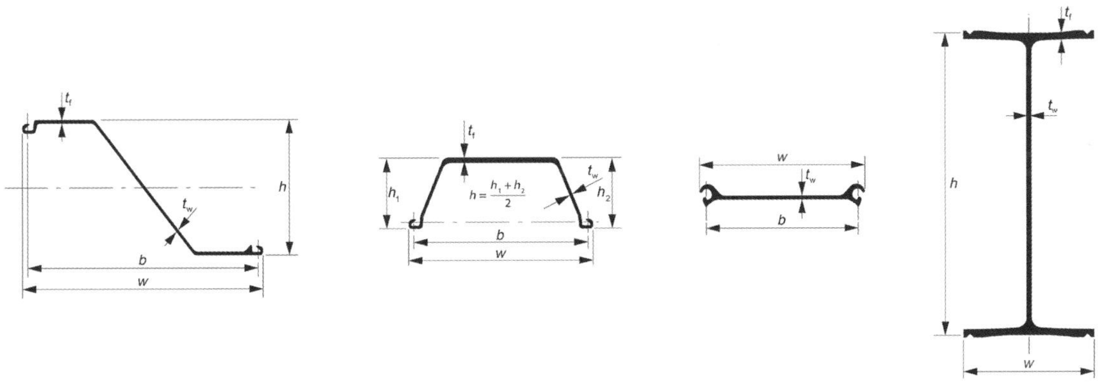

Height				
Z piles	$h \leq 200$ mm ± 5 mm		200 < h < 300 ± 6 mm	$h \geq 300$ mm ± 7 mm
U piles	$h \leq 200$ mm ± 4 mm		$h > 200$ mm ± 5 mm	
H piles	$h \leq 500$ mm ± 5 mm		$h > 500$ mm ± 7 mm	
Wall thickness[a]				
	Flange		**Web**	
Z and U piles	$t_f \leq 8.5$ mm ± 0.5 mm	$t_f > 8.5$ mm ± 6 % t_f	$t_w \leq 8.5$ mm ± 0.5 mm	$t_w > 8.5$ mm ± 6 % t_w
H piles	$t_f \leq 12.5$ mm +2.0 /−1.0 mm	$t_f > 12.5$ mm +2.5 /−1.5 mm	$t_w \leq 12.5$ mm +2.0 /−1.0 mm	$t_w > 12.5$ mm +2.5 /−1.5 mm
Straight web sheet piles			$t_w \leq 8.5$ mm ± 0.5 mm	$t_w > 8.5$ mm ± 6 % t_w

[a] For Z-shaped, U-shaped and straight web sheet piles, the positive tolerance shall be at the discretion of the manufacturer. At the time of the enquiry and order, a limitation on the positive tolerance can be agreed (option). The recommended positive tolerances are shown in the table.

Figure 85.19 Manufacturing tolerances for steel sheet piles (Reproduced with permission from ArcelorMittal; all rights reserved) (continued on next page)

Width		
	Single piles	**Double and triple piles**
Z, U and H piles	± 2% *w*	± 3% *w*
Straight web sheet piles	± 2% *w*	

Straightness, length, mass			
	Straightness	**Length**	**Mass**
All sections	0.2% *L*	±200 mm	±5 %

Squareness of end of profiles			
Z, U and straight web sheet piles	Single piles	Width	±2% *w*
	Double piles	Width	±1% *w*
H piles	Single piles	Height	±2% *h*
	Single piles	Width	±2% *w*

where

w = width of a single pile, respectively double pile or triple pile

L = pile length

Notes

- Tolerances on the height of U-shaped sheet pile sections made of a pair of crimped or welded piles may be agreed at the time of the enquiry and order.
- The width *w* is the overall width of the section. The system width *b* is a theoretical width between fictitious interlocking points inside the interlocks.

Figure 85.19 Continued

should also be considered when detailing the pile layout, particularly for complex geometries or over long plan lengths of wall where the cumulative effect may lead to a discrepancy in the number of piles required.

85.5.7 Interfaces, joints and connections

One of the major benefits of utilising a sheet pile wall is that each element is manufactured in factory conditions and is supplied with a designed connection detail, which is capable of providing the full section modulus. This interlock is complemented by a series of special sections, such as ‘corners and junctions’, which allow a sheet pile wall to be installed in numerous configurations. Some of the more common corner products available are shown in **Figure 85.20**.

As with all wall types, the issue of watertightness is most critical at the location of joints and interfaces. For sheet pile

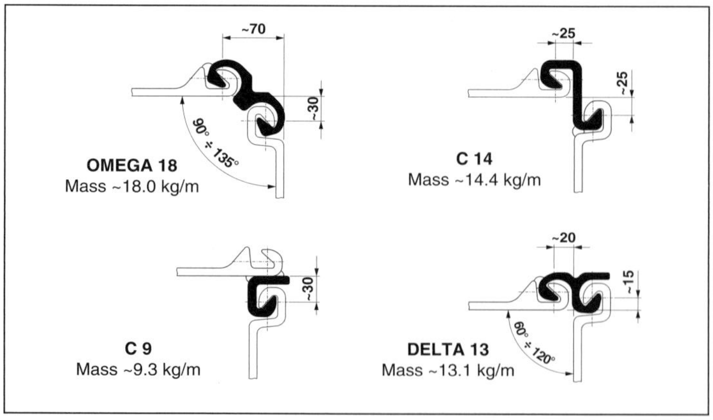

Figure 85.20 Range of corner profiles used to change wall direction (Reproduced with permission from ArcelorMittal; all rights reserved)

Figure 85.21 Detail of floor/wall connection in a permanent steel basement

Figure 85.22 Driving of combi-wall elements using a hydraulic impact hammer

walls, where watertightness is important, a number of methods can be utilised to exclude or minimise water penetration. There are a number of clutch sealants on the market, which can be placed within the clutch prior to pile installation. The highest level of impermeability is provided by welding the clutches after installation, during basement construction. This provides a permanent seal, which will not degrade with time. As the seal is created post-installation, it can normally only be carried out over the exposed length of pile, from the surface level down to the formation level of the basement or excavation. This may mean that some sort of flange is needed to link the sheet pile wall with a basement slab to ensure that water cannot penetrate through the base. As sheet pile walls are made from steel, there are more possibilities for carrying out post-installation work to improve or maintain watertightness, such as welding at clutch locations or installing additional steel products over damaged joints. **Figure 85.21** shows a typical steel flange connection at the base of a permanent sheet pile basement construction.

85.6. High-modulus steel walls (combi-walls)

85.6.1 Description and use

The main characteristic of a high-modulus (combi-)wall is that the design approach relies upon any load being retained on individual primary elements, which are separated by infill panels (secondary elements), usually consisting of steel sheet piles. As the retained height is generally greater, much larger plant and equipment are required to install the heavier sections. Some common configurations for various steel combi walls are given in BS EN 1993-5 (BSI, 2009).

Common uses for combi steel pile retaining walls include

- flood defences
- quay walls
- deep basements
- retaining walls
- cofferdams.

85.6.2 Plant

Due to the size of the wall elements it is common for combi-walls to be installed using traditional crane-hung vibratory and impact hammers (**Figure 85.22**). As the elements are relatively large and, thus, heavy, greater amounts of energy are required to drive the piles to the required depth. A further advantage of using crane-hung equipment is that when working near water, the heavy plant and equipment can be set back from the edge, which minimises any surcharge loading on the unprotected slopes.

In some cases, such as deep basements within city centre locations, it may not be possible to use either vibro- or impact-driving methods due to existing structures. In recent years, pile-pressing methods have been developed to allow the installation of combi-walls with low levels of noise and no vibration. These techniques can be slower and, therefore, more expensive than traditional methods, but have extended considerably the range of locations where combi steel walls can be utilised.

85.6.3 Materials

Most materials used will conform to relevant standards such as BS EN 10248 (BSI, 2023b) for hot-rolled sheet piling. HZM

Figure 85.23 Typical combi-wall arrangement showing primary tubular piles and infill sheet pile elements

and box piles as primary elements should also conform to BS EN 10248. Tubular steel piles may be produced to a relevant standard such as BS EN 10219 (BSI, 2006); however, tubes may be sourced as used or repurposed from other industries such as line pipe. If this is the case, then it is essential that all materials used are certified and if necessary tested to ensure they comply with the design and project requirements.

85.6.4 Tolerances

Due to the greater retained heights leading the design towards a combi-wall solution, the weight and physical size of the wall elements will be much larger than for sheet pile walls. In order to ensure that the various elements are installed accurately it is generally found that the position and verticality will be controlled through the use of gates. These consist of a framework of temporary steelwork set at two levels, which ensures that the wall elements are installed accurately (**Figure 85.23**). The use of Z-piles as infill elements allows for adaptation to as-built geometries both in plan position and verticality. The width of a pair of Z-piles adapts through rotation at the interlocks.

85.6.5 Interfaces, joints and connections

In order to join the various elements of a combi steel wall together, the primary elements should be produced or prefabricated with suitable interlocks. This allows them to be clutched directly to the steel sheet piles, which will act as the infill panels. See **Figure 85.20** for typical details of corner and junction profiles, as detailed in the *Piling Handbook* (ArcelorMittal, 2022).

85.7. Soldier pile walls (king post or Berlin walling)

The soldier pile, usually a steel or concrete member, is driven or set in a pre-excavated hole at about 2–3 m centres. Excavation then proceeds in approximately 1 m 'steps' and horizontal sheetings or 'laggings' (timber, steel or precast concrete) are

Figure 85.24 Typical soldier pile wall showing use of timber laggings

placed to restrain the ground between the soldier pile or king post. The system is illustrated in **Figure 85.24**.

The lagging encourages horizontal arching and the king posts carry the major thrust of the ground. By incorporating walings and tiebacks, significant depths of excavation can be achieved. In very competent soil (stiff clays and soft rocks), lagging may not be necessary and simple local safety measures such as mesh pinning or guniting may be sufficient. The works are not considered permanent, although the soldier piles may be incorporated into the final structure.

King post walls are most suited to competent ground, typically dense granular material (with lowered groundwater) and stiff to very stiff clays. Conventional soldier piles are not used in soft clays and loose granular material, which may run and contribute to loss of ground. Such works are comparatively cheap to construct in suitable ground. Of all the wall types, they are the most associated with 'failure' (usually excess deformation and local movement), probably caused by attempting to use them in the wrong conditions.

King post walls are not often employed in the UK (because of ground conditions) but are much employed in the USA, Asia and in Central Europe. Local practices and useful semi-empirical processes and rules have evolved to suit local ground. Examples of such are

- the German wedging practice, which encourages arching in poorer soils (**Figure 85.25**)
- the 'louvre' lagging gaps to permit back-packing of the ground in timber and concrete lagging practice (**Figure 85.26**)
- the empirical evolution of timber capacity to loading – illustrated for competent and difficult soils.

The range of king post walling is typically up to three or four basements, although in good conditions this is only constrained by the depth to which the structural king post itself

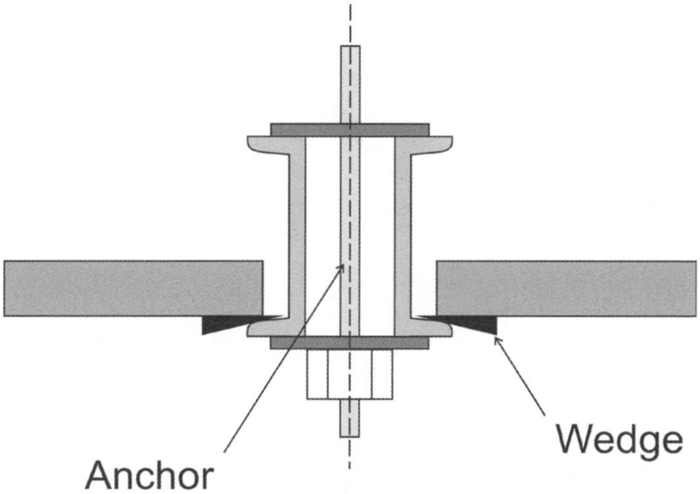

Figure 85.25 Plan of anchored retained king post with lagging wedges

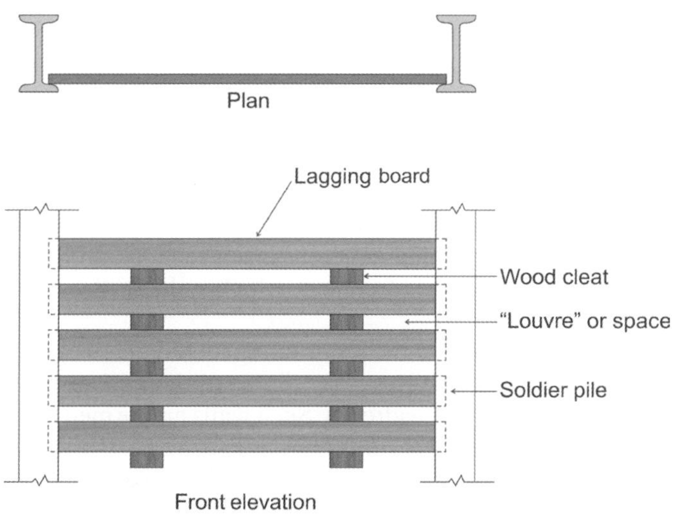

Figure 85.26 Typical simple king post wall with wood lagging

can be sensibly placed or driven and the propping arrangements. Shallow walls may be able to cope with larger lagging spans. The arching effect means that a fairly modest lagging provision can be used at depth, provided the horizontal spans are not excessive (i.e. about 2 m).

Plant used for the king post or soldier pile is dependent upon the construction system, but is similar to the driven or bored pile techniques described above. Where bored piles are used, below the dredge level, larger-diameter piles can be employed to increase bending capacity with an appropriate steel or concrete soldier pile plunged (or placed) into the bored pile.

Soldier pile walls have only the king post as their embedded component. The majority of their exposed area is dependent on the excavation, lagging and packing process, and this needs much care, discipline and good working practice and interface consideration in construction. Suitable local drainage arrangements and alert observational techniques are required for the successful completion of the system. Particular attention has to be paid if the works are adjacent to sensitive structures, since local 'blowouts', squeezing or runs can be catastrophic; this process is more prone to systemic failure than other walling types.

Despite having to cope with many significant interfaces, in the correct ground conditions and with careful processes, soldier pile walls (and their derivatives) can give very effective and economic solutions to earth-retaining problems.

85.8. Other wall types

Although categorised as 'other' wall types, the remaining types of embedded retaining wall are in essence variations or hybrids of those mentioned previously. They are generally used in niche markets in areas where local ground conditions are appropriately favourable.

Five such 'other walls' are those employing

- concrete sheeting and plastic/vinyl sheet pile walls
- soft walls with precast walls or steel sheet insertions
- unreinforced mix-in-place walls
- reinforced mix-in place walls
- jet grouted walls.

Concrete sheeting is much less robust than its steel precursor but nevertheless can be competitive if used in deep and soft soil deposits – such as the clays of Mexico City – and deep soft marine alluvium. Other than specialist head protection, its process and provision follow very similar patterns to its steel counterpart. In the correct situations it is a real and useful alternative and it can give, for example, more robust protection in the aerobic zone. It tends to be limited in depth because of joining and transportation issues for the precast panels. It has been used to support dredge depths up to typically 5 m. Plastic or vinyl sheet piles have also found some application, although their use is limited to modest retained heights.

'Soft' walls have steel sheeting or precast concrete inserts. Normally when a trench with support fluid is constructed the most economical solution is to move on to a full diaphragm wall. Occasionally (often because of space constraints) it is necessary to provide precast solutions, which can take the form of concrete panels or sheet piles. This technique has been used successfully in Germany on sites where a deep cut-off is required.

Plant and construction techniques are identical to diaphragm techniques. The 'soft' mix, which surrounds the inserts, is often a cement bentonite, which can be removed from the front of the wall and which aids 'watertightness' at the joints.

Unreinforced mix-in-place walls are employed where the ground is suitably granular, such that a reasonably consistent and homogeneous mixing of cement can guarantee a sensibly 'strong' compressive strength material. This can be employed, for example, in shafts. The walls will be relatively thick since they must act in compression, and relatively shallow (< 10 m). The required plant is highly specialised for an effective mix-in-place (**Figure 85.27**). The technique is more often encountered in Asia and increasingly in the USA.

Reinforced mix-in-place or 'firm' walls, where the reinforcement is essentially a series of king posts, have grown in popularity in areas with suitable (again mainly granular) ground. Steel and concrete precast sections carry the majority of the structural loading with the unreinforced mix-in-place material containing the ground between the soldiers. King posts are set at a spacing sufficiently close to encourage arching or to cope with the applied loading (**Figure 85.28**).

Unlike soldier pile walls, such walls are fully embedded and do restrain local ground and groundwater at the excavation stage. The mix-in-place material requires considerable attention to its formation to ensure that the king posts can be accurately plunged, yet the mix will 'firm up' (in time) to sufficient strength to cope with local anomalies and flexure. Such walls are not encountered in the UK but are increasingly popular in Asia and the USA where ground conditions are suitable.

Plant for mix-in-place applications is continually developing and can vary from drumhead rollers to multi-armed auger systems. Both reinforced and unreinforced mix-in-place walls are gaining wider application as soil is embedded in the final material and there is consequently much reduced production of spoil and utilisation of cement which reduces the overall carbon footprint.

Figure 85.28 Reinforced mix-in-place or 'firm' walls

Figure 85.27 Plant for mix-in-place walls

Jet grouted walls use the jet grouting technique, which can be carried out with compact rigs in conditions of restricted working space and/or headroom. Such walls are formed from a series of overlapping jet grout columns and are normally unreinforced and are typically used in underpinning applications.

References

ArcelorMittal (2022). *Piling Handbook*, 9th edn. ArcelorMittal, London, UK.

BSI (British Standards Institution) (2006) BS EN 10219:2006: Cold formed welded structural hollow sections. BSI, London, UK.

BSI (2009) BS EN 1993-5:2009: Eurocode 3: Design of steel structures – Part 5: Piling. BSI, London, UK.

BSI (2015) BS EN 1538:2015: Execution of special geotechnical works – Diaphragm walls. BSI, London, UK.

BSI (2022) BS 8102:2022: Protection of below ground structures against water ingress. BSI, London, UK.

BSI (2023a) BS EN 10249:2023: Cold formed sheet piling of non alloy steels. BSI, London, UK.

BSI (2023b) BS EN 10248:2023: Hot rolled sheet piling of non alloy steels. BSI, London, UK.

BSI (2023c) BS EN 12063:2023: Execution of special geotechnical works – Sheet pile walls. BSI, London, UK.

EFFC/DFI (2016) *Best Practice Guide to Tremie Concrete for Deep Foundations*. EFFC, Faversham, UK.

EFFC/DFI (2019) *Guide to Support Fluids for Deep Foundations*. EFFC, Faversham, UK.

Gaba AR, Hardy S, Doughty LD, Powrie P and Selemetas D (2017) *Guidance on Embedded Retaining Wall Design*. Report C760. CIRIA, London, UK.

ICE (2017) *Specification for Piling and Embedded Retaining Walls*, 3rd edn. ICE, London, UK.

Maloney M, Skinner H, Vaziri M and Windle J (2009) *Reducing the Risk of Leaking Substructures: A Client's Guide*. ICE, London, UK.

Yanzio E and Biddle AR (2001) *Steel Intensive Basements*. SCI Publication 275. Steel Construction Institute, Ascot, UK.

It is recommended this chapter is read in conjunction with

- Chapter 100 *Observational method*
- Section 6 *Design of retaining structures*

All chapters in this book rely on the guidance in Sections 1 *Context* and 2 *Fundamental principles*. A sound knowledge of ground investigation is required for all geotechnical works, as set out in Section 4 *Site investigation*.

doi: 10.1680/icemge.66830.1405

CONTENTS

Chapter 86
Soil reinforcement construction

Chaido Doulala-Rigby (Yuli) Tensar International Ltd, Blackburn, UK
Dave Woods Huesker UK Ltd, Warrington, UK

Reinforced soil structures have become established as reliable alternatives to conventional mass and cantilever earth-retaining structures. The combination of tensile reinforcement within the soil fill creates an internally stabilised hybrid mass as opposed to most conventional externally stabilised retaining structures. In many situations the use of reinforcement within the fill has opened up possibilities for the construction of extraordinary retaining structures that would not otherwise be feasible or would be overly expensive, enabling retaining structures in excess of 60 m in height and facilitating rapid construction while providing earth-retaining solutions resulting in attractive, stable, cost-effective and maintenance-free structures, such as bridge abutments, for their 120-year design lives. The two main components of soil reinforced structures are soil and reinforcement – hence the soil itself has a very important part to play in the success of the final result. The reinforcing materials and the available soil should have had their assessment completed before the installation process begins, while the combination of the elements can only be achieved in situ. The importance of adhering to the design specifications and materials, as well as the placement and compaction of the soil materials to achieve the adopted properties used in the design, and hence a strong and stable reinforced soil block, must be clearly communicated and appreciated by the construction team.

86.1. Introduction

Soil reinforced structures are composite structures which combine the properties of soil (earth) and soil reinforcement materials. The applications can be divided into two main areas.

- *Vertical and near-vertical walls, and steep slopes.* Walls and steep slopes are essentially soil-retaining structures (**Figure 86.1**). The composite material of soil and reinforcement is designed to have the mass and geometry necessary to prevent the retained unreinforced soil from failing. Vertical and near-vertical walls are generally faced with some form of hard facing material, while steep slopes are normally vegetated to provide a green and environmentally pleasing face. The function of the facing is essentially erosion protection and aesthetics. In terms of the design, the contribution of the facing is relatively small, but the consequences of the facing not providing the appropriate protection could be quite severe. The importance of the soil element of the composite material must be appreciated. Its adequacy is defined by both its inherent quality (i.e. its shear strength properties, chemical/toxic/pH composition, degradation and angularity) as well as the quality of the construction process, which it is highly dependent on.
- *Embankment foundations.* Embankment foundations (**Figure 86.2**) applications are generally used when the existing soils beneath a proposed embankment, which is stable within itself, are not capable of supporting the new construction. Embankment foundation applications can range from relatively light reinforcement materials providing a working platform through to very robust, piled embankment solutions with very high-strength geotextile reinforcement sheet materials. There are two British Standards publications that are relevant to the construction phase of a project, BS EN 14475:2006 (BSI, 2006) and BS 8006-1:2010+A1:2016 (BSI, 2010).

Figure 86.1 60 m high, three-tier reinforced soil wall (Courtesy of Tensar International Ltd)

The construction process can be divided into three distinct phases: pre-construction, construction and post-construction, which are discussed below.

86.2. Pre-construction

The gathering of information before construction begins is a very important part of the process and can be the key to avoiding problems later. This is particularly important if the contractor who is actually carrying out the construction is outside the original contract team. Information regarding existing

Figure 86.2 Reinforced soil load-bearing bridge abutment (Courtesy of Tensar International Ltd)

site conditions, services and any restrictions that may apply to the work should be supplied to that contractor at a very early stage. A comprehensive list of particular information required is given in BS EN 14475:2006, but there are some areas that warrant additional emphasis.

86.2.1 Roles and responsibilities

The roles and responsibilities of all the parties involved in the reinforced soil construction should be defined very clearly at an early stage of the process. Good definition at this early stage can ensure that any problems that occur during the construction process are dealt with quickly and efficiently by the responsible party. Poor definition inevitably leads to confusion and a much slower response to a problem, which may have escalated in the interim. The split of the responsibilities between the engineer, the reinforced soil designer, the reinforcing material supplier, the main contractor and the reinforced soil subcontractor could be quite complex, and the contractual links need to be well defined. In many cases one party will carry out a number of these tasks, but there could be situations where all the functions are separate.

Regardless of the type of reinforced soil structure, the first vital activity that needs to be carefully handled at the onset of every contract is a clear, unambiguous and honest conversation between all stakeholders, and in particular between the reinforced soil designer and the constructor who will build it. It is therefore vital that all parties understand the scope and importance of the various components specified in a project, 'specialised' or 'standard', to avoid, for example, unauthorised material replacements or indeed omissions that could lead to catastrophic consequences. A typical example is the unauthorised replacement of a high-spec reinforced fill with one of lesser quality due to the lack of understanding of the importance of the fill quality in the performance of a reinforced soil structure. Another typical example is the responsibility for the assessment of global stability of a reinforced soil structure often falling between the overall project designer and the designer/supplier of the proprietary reinforced soil-retaining structure, which if not clearly defined at the beginning of the project can lead to painful delays and indeed project failures. Another typical example is that, often, the responsibility of designing and coordinating temporary works vital to support the excavated retained fill prior to the installation of the reinforced soil structure is overlooked, becoming an oversight which could lead to a potentially major collapse, possibly with catastrophic consequences. An all-stakeholder 'kick-off' project meeting to ensure everybody understands the scheme and the importance of its components is vital to the successful execution of any project prior to work beginning.

86.2.1.1 Reporting channels

The arrangements for change management need to be understood by all parties to deal with any unforeseen circumstances in soil conditions, for example, and the inevitable alterations to geometry and detail that occur during the life of a project. The definition of the line of communication is vital to the smooth running of the process for standard design-generated structures, and becomes even more important when the observational approach to design is adopted.

86.2.2 Site investigation

An adequate site investigation should be carried out to determine the existing soil conditions and also the possibility of using existing, site-won soils in the reinforced soil structures. This is particularly relevant as the industry moves forward, and recycling of materials is becoming of paramount importance in efforts to offer the most sustainable solutions and to work towards achieving net-zero in construction by 2030. There may be situations where the reinforced soil contractor needs to design and install temporary works in order to carry out construction of the permanent works. Either way, the contractor will require appropriate soil testing and classification to be able to assess the suitability of locally available, site-won fills and soil properties for both permanent and temporary works design (see Chapter 49 *Sample and laboratory testing*).

Of particular importance, when choosing the reinforcement material, the chemical properties of the existing soils and the fill should be checked for any chemical degradation effects, which could be important in the selection process. Similarly, the grading of the fill should be checked to ensure material damage and soil reinforcement interaction are predictable by interpolating from existing test results, or are verified by additional investigation ahead of final design. Issues such as fill selection, compactability, global stability of the structure and temporary works staging all need to be considered holistically.

Guidance on the suitable fill types and durability data for different materials is given in BS 8006-1:2010+A1:2016.

86.2.3 Planning

The reinforced soil element of a project may be very large – in some situations being the whole construction; in others it may be only a very small part of the complete project. Additionally, the reinforced soils part of a project might be in the critical path of the project (e.g. temporary slopes/walls forming the working platforms for permanent bridge deck replacement). It is very important, therefore, to plan for the reinforced soil construction and to ensure that all the components are available in the right place at the right time.

86.2.3.1 Materials

The design and construction drawings for the reinforced soil element of the works will include reinforcing and soil material properties and facing details as appropriate. The importance of supplying the exact soil material and polymer geogrid specified in the drawings must be fully communicated to the contractor together with the implications in case the contractor fails to do so. It should also be made clear that should the contractor wish to use alternative fill material and/or reinforcement, then a full redesign must be carried out rather than simply substituting individual components (i.e. soil and/or reinforcement material 'equivalents').

In some cases, the required facing appearance and the reinforcement type will be linked by a proprietary connection detail, which must be adhered to unless an alternative reinforcement type and design are proposed altogether. Information on suitable fill types for particular applications is given in BS EN 14475:2006.

Considerations when Using Non-standard Soil Fill

Some practical but equally important considerations, specific to when using 'non-standard' fill, are listed below.

- Non-standard, site-won fill is often highly variable, so appropriate mixing and adequate numbers and types of soil samples must be taken for laboratory testing to be representative.
- Non-standard fill needs to be appropriately classified and the use of appropriate testing needs to be assigned to derive its required design properties. The most usual laboratory testing to derive shear strength parameters to be used in the design of any reinforced soil structure is a shear box test; use of an appropriately sized shear box is vital, and in cases where stones are present in the site-won fill, careful consideration of the amount of stone removal must take place together with the effect of such stone removal on the behaviour of the soil. If too many stones are removed then the derived shear strength properties will not be representative as the reduced-stone soil samples are likely to behave in a different manner to the actual site-won fill from the available raw stockpile.
- Appropriate screening of site-won material and removal of extra-large cobbles (>125 mm) must take place, preferably prior to construction beginning or during construction, if necessary.
- Unlike the reasonably fast shearing rate in the order of 0.8 mm/min that is expected for coarser gravelly sandy fills, if finer/clay-rich materials are proposed as reinforced fill, the rate of shearing in the shear box testing needs to be carefully specified. For a fine, sandy clay, for example, a very low shearing rate in the order of 0.001–0.009 mm/min is expected to be used after the specimens are submerged/saturated and then consolidated, in which case testing would take at least 6 days! This is the only correct way for slow-draining clay fills to be tested to produce true, representative 'drained' shear strength parameters to be used in design rather than assigning the 'typical' fast rate of shearing, which would lead to false, non-representative shear strength design properties and consequently to unsafe design.
- Depending on the specified reinforced fill material nature, whether imported, site-won or man-made, especially if it is particularly angular, unusually fine, rounded or in general 'non-standard', and has not been used with the specific geogrid before, bespoke, project-specific laboratory shear and pull-out testing, and maybe even damage or abrasion testing, will need to take place preferably prior to the start of design, but definitely prior to construction, to investigate the effect it will have on the geogrid used to reinforce, as well as to derive representative interaction parameters between geogrid and fill, such as pull-out and sliding factors to be used in the design.
- Careful chemical analysis of 'non-standard' fill, especially when elevated pH levels are suspected, must be carried out. HDPE is largely inert to chemical attack and to environments with pH4 to pH12.5, but not all soil reinforcement geogrids are, so added FoS should be incorporated in the design as required, or avoidance of specific polymer geogrids altogether if they seem overly susceptible, based on the findings of the chemical analysis.
- Likewise, in projects where the 'non-standard' fill comprises recycled waste fill rich in contaminants like bisphenol A (BPA) or polychlorinated biphenyl (PCB) and/or similar, apart from added personal protection equipment (PPE) considerations, special consideration must also be given to the design of the capping layer over the reinforced soil structure to effectively contain the contaminants and prevent future contamination through leaching or even through airborne paths.
- For recycled site-won fills, especially fills that are derived from demolition waste rich in brick fragments, for example, additional abrasion testing may be necessary to establish crushability under compaction so that appropriate plant and method of compaction can be used that will not lead to fill particles breaking down under compaction.
- Similar considerations should be taken into account when other 'non-standard' fills, such as lightweight aggregate (LWA), is used as reinforced fill, in which case close cooperation with the LWA provider is imperative, and their advice for handling and compacting LWA must be strictly followed.
- In situ specific testing and site controls should be prepared, such as bespoke, site-specific schedules of contingency measures to be followed on site for projects where highly cohesive soil is used and challenging climatic conditions are expected.

86.2.3.2 Material testing

The design will include the specified properties of the fill materials that have been used to derive the reinforcement layout. The actual fill to be used for construction on site should be tested or verified to ensure that it satisfies the design value (see also Chapter 99 *Materials and material testing for foundations*).

The reinforcement materials are the subject of the Construction Products Regulation (CPR, 2011), which replaced the Construction Products Directive (CPD, 1988), and as such will have independent CE marking or equivalent information to verify the properties – providing that the harmonised properties are relevant to the application for which the reinforcement is being used.

Once the fill material and the reinforcement have been selected, the interaction parameters between the two materials need to be determined to verify whether the design parameters are satisfied. A lot of interaction factors are freely available for the most commonly used fills. Where bespoke new fill materials are used, such as LWA, a reinforcement material-specific interaction testing must be carried out for the derivation of representative interaction parameters. More comprehensive details of the information that would be expected to be in a design output are given in BS EN 14475:2006.

86.3. Construction

86.3.1 Delivery and storage of materials

Identification of the different components involved in the construction, separation and isolation of apparently similar materials is very important in preventing mistakes. Facing components should be stored carefully as they are the face of the structure on completion. Some reinforcing materials may require specific storage conditions, such as being protected from the sunlight (UV exposure) and/or remaining dry; storage guidance should be obtained from the material supplier and strictly adhered to.

86.3.2 Preparation of the foundation

The foundation level and extent will be indicated in the design and, as with any structure, the accuracy in defining foundation layer strength is vital in the production of a serviceable end result. In situations where the foundation soils are weak, it is important to limit the disturbance that they have to endure during the construction process. The parameters used in the design will generally be those of the foundation soils in their pre-construction condition using the parameters from the ground investigation report (see Chapter 50 *Geotechnical reporting*) and hence any major disturbance of those soils will have a significant effect on the final result. Foundation soil may need to be remediated to be capable of supporting the proposed reinforced soil structure above. The depth of the founding layer for the reinforced soil is typically governed by the minimum depth for frost heave of 450–500 mm, but may be deeper to improve global stability or to provide additional protection against future excavation for service installation or maintenance in front of the reinforced soil structure.

When hard facing materials are used in the construction of reinforced soil walls, there is generally a requirement for some form of levelling pad, concrete or sometimes gravel, on which to place the first face elements. This pad does not have a structural purpose, but it does provide a firm surface from which the wall face rises. As with any engineering structure, errors in line or level at the foundation level will be very difficult, if not impossible, to rectify later. Soft-faced walls and slopes, such as vegetated faces, do not generally require a levelling pad.

When constructing reinforced soil structures over variable foundation soils, the inclusion of a 300–500 mm granular mattress stabilised with one or two layers of geogrids may be considered as a good measure to mitigate differential settlements. Such a granular layer may also serve as a drainage blanket layer at the base of the reinforced soil structures, as further captured in the section below.

86.3.3 Installation of drainage

Drainage of structures is important for two reasons. First, there is a design aspect regarding the possible presence of water within or outside a structure. Reinforced soil structures are typically designed to remain reasonably dry during construction, but more importantly also during their design life. Water that is not allowed to drain away from the reinforced soil structures can have a significant effect, so if there is a possibility that the materials being used are not free-draining, then drainage layers should be included at the back and underside of the reinforced soil structure to prevent any groundwater entering the reinforced soil block and linked into the drainage system on the site. Similarly, appropriate measures should be taken to make the crest reasonably impermeable and with an appropriate inclination for the surface water to be drained/collected and diverted away from the wall face, and to also prevent excessive surface water infiltration. Some particular fill materials, such as pulverised fuel ash (PFA), require particular drainage arrangements to ensure isolation of the PFA block (BS 8006:2010 Cl. 6.10.5.2, Cl. 6.10.5.3 and Cl. 6.10.2.6.3).

Drainage around the reinforced soil should be viewed in conjunction with the site drainage as a whole to ensure that excess water from elsewhere on site is not drained inadvertently into the reinforced fill. Sequencing of any site drainage installation and potential for run-off from partially completed stages should be reviewed in conjunction with the reinforced soil design to ensure the structure is not adversely affected in the event of heavy rainfall during construction.

The second important reason for drainage provision is aesthetic. A number of facing block finishes have open joints and staining can occur if there is no drainage layer behind the block face to collect and transfer any water seepage down to the toe and away from the wall. It is very important that the drainage layer behind the block face must not be exposed to daylight at the wall crest but must be terminated at least

500 mm below the crest to avoid the drain acting as a funnel for the surface water into the reinforced soil block.

86.3.4 Placement of facings

General information on the placement of facings is given in BS EN 14475:2006. It must be noted that while the facing does not have a true structural function, it does provide erosion protection and it is the visible result of the construction. The accuracy of placement of the facing and the care with which the fill is placed and compacted behind it define the quality of the final structure. A good understanding of these principles by the reinforced soil contractor's team will result in sound and serviceable structures.

The flexibility of modular block facings for reinforced soil walls enables complex geometric shapes to be constructed. Care should be taken to ensure that the heights of free-standing blocks without reinforcement are limited to those advised by the supplier or manufacturer. Exceeding the recommended heights can cause great difficulties in keeping the line of the wall within acceptable tolerances or future failures. Indicative tolerances are given in Annex C of BS EN 14475:2006 (see **Figure 86.3**).

Connection details between the reinforcement layers and the facing must be reflected in the design and must comply with the manufacturers' specifications.

Lifting and placing of the larger panel facings can be carried out using craneage, whereas the modular block facings can generally be handled manually. The limitations of block size and weight to comply with the health and safety requirements for manual lifting must always be observed.

86.3.5 Placement of fill materials

86.3.5.1 Walls and slopes

While all the tasks that are involved in the construction of a reinforced soil structure are important to produce a satisfactory and pleasing result, the placement and compaction of the fill is probably the most important operation of the whole process, assuming that the project-specific fill that was been specified is used. The development of a true composite mass of soil and reinforcement depends on the two materials being intimately related to each other by the appropriate mechanism. Strip-type reinforcement materials generally rely on a frictional interaction on the surface of the reinforcement, while grid-type materials interact through an interlocking of the soil particles into the geogrid apertures. Whichever mechanism is appropriate, it will only be mobilised effectively if the soil is placed and compacted to standard specification levels. The design of structures, vertical walls and steep slopes to BS 8006-1:2010+A1:2016 uses the peak friction angle for the reinforced fill and for the unreinforced backfill. The peak friction angle will only be achieved if the compaction of the soil is carried out to specification. The case study in **Box 86.1** shows how a well-compacted, good-quality fill material can produce

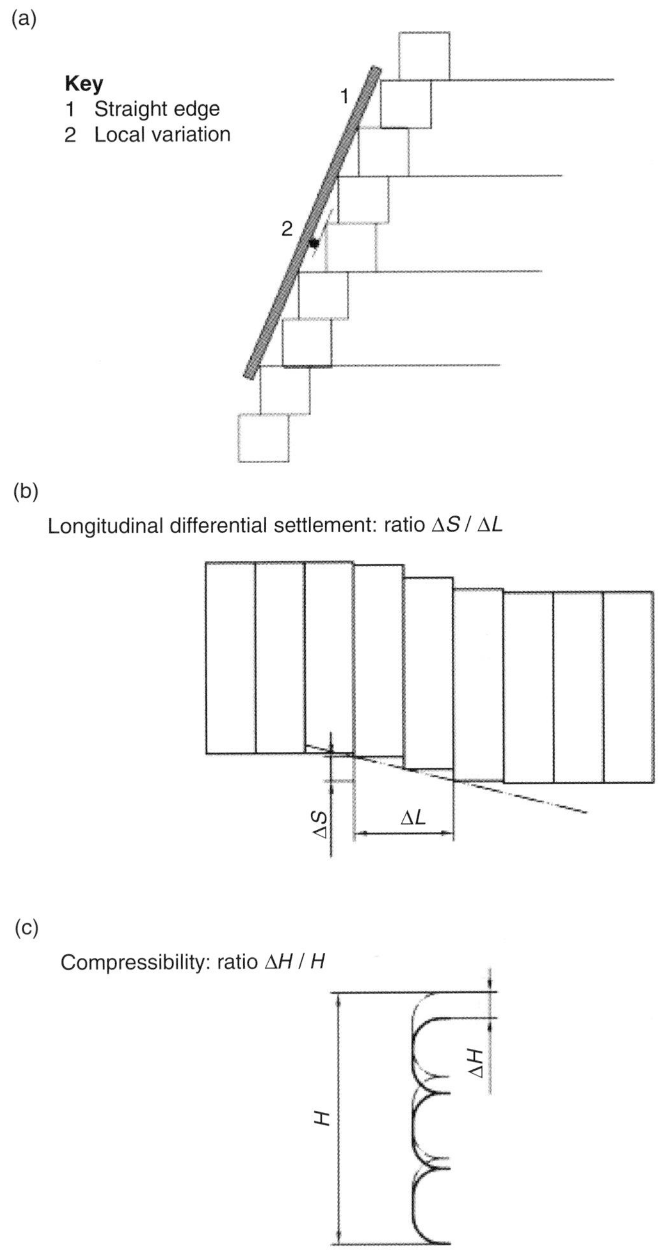

Figure 86.3 Indicative construction tolerances from BS EN 14475. (a) Cross-section through wall; (b) front elevation of wall; (c) cross-section through wall (Reproduced with permission from BS EN 14475, Annex C (2006) © BSI)

a completely stable and durable soil reinforcement composite mass even under severe circumstances. BS EN 14475:2006 gives a large amount of guidance on this aspect of the construction, recognising its importance.

Particular care should be taken to follow the guidance regarding compaction close to the face; see **Figure 86.4**. Incorrect operations in this area can cause face deformations which may be irretrievable and result in an unsatisfactory appearance.

Figure 86.4 Limitations of compaction plant close to the structure face (Courtesy of Tensar International Ltd)

86.3.6 Placement of reinforcement

86.3.6.1 Walls and slopes

Guidance on reinforcement placement is generally available from the manufacturer or supplier, and BS EN 14475:2006 also gives helpful information. The main points to bear in mind are that the reinforcement should not be damaged in the construction process and that any flexible reinforcements should be pulled tight, not tensioned, to ensure that any slack is removed from the system. Construction equipment should not run on the reinforcement, regardless of type – there should always be a layer of at least 150 mm of soil above the reinforcement before any traffic is allowed, whether by the compaction plant or other site equipment.

Connection of the reinforcement to the face unit is a critical aspect of reinforced soil structures and can be achieved in many ways. Steel strip reinforcement may be bolted to anchor plates embedded within the facing panel, while there are many proprietary fixing devices for the connection of geosynthetic reinforcement to modular block facing systems. Some polymer reinforcement soil systems rely on mechanical connection between the reinforcement and the modular facing blocks, while others rely purely on frictional connection. Reference should be made to manufacturers' literature for full details.

Pulling the reinforcement tight minimises the possibility of post-construction face movement. This may happen if the face connection between the reinforcement and a facing block, or a joint between two pieces of reinforcement, is left with slack. A typical example of removing the slack is shown in **Figure 86.7**.

Protection of the connections during the construction process, particularly the placing and compacting of the fill materials, is very important. The restrictions on compaction equipment weight and thickness of fill over the reinforcement layers must be followed. Long-term durability requirements of the connection should also be carefully implemented in accordance with the manufacturer's specification to avoid possible post-construction problems.

Box 86.1 Case study

A 19 m high reinforced soil wall constructed in a very dry climate was subjected to a flash flood before the contractor had placed sufficient scour protection at the toe of the structure. Consequently, over a short length, the levelling pad was undermined and the facing blocks fell away from the reinforced soil block. **Figure 86.5** illustrates a flash flood that occurred later at the site.

Figure 86.5 An example of severe flood water flow (Courtesy of Tensar International Ltd)

However, the fill material was of very good quality and the contractor was careful to ensure that the compaction matched the specification required, with the result that the main structural element, the reinforced soil composite block, remained completely serviceable throughout this very severe encounter. During the face reconstruction, the reinforced soil block did not require any treatment and the road remained open and serviceable with no signs of cracking or damage. **Figure 86.6** illustrates the stable reinforced soil block.

Figure 86.6 Result of lack of scour protection, stable reinforced soil block after critical event (Courtesy of Tensar International Ltd)

The wall face was reconstructed with the correct scour protection in place to prevent a recurrence of the problem.

(a)

(b)

(c)

Figure 86.7 Typical procedure for the removal of any slack in flexible geosynthetic reinforcement. (a) A length of geogrid reinforcement having a nominal manual tension applied through a hooked beam to remove any slack in the system. (b) The nominal manual tension being applied to the connection between the geosynthetic and the facing blocks. (c) Temporary fixing of the nominal tension with anchor pins while the fill is placed onto the reinforcement (Courtesy of Tensar International Ltd)

Reinforcement geogrids can be cut carefully and locally to accommodate reasonable-size tree pits and/or isolated fence footings. However, reinforcement geogrids cannot be cut longitudinally to accommodate continuous footings. Such continuous foundations to accommodate fences or services need to be contained as independent structures on top of or within the reinforced soil structure. Water pipes can be accommodated within reinforced soil structures but need to be installed concurrently with the reinforced soil structure construction and must be made watertight to avoid any future leakage. Reinforcement geogrids cannot be cut through for future maintenance, and if access to the services is required for future maintenance, an independent service trench must be designed and installed within the reinforced soils structure concurrently during construction.

86.3.6.2 Embankment foundations

The principles of reinforcement placement in embankment foundations are essentially the same as for walls and slopes, except there will not be a face against which to take any slack out of the system. Joints or overlaps will probably be required, and these are the areas that require the most care. Working below water brings its own special requirements when using materials which actually float; weighting down may be required.

86.3.7 Vegetation

A very common facing for steep slopes is vegetation, where the nutrient soil for the vegetation is placed behind some sort of open facing and the vegetation grows through to provide a soft face to the structure. There is a very comprehensive section on vegetation in the revised BS 8006-1:2010+A1:2016, but the main requirements for survival of the face are sun-facing exposure, water and nutrients in the face soil. To ensure the successful establishment and continued success of any vegetative facing, either hand-planted or seeded, a number of points should be considered.

- Topsoil should be clean and free of any potential phytotoxic substances.
- Steeper face angles will receive less rainfall and suffer faster run-off; steps in the facing can increase infiltration.
- Face orientation can be critical to vegetation growth, with north-facing slopes receiving very little direct sunlight and south-facing slopes receiving full sunlight for many hours during summer, both of which can hinder or prevent the growth of certain species of shrubs and grasses.
- For slopes adjacent to highways, the lower portion of the wall may be subjected to road de-icing salts during the winter, with the raised salinity levels posing a problem for some species of vegetation.
- Maintenance of the slope surface will depend upon the vegetation species selected, but some seasonal cutting may be required to maintain sight lines on roads and to rejuvenate overgrown faces. Timing of the introduction of vegetation is a planning issue and careful thought can improve its establishment.

86.3.8 Monitoring and supervision

86.3.8.1 Walls and slopes

The construction of reinforced soil walls and slopes is a relatively straightforward operation using well-accepted engineering earthworks principles. The main areas of interest – the supervising and monitoring of the construction – can be summarised as follows.

- Compaction of the fill. Ensuring and verifying that the compaction meets the specification is of major importance with regard to the development of the design soil properties and the true composite block performance.
- Diligent observation of the line of any facing during construction will ensure an early warning of any tendency to veer out of tolerance.
- Monitoring any reinforcement connection details – reinforcement to face or reinforcement to reinforcement – and ensuring that any slack is taken out, particularly of flexible reinforcement materials, will benefit the final appearance.
- Good temporary drainage provision during construction to avoid the effects of excessive infiltration and residual groundwater within the reinforced soil. A good way of doing this is by creating a slight inclination at the reinforced soil structure level reached at the end of each working day and also covering it with tarpaulin sheets or similar to avoid water inundating the structure.
- In critical structures, like load-bearing bridge abutments, post-construction monitoring of both vertical and horizontal movement is recommended to confirm the design predictions and to also increase confidence in these relatively novel, sustainable and cost-effective reinforced soil bridge abutment structure alternatives to reinforced concrete.

86.4. Embankment foundations, basal mattresses, reinforced working platforms and haul roads

The construction of reinforced embankment foundation projects is not as formalised as for walls and slopes, and each project will have its own specific characteristics. A construction sequence, specific to the project, should be part of the design and any monitoring and supervision requirements will be specified in that document. General earthworks principles will, of course, still apply.

86.4.1 Embankment foundations and basal mattresses

BS EN 14475:2006 and BS 8006-1:2010+A1:2016 give good guidance on this quite specific application area. The reinforcement of embankment foundations generally involves the requirement to place fill over soft soils to provide a higher-strength reinforced layer which resists potential slip failures passing through the embankment fill into the soft soils beneath.

Where the founding soils are insufficient to carry the weight of the embankment, rigid inclusions such as piles, vibro stone columns or constant modulus columns can be used to carry the embankment load to a suitable bearing stratum. This combination of rigid inclusions in a very soft soil can lead to uneven deformation of the embankment and cracking at the finished ground surface. In this instance, high-strength reinforced soil mattresses may be designed to span between the rigid inclusions, providing a catenary suspension of the embankment weight between the support points to minimise reflected deformations at the embankment surface (**Figure 86.8**).

Figure 86.8 High-strength geosynthetic spanning between rigid inclusions (Courtesy of Huesker)

In situations where soft soils of relatively low permeability underlie a new embankment and pose problems in terms of potential lengthy consolidation periods, vertical sand drains or geosynthetic preformed vertical drains are often used to shorten the drainage path and facilitate rapid dissipation of the elevated pore pressures generated by the placement of the new embankment fill, such that the consolidation settlement occurs within the construction period of the embankment. It may be necessary to use geosynthetic reinforcement at the base of the new embankment to resist slip circles forming through the underlying soft soils as the embankment height increases. Care should be taken in the sequencing of the embankment construction to ensure adequate dissipation of the pore pressure, with the reinforced global stability design being checked for all potential stress conditions and if necessary a system of pore pressure measurement being installed with critical trigger levels set to ensure the embankment is stable throughout all stages of its construction.

The edge detailing of all reinforced embankment foundations and basal mattresses needs to be carefully considered to ensure there is sufficient cover to the reinforcement to generate the required tension and pull-out resistance and to prevent the risk of lateral extrusion of the soft soil beneath the edges of the embankment. Details such as anchor trenches, wrap-around returns or anchoring the reinforcement around a dead weight such as a filled gabion or soil bag might be necessary to prevent edge failure (see **Figure 86.9**).

Figure 86.9 Anchor trenches necessary to prevent edge failure (Courtesy of Huesker)

In cases where the foundation soils in between the piles are of firmer nature and in order to induce high-strength reinforcement straining during construction, a layer of soft soil may be necessary between the pile caps/pile heads (**Figure 86.10**).

86.4.2 Reinforced working platforms and haul roads

In the temporary condition, sites over poor soil will require adequate working platforms and haul roads to facilitate site access and the safe movement and operation of heavy plant such as piling rigs and cranes. With permission from the Temporary Works Forum, this section has largely been extracted from their recently published 'Guide to good practice' of 2019.

Historically, the design of granular working platforms for construction plant has not been carried out in a consistent manner across the industry. In the past, the methods have generally consisted of what might loosely be described as 'empirical', and have largely been based on previous experience of suitable materials and thickness. Additionally, formal design methods have been used such as classical bearing capacity methods, for crane and piling platforms, or the TRRL LR1132 method, for haul roads. One of the alternative approaches that has been adopted is the use of plate loading tests to prove platform capacity. Another is the use of design methods developed by specialist geosynthetic manufacturers. The publication of CIRIA SP123 and BRE BR470 introduced new analytical design procedures for the design of both unreinforced and reinforced granular platforms. Both methods include the ability to incorporate geosynthetic reinforcement to reinforce the working platform and allow an effective reduction in the depth of granular fill required to ensure the platform is secure.

Figure 86.10 A layer of soft soil placed between the pile caps/pile heads to facilitate high-strength geotextile straining during construction (Courtesy of Huesker)

It is stated in BR470 that alternative methods may be adopted and the use of geosynthetic manufacturer's design methods was expanded on in a supplement, 'Use of "structural geosynthetic reinforcement" – a BRE review seven years on' (issued by the BRE in 2011). In addition to the standards and guidance documents that are available, a great deal of product design and development have also been undertaken over the years by geosynthetic manufacturers. This has resulted in the development of various design methodologies that are bespoke to their products. Such alternative methods are deemed acceptable for the design of temporary working platforms, provided that the manufacturers can produce sufficient evidence that their proprietary methods have been validated by extensive past experience and by appropriate experimental testing.

For temporary haul roads subjected to repeated trafficking and the subsequent risk of rutting of the granular surface but less likely to be subjected to the higher loads applied to working platforms through crane outriggers and tracks there are a number of design methods used by different geosynthetic manufacturers, these typically rely on either a tension membrane effect to support the area as a rut begins to appear or the lateral restraint of the granular fill in the haul road through friction, interlock or confinement. The type of geosynthetic selected usually determines which method is the more dominant with the two effects generally being separated into different geosynthetic functions of reinforcement for the tensioned membrane effect where the ultimate tensile strength of the geosynthetic resists the loading generated by rut formation at a strain of 2% to 6% typically, or stabilisation where lateral restraint achieved by confinement and interlock of the granular fill at the level of the geosynthetic occurs at very low strain, typically in the order of 0.5%.

In all cases, it is important to note that the contribution of the geosynthetic to the performance of a granular working platform varies between the different types of geosynthetics available in the market. Direct substitution of alternative geosynthetics in proprietary geosynthetic working platform designs must be avoided due to the specific differences in product development. It is, therefore, strongly recommended that when alternative geosynthetic is considered for a design, it should be accompanied by a corresponding alternative, geosynthetic specific design.

For construction, the following points are recommended as guidance and good practice.

- For reinforced/stabilised platforms the minimum platform thickness should be 300 mm; the maximum thickness is 1.0 × track width.
- Minimum cover over geosynthetic reinforcement should be 300 mm. It should be noted that the minimum cover over geosynthetic reinforcement may be reduced to 150 mm where further advice is obtained from a geosynthetic manufacturer.

Spacing of geosynthetic reinforcement. Under certain circumstances more than one layer of geosynthetics may be necessary or prove beneficial. While the positioning of the layers within the platform can be relatively arbitrary, it is recommended that

- they are evenly spaced
- vertical spacing should not exceed 450 mm
- vertical spacing should not be less than 150 mm.

Edge detailing for all working platforms must be carefully considered and designed for, depending on each sites individual soil and loading conditions. The edge of the working area must be clearly marked to define an exclusion zone between the working area and the edge of the platform.

It is important to maintain both the minimum thickness of platforms and the engineering properties of the fill. If either becomes compromised, the structural integrity of the platform may be in doubt.

It is also important to ensure that the crushing strength of the aggregate, particularly where recycled brick is used, is suitable. Where 'house brick' may be included in the mix, for example, it may be necessary to specify a maximum proportion to be mixed with crushed concrete.

The specification for construction of the platform should cover materials, workmanship and use. It should generally be based upon or refer to standard specifications (e.g. Specification for Highway Works), standards, guidance documents and/or supplier literature

Upon completion of the working platform and/or haul road construction, inspections and appropriate tests should be carried out to confirm the adequacy of the construction. Such testing should be specified and included in the specs by the designer. This may include separate testing for both the formation and finished platform to confirm that strength and deformation parameters are within acceptable margins and/or comply with design assumptions. The type, frequency and acceptance criteria should be provided for all specified tests. All applicable standards and/or equipment suppliers' guidance for execution of tests should also be referenced.

The most common test used for on-site working platform adequacy verification is the plate load test. In order to be representative of the actual site loading conditions, the plate diameter should be similar to the diameter of the piling rig/crane track width to cover the track loading influence zone. The frequency of plate bearing tests should be as follows.

- For the working platform
  - a minimum of three tests per site
  - at least one test per 500 m^2 for sites up to 3000 m^2
  - at least one additional test per 1000 m^2 for areas over and above 3000 m^2
- For the subgrade
  - as for the platform but this should be increased if the subgrade is known to be significantly variable over the site.

In all cases, additional tests should be carried out if initial results are variable.

86.5. Post-construction

86.5.1 Records

Records of weather conditions, for example, during reinforced soil construction should be the same as those conventionally kept on construction sites. A particularly important area of record keeping is the location of the reinforced soil structure itself. The fact that a structure is constructed of reinforced soil may not be obvious; a concrete-faced wall may look the same as a reinforced concrete wall, and hence an indication that soil reinforcement is present is important in the event of further construction in the future to make sure that the reinforcement elements will not be damaged, which will compromise the stability of such a structure.

86.5.2 Maintenance

In general, reinforced soil structures are low-maintenance or maintenance-free. The usual requirements of ensuring that drainage is maintained and excessive vegetation, if any, is managed still apply, and in extreme conditions can be very important. Maintenance of vegetated faces is an area that may need some consideration in terms of irrigation and plant species selection.

References

BSI (British Standards Institution) (2006) BS EN 14475:2006: Execution of special geotechnical works – Reinforced fill. BSI, London, UK.

BSI (2010) BS 8006-1:2010: Code of practice for strengthened/reinforced soils and other fills. BSI, London, UK.

Temporary Works Forum (2019) *Working Platforms: Design of Granular Working Platforms for Construction Plant – A Guide to Good Practice*. Temporary Works Forum, London, UK.

Further reading

European Federation of Foundation Contractors (2020) *Guide to Working Platforms*. EFFC, Faversham, UK.

Forde MC (ed) (2009) *ICE Manual of Construction Materials*. Thomas Telford, London, UK.
Jones CJFP (1996) *Earth Reinforcement and Soil Structures*. Thomas Telford, London, UK.
Koerner RM (2005) *Designing with Geosynthetics*, 5th edn. Prentice Hall, Upper Saddle River, NJ.
McAleenan C and Oloke D (2010) *ICE Manual of Health and Safety in Construction*. Thomas Telford, London, UK.
Raymond GP and Giroud JP (eds) (1993) *Geosynthetics Case Histories*. BiTech Publishers, Richmond, BC.

Useful websites

International Geosynthetics Society (IGS): http://www.geosyntheticssociety.org
Temporary Works Forum Working Platforms Guide to Good Practice: https://www.twforum.org.uk/viewdocument/working-platforms-design-of-granu
UK chapter of the IGS: http://www.igs-uk.org

It is recommended this chapter is read in conjunction with

- Chapter 72 *Slope stabilisation methods*
- Chapter 73 *Design of soil reinforced slopes and structures*

All chapters in this book rely on the guidance in Sections 1 *Context* and 2 *Fundamental principles*. A sound knowledge of ground investigation is required for all geotechnical works, as set out in Section 4 *Site investigation*.

doi: 10.1680/icemge.66830.1417

Chapter 87
Rock stabilisation

Richard Nicholson AECOM Ltd, Chesterfield, UK

CONTENTS

Rock faces can be natural or man-made features of the landscape that can arouse significant local or national interest. A designer needs to be sensitive to the position and location of a rock face and give careful consideration to the likelihood and consequences of a failure. This is likely to determine the conceptual approach to any stabilisation works. Risk mitigation measures can comprise a management approach where failure of the rock face is permitted, but the consequences of a rockfall are managed. This can be achieved through the installation of warning systems, rockfall netting or catch fences. Alternatively, an engineered solution can be sought, typically comprising the removal of the hazard or a reduction in the likelihood of a rockfall. Examples of remedial works adopting this approach would include scaling, installation of grouted anchors or the application of sprayed concrete.

Installation and subsequent maintenance of any remedial works should be undertaken by a specialist with experience of working in potentially exposed locations on an unstable rock face. Careful consideration needs to be given to access requirements, the safety of the workforce and third parties. It is therefore essential that a full risk assessment is undertaken at the design stage and reviewed throughout construction.

87.1. Introduction
87.1.1 Principles

Rock faces can be formed naturally or as a result of construction activities. The stabilisation of rock faces can therefore be undertaken on existing rock slopes (often on a reactive basis) or newly formed faces, either during or after their excavation. Landowners and asset operators owe a duty of care to ensure that the safety of the site user is not compromised.

In situ rock is unlike the majority of materials encountered by engineers. Rock can be highly variably within a short distance and usually only the front face is visible. When working with in situ rock it is critical to have a clear understanding of the geological context and the engineering characteristics of the rock mass, as well as the forces that are acting upon it. The design process is likely to require an inspection of the exposed face and an intrusive ground investigation to obtain an understanding of not only the engineering properties of the rock, but also the quality of the rock mass as a whole. The design of stabilisation measures should therefore be undertaken by a suitably qualified and experienced specialist.

When selecting the design approach to be adopted, the designer will need to consider both the likelihood and consequences of a failure of the rock face. The land use at the toe and crest of the rock face is likely to determine whether risk mitigation measures comprise a management approach, where failure of the rock face is permitted, or whether a fully engineered solution is required.

87.1.2 Access considerations

The design of any remedial measures will need to give careful consideration to the access requirements, both for construction and subsequent maintenance of the works (**Figure 87.1**). Newly cut faces can be stabilised during excavation in a top-down manner as long as there is a clear understanding of the scope of the whole works.

Existing rock faces in remote locations may be difficult to access from the nearest highway. Alternatively, some rock faces will preclude working from the toe or crest (e.g. due to the presence of infrastructure or water). The inherent nature of existing steep and irregular rock faces may also preclude the use of traditional plant and scaffolding. Consideration may need to be given to employing a contractor specialising in difficult access techniques. This will typically involve specialist plant reaching up from the toe or reaching down from the crest of the rock face. Alternatively, equipment may be suspended by steel cables attached to a temporary access system located behind the crest and winched into position on the face.

Figure 87.1 Different access solutions (© CAN Geotechnical Ltd)

Personnel undertaking the works must therefore not only be competent in undertaking the specified work tasks, but also be competent in accessing the work site. Rope access methods in accordance with BS 7985 (BSI, 2013b) have become increasingly accepted as offering a safe, cost-effective access solution.

87.1.3 Safety considerations

Given the nature of rock stabilisation works, they are often undertaken on faces that pose a risk of injury due to rockfall. In assessing the stability of the face, careful consideration must be given to the safety of the workforce, other contractors and members of the public. The works should be planned such that a safe working environment can be provided at all times. Consideration may therefore need to be given to the implementation of the following mitigation measures

- temporary works including removal of loose rock or strapping of blocks
- implementation of an inspection and monitoring regime during the works
- a method of working such that construction is undertaken in a controlled, systematic manner (e.g. top-down)
- instigation of exclusion areas at the toe or crest to protect the workforce and third parties that in turn may necessitate the temporary suspension of infrastructure such as road or rail services.

It is therefore critical that a full risk assessment is undertaken at the design stage and reviewed throughout the construction stage. It is also important that only competent contractors are employed, with a proven track record for the safe delivery of rock stabilisation works.

87.1.4 Environmental considerations

Existing rock faces are natural features of the landscape that often arouse significant local or national interest from various organisations. When designing or undertaking any stabilisation works, due regard for the conservation, recreation and access requirements for the rock face and the surrounding area should be given. Environmental considerations may include

- setting and location
- fauna and flora
- geological interest
- historical interest
- recreational amenity (walking/climbing).

Each of the above may affect the design approach and the scope of works or the time of year in which the works are undertaken. In addition, the rock face or the surrounding area may have statutory protection (e.g. RIGS, SSSI); where this is the case, consultation with regulatory bodies will be required.

87.2. Management solutions

Risk mitigation measures can comprise a management approach where failure of the rock face is permitted, but the consequence of a rockfall is managed to minimise the potential impact of a failure.

87.2.1 Warning systems

Warning systems can be quick and easy to install and provide a reactive solution while permanent works are designed, or alternatively can be installed as part of the ongoing management regime for the site.

Warning systems can take the approach of simple signage to warn site users in areas of known rockfall (e.g. as typically seen on highways). More sophisticated warning systems include the regular inspection and/or monitoring of rock faces where the consequences of a failure may be considered more significant (e.g. as used on railway cuttings). This may include the installation of simple trip wires or instrumentation such as load sensors and tiltmeters that can be monitored remotely by a data logger and modem. If agreed trigger values are reached, then alarms can be activated, and appropriate action taken (such as activation of automated warning signs or signals). While warning systems do not reduce the likelihood of a rockfall occurring, they can reduce the potential consequence and therefore the risk associated with a rockfall event.

87.2.2 Rockfall netting

Rockfall netting is a cost-effective method of controlling the fall of rock debris and discrete blocks (**Figure 87.2**). There are several products readily available, including composite and PVC-coated steel wire meshes. These products have a variety of aperture sizes and tensile strengths, and are manufactured to suit different corrosive environments. The netting is typically hung as a drapery with rock allowed to fall to the toe of the face.

Consideration must be given to the typical block size, mass and trajectory to ensure that it will be contained by the

Figure 87.2 Typical draped rockfall netting installation (© CAN Geotechnical Ltd)

netting. It is important to realise that the netting will only be as effective as the weakest component within the system. The netting must therefore be secured to a robust anchor system installed behind the crest. This typically comprises a crest cable fixed to rock bolts with a bond length within competent rock. Should soil materials be present at the crest then short soil nails or sometimes driven earth anchors are used to provide the necessary restraint.

Adjacent rolls of netting must be securely connected in accordance with the manufacturer's recommendations. This typically comprises the use of tie wire or steel hog rings fixed along the seam, or selvedge wire at regular centres. This join is often the weakest component of the system. Steel cables can also be woven or fixed to the selvedge to act as reinforcing elements.

Consideration needs to be given to the proximity of infrastructure to the rock face, as accumulations of trapped debris may extend well beyond the toe of the face. Careful detailing of the toe fixings is also important to ensure that rock debris can be easily and safely released for periodic uplift and clearance.

87.2.3 Catch fences and ditches

Where there is room at the toe of a rock face, consideration can be given to the construction of catch ditches (also known as rock traps) and/or bunds. The purpose of a catch ditch is to contain rock debris within a zone close to the toe of the face. Simple earthwork solutions can be cost-effective, particularly where the height or condition of a face would require extensive remedial measures to be installed. For example, bunds are commonly used in working quarries where the location of rock faces can be temporary and earthmoving plant is readily available.

Ditches and bunds can be constructed in combination where there are soil materials at the toe of the face (e.g. scree slope). Ditches can be lined with uniformly graded aggregate to dissipate the impact energy from falling rocks to reduce the bounce height and further mitigate the risks associated with rockfall. Bunds can also take the form of walls (typically gabion structures), as often seen in alpine areas to contain or divert debris flows away from infrastructure.

Catch fences may be installed where trajectory analysis suggests that bounce heights would require substantial bunds or ditches to be constructed (**Figure 87.3**). Catch fences can be sacrificial, designed to slow down and reduce the kinetic energy of a falling rock, or may be permanent, being designed to withstand repeated impacts. The location, extent, height and capacity of the fence are critical to contain rockfall. Rigorous trajectory analysis should be undertaken utilising one of the specialist software products that are widely available. Catch fences can be cantilevered from the face (e.g. above a tunnel portal) or free-standing structures located downslope from the toe of the rock face, often on scree slopes or areas of unconsolidated ground.

Various proprietary products are available, typically tested and certified from 100 to 10 000 kJ. While the fence and individual components are likely to have been subjected to rigorous testing within a controlled environment, considerable uncertainty may exist in relation to the foundation detail for the fence posts. Fence posts are a critical component of the system, transferring impact energy to the ground. The foundations may comprise concrete pads with compression/tension piles securing the fence to underlying competent strata. Additional raking and lateral anchors may also be required to secure stay and end fence cables, depending on whether the fence is a dynamic or static system. **Figure 87.4** summarises the typical management solutions.

Figure 87.3 Construction of a 3000 kJ catch fence in Gibraltar (© CAN Geotechnical Ltd)

An important consideration when designing catch fences and ditches is maintenance. The effectiveness of these structures reduces as rockfall material accumulates and capacity is reduced. Consideration therefore needs to be given at the design stage to the ease of safely undertaking maintenance.

87.2.4 Case study: Merstham cutting

Failure of a steep-sided chalk cutting near Merstham, south London, resulted in the reactive installation of remedial works. The cutting is located on the northern approach to the South Downs tunnel on the London to Brighton railway. The first phase of works comprised the installation of rockfall netting over approximately 40 000 m^2 to provide an effective

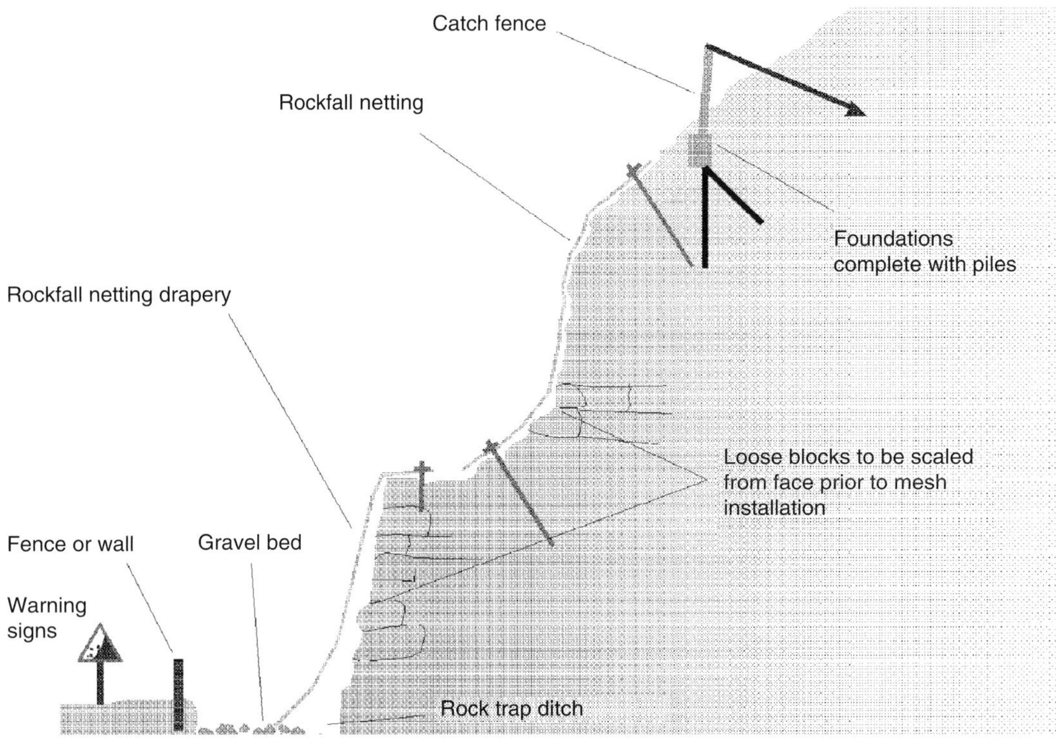

Figure 87.4 Summary of typical management solutions (Reproduced from Fookes and Sweeney (1976) © The Geological Society)

Figure 87.5 During and after installation of rockfall netting (© CAN Geotechnical Ltd)

containment system (**Figure 87.5**). This netting was installed predominantly with the line open to trains within a 6 week period. A second phase comprised the construction of an additional king post wall at the toe of the cutting towards the tunnel portal, utilising temporary possessions of the railway.

87.3. Engineered solutions

Engineered solutions typically comprise the removal of the hazard or reducing the likelihood of rockfall. Stabilisation of the rock face is often achieved by reducing pore water pressures or changing the force regime acting on the face (e.g. by the provision of structural support or re-profiling the face).

87.3.1 Scaling, block removal and re-profiling

Scaling comprises the removal of rock debris and loose blocks from the face, typically utilising light hand tools (**Figure 87.6**). Care should be taken not to dislodge material that could result in the stability of the face being compromised. It should also be noted that scaling is a palliative treatment,

Figure 87.6 Controlled lowering of a block (© CAN Geotechnical Ltd)

often undertaken at the start of works to mitigate against the short-term risk of rock fall only.

Block removal is the controlled dislodgement of discrete blocks, often by the application of pneumatic or hydraulic breakers or jacks. Pyrotechnic capsules, expansive grouts and gas systems can also be used to undertake rock breaking.

Where land is available at the crest, wholesale re-profiling of the rock face to a stable angle will provide a long-term, low-maintenance solution. However, consideration needs to be given to the method of excavation as uncontrolled blasting can dilate joints, resulting in further instability. Alternatively benches can be formed to reduce the face height, contain rockfall and/or provide access to undertake further remedial works.

87.3.2 Grouted anchors

Further details of grouted anchors are provided within Chapter 89 *Grouted anchor construction*; however they are also discussed within this section for completeness (**Figure 87.7**).

Significant rock features can be restrained by the introduction of reinforcing elements. Monobar or multi-strand grouted anchors are installed into pre-drilled, large-diameter holes. Substantial compressive loads can then be applied to ensure the stability of the rock mass.

A grouted anchor can therefore be considered as an installation capable of transmitting an applied tensile load to a load-bearing stratum and comprises an anchor head, free length and bond length.

Given the high capacity and safety critical nature of grouted anchors, they are manufactured to specific requirements with regard to corrosion protection, in accordance with BS EN 1537:2013 (BSI, 2013a). Care must be taken to ensure that the

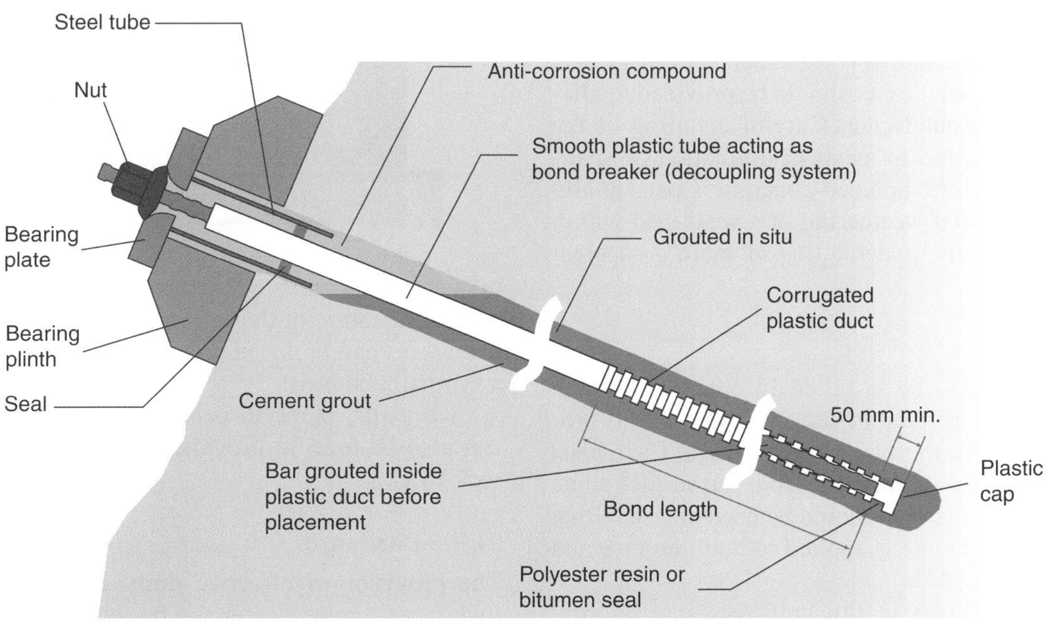

Figure 87.7 Grouted anchor detail (© CAN Geotechnical Ltd)

integrity of this corrosion protection is not compromised during anchor installation. Rigorous testing and commissioning of the anchor must also be undertaken to demonstrate the efficacy of the installation.

Low-capacity grouted anchors that are not safety critical are often referred to as rock bolts. They are available in a range of materials (stainless steel, glass-reinforced plastic, epoxy-coated and/or galvanised high-yield steel) and can be grout- or resin-bonded. In addition, dowels are passive installations relying primarily on the shear strength characteristics of the bar being used. They are usually installed into small-diameter drill holes and resin-bonded.

Ground conditions will determine the method of drilling and the need for any additional ground improvement or grouting to ensure the integrity of the anchor bond.

87.3.3 Surface protection

Often the stability of a rock face can be compromised by the effects of ongoing weathering, particularly where the rock mass is weak or fissile. Protection of the face can be afforded by the provision of 'soft' or 'hard' facing systems.

Soft facing systems typically comprise rockfall netting that can be tensioned using an array of pre-installed rock bolts. The rock bolts are installed within surface hollows so that the netting will conform tightly to the profile of the rock face. This method of installation may therefore be appropriate for the stabilisation of the near-surface materials, and can in turn encourage indigenous vegetation to become established. This can provide an aesthetically pleasing finish, but it should also be noted that vegetation may become a maintenance liability.

Hard facing systems typically comprise concrete pneumatically sprayed to suit the profile of the rock face. The concrete can comprise 'wet' or 'dry' mix and may be additionally reinforced with glass fibres or steel staples. The mix can also include pigment in an attempt to complement the rock face. Weepholes or drainage membranes should be provided to alleviate any build up of groundwater. Careful detailing of the tie-in detail is also required to avoid delamination or spalling of the finished concrete. Sprayed concrete is particularly effective where preferential weathering of a weak bed within the face has resulted in the undercutting of more competent strata above.

87.3.4 Dentition and buttressing

Sprayed concrete can also be installed in conjunction with steel reinforcement (either fixed in situ or weldmesh) to provide structural support to the face (**Figure 87.8**). Care must be taken to remove any loose material from the surface to be treated and depth gauge markers should be used to ensure that the required depth of concrete is applied to both the face and any reinforcement.

For larger structures, cast in situ buttresses can be constructed. These will typically be tied back into the rock face by grouted anchors with the head blocks either concealed within the fabric of the structure or exposed on the front face. Buttresses can be faced with locally won masonry to provide a sympathetic finish.

(a)

(b)

Figure 87.8 Application of sprayed concrete and dentition works (© CAN Geotechnical Ltd)

Crib walls, pier and beam structures and reno mattresses can also be used to provide toe support or protection of the rock face.

87.3.5 Drainage

The provision of effective drainage is often overlooked for rock faces. Groundwater can be the controlling factor in rock stability, particularly where there are interbedded or steeply

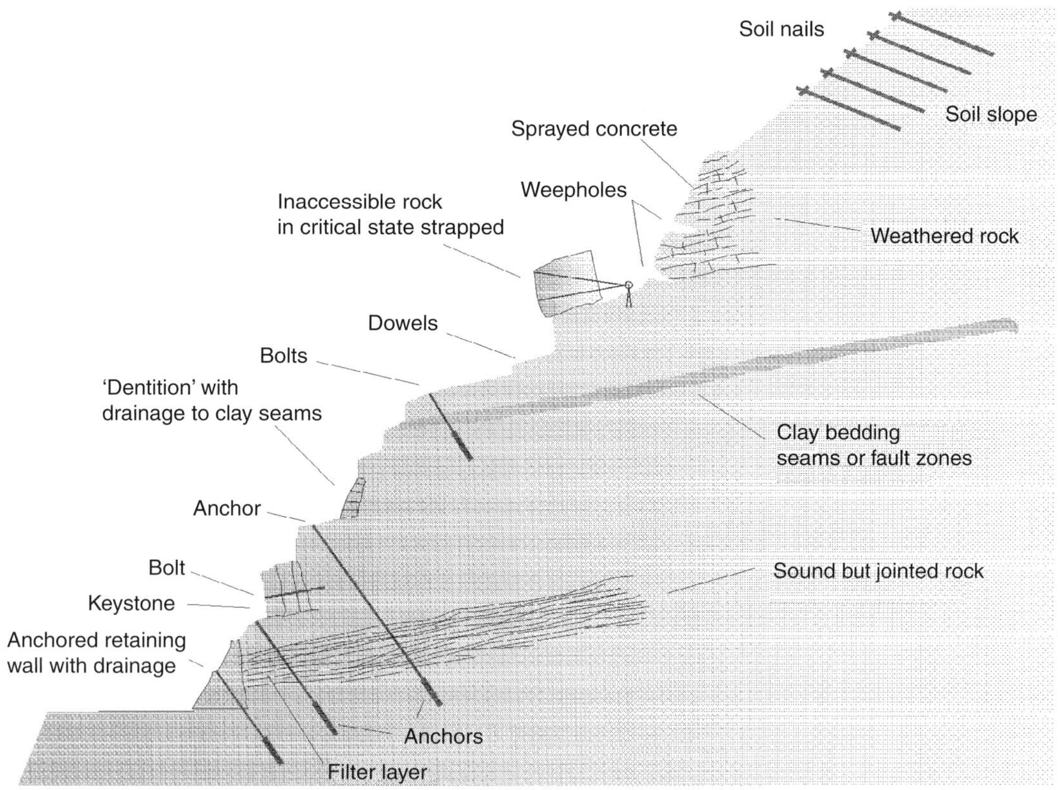

Figure 87.9 Summary of typical engineered solutions (Reproduced from Fookes and Sweeney (1976) © The Geological Society)

dipping strata, or where outdipping clay seams are present. An understanding of the hydrogeological characteristics of the rock mass should therefore be obtained if the implementation of an effective drainage regime is being considered as a stabilisation solution.

Adits have long been used for the deep drainage of mines and rock faces. Face drainage such as the installation of inclined drains can also reduce groundwater levels, particularly within cuttings. Cut-off or interceptor drains can be installed behind the crest of the rock face to divert surface run-off away from the rock face. This will reduce potential infiltration of joints and consequent build up of pore water pressures behind the face. **Figure 87.9** summarises the typical engineered solutions.

87.3.6 Case study: Camp Bay, Gibraltar

A 15 000 m^3 failure of the former sea cliff above the main beach area in Gibraltar resulted in the multiphased stabilisation of a 100 m high rock face. The works comprised the installation of over 250 typically 100 t multi-strand rock anchorages up to 38 m in length. The fault brecciated, karstic limestone resulted in the need to pre-grout the drill holes and undertake rigorous testing of all installed anchors. Other engineered solutions included the construction of cast in situ buttresses, application of sprayed concrete and the installation of rock bolts, dowels and rockfall netting. The works required bespoke drill rigs to facilitate both down-the-hole (DTH) and drifter drilling in sub-vertical locations (**Figure 87.10**).

87.4. Maintenance requirements

As part of the management of any asset, a maintenance regime should be developed that is specific to the installed stabilisation works. However, rock stabilisation works typically require minimal maintenance.

87.4.1 Inspection

It is prudent to instigate a monitoring and inspection regime to ensure that any issues are identified at an early stage. A bespoke inspection regime commensurate with the residual risk level should be established. As a guide, annual inspections should be undertaken by maintenance staff and comprise a visual inspection from the toe and crest of the rock face. Principal inspections should be undertaken 1 year following completion of the works and then at 3–5 year intervals by a geotechnical specialist. These inspections may entail tactile inspection of the rock face and the installed works.

Figure 87.10 DTH drilling for 100 t capacity multi-strand anchors in Gibraltar (© CAN Geotechnical Ltd)

The inspections should identify any significant changes or defects in the face and installed works compared to the as-built photographs and drawings. A written record and appropriate photographs of the inspection and findings should be made and kept in a health and safety/operation and maintenance file. Recommendations should be provided for proactive maintenance works, with timescales for completion, to ensure that the installed system offers continued risk reduction.

87.4.2 Scaling and rock removal

As part of the principal inspection, it may be appropriate to scale the rock face, removing any rock debris that may have the potential to cause harm to people or infrastructure below. Rock traps and rockfall netting should be cleared of any arisings to avoid accumulation of rock debris that may damage netting or negate the benefit of the rock ditch.

87.4.3 Vegetation control

Trees and vegetation (including creepers such as ivy) should in general be prevented from becoming established on the face or crest. This is due to the detrimental effect of root action and potential concealment of any hazard during maintenance inspections. Vegetation should be removed, and any stumps or roots injected with systemic herbicide to prevent regrowth.

Where it is desired that vegetation does become established, careful consideration of the aspect and drainage of the rock face needs to be given. Only very shallow-rooted and drought-tolerant species should be promoted.

Shrubs (particularly prickly species such as hawthorn and blackthorn) should be encouraged at the toe of the face as they will deter access and also act as a natural catch system for any spalling debris weathering from the face.

87.4.4 Stabilisation works

The maintenance regime should be bespoke to the works installed, with a checklist to ensure a systematic record of repairs is kept. The list below is generic, covering typical areas of maintenance only and not intended for site use.

- Catch fences and rockfall netting should be inspected for signs of impact damage to posts, fraying of cables or damage to the netting, as well as corrosion.
- Grouted anchorages should be inspected for signs of corrosion and the tightness of anchor nuts/plates checked. Anchor caps, if installed, should be re-greased and new gaskets fitted. Periodic lift-off tests should be undertaken on anchors to monitor the residual load applied.
- The condition of sprayed concrete should be monitored so that any damage (concrete spalling/cracking) can be repaired.
- Inclined drains should be flushed and weepholes rodded to ensure they are free-flowing.

References

BSI (British Standards Institution) (2013a) BS EN 1537:2013: Execution of special geotechnical work – Ground anchors. BSI, London, UK.

BSI (2013b) BS 7985:2013: Code of practice for the use of rope access methods for industrial purposes – Recommendations and guidance supplementary to BS ISO 22846. BSI, London, UK.

Fookes PG and Sweeney M (1976) Stabilisation and control of local rockfalls and degrading of slopes. *Quarterly Journal of Engineering Geology* **9**: 37–55.

Further reading

Agnostini R, Mazzalai P and Papetti A (1988) *Hexagonal Wire Mesh for Rock-Fall and Slope Stabilization*. Maccaferri, Bologna, Italy.

Cata M, Flum D, Roduner A, Rüegger R and Wartmann S (2020) *Tecco® Slope Stabilization System and Ruvolum® Dimensioning Method*. Geobrugg, Romanshorn, Switzerland.

Goodman RE (1989) *Introduction to Rock Mechanics*, 2nd edn. Wiley, New York, NY.

Highways England (2020) *HE-DMRB-G CS 641 Managing the Maintenance of Highway Geotechnical Assets (formerly HD 41/95). Revision 1.* Highways England, Birmingham, UK.

Hoek E and Bray JW (1981) *Rock Slope Engineering*, 3rd edn. E. & F.N. Spon, London, UK.

Koe A, Murphy W and Nicholson R (2018) *Rock Netting Systems – Design, Installation and Whole-Life Management (C775).* CIRIA, London, UK.

Perry J, Pedley M and Brady K (2003) *Infrastructure Cuttings Condition Appraisal and Remedial Treatment (C591).* CIRIA, London, UK.

Simons N, Menzies B and Matthews M (2001) *A Short Course in Soil and Rock Slope Engineering.* Thomas Telford, London, UK.

Sprayed Concrete Association (2016) *An Introduction to Sprayed Concrete.* Sprayed Concrete Association, Bordon, UK.

Ulusay R and Hudson JA (2007) *The Complete ISRM Suggested Methods for Rock Characterization, Testing and Monitoring: 1974–2006.* ISRM, Ankara, Turkey.

It is recommended this chapter is read in conjunction with

- Chapter 18 *Rock behaviour*
- Chapter 72 *Slope stabilisation methods*
- Chapter 89 *Grouted anchor construction*

All chapters in this book rely on the guidance in Sections 1 Context and 2 Fundamental principles. A sound knowledge of ground investigation is required for all geotechnical works, as set out in Section 4 Site investigation.

doi: 10.1680/icemge.66830.1427

Chapter 88
Soil nailing construction

Matthew Large Keller, Chesterfield, UK
Richard Garland BAM Nuttall Ltd, Camberley, UK
Philip Ball SOCOTEC UK, UK
Michael R Gavins Atkins Global, UK

Soil nailing is an established method of providing efficient slope stabilisation and structural retaining wall stability. Soil nails are utilised on a wide range of slope geometries and in variable ground conditions; this has resulted in the development of a variety of innovative materials and construction methods.

The technique involves the installation of a reinforcement bar within a drilled hole which is usually filled with a cementitious grout. The grout can be placed before, during or after the reinforcement is inserted. The soil nails are usually arranged on regular grid spacing depending on ground conditions, slope geometry and the nature of the facing between the nails. The surface of the slope is also usually covered with either a hard or flexible facing, with a head plate being fixed to the nail reinforcement and bearing on the facing. Soil nail projects also involve the excavation or preparation of the slope face, construction of crest and toe drainage and the installation of drainage holes in the slope face. Quality control includes sampling and testing of grout and the load testing of nails.

CONTENTS

88.1. Introduction

The soil nail technique may be used to improve the stability of a recently cut slope, an existing unstable slope subject to landslips (or to reduce the risk of future landslip by reinforcing the slope) or a retaining wall formed before or during the excavation process. Soil nails are installed within the soil (or weak rock) of the slope, providing reinforcement to resist destabilising forces. This chapter outlines the various stages and methods required to construct a soil nail scheme. The design of soil nails is covered in Chapter 74 *Design of soil nails.*

Detailed planning of the scheme is required to ensure a safe and economical site and construction sequence is established using the appropriate plant. Following the provision of suitable access, the slope needs to be carefully prepared before the nails are installed. Soil nails consist of slender reinforced elements, usually steel bars, and cement grout installed on a regular grid within the slope. There are various techniques available for the drilling of the nails, placing of reinforcement and grouting. The optimum combination depends on slope geometry, access and ground conditions. A surface facing is usually placed to distribute forces between the soil nails, and the design will often incorporate construction of drainage for which there are also several possible solutions. Sampling and testing of grout and the load testing of the soil nails form part of the quality control and design verification and require careful on-site practice.

CIRIA has published a comprehensive report, *Soil Nailing: Best Practice Guidance* (C637) (Phear *et al.*, 2005), which provides detailed information on the design, construction, testing and maintenance of soil-nailed walls and slopes and can be read in conjunction with this chapter.

General principles for the execution, testing, supervision and monitoring of soil nailing are provided in 'BS EN 14490:2010: Execution of special geotechnical works – Soil nailing'.

88.2. Planning

88.2.1 Health and safety

The environments in which soil nails are installed are often associated with a range of hazards that need to be considered and managed: working at height, slippery slopes, manual handling, haul roads close to the face and rotating drill strings. It is essential that the planning phase considers all the concurrent activities and how they impact on one another. Invariably, several different trades are involved and so coordination is required. It is good practice to restrict access for personnel and equipment not directly involved in operations from the working area to reduce the risks associated with plant movements (**Figure 88.1**).

On new cutting slopes it is typical for the site/haul road to pass along the face of the works at the toe of the slope. The drilling rig, and any mobile elevated working platforms (MEWPs), will often need to be positioned close to passing site traffic, with the potential for clashes. On road-widening schemes this is exacerbated by public traffic alongside generating noise and distractions. Adequate separation and management of site traffic is therefore required and personnel must be segregated from the movements as much as possible. Soil nailing is commonly used to stabilise cuttings and embankments in the railway environment, which brings particular hazards and requirements to ensure the safety of operations.

Figure 88.1 Example of signage used to segregate soil nailing work area during construction (Image courtesy of BAM Ritchies)

Cut slopes will vary widely but often present wet slippery surfaces that lie between 30 and 70 degrees from horizontal, and in some cases vertical slopes may be formed using soil nails (usually with a hard facing). These are sources of considerable risk if access to nail positions and the crest are not planned adequately. There may be several phases of access required to any one part of the slope. If the slope is excavated in a top-down manner, access is required near the crest to install the top row of nails and again at intermediate rows. To test sample nails, access may be required at later stages of construction, which may require work to be carried out at height to set up testing equipment and reaction frames on the slope face. To fit the facing, access will be required from behind/above the crest, down the face; to fit the face plates and tighten them up, access will again be required. On each occasion the means of access, the slope surface and the type of work change. Preparation is vital to ensure these activities are well managed.

When soil nails are installed on steep and/or high slopes, MEWPs and industrial rope access methods are commonly used to provide safe systems of work. Each of these requires careful planning to ensure the safety of personnel involved in soil nail installation and adjacent activities.

Where existing faces are to be strengthened or over-steepened, similar risks are likely to be present but less severe. Nevertheless, each activity needs to be considered.

Selection of the most appropriate methods depends on local conditions and constraints and the details of the soil nails to be installed.

Digital construction methods may be used to model the specific plant to be used during nailing operations (**Figure 88.2**) to assess the feasibility of the proposed methods and equipment. This can be especially valuable when planning work in railway or highway possessions to confirm that all nail positions can be accessed and that drill masts, for example, can operate effectively in areas of restricted space.

88.2.2 Access

Soil nailing on new cutting slopes is usually best suited to a sequence of tasks from one end of the slope to the other.

Figure 88.2 Example of a digital model used to confirm the suitability of equipment proposed for use in a railway possession (Image courtesy of BAM Ritchies)

Consequently, access along the slope should ideally be continuous and have regard for the constraints (if any) imposed by the designers. For instance, in top-down construction it may only be permissible to excavate for one or two rows of nails at a time, which may require a considerable length of slope to be accessible for one continuous shift of work. To allow for curing of the grout before the next level of excavation, several days' production may need to be available along the length of the slope to permit efficient working. Therefore, the sequencing, access and deliveries need to be considered when planning the installation to balance access arrangements with plant productivity. The key to maximising productivity is clear planning and coordination between the groundworker carrying out the excavation activity and the soil nail contractor to ensure continuity for both operations, with sufficient space between operations to maintain safe working conditions.

Access up the face of the slope for personnel may be achieved using MEWPs. These are available with significant reach and can provide an efficient method of installation where the benches are deep or to gain access for the 'finishing' activities once the slope is cut (**Figure 88.3**). Industrial rope access methods are also commonly employed for soil nailing activities (**Figure 88.4**).

In the case of a high slope where benches are not being cut, it may be necessary to resort to specialist drilling equipment that can access the face from the toe or crest of the slope, or cable-supported/-suspended rigs anchored at the top of the slope. Examples are shown in **Figures 88.5** and **88.6**.

Consideration must be given to the means of access for fitting plates, meshing or similar surface covering and testing. Modern test frames need to be located on the slope surface, with independent reference beams and access for the technician. The selection of the location of load test and access to it should be planned and agreed early (see Section 88.9).

Figure 88.3 Example of long-reach personnel access

Figure 88.4 Drill mast on rail-mounted excavator in combination with rope access for personnel (Image courtesy of BAM Ritchies)

Figure 88.5 Drill mast mounted on long-reach excavator (Image courtesy of Keller Geotechnique)

88.2.3 Rig types

There are numerous types of drill rig that can be used for soil nailing, including

- tracked hydraulic drilling/mini-piling machines
- excavator-based drill masts (**Figures 88.4** and **88.5**)
- telehandler-mounted drill masts
- cable-supported drill rigs (**Figure 88.6**)
- air-powered drills – sometimes used with rope access systems.

Provided a hole can be formed that is suitably accurate, stable and straight, any drilling machine that can reach the installation position can be used. Holes can be drilled with purely rotary methods (tri-cone rollers, drag bits or similar) or augered. Where harder ground is encountered, percussion is frequently used where the impact energy and a flushing medium are used to penetrate. This is the most common style of drilling for self-drilled hollow bars (see Section 88.4.4).

The decisions on the type of drilling and type of reinforcement are linked and are sensitive to cost and programme. Occasionally, access constraints will also determine the selection. The relatively high cost of self-drilling hollow bars must outweigh the cost of installing solid bars in pre-drilled holes. With the former, the hole is drilled with the bar that will form the soil nail, with a sacrificial drill bit. On reaching depth it is left in the hole and grouted. Therefore, the speed at which

Figure 88.6 Cable-supported drill rig (Image courtesy of BAM Ritchies)

these holes can be drilled is high – several hundred linear metres per shift in favourable conditions with good access. In the case where the hole is pre-drilled, the drilling tools have to reach depth and then be removed before the reinforcing bar is positioned. The extra time is the key factor, but the stability of the hole can present a risk. All of the above will influence the choice of drill rigs and equipment.

Rotating drill strings present a significant hazard, and drill masts should be fitted with interlocking guards to isolate rotating parts.

Where hand-held percussive drills are used, particular consideration must be given to hazards associated with vibrating tools, and measures included to prevent harmful exposure leading to musculoskeletal disorders and long-term health impacts. With advances in modern drilling techniques and equipment the use of hand-held drills should be considered a last resort, and where used they require adoption of control measures and monitoring.

Where excavator- or telehandler-mounted drill masts are used, consideration must be given to the personnel working in close proximity to the mast or adjacent infrastructure, public vehicles or train movements and adjacent line open (ALO) working (e.g. slew restrictors).

88.2.4 Working platforms

The working surfaces for new slopes in cuttings tend to be the most difficult to manage. Because the benches cut to provide access to the soil nail entry positions are also the working platforms for the rigs – and are temporary – they can be a source of problems if not properly designed, constructed and maintained. Trafficking and uncontrolled surface water will soon cause deterioration that poses risks and can adversely affect the quality of the platform and the stability of plant using it.

The activities of other trades near the slope have to be allowed for in terms of access and impact on the slope. For instance, if the temporary slope is only stable over a specific height, the excavation of a trench along the toe to provide for services would not be acceptable. It is quite common for communications cables and drainage to run parallel to the toe of the slope and also for gantry positions to require special consideration.

88.2.5 Sequencing

This issue has been referred to above with regard to safety considerations. Each project will need to be considered separately as there are no ‘fixed’ approaches. When there are repeat visits to a section of slope it can extend the period until the slope is completed. This is particularly true for slopes where the nail installation is followed by acceptance testing, mesh, soil-retention geofabrics, topsoiling and then final geofabric covering. In such circumstances the overall stability of the slope and the risk of surface erosion need to be considered. A temporary cover may be beneficial but is obviously extra work.

The application of a mesh, where required, on the surface is a key step. The mesh has to be in place before the nail heads are fitted; therefore, the period between the nails being drilled and the mesh being placed can be a ‘hold point’ preventing further excavation.

In top-down excavation the designer should consider the stability of the cut slope in the period between excavation and completion of soil nail installation and fixing of the facing. In granular soils or weathered rock masses containing adversely orientated discontinuities it may not be feasible to cut a steep slope face which can stand unsupported over this period and additional (temporary or permanent) measures may be needed to maintain stability until the permanent facing is constructed.

88.2.6 Spoil/drill arisings

There tends to be little spoil produced from the drilling of soil nails unless formed in soft clays. Typically bores will be 100–150 mm dia. and so produce a small volume of waste.

When liquid flush is used to drill the holes there can be a continuous flow from the top of the hole throughout the drilling process, requiring control and disposal or recycling. In some cases (i.e. where hollow bar self-drilling nails are installed) that flush could be grout but frequently is water. Either way, the final grouting of the hole will cause overspill and so needs to be cleaned off the slope. Such waste needs to be segregated and disposed of appropriately.

88.2.7 Environmental considerations

Soil nail applications may be carried out on slopes with freshly cut faces or on existing slopes where vegetation has become established. Vegetation clearance and ecological mitigation (e.g. reptile control) would normally all be in place in advance of soil nailing. Some slopes may need enhanced stabilisation, and so existing trees, shrubs and surface vegetation may be retained and the nails and facing installed around them. Each case needs to be considered on its merit to minimise environmental impact, and consideration should be given to the aesthetics of the end product.

The issue of spoil from the drilling and grouting is described in Section 88.2.6.

Some fibrous geofabrics – coir matting in particular – are easily set alight. This can be accidental, deliberate vandalism or inadvertent when cutting the protruding bars from the finished slope. In hot, dry periods this has to be considered as it will result in damage to the other facing materials.

88.2.8 Pre-construction approvals

The nature of soil nailing projects means that often preliminary testing is difficult. This may be because access is not available, the slope cannot be cut or the site is not available.

However, the nature of this type of work means that early agreement as to the size, length and construction method is a significant advantage that will avoid delays and aborted work when the main phase starts. Where the lack of preliminary testing necessitates conservative design assumptions to be made, the benefits of testing preliminary nails before the design is finalised may outweigh the costs.

Refer to Section 88.10.1 for details of preliminary testing.

88.3. Slope/site preparation

The site access roads, working platform and slope face should be planned in consultation with the soil nail contractor and designer and then constructed and maintained to provide a safe and efficient site for the installation of soil nails, taking into account the proposed installation equipment and methods.

88.3.1 Topographic accuracy

An accurate topographic survey will be required for the soil nail designer, the setting out engineer and the soil nail and facing installation contractor. The survey should detail the overall slope geometry and any localised variations as these can affect nail design, spacing, the setting out and difficulties in placing head facing materials. The area outside the location of the slope should be included to identify working areas, land boundaries and access constraints.

88.3.2 Haul roads/platforms

Haul roads to the site should be designed and prepared to allow the safe passage to and from the slope for all the relevant plant that is required to excavate the slope and install the soil nails. The haul roads may have to be frequently moved due to the excavation and construction sequences.

Working platforms should be designed and constructed to ensure the safe operation of the rigs during nail installation. Soil nail rigs can vary in weight from 4 to 40 tonnes for long-reach machines. It is therefore recommended that platforms are designed, constructed and maintained in line with BR470 *Working Platforms for Tracked Plant* guidelines (Skinner, 2004) or other approved design methodology. The working platforms should be placed to ensure that the rigs provided can reach all the soil nail positions and allow access for nail testing.

Soil nails can also be installed using rope access techniques where the slope topography and access dictate. A contractor experienced in this method of installation should be consulted during the initial planning phase to ensure the feasibility of this methodology where it is considered.

88.3.3 Slope surface preparation

The method of slope surface preparation depends on whether the slope is to be excavated and cut to a steeper angle or whether the existing slope is to be soil nailed at its existing angle.

Where an existing slope is heavily vegetated and impedes the safe and efficient installation of the soil nails and slope facing, the vegetation should be removed. It should be noted that on some slopes the vegetation is providing a beneficial effect on the stability of the slope, and consideration should be given to ensure its removal will not cause instability. It is often unnecessary to remove grass or other low-lying vegetation prior to nail and facing installation.

Following removal of vegetation, the slope should be reinstated to provide a uniform surface suitable for the placing of the facing and head plates. Excavated slopes should be cut as uniformly as possible to reduce problems with the placing of facing and head plates. Any very loose material or surface anomalies should be removed.

Guidelines for the excavation tolerances of the slope surface are available and given in CIRIA Report C637 (Phear *et al.*, 2005). The quality of the finished soil-nailed slope will reflect the accuracy and uniformity of the cut slope surface, so consideration should be given to minimising the period between excavation, final trimming and soil nailing to reduce the risk of local failures and slumping before the nails are installed.

88.3.4 Setting out

Sufficient data for the level and position of the soil nail heads, angle and length of the soil nails should be provided by the soil nail designer to enable accurate setting out and construction. The entry position of the soil nail should be marked accurately on the slope. The slope topography can vary significantly compared to the theoretical design profile (especially where nails are installed on an existing slope rather than a newly cut slope): in these cases the setting out of the nails can be difficult and the designer should consider the effect of potential variations and be consulted during construction to ensure any variations satisfy the design requirements.

88.3.5 Services/land drains

A full survey to establish the existence of services, land drains and existing infrastructure should be undertaken before the start of the works. This is to ensure the safety of personnel and that the services or drains are not damaged during excavation and drilling, or filled with grout during the works.

88.4. Drilling

Recommended tolerances for setting up and drilling soil nail bores is set out in section 10.6 of *Soil Nailing: Best Practice Guidance* (Phear *et al.*, 2005). Recommendations given there are summarised as

- slope face angle deviation: ±2.5°
- localised tolerance of the slope face: +150 mm/−300 mm positional tolerance at the installation ground surface: ±100 mm horizontally and vertically
- orientation at the head of the completed soil nail: ±5°
- orientation of the soil nail along its length: ±2°.

Care is needed in applying these tolerances. Unduly stringent demands can significantly affect the costs and programme but do not provide any benefits to the client. For instance, the accuracy of the cut face in a new slope may not need to be so tightly specified if the design is not too sensitive to local variation and the nail head plate/washer assembly is designed to cope with some variability. Conversely, if a flat sprayed concrete finish is required a variation in the face of up to 450 mm would consume a lot of material. What is essential is that each aspect is considered in the context of the project and specified accordingly.

Aligned to the issue of setting up rigs is the risk of deflection. Thin self-drill bars can readily be deflected on boulders or tree roots and so may not be appropriate if such a deflection were to endanger services or other features. Casings or stiffer bars would have to be used to mitigate the risk.

88.4.1 Open holes (unsupported)

In stable clayey soils or weak rock the bores can usually be formed by standard drilling rotary techniques or augers. The hole would be drilled to depth, cleaned out and the drill string withdrawn. The bar can then be placed in the hole with requisite centralisers and couplers with a grout tremie pipe attached, and finally the hole grouted from the bottom to the top (Phear *et al.*, 2005: section 4.0). Alternatively, the bore can be filled with grout first (using a tremie pipe inserted to the base of the hole) and the bar plunged into it. This is the preferred method if there is any risk of instability. Grout should not be pumped into the top of the hole due to the risk of collapse of the hole at depth and/or the potential for dilution of the grout by groundwater.

88.4.2 Cased holes (temporarily supported)

Temporary casing is often used where there is a risk of unstable ground interfering with the installation of the reinforcement (and occasionally where extra accuracy is required). The casing is advanced to the required depth, which may only be part way down the hole if the risk of instability is confined to the near-surface soils. The bore is drilled to depth and then the installation of the nail bar is undertaken as described above. However, the casing then needs to be removed using the drill rig, so productivity is adversely affected.

88.4.3 Drilling with fluids

Some ground will lend itself to drilling using a fluid rather than a casing to support the bore. The fluid supports the bore's sides but also clears the arisings by flushing them to the surface. The choice of fluid is usually either neat grout or water. Occasionally a polymer may be added. Whatever the fluid, it must not be detrimental to the bond of the grout to the ground.

88.4.4 Self-drill hollow bars

Usually drilled with grout flush (or sometimes water, noting the potential for water to adversely affect the friction between the soil nail and surrounding ground as described above), the hollow bars form the reinforcement as well as the drill tool. Each bar is drilled into the hole and the subsequent bar added using an external coupler. Drill bits can be attached to cater for a variety of soil types and are available in different diameters. The flush is injected through a specially fabricated flushing coupler/swivel at the drill head and flows down through the central bore of the hollow bar, out from ports in the drill bit and back up the borehole, filling the annulus between the bar and the soil. The bar diameter and wall thickness (therefore torsion capacity) determines the size (diameter) of bit that can be used in any one ground condition. On reaching depth the drill string should be surged up and down the hole a few times while grout flushing continues to ensure a clear bore. The drill bit is sacrificial in this operation and remains in the hole with the hollow bar.

88.5. Placing the soil nail reinforcement

The essential element for a soil nail is the reinforcing bar. It provides the tension capacity in the ground by transferring load through the grout to the surrounding ground, countering the strain along its length and resisting the loads on the head plates. To do this over the design life the grout–bar–soil bond must be effective, and the bar must retain sufficient tensile strength to meet the design requirements after allowing for any corrosion. The head plate and the connection details must be equally resilient.

It is essential that the bore is clean and free of debris and water. The grout, however placed, must form a constant column in the ground completely surrounding the bar. It is not critical that the bar lies on the axis of the bore, but it must be surrounded by a sufficient thickness of good grout to provide a bond to the soil and in some cases corrosion protection. Galvanised steel, stainless steel or composite nails may not rely on the grout for corrosion protection. They still need the bond provided by the grout, but not its protection. Refer to Chapter 74 *Design of soil nails.*

Given the variables described above, judgement can be made on which elements of the nail are critical and which are not. If it is essential to maintain the bar on the axis of the hole, adequate centralisers must be provided at appropriate intervals. In the case of self-drilled hollow bars, it is difficult to achieve a consistent result due to the installation method. A series of radial splines arranged around a circular collar are typically used at each coupling. These are loose-fitting and allow the bar to rotate inside the collar. When drilling down, the centralisers are pushed up against the coupler. However, on completion of the drilling, any surging of the bar to flush the hole or any particularly soft areas may allow the centralisers to move and/or dig in, thus compromising the grout cover. Consequently, consideration needs to be given to the finished product and whether the method of installation is entirely appropriate.

Where bars are not self-drilled, standard centralisers can be used so that centralising is generally more reliable.

In the event that bars have been specifically coated for their durability, care must be taken not to damage that protective system during installation.

88.6. Grouting

In almost all cases soil nails are grouted with a conventional cement–water grout. The type of cement used generally reflects the design sulfate class of the ground. Commercially available cement grouts in the UK will typically conform to a maximum DC3 sulfate classification due to the water–cement ratios required to pump the grout, which requires 0.45 water–cement ratios or higher. The characteristic strength rarely needs to exceed an unconfined compressive strength (UCS) of 30 N/mm^2 because soil nails are usually used in weak ground. The composition can vary, but a water–cement ratio between 0.4 and 0.5 is commonly used for the final product. Where grout flush is used during installation, a 'lighter' grout mix may sometimes be used to support the bores and flush the arisings effectively, with a thicker grout injected after drilling to displace the lighter grout and fill the bore. Where this is carried out, careful quality control is essential to ensure that the final grout placed is of the design mix and strength.

The mixing of the grout is important. A properly blended grout needs to be mixed in a high-shear mixer for about 3 minutes before use. Other types of mixer, such as those that mix dry cement immediately before pumping, must be verified before use to ensure thorough blending is achieved. This can be done by passing the mix over a fine mesh filter and looking for 'lumps' of dry cement.

Bleed, the loss of excess water, has the potential to be a significant problem. When grout bleeds, the particles of cement settle, the free water separates and there is a loss of volume. The problem is not at the surface where the effects can be seen (the column of grout recedes and may crack on the surface) but below ground, where the effects cannot be seen. The results of excessive bleed will include the loss of grout cover, deterioration of its durability and at worst loss of cross-section. Regular bleed testing should be an integral part of the quality control on a project.

The portion of the hole immediately behind the head plate is often the main cause for concern. The grout may settle or shrink at the face while curing. This can be due to bleed as described, but also filtration into the host ground. The grout surface is usually sub-horizontal in its liquid form and so a separate operation is required to top up the grout to form a surface parallel with the ground surface (**Figure 88.7**). In this location the plate and grout need to be in close contact and the bar must be adequately protected against near-surface moisture that would readily corrode an unprotected bar. Obviously, as discussed above, the degree to which this issue becomes critical depends on the environment, the protection system and the design life.

Representative samples of the grout used should be taken throughout installation and tested to verify that the compressive strength meets the specification requirements. Where an unproven grout mix or product is proposed, trials should be carried out in advance of installation to verify that the proposed mix design will meet the design requirements.

Figure 88.7 Finished grout around a plain solid bar nail

88.7. Completion/finishing

The issue of the final preparation of the grout column near the surface has been covered above.

The head plates form an integral part of the soil nail and so need to be placed and positioned properly to work well. There are few problems with plates (provided the protection is not compromised) but frequently the slope surface is not trimmed well enough to permit a flush fitting of the plates. This is particularly true where weak rock or mixed ground cause an uneven surface to be cut. In such instances the plate is only in partial contact and may put the bar into bending. Hemispherical nuts and tapered washers can provide some flexibility, but it is limited. Where necessary the face should be prepared after trimming and patched to provide a suitable bearing surface for the plates. In the worst cases this could require the use of mortar packing placed around the nail prior to installing the head plate. However, this process would be costly and should be avoided where possible.

88.8. Slope facing

Slope facing is required to provide resistance to any localised failure or ravelling between the soil nails and also to provide additional bearing support for the nail head plates. In addition, slope facing can provide topsoil retention and promote vegetation growth. The slope facing should be specified by the soil nail designer. Various types of facing and methods of installation are available and are dependent on the site topography and design requirements.

88.8.1 Types

The three main categories of slope facing are as follows.

- *Flexible structural facings.* These are generally composed of a metallic mesh which may be coated for corrosion protection and also may incorporate or be bonded to a fibrous matting to assist in the growth of vegetation (**Figure 88.8**).
- *Hard structural facings.* Hard facings may comprise reinforced concrete which can be formed using precast panels or more often utilising the sprayed concrete method into which the soil nail heads are incorporated or fixed on the external face (**Figure 88.9**). Alternatives to concrete may be adopted to prevent soil erosion or for purely aesthetic reasons. These may consist of solutions such as timber crib walls or stone-filled gabion-like baskets.
- *Soft facings.* Soft facings are rarely used on their own in modern practice because they do not provide positive support or retention between soil nails. Where they are used, common forms of soft facings are coir mats or geotextiles which promote the growth of vegetation and help to prevent soil erosion. In addition, the use of cellular topsoil retention facings is becoming frequent as they have the aesthetic benefit of completely covering the nail head plates and providing a good depth of topsoil to promote vegetation.

Figure 88.9 Near-vertical hard facing (note the cored UCS core sample holes)

(a)
(b)

Figure 88.8 (a) Nearly finished slope showing nail plate and fibrous matting under mesh; (b) galvanised hollow bar with plate over mesh with coir matting below

88.8.2 Installation and joining details

88.8.2.1 Flexible structural facings

The most widely used flexible structural facings consist of a coated wire mesh. The mesh is usually placed following the installation of the soil nail bar reinforcement and grouting. In some cases the mesh may be laid prior to nail installation, in which case the installation methods must include measures to avoid damage to the mesh. The mesh is usually supplied rolled and installed in a top-down sequence. Consideration must be given to how the mesh will be transported to the crest of the slope prior to placement. It should be initially fixed at the crest of the slope above the top row of nails and then rolled vertically down the slope. The mesh may be fixed to the slope using hooked pins at regular spacings between the nails. At the top of the slope it is good practice to provide sufficient facing so it can be anchored above the crest of the slope. The spacing of the pins and pin length (typically in the range 400–600 mm long) should be specified by the designer or specialist contractor and will depend on slope geometry and soil conditions in the near-surface. The soil nail head plates will then be placed over the mesh onto the nail reinforcement and secured by a locking nut. This helps to fix the mesh.

In many cases the locking nuts are tightened using hand tools, but some high-strength meshes and proprietary systems permit an 'active load' to be imparted to the mesh by applying a specified load to the nail head. This is usually achieved with a calibrated torque wrench and may require a shallow 'hollow' to be formed on the slope surface immediately below the head plate to allow the tightening of the nut to pull the mesh into the hollow.

Figure 88.10 Installation of flexible facing mesh using rope access methods (Image courtesy of BAM Ritchies)

Figure 88.11 Applying sprayed concrete to the vertical face (note the double mesh and drainage holes)

Where the slope is to be cut top-down in staged excavations it may be necessary to install the facing in the same sequence, leaving the roll of mesh securely fastened above the next row of nails to be installed.

At joints to adjacent strips of mesh a suitable overlap should be provided and the meshes joined using ties or clips in accordance with the manufacturer's recommendations to provide a structural join.

The use of MEWPs can simplify the procedure and provide a safe access platform for the personnel installing the meshing and head plates (**Figure 88.3**) or in the case of high existing slopes, or where access at the toe is restricted, rope access methods are commonly employed (**Figure 88.10**).

88.8.2.2 Hard facings: sprayed concrete

Sprayed concrete involves the placing of fresh concrete on a slope surface using a high-pressure spraying technique. The slope surface is first cut to the required angle. This should be cut as accurately as possible as filling voids with sprayed concrete can be uneconomical. The permanent sprayed concrete facing is usually applied after installation of the nails, but in some cases where the cut slope is steep and liable to ravelling or slumping an initial layer of sprayed concrete may be applied to protect the face prior to installation of nails. Following installation of the soil nails, the facing material, usually steel mesh (although steel fibre-reinforced sprayed concrete may also be used), is placed against the slope surface, ensuring spacers are placed at sufficient intervals to provide the required cover. The soil nail head plates can be installed before or after the sprayed concrete is placed, depending on whether they are to be incorporated into the concrete or left exposed on the surface.

Concrete is then sprayed onto the surface around and through the reinforcement and head plates (**Figures 88.11** and **88.12**). This can be done in one or more layers, depending on the thickness of the concrete facing specified. Sprayed concrete can be placed using wet or dry mix methods. It is recommended that a specialist sprayed concrete contractor is consulted during both the design and planning stages to ensure a practical solution is specified. Consideration must be given to the density of reinforcement within the facing to ensure that it does not create a barrier to the application of concrete behind the reinforcement, resulting in voids or 'shadowing'.

Sprayed concrete is generally used on steeper or even vertical cuttings (**Figures 88.9** and **88.12**). The stability of the slope is therefore often more critical, and the sequencing of the excavation, nailing and sprayed concrete should be given careful consideration. It may be necessary to cut the slope in discrete panels using a hit-and-miss basis and install the soil nails and sprayed concrete before excavating the adjacent section. In this case an allowance for the lapping of the sprayed concrete reinforcement should also be made.

Samples of sprayed concrete for compressive strength testing may be taken directly from the permanent facing by coring or by spraying test panels into wooden formers placed adjacent to the works. The use of test panels avoids the need for reinstatement of core holes in the final facing, but care must be taken to ensure that the test panels are representative of the permanent facing.

88.8.2.3 Hard facings: alternatives to concrete

Various facings can be installed in conjunction with soil nails, including timber crib walls, modular blocks and gabion basket facings, and these will usually be constructed following the completion of the excavation and soil nails. The facing is normally connected to the soil nails and is installed from the base of the slope. In some cases, a temporary flexible facing

Figure 88.12 Sprayed concrete applied to a steep face using a MEWP. Note the use of plastic pipes to sleeve nail bars to enable head plates and nuts to be installed on the finished surface, and use of guide rails to control thickness and profile (Image courtesy of BAM Ritchies)

is installed top-down as the slope is excavated to ensure the stability of the slope until the hard facing is constructed. Specialist suppliers and contractors will provide facing materials, guidance on installation and design, and if required full construction for these types of facings (**Figure 88.13**).

(a)
(b)

Figure 88.13 (a) Construction of timber crib facing; (b) soil or stone panel facing (Courtesy of Phi Group)

88.8.2.4 Soft facings

Soft facings are installed in a similar manner to the flexible facings described in Section 88.8.2.1.

88.8.3 Soil-retention systems

Where the design requires a layer of soil to be placed on top of the facing, cellular soil-retention systems may be installed following the completion of the soil nail head plates and may be placed over flexible structural facings. They are fixed to the slope using steel pins in a similar fashion to the facing (**Figure 88.14**). Alternatively, a narrow steel mesh cage, or soil panel, can be used to contain the topsoil. Topsoil can then be placed by normal excavators before hydro seeding.

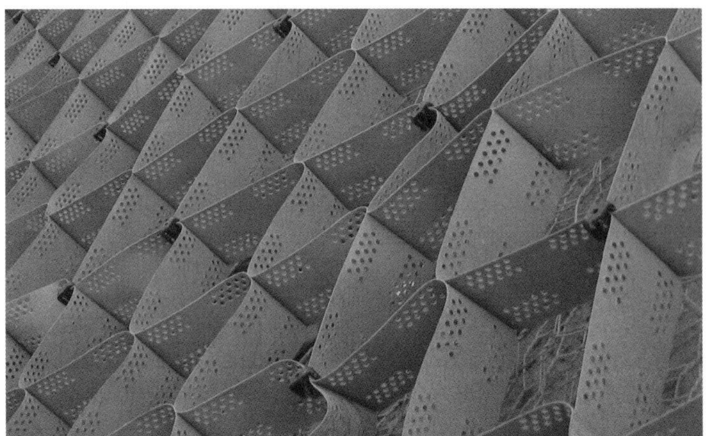

Figure 88.14 Soil-retention system on a 60° slope (note the holding-down pins)

88.8.4 Vegetation: hydro seeding

Hydro seeding (the placement of a seeded solution by spraying) is generally carried out by specialist landscaping companies. This operation should be carried out following the placement of all the slope facings and topsoil if required.

88.9. Drainage

The stability of a slope is usually highly dependent on the groundwater levels within the slope strata, and it is therefore essential that any drainage systems specified are constructed properly to ensure they are fully operational throughout the operational life of the slope.

88.9.1 Crest and toe drainage

If crest or toe drainage channels or trenches are specified, they are usually installed using standard excavation and drainage methods. There are, however, increased risks working above and below a slope while excavating, and precautions should be taken to ensure the slope stability is not compromised. Care must be taken to ensure that any excavations at the toe of the slope do not result in undermining of the soil-nailed slope and facing.

88.9.2 Slope surface drainage

For most soil nail schemes the slope surface drainage occurs naturally through any meshing and vegetation. Where an impermeable hard facing such as sprayed concrete is specified, drainage measures can consist of weep holes and/or a drainage fabric placed between the facing and soil on the slope. Weep holes can be formed using plastic pipe and a filter membrane and fixed to the mesh before sprayed concrete is placed. Weep holes can also be drilled through the sprayed concrete after it has been constructed and hardened.

The drainage fabric should be placed and pinned to the slope surface before the placing of any mesh reinforcement and sprayed concrete.

88.9.3 Sub-horizontal or subsurface drainage

The design of the slope may also require the installation of sub-horizontal inclined drains which can assist in the control of groundwater pressures. The drains usually consist of a slotted uPVC pipe with a geotextile membrane which is placed inside a pre-drilled borehole. The drains should be inclined 5° to 10° upward to allow the water to drain to the slope face. The drains can be installed by the same rigs and personnel as the soil nails, especially for cut slopes where the excavation is undertaken in stages (**Figure 88.15**).

Figure 88.15 Installing deep drains into the slope face

88.10. Testing

Soil nail performance can be affected by variations in ground conditions and soil nail installation methods. Soil nail testing is therefore required to verify that the nail–grout–ground interaction performance is in accordance with the design assumptions.

88.10.1 Preliminary and acceptance testing

The type, frequency, location and loading schedule of the load test should be specified by the soil nail designer (see Chapter 74 *Design of soil nails*). Guidance on this can also be found in section 11.3 of CIRIA 637 and reference should be made to BS EN 14490:2010.

Preliminary soil nail tests may be carried out in advance of the main works usually on sacrificial nails installed using the same technique as the proposed working test nails. Depending on the site conditions and topography, it may be necessary and more efficient to undertake vertical nail tests which can still demonstrate the nail–ground bond in the same strata as the working nails.

Acceptance or working nail tests are undertaken on production or working nails which form part of the permanent works. These tests are carried out during or after the main working nails are installed, using similar testing methods as for the preliminary nails.

88.10.2 Reaction system

A suitable reaction frame should be provided to transfer the stressing loads into the slope or ground adjacent to the nail. The actual detail of the reaction frame will depend on the slope topography, ground conditions and test load to be applied.

These should be considered carefully for each project. A typical beam test set up is shown in **Figure 88.16**.

The reaction system design should take into account the test load and condition of the slope surface. It is essential to ensure the reaction system is safely secured to the slope.

Testing of soil nails is a potentially hazardous activity requiring experience and careful planning to ensure that a safe system of work is planned and implemented. Risks associated with the failure of soil nails or testing equipment while

Figure 88.16 Load testing frame mounted on a slope

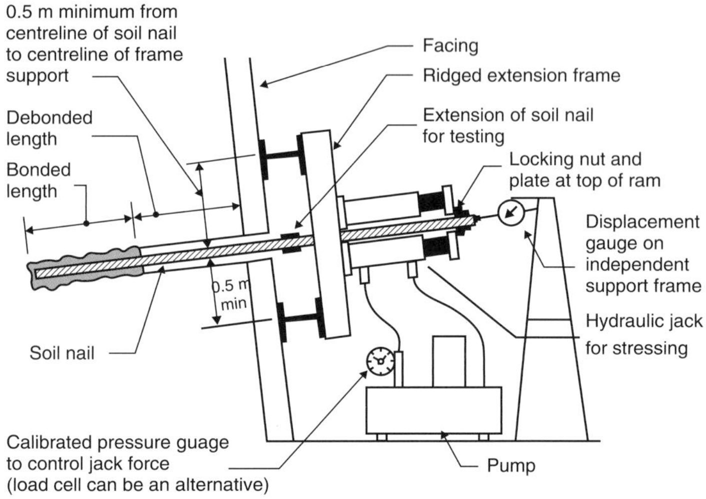

Figure 88.17 Schematic layout of a soil nail testing system (Reproduced with permission from CIRIA C637, Phear *et al.*, 2005; www.ciria.org (originally developed from Pr EN14490:2002))

operatives are conducting tensioning operations need to be carefully considered, with the use of automated monitoring methods a preferred option to remove the operatives from the immediate area; an exclusion zone around the test nail and the reaction system must be established during tensioning operations.

88.10.3 Monitoring

The extension of the soil nail during the load test should be accurately monitored. The actual method of measuring the extension will vary depending on the load test location and the topography of the slope (**Figure 88.17**). Ideally a displacement (dial) gauge or transducer should be placed on the exposed soil nail bar end and supported on a frame which is not subject to movement from the stressing operation or other factors (**Figure 88.17**). The stressing load should be measured using a calibrated pressure gauge connected to the jack.

88.10.4 Free length formation

The designer may specify that the soil nail to be tested should be debonded over part of its upper length to prove the nail capacity in a certain strata or level in the resistant zone. The debonding can be carried out using a plastic sheathing between couplers which should be sealed against grout intrusion. Effective debonding of self-drilling hollow bar nails can be very difficult to achieve due to the installation method, particularly where the specified free length extends beyond the top coupler. Where possible the free length for self-drilled nails should be restricted to the uppermost segment of the bar.

88.10.5 Approvals

For preliminary tests sufficient time must be allowed in the construction programme for the contractor and nail designer to be able to review the soil nail testing and make any required changes to the design before the main works begin. The working test results should be reported, reviewed and approved as soon as practically possible after the tests so any consequential actions can be implemented, minimising possible delays and costs.

References

BSI (2010) BS EN 14490:2010: Execution of special geotechnical works – Soil nailing. BSI, London, UK.

Phear A, Dew C, Ozsoy B, *et al.* (2005) *Soil Nailing: Best Practice Guidance. CIRIA Report C637.* CIRIA, London, UK.

Skinner H (2004) *Working Platforms for Tracked Plant: Good Practice Guide to the Design, Installation, Maintenance and Repair of Ground-Supported Working Platforms.* Report BR470. BRE Press, London, UK.

Further reading

BSI (British Standards Institution) (1995) BS 8006:1995: Code of practice for strengthened/reinforced soils and other fills [partially replaced]. BSI, London, UK.

BSI (2010) BS 8006-1:2010: Code of practice for strengthened/reinforced soils and other fills. BSI, London, UK.

Bye G, Livesey P and Struble L (2011) *Portland Cement*, 3rd edn. ICE, London, UK.

Useful websites

Deep Foundations Institute (DFI), USA: http://www.dfi.org

Federation of Piling Specialists (FPS), UK: http://www.fps.org.uk

It is recommended this chapter is read in conjunction with

- Chapter 72 *Slope stabilisation methods*
- Chapter 74 *Design of soil nails*

All chapters in this book rely on the guidance in Sections 1 *Context* and 2 *Fundamental principles.* A sound knowledge of ground investigation is required for all geotechnical works, as set out in Section 4 *Site investigation.*

doi: 10.1680/icemge.66830.1439

Chapter 89
Grouted anchor construction

Bob Thompson Keller, Coventry, UK

Grouted anchors are an important geotechnical component in the resisting of destabilising forces in structures and slopes. The construction process is of fundamental importance to the performance of the grouted anchors and the industry offers many proprietary contractor-designed systems. The history and experience of anchoring contractors is critical in terms of the use of drilling and grouting systems and their effect on the surrounding ground conditions. The long-term corrosion resistance of the anchor system is important, together with the long-term monitoring and maintenance of the anchor heads.

CONTENTS

89.1. Introduction

Grouted anchors are tensile members that allow load transfer from a structural face to a discrete zone of ground. The particular ground where the load transfer is carried has been previously examined and defined through good ground investigation and survey works and targeted as part of the preliminary anchor design process. This transfer of load to a particular stratum of ground makes the grouted anchor approach different from systems such as the rock bolt, soil nail and tension micropile, where load is generally carried along the full length of the structural member.

Grouted anchors can be installed for loads from 10 kN to in excess of 5000 kN. However, the more typical range of loads is 100–1000 kN for permanent systems. The grouted anchor is commonly used to support retaining structures, resist anti-flotation and earthquake loads and for slope stability schemes. The grouted anchor is also used to improve rock face stability with similar techniques as used for soil nail installation (see Chapter 88 *Soil nailing construction*). The design of these systems is closely linked to the construction process; hence good understanding is required between the designer, contractor and specialist anchor contractor when using this system within a project design.

This chapter reviews

- applications of grouted anchors
- types of grouted anchors
- grouted anchor tendons
- construction methods in various ground types: cohesive and granular soils, and weak and strong rock
- grouted anchor testing and long-term maintenance of anchor systems.

89.2. Applications of grouted anchors

Grouted anchors in the UK are generally installed based upon the requirements of 'BS 8081:2015+A2:2018: Code of practice for grouted anchors' (BSI, 2018). However, 'BS EN 1537:2013: Execution of special geotechnical work – Ground anchors' (BSI, 2013) also reviews construction practice.

Grouted anchors (or anchorages) consist of an anchor head, free anchor length and fixed length (see **Figure 89.1**).

Each part requires a detailed design based upon the structure type and the expected ground conditions (see Chapter 66 *Geotechnical design of grouted anchors*). The design of the fixed anchor length (bonded length) where the load transfer occurs has a particularly important link to the construction process.

One of the key aspects of grouted anchor design is that specialist grouted anchorage contractors must be consulted at an early stage to advise on construction issues and how they can influence the proposed design. In addition, where specialist grouted anchorage contractors are proposed, the design engineer and main contractor must be satisfied with the experience of the installer. Failure to follow this approach may lead to construction and performance problems with the grouted anchors.

The design of fixed lengths is summarised within BS 8081:2015+A2:2018 (BSI, 2018) and is covered in detail in Chapter 66 *Geotechnical design of grouted anchors*. It is dependent upon the construction process adopted, summarised as follows.

- **Type A anchor:** tremie (gravity displacement), packer or cartridge-grouted, straight-shaft boreholes, which may be temporarily lined or unlined, depending on borehole stability.
- **Type B anchor:** low-pressure grouted anchors (typically, grout injection pressure $p_i < 1000$ kPa) grouted using a lining tube or in situ packer.
- **Type C anchor:** boreholes grouted to high pressure (typically $p_i > 2000$ kPa), using a lining tube or in situ packer.
- **Type D anchor:** tremie-grouted boreholes, in which a series of enlargements, either bells or under-reams, have previously been formed.

The construction of the grouted anchor involves one or more of these processes, which are referenced throughout this chapter.

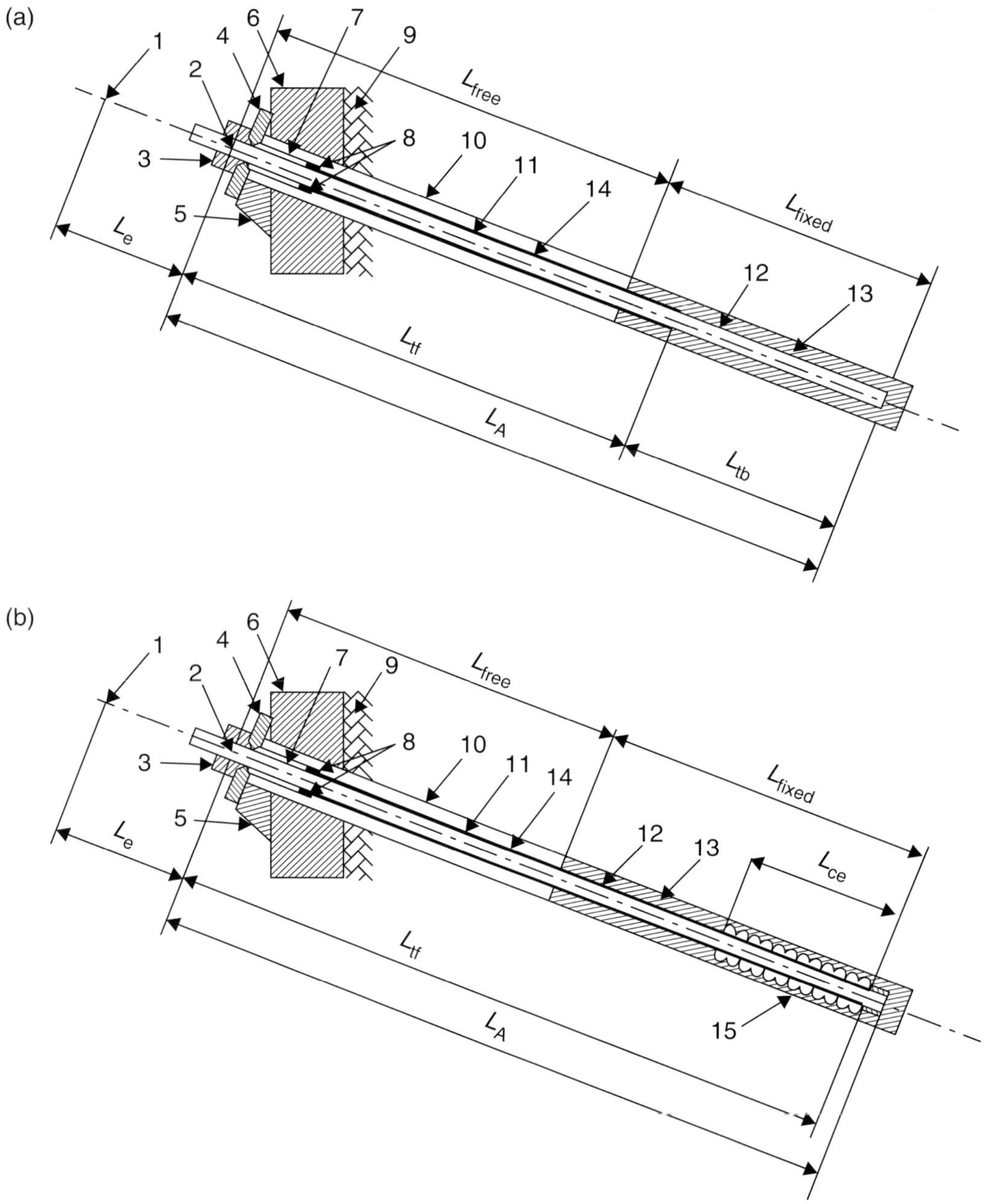

Key

1 Anchorage point at jack during stressing
2 Anchorage point at anchor head in service
3 Tensioning element at anchor head
4 Bearing plate
5 Load transfer block
6 Structural element
7 Trumpet or anchor head tube
8 O-ring
9 Soil/rock
10 Borehole
11 Debonding sleeve
12 Tendon
13 Fixed-length grout body
14 Free-length filling, where appropriate
15 Compression element

Figure 89.1 (a) Bond-type grouted anchor and (b) Compression-type grouted anchor. Details of anchor head and head protection omitted (Reproduced with permission from BS 8081:2018)

89.3. Types of grouted anchors

Ground anchors can be divided into either mechanical or grouted-type installations.

89.3.1 Mechanical ground anchors

Mechanical anchors are generally used for low-capacity anchors. The resistance is generated by driving the anchor tendon to depth and forming a load transfer mechanism with a plate system; two examples are the Platipus system (**Figure 89.2**) and the Duckbill system.

The advantage of these systems is fast installation, no requirement for curing of grout and instant load application. However, the design method for these systems is semi-empirical, with the load capacity based upon the expected ground conditions and the structural loads required by the designer. These systems are suitable for small-scale installations such as masonry wall repair, foundation anchors and shallow slope support.

Figure 89.2 Platipus mechanical anchor system (Courtesy of Platipus Anchors Ltd)

Caution needs to be exercised with these systems as confirmation of acceptance does not conform to BS 8081:2015+A2:2018 (BSI, 2018) in terms of testing and corrosion protection requirements. The use of such systems in permanent works therefore needs careful consideration with regard to the long-term performance and the factor of safety involved. Typical systems are designed for working loads of 10–120 kN; however, it is rare that these are tested to the full three times working load required by table 2 of BS 8081:2015+A2:2018 (BSI, 2018). Actual factors of safety on the system would therefore require preliminary tests on site to give confirmation of the levels of safety achieved.

Mechanical anchors are generally driven into the ground to the required length using a pneumatic hammer system. Due to this method of installation, they are particularly susceptible to obstructions within the ground. The key difference with other geotechnical products is that the only section of the ground anchor that can be shortened is the fixed anchor length (following detailed site trials). The ground anchor must achieve the free length detailed by the designer to maintain overall stability of the structure. Hence, if obstructions are encountered, pre-drilling works may be required to allow installation. Due to these problems, these systems tend to be installed in relatively low-strength, fine-grained soils of a homogeneous nature. It is important that specialist suppliers and installers of these systems are consulted if this method is to be employed.

89.3.2 Drilled and grouted anchors

Drilled and grouted anchors are generally the most common form of grouted anchor installation for large civil engineering projects. These systems allow the designer to use developments in drilling technology as well as specialist grouting techniques. Both of these areas of construction have a significant influence on the load-carrying capacity of the grouted anchor in its final condition (and to achieve the factors of safety required by the Design Codes).

The drilled and grouted system generally involves the formation of the anchor borehole using the following: open hole drilling techniques, rotary duplex cased systems, rotary percussive cased systems or down-the-hole hammer systems. Significant variation and overlap of these systems can be used, and the type of drilling system generally depends on the type and strength of ground expected over the anchor length. These systems are explored further within this chapter in relation to different types of ground encountered.

On completion of the borehole, an anchor tendon is installed to the base of the borehole and grout introduced using a tremie system. This allows full grouting of the borehole and prevents the occurrence of voids within the anchor bore, which could lead to failure of the anchor. Where temporary casing has been utilised, this is gradually withdrawn while continually topping up the grout level.

Based upon the tendon arrangement and the drilling and grouting methods, a number of grouted anchor systems have been provided by specialist contractors and suppliers. Examples of these types of developments are single-bore multiple-anchor (SBMA) systems, multi-stage multi-bell/under-reamed grouted anchors and tube-à-manchettes. All these systems allow modification and efficiencies in the load transfer mechanism compared to conventional anchors, and in particular allow high anchor loads to be obtained in otherwise unsuitable ground conditions.

89.4. Grouted anchor tendons

The anchor tendon generally consists of either a high-strength steel bar or woven wire prestressing strands (**Figure 89.3**). The tendon system is typically factory-fabricated to avoid contamination of the units on site. This is especially important with regard to the corrosion protection system (for permanent anchors) and the greasing and free movement over the anchor free length.

The factory-fabricated anchor tendon is delivered to site in a secure storage system to allow installation (**Figure 89.4**). Any corrosion damage to the anchor tendon during transportation or installation could affect its long-term performance.

89.5. Construction methods in various ground types

89.5.1 Construction issues and risks

The key area to consider in grouted anchor construction is the effect the installation method will have on the load-carrying capacity of the grouted anchor. The preference with most grouted anchors is to use a construction method that allows formation of the fixed length to the diameter of the borehole. Although enlargement of the fixed length (by techniques such as grout enhancement or under-reaming) is accepted, control of these processes is required.

Figure 89.3 Factory-fabricated double-plastic corrosion-protected grouted anchors (Courtesy of Keller Ltd)

Figure 89.4 Storage frame (Courtesy of Keller Ltd)

Where ground conditions are such that actual grout loss can occur, this can be detrimental to the construction and performance of the grouted anchor. Grout loss should be carefully considered within materials such as chalk (where solution features are expected) and Coal Measures (due to past mining activity and fissured bedrock). Methods to deal with this problem include the use of thixotropic grouts or pre-grouting works prior to construction (i.e. a two-stage process).

The effect of the flushing medium on the ground and the knock-on effect on the load transfer mechanism can also be detrimental. Careful consideration of the flushing medium and its control is required – not only from a design point of view, but also from an environmental and control perspective. In certain situations, poor control of flushing systems can cause severe effects on adjacent structures and ground types.

Careful consideration is required regarding the tolerances and location of as-built anchors. Grouted anchor construction is now routinely used in areas where retaining walls support structures that are supported on piles. The grouted anchor system can be designed to pass between piles and be installed at inclinations (both vertical and horizontal) that pass with safe clearance to the deep foundations. The installation tolerances must conform to BS EN 1537:2013, Clause 8.1.2.

The location of as-built anchors is also important in terms of construction design and management (CDM) requirements, and for future construction works causing disturbance and possible failure of the grouted anchors. This has been especially common in the Docklands area of London where new developments utilising piling techniques have had to be carefully monitored in relation to the effect on existing grouted anchors to the river walls. Where concern is raised regarding anchor location and possible interference, accurate borehole alignment monitoring techniques can be used as part of the construction works. A precise understanding of the accuracy of these techniques is fundamental in the choice of system and data management.

89.5.2 Construction of fixed anchor lengths within clays

Grouted anchor construction in clays should be viewed with some caution, and detailed assessment of strength development with depth should be carefully considered at the design stage. Design of fixed lengths within clay deposits tend to start with the Type A anchor system approach. This is very common within overconsolidated clays such as London Clay, Oxford Clay and Gault Clay, found in the south-east of the UK.

To support loads in excess of 100 kN, the clay needs to be of firm to stiff consistency according to BS 5930:2015 (BSI, 2015). Clay strengths less than 75 kN/m^2 should be treated with caution in terms of anchor design and construction, although some success has been achieved in softer soils with membrane expansion-type anchors. The limitations on fixed anchor lengths due to the effect of progressive debonding mean that they are generally limited to less than 10 m. This limits the load-carrying capacity of the grouted anchor, hence the development of the under-reamed system. Littlejohn (1977) details the under-reamed anchor (or multi-bell) system, which gives a load-carrying capacity of up to five times more than a traditional anchor of similar length and diameter.

Even with this increase, under-reamed anchors have limited use outside the London area. This is due to the requirement of careful quality control of the under-ream and demonstration of good construction. The influence of granular material within clays or low-strength clays can cause instability of the bore during construction. However, the system should not be discounted as it has major advantages where issues with land take and boundary constraints on anchor length exist.

One system that has allowed higher capacities to be achieved with clay deposits is the SBMA system (Barley, 1997). This system allows the fixed-length construction to exceed the maximum of 10 m used by single encapsulated lengths, and hence to fully utilise the increase in strength with depth of clay deposits. The disadvantage of this system in clays is the large anchor lengths and the effects on boundary constraints. A schematic of the differences between conventional and SBMA fixed length load distribution is shown in **Figure 89.5**. Another similar system is the DYWIDAG (an acronym for Dyckerhoff & Widmann AG) multi-stage anchor tendon.

The most common form of anchor construction in clay is the use of auger drilling systems (**Figure 89.6**), where the drill string helical causes the material to pass to the surface for disposal. This system is quick and allows the borehole to be constructed with minimal use of temporary casing. However, problems can develop where groundwater is encountered within clays, an example being mudstone bands in London Clay. The drilling system does not allow for removal of the water and this can create a slurry that can rapidly reduce the clay strength and grout–ground bond stress

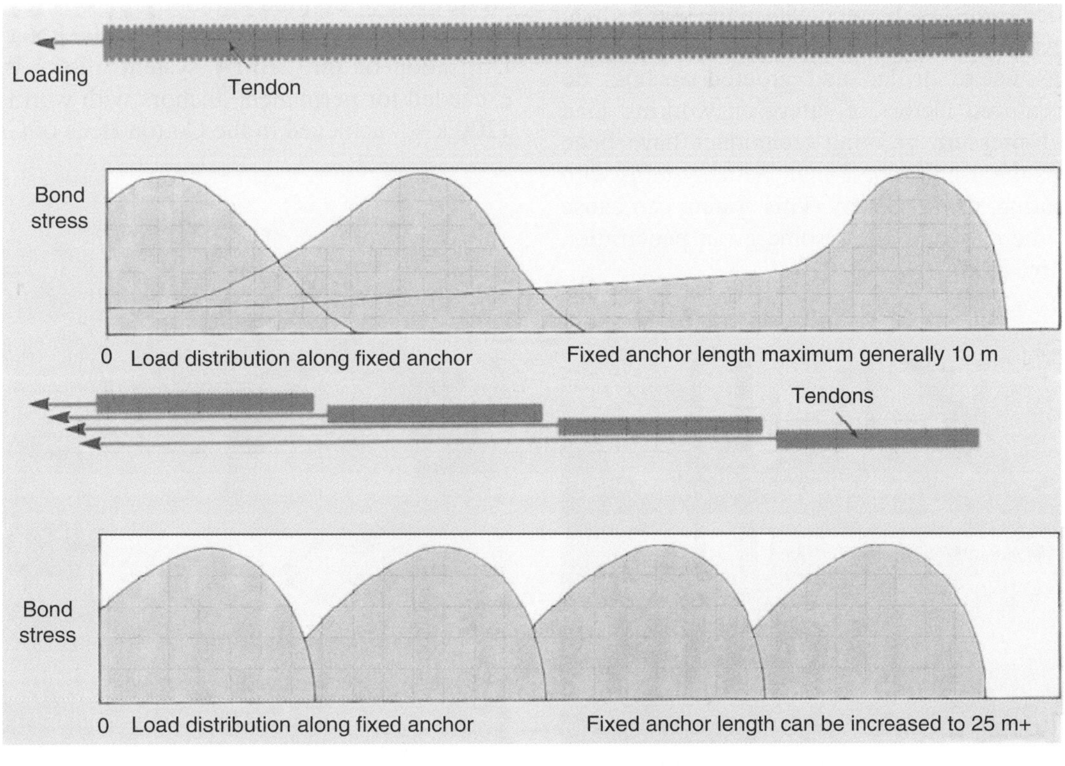

Figure 89.5 Load transfer of conventional and SBMA systems (Courtesy of A.D. Barley/Geoserve Global Ltd)

Figure 89.6 Auger drilling in London Clay (Courtesy of A.D. Barley/Geoserve Global Ltd)

properties. This can also prevent installation of the grouted anchor as the borehole cannot be cleaned to allow anchor tendon installation.

For clays that can contain granular material and possible groundwater, rotary duplex or rotary percussive systems are often employed (**Figure 89.7**). These are commonly used in the boulder clays and glacial tills encountered within the north-east of the UK, where full-length casing can be employed, together with a water flush to maintain stability of the borehole. Water-flushed drilling techniques ensure that a positive hydrostatic head is maintained within the bore, which will generally prevent 'blowing' conditions.

For applications where drilled and grouted anchors do not achieve the required factors of safety on working load requirements, high-pressure grouting techniques have been used. This involves the use of tube-à-manchette systems with high-pressure grouting, using packers. This system can cause hydro-fracture of the clay and allow some grout penetration into the clay (**Figure 89.8**).

Figure 89.7 Rotary duplex drilling (Courtesy of Keller Geotechnique)

Figure 89.8 Post-grouted fissures in stiff clay sands (Courtesy of A.D. Barley/Geoserve Global Ltd)

89.5.3 Construction of fixed anchor lengths within sands and gravels

The construction of fixed anchor lengths within sands and gravels is almost exclusively based upon low-pressure grouting techniques (Type B anchors). Low-pressure grouting of the high-strength grout uses the fully cased system to increase the borehole diameter of fixed lengths through permeation and compaction of the granular material (**Figure 89.9**). With limitations to fixed lengths of up to 10 m due to progressive debonding, permanent anchor loads of 350–450 kN can be achieved within marine sands commonly found within the coastal areas of the UK. However, for coarser deposits which are proven to be predominantly gravels (e.g. terrace gravel), permanent anchor loads of 600–800 kN can be achieved. Utilisation of the SBMA system allows these loads to be exceeded for permanent anchors with working loads of up to 1200 kN – achieved in the Cotton Beds of Great Yarmouth.

Figure 89.9 Low-pressure grouted anchor. The casing shoe on the left was used to drill the section on the right using anchor tension. The exhumed grout body on the right shows the expansion of the grout body through pressure grouting (Courtesy of A.D. Barley/Geoserve Global Ltd)

The key installation method for grouted anchors within granular materials is to have the casing closely follow the drill bit, with a good control on the flush system. The use of water flush is common as this minimises disturbance to the material and is a safe flushing system in terms of recirculation. However, careful monitoring is required to stop 'water jetting' where the soils are loosened and allowed to collapse, creating loose zones. This can cause settlement of the ground surface and affect adjacent structures. The use of air flush is not recommended in granular soils due to the aggressive nature of the flush and the migration of the air through the ground – which again may affect adjacent structures.

Grouted anchor construction within alluvial and marine sand deposits generally involves river wall and harbour wall support. Typical installation plant used is illustrated in **Figure 89.10**.

89.5.4 Construction of fixed anchor lengths within chalk

The UK contains large areas of chalk where major infrastructure works have been carried out. This generally is around the south-east of England, Norwich and Humberside. Anchoring in chalk carries a greater degree of risk compared to soils due to the high variation in weathering, and the strength and presence of solution-related features (causing excessive grout take).

Barley *et al.* (1992) reviewed works completed at Castle Mall in Norwich, where pre-grouting works were carried out using a sand–cement grout prior to re-drilling and anchor installation. Grout volumes of 3–7 times the theoretical borehole volume were recorded for the test anchors using this technique, due to the existence of solution features. Careful drilling and pre-grouting is required within the chalk as the use of water flush can trigger ground subsidence from collapse of the solution feature.

Construction methods used within chalk are generally end-of-casing grouting techniques (Type B), utilising rotary duplex or rotary percussive drilling methods. The use of these techniques, the structure of the chalk and the increased grout take cause expansion of the fixed-length diameter (**Figure 89.11**). Auger techniques can also be utilised but are less common.

89.5.5 Construction of fixed anchor lengths within rock

Grouted anchors are commonly associated with rock anchors used within rock cutting support or anti-flotation anchors. The question 'when is a soil classified as a rock?' is very much open to debate, especially in grouted anchor construction. This is always a borderline discussion in terms of chalk type, based upon the weathering profile of the material. Grouted anchors within the UK can be subdivided by strength and geological history.

Problems of anchor construction within weak bedrock in the UK generally revolve around mudstones and siltstones such as Coal Measure deposits, Mercia Mudstones and Lias Clay deposits, which tend to cross the country from the Severn Estuary to the north of England. For very weak mudstones, the drilling process can have a major impact on the bond stress properties of the material, hence caution should be used with regard to overestimated designs. It is recommended that very weak mudstones are designed based upon a clay soil approach.

Because of the unpredictable nature of weak rocks, the use of water, air or auger drilling techniques should be carefully considered regarding the effect they have on the structure and bond properties of the rock type. Auger drilling can create a smear

Figure 89.10 Caterpillar machine modified for under-slung drilling (Courtesy of Keller Geotechnique)

Figure 89.11 Exhumed grout body within chalk using Type B installation – showing grout penetration (Courtesy of Geoserve Global Ltd)

effect on the ground interface (especially in mudstones) which can rapidly reduce the load-carrying capacity of the grouted anchor. Air flush can cause migration along open fractures and affect adjacent structures. The use of air can also cause disintegration of the bedrock, reducing bond stress properties. Water flush drilling can cause localised softening, especially where boreholes are left open for a period of time. The use of anchor contractor experience and trial anchor testing are important parts in demonstrating constructability in weak rocks.

One major problem with forming grouted anchors within weak, weathered rock is the interbedded nature with stronger, more competent materials. This is especially common with Coal Measures, where bands of strong sandstone can be encountered. Also, the Mercia Mudstones can be interbedded with strong bands of limestone. The drilling system in these conditions is generally chosen for the strongest element; hence when anchors are formed within these materials, variation in the positioning of the fixed length in the rock type can be a problem. In these materials, careful review of construction methods compared to design is required. Detailed review of testing within mudstones is covered by Barley (1988).

For stronger, more competent bedrock (e.g. igneous deposits in Scotland, Wales and south-east England, limestone, Sherwood Sandstone and metamorphic rocks), down-the-hole hammer systems are generally used to form the anchor boreholes. These systems are efficient because the energy of the drill is directly applied at the drill bit (and not at the drill head, as with rotary percussive drilling). This is especially important in drilling strong rocks as the rate of penetration is quicker than top-drive systems.

Down-the-hole hammer systems use air flush to return the drill spoil to the surface, hence the calculation of uplift velocity compared to depth drilled is extremely important. These drilling systems rely on air compressors to have the capacity to return the spoil, so the diameter of hole compared to the rod size needs careful review. If the air gap between the rods and the hole diameter is too great, the returns will not be expelled at the surface and will affect the grouted anchor construction. These systems have one key problem compared to water flush systems – control of the spoil is difficult due to the airborne nature of the returns. Head collecting systems or reverse circulation systems can be used to control the material. Symmetrix and MaxBit down-the-hole hammer overburden drilling systems are becoming increasingly popular due to their speed and ability to bore through difficult ground.

Drilling within igneous deposits that have a high quartz content can cause extreme wear on the drill bits, which is accelerated when drilling in rock that contains saline groundwater. Careful review of these systems is required. Large, high-capacity grouted anchors are used within dam construction and dock wall remediation where earthquake loading is required; such systems are very common as these structures are often constructed on bedrock (**Figure 89.12**). High tolerances of construction are required for these structures, and hammer and rod systems can be flexible in their nature (similar to auger drilling). In these situations, the hammer is often connected to a strong casing member over the initial 10 m to give a stiff drill string. Using this method can give a high degree of drilling accuracy and tolerance. These techniques are reviewed in Cavill (1997) and Shields (1997). Centralising rods can also be used, but again, trial drilling and

Figure 89.12 Installation of permanent multi-strand anchors at Clunie Dam, Perthshire (Courtesy of BAM Ritchies)

measurement should be carried out to confirm that the tolerances can be achieved as required by the designer.

89.6. Grouted anchor testing and maintenance

Grouted anchors are probably the most tested component in geotechnical works. Many grouted anchors that are installed are active (carry prestress load). Some grouted anchors are considered passive – where a nominal load is introduced, based upon the use of a torque wrench or jack system. These passive anchors are always bar systems and are often used to resist hydrostatic uplift forces.

Grouted anchor testing is covered in some detail by BS 8081:2015+A2:2018 (BSI, 2018) and BS EN 1537:2013 (BSI, 2013), and the reader is directed to these references. The testing regime is covered by preliminary trial anchor tests, on-site acceptance tests and suitability tests. The anchor testing and lock-off is the most important part of the anchor commissioning process. The designer and contractor should make sure any specialist grouted anchor contractor has the necessary experience and history of installing grouted anchors. Indeed, if there is concern about the contractor's experience, a contract should be let on a preliminary trial anchor programme to confirm the contractor's competency and design proposals.

Anchor testing is generally carried out using a mono- or multi-jack system (**Figure 89.13**); hence strict quality control and calibration certificates for the jacks are of paramount importance. Also, the necessary access is required for the stressing crews in terms of anchor locations and temporary platforms, as the stressing jacks can be large and heavy. Risks associated with the failure of anchorages or testing equipment while operatives are conducting tensioning operations need to be carefully considered, with the use of automated monitoring methods being a preferred option to remove the operatives from the immediate area; an exclusion zone around anchors and the reaction system must be established during tensioning operations.

Figure 89.13 Hydraulic multi-jack system for SBMA tendons (Courtesy of A.D. Barley/Geoserve Global Ltd)

One of the most important areas of grouted anchor construction, which is commonly overlooked, is the long-term maintenance of the anchors. These tendons are often critical in terms of the performance of the structure; hence it is important to include a long-term maintenance programme for the duration of the design life. Strong evidence exists that long-term anchor failure and/or poor performance is due to problems at the anchor head, so load checks and visual inspection by an anchor specialist are very important. Some engineers consider the use of load cells for long-term monitoring as an option as remote data collection is becoming more popular. However, caution should be exercised as load cells can have long-term performance issues and the use of load cells enlarges the anchor head. Their use does not review the state of the anchor locking system and corrosion protection compound in the anchor head. Therefore a proper anchor maintenance programme must be considered when designing and constructing grouted anchors.

References

Barley AD (1988) Ten thousand anchorages in rock. *Ground Engineering*, Sept.–Nov.

Barley AD (1997) The single bore multiple anchor system. In *Ground Anchorages and Anchored Structures* (Littlejohn GS (ed)). Thomas Telford, London, UK, pp. 65–75.

Barley AD, Eve R and Twine D (1992) Design and construction of temporary ground anchors at Castle Mall development, Norwich. *Conference on Retaining Structures, Cambridge, UK.*

BSI (British Standards Institution) (2013) BS EN 1537:2013: Execution of special geotechnical work – ground anchors. BSI, London, UK.

BSI (2015) BS 5930:2015: Code of practice for site investigation. BSI, London, UK.

BSI (2018) BS 8081:2015+A2 2018: Code of practice for grouted anchorages. BSI, London, UK.

Cavill BA (1997) Very high capacity ground anchors used in strengthening concrete gravity dams. In *Ground Anchorages and Anchored Structures* (Littlejohn GS (ed)). Thomas Telford, London, UK, pp. 262–271.

Littlejohn GS (1997) *Ground Anchorages and Anchored Structures.* Thomas Telford, London, UK.

Shields JG (1997) Post-tensioning Mullardoch Dam in Scotland. In *Ground Anchorages and Anchored Structures* (Littlejohn GS (ed)). Thomas Telford, London, pp. 206–216.

It is recommended this chapter is read in conjunction with

- Chapter 66 *Geotechnical design of grouted anchors*
- Chapter 72 *Slope stabilisation methods*
- Chapter 79 *Sequencing of geotechnical works*

All chapters in this book rely on the guidance in Sections 1 *Context* and 2 *Fundamental principles.* A sound knowledge of ground investigation is required for all geotechnical works, as set out in Section 4 *Site investigation.*

doi: 10.1680/icemge.66830.1449

CONTENTS

Chapter 90
Geotechnical grouting and soil mixing

Paul Marsden Keller, Ryton-on-Dunsmore, UK
Andrew Heathcote Keller, Ryton-on-Dunsmore, UK

Grouting and soil mixing are a range of processes that involve the injection of wet or dry materials into the ground to provide improved engineering properties. Common aims are to increase strength or stiffness or to reduce permeability within the mass of ground treated. The simplest process in concept is the permeation of the pore spaces with a fluid grout which then sets and provides the desired properties. Soil fracture and compaction grouting methods overcome some of the limitations of permeation and employ displacement to insert volumes of grout into the ground. Compaction grouting can be used to densify cohesionless soils, and soil fracture can be used to improve cohesive soils by consolidation. Both compaction and soil fracture processes can be used to arrest or even control ground movements. Jet grouting employs erosion and mixing by high-energy jets, making it applicable to a wide-ranging set of soils and applications. Soil mixing employs powerful mixing tools to mix soils with grouts or dry binders in situ. A range of shapes and applications are in use. Verification is necessary to ensure that the intended performance has been achieved. The main methods are briefly described in this chapter.

90.1. Introduction and background

This chapter covers methods using the injection or mixing of wet or dry materials to improve the performance of the ground. Mechanical improvement employing deep vibratory and dynamic compaction methods is covered separately in Chapter 84 *Ground improvement*.

Geotechnical grouting is the controlled injection of a fluid grout mix into soil or rock. The grout is typically designed to harden with time to secure improvement in the strength, stiffness or watertightness of the ground so that specified performance can be realised. Special applications of grouting include the restoration of levels to foundations that have settled excessively and the control of ground movements caused by underground works such as tunnelling.

Soil mixing, as the term implies, is a related set of processes whereby the ground is improved by mixing grouts in the ground, forming a new material which hardens to deliver the desired properties.

Success in these processes depends on four crucial and related issues

- the type, state, properties, profile and variability of the soil or rock to be treated
- the formulation of the grout and its preparation
- the method for placing or injecting the grout into the ground
- the intended properties or performance required of the treated ground and how these will be verified.

These issues will determine the selection of the process to use, the level of improvement possible and the costs.

Early attempts at direct injection of simple cement grouts into the ground were often hampered by failure to permeate the ground, leading to an appreciation of the limits of this grout type for permeation, and the development of silicates and more complex chemical formulations to extend the range of treatable soils, but with significant limitations.

Subsequent grouting development concentrated more on the means of delivery of simple cementitious grouts. This has opened up the range of soils and processes available and has considerably broadened the range of applications. By the early 1990s jet grouting, compaction grouting, soil-fracture grouting and soil mixing methods were all being widely and successfully applied in addition to permeation grouting (see Bell, 1994; Rawlings *et al.*, 2000). Subsequently there has been increased use of soil mixing, and continued development of other grouting methods particularly in relation to electronic monitoring and control of the processes on site.

Today, many of the processes require only fairly simple grouts consisting of suspensions of water and cement, or microfine cement, with the water to cement ratio being a key variable. Additives can be used to modify either the temporary fluid grout behaviour, such as longer gel times, or the hardened properties, such as unconfined compressive strength. Reactive additives such as PFA (pulverised fuel ash), bentonite, lime, GGBS (blast-furnace slag) or gypsum may be included in the grout mix. Indeed, some of these materials are sometimes used without cement. Inert materials such as sand, rock flour or talc can be added as fillers. Chemical solutions and emulsions may also be used as grouts. Silicates are used for

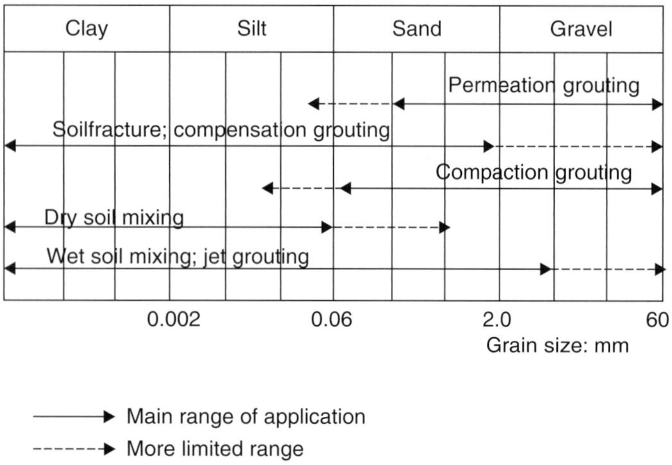

Figure 90.1 Grouting processes applicability according to grain size

permeating finer-grained sands, and epoxies are sometimes used in sealing (e.g. for fine rock fissures).

Figure 90.1 shows the main methods and their usual ranges of application in soils as a function of grain size. However, this figure does not allow for the effects of layered soils, highly variable or mixed-grain soils, and the soil strength, density or stiffness, and is for general guidance only. A summary of the key physical effects and property changes of the various grouting and soil mixing processes is included in **Table 90.1**.

The methods are extensively used to solve a range of geotechnical and foundation problems. All have been used for improving foundation-bearing capacity and for reducing total or differential settlements. Permeation, jet grouting and soil mixing have been used in the formation of barriers to minimise the flow of water or waterborne pollution. Soilfracture and compaction grouting have been used to minimise foundation movement or surface displacements caused by adjacent excavations or tunnelling and for re-levelling previously tilted or distorted foundations. However, matching the problem to be solved with the appropriate technique requires careful consideration.

It can be seen from the above that grouting and soil mixing are a range of processes that depend on the ground profile, the type of grout selected, the method employed and the end objective. They do not produce homogeneous volumes of improved ground, but depend instead on the mass properties produced, with necessary variations in properties within the mass. This reality must be accepted in design and verification approaches to ensure technical and economic success.

Early in the project planning process it is important to ensure that the ground investigation is designed to include relevant ground properties for the potential methodologies. Precontract or preliminary trials may be necessary to optimise the approach, or in some cases to confirm the viability of the process. Use of an observational approach, requiring changes to the quantities or injection parameters as the project proceeds, has also been found to be helpful for difficult sites. However, a detailed consideration of the limitations and considerations in grouting and soil mixing is beyond the scope of the present chapter, and those requiring further information should refer to the references and reading lists, notably CIRIA reports C572 (Charles and Watts, 2002), C573 (Mitchell and Jardine, 2002), C514 (Rawlings *et al.*, 2000), and C774 (Hughes *et al.*, 2018). Specialist advice needs to be sought in all cases.

90.2. Permeation grouting in soils

Permeation grouting is achieved by grouting to partially fill the pores in cohesionless soils, largely displacing the water or air that is present naturally in these spaces, and without altering the structure of the ground. Groutability is a key issue, ensuring that permeation is possible in the target soil or rock mass. In soil it is influenced by the grout type and grain size distribution, and the soil permeability, grain size and grading, fines content and variability.

Soil pore size is theoretically more relevant than grain size, but establishing the pore size, shape and distribution, even in cohesionless soils, is difficult. An additional issue is that small grout particles become wedged together to form a filter cake, thus effectively preventing further penetration. Groutability based on soil and grout particle gradings can be developed, and a rough guide for acceptability is given by Mitchell and Katti (1981) as

$$\begin{aligned} & \mathrm{D}_{15(\mathrm{soil})}/\mathrm{D}_{85(\mathrm{grout})} > 24 \\ \text{or} \quad & \mathrm{D}_{10(\mathrm{soil})}/D_{95(\mathrm{grout})} > 11 \end{aligned} \qquad (90.1)$$

Method	Major physical effect	Key properties for application
Permeation grouting	Infills voids and fissures with grout	Reduces permeability; increases strength or stiffness
Soilfracture grouting	Opens network of fissures using grout	Increases strength or stiffness; controls ground movement
Compaction grouting	Forces a body of grout into the ground	Increases density; controls ground movement
Jet grouting	Erodes the ground, mixes the soil with grout in situ	Increases strength or stiffness; reduces permeability
Wet soil mixing	Mixes grout with the ground in situ	Increases strength or stiffness; reduces permeability
Dry soil mixing	Mixes dry binders with the ground in situ	Increases strength or stiffness

Table 90.1 Summary of physical effects and properties for application

where the D values relate to the percentages passing a given size. This kind of approach is useful in a general sense, but care is needed as grout particles flocculate after hydration, changing the apparent particle sizes. Decisions about groutability are usually taken on the basis of experience coupled with detailed information on the variability and true grading range of the soils being considered. It should be appreciated that many ground investigation sampling methods may not recover all of the fines in cohesionless soils, and this may distort the view taken on groutability or variability of the ground. Advice on the number and quality of samples for grading, and the interpretation of gradings, and similar care with rock discontinuity assessment, must be sought as part of the early planning process. Greenwood (1994) provides a more detailed consideration of the issue.

Generally, experience indicates that low-pressure permeation grouting using simple cement grouts is usually limited at best to gravels containing perhaps some coarse sands, or to fissures greater than *c*.0.2 mm in rock. With microfine or ultrafine cements (more finely ground cements) and/or silicates, described below, it is possible to extend permeation into the coarse sands and even fine to medium sands with silicates.

Key properties of simple cement grouts as a function of water–cement ratio are shown in **Figure 90.2**.

Sandy cohesionless soils may be treated using sodium silicate – the process developed in 1926 by Dr Hugo Joosten using a two-stage approach to combine waterglass and calcium chloride to form precipitate silicates in situ. By the late 1950s a single-shot approach was developed by mixing an organic hardener and waterglass before injection, which then gels after predetermined periods in the ground. This principle

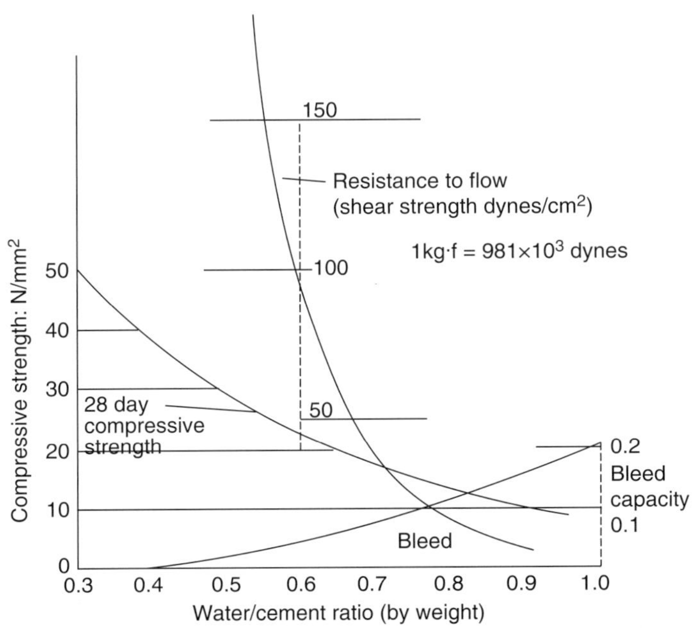

Figure 90.2 Properties of simple cement grouts (Reproduced from Littlejohn, 1982)

remains applicable in soil permeation, and various proprietary reagents are available as hardeners, with widely differing properties and durabilities.

Other single-fluid low-viscosity chemical grouts capable of penetrating finer-grained soils such as silts have also been developed, but these are expensive, and the strength, stiffness and durability often more limited than with simple cements. Several are also highly toxic, leading to restrictions or bans on their use, and are not considered further here as they now see little use and are for very specialist applications only.

The mass properties of permeated ground are quite different from the neat grout properties, and **Table 90.2** provides an indication of the strength and permeability limits expected. The expected properties of treated ground depend on the grout type, the grading and state of the ground and the details of the permeation grouting method employed. Considerable variation, particularly in strength and stiffness, is normal.

Equipment employed in permeation grouting is wide-ranging and to some extent dependent on the grout selected and the method of injection. Drilling boreholes is the normal precedent to grouting, necessary in order to get access to the ground to be grouted. A description of drilling processes is beyond the scope of this chapter (see Warner, 2004, and the suggested reading list). Only relatively small-diameter holes are needed, less than 100 mm dia., and while vertical holes with long masts in open sites are cheapest, inclined holes, including drilling from shafts, galleries and tight corners, are perfectly feasible if the application demands it.

For low-cost treatment in highly porous ground, grouting can be performed directly from the end of borehole casings or lances or through slotted pipes installed in the boreholes. High pressures are not effective due to grout surfacing, and control of permeation is relatively crude. However, such methods can be relevant for improving marginal situations in open ground, and particularly for temporary works.

Soil grouting was considerably improved with the invention of the tube-à-manchette (TAM) pipe by Ischy in 1933 (**Figure 90.3**). These pipes, typically 25–50 mm OD are installed in lengths that screw together in a borehole to attain the depths required. They are grouted in place using a relatively weak sleeve grout. Grout ports closed with cylindrical rubber seals are spaced at regular intervals, ranging from 250 to 500 mm. A packer placed inside the pipe isolates one set of ports at a time. Grout is injected at sufficient pressure to expand the rubber seal, break the sleeve grout and then flow into the surrounding ground.

	Cements	Microfine	Silicates
Unconfined compression strength: N/mm^2	<15	<10	<7
Coefficient of permeability *k*: m/s	>10^{-6}	>10^{-7}	>10^{-7}

Table 90.2 Properties of cohesionless soils treated by permeation grouting

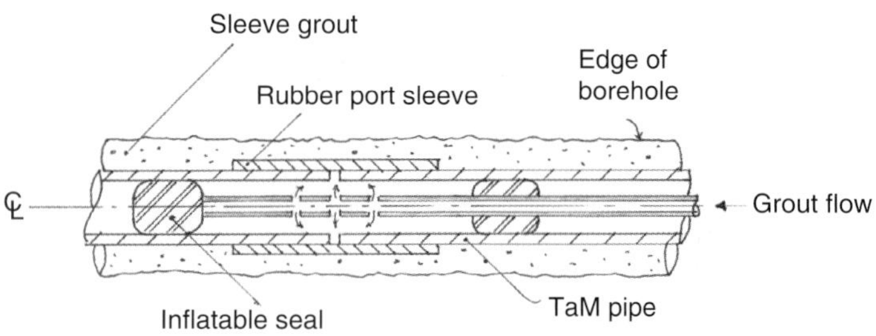

Figure 90.3 Principle of operation of a tube-à-manchette pipe

The TAM pipe method provides much more control for permeation and related forms of soil grouting because the injection position can be selected, repeated injections from any port location can be made, and grout pressure or volume can be measured or controlled. The TAM method is now almost invariably used in all but the least demanding soil permeation applications. Borehole spacings for TAM installation depend on the grout and permeability of the soil, and the application, but range typically from 0.6 to 2 m.

Permeation grouting is a flexible process because discrete layers or zones of soil can usually be treated, even when beneath buildings and structures as only small-diameter drilling is needed to reach them, and the shape of the treated volume can be made flexible to suit. Minimal spoil is generated, and ground movements likewise, if the grouting is properly selected, specified and executed (Greenwood, 1994).

Major applications include

- underpinning foundations
- sealing cracks, joints and other rock discontinuities
- temporary support for excavations, shafts or tunnel construction
- temporary zones of reduced permeability
- reducing permeability below dams
- improving stability of retained ground
- support for services, such as pipelines on poor ground.

Perhaps the major use of permeation grouting worldwide is to reduce soil or rock permeability. Many dams have reduced-permeability grouted 'curtains' formed in this way (**Figure 90.4**). Another very broad range of applications concerns the provision of temporary stability to facilitate the construction of tunnels, shafts and excavations. Underpinning has also been a key use of permeation, where ground conditions allow (e.g. **Figure 90.5**).

90.2.1 Permeation grouting in rock

Permeation grouting in rock is usually applied to infill the discontinuities in otherwise sound rock. It is mainly used for sealing, or less frequently for strength or stiffness, in fissured or jointed rock: for example, reduced-permeability curtains or blankets beneath dams; sealing rock near nuclear installations; to enable tunnel construction; or for permanent improvement around or near tunnels. Site investigation needs to identify the volume and size of the rock fissures and voids, so cores and in situ tests will be needed to properly explore the rock mass.

Grout hole diameters are small, typically 35–75 mm, and spacings typically 1–2 m with the opportunity for infill holes if needed. For curtain grouting the holes may be spaced in two or more rows, depending on the specification and rock openness, to improve the effective thickness of the curtain. The rock is often grouted within the bores in stage lengths of 3–5 m. Each stage is isolated by down-the-hole packers which seal the boundaries of the stage. Depending on the rock mass

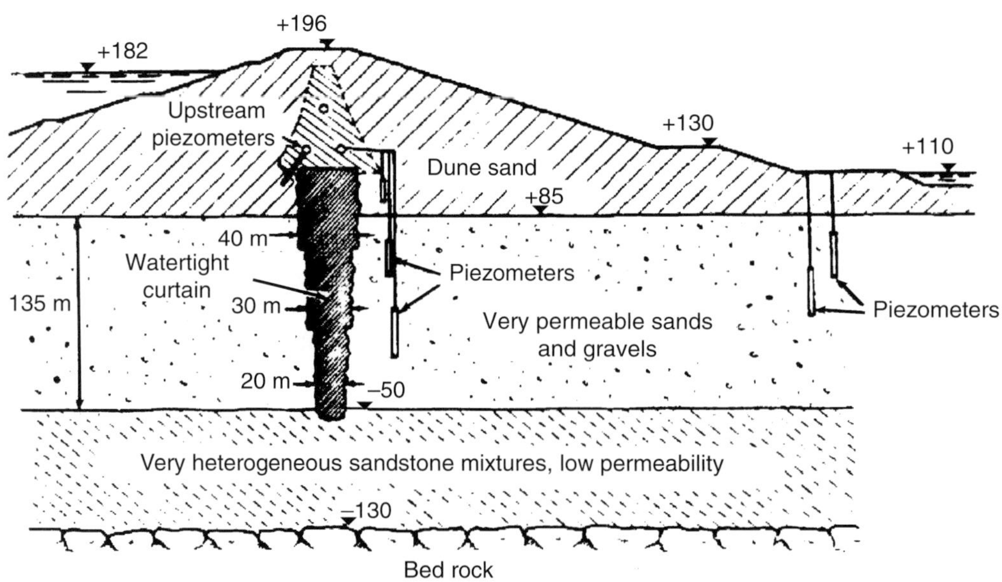

Figure 90.4 Grout curtain for Aswan High Dam (Reproduced from Littlejohn, 1983)

state, grouting may proceed from the top (descending stage) or from the base of the borehole (ascending stage).

Seepage through grouted rock can be determined using the Lugeon unit (1 Lugeon is the volume of water (litres) lost per metre length of hole per minute at 10 bars pressure). However, this does not necessarily relate well to groutability, so the GIN (grouting intensity number) method has been developed, which avoids the need for Lugeon testing. The method was introduced by Lombardi and Deere (1993) and is described by Warner (2004).

Cavities in rock are also frequently backfilled with grouts, particularly in areas of old underground mine workings in the UK. Low-cost grouts are used to fill old shafts and working tunnels to prevent migration of voids to the surface and thus to prevent or minimise any consequent surface subsidence. Methods are described in CIRIA SP32 (Healy and Head, 2002) and on the websites of specialist contractors.

(a)

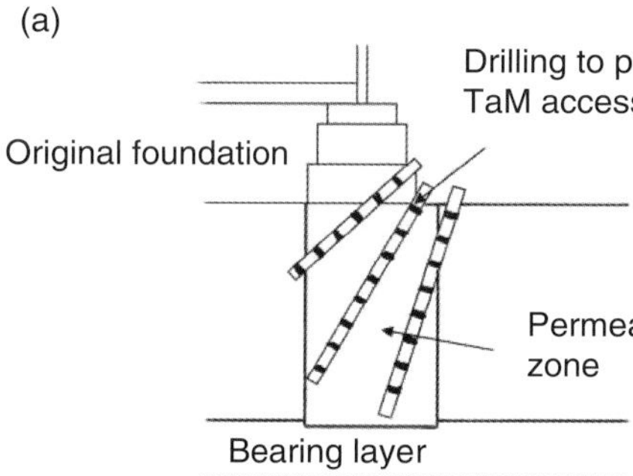

(b)

Figure 90.5 (a) Underpinning using permeation grouting; (b) Thames Gravel section treated using silicate grout permeation (Courtesy of Keller)

90.3. Soilfracture and compensation grouting

Fracture grouting, or soilfrac grouting, employs similar equipment and approach as permeation grouting in soils, but uses higher pressures to deliberately cause 'hydrofracture', which opens up a network of fissures in the ground, thus enabling the injection of grout. This widens the range of applicable soils and enables treatment to extend even into the clay range. Compensation grouting is a particular application of fracture grouting which enables the control of ground movements to compensate for deformations arising from construction.

Fracture grouting was often correctly seen as an undesirable effect of overpressurising in soil permeation grouting, but was first developed in France as a positive measure and is thus sometimes referred to as 'claquage' grouting. Controlled hydrofracture is achieved by injecting grout under high pressures, so inducing cracks in the ground which are opened up by the grout flowing into them. By repeated injections of relatively small amounts of grout, a network of veins and lamellae form a dendritic structure in the ground (**Figure 90.6**). Steel TAM pipes are often used to ensure the repeatability of the injections, especially for compensation grouting applications, which can go on for many months and sometimes years. Grouts employed are commonly cements but often contain additives to assist in achieving early strength or other required characteristics.

It is important with hydrofracture grouting that experienced specialists are employed who understand the process well, as uncontrolled hydrofracture at best will be ineffective and at worst can create deleterious effects for the ground and may cause grout to travel into such structures as drains, sewerage systems or basements, creating blockages or damage.

The technique can be applied in mixed and cohesive soils for straightforward ground improvement for foundations, although applications are limited due to time and cost. In most cases the process causes consolidation or tightening between the grout veins and stiffness increases by a factor of 2–5 are possible.

However, perhaps of more interest is that, particularly in cohesive or more dense granular ground, the volume of the soil mass has to increase to accommodate a significant proportion

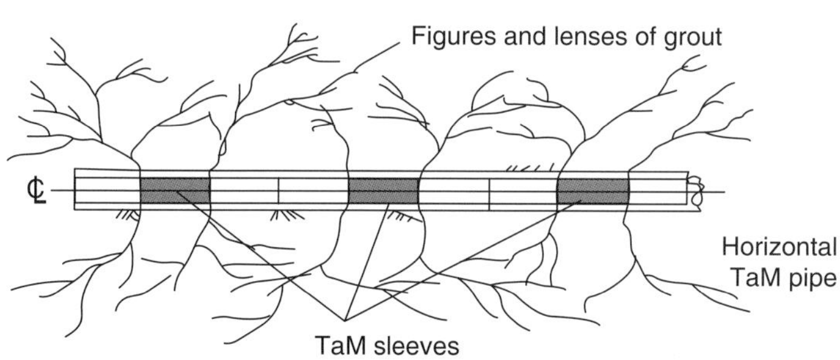

Figure 90.6 Formation of dendritic structure of grout inclusions using soilfracture grouting

of the extra volume of the injected grout. Consequently, a powerful application is to employ the technique to arrest excessive total or differential settlement beneath existing buildings and structures.

In such cases it is advantageous to have the pipes drilled in at low angles to the horizontal or even from specially constructed shafts in order to get the array of pipes installed. Since the movements induced in the ground are locally very small and are attenuated over a larger area when transmitted underneath a foundation, the technique is kind to the structure when properly managed. In any case the existing state of the structure and its foundations, and any adjacent structures or buried services, must be well understood prior to design and grouting operations.

The volume increase caused by soilfracture, if injection persists, can migrate to the surface as heave. This effect, when controlled by the layout of the TAM pipes, the grout type and the sequence and intensity of injection, can be used not only to arrest settlement, but to take the process further in order to re-level previously settled buildings and foundations – a major application of the technique (**Figure 90.7**).

This application was first employed in Germany, and not only led to high accuracy in many cases, but also to the development and application of soilfracture grouting to compensate in real time for ground loss (Raabe and Esters, 1990). This takes concepts of re-levelling even further and attempts to balance the ground loss arising from tunnelling with the volume increases in the ground caused by the soilfracturing grout injection. Basic principles of compensation grouting are outlined in **Figure 90.8**.

The technique has been applied in the UK, where it is known as compensation grouting, and the first major applications were in connection with protection of many key structures during construction of the running tunnels for several of the Jubilee Line Extension contracts in London (Linney and Essler, 1994). It is now a common technique across Europe and further afield (Moseley and Kirsch, 2004; Mair, 2008).

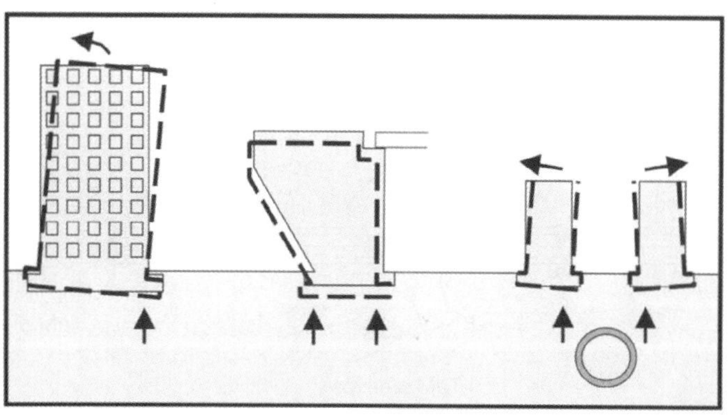

Figure 90.7 Applications for soilfracture grouting (Modified from Raabe and Esters, 1990)

90.4. Compaction grouting

Developed in the USA in the late 1950s, compaction grouting has now been adopted in the UK (Crockford and Bell, 1996) and increasingly in Europe and Asia. It offers a very different approach from permeation or soilfracture. The method employs low-slump mortar-like grout (slump typically 25–75 mm, but can be more), which is forced into the ground under high pressures to form bulb-shaped masses. These can be employed to compact loose granular soils or to displace cohesive and mixed soils to create heave, close voids or control ground movements during construction. It should be noted that in the USA recent terminology distinguishes between 'compaction grouting', used only in applications to compact soils, and other applications of the technique, referred to as 'low mobility grouting'. In Europe all applications come under the 'compaction grouting' umbrella.

A key requirement of the grout mix is that it does not permeate or mix with the soil, or cause hydrofracture as a fluid, but remains a plastic growing mass during injection, with enough internal friction to allow flow but not so much that segregation or filter blocking occurs. Mixes consist of a wide range of aggregate sizes, often a silty sand, with silt sizes accounting for 10–25% of the total by weight, with cement or cement substitutes such as fly ash. Further details are included in Rubright and Bandimere (2004) and Warner (2004). The grout is often batch-mixed on site in simple pugmills or is supplied from ready-mix plants. More expensive continuous batching set-ups can be effective for large or high-volume demand sites.

The grout bulbs are generally formed from the end of casings installed in boreholes or pipes of about 50–100 mm dia. driven directly. Gaining access under structures can be facilitated by the use of inclined borings. A typical set-up for ground improvement using compaction grouting is shown in **Figure 90.9**. Typical pressures needed are in the 40–60 bar range, and concrete pumps with capabilities of 80 bar or more are typically specified. Pressures at which further grout intrusion refuses are not usually obtained before surface heave is reached, which is helpful for applications in which surface movements are required.

Lateral spacings between the pipes, forming the bulbs upstage or downstage, the vertical spacing between bulbs and the order in which adjacent casings are injected are interrelated and depend on the objective of the process, the ground conditions and the site environment. Spacings between pipes vary for compaction purposes from about 1 m where overburden pressures are low, to about 4.5 m or so where deeper layers at 10 m or more are being addressed.

To enable efficient compaction, pipes are often installed in primary and secondary grid arrangements. Upstage grouting is the most common and least costly approach, with each bulb injected at intervals as the pipe is withdrawn from the maximum required depth. Downstage grouting can be useful where compaction is needed only at shallow depths to provide some near-surface confinement for subsequent grouting

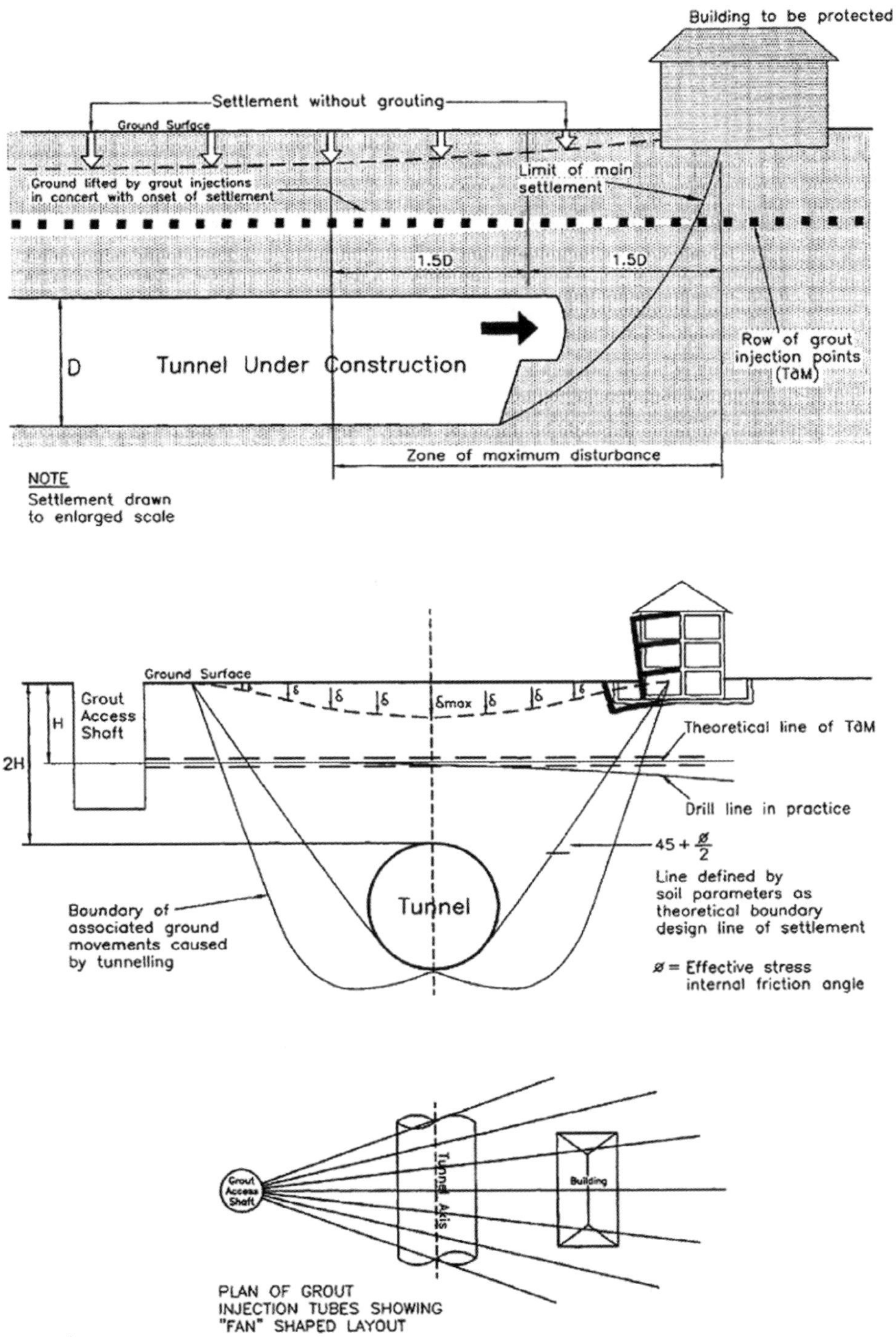

Figure 90.8 Principle of compensation grouting for tunnels (Reproduced with permission from CIRIA C514, Rawlings *et al.*, 2000; www.ciria.org)

beneath. Compaction grouting applications usually proceed by injecting target volumes, to ensure relatively even distributions of grout in the ground. Experience indicates grout takes of 3–12% of ground volume, more frequently at the upper end, are needed to achieve intended levels of densification or movement in non-voided soils.

Compaction grouting can be used as a ground improvement technique to densify loose natural or voided filled ground (e.g. **Figure 90.9**) to be employed as bearing material for new or adjacent construction; or to deal with features such as loose material in chalk cavities. The latter is not uncommon in southern England, and compaction grouting has been used

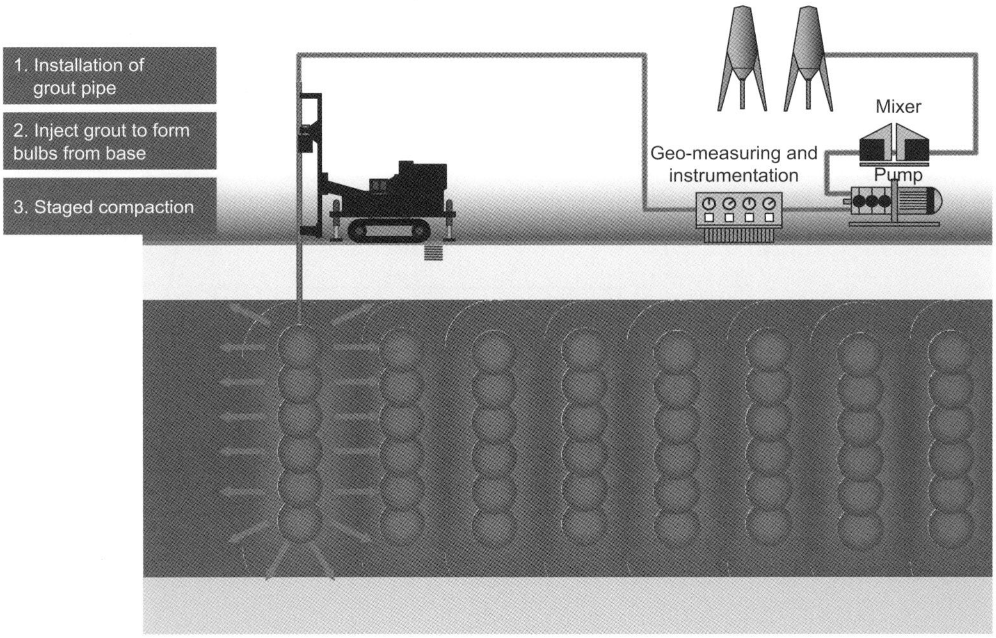

Figure 90.9 Equipment and procedure for ground improvement using compaction grouting (Courtesy of Keller Grundbau)

for several foundations for structures and the preparation of waste disposal sites, as well as for remedial works beneath existing buildings. It has also been applied in the densification of potentially liquefiable zones for earthquake protection and has been applied retrospectively to provide seismic protection for pre-existing structures. Warner (2004) and Moseley and Kirsch (2004) provide many examples.

However, because it can create heave, like soilfracture grouting, a key application area is in re-levelling structures (**Figure 90.10**) and to compensate for ground movements caused by tunnelling or other excavations. Indeed, the first compensation grouting in the USA employed compaction grouting and pre-dates the use of soilfracture grouting in Europe for the same purpose (Baker *et al*., 1983).

90.5. Jet grouting

Jet grouting uses erosion to enable grouting to proceed across a very wide range of soil types. High-pressure jets of water or grout break down the soil structure and remove varying proportions of the resultant soil particles. The grout replaces the eroded material and becomes mixed with the residue, which can be a very small proportion depending on the method employed and its ability to discharge spoil to the surface.

There are now several forms of jet grouting in use, and the majority stem from three basic variants: single, double or triple systems (**Figure 90.11**). Examples of the drilling equipment and high-pressure pumps required are shown in **Figure 90.12**. Compact crawler rotary drills with rotary heads are most often used to reach the zone identified for treatment by jet grouting. Where possible, long masts should be used to enable the jet grouting for each element to be completed in one lift, so avoiding costly re-entry and joints in the treatment. The erosive pumps (grout for the single and double, and water for triple) need to operate consistently at high pressures of up to 600 bar or more. Grout is usually mixed and batched on site and held in agitation tanks before injection.

In all cases the work begins by drilling a small borehole of *c*.150 mm dia. to take the hollow jetting rods containing the relevant jetting fluids to the maximum treatment depth. A column is formed by simultaneously withdrawing and rotating the rods at constant predetermined rates, while the erosive jets cut the ground and the grout is placed at predetermined volumetric injection rates. In some cases panels of grout are formed as this geometry can be useful in sealing applications. The jetting parameters (i.e. lift and rotation speeds, fluid flows and pressures) are monitored closely and are generally controlled automatically to ensure highest quality. Columns are the most commonly required shape and are usually intersected to form a treated mass of ground with properties determined in part by the nature of the ground being eroded, as well as the injected grout. Some examples of column and panel arrangements for various applications are shown in **Figures 90.13**, **90.14** and **90.15**. The grout mix is typically a simple water–cement grout, sometimes with additives such as bentonite. It must be free from impurities and sufficiently fluid to flow easily through the jetting nozzles.

The single and double systems use a grout jet to erode and place the grout, shrouded with compressed air in the case of the double system to improve efficiency. The triple system employs an efficient air-shrouded water jet for erosion and the grout is separately placed below the erosive jet. Precutting using water or weak low-density grout may be useful in stiffer

Figure 90.10 Re-levelling and protection of old foundations (Courtesy of Keller Grundbau)

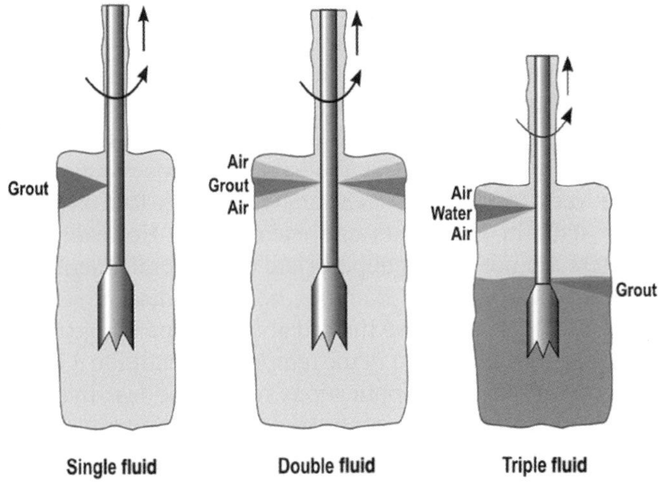

Figure 90.11 The main variants of jet grouting processes

Figure 90.12 (a) Two jet grout rigs; (b) high pressure pump used for jet grouting

cohesive layers or with mixed soils, to improve the efficiency and uniformity of the final column.

It is important to ensure the spoil has an unhindered passage to the surface to prevent excess ground pressure, which can result in unwanted heave at the surface. Similarly, planned and efficient spoil handling and disposal, possibly by recycling, will be necessary. In many applications, notably underpinning, a thorough survey of the existing structural integrity of any structures likely to be influenced will be needed. Monitoring of structural movements and ground heave will be necessary, and proficient real-time systems of communication established between the monitoring and the jetting process and personnel.

Since jet grouting can erode large volumes of soil, attention also needs to be paid to temporary stability, the gain of grout strength with time and the sequence of forming interlocking columns. Diameters of up to 6 m or more are possible with higher-energy air-shrouded grout jet systems using double opposed nozzles (i.e. the 'superjet' process) and these can be very efficient in such applications as forming basal cut-off slabs in basements and excavations. However, large diameters are not always appropriate or economic, depending on the application – for example, beneath adjacent sensitive structures which may have limited ability to span large unsupported volumes of ground in the temporary condition. Column diameters in most UK applications tend to be less than 3 m. Erodibility of the soils being treated is also highly relevant to the diameters that can realistically be achieved, and variation in soil type or state in the treatment zone must be understood in advance to ensure success.

Table 90.3 provides a rough guide to the range of mass properties expected, but actual properties will depend on the ground properties and variability, the process and jetting parameters used and the properties of the grout injected.

Jet grouting has the ability to treat discrete layers in the ground, along with other forms of grouting, but is able to treat a wider range of soil and rock and has additional advantages in that the geometry can be more flexible, the properties more predictable and able to meet demanding specifications, with good performances available in a wide range of soils. Indeed, jet grouting enjoys wide application due to its flexibility of geometry in the ground, and the broad range of treatable ground conditions.

A key further advantage is the intimate contact which jet grouting can make with overlying structural foundations, or with sealing sheet piles, particularly with the double and triple systems as the erosive jet cleans existing structural surfaces prior to the grout being placed (**Figure 90.13**). This makes jet grouting suitable, with appropriate consideration, for sealing and underpinning applications. It can also be used in weak rock, in cleaning out softer zones in weathered rock, in debris-filled fault zones, fractures, fissures or joints.

Underpinning of existing buildings needed for new adjacent construction is one of the most common applications of jet grouting. This is frequently done in conjunction with basement construction, as the jet grouted mass not only provides support of the foundations of the existing building, but also support for the excavation and formation of the new basement, which can be taken close to the existing building line thus saving space over other methods. It has been used frequently in many cities around the world, including London. Typical arrangements for jet grouting underpinning are shown in **Figure 90.14**.

As well as underpinning, jet grouting has been widely used in both permanent and temporary cases for providing basal seals and plugs and in forming new 'cut-offs' (reduced-permeability zones) or extending or remediating existing 'cut-offs'

UCS (MPa)	3–28	Sands and gravels
	2–6	Silts and silty sands
	1–5	Inorganic clays
	<2	Organic clays, silts and peats
Permeability *k*	>10^{-8} m/s	Granular and cohesive

Table 90.3 Typical properties for jet grouted soil masses

(a)

(b)

Figure 90.13 (a) Intimate contact between the underside of the footing and the jet grout; (b) exposed and trimmed jet grout columns with anchor plates

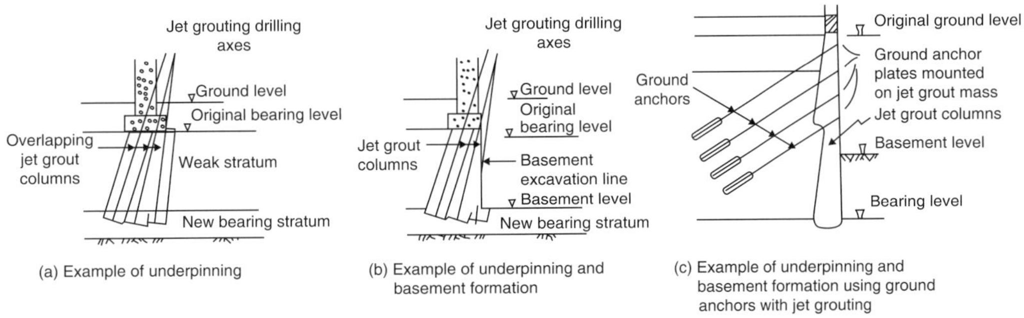

Figure 90.14 Examples of underpinning and basement formation using jet grouting

beneath dams and other water-retaining structures. Due to its capacity to be bonded to existing structures in situ, it has been used to seal leaking cofferdams and shafts. It is also used directly and indirectly to provide support – for example, improving the capacity of existing foundations and abutments. It is also extensively employed in creating temporary stability for tunnel construction or for break-in/break-out blocks. Some examples of such applications are included in **Figure 90.15**.

Figure 90.16 shows a sheet-piled excavation supported by a jet grout base plug. After driving piles to support the eventual structure, the base plug is formed using interlocking jet grout columns, drilling from the surface and jetting only over the depth range needed for the plug. Sheet piles to support the sides of the excavation are then driven, and a structural high-level propping system installed during initial excavation. Further excavation proceeds to final excavation level with no need to install further support as the jet grout seals against water ingress, stabilises against heave and acts as a basal prop. This approach to constructing excavations, with some variations in sequencing of the elements illustrating the flexibility of jet grout construction, is widely applied and has been employed on several projects in the UK.

90.6. Soil mixing

Deep soil mixing has developed in two major ways. Dry soil mixing uses dry powders, often cement or lime-cement, which is usually blown with compressed air into the mixing train, and the soil moisture is sufficient for the properties of the mix. Dry soil mixing was developed in Sweden in the 1960s to address the compressible soft soils found there. The method, which initially employed lime binders, is now also widespread in Finland and Norway, and more recently in other parts of Europe, including the UK, and in the USA. Cement and cement-lime binders are now more common.

Wet soil mixing uses wet grout, usually a cement suspension, prepared and mixed at the surface, before being pumped into the ground through the mixing tool for subsequent mixing with the soil. Water in the grout mix and water in the ground contribute to the final mix. Wet soil mixing was introduced in the USA in the 1950s and has been developed steadily since the 1970s, notably in Japan with depths of 50 m or more having been treated using some very sophisticated equipment developed for near- and offshore work. The method is now well accepted in the USA and the Far East, and is increasingly being employed across Europe, notably in Poland.

Soil mixing is being applied in a wide range of applications, often using columns to provide an improved mass of ground, similar to jet grouting. Soil mixing needs to be performed from the surface in each case, lacking some of the flexibility of jet grouting, but is nevertheless usually competitive due to its generally lower cost per cubic metre. Applications of soil mixing include

- improvement of bearing capacity
- reduction of total or differential settlement
- improvement of slope stability
- support in temporary or permanent cases for excavations or tunnels
- reduction of earth pressures behind retaining structures
- cut-off barriers (reduction of permeability)
- immobilisation or confinement of in situ soil contaminants
- improvement of seismic (liquefaction) resistance in new and retrofit construction.

A more complete description of the methods may be found in Topolnicki (2004, 2009), and in the proceedings from the Deep Foundations Institute (DFI) international conferences at San Francisco in 2015 and virtually in 2021 (see references below).

90.6.1 Dry soil mixing

Dry soil mixing is typically applied to soft soils in which the natural moisture content is close to or above the liquid limit. It is increasingly being used for more organic soils and even some peats. Typical dosage rates are 75–200 kg cement per cubic metre of natural soil, or more for very soft or organic soils, but depend on the soil and binder types, the intended performance of the mixed soils and the capability of the equipment. Undrained shear strengths attainable are

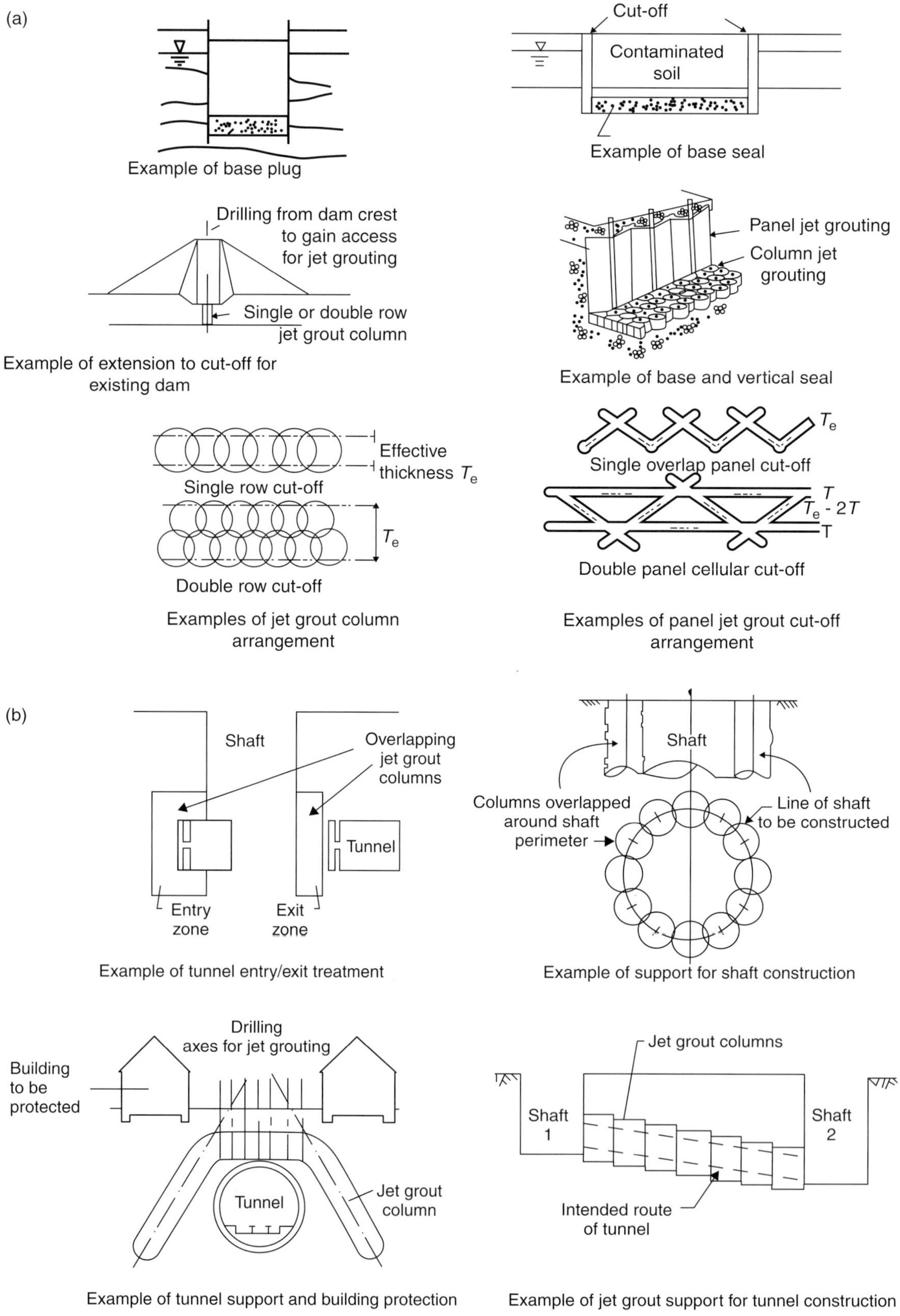

Figure 90.15 Some examples of jet grouting (a) in sealing applications; (b) for shafts and tunnels (Bell, 1993)

Figure 90.16 Jet grouted base slab (Courtesy of Keller)

up to 1000 kPa or better at 28 days in inorganic silts, but are typically much lower, and especially so in the case of organic clays and peats.

In any case, it is normal to conduct pre-contract laboratory mixing tests with a range of prospective binder dosages on representative soil samples to confirm the expected level of improvement. Often, and in particular where experience with the soil type or area is minimal, it will be necessary to conduct preliminary site trials as well, as laboratory mixing can never be truly representative of the ground profile or the mixing efficiencies achieved in the field.

The mass of ground to be improved is often formed using columns of mixed soil that are designed to interlock. After the initial set of the columns, a small surcharge can be applied to optimise the settlement performance in embankment and foundation applications. Mixing tools to form columns range in size from 500 to 1000 mm dia. and are commonly 600–800 mm dia., to suit the binder injection equipment. Most designs employ vertical columns, with construction tolerances of up to 1 in 100, but small inclinations of up to 10V:1H may also be used, especially beneath embankments.

Equipment (**Figure 90.17**) consists of a relatively lightweight hydraulic base machine of up to 40 tonnes or so, with wide tracks to assist with operating on working platforms formed over soft ground. The machine operates in tandem

Figure 90.17 Dry soil mixing equipment: dry binder shuttle and mixing machine (Courtesy of Keller)

with a self-propelled binder delivery unit, a pressurised air vessel known as a ‘shuttle’. The base machine rotates a kelly bar with a mixing tool attached to its base into the ground to the desired depth. It is then withdrawn at a steady rate and the tool rotated at a high rate, typically 150–200 rpm, as the binder is injected using compressed air, so forming the mixed column. Columns can be constructed to about 20 m depth or beyond, given due consideration of machine stability and enhanced air pressure. Dosage rates can be modified to some extent during withdrawal to suit significantly differing layers in the ground. However, the horizons of such layers must be well established in advance of construction using, for example, cone penetration testing (CPT) equipment, as the dry soil mixing process is not intended for ground investigation. Binder wastage is under 5% in most cases.

It is important to ensure even distribution and mixing of the binder into the ground. The shape, numbers of blades and nozzle spacing on the tool are set to optimise binder injection across the diameter of the column. Distribution with depth is dependent on steady binder delivery and tool rotation and withdrawal. It is common to automatically measure the binder delivery and to display this against depth and other tool parameters in the cab of the mixing machine. Some machines are now using such information to not only measure but also control binder delivery, thus improving the quality of the mixing and minimising waste.

A useful broad indication of mixing efficiency is provided by the blade rotation number (BRN), defined as the total number of mixing blades passing during mixing over a 1 m depth interval for a single shaft, whence

$$\mathrm{BRN} = (NR)/V \qquad (90.2)$$

where N is the number of blades, R is the rpm of the tool and V is the withdrawal velocity (m/min). Experience from Sweden indicates that BRN values for adequate dry mixing in normal soft soils range from 220 upwards, and in practice are often 300 or more.

At relatively shallow depths, typically 3–5 m, mass mixing rather than column mixing may be used. Here, tools rotating on a horizontal axis, or larger than normal vertical axis tools, are mounted on the booms of basic hydraulic diggers. These are used to form large volumes of improved ground relatively quickly by repeated mixing within designated zones. Care is needed with site control to ensure appropriate binder distribution and mixing, as there is no means of accurately measuring binder dosage due to the large overlaps used. However, provided the applications for mass mixing, such as highway sub-grades, are carefully selected to match this approach it can represent a cost-effective solution.

Like jet grouting, blocks of treated ground can be formed by interlocking dry soil mixed columns together to attack a wide range of geotechnical problems. Dry soil mixing has been extensively applied to support road and rail embankments over soft ground (**Figure 90.18**). It has also been employed beneath housing and other structures; to provide side and base support to excavations; to buttress retaining walls; and in shaft and tunnel construction.

90.6.2 Wet soil mixing

Wet soil mixing differs from dry soil mixing, not only in that wet grout is injected, but in that the equipment, range of applications and soil types treatable are very much broader. Columns can be formed using single vertical axis systems, or with double, triple or more axes constructing interlocking columns in one cycle (**Figure 90.19**). The multi-axis systems tend to be more efficient when constructing linear structures in the ground, such as cut-off walls or long earth-retaining structures. Single-axis soil mixing is used to construct columns typically of 600–2500 mm dia., and multi-axis systems have a wide range of geometry, with nominal diameters again of 600 mm and greater. Japanese development has gone even further and large multi-axis mixing machines are available for working off barges to depths of 60 m or more from sea level.

The key equipment for wet soil mixing consists of silos for storage of the cement and other materials, high-shear mixers, water supply and pumps capable of delivering sufficient slurry grout to ensure that each cycle of mixing in the ground is not held up – for quality as well as production reasons. The grout is pumped through delivery lines into the hollow drilling stem and thence through ports placed at the end of the mixing tool. The stem or kelly bar is normally mounted on the masts or leaders of cranes and rotated using high-torque heads. The base machines tend to be heavier than with dry soil mixing, reflecting the wider range of soil types and larger column diameters, and multi-axis approaches that are employed, and additional consideration for good working platforms needs to be made. Mixing tools vary widely in design, depending mainly on the soil type being treated, and most have multi-level blades.

Construction for single-axis and many multi-axis cycles begins with penetration to the maximum treatment depth, and unlike dry mixing, grout is usually injected both during this phase and the withdrawal phase. The wet grout assists penetration in drier cohesive soils and in sandy and mixed soils and in the uniformity of the mixing. Penetration speeds and withdrawal speeds are normally different, as are the injection rates during penetration and withdrawal. Some approaches employ a further complete cycle or partial re-penetration to ensure a higher mixing efficiency. Rotation rates are lower than for dry soil mixing, but the BRN has been found to be helpful also for wet soil mixing. Topolnicki (2009) showed that the BRN should exceed 430 for a reasonable consistency in relation to the variation in strength of the mixed ground in single-axis mixing using cement grout in clay soils. It is expected this should also apply for multi-axis methods. In practice most wet mixing approaches in cohesive soils provide at least this level of BRN. Different considerations apply for wet soil mixing in cohesionless soils such as sands, and lower BRN values may apply.

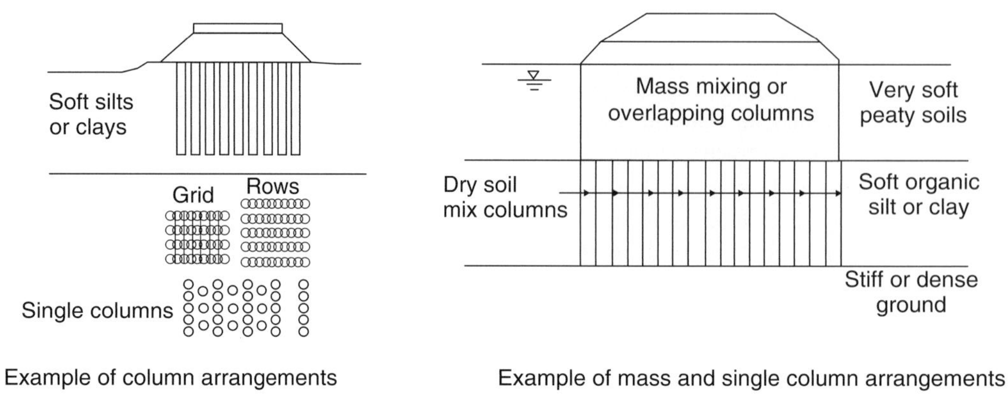

Figure 90.18 Methods for supporting embankments on soft ground using dry soil mixing

As with dry soil mixing, electronic measurement and reporting are commonplace, and are also now increasingly being used to control the mixing cycles. The reporting and control provide confidence in the dosage rates with depth and assist in minimising variability and waste.

Dosage rates vary, typically from 100 to 350 kg cement or more per cubic metre of natural soil, with the higher dosage rates used in the softer and more organic soils, or where very high strengths are required by the particular application (**Table 90.4**). Stiffness is less often required, but where measured is typically in the range of 75–100×UCS (Filz, 2009). The mass coefficient of permeability, *k*, may be as low as 10^{-8} m/s or better where the grout mix, method and application demand, as shown below, but can be at least one or two orders of magnitude higher.

Wet soil mixing, like jet grouting, has found applications in a broad range of geotechnical problems. Given that relatively high strengths and consistency can be achieved, wet soil mixing can be used to provide foundations not only for embankments and the like, but also for relatively large loaded structures (see **Figure 90.20**). Wet mixed columns can be arranged to form gravity walls or alternatively slimmer forms can be reinforced with H-section steel beams or the like to provide additional bending capacity and stiffness. An excellent example is shown in **Figure 90.21**. In suitable ground conditions wet soil mixing can provide low permeability and has been used to provide new or remedial cut-off elements

(a)

(b)

Figure 90.19 (a) Single-axis mixing; (b) double-axis mixing (Courtesy of Topolnicki)

Soil type	Cement factor: kg/m³	UCS at 28 days: Mpa	Permeability *k*: m/s
Sludge	250–400	0.1–0.4	1×10^{-8}
Peat, organic silts, clays	150–300	0.2–1.2	5×10^{-9}
Soft clays	150–300	0.5–1.7	5×10^{-9}
Medium/hard clays	120–300	0.7–2.5	5×10^{-9}
Silts and silty sands	120–300	1.0–3.0	1×10^{-8}
Fine to medium sands	120–300	1.5–5.0	5×10^{-8}
Coarse sands and gravels	120–250	3.0–7.0	1×10^{-7}

Table 90.4 Typical field strength and permeability data for wet soil mixing (Data taken from Topolnicki, 2004)

Figure 90.20 Prepared foundation excavation showing 0.9 m dia. wet soil mix columns (Reproduced from Topolnicki, 2004)

Figure 90.21 Reinforced soil mix columns to form retaining structure; Music Academy, Posnan, Poland (Courtesy of Topolnicki)

for dams and other barriers to water or contaminated flows. Triple-axis mixing is often more efficient than single-axis for this range of applications.

90.6.3 Soil mixing using trench cutting

Another approach to soil mixing has been provided by the development of equipment which enables mixing of wet grout while cutting trench structures in the ground. Originally developed and used in Japan, there are now several approaches available and in use in Japan, the USA and Europe. The major applications are in forming linear in situ structures such as cut-off barriers and retaining walls. Reinforcement in the form of steel sections can be used to provide additional structural strength where needed. Advantages of these systems, given the right conditions, are a reduction in the number of joints over competing methods and less waste for any required wall thickness as overlap is minimised, thus leading to reduced costs and improved quality. Wet grout mixes are generally used, with similar mix proportions and final in situ mixed characteristics as with other methods of wet soil mixing.

The TRD (trench cutting and remixing deep) equipment (**Figure 90.22**) consists of a large machine of about 100 tonnes and 8 m tall which effectively enables cutting and mixing by means of a chainsaw concept. An initial starter trench is formed into which a post is assembled and lowered to the required depth. Depths of 30 m or more have been treated. The post holds the cutting chain and injection proceeds as the machine crawls forward, cutting a full depth face and providing a uniform mix-in-place material. Joints in the soil mixed trench material only occur if production is stopped – for example, if only day shifts are being used. In this way joints are few and the system is particularly efficient for long cut-off barriers.

Figure 90.22 TRD CSM machine (Courtesy of Hayward Baker)

The CSM (cutter soil mixing) method (Stoetzer *et al.*, 2006) is increasingly being used worldwide. It uses combination cutter wheel arrangements attached to kelly bars or suspended from ropes (**Figure 90.23**) to form panels up to 2.8 m in length or greater. An advantage of the method is that both fresh-in-fresh and fresh-to-hard panel construction joints can be facilitated. Depths up to 55 m can be achieved with the rope-suspended CSM. In relatively uniform soils, or for retaining structures up to about 20 m deep, mixing is conducted during cutting (penetration) and withdrawal. For deeper cut-off walls and for less uniform soils a two-phase approach is adopted using bentonite for temporary trench support during cutting. As with conventional diaphragm wall construction, the bentonite slurry is recirculated and cleaned by passing through de-sanding equipment. Grout is injected during the withdrawal phase and mixed with the remaining soils.

Figure 90.23 (a) Mixing wheels and housing; (b) kelly bar-guided CSM unit (Reproduced from Stoetzer *et al.*, 2006)

90.7. Verification for grouting and soil mixing

It is important to verify the mass properties of the grouted or mixed ground in some way to ensure that the treatment has delivered the intended specification.

Verification begins with the careful execution of the work within the specified tolerances to an appropriate level of quality control. Further verification using field trials, laboratory or in situ tests or some other means is usually necessary. Attention should be paid to guidance available in the CIRIA reports C572 (Charles and Watts, 2002), C573 (Mitchell and Jardine, 2002), C514 (Rawlings *et al.*, 2000), and C774 (Hughes *et al.*, 2018), and the European execution codes listed in the further reading section. As with the processes themselves, it is essential that advice from experienced specialists is obtained.

Considerable care is needed in the selection, execution and interpretation of these tests, and all parties to the work need to be clear on the verification strategy and the interpretation of the results before testing begins. It is clearly wise to select a verification programme that has a proven track record. **Table 90.5** indicates the main means of verifying treatment, but the reader should be aware that there are several other means, some specific to individual processes, which may be particularly relevant to their project.

Pre-contract trials at full scale are often useful, and sometimes essential, to establish the properties or even the viability of the prospective process. In jet grouting, for example, evidence of the diameters and associated properties may require the formation of full-scale test columns, which can then be examined in detail to provide confidence in conditions for which previous experience is small or unavailable (**Figure 90.24**).

For both dry and wet soil mixing, laboratory trial mixes are particularly useful, but cannot directly take account of variability in the ground, and lab mixing efficiencies will inevitably be different from in situ mixing. Careful interpretation of such tests is therefore necessary.

Tests on cores, or more rarely on block samples, recovered from the treated zone are often considered and can be good guides to performance. However, low sample disturbance, accuracy in aligning the coring tool and high core recovery to properly represent the mass properties are needed. Realism in assessing the prospects of achieving good-quality cores, together with the employment of specialist crews experienced in verification coring, is essential to achieve a successful outcome. In wet soil mixing and in jet grouting, triple tube coring methods have shown good efficiency where the treated mass is suitable. In grouted soils containing gravel or cobbles, it is often difficult or impractical to recover good-quality cores.

Another option possible in some situations for soil mixing and jet grouting consists of the recovery of 'wet' samples soon after completion of the process and before setting takes place. These samples can be placed in cylindrical or other approved moulds and tested at the specified time intervals.

A range of tests carried out from within boreholes can be conducted. It is important to ensure the positional tolerance of these and take account of the possible effects of disturbance on the properties being measured. They are particularly helpful in measuring permeability as some feel for the mass behaviour is gained. Other common in situ tests such as SPT or CPT are commonly considered for verification tests. It should be appreciated that these were developed for soil investigation and they can be unpredictable in assessing the properties of what in effect are artificial cemented materials produced by one of a range of grouting and soil mixing

	Permeation	Jet grouting	Soilfracture	Compaction	Dry soil mixing	Wet soil mixing
Permeability						
Pre-contract full-scale trials	Possible	Possiblè	N/A	N/A	Possible	Possible
Laboratory tests on trial mixes	Unlikely	Unlikely	N/A	N/A	Possible	Possible
Laboratory tests on in situ cored or block samples	Possible	Possible	N/A	N/A	Possible	Possible
Laboratory tests on in situ wet samples	Unlikely	Possible	N/A	N/A	Possible	Possible
Tests from boreholes in the treated ground	Possible	Possible	N/A	N/A	Possible	Possible
Full-scale assessment from piezometers	Possible	Possible	N/A	N/A	Possible	Possible
Unconfined compressive strength						
Pre-contract full-scale trials	Possible	Possible	Possible	N/A	Possible	Possible
Laboratory tests on trial mixes	Unlikely	Unlikely	N/A	N/A	Possible	Possible
Laboratory tests on in situ cored or block samples	Possible	Possible	Possible	N/A	Possible	Possible
Laboratory tests on in situ wet samples	Unlikely	Possible	Unlikely	N/A	Unlikely	Possible
In situ standard penetration testing (SPT) or static CPT tests	These can only be recommended for situations in which good experience and calibration exist. Note: they are useful for assessment of density change in the ground around compaction grouting bulbs.					
Stiffness						
Pre-contract full-scale trials	Possible	Possible	Possible	Possible	Possible	Possible
Laboratory tests on trial mixes	Unlikely	Unlikely	N/A	N/A	Possible	Possible
Laboratory tests on in situ cored or block samples	Possible	Possible	Possible	Possible	Possible	Possible
Laboratory tests on in situ wet samples	Possible	Possible	N/A	N/A	Unlikely	Possible
In situ SPT or static CPT tests	These can only be recommended for situations in which good experience and calibration exist.					
Pressuremeter or dilatometer tests	These can only be recommended for situations in which good experience and calibration exist.					
Full-scale trial loading tests	Possible	Possible	Possible	Possible	Possible	Possible

Table 90.5 Guide to the main verification methods for grouting and soil mixing

Figure 90.24 Example of jet grouted column formed as part of a pre-contract trial

processes. The authors are aware of sites for which the CPT grossly underpredicted the apparent in situ shear strength of cement bentonite injections, jet grouted columns and soil mixed masses, by comparison with tests on cores and wet samples. These systems can therefore only be recommended when there exists a good body of relevant information for the processes and specific soils to be treated, or where sufficient pre-contract trials are specified in order to calibrate such tests with the required properties.

Finally, testing at or near full scale can be a powerful means of verifying the performance of the treated mass. However, this is often expensive and can suffer from lack of coverage of the whole volume of treatment. In assessing permeation, for example, such as for cut-off walls, piezometers suitably placed either side at the key levels can be used to check the overall cut-off performance. However, the timescales involved in assessment may be considerable and well beyond the construction programme. Similarly, foundation settlement under load can be assessed to some extent by full-scale loading on the improved ground, although usually this needs further interpretation to cover the likely timescales and loading complexities imposed during the design life of the structure being supported.

References

Baker WH, Cording E and McPherson H (1983) Compaction grouting to control ground movements during tunnelling. *Underground Space*, vol. 7. Pergamon Press, Oxford, UK, pp. 205–212.

Bell AL (1993) Jet grouting. In *Ground Improvement*, 1st edn. Spon, Abingdon, UK.

Bell AL (ed) (1994) *Grouting in the Ground*. Thomas Telford, London, UK.

Charles JA and Watts KS (2002) *Treated Ground: Engineering Properties and Performance*. CIRIA Report C572. CIRIA, London, UK.

Crockford RM and Bell AL (1996) Compaction grouting in the UK. *Proceedings of the International Conference on Grouting and Deep Mixing, Tokyo, Japan* (Yonekura R, Shibazaki M and Terashi M (eds)). Balkema, Rotterdam, the Netherlands, pp. 279–285.

Deep Foundations Institute (2015) *Proceedings of the International Conference on Deep Mixing, San Fransisco, USA*.

Deep Foundations Institute (2021) Best practice and legacy. *Proceedings of the International Conference on Deep Mixing* [virtual conference].

Filz GM (2009) Design of deep mixing support for embankments and levees. Keynote lecture. *Proceedings of the International Symposium on Deep Mixing and Admixture Stabilization, Okinawa, Japan*.

Greenwood DA (1994) Permeation grouting. In *Grouting in the Ground* (Bell, AL (ed)). Thomas Telford, London, UK, pp. 72–94.

Healy PR and Head JM (2002) *Construction over Abandoned Mineworkings*. CIRIA Report SP32. CIRIA, London, UK.

Hughes AK, Bruggemann DA and Kettle C (2018) *Grouting for Reservoir Dams: A Guide to Good Practice*. CIRIA Report C774). CIRIA, London, UK.

Linney LF and Essler RD (1994) Compensation grouting trial works at Redcross Way. In *Grouting in the Ground* (Bell AL (ed)). Thomas Telford, London, UK, pp. 313–327.

Littlejohn S (1982) Design of cement based grouts. *Proceedings of the Conference on Grouting in Geotechnical Engineering, New Orleans, LA* (Baker WH (ed)). ASCE, Reston, VA.

Littlejohn S (1983) *Chemical Grouting*. University of Witwatersrand, Johannesburg, South Africa. Reprint from the South African Institution of Civil Engineers.

Lombardi G and Deere D (1993) Grouting design and control using the GIN principle. *International Water Power and Dam Construction*, June: pp. 15–22.

Mair RJ (2008) Tunnelling and geotechnics: new horizons. Rankine Lecture. *Géotechnique* **58(9)**: 695–736.

Mitchell JK and Katti RK (1981) Soil improvement: state of the art report. *Proceedings of the 10th International Conference on Soil Mechanics and Foundation Engineering, Stockholm, Sweden*, pp. 509–565.

Mitchell JM and Jardine F (2002) *A Guide to Ground Treatment*. CIRIA Report C573. CIRIA, London, UK.

Moseley MP and Kirsch K (eds) (2004) *Ground Improvement*, 2nd edn. Spon, Abingdon, UK.

Raabe EW and Esters K (1990) Soil fracturing techniques for terminating settlements and restoring levels of buildings and structures. *Ground Engineering*, May: pp. 33–45.

Rawlings CG, Hellawell EE and Kilkenny WM (2000) *Grouting for Ground Improvement*. CIRIA Report C514. CIRIA, London, UK.

Rubright R and Bandimere S (2004) Compaction grouting. In *Ground Improvement*, 2nd edn (Moseley MP and Kirsch K (eds)). Spon, Abingdon, UK, chapter 6.

Stoetzer E, Gerressen FW and Schoepf M (2006) CSM cutter soil mixing: a new technique for the construction of subterranean walls. *Proceedings of the DFI Conference, Amsterdam, the Netherlands*.

Topolnicki M (2004) In-situ soil mixing. In *Ground Improvement*, 2nd edn (Moseley MP and Kirsch K (eds)). Spon, Abingdon, UK, chapter 9.

Topolnicki M (2009) Design and execution practice of wet soil mixing in Poland. *Proceedings of the International Symposium on Deep Mixing and Admixture Stabilization, Okinawa, Japan*.

Warner J (2004) *Practical Handbook of Grouting*. Wiley, New York, NY.

Further reading

Anon (2009) *Proceedings of the International Symposium on Deep Mixing and Admixture Stabilization, Okinawa, Japan*.

BSI (British Standards Institution) (2000) BS EN 12715:2020: Execution of special geotechnical work – Grouting. BSI, London, UK.

BSI (2001) BS EN 12716:2018: Execution of special geotechnical work – Jet grouting. BSI, London, UK.

BSI (2005) BS EN 14679:2005: Execution of special geotechnical work – Deep mixing. BSI, London, UK.

Terashi M and Kitazume M (2009) Current practice and future perspective of QA/QC for deep mixed ground. Keynote lecture. *Proceedings of the International Symposium on Deep Mixing and Admixture Stabilization, Okinawa, Japan*.

Yonekura R, Shibazaki M and Terashi M (eds) (1996) *Proceedings of the International Conference on Grouting and Deep Mixing, Tokyo, Japan*. Balkema, Rotterdam, the Netherlands.

It is recommended this chapter is read in conjunction with

- Chapter 25 *The role of ground improvement*
- Chapter 59 *Design principles for ground improvement*
- Chapter 84 *Ground improvement*
- Chapter 94 *Principles of geotechnical monitoring*

All chapters in this book rely on the guidance in Sections 1 *Context* and 2 *Fundamental principles.* A sound knowledge of ground investigation is required for all geotechnical works, as set out in Section 4 *Site investigation.*

doi: 10.1680/icemge.66830.1469

Chapter 91
Modular foundations and retaining walls

Simon Bullivant Taranto Ltd UK, Castle Donington, UK

The benefits of using modular foundations are becoming more environmentally obvious and financially pronounced. With the trend for 'off-site superstructure fabrication' continuing to progress, the use of such sustainable foundation options will only increase. The leading ground engineering specialists continue to develop new systems that assist building contractors to achieve an increase in efficiency and sustainability that cannot be realised by using traditional cast in situ foundations.

CONTENTS

91.1. Introduction

Structures that utilise a form of modular foundation have been in use for decades. The British landscape has many excellent examples of just how forward-thinking our ancestors were. Sustainability was not high on their agenda. They had not heard of lean construction, Egan, or Latham, and they did not appreciate the effects of climate change and the lasting legacy of their actions. They built foundations for the structures like those shown in **Figures 91.1** and **91.2** because they had practical reasons to do so.

The methods they employed depended on the skills and equipment of the time but were always tried-and-tested techniques. The drivers for their construction were purely practical. In the case of **Figure 91.1**, the requirement was to store grain off the floor in a dry environment away from rodents and other vermin; in the case of **Figure 91.2** the need was to live above water. They were not designed using mathematical formulae and did not employ the concepts of soil mechanics or geotechnical engineering. They were built using techniques passed down from generation to generation and, more importantly, the methods they employed worked.

Engineers seek to improve on the 'traditional' methods employed by our ancestors. The Egan report *Rethinking Construction* (Construction Task Force, 1998) brought to the attention of the construction industry the concept of 'lean construction'. One of the cornerstones of this concept is the use of modularisation, off-site techniques and 'just-in-time' delivery. The concept was not really new to the foundations industry. It had been installing factory-produced and 'just-in-time' delivered precast driven piles for decades and had been fabricating precast concrete walls and foundation beams since the early 1930s. Many of these early modular buildings still survive and are in good, serviceable condition (BRE, 1984). However, the principles behind *Rethinking Construction* require the piling and foundations industry to take a more holistic view and move towards the use of a truly modular foundation. Generally, for civil engineering and high-rise structures this is not a practical solution, but for house foundations and lightly loaded structures there have been numerous contractor-led developments in the use of modular off-site foundation systems.

Figure 91.1 Sixteenth-century grain store foundation and staddle stones (Courtesy of Nigel Rake, all rights reserved)

Figure 91.2 House on stilts (© Mikhail Nekrasov)

91.2. Modular foundations

A *module* can be defined as being 'a separable component, frequently one that is interchangeable with others, for assembly into units of differing size, complexity or function', and *modular* as being 'composed of standardised units or sections for easy construction or flexible arrangement'. What is therefore meant by a modular foundation? The historic sixteenth-century grain store foundation shown in **Figure 91.1** provides us with an excellent example of how a simple practical design solution of the past can be translated into modern concepts. The foundation *system* is modular; the staddle stones provided the support mechanism in a similar way to a foundation system patented some 450 years later.

Perhaps the most well-known example of a modular foundation is the precast pile, a concrete unit manufactured in a factory environment and delivered to site, where it is subsequently driven into the ground. Individual precast concrete sections, or modules, can be joined together during the driving process and the final length of pile is generally determined during the actual installation process and verified by static design or testing. In the strictest sense the system is truly modular, with significant efficiency advantages being gained by the ability to manufacture and stock standard-sized modules, as shown in **Figure 91.3**. A more in-depth review of modular piling systems is covered elsewhere in this publication: Chapter 81 *Types of bearing piles* provides further information on the construction of these and other displacement pile types; Chapter 82 *Piling problems* outlines some of the commoner problems that may be encountered; and suitable design methods are discussed in Chapter 54 *Single piles*.

In house building the trend for 'off-site superstructure fabrication' continues to progress. There is now a diversity of approaches based on steel or concrete systems, as well as timber frame or panels. Timber-framed houses had a poor reputation in the 1980s due to a lack of understanding and numerous construction issues, but they have recently improved due to better product quality control and environmental credentials. The increased use of light steel and other panel systems has implications for traditional cladding and roofing products, as well as improving site practices. This applies across the whole housing and offices sector. Designs based on modular foundation units fabricated in a factory environment and assembled on site have been around for many years, but it is only recently that they have been viewed in a more holistic way, with the substructure becoming an integral part of the superstructure design. The holistic approach should look at the structural interfaces. A key component is the ground floor design. The relationship between the design of the ground floor and the foundation has a bigger impact than between the floor and the superstructure. The design integration of floors and foundations together can lead to significant optimisation and savings.

Figure 91.3 Precast piles (Courtesy of Roger Bullivant Ltd, all rights reserved)

91.3. Off-site manufactured solutions: the rationale

'It is unequivocal that human influence has warmed the atmosphere…' (IPCC, 2021). The 28 million households in the UK account for over 27% of the UK's energy consumption, and approximately 15% of the UK's greenhouse gas emissions are attributed to heating (Department for Business Energy and Industrial Strategy, 2019a, 2019b; ONS, 2021). The need for contractors and engineers to adopt a responsible approach to construction has never been greater. Chapter 11 *Sustainable geotechnics* provides a holistic overview of sustainable geotechnics and also contains some excellent references that outline the subject in greater detail.

Recent changes in 2021 to the Building Regulations Part L 2010 and the issuance of the Future Homes Standard for 2025 have already had an impact on design parameters, particularly in the superstructure, with a *fabric first* approach to ensure radical improvements to efficiency and huge reductions in the production of greenhouse gases. These improved standards have also required changes to modular foundation design, with a particular focus on the elimination of cold bridging wherever possible.

Several proprietary examples of modular foundation systems are currently marketed in the UK, and all offer differing but significant advantages to prospective developers. The most influential advantage in the commercial environment is that of cost. Modular foundation systems were initially only competitive in 'poor ground' conditions. This was regarded as ground that demanded piling due to deep fill, soft soils or high water table conditions. As time has passed, the financial comparison with cast in situ foundations has inexorably swung towards the modular foundation due to reduced labour and material content. Now the major UK house builders regularly utilise piled modular foundations in 'marginal ground' conditions where fill, soft or wet ground may be only 1.2 m. Another key feature that is often overlooked in these comparisons is that if there is any variation in soils that require deeper digging, then the cost implications are exponential as risks of collapsing sides and groundwater increase. However, the cost increase risk with longer piles is generally linear.

A significant key growth area of modular piled foundations is where soil heave conditions exist. It is cost-effective to have a modular piled foundation constructed and fitted with a proprietary void-maintaining product to prevent the migration of backfill, as illustrated in **Figure 91.4.**

The most efficient modular foundation systems provide major sustainable benefits (NHBC Foundation, 2010). In the marginal ground case above there is approximately 0.75 m^3/linear metre of in situ concrete in a traditional trench-fill foundation, compared with 0.19 m^3 of precast concrete in a modular foundation system consisting of a 6.0 m long 200 mm section pile, pile cap and precast beam. There is the added security of having an engineered factory-produced product and significant reduction in dumping of excavated material.

Given the major impact that construction, and house building in particular, has on the environment, the changes in Building Regulations Part L and the Future Homes Standard for 2025 is driving new homes towards net zero in carbon emissions. This has an effect immediately of making most modular foundation beam profiles flat-topped rather than the inverted-T profile that dominated the modular foundation sector for many years. This is to avoid cold bridges and provides clear vertical separation between the foundation and floor wall intersection. The secondary effect of these increased performance standards is that the superstructure will inevitably become more lightweight with timber and light gauge steel becoming the predominant skeletal materials. This change makes modular foundations even more competitive, with longer spans achievable from smaller-section beams.

Another key area of carbon emissions that is coming increasingly to the forefront, particularly as building-use emissions reduce, is the embodied carbon in foundations. Calculations for the embodied carbon in superstructures are now widespread, with BREEAM becoming the norm for commercial buildings. However, assessment of foundations has proved difficult due to the significant variations in ground conditions and foundation solutions (Livesey and Mace, 2016). The manufacture of cement and steel have significant amounts of embodied carbon, but their versatility and durability below ground are very difficult to match. Modular foundations ensure that with factory control conditions and tight design, huge steps can be taken in reducing embodied carbon within foundations by optimising the materials used.

Specialist manufacturers will see the opportunity for further market growth by providing reliable calculations of the embodied carbon in their products, which will not so easily be replicated with the variables associated with traditional cast in situ methods. These long-term benefits could have significant implications for the building industry as a whole.

Most modular foundation systems require no trench excavation, thus avoiding the hazardous practice of working in excavated ground. Components are manufactured off site (**Figure 91.5**) and installed on site (**Figure 91.6**), with a significant time saving on traditional house foundation construction methods. These advantages are more pronounced when the systems are installed in large volumes. The dry placement of preformed beam sections provides a high degree of accuracy, allowing beam width to closely match wall width. Some manufacturers provide a suite of standard beam widths to minimise unnecessary waste.

Significantly, most of these proprietary systems have been developed by specialist geotechnical contractors rather than general building contractors. The most common types are detailed below.

Figure 91.4 Precast beams fitted with compressible polystyrene void formers (Courtesy of Mawson Piling Ltd, all rights reserved)

Figure 91.5 Factory manufacture of foundation components

Figure 91.6 Precast concrete beams for a modular foundation system (Courtesy of Roger Bullivant Ltd, all rights reserved)

91.4. Precast concrete systems

Modular precast concrete foundation systems have been used successfully for many years, particularly in the housing sector where they form one of the larger component sub-assemblies for either conventionally built or modern methods of construction dwellings. This type of modular foundation forms the overwhelming majority of systems sold in the UK and can be broken down into a number of subgroups that are offered by a variety of specialists. These systems can be supported by either precast or cast in situ piles, and provide support either directly under the foundation beams or in conjunction with individual pile caps, depending upon the particular modular system.

All of the systems detailed have the advantages of factory production and quality control together with clean, fast and non-weather-dependent installation at typical rates of 250–350 m of beams per week.

91.4.1 Interlocking reinforced concrete beams systems

This system was adopted from designs imported from the Benelux countries, where deep precast driven piles are the predominant support method. The system is generally supported directly on top of the piles with a single starter bar drilled and grouted into the top of the pile. This protruding bar is then grouted into a preformed hole through the precast concrete beam once lowered into place. The beams are jointed with half depth laps into pockets in the primary beams and secured by way of another grouted single locator bar. The system is shown in **Figure 91.7**.

The reinforced precast concrete beams are individually designed and casting requires a high degree of accuracy to ensure the system fits together correctly. The lap jointing system results in primary and secondary beams, which dictates a specific laying order. This means that reach over a full plot is necessary and consequently the system is installed with a crane rather than the 13 t excavator that most other precast concrete systems utilise.

Figure 91.7 Interlocking reinforced concrete beam system (Courtesy of Mawson Piling Ltd, all rights reserved)

The system is very effective, but generally has a higher degree of reinforcement than other precast concrete systems due to the high shear stresses at the lap joints. The system is particularly advantageous when ground conditions require very deep piles as the beams can accommodate joints independent of pile positions. Therefore, a beam support arrangement for a given house design may have a reduced number of piles but at the expense of heavier beams.

91.4.2 Simply supported concrete beams systems

This type of system is supported on either precast driven or bored piles and generally utilises an individual precast or cast in situ pile cap on which to land the ends of the precast beams. The most prevalent proprietary system until recently was the inverted-T beam system, which was able to support the outer brick skin on the outer toe and the floor on the inner toe. The spine then supported the inner block skin. However, changes in the Building Regulations Part L rendered this system obsolete as the spine encroached too near the finished floor level to satisfy the linear cold bridge requirements. This can be overcome by lowering the system, but this results in additional under-build. As a result, most of the modular foundation systems are produced with a level top surface from which to build the walls. A detail of this is shown in **Figure 91.8**.

The modular beams may be simple rectangular shapes or may utilise more sophisticated shapes below the level top surface to minimise the use of concrete. Systems may be either normally reinforced or prestressed. The beams are either jointed on the pile caps by means of chamfered ends and cast in situ corners, or may be butt-jointed. In either case there is a nominal amount of reinforcement required for continuity around the corner and in some cases to the pile cap.

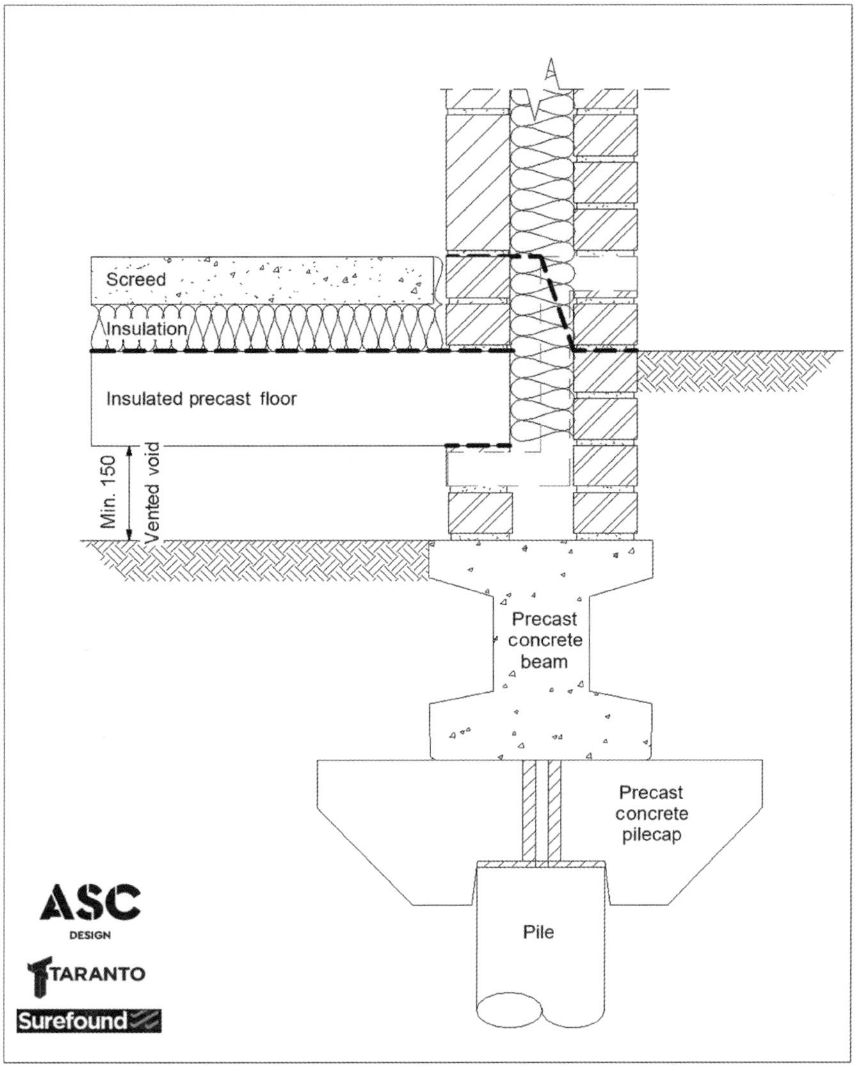

Figure 91.8 Detail illustrating a typical outer wall foundation section (Courtesy of Surefound by Taranto Ltd, all rights reserved)

These systems are very flexible, often taking the modularisation deeper to use standard reinforcement and beam sizes in order to take advantage of production for stock rather than custom manufacture. The generally smaller beam sections lend themselves to being positioned by a 13 t excavator and have the added flexibility to easily deal with stepped sites, as illustrated in **Figure 91.9**.

Figure 91.9 Example of the flexible nature of modular foundations on a stepped site (Courtesy of Roger Bullivant Ltd, all rights reserved)

91.4.3 Post-tensioned concrete modular systems

A post-tensioned precast concrete modular foundation system offers similar benefits to the systems already outlined above for a wide range of structures, including traditional build, modular structures, housing, hotels, prisons and schools (**Figure 91.10**).

The beam modules are rectangular in section and cast with a central longitudinal hole. The beams are supported on pile caps and simply butt-jointed. Once in position a stressing cable is passed through the central hole and then post-tensioned.

Figure 91.10 Post-tensioned concrete modular foundation – Smartfoot® (Courtesy of Van Elle Ltd, all rights reserved)

The advantage of this is that other than the grouting material there is minimal cast in situ concrete used.

Like the preceding modular systems, this type of foundation provides many benefits over traditional strip or trench-fill foundations. It is cost-effective, quick and can be installed to high tolerances.

91.4.4 Modular precast flooring systems

Modular ground floor systems can be an integral part of the structural support of the superstructure. However, the floor is often disregarded in the overall design of the foundation. This is generally because the modular foundation is designed and manufactured by a geotechnical specialist and the modular floors are manufactured by flooring specialists. Where a project relationship provides the modular foundation specialist the opportunity to influence the choice of floor, significant optimisation may be provided by integrating the modular floor and foundation designs.

A variety of proprietary precast suspended flooring systems are available and utilised on conventional strip, trench-fill foundations or on piled modular foundations.

The more common types of system are as follows.

(*a*) *Beam and block*, which is suitable for use for both ground and upper floors and is used in conjunction with standard format building blocks to provide a fast and robust floor deck. The spacing of beams can be varied by using the block to span either in length or width to suit the required span and loading requirements. Beams may also be positioned together for point or wall loads. Typical spans for domestic loading are *c*.5 m for a typical 150 mm beam, but this can be extended up to 8 m for deeper beam sections. This system has been the most widely used suspended ground floor method for many years. Once installed with blocks and grouted, the platform provided can be used as the ongoing working platform. Insulation sheets are generally placed over the whole construction and topped with screed or dry boards.

(*b*) *Beam and insulation block* utilises prestressed beams in conjunction with expanded polystyrene blocks, rather than standard building blocks, to provide inherent thermal insulation for ground floors. Early methods encompassed shaped polystyrene that spanned between the beams and also wrapped under the beam. The screed would then sit directly onto the top of the beams. This system was prone to cold bridging if not carefully installed and now is being superseded by using higher-grade polystyrene units that span between and a further high-grade polystyrene on top of the beams. This ensures a clear and easily checked thermal break is formed between the beams and the screed. These systems require the design consideration of wall loadings that sit directly on the screed to ensure that the strain limit of the polystyrene sitting on the narrow top of the floor beams is not exceeded.

(*c*) *Hollowcore*, comprising prestressed and reinforced hollow-cored slab elements which easily accommodate the longest spans for domestic applications. In situ concrete is placed between the units to fill the shaped shear keys. The top is covered with insulation boards and finished with screed. The drawback of this large unit type floor is the accuracy required for preformed service entries. While not widely used for the ground floors in housing, their superior span and load capabilities make them ideal for apartments.

(*d*) *Insulated reinforced precast floor modules*, comprising shaped polystyrene units to form integral beam ribs that are then filled with reinforced concrete in the factory and arrive as custom-designed modules up to 1.2 m wide. Once positioned a screed is laid directly on top of the units.

91.4.5 Composite reinforced concrete and galvanised steel

This type of foundation system is a lightweight, modular, composite, insulated foundation and integral ground floor system for low-rise buildings. The system comprises fibre-reinforced screed on expanded polystyrene insulation which is incorporated into galvanised steel channel section floor beams. This floor is placed on composite reinforced concrete and galvanised steel primary support beams. Manufactured predominantly off site in a factory environment, the system is assembled on site by hand, and is connected together with an appropriate amount of reinforced steel and fibre-reinforced in situ concrete, as shown in **Figure 91.11**.

As originally designed and patented the system was not cost-effective compared to the precast concrete modular systems in the mainstream market, but the lightweight nature and

Figure 91.11 Composite foundation floor assembly – SystemFirst™ (Courtesy of Roger Bullivant Ltd, all rights reserved)

build-up by hand made it effective in specialist restricted areas for holiday forest lodges and extensions. The system requires further development to meet the insulation requirements of the current Building Regulations Part L, but the structural modular elements warrant inclusion of the system in this chapter.

91.5. Modular retaining structures

A retaining structure or retaining wall is used to provide lateral support for a soil mass and may be used to carry vertical load – for example, bridge abutments or basement walls. The varieties of structural arrangement that may serve as a retaining wall are almost unlimited. Their classification is based on their method of achieving stability; the most common are the gravity wall, cantilever wall, counterfort wall, buttress wall, diaphragm wall, sheet pile wall and reinforced earth wall (Clayton *et al.*, 1993). For a more in-depth review of retaining walls, reference should also be made to the chapters within Section 6 *Design of retaining structures*.

The use of modularisation is beneficial in most retaining structures where a modular solution is possible. The modular approach provides benefits of cost, quality, appearance and productivity. For these reasons, the use of repeating modular units to form structural retaining systems has long been favoured by foundation and groundwork specialists. Gravity walls, for example, which rely on their own weight and rigidity for stability, can be constructed using smaller interlocking modular units. Examples include gabion walls, crib walls and precast reinforced concrete walls.

Gabion walls are formed from rectangular baskets made from galvanised steel mesh or woven strips filled with stone rubble or cobbles. The stone can be hand placed for a particular architectural style, or machine filled. They are aesthetically pleasing and provide a quick method of installing free-draining wall units, as shown in **Figure 91.12**. Due to the flexible nature and ease of construction, gabion walls are often formed without a solid foundation and thus can be particularly useful for river bank protection.

Figure 91.12 Gabion wall (Courtesy of Taranto Ltd, all rights reserved)

Crib walls are formed with interlocking precast concrete or treated timber units. Stretchers run parallel to the wall face and headers are laid perpendicular to the stretchers. The space formed by the 'cribs' is then filled with free-draining material such as rock, cobble or gravel.

Small module precast concrete gravity walls form a large group of architectural gravity walls with a wide array of surface finishes. These smaller units can be combined with soil reinforcement to form high walls and can be particularly useful in confined access areas. In these cases units are sized to be placed by hand. The example shown in **Figure 91.13** has a thin precast concrete stone-effect facing which links to a separate anchor unit using a galvanised steel loop. This then integrates with soil reinforcement mesh where retention height demands.

Precast reinforced concrete gravity walls are the commonest modern form of larger utility retaining wall. Standard stock L walls are produced to retain heights of up to 5 m, although at this height the units usually rely on a cast in situ base slab to provide additional stability. Beyond this height precast concrete counterfort wall units can be designed for retention heights of *c*.11 m. The stems are generally manufactured in standard modules of *c*.2 m wide with twin counterfort ribs. These are then craned onto a prepared reinforced cast in situ concrete base with large holding down bolts. An example is shown in **Figure 91.14**.

Interlocking sheet piles are commonly used for embedded retaining walls, both for temporary works and permanent structures. A factory-produced product, these steel modules

Figure 91.13 Georoc units under construction (Courtesy of Retaining UK Ltd, all rights reserved)

Figure 91.14 Precast counterfort retaining wall (Courtesy of Taranto Ltd, all rights reserved)

are designed to provide the maximum strength and durability at the lowest possible weight. The form of the sheet pile section interlocks with clutches between modules, facilitating pitching and driving, and results in a continuous wall with a series of closely fitting joints (ArcelorMittal, 2022). Steel sheet piling has traditionally been used for river control and flood defence structures and is a tried-and-tested material for the construction of quay walls for ports and harbours. It is also an ideal material for constructing basement retaining walls as it requires minimal construction width. One example where steel sheet piling has been found to be particularly effective is for the creation of underground car parks (SCI, 2001). Profiled interlocking plastic sheet piles, with their excellent durability, have gained popularity particularly in river and canal bank protection where retention heights are limited and driving conditions allow.

The reader is directed to the following references which provide further detail on many aspects of the use of modular foundations.

References

ArcelorMittal (2022) *Piling Handbook*, 9th edn. ArcelorMittal Commercial RPS, Luxembourg, Luxembourg. See http://www.arcelormittal.com/sheetpiling/downloadcentre/piling-handbook (accessed 18/04/2023).

BRE (1984) *The Structural Condition of Prefabricated Reinforced Concrete Houses Designed before 1960*. BRE, Watford, UK.

Clayton CRI, Milititsky J and Woods RI (1993) *Earth Pressure and Earth-Retaining Structures*, 2nd edn. Blackie, London, UK.

Construction Task Force (1998) *Rethinking Construction*. Department of the Environment, Transport and the Regions, London, UK.

Department for Business Energy and Industrial Strategy (2019a) Energy Flow Chart 2019. https://www.gov.uk/government/statistics/energy-flow-chart-2019 (accessed 18/04/2023).

Department for Business Energy and Industrial Strategy (2019b) Energy UK Greenhouse Gas Emissions 2019. https://assets.publishing.service.gov.uk/government/uploads/system/uploads/attachment_data/file/957687/2019_Final_emissions_statistics_one_page_summary.pdf (accessed 18/04/2023).

IPCC (Intergovernmental Panel on Climate Change) (2021) *Sixth Assessment Report (AR6)*. Cambridge University Press, Cambridge, UK.

Livesey K and Mace E (2016) *The Challenge of Measuring and Mitigating the Environmental Performance of Foundations and Substructures*. Report. BRE, Watford, UK.

NHBC Foundation (2010) *Efficient Design of Piled Foundations for Low-Rise Housing*. IHS BRE Press, London, UK.

ONS (Office of National Statistics) (2021) Families and Households in the UK: 2021. https://www.ons.gov.uk/peoplepopulationandcommunity/birthsdeathsandmarriages/families/bulletins/familiesandhouseholds/2021 (accessed 18/04/2023).

SCI (Steel Construction Institute) (2001) *Steel Intensive Basements*. SCI, Ascot, UK.

Further reading

BRE (2003) *Off-Site Construction: An Introduction.* IHS BRE Press, London, UK.

Department for Communities and Local Government (2008) *Innovative Construction Products and Techniques.* DCLG, London, UK.

Department for Levelling Up, Housing and Communities (2021) Building Regulations 2010 Approved Document (Part) L Conservation of Fuel and Power 2021 Edition – for use in England. Department for Levelling Up, Housing and Communities, London, UK.

Latham, M., Sir (1994) *Constructing the Team.* HMSO, London, UK.

National Audit Office (2005) *Using Modern Methods of Construction to Build Homes More Quickly and Efficiently.* NAO, London, UK.

NHBC Foundation (2006) *A Guide to Modern Methods of Construction.* IHS BRE Press, London, UK.

Stern N (2007) *The Economics of Climate Change: The Stern Review.* Cambridge University Press, Cambridge, UK.

Tomlinson MJ (2001) *Foundation Design and Construction*, 7th edn. Pearson Education, Harlow, UK.

Useful websites

Association of Specialist Underpinning Contractors: http://www.asuc.org.uk

BRE Group BREEM: http://www.bregroup.com

Federation of Piling Specialists: http://www.fps.org.uk

NHBC: http://www.nhbc.co.uk

NHBC Foundation: http://www.nhbcfoundation.org.

Precast Flooring Federation: http://www.mpaprecast.org.uk

It is recommended this chapter is read in conjunction with

- Chapter 11 *Sustainable geotechnics*
- Chapter 54 *Single piles*
- Chapter 81 *Types of bearing piles*
- Chapter 82 *Piling problems*
- Chapter 99 *Materials and material testing for foundations*
- Section 5 *Design of foundations*
- Section 6 *Design of retaining structures*

All chapters in this book rely on the guidance in Sections 1 *Context* and 2 *Fundamental principles.* A sound knowledge of ground investigation is required for all geotechnical works, as set out in Section 4 *Site investigation.*

Section 9: Construction verification

Section editors: **Michael Brown and Michael Devriendt**

doi: 10.1680/icemge.66830.1481

Chapter 92
Introduction to Section 9

Michael Devriendt Arup, London, UK
Michael Brown University of Dundee, Dundee, UK

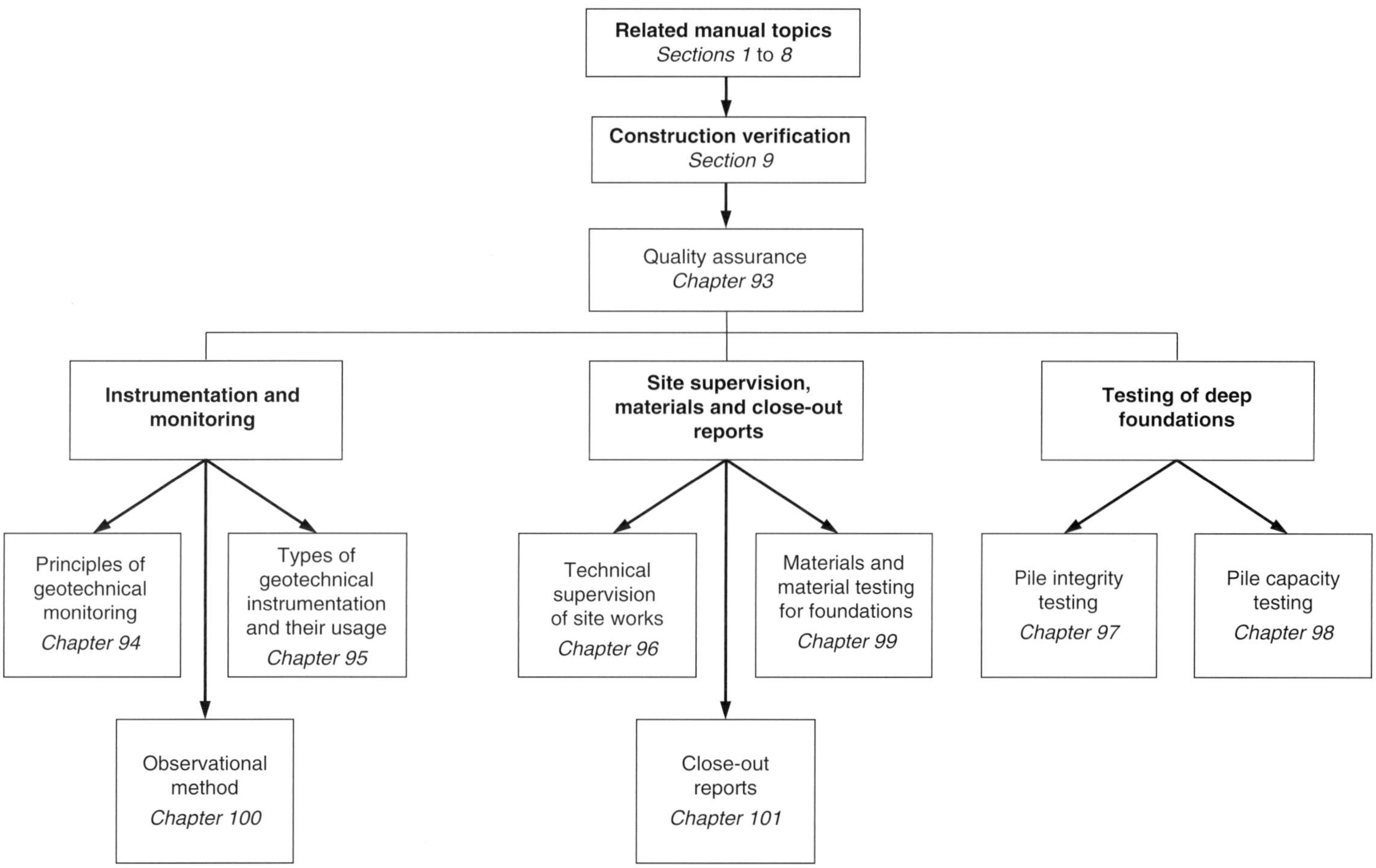

Figure 92.1 Layout of chapters in Section 9

Construction verification can be described as the process of providing test or monitoring results to contribute towards design or management of the works, or to confirm that construction work is being carried out in accordance with the proposed design, specification and appropriate standards.

In no other area of construction is the interrelationship between design and construction so pronounced. Ground is inherently variable so the main construction material can never be known in advance with enough precision. Additionally, design parameters are often dependent on the methods of construction, so can be easily invalidated by minor changes in technique or in soils encountered. Hence construction verification has a much higher significance in geotechnical design than it does in most other areas of construction. Additionally, there is much to be learned from observing construction, and most innovations in geotechnical engineering have their roots in watching and understanding construction taking place. Hence there is a long-term virtuous circle in improving the links between construction and design. Designers must understand how geotechnical works will be enacted and constructors must understand the significance of the design assumptions underpinning the methods they are using.

This section is composed of nine chapters (see **Figure 92.1**) and begins with Chapter 93 *Quality assurance* (QA), which outlines the processes and management systems that may be put in place to ensure a minimum quality of the final construction product. One of the most important elements of a QA process is construction verification, which is particularly important for geotechnical works as completed elements are usually buried and difficult to inspect subsequently. They are also among the most heavily loaded structural elements and their construction is invariably on the project's critical path. This results in a need for early identification and remediation of non-conformances.

The remaining chapters in the section can then be divided into three general themes:

Instrumentation and monitoring

- Chapter 94 *Principles of geotechnical monitoring*
- Chapter 95 *Types of geotechnical instrumentation and their usage*
- Chapter 100 *Observational method*

Construction verification and close-out reports

- Chapter 96 *Technical supervision of site works*
- Chapter 99 *Materials and material testing for foundations*
- Chapter 101 *Close-out reports*

Testing of deep foundations

- Chapter 97 *Pile integrity testing*
- Chapter 98 *Pile capacity testing*

Geotechnical monitoring provides confirmation that works are being carried out in a safe, controlled and efficient manner. By setting pre-agreed limits, monitoring also provides information on when and how contingency measures should be adopted. Comparisons of the monitoring data can also be made with design calculations, or results can be back-analysed to improve calculations on future projects with the potential for significant cost savings. Although monitoring may be viewed as being beneficial, it is important that systems are installed with a clear objective and a plan of how the measured data will be implemented. The application of the observational method is a particular example of this, and the approach can be used to achieve construction efficiencies, or if unexpected behaviour is observed it can provide a process for intervention to prevent failure occurring. Monitoring is also very useful for avoiding claims from third parties on the effects of construction on them and their activities – movement, noise and vibration monitoring can be particularly valuable for refuting claims which could otherwise result in long and costly legal procedures.

Site supervision, close-out reports and materials testing provide confidence that the site works are being carried out and documented in an appropriate manner. Improvements in data recording, management and storage have increased the importance of these subjects, particularly when modifications are carried out to existing structures or where sites are redeveloped. For example, there has been more interest in the reuse of foundation elements, especially deep foundations in congested urban environments. Knowledge captured as part of the original construction (as covered in all of the chapters in this section) will form a valuable source of data for future construction teams which may be faced with a new set of challenges.

The testing of deep foundations allows efficiencies to be made in the design of foundations and can confirm that the piles have been constructed in accordance with the design intent. Deep foundation testing has seen recent changes in techniques (e.g. automated static testing, rapid load testing) and codification as well as new applications through the reuse of piles. This means that the engineer needs to decide

- when testing is necessary
- how many piles to test
- the most appropriate techniques to use
- how to assess and interpret the findings of testing.

The publishing and dissemination of pile test data are also invaluable for improving the collective knowledge for the design and construction of piles in similar materials in the future.

All of the chapters covered in this section contribute significantly towards the centre of the geotechnical triangle (Burland, 1987; and Chapter 4 *The geotechnical triangle*). The geotechnical triangle is used as a framework for geotechnical engineering; its centre focuses on

- *establishing precedent* – for example, the use of instrumentation, monitoring and the observational method for new construction techniques, or where ground conditions are unfamiliar, have uncertain behaviour or are difficult
- *the use of empiricism* – for example, pile capacity testing to refine design assumptions and increase efficiency
- *forming well-winnowed experience* – for example, technical site supervision, selection of appropriate materials and dealing with non-conformances.

The range of subject matter in this section has been chosen to assist an engineer in making appropriate decisions to manage and verify construction processes. It is important, however, to realise that such techniques are only one component in the overall strategy to successfully complete a set of foundations or other geotechnical works. Construction verification must also consider the influence of the ground profile, soil behaviour and the need for an appropriate model to compare experience with. In this respect, the following quote from Terzaghi (1935) summarises the importance of having knowledge from both theory and experience: 'Experience alone leads to a mass of incoherent facts. But theory alone is equally worthless in the field of foundation engineering, because there are too many factors whose relative importance can be learned only from experience.'

As a profession, geotechnical engineers can collectively make improvements to existing knowledge by utilising best practice, some of which is contained in this section. This will enable improvements and efficiencies to be made on current and future projects.

References

Burland JB (1987) The teaching of soil mechanics: a personal view. *Proceedings of the 9th European Conference SMFE, Dublin, Ireland*, vol. 3, pp. 1427–1447.

Terzaghi K (1935) The actual factor of safety in foundations. *The Structural Engineer* **13(3)**: 126–160.

doi: 10.1680/icemge.66830.1485

Chapter 93
Quality assurance

Martin Stanley Bachy Soletanche Ltd, Camberley, UK
David Corke previously at DC Project Solutions, UK
Tony P Suckling A-squared Studio Engineers Ltd, London, UK

Quality assurance is an essential part of the design and construction process and provides confidence to the other parties to the contract that the end product will be built to at least a minimum standard. The specification defines this minimum standard – in geotechnical engineering there are many different model specifications available to suit different construction processes. An additional complementary project-specific specification is needed considering the unique nature of the project and its ground conditions. Workmanship is verified either by a resident engineer, or similar, or by the contractor undertaking self-certification. A key component of quality assurance is the identification and reparation, if necessary, of non-conformances. If any workmanship is thought to be faulty, or if there is a failure, then a forensic investigation will need to be carried out. The records produced by the quality assurance procedure provide factual data for use as part of this investigation.

CONTENTS

93.1. Introduction

Geotechnical works are often supplied by different parties undertaking the choice of the type, design, procurement and construction of the works. However, all of these are interdependent and a failure in one will likely lead to a failure of the whole works.

Quality assurance (QA) is the management system that will ensure that all key parts of the geotechnical design and construction process are undertaken consistently in order to provide at least a minimum standard. Certification of the QA system by an accredited independent body is the evidence that clients, consultants or contractors can rely upon so that each party does not need to audit the other organisations.

A designer will use quality control (QC) systems to ensure that the design and checking process is undertaken to meet the QA requirements. The contractor will use QC systems to ensure that the construction is in accordance with the specification and design. Dependent upon the contractual arrangements on any particular project, the contractor will either have their workmanship verified by an external party, such as a resident engineer (RE) or client representative, or will be required to self-certify their own work.

Identifying and rectifying non-conformant work is a key part of the QA process. For geotechnical work it is paramount that a thorough forensic investigation is carried out at the earliest opportunity so that these issues are closed out early on in the construction process to avoid repeated issues or consequential problems or delays for the follow-on works.

93.2. Quality management systems

For background information on quality management systems, refer to 'ISO 9001: Quality management systems', 'ISO 14001: Environmental management systems', 'ISO 27001: Information security management systems' and 'ISO 45001: Occupational health & safety management systems'.

The design and construction of geotechnical work need to produce an end product that meets three key criteria

- to have an adequate safety margin against failure
- to have acceptable displacements over the likely range of applied loads
- to be durable for the stated design life.

For UK design work, British Standards and Codes of Practice define methods of achieving these criteria. Execution codes, as listed in Chapter 78 *Procurement and specification*, also provide recommendations to help the designer achieve the criteria.

For geotechnical work it is essential that the method of construction is compatible with the design or, if a particular method is prescribed, that the design is cognisant of the methodology to be employed, and that the workmanship meets at least the minimum requirements of the design. The contractor will use QC systems to ensure that these are achieved. QC systems are important in all aspects of construction, but are especially so in geotechnical work as there may be variation in the ground conditions not shown by a site investigation – which only provides information from a limited and discrete number of locations. This is alongside the fact that the majority of works are undertaken below ground level and cannot be visually inspected during most of the construction process for those conducting the works to be built and thus data recording of the execution is critical in verifying the correct execution or gathering information for analysis when issues arise. In addition there is the potential value for clients to capture detailed records of construction to enable the potential reuse of foundations at some stage if modification or demolition of the structure is proposed. This is becoming an increasingly important issue for both sustainability and change-of-use reasons. Many geotechnical processes, such as piling, have to be carried out without necessarily having the knowledge of specific ground information at the particular location. Consequently, continuous monitoring of these processes is a key aspect of the QA management system.

93.3. Geotechnical specifications

The specification defines the minimum standard of work necessary to meet the requirements of the design and of the employer. Because there are many geotechnical processes and solutions, there are many model specifications available; some of these are listed in Chapter 78 *Procurement and specification*.

Model specifications should be used because these have been written by experts in the field and will have been subjected to critical scrutiny by the geotechnical community. However, geotechnical processes can be undertaken with variations, and so systems and methods may vary from those stated in a model specification. Such variations in the method of construction or equipment must be clearly identified in the specialist subcontractor's method statement and be compatible with the design.

Geotechnical engineering is a very innovative part of the construction industry, and both design methods and construction processes are improving all the time. This means that model specifications, if not recently updated, can become outdated and may even be contradictory to current practice. It is essential that experienced geotechnical engineers are used to assess any proposals that would vary the requirements of these specifications.

Early contractor involvement (ECI) or preconstruction services agreements (PCSAs) can be used by the employer and their design teams to help inform the development of the design and specification to suit the potential methods to be used by contractors where works are complex or require uncommon or bespoke solutions. In addition to model specifications, a project-specific specification is very important for geotechnical work because ground conditions are unique for any site. The particular aspects that may affect the geotechnical work need to be stated so that all concerned can be informed. Project-specific specifications should include all available factual information on the ground conditions, including the groundwater conditions. Clause B1.2 of the ICE *Specification for Piling and Embedded Retaining Walls*, 3rd edition (2017), commonly referred to as SPERW, provides a checklist of the matters that should be considered for inclusion in a project specification, and Appendix 1 of the SPERW contains an example of what needs to be included in a project-specific specification.

93.4. Role of the resident engineer

The detailed responsibilities of the RE on any site will probably not be defined in the contract documents. Normally the role of the RE is to supervise the works on behalf of the employer and to make sure that the employer's project is delivered as intended – see Chapter 96 *Technical supervision of site works*. If the engineer is also the designer of the geotechnical work, then the RE will also be supervising the works to ensure that the standard of construction is in accordance with the design and specification requirements.

To save costs, it has become common for there to be no RE at all, or for the RE to be on site on a part-time basis with no formal power under the contract. This is not recommended except for the most straightforward of geotechnical work. Geotechnical engineering requires experience of a wide range of ground conditions and of complex construction processes, and it can only be of benefit to the project, and ultimately the employer, to have an RE present. It is preferable that the RE supervising geotechnical work should be a geotechnical engineer experienced in the prevailing ground conditions and construction process being used. There are many specialist geotechnical consultancies that can provide this service if the employer's professional team does not have the appropriate experience.

93.5. Self-certification

Most geotechnical work is carried out on a design and construct basis by specialist organisations that are often contracted to self-certify their work and provide a warranty for it. Self-certification by the trade contractor or specialist subcontractor requires them to apply the QA processes on site in relation to their own works. If a project is self-certified there may not necessarily be an RE on site; if there is, they will have little or no contractual responsibility for the self-certified works. The process is often associated with the use of the new engineering contract (NEC), especially on large civil engineering projects, although the NEC itself does not impose self-certification, nor does it actually impose QA. Consequently, any requirement for self-certification will either be defined in specific contract amendment clauses ('Z' clauses in NEC), the *project specification* or in *works information*.

Self-certification by the contractor reduces the burden on the employer and the professional team, particularly where new technologies are involved. If the process is carried out efficiently, the specialist contractor is best placed to coordinate the construction and the checking of the construction process. Any faults or non-conformances should be found and dealt with at the earliest opportunity, minimising the cost of rectification and any consequences.

93.5.1 The self-certification process

The fundamental requirements of the self-certification process are that every part of the construction process should be monitored for compliance and performance. The basic questions are

- Does the work satisfy the specification?
- Is the work being implemented as planned?
- Are the construction and design requirements being met?

The contractor's staff on site should be competent to self-check their own work and supervisory staff made responsible for review and inspection.

The employer will need to approve the contractor's quality management system and will monitor that it is being applied correctly, and carry out spot-checks on finished work. Under NEC options C and D (target cost), defects identified prior to completion are paid for by the client. However, if they are found after completion the contractor pays – which provides an incentive to the contractor to ensure that their QC processes are applied consistently. If the process is to be carried out with confidence, the contractor must set out in their quality management system a detailed description of how the construction standards will be monitored, checked and maintained at all times.

Clause 41.3 in the NEC4 describes how self-certification works best

> The Contractor and the Supervisor informs the other of each of their tests and inspections before the test or inspection starts and afterwards informs the other of the results. The Contractor informs the Supervisor in time for a test or inspection to be arranged and done before doing work which would obstruct the test or inspection. The Supervisor may watch any test done by the Contractor.

The Construction (Design and Management) Regulations 2015 (CDM) require the contractor to

> Plan, manage and monitor construction work carried out by him or under his control in a way which ensures that, so far as is reasonably practicable, it is carried out without risks to health and safety.

These CDM requirements must be satisfied in addition to any self-certification requirements, including monitoring and testing.

The records that need to be kept and submitted as part of the self-certification process need to be very detailed, and are likely to be incorporated into the contractor's method statements. The contractor's quality management system needs to define who will be responsible for the implementation of each element of the works, who will be responsible for independent inspection and, if appropriate, who will be responsible for auditing the whole process.

An important part of the self-certification process is the ability to react quickly to any revealed non-conformances. The processes and responsibilities for dealing with these situations should also be defined by the contractor's quality management system.

93.6. Finding non-conformances

Undetected problems with work carried out in the ground can have severe consequences later on, including major programme and/or financial implications for all involved. For example, a defective pile can normally be easily replaced during the piling works, but once the substructure and superstructure are in place it can be very difficult to do so. (Also refer to Chapter 82 *Piling problems*.)

The methods for finding non-conformances vary depending on the construction process, but would normally involve a combination of some or all of the following: design review, observation, measurement, testing and investigation. The most common non-conformances in geotechnical work are described below.

Under NEC Clause 43, searching for and notifying defective works is managed as follows. Until the defects date, the supervisor may instruct the contractor to search for a defect. The supervisor gives reasons for the search with the instruction. Searching may include uncovering, dismantling, re-covering and re-erecting work, providing facilities, materials and samples for tests and inspections done by the supervisor and doing tests and inspections which the scope does not require.

93.6.1 Concrete in the ground

Concrete defects can be numerous in nature. The classic example is inadequate strength compared to that required by the design or specification. Further issues such as lack of cover or inclusions are covered elsewhere in this section.

Concrete placed in the ground is particularly at risk from expensive investigation and remedial measures if the concrete strength test results prove to be unacceptable. The first point to consider is the source of the concrete. Ready mixed concrete is often delivered to the site, but on larger projects the concrete may be batched and mixed on site. The quality scheme for ready mixed concrete (QSRMC) certification for the batching plant should be checked, along with the date of certification and any changes that may have been made at the batching plant since then.

Batch records for concrete production can be analysed to review matters such as adequate dosing of mix components and the correct water–cement ratio. An important aspect to consider is the moisture content of aggregates in particular, which can vary throughout the day and year on the basis of prevailing weather conditions and the nature of their storage. This has the potential to amend the water content of the mix, affecting both fresh workability and hardened cast-in-place properties detrimentally.

The strength of blended mixed concretes can be particularly sensitive to weighing accuracy. These are mixes which use either pulverised fuel ash (PFA) or ground granulated blast furnace slag (GGBS) as partial replacement for the cement content. The higher the cement replacement ratio (i.e. the less cement the mix contains) the more sensitive the ultimate mix strength becomes to the batch weight of the cement. The use of cement replacement mixes is high in the geotechnical sector due to the need to provide high workability, controlled strength gain and high ultimate strength. This will increase as more sustainable lower-cement concretes are developed for environmental reasons.

Different types of concrete have different characteristic rates of gain of strength that are well documented; they can be used to assess the likely 28- or 56-day strength from earlier testing. Careful tabulation, analysis and plotting of early age (e.g. 3- or 7-day) testing maximises the chance of early identification of any problematic concrete. The presence of low-strength concrete may affect its durability, the protection of reinforcement and the structural capacity of the element.

It is always advisable to review the early cube strength test results to check for any indications that the specified strength may not be achieved. If low-strength concrete results occur, with a knowledge of the typical rate of gain of strength characteristics of the particular type of mix, it may be possible to predict whether or not the required strength will be achieved – even if at a later date than specified. Alternatively, cores can be taken and tested to establish the actual in situ strength of the concrete. For low-strength concretes, the method of testing and the standard to which it is executed should also be considered to ensure representative results are obtained.

Upfront concrete trials, particularly for those mixes with tight specification requirements or complex performance requirements, are highly recommended to be conducted by the contractor in advance of the works to verify the mix design prior to starting and then ongoing compliance testing and results monitoring should be maintained throughout the works. If the results are observed to be non-compliant, action should be taken as soon as possible to avoid ongoing issues until completion of the project.

Concrete durability is an increasingly important subject, as well as concrete strength, and for projects with extended design lives it is not uncommon for advanced and ongoing works testing to include accelerated carbonation and chloride content testing. The lead time associated with these tests should be considered by all parties to ensure sufficient time is allowed to gain satisfactory results.

Concrete placement is a key factor in the success of geotechnical works construction, particularly when the concrete is placed by way of tremie methods. The European Federation of Foundation Contractors (EFFC, 2018) has published guidance on tremied concrete for deep foundations in its *Guide to Tremie Concrete for Deep Foundations*.

Where support fluid is used for the excavation of piles, barrettes or diaphragm walls, consideration should also be given to the recommendations in the EFFC's (2019) *Guide to Support Fluids for Deep Foundations*.

93.6.2 Pile load testing

One risk associated with load testing is that in certain ground conditions the ultimate load capacity may increase with time (see Chapter 98 *Pile capacity testing*). Consequently, testing foundation elements such as piles too early may result in underestimating their load capacity, even if the concrete strength has developed sufficiently to carry the load. If early testing of a pile, typically 7–14 days after installation, does not verify the required load capacity, retesting at a later date may show some improvement. This behaviour can affect the performance of bored and driven displacement piles in fine-grained soils and rocks.

If load testing cannot verify the required pile load capacity, the most common solution is to down-rate capacity and install additional piles, which may require modifications to the substructure.

93.6.3 Integrity testing of piles

There are three types of integrity testing for piles – see Chapter 97 *Pile integrity testing*. The sonic echo method, covered by ASTM D5882-00, is usually restricted to small- to medium-diameter piles due to the limited energy that can be imparted with a small hammer to the pile head at cut-off level. The potential difficulty with this testing is that although the measurement of the response to the applied impulse is common to all practitioners, the signal processing that is applied to produce the test response can vary. This may result in testing by one organisation showing an anomaly in a pile, whereas testing by another may not show the anomaly. Integrity problems are least likely to occur in relatively uniform ground conditions and are most likely to occur in highly variable ground conditions, particular those with less stable granular strata which can be drawn into the pile during the concreting phase through poor workmanship or method execution.

Reported anomalies can also result from reflections at the bottom of a reinforcement cage or from a bulge in the concrete in soft ground, as well as from defects such as cracks and necking of the concrete section.

If integrity testing does indicate an anomaly and retesting is carried out, the reported results should include the recorded data (i.e. before processing) so that it is available for an independent re-analysis. Ultimately it is often necessary to excavate around the pile to investigate the true nature of a reported anomaly if it is shallow enough to do so (**Figure 93.1**). Otherwise concrete coring can be conducted to recover samples for inspection and testing. Alternatively, replacement pile(s) may need to be installed. Small-diameter piles are particularly susceptible to lateral impact or ground-induced loading, such as from a truck, piling rig or crane passing by too close following construction, which can lead to problems with integrity. These piles need to be protected from such events either by increasing reinforcement provision (using high-visibility marking on any protruding reinforcement) or by sectioning off completed areas with barriers or similar. Trimming piles to cut-off level from working platform level for onward capping beam/pile cap construction can also induce pile damage, which will not be identified by integrity testing. To that end testing should take place after the pile is trimmed down to its cut-off level.

Cross-hole sonic logging (see Chapter 97 *Pile integrity testing*) is another method to investigate the integrity of piles/barrettes/diaphragm wall panels. This is typically completed

Figure 93.1 Temporary casing withdrawn from pile after initial setting of the concrete

in elements with cages, which are placed prior to the concreting process. The first common issue with this method is that the pipes that are placed in the reinforcement cages to provide access for the measuring sondes become blocked either by concrete/grout ingress through the pipe joints or by way of material entering the tube at the top during concreting or during the subsequent element breaking down process to cut-off level. Therefore sufficient protection should be provided at the joints and tops of the tubes to avoid this; a good practice would be handover records between trade contractors following completion of geotechnical works validating that, for example, prior to element breakdown, all tubes were verified as being clear for testing.

In terms of the testing results themselves, much like pile load testing, if an element is tested too soon the result can identify flaws which are not real. The main apparent anomalies identified occur if the tubes are locally debonded from the adjacent concrete or if there is a small volume of no-fines concrete close to the tube. An important point to check is the competence of technicians undertaking the testing on site and those analysing the results. Recent experience also suggests that suction excavation methods being developed by follow-on contractors may also affect results. There are health and safety implications with installing the tubes for sonic logging into piles that require cages made up of more than one element. As such the specifier should consider if the most appropriate method of integrity testing has been listed. Cross-hole sonic logging is covered by ASTM D6760-16.

A third method of integrity testing is now gaining greater use. This is known as thermal integrity profiling and is a technology that utilises the heat generated by curing of concrete to assess the integrity and quality of concrete foundations. From the test results, necks, bulges, the as-built shape of the shaft and variations in concrete cover, including steel cage eccentricities, can be characterised. Piles are prepared for testing during their construction by attaching instrumentation cables to the reinforcing cage or centre bar. Temperatures are then sampled throughout the concrete hydration process starting immediately after casting. These results can be evaluated for shaft shape and integrity, concrete cover, concrete quality and for location of the reinforcing cage. If the measured average temperature against depth is consistent, the shaft is considered to be uniform in shape and quality. Bulges can be identified as localised increases in average temperature, while insufficient concrete quality or cross-section reductions can be identified as localised decreases in average temperature. Because soil and/or slurry pockets produce no heat, areas of soil intrusion or inclusion are indicated by lower local temperatures. Thermal integrity profiling is covered by 'ASTM D7949: Standard test methods for thermal integrity profiling of concrete deep foundations'.

93.6.4 Retaining wall deflections

Retaining wall deflections are often specified as a means of controlling a wide range of performances, as they are relatively simple to measure. However, care should be taken to specify deflections that are appropriate to the required performance and that should realistically be achieved, to ensure that economy in design is maintained. Any behaviour which is not 'as predicted' must be observed and promptly reported to experienced geotechnical engineers.

93.6.5 Prop loads

In major excavations it is becoming more common to monitor the loads in props, the largest of which may carry loads of several hundred kilonewtons. Monitoring of the loads, which may include an initial prestress load, should be carried out from installation, and the measured loads compared with the predicted ones for each relevant stage in the construction. If the monitoring shows that there are early signs that the prop load is increasing faster than predicted, this may indicate not only that the final load in the prop may exceed the allowable value, but also that the loads on the supported wall are higher than anticipated – which may result in overstressing of the wall itself.

The first thing to establish is whether the prop overload is indicating that the total load from the wall may have been underestimated, or the distribution of load from the wall to the prop(s) is not as predicted. Increased loads are often due to temperature effects or to construction imperfections, or they may indicate a departure from the design construction sequence or a change in the construction itself.

In order to be able to redesign the construction process, the design must be reassessed and parameters and loadings reviewed and adjusted until the design model predicts the measured prop load(s) and the observed wall deflection. Then, using the revised parameters and loading, additional props and/or berms can be considered in a revised design and construction sequence.

93.6.6 Lack of cover to reinforcement

The five main reasons for a pile, barrette or diaphragm wall panel losing cover to the reinforcement cage are as follows.

- *Concrete.* If the concrete mix is too stiff to flow and fill the annulus outside the reinforcement, then the thickness of the concrete cover is likely to be variable, and typically be thinnest in front of the reinforcing bars. Upfront concrete workability testing should be conducted in advance of mobilisation to site to conduct the works to mitigate this risk. Tests on slump (BS EN 12350-2), flow (BS EN 12350-5) or slump flow (BS EN 12350-8 and ASTM C1611) should be conducted and the retention of these parameters with time recorded and analysed in relation to the likely construction process time. The BS EN 12350-10 L-box test should also be considered to ascertain the ability of the mix to travel between the reinforcement. Concrete segregation should be tested in line with the ASTM C1610 standard, with an extended standing period of 2 hours as per SPERW (ICE, 2017) Clause 21.5.4.
- *Reinforcement spacing.* The project specification and appropriate design standards should state the minimum spacing for both vertical and horizontal reinforcement so that there is adequate space for the concrete to flow between the reinforcement and completely fill the annulus beyond it (**Figures 93.2** and **93.3**). The execution codes BS EN 1536 and BS EN 1538 (BSI, 2000a, 2000b) give the minimum reinforcement spacing for various types of piles and for diaphragm walls. Specialist geotechnical contractor involvement early in the design and reinforcement detailing process can help reduce this risk.
- *Tremie concreting.* Control of the flow of fresh concrete into a pile, barrette or diaphragm wall panel is essential in order to ensure that no segregation of the concrete takes place, and that older hardening and stiffening concrete is not continually pushed vertically upwards where it could impede the flow of the fresh concrete around and to the outside of the reinforcement. The specification combined with the contractor's method statement should describe the acceptable procedures for placing concrete by tremie pipes. The EFFC (2018) has published guidance on tremied concrete for deep foundations: *Guide to Tremie Concrete for Deep Foundations.* Where support fluid is used for the excavation of piles, barrettes or diaphragm walls, consideration should also be given to the recommendations in the EFFC's (2019) *Guide to Support Fluids for Deep Foundations.*
- *Concrete spacers.* The use and appropriate placement of suitable concrete spacers is important in ensuring adequate cover is provided. The type of spacer (plastic wheel, plastic-coated weld-on metal skid or precast concrete) should be considered in relation to the ground conditions, cage type (including its size and rigidity) and the durability requirements of the project. Weld-on skid spacers are the most suitable to maintain cover. When cages are stored on site care should be taken to avoid crushing spacers resulting in the loss of cover. Consideration should be given to varying the radial and longitudinal frequency of spacers around the cage. For example, at the top of cages and in soft soils additional spacers both around and up and down the cage should be considered.

Figure 93.2 Example of well-spaced steel in a diaphragm wall cage

Figure 93.3 Example of heavily congested steel in a large pile cage which could result in issues with concrete cover

- *Additional cage items.* There is an increasing desire to include provisions in prefabricated reinforcement cages for coupler connections to future slabs as well as reservations for anchors/inclinometers and so on to avoid the drilling and fixing of bars, for example. Consideration of concrete flow and respecting standards on bar spacing should be taken into account when including these additional steel items. Remedial measures to address a lack of concrete cover can include breaking of exposed concrete back to the reinforcement (using hydro-demolition or mechanical methods) and then re-concreting the surface to make good the cover. The difficulty with this method is that if the cover has been compromised on the exposed face of an element, it is also likely to be compromised on unseen and relatively inaccessible faces cast against the ground. In other cases chemical treatments can be applied to the face of the concrete. Advice from specialist suppliers and contractors should be sought in this instance and the durability requirements of the projects understood in terms of what the materials are certified for. In extreme cases, measures such as jet grouting adjacent to the ground face may be required.

93.6.7 Water ingress

The specification should set out the minimum standard for water ingress into any excavation below the water table (**Figure 93.4**) and how this is to be measured, although this may be modified to suit particular usage requirements. If water ingress does not satisfy the specified requirements, then it is most common to resort either to treatment of the ground by drilling from an exposed face or, in some situations, to drilling from the ground surface behind the retaining element. The treatment itself can use a number of products, including conventional cements, special cements or chemical resins to exclude the water. Microcements and mineral grouts, known as colloidal silica, provide extremely effective penetration into fine cracks in rock and fine-grained soils to give efficient water tightness, stability and durability. Rapidly foaming, water-sensitive polyurethanes are ideally suited to combat water ingress quickly and effectively. Quick-reacting and non-water-sensitive polyurea silicate systems can be used where there is an additional need to stabilise ground. Expanding foaming systems can be used where there is a need for void filling, while acrylic resins are particularly suitable for rehabilitation injections for concretes. Advice from suppliers and specialist contractors should be sought when specifying methods and products to resolve water ingress issues.

Figure 93.4 Example of a damp water patch

There are a number of risks associated with this particular problem. First, there is an inevitable delay while remedial grouting takes place. Second, and potentially more seriously, water flow may have an impact on the integrity of the wall itself or induce ground settlements that may cause damage to adjacent structures. Finally, there may be a need to obtain third-party consents for the use of chemical products where groundwater and/or surface water bodies are involved.

93.6.8 Position and verticality of elements

Many non-conformances can be attributed to the lack of positional control at the starting level and during execution of the works. Surveying methods are becoming increasingly accurate for setting out and there are numerous systems available to monitor the as-built position of a pile/panel excavation, including open-market systems from Koden (Ultrasonic Drilling Monitor), Jean Lutz (PRAD/TAUPE/NEMO), Fugro (SoniCaliper) and PDI (Shape), as well as specialist geotechnical contractors' proprietary systems. Adequate consideration should be given at design stage to the method(s) being employed and their inherent accuracies, and in turn adequate time and effort should be deployed on site during the works to supervise and control works within tolerances. These tools can aid in the process, either identifying an issue at a point in construction when rectification is possible or at a stage when the overall scheme design can be reviewed and amended. Out-of-position elements often require additional works to be installed to supplement or replace the lost capacity, with corresponding time and cost for the project.

93.7. Forensic investigations

The definition of a failure is 'an unacceptable difference between expected and observed performance' (Leonards, 1982). When geotechnical failures occur, they are often a surprise event and can happen very quickly. Having said that, it is also common to find that there were earlier warning signs that were either unseen or seen but unheeded. It is common to find that when first called upon to investigate a failure there is a lack of relevant information.

Forensic engineering is becoming increasingly important since foundation and other geotechnical failures can lead to litigation and possibly criminal action. The causes of a failure may be complex, but it is important to be able to explain the failure in simple terms to those charged with resolving any contractual or legal matters arising from the failure.

Failures can occur, for example, due to a fundamental flaw in the design assumptions, or due to a small defective detail. There is no unique recipe for failure and no definitive exhaustive checklist of possible causes.

Attention to any possible warning signs may give some advance notice of an impending failure, and may save money and ultimately lives. Simple signs, such as cracking of concrete sections, cracks around bolted or welded connections in steel structures or unexpected deflections, should be noted and reported to experienced engineers, and reacted to promptly. To ignore any such signs may lead to expensive or even fatal consequences.

93.7.1 The process

Following a failure, information gathering is often a cyclical process. Initial documents provided are reviewed, collated and assessed. This invariably leads to requests for other documents that were mentioned but not provided, or there may be existing documents that have not yet been referred to. Much of the documentation presented may ultimately prove to be irrelevant, but everything should be considered. Small pieces of information contained in large documents can turn out to be vital.

It is necessary to be completely methodical with all the information provided, which should be catalogued and arranged chronologically. The precise order and timing of events leading up to a failure are often important.

Cross-checking information is always an important part of the forensic process. For example, do the 'as constructed' records agree with the approved design and construction sequence? Are the ground investigation data (including field and laboratory test results) as would be expected, taking into account the local geology? Are any materials tests consistent within themselves – for example, are variations in concrete strength reflected in variations of the measured concrete density?

When reviewing the information, note all the important points as they are found. It is frustrating to remember seeing something important, but not being able to remember which document/page/line it came from.

93.7.2 Design

It is often the case that the fundamental cause of a failure is that the geotechnical or structural behaviour assumed for design purposes is not appropriate and does not reflect reality.

The following quotations from eminent geotechnical practitioners relate specifically to the topic of geotechnical design, and have much to commend them

> Simple calculations based on a sound conception are far more meaningful than elaborate calculations which ignore controlling factors.
>
> *Ralph B. Peck*, 1973

> Recent examples of inappropriate analysis … suggest modern engineers could benefit from a wider appreciation of elementary principles.
>
> *Sir Alan Muir Wood*, 2004

> I express much concern over the extent to which engineers, presented with a technical problem, tend to fly directly to numerical solutions without pausing to understand the physical basis of the problem. Too often, the numerical model does not address the actual conditions, or provides output in a form that confuses rather than informs. Almost universally, the apparent precision of a solution bears no relationship to the uncertainty of the data. The ability to make a first estimate on the basis of a sketch and the formulation of equations for simple analysis continues to provide an essential tool to avoid expensive errors and misunderstandings.
>
> *Sir Alan Muir Wood*, 2004

When considering the design of the elements relevant to a failure, some basic questions need to be asked.

- Does the analytical model represent the likely actual mechanical behaviour of the construction in the ground?
- How were the design parameters derived, and are they representative of the physical behaviour of the structure and the ground?
- Have design parameters been used that are appropriate to the particular situation? Short- and long-term parameters, small- and high-strain parameters: were they appropriate to the structural mode in compression, tension, shear and bending?

Finally, does the end result of the analysis appear reasonable? The use of simplistic hand calculations to check the outcome is invaluable – it is too easy for the effect of a slipped decimal place to become lost in a complex numerical model analysis.

At the same time as considering the design, construction and workmanship must also be investigated. The construction records must be checked to confirm that the correct sequence was followed. Webcams that continuously monitor and record site activities have been found to be invaluable when checking against contractors' records.

Test results and monitoring records need to be interrogated for clues regarding the possible causes of failure.

Observations can take many forms, from anecdotal evidence to measurements of deflections, monitored stresses and strains. Some of this evidence may prove to be correct and true, some may prove to be untrue, and for some it may not be possible to verify either way. As with ground investigation information, the task of forensic investigation is to attempt to identify what is reliable, and why this should be so. Conversely, it must also attempt to identify what is not reliable, and for what reasons.

Having considered all the available evidence, one or more hypotheses must be formulated and tested for consistency with the observed behaviour. It is important to consider that there may be more than one explanation for what has occurred. All possibilities should be considered, and rejected only when they have been thoroughly examined. It is never sufficient to consider just one explanation and look no further, even if that one explanation happens to be consistent with the observed behaviour.

93.7.3 Reporting

Forensic investigation usually has two functions: (*a*) to find a solution to a problem; and (*b*) to document the investigation, the findings and the conclusions. If there has been a significant problem resulting in a dispute, then the matter may be resolved directly under the project contract, or indirectly by such means as set out in the contract. Resolution may involve adjudication, mediation, arbitration or, ultimately, litigation. In case the matter eventually has to be resolved by litigation, great care should be taken when reporting on a forensic investigation to ensure the accuracy and validity of all statements made.

If it is possible that a report may be used in litigation, then Civil Procedure Rules Part 35 (CPR35) 'Experts and assessors', and the accompanying 'Practice direction', include the requirements for expert evidence to be given in a written report (Ministry of Justice, 2021). It is important to distinguish between the evidence considered (the input data), the assessment of the evidence and opinions expressed, and the conclusions reached, and present a report with clearly defined sections dealing with all aspects required by CPR35. Thus the forensic report should be clearly divided into the following sections

- evidence
- assessment of the evidence
- opinions and conclusions.

Most reports on a forensic investigation will be read by people who will be less qualified than the author. It is therefore helpful to minimise the use of technical terminology and to add explanations where necessary. Judgements may be made based on the report, so it has to be easily understood by the reader. A style that works well is to minimise the content of the main body of the report so that it is easy to read and absorb, and to use appendices for reference information. Conclusions should be as succinct as possible.

A one-page summary is a useful inclusion in a report on a forensic investigation, both for the author and the reader. For the author it ensures that the essence of the problem and the conclusions are presented logically, and for the reader it provides an easily absorbable overview.

93.8. Conclusions

QA is the management system that will ensure that all key parts of the geotechnical design and construction process are undertaken consistently in order to provide at least a minimum standard. A designer will use QC systems to ensure that the design and checking process is undertaken to meet the QA requirements. The contractor will use QC systems to ensure that the construction is in accordance with the specification and design.

Dependent upon the contractual arrangements on any particular project, the contractor will either have their workmanship verified by an external party such as an RE, or will be required to self-certify their own work.

Identifying and rectifying non-conformant work is a key part of the QA process. For geotechnical work it is paramount that a thorough forensic investigation is carried out at the earliest opportunity so that these issues are closed out early in the construction process. With the increasing use of digital data to input and export from the geotechnical design and construction process, robust review of the outputs from calculation and construction records must be undertaken to confirm that these are realistic and correct in terms of informing decisions.

Disclaimer

The example non-conformances included in this chapter are for illustrative purposes only and are not associated with, or reflective of, the authors or their employers.

References

ASTM International (2021a) C1611: Standard test method for slump flow of self-consolidating concrete. ASTM International, West Conshohocken, PA.

ASTM International (2021b) C1610: Standard test method for static segregation of self-consolidating concrete using column technique. ASTM International, West Conshohocken, PA.

BSI (British Standards Institution) (2000a) BS EN 1538:2000: Execution of specialist geotechnical work – Diaphragm walls. BSI, London, UK.

BSI (2000b) BS EN 1536:2000: Execution of specialist geotechnical work – Piles. BSI, London, UK.

BSI (2009) BS EN 12350-2:2009: Testing fresh concrete – Slump-test. BSI, London, UK.

BSI (2010) BS EN 12350-10:2010: Testing fresh concrete – Self-compacting concrete: L box test. BSI, London, UK.

BSI (2014) ISO 14001:2014: Environmental management systems – Requirements with guidance for use. BSI, London, UK.

BSI (2015) ISO 9001:2015: Quality management systems – Requirements. BSI, London, UK.

BSI (2018) BS ISO 45001:2018: Occupational health and safety management systems – Requirements with guidance for use. BSI, London, UK.

BSI (2019a) BS EN 12350-5:2019: Testing fresh concrete – Flow table test. BSI, London, UK

BSI (2019b) BS EN 12350-8:2019: Testing fresh concrete – Self-compacting concrete: Slump-flow test. BSI, London, UK.

BSI (2022) ISO 27001:2022: Information security, cybersecurity and privacy protection – Information security management system: Requirements. BSI, London, UK.

EFFC (European Federation of Foundation Contractors) (2018) *Guide to Tremie Concrete for Deep Foundations*, 2nd edn. EFFC, Faversham, UK.

EFFC (2019) *Guide to Support Fluids for Deep Foundations*, 1st edn. EFFC, Faversham, UK.

HMG (Her Majesty's Government) (2015) The Construction (Design and Management) Regulations. The Stationery Office, London, UK.

ICE (Institution of Civil Engineers) (2017) *ICE Specification for Piling and Embedded Retaining Walls*, 3rd edn. Thomas Telford, London, UK.

Leonards G (1982) Investigation of failures. *Journal of the Geotechnical Engineering Division* **108(2)**: 185–246.

Ministry of Justice (2021) Civil procedure rules part 35 – Experts and assessors and practice direction. Ministry of Justice, London, UK. See www.justice.gov.uk/courts/procedure-rules/civil/rules/part35/pd_part35 (accessed 29 March 2023).

Muir Wood A (2004) *Civil Engineering in Context*. Thomas Telford, London, UK.

Peck R (1973) Opening address. *8th International Conference on Soil Mechanics and Foundation Engineering, Moscow, Russia*, p. 158.

Useful websites

Get It Right Initiative: https://getitright.uk.com

Quality Scheme for Ready Mixed Concrete (QSRMC): http://www.qsrmc.co.uk

Quality Scheme for Reinforcement (CARES): http://www.carescertification.com

It is recommended this chapter is read in conjunction with

- Chapter 78 *Procurement and specification*
- Chapter 82 *Piling problems*
- Chapter 96 *Technical supervision of site works*

All chapters in this book rely on the guidance in Sections 1 *Context* and 2 *Fundamental principles*. A sound knowledge of ground investigation is required for all geotechnical works, as set out in Section 4 *Site investigation*.

doi: 10.1680/icemge.66830.1495

CONTENTS

Chapter 94
Principles of geotechnical monitoring

W Allen Marr Geocomp, Inc, Acton, MA, USA
Jamie Standing Imperial College London, London, UK

Almost every construction project involving soil or rock runs some risk of encountering surprises, in particular 'unforeseen ground conditions'. Compared with other branches of civil engineering where there is greater control over the materials used, monitoring is vital to the practice of geotechnical design and construction. For this reason geotechnical engineers, unlike their colleagues in many other fields, must have more than a casual knowledge of geotechnical monitoring and instrumentation: it is an essential working tool.

The benefits of geotechnical monitoring are outlined in this chapter. These are followed by a systematic approach to planning monitoring programmes using geotechnical instrumentation, an example of systematic planning and general guidelines on the execution of monitoring programmes.

94.1. Introduction

Every geotechnical design has to cater for some uncertainty, and every construction project involving soil or rock runs some risk of encountering surprises. These circumstances are the inevitable result of working with materials that are a product of nature, created by processes that seldom result in uniform conditions. The limitation of ground investigation procedures' ability to detect all significant properties and conditions of natural materials in advance requires the designer to make assumptions that may not reflect the overall ground conditions. The construction contractor may choose equipment and construction procedures without full knowledge of what might be encountered. Despite the limitations of not being able to explore every point within a site, geotechnical monitoring, including quantitative measurements obtained from field instrumentation, provides the means by which the geotechnical engineer can design a project to be safe and efficient, and the contractor can execute the work with safety and economy. However, due to uncertainties in subsurface conditions, actual performance may deviate from expected performance of the design in ways that create substantial risk. Thus, geotechnical monitoring is vital to the practice of geotechnical design and construction. For this reason, geotechnical engineers frequently need a comprehensive knowledge of geotechnical monitoring and instrumentation.

Geotechnical monitoring is not merely the selection of instruments, but a comprehensive step-by-step engineering process, beginning with a definition of the objective(s) and ending with implementation of contingency measures if the data warrant. Each step is critical to the success (or failure) of the monitoring programme, and the monitoring process involves combining the capabilities of instruments and people.

Peck (1984) states 'The legitimate uses of instrumentation are so many, and the questions that instruments and observation can answer so vital, that we should not risk discrediting their value by using them improperly or unnecessarily.' Clauses 4.5, 11.7, 12.7 and Annex J in Eurocode 7, Part 1 (BS EN 1997-1:2004; BSI, 2004) provide guidelines on geotechnical monitoring.

The benefits of geotechnical monitoring are outlined in the following sections. These are followed by a systematic approach to planning monitoring programmes using geotechnical instrumentation, an example of systematic planning and general guidelines on the execution of monitoring programmes.

Instruments that are used for geotechnical monitoring are described in Chapter 95 *Types of geotechnical instrumentation and their usage.*

Primary sources of text for this chapter have been Dunnicliff (1993, 1998) and, with permission from the American Society of Civil Engineers (ASCE), Marr (2007).

94.2. Benefits of geotechnical monitoring

The principal technical reasons for recommending a geotechnical monitoring programme for a project are

- minimising damage to adjacent structures
- implementing the observational method
- revealing unknowns
- assessing a contractor's construction methods
- devising remedial measures to address problems
- improving performance
- documenting performance for assessment of damages

- showing that everything is satisfactory and as expected
- warning of impending failure
- advancing the state of knowledge.

Each of these is discussed next in the context of today's practice of geotechnical engineering. In general, two common features of these technical reasons are that monitoring programmes help manage risk and they save money.

94.2.1 Minimising damage to adjacent structures

Geotechnical construction may affect adjacent property with undesirable results, such as expensive repairs, bad relations and protracted litigation. Geotechnical monitoring can be used to warn of unacceptable movements of adjacent property. By doing this, the costs required to repair the damage are likely to be avoided. **Figure 94.1** shows a motorised (robotic) total station for monitoring any movement of overlying structures while tunnelling beneath a road. Such instruments can work day and night every day to provide detailed facts on project performance.

94.2.2 Implementing the observational method

The complete application of the observational method (see also Chapter 100 *Observational method*) embodies the following steps (Peck, 1969)

- Exploration sufficient to establish at least the general nature, pattern and properties of the deposits, but not necessarily in detail.
- Assessment of the most probable conditions and the most unfavourable conceivable deviations from these conditions.
- Establishment of the design based on a working hypothesis of behaviour anticipated under the most probable conditions.

Figure 94.1 Motorised total station for monitoring any movement of structures (Photograph courtesy of SolData; all rights reserved)

- Selection of quantities to be observed as construction proceeds and calculation of their anticipated values on the basis of the working hypothesis.
- Calculation of values of the same quantities under the most unfavourable conditions compatible with the available data concerning the subsurface conditions.
- Selection in advance of a course of action or modification of design for every foreseeable significant deviation of the observational findings from those predicted on the basis of the working hypothesis.
- Measurement of quantities to be observed and evaluation of actual conditions.
- Modification of design to suit actual conditions.

The degree to which all these steps can be followed depends on the nature and complexity of the work, but it is clear that in all cases a geotechnical monitoring programme is essential.

For example, it is possible to build an embankment over a soft soil stratum by constructing it in stages. Placed all at once, the embankment might cause the foundation to fail. Placing the embankment in stages at suitable time intervals allows the soft soil to strengthen by consolidating between each stage. Instruments to measure settlements and pore water pressures can be used to determine when enough consolidation of the clay has occurred, so that the next stage of fill can safely be added. A balance is usually sought between adding the next stage as quickly as possible to minimise construction time – but not so fast as to cause foundation failure. This example is described in more detail in Section 94.4.

Section 2.7 in BS EN 1997-1 (BSI, 2004) indicates that it is appropriate to apply the observational method when prediction of geotechnical behaviour is difficult or involves considerable uncertainties. It includes: requirements to be met before construction is started; requirements to be met during construction; assessment of the results of monitoring; replacement of monitoring equipment.

94.2.3 Revealing unknowns

Using procedures which reveal unknown conditions and help engage remedial work as early as possible will lead to the lowest project cost. A good geotechnical monitoring programme is vital to this approach. The alternative of delay, denial and blame almost always costs more.

94.2.4 Assessing a contractor's construction methods

The outcome of some geotechnical projects depends on the construction methods used and the ability of the construction contractor. Geotechnical monitoring is used to determine whether these meet the specified performance requirements.

94.2.5 Devising remedial methods to address problems

When problems occur in geotechnical construction, the cause must be identified as far as possible, and remedied. Finding

Figure 94.2 Instruments in an unstable slope

Figure 94.3 Vibrating wire crack gauge on the side of a building

the best remedy requires an understanding of what went wrong. Data from geotechnical monitoring can help engineers determine what caused the problem. A remedial action can then be devised that focuses on the specific cause rather than addressing several potential causes in the light of uncertainty. **Figure 94.2** shows an unstable slope that is being monitored with subsurface deformation gauges to determine the depth of the failure surface so that an appropriate remedial action can be implemented.

94.2.6 Improving performance

Modern concepts of business management stress continual improvement and the need for measurements to gauge success. A common saying in business practice is 'that which is measured improves, while things not measured eventually fail'. The mere process of measuring performance coupled with normal human behaviour leads to improved performance.

94.2.7 Documenting performance for assessment of damages

Damage claims by third parties represents one of the substantial risks encountered in geotechnical projects. Claims sometimes include charges for damage unrelated to the construction in question. Others may be inflated, such as a claim for structural damage when only minor architectural damage has occurred. Data from geotechnical monitoring can help establish the validity of such claims, thereby satisfying stakeholders and regulators, and reducing litigation. For example, if instrumentation shows that an adjacent building has not moved during construction, it becomes more difficult for the owner to claim that cracks in the building resulted from the construction activity. **Figure 94.3** shows a vibrating wire crack gauge on the side of a building to determine whether nearby construction is causing changes in crack width.

94.2.8 Showing that everything is satisfactory and as expected

Increasingly, geotechnical monitoring programmes are used to demonstrate that the actual performance is within the bounds anticipated by the designers. The presumption is that there will be no surprises or unexpected consequences to the overall cost and schedule, and that unexpected behaviour can be identified early enough to maintain control of the project cost and schedule.

Data from a geotechnical monitoring programme help maintain the various parties' confidence in the performance of the work, thus allowing them to focus on other issues. Increasingly, project owners desire performance monitoring systems that are comprehensive and robust, but with instant reporting as simple as a green light to indicate that everything is progressing acceptably.

Figure 94.4 shows a tied-back retaining wall instrumented with load cells on some of the tieback anchors to monitor any changes in load.

94.2.9 Warning of impending failure

Geotechnical structures can fail with catastrophic consequences to life and property. Such failures may be the result of excessive loads, design errors, construction deficiencies, unknown or unexpected conditions, deterioration, operational errors or intentional action. Geotechnical monitoring has been widely used to detect the onset of failure in dams, slopes, embankments and excavations. **Figure 94.5** shows vibrating wire strain gauges mounted on a pipe strut across an internally braced excavation to warn of excessive load in the strut. In practice, monitoring of a single strut in this way is insufficient – a significant number of struts need to be monitored to provide representative data.

Instrumentation and monitoring are expected to take on an increasingly important role in informing early warning and emergency warning systems, particularly in relation to dam safety, but also for other facilities such as tunnels and excavations. As dams become older and the downstream risks grow with more development, warning systems become an increasingly important component of risk management. Society is also demanding a higher level of protection from risks from sources that are beyond their control. Lives can be saved if people can be given a warning to immediately move out of harm's way. Equipment can be saved if an impending landslide can be detected in time to move it to a safe location. With modern automated data acquisition systems that are increasingly reliable and lower cost, and with immediate communications from the site to anywhere, warning systems are becoming an important application of geotechnical instrumentation.

Figure 94.4 Load cells on tieback anchors

Figure 94.5 Vibrating wire strain gauges mounted on a pipe strut across an internally braced excavation

94.2.10 Advancing the state of knowledge

Many of the advances in understanding ground and structural responses in geotechnical engineering have their roots in data from geotechnical monitoring of full-scale projects. The data give us insight into how the ground/structure is responding to construction and about causal relationships. A significant amount of geotechnical instrumentation is currently used and has been used historically as part of a research effort to improve our knowledge. Much of this has been paid for by governmental agencies with a mission to improve practice. **Figure 94.6** shows a long-stroke vibrating wire strain gauge mounted on a geotextile for advancing the state of knowledge about geotextile performance used as reinforcement.

94.2.11 An overview of the benefits of geotechnical monitoring

Ten principal technical reasons (or benefits) for recommending a geotechnical monitoring programme for a project have been identified and discussed in the context of today's practice of geotechnical engineering. As stated earlier, a common feature of these technical reasons is that monitoring programmes generally help manage risk and save money. It is probable that the first reason, minimising damage to adjacent structures, is the most common. However, it is strongly recommended that designers of construction projects should study the other

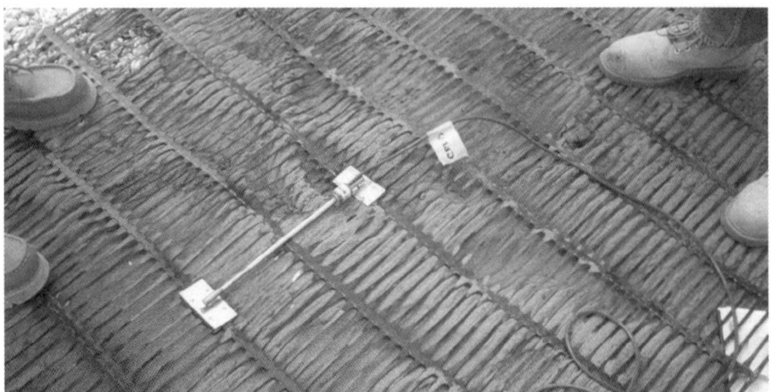

Figure 94.6 Long-stroke vibrating wire strain gauge mounted on a geotextile

nine potential reasons in the context of the project specifics, and endeavour to persuade their clients to adopt geotechnical monitoring if it can be shown that there is a clear technical or financial benefit.

94.3. Systematic approach to planning monitoring programmes using geotechnical instrumentation

Planning a monitoring programme using geotechnical instrumentation should begin with defining the objective and end with planning how the measurement data will be implemented. Planning should proceed through the steps listed in **Table 94.1** and outlined below. More detail is given in Dunnicliff (1993). All of these planning steps should, if possible, be completed before geotechnical monitoring work begins in the field.

94.3.1 Step 1: Define the project conditions

The person responsible for planning a monitoring programme must become very familiar with the project type and layout, subsurface stratigraphy and engineering properties of subsurface materials, groundwater conditions, status of nearby structures or other facilities, environmental conditions and the planned construction method.

Step	Task
1	Define the project conditions
2	Predict mechanisms that control behaviour
3	Define the geotechnical questions that need to be answered
4	Identify, analyse, allocate and plan for control of risks
5	Select the variables to be monitored
6	Predict magnitudes of change
7	Devise remedial action
8	Assign tasks for the construction phase
9	Select instruments
10	Select instrument locations
11	Plan documentation of factors that may influence measured data
12	Establish procedures for ensuring data correctness
13	List the specific purpose of each instrument
14	Prepare the budget
15	Prepare the instrumentation system design report
16	Plan installation
17	Plan regular calibration and maintenance
18	Plan data collection and data management
19	Prepare contract documents
20	Update the budget

Table 94.1 Steps in a systematic approach to planning monitoring programmes using geotechnical instrumentation

94.3.2 Step 2: Predict mechanisms that control behaviour

Prior to developing a monitoring programme, one or more working hypotheses must be developed for mechanisms that are likely to control behaviour of the monitored facility. This step should consider potential failure mechanisms and define the possible risks were each of these to develop.

94.3.3 Step 3: Define the geotechnical questions that need to be answered

Every instrument on a project should be selected and placed to assist in answering a specific question: if there is no question, there should be no instrumentation. This is the first golden rule. The questions come directly from the potential mechanisms developed in Step 2. They mostly result from uncertainties about loads, site conditions, material parameters, construction procedures and other unknowns.

94.3.4 Step 4: Identify, analyse, allocate and plan for control of risks

All risks associated with construction should be identified, and each 'geotechnical question' should be prioritised based on risk. Responsibility for each risk may be allocated to a single party or to more than one party. Risk responsibility allocation should be included in the construction contract documents.

Risk analysis embodies a wide range of scientific theories and engineering analyses to identify potential sources of risk, determine the probability of occurrence for each source, and estimate the consequences from each source of risk. Total risk is the summation of the probability of each source of risk occurring multiplied by the consequences of that occurrence. Risk can be decreased by actions that reduce the probability of a source of risk or its adverse consequences occurring.

Van Staveren (2006) provides comprehensive guidance on identification, analysis, allocation and planning for control of risks, and Marr (2007) provides an approximate method for quantifying the benefits that have been described above, based on concepts of decision theory and risk analysis. Risks are also discussed in Chapter 7 *Geotechnical risks and their context for the whole project.*

From the considerations of risk and unknowns, one can proceed to identify what measurements might be taken to indicate the emergence of each significant risk and track its evolution.

94.3.5 Step 5: Select the variables to be monitored

Typical geotechnical variables include

- pore water pressure
- deformation
- tilt
- total stress
- load and strain in structural members
- temperature.

The question 'which variables are most relevant?' should be answered. The answer comes by considering the results of Step 4.

Clauses 4.4, 11.7, 12.7 and Annex J in BS EN 1997-1 (BSI, 2004) include guidance on selecting the variables to be monitored.

94.3.6 Step 6: Predict magnitudes of change

Manufacturers of geotechnical instruments will define their range and accuracy. In order that appropriate instruments are selected, there must be a prediction of the maximum change that might occur, hence determining the needed range for the instrument. For the same reason, the needed accuracy must also be determined.

If measurements are for construction control or safety purposes, a predetermination should be made of numerical values that indicate the need for decisive mitigation measures. These values are often referred to as trigger levels. The concept of green, amber and red trigger levels is useful.

- Green indicates that all is well.
- Amber indicates the need for cautionary measures, including an increase in monitoring frequency.
- Red indicates the need for timely remedial actions and being prepared to implement them quickly.

The following are guidelines for selection of trigger levels.

- Early trigger levels can be based on calculated changes, whereas later levels can be based on (unrelated) tolerable changes.
- Trigger levels must recognise the changes that can occur from causes other than construction.
- Trigger levels should be several times larger than the accuracy of measured changes (those last four words are very carefully chosen).
- Provisions should be included to revise trigger levels if needed as the project proceeds and more information about uncertainties becomes known.

94.3.7 Step 7: Devise remedial action

Inherent in the use of instrumentation for construction purposes is the absolute necessity for deciding, in advance, a positive means for dealing with any problem that may become evident from the results of the observations. This is having contingency measures ready to put in place when needed. If the observations should demonstrate that remedial action is needed, that action must be based on appropriate, previously anticipated plans. Arrangements should be made to determine how all parties will be forewarned of the planned remedial actions.

94.3.8 Step 8: Assign tasks for the construction phase

The second golden rule (remember that the first golden rule has been given in Step 3) is: *Tasks should be assigned to the people who have the greatest motivation to provide high-quality data and results.* Many geotechnical monitoring programmes have been unsuccessful because planners of the programmes have assigned key tasks to people who have inadequate motivation or training. Hence this step is of the utmost importance and is covered in some detail in this section. The tasks include

- buying instruments
- installing instruments
- collecting data
- evaluating data
- interpreting data
- implementing actions resulting from the data

and it is crucial to ensure that these tasks are assigned to the people who are most likely to maximise quality.

Data evaluation means to examine the data to remove any suspect or wrong information. The instrumentation team is most knowledgeable in this respect and so it is appropriate that they do this. Their goal should be to pass on data of high quality and that is not suspect, or else clearly identify questionable or suspect data. Data interpretation should be the responsibility of the people who understand how the facility is expected to perform working with those on the monitoring team who understand facility performance. Implementation should be by construction personnel.

If principal construction contractors, temporary works contractors, specialist geotechnical subcontractors or design/build contractors have initiated the monitoring programme, clearly they have the greatest motivation, and tasks 1, 2 and 3 should be assigned to them. However, if the programme has been initiated by the designer of the project, there are four reasons for not assigning tasks 1, 2 and 3 to the principal construction contractor.

- Principal construction contractors may not have enough motivation to ensure quality. Use of the conventional lowest tender procedure selected by the lead bid estimator, whereby these tasks are included as items in the principal construction contactor's contract, has often led to poor-quality data.
- Monitoring cannot start until after the award of the principal construction contract, hence adequate pre-construction (baseline) data are usually unavailable. Structures move and groundwater regimes often change from season to season, and geotechnical monitoring data cannot be interpreted correctly if baseline data are not established. Critical data for evaluating differing site condition claims may not be available. Whenever practicable, it is highly preferable to establish at least one year of baseline data.
- It costs the project owner more. Potential monitoring subcontractors give prices to principal construction contractors prior to the latter tendering. After contract award, there is normally a bargaining process, and the project owner pays £1 for work that costs the principal contractor 90 or 80 or 70 pence.

- For multi-principal contract projects there would be one monitoring subcontractor for each principal construction contract. It is desirable on large contracts to have one principal instrumentation contractor collecting and providing instrumentation data for the entire project so that effects of concurrent works can be determined and lessons learned from one contract can be passed to others.

The following assignments are recommended for tasks 1, 2 and 3.

- *Pre-construction baseline data*: specialist firm under contract with the project owner, using a qualifications-based selection procedure.
- *During construction, outside principal construction contractor's work area*: same as above.
- *During construction, within principal construction contractor's work area*
  - owner's representative, with assistance from principal construction contractor for access as necessary, or
  - specialist firm, as a nominated (assigned) specialist subcontractor, selected using a qualifications-based procedure, with instruments selected on the basis of proven performance, or
  - principal construction contractor, with partnering and rigorous and *enforced* specifications.

A hybrid model where the construction contractor subcontracts the procurement and installation of instruments to a qualified instrumentation subcontractor and an instrumentation data specialist working for the owner's representative collects, evaluates and reports project-wide data and related context information has worked well on large projects.

For further information on assignment of tasks and for details of the above recommendations, see Dail and Volterra (2009), Dunnicliff (2009), Dunnicliff and Powderham (2001a, 2001b) and Klingler (2001). Additional information on writing specifications is given in Chapter 78 *Procurement and specification.*

94.3.9 Step 9: Select instruments

The preceding eight steps should be completed before instruments are selected. When selecting instruments, the overriding desirable feature is *reliability*. Selection of instruments and read-out equipment depends directly on the methods used for data collection – see Section 94.5.5 for general guidelines. The scale of the geotechnical monitoring programme should match the identified risks and the scale and complexity of the geotechnical questions.

Details of some instruments, together with their applications, are given in Chapter 95 *Types of geotechnical instrumentation and their usage.*

The lowest cost to purchase an instrument should never dominate the selection. The least expensive instrument is unlikely to result in minimising project cost. In evaluating the economics of alternative instruments, the overall cost of procuring, calibration, installation, maintenance, monitoring, processing and evaluating the data should be compared.

94.3.10 Step 10: Select instrument locations

A practical approach to selecting instrument locations entails three steps.

- Establish the zones where the risks are highest. Instrument these thoroughly.
- Zones are identified (normally cross-sections) where predicted behaviour is considered representative of behaviour as a whole. These cross-sections are regarded as *primary instrumented sections*, and instruments are located to provide comprehensive information on ground/structural responses and construction performance.
- Because unknown factors may cause poorer performance at other locations, instrumentation should be installed at a number of *secondary instrumented sections*, to serve as indices of comparative behaviour. Instruments at these secondary sections should be as simple as the requirements allow and be installed at the primary sections so that comparisons can be made. If, in fact, the behaviour at a secondary section appears to be significantly different from the behaviour at the primary sections, additional instrumentation may be installed at the secondary section as construction progresses.

When selecting locations it should be recognised that some instruments will probably cease to function when data are still required, hence some duplication may be needed. Some limited duplication of measurements leads to increased confidence in data, especially in areas of uncertainty or greatest change. Also provisions should be made to replace any priority instruments should they stop working.

94.3.11 Step 11: Plan documentation of factors that may influence measured data

Measurements by themselves are rarely sufficient to provide useful conclusions. Geotechnical monitoring normally involves relating measurements to causes, and therefore records and diaries must be maintained of all factors that might cause changes in the measured parameters, including construction details and progress. Including time lapse cameras, drone surveys and InSAR data may be beneficial to the project.

94.3.12 Step 12: Establish procedures for ensuring data correctness

Personnel responsible for geotechnical monitoring must be able to answer the question: *Are the data reliable?* The ability to answer depends on the availability of good evidence, for which planning is required. In critical situations, duplicate instruments can be used. A backup system is often useful and will frequently provide an answer even when its accuracy is significantly less than that of the primary system. Key to the success of this step is to have people well experienced with evaluation of instrumentation data available with sufficient time to do the evaluation.

94.3.13 Step 13: List the specific purpose of each instrument

At this point in the planning it is wise to question whether all planned instruments are justified. Each planned instrument should be numbered and its purpose listed. If no viable specific purpose can be found for a planned instrument, it should be deleted.

94.3.14 Step 14: Prepare the budget

Even though the planning steps are not complete, a budget should be prepared at this stage for all tasks that are included in Section 94.5 to ensure that sufficient funds are available. A frequent error in budget preparation is to underestimate the duration of the project and the total costs for data collection and interpretation. If insufficient funds are available, the monitoring programme may have to be curtailed or more funds obtained.

94.3.15 Step 15: Prepare the instrumentation system design report

This report should summarise the results of planning steps 1–14. It should then be reviewed by others to ensure that everything is consistent, that the plan is a good one and that it covers the needs of the project.

94.3.16 Step 16: Plan installation

Installation procedures should be planned well in advance of scheduled installation dates. Written step-by-step procedures should be prepared, making use of the manufacturer's instruction manual and the designer's knowledge of specific site geotechnical conditions. The written procedures should include a detailed listing of required materials and tools, and installation record sheets should be prepared for documenting factors that may influence measured data. Installation plans should be coordinated with the construction contractor and arrangements made for access and protection from damage of installed instruments. A review of the needs and requirements for qualified staff should be made. An installation schedule should be prepared, consistent with the construction schedule.

94.3.17 Step 17: Plan regular calibration and maintenance

Calibration and maintenance should be planned well before fieldwork starts; calibrations are described in Section 94.5.2. Maintenance planning should include read-out units, field terminals and embedded components.

94.3.18 Step 18: Plan data collection and data management

Written procedures for data collection and data management should be prepared well before fieldwork starts. These are significant tasks, and the effort required for them should not be underestimated. Guidelines are given by Dunnicliff (1993) and in Section 94.5.5.

At this stage in the planning it should be verified that field staff are appropriately trained, remedial actions have been planned, personnel responsible for interpretation of monitoring data have the contractual authority to signal remedial action, communication channels between design and construction personnel are open and arrangements have been made to forewarn all parties of the planned remedial actions.

94.3.19 Step 19: Prepare contract documents

Any necessary contract documents should be prepared, to match the decisions made during Step 8.

94.3.20 Step 20: Update the budget

With planning complete, the budget for all tasks listed in planning Step 8 should be updated in light of all planning steps.

94.3.21 An overview of the most common failures in implementing a systematic approach to planning monitoring programmes

All of the above 20 steps are important. In the experience of the authors, the following steps are the most crucial.

- *Step 3: Define the geotechnical questions that need to be answered.* As indicated above, every instrument on a project should be selected and placed to assist in answering a specific question: if there is no question, there should be no instrumentation. This is the first golden rule, and is commonly ignored.
- *Step 4: Identify, analyse, allocate and plan for control of risks.* When a risk analysis is made, it is often not comprehensive enough and risk responsibility allocation is commonly not adequately covered in the construction contract documents. Define in writing how the instrumentation programme fits into the project risk management plan.
- *Step 7: Devise remedial actions through contingency planning.* All too often instrumentation programmes are planned without ensuring that viable remedial actions are available in the event that trigger levels are reached.
- *Step 8: Assign tasks for the construction phase.* As indicated above, the second golden rule is that tasks should be assigned to the people who have the greatest motivation to achieve high-quality data. Many geotechnical monitoring programmes have been unsuccessful because planners of the programmes have assigned key tasks to people who have inadequate motivation or lack the necessary experience to be successful.

Singling out these four steps in this overview in no way implies that the other 16 steps are unimportant.

94.4. Example of a systematic approach to planning a monitoring programme: using geotechnical instrumentation for an embankment on soft ground

In this example, the geotechnical questions are associated only with the soft ground below the embankment, and not with the embankment itself. Only selected planning steps are included in the example (i.e. not all 20 steps that have been detailed in the previous section). Ladd (1991) presents valuable guidelines for geotechnical monitoring of embankments on soft ground.

Examples of Steps 3, 5 and 9 for 12 additional project types are given in Chapter 95 *Types of geotechnical instrumentation and their usage*. These additional project types are

- internally braced excavations
- externally braced excavations
- deep excavations – basement heave
- clay embankments
- embankment dams
- waste retention dams (tailings dams)
- cut slopes in soil
- landslides in soil
- cut slopes in rock
- landslides in rock
- tunnels
- driven piles
- bored piles.

94.4.1 Step 1: Define the project conditions

- **Figure 94.7** illustrates the project conditions for this example of planning.
- The embankment is taken to be wide compared with the thickness of the very soft clay layer.
- Staged construction and vertical drains are planned so that the roadway on top of the embankment can be completed as soon as possible.
- **Figure 94.8** shows the installation of vertical drains through very soft clay.
- There are some silt layers in the very soft clay.

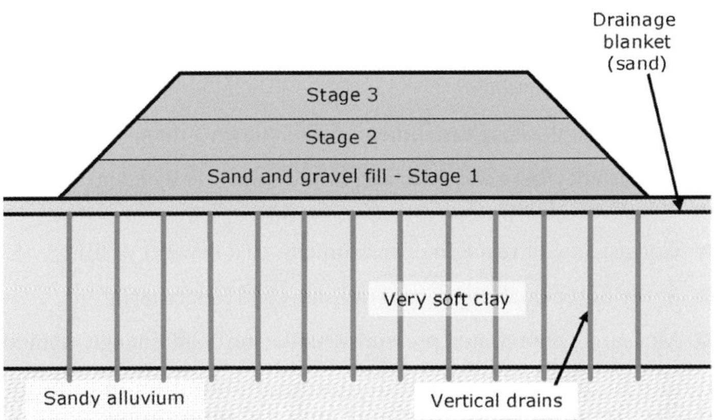

Figure 94.7 Project conditions for an embankment on soft ground

Figure 94.8 Installation of vertical drains through soft clay

- Initial groundwater conditions are hydrostatic.
- There are no nearby structures.
- The predicted total vertical compression of the clay is 15% strain.

94.4.2 Step 2: Predict mechanisms that control behaviour

- Settlement (consolidation of very soft clay).
- Bulging displacements of the clay from lateral squeeze, causing heave either side of the embankment.
- A rotational slide, causing failure.

94.4.3 Step 3: Define the geotechnical questions that need to be answered

(See Chapter 95 *Types of geotechnical instrumentation and their usage*.)

- What are the initial site conditions in the soft ground?
- Is the embankment stable in the short term?
- So that decisions can be made as to when to place the next stage of fill, what is the progress of consolidation and associated gain of strength of the soft ground?

94.4.4 Step 4: Identify, analyse, allocate and plan for control of risks

- Risks are that (*a*) an embankment failure may occur and (*b*) the completion of the roadway is delayed.
- The project owner's construction manager will be on site to make decisions about placing subsequent stages of fill, and the project owner accepts responsibility for risks (*a*) and (*b*).
- Strength gain will occur more slowly than predicted, thereby slowing construction, delaying completion and creating a possibility for the construction contractor to submit a claim for extra money.

94.4.5 Step 5: Select the parameters to be monitored

(See **Table 95.3**.)

- Pore water pressure in the clay at appropriately spaced lateral and vertical locations throughout the clay.
- Vertical deformation of ground surface.
- Horizontal deformation of the side slopes.

94.4.6 Step 6: Predict magnitudes of change

- Pore water pressure in the clay
  - range – initial pressure plus additional pressure caused by the weight of the embankment
  - accuracy – consider the time plot that shows dissipation of pore water pressure under a single stage of fill, and the need to extrapolate from that to decide when the next stage can be placed.
- Vertical deformation
  - range – as the predicted vertical compression is given as 15%, surface settlement can be calculated
  - accuracy – same logic as for pore water pressure.
- Horizontal deformation
  - range – base this on an estimate of the maximum deformation that could occur
  - accuracy – no general recommendation can be given, as required accuracy depends on project details.
- Trigger levels need to be determined.

94.4.7 Step 7: Devise potential remedial actions

- Do not place any more fill for some rest time.
- Remove fill.
- Place a berm at the toe of the embankment.

94.4.8 Step 8: Assign tasks for the construction phase

- The monitoring programme has been initiated by the project designers.
- Responsibility for pre-construction baseline data is assigned to a specialist firm under contract with the project owner, selected using a qualifications-based procedure.
- During construction, the tasks of buying and installing instruments and collecting data are assigned to a nominated specialist subcontractor, selected using a qualifications-based procedure. Instruments are selected on the basis of proven performance.
- The specialist subcontractor is responsible for evaluating the data and delivering reliable data and related information.
- The project designers are responsible for interpreting monitoring data and advising on any need for remedial action.
- The construction contractor is responsible for implementing actions resulting from the data.

94.4.9 Step 9: Select instruments

(See **Table 95.3**.)

- Pore water pressure in the clay
  - vibrating wire piezometers.
- Vertical deformation
  - surface monuments with surveying – probably by total station or precise levelling if greater accuracy required
  - probe extensometers – probably magnet/reed switch type.
- Horizontal deformation
  - surface monuments with surveying – probably by total station
  - vertical inclinometers.

94.4.10 Step 10: Select instrument locations

- This would depend on the length of the new roadway.
- Perhaps a minimum of three primary instrumented sections and three times as many secondary instrumented sections.
- Within each of these sections, decisions would have to be made regarding the position of instruments (e.g. piezometers midway between vertical drains, locating them to avoid damage by construction activities).

94.4.11 Step 11: Plan recording of factors that may influence measured data

- Fill level.
- Weather (e.g. amount of precipitation, sun, wind).
- Temperature.
- Barometric pressure (if vibrating wire piezometers have been selected).
- Visual observations of unexpected or unusual behaviour.
- Construction activities.

94.4.12 Step 12: Establish procedures for ensuring data correctness

- Visual observations.
- Consistency among instruments for monitoring the same variable.
- Consistency between pore pressure dissipation data and vertical compression data.
- Consistency of reaction of instruments to a new lift of fill.
- Study of repeatability, over both short and longer term.
- Measured pore water pressures, under no load change, should extrapolate to initial pressures.
- Regular calibrations of read-out units and survey equipment.

94.5. General guidelines on execution of monitoring programmes

94.5.1 Introduction

This section includes general guidelines for

- calibration
- installation
- maintenance
- data collection
- processing and presentation of monitoring data
- evaluation of monitoring data
- interpretation of monitoring data
- reporting of conclusions
- implementation.

Additional guidelines are given by Dunnicliff (1993).

94.5.2 Calibration

An instrument reading is only useful if the correct calibration is applied. Changes in calibration factors occur due to wear and tear, misuse, creep, moisture ingress and corrosion. If these changes are not accounted for, the entire monitoring programme can become worthless or lead to wrong conclusions and costly recommendations. To maximise their effectiveness, all instruments must be calibrated and checked properly.

Instrument calibrations and/or function checks are generally required at the following stages:

- Calibrations prior to shipment. These are the responsibility of the manufacturer, and are referred to as *factory calibrations.* Calibration certificates should be provided to the user. **Figure 94.9** shows factory calibrations of transducers for vibrating wire piezometers.
- Function checks by the user before installation. These are referred to as *pre-installation acceptance tests*, their purpose being to verify that no damage has occurred during transit. Manufacturers will provide guidelines on what checks should be made.
- Function checks after installation. These are referred to as *post-installation acceptance tests.* Try to apply/identify some known change to test whether the instrument continues to function as expected.
- Calibrations during service life. Calibrations or function checks of instruments and read-out units are required during the service life. These calibrations are usually performed by personnel responsible for data collection.

94.5.3 Installation

Each instrument must be installed before its needed date. The schedule will be based on the need to establish pre-construction (baseline) conditions, construction activity, availability of installation personnel, instrument delivery dates and other relevant factors.

Figures 94.10 to **94.12** show installations of a borehole extensometer, inclinometer casing and electrical resistance strain gauges, respectively.

Surface exposures of instruments should usually be protected with robust cover plates, welded, bolted or otherwise attached to the surface. Exposed tubes and cables are extremely vulnerable to damage or vandalism and should usually be protected by conduit or armouring. **Figure 94.13** shows precast concrete pipe segments used as surface protection for borehole instruments.

The presence of a measuring instrument should cause minimum alteration to the value of the variable being measured. If the instrument alters the value significantly, it is said to have *poor conformance.*

As indicated in Step 17, installation record sheets should be prepared during the planning phase and completed during the installation. These records serve two purposes. First, when installation personnel are required to enter data/information in the blank spaces on a field form they are more likely to follow the installation procedure with care. Second, 'as-built' data are required both for record purposes and for use during evaluation of data. These records form an important part of any quality assurance programme. Alternatively, the data may be entered into a form on a hand-held devicc for automated processing. Such devices must be able to save and produce the data in their original raw form.

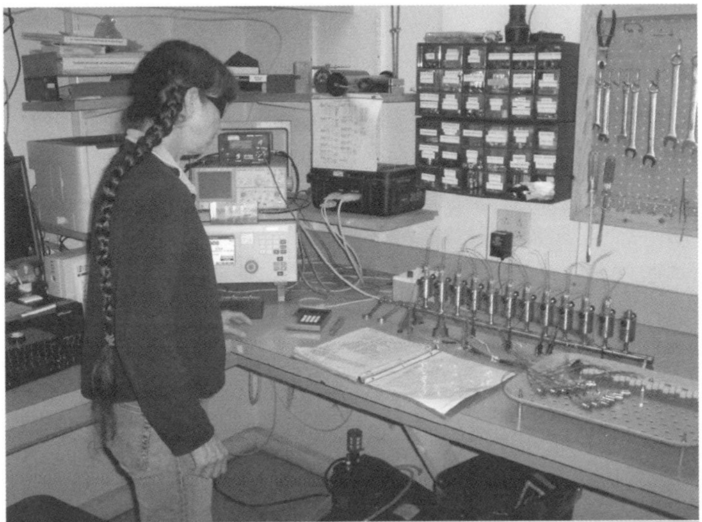

Figure 94.9 Factory calibration of transducers for vibrating wire piezometers (Courtesy of Geokon Inc., Lebanon, New Hampshire, USA)

Figure 94.10 Installation of borehole extensometer above future tunnel. Photo supplied by Dr Michael Wan

Figure 94.11 Installation of inclinometer casing

On completion of installation, an installation report should be produced to provide a complete compilation of information needed by personnel responsible for data collection, processing, presentation and interpretation. The installation report should contain as-built information, descriptions of instruments and read-out units, results for calibrations and checks, details of installation procedures, initial readings and a copy of each installation record sheet.

94.5.4 Maintenance

Regular maintenance required during the service life is usually performed by personnel responsible for data collection. They should always be on the lookout for damage, potential for damage and deterioration or malfunction.

Figure 94.12 Installation of electrical resistance strain gauges on segment of cast iron tunnel lining

Figure 94.13 Borehole instruments protected from damage by precast concrete pipe segments

94.5.5 Data collection

There are three primary means of collecting data.

- Manual reading involves people going to the instrument read-out location and using some device to read the instrument. The reading is manually recorded on paper or perhaps in an electronic device. This approach may be the most cost-effective when readings are required less than once per week and the distance to the instrument is not too far, and the read-out location can be safely reached. Instruments for manual readings do not have to be electronic and may be of low cost and easy to maintain. **Figure 94.14** shows the manual reading of piezometric level in an open standpipe.

Figure 94.14 Manual reading of piezometric level in an open standpipe

- Data loggers with periodic manual data download may be used when the reading interval is less than one week or it is difficult to get to the instrument location and there is no immediate need to know the readings. Readings are stored in the memory of the data logger, where they are periodically downloaded manually to a portable electronic device for transport back to the office. Someone travels to the data logger once per week or month or quarter to collect data recorded since the last visit. This approach requires the use of electronic instruments that are read by the data logger. Hardware for this approach may cost more than for the manual reading approach, but the cost of collecting data may be less, and more frequent readings can be obtained for essentially no added cost.
- Data loggers with remote data access are used when there is an immediate need for the instrument readings and when reading frequency is more than once per day. In this approach, the data logger is connected to a remote computer system by landline or mobile phone modem, satellite modem, radio or the internet. These systems allow users to see data within seconds of when it is recorded. They also allow alarm systems to be set up that can inform project staff whenever an instrument indicates a value that exceeds a pre-established trigger level. An example of such a data logger, with solar panels for battery charging, is shown in **Figure 94.15**.

Figure 94.15 Data logger with remote data access for monitoring lateral movement of a lock wall

Each of these approaches has its advantages and limitations which directly influence how data are collected, reduced and communicated – hence the importance of selecting the data collection approach during the planning phases of the monitoring programme.

The frequency of data collection should be related to construction activity, to the rate at which the readings are changing and to the requirements of data interpretation. Clause 4.5 in BS EN 1997-1 (BSI, 2004) includes guidance on the duration of monitoring.

When construction begins and approaches the instrument location, readings should be taken frequently; for example, once per week, once per day, once per shift or even more frequently in relation to construction activity (such as before and after each blast, during pile driving or during the placement or removal of a preload). It is often wise to increase the frequency of readings during heavy precipitation. As construction activity moves away from the instrument location or ceases altogether, and when readings have stabilised, the frequency may then be decreased. It is generally advisable to continue the monitoring effort until near completion of all construction activities because some effects of construction take time to develop fully and there may be activities by others that affect the project.

94.5.6 Processing and presentation of monitoring data

The first aim of data processing and presentation is to provide a rapid assessment of data in order to evaluate them (removing erroneous readings, leaving only reliable data – see next section) and to detect changes that require immediate action. The second aim is to summarise and present the data in order to show trends and to compare observed with predicted performance so that any necessary action can be taken.

After calculations have been made, graphs of data should always be prepared. Various types of graphs are described by Dunnicliff (1988, 1993)

- graphs to assist with data screening (evaluation)
- graphs of data against time – an example of this type of graph is shown in **Figure 94.16**
- graphs for comparing observed and predicted behaviour
- graphs for comparison of measurements with construction events – **Figure 94.16** includes construction events on the time axis
- graphs to examine cause-and-effect relationships – **Figure 94.17** is an example of this type of graph, showing load as 'cause' and displacement as 'effect' during a pile load test
- summary graphs.

Increasingly, project management staff want to know the status of the monitoring programme without being involved with the details, unless unacceptable values are imminent. A one-page summary report of important information is useful for this purpose.

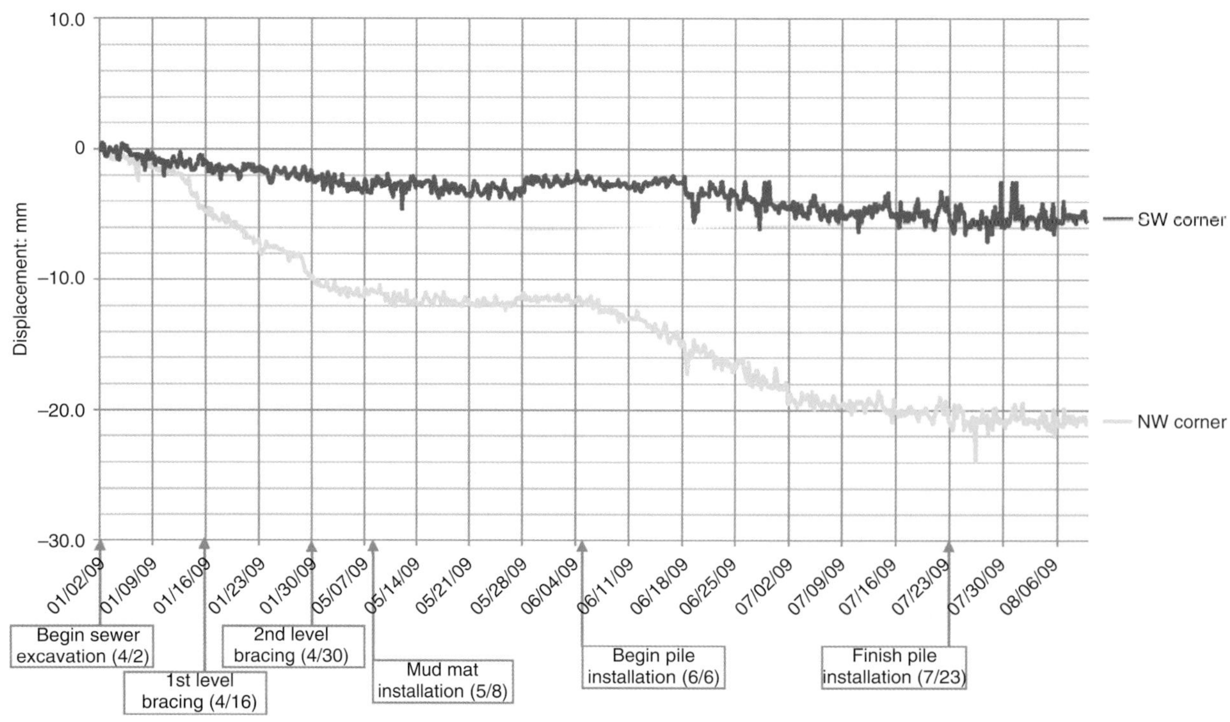

Figure 94.16 Graph of building settlement adjacent to excavation over time

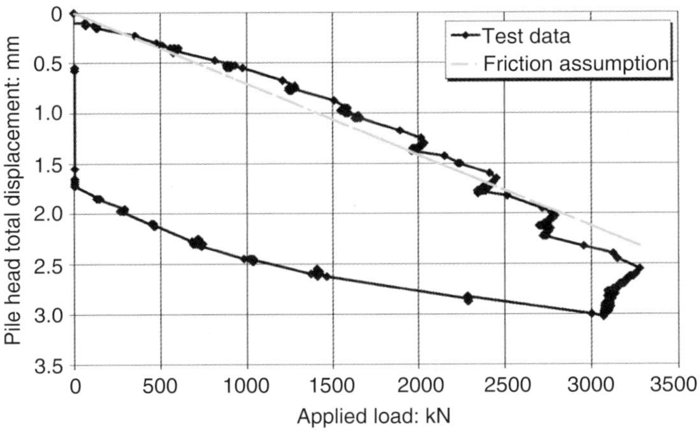

Figure 94.17 Graph of pile load test data

94.5.7 Evaluating monitoring data

Having produced various graphs, as suggested in the previous section, the data need to be evaluated to check factors such as: whether they look sensible; whether there are clear erroneous data that can be removed; assessment of scatter and accuracy; and trends (e.g. to investigate whether a re-calibration is necessary). In a sense, this stage is to check the quality and reliability of the data prior to making interpretations. When evaluating monitoring data, it is necessary to consider many aspects of the monitoring device itself, such as what it is measuring, its nature (mechanical, electrical), what might be expected from it, the details of where it is installed, potential seasonal and diurnal changes, expected responses, human error and other factors that will vary depending on the specific installation situation. The goal of the data evaluation step is to minimise any suspicion about the reliability and quality of the data from those who must interpret the data.

Historically and still today the evaluation is largely based on knowledge and experience, looking for anomalies such as sudden changes without reason, drifting responses under stable conditions and values or trends that do not make sense. There is an increasing trend in many disciplines to use statistical/machine learning/big data approaches which help remove subjectivity in this step and can help detect anomalies that might otherwise be missed. Although a detailed review of these approaches is beyond the scope of this chapter, they are developing rapidly and can also help with calibrating and validating numerical models used for the construction design.

94.5.8 Interpretation of monitoring data

Many monitoring programmes have failed because the collected data were never used. If there is a clear sense of purpose for a monitoring programme, the method of data interpretation will be guided by that sense of purpose.

The first data interpretation step should be to provide a rapid assessment of data in order to detect changes that require immediate action. The essence of subsequent data interpretation steps is to correlate the instrument readings with other factors (cause-and-effect relationships) and to evaluate the deviation of the readings from the predicted behaviour.

Trigger levels (see Section 94.3.6) should be watched very closely when interpreting monitoring data. Do not ignore changes during the green trigger level period by simply waiting for the green flag to change to amber. Trends during the green period can give useful forewarning. It may be appropriate to change trigger levels up or down as the project progresses and more is learned about how the site is performing.

Interpretation is an ongoing process. Interpretations should be made as frequently as change is expected; those from early data will usually be revised or refined as more data become available, and as a clearer understanding of real behaviour is developed.

An open communication channel should be maintained between design and construction personnel, so that discussions can be held between design engineers (who raised the questions that caused the instrumentation to be used), field engineers (who provide the data) and construction or operations personnel (who use the data). It is important to establish at the start of the project the channel of communications for transmitting reports and alarm messages and for deciding and communicating what response actions are needed. Clause 4.5 of BS EN 1997-1 (BSI, 2004) includes guidance on evaluation of monitoring data.

94.5.9 Reporting of conclusions

After each set of data has been interpreted, conclusions should be reported in the form of an interim monitoring report and submitted to the personnel responsible for implementation of contingency measures. These interim reports may be daily, weekly or monthly, depending on the specific needs of the project.

94.5.10 Implementation

Implementation includes carrying out all the steps described above. For the geotechnical monitoring to be effective there must be actions that implement the data in order to manage risk. These actions include

- executing remedial actions where required
- revising design as necessary
- changing contractors' construction methods
- evaluating measured effects on the remaining work
- synthesising lessons learned and modifying standard procedures where required.

94.6. Summary

Ten principal technical reasons for recommending a geotechnical monitoring programme for a project have been outlined. Each of these has been discussed in the context of today's practice of geotechnical engineering. In general, a common feature of these technical reasons is that monitoring programmes help manage risk and save money.

There is a tendency among some users of instrumentation to proceed in an illogical manner, often first selecting an instrument, making measurements and then wondering what to do with the resulting data. Planning a monitoring programme using geotechnical instrumentation should begin with defining the objective and end with planning how the measurement data will be implemented. Twenty steps in a systematic planning process have been outlined, and an example has been given.

Finally, guidance is given for maximising the quality of field tasks associated with monitoring programmes.

References

BSI (British Standards Institution) (2004) BS EN 1997-1:2004+A1:2013: Eurocode 7 – Geotechnical design – General rules. BSI, London, UK.

Dail EB and Volterra JL (2009) Instrumentation and monitoring trends in New York City and beyond. *Geotechnical News* **27(3)**: 31–34.

Dunnicliff J (1993) *Geotechnical Instrumentation for Monitoring Field Performance*. Wiley, New York, NY.

Dunnicliff J (1998) *Geotechnical Instrumentation Reference Manual*. Training course in geotechnical and foundation engineering, NHI Course 13241 – Module, FHWA HI-98-034. National Highway Institute, Arlington, VA.

Dunnicliff J (2009) A designer's dilemma. *Geotechnical News* **27(3)**: 30.

Dunnicliff J and Powderham AJ (2001a) Recommendations for procurement of geotechnical instruments and field instrumentation services. *Geotechnical News* **19(3)**: 30–35, 37.

Dunnicliff J and Powderham AJ (2001b) Recommendations for procurement of geotechnical instruments and field instrumentation services. *Proceedings of the International Conference on Response of Buildings to Excavation-Induced Ground Movements, London, UK* (Jardine FM (ed)). CIRIA, London, UK, pp. 267–276.

Klingler FJ (2001) Discussion: recommendations for procurement of geotechnical instruments and field instrumentation services. *Geotechnical News* **19(3)**: 36–37.

Ladd CC (1991) Stability evaluation during staged construction. 1986 Terzaghi Lecture, American Society of Civil Engineers. *Journal of Geotechnical Engineering* **117(4)**: 593–604.

Marr WA (2007) *Why Monitor Performance?* ASCE Geotechnical Special Publication 117. Keynote paper. 7th International Symposium on Field Measurements in GeoMechanics, Boston, MA. ASCE, New York, NY.

Peck RB (1969) Advantages and limitations of the observational method in applied soil mechanics. *Géotechnique* **19(2)**: 171–187.

Peck RB (1984) Observation and instrumentation, some elementary considerations. In *Judgment in Geotechnical Engineering: Professional Legacy of Ralph B. Peck* (Dunnicliff J and Deere DU (eds)). Wiley, New York, NY, pp. 128–130.

Van Staveren M (2006) *Uncertainty and Ground Conditions: A Risk Management Approach*. Butterworth, Oxford, UK.

Useful websites

American Society of Civil Engineers (ASCE): www.asce.org

Geotechnical Instrumentation News (GIN) articles: www.geotechnicalnews.com/instrumentation_news.php. Articles published since 2001 can be downloaded from this site. At the end of each year, that year's articles will be posted on the site.

International Society of Soil Mechanics and Geotechnical Engineering (ISSMGE) Technical Committee 220 (TC220) covering Field Monitoring in Geomechanics: https://www.issmge.org/committees/technical-committees/applications/field-monitoring

It is recommended this chapter is read in conjunction with

- Chapter 78 *Procurement and specification*
- Chapter 79 *Sequencing of geotechnical works*
- Chapter 95 *Types of geotechnical instrumentation and their usage*
- Chapter 100 *Observational method*

All chapters in this book rely on the guidance in Sections 1 *Context* and 2 *Fundamental principles.* A sound knowledge of ground investigation is required for all geotechnical works, as set out in Section 4 *Site investigation.*

doi: 10.1680/icemge.66830.1511

Chapter 95
Types of geotechnical instrumentation and their usage

W Allen Marr Geocomp Corporation, Massachusetts, USA
Jamie Standing Imperial College, London, UK

Instruments for the monitoring of groundwater pressure, ground deformation, load and strain in structural members and total stress are described, together with a brief explanation of how they work, schematic diagrams of selected instruments and other important issues. Applications for the instruments are included. The general role of instrumentation for various types of projects is followed by tabular listings of possible geotechnical questions that need to be answered and that may lead to the use of instrumentation, with summaries to indicate some of the types of instruments that can be considered for helping to provide answers to these geotechnical questions.

CONTENTS

95.1. Introduction

In the first part of this chapter, instruments for the monitoring of groundwater pressure, ground deformation, load and strain in structural members and total stress are described, together with a brief explanation of how they work, schematic diagrams of selected instruments and other important issues. Applications for the instruments are included. In recent times there has been an increasing trend towards the development of instruments that allow for electronic data acquisition and wireless data transmission, thereby reducing costs of data collection, transmission, reduction and reporting. There have also been developments to make these systems more robust against damage and early failure.

In the second part of this chapter, the general role of instrumentation is given for various types of projects. This is followed by tabular listings of possible 'geotechnical questions that need to be answered' (see Chapter 94 *Principles of geotechnical monitoring*) that may lead to the use of instrumentation, with summaries to indicate some of the types of instruments that can be considered for helping to provide answers to these geotechnical questions.

When references are cited, priority is given to those that are available online. A significant part of the contents of this chapter is based on Dunnicliff (1993, 1998).

The associated Chapter 94 *Principles of geotechnical monitoring* includes the benefits of geotechnical monitoring, a systematic approach to planning monitoring programmes using geotechnical instrumentation, and general guidelines on the execution of monitoring programmes.

95.2. Instruments for monitoring groundwater pressure

95.2.1 Introduction

In this chapter the term 'piezometer' is used to indicate a closed system device that is hydraulically sealed within the ground so that it responds only to groundwater pressure around itself and not to groundwater pressures at other elevations. Piezometers are used to monitor positive (and sometimes negative) pore water pressure in soil and joint water pressure in rock. An observation well is a different type of instrument: it has no subsurface seals, and creates a vertical connection between strata.

Applications for piezometers fall into three general categories.

- First, for monitoring the pattern of water flow; second, to provide an index of soil or rock mass strength; and third, for measuring groundwater pressure at a specific point. Examples in the first category include monitoring subsurface water flow during large-scale pumping tests to determine permeability in situ, and monitoring the long-term seepage pattern in slopes and dams.
- Monitoring pore water pressure (positive or negative) or joint water pressure allows effective stresses and shear strength to be calculated. Examples include assessing the strength along

a potential failure plane behind a cut slope in soil or rock, and monitoring pore water pressure in soft clay foundations to control staged construction of an embankment.

- When readings are taken from multiple piezometers at different elevations, a piezometric profile can be inferred to indicate the presence of vertical flow. This is of significance for geotechnical design in many cases.

95.2.2 Observation wells

As shown in **Figure 95.1**, an observation well is an open system that consists of a perforated section of pipe attached to a riser pipe, installed in a sand- or gravel-filled borehole. Typical diameters are: pipe 25–50 mm; borehole 75–250 mm. A surface seal, with cement mortar or other material, is needed to prevent surface run-off from entering the borehole, and a vent is required in the pipe cap so that water is free to flow into and out of the perforated pipe section. The elevation of the water surface in the observation well is determined by sounding with an electrical probe attached to a graduated cable (sometimes an integral survey tape and cable), commonly referred to as a dipmeter. When the probe reaches the water surface in the riser pipe, an electrical circuit is completed and this is indicated by a light or buzzer in the read-out unit.

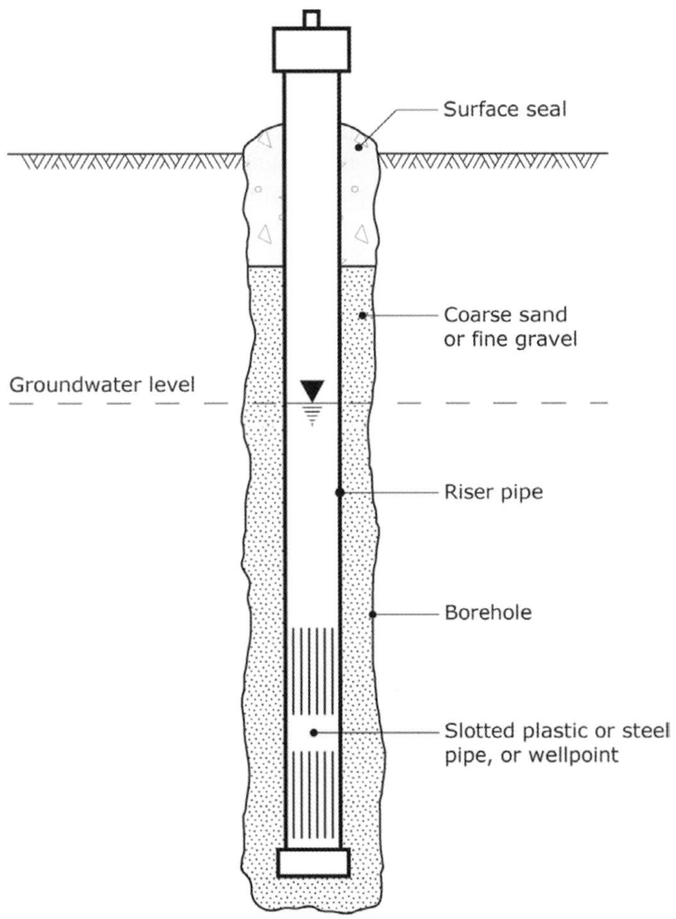

Figure 95.1 Schematic of an observation well

Appropriate applications for observation wells are very limited. In current practice they are frequently installed in boreholes during the site investigation phase of a project, ostensibly to define existing groundwater pressures and seasonal fluctuations. However, because observation wells create a vertical connection between strata, their *only* valid application is in homogeneous permeable ground in which there is no component of vertical flow and groundwater pressure increases uniformly with depth. This condition can rarely be assumed. There is a specific caution with respect to specifying observation wells with screens bridging between 'contaminated' and unadulterated strata separated by relatively impermeable strata – this provides a preferential route for contaminant flow, hence is improper practice.

95.2.3 Open standpipes

The components are identical in principle to components of an observation well, and so a standpipe is still an open system, with the addition of a subsurface seal just above the porous section of pipe, so that the instrument responds only to groundwater pressure around the filter element and not to groundwater pressures at other elevations. **Figure 95.2** shows an installation in a borehole. Typical diameters of standpipe and borehole are

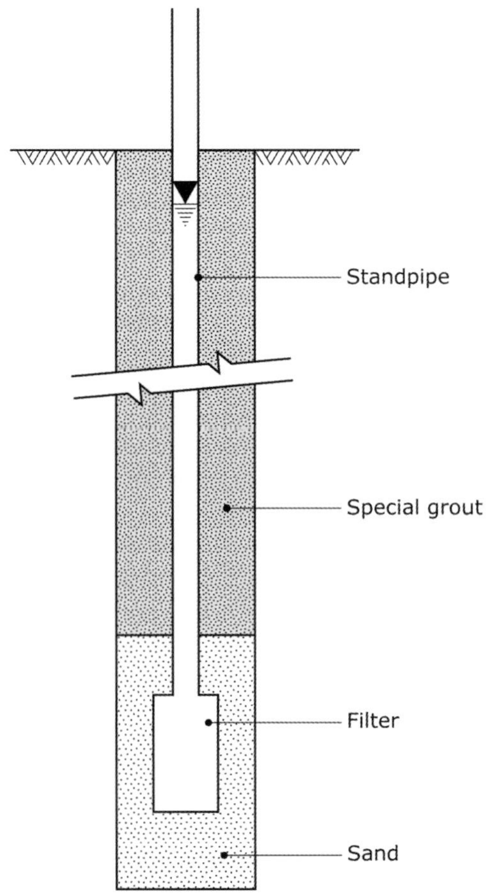

Figure 95.2 Schematic of open standpipe piezometer installed in a borehole

the same as for observation wells. Sometimes they are referred to as a Casagrande piezometer (Casagrande, 1949); however, as noted at the start of Section 95.2.1, the term piezometer is reserved for closed systems in modern usage. Traditionally, following Casagrande's installation procedure, a seal of bentonite chips or pellets was installed between the sand filter and the special grout to improve the response time of the instrument. However, this is no longer necessary (Dunnicliff, 2008). The special grout (Contreras *et al.*, 2008; Mikkelsen, 2002) can be tremied directly onto the sand filter. To avoid jetting of the sand, the bottom of the grout pipe should be plugged and the grout pipe should have side-discharge holes with either grouting under gravity or pumping very slowly. As with observation wells, the elevation of the water surface in the standpipe is determined by sounding with an electrical probe.

Open standpipes have a long successful performance record, and can be converted to remote-reading systems by insertion of pressure transducers inside the standpipe below the piezometric level (water table). They can also be used to perform either rising or falling head tests to determine the permeability of the soil surrounding the filter element. Details of the procedures and calculation method are given in BS 5930 (BSI, 2015). However, standpipes can have a long time lag (response time), are subject to damage by construction equipment and by vertical compression of soil around the standpipe and may experience freezing problems if the piezometric level rises above the frost line.

The simplest open standpipe is an open borehole. It is common practice to leave a borehole open overnight (properly secured) and record the groundwater level the next day. Many take this reading as indicative of the surface of a hydrostatic groundwater condition. This can be very misleading for a number of reasons. If the boring passes through multiple soil layers of different permeability, the measured water level may not indicate the actual pore pressure at any location. If there is a perched water condition and one or more of the lower soil layers is more permeable, the data will not reveal the perched water condition. If the soils are all low-permeability soils there will not have been enough time for the measured water level in the boring to have come to equilibrium as this can take days to weeks. If artesian conditions exist within some layer they may not be detected as the inflow from the artesian layer may flow out through other layers. For these reasons, unless one can demonstrate that an observation well or open borehole will give a meaningful indicator of the pore water pressures in the soil, sealed (closed system) piezometers are recommended.

95.2.4 Vibrating wire piezometers

As shown in **Figure 95.3**, this electronic pressure sensor has a thin metallic diaphragm separating the pore water from the measuring system. A tensioned wire is attached to the midpoint of the diaphragm such that deflection of the diaphragm causes changes in wire tension. As with a piano string, the frequency of vibration varies with the tension in the wire. The wire is plucked magnetically by an electrical coil attached near the wire at its midpoint, and either this same coil or a second coil is used to measure the frequency of vibration by using a frequency counter. The manufacturer supplies a calibration between frequency and pressure. In **Figure 95.3**, P is the current pore water pressure, K is a constant that is provided by the manufacturer, f_0 is the pre-installation frequency of vibration and f is the current frequency of vibration. Vibrating wire piezometers respond to changes of atmospheric pressure so that, if these changes need to be removed from monitored data, atmospheric pressure must be monitored separately. Attempts to circumvent this issue by using 'vented' piezometers have generally been unsuccessful, as the small-diameter vent tube may not allow for full pressure equalisation.

Figure 95.3 shows the piezometer sensor installed by the 'fully grouted method' (Contreras *et al.*, 2008; Weber, 2009). As mentioned above, following Casagrande's installation procedure for open standpipe piezometers (Casagrande, 1949), a seal of bentonite chips or pellets used to be installed between the sand and the special grout. Several decades ago, Vaughan (1969) showed that it is acceptable to use the fully grouted method, but this practice was slow to be adopted. It is not necessary to use sand and bentonite chips or pellets (Contreras *et al.*, 2008). Dunnicliff (2008) summarises some worldwide experiences with the fully grouted method, providing contact information for eight practitioners who have adopted the method as their standard procedure. Typical borehole diameter is 100–150 mm. In recent times, multiple piezometer installations within a single fully grouted borehole have been successfully used to establish pore water pressure profiles and whether they are hydrostatic or not (e.g. Wan and Standing, 2014). Alternatively, the pressure transducer may be placed in a housing and pushed into place in soft soils, negating the need for a borehole. However, steps must be taken to avoid leakage of water along the interface between the housing and the soil.

Wireless data collection and transmission are available for vibrating wire piezometers. The components require power and should be fully grounded to minimise chance of damage from electrical surges. As mentioned above, because the pressure transducer is sealed, changes in atmospheric pressure affect the readings. Hence the readings should be corrected for changes in atmospheric pressure which can be measured with a separate pressure transducer placed on site and read each time the piezometers are read.

Vibrating wire piezometers are very often the instruments of choice for monitoring groundwater pressure. They are easy to read and to data log, have a short time lag, cause minimum interference to construction and lead wire effects are minimal. They have long-term stability and can be used with long signal cables which in other sensing types may result in lead wire effects. However, lightning protection is required in locations where electrical storms occur.

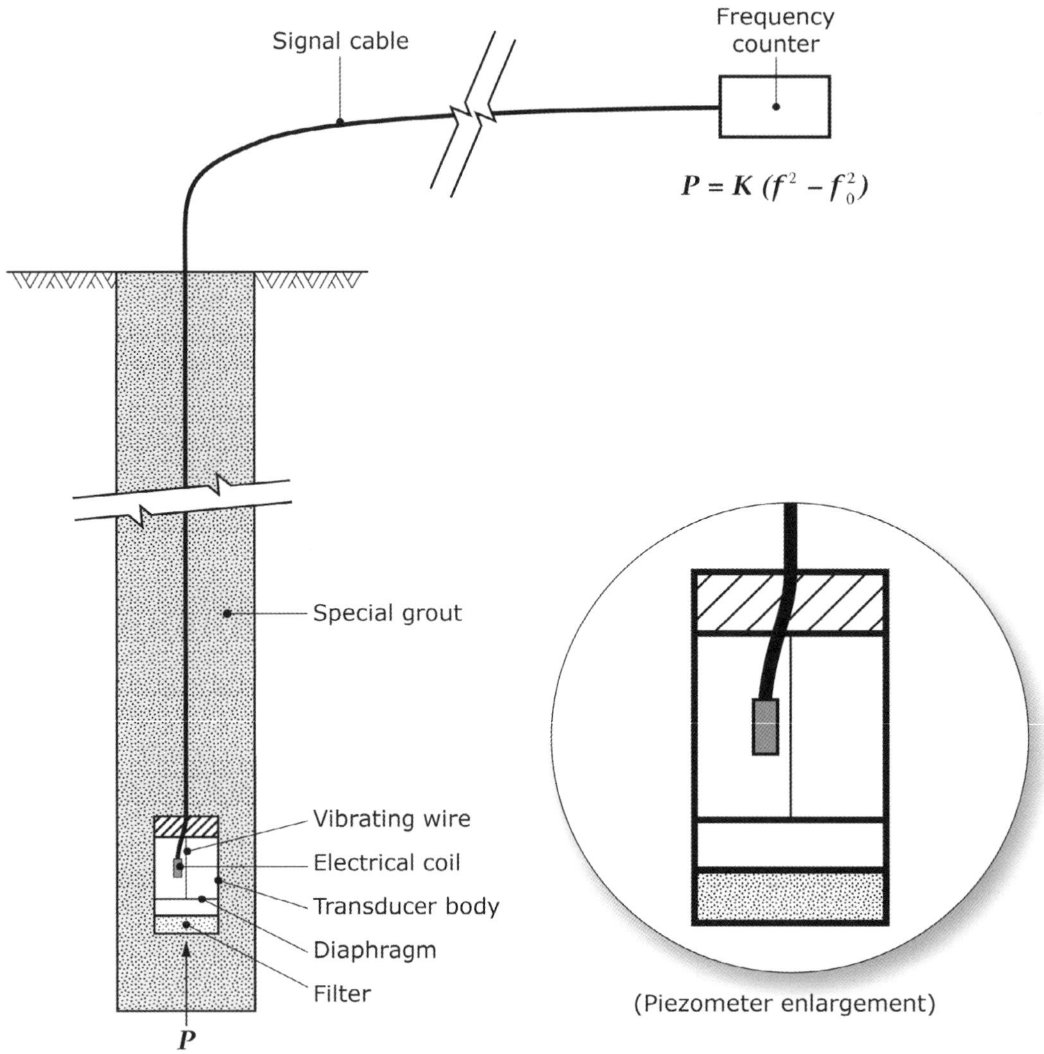

Figure 95.3 Schematic of vibrating wire piezometer installed in a borehole by the fully grouted method

95.2.5 Pneumatic piezometers

Figure 95.4 shows the basic arrangement of a pneumatic piezometer. An increasing gas pressure is applied to the inlet tube and, while the gas pressure is less than the pore water pressure, it merely builds up in the inlet tube. When the gas pressure exceeds the pore water pressure, the diaphragm deflects, allowing gas to circulate behind the diaphragm into the outlet tube, and flow is recognised either by noting a peak on the pressure gauge, or by hearing the gas flow emerging from the outlet tube. The gas supply is then shut off at the inlet valve, and any excess pressure in the tubes bleeds away, such that the diaphragm returns to its original position when the pressure in the inlet tube equals the pore water pressure. This pressure is read on a Bourdon tube or electrical pressure gauge. As with vibrating wire piezometers, pneumatic piezometers respond to changes of atmospheric pressure so that, if these changes need to be isolated from monitored data, atmospheric pressure must be monitored separately. As with open standpipes, there is no need for a bentonite chip or pellet seal between the sand and special grout. The fully grouted method is not recommended for installation of pneumatic piezometers. Typical borehole diameter is 100–150 mm.

Pneumatic piezometers are also manufactured for installation in soft soils by the push-in method.

As with vibrating wire piezometers, they have a short time lag and cause minimum interference to construction. There is no need for lightning protection, but they cannot readily be data-logged and readings are somewhat operator-dependent. Long-term measurements, where there are long periods (e.g. years) between readings, may be hindered by the diaphragm becoming stuck, requiring excessive pressures to release it, hence producing overestimates of pore water pressures. Also in situations where large displacements occur the tubing may become crimped, rendering the instrument useless.

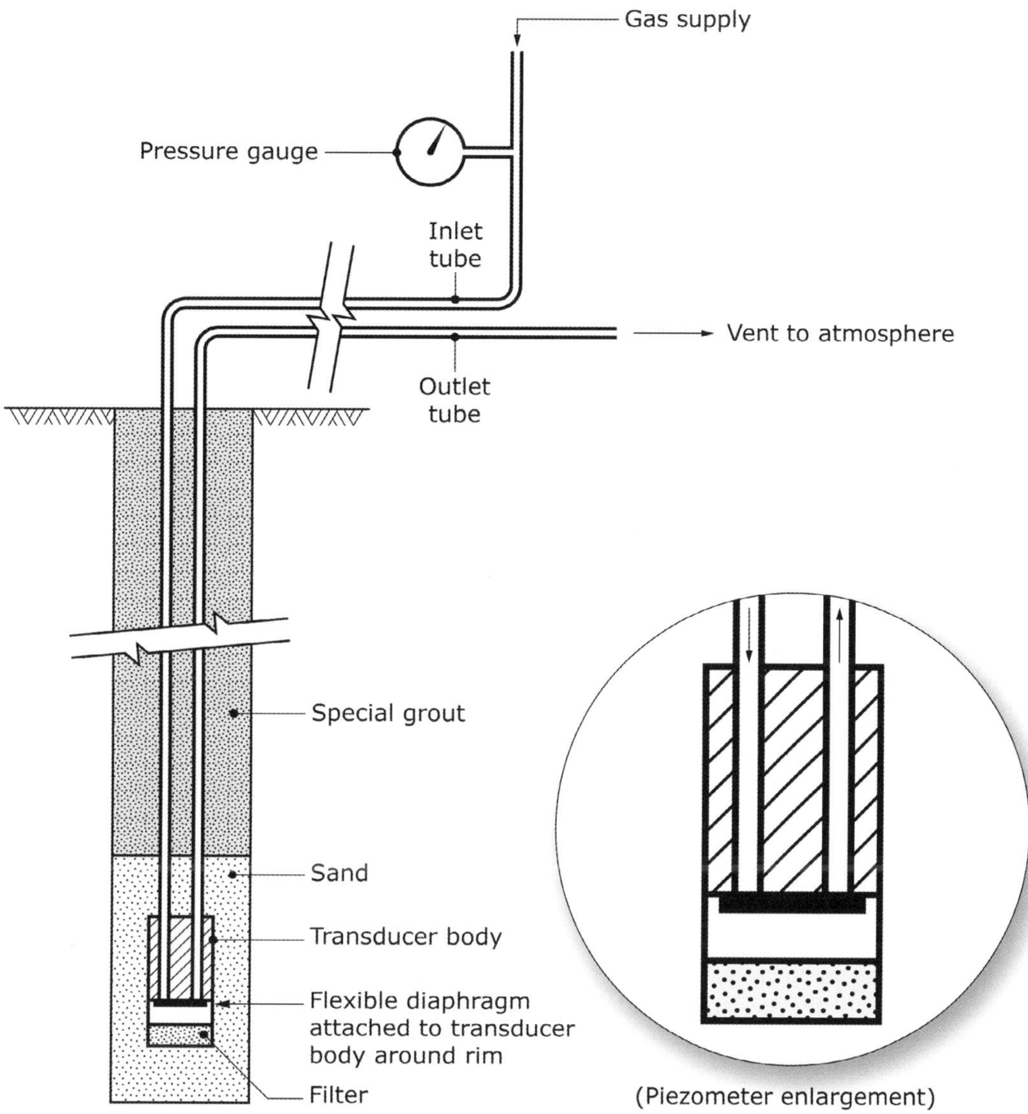

Figure 95.4 Schematic of pneumatic piezometer installed in a borehole

95.2.6 Twin-tube hydraulic piezometers

Twin-tube hydraulic piezometers were developed for installation in the foundations and fill during construction of embankment dams where the pressure head is higher than the elevation of the measuring elevation. Each consists of a porous filter element connected to two plastic tubes, with a Bourdon tube pressure gauge or pressure transducer on the top end of one tube and the other tube closed with a valve. The tubes are filled and flushed with de-aired water and piezometric elevation is determined from the pressure readings. When used in the unsaturated cores of embankment dams attention must be paid to many details to ensure that pore water pressure (as opposed to pore gas pressure) is measured, and that the tubes remain continuously filled with water (Dunnicliff, 1993). This requires flushing, which is why two tubes are used.

95.2.7 Flushable piezometers for measuring negative pore water pressures

Flushable piezometers (Ridley, 2003; Ridley *et al.*, 2003) have been developed primarily for measuring negative pore water pressures. Appropriate applications are slopes and excavations in clays. Negative pore water pressures will transmit tension to the water in a piezometer. This is likely to lead to cavitation, the formation of air bubbles, in a piezometer. If air is present in a piezometer it will not record the correct pore water pressures. Therefore the air must be removed and replaced with fresh de-aired water.

Twin-tube hydraulic piezometers are inherently flushable, but suffer from the principal disadvantage that the hydraulic tubes that are used for flushing water through them have to be laid horizontally to maximise the range of negative pore

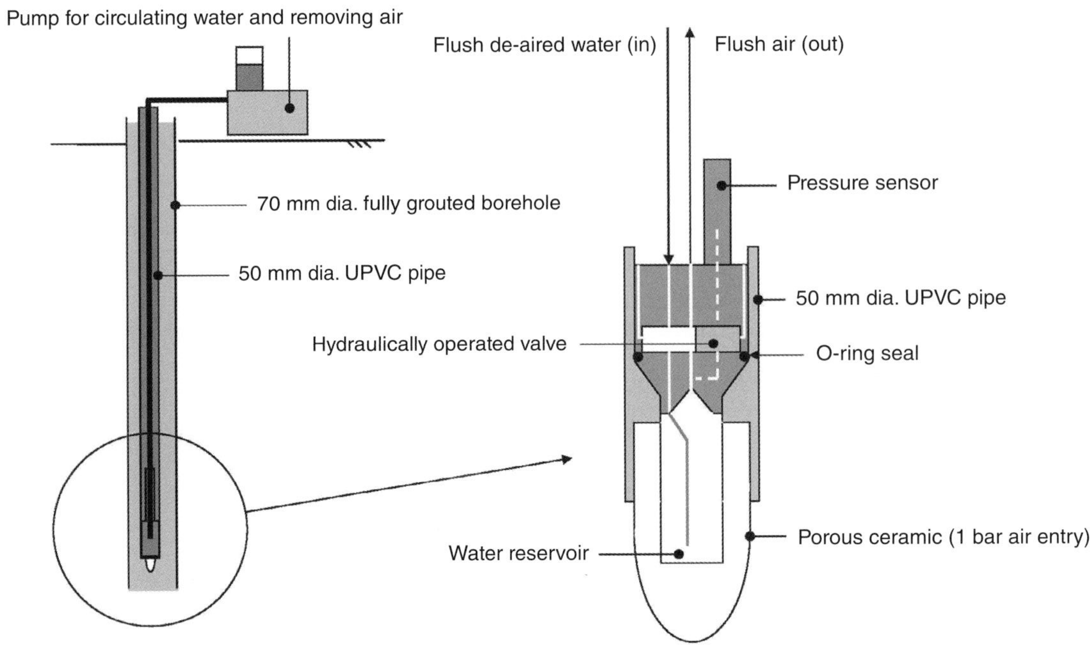

Figure 95.5 Schematic of flushable piezometer (Courtesy of Geotechnical Observations Ltd, Weybridge, Surrey)

water pressures that can be recorded and to avoid air becoming trapped within the tubes. The flushable piezometer shown in **Figure 95.5** overcomes this problem by incorporating a hydraulically operated shuttle valve that enables the piezometers to be installed in boreholes, typically 70 mm dia. The valve is used to isolate the sensor from the flushing tubes, thereby enabling the piezometers to measure positive pore water pressures and negative pore water pressures to approximately –95 kPa, irrespective of the depth of installation. Using a high air entry porous filter also inhibits the formation of air in the piezometer, and adopting the fully grouted method of installation allows hydraulic continuity to be maintained between the pore water and the water in the piezometer.

95.2.8 Tensiometers for measuring negative pore water pressures

Tensiometers are similar to open standpipes, except for three primary differences. The ceramic filter at the base has a rounded end and is installed to be in intimate contact with the soil (often achieved using a boring tool that creates a hole into which the filter fits snugly). The upper end of the standpipe is sealed rather than being open and the standpipe is filled with de-aired water. As the pore water in the soil surrounding the filter is negative, it will try to draw water from the tensiometer, resulting in a negative pressure in the sealed standpipe. This negative pressure is measured using a Bourdon gauge or electrical pore water pressure transducer mounted at the top of the sealed standpipe. As with the flushable piezometer described above, the limit of negative pressure that can be measured is usually about –90 to –70 kPa. Cavitation occurs at greater negative pressures which creates air bubbles in the tubing and leads to inaccurate readings. Flushable tensiometers are commercially available, allowing any air that might have accumulated in the sealed standpipe to be removed.

95.2.9 Fibre-optic piezometers

Fibre-optic piezometers are an alternative to vibrating wire piezometers, which offer the benefit of being unaffected by electrical, radio or magnetic interference. The instrument design is similar to that of a vibrating wire piezometer, with the exception that the deflection of the metallic diaphragm during changes in pressure are measured by changes in light transmitted by a fibre-optic cable.

Two point-based (discrete) fibre-optic sensing technologies have been applied for pore pressure monitoring. The first type of fibre-optic piezometer uses Fabry–Pérot interferometry (Glišić and Inaudi, 2007; Inaudi and Glišić, 2007a). This technology relies upon an air gap between two partially mirrored optical fibres facing each other. When light is coupled into one of the fibres, a back-reflected interference signal is obtained due to the reflection of the incoming light on the two mirrors. By packaging the diaphragm at one side of the air cavity, deflections of the diaphragm result in changes in the reflected signal. The second type of fibre-optic piezometers function using fibre Bragg grating (FBG) sensors. These sensors are prepared by slightly changing the refractive index of the fibre-optic core in a series of controlled evenly spaced intervals, referred to as a grating. When a broadband light pulse is passed through the fibre-optic cable, the grating causes a large reflection at a signature wavelength. Changes in the interval spacing due to positive or negative longitudinal strain on the grating result in a proportional shift in the reflected wavelength. For a piezometer the grating is suspended within the piezometer housing, with one end of the fibre attached to the diaphragm. Deflections in the diaphragm caused by changes in pressure result in a proportional change in the wavelength of the FBG reflection that is measured with a fibre-optic interrogator.

95.3. Instruments for monitoring deformation

95.3.1 Surveying methods

Surveying methods are used to monitor the magnitude and rate of horizontal and vertical deformations of structures, the ground surface and accessible parts of subsurface instruments in a wide variety of construction situations. In general, whenever geotechnical instruments are used to monitor displacements, surveying methods are also used to relate measurements to a reference datum: a benchmark for vertical displacement monitoring or a horizontal control station for horizontal displacement monitoring.

Surveyors who work on construction sites often have little experience with the accuracies required for deformation monitoring, and a well-trained instrumentation crew is essential to achieve maximum accuracy. Precise levelling techniques are usually adopted where measurements are made to sub-millimetre accuracy. Measurement accuracy is controlled by the choice and quality of surveying technique and by characteristics of reference data and measuring points (Cheney, 1973). Survey instrument technology is well established, and most reputable manufacturers include a statement of accuracy in their instrument specifications, which can be relied on if the instrument is calibrated, maintained and operated in accordance with instructions.

A detailed discussion of surveying methods is beyond the scope of this chapter, and readers should make use of information provided by manufacturers of surveying equipment. However, there is a current trend to make wide use of both manual and motorised (robotic) total stations (Beth 2007; Cook, 2006; Hope and Chaqui, 2008; Kontogianni, 2007; Marr, 2008; Volterra, 2008). Total stations are survey instruments combining a theodolite and electronic distance measurement. Retro-reflective targets or prisms are mounted at the locations of interest and their displacements are monitored in 3D space. When a motor is added to the theodolite and automatic target recognition included, monitoring can be achieved automatically and remotely

95.3.2 Mounted distometers

Distometers consist of a sensor to measure the distance between the device and a reflective surface using a laser beam. Differences between successive readings indicate the change in the distance to a resolution of about 2 mm. They are used to measure displacement between two points. A portable unit placed on a seat may be used, but the positioning must be repeatable to minimise error. The distometer can be permanently mounted and read with a data logger. Cost increases and accuracy decreases with increases in the distance between the two points. They have been used to measure displacements in tunnels, open-pit mines, retaining walls, structures and other applications.

95.3.3 Probe extensometers

Probe extensometers are defined in this chapter as devices for monitoring the changing distance between two or more points along a common axis, by passing a probe through a pipe. Measuring points along the pipe are identified by the probe, and the distance between points is determined by measurements of probe position. Typical applications of probe extensometers are monitoring vertical strain within embankments or embankment foundations, and settlement alongside excavations and above tunnels.

Various types of transducers are used in probe extensometers, including induction coils and magnet/reed switch transducers. A magnet/reed switch transducer is an on/off position detector, arranged to indicate when the reed switch is in a certain position with respect to a magnet. The switch contacts are normally open and one of the reeds must be magnetically susceptible. When the switch enters a sufficiently strong magnetic field, the reed contacts snap closed and remain closed as long as they stay in the magnetic field. The closed contacts actuate a buzzer or indicator light in a portable read-out unit (Burland *et al.*, 1972). A schematic of a borehole installation is shown in **Figure 95.6**.

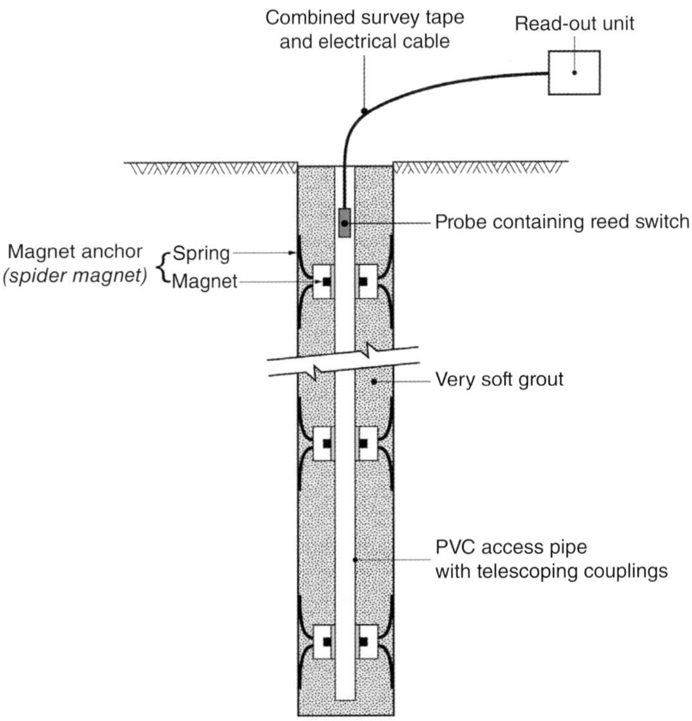

Figure 95.6 Schematic of probe extensometer with magnet/reed switch transducer

The spider magnets shown in **Figure 95.6** are suitable for installation in soft ground. The three springs above the magnets and the three below are held back while being lowered into the borehole, and tripped once at the required depth. To ensure that a spider magnet moves with the ground around it, the stiffness of the six springs should be such that a force of at least 20 N is required to move a 200 mm long spring from its extended position to the longitudinal axis of the magnet housing. The tip-to-tip diameter of the springs when in their extended positions should be at least 220 mm. For installation in stiff ground it is preferable to have only the three upper springs and to push them to the required depth with an insertion pipe that surrounds the access pipe, followed by an upward pull to ensure that they 'bite' into the ground. A bayonet disconnect between the bottom of the insertion pipe and the magnet housing allows removal of the insertion pipe.

The diameter of the access pipe is typically 33 mm, and the borehole diameter typically 65–225 mm. A telescoping access pipe is needed when the predicted vertical strain is greater than about 1%.

Some users have installed this instrument surrounding inclinometer casing, hoping to be able to monitor deformation in three dimensions. However, because the criteria for grout properties are different – high compressibility for vertical deformation measurements and significantly less compressibility for horizontal deformation measurements – data have often been unsatisfactory.

95.3.4 Fixed borehole extensometers

Fixed borehole extensometers are defined in this chapter as devices installed in boreholes in soil or rock for monitoring the changing distance between two or more points along the axis of a borehole, without use of a movable probe. When the location of a measurement point is determined with respect to a fixed reference datum (e.g. by precise levelling), the devices also provide absolute deformation data. The operating principle is shown in **Figure 95.7**. Typical applications are monitoring deformations around underground excavations in rock, tunnels under construction in soil and behind the faces of excavated slopes.

Changes in the distance from the face of the collar anchor, sometimes referred to as the reference head, to the end of the rod is measured using either a mechanical dial gauge or an electrical displacement transducer. The device shown is a single-point borehole extensometer (SPBX), but several down-hole anchors can be located in a single borehole, each with an attached rod from the down-hole anchor to the collar anchor, to create a multipoint borehole extensometer (MPBX). The normal maximum number of anchors and rods for an MPBX is eight, and typical borehole diameter is 150 mm.

Many types of fixed borehole extensometers are available, the primary variables being choices of anchor type, SPBX or MPBX, transducer type and extensometer reference head. Wireless data collection and transmission are available for fixed borehole extensometers.

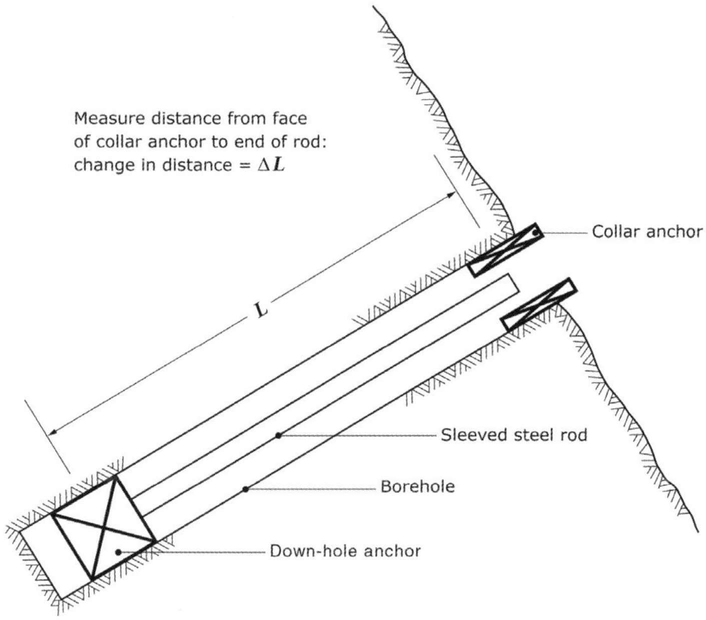

Figure 95.7 Operating principle of fixed borehole extensometer

Distributed fibre-optic sensing has also been successfully deployed as a continuous extensometer to measure and localise vertical deformation within the subsurface. In this application, a strain-sensing fibre-optic cable is installed within a borehole and backfilled with a cement–bentonite grout. A pre-tension is then applied to the cable which allows the measurement of both extension and compression without the fibre-optic cable going slack. The continuous nature of the readings allows for the integration of the strain data to calculate local and cumulative displacement along the borehole. Thermal compensation is necessary for fibre-optic strain data, and as such it is recommended to co-deploy a temperature-sensing cable with the strain cable. Examples of this approach have been presented by Bell *et al.* (2013), Wan (2014), where optical fibres were installed within a rod extensometer borehole and readings from the two independent systems compared, and more recently by Yeskoo *et al.* (2023).

95.3.5 Tiltmeters

Tiltmeters are used to monitor the change in inclination (rotation) of points on or in the ground or a structure. A tiltmeter consists of a gravity-sensing transducer within an appropriate housing, either fixed-in-place or arranged as a portable device by mating with reference points permanently attached to the surface of the ground or structure. Various types of transducers are used in tiltmeters, including force balance servo-accelerometers, electrolytic levels (electrolevels), vibrating wire transducers and micro-electrical mechanical systems (MEMS). Wireless data collection and transmission are available for tiltmeters.

Force balance servo-accelerometers were traditionally used in inclinometers due to their high precision and stability. The transducer is based on a flexure-suspended proof mass which is deflected as the component of gravity changes with tilt angle.

An electrolytic level consists of a sealed glass vial similar to the vial on a conventional spirit level, partly filled with a conductive liquid. Electrical conductors inserted through the vial enable tilt to be monitored remotely. Although historically an inexpensive and often-used transducer, the sensor tends to be very temperature-sensitive. Precise monitoring requires that temperature be measured with each tilt measurement if using this device. Electrolevels have now largely been superseded by MEMS devices, but these are also quite temperature-sensitive.

A vibrating wire tilt sensor consists of a pendulous mass which is supported by a vibrating wire strain gauge and an elastic hinge.

MEMS (Sellers and Taylor, 2008; Sheahan *et al.*, 2008) are now extensively used as tilt transducers for the geotechnical instrumentation community. They are the integration of mechanical elements, sensors, actuators and electronics on a common silicon substrate through microfabrication technology. There are numerous types of MEMS, which is still a large and rapidly growing high-technology area. Similar to the force balance servo-accelerometer, a MEMS tilt sensor is based on a flexure-suspended proof mass which is deflected as the component of gravity changes with tilt angle. A position sensor senses this movement by differential capacitance sensors of high sensitivity. The proof mass, flexible mounting, position sensor and supporting electronics are all constructed from a single wafer of non-metallic material, usually silicon.

95.3.6 Inclinometers

Inclinometers are devices for monitoring horizontal deformation below the ground surface. Typical applications include determining the shear zone in landslide movement, monitoring the extent and rate of horizontal movement of embankments on soft ground and alongside open-cut excavations, and monitoring the deflection of bulkheads, piles or retaining walls.

Inclinometers have four major components. First, a permanently installed guide casing, usually made of plastic or fibreglass, is installed in a near-vertical alignment. The guide casing has tracking grooves (keyways) for controlling the orientation of the probe. Second is a portable probe containing two tilt transducers mounted 90° apart for monitoring horizontal deformations in two directions. Third is a portable read-out unit for power supply and indication of probe inclination. Fourth is a graduated electrical cable linking the probe to the read-out unit. Care needs to be taken when installing the guide casing to make sure that the keyways are orientated in the right direction (e.g. perpendicular to a retaining wall) and that they do not become twisted. Usually the casing is weighted during grouting to avoid it rising from buoyancy effects. The weight may be a drilling rod placed within the casing to hold it down until the grout sets.

Figure 95.8 shows the normal principle of inclinometer operation. The probe measures rotations (tilt) at specific depths, as indicated in the schematic on the right, and these are integrated along the length of the casing, as shown, to obtain horizontal displacements. This is usually done automatically within the probe read-out unit. After installing the casing and set-up of the surrounding grout, the probe is lowered to the bottom and an inclination reading is made. Additional readings are made as the probe is raised incrementally to the top of the casing, providing data for determination of initial casing alignment. The differences between these initial readings and a subsequent set define any change in alignment. Provided that the lower end of the casing is fixed from translation by installing it in material such as bedrock that will not move laterally, these differences allow calculation of absolute horizontal deformation at any point along the casing. Otherwise, to obtain absolute readings, the top of the inclinometer tube should be monitored as well using position-measuring equipment such as a total station. Since one may not know for sure if the bottom of the casing is fixed, good practice involves measuring the precise position of a point marked on the top of the casing after it is installed, then making this measurement again at different times to prove the bottom of the casing is fixed or to measure the change in top position and make a correction to the inclinometer data.

Standard inclinometer casing diameters are 70 mm and 85 mm. Most commercially available inclinometers make use of a force balance servo-accelerometer transducer or increasingly a MEMS transducer. Most have portable read-out units that store data in on-board memory. Inclinometer casings can also be installed horizontally, such that vertical deformation data are monitored – for example, below embankments on soft ground.

Valuable guidelines on inclinometer data analysis are given by Mikkelsen (2003).

95.3.7 In-place inclinometers

In-place inclinometers are typically used for monitoring subsurface deformations around excavations or within slopes, when rapid or automated monitoring is required.

An in-place inclinometer is generally designed to operate in a near-vertical borehole and provides essentially the same data as a conventional inclinometer. The device, shown schematically in **Figure 95.9**, consists of a series of tilt transducers joined by articulated rods, but sometimes mounted individually on carriages. The transducers are the same as used in tiltmeters – force balance servo-accelerometers, electrolytic levels (electrolevels), vibrating wire transducers or MEMS tilt sensors. Uniaxial or biaxial transducers can be used. The transducers are positioned at fixed intervals along the borehole axis and can be concentrated in zones of expected movement. It is important that the spacing between the transducers

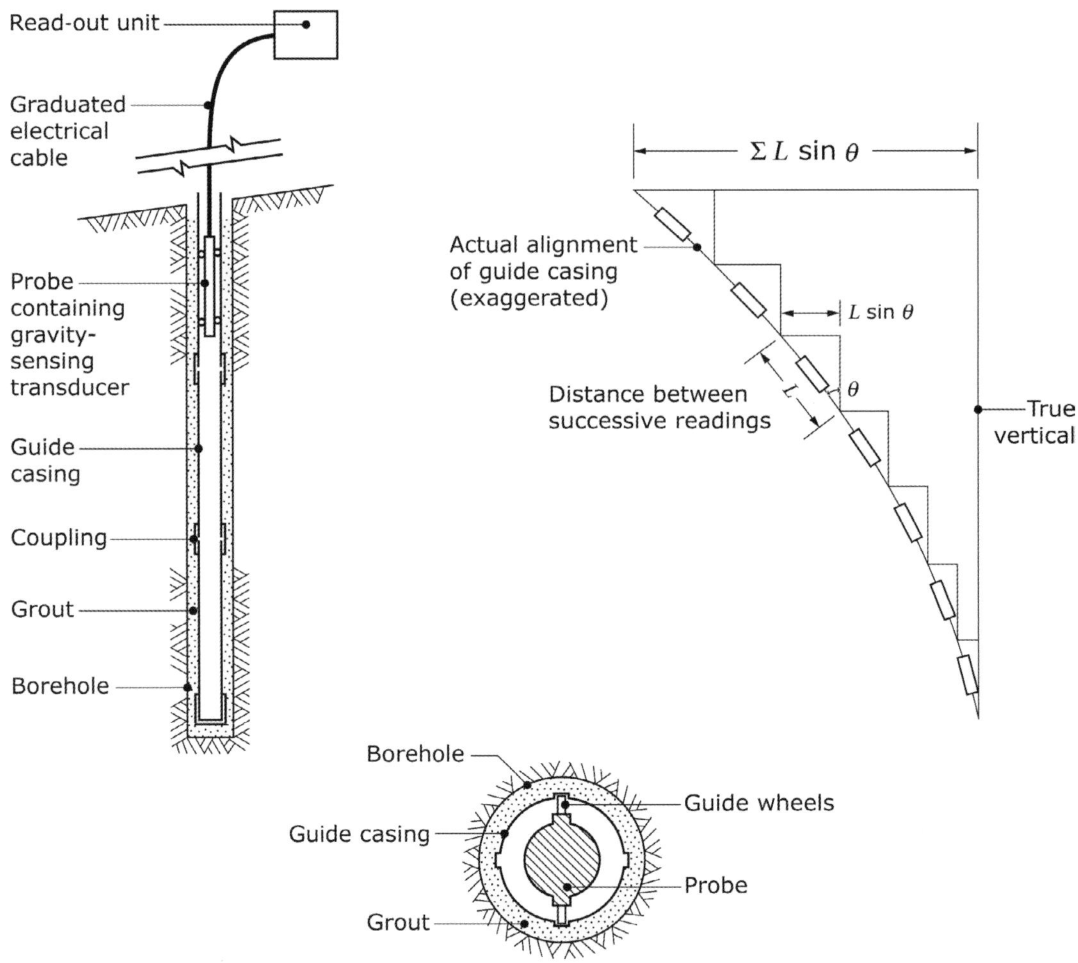

Figure 95.8 Principle of inclinometer operation

is not too great in order to improve accuracy, to minimise the possibility of the connecting rods interacting with the casing and to locate a shear zone more precisely. Usually a maximum spacing of 2 m is specified (BSI, 2017); however, it is often preferable to have much shorter spacings to improve overall accuracy. The appropriate spacing should also be selected after determining the initial alignment of the guide casing (e.g. by performing a probe survey) to assess how straight it is; if there is excessive curvature reduced spacings should be used. Movement data are calculated using the same methods as for conventional inclinometers. Wireless data collection and transmission are available for in-place inclinometers.

In-place inclinometers can be used effectively in combination with a conventional inclinometer. An in-place version can first be installed to define the location of any horizontal deformation, with minimal labour costs for reading. If deformation occurs, the in-place system can be removed and the moving zone monitored with a conventional inclinometer. Alternatively, a conventional inclinometer can be used first to indicate any deformation and an in-place version later installed across a critical deforming zone to minimise subsequent effort and perhaps to provide an alarm trigger.

A recently developed system that has great flexibility consists of a series of 300 mm long rigid segments connected by composite joints that prevent torsion but allow flexibility of segments in two degrees of freedom (Abdoun and Bennett, 2008; Barendse, 2008; Mikkelsen and Dunnicliff, 2008). MEMS tilt transducers are contained in each segment, such that when installed in a vertical borehole horizontal deformation data are obtained in real time. The instrument is typically installed within a 25 mm pipe and, in contrast to inclinometers and conventional in-place inclinometers, tracking grooves are not required for controlling orientation. This system can also be installed horizontally to monitor vertical deformation. The instrument can be recovered and reused. If installed within a conventional inclinometer casing, the device can be removed and monitoring continued using a conventional probe inclinometer.

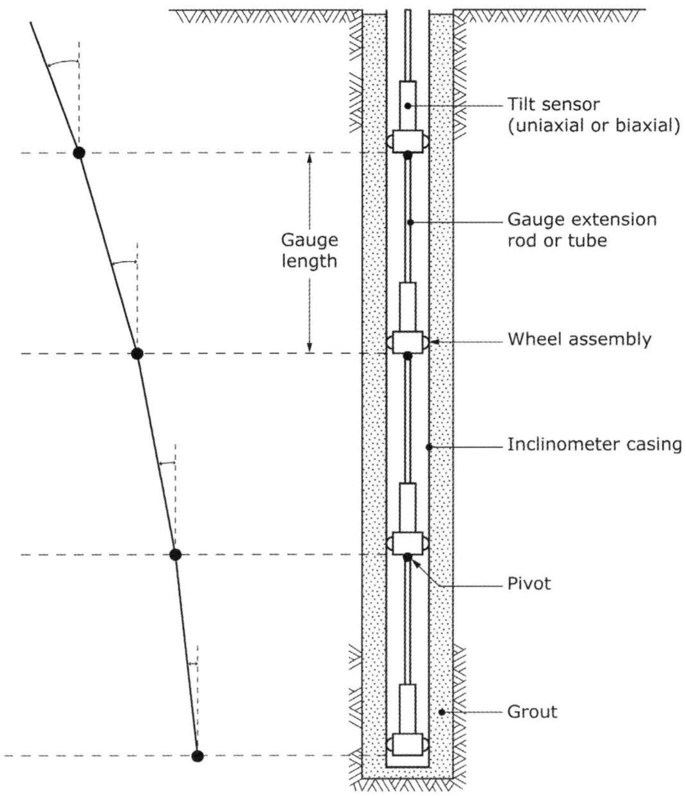

Figure 95.9 Schematic of in-place inclinometer

95.3.8 Liquid-level gauges

Liquid-level gauges are defined in this chapter as instruments that incorporate a liquid-filled tube or pipe for determination of relative vertical deformation. Relative elevation is determined either from the equivalence of liquid level in a manometer or from the pressure transmitted by the liquid.

The primary application for liquid-level gauges is monitoring settlements within embankments or embankment foundations (although recently they have also been used to monitor real-time tunnelling-induced building settlements). They allow installation to be made without frequent interruption to normal fill placement and compaction, and minimise the potential for instrument damage. They may also be mounted on different points on a structure, such as structural columns or tunnel lining, to measure differences in vertical deformation with time. Most liquid-level gauges also allow measurements to be taken at a central reading location. The gauges only provide a means of measuring *relative* elevations between two or more points. If *absolute* settlement or heave is required, as is usually the case, data must be referenced to a benchmark. If one end of the gauge cannot be mounted directly on a benchmark, a surveying method will normally be used (e.g. precise levelling), and accuracy may be dependent on the accuracy of the surveying method.

The most common type of liquid-level gauge is shown in **Figure 95.10**. The pressure transducer is usually a vibrating wire transducer. The upper surface of the liquid column is at a

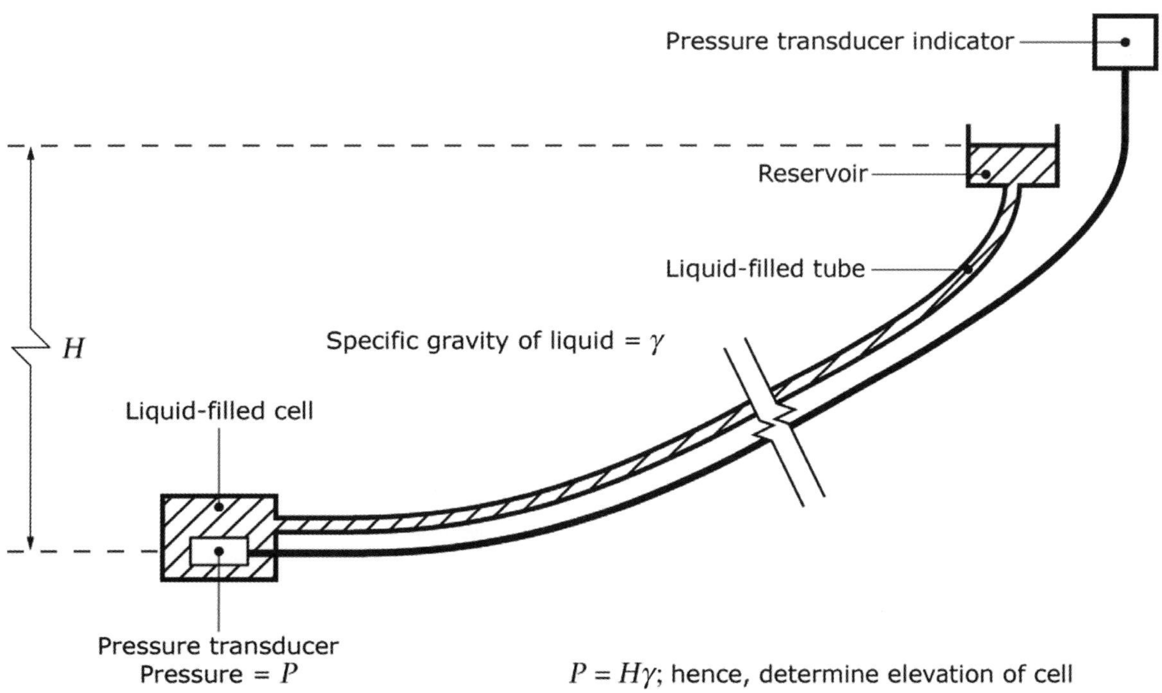

Figure 95.10 Schematic of liquid-level gauge

known elevation at the read-out location, and therefore relative elevation of the transducer and reservoir can be determined from the pressure measurement and liquid density. **Figure 95.10** shows only one liquid-filled tube, but in practice it is preferable to have two tubes so that flushing is possible and independent measurements can be made on each tube as a check. The greatest potential source of error is discontinuity of liquid caused by the presence of gas, and this is a frequent problem despite use of de-aired liquid. A much better arrangement is shown in **Figure 95.11** (McRae, 2000). Sufficient backpressure is applied to force any gas into solution, and any increase in backpressure will not change the transducer reading.

The same arrangement can be packaged as a probe that is pulled along a buried pipe, thereby providing vertical deformation data at any point along the pipe. This is called a full-profile liquid-level gauge.

Various liquid-level gauges (sometimes called liquid levelling systems) are available with interconnected sensors, for monitoring settlement at a series of locations. The sensors are connected together with both a liquid-filled tube and a tube for equalising air pressure. However, the small diameter of the liquid-level tube normally results in discontinuities in the liquid and a consequent major maintenance effort.

95.3.9 Crack gauges

Crack gauges (sometimes called joint-meters) are typically used for monitoring tension cracks behind slopes and for monitoring cracks in concrete or other structures, or joints or faults in rock. There are numerous versions, including portable or fixed-in-place mechanical gauges, and portable or fixed-in-place gauges that incorporate a locally or remotely read displacement transducer. Versions exist to measure relative movements in one, two and three directions. Wireless data collection transmission is available for crack gauges with electrical transducers.

95.3.10 Convergence gauges

Convergence gauges have typically been used for monitoring convergence across braced excavations and tunnels. While they are still used for these applications today, in most cases the measurements are now made by using total stations.

A typical convergence gauge, often called a tape extensometer, has a spring to control tension in the tape, and each end of the tape is attached to an anchor installed at an end of the span to be monitored. After attachment of the extensometer to the anchors and standardising the tension, readings of distance are made by adding the dial or digital indicator reading to the tape reading. Modern digital devices can read to a resolution of 0.01 mm.

It is important to measure temperature at the time of readings to apply corrections, as the tape is usually made of steel and thermal contraction/expansion of the tape itself can be significant. Temperature should be measured for each span to within 0.1°C using a portable digital thermometer placed

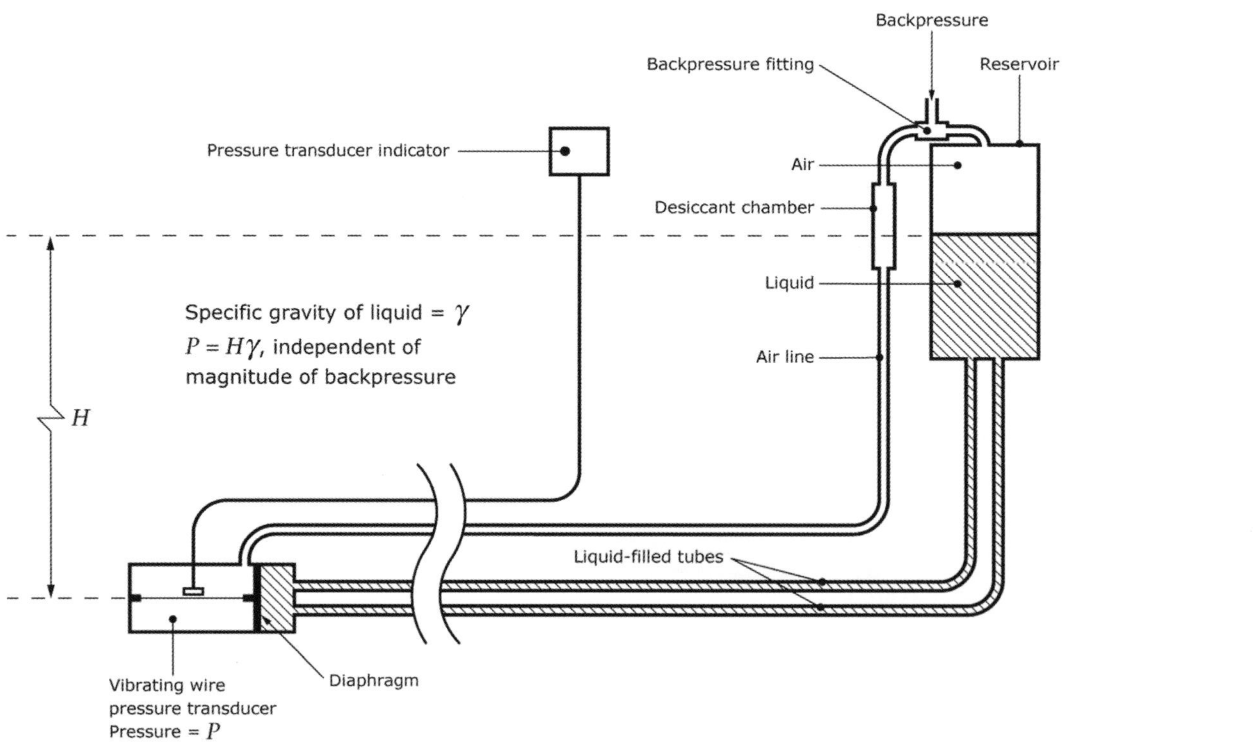

Figure 95.11 Schematic of closed-loop liquid-level gauge (Courtesy of Geokon, Inc., Lebanon, NH, USA)

roughly midway along the length of the steel tape. When using tape extensometers outdoors, flutter from wind can also be an issue. Factors influencing tape extensometer measurements and how to correct for them are provided by Standing *et al.* (2001).

95.3.11 Settlement platforms

Settlement platforms are typically used for monitoring settlement below embankments on soft ground. A settlement platform consists of a square plate of steel, wood or concrete placed on the original ground surface, to which a stiff riser pipe is attached. Optical levelling measurements to the top of the riser pipe provide a record of plate elevations. The riser pipe is extended as embankment filling proceeds or shortened if inside an excavation as it gets deeper. Recommended practice is to include a casing surrounding the riser pipe but not connected to the platform to reduce downdrag from the compressing surrounding soil. These devices are prone to construction damage and errors in readings if monitored by a conventional construction survey team.

95.3.12 Fibre-optic instruments

Fibre-optic instruments for monitoring deformation have been in development since the mid-1990s. Fibre-optic sensing can be divided into point-based technologies that provide measurement data at specially prepared measurements points, or continuous technologies that provide data at closely spaced read-out intervals along the sensing fibre. The light signal is processed to indicate deformation between fixed distances along the fibre. Glišić and Inaudi (2007) and Inaudi and Glišić (2007a) describe four types of fibre-optic sensing techniques: Fabry–Pérot interferometric point sensors for monitoring distance across a discontinuity such as a crack; FBG sensors that allow deformation measurements at multiple points along a single fibre line; SOFO interferometric sensors that are used for integrating deformation measurement over a long measurement base; and distributed fibre-optic sensors that are able to sense deformation at any point along a single fibre line. This last technique has very high potential for monitoring deformation over large areas, such as on slopes. Inaudi and Glišić (2007b) describe the fourth technique in more detail.

Distributed fibre-optic sensing (DFOS) techniques rely on the fibre-optic cable itself as the sensing instrument. When a light pulse is transmitted through a standard fibre-optic cable, a small subportion of light is reflected back towards the source at every location along the fibre. This phenomenon is referred to as backscattering. Changes in the backscattered light are proportional to changes in physical phenomena along the fibre-optic cable, with the location of measurement points determined by time-of-flight of the light within the cable. The types of DFOS technologies are separated by the type of scattering that they rely upon. These are divided into Raman scattering, which is proportional to temperature; Rayleigh scattering, which is proportional to vibration or dynamic rate-of-strain; and Brillouin scattering, which is proportional to strain and temperature change. Of these, Brillouin-based sensing systems are the most common for deformation monitoring in civil engineering applications. Soga and Luo (2018) summarise the measurement principles of distributed strain/temperature/acoustic sensing and their state-of-the-art status in terms of their capabilities. Soga (2014) has demonstrated use of the distributed fibre-optic displacement technique to monitor strains in bored piles. For a bored pile, a single fibre is run down and back up at 4–8 locations around the rebar cage. Average strain can be read in the fibre at sub-metre spacings along its entire length. These readings allow axial load and moment for the entire length of the pile shaft to be computed (e.g. Pelecanos *et al.*, 2018). Other examples of where this technology has been used are reported by Klar *et al.* (2014) and Hauswirth *et al.* (2014) for tunnelling-induced ground movements and an overall practical guide is provided by Kechavarzi et al. (2016).

95.3.13 Time-domain reflectometry

Time-domain reflectometry (TDR) (O'Connor and Dowding, 1999) is a remote sensing electrical technique, similar in concept to radar along a cable. A voltage pulse, produced by a TDR pulser, travels along a two-conductor coaxial metallic cable until it is partially reflected by a significant change in cross-sectional area of the cable. The distance to the change can be calculated knowing the propagation velocity of the signal in the cable and the time of travel of the voltage pulse from the disruption to the cable tester. TDR is used for monitoring unstable slopes, and for other applications for which an inclinometer might be considered – this application is described by Dowding *et al.* (2003). O'Connor (2008) describes geotechnical alarm systems based on TDR technology, and Lin (2009) presents some details of various types of measurements. Wireless data collection and transmission are available for TDR.

95.3.14 Acoustic emission monitoring

Acoustic emission (AE) is a non-destructive technique for providing an early warning of developing failures. The primary application is for monitoring slopes in soil and rock. AE is a natural phenomenon that occurs when a solid is subjected to stress. This stress, from an external source, causes a sudden release of sound waves resulting in microseismic activity, which can be detected by transducers. Within a slope the stress induced by destabilising forces causes a rearrangement of particles along developing shear surfaces. This interparticle friction results in the release of AE, and is an indication of straining within a soil body.

Although historically used as a qualitative monitoring technique, Spriggs and Dixon (2005) describe a quantitative approach to using AE data as an early warning system. This technique went dormant for many years largely due to background noise overwhelming the signals from the soil. It is experiencing a resurgence with fibre-optic acoustic monitoring that has higher sensitivity and expanded signal processing to reduce noise effects.

95.3.15 Photogrammetry techniques

Historically photogrammetry was used for tracking displacements using high-resolution photographic images of a structure taken at different times. The photographs are usually taken using a large-format Fresnel lens to improve the quality of the images, which are taken from various angles and elevations, often with some form of scale in the picture. The images were digitised and coordinates assigned from which displacements could be determined relating to movements that occurred between monitoring episodes.

Recently there has been a focus on implementing PIV (particle image velocimetry) techniques in photogrammetry applications. This involves mounting digital cameras on a stable point and taking images at a relevant frequency and processing the data using PIV software. An example is provided by Take *et al.* (2006), where the data were processed in real time. Now the digital camera may be placed on a drone, airplane or satellite and the data processed by computer to determine changes in successive images from which line-of-sight displacements can be determined.

95.3.16 Global positioning system

The global positioning system (GPS) consists of three parts: satellites, a ground control network and fixed-point ground receivers. Radio signals are used in an interferometric mode. Two or more GPS receivers simultaneously receive signals from the same set of satellites, and the resulting observations are subsequently processed to obtain the inter-station difference in position. Connection with at least three satellites is needed to locate the ground-based receiver, but five more are typically used. If one of the receivers is placed at a known position, the three-dimensional x, y, z position of the second receiver may be determined to a typical accuracy of 1 mm. The number of stations determined simultaneously is limited only by the number of receivers available. GPS is proving to be very useful for long-term performance monitoring of dams and other large structures (Rutledge and Meyerholtz, 2005).

95.3.17 Satellite monitoring systems

In recent years the applicability of satellite monitoring systems has improved greatly with increasing numbers of satellites orbiting the Earth. Currently, the most commonly used systems involve InSAR measurements (interferometric synthetic aperture radar) and permanent scatterer interferometry (a measurement method relying on the phenomenon of interference of radio waves). Observations within the microwave spectrum allow changes in the position of permanent scatterers to be determined from the phase difference between two SAR passes, taken from different sensor positions (ascending and descending paths). The permanent scatterers can be buildings, well-defined points on the ground surface or specifically designed reflectors. Historic records from satellites are available, allowing the development of movements to be tracked back by 1–3 decades, and many of the data are freely available. Typical frequency of satellite orbits and hence measurements is now weekly. The technology for these systems is developing rapidly both in terms of the hardware and software and increased accuracy. Current accuracy of line-of-sight measurements can be ±2 mm in the better cases. Vegetation and sharp irregularities in the ground surface considerably reduce this value. A recent guide to satellite monitoring has been prepared by Mason *et al.* (2022).

InSAR technology has been used successfully to monitor the deformations of dams, landslides, buildings, glacier movements, subsidence and volcanoes. There are three common frequency bands available. L uses a wavelength of 300 to 150 mm, with better penetration of vegetation and measurement precision to a few tens of millimetres. C uses a wavelength of 75 to 38 mm, with low to moderate penetration of vegetation and a few millimetres precision. X uses a wavelength of 38 to 24 mm, with little penetration of vegetation but better precision. InSAR data are usually obtained from service providers, who deliver point clouds of changes in line-of-site distances. These data reflect mostly up–down vertical motion of points on the surface, with a small east–west component. North–south displacement is not obtainable with InSAR.

95.3.18 LiDAR monitoring systems

Displacements can be measured using LiDAR (light or laser image detection and ranging). This is a ground-based or airborne method for determining ranges (variable distance) by targeting an object with a laser and measuring the time for the reflected light to return to the receiver. LiDAR sensors are able to achieve a range of accuracy from 0.5 to 10 mm relative to the sensor and a mapping accuracy of up to 10 mm horizontal (x, y) and 20 mm vertical (z). This makes them particularly useful as a remote sensing tool for mobile mapping. They can scan a large area with closely spaced points to deliver a point cloud of data. The data from successive scans are processed to give a point cloud of change in line-of-sight distance between the radar unit and each point. Ground-based systems are used to monitor large slopes where movements can grow rapidly and an early warning of impending failure has high value, such as in open-pit mines.

95.3.19 Ground-based radar system

This system uses a scanning radar positioned on the ground to scan a slope and detect movements on a millimetre scale. It is mostly used in open-pit mining to detect and warn of impending slope instabilities so people and equipment can be moved to safe locations (Sharon and Eberhardt, 2020). The equipment can be prohibitively expensive but there have been some applications to monitor earth dams while repair work is done.

95.4. Instruments for monitoring load and strain in structural members

95.4.1 Introduction

Instruments for measuring load and strain in structures fall into two groups: load cells and strain gauges. In each case, the transducers are used to measure small extensions and compressions. Load cells are interposed in the structure in such a way that structural forces pass through the cells, and strain gauges are attached directly to the surface of the structure or are embedded within the structure to sense the extensions and compressions in the structure itself.

Typical load cell applications include load testing of driven and bored piles, tiebacks and rockbolts and long-term performance monitoring of tiebacks and end-anchored rockbolts. Calibrated hydraulic jacks are designed for application of load, and not for measurement of load. When used for load measurement in the field, errors caused by friction between the cylinder and the piston can cause underestimates of load of up to 20%. Calibration errors from mishandling of the equipment can further increase this error.

Strain gauges are used where load cells cannot be interposed in the structure for reasons of geometry, capacity or economy, and where load and stress can be calculated with adequate accuracy from knowledge of the relationship between strain and stress. For example, strain gauges are used together with the modulus of elasticity of the structural member to determine temporary or permanent stresses or loads in struts across braced excavations, in rockbolts, retaining walls, diaphragm walls and deep foundations.

95.4.2 Load cells

Most load cells have either electrical resistance strain gauge or vibrating wire transducers. As shown in **Figure 95.12**, a cell with electrical resistance strain gauge transducers consists of a cylinder of steel or aluminium alloy, with electrical resistance strain gauges bonded to the outer periphery of the cylinder at its mid-section. Several strain gauges are used at regular intervals around the periphery, thereby reducing errors that result from load misalignment and off-centre loading. However, in all cases adequately thick bearing plates should be used at both ends of the cell, ground flat, smooth and parallel. Typical thicknesses are 38, 63 and 76 mm for cell capacities of 0.7, 1.8 and 2.2 MN, respectively.

In most vibrating wire load cells, deformation of the load-bearing member is measured by using three or more vibrating

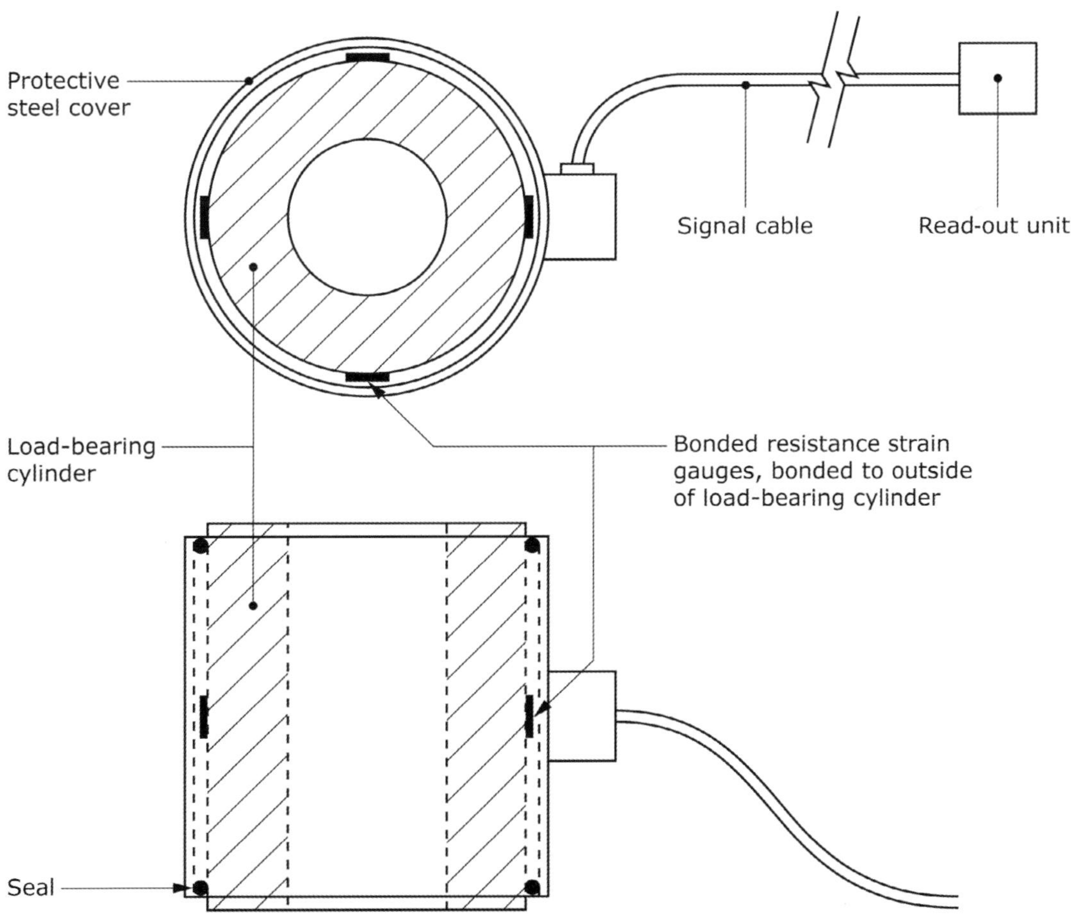

Figure 95.12 Schematic of load cell with electrical resistance transducers

wire transducers, and outputs from each transducer must be measured separately and averaged. The arrangement is similar to the electrical resistance load cell shown in **Figure 95.12**, but with vibrating wire transducers instead of bonded resistance strain gauges. However, the vibrating wire strain gauges are installed at a diameter mid-way between the inside and outside diameters of the cylindrical load-bearing member, thereby minimising errors caused by size mismatch between the cell and adjacent components. Wireless data collection and transmission are available for load cells.

Load cells with outside diameters of 80–320 mm are available, with the size depending on the load range.

Sellers (1994) reports on tests made to examine the error caused by a size mismatch between electrical resistance load cells and adjacent hydraulic jacks. If the hydraulic jack is larger than the load cell, there is a tendency for it to try to wrap the intervening bearing plate around the load cell and to bend the cylinder walls inward. If the hydraulic jack is smaller than the load cell it will try to push the intervening bearing plate through the hole in the load cell and to bend the cylinder walls outward. Sellers reports on errors of up to 8%. Dunnicliff (1995) reports on tests made by Sellers, subsequent to the tests referred to above, to examine the errors caused by a size mismatch between vibrating wire load cells and adjacent hydraulic jacks, where maximum errors were only 1%. In summary, it appeared that cells with vibrating wire transducers are preferable to cells with electrical resistance strain gauge transducers.

Calibrated hydraulic jacks are used for the application of load to rockbolts, tieback anchors, piles, struts across braced excavations and other structural elements. However, reliance on the measured pressure in the jack fluid as the sole method for load determination can often lead to significant inaccuracies. Off-centre and/or misaligned loading causes friction between the piston and cylinder, so use of the hydraulic pressure as an indication of load will indicate that the load is more than it truly is – that is, a measurement on the unsafe side. Good practice when load testing piles and tiebacks includes use of a load cell in series with the hydraulic jack.

95.4.3 Surface-mounted strain gauges

Historically, three types of strain gauges have been used for measurements on structural surfaces: mechanical, electrical resistance and vibrating wire, but most applications today make use of vibrating wire gauges. The recent development of fibre-optic instruments provides a fourth method.

A schematic of a surface-mounted vibrating wire gauge is shown in **Figure 95.13**. The operating principle is described above in the section on vibrating wire piezometers (see Section 95.2.4). The gauge is available in two basic configurations: typically 100–200 mm long for installation by arc welding and typically 50 mm long for installation by spot welding. When compared with the arc-welded gauge, advantages of the spot-welded gauge are small size, an installation procedure not requiring arc welding and minimum errors that result from bending of the structure because of the close proximity between the vibrating wire and the structure surface. However, if adequate space is available the larger arc-welded version is generally preferred. Wireless data collection and transmission are available for these gauges.

Fibre-optic gauges are described above for monitoring deformation, and references are given (see Section 95.3.12). When used for monitoring strain, all the same techniques are applicable. For distributed fibre-optic systems, special care

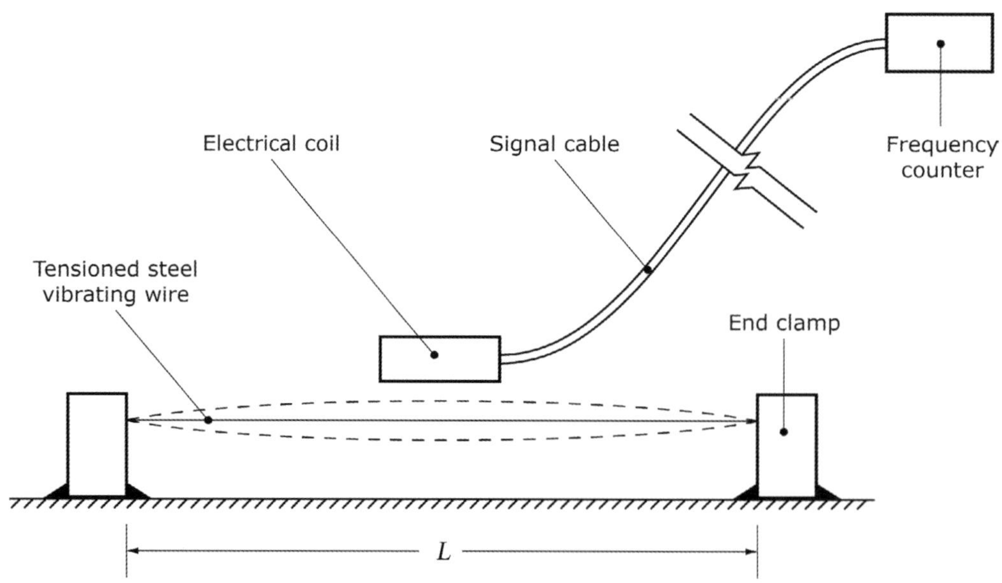

Figure 95.13 Schematic of surface-mounted vibrating wire strain gauge

must be taken to ensure continuous attachment of the fibre-optic cable to the monitored structure if the strain measurement is intended to be continuous. Gue *et al.* (2015) describe an application of distributed fibre-optic sensors for monitoring strain in an existing tunnel.

Strain data are rarely of interest by themselves; the data are merely a step in the determination of stress. When taking measurements on steel, for which the modulus is well known, strict attention must be paid to the influence of temperature (Boone and Bidhendi, 2001; Boone and Crawford, 2000; Daigle, 2005; Druss, 2000; Hashash and Marulanda, 2003; Osborne and Tan, 2009).

Stress determination in concrete is by no means straightforward, and accurate results should not be expected. Strain in concrete can be caused by several factors other than stress change, and a strain gauge responds to all causes of strain. Strain other than that caused by stress may be due to creep (strain under constant stress), shrinkage and swelling (moisture content change), temperature change, and the progress of autogenous volume change (dimensional change that is self-generated and not due to temperature, moisture change or stress). Other strain may also occur as concrete cures. Strain due to creep is generally more extensive in highly stressed prestressed concrete than in conventional reinforced concrete; therefore, for structural members such as prestressed concrete piles, creep strains are particularly significant. Local discontinuities in strain due to the larger aggregates in concrete can also affect interpretation of the readings. A special effort must therefore be made to determine the relationship between strain and stress, as described by Dunnicliff (1993) and Acerbis *et al.* (2011).

95.4.4 Embedment strain gauges

The primary application for embedment strain gauges is measurement of strain in concrete. Typical applications are driven concrete piles, bored piles, diaphragm walls, concrete tunnel linings and concrete dams. As with surface-mounted versions, vibrating wire gauges are used in most applications today. Fibre-optic sensing, both point-based and distributed, have also been successfully deployed in embedded applications for numerous geostructural monitoring applications. When using DFOS, embedded installation has the added benefit of providing continuous coupling along the fibre length, providing continuous strain data which can be integrated to calculate displacements within the structure. DFOS has been deployed in a range of embedded structures, including diaphragm walls (Li *et al.*, 2018) and bored piles (Pelecanos *et al.*, 2018).

There are two basic types of vibrating wire gauge. The first is similar to the arrangement shown in **Figure 95.13**, with circular end flanges instead of the end blocks, typically 100–200 mm long. Most gauges are supplied with a preset tension, which can be specified by the user. The second type consists of a vibrating wire transducer mounted in the central portion of a length of steel bar, such as reinforcing steel. The use of a length of steel bar, embedded in concrete to measure strain, is sometimes referred to as a sister bar, or a rebar strain meter. **Figure 95.14** shows a sister bar.

The first type cannot be installed without protection prior to concreting because of the likely damage when the concrete is vibrated. It must therefore be encapsulated in a briquette of fine-aggregate concrete, and the briquette must not be more than 24 hours old at the time that the primary concrete is placed – if it is older than this there will be an inclusion effect that will distort the gauge reading. The second type can be installed directly in concrete without protection and is therefore preferred if sufficient space is available.

As with gauges surface-mounted on concrete, a special effort must be made to determine the relationship between strain and stress, as described by Dunnicliff (1993) and Acerbis *et al.* (2011).

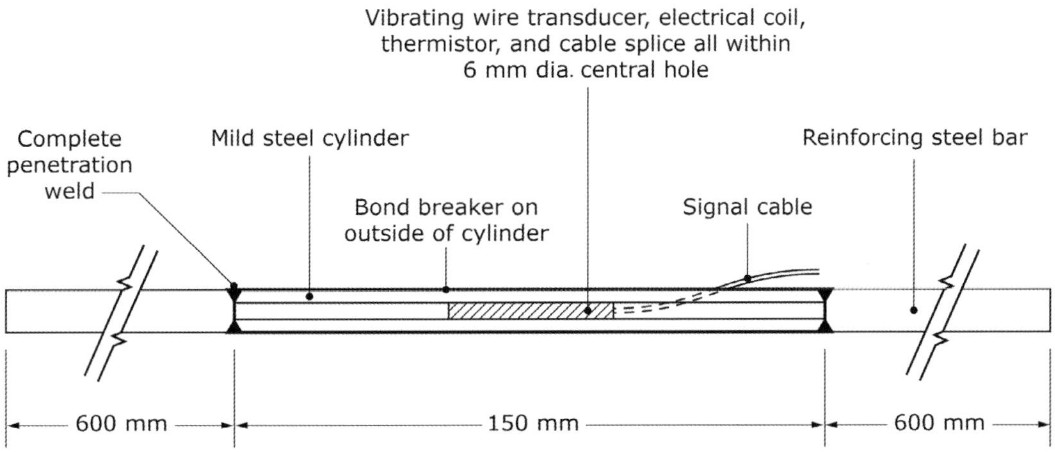

Figure 95.14 Schematic of sister bar with vibrating wire transducer

95.5. Instruments for monitoring total stress

95.5.1 Introduction

Total stress measurements in soil fall into two basic categories: measurements within a soil mass and measurements at the face of a structural element. Instruments are referred to as earth pressure cells, soil stress cells and soil pressure cells, and in this chapter the terms 'embedment earth pressure cells' and 'contact earth pressure cells' will be used for the two basic categories.

Embedment earth pressure cells are installed within fill – for example, to determine the distribution, magnitude and direction of total stress within an embankment. Applications for contact earth pressure cells include measurement of total stress against retaining walls, culverts, piles and diaphragm walls.

The primary uses of earth pressure cells are to confirm design assumptions and to provide information for the improvement of future designs. They are less commonly used for construction control or other reasons. When concerned with stresses acting on a structure during construction or after construction is complete, it is usually preferable to isolate a portion of the structure and to determine stresses by use of load cells and strain gauges within the structure. For example, this approach has been used successfully to determine earth pressures in braced excavation from measurements of support loads.

95.5.2 Embedment earth pressure cells

Attempts to measure total stress within a soil mass can be problematic because both the presence of the cell and the installation method generally create significant changes in the free-field stress. Therefore, it is usually not possible to measure total stress with great accuracy.

Most available embedment earth pressure cells are of the type shown in **Figure 95.15**. A cell consists of two circular or rectangular steel plates, welded together around their periphery, with liquid filling the intervening cavity and a length of high-pressure steel tubing connecting the cavity to a nearby pressure transducer. Total stress acting on the outside of the cell is balanced by an equal pressure induced in the internal liquid. It is essential that the cell is filled with de-aired liquid with low compressibility and that no gas bubbles are trapped within the cavity during filling. Typical cell diameter is 200 mm, but larger-diameter cells are available.

Push-in spade-type earth pressure cells operate on the same principle as described in **Figure 95.15**, but are installed from the base of a pre-drilled borehole. The spade cell is connected to a string of rods and lowered to the base of the borehole (typically at a depth about 1 m above the target level) and then pushed to the required depth using a hydraulic jack. Care must be taken when lowering the rods to make sure that the correct orientation of the spade cell is achieved, as it measures stresses perpendicular to its flat face. Many versions of spade cells also have a small piezometer incorporated within them to allow pore water pressures to be measured and hence effective stresses to be determined. As mentioned above, installation effects can strongly influence absolute total stress measurements: Tedd *et al.* (1989) provide recommendations for correcting readings. Changes in total stress can be interpreted with greater confidence than absolute stress measurement. An example where the output from such installations have been used to interpret mechanisms of ground response to tunnelling is given by Wan *et al.* (2019).

95.5.3 Contact earth pressure cells

Measurements of total stress against a structure are not plagued by so many of the errors associated with measurements within a soil mass, and it is possible to measure total stress at the face of a structural element with greater accuracy than within a soil mass. This is because both the stress cell and the surrounding structure are of comparable stiffness, whereas the stiffness of an embedded earth pressure cell is much higher than the surrounding soil, causing changes to the stress applied to the earth pressure cell (often referred to as cell action effects).

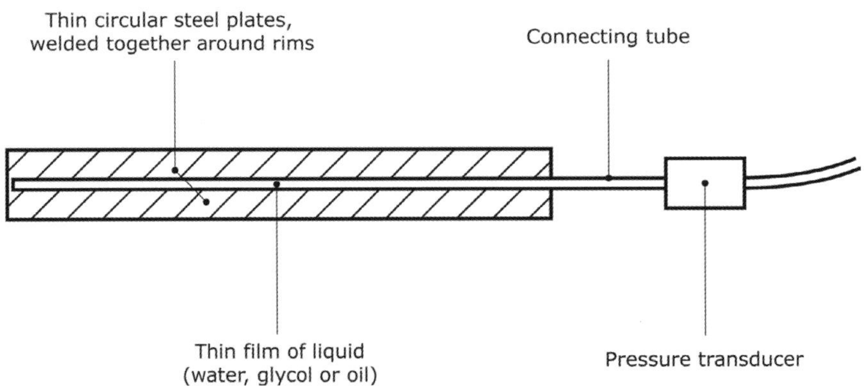

Figure 95.15 Schematic of embedment earth pressure cell

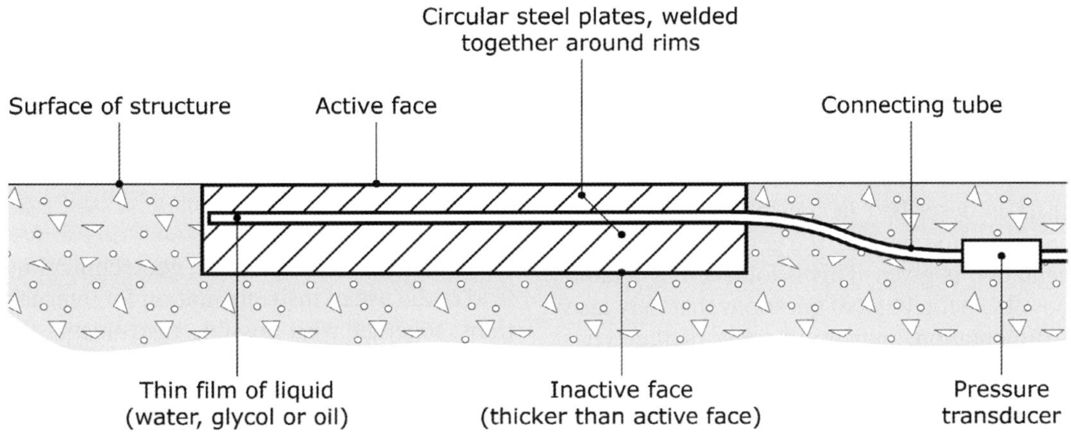

Figure 95.16 Schematic of contact earth pressure cell

Figure 95.16 shows a typical contact earth pressure cell. This is similar to the embedment cell shown in **Figure 95.15**, with one of the active faces replaced by a thick inactive face. Cell diameters are the same as for embedment cells.

95.6. Instruments for monitoring temperature

Measuring temperature can provide numerous benefits when performing any form of surface monitoring. Temperature changes can affect the measuring system and its components, the ground or structure being monitored and those taking the measurements. For instance, changes in loads in struts used to support open excavations can result from thermal changes to the struts. Being aware of temperature changes can provide insight into observed responses and facilitate the interpretation of the field data, especially when measurements are made when there are large diurnal or seasonal variations. Corrections can often be made to data using appropriate coefficients of thermal expansion, applied in conjunction with the observed temperature change. Ideally temperatures should be monitored to a resolution of 0.1°C at the same time that other measurements are being made. In some systems thermal changes can be minimised by using temperature-insensitive materials such as invar. For electrical resistance strain gauges the backing material can be selected such that it has a similar coefficient of thermal expansion as the element to which it is to be attached (e.g. cast iron tunnel lining), so that it is temperature self-compensating.

Thermal effects have historically not been treated as important for subsurface monitoring where the temperature tends to be quite stable. However, for instruments that require thermal compensation such as some strain gauges and DFOS applications, thermal measurements should be taken so that appropriate corrections can be applied. This is particularly important for instruments extending to the ground surface as fluctuations in the subsurface temperature associated with diurnal and seasonal changes at the surface have been observed to depths of several metres (Yeskoo *et al.*, 2023).

There are a variety of devices to measure temperature for geotechnical applications. These include thermometers, thermistors, thermocouples and resistive temperature devices (RTDs), with the last three used with data acquisition systems.

95.7. Monitoring visual change

Dr Ralph Peck often remarked that 'an instrument too often overlooked in our technical world is a human eye connected to the brain of an intelligent human being' (Dunnicliff and Deere, 1984). The importance of this powerful statement never fades. However, someone with a trained eye cannot watch over a facility constantly, but many adverse performance events occur when not expected and therefore are not seen in real time by a human. Modern cameras for taking photographs and videos play an increasingly useful role in this regard. These technologies are increasingly powerful, can be remotely operated and are inexpensive. Images and videos can be transmitted by communications networks almost instantly. These technologies can allow a human with a trained eye to see the facility from anywhere and at any time, as well as give the ability to pan, tilt and zoom the camera to zones of interest. The images can be examined with software to detect changes that might indicate an unravelling situation, thus warranting a warning message to be issued. Images and videos can be recorded for later examination of evidence of causes and documentation of consequences.

Real-time photographic information is rapidly becoming an integral part of performance monitoring programmes. It helps provide context information about activities on the project that may affect the readings from instrumentation. But perhaps more importantly, it provides complementary and supplemental information about changes in performance that instrumentation may not detect. Of course, photography is limited to information on exposed surfaces within the line of sight of the camera and so does not provide a complete monitoring system by itself. It becomes an increasingly important tool as part of an integrated monitoring programme for many applications.

95.8. General role of instrumentation, and summaries of instruments to be considered for helping to provide answers to various geotechnical questions

95.8.1 Introduction

Instrumentation that may be considered for various types of construction projects is described in the following sections. For each type of construction project, the general role of geotechnical instrumentation is given. This is followed by a tabular listing of the possible geotechnical questions that may lead to the use of instrumentation, indicating some of the types of instruments that can be considered for helping to provide answers to those questions. It is recognised that there may be additional geotechnical questions and also additional instruments that are not described in this chapter.

The existence of significant questions indicates the presence of uncertainty. Uncertainty introduces risk. Therefore, the use of instrumentation can be viewed as a tool to help identify and manage emerging risk (Marr, 2007). Data from instrumentation help answer questions, thereby reducing uncertainties and resulting in reduced risk. Well-executed instrumentation and monitoring can play an important role in risk management of structural works.

The sequence of geotechnical questions is intended to match the time sequence in which the question may be addressed during the design, construction and performance process, and does not indicate any rating of importance. The observational method (Chapter 100 *Observational method*) relies primarily on real-time feedback from field monitoring and so the following scenarios can facilitate the choice of instrumentation to incorporate within the works when relevant.

The suggestions for types of instruments are not intended to be dogmatic, because the selection always depends on issues specific to each project and is influenced by the personal experience of the person making the selection. In the tables that follow some of the most likely instruments to consider are listed, with other possible types in brackets.

95.8.2 Internally braced excavations

95.8.2.1 General role of instrumentation

The design of internally braced (strutted or propped) excavations is based for the most part on empirical procedures and past experience. The consequences of poor performance can be severe and may on occasion be catastrophic (e.g. the Nicoll Highway collapse; Davies *et al.*, 2006). A monitoring programme may not be required if the design is very conservative with few uncertainties, if there is previous experience with design and construction of similar facilities under similar conditions or if the consequences of poor performance will be acceptable. However, under other circumstances a monitoring programme will normally be required to demonstrate that the excavation is stable and that nearby structures are not adversely affected. Depending on the specific needs of each case, the monitoring programme may apply to the wall and struts, to the ground beneath or surrounding the excavation and/or to adjacent structures or utilities, and to the pore water pressures outside and below the bottom of the excavation.

95.8.2.2 Summary of instruments to be considered for helping to provide answers to various geotechnical questions

Table 95.1 lists the possible geotechnical questions that may lead to the use of instrumentation for internally braced excavations, together with possible instruments that can be considered for helping to provide answers to those questions.

95.8.3 Externally braced excavations using anchors or tiebacks

95.8.3.1 General role of instrumentation

The general role of instrumentation for externally braced excavations is the same as for internally braced excavations. However, it is possible to make regular visual inspections of internal bracing, but external bracing cannot be seen. Although confidence in the performance of an externally braced excavation is increased by conducting a proof test on every tieback anchor, if an anchor subsequently fails the failure may be progressive and catastrophic. In general, therefore, instrumentation plays a role in three phases of external bracing that are not applicable to internal bracing: testing of *test anchors* during the design phase or at the start of construction, as input to design of the project anchors; *performance* and *proof testing* of anchors during construction; and subsequent *monitoring* of selected representative anchors. The third phase may be omitted if a conservative design has been used.

95.8.3.2 Summary of instruments to be considered for helping to provide answers to various geotechnical questions

Table 95.2 lists the possible geotechnical questions that may lead to the use of instrumentation for externally braced excavations, together with possible instruments that can be considered for helping to provide answers to those questions.

95.8.4 Deep excavations: basement heave

95.8.4.1 General role of instrumentation

Basement heave is an important consideration when making a deep excavation, either within retaining walls (as discussed above) or between cut slopes (e.g. for railway cuttings). There are short- and long-term components of base heave, the latter often taking decades to develop. Understanding the magnitude of heave and swelling pressure is necessary if basement slabs are to be cast directly onto excavated ground (although sometimes a gap is left or a void former used to allow free swelling to take place) or if a cutting has been formed as part of the route of a railway line (especially for high-speed trains where high tolerances need to be maintained to avoid

Some geotechnical questions	Measurement	Some instruments to consider
What are the initial site conditions?	Groundwater pressure	Open standpipes Vibrating wire piezometers installed by the fully grouted method (Pneumatic piezometers)
	Vertical deformation	Displacement monitoring points with surveying methods
	Widths of cracks in structures	Crack gauges
Are the struts being installed correctly?	Load in struts	Calibrated hydraulic jacks, load cells
Is the excavation stable, and are nearby structures being affected adversely by ground movements?	Settlement of ground surface, structures and top of supporting wall	Displacement monitoring points with surveying methods
	Horizontal deformation of ground surface, structures and exposed part of supporting wall	Displacement monitoring points with surveying methods (Convergence gauges)
	Change in width of cracks in structures and utilities	Crack gauges
	Subsurface horizontal deformation of ground	Inclinometers In-place inclinometers (Fixed borehole extensometers, fibre-optic instruments)
	Subsurface settlement of ground and utilities	Probe extensometers (Fixed borehole extensometers)
	Load in struts	Surface-mounted strain gauges, load cells, temperature
	Groundwater pressure	Open standpipes Vibrating wire piezometers installed by the fully grouted method (Pneumatic piezometers)
	Base heave	Probe extensometers, heave points
Is an individual strut being overloaded?	Load in strut	Surface-mounted strain gauges, load cells, temperature
Is the groundwater table being lowered?	Groundwater pressure	Open standpipes Vibrating wire piezometers installed by the fully grouted method (Pneumatic piezometers)
Is excessive base heave occurring?	Base heave	Probe extensometers, heave points, piezometers
	Subsurface horizontal deformation	Inclinometers In-place inclinometers

Table 95.1 Some instruments to consider for monitoring internally braced excavations

differential movements). Heave-reducing piles are sometimes installed to minimise upward displacements of the ground. Predicting and monitoring basement heave is difficult, especially for long-term conditions.

A scenario frequently encountered in the urban environment is where a deep excavation takes place above an existing tunnel or sensitive services. In these cases, deformations of the existing structures will start along with the beginning of excavation, and it might be prudent to monitor the ground response around them. Generally, it is vertical displacements that are of interest, and fixed borehole extensometers would be appropriate. The challenge is that they need to be dismantled as excavation progresses and reset at each stage. As stated in Section 95.3.4, if absolute movements are required, precise levelling of the reference head is necessary (taken back to a stable datum) and this too is challenging. Another alternative, if possible, is to undertake monitoring within the existing structure itself. This is often best achieved with robotic total station measurements and/or continuous in-place inclinometers (as described at the end of Section 95.3.7). A good

Some geotechnical questions	Measurement	Some instruments to consider
What are the initial site conditions?	As in **Table 95.1**	As in **Table 95.1**
What is a suitable design for tieback anchors (by constructing and testing *test anchors*)?	Load in tieback	Load cells; strain gauges attached to the anchors
	Deformation at head	Dial indicators, prisms for total station
	Load transfer in grouted zone	Surface-mounted strain gauges on anchor
Are the tiebacks being installed correctly (by *performance* and *proof testing*)?	Load in tieback	Calibrated hydraulic jacks (load cells), surface-mounted strain gauges
	Deformation at head	Dial indicators
Is the excavation stable, and are nearby structures being affected adversely by ground movement?	As in **Table 95.1**, except for load in struts	As in **Table 95.1**, except for load in struts
	Load in tieback	Load cells (Calibrated hydraulic jacks and load cells: *lift-off tests*) Surface-mounted strain gauges
	Monitor displacements of the wall and adjacent structures	Displacement monitoring points and survey methods
Is the groundwater table being lowered?	As in **Table 95.1**	As in **Table 95.1**
Is excessive bottom heave occurring?	As in **Table 95.1**	As in **Table 95.1**

Table 95.2 Some instruments to consider for monitoring externally braced excavations

example of long-term monitoring of existing tunnels beneath a deep excavation is given by Burford (1988). **Table 95.1** can be used as a guide for choosing instrumentation for basement heave. Primary items to monitor in deep excavations include vertical heave at representative points within the excavation, pore water pressure with depth below the bottom of the excavation, pore water pressure outside the retaining walls in each soil layer, lateral movement of the retaining walls, vertical movement of the retaining walls if bottoms are in or over soft soils and vertical and horizontal movements of any nearby buildings within a distance of one to two times the depth of the excavation.

95.8.5 Embankments on soft ground

95.8.5.1 General role of instrumentation

This section relates to the use of geotechnical instrumentation where all the geotechnical questions are associated with the soft ground itself, and not with the embankment.

In many cases, selection of soil parameters for the foundation soil is reliably conservative. The embankment is therefore designed with confidence that performance will be satisfactory, and 'comfortable' factors of safety are used. In such cases many projects will proceed without the use of instrumentation. However, some uncertainties always exist. Where design uncertainties are great, factors of safety small or the consequences of poor performance severe, a prudent designer will include a performance monitoring programme in the design. This is particularly true for staged embankment construction.

In spite of a long record of successful embankment construction throughout the history of civil engineering, embankments that are designed with a factor of safety greater than unity can fail and have disastrous consequences to the project. Thus, it is not surprising that instrumentation plays a significant role in design and construction of embankments on soft ground.

The most frequent uses of instrumentation for embankments on soft ground are to monitor the progress of consolidation and strength gain of the soft ground foundation and to determine whether the embankment is stable. If the calculated factor of safety may drop below 1.5, instrumentation will generally be installed to provide a warning of any instability, thereby allowing remedial measures to be implemented before critical situations arise.

Chapter 70 *Design of new earthworks* provides additional information on embankments on soft ground.

95.8.5.2 Summary of instruments to be considered for helping to provide answers to various geotechnical questions

Table 95.3 lists the possible geotechnical questions that may lead to the use of instrumentation for embankments on soft ground, together with possible instruments that can be considered for helping to provide answers to those questions. If the project involves staged construction of the embankment, piezometers or settlement gauges at multiple levels and multiple locations are recommended to help determine the progress of consolidation over space and time.

95.8.6 Clay embankments

95.8.6.1 General role of instrumentation

This section relates to the use of geotechnical instrumentation where the primary geotechnical questions are associated with the embankment, and not necessarily with the embankment's foundation.

Some geotechnical questions	Measurement	Some instruments to consider
What are the initial site conditions in the soft ground?	Pore water pressure	Vibrating wire piezometers installed by the fully grouted method (Open standpipes, pneumatic piezometers)
	Vertical deformation	Displacement monitoring points with surveying methods
Is the embankment stable?	Horizontal deformation	Displacement monitoring points with surveying methods Inclinometers (In-place inclinometers)
What is the progress of consolidation of the soft ground?	Vertical deformation of embankment surface on crest and ground surface at and beyond toe of embankment	Displacement monitoring points with surveying methods
	Vertical deformation of original ground surface and within foundation below embankment	Probe extensometers (Single-point and full-profile liquid-level gauges, settlement platforms, settlement points, horizontal inclinometers)
	Vertical deformation and compression of subsurface	Probe extensometers
	Pore water pressure	Vibrating wire piezometers installed into foundation by the push-in method or grouted-in-place piezometers at different depths

Table 95.3 Some instruments to consider for monitoring embankments on soft ground

All clay embankments begin life with negative pore water pressures present in parts of the embankment, which should provide enhanced stability. With time, increases of the pore water pressure in the fill from added load or groundwater inflow could lead to the development of instability. In some cases such failures will occur shortly after construction. Longer-term failures can occur in embankments that have an otherwise stable geometry, caused by progressive deformations induced by shrinkage and swelling that result from the presence of vegetation or by the slow long-term in-flow of water. It is these that give most cause for concern. Seasonal variations of pore water pressure giving rise to shrinkage and swelling also cause problems with the serviceability of clay embankments, particularly where they carry railway infrastructure. Identifying the presence of seasonal pore water pressure changes or shrinkage and swelling is therefore an important factor in determining the need for further assessment (an example of monitoring such changes is given by Smethurst *et al.*, 2015).

Once identified, the magnitude of the stress changes can be used in an analysis of the embankment, to determine if a failure surface is likely to develop. However, it is important to recognise that the analysis will be imprecise in its ability to predict the timing of a failure because the seasonal changes will not be consistent from year to year, and the history of stress changes will be unknown. Therefore long-term monitoring of the pore water pressures is likely to be required and suitable instrumentation should be selected. Inclinometers can be used to detect acceleration in the movement along discontinuities near the toe of the embankment, but these measurements must also be long-term.

Ultimate stability may be influenced by the underlying foundation. Low-permeability soils could prolong the increase of pore water pressures within the embankment during extreme environmental conditions. Higher-permeability soils in the foundation can provide underdrainage and help to maintain lower pore water pressures within the embankment. Deep open standpipes are therefore useful for identifying the underlying groundwater regime.

95.8.6.2 Summary of instruments to be considered for helping to provide answers to various geotechnical questions

Table 95.4 lists the possible geotechnical questions that may lead to the use of instrumentation for clay embankments, together with possible instruments that can be considered for helping to provide answers to those questions.

95.8.7 Embankment dams

95.8.7.1 General role of instrumentation

The main purpose of instrumentation installed within an embankment dam is to study whether or not the dam is behaving according to design predictions during construction and operation. This general statement can be subdivided into two categories: first, the study of special problems at individual sites that can occur during construction due to special foundation conditions or uncommon design features; and second, the study of behaviour when the dam is in its operational state especially during extreme loading.

When instrumentation is used for studying special problems that are related to special foundation conditions or uncommon design features, the design of the monitoring programme is

Some geotechnical questions	Measurement	Some instruments to consider
Is the embankment stable?	Negative and positive pore water pressure	Flushable piezometers
	Horizontal deformation	Surveying methods for settlement of embankment and heave outside toe. Inclinometers (In-place inclinometers)
What are pore pressures and the internal drainage condition? How fast are excess pore pressures dissipating?	Pore water pressure	Piezometers
Is there any shrink or swell?	Pore water pressure	Flushable piezometers
	Horizontal deformation	Inclinometers (In-place inclinometers)
	Vertical deformation	Probe extensometers Surveying methods

Table 95.4 Some instruments to consider for monitoring clay embankments

tuned directly for the special conditions and features and, because the designers of the dam know the weaknesses of the particular site and the sensitive features of the design, they must play a leading role in the choice of type and location of instruments.

The practice of placing instruments routinely in embankment dams where there is no special problem to be studied – for example, a conventional dam of moderate height on a good foundation – has seen several ups and downs in the past 60 years. In the 1960s many important dams were constructed with essentially no internal instruments. Each failure or major problem tends to be widely publicised, and dam engineers feel pressure to install instruments routinely for their own protection, even if they do not believe it absolutely necessary.

Peck (1985) comments:

> Instrumentation, vital for obtaining quantitative answers to significant questions, is too often misused, especially in earth and rockfill dams. In some countries regulations concerning the safety of dams demand the incorporation of inclinometers, settlement indicators, and piezometers in the cores of virtually all new dams, but for what purpose? Not for research, because the patterns of deformation and pore pressure development for ordinary geometries and materials are now well known and can be predicted by calculation. Only under unusual circumstances can it be said that design assumptions in these regards require verification. Yet, installation of instruments, even under the best of circumstances, introduces inhomogeneities into the cores, and occasionally is the direct cause of such local defects as sinkholes. The potential weakness introduced by an installation should be balanced against the potential benefit from the observations. In contrast to those located in cores, piezometers in foundation materials near the downstream toes detect upward seepage pressures that cannot be predicted reliably, and can thus give timely warning if measures are needed to ensure safety. There is a danger that instrumentation may be discredited because of indiscriminate use. Of all the internal instruments in embankment dams, piezometers are by far the most common, and an installation consisting only of piezometers can serve two purposes. First, piezometers in a clay core indicate rates of dissipation of pore water pressure and the approach of equilibrium conditions after impounding. Second, piezometers may be appropriate where it is desired to avoid criticism when designing a dam of conventional design and moderate height on a good foundation.

Although written in 1985, these comments are still applicable today.

Peck (2006) puts instrumentation into perspective: 'Monitoring of every dam is mandatory, because dams change with age and may develop defects. There is no substitute for systematic and intelligent surveillance. But monitoring and surveillance are not synonymous with instrumentation.'

A helpful way to consider instrumentation of a dam is to tie its design to a potential failure modes analysis (Marr, 2015a, 2015b). The project team reviews the more significant potential failure modes and determines what measurements might be used to reveal the emergence and development of each. With this as a starting point the instrumentation design team can then choose the types and locations of instruments appropriate to each significant potential failure mode.

95.8.7.2 Instruments to be considered for helping to provide answers to various geotechnical questions

Among the possible geotechnical questions are the following.

- What are the initial site conditions?
- Is performance satisfactory during construction?
- Is performance satisfactory during first filling?
- Is performance satisfactory during drawdown?

- What was the performance during an extreme event?
- Is the performance deteriorating over time and likely to become unsatisfactory?

The types of instruments to be considered for helping to provide answers to these questions are too numerous to be tabulated here, and readers are referred to Dunnicliff (1993). The most important methods for addressing the last question are visual observations of the entire structure by appropriately trained people and monitoring of leakage emerging downstream by using leakage weirs. Monitoring methods that have been developed or improved since the above reference was written, and which may be applicable for addressing the last question, include GPS, DFOS and robotic total stations for monitoring deformation. Distributed fibre-optic sensors also have potential for monitoring the temperature of seepage water (Inaudi and Glišić, 2007b), thereby indicating areas where water flow is concentrated.

95.8.8 Waste retention dams (tailings dams)

95.8.8.1 General role of instrumentation

Tailings is a term used for waste products produced from mining and ore extraction processes. Usually in their initial raw state they are liquids (or slurries) containing varying proportions of ground solids and dissolved chemicals (e.g. salts). As these slurries are pumped into the storage area, the coarser particles settle first and 'beaches' are formed close to where they are pumped out. The finer particles in suspension flow much further, often into a 'pond' where deposition is much slower. These then consolidate and may take decades to come to equilibrium. Impoundments for coal combustion residuals and other industrial solid waste products involve similar concepts as discussed here but some can have unique characteristics that make their performance complex.

Tailings dams have many similarities to embankment dams in that they provide a means of retaining a liquid (the tailings slurry that may evolve into a soil-like material). The main differences are that tailings dams are usually constructed continuously over time as the waste material is produced, and the tailings themselves have their own inherent behaviour, rather than being just water; in particular, they are prone to liquefaction. Construction of the dams can be inwards towards the stored waste material (upstream method), upwards (centre-line method) or outwards (downstream method). More detailed descriptions of these and discussion of their advantages and disadvantages is beyond the scope of this chapter. Much of the structural component of the dam may be placed without compaction or good quality control over the material placement. Additionally, the increase in stress level and pore pressures within the bottom core of the dam can adversely affect the strength behaviour of components of the dam.

The fact that the dam is being raised over prolonged periods of years and/or decades means that it is far more critical to monitor the response of (*a*) the foundation, (*b*) the dam itself (which after construction of the initial starter dam is often made from the coarser tailings) and (*c*) the stored waste itself. In view of this it is necessary to monitor from the outset of dam construction and continue installing monitoring systems as the dam is raised.

The geotechnical questions and how to address them are largely the same as those for embankment dams. The Želazny Most dam in Poland is an example where comprehensive monitoring was undertaken as described by Jamiolkowski (2013). In this case, the primary monitoring undertaken was of surface displacements at various levels and positions on the dam using conventional surveying techniques and also GPS, subsurface horizontal displacements using inclinometer systems (in particular to assess existing shear planes) and pore water pressures within the foundation soils. Other additional measurements such as flow rates into the drainage systems, salinity levels and concentrations of suspended solids can provide vital information about how the dam is functioning.

It is important to recognise that many tailings dams have elements that exhibit brittle behaviour whereby the tailings may reach a certain shear stress level with additional construction that matches the material's peak strength and thereafter that element may experience a major loss in strength through liquefaction. Strains before this point can be very small and pore water pressures can appear to be acceptable. But once strength loss begins, a progressive failure can develop very rapidly in which large zones lose strength, large pore water pressures develop and large downward and outward deformations occur in seconds. A monitoring programme to detect the onset of liquefaction with the objective of avoiding a failure is unlikely to work because there will be little change in instrument readings leading up to the failure and no time to respond when instrument readings do suddenly start to change. Instead, one needs to look for the precursors to a liquefaction failure that might be monitored to reveal a warning. Monitoring of tailings dams is currently a very active area of research.

At this time the most useful measurements for tailings dams which have liquefiable materials within their structural zone are pore water pressures at many points throughout the dam to develop a full understanding of how flow is occurring within the dam and to provide pore water pressures to use in detailed stability analyses. Stability analyses with conservative assumptions, measured pore water pressures, and an appropriate minimum factor of safety presently provide the most reliable application of instrumentation to tailings dams.

Deformation monitoring of the surface of the dam using techniques like InSAR and LiDAR might also help identify local anomalies in performance that need more investigation and perhaps additional instrumentation (see Mazzanti *et al.* (2021) for an InSAR assessment of the Želazny Most dam). But in and of themselves they cannot be counted on to detect an impending liquefaction failure with high reliability.

95.8.9 Cut slopes in soil

95.8.9.1 General role of instrumentation

It is imperative that, prior to planning an instrumentation programme for a cut slope in soil, an engineer first develops one or more working hypotheses for a potential behaviour mechanism. The hypotheses must be based on a comprehensive knowledge of the locations and properties of stratigraphic discontinuities.

Instrumentation can be used to define the groundwater regime prior to excavating a slope. Results of measurements during excavation can be used as a basis for modification of the designed slope angle. Measurements of ground movement and positive or negative groundwater pressure can assist in documenting whether or not performance during and after excavation is in accordance with predicted behaviour. Measurements can also be used to document whether short- and long-term surface and/or subsurface drainage measures are performing effectively. If evidence of instability appears during or after construction, instrumentation plays a role in defining the characteristics of the instability, thus permitting selection of an appropriate remedy.

A very important subset is the case of a cut slope in clay. Here, negative pore water pressures generated during excavation can give rise to temporary stability, the lifetime of which will be related to the height of the slope and the slope angle. Therefore, monitoring the negative pore water pressures is an effective way of assessing the stability of a cut slope in clay. In some instances the stability may be maintained for long enough to undertake temporary works within the excavation and thereby save on expensive stabilisation measures.

Another key point to consider is that a cut slope produces an unloading of the remaining soil. Unloading of most soils produces negative pore water pressures and small strains. The negative pore water pressures impart strength to the soil, but these may decrease over time. Therefore strength may decrease over time and an apparently stable slope with no signs of distress suddenly fail without much warning. Many of these failures can involve metres of very rapid mass movement. Extra consideration must be given to what to measure where and what is the expected benefit.

Chapter 72 *Slope stabilisation methods* provides additional information on cut slopes in soil.

95.8.9.2 Summary of instruments to be considered for helping to provide answers to various geotechnical questions

Table 95.5 lists the possible geotechnical questions that may lead to the use of instrumentation for cut slopes in soil, together with possible instruments that can be considered for helping to provide answers to those questions.

Some geotechnical questions	Measurement	Some instruments to consider
What are the initial site conditions?	Pore water pressure	Open standpipes Vibrating wire piezometers installed by the fully grouted method Flushable piezometers (Pneumatic piezometers)
	Surface deformation	Surveying methods (Tiltmeters, fibre-optic instruments, GPS)
	Subsurface deformation	Inclinometers In-place inclinometers (TDR, fibre-optic instruments)
Is the slope stable during excavation?	Surface deformation	Surveying methods (Tiltmeters, TDR, fibre-optic instruments, GPS)
	Subsurface deformation	Inclinometers In-place inclinometers (TDR, fibre-optic instruments)
	Pore water pressure	Vibrating wire piezometers installed by the fully grouted method Flushable piezometers
Is the slope stable in the long term?	As for 'Is the slope stable during excavation?'	As for 'Is the slope stable during excavation?'
	Rainfall, for possible correlation with any deformation	Rain gauges
	Load in tiebacks	Load cells, strain gauges on the tie

Table 95.5 Some instruments to consider for monitoring cut slopes in soil

95.8.10 Landslides in soil

95.8.10.1 General role of instrumentation

If there is evidence of slope instability, its characteristics must be defined so that any necessary remedial measures may be taken. The question of *how much ground is moving* can be answered by use of instrumentation. The question of *why the ground is moving* will not be answered by instrumentation alone: the answer requires a complete geotechnical investigation and analysis. Instrumentation also plays a role in monitoring the long-term stability of the slope to determine the effectiveness of any remedial measures that have been taken. Almost all slope failures involve increase of pore water pressures on the failure surface, but these are usually not measured or known. Piezometers located on or near the expected failure surface should be used to monitor active or suspected landslide areas where the consequences of a failure are significant.

Chapter 72 *Slope stabilisation methods* provides additional information on landslides in soil.

95.8.10.2 Summary of instruments to be considered for helping to provide answers to various geotechnical questions

Table 95.6 lists the possible geotechnical questions that may lead to the use of instrumentation for landslides in soil, together with possible instruments that can be considered for helping to provide answers to those questions.

95.8.11 Cut slopes in rock

95.8.11.1 General role of instrumentation

The general role of instrumentation is identical to the role for cut slopes in soil, as discussed above. However, when planning to monitor the stability of rock slopes, it is important to recognise that if the slope is subject to a brittle failure mode, movement will be sudden. In such cases, geotechnical instrumentation may not be appropriate to forewarn of instability. It may be more appropriate to develop an area-wide correlation between rainfall intensity and slope instability, and to use rainfall measurements to warn of potential problems. Also, many slopes in rock move in a stick–slip fashion with small deformation rates followed by rapid movement that then stops. An automated system to obtain several readings per day may be warranted to obtain meaningful data over time if the consequences of a failure are significant.

Chapters 72 *Slope stabilisation methods* and 87 *Rock stabilisation* provide additional information on cut slopes in rock.

95.8.11.2 Summary of instruments to be considered for helping to provide answers to various geotechnical questions

Table 95.7 lists the possible geotechnical questions that may lead to the use of instrumentation for cut slopes in rock, together with possible instruments that can be considered for helping to provide answers to those questions.

95.8.12 Landslides in rock

95.8.12.1 General role of instrumentation

The general role of instrumentation is identical to the role for landslides in soil, as discussed above.

95.8.12.2 Summary of instruments to be considered for helping to provide answers to various geotechnical questions

Table 95.8 lists the possible geotechnical questions that may lead to the use of instrumentation for landslides in rock,

Some geotechnical questions	Measurement	Some instruments to consider
What are the post-landslide conditions?	Pore water pressure	Open standpipe Vibrating wire piezometers installed by the fully grouted method Flushable piezometers (Pneumatic piezometers)
	Surface deformation	Surveying methods (Tiltmeters, fibre-optic instruments, GPS)
	Subsurface deformation	Inclinometers In-place inclinometers TDR (Fibre-optic instruments)
Is the slope stable in the long term?	As for 'What are the post-landslide conditions?'	As for 'What are the post-landslide conditions?'
	Rainfall, for possible correlation with any deformation	Rain gauges
	Load in tiebacks	Load cells, strain gauges on ties

Table 95.6 Some instruments to consider for monitoring landslides in soil

Some geotechnical questions	Measurement	Some instruments to consider
What are the initial site conditions?	Joint water pressure	Open standpipes Vibrating wire piezometers installed by the fully grouted method (Pneumatic piezometers)
	Surface deformation	Surveying methods Crack gauges (Tiltmeters, fibre-optic instruments, GPS)
	Subsurface deformation	Fixed borehole extensometers In-place inclinometers (AE monitoring, TDR, fibre-optic instruments)
Is the slope stable during excavation?	Surface deformation	Surveying methods Crack gauges (Tiltmeters, TDR, fibre-optic instruments, GPS)
	Subsurface deformation	Fixed borehole extensometers In-place inclinometers (AE monitoring, TDR, fibre-optic instruments)
	Joint water pressure	Vibrating wire piezometers installed by the fully grouted method
Is the slope stable in the long term?	As for 'Is the slope stable during excavation?'	As for 'Is the slope stable during excavation?'
	Rainfall, for possible correlation with any deformation	Rain gauges
	Load in tiebacks	Load cells, strain gauges

Table 95.7 Some instruments to consider for monitoring cut slopes in rock

Some geotechnical questions	Measurement	Some instruments to consider
What are the post-landslide conditions?	As in **Table 95.7** for 'What are the initial site conditions?'	As in **Table 95.7** for 'What are the initial site conditions?'
Is the slope stable in the long term?	As in **Table 95.7** for 'Is the slope stable in the long term?'	As in **Table 95.7** for 'Is the slope stable in the long term?'
	Rainfall, for possible correlation with any deformation	Rain gauges
	Load in tiebacks	Load cells, strain gauges

Table 95.8 Some instruments to consider for monitoring landslides in rock

together with possible instruments that can be considered for helping to provide answers to those questions.

95.8.13 Tunnels

95.8.13.1 General role of instrumentation

The consequence of poor performance of a tunnel can be severe and may on occasion be catastrophic. A monitoring programme may not be required if the design is very conservative, if the ground conditions are good, if there is previous experience with design and construction of similar facilities under similar conditions or if the consequences of poor performance will not be severe. However, under other circumstances a monitoring programme will normally be required to demonstrate that the tunnel is stable and that nearby structures are not adversely affected. Monitoring programmes for tunnels focus on three primary factors: (*a*) Is the tunnelling operation likely to become unstable? (*b*) Are there undesirable impacts on the adjacent structures caused by tunnelling? And (*c*) can the tunnelling operations be made more efficient? Much of a monitoring programme for tunnels focuses on effects of tunnelling on groundwater pressures and deformations of the tunnel lining and the adjacent structures.

Some geotechnical questions	Measurement	Some instruments to consider
What are the initial site conditions?	As in **Table 95.1**	As in **Table 95.1**
Is the tunnel stable, and are overlying structures being affected adversely by ground movements?	Settlement of ground surface and structures	Surveying methods
	Horizontal deformation of ground surface and structures	Surveying methods
	Change in width of cracks in structures and utilities	Crack gauges
	Subsurface horizontal deformation of ground	Inclinometers In-place inclinometers (Fibre-optic instruments)
	Subsurface settlement of ground and utilities	Probe extensometers Fixed borehole extensometers
	Deformation within tunnel	Surveying methods (Fibre-optic instruments, tape extensometers, robotic total stations, distometers)
	Stresses within tunnel lining	Embedded stress gauges, strain gauges on rebar

Table 95.9 Some instruments to consider for monitoring tunnels

95.8.13.2 Summary of instruments to be considered for helping to provide answers to various geotechnical questions

Table 95.9 lists the possible geotechnical questions that may lead to the use of instrumentation for tunnels, together with possible instruments that can be considered for helping to provide answers to those questions.

95.8.14 Driven piles

95.8.14.1 General role of instrumentation

The subsurface length of a driven pile cannot usually be inspected after driving; thus, its physical condition and alignment are unknown. Subsurface geotechnical conditions are rarely known with certainty, and therefore the design of driven piles involves assumptions and uncertainties that are often addressed by conducting instrumented full-scale tests. Tests may examine the behaviour of the pile under load applied to the pile head or under load caused by settlement of soil with respect to the pile.

Defects in piles can be created during driving, and inspection procedures are available for examining the condition and alignment after driving. Certain types of driven pile cause large displacements and changes of pore water pressure in the surrounding soil, and these may in turn have a detrimental effect on neighbouring piles or on the stability of the site as a whole. Instrumentation can be used to quantify the consequences of pile driving and thus to assist in planning any necessary action.

The following chapters provide additional information on driven piles: Chapter 54 *Single piles*, 55 *Pile-group design*, 81 *Types of bearing piles* and 82 *Piling problems.*

95.8.14.2 Summary of instruments to be considered for helping to provide answers to various geotechnical questions

Table 95.10 lists the possible geotechnical questions that may lead to the use of instrumentation for driven piles, together with possible instruments that can be considered for helping to provide answers to those questions.

Some geotechnical questions	Measurement	Some instruments to consider
What is the load–movement relationship of the pile? (See Chapter 98 *Pile capacity testing*)	Deformation at head	Dial indicators with reference beams Wire/mirror/scale Surveying methods
	Load at head	Load cell
	Deformation at toe	Tell-tales
	Stress along pile	Embedment or surface-mounted strain gauges (Fibre-optic instruments)
Has the capacity of the pile been reduced by defects caused during driving?	Curvature of pile	Inclinometer
	Condition of pile	Integrity testing (See Chapter 97 *Pile integrity testing)*

Table 95.10 Some instruments to consider for monitoring driven piles

95.8.15 Bored piles

95.8.15.1 General role of instrumentation

Many uncertainties exist during design of bored piles, and instrumentation plays a role in determining the load–movement relationship by conducting load tests. Concrete integrity is often uncertain during construction, particularly when bored piles are constructed in granular soils below the water table or in softer, squeezing clays, when concrete slump is inadequate or when concrete placement practices are inferior.

Some geotechnical questions	Measurement	Some instruments to consider
What is the load–movement relationship of the pile? (See Chapter 98 *Pile capacity testing*)	As in **Table 95.10** Load at toe or in pile	As in **Table 95.10** Osterberg load cell
What is the integrity of the concrete?	Condition of pile	Integrity testing (See Chapter 97 *Pile integrity testing*)

Table 95.11 Some instruments to consider for monitoring bored piles

Instrumentation can be used to examine the integrity of the concrete. For piles cast using support fluid, concrete integrity at the pile toe is particularly important.

95.8.15.2 Summary of instruments to be considered for helping to provide answers to various geotechnical questions

Table 95.11 lists the possible geotechnical questions that may lead to the use of instrumentation for bored piles, together with possible instruments that can be considered for helping to provide answers to those questions.

Acknowledgements

The authors are indebted to Dr Andrew Ridley, Managing Director, Geotechnical Observations Limited, Weybridge, Surrey for the text on the description and applications of flushable piezometers for measuring negative pore water pressure and discussions concerning in-place inclinometers. We also appreciate the added information on fibre-optic sensors provided by Professor Kenichi Soga of University of California at Berkeley and Dr Andrew Yeskoo of Geocomp.

References

Abdoun T and Bennett V (2008) A new wireless MEMS-based system for real-time deformation monitoring. *Geotechnical News* **26(1)**: 36–40.*

Acerbis R, Asche H, Barbieri G and Collotta T (2011) Recommendations for converting strain measured in concrete to stress. *Geotechnical News* **29(1)**: 29–33.*

Barendse MB (2008) Field evaluation of a MEMS-based real-time deformation monitoring system. *Geotechnical News* **26(1)**: 41–44.*

Bell A, Soga K, Ouyang Y, Yan J and Wang F (2013) The role of fibre optic instrumentation in the re-use of deep foundations. *Proceedings of the 18th International Conference on Soil Mechanics and Geotechnical Engineering, Paris, France*, pp. 1863–1866.

Beth M (2007) Discussion of robotic total stations and remote data capture: challenges in construction. *Geotechnical News* **25(1)**: 33–38.*

Boone SJ and Bidhendi H (2001) Strain gauges, struts and sunshine. *Geotechnical News* **19(1)**: 39–41.

Boone SJ and Crawford AM (2000) The effects of temperature and use of vibrating wire strain gauges for braced excavations. *Geotechnical News* **18(3)**: 24–28.

BSI (British Standards Institution) (2015) BS 5930+A1:2020: Code of practice for ground investigations. BSI, London, UK.

BSI (2017) BS EN ISO 18674-3:2017: Geotechnical investigation and testing – Geotechnical monitoring by field instrumentation. Part 3: Inclinometers. BSI, London, UK.

Burford D (1988) Heave of tunnels beneath the Shell Centre, London, 1959–1986. *Géotechnique* **38(1)**: 135–137.

Burland JB, Moore JFA and Smith PDK (1972) A simple and precise borehole extensometer. *Géotechnique* **22(1)**: 174–177.

Casagrande A (1949) Soil mechanics in the design and construction of the Logan airport. *Journal of the Boston Society of Civil Engineers* **36(2)**: 192–221.

Cheney JE (1973) Techniques and equipment using the surveyor's level for accurate measurement of building movement. *Proceedings of the Symposium on Field Instrumentation in Geotechnical Engineering, London, UK. British Geotechnical Society, London, UK*, pp. 85–99.

Contreras IA, Grosser AT and Ver Strate RH (2008) The use of the fully-grouted method for piezometer installation. *Geotechnical News* **26(2)**: 30–37, 40.*

Cook D (2006) Robotic total stations and remote data capture: challenges in construction. *Geotechnical News* **24(3)**: 42–45.*

Daigle L (2005) Temperature influences on earth pressure cell readings. *Geotechnical News* **23(4)**: 32–36.*

Davies RV, Fok P, Norrish A and Poh ST (2006) The Nicoll Highway collapse: field measurements and observations. *Proceedings of the International Conference on Deep Excavations, Singapore*, pp. 1–15.

Dowding CH, Dussud ML, Kane WF and O'Connor KM (2003) Monitoring deformation in rock and soil with TDR sensor cables. *Geotechnical News* **21(2)**: 51–59.*

Druss DL (2000) Discussion: the effects of temperature and use of vibrating wire strain gauges for braced excavations. *Geotechnical News* **18(4)**: 24.

Dunnicliff J (1993) *Geotechnical Instrumentation for Monitoring Field Performance*. Wiley, New York, NY.

Dunnicliff J (1995) Load cell calibrations. *Geotechnical News* **13(1)**: 35.

Dunnicliff J (1998) *Geotechnical Instrumentation Reference Manual*. Training course in geotechnical and foundation engineering, NHI Course 13241 – Module, FHWA HI-98-034. National Highway Institute, Arlington, VA.

Dunnicliff J (2008) Discussion: the use of the fully-grouted method for piezometer installation. *Geotechnical News* **26(2)**: 38–39.*

Dunnicliff J and Deere DU (eds) (1984) *Judgment in Geotechnical Engineering: The Professional Legacy of Ralph B. Peck*. Wiley, New York, NY.

Glišić B and Inaudi D (2007) *Fibre Optic Methods for Structural Health Monitoring*. Wiley, Chichester, UK.

Gue CY, Wilcock M, Alhaddad MM, *et al.* (2015) The monitoring of an existing cast iron tunnel with distributed fibre optic sensing (DFOS). *Journal of Civil Structural Health Monitoring* **5**: 573–586, 10.1007/s13349-015-0109-8

Hashash MA and Marulanda C (2003) Temperature correction and strut loads interpretation in central artery excavations. *Geotechnical News* **21(4)**: 30–31.*

Hauswirth D, Puzrin AM, Carrera A, Standing JR and Wan SPM (2014) Application of fibre optic sensors for simple assessment of ground surface displacements during tunnelling. *Géotechnique* **64(10)**: 837–842, 10.1680/geot.14.T.0009.

Hope C and Chaqui M (2008) Manual total station monitoring. *Geotechnical News* **26(3)**: 28–30.*

Inaudi D and Glišić B (2007a) Overview of fiber optic sensing technologies for geotechnical instrumentation and monitoring. *Geotechnical News* **25(3)**: 27–31.*

Inaudi D and Glišić B (2007b) Distributed fiber optic sensors: novel tools for the monitoring of large structures. *Geotechnical News* **25(3)**: 31–35.*

Jamiolkowski M (2013) Soil mechanics and the observational methods: challenges at the Zelazny Most tailings disposal facility. *Géotechnique* **64(8)**: 590–619.

Kechavarzi C, Soga K, de Battista N, *et al.* (2016) *Distributed Fibre Optic Strain Sensing for Monitoring Civil Infrastructure: A Practical Guide*. ICE Publishing, London, UK.

Klar A, Dromy I and Linker R (2014) Monitoring tunnelling induced ground displacements using distributed fiber-optic sensing. *Tunnelling and Underground Space Technology* **40**: 141–150.

Kontogianni V, Kornarou S and Stiros S (2007) Monitoring with electronic total stations: performance and accuracy of prismatic and non-prismatic reflectors. *Geotechnical News* **25(1)**: 30–33.*

Li Z, Soga K and Kechavarzi C (2018) Distributed fibre optic sensing of a deep excavation adjacent to pre-existing tunnels. *Géotechnique Letters* **8**: 171–177.

Lin C-P (2009) TDR as a geo-nerve: a slope monitoring system example. *Geotechnical News* **27(1)**: 38–40.*

Marr WA (2007) *Why Monitor Performance?* ASCE Geotechnical Special Publication 117. Keynote paper. 7th International Symposium on Field Measurements in GeoMechanics, Boston, MA. ASCE, New York, NY.

Marr WA (2008) Monitoring deformations with automated total stations. *Geotechnical News* **26(3)**: 30–33.*

Marr WA (2015a) Performance monitoring to help manage risk for dams and impoundments. *Proceedings of the 35th Annual USSD: Managing Risk and Uncertainty – Ensuring a Sustainable Future for Dams, Levees and Reservoirs, Louisville, KY*, pp 1149–1166.

Marr WA (2015b) Performance monitoring as a risk management tool in dam safety. *Proceedings of the Ninth International Symposium on Field Measurements in Geomechanics, Sydney, Australia*, pp. 63–88.

Mason PJ, Devanthéry N, Camafort M, *et al.* (2022) *InSAR and Earth Observation Techniques for Infrastructure*. CIRIA Report C805D. CIRIA, London, UK.

Mazzanti P, Antonielli B, Sciortino A, Scancella S and Bozzano F (2021) Tracking deformation processes at the Legnica Glogow Copper District (Poland) by satellite InSAR – II: Zelazny Most tailings dam. *Land* **10**: 654, 10.3390/land10060654

McRae J (2000) Vibrating wire settlement cells: an alternative technique. *Geotechnical News* **18(1)**: 40.

Mikkelsen PE (2002) Cement–bentonite grout backfill for borehole instruments. *Geotechnical News* **20(4)**: 38–42.*

Mikkelsen PE (2003) Advances in inclinometer data analysis. *Proceedings of the Symposium on Field Measurements in Geomechanics, Oslo, Norway*, pp. 555–567.

Mikkelsen PE and Dunnicliff J (2008) Some views on a recent addition to our instrumentation tool box. *Geotechnical News* **26(4)**: 28–30.*

O'Connor KM (2008) Geotechnical alarm systems based on TDR technology. *Geotechnical News* **26(2)**: 40–44.*

O'Connor KM and Dowding CH (eds) (1999) *GeoMeasurements by Pulsing TDR and Probes*. CRC Press, Boca Raton, FL.

Osborne N and Tan GH (2009) Factors influencing the performance of strain gauge monitoring systems. *Geotechnical News* **27(2)**: 34–37.*

Peck RB (1985) The last sixty years. *Proceedings of the 11th International Conference on Soil Mechanics and Foundation Engineering, San Francisco, CA*. Balkema, Rotterdam, the Netherlands, pp. 123–133.

Peck RB (2006) Embankment dams: instrumentation versus monitoring. In *Ralph B. Peck, Educator and Engineer: The Essence of the Man* (Dunnicliff J and Young N (eds)). BiTech Publishers, Vancouver, BC, pp. 253–254.

Pelecanos L, Soga K, Elshafie MZEB, *et al.* (2018) Distributed fiber optic sensing of axially loaded bored piles. *Journal of Geotechnical and Geoenvironmental Engineering* **144(3)**, 10.1061/(ASCE)GT.1943-5606.0001843.

Ridley AM (2003) Recent developments in the measurement of pore water pressure and suction. *Geotechnical News* **21(1)**: 47–50.*

Ridley AM, Dineen K, Burland JB and Vaughan PR (2003) Soil matrix suction: some examples of its measurement and application in geotechnical engineering. *Géotechnique* **53(2)**: 241–253.

Rutledge DR and Meyerholtz SZ (2005) Using the global positioning system (GPS) to monitor the performance of dams. *Geotechnical News* **23(4)**: 24–28.*

Sellers JB (1994) Load cell calibrations. *Geotechnical News* **12(3)**: 65–66.

Sellers JB and Taylor R (2008) MEMS basics. *Geotechnical News* **26(1)**: 32–33.*

Sharon R and Eberhardt E (2020) *Guidelines for Slope Performance Monitoring*. CSIRO Publishing, Canberra, Australia.

Sheahan TC, Mazzei D and McRae J (2008) Performance testing of MEMS-based tilt sensors. *Geotechnical News* **26(1)**: 33–36.*

Smethurst JA, Briggs KM, Powrie W, Ridley AM and Butcher DJE (2015) Mechanical and hydrological impacts of tree removal on a clay fill railway embankment. *Géotechnique* **65(11)**: 869–882.

Soga K (2014) Understanding the real performance of geotechnical structures using an innovative fibre optic distributed strain measurement technology. *Rivista Italiana di Geotechnica* **4**: 7–48.

Soga K and Luo L (2018) Distributed fiber optics sensors for civil engineering infrastructure sensing. *Journal of Structural Integrity and Maintenance* **3(1)**: 1–21.

Spriggs M and Dixon N (2005) The instrumentation of landslides using acoustic emission. *Geotechnical News* **23(3)**: 27–31.*

Standing JR, Withers AD and Nyren RJ (2001) Measuring techniques and their accuracy. *Building Response to Tunnelling. Case Studies from the Jubilee Line Extension, London, Volume 1, Projects and Methods* (Burland JB, Standing JR and Jardine FM (eds)). CIRIA Special Publication 200. CIRIA and Thomas Telford, London, UK, pp. 273–299.

Take WA, White DJ, Bowers KH and Moss NA (2006) Remote real-time monitoring of tunnelling-induced settlement using image analysis. *Proceedings of the International Symposium on Geotechnical Aspects of Underground Construction in Soft Ground, Amsterdam, the Netherlands* (Bakker K, Bezuijen A, Broere W and Kwast EA (eds)). Taylor & Francis, London, pp. 771–777.

Tedd P, Powell JJ, Charles JA and Uglow IM (1989) In-situ measurements of earth pressures using push-in spade-shaped pressure cells: 10 years' experience. In *Geotechnical Instrumentation in Practice: Purpose, Performance and Interpretation*. Thomas Telford, London, UK, pp. 701–715.

Vaughan PR (1969) A note on sealing piezometers in bore holes. *Géotechnique* **19(3)**: 405–413.

Volterra JL (2008) Monitoring by manual and/or automated optical survey. *Geotechnical News* **26(4)**: 26–27.*

Wan MSP (2014) *Field Monitoring of Ground Response to EPBM Tunnelling Close to Existing Tunnels in London Clay*. PhD thesis, Imperial College London, UK.

Wan MSP and Standing JR (2014) Field measurement by fully grouted vibrating wire piezometers. *Proceedings of the Institution of Civil Engineers: Geotechnical Engineering* **167(GE6)**: 547–564, 10.1680/geng.13.00153.

Wan MSP, Standing JR, Potts DM and Burland JB (2019) Pore water pressure and total horizontal stress response to EPBM tunnelling in London Clay. *Géotechnique* **69(5)**: 434–457, 10.1680/jgeot.17.P.309.

Weber DS (2009) In support of the fully-grouted method for piezometer installation. *Geotechnical News* **27(2)**: 33–34.*

Yeskoo RA, Murphy JW, Hubbard P, *et al.* (2023) Application of distributed fiber optic sensing for subsurface levee monitoring. *Proceedings of Geo-Congress 2023: Geotechnical Data Analysis and Computation, Los Angeles, CA*, pp. 367–376.

* Reference can be downloaded from www.geotechnicalnews.com/instrumentation_news.php

Useful websites

Geotechnical Instrumentation News (GIN) articles: https://cgs.ca/instrumentation_news.html

International Society of Soil Mechanics and Geotechnical Engineering (ISSMGE) Technical Committee 220 (TC220) covering field monitoring in geomechanics: https://www.issmge.org/committees/technical-committees/applications/field-monitoring

Papers presented as some other FMGM conferences can be found at https://www.field-monitoring.org/symposia

Proceedings of the Symposium on Field Measurements in Geomechanics (FMGM 2011): http://www.fmgm2011.org

It is recommended this chapter is read in conjunction with

- Chapter 94 *Principles of geotechnical monitoring*
- Chapter 100 *Observational method*
- Section 8 *Construction processes*

All chapters in this book rely on the guidance in Sections 1 *Context* and 2 *Fundamental principles*. A sound knowledge of ground investigation is required for all geotechnical works, as set out in Section 4 *Site investigation*.

doi: 10.1680/icemge.66830.1543

Chapter 96
Technical supervision of site works

Sarah Glover Arup, London, UK
Chris Brewis Arup, London, UK

This chapter explains the need for geotechnical site supervision and the general tasks expected of an engineer responsible for supervising site work. It explains how an engineer's site role can vary dependent on the contract form and procurement route and gives guidance to help an engineer prepare for and carry out a technical site role. Particular attention is paid to the documentation process and typical engineer's responsibilities – in general, to health and safety and for the main types of geotechnical site supervision.

CONTENTS

Box 96.1 Note on terminology

The titles in this section are used in their general sense, not as contractual terms. Particularly, the term 'resident engineer' (RE) is used to describe the general role of an engineer responsible for independent overviewing of site works.

96.1. Introduction

There are many different end stages within civil engineering projects, including asset management, maintenance, renewal and so on. However, a large proportion of civil engineering projects culminate in traditional construction site works. For this reason, it is likely that every civil engineer will spend at least some part of their career involved in site work, typically called a resident engineer (RE) role. Perhaps more than any other engineering discipline, the role of site supervision in geotechnical engineering is essential to understanding the pre-design ground conditions and later verifying construction quality prior to the works being covered.

In the early stages of a project, investigation of the ground and groundwater conditions is required to develop the geological model and parameters for use in design. This work will normally be carried out by a specialist contractor; however, the engineer who also attends the site works will gain a first-hand knowledge of the ground conditions over and above what can be gained through study of logs and laboratory test data alone. The engineer will also gain a thorough appreciation of the site constraints and key ground risks, which can then be addressed in their design.

Site supervision during the construction phase is usually focused on monitoring the quality and progress of the works and ensuring that the contractor's working methods are in line with the design assumptions. Unlike most other forms of construction, for the majority of geotechnical works it is either very difficult or impossible to check the construction works once completed as they are usually buried in the ground. Problems with geotechnical construction are likely to come to light at a much later stage in the project – for example, in the form of cracking in a building with poorly constructed foundations, where the defects will only become apparent upon construction of most of the building (and therefore application of most of the load). Investigation of problems and the subsequent remedial works are usually expensive and time-consuming, resulting in claims, disputes and general dissatisfaction. While the presence of an engineer on site can never guarantee that all defects are eliminated, their role during construction works can provide a valuable check on workmanship. For the client, the costs involved in correcting defects are usually disproportionately high compared to the cost of providing site supervision. Refer to Chapter 93 *Quality assurance*, for further examples of potential construction problems and also discussion of the different forms of technical site supervision – for example, the client employing an RE or the contractor using a self-certification process. It is recognised that today's engineering site roles can vary greatly from a site engineer acting as the conduit between the design team and the site, to an engineer working directly for the client or even a contractor. This chapter specifically provides guidance to engineers responsible for independently reviewing contractors' work. However, much of the guidance applies equally to other engineering site roles. The chapter primarily refers to technical supervision and ensuring technical compliance. It is not intended to give a detailed description of site roles that primarily involve contract administration.

Recent developments in digital innovation, such as the use of pile log input on handheld tablets, have created quicker methods of quality assurance (QA) checks for the contractor's as-built records and engineer checks, to create lasting as-built records which are critical for development insurers and future reuse projects. In addition to the checking of physical works, the RE can bring value to the client by being on site and reviewing movement monitoring readings and highlighting potential trends in a traffic-light management system, for example, where green movement readings (green values are as predicted) move towards amber and red alert levels (which require contingency actions to be implemented) – refer to Chapter 100 *Observational method* for further discussion of movement management processes on site.

For the engineer, site work can be a very enjoyable and rewarding experience. An engineer that has spent time on site will generally have a better understanding of construction practicalities and the difficulties encountered by contractors. This knowledge of buildability will be invaluable to the engineer throughout the design process and should result in more workable designs and better construction details.

However, site supervision can also be a difficult experience if the RE is not suitably prepared and supported throughout the site works, as the RE may not have been part of the project's design stages (although, if possible, it is beneficial if the RE was a member of the design team before the construction phase). For all but the largest projects, the RE may be the only person from their organisation on site; in this case, regular communication with the office-based team becomes critical. It is important that the design team provide adequate training and a thorough briefing to the RE before the start of the site role and that they continue to do so throughout the site works. Additionally, it is important to recognise that supervision of site works requires a different skill set to that developed in the design office, with much more emphasis needed on communication, prompt decision-making and pragmatism.

This chapter aims to give guidance to engineers new to site work on

- how to prepare for a site role (Section 96.3)
- what to expect while on site, including common difficulties and guidance for dealing with them (Section 96.4)
- the engineer's responsibilities with respect to health and safety law (Section 96.5)
- various examples of geotechnical site work (Sections 96.6–96.8).

It should also act as an aide-memoire to the more experienced engineer who is preparing to send a junior engineer to site.

96.2. Reasons for supervision of geotechnical works

96.2.1 General

Supervision of site works is relatively common practice in geotechnical engineering. It is usually required to ensure that the planned works are carried out in accordance with the contract and as an independent check on the quality of workmanship.

Site supervision is particularly important in geotechnics, as in most cases the works are covered and are neither visible nor accessible for checking post-construction. For example, where the design of bored piles relies on a dry pile shaft with a base free from debris, it is important to observe the works and to visually check the pile bore prior to concreting (**Figure 96.1**). If the pile does not perform as designed, it will be very difficult to check these details retrospectively. Supervision therefore becomes a key quality control (QC) and risk management tool that can be used to help eliminate the need for remedial works and their associated disproportionately high costs (refer to Chapter 7 *Geotechnical risks and their context for the whole project*).

Figure 96.1 Inspecting a pile bore before concreting (© Arup/Daniel Clements; all rights reserved)

The provision of site supervision is linked to the foundation design responsibility – for example, where the works are contractor-designed, often the contractor's self-certification will be adopted and the client may require some additional level of oversight from an independent RE. If the foundations are engineer-designed, it is usual for an RE to be commissioned for the purpose of checking the work is carried out in accordance with the engineer's design. It should be noted that the client, and particularly potentially affected third parties, may still have some oversight role in both of these scenarios too. The site supervision of ground investigations (GI) is slightly different and can be seen as a 'special' or additional case, in that the GI information will be used by the designer and thus technical supervision will offer benefits for this. For example, a site RE role during the GI will offer additional confidence to the design team in the reliability of the factually reported data; observations on the reliability of the data should be captured in the geotechnical design reports.

The provision of site supervision is often a contractual requirement of the engineer. Even if it is not, the engineer generally has a duty to advise the client on the appropriate level of supervision for a particular project. This advice should include the level of expertise and skills of the supervisor and the proportion of time that should be spent on site. It should be based on professional judgement and knowledge of the project and should take account of

- the complexity of the project
- the degree to which the design is dependent on good workmanship

- the extent of work that will be covered up post-construction (and subsequently not visible or accessible for checking) and the time frame for it
- the degree of completeness of the design
- the method of procurement and the responsibility of the contractor
- the level of experience of the contractor
- the engineer's responsibilities as defined in their terms of appointment.

These recommendations should always be provided in writing to the client, and the final agreed brief clearly documented.

96.2.2 Core aims of supervision

Regardless of the particular project, most supervisory roles will endeavour to meet one or more of the following core aims.

96.2.2.1 Provide an independent check on quality

The contractor will generally have a duty to ensure their workmanship is of the required quality. To demonstrate this, the contractor should put in place their own QA procedures, which will normally involve checking by members of their site team, as detailed in their inspection and test plan (ITP), linked to their works method statement. Refer to Chapter 93 *Quality assurance* for further background on quality management systems for both designs and construction phases.

However, it can be reassuring for the client to have an independent check on the quality of the contractor's work, particularly if the work is unusual or if the contractor is relatively inexperienced. The presence of the engineer should never relieve the contractor of their contractual obligations, but a second set of eyes can assist with identifying potential quality issues early and agreeing methods for their resolution. Additionally, the engineer is less likely to be influenced by other external factors, such as programme deadlines, and can therefore ensure that these factors do not adversely influence the quality of the work.

For the particular case of site investigation work, the engineer will normally need to rely on the results of the investigation in later design work. Observation of the work can identify if the contractor's drillers/site team are recording the specified level of detail correctly and also identify any problems that may affect the soil parameters (for example, a sticking catch on a standard penetration test hammer invalidating the use of correlations with standard penetration test (SPT) 'N' values) and will give confidence that the related design is based on reliable data. Doubts as to the quality of the data may necessitate an over-conservative approach, which will have a knock-on effect on the efficiency of the design and therefore project costs. If the engineer is unaware of problems with the data, then this could equally result in an unconservative design and, in an extreme case, potential failures.

96.2.2.2 Ensure compliance with technical requirements

Technical requirements for site works will be communicated to the contractor through the contract documents and will normally comprise a specification and drawings. Through preparation of these documents the engineer should have identified any key design assumptions that could be affected by the contractor's working method. For example, the soil properties required from the fill material in an earthworks operation and the associated method and type of compaction plant.

The contractor will be responsible for ensuring that their work meets the specified technical requirements and should do so through agreed method statements, rigorous checking and record keeping. Again, the presence of an independent RE will provide a check that the technical requirements are being met and that the contractor is working in accordance with their method statements. Problems can, in this way, be identified early and remediated, and repeat occurrences can be eliminated rather than being left to manifest later in the project, when their cause will be more difficult to determine and their resolution could have a major impact on the project.

The importance of the contractor producing good as-built records has drawn much attention in recent years due to digital innovations in record keeping, use of as-built information in BIM modelling, monitoring carbon against targets during construction and longer-term drives such as promoting the reuse of previous buildings' foundations. For example, if the building has a change of use or additional storeys are considered to increase building occupancy, good as-built records are critical for reuse redesigns of foundations and justifying designs to insurers. The RE can comment upon the as-built records during the works rather than towards the end of construction phases by checking against their independent measurements and observations. Refer to the further reading section at the end of the chapter for suggested guidance documents relating to reuse of foundations, lean construction, BIM and carbon management.

96.2.2.3 Ensure design-related issues are raised and resolved promptly

Where observations give cause for concern, the RE will be responsible for ensuring that problems are raised promptly by the contractor through the predefined formal channels (usually non-conformance reports (NCRs)). The RE can then liaise with the design team to ensure that responses to the non-conformances and suggested resolutions are tackled and reported back to the construction team in a timely manner. The RE should use their knowledge of the design to confirm immediate redesign changes, such as suitable deepening of pile toe levels in response to extended bore opening times, and use their judgement when larger issues/impacts on the design intent will require more detailed redesigns in the design office. It is important for issues and changes to be well documented in NCRs. The RE is well placed to comment on these documents if the non-conforming works have been observed first-hand or if they have a good understanding from observations of similar site works.

This aspect of the role becomes particularly important towards the end of the site works as it is ideally required that any identified problems are resolved and remediated before the contractor leaves site.

96.2.2.4 Provide a communication link between design and construction teams

A defining factor of many successful civil engineering projects is clear and timely communication between all of the project parties, and particularly the design and construction teams. Part of the RE's role will be to assist the contractor in understanding the contract documents and design intentions and providing clarification where there are questions or conflicts within the documents.

The RE should ideally have a good understanding of the design and any site constraints such as the site history, buried structures and ground and groundwater conditions (typically documented in the desk study, interpretation of GI and other design reports), and bring value by sharing or highlighting key project and site aspects. The RE may also have good relationships/knowledge of other parties which influence the contractor's work and can offer assistance/insight in terms of, for example, movement monitoring requirements that are often carried out by monitoring contractors or specific neighbouring building requirements to ensure potential impacts are limited.

Through the review of the contractor's method statements, ITPs, concrete mix designs and drawings against the design intent, specification and engineer's drawings, the RE can early in the process help to identify issues and any conflicts to allow optimal time for their resolution (ideally before the works have started). The RE can also provide regular updates to the design team, which can help them to plan and prioritise their activities to meet changing site requirements. Good record keeping by the RE is essential – the design team may request information about the workmanship of an item constructed several weeks or months ago.

96.2.2.5 Contract administration

For some projects the RE will also be responsible for administration of the construction contract, but this will depend on the form of contract and procurement route used for the particular project. For example, civil engineering projects which use the ICE conditions of contract can be administered by the engineer, whereas building projects using the JCT contract forms will normally be administered by an architect or project manager. For many site investigation projects the engineer will be called upon to provide both a technical and contract administration role on site.

The engineer's duties relating to contract administration should be established and agreed early in the project.

96.2.3 Additional benefits of site supervision

In addition to carrying out a key project role, the engineer's site experience should improve their skills as a designer.

Through observation of the construction process the engineer will be able to understand elements of their design work well and which ones cause practical problems for the contractor. This should help to develop a pragmatic approach and result in the production of more effective and buildable designs in the future. For example, by witnessing the reinforcement cages of bored piles being spliced together on site, the engineer will appreciate how the design should consider the density and thickness of the bars to facilitate buildability, and also offer appreciation of health and safety (H&S) and CDM aspects of the designs more broadly which are to be built by the contractors.

Observation of the number of interrelated activities on site should give the engineer an appreciation of the complexity of coordinating site work (see **Figure 96.2**), which should help with the design of future projects in early feasibility and concept stages. It will also help in discussions with contractors on future projects.

Through working with the contractor to identify and resolve problems, the engineer will gain valuable experience that should allow those same problems to be eliminated from future designs but also to resolve other problems more quickly and effectively in the future. The engineer may also be working directly with a larger range of different types of individuals in comparison to only office-based design roles (e.g. site supervisors, rig drivers, testing specialists, project managers, client programme and cost representatives) and this offers a great opportunity to build personal networks within the industry. The engineer may be exposed to industry innovations and best practice in site work techniques and should collate case study data and share knowledge and improvements with colleagues, and seek opportunities to assist and write technical papers.

Figure 96.2 Deep basement construction involves a number of interrelated activities (© Arup; all rights reserved)

Box 96.2 Key points: site supervision

- Site supervision is a key QC and risk management tool that can be used to help eliminate the need for remedial works.
- The core aims of site supervision are to
  - provide an independent check on quality and use digital tools where possible
  - ensure compliance with technical requirements and assist in the production of good as-built records for foundation reuse
  - ensure design-related issues are raised and resolved promptly
  - provide a communication link between design and construction teams and contract administration.
- An engineer's site experience should improve their skills as a designer, give opportunities to build personal networks within the industry and assist in developing improvements and innovation which can be shared within new technical papers.

96.3. Preparing for a site role

96.3.1 Importance of preparation

Every site and corresponding site role is different and therefore requires a tailored approach based on a clear understanding of the project parties, the procurement route and contract conditions and the particular technical requirements. It is therefore essential that the RE prepares thoroughly for each and every site role so that they can, from the outset, operate effectively with all project parties to fulfil their supervisory obligations.

The first few weeks of site work can often be the most challenging as several project parties, who have not usually worked together before, start to establish working processes and relationships. The RE's time will be filled with carrying out initial site inspections and answering technical queries from the construction team. It is important that the RE has a thorough understanding of the technical requirements of the work so that they can set a clear expectation for the contractor. Once established, this should allow the contractor to settle quickly into a routine and will help to ensure the smooth running of the project.

In addition to setting the technical expectation, it is also essential that the RE establishes clear lines of communication with the various project parties. In order to do this the RE needs to understand the roles and responsibilities of the individuals working on the project and ensure that they in return are clear on those of the RE.

The amount of work required when preparing for a site role should not be underestimated, particularly where the RE is new to the project or to site work in general. It is the responsibility of the design team to ensure that the RE is suitably experienced in the type of fieldwork they are about to supervise or is trained and briefed before starting a site role and always remains well supported throughout. The benefits of this preparation will be felt by all involved in the project.

The RE should use the other chapters within this manual to gain a quick understanding of different aspects of site works (e.g. Chapters 47–49 detail GI, Chapters 52–91 detail different structural elements such as shallow foundations, retaining walls, piles and ground anchors, Chapter 93 discusses QA, Chapter 94 discusses monitoring and Chapter 99 discusses material testing). The RE will also be exposed to testing on site and should be aware of correct test procedures as detailed in British Standards. Another quick method of understanding similar site works is to read previously written geotechnical close-out reports or formal papers which should detail key aspects of thc construction works, issues encountered and remedial actions taken (see Chapter 101 *Close-out reports* for further details). This preparation work is in addition to the RE gaining familiarity with the project specifics, such as design reports, specifications and drawings (see Section 96.3.4).

96.3.2 Defining roles and responsibilities

The roles and responsibilities of the RE should be agreed with the client well in advance of the start of site works. These should be recorded in a clear written statement of the brief that is incorporated into the design team's contract.

For some construction contract forms the engineer is a named party with specific contractual duties clearly defined in the contract. Other contract forms may refer instead to a supervising officer or contract administrator and may not define the role of the engineer specifically. Under these types of arrangement, the role of the contract administrator may be carried out by another project party (e.g. the architect, project manager or construction manager) and the engineer will usually act as a technical adviser to the contract administrator. In this instance the engineer's role will not be clearly defined in the main contract and it will be necessary for the design team to agree the extent of the technical supervision role (refer to Chapter 78 *Procurement and specification*).

The RE's brief will need to be tailored to suit the type of works being constructed, the procurement route and contract form, but should typically describe

- the anticipated duration of the site work and the level of attendance (full-time, part-time, occasional visits)
- the elements of work that the RE will be responsible for witnessing and those for which the RE will not be responsible – this should also include the role associated with reviewing movement monitoring readings
- the checks that are to be made and the records that must be kept, including any procedures for web-based document management systems being used (and formats of the engineer's posted comments and sign-off status)
- the format, frequency and distribution for progress reports
- the RE's duties in relation to ensuring the contractor is carrying out their role to a satisfactory standard

- the extent of the RE's authority, including details of any limitations under the contract
- the project change management system
- the procedure for issuing instructions
- the procedure for responding to requests for information and NCRs
- the interfaces and lines of communication with other members of the team.

Preparation of this brief will be a useful prompt in defining the boundaries between the RE's responsibilities and those of other project team members. Through the brief, it is likely that areas of conflict or split responsibility will be identified, and early preparation should enable the resolution of areas of conflict prior to start of work on site.

96.3.3 Understanding the construction team

For the RE to operate effectively in the site environment, it is important that a good knowledge of the various project parties and their organisation is gained. To develop this understanding, it can be useful to draw an organogram of the project team to clearly indicate the lines of communication between the various project parties and summarising their main responsibilities (**Figure 96.3**). This could prove a very simple task where the construction work is procured through a traditional route, with the client separately appointing an architect, engineer and main contractor; or, it could be a more complicated arrangement where there are many more contracts and parties involved in the construction works.

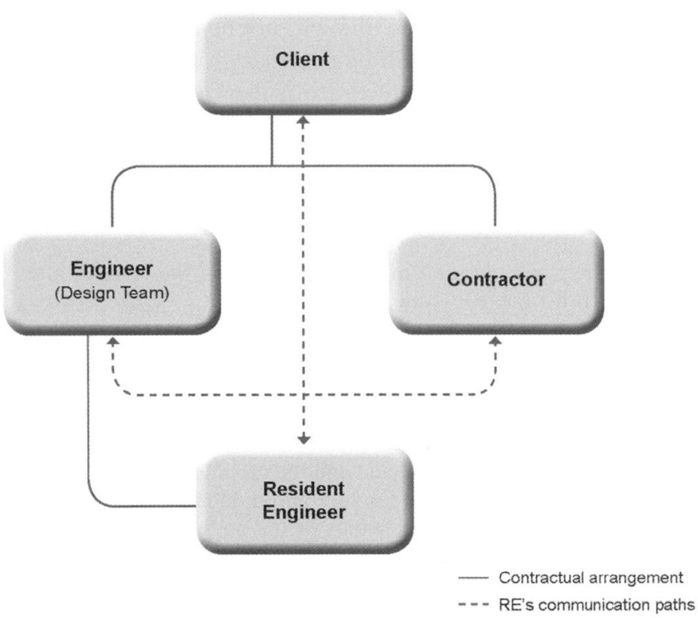

Figure 96.3 Typical lines of communication for a resident engineer under a traditional project arrangement (© Arup; all rights reserved)

Having established the general project organisation, the RE should be able to identify named individuals with the authority to make project decisions, issue instructions and tackle problems. The RE should be able to identify who to communicate with on a day-to-day basis and can then confirm this with the project team.

A major part of the RE's role will normally be to ensure that the contractor fulfils their obligations under the contract. These should be set out clearly in the contract documents for the construction works, which the RE should review and check for compatibility with their own duties.

If, following this assessment of the various site parties' responsibilities, there remain any uncertainties or conflicts, the RE should proactively raise these issues with the project team and ensure that they are resolved prior to start of the site role.

96.3.4 Preparatory work of the project specifics

The RE should be thoroughly briefed on the organisational and technical aspects of the site role by a senior member of the design team (Section 96.3.2).

Additionally, the design team and RE should gather together a set of base information for the project to refer to while on site. This should include the items in **Checklist 96.1**.

Generally, it is advantageous for the RE to have had some experience in carrying out the design before going to site. However, this is not always possible and therefore the RE should have a good knowledge of the design prior to arriving on site. This will allow the RE to understand the criticality of certain aspects of the design and also reduce their dependence on the design team to resolve technical queries. A junior engineer

Checklist 96.1 Preparatory documents

- ✓ The written statement of brief including the schedule of RE duties
- ✓ An overview of the project
- ✓ The design basis and, if appropriate, calculations
- ✓ Identification of critical elements of design that require particular checking procedures
- ✓ Specifications for the elements of the work for which the RE is responsible
- ✓ Construction drawings
- ✓ Proforma check sheets for each element of the work that requires inspection
- ✓ Method statements, inspection and test plans and fabrication drawings prepared by the contractor and agreed with the project team
- ✓ The project programme
- ✓ Health and safety documentation and records of briefings (refer to Section 96.5)
- ✓ Copies of relevant codes of practice, handbooks and guidance notes
- ✓ Contact details for key project individuals

should seek out opportunities to visit construction sites with colleagues prior to starting the RE role and gain familiarity of the site environment and types of construction works.

It is advisable to involve the RE in the initial review and checking of the contractor's method statements, ITPs, mix designs and drawings and any pre-start contractor meetings. This will allow the RE to become familiarised with the site work and to identify the major issues in advance of the start of the works. It will also allow the RE to be introduced to the key site parties so they are already a familiar face before arriving on site. If a web-based document management system is set up at this time, the RE should gain experience in posting review comments and assigning a document status.

The RE should insist that the contractor prepares the information required by the contract documents well in advance of the start of site works. These will generally include the items in **Checklist 96.2**, as well as requirements for movement, noise, vibration and dust monitoring.

The RE should ensure that information received is complete, and that their comments are issued promptly and addressed by the contractor. Tracking of comments until closure is a key part of the RE role.

Checklist 96.2 Initial information prepared by contractor

- ✓ Programme and sequence of work
- ✓ Site organisation chart
- ✓ Method statements
- ✓ Inspection and testing plan
- ✓ Typical proforma sheets for inspection requests, contractor's submissions, requests for information, non-conformance reports

Box 96.3 Key points: preparatory work

- Thorough preparation is essential to a successful site role.
- A clear brief and schedule of duties will help the RE understand their site role.
- The RE should undertake appropriate training prior to starting on site.
- The RE needs to understand the roles and responsibilities of others working on the project and know who to communicate with on a day-to-day basis.
- The RE should ideally be involved in the project for some time prior to the start of a site role and have experience of the type of works they will be responsible for.

96.4. Managing the site works

96.4.1 Setting up on site

The RE should aim to arrive on site slightly in advance of the start of site works to allow time to get organised before the works begin in earnest. This time should be occupied with setting up on site and making contact with the key site individuals. Importantly, the RE should check that all H&S requirements are in place and, if this is not the case, these should be addressed as a priority.

The contractor will normally be required to provide office space for the RE. This may be separate from or (as is becoming more common) contained within the contractor's own site office. The contract documents should contain a list of the facilities the contractor must provide, but these will normally include

- electric power (ideally internet access as well)
- lighting
- air conditioner/heater
- a desk with lockable drawers
- a chair.

The contractor is also legally obliged to provide adequate welfare provisions. If reasonable welfare facilities are not provided, then the contractor should rectify this immediately.

The RE will normally be responsible for arranging provision of other facilities that are needed, such as a laptop, printer and phone, possibly internet access and personal protective equipment. The RE should also ensure that the relevant project documents and reference material are taken to site so they are instantly accessible (see Section 96.3.4 and summary in **Figure 96.4**).

96.4.2 Documentation and record keeping

The importance of clearly documented, timely and well-organised site records must not be underestimated. Comprehensive and accurate site records can contribute significantly to the smooth running of a project and can be used to demonstrate that the RE is carrying out their role effectively.

Good record keeping can also become critically important in the event of claims, particularly when these are raised long after the project has been completed. For example, if a pile integrity test is undertaken many days after a pile has been completed and shows the pile to have inclusions, then detailed site records may be essential in understanding the cause of the problem and informing the subsequent investigation and remedial works design.

The volume of records that will be generated for even a short site role can be significant. For this reason, it is imperative that the RE is well organised and establishes a workable system that allows information to be stored neatly and in such a way that it can be found again quickly. Some projects may have clearly defined processes for doing this, but if not the RE should choose a method that suits the particular project and their style of working.

The RE will normally be responsible for both keeping and maintaining their own records and for collating and commenting on those produced by the contractor. **Figure 96.4** provides a schedule of the typical records that should be collated by the RE before, during and at the end of the site works, in addition to the parties that would normally prepare the records.

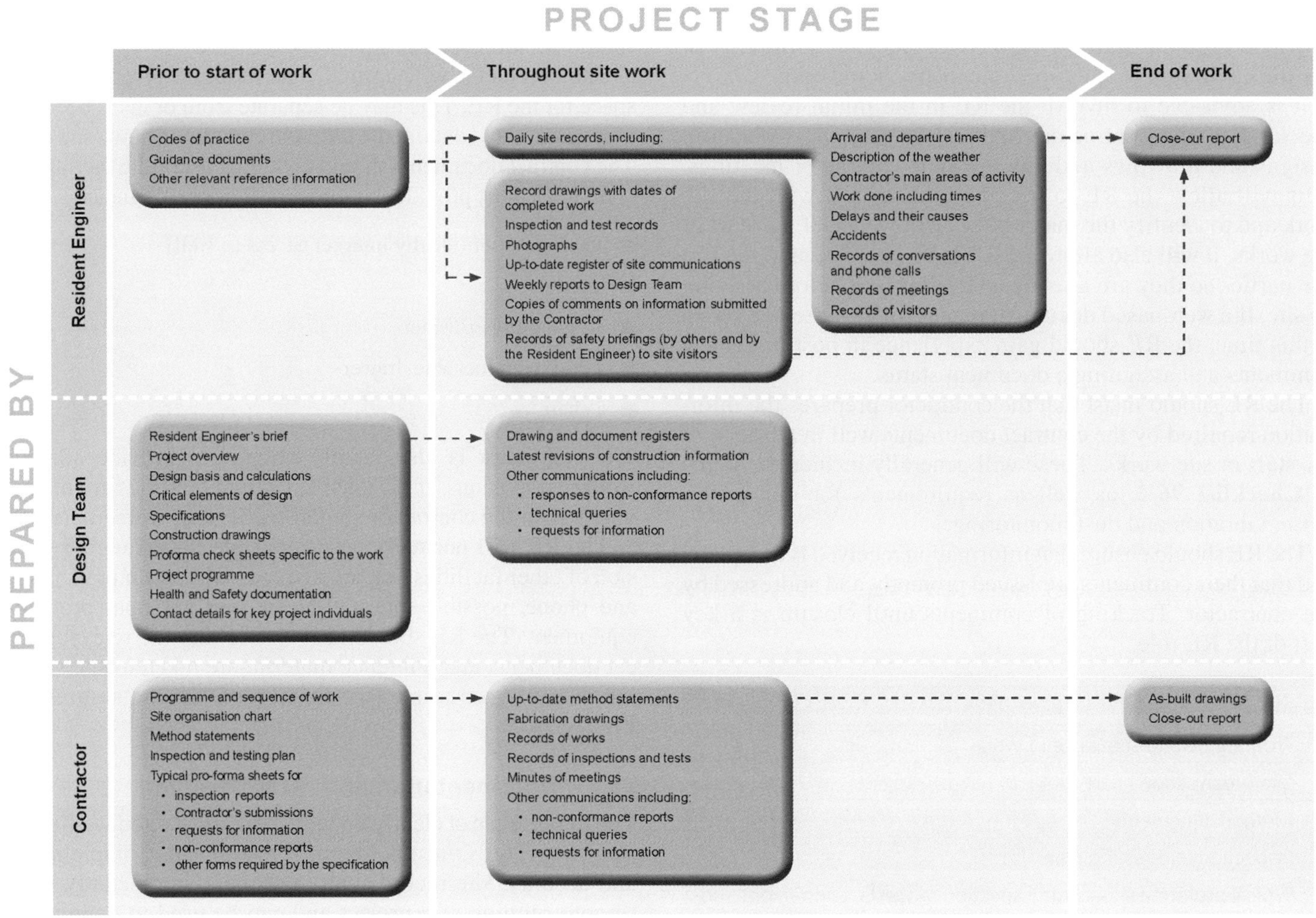

Figure 96.4 Schedule of site work documentation to be collated by the resident engineer (© Arup; all rights reserved)

Possibly the RE's most important record is their diary. This should be written daily and should include the information listed in **Figure 96.4**.

Recent digital innovations in recording site measurements and observations, such as use of tablets with predefined input boxes linked to databases, can speed up the paperwork tasks. These databases can also be used to auto-generate weekly summary reports and for tracking graphs of key aspects of the construction works. These graphs can be used during the works to visually see trends in the collated data and give an indication of workmanship performance over time, and to highlight particular areas of concern for more focused supervision. Having records digitally rather than handwritten also assists in linking construction activities to movement monitoring readings, in close-out reporting and developing case study data for potential use in writing technical papers.

Digital records produced by contractors in issued databases (rather than PDFs) are also being introduced as requirements for BIM modelling of as-built information, and they bring benefits for potential future reuse of foundations.

Project changes can often be the cause of confusion and problems on site, particularly if these need to be justified retrospectively. It is essential the RE clarifies the project change procedure prior to starting the works and follows it precisely should a change be needed. On some projects the change procedure will be clearly defined; however, on other projects it can be useful for the RE to prompt the site and design teams to clarify the change procedure prior to the site works commencing. For example, if during a site investigation it is necessary to introduce an additional borehole to provide clarity on conflicting information, the RE should follow the project change procedure and clearly state the reason for the addition so that payment can be easily certified later in the work when the contractor submits their application for payment.

96.4.3 Inspection and checking

Probably the most important aspect of the RE's role is the witnessing and selective checking of the contractor's work. The RE should ensure that checks are carried out rigorously and that each check is recorded by the contractor and separately by the RE using the pre-agreed proforma check sheets. It is important not to get behind on this paperwork as it can be difficult or impossible to fill out check sheets even a short time after the check has been carried out. Good-quality as-built records are a key sustainability driver and allow for future assessments for schemes relying on the reuse of foundations.

The contractor should normally be present at the time the RE carries out their independent checks. If the RE is unhappy with any element of the check, they should raise it with the contractor immediately to give the contractor the opportunity to rectify the problem before the work is completed. In the unlikely instance that the contractor refuses to rectify the defect, then the RE should make the contractor aware of the consequences of not rectifying the defect and then follow up this conversation in writing as soon as possible after the event. The RE should focus on developing a clear communication style and good working relationships with the contractor to ensure joint inspections run smoothly and respect is fostered for all individuals involved.

The contractor will normally provide their own records of the works, copies of which should be sent to the RE in addition to the contract administrator. The RE should check the contractor's records for compatibility with their own and raise any concerns or incompatibilities to the contractor and the project team. This information is best transferred digitally to avoid duplication of data input from PDF records for the comparison check. In advance of start of site works, the RE should obtain proforma check sheets from the contractor within their ITP to check that these cover the elements required by the contract documents.

Precise details of the checks to be made, their frequency and the records to be taken will depend on the nature of the work being inspected and the RE's brief.

96.4.4 Communication

This will depend upon the form of procurement and the construction contract. However, the following sections contain general advice that can be modified to suit the particular project requirements.

96.4.4.1 Communicating with the contractor

The RE's most frequent contact will be with the contractor. Communication with the contractor will range from verbal discussions to formal written site instructions. At first it may be difficult to judge the most appropriate means of communication for a given situation.

The RE's first aim should be to establish a good working relationship with the contractor's staff. In doing this the RE must recognise that the contractor's staff will be working to time and cost deadlines; that their jobs may depend on the project's progress and profitability, and that in these circumstances the contractor may seek cost-driven, rather than quality-driven, solutions. However, while the RE should acknowledge that the contractor is in business to make a profit, they should also be clear that the contractor is obliged to work within the contract terms.

In all site dealings the RE should aim to adopt a fair, straightforward and consistent approach. Where the RE observes something going wrong, they should raise the issue as early as possible with the appropriate individual (typically the site supervisor and project managers) and should not wait until it is completed and they are formally invited to inspect it. The RE should try not to reverse earlier decisions or to find further faults in work that they have already accepted. For more junior REs developing a straightforward approach, note that the contractor is looking for clear statements on their works from the RE and they will be responsive if the RE says that they need five minutes to double-check with the design team on more complex issues. However, once said, the RE will need to return to the contractor and not leave issues unresolved, as they will be expecting a response. A consistent approach is essential.

In this way the RE will gain increasing respect and authority on site and typically this will result in the RE being listened to and being well informed of the works. Developing a good reputation with the contractor will widen the RE's personal network within the industry and be valuable for future projects where early contractor advice/engagement could be gained from the contacts established during their time on site.

It is usual for the RE to have daily discussions with the contractor, and these discussions should be summarised in the REs' site diary. If the discussions are informal, such as comments during site inspections or suggestions for improved working, then it should not be necessary to record and circulate them more formally provided the contractor acts on them. The RE should give comments only to pre-identified individuals on the contractor's team and not directly to the contractor's site operatives, or to their subcontractors.

Where discussions are more formal, the RE should send a written record to the contractor's team and distribute to other affected site parties. These communications should use a pre-agreed form such as a site memo. Examples of when a written record may be required are when the contractor has not responded to the RE's verbal comments during inspections, when safety matters are ignored or when requesting written information such as method statements.

Site instructions should be reserved for formal contractual matters such as: issuing sketches and information for construction; instructing the contractor to rectify breaches of the specification; dealing with safety issues (when the initial correspondence has not received a response); or for requiring the contractor to issue remedial work proposals. Remember that site instructions usually need confirming by the party acting as contract administrator. To aid the understanding of all parties involved, the RE should clearly state the reason for the issue of an instruction as this information can be of use later. If the RE is uncertain whether a site instruction is warranted, they should discuss with their design team.

It is important to recognise that there is no such thing as a 'verbal instruction'. Any instruction given by the RE must be confirmed in writing so that it is on record.

The contractor will send the RE many communications, including method statements, requests for inspection, requests for information, site memos and NCRs. The RE should insist that all requests are in writing and that they follow the agreed site procedures. The RE will normally be contractually obliged to respond to each communication within a defined time period and should ensure that this period is known and adhered to. All responses should be in writing. In instances where a large number of requests are received, the RE may find it useful to set up a register to keep a check on which have been answered. The RE will need to liaise closely with other members of the design team to seek input where required and ensure a coordinated response.

96.4.4.2 Communicating with the design team

For the benefit of all, the RE will need to keep in close contact with their design team. The RE should speak regularly to a designated member of the design team to give details of site progress, to pass on queries and to receive information. In the early days of a project it is a recommended to make contact at least daily, even if there is little to report. The RE should make a brief note in their diary to record these communications.

It is a good idea for members of the design team to visit the site, especially at the beginning of the works. This will provide support to the RE and will give the design team an appreciation of the site conditions.

The RE should also provide written summaries of site events. The format and frequency of reporting should be agreed in advance with the design team. For most projects a weekly report summarising the following issues should be sufficient

- project progress against programmed progress
- major problems and delays
- technical issues encountered and their resolution
- instructions issued
- communications received and responses outstanding
- test results received
- non-conformances identified.

The RE should ensure that the design team is clear on the information that is required on site and that a designated person is responsible for ensuring that the RE is supplied with the latest versions of specifications and drawings in a timely manner. Additionally, the design team will be responsible for keeping the RE apprised of project changes that could affect their work. Increasingly, web-based systems are being used to manage this documentation process and the RE must have the means to access such systems efficiently while on site and be provided notification of new uploads. However, due to web-based systems potentially having a finite period for access/subscriptions, it is recommended that the RE keeps copies of their own independent records as it may be problematic to resolve issues if access is lost to the web-based system a number of years after the works and still within defect correction/liability periods.

The RE should not be afraid to ask the design team for advice, especially in the early days on site. As the RE becomes more familiar with site work they will become more confident and independent. However, they should recognise that in unfamiliar or unusual situations they should seek clarification from the design team. Maintaining a communication line to a designated person within the design team is critical as issues may occur late at night or at weekends, particularly for responses to more complex issues.

96.4.4.3 Communicating with other parties

Dependent on their particular brief, the RE may be required to interface with other site parties such as contract administrators, project managers, the client, quantity surveyors, clerks of works, H&S teams, members of the public and so on. The level and form of communication will vary from project to project and should be clearly defined prior to start of site work.

While the RE should always be reasonably helpful and informative, it is important to ensure that the level of communication provided is appropriate to the role and authority of the party. Particularly, the RE should only accept instructions from parties that have authority to issue them. Also, interactions with the public may be sensitive on certain projects, and a briefing on how to act should be carried out prior to starting on site.

Much of the communication with other site parties will be verbal but should always be recorded in site diaries or by more formal means if deemed necessary.

96.4.5 Site meetings

The contract administrator should arrange regular site meetings with the contractor to review progress and discuss problems. These should be attended by the RE (unless critical works are ongoing and checks are required by the RE) and representatives of the client and design team should also be invited.

A typical agenda for the meeting should as a minimum include

- progress against programme and issues delaying progress
- design-related issues
- status of requests for information
- status of NCRs
- problems encountered and a means for resolution
- H&S issues

- proposed works before the next meeting
- any other business.

Minutes should be prepared for all meetings, and these should be distributed to the attendees as soon as possible and certainly before the time of the next meeting. The RE should thoroughly review all minutes to ensure that they are a fair representation of the meeting. Any issues with the minutes should be raised at the start of the next meeting, or earlier if the need dictates. The RE should ensure that action items identified in the minutes are dealt with promptly.

In advance of the meeting the RE should make a note of points they wish to raise, such as quality of work (use an evidence-based approach – e.g. use photos of measurements or videos), matters of safety or information required from the contractor. The RE should also take some time before the meeting to brief the design team representative on current topics.

96.4.6 Dealing with problems

In spite of the RE's best efforts, there will be times when problems arise and the RE will need to deal with these as proactively as possible. This is one of the key benefits of the RE being based on site and observing the works.

To get an unwilling contractor to improve quality or correct mistakes is one of the hardest jobs on site. It is therefore wise to try to avoid head-on confrontation. The RE should initially try to avoid defective work by making the required standard clear the first time that any particular element is constructed or each time a new team arrives on site.

The RE's aim should be to establish a good working relationship with the contractor. If things are wrong or done poorly, the RE should explain the technical reasons behind the requirements. The RE should raise issues as soon as possible to give the contractor the opportunity to correct problems quickly. Communication should be straightforward and the RE should use an evidence-based approach, such as using photos of measurements or videos to explain points further.

If the contractor does not act on initial verbal warnings, the RE must write a site memo to the contractor recording their verbal comments and requesting action is taken. This is often enough to resolve issues. It also provides a written record of the request and the date of it, for future reference. The RE should be careful not to tell the contractor how to correct the fault; it is better to ask for their proposals and for the RE to provide comments on their proposals (sometimes for complex issues the design team will need to provide input or carry out redesigns).

Sometimes the contractor will fail to act on the RE's request. If the RE is satisfied that they have given the contractor reasonable opportunity to correct the issue, then the RE should not be afraid to take further action. In the case of an unresponsive individual within the contractor's organisation, the RE could contact one of their superiors; in the case of the contractor's organisation generally being unresponsive, the RE should contact the contract administrator or project manager. In either case the RE should seek the support of their design team, record their concerns in writing and refer to previous discussions and written communication. The RE should set out the reasons for their concern and the consequences of not resolving the issue promptly (for example, repeating the faulty work).

Such drastic measures should not be taken for something trivial. The RE will need to judge the severity of the fault initially and may need to seek the advice of the design team. If the defective work has since been covered, the RE will need to be able to support their claim rigorously, particularly if it contradicts the contractor's own records. In such circumstances, clear record photographs can be very convincing evidence of whether the alleged fault ever existed and the degree of its severity.

With one exception, the RE should never issue a stop-work instruction without first consulting the design team and the project manager or contract administrator. It is a very serious step leading to a difficult contractual situation. The one exception would be when there is immediate danger to human life (e.g. an unsupported and unstable excavation), but even then it is vital that the design team and project manager are informed immediately. Refer to Chapter 93 *Quality assurance* for more detailed information on avoiding defective work.

The RE will gain awareness of potential problems by their design team giving them good initial briefings and then by experiencing issues during their supervision which they have identified or by the contractor informing them. The RE will quickly build a knowledge base of likely issues, their seriousness and how they were previously resolved.

Box 96.4 Key points: managing the site works

- Preparing clearly documented, timely and well-organised site records is an essential part of the RE's duties.
- Numerous records will need to be produced by the RE and other site parties (refer to **Figure 96.4**).
- The RE needs to be clear on the project change procedure and follow it precisely if a change is needed.
- The RE must ensure that inspections and checks are carried out thoroughly and recorded using the pre-agreed forms.
- The RE should communicate in a clear and timely manner with the site teams and should vary the form of communication for each particular situation.
- The RE should not be afraid to ask their team for advice, especially in the early days on site.
- Regular site meetings provide an essential forum for discussing issues relating to the site works.
- When problems arise, the RE will need to deal with them proactively, raising issues as soon as possible to give the contractor the opportunity to correct problems quickly.

96.5. Health and safety responsibilities

96.5.1 The law

H&S practice in the UK is primarily governed by the Health & Safety at Work Act 1974 (HSAWA). This Act outlines the general duties imposed upon employers and other professionals to safeguard the health, safety and welfare at work of their employees and others who might be affected by their activities.

Implementation of the Act is through Regulations. These are in turn supplemented by Approved Codes of Practices, which contain detailed provisions and give practical guidance. Numerous regulations exist governing all aspects of construction work, with those that are most commonly used listed below (note specific regulations may be required for your site, such as confined space training and procedures or nuclear sites)

- The Construction (Design & Management) Regulations 2015 ('The CDM Regulations')
- Reporting of Injuries Diseases and Dangerous Occurrences Regulations (RIDDOR)
- Control of Substances Hazardous to Health Regulations (COSHH)
- Lifting Operations and Lifting Equipment Regulations (LOLER)
- Provision and Use of Work Equipment Regulations (PUWER)

Before embarking on a site role, the RE should ensure they are familiar with the duties imposed by the Act and also the principal sets of Regulations under the Act. Refer to Chapter 8 *Health and safety in geotechnical engineering* for more detailed information on H&S law and its application to geotechnical engineering.

96.5.2 Personal safety

While the contractor is legally responsible for the safe execution of the works and supervision of the workforce, the RE also has a duty to take reasonable care of their own safety. The RE should be aware that other construction works may be ongoing alongside the works they are supervising (e.g. demolition and excavation) and communication with the contractor is critical. H&S should form an essential part of the engineer's preparation for site works and will probably comprise a mix of formal training, briefings and reading of literature, as described in **Checklist 96.3**.

Checklist 96.3 Preparatory health and safety requirements

- ✓ Appropriate general training in site safety including a general briefing from their own employer
- ✓ Health and safety certification required by the contractor (for example a Construction Skills Certification Scheme card)
- ✓ Briefing on the risks associated with the particular site and construction works by both their own team and the contractor
- ✓ Access to and understanding of the necessary health and safety documentation including the construction phase plan, designer's risk assessments and contractor's risk assessments

The RE should keep up to date with changes to the site H&S procedures throughout the site works and should seek to contribute proactively to the management of H&S on site.

During the site works the RE should ensure they wear appropriate personal protective equipment (PPE) at all times and do not take unnecessary risks. If the RE is asked to undertake work they do not consider safe – for example to enter an unsupported excavation – then they should insist the contractor rectifies the problem before carrying out the work. More typical good H&S practices should be discussed with the RE's design team by undertaking a site-specific risk assessment (SSRA) (e.g. avoid standing under crane lifts, ensure understanding of piling rig blind spots).

In the event that the RE is injured, they should report the injury to the site team and to their design team immediately. As a minimum, the injury should be reported in the site accident book, with further reporting to the Health and Safety Executive (HSE) if warranted by the nature of the injury.

96.5.3 Safety of others

Under the Health and Safety at Work Act the RE has a duty to 'take reasonable care for the health and safety of themselves and of other persons who may be affected by their acts or omissions at work'. The RE should raise any observations relating to unsafe working practice promptly with the contractor. If these are not addressed in good time, the RE should refer the observations to a higher level within the contractor's and/or their own organisation as appropriate.

During their time on site the RE will receive visitors and they will need to take responsibility for ensuring their H&S while on site. The RE should carry out an SSRA for each visitor and provide a briefing before the site visit. The contractor should always be notified before the RE receives visitors on site and may require visitors to also attend a safety briefing given by the contractor.

Box 96.5 Key points: health and safety responsibility

- Before embarking on a site role, the RE should ensure they are familiar with the duties imposed by the Health & Safety at Work Act 1974 and also the principal sets of Regulations under the Act (e.g. CDM).
- The RE has a duty to take reasonable care of their own safety. They must ensure they
  - have received appropriate training
  - have received SSRA briefings
  - are familiar with the key site safety documents, including the construction phase plan and risk assessments.
- The RE also has a duty to take reasonable care for the H&S of others.

96.6. Supervision of site investigation works

Site investigation works aim to provide information on the ground and groundwater conditions and sometimes conditions of existing structures. The design of, and methods for, site investigation are discussed in detail in Section 4 *Site investigation*.

Each site investigation will be bespoke, designed to target the issues relating to the particular site and the proposed development (**Figures 96.5** and **96.6** present photos from examples of common GI techniques). However, the engineer can never really know what the site investigation will reveal, and on some sites the results can show ground conditions significantly different to those expected.

The RE for site investigation works therefore needs to be prepared to respond to the findings of the investigation as they are observed, check that they align with the anticipated results and, if not, adjust the investigative works to ensure that information suitable to the needs of the project is obtained. In order to do this the RE needs to fully understand the overall aims of the site investigation and the specific aim of each exploratory hole. Ideally the RE should be involved with the preliminary design and scoping of the site investigation.

The RE needs to collate and assess the results of the exploratory hole information as they are produced and be prepared to instruct changes to respond to unexpected results. In doing so, the RE also needs to be aware of the cost implications of the changes they instruct. Where these are large, the RE should get written agreement of the change and the cost implication from the contractor and the contract administrator before the change is implemented. Often the RE is involved with certifying payment of the site investigation items; recording data on activity dates and delays is critical.

Sometimes the contractor will suggest alternative ways for progressing holes or carrying out tests. The RE should be open to alternatives, but should always question the need to depart from the specification and ensure that the proposed new method will produce data of the same or better quality than the specified method.

A site investigation is only of use to a project if the data obtained are of good quality and subsequent phases of GI may be required. The RE should be familiar with the codes and standards governing the site investigation operations. Documents relevant to most site investigations in the UK are listed below; however, the RE should be aware of other codes or guidance documents relating to more unusual site investigations or works in other countries, plus ongoing developments in site investigation guidance

- BS 5930:2015+A1:2020: Code of practice for site investigations (BSI, 2015a)
- BS EN 1997-2:2007: Eurocode 7: Geotechnical design – Ground investigation and testing (BSI, 2007a)
- NA to BS EN 1997-2:2007: UK National Annex to Eurocode 7: Geotechnical design – Ground investigation and testing (BSI, 2009a)
- individual British Standards for correct procedures for taking samples, carried out in situ and for laboratory testing.

Figure 96.5 Cable percussive drilling rig in use (© Arup; all rights reserved)

Figure 96.6 Logging core recovered from rotary boreholes (© Arup; all rights reserved)

The RE should observe the works carefully to ensure they are confident that these standards are being achieved by the contractor. The RE should also check early on that the specified sampling and testing frequencies are being achieved.

There will often be significant pressure on the RE to supervise the site works in a part-time capacity. In this instance the RE needs to time their site visits carefully to ensure that they are present at the most critical times. This will vary from site to site, but as a general rule the RE should be present to agree all exploratory hole locations and to observe the early works to gain confidence in the capability of the contractor. Where numerous crews are working on the site, the RE should ensure they observe the early works of each crew to check for variations in quality and working practices.

If the RE is on site as a person who will later prepare the geotechnical interpretative report (as is preferable), then they also need to observe and understand the nature of the soils exposed and should time their site visits so that soils from each stratum are observed and independent soil logging and photography can be carried out by the RE so checks of the contractor's logs are possible. The RE should also take independent measurements of structures exposed in trial pits if safe to do so. The independent records are key to checking contractor logs, in both situations where the RE is doing the interpretation or is passing it on to a colleague or third party.

Other key inputs required from the RE during site investigation works are checking locations of trial pits investigating existing foundations (e.g. piles under columns); making decisions on extending pit dimensions and depths in a timely manner; checking cone penetration test (CPT) refusals and taking timely decisions on any retesting; plus more standard checks on the condition of sample cutting shoes, SPT hammer certifications, counting SPT blows, considering checks on water levels/water strikes recorded by drillers and within trial pits, and more.

Because site investigations often occur in the early stages of a project and because the site teams are normally relatively small, welfare provisions and security for site investigations can often be less well established than for other construction works. The RE should nevertheless ensure that adequate welfare, security and H&S provisions are made on site and should insist that inadequate conditions are rectified immediately.

Box 96.6 Key points: supervision of site investigation works

The RE should:

- observe the works carefully to ensure that the required quality and quantity of data are being achieved in line with the specification and relevant codes of practice
- check the findings of the site investigation throughout the fieldwork and adjust the scope of work if the ground conditions are shown to be significantly different to what was expected
- ensure they are on site at critical times if their role is part-time.

96.7. Supervision of piling works

The precise nature of the RE's supervision activities will depend on the type of piling being witnessed and the design requirements (**Figure 96.7** presents an example of a common piling site). Sections 5 *Design of foundations* and 8 *Construction processes* describe the many types of piling that are commonly used in geotechnical engineering today. These include standard rotary or continuous flight auger (CFA) bearing piles, support fluid piles, secant piled walls, mini-piles, diaphragm walls, displacement piles, base-grouted piles, sheet piling, king post wall piles, pile removal and so on. Relevant codes and guidance documents include

- BS EN 1997-1:2004: Eurocode 7: Geotechnical design – General rules (BSI, 2004)
- NA to BS EN 1997-1:2004: UK National Annex to Eurocode 7: Geotechnical design – General rules (BSI, 2007b)
- BS EN 1536: Execution of special geotechnical work – Bored piles (BSI, 2010a)
- BS EN 1583: Execution of special geotechnical work – Diaphragm walls (BSI, 2010b)
- BS EN 12063: Execution of special geotechnical work – Sheet pile walls (BSI, 2022)
- BS EN 14199: Execution of special geotechnical works – Micropiles (BSI, 2015b)
- BS EN 12699: Execution of special geotechnical works – Displacement piles (BSI, 2015c)
- *Pile Design and Construction Practice* (Tomlinson and Woodward, 2008)
- various CIRIA guides – see the CIRIA website.

The RE should be clear on the level of checking required, the construction details to be checked and the criteria for acceptance. This will depend largely on the design. For example, if single piles are used to support critical elements of the

Figure 96.7 Typical bored piling site in Central London (© Arup; all rights reserved)

structure, then it would be expected that every such pile is rigorously checked due to the serious consequences of failure of a defective pile. However, it may be acceptable to check a proportion of piles for a design using pile groups to support a structure due to the inherent redundancy in the pile group. The complexity of the construction technique and relative experience of the contractor will also be factors – for example, more rigorous checking of new or complex piling methods would be expected over proven and relatively straightforward piling methods. The requirements for contractors to validate their methodology in pre-start trials and testing should be written into the piling specification. All the contractor's checks and acceptance criteria should be included in their ITP and match the piling specification requirements.

It is also important that the RE is familiar with the practical elements of the piling technique before they take sole responsibility for the site supervision. Ideally this would be achieved by visiting another site using the same piling technique and assisting the RE there. Otherwise, an experienced engineer should initially accompany the RE to site to assist in understanding the piling process and gaining confidence in the checking procedure.

The RE should ensure that they clearly understand the contractor's piling method and also the consequences of varying that method on the pile design.

There will be times on site when circumstances beyond the contractor's control disrupt the piling works, such as breakdown of plant, unforeseen ground conditions or severe weather. At such times the contractor and the RE will come under pressure to resolve the situation as quickly as possible and with minimal effect on the pile design. It is therefore useful for the RE in conjunction with the contractor to consider the possible situations where things can go wrong and plan their actions in advance (best practice is for this to be written into the contractor's method statement prior to works beginning).

Where the piling works are repetitive, it is even more important than usual for the RE to establish the required quality standard at the start of the works. The RE also needs to ensure that standards are maintained throughout the work, particularly on long projects where personnel may change over the course of the contract. The RE will often be a junior engineer compared to the site operatives and should recognise that these people can be very knowledgeable. The RE should consider their suggestions carefully but should also be aware that just because the crew have 'always done it like that' does not mean it is the right solution for the project. Particular awareness of industry innovations and lessons learned from previous projects is important here.

On piling projects, particularly those with high pile production rates, it can be tempting for the RE and contractor to fall behind with their record keeping. It is important that the RE does not let this happen and insists on records being produced in the time frames specified in the contract. Note that further inspections may be required after the piling works have been completed (e.g. wall watertightness inspections, remedial works inspections or movement monitoring checks), which may require return visits to site.

Timely and accurate production of close-out reports is increasingly important on piling projects (refer to Chapter 101 *Close-out reports*). Close-out reports should be comprehensive and detail not only the as-built pile details but also record any problems encountered during the works and the chosen resolution. This will aid in feasibility studies for the reuse of the piles in the future.

Box 96.7 Key points: supervision of piling works

The RE should

- ensure the documents/checks are suitable for the type of piling
- be familiar with the practical elements of the piling technique (through previous experience or visits to other sites using the same technique)
- be clear on the level of checking required, construction details to be checked and the criteria for acceptance
- keep up-to-date records, especially if production rates are high
- have considered the possible situations where things can go wrong and (with the contractor) plan their actions in advance
- ensure that standards are maintained throughout the work, particularly on long projects where personnel may change over the course of the contract.

96.8. Supervision of earthworks

Earthworks construction projects can range from the very small (excavation of pad foundations for a low-level building), to formation inspections for raft formations to the very large (major road infrastructure construction). However, the level of supervision required should be decided based on the complexity of the project, as well as its scale (**Figure 96.8** presents an example common earthworks site).

For detailed information on design of earthworks, refer to Section 7 *Design of earthworks, slopes and pavements*. Relevant codes and guidance documents include

- BS EN 1997-1:2004+A1:2013: Eurocode 7: Geotechnical design – General rules (BSI, 2004)
- NA to BS EN 1997-1:2004: UK National Annex to Eurocode 7: Geotechnical design – General rules (BSI, 2007a)
- BS 6031: Code of practice for earthworks (BSI, 2009b)
- various CIRIA guides – see the CIRIA website.

First, it is important that the RE understands the intended use of the earthwork structure as this will influence the level of inspection and the decisions they are required to make. For example, criteria for a structure required to carry heavy loads or plant with strict movement tolerances will be much more stringent than for a non-structural fill providing a platform for soft landscaping.

Figure 96.8 Typical earthworks site (© Arup; all rights reserved)

Prior to the site works beginning it is important to pre-classify the materials on site through site investigation. This allows the design and site teams to identify the appropriate type of plant to use for excavation, the angle of slopes that can be achieved and the level of compaction that can be achieved for filling. In addition, the RE should make themselves familiar with the types of compaction plant that may be suitable for the material on site and how they function.

With the exception of the smallest earthworks projects, it is usually only possible to inspect a selection of the works. It is important that the RE ensures that the required frequency of testing (either in situ or laboratory-based) is obtained and that the samples are representative of the site as a whole and not, for example, just a limited area of the site or soil type. It is normally preferable for the RE to choose the testing locations to ensure these are independent of the contractor and focused on the key activities. Alternatively, the contractor may propose a sampling and testing schedule for agreement with the RE. In addition, procedures for dealing with contaminated material should be agreed prior to testing in case any such material is encountered during the works.

Where testing of the samples is required, the RE must ensure that the laboratories have achieved the specified certification for the particular tests that have been carried out. This is particularly important for site-based laboratories. For off-site laboratories it is important to ensure that samples are stored and transported in such a way as to have minimal effect on their behaviour.

Detailed daily recording of the weather is essential to earthworks operations as it could affect the performance of the soils used. Additionally, the possible effects of inclement weather on the performance of the materials should be considered by the design team and understood by both the RE and the contractor. The RE should be aware of the contractual implications of inclement weather and if it is counted as 'extraordinary', in which case the client could be liable for any costs incurred, or if it is within the contractor's liability.

Where earthworks are inspected, the RE should keep good-quality annotated photographic records complemented by detailed descriptions of the works they have observed and the type and location of compliance testing (e.g. hand vane testing of raft clay formations or plate load testing for piling platforms). The contractor will often have to produce a validation report at the end of the works, which the RE may be asked to comment on. The RE should ensure that the contractor is providing elements of this report throughout the site works to prevent problems occurring at the end of the works if the information is inconsistent or incomplete. The RE should also produce a comprehensive and timely close-out report (refer to Chapter 101: *Close-out reports*) to summarise the works and to collate their independent records.

Box 96.8 Key points: supervision of earthworks

The RE should:

- understand the intended use of the earthworks and how that influences their supervisory role
- have a thorough understanding of the site investigation and therefore the site soils
- ensure that the sampling frequency and quality of testing is achieved and that samples are representative of the site as a whole
- keep detailed records of the weather and understand how conditions could affect the performance of the soils
- keep good-quality annotated photographic records.

References

BSI (British Standards Institution) (2004) BS EN 1997-1:2004: Eurocode 7: Geotechnical design – General rules. BSI, London, UK.

BSI (2007a) BS EN 1997-2:2007: Eurocode 7: Geotechnical design – Ground investigation and testing. BSI, London, UK.

BSI (2007b) NA to BS EN 1997-1:2004:2007: UK National Annex to Eurocode 7: Geotechnical design – General rules. BSI, London, UK.

BSI (2009a) NA to BS EN 1997-2:2007:2009: UK National Annex to Eurocode 7: Geotechnical design – Ground investigation and testing. BSI, London, UK.

BSI (2009b) BS 6031:2009: Code of practice for earthworks. BSI, London, UK.

BSI (2010a) BS EN 1536:2010: Execution of special geotechnical work – Bored piles. BSI, London, UK.

BSI (2010b) BS EN 1583:2010: Execution of special geotechnical work – Diaphragm walls. BSI, London, UK.

BSI (2015a) BS 5930:2015: Code of practice for site investigations. BSI, London, UK.

BSI (2015b) BS EN 14199:2015: Execution of special geotechnical works – Micropiles. BSI, London, UK.

BSI (2015c) BS EN 12699:2015 Execution of special geotechnical works – Displacement piles. BSI, London, UK.
BSI (2022) BS EN 12063:2022: Execution of special geotechnical work – Sheet pile walls. BSI, London, UK.
HMG (Her Majesty's Government) (1974) Health & Safety at Work Act. Elizabeth II. Chapter 37. The Stationery Office, London, UK.
HMG (1998) Lifting Operating and Lifting Equipment Regulations. The Stationery Office, London, UK.
HMG (1998) Provision and Use of Work Equipment Regulations. The Stationery Office, London, UK.
HMG (2002) Control of Substances Hazardous to Health Regulations. The Stationery Office, London, UK.
HMG (2013) Reporting of Injuries Diseases and Dangerous Occurrences Regulations. The Stationery Office, London, UK.
HMG (2015) Construction (Design & Management) Regulations. The Stationery Office, London, UK.
Tomlinson M and Woodward J (2008) *Pile Design and Construction Practice*, 5th edn. Taylor & Francis, Abingdon, UK.

Further reading

BSI (2016) PAS 2080:2016: Carbon management in infrastructure. BSI, London, UK.
Butcher AP, Powell JJM and Skinner HD (eds) (2006) *Reuse of Foundations for Urban Sites: Proceedings of the International Conference*. BRE, Watford, UK.
Chapman T, Anderson S and Windle J (2007) *Reuse of Foundations*. CIRIA Publication C653. CIRIA, London, UK.
CIRIA (Construction Industry Research and Information Association) (1996) *Site Guide to Foundation Construction*. Special Publication 136. CIRIA, London, UK.
CIRIA (2001) *Site Safety Handbook*, 3rd edn. Special Publication 151. CIRIA, London, UK.
CIRIA (2013) *Implementing Lean in Construction: Lean Construction and BIM*. Publication C725. CIRIA, London, UK.
CIRIA (2015) *CDM 2015: Workplace 'In-Use' Guidance for Designers*. CIRIA, London, UK.
Clarke RH (1984) *Site Supervision*. Thomas Telford, London, UK.
Twort AC and Rees JG (2020) *Civil Engineering Project Management*, 4th edn. Elsevier Butterworth-Heinemann, Oxford, UK.

Useful websites

British Standards Institution (BSI): http://www.bsigroup.co.uk
Construction Industry Research and Information Association (CIRIA): http://www.ciria.org
Construction Skills Certification Scheme: http://www.cscs.uk.com
Health and Safety Executive (HSE): http://www.hse.gov.uk/construction

It is recommended this chapter is read in conjunction with:

- Chapter 8 *Health and safety in geotechnical engineering*
- Chapter 78 *Procurement and specification*
- Chapter 93 *Quality assurance*
- Chapter 100 *Observational method*
- Chapter 101 *Close-out reports*

All chapters in this book rely on the guidance in Sections 1 *Context* and 2 *Fundamental principles.* A sound knowledge of ground investigation is required for all geotechnical works, as set out in Section 4 *Site investigation.*

doi: 10.1680/icemge.66830.1561

Chapter 97
Pile integrity testing

Simon French BES Group Ltd, Warrington, UK
Alex Whittaker BES Group Ltd, Warrington, UK

A pile integrity test is defined as a non-destructive test (NDT) which examines the response of a pile to some form of indirect physical scanning technique; such a test may employ acoustic shock waves, electrical energy, nuclear radiation or other input to excite the pile. Analysis of the response of the pile may allow an interpretation of the 'integrity' of construction of the pile body, typified by the homogeneity of the construction concrete or the uniformity of its external shape. Such tests are commonly used as part of a pre-planned quality control programme, as a retrospective testing programme or as an aid to the evaluation of existing foundations.

The most common non-destructive pile testing techniques are described with particular reference to low-strain integrity tests, cross-hole sonic logging and thermal integrity profiling. Guidance is given on the analysis, interpretation and reporting of the results of such tests, including computer-aided signal matching and modelling techniques to enhance the interpretation of the signal responses obtained.

Parallel seismic testing and high-strain integrity testing are also reviewed. The former is commonly used for retrospective testing of foundation piles when they are to be reused and the latter is a by-product of dynamic load testing.

CONTENTS

97.1. Introduction

CIRIA Report 144 (Turner, 1997) defines a pile integrity test as a non-destructive test (NDT) which examines the response of a pile to some form of indirect physical scanning technique. Integrity testing is most commonly used as part of a pre-planned quality control programme for a particular site. It can also be used as part of a retrospective testing programme, where a problem has become apparent during or subsequent to the construction of the piled foundations, or as an aid in the evaluation of existing foundations – where they might be required to be reused as part of a new development, for example.

Non-destructive, or integrity, testing of foundation piles has become routine on most UK construction sites but is often considered something of a 'black art' by many engineers and industry professionals. Such tests employ indirect testing techniques to examine the structure of the pile. These techniques include the use of acoustic shock waves from a hammer blow; ultrasonic pulses; electrical resistivity; nuclear radiation such as gamma or neutron scanning; or other input to excite the pile. The response of the pile allows interpretation to be made and conclusions to be drawn on the 'integrity' of construction of the pile body, typically related to its internal construction or the uniformity of its external shape.

Pile integrity testing and its principal methods have evolved since their introduction in the 1970s to become an important part of the quality control process. It can provide additional information on the as-built construction of piles, as well as enabling further evaluation of piling systems. Recent advances in computer technology have provided convenient solutions for compact and portable site acquisition equipment, and developments of sophisticated analysis programs have led to their widespread use by the construction industry.

In many respects, and in particular since publication of CIRIA R144 and the previous version of this chapter, little has changed with the 'traditional' testing methods. Testing hardware has improved marginally; with advances in computer and battery technology the emergence of 'new' technology in the form of thermal integrity testing is beginning to gain traction with many practitioners. Updated information and a review of testing methods is published by the Federation of Piling Specialists in their *Guidance Note on Current Integrity Testing Methods* (FPS, 2017).

This chapter provides an insight into the main pile testing methods currently in use, together with background theory and guidelines on the choice of the most appropriate method. Specifying the most appropriate test is not universally understood and the test method applied is often driven by varying factors. A wider understanding is therefore essential, and all engineers with responsibility in this area should have a working knowledge of the appropriateness of the various test methods, their limitations and the interpretation of the data obtained from such tests.

The main test techniques currently in use in the UK, which will be discussed in this chapter, can be identified as

- low-strain integrity testing
- cross-hole sonic logging

- thermal integrity profiling
- parallel seismic testing
- high-strain dynamic testing.

Electrical and nuclear, or radiometric, tests are not commonly used in general site practice. Further information on these and other test techniques can be found in Fleming *et al.* (2009) and CIRIA R144 (Turner, 1997).

The principal features of these four main test techniques can be summarised as follows.

(*a*) *Low-strain integrity testing* is the examination of the response of the pile to an external impulsive force: usually imparted by striking the head of the pile with a light, hand-held hammer, generally fitted with a hard plastic tip. The shock wave from the hammer blow travels down the pile. Part or all of the energy from this blow may be reflected or affected by a change in pile properties – the pile toe being an obvious source of such an effect. An analysis of the response of the pile (measured by a receiving sensor at the pile head) to both the initial blow and any returning reflections from depth allows inferences or conclusions to be drawn on the properties of the buried pile.

(*b*) *Cross-hole sonic logging* (*CSL*) is the examination of the response of the pile to a sonic or ultrasonic pulse travelling between a number of vertical ducts cast into the pile concrete. A signal transmitter or emitter is lowered to the bottom of one duct, and a receiver is lowered to the bottom of an adjacent duct. Both transducer and receiver are then raised in tandem at a predetermined rate. Changes in the characteristics of the pile or its constituent materials can affect the travel time or quality of the signal arriving at the receiver.

(*c*) *Thermal integrity profiling* is the evaluation of the entire cross-section over the entire length of the pile by measuring heat generated during cement hydration. The exothermic chemical reaction in which the major compounds in cement form chemical bonds with water molecules generates heat, which may be assessed to determine the quality of concrete within the piles.

(*d*) *Parallel seismic testing* is a method of estimating the acoustic length of a pile where no direct access can be gained to the pile head (a pile within an existing foundation, for example). A vertical tube, such as a borehole casing, is driven or bored into the ground alongside the pile. A sensor is lowered sequentially down the tube while the foundation above the pile is struck with a small hand-held hammer. The transit time of the shock wave to travel through the pile and the intervening ground to the sensor is measured. Once the sensor passes the end of the pile, the shock wave must pass through an increasing distance of soil to reach the sensor and the transit time will increase markedly.

(*e*) *High-strain integrity testing* is an ancillary use of dynamic pile testing systems such as those described in Chapter 98 *Pile capacity testing*. The test involves striking the pile with a large falling weight, such as a piling hammer. Sensors attached to the pile measure the impulsive force imparted into the pile by the falling weight and the resulting acceleration and velocity of the pile as it is driven downwards. Analysis of the signals from these sensors can allow the identification of pile damage such as a broken pile.

97.2. The history and development of non-destructive pile testing

97.2.1 Low-strain integrity testing

Most of the integrity testing of piles in the UK involves measuring the acoustic properties of the pile by striking it or causing some other acoustic vibration within it. This mechanical impulse generates ‘shock’ or ‘acoustic’ waves, which travel along or through the pile at a velocity determined by the characteristics of the pile material. The transmission of such waves through the pile is most readily represented or analysed by stress wave propagation theory using the principles of one-dimensional wave mechanics in elastic rods.

During the 1960s, low-strain integrity testing research was pursued in France by Paquet, working at the Centre Expérimental de Recherches et d’Études du Bâtiment et des Travaux Publics (CEBTP). This centred upon the response of a pile to an external impulse in both the time (‘sonic echo’) and frequency (‘vibrational’) domains. This work led to the conclusion by CEBTP (Paquet, 1968) that there were limits to the time-based ‘echo’ methods and that the assessment of piles by the ‘vibration method’ was preferable, particularly when defects in the upper few metres of the pile were a concern. These developments are discussed in detail in Hertlein and Davis (2006). Contemporary technology made practical on-site application of the theories discussed in the paper difficult to apply, and it was not until the early 1970s that appropriate equipment became commercially available.

At that time, Davis and Dunn (1974) described the application, theory and interpretation of results gathered by the frequency-based vibration method, using an electro-dynamic vibrator or ‘shaker’ attached to the pile head and measuring the response of the pile to the spectrum of vibration frequencies. The significant downside to the method at this time was the extensive pile head preparation necessary to perform the test and the cumbersome equipment required.

Over the same period research was also being undertaken at the Institute TNO in the Netherlands, and Case Western University in Texas, for example. These projects were generally focused on pile-driving analysis using the wave equation

(Smith, 1960). These researchers also investigated the use of the wave equation to examine pile integrity based upon 'low-strain' test techniques using a small hand-held test hammer. These forms of analysis concentrated upon time-based 'sonic echo'-type techniques.

During the early 1980s the CEBTP 'vibration' test method was further developed by substituting the vibrator with a broad-spectrum load cell which was struck with a small hand-held hammer to generate the necessary range of frequencies to accommodate the test. This was further developed into an integrated hammer which incorporated the load cell (Stain, 1982). After 1982 the test method was characterised by CEBTP as the Transient Dynamic Response (TDR) method. Other specialists use terms such as the impulse response method. In the following text, such tests are described under the generic name of frequency response tests.

Parallel developments in the 1980s by both research and commercial organisations saw the development from analogue to digital processing, recording and analytical techniques and, by the mid-1990s, site equipment that had become truly portable, being both compact and battery-powered. Such systems can store several hundred pile records and, if necessary, transfer them using a mobile telecoms link to an office base for speedier processing and report production. Advances in software and analysis programming with the advent of greater computing power have also enabled result simulation and modelling techniques to be used to aid result interpretation, as described later.

97.2.2 Cross-hole sonic logging and parallel seismic testing

The history of the development of CSL and the parallel seismic test broadly follows a similar path to low-strain integrity testing, as outlined above. CSL, in particular, was developed by CEBTP in the late 1960s (Paquet, 1969) to counter flaws with low-strain integrity testing, which was that the response of the pile to the pile head excitation is depth-limited. Cross-hole logging utilises sonic transmitter and receiver probes which are lowered down ducts within the pile and allow the construction characteristics of the pile to be examined at depth.

Parallel seismic testing (Paquet and Briard, 1976) utilises a duct placed outside a previously constructed pile (which might have been already incorporated into the as-built structure, for example) to examine the length or characteristics of the pile.

As with low-strain testing, the use of these systems was limited by the technology available at that time. Advances in portable computers throughout the 1980s and the introduction of improved analysis equipment saw widespread use of CSL on major projects around the world.

97.2.3 Thermal integrity profiling

This testing method was first developed at the University of South Florida between 2004 and 2007 by Mullins *et al.* (2007) as part of a research project. It relates the heat generated by curing of cement to the integrity and quality of cast in situ piles and other concrete foundations.

The project investigated the merits of assessing shaft integrity using infrared thermal detection methods to assess the presence or absence of intact concrete both outside and inside the reinforcing cage. The objectives included temperature measurement and numerical modelling to verify anticipated temperature response within the body of the pile. Using thermal modelling software, the physical dimensions of the pile shaft and the locations of known anomalies were signal-matched against the measured temperature traces. Correlations obtained enable prediction of the size and location of anomalies.

In addition, it is proposed that the data are analysed in conjunction with heat transfer process modelling considering the thermal properties of the ground strata in an attempt to improve the reliability of the thermal integrity testing method. More recently an interpretation approach has been developed in which thermal integrity profiling (TIP) constitutes the first step. An investigative staged process is then followed to establish and assess the anomalies in the problematic regions along the pile employing the combined use of detailed finite element simulations, the actual temperature data from TIP and genetic evolution algorithms (Sun *et al.*, 2022).

The use of distributed fibre-optic sensing for structural health monitoring has gained traction since 2020. This methodology has promising advantages over discrete instrumentation because it provides thousands of temperature points along the embedded fibre. The ease of installation of unobtrusive fibre-optic cables and the low cost of fibre, together with the very low failure rate of such sensors, mean it is emerging as a viable alternative to conventional thermal probes.

97.3. A review of defects in piles in the context of NDT

97.3.1 Introduction

The foundation system is selected by the designer of the piles and/or the overall scheme designer from a range of possible alternatives to suit the ground conditions and the imposed structural loads. In selecting the system both parties need to consider the means by which the as-constructed works are checked against the design intentions and assumptions. Ideally, these design and performance requirements should be outlined within the project specification.

To a large extent, each pile is unique and, because it is buried in the ground, it cannot be examined in the same way as the structure it supports. Additionally, the foundation system cannot normally be subjected to preloading to confirm its adequacy before being put into service (except in a case such as a storage tank or silo). Hence, the evaluation of the foundation pile system usually requires inference and extrapolation from a combination of direct and indirect testing and, if necessary, the targeted examination of individual piles.

It should be understood that a key point in any such evaluation process is the accurate recording of construction information for each pile, usually by the pile installer. Guidance on essential construction records is listed, for example, in the ICE *Specification for Piling and Embedded Retaining Walls* (SPERW), third edition (ICE, 2016).

NDT is therefore undertaken as part of the contractual evaluation system, to provide additional information on aspects of the pile construction, as an aid to the evaluation of the foundation system.

97.3.2 An overview of pile defects and NDT

Problems that may arise in the construction of various types of pile are reviewed in Chapter 82 *Piling problems*. In the context of the use of NDT, the features arising from the problems, or potential problems, identified in Chapter 82 *Piling problems* fall into four general categories (Turner, 1997):

- type A: changes to the intended shape of the pile body itself
- type B: total rupture of the pile in the form of a transverse crack or break across the body of the pile
- type C: changes in the internal properties of the pile
- type D: features that affect the interaction of the pile with its environment and, consequently, its ability to transfer the design load into the surrounding soil or rock.

Type A features are usually associated with some aspect of the pile-forming process, either in the pile construction or post-construction phase. Features that affect the shape of the pile body were generally identified in CIRIA R144 (Turner, 1997) by the terms necks, waists, bulbs, expansions, steps or bites, as illustrated in **Figure 97.1**.

Type B features are linked to the action of unplanned external tensile or lateral forces acting upon the hardened pile material, before (as in the case of preformed or precast piles), during or after installation. The effect of such an external force is typically to produce a transverse fracture perpendicular or at an acute angle to the long axis of the pile, as illustrated in **Figure 97.2**. Transverse cracks can also be formed by shrinkage of the concrete or by the effects of ground heave.

Type C features might be caused by faulty concrete or concreting processes, or the use of substandard materials. They therefore reflect internal changes in the properties of the pile, as illustrated in **Figure 97.3**. The changes in properties, such as concrete strength, could be gradational or sharp. Other examples are where some of the pile concrete is contaminated by suspended spoil (e.g. piles formed under bentonite) or variations in concrete cover to steel reinforcement.

Type D features include poor toe conditions and softened or degraded bore sides due to relaxation or water ingress. Type D features are therefore typically associated with deficiencies in the pile construction process at the boring or drilling stage, prior to concreting, or, in the case of driven piles, with pile heave in congested groups, or with relaxation effects after driving. Such factors can therefore directly affect the load-carrying capacity of the pile.

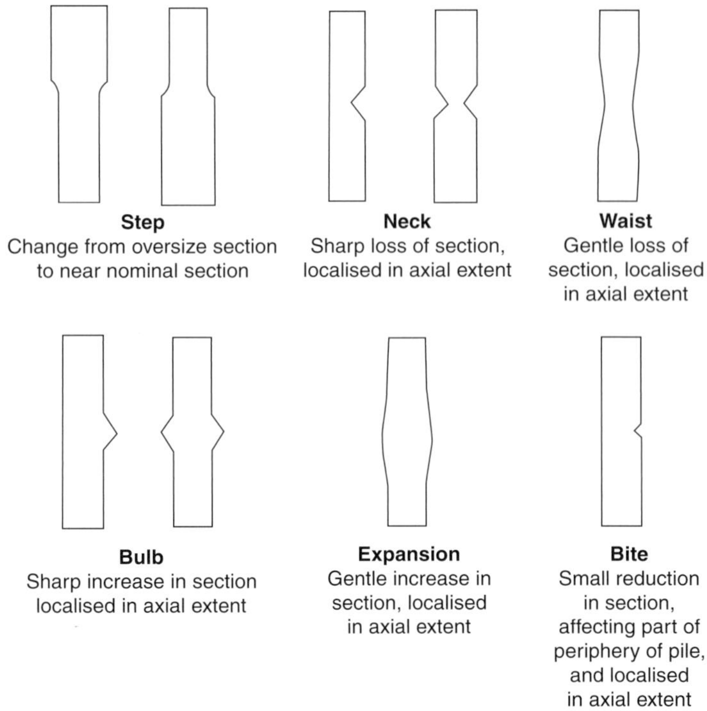

Figure 97.1 Typical type 'A' features in a pile (Reproduced with permission from CIRIA R144 (Turner, 1997); www.ciria.org)

It is important to understand, however, that these 'features' are not necessarily defects: in the sense that a 'defect' might be considered to be something that would significantly affect the short- or long-term performances or the load-carrying capacity of the pile. The identification of a particular feature is not a judgement of whether a pile is defective, sub-standard

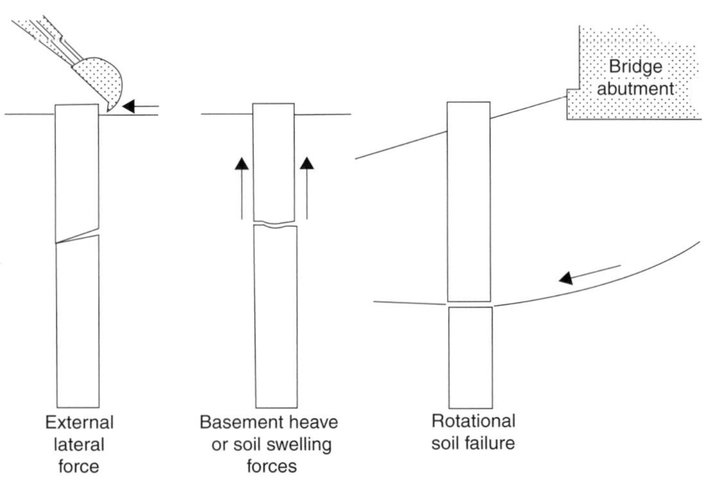

Figure 97.2 Typical external causes resulting in type 'B' fractures (Reproduced with permission from CIRIA R144 (Turner, 1997); www.ciria.org)

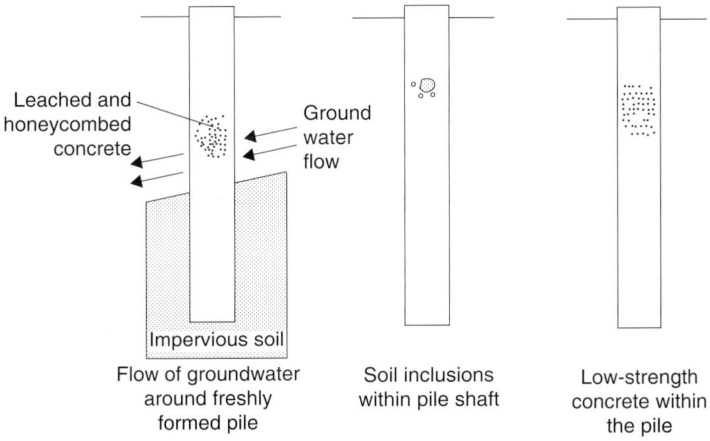

Figure 97.3 Typical type 'C' features within a pile (Reproduced with permission from CIRIA R144 (Turner, 1997); www.ciria.org)

or non-compliant. Further evaluation and, possibly, further investigation may need to be undertaken to determine whether the feature is an unacceptable 'defect'.

CIRIA R144 (Turner, 1997) proposed the terms 'anomaly' and 'defect' where an anomaly was an irregular or unexpected response that might or might not represent a real feature of the pile shaft. In contrast, a defect was categorised as a feature in the pile that was not in accordance with its specified construction, and which might or might not affect the ability of the pile to perform as required.

In terms of the latter, Amir (2001) proposed further that a defect which was not significant in terms of its effect on pile capacity or durability should be termed a 'flaw', reserving the term 'defect' for a feature that would affect the bearing capacity or likely durability of the pile. A 'defect' would require engineering evaluation: perhaps the pile could be accepted at a reduced capacity, for instance.

Integrity testing therefore inherently requires a step-by-step process leading from the identification of a 'feature' or an 'anomaly' by the test to the evaluation of its significance and the decision on any further action required.

Section 97.4.6 discusses in more detail these analytical and interpretative processes associated with the evaluation of such 'features'.

It is important that the method of testing and evaluation should be chosen with reference to the pile type and construction method, pile layout, piling programme, subsoil conditions and level of site control, both during and after pile installation, to ensure it is suited to the detection of the type of fault which could exist.

A further point to understand is that, in general, non-destructive tests are better at highlighting sharp changes in pile properties, as opposed to gradational ones. Hence in **Figure 97.1** small changes such as 'waists' or 'expansions' or gradual changes in pile diameter, rather than 'steps', would typically be difficult or impossible to identify in practice.

97.4. Low-strain integrity testing

97.4.1 Time-based analysis

The general arrangements for a time-based integrity test are illustrated in **Figure 97.4**. These tests are often known as 'echo' or 'sonic echo' (time domain) tests.

To understand the essence of the test technique it is helpful to initially consider the case of a perfect free pile (i.e. a pile which has no lateral soil restraint or base support) of length L. In this case, the toe of the pile is said to be a 'free' end, as it is free to move downwards in response to the downward-travelling stress wave from the hammer blow (Equation 97.1). Applying a short-duration impact to the pile head, such as a blow from a hand-held hammer, causes a compression wave to propagate down the pile in the form of a longitudinal or bar wave (commonly termed a stress wave). Upon reaching the pile toe the wave reflects back to the pile head as a tension wave. The fundamental propagation velocity, c, for longitudinal waves is given by one-dimensional wave theory as

$$c = \sqrt{E/\rho} \tag{97.1}$$

where E is the material Young's modulus and ρ is the material density.

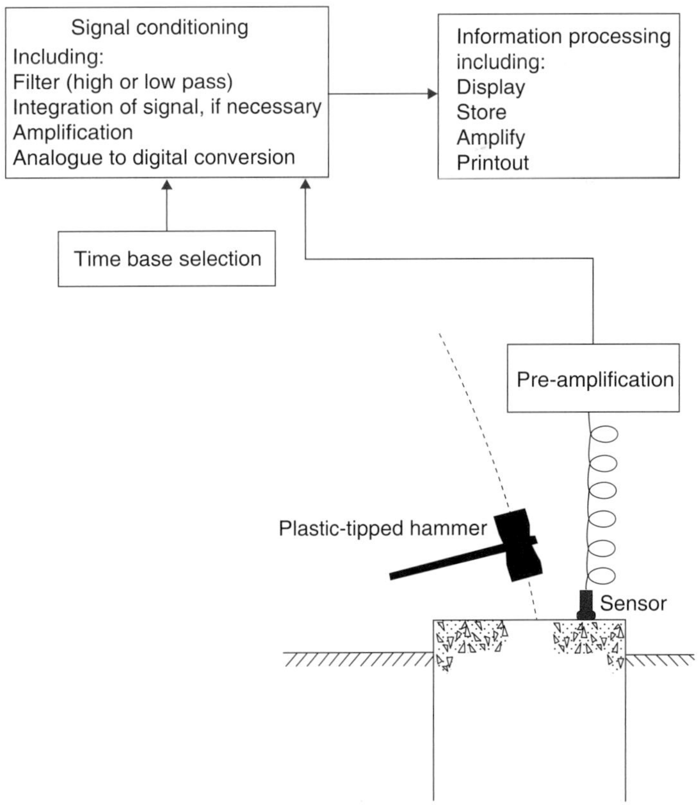

Figure 97.4 Schematic of a typical time-based 'sonic echo' test layout (Reproduced with permission from CIRIA R144 (Turner, 1997); www.ciria.org)

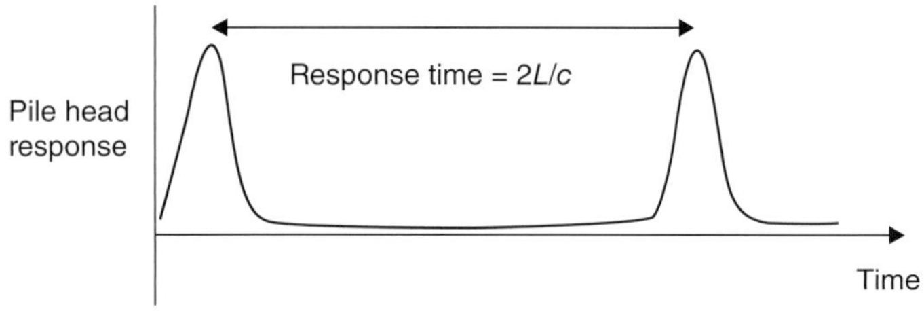

Figure 97.5 Graphical representation of hammer blow and 'echo' reflection observed at the pile head from a 'free' end at the pile toe

It can be seen that the time, t, taken for this wave to propagate from the pile head to the pile toe and back up to the head will be given by

$$t = 2L/c \tag{97.2}$$

This is illustrated in **Figure 97.5**, where the first peak is generated by the hammer blow and the second peak is the reflection or 'echo' of the wave returning from the 'free end' at the pile toe. It should be noted that on most of the illustrations of time-based (sonic echo) testing in this chapter the initial hammer blow is represented as an upward peak on the signal response graph. It will be realised that it could equally be represented as a downward-facing peak, and hence all other peaks or troughs will be similarly reversed, and other figures show this mode. Commercially available testing systems usually adopt one or the other method of data presentation.

Now consider the case of a similar pile, without skin friction but this time supported on an infinitely rigid base. The applied compression wave from the hammer blow upon reaching the supported base reflects in the form of a compression wave, but one with an opposite polarity to the downward-travelling wave, as shown in **Figure 97.6**.

In practice, of course, the pile will be embedded in the ground. This has the effect of dissipating the energy from the stress wave as it propagates down the pile shaft. The degree of signal attenuation is a function of the pile diameter, the pile length and the stiffness of the surrounding soils. Generally

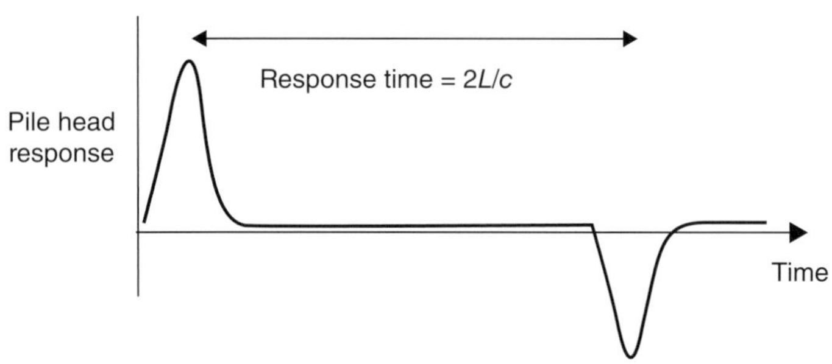

Figure 97.6 Hammer blow and 'echo' reflection observed at the pile head from a 'fixed' end at the pile toe, showing an upward-travelling compression wave, with its polarity opposite to the downward-travelling wave

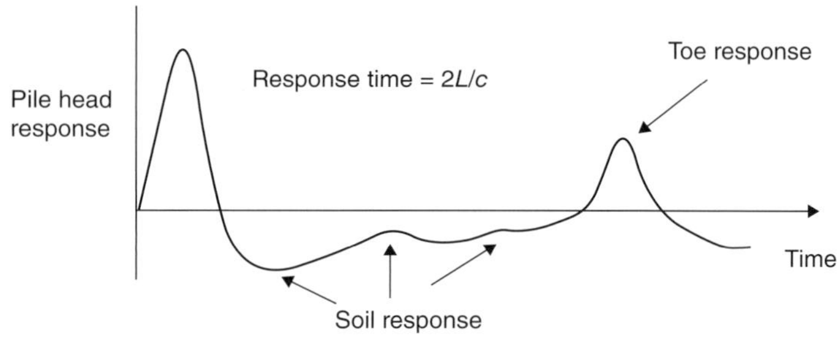

Figure 97.7 Pile head response in multiple soil horizons

speaking, piles with a length to diameter ratio in excess of 30:1, embedded in stiff to very stiff clays, exhibit 100% signal attenuation so that effectively no return signal reflected from the pile toe reaches the pile head. In such a situation it can be seen that the test may be of diminishing use in identifying the location of the toe of the pile, if this was of concern to the engineer. However, it must be appreciated that the test will still have application in examining the upper section of such a pile, which is more susceptible to mechanical damage during construction and pile cap preparation.

The attenuation of the low-strain integrity test signal from longer piles was also the reason that CSL was developed (see Section 97.4.5).

Changes in soil characteristics along the shaft will also generate partial reflections of the downward-travelling stress wave. At each of these soil horizons a proportion of the stress wave will be reflected back to the pile head, the remaining portion continuing to the pile toe. Multiple soil horizons along the shaft can therefore have a major effect on the theoretical pile response, so that interpretation of the return signal becomes more complex, as shown in **Figure 97.7**.

Inevitably, the resultant response from an embedded pile is often not as straightforward to interpret as the theory may suggest. A more detailed explanation of wave propagation in piles is given in CIRA R144 (Turner, 1997).

97.4.2 Typical pile and soil properties

As discussed in Section 97.4.1, in order to evaluate the response of a pile to a low-strain integrity test impact it is necessary to consider the properties of both the material from which the pile is constructed and the soil surrounding the shaft. It can be appreciated that a sound pile shaft constructed in a very stiff soil, or socketed into rock, would produce a response significantly different to a pile floating in space with no skin friction or base support. The stiffness of the soil surrounding the embedded pile contributes to damping of the stress wave as it propagates down the shaft. This damping factor was first recognised by Briard (1970) and can be measured by the attenuation of the response from the pile base. Typical values of material properties are shown in **Table 97.1**

Velocity of plane wave propagation in concrete	3500–4000 m/s
Density/unit weight of concrete	23–24 kN/m^3
Young's modulus of concrete	Approximately 38 GN/m^2
Soil shear wave velocity	100–300 m/s
Soil density/unit weight	16–21 kN/m^3

Table 97.1 Typical values of material properties

97.4.3 Pile impedance

Any changes in pile properties along the shaft will also generate partial reflections of the stress wave as it passes the boundary at which the change occurs. Such property changes within the pile are defined by a parameter termed the impedance, Z, of the pile. The value of Z at any given location along the shaft is given by

$$Z = \rho \times c \times A \tag{97.3}$$

An increase in pile impedance, such as an increase in the cross-sectional area, A, of the pile, will generate a compression wave at the level of the change. This will propagate back up to the pile head and produce a pile head response in the *opposite* direction to the hammer impact response, in a similar manner to the rigid toe reflection illustrated in **Figure 97.6**. Conversely, a reduction in pile impedance, such as a reduction in cross-sectional area, or a reduction in the density of the concrete forming the pile, will generate a tension wave, producing a response at the pile head in the same direction as the hammer impact response, in a similar manner to the 'free end' toe reflection in **Figure 97.5**.

It can be seen, therefore, that the reflection of the signal from the toe of the pile, and its interpretation as either a free end or a fixed end effect, would also be produced by other features within the pile, such as a change of section, for example. Typical features that might be expected to produce a free end response are illustrated in **Figure 97.8(a)**. Similarly, features which would be expected to produce fixed end responses are illustrated in **Figure 97.8(b)**.

In addition, the impedance change associated with a particular feature might only cause a partial reflection of the downward-travelling stress wave from the initial hammer blow on the pile head, while the remainder of the signal continues downwards towards the toe of the pile. An example of such a feature might be a bulge or a neck in the pile section. It will be appreciated that, in such a case, the part of the signal that reaches the pile toe and is reflected back towards the pile head will again suffer partial reflections, and hence attenuation or degradation, as it passes through the feature, so that it might be barely detectable at the pile head. The practical significance of such partial reflections is that the pile head response becomes more complex as a consequence of the effect of more than one impedance change upon the transmitted and reflected stress waves. **Figure 97.8(c)** illustrates the effect of the possible 'combined' response of piles with a neck or a bulb feature within the pile section. In addition, **Figure 97.8(d)** illustrates the effect of the attenuation of the signal response as a result of soil skin

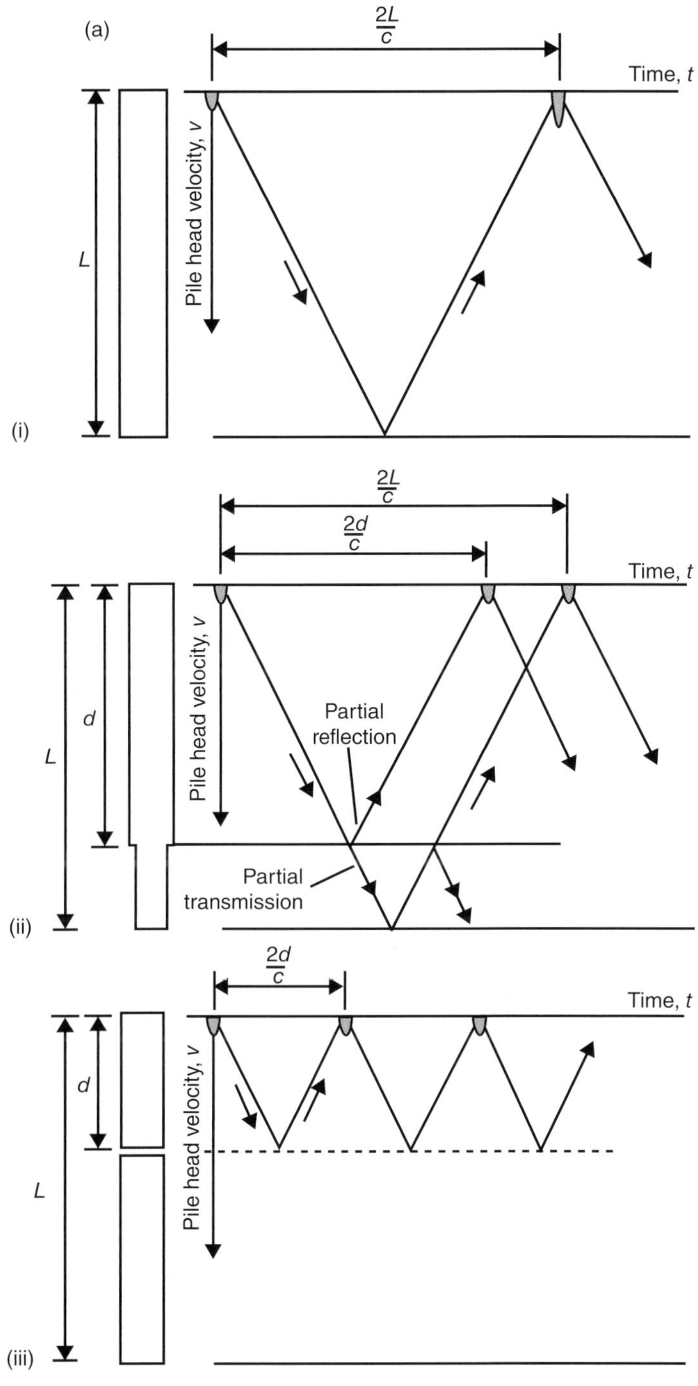

Figure 97.8 (a) Typical simplified free end responses for time-based tests: (*i*) toe of pile (note: repeated reflection at time intervals of 2*L*/*c*); (*ii*) intermediate decrease in cross-section (partial reflection at change of section, reduced toe reflection); (*iii*) broken pile/complete loss of section/crack (free end reflection from break, repeated at time intervals of 2*d*/*c*) (Reproduced with permission from CIRIA R144 (Turner, 1997); www.ciria.org)

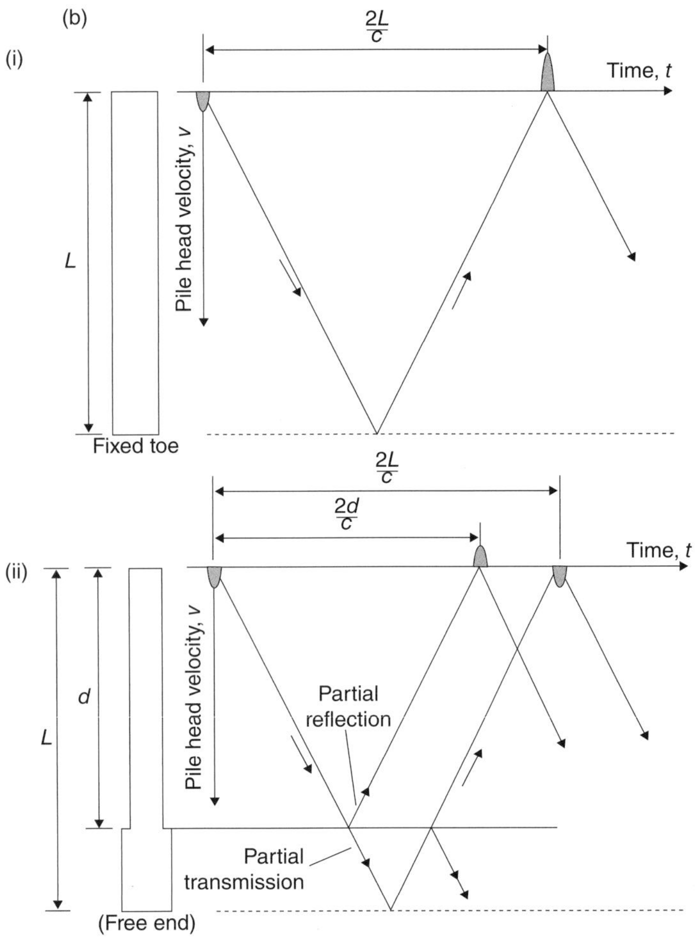

Figure 97.8(b) Typical simplified fixed end responses for time-based tests: (*i*) toe of pile (note: repeated reflection at time intervals of 2*L*/*c*); (*ii*) intermediate increase in cross-section (partial reflection at change of section, reduced toe reflection) (Reproduced with permission from CIRIA R144 (Turner, 1997); www.ciria.org)

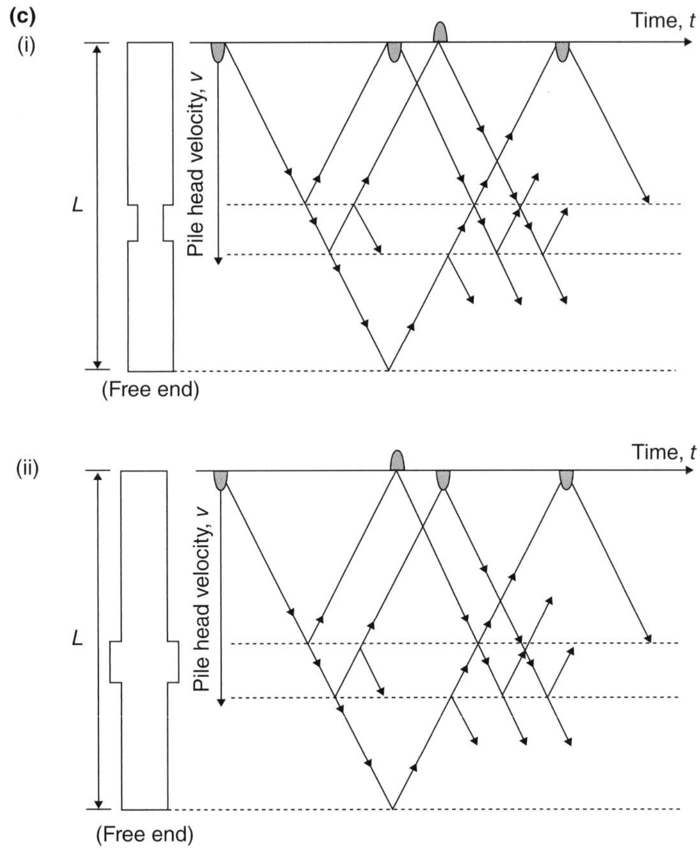

Figure 97.8(c) Typical 'combined' responses for time-based tests: (*i*) free end with neck (partial reflections at both changes in section, reduced toe reflection); (*ii*) free end with bulb (partial reflection at both changes in section, reduced toe reflection) (Reproduced with permission from CIRIA R144 (Turner, 1997); www.ciria.org)

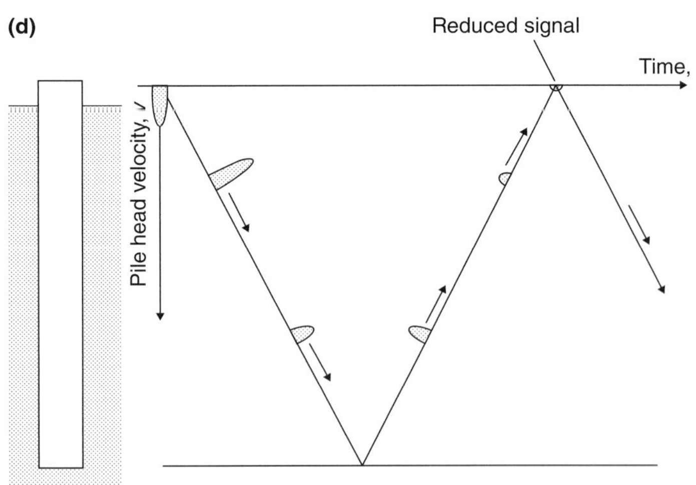

Figure 97.8(d) Attenuation of stress wave due to skin friction

friction upon the pile. At the limit, for a long pile embedded in a stiff soil, no toe reflection would return to the sensor at the pile head, and only the initial impulse from the hammer blow would be recorded, as illustrated in **Figure 97.8(e)**.

It will be appreciated that a relatively small impedance change would only reflect a small percentage of the downward-travelling wave, so that most of the signal would continue downward towards the pile toe. As the impedance change becomes greater, an increasing proportion of the downward-travelling wave would be reflected. Ellway (1987) suggested that when the ratio Z_1/Z_2 or Z_2/Z_1 is greater than 4, the downward-travelling wave will be effectively completely reflected at the interface between Z_1 and Z_2, so that no return signal would be received from below this level.

97.4.4 Frequency-based analysis

Initially, as discussed in Section 97.2.1, for the 'vibration' test method employed by Paquet (1968) and Davis and Dunn

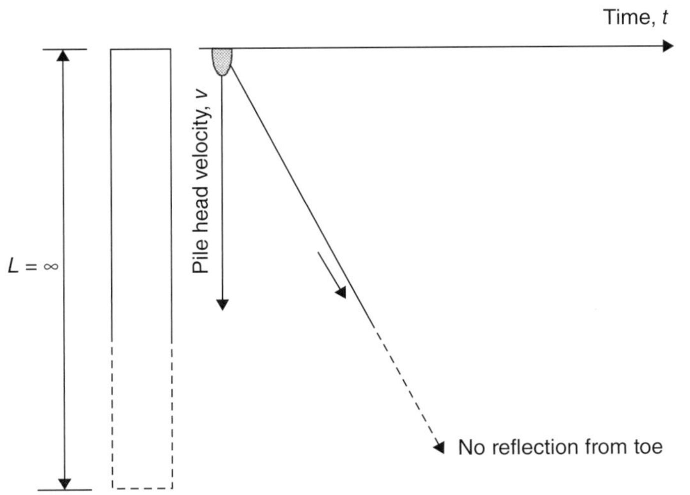

Figure 97.8(e) Signal response for an infinite pile (no toe reflection) (Reproduced with permission from CIRIA R144 (Turner, 1997); www.ciria.org)

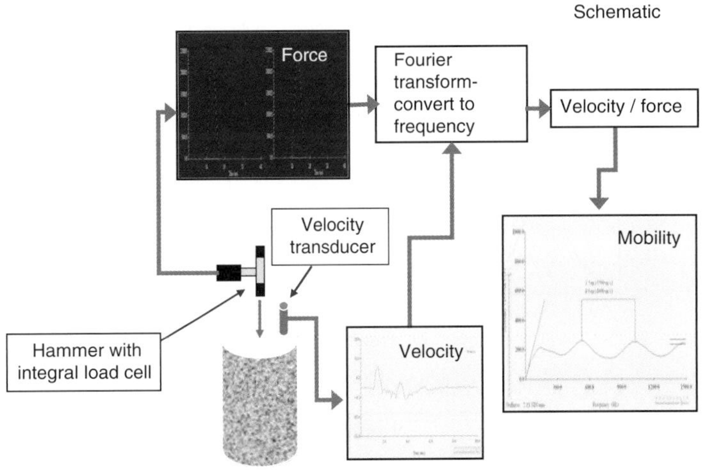

Figure 97.9 Schematic of general arrangements for the frequency response method (Courtesy of British Engineering Services Limited (formerly Testconsult Limited))

(1974), the pile head was subjected to a cyclically varying excitation force using an electro-dynamic 'shaker' attached to the pile head. By this means the pile head force was applied in discrete frequency steps, gradually sweeping through a frequency range from 0 to 1000 Hz. By applying this vibration at constant amplitude and force F_o and observing the maximum velocity V_o at the head of the pile within this frequency range it will be seen that the discrete frequencies at which resonance occurs are equally spaced along the frequency spectrum. The first resonating peak in the series is the natural frequency of the pile, and subsequent peaks are harmonics of the natural frequency.

As described earlier, later developments utilised a load cell attached to the pile head, which was then struck with a small hand-held hammer fitted with a suitable impact head capable of generating the same bandwidth of frequencies used for the vibration tests. Current systems use a hammer incorporating the load cell as a single unit.

The current frequency-based test techniques are commonly known as 'impulse' response and by such proprietary names as Transient Dynamic Response (TDR) methods. Within this text they will be described under the generic term of 'frequency response' tests.

An important difference between the current time- and frequency-based techniques is that the latter always measures the force applied to the pile head by the hand-held test hammer. A schematic of the general arrangement of this test method is shown in **Figure 97.9**.

This method can best be understood by referring back to the original 'vibration' test technique using an electro-dynamic shaker. For any discrete frequency of excitation generated at the pile head, the returning wave and consequential pile head velocity will be the vector sum of the phase of the returning wave and the original input wave. For example, if the returning wave is in phase with the input wave, then the pile head velocity will be at a maximum; if out of phase by 180 degrees the resultant pile head velocity will be at a minimum.

From standard wave theory relating to vibrations in long slender rods, the distance between resonating peaks, Δf, is given by

$$\Delta f = c/2L \qquad (97.4)$$

where c is the velocity of plane wave propagation along the pile and L is the length of the pile or depth to an acoustic feature.

To 'normalise' the response from the pile impact and produce a unique pile signature, the output from the pile head velocity transducer (V_o) is divided by the input force (F_o), where V_o/F_o is the mobility, M_o (also termed the mechanical admittance), of the pile at the particular frequency. The units for mobility are therefore m/s/N. The pile head response is thus plotted in terms of pile head mobility against frequency.

In the case where the elastic base is infinitely rigid (see Section 97.4.1), the lowest frequency of resonance has a value of $c/4L$, as shown in **Figure 97.10(a)**.

By contrast, when a pile rests on an infinitely compressible base, resonance first occurs at a very low frequency (**Figure 97.10(b)**). When the base is an elastic one of normal compressibility the lowest frequency of resonance lies in an intermediate position between that of the rigid and infinitely compressible bases, as shown in **Figure 97.10(c)**.

In a similar manner to that described above for time-based analyses, a free or fixed end effect could also be produced by other features within the pile, such as a reduction or increase of the pile section, for example, or a broken or cracked pile.

As described by Ellway (1987), the sharpness of the pile shaft resonances depends upon the relative amount of energy of the stress wave that is either transmitted or dissipated each time it is reflected from a boundary layer, such as the pile toe. A short pile in free air would have little energy loss and would

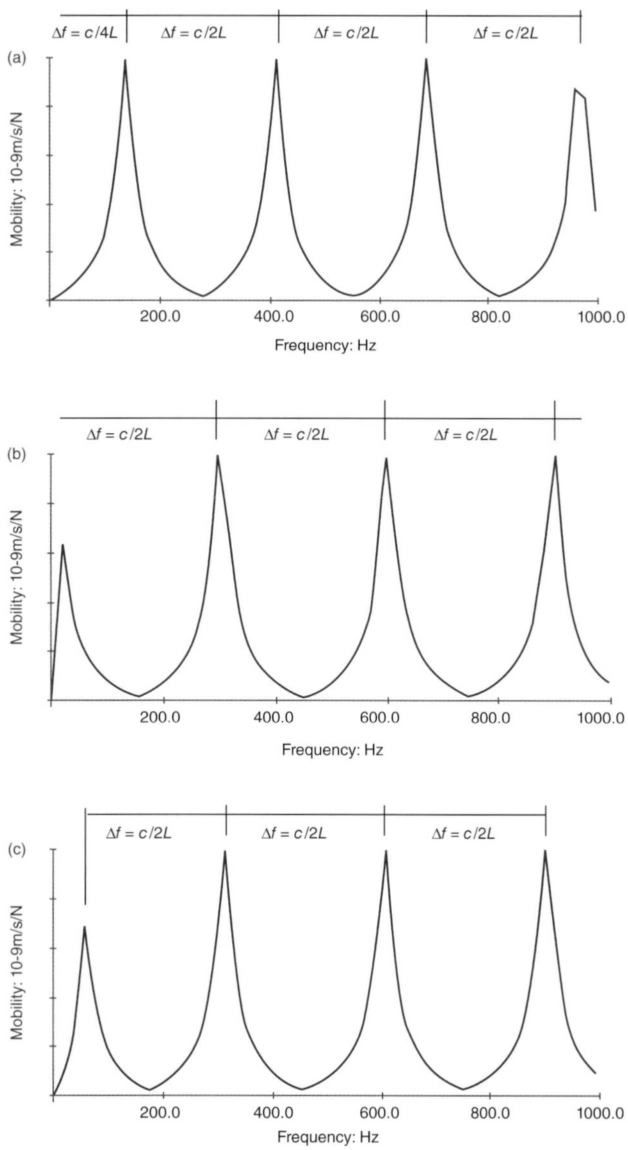

Figure 97.10 Pile with (a) rigid base support; (b) no base support; (c) normal base support

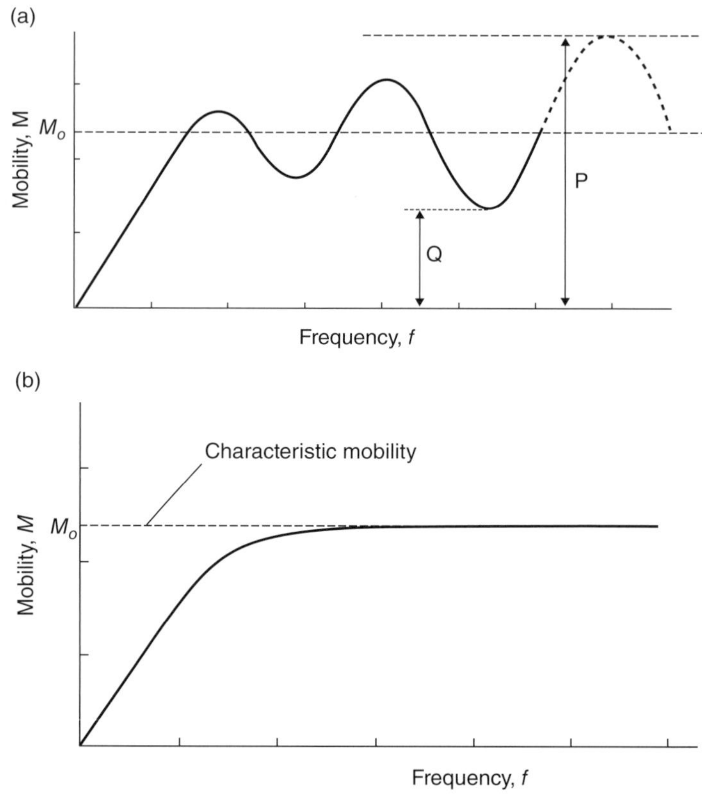

Figure 97.11 (a) Effect of soil damping on signal response curve; (b) idealised frequency response curve for an infinitely long pile (Reproduced with permission from CIRIA R144 (Turner, 1997); www.ciria.org)

be expected to exhibit the sharp resonant peaks illustrated in **Figures 97.10(a)** to **97.10(c)**. However, where a pile is embedded within a soil then the propagation of the stress wave will be dissipated or *damped* by the soil. The idealised signal response curve of **Figure 97.10** would therefore be modified to the form shown in **Figure 97.11(a)**. This form is typical of most frequency response signal curves met in practice.

In an exactly similar way to the 'echo' test, if the soil was sufficiently stiff, all of the stress wave would be attenuated and no reflection would be received from the toe of the pile by the sensor at the pile head. Hence, no resonances or frequency peaks will be registered, and the signal response will tend towards a single value of mobility, as illustrated in **Figure 97.11(b)**.

The stress wave can be attenuated not only by the surrounding soil, but also by a very long pile, where the wave has to travel through the pile to the toe before returning to the sensor at the pile head. It can be seen that an infinitely long pile of uniform cross-section would produce no reflection from the toe, and there would be no intermediate impedance changes to cause reflections of the signal. Thus, the frequency response curve of a very long pile would therefore also tend to that illustrated in **Figure 97.11(b)**, and the value of the mobility would reach a constant value. This value is termed the characteristic mobility, M_o, of the pile, and is given by

$$M_o = L/\rho c A \tag{97.5}$$

The characteristic mobility, M_o, for a particular pile therefore depends only upon its internal properties and is a unique calculable value. The calculated theoretical value of M_o for a pile tested in the field is often annotated onto commercial signal response curves as an aid to interpretation and comparison.

From a consideration of Equation 97.5, it can be seen that a test value of M_o greater than its theoretical value could be caused by one or more of the pile parameters σ, c or A being lower than its design or intended value. Conversely, a lower value of M_o would imply σ, c or A being greater than their design value – a larger-diameter pile, for example. It is also of use to the equipment operator to signify poor pile head preparation for testing of poor-quality concrete at the pile head.

The above text and figures also clarify that the attenuation of the frequency response curve can be the result of either a shorter pile embedded within a stiffer soil, or a longer pile in a weaker soil. In both cases the difference in amplitude between maxima and minima (indicated by P and Q in **Figure 97.11(a)**) is reduced, so that the response curve gradually approaches the value of the characteristic mobility of the pile.

In addition, the initial straight line portion of the mobility–frequency curve is a measure of the *dynamic stiffness*, *E*, of the pile head. *E* would be expected to be a specific value for a group or 'class' of piles on a site having the same cross-sectional area, length and unit weight, installed into the same soil stratum. Hence, a pile with a lower dynamic stiffness than its fellows might be considered for further examination or test. Further description of this feature is given in Hertlein and Davis (2006) and CIRIA R144 (Turner, 1997).

97.4.4.1 Pile impedance and partial reflection

In a similar manner to analysis in the time domain discussed in Sections 97.4.2 and 97.4.3, it would be expected that an intermediate feature in the pile, such as a reduction or increase in diameter, which caused only a partial reflection of the stress wave, would be identifiable by its characteristic resonant frequency in addition to the frequency peaks due to, say, the toe of the pile. Hence, the two signals might be superimposed onto the response registered for the pile. This is illustrated in **Figure 97.12**, taken from CIRIA R144 (Turner, 1997), after Ellway (1987). Ellway noted that the reflection coefficient (the proportion of the signal either reflected or transmitted) across the change of impedance (shown as a bulb in **Figure 97.12**) is also governed by the frequency of the vibration wave. In general, this coefficient increases with frequency, so that a greater proportion of the vibration energy is reflected from the change in impedance interface at higher frequencies. Thus, as seen in **Figure 97.12**, at lower frequencies the full length of the pile can be discerned. At higher frequencies the intermediate impedance change becomes increasingly discernible. Ellway suggested that, for this reason, a frequency-based test should examine the response of the pile over a minimum frequency range of 0–2500 Hz.

Figure 97.13 further illustrates the return signal from a pile with a major impedance change at shallow depth. Section 97.4.5 discusses the interpretation of site data in further detail.

97.4.5 Features detected by low-strain integrity tests

97.4.5.1 Features detected using time-based sonic echo test methods

Figure 97.14 illustrates a typical response from a pile with a previously placed 'neck' defect (from a test site at Blyth, in north-east England) as described by Lilley *et al.* (1987). The first upwards peak is from the hammer impact and the second from the reflection or 'echo' from a change in pile or soil impedance. As the echo is of positive polarity (relative to the hammer impact peak) the returning wave is a tension wave and is therefore a result of a reduction in impedance due to a change in pile or soil properties. The magnitude of the reflection would suggest a shaft feature rather than a soil feature and this is most likely to take the form of a reduction in section or 'necking' of the pile shaft.

The separation of the two peaks is 1.4 ms, which, based upon a wave speed within the concrete pile of between 3500 and 4000 m/s, is equivalent to a depth of 2.4–2.8 m below pile

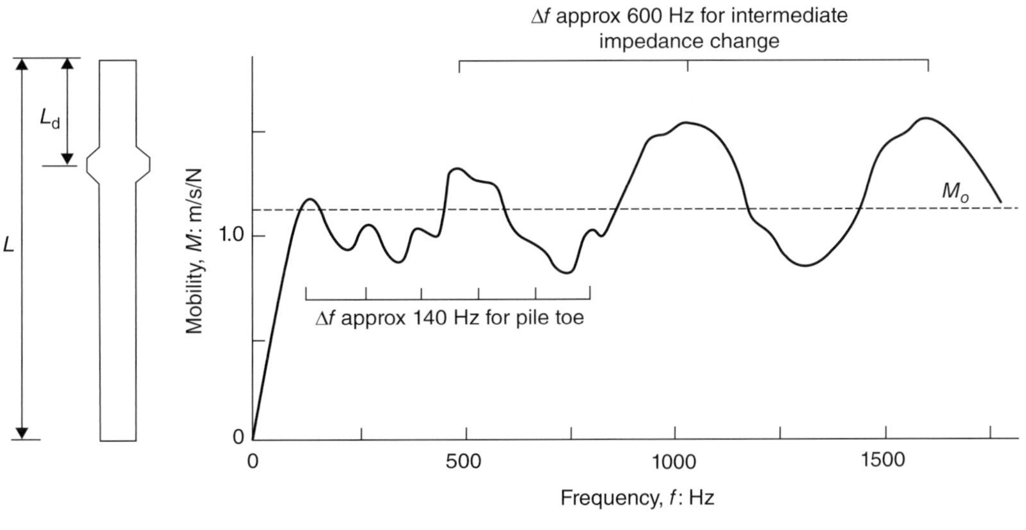

Figure 97.12 Frequency response test, illustrating the effect of an intermediate impedance change (Reproduced with permission from CIRIA R144 (Turner, 1997); www.ciria.org; after Ellway, 1987)

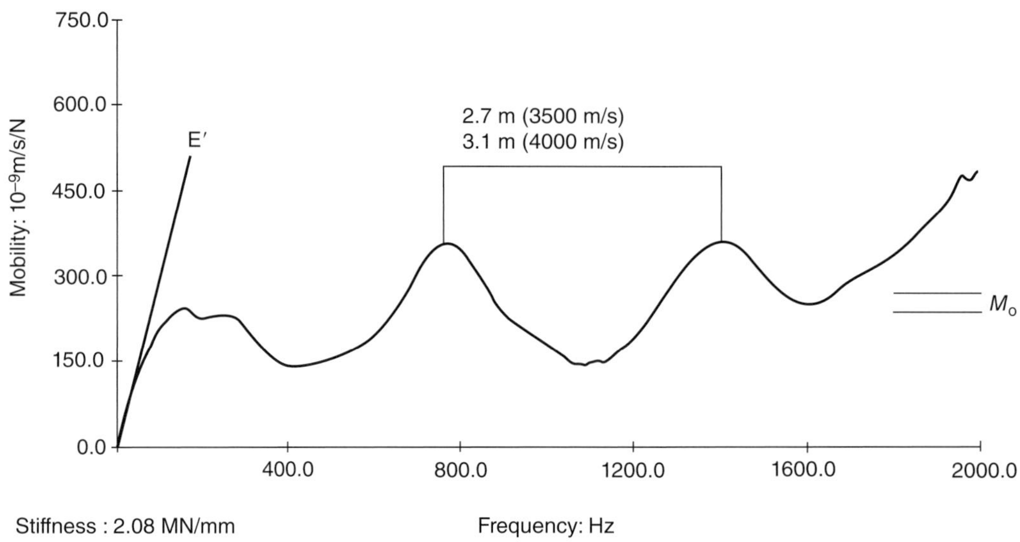

Figure 97.13 Typical site trace of a frequency response test on a pile with a neck defect, showing interpretation of pile head dynamic stiffness, *E′*, and characteristic mobility, M_o (Courtesy of British Engineering Services Limited (formerly Testconsult Limited))

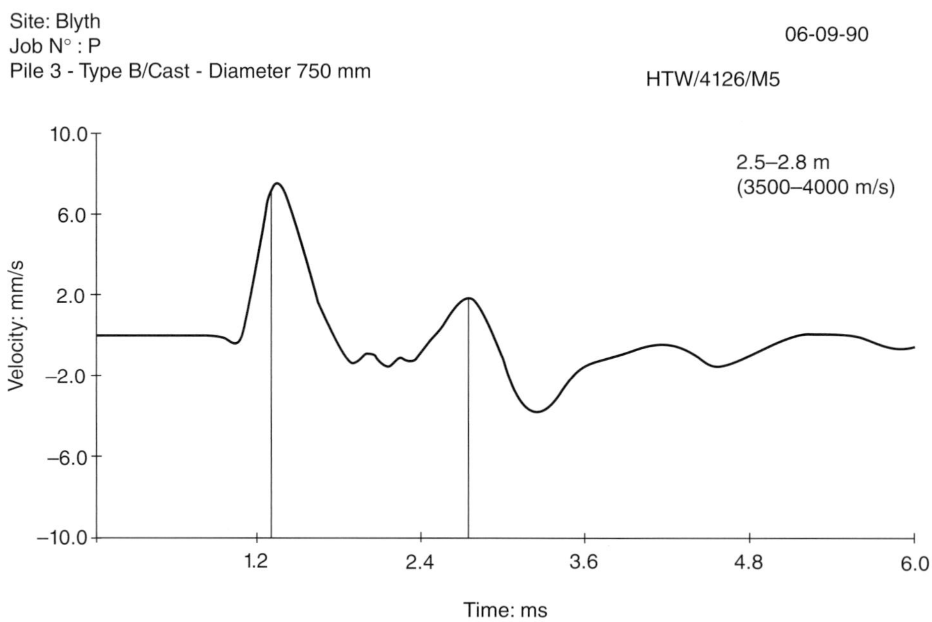

Figure 97.14 Sonic echo response from an anomaly at approximately 2.4–2.8 m (Courtesy of British Engineering Services (formerly Testconsult Limited))

head level. The signal response may also suggest an increase in impedance very shortly after (probably as the pile returns to normal section).

There is no obvious toe reflection from the pile, which was recorded as 11.3 m long and installed within very stiff to hard glacial till (boulder clay). Again, based on a wave speed of between 3500 and 4000 m/s the pile toe would be expected on the signal trace at around 5.5–6.5 m/s from the hammer peak. The toe is not clearly discernible because of attenuation of the signal by the strong soil.

Caution should always be taken with this type of result as the reduction in pile impedance suggested may also be associated with an oversize pile section reducing to nominal diameter at the indicated depth. If this is suspected, then reference should be made to pile construction records, concrete volume and geotechnical conditions.

The normal course of action with this pile result (once reference to construction records has eliminated an oversize upper section) would be to excavate to the level of the anomaly to inspect the pile shaft or undertake a performance test on the pile. This may take the form of either dynamic, rapid or static load tests.

97.4.5.2 Features detected using frequency response test methods

The site analyser processes the data from the hammer and velocity transducer and produces a signature response curve showing mobility against frequency. The response curve from the same 'defective' pile is shown in **Figure 97.15**. This also allows a direct comparison of the results when displayed in the time domain and frequency domain.

Interpretation of the response curve would initially be undertaken by measuring the frequency difference between adjacent peaks. This shows a Δf frequency of approximately 650 Hz. By applying the formula $L = C/2\,\Delta f$, length measurements of approximately 2.7 and 3.1 m can be calculated for the wave propagation speeds (C) of 3500 and 4000 m/s. The phasing of the first harmonic peak is close to Δf frequency (650 Hz), which suggests a reduction in pile impedance. In addition, the pile head dynamic stiffness (E) is calculated to be 2.08 MN/mm and the characteristic mobility (M) calculated to be approximately 300×10^{-9} m/s per N, which suggests that the pile has a comparable pile head stiffness to similar pile/soil installations and that pile section, concrete density and wave propagation velocity are within the normal range for the given pile diameter and concrete properties.

Further interpretation of the response curve can be undertaken using numeric simulation techniques. **Figure 97.16** shows a numeric simulation performed on the impulse response curve discussed above: proprietary software is, however, required for this. The simulation includes a reduction in pile section at 2.9 m and shows a close match between the real and simulated response following manual adjustment to both soil and pile variables. The resultant graphics confirm that the inputted localised necking of the pile shaft at approximately 3.0 m is the correct interpretation of the response, as any mismatch of the response curve would require differing pile and soil properties.

Further modelling of the pile section using impedance analysis of the impulse response was first postulated by Paquet (1992). Advances in computer technology have enabled this to become a reality, and several equipment manufacturers now include modelling of the pile section in this way. The resulting pile model is termed an impedance-log. The impedance-log technique essentially produces a one-dimensional graphic representing changes in pile impedance.

97.4.5.3 Impedance-log analysis

As noted by Hertlein and Davis (2006), it has always been the ultimate aim of low-strain integrity test specialists to create an image of the as-constructed shape of the pile shaft 'in the ground'. They suggested that the closest to this ideal at that date is the impedance-log analysis method developed by Paquet (1991, 1992). This technique combines the amplified time-domain response from the echo-type test with the characteristic pile shaft impedance measured in the frequency response test. In common with other analytical tests, however, the method is based upon a one-dimensional model, so that

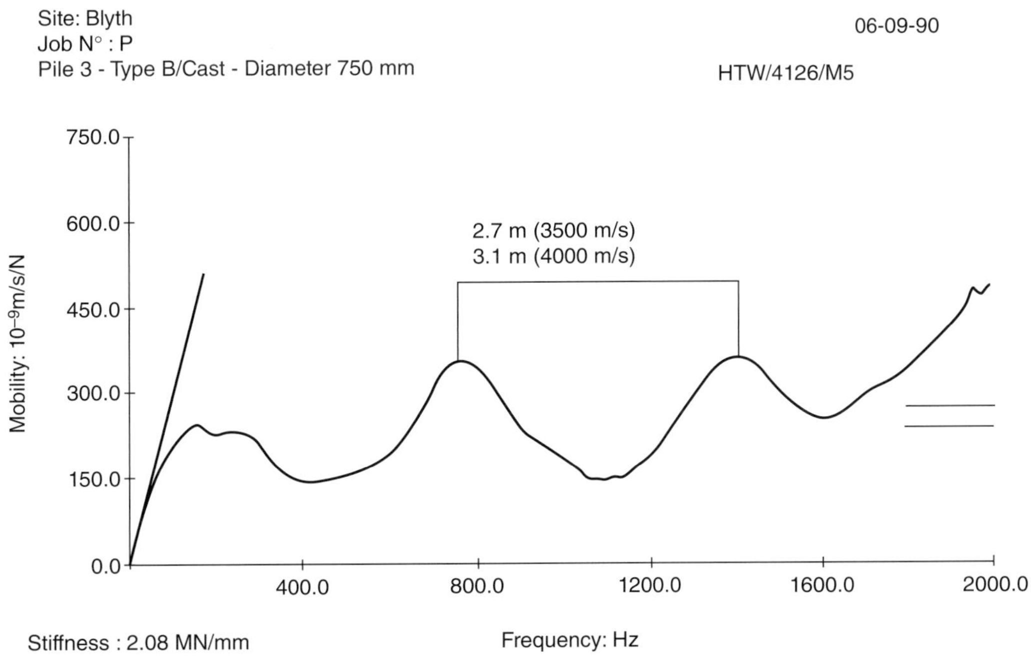

Figure 97.15 Impulse response curve from anomaly at approximately 2.7–3.1 m (Courtesy of British Engineering Services (formerly Testconsult Limited))

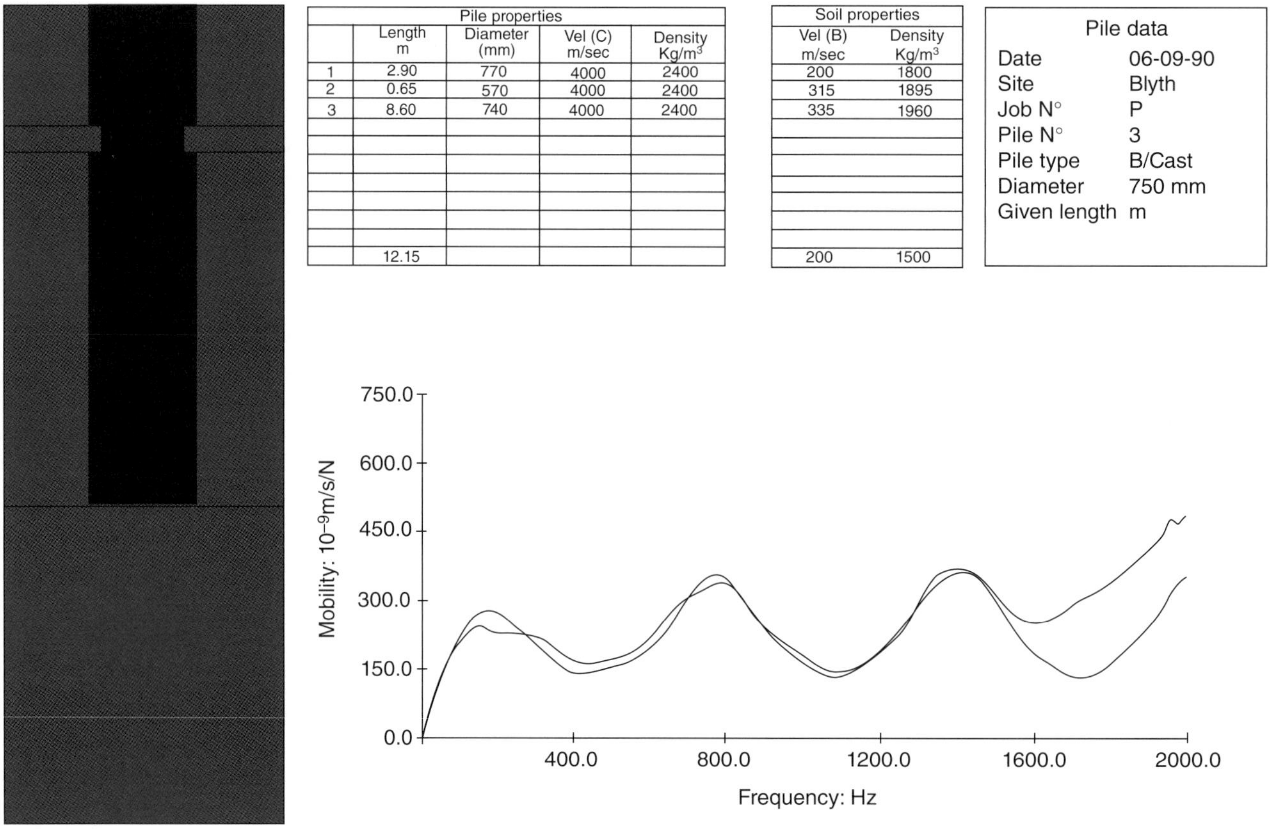

Figure 97.16 Simulation of mobility response and resultant pile section (Courtesy of British Engineering Services (formerly Testconsult Limited))

an asymmetrical feature such as a partial neck on one side of the pile would not be identified as such. The technique utilises the data from both the time and frequency domains to remove the motion of the pile head caused by the initial hammer blow and the effect of impedance changes due to the surrounding soil. In principle, this is obtained by calculating a theoretical mobility plot for an infinitely long defect-free pile of the given diameter and predicted ground conditions from the available ground investigation information. In practice this model is typically derived from a simulation analysis on a pile at the site which is considered to be satisfactory. The resulting calculated soil characteristics are then used in the impedance-log model for other piles.

This idealised mobility plot is subtracted from the test response, resulting in a 'reflected' mobility response which contains information about changes in the pile shaft geometry and the surrounding soil. The response in the time domain is then calculated from this plot to produce a 'relative reflectogram', as illustrated in **Figure 97.17(b)**.

Further analysis allows the removal of the soil effects from the model, and the calculation of the impedance down the pile shaft as a function of time, which is equivalent to depth. As described in earlier parts of this chapter, if the density and wave speed of the pile material are known, or can be reasonably approximated, changes in impedance correspond to changes in the cross-sectional area of the pile, and the resulting impedance–depth relationship can be plotted in terms of impedance against either the diameter or cross-sectional area of the pile.

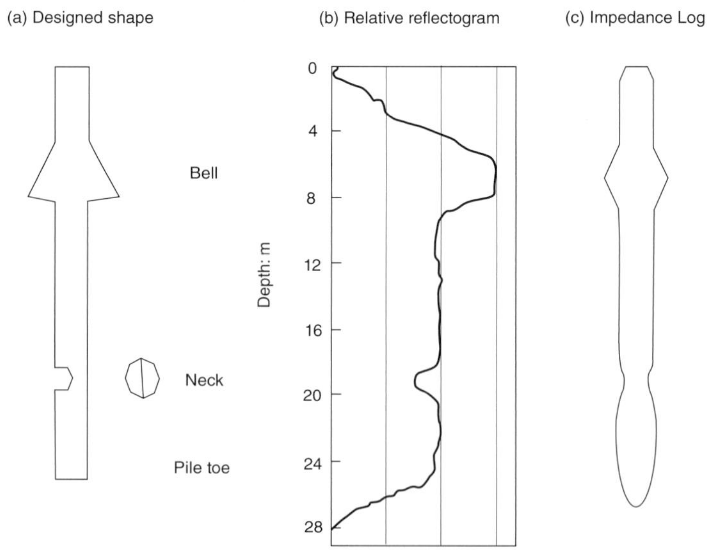

Figure 97.17 The impedance-log and relative reflectogram calculated from the velocity–time reflections for a pile constructed with artificial pile defects (Reproduced from Hertlein and Davis, 2006; Wiley)

A resulting pile impedance-log analysis is shown in **Figure 97.17(c)**. This analysis was undertaken on a pile on a test site at San Jose, California, which was constructed with known defects. It can be appreciated that for such analyses it is most important to have full pile construction details and soil data.

Further description of the technique is given by Hertlein and Davis (2006).

97.4.6 Analysis, interpretation and reporting of low-strain integrity tests

CIRIA R144 (Turner, 1997) identified two main fields of use for low-strain integrity testing.

- Routine control testing, as part of a pre-planned site quality control regime.
- Retrospective investigations, where a problem has become apparent during the works: often while piles are being cut down ready for incorporation into the foundation system. A further use of such testing would be where it was proposed to reuse existing piles following demolition or alterations to an older structure (e.g. Butcher *et al.*, 2006; Chapman *et al.*, 2007), or where the testing formed part of a forensic investigation into a possible foundation failure.

In the case of routine control testing, CIRIA R144 (Turner, 1997) suggests that it should be reasonable to expect that most of the signals from such integrity tests should be understandable to other non-specialist engineers involved with the project. Since the results are of interest and importance to a number of contracting parties within the project, CIRIA R144 suggested that, for a testing programme to be useful, it was also important that the engineering advisers to those parties were able, with proper guidance, to agree to the validity of the basic information.

It is also imperative that such engineers are aware that the results of integrity testing should not serve as the sole basis for the acceptance or rejection of any particular pile. Other information, such as ground conditions, construction records and on-site observations, should also be considered. Crucially, if the interpretation of the test data is perceived by others to be subjective, or even whimsical, then there will be a corresponding decrease of confidence in its use as a quality control tool.

Again, it is important that the non-specialist engineer should be aware that, in practical terms, in some situations the data will allow no clear interpretation.

In an attempt to resolve this problem CIRIA R144 proposed a signal classification system for both sonic echo and frequency response tests. This was intended to differentiate the simpler signal responses from those that are more complex.

97.4.6.1 Classification of signal responses

CIRIA R144 classification

To attempt to resolve this problem, CIRIA R144 (Turner, 1997) suggested a threefold classification of signal responses:

- *Type 0:* no distinct return signal received – that is, no significant impedance change within the pile, for the depth penetrated by the test
- *Type 1:* one clear major response – that is, one significant impedance change within the pile
- *Type 2:* more than one major response – that is, two or more impedance changes within the pile.

Type 0 signal

A Type 0 signal would be one in which the damping effect of the surrounding soil has attenuated any return signals to such an extent that the toe cannot be discerned. Therefore, there is no significant impedance change within the effective penetration depth of the system. The non-specialist engineer would easily appreciate the reason for this, provided that the basic principle of the test was understood. **Figure 97.18** illustrates typical examples of Type 0 signals for time-based and frequency-based systems.

Type 1 signal

A Type 1 signal would contain one clear, major response, indicating that the shaft was responding as a single acoustic unit. This would represent a pile containing a single major impedance change, either from the pile toe or

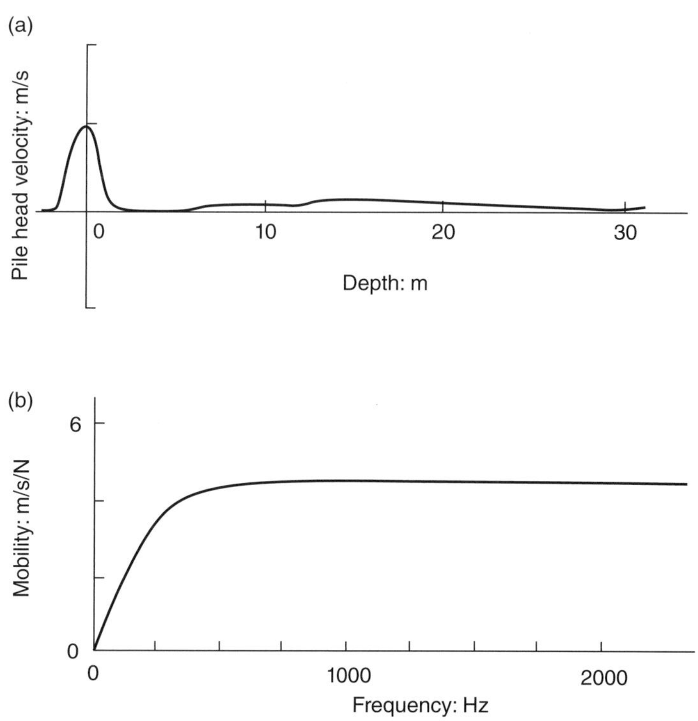

Figure 97.18 Examples of Type 0 signal response: (a) time-based (sonic echo) test; (b) frequency-based (frequency response) test (Reproduced with permission from CIRIA R144 (Turner, 1997); www.ciria.org)

some significant intervening feature. No other significant response would be visible on the recorded trace. Type 1 signal responses should be very similar to the theoretical simple signal expected from the test and be easily recognisable. Hertlein and Davis (2006) also suggested that such a signal response would be easily simulated by computer modelling. Typical Type 1 signal responses would be like those shown in **Figure 97.19**.

Type 2 signal

A Type 2 signal would be one containing more than one major response, so that the interaction of overlapping responses from different levels within the shaft would make interpretation of the resulting response a complex matter. At one extreme, Type 2 signals might display a clear major response from the length of the shaft responding as a major acoustic unit, but with intermediate responses due to local changes in shaft impedance within that acoustic unit, as shown in **Figures 97.20** and **97.12**. At the other extreme, Type 2 signals might contain no clear major response to indicate if part of the shaft is responding as a single acoustic unit, as indicated in **Figure 97.21**.

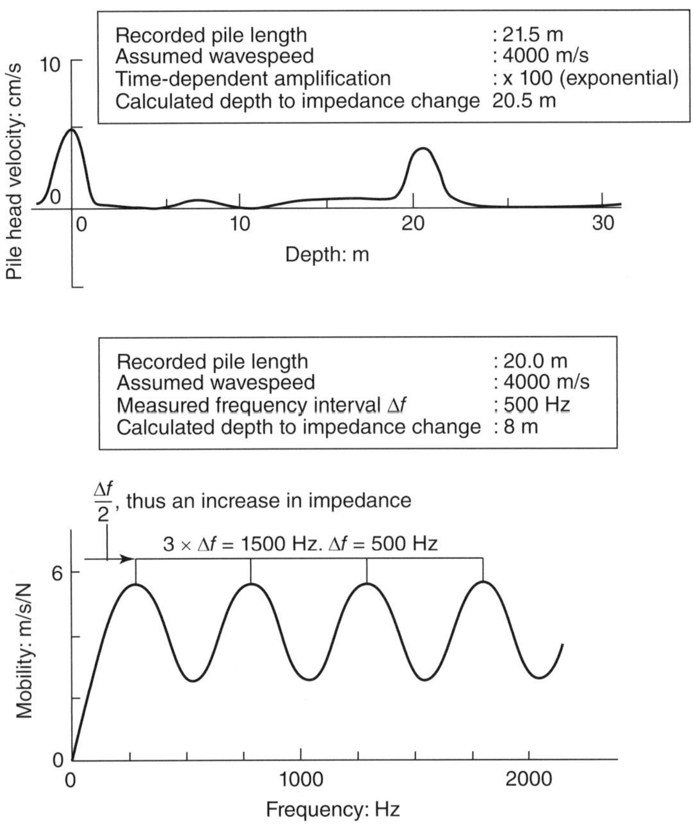

Figure 97.19 Examples of Type 1 signal response: (a) time-based (sonic echo) test; (b) frequency-based (frequency response) test (Reproduced with permission from CIRIA R144 (Turner, 1997); www.ciria.org)

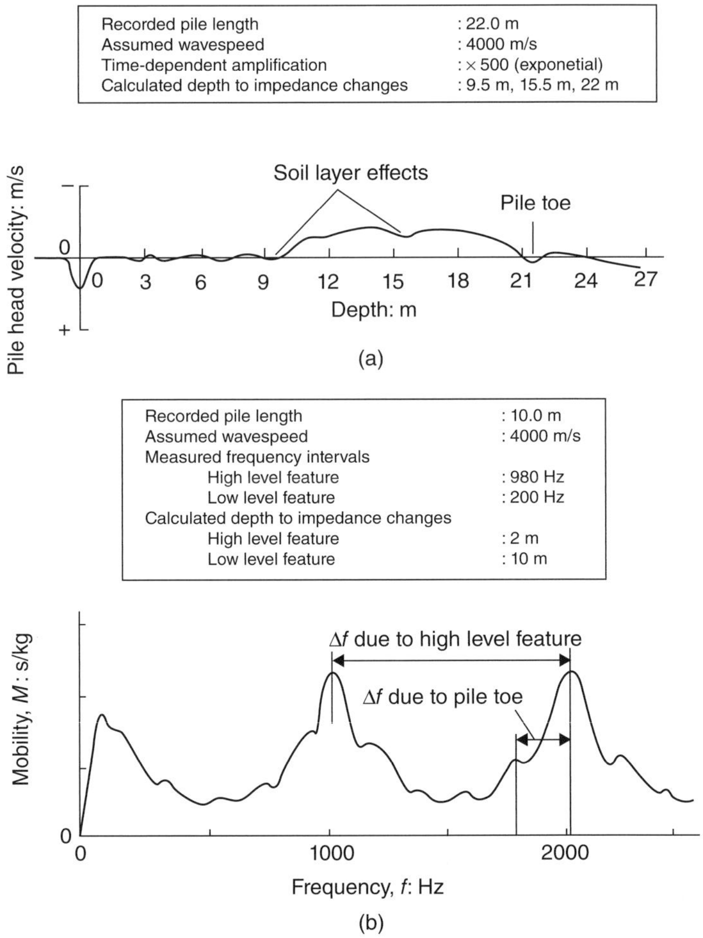

Figure 97.20 Examples of 'clear' Type 2 signal response, clear toe response and additional secondary response: (a) time-based (sonic echo) test; (b) frequency-based (frequency response) test (note: sonic echo test shows inversion of *y*-axis, so that initial hammer blow response is shown pointing downwards. This is a feature of the particular test system.) (Reproduced with permission from CIRIA R144 (Turner, 1997); www.ciria.org)

A Type 2 signal would require interpretation by a specialist, because simple models do not easily explain the response. A summary of the CIRIA R144 (Turner, 1997) classification is provided in **Table 97.2**.

Classification of the signal responses obtained from the piles on an individual site in this way will help to assess how much weight should be given to the test results for the piles on that site. The quality of the test results will depend upon

- the characteristics of the test system, particularly its dynamic range (the ratio of the specified maximum value and the minimum detectable value of the parameter being measured), its resolution and its signal-to-noise ratio
- the pile characteristics, especially the length/diameter ratio, the quality of the pile material and the shape of the pile
- the nature of the surrounding soil: the stiffer the soil, the greater the signal attenuation. In addition, a boundary between soils of

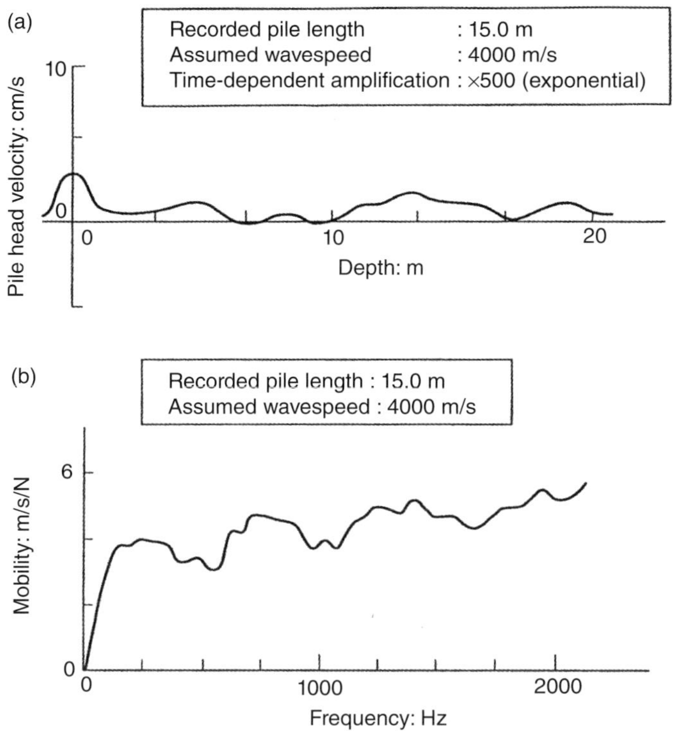

Figure 97.21 Type 2 signal response, no clear response

different relative stiffness acts as a reflective layer or impedance change within the shaft–soil system

- the quality and standard of the pile head preparation for the test.

It is important to note that CIRIA R144 highlights that there is a distinction between the analysis of the signal (i.e. what it is showing or saying) and its interpretation (i.e. what it means). The CIRIA classification 0, 1 or 2 is intended as a preliminary *analysis* of the characteristics of the signal, so that the tester can identify to the non-specialist engineer what are the general features of the signal response for each pile. The next step in the process is the tester's *interpretation* of the signal (what it means, and its significance).

An example of this *interpretative* aspect of the analysis process, which was developed by Testconsult Ltd, in the UK, is outlined below.

97.4.6.2 Interpretative classification

The interpretative classification system seeks to group results from frequency response testing into seven categories

- CAT 1 – full pile length measured
- CAT 2 – damped response
- CAT 3 – pile with over-break
- CAT 4 – oversize pile reducing to nominal diameter
- CAT 5 – pile with crack in upper shaft
- CAT 6 – pile with reduction in properties (i.e. necking)
- CAT 7 – invalid result due to poor head condition.

It can be seen that a combined CIRIA/Testconsult classification might be, for instance, a CIRIA 1 + Cat 1 signal response.

97.4.6.3 Analysis and interpretation

As outlined above, CIRIA R144 (Turner, 1997) emphasises that there are two stages to the interpretation process. These can be summarised as

- analysis of the acoustic data
- interpretation of the significance of this analysis, taking account of all other relevant information for the pile.

Signal type	Characteristics of signal	Signal responses obtained from: Remote end	Intermediate sources	Remarks
0	No impedance change within depth of penetration of signal	Not distinguishable because of soil damping or other effects	Not distinguishable because of soil damping or other effects, or not present	The reason for the response is readily understandable to the non-specialist engineer. Caution required in evaluating depth of penetration of signal
1	One clear impedance change	Clear major signal, indicating the pile system is responding as a single acoustic unit	Absent or very subordinate. Less than 50% of relative magnitude of remote-end signal	The response is readily understandable to the non-specialist engineer
2	More than one impedance change	(1) Clearly discernible, reasonably prominent signal	Moderate to strong signals, but not completely obscuring the remote-end response	The response is understandable to a non-specialist engineer only with expert assistance
		(2) Not clearly discernible, because it is similar to or weaker than other parts of signal from intermediate levels	Signals of similar strength to or stronger than remote-end response	The response is not capable of interpretation by a non-specialist engineer

Table 97.2 Summary of CIRIA classification of low-strain integrity test signals (Reproduced with permission from CIRIA R144 (Turner, 1997); www.ciria.org)

The analysis of the acoustic data includes the identification and depth of impedance changes within the signal response curve, and the evaluation of these to determine whether they are relative increases or decreases in impedance.

The interpretation should take account of the site geology and ground conditions, the pile construction records, and particular features of the piling system (such as the use of permanent casing to the upper section of the pile) or the like.

CIRIA R144 (Turner, 1997) provides further insight into the way this information should be used by the interpreter to provide a reasoned explanation of the condition of the piles. It emphasises that the purpose of such testing is to help evaluate the pile, not the testing system, so that as much information as possible should be given to the specialist to assist in interpretation.

97.4.7 Testing within the contract

CIRIA R144 provides extensive guidance on the use of control testing within the framework of the contract and the constraints of an active working site. It emphasises that many parties have an interest in the results of such testing, but the lines of responsibility and communication are not always clear. This applies especially to low-strain integrity testing but has application to CSL also.

Hertlein and Davis (2006) also highlight this problem from US experience, and point out that NDT, when used in quality assurance testing of piles and other foundation systems, can be viewed from two extreme viewpoints. They suggest that NDT is either

- to assist *engineers* to confirm that the design and performance criteria have been met, or
- to check that the *contractor* has supplied the *owner* with the product that the owner has paid for (material quality, minimal geometric requirements, etc.).

From the first point of view, certain deviations from construction specifications can be tolerated and compensated for, provided that there is strong interaction between *engineer*, *contractor* and *tester*. In the second case, no margin for error outside the contract specifications can be allowed, since they are effectively 'set in stone'.

Davis (1998) contends that the NDT methods now available are more suited to the first scenario (performance specification). This is because the variables inherent in drawing conclusions from NDT on piles are numerous, and it is not always possible to clearly define these variables to reach a complete answer to satisfy the materials specification.

In UK practice there is typically a view that such testing is intended to 'verify' that the construction of the piles is acceptable. However, there is almost always no contractual acceptance criterion against which 'verification' can be judged. Hence, acceptability or otherwise of a particular pile or piles becomes a matter of argument or 'discussion'.

UK practice has also tended towards the specifier of the test being the technical representatives of the employer, owner, developer or funder of the project. Unfortunately, in the authors' opinion, the NDT contract is almost always included within the piling contract. This means that the test contractor must report test results directly to the entity most likely to suffer monetary cost and disruption if adverse results are reported. It is not unusual that the test contractor is specifically instructed not to report to or discuss the test results directly with the employer's technical representative – either by the piling contractor or the general contractor to whom the piling contractor is subcontracted.

This procurement method has led not only to a 'least cost' rather than a 'best specialist' approach, but also to the production of specialist reports which tend to give as little information and comment as possible about the results obtained.

97.4.8 Guidance on the specification of low-strain integrity testing

In the UK, the ICE (2016) SPERW document provides some useful guidance on the specification for the integrity testing of piles and embedded retaining walls, together with some accompanying guidance notes. Unfortunately, there is some confusion in the document on the identification of the various test methods, partly due to the use of proprietary names rather than the generic terms adopted in CIRIA R144 (Turner, 1997). In addition, practice, equipment and techniques have advanced further since CIRIA R144 was first drafted.

Both CIRIA R144 and Hertlein and Davis (2006) provide guidance on the specification of low-strain testing. Hertlein and Davis provide sample specifications for both low-strain testing and CSL. It is not appropriate in this particular chapter to provide a detailed specification, but it is probably useful to highlight crucial points to be borne in mind when specifying such tests.

ASTM D5882-16 covers the procedure for determining the integrity of individual vertical or inclined piles by measuring and analysing the velocity (required) and force (optional) response of the pile induced by an impact device (hand-held hammer or other similar type) usually applied axially and perpendicularly to the pile head surface. The standard provides minimum requirements for low-strain impact testing of piles. Plans, specifications and/or provisions prepared by a qualified engineer, and approved by the agency requiring the test(s), may provide additional requirements and procedures as needed to satisfy the objectives of a particular test programme.

97.4.8.1 Type of test

It can be seen from the earlier discussions that there is a general hierarchy of technical quality as a result of the methodology and techniques employed. In order of increasing sophistication these can be identified as

- time-based 'echo' techniques
- frequency-based techniques
- frequency-based techniques with simulation modelling, including impedance-log analysis.

In this context it should be appreciated that time-based techniques only examine data in the time domain. Frequency-based techniques can examine the data in both the time and frequency domains and also monitor the input force from the test hammer. Simulation modelling techniques attempt to remove extraneous information from the signal, such as that caused by the influence of the ground within which the pile is embedded.

It is important that the specifier is aware of and appreciates these differences. The more knowledgeable the specifier, the better they can target the testing. It is also important to realise that a more sophisticated test, with a correspondingly more complex analysis, should be expected to be more expensive. Hence, the selection of a test organisation on price alone may not be appropriate in many cases.

An important difference between time-based and frequency-based techniques, which may be of major interest to the end-user, is that the examination of the data in the frequency domain can assist in examining anomalies close to the pile head. This is because the stress wave imparted by the hammer blow would typically be 1.0–1.5 ms long, which is equivalent to a wavelength of around 2–3 m in a concrete pile. Hence, reflections returning from a near-surface anomaly can overlap with part of the downward-travelling wave from the initial hammer blow. The resulting complex signal can be difficult to differentiate and interpret in the time domain. By comparison, in the frequency domain such a feature would be expected to be visible at a frequency interval of 666–1000 Hz and might be more readily discernible (see **Figures 97.12** or **97.20(b)**, for example).

As a generality, a defect close to the pile head would often be of greater concern to the design engineer, especially if the bulk of the pile load is carried by shaft friction rather than in end-bearing.

97.4.8.2 Experience of testing specialists

Research testing in the USA (e.g. Baker *et al.*, 1993; Iskander *et al.*, 2001) indicated that integrity testing methods were quite dependent upon the skill and experience of the test operator. Hertlein and Davis (2006) suggest that field testing and preliminary interpretation should be undertaken by an experienced operator with at least one year's experience with the particular method employed. They further suggest that final interpretation and reporting should be performed by, or under the direct supervision of, an engineer or senior technician with at least three years' experience with the specific method being employed.

The testing specialist should be expected to have such experienced and qualified personnel on its staff. It should also be able to provide computer simulations of the anticipated response of the piles at the subject site, given details of pile characteristics (length, diameter, concrete and reinforcement details), cut-off levels and ground conditions from test level.

It should also be able to provide a sample test report, with full explanatory annotation as necessary. The selection of a suitable test specialist is of paramount importance to all parties, and the depth and coverage of a sample test report can be a guide to the test organisation's understanding and ability in this highly specialist field.

97.4.8.3 Pile head preparation

Pile head preparation is most important to the quality of the data acquired for subsequent analysis. It is to be expected that most integrity test houses can provide guidance on pile head preparation, and for best results this advice should normally be followed.

97.4.8.4 Equipment

Current recommendations suggest that the measuring and data acquisition systems should have certified dynamic ranges of at least 60 dB and flat frequency responses over a minimum bandwidth of 20–2500 Hz, and the analogue to digital resolution should be no less than 16 bits.

All equipment should be provided with calibration certificates traceable to UKAS standards. Impulse hammers, velocity transducers and the data acquisition system should have recalibration checks at intervals of no more than 12 months.

97.4.8.5 Reporting of test results

A reporting timescale should be specified. This would typically include preliminary 'overview' reporting at the time of test or within 24 hours, with the issue of a formal analysis and evaluation report within 3–5 working days.

The testing specialist's report should include signal response graphs for each pile, as applicable, and an analysis of the response of each pile.

Analysis and interpretation of the test data for each pile by the testing specialist should include

- whether a clear response has been detected consistent with the level of the pile toe, or any other level within the pile
- whether any secondary responses have been detected
- whether the responses are due to an increase or decrease in pile impedance, and a view of their magnitude and importance
- an interpretation of the cause and significance of these responses or 'features'
- recommendations for any further investigation or measures if required.

97.4.8.6 The review of acceptance or rejection criteria for the piles, based upon integrity testing

It is important that the specifying engineer, the main contractor and the piling contractor have in mind how they are going to deal with the results obtained from the testing, and the specification should address this point. It should be appreciated that this is not a minor matter. There is often a major cost implication in terms of delay and disruption to the contract progress, quite apart from the costs of remedial works to the pile(s).

Hertlein and Davis (2006) suggest the following approach.

- If a clear response consistent with the level of the pile toe is detected, with no significant secondary responses from levels above this, then the pile would be deemed acceptable in terms of the integrity testing.
- If no clear response from the pile toe is detected, with no significant secondary intermediate responses above the level of the pile toe, then the supervising engineer of the test house as identified in Section 97.4.8.2 (the experience of the testing specialist) should report what length of the foundation pile may be considered to be proven, based on their interpretation of the test data.
- Reflections interpreted to be from any level above the level of the toe of the pile should be analysed by the testing specialist to determine whether the reflection is from a significant reduction or increase in impedance, since these are consistent, in the main, with a corresponding reduction or increase in the pile cross-section. The probable/possible nature and significance of the reflector should then be assessed, taking into consideration the intent behind the design of the pile. The decision to accept or reject the pile or to require any further investigative works should be made by the engineer responsible to the client for the overall design of the structure.

If the test result is deemed to be inconclusive, the pile head may be trimmed back, and the pile re-tested to try to provide more conclusive data. If this is impractical or is not successful, additional testing by other means may be instructed. Such measures would include excavation to expose the area of concern, coring the pile or undertaking dynamic or static load testing. Such measures are reviewed in CIRIA R144 (Turner, 1997).

97.4.8.7 Timing of the testing

CIRIA R144 discusses the potential clash of interests between the contractor, who might prefer to test early, and the client, who is really most concerned about the pile immediately before it is incorporated into the works. However, the later the test the greater is the potential disruption to the overall contract if problems are found. The specifier should address this point, and the specification should be clear on the requirements. As a guide, it is not unusual to require that the testing of any pile should be undertaken within 28 days of installation, and that all testing should be complete within 14 days of the installation of the last pile. For cast-in-place piles, it is usual to specify a minimum curing period before testing. This would typically be 3–7 days, depending on the characteristics of the concrete mix.

97.5. Cross-hole sonic logging

97.5.1 Introduction

The method is a development of the ultrasonic pulse velocity (UPV) test method commonly used to determine ultrasonic transit time in concrete. The UPV test method is described in detail in BS EN 12504-4:2021.

Vertical ducts are cast into the pile over its full length. Two probes, one a transmitter and one a receiver, are lowered to the bottom of adjacent ducts. An ultrasonic pulse from the transmitter is detected by the receiver and the signal is captured by the data acquisition system, as illustrated in **Figure 97.22**. A continuous series of such measurements is taken as the probes are simultaneously raised up the ducts, so that a continuous profile of the transit time and characteristics of the wave-train generated by each pulse is built up at successive levels in the pile.

97.5.2 Description of the technique

CSL is usually used on piles with a diameter of 600 mm or greater. Steel or, less usually, plastic access tubes are cast into the pile during the construction phase. Plastic tubes can often suffer from debonding from the concrete, rendering the test ineffective; hence steel tubes are preferred in most cases. The tubes are typically attached to the inside of the reinforcing cage over the full length of the pile and extend above ground level by approximately 500 mm. A minimum of three tubes are used, usually increasing to a minimum of four for pile diameters of 750 mm or more. The tubes are usually

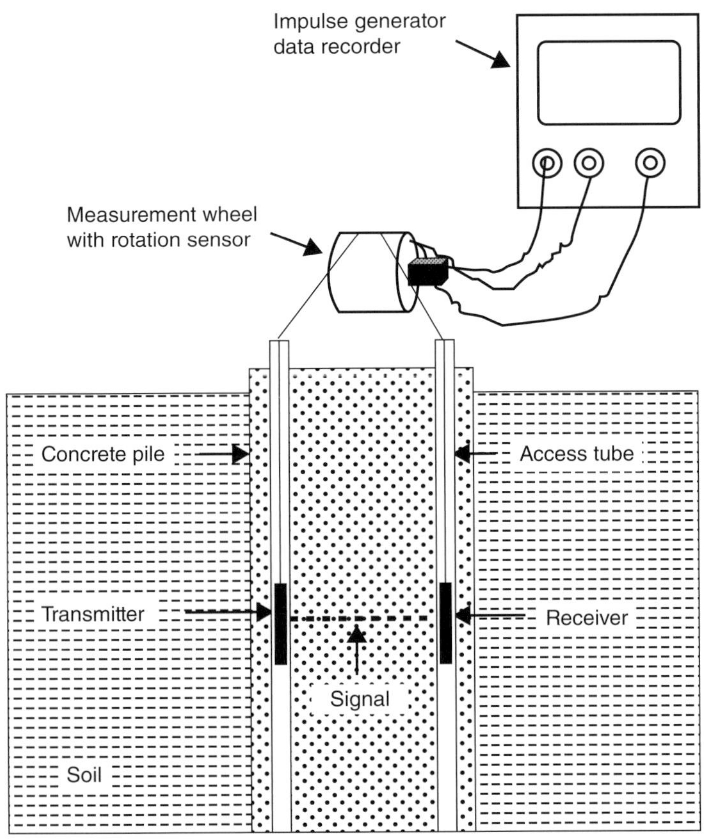

Figure 97.22 Schematic of a CSL system (Reproduced from Hertlein and Davis, 2006; Wiley)

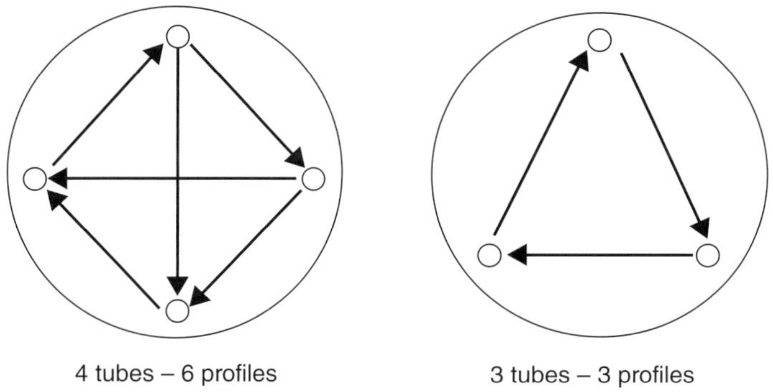

Figure 97.23 Plan view of CSL access tube configuration

equispaced around the circumference of the reinforcing cage to provide a minimum of three longitudinal 'profiles' of the concrete between the tubes for a three-tube configuration, and up to six for a four-tube configuration, as illustrated in **Figure 97.23**. The steel tubes are generally approximately 40 mm in internal diameter to facilitate the passage of transmitter and receiver transducers.

Prior to the test, the tubes are filled with clean water to act as a coupling medium for the transducers.

The technique is also suitable for use in diaphragm wall panels or barrettes. A typical layout of ducts within diaphragm wall panels is illustrated in **Figure 97.24**.

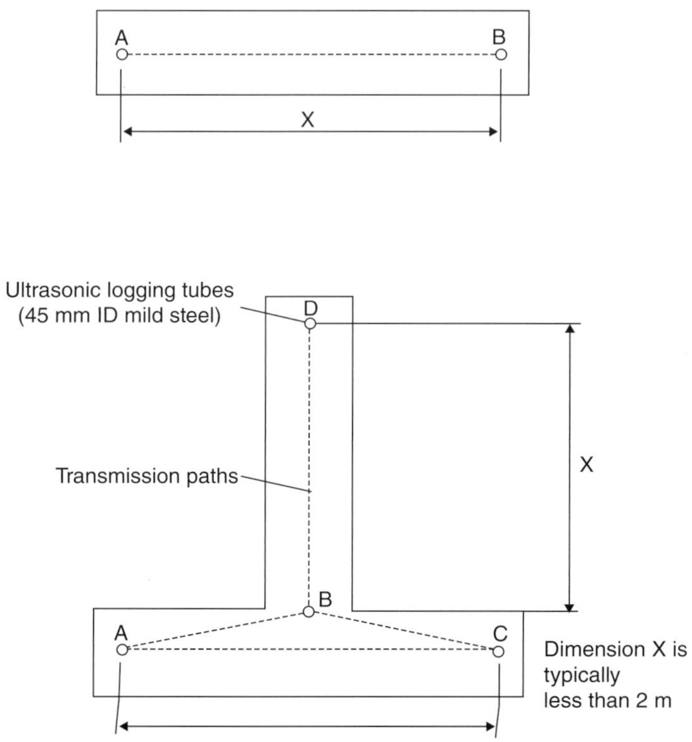

Figure 97.24 Plan view of sonic logging ducts within diaphragm wall panels (Reproduced from Stain and Johns, 1987; Deep Foundations Institute)

The transit time (t) of the wave-train from each pulse is a function of the distance between the tubes (L) and the propagation velocity (c) of the concrete between the piles. In turn, the propagation velocity of the pulse is a function of the modulus, E, density, ρ, and Poisson's ratio, ν, of the pile concrete. Hence anomalies within the pile, such as voids, soil inclusions or low-modulus concrete, would be expected to give an increase in the transit time, provided that the access tubes remain at a constant relative position apart (i.e. L remains constant). In addition, if the anomaly was caused by voided or poorly compacted material, for instance, this might also lead to an attenuation of the signal.

The two probes (transmitter and receiver) are slowly raised up the tubes, and a reading is taken every 20 mm. Each received ultrasonic pulse is collected and the wave-train would have the form illustrated in **Figure 97.25(a)**.

With the system developed by CEBTP, each received ultrasonic pulse is processed by the data acquisition system so that each 'positive' peak of the wave-train is printed as a black line with a width matching that of the original peak; conversely, each 'negative' trough is printed as a gap. Hence a dashed line is created as illustrated in **Figure 97.25(b)**, which contains all of the timing and much of the amplitude information of the original record. Each consecutive pulse is printed contiguously to form a vertical profile of the data for that particular pair of tubes, as illustrated in **Figure 97.25(c)**. This profile is often known as a 'waterfall plot' or a 'sonic profile'. In this way any changes in transit time can readily be distinguished from the mean value and identified for further investigation.

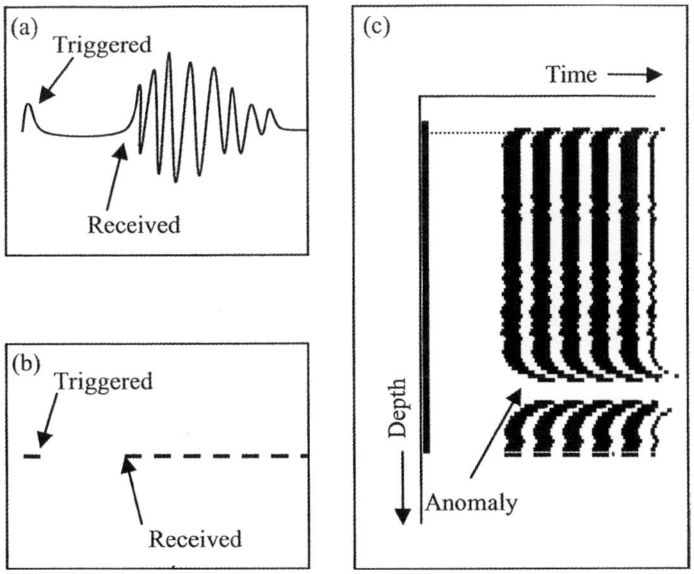

Figure 97.25 Compilation of a CSL plot: (a) original single pulse; (b) pulse data modulated to form a single positive/negative dashed line; (c) modulated pulses stacked to form the complete 'waterfall' plot (Reproduced from Hertlein and Davis, 2006; Wiley)

Other CSL systems measure the arrival time of the first peak in the ultrasonic pulse wave-train, known as the first arrival time (FAT), together with the overall amplitude of the early part of the pulse, as illustrated in **Figure 97.26**.

Hertlein and Davis (2006) suggest that a drawback of this method is that the system has to detect the first arrival peak in a digitised wave-train and has to be able to differentiate between background noise and the arriving signal. Thus, a threshold level has to be set that eliminates the background noise, as illustrated in **Figure 97.27**. Unfortunately, the signal pulse is a relatively small ‘noise’, and construction sites tend to be noisy. Hence, if the first or second peak in the arriving wave-train falls below the threshold level, the computer will identify the ‘first arrival’ as the first peak that exceeds the specified threshold, be it the second, third or even later arrival.

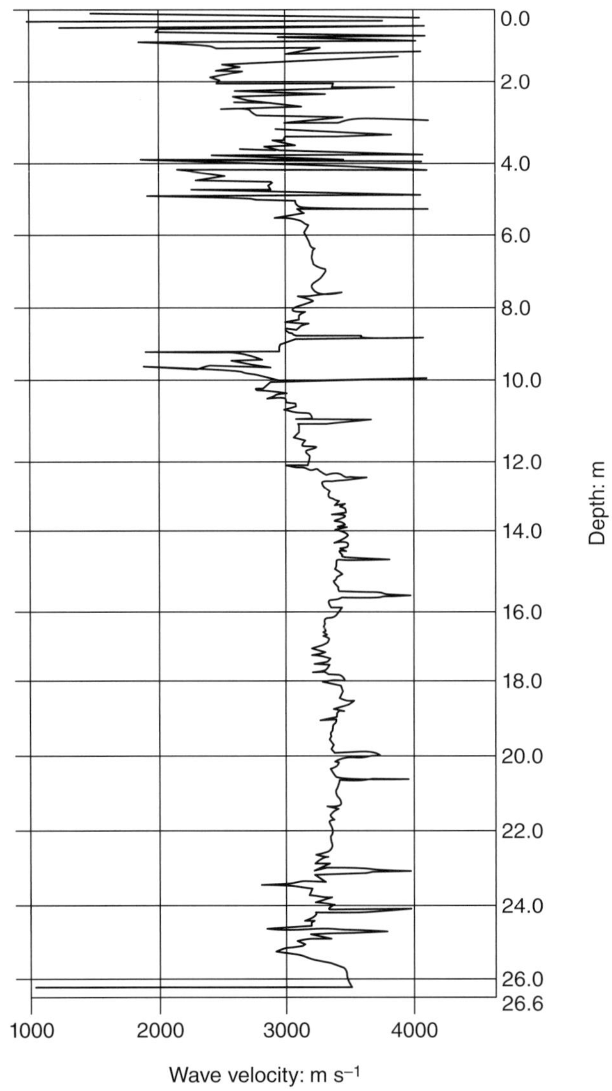

Figure 97.26 Typical FAT (first arrival time) CSL plot (Reproduced from Hertlein and Davis, 2006; Wiley)

97.5.3 The detection of defects

As noted in CIRIA R144 (Turner, 1997), the test identifies the shortest acoustic path between the transmitter and receiver. With a three-tube system, for example, it is possible that a feature at the centre of the pile would not be detected, as illustrated in **Figure 97.28(a)**. Similarly, since the access ducts are usually attached to the pile reinforcement, a defect such as loss of concrete cover would not typically be detectable as in **Figure 97.28(b)**. The method is also unsuitable for detecting transverse defects unless a staggered probe method is used.

97.5.4 Features detected using the CSL method

A typical test result from a defective diaphragm wall section is given in **Figure 97.29**. Interpretation of the results is generally straightforward when signal loss increases dramatically, but the threshold for what is considered to be anomalous concrete is less clear. In practice, an increase in transit time of less than 10% is not considered to be significant unless all profiles are affected. If this is the case then further investigation may be required, which could include coring the pile, reference to the pile construction records, test cube analysis and further clarification of axial loading and bending moments on the pile.

The association between ultrasonic propagation velocity and concrete strength has been advocated in several technical papers, and an argument exists to suggest that this may be applied to data collected from CSL data. Tube spacing is, however, an additional variable which needs to be taken into consideration. For further reading on the correlation between concrete strength and ultrasonic pulse velocity, refer to Turgut (2004).

Signal damping is often a feature of the CSL results, and this may be due to debonding of the access tubes or a thin layer of contaminated concrete surrounding one or more of the tubes. Signal damping with no associated increase in transit time is not considered to be significant.

Three-dimensional computer tomography has been introduced by several equipment manufacturers since 2000. As noted by Fleming *et al.* (2009), this can model the extent and form of a defect or anomaly within the pile in 3D form by utilising and combining, for example, the six 2D profiles obtained from a four-tube layout. The vertical position of such anomalies, together with the significance of relative increases in transit time between adjacent pairs of tubes, is colour-weighted to produce a 3D colour-coded graphical representation of the pile and to help visualise the potential anomalies.

The nature of these anomalies and their structural significance is, however, still subject to the analysis and interpretation of the test specialist. An example of computer tomography is presented in **Figure 97.30**.

97.5.5 Health and safety considerations

Over recent years safety concerns have been raised regarding the potential for serious injury during splicing of CSL access tubes in long piles. Careful consideration of the benefits of

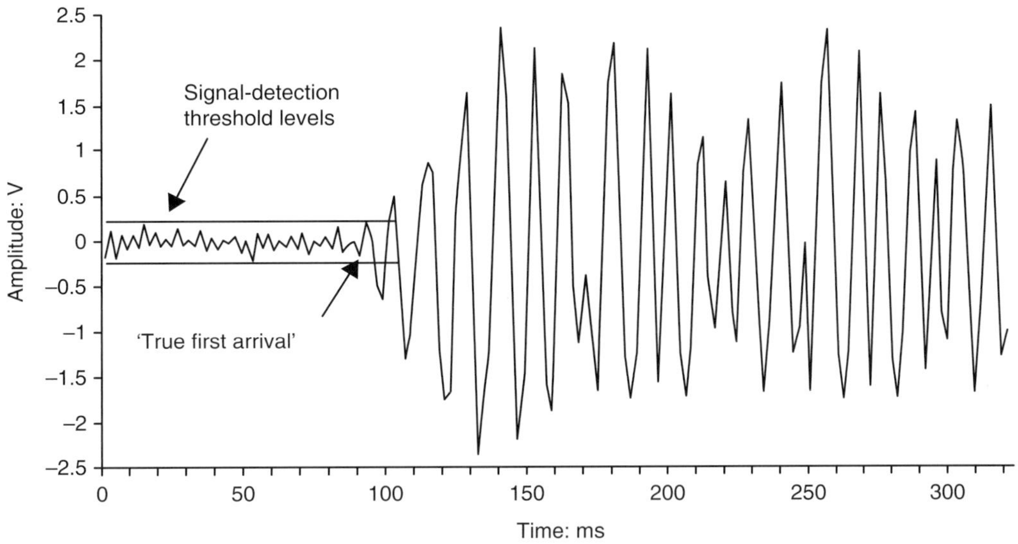

Figure 97.27 Typical CSL pulse, with illustration of signal threshold levels and 'first arrival time' (Reproduced from Hertlein and Davis, 2006; Wiley)

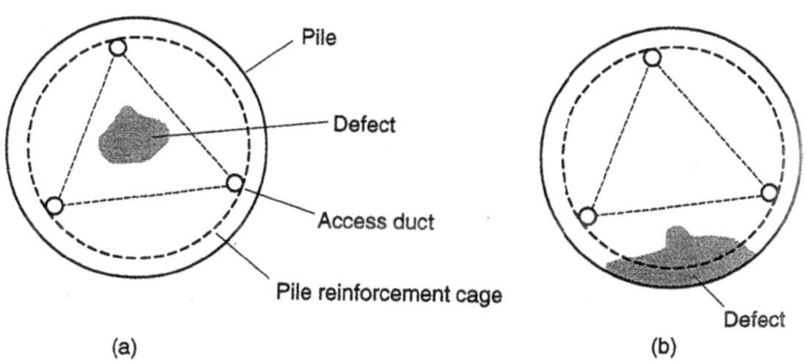

Figure 97.28 Cross-hole sonic logging. Examples of possible defects off the shortest direct path between probes: (a) defect in centre; (b) defect on perimeter (Reproduced with permission from CIRIA R144 (Turner, 1997); www.ciria.org)

CSL against risk of injury in all circumstances must therefore be undertaken and robust risk assessment and access tube installation method statement must be undertaken.

When specifying CSL, due consideration should be given to prevailing ground conditions. Piles installed using dry, open-bore methods in stable ground such as stiff clay and rock sockets and where visual inspection to confirm shaft integrity prior to concreting can be safely undertaken renders CSL unnecessary. Installation of CSL tubes in CFA (continuous flight auger) piles below 600 mm dia. where the overall steel congestion may restrict adequate concrete flow should also be avoided.

Where associated risks with installation of CSL access tubes cannot be fully mitigated, then alternative testing should be

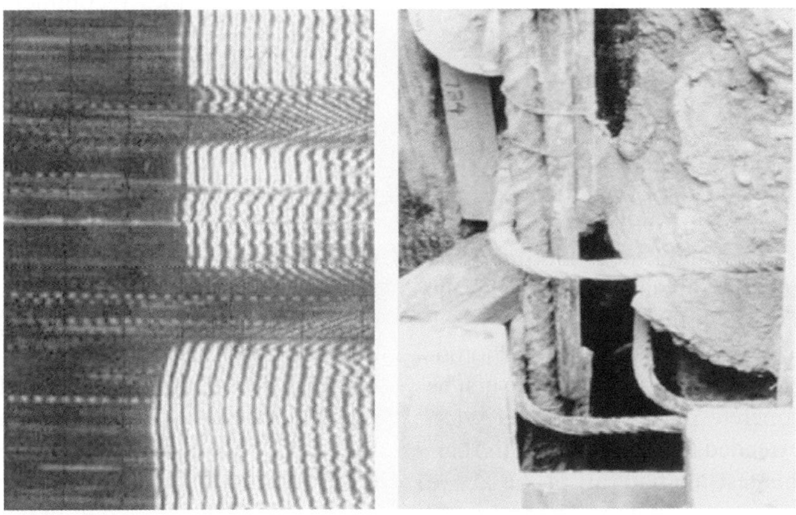

Figure 97.29 CSL profile and excavated diaphragm wall (Reproduced from Stain and Williams, 1991; Deep Foundations Institute)

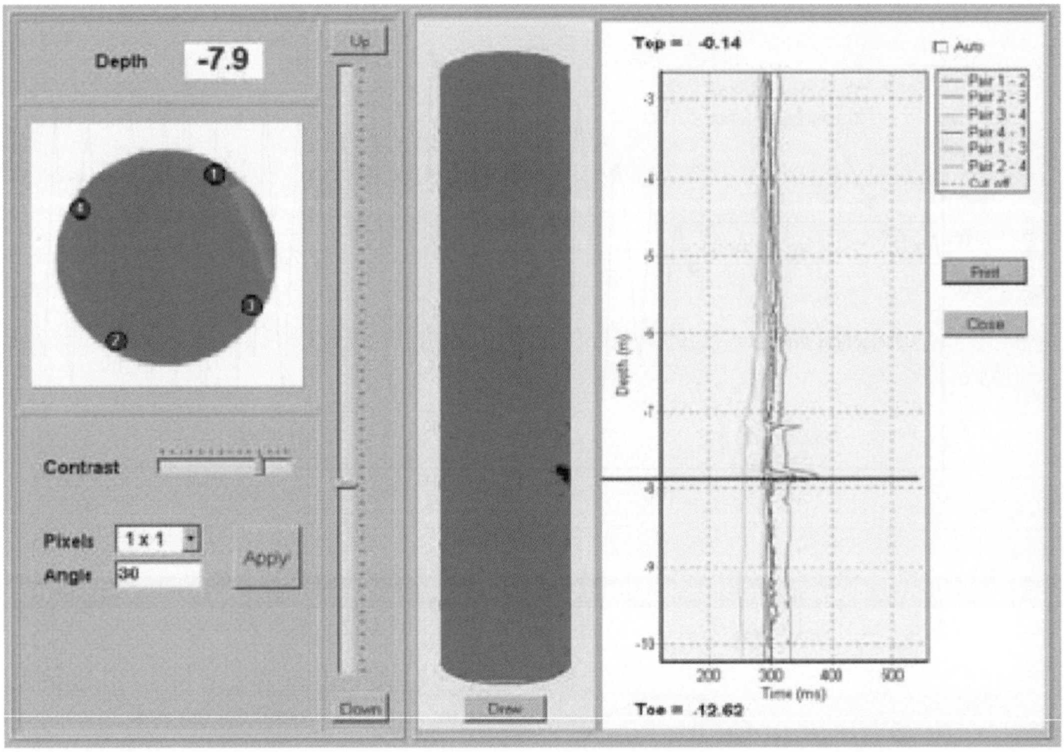

Figure 97.30 Example of computer tomography (Courtesy of British Engineering Services Limited (formerly Testconsult Limited))

considered. Conversely, CSL may be the only method for end-bearing piles where pile toe support is critical or where base grouting is necessary, particularly where drilling using supporting liquid (water, bentonite, polymers, etc.) is employed.

Advances in hardware and computer modelling and mainstream adoption of temperature integrity profiling provide an alternative to CSL testing – Section 97.6 describes the method in more detail. Although risks are reduced (cable splicing will still be required), little information will be provided on the pile toe interface. Further advances with modelling of the pile toe/soil interface may improve understanding of concrete quality in this critical area.

97.5.6 The specification of CSL

Similar considerations to those highlighted in Section 97.4.8 for low-strain integrity testing would also apply to the specification of CSL testing. Again, ICE's (2016) SPERW document, and both Turner (1997) and Hertlein and Davis (2006) provide guidance on this issue. The latter document also references the French AFNOR and US ASTM D6760 standards.

ASTM D6760 covers procedures for checking the homogeneity and integrity of concrete in deep foundations such as bored piles, drilled shafts, concrete piles or augercast piles. The test method can also be extended to diaphragm walls, barrettes, dams and so on. In this test method, all of the above will be designated 'deep foundation elements'. The test measures the propagation time and relative energy of an ultrasonic pulse between parallel access ducts (cross-hole) or in a single tube (single-hole) installed in the deep foundation element. This method is most applicable when performed in tubes that are installed during construction.

In a similar manner to that proposed for low-strain testing, Hertlein and Davis (2006) suggest that the person responsible for analysis and reporting on the CSL test data should be able to demonstrate a minimum of, say, five years' experience in CSL testing, with at least three CSL projects in each of those years. Similarly, they recommend that field personnel responsible for performing or supervising the performance of the CSL test shall be able to demonstrate at least two years' experience with CSL testing, with a minimum of three projects in each of those years.

Hertlein and Davis also recommend that the text of the CSL report should include a description of any anomalous zones identified by the test data and a discussion of the apparent difference in pulse velocity. Account should be taken of the sources of possible variation. One particular point that is highlighted is that when assessing the likely significance of an anomaly, any available information concerning test strengths or other actual measurements of concrete properties should be included in the report. These authors highlight that it is not unusual for actual concrete strength to significantly exceed the design strength, in which case the concrete in an anomaly that shows a 15–20% reduction in pulse velocity may still meet or exceed project specifications.

With regard to the review and acceptance/rejection of the piles, a time period (typically three working days) should be specified for the engineer to evaluate the CSL report and determine whether or not the tested piles are acceptable. If the pile is not acceptable, the engineer should decide what additional testing, investigation or analysis is necessary to properly characterise the anomaly and determine whether or not the pile needs any repair or remediation. In such an eventuality the repair technique to provide satisfactory remediation must be proposed and agreed between the parties. Once the testing regime is completed, access tubes should be fully grouted to eliminate any potential internal corrosion activity.

97.6. Thermal integrity profiling (TIP)

97.6.1 Introduction

Following development in the late 1970s and early 1980s, cone penetrometers were being used with various sensors (e.g. pore pressure, resistivity, cameras, etc.). At that time, faculty researchers at the University of South Florida gave serious consideration to taking soil temperature measurements around freshly cast shafts using the cone as the means to gain access to these regions.

Two hurdles seemed insurmountable: (*a*) the time to achieve thermal equilibrium between a cone-based temperature sensor and the soil (without creating thermal disturbances) was too long to be practical and (*b*) the inability to penetrate rock or stiff soil.

A research project to investigate the merits of assuring shaft integrity using infrared thermal detection methods was awarded to the University of South Florida by the Florida Department of Transport. The objective included field temperature measurement of mass concrete structures with specific interest in drilled shafts and numerical modelling to verify the anticipated temperature response within a drilled shaft or mass concrete structure. The project report describes the results and conclusions of the research and has become one of the defining reference documents on the subject (see Mullins *et al.*, 2007).

The cost of throw-away embedded instrumentation (e.g. thermocouples or similar) in the reinforcing cage or in small boreholes surrounding the excavation was exorbitant at the time. However, as instrumented load tests have become more common, so have embedded inclinometer casings, which opened the door to measurements from reusable down-hole devices capable of monitoring inclination, lateral acceleration, axial strain, density, wave speed and temperature.

The first full-scale versions of thermal integrity profilers used inclinometer wheel bodies with larger infrared sensors rather than the embedded sensors used today. By the turn of the twenty-first century, several versions of the equipment had evolved progressively smaller probes to provide access in smaller-diameter tubes, staying abreast of the trend towards smaller CSL devices. Smaller access tubes reduce cage congestion and aid in providing better concrete flow through the cage openings.

TIP uses the measured temperature generated during concrete curing to assess the quality of cast in situ concrete foundations. In general, the absence of competent concrete is registered by cool regions (necks or inclusions) relative to the shaft norm; additional concrete volume is signified by relatively warm regions (e.g. bulbs into soft strata). Relative changes in concrete volume at any specific depth both inside and outside the reinforcing cage moderate the normal temperature signature detected by the nearest temperature sensor. Anomalies (inclusions, necks, bulges, etc.) are detected by more distant tubes, but with progressively less effect.

To assess the quality of a cast in situ concrete foundation (i.e. bored, cast in situ or CFA piles), the relative concrete temperature which is interdependent of the pile diameter, mix design and the time of measurement with respect to hydration process (concrete maturity) is considered. **Figure 97.31** shows concrete temperature during the five phases of hydration.

Whereas other methods of integrity testing have limits in assessing the full cross-section or length, TIP measurements evaluate the concrete quality from all portions of the cross-section along the entire length.

97.6.2 Description of the technique

The TIP method generally measures in situ temperatures using low-cost PT100 or thermocouple sensors

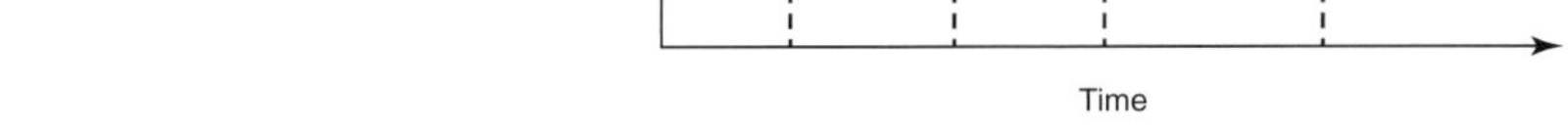

Figure 97.31 Five phases of the hydration process (Courtesy Maturix)

interconnected with signal cable and attached to the reinforcing cage prior to casting the shaft. The temperature sensors are evenly spaced approximately every 300 mm along the length of the shaft. Generally, one temperature sensor is installed for each 300 mm of drilled shaft diameter, evenly spaced around the reinforcing cage and rounded to the nearest whole number.

The temperature sensors are connected to a logging system which begins collecting data immediately after the shaft is cast. Procedures for performing the test are further described in ASTM Standard D7949. **Figure 97.32** shows the thermal wire cables extending above the top of a concreted shaft with the end of each cable attached to a wireless data logger module.

As an alternative, distributed fibre-optic sensing (DFOS) is becoming popular where reliability, high sensor density and ease of installation are proving advantageous. DFOS is now regularly used for preliminary pile load tests. Another advantage is that installed fibre may be used for TIP during the curing phase and to measure shaft axial strain during subsequent load testing.

The particular DFOS technique emerging as most appropriate is known as Brillouin optical time-domain reflectometry (BOTDR). Brillouin scattering defines the spatial resolution (the smallest distance over which strain or temperature can be measured with full accuracy), which is limited to 0.5–1.0 m for most commercial BOTDR equipment. The procedure is described in detail by Rui *et al.* (2017).

Figure 97.32 Temperature probes attached to loggers (Courtesy Pile Dynamics Inc.)

Concrete temperatures are either sampled throughout the concrete hydration process or at a time prior to the peak curing temperature being reached (phase three curing region). Relative temperature evaluation or profiling is consequently available relatively soon after pouring (6–72 hours), earlier than when other integrity testing methods such as CSL and low-strain integrity testing can be performed.

Temperatures are measured along the entire depth of the foundation. If records at a certain depth show regions with cooler temperatures (when compared to the average temperature at that depth), a concrete deficiency or defect may be present. An average temperature at a certain depth that is significantly lower than the average temperatures at other depths may also be an indication of a potential problem. It is also possible to estimate the effective area of the foundation and to assess whether the reinforcing cage is properly aligned and centred.

97.6.3 The detection of defects

TIP provides various details of shaft integrity, which include effective shaft size (diameter and length), anomaly detection inside and outside the reinforcement cage, cage alignment and proper hydration of the concrete. The ability to detect concrete volumes outside the reinforcing cage is perhaps its strongest feature. Defects are best observed during the early curing stage, at a time when the shaft temperature is one-half the peak temperature.

Each sensor on each string is scanned for local reductions in temperature. The overall average temperature of all temperature probe readings of a given foundation element over the embedded depths can be directly related to the overall volume of concrete installed. In short, the average temperature profile (from all wires or tubes) defines the shape of the shaft and local variations from the average indicate cage offsets from centre.

For bored piles, the pile integrity can be assessed based on the average temperature measurements from each temperature probe at each depth increment. If the measured average temperature is consistent over the monitored range of depths, the pile is considered to be of uniform shape and quality. Bulges can be identified as localised increases in average temperature, while insufficient concrete quality or section reductions can be identified as localised decreases in average temperature. Anomalies present over more than 10% of the effective cross-sectional area are generally indicated in multiple temperature probes at the same depth. Because soil and slurry pockets produce no heat, areas of soil intrusion or inclusion are indicated by lower local temperatures.

97.6.4 Features detected using the TIP method

Reinforcing cage location can be estimated based on the relative temperature difference between an individual temperature probe and the average of all probes. Higher individual sensor temperatures indicate that the cable is closer to the centre of the bored pile, or near a local bulge, while lower individual temperatures indicate that the sensor is closer to the soil–pile interface, or to a local defect. Lateral shift of the reinforcing cage can be determined if one cable temperature is higher than average, and the diametrically opposite cable temperature is lower than average.

Figure 97.33 presents TIP results for a 1524 mm dia. bored pile. The leftmost plot presents the measured temperatures against depth. Note that the shaft has a relatively uniform temperature against depth, with the exception of the top and bottom as well as near a depth of 11.5 m. The top and bottom temperature variations are normal, where the shaft temperature reduces to the air temperature at the top and the soil temperature at the base. These environmental influences can be modelled and test results adjusted for their effects, as described by Sun *et al.* (2020).

For example, the centre plot in **Figure 97.33** shows the average temperature of all temperature readings over the embedded depths. These are 'normalised' by cross correlation to allow for the overall volume of concrete installed to yield the pile radius against depth and the concrete cover against depth. The significant drop in temperature near 11.5 m indicates an integrity concern. The temperature recorded by sensors is below the 6% reduction criteria as defined by Piscsalko *et al.* (2016), warranting further evaluation. Subsequent investigation by concrete coring encountered a 150 mm thick void at this location, necessitating pile remediation. A 3D representation of the pile and cage overlain on the soil description is presented in the rightmost plot.

Further analysis of the collected data may be undertaken by modelling of the surrounding soils. For a perfect cylindrical pile without any defect, the longitudinal temperature distribution is nearly constant within the same soil layer. The change of soil thermal properties between different strata affects the heat dissipation rate from the concrete body to the surrounding ground and leads to longitudinal temperature variation, as shown in **Figure 97.34**. The soil with higher thermal conductivity allows faster heat transfer; thus, lower temperature would be recorded within the corresponding strata on the longitudinal temperature profile. Generally, a significant 'roll-off' would exist on both ends of the longitudinal temperature profile, as heat dissipates not only radially outwards but also vertically through the ends to the ambient environment (air) on the top or to the soil layer at the bottom.

97.6.5 The specification of TIP

ASTM D7949-14 provide procedures for measuring the temperature profile within a deep foundation element constructed using cast in situ concrete, such as bored piles, drilled shafts, augered piles, diaphragm walls, barrettes and dams. Two alternative procedures are suggested.

- Method A uses a thermal probe lowered into access ducts installed in the deep foundation element during construction.
- Method B uses multiple embedded thermal sensors attached to the reinforcing cage installed in the deep foundation element during construction.

All observed and calculated values shall conform to the guidelines for significant digits and rounding established in ASTM D6026-21 unless superseded by this standard.

The standard provides minimum requirements for thermal profiling of concrete deep foundation elements. Plans, specifications and/or provisions prepared by a qualified engineer, and approved by the agency requiring the test, may provide additional requirements and procedures as needed to satisfy the objectives of a particular test programme.

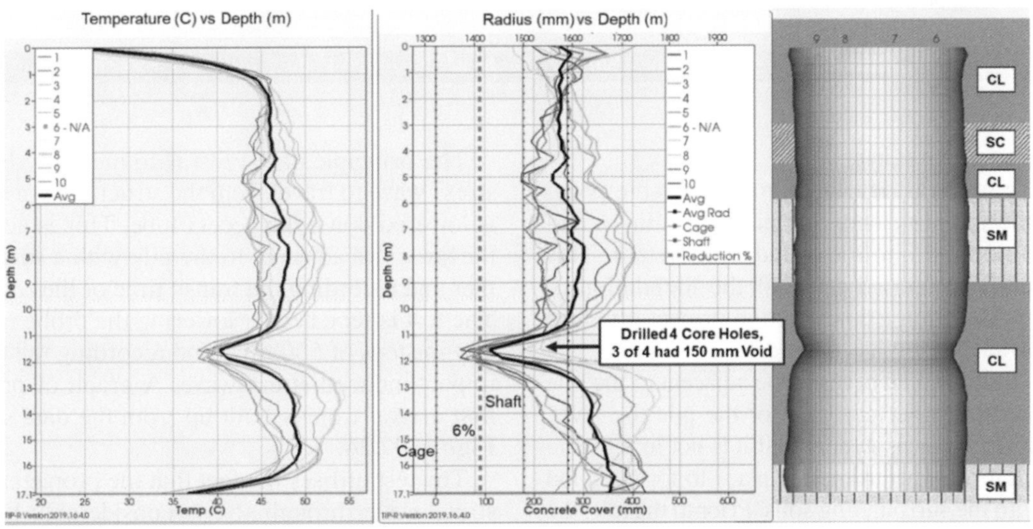

Figure 97.33 TIP results on a bored pile (Reproduced from Hannigan and Piscsalko, 2020)

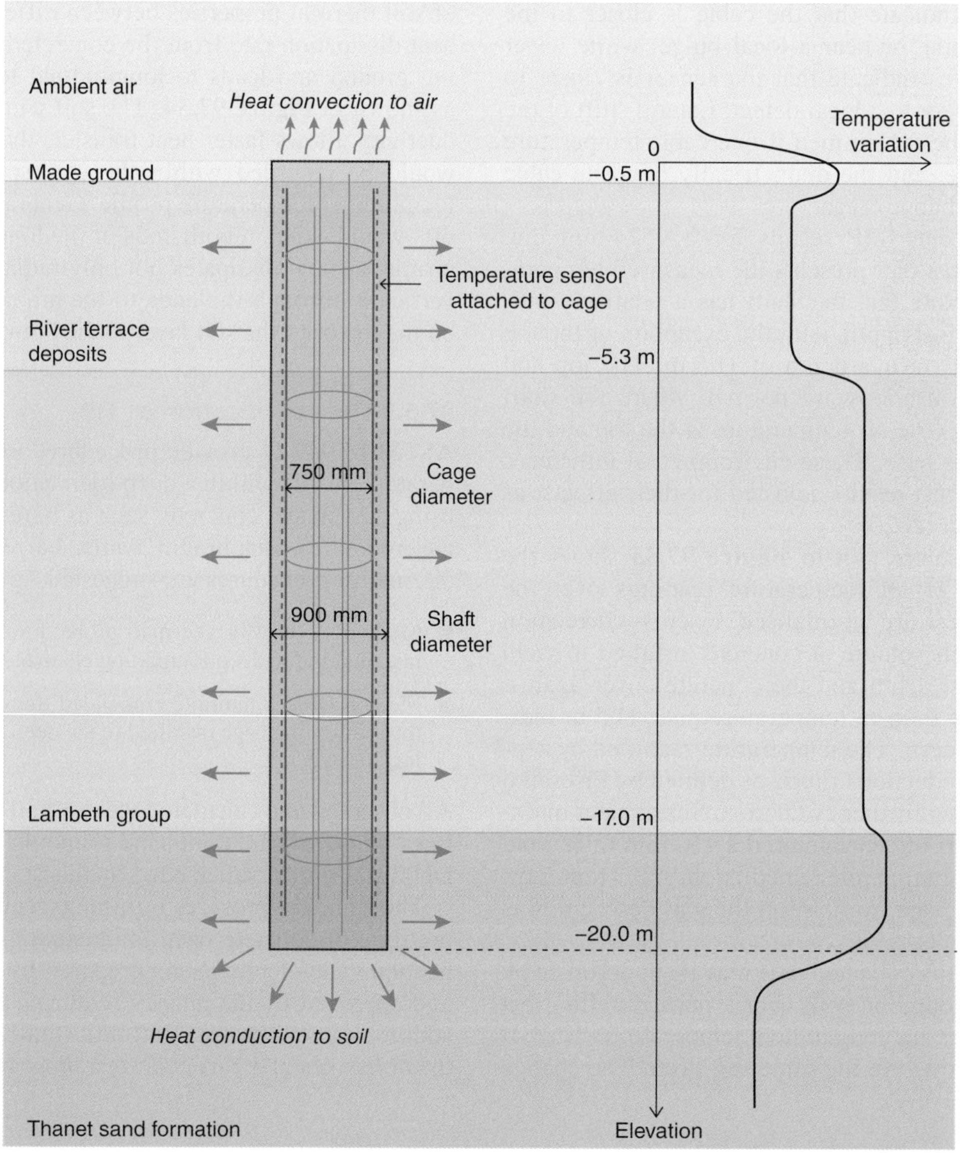

Figure 97.34 Longitudinal temperature variation profile for a perfect cylindrical pile without any defects

97.7. Parallel seismic testing

The parallel seismic test was developed to address the problem of assessing pile integrity once the pile under investigation had been incorporated into the foundation system. This might be as a result of proposed reuse of the foundation or to augment quality control on a construction project after the option for more normal testing has passed.

The test involves installing an access tube adjacent to the pile to a depth exceeding the expected depth of the pile toe level. The tube can either be plastic or steel and, if it is not in intimate contact with the soil, it must be grouted in place to form a good acoustic coupling with the surrounding soils. Prior to the test, the tube is filled with water – again to ensure an acoustic coupling. The arrangements for the test are illustrated in **Figure 97.35.**

The principle of the test is to measure the time taken for a stress wave to travel from the structure to a receiver probe as it is lowered down the access tube. This is achieved by striking the structure adjacent to the pile with a small hand-held hammer and recording the transit time of the resultant stress wave. The test is repeated by lowering the probe incrementally (usually in steps of 500 mm) and recording the transit time for the impact-induced stress wave. A graph of probe depth against first arrival time is built up from the data series, as shown in **Figure 97.36**.

The test utilises the fact that the propagation velocity of the stress wave through soil is considerably lower than through concrete or steel. As the probe passes the toe of the pile, or a defect such as a discontinuity in the pile shaft, the path length

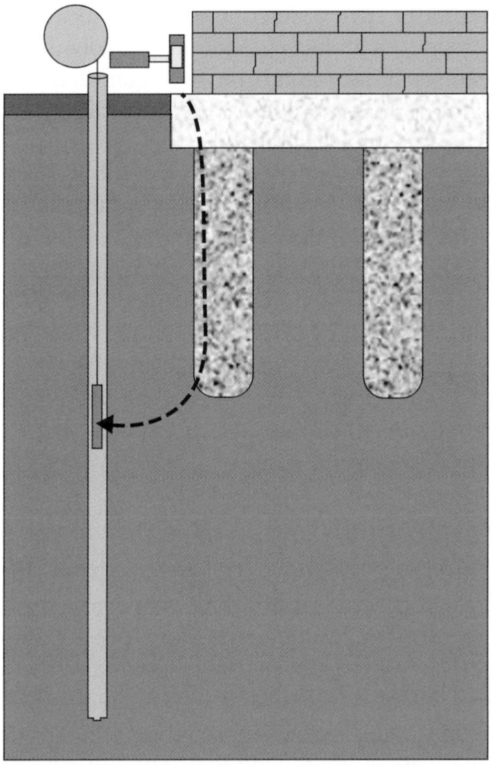

Figure 97.35 Arrangements for parallel seismic test (Courtesy of British Engineering Services Limited (formerly Testconsult Limited))

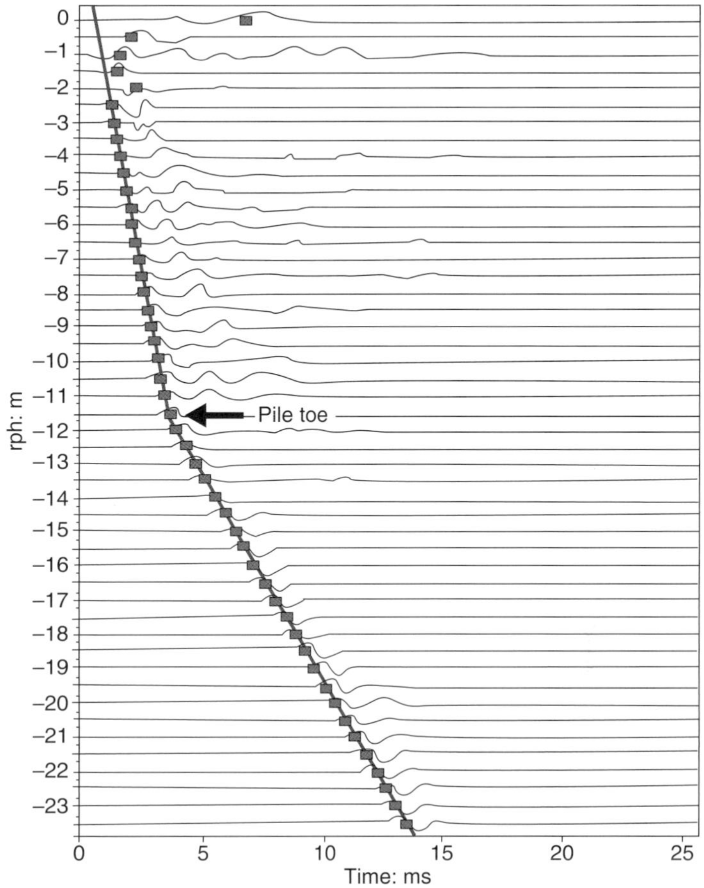

Figure 97.36 Typical parallel seismic test result (Courtesy of British Engineering Services Limited (formerly Testconsult Limited))

of the stress wave will be increasingly through the soil, and the transit time will correspondingly increase at a greater rate, as indicated in **Figure 97.36**. A change in gradient of the first arrival time at the probe depth should therefore indicate the base of the pile (or, alternatively, it could be indicative of a discontinuity in the pile shaft).

97.7.1 Limitations

A limitation of the test is that a borehole must be drilled alongside the pile to provide a duct for the receiver. This needs to be sensibly parallel to the pile (usually vertical) and typically within 1 m of the pile under test. The method is ineffective in dry or loose granular material or unconsolidated fill.

The impact point on the structure for the test hammer must have a good mechanical coupling to the top of the foundation and be as close as possible to the axis of the pile.

97.8. High-strain integrity testing

High-strain dynamic load testing is described in Chapter 98 *Pile capacity testing*. Incidental to predictions of the load-carrying capacity of the pile is the possibility of detecting significant flaws in the pile which may be as a result of overdriving, or construction defects in the case of a cast in situ pile.

The dynamic load test (DLT) involves instrumenting the pile head with strain gauges and accelerometers to measure both upward and downward forces in the pile shaft. The raw data from the instrumentation take the form of total force (F) from the strain gauges (F = strain × cross-sectional area × Young's modulus) and force measured by the processed accelerometer data (F = pile impedance, Z, times pile head velocity, V). If these two sets of data are plotted on the same graph, as illustrated in **Figure 97.37**, any divergence between the two data series will be as a result of changes in pile support from the surrounding ground or the impedance of the pile. In the example shown in **Figure 97.37** the divergence is shown as an increase in force derived from strain measurements (F_t) and a decrease in force based upon velocity measurements ($Z \times V$). This has been generated as a result of shaft skin friction.

If the pile shaft has been damaged due to overdriving or constructed with a shaft defect, then this will present itself as a decrease in the F_t force and a corresponding increase in $Z \times V$ force at the point on the timescale representing the depth to the anomaly.

97.8.1 Features detected using high-strain dynamic load testing

The detection of pile defects using the high-strain DLT should be considered a by-product of the load test and should

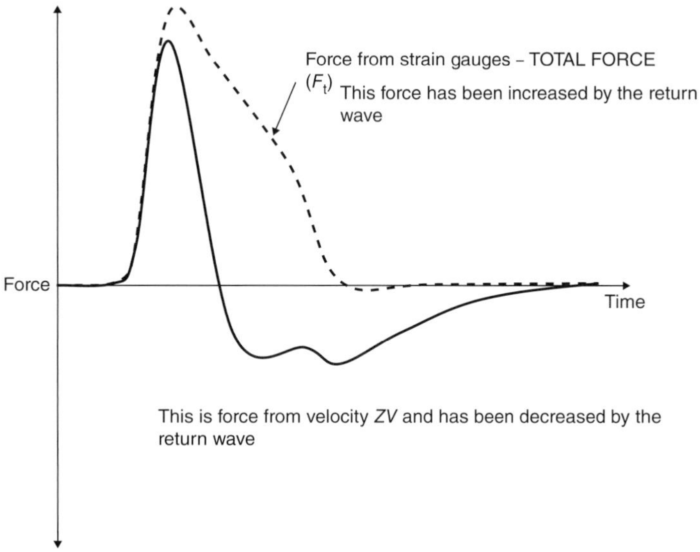

Figure 97.37 Force diagram from a dynamic load test

not be relied upon as the principal method for confirming pile integrity. The reason for this is that the duration of the hammer impact overlaps any return signal from defects close to the pile head, effectively masking them. In addition, if a defect is limited to a partial reduction in section, as may be the case with a bored cast in situ pile, the stress wave will pass the anomaly and continue towards the pile toe. The partial returning signal under these circumstances may be mistakenly identified as a change in soil characteristics at the indicated depth.

This method is, however, useful in identifying damage to precast driven piles which may be due to overdriving. A crack or fracture along the shaft, and in particular close to the pile toe, will exhibit a recognisable change in both total force and $Z \times V$ data. CIRIA R144 (Turner, 1997) provides further information on interpretation and limitations of this method.

97.9. The reliability of pile integrity testing

97.9.1 Introduction

The reliability and even 'trustworthiness' of integrity testing as a quality control or quality assurance tool is a question that constantly arises within any discussion or dispute on the results being found on a particular site. Such a question is linked not only to the ability or reliability of the test method to detect an anomaly or defect within the pile or group of piles, but also to the issue of whether, as discussed in Section 97.3.2, such an 'anomaly' represents a 'defect' and whether such a defect could be

- minor and not significant to the performance of the pile (a 'flaw' as described by Amir, 2001; see Section 97.3.2), or
- major and potentially detrimental to the pile (a true 'defect' according to Amir, 2001).

In the latter case, such a defect could lead to the rejection of a particular pile, or its acceptance at a reduced load or after some agreed secondary repair or investigations.

The following section reviews the implications of the 'reliability' of an integrity test method, and how this is linked to the question of how large a percentage of the total pile population should be tested to provide an acceptable level of assurance of the adequacy of the piled foundations. It also discusses the corresponding implications of providing for a degree of redundancy within the foundation design.

The planning and implementation of a test programme should also consider how many unacceptable defects might, on average and on past experience, be anticipated within the pile population installed at a particular site; this is also reviewed in this section.

97.9.2 Considerations on the reliability of the test

As highlighted by Cameron and Chapman (2004), the methods of integrity testing are not infallible. An erroneous classification of the quality or characteristics of a pile by an inaccurate integrity test, or an inaccurate interpretation of the test result, can lead to a defective pile not being detected (a false negative result) and consequently incorporated into the foundation system. Alternatively, it could lead to a sound pile being wrongly condemned (a false positive result) and subjected to unnecessary inspection or remediation work, or even rejection and replacement.

Hertlein and Davis (2006) identify the following 'links' in the chain of reliability for low-strain integrity testing methods

(*a*) adequacy of pile head preparation for the pile head sensor, including trimming the pile to its test level

(*b*) correct and recent calibration of load cells and sensors

(*c*) suitable data acquisition, signal filtering and processing systems

(*d*) trained and experienced site operatives

(*e*) the degree of signal damping of the signal response: high length/diameter (*l*/*d*) ratios, stiff confining soils and bulges in the upper portion of the shaft all cause high signal damping

(*f*) the presence or absence of multiple anomalies down the pile length (where an anomaly can be a bulge, a neck, cracking, honeycombing or soil/laitance inclusions)

(*g*) the experience of the testing engineer in test data interpretation (the 'personal' component).

They suggest that these factors can be combined into a 'lumped' reliability model.

It can be appreciated that the breakdown of a single component in this model can throw the reliability of the system into question. Problems with links (*a*)–(*d*) can be improved by the test organisation and the site management. Problems associated with links (*e*)–(*g*) can be minimised by improvements in both the testing hardware and software. Of particular note is the use of equipment capable of improved data analysis, using methods such as the impedance-log analysis described above.

Hertlein and Davis suggest that the present state-of-the-art test methods are very reliable if

(*a*) *l*/*d* ratios are less than 30:1 in relatively soft soils

(*b*) significant defects are limited to the upper two-thirds of the pile length.

Conversely, reliability decreases rapidly with

(*a*) high *l*/*d* ratios in 'stiff' soils and

(*b*) when defects are in the bottom one-third of the shaft, or

(*c*) there are multiple defects.

Further insight into these features has been given in the preceding text or can be seen in Hertlein and Davis (2006).

For CSL the 'chain of reliability' can be identified as

(*a*) quality of the bond between the access tubes and the surrounding concrete

(*b*) trained and experienced site operatives

(*c*) distance between access tubes

(*d*) number of tubes per unit area of shaft cross-section

(*e*) distance of perimeter access tubes from the perimeter of the pile

(*f*) experience of the testing engineer in the interpretation of test data (personal factor).

Item (*b*) was considered less important by Hertlein and Davies, but, as noted in CIRIA R144 (Turner, 1997), cross-hole testing is also reliant upon the field expertise of the site operative – particularly if there is a need to examine an anomaly in greater detail.

97.9.3 The frequency of occurrence of potential flaws or defects within piles

Cameron and Chapman (2004) reviewed the frequency of defective bored piles identified by low-strain integrity testing, as published in the technical literature. This review noted that within six published papers a range of 1.5–10% of such tests identified piles as being defective, or perhaps more strictly potentially defective. One of these references, Fleming *et al.* (2009), which covered UK practice, reviewed the results of time-based 'sonic echo'-type pile integrity testing undertaken in 1981 and 1982. The data suggested that defects were identified in 1.5% and 1.9% of piles tested (for two investigations of 5000 and 4550 piles per year, respectively). Of these, the percentage due to pile construction defects was recorded at 0.4–0.6%. The balance, amounting to 60–80% of the defects detected by the test, was due to damage after the construction of the pile.

A further reference was Ellway (1987), who reported on the results of frequency-based integrity testing in the UK on some 4400 small-diameter bored piles in 1985. He suggested that just over 4% of the piles showed signs of 'potentially significant structural faults'. Some one-third of these were attributed to post-construction damage from mechanical plant or pile trimming techniques. A further one-third were due to weak or contaminated concrete within the upper 2 m of the pile. The remaining one-third (1.4% of the piles) suggested evidence of soil contamination or loss of section below 2 m.

These two sets of data are now over 25 years old, and it appears to be the general view of the industry that the quality of pile construction has improved greatly since that time – particularly in the case of CFA piles. The improvement in construction quality in the latter technique is largely driven by the now widespread use of instrumentation to monitor and record the pile construction process. The development of these systems also enabled users to better understand, and gave a greater insight into, the pile construction process.

Taking these factors into account, the limited data probably suggest that perhaps 0.5–1.0% of bored cast in situ piles might be expected to display imperfections within their acoustic response which *might* have some structural significance to the load-carrying capacity or durability of the pile. Although this is a relatively low frequency, the implication is still that, on a site containing a large number of piles, it is to be expected that some piles will be identified as requiring further engineering investigation and evaluation.

97.9.4 How many piles to test?

When electing to use low-strain integrity testing on a particular site it is necessary to also consider what should be the minimum number of tests to be undertaken on the piles. A difficulty is that defects or problems with piles are not always randomly distributed but might be associated with more difficult ground conditions in a part of the site, or the poor workmanship of a particular rig operative or some other factor.

CIRIA R144 (Turner, 1997) suggested that it was the considered view of most practitioners that a statistical approach was often difficult to apply in practice. Hence the decision on whether to test or not on a particular site had to take account of all the circumstances at the site, such as the availability of other construction records or data, the consistency or variability of the ground conditions across the site and the type of pile and piling systems. As a general recommendation, CIRIA R144 suggests that in situations where low-strain integrity testing is considered to be worth doing, all the piles on the site or in a particular area of concern should be tested. However, if the picture emerging from the test results was sufficiently consistent, the designer might consider reducing the number of tests. At the same time, if no meaningful information was being obtained from the tests, the designer might choose to dispense with the testing. This advice was intended to highlight that the designer and the project team should be able to make informed decisions on the applicability and/or usefulness of the testing on a particular site, and to modify the test programme if this was considered appropriate.

Williams and Stain (1987) outlined a testing strategy or 'decision tree', as illustrated in **Figure 97.38**, that also served to give pragmatic guidance on 'when' and 'how many' to test. This suggested, as a guide, that a minimum of around 30 tests would give a useful guide to the condition of the piles on a site.

Cameron and Chapman (2004) have considered the statistical approaches that may be adopted in selecting the number of piles that should be tested to achieve predetermined acceptable confidence limits. Such methods require an estimate of whether a percentage of defective piles can be tolerated within the foundation (in the case of a piled raft foundation, for example). Importantly, these authors also included within their review a consideration of the reliability of the testing system.

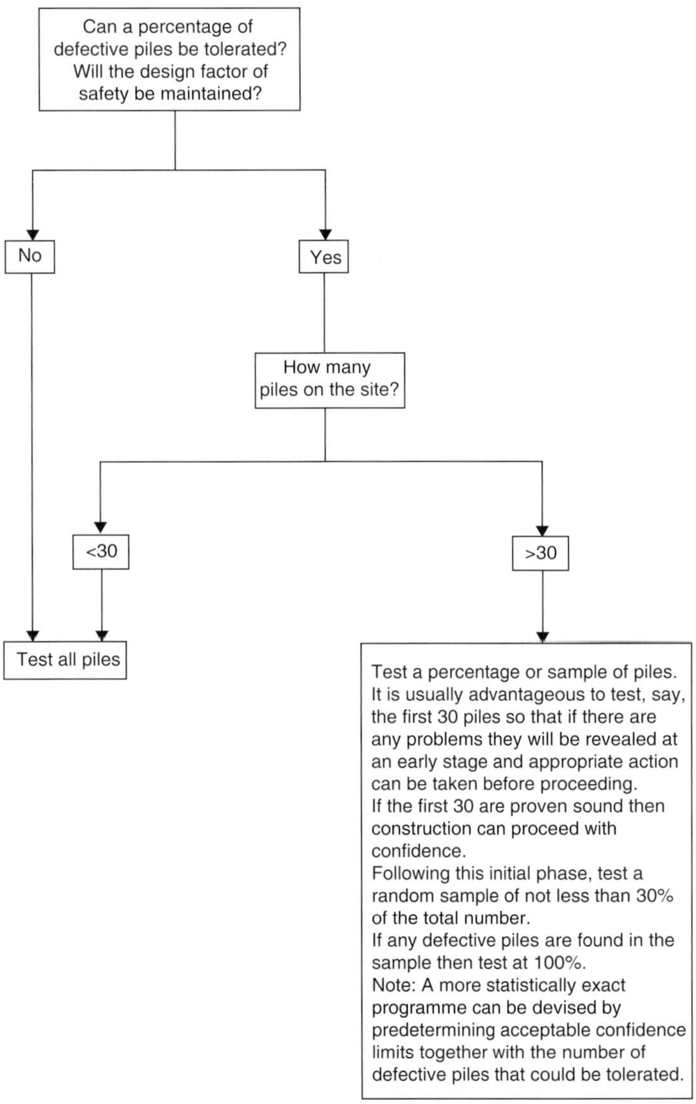

Figure 97.38 Evaluation of numbers of piles to test on a site (Reproduced from Williams and Stain, 1987; Engineering Technics Press)

Clearly, the options available when planning an integrity testing programme are to examine all of the piles or none of them, or a selected sample. Cameron and Chapman (2004) give particular attention to the last option, where conventional sampling theory can allow conclusions to be drawn about the pile population on a particular site, based upon testing a sample of the piles. These authors also highlight the effects of planning the foundation design based upon a degree of redundancy within the overall foundation design. They compare the relative advantages and disadvantages of using either single piles or multiple piles in groups to support individual loads. In the former case, the occurrence of defects is clearly more critical to the structural performance of the foundation (**Figure 97.39**). A similar consideration would apply to piles constructed on strip or grid formations, as illustrated in **Figure 97.40**. Thus, if the interconnecting ground beams are capable of spanning to transfer the reduction in capacity within a structure due to a potentially faulty or unacceptable pile onto the adjacent piles, then such a robust foundation solution may justify a reduced integrity testing regime. At the least it can allow a reasoned view to be taken of the effect of such a predicted shortfall of capacity.

Cameron and Chapman (2004) also point out that such a comparative exercise can often highlight the perceived relative costs and savings of using single piles to support individual high column loads, compared with a notionally more expensive solution using multiple piles, providing a relatively high degree of redundancy. In the former case, the consequential programme risks of the failure or rejection of such a pile, including delays from re-testing or repairing such a pile, have to be considered. It is often not appreciated how, on a major fast-track project, the overall costs of the piled foundations are often small compared with overall project and programme costs.

As noted earlier, however, if the reliability of the integrity testing system is also factored into the picture, then even with 100% testing it is not possible to be 100% certain that all the piles are defect-free.

Cameron and Chapman highlight a statistical approach developed by Preiss and Shapiro (1979), which gave quantitative guidance on the level of testing necessary to achieve a required degree of 'assurance' in the foundations. **Figure 97.41** shows the relationship between sample size, the percentage of defective piles which are acceptable within the pile population on that particular site and the statistical implications of the number of defective piles detected in the test sample.

As an example, consider that the foundation designer requires a confidence level of at least 90% that there will be less than a certain number of 'defective' piles among both the tested and non-tested piles in the population. Consider that

	Non-redundant	Working load is equal to failure load Extremely unsafe with potentially very large settlements	Total loss of pile capacity Ultimate failure
	Non-redundant	Deficient pile on the verge of failure Very unsafe	Failure is likely
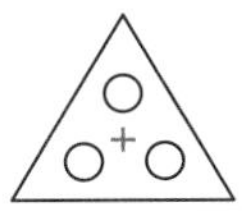	Non-redundant	Just sufficient capacity to avoid failure by much tilting	Failure still likely as centre of support offset from centre of load
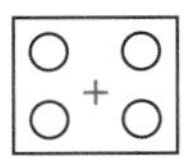	Logically non-redundant	Support reduced to two piles providing full and two piles providing half support Likely to be extra settlement and reduced factor safety	Support reduced to just two piles Factor of safety equals 1.0, therefore stability is marginal
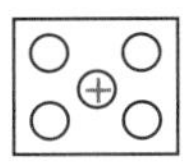	Clearly redundant	Minimal effect	Support reduced to potentially three or four piles, depending on location of deficient pile. Lowered factor of safety but performance may be adequate
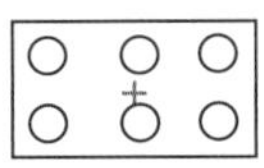	Clearly redundant	Minimal effect	Minimal effect

Legend: ○ Pile ▭ Pilecap + Position of column load

Figure 97.39 Degree of redundancy offered by different pile cap arrangements for a factor of safety of 2 (Reproduced from Cameron and Chapman, 2004)

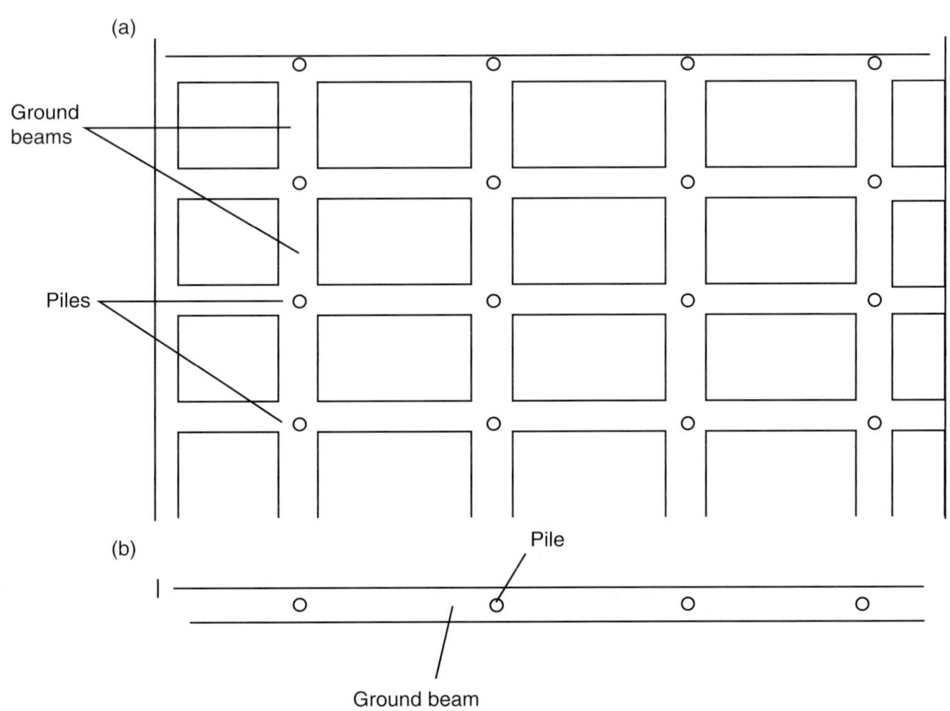

Figure 97.40 Potential redundancy of (a) grid and (b) piled strip foundations

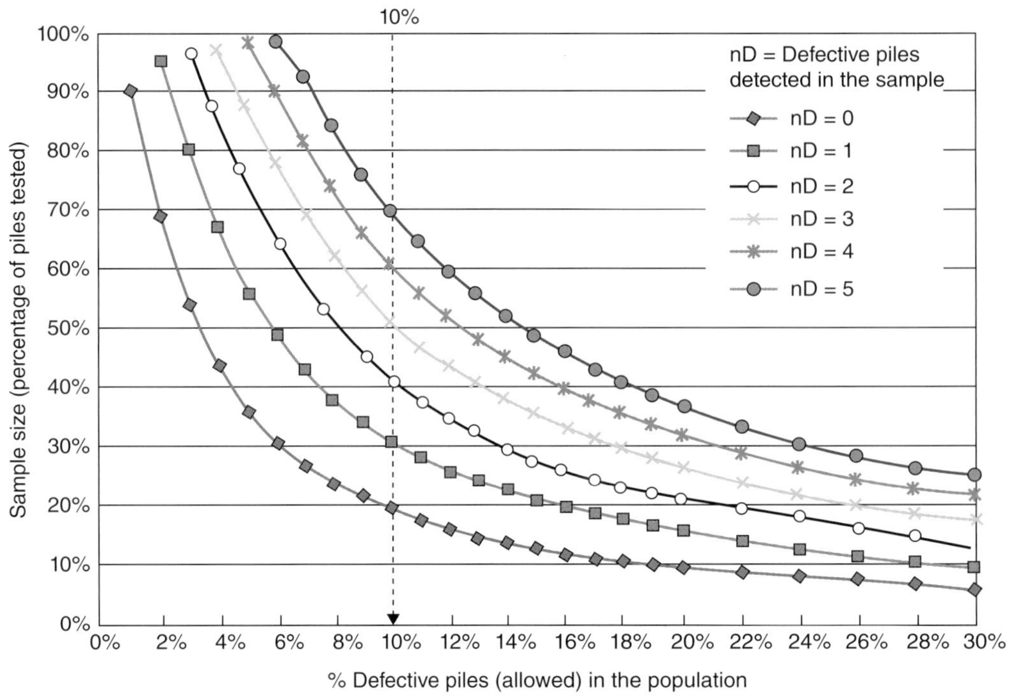

Figure 97.41 Number of piles to be integrity tested for 90% confidence within a population of piles (Reproduced from Cameron and Chapman, 2004)

10% of the piles might be allowed to be defective. In such a case, **Figure 97.41** indicates the following.

- If 20% of the piles are tested and no defective piles are detected, then the 90% confidence limit is satisfied.
- If, however, one defective pile is identified, then 31% of the piles must be tested, with no further 'defectives'.
- If two defective piles are detected, the sample must be increased to over 40% of the piles, with no further 'defectives'.
- If five defective piles were detected the sample size would need to increase again, to around 70% of the piles.

The exercise would clearly continue until the number of tested piles, the sample size and the number of defective piles identified within the sample size complied with the required degree of confidence, or all the piles had been tested.

However, Cameron and Chapman also highlight a further complication in the Preiss and Shapiro analysis: it is based upon the test being 100% accurate. If the reliability of the test is less than 100%, then this affects the confidence level of an accurate prediction. If a system was only 80% reliable, then even testing all the piles would not give an overall confidence level in the test results of greater than 80%.

The practical outcome from a less reliable test is that a greater number of false positive, false negative or 'inconclusive' results are likely to be obtained, which might have to be followed up by some other examination or review – such as coring the pile or excavating alongside it to determine its true condition.

As an example, **Figure 97.42** illustrates the increase in confidence gained for every additional 10% of piles which are integrity tested within a pile population, using a test system which has an 80% 'reliability'. In this case the results are shown for a foundation system that can tolerate 0–4% defective piles. **Figure 97.42** illustrates that, where the foundation system can tolerate up to 4% defective piles, the greatest

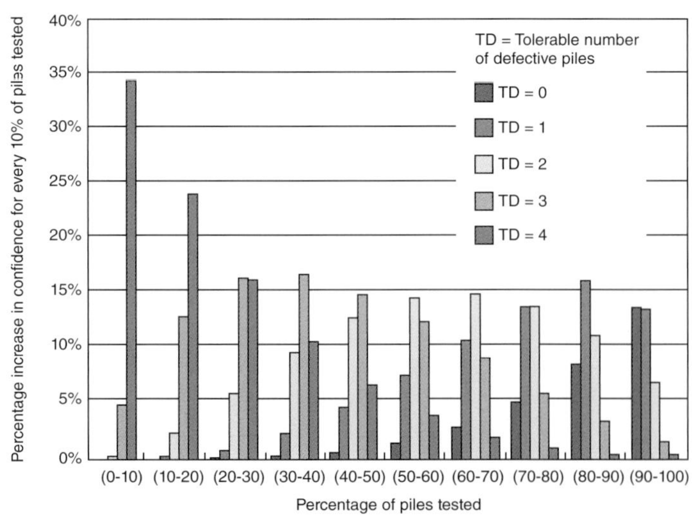

Figure 97.42 Additional increase in confidence by testing each additional 10% of the pile population, using a test system with 80% reliability (Reproduced from Cameron and Chapman, 2004)

benefit comes from the initial tests on the pile population. However, where the foundation system cannot tolerate any defective piles, then the full benefit only comes when testing the last 20% or so of the piles. Hence, the greater the reliance on each individual pile within the pile population, the more important it is to check the integrity of every single pile.

It should be borne in mind that, although statistically in most practical cases the most important piles are the first 20–40% of those tested, these do not necessarily equate to the first piles installed on the project. The sample must be representative of the whole population, and not skewed in some way, such as by different ground conditions, different piling operatives or equipment or other considerations. This has particular relevance where there is particular pressure in terms of cost and programme to omit some piles from testing.

Cameron and Chapman point out that in some cases the added value of testing the last piles (**Figure 97.42**) could be judged to not be worth the time and consequential cost.

Test method	Low-strain integrity tests	Cross-hole sonic logging	Temperature integrity profiling	Parallel seismic tests	High-strain integrity tests
Property measured	Characteristics of the behaviour of acoustic shock waves or stress waves travelling through the pile	Transmission time of an ultrasonic wave through the pile material	Relative temperature in concrete over full depth and around reinforcing cage	Transmission time of acoustic shock waves or stress waves through the pile and intervening soil to a detector	Characteristics of the behaviour of stress waves travelling through the pile from a heavy impact
Pre-planning required?	None	Access ducts have to be cast into preselected piles	Temperature sensors must be attached to reinforcing into preselected piles	Sinking of measurement bore alongside pile	Not strictly necessary, but access for heavy plant may have to be provided
When can test be carried out?	After concrete has achieved design strength (usually 5–7 days min.)	After concrete has achieved design strength (usually 5–7 days min.)	48–72 hours after construction	After construction	After concrete has achieved design strength and typically 7 days min. after construction
Type of pile suited to test method	All types	Large-diameter cast in place typically (usually 600 mm dia. or larger)	Large-diameter cast in place typically (usually 600 mm dia. or larger)	Any pile	All types
Approximate relative cost	Low	Low to medium	Low to medium	Medium to very high	Medium to high
Relative frequency of use:	High	Medium	Low to medium – gaining traction	Low	Low
Control testing	5	4–5	1–2	Not applicable	1–2
Retrospective investigation	3–5	Not applicable	Not applicable	0–1	1–2
Availability	Readily available from specialist testing houses	Available from specialist testing houses	Available from specialist testing houses	From specialist testing houses	Readily available from piling contractor and specialist test house
Test affected by pile length?	Yes, signals increasingly attenuated with depth	No	No	Yes	No, not within normal pile depth
Comments	Very common technique. Pile response is investigated in terms of time and/or frequency	Mainly used for large-diameter cast-in-place piles, piers and barrettes. Especially large single piles supporting high column loads. Not usually suitable for retrospective investigation because of necessity to install access ducts	Provides information on concrete quality outside of the reinforcing cage	Used for retrospective investigation only	Not commonly used in routine testing. Typically, may be used to investigate a post-installation problem such as pile damage

Relative frequency of use: scale 0–5: 0: very rare; 1: rare; 2: occasional; 3: sometimes; 4: common; 5: very common.
Approximate relative cost: scale low–very high: low <10% of pile cost; medium: 10–50%; high: 50–100%; very high: >100%.
Relative cost excludes mobilisation costs.

Table 97.3 Summary of the suitability and applicability of the test methods (Reproduced with permission from CIRIA R144 (Turner, 1997); www.ciria.org)

Figure 97.42 also demonstrates that such an omission may be possible only if the tolerable proportion of defective piles is high. Hence there would be programme advantages in selecting a more redundant foundation system.

97.10. Selection of a suitable test method

Table 97.3 summarises the features of the four integrity test techniques outlined in this chapter.

For each technique the table outlines the property measured by the test, details of any pre-planning requirements and the type of pile suited to the test.

In addition, the table gives a guide to the approximate relative cost of the particular test, its availability and its relative frequency of use. Any other relevant points are noted in the comments row.

References

Amir J (2001) Reflections on pile integrity testing. *Proceedings of the Deep Foundations Institute Specialty Seminar on Nondestructive Testing for Drilled Shafts, St Louis, MO.* DFI, Hawthorne, NJ.

ASTM (American Society for Testing Materials) (2014) D7949: Standard test methods for thermal integrity profiling of concrete deep foundations. ASTM. West Conshohocken, PA.

ASTM (2016) D5882: Standard test method for low strain impact integrity testing of deep foundations. ASTM, West Conshohocken, PA.

ASTM (2017) D6760: Standard test method for integrity testing of deep foundations by cross-hole testing. ASTM, West Conshohocken, PA.

ASTM (2021) D6026: Standard practice for using significant digits and data records in geotechnical data. ASTM, West Conshohocken, PA.

Baker CN, Drumwright EE, Briaud J-L, Mensah-Dumwah F and Parikh G (1993) *Drilled Shafts for Bridge Foundations.* Federal Highway Administration, Washington, DC.

Briard M (1970) Contrôle des pieux par la méthode des vibrations. *Annales de l'Institut Technique du Bâtiment et des Travaux Publics* **270**: 105–107.

BSI (British Standards Institution) (2021) BS EN 12504-4:2021: Testing concrete: Determination of ultrasonic pulse velocity. BSI, London, UK.

Butcher AP, Powell JM and Skinner HD (2006) *Proceedings of an International Conference on Reuse of Foundations for Urban Sites, London, UK.* BRE Press, Watford, UK.

Cameron G and Chapman T (2004) Quality assurance of bored pile foundations. *Ground Engineering* **37(2)**: 35–40.

Chapman T, Anderson S and Windle J (2007) *Reuse of Foundations.* CIRIA Report C653. CIRIA, London, UK.

Davis AG (1998) Assessing the reliability of drilled shaft integrity testing. In *Transportation Research Record 1633.* Transportation Research Board, Washington, DC, pp. 108–116.

Davis AG and Dunn CS (1974) From theory to field experience with the non-destructive vibration testing of piles. *Proceedings of the Institution of Civil Engineers* **57(2)**: 571–593.

Ellway K (1987) Practical guidance on the use of integrity tests for the quality control of cast-in-situ piles. *Proceedings of the International Conference on Foundation and Tunnels, London, UK*, pp. 228–234.

Fleming K, Weltman A, Randolph M and Elson K (2009) *Piling Engineering*, 3rd edn. Taylor & Francis, London, UK.

FPS (Federation of Piling Specialists) (2017) *Guidance Note on Current Integrity Testing Methods.* FPS, Faversham, UK.

Hannigan PJ and Piscsalko G (2020) Advances in quality control methods for bored pile and diaphragm wall foundations with case histories. *14th Baltic Sea Region Geotechnical Conference [virtual conference]*, 10.1088/1755-1315/727/1/012030

Hertlein B and Davis A (2006) *Nondestructive Testing of Deep Foundations.* Wiley, New York, NY.

ICE (Institution of Civil Engineers) (2016) *ICE Specification for Piling and Embedded Retaining Walls*, 3rd edn. Thomas Telford, London, UK.

Iskander M, Roy D, Earley C and Kelley S (2001) Class-A prediction of construction defects in drilled shafts. In *Transportation Research Record 1772.* Transportation Research Board, Washington, DC, pp. 73–83.

Lilley DM, Kilkenny WM and Ackroyd RF (1987) Investigation of structural integrity of pile foundations using a vibration method. *Proceedings of the International Conference on Foundations and Tunnels, London, UK*, pp. 177–183.

Mullins G, Kranc S, Johnson K, *et al.* (2007) *Thermal Integrity Testing of Drilled Shafts: Final Report.* Florida Department of Transportation, Tallahassee, FL.

Paquet J (1968) Étude vibratoire des pieux en béton; réponse harmonique et impulsionnelle: application et contrôle. *Annales de l'Institut Technique du Bâtiment et des Travaux Publics* **245**: 789–803.

Paquet J (1991) A new method for testing integrity of piles by dynamic impulse: the impedance log. *Proceedings of the International Colloquium on Deep Foundations, Paris, France*, pp. 1–10.

Paquet J (1992) Pile integrity testing: the CEBTP reflectogram. *Piling European Practice and Worldwide Trends, London, UK.* ICE, London, UK, pp. 177–188.

Paquet J and Briard M (1976) Contrôle non destructif des pieux en béton. *Annales de l'Institut Technique du Bâtiment et des Travaux Publics. Série: Sols et Fondations* **128**(Suppl. 337).

Piscsalko G, Likins G and Mullins G (2016) Drilled shaft acceptance criteria based upon thermal integrity profiling, *DFI 41st Annual Conference on Deep Foundation, New York, NY*, pp. 1–10

Preiss K and Shapiro J (1979) *Statistical Estimation of the Number of Piles to be Tested on a Project.* RILEM Commission on Non-Destructive Testing, Stockholm, Sweden.

Rui Y, Kechavarzi C, O'Leary F, *et al.* (2017) Integrity testing of pile cover using distributed fibre optic sensing. *Sensors (Basel)* **17(12)**: 2949.

Smith EAL (1960) Pile-driving analysis by the wave equation. *Journal of the Soil Mechanics and Foundations Division* **86**: 36–61.

Stain RT (1982) Integrity testing. *Civil Engineering*, Apr./May.

Stain RT and Johns D (1987) Integrity testing of deep foundations. *Proceedings of the Second International Symposium of the Deep Foundations Institute, Luxembourg, Luxembourg.*

Stain RT and Williams HT (1991) Interpretation of sonic coring results: a research project. *Proceedings of the 4th International*

DFI Conference on Piling and Deep Foundations, Stresa, Italy. Balkema, Rotterdam, the Netherlands, pp. 633–640.

Sun Q, Elshafie M, Barker C, *et al.* (2020) Thermal integrity testing of cast in situ piles: an alternative interpretation approach. *Structural Health Monitoring* **20(5)**: 2493–2512

Turgut P (2004) Research into the correlation between concrete strength and UPV values. *NDT.net* **12(12)**.

Turner MJ (1997) *Integrity Testing in Piling Practice*. CIRIA Report 144. CIRIA, London, UK.

Williams HT and Stain RT (1987) Pile integrity testing: horses for courses. *Proceedings of the International Conference on Foundation and Tunnels, London, UK*. Engineering Technics Press, London, UK, pp. 184–191.

Further reading

Although cited in the reference list above, the following publications are recommended for further reading and background information.

Fleming K, Weltman A, Randolph M and Elson K (2009) *Piling Engineering*, 3rd edn. Taylor & Francis, London, UK.

Hertlein B and Davis A (2006) *Nondestructive Testing of Deep Foundations*. Wiley, New York, NY.

ICE (2016) *ICE Specification for Piling and Embedded Retaining Walls*, 3rd edn. Thomas Telford, London, UK.

Turner MJ (1997) *Integrity Testing in Piling Practice*. CIRIA Report 144. CIRIA, London, UK.

Useful websites

Corporate market research procurement services: for example, http://www.profound.com

Testing, monitoring, analysis and consulting services for deep foundations: Pile Dynamics, Inc. and GRL Engineers, Inc.: http://www.pile.com

It is recommended this chapter is read in conjunction with

- Chapter 82 *Piling problems*
- Chapter 93 *Quality assurance*

All chapters in this book rely on the guidance in Sections 1 *Context* and 2 *Fundamental principles*. A sound knowledge of ground investigation is required for all geotechnical works, as set out in Section 4 *Site investigation*.

Dedication: To my friend and colleague Bernie Hertlein who gave so much to me as a young engineer and to the industry with his relentless pursuit of advances in NDT.

Chapter 98
Pile capacity testing

Michael Brown University of Dundee, Dundee, UK
Jon Ball Roger Bullivant Limited, Swadlincote, UK

doi: 10.1680/icemge.66830.1599

CONTENTS

As there is still uncertainty in accurately predicting the performance of piled foundations based upon design calculations, there remains a need to test piles. Several methods of pile testing are available, some of which have been in common usage for many years, while others are relatively recent developments. Static pile testing is a well understood and simple test technique that has been in use for many years. This approach has the benefit of directly producing test results but is hindered by the increasing size of associated testing infrastructure as pile capacity increases. A recent variation of classic top-down static testing is bi-directional testing which relies on the incorporation of a specialised loading jack or jacks in the pile shaft at some depth below the ground surface. This technique has the ability to apply test loads that greatly exceed those possible in other pile test types by effectively using one portion of the pile capacity to test against the other or others. Alternative pile testing techniques come in the form of rapid and dynamic load tests that have the benefit of quick testing and reduced testing infrastructure, but require more complicated analysis and interpretation techniques due to the rapid/dynamic nature of the tests.

98.1. An introduction to pile testing

Although pile design has advanced in recent decades, the determination of axial pile capacity is still dependent on the use of empirical correlations. As a result, it is only possible to estimate capacity to ±30% in many soil types (Randolph, 2003), whereas in rock capacity may be significantly underpredicted. Owing to the uncertainty associated with predicting load–settlement behaviour using existing design methods, it is common practice to carry out pile load tests for verification.

The information obtained from pile load testing may be used in a number of ways (Poulos, 2000) including

- construction and quality verification
- verification of design information
- to allow for a more refined or confident design with potential cost savings for subsequent piling works.

As pile testing may prove expensive (especially on small contracts), the need for pile testing may be considered in terms of risk reduction. On larger sites, where significant numbers of piles are being installed, pile testing may be considered from the perspective of potential cost saving. Through testing, improved design parameters may be determined, resulting in optimisation of the piles through a reduction in length, for example. The pile testing strategy recommended by the Federation of Piling Specialists (FPS, 2006) in their *Handbook on Pile Load Testing* is shown in **Table 98.1**. Similar pile testing recommendations can be found in BS 8004:2015 for preliminary and working piles in low- and medium-risk situations.

Eurocode 7 allows pile design based upon a pile load test(s) and states that pile tests should be undertaken in the high- to medium-risk situations highlighted in **Table 98.1**. Eurocode 7 also allows the characteristic resistance of axially loaded piles to be derived directly from pile load tests, with the level of correlation increasing with increasing number of pile tests (from one to five). For five or more pile tests direct correlation is allowed with the measured test results, which in turn are reduced by a partial factor to obtain the design resistance (see the relevant sections in BS EN 1997 (BSI, 2004), otherwise known as EC7, and the relevant National Annex for further detail). This prescriptive approach to factoring may be considered an improvement over the ad hoc manner in which pile load tests have previously been used to influence safety factors. Further information with respect to pile testing as part of the Eurocode framework is available in 'BS EN 1536:2010+A1:2015, Special geotechnical works: Bored piles'. Piles may be tested at different stages throughout a project, depending on how the outcome of the test will be used. For instance, preliminary or trial piles are tested prior to the main works or pre-contract and are designed to validate design and achievable performance along with checking construction techniques in certain ground types. It is important that preliminary piles are tested sufficiently in advance of the main works such that findings from the tests can be incorporated in working pile designs. Piles tested that will form part of the final structure are referred to as working piles.

As well as the stage of the project at which piles should be tested, it is also important to consider the time after installation at which piles should be tested. For instancc, driven or displacement piles installed in clay may cause significant disturbance of the ground. This results in changes to the local effective stress regime, which with time may recover depending on the soil's permeability and the size of the pile (Jardine *et al*., 2005). For instance, the ICE *Specification for Piles and Embedded Retaining Walls* (ICE, 2017) (often referred to as SPERW) suggests a minimum of 4 days between installation and testing for cast in situ piles to avoid damage and achieve suitable cube strength, although it does go on to say that a 12-hour delay between installation and testing is adequate for driven piles, which appears to ignore the potential for set-up (gain in strength with time) or set-down (reduction in strength with time) in different soil types caused by the installation process. In contrast, Fleming *et al.* (2009) suggest 1–3 weeks between installation and testing, depending on soil type and experience, where up to 28 days may be adopted in practice. BS EN ISO 22477-1 recommends 7 days between installation and testing for coarse soils and 21–28 days in fine-grained soils, dependent on whether the pile is of bored or displacement type.

Pile load testing methods include static tests, dynamic tests and kinematic or rapid load tests. There are also alternative static load test methods such as bi-directional testing in the form of the Osterberg Cell (O-Cell). Generally, static pile testing methods are expensive and time consuming (Fleming *et al.*, 2009), becoming more so with increased load requirements, but have the advantage of simple analysis and interpretation. Conversely, dynamic and rapid load testing methods are quick to carry out and become cheaper than static tests with increased load requirements but require more specialised equipment and analysis. Comparisons of some of the typical pile test characteristics are shown in **Table 98.2**. Note that this table is based upon current typical approaches to testing and does not dictate use of the techniques; for instance, dynamic tests may be undertaken on cast in situ piles but this may not be a frequent occurrence in practice or in some parts of the world.

Characteristics of the piling works	Risk level	Pile testing strategy
Complex or unknown ground conditions	High	Both preliminary and working pile tests essential
No previous pile test data		1 preliminary pile test per 250 piles
New piling technique or very limited relevant experience		1 working pile test per 100 piles
Consistent ground conditions	Medium	Pile tests essential
No previous pile test data		Either preliminary and/or working pile tests can be used
Limited experience of piling in similar ground		1 preliminary pile test per 500 piles
		1 working pile test per 100 piles
Previous pile test data available	Low	Pile tests not essential
Extensive experience of piling in similar ground		If using pile tests, either preliminary and/or working pile tests can be used
		1 preliminary pile test per 500 piles
		1 working pile test per 100 piles

Table 98.1 Typical pile testing strategy based upon risk levels (Reproduced from FPS, 2006)

	Type of pile test		
Characteristic	Static[a]	Rapid	Dynamic
Load duration	1–24 hours	100 ms	7 ms
Tests per day	1	2–6	8
Reaction mass required (as a percentage of pile capacity)	120%	5–10%	1.5–2%
Time needed for results	Directly	10 minutes	4 hours
Pile types tested			
precast	Yes	Yes	Yes
cast in situ	Yes	Yes	Yes
tubular steel	Yes	Yes	Yes
Perceived reliability	High	Experience too limited to assess	Intermediate
Cost[b] (£ per kN of pile capacity)	0.42–0.75	0.45	0.05–0.1
Cost per pile (£)	850–1500	400	50–200

[a] Static definition here does not include O-Cell (top-down).
[b] Based upon testing at 1000–2000 kN capacity piles on the same site; note dynamic test costs assume a pile-driving rig is present on site. Costs given in GBP.

Table 98.2 Comparison of typical features of pile testing techniques (Data taken from Hoelscher and van Toll, 2009)

98.2. Static pile testing

98.2.1 What is static pile testing and what types of test are there?

The most common method of pile load testing – static testing – may take one of two forms. These are the maintained load test (MLT) and the constant rate of penetration test (CRP). The names of the two tests are derived from their methodology. Since the invention of tests such as the O-Cell method, these types of tests are often referred to as top-down as the loading is applied at the head of the pile.

98.2.2 Maintained load compression test

This type of test is often referred to using the acronyms ML or MLT. MLT works by applying and maintaining increments of load to the head of the pile for a minimum specified time (**Table 98.3**) and until a specified rate of settlement criterion is satisfied (**Table 98.4**), at which point the load is either increased or reduced (Tomlinson and Woodward, 2008). According to SPERW (ICE, 2017) the minimum time for holding an increment typically varies from 30 minutes to 6 hours, with unloading increments typically held for 10 minutes. The approach in SPERW (ICE, 2017) varies from that in BS EN ISO 22477-1:2018, but it is noted that minimum hold periods are recommended rather than prescribed. An example of the pile response seen during an ML test is shown in **Figure 98.1**.

Generally, this test method is not always used to prove the pile ultimate load capacity or generate 'plunge' as it is difficult to maintain constant load at high settlement rates. Additionally, it is typical to increase the pile load in 25% increments (**Table 98.3**) of the working design verification load (DVL; see definition in **Table 98.3** and specification section later), which may mean the application of many hundreds of kilonewtons between load increments. It is then possible for the actual ultimate load to be missed, resulting in an underestimation of ultimate capacity as the pile plunges at a higher load increment. To avoid the underestimation, the load increment may be reduced to 10% DVL throughout the test or when load stability issues occur, as is recommended for testing of preliminary piles (ICE, 2017). SPERW (ICE, 2017) also includes additional guidance on short-duration MLT proof testing for steel screw piles which differs from other pile types.

98.2.3 Constant rate of penetration testing

The CRP test varies from the MLT in that a varying load is applied to the pile to maintain a constant rate of penetration. The rate of penetration is typically chosen to reflect the main

Load[a]	Minimum time of holding load for a single-cycle pile test	Minimum time of holding for a multi-cyclic pile test
25% DVL	30 minutes	30 minutes
50% DVL	30 minutes	30 minutes
75% DVL	30 minutes	30 minutes
100% DVL	6 hours	6 hours
75% DVL	n/a	10 minutes
50% DVL	n/a	10 minutes
25% DVL	n/a	10 minutes
0% DVL	n/a	1 hour
100% DVL	n/a	1 hour
100% DVL+25% F_{rep}	1 hour	1 hour
100% DVL+50% F_{rep}	6 hours	6 hours
100% DVL+25% F_{rep}	10 minutes	10 minutes
100% DVL	10 minutes	10 minutes
75% DVL	10 minutes	10 minutes
50% DVL	10 minutes	10 minutes
25% DVL	10 minutes	10 minutes
0% DVL	1 hour	1 hour

[a] F_{reb} denotes the representative action, previously referred to as SWL (specified working load). DVL, design verification load.

Table 98.3 Minimum loading times for a maintained load compression proof test (Reproduced from ICE, 2017)

Pile head displacement range	Rate of settlement criteria
<10 mm	≤0.1 mm/hour
>10 mm	≤0.2 mm/hour

Table 98.4 Rate of settlement to be used along with minimum hold times

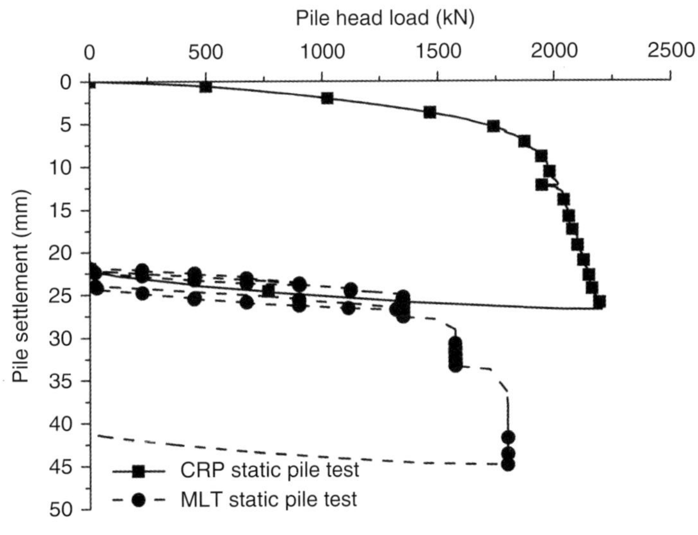

Figure 98.1 Example of CRP and MLT static load tests on a 12 m long, 600 mm dia. cast in situ bored pile in glacial till (see Brown *et al.*, 2006 for more information on the pile installation and testing) (Data taken from Brown *et al.*, 2006)

soil type that the pile installation encounters (**Table 98.5**). Due to these penetration rates, tests are completed relatively quickly. For instance, a 600 mm dia. pile installed in clay pile can be taken to a penetration equal to 15% of the pile diameter (90 mm) in 2.5 hours. It can be seen in **Table 98.5** that the rates used for CRP in US practice may be 50% slower or faster than those specified for UK use. It should also be noted when comparing tests or correlations from the USA that there is also a different MLT referred to as the quick load test method (QLT), where load increments are only held for 2.5 minutes (ASTM D1143/D1143M-20).

The CRP test is not considered appropriate for most testing situations and is typically reserved for research purposes, and has more recently been removed from revisions of codification (as has guidance on pile penetration rates adopted during a CRP test). Although testing is faster, it may require greater capacity from the loading and reaction systems to produce plunge. There are also reservations about the relatively high penetration rates and short test duration (rate effects), especially where piles are installed in clay. It has been shown that as the penetration rates in CRP increase so does the ultimate pile capacity and stiffness (**Figure 98.1**). In terms of representing meaningful structural loading rates, it may be argued that the loading rate should be selected based upon the permeability of the soil specific to the pile installation and effective drainage path length (often related to pile diameter). It should also reflect whether the desired behaviour to be proven is drained or undrained (Brown and Hyde, 2008a).

	Rate of penetration mm/s (mm/min)		
Major soil type	SPERW 2017	BS 8004:1986	ASTM D1143
Fine-grained soils (e.g. clay)	0.01 (0.6)	0.0125 (0.75)	0.0042–0.021 (0.25–1.25)
Coarse-grained soils (e.g. sand or gravel)	0.02 (1.2)	0.025 (1.5)	0.0125–0.042 (0.75–2.5)

Table 98.5 Examples of different pile penetration rates specified for CRP testing; note that BS 8004:1986 has now been superseded/withdrawn and this information is not included in the latest revision

98.2.4 How it works and the various test set-ups

Static pile testing systems typically require a structure to react against to allow the application of load to the pile. A typical reaction or anchor pile type arrangement for pile testing is shown in **Figure 98.2**. Alternatively, the reaction to the hydraulic jack can be provided by placing kentledge (reaction mass) above the jacking arrangement (**Figure 98.3**). If the pile is a preliminary pile – that is, one required to validate construction performance prior to construction of the working piles – then either arrangement of pile test would be suitable. Where load tests are required on piles that will form part of the final structure (working piles) then the test arrangement using kentledge would appear more appropriate, but often reaction piles are used and greater care is required with respect to the spacing between the reaction piles and those under test. Unfortunately, if several working pile tests are needed, multiple individual test arrangements may be required to avoid time delays. Greater detail regarding pile testing procedure and equipment arrangements is given by Weltman (1980) and FPS (2006).

Application of the load to the test pile is usually undertaken using a large hydraulic jack with manual or automated control of hydraulic oil flow to allow either a maintained load to be applied or varying load for constant penetration rates (**Figure 98.4**).

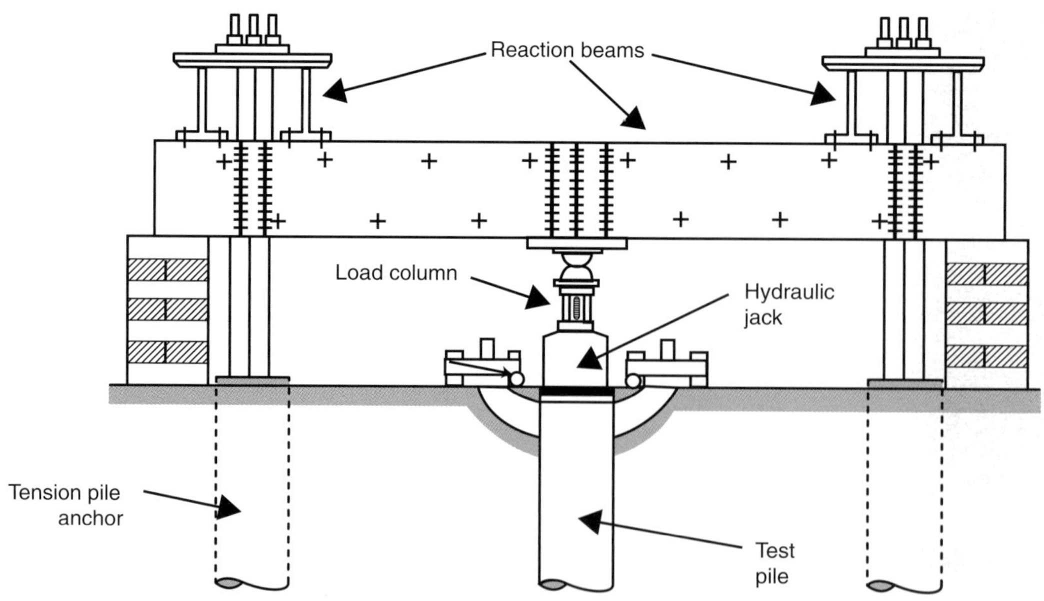

Figure 98.2 Top-down static pile testing arrangement utilising a tension pile reaction system (Modified with permission from CIRIA PG7, Weltman (1980); www.ciria.org)

Figure 98.3 Top-down static pile testing arrangement utilising water-filled tanks as a reaction system; the testing arrangement is completely obscured by the tanks (Photograph courtesy of Deltares; all rights reserved)

Figure 98.4 Pile head loading arrangement showing loading jack placed on the pile with calibrated load column above with spacers to the reaction frame; note the dial gauges for settlement measurement in contact with the head of the pile

Historically load has been measured with a manually read calibrated load column placed between the jack and the reaction system (**Figure 98.4**). It is now more likely that an automated strain gauge-based load cell will be used which produces readings that can automatically be logged by a computer or datalogger. Similarly, the piles' deflection has been measured using manually read dial gauges (mounted on a reference beam placed three diameters or 2 m from the test pile), whereas these have been typically replaced by displacement transducers that can be logged by a computer or datalogger (Fleming *et al.*, 2009). Measurement by displacement transducers or dial gauges should also be verified by optical levelling to an accuracy of 0.2 mm. Developments in monitoring equipment along with reliable and cost-effective servo hydraulic control systems (which allow true stress and displacement controlled tests to be undertaken) have led to the development of fully automated test rigs that can be operated remotely (England, 2002). Automating such a system means that staff spend less time on an individual site and less time travelling between sites.

Another advantage of automation is that the safety of site operatives is enhanced as they do not have to spend long periods close to highly stressed structures which in extreme cases have failed, causing injury. There is also less need for shift working that may result in operatives working on deserted sites during the hours of darkness.

98.2.5 Static tension tests

Tension or uplift testing of piles (**Figure 98.5**) utilises similar equipment and testing procedures to compression testing, although a single hollow jack or two jacks may be used to avoid inducing bending stresses in the pile. Dependent upon beam arrangement and surface soil strength in compression, resisting piles may be required. This will not be the case where the surface is stiff enough to accommodate the compressive force imparted, in which case trestles and spreader plates may be sufficient. Connections to cast in situ piles are similar to those employed in tensile reaction piles, whereas preformed steel

Figure 98.5 Tension pile test; note the loading jack above the reaction beam (Photograph courtesy of Cementation Foundations Skanska Ltd; all rights reserved)

piles may require additional welded brackets (Fleming *et al.*, 2009; Tomlinson and Woodward, 2008). It is normal to test more than one pile that may be subject to tension and at least 2% of piles where a large number are subject to tension.

98.2.6 Static lateral load testing

As well as tensile load, piles are often subject to combined lateral, vertical and moment loading, which in certain circumstances, such as foundations for offshore structures, may be cyclic in nature. It is not easy to achieve such complicated load paths in full-scale pile tests, but both monotonic and cyclic lateral tests may be specified (Burd *et al.*, 2020). Typically, an adjacent pile(s) is used as reaction, with a jack placed between one or more reaction pile(s) and the test pile (**Figure 98.6**). Where cyclic loading is required, automated jacks and monitoring systems are recommended. Similar load/deflection measuring techniques may be used as per axial pile load tests, but it is also advisable to measure relative movement of the test pile to the reaction piles and pile head rotations. In addition, valuable information for test analysis can be obtained by instrumenting the pile (Reese and van Impe, 2011) to allow determination of moments with depth. When selecting the number of lateral piles to be tested, close attention should be given to the variability of the ground over the top few metres of the test pile as this zone has significant effect on the lateral pile behaviour.

98.2.7 Advantages and disadvantages of static testing

The advantages of static testing lie in the simplicity of the test and it having been a long accepted technique. Generally, interpretation is simple and results can be produced quickly, without the need for specialist interpretation or in some cases with only limited knowledge of the ground conditions.

Figure 98.6 Lateral load pile tests showing piles being jacked off each other; displacement transducers (mounted on the white reference frame) to monitor lateral deflections can be seen against the test pile on the right-hand side of the image (Photograph courtesy of Cementation Foundations Skanska Ltd; all rights reserved)

Static tests are relatively slow. For instance, a typical MLT generally takes a minimum of 19 hours but may take much longer depending on the particular test specification and the settlement rate characteristics of the pile under test. This also neglects the time required for setting up the test equipment. This has programme impacts and implications regarding site safety, with the need in some cases for 24-hour working, although system automation can further reduce risks.

The main disadvantages of static testing stem from the need to have enough reaction at ground level to apply loading. This results in infrastructure-intensive reaction frames connected to tension piles or kentledge (**Figures 98.2** and **98.3**). Such systems are time consuming to construct, expensive and may have significant space requirements both in terms of footprint and materials handling. The need for such large reaction (120% of anticipated ultimate capacity; **Table 98.2**) also means that the reaction system is highly stressed during testing, resulting in potential safety concerns. This may result in the testing of reduced-diameter preliminary piles or model piles compared to those proposed as contract piles to further understand the components of pile resistance and verify design approaches. More recently, for very large piles it has been proposed to undertake multi-suite testing where a series of preliminary piles with different elements of the pile capacity that are isolated or significantly reduced in certain strata through the use of sleeving (Hardy *et al.*, 2019), low-friction bitumen coatings or including compressible bases. In this way the pile diameter may be kept as designed and the limiting shaft friction for each stratum and base capacity established from the different tests. The multi-suite approach can then allow valuable information to be gained on pile performance in different layers while adopting a smaller reaction frame and testing system.

One of the possible drawbacks of the anchor pile or kentledge-based reaction systems is that they may interfere with the behaviour of the pile under test. Poulos (2000) suggests that the use of kentledge may cause an increase in test pile capacity and stiffness. Predictions in sand suggest that pile capacity and stiffness might be increased by 10–20% due to the presence of kentledge (Poulos and Davis, 1980). With this in mind, SPERW (ICE, 2017) suggests that there should be a minimum separation of 2 m between the kentledge supporting arrangement or three times the test pile diameter (BS EN ISO 22477-1:2018 suggests a maximum of 2.5D or 2.5 m). For a reaction pile arrangement where the reaction piles are shorter than the test pile there should be a minimum separation of 2–3 m (depending on guidance consulted) or 2.5–3 times the test pile diameter. If the reaction piles are longer than the test pile (or inclined) the spacing should be increased to five times the diameter of the largest pile (test or reaction) unless the base capacity of the test pile is 20% less than that of the reaction piles. In the case of tension tests, any ground beams or adjacent compression piles used to support the reaction system should be at least three times the test pile diameter away from the test pile.

98.2.8 Interpretation of static tests

Inspection of the load–settlement curves determined from pile testing can be used in several ways. First, by inspection the shapes of the curves may hint at the adequacy or integrity of the installation and highlight problems occurring during pile formation. Additionally, the result may be used to check performance criteria.

98.2.8.1 Settlement criteria

Pile design is typically based upon determining both ultimate and serviceability limit cases or states. When assessing performance from pile tests, it seems that settlement criteria at working loads may be given more attention (in an onshore setting) than ultimate pile capacity. This is probably a result of ultimate capacity being associated with settlements that would cause serviceability damage to structures.

For working load tests, historically SPERW suggested that settlement criteria should be in the range of 5–10 mm at the design verification load (DVL) and 15–25 mm DVL + $0.5F_{rep}$ (F_{rep} is the representative action, previously referred to as the specified working load, or SWL), and that this should be assessed based on the specific building type. These values have been removed in the most recent version of SPERW, which may be an attempt to reinforce the need for them to be determined on a structure-specific basis, but there may still be a need to specify limits.

98.2.8.2 Determination of ultimate capacity and test termination

Another potential barrier to using ultimate axial pile capacity to define performance seems to lie in the various definitions of what constitutes the ultimate pile load or capacity. For example, Tomlinson and Woodward (2008) list seven different recognised techniques. These may be based upon the load at a certain settlement defined by the pile diameter (for instance, 10–15% of the test pile diameter, BS 8004:2015) or a feature related to the shape of the load–settlement curve, such as the load at which settlement continues to increase without further increase in load. Definitions such as the latter may be problematic; for instance, in granular soils it is common to see gradual increases in capacity with increasing settlement – that is, the criterion is never met.

EC7 seems relatively vague on this point but does state that where there is difficulty in determining the ultimate capacity, a value equal to 10% of the diameter should be assumed. SPERW (ICE, 2017) avoids the use of a specific settlement criterion and defines the ultimate capacity in MLT as the maximum load that can be applied while achieving the specific settlement rate criteria (**Table 98.4**). Conversely, CRP tests in SPERW (ICE, 2017) are terminated on the basis of where loads are constant or reducing for at least 10 mm or a settlement equal to 15% of diameter is achieved.

If a preliminary pile does not reach ultimate capacity, then the results may need to be extrapolated to give an indication of the expected capacity. Methods proposed by Chin (1972) and Fleming (1992), referred to as hyperbolic methods, are often used for this with varying degrees of performance for particular pile and soil types (BS 8004:2015). The Chin method is based on the assumption that measured pile load–settlement behaviour can be represented by a single hyperbolic function such that the ultimate capacity (U) of a rigid pile can be represented by

$$U = \frac{\Delta}{(\Delta / P) - K} \tag{98.1}$$

where Δ represents the pile head settlement at any pile head load (P), and K is the intercept on the horizontal axis (Fleming, 1992). Chin observed that if Δ/P was plotted against settlement (Δ), a straight line was obtained (**Figure 98.7**). The gradient of the straight line is the ultimate capacity of the pile (Equation 98.1). **Figure 98.7** also shows the CRP test previously presented in **Figure 98.1**, with the ultimate pile capacity predicted from the Chin analysis shown as an asymptote at 2323 kN.

The form of the hyperbolic function was then improved upon by Fleming (1992) through incorporating parameters from back-analysis of a large pile testing database and the effects of pile shortening by recognising that there are potentially two separate behaviours and consequently two hyperbolic functions. Fleming (1992) separates consideration of pile shaft (denoted subscript s) and base (denoted subscript b) components such that

$$\Delta_s = \frac{M_s D_s P_s}{U_s - P_s} \tag{98.2}$$

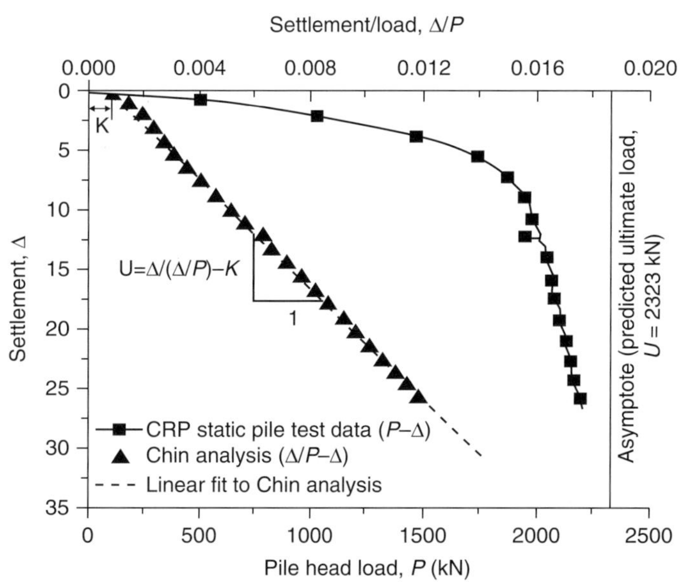

Figure 98.7 Chin analysis of a CRP static load test on a 12 m long, 600 mm dia. cast in situ bored pile in glacial till

where the pile diameter (D_s) has been incorporated to recognise its influence on load–settlement behaviour and M_s is a dimensionless flexibility factor with low variance with typical values between 0.001 and 0.0015 (Azizi, 2000). Pile base settlement is based upon the assumptions made for linear-elastic settlement of a footing and assessment of soil secant modulus (E_b) from a real load–settlement relationship under a footing at one-quarter of the ultimate stress (U_b /4).

$$\Delta_b = \frac{0.6 U_b P_b}{D_b E_b \left(U_b - P_b\right)} \quad (98.3)$$

Note that Equations 98.2 and 98.3 are for consideration of a rigid pile only. See Fleming (1992) for inclusion of elastic shortening. A useful summary of typical soil secant modulus values (E_b) can be found in Azizi (2000).

More detail of these techniques and interpretation of the results of static testing can be found in Fleming *et al.* (2009) and Tomlinson and Woodward (2008). BS 8004:2015 also refers to other methods, such as Davisson's (1972) offset limit and Butler and Hoy's (1976) slope and tangent method.

One of the aims of pile testing may be to verify or update design parameters and techniques. Where this requires specific knowledge of separate shaft and tip resistance and mobilised shaft frictional resistance with depth and settlement, it is recommended that the test pile is appropriately instrumented rather than attempting to derive such information based upon pile head measurements alone.

98.2.9 Specifying a static pile test

The magnitudes of the applied load increments for an MLT (**Table 98.3**) are chosen to verify the ability of the pile to carry the design loads associated with the structure and the resulting settlement. Typically, up to 10 mm of settlement is considered acceptable at the DVL for insensitive buildings. The designer's expectations of an acceptable test result should be set out in advance of the works. The DVL referred to in SPERW is typically 40–50% of the anticipated or previously measured ultimate bearing capacity. The load a pile will actually carry in service is referred to as the representative action (F_{rep}) or specified working load (SWL). The DVL should account for the F_{rep} and any specific differences that may occur between the pile load test and the pile in service – for example, pile downdrag or variations in final levels of the site. Working piles are typically loaded to the DVL plus 50% of their F_{rep} (proof load test). Preliminary piles may be further subject to increments of 25% of their F_{rep}. When calculating the anticipated static capacity, it should be remembered that the calculated capacity should be as realistic as possible and avoid the adoption of unnecessarily conservative soil parameters to allow specification of adequate loading systems. The equipment required to carry out the testing should be specifically designed for the purpose, include careful consideration of the peak load required, and the potential for change should be made at an early stage of the process. Once the equipment is on site, increases in load application beyond that first envisaged cannot be easily accommodated.

98.3. Bi-directional pile testing

98.3.1 What is bi-directional load testing?

The bi-directional static load test is an alternative to the top-down static load tests described in the previous sections (England, 2008). The method varies from the top-down load tests in that the major component of the system is a purpose-built high-capacity jack (or jacks) cast in the pile length. The most common form of this type of system is referred to as the O-Cell, which is receiving increased use and acceptance. SPERW (ICE, 2017) refers to the bi-directional method as another form of the MLT and it is included in BS EN ISO 22477-1:2018, which focuses on compressive static load testing.

The bi-directional O-Cell method of testing was originally developed by Professor Jorj Osterberg to load the pile from the pile base rather than from the head and more specifically for the evaluation of skin friction and end-bearing in rock sockets. On pressurising the jack(s), reaction is provided by the pile end-bearing capacity to mobilise the pile's skin resistance and vice versa until the capacity of either the jack or the upper or lower components of resistance are exceeded. More recently, individual jacks or multiple jacks have been installed at various levels within cast in situ piles and diaphragm walls (barrettes) to allow testing of different sections of the foundation length (England, 2008; England and Cheesman, 2010; Randolph, 2003).

98.3.2 How does bi-directional testing work?

The system typically works by incorporating a purpose-built, low-friction jacking device within a cast in situ pile (**Figure 98.8**). The loading arrangement is attached to the reinforcing steel or other support structure to ensure precise location and depth (**Figure 98.9**). When pile construction is complete and the concrete has attained sufficient strength, the jack is pressurised using a hydraulic pump, with the pressure monitored by a pressure transducer attached to the hydraulic return line. The foundation is then effectively separated into two elements which are subjected to simultaneous loading.

During testing, the separation of the jack is monitored by displacement transducers (linear variable differential transformers – LVDTs or LVWDTs) mounted between the two faces of the jack. Telltales are also attached to the top of the jack that extend up to the head of the pile, which allow the compression of the pile shaft to be monitored and the location of the top and bottom of the jack(s) to be determined. The movement of the pile head is also monitored along with

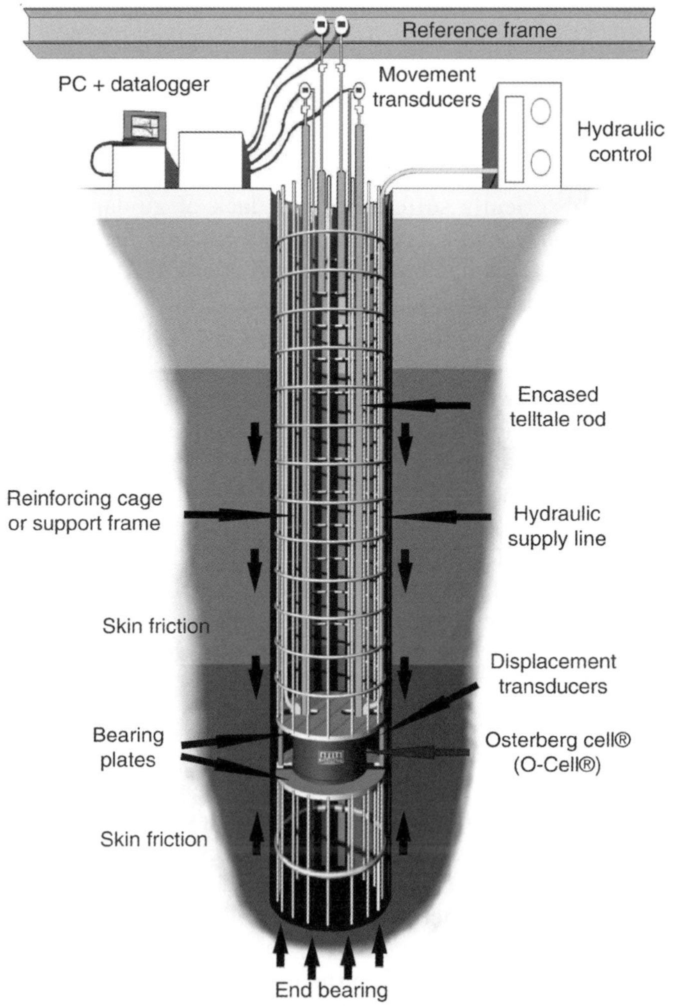

Figure 98.8 Schematic of a single-level O-Cell testing arrangement (Image supplied courtesy of Fugro Loadtest; all rights reserved)

Figure 98.9 Insertion of the O-Cell in the reinforcing cage of a diaphragm wall (Photograph courtesy of Fugro Loadtest; all rights reserved)

compression of the pile. On completion of testing, the jack(s) and the annulus around the jack(s) can be grouted up through the hydraulic circuit to allow the pile to be integrated into the structure as a working pile. The bi-directional cells come in various diameters from 180 to 870 mm, with capacities from 0.9 to 27 MN. Test loads as high as 320 MN have been applied using several bi-directional cells installed at the same level in a pile socketed in rock, and up to 360 MN mobilised in a barrette (Fugro Loadtest, 2023).

98.3.3 The advantages and disadvantages of bi-directional testing

The bi-directional type of load test has several obvious advantages over static top-down methods. The systems require no large surface reaction such as kentledge or anchored steel reaction frames, thus reducing space requirements, set-up time and transportation costs. The system is also safer, with the loads being applied at depth. There are reported cost savings, with bi-directional load testing being comparable in cost with top-down static load tests at 5–10 MN but then becoming much more cost-effective at higher loads. There is also the potential with the bi-directional cells to apply loads that cannot be achieved by top-down reaction or kentledge systems. Although the design of bi-directional cells lends itself to use in auger-bored cast in situ piles, it has also been successfully deployed in continuous flight auger (CFA) piles, Fundex piles, drilled steel piles with rock sockets, steel and concrete driven piles and barrettes. Another advantage is the ability to use a bi-directional test as a substitute for tension tests at high loads and push the pile up instead of attempting to pull the pile upwards from ground level.

One specific disadvantage associated with the bi-directional approach is that the jacking system needs to be pre-installed. This means it is not possible to select a random working pile for testing. The other disadvantage of the system is that the components installed within the pile are not recoverable after testing. This means that the bi-directional cell and some of the instrumentation is sacrificial, which has a cost implication, although when testing high loads the cost per unit load decreases with increasing magnitude (contrary to traditional top-down load tests).

98.3.4 Interpreting the data from bi-directional testing

Typical results from a single O-Cell installation similar to that shown in **Figure 98.8** are shown in **Figures 98.10** and **98.11**. Several techniques can be used to estimate the load–displacement behaviour at the head of the pile. The two most direct methods are to sum the measured responses and include the additional elastic shortening or alternatively to model the response of each of the pile elements using Cemsolve (Fleming, 1992) and then add together the geotechnical behaviour and corresponding total anticipated elastic shortening.

The total load–displacement responses can be obtained by adding together the loads measured above and below the bi-directional cell (at the same level of displacement). Where one section has mobilised less displacement than the other – for instance in the case of the movement of the upper section (**Figure 98.10**) – it may be necessary to extrapolate the obtained results using methods such as the Chin approach (Chin, 1972) for a single hyperbolic function or the Cemsolve approach (Fleming, 1992) to obtain the sum of the measured response as shown in **Figure 98.11**.

During the bi-directional loading some elastic compression is contained within the results. During top-down static loading, where the full load is applied to the pile head, additional elastic shortening would occur. This feature can be modelled and added to the results of bi-directional loading. Alternative approaches to analysis are discussed by England (2008).

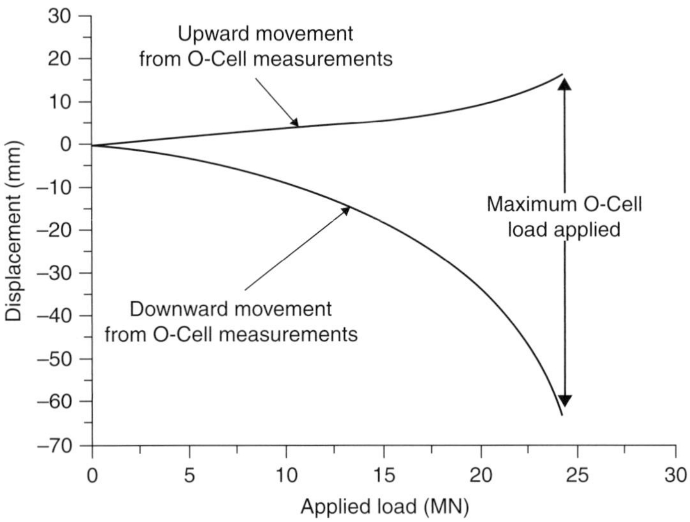

Figure 98.10 Example of results from a typical bi-directional load test (Data supplied courtesy of Fugro Loadtest)

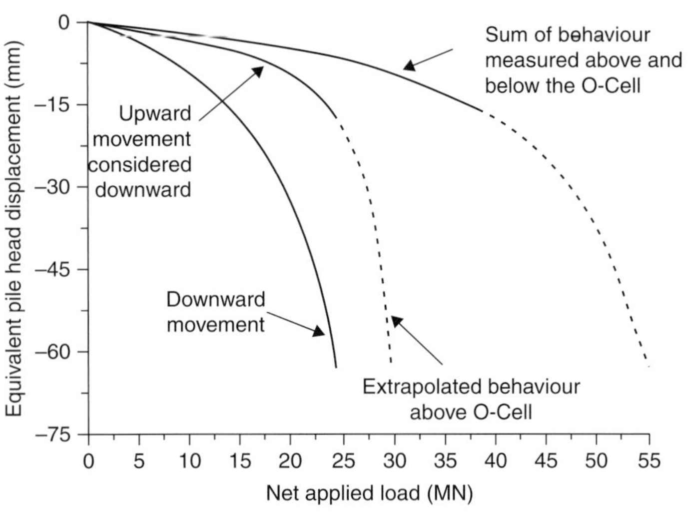

Figure 98.11 Sum of the measured responses from a bi-directional load test (Data supplied courtesy of Fugro Loadtest)

98.3.5 Standardisation and guidance for bi-directional testing

As a relatively new method of pile testing, bi-directional testing has historically suffered from a lack of guidance to aid specification; for instance, it is not specifically mentioned in Eurocode 7. This has been addressed in the UK by the second and third editions of SPERW (ICE, 2017). The FPS has also published guidance on the use of the technique as part of a general pile testing guide. A more critical appraisal of the technique with case study comparisons can be found in a review of innovative pile testing techniques by Paikowsky (2004). Codification was produced in the USA in 2018 with the publication of ASTM D8169/D8169M-18 and more recently as per other pile testing methods it is included in the informative annex of BS EN ISO 22477-1:2018. For Eurocode 7 purposes (and other codes and specifications where bi-directional testing is not specifically mentioned), it may be adequate to class the bi-directional test as a full-scale static load test, as suggested in SPERW (ICE, 2017).

98.3.6 Specifying a bi-directional test

Care needs to be taken when specifying systems where the jack is to be installed in a preliminary test as the optimum level will be where the frictional capacity upwards (above the cell) matches the available reaction below the cell. In an ideal test, the skin friction below the cell (downwards) is fully mobilised with sufficient of the end-bearing mobilised to allow characterisation before the ultimate skin friction above the cell (upwards) is mobilised. Where elements of the pile are not fully mobilised, back-analysis to reveal ultimate capacity may be used (as discussed above) where sufficient movement has occurred.

If the required test loads cannot be accommodated at a single level due to limitations on pile diameter, it may be appropriate to consider multiple installation levels so that the entire foundation element may be mobilised in turn during phased testing. A specific advantage of using multiple-level jack installations is that it allows individual pile sections to be mobilised separately. Cells placed close to the pile tip should also give due regard to the quality of the base and concrete at this level, with cells typically placed 1–2 diameters above the base.

As the bi-directional cell applies load using jacking systems it is possible to specify load increments, hold periods and loading rates as per static tests.

98.4. High-strain dynamic pile testing

98.4.1 What is high-strain dynamic testing?

The previously discussed load application techniques used to measure pile capacity typically have load application durations in terms of hours. Dynamic pile testing differs in that loading the pile occurs by a drop weight applying a very short-duration impact (≈5 ms) load at the head of the pile. The impact of the weight produces a stress wave that travels down the pile. Where the movement of the pile is resisted or there is a change in impedance, such as at the pile tip, a wave will be reflected back up the pile. Based upon comparing the waves travelling up and down the pile it is possible to assess the dynamic pile reaction. Through further analysis it is possible to derive the equivalent static pile load–displacement behaviour. Due to phenomena such as rate effects (including creep), excess pore water pressures and inertia effects (due to acceleration), dynamic or rapid load test results might differ from what would be expected from an equivalent static load test (BS 8004:2015). Further detailed guidance on dynamic pile test specification and analysis approaches can be found in BS EN ISO 22477-4:2018.

98.4.2 Methods available for dynamic testing

Dynamic testing has the advantage that the main equipment required for monitoring the test is relatively compact, and where the piles are being driven on site the normal pile-driving hammer can be used to produce the required stress wave. Piles are typically tested some time after installation, which is referred to as re-strike. This involves the piling rig returning briefly to the test pile to deliver a series of hammer blows (typically five, depending on the approach taken). If the tests are being carried out on a cast in situ pile or are re-strike tests of driven piles where pile driving for installation is complete, then it may be necessary to mobilise a separate drop-weight system (**Figure 98.12**). Alternative dynamic testing methods are also available which may use dedicated testing equipment and utilise alternative approaches to testing and analysis, such as the multi-blow dynamic testing technique referred to in BS EN ISO 22477-4:2018 and shown in **Figure 98.12**.

98.4.3 How does dynamic testing work?

Measurements are typically taken during the hammer impact from a pair of accelerometers and strain gauges that are attached above ground level to the pile head (BS EN ISO 22477-4:2018) either during or after driving (**Figure 98.13**). The response of this instrumentation is logged during and after the weight impact on a specially designed logging and analysis device, for example the Pile Driving Analyzer (PDA) produced by Pile Dynamics Inc. Data from the instrumentation are used to derive the force (F) applied to the pile by multiplying the measured strain by the cross-sectional rigidity of the pile and the velocity (v) from integration of the accelerometer readings.

The force measured by the strain gauges at the pile head due to the hammer impact is compared with the equivalent force derived from measurements made by the accelerometers (**Figure 98.14**). The data shown are for a re-strike test on a 250 mm square pre-cast concrete pile where the pile tip was 23.2 m below the strain gauges.

Figure 98.12 SIMBAT mini dynamic drop-weight system (1000 kg hammer); note the pile head instrumentation installed on the exposed concrete after local removal of the temporary casing (Photograph courtesy of James Fisher Testing Services Ltd; all rights reserved)

It is assumed in stress wave theory that force and velocity are proportional as the wave travels through a material. Where

$$v = \frac{F}{Z} \tag{98.4}$$

or

$$F = v \times Z \tag{98.5}$$

the pile impedance is

$$Z = \frac{E_p A_p}{c} \tag{98.6}$$

where E_p is Young's modulus of the pile, A_p is the pile's cross-sectional area and c is the wave velocity in the pile, which can be found based upon the stiffness and density of the pile (ρ_p)

$$c = \sqrt{\frac{E_p}{\rho_p}} \tag{98.7}$$

Figure 98.13 Installation of instrumentation (strain gauge and accelerometer) on a pre-cast concrete driven pile (pile hammer visible at the top of the image)

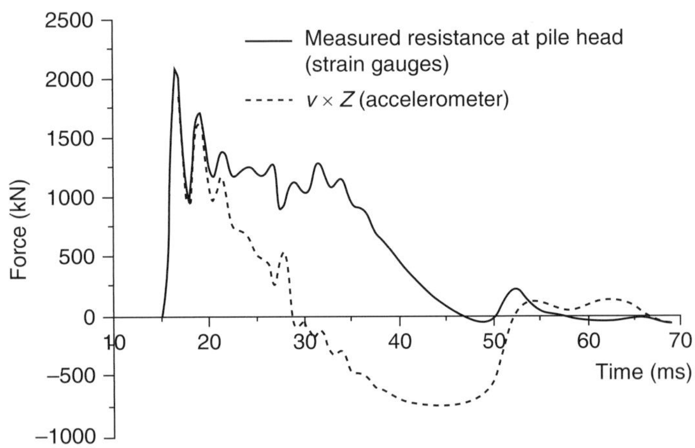

Figure 98.14 An example of measured pile stress wave data (Data provided by Technical Services Ltd)

Where the movement of the pile is resisted, or there is a change in impedance such as at the pile tip, a wave will be reflected back up the pile (**Figure 98.15**). The total dynamic resistance of the pile (R) to the stress wave passing up and down the pile has been shown to equal the sum of the downward-travelling force ($F_{(t_0)}$), force measured at maximum pile head velocity ($v_{(t_0)}$) plus the upward-travelling force that arrives back at the pile head ($F_{(t_0+2L/c)}$) at approximately $2L/c$ (where L is the pile length, **Figure 98.15**) after the initial peak load (Randolph, 2003; Rausche *et al.*, 1985)

$$R = \frac{F_{(t_0)} + F_{(t_0+2L/c)}}{2} + Z\left(v_{(t_0)} + v_{(t_0+2L/c)}\right) \qquad (98.8)$$

98.4.4 Advantages and disadvantages of dynamic testing

The advantages of dynamic testing come from the simplicity of the equipment and the speed of testing. If driven piles are being tested it is typical to test at re-strike at some time after driving. The on-pile strain gauge and accelerometer are quickly bolted to the test pile (**Figure 98.13**), with logging undertaken on a small hand-held logging and analysis system (**Figure 98.16**). This means the system can be operated by a single operator with quick instrument installation and causing little disruption to construction. As the piles can be tested during installation, many working piles can also be tested without concerns over testing altering working performance. This data collection process has been improved recently with the development of wireless instrumentation and the option of replacing strain gauge load determination with direct pile head load cell readings and direct measurement of pile head displacement with the use of high-speed optical scanning theodolites.

If the piles are cast in situ piles or driven piles tested after the original pile-driving rig has been demobilised, then it is possible to mobilise a drop-weight system. Tests up to 36 MN have been undertaken by dropping 36 tonne masses. Typical

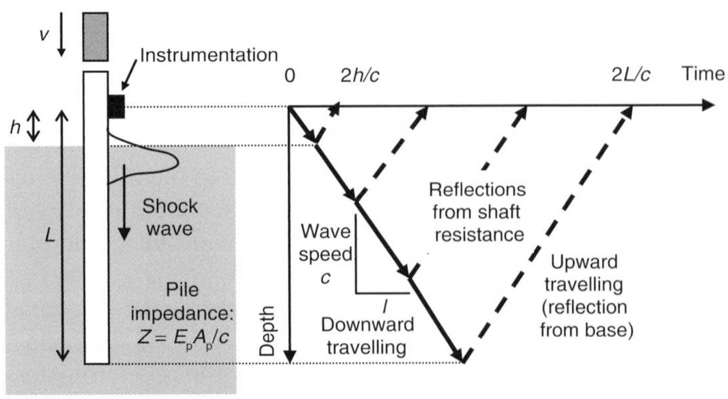

Figure 98.15 Schematic of stress wave travel in a pile (Reproduced from Randolph, 2003)

Figure 98.16 Pile Driving Analyzer capturing data from wireless instrumentation during pile driving (Photograph courtesy of Technical Services Ltd; all rights reserved)

loads achieved are 1 MN for 1000–1500 kg drop-weight systems (typically 1.5–2% of the applied load is required) and 3 MN for 4000 kg drop-weight systems. The ability to produce large loads with limited drop mass significantly reduces cost (**Figure 98.12**).

Other benefits associated with the nature of the test mean that it may be possible to verify the integrity of the pile during normal testing (see Chapter 97 *Pile integrity testing*). Through analysis it is also possible to investigate the distribution of resistance down the pile and the split of base to shaft resistance without the need for additional instrumentation as is required in techniques where only pile head measurements are made.

Due to the nature of using the test during pile driving the test only gives an indication of the pile resistance during the installation process. Depending on the soil type, the capacity of the pile may change significantly after driving due to, for instance, the dissipation of pore pressures which may lead to increases in capacity with time, which is often referred to as set up. This may lead to the need for re-strike testing at various times after driving to verify pile capacity. This is not a problem specific to dynamic testing but may need to be considered for any pile testing method where displacement piles are being tested.

98.4.5 Interpretation of dynamic testing

It has previously been briefly explained how the dynamic pile resistance can be derived from pile head measurements during dynamic testing. It is then necessary to determine the ultimate static pile resistance from these measurements. Several methods are available to do this, with the most common approaches based upon signal processing and numerical models (Holeyman, 1992).

98.4.5.1 CASE method

One of the earlier methods developed for signal processing is referred to as the CASE method (Rausche *et al.*, 1985). This simple technique may be used to quickly assess capacity in the field based upon an individual blow. It requires the determination of a damping force (R_d) which is assumed to be proportional to the pile tip velocity (v_{tip}) that can be removed from the total dynamic resistance to give the equivalent static resistance. Simplification lies in the assumption that dynamic pile resistance is proportional to pile tip velocity only

$$R_d = JZv_{tip} = J\left(F_{(t_0)} + Zv_{(t_0)} - R\right) \tag{98.9}$$

The performance of this technique is dependent on the soil-dependent CASE damping factor (J). Suggested values for the damping factor are shown in **Table 98.6**. Depending on the exact time point at which dynamic pile capacity is determined when analysing an individual blow, some variation in the value of the damping parameters may occur. Further guidance is given in Rausche *et al.* (1985), EA Pfähle (2013) and BS EN ISO 22477-4:2018.

Based upon the range of potential damping parameters shown in **Table 98.6** it is recommended that the actual damping factors adopted are selected based upon the results of signal matching or experience of testing of similar piles in similar ground conditions.

As an alternative, the TNO closed form solution method, includes both skin friction and pile tip-related terms, which results in different damping constants for both components. More detail on the TNO method can be found in Foeken *et al.* (1996) and BS EN ISO 22477-4:2018.

98.4.5.2 Signal matching

The most common methods of analysing dynamic load tests presently used are based upon lumped parameter finite difference or finite element techniques where the pile is modelled as an assembly of interconnected masses with varying properties. These properties, predominantly soil parameters, are varied until computer-simulated pile head forces and velocities match those measured. Several computer packages have been developed that utilise this 'signal matching' method, such as

Soil type	Damping factor, J (s/m)
Sand	0.05–0.20
Silty sand/sandy silt	0.15–0.30
Silt	0.20–0.45
Silty clay/clayey silt	0.40–0.70
Clay	0.60–1.10

Table 98.6 Values for CASE damping coefficient (Data taken from Rausche *et al.*, 1985)

CAPWAP (Case Pile Wave Analysis Program), TNOWAVE and SIMBAT (Stain, 1992). Further detail on signal matching is given in BS EN ISO 22477-4:2018.

Common issues that may affect analysis are

- insufficient pile settlement mobilised during testing to allow determination of ultimate pile capacity
- variation in material properties or cross-section down the pile length
- stresses locked in the pile as a result of pile driving.

It should be noted that these issues are not specific to dynamic testing and may cause problems to both static and rapid pile testing, especially where pile instrumentation is being used to supplement testing.

To allow the ultimate behaviour of a pile to be assessed from dynamic analysis, it is necessary to mobilise adequate settlement in the pile and allow appropriate strain levels in the surrounding soil. Dynamic analysis based upon low settlement levels is likely to lead to a conservative assessment of pile performance (i.e. an underestimation of pile capacity). Where insufficient settlement is mobilised, analysis is not compromised but reflects the mobilised soil resistance rather than ultimate pile capacity. Unfortunately, in the case of rapid load testing insufficient settlement may render analysis difficult. The situation of low mobilisation may come about due to ignorance of the importance of mobilising adequate settlement, but more commonly it is a result of specifying rigs that are unable to apply the required load levels. Another reason for low mobilisation may be a limit on load–stress levels due to concerns over pile damage.

As dynamic test analysis is dependent on knowledge of the pile cross-sectional stiffness, it is necessary to accurately know the cross-sectional profile of the pile and the stiffness of the pile material. These properties may be easy to determine for a steel or pre-cast concrete pile that is constructed to tight size tolerances and is cured in a factory environment. Where cast in situ piles are formed, the profile of the pile may be less certain or complex – for instance in the case of a cast in situ screw pile or under-reamed piles. There are also material issues such as how the inclusion of reinforcement affects the pile stiffness and whether site-cured concrete has the same properties as laboratory-cured samples. Where piles have complex geometry this can easily be incorporated in dynamic analysis if the geometry is known. Where less information is available for the test, for instance in the case of pile reuse assessment where there is little information about the length, cross-section and quality of historic piles (Butcher *et al.*, 2006), dynamic testing may not be appropriate.

98.4.5.3 Multi-blow analysis

Further detailed guidance on the analysis of multi-blow testing can be found in BS EN ISO 22477-4:2018. This approach differs in that the aim is to determine a site-specific damping and soil-dependent factors which reduces measured dynamic resistance to static resistance for a series of blows. The method consists of applying multiple blows of different drop height to the top of an instrumented pile, such that both high and low pile head displacements and hence pile velocities are produced. Strain and acceleration are measured using sensors attached to the pile as described above, with the pile head displacement signal measured using a high-speed optical scanning theodolite. Total dynamic pile resistance is determined in a similar manner to that described above (and described in more detail in BS EN ISO 22477-4:2018) and this resistance is compared for the different drop heights to determine the soil or site-specific correction factors due to controls such as soil-dependent rate effects. The dynamic resistance is then corrected on this basis to obtain a static equivalent pile resistance. As per other dynamic or rapid methods, this approach cannot capture the long-term behaviour of a pile where consolidation, pore pressure dissipation or creep will change or affect pile capacity with time.

98.4.6 Standardisation of dynamic testing

The dynamic pile testing method has been in common use for many years, with BS EN ISO 22477-4:2018 giving useful recent coverage. In the UK it is well documented in SPERW (ICE, 2017) and the FPS general pile testing guide (FPS, 2006). SPERW (ICE, 2017) also refers to both dynamic and rapid load pile testing as being well established. In the USA there is an ASTM standard test method for high-strain dynamic testing of piles (ASTM D4945-17). One common thread throughout the guidance documents is the need for experience of the dynamic load testing in similar soils and on similar pile types (EC7). Where such experience does not exist, or certain ground conditions prevail such as fine-grained and laminated soils, it is recommended that the tests are calibrated against a site-specific static test (ICE, 2017).

98.4.7 Specifying a dynamic test

It is important that the installation equipment being used for dynamic pile testing is first capable of driving the piles to the required depth or capacity and that it can mobilise the pile adequately during testing. Where time allows, details of the piles being driven and the driving equipment should be analysed prior to installation by the dynamic pile testing specialist. This wave equation analysis can be used to assess the ability of the proposed driving system to install the pile to the required capacity and desired penetration without exceeding allowable driving stresses in the pile material.

The other important decision to be made during dynamic testing is when to test, due to set up and change of capacity with time. As mentioned, testing can be undertaken at the end of initial pile installation and through re-strike at any time after installation.

The piles to be tested dynamically should be designed to resist the driving stresses. They should also be designed to have at least 300 mm sticking up above ground at the end of pile testing to avoid sensor damage. To allow sensor installation it is important that there is adequate working room near the pile heads for sensor mounts to be installed prior to lifting piles. It is normal practice to have at least 1.5 diameter stick-up above any attached instrumentation as a minimum. For cast in situ piles it may be necessary to include an extension to the top of the pile to approximately three pile diameters above ground level and at least 100 mm above the existing pile reinforcement. This may be achieved by incorporating a section of casing above the pile head and filling with concrete while pouring the main pile concrete. Once the concrete has set, windows may be cut in the casing to allow instruments to be attached to the concrete (**Figure 98.12**). Instrumentation is typically installed at a minimum of 1.5–2 pile diameters below the pile head. It is noted that dynamic testing and analysis is a relatively specialist area of pile testing and it is recommended that further guidance on appropriate analysis techniques and testing methodology should be sought through early engagement with a specialist pile testing contractor.

98.5. Rapid load testing

98.5.1 What is rapid load testing?

The main difference between rapid load pile testing and static and dynamic testing is the duration of application of load. A rapid load test usually has a load duration of 90–250 ms, which is approximately 30–40 times that of a dynamic pile load test (ASTM D7383-19; Hoelscher and van Tol, 2009). The duration of the loading has been designed such that piles less than 40 m in length remain in compression throughout the loading event, resulting in negligible stress wave effects (BS EN ISO 22477-10:2016) and simpler analysis (**Figure 98.17**). Generally, stress wave analysis cannot be utilised with this type of test.

98.5.2 Methods available for rapid load testing

Several different techniques are available to produce rapid loading events. Statnamic testing works by rapid burning of a solid fuel that produces gas in a pressure chamber (**Figure 98.18**). The venting of this gas is used to accelerate a mass upward that in turn imparts a load onto the foundation pile below the Statnamic device. The load is applied and removed smoothly through the controlled venting of the gas. Alternatives to this approach typically take the form of a drop mass system where the load duration is increased through various cushioning techniques, for instance a spring or springs in the pseudo-static pile load tester (Schellingerhout and Revoort, 1996) and spring hammer systems (Matsuzawa *et al.*, 2008) and specialised cushioning in the hybrid dynamic test (Miyasaka *et al.*, 2008) or tuneable cushioning in the StatRapid system (**Figure 98.19**) (Middendorp *et al.*, 2013).

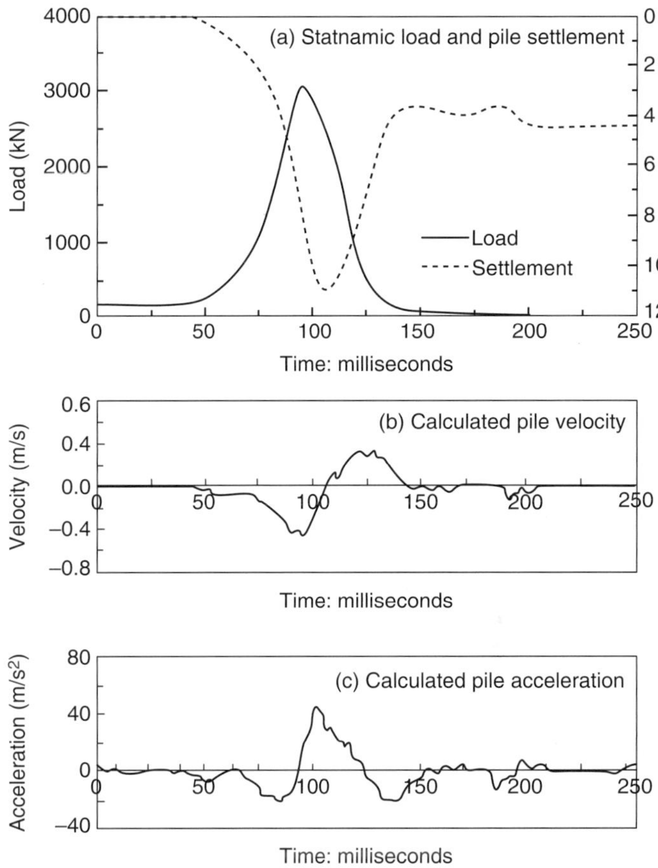

Figure 98.17 Measured and calculated results from a 3000 kN Statnamic load cycle (Reproduced from Brown and Hyde, 2006)

98.5.3 What are the advantages and disadvantages of rapid load testing?

The main advantage of rapid load testing devices comes about as a result of the method of load application. For example, in the rapid load testing method known as Statnamic, a reaction mass is accelerated upwards resulting in a load application of 3.5 MN from 18 tonnes. Typically, for Statnamic testing the reaction mass is only 5% of the equivalent static load reaction mass required to produce the same load. This results in rapid load testing devices being significantly smaller and lighter than static equivalent set-ups. This means that they can be used on sites with space restrictions, are easier to mobilise and quicker to set up and test. For example, a 3 MN Statnamic rig (**Figure 98.20**) can be mobilised with one articulated truck and a 70 tonne crane. If the rig incorporates a hydraulic catch mechanism (catches the reaction mass) it can test up to ten individual piles per day or allow multiple load cycles to be carried out on an individual pile with minutes between cycles.

Typically disadvantages of rapid load testing methods stem from the rapid nature of the load application, which may result in piles displacing at rates measured in metres per second, resulting in very high strain rates in the soil (**Figure 98.17**). In

Figure 98.18 Statnamic pressure chamber with a load cell at its base shown mounted on a test pile

Figure 98.19 A 1.5–10 MN StatRapid rig with remote optical displacement tracking system for displacement measurement in the foreground (Photograph courtesy of Allnamics; all rights reserved)

fine-grained soils such as clay such high strain rates have been shown to significantly increase the strength of the clay (Brown and Hyde, 2008a). This may result in pile ultimate capacity up to three times the measured or predicted static pile capacity (Holeyman, 1992). This effect appears to become more pronounced as the liquidity index of a clay increases (Robinson 2019; Robinson *et al.*, 2023). It is important to be aware of this effect to avoid specifying rapid load testing devices that are incapable of adequately mobilising a pile.

Another major drawback is that certain techniques such as Statnamic may operate over a very discrete loading range; for instance, a 3 MN Statnamic rig may apply loads as high as 3.5 MN, but would not be able to apply loads below 700 kN.

98.5.4 How does StatRapid work?

Historically in Europe Statnamic was the most widely used form of rapid load testing, but has been overtaken more recently by StatRapid testing. As StatRapid is the most widely deployed method of rapid load testing in Europe it is described in more detail. The StatRapid test method uses a drop weight to axially load the pile by dropping a mass onto a specifically designed polymer-type cushion which has modular elements in terms of the number of spring elements and stiffness of each spring element that make up the cushion. This allows optimisation of the loading event, which is numerically modelled prior to testing to choose the spring modules that control the load pulse or 'push' applied and to maintain pile compression throughout the loading event. The drop mass or weight is guided within a frame that includes hydraulics for lifting and catching the weight after impact to avoid multiple uncontrolled impacts on the pile. The test mass is also modular, and weights can be added or removed in order to optimise the set-up for each pile, depending on soil conditions and on the required test load.

Statnamic varies from StatRapid in that load is applied vertically down a pile by burning a rapid-burning fuel which produces significant volumes of gas within a piston and accelerates a reaction mass upwards. The duration and shape of the load pulse (**Figure 98.17**) are controlled by venting of the gas. The reaction mass is forced upwards and is caught by either gravel in the case of large load tests (20–60 MN) or by hydraulic catch mechanisms for smaller rigs (3–20 MN)

(**Figure 98.20**). As Statnamic load application is independent of gravity, the device can easily be used to test raked piles and has been deployed horizontally to test piles laterally and simulate impacts on structures.

During rapid load tests, load is measured directly by a calibrated load cell mounted below the cushioning system, or incorporated in the base of the combustion piston for Statnamic, which sits on the pile head (**Figure 98.18**). Displacement is also measured directly using a non-contact system either by a photovoltaic sensor mounted in the piston (Statnamic) which is excited by a remote laser beam or the movement of targets mounted on the pile are recorded by means of a remote optical displacement tracking system such as a high-speed optical scanning theodolite (**Figure 98.19**). As the rapid load test event may induce stress waves in the ground it is important that the remote component of the measuring device is located far enough away from the pile to avoid disturbance during the duration of the loading event (Brown and Hyde, 2006). The recording of displacement–time history can then be differentiated once or twice to determine pile velocity and acceleration, both of which are required for test analysis.

Figure 98.20 A 3.5 MN Statnamic rig with hydraulic catch mechanism ready for testing

It is also commonplace for the system to incorporate an accelerometer which can be used for direct recording of acceleration. The acceleration–time history can also be integrated to determine velocity, which may be less 'noisy' than that found by differentiation, and for a second time to verify optical or laser-based displacement measurements.

98.5.5 How to interpret the data

Rapid load tests must be analysed to remove both the effects of inertia associated with the pile/soil and strain rate-dependent response of the ground. The most common approach for this is referred to as the unloading point method (UPM).

Based on inspection of load–displacement curves from Statnamic tests, Middendorp (2000) observed that during the unloading the pile velocity reached zero, which in coarse-grained soils corresponded approximately to the ultimate static resistance of the pile. This point (the unloading point, UP) was then used to determine a constant damping coefficient to correct the measured Statnamic data for the velocity-dependent resistance of the soil. Unfortunately, when this technique is applied to piles installed in fine-grained soils there is a tendency for the ultimate pile capacity to be significantly overpredicted. In order to correct for this effect, a series of soil-dependent average correction factors were developed by Paikowsky (2004) (**Table 98.7**), which the derived static load was then multiplied by to obtain a corrected UPM analysis (**Figure 98.21**). More recently, it has been proposed that much greater average correction factors in clay are required, resulting in a μ value of 0.47 (Weaver and Rollins, 2010). BS EN ISO 22477-10:2016 gives similar correction factors based upon recent experience in Europe and as collated in Hoelscher *et al.* (2011).

In response to the known shortcomings of the UPM technique, a series of alternative methods have been developed which have their basis in the velocity or strain rate-dependent behaviour of soils. One nonlinear velocity-dependent method of analysis was developed by Brown (2004). The main conceptual difference from UPM is that the method is reliant on the user input of soil-specific rate parameters

$$F_u = \frac{F_{STN} - Ma}{1 + \left(\frac{F_{STN}}{F_{STNpeak}}\right)\alpha\left(\frac{\Delta v}{v_0}\right)^{\beta} - \left(\frac{F_{STN}}{F_{STNpeak}}\right)\alpha\left(\frac{v_{min}}{v_0}\right)^{\beta}} \quad (98.10)$$

Soil type	Rate effect factor (μ)	FOS without μ	FOS with μ
Rock	0.96	2.0	2.0
Sand	0.91	2.1	2.0
Silt	0.69	2.8	2.0
Clay	0.65	3.0	2.0

FOS, factor of safety.

Table 98.7 Correction factors for UPM analysis (Information taken from Paikowsky, 2004)

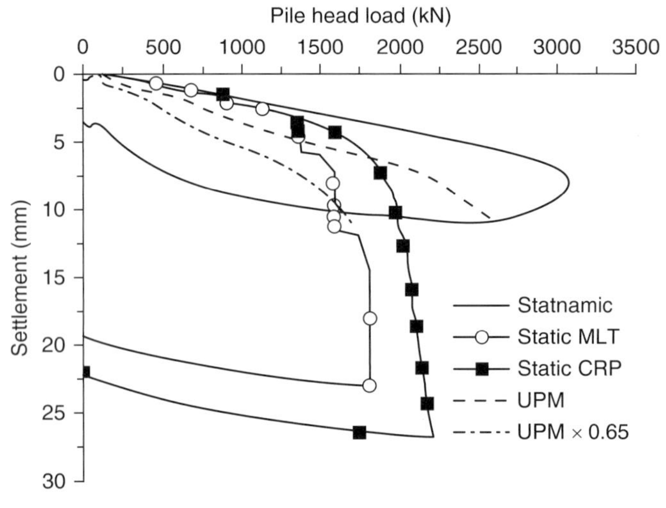

Figure 98.21 Performance of UPM correction of Statnamic data; the pile tests have been reset to zero settlement to allow comparison

where F_u is the derived static pile resistance, F_{STN} is the measured Statnamic load, Ma is the pile inertia, Δv is the pile's velocity relative to the soil and v_{min} is the velocity of the static CRP pile test used to define the soil-specific rate parameters α and β (β is normally set to 0.2 in clay). Both Δv and v_{min} are normalised by v_0, which is assumed to be 1 m/s. This form of analysis has evolved from the analysis of dynamic tests where the majority of the pile capacity is developed through skin friction (Randolph, 2003). The value of α may be selected based on the tentative relationship with plasticity index (PI) as proposed by Powell and Brown (2006) and Brown and Powell (2013)

$$\alpha = 0.03PI(\%) + 0.5 \tag{98.11}$$

More recent work looking at rate effects and influencing factors in clays using high-speed triaxial testing (Robinson, 2019) suggests that rate effects in clays are influenced by the state of the soil, which could be determined based upon the in situ moisture content (Brown and Hyde, 2008b) and the clay particle surface area. Better correlations were found using the liquidity index (LI) and there appeared to be a strong correlation with the slope of the critical state line, which itself is thought to be influenced by a clay's particle surface area (Muir Wood, 1990).

One additional benefit of rapid load testing methods that incorporate hydraulic catching and lifting mechanisms is the ability to quickly recycle the testing equipment and apply multiple cycles of loading to a single pile. BS EN ISO 22477-10:2016 outlines how the UP method of analysis can be modified to accommodate multiple load cycles.

98.5.6 Guidance and standardisation of rapid load testing

Like bi-directional testing, rapid load pile testing has historically suffered from a lack of guidance to aid specification. This has again been addressed in the UK by SPERW (ICE, 2017) and the FPS pile testing guide (FPS, 2006). In the USA there is an ASTM standard for rapid load testing of deep foundations (ASTM D7383-19) along with guidance produced for the federal and state highways agencies (McVay *et al.*, 2003; Paikowsky, 2004). The Japanese Geotechnical Society has also produced a testing specification, an English-language draft of which was published in 2000. A European testing standard within the Eurocode/ISO framework (which has been adopted by BSI) was developed and published in 2016 (EN ISO 22477-10:2016). This document was initially outlined by Hoelscher *et al.* (2010) and followed the publication of Dutch guidance (Hoelscher *et al.*, 2011) which contains a useful database of rapid load testing in different soils. The publication of EN ISO 22477-10:2016 makes a clear statement that rapid load testing is its own separate entity and not a subset of dynamic pile testing. This standard is predominantly concerned with test definition and guidance on how to undertake testing to achieve the best results, but it does outline potential methods of test analysis based upon the UPM approach for single and multicycle testing. EN ISO 22477-10:2016 does not consider model or partial factors for design as these were considered the remit of an evolved Eurocode 7. Inspection of the evolved Eurocode 7 as part of public consultation would suggest that both rapid and dynamic load testing of piles will see greater independent recognition and guidance on how these techniques can be utilised in a Eurocode design framework in the future.

98.5.7 Specifying a rapid load test

Prior to specifying a rapid load test, it is necessary to have access to the results of geotechnical investigation and laboratory testing to allow selection of adequate testing equipment and loading levels. This should really be the case prior to undertaking any type of pile test. The encountered ground will also influence the analysis approach. For analysis of tests in fine-grained soils, ideally regular determination of moisture content and plastic and liquid limits should be undertaken down to at least the depth of the toe of the pile.

In granular soils and piles installed in rock, the applied loads will be similar to the anticipated ultimate pile capacity, with additional loading capacity required for the inertial forces generated by the pile and soil/rock. In addition to the inertial resistance, load testing devices used in fine-grained soils such as clays will need to be capable of applying additional loads due to strain rate effects. For example, in clay the ultimate capacity measured during rapid loading may be 1.8–3.5 times greater than the static capacity, or higher (Brown and Powell, 2013; **Figure 98.22**). The need to apply such high loads should be considered when designing and specifying test piles. Further detailed guidance can be found in 'BS EN ISO 22477-10 Testing of piles: Rapid load testing'. There is also further useful commentary in SPERW (ICE, 2017).

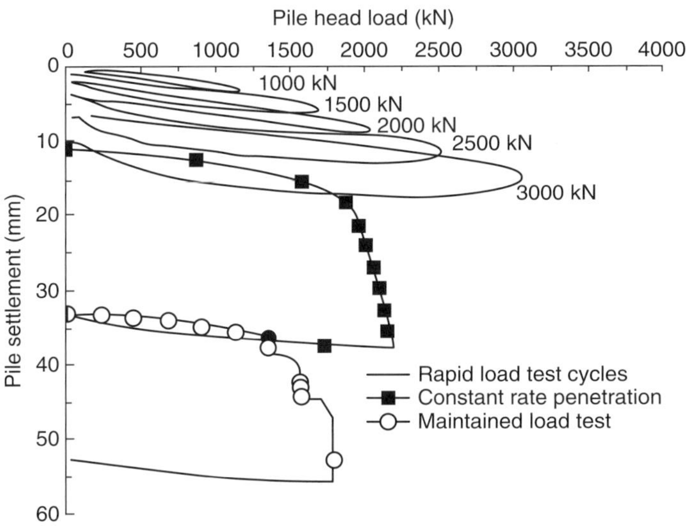

Figure 98.22 Cycles of Statnamic loading prior to analysis compared with CRP and MLT static tests (see Figure 98.1) (Reproduced from Brown *et al.*, 2006)

98.6. Pile testing safety

98.6.1 Why is there a need to be careful?

The testing of piles poses a safety hazard for several reasons. The most obvious of these are the high tensile and compressive loads generated in both the pile and reaction system during testing. Failure in any part of the system may lead to rapid release of energy and collapse of the testing arrangement. Other issues may arise due to the specific reaction system; for instance, kentledge relies on placing weights above a supporting frame which needs to be stable throughout erection, and testing thus requires appropriate structural design and checks on bearing capacity.

98.6.2 Good practice

As relatively high loads are applied over the relatively small area of the pile head it is important that all attempts are made to reduce eccentricity of loading on the pile and in other areas of the reaction system, such as connections to anchor piles. Excessive eccentricity can lead to large moments that the load or reaction system has not been designed for. Eccentricity can be minimised by accurate setting out of reaction piles relative to the test pile, correct alignment of the load column and proper mating of contact surfaces and levelling of reaction beams.

An obvious means of reducing risk to site and testing operatives is to restrict access to the load testing set-up and reduce the need to be close to the heavily loaded elements or reduce the need to work under kentledge. This can be achieved by restricting access only to relevant personnel. The need for direct access to the load column can also be reduced or removed through the use of computer-controlled loading jacks and displacement transducers. Fully automated testing systems have been developed which can be controlled remotely (England, 2002), although it is generally advised that tests should not be left unattended while loading.

Good practice can also be incorporated in design of the loading system. For instance, where the testing system uses reaction piles, this should ideally incorporate a minimum of three anchor piles as systems based upon two reaction piles are significantly less stable. The use of a two-pile reaction system should be reserved for relatively low loading levels with high tolerance pile and reaction system positioning.

98.6.3 Loading and inspection

It is important that the maximum test load to be applied to the pile is agreed in advance of the test so that the test pile/pile cap and all associated elements of the load testing equipment can be designed and specified to apply the maximum test load safely. This information must be communicated to the relevant operatives and, where possible, relevant limiting capacities should be marked on individual elements of the testing equipment (see **Figures 98.4–98.6**).

It may be the intention during testing to take the pile to its ultimate capacity and cause significant permanent settlement. In some cases a pile's capacity during testing may exceed that anticipated prior to testing and require additional loads to prove ultimate behaviour. Although the goal may be to prove ultimate behaviour, this should never be used to justify exceeding the maximum test load the testing system has been designed for.

During the course of the load test the whole system should be monitored for eccentricities and excessive deflections, with appropriate action taken if this occurs. Typical signs of distress may include

- excessive deflection of the reaction beams
- upward movement of tensile bars cast into reaction piles
- horizontal deflection of the reaction system
- movement of kentledge
- difficulty in maintaining test loads.

Further key safety issue guidance is given by FPS (2006).

98.7. Simple overview of pile testing methods

Selection of the most appropriate pile testing technique will vary from one situation to the next. Criticism is often levelled at the different techniques for their shortcomings, whereas any form of pile test (if correctly undertaken and analysed) should be viewed as a tool to give confidence in the design assumptions made, improve efficiency in construction and verify as-built performance. For instance, where a static load test identifies problems with trial pile performance during working pile construction, rapid or dynamic testing techniques can be rapidly deployed to assess the effect on the

Test type	Advantages	Limitations	Potential deployment
Static	• Simple test • Simple and quick interpretation • Well understood and accepted	• Slow tests • Significant infrastructure (especially as loads increase) • High space requirements • Safety concerns with increasing loads	• Lower loading (<20 MN) • Low pile numbers • Trial piles/limited number of working piles
Bi-directional	• Capable of very high test loads (higher than other techniques) • Low infrastructure • Low space requirement • Cost-effective as load magnitude increases	• Pile for testing needs to be preselected • Less experience • Analysis needs to take account of different surface boundary conditions • Specialised analysis and interpretation	• Medium to very high loading (2–320 MN) • Low pile numbers • Trial piles/limited working piles
Dynamic	• Low infrastructure • Low space requirement • Fast tests • Quick repeat testing • Mature technique	• Pile damage may be a concern • Perceived reliability: intermediate • Tests may be influenced by pile material and geometry • Specialised analysis and interpretation	• Medium to high loading (1–3 MN) • Low to high pile numbers • Trial and working piles • Where pile-driving equipment already on site
Rapid	• Low infrastructure • Low space requirement • Rapid testing • Quick repeat testing	• Less experience • Analysis techniques still evolving • Less case study experience • Availability of high-capacity equipment	• Medium to high loading (0.6–40 MN) • Medium to high pile numbers • Trial and working piles • Quality control of working piles • Problems with working pile performance

Table 98.8 Summary of pile test characteristics and potential deployment criteria

constructed working piles. This may then allow modifications to pile design and construction that can be used to mitigate problems during construction without significant effect on the construction programme. **Table 98.8** summarises some of the pros and cons of the methods described in this chapter and offers suggestions for potential deployment considerations. This table only serves as a quick summary and should not be used for definitive selection of a test. Decisions on the most appropriate testing method are influenced by many factors – for example, if there is a pile-driving rig on site it will likely be more cost-effective to undertake dynamic testing than to mobilise another technique. Definitive guidance should be sought from the specific providers of such techniques, who can offer cost-effective and bespoke solutions both for deployment and subsequent analysis and interpretation.

Acknowledgements

The authors would like to acknowledge the input of Melvin England (Fugro Loadtest UK) and Mike Kightley (Pile Analysis (WEAP) Limited) on the Osterberg Cell and dynamic testing sections, respectively. The authors are grateful to Andrew Bell of Cementation Skanska Foundations, Simon French of James Fisher Testing Services Ltd, Paul Hoelscher of Deltares and Rob van Dorp of Allnamics.

References

ASTM (2017) D4945-17: Standard test method for high-strain dynamic testing of piles. ASTM, West Conshohocken, PA.

ASTM (2018) D8169/D8169M-18: Standard test methods for deep foundations under bi-directional static axial compressive load. ASTM, West Conshohocken, PA.

ASTM (2019) ASTM D7383-19: Standard test methods for axial rapid load (compressive force pulse) testing of deep foundations. ASTM, West Conshohocken, PA.

ASTM (2020) ASTM D1143/D1143M-20: Standard test method for deep foundations under static axial compressive load. ASTM, West Conshohocken, PA.

Azizi F (2000) *Applied Analysis in Geotechnics*. Spon, London, UK.

Brown MJ (2004) *The Rapid Load Testing of Piles in Fine Grained Soils*. PhD thesis, University of Sheffield, Sheffield, UK.

Brown MJ and Hyde AFL (2006) Some observations of Statnamic pile testing. *Proceedings of the Institution of Civil Engineers: Geotechnical Engineering Journal* **159(GE4)**: 269–273.

Brown MJ and Hyde AFL (2008a) Rate effects from pile shaft resistance measurements. *Canadian Geotechnical Journal* **45(3)**: 425–431.

Brown MJ and Hyde AFL (2008b) High penetration rate CPT to determine damping. *3rd International Conference on Geotechnical and Geophysical Site Characterisation, Taipei, Taiwan* (Huang A-B and Mayne PW (eds)). Routledge, London, UK, pp. 657–663.

Brown MJ and Powell JJM (2013) Comparison of rapid load test analysis techniques in clay soils. *ASCE Journal of Geotechnical & Geoenvironmental Engineering* **139(1)**: 152–161.

Brown MJ, Hyde AFL and Anderson WF (2006) Instrumented rapid load pile tests in clay. *Géotechnique* **56(9)**: 627–638.

BSI (British Standards Institution) (1986) BS 8004:1986: Code of practice for foundations. BSI, London, UK (superseded/withdrawn).

BSI (2004) BS EN 1997-1:2004: Eurocode 7: Geotechnical design – Part 1: General rules. BSI, London, UK. See also corrigendum and national annexes.

BSI (2010) BS EN 1536-2010+A1:2015: Execution of special geotechnical works – Bored piles. BSI, London, UK.

BSI (2015) BS 8004:2015+A1:2020: Code of practice for foundations. BSI, London, UK.

BSI (2016) EN ISO 22477-10:2016: Geotechnical investigation and testing – Testing of geotechnical structures – Part 10: Testing of piles: Rapid load testing. BSI, London, UK.

BSI (2018a) BS EN ISO 22477-1:2018: Geotechnical investigation and testing: Testing of geotechnical structures – Part 1: Testing of piles: Static compression load testing. BSI, London, UK.

BSI (2018b) EN ISO 22477-4:2018: Geotechnical investigation and testing: Testing of geotechnical structures – Part 4: Testing of piles: Dynamic load testing. BSI, London, UK.

Burd HJ, Beuckelaers WJAP, Byrne BW, *et al.* (2020) New data analysis methods for instrumented medium-scale monopile field tests. *Géotechnique* **56(11)**: 961–969.

Butcher AP, Powell JJM and Skinner HD (2006) *Reuse of Foundations for Urban Sites: A Best Practice Handbook.* IHS BRE Press, Watford, UK.

Butler HD and Hoy HE (1976) *The Texas Quick-Load Method for Foundation Load Testing: User's Manual.* Report FHWA-IP-77-8. Federal Highways Administration, Office of Development, Washington, DC.

Chin FK (1972) The inverse slope as a prediction of ultimate bearing capacity of piles. *Proceedings of the 3rd Southeast Asian Conference on Soil Engineering* (Lumb P (ed)). Southeast Asian Society of Soil Engineering, Hong Kong, China.

Davisson MT (1972) High capacity piles. *Proceedings in the Lecture Series on Innovations in Foundation Construction, Chicago, IL.* American Society of Civil Engineers, Reston, VA, pp. 81–112.

EA Pfähle (2013) *Recommendations on Piling*, 1st edn. Wiley, Oxford, UK.

England M (2002) Easy static load tests: expert results. *Proceedings of the 9th International Conference on Piling and Deep Foundations, Nice, France*, pp. 657–662.

England M (2008) Review of methods of analysis of test results from bi-directional static load tests. *Proceedings of the 5th International Seminar on Deep Foundations on Bored and Auger Piles (BAP 2008), Ghent, Belgium* (Van Impe WF and Van Impe PO (eds)). Taylor & Francis, London, UK, pp. 235–239.

England M and Cheesman P (2010) Design benefits of bi-directional load testing of barrettes. *Proceedings of the 11th International Conference on Piling and Deep Foundations, London, UK.*

Fleming WKG (1992) A new method for single pile settlement prediction and analysis. *Géotechnique* **42(3)**: 411– 425.

Fleming WKG, Weltman AJ, Randolph MF and Elson K (2009) *Piling Engineering*, 3rd edn. Taylor & Francis, London, UK.

Foeken van RJ, Daniels B and Middendorp P (1996) An improved method for the real time calculation of soil resistance during driving. *Proceedings of the Fifth International Conference on the Application of Stress-Wave Theory to Piles, Orlando, FL*, pp. 1132–1143.

FPS (Federation of Piling Specialists) (2006) *Handbook on Pile Load Testing.* FPS, Faversham, UK

Fugro Loadtest (2023) http://www.loadtest.com (accessed 17/03/2023).

Hardy S, Nicholson D, Srisakthivel S, Olivier L and Disley T (2019) Preliminary test pile results for the South Bank Place development. *Ground Engineering*, Dec.: pp. 26–33.

Hoelscher P and van Toll F (2009) *Rapid Load Testing on Piles.* CRC Press/Balkema, Leiden, the Netherlands.

Hoelscher P, van Toll AF and Brown MJ (2010) European standards and guidelines for rapid load testing on piles. *Proceedings of the 11th International Conference on Piling and Deep Foundations, London, UK.*

Hoelscher P, Brassinga H, Brown MJ, *et al.* (2011) *Rapid Load Testing in Piles: Interpretation Guideline.* CRC Press/Balkema, Leiden, the Netherlands.

Holeyman AE (1992) Technology of pile dynamic testing. *Proceedings of the International Conference on the Application of Stress Wave Theory to Piles, The Hague, the Netherlands* (Barends FBJ (ed)). Balkema, Rotterdam, the Netherlands, pp. 195–215.

ICE (Institution of Civil Engineers) (2017) *ICE Specification for Piling and Embedded Retaining Walls* (SPERW), 3rd edn. Thomas Telford, London, UK.

Jardine R, Chow F, Overy R and Standing J (2005) *ICP Design Methods for Driven Piles in Sands and Clays.* Thomas Telford, London, UK.

Matsuzawa K, Nakashima Y and Matsumoto T (2008) Spring hammer rapid load test method and its validation. *Proceedings of the 2nd International Conference on Foundations, Dundee, UK* (Brown M, Bransby M, Brennan A and Knappett J (eds)). IHS BRE Press, Watford, UK, pp. 223–234.

McVay MC, Kuo CL and Guisinger AL (2003) *Calibrating Resistance Factor in Load and Resistance Factor Design of Statnamic Load Testing.* Research Report 4910-4504-823-12. Florida Department of Transportation, Tallahassee, FL.

Middendorp P. (2000) Statnamic the engineering of art. *Proceedings of the 6th International Conference on the Application of Stress Wave Theory to Piles, São Paulo, Brazil* (Niyama S and Beim J (eds)). Balkema, Rotterdam, the Netherlands, pp. 551–562.

Middendorp P, de Neef L, van Dorp R and Lambo A (2013) StatRapid, eine Neuentwicklung für Rapid-Load-Testing. *Pfahlsymposium, Braunschweig, Germany.* (In German.)

Miyasaki T, Kuwabara F, Likins G and Rausche F (2008) Rapid load test on high bearing capacity piles. *Proceedings of the 8th International Conference on Application of Stress-Wave Theory to Piles, Lisbon, Portugal* (Santos JA (ed)). IOS Press, Amsterdam, the Netherlands, pp. 501–506.

Muir Wood D (1990) *Soil Behaviour and Critical State Soil Mechanics.* Cambridge University Press, Cambridge, UK.

Paikowsky SG (2004) *Innovative Load Testing Systems.* Geosciences Testing and Research, North Chelmsford, MA.

Poulos HG (2000) Pile testing: from the designer's viewpoint. *Proceedings of the Second International Statnamic Seminar, Tokyo, Japan* (Kusakabe O, Kuwabara F and Matsumoto T (eds)). Rotterdam, Balkema, the Netherlands, pp. 3–21.

Poulos HG and Davis EH (1980) *Pile Foundation Analysis and Design.* Wiley, New York, NY.

Powell JJM and Brown MJ (2006) Statnamic pile testing for foundation re-use. *Proceedings of the International Conference on the Re-use of Foundations for Urban Sites, Watford, UK* (Butcher AP, Powell JJM and Skinner HD (eds)). IHS BRE Press, Watford, UK, pp. 223–236.

Randolph MF (2003) Science and empiricism in pile foundation design. *Géotechnique* **53(10)**: 847–875.

Rausche F, Goble GG and Likins GE (1985) Dynamic determination of pile capacity. *ASCE Journal of Geotechnical Engineering* **111(3)**: 367–383.

Reese LC and van Impe WF (2011) *Single Piles and Pile Groups Under Lateral Loading*, 2nd edn. Taylor & Francis, London, UK.

Robinson S (2019) *Rate Effect Behaviour of Different Clays from High Speed Triaxial Element Testing.* PhD thesis, University of Dundee, Dundee, UK.

Robinson S, Brown MJ, Knappett JA and Brennan A (2023) Impacts of reference strain rate and strain level dependency on rate effects in fine grained soils. *Géotechnique Letters* **13(1)**: 1–6.

Schellingerhout AJG and Revoort E (1996) Pseudo static pile load tester. *Proceedings of the 5th International Conference on Application of Stress-Wave Theory to Piles, Orlando, FL*, pp. 1031–1037.

Stain RT (1992) SIMBAT: A dynamic load test for bored piles. In *Piling: European Practice and Worldwide Trends* (Sands MJ (ed)). Thomas Telford, London, UK, pp. 198–205.

Tomlinson M and Woodward J (2008) *Pile Design and Construction Practice*, 5th edn. Taylor & Francis, London, UK.

Weaver TJ and Rollins KM (2010) Reduction factor for the unloading point method at clay soil sites. *Journal of Geotechnical and Geoenvironmental Engineering, ASCE* **136(4)**: 643–646.

Weltman AJ (1980) *Pile Load Testing Procedures.* DOE and CIRIA Pile Development Group Report PG7. CIRIA, London, UK.

Useful websites

Allnamics: allnamics.com
Federation of Piling Specialists (FPS): www.fps.org.uk
Fugro Loadtest: www.loadtest.co.uk
Pile Dynamics Inc. and GRL Engineers Inc.: www.pile.com

It is recommended this chapter is read in conjunction with

- Chapter 8 *Health and safety in geotechnical engineering*
- Chapter 22 *Behaviour of single piles under vertical loads*
- Chapter 54 *Single piles*
- Chapter 81 *Types of bearing piles*
- Chapter 82 *Piling problems*
- Chapter 97 *Pile integrity testing*

All chapters in this book rely on the guidance in Sections 1 *Context* and 2 *Fundamental principles.* A sound knowledge of ground investigation is required for all geotechnical works, as set out in Section 4 *Site investigation.*

doi: 10.1680/icemge.66830.1621

CONTENTS

Chapter 99
Materials and material testing for foundations

Alice Berry Arup, London, UK
Chris Brewis Arup, London, UK
Stuart Pennington Arup, Melbourne, Australia
Sima Jovanovic Arup, London, UK

This chapter presents an introduction to a selection of construction materials and their verification, at construction stage, for the geotechnical engineer. Verification is discussed in terms of its forms, levels of complexity, documentation and appropriate tests, with an emphasis on UK practice. An introduction to the verification of existing foundations for reuse is also presented.

99.1. Introduction

The geotechnical engineer is faced with a variety of construction materials as most geotechnical structures involve more than just soil and rock. There is often an overlap with other disciplines, such as structural engineering or materials technology. Verification of these materials during construction requires an understanding of the designer's intent, the characteristics of the material being verified and practical experience of the verification process.

This chapter introduces verification and presents a selection of common materials encountered in the construction of geotechnical structures, with an emphasis on UK practice and standards to illustrate the verification process. Verification by testing is addressed in terms of the general requirements for each material rather than the requirements for a material in a specific structure (e.g. grout for anchors). The chapter concludes with an introduction to foundation reuse. Discussion of pavements and contamination is not included in this chapter.

At the end of this chapter a selected list of documents and websites provides more specific information on the topics addressed in the chapter.

There are numerous overlaps with other parts of this manual for the issues discussed in this chapter, and the reader should peruse the contents list to identify potential sources of more detail. This is particularly relevant to topics such as earthworks (e.g. Section 7 *Design of earthworks, slopes and pavements*), integrity testing (Chapter 97 *Pile integrity testing*), site investigation (Section 4 *Site investigation*), ground improvement (Chapters 25 *The role of ground improvement*, 59 *Design principles for ground improvement*, 84 *Ground improvement*), specification (Chapters 75 *Earthworks material specification, compaction and control*, 78 *Procurement and specification*, 93 *Quality assurance*), site supervision (Chapter 96 *Technical supervision of site works*), pavements (Chapter 76 *Pavement foundation design*) and reinforced soil (Chapters 73 *Design of soil reinforced slopes and structures*, 74 *Design of soil nails*, 86 *Soil reinforcement construction*, 88 *Soil nailing construction*).

99.2. Eurocodes

European Standards have made construction verification a requirement for designs under their jurisdiction. Documents relevant to material verification include

- EN Eurocodes (e.g. BS EN 1997:1-2004)
- National Annexes (e.g. NA to BS EN 1997:1-2004)
- investigation and testing standards (e.g. BS EN 12620: 2002+A1:2008)
- execution standards (e.g. BS EN 12715:2020)
- non-contradictory complementary information (e.g. BS 8004:2015).

There are also specific practitioners' guides developed by the foundations industry, such as the EFFC/DFI *Guide to Tremie Concrete for Deep Foundations* (2018), the EFFC/DFI *Guide to Support Fluids for Deep Foundations* (2019) and CIRIA C653 *Reuse of Foundations* (Chapman *et al.*, 2007). The consequence of this array of documentation is that there are many scattered references to construction verification of materials that an engineer is required to consider.

99.3. Materials

Geotechnical structures often involve a variety of construction materials in addition to soil and rock, such as polymers, timber, steel, concrete and grout. The processes of production, transport, placement and use will influence the method of verification. For example, materials can be

- relatively unprocessed single materials (e.g. timber)
- processed composite materials (e.g. concrete)

- imported to a particular site (e.g. fresh concrete)
- transported within a particular site (e.g. excavated soil)
- used as-is (e.g. in situ soil)
- used with other materials (e.g. geosynthetics)
- in a ready-to-use format (e.g. reinforcement cages)
- placed under water (e.g. fresh concrete)
- used under water (e.g. piles)
- in need of specific on-site management (e.g. fresh concrete)
- reused (e.g. existing piles).

The practical application of verification can be undertaken once a more general understanding of a material's composition, potential problems and suitable methods of testing is obtained. Specialist understanding of a material is often necessary to determine the appropriate methods of verification.

99.4. Verification

Verification of materials during construction is the process of checking (by an appropriate or specified method) that the materials brought to site and used in a structure are what the designer intended. While the designer's intent and requirements should be clear in the construction specifications and drawings, a face-to-face meeting with the designer will be invaluable. It should be recognised that additional verification beyond that specified by the designer may be necessary or prudent.

Where appropriate, verification should be planned in advance, for instance by creating an Inspection and Test Plan (ITP), which should include example blank record sheets and acceptance criteria.

When one thinks of verification of materials, testing in a laboratory may spring to mind. However, this is only one part of the process and should be complemented by other verification procedures such as

- visual inspection
- collection and review of records and certificates (e.g. concrete lorry tickets)
- source quality control and certification (e.g. CE (Conformité Européenne) marking)
- review of quality control procedures at the point of manufacture (e.g. geosynthetics)
- engineering judgement
- appropriate sampling
- field testing (e.g. flow table for concrete)
- comparable experience.

The verifier should

- understand the material being verified
- know when to request additional testing
- be prepared to say that a material is not satisfactory.

The level, quantity, complexity of verification and responsibility for verification will and should depend upon factors such as

- the category of the structure in which it is used
- the level of quality control provided at source
- the type of material
- the parameters used in design and the extent to which the designer is relying on the testing of materials during construction
- the importance or size of the structure
- material origin
- the construction contract.

Guidance is included in standards such as BS EN 206 and BS EN 1990.

In considering what verification is required, the following questions should provide the engineer with a starting point for further thought.

- What are the verification requirements in the designer's specification?
- What are the referenced standards in the specification and in the construction contract?
- Who is responsible for the various aspects of verification?
- Is the material covered by a manufacturing standard (e.g. CE marking)?
- What can go wrong with the material once it arrives on site?
- What level of responsibility do you have as a geotechnical engineer on site?
- Is on-site testing required to progress or verify the design?
- Does the specification adequately address verification requirements?
- What level of verification control is required?
- Are the records being provided adequate?
- Are there materials not covered by standards?
- Under what conditions has the material been transported to site?
- What has happened to the material since it arrived on site?
- Is the material being stored and handled correctly?
- What are the limitations of any controlling standards?
- Are there any hold points during construction for specific items of verification?
- What is the process for rejecting a material?
- What are the requirements for personnel or companies undertaking material testing?
- Has a plan for verification been created?
- Is the quality control at the material source satisfactory?

- How are the results of verification to be communicated to the designer and client?

99.5. Concrete

99.5.1 General

The use of concrete in geotechnical applications is extensive – for example, piles, shallow foundations, retaining walls, facing units, tunnel linings and masonry blocks.

In simple form, concrete is composed of cement, fine aggregate, coarse aggregate and water. Typically, by volume cement is 6–16%, fine aggregate is 20–30%, coarse aggregate is 40–55% and water is 12–20%. The water–cement ratio can range from 0.3 to 1.0 (Illston and Domone, 2010). A more complex mix can include admixtures and cement replacement materials which enhance its performance. Admixtures can include plasticisers, superplasticisers, accelerators, retarders, air-entraining agents and foaming agents. Cement replacement materials can include pulverised fuel ash/fly ash, ground granulated blast furnace slag (GGBS), silica fume, metakaolin and natural pozzolans; these are often included to address sustainability targets associated with reducing embodied carbon (Low Carbon Concrete Group, 2022).

Concrete is batched using hopper dispensers and mixing drums which combine the constituents (**Figure 99.1**). Concrete is either poured or pumped into the ground (e.g. a pile) or formwork (e.g. a retaining wall or precast unit). The use of a tremie may be required, particularly if concreting is undertaken below water, to prevent segregation (**Figure 99.2**). Concrete placed in formwork is typically vibrated when fresh to expel air and compact it. Concrete hydrates and typically achieves an initial set after 2–4 hours, a final set after many hours and then hardens over weeks to months, although this can be altered through the use of admixtures. Concrete is typically delivered to site as ready-mix or as precast units, although on-site batching may be undertaken in special circumstances.

Figure 99.1 Concrete batching plant (Courtesy of Texnokat)

Figure 99.2 Tremie being used to pour concrete into a bored pile

99.5.2 Potential problems

Problems with concrete can occur when it is fresh, during setting or when it has hardened, as a result of the mix proportions, the mix constituents, the environment in which it is used or how it is handled during placement.

While concrete is fresh it can suffer with respect to its consistency, compactability and cohesiveness. For example, if workability is reduced it may be difficult to plunge a reinforcement cage for a continuous flight auger (CFA) pile (**Figure 99.3**). Alternatively, if it is dropped through water that has collected at the base of a pile it may segregate.

When concrete is setting it can suffer from bleed, plastic settlement and plastic shrinkage. Once concrete has hardened it can suffer from durability issues caused by internal or external attack and by physical or chemical attack, such as sulfate attack, sea water attack, acid attack, alkali–aggregate reactions, abrasion and freeze–thaw.

During the time that concrete is fresh or setting the problems it may encounter are mainly associated with mix design and handling, whereas problems during its design life are more typically associated with mix constituents and quality of construction.

99.5.3 Verification

Verification of concrete starts with the mix design, which can be specified in various manners. For example, designed – where the concrete supplier chooses the mix – or prescribed – where the designer chooses the mix. Mix constituents and proportions are covered by standards and are typically controlled by requirements for workability (short-term performance), durability (long-term performance) and strength. If the mix design

Figure 99.3 Partially plunged CFA cage

(a)

(b)

(c)

Figure 99.4 (a) Flow table test: (a) compaction of concrete into a cone; (b) removal of the cone; (c) measurement of flow width

is not typical, then trial mixes in a laboratory and subsequently in the field may be required, as described in the EFFC/DFI *Guide to Tremie Concrete for Deep Foundations* (2018).

For concrete batching plants supplying ready-mix concrete or factories manufacturing precast concrete, there should be quality controls in place that monitor performance as detailed

Box 99.1 Case study: flow table (slump flow) test

The flow table (slump flow) test is similar to the slump test in that it measures the consistence of fresh concrete. The flow table test is suited to high-consistence mixes, such as those used in CFA piles, which would be difficult to measure using the slump test. The test procedure is detailed in BS EN 12350-5:2019; in simple terms it involves

- compaction of concrete into a cone (**Figure 99.4(a)**)
- removal of the cone to release the concrete (**Figure 99.4(b)**)
- jolting of the concrete on a flow table
- measurement of the spread of the concrete (see **Figure 99.4(c)**).

in the standards. Ready-mix suppliers may subscribe to QSRMC (the UK Quality Scheme for Ready-Mixed Concrete) and precast concrete may be CE marked.

The first check for fresh concrete should be a review of the delivery ticket, the contents of which are specified by standards. The ticket includes information such as time of batching, volume, strength class, mix description and consistence. No materials (including water) should be added to ready-mix concrete unless required by the producer and this should be recorded on the delivery ticket. On-site verification tests prior to use, which are covered by standards, address issues of consistency, compactability and cohesiveness. Consistency can be assessed by the slump test, vebe test or flow table test (slump flow); compactability can be assessed by the compacting factor test; and cohesiveness can be assessed visually. Further verification typically involves the preparation of cube or cylinder samples which are then tested for strength (typically after 7, 28 and 56 days of curing). When referring to concrete strength grade under Eurocode terminology, for example C32/40, the '32' refers to the cylinder strength and the '40' to the cube strength.

With the introduction of an authoritative guide to tremie concrete (EFFC/DFI, 2018) steps have been taken to improve the quality of concrete used in foundations through the introduction of additional testing recommendations.

Subsequent to the checks, testing and sampling undertaken at delivery time, the quality control of ready-mix concrete until it has hardened relates primarily to execution rather than material verification, and generally involves inspection of placement techniques, compaction and curing.

Verification of hardened concrete can be undertaken indirectly, destructively or non-destructively. Indirect verification may involve inspection of surface quality or dimensional tolerance in the case of concrete cast using formwork. Destructive verification may involve full-scale load testing of a specially prepared member or small-scale testing of a core sample (see **Figure 99.5** for an example of a concrete core from a pile with defects). Non-destructive verification can include: rebound hammer testing, resonant frequency testing and ultrasonic pulse testing (Illston and Domone, 2010); however, these tests need to be correlated with core sample testing. Non-destructive verification for piles is generally termed 'integrity testing' and takes several forms, such as low-strain 'hammer-type' tests (CIRIA, 1997) and thermal integrity profiling using embedded thermistors.

99.6. Steel and cast iron

99.6.1 General

The use of steel in geotechnical applications is extensive. For example, structural sections can be used as piles (e.g. H-piles or sheet piles) or as temporary props. **Figure 99.6** illustrates the use of sheet piles to form a cofferdam. Reinforcing bars (generally 8–40 mm dia.) are used in reinforced concrete, soil nailing and anchors. Steel tendons are used in anchors. Cast

Figure 99.5 Concrete core showing defects

Figure 99.6 Steel sheet pile cofferdam

iron was used historically in tunnels in the form of grey cast iron lining segments; however, in more modern tunnels spheroidal graphite cast iron lining segments are used. Stainless steel can be used in the same manner as non-stainless steel where additional protection against corrosion is required, albeit at higher cost and embodied carbon. Steel fibres are used in segmental concrete tunnel linings and shotcrete.

Steel and cast iron primarily consist of iron alloyed with carbon and to a lesser extent other elements such as silicon (for cast iron), chromium and manganese. Steel is typically formed by hot rolling or cold working and then bent/welded to form more complex sections (reinforcement cages, girders, hollow tubes, etc.); the length of any section is generally limited to 12 m (the length of delivery truck). Cast iron such as that used in tunnel lining segments is moulded and then bolted together.

In the UK, structural steel is typically grade S275 or S355 (where 'S' denotes structural steel and the number represents the yield strength in MPa). Reinforcing bars are typically grade B500A, B500B or B500C (where the initial 'B' denotes reinforcing steel, the number represents the yield strength in MPa and the final letter represents the ductility rating).

Structural steel is typically formed in a basic oxygen furnace using approximately 20% recycled content in the UK, while reinforcing bars are typically formed with very high recycled content (>90%). The percentage of recycled content greatly influences the embodied carbon of steel – the limiting factor is often the availability of scrap steel for recycling, so care should be taken to select steel based on embodied carbon for a particular, realistic, recycled percentage content, rather than simply demand low embodied carbon steel, which could incentivise higher recycling rates in some steels at the expense of others.

Figure 99.7 Variability in steel corrosion near the waterline

99.6.2 Potential problems

The primary issue with steel and cast iron is its durability, as its strength and stiffness properties are closely controlled by factory production. It can be degraded by dry oxidation but more so by wet corrosion, which may be chemically or biologically based. However, it should also be noted that steel can be damaged mechanically once delivered to site if it is subjected to excessive bending or poorly controlled welding.

The extent to which steel in reinforced concrete corrodes is very much dependent upon the density, quality and thickness of concrete cover (generally 75 mm or more for concrete cast against the ground) and if there are any chemicals in the concrete that may attack the steel. In applications where steel or cast iron is exposed to the environment (e.g. sheet piles) alternative measures are necessary, such as good design/detailing, coatings, cathodic protection or sacrificial thickness. **Figure 99.7** illustrates how corrosion can vary over relatively short distances – in this case, the steel in the splash zone (the central part of the photograph) has corroded significantly more than that above and below.

99.6.3 Verification

For mills supplying steel and cast iron there should be quality controls in place that monitor performance as detailed in the standards. CE marking can be used to guarantee certain characteristics, and producers may subscribe to CARES (Certification Authority for Reinforcing Steels).

Verification of steel and cast iron typically starts with its delivery to site, where inspection of delivery records and the material should take place. Pertinent material characteristics should be checked on identification labels. Reinforcing bars produced by a CARES-certified body may have identifying rib patterns, and may come with a 'material passport' that can be linked to as-built models, to facilitate future reuse. The material should be free of scale rust, although a dusting of iron oxide is acceptable. Dimensions should be as specified, subject to allowable tolerances.

Samples can be taken for laboratory testing as there is not usually scope for on-site testing to verify material properties such as strength or composition. However, an example of where steel can be verified on site (albeit indirectly) is the case

Box 99.2 Case study: reinforcement cage check

Steel reinforcement used in piles or barrettes typically arrives on site as a cage (**Figure 99.8**). Verification against construction drawings should comprise checking

- overall dimensions (width, length, breadth, diameter)
- number of bars, bar spacing, bar diameter, lap lengths
- integrity testing installations (e.g. sonic logging ducts)
- cleanliness (absence of rust and soil)
- absence of any indication of uncontrolled joining processes, such non-factory welds
- any box-outs
- steel grade by way of identification tags and rib patterns (**Figure 99.9**)
- robustness for lifting purposes
- appropriate spacers attached at agreed locations
- interface/couplers for multiple cages.

Figure 99.8 Reinforcement cage for barrette

Figure 99.9 Reinforcing bars showing a rib pattern

of anchors, where steel tendons are tested non-destructively during the performance of load tests.

Where a protective coating is specified, for example on sheet piles, this should ideally be applied under factory conditions, and once delivered to site should be inspected for composition, integrity, quality and thickness.

Verification of welding is not addressed here. However, it is worth noting that high yield strength bars should not be heated (e.g. by welding) as this can lead to them becoming brittle.

99.7. Timber

99.7.1 General

Timber can be used for driven piles, retaining walls, shoring/falsework, temporary support for tunnelling and underpinning and as formwork. Timber can be described as a low-density, cellular, polymeric composite (Illston and Domone, 2010). At the microscopic level timber is classed as

- softwood (cells are primarily vertical and density is generally lower, e.g. pine)
- hardwood (cells are more anisotropic, e.g. oak).

On a larger scale, timber can be subdivided into

- sapwood (outer layer, growing later)
- heartwood (dead centre section)
- latewood (formed late in the growing season)
- earlywood (formed early in the growing season).

Timber should be obtained from certified sustainable sources and is typically dried before use. Natural timber is available as whole logs or as a sawn product. Timber can also be processed to produce products such as plywood for formwork applications, and cross-laminated timber (although these applications are not discussed here).

99.7.2 Potential problems

Natural timber cannot be controlled in the same way that, for example, steel production is. Its properties vary according to, for example, density, grain structure, temperature, knots, ring width, cell structure, moisture content, time and splits. There is a need to consider the small-scale structure as well as gross features. **Figure 99.10** illustrates the effect of a knot on timber structure.

Timber needs to be seasoned before use, otherwise it can be attacked by fungi and its strength and stiffness will be lower. Timber (particularly heartwood) has a natural resistance to deterioration provided its moisture content is kept low; ideally below 20% (Illston and Domone, 2010). However, it should be remembered that timber is hygroscopic (it has the ability to absorb water from the air).

Timber is mainly susceptible to attack by biological organisms and fire. Other forms of attack, such as chemical and mechanical, are of lesser concern. When timber is placed below groundwater level is it essentially immune from deterioration as there is very little air. However, where water can

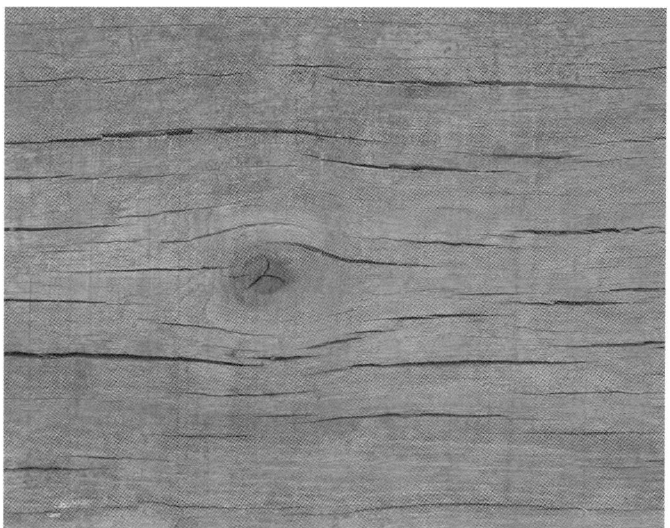

Figure 99.10 Timber is not a uniform material

be oxygenated and in marine environments where borers (animals that eat wood) are present, decay can occur.

While timber has a natural resistance to deterioration due to natural chemicals within it, these vary between species and the part of the tree used. Chemical processing typically provides more resistance. However, its success depends on the part of the tree (e.g. sapwood or heartwood), the tree species and the moisture content of the timber. Preservatives can be tar oil-based, water-based or solvent-based. Tar oil- and water-based products are typically used for in-ground applications. It is noted that EU regulations restrict/forbid the use of many traditional preservatives such as copper–chromium–arsenic (CCA) and creosote (Illston and Domone, 2010).

99.7.3 Verification

Timber for structural use should be graded into strength classes according to its anticipated performance. Grading is either by visual inspection or by machine inspection and a strength class is assigned to each piece of timber on this basis. In terms of durability, timber is classified according to its life when in contact with the ground. CE marking can be used to guarantee certain characteristics.

Verification of timber on site should start with an inspection of the delivery records and the grade stamps/marks. Samples can be taken for testing to verify various material properties. However, it should be remembered that the size of sample will affect the measured properties due to the presence of gross features, and the testing of timber samples usually only informs the values presented in standards.

Grade stamps/marks (an example of which is shown in **Figure 99.11**) should record information such as

- strength class (C14 to C50 for softwood and D30 to D70 for hardwood – the higher the number the higher the strength)
- species
- condition (e.g. dry)
- origin (e.g. FSC or PEFC for sustainable sourcing)
- supplier name
- certifying body
- standard used
- grading method (i.e. visual or machine).

Dimensions should be verified against construction drawings subject to allowable tolerances, and checks should be made on straightness and for imperfections. Timber should be stored in a sheltered location off the ground so as to avoid changes in moisture content, and handled appropriately to avoid damage (**Figure 99.12**). The moisture content can be checked with an electrical moisture meter if required.

Where preservatives are specified, these should be applied under factory conditions; once delivered to site, the integrity and quality of the preservative application should be inspected. If timber is subsequently cut or drilled, site treatment will be needed.

Figure 99.12 Storage of timber

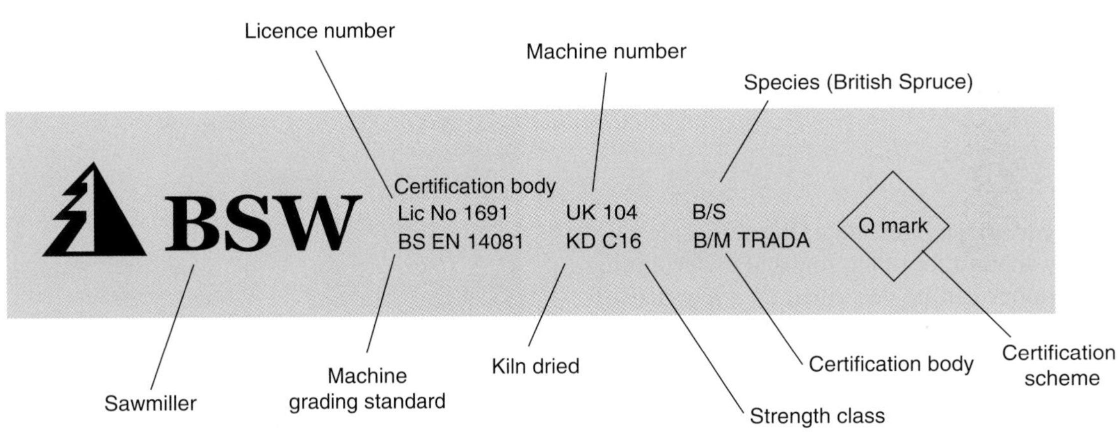

Figure 99.11 Example of a grading stamp (Courtesy of BSW Timber Ltd)

Box 99.3 Case study: timber decay at wharf

The importance of verification with respect to characteristics such as species, heartwood or sapwood, type and application of preservative is illustrated in **Figure 99.13**. The timber in **Figure 99.13(a)** has not been immersed in water, and apart from some surface discolouration is essentially fresh. However, just below, in the splash zone (**Figure 99.13(b)**), the timber has decayed and crumbles in the hand.

(a)

(b)

Figure 99.13 Variability in timber decay: (a) above the waterline; (b) below the waterline

99.8. Geosynthetics

99.8.1 General

Geosynthetics in geotechnical applications can be used to separate materials, reinforce soil (see Chapter 73 *Design of soil reinforced slopes and structures*), permit drainage, provide filtration or act as an impermeable barrier. Geosynthetics can be subdivided into various forms, such as geotextiles, geogrids, geomembranes, geocomposites and geonets. **Figure 99.14** illustrates a geogrid and geocomposite.

Geosynthetics are typically composed of polymers such as polyester, polyethylene, polypropylene and polyamide, and can be formed by processes such as extrusion, spinning, stretching, weaving and bonding. Geosynthetics may be reinforced with fibres or they can be combined with other materials such as bentonite.

99.8.2 Potential problems

Issues for geosynthetics during construction include durability and interface friction. The durability of geosynthetics can be affected on site by acids, alkalis, heat, abrasion, fungus,

(a)

(b)

Figure 99.14 (a) Example of a geogrid; (b) example of a geocomposite oxidising agents, UV light and construction plant. While a designer should take account of interface friction and stress–strain behaviour, poor construction practice can adversely affect assumptions made during design.

99.8.3 Verification

For factories supplying geosynthetics, there should be quality controls in place that monitor performance as detailed in the standards. CE marking can be used to guarantee certain characteristics.

On-site verification of geosynthetics should start with inspection of delivery records. Geosynthetics should be clearly labelled, stored in the shade and handled with care. Geosynthetic products should be inspected for quality, consistency, integrity and conformity with design specifications. Samples can be taken for off-site testing to verify various material properties such as tensile strength; if tests are undertaken they should be consistent with service conditions.

The surface onto which geosynthetics are to be placed should be appropriately prepared (e.g. level, free of protruding objects) and then inspected to verify that the potential for damage to the geosynthetic to be laid is minimised. While most testing is laboratory-based, it may be specified that on-site testing be undertaken to assess the impact of damage from construction plant on a particular geosynthetic being used.

Once a geosynthetic is laid, inspection relating to aspects such as lapping or welded joints, damage, orientation, spacing and flatness (i.e. lack of ripples) should be undertaken and fill promptly placed to avoid damage. **Figure 99.15** illustrates the verification of welded joints by pressure testing.

Figure 99.15 Pressure testing of welded laps (Courtesy of Escambia)

99.9. The ground

99.9.1 General

The ground (soil/rock) is primarily used to support structures in its natural state. However, when excavated, processed, transported or emplaced (fill/made ground) it can be used, for example, to raise surface levels, form part of concrete, act as a drainage filter or provide protection against erosion. Its characteristics generally vary according to its origin and composition.

99.9.2 Potential problems

Natural ground can hide many potentially problematic issues; while undisturbed these issues are generally not of concern. However, once humans interact with the ground then these issues often become problems. Soft soils can settle, erosion features may collapse, excess water pressures can cause collapse, aggressive chemicals may attack concrete, sensitive soils are susceptible to landslide, hard rock can cause excavation problems and clay can expand. Made ground can hide additional issues such as collapsible soil, poor compaction, deleterious material and contaminants.

99.9.3 Verification

The ground can be divided into four groups for the purposes of verification

- in situ ground (existing natural or made)
- engineered ground (engineered fill)
- treated ground (improved existing ground)
- excavated material.

Verification of in situ ground during construction should be an extension to earlier desk study work and intrusive ground investigation. It should be used to validate design assumptions but may be used to fill gaps in knowledge or investigate features that were previously inaccessible. The simplest form of verification could involve recording strata and groundwater in cuttings, excavations or pile bores (**Figure 99.16**). Simple portable assessment tools may be used (e.g. handheld shear vane – **Figure 99.17**).

Existing groundwater installations may be used to monitor water levels. More complex verification could involve intrusive sampling and testing through boreholes or field experiments such as load testing, pumping trials or plate load tests.

Verification of engineered ground that is emplaced during construction requires additional measures such as density testing and classification of materials with respect to their ultimate use and compaction characteristics, although the techniques above may also be used (see also Section 7 *Design of earthworks, slopes and pavements*).

Verification of ground that is treated during construction to improve its geotechnical properties also requires additional verification specific to the properties being improved, although

Figure 99.16 Soil exposure in an excavation

Figure 99.17 Shear vane test

again the techniques above may also be used. The Mackintosh Probe (**Figure 99.18**) is a simple hand tool that can be used relatively quickly to assess improvement in strength over the top few metres of ground.

Verification of excavated material relates to ground that is going to be moved from its source location for use elsewhere. Verification of this type of ground will be specific to its end use and generally involves laboratory testing and visual inspection. For example, visual assessment that it is free of deleterious materials, maximum dry density testing for clay fill or durability testing of rock for armourstone.

99.10. Aggregates

99.10.1 General

Aggregates have a number of geotechnical uses – for example, they can be used in concrete, as drainage filters, as protection against scour, as fill for embankments, in road pavements or in ground improvement (e.g. stone columns). Aggregates range from sand to boulder in particle size; they can be angular to smooth and their strength and durability will vary according to their origin. They can be heavyweight, such as those sourced from igneous rocks, or lightweight, such as pumice. Natural aggregates can be collected from locations such as buried channels, riverbeds, river terraces and glacial outwash; however, they are also often produced by excavation and crushing of rock (**Figure 99.19**). Aggregates can also be manmade, in that they are created by recycling materials such as concrete, asphalt, GGBS or plastic.

99.10.2 Potential problems

Aggregate problems typically relate to their formation process and constituents. Strength is usually only important in high-strength concrete. Of more importance are durability, participation in chemical reactions and physical characteristics.

When some rocks are excavated, exposed to air, mixed with other materials or loaded, they may react to produce undesirable side effects. Issues that need to be addressed by selection and testing include fines content, shell content, resistance to fragmentation, resistance to polishing, resistance to wear,

Figure 99.18 Mackintosh Probe test

Figure 99.19 Crushing plant

freeze–thaw, drying shrinkage, chloride content, alkali–silica reactivity, carbonate content, shape, size, density, porosity and impurities.

99.10.3 Verification

The production or collection of natural aggregates should be controlled at source. While verification typically starts when aggregates are delivered to site, visits to the proposed source quarry will be beneficial, particularly if suitability has not been addressed during ground investigations.

Once on site, delivery records should be inspected and features such as origin, mass distribution, shape and integrity verified to be satisfactory. Samples can be taken for testing to verify material properties such as particle size grading, abrasion value (resistance to wear), magnesium sulfate soundness (resistance to weathering) and 10% fines value (resistance to crushing). An example of aggregate testing is illustrated in **Figure 99.20**, which shows the test apparatus used to undertake the Los Angeles abrasion test.

Figure 99.20 Los Angeles abrasion test apparatus (Source: Pavement Interactive)

Man-made materials such as recycled concrete may be produced on site. Depending upon their intended use, they should be subjected to similar suites of tests as undertaken for natural aggregates. Recycled material may be required to undergo contaminant testing.

99.11. Grout

99.11.1 General

Grout can be used in geotechnical applications to reduce permeability (e.g. rock grouting and grouting of granular soils), stiffen soils (e.g. compaction grouting), stabilise ground (e.g. permeation grouting), provide support (e.g. jet grouting), compensate for ground movement (e.g. compensation grouting) and fill cavities (see Chapter 90 *Geotechnical grouting and soil mixing*).

Grouts are commonly cement-based, silicate-based or resin-based, and can be classified in terms of their particle size as mortars, pastes, suspensions, colloids and solutions (Rawlings *et al.*, 2000). Depending upon the application, they can contain other materials such as cement replacements, clay, sand, fillers, chemicals, admixtures and water. They can be thick and viscous, thin and fluid, weak or very strong.

99.11.2 Potential problems

Problems occur with grouting when the ground being treated is not fully understood, when the grout or grouting technique being used is not appropriate to the situation at hand, when the mix is poorly designed, when control over the procedure is inadequate or when the grout behaviour is not understood by those using it.

99.11.3 Verification

As grout is often mixed on site, verification of the ingredients and the mixed product is necessary. There are many possible ingredients, although the verification process is similar in that characteristics such as brand name, model/type, remaining shelf life and evidence of correct storage are checked. In terms of the mixed product, testing should be undertaken at predetermined intervals appropriate to the product during construction to ensure it continues to have the expected properties, and that it is only used within its pot life.

Taking cement grout as an example, typical on-site tests for freshly mixed grout include: rheology (viscosity, gel strength, penetrability, thixotropy, workability, bleed) and setting time. For cement grouts where strength is important, grout cubes prepared from fresh grout should be prepared. **Figure 99.21** shows a group of freshly prepared cement grout cubes and a cylinder of grout being measured for bleed.

Field trials (appropriate to the scale of the project) should be undertaken to demonstrate that the grout is achieving the aims of the designer, especially where there is no prior experience. This may involve intrusive testing or exhumation (for example, to ascertain how far or how well a grout has permeated a particular stratum) for verification.

Figure 99.21 Fresh grout cubes and cylinder bleed test

Post-grouting verification of works can include laboratory testing of hardened grout samples or field techniques similar to those described for trials.

99.12. Support fluid

99.12.1 General

Bentonite and polymer support fluids are used in the construction of bored piles and diaphragm walls to provide temporary stability to the excavation prior to being replaced by concrete,

Box 99.4 Case study: cone test

The cone test is used to provide a viscosity measurement for cement grout. A common cone used for this test is the Marsh cone; however, there are others, such as the one presented in BS EN 445:2007. It is important to understand which cone is being specified as different cones may have different volumes and discharge orifice sizes. However, the principles of the test are generally the same and comprise

- filling the cone with a set volume of grout while blocking the discharge orifice (**Figure 99.22(a)**)
- timing how long it takes a set volume (typically less than that input) of grout to drain from the cone (**Figure 99.22(b)**)
- using the same procedure to calibrate the cone with water instead of grout.

(a)

(b)

Figure 99.22 Cone test: (a) filling of the cone; (b) timing of the discharge duration

and can also be used to assist in the removal of cuttings. Their key performance indicators are

- retaining properties over time
- limiting fluid loss into the surrounding ground
- controlling bore stability
- managing soil particles in the fluid
- limiting impacts to shaft friction
- displacement during concreting.

Bentonite and polymer behave differently and require separate verification test procedures and acceptance criteria.

Bentonite typically comprises natural or activated sodium bentonite (a type of clay) and has been used since the 1960s. Bentonite controls soil sedimentation through its viscosity. It develops a filter cake at the soil surface along the bore's shaft wall to restrict fluid loss and needs to be cleaned/exchanged after digging, before the concreting.

Polymer use has been developed more recently. It comprises synthetic long-chain molecules which interact with each other and soil in the fluid, increasing its viscosity. Polymers control fluid loss by forming a membrane at the soil surface along the bore's shaft wall. Polymer is not typically replaced after digging and concreting. It is typically modified by adding further product before reuse.

Both bentonite and polymer are mixed with water to supplier- and contractor-specified percentages and additional products are added to assist with the performance of the support fluid. For bentonite this can be sodium carbonate, while for polymer the supplier's design typically includes one or two admixtures, potentially tailored for different soil types (clay, sands, gravels, etc.).

Typically, bentonite is cleaned on site by the use of shaker separating screens and de-sanding hydrocyclones, while polymer is returned to mixing tanks to be 'reboosted' – a process of adding more base polymer product and admixtures to the returned polymer to boost its properties before reuse. Both bentonite and polymer require large area production farms on site (storage tanks, mixing tanks, pipelines, cleaning units, etc.); polymer farms are slightly smaller in area. Specialist digging tools and procedures typically result in extended construction periods – for example, deep support fluid piles can take 48 hours to construct.

Key properties of the support fluids to understand for verification testing are density, sand content percentage, viscosity, yield point, flowability/rheology, sorption (physical and chemical processes of attaching substances together), fluid loss, gel strength/shear stress, pH, filter cake (at the transition zone with soil) and interface layer (which accumulates between displaced support fluid and concrete). Further background and guidance can be found in the EFFC/DFI *Guide to Support Fluids for Deep Foundations*.

99.12.2 Potential problems

Without adequate site control by a specialist contractor support fluids can easily affect the performance of piles or walls in an adverse manner. Problems can include reduced concrete–soil interface friction, reduced reinforcement bond and excavation collapse. Further examples of issues are highlighted below:

Bentonite: The exchange of sodium in clays and calcium in water can reduce bentonite performance; however, treating the water with sodium carbonate should reduce issues. Bentonite contamination with cement can also adversely impact the performance. Sand content percentage impacts the ability to displace the bentonite out of the bore ahead of tremied concrete; thus bentonite cleaning procedures are important. The bentonite supply source (country) can also create performance issues, thus pre-start testing of performance is critical.
Polymers: The case history of performance and the understanding of recent product changes requires ongoing research and data collection to develop further industry confidence in polymer use. Records of polymer level drops in bores and resupply volumes are also important to check fluid loss rates. The 'reboosting' process is not well documented and further testing records should be implemented. Currently the polymer mix is heavily controlled on site by Marsh funnel results, which is the key indicator for 'reboosting'/adding polymer and admixtures. Further control and testing developments based on material/soil permeability/pore size are ongoing, currently led by suppliers, contractors and working groups such as the EFFC (referenced previously).

99.12.3 Verification

As support fluids are often mixed on site, verification of the ingredients and the mixed product is necessary; however, only the mixed product is considered here. In terms of test methods and acceptance criteria values, bentonite testing is better defined in current industry guidance, whereas polymer products and admixtures are being developed and best practice/research studies are ongoing.

Verification should start with checking that mixing of the constituents with water is thorough and in the correct proportions. The chemical composition of the support fluid should be noted. Once in use, testing should take place at various stages throughout excavation and until concrete is placed. The timings of these stages are slightly different for bentonite and polymer; a typical full regime of testing can comprise the following:

Test stages for bentonite: (1) fresh mix; (2) ready to reuse (mixture of fresh mix and cleaned bentonite); (3) dirty bentonite before and after base cleaning; (4) exchanged bentonite in bore prior to concreting; and ongoing measurement of bentonite level drop and resupply volumes to establish fluid loss.
Test stages for polymer: (1) fresh mix; (2) return polymer from previous use; (3) after reboosting/ready to reuse; (4) dirty polymer before and after base cleaning; (5) before concreting; and ongoing measurement of polymer level drop and resupply volumes to establish fluid loss. The volume of polymer product and admixtures added during reboosting should be recorded and linked to the piles/walls where used.
Testing should comprise the measurement of density, viscosity, pH and sand content. For bentonite, additional tests for setting time, gel strength and fluid loss/filter cake thickness are necessary. For polymer, suitable additional tests are being developed – further guidance is expected to be published soon in the EFFC/DFI *Guide to Support Fluids for Deep Foundations*. Samples of the support fluids should be taken from an appropriate point so as to obtain a representative sample – for example, from near the base of a pile bore for sand

(a)

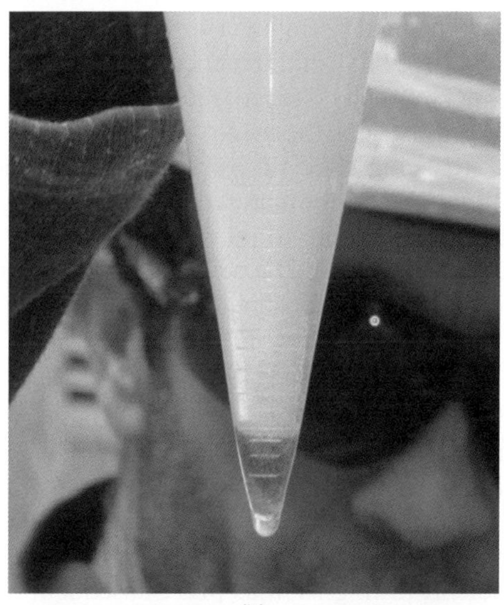

(b)

Figure 99.23 Sand content test:
(a) washing sand from the sieve into a vial;
(b) measurement of the sand content

Figure 99.24 Density testing of bentonite

Figure 99.25 Vegetation planted in geosynthetic cells on a steep slope (Courtesy of David B. Andrews, P.E., Atlanta, GA, USA; www.cabeceo.net (personal collection))

content prior to concreting. **Figure 99.23** illustrates parts of the sand content test: final washing of the sieve and measuring the content. **Figure 99.24** illustrates the apparatus used to determine bentonite density.

99.13. Miscellaneous materials

99.13.1 Vegetation

Vegetation is often used as a cover for slopes or retaining walls, but does not always successfully survive – for example, if the vegetation type is inappropriate for the slope, shading and moisture conditions. Use on soil/rock slopes can help to reduce erosion and provide an increased level of stability. However, on steep slopes vegetation is likely to require temporary support until it is established, and in some cases it (or the soil) may need permanent support. The effect of vegetation on biodiversity, maintenance and longer-term slope stability under climate change should also be considered. **Figure 99.25** illustrates the use of vegetation on a slope.

If seeds are being used they should be sampled for testing. If germinated seeds or matured plants are being used, test beds can be set up on site to provide a control. If plants are delivered to site they should be covered while in transport and acclimatised on site. Roots should not be exposed for longer than necessary.

99.13.2 Water

Water is used in many applications on site. Generally, potable water is fine to use without further testing, whereas sewage water is not. Non-potable water should be subjected to a preliminary inspection for visual signs of contamination and simple handheld testing such as pH. Subsequent testing may involve laboratory testing for chemicals. However, the extent of testing should be appropriate to the end use of the water.

99.13.3 Polystyrene blocks

In a similar manner to lightweight aggregates, polystyrene blocks can be used to reduce the impact of embankments on soft soils or earth pressures on retaining walls. Problems that may be encountered during construction include hydrocarbon (e.g. petroleum) attack, heat damage and UV light damage, although these issues are generally easy to manage with appropriate on-site storage. **Figure 99.26** illustrates the use of polystyrene blocks behind a retaining wall.

Testing is generally undertaken prior to delivery to site; however, samples may be collected to determine properties such as density, compressive strength and shear strength. Delivery records should be reviewed, dimensions checked (subject to allowable tolerances) and checks made for imperfections.

Figure 99.26 Polystyrene blocks used as filling behind a retaining wall

99.14. Reuse of foundations

The reuse of foundations has drawn much attention in recent years due to concerns over future use of sites and conservation of resources, and clients seeking opportunities to change buildings' use or add additional storeys to existing buildings. The appraisal of the viability of reusing foundations should be started at the earliest project stage and a reuse strategy written to inform the client's insurers (for latent defects insurance policies). This strategy should include a responsibility matrix (across the client, structural and geotechnical engineers and contractors), summaries of existing as-built information, risk registers of potential defects and unknowns, surveys and investigations required, assessments of existing and proposed loads and so on. The reuse decision process throughout the design stages towards construction is based on principles of increasing the certainty of the collated information and validating initial assumptions through assessments, surveys, testing and trials (as defined in CIRIA C653 – Chapman *et al.*, 2007).

Assessments of existing foundations for reuse involves several investigation steps

- **desk study:** records of construction, historical changes in condition, loading, information gap analysis
- **physical:** geometry and integrity (i.e. direct measurement of foundation locations, dimensions, depths, rebar cover, ground conditions and so on; and material testing: pile concrete strength and chemical testing, reinforcement exposure for assessment of condition and signs of defects or deterioration e.g. concrete sulfate, acid attack, rebar corrosion, timber rot, etc.)
- **performance of existing:** load/settlement testing and historical behaviour for load capacities
- **pre-start site trials** of any proposed strengthening works (e.g. additional piling adjacent to existing foundations) to test impacts on reused structures and validate construction methodologies to ensure that the design intent is satisfied.

Following investigation steps and any resultant strengthening designs, it is important for rigorous site management and control procedures to be established, supported by instrumentation and monitoring, to minimise the impact on the existing and proposed structures.

In terms of verification of materials, the 'physical' step is most relevant to this chapter. The aims of physical investigation are to determine the geometry, integrity, strength and durability of the ground and foundation elements. The extent of the site investigation is dependent on the reliance to be placed on the old foundation to support the new structure, the costs, the time allocated for the investigation and the potential for damaging existing foundations. Investigation of the elements can be intrusive or non-intrusive. Intrusive testing comprises coring to obtaining samples for laboratory testing (e.g. concrete cube strength, concrete petrography, chemical and carbonation analysis, reinforcement chloride ion content profiles, etc.). Trial pitting to expose parts of the foundations for visual inspection, direct measurement and sampling is also a typical investigation task which requires the impact on the existing foundation to be carefully considered. Non-intrusive testing comprises external field testing. It is important to understand the limitations of the various non-intrusive tests and their needs in terms of calibration and correlation. Non-intrusive tests include: ultrasonic echo testing (to derive the thickness and location of objects – **Figure 99.27**), radar testing (to locate reinforcement), low-strain testing (to derive pile length and integrity), parallel seismic testing (down a parallel borehole, to derive pile length), mise-à-la-masse (to locate reinforcement and the length of steel piles), rebound hammer (to derive concrete strength – **Figure 99.28**) and cover meters (to derive the thickness of concrete cover, number of reinforcement bars

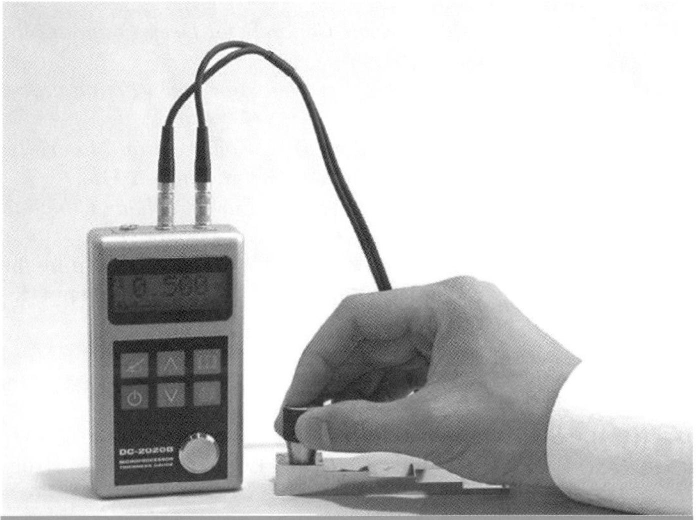

Figure 99.27 Ultrasonic echo tester – used to measure thickness

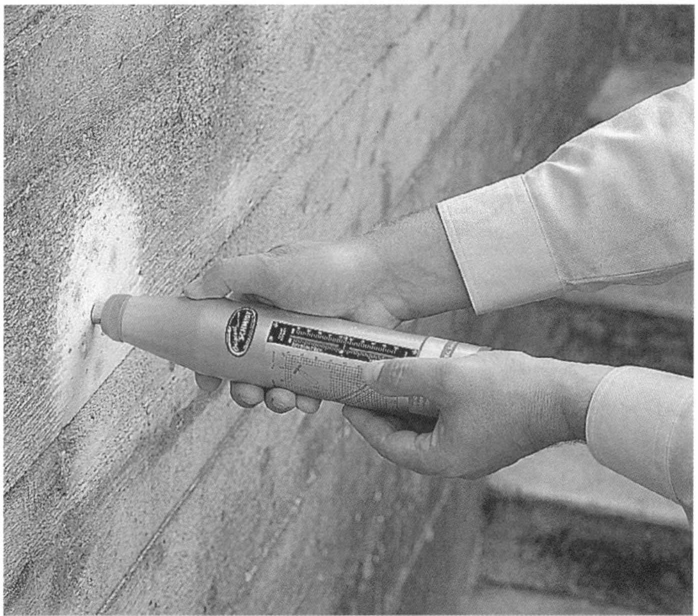

Figure 99.28 Rebound hammer – used to measure concrete strength

and options to carry out magnetometer testing down a parallel borehole to establish length of reinforcement).

Useful websites, references and further reading

There are numerous sources of information relating to materials in books, CIRIA guides, British Standards, websites and the like. Some of these documents and websites are listed below. Readers should refer to the bibliographies of these documents and document databases to find further information.

References

Chapman T, Anderson S and Windle J (2007) *Reuse of Foundations*. CIRIA Report C653. CIRIA, London, UK.

CIRIA (1997) *Integrity Testing in Piling Practice*. CIRIA Report R144. CIRIA, London, UK.

EFFC/DFI (2018) *Guide to Tremie Concrete for Deep Foundations*. EFFC/DFI, Faversham, UK.

EFFC-DFI (2019) *Guide to Support Fluids for Deep Foundations*. EFFC/DFI, Faversham, UK.

Illston JM and Domone PLJ (eds) (2010) *Construction Materials: Their Nature and Behaviour*, 4th edn. Spon, London, UK.

Low Carbon Concrete Group (2022) *Low Carbon Concrete Routemap*. Green Construction Board/ICE, London, UK.

Rawlings CG, Hellawell EE and Kilkenny WM (2000) *Grouting for Ground Engineering*. CIRIA Report C514. CIRIA, London, UK.

Further reading

General

BSI (British Standards Institute): http://www.bsigroup.co.uk

BSI (2002) BS EN 1990:2002+A1:2005: Eurocode: Basis of structural design. BSI, London, UK.

CIRIA (Construction Industry Research and Information Association): http://www.ciria.org

Dean Y (1996). *Mitchell's Materials Technology*. Longman, Harlow, UK.

Forde MC (ed) (2009) *ICE Manual of Construction Materials*. Thomas Telford, London, UK.

IHS (Construction Information Service): http://uk.ihs.com

Institution of Civil Engineers (2016) *Specification for Piling and Embedded Retaining Walls*, 3rd edn. Thomas Telford, London, UK.

Tomlinson M and Woodward J (2015) *Pile Design and Construction Practice*, 6th edn. Taylor & Francis, Abingdon, UK.

Concrete

BRE (2017) *Concrete in Aggressive Ground*. BRE, Watford, UK.

BSI (2000) BS EN 206-1:2000: Concrete – Part 1: Specification, performance, production and conformity. BSI, London, UK.

BSI (2005) BS EN 12794:2005: Precast concrete products – Foundation piles. BSI, London, UK.

BSI (2014) BS EN 1992-1.1:2014 +A1:2014: Eurocode 2: Design of concrete structures – Part 1-1: General rules and rules for buildings. BSI, London, UK.

BSI (2019) BS EN 12350-5:2019: Testing fresh concrete – Part 5: Flow table test. BSI, London, UK.

Concrete Centre (2020) *Specifying Sustainable Concrete*. MPA Concrete Centre, London, UK.

Henderson NA (2002) *Concrete Technology for Cast In-Situ Foundations*. CIRIA Report C569. CIRIA, London, UK.

QSRMC (Quality Scheme for Ready-Mixed Concrete): http://www.qsrmc.co.uk

Steel

ArcelorMittal (steel products): http://www.arcelormittal.com

BSI (1999) BS EN 12063:1999: Execution of special geotechnical work – Sheet pile walls. BSI, London, UK.

BSI (2005) BS 4449:2005+A3:2016: Steel for the reinforcement of concrete – Weldable reinforcing steel. BSI, London, UK.

BSI (2007) BS EN 1993-5:2007: Eurocode 3: Design of steel structures – Part 5: Piling. BSI, London, UK.

CARES (Certification Authority for Reinforcing Steels): http://www.ukcares.com/

Corus (steel products): http://www.corusgroup.com

Timber

BSI (2004) BS EN 1995-1-1:2004: Eurocode 5: Design of timber structures – General – Part 1-1: Common rules and rules for buildings. BSI, London, UK.

TRADA (Timber Research and Development Association): http://www.trada.co.uk

Geosynthetics

BSI (2006) BS EN 14475:2006: Execution of special geotechnical works – Reinforced fill. BSI, London, UK.

BSI (2019) BS EN ISO 13437:2019: Geosynthetics – Installing and retrieving samples in the field for durability assessment. BSI, London, UK.

Jewell RA (ed) (1996) *Soil Reinforcement with Geotextiles*. CIRIA Special Publication 123. CIRIA, London, UK.

Ground

BSI (2004) BS EN 1997-1:2004 Eurocode 7: Geotechnical design – Part 1: General rules. BSI, London, UK.

BSI (2004) NA to BS EN 1997-1:2004: UK National Annex to Eurocode 7: Geotechnical design – Part 1: General rules. BSI, London, UK.

BSI (2009) BS 6031:2009 Code of practice for earthworks. BSI, London, UK.

BSI (2015) BS 8004:2015+A1:2020: Code of practice for foundations. BSI, London, UK.

Aggregates

BSI (2008) BS EN 12620-2:2008 Aggregates for concrete. BSI, London, UK.

CIRIA (2016) *The Rock Manual: The Use of Rock in Hydraulic Engineering*, 2nd edn. CIRIA Report C683. CIRIA, London, UK.

Pavement Interactive: https://pavementinteractive.org

Smith MR and Collis L (eds) (1993) *Aggregates*, 2nd edn. Geological Society, London, UK.

Grout

BSI (2007) BS EN 445:2007 Grout for prestressing tendons – Test methods. BSI, London, UK.

BSI (2020) BS EN 12715:2020 Execution of special geotechnical work – Grouting. BSI, London, UK.

Support fluids

Fleming WK and Sliwinski ZJ (eds) (1977) *The Use and Influence of Bentonite in Bored Pile Construction*. CIRIA, London, UK.

FPS (2006) *Bentonite Support Fluids in Civil Engineering*, 2nd edn. FPS, Faversham, UK.

Lam C and Jefferis S (2017) *Polymer Support Fluids in Civil Engineering*. ICE, London, UK.

Vegetation

Coppin NJ and Richards IG (eds) (2007) *Use of Vegetation in Civil Engineering*. CIRIA Report C708. CIRIA, London, UK.

Leung AK, Boldrin D, Liang T, *et al.* (2018) Plant age effects on soil infiltration rate during early plant establishment. *Géotechnique* **68(7):** 646–652.

Meijer GJ, Bengough AG, Knappett JA, Loades KW and Nicoll BC (2016) New in situ techniques for measuring the properties of root-reinforced soil: laboratory evaluation. *Géotechnique* **66(1):** 27–40.

Water

BSI (2002) BS EN 1008:2002: Mixing water for concrete – Specification for sampling, testing and assessing the suitability of water, including water recovered from processes in the concrete industry, as mixing water for concrete. BSI, London, UK.

Polystyrene

Sanders RL and Seedhouse RL (1994) *Use of Polystyrene for Embankment Construction*. Contractor Report 356. Transport Research Laboratory, Wokingham, UK.

Foundation reuse

Butcher AP, Powell JJM and Skinner HD (eds) (2006) *Reuse of Foundations for Urban Sites: Proceedings of the International Conference*. BRE, Watford, UK.

Coventry S, Woolveridge C and Hillier S (eds) (1999) *The Reclaimed and Recycled Construction Materials Handbook*. CIRIA Report C513. CIRIA, London, UK.

It is recommended this chapter is read in conjunction with

- Chapter 75 *Earthworks material specification, compaction and control*
- Chapter 78 *Procurement and specification*
- Chapter 96 *Technical supervision of site works*

All chapters in this book rely on the guidance in Sections 1 *Context* and 2 *Fundamental principles*. A sound knowledge of ground investigation is required for all geotechnical works, as set out in Section 4 *Site investigation*.

Chapter 100
Observational method

doi: 10.1680/icemge.66830.1641

CONTENTS

Dinesh Patel Arup, London, UK

This chapter describes the use of the observational method (OM) on engineering projects. The latest thinking on the two approaches to OM, the *ab initio* and the *ipso tempore* is also described with reference to CIRIA C760 (2017). Traditional ground engineering projects are usually based on a single, fully developed, robust design. This can lead to conservative but costly designs. The objective of OM is to achieve greater overall economy by having less conservative designs without compromising safety. Successful implementation of the *ab initio* OM relies on a structured approach to both design and construction, an assessment of the most realistic design parameters, comparing these with the cautious design parameters used in traditional designs, having a rigorous monitoring and observation strategy, having a design/construction which in the light of the monitoring data can be modified in a timely way, and having a strong management and reporting structure with well-defined contingency plans in case problems occur. These key ingredients to the successful implementation of OM are described with the help of some case examples. OM can also be used when a project, designed by traditional methods, is in difficulty and the *ipso tempore* or 'best way out' approach is required; this is also discussed. The *ipso tempore* approach can also be applied to traditionally designed installed structures where its performance during construction exceeds the designer's expectations and programme savings can be realised by the project team. Eurocode 7 allows the use of OM in design, but has shortcomings which are also highlighted.

100.1. Introduction

This chapter aims to provide guidance to engineers on the use of the observational method (OM) on engineering projects. It describes the important differences between carrying out designs based on the traditional approach, using 'characteristic values' of parameters (defined in Eurocode 7 (EC7; BSI, 2004)) and those developed using realistic parameters ('most probable') when using OM. It also explains that traditional designs are by their nature conservative and not flexible to change, but that OM allows the project team to produce an integrated approach to both the design and construction, which can yield significant cost and programme savings. It also fosters good working relationships between all the team members and can be very satisfying, but only if it is implemented properly.

The use of OM is permitted in EC7, but engineers should be aware that there are some shortcomings that need to be understood before applying EC7. The use of OM reduces the safety margins on design compared with traditional designs, but this can be managed within a rigorous monitoring and observation strategy, implemented within a coordinated team, to still give safe designs. Before applying OM, the management of ground risks and pre-agreed contingency plans (should things not go according to plan) need to be carefully evaluated and explained to the client/stakeholders. Approval for implementation of OM should be agreed with the client in the knowledge that it provides benefits and that there may be some drawbacks (e.g. implementing pre-agreed contingency plans), even if the risks are low. Approval from other third-party checking engineers may also be required before implementing OM.

While the OM approach is intended to provide overall economy, obviously there are higher associated costs for increased instrumentation and monitoring, for implementing a stronger management team to implement the OM, for increased design services and for reviewing of the monitoring data. However, on complex projects much of this cost may already be accounted for as instruments and monitoring are required for other reasons, and usually there is already a strong contractor's management team established on site.

The OM approach is not intended to be used where there is a risk-averse client, where there is likely to be a lack of total commitment from any member of the project team, where there is no thorough investigation of the ground and water conditions and where there is a likelihood of a rapid or brittle mode of failure of the ground (including temporary structural elements).

This chapter guides the engineer through the principles of the use of OM on projects from inception (referred to as the *ab initio* approach) and also on projects under construction which for unexpected reasons are running into difficulty (referred to as the 'best way out' approach). It relies heavily on the work of some key authors, including Nicholson *et al.* (1999), Peck (1969), Powderman and Nicholson (1996) and Gaba *et al.* (2017: chapter 7.4). These authors provide plenty of examples of the structured implementation of OM on engineering projects. Additional examples from Europe are also given in a recent study carried out by a working party on the use of OM in Europe (GeoTechNet, 2005); this includes seven case examples of the 'structured' use of OM on both building and civil engineering projects and also highlights the overall cost and programme savings made, which the reader might find useful.

100.1.1 Differences between Peck's and the CIRIA R185 approach to OM

The OM has developed over the last 60 years and was initially applied on a 'trial and error' basis to improve designs. It was not until Peck's Rankine lecture (1969) that an integrated process for predicting, monitoring, reviewing and modifying designs was advocated, operating within a framework of OM and without compromising the safety of the structure.

To successfully implement OM, Peck (1969) identified that it was necessary to have two designs compared with the traditional single-design approach in geotechnics. A range of foreseeable conditions needed to be considered, which Peck associated with the most likely condition to happen in practice ('most probable') and the least likely condition to happen ('most unfavourable') (refer to Section 100.3.2 for definitions, also illustrated in **Figures 100.4** and **100.5**). He suggested a design starting with the most probable (best estimate) condition and varying the design and/or construction to the planned most unfavourable condition, should observed behaviour be worse than that predicted based on best-estimate parameters. Peck's (1969) eight key ingredients for successful use of OM are given in **Table 100.1**.

This approach is fundamentally different to the recent work on OM published by CIRIA (Nicholson *et al.*, 1999). The approach advocated by Nicholson *et al.* (1999) starts with initial moderately conservative parameters (the same as 'characteristic' parameters in EC7), to be relaxed to a likely real situation (i.e. most probable condition) during construction, should the observed behaviour warrant it. This approach is also known as *progressive modification* to the design and was first suggested by Powderham (1994). The setting up of a rigorous traffic-light trigger system (red, amber and green) to deal with uncertainties in the ground during construction was also established after Peck's work. Both of these modifications result in improvement in the use of OM on projects, leading to safer designs.

1	There must be sufficient site investigation
2	Design is developed on most probable (best estimates) to predict behaviour
3	Develop monitoring strategy on calculated values for best case
4	Perform calculations on most unfavourable conditions
5	Identify contingency plans for most unfavourable conditions
6	Monitor and evaluate actual conditions
7	Modify design to suit actual conditions if triggers are exceeded
8	OM can only be done if there is adequate time to make decisions and implement

Table 100.1 Peck's (1969) eight key ingredients for OM

100.1.2 Definition of OM

The best definition of the OM approach is given in CIRIA R185 (Nicholson *et al.*, 1999) (see **Box 100.1**). Often engineers mistake monitoring instruments on a project as following an OM approach, and **Box 100.1** shows that there is a proper operational framework for carrying out OM and monitoring is just part of this process.

Box 100.1 Definition of OM (Nicholson *et al.*, 1999)

'The Observational Method in ground engineering is a continuous, managed, integrated process of design, construction control, monitoring and review that enables previously defined modifications to be incorporated during or after construction as appropriate. All these aspects have to be demonstrably robust. The objective is to achieve greater overall economy without compromising safety.'

100.1.3 Comparison between traditional designs and OM design

Traditional ground engineering projects are usually based on a single, fully developed, robust design and there is no intention to vary the design during construction. Instrumentation and monitoring may also be carried out, but this plays a very *passive* role, to check original predictions are still valid and provide confidence to third-party checkers, such as designers for adjacent building owners affected by a development. In Nicholson *et al.* (1999) this traditional design is termed 'predefined design'. In comparison, in OM the monitoring plays a very *active* role in both the design and construction, allowing planned modifications to be carried out within an agreed contractual framework that involves all the main parties (client, designer and contractor). The differences in the two design approaches are illustrated in **Table 100.2**. Peck (1969) defined two OM approaches

Predefined design process (traditional design)	The OM process
■ Permanent works	■ Temporary works
■ One set of parameters	■ Two sets of parameters
■ One design/predictions	■ Two designs and predictions
■ Outline of construction method	■ Integrated design and construction methods
■ Contractor's temporary works design/method statement	■ Methods relate to triggers
■ Monitoring checks predictions not exceeded	■ Comprehensive and robust monitoring system
■ If checks are exceeded, consider (a) best way out approach to design or (b) redefine the predefined design approach, reassessing the geotechnical uncertainties in the ground (see **Table 100.5**)	■ Review and modify process – contingency plan – improvement plan ■ Emergency plan
■ Emergency plan	

Table 100.2 Comparison of the predefined design process and the observational method

- the *ab initio* approach, adopted from inception of the project
- the 'best way out' approach (Peck, 1969), adopted after the project has begun and some unexpected event has occurred that is different to the predefined design, or failure occurs, and where OM is required to establish a way of getting out of a difficulty.

Case examples presented on the GeoTechNet website give examples of both the *ab initio* and 'best way out' OM approaches. A paper by Nicholson *et al.* (2006) suggests a structural framework for operating the 'best way out' approach for recovery of deep, multi-stage excavation projects when problems occur during construction, like that which occurred at Nicoll Highway, Singapore (COI, 2005). This operational framework for carrying out the 'best way out' is also described in Section 100.5.

100.1.4 *Ipso tempore* OM definition (CIRIA C760)

The application of the OM method is also described in the latest CIRIA guidance (C760; Gaba *et al.*, 2017) on embedded retaining wall design. As well as recognising the *ab initio* OM method of design applied from project inception, it refers to a new definition – the *ipso tempore* method, not previously defined by earlier researchers, but which is actually the most common form in practice when there is a better than expected result or an unplanned event requiring the use of OM.

The two scenarios when *ipso tempore* is possible for retaining walls are

- the wall is performing much better than the predicted behaviour and the project team decide on modifying the construction sequence to achieve programme savings
- the wall is performing unpredictably during construction and active intervention is required to prevent a service or ultimate limit state (SLS or ULS) failure occurring ; this approach then defaults to the 'best way out' method proposed by Peck (1969).

Box 100.2 Eurocode 7, clause 2.7 (BSI, 2004)

Observational method

(1) When prediction of geotechnical behaviour is difficult, it can be appropriate to apply the approach known as 'the observational method', in which the design is reviewed during construction.

(2) The following requirements shall be met before construction is started:

- acceptable limits of behaviour shall be established;
- the range of possible behaviour shall be assessed and it shall be shown that there is an acceptable probability that the actual behaviour will be within the acceptable limits;
- a plan of monitoring shall be devised, which will reveal whether the actual behaviour lies within the acceptable limits (the monitoring shall make this clear at a sufficiently early stage, and with sufficiently short intervals to allow contingency actions to be undertaken successfully);
- the response time of the instruments and the procedures for analysing the results shall be sufficiently rapid in relation to the possible evolution of the system;
- a plan of contingency actions shall be devised, which may be adopted if the monitoring reveals behaviour outside acceptable limits.

(3) During construction, the monitoring shall be carried out as planned.

(4) The results of the monitoring shall be assessed at appropriate stages and the planned contingency actions shall be put into operation if the limits of behaviour are exceeded.

(5) Monitoring equipment shall either be replaced or extended if it fails to supply reliable data of appropriate type or in sufficient quantity.

100.2. Fundamentals of OM implementation and pros and cons of its use

100.2.1 General

EC7 does not clearly define the framework to be followed when adopting OM, and there are other drawbacks which can be strengthened if use of the CIRIA approach (Nicholson *et al.*, 1999) is also made on OM projects, as described below.

100.2.2 OM drawbacks in Eurocode 7 (BSI, 2004)

The OM method described in clause 2.7 of EC7 has been reproduced in **Box 100.2** to illustrate the main drawbacks, which are as follows.

- It is primarily aimed at the *ab initio* approach to OM, although it does not exclude the 'best way out' application of OM.
- While it refers to 'acceptable limits of behaviour' it does not define how these may be derived, since EC7's premise for design is based on use of 'characteristic values' which present a lower cautious limit in design, but not the upper limit to represent the most likely behaviour (most probable case), needed to implement the OM approach.
- No trigger limits are defined to establish planned contingency actions to check behaviour.
- There is no operational framework described for management of the OM within a contract, either within national policy or in a project organisation.

100.2.3 CIRIA C760 definition

The two categories of the OM method which have been historically described in various published papers/reports are more clearly defined in CIRIA C760 (Gaba *et al.*, 2017). For retaining walls the two categories are

- *ab initio* (from the start) where the use of the OM is planned from the start of the project
- *ipso tempore* (in the moment) where the design of the embedded retaining wall is reassessed during construction. It is not planned from the beginning, but rather adopted during construction after the wall has been installed. It either defaults to Peck's (1969) 'best way out' method if anything unexpected happens during construction, or in the case of a better wall performance it can be used to achieve programme savings.

100.2.4 Operational framework for following OM

The operational framework for implementing OM is illustrated in **Figure 100.1** for the *ab initio* method. The OM has to be carried out within the framework of any national and corporate policies governing design codes, specifications, quality management systems and health and safety regulations (e.g. in the UK this is the Health and Safety (HSE) Regulations 1994 and the Construction Design and Management (CDM) Regulations 2015). This is represented by the upper box in **Figure 100.1** (see also Section 100.3.5 for details).

The second box defines the structure of the key players in the stakeholder's organisation (i.e. the client, designer, contractors, third-party checkers and other inspectors), their roles and responsibilities and the relationship between organisations and the individuals. This needs to address the culture of each organisation, the level of staff training, experience, openness to communication and management commitment to implementing the OM approach. The stakeholders also need to 'buy into' the technical and commercial risks should any planned contingency or emergency measures need to be implemented, even if this risk is considered low.

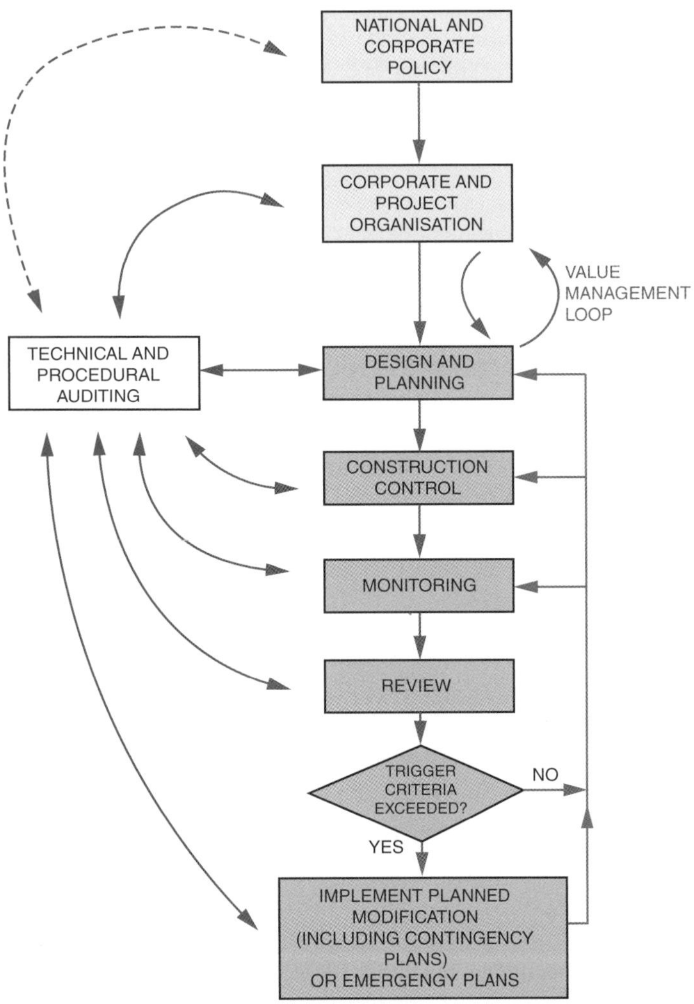

Figure 100.1 The observational method (Modified with permission from CIRIA R185, Nicholson *et al.*, 1999 ; www.ciria.org)

Once the OM is agreed at 'project organisation' level, the remaining boxes describe the robust management structure required to implement OM at both design and construction stage and to control the monitoring and reviewing aspects of the OM when the works are on site. The works have to progress to an agreed plan, with risks being recognised at each construction phase. Daily construction progress has to be under the control of a management structure that ensures any deviation from the method is fully thought through by all members of the project team and covered by an amendment to the plan. A monitoring regime has to be in place, with competent staff made available to check, review and respond to all monitoring results within a given timescale from when they become available. There then needs to be clear instructions to all involved for all foreseeable situations. Finally, contingency plans need to be in place that can be rapidly implemented should preset 'trigger' limits be breached or if any other unforeseen situation develops.

'Auditing', preferably by an independent geotechnical firm, is essential as it checks that the OM designer and project team are following established procedures and reaching the correct technical interpretations. Ideally this should be carried out by a designer who is unconnected with the OM process.

100.2.5 OM management process on site

At site level there are usually many layers of contracting organisations involved in a project and all the main players need to (*a*) buy into the OM process and (*b*) have clearly defined responsibility levels. An example of the interaction between these organisations and the *managed* reporting of the construction and monitoring process is given in a paper by Chapman and Green (2004) for a deep basement project in central London; this structure chart is illustrated in **Figure 100.2**. On this project successful implementation of the OM relied on a clear understanding of the process, roles and responsibilities between the main contractor (HBG), the groundworks contractor (McGee), the concreting contractor (Byrne Bros), the instrumentation contractor (Soil Instruments) and the designer/reviewer (Arup). The management and reporting structure was defined under a traffic-light system of green, amber and red trigger levels, so that all parties were clear on their responsibilities should pre-agreed contingency or emergency plans need to be implemented (see Section 100.4 for details). These plans consisted of increased monitoring, stopping work and implementing additional propping or berms and/or reverting to the predefined design. In the event, the measured movements of the walls during construction were within acceptable limits, thus allowing considerable savings to be made as the alternative design allowed the excavation of double-height basements to be sequenced for faster top-down construction.

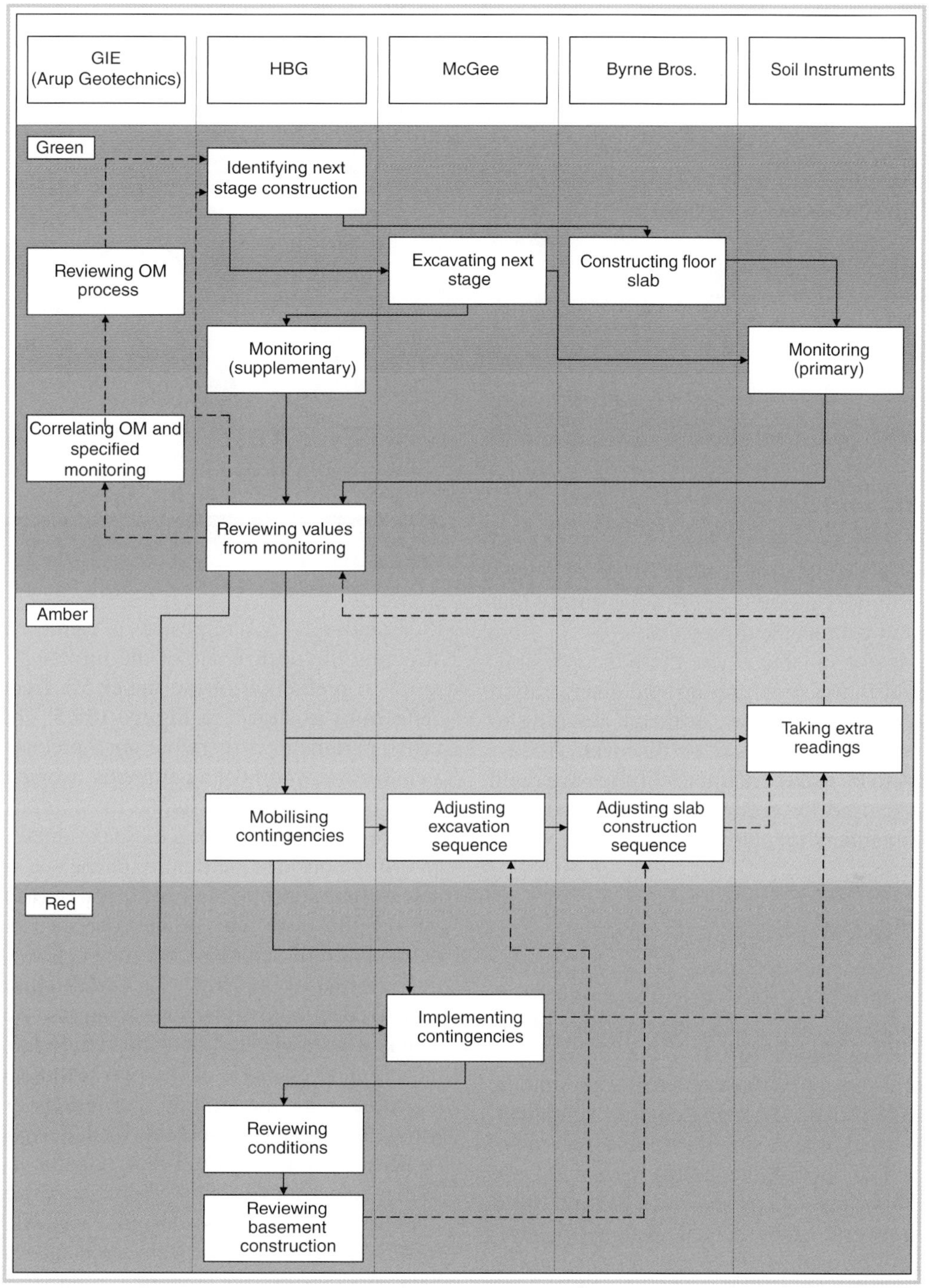

Figure 100.2 OM management and reporting structure on site (Data taken from Chapman and Green, 2004)

100.2.6 Pros and cons of OM

OM offers potential savings of time and money, and the monitoring provides the needed assurance concerning safety. Some potential benefits of OM are illustrated in **Figure 100.3**, and seven detailed case examples of the benefits provided to clients are described in GeoTechNet (2005) (see also Nicholson *et al.*, 1999).

However, while there are significant advantages associated with the OM application, there are extra costs associated with prescribing a higher level of management and control on site, more instrumentation, preparing for contingency plans and readiness to use back-up plans (e.g. extra propping) and reporting compared with a conventional design situation.

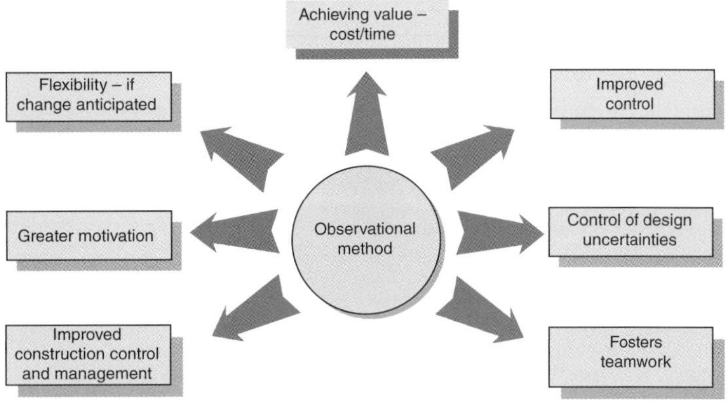

Figure 100.3 Some potential benefits of OM (Modified with permission from CIRIA R185, Nicholson *et al.*, 1999; www.ciria.org)

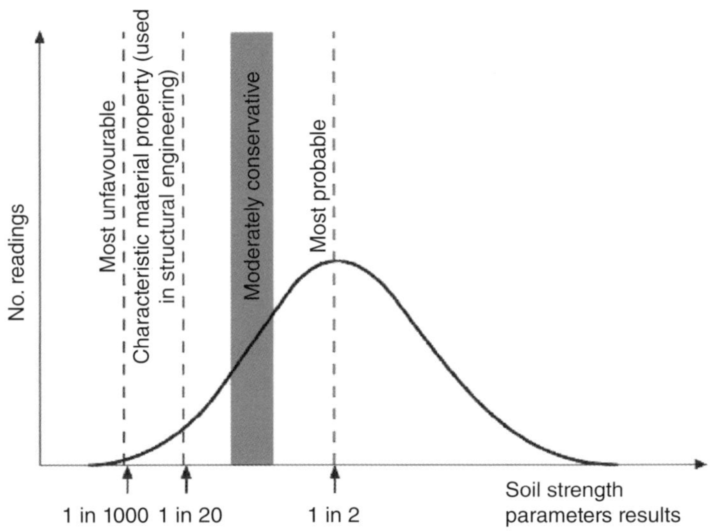

Figure 100.4 Types of soil strength parameters (Modified with permission from CIRIA R185, Nicholson *et al.*, 1999; www.ciria.org)

100.3. OM concepts and design

100.3.1 Uncertainty and serviceability

OM is most effective where there is a wide range of uncertainty. **Table 100.3** summarises the types of uncertainty that are often encountered in geotechnical projects.

The OM approach is not suitable where there is a possibility of 'brittle' behaviour in the structure or rapid deterioration in the materials which does not allow sufficient warning to implement any planned modifications (e.g. 'discovery–recovery' contingency plans to be used). Examples of such are rapid deterioration of soils caused by groundwater or non-ductile failures of structural members (struts/walling connections) in multi-propped basements.

100.3.2 Selection of design soil parameters

Where there is a wide range of uncertainty in the soil parameters the OM process in CIRIA R185 uses the terms 'most probable' and 'most unfavourable' to describe the range of soil conditions, as illustrated in **Figure 100.4**.

The 'most probable' is a set of parameters that represent the probabilistic mean of all the data, although a degree of engineering judgement must be used in assessing this to take account of the quality of the data. The 'most unfavourable' parameter represents the 0.1% fractile of the data as shown in **Figure 100.4**, and this represents the worst value that the designer believes might occur in practice. The moderately conservative parameter or 'characteristic value' of geotechnical parameters (defined in EC7, clause 2.4.5.2) represents a 'cautious estimate of the value affecting the occurrence of the limit state', and should ideally result in prediction of the upper 5% fractile of the *measured deflections* as shown in **Figure 100.5**. The moderately conservative parameter is therefore not a precisely defined value. It is a *cautious estimate* of a parameter, worse than the probabilistic mean but not as severe as the most unfavourable, as shown in **Figure 100.4**. In assessing these parameters the designer should carefully consider the quality of the site investigation data and assess their appropriateness for use in the OM approach. Often the original data may be appropriate for a more robust 'predefined design' approach but may not be of a higher quality for the purposes of implementing OM. In this case it may be necessary to carry out further investigations – for instance, if there is a lack of groundwater table information.

Geotechnical uncertainty	Example
Geological	Complex geology and hydrogeology
Parameter and modelling	Undrained soil vs drained behaviour
Ground treatment	Grouting, dewatering
Construction	Complex temporary work

Table 100.3 Examples of uncertainty in the ground

A typical example of the two sets of OM parameters (most probable and moderately conservative) which were used for a 20-m-deep basement wall design into London Clay is illustrated in **Table 100.4** (Chapman and Green, 2004). When selecting the most probable parameter for use in OM it is important that the designer can justify the choice – for instance, this may be proven experience from other projects or back-analysis of case studies.

100.3.3 SLS and ULS prediction

When designing to EC7, checks are required to ensure that the following ULS are not exceeded

- loss of equilibrium of the structure or the ground
- internal failure or excessive deformations of the structure or structural elements
- failure or excessive deformations of the ground due to loss of strength

Parameter	Moderately conservative Conventional design	Most probable (OM)	Reference
Made ground	$\phi' = 25°$	$\phi' = 38°$	OAP – Broadgate
Taplow Gravel	$\phi' = 25°$	$\phi' = 40°$	Lehane *et al.*
London Clay	$\phi' = 36°$	$\phi' = 24°$ $c' = 1.0$ kPa	OAP, Crossrail
Lambeth Clay	$\phi' = 23°$ $c' = 0$	$\phi' = 28°$ $c' = 25$ kPa	OAP, Crossrail
Water pressure in Taplow Gravel	None		Geotechnical report
MEFP	Applied from GL	None	Observations
Softening	Passive Soil	None	Observations – OAP Horseferry Road
Surcharge	20 kPa	20 kPa (Perm) 10 kPa (Temp)	–
Overdig	0.5 m	None	–
Undrained shear strength London Clay	$70 + 7.5z^*$ kPa	$+5.5 < d > -11.0$ Mod $112 + 5.19z +$ kPa $-11.0 \leq d \geq -17.5$ Mod 200 kPa $-17.5 \leq d \geq -29.2$ Mod 400 kPa	Results – see **Figure 100.4**
Undrained shear strength Lambeth Clay	300 kPa	400 kPa	Results – see **Figure 100.4**
Stiffness of clay (E_q)	1000 c_u	1500 c_u	Back-analysis

Table 100.4 Two sets of design parameters used in OM (Reproduced from Chapman and Green, 2004)

- loss of equilibrium of the structure or ground from uplift water pressures
- hydraulic heave, internal erosion and piping of ground caused by hydraulic gradients.

EC7 also require checks on SLS – states that are less serious than ULS but are nevertheless undesirable and would need intervention or repair. In (traditional) predefined designs, calculations made to check these states use 'characteristic values'.

In OM, the acceptable limit of behaviour is a 'serviceability' calculation, made using both the 'most probable' and 'characteristic' parameters and conditions. These provide the predictions against which the field performance can be monitored and reviewed. Trigger values can be established and contingency plans introduced as illustrated in **Figure 100.5** and **Table 100.5**.

The OM approach illustrated in **Figure 100.5** is for a cantilever wall design, but the principle applies to other examples. The green, amber and red zones represent the trigger limits or traffic-light control system used in OM. The precise deflections set for the trigger values will depend on the 'discovery–recovery' contingency plans being used and not simply on the calculated values of predictions made. For

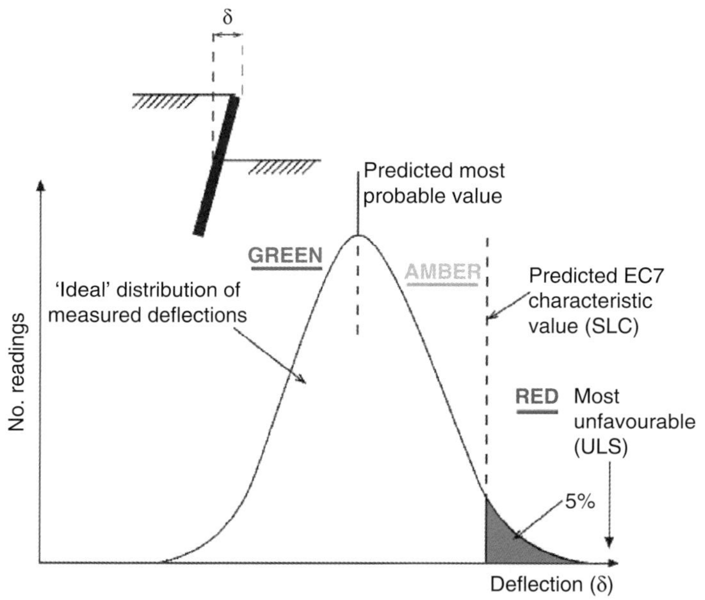

Figure 100.5 Ideal EC7 predicted against measured performance (Modified with permission from CIRIA R185, Nicholson *et al.*, 1999; www.ciria.org)

Most probable	50% likelihood of movement predictions being exceeded
Characteristic values (EC7) or moderately conservative (CIRIA R185)	5% likelihood of movement predictions being exceeded
Most unfavourable (CIRIA R185)	0.1% likelihood of movement predictions being exceeded

Table 100.5 Definitions of most probable, characteristic values and most unfavourable (Data taken from CIRIA R185, Nicholson *et al.*, 1999; www.ciria.org)

instance, if the contingency plans involve a berm in front of the cantilever wall or introduction of a raking prop, then the time taken to implement these measures will influence the setting of these trigger limits. The triggers may then be based on two criteria: the first is wall deflection and the second is based on the rate of wall movement to ensure that sufficient time remains to implement the contingency measure. The SLS wall deflection limit is sometimes used as an easily measured proxy for a range of undesirable outcomes, including bending moment failure and buckling of props. It should be noted that OM should not result in excessive movements and that, if limits were breached, the risk of a collapse would still be remote.

In respect to ULS predictions, EC7 identifies three sets of partial factors to apply when assessing the ultimate limit case. These partial factors are applied to 'characteristic values' of the ground, but in essence the ULS design values are then similar to the most unfavourable conditions (see **Figure 100.5**). Although, for example in retaining walls, the ULS predictions are used for assessing structural forces, moments and shear, the ULS deformations of the wall can also be a useful guide to determining the maximum predicted movements in the red zone (**Figure 100.5**). This 'upper limit' of wall deformation (or curvature) can provide a useful input when developing emergency plans of unexpected behaviour in OM and also with the 'best way out' approach in OM, provided the problem has been identified in time to implement a disaster and recovery plan before a ULS condition occurs, and with due regard to safety.

100.3.4 Factors of safety

The design values for ULS (stability) calculations are chosen so that the probability of failure will be acceptably small. It should be noted that the intent of OM is to take out uncertainties in the ground (see **Box 100.2**), *not reduce factors of safety*, when assessing the ULS condition in design.

UK standards and those of some other European countries have traditionally applied 'factors of safety' on the design soil parameter, which can vary depending on the type of foundation and type of redistribution of load in the ground (e.g. piled rafts operate differently from single piles per column). EC7, on the other hand, applies partial factors on the characteristic value of the ground parameters. **Figure 100.6** illustrates

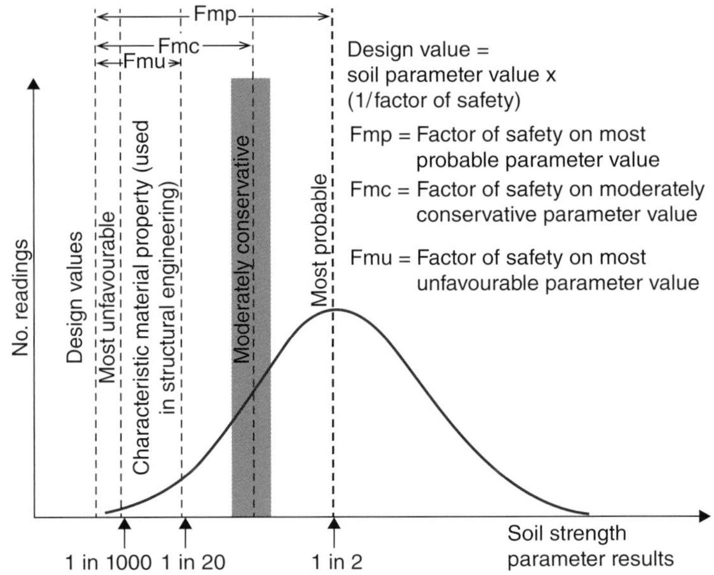

Figure 100.6 Application of factors of safety to different types of design soil parameters (Modified with permission from CIRIA R185, Nicholson *et al.*, 1999; www.ciria.org)

how the two approaches can produce different factors of safety and how they can vary for different assumed design soil parameters.

100.3.5 Setting OM within the context of a contractual model and safe design

In the UK there are essentially two forms of main design contracts.

- The client appoints a consulting practice to carry out the permanent design ('engineer design') and the contractor is responsible for carrying out the specified works. In this form of contract the contractor is only responsible for any temporary works design required to complete the permanent works.
- The client appoints a 'design and build' contractor to complete the design based generally on an outline or scheme design performed by a consulting engineering firm.

Other variations to this also occur when a construction manager or project manager is appointed by a client to manage the overall contract. It should be noted that in some European countries the contractual model is very different, responsibilities between parties are less clear and the legal obligations on parties may be severe. The application of OM needs to consider these factors.

The OM process can be applied to both forms of contract in the UK model described above. In both cases the design product comprises

- drawings
- work specifications and bills of quantities
- calculations.

In addition to this, the UK Construction (Design and Management) (CDM) Regulations 2015 place new duties on the client, designers and contractors to take health and safety into account in both the design and construction of a project. For the designer this means that the design is no longer a set of calculations but must also

- address buildability issues
- identify hazards and risks in respect to safety
- eliminate hazards through good design or, where it is not possible, reduce the risk to a low level
- show how this process has evolved in the design by producing a 'risk register'
- address impact on adjacent structures (above or below ground).

These regulations are intended to produce stronger links between the designer and the contractor and minimise risk of failures by way of

- production of 'health and safety plans' by both the engineer and contractor
- appointment of a planning supervisor by the client, who vets these plans before and during construction
- seeking approval from third-party checkers.

The CDM Regulations are in line with the OM objective of integrating the design and construction process to produce safe designs and construction practices. From work carried under Work Package 3 of GeoTechNet funded by the European Commission, the UK contractual model with appropriate CDM Regulations and risk assessments does not appear to exist in all countries in Europe. Again, when carrying out OM in these countries, the national polices and regulations need to be fully understood and incorporated within the OM process.

On NATM tunnelling projects, HSC (1996) also describes the process of managing risk through the use of a 'discovery–recovery' model before an unacceptable failure scenario is reached, and this approach is entirely aligned with the OM approach.

100.3.6 Rapid deterioration

In certain engineering situations rapid deterioration can be controlled by modifying the construction sequence as follows.

- **Using the multi-stage construction process** – for instance, an example may be an embankment construction over soft clays. In this situation a rapid deterioration in the factor of safety can be controlled by 'staged' filling, between rest periods, using monitoring to control when the next lift is done (an example of this is given in **Figure 100.7**).
- **Using the incremental construction process** – for instance, an example may be in NATM tunnelling work, where the rate of advancing the tunnel face and controlling face loss is a critical component in determining how the ground movements are controlled in the discovery–recovery programme (**Figure 100.8**) using the traffic-light system described above. This figure shows that the later the problem is discovered, the higher the risk and the longer the structure remains in a state of reduced stability (red zone). Late instigation of decision-making and recovery would have the same effect. In such instances, trigger limits can be set as both absolute values and/or rate of movements.

The importance of early decision-making to instigate actions for recovery is an important feature of the UK Health and Safety Executive (HSC, 1996) discovery–recovery model, and is a legal requirement for use on all UK construction sites. Trigger values described below, an essential feature of OM, can also be used in the 'predefined designs' (see Section 100.1.3) to allow sufficient time for implementation of emergency measures when monitoring is being used.

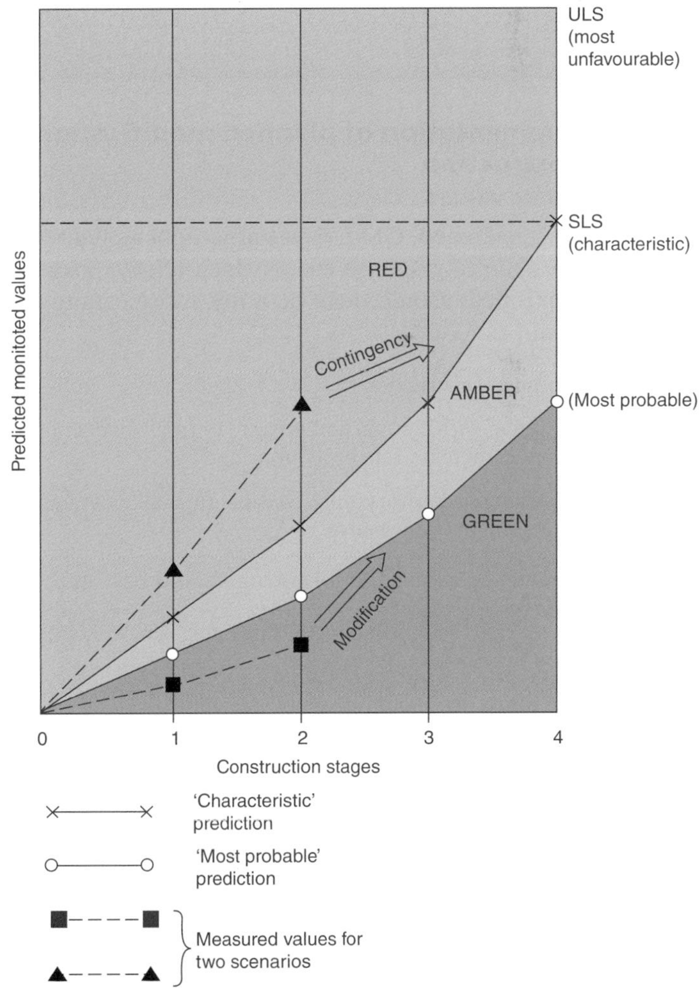

Figure 100.7 Multi-stage construction trigger values (Modified with permission from CIRIA R185, Nicholson *et al.*, 1999; www .ciria.org)

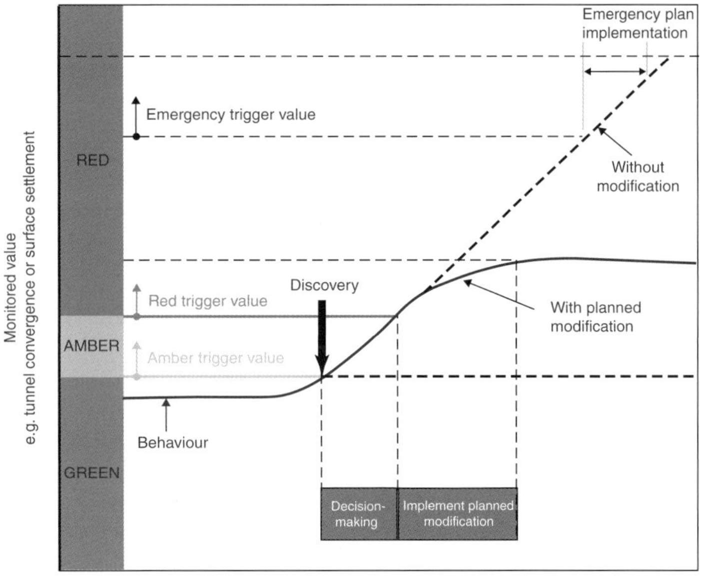

Figure 100.8 Traffic-light system for an incremental excavation (tunnels) process (Modified with permission from CIRIA R185, Nicholson *et al.*, 1999; www .ciria.org)

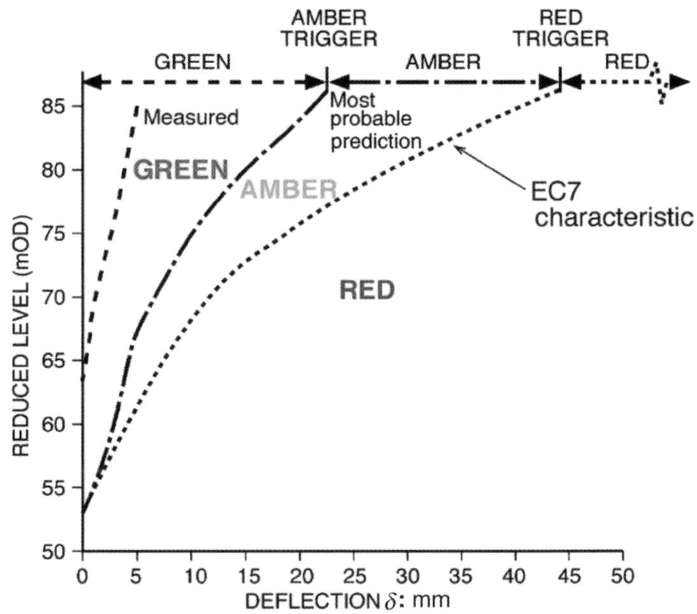

Figure 100.9 Example of trigger limits in retaining walls (Modified with permission from CIRIA R185, Nicholson *et al.*, 1999; www .ciria.org)

100.4. Implementation of planned modifications during construction

100.4.1 Trigger values

As previously mentioned, OM uses a traffic-light system with green, amber and red response zones which allow construction to be controlled, should there be a risk of exceeding the safe green limit, as follows

- green – continue construction
- amber – continue with caution and prepare to implement contingency measures, increase rate of monitoring
- red – stop progress, do everything possible to slow movements, implement contingency measures.

In setting the trigger values the following should be noted.

- The values set may be absolute values or rates of movements or both.
- The trigger limits set should also consider the accuracy of the instrument and whether it is practically measurable.
- The choice of instruments to measure movements should therefore be appropriate for the project and not based simply on lowest cost.
- The amber/green limits should be sensibly set based on the design so that the likelihood of breaching this limit is small (e.g. values may occasionally stray outside the green zone) but also treated seriously if it is breached.
- The contingency/implementation plans should be taken seriously if the trigger limits are being breached.
- The choice of the instrumentation should be treated seriously and the monitoring contractor should be experienced in this type of work.

The trigger limits should be linked to the 'most probable' (green/ amber limit) and the 'characteristic values' (amber/red limit) for implementation of the planned modifications in OM, as illustrated in **Figures 100.5**, **100.7** and **100.8**. A further example, for use with cantilever retaining wall movements, is illustrated in **Figure 100.9**. It can be seen that in this example the measured movements were also well below the most probable conditions.

100.4.2 Monitoring systems

Monitoring systems will vary depending on the type of construction project in which OM is implemented. It is also very important to define both 'primary' and 'secondary' monitoring systems in OM. For example, in multi-prop deep basements, the primary system (e.g. inclinometers and wall-mounted settlement gauges on neighbouring party wall structures) may be the main instruments relied upon to allow implementation of any contingency measures in OM, while the secondary system (e.g. 3D targets at the tops of walls or levelling surrounding ground) might be a more frequent and fast monitoring system to quickly assess the progress of the excavation works and aid a broader understanding of the pattern of ground movements on a site. When considering the amount of instrumentation for use on a project, considerations should be made in respect to whether to monitor using remote methods (can be expensive) or by manual methods (labour-intensive), and if the latter, how long it will take to carry out a round of monitoring in a single day. This will then allow assessments of the most important instruments to be monitored to be made as part of the overall OM strategy.

An essential part of OM is that the primary system needs to be immediately repaired if damaged on site, to ensure that OM can be continued. Chapman and Green (2004) explain the use of primary and secondary instrumentation for a deep basement in London, in the context of the trigger limits and the process owner (see **Figure 100.2**).

100.4.3 OM quality plans

It is essential to have a quality plan before implementing OM on a site. This plan would present the designer's movement predictions based on the defined construction sequence. Each stage of the construction sequence should also show the acceptable limits of predicted behaviour using the traffic-light system described above.

Simple graphical outputs that the whole project team can understand are essential, and an example of this is illustrated in **Figure 100.10** for a 20-m-deep basement in London, excavated in a top-down manner. This quality plan shows the predicted green and red limits for wall deflection at each dig stage for a planned double-height excavation technique to allow floors to be built quickly. The actual movements recorded from inclinometers can then be plotted for comparison with predicted behaviour. These graphs allow the OM reviewer to make informed judgements and the whole project team is brought into the decision-making process: whether to continue to the next dig level or to implement contingency measures before continuing. For the example illustrated, contingency measures involved additional propping, use of natural berms in front of the wall before proceeding to the next level, and changing the construction sequence to the original 'predefined design' which involved progressing excavation/floors one after another.

100.4.4 Construction control

Successful construction control is a vital part of OM; the main process is as follows.

- A construction control proforma is used to record all details of construction operations, strengths of materials exposed during staged excavations, the fabric and structure of exposed materials and the deterioration of surfaces exposed to water (see the example of cutting in **Figure 100.11**).
- This control has to be fully integrated within the project team, simple to use and the data easy to read; graphical outputs are essential for making informed decisions (see **Figure 100.10**).
- Each process has to have a process owner, with certain levels of responsibilities and implementing of actions, as illustrated

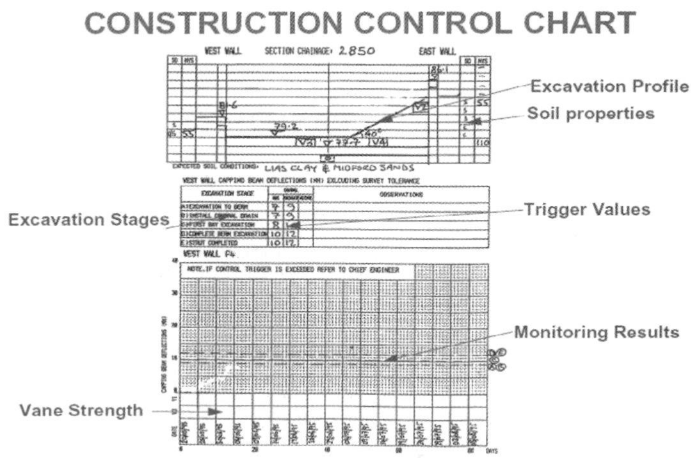

Figure 100.11 Example of construction control proforma sheets on site prepared by a contractor (Reproduced from Nicholson *et al.*, 2006)

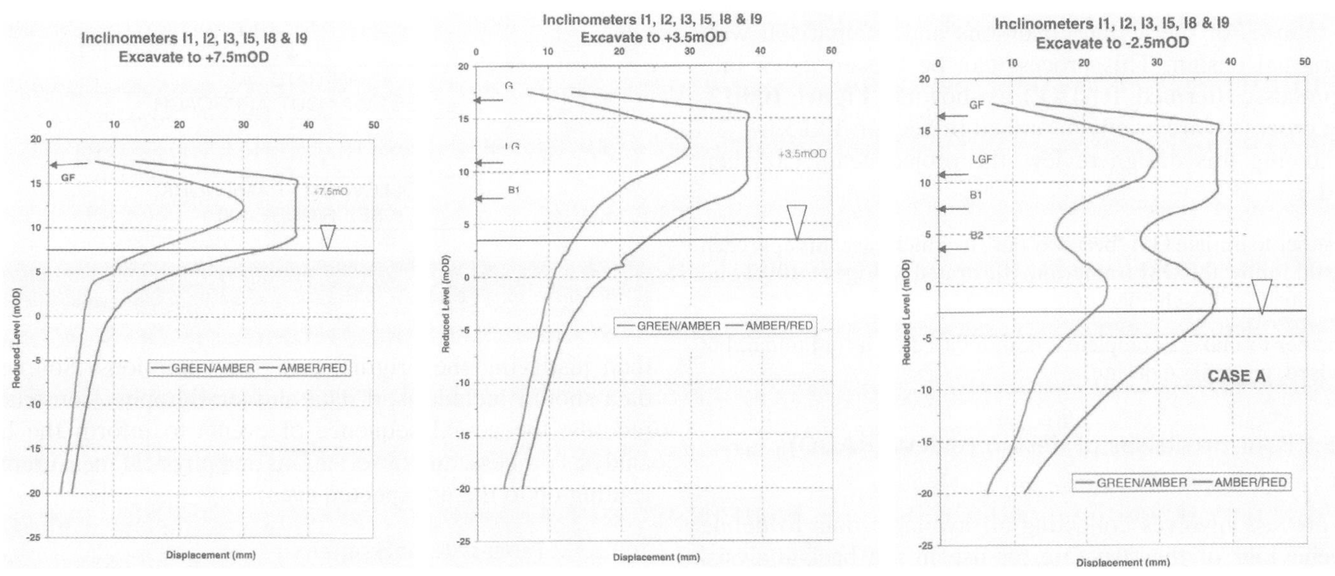

Figure 100.10 Typical quality plan developed for a phased top-down basement excavation (Reproduced from Nicholson *et al.*, 2006)

in **Figure 100.2** and described in Chapman and Green (2004). Where there are many trade subcontractors involved in a project, the key organisation for making the decision for reviewing the values from the site monitoring and implementing planned contingencies is often the main contractor, who has greater control and powers to immediately stop works on site.

100.5. *Ipso tempore* or 'best way out' approach in OM

The 'best way out' approach is used when monitoring checks in a predefined design (**Table 100.1**) exceed, for unexpected reasons, predicted values but before an emergency condition is reached. Peck (1969) gives some good examples of the 'best way out' approach in which OM is adopted in response to some unexpected events or failure to establish a way out of a difficulty. Nicholson *et al.* (2006) present a more structured approach for recovery of deep, multi-stage excavation projects when problems occur during excavation. These authors studied the post-inquiry reports following the collapse of Nicoll Highway (COI, 2005), and to avoid such failures occurring again suggested a 'best way out' solution which is described below.

In the event of a 'discovery', the 'best way out' approach would trigger an '*initial recovery decision-making*' stage, as shown in **Figure 100.12**.

In all cases this will result in stopping work and/or implementing emergency planned measures to secure the safety of site staff and the general public while the unexpected event is fully investigated. This assessment will inevitably be somewhat qualitative rather than quantitative, as decisions need to be made rapidly at this initial stage.

Once the safety of the site has been secured, the project team can then turn their attention to recovery of the project back to a fully stable condition, which means first carrying out a 'design review process' of the unexpected event by back-analysis of the actual conditions and comparison with the original design. This process can be broken down into four processes (termed 'RADO'), as shown in **Figure 100.12**. These processes are briefly described in Section 100.5.1.

Following this design review, the project team can then consider

- whether to initiate OM 'best way out', in which case this approach would follow the OM framework illustrated in **Figure 100.1** and described in this chapter, or
- whether to make a complete redesign based on a traditional, but revised, predefined design.

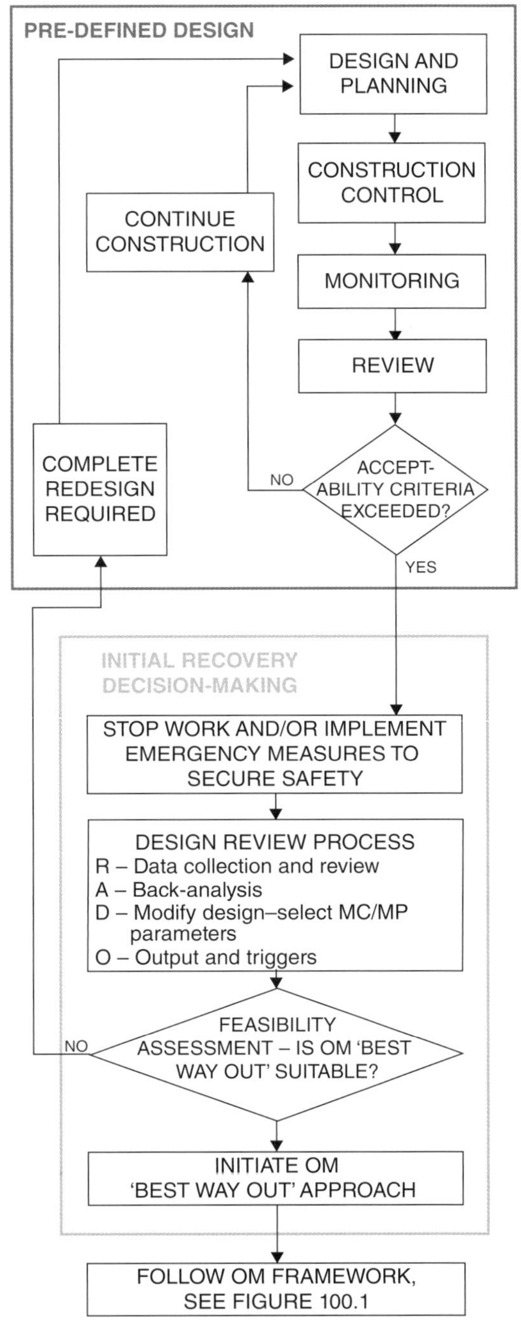

Figure 100.12 'Best way out' operational framework (Reproduced from Nicholson *et al.*, 2006)

100.5.1 Four processes of design review (RADO)

100.5.1.1 Process R: data collection and review

This process involves collecting all available data to define the behaviour of the structure for use in the back-analysis. Particular emphasis should be placed on understanding the actual conditions and behaviour operating in the field, rather than justifying the original design assumptions. Sources of data should include: soil data and stratigraphy; construction records; the actual sequence of events to inform the back-analysis process; and observations and physical measurements leading up to the unexpected event.

100.5.1.2 Process A: back-analysis

The purpose of this process is to refine the designer's understanding of the actual behaviour of the structure and reduce uncertainty in the design. The process involves: establishing most probable parameters; developing a satisfactory model using most probable parameters; comparing results with monitoring data and field observations; reviewing/revising parameters if good agreement is not achieved; and once a reliable model has been produced, proceeding to design.

100.5.1.3 Process D: verify modified design

This process involves predicting the future behaviour using the realistic model and set of parameters developed from back-analysis for the remaining construction stages, but adopting a level of conservatism into the model. The structure behaviour should adopt moderately conservative (characteristic) parameters for the serviceability design and 'worst credible' or factored parameters for stability checks (see Sections 100.3.3 and 100.3.4).

100.5.1.4 Process O: Output plans and triggers

If the OM 'best way out' is to be used, then the process of OM as described in this chapter has to be agreed with all stakeholders in the project, with appropriate contingency and monitoring plans and setting up of trigger values and management teams.

100.6. Concluding remarks

This chapter provides an overview of the use of OM in engineering projects. It provides a structural framework for carrying out OM for both the *ab initio* (projects from initial conception) and *ipso tempore* or 'best way out' approaches. *The ipso tempore* OM method can be used on traditionally designed projects running into difficulty during construction or where programme savings can be realised if wall performance exceeds the designer's expectations. An extensive guide to OM is CIRIA R185 (Nicholson *et al.*, 1999), and the two OM categories described above are clearly presented in CIRIA 760 (Gaba *et al.*, 2017) for retaining wall design. Both documents should be considered, as they set out a proper framework for operating OM and are more rigorous than currently described in EC7. Other case examples quoted in this chapter also provide an excellent history of the use of OM on more recent projects since CIRIA R185. The design and installation of a robust instrumentation and monitoring system is absolutely crucial to the application of OM, so that the actual performance of the structure can be tracked, with an appropriate feedback loop to the project team during construction.

Acknowledgement

Thanks to Stuart Hardy – Technical Leader , Laing O'Rourke – for proofreading this chapter.

References

BSI (British Standards Institution) (2004) BS EN 1997-1:2004: Eurocode 7: Geotechnical design – Part 1: General rules. BSI, London, UK.

Chapman T and Green G (2004) Observational method looks set to cut city building costs. *Civil Engineering* **157(3)**: 125–133.

COI (Committee of Inquiry) (2005) *Report of the Committee of Inquiry into the Incident at the MRT Circle Line Worksite that Led to the Collapse of the Nicoll Highway on 20 April 2004.* Ministry of Manpower, Singapore.

Gaba A, Hardy S, Doughty L, Powrie W and Selemetas D (2017) *Guidance on Embedded Retaining Wall Design.* CIRIA Report C760. CIRIA, London, UK.

GeoTechNet *(European Geotechnical Thematic Network) (2005) Work Package, WP3: Innovative Design Tools in Geotechnics – Observational Method and Finite Element Method.* Project with financial support of the European Commission under the 5th Framework, Project GTC2-2000-33033.

HSC (Health and Safety Commission) (1994) *Managing Construction for Health and Safety: Construction (Design and Management) Regulations 1994.* HSE Books, Sudbury, UK.

HSC (1996) *Safety of the New Austrian Tunnelling Method (NATM) Tunnels: A Review of Sprayed Concrete Lined Tunnels with Particular Reference to London Clay.* HSE Books, Sudbury, UK.

HSE (Health and Safety Executive) (2015) Construction (Design and Management) Regulations 2015. HSE, London, UK.

Nicholson D, Tse C and Penny C (1999) *The Observational Method in Ground Engineering: Principles and Applications.* CIRIA Report R185. CIRIA, London, UK.

Nicholson D, Dew CE and Grose WJ (2006) A systematic 'best way out' approach using back analysis and the principles of the observational method. *Proceedings of the International Conference on Deep Excavations, Singapore.*

Peck RB (1969) Advantages and limitations of the observational method in applied soil mechanics. *Géotechnique* **19(2)**: 171–187.

Powderham AJ (1994) The value of the observational method: development in cut and cover bored tunnelling projects. *Géotechnique* **44(4)**: 619–636.

Powderham AJ and Nicholson DP (1996) *The Observational Method in Geotechnical Engineering.* ICE and Thomas Telford, London, UK.

Useful websites

GeoTechNet (European Geotechnical Thematic Network): http://www.geotechnet.org

Health and Safety Executive (HSE): http://www.hse.gov.uk

It is recommended this chapter is read in conjunction with

- Chapter 78 *Procurement and specification*
- Chapter 79 *Sequencing of geotechnical works*
- Chapter 94 *Principles of geotechnical monitoring*
- Chapter 96 *Technical supervision of site works*

All chapters in this book rely on the guidance in Sections 1 *Context* and 2 *Fundamental principles.* A sound knowledge of ground investigation is required for all geotechnical works, as set out in Section 4 *Site investigation.*

Acknowledgements

References

Chapter 101
Close-out reports

Roger Lindsay Atkins, Epsom, UK
Martin Kemp Atkins, Epsom, UK

The geotechnical close-out report is intended to describe, collate and summarise the geotechnical works that occurred during the construction phase of the works, together with any post-construction monitoring. The report forms an essential part of the construction verification process, and it provides information relevant for maintenance, demolition or reuse purposes by future owners of the asset. The close-out report should represent the final step in the geotechnical reporting cycle.

doi: 10.1680/icemge.66830.1655

CONTENTS

101.1. Introduction

This chapter describes the relevance and importance of producing a close-out report. Although it is often a contractual requirement of major civil infrastructure projects, the close-out report rarely attracts the right level of attention to do it justice. Typically, this report is produced in a rush, after the event, without the continuity of staff and during a period when the emphasis is on staff leaving site and demobilisation of the design team.

The importance of the close-out report during its production is not always immediately apparent to the author, as it is not until some future event (usually involving others) that the recorded information and specific details are required. In recognition of this, the report should be prepared on the basis that it forms an essential and valuable part of the design and construction process rather than a contractual compliance.

The geotechnical close-out report is the culmination of the geotechnical reporting cycle, which includes preliminary studies and geotechnical reporting of the factual, interpretative and design data (see Chapter 50 *Geotechnical reporting* and Eurocode 7 (BSI, 2004)). The report may also be known as a feedback, completion, verification or validation report.

101.2. Reasons for writing close-out reports

It is good practice to record and document the geotechnical elements of the construction phase of a project as part of the construction verification in a close-out report. In some instances, as described below, it is a requirement to produce one. The report is an essential component of assessing reuse of the geotechnical elements in future projects (see BRE, 2006a, 2006b; Chapman *et al.*, 2007). For example, faced with 100 30-year-old piles, there can be much greater confidence if there is a final as-built plan and knowledge that only piles numbered 23 and 76 had non-conformance reports (NCRs) and that they were satisfactorily resolved, together with full details of installation and testing.

The close-out report provides a means of demonstrating the verification and implementation of the geotechnical design (from the geotechnical reporting phase) and records any changes to the design during the construction phase. As-built works frequently differ from those shown on the 'design' or 'for construction' drawings. **Figure 101.1** shows a chalk slope cut to the 'for construction' requirement, while **Figure 101.2** shows the additional remedial works added during site works to maintain stability of the infill deposits prior to topsoiling the slope.

A key element of the report is feedback on constructability issues which can then lead to more cost-effective buildable designs. This is an important feature for infrastructure owners, such as UK National Highways, where feedback can be incorporated into future improvements to their standard specifications. An example would be reporting on how earthworks material behaved when placed and compacted just outside the specified grading envelope and whether the standard compaction criterion had to be altered to achieve a compliant compaction.

Figure 101.1 Cut slope in chalk formed to the 'for construction' slope angle

Figure 101.2 Cut slope to be recorded on the 'as-built' drawing after remedial works

Figure 101.3 Historical foundations and a well encountered during site clearance

The close-out report also provides information on the geotechnical works for ongoing maintenance, demolition, reuse and on rare occasions for forensic investigations when problems had occurred, such as excessive foundation settlement to the works.

Projects designed under the Eurocodes require the production of a Geotechnical Design Report (see Chapter 50 *Geotechnical reporting*). This design report, which is detailed in Eurocode 7 (BSI, 2004) as a principle, has a requirement to provide a 'plan of supervision and monitoring' for the works. This plan identifies items to be appropriately checked and monitored both during and after construction of the works. When the checks and monitoring are completed, Eurocode 7 requires the checks and monitoring results to be recorded as an 'addendum to the [Geotechnical Design] Report'. The close-out report can be used as a means of recording and providing the 'addendum'.

Box 101.1 Illustration of the benefit of a close-out report

The designed foundation solution for a ten-storey development in central London was for pile groups supporting the column loadings. During construction the following were noted by the supervision team: (*a*) some of the installed continuous flight auger (CFA) pile depths were shortened due to the reassessment of pile capacities using the results from site pile tests; (*b*) the installed pile reinforcement was of varying depth (10 m design length, 8–10 m installed lengths) due to difficulties in pushing the cage into the fresh concrete; and (*c*) the presence of historical mass concrete footings and a well (see **Figure 101.3**), which meant that a number of pile cap arrangements were redesigned to avoid the footing. Recording all these variations in the design in a close-out report means that in the future these piles could be assessed for reuse in a future redevelopment. Additionally, the report can provide information on the piles as an obstruction to tunnelling projects, or the foundation information can be used to assess the effects that ground movement from a tunnelling project would have on the piles.

An important justification for producing a close-out report is in satisfying the requirements of the UK CDM 2015 regulations (HSE, 2015). When these regulations apply to a project, there is a requirement for a 'Health and Safety File' (see Chapter 42 *Roles and responsibilities*) to be produced on completion of the project. This file is intended to provide information to future users, designers and contractors. A close-out report can be used as the mechanism for recording the relevant information for inclusion into the file. It should be noted that other parties to the construction works also input to the 'Health and Safety File' and that agreement should be reached between these parties as to the scope of their individual input. This agreement reduces the potential of either overlap and conflicting information or gaps in the information supplied.

Previously, Highways England (now National Highways) has recognised the importance of recording the construction activities on its projects, and as part of its design standards has a requirement for a feedback report which is described in CD 622 – *Managing Geotechnical Risk* (Highways England, 2020). The piling industry has also understood the need for recording its on-site construction activities and has included the requirement for an 'as-built' element in the widely used industry standard *Specification for Piling and Embedded Retaining Walls* (ICE, 2016).

The 'Reuse of Foundations for Urban Sites' (RuFUS) international conference held at the Building Research Establishment in 2006 highlighted the problem of verification of existing foundations from available information for reuse in new works. The proceedings (BRE, 2006a) and its associated best-practice handbook (BRE, 2006b) provide useful background information on what information is required for assessing the reuse of foundations. This was further augmented in 2007 by CIRIA C653 *Reuse of Foundations* (Chapman *et al.*, 2007) which provides greater detail on the information required for assessing the reuse of piled foundations, although the principles described could be applied to other foundation solutions.

Important issues relating to insurance and liability of foundation reuse are described by Chapman *et al.* (2001, 2007). Examples of the benefits of foundation reuse are provided in an article in *The Structural Engineer* by Tayler (2020).

The AGS (2005) report *AGS Guidelines for the Preparation of the Ground Report* includes reasons for producing a close-out report, of which the main points have been summarised in this chapter.

The BRE has developed a methodology, 'Design for deconstruction: helping construction unlock the benefits of the circular economy' (DfD), against a backdrop of buildings, assets and infrastructure having specific design lives. The DfD enables these resources to be reused and recycled in the most efficient way. Although this approach is driven largely at the design stage through the choice of materials and components and how they are put together, it complements the information contained in the close-out report (BRE, 2023).

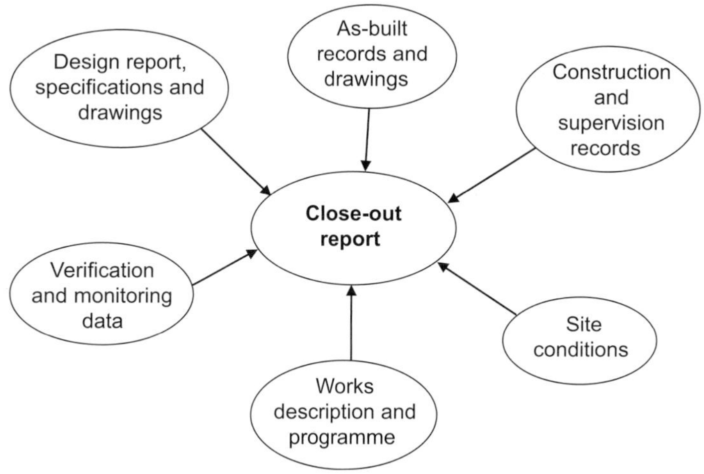

Figure 101.4 Major inputs to close-out reports

101.2.1 Key learning points

- A close-out report can demonstrate the verification and implementation of the geotechnical design in accordance with the design specification.
- The report brings together and records design changes and how they were implemented during the construction phase.
- The report provides useful information for the future reuse or deconstruction of the works.

101.3. Contents of close-out reports

The close-out report should be a stand-alone report that may cross-reference the design reports or the design specifications (including drawings). It does not completely replace them but should add to the geotechnical knowledge bank of the works. Therefore, the report should describe the site, the geotechnical works, verification and monitoring data in succinct text and be supported by material such as drawings, figures and photographs (see **Figure 101.4**). The source material should accompany the report, either as appendices to the report or as stand-alone reports, to which cross-reference is made. 'As-built' drawings are also an essential component of a close-out report detailing the actual constructed works. Photographs properly referenced (e.g. with date, orientation, subject) also provide a good record of the works on site. A key element of the report is the inclusion of the site documentation, construction and supervision records (see Chapter 96 *Technical supervision of site works*).

The close-out report is not intended to be verification/validation of any geo-environmental works carried out as part of the construction works. Geo-environmental works include works such as testing of samples for contamination and remediation of contaminated ground. However, salient factors from the geo-environmental works, such as contamination barriers, should be included in the close-out report. A separate report dealing with the verification/validation of any geo-environmental works is generally required for issuing to the relevant authorities.

The majority of projects will generate large quantities of data that can be difficult to present or understand in the context of the overall project. The close-out report should present this information clearly within a logical framework. Summary data can often be presented more effectively in graphical format, such as time series graphs or as spatial plots (**Figure 101.5**). In common with all reporting, the presentation and clarity are important. Guidance on good practice in writing ground reports is given by the AGS (2015) *Guide to Good Practice in Writing Ground Reports*.

As a minimum, the contents of a close-out report should be tailored to the size and complexity of the works, and for the geotechnical works should include details of

- project organisational structure involved (e.g. client, consultant, main works contractor, groundworks contractor)
- description of the works (main and geotechnical) and other relevant works (e.g. geo-environmental remediation)
- works programme
- works location related to a nationally recognised survey grid
- description of the site background and original site conditions
- summary of the relevant site investigation reports (including design reports)
- detail of additional site investigation works
- site documentation, construction and supervision records
- description of the ground and groundwater conditions encountered
- geotechnical and compliance testing (including calibration certificates)
- monitoring of the works and its effects on adjacent buildings, structures and utilities

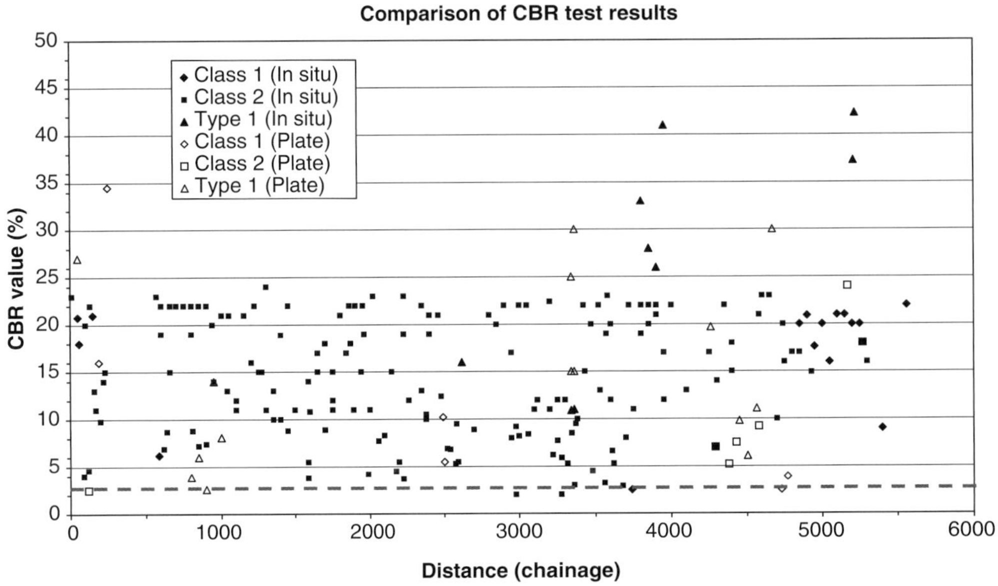

Figure 101.5 Example of graphical presentation of data

- temporary works (including as-built details if significant)
- as-built and quality assurance records including reports, drawings, schedules and registers (e.g. piles, soil nails)
- NCRs or similar, together with remedial actions
- site and progress photographs
- environmental monitoring (e.g. vibration from piling or plant movements)
- details of communications with regulators (e.g. the UK Environment Agency)
- post-works monitoring arrangements and maintenance requirements
- critical requirements for decommissioning of the works.

Highways England, in their design standard CD622, provide a detailed format for their feedback report. This format and content listing could form the starting point for close-out reports for other infrastructure-based projects. Further advice on the preparation and content of a close-out report has been published by the AGS (2005).

101.3.1 Key learning points

- The purpose of the close-out report is to succinctly describe the site, the geotechnical works, verification and monitoring data, supported by drawings, figures and photographs.
- The report should clearly record any changes made on site from that detailed in the Geotechnical Design Report, specifications and construction drawings.

101.4. Reporting on quality issues

The quality system in place for the construction of the main works and geotechnical works (see Chapter 93 *Quality assurance*) should be described in the close-out report. The quality assurance strategy for the supervision of the works should also be described in the report. Relevant records relating to quality issues for the geotechnical works should be included in the close-out report together with a summary of the key quality issues and their resolution.

NCRs or similar are to be expected on geotechnical works due mainly to the variability of the ground (conditions and response) and the limitation on the extent that the ground can be assessed by intrusive ground investigations. The resolution of NCRs should be described in the close-out report together with the remedial action taken and its effects on the design, construction and use of the geotechnical works. The fact that an NCR has been raised often means that there has been a variation in the construction of the geotechnical works, and this would require additional design considerations. Therefore, the close-out report needs to show that the NCR has been rationally considered and the appropriate action taken to close out the NCR.

101.4.1 Key learning points

- Inclusion of quality-related records aids the verification process of the works.
- Reporting on the resolution of construction-related non-conformances is a key part of proving the quality of the works.

101.5. Reporting on health and safety issues

The designers of the geotechnical works will have identified the key health and safety issues and risks relating to their design. The preferable method of communicating these issues and risk is for them to be shown on the 'for construction' drawings, typically as notes. These issues/risks should relate to the designer's significant assumptions on construction, maintenance and decommissioning. On completion of the construction activities the designer's significant assumptions on maintenance and decommissioning should be included on the 'as-built' drawings and summarised in the close-out report.

At the construction stage, where a significant issue has been detailed by the designer – such as a suggested construction sequence – the contractor may adopt this sequence or develop their own alternative approach. This may significantly alter the original designer's assumptions relating to the maintenance and decommissioning of the works. Therefore, these issues or risks would need reassessment by the contractor and/or designer before being included on the 'as-built' drawings.

The majority of geotechnical works once constructed are buried and not readily accessible for inspection/survey, and therefore accurate and detailed recording of exposed features and their treatment during the construction works are necessary. The following are examples of features that could have health and safety issues relating to maintenance, decommissioning or reuse for inclusion in the close-out report

- observations on founding stratum (e.g. replacement areas of poor founding material)
- groundwater in excavations (e.g. amount, quality and from which stratum)
- areas where contaminated ground was encountered
- historic man-made obstructions (e.g. foundations, archaeological remains, wells)
- newly created man-made obstructions (e.g. test piles, sheet piles or dewatering wells left in place)
- utility services, especially on private land
- temporary works (e.g. sheet piles, soil nails, extent of excavations/slopes, dewatering)
- areas of slope/excavation failures
- details of adjacent structure foundations
- areas of in-ground testing (e.g. piles, ground treatment, soil nails)
- locations of basements and voids.

Further information is described in the CIRIA SP136 *Site Guide to Foundation Construction* (1996), which contains useful information relating to observations to make during the construction works.

101.5.1 Key learning points

- Significant health and safety issues where appropriate should be included on the as-built drawings.
- Observations made during construction works are important and need to be recorded even if they do not impact on the current design.

101.6. Documentation systems and preserving data

Consideration needs to be given to the maintenance and accessibility of close-out reports. To facilitate the handling and storage of reports it is increasingly common for such reports and supporting data to be stored in digital form. Various international standards are available that relate to the indexing, handling and transferring of digital information. Increasingly the ISO 19650 series is used as a basis for managing this information through whole life (BSI, 2018; UK BIM Framework, 2023). Digital storage is subject to fast-moving technological changes in storage media type and availability of software, and due consideration should be given to backward compatibility.

Where it is necessary for a paper print-out of a digital file (typically a spreadsheet) to be provided, it is important that the format of the print-out visibly includes all the computational steps that were undertaken within the document or spreadsheet to produce the end result. For example, the steps in conversion of a frequency reading from a vibrating wire piezometer into a groundwater level should be shown together with cross-reference to the appropriate calibration data sheets.

In addition to providing the close-out report to their client, the consultant or contractor should hold such reports, information and data for the period required by their contract (typically 6 or 12 years). Eurocode 7 recommends ten years, with the more important documents stored for the lifetime of the relevant works. There should be mechanisms in place to ensure the ability to read the digital data throughout this period.

Finally, consideration should also be given to the safeguarding of the documentation storage site with off-site back-up of digital data being a benefit against total loss of the main storage site. With the rising cost of storage there will be continuing pressure on destruction of the original design information. Consequently, the close-out report will become an invaluable reference for future works, be they ongoing maintenance, demolition or reuse.

101.6.1 Key learning points

- Comply with the digital requirements of the client as well as digital best practice and standards.
- Maintain a digital copy of the close-out report.
- Regularly review and maintain backward compatibility.

101.7. Summary

The close-out report, the final part of the geotechnical cycle, is intended to describe, collate and summarise all relevant geotechnical works that occur during the construction phase of the works, together with any post-construction monitoring. It forms an essential part of the construction verification process. The report also provides valuable information on the geotechnical works, allowing the future reuse of these works and hence increasing the sustainability profile of the initial works.

References

AGS (2005) *Guidelines for the Preparation of the Ground Report.* Association of Geotechnical and Geoenvironmental Specialists, Faversham, UK.

AGS (2015) *Guide to Good Practice in Writing Ground Reports.* Association of Geotechnical and Geoenvironmental Specialists, Faversham, UK.

BRE (Building Research Establishment) (2006a) *Reuse of Foundations for Urban Sites: Proceedings of the 2nd International Conference of the BRE, Watford, UK.* IHS BRE Press, Watford, UK.

BRE (2006b) *Reuse of Foundations for Urban Sites: A Best Practice Handbook.* IHS BRE Press, Watford, UK.

BRE (2023) Helping construction unlock the benefits of the circular economy. https://bregroup.com/buzz/design-for-deconstruction-helping-construction-unlock-the-benefits-of-the-circular-economy/construction (accessed 07/05/2023).

BSI (British Standards Institution) (2004) BS EN1997-1:2004+A1:2013: Eurocode 7: Geotechnical design – Part 1: General rules. BSI, London, UK.

BSI (2018) ISO 19650: Building information modelling (BIM). BSI, London, UK.

Chapman M, Marsh B and Foster A (2001) Foundations for the future. *Proceedings of the ICE Civil Engineering* **144**: 36–41.

Chapman T, Anderson S and Windle J (2007) *Reuse of Foundations.* CIRIA Report C653. CIRIA, London, UK.

CIRIA (1996) *Site Guide to Foundation Construction. CIRIA Special Publication SP136.* CIRIA, London, UK.

Highways England (2020) *Managing Geotechnical Risk. Report CD 622.* The Stationery Office, London, UK.

HSE (Health and Safety Executive) (2015) *Managing Health and Safety in Construction: Construction (Design and Management) Regulations 2015.* HSE, London, UK.

ICE (Institution of Civil Engineers) (2016) *Specification for Piling and Embedded Retaining Walls*, 3rd edn. ICE, London, UK.

Tayler H (2020) A short guide to reusing foundations. *The Structural Engineer* **98(11)**, 10.56330/GYPR7761.

UK BIM Framework (2023) https//www.ukbimframework.org (accessed 07/05/2023).

Useful websites

Association of Geotechnical and Geoenvironmental Specialists (AGS): http://www.ags.org.uk

The Structural Engineer: https//www.thestructuralengineer.org

It is recommended this chapter is read in conjunction with

- Chapter 78 *Procurement and specification*
- Chapter 93 *Quality assurance*
- Chapter 96 *Technical supervision of site works*

All chapters in this book rely on the guidance in Sections 1 *Context* and 2 *Fundamental principles.* A sound knowledge of ground investigation is required for all geotechnical works, as set out in Section 4 *Site investigation.*

Index

Page numbers in **bold** refer to figures and *italics* refer to tables.